Summary and Review Tools

Mechanism Review: Substitution versus Elimination

S_N2

Primary substrate
Back-side attack of Nu: with respect to LG
Strong/polarizable unhindered nucleophile

Bimolecular in rate-determining step
Concerted bond forming/bond breaking
Inversion of stereochemistry
Favored by polar aprotic solvent

$Nu/B^{\delta-}$ ····· $LG^{\delta-}$

S_N1 and E1

Tertiary substrate
Carbocation intermediate
Weak nucleophile/base (e.g., solvent)

Unimolecular in rate-determining step
Racemization if S_N1
Removal of β-hydrogen if E1
Protic solvent assists ionization of LG
Low temperature (S_N1) / high temperature (E2)

Nu:/B:⁻ — LG, H

S_N2 and E2

Secondary or primary substrate
Strong unhindered base/nucleophile leads to S_N2
Strong hindered base/nucleophile leads to E2
Low temperature (S_N2) / high temperature (E2)

E2

Tertiary or secondary substrate
Concerted anti-coplanar TS

Bimolecular in rate-determining step
Strong hindered base
High temperature

$Nu/B^{\delta-}$ ····· H ····· $LG^{\delta-}$

TABLE 3.1 Relative Strength of Selected Acids and Their Conjugate Bases

	Acid	Approximate pK_a	Conjugate Base	
Strongest acid	$HSbF_6$	< −12	SbF_6^-	Weakest base
	HI	−10	I^-	
	H_2SO_4	−9	HSO_4^-	
	HBr	−9	Br^-	
	HCl	−7	Cl^-	
	$C_6H_5SO_3H$	−6.5	$C_6H_5SO_3^-$	
	$(CH_3)_2\overset{+}{O}H$	−3.8	$(CH_3)_2O$	
	$(CH_3)_2C{=}\overset{+}{O}H$	−2.9	$(CH_3)_2C{=}O$	
	$CH_3\overset{+}{O}H_2$	−2.5	CH_3OH	
	H_3O^+	−1.74	H_2O	
	HNO_3	−1.4	NO_3^-	
	CF_3CO_2H	0.18	$CF_3CO_2^-$	
	HF	3.2	F^-	
	CH_3CO_2H	4.75	$CH_3CO_2^-$	
	H_2CO_3	6.35	HCO_3^-	
	$CH_3COCH_2COCH_3$	9.0	$CH_3CO\overset{-}{C}HCOCH_3$	
	NH_4^+	9.2	NH_3	
	C_6H_5OH	9.9	$C_6H_5O^-$	
	HCO_3^-	10.2	CO_3^{2-}	
	$CH_3NH_3^+$	10.6	CH_3NH_2	
	H_2O	15.7	OH^-	
	CH_3CH_2OH	16	$CH_3CH_2O^-$	
	$(CH_3)_3COH$	18	$(CH_3)_3CO^-$	
	CH_3COCH_3	19.2	$^-CH_2COCH_3$	
	$HC{\equiv}CH$	25	$HC{\equiv}C^-$	
	H_2	35	H^-	
	NH_3	38	NH_2^-	
	$CH_2{=}CH_2$	44	$CH_2{=}CH^-$	
Weakest acid	CH_3CH_3	50	$CH_3CH_2^-$	Strongest base

Increasing acid strength (upward arrow); Increasing base strength (downward arrow)

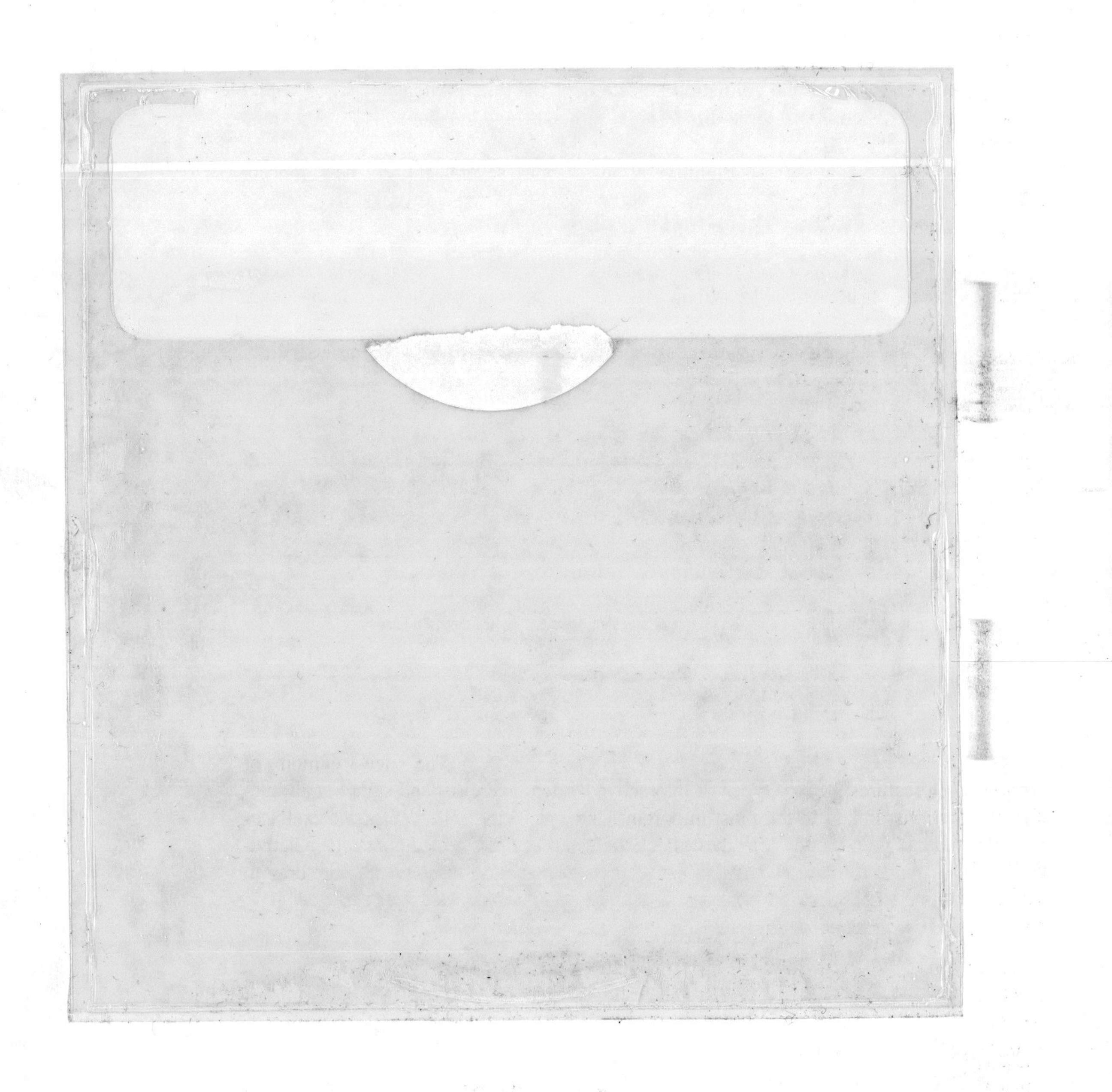

OrganicView CD

The CD packaged with this text includes the following components to help students visualize and understand the basic concepts of organic chemistry.

1. CONCEPT UNITS The *OrganicView CD* contains some 50 animated 3D "Concept Units," from the Science Teaching Graphics collection of Darrell J. Woodman (University of Washington). These are presentations, with audio, on key topics in organic chemistry where computer graphics and animation help to depict: 1) the particulate (microscopic) view of matter; 2) complex three-dimensional structure and relationships; 3) dynamic processes.

The CD accompanying the Eighth Edition includes several entirely new concept units on reaction mechanisms such as Electrophilic Aromatic Substitution, Acyl Substitution, Enolate Anions, and Epoxidation.

2. 3D MODELS A library of more than 400 3D molecular models is included on the *OrganicView CD*, linked to where the particular molecules are discussed in the text. These open automatically for student exploration, using either the well-known Chem3D® plugin or Rasmol® scientific visualization software (included by permission of the author, Roger Sayle). The software and the special model files allow the student to switch between various model types (wireframe, ball-stick, space-filling) with special color schemes and display options (backbone, strand, ribbon) for biomolecules. Of particular significance in this category are 3D versions of many new color molecular graphics in ***Organic Chemistry, Eighth Edition,*** based on quantum mechanical calculations with Chem3D® and Gaussian® software, prepared by Craig Fryhle. These new graphics feature accurate molecular orbital, electron density surface, and electrostatic potential displays, bringing a higher level of scientific sophistication to this edition.

3. ANIMATED GRAPHICS The *OrganicView CD* provides over 60 additional interactive, animated presentations of other specific text graphics and structures, including rotational or other 3D movies of many molecules of special interest, for example, the enzyme carbonic anhydrase, complexation by a crown ether, and many of the electrostatic potential maps shown in the book.

4. INTERACTIVE EXERCISES AND ASSESSMENT The new edition of *OrganicView* features 30 new types of interactive student exercises and sample multiple-choice questions, linked to the most important Concept Units. The interactive exercises provide students with feedback as they check their mastery of selected concepts in activities such as "guided drawing" (selected structures and reaction mechanisms) or "drag to assemble" complex structures. Many of the exercises utilize Shockwave® 3D, which enables students to manipulate complex chemical structures in 3D space.

5. DRILL/REVIEW AND MECHANISM PRACTICE Designed to help students practice the skills needed to be successful in this course, the Drill/Review units focus on remembering reagents for chemical transformations and applying the basic patterns of reactions either to predict the main product or to deduce the needed starting material.

6. IR TUTOR The popular IR Tutor software by Charles Abrams (Beloit College) is also included on the *OrganicView CD.* IR Tutor provides the student with animated tutorials on the theory and practice of infrared spectroscopy, as well as illustrative spectra of representative compounds. The illustrative spectra show both an animation and an assignment of the vibrational mode when the user clicks on a peak. Comparison overlays and tips help the student quickly develop the ability to analyze infrared spectra.

eGrade

Wiley's eGrade is web-based software that automates the process of assigning and grading homework, quizzes, and exams through a mathematically "smart" grading engine. eGrade *provides professors and students with numerous benefits; it increases students' time-on-task while providing them with immediate feedback and scoring on their work.*

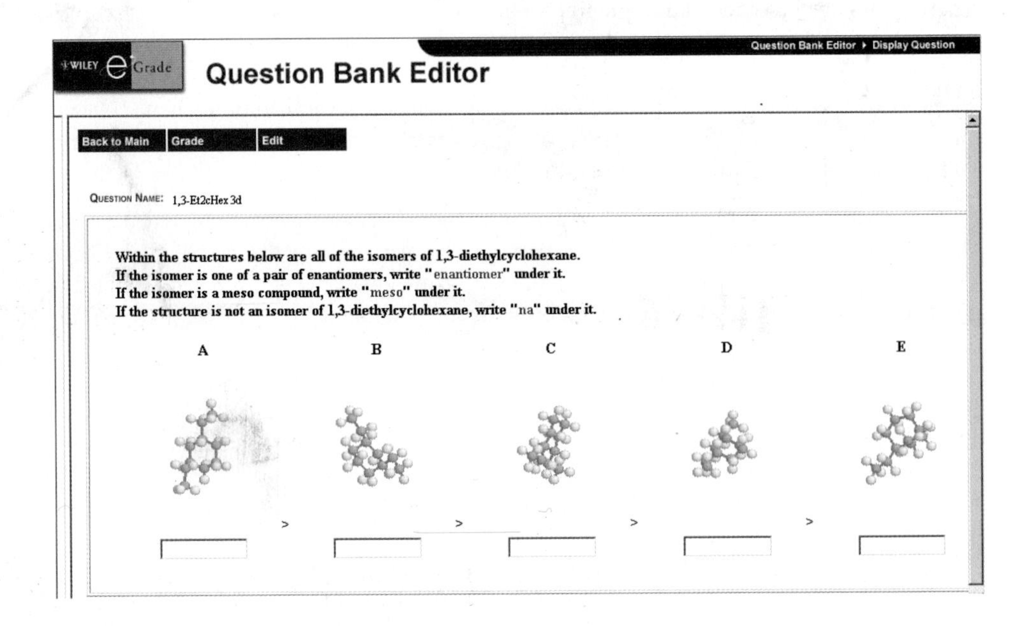

1. HOMEWORK PROBLEMS Developed by K. C. Russell (Northern Kentucky University), the *eGrade* for Solomons/Fryhle **Organic Chemistry, Eighth Edition**, features 15–20 problems per chapter, derived from the end-of-chapter problems in the textbook. Many problems feature 3D molecules which students can manipulate as they work out answers. Each *eGrade* homework problem has multiple parts, each part an individual numeric response, multiple choice or fill-in-the-blank question. Most homework problems feature randomized variables, offering each student a different version of the same problem.

2. TEST BANK QUESTIONS The Solomons/Fryhle **Organic Chemistry, Eighth Edition** Test Bank material is also available in *eGrade,* allowing for online testing and quizzing opportunities. Instructors making use of both homework and Test Bank question files will be able to take full advantage of *eGrade*'s robust gradebook and assignment management capabilities.

Organic Chemistry

Eighth Edition

T.W. Graham Solomons
University of South Florida

Craig B. Fryhle
Pacific Lutheran University

John Wiley & Sons, Inc.

For Judith, Allen, Graham, Jennie, Guido, and Cory. **TWGS**
For Deanna, Lauren, and Heather. **CBF**

Executive Editor	*Deborah Brennan*
Acquisition Editor	*Kevin Molloy*
Senior Marketing Manager	*Robert Smith*
Senior Production Editor	*Elizabeth Swain*
Senior Designer	*Karin Gerdes Kincheloe*
Illustration Editor	*Sandra Rigby*
Photo Editor	*Lisa Gee*
Associate Director of Development	*Johnna Barto*
Cover Image	*Image courtesy of Anthony Pease*
Text Design	*Delgado Design, Inc.*
Cover Design	*David Levy*

Cover
The cover shows a fanciful representation of nanoscale molecular switches on the surface of a droplet. Molecules such as these, called [2]rotaxanes and consisting of a cyclic molecule through which a molecular shaft is threaded (see Section 4.12), have the potential to behave like nanoscale transporters or shuttles. Changing the oxidation state of a group on the shaft leads to electrostatic forces that drive the ring along the shaft from one position to another. Research with organic molecules like these, synthesized by J. Fraser Stoddart and colleagues at UCLA, holds great promise for revolutionizing our world with new nanoscale machines and devices. Nanotechnology is indeed a burgeoning and exciting area, and it is one of the dimensions of chemistry we highlight in this edition. Image courtesy of Theresa Chang.

This book was set in Times Roman by Progressive Information Technologies
and printed and bound by Von Hoffman Press. The cover was printed by Von Hoffman.

This book is printed on acid-free paper. ♾

Library of Congress Cataloging-in-Publication Data
Solomons, T. W. Graham.
Organic chemistry / T. W. Graham Solomons, Craig B. Fryhle.—8th ed.
p. cm.
Includes index.
ISBN 0-471-41799-8 (cloth : acid-free paper)
Wiley International Edition 0-471-44890-7
1. Chemistry, Organic. I. Fryhle, Craig B. II. Title.

QD253.2.S65 2003
547—dc21 2003041110

Printed in the United States of America

10 9 8 7 6 5

Preface

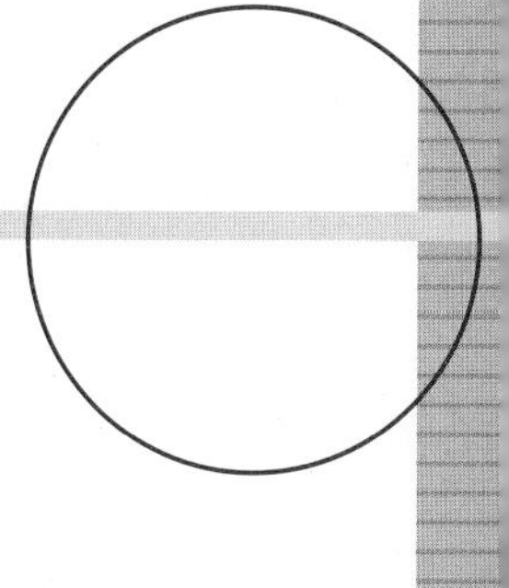

The goal of our book is to bring organic chemistry to students in the most interesting and comprehensible way possible. We believe that the Eighth Edition offers our strongest pedagogy yet for achieving these goals. The text includes many visual tools for learning, including Concept Maps, details of reaction mechanisms and thematic Mechanism Review summaries, Synthetic Connections, study tips and tool box notations, enlightening illustrations, and abundant problems. We have included chapter opening vignettes and "The Chemistry of..." boxes to help students relate organic chemistry to everyday life. The CD packaged with the text is rich with opportunities for students to review and practice the concepts of organic chemistry. Students who use the in-text learning aids, work the problems, regularly employ resources on the CD, and use the supplementary modeling kit, will be assured of success in organic chemistry.

Organization

A central theme of our approach to organic chemistry is to emphasize the *relationship between structure and reactivity.* To accomplish this, we have chosen an organization that combines the most useful features of the traditional functional group approach with one based on reaction mechanisms. Our philosophy is to emphasize mechanisms and their common aspects as often as possible, and at the same time to use the unifying features of functional groups as the basis for most chapters. The structural aspects of our approach show students *what organic chemistry is.* Mechanistic aspects of our approach show students *how it works.* And wherever an opportunity arises, we show them *what it does* in living systems and the physical world around us.

Most important is for students to have a solid understanding of structure—of hybridization and geometry, steric hindrance, electronegativity, polarity, and formal charges—so that they can make intuitive sense of mechanisms. It is with these topics that we begin in Chapter 1. In Chapter 2 we introduce all of the important functional groups, intermolecular forces, and a key tool for identifying functional groups–infrared spectroscopy. Throughout the book we have updated our calculated models of molecular orbitals, electron density surfaces, and maps of electrostatic potential. These models enhance students' appreciation for the role of structure in properties and reactivity.

We begin our study of mechanisms in Chapter 3 in the context of acid-base chemistry. Why? Because acid-base reactions are fundamental to organic chemistry. When looked at from the point of view of Lewis acid-base theory, the steps of most organic reaction mechanisms are acid-base reactions. Acid-base reactions, moreover, are relatively simple and they are reactions that students will find familiar. Acid-base reactions also lend themselves to an introduction of several important topics that students need to know about early in the course: (1) the curved arrow notation for illustrating mechanisms, (2) the relationship between free-energy changes and equilibrium constants, (3) how enthalpy and entropy changes affect reactions under equilibrium control, and (4) the importance of inductive and resonance effects and of solvent effects. In Chapter 3, we also begin to show students how organic chemistry works by presenting the first of many boxes called "A Mechanism for the Reaction." All through the book, these boxes highlight and bring forth the details of important reaction mechanisms.

Throughout our study we use various opportunities to show what organic chemistry *does* in life, both in biological terms and in our physical environment, through real world applications highlighted by chapter opening vignettes and "The Chemistry of..." boxes. As students come to realize that life and much of the world around us involves organic chemistry, their fascination with the subject cannot help but increase.

Key Features of the Eighth Edition

Some major highlights of the Eighth Edition include:

- *Concept Maps, Mechanism Reviews, and Synthetic Connections:* comprehensive new summary and review tools to enhance student learning.
- New material relating to environmentally benign ("green") chemistry, nanotechnology, and biochemistry introduces exciting new frontiers of organic chemistry.
- All of the graphics and illustrations from the previous edition have been modernized, revised, and updated. Stylized orbital representations have been redrawn with improved shapes, and an attractive shine and color scheme permeates the presentation of models and other graphics.
- New electrostatic potential maps and electron density surfaces prepared with Gaussian software help students visualize polarity and electron distribution.
- Several new chapter-opening vignettes and boxes relate concepts to the real world.
- Highlights of Nobel Prize-winning chemistry are integrated into the text.
- Organic examples of biological and other real-world chemistry are highlighted in "The Chemistry of …" boxes.
- Early introduction of spectroscopy gives students evidence for functional groups and structure, and supports their use of instrumentation in laboratory classes.
- Pedagogical margin notes focus attention on key tools and tips for learning.
- Learning Group Problems provide active integration of concepts and opportunities for peer-led teaching.
- New problems employ calculated molecular models created with Gaussian for viewing with the Chem3D plugin.
- OrganicView CD and web site offer technology-based learning–Woodman Concept Unit Tutorials, Shockwave® Interactive Web Exercises, Reaction Animations, and 3D models provide reinforcement of basic concepts.
- eGrade self-assessment software allows students to practice problems from the text and receive immediate feedback on their progress.

Concept Maps, Mechanism Summaries, and Synthetic Connections

We introduce, for the first time with this edition, comprehensive visual summary and review tools for students. These tools come in three forms: Concept Maps, Mechanism Summaries, and Synthetic Connections. **Concept Maps** are hierarchical flowcharts that join one key concept to the next with a linking phrase. Our Concept Maps help students summarize, review, and organize the material in a chapter. Our new **Mechanism Summaries** tie together common themes and highlight key attributes of important mechanisms. They highlight factors influencing the type of mechanism by which a molecule will react, as well as show regiochemical and stereochemical aspects of mechanisms. **Synthetic Connections** are roadmaps that show pathways for converting molecules from one type to another. Using our Synthetic Connections, students can see how reactions they have learned are part of their growing repertoire for synthesis.

Green Chemistry, Biochemistry, and Nanotechnology

The Eighth Edition brings new material on green chemistry, biochemistry, and nanotechnology. Environmentally-benign chemical methods are increasingly important in our world, and we have highlighted examples and new directions in 'green chemistry' whenever possible. Students

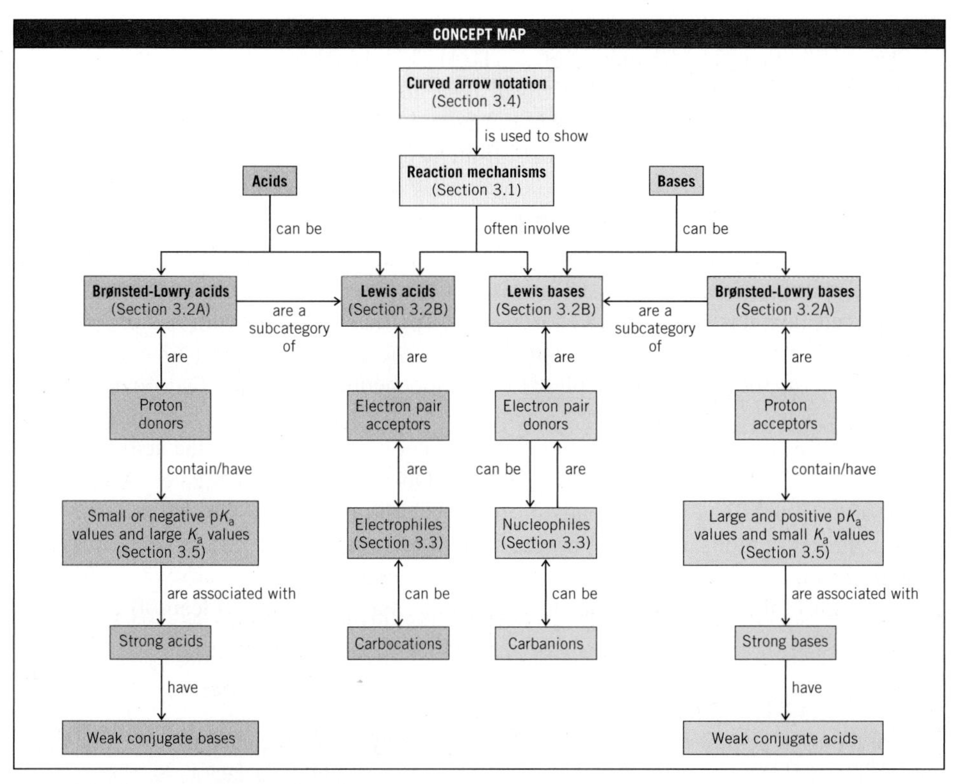

Part of the Chapter 3 Concept Map

need to know that environmentally-benign chemical methods are in use, and that opportunities exist to develop new methods. We encourage students and instructors to consult the American Chemical Society (www.chemistry.org) and the Environmental Protection Agency (www.epa.gov) web sites for further resources on green chemistry. In addition, an excellent resource is "Real World Cases in Green Chemistry," by Michael C. Cann (American Chemical Society Publications: Washington, D.C., 2000).

Green chemistry opening vignette

Biochemistry and biotechnology are always areas of high interest for students of organic chemistry, and dramatic changes are occurring in these areas. Recognizing the rapid pace of developments, we have substantially updated the biochemistry chapters in the Eighth Edition. We have included new sections on proteomics and genomics. We have updated information about key chemical tools for biotechnology, including solid-phase carbohydrate synthesis, mass spectrometry of proteins and nucleic acids, and the dideoxy method for sequencing nucleic acids. Research advances, such as the recently solved crystal structure of the ribosome, have opened many doors for revisions in our coverage. As always, we draw connections to biochemistry throughout the book in chapter-opening vignettes, "The Chemistry of…" boxes, and Special Topics.

Nanotechnology and materials science are among the most exciting areas in chemistry today. We have incorporated several new vignettes and boxes pertaining to nanotechnology, materials science, and bioengineering. Among these are the Chapter 2 opening vignette "Structure and Function: Organic Chemistry, Nanotechnology, and Bioengineering," a box in

Chapter 4 titled "The Chemistry of... Nanoscale Motors and Molecular Switches," and a box in Chapter 23 titled "The Chemistry of... STEALTH® Liposomes for Drug Delivery." *The cover of our book shows a spherical array of molecules like the proposed molecular switches and motors discussed in the Chapter 4 box.* Advances in nanotechnology are literally revolutionizing our lives, with pervasive influence from medicine to personal technology.

Chapter-Opening Vignettes

Students who find a subject interesting will be motivated to learn it. This idea was an innovation that made the first edition of this text the success that it became, and it is one that has been emphasized in every edition since. As in the previous edition, we open each chapter with a vignette that shows students how the chapter's subject matter relates to "real world" applications—to applications of biochemical, medical and environmental importance. Chapter 3, on acid-base chemistry, begins with an essay describing the role of the enzyme carbonic anhydrase in regulating blood acidity through the acid-base reactions involved in the mechanism of this enzyme. Opening Chapter 4 is an essay on the rotation of carbon-carbon single bonds in the muscle protein actin, one that sets the stage for the chapter's emphasis on conformational analysis. Chapter 13, where we discuss benzene in detail, begins with an essay about green chemistry and alternatives to using benzene in industrial processes. Other topics of similar engaging interest lead off each chapter in the text.

Electrostatic Potential Maps

Two of the most helpful concepts that students can apply in organic chemistry are that opposite charges attract and that delocalization of charge is a stabilizing factor. As chemists, we know that many reactions occur because molecules with opposite charges are attracted to each other. We also know that reaction pathways are favored or disfavored partly on the basis of relative stability of charged intermediates. To fully utilize this pedagogy, we use maps of electrostatic potential at the van der Waals surface of molecules in which colors indicate the charge distribution in the various regions of a molecule or ion.

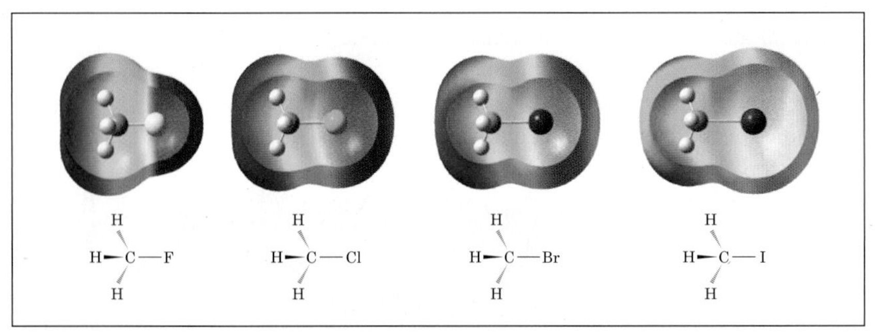

Electrostatic potential maps for the methyl halides

Electrostatic potential maps help students visualize basic principles of structure and reactivity. For example, because an understanding of Brønsted-Lowry and Lewis acid-base reactivities is essential for success in organic chemistry, in the Seventh Edition we improved Chapter 3 (on acids and bases) by including a number of calculated electrostatic potential maps to illustrate how charge distribution influences the relative acidity of an acid and how it affects the relative stability of the conjugate base. These graphics greatly assist visualization of charge separation, localization, and dispersal. Other examples in this chapter include illustrations of the acidity of terminal alkynes, of the charge distribution in acetate anion versus that of ethoxide anion, and in the Lewis acid-base reaction of boron trifluoride and ammonia.

We also use electrostatic potential maps to focus attention on the complementary charges in nucleophiles and electrophiles, to show the relative charge distribution in asymmetric bromonium ions, to compare the relative stabilities of arenium ion intermediates in electrophilic aromatic substitution, and to illustrate the electrophilic nature of carbonyl groups and the β carbon in α,β-unsaturated carbonyl compounds. In one of the early in-chapter boxes we also show how the LUMO of one reactant and the HOMO of another are important in reactions.

We generated most of the electron density surfaces and electrostatic potential maps in this book using ab initio quantum mechanical methods at the 6-31G level using Gaussian® software. Molecules that we compare within a series are depicted over the same charge range to insure that comparisons are accurate and meaningful in relative terms. Structures are energy-minimized except for those where a particular higher energy conformation is desired.

Orbital Hybridization and the Structure of Organic Molecules

Students must develop a sound understanding of the structure and shape of organic molecules. We build the foundation for their understanding of structure by introducing orbital hybridization and VSEPR theory in Chapter 1. We begin with methane for sp^3 hybridization, move directly to ethene for sp^2 hybridization, and then to ethyne for sp hybridization. We also use calculated molecular orbitals, and electron density surfaces to illustrate regions of bonding electron density and overall molecular shape. In this edition we improved our presentation of resonance in Chapter 1 by including some additional rules for drawing proper resonance structures. Students will consider these rules further when they study conjugated systems in Chapter 13.

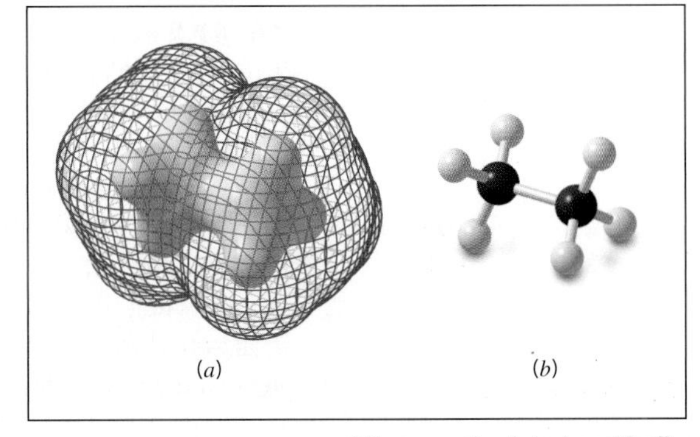

Ethane: calculated and ball-and-stick models

Molecular Models: Hand-held and Computer-based

We emphasize the importance of three-dimensional structure by frequently encouraging students to use hand-held molecular models. We feel that the tactile experience of manipulating physical models is key to students' understanding that organic molecules have shape and occupy space. To facilitate this effort, we have arranged with the Darling Company to bundle inexpensive Molecular Visions™ model kits with the book (for those who choose that option). The model kit has been specially assembled to accompany this text.

We also emphasize structure by providing several hundred 3D computer-based molecular models on the CD accompanying the book. We use the Chem3D plugin for models that have molecular orbitals and isosurfaces (calculated using Chem3D and Gaussian), and Rasmol for models that do not have calculated properties associated with them. To reinforce students' learning through use of these models, we have included problems at the end of some chapters that call upon students to view and manipulate computer models stored on the CD or Web site.

Learning Group Problems

Active Integration of Concepts

To facilitate active and collaborative student involvement in learning we have included problems at the end of each chapter that are intended to be solved by students working in small "Learning Groups." Each problem, called a Learning Group Problem (LGP), integrates concepts and requires gathering of information from the chapter for students to arrive at a complete solution. The problems can be worked inside or outside of class, with three to six students per group as a desirable size. The LGPs are a useful culminating activity to help students draw together what they have learned from each chapter, and to integrate this knowledge with ideas they have learned earlier.

Student-Led Teaching

The nature of the Learning Group Problems makes them useful as a vehicle for students to *teach* organic chemistry to their peers, as well. For example, because solutions to the problems draw out a variety of important concepts from each chapter, a group's classroom presentation of their Learning Group Problem can be the teaching mode for a given day in class. In this way, students can have the powerful experience of learning through teaching (a wonderful experience that we teachers already know first-hand). The instructor can coach the student presentations from the side of the classroom to insure that all the desired ideas are brought forth and articulated. Detailed suggestions for orchestrating a class involving Learning Group Presentations are available with information for instructors using this text.

Early Introduction to Retrosynthetic Analysis and Organic Synthesis

We use the alkylation of alkynide anions in Chapter 4 to introduce organic synthesis and retrosynthetic analysis. One advantage of using alkynide anions to introduce synthesis is that the reactivity of terminal alkynes and alkyl halides is readily understood on the basis of concepts students have learned in the beginning chapters of the book, namely acid-base chemistry and polarity. Students will use Brønsted-Lowry acid-base chemistry to outline the preparation of alkynide anions from alkynes, and they will recall Lewis acid-base concepts when they consider the reaction of an alkynide anion with an alkyl halide. They will also find reinforcement of a common theme in many organic reactions—the interaction between molecules or groups bearing opposite charges.

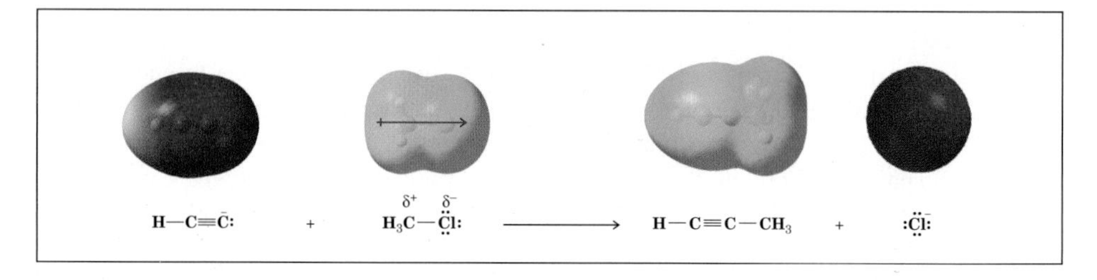

The alkylation of alkynide anions also gives students a method for carbon-carbon bond formation very early in their study of organic chemistry. And, it gives them a product that contains a functional group from which they can make many other compounds as their synthetic repertoire grows. Finally, because 'better' or 'worse' retrosynthetic pathways using alkynide-anion alkylation are conceivable for a given molecule, this reaction is a realistic vehicle for teaching the logic of retrosynthetic analysis.

Substitution and Elimination Reactions

Substitution and elimination reactions provide an opportunity for students to encounter one of the important realities of organic chemistry. Reactions almost never follow exclusively one path, much as we would like them to. As chemists, we know very well how frequently one kind of reaction competes with another to complicate our synthetic plans. Therefore, bringing students to the point where they can propose reasonable syntheses using substitutions or eliminations requires a careful orchestration of topics. In this edition, as in the last, Chapter 6 focuses on substitution, but it also briefly introduces elimination reactions. We have done this because the two reactions nearly always occur together, and it is vitally important that students gain a chemically accurate understanding of this. Chapter 7, then, rounds out the discussion by giving detailed treatment of E2 and E1 reactions, including stereochemistry, regiochemistry, and overall product distribution. The Eighth Edition includes new Mechanism

Review tools for summarizing the factors that favor substitution or elimination and unimolecular or bimolecular reaction. Due to the importance of these reaction types, we have placed a copy of this Mechanism Review inside the front cover of the book.

Synthesis Updates

The 2001 Nobel Prize in Chemistry was awarded to K. Barry Sharpless, William Knowles, and Ryoji Noyori for their work on catalytic asymmetric methods for oxidation and reduction. We have highlighted these powerful synthesis tools in several new or revised boxes in the Eighth Edition, including mention of their relevance to well-known compounds such as Naproxen™ and L-DOPA. The Sharpless asymmetric epoxidation was already a boxed topic in Chapter 11 of the Seventh Edition, and we have now included the Sharpless asymmetric dihydroxylation. We have also included information on "green" catalytic methods such as the oxidation of alkenes using catalytic rather than stoichiometric amounts of osmium, and oxidation using Jacobsen's catalyst.

In the previous edition we updated a number of sections so that other important tools for synthesis were illustrated. These prior updates included sections on the use of lithium enolates, silyl ether protecting groups, and silyl enol ethers in Chapter 17. We also added enantioselective carbonyl reduction methods to Chapter 12, including the use of enzymes in organic synthesis (e.g., use of extremozymes from thermophilic bacteria for reduction reactions), and we substantially updated discussion of the Diels-Alder reaction in Chapter 13. In addition, we mentioned development of catalytic antibodies for aldol condensations (in relation to the Robinson annulation, Diels-Alder reactions, and ester hydrolyses). The latter are found in Chapter 24 regarding proteins.

In this edition we moved oxymercuration-demercuration and hydroboration-oxidation from Chapter 11 to Chapter 8 because these reactions belong with other addition reactions of alkenes. This change also groups these reactions with acid-catalyzed hydration of alkenes as a collection of methods for alcohol synthesis from alkenes. Then, when alcohols and ethers are discussed in Chapter 11, we briefly review the ensemble of methods given in Chapter 8 for synthesis of alcohols from alkenes.

We have also deleted some sections that, although dear to us as chemists, provided reasonable opportunities to trim material from the book. We removed our coverage of the Hell-Vollhard-Zelinski reaction, the Wolff-Kishner reduction, and the debromination of vicinal dibromides, and we moved the Special Topic on Reactions and Synthesis of Heterocyclic Amines to the CD as an electronic rather than a print resource. Though we were tempted to delete it due to our emphasis on IR, we retained the Hinsberg classification test for amines. We welcome your suggestions for other topics that could be shortened or removed.

Advantages of Introducing Spectroscopy Early

Our book gives instructors the opportunity, if they desire, to use spectroscopy as an early and integral part of their course. We present infrared (IR) spectroscopy in Chapter 2, immediately after the introduction of functional groups. Placement of IR here shows students one of the important ways chemists obtain evidence about functional groups. It also supports early experiences students may have using IR in their laboratory experiments. Thus, as students study functional groups for the first time in Chapter 2, they also learn about the best method for detecting the presence of functional groups in a molecule. In addition, IR gives evidence for intermolecular forces, such as hydrogen bonding, which is also discussed in Chapter 2.

The CD that accompanies the book also includes ***IR Tutor,*** *a widely acclaimed computerized tutorial on IR spectroscopy created at Columbia University. It has been said that students cannot help but learn IR spectroscopy if they use the IR Tutor program.*

We also give nuclear magnetic resonance (NMR) and mass spectrometry (MS) prominence by placing them together in Chapter 9, relatively early in the book. These two methods are among the most powerful structure elucidation tools in organic chemistry and biochemistry, and for this reason they deserve early and substantial coverage. We also use spectroscopy with problems in almost every chapter after the introduction of each spectroscopic method, and we give details about the spectroscopic characteristics of each functional group as we study them in later chapters. To enhance students' appreciation for ultraviolet-visible (UV-Vis) spectroscopy, we place it in the context of conjugated unsaturated systems (Chapter 13).

We also briefly introduce gas chromatography (GC) before mass spectrometry so that we can describe GCMS as a tool for obtaining structural information on compounds in mixtures. In Chapter 9 and again in Chapter 24 we discuss electrospray ionization mass spectrometry (ESI-MS) because it is key to the analysis of biomolecules.

The essence of our rationale for making all of these changes is that modern instrumental methods are central to the way chemists and biochemists elucidate molecular structures. Early introduction of spectroscopic methods provides important support for the laboratory experiences of organic chemistry students, where instrumental methods play an increasing and early role in their training, and NMR and mass spectrometry, together with IR, complete the typical analytical ensemble used by many of today's organic chemists.

User-friendly Spectra and Interpretation Tools

All of the ^{1}H and ^{13}C NMR spectra in the book are 300 MHz Fourier transform NMR spectra. NMR spectra used to teach spectral interpretation are clearly annotated to show which atoms are responsible for producing each signal in the spectrum. Offset zoom expansions of many ^{1}H spectra are provided for clarity, and integral curves are shown. ^{13}C NMR data is given with DEPT information to indicate the number of hydrogen atoms bonded to each unique carbon atom. Two-dimensional NMR spectroscopy (COSY and HETCOR) is discussed in Chapter 9, as well.

In this edition we have included new graphical chemical shift correlation charts for ^{1}H and ^{13}C NMR spectra. (There is also a new IR frequency chart.) These figures are found inside the covers of the book for easy reference.

All of the 1D NMR data files are provided on the World-Wide Web via a link from the book's website in both JCAMP format and NUTS format (an NMR software program from Acorn NMR, Inc.). The JCAMP format allows direct viewing of the NMR spectra using a Web browser (but does not provide for extensive data manipulation). The NUTS format allows users to manipulate the data and prepare it for presentation in whatever manner they desire.

"The Chemistry of ..." Boxes

In most chapters, we use one or more boxes called "The Chemistry of ..." to provide enhanced coverage of a chapter topic, to supply a relevant biological, environmental, or materials science example, or to expand upon concepts from the opening vignettes. Some examples are:

- The Chemistry of ... The Bombardier Beetle's Noxious Spray
- The Chemistry of ... Organic Templates Engineered to Mimic Bone Growth
- The Chemistry of ... Nanoscale Motors and Molecular Switches
- The Chemistry of ... Radicals in Biology, Medicine, and Industry
- The Chemistry of ... The Sharpless Asymmetric Epoxidation
- The Chemistry of ... Epoxides, Carcinogens, and Biological Oxidation
- The Chemistry of ... Sunscreens (Catching the Sun's Rays and What Happens to Them)

The Chemistry of...

The Bombardier Beetle's Noxious Spray

The bombardier beetle defends itself by spraying a jet stream of hot (100°C), noxious *p*-benzoquinones at an attacker. The beetle mixes *p*-hydroquinones and hydrogen peroxide from one abdominal reservoir with enzymes from another reservoir. The enzymes convert hydrogen peroxide to oxygen, which in turn oxidizes the *p*-hydroquinones to *p*-benzoquinones and explosively propels the irritating spray at the attacker. Photos by T. Eisner and D. Aneshansley (Cornell University) have shown that the amazing bombardier beetle can direct its spray in virtually any direction, even parallel over its back, to ward off a predator.

Bombardier beetle in the process of spraying. From Eisner, T.; Aneshansley, D. J. *Proc. Natl. Acad. Sci. USA* **1999,** *96,* 9705–9709.

- The Chemistry of ... Vaccines Against Cancer
- The Chemistry of ... Antibody-Catalyzed Aldol Condensations
- The Chemistry of ... A Suicide Enzyme Substrate
- The Chemistry of ... Artificial Sweeteners

These and other boxes within the text show students the many ways in which organic chemistry is central to life and the world around us.

Relating Organic Chemistry to Biosynthesis

When an aspect of organic chemistry arises that has a biosynthetic counterpart we have juxtaposed that topic with coverage of the relevant fundamental organic chemistry. An example is the biosynthesis of lanosterol from 2,3-oxidosqualene, a step along the pathway to cholesterol. Because this biosynthetic transformation is such a beautiful example of enzyme-mediated epoxide ring opening, alkene addition steps, and skeletal migrations, we placed "The Chemistry of ... Cholesterol Biosynthesis" in Chapter 8, directly in relation to alkene addition reactions and shortly after students' acquaintance with hydride and methanide migrations. Another example of this approach is placement of "The Chemistry of ... Polyketide Antibiotic Biosynthesis" in Chapter 19, directly in relation to the malonic ester synthesis and the Claisen condensation. We believe that a student's appreciation of organic chemistry can be enhanced tremendously by showing the elegance of organic chemical reactions that take place in nature. Some other examples are:

- The Chemistry of ... Biochemical Nucleophilic Substitution
- The Chemistry of ... Pyridoxal Phosphate (Vitamin B_6)
- The Chemistry of ... Thiamine Pyrophosphate (Vitamin B_1)
- The Chemistry of ... Biological Methylation

Other Pedagogical Features

We use the margins of the wide-format pages to note the central importance of certain topics, to provide study aids or practical tips for students where appropriate, and to add brief notes of a practical or historical nature where appropriate.

Concept Maps, Mechanism Reviews, and Synthetic Connections These new features for the Eighth Edition appear at the ends of chapters as summary and review tools.

Study Tip Icons We have used "Study Tip" icons to highlight places in the text where a point is made that can be particularly helpful for a student learning organic chemistry. An example is the Study Tip icon in Section 1.7A regarding formal charges, pointing out that it will be necessary for students to keep track of formal charges later when they learn organic reactions. Another example is the Study Tip icon used to emphasize conventions used by chemists when we draw reaction and electron movement arrows (Section 1.8). Study Tip icons occur throughout the book.

Tool Box Icons We have used "Tool Box" icons in the margin to emphasize concepts that are fundamental "tools" in organic chemistry. Some examples where the Tool Box icon is used are in Chapter 1, when we introduce the hybridization states of carbon and when we introduce VSEPR theory. These concepts are among the many essential "tools" for success in learning organic chemistry. We also use the Tool Box icon when a key process or reaction is described, such as the Robinson annulation for synthesis of carbocylic rings. Tool Box icons occur throughout the book.

"A Mechanism for the Reaction" These specially designed mechanism presentations give detailed explanations for every key mechanism in the book. Curved arrows show precisely and unambiguously how electrons flow in each step of a mechanism. Steps are identified and annotated in the mechanisms to further explain each transformation.

Solved Problems Sample problems with solutions are included at key points to show students how to approach problems in organic chemistry.

In-Chapter Problems Numbered problems appear at the end of sections to reinforce students' learning immediately after each topic is introduced.

Key Terms and Concepts A list of key terms and concepts with section references at the end of each chapter allows students to test their memory regarding key ideas and if necessary to easily refer back to a full presentation of the concept in the chapter. The key terms and concepts listed are also defined in the glossary.

Library of 3D Computer Models All ball-and-stick, space filling, and many other molecules in the book are included in a library of 3D molecules in Chem3D or Rasmol format on the CD. Downloadable JPEG files for all illustrations and selected photographs are available at the Web site.

Use of Color in Mechanisms Consistent color schemes are used in presentation of reaction mechanisms. Changes in the bonding of atoms from reactants to products are highlighted on the basis of the color used for atoms and bonds.

Molecular Modeling Exercises Probe Structure and Properties Problems at the end of some chapters call upon students to answer questions where viewing is required of molecules from the CD or Web site in Chem3D or Rasmol format.

Supplements

OrganicView CD and Web Site

The CD included with the text includes "Chemistry Teaching Graphics for Organic Chemistry" by Darrell Woodman (University of Washington). The Woodman graphics use multimedia and animation to amplify many concepts in the text with compelling interactive tutorials and problems. The CD also features:

- A full multimedia approach that includes Shockwave animations and Interactive Exercises
- Concept Units comprised of animations and simulations accompanied by audio explanations

- An intuitive, browser-based graphical interface that makes it easy to seamlessly use the Web and CD together
- A library of 3D computer molecular models in Chem3D and Rasmol formats
- Bookmarks and a printable notepad
- Quick and easy access to all program features through tool bars and buttons

Web Site

A Web site (www.wiley.com/college/solomons) is provided to support faculty and students who use the text. One aspect of the Web site is a complete collection of the Fourier transform NMR spectra used in the text. The NMR spectra are available as JCAMP or NUTS format data files.

Study Guide

The Study Guide for **Organic Chemistry, Eighth Edition** contains explained solutions to all of the problems in the text. The Study Guide also contains:

- An introductory essay "Solving the Puzzle—or—Structure is Everything" that serves as a bridge from general to organic chemistry
- Summary tables of reactions by mechanistic type and functional group
- A review quiz for each chapter
- A section describing the calculation of molecular formulas
- A set of hands-on molecular model exercises

Molecular Visions™ Model Kits

We believe that the tactile experience of manipulating physical models is key to students' understanding that organic molecules have shape and occupy space. To support our pedagogy, we have arranged with the Darling Company to bundle a special ensemble of Molecular Visions™ model kits with our book (for those who choose that option). We use Study Tip icons and margin notes to frequently encourage students to use hand-held models to investigate the three-dimensional shape of molecules we are discussing in the book.

Instructor Resources

Test Bank. The Test Bank includes over 1600 questions to aid instructors and is available in both a printed and computerized version.

Instructor's Resource CD. This CD includes slides, with text photos and illustrations for use in lecture presentations.

eGrade Course Management. eGrade is an online quizzing and homework management program that allows students to take practice tests and email homework assignments directly to the professor.

—Craig B. Fryhle
—T. W. Graham Solomons

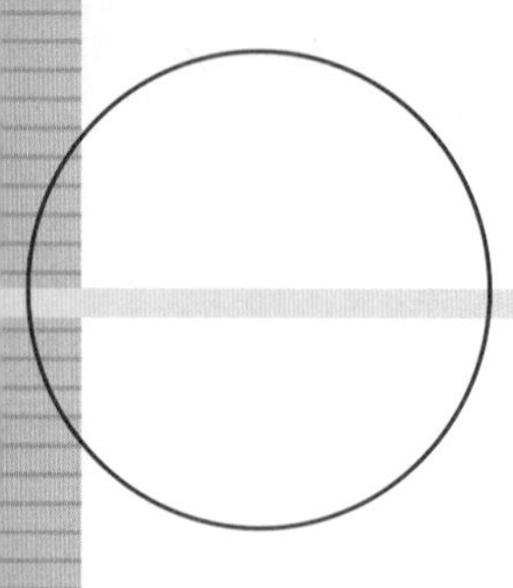

Acknowledgments

We are especially grateful to the following people who provided detailed reviews that helped us prepare this new addition of Organic Chemistry.

Merritt B. Andrus
Brigham Young University

Eric Bosch
Southwest Missouri State University

Christine Brzezowski
University of Alberta

Jeff Charonnat
California State University, Northridge

Roman Dembinski
Oakland University

Trudy Dickneider
University of Scranton

George Fisher
Barry University

Mark Forman
Saint Joseph's University

Steven A. Hardinger
University of California at Los Angeles

John L. Isidor
Montclair State University

Robert G. Johnson
Xavier University

Thomas Lectka
Johns Hopkins University

Eugene Losey
Elmhurst College

Andrew Morehead
University of Maryland

Michael J. Panigot
Arkansas State University, Jonesboro

Richard Steiner
University of Utah

Joseph J. Tufariello
State University of New York, Buffalo

Rueben Walter
Tarleton State University

Kraig Wheeler
Delaware State University

We are also grateful to the many people who provided reviews that guided preparation of the earlier editions of our book.

Chris Abelt, *College of William and Mary;* James Ames, *University of Michigan, Flint;* Winfield M. Baldwin, *University of Georgia;* David Ball, *California State University, Chico;* George Bandik, *University of Pittsburgh;* Paul A. Barks, *North Hennepin State Junior College;* Ronald Baumgarten, *University of Illinois at Chicago;* Harold Bell, *Virginia Polytechnic Institute and State University;* Kenneth Berlin, *Oklahoma State University;* Stuart R. Berryhill, *California State University, Long Beach;* Edward V. Blackburn, *University of Alberta;* Newell S. Bowman, *The University of Tennessee;* Bruce Branchaud, *University of Oregon;* Wayne Brouillette, *University of Alabama;* Ed Brusch, *Tufts University;* Edward M. Burgess, *Georgia Institute of Technology;* Robert Carlson, *University of Minnesota;* Lyle W. Castle, *Idaho State University;* George Clemans, *Bowling Green State University;* William D. Closson, *State University of New York at Albany;* Sidney Cohen, *Buffalo State College;* Randolph Coleman, *College of William & Mary;* David Collard, *Georgia Institute of Technology;* Brian Coppola, *University of Michigan;* Phillip Crews, *University of California, Santa Cruz;* James Damewood, *University of Delaware;* O. C. Dermer, *Oklahoma State University;* Phillip DeShong, *University of Maryland;* John DiCesare, *University of Tulsa;* Marion T. Doig III, *College of Charleston;* Paul Dowd, *University of Pittsburgh;* Robert C. Duty, *Illinois State University;* Eric Edstrom, *Utah State University;* James Ellern, *Consulting Chemist;* Stuart Fenton, *University of Minnesota;* Gideon Fraenkel, *The Ohio State University;* Jeremiah P. Freeman, *University of Notre Dame;* Peter Gaspar, *Washington University, St. Louis;* M. K. Gleicher, *Oregon State University;* Brad Glorvigen, *University of St. Thomas;* Roy Gratz, *Mary Washington College;* Wayne Guida, *Eckerd College;* Frank Guziec, *New Mexico State University;* Dennis Hall, *University of Alberta;* Philip L. Hall, *Virginia Polytechnic Institute and State University;* Lee Harris, *University of Arizona;* Kenneth Hartman, *Geneva College;* Michael Hearn, *Wellesley College;* John Helling, *University of Florida;* William H. Hersh, *Queens College;* Jerry A. Hirsch, *Seton Hall University;* John Hogg, *Texas A & M University;* John Holum, *Augsburg College;* John Jewett, *University of Vermont;* A. William Johnson, *University of North Dakota;* Robert G. Johnson, *Xavier University;* Stanley N. Johnson, *Orange Coast College;* John F. Keana, *University of Oregon;* David H. Kenny, *Michigan Technological University;* Robert C. Kerber, *State*

University of New York at Stony Brook; Karl R. Kopecky, *The University of Alberta;* Paul J. Kropp, *University of North Carolina at Chapel Hill;* Michael Kzell, *Orange Coast College;* John A. Landgrebe, *University of Kansas;* Paul Langford, *David Lipscomb University;* Allan K. Lazarus, *Trenton State College;* James Leighton, *Columbia University;* Philip W. LeQuesne, *Northeastern University;* Robert Levine, *University of Pittsburgh;* Samuel G. Levine, *North Carolina State University;* James W. Long, *University of Oregon;* Patricia Lutz, *Wagner College;* Frederick A. Luzzio, *University of Louisville;* Ronald M. Magid, *University of Tennessee;* John Mangravite, *West Chester University;* Jerry March, *Adelphi University;* Przemyslaw Maslak, *Pennsylvania State University;* James McKee, *University of the Sciences, Philadelphia;* Mark C. McMills, *Ohio University;* John L. Meisenheimer, *Eastern Kentucky University;* Gerado Molina, *Universidad de Puerto Rico;* Renee Muro, *Oakland Community College;* Everett Nienhouse, *Ferris State College;* John Otto Olson, *Camrose Lutheran College;* Kenneth R. Overly, *Richard Stockton College, NJ;* Paul Papadopoulos, *University of New Mexico;* Cyril Parkanyi, *Florida Atlantic University;* James W. Pavlik, *Worcester Polytechnic Institute;* William A. Pryor, *Louisiana State University;* Shon Pulley, *University of Missouri, Columbia;* Eric Remy, *Virginia Polytechnic Institute;* Michael Richmond, *University of North Texas;* Thomas R. Riggs, *University of Michigan;* Frank Robinson, *University of Victoria, British Columbia;* Stephen Rodemeyer, *California State University, Fresno;* Alan Rosan, *Drew University;* Christine Russell, *College of DuPage;* Tomikazu Sasaki, *University of Washington;* Yousry Sayed, *University of North Carolina at Wilmington;* Adrian L. Schwan, *University of Guelph;* Jonathan Sessler, *University of Texas at Austin;* John Sevenair, *Xavier University of Louisiana;* Warren Sherman, *Chicago State University;* Don Slavin, *Community College of Philadelphia;* Chase Smith, *Ohio Northern University;* Doug Smith, *University of Toledo;* John Sowa, *Seton Hall University;* Jean Stanley, *Wellesley College;* Ronald Starkey, *University of Wisconsin—Green Bay;* Robert Stolow, *Tufts University;* Frank Switzer, *Xavier University;* Richard Tarkka, *George Washington University;* James G. Traynham, *Louisiana State University;* Daniel Trifan, *Fairleigh Dickinson University;* Kay Turner, *Rochester Institute of Technology;* Rik R. Tykwinski, *University of Alberta;* James Van Verth, *Canisius College;* George Wahl, *North Carolina State University;* Darrell Watson, *GMI Engineering and Management Institure;* Arthur Watterson, *U. Massachusetts-Lowell;* Donald Wedegaertner, *University of the Pacific;* Mark Welker, *Wake Forest University;* Desmond M. S. Wheeler, *University of Nebraska;* James K. Whitesell, *The University of Texas at Austin;* David Wiedenfeld, *University of North Texas;* Carlton Willson, *University of Texas at Austin;* Joseph Wolinski, *Purdue University;* Darrell J. Woodman, *University of Washington;* Stephen A. Woski, *University of Alabama;* Linfeng Xie, *University of Wisconsin, Oshkosh;* Viktor V. Zhdankin, *University of Minnesota, Duluth;* Regina Zibuck, *Wayne State University;* Herman E. Zieger, *Brooklyn College*

We owe great thanks to yet a number of other people for their help with this edition. We would especially like to thank Robert G. Johnson (Professor Emeritus, Xavier University) for proofreading the manuscript through all of its stages. Bob has an uncanny ability to spot the minutest inconsistency or error. He also made many valuable suggestions along the way, and he coordinated revision of the 8th edition Study Guide and Solutions Manual. We would also like to thank Professor Steve Hardinger (UCLA) for his thoughtful comments about pedagogy and for proofreading the galleys.

Professor Michael Cann (University of Scranton) offered a variety of helpful ideas relating to green chemistry. We are grateful to him for conversations and suggestions regarding this important dimension of chemistry in our world. In addition we would like to thank Professors Muhammed Ali (Howard University), Charles Anderson (Pacific Lutheran University), Roman Dembinski (Oakland University), Michael Gelb (University of Washington), and Andreas Zavitsas (Long Island University). Each of them offered specific suggestions and advice that enriched the 8th edition manuscript. We are grateful to Alan Shusterman (Reed College) and Warren Hehre (Wavefunction, Inc.) for earlier assistance regarding explanations of electrostatic potential maps and other calculated molecular models. We would also like to thank those scientists who allowed us to use or adapt figures from their research as illustrations for a number of the new topics in the 8th edition of our book.

Special thanks goes to Professor Darrell Woodman (University of Washington) for creating outstanding computer-based pedagogical tools that accompany the 8th edition. His Interactive Exercises, Concept Units, and other materials on the OrganicView CD and website are first rate. Students and faculty will find Professor Woodman's new and preexisting resources enjoyable and effective tools for learning. We would also like to thank Professor K.C. Russell (Northern Kentucky University) for converting end-of-chapter problems to eGrade format for web delivery.

Without the excellent work of the people at Wiley it would not be possible to produce a book of this scope. Johnna Barto, Associate Director of Development, provided a steadfast eye for detail and consistency, as well as background support in a variety of other key ways. Karin Kincheloe created the striking new design of the 8th edition. Photo Editor Lisa Gee helped obtained stunning images for the book. Illustration Editor Sandra Rigby guided the massive task of completely recasting the art program for the 8th edition. David Harris began editorial oversight of the book, which was later adroitly assumed by Debbie Brennan and Kevin Molloy. Production Editor Elizabeth Swain orchestrated production of the 8th edition with her characteristic skill and efficiency. Martin Batey oversaw production of the OrganicView CD, and Tom Kulesa supported development of eGrade resources for the book. We are thankful to all of these people at Wiley for the skills and dedication they provided to bring this book to fruition.

CBF would like to thank his faculty and administrative colleagues at Pacific Lutheran University for their support during preparation of the 8th edition. He is also grateful to students and former mentors for what they have taught him about teaching. He thanks his parents for decades of support. And, last yet first, CBF would like to thank his wife Deanna, and daughters Lauren and Heather. Their patience during preparation of the 8th edition was immeasurable, and their encouragement essential.

T. W. Graham Solomons
Craig B. Fryhle

Contents

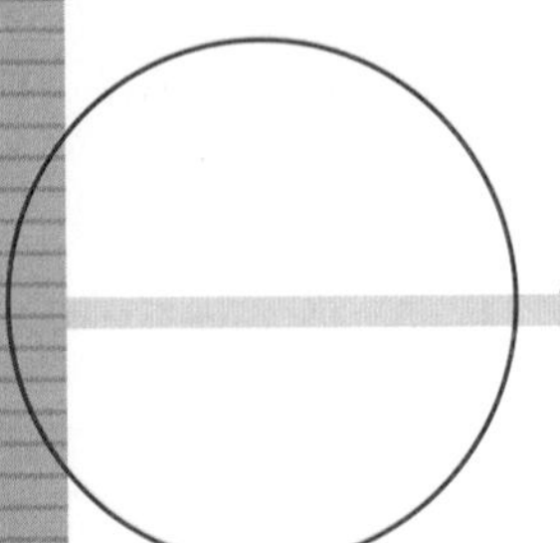

About the Authors

T. W. Graham Solomons

T. W. Graham Solomons did his undergraduate work at The Citadel and received his doctorate in organic chemistry in 1959 from Duke University where he worked with C. K. Bradsher. Following this he was a Sloan Foundation Postdoctoral Fellow at the University of Rochester where he worked with V. Boekelheide. In 1960 he became a charter member of the faculty of the University of South Florida and became Professor of Chemistry in 1973. In 1992 he was made Professor Emeritus. In 1994 he was a visiting professor with the Faculté des Sciences Pharmaceutiques et Biologiques, Université René Descartes (Paris V). He is a member of Sigma Xi, Phi Lambda Upsilon, and Sigma Pi Sigma. He has received research grants from the Research Corporation and the American Chemical Society Petroleum Research Fund. For several years he was director of an NSF-sponsored Undergraduate Research Participation Program at USF. His research interests have been in the areas of heterocyclic chemistry and unusual aromatic compounds. He has published papers in the *Journal of the American Chemical Society,* the *Journal of Organic Chemistry,* and the *Journal of Heterocyclic Chemistry.* He has received several awards for distinguished teaching. His organic chemistry textbooks have been widely used for 20 years and have been translated into French, Japanese, Chinese, Korean, Malaysian, Arabic, Portuguese, Spanish, Turkish, and Italian. He and his wife Judith have a daughter who is a building conservator, a son who is an artist, and another son who is a graduate student studying biochemistry.

Craig Barton Fryhle

Craig Barton Fryhle is Chair and Professor of Chemistry at Pacific Lutheran University. He earned his B.A. degree from Gettysburg College and Ph.D. from Brown University. His experiences at these institutions shaped his dedication to mentoring undergraduate students in chemistry and the liberal arts, which is a passion that burns strongly for him. His research interests have been in areas relating to the shikimic acid pathway, including molecular modeling and NMR spectrometry of substrates and analogues, as well as structure and reactivity studies of shikimate pathway enzymes using isotopic labeling and mass spectrometry. He has mentored many students in undergraduate research, a number of whom have later earned their Ph.D. degrees and gone on to academic or industrial positions. He has participated in workshops on fostering undergraduate participation in research, and has been an invited participant in efforts by the National Science Foundation to enhance undergraduate research in chemistry. He has received research and instrumentation grants from the National Science Foundation, the M J. Murdock Charitable Trust, and other private foundations. His work in chemical education, in addition to textbook co-authorship, involves incorporation of student-led teaching in the classroom and technology-based strategies in organic chemistry. He has also developed experiments for undergraduate students in organic laboratory and instrumental analysis courses. He has been a volunteer with the hands-on science program in Seattle public schools, and Chair of the Puget Sound Section of the American Chemical Society. He lives in Seattle with his wife and two daughters.

To the Student

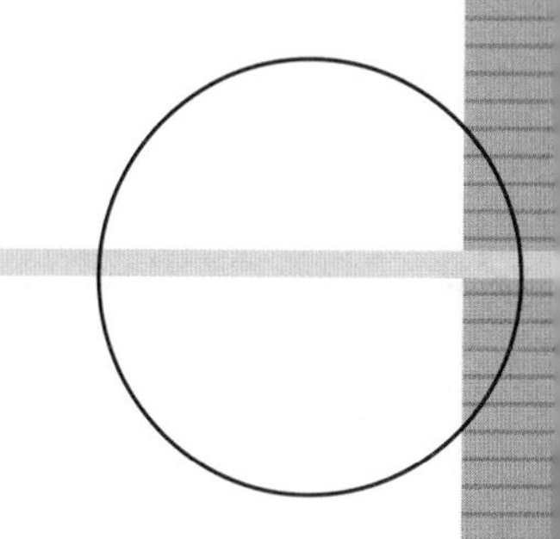

Contrary to what you may have heard, organic chemisty does not have to be a difficult course. It will be a rigorous course, and it will offer a challenge. But you will learn more in it than in almost any course you will take—and what you learn will have a special relevance to life and the world around you. However, because organic chemistry can be approached in a logical and systematic way, you will find that with the right study habits, mastering organic chemistry can be a deeply satisfying experience. Here, then, are some suggestions about how to study:

1. **Keep up with your work from day to day—never let yourself get behind.** Organic chemistry is a course in which one idea almost always builds on another that has gone before. It is essential, therefore, that you keep up with, or better yet, be a little ahead of your instructor. Ideally, you should try to stay one day ahead of your instructor's lectures in your own class preparations. The lecture, then, will be much more helpful because you will already have some understanding of the assigned material. Your time in class will clarify and expand ideas that are already familiar ones.

2. **Study material in small units, and be sure that you understand each new section before you go on to the next.** Again, because of the cumulative nature of organic chemistry, your studying will be much more effective if you take each new idea as it comes and try to understand it completely before you move on to the next concept. Many key concepts are emphasized by *Toolbox icons* in the margin and their accompanying captions. These concepts, once you learn them, will be part of your toolbox for success in organic chemistry. Similarly, tips or suggestions for studying and thinking about organic chemistry are highlighted by *Study Tip icons* and captions. Whether or not a concept is highlighted by a Toolbox or Study Tip icon, be sure you understand it before moving on.

3. **Work all of the in-chapter and assigned problems.** One way to check your progress is to work each of the in-chapter problems when you come to it. These problems have been written just for this purpose and are designed to help you decide whether or not you understand the material that has just been explained. If you can work the in-chapter problem, then you should go on; if you cannot, then you should go back and study the preceding material again. Work all of the problems assigned by your instructor from the end of the chapter, as well. Do all of your problems in a notebook and bring this book with you when you go to see your instructor for extra help.

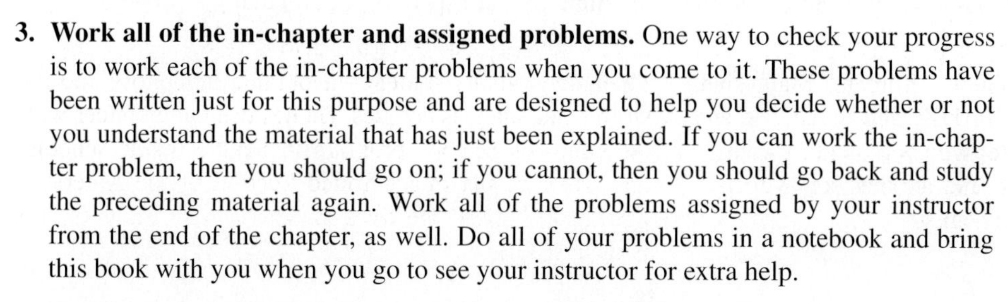

4. **Write when you study.** Write the reactions, mechanisms, structures, and so on, over and over again. Organic chemistry is best assimilated through the fingertips by writing, and not through the eyes by simply looking, or by highlighting material in the text, or by referring to flash cards. There is a good reason for this. Organic structures, mechanisms, and reactions are complex. If you simply examine them, you may think you understand them thoroughly, but that will be a misperception. The reaction mechanism may make sense to you in a certain way, but you need a deeper understanding than this. You need to know the material so thoroughly that you can explain it to someone else. This level of understanding comes to most of us (those of us without photographic memories) through writing. Only by writing the reaction mechanisms do we

pay sufficient attention to their details, such as which atoms are connected to which atoms, which bonds break in a reaction and which bonds form, and the three-dimensional aspects of the structures. When we write reactions and mechanisms, connections are made in our brains that provide the long-term memory needed for success in organic chemistry. We virtually guarantee that your grade in the course will be directly proportional to the number of pages of paper that your fill with your own writing in studying during the term.

5. **Learn by teaching and explaining.** Study with your student peers and practice explaining concepts and mechanisms to each other. Use the *Learning Group Problems* and other exercises your instructor may assign as vehicles for teaching and learning interactively with your peers.

6. **Use the answers to the problems in the *Study Guide* in the proper way.** Refer to the answers only in two circumstances: (1) When you have finished a problem, use the Study Guide to check your answer. (2) When, after making a real effort to solve the problem, you find that you are completely stuck, then look at the answer for a clue and go back to work out the problem on your own. The value of a problem is in solving it. If you simply read the problem and look up the answer, you will deprive yourself of an important way to learn.

7. **Use the introductory material in the *Study Guide* entitled "Solving the Puzzle—or—Structure Is Everything (Almost)"** as a bridge from general chemistry to your beginning study of organic chemistry. You might find this section helpful as a way to see the relevance to organic chemistry of some of the concepts you learned in general chemistry, while at the same time refreshing these ideas in your mind and gearing you up to study organic chemistry. It is also meant to help you see that an understanding of certain fundamental principles, largely having to do with structure, can help immensely to reduce the complexity of the puzzle you may feel lies ahead of you. Indeed, once you have a firm understanding of structure, the puzzle of organic chemistry can become one of very manageable size and comprehensible pieces.

8. **Use molecular models when you study.** Because of the three-dimensional nature of most organic molecules, molecular models can be an invaluable aid to your understanding of them. When you need to see the three-dimensional aspect of a particular topic, use the Molecular Visions™ model set that may have been packaged with your textbook, or buy a set of models separately. An appendix to the *Study Guide* that accompanies this text provides a set of highly useful molecular model exercises.

Carbon Compounds and Chemical Bonds

"We are Stardust," J. Mitchell

We have stardust in our eyes and, for that matter, in every part of our bodies. According to the Big Bang Theory, all matter in the universe is derived from one gigantic explosion that released a sea of subatomic particles and radiation. In the seconds and minutes that followed, these raw materials coalesced to form protons, neutrons, and electrons. From these were formed hydrogen and deuterium atoms and then atoms of helium. Eventually interaction between these atomic and subatomic particles created heavier elements, and gravitational forces, collisions, and cosmic energy sources led to combination of atoms into simple molecules and of molecules into great molecular clouds.

Even though the interstellar medium consists of approximately 90% hydrogen and 10% helium, simple organic molecules formed from the remaining 0.1% carbon, nitrogen, and oxygen. Cooler, dense molecular clouds contain a variety of common classes of organic compounds, including alcohols, aldehydes, amines, ethers, ketones, nitriles, and long, linear molecules called cyanopolyynes. Over 100 molecular species have been identified, including familiar compounds such as acetylene (C_2H_2), ethylene (C_2H_4), hydrogen cyanide (HCN), methanol, and diethyl ether.* Meanwhile, hotter, diffuse molecular clouds produced even more complex species, including graphite, fullerenes, and perhaps even diamonds.

Comets, which are aggregates of frozen gases and dust, carry a large variety of organic molecules. Scientists studying Comet Halley have found evidence for most of the classes of organic compounds mentioned above, as well as more highly evolved com-

---

*Ehrenfreund, P.; Charnley, S. B. Organic Molecules in the Interstellar Medium, Comets and Meteorites: A Voyage from Dark Clouds to the Early Earth. *Ann. Rev. Astron. Astrophys.* **2000,** *38,* 427–483.

pounds related to those in nucleic acids, the building blocks of deoxyribonucleic acid (DNA) and ribonucleic acid (RNA). Meteorites have also been shown to contain a diverse collection of organic molecules, including simple amino acids, the building blocks of proteins. Of course, it is not likely that life or the more complex molecules required for life evolved on comets or meteors, but there has been speculation that comet and meteorite impacts on the primordial Earth delivered raw organic materials that may have facilitated the evolution of life on Earth. Whether or not comets and meteorites delivered the materials, there is little doubt that all of the atoms that comprise our bodies and the world around us ultimately have their origin in stars. We need only gaze at the stars to see the origin of chemistry on Earth.

1.1 Introduction

Organic chemistry is the chemistry of the compounds of carbon. The compounds of carbon are central to life on this planet. Carbon compounds include DNAs, the giant helical molecules that contain all of our genetic information. They include the proteins that catalyze all of the reactions in our body and that constitute the essential compounds of our blood, muscle, and skin. Together with oxygen from the air we breathe, carbon compounds furnish the energy that sustains life.

Methane

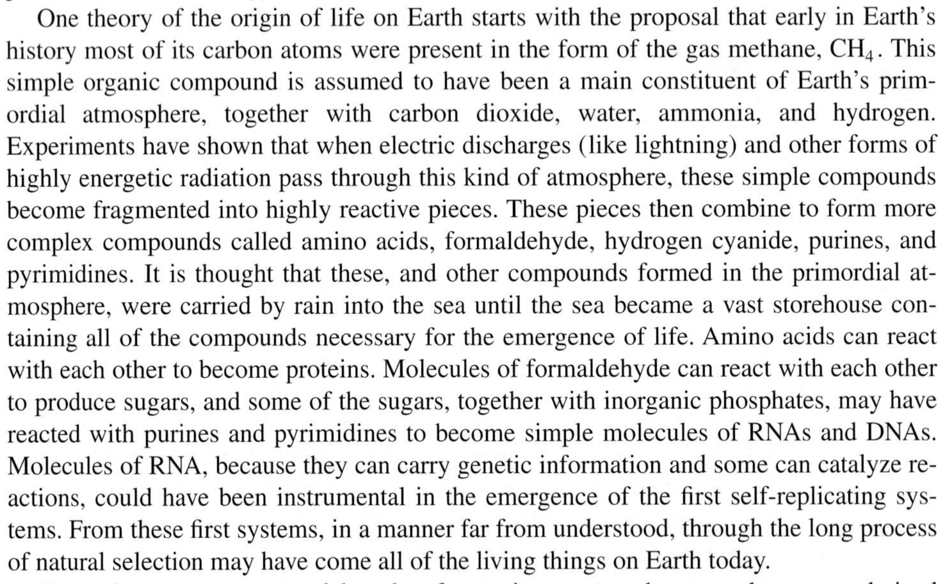

One theory of the origin of life on Earth starts with the proposal that early in Earth's history most of its carbon atoms were present in the form of the gas methane, CH_4. This simple organic compound is assumed to have been a main constituent of Earth's primordial atmosphere, together with carbon dioxide, water, ammonia, and hydrogen. Experiments have shown that when electric discharges (like lightning) and other forms of highly energetic radiation pass through this kind of atmosphere, these simple compounds become fragmented into highly reactive pieces. These pieces then combine to form more complex compounds called amino acids, formaldehyde, hydrogen cyanide, purines, and pyrimidines. It is thought that these, and other compounds formed in the primordial atmosphere, were carried by rain into the sea until the sea became a vast storehouse containing all of the compounds necessary for the emergence of life. Amino acids can react with each other to become proteins. Molecules of formaldehyde can react with each other to produce sugars, and some of the sugars, together with inorganic phosphates, may have reacted with purines and pyrimidines to become simple molecules of RNAs and DNAs. Molecules of RNA, because they can carry genetic information and some can catalyze reactions, could have been instrumental in the emergence of the first self-replicating systems. From these first systems, in a manner far from understood, through the long process of natural selection may have come all of the living things on Earth today.

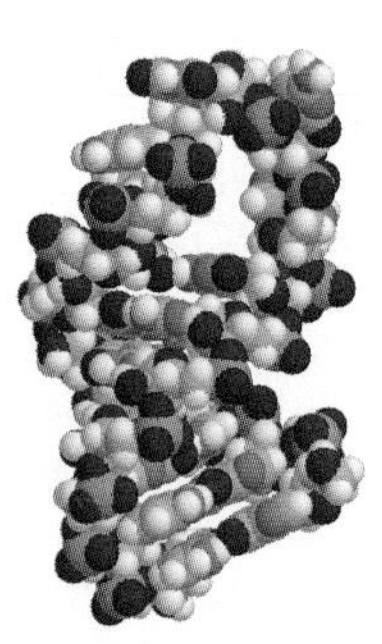

An RNA molecule

Not only are we composed largely of organic compounds, not only are we derived from and nourished by them, we also live in a time when the central role of organic chemistry in medicine, bioengineering, nanotechnology, and other disciplines is more apparent than ever. The clothing we wear, whether a natural substance such as wool or cotton or a synthetic such as nylon or a polyester, is made of carbon compounds. Many of the materials that go into the houses that shelter us are organic. The gasoline that propels our automobiles, the rubber of their tires, and the plastic of their interiors are all organic. Most of the medicines that help us cure diseases and relieve suffering are organic.

Organic chemicals are also factors in some of our most serious problems. Many of the organic chemicals introduced into the environment in years past have had consequences far beyond those originally intended. In a number of these situations, however, methods that are more environment-friendly are being developed. For example, natural insect attractants called pheromones are now often used to lure insects into traps, instead of widespread pesticide application. International negotiations to replace with less reactive compounds organic refrigerants and aerosol propellants that destroy Earth's ozone layer hold promise for a worldwide rescue of our outer atmosphere. Cars with more efficient combustion engines use less gasoline and help reduce the massive amount of automobile pollutants that foul the air and contribute to the greenhouse effect.

Inasmuch as this is an age where many disciplines are linked by organic chemistry, it is also an age where our working philosophy is reduce, reuse, and recycle. Organic compounds play a big part in this effort. Plastics that were once bottles for soft drinks and milk are recycled into fabric and carpet. Paper recycling allows harvesting of fewer trees for pulp. Engine oils, paints, and solvents are collected by environmental agencies and recycled. Chemistry laboratories conduct experiments on smaller and smaller scales, thereby using less material and generating less waste. Recognizing and encouraging these efforts, the U.S. Environmental Protection Agency honors the most effective innovators with Presidential Green Chemistry Awards. On all fronts, chemists are developing environment-friendly procedures for the world's benefit.

Clearly, organic chemistry is associated with nearly every aspect of our existence. We would be wise to understand it as best we can.

1.2 The Development of Organic Chemistry as a Science

Humans have used organic compounds and their reactions for thousands of years. Their first deliberate experience with an organic reaction probably dates from their discovery of fire. The ancient Egyptians used organic compounds (indigo and alizarin) to dye cloth. The famous "royal purple" used by the Phoenicians was also an organic substance, obtained from mollusks. The fermentation of grapes to produce ethyl alcohol and the acidic qualities of "soured wine" are both described in the Bible and were probably known earlier.

As a science, organic chemistry is less than 200 years old. Most historians of science date its origin to the early part of the nineteenth century, a time in which an erroneous belief was dispelled.

1.2A Vitalism

During the 1780s scientists began to distinguish between **organic compounds** and **inorganic compounds.** Organic compounds were defined as compounds that could be obtained from *living organisms.* Inorganic compounds were those that came from *nonliving sources.* Along with this distinction, a belief called "vitalism" grew. According to this idea, the intervention of a "vital force" was necessary for the synthesis of an organic compound. Such synthesis, chemists held then, could take place only in living organisms. It could not take place in the flasks of a chemistry laboratory.

Between 1828 and 1850 a number of compounds that were clearly "organic" were synthesized from sources that were clearly "inorganic." The first of these syntheses was accomplished by Friedrich Wöhler in 1828. Wöhler found that the organic compound urea (a constituent of urine) could be made by evaporating an aqueous solution containing the inorganic compound ammonium cyanate:

$$\underset{\textbf{Ammonium cyanate}}{NH_4^+NCO^-} \xrightarrow{\text{heat}} \underset{\textbf{Urea}}{H_2N-\overset{\overset{\displaystyle O}{\|}}{C}-NH_2}$$

Although vitalism disappeared slowly from scientific circles after Wöhler's synthesis, its passing made possible the flowering of the science of organic chemistry that has occurred since 1850.

Despite the demise of vitalism in science, the word "organic" is still used today by some people to mean "coming from living organisms" as in the terms "organic vitamins" and "organic fertilizers." The commonly used term "organic food" means that the food was grown without the use of synthetic fertilizers and pesticides. An "organic vitamin" means to these people that the vitamin was isolated from a natural source and not synthesized by a chemist. While there are sound arguments to be made against using food contaminated with certain pesticides, while there may be environmental benefits to be obtained from organic farming, and while "natural" vitamins may contain beneficial substances not present in synthetic vitamins, it is impossible to argue that pure "natural" vitamin C, for example, is healthier than pure "synthetic" vitamin C, since the two substances are identical in all respects. In science today, the study of compounds from living organisms is called natural products chemistry.

Vitamin C

Vitamin C is found in various citrus fruits.

1.2B Empirical and Molecular Formulas

In the eighteenth and nineteenth centuries extremely important advances were made in the development of qualitative and quantitative methods for analyzing organic substances. In 1784 Antoine Lavoisier first showed that organic compounds were composed primarily of carbon, hydrogen, and oxygen. Between 1811 and 1831, *quantitative* methods for determining the composition of organic compounds were developed by Justus Liebig, J. J. Berzelius, and J. B. A. Dumas.

A great confusion was dispelled in 1860 when Stanislao Cannizzaro showed that the earlier hypothesis of Amedeo Avogadro (1811) could be used to distinguish between **empirical** and **molecular formulas.** As a result, many molecules that had appeared earlier to have the same formula were seen to be composed of different numbers of atoms. For example, ethene, cyclopentane, and cyclohexane all have the same empirical formula: CH_2. However, they have molecular formulas of C_2H_4, C_5H_{10}, and C_6H_{12}, respectively. Appendix A of the Study Guide that accompanies this book contains a review of how empirical and molecular formulas are determined and calculated.

1.3 The Structural Theory of Organic Chemistry

Between 1858 and 1861, August Kekulé, Archibald Scott Couper, and Alexander M. Butlerov, working independently, laid the basis for one of the most fundamental theories in chemistry: the **structural theory.**

Two central premises are fundamental:

1. The atoms in organic compounds can form a fixed number of bonds. The measure of this ability is called **valence.** Carbon is *tetravalent;* that is, carbon atoms form four bonds. Oxygen is *divalent,* and hydrogen and (usually) the halogens are *monovalent:*

Carbon atoms are tetravalent **Oxygen atoms are divalent** **Hydrogen and halogen atoms are monovalent**

Knowing the number of bonds an atom typically forms is a basic tool for learning organic chemistry.

2. A carbon atom can use one or more of its valences to form bonds to other carbon atoms:

Carbon–carbon bonds

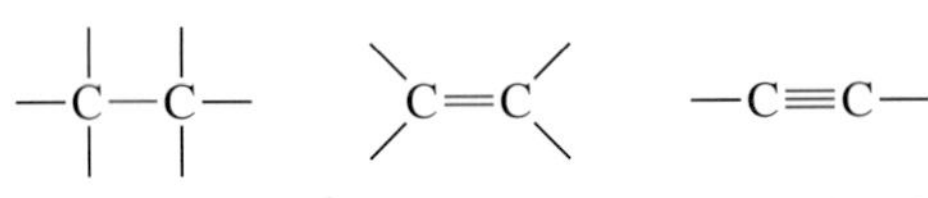

Single bond **Double bond** **Triple bond**

TABLE 1.1 Properties of Ethyl Alcohol and Dimethyl Ether

	Ethyl Alcohol C_2H_6O	Dimethyl Ether C_2H_6O
Boiling point (°C)	78.5	−24.9
Melting point (°C)	−117.3	−138

In his original publication Couper represented these bonds by lines much in the same way that most of the formulas in this book are drawn. In his textbook (published in 1861), Kekulé gave the science of organic chemistry its modern definition: ***a study of the compounds of carbon.***

1.3A Isomers: The Importance of Structural Formulas

The structural theory allowed early organic chemists to begin to solve a fundamental problem that plagued them: the problem of **isomerism.** These chemists frequently found examples of **different compounds that have the same molecular formula.** Such compounds are called **isomers.**

STUDY TIP!

Terms and concepts that are fundamentally important to your learning organic chemistry are set in bold **green** type. You should learn them as they are introduced. These terms are also defined in the glossary.

Let us consider an example. There are two compounds with the molecular formula C_2H_6O that are clearly different because they have different properties (see Table 1.1). These compounds, therefore, are classified as being isomers of one another; they are said to be **isomeric.** Notice that these two isomers have different boiling points, and, because of this, one isomer, called *dimethyl ether,* is a gas at room temperature; the other isomer, called *ethyl alcohol,* is a liquid at room temperature. The two isomers also have different melting points.

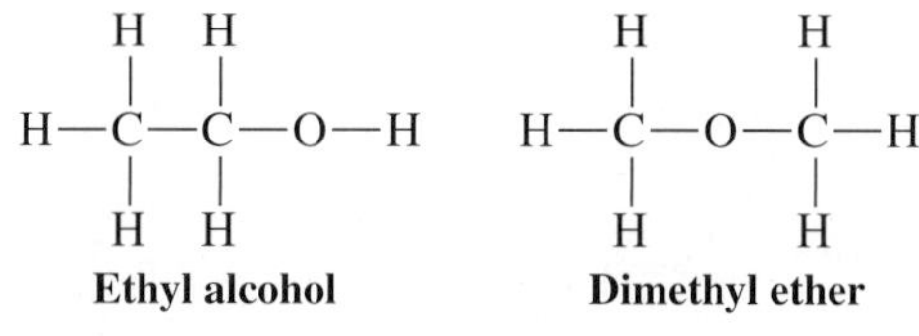

Ethyl alcohol **Dimethyl ether**

Build hand-held models of these compounds and compare their structures.

Because the molecular formula (C_2H_6O) for these two compounds is the same, it gives us no basis for understanding the differences between them. The structural theory remedies this situation, however. It does so by giving us different **structures** (Fig. 1.1) and different **structural formulas** for the two compounds.

One glance at the structural formulas for these two compounds reveals their difference. The two compounds differ in the **connectivity** of their atoms: The atoms of ethyl alcohol are connected in a way that is different from those of dimethyl ether. In ethyl alcohol there is a C—C—O linkage; in dimethyl ether the linkage is C—O—C. Ethyl

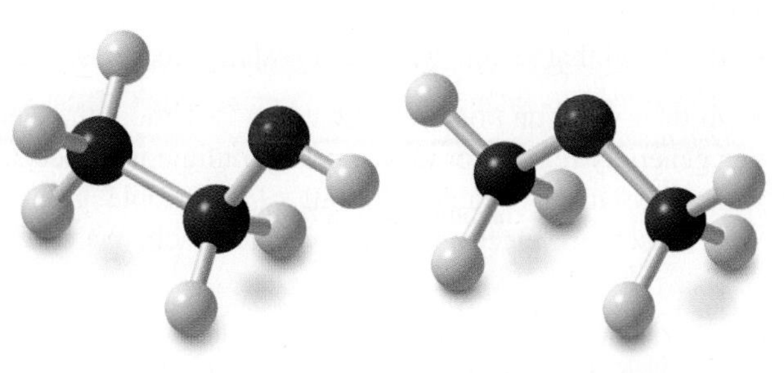

FIGURE 1.1 Ball-and-stick models show the different structures of ethyl alcohol and dimethyl ether.

Chem3D Models

This icon denotes molecular models or other features that are provided on the CD with this book.

alcohol has a hydrogen atom attached to oxygen; in dimethyl ether all of the hydrogen atoms are attached to carbon. It is the hydrogen atom covalently bonded to oxygen in ethyl alcohol that accounts for the fact that ethyl alcohol is a liquid at room temperature. As we shall see in Section 2.14C, this hydrogen atom allows molecules of ethyl alcohol to form hydrogen bonds to each other and gives ethyl alcohol a boiling point much higher than that of dimethyl ether.

Constitutional isomers

Ethyl alcohol and dimethyl ether are examples of what are now called **constitutional isomers.*** ***Constitutional isomers are different compounds that have the same molecular formula but differ in their connectivity, that is, in the sequence in which their atoms are bonded together.*** Constitutional isomers usually have different physical properties (e.g., melting point, boiling point, and density) and different chemical properties. The differences, however, may not always be as large as those between ethyl alcohol and dimethyl ether.

1.3B The Tetrahedral Shape of Methane

In 1874, the structural formulas originated by Kekulé, Couper, and Butlerov were expanded into three dimensions by the independent work of J. H. van't Hoff and J. A. Le Bel. van't Hoff and Le Bel proposed that the four bonds of the carbon atom in methane, for example, are arranged in such a way that they would point toward the corners of a regular tetrahedron, the carbon atom being placed at its center (Fig. 1.2). The necessity for knowing the arrangement of the atoms in space, taken together with an understanding of the order in which they are connected, is central to an understanding of organic chemistry, and we shall have much more to say about this later, in Chapters 4 and 5.

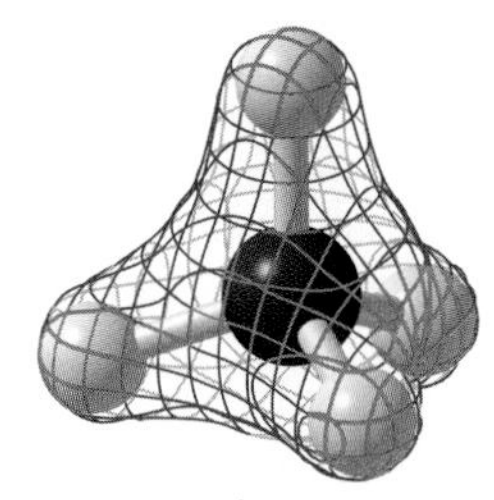

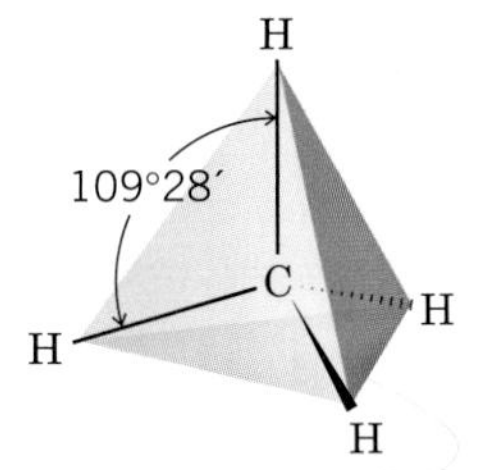

FIGURE 1.2 The tetrahedral structure of methane. Bonding electrons in methane principally occupy the space within the wire mesh.

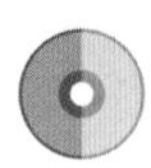

CD Tutorial

Methane models

Chem3D Model, Animated Graphics

1.4 Chemical Bonds: The Octet Rule

G. N. Lewis

The first explanations of the nature of chemical bonds were advanced by G. N. Lewis (of the University of California, Berkeley) and W. Kössel (of the University of Munich) in 1916. Two major types of chemical bonds were proposed.

1. the **ionic** (or electrovalent) bond, formed by the transfer of one or more electrons from one atom to another to create ions and
2. the **covalent** bond, a bond that results when atoms share electrons.

The central idea in their work on bonding is that atoms without the electronic configuration of a noble gas generally react to produce such a configuration because these configurations are known to be highly stable. For all of the noble gases except helium, this means achieving an octet of electrons in the valence shell. We call the tendency for

*An older term for isomers of this type was **structural isomers.** The International Union of Pure and Applied Chemistry (IUPAC) now recommends that use of the term "structural" when applied to isomers of this type be abandoned.

an atom to achieve a configuration where its valence shell contains eight electrons the **octet rule.**

The concepts and explanations that arise from the original propositions of Lewis and Kössel are satisfactory for explanations of many of the problems we deal with in organic chemistry today. For this reason we shall review these two types of bonds in more modern terms.

1.4A Ionic Bonds

Atoms may gain or lose electrons and form charged particles called *ions.* An ionic bond is an attractive force between oppositely charged ions. One source of such ions is a reaction between atoms of widely differing electronegativities (Table 1.2). ***Electronegativity is a measure of the ability of an atom to attract electrons.*** Notice in Table 1.2 that electronegativity increases as we go across a horizontal row of the periodic table from left to right and that it increases as we go up a vertical column.

We will use electronegativity frequently as a tool for understanding the properties and reactivity of organic molecules.

TABLE 1.2 Electronegativities of Some of the Elements

Increasing electronegativity →

			H				
			2.1				
Li	Be	B	C	N	O	F	↑ Increasing electronegativity
1.0	1.5	2.0	2.5	3.0	3.5	4.0	
Na	Mg	Al	Si	P	S	Cl	
0.9	1.2	1.5	1.8	2.1	2.5	3.0	
K						Br	
0.8						2.8	

An example of the formation of an ionic bond is the reaction of lithium and fluorine atoms:

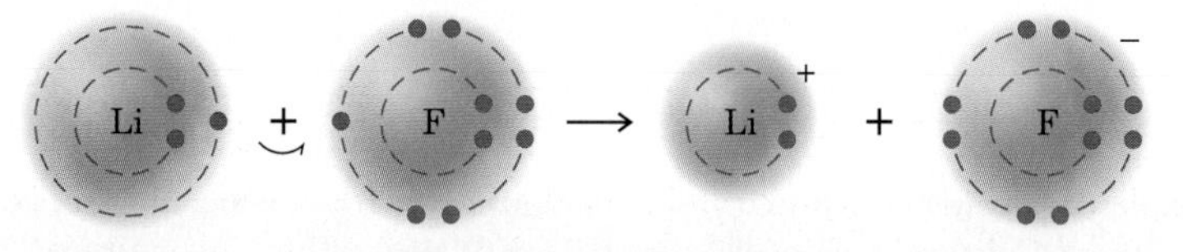

CD Tutorial

Ionic and covalent bonding

Lithium, a typical metal, has a very low electronegativity; fluorine, a nonmetal, is the most electronegative element of all. The loss of an electron (a negatively charged species) by the lithium atom leaves a lithium cation (Li^+); the gain of an electron by the fluorine atom gives a fluoride anion (F^-). Why do these ions form? In terms of the Lewis–Kössel theory, both atoms achieve the electronic configuration of a noble gas by becoming ions. The lithium cation with two electrons in its valence shell is like an atom of the noble gas helium, and the fluoride anion with eight electrons in its valence shell is like an atom of the noble gas neon. Moreover, crystalline lithium fluoride forms from the individual lithium and fluoride ions. In this process negative fluoride ions become surrounded by positive lithium ions, and positive lithium ions by negative fluoride ions. In this crystalline state, the ions have substantially lower energies than the atoms from which they have been formed. Lithium and fluorine are thus "stabilized" when they react to form crystalline lithium fluoride.

We represent the formula for lithium fluoride as LiF, because that is the simplest formula for this ionic compound.

Ionic substances, because of their strong internal electrostatic forces, are usually very high melting solids, often having melting points above 1000°C. In polar solvents, such as water, the ions are solvated (see Section 2.14E), and such solutions usually conduct an electric current.

Ionic compounds, often called **salts,** form only when atoms of very different electronegativities transfer electrons to become ions.

1.4B Covalent Bonds

When two or more atoms of the same or similar electronegativities react, a complete transfer of electrons does not occur. In these instances the atoms achieve noble gas configurations by *sharing electrons. Covalent* bonds form between the atoms, and the products are called *molecules.* Molecules may be represented by electron-dot formulas or, more conveniently, by dash formulas, where each dash represents a pair of electrons shared by two atoms. Some examples are shown here:

$$H_2 \quad H\cdot + \cdot H \longrightarrow H{:}H \quad \text{or} \quad H{-}H$$

$$Cl_2 \quad :\ddot{\underset{..}{Cl}}\cdot + \cdot\ddot{\underset{..}{Cl}}: \longrightarrow :\ddot{\underset{..}{Cl}}{:}\ddot{\underset{..}{Cl}}: \quad \text{or} \quad :\ddot{\underset{..}{Cl}}{-}\ddot{\underset{..}{Cl}}:$$

$$CH_4 \quad \cdot\dot{\underset{\cdot}{C}}\cdot + 4\,H\cdot \longrightarrow H{:}\overset{H}{\underset{H}{\ddot{\underset{..}{C}}}}{:}H \quad \text{or} \quad H{-}\overset{H}{\underset{H}{\overset{|}{\underset{|}{C}}}}{-}H$$

These formulas are often called Lewis structures; in writing them we show only the electrons of the valence shell.

In certain cases, multiple covalent bonds are formed; for example,

$$N_2 \quad :N\vdots\vdots N: \quad \text{or} \quad :N{\equiv}N:$$

and ions themselves may contain covalent bonds:

$$NH_4^+ \quad H{:}\overset{H}{\underset{H}{\ddot{\underset{..}{N}}}}{}^{+}{:}H \quad \text{or} \quad H{-}\overset{H}{\underset{H}{\overset{|}{\underset{|}{N^{+}}}}}{-}H$$

1.5 Writing Lewis Structures

When we write **Lewis structures** (electron-dot formulas) we assemble the molecule or ion from the constituent atoms showing only the valence electrons (i.e., the electrons of the outermost shell). By having the atoms share or transfer electrons, we try to give each atom the electronic configuration of the noble gas in the same horizontal row of the periodic table. For example, we give hydrogen atoms two electrons because by doing so we give them the structure of helium. We give carbon, nitrogen, oxygen, and fluorine atoms eight electrons because by doing this we give them the electronic configuration of neon and satisfy the octet rule. **The number of valence electrons of an atom can be obtained from the periodic table because it is equal to the group number of the atom.** (A periodic table is provided inside the back cover of this book.) Carbon, for example, is in Group **4A** and it has four valence electrons; fluorine, in Group **7A,** has seven; hydrogen, in Group **1A,** has one. **If the structure is an ion, we add or subtract electrons to give it the proper charge.**

The ability to write proper **Lewis structures** is one of the most important tools for learning organic chemistry.

SOLVED PROBLEM

Write the Lewis structure of CH_3F.

ANSWER

1. We find the total number of valence electrons of all the atoms:

$$\underset{\text{C}}{4} + \underset{\text{3 H}}{3(1)} + \underset{\text{F}}{7} = 14$$

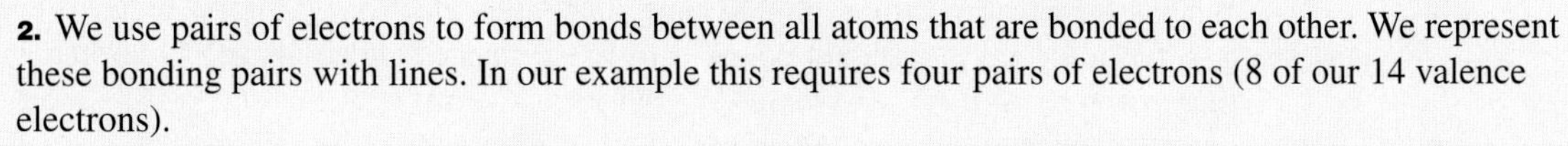

2. We use pairs of electrons to form bonds between all atoms that are bonded to each other. We represent these bonding pairs with lines. In our example this requires four pairs of electrons (8 of our 14 valence electrons).

```
     H
     |
  H—C—F
     |
     H
```

3. We then add the remaining electrons in pairs so as to give each hydrogen 2 electrons (a duet) and every other atom 8 electrons (an octet). In our example, we assign the remaining 6 valence electrons to the fluorine atom in three nonbonding pairs.

```
     H
     |    ..
  H—C—F:
     |    ..
     H
```

SOLVED PROBLEM

Write the Lewis structure for the chlorate ion (ClO_3^-). (*Note:* The chlorine atom is bonded to all three oxygen atoms.)

ANSWER

1. We find the total number of valence electrons of all the atoms including the extra electron needed to give the ion a negative charge:

$$\underset{\text{Cl}}{7} + \underset{\text{3 O}}{3(6)} + \underset{e^-}{1} = 26$$

2. We use three pairs of electrons to form bonds between the chlorine atom and the three oxygen atoms:

```
      O
      |
  O—Cl—O
```

3. We then add the remaining 20 electrons in pairs so as to give each atom an octet:

```
 [      ..       ]-
 [     :O:       ]
 [  ..  |   ..   ]
 [ :O—Cl—O:      ]
 [  ..  ..  ..   ]
```

If necessary, we use multiple bonds to satisfy the octet rule (i.e., give atoms the noble gas configuration). The carbonate ion (CO_3^{2-}) illustrates this:

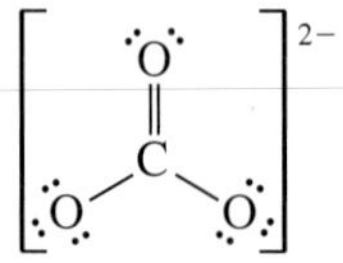

$$\left[\begin{array}{c} \ddot{\text{O}} \\ \| \\ \text{C} \\ :\ddot{\text{O}}\diagup \quad \diagdown\ddot{\text{O}}: \end{array}\right]^{2-}$$

The organic molecules ethene (C_2H_4) and ethyne (C_2H_2) have a double and triple bond, respectively:

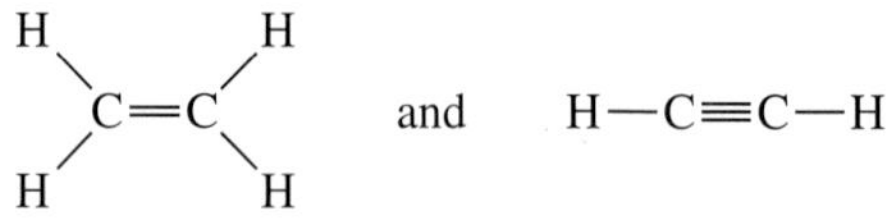

$$\begin{array}{c} \text{H} \quad\quad \text{H} \\ \diagdown \quad \diagup \\ \text{C}{=}\text{C} \\ \diagup \quad \diagdown \\ \text{H} \quad\quad \text{H} \end{array} \quad \text{and} \quad \text{H}{-}\text{C}{\equiv}\text{C}{-}\text{H}$$

1.6 EXCEPTIONS TO THE OCTET RULE

Atoms share electrons, not just to obtain the configuration of an inert gas, but because sharing electrons produces increased electron density between the positive nuclei. The resulting attractive forces of nuclei for electrons is the "glue" that holds the atoms together (cf. Section 1.10). Elements of the second period of the periodic table can have a maximum of four bonds (i.e., have eight electrons around them) because these elements have only one 2*s* and three 2*p* orbitals available for bonding. Each orbital can contain two electrons, and a total of eight electrons fills these orbitals (Section 1.10). The octet rule, therefore, only applies to these elements, and even here, as we shall see in compounds of beryllium and boron, fewer than eight electrons are possible. Elements of the third period and beyond have *d* orbitals that can be used for bonding. These elements can accommodate more than eight electrons in their valence shells and therefore can form more than four covalent bonds. Examples are compounds such as PCl_5 and SF_6.

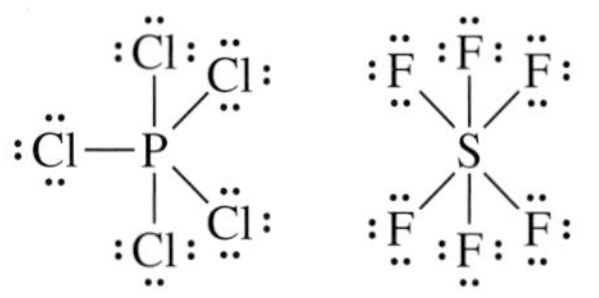

[Lewis structures of PCl_5 (P bonded to five :Cl: atoms) and SF_6 (S bonded to six :F: atoms)]

SOLVED PROBLEM

Write a Lewis structure for the sulfate ion (SO_4^{2-}). (*Note:* The sulfur atom is bonded to all four oxygen atoms.)

ANSWER

1. We find the total number of valence electrons including the extra 2 electrons needed to give the ion the double negative charge:

$$\begin{array}{ccccccc} 6 & + & 4(6) & + & 2 & = & 32 \\ \uparrow & & \uparrow & & \uparrow & & \\ \text{S} & & 4\,\text{O} & & 2e^- & & \end{array}$$

2. We use four pairs of electrons to form bonds between the sulfur atom and the four oxygen atoms:

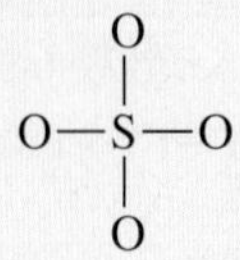

$$\begin{array}{c} \text{O} \\ | \\ \text{O}{-}\text{S}{-}\text{O} \\ | \\ \text{O} \end{array}$$

3. We add the remaining 24 electrons as unshared pairs on oxygen atoms and as double bonds between the sulfur atom and two oxygen atoms. This gives each oxygen 8 electrons and the sulfur atom 12:

$$\left[\begin{array}{c} \ddot{O} \\ \| \\ :\ddot{O}-S-\ddot{O}: \\ \| \\ \ddot{O} \end{array} \right]^{2-}$$

Some highly reactive molecules or ions have atoms with fewer than eight electrons in their outer shell. An example is boron trifluoride (BF_3). In a BF_3 molecule the central boron atom has only six electrons around it:

$$:\ddot{F}-B(-\ddot{F}:)-\ddot{F}:$$

Finally, one point needs to be stressed: **Before we can write some Lewis structures, *we must know how the atoms are connected to each other.*** Consider nitric acid, for example. Even though the formula for nitric acid is often written HNO_3, the hydrogen is actually connected to an oxygen, not to the nitrogen. The structure is $HONO_2$ and not HNO_3. Thus the correct Lewis structure is

$$H-\ddot{O}-N(=\ddot{O}:)-\ddot{O}: \quad \textbf{and not} \quad H-N(=O)-\ddot{O}-\ddot{O}:$$

This knowledge comes ultimately from experiments. If you have forgotten the structures of some of the common inorganic molecules and ions (such as those listed in Problem 1.1), this may be a good time for a review of the relevant portions of your general chemistry text.

PROBLEM 1.1

Write a Lewis structure for each of the following:

(a) HF **(c)** CH_3F **(e)** H_2SO_3 **(g)** H_3PO_4 **(i)** HCN

(b) F_2 **(d)** HNO_2 **(f)** BH_4^- **(h)** H_2CO_3

1.7 Formal Charge

When we write Lewis structures, it is often helpful for understanding properties and reactivity to assign unit positive or negative charges, called **formal charges,** to certain atoms in the molecule or ion. The determination of formal charges is nothing more than a bookkeeping method for electrical charges, because ***the arithmetic sum of all of the formal charges equals the total charge on the molecule or ion.***

The proper assignment of **formal charges** is another essential tool for learning organic chemistry.

We calculate formal charges on individual atoms **by subtracting the number of valence electrons assigned to an atom in its bonded state from the number of valence electrons it has as a neutral free atom.** (Recall that the number of valence electrons in a neutral free atom is equal to its **group number** on the periodic table.)

We assign valence electrons to atoms in the bonded state by apportioning them. **We divide electrons in covalent bonds equally between the atoms that share them and we assign unshared pairs to the atom that possesses them.**

Consider first the ammonium ion, an ion that has no unshared pairs. We divide all of the electrons in bonds equally between the atoms that share them. Each hydrogen is assigned *one electron* (e^-) and we subtract this from *one* (the number of valence electrons in a neutral hydrogen atom) to give a formal charge of 0 for each hydrogen atom. The nitrogen atom is assigned *four electrons* (one from each bond). We subtract this from *five* (the number of valence electrons in a neutral nitrogen atom) to give a formal charge of + 1. In effect, we say that because the nitrogen atom in the ammonium ion lacks one electron when compared to a neutral nitrogen atom (in which the number of protons and electrons are equal), it has a formal charge of + 1.*

$$\begin{matrix} & H & + \\ H : & \ddot{N} & : H \\ & \ddot{H} & \end{matrix}$$

For hydrogen: valence electrons of free atom = 1
subtract assigned electrons = −1
Formal charge = 0

For nitrogen: valence electrons of free atom = 5
subtract assigned electrons = −4
Formal charge = +1

Charge on ion = 4(0) + 1 = +1

Let us next consider the nitrate ion (NO_3^-), an ion that has oxygen atoms with unshared electron pairs. Here we find that the nitrogen atom has a formal charge of + 1, that two oxygen atoms have formal charges of − 1, and that one oxygen has a formal charge equal to 0.

Formal charge = 6 − 7 = −1

Formal charge = 5 − 4 = +1

Formal charge = 6 − 6 = 0

Charge on ion = 2(−1) + 1 + 0 = −1

Molecules, of course, have no net electrical charge. Molecules, by definition, are neutral. Therefore, the sum of the formal charges on each atom making up a molecule must be zero. Consider the following examples:

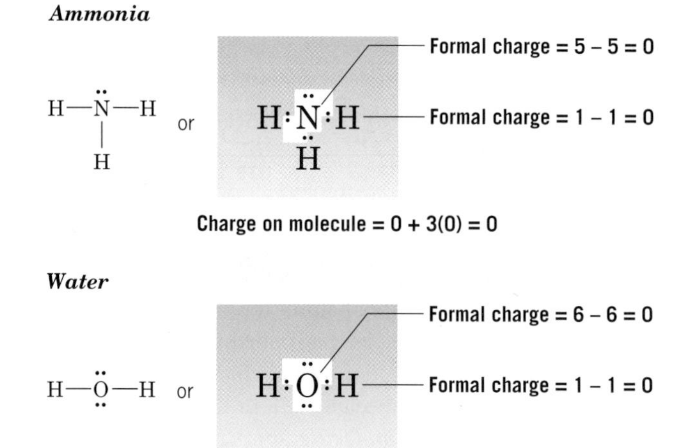

Charge on molecule = 0 + 2(0) = 0

*An alternative method for calculating formal charge is to use the equation

$$F = Z - S/2 - U$$

where F is the formal charge, Z is the group number, S equals the number of shared electrons, and U is the number of unshared electrons.

PROBLEM 1.2

Write a Lewis structure for each of the following negative ions, and assign the formal negative charge to the correct atom:

(a) NO_2^- (c) CN^- (e) HCO_3^-
(b) NH_2^- (d) HSO_4^- (f) HC_2^-

1.7A Summary of Formal Charges

With this background, it should now be clear that each time an oxygen atom of the type —Ö: (with three unshared pairs) appears in a molecule or ion, it will have a formal charge of − 1, and that each time an oxygen atom of the type =Ö: or —Ö— appears, it will have a formal charge of 0. Similarly, —N— (with four bonds) will be + 1, and —N̈— (three bonds, one unshared pair) will be zero. These and other common strutures are summarized in Table 1.3.

TABLE 1.3 A Summary of Formal Charges

Group	Formal Charge of +1	Formal Charge of 0	Formal Charge of −1
3A		>B— (three bonds)	—B⁻— (four bonds)
4A	>C⁺— (three bonds) =C⁺— ≡C⁺	—C— (four bonds) =C< ≡C—	—C̈⁻— (three bonds) =C̈⁻— ≡C:⁻
5A	—N⁺— (four bonds) =N⁺< ≡N⁺—	—N̈— (three bonds) =N̈— ≡N:	—N̈⁻— =N̈:⁻
6A	—Ö⁺— (three bonds) =Ö⁺—	—Ö— =Ö:	—Ö:⁻
7A	—Ẍ⁺—	—X: (X = F, Cl, Br, or I)	:X:⁻

In later chapters, when you are evaluating how reactions proceed and what products form, you will find it essential to keep track of formal charges.

PROBLEM 1.3

Assign the proper formal charge to the colored atom in each of the following structures:

(a) $H_3C—CH_2$ (H—C(H)(H)—C(H)(H))
(b) H—Ö—H with H on O (H₃O)
(c) H—C(=Ö:)—Ö:
(d) H—C(—Ö:)(H)—H
(e) H—C(H)(H)—N(H)(H)—H
(f) H—C(H)(H)—Ö(H)—H

1.8 Resonance

One problem with Lewis structures is that they impose an *artificial* **location** on the electrons. As a result, more than one *equivalent* Lewis structure can be written for many molecules and ions. Consider, for example, the carbonate ion (CO_3^{2-}). We can write three *different* but *equivalent* structures, **1–3**:

1 2 3

Notice two important features of these structures. First, each atom has the noble gas configuration. Second, *and this is especially important,* we can convert one structure into any other by *changing only the positions of the electrons.* We do not need to change the relative positions of the atomic nuclei. For example, if we move the electron pairs in the manner indicated by the **curved arrows** in structure **1**, we change structure **1** into structure **2**:

Curved arrows (Section 3.4) show movement of electron pairs, *not atoms.* The *tail* of the arrow begins at the current position of the electron pair. The *head* of the arrow points to the location where the electron pair will be in the next structure. Curved-arrow notation is one of the most important tools that you will use to understand organic reactions.

1 becomes 2

In a similar way we can change structure **2** into structure **3**:

2 becomes 3

Structures **1–3,** although not identical on paper, *are equivalent.* None of them alone, however, fits important data about the carbonate ion.

X-ray studies have shown that carbon–oxygen double bonds are shorter than single bonds. The same kind of study of the carbonate ion shows, however, that all of its carbon–oxygen bonds *are of equal length.* One is not shorter than the others as would be expected from the representations **1**, **2**, and **3**. Clearly none of the three structures agrees with this evidence. In each structure, **1–3**, one carbon–oxygen bond is a double bond and the other two are single bonds. None of the structures, therefore, is correct. How, then, should we represent the carbonate ion?

One way is through a theory called **resonance theory.** This theory states that whenever a molecule or ion can be represented by two or more Lewis structures *that differ only in the positions of the electrons,* two things will be true:

Resonance structures are a tool we will use for understanding structure and reactivity.

1. None of these structures, which we call **resonance structures** or **resonance contributors,** will be a correct representation for the molecule or ion. None will be in complete accord with the physical or chemical properties of the substance.
2. The actual molecule or ion will be better represented by a *hybrid (average) of these structures.*

Resonance structures, then, are not structures for the actual molecule or ion; they exist only on paper. As such, they can never be isolated. No single contributor adequately represents the molecule or ion. In resonance theory we view the carbonate ion, which is, of course, a real entity, as having a structure that is a **hybrid** of these three **hypothetical** resonance structures.

What would a hybrid of structures **1–3** be like? Look at the structures and look especially at a particular carbon–oxygen bond, say, the one at the top. This carbon–oxygen bond is a double bond in one structure (**1**) and a single bond in the other two (**2** and **3**). The actual carbon–oxygen bond, since it is a hybrid, must be something in between a double bond and a single bond. Because the carbon–oxygen bond is a single bond in two of the structures and a double bond in only one, it must be more like a single bond than a double bond. It must be like a one and one-third bond. We could call it a partial double bond. And, of course, what we have just said about any one carbon–oxygen bond will be equally true of the other two. Thus all of the carbon–oxygen bonds of the carbonate ion are partial double bonds, and *all are equivalent.* All of them *should be* the same length, and this is exactly what experiments tell us. The bonds are all 1.28 Å long, a distance which is intermediate between that of a carbon–oxygen single bond (1.43 Å) and that of a carbon–oxygen double bond (1.20 Å). One angstrom equals 1×10^{-10} meter.

One other important point: By convention, when we draw resonance structures, we connect them by double-headed arrows to indicate clearly that they are hypothetical, not real. For the carbonate ion we write them this way:

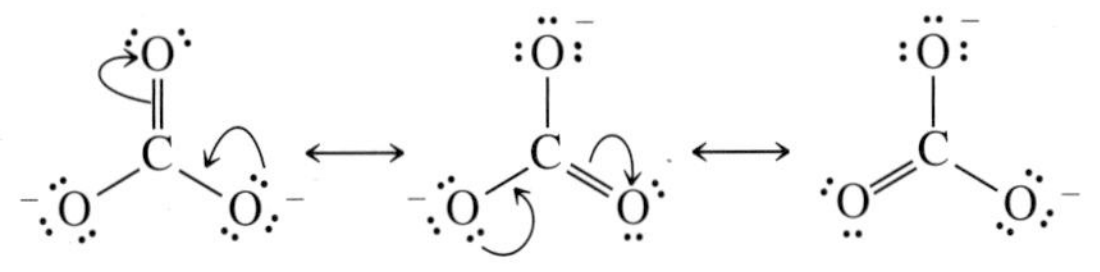

STUDY TIP!

Each type of arrow in organic chemistry (e.g., ⌒, ⇌ and ↔) has a specific meaning. It is important that you use each type of arrow only for the purpose for which it is defined.

We should not let these arrows, or the word "resonance," mislead us into thinking that the carbonate ion fluctuates between one structure and another. These structures exist only on paper; therefore, the carbonate ion cannot fluctuate among them because it is a hybrid of them. It is also important to distinguish between resonance and an **equilibrium.** In an equilibrium between two, or more, species, it is quite correct to think of different structures and moving (or fluctuating) atoms, *but not in the case of resonance* (as in the carbonate ion). Here the atoms do not move, and the "structures" exist only on paper. **An equilibrium is indicated by ⇆ and resonance by ↔.**

How can we write the structure of the carbonate ion in a way that will indicate its actual structure? We may do two things: we may write all of the resonance structures as we have just done and let the reader mentally fashion the hybrid, or we may write a non-Lewis structure that attempts to represent the hybrid. For the carbonate ion we might do the following:

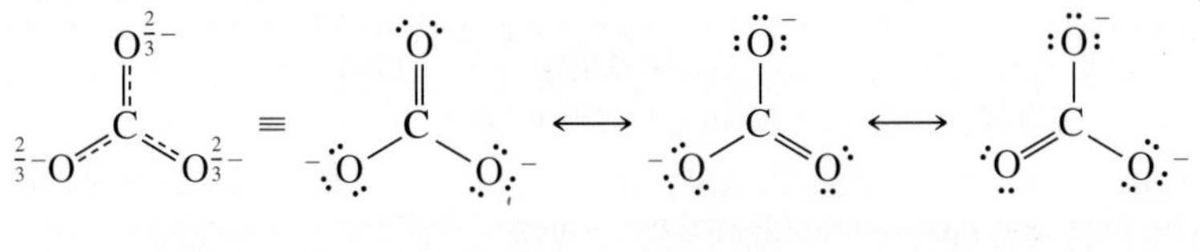

The bonds in the structure on the left are indicated by a combination of a solid line and a dashed line. This is to indicate that the bonds are something in between a single bond and a double bond. As a rule, we use a solid line whenever a bond appears in all structures, and a dashed line when a bond exists in one or more but not all. We also place a $\delta-$ (read partial minus) beside each oxygen to indicate that something less than a full negative charge resides on each oxygen atom. (In this instance each oxygen atom has two-thirds of a full negative charge.)

FIGURE 1.3 A calculated electrostatic potential map for the carbonate anion, showing the equal charge distribution at the three oxygen atoms. In electrostatic potential maps like this one, colors trending toward red mean increasing concentration of negative charge, while those trending toward blue mean less negative (or more positive) charge.

Animated Graphics

Electrostatic potential maps are useful models for visualizing molecular charge distribution.

Calculations from theory show the equal charge density at each oxygen in the carbonate anion. Figure 1.3 shows a calculated **electrostatic potential map** of the electron density in the carbonate ion. In an electrostatic potential map, regions of relatively more negative charge are red, while more positive regions (i.e., less negative regions) are indicated by colors trending toward blue. Equality of the bond lengths in the carbonate anion (partial double bonds as shown in the resonance hybrid above) is also evident in this model.

We shall make use of electrostatic potential maps and other types of structures calculated from theory at various points in this book to help explain aspects of molecular properties and reactivity. Further explanation of calculated structures will be given in Section 1.12.

1.8A Summary of Rules for Resonance

1. Resonance structures exist only on paper. Although they have no real existence of their own, **resonance structures** are useful because they allow us to describe molecules and ions for which a single Lewis structure is inadequate. We write two or more Lewis structures, calling them resonance structures or resonance contributors. We connect these structures by double-headed arrows (↔), and we say that the real molecule or ion is a hybrid of all of them.

2. In writing resonance structures we are only allowed to move electrons. The positions of the nuclei of the atoms must remain the same in all of the structures. Structure **3** is not a resonance structure of **1** or **2**, for example, because in order to form it we would have to move a hydrogen atom and this is not permitted:

$$\underset{\mathbf{1}}{CH_3-\overset{+}{C}H-CH=CH_2} \longleftrightarrow \underset{\mathbf{2}}{CH_3-CH=CH-\overset{+}{C}H_2} \qquad \underset{\mathbf{3}}{\overset{+}{C}H_2-CH_2-CH-CH_2}$$

These are resonance structures. (1 and 2)

This is not a proper resonance structure of 1 or 2 because a hydrogen atom has been moved. (3)

Generally speaking, when we move electrons, we move only those of π bonds (as in the example above) and those of nonbonding electron pairs.

3. All of the structures must be proper Lewis structures. We should not write structures in which carbon has five bonds, for example:

$$H-\overset{H}{\underset{H}{\overset{|}{\underset{|}{\overset{-}{C}}}}}=\overset{+}{\underset{\cdot\cdot}{O}}-H$$

This is not a proper resonance structure for methanol because carbon has five bonds. Elements of the first major row of the periodic table cannot have more than eight electrons in their valence shell.

4. **The energy of the actual molecule is lower than the energy that might be estimated for any contributing structure.** Chemists often call this kind of stabilization *resonance stabilization.*

5. **Equivalent resonance structures make equal contributions to the hybrid, and a system described by them has a large resonance stabilization.** The following cation, for example, is more stable than either contributing resonance structure taken separately:

$$\underset{\mathbf{4}}{CH_2{=}CH{-}\overset{+}{C}H_2} \longleftrightarrow \underset{\mathbf{5}}{\overset{+}{C}H_2{-}CH{=}CH_2} \qquad \overset{\frac{1}{2}+}{C}H_2{\overset{\text{- - -}}{=}}CH{\overset{\text{- - -}}{=}}\overset{\frac{1}{2}+}{C}H_2$$

Contributing resonance structures **Resonance hybrid**

In Chapter 14 we shall find that benzene is highly resonance stabilized because it is a hybrid of the two equivalent forms that follow:

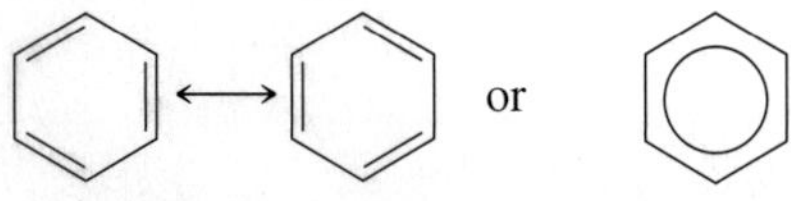

or

Resonance structures for benzene **Representation of hybrid**

6. **The more stable a structure is (when taken by itself), the greater is its contribution to the hybrid.**

The following rules will help us in making decisions about the relative stabilities of resonance structures.

a. **The more covalent bonds a structure has, the more stable it is.** This is exactly what we would expect because we know that forming a covalent bond lowers the energy of atoms. This means that of the following structures for 1,3-butadiene, **6** is by far the most stable and makes by far the largest contribution because it contains one more bond. (It is also more stable for the reason given under the rule **c.**)

$$\underset{\mathbf{6}}{CH_2{=}CH{-}CH{=}CH_2} \longleftrightarrow \underset{\mathbf{7}}{\overset{+}{C}H_2{-}CH{=}CH{-}\overset{\cdot\cdot}{C}H_2^{-}} \longleftrightarrow \underset{\mathbf{8}}{\overset{\cdot\cdot}{C}H_2^{-}{-}CH{=}CH{-}\overset{+}{C}H_2}$$

This structure is the most stable because it contains more covalent bonds.

b. **Structures in which all of the atoms have a complete valence shell of electrons (i.e., the noble gas structure) are especially stable and make large contributions to the hybrid.** Again, this is what we would expect from what we know about bonding. This means, for example, that **10** makes a larger stabilizing contribution to the cation below than **9** because all of the atoms of **10** have a complete valence shell. (Notice too that **10** has more covalent bonds than **9**; see rule **a.**)

$$\underset{\mathbf{9}}{\overset{+}{C}H_2{-}\ddot{\underset{\cdot\cdot}{O}}{-}CH_3} \longleftrightarrow \underset{\mathbf{10}}{CH_2{=}\overset{+}{\underset{\cdot\cdot}{O}}{-}CH_3}$$

Here this carbon atom has only six electrons. **Here the carbon atom has eight electrons.**

c. **Charge separation decreases stability.** Separating opposite charges requires energy. Therefore, structures in which opposite charges are separated have greater energy (lower stability) than those that have no charge separation. This means that of the following two structures for vinyl chloride, structure **11** makes a larger contribution because it does not have separated charges. (This does not mean that structure **12** does not contribute to the hybrid; it just means that the contribution made by **12** is smaller.)

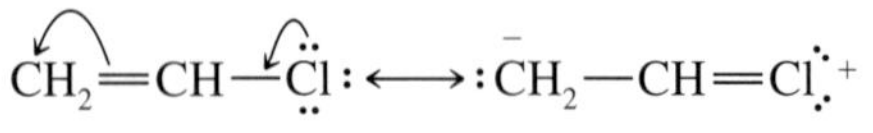

$$\underset{\mathbf{11}}{CH_2{=}CH{-}\ddot{\underset{\cdot\cdot}{Cl}}:} \longleftrightarrow \underset{\mathbf{12}}{:\bar{C}H_2{-}CH{=}\ddot{Cl}:^{+}}$$

d. Resonance contributors with negative charge on highly electronegative atoms are more stable than ones with negative charge on less or nonelectronegative atoms. Conversely, resonance contributors with positive charge on highly electronegative atoms are less stable than ones with positive charge on nonelectronegative atoms.

SOLVED PROBLEM

The following is one way of writing the structure of the nitrate ion:

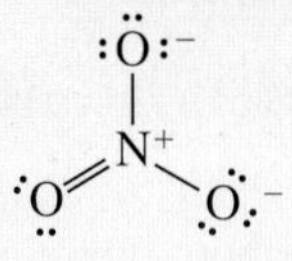

However, considerable physical evidence indicates that all three nitrogen–oxygen bonds are equivalent and that they have the same length, a bond distance between that expected for a nitrogen–oxygen single bond and a nitrogen–oxygen double bond. Explain this in terms of resonance theory.

ANSWER We recognize that if we move the electron pairs in the following way, we can write three *different* but *equivalent* structures for the nitrate ion:

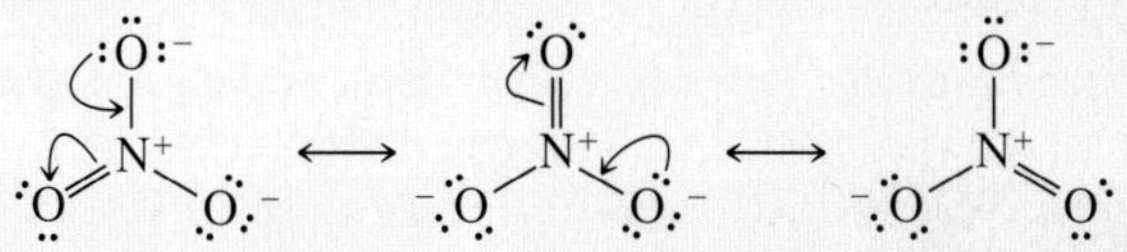

Since these structures differ from one another *only in the positions of their electrons,* they are *resonance structures* or *resonance contributors.* As such, no single structure taken alone will adequately represent the nitrate ion. The actual molecule will be best represented by a *hybrid of these three structures.* We might write this hybrid in the following way to indicate that all of the bonds are equivalent and that they are more than single bonds and less than double bonds. We also indicate that each oxygen atom bears an equal partial negative charge. This charge distribution corresponds to what we find experimentally.

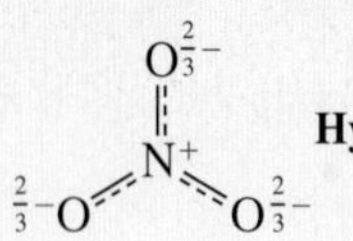

Hybrid structure for the nitrate ion

PROBLEM 1.4 **(a)** Write two resonance structures for the formate ion HCO_2^-. (*Note:* The hydrogen and oxygen atoms are bonded to the carbon.) Explain what these structures predict for **(b)** the carbon–oxygen bond lengths of the formate ion and **(c)** for the electrical charge on the oxygen atoms.

PROBLEM 1.5

Write the contributing resonance structures and resonance hybrid for each of the following:

(a) $CH_3CH{=}CH{-}CH{=}\overset{+}{\ddot{O}}H$

(b) $CH_2{=}CH{-}\underset{+}{C}H{-}CH{=}CH_2$

(c) cyclohexadienyl cation (ring with + charge)

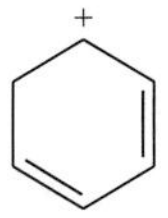

(d) $CH_2{=}CH{-}Br$

(e) $C_6H_5CH_2^+$ (benzene ring bearing CH_2^+)

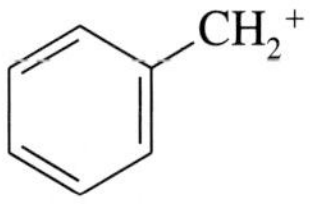

(f) $^-{:}CH_2{-}\overset{O}{\overset{\|}{C}}{-}CH_3$

(g) $CH_3{-}S{-}CH_2^+$

(h) $CH_3{-}NO_2$

PROBLEM 1.6

From each set of resonance structures that follow, designate the one that would contribute most to the hybrid and explain your choice:

(a) $\overset{+}{C}H_2{-}\ddot{N}(CH_3)_2 \longleftrightarrow CH_2{=}\overset{+}{N}(CH_3)_2$

(b) $CH_3{-}\overset{\ddot{O}:}{\overset{\|}{C}}{-}\ddot{O}{-}H \longleftrightarrow CH_3{-}\overset{:\ddot{O}:^-}{\overset{|}{C}}{=}\overset{+}{\ddot{O}}{-}H$

(c) ${:}NH_2{-}C{\equiv}N{:} \longleftrightarrow \overset{+}{N}H_2{=}C{=}\ddot{N}{:}^-$

1.9 Quantum Mechanics

In 1926 a new theory of atomic and molecular structure was advanced independently and almost simultaneously by three men: Erwin Schrödinger, Werner Heisenberg, and Paul Dirac. This theory, called **wave mechanics** by Schrödinger and **quantum mechanics** by Heisenberg, has become the basis from which we derive our modern understanding of bonding in molecules.

The formulation of quantum mechanics that Schrödinger advanced is the form that is most often used by chemists. In Schrödinger's publication the motion of electrons is described in terms that take into account the wave nature of the electron.* Schrödinger developed a way to convert the mathematical expression for the total energy of the system consisting of one proton and one electron—the hydrogen atom—into another expression called a **wave equation.** This equation is then solved to yield not one but a series of solutions called **wave functions.**

Wave functions are most often denoted by the Greek letter psi (ψ), and each wave function (ψ function) corresponds to a different state for the electron. Corresponding to each state, and calculable from the wave equation for the state, is a particular energy.

Each state is a sublevel where one or two electrons can reside. *The solutions to the wave equation for a hydrogen atom can also be used* (with appropriate modifications) *to give sublevels for the electrons of higher elements.*

A wave equation is simply a tool for calculating two important properties: (a) the energy associated with the state of the electron, and (b) the relative probability of an electron residing at particular places in the sublevel (Section 1.10). When the value of a wave equation is calculated for a particular point in space relative to the nucleus, the result may be a positive number or a negative number (or zero). These signs are sometimes called **phase signs.** They are characteristic of all equations that describe waves. We do not need

*The idea that the electron has the properties of a wave as well as those of a particle was proposed by Louis de Broglie in 1923.

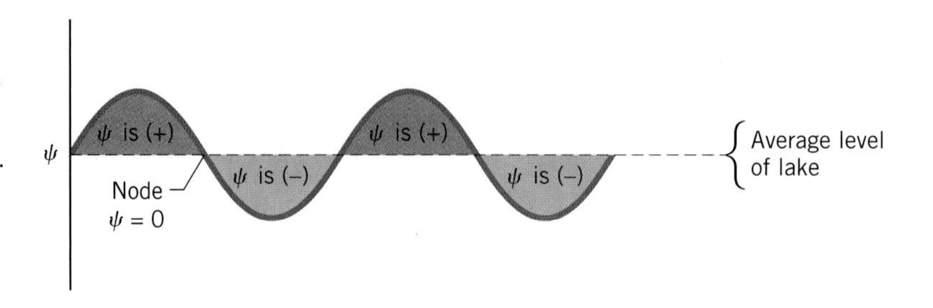

FIGURE 1.4 A wave moving across a lake is viewed along a slice through the lake. For this wave the wave function, ψ, is plus (+) in crests and minus (−) in troughs. At the average level of the lake it is zero; these places are called nodes.

to go into the mathematics of waves here, but a simple analogy will help us understand the nature of these phase signs.

Imagine a wave moving across a lake. As it moves along, the wave has crests and troughs; that is, it has regions where the wave rises above the average level of the lake or falls below it (Fig. 1.4). Now, if an equation were to be written for this wave, the wave function (ψ) would be plus (+) in regions where the wave is above the average level of the lake (i.e., in crests) and it would be minus (−) in regions where the wave is below the average level (i.e., in troughs). The relative magnitude of ψ (called the amplitude) will be related to the distance the wave rises above or falls below the average level of the lake. At the places where the wave is exactly at the average level of the lake, the wave function will be zero. Such a place is called a **node.**

One other characteristic of waves is their ability to reinforce each other or to interfere with one another. Imagine two waves approaching each other as they move across a lake. If the waves meet so that a crest meets a crest, that is, so that *waves of the same phase sign meet each other,* the waves **reinforce** each other, they add together, and the resulting wave is larger than either individual wave. On the other hand, if a crest meets a trough, that is, if waves of opposite sign meet, the waves **interfere** with each other, they subtract from each other, and the resulting wave is smaller than either individual wave. (If the two waves of opposite sign meet in precisely the right way, complete cancellation can occur.)

The wave functions that describe the motion of an electron in an atom or molecule are, of course, different from the equations that describe waves moving across lakes. And when dealing with the electron we should be careful not to take analogies like this too far. Electron wave functions, however, are like the equations that describe water waves in that they have phase signs and nodes, and *they undergo reinforcement and interference.*

1.10 Atomic Orbitals

For a short time after Schrödinger's proposal in 1926, a precise physical interpretation for the electron wave function eluded early practitioners of quantum mechanics. It remained for Max Born, a few months later, to point out that the square of ψ *could* be given a precise physical meaning. According to Born, ψ^2 for a particular location (x,y,z) expresses the **probability** of finding an electron at that particular location in space. If ψ^2 is large in a unit volume of space, the probability of finding an electron in that volume is great—we say that the **electron probability density** is large. Conversely if ψ^2 for some other unit volume of space is small, the probability of finding an electron there is low.* **Plots of ψ^2 in three dimensions generate the shapes of the familiar *s*, *p*, and *d* atomic orbitals (AOs), which we use as our models for atomic structure.**

The *f* orbitals are practically never used in organic chemistry, and we shall not concern ourselves with them in this book. The *d* orbitals will be discussed briefly later when we discuss compounds in which *d*-orbital interactions are important. The *s* and *p* orbitals are,

*Integration of ψ^2 over all space must equal 1; that is, the probability of finding an electron somewhere in all of space is 100%.

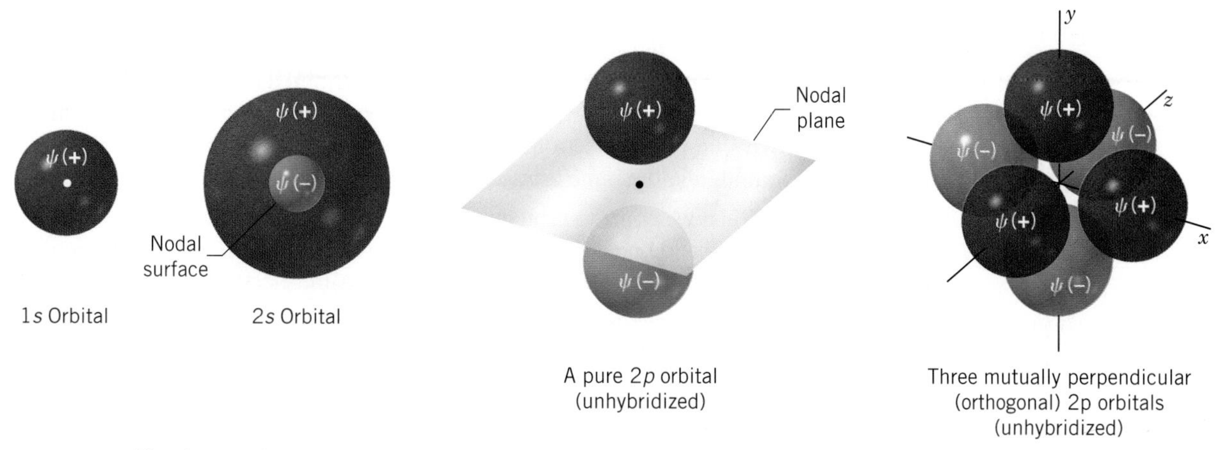

FIGURE 1.5 The shapes of some *s* and *p* orbitals. Pure, unhybridized *p* orbitals are almost-touching spheres. The *p* orbitals in hybridized atoms are lobe-shaped (Section 1.14).

CD Animated Graphics

by far, the most important in the formation of organic molecules and, at this point, we shall limit our discussion to them.

An orbital is a region of space where the probability of finding an electron is large. The shapes of *s* and *p* orbitals are shown in Fig. 1.5. There is a finite, but very small, probability of finding an electron at greater distances from the nucleus. The volumes that we typically use to illustrate an orbital are those volumes that would contain the electron 90–95% of the time.

Both the 1*s* and 2*s* orbitals are spheres, as are all higher *s* orbitals (Fig. 1.5). The 2*s* orbital contains a nodal surface, that is, an area where $\psi = 0$. In the inner portion of the 2*s* orbital, ψ_{2s} is negative.

The 2*p* orbitals have the shape of two almost-touching spheres. The phase sign of the 2*p* wave function, ψ_{2p}, is positive in one lobe (or sphere) and negative in the other. A nodal plane separates the two lobes of a *p* orbital, and the three *p* orbitals are arranged in space so that their axes are mutually perpendicular.

You should not associate the sign of the wave function with anything having to do with electrical charge. As we said earlier the (+) and (−) signs associated with ψ are simply the arithmetic signs of the wave function in that region of space. The (+) and (−) signs **do not** imply a greater or lesser probability of finding an electron either. The probability of finding an electron is ψ^2, and ψ^2 is always positive. (Squaring a negative number always makes it positive.) Thus the probability of finding the electron in the (−) lobe of a *p* orbital is the same as that of the (+) lobe. The significance of the (+) and (−) signs will become clear later when we see how atomic orbitals combine to form molecular orbitals and when we see how covalent bonds are formed.

There is a relationship between the number of nodes of an orbital and its energy. ***The greater the number of nodes, the greater the energy.*** We can see an example here; the 2*s* and 2*p* orbitals have one node each and they have greater energy than a 1*s* orbital, which has no nodes.

The relative energies of the lower energy orbitals are as follows. Electrons in 1*s* orbitals have the lowest energy because they are closest to the positive nucleus. Electrons in 2*s* orbitals are next lowest in energy. Electrons of the three 2*p* orbitals have equal but still higher energy. (Orbitals of equal energy are said to be **degenerate orbitals.**)

We can use these relative energies to arrive at the electronic configuration of any atom in the first two rows of the periodic table. We need only follow a few simple rules.

1. **Aufbau principle:** Orbitals are filled so that those of lowest energy are filled first. (*Aufbau* is German for "building up.")

2. Pauli exclusion principle: A maximum of two electrons may be placed in each orbital *but only when the spins of the electrons are paired.* An electron spins about its own axis. For reasons that we cannot develop here, an electron is permitted only one or the other of just two possible spin orientations. We usually show these orientations by arrows, either ↿ or ⇂. Thus two spin-paired electrons would be designated ⥮. Unpaired electrons, which are not permitted in the same orbital, are designated ↿ ↿ (or ⇂ ⇂).

3. Hund's rule: When we come to orbitals of equal energy (degenerate orbitals) such as the three *p* orbitals, we add one electron to each *with their spins unpaired* until each of the degenerate orbitals contains one electron. (This allows the electrons, which repel each other, to be farther apart.) Then, we begin adding a second electron to each degenerate orbital so that the spins are paired.

If we apply these rules to some of the second-row elements of the periodic table, we get the results shown in Fig. 1.6.

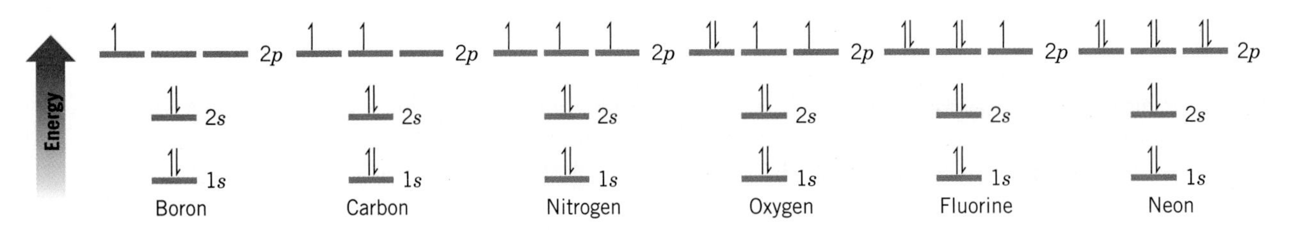

FIGURE 1.6 The electron configurations of some second-row elements.

1.11 MOLECULAR ORBITALS

For the organic chemist the greatest utility of atomic orbitals is in using them as models for understanding how atoms combine to form molecules. We shall have much more to say about this subject in subsequent chapters, for, as we have already said, covalent bonds are central to the study of organic chemistry. First, however, we shall concern ourselves with a very simple case: the covalent bond that is formed when two hydrogen atoms combine to form a hydrogen molecule. We shall see that the description of the formation of the H—H bond is the same as, or at least very similar to, the description of bonds in more complex molecules.

Let us begin by examining what happens to the total energy of two hydrogen atoms with electrons of opposite spins when they are brought closer and closer together. This can best be shown with the curve in Fig. 1.7.

FIGURE 1.7 The potential energy of the hydrogen molecule as a function of internuclear distance.

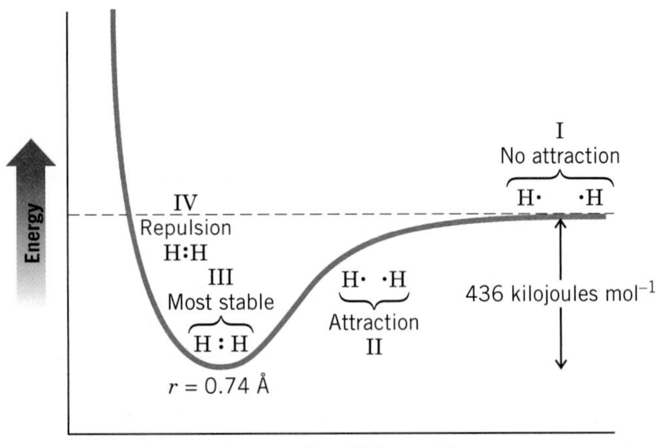

When the atoms of hydrogen are relatively far apart **(I)**, their total energy is simply that of two isolated hydrogen atoms. As the hydrogen atoms move closer together **(II)**, each nucleus increasingly attracts the other's electron. This attraction more than compensates for the repulsive force between the two nuclei (or the two electrons), and the result of this attraction *is to lower the energy of the total system.* When the two nuclei are 0.74 Å apart **(III)**, the most stable (lowest energy) state is obtained. This distance, 0.74 Å, corresponds to the *bond length* for the hydrogen molecule. If the nuclei are moved closer together **(IV)** the repulsion of the two positively charged nuclei predominates, and the energy of the system rises.

There is one serious problem with this model for bond formation. We have assumed that the electrons are essentially motionless and that as the nuclei come together they will be stationary in the region between the two nuclei. Electrons do not behave that way. Electrons move about, and according to the **Heisenberg uncertainty principle,** we cannot know simultaneously the position and momentum of an electron. That is, we cannot pin the electrons down as precisely as our explanation suggests.

We avoid this problem when we use a model based on quantum mechanics and *orbitals,* because now we describe the electron in terms of probabilities (ψ^2) of finding it at particular places. By treating the electron in this way we do not violate the uncertainty principle, because we do not talk about where the electron is precisely. We talk instead about where the *electron probability density* is large or small.

Thus an orbital explanation for what happens when two hydrogen atoms combine to form a hydrogen molecule is the following: As the hydrogen atoms approach each other, their 1*s* orbitals (ψ_{1s}) begin to overlap. As the atoms move closer together, orbital overlap increases until the **atomic orbitals (AOs)** combine to become **molecular orbitals (MOs).** The molecular orbitals that are formed encompass both nuclei, and, in them, the electrons can move about both nuclei. The electrons are not restricted to the vicinity of one nucleus or the other as they were in the separate atomic orbitals. Each molecular orbital, like each atomic orbital, *may contain a maximum of two spin-paired electrons.*

When atomic orbitals combine to form molecular orbitals, ***the number of molecular orbitals that results always equals the number of atomic orbitals that combine.*** Thus in the formation of a hydrogen molecule the *two* atomic orbitals combine to produce *two* molecular orbitals. Two orbitals result because the mathematical properties of wave functions permit them to be combined by either *addition* or *subtraction.* That is, they can combine either *in* or *out of* phase. What are the natures of these new molecular orbitals?

One molecular orbital, called the **bonding molecular orbital** (ψ_{molec}) contains both electrons in the lowest energy state, or *ground* state, of a hydrogen molecule. It is formed when the atomic orbitals combine in the way shown in Fig. 1.8. Here atomic orbitals combine by *addition,* and this means that atomic *orbitals of the same phase sign overlap.* Such overlap leads to *reinforcement* of the wave function in the region between the two

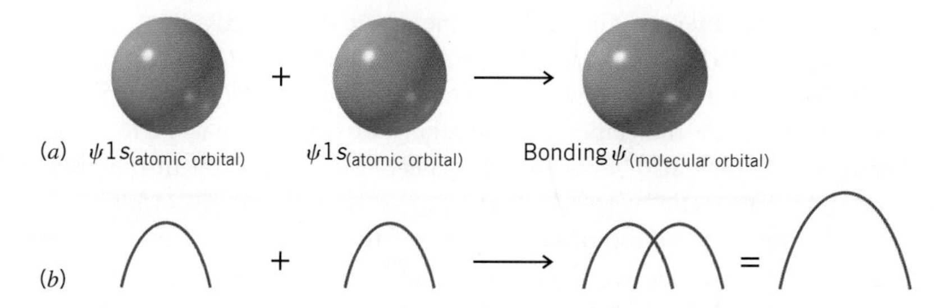

FIGURE 1.8 *(a)* The overlapping of two hydrogen 1*s* atomic orbitals with the same phase sign (indicated by their identical color) to form a bonding molecular orbital. *(b)* The analogous overlapping at two waves with the same phase, resulting in constructive interference and enhanced amplitude.

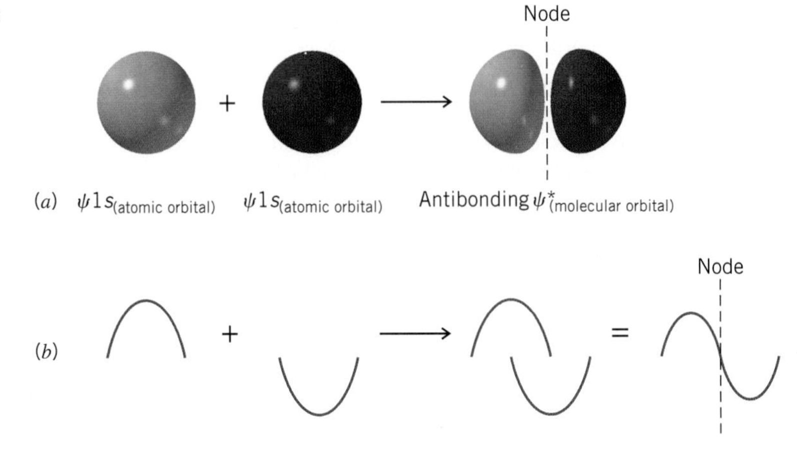

FIGURE 1.9 *(a)* The overlapping of two hydrogen 1*s* atomic orbitals with opposite phase signs (indicated by their different colors) to form an antibonding molecular orbital. *(b)* The analogous overlapping of two waves with the opposite sign, resulting in destructive interference and decreased amplitude. A node exists where complete cancellation by opposite phases makes the value of the combined wave function zero.

nuclei. Reinforcement of the wave function not only means that the value of ψ is larger between the two nuclei, it means that ψ^2 is larger as well. Moreover, since ψ^2 expresses the probability of finding an electron in this region of space, we can now understand how orbital overlap of this kind leads to bonding. It does so by increasing the electron probability density in exactly the right place—in the region of space between the nuclei. When the electron density is large here, the attractive force of the nuclei for the electrons more than offsets the repulsive force acting between the two nuclei (and between the two electrons). This extra attractive force is, of course, the "glue" that holds the atoms together.

The second molecular orbital, called the **antibonding molecular orbital** (ψ^*_{molec}), contains no electrons in the ground state of the molecule. It is formed by interaction of orbitals with opposite phase signs. As shown in Fig. 1.9, the wave functions *interfere* with each other in the region between the two nuclei and a node is produced. At the node $\psi = 0$, and on either side of the node ψ is small. This means that in the region between the nuclei ψ^2 is also small. Thus if electrons were to occupy the antibonding orbital, the electrons would avoid the region between the nuclei. There would be only a small attractive force of the nuclei for the electrons. Repulsive forces (between the two nuclei and between the two electrons) would be greater than the attractive forces. Having electrons in the antibonding orbital would not tend to hold the atoms together; it would tend to make them fly apart.

What we have just described has its counterpart in a mathematical treatment called the **LCAO (linear combination of atomic orbitals)** method. In the LCAO treatment, wave functions for the atomic orbitals are combined in a linear fashion (by addition or subtraction) in order to obtain new wave functions for the molecular orbitals.

Molecular orbitals, like atomic orbitals, correspond to particular energy states for an electron. Calculations show that the relative energy of an electron in the bonding molecular orbital of the hydrogen molecule is substantially less than its energy in a ψ_{1s} atomic orbital. These calculations also show that the energy of an electron in the antibonding molecular orbital is substantially greater than its energy in a ψ_{1s} atomic orbital.

An energy diagram for the molecular orbitals of the hydrogen molecule is shown in Fig. 1.10. Notice that electrons are placed in molecular orbitals in the same way that they were in atomic orbitals. Two electrons (with their spins opposed) occupy the bonding molecular orbital, where their total energy is less than in the separate atomic orbitals. This is, as we have said, the *lowest electronic state* or *ground state* of the hydrogen molecule. (An electron may occupy the antibonding orbital in what is called an *excited state* for the mol-

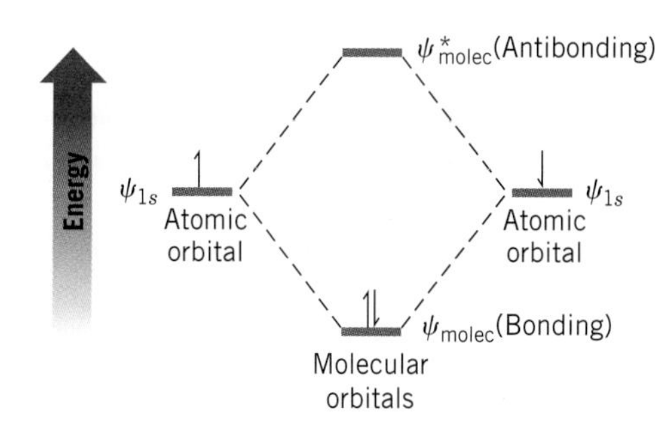

FIGURE 1.10 Energy diagram for the hydrogen molecule. Combination of two atomic orbitals, ψ_{1s}, gives two molecular orbitals, ψ_{molec} and ψ^*_{molec}. The energy of ψ_{molec} is lower than that of the separate atomic orbitals, and in the lowest electronic energy state of molecular hydrogen it contains both electrons.

CD Tutorial

Molecular orbitals

ecule. This state forms when the molecule in the ground state absorbs a photon of light of proper energy.)

1.12 The Structure of Methane and Ethane: sp^3 Hybridization

The *s* and *p* orbitals used in the quantum mechanical description of the carbon atom, given in Section 1.10, were based on calculations for hydrogen atoms. These simple *s* and *p* orbitals do not, when taken alone, provide a satisfactory model for the *tetravalent–tetrahedral* carbon of methane (CH_4, see Problem 1.7). However, a satisfactory model of methane's structure that is based on quantum mechanics *can* be obtained through an approach called **orbital hybridization.** Orbital hybridization, in its simplest terms, is nothing more than a mathematical approach that involves the combining of individual wave functions for *s* and *p* orbitals to obtain wave functions for new orbitals. The new orbitals have, *in varying proportions,* the properties of the original orbitals taken separately. These new orbitals are called **hybrid atomic orbitals.**

According to quantum mechanics, the electronic configuration of a carbon atom in its lowest energy state—called the *ground state*—is that given here:

C	⇅	⇅	↑	↑	
	$1s$	$2s$	$2p_x$	$2p_y$	$2p_z$

Ground state of a carbon atom

The valence electrons of a carbon atom (those used in bonding) are those of the *outer level,* that is, the $2s$ and $2p$ electrons.

1.12A The Structure of Methane

Hybrid atomic orbitals that account for the structure of methane can be obtained by combining the wave functions of the $2s$ orbital of carbon with those of the three $2p$ orbitals. The mathematical procedure for hybridization can be approximated by the illustration that is shown in Fig. 1.11.

sp^3 Hybridization

In this model, four orbitals are mixed—or hybridized—and four new hybrid orbitals are obtained. The hybrid orbitals are called sp^3 orbitals to indicate that they have one part the character of an *s* orbital and three parts the character of a *p* orbital. The mathematical treatment of orbital hybridization also shows that *the four sp^3 orbitals should be oriented at angles of 109.5° with respect to each other.* This is precisely the spatial orientation of the four hydrogen atoms of methane.

If, in our imagination, we visualize the hypothetical formation of methane from an sp^3-hybridized carbon atom and four hydrogen atoms, the process might be like that shown in Fig. 1.12. For simplicity we show only the formation of the *bonding molecular orbital* for each carbon–hydrogen bond. We see that an sp^3-hybridized carbon gives a *tetrahedral structure for methane, and one with four equivalent C—H bonds.*

FIGURE 1.11 Hybridization of pure atomic orbitals of a carbon atom to produce sp^3 hybrid orbitals.

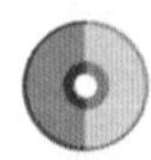

CD Tutorial

sp^3 Hybridization

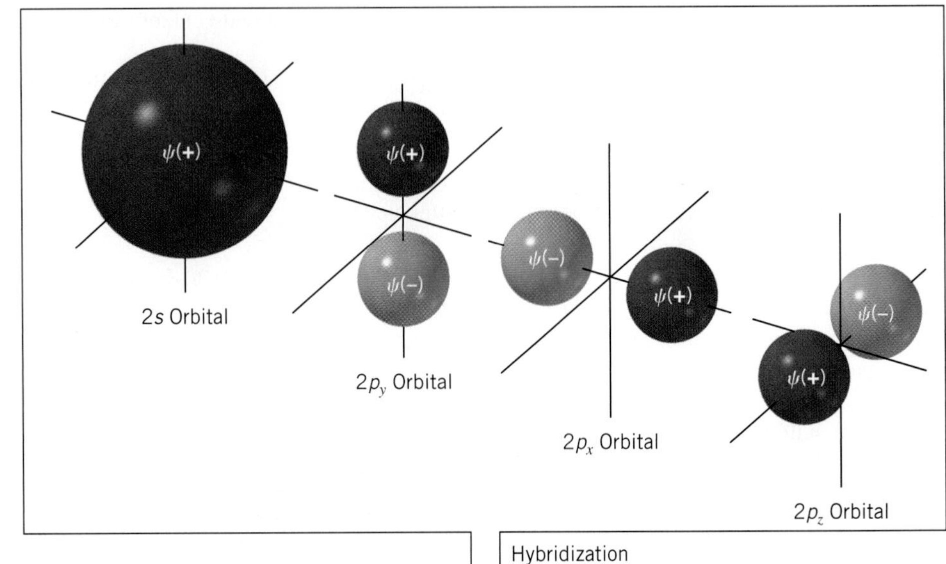

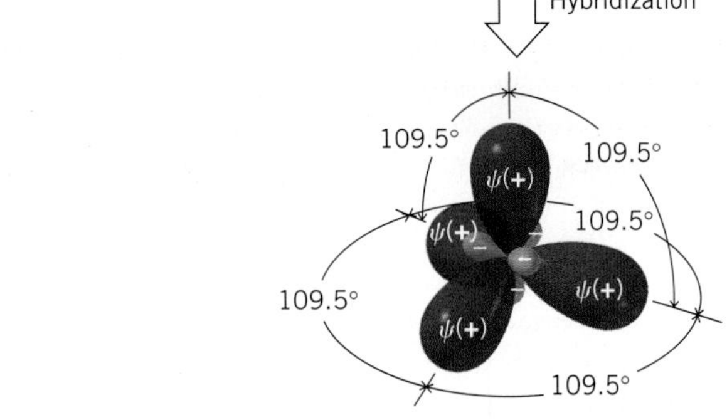

CD Tutorial

Methane bonding

FIGURE 1.12 The hypothetical formation of methane from an sp^3-hybridized carbon atom and four hydrogen atoms. In orbital hybridization we combine orbitals, *not* electrons. The electrons can then be placed in the hybrid orbitals as necessary for bond formation, but always in accordance with the Pauli principle of no more than two electrons (with opposite spin) in each orbital. In this illustration we have placed one electron in each of the hybrid carbon orbitals. In addition, we have shown only the bonding molecular orbital of each C—H bond because these are the orbitals that contain the electrons in the lowest energy state of the molecule.

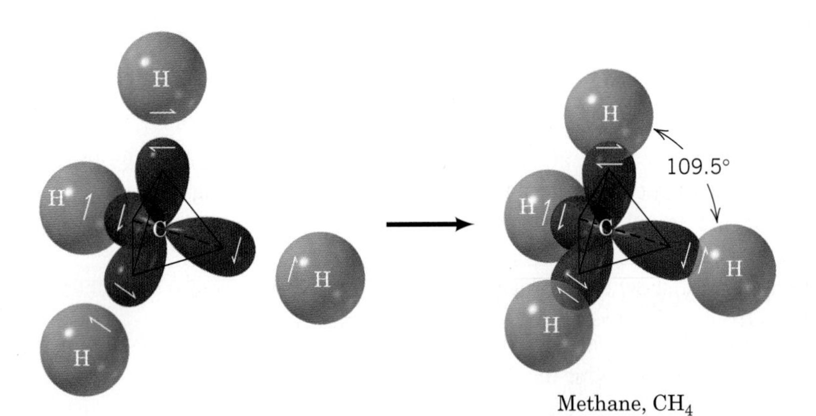

PROBLEM 1.7

(a) Consider a carbon atom in its ground state. Would such an atom offer a satisfactory model for the carbon of methane? If not, why not? (*Hint:* Consider whether or not a ground state carbon atom could be tetravalent, and consider the bond angles that would result if it were to combine with hydrogen atoms.)

(b) What about a carbon atom in the excited state:

C	↿⇂	↿	↿	↿	↿
	$1s$	$2s$	$2p_x$	$2p_y$	$2p_z$

Excited state of a carbon atom

Would such an atom offer a satisfactory model for the carbon of methane? If not, why not?

In addition to accounting properly for the shape of methane, the orbital hybridization model also explains the very strong bonds that are formed between carbon and hydrogen. To see how this is so, consider the shape of the individual sp^3 orbital shown in Fig. 1.13. Because the sp^3 orbital has the character of a p orbital, the positive lobe of the sp^3 orbital is large and extends relatively far from the carbon nucleus.

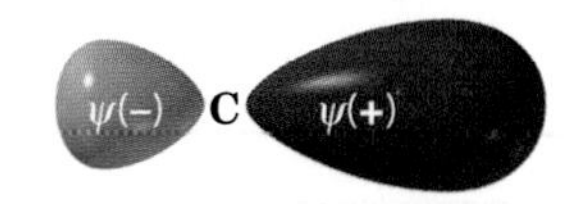

FIGURE 1.13 The shape of an sp^3 orbital.

It is the positive lobe of the sp^3 orbital that overlaps with the positive $1s$ orbital of hydrogen to form the bonding molecular orbital of a carbon–hydrogen bond (Fig. 1.14). Because the positive lobe of the sp^3 orbital is large and is extended into space, the overlap between it and the $1s$ orbital of hydrogen is also large, and the resulting carbon–hydrogen bond is quite strong.

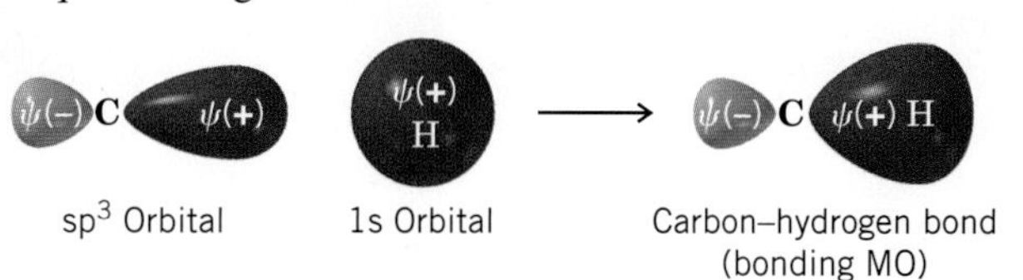

FIGURE 1.14 Formation of a C—H bond.

The bond formed from the overlap of an sp^3 orbital and a $1s$ orbital is an example of a **sigma (σ) bond** (Fig. 1.15). The term *sigma bond* is a general term applied to those bonds in which orbital overlap gives a bond that is *circularly symmetrical in cross section when viewed along the bond axis.* ***All purely single bonds are sigma bonds.***

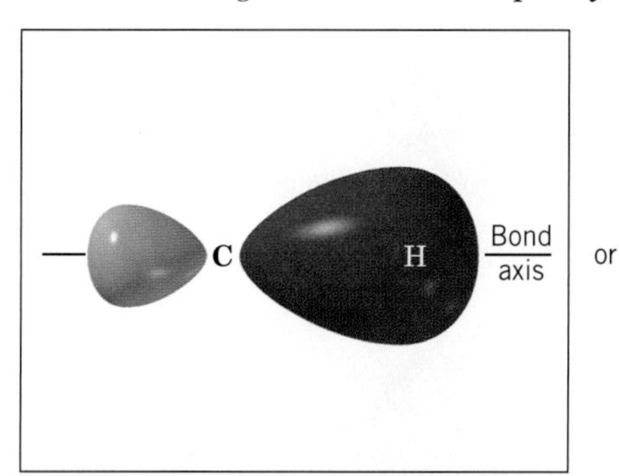

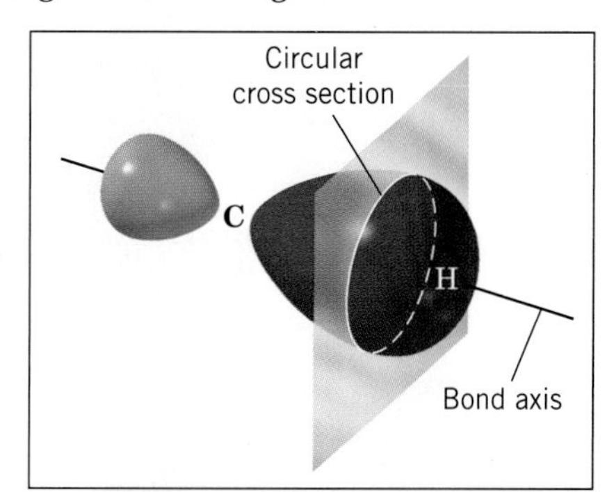

FIGURE 1.15 A σ (sigma) bond.

From this point on we shall often show only the bonding molecular orbitals because they are the ones that contain the electrons when the molecule is in its lowest energy state. Consideration of antibonding orbitals is important when a molecule absorbs light and in explaining certain reactions. We shall point out these instances later.

In Fig. 1.16 we show a calculated structure for methane where the tetrahedral geometry derived from orbital hybridization is clearly apparent.

Chem3D Model, Animated Graphics

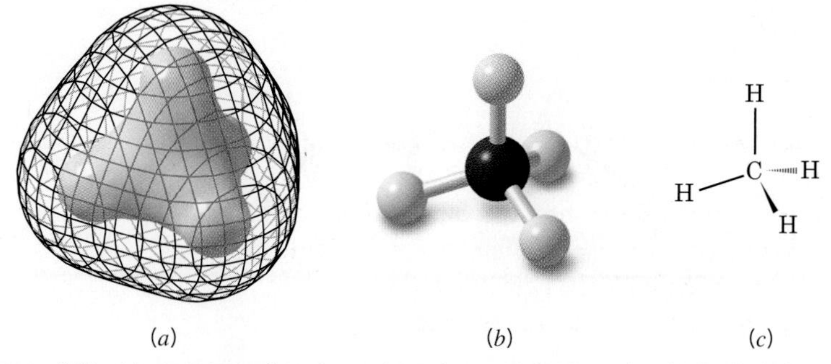

(a) (b) (c)

FIGURE 1.16 *(a)* In this structure of methane, based on quantum mechanical calculations, the inner solid surface represents a region of high electron density. High electron density is found in each bonding region. The outer mesh surface represents approximately the furthest extent of overall electron density for the molecule. *(b)* This ball-and-stick model of methane is like the kind you might build with a molecular model kit. *(c)* This structure is how you would draw methane. Ordinary lines are used to show the two bonds that are in the plane of the paper, a solid wedge is used to show the bond that is in front of the paper, and a dashed wedge is used to show the bond that is behind the plane of the paper.

1.12B The Structure of Ethane

The bond angles at the carbon atoms of ethane, and of all alkanes, are also tetrahedral like those in methane. A satisfactory model for ethane can be provided by sp^3-hybridized carbon atoms. Figure 1.17 shows how we might imagine the bonding molecular orbitals of an ethane molecule being constructed from two sp^3-hybridized carbon atoms and six hydrogen atoms.

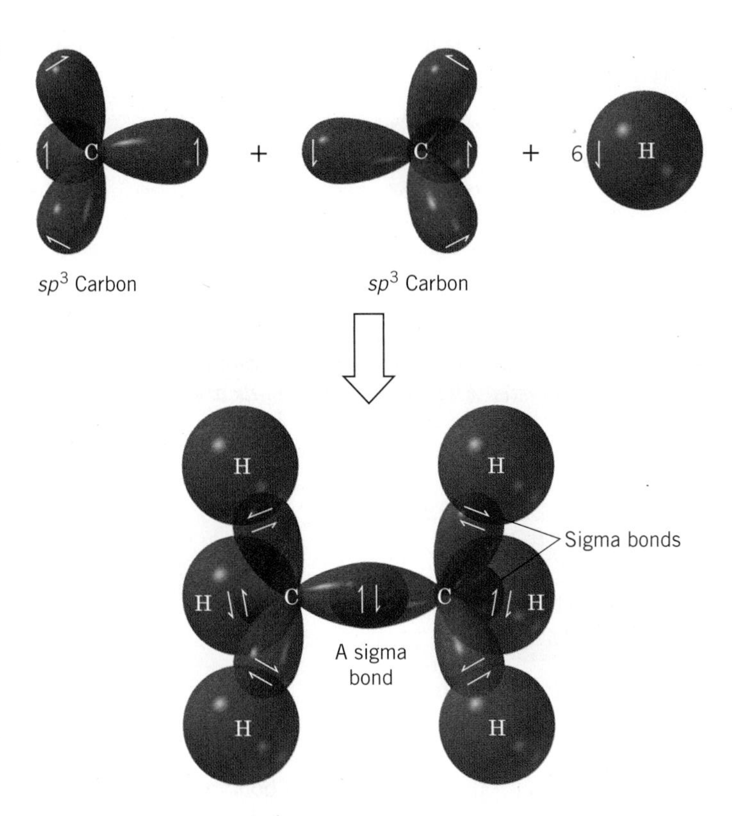

FIGURE 1.17 The hypothetical formation of the bonding molecular orbitals of ethane from two sp^3-hybridized carbon atoms and six hydrogen atoms. All of the bonds are sigma bonds. (Antibonding sigma molecular orbitals—called σ^* orbitals—are formed in each instance as well, but for simplicity these are not shown.)

The carbon–carbon bond of ethane is a *sigma bond* formed by two overlapping sp^3 orbitals. (The carbon–hydrogen bonds are also sigma bonds. They are formed from overlapping carbon sp^3 orbitals and hydrogen *s* orbitals.)

Chem3D Model, Animated Graphics

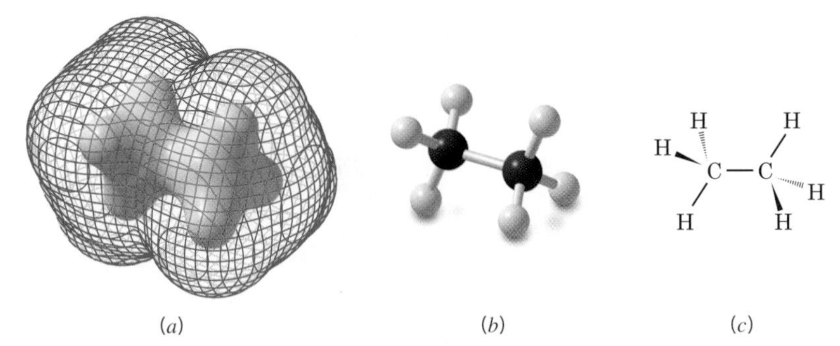

FIGURE 1.18 *(a)* In this structure of ethane, based on quantum mechanical calculations, the inner solid surface represents a region of high electron density. High electron density is found in each bonding region. The outer mesh surface represents approximately the furthest extent of overall electron density for the molecule. *(b)* A ball-and-stick model of ethane, like the kind you might build with a molecular model kit. *(c)* A structural formula for ethane as you would draw it using lines, wedges, and dashed wedges to show in three dimensions its tetrahedral geometry at each carbon.

Because a sigma bond (i.e., any nonmultiple bond) has cylindrical symmetry along the bond axis, *rotation of groups joined by a single bond does not usually require a large amount of energy.* Consequently, groups joined by single bonds rotate relatively freely with respect to one another. (We discuss this point further in Section 4.8.) In Fig. 1.18 we show a calculated structure for ethane in which the tetrahedral geometry derived from orbital hybridization is clearly apparent.

The Chemistry of...

Calculated Molecular Models: Electron Density Surfaces

We will make frequent use in this book of molecular models derived from quantum mechanical calculations. These models will help us visualize the shapes of molecules as well as understand their properties and reactivity. A useful type of model is one that shows a calculated three-dimensional surface at which a chosen value of electron density is the same all around a molecule, called an **electron density surface.** If we make a plot where the value chosen is for low electron density, the result is a van der Waals surface, the surface that represents approximately the overall shape of a molecule as determined by the furthest extent of its electron cloud. On the other hand, if we make a plot where the value of electron density is relatively high, the resulting surface is one that approximately represents the region of covalent bonding in a molecule. Surfaces of low and high electron density are shown in this box for dimethyl ether. Similar models are shown for methane and ethane in Figs. 1.16 and 1.18.

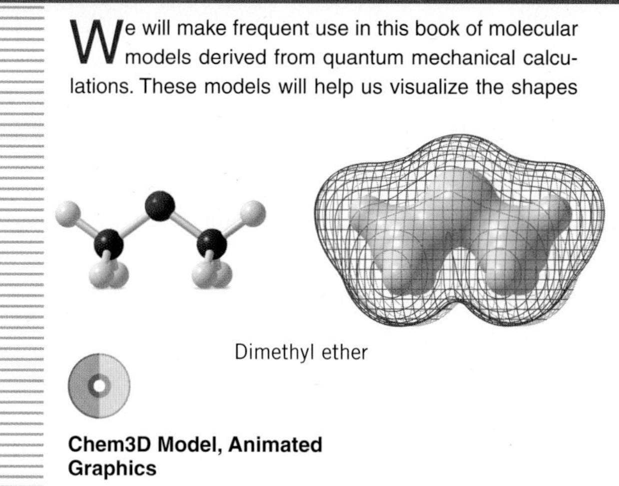

Dimethyl ether

Chem3D Model, Animated Graphics

1.13 The Structure of Ethene (Ethylene): sp^2 Hybridization

The carbon atoms of many of the molecules that we have considered so far have used their four valence electrons to form four single covalent (sigma) bonds to four other atoms. We find, however, that many important organic compounds exist in which carbon atoms share more than two electrons with another atom. In molecules of these compounds some bonds that are formed are multiple covalent bonds. When two carbon atoms share two pairs of electrons, for example, the result is a carbon–carbon double bond:

$$\dot{\underset{\cdot}{C}}\!:\,:\!\dot{\underset{\cdot}{C}} \quad \text{or} \quad \gt C{=}C \lt$$

Hydrocarbons whose molecules contain a carbon–carbon double bond are called **alkenes.** Ethene (C_2H_4) and propene (C_3H_6) are both alkenes. (Ethene is also called ethylene, and propene is sometimes called propylene.)

$$(H)(H)C{=}C(H)(H) \qquad (H)(H_3C)C{=}C(H)(H)$$

Ethene **Propene**

In ethene the only carbon–carbon bond is a double bond. Propene has one carbon–carbon single bond and one carbon–carbon double bond.

The spatial arrangement of the atoms of alkenes is different from that of alkanes. The six atoms of ethene are coplanar, and the arrangement of atoms around each carbon atom is triangular (Fig. 1.19).

FIGURE 1.19 The structure and bond angles of ethene. The plane of the atoms is perpendicular to the paper. The dashed wedge bonds project behind the plane of the paper, and the solid wedge bonds project in front of the paper.

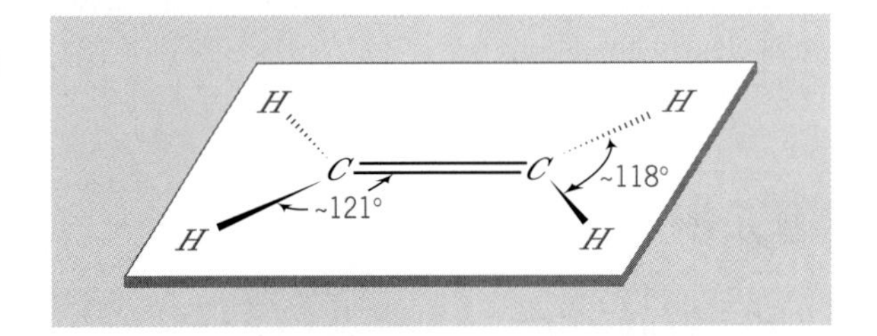

A satisfactory model for the carbon–carbon double bond can be based on sp^2-hybridized carbon atoms.* The mathematical mixing of orbitals that furnish the sp^2 orbitals for our model can be visualized in the way shown in Fig. 1.20. The $2s$ orbital is mathematically mixed (or hybridized) with two of the $2p$ orbitals. (The hybridization procedure applies only to the orbitals, not to the electrons.) One $2p$ orbital is left unhybridized. One electron is then placed in each of the sp^2 hybrid orbitals and one electron remains in the $2p$ orbital.

FIGURE 1.20 A process for obtaining sp^2-hybridized carbon atoms.

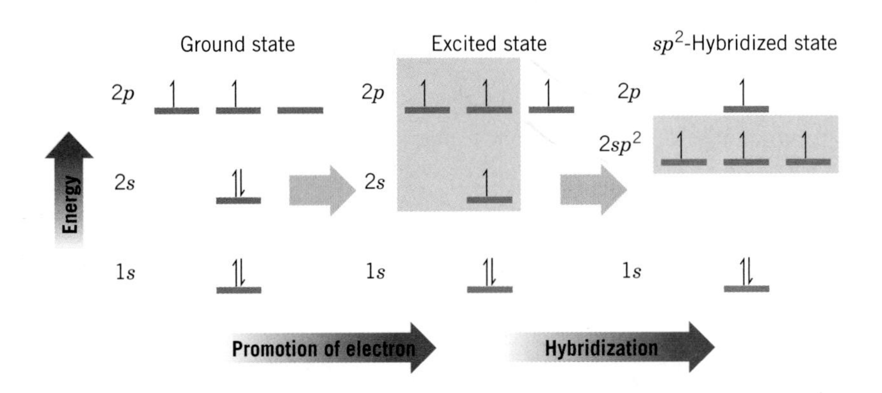

The three sp^2 orbitals that result from hybridization are directed toward the corners of a regular triangle (with angles of 120° between them). The carbon p orbital that is not hybridized is perpendicular to the plane of the triangle formed by the hybrid sp^2 orbitals (Fig. 1.21).

FIGURE 1.21 An sp^2-hybridized carbon atom.

CD Tutorial
sp^2 Hybridization

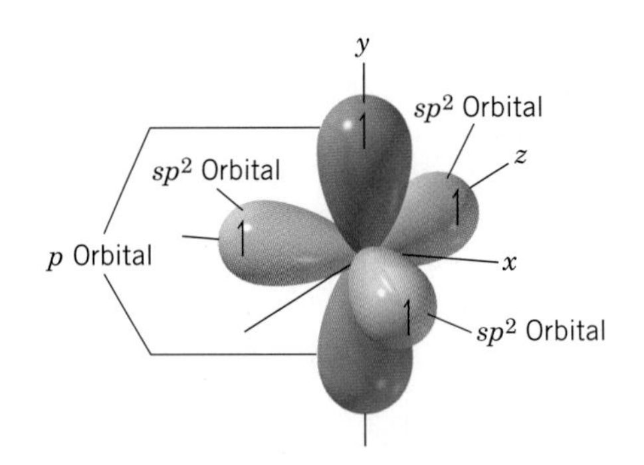

*An alternative model for the carbon–carbon double bond is discussed in an article by W. E. Palke, *J. Am. Chem. Soc.* **1986,** *108,* 6543–6544.

In our model for ethene (Fig. 1.22) we see that two sp^2-hybridized carbon atoms form a sigma (σ) bond between them by the overlap of one sp^2 orbital from each. The remaining sp^2 orbitals of the carbon atoms form σ bonds to four hydrogen atoms through overlap with the 1*s* orbitals of the hydrogen atoms. These five bonds account for 10 of the 12 bonding electrons of ethene, and they are called the **σ-bond framework.** The bond angles that we would predict on the basis of sp^2-hybridized carbon atoms (120° all around) are quite close to the bond angles that are actually found (Fig. 1.19).

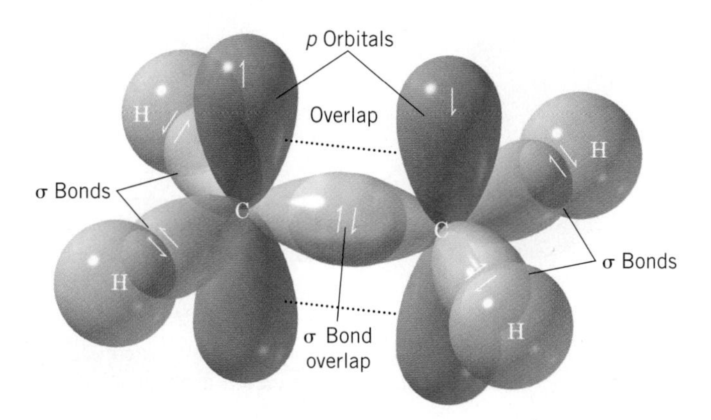

FIGURE 1.22 A model for the bonding molecular orbitals of ethene formed from two sp^2-hybridized carbon atoms and four hydrogen atoms.

CD Tutorial
Ethene

The remaining two bonding electrons in our model are located in the *p* orbitals of each carbon atom. The overlap of these *p* orbitals is shown schematically in Fig. 1.22. We can better visualize how these *p* orbitals interact with each other if we view a structure showing calculated molecular orbitals for ethene (Fig. 1.23). We see that the parallel *p* orbitals *overlap above and below the plane of the σ framework.* This sideways overlap of the *p* orbitals results in a new type of covalent bond, known as a **pi (π) bond.** Note the difference in shape of the bonding molecular orbital of a π bond as contrasted to that of a σ bond. A σ bond has cylindrical symmetry about a line connecting the two bonded nuclei. A π bond has a nodal plane passing through the two bonded nuclei and between the π molecular orbital lobes.

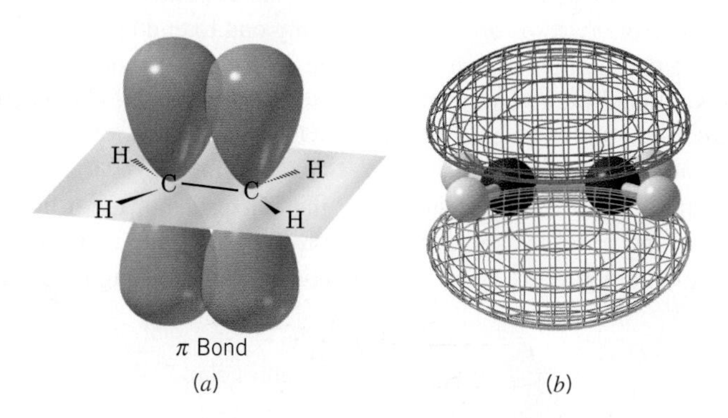

FIGURE 1.23 *(a)* A wedge–dashed wedge formula for the sigma bonds in ethene and a schematic depiction of the overlapping of adjacent *p* orbitals that form the π bond. *(b)* A calculated structure for ethene. The blue and red colors indicate opposite phase signs in each lobe of the π molecular orbital. A ball-and-stick model for the σ bonds in ethene can be seen through the mesh that indicates the π bond.

Chem3D Model, Animated Graphics

According to molecular orbital theory, both bonding and antibonding π molecular orbitals are formed when *p* orbitals interact in this way to form a π bond. The bonding π orbital results when *p*-orbital lobes of like signs overlap; the antibonding π orbital is formed when *p*-orbital lobes of opposite signs overlap (Fig. 1.24).

The bonding π orbital is the lower energy orbital and contains both π electrons (with opposite spins) in the ground state of the molecule. The region of greatest probability of finding the electrons in the bonding π orbital is a region generally situated above and below the plane of the σ-bond framework between the two carbon atoms. The antibonding

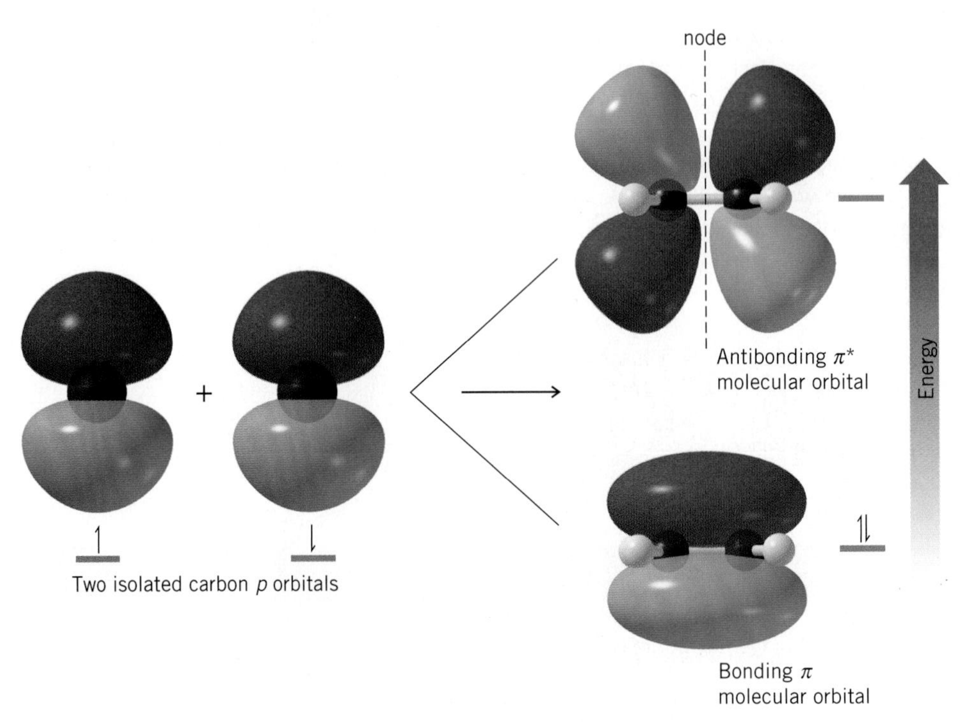

FIGURE 1.24 How two isolated carbon p orbitals combine to form two π (pi) molecular orbitals. The bonding MO is of lower energy. The higher energy antibonding MO contains an additional node. (Both orbitals have a node in the plane containing the C and H atoms.)

π^* orbital is of higher energy, and it is not occupied by electrons when the molecule is in the ground state. It can become occupied, however, if the molecule absorbs light of the right frequency and an electron is promoted from the lower energy level to the higher one. The antibonding π^* orbital has a nodal plane between the two carbon atoms.

sp^2 Hybridization

To summarize: In our model based on orbital hybridization, the carbon–carbon double bond is viewed as consisting of two different kinds of bonds, *a σ bond and a π bond.* The σ bond results from two sp^2 orbitals overlapping end to end and is symmetrical about an axis linking the two carbon atoms. The π bond results from a sideways overlap of two p orbitals; it has a nodal plane like a p orbital. In the ground state the electrons of the π bond are located between the two carbon atoms but generally above and below the plane of the σ-bond framework.

Electrons of the π bond have greater energy than electrons of the σ bond. The relative energies of the σ and π molecular orbitals (with the electrons in the ground state) are shown in the following figure. (The σ^* orbital is the antibonding sigma orbital.)

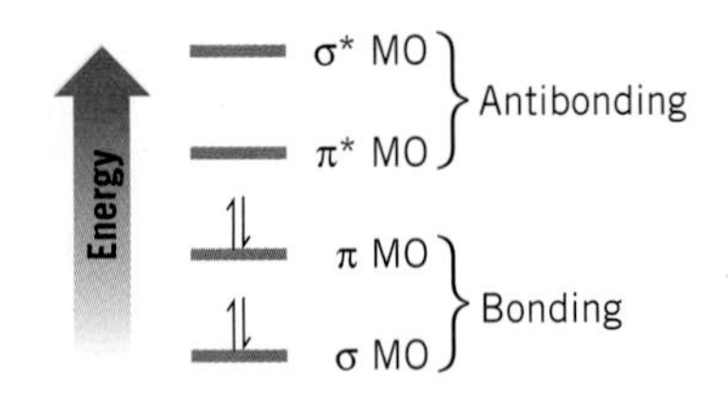

1.13A Restricted Rotation and the Double Bond

The σ–π model for the carbon–carbon double bond also accounts for an important property of the double bond: *There is a large energy barrier to rotation associated with*

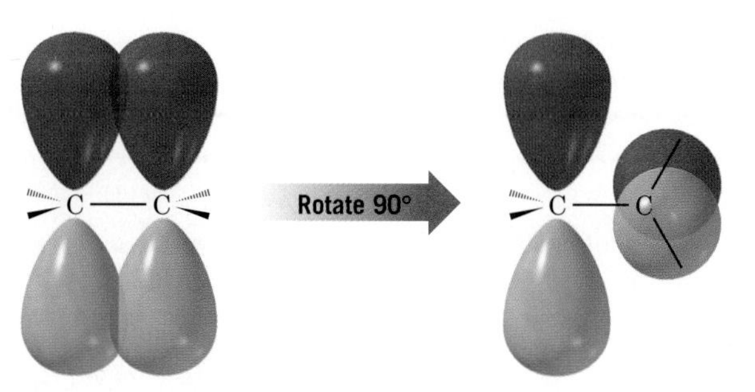

FIGURE 1.25 A stylized depiction of how rotation of a carbon atom of a double bond through an angle of 90° results in breaking of the π bond.

CD Tutorial
Ethene rotation barrier

groups joined by a double bond. Maximum overlap between the *p* orbitals of a π bond occurs when the axes of the *p* orbitals are exactly parallel. Rotating one carbon of the double bond 90° (Fig. 1.25) breaks the π bond, for then the axes of the *p* orbitals are perpendicular and there is no net overlap between them. Estimates based on thermochemical calculations indicate that the strength of the π bond is 264 kJ mol^{-1}. This, then, is the barrier to rotation of the double bond. It is markedly higher than the rotational barrier of groups joined by carbon–carbon single bonds (13–26 kJ mol^{-1}). While groups joined by single bonds rotate relatively freely at room temperature, those joined by double bonds do not.

1.13B Cis–Trans Isomerism

Restricted rotation of groups joined by a double bond causes a new type of isomerism that we illustrate with the two dichloroethenes written as the following structures:

```
Cl   H            H   Cl
  \ /              \ /
   C                C
   ‖                ‖
   C                C
  / \              / \
Cl   H            Cl   H
```

***cis*-1,2-Dichloroethene** ***trans*-1,2-Dichloroethene**

These two compounds are isomers; they are different compounds that have the same molecular formula. We can tell that they are different compounds by trying to place a model of one compound on a model of the other so that all parts coincide, that is, to try to **superpose** one on the other. We find that it cannot be done. Had one been **superposable** on the other, all parts of one model would correspond in three-dimensions exactly with the other model. (*The notion of superposition is different from simply superimposing one thing on another.* The latter means only to lay one on the other without the necessary condition that all parts coincide.)

We indicate that they are different isomers by attaching the prefixes cis or trans to their names (cis, Latin: on this side; trans, Latin: across). *cis*-1,2-Dichloroethene and *trans*-1,2-dichloroethene are not constitutional isomers because the connectivity of the atoms is the same in each. The two compounds ***differ only in the arrangement of their atoms in space.*** Isomers of this kind are classified formally as **stereoisomers,** but often they are called simply cis–trans isomers. (We shall study stereoisomerism in detail in Chapters 4 and 5.)

The structural requirements for cis–trans isomerism will become clear if we consider a few additional examples. 1,1-Dichloroethene and 1,1,2-trichloroethene do not show this type of isomerism.

```
Cl      H          Cl      Cl
  \    /             \    /
   C=C                C=C
  /    \             /    \
Cl      H          Cl      H
```

1,1-Dichloroethene (no cis–trans isomerism) **1,1,2-Trichloroethene (no cis–trans isomerism)**

1,2-Difluoroethene and 1,2-dichloro-1,2-difluoroethene do exist as cis–trans isomers. Notice that we designate the isomer with two identical groups on the same side as being cis:

cis-**1,2-Difluoroethene** *trans*-**1,2-Difluoroethene**

cis-**1,2-Dichloro-1,2-difluoroethene** *trans*-**1,2-Dichloro-1,2-difluoroethene**

Clearly, then, ***cis–trans isomerism of this type is not possible if one carbon atom of the double bond bears two identical groups.***

PROBLEM 1.8 Which of the following alkenes can exist as cis–trans isomers? Write their structures. Build hand-held models to prove that one isomer is not superposable on the other.

(a) $CH_2{=}CHCH_2CH_3$ **(c)** $CH_2{=}C(CH_3)_2$
(b) $CH_3CH{=}CHCH_3$ **(d)** $CH_3CH_2CH{=}CHCl$

1.14 The Structure of Ethyne (Acetylene): *sp* Hybridization

Hydrocarbons in which two carbon atoms share three pairs of electrons between them, and are thus bonded by a triple bond, are called **alkynes.** The two simplest alkynes are ethyne and propyne.

$$H{-}C{\equiv}C{-}H \qquad CH_3{-}C{\equiv}C{-}H$$

Ethyne (acetylene) (C_2H_2) **Propyne (C_3H_4)**

Ethyne, a compound that is also called acetylene, consists of a linear arrangement of atoms. The H—C≡C bond angles of ethyne molecules are 180°:

$$H{-}C{\equiv}C{-}H$$
180° 180°

We can account for the structure of ethyne on the basis of orbital hybridization as we did for ethane and ethene. In our model for ethane (Section 1.12B) we saw that the car-

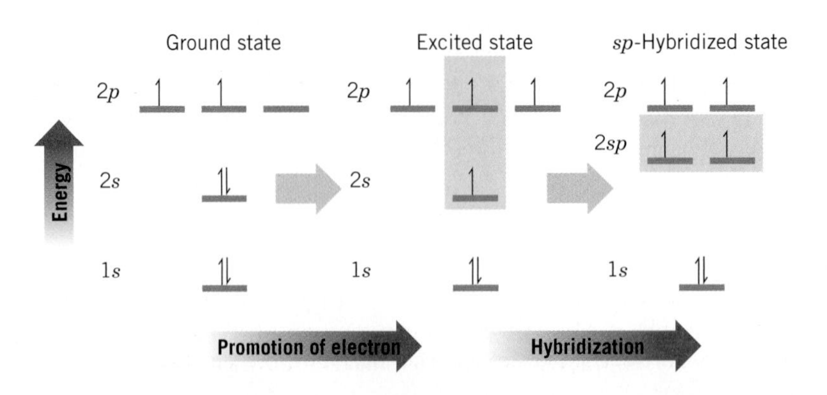

FIGURE 1.26 A process for obtaining *sp*-hybridized carbon atoms.

bon orbitals are sp^3 hybridized, and in our model for ethene (Section 1.13) we saw that they are sp^2 hybridized. In our model for ethyne we shall see that the carbon atoms are *sp hybridized.*

The mathematical process for obtaining the *sp* hybrid orbitals of ethyne can be visualized in the following way (Fig. 1.26). The 2*s* orbital and one 2*p* orbital of carbon are hybridized to form two *sp* orbitals. The remaining two 2*p* orbitals are not hybridized. Calculations show that the *sp* hybrid orbitals have their large positive lobes oriented at an angle of 180° with respect to each other. The two 2*p* orbitals that were not hybridized are each perpendicular to the axis that passes through the center of the two *sp* orbitals (Fig. 1.27).

σ and π bonds

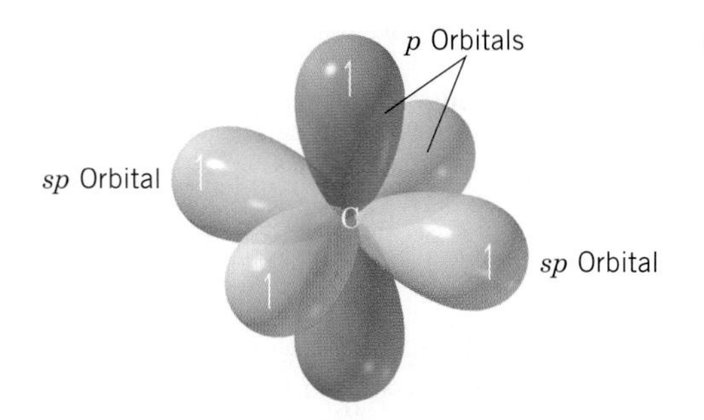

FIGURE 1.27 An *sp*-hybridized carbon atom.

CD Tutorial
sp Hybridization

We envision the bonding molecular orbitals of ethyne being formed in the following way (Fig. 1.28). Two carbon atoms overlap *sp* orbitals to form a sigma bond between them (this is one bond of the triple bond). The remaining two *sp* orbitals at each carbon atom overlap with *s* orbitals from hydrogen atoms to produce two sigma C—H bonds. The two *p* orbitals on each carbon atom also overlap side to side to form two π bonds. These are the other two bonds of the triple bond. Thus we see that **the carbon–carbon triple bond consists of two π bonds and one σ bond.**

sp Hybridization

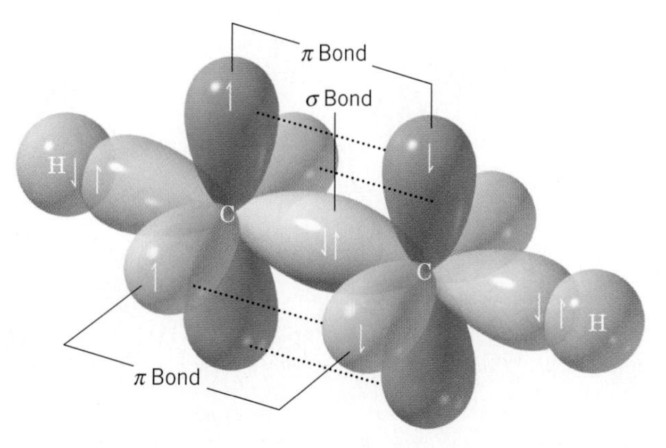

FIGURE 1.28 Formation of the bonding molecular orbitals of ethyne from two *sp*-hybridized carbon atoms and two hydrogen atoms. (Antibonding orbitals are formed as well, but these have been omitted for simplicity.)

CD Tutorial
Ethyne

Structures for ethyne based on calculated molecular orbitals and electron density are shown in Fig. 1.29. Circular symmetry exists along the length of a triple bond (Fig. 1.29*b*). As a result, there is no restriction of rotation for groups joined by a triple bond (as compared with alkenes), and if rotation would occur, no new compound would form.

1.14A Bond Lengths of Ethyne, Ethene, and Ethane

The carbon–carbon triple bond is shorter than the carbon–carbon double bond, and the carbon–carbon double bond is shorter than the carbon–carbon single bond. The carbon–hydrogen bonds of ethyne are also shorter than those of ethene, and the carbon–

Chem3D Model, Animated Graphics

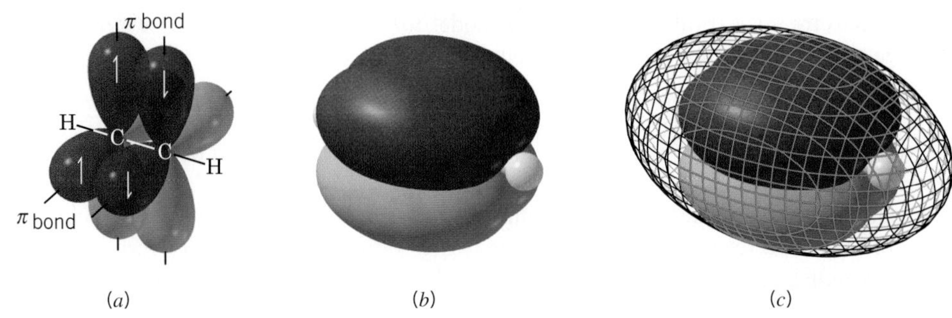

FIGURE 1.29 *(a)* The structure of ethyne (acetylene) showing the sigma-bond framework and a schematic depiction of the two pairs of *p* orbitals that overlap to form the two π bonds in ethyne. *(b)* A structure of ethyne showing calculated π molecular orbitals. Two pairs of π molecular orbital lobes are present, one pair for each π bond. The red and blue lobes in each π bond represent opposite phase signs. The hydrogen atoms of ethyne (white spheres) can be seen at each end of the structure (the carbon atoms are hidden by the molecular orbitals). *(c)* The mesh surface in this structure represents approximately the furthest extent of overall electron density in ethyne. Note that the overall electron density (but not the π-bonding electrons) extends over both hydrogen atoms.

hydrogen bonds of ethene are shorter than those of ethane. This illustrates a general principle: *The shortest C—H bonds are associated with those carbon orbitals with the greatest s character.* The *sp* orbitals of ethyne—50% *s* (and 50% *p*) in character—form the shortest C—H bonds. The sp^3 orbitals of ethane—25% *s* (and 75% *p*) in character—form the longest C—H bonds. The differences in bond lengths and bond angles of ethyne, ethene, and ethane are summarized in Fig. 1.30.

FIGURE 1.30 Bond angles and bond lengths of ethyne, ethene, and ethane.

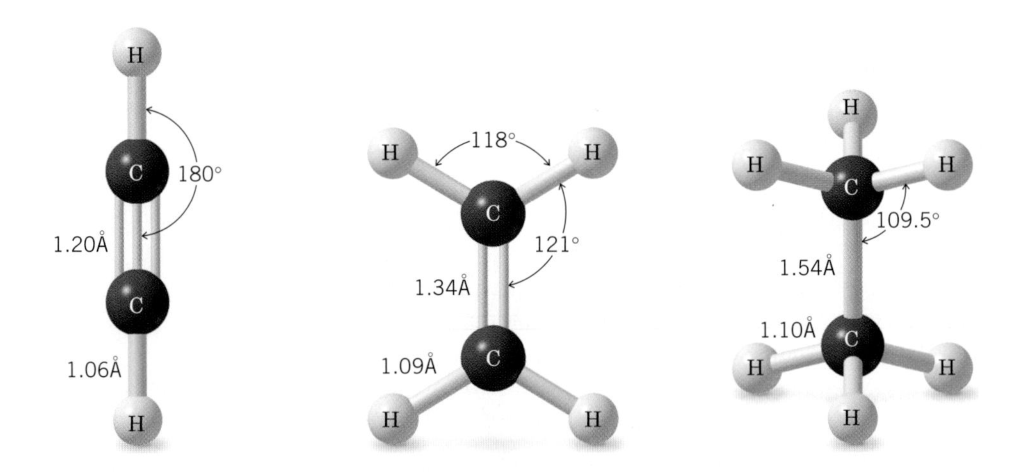

1.15 A Summary of Important Concepts That Come from Quantum Mechanics

1. An **atomic orbital (AO)** corresponds to a region of space about the nucleus of a single atom where there is a high probability of finding an electron. Atomic orbitals called *s* orbitals are spherical; those called *p* orbitals are like two almost-tangent spheres. Orbitals can hold a maximum of two electrons when their spins are paired. Orbitals are described by a wave function, ψ, and each orbital has a characteristic energy. The phase signs associated with an orbital may be + or −.

2. When atomic orbitals overlap, they combine to form **molecular orbitals (MOs).** Molecular orbitals correspond to regions of space encompassing two (or more) nuclei

where electrons are to be found. Like atomic orbitals, molecular orbitals can hold up to two electrons if their spins are paired.

3. When atomic orbitals with the same phase sign interact, they combine to form a **bonding molecular orbital:**

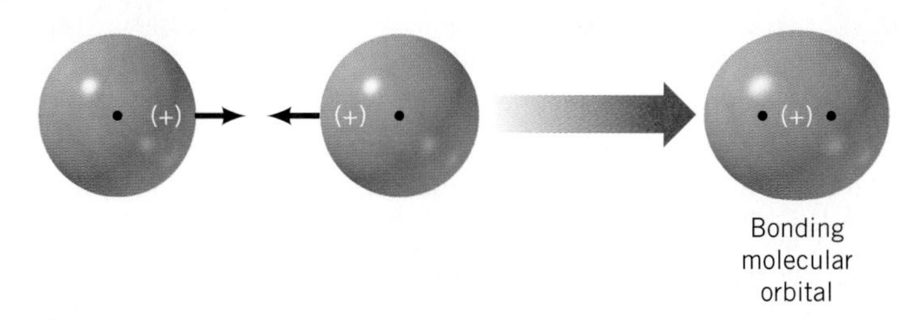

The electron probability density of a bonding molecular orbital is large in the region of space between the two nuclei where the negative electrons hold the positive nuclei together.

4. An **antibonding molecular orbital** forms when orbitals of opposite phase sign overlap:

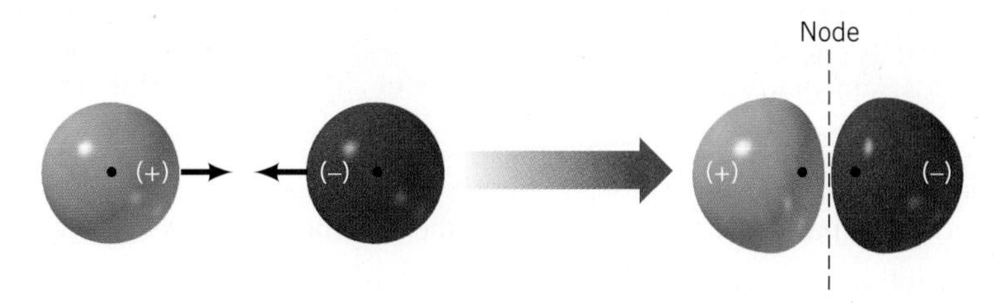

An antibonding orbital has higher energy than a bonding orbital. The electron probability density of the region between the nuclei is small and it contains a **node**—a region where $\psi = 0$. Thus, having electrons in an antibonding orbital does not help hold the nuclei together. The internuclear repulsions tend to make them fly apart.

5. The **energy of electrons** in a bonding molecular orbital is less than the energy of the electrons in their separate atomic orbitals. The energy of electrons in an antibonding orbital is greater than that of electrons in their separate atomic orbitals.

6. The **number of molecular orbitals** always equals the number of atomic orbitals from which they are formed. Combining two atomic orbitals will always yield two molecular orbitals—one bonding and one antibonding.

7. Hybrid atomic orbitals are obtained by mixing (hybridizing) the wave functions for orbitals of different types (i.e., *s* and *p* orbitals) but from the same atom.

A summary of sp^3, sp^2, and *sp* hybrid orbital geometries.

8. Hybridizing three *p* orbitals with one *s* orbital yields four $\boldsymbol{sp^3}$ **orbitals.** Atoms that are sp^3 hybridized direct the axes of their four sp^3 orbitals toward the corners of a tetrahedron. The carbon of methane is sp^3 hybridized and **tetrahedral.**

9. Hybridizing two *p* orbitals with one *s* orbital yields three $\boldsymbol{sp^2}$ **orbitals.** Atoms that are sp^2 hybridized point the axes of three sp^2 orbitals toward the corners of an equilateral triangle. The carbon atoms of ethene are sp^2 hybridized and **trigonal planar.**

10. Hybridizing one *p* orbital with one *s* orbital yields two ***sp* orbitals.** Atoms that are *sp* hybridized orient the axes of their two *sp* orbitals in opposite directions (at an angle of 180°). The carbon atoms of ethyne are *sp* hybridized and ethyne is a **linear** molecule.

11. A **sigma bond** (a type of single bond) is one in which the electron density has circular symmetry when viewed along the bond axis. In general, the skeletons of organic molecules are constructed of atoms linked by sigma bonds.

12. A **pi bond,** part of double and triple carbon–carbon bonds, is one in which the electron densities of two adjacent parallel *p* orbitals overlap sideways to form a bonding pi molecular orbital.

1.16 Molecular Geometry: The Valence Shell Electron Pair Repulsion Model

The VSEPR model is a useful tool for predicting approximate molecular geometry.

We have been discussing the geometry of molecules on the basis of theories that arise from quantum mechanics. It is possible, however, to predict the arrangement of atoms in molecules and ions on the basis of a theory called the **valence shell electron pair repulsion (VSEPR) theory.**

We apply **VSEPR** theory in the following way:

1. We consider molecules (or ions) in which the central atom is covalently bonded to two or more atoms or groups.

2. We consider all of the valence electron pairs of the central atom—both those that are shared in covalent bonds, called **bonding pairs,** and those that are unshared, called **nonbonding pairs** or **unshared pairs.**

3. Because electron pairs repel each other, the electron pairs of the valence shell tend to stay as far apart as possible. The repulsion between nonbonding pairs is generally greater than that between bonding pairs.

4. We arrive at the *geometry* of the molecule by considering all of the electron pairs, bonding and nonbonding, but we describe the *shape* of the molecule or ion by referring to the positions of the nuclei (or atoms) and not by the positions of the electron pairs.

Consider the following examples found in Sections 1.16A–F.

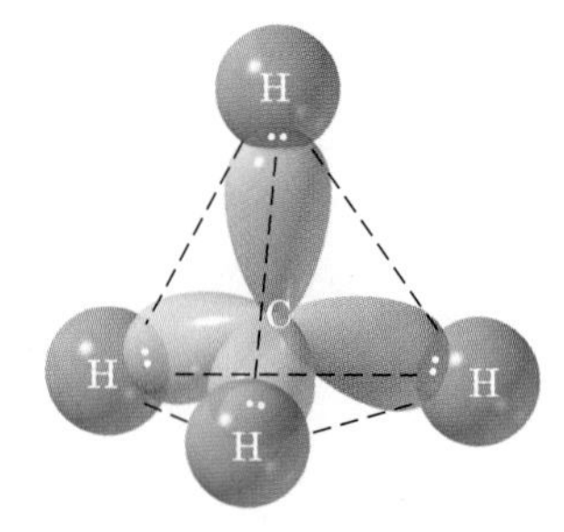

FIGURE 1.31 A tetrahedral shape for methane allows the maximum separation of the four bonding electron pairs.

1.16A Methane

The valence shell of methane contains four pairs of bonding electrons. Only a tetrahedral orientation will allow four pairs of electrons to have the maximum possible separation (Fig. 1.31). Any other orientation, for example, a square planar arrangement, places the electron pairs closer together.

Thus, in the case of methane, the VSEPR model accommodates what we have known since the proposal of van't Hoff and Le Bel (Section 1.3B): The molecule of methane has a tetrahedral shape.

Problem 1.9

Part of the reasoning that led van't Hoff and Le Bel to propose a tetrahedral shape for molecules of methane was based on the number of isomers possible for substituted methanes. For example, only one compound with the formula CH_2Cl_2 has ever been observed (i.e., there is no isomeric form). Consider both a square planar structure and a tetrahedral structure for CH_2Cl_2 and explain how this observation supports a tetrahedral structure.

The bond angles for any atom that has a regular tetrahedral structure are 109.5°. A representation of these angles in methane is shown in Fig. 1.32.

FIGURE 1.32 The bond angles of methane are 109.5°.

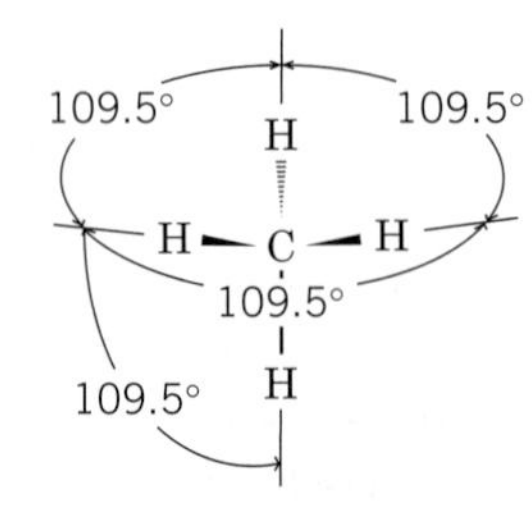

1.16B Ammonia

The shape of a molecule of ammonia (NH_3) is a **trigonal pyramid.** There are three bonding pairs of electrons and one nonbonding pair. The bond angles in a molecule of ammonia are 107°, a value very close to the tetrahedral angle (109.5°). We can write a general tetrahedral structure for the electron pairs of ammonia by placing the nonbonding pair at one corner (Fig. 1.33). A *tetrahedral arrangement* of the electron pairs explains the *trigonal pyramidal* arrangement of the four atoms. The bond angles are 107° (not 109.5°) because the nonbonding pair occupies more space than the bonding pairs.

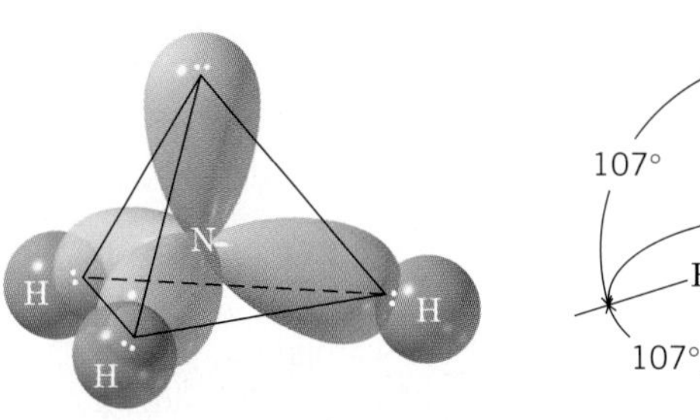

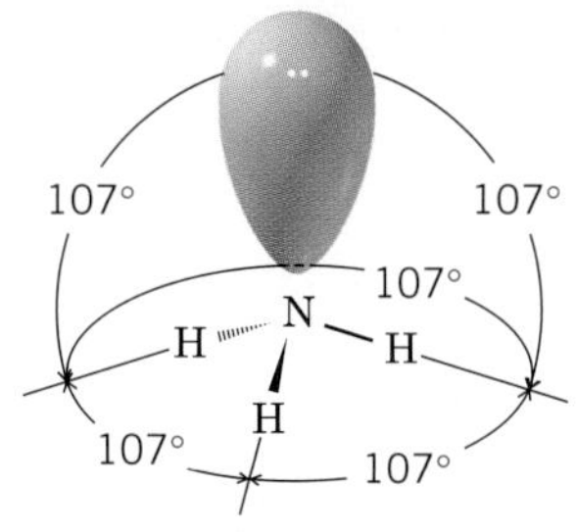

FIGURE 1.33 The tetrahedral arrangement of the electron pairs of an ammonia molecule that results when the nonbonding electron pair is considered to occupy one corner. This arrangement of electron pairs explains the trigonal pyramidal shape of the NH_3 molecule.

1.16C Water

A molecule of water has an **angular** or **bent** shape. The H—O—H bond angle in a molecule of water is 105°, an angle that is also quite close to the 109.5° bond angles of methane.

We can write a general tetrahedral structure for the electron pairs of a molecule of water *if we place the two bonding pairs of electrons and the two nonbonding electron pairs at corners of the tetrahedron.* Such a structure is shown in Fig. 1.34. A *tetrahedral arrangement* of the electron pairs accounts for the *angular arrangement* of the three atoms. The bond angle is less than 109.5° because the nonbonding pairs are effectively "larger" than the bonding pairs and, therefore, the structure is not perfectly tetrahedral.

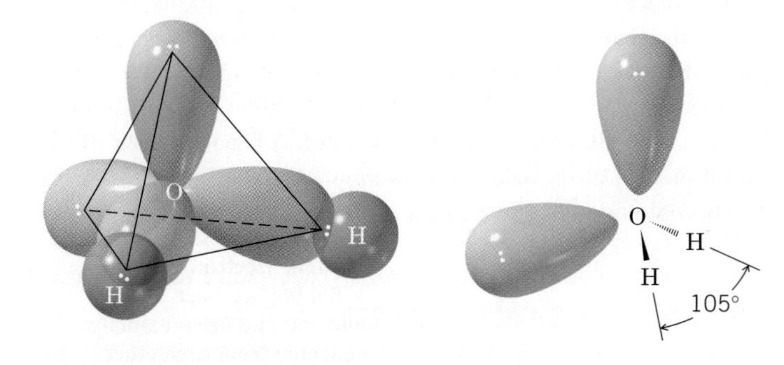

FIGURE 1.34 An approximately tetrahedral arrangement of the electron pairs of a molecule of water that results when the pairs of nonbonding electrons are considered to occupy corners. This arrangement accounts for the angular shape of the H_2O molecule.

1.16D Boron Trifluoride

Boron, a Group 3A element, has only three outer-level electrons. In the compound boron trifluoride (BF_3) these three electrons are shared with three fluorine atoms. As a result, the boron atom in BF_3 has only six electrons (three bonding pairs) around it. Maximum separation of three bonding pairs occurs when they occupy the corners of an equilateral triangle. Consequently, in the boron trifluoride molecule the three fluorine atoms lie in a plane at the corners of an equilateral triangle (Fig. 1.35). Boron trifluoride is said to have a *trigonal planar structure.* The bond angles are 120°.

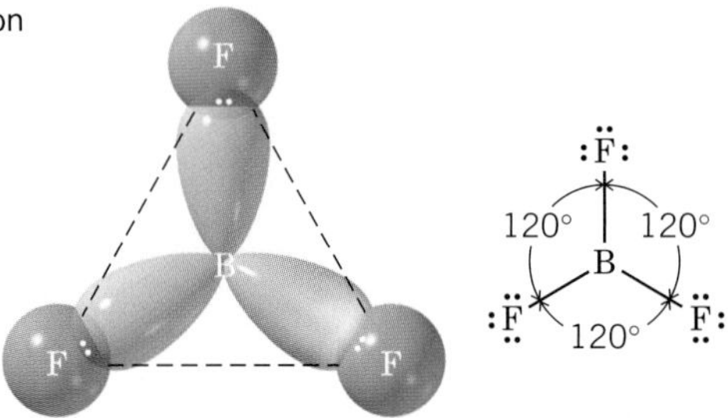

FIGURE 1.35 The triangular (trigonal planar) shape of boron trifluoride maximally separates the three bonding pairs.

1.16E Beryllium Hydride

CD Tutorial
Beryllium hydride

The central beryllium atom of BeH_2 has only two electron pairs around it; both electron pairs are bonding pairs. These two pairs are maximally separated when they are on opposite sides of the central atom, as shown in the following structures. This arrangement of the electron pairs accounts for the *linear geometry* of the BeH_2 molecule and its bond angle of 180°.

H:Be:H or H—Be—H (180°)

Linear geometry of BeH_2

PROBLEM 1.10 Use VSEPR theory to predict the geometry of each of the following molecules and ions:
(a) BH_4^- **(c)** NH_4^+ **(e)** BH_3 **(g)** SiF_4
(b) BeF_2 **(d)** H_2S **(f)** CF_4 **(h)** $:CCl_3^-$

1.16F Carbon Dioxide

The VSEPR method can also be used to predict the shapes of molecules containing multiple bonds if we assume that *all of the electrons of a multiple bond act as though they were a single unit* and, therefore, are located in the region of space between the two atoms joined by a multiple bond.

This principle can be illustrated with the structure of a molecule of carbon dioxide (CO_2). The central carbon atom of carbon dioxide is bonded to each oxygen atom by a double bond. Carbon dioxide is known to have a linear shape; the bond angle is 180°.

:Ö=C=Ö: or :Ö::C::Ö: (180°)

The four electrons of each double bond act as a single unit and are maximally separated from each other.

Such a structure is consistent with a maximum separation of the two groups of four bonding electrons. (The nonbonding pairs associated with the oxygen atoms have no effect on the shape.)

PROBLEM 1.11 Predict the bond angles of **(a)** $F_2C{=}CF_2$, **(b)** $CH_3C{\equiv}CCH_3$, **(c)** $HC{\equiv}N$.

TABLE 1.4 Shapes of Molecules and Ions from VSEPR Theory

Number of Electron Pairs at Central Atom			Hybridization State of Central Atom	Shape of Molecule or Ion[a]	Examples
Bonding	Nonbonding	Total			
2	0	2	*sp*	Linear	BeH_2
3	0	3	sp^2	Trigonal planar	BF_3, CH_3^+
4	0	4	sp^3	Tetrahedral	CH_4, NH_4^+
3	1	4	~ sp^3	Trigonal pyramidal	NH_3, CH_3^-
2	2	4	~ sp^3	Angular	H_2O

[a]Referring to positions of atoms and excluding nonbonding pairs.

The shapes of several simple molecules and ions as predicted by VSEPR theory are shown in Table 1.4. In this table we have also included the hybridization state of the central atom.

1.17 Representation of Structural Formulas

Organic chemists use a variety of ways to write structural formulas. The most common types of representations are shown in Fig. 1.36. The **dot structure** shown further below shows all of the valence electrons, but writing it is tedious and time-consuming. The other representations are more convenient and are, therefore, more often used.

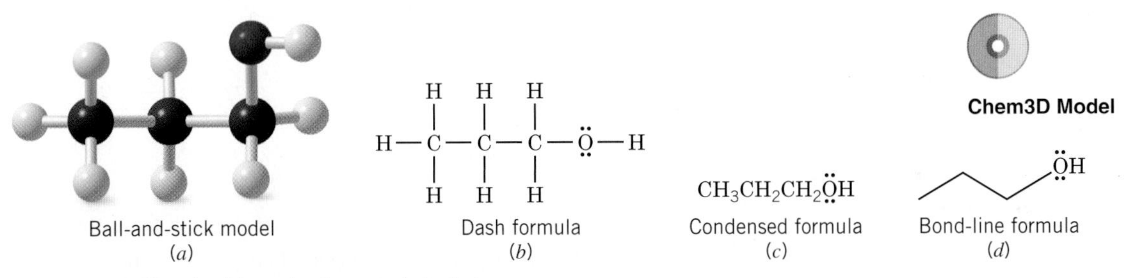

FIGURE 1.36 Structural formulas for propyl alcohol.

Sometimes we even omit unshared pairs when we write formulas. However, when we write chemical reactions, we see that it is necessary to include the unshared electron pairs when they participate in the reaction. It is a good idea, therefore, to get into the habit of writing the unshared (nonbonding) electron pairs in the structures you draw.

$$\mathrm{H{:}\overset{H}{\underset{H}{C}}{:}\ddot{\underset{\cdot\cdot}{O}}{:}\overset{H}{\underset{H}{C}}{:}H} = \mathrm{H{-}\overset{H}{\underset{H}{C}}{-}\ddot{\underset{\cdot\cdot}{O}}{-}\overset{H}{\underset{H}{C}}{-}H} = \mathrm{CH_3\ddot{\underset{\cdot\cdot}{O}}CH_3}$$

Dot structure **Dash formula** **Condensed formula**

1.17A Dash Structural Formulas

If we look at the model for propyl alcohol given in Fig. 1.36*a* and compare it with the dash and condensed formulas in Figs. 1.36*b,c,* we find that the chain of atoms is straight in those formulas. In the model, which corresponds more accurately to the actual shape of the molecule, the chain of atoms is not at all straight. Also of importance is this: ***Atoms joined by single bonds can rotate relatively freely with respect to one another.*** (We discussed this point briefly in Section 1.12B.) This relatively free rotation means that the chain of atoms in propyl alcohol can assume a variety of arrangements like those that follow:

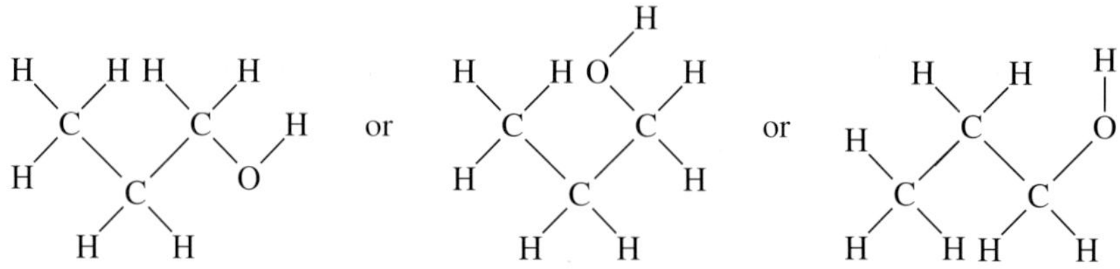

Equivalent dash formulas for propyl alcohol

STUDY TIP!

It is important that you be able to recognize when a set of structural formulas has the same connectivity versus when they are constitutional isomers.

It also means that all of the dash structures above are *equivalent* and all represent propyl alcohol. (Notice that in these formulas we represent the bond angles as being 90°, not 109.5°. This convention is followed simply for convenience in printing.)

Structural formulas such as these indicate the way in which the atoms are attached to each other and *are not* representations of the actual shapes of the molecule. They show what is called the **connectivity** of the atoms. *Constitutional isomers (Section 1.3A) have different connectivity and, therefore, must have different structural formulas.*

Consider the compound called isopropyl alcohol, whose formula we might write in a variety of ways:

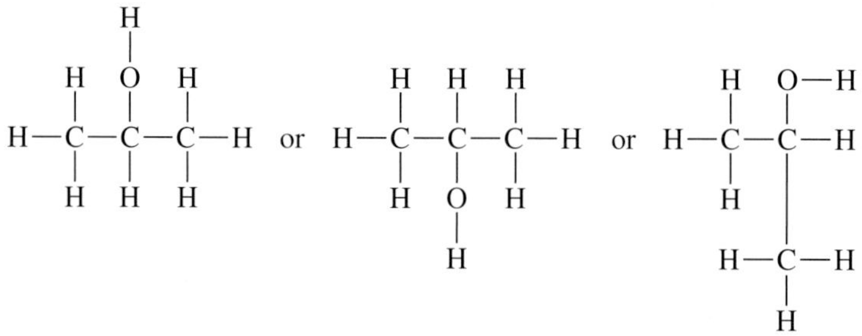

Equivalent dash formulas for isopropyl alcohol

Isopropyl alcohol is a constitutional isomer (Section 1.3A) of propyl alcohol because its atoms are connected in a different order and both compounds have the same molecular formula, C_3H_8O. In isopropyl alcohol the OH group is attached to the central carbon; in propyl alcohol it is attached to an end carbon.

One other point: In problems you will often be asked to write structural formulas for all the isomers with a given molecular formula. Do not make the error of writing several equivalent formulas, like those that we have just shown, mistaking them for different constitutional isomers.

PROBLEM 1.12 There are actually three constitutional isomers with the molecular formula C_3H_8O. We have seen two of them in propyl alcohol and isopropyl alcohol. Write a dash formula for the third isomer.

1.17B Condensed Structural Formulas

Proper use of condensed structural formulas is an essential tool in organic chemistry.

Condensed structural formulas are easier to write than dash formulas and, when we become familiar with them, they will impart all the information that is contained in the dash structure. In condensed formulas all of the hydrogen atoms that are attached to a particular carbon are usually written immediately after the carbon. In fully condensed formulas, all of the atoms that are attached to the carbon are usually written immediately after that carbon, listing hydrogens first. For example,

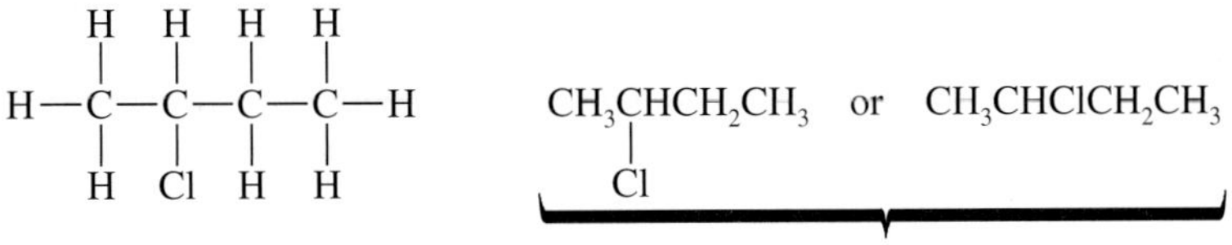

```
   H  H  H  H
   |  |  |  |
H—C—C—C—C—H          CH3CHCH2CH3   or   CH3CHClCH2CH3
   |  |  |  |               |
   H  Cl H  H               Cl
```

Dash formula — $H{-}C(H)(H){-}C(H)(Cl){-}C(H)(H){-}C(H)(H){-}H$

Condensed formulas — $CH_3CH(Cl)CH_2CH_3$ (Cl drawn below the second C) or $CH_3CHClCH_2CH_3$

The condensed formula for isopropyl alcohol can be written in four different ways:

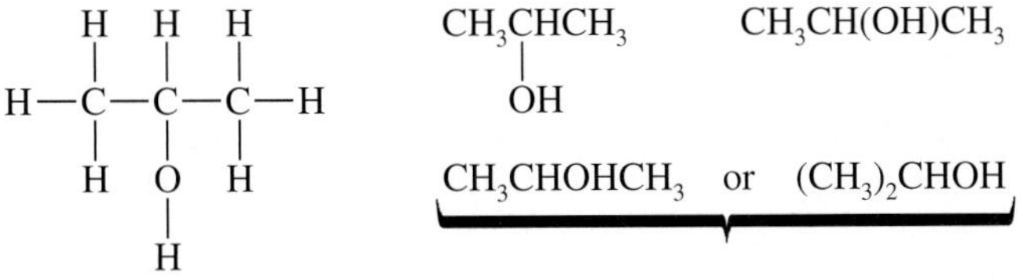

```
   H  H  H
   |  |  |                CH3CHCH3        CH3CH(OH)CH3
H—C—C—C—H                    |
   |  |  |                   OH
   H  O  H
      |                   CH3CHOHCH3  or  (CH3)2CHOH
      H
```

Dash formula

Condensed formulas — $CH_3CH(OH)CH_3$ (OH drawn below the second C), $CH_3CH(OH)CH_3$, $CH_3CHOHCH_3$ or $(CH_3)_2CHOH$

SOLVED PROBLEM

Write a condensed structural formula for the compound that follows:

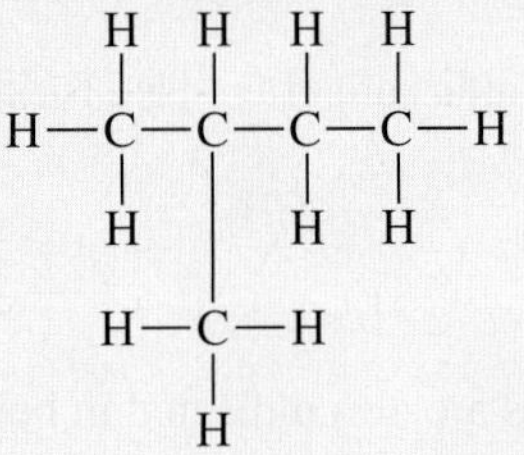

```
   H  H  H  H
   |  |  |  |
H—C—C—C—C—H
   |  |  |  |
   H  |  H  H
      |
   H—C—H
      |
      H
```

ANSWER

$CH_3CHCH_2CH_3$ (with CH_3 drawn below the second C) or $CH_3CH(CH_3)CH_2CH_3$ or $(CH_3)_2CHCH_2CH_3$

or $CH_3CH_2CH(CH_3)_2$ or $CH_3CH_2CHCH_3$ (with CH_3 drawn below the third C)

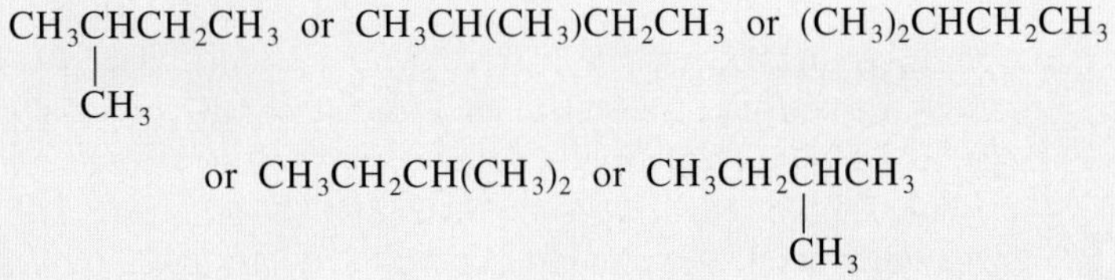

1.17C Cyclic Molecules

Organic compounds not only have their carbon atoms arranged in chains, they can also have them arranged in rings. The compound called cyclopropane has its carbon atoms arranged in a three-membered ring:

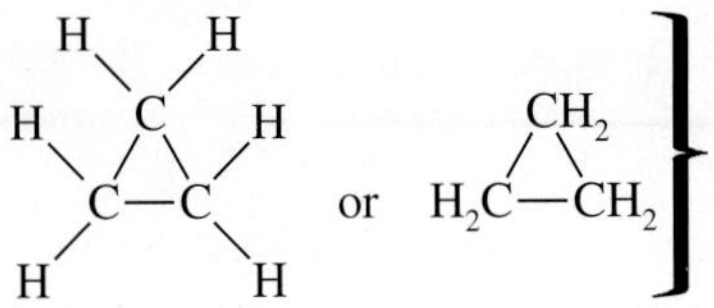

```
  H   H
   \ /
H   C   H              CH2
 \ / \ /               /  \
  C—C       or     H2C—CH2
 /    \
H      H
```

Formulas for cyclopropane

1.17D Bond-Line Formulas

Many organic chemists use a very simplified formula called a **bond-line formula** to represent structural formulas. The bond-line representation is the quickest of all to write be-

As you become more familiar with organic molecules, you will find bond-line formulas to be very useful tools for representing structures.

cause it shows only the carbon skeleton. The number of hydrogen atoms necessary to fulfill the carbon atoms' valences are assumed to be present, but we do not write them in. Other atoms (e.g., O, Cl, N) *are* written in. Each intersection of two or more lines and the end of a line represent a carbon atom unless some other atom is written in:

$CH_3CHClCH_2CH_3$ = 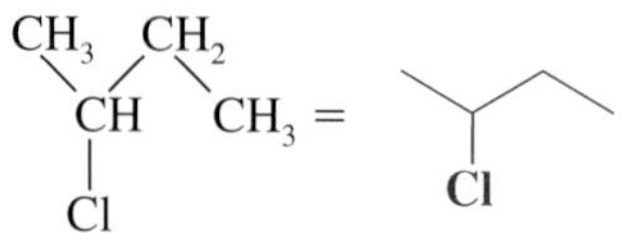

CD Tutorial

Bond-line formulas

$CH_3CH(CH_3)CH_2CH_3$ = 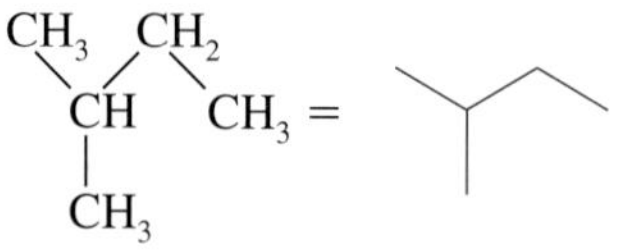

$(CH_3)_2NCH_2CH_3$ = 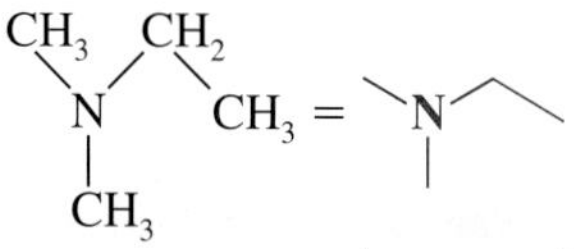

Bond-line formulas

Bond-line formulas are often used for cyclic compounds:

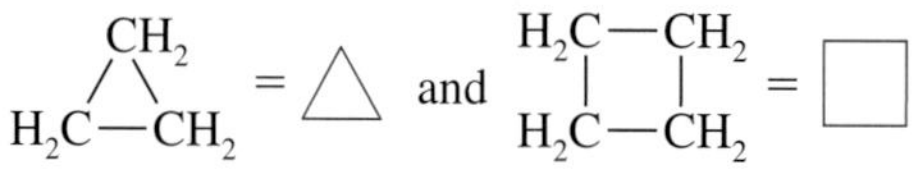

Multiple bonds are also indicated in bond-line formulas. For example:

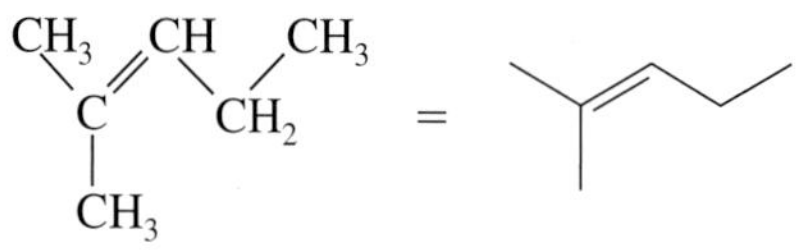

$CH_2{=}CHCH_2OH$ = 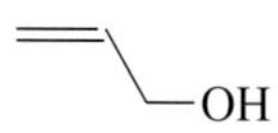

SOLVED PROBLEM

Write the bond-line formula for

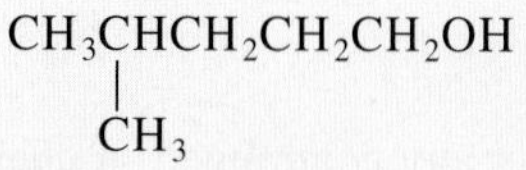

ANSWER First, we outline the carbon skeleton, including the OH group, as follows:

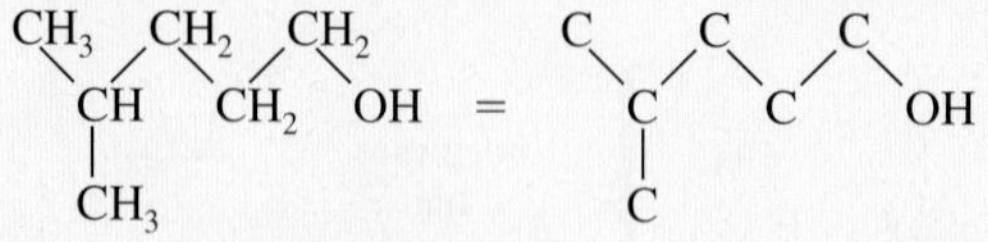

Thus, the bond-line formula is

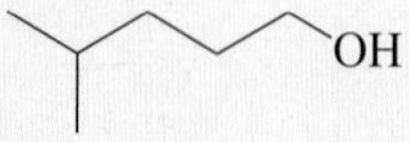

PROBLEM 1.13

Outline the carbon skeleton of the following condensed structural formulas and then write each as a bond-line formula:

(a) $(CH_3)_2CHCH_2CH_3$
(b) $(CH_3)_2CHCH_2CH_2OH$
(c) $(CH_3)_2C{=}CHCH_2CH_3$
(d) $CH_3CH_2CH_2CH_2CH_3$
(e) $CH_3CH_2CH(OH)CH_2CH_3$
(f) $CH_2{=}C(CH_2CH_3)_2$
(g) $CH_3\overset{O}{\overset{\|}{C}}CH_2CH_2CH_2CH_3$
(h) $CH_3CHClCH_2CH(CH_3)_2$

PROBLEM 1.14

Which molecules in Problem 1.13 form sets of constitutional isomers?

PROBLEM 1.15

Write a dash formula for each of the following bond-line formulas:

(a) 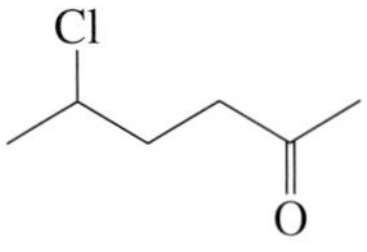 Cl, O (b) 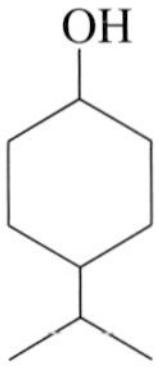 OH (c)

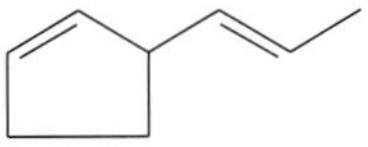

1.17E Three-Dimensional Formulas

None of the formulas that we have described so far conveys any information about how the atoms of a molecule are arranged in space. There are several types of representations that do this. The type of formula that we shall use is shown in Fig. 1.37. In this representation, bonds that project upward out of the plane of the paper are indicated by a wedge (◀), those that lie behind the plane are indicated with a dashed wedge (······), and those bonds that lie in the plane of the page are indicated by a line (—). For tetrahedral atoms, notice that we draw the two bonds that are in the plane of the page with an angle of approximately 109° between them and that proper three-dimensional perspective then requires the wedge and dashed-wedge bonds to be drawn near each other on the page (i.e., the atom in front nearly eclipses the atom behind). We can draw trigonal planar atoms either with all bonds in the plane of the page separated by approximately 120° or with one of the three bonds in the plane of the page, one behind, and one in front (as in Fig. 1.19). Atoms with linear bonding geometry are best drawn with all bonds in the plane of the page. Generally we only use three-dimensional formulas when it is necessary to convey information about the shape of the molecule.

Wedge and dashed-wedge formulas are a tool for unambiguously showing three dimensions.

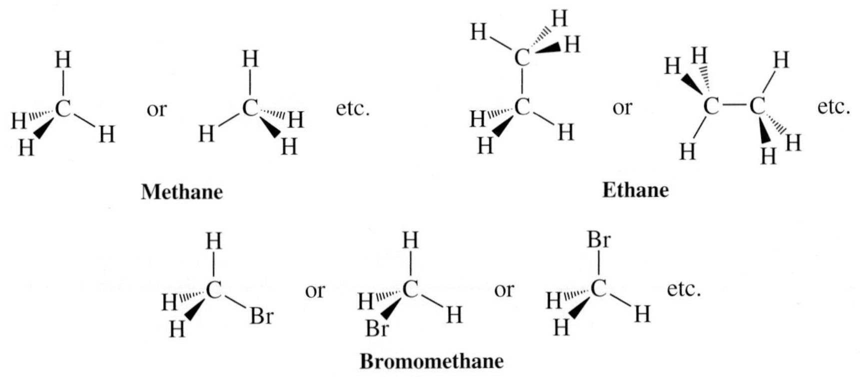

Methane: H–C(H)(H)H or ... etc.

Ethane: ... or ... etc.

Bromomethane: ... or ... or ... etc.

FIGURE 1.37 Three-dimensional formulas using wedge–dashed wedge–line formulas.

PROBLEM 1.16

Write three-dimensional (wedge–dashed wedge–line) representations for each of the following: **(a)** CH_3Cl, **(b)** CH_2Cl_2, **(c)** CH_2BrCl, **(d)** CH_3CH_2Cl.

Key Terms and Concepts

Atomic orbitals	Sections 1.10, 1.11, 1.15
Bond-line formulas	Section 1.17D
Cis–trans isomerism	Section 1.13B
Condensed structural formulas	Section 1.17B
Constitutional isomers	Section 1.3A
Connectivity	Sections 1.3A, 1.17A
Covalent bonds	Section 1.4B
Curved arrows	Section 1.8
Dash structural formulas	Section 1.17A
Double bonds	Sections 1.3, 1.13
Electron density surfaces	Section 1.12B
Electronegativity	Section 1.4A
Electron probability density	Section 1.10
Electrostatic potential maps	Section 1.8
Formal charge	Section 1.7
Hybrid atomic orbitals	Sections 1.12, 1.15
Ionic bonds	Section 1.4A
Isomers	Section 1.3A
LCAO (linear combination of atomic orbitals)	Section 1.11
Lewis structures	Section 1.5
Molecular orbitals	Sections 1.11, 1.15
Octet rule	Sections 1.4, 1.6
Orbital	Section 1.10
Resonance structures	Sections 1.8, 1.8A
Salts	Section 1.4A
Single bonds	Sections 1.3, 1.12
Stereoisomers	Section 1.13B
Structural formulas	Sections 1.3A, 1.17
Superposable	Section 1.13B
Triple bonds	Sections 1.3, 1.14
Wave function (ψ)	Section 1.9
VSEPR model	Section 1.16

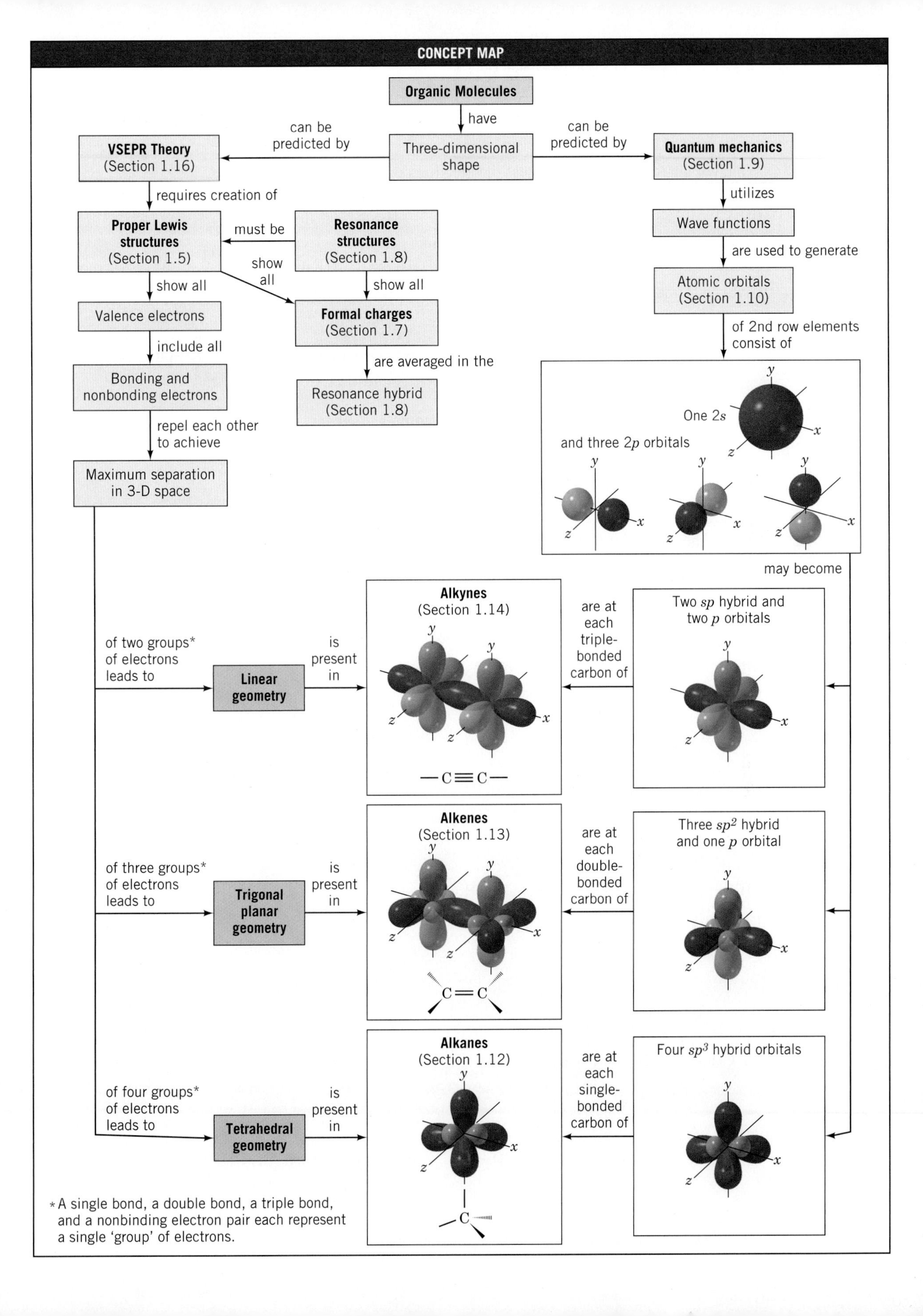

CONCEPT MAP
Organic Molecules
have
Three-dimensional shape
can be predicted by
VSEPR Theory (Section 1.16)
can be predicted by
Quantum mechanics (Section 1.9)
requires creation of
Proper Lewis structures (Section 1.5)
must be
Resonance structures (Section 1.8)
show all
Valence electrons
show all
Formal charges (Section 1.7)
show all
include all
Bonding and nonbonding electrons
are averaged in the
Resonance hybrid (Section 1.8)
repel each other to achieve
Maximum separation in 3-D space
utilizes
Wave functions
are used to generate
Atomic orbitals (Section 1.10)
of 2nd row elements consist of
One 2s
and three 2p orbitals
may become
of two groups* of electrons leads to
Linear geometry
is present in
Alkynes (Section 1.14)
are at each triple-bonded carbon of
Two sp hybrid and two p orbitals
of three groups* of electrons leads to
Trigonal planar geometry
is present in
Alkenes (Section 1.13)
are at each double-bonded carbon of
Three sp2 hybrid and one p orbital
of four groups* of electrons leads to
Tetrahedral geometry
is present in
Alkanes (Section 1.12)
are at each single-bonded carbon of
Four sp3 hybrid orbitals
*A single bond, a double bond, a triple bond, and a nonbinding electron pair each represent a single 'group' of electrons.

ADDITIONAL PROBLEMS

1.17 Which of the following ions possess the electron configuration of a noble gas?

(a) Na^+ **(c)** F^+ **(e)** Ca^{2+} **(g)** O^{2-}
(b) Cl^- **(d)** H^- **(f)** S^{2-} **(h)** Br^+

1.18 Write a Lewis structure for each of the following:

(a) $SOCl_2$ **(b)** $POCl_3$ **(c)** PCl_5 **(d)** $HONO_2$ (HNO_3)

1.19 Give the formal charge (if one exists) on each atom of the following:

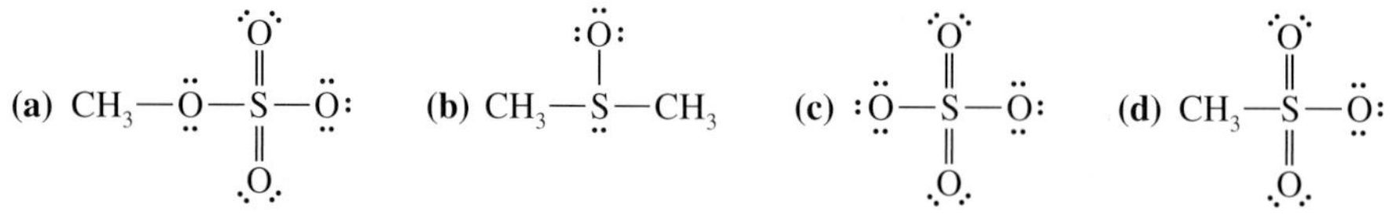

(a) $CH_3—\ddot{O}—S(=O)(=O)—\ddot{O}:$ **(b)** $CH_3—\ddot{S}(—\ddot{O}:)—CH_3$ **(c)** $:\ddot{O}—S(=O)(=O)—\ddot{O}:$ **(d)** $CH_3—S(=O)(=O)—\ddot{O}:$

1.20 Write a condensed structural formula for each compound given here.

(a) 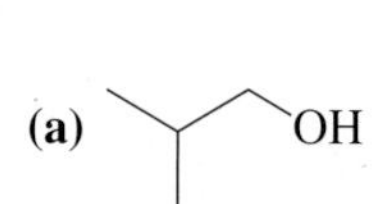 **(b)** 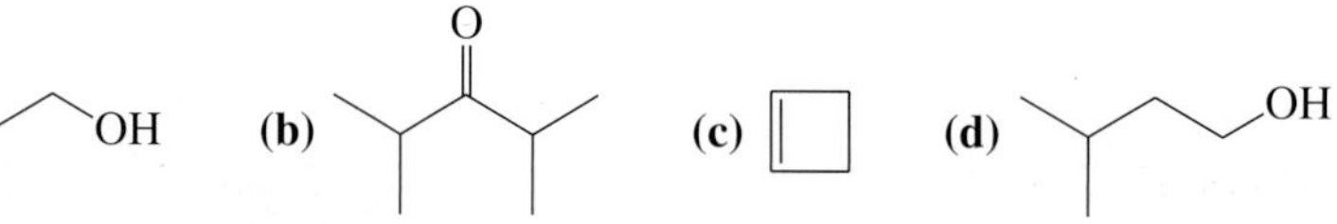 **(c)** 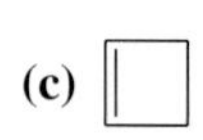 **(d)**

1.21 What is the molecular formula for each of the compounds given in Problem 1.20?

1.22 Consider each pair of structural formulas that follow and state whether the two formulas represent the same compound, whether they represent different compounds that are constitutional isomers of each other, or whether they represent different compounds that are not isomeric.

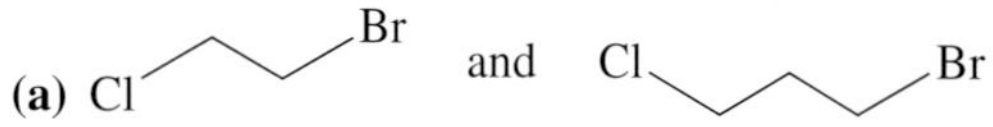

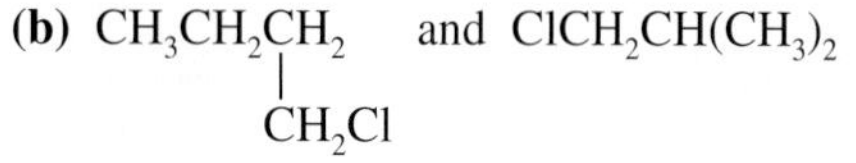

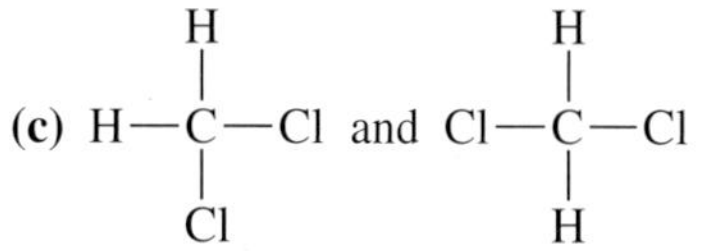

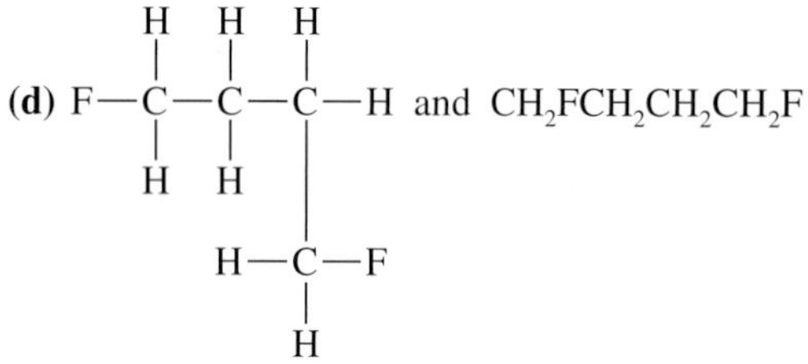

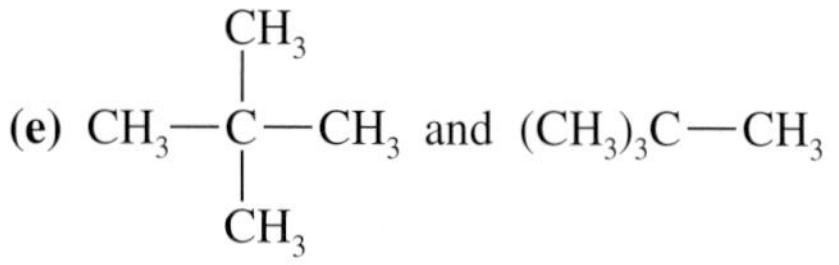

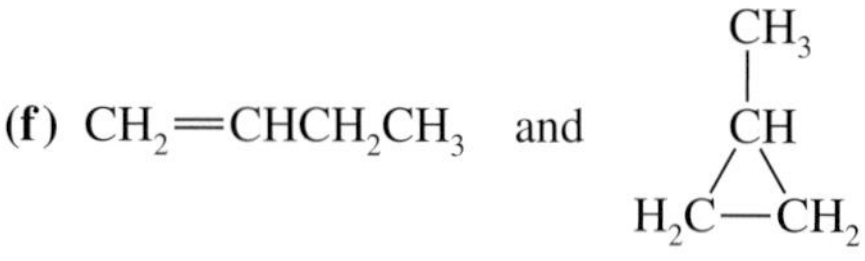

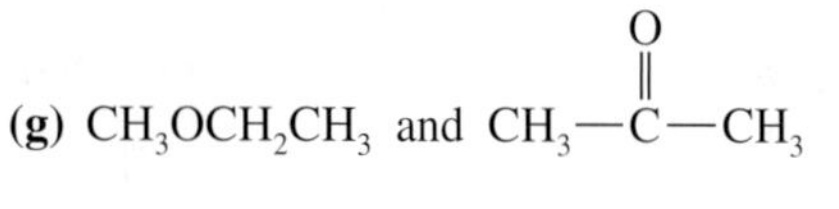

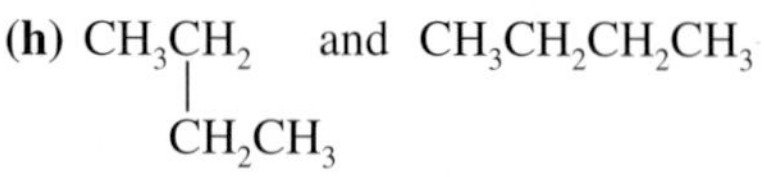

(a) Cl‒CH₂CH₂CH₂‒Br (bond-line) and Cl‒CH₂CH₂CH₂CH₂‒Br (bond-line)

(b) $CH_3CH_2CH_2$ (with CH_2Cl attached) and $ClCH_2CH(CH_3)_2$

(c) $H—C(H)(Cl)—Cl$ and $Cl—C(H)(H)—Cl$

(d) $F—CH_2—CH_2—CH(—CH_2F)—H$ and $CH_2FCH_2CH_2CH_2F$

(e) $CH_3—C(CH_3)_2—CH_3$ and $(CH_3)_3C—CH_3$

(f) $CH_2{=}CHCH_2CH_3$ and $CH_3—CH$ (ring with $H_2C—CH_2$)

(g) $CH_3OCH_2CH_3$ and $CH_3—C(=O)—CH_3$

(h) CH_3CH_2 (with CH_2CH_3 attached) and $CH_3CH_2CH_2CH_3$

Note: Problems marked with an asterisk are "challenge problems."

(i) $CH_3OCH_2CH_3$ and $H_2C{-}C({=}O){-}CH_2$ (three-membered ring: C=O above, H_2C—CH_2)

(j) $CH_2ClCHClCH_3$ and $CH_3CHClCH_2Cl$

(k) $CH_3CH_2CHClCH_2Cl$ and CH_3CHCH_2Cl with CH_2Cl on C-2

(l) CH_3CCH_3 (with C=O) and $H_2C{-}C({=}O){-}CH_2$ (three-membered ring)

(m) H—C(Cl)(H)—Br and Cl—C(H)(H)—Br

(n) CH_3—C(CH_3)(H)—H and CH_3—C(H)(H)—CH_3

(o) H, H, F on C—C with H, H, F (wedge–dash formula) and H, H, F on C—C with H, F, H

(p) H, H, F on C—C with H, H, F (wedge–dash formula) and F, H, F on C—C with H, H, H

1.23 Rewrite each of the following using bond-line formulas:

(a) $CH_3CH_2CH_2CCH_3$ (with C=O)

(b) $CH_3CHCH_2CH_2CHCH_2CH_3$ with CH_3 on C-2 and CH_3 on C-5

(c) $(CH_3)_3CCH_2CH_2CH_2OH$

(d) $CH_3CH_2CHCH_2COH$ (with C=O) and CH_3 on C-3

(e) $CH_2{=}CHCH_2CH_2CH{=}CHCH_3$

(f) Six-membered ring: C(=O), CH_2, CH_2, C(H_2), HC=HC

1.24 Write a dash formula for each of the following showing any unshared electron pairs:

(a) 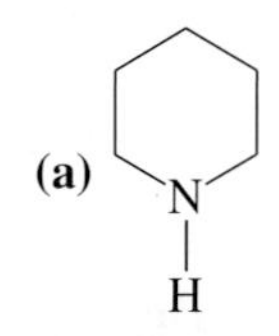

(b) 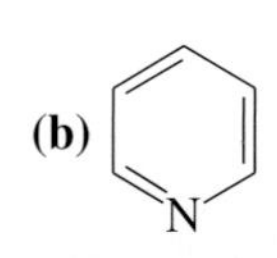

(c) $(CH_3)_2NCH_2CH_3$

(d) Five-membered ring containing O

1.25 Write structural formulas of your choice for all of the constitutional isomers with the molecular formula C_4H_8.

1.26 Write structural formulas for at least three constitutional isomers with the molecular formula CH_3NO_2. (In answering this question you should assign a formal charge to any atom that bears one.)

*1.27 Cyanic acid (H—O—C≡N) and isocyanic acid (H—N=C=O) differ in the positions of their electrons but their structures do not represent resonance structures. **(a)** Explain. **(b)** Loss of a proton from cyanic acid yields the same anion as that obtained by loss of a proton from isocyanic acid. Explain.

1.28 Consider a chemical species (either a molecule or an ion) in which a carbon atom forms three single bonds to three hydrogen atoms and in which the carbon atom possesses no other valence electrons. **(a)** What formal charge would the carbon atom have? **(b)** What total charge would the species have?

(c) What shape would you expect this species to have? (d) What would you expect the hybridization state of the carbon atom to be?

1.29 Consider a chemical species like the one in the previous problem in which a carbon atom forms three single bonds to three hydrogen atoms, but in which the carbon atom possesses an unshared electron pair. **(a)** What formal charge would the carbon atom have? **(b)** What total charge would the species have? **(c)** What shape would you expect this species to have? **(d)** What would you expect the hybridization state of the carbon atom to be?

1.30 Consider another chemical species like the ones in the previous problems in which a carbon atom forms three single bonds to three hydrogen atoms but in which the carbon atom possesses a single unpaired electron. **(a)** What formal charge would the carbon atom have? **(b)** What total charge would the species have? **(c)** Given that the shape of this species is trigonal planar, what would you expect the hybridization state of the carbon atom to be?

1.31 Ozone (O_3) is found in the upper atmosphere where it absorbs highly energetic ultraviolet (UV) radiation and thereby provides the surface of Earth with a protective screen (cf. Section 10.11E). One possible resonance structure for ozone is the following:

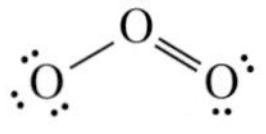

:Ö–Ö=Ö:

(a) Assign any necessary formal charges to the atoms in this structure. **(b)** Write another equivalent resonance structure for ozone. **(c)** What do these resonance structures predict about the relative lengths of the two oxygen–oxygen bonds of ozone? **(d)** The structure above, and the one you have written, assume an angular shape for the ozone molecule. Is this shape consistent with VSEPR theory? Explain your answer.

1.32 Write resonance structures for the azide ion, N_3^-. Explain how these resonance structures account for the fact that both bonds of the azide ion have the same length.

1.33 Write structural formulas of the type indicated: **(a)** bond-line formulas for seven constitutional isomers with the formula $C_4H_{10}O$; **(b)** condensed structural formulas for two constitutional isomers with the formula C_2H_7N; **(c)** condensed structural formulas for four constitutional isomers with the formula C_3H_9N; **(d)** bond-line formulas for three constitutional isomers with the formula C_5H_{12}.

***1.34** What is the relationship between the members of the following pairs? That is, are they constitutional isomers, the same, or something else (specify).

(a) 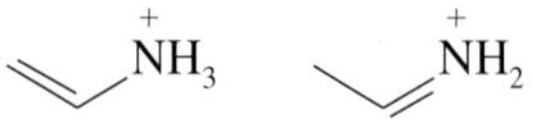 $\overset{+}{N}H_3$ $\overset{+}{N}H_2$

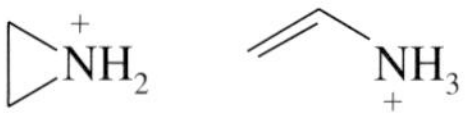

(b) 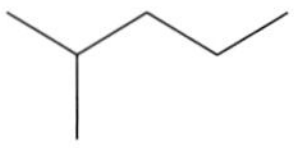 (bond-line structure) $CH_3CH_2CH_2CH(CH_3)_2$

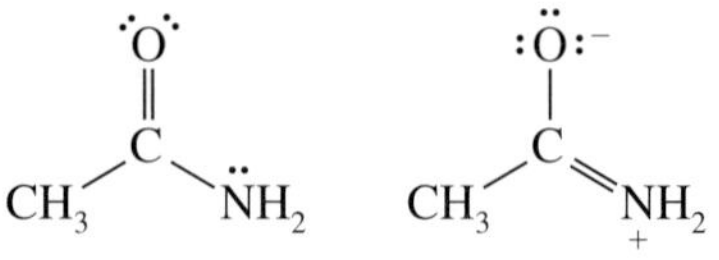

(c) 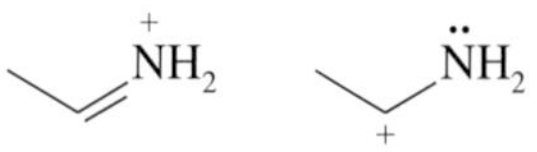 $\overset{+}{N}H_2$ $\ddot{N}H_2$ (+)

(d) $\overset{+}{N}H_2$ (ring) NH_3^+

(e) $CH_3C(=\ddot{O})\ddot{N}H_2$ $CH_3C(-\ddot{O}:^-)=\overset{+}{N}H_2$

(f) 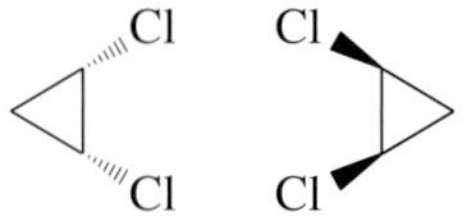 Cl, Cl (cyclopropane)   Cl, Cl (cyclopropane)

***1.35** In Chapter 15 we shall learn how the nitronium ion, NO_2^+, forms when concentrated nitric and sulfuric acids are mixed. **(a)** Write a Lewis structure for the nitronium ion. **(b)** What geometry does VSEPR theory predict for the NO_2^+ ion? **(c)** Give a species that has the same number of electrons as NO_2^+.

***1.36** Given the following sets of atoms, write bond-line formulas for all of the possible constitutionally isomeric compounds or ions that could be made from them. Show all unshared electron pairs and all formal charges, if any.

Set	C atoms	H atoms	Other
A	3	6	2 Br atoms
B	3	9	1 N atom and 1 O atom (not on same C)
C	3	4	1 O atom
D	2	7	1 N atom and 1 proton
E	3	7	1 extra electron

1.37 Open computer molecular models for dimethyl ether, dimethylacetylene, and *cis*-1,2-dichloro-1,2-difluoroethene from the 3D Molecular Models section of the CD. By interpreting the computer molecular model for each one, draw **(a)** a dash formula, **(b)** a bond-line formula, and **(c)** a three-dimensional dashed-wedge formula. Draw the models in whatever perspective is most convenient—generally the perspective in which the most atoms in the chain of a molecule can be in the plane of the paper.

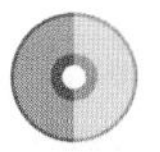

Chem3D Model

1.38 Boron is a Group 3A element. Open the molecular model for boron trifluoride from the 3D Molecular Models section of the CD. Near the boron atom, above and below the plane of the atoms in BF_3, are two relatively large lobes. Considering the position of boron in the periodic table and the three-dimensional and electronic structure of BF_3, what type of orbital does this lobe represent? Is it a hybridized orbital or not?

Chem3D Model

1.39 There are two contributing resonance structures for an anion called acetaldehyde enolate, whose condensed molecular formula is CH_2CHO^-. Draw the two resonance contributors and the resonance hybrid, then consider the map of electrostatic potential (MEP) shown below for this anion. Comment on whether the MEP is consistent or not with predominance of the resonance contributor you would have predicted to be represented most strongly in the hybrid.

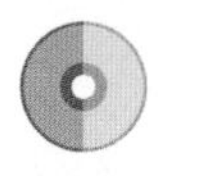

Chem3D Model

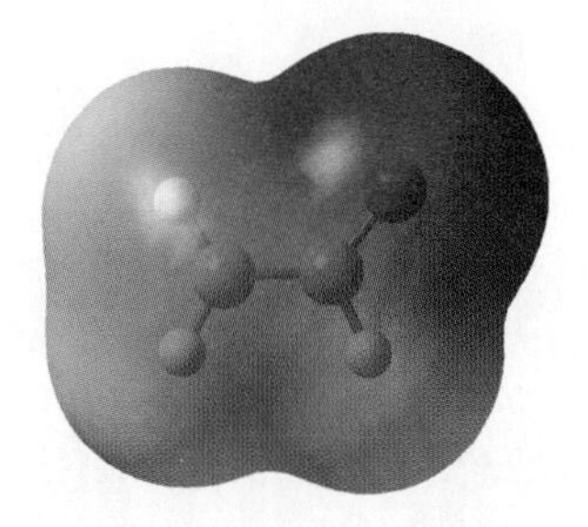

LEARNING GROUP PROBLEM

Your instructor will tell you how to work these problems as a Learning Group.

Consider the compound with the following condensed molecular formula:

$$CH_3CHOHCH{=}CH_2$$

1. Write a full dash structural formula for the compound.
2. Show all nonbonding electron pairs on your dash structural formula.
3. Indicate any formal charges that may be present in the molecule.
4. Label the hybridization state at every carbon atom and the oxygen.
5. Draw a three-dimensional perspective representation for the compound showing approximate bond angles as clearly as possible. Use ordinary lines to indicate bonds in the plane of the paper, bold wedges for bonds in front of the paper, and dashed wedges for bonds behind the paper.
6. Label all the bond angles in your three-dimensional structure.
7. Draw a bond-line formula for the compound.
8. Devise two structures, each having two *sp*-hybridized carbons and the molecular formula C_4H_6O. Create one of these structures such that it is linear with respect to all carbon atoms. Repeat parts 1–7 above for both structures.

CHAPTER 2

Representative Carbon Compounds: Functional Groups, Intermolecular Forces, and Infrared (IR) Spectroscopy

Structure and Function: Organic Chemistry, Nanotechnology, and Bioengineering

Molecules that move like machines and others that mimic natural templates for bone growth seem far-fetched, but research in these areas is at the core of exciting fields like nanotechnology and bioengineering. Indeed, we live in an age where nanotechnology, bioengineering, and molecular-level innovations promise to revolutionize our world. Organic chemistry, of course, is central to the advances being made in these fields. Organic chemists have synthesized molecules that behave like motors, molecules that mimic nature's prowess in bioengineering, and others that hold promise for new molecular devices in computing. For example, chemists have synthesized a threaded assembly of molecular rings and rods that can be made to slide back and forth in pistonlike fashion, mimicking the molecular sliding motion of proteins in muscle. Other chemists have mimicked nature's template for bone growth by designing molecules that self-assemble into macroscopic tubules and then promote calcium crystallization. Chemists have also synthesized molecules that coil and uncoil into mirror image forms under the influence of light, a property that could find application in the computer industry. Still others are designing molecules whose three-dimensional structure and properties act to fight diseases by binding to enzymes or receptor sites. Nanotechnology and bioengineering are fields that capture our imagination, and amazing accomplishments are being reported at an increasing rate.

Key to all of the advances described above is molecular structure, because as you shall see, molecular structure determines function. In the first chapter we learned about fundamental aspects of bonding and three-dimensional structure in organic molecules. In this chapter we learn about structural families of organic molecules characterized by groups of atoms called functional groups. Functional groups impart specific properties and reactivity to molecules. In this chapter we shall also learn about forces between molecules or parts of molecules that cause them to attract or repel each other, including forces like those that cause motion in molecular machines and the self-assembly of bioengineered fibers. Lastly, we shall learn about a rapid instrumental method called infrared spectroscopy that can give us information about the structure of organic molecules and the functional groups present in them.

2.1 Carbon–Carbon Covalent Bonds

Carbon's ability to form strong covalent bonds to other carbon atoms is the single property of the carbon atom that—more than any other—accounts for the very existence of a field of study called organic chemistry. It is this property too that accounts in part for carbon being the element around which most of the molecules of living organisms are constructed. Carbon's ability to form as many as four strong bonds to other carbon atoms, and to form strong bonds to hydrogen, oxygen, sulfur, and nitrogen atoms as well, provides the necessary versatility of structure that makes possible the vast number of different molecules required for complex living organisms. In this chapter we shall see how organic compounds can be organized into families of compounds on the basis of certain groupings of atoms that their molecules may contain. Organic chemists call these groupings of atoms *functional groups,* and it is these functional groups that determine most of the chemical and physical properties of the family.

We shall also study here an instrumental technique called *infrared spectroscopy* that can be used to demonstrate the presence of particular functional groups in the molecules of a compound.

2.2 Hydrocarbons: Representative Alkanes, Alkenes, Alkynes, and Aromatic Compounds

Hydrocarbons, as the name implies, are compounds whose molecules contain only carbon and hydrogen atoms. Methane (CH_4) and ethane (C_2H_6) are hydrocarbons. They also belong to a subgroup of hydrocarbons known as **alkanes** whose members do not have multiple bonds between carbon atoms. ***Hydrocarbons whose molecules have a carbon–carbon double bond*** are called **alkenes,** and ***those with a carbon–carbon triple bond*** are called **alkynes.** Hydrocarbons that contain a special ring that we shall introduce in Section 2.2D and study in Chapter 14 are called **aromatic hydrocarbons.**

Generally speaking, compounds such as the alkanes, whose molecules contain only single bonds, are referred to as **saturated compounds** because these compounds contain the maximum number of hydrogen atoms that the carbon compound can possess. Compounds with multiple bonds, such as alkenes, alkynes, and aromatic hydrocarbons, are called **unsaturated compounds** because they possess fewer than the maximum number of hydrogen atoms, and they are capable of reacting with hydrogen under the proper conditions. We shall have more to say about this in Chapter 7.

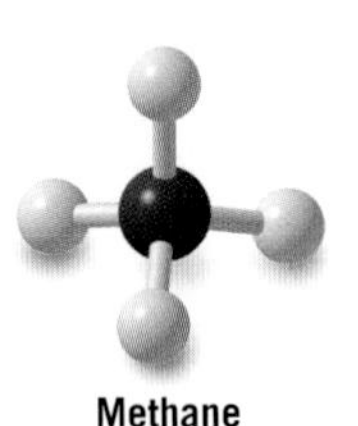

Methane

2.2A Alkanes

The principal sources of alkanes are natural gas and petroleum. The smaller alkanes (methane through butane) are gases under ambient conditions. The higher molecular weight alkanes are obtained largely by refining petroleum. Methane, the simplest alkane, was one major component of the early atmosphere of this planet. Methane is still found in the atmosphere of Earth, but no longer in appreciable amounts. It is, however, a major component of the atmospheres of Jupiter, Saturn, Uranus, and Neptune. Titan, Saturn's largest moon, contains an unexpectedly large amount of methane in its atmosphere. Scientists believe Titan's methane atmosphere is continuously being replenished from methane on or near the moon's surface, where it is found in a form with water called methane hydrate.

On Earth, methane is the major component of natural gas. The United States is currently using its large reserves of natural gas at a very high rate. Because the components of natural gas are important in industry, efforts are being made to develop coal gasification processes to provide alternative sources.

Some living organisms produce methane from carbon dioxide and hydrogen. These very primitive creatures called *methanogens* may be Earth's oldest organisms, and they may represent a separate form of evolutionary development. Methanogens can survive only in an anaerobic (i.e., oxygen-free) environment. They have been found in ocean trenches, in mud, in sewage, and in cows' stomachs.

2.2B Alkenes

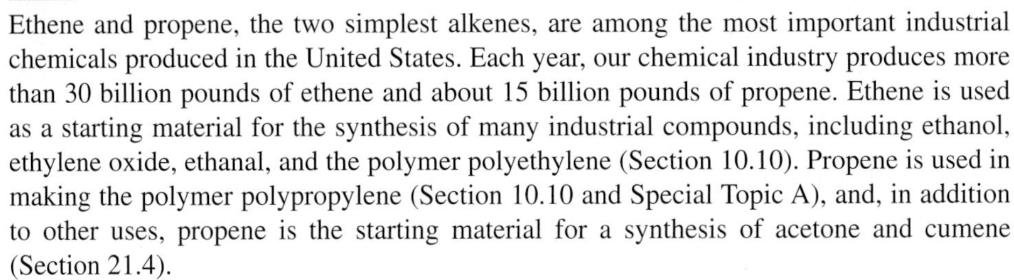

Ethene

Ethene and propene, the two simplest alkenes, are among the most important industrial chemicals produced in the United States. Each year, our chemical industry produces more than 30 billion pounds of ethene and about 15 billion pounds of propene. Ethene is used as a starting material for the synthesis of many industrial compounds, including ethanol, ethylene oxide, ethanal, and the polymer polyethylene (Section 10.10). Propene is used in making the polymer polypropylene (Section 10.10 and Special Topic A), and, in addition to other uses, propene is the starting material for a synthesis of acetone and cumene (Section 21.4).

Ethene also occurs in nature as a plant hormone. It is produced naturally by fruits such as tomatoes and bananas and is involved in the ripening process of these fruits. Much use is now made of ethene in the commercial fruit industry to bring about the ripening of tomatoes and bananas picked green because the green fruits are less susceptible to damage during shipping.

There are many naturally occurring alkenes. Two examples are the following:

β-Pinene (a component of turpentine)

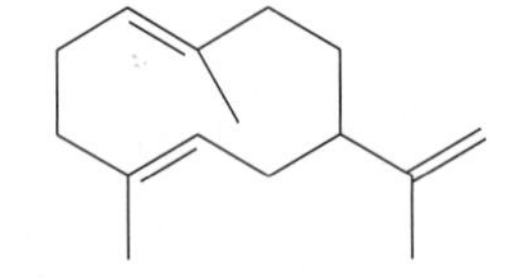

An aphid alarm pheromone

Ethyne

2.2C Alkynes

The simplest alkyne is ethyne (also called acetylene). Alkynes occur in nature and can be synthesized in the laboratory. Friedrich Wöhler synthesized ethyne in 1862 by the reaction of calcium carbide and water. Ethyne produced this way was burned in carbide lamps, such as those formerly used on miners' headlamps. Because it burns at a high temperature, ethyne is used in welding torches.

Alkynes whose molecules have multiple triple bonds exist in the atmosphere of the outer planets in our solar system. H. W. Kroto (University of Sussex, England) has identified compounds called cyanopolyynes in intersteller space. (Kroto shared the 1996 Nobel Prize in Chemistry for the discovery of buckminsterfullerenes, Section 14.8C.)

Two examples of alkynes among thousands that have a biosynthetic origin are capillin, an antifungal agent, and dactylyne, a marine natural product that is an inhibitor of pentobarbital metabolism. Ethinyl estradiol is a synthetic alkyne whose estrogen-like properties have found use in oral contraceptives.

Capillin

Dactylyne

Ethinyl estradiol
[17α-ethynyl-1,3,5(10)-estratriene-3,17β-diol]

2.2D Benzene: A Representative Aromatic Hydrocarbon

In Chapter 14 we shall study in detail a group of unsaturated cyclic hydrocarbons known as aromatic compounds. The compound known as **benzene** is the prototypical aromatic compound. Benzene can be written as a six-membered ring with alternating single and double bonds, called a Kekulé structure after August Kekulé (Section 1.3), who first conceived of this representation:

Kekulé structure for benzene or **Bond-line representation of Kekulé structure**

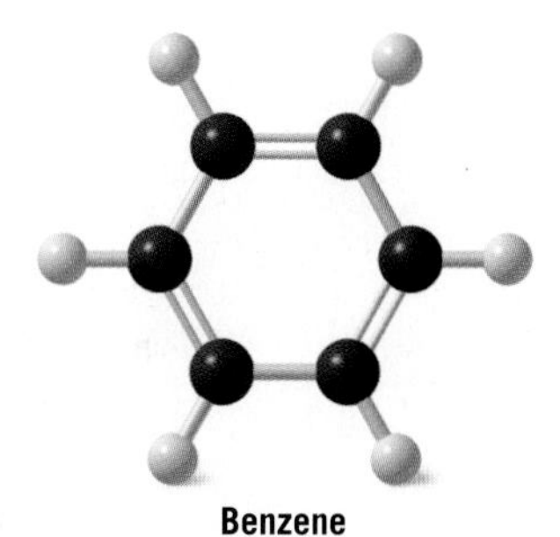

Benzene

Even though the Kekulé structure is frequently used for benzene compounds, there is much evidence that this representation is inadequate and incorrect. For example, if benzene had alternating single and double bonds as the Kekulé structure indicates, we would expect the lengths of the carbon–carbon bonds around the ring to be alternately longer and shorter, as we typically find with carbon–carbon single and double bonds (Fig. 1.30). In fact, the carbon–carbon bonds of benzene are all the same length (1.39 Å), a value in between that of a carbon–carbon single bond and a carbon–carbon double bond. There are two ways of dealing with this problem: with resonance theory or with molecular orbital theory.

If we use resonance theory, we visualize benzene as being represented by either of two equivalent Kekulé structures:

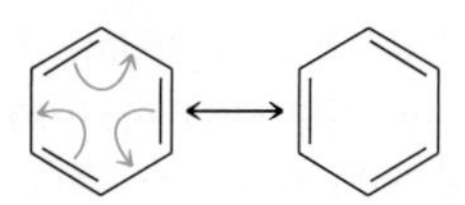

Two contributing Kekulé structures for benzene

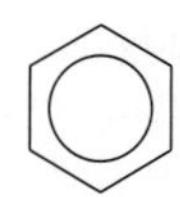

A representation of the resonance hybrid

CD Tutorial
Benzene resonance

Based on the principles of resonance theory (Section 1.8) we recognize that benzene cannot be represented adequately by either structure, but that, instead, *it should be visualized as a hybrid of the two structures.* We represent this hybrid by a hexagon with a circle in the middle. Resonance theory, therefore, solves the problem we encountered in understanding how all of the carbon–carbon bonds are the same length. According to resonance theory, the bonds are not alternating single and double bonds, they are a resonance hybrid of the two: Any bond that is a single bond in the first contributor is a double bond in the second, and vice versa. Therefore, we should expect *all of the carbon–carbon bonds to be the same,* to be one and one-half bonds, and to have a bond length in between that of a single bond and a double bond. That is, of course, what we actually find.

In the molecular orbital explanation, which we shall describe in much more depth in Chapter 14, we begin by recognizing that the carbon atoms of the benzene ring are sp^2 hybridized. Therefore, each carbon has a *p* orbital that has one lobe above the plane of the ring and one lobe below.

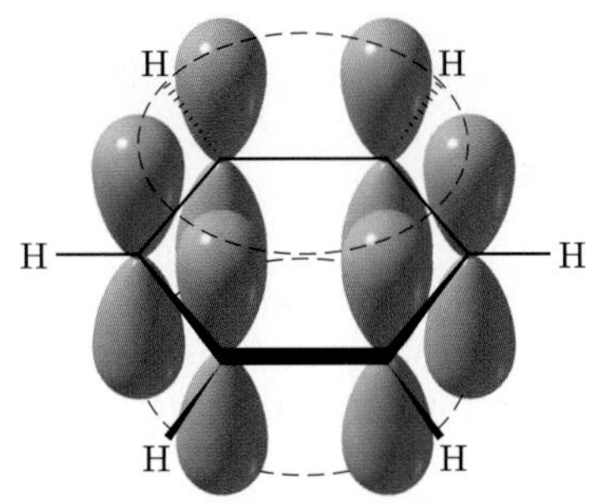

Schematic representation of benzene *p* orbitals

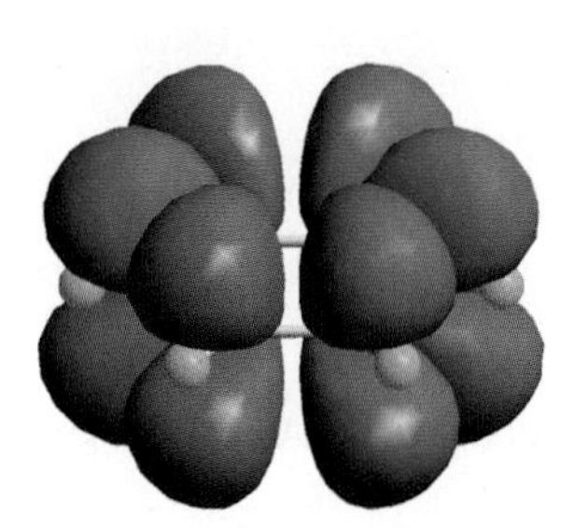
Calculated *p* orbital shapes in benzene

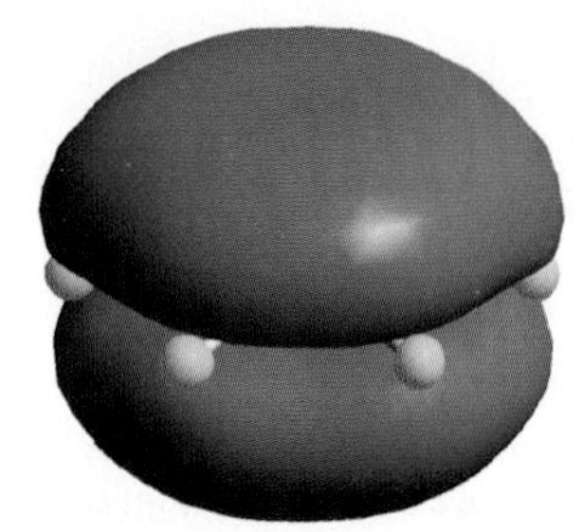
Calculated benzene molecular orbital resulting from favorable overlap of *p* orbitals above and below plane of benzene ring

Although the schematic depiction of *p* orbitals in our illustration does not show this, each lobe of each *p* orbital, above and below the ring, overlaps with the lobe of a *p* orbital on each atom on either side of it. This kind of overlap of *p* orbitals leads to a set of bonding molecular orbitals that encompass all of the carbon atoms of the ring, as shown in the calculated molecular orbital. Therefore, the six electrons associated with these *p* orbitals (one from each) are **delocalized** about all six carbon atoms of the ring. This delocalization of electrons explains how all the carbon–carbon bonds are equivalent and have the same length. In Section 14.7B, when we study nuclear magnetic resonance spectroscopy, we shall present convincing physical evidence for this delocalization of the electrons.

2.3 Polar Covalent Bonds

Thus far we have principally been studying molecules having only carbon–carbon and carbon–hydrogen bonds, that is, bonds having little or no difference in electronegativity between the attached atoms. (Electronegativity was first discussed in Section 1.4.) Shortly we shall begin to study groups of atoms called functional groups (Section 2.5). Many functional groups contain atoms that have different electronegativities and unshared electron pairs. (Atoms that form covalent bonds and have unshared electron pairs are called **heteroatoms.**) First, however, it is useful to consider some properties of bonds between atoms that have different electronegativities.

When two atoms of different electronegativities form a covalent bond, the electrons are not shared equally between them. The atom with greater electronegativity draws electron density closer to it, and a **polar covalent bond** results. (One definition of **electronegativity** is *the ability of an element to attract electrons that it is sharing in a covalent bond;* see Section 1.4A.) An example of such a polar covalent bond is the one in hydrogen chloride. The chlorine atom, with its greater electronegativity, pulls the bonding electrons closer to it. This makes the hydrogen atom somewhat electron deficient and gives it a *partial* positive charge ($\delta+$). The chlorine atom becomes somewhat electron rich and bears a *partial* negative charge ($\delta-$):

The concepts of electronegativity and bond polarity will help greatly in understanding molecular properties and reactivity.

$$\overset{\delta+}{\mathrm{H}} : \overset{\delta-}{\ddot{\underset{..}{\mathrm{Cl}}}}:$$

Because the hydrogen chloride molecule has a partially positive end and a partially negative end, it is a **dipole,** and it has a **dipole moment.**

The direction of polarity of a polar bond can be symbolized by a vector quantity $\longmapsto$. The crossed end of the arrow is the positive end and the arrow head is the negative end:

$$(\text{positive end}) \longmapsto (\text{negative end})$$

In HCl, for example, we would indicate the direction of the dipole moment in the following way:

$$\underset{\longmapsto}{\mathrm{H{-}Cl}}$$

The dipole moment is a physical property that can be measured experimentally. It is defined as the product of the magnitude of the charge in electrostatic units (esu) and the distance that separates them in centimeters (cm):

$$\text{Dipole moment} = \text{charge (in esu)} \times \text{distance (in cm)}$$
$$\mu = e \times d$$

The charges are typically on the order of 10^{-10} esu and the distances are on the order of 10^{-8} cm. Dipole moments, therefore, are typically on the order of 10^{-18} esu cm. For convenience, this unit, 1×10^{-18} esu cm, is defined as one **debye** and is abbreviated D. (The unit is named after Peter J. W. Debye, a chemist born in the Netherlands, who taught at Cornell University from 1936 to 1966. Debye won the Nobel Prize in 1936.)

If necessary, the length of the arrow can be used to indicate the magnitude of the dipole moment. Dipole moments, as we shall see in Section 2.4, are very useful quantities in accounting for physical properties of compounds.

PROBLEM 2.1

Write $\delta+$ and $\delta-$ by the appropriate atoms and draw a dipole moment vector for any of the following molecules that are polar:
(a) HF, **(b)** IBr, **(c)** Br_2, **(d)** F_2.

2.3A Maps of Electrostatic Potential

One way to visualize the distribution of charge in a molecule is with a **map of electrostatic potential (MEP).** Regions of an electron density surface that are more negative than others in an MEP are colored red. These regions would attract a positively charged species (or repel a negative charge). Regions in the MEP that are less negative (or are positive) are blue. Blue regions are likely to attract electrons from another molecule. The spectrum of colors from red to blue indicates the trend in charge from most negative to least negative (or positive).

Figure 2.1 shows a map of electrostatic potential for the low-electron-density surface of hydrogen chloride. We can see clearly that negative charge is concentrated near the chlorine atom and that positive charge is localized near the hydrogen atom, as we predict based on the difference in their electronegativity values. Furthermore, because this MEP is plotted at the low-electron-density surface of the molecule (the van der Waals surface, Section 2.14D), it also gives an indication of the molecule's overall shape.

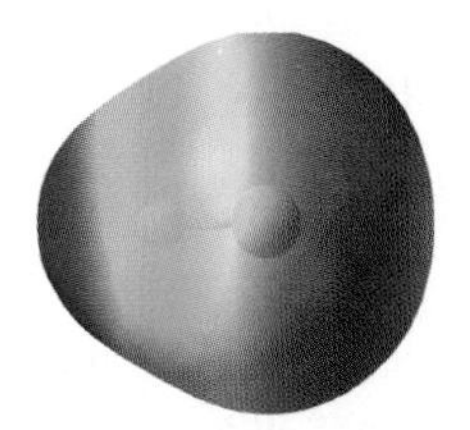

FIGURE 2.1 A calculated map of electrostatic potential for hydrogen chloride showing regions of relatively more negative charge in red and more positive charge in blue. Negative charge is clearly localized near the chlorine, resulting in a strong dipole moment for the molecule.

Animated Graphics

The Chemistry of...

Calculated Molecular Models: Maps of Electrostatic Potential

A map of electrostatic potential is prepared by carrying out a quantum mechanical calculation that involves moving an imaginary positive point charge at a fixed distance over a given electron density surface of a molecule. As this is done, the varying potential energy in the attraction between the electron cloud and the imaginary positive charge is plotted in color-coded fashion. Red in the MEP indicates strong attraction between the electron density surface at that location and the probing positive charge—in other words, greater negative charge at that part of the surface. Blue regions in the map indicate weaker attraction between the surface and the positive charge probe. The overall distribution of charge is indicated by the trend from blue (most positive or least negative) to green or yellow (neutral) to red (most negative). Most often we plot the MEP at the van der Waals surface of a molecule since that represents approximately the furthest extent of a molecule's electron cloud and therefore its overall shape. The molecular model in this box is for dimethyl ether. The MEP shows the concentration of negative charge where the unshared electron pairs are located on the oxygen atom.

It is important to note that when directly comparing the MEP for one molecule to that of another, the color scheme used to represent the charge scale in each model must be the same. When we make direct comparisons between molecules, we will plot their MEPs on the same scale. We will find that such comparisons are especially useful because they allow us to compare the electron distribution in one molecule to another and predict how one molecule might interact with the electrons of another molecule.

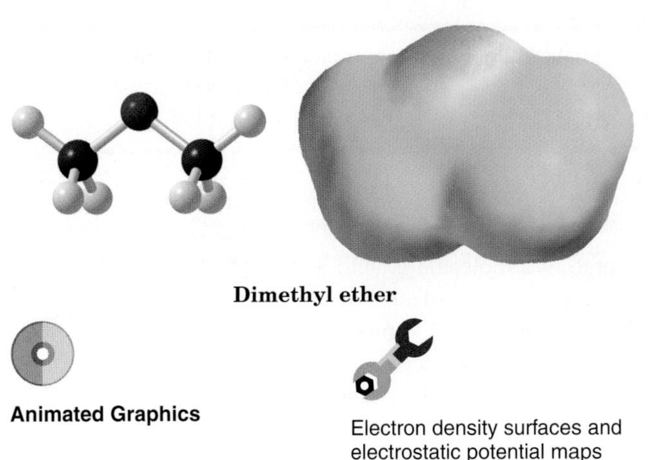

Most positive (Least negative)

Most negative

Animated Graphics

Electron density surfaces and electrostatic potential maps

2.4 Polar and Nonpolar Molecules

In the discussion of dipole moments in the previous section, our attention was restricted to simple diatomic molecules. Any *diatomic* molecule in which the two atoms are *different* (and thus have different electronegativities) will, of necessity, have a dipole moment. If we examine Table 2.1, however, we find that a number of molecules (e.g., CCl_4, CO_2) consist of more than two atoms, have *polar* bonds, *but have no dipole moment.* With our understanding of the shapes of molecules (Sections 1.12–1.16) we can understand how this can occur.

TABLE 2.1 Dipole Moments of Some Simple Molecules

Formula	μ (D)	Formula	μ (D)
H_2	0	CH_4	0
Cl_2	0	CH_3Cl	1.87
HF	1.91	CH_2Cl_2	1.55
HCl	1.08	$CHCl_3$	1.02
HBr	0.80	CCl_4	0
HI	0.42	NH_3	1.47
BF_3	0	NF_3	0.24
CO_2	0	H_2O	1.85

Consider a molecule of carbon tetrachloride (CCl_4). Because the electronegativity of chlorine is greater than that of carbon, each of the carbon–chlorine bonds in CCl_4 is polar. Each chlorine atom has a partial negative charge, and the carbon atom is considerably positive. Because a molecule of carbon tetrachloride is tetrahedral (Fig. 2.2), however, *the center of positive charge and the center of negative charge coincide, and the molecule has no net dipole moment.*

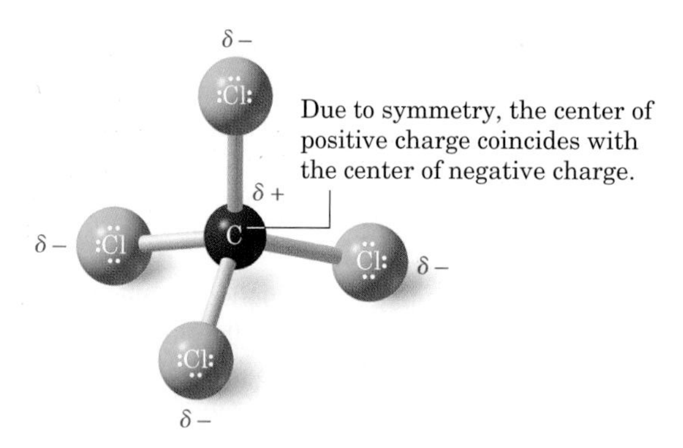

FIGURE 2.2 Charge distribution in carbon tetrachloride. The molecule has no net dipole moment.

This result can be illustrated in a slightly different way: If we use arrows (⟼) to represent the direction of polarity of each bond, we get the arrangement of bond moments shown in Fig. 2.3. Since the bond moments are vectors of equal magnitude arranged tetra-

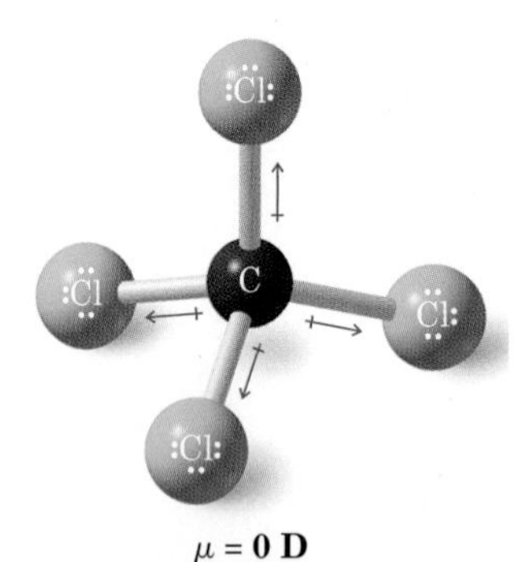

FIGURE 2.3 A tetrahedral orientation of equal bond moments causes their effects to cancel.

hedrally, their effects cancel. Their vector sum is zero. The molecule has *no net dipole moment.*

The chloromethane molecule (CH_3Cl) has a net dipole moment of 1.87 D. Since carbon and hydrogen have electronegativities (Table 1.2) that are nearly the same, the contribution of three C—H bonds to the net dipole is negligible. The electronegativity difference between carbon and chlorine is large, however, and this highly polar C—Cl bond accounts for most of the dipole moment of CH_3Cl (Fig. 2.4).

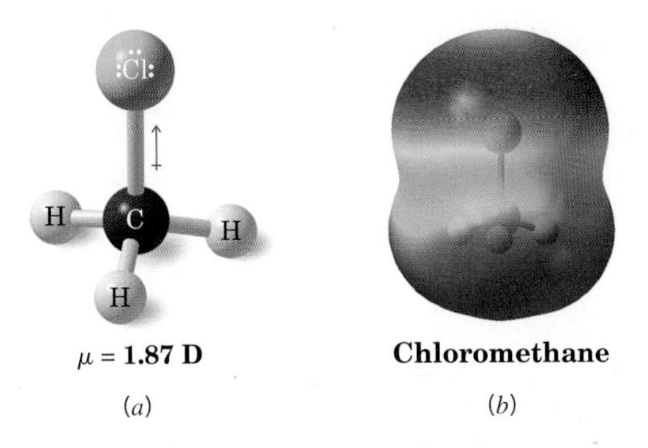

FIGURE 2.4 *(a)* The dipole moment of chloromethane arises mainly from the highly polar carbon–chlorine bond. *(b)* A map of electrostatic potential illustrates the polarity of chloromethane.

PROBLEM 2.2 Boron trifluoride (BF_3) has no dipole moment ($\mu = 0$ D). Explain how this observation confirms the geometry of BF_3 predicted by VSEPR theory.

PROBLEM 2.3 Tetrachloroethene ($CCl_2{=}CCl_2$) does not have a dipole moment. Explain this fact on the basis of the shape of $CCl_2{=}CCl_2$.

PROBLEM 2.4 Sulfur dioxide (SO_2) has a dipole moment ($\mu = 1.63$ D); on the other hand, carbon dioxide (CO_2) has no dipole moment ($\mu = 0$ D). What do these facts indicate about the geometries of the two molecules?

Unshared pairs of electrons make large contributions to the dipole moments of water and ammonia. Because an unshared pair has no other atom attached to it to partially neutralize its negative charge, an unshared electron pair contributes a large moment directed away from the central atom (Fig. 2.5). (The O—H and N—H moments are also appreciable.)

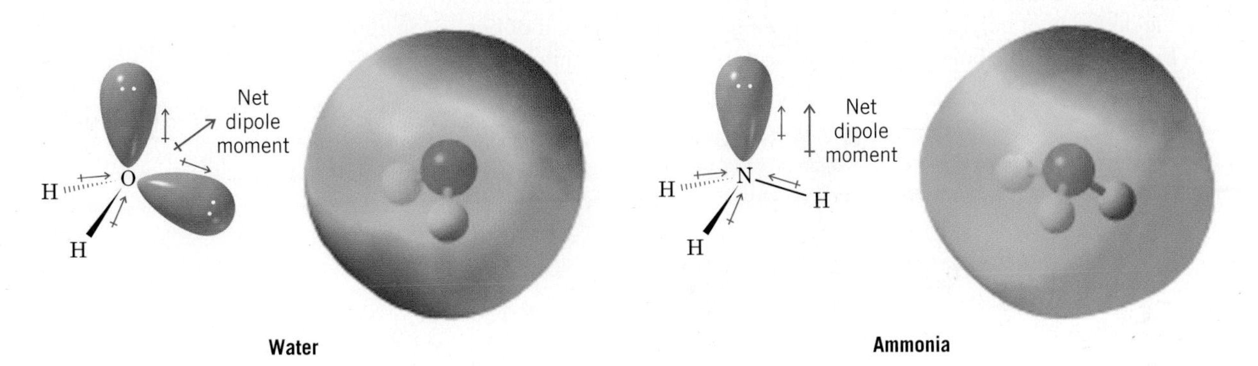

FIGURE 2.5 Bond moments and the resulting dipole moments of water and ammonia.

PROBLEM 2.5

Using a three-dimensional formula, show the direction of the dipole moment of CH_3OH. Write $\delta+$ and $\delta-$ signs next to the appropriate atoms.

PROBLEM 2.6

Trichloromethane ($CHCl_3$, also called *chloroform*) has a larger dipole moment than $CFCl_3$. Use three-dimensional structures and bond moments to explain this fact.

2.4A Dipole Moments in Alkenes

Cis–trans isomers of alkenes (Section 1.13B) have different physical properties. They have different melting points and boiling points, and often cis–trans isomers differ markedly in the magnitude of their dipole moments. Table 2.2 summarizes some of the physical properties of two pairs of cis–trans isomers.

TABLE 2.2 Physical Properties of Some Cis–Trans Isomers

Compound	Melting Point (°C)	Boiling Point (°C)	Dipole Moment (D)
cis-1,2-Dichloroethene	−80	60	1.90
trans-1,2-Dichloroethene	−50	48	0
cis-1,2-Dibromoethene	−53	112.5	1.35
trans-1,2-Dibromoethene	−6	108	0

PROBLEM 2.7

Indicate the direction of the important bond moments in each of the following compounds (neglect C—H bonds). You should also give the direction of the net dipole moment for the molecule. If there is no net dipole moment, state that $\mu = 0$ D.

(a) *cis*-CHF=CHF
(b) *trans*-CHF=CHF
(c) $CH_2=CF_2$
(d) $CF_2=CF_2$

PROBLEM 2.8

Write structural formulas for all of the alkenes with **(a)** the formula $C_2H_2Br_2$ and **(b)** the formula $C_2Br_2Cl_2$. In each instance designate compounds that are cis–trans isomers of each other. Predict the dipole moment of each one.

2.5 Functional Groups

One great advantage of the structural theory is that it enables us to classify the vast number of organic compounds into a relatively small number of families based on their structures. Table 2.3 in Section 2.12 summarizes the most important of these families. The molecules of compounds in a particular family are characterized by the presence of a certain arrangement of atoms called a **functional group.**

Functional groups will help us organize our knowledge about the properties and reactivity of organic molecules.

A functional group is the part of a molecule where most of its chemical reactions occur. It is the part that effectively determines the compound's chemical properties (and many of its physical properties as well). The functional group of an alkene, for example, is its carbon–carbon double bond. When we study the reactions of alkenes in greater detail in Chapter 8, we shall find that most of the chemical reactions of alkenes are the chemical reactions of the carbon–carbon double bond.

The functional group of an alkyne is its carbon–carbon triple bond. Alkanes do not have a functional group. Their molecules have carbon–carbon single bonds and carbon–hydrogen bonds, but these bonds are present in molecules of almost all organic molecules, and C—C and C—H bonds are, in general, much less reactive than common functional groups.

2.5A Alkyl Groups and the Symbol R

CD Tutorial
Alkyl groups

Alkyl groups are the groups that we identify for purposes of naming compounds. They are groups that would be obtained by removing a hydrogen atom from an alkane:

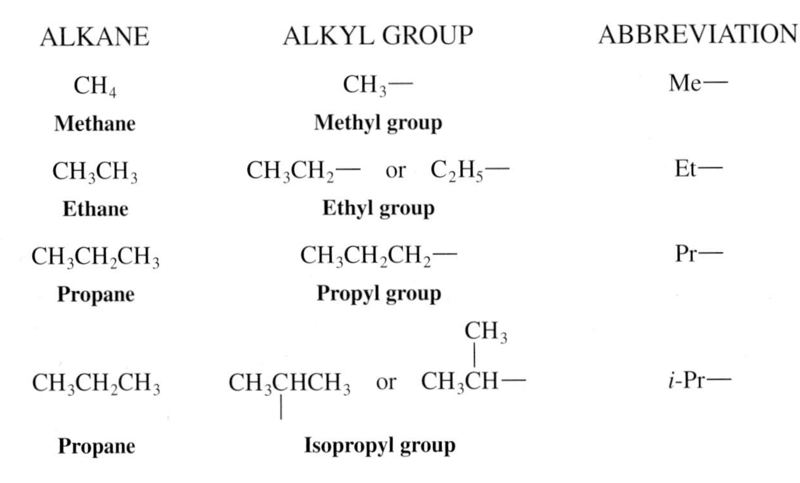

ALKANE	ALKYL GROUP	ABBREVIATION
CH_4 **Methane**	$CH_3—$ **Methyl group**	Me—
CH_3CH_3 **Ethane**	$CH_3CH_2—$ or $C_2H_5—$ **Ethyl group**	Et—
$CH_3CH_2CH_3$ **Propane**	$CH_3CH_2CH_2—$ **Propyl group**	Pr—
$CH_3CH_2CH_3$ **Propane**	$CH_3\underset{\vert}{C}HCH_3$ or $CH_3\overset{CH_3}{\overset{\vert}{C}}H—$ **Isopropyl group**	*i*-Pr—

While only one alkyl group can be derived from methane or ethane (the **methyl** and **ethyl** groups, respectively), two groups can be derived from propane. Removal of a hydrogen from one of the end carbon atoms gives a group that is called the **propyl** group; removal of a hydrogen from the middle carbon atom gives a group that is called the **isopropyl** group. The names and structures of these groups are used so frequently in organic chemistry that you should learn them now. See Section 4.3C for names and structures of branched alkyl groups derived from butane and other hydrocarbons.

We can simplify much of our future discussion if, at this point, we introduce a symbol that is widely used in designating general structures of organic molecules: the symbol R. **R *is used as a general symbol to represent any alkyl group.*** For example, R might be a methyl group, an ethyl group, a propyl group, or an isopropyl group:

$CH_3—$	Methyl	**These and others can be designated by R.**
$CH_3CH_2—$	Ethyl	
$CH_3CH_2CH_2—$	Propyl	
$CH_3\underset{\vert}{C}HCH_3$	Isopropyl	

Thus, the general formula for an alkane is R—H.

2.5B Phenyl and Benzyl Groups

When a benzene ring is attached to some other group of atoms in a molecule, it is called a **phenyl group,** and it is represented in several ways:

or   or $C_6H_5—$ or Ph— or $\phi—$ or Ar— (if ring substituents are present)

Ways of representing a phenyl group

The combination of a phenyl group and a —CH_2— group is called a **benzyl group:**

$C_6H_5CH_2$— (phenyl ring)—CH_2— or (ring with circle)—CH_2— or $C_6H_5CH_2$— or Bn—

Ways of representing a benzyl group

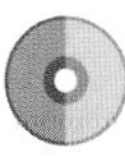

CD Drill Review

Group Names and Structure Study Aid

2.6 Alkyl Halides or Haloalkanes

Alkyl halides are compounds in which a halogen atom (fluorine, chlorine, bromine, or iodine) replaces a hydrogen atom of an alkane. For example, CH_3Cl and CH_3CH_2Br are alkyl halides. Alkyl halides are also called **haloalkanes.**

Alkyl halides are classified as being primary (1°), secondary (2°), or tertiary (3°). ***This classification is based on the carbon atom to which the halogen is directly attached.*** If the carbon *atom* that bears the halogen is attached to only one other carbon, the carbon atom is said to be a **primary carbon atom** and the alkyl halide is classified as a **primary alkyl halide.** If the carbon that bears the halogen is itself attached to two other carbon atoms, then the carbon is a **secondary carbon** and the alkyl halide is a **secondary alkyl halide.** If the carbon that bears the halogen is attached to three other carbon atoms, then the carbon is a **tertiary carbon** and the alkyl halide is a **tertiary alkyl halide.** Examples of primary, secondary, and tertiary alkyl halides are the following:

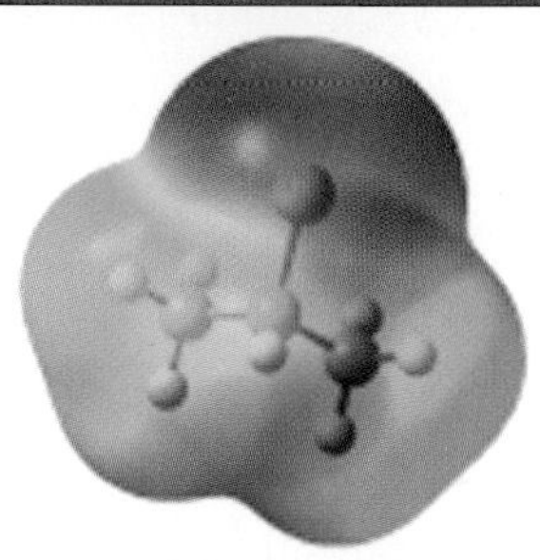

2-Chloropropane

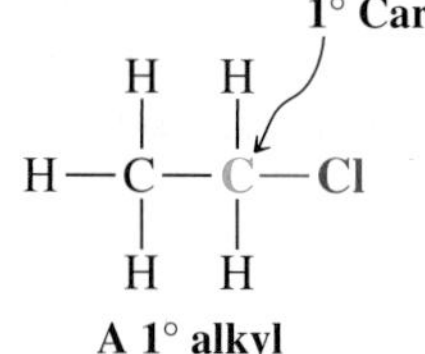

1° Carbon — $H_3C{-}CH_2Cl$ — **A 1° alkyl chloride**

2° Carbon — $CH_3CHClCH_3$ — **A 2° alkyl chloride**

3° Carbon — $(CH_3)_3C{-}Cl$ — **A 3° alkyl chloride**

Although we use the symbols 1°, 2°, 3°, we do not *say* first degree, second degree, and third degree; we say *primary, secondary,* and *tertiary.*

PROBLEM 2.9

Write structural formulas for **(a)** two constitutionally isomeric primary alkyl bromides with the formula C_4H_9Br, **(b)** a secondary alkyl bromide, and **(c)** a tertiary alkyl bromide with the same formula. Build hand-held molecular models for each structure and examine the differences in their connectivity.

PROBLEM 2.10

Although we shall discuss the naming of organic compounds later when we discuss the individual families in detail, one method of naming alkyl halides is so straightforward that it is worth describing here. We simply name the alkyl group attached to the halogen and add the word *fluoride, chloride, bromide,* or *iodide.* Write formulas for **(a)** ethyl fluoride and **(b)** isopropyl chloride. What are the names for **(c)** $CH_3CH_2CH_2Br$, **(d)** CH_3CHFCH_3, and **(e)** C_6H_5I?

2.7 Alcohols

Methyl alcohol (more systematically called methanol) has the structural formula CH_3OH and is the simplest member of a family of organic compounds known as **alcohols.** The characteristic functional group of this family is the hydroxyl (OH) group attached to an

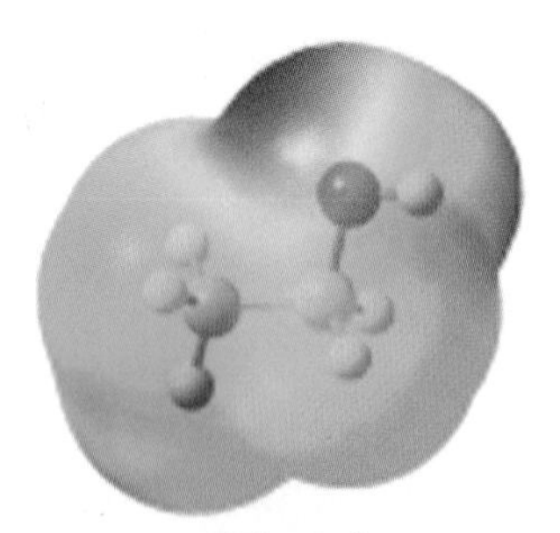

Ethanol

sp^3-hybridized carbon atom. Another example of an alcohol is ethyl alcohol, CH_3CH_2OH (also called ethanol).

$$-\overset{|}{\underset{|}{C}}-\ddot{\underset{..}{O}}-H$$

This is the functional group of an alcohol.

Alcohols may be viewed in two ways structurally: (1) as hydroxyl derivatives of alkanes and (2) as alkyl derivatives of water. Ethyl alcohol, for example, can be seen as an ethane molecule in which one hydrogen has been replaced by a hydroxyl group or as a water molecule in which one hydrogen has been replaced by an ethyl group:

Ethyl group

CH_3CH_3 — CH_3CH_2 / 109° / $\ddot{O}$: / H — H / 105° / $\ddot{O}$: / H

Hydroxyl group

Ethane **Ethyl alcohol (ethanol)** **Water**

As with alkyl halides, alcohols are classified into three groups: primary (1°), secondary (2°), or tertiary (3°) alcohols. ***This classification is based on the degree of substitution of the carbon to which the hydroxyl group is directly attached.*** If the carbon has only one other carbon attached to it, the carbon is said to be a **primary carbon** and the alcohol is a **primary alcohol:**

Geraniol is a major component of the oil of roses.

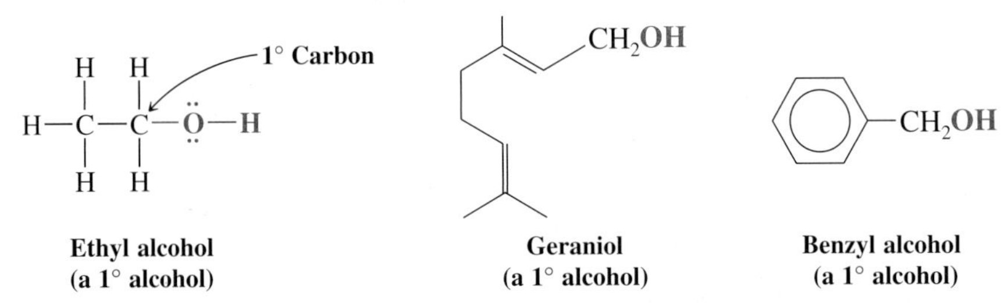

Ethyl alcohol (a 1° alcohol) **Geraniol (a 1° alcohol)** **Benzyl alcohol (a 1° alcohol)**

If the carbon atom that bears the hydroxyl group also has two other carbon atoms attached to it, this carbon is called a secondary carbon, and the alcohol is a secondary alcohol:

Practice with hand-held molecular models by building models of as many of the compounds on this page as you can.

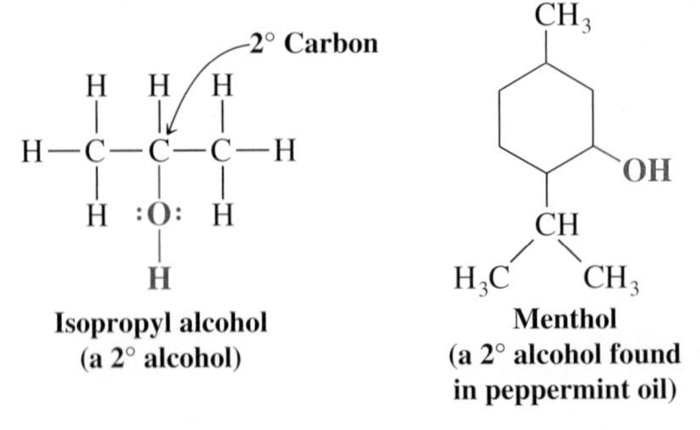

Isopropyl alcohol (a 2° alcohol) **Menthol (a 2° alcohol found in peppermint oil)**

If the carbon atom that bears the hydroxyl group has three other carbons attached to it, this carbon is called a tertiary carbon, and the alcohol is a tertiary alcohol:

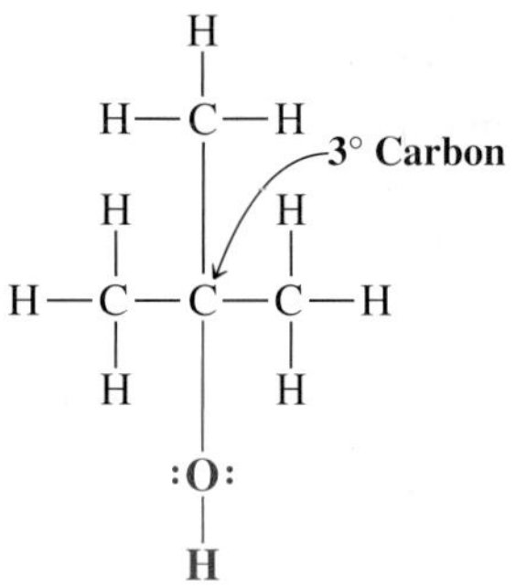

3° Carbon

tert-**Butyl alcohol**
(a 3° alcohol)

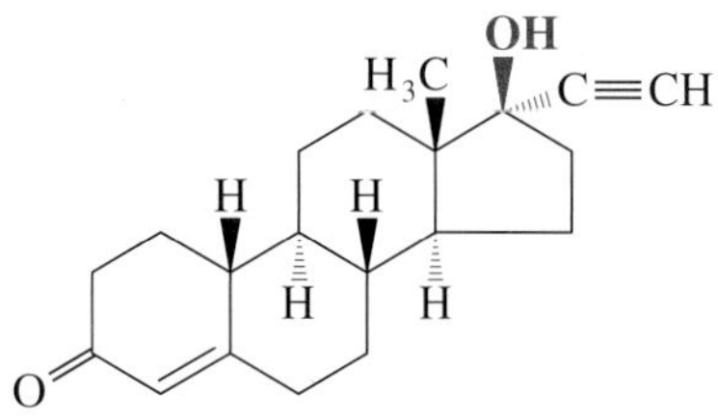

Norethindrone
(an oral contraceptive that contains a 3° alcohol group as well as a ketone group and carbon–carbon double and triple bonds)

PROBLEM 2.11

Write structural formulas for **(a)** two primary alcohols, **(b)** a secondary alcohol, and **(c)** a tertiary alcohol—all having the molecular formula $C_4H_{10}O$.

PROBLEM 2.12

One way of naming alcohols is to name the alkyl group that is attached to the —OH and add the word *alcohol*. Write the structures of **(a)** propyl alcohol and **(b)** isopropyl alcohol.

2.8 Ethers

Ethers have the general formula R—O—R or R—O—R′, where R′ may be an alkyl (or phenyl) group different from R. Ethers can be thought of as derivatives of water in which both hydrogen atoms have been replaced by alkyl groups. The bond angle at the oxygen atom of an ether is only slightly larger than that of water:

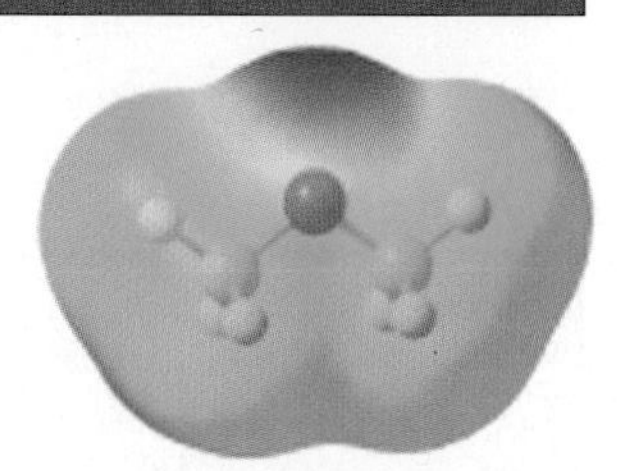

Dimethyl ether

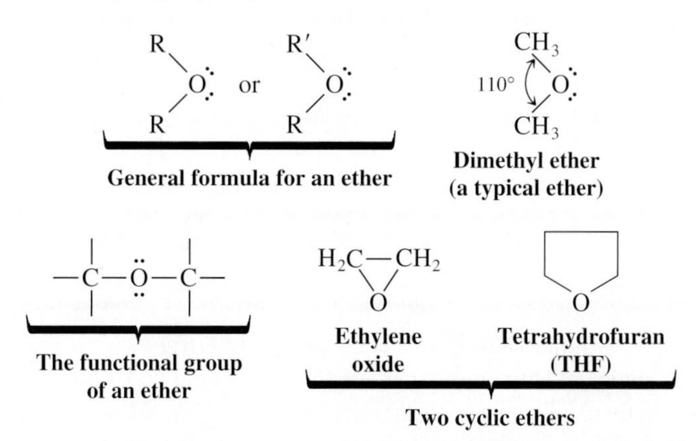

PROBLEM 2.13

One way of naming ethers is to name the two alkyl groups attached to the oxygen atom in alphabetical order and add the word *ether.* If the two alkyl groups are the same, we use the prefix *di-*, for example, as in *dimethyl ether.* Write structural formulas for **(a)** diethyl ether, **(b)** ethyl propyl ether, and **(c)** ethyl isopropyl ether. What name would you give to **(d)** $CH_3OCH_2CH_2CH_3$, **(e)** $(CH_3)_2CHOCH(CH_3)_2$, and **(f)** $CH_3OC_6H_5$?

2.9 Amines

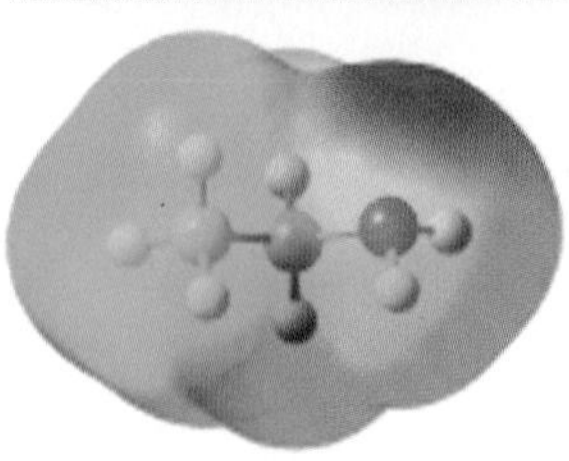

Ethylamine

Just as alcohols and ethers may be considered as organic derivatives of water, amines may be considered as organic derivatives of ammonia:

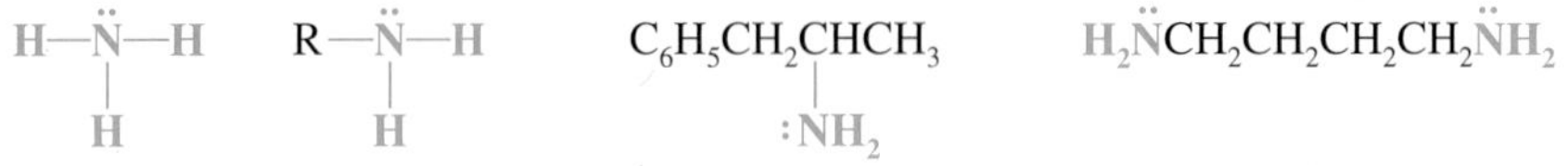

Ammonia **An amine** **Amphetamine (a dangerous stimulant)** **Putrescine (found in decaying meat)**

Amines are classified as primary, secondary, or tertiary amines. **This classification is based on *the number of organic groups that are attached to the nitrogen atom:***

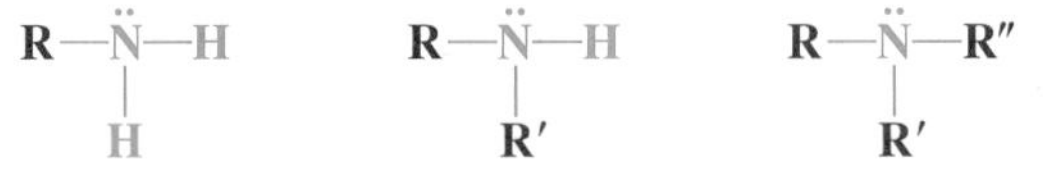

A primary (1°) amine **A secondary (2°) amine** **A tertiary (3°) amine**

Notice that this is quite different from the way alcohols and alkyl halides are classified. Isopropylamine, for example, is a primary amine even though its —NH_2 group is attached to a secondary carbon atom. It is a primary amine because only one organic group is attached to the nitrogen atom:

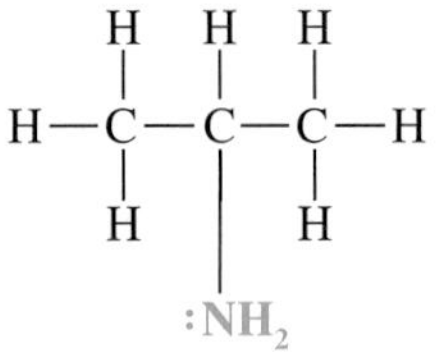

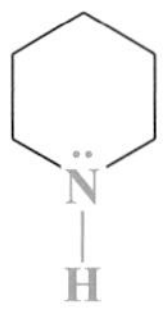

Isopropylamine (a 1° amine) **Piperidine (a cyclic 2° amine)**

Problem 2.14 One way of naming amines is to name in alphabetical order the alkyl groups attached to the nitrogen atom, using the prefixes *di-* and *tri-* if the groups are the same. An example is *isopropylamine* for $(CH_3)_2CHNH_2$. Write formulas for **(a)** propylamine, **(b)** trimethylamine, and **(c)** ethylisopropylmethylamine. What are names for **(d)** $CH_3CH_2CH_2NHCH(CH_3)_2$, **(e)** $(CH_3CH_2CH_2)_3N$, **(f)** $C_6H_5NHCH_3$, and **(g)** $C_6H_5N(CH_3)_2$? Build hand-held molecular models for the compounds in parts (d)–(f).

Problem 2.15 Which amines in Problem 2.14 are **(a)** primary amines, **(b)** secondary amines, and **(c)** tertiary amines?

Amines are like ammonia (Section 1.16B) in having a trigonal pyramidal shape. The C—N—C bond angles of trimethylamine are 108.7°, a value very close to the H—C—H bond angles of methane. Thus, for all practical purposes, the nitrogen atom of an amine can be considered to be sp^3 hybridized with the unshared electron pair occupying one orbital (see page 67). This means that the unshared pair is relatively exposed, and as we shall see this is important because it is involved in almost all of the reactions of amines.

Problem 2.16 Amines are like ammonia in being weak bases. They do this by using their unshared electron pair to accept a proton. **(a)** Show the reaction that would take place between trimethylamine and HCl. **(b)** What hybridization state would you expect for the nitrogen atom in the product of this reaction?

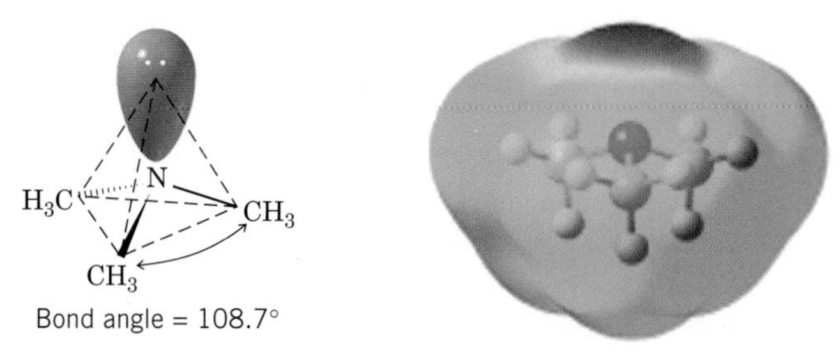

Trimethylamine

2.10 Aldehydes and Ketones

Aldehydes and ketones both contain the **carbonyl group**—a group in which a carbon atom has a double bond to oxygen:

$$>C=\ddot{O}:$$

The carbonyl group

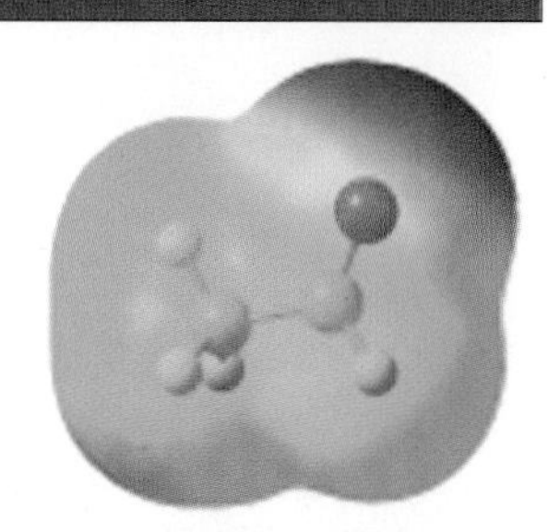

Acetaldehyde

The carbonyl group in aldehydes is bonded to at least one *hydrogen atom,* and in ketones it is bonded to *two carbon atoms.* Using R, we can designate the general formula for an aldehyde as

$$R-C(=\ddot{O})-H \quad \text{or} \quad RCHO \quad \text{R may also be H}$$

and the general formula for a ketone as

$$R-C(=\ddot{O})-R \quad \text{or} \quad RCOR; \quad R-C(=\ddot{O})-R' \quad \text{or} \quad RCOR'$$

(where R′ is an alkyl group different from R).

Some examples of aldehydes and ketones are the following:

ALDEHYDES

$H-C(=\ddot{O})-H$ **Formaldehyde**

$CH_3-C(=\ddot{O})-H$ **Acetaldehyde**

$C_6H_5-C(=O)-H$ **Benzaldehyde**

$C_6H_5(H)C=C(H)-C(=\ddot{O})-H$ ***trans*-Cinnamaldehyde (present in cinnamon)**

KETONES

$CH_3-C(=\ddot{O})-CH_3$ **Acetone**

$CH_3CH_2-C(=\ddot{O})-CH_3$ **Ethyl methyl ketone**

CH_3, $\ddot{O}$:, $H_3C-C=CH_2$ **Carvone (from spearmint)**

STUDY TIP

Computer molecular models can be found in the 3D Models section of the CD for these and many other compounds we discuss in this book.

Aldehydes and ketones have a trigonal planar arrangement of groups around the carbonyl carbon atom. The carbon atom is sp^2 hybridized. In formaldehyde, for example, the bond angles are as follows:

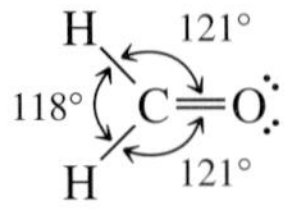

H 121° 118° C=O H 121°

2.11 Carboxylic Acids, Esters, and Amides

Carboxylic acids, esters, and amides all contain a carbonyl group that is bonded to an oxygen or nitrogen atom. Carboxylic acids have a hydroxyl group bonded to a carbonyl group. Esters have an alkoxyl group (—OR) bonded to a carbonyl group and are related to carboxylic acids by replacement of the hydroxyl group with the alkoxyl group. Amides have a nitrogen atom bonded to a carbonyl group and are related to acids and esters by replacement of their hydroxyl or alkoxyl group with a substituted or unsubstituted nitrogen atom. As we shall learn in later chapters, all of these functional groups are interconvertible by appropriately chosen reactions.

2.11A Carboxylic Acids

Carboxylic acids have the general formula R—C(=O)—O—H. The functional group, —C(=O)—O—H, is called the **carboxyl group** (**carbo**nyl + hydro**xyl**):

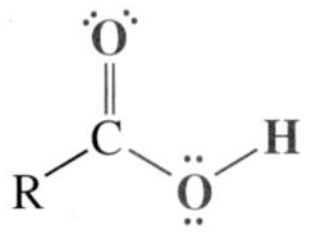

R—C(=O)—O—H or RCO_2H or RCOOH

A carboxylic acid

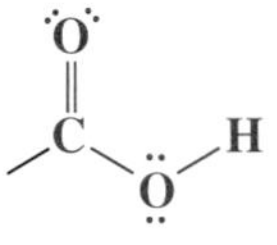

—C(=O)—O—H or $—CO_2H$ or —COOH

The carboxyl group

Examples of carboxylic acids are formic acid, acetic acid, and benzoic acid:

H—C(=O)—O—H or HCO_2H or HCOOH

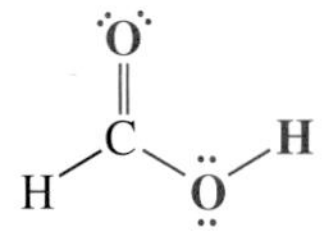

Formic acid

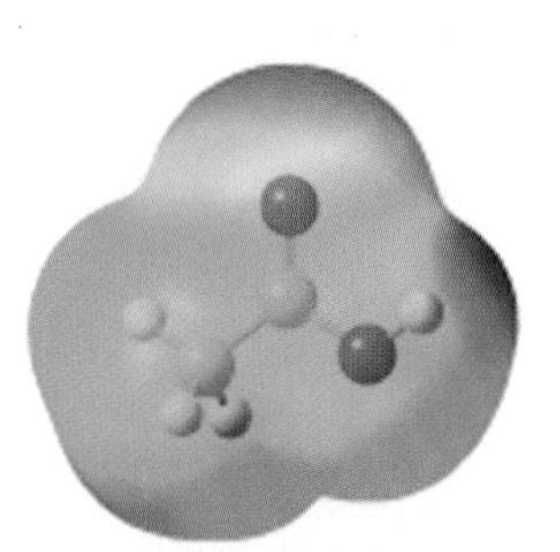

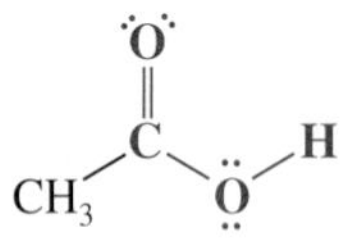

Acetic acid

CH_3—C(=O)—O—H or CH_3CO_2H or CH_3COOH

Acetic acid

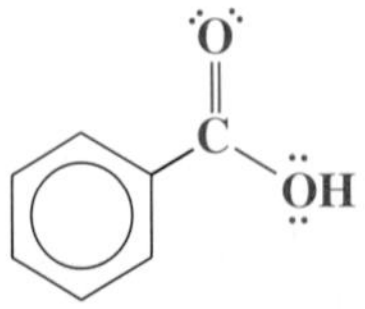

C_6H_5—C(=O)—OH or $C_6H_5CO_2H$ or C_6H_5COOH

Benzoic acid

Formic acid is an irritating liquid produced by ants. (The sting of the ant is caused, in part, by formic acid being injected under the skin.) Acetic acid, the substance responsible for the sour taste of vinegar, is produced when certain bacteria act on the ethyl alcohol of wine and cause the ethyl alcohol to be oxidized by air.

2.11B Esters

Esters have the general formula RCO_2R' (or RCOOR′), where a carbonyl group is bonded to an alkoxyl group:

$$R-C(=\ddot{O}:)-\ddot{O}-R' \quad \text{or} \quad RCO_2R' \quad \text{or} \quad RCOOR'$$

General formula for an ester

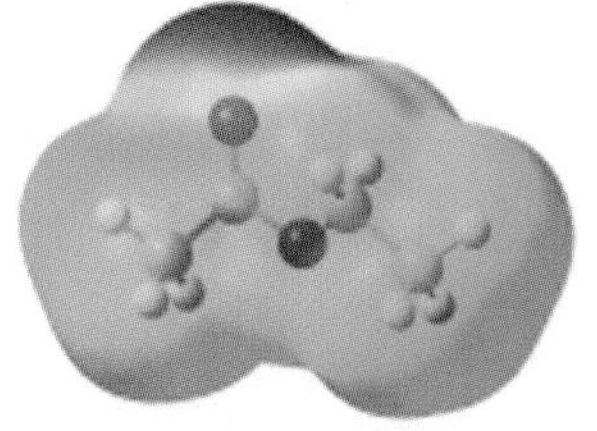

Ethyl acetate

$$CH_3-C(=\ddot{O}:)-\ddot{O}CH_2CH_3 \quad \text{or} \quad CH_3CO_2CH_2CH_3 \quad \text{or} \quad CH_3COOCH_2CH_3$$

A specific ester called ethyl acetate

Esters can be made from a carbonxylic acid and an alcohol through the acid-catalyzed loss of a molecule of water. For example:

$$\underset{\textbf{Acetic acid}}{CH_3-C(=O)-OH} + \underset{\textbf{Ethyl alcohol}}{HOCH_2CH_3} \xrightarrow{\text{acid-catalyzed}} \underset{\textbf{Ethyl acetate}}{CH_3-C(=O)-OCH_2CH_3} + H_2O$$

2.11C Amides

Amides have the formulas $RCONH_2$, RCONHR′, or RCONR′R″ where a carbonyl group is bonded to a nitrogen atom bearing hydrogen and/or alkyl groups. Specific examples are the following:

$$\underset{\textbf{Acetamide}}{CH_3-C(=\ddot{O}:)-\ddot{N}H_2} \qquad \underset{\textbf{\textit{N}-Methylacetamide}}{CH_3-C(=\ddot{O}:)-\ddot{N}HCH_3} \qquad \underset{\textbf{\textit{N,N}-Dimethylacetamide}}{CH_3-C(=\ddot{O}:)-\ddot{N}(CH_3)-CH_3}$$

Acetamide

The *N*- and *N,N*- indicate that the substituents are attached to the nitrogen atom.

2.12 Nitriles

A nitrile has the formula R—C≡N: (or R—CN). The carbon and the nitrogen of a nitrile are *sp* hybridized. In IUPAC systematic nomenclature, acyclic nitriles are named by adding the suffix *nitrile* to the name of the corresponding hydrocarbon. The carbon atom of the —C≡N group is assigned number 1. The name acetonitrile is an acceptable common name for CH_3CN, and acrylonitrile is an acceptable common name for $CH_2{=}CHCN$:

TABLE 2.3 Important Families of Organic Compounds

	Family						
	Alkane	**Alkene**	**Alkyne**	**Aromatic**	**Haloalkane**	**Alcohol**	**Ether**
Functional group	C—H and C—C bonds	$>C{=}C<$	$—C{\equiv}C—$	Aromatic ring	$—\overset{\mid}{\underset{\mid}{C}}—\ddot{\underset{\cdot\cdot}{X}}:$	$—\overset{\mid}{\underset{\mid}{C}}—\ddot{\underset{\cdot\cdot}{O}}H$	$—\overset{\mid}{\underset{\mid}{C}}—\ddot{\underset{\cdot\cdot}{O}}—\overset{\mid}{\underset{\mid}{C}}—$
General formula	RH	$RCH{=}CH_2$ $RCH{=}CHR$ $R_2C{=}CHR$ $R_2C{=}CR_2$	$RC{\equiv}CH$ $RC{\equiv}CR$	ArH	RX	ROH	ROR
Specific example	CH_3CH_3	$CH_2{=}CH_2$	$HC{\equiv}CH$	(benzene ring)	CH_3CH_2Cl	CH_3CH_2OH	CH_3OCH_3
IUPAC name	Ethane	Ethene	Ethyne	Benzene	Chloroethane	Ethanol	Methoxymethane
Common name[a]	Ethane	Ethylene	Acetylene	Benzene	Ethyl chloride	Ethyl alcohol	Dimethyl ether

[a]These names are also accepted by the IUPAC.

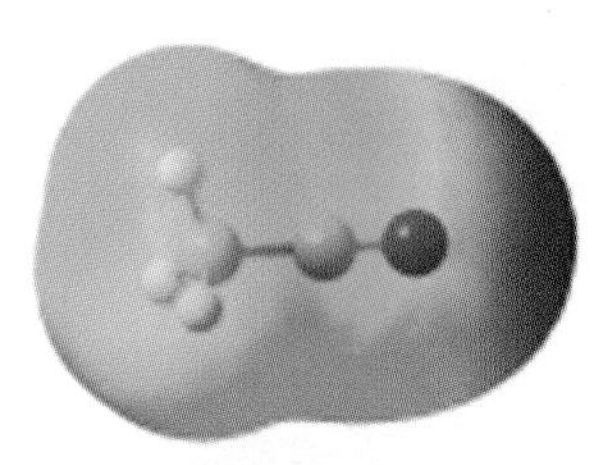

Acetontrile

$$\overset{2}{CH_3}—\overset{1}{C}{\equiv}N: \qquad \overset{4}{CH_3}\overset{3}{CH_2}\overset{2}{CH_2}—\overset{1}{C}{\equiv}N:$$

Ethanenitrile (acetonitrile) **Butanenitrile**

$$\overset{3}{CH_2}{=}\overset{2}{CH}—\overset{1}{C}{\equiv}N: \qquad \overset{5}{CH_2}{=}\overset{4}{CH}\overset{3}{CH_2}\overset{2}{CH_2}—\overset{1}{C}{\equiv}N:$$

Propenenitrile (acrylonitrile) **4-Pentenenitrile**

Cyclic nitriles are named by adding the suffix *carbonitrile* to the name of the ring system to which the —CN group is attached. Benzonitrile is an acceptable common name for C_6H_5CN:

$$C_6H_5—C{\equiv}N: \qquad C_6H_{11}—C{\equiv}N:$$

Benzenecarbonitrile (benzonitrile) **Cyclohexanecarbonitrile**

CD Tutorial
Functional groups

2.13 Summary of Important Families of Organic Compounds

A summary of the important families of organic compounds is given in Table 2.3. You should learn to identify these common functional groups as they appear in other more complicated molecules.

2.14 Physical Properties and Molecular Structure

So far, we have said little about one of the most obvious characteristics of organic compounds, that is, *their physical state or phase.* Whether a particular substance is a solid, or a liquid, or a gas would certainly be one of the first observations that we would note in

TABLE 2.3 Important Families of Organic Compounds (cont.)

Family						
Amine	**Aldehyde**	**Ketone**	**Carboxylic Acid**	**Ester**	**Amide**	**Nitrile**
$-\overset{\vert}{\underset{\vert}{C}}-\ddot{N}<$	$-C(=\ddot{O})-H$	$-\overset{\vert}{\underset{\vert}{C}}-C(=\ddot{O})-\overset{\vert}{\underset{\vert}{C}}-$	$-C(=\ddot{O})-\ddot{O}H$	$-C(=\ddot{O})-\ddot{O}-\overset{\vert}{\underset{\vert}{C}}-$	$-C(=\ddot{O})-\ddot{N}<$	$-C\equiv N:$
RNH_2 R_2NH R_3N	$RCH(=O)$	$RCR'(=O)$	$RCOH(=O)$	$RCOR'(=O)$	$RCNH_2(=O)$ $RCNHR'(=O)$ $RCNR'R''(=O)$	RCN
CH_3NH_2	$CH_3CH(=O)$	$CH_3CCH_3(=O)$	$CH_3COH(=O)$	$CH_3COCH_3(=O)$	$CH_3CNH_2(=O)$	$CH_3C\equiv N$
Methanamine	Ethanal	Propanone	Ethanoic acid	Methyl ethanoate	Ethanamide	Ethanenitrile
Methylamine	Acetaldehyde	Acetone	Acetic acid	Methyl acetate	Acetamide	Acetonitrile

any experimental work. The temperatures at which transitions occur between phases, that is, melting points (mp) and boiling points (bp), are also among the more easily measured **physical properties.** Melting points and boiling points are also useful in identifying and isolating organic compounds.

CD Drill Review

Functional group study aid

Suppose, for example, we have just carried out the synthesis of an organic compound that is known to be a liquid at room temperature and 1 atm pressure. If we know the boiling point of our desired product and the boiling points of byproducts and solvents that may be present in the reaction mixture, we can decide whether or not simple distillation will be a feasible method for isolating our product.

In another instance our product might be a solid. In this case, in order to isolate the substance by crystallization, we need to know its melting point and its solubility in different solvents.

The physical constants of known organic substances are easily found in handbooks and other reference books.* Table 2.4 lists the melting and boiling points of some of the compounds that we have discussed in this chapter.

Often in the course of research, however, the product of a synthesis is a new compound—one that has never been described before. In these instances, success in isolating the new compound depends on making reasonably accurate estimates of its melting point, boiling point, and solubilities. Estimations of these macroscopic physical properties are based on the most likely structure of the substance and on the forces that act between molecules and ions. The temperatures at which phase changes occur are an indication of the strength of these intermolecular forces.

Understanding how molecular structure influences physical properties is very useful in practical organic chemistry.

2.14A Ion–Ion Forces

The **melting point** of a substance is the temperature at which an equilibrium exists between the well-ordered crystalline state and the more random liquid state. If the substance

*Two useful handbooks are *Handbook of Chemistry;* Lange, N. A., Ed.; McGraw-Hill: New York, and *CRC Handbook of Chemistry and Physics;* CRC: Boca Raton, FL.

TABLE 2.4 Physical Properties of Representative Compounds

Compound	Structure	mp (°C)	bp (°C) (1 atm)[a]
Methane	CH_4	−182.6	−162
Ethane	CH_3CH_3	−183	−88.2
Ethene	$CH_2{=}CH_2$	−169	−102
Ethyne	$HC{\equiv}CH$	−82	−84 subl
Chloromethane	CH_3Cl	−97	−23.7
Chloroethane	CH_3CH_2Cl	−138.7	13.1
Ethyl alcohol	CH_3CH_2OH	−115	78.5
Acetaldehyde	CH_3CHO	−121	20
Acetic acid	CH_3CO_2H	16.6	118
Sodium acetate	CH_3CO_2Na	324	dec
Ethylamine	$CH_3CH_2NH_2$	−80	17
Diethyl ether	$(CH_3CH_2)_2O$	−116	34.6
Ethyl acetate	$CH_3CO_2CH_2CH_3$	−84	77

[a]In this table dec = decomposes and subl = sublimes.

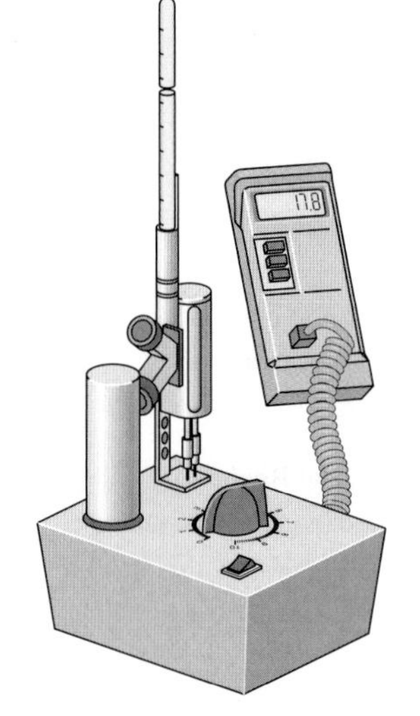

An instrument used to measure melting point

is an ionic compound, such as sodium acetate (Table 2.4), the forces that hold the ions together in the crystalline state are the strong electrostatic lattice forces that act between the positive and negative ions in the orderly crystalline structure. In Fig. 2.6 each sodium ion is surrounded by negatively charged acetate ions, and each acetate ion is surrounded by positive sodium ions. A large amount of thermal energy is required to break up the orderly structure of the crystal into the disorderly open structure of a liquid. As a result, the temperature at which sodium acetate melts is quite high, 324°C. The **boiling points** of ionic compounds are higher still, so high that most ionic organic compounds decompose (are changed by undesirable chemical reactions) before they boil. Sodium acetate shows this behavior.

2.14B Dipole–Dipole Forces

Most organic molecules are not fully ionic but have instead a *permanent dipole moment* resulting from a nonuniform distribution of the bonding electrons (Section 2.4). Acetone and acetaldehyde are examples of molecules with permanent dipoles because the car-

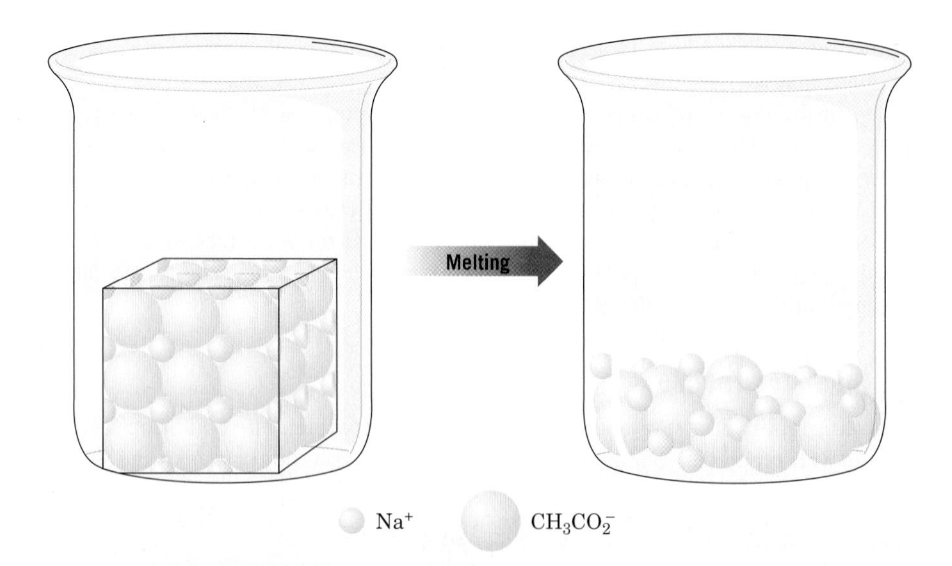

FIGURE 2.6 The melting of sodium acetate.

CD Tutorial
Ion–ion forces

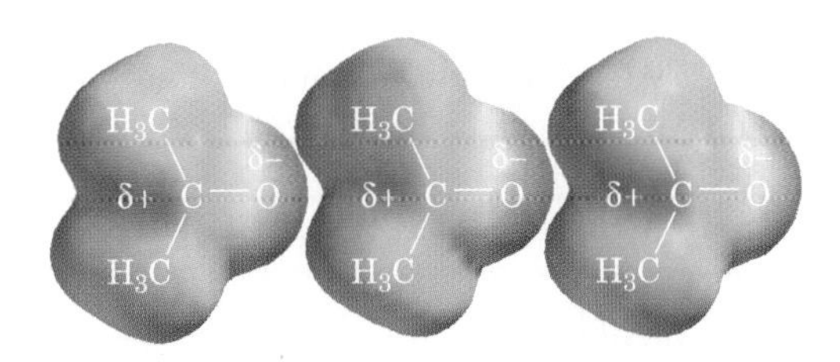

FIGURE 2.7 Electrostatic potential models for acetone molecules that show how acetone molecules might align according to attractions of their partially positive regions and partially negative regions (dipole–dipole interactions).

Animated Graphics

bonyl group that they contain is highly polarized. In these compounds, the attractive forces between molecules are much easier to visualize. In the liquid or solid state, **dipole–dipole** attractions cause the molecules to orient themselves so that the positive end of one molecule is directed toward the negative end of another (Fig. 2.7).

2.14C Hydrogen Bonds

Very strong dipole–dipole attractions occur between hydrogen atoms bonded to small, strongly electronegative atoms (O, N, or F) and nonbonding electron pairs on other such electronegative atoms. This type of intermolecular force is called a **hydrogen bond.** The hydrogen bond (bond dissociation energy about 4–38 kJ mol^{-1}) is weaker than an ordinary covalent bond but is much stronger than the dipole–dipole interactions that occur in acetone:

CD Tutorial

Hydrogen bonds and dipole–dipole attractions

$$\overset{\delta-}{:Z}—\overset{\delta+}{H}\cdots\cdots\overset{\delta-}{:Z}—\overset{\delta+}{H}$$

A hydrogen bond (shown by red dots)

Z is a strongly electronegative element, usually oxygen, nitrogen, or fluorine.

Hydrogen bonds hold the base pairs of double-stranded DNA together (see Section 25.4). Thymine hydrogen bonds with adenine. Cytosine hydrogen bonds with guanine.

DNA backbone | DNA backbone | DNA backbone | DNA backbone

Thymine **Adenine** **Cytosine** **Guanine**

Hydrogen bonding accounts for the fact that ethyl alcohol has a much higher boiling point (+ 78.5°C) than dimethyl ether (− 24.9°C) even though the two compounds have the same molecular weight. Molecules of ethyl alcohol, because they have a hydrogen atom covalently bonded to an oxygen atom, can form strong hydrogen bonds to each other.

$$CH_3CH_2\overset{\delta-}{\ddot{O}}—\overset{\delta+}{H}\cdots:\overset{\delta-}{\ddot{O}}(\overset{\delta+}{H})CH_2CH_3$$

The dotted bond is a hydrogen bond. Strong hydrogen bonding is limited to molecules having a hydrogen atom attached to an O, N, or F atom.

Molecules of dimethyl ether, because they lack a hydrogen atom attached to a strongly electronegative atom, cannot form strong hydrogen bonds to each other. In dimethyl ether the intermolecular forces are weaker dipole–dipole interactions.

PROBLEM 2.17 The compounds in each part below have the same (or similar) molecular weights. Which compound in each part would you expect to have the higher boiling point? Explain your answers. **(a)** $CH_3CH_2CH_2CH_2OH$ or $CH_3CH_2OCH_2CH_3$, **(b)** $(CH_3)_3N$ or $CH_3CH_2NHCH_3$, **(c)** $CH_3CH_2CH_2CH_2OH$ or $HOCH_2CH_2CH_2OH$.

A factor (in addition to polarity and hydrogen bonding) that affects the *melting point* of many organic compounds is the compactness and rigidity of their individual molecules. Molecules that are symmetrical generally have abnormally high melting points. *tert*-Butyl alcohol, for example, has a much higher melting point than the other isomeric alcohols shown here:

$(CH_3)_3C{-}OH$	$CH_3CH_2CH_2CH_2OH$	$CH_3CH(CH_3)CH_2OH$	$CH_3CH_2CH(CH_3)OH$
***tert*-Butyl alcohol** (mp 25°C)	**Butyl alcohol** (mp −90°C)	**Isobutyl alcohol** (mp −108°C)	***sec*-Butyl alcohol** (mp −114°C)

PROBLEM 2.18 Which compound would you expect to have the higher melting point, propane or cyclopropane? Explain your answer.

2.14D van der Waals Forces

If we consider a substance like methane where the particles are nonpolar molecules, we find that the melting point and boiling point are very low: − 182.6°C and − 162°C, respectively. Instead of asking, "Why does methane melt and boil at low temperatures?" a more appropriate question might be "Why does methane, a nonionic, nonpolar substance, become a liquid or a solid at all?" The answer to this question can be given in terms of attractive intermolecular forces called **van der Waals forces** (or **London forces** or **dispersion forces**).

CD Tutorial
van der Waals (London) forces

An accurate account of the nature of van der Waals forces requires the use of quantum mechanics. We can, however, visualize the origin of these forces in the following way. The average distribution of charge in a nonpolar molecule (such as methane) over a period of time is uniform. At any given instant, however, *because electrons move,* the electrons and therefore the charge may not be uniformly distributed. Electrons may, in one instant, be slightly accumulated on one part of the molecule, and, as a consequence, *a small temporary dipole will occur* (Fig. 2.8). This temporary dipole in one molecule can induce opposite (attractive) dipoles in surrounding molecules. It does this because the negative (or positive) charge in a portion of one molecule will distort the electron cloud of an adjacent portion of another molecule, causing an opposite charge to develop there. These temporary dipoles change constantly, but the net result of their existence is to produce attractive forces between nonpolar molecules and thus make possible the existence of their liquid and solid states.

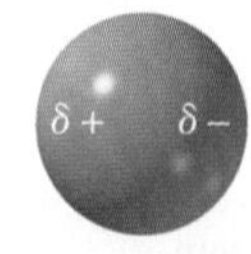

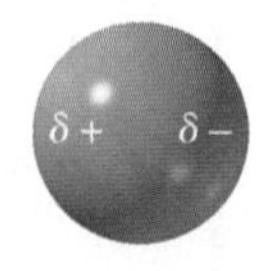

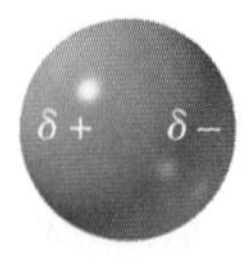

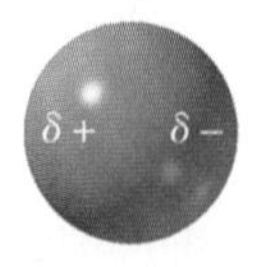

FIGURE 2.8 Temporary dipoles and induced dipoles in nonpolar molecules resulting from an uneven distribution of electrons at a given instant.

One important factor that determines the magnitude of van der Waals forces is the relative **polarizability** of the electrons of the atoms involved. By polarizability we mean *the ability of the electrons to respond to a changing electric field.* Relative polarizability de-

TABLE 2.5 Attractive Energies in Simple Covalent Compounds

Molecule	Dipole Moment (D)	Attractive Energies (kJ mol^{-1}) Dipole–Dipole	van der Waals	Melting Point (°C)	Boiling Point (°C)
H_2O	1.85	36[a]	8.8	0	100
NH_3	1.47	14[a]	15	−78	−33
HCl	1.08	3[a]	17	−115	−85
HBr	0.80	0.8	22	−88	−67
HI	0.42	0.03	28	−51	−35

[a]These dipole–dipole attractions are called hydrogen bonds.

pends on how loosely or tightly the electrons are held. In the halogen family, for example, polarizability increases in the order $F < Cl < Br < I$. Fluorine atoms show a very low polarizability because their electrons are very tightly held; they are close to the nucleus. Iodine atoms are large and hence are more easily polarized. Their valence electrons are far from the nucleus. Atoms with unshared pairs are generally more polarizable than those with only bonding pairs. Thus a halogen substituent is more polarizable than an alkyl group of comparable size. Table 2.5 gives the relative magnitude of van der Waals forces and dipole–dipole interactions for several simple compounds. Notice that except for the molecules where strong hydrogen bonds are possible, van der Waals forces are far more important than dipole–dipole interactions.

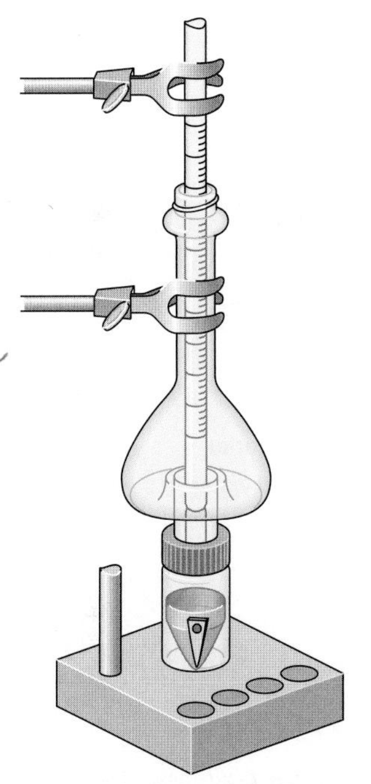
A microscale distillation apparatus

The *boiling point* of a liquid is the temperature at which the vapor pressure of the liquid equals the pressure of the atmosphere above it. For this reason, the boiling points of liquids are *pressure dependent,* and boiling points are always reported as occurring at a particular pressure, at 1 atm (or at 760 torr), for example. A substance that boils at 150°C at 1 atm pressure will boil at a substantially lower temperature if the pressure is reduced to, for example, 0.01 torr (a pressure easily obtained with a vacuum pump). The normal boiling point given for a liquid is its boiling point at 1 atm.

In passing from a liquid to a gaseous state, the individual molecules (or ions) of the substance must separate considerably. Because of this, we can understand why ionic organic compounds often decompose before they boil. The thermal energy required to completely separate (volatilize) the ions is so great that chemical reactions (decompositions) occur first.

Nonpolar compounds, where the intermolecular forces are very weak, usually boil at low temperatures even at 1 atm pressure. This is not always true, however, because of other factors that we have not yet mentioned: the effects of molecular weight and molecular size. Heavier molecules require greater thermal energy in order to acquire velocities sufficiently great to escape the liquid surface, and because the surface areas of heavier molecules are usually much greater, intermolecular van der Waals attractions are also much larger. These factors explain why nonpolar ethane (bp −88.2°C) boils higher than methane (bp −162°C) at a pressure of 1 atm. It also explains why, at 1 atm, the even heavier and larger nonpolar molecule decane ($C_{10}H_{22}$) boils at +174°C.

Fluorocarbons (compounds containing only carbon and fluorine) have extraordinarily low boiling points when compared to hydrocarbons of the same molecular weight. The fluorocarbon C_5F_{12}, for example, has a slightly lower boiling point than pentane (C_5H_{12}) even though it has a far higher molecular weight. The important factor in explaining this behavior is the very low polarizability of fluorine atoms that we mentioned earlier, resulting in very small van der Waals forces. The fluorocarbon polymer called *Teflon* [$\text{+}CF_2CF_2\text{+}_n$, see Section 10.10] has self-lubricating properties, which are exploited in making "nonstick" frying pans and lightweight bearings.

2.14E Solubilities

Your ability to make qualitative predictions regarding solubility will prove very useful in the organic chemistry laboratory.

Intermolecular forces are of primary importance in explaining the **solubilities** of substances. Dissolution of a solid in a liquid is, in many respects, like the melting of a solid. The orderly crystal structure of the solid is destroyed, and the result is the formation of the more disorderly arrangement of the molecules (or ions) in solution. In the process of dissolving, too, the molecules or ions must be separated from each other, and energy must be supplied for both changes. The energy required to overcome lattice energies and intermolecular or interionic attractions comes from the formation of new attractive forces between solute and solvent.

Consider the dissolution of an ionic substance as an example. Here both the lattice energy and interionic attractions are large. We find that water and only a few other very polar solvents are capable of dissolving ionic compounds. These solvents dissolve ionic compounds by **hydrating** or **solvating** the ions (Fig. 2.9).

Water molecules, by virtue of their great polarity as well as their very small, compact shape, can very effectively surround the individual ions as they are freed from the crystal surface. Positive ions are surrounded by water molecules with the negative end of the water dipole pointed toward the positive ion; negative ions are solvated in exactly the opposite way. Because water is highly polar, and because water is capable of forming strong hydrogen bonds, the *dipole–ion* attractive forces are also large. The energy supplied by the formation of these forces is great enough to overcome both the lattice energy and interionic attractions of the crystal.

CD Tutorial
Oil and water

A rule of thumb for predicting solubilities is that "like dissolves like." Polar and ionic compounds tend to dissolve in polar solvents. Polar liquids are generally miscible with each other. Nonpolar solids are usually soluble in nonpolar solvents. On the other hand, nonpolar solids are insoluble in polar solvents. Nonpolar liquids are usually mutually miscible, but nonpolar liquids and polar liquids "like oil and water" do not mix.

Methanol and water are miscible in all proportions; so too are mixtures of ethanol and water and mixtures of both propyl alcohols and water. In these cases the alkyl groups of the alcohols are relatively small, and the molecules therefore resemble water more than they do an alkane. Another factor in understanding their solubility is that the molecules are capable of forming strong hydrogen bonds to each other:

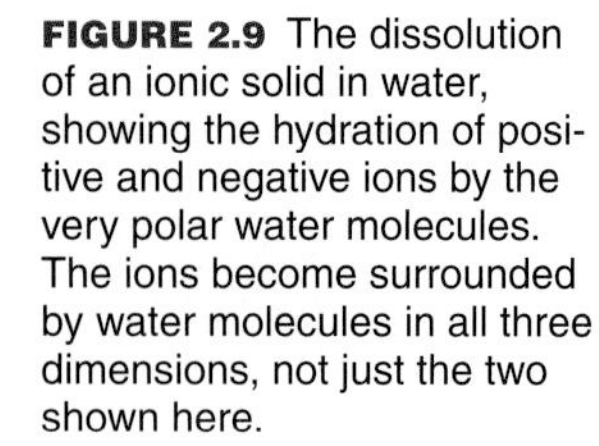

FIGURE 2.9 The dissolution of an ionic solid in water, showing the hydration of positive and negative ions by the very polar water molecules. The ions become surrounded by water molecules in all three dimensions, not just the two shown here.

CH_3CH_2 $\overset{\delta+}{H}$

$\overset{\delta-}{O}$ ⋯ $H^{\delta+}$ $H^{\delta+}$ ⟵ **Hydrogen bond**

$O^{\delta-}$

If the carbon chain of an alcohol is long, however, we find that the alcohol is much less soluble in water. Decyl alcohol (see following structure) with a chain of 10 carbon atoms is only very slightly soluble in water. Decyl alcohol resembles an alkane more than it does water. The long carbon chain of decyl alcohol is said to be **hydrophobic** (hydro, water; *phobic,* fearing or avoiding—"water avoiding"). Only the OH group, a rather small part of the molecule, is **hydrophilic** (*philic,* loving or seeking—"water seeking"). (On the other hand, decyl alcohol is quite soluble in less polar solvents, such as chloroform.)

Hydrophobic portion — **Hydrophilic group**

$CH_3CH_2CH_2CH_2CH_2CH_2CH_2CH_2CH_2CH_2OH$

Decyl alcohol

An explanation for why nonpolar groups such as long alkane chains avoid an aqueous environment, that is, for the so-called **hydrophobic effect,** is complex. The most important factor seems to involve an **unfavorable entropy change** in the water. Entropy changes (Section 3.9) have to do with changes from a relatively ordered state to a more disordered one or the reverse. Changes from order to disorder are favorable, whereas changes from disorder to order are unfavorable. For a nonpolar hydrocarbon chain to be accommodated by water, the water molecules have to form a more ordered structure around the chain, and for this, the entropy change is unfavorable.

2.14F Guidelines for Water Solubility

Organic chemists usually define a compound as water soluble if at least 3 g of the organic compound dissolves in 100 mL of water. We find that for compounds containing one hydrophilic group—and thus capable of forming strong hydrogen bonds—the following approximate guidelines hold: Compounds with one to three carbon atoms are water soluble, compounds with four or five carbon atoms are borderline, and compounds with six carbon atoms or more are insoluble.

CD Animated Graphics

When a compound contains more than one hydrophilic group, these guidelines do not apply. Polysaccharides (Chapter 22), proteins (Chapter 24), and nucleic acids (Chapter 25) all contain thousands of carbon atoms *and many are water soluble.* They dissolve in water because they also contain thousands of hydrophilic groups.

2.14G Intermolecular Forces in Biochemistry

Later, after we have had a chance to examine in detail the properties of the molecules that make up living organisms, we shall see how **intermolecular forces** are extremely important in the functioning of cells. **Hydrogen bond** formation, the hydration of polar groups, and the tendency of nonpolar groups to avoid a polar environment all cause complex protein molecules to fold in precise ways—ways that allow them to function as biological catalysts of incredible efficiency. The same factors allow molecules of hemoglobin to assume the shape needed to transport oxygen. They allow proteins and molecules called lipids to function as cell membranes. Hydrogen bonding gives certain carbohydrates a globular shape that makes them highly efficient food reserves in animals. It gives molecules of other carbohydrates a rigid linear shape that makes them perfectly suited to be structural components in plants.

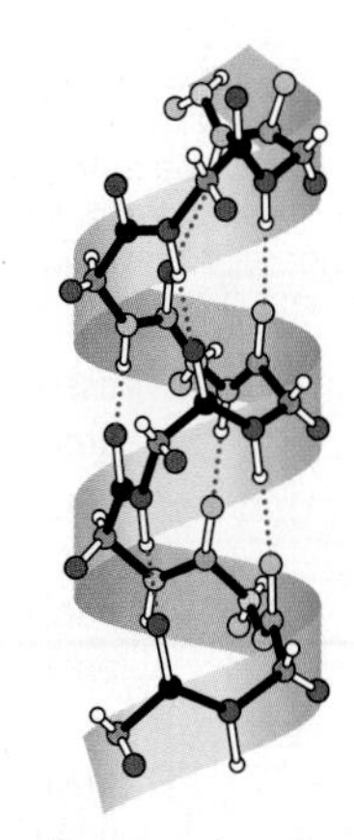

Hydrogen bonding (red dotted lines) in the α - helix structure of proteins

2.15 Summary of Attractive Electric Forces

CD Tutorial

Intermolecular attractive forces

The attractive forces occurring between molecules and ions that we have studied so far are summarized in Table 2.6.

TABLE 2.6 Attractive Electric Forces

Electric Force	Relative Strength	Type	Example
Cation–anion (in a crystal)	Very strong	⊕ ⊖	Lithium fluoride crystal lattice
Covalent bonds	Strong (140–523 kJ mol^{-1})	Shared electron pairs	H—H (436 kJ mol^{-1}) CH_3—CH_3 (378 kJ mol^{-1}) I—I (151 kJ mol^{-1})
Ion–dipole	Moderate	⊕ surrounded by dipoles ($\delta+\delta-$, $\delta-\delta+$)	Na^+ in water (see Fig. 2.9)
Dipole–dipole (including hydrogen bonds)	Moderate to weak (4–38 kJ mol^{-1})	$\overset{\delta-}{—Z}:\cdots\overset{\delta+}{H}—$	R—$\overset{\delta-}{O}$(—$^{\delta+}$H):···$\overset{\delta+}{H}$—$\ddot{O}^{\delta-}$—R and $\overset{\delta+}{CH_3}$—$\overset{\delta-}{Cl}$ $\overset{\delta+}{CH_3}$—$\overset{\delta-}{Cl}$
van der Waals	Variable	Transient dipole	Interactions between methane molecules

The Chemistry of...

Organic Templates Engineered to Mimic Bone Growth

Intermolecular forces play a myriad of roles in life and the world around us. Intermolecular forces hold together the strands of our DNA, provide structure to our cell membranes, cause the feet of gecko lizards to stick to walls and ceilings, keep water from boiling at room temperature and ordinary pressure, and literally provide the adhesive forces that hold our cells, bones, and tissues together. As these examples show, the world around us provides exquisite instruction in nanotechnology and bioengineering, and scientists throughout the ages have been inspired to create and innovate based on nature. One target of recent research in bioengineering is the development of synthetic materials that mimic nature's template for bone growth. A synthetic material with bone-promoting properties could be used to help repair broken bones, offset osteoporosis, and treat bone cancer.

Both natural bone growth and the synthetic system under development depend strongly on intermolecular forces. In living systems, bones grow by adhesion of specialized cells to a long fibrous natural template called collagen. Certain functional groups along the collagen promote the binding of bone-growing cells, while other functional groups facilitate calcium crystallization. Chemists at Northwestern University (led by S. I. Stupp) have engineered a molecule that can be made in the laboratory and that mimics this process. The molecule (shown on page 79) spontaneously self-assembles into a long tubular aggregate, imitating the fibers of collagen. van der Waals forces between hydrophobic alkyl tails on the molecule cause self-assembly of the molecules into tubules. At the other end of the molecule, the researchers included functional groups that promote cell binding and still other functional groups that encourage calcium crystallization. Lastly, they included functional groups that allow one molecule to be covalently linked to its neighbors after the van der Waals self-assembly process has occurred, thus adding further stabilization to the initially noncovalent structure. Designing all of these features into the molecular structure has paid off, because the self-assembled fiber promotes calcium crystallization along its axis, much like nature's collagen template. This example of molecular design is just one exciting development at the intersection of nanotechnology and bioengineering.

Hydrophobic alkyl region

Flexible linker region

Hydrophilic cell adhesion region

2.16 Infrared Spectroscopy: An Instrumental Method for Detecting Functional Groups

Infrared (IR) spectroscopy provides a simple and rapid instrumental technique that can give evidence for the presence of various functional groups. If you had a sample of unknown identity, for example, among the first things you would do is obtain an infrared spectrum, along with determining its solubility in common solvents and its melting and/or boiling point.

CD Tutorial
IR tutor

Infrared spectroscopy, as all forms of spectroscopy, depends on the interaction of molecules or atoms with electromagnetic radiation. We shall have more to say about the detailed properties of electromagnetic radiation in Chapter 9, but for now the following will suffice to describe the interaction of IR radiation and organic molecules.

Infrared radiation causes atoms and groups of atoms of organic compounds to vibrate with increased amplitude about the covalent bonds that connect them. (Infrared radiation is not of sufficient energy to excite electrons, as is the case when some molecules interact with visible, ultraviolet, or higher energy forms of light.) Since the functional groups of organic molecules include specific arrangements of bonded atoms, absorption of IR energy by an organic molecule will occur in a manner characteristic of the types of bonds and atoms present in the specific functional groups of that molecule. These vibrations are *quantized,* and as they occur, the compounds absorb IR energy in particular regions of the IR portion of the spectrum.

An infrared spectrometer (Fig. 2.10) operates by passing a beam of IR radiation through a sample and comparing the radiation transmitted through the sample with that transmitted in the absence of the sample. Any frequencies absorbed by the sample will be apparent by the difference. The spectrometer plots the results as a graph showing absorbance versus frequency or wavelength.

The location of an IR absorption band (or peak) can be specified in **frequency-related units** by its **wavenumber** ($\bar{\nu}$) measured in reciprocal centimeters (cm^{-1}) or by its **wavelength** (λ) measured in micrometers (μm; old name micron, μ). The wavenumber is the number of cycles of the wave in each centimeter along the light beam, and the wavelength is the length of the wave, crest to crest:

$$\bar{\nu} = \frac{1}{\lambda} \text{ (with } \lambda \text{ in cm)} \quad \text{or} \quad \bar{\nu} = \frac{10{,}000}{\lambda} \text{ (with } \lambda \text{ in } \mu\text{m)}$$

In their vibrations covalent bonds behave as if they were tiny springs connecting the atoms. When the atoms vibrate, they can do so only at certain frequencies, as if the bonds were "tuned." Because of this, covalently bonded atoms have only particular vibrational energy levels; that is, the levels are quantized.

The excitation of a molecule from one vibrational energy level to another occurs only when the compound absorbs IR radiation of a particular energy, meaning a particular wavelength or frequency (since $\Delta E = h\nu$).

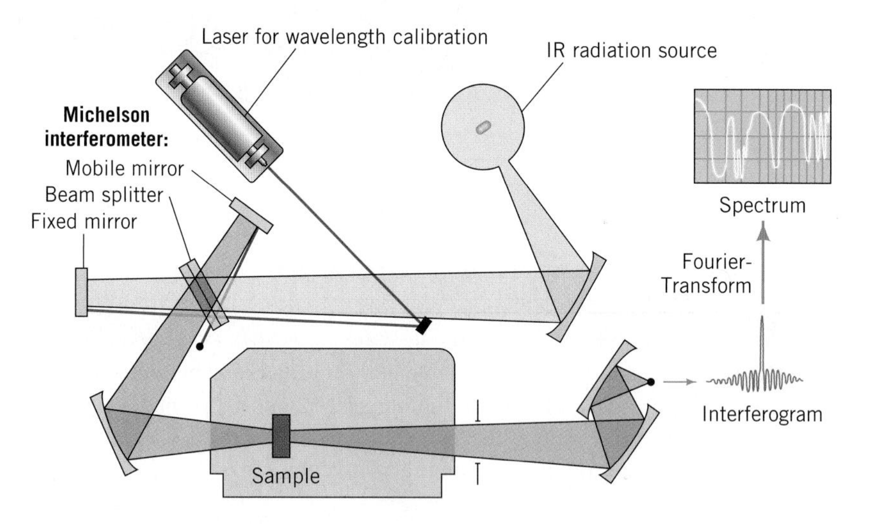

FIGURE 2.10 A diagram of a Fourier transform infrared (FTIR) spectrometer. FTIR spectrometers employ a Michelson interferometer, which splits the radiation beam from the IR source so that it reflects simultaneously from a moving mirror and a fixed mirror, leading to interference. After the beams recombine, they pass through the sample to the detector and are recorded as a plot of time versus signal intensity, called an interferogram. The overlapping wavelengths and the intensities of their respective absorptions are then converted to a spectrum by applying a mathematical operation called a Fourier transform.

The FTIR method eliminates the need to scan slowly over a range of wavelengths, as was the case with older types of instruments called dispersive IR spectrometers, and therefore FTIR spectra can be acquired very quickly. The FTIR method also allows greater throughput of IR energy. The combination of these factors gives FTIR spectra strong signals as compared to background noise (i.e., a high signal to noise ratio) because radiation throughput is high and rapid scanning allows multiple spectra to be averaged in a short period of time. The result is enhancement of real signals and cancellation of random noise. (Diagram adapted from the computer program IR Tutor, Columbia University.)

Molecules can vibrate in a variety of ways. Two atoms joined by a covalent bond can undergo a stretching vibration where the atoms move back and forth as if joined by a spring. Three atoms can also undergo a variety of stretching and bending vibrations.

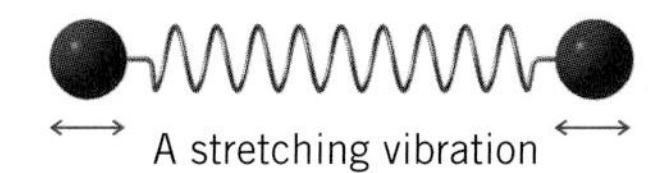

CD Tutorial
Methane dynamics

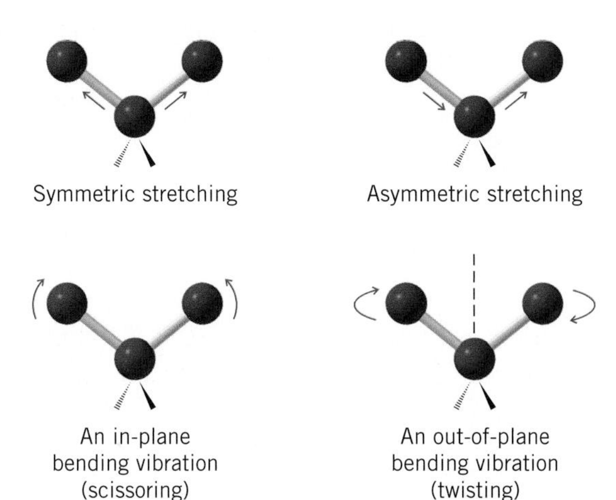

The *frequency* of a given stretching vibration *in an IR spectrum* can be related to two factors. These are *the masses of the bonded atoms*—light atoms vibrate at higher frequencies than heavier ones—*and the relative stiffness of the bond.* (These factors are accounted for in Hooke's law, a relationship you may study in introductory physics.) Triple bonds are stiffer (and vibrate at higher frequencies) than double bonds, and double bonds are stiffer (and vibrate at higher frequencies) than single bonds. We can see some of these effects in Table 2.7. Notice that stretching frequencies of groups involving hydrogen (a light atom) such as C—H, N—H, and O—H all occur at relatively high frequencies:

GROUP	BOND	FREQUENCY RANGE (CM^{-1})
Alkyl	C—H	2853–2962
Alcohol	O—H	3590–3650
Amine	N—H	3300–3500

Notice, too, that triple bonds vibrate at higher frequencies than double bonds:

BOND	FREQUENCY RANGE (CM^{-1})
C≡C	2100–2260
C≡N	2220–2260
C=C	1620–1680
C=O	1630–1780

Not all molecular vibrations result in the absorption of IR energy. ***In order for a vibration to occur with the absorption of IR energy, the dipole moment of the molecule must change as the vibration occurs.*** Thus, when the four hydrogen atoms of methane vibrate symmetrically, methane does not absorb IR energy. Symmetrical vibrations of the car-

TABLE 2.7 Characteristic Infrared Absorptions of Groups

Group		Frequency Range (cm^{-1})	Intensity[a]
A. Alkyl			
C—H (stretching)		2853–2962	(m–s)
Isopropyl, —$CH(CH_3)_2$		1380–1385	(s)
	and	1365–1370	(s)
tert-Butyl, —$C(CH_3)_3$		1385–1395	(m)
	and	~1365	(s)
B. Alkenyl			
C—H (stretching)		3010–3095	(m)
C═C (stretching)		1620–1680	(v)
R—CH═CH_2 (out-of-plane C—H bendings)		985–1000	(s)
	and	905–920	(s)
R_2C═CH_2 (out-of-plane C—H bendings)		880–900	(s)
cis-RCH═CHR (out-of-plane C—H bendings)		675–730	(s)
trans-RCH═CHR (out-of-plane C—H bendings)		960–975	(s)
C. Alkynyl			
≡C—H (stretching)		~3300	(s)
C≡C (stretching)		2100–2260	(v)
D. Aromatic			
Ar—H (stretching)		~3030	(v)
Aromatic substitution type (C—H out-of-plane bendings)			
Monosubstituted		690–710	(very s)
	and	730–770	(very s)
o-Disubstituted		735–770	(s)
m-Disubstituted		680–725	(s)
	and	750–810	(very s)
p-Disubstituted		800–860	(very s)
E. Alcohols, Phenols, and Carboxylic Acids			
O—H (stretching)			
Alcohols, phenols (dilute solutions)		3590–3650	(sharp, v)
Alcohols, phenols (hydrogen bonded)		3200–3550	(broad, s)
Carboxylic acids (hydrogen bonded)		2500–3000	(broad, v)
F. Aldehydes, Ketones, Esters, Carboxylic Acids and Amides			
C═O (stretching)		1630–1780	(s)
Aldehydes		1690–1740	(s)
Ketones		1680–1750	(s)
Esters		1735–1750	(s)
Carboxylic acids		1710–1780	(s)
Amides		1630–1690	(s)
G. Amines			
N—H		3300–3500	(m)
H. Nitriles			
C≡N		2220–2260	(m)

[a]Abbreviations: s = strong, m = medium, w = weak, v = variable, ~ = approximately.

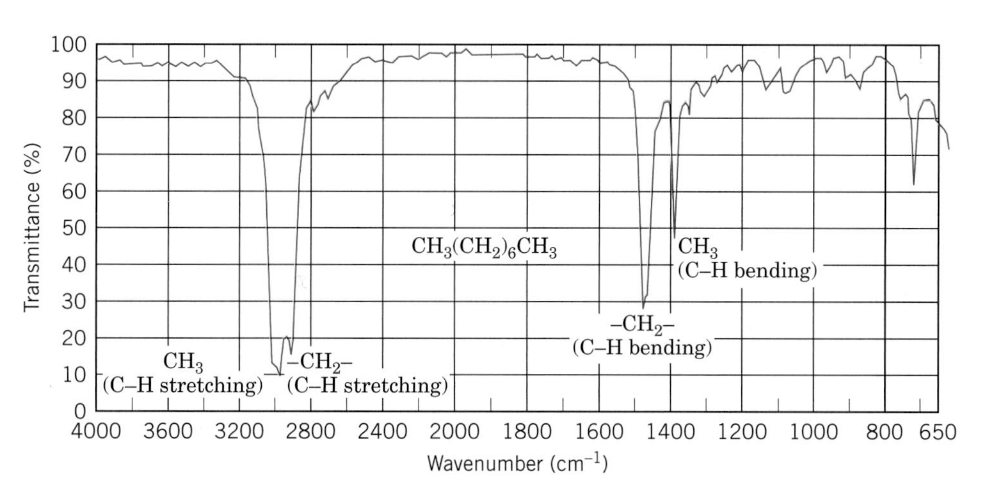

FIGURE 2.11 The IR spectrum of octane. (Notice that, in IR spectra, the peaks are usually measured in % transmittance. Thus, the peak at 2900 cm^{-1} has 10% transmittance, that is, an absorbance, A, of 0.90).

bon–carbon double and triple bonds of ethene and ethyne do not result in the absorption of IR radiation, either.

Vibrational absorption may occur outside the region measured by a particular IR spectrometer, and vibrational absorptions may occur so closely together that peaks fall on top of peaks.

Other factors bring about even more absorption peaks. Overtones (harmonics) of fundamental absorption bands may be seen in IR spectra even though these overtones occur with greatly reduced intensity. Bands called combination bands and difference bands also appear in IR spectra.

Because IR spectra of even relatively simple compounds contain so many peaks, the possibility that two different compounds will have the same IR spectrum is exceedingly small. It is because of this that an IR spectrum has been called the "fingerprint" of a molecule. Thus, with organic compounds, if two pure samples give different IR spectra, one can be certain that they are different compounds. If they give the same IR spectrum, then they are very likely to be the same compound.

In the hands of one skilled in their interpretation, IR spectra contain a wealth of information about the structures of compounds. We show some of the information that can be gathered from the spectra of octane and methylbenzene (commonly called toluene) in Figs. 2.11 and 2.12. We have neither the time nor the space here to develop the skill that would lead to complete interpretations of IR spectra, but we can learn how to recognize the presence of absorption peaks in the IR spectrum that result from vibrations of characteristic functional groups in the compound. By doing only this, however, we shall be able

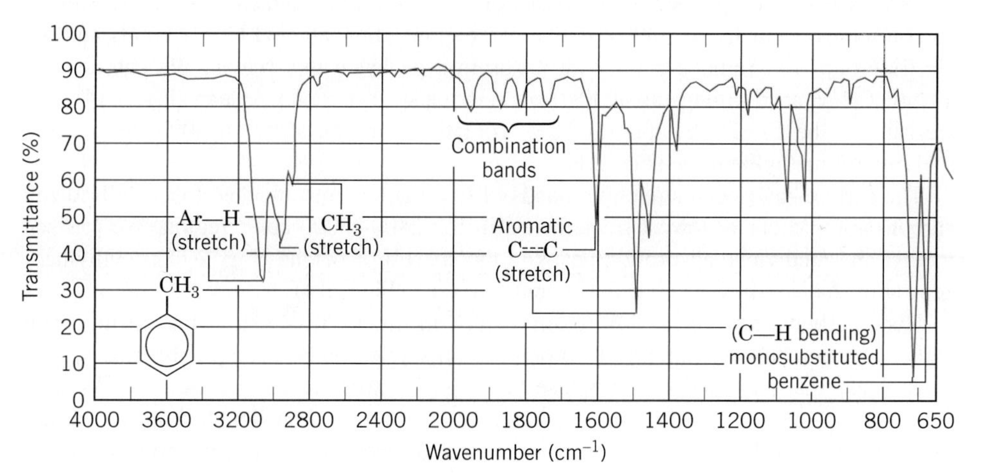

FIGURE 2.12 The IR spectrum of methylbenzene (toluene).

to use the information we gather from IR spectra in a powerful way, particularly when we couple it with the information we gather from NMR (nuclear magnetic resonance) and mass spectra in Chapter 9.

Let us now see how we can apply the data given in Table 2.7 to the interpretation of IR spectra.

2.16A Hydrocarbons

All hydrocarbons give absorption peaks in the 2800–3300-cm^{-1} region that are associated with carbon–hydrogen stretching vibrations. We can use these peaks in interpreting IR spectra because the exact location of the peak depends on the strength (and stiffness) of the C—H bond, which in turn depends on the hybridization state of the carbon that bears the hydrogen. The C—H bonds involving *sp*-hybridized carbon are strongest and those involving sp^3-hybridized carbon are weakest. The order of bond strength is

$$sp > sp^2 > sp^3$$

This too is the order of the bond stiffness.

The carbon–hydrogen stretching peaks of hydrogen atoms attached to *sp*-hybridized carbon atoms occur at highest frequencies, about 3300 cm^{-1}. Thus, ≡C—H groups of terminal alkynes give peaks in this region. We can see the absorption of the acetylenic C—H bond of 1-hexyne at 3320 cm^{-1} in Fig. 2.13.

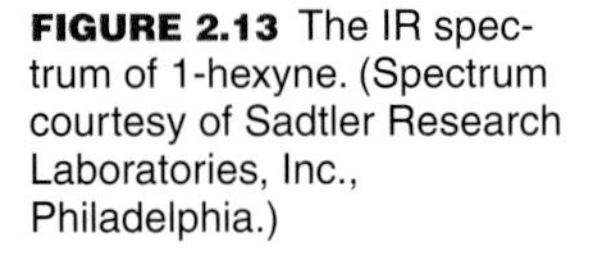

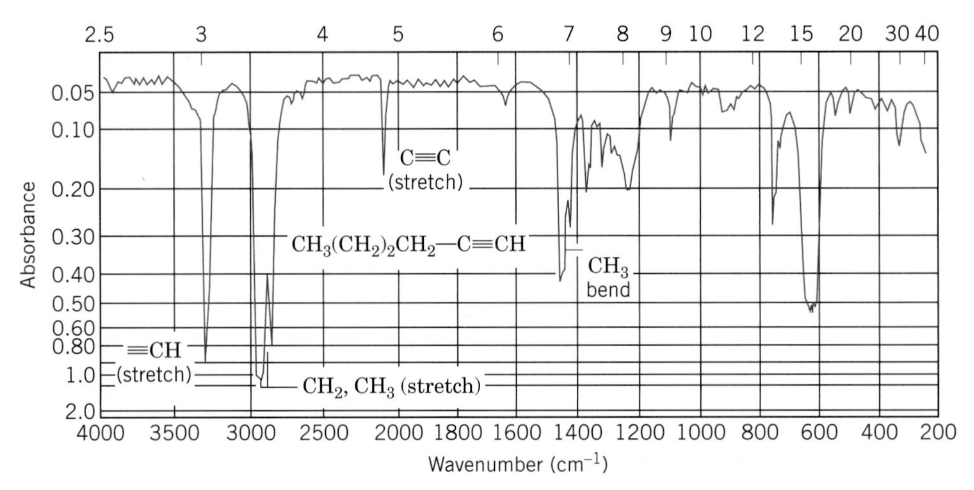

FIGURE 2.13 The IR spectrum of 1-hexyne. (Spectrum courtesy of Sadtler Research Laboratories, Inc., Philadelphia.)

The carbon–hydrogen stretching peaks of hydrogen atoms attached to sp^2-hybridized carbon atoms occur in the 3000–3100-cm^{-1} region. Thus, alkenyl C—H bonds and the C—H groups of aromatic rings give absorption peaks in this region. We can see the alkenyl C—H absorption peak at 3080 cm^{-1} in the spectrum of 1-hexene (Fig. 2.14), and we can see the C—H absorption of the aromatic hydrogen atoms at 3090 cm^{-1} in the spectrum of methylbenzene (Fig. 2.12).

The carbon–hydrogen stretching bands of hydrogen atoms attached to sp^3-hybridized carbon atoms occur at lowest frequencies, in the 2800–3000-cm^{-1} region. We can see methyl and methylene absorption peaks in the spectra of octane (Fig. 2.11), methylbenzene (Fig. 2.12), 1-hexyne (Fig. 2.13), and 1-hexene (Fig. 2.14).

Hydrocarbons also give absorption peaks in their IR spectra that result from carbon–carbon bond stretchings. Carbon–carbon single bonds normally give rise to very weak peaks that are usually of little use in assigning structures. More useful peaks arise from carbon–carbon multiple bonds, however. Carbon–carbon double bonds give absorption peaks in the 1620–1680-cm^{-1} region, and carbon–carbon triple bonds give ab-

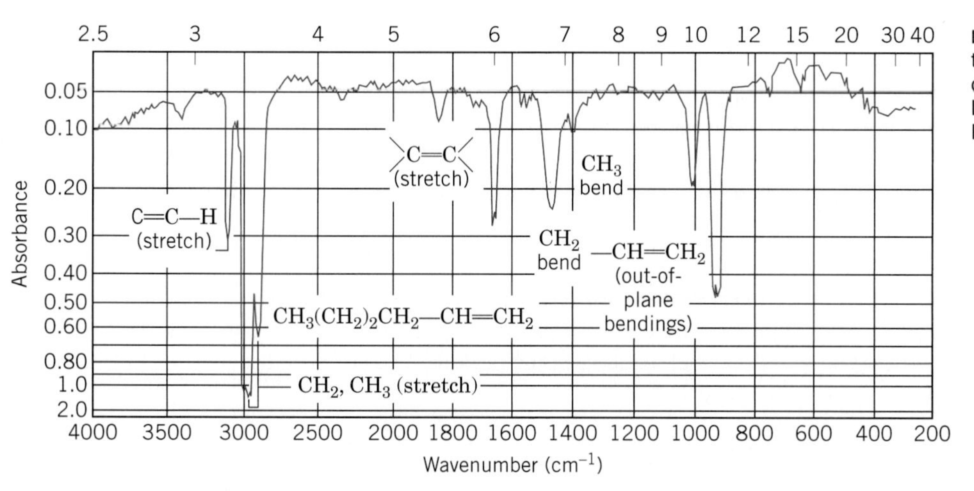

FIGURE 2.14 The IR spectrum of 1-hexene. (Spectrum courtesy of Sadtler Research Laboratories, Inc., Philadelphia.)

sorption peaks between 2100 and 2260 cm^{-1}. These absorptions are not usually strong ones, and they are absent if the double or triple bond is symmetrically substituted. (No dipole moment change will be associated with the vibration.) The stretchings of the carbon–carbon bonds of benzene rings usually give a set of characteristic sharp peaks in the 1450–1600-cm^{-1} region.

Absorptions arising from carbon–hydrogen bending vibrations of alkenes occur in the 600–1000-cm^{-1} region. The exact location of these peaks can often be used to determine the ***substitution pattern of the double bond and its configuration.***

Figure 2.15 shows the regions of characteristic absorption for alkenes with different substitution patterns. These ranges can be used with fair reliability for alkenes that do not have an electron-releasing or electron-withdrawing substituent (other than an alkyl group) on one of the carbon atoms of the double bond. When electron-releasing or electron-withdrawing substituents are present on a double-bond carbon, the bending absorption peaks may be shifted out of the regions we have given.

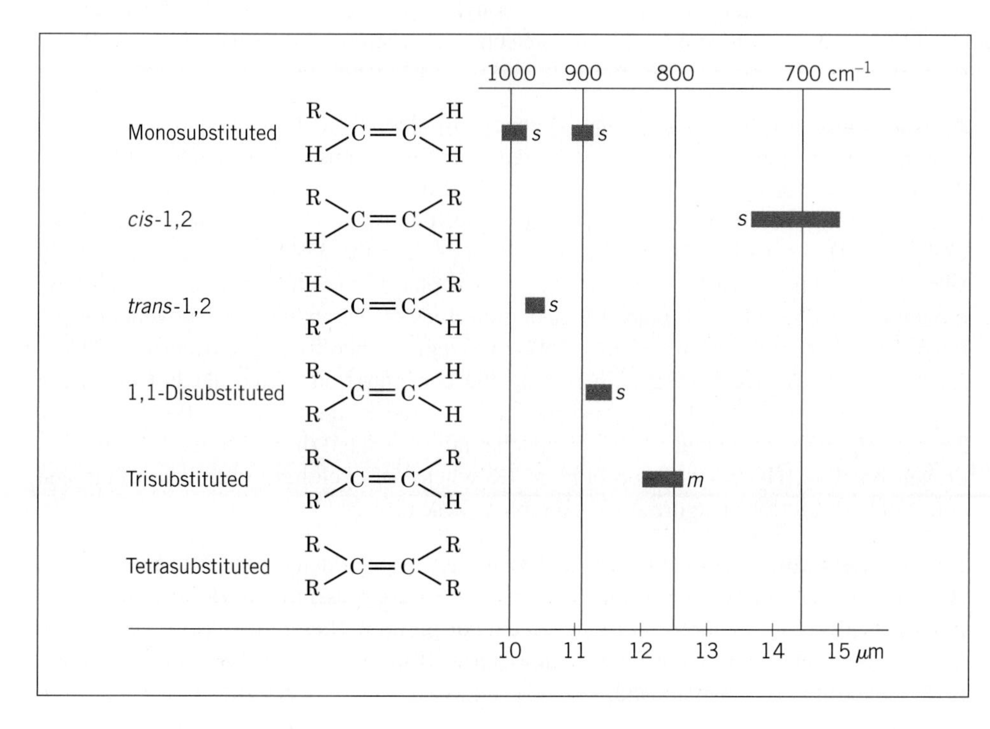

FIGURE 2.15 The C—H out-of-plane bending vibrations for substituted alkenes. (From Pavia, D. L.; Lampan, G. M.; Kriz, G. S.; Engel, R. G. *Introduction to Organic Laboratory Techniques,* 3rd ed.; Saunders College Publishing: Philadelphia, 1999; p A22.)

IR is an exceedingly useful tool for detecting functional groups.

2.16B Other Functional Groups

Infrared spectroscopy gives us an invaluable method for recognizing quickly and simply the presence of certain functional groups in a molecule.

Carbonyl Functional Groups One important functional group that gives a prominent absorption peak in IR spectra is the **carbonyl group** $\rangle C{=}O$. This group is present in aldehydes, ketones, esters, carboxylic acids, amides, and so forth. The carbon–oxygen double-bond stretching frequency of all these groups gives a strong peak between 1630 and 1780 cm^{-1}. The exact location of the peak depends on whether it arises from an aldehyde, ketone, ester, and so forth.

$$\underset{\text{Aldehyde}}{R{-}C({=}O){-}H} \qquad \underset{\text{Ketone}}{R{-}C({=}O){-}R} \qquad \underset{\text{Ester}}{R{-}C({=}O){-}OR}$$

Aldehyde **1690–1740 cm^{-1}** | **Ketone** **1680–1750 cm^{-1}** | **Ester** **1735–1750 cm^{-1}**

$$\underset{\text{Carboxylic acid}}{R{-}C({=}O){-}OH} \qquad \underset{\text{Amide}}{R{-}C({=}O){-}NH_2}$$

Carboxylic acid **1710–1780 cm^{-1}** | **Amide** **1630–1690 cm^{-1}**

PROBLEM 2.19

Use arguments based on resonance and electronegativity effects to explain the trend in carbonyl IR stretching frequencies from higher frequency for esters and carboxylic acids to lower frequencies for amides. (*Hint:* Use the range of carbonyl stretching frequencies for aldehydes and ketones as the "base" frequency range of an unsubstituted carbonyl group and consider the influence of electronegative atoms on the carbonyl group and/or atoms that alter the resonance hybrid of the carbonyl.) What does this suggest about the way the nitrogen atom of an amide influences electronic structure of the carbonyl group?)

Alcohols and Phenols The **hydroxyl groups** of alcohols and phenols are also easy to recognize in IR spectra by their O—H stretching absorptions. These bonds also give us direct evidence for hydrogen bonding (Section 2.14C). If an alcohol or phenol is present as a very dilute solution in a solvent that cannot contribute to hydrogen bonding (e.g., CCl_4), O—H absorption occurs as a very sharp peak in the 3590–3650-cm^{-1} region. In very dilute solution in such a solvent or in the gas phase, formation of intermolecular hydrogen bonds does not take place because molecules of the analyte are too widely separated. The sharp peak in the 3590–3650-cm^{-1} region, therefore, is attributed to "free" (unassociated) hydroxyl groups. Increasing the concentration of the alcohol or phenol causes the sharp peak to be replaced by a broad band in the 3200–3550-cm^{-1} region. This absorption is attributed to OH groups that are associated through intermolecular hydrogen bonding. Hydroxyl absorptions in IR spectra of cyclohexanol run in dilute and concentrated solutions (Figure 2.16) exemplify these effects.

Carboxylic Acids The **carboxylic acid group** can also be detected by IR spectroscopy. If both carbonyl and hydroxyl stretching absorptions are present in an IR spectrum, there is good evidence for a carboxylic acid functional group (although it is possible that isolated carbonyl and hydroxyl groups could be present in the molecule). Figure 2.17 shows the IR spectrum of propanoic acid.

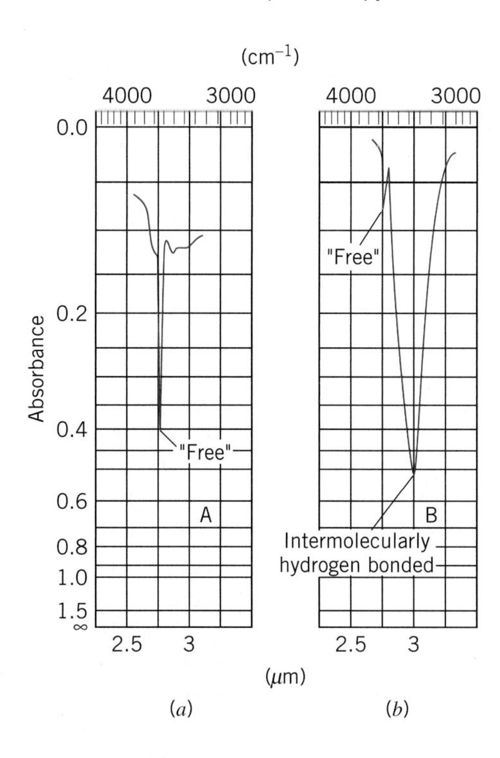

FIGURE 2.16 *(a)* The IR spectrum of cyclohexanol in a dilute solution shows the sharp absorption of a "free" (non-hydrogen-bonded) hydroxyl group at 3600 cm^{-1}. *(b)* The IR spectrum of cyclohexanol as a concentrated solution shows a broad hydroxyl group absorption at 3300 cm^{-1} due to hydrogen bonding. (From Silverstein, R. M.; Webster, F. X. *Spectrometric Identification of Organic Compounds,* 6th ed.; Wiley: New York, 1998; p 89.)

Amines Very dilute solutions of 1° and 2° **amines** also give sharp peaks in the 3300–3500-cm^{-1} region arising from free N—H stretching vibrations. Primary amines give two sharp peaks arising from symmetric and asymmetric stretching of the two N—H bonds; secondary amines give only one stretching absorption. Tertiary amines, because they have no N—H bond, do not absorb in this region.

RNH_2	R_2NH
1° Amine	**2° Amine**
Two peaks in	**One peak in**
3300–3500-cm^{-1}	**3300–3500-cm^{-1}**
region	**region**

Hydrogen bonding causes N—H stretching peaks of 1° and 2° amines to broaden. The NH groups of **amides** give similar absorption peaks and include a carbonyl absorption as well.

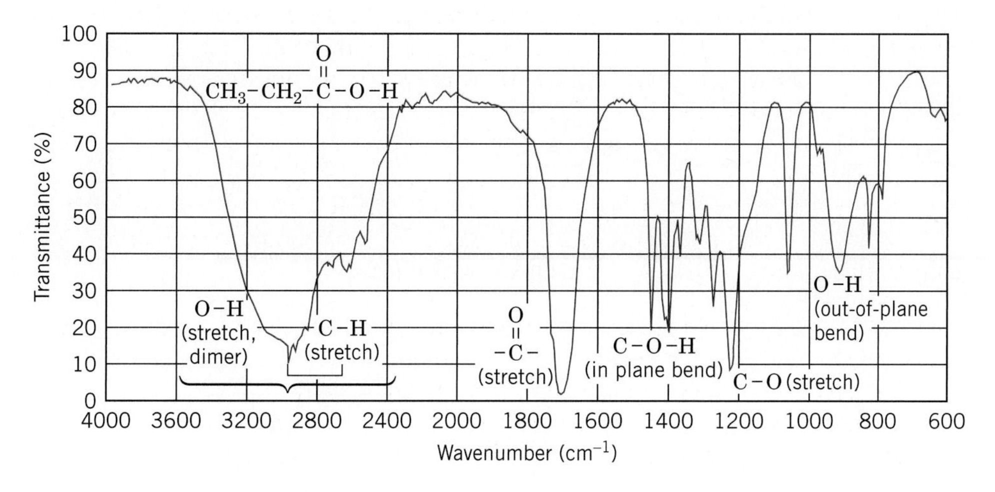

FIGURE 2.17 The IR spectrum of propanoic acid.

Key Terms and Concepts

Alkanes	Sections 2.2, 2.2A
Alkenes	Sections 2.2, 2.4A
Alkyl groups	Section 2.5A
Alkynes	Section 2.2
Aromatic hydrocarbons	Section 2.2
Boiling point	Sections 2.14A, 2.14D
Dipole–dipole forces	Section 2.14B
Dipole moment	Section 2.3
Electronegativity	Sections 1.4, 2.3
Electrostatic potential maps	Sections 1.8, 2.3A
Functional groups	Section 2.5
Heteroatoms	Section 2.3
Hydrocarbons	Section 2.2
Hydrogen bonds	Sections 2.14C, 2.14F, 2.14G
Hydrophilic groups	Section 2.14E
Hydrophobic groups	Section 2.14E
Infrared spectroscopy	Section 2.16
Intermolecular forces	Section 2.14G
Ion–ion forces	Section 2.14A
Melting point	Section 2.14A
Physical properties	Section 2.14
Polar covalent bond	Section 2.3
Saturated/unsaturated compounds	Section 2.2
Solubility	Sections 2.14E, 2.14F
van der Waals forces	Section 2.14D

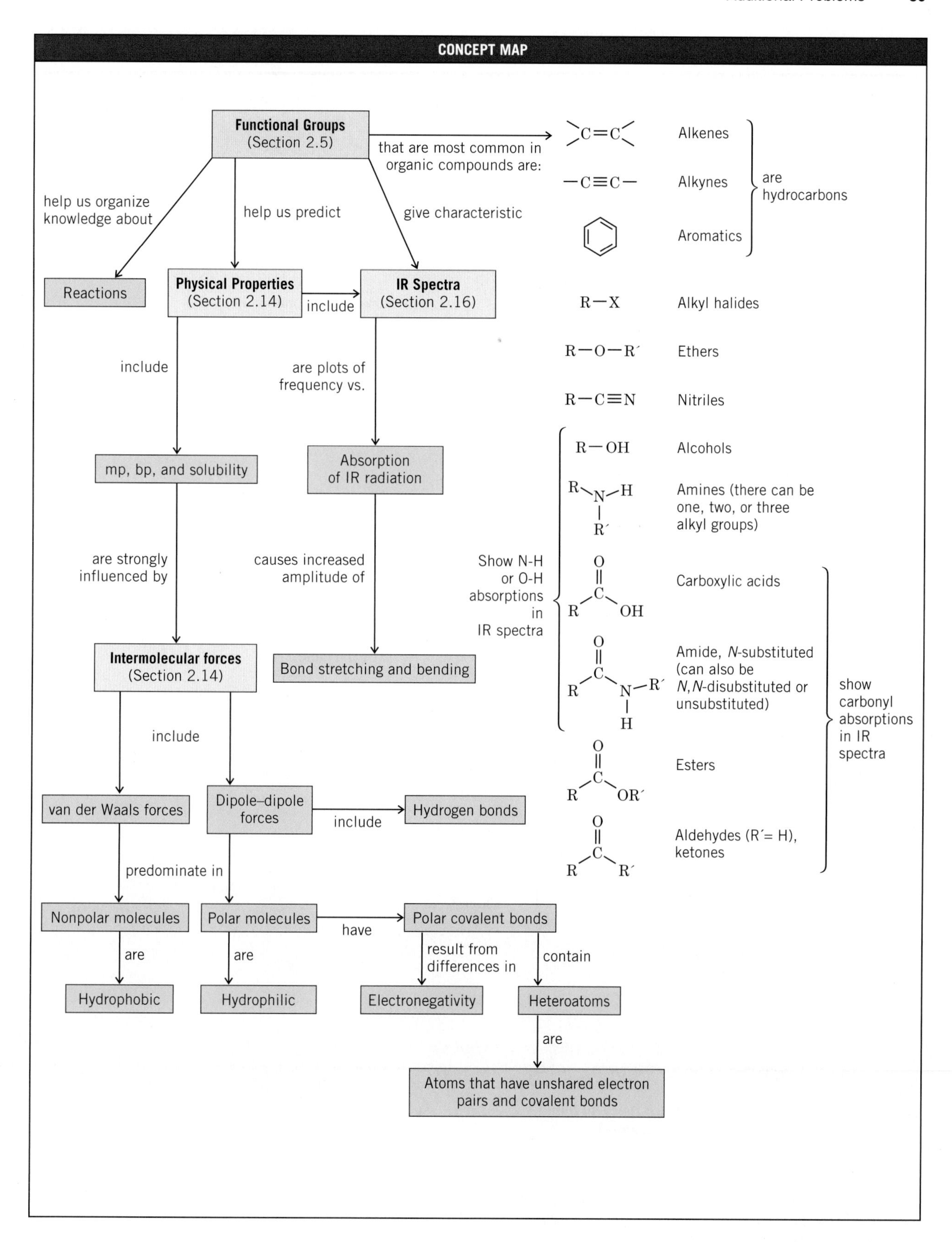

CONCEPT MAP
Functional Groups (Section 2.5)
that are most common in organic compounds are:
Alkenes
Alkynes
Aromatics
are hydrocarbons
R—X
Alkyl halides
R—O—R′
Ethers
R—C≡N
Nitriles
R—OH
Alcohols
Amines (there can be one, two, or three alkyl groups)
Carboxylic acids
Amide, N-substituted (can also be N,N-disubstituted or unsubstituted)
Esters
Aldehydes (R′= H), ketones
Show N-H or O-H absorptions in IR spectra
show carbonyl absorptions in IR spectra
help us organize knowledge about
help us predict
give characteristic
Reactions
Physical Properties (Section 2.14)
include
IR Spectra (Section 2.16)
include
are plots of frequency vs.
mp, bp, and solubility
Absorption of IR radiation
are strongly influenced by
causes increased amplitude of
Intermolecular forces (Section 2.14)
Bond stretching and bending
include
van der Waals forces
Dipole–dipole forces
include
Hydrogen bonds
predominate in
Nonpolar molecules
Polar molecules
have
Polar covalent bonds
are
are
result from differences in
contain
Hydrophobic
Hydrophilic
Electronegativity
Heteroatoms
are
Atoms that have unshared electron pairs and covalent bonds

Additional Problems

2.20 Classify each of the following compounds as an alkane, alkene, alkyne, alcohol, aldehyde, amine, and so forth.

(a) 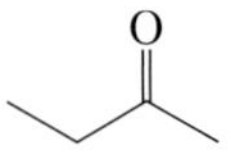

(b) $CH_3—C≡CH$

(c) 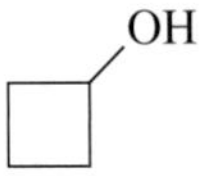

(d) 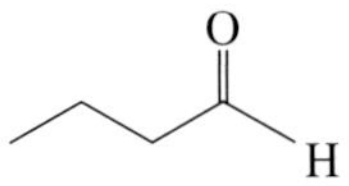

(e) 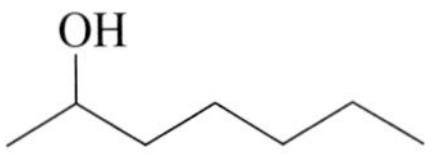

(obtained from oil of cloves)

(f) 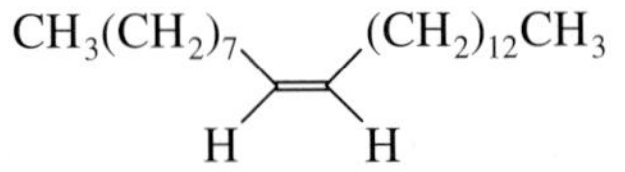

(sex attractant of common housefly)

2.21 Identify all of the functional groups in each of the following compounds:

(a) 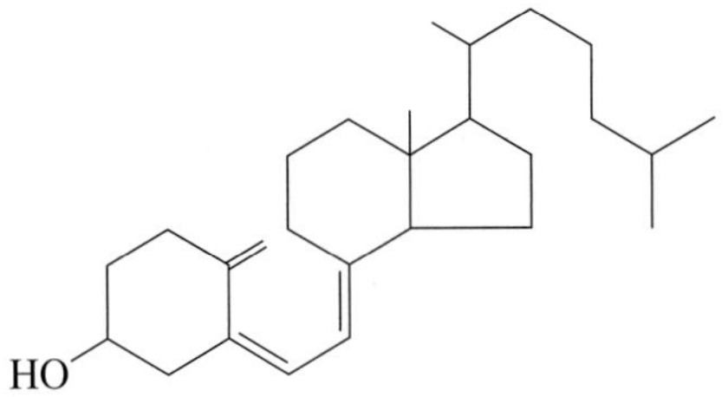

Vitamin D_3

(b) 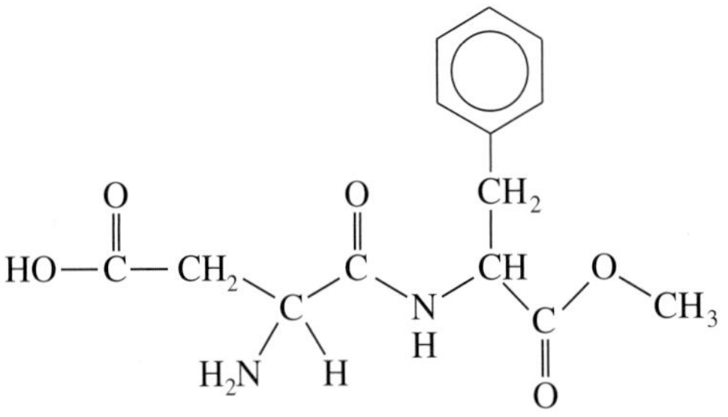

Aspartame

(c) 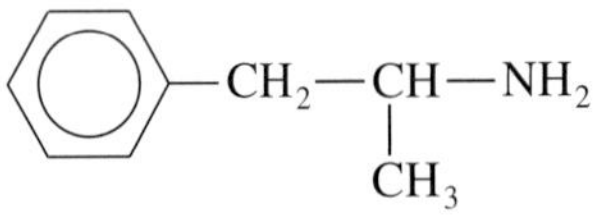

Amphetamine

(d) 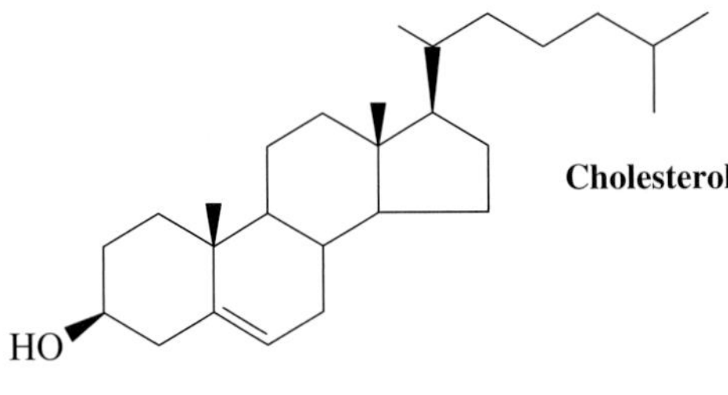

Cholesterol

(e) 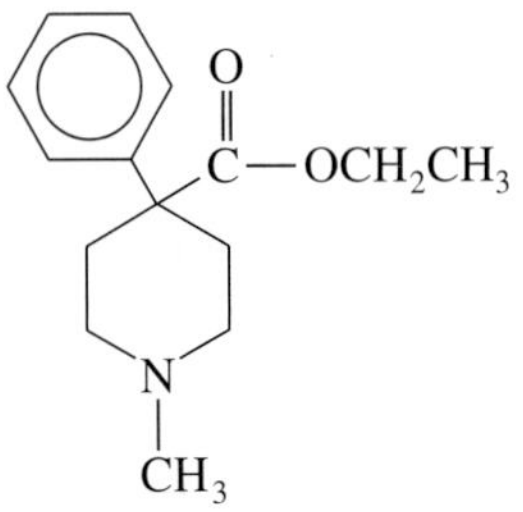

Demerol

(f) 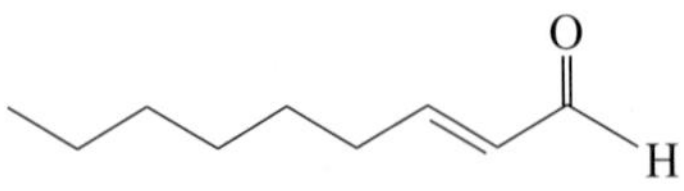

A cockroach repellent found in cucumbers

(g) 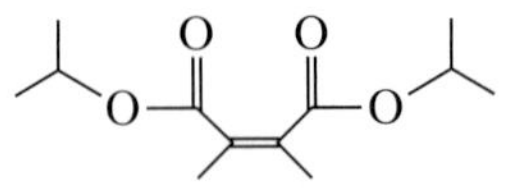

A synthetic cockroach repellent

Note: Problems marked with an asterisk are "challenge problems."

2.22 There are four alkyl bromides with the formula C_4H_9Br. Write their structural formulas and classify each as to whether it is a primary, secondary, or tertiary alkyl bromide.

2.23 There are seven isomeric compounds with the formula $C_4H_{10}O$. Write their structures and classify each compound according to its functional group.

2.24 Write structural formulas for four compounds with the formula C_3H_6O and classify each according to its functional group. Predict IR absorption frequencies for the functional groups you have drawn.

2.25 Classify the following alcohols as primary, secondary, or tertiary:

(a) $(CH_3)_3CCH_2OH$

(b) $CH_3CH(OH)CH(CH_3)_2$

(c) $(CH_3)_2C(OH)CH_2CH_3$

(d) OH

(e) OH

2.26 Classify the following amines as primary, secondary, or tertiary:

(a) $CH_3NHCH(CH_3)_2$
(b) $CH_3CH_2CH(CH_3)CH_2NH_2$
(c) $(CH_3CH_2)_3N$
(d) $(C_6H_5)_2CHCH_2NHCH_3$

(e) HN

(f) N

2.27 Write structural formulas for each of the following:
(a) Three ethers with the formula $C_4H_{10}O$.
(b) Three primary alcohols with the formula C_4H_8O.
(c) A secondary alcohol with the formula C_3H_6O.
(d) A tertiary alcohol with the formula C_4H_8O.
(e) Two esters with the formula $C_3H_6O_2$.
(f) Four primary alkyl halides with the formula $C_5H_{11}Br$.
(g) Three secondary alkyl halides with the formula $C_5H_{11}Br$.
(h) A tertiary alkyl halide with the formula $C_5H_{11}Br$.
(i) Three aldehydes with the formula $C_5H_{10}O$.
(j) Three ketones with the formula $C_5H_{10}O$.
(k) Two primary amines with the formula C_3H_9N.
(l) A secondary amine with the formula C_3H_9N.
(m) A tertiary amine with the formula C_3H_9N.
(n) Two amides with the formula C_2H_5NO.

2.28 Which compound in each of the following pairs would have the higher boiling point? Explain your answers.

(a) $CH_3CH_2CH_2OH$ or $CH_3CH_2OCH_3$
(b) $CH_3CH_2CH_2OH$ or $HOCH_2CH_2OH$

(c)

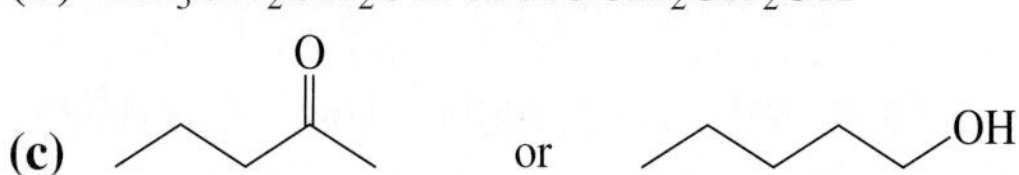

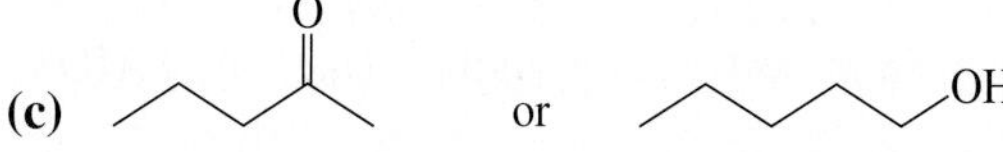

(d)

(e) NH or N—CH_3

(f) F, F or F, F

(g)

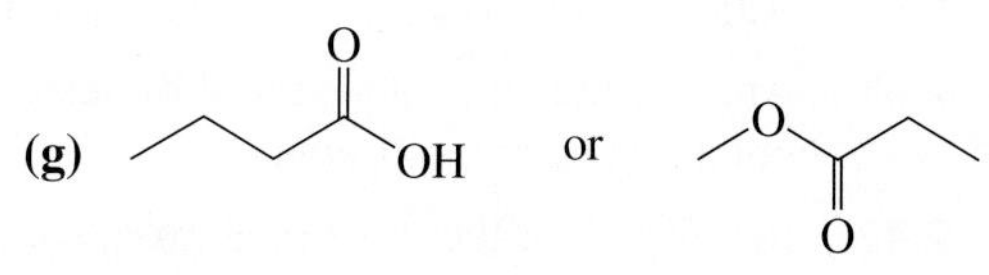

(h) Hexane $CH_3(CH_2)_4CH_3$ or nonane $CH_3(CH_2)_7CH_3$

(i) or =O

2.29 Predict the key IR absorption bands whose presence would allow each compound in pairs (a), (c), (d), (e), (g), and (i) from Problem 2.28 to be distinguished from each other.

2.30 There are four amides with the formula C_3H_7NO. **(a)** Write their structures. **(b)** One of these amides has a melting and a boiling point that are substantially lower than those of the other three. Which amide is this? Explain your answer. **(c)** Explain how these amides could be differentiated on the basis of their IR spectra.

2.31 Write structures for all compounds with molecular formula C_4H_6O that would not be expected to exhibit infrared absorption in the 3200–3550-cm^{-1} and 1620–1780-cm^{-1} regions.

2.32 Cyclic compounds of the general type shown here are called lactones. What functional group does a lactone contain?

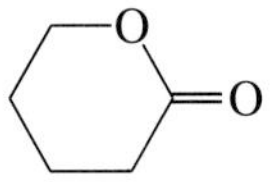

2.33 Hydrogen fluoride has a dipole moment of 1.82 D; its boiling point is 19.34°C. Ethyl fluoride (CH_3CH_2F) has an almost identical dipole moment and has a larger molecular weight, yet its boiling point is − 37.7°C. Explain.

2.34 Why does one expect the cis isomer of an alkene to have a higher boiling point than the trans isomer?

2.35 Cetylethyldimethylammonium bromide is the common name for $(CH_3)_2N^+(CH_2CH_3)(CH_2)_{15}CH_3\ Br^-$ (with CH_2CH_3 attached to N), a compound with antiseptic properties. Predict its solubility behavior in water and diethyl ether.

2.36 Which of the following solvents should be capable of dissolving ionic compounds?
(a) Liquid SO_2 **(b)** Liquid NH_3 **(c)** Benzene **(d)** CCl_4

2.37 Write a three-dimensional formula for each of the following molecules using the wedge–dashed wedge–line formalism. If the molecule has a net dipole moment, indicate its direction with an arrow, ⟼. If the molecule has no net dipole moment, you should so state. (You may ignore the small polarity of C—H bonds in working this and similar problems.)
(a) CH_3F **(c)** CHF_3 **(e)** CH_2FCl **(g)** BeF_2 **(i)** CH_3OH
(b) CH_2F_2 **(d)** CF_4 **(f)** BCl_3 **(h)** CH_3OCH_3 **(j)** CH_2O

2.38 The infrared spectrum of 1-hexyne exhibits a sharp absorption peak near 2100 cm^{-1} due to C≡C stretching. However, 3-hexyne shows no absorption in that region. Explain.

2.39 Consider each of the following molecules in turn: **(a)** dimethyl ether, $(CH_3)_2O$; **(b)** trimethylamine, $(CH_3)_3N$; **(c)** trimethylboron, $(CH_3)_3B$; and **(d)** dimethylberyllium, $(CH_3)_2Be$. Describe the hybridization state of the central atom (i.e., O, N, B, or Be) of each molecule, tell what bond angles you would expect at the central atom, and state whether the molecule would have a dipole moment.

2.40 Alkenes can interact with metal ions such as Ag^+. What is the nature of this interaction?

2.41 Analyze the statement: For a molecule to be polar, the presence of polar bonds is necessary, but it is not a sufficient requirement.

2.42 Identify all of the functional groups in Crixivan, an important drug in the treatment of AIDS.

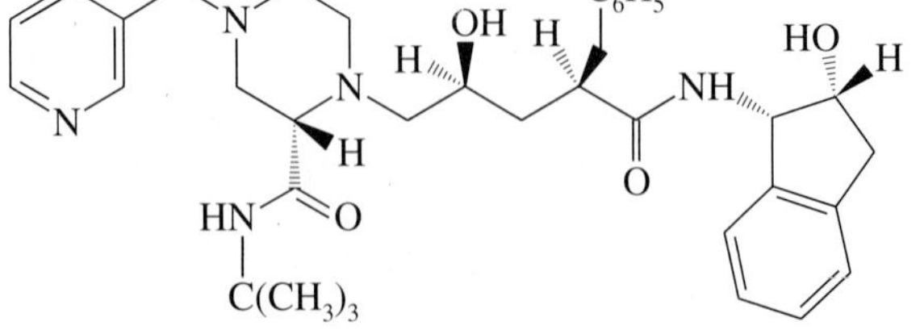

Crixivan (an HIV protease inhibitor)

2.43 The IR spectrum of propanoic acid (Fig. 2.17) indicates that the absorption for the O—H stretch of the carboxylic acid functional group is due to a hydrogen-bonded form. Draw the structure of two propanoic acid molecules showing how they could dimerize via hydrogen bonding.

2.44 In infrared spectra, the carbonyl group is usually indicated by a single strong and sharp absorption. However, in the case of carboxylic acid anhydrides, $R-\overset{\overset{O}{\|}}{C}-O-\overset{\overset{O}{\|}}{C}-R$, two peaks are observed even though the two carbonyl groups are chemically equivalent. Explain.

***2.45** Two isomers having molecular formula C_4H_6O are both symmetrical in structure. In their infrared spectra, neither isomer when in dilute solution in CCl_4 (used because it is nonpolar) has absorption in the 3600-cm^{-1} region. Isomer A has absorption bands at approximately 3080, 1620, and 700 cm^{-1}. Isomer B has bands in the 2900-cm^{-1} region and at 1780 cm^{-1}. Propose a structure for A and two possible structures for B.

***2.46** When two substituents are on the same side of a ring skeleton, they are said to be cis, and when on opposite sides, trans (analogous to use of those terms with 1,2-disubstituted alkene isomers). Consider stereoisomeric forms of 1,2-cyclopentanediol (compounds having a five-membered ring and hydroxyl groups on two adjacent carbons that are cis in one isomer and trans in the other). At high dilution in CCl_4, both isomers have an infrared absorption band at approximately 3626 cm^{-1} but only one isomer has a band at 3572 cm^{-1}. **(a)** Assume for now that the cyclopentane ring is coplanar (the interesting actuality will be studied later) and then draw and label the two isomers using the wedge–dashed wedge method of depicting the OH groups. **(b)** Designate which isomer will have the 3572-cm^{-1} band and explain its origin.

***2.47** Compound C is asymmetric, has molecular formula $C_5H_{10}O$, and contains two methyl groups and a 3° functional group. It has a broad infrared absorption band in the 3200–3550-cm^{-1} region and no absorption in the 1620–1680-cm^{-1} region. **(a)** Propose a structure for C. **(b)** Is your suggested structure capable of stereoisomerism? If so, draw the stereoisomers using the wedge–dashed wedge method.

2.48 Open the alpha-helix computer model from the Animated Graphics section of the CD. Between what specific atoms and of what functional groups are the hydrogen bonds formed that give the molecule its helical structure? Sketch a dash molecular formula for the model and show the hydrogen bonds.

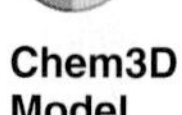
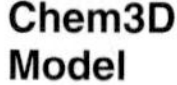

Chem3D Model

Learning Group Problem

Consider the molecular formula $C_4H_8O_2$.

1. Write structures for at least 15 different compounds that all have the molecular formula $C_4H_8O_2$.
2. Provide at least one example each of a structure written using the dash format, the condensed format, the bond-line format, and the full three-dimensional format. Use your choice of format for the remaining structures.
3. Identify four different functional groups from among your structures. Circle and name them on the representative structures.
4. Predict approximate frequencies for IR absorptions that could be used to distinguish the four compounds representing these functional groups.
5. If any of the 15 structures you drew have atoms where the formal charge is other than zero, indicate the formal charge on the appropriate atom(s) and the overall charge for the molecule.
6. Identify which types of intermolecular forces would be possible in pure samples of each of these compounds.
7. Pick five structures you have drawn and predict their order with respect to trend in increasing boiling point.
8. Explain the order of boiling points on the basis of intermolecular forces and polarity.

CHAPTER 3

An Introduction to Organic Reactions: Acids and Bases

Shuttling the Protons

An enzyme (a biological catalyst) called carbonic anhydrase regulates the acidity (or pH) of blood and the physiological conditions relating to blood pH. The reaction that carbonic anhydrase catalyzes is the following:

$$HCO_3^- + H^+ \rightleftharpoons H_2CO_3 \xrightleftharpoons{\text{carbonic anhydrase}} H_2O + CO_2$$

The rate at which one breathes, for example, is influenced by one's relative blood acidity. Mountain climbers going to high elevations sometimes take a drug called Diamox (acetazolamide, whose structure is by the photo above) to prevent altitude sickness. Diamox inhibits carbonic anhydrase, and this, in turn, increases blood acidity. This increased blood acidity stimulates breathing and thereby decreases the likelihood of altitude sickness.

Carbonic anhydrase

Animated Graphics

The overall reaction shown above is clearly an acid–base reaction because acids and bases appear in it. The details of *how* carbonic anhydrase catalyzes this reaction (the reaction mechanism) also require acid–base steps that occur inside the enzyme itself. These details are provided in "The Chemistry of . . . Carbonic Anhydrase" box later in this chapter. For now, suffice it to say that these steps in the mechanism involve two operating definitions of acid–base chemistry. One is the Brønsted–Lowry definition of acids and bases, the other the Lewis definition. In this chapter, we shall consider both definitions, and you will see that both are essential to an understanding of organic and biological chemistry.

3.1 Reactions and Their Mechanisms

Virtually all organic reactions fall into one of four categories: *substitutions, additions, eliminations,* or *rearrangements.*

Substitutions are the characteristic reactions of saturated compounds such as alkanes and alkyl halides and of aromatic compounds (even though they are unsaturated). In a substitution, *one group replaces another.* For example, methyl chloride reacts with sodium hydroxide to produce methyl alcohol and sodium chloride:

$$H_3C{-}Cl + Na^+OH^- \xrightarrow{H_2O} H_3C{-}OH + Na^+Cl^-$$

A substitution reaction

In this reaction a hydroxide ion from sodium hydroxide replaces the chlorine of methyl chloride. We shall study this reaction in detail in Chapter 6.

Additions are characteristic of compounds with multiple bonds. Ethene, for example, reacts with bromine by an addition. In an addition *all parts of the adding reagent appear in the product; two molecules become one:*

$$H_2C{=}CH_2 + Br{-}Br \xrightarrow{CCl_4} H{-}CHBr{-}CHBr{-}H$$

An addition reaction

Eliminations are the opposite of additions. *In an elimination one molecule loses the elements of another small molecule.* Elimination reactions give us a method for preparing compounds with double and triple bonds. In Chapter 7, for example, we shall study an important elimination, called *dehydrohalogenation,* a reaction that is used to prepare alkenes. In dehydrohalogenation, as the word suggests, the elements of a hydrogen halide are eliminated. An alkyl halide becomes an alkene:

$$H{-}CH_2{-}CHBr{-}H \xrightarrow[(-HBr)]{KOH} H_2C{=}CH_2$$

An elimination reaction

In a **rearrangement** *a molecule undergoes a reorganization of its constituent parts.* For example, heating the following alkene with a strong acid causes the formation of another isomeric alkene:

$$(CH_3)_3C{-}CH{=}CH_2 \xrightarrow{\text{acid cat.}} (H_3C)_2C{=}C(CH_3)_2$$

A rearrangement

In this rearrangement not only have the positions of the double bond and a hydrogen atom changed, but a methyl group has moved from one carbon to another.

In the following sections we shall begin to learn some of the principles that explain how these kinds of reactions take place.

A Mechanism for the Reaction

To the uninitiated, a chemical reaction must seem like an act of magic. A chemist puts one or two reagents into a flask, warms them for a period of time, and then takes from the flask one or more completely different compounds. It is, until we understand the details of the reaction, like a magician who puts apples and oranges in a hat, shakes it, and then pulls out rabbits and parakeets.

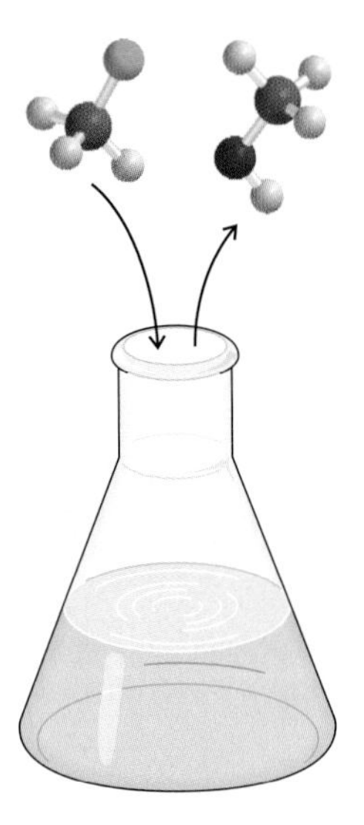

One of our goals in this course will be, in fact, to try to understand how this chemical magic takes place. We will want to be able to explain *how the products of the reaction are formed.* This explanation will take the form of **a mechanism for the reaction—a description of the events that take place on a molecular level as reactants become products.** If, as is often the case, the reaction takes place in more than one step, we will want to know what chemical species, called **intermediates**, intervene between each step along the way.

By postulating a mechanism, we may take some of the magic out of the reaction, but we will put rationality in its place. Any mechanism we propose must be consistent with what we know about the reaction and with what we know about the reactivity of organic compounds generally. In later chapters we shall see how we can glean evidence for or against a given mechanism from studies of reaction rates, from isolating intermediates, and from spectroscopy. We cannot actually see the molecular events because molecules are too small, but from solid evidence and from good chemical intuition, we can propose reasonable mechanisms. If at some later time an experiment gives results that contradict our proposed mechanism, then we change it, because in the final analysis our mechanism must be consistent with all our experimental observations.

One of the most important things about approaching organic chemistry mechanistically is this: It helps us organize what otherwise might be an overwhelmingly complex body of knowledge into a form that makes it understandable. There are millions of organic compounds now known, and there are millions of reactions that these compounds undergo. If we had to learn them all by rote memorization, then we would soon give up. But, we don't have to do this. In the same way that functional groups help us organize compounds in a comprehensible way, mechanisms help us organize reactions. Fortunately, too, there is a relatively small number of basic mechanisms.

3.1A Homolysis and Heterolysis of Covalent Bonds

Reactions of organic compounds always involve the making and breaking of covalent bonds. A covalent bond may break in two fundamentally different ways. The bond may break so that one fragment takes away both electrons of the bond, leaving the other fragment with an empty orbital. This kind of cleavage, called **heterolysis** (Gr: *hetero-*, different, + *lysis,* loosening or cleavage), produces charged fragments or **ions.** The bond is said to have broken *heterolytically:*

Notice in these illustrations that we have used curved arrows to show the movement of electrons. We will have more to say about this convention in Section 3.4, but for the moment notice that we use a double-barbed curved arrow to show the movement of a pair of electrons and a single-barbed curved arrow to show the movement of a single electron.

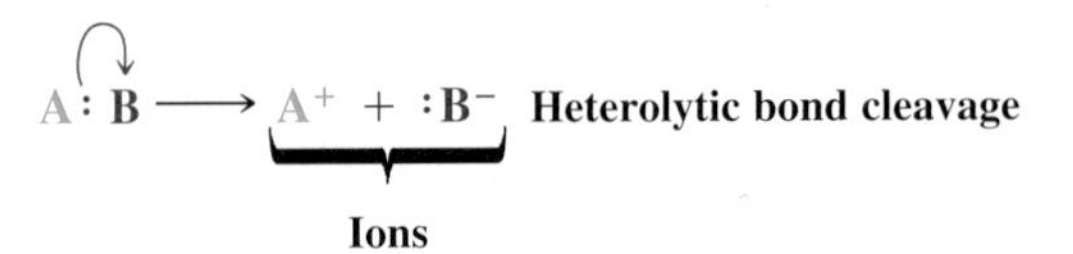

The other possibility is that the bond breaks so that each fragment takes away one of the electrons of the bond. This process, called **homolysis** (Gr. *homo-*, the same, + *lysis*), produces fragments with unpaired electrons called **radicals.**

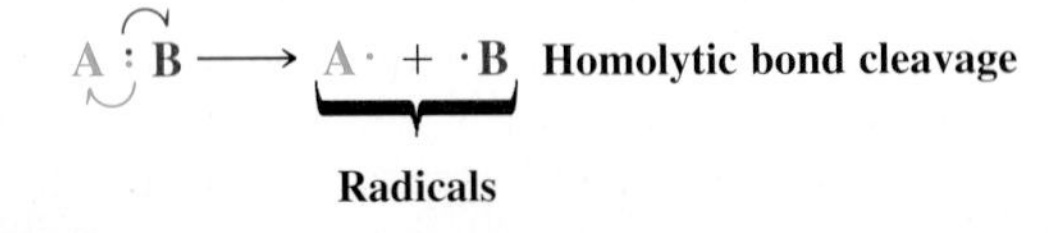

We shall postpone further discussions of reactions involving radicals and homolytic bond cleavage until we reach Chapter 10. At this point we shall focus our attention on reactions involving ions and heterolytic bond cleavage.

Heterolysis of a bond normally requires that the bond be polarized:

$$^{\delta+}A:B^{\delta-} \longrightarrow A^{+} + :B^{-}$$

Polarization of a bond usually results from differing electronegativities (Section 2.3) of the atoms joined by the bond. The greater the difference in electronegativity, the greater the polarization. In the instance just given, atom B is more electronegative than A.

Even with a highly polarized bond, heterolysis rarely occurs without assistance. The reason: *Heterolysis requires separation of oppositely charged ions.* Because oppositely charged ions attract each other, their separation requires considerable energy. Often, heterolysis is assisted by a molecule with an unshared pair that can form a bond to one of the atoms:

$$Y: + {}^{\delta+}A{-}B^{\delta-} \longrightarrow \overset{+}{Y}{-}A + :B^{-}$$

Formation of the new bond furnishes some of the energy required for the heterolysis.

3.2 Acid–Base Reactions

We begin our study of chemical reactions by examining some of the basic principles of acid–base chemistry. There are several reasons for doing this: Many of the reactions that occur in organic chemistry are either acid–base reactions themselves or they involve an acid–base reaction at some stage. An acid–base reaction is also a simple fundamental reaction that will enable you to see how chemists use curved arrows to represent mechanisms of reactions and how they depict the processes of bond breaking and bond making that occur as molecules react. Acid–base reactions also allow us to examine important ideas about the relationship between the structures of molecules and their reactivity and to see how certain thermodynamic parameters can be used to predict how much of the product will be formed when a reaction reaches equilibrium. Acid–base reactions also provide an illustration of the important role solvents play in chemical reactions. They even give us a brief introduction to organic synthesis. Finally, acid–base chemistry is something that you will find familiar because of your studies in general chemistry. We begin, therefore, with a brief review.

3.2A The Brønsted–Lowry Definition of Acids and Bases

According to the Brønsted–Lowry theory, an acid is a substance that can donate (or lose) a proton, and a base is a substance that can accept (or remove) a proton. Let us consider, as an example of this concept, the reaction that occurs when gaseous hydrogen chloride dissolves in water:

$$H_2\ddot{O}: + H{-}\ddot{C}l: \longrightarrow H{-}\overset{+}{\ddot{O}}{-}H + :\ddot{C}l:^{-}$$

Base (proton acceptor)	Acid (proton donor)	Conjugate acid of H_2O	Conjugate base of HCl

Hydrogen chloride, a very strong acid, transfers its proton to water. Water acts as a base and accepts the proton. The products that result from this reaction are a hydronium ion (H_3O^+) and a chloride ion (Cl^-).

We shall see that an understanding of acid and base conjugates is very helpful for evaluating relative acid–base strength.

The molecule or ion that forms when an acid loses its proton is called the **conjugate base** of that acid. The chloride ion, therefore, is the conjugate base of HCl. The molecule

The extent to which an acid transfers protons to a base, such as water, is a measure of its strength as an acid. Acid strength is therefore a measure of the percentage of ionization and *not* of concentration.

or ion that forms when a base accepts a proton is called the **conjugate acid** of that base. The hydronium ion, therefore, is the conjugate acid of water.

Other strong acids that completely transfer a proton when dissolved in water are hydrogen iodide, hydrogen bromide, and sulfuric acid:

$$HI + H_2O \longrightarrow H_3O^+ + I^-$$
$$HBr + H_2O \longrightarrow H_3O^+ + Br^-$$
$$H_2SO_4 + H_2O \longrightarrow H_3O^+ + HSO_4^-$$
$$HSO_4^- + H_2O \rightleftharpoons H_3O^+ + SO_4^{2-}$$

Because sulfuric acid has two protons that it can transfer to a base, it is called a diprotic (or dibasic) acid. The proton transfer is stepwise; the first proton transfer occurs completely, the second only to the extent of ~ 10%.

Hydronium ions and hydroxide ions are the strongest acids and bases that can exist in aqueous solution in significant amounts. When sodium hydroxide (a crystalline compound consisting of sodium ions and hydroxide ions) dissolves in water, the result is a solution containing solvated sodium ions and solvated hydroxide ions:

$$Na^+OH^-_{(solid)} \longrightarrow Na^+_{(aq)} + OH^-_{(aq)}$$

Sodium ions (and other similar cations) become solvated when water molecules donate unshared electron pairs to their vacant orbitals. Hydroxide ions (and other anions with unshared electron pairs) become solvated when water molecules form hydrogen bonds to them:

Solvated sodium ion

Solvated hydroxide ion

When an aqueous solution of sodium hydroxide is mixed with an aqueous solution of hydrogen chloride (hydrochloric acid), the reaction that occurs is between hydronium and hydroxide ions. The sodium and chloride ions are called **spectator ions** because they play no part in the acid–base reaction:

Total Ionic Reaction

$$H_3O^+ + Cl^- + Na^+ + {}^-OH \longrightarrow 2\,H_2O + Na^+ + Cl^-$$

Spectator ions

Net Reaction

$$H_3O^+ + {}^-OH \longrightarrow 2\,H_2O$$

What we have just said about hydrochloric acid and aqueous sodium hydroxide is true when solutions of all aqueous strong acids and bases are mixed. The net ionic reaction is simply

$$H_3O^+ + OH^- \longrightarrow 2\,H_2O$$

3.2B The Lewis Definition of Acids and Bases

Acid–base theory was broadened considerably by G. N. Lewis in 1923. Striking at what he called "the cult of the proton," Lewis proposed that **acids be defined as electron pair acceptors** and **bases be defined as electron pair donors.** In the **Lewis acid–base theory,** proton donors are not the only acids; many other species are acids as well. Aluminum chloride, for example, reacts with ammonia in the same way that a proton donor does. Using curved arrows to show the donation of the electron pair of ammonia (the Lewis base), we have the following examples:

Most of the reactions we shall study involve Lewis acid–base interactions. A sound understanding of Lewis acid–base chemistry will help greatly as you learn organic chemistry.

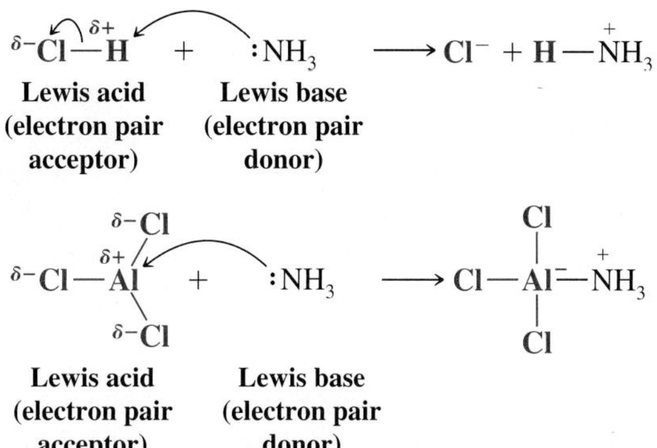

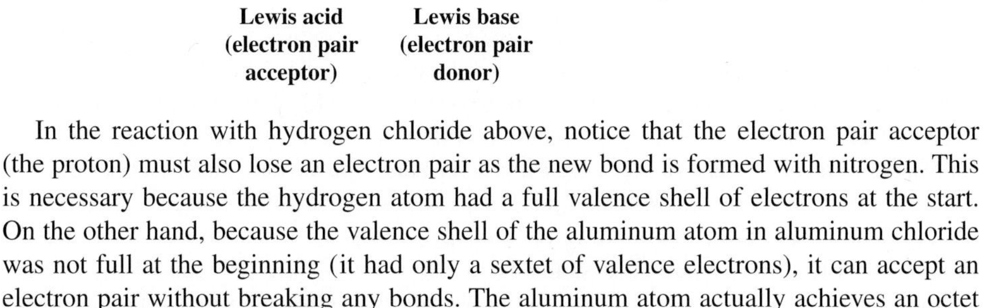

Verify for yourself that you can calculate the formal charges in these structures.

In the reaction with hydrogen chloride above, notice that the electron pair acceptor (the proton) must also lose an electron pair as the new bond is formed with nitrogen. This is necessary because the hydrogen atom had a full valence shell of electrons at the start. On the other hand, because the valence shell of the aluminum atom in aluminum chloride was not full at the beginning (it had only a sextet of valence electrons), it can accept an electron pair without breaking any bonds. The aluminum atom actually achieves an octet by accepting the pair from nitrogen, although it gains a formal negative charge. When it accepts the electron pair, aluminum chloride is, in the Lewis definition, *acting as an acid.*

Bases are much the same in the Lewis theory and the Brønsted–Lowry theory, because in the Brønsted–Lowry theory a base must donate a pair of electrons in order to accept a proton.

The Lewis theory, by virtue of its broader definition of acids, allows acid–base theory to include all of the Brønsted–Lowry reactions and, as we shall see, a great many others.

Any *electron-deficient atom* can act as a Lewis acid. Many compounds containing Group 3A elements such as boron and aluminum are Lewis acids because Group 3A atoms have only a sextet of electrons in their outer shell. Many other compounds that have atoms with vacant orbitals also act as Lewis acids. Zinc and iron(III) halides (ferric halides) are frequently used as Lewis acids in organic reactions. Two examples that we shall study later are the following:

Carbonic anhydrase

A zinc ion acts as a Lewis acid in the mechanism of the enzyme carbonic anhydrase. See "The Chemistry of . . . Carbonic Anhydrase" later in this chapter.

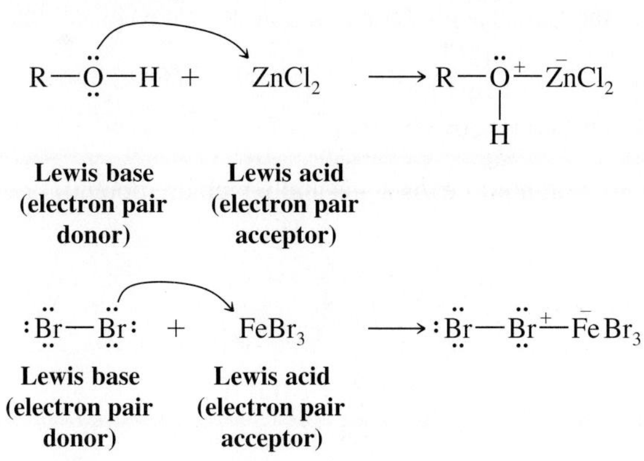

3.2C Opposite Charges Attract

In Lewis acid–base theory, as in many organic reactions, the attraction of oppositely charged species is fundamental to reactivity. As one further example, we consider boron trifluoride, an even more powerful Lewis acid than aluminum chloride, and its reaction with ammonia. The calculated structure for boron trifluoride in Fig. 3.1 shows **electrostatic potential** at its van der Waals surface (like that in Section 2.3 for HCl). It is obvious from this figure (and you should be able to predict this) that BF_3 has substantial posi-

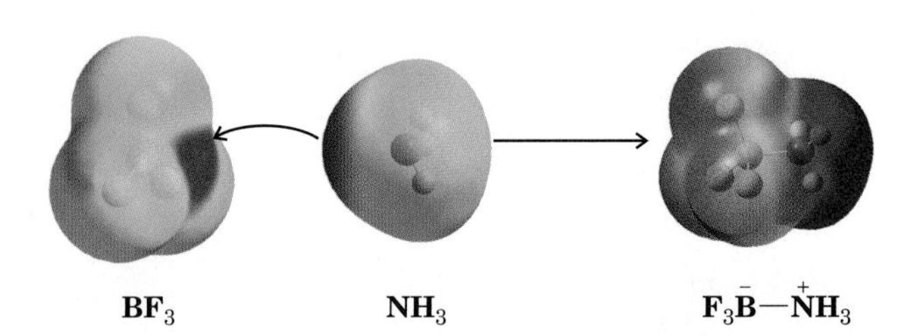

FIGURE 3.1 Electrostatic potential maps for BF_3 and NH_3 and the product that results from reaction between them. Attraction between the strongly positive region of BF_3 and the negative region of NH_3 causes them to react. The electrostatic potential map for the product shows that the fluorine atoms draw in the electron density of the formal negative charge, and the nitrogen atom, with its hydrogens, carries the formal positive charge.

Animated Graphics

tive charge centered on the boron atom and negative charge located on the three fluorines. (The convention in these structures is that blue represents relatively positive areas and red represents relatively negative areas.) On the other hand, the surface electrostatic potential for ammonia shows (as you would expect) that substantial negative charge is localized in the region of ammonia's nonbonding electron pair. Thus, the electrostatic properties of these two molecules are perfectly suited for a Lewis acid–base reaction. When the expected reaction occurs between them, the nonbonding electron pair of ammonia attacks the boron atom of boron trifluoride, filling boron's valence shell. The boron now carries a formal negative charge and the nitrogen carries a formal positive charge. This separation of charge is borne out in the electrostatic potential map for the product shown in Fig. 3.1. Notice that substantial negative charge resides in the BF_3 part of the molecule, and substantial positive charge is localized near the nitrogen.

The need for a firm understanding of structure, formal charges, and electronegativity can hardly be emphasized enough as you build a foundation of knowledge for learning organic chemistry.

Although calculated electrostatic potential maps like these illustrate charge distribution and molecular shape well, it is important that you are able to draw the same conclusions based on what you would have predicted about the structures of BF_3 and NH_3 and their reaction product using orbital hybridization (Sections 1.12–1.14), VSEPR models (Section 1.16), consideration of formal charges (Section 1.7), and electronegativity (Section 2.3).

PROBLEM 3.1

Write equations showing the Lewis acid–base reaction that takes place when:

(a) Methanol (CH_3OH) reacts with BF_3.
(b) Chloromethane (CH_3Cl) reacts with $AlCl_3$.
(c) Dimethyl ether (CH_3OCH_3) reacts with BF_3.

PROBLEM 3.2

Which of the following are potential Lewis acids and which are potential Lewis bases?

(a) $CH_3CH_2—\ddot{N}(CH_3)—CH_3$ **(b)** $H_3C—C^+(CH_3)_2$

(c) $(C_6H_5)_3P:$ **(d)** $:\ddot{Br}:^-$ **(e)** $(CH_3)_3B$ **(f)** $H:^-$

The Chemistry of...

HOMOs and LUMOs in Reactions

The calculated lowest unoccupied molecular orbital (LUMO) for BF_3 is shown by solid red and blue lobes. Most of the volume represented by the LUMO corresponds to the empty *p* orbital in the sp^2-hybridized state of BF_3 (located perpendicular to the plane of the atoms). This orbital is where electron density fills (bonding occurs) when BF_3 is attacked by NH_3. The van der Waals surface electron density of BF_3 is indicated by the mesh. As the structure shows, the LUMO extends beyond the electron density surface, and hence it is easily accessible for reaction.

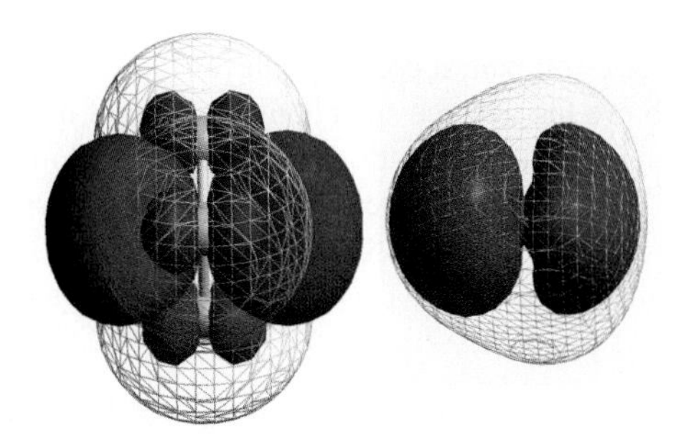

The LUMO of BF_3 (left) and the HOMO of NH_3 (right).

Chem3D Model

The highest occupied molecular orbital (HOMO) of ammonia, where the nonbonding pair resides, is shown by red and blue lobes in its structure. When the reaction occurs, the electron density from the HOMO of ammonia is transferred to the LUMO of boron trifluoride. This interaction involving the HOMO of one molecule with the LUMO of another is, from a molecular orbital perspective, the way reactions occur.

3.3 Heterolysis of Bonds to Carbon: Carbocations and Carbanions

Heterolysis of a bond to a carbon atom can lead to either of two ions: either to an ion with a positive charge on the carbon atom, called a **carbocation,*** or to an ion with a negatively charged carbon atom, called a **carbanion:**

$$\overset{\delta+}{-\mathrm{C}}-\mathrm{Z}^{\delta-} \xrightarrow{\text{heterolysis}} -\mathrm{C}^{+} + :\mathrm{Z}^{-}$$

Carbocation

$$\overset{\delta-}{-\mathrm{C}}-\mathrm{Z}^{\delta+} \xrightarrow{\text{heterolysis}} -\mathrm{C}:^{-} + \mathrm{Z}^{+}$$

Carbanion

Carbocations are electron deficient. They have only six electrons in their valence shell, and because of this, carbocations are Lewis acids. In this way they are like BF_3 and $AlCl_3$. Most carbocations are also short-lived and highly reactive. They occur as intermediates in some organic reactions. Carbocations react rapidly with Lewis bases—with molecules or ions that can donate the electron pair that they need to achieve a stable octet of electrons (i.e., the electronic configuration of a noble gas):

$$-\mathrm{C}^{+} + :\mathrm{B}^{-} \longrightarrow -\mathrm{C}-\mathrm{B}$$

Carbocation (a Lewis acid) **Anion (a Lewis base)**

*Some chemists refer to carbocations as **carbonium ions.** This older term, however, no longer has wide usage; therefore, we shall use the term *carbocation.* Its meaning is clear and distinct.

$$-\overset{|}{\underset{|}{C}}{}^{+} + :\ddot{O}-H \longrightarrow -\overset{|}{\underset{|}{C}}-\ddot{O}^{+}-H$$

Carbocation (a Lewis acid) **Water (a Lewis base)**

Because carbocations are electron-seeking reagents, chemists call them electrophiles. ***Electrophiles*** *are reagents which in their reactions seek the extra electrons that will give them a stable valence shell of electrons.* All Lewis acids, including protons, are electrophiles. By accepting an electron pair, a proton achieves the valence shell configuration of helium; carbocations achieve the valence shell configuration of neon.

Carbanions are Lewis bases. In their reactions they seek a proton or some other positive center to which they can donate their electron pair and thereby neutralize their negative charge. *Reagents, like carbanions, that seek a proton or some other positive center are called* ***nucleophiles*** (because the nucleus is the positive part of an atom):

$$-\overset{|}{\underset{|}{C}}:^{-} + \overset{\delta+}{H}-A^{\delta-} \longrightarrow -\overset{|}{\underset{|}{C}}-H + :A^{-}$$

Carbanion **Lewis acid**

$$-\overset{|}{\underset{|}{C}}:^{-} + -\overset{|}{\underset{|}{C}}{}^{\delta+}-L^{\delta-} \longrightarrow -\overset{|}{\underset{|}{C}}-\overset{|}{\underset{|}{C}}- + :L^{-}$$

Carbanion **Lewis acid**

3.4 The Use of Curved Arrows in Illustrating Reactions

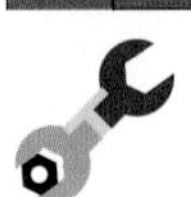

Curved arrows are one of the most important tools you will use to learn organic chemistry.

In the preceding sections we have shown the movement of an electron pair with a **curved arrow.** This type of notation is commonly used by organic chemists to show *the direction of electron flow in a reaction. The curved arrow does not show the movement of atoms,* however. The atoms are assumed to follow the flow of electrons. Consider as an example the reaction of hydrogen chloride with water.

A Mechanism for the Reaction

Reaction:

$$H_2O + HCl \longrightarrow H_3O^+ + Cl^-$$

Mechanism:

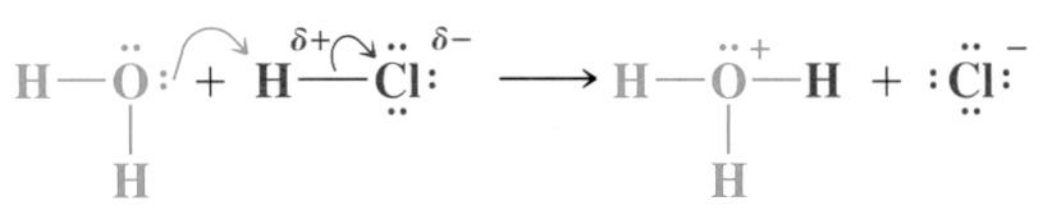

A water molecule uses one of the nonbonding electron pairs to form a bond to a proton of HCl. The bond between the hydrogen and chlorine breaks with the electron pair going to the chlorine atom.

This leads to the formation of a hydronium ion and a chloride ion.

Curved arrows point from electrons to the atom receiving the electrons.

Sir Robert Robinson

Sir Robert Robinson (1885–1975) is credited with publishing the first paper to use the "curly arrow" to show electron movements (*J. Chem Soc.* **1922,** *121,* 427–440). Robinson received the 1947 Nobel Prize in Chemistry for his work on the synthesis and biogenesis of natural products. Among his other accomplishments is elucidation of the structure of strychnine (Special Topic E).

The curved arrow begins with a covalent bond or unshared electron pair (a site of higher electron density) and points toward a site of electron deficiency. We see here that as the water molecule collides with a hydrogen chloride molecule, it uses one of its unshared electron pairs (shown in blue) to form a bond to the proton of HCl. This bond forms because the negatively charged electrons of the oxygen atom are attracted to the positively charged proton. As the bond between the oxygen and the proton forms, the hydrogen–chlorine bond of HCl breaks, and the chlorine of HCl departs with the electron pair that formerly bonded it to the proton. (If this did not happen, the proton would end up forming two covalent bonds, which, of course, a proton cannot do.) We, therefore, use a curved arrow to show the bond cleavage as well. By pointing from the bond to the chlorine, the arrow indicates that the bond breaks and the electron pair leaves with the chloride ion.

The following acid–base reactions give other examples of the use of the curved-arrow notation:

$$\underset{\text{Acid}}{H_2\ddot{O}^+{-}H} + \underset{\text{Base}}{{}^-:\ddot{O}{-}H} \longrightarrow H_2\ddot{O}: + H{-}\ddot{O}{-}H$$

$$\underset{\text{Acid}}{CH_3C(=O)\ddot{O}{-}H} + \underset{\text{Base}}{:\ddot{O}H_2} \rightleftharpoons CH_3C(=O)\ddot{O}:^- + H{-}\ddot{O}^+{-}H_2$$

$$\underset{\text{Acid}}{CH_3C(=O)\ddot{O}{-}H} + \underset{\text{Base}}{{}^-:\ddot{O}{-}H} \longrightarrow CH_3C(=O)\ddot{O}:^- + H{-}\ddot{O}{-}H$$

PROBLEM 3.3

Use the curved-arrow notation to write the reaction that would take place between dimethylamine and boron trifluoride. Identify the Lewis acid and Lewis base and assign appropriate formal charges.

3.5 The Strength of Acids and Bases: K_a and pK_a

In contrast to the strong acids, such as HCl and H_2SO_4, acetic acid is a much weaker acid. When acetic acid dissolves in water, the following reaction does not proceed to completion:

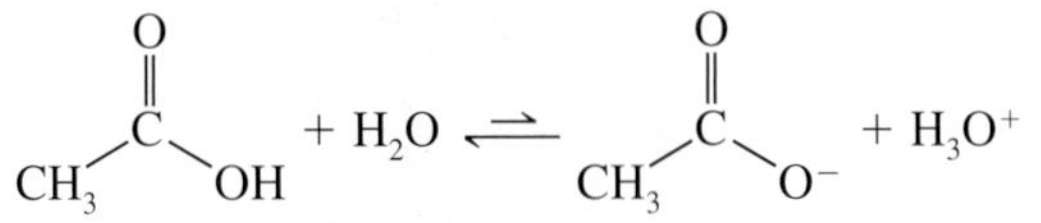

$$CH_3COOH + H_2O \rightleftharpoons CH_3COO^- + H_3O^+$$

Experiments show that in a 0.1*M* solution of acetic acid at 25°C only about 1% of the acetic acid molecules ionize by transferring their protons to water.

3.5A The Acidity Constant, K_a

Because the reaction that occurs in an aqueous solution of acetic acid is an equilibrium, we can describe it with an expression for the equilibrium constant:

$$K_{eq} = \frac{[H_3O^+][CH_3CO_2^-]}{[CH_3CO_2H][H_2O]}$$

For dilute aqueous solutions, the concentration of water is essentially constant (~55.5*M*), so we can rewrite the expression for the equilibrium constant in terms of a new constant (K_a) called the **acidity constant:**

$$K_a = K_{eq}[H_2O] = \frac{[H_3O^+][CH_3CO_2^-]}{[CH_3CO_2H]}$$

At 25°C, the acidity constant for acetic acid is 1.76×10^{-5}.

We can write similar expressions for any weak acid dissolved in water. Using a generalized hypothetical acid (HA), the reaction in water is

$$HA + H_2O \rightleftharpoons H_3O^+ + A^-$$

and the expression for the acidity constant is

$$K_a = \frac{[H_3O^+][A^-]}{[HA]}$$

Because the concentrations of the products of the reaction are written in the numerator and the concentration of the undissociated acid in the denominator, **a large value of K_a means the acid is a strong acid and a small value of K_a means the acid is a weak acid.** If the K_a is greater than 10, the acid will be, for all practical purposes, completely dissociated in water.

Problem 3.4 Formic acid (HCO_2H) has a $K_a = 1.77 \times 10^{-4}$. **(a)** What are the molar concentrations of the hydronium ion and formate ion (HCO_2^-) in a 0.1*M* aqueous solution of formic acid? **(b)** What percentage of the formic acid is ionized?

3.5B Acidity and pK_a

Chemists usually express the acidity constant, K_a, as its negative logarithm, $\mathbf{pK_a}$**:**

$$pK_a = -\log K_a$$

This is analogous to expressing the hydronium ion concentration as pH:

$$pH = -\log[H_3O^+]$$

For acetic acid the pK_a is 4.75:

$$pK_a = -\log(1.76 \times 10^{-5}) = -(-4.75) = 4.75$$

Notice that there is an inverse relationship between the magnitude of the pK_a and the strength of the acid. **The larger the value of the pK_a, the weaker is the acid.** For example, acetic acid with a $pK_a = 4.75$ is a weaker acid than trifluoroacetic acid with a $pK_a = 0$ ($K_a = 1$). Hydrochloric acid with a $pK_a = -7$ ($K_a = 10^7$) is a far stronger acid than trifluoroacetic acid. (It is understood that a positive pK_a is larger than a negative pK_a.)

K_a is an indicator of acid strength.

$$CH_3CO_2H \quad < \quad CF_3CO_2H < HCl$$

$$pK_a = 4.75 \qquad pK_a = 0 \qquad pK_a = -7$$

Weak acid **Very strong acid**

Increasing acid strength →

pK_a is an indicator of acid strength.

Table 3.1 lists pK_a values for a selection of acids relative to water as the base. The values in the middle pK_a range of the table are the most accurate because they can be meas-

TABLE 3.1 Relative Strength of Selected Acids and Their Conjugate Bases

	Acid	Approximate pK_a	Conjugate Base	
Strongest acid	$HSbF_6$	<-12	SbF_6^-	Weakest base
↑ Increasing acid strength	HI	-10	I^-	Increasing base strength ↓
	H_2SO_4	-9	HSO_4^-	
	HBr	-9	Br^-	
	HCl	-7	Cl^-	
	$C_6H_5SO_3H$	-6.5	$C_6H_5SO_3^-$	
	$(CH_3)_2\overset{+}{O}H$	-3.8	$(CH_3)_2O$	
	$(CH_3)_2C{=}\overset{+}{O}H$	-2.9	$(CH_3)_2C{=}O$	
	$CH_3\overset{+}{O}H_2$	-2.5	CH_3OH	
	H_3O^+	-1.74	H_2O	
	HNO_3	-1.4	NO_3^-	
	CF_3CO_2H	0.18	$CF_3CO_2^-$	
	HF	3.2	F^-	
	CH_3CO_2H	4.75	$CH_3CO_2^-$	
	H_2CO_3	6.35	HCO_3^-	
	$CH_3COCH_2COCH_3$	9.0	$CH_3CO\overset{-}{C}HCOCH_3$	
	NH_4^+	9.2	NH_3	
	C_6H_5OH	9.9	C_6H_5O-	
	HCO_3^-	10.2	CO_3^{2-}	
	$CH_3NH_3^+$	10.6	CH_3NH_2	
	H_2O	15.7	OH^-	
	CH_3CH_2OH	16	$CH_3CH_2O^-$	
	$(CH_3)_3COH$	18	$(CH_3)_3CO^-$	
	CH_3COCH_3	19.2	$^-CH_2COCH_3$	
	$HC{\equiv}CH$	25	$HC{\equiv}C^-$	
	H_2	35	H^-	
	NH_3	38	NH_2^-	
	$CH_2{=}CH_2$	44	$CH_2{=}CH^-$	
Weakest acid	CH_3CH_3	50	$CH_3CH_2^-$	Strongest base

ured in aqueous solution. Special methods must be used to estimate the pK_a values for the very strong acids at the top of the table and for the very weak acids at the bottom.* The pK_a values for these very strong and weak acids are therefore approximate. All of the acids that we shall consider in this book will have strengths in between that of ethane (an extremely weak acid) and that of $HSbF_6$ (an acid that is so strong that it is called a "superacid"). As you examine Table 3.1, take care not to lose sight of the vast range of acidities that it represents (a factor of 10^{62}).

PROBLEM 3.5 **(a)** An acid (HA) has a $K_a = 10^{-7}$. What is its pK_a? **(b)** Another acid (HB) has a $K_a = 5$; what is its pK_a? **(c)** Which is the stronger acid?

Water, itself, is a very weak acid and undergoes self-ionization even in the absence of acids and bases:

$$H{-}\ddot{O}{:}(H) + H{-}\ddot{O}{:}(H) \rightleftharpoons H{-}\ddot{O}^{+}(H){-}H + {}^{-}{:}\ddot{O}{-}H$$

In pure water at 25°C, the concentrations of hydronium and hydroxide ions are equal to $10^{-7}M$. Since the concentration of water in pure water is 55.5*M*, we can calculate the K_a for water.

$$K_a = \frac{[H_3O^+][OH^-]}{[H_2O]} \qquad K_a = \frac{(10^{-7})(10^{-7})}{(55.5)} = 1.8 \times 10^{-16} \qquad pK_a = 15.7$$

PROBLEM 3.6 Show calculations proving that the pK_a of the hydronium ion (H_3O^+) is -1.74 as given in Table 3.1.

3.5C Predicting the Strength of Bases

The relationship of acid–base conjugates is a very useful tool for predicting base strength.

In our discussion so far we have dealt only with the strengths of acids. Arising as a natural corollary to this is a principle that allows us to estimate the strengths of bases. Simply stated, the principle is this: **The stronger the acid, the weaker will be its conjugate base.**

We can, therefore, **relate the strength of a base to the pK_a of its conjugate acid. The larger the pK_a of the conjugate acid, the stronger is the base.** Consider the following as examples:

Increasing base strength →

Cl^-	$CH_3CO_2^-$	OH^-
Very weak base	**Weak base**	**Strong base**
pK_a of conjugate acid (HCl) = −7	**pK_a of conjugate acid (CH_3CO_2H) = 4.75**	**pK_a of conjugate acid (H_2O) = 15.7**

We see that the hydroxide ion is the strongest in this series of three bases because its conjugate acid, water, is the weakest acid. (We know that water is the weakest acid because it has the largest pK_a.)

*Acids that are stronger than a hydronium ion and bases that are stronger than a hydroxide ion react completely with water (a phenomenon called the leveling effect, see Sections 3.2A and 3.14). Therefore, it is not possible to measure acidity constants for these acids in water. Other solvents and special techniques are used, but we do not have the space to describe these methods here.

Amines are like ammonia in that they are weak bases. Dissolving ammonia in water brings about the following equilibrium:

$$\ddot{N}H_3 + H-\ddot{\underset{\cdot\cdot}{O}}-H \rightleftharpoons H-\overset{H}{\underset{H}{\overset{|}{\underset{|}{N^+}}}}-H + {}^-:\ddot{\underset{\cdot\cdot}{O}}-H$$

Base	Acid	Conjugate acid $pK_a = 9.2$	Conjugate base

Dissolving methylamine in water causes the establishment of a similar equilibrium.

$$CH_3\ddot{N}H_2 + H-\ddot{\underset{\cdot\cdot}{O}}-H \rightleftharpoons CH_3-\overset{H}{\underset{H}{\overset{|}{\underset{|}{N^+}}}}-H + {}^-:\ddot{\underset{\cdot\cdot}{O}}-H$$

Base	Acid	Conjugate acid $pK_a = 10.6$	Conjugate base

Again we can relate the basicity of these substances to the strength of their conjugate acids. The conjugate acid of ammonia is the ammonium ion, NH_4^+. The pK_a of the ammonium ion is 9.2. The conjugate acid of methylamine is the $CH_3NH_3^+$ ion. This ion, called the methylaminium ion, has a $pK_a = 10.6$. Since the conjugate acid of methylamine is a weaker acid than the conjugate acid of ammonia, we can conclude that methylamine is a stronger base than ammonia.

PROBLEM 3.7

The pK_a of the anilinium ion ($C_6H_5\overset{+}{N}H_3$) is equal to 4.6. On the basis of this fact, decide whether aniline ($C_6H_5NH_2$) is a stronger or weaker base than methylamine.

3.6 Predicting the Outcome of Acid–Base Reactions

Table 3.1 gives the approximate pK_a values for a range of representative compounds. While you probably will not be expected to memorize all of the pK_a values in Table 3.1, it is a good idea to begin to learn the general order of acidity and basicity for some of the common acids and bases. The examples given in Table 3.1 are representative of their class or functional group. For example, acetic acid has a $pK_a = 4.75$, and carboxylic acids generally have pK_a values near this value (in the range $pK_a = 3–5$). Ethyl alcohol is given as an example of an alcohol, and alcohols generally have pK_a values near that of ethyl alcohol (in the pK_a range 15–18), and so on. (There are exceptions, of course, and we shall learn what these exceptions are as we go on.)

By learning the relative scale of acidity of common acids now, you will be able to predict whether or not an acid–base reaction will occur as written. The general principle to apply is this: **Acid–base reactions always favor the formation of the weaker acid and the weaker base.** The reason for this is that the outcome of an acid–base reaction is determined by the position of an equilibrium. Acid–base reactions are said, therefore, to be **under equilibrium control,** and reactions under equilibrium control always favor the formation of the most stable (lowest potential energy) species. The weaker acid and weaker base are more stable (lower in potential energy) than the stronger acid and stronger base.

A general principle for predicting the outcome of acid–base reactions.

Using this principle, we can predict that a carboxylic acid (RCO_2H) will react with aqueous NaOH in the following way because the reaction will lead to the formation of the weaker acid (H_2O) and weaker base (RCO_2^-):

$$RCO_2H + Na^+ \, {}^-OH \longrightarrow RCO_2^- \, Na^+ + H{-}O{-}H$$

Stronger acid $pK_a = 3–5$ | Stronger base | Weaker base | Weaker acid $pK_a = 15.7$

Because there is a large difference in the value of the pK_a of the two acids, the position of equilibrium will greatly favor the formation of the products. In instances like these we commonly show the reaction with a one-way arrow even though the reaction is an equilibrium.

3.6A Water Solubility as the Result of Salt Formation

Although acetic acid and other carboxylic acids containing fewer than five carbon atoms are soluble in water, many other carboxylic acids of higher molecular weight are not appreciably soluble in water. Because of their acidity, however, *water-insoluble carboxylic acids dissolve in aqueous sodium hydroxide;* they do so by reacting to form water-soluble sodium salts:

$$C_6H_5CO_2H + Na^+ \, {}^-OH \longrightarrow C_6H_5CO_2^- \, Na^+ + H{-}O{-}H$$

Insoluble in water | Soluble in water (due to its polarity as a salt)

We can also predict that an amine will react with aqueous hydrochloric acid in the following way:

$$R{-}NH_2 + H{-}OH_2^+ \, Cl^- \longrightarrow R{-}NH_3^+ \, Cl^- + H_2O$$

Stronger base | Stronger acid $pK_a = -1.74$ | Weaker acid $pK_a = 9–10$ | Weaker base

While methylamine and most amines of low molecular weight are very soluble in water, amines with higher molecular weights, such as aniline ($C_6H_5NH_2$), have limited water solubility. However, these *water-insoluble amines dissolve readily in hydrochloric acid* because the acid–base reactions convert them into soluble salts:

$$C_6H_5{-}NH_2 + H{-}OH_2^+ \, Cl^- \longrightarrow C_6H_5{-}NH_3^+ \, Cl^- + H_2O$$

Water insoluble | Water-soluble salt

3.7 The Relationship between Structure and Acidity

The strength of a Brønsted–Lowry acid depends on the extent to which a proton can be separated from it and transferred to a base. Removing the proton involves breaking a bond to the proton, and it involves making the conjugate base more electrically negative.

When we compare compounds in a single column of the periodic table, the strength of the bond to the proton is the dominating effect. Bond strength to the proton decreases as we move down the column, a phenomenon mainly due to decreasing effectiveness of orbital overlap between the hydrogen 1*s* orbital and the orbitals of successively larger elements in the column. The less effective the orbital overlap, the weaker the bond, and the stronger is the acid. The acidities of the hydrogen halides furnish an example:

Acidity increases as we descend a column in the periodic table.

pK_a	
3.2	H—F
−7	H—Cl
−9	H—Br
−10	H—I

Acidity increases ↓

Comparing the hydrogen halides with each other, H—F is the weakest acid and H—I is the strongest. This follows from the fact that the H—F bond is by far the strongest and the H—I bond is the weakest.

Because HI, HBr, and HCl are strong acids, their conjugate bases (I^-, Br^-, Cl^-) are all weak bases. HF, however, which is less acidic than the other hydrogen halides by 10–13 orders of magnitude (compare their pK_a values), has a conjugate base that is correspondingly more basic than the other halide anions. The fluoride anion is still not nearly as basic as other species we commonly think of as bases, such as the hydroxide anion, however. A comparison of the pK_a values for HF (3.2) and H_2O (15.7) illustrates this point.

The same trend of acidities and basicities holds true in other columns of the periodic table. Consider, for example, the column headed by oxygen:

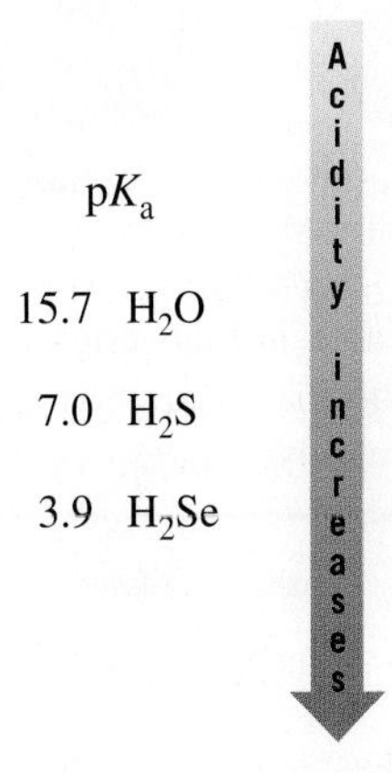

Here the strongest bond is the O—H bond and H_2O is the weakest acid; the weakest bond is the Se—H bond and H_2Se is the strongest acid.

Acidity increases from left to right in a given row of the periodic table.

Acidity increases from left to right when we compare compounds in a given row of the periodic table. Bond strengths vary somewhat, but the predominant factor becomes the electronegativity of the atom bonded to the hydrogen. The electronegativity of the atom in question affects acidity in two related ways. It affects the polarity of the bond to the proton and it affects the relative stability of the anion (conjugate base) that forms when the proton is lost.

We can see an example of this effect when we compare the acidities of the compounds CH_4, NH_3, H_2O, and HF. These compounds are all hydrides of first-row elements, and electronegativity increases across a row of the periodic table from left to right (see Table 1.2):

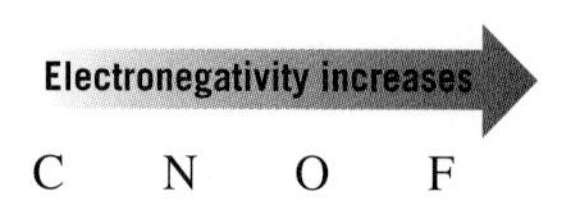

Because fluorine is the most electronegative, the bond in H—F is most polarized, and the proton in H—F is the most positive. Therefore, H—F loses a proton most readily and is the most acidic in this series:

CD Tutorial

pK_as and an acid–base reaction.

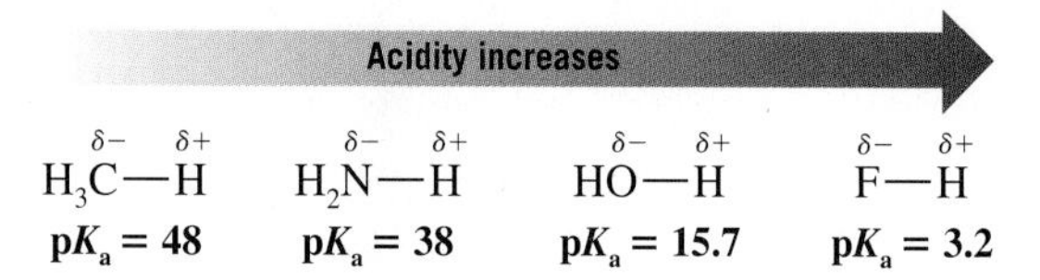

Electrostatic potential maps for these compounds directly illustrate this trend based on electronegativity and increasing polarization of the bonds to hydrogen (Fig. 3.2). Almost no positive charge (indicated by extent of color trending toward blue) is evident at the hydrogens of methane. Very little positive charge is present at the hydrogens of ammonia. This is consistent with the weak electronegativity of both carbon and nitrogen and hence with the behavior of methane and ammonia as exceedingly weak acids (pK_as of 48 and 38, respectively). Water shows significant positive charge at its hydrogens (pK_a more than 20 units lower than ammonia), and hydrogen fluoride clearly has the highest amount of positive charge at its hydrogen (pK_a of 3.2), resulting in strong acidity.

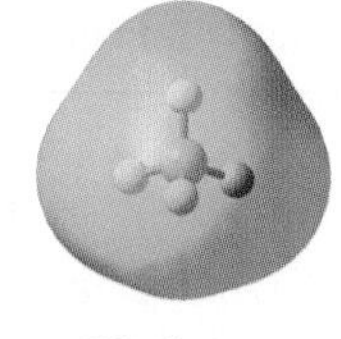
Methane

Ammonia

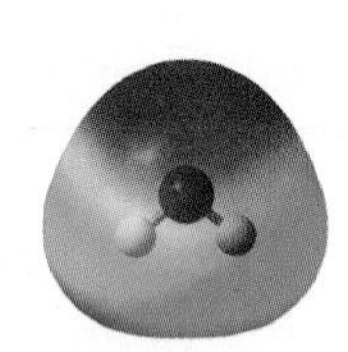
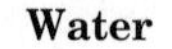
Water

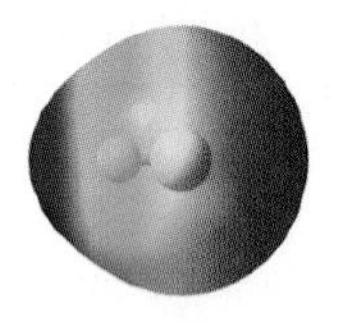
Hydrogen fluoride

FIGURE 3.2 The effect of increasing electronegativity among elements from left to right in the first row of the periodic table is evident in these maps of electrostatic potential for methane, ammonia, water, and hydrogen fluoride.

Animated Graphics

Because H—F is the strongest acid in this series, its conjugate base, the fluoride ion (F^-), will be the weakest base. Fluorine is the most electronegative atom and it accommodates the negative charge most readily:

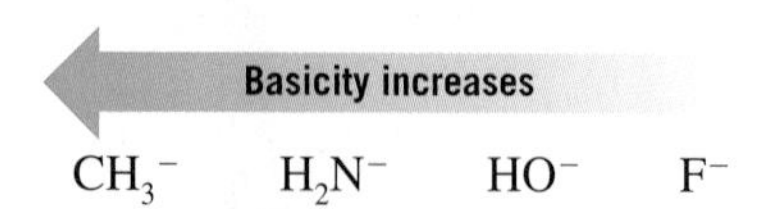

The methanide ion (CH_3^-) is the least stable anion of the four, because carbon being the least electronegative element is least able to accept the negative charge. The methanide ion, therefore, is the strongest base in this series. [The methanide ion, a **carbanion,** and

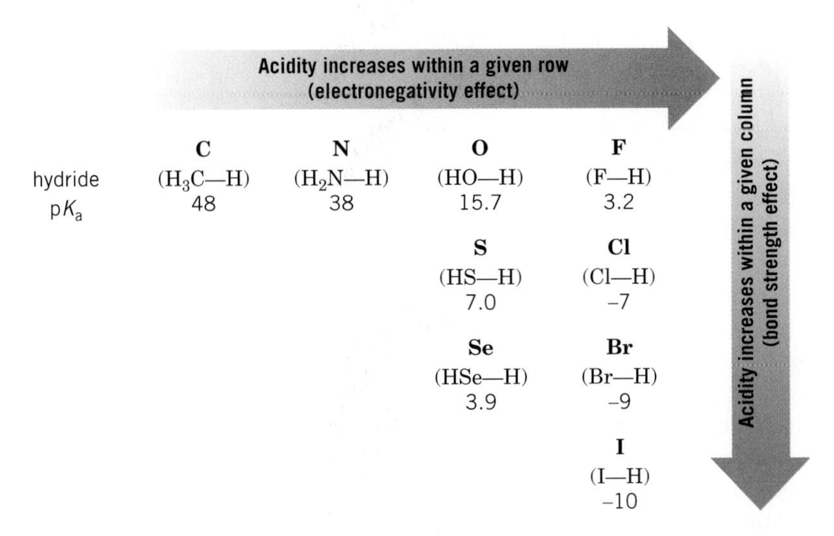

FIGURE 3.3 A summary of periodic trends in relative acidity. Acidity increases from left to right across a given row (electronegativity effect) and from top to bottom in a given column (bond strength effect) of the periodic table.

the amide ion (NH_2^-) are exceedingly strong bases because they are the conjugate bases of extremely weak acids. We shall discuss some uses of these powerful bases in Section 3.14.]

Trends in acidity within the periodic table are summarized in Fig. 3.3.

3.7A The Effect of Hybridization

The protons of ethyne are more acidic than those of ethene, which in turn are more acidic than those of ethane:

H—C≡C—H

Ethyne
pK_a = 25

Ethene
pK_a = 44

Ethane
pK_a = 50

We can explain this order of acidities on the basis of the hybridization state of carbon in each compound. Electrons of 2*s* orbitals have lower energy than those of 2*p* orbitals because *electrons in 2s orbitals tend, on the average, to be much closer to the nucleus than electrons in 2p orbitals.* (Consider the shapes of the orbitals: 2*s* orbitals are spherical and centered on the nucleus; 2*p* orbitals have lobes on either side of the nucleus and are extended into space.) With hybrid orbitals, therefore, **having more *s* character means that the electrons of the anion will, on the average, be lower in energy, and the anion will be more stable.** The *sp* orbitals of the C—H bonds of ethyne have 50% *s* character (because they arise from the combination of one *s* orbital and one *p* orbital), those of the sp^2 orbitals of ethene have 33.3% *s* character, while those of the sp^3 orbitals of ethane have only 25% *s* character. This means, in effect, that the *sp* carbon atoms of ethyne act as if they were the most electronegative when compared to the sp^2 carbon atoms of ethene and the sp^3 carbon atoms of ethane. (Remember: Electronegativity measures an atom's ability to hold bonding electrons close to its nucleus, and having electrons closer to the nucleus makes it more stable.)

The effect of hybridization on acidity is borne out in the calculated electrostatic potential maps for ethyne, ethene, and ethane shown in Fig. 3.4. Some positive charge (indicated by blue color) is clearly evident on the hydrogens of ethyne (pK_a = 25), but almost

FIGURE 3.4 Electrostatic potential maps for ethyne, ethene, and ethane.

Animated Graphics

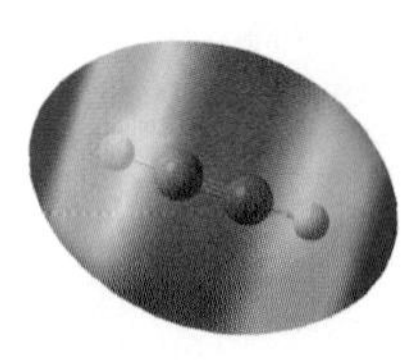
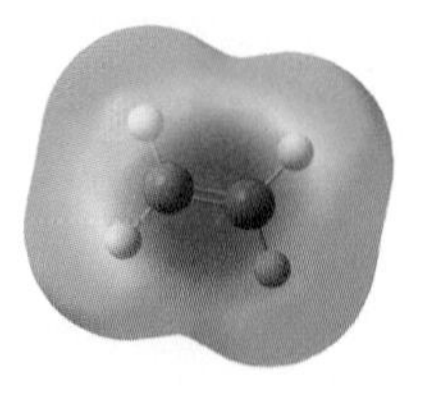

Ethyne **Ethene** **Ethane**

no positive charge is present on the hydrogens of ethene and ethane (both having pK_as more than 20 units greater than ethyne). This is consistent with the effectively greater electronegativity of the *sp* orbitals in ethyne, which have more *s* character than the sp^2 and sp^3 orbitals in ethene and ethane. [Also evident in Fig. 3.4 is the negative charge resulting from electron density in the π bonds of ethyne and ethene (indicated by red in the region of their respective π bonds). Note the cylindrical symmetry of π electron density in the triple bond of ethyne. In the π bond of ethene there is a region of high electron density on its underneath face complementary to that visible on the top face of its double bond.]

Now we can see how the order of relative acidities of ethyne, ethene, and ethane parallels the effective electronegativity of the carbon atom in each compound:

Relative Acidity of the Hydrocarbons

$$HC{\equiv}CH > H_2C{=}CH_2 > H_3C{-}CH_3$$

Being the most electronegative, the *sp*-hybridized carbon atom of ethyne polarizes its C—H bonds to the greatest extent, causing its hydrogens to be most positive. Therefore, ethyne donates a proton to a base more readily. And, in the same way, the ethynide ion is the weakest base because the more electronegative carbon of ethyne is best able to stabilize the negative charge.

Relative Basicity of the Carbanions

$$H_3C{-}CH_2{:}^- > H_2C{=}CH{:}^- > HC{\equiv}C{:}^-$$

Notice that the explanation given here is the same as that given to account for the relative acidities of HF, H_2O, NH_3, and CH_4.

3.7B Inductive Effects

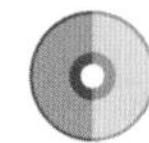

Animated Graphics

The carbon–carbon bond of ethane is completely nonpolar because at each end of the bond there are two equivalent methyl groups:

$$CH_3{-}CH_3$$

Ethane

The C—C bond is nonpolar.

This is not the case with the carbon–carbon bond of ethyl fluoride, however:

$$\overset{\delta+}{CH_3}{\rightarrowtail}\overset{\delta+}{CH_2}{\rightarrowtail}\overset{\delta-}{F}$$

2 **1**

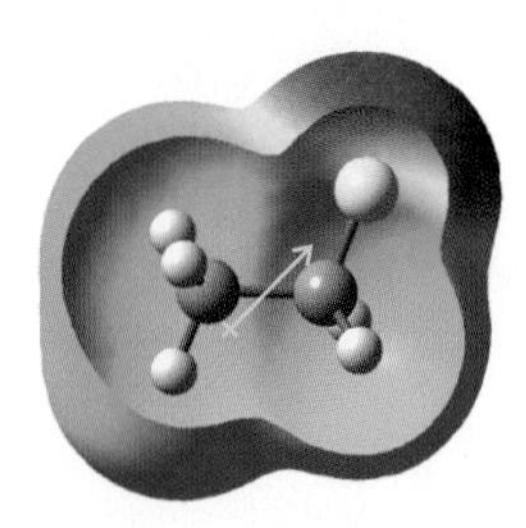

FIGURE 3.5 Ethyl fluoride, showing its dipole moment inside a cut-away view of the electrostatic potential at its van der Waals surface.

One end of the bond, the one nearer the fluorine atom, is more positive than the other. This polarization of the carbon–carbon bond results from an intrinsic electron-attracting ability of the fluorine (because of its electronegativity) that is transmitted *through space* and *through the bonds of the molecule.* Chemists call this kind of effect an **inductive effect.** The inductive effect here is **electron attracting** (or **electron withdrawing**), but we shall see later that inductive effects can also be **electron releasing.** *Inductive effects* *weaken as the distance from the substituent increases.* In this instance, the positive charge that the fluorine imparts to **C1** is greater than that imparted to **C2** because the fluorine is closer to **C1**.

Figure 3.5 shows the dipole moment for ethyl fluoride (fluoroethane). The distribution of negative charge around the electronegative fluorine is plainly evident in the calculated electrostatic potential map.

3.8 Energy Changes

Since we will be talking frequently about the energies of chemical systems and the relative stabilities of molecules, perhaps we should pause here for a brief review. *Energy* is defined as the capacity to do work. The two fundamental types of energy are **kinetic energy** and **potential energy.**

Kinetic energy is the energy an object has because of its motion; it equals one-half the object's mass multiplied by the square of its velocity (i.e., $\frac{1}{2}mv^2$).

Potential energy is stored energy. It exists only when an attractive or repulsive force exists between objects. Two balls attached to each other by a spring (an analogy we used for covalent bonds when we discussed infrared spectroscopy in Section 2.16) can have their potential energy increased when the spring is stretched or compressed (Fig. 3.6). If the spring is stretched, an attractive force will exist between the balls. If it is compressed, a repulsive force will exist. In either instance releasing the balls will cause the potential energy (stored energy) of the balls to be converted into kinetic energy (energy of motion).

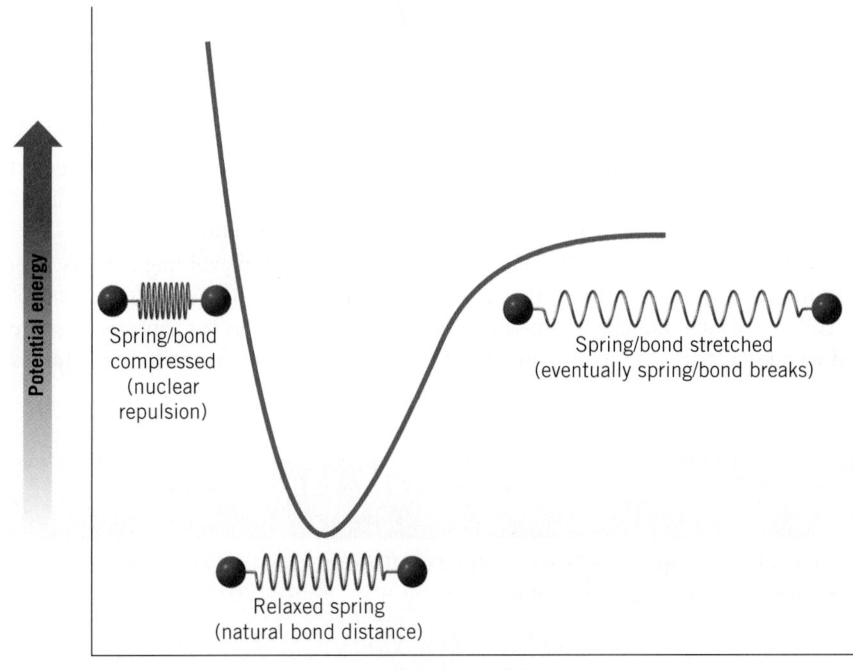

FIGURE 3.6 Potential energy exists between objects that either attract or repel each other. In the case of atoms joined by a covalent bond, or objects connected by a spring, the lowest potential energy state is when atoms are at their ideal internuclear distance (bond length), or when a spring between objects is relaxed. Lengthening or shortening the bond distance, or compressing or stretching a spring, raises the potential energy.

Chemical energy is a form of potential energy. It exists because attractive and repulsive electrical forces exist between different pieces of the molecules. Nuclei attract electrons, nuclei repel each other, and electrons repel each other.

It is usually impractical (and often impossible) to describe the *absolute* amount of potential energy contained by a substance. Thus we usually think in terms of its *relative potential energy.* We say that one system has *more* or *less* potential energy than another.

Another term that chemists frequently use in this context is the term **stability** or **relative stability.** *The relative stability of a system is inversely related to its relative potential energy.* **The *more* potential energy an object has, the *less stable* it is.** Consider, as an example, the relative potential energy and the relative stability of snow when it lies high on a mountainside and when it lies serenely in the valley below. Because of the attractive force of gravity, the snow high on the mountain *has greater potential energy and is much less stable* than the snow in the valley. This greater potential energy of the snow on the mountainside can become converted into the enormous kinetic energy of an avalanche. By contrast, the snow in the valley, with its lower potential energy and with its greater stability, is incapable of releasing such energy.

3.8A Potential Energy and Covalent Bonds

Atoms and molecules possess potential energy—often called chemical energy—that can be released as heat when they react. Because heat is associated with molecular motion, this release of heat results from a change from potential energy to kinetic energy.

From the standpoint of covalent bonds, the state of greatest potential energy is the state of free atoms, the state in which the atoms are not bonded to each other at all. This is true because the formation of a chemical bond is always accompanied by the lowering of the potential energy of the atoms (cf. Fig. 1.7). Consider as an example the formation of hydrogen molecules from hydrogen atoms:

$$H\cdot + H\cdot \longrightarrow H{-}H \qquad \Delta H^\circ = -436 \text{ kJ mol}^{-1}*$$

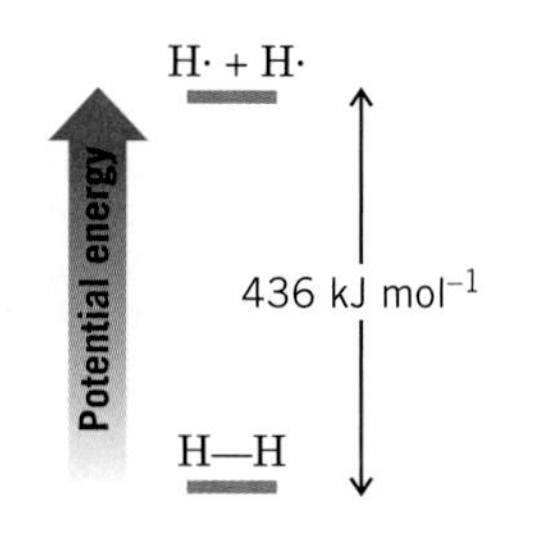

FIGURE 3.7 The relative potential energies of hydrogen atoms and a hydrogen molecule.

The potential energy of the atoms decreases by 436 kJ mol^{-1} as the covalent bond forms. This potential energy change is illustrated graphically in Fig. 3.7.

A convenient way to represent the relative potential energies of molecules is in terms of their relative **enthalpies,** or **heat contents,** H. (*Enthalpy* comes from *en* + Gk: *thalpein,* to heat.) The difference in relative enthalpies of reactants and products in a chemical change is called the enthalpy change and is symbolized by ΔH°. [The Δ (delta) in front of a quantity usually means the difference, or change, in the quantity. The superscript ° indicates that the measurement is made under standard conditions.]

By convention, the sign of ΔH° for **exothermic** reactions (those evolving heat) is negative. **Endothermic** reactions (those that absorb heat) have a positive ΔH°. The heat of reaction, ΔH°, measures the change in enthalpy of the atoms of the reactants as they are converted to products. For an exothermic reaction, the atoms have a smaller enthalpy as products than they do as reactants. For endothermic reactions, the reverse is true.

3.9 The Relationship Between the Equilibrium Constant and the Standard Free-Energy Change, ΔG°

An important **relationship exists between the equilibrium constant (K_{eq}) and the standard free-energy change† (ΔG°)** for a reaction.

$$\Delta G^\circ = -RT \ln K_{eq}$$

*The unit of energy in SI units is the joule, J, and 1 cal = 4.184 J. (Thus 1 kcal = 4.184 kJ.) A kilocalorie of energy (1000 cal) is the amount of energy in the form of heat required to raise by 1°C the temperature of 1 kg (1000 g) of water at 15°C.

†By standard free-energy change (ΔG°), we mean that the products and reactants are taken as being in their standard states (1 atm of pressure for a gas and 1M for a solution). The free-energy change is often called the **Gibbs free-energy change,** to honor the contributions to thermodynamics of J. Willard Gibbs, a professor of mathematical physics at Yale University in the latter part of the nineteenth century. Gibbs ranks as one of the greatest scientists produced by the United States.

where R is the gas constant and equals 8.314 J K^{-1} mol^{-1} and T is the absolute temperature in kelvin (K).

It is easy to show with this equation that **a negative value of ΔG° is associated with reactions that favor the formation of products when equilibrium is reached** and for which the equilibrium constant is greater than 1. Reactions with a ΔG° more negative than about 13 kJ mol^{-1} are said *to go to completion,* meaning that almost all (>99%) of the reactants are converted into products when equilibrium is reached. Conversely, **a positive value of ΔG° is associated with reactions for which the formation of products at equilibrium is unfavorable** and for which the equilibrium constant is less than 1. Inasmuch as K_a is an equilibrium constant, it is related to ΔG° in the same way.

The free-energy change (ΔG°) has two components, the enthalpy change (ΔH°) and the entropy change (ΔS°). The relationship between these three thermodynamic quantities is

$$\Delta G^\circ = \Delta H^\circ - \mathrm{T}\,\Delta S^\circ$$

We have seen (Section 3.8) that ΔH° is associated with changes in bonding that occur in a reaction. If, collectively, stronger bonds are formed in the products than existed in the starting materials, then ΔH° will be negative (i.e., the reaction is *exothermic*). If the reverse is true, then ΔH° will be positive (the reaction is *endothermic*). A negative value for ΔH°, therefore, will contribute to making ΔG° negative and will consequently favor the formation of products. For the ionization of an acid, the less positive or more negative the value of ΔH°, the stronger the acid will be.

Entropy changes have to do with *changes in the relative order of a system.* **The more random a system is, the greater is its entropy.** Therefore, a positive entropy change ($+\Delta S^\circ$) is always associated with a change from a more ordered system to a less ordered one. A negative entropy change ($-\Delta S^\circ$) accompanies the reverse process. In the equation $\Delta G^\circ = \Delta H^\circ - T\,\Delta S^\circ$, the entropy change (multiplied by T) is preceded by a negative sign; this means that *a positive entropy change (from order to disorder) makes a negative contribution to ΔG° and is energetically favorable for the formation of products.*

For many reactions in which the number of molecules of products equals the number of molecules of reactants (e.g., when two molecules react to produce two molecules), the entropy change will be small. This means that except at high temperatures (where the term $T\,\Delta S^\circ$ becomes large even if ΔS° is small) the value of ΔH° will largely determine whether or not the formation of products will be favored. If ΔH° is large and negative (if the reaction is exothermic), then the reaction will favor the formation of products at equilibrium. If ΔH° is positive (if the reaction is endothermic), then the formation of products will be unfavorable.

PROBLEM 3.8

State whether you would expect the entropy change, ΔS°, to be positive, negative, or approximately zero for each of the following reactions. (Assume the reactions take place in the gas phase.)

(a) $A + B \longrightarrow C$ **(b)** $A + B \longrightarrow C + D$ **(c)** $A \longrightarrow B + C$

PROBLEM 3.9

(a) What is the value of ΔG° for a reaction where $K_{eq} = 1$? **(b)** Where $K_{eq} = 10$? (The change in ΔG° required to produce a 10-fold increase in the equilibrium constant is a useful term to remember.) **(c)** Assuming that the entropy change for this reaction is negligible (or zero), what change in ΔH° is required to produce a 10-fold increase in the equilibrium constant?

3.10 The Acidity of Carboxylic Acids

Both carboxylic acids and alcohols are weak acids. However, *carboxylic acids are much more acidic than the corresponding alcohols.* Unsubstituted carboxylic acids have pK_as in the range of 3–5; alcohols have pK_as in the range of 15–18. Consider as examples two compounds of about the same molecular proportions but with very different acidities, acetic acid and ethanol:

$$CH_3C(=O)OH \qquad CH_3—CH_2—OH$$

Acetic acid	**Ethanol**
pK_a = 4.75	**pK_a = 16**
$\Delta G°$ = 27 kJ mol^{-1}	**$\Delta G°$ = 90.8 kJ mol^{-1}**

From the pK_a for acetic acid (pK_a = 4.75) one can calculate (Section 3.9) that the free-energy change for the ionization of acetic acid is positive and equal to 27 kJ mol^{-1}. For ethanol (pK_a = 16), the free-energy change for the ionization is much larger and is equal to 90.8 kJ mol^{-1}. These values (see Fig. 3.8) show that whereas both compounds are weak acids, ethanol is a much weaker acid than acetic acid.

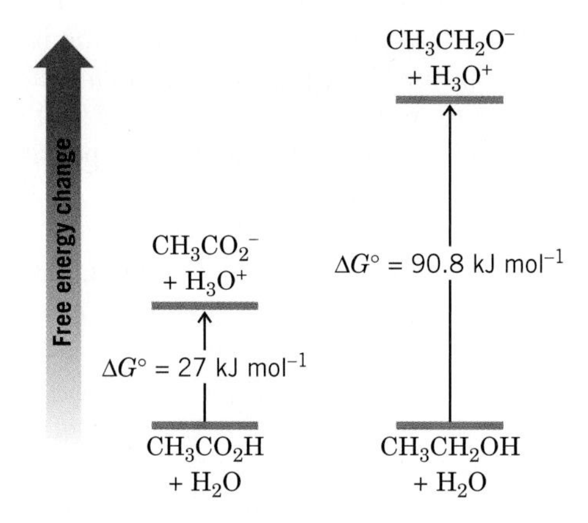

FIGURE 3.8 A diagram comparing the free-energy changes that accompany ionization of acetic acid and ethanol. Ethanol has a larger positive free-energy change and is a weaker acid because its ionization is more unfavorable.

How do we explain the greater acidity of carboxylic acids relative to alcohols? Two explanations have been proposed, one based on effects arising from resonance theory (and called, therefore, **resonance effects**) and another explanation based on **inductive effects** (Section 3.7B). Although both effects make a contribution to the greater acidity of carboxylic acids, the question as to which effect is more important, as we shall see, has not been resolved.

3.10A An Explanation Based on Resonance Effects

For many years, the greater acidity of carboxylic acids was attributed primarily to **resonance stabilization of the carboxylate ion.** This explanation invokes a principle of resonance theory (Section 1.8) that states that molecules and ions *are stabilized by resonance* especially *when the molecule or ion can be represented by two or more* **equivalent** *resonance structures (i.e., by resonance structures of equal stability).*

Two resonance structures can be written for a carboxylic acid and two for its anion (Fig. 3.9), and it can be assumed that *resonance stabilization of the anion is greater* because the resonance structures for the anion are equivalent and because no separation of opposite charges occurs in them. [Separation of opposite charges requires energy, and resonance structures that have separated charges are assumed to be less important in providing stability than those that do not (Section 1.8A).] The greater stabilization of the anion of a carboxylic acid (relative to the acid itself) lowers the free energy of the anion and

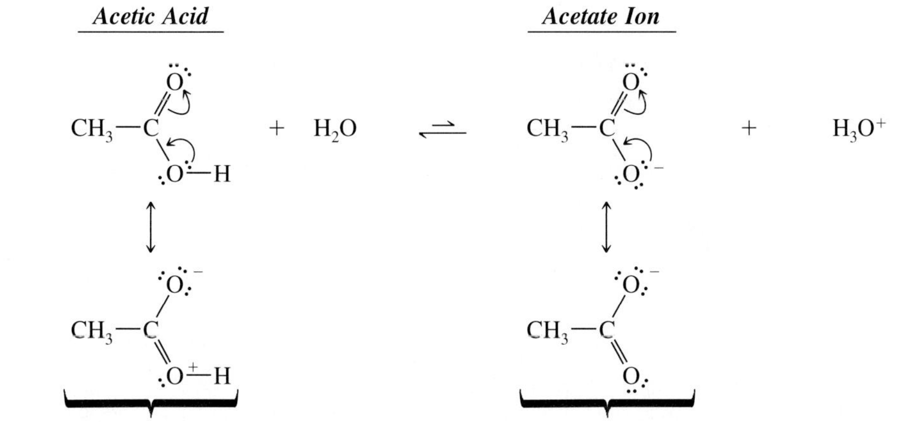

Acetic Acid | Acetate Ion

$$CH_3-C(=O)-O-H + H_2O \rightleftharpoons CH_3-C(=O)-O^- + H_3O^+$$

$$\updownarrow \qquad\qquad \updownarrow$$

$$CH_3-C(-O^-)=O^+-H \qquad\qquad CH_3-C(-O^-)=O$$

Small resonance stabilization
(The structures are not equivalent and the lower structure requires charge separation.)

Larger resonance stabilization
(The structures are equivalent and there is no requirement for charge separation.)

FIGURE 3.9 Two resonance structures that can be written for acetic acid and two that can be written for the acetate ion. According to a resonance explanation of the greater acidity of acetic acid, the equivalent resonance structures for the acetate ion provide it greater resonance stabilization and reduce the positive free-energy change for the ionization.

thereby decreases the positive free-energy change required for the ionization. Remember: *Any factor that makes the free-energy change for the ionization of an acid less positive (or more negative) makes the acid stronger* (Section 3.9).

No stabilizing resonance structures are possible for an alcohol or its anion:

$$CH_3-CH_2-\ddot{O}-H + H_2O \rightleftharpoons CH_3-CH_2-\ddot{O}:^- + H_3O^+$$

No resonance stabilization **No resonance stabilization**

The positive free-energy change for the ionization of an alcohol is not reduced by resonance stabilization, and this explains why the free-energy change is much larger than that for a carboxylic acid. An alcohol, consequently, is much less acidic than a carboxylic acid.

3.10B An Explanation Based on Inductive Effects

In 1986 an alternative explanation for the greater acidity of carboxylic acids stated that the inductive effect of the carbonyl group is the most important factor.* To understand this explanation based on inductive effects, let us consider the same two compounds:

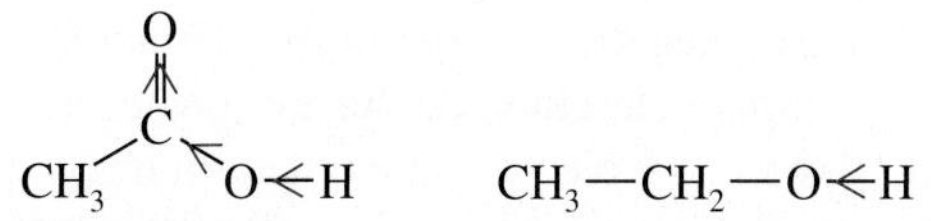

$$CH_3-C(=O)-O \leftarrow H \qquad\qquad CH_3-CH_2-O \leftarrow H$$

Acetic acid **(stronger acid)** | **Ethanol** **(weaker acid)**

In both compounds the O—H bond is highly polarized by the greater electronegativity of the oxygen atom. The key to the much greater acidity of acetic acid, in this explanation, is the powerful electron-attracting inductive effect of its carbonyl group (C=O group) when compared with the CH_2 group in the corresponding position of ethanol. The carbonyl group is highly polarized, as can be seen in calculated maps of electrostatic po-

*See Siggel, M. R. F.; Thomas, T. D. *J. Am. Chem. Soc.* **1986,** *108,* 4360–4362. Siggel, M. R. F.; Streitwieser, A. R., Jr.; Thomas, T. D. *J. Am. Chem. Soc.* **1988,** *110,* 8022–8028.

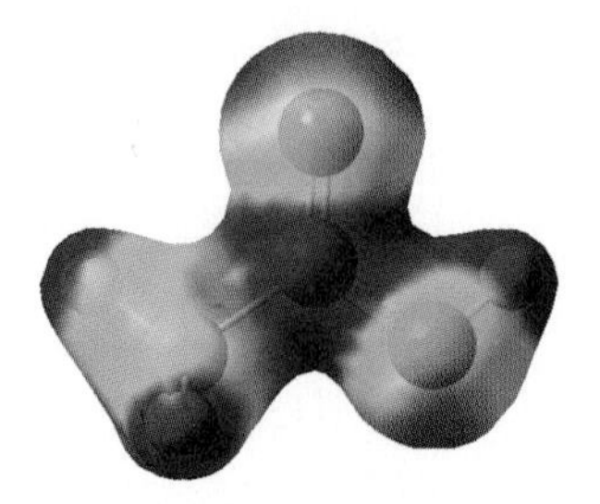

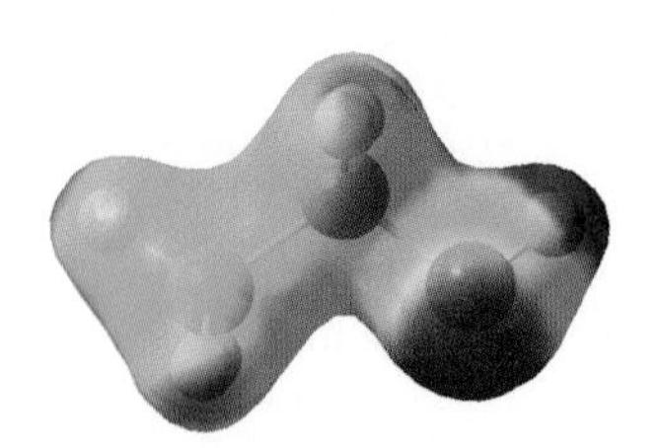

Acetic acid **Ethanol**

FIGURE 3.10 Maps of electrostatic potential at approximately the bond density surface for acetic acid and ethanol. The positive charge at the carbonyl carbon of acetic acid is evident in the blue color of the electrostatic potential map at that position, as compared to the hydroxyl carbon of ethanol. The inductive electron-withdrawing effect of the carbonyl group in carboxylic acids contributes to the acidity of this functional group.

tential at approximately the bond density surface for acetic acid and ethanol (Fig. 3.10). These models show clearly the significant positive charge at the carbonyl carbon of acetic acid as compared to the CH_2 carbon of ethanol.

Because the carbon atom of the carbonyl group of acetic acid bears a large positive charge, it adds its electron-attracting inductive effect to that of the oxygen atom of the hydroxyl group attached to it; *these combined effects make the carboxylic acid proton more positive than the proton of the alcohol.* This greater positive charge on the proton of the carboxylic acid helps explain why the proton separates more readily.

Further explanation of the acidity of carboxylic acids relates to the relative stability of the conjugate bases of carboxylic acids and alcohols. The electron-attracting inductive effect of the carbonyl group stabilizes the acetate ion that forms from acetic acid, and therefore the acetate ion is a weaker base than the ethoxide ion:

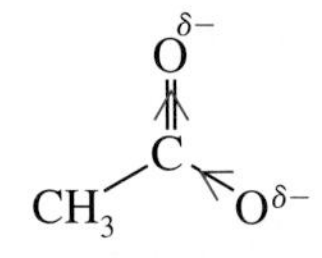

$CH_3{-}CH_2{\leftarrow}O^-$

Acetate anion
Weaker base

Ethoxide anion
Stronger base

Evaluation of an acid's conjugate base is a tool for predicting acid strength.

The ability of acetate to stabilize the negative charge better than ethoxide is evident in electrostatic potential maps at a surface approximating the bonding electron density for the two anions (Fig. 3.11). The negative charge in the acetate anion is evenly distributed over the two oxygens, as compared to ethoxide, where the negative charge is localized on its oxygen (as indicated by red in the electrostatic potential map). The ability to better sta-

FIGURE 3.11 Calculated electrostatic potential maps at a surface approximating the bonding electron density for acetate anion and ethoxide anion. Although both molecules carry the same −1 net charge, acetate stabilizes the charge better by dispersing it over both oxygen atoms.

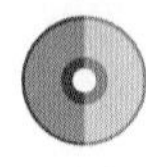

Animated Graphics

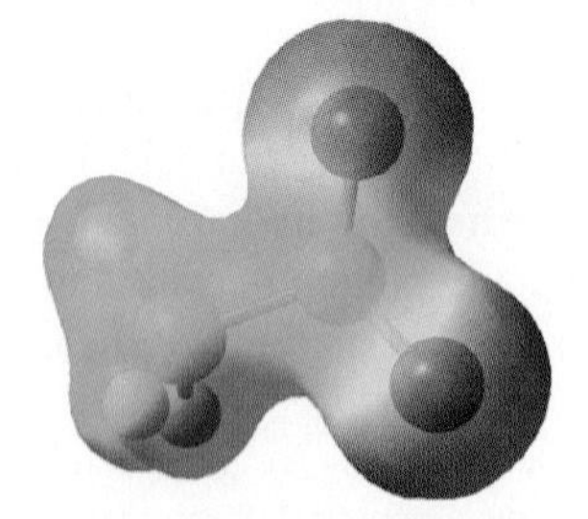

Acetate anion **Ethoxide anion**

bilize the negative charge is what makes acetate a weaker base than ethoxide (and hence its conjugate acid stronger than ethanol).

G. W. Wheland, who many years ago formulated the resonance explanation for the greater acidity of carboxylic acids, recognized that the inductive effect of the carbonyl group could be important and saw that deciding which effect, resonance or inductive, was more important would be difficult. Today this difficulty persists. A vigorous and interesting debate on the relative importance of the two effects is still being carried on.*

PROBLEM 3.10

Resonance theory may or may not provide the best explanation for the acidity of a carboxylic acid; however, it does provide an appropriate explanation for two related facts: The carbon–oxygen bond distances in the acetate ion are the same, and the oxygens of the acetate ion bear equal negative charges. Provide the resonance explanation for these facts.

3.10C Inductive Effects of Other Groups

The acid-strengthening effect of other electron-attracting groups (other than the carbonyl group) can be shown by comparing the acidities of acetic acid and chloroacetic acid:

CH_3COOH (O→H): **pK_a = 4.75** $ClCH_2COOH$ (Cl←CH_2, O→H): **pK_a = 2.86**

The greater acidity of chloroacetic acid can be attributed, in part, to the extra electron-attracting inductive effect of the electronegative chlorine atom. By adding its inductive effect to that of the carbonyl group and the oxygen, it makes the hydroxyl proton of chloroacetic acid even more positive than that of acetic acid. It also stabilizes the chloroacetate ion that is formed when the proton is lost *by dispersing its negative charge* (Fig. 3.12):

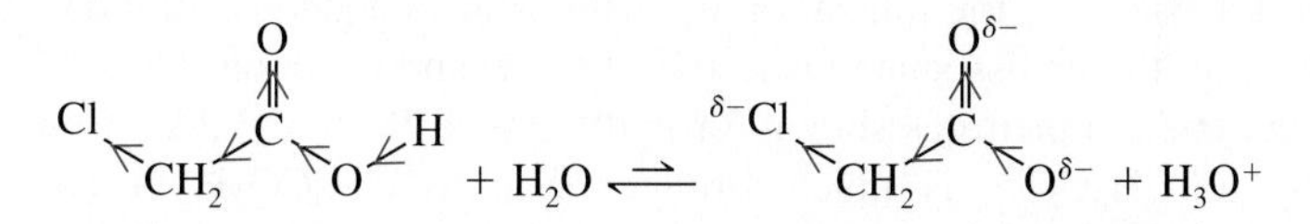

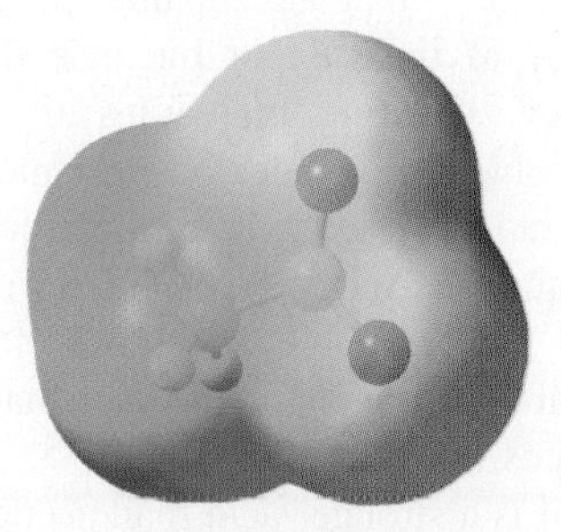

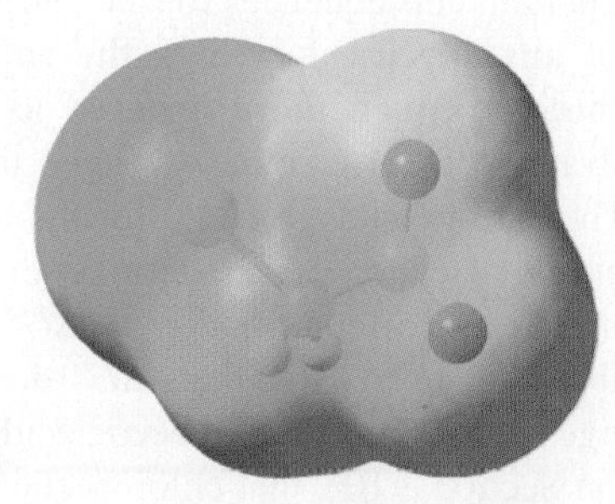

Acetate anion **Chloroacetate anion**

FIGURE 3.12 The electrostatic potential maps for acetate and chloroacetate ions show the relatively greater ability of chloroacetate to disperse the negative charge.

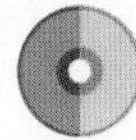

Animated Graphics

*The explanation based on the greater importance of inductive effects has been challenged. Those who may be interested in pursuing this debate further should consult the following article and the references given in it: Bordwell, F. G.; Satish, A. V. *J. Am. Chem. Soc.* **1994,** *116,* 8885–8889.

Dispersal of charge always makes a species more stable, and, as we have seen now in several instances, **any factor that stabilizes the conjugate base of an acid increases the strength of the acid.** (In Section 3.11, we shall see that entropy changes in the solvent are also important in explaining the increased acidity of chloroacetic acid.)

PROBLEM 3.11

Which would you expect to be the stronger acid? Explain your reasoning in each instance.

(a) CH_2ClCO_2H or $CHCl_2CO_2H$
(b) CCl_3CO_2H or $CHCl_2CO_2H$
(c) CH_2FCO_2H or CH_2BrCO_2H
(d) CH_2FCO_2H or $CH_2FCH_2CO_2H$

3.11 The Effect of the Solvent on Acidity

In the absence of a solvent (i.e., in the gas phase), most acids are far weaker than they are in solution. In the gas phase, for example, acetic acid is estimated to have a pK_a of about 130 (a K_a of $\sim 10^{-130}$)! The reason is this: When an acetic acid molecule donates a proton to a water molecule in the gas phase, the ions that are formed are oppositely charged particles and these particles must become separated:

$$CH_3C(=O)OH + H_2O \rightleftharpoons CH_3C(=O)O^- + H_3O^+$$

In the absence of a solvent, separation is difficult. In solution, solvent molecules surround the ions, insulating them from one another, stabilizing them, and making it far easier to separate them than in the gas phase.

In a solvent such as water, called a **protic solvent,** solvation by hydrogen bonding is important (Section 2.14C). **A protic solvent is one that has a hydrogen atom attached to a strongly electronegative element such as oxygen or nitrogen.** Molecules of a protic solvent, therefore, can form hydrogen bonds to the unshared electron pairs of oxygen (or nitrogen) atoms of an acid and its conjugate base, but they may not stabilize both equally.

Consider, for example, the ionization of acetic acid in aqueous solution. Water molecules solvate both the undissociated acid (CH_3CO_2H) and its anion ($CH_3CO_2^-$) by forming hydrogen bonds to them (as shown for hydroxide in Section 3.2A). However, hydrogen bonding to $CH_3CO_2^-$ is much stronger than to CH_3CO_2H because the water molecules are more attracted by the negative charge. This differential solvation, moreover, has important consequences for the entropy change that accompanies the ionization. **Solvation of any species decreases the entropy of the solvent because the solvent molecules become much more ordered as they surround molecules of the solute.** Because solvation of $CH_3CO_2^-$ is stronger, the solvent molecules become more orderly around it. The entropy change ($\Delta S°$) for the ionization of acetic acid, therefore, is negative. This means that the $-T\Delta S°$ term in the equation $\Delta G° = \Delta H° - T\Delta S°$, makes an acid-weakening positive contribution to $\Delta G°$. In fact, as Table 3.2 shows, the $-T\Delta S°$ term contributes more to $\Delta G°$ than $\Delta H°$ does and accounts for the fact that the free-energy change for the ionization of acetic acid is positive (unfavorable).

We saw in Section 3.10C that chloroacetic acid is a stronger acid than acetic acid, and we attributed this increased acidity to the presence of the electron-withdrawing chlorine atom. Table 3.2 shows us that both $\Delta H°$ and $-T\Delta S°$ are more favorable for the ionization of chloroacetic acid ($\Delta H°$ is more negative by 4.2 kJ mol^{-1}, and $-T\Delta S°$ is less positive by 7 kJ mol^{-1}). The larger contribution is clearly in the entropy term. Apparently, by stabilizing the chloroacetate anion, the chlorine atom makes the chloroacetate ion less prone to cause an ordering of the solvent because it requires less stabilization through solvation.

TABLE 3.2 Thermodynamic Values for the Dissociation of Acetic and Chloroacetic Acids in H_2O at 25°C[a]

Acid	pK_a	$\Delta G°$ (kJ mol^{-1})	$\Delta H°$ (kJ mol^{-1})	$-T\Delta S°$ (kJ mol^{-1})
CH_3CO_2H	4.75	+27	−0.4	+28
$ClCH_2CO_2H$	2.86	+16	−4.6	+21

[a]Table adapted from March, J. *Advanced Organic Chemistry,* 3rd ed.; Wiley: New York, 1985; p 236.

3.12 Organic Compounds as Bases

If an organic compound contains an atom with an unshared electron pair, it is a potential base. We saw in Section 3.5C that compounds with an unshared electron pair on a nitrogen atom (i.e., amines) act as bases. Let us now consider several examples in which organic compounds having an unshared electron pair on an oxygen atom act in the same way.

Dissolving gaseous HCl in methanol brings about an acid–base reaction much like the one that occurs with water (Section 3.2A):

$$H_3C{-}\ddot{O}(H){:} + H{-}\ddot{Cl}{:} \longrightarrow H_3C{-}\ddot{O}^{+}(H){-}H + {:}\ddot{Cl}{:}^{-}$$

Methanol → **Methyloxonium ion (a protonated alcohol)**

The conjugate acid of the alcohol is often called a **protonated alcohol,** although more formally it is called an **alkyloxonium ion.**

Alcohols, in general, undergo this same reaction when they are treated with solutions of strong acids such as HCl, HBr, HI, and H_2SO_4:

$$R{-}\ddot{O}(H){:} + H{-}A \longrightarrow R{-}\ddot{O}^{+}(H){-}H + {:}A^{-}$$

Alcohol, **Strong acid** → **Alkyloxonium ion**, **Weak base**

So, too, do ethers:

$$R{-}\ddot{O}(R){:} + H{-}A \longrightarrow R{-}\ddot{O}^{+}(R){-}H + {:}A^{-}$$

Ether, **Strong acid** → **Dialkyloxonium ion**, **Weak base**

Compounds containing a carbonyl group also act as bases in the presence of a strong acid:

$$R_2C{=}\ddot{O}{:} + H{-}A \rightleftharpoons R_2C{=}\dot{O}^{+}{-}H + {:}A^{-}$$

Ketone, **Strong acid** ⇌ **Protonated ketone**, **Weak base**

Proton transfers are a common first step in many reactions we shall study.

Proton transfer reactions like these are often the first step in many reactions that alcohols, ethers, aldehydes, ketones, esters, amides, and carboxylic acids undergo. The pK_as for some of these protonated intermediates are given in Table 3.1.

An atom with an unshared electron pair is not the only locus that confers basicity on an organic compound. The π bond of an alkene can have the same effect. Later we shall study many reactions in which, as a first step, alkenes react with a strong acid by accepting a proton in the following way:

The π bond breaks

This bond breaks

This bond is formed

$$\mathrm{>C{=}C<} + \mathrm{H{-}A} \rightleftharpoons \mathrm{>\overset{+}{C}{-}\overset{|}{\underset{|}{C}}{-}H} + :\mathrm{A}^-$$

Alkene **Strong acid** **Carbocation** **Weak base**

In this reaction the electron pair of the π bond of the alkene is used to form a bond between one carbon of the alkene and the proton donated by the strong acid. Notice that two bonds are broken in this process: the π bond of the double bond and the bond between the proton of the acid and its conjugate base. One new bond is formed: a bond between a carbon of the alkene and the proton. This process leaves the other carbon of the alkene trivalent, electron deficient, and with a formal positive charge. A compound containing a carbon of this type is called a **carbocation** (Section 3.3). As we shall see in later chapters, carbocations are unstable intermediates that react further to produce stable molecules.

PROBLEM 3.12 It is a general rule that any organic compound containing oxygen, nitrogen, or a multiple bond will dissolve in concentrated sulfuric acid. Explain the basis of this rule.

3.13 A Mechanism for an Organic Reaction

In Chapter 6 we shall begin our study of mechanisms of organic reactions in earnest. Let us consider now one mechanism as an example, one that allows us to apply some of the chemistry we have learned in this chapter and one that, at the same time, will reinforce what we have learned about how curved arrows are used to illustrate mechanisms.

Dissolving *tert*-butyl alcohol in concentrated (concd) aqueous hydrochloric acid soon results in the formation of *tert*-butyl chloride. The reaction is a substitution reaction:

$$(CH_3)_3C{-}OH + H{-}\ddot{O}^{+}{-}H\,(H) + :\ddot{Cl}:^{-} \xrightarrow{H_2O} (CH_3)_3C{-}Cl + 2\,H_2O$$

***tert*-Butyl alcohol (soluble in H_2O)** **Concd HCl** ***tert*-Butyl chloride (insoluble in H_2O)**

That a reaction has taken place is obvious when one actually does the experiment. *tert*-Butyl alcohol is soluble in the aqueous medium; however, *tert*-butyl chloride is not, and consequently it separates from the aqueous phase as another layer in the flask. It is easy to remove this nonaqueous layer, purify it by distillation, and thus obtain the *tert*-butyl chloride.

Considerable evidence, described later, indicates that the reaction occurs in the following way.

A Mechanism for the Reaction

Reaction of *tert*-Butyl Alcohol with Concentrated Aqueous HCl

Step 1

$$H_3C-C(CH_3)_2-\ddot{O}-H + H-\overset{+}{\ddot{O}}(H)-H \rightleftharpoons H_3C-C(CH_3)_2-\overset{+}{\ddot{O}}(H)-H + :\ddot{O}(H)-H$$

***tert*-Butyloxonium ion**

tert-Butyl alcohol acts as a base and accepts a proton from the hydronium ion. (Chloride anions are spectators in this step of the reaction.)

The products are a protonated alcohol and water (the conjugate acid and base).

Step 2

$$H_3C-C(CH_3)_2-\overset{+}{\ddot{O}}(H)-H \rightleftharpoons H_3C-C^{+}(CH_3)_2 + :\ddot{O}(H)-H$$

Carbocation

The bond between the carbon and oxygen of the *tert*-butyloxonium ion breaks heterolytically, leading to the formation of a carbocation and a molecule of water.

Step 3

$$H_3C-C^{+}(CH_3)_2 + :\ddot{Cl}:^{-} \rightleftharpoons H_3C-C(CH_3)_2-\ddot{Cl}:$$

***tert*-Butyl chloride**

The carbocation, acting as a Lewis acid, accepts an electron pair from a chloride ion to become the product.

Notice that **all of these steps involve acid–base reactions.** Step 1 is a straightforward Brønsted acid–base reaction in which the alcohol oxygen removes a proton from the hydronium ion. Step 2 is the reverse of a Lewis acid–base reaction. In it, the carbon–oxygen bond of the protonated alcohol breaks heterolytically as a water molecule departs with the electrons of the bond. This happens, in part, because the alcohol is protonated. The presence of a formal positive charge on the oxygen of the protonated alcohol weakens the carbon–oxygen bond by drawing the electrons in the direction of the positive oxygen. Step 3 is a Lewis acid–base reaction, in which a chloride anion (a Lewis base) reacts with the carbocation (a Lewis acid) to form the product.

A question might arise: Why doesn't a molecule of water (also a Lewis base) instead of a chloride ion react with the carbocation? After all, there are many water molecules around, since water is the solvent. The answer is that this step does occur sometimes, but it is simply the reverse of step 2. That is to say, not all of the carbocations that form go on directly to become product. Some react with water to become protonated alcohols again. However, these will dissociate again to become carbocations (even if, before they do, they lose a proton to become the alcohol again). Eventually, however, most of them are converted to the product because, under the conditions of the reaction, the equilibrium of the last step lies far to the right (and draws the reaction to completion).

The Chemistry of...

Carbonic Anhydrase

Carbonic anhydrase is an enzyme (a biological catalyst) that is involved in regulating blood acidity. As mentioned in the chapter opening vignette, breathing rate is among the physiological functions influenced by blood acidity. The reaction catalyzed by carbonic anhydrase is the equilibrium conversion of water and carbon dioxide to carbonic acid (H_2CO_3):

$$H_2O + CO_2 \xrightleftharpoons{\text{carbonic anhydrase}} H_2CO_3 \rightleftharpoons HCO_3^- + H^+$$

Carbonic anhydrase is a protein consisting of a chain of 260 subunits (amino acids) that naturally folds into a globule of a specific shape. Included in its structure is a cleft or pocket, called the active site, where the reactants are converted to products. The protein chain of carbonic anhydrase is shown here as a blue ribbon.

At the active site of carbonic anhydrase a water molecule loses a proton to form a hydroxide (OH^-) ion. This proton is removed by a part of carbonic anhydrase that acts as a base. Ordinarily the proton of a water molecule is not very acidic. However, the Lewis acid–base interaction between a zinc cation at the active site of carbonic anhydrase and the oxygen atom of a water molecule leads to positive charge on the water oxygen atom. This makes the protons of the water molecule more acidic. Removal of one of the protons of the water molecule forms hydroxide, which reacts with a carbon dioxide molecule at the active site to form HCO_3^- (hydrogen carbonate, or bicarbonate). In the structure of carbonic anhydrase shown here (based on X-ray crystallographic data), a bicarbonate ion at the active site is shown in red, the zinc cation at the active site is green, a water molecule is shown in blue, and the basic sites that coordinate with the zinc cation (as Lewis bases) or remove the proton from water to form hydroxide (as Brønsted–Lowry bases) are magenta (these bases are nitrogen atoms from histidine imidazole rings). No hydrogen atoms are shown in any of these species. As you can see, a remarkable orchestration of Lewis and Brønsted–Lowry acid–base reactions is involved in catalysis by carbonic anhydrase.

Chem3D Model, Animated Graphics

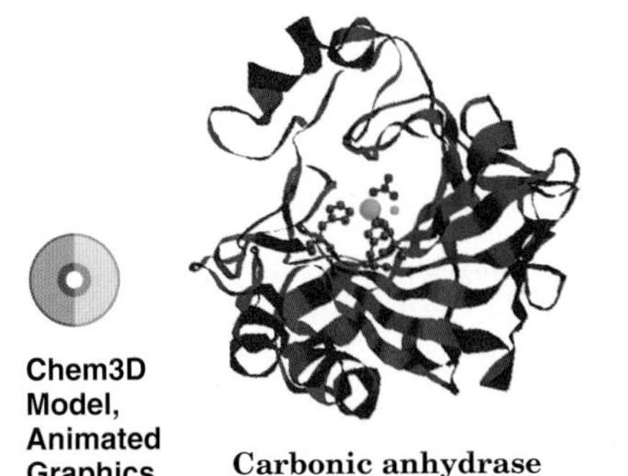

Carbonic anhydrase

3.14 Acids and Bases in Nonaqueous Solutions

If you were to add sodium amide ($NaNH_2$) to water in an attempt to carry out a reaction using the very powerful base, the amide ion (NH_2^-), the following reaction would take place immediately:

$$\underset{\substack{\textbf{Stronger acid} \\ \mathbf{p}K_a = \mathbf{15.7}}}{H{-}\ddot{\underset{..}{O}}{-}H} + \underset{\textbf{Stronger base}}{:\ddot{N}H_2^-} \longrightarrow \underset{\textbf{Weaker base}}{H{-}\ddot{\underset{..}{O}}:^-} + \underset{\substack{\textbf{Weaker acid} \\ \mathbf{p}K_a = \mathbf{38}}}{\ddot{N}H_3}$$

The amide ion would react with water to produce a solution containing hydroxide ions (a much weaker base) and ammonia. This example illustrates what is called the **leveling effect** of the solvent. The solvent here, *water, donates a proton to any base stronger than a* *hydroxide ion.* Therefore, *it is not possible to use a base stronger than hydroxide ion in aqueous solution.*

We can use bases stronger than hydroxide ion, however, by using solvents that are weaker acids than water. We can use amide ion (e.g., from $NaNH_2$) in a solvent such as hexane, diethyl ether, or liquid NH_3 (the liquified gas, bp $-33°C$, not the aqueous solution that you may have used in your general chemistry laboratory). All of these solvents are very weak acids, and therefore they will not donate a proton even to the strong base NH_2^-.

We can, for example, convert ethyne to its conjugate base, a carbanion, by treating it with sodium amide in liquid ammonia:

We shall use this reaction as part of our introduction to organic synthesis in Chapter 4.

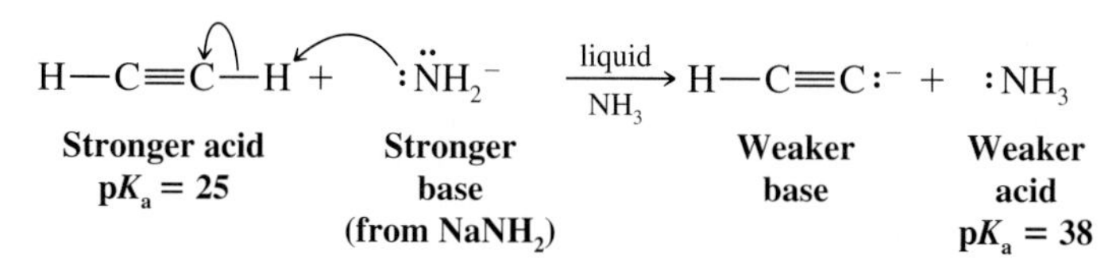

Most alkynes with a proton attached to a triply bonded carbon (called **terminal alkynes**) have pK_a values of about 25; therefore, all react with sodium amide in liquid ammonia in the same way that ethyne does. The general reaction is

$$R{-}C{\equiv}C{-}H + :\ddot{N}H_2^- \xrightarrow[NH_3]{\text{liquid}} R{-}C{\equiv}C:^- + :NH_3$$

Stronger acid $pK_a \cong 25$	Stronger base	Weaker base	Weaker acid $pK_a = 38$

Alcohols are often used as solvents for organic reactions because, being somewhat less polar than water, they dissolve less polar organic compounds. Using alcohols as solvents also offers the advantage of using RO^- ions (called **alkoxide ions**) as bases. Alkoxide ions are somewhat stronger bases than hydroxide ions because alcohols are weaker acids than water. For example, we can create a solution of sodium ethoxide (CH_3CH_2ONa) in ethyl alcohol by adding sodium hydride (NaH) to ethyl alcohol. We use a large excess of ethyl alcohol because we want it to be the solvent. Being a very strong base, the hydride ion reacts readily with ethyl alcohol:

$$CH_3CH_2\ddot{\underset{..}{O}}{-}H + :H^- \xrightarrow{\text{ethyl alcohol}} CH_3CH_2\ddot{\underset{..}{O}}:^- + H_2$$

Stronger acid $pK_a = 16$	Stronger base (from NaH)	Weaker base	Weaker acid $pK_a = 35$

The *tert*-butoxide ion, $(CH_3)_3CO^-$, in *tert*-butyl alcohol, $(CH_3)_3COH$, is a stronger base than the ethoxide ion in ethyl alcohol, and it can be prepared in a similar way:

$$(CH_3)_3C\ddot{\underset{..}{O}}{-}H + :H^- \xrightarrow[\text{alcohol}]{\textit{tert}\text{-butyl}} (CH_3)_3C\ddot{\underset{..}{O}}:^- + H_2$$

Stronger acid $pK_a = 18$	Stronger base (from NaH)	Weaker base	Weaker acid $pK_a = 35$

Although the carbon–lithium bond of an alkyllithium (RLi) has covalent character, it is polarized so as to make the carbon negative:

$$\overset{\delta-}{R}{\leftarrow\!\!\!+}\overset{\delta+}{Li}$$

Alkyllithium reagents react as though they contain alkanide ($R:^-$) ions and, being the conjugate bases of alkanes, alkanide ions are the strongest bases that we shall encounter. Ethyllithium (CH_3CH_2Li), for example, acts as though it contains an ethanide ($CH_3CH_2:^-$) carbanion. It reacts with ethyne in the following way:

$$H{-}C{\equiv}C{-}H + {}^{-}{:}CH_2CH_3 \xrightarrow{\text{hexane}} H{-}C{\equiv}C{:}^{-} + CH_3CH_3$$

Stronger acid $\mathbf{pK_a = 25}$ | **Stronger base (from CH_3CH_2Li)** | **Weaker base** | **Weaker acid** $\mathbf{pK_a = 50}$

Alkyllithiums can be easily prepared by allowing an alkyl bromide to react with lithium metal in an ether solvent (such as diethyl ether). See Section 12.6.

PROBLEM 3.13

Write equations for the acid–base reaction that would occur when each of the following compounds or solutions are mixed. In each case label the stronger acid and stronger base, and the weaker acid and weaker base, by using the appropriate pK_a values (Table 3.1). (If no appreciable acid–base reaction would occur, you should indicate this.)

(a) NaH is added to CH_3OH.
(b) $NaNH_2$ is added to CH_3CH_2OH.
(c) Gaseous NH_3 is added to ethyllithium in hexane.
(d) NH_4Cl is added to sodium amide in liquid ammonia.
(e) $(CH_3)_3CONa$ is added to H_2O.
(f) $NaOH$ is added to $(CH_3)_3COH$

3.15 Acid–Base Reactions and the Synthesis of Deuterium- and Tritium-Labeled Compounds

Chemists often use compounds in which deuterium or tritium atoms have replaced one or more hydrogen atoms of the compound as a method of "labeling" or identifying particular hydrogen atoms. Deuterium (^2H) and tritium (^3H) are isotopes of hydrogen with masses of 2 and 3 atomic mass units (amu), respectively.

For most chemical purposes, deuterium and tritium atoms in a molecule behave in much the same way that ordinary hydrogen atoms behave. The extra mass and additional neutrons associated with a deuterium or tritium atom often make its position in a molecule easy to locate by certain spectroscopic methods that we shall study later. Tritium is also radioactive, which makes it very easy to locate. (The extra mass associated with these labeled atoms may also cause compounds containing deuterium or tritium atoms to react more slowly than compounds with ordinary hydrogen atoms. This effect, called an "isotope effect," has been useful in studying the mechanisms of many reactions.)

One way to introduce a deuterium or tritium atom into a specific location in a molecule is through the acid–base reaction that takes place when a very strong base is treated with D_2O or T_2O (water that has deuterium or tritium in place of its hydrogens). For example, treating a solution containing $(CH_3)_2CHLi$ (isopropyllithium) with D_2O results in the formation of propane labeled with deuterium at the central atom:

$$CH_3\overset{\overset{CH_3}{|}}{C}H{:}^{-}Li^{+} + D_2O \xrightarrow{\text{hexane}} CH_3\overset{\overset{CH_3}{|}}{C}H{-}D + OD^{-}$$

Isopropyllithium *(stronger base)* | *(stronger acid)* | **2-Deuteriopropane** *(weaker acid)* | *(weaker base)*

SOLVED PROBLEM

Assuming you have available propyne, a solution of sodium amide in liquid ammonia, and T_2O, show how you would prepare the tritium-labeled compound $CH_3C{\equiv}CT$.

ANSWER: First add propyne to sodium amide in liquid ammonia. The following acid–base reaction will take place:

$$\underset{\text{Stronger acid}}{CH_3C{\equiv}CH} + \underset{\text{Stronger base}}{NH_2^-} \xrightarrow{\text{liq. ammonia}} \underset{\text{Weaker base}}{CH_3C{\equiv}C{:}^-} + \underset{\text{Weaker acid}}{NH_3}$$

Then adding T_2O (a much stronger acid than NH_3) to the solution will produce $CH_3C{\equiv}CT$:

$$\underset{\text{Stronger base}}{CH_3C{\equiv}C{:}^-} + \underset{\text{Stronger acid}}{T_2O} \xrightarrow{\text{liq. ammonia}} \underset{\text{Weaker acid}}{CH_3C{\equiv}CT} + \underset{\text{Weaker base}}{OT^-}$$

PROBLEM 3.14

Complete the following acid–base reactions:

(a) $HC{\equiv}CH + NaH \xrightarrow{\text{hexane}}$

(b) The solution obtained in (a) + $D_2O \longrightarrow$

(c) $CH_3CH_2Li + D_2O \xrightarrow{\text{hexane}}$

(d) $CH_3CH_2OH + NaH \xrightarrow{\text{hexane}}$

(e) The solution obtained in (d) + $T_2O \longrightarrow$

(f) $CH_3CH_2CH_2Li + D_2O \xrightarrow{\text{hexane}}$

KEY TERMS AND CONCEPTS

Acid–base conjugate pairs	Sections 3.2A, 3.5C
Acid–base strength, K_a, and pK_a	Section 3.5
Brønsted–Lowry theory	Section 3.2A
Carbocations and carbanions	Section 3.3
Curved-arrow (⌒) notation	Section 3.4
Electrostatic potential maps	Sections 1.8, 2.14B, 3.2C
Enthalpy	Section 3.8A
Inductive effects	Sections 3.7B, 3.10
Leveling effect	Section 3.14
Lewis acid–base theory	Section 3.2B
Mechanisms for reactions	Sections 3.1, 3.13
Nucleophiles and electrophiles	Section 3.3
Predicting the outcome of acid–base reactions	Section 3.6
Protic solvents	Section 3.11
Reaction equilibria and the standard free-energy change ($\Delta G°$)	Section 3.9
Resonance effect	Section 3.10

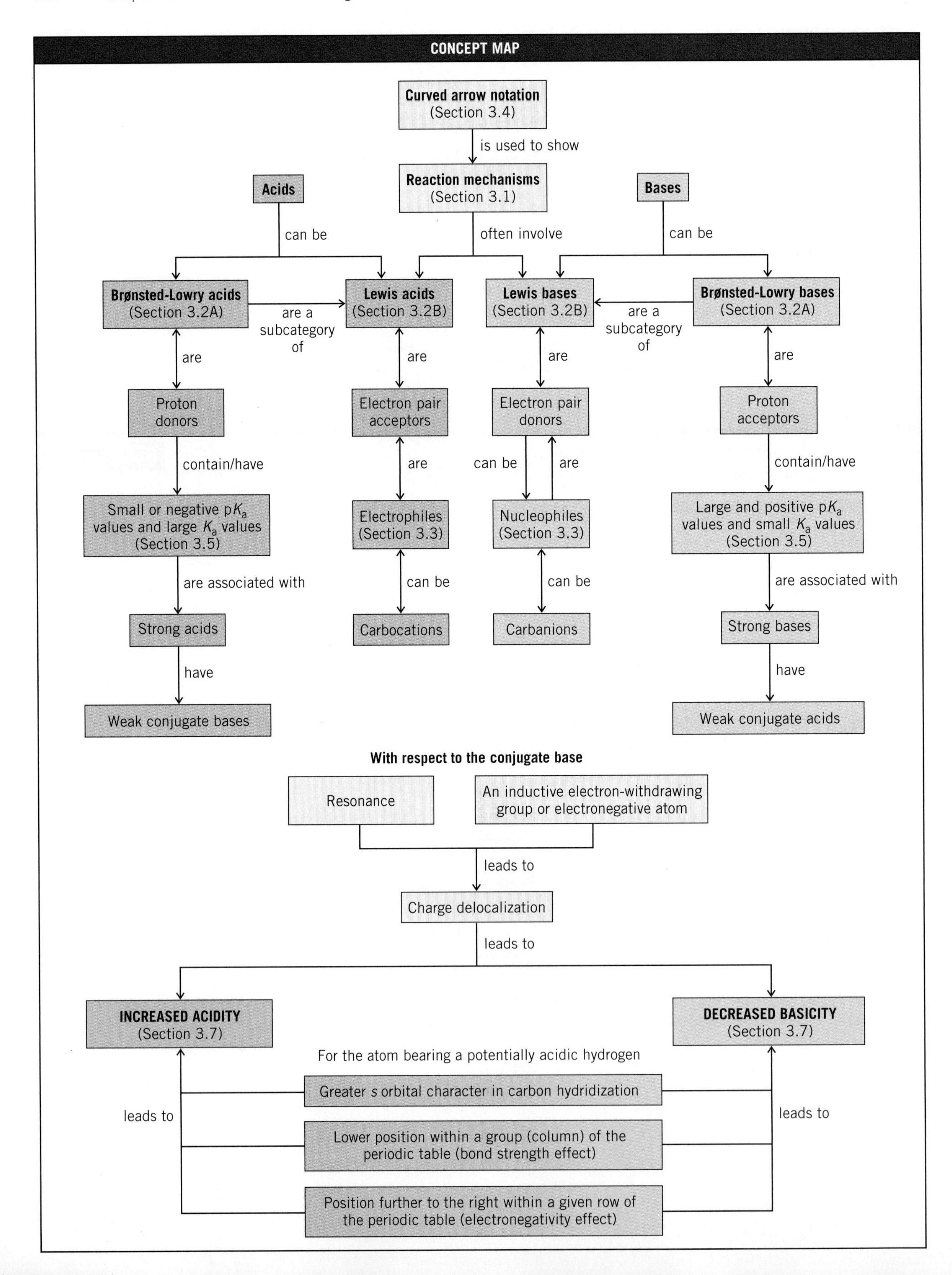
CONCEPT MAP
Curved arrow notation (Section 3.4)
is used to show
Reaction mechanisms (Section 3.1)
often involve
Acids
can be
Bases
can be
Brønsted-Lowry acids (Section 3.2A)
are a subcategory of
Lewis acids (Section 3.2B)
Lewis bases (Section 3.2B)
are a subcategory of
Brønsted-Lowry bases (Section 3.2A)
are
Proton donors
contain/have
Small or negative pK_a values and large K_a values (Section 3.5)
are associated with
Strong acids
have
Weak conjugate bases
are
Electron pair acceptors
are
Electrophiles (Section 3.3)
can be
Carbocations
are
Electron pair donors
can be
are
Nucleophiles (Section 3.3)
can be
Carbanions
are
Proton acceptors
contain/have
Large and positive pK_a values and small K_a values (Section 3.5)
are associated with
Strong bases
have
Weak conjugate acids
With respect to the conjugate base
Resonance
An inductive electron-withdrawing group or electronegative atom
leads to
Charge delocalization
leads to
INCREASED ACIDITY (Section 3.7)
DECREASED BASICITY (Section 3.7)
For the atom bearing a potentially acidic hydrogen
Greater *s* orbital character in carbon hydridization
Lower position within a group (column) of the periodic table (bond strength effect)
Position further to the right within a given row of the periodic table (electronegativity effect)
leads to
leads to

ADDITIONAL PROBLEMS

3.15 What is the conjugate base of each of the following acids?
(a) NH_3 **(c)** H_2 **(e)** CH_3OH
(b) H_2O **(d)** $HC\equiv CH$ **(f)** H_3O^+

3.16 List the bases you gave as answers to Problem 3.15 in order of decreasing basicity.

3.17 What is the conjugate acid of each of the following bases?
(a) HSO_4^- **(c)** CH_3NH_2 **(e)** $CH_3CH_2^-$
(b) H_2O **(d)** NH_2^- **(f)** $CH_3CO_2^-$

3.18 List the acids you gave as answers to Problem 3.17 in order of decreasing acidity.

3.19 Designate the Lewis acid and Lewis base in each of the following reactions:

(a) 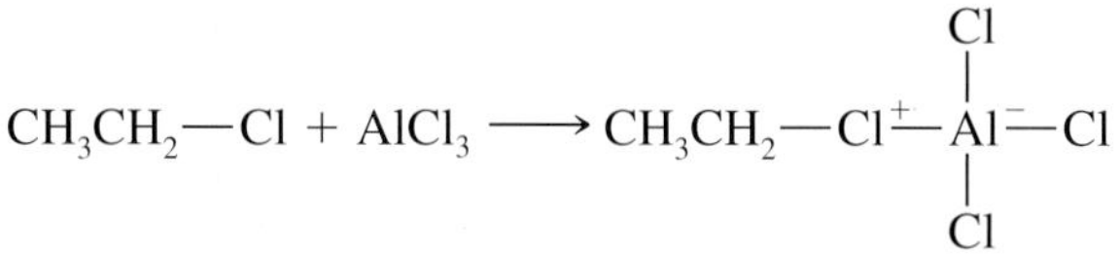 $CH_3CH_2{-}Cl + AlCl_3 \longrightarrow CH_3CH_2{-}\overset{+}{Cl}{-}\overset{-}{Al}(Cl)_2{-}Cl$

(b) 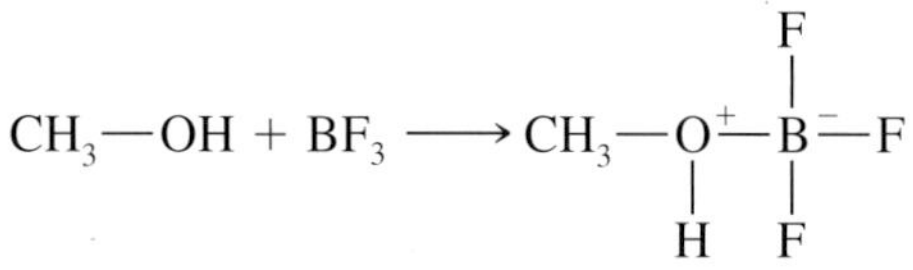 $CH_3{-}OH + BF_3 \longrightarrow CH_3{-}\overset{+}{O}(H){-}\overset{-}{B}(F)_2{-}F$

(c) 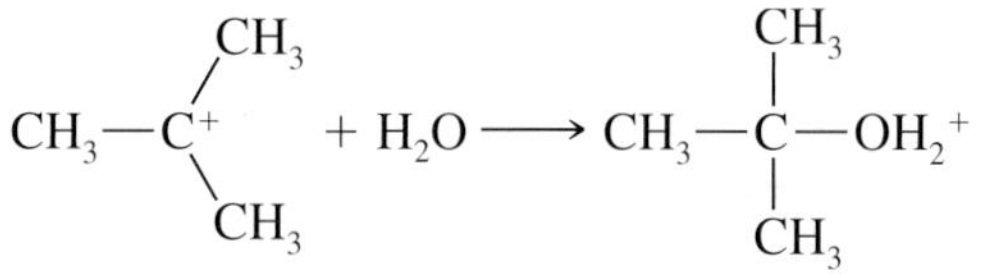 $CH_3{-}\overset{+}{C}(CH_3)_2 + H_2O \longrightarrow CH_3{-}C(CH_3)_2{-}OH_2^+$

3.20 Rewrite each of the following reactions using curved arrows and show all nonbonding electron pairs:
(a) $CH_3OH + HI \longrightarrow CH_3OH_2^+ + I^-$
(b) $CH_3NH_2 + HCl \longrightarrow CH_3NH_3^+ + Cl^-$
(c) 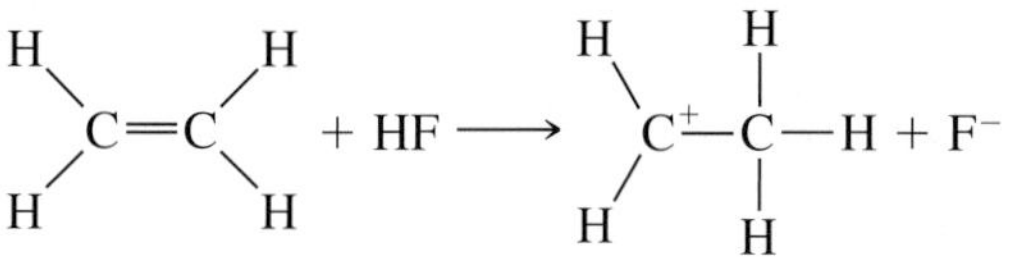 $H_2C{=}CH_2 + HF \longrightarrow H_2\overset{+}{C}{-}CH_3 + F^-$

3.21 When methyl alcohol is treated with NaH, the product is $CH_3O^-Na^+$ (and H_2) and not $Na^+\ {}^-CH_2OH$ (and H_2). Explain why this is so.

3.22 What reaction will take place if ethyl alcohol is added to a solution of $HC\equiv C{:}^-Na^+$ in liquid ammonia?

3.23 **(a)** The K_a of formic acid (HCO_2H) is 1.77×10^{-4}. What is the pK_a?
(b) What is the K_a of an acid whose $pK_a = 13$?

3.24 Acid HA has a $pK_a = 20$; acid HB has a $pK_a = 10$.
(a) Which is the stronger acid?
(b) Will an acid–base reaction with an equilibrium lying to the right take place if Na^+A^- is added to HB? Explain your answer.

*_Note:_ Problems marked with an asterisk are "challenge problems."

3.25 Write an equation, using the curved-arrow notation, for the acid–base reaction that will take place when each of the following are mixed. If no appreciable acid–base reaction takes place, because the equilibrium is unfavorable, you should so indicate.
(a) Aqueous NaOH and $CH_3CH_2CO_2H$
(b) Aqueous NaOH and $C_6H_5SO_3H$
(c) CH_3CH_2ONa in ethyl alcohol and ethyne
(d) CH_3CH_2Li in hexane and ethyne
(e) CH_3CH_2Li in hexane and ethyl alcohol

3.26 Starting with appropriate unlabeled organic compounds, show syntheses of each of the following:
(a) $C_6H_5{-}C{\equiv}C{-}T$ **(b)** $CH_3{-}CH(CH_3){-}O{-}D$ **(c)** $CH_3CH_2CH_2OD$

3.27 **(a)** Arrange the following compounds in order of decreasing acidity and explain your answer: $CH_3CH_2NH_2$, CH_3CH_2OH, and $CH_3CH_2CH_3$. **(b)** Arrange the conjugate bases of the acids given in part (a) in order of increasing basicity and explain this answer.

3.28 Arrange the following compounds in order of decreasing acidity:
(a) $CH_3CH{=}CH_2$, $CH_3CH_2CH_3$, $CH_3C{\equiv}CH$
(b) $CH_3CH_2CH_2OH$, $CH_3CH_2CO_2H$, $CH_3CHClCO_2H$
(c) CH_3CH_2OH, $CH_3CH_2OH_2{}^+$, CH_3OCH_3

3.29 Arrange the following in order of increasing basicity:
(a) CH_3NH_2, $CH_3NH_3{}^+$, CH_3NH^-
(b) CH_3O^-, CH_3NH^-, $CH_3CH_2{}^-$
(c) $CH_3CH{=}CH^-$, $CH_3CH_2CH_2{}^-$, $CH_3C{\equiv}C^-$

3.30 Whereas H_3PO_4 is a triprotic acid, H_3PO_3 is a diprotic acid. Draw structures for these two acids that account for this difference in behavior.

3.31 Supply the curved arrows necessary for the following reactions:

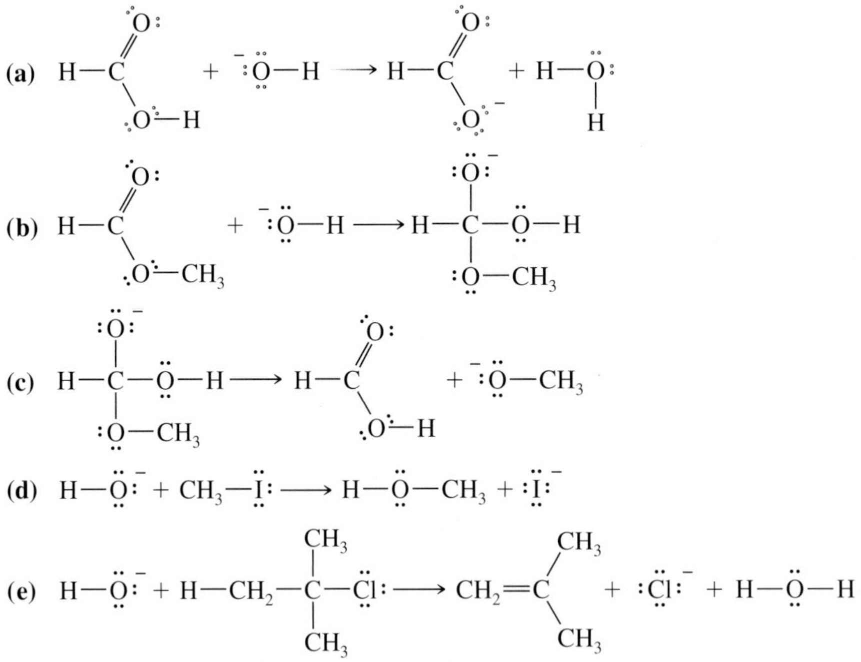

(a) $H{-}C({=}\ddot{O}{:}){-}\ddot{O}{-}H + {}^-{:}\ddot{O}{-}H \longrightarrow H{-}C({=}\ddot{O}{:}){-}\ddot{O}{:}^- + H{-}\ddot{O}{:}{-}H$

(b) $H{-}C({=}\ddot{O}{:}){-}\ddot{O}{-}CH_3 + {}^-{:}\ddot{O}{-}H \longrightarrow H{-}C({:}\ddot{O}{:}^-)({:}\ddot{O}{-}CH_3){-}\ddot{O}{-}H$

(c) $H{-}C({:}\ddot{O}{:}^-)({:}\ddot{O}{-}CH_3){-}\ddot{O}{-}H \longrightarrow H{-}C({=}\ddot{O}{:}){-}\ddot{O}{-}H + {}^-{:}\ddot{O}{-}CH_3$

(d) $H{-}\ddot{O}{:}^- + CH_3{-}\ddot{I}{:} \longrightarrow H{-}\ddot{O}{-}CH_3 + {:}\ddot{I}{:}^-$

(e) $H{-}\ddot{O}{:}^- + H{-}CH_2{-}C(CH_3)_2{-}\ddot{Cl}{:} \longrightarrow CH_2{=}C(CH_3)_2 + {:}\ddot{Cl}{:}^- + H{-}\ddot{O}{-}H$

3.32 Glycine is an amino acid that can be obtained from most proteins. In solution, glycine exists in equilibrium between the two forms:

$$H_2NCH_2CO_2H \rightleftharpoons H_3\overset{+}{N}CH_2CO_2^-$$

(a) Consult Table 3.1 and state which form is favored at equilibrium.
(b) A handbook gives the melting point of glycine as 262°C (with decomposition). Which structure better represents glycine?

3.33 Malonic acid, $HO_2CCH_2CO_2H$, is a diprotic acid. The pK_a for the loss of the first proton is 2.83; the pK_a for the loss of the second proton is 5.69. **(a)** Explain why malonic acid is a stronger acid than acetic acid ($pK_a = 4.75$). **(b)** Explain why the anion, $^{-}O_2CCH_2CO_2H$, is so much less acidic than malonic acid itself.

3.34 The free-energy change, $\Delta G°$, for the ionization of acid HA is 21 kJ mol^{-1}; for acid HB it is -21 kJ mol^{-1}. Which is the stronger acid?

3.35 At 25°C the enthalpy change, $\Delta H°$, for the ionization of trichloroacetic acid is $+6.3$ kJ mol^{-1} and the entropy change, $\Delta S°$, is $+0.0084$ kJ mol^{-1} K^{-1}. What is the pK_a of trichloroacetic acid?

3.36 The compound below has (for obvious reasons) been given the trivial name **squaric acid.** Squaric acid is a diprotic acid, with both protons being more acidic than acetic acid. In the dianion obtained after the loss of both protons, all of the carbon–carbon bonds are the same length as well as all of the carbon–oxygen bonds. Provide a resonance explanation for these observations.

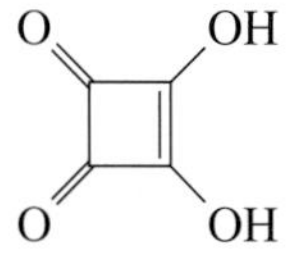

O OH

O OH

Squaric acid

***3.37** $CH_3CH_2SH + CH_3O^- \longrightarrow$ A (contains sulfur) + B
A + $CH_2—CH_2$ (bridged by O) $\longrightarrow$ C (which has the partial structure A—CH_2CH_2O)
C + $H_2O \longrightarrow$ D + E (which is inorganic)
(a) Given the above sequence of reactions, draw structures for A through E.
(b) Rewrite the reaction sequence, showing all nonbonding electron pairs and using curved arrows to show electron pair movements.

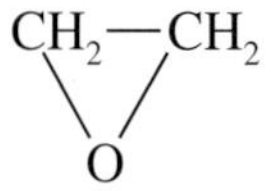

***3.38** First, complete and balance each of the equations below. Then, choosing among ethanol, hexane, and liquid ammonia, state which (there may be more than one) might be suitable solvents for each of these reactions. Disregard the practical limitations that come from consideration of "like dissolves like" and base your answers only on relative acidities.
(a) $CH_3(CH_2)_8OD + CH_3(CH_2)_8Li \rightarrow$
(b) $NaNH_2 + CH_3C{\equiv}CH \rightarrow$
(c) HCl + C_6H_5—$NH_2 \longrightarrow$
(The conjugate acid of this amine, aniline, has pK_a 4.6.)

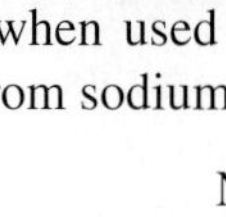

***3.39** Dimethylformamide (DMF), $HCON(CH_3)_2$, is an example of a polar aprotic solvent, aprotic meaning it has no hydrogen atoms attached to highly electronegative atoms.
(a) Draw its dash-type structural formula, showing unshared electron pairs.
(b) Draw what you predict to be its most important resonance forms [one is your answer to part (a)].
(c) DMF, when used as the reaction solvent, greatly enhances the reactivity of nucleophiles (e.g., CN^- from sodium cyanide) in reactions like this:

$$NaCN + CH_3CH_2Br \longrightarrow CH_3CH_2C{\equiv}N + NaBr$$

Suggest an explanation for this effect of DMF on the basis of Lewis acid–base considerations. (*Hint:* While water or an alcohol solvates both cations and anions, DMF is only effective in solvating cations.)

***3.40** As noted in Table 3.1, the pK_a of acetone, CH_3COCH_3, is 19.2.

(a) Draw the bond-line formula for acetone and of any other contributing resonance form.

(b) Predict and draw the structure of the conjugate base of acetone and of any other contributing resonance form.

(c) Write an equation for a reaction that could be used to synthesize CH_3COCH_2D.

3.41 The chapter opening vignette and "The Chemistry of. . ." box in this chapter discussed carbonic anhydrase, an enzyme that is important for regulation of blood pH and other physiological functions. Open the 3D Molecular Model titled "Carbonic Anhydrase Active Site" from the CD that accompanies this book and examine the active-site functional groups. The model shows a centrally located zinc cation, a carbonate anion, and three five-membered rings containing two nitrogen atoms each (called imidazole groups) as well as atoms that connect these rings to the backbone of the enzyme (although the backbone itself is not shown in this model). The carbonate anion and the three five-membered rings engage in Lewis acid–base interactions with the zinc ion.

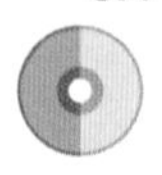

Chem3D Model

(a) Draw a Lewis structure for one of the imidazole rings. What are the hybridization states of all the atoms in the imidazole ring? In what type of orbital are the electrons that coordinate from the imidazole nitrogen to the zinc cation?

(b) The carbonate anion at the active site is formed by reaction of a hydroxide anion coordinated to the zinc with a carbon dioxide molecule that binds at the active site. Draw a reaction mechanism with curved arrows to show the attack of a hydroxide anion on a carbon dioxide molecule and the resulting formation of a hydrogen carbonate anion.

3.42 Formamide ($HCONH_2$) has a pK_a of approximately 25. Predict, based on the map of electrostatic potential for formamide shown here, which hydrogen atom(s) have this pK_a value. Support your conclusion with arguments having to do with the electronic structure of formamide.

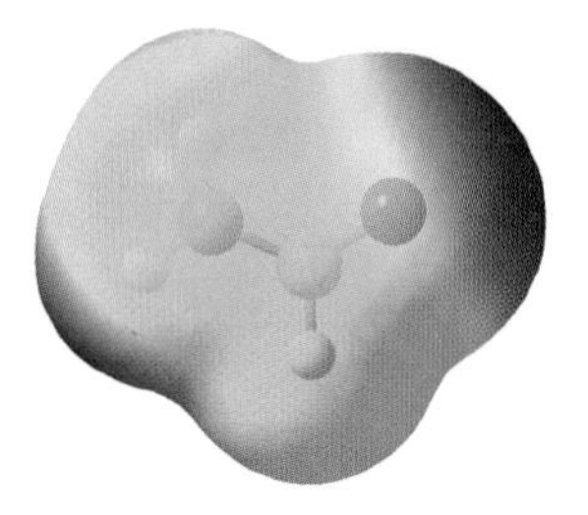

LEARNING GROUP PROBLEM

Suppose you carried out the following synthesis of 3-methylbutyl ethanoate (isoamyl acetate):

$$CH_3COOH \text{ (Ethanoic acid, excess)} + HOCH_2CH_2CH(CH_3)_2 \text{ (3-Methyl-1-butanol)} \xrightleftharpoons{H_2SO_4 \text{ (trace)}} CH_3COOCH_2CH_2CH(CH_3)_2 \text{ (3-Methylbutyl ethanoate)} + H_2O$$

As the chemical equation shows, 3-methyl-1-butanol (also called isoamyl alcohol or isopentyl alcohol) was mixed with an excess of acetic acid (ethanoic acid by its systematic name) and a trace of sulfuric acid (which serves as a catalyst). This reaction is an equilibrium reaction, so it is expected that not all of the starting materials will be consumed. The equilibrium should lie quite far to the right due to the excess of acetic acid used, but not completely.

After an appropriate length of time, isolation of the desired product from the reaction mixture was begun by adding a volume of 5% aqueous sodium bicarbonate ($NaHCO_3$ has an effective pK_a of 7.)

roughly equal to the volume of the reaction mixture. Bubbling occurred and a mixture consisting of two layers resulted—a basic aqueous layer and an organic layer. The layers were separated and the aqueous layer was removed. The addition of aqueous sodium bicarbonate to the layer of organic materials and separation of the layers were repeated twice. Each time the predominantly aqueous layers were removed, they were combined in the same collection flask. The organic layer that remained after the three bicarbonate extractions was dried and then subjected to distillation in order to obtain a pure sample of 3-methylbutyl ethanoate (isoamyl acetate).

1. List all the chemical species likely to be present at the end of the reaction but before adding aqueous $NaHCO_3$. Note that the H_2SO_4 was not consumed (since it is a catalyst), and is thus still available to donate a proton to atoms that can be protonated.
2. Use a table of pK_a values, such as Table 3.1, to estimate pK_a values for any potentially acidic hydrogens in each of the species you listed in part 1 (or for the conjugate acid).
3. Write chemical equations for all the acid–base reactions you would predict to occur (based on the pK_a values you used) when the species you listed above encounter the aqueous sodium bicarbonate solution. (*Hint:* Consider whether or not each species might be an acid that could react with $NaHCO_3$.)
4. **(a)** Explain, on the basis of polarities and solubility, why separate layers formed when aqueous sodium bicarbonate was added to the reaction mixture. [*Hint:* Most sodium salts of organic acids are soluble in water, as are neutral oxygen-containing organic compounds of four carbons or less.]
   (b) List the chemical species likely to be present after the reaction with $NaHCO_3$ in (i) the organic layer and (ii) the aqueous layer.
   (c) Why was the aqueous sodium bicarbonate extraction step repeated three times?

CHAPTER

4

Alkanes: Nomenclature, Conformational Analysis, and an Introduction to Synthesis

To Be Flexible or Inflexible—Molecular Structure Makes the Difference

When your muscles contract, it is largely because many carbon–carbon sigma (single) bonds are undergoing rotation (conformational changes) in a muscle protein called myosin (see the electron micrograph above). But when you etch glass with a diamond, the carbon–carbon single bonds that make up diamond are resisting all the forces brought to bear on them such that the glass is scratched and not the diamond (a partial molecular structure for diamond is shown above). Nanotubes, a new class of carbon-based materials with strength roughly one hundred times that of steel, also have an exceptional toughness. (Nanotubes are related to buckminsterfullerenes, or "buckyballs," for which the Nobel Prize in Chemistry was awarded in 1996; see Section 14.8). The properties of these materials—muscle protein, diamond, and nanotubes—depend on many things, but central to them is whether or not rotation is possible around carbon–carbon bonds.

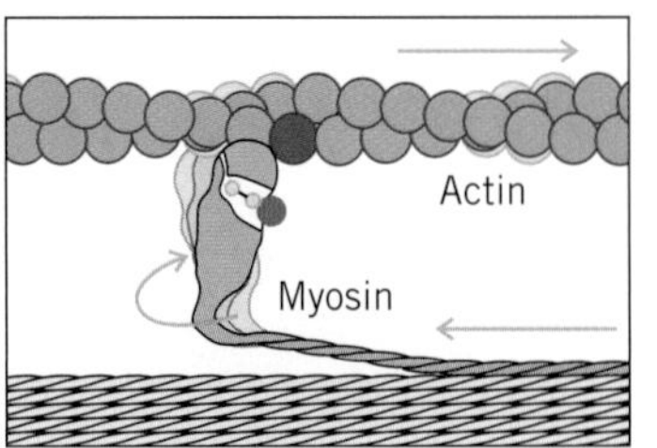

Power stroke in muscle

Muscle proteins are essentially very long linear molecules (folded into a compact shape) whose atoms are connected by single bonds in a chain-like fashion. As we shall learn in this chapter, relatively free rotation is possible about atoms joined by single bonds. In muscles, the cumulative effect of rotations about many single bonds is to move the tail of each myosin molecule 60 Å along the adjacent protein (called actin) in a step called the "power stroke." This process occurs over and over again as part of a ratcheting mechanism between many myosin and actin molecules for each muscle movement.

In diamond and nanotubes, the molecules contain a network of interconnecting carbon rings rather than long chains. Their molecular structure is such that very little rotation is possible about carbon–carbon bonds without breaking carbon–carbon bonds, something that is difficult. It is this lack of bond rotation that gives these molecules their hardness and strength.

4.1 Introduction to Alkanes and Cycloalkanes

We noted earlier that the family of organic compounds called hydrocarbons can be divided into several groups on the basis of the type of bond that exists between the individual carbon atoms. Those hydrocarbons in which all of the carbon–carbon bonds are single bonds are called **alkanes,** those hydrocarbons that contain a carbon–carbon double bond are called **alkenes,** and those with a carbon–carbon triple bond are called **alkynes.**

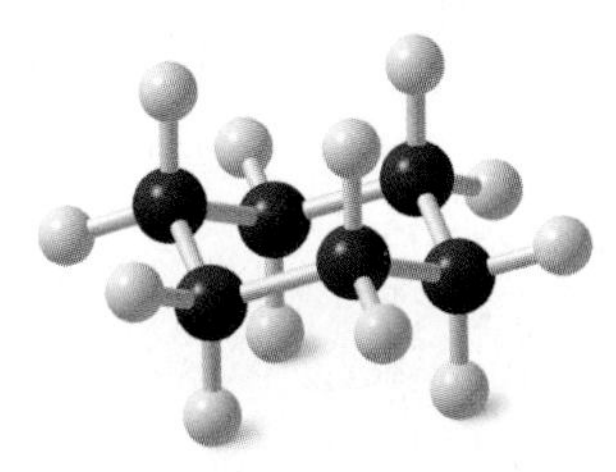
Cyclohexane

Cycloalkanes are alkanes in which all or some of the carbon atoms are arranged in a ring. Alkanes have the general formula C_nH_{2n+2}; cycloalkanes containing a single ring have two fewer hydrogen atoms and thus have the general formula C_nH_{2n}.

Alkanes and cycloalkanes are so similar that many of their properties can be considered side by side. Some differences remain, however, and certain structural features arise from the rings of cycloalkanes that are more conveniently studied separately. We shall point out the chemical and physical similarities of alkanes and cycloalkanes as we go along.

4.1A Sources of Alkanes: Petroleum

The primary source of alkanes is petroleum. Petroleum is a complex mixture of organic compounds, most of which are alkanes and aromatic hydrocarbons (cf. Chapter 14). It also contains small amounts of oxygen-, nitrogen-, and sulfur-containing compounds.

Some of the molecules in petroleum are clearly of biological origin. The natural origin of petroleum is still under debate, however. Many scientists believe petroleum originated with decay of primordial biological matter. Recent theories suggest, however, that organic molecules may have been included as Earth formed by accretion of interstellar materials. Analysis of asteroids and comets has shown that they contain a significant amount and variety of organic compounds. Methane and other hydrocarbons are found in the atmospheres of Jupiter, Saturn, and Uranus. Saturn's moon Titan has a solid form of methane–water ice at its surface and an atmosphere rich in methane. Earth's petroleum may therefore have originated similarly to the way methane became part of these other bodies in our solar system. The discovery of microbial life in high-temperature ocean vents and the growing evidence for a deep, hot biosphere within Earth suggest that compounds in petroleum of biological origin may simply be contaminants introduced by primitive life into a nonbiologically formed petroleum reserve that was present from Earth's beginning.

Petroleum is a finite resource whose origin is under debate (see Section 4.1A). The La Brea Tar Pits in Los Angeles is a place where many prehistoric animals perished in a natural vat containing hydrocarbons.

4.1B Petroleum Refining

The first step in refining petroleum is distillation; the object here is to separate the petroleum into fractions based on the volatility of its components. Complete separation into

A petroleum refinery. The tall towers are fractioning columns used to separate components of crude oil according to their boiling point.

fractions containing individual compounds is economically impractical and virtually impossible technically. More than 500 different compounds are contained in the petroleum distillates boiling below 200°C, and many have almost the same boiling points. Thus the fractions taken contain mixtures of alkanes of similar boiling points (cf. Table 4.1). Mixtures of alkanes, fortunately, are perfectly suitable for uses as fuels, solvents, and lubricants, the primary uses of petroleum.

4.1C Cracking

The demand for gasoline is much greater than that supplied by the gasoline fraction of petroleum. Important processes in the petroleum industry, therefore, are concerned with converting hydrocarbons from other fractions into gasoline. When a mixture of alkanes from the gas oil fraction (C_{12} and higher) is heated at very high temperatures (~ 500°C) in the presence of a variety of catalysts, the molecules break apart and rearrange to smaller, more highly branched alkanes containing 5–10 carbon atoms (see Table 4.1). This process is called **catalytic cracking.** Cracking can also be done in the absence of a

TABLE 4.1 Typical Fractions Obtained by Distillation of Petroleum

Boiling Range of Fraction (°C)	Number of Carbon Atoms per Molecules	Use
Below 20	C_1–C_4	Natural gas, bottled gas, petrochemicals
20–60	C_5–C_6	Petroleum ether, solvents
60–100	C_6–C_7	Ligroin, solvents
40–200	C_5–C_{10}	Gasoline (straight-run gasoline)
175–325	C_{12}–C_{18}	Kerosene and jet fuel
250–400	C_{12} and higher	Gas oil, fuel oil, and diesel oil
Nonvolatile liquids	C_{20} and higher	Refined mineral oil, lubricating oil, and grease
Nonvolatile solids	C_{20} and higher	Paraffin wax, asphalt, and tar

Adapted with permission from Holum, J. R. *Elements of General, Organic, and Biological Chemistry,* 9th ed; Wiley: New York, 1995; p 213.

catalyst—called **thermal cracking**—but in this process the products tend to have unbranched chains, and alkanes with unbranched chains have a very low "octane rating."

The highly branched compound 2,2,4-trimethylpentane (called "isooctane" in the petroleum industry) burns very smoothly (without knocking) in internal combustion engines and is used as one of the standards by which the octane rating of gasolines is established.

$$CH_3-C(CH_3)_2-CH_2-CH(CH_3)-CH_3 \quad \text{or} \quad \text{(bond-line formula)}$$

2,2,4-Trimethylpentane
("isooctane")

According to this scale, 2,2,4-trimethylpentane has an octane rating of 100. Heptane, $CH_3(CH_2)_5CH_3$, a compound that produces much knocking when it is burned in an internal combustion engine, is given an octane rating of 0. Mixtures of 2,2,4-trimethylpentane and heptane are used as standards for octane ratings between 0 and 100. A gasoline, for example, that has the same characteristics in an engine as a mixture of 87% 2,2,4-trimethylpentane and 13% heptane would be rated as 87-octane gasoline.

4.2 Shapes of Alkanes

A general tetrahedral orientation of groups—and thus sp^3 hybridization—is the rule for the carbon atoms of all alkanes and cycloalkanes. We can represent the shapes of alkanes as shown in Fig. 4.1.

Butane and pentane are examples of alkanes that are sometimes called "straight-chain" alkanes. One glance at three-dimensional models however, shows that because of their tetrahedral carbon atoms the chains are zigzagged and not at all straight. Indeed, the structures that we have depicted in Fig. 4.1 are the straightest possible arrangements of the chains because rotations about the carbon–carbon single bonds produce arrangements

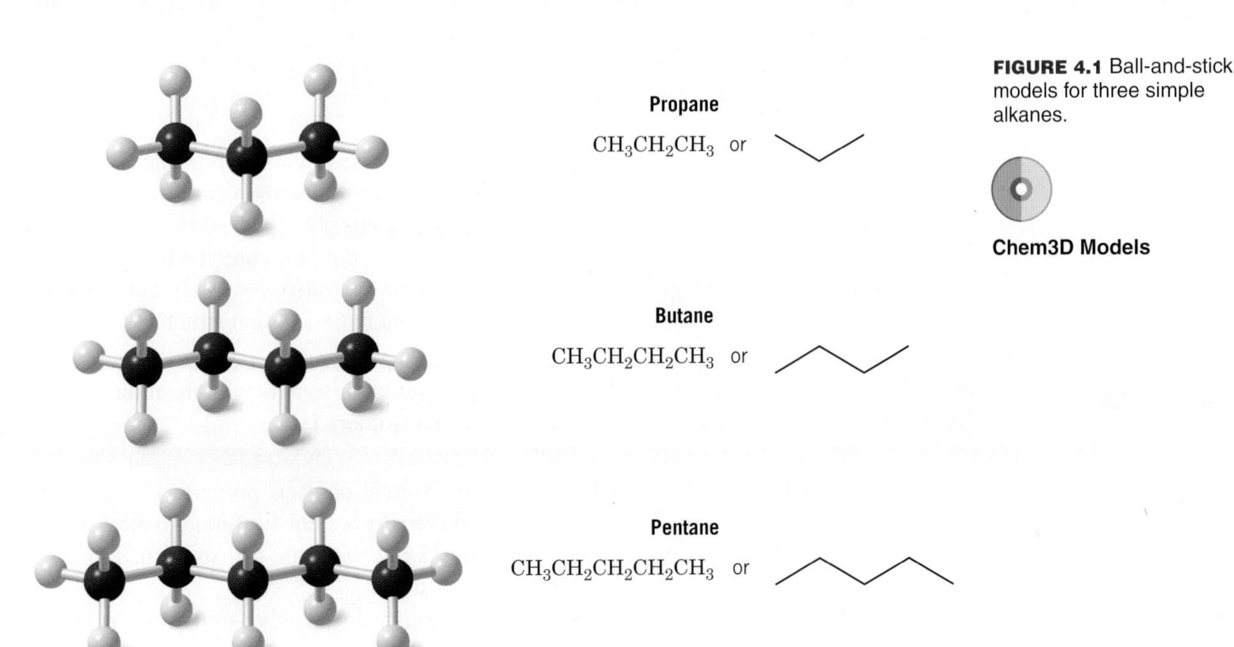

FIGURE 4.1 Ball-and-stick models for three simple alkanes.

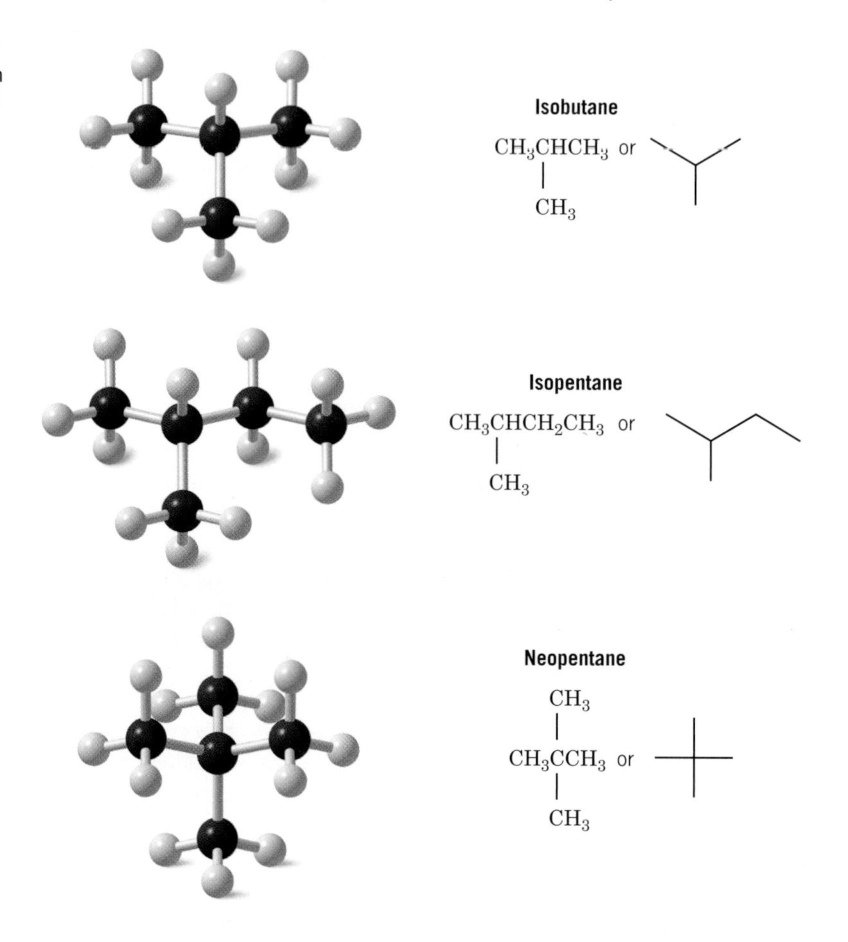

FIGURE 4.2 Ball-and-stick models for three branched-chain alkanes. In each of the compounds one carbon atom is attached to more than two other carbon atoms.

You should build your own molecular models of the compounds in Figs. 4.1 and 4.2. View them from different perspectives and experiment with how their shapes change when you twist various bonds. Make drawings of your structures.

that are even less straight. A better description is **unbranched.** This means that each carbon atom within the chain is bonded to no more than two other carbon atoms and that unbranched alkanes contain only primary and secondary carbon atoms. Primary, secondary, and tertiary carbon atoms were defined in Section 2.6. (Unbranched alkanes used to be called "normal" alkanes or *n*-alkanes, but this designation is archaic and should not be used now.)

Isobutane, isopentane, and neopentane (Fig. 4.2) are examples of branched-chain alkanes. In neopentane the central carbon atom is bonded to four carbon atoms.

Butane and isobutane have the same molecular formula: C_4H_{10}. The two compounds have their atoms connected in a different order and are, therefore, **constitutional isomers** (Section 1.3A). Pentane, isopentane, and neopentane are also constitutional isomers. They, too, have the same molecular formula (C_5H_{12}) but have different structures.

PROBLEM 4.1 Write condensed structural formulas for all of the constitutional isomers with the molecular formula C_7H_{16}. (There is a total of nine constitutional isomers.)

Constitutional isomers, as stated earlier, have different physical properties. The differences may not always be large, but constitutional isomers are always found to have different melting points, boiling points, densities, indexes of refraction, and so forth. Table 4.2 gives some of the physical properties of the C_6H_{14} isomers.

As Table 4.3 shows, the number of constitutional isomers that is possible increases dramatically as the number of carbon atoms in the alkane increases.

TABLE 4.2 Physical Constants of the Hexane Isomers

Molecular Formula	Structural Formula	mp (°C)	bp (°C)[a] (1 atm)	Density[b] (g mL^{-1})	Index of Refraction[c] (n_D 20°C)
C_6H_{14}	$CH_3CH_2CH_2CH_2CH_2CH_3$	−95	68.7	0.6594^{20}	1.3748
C_6H_{14}	$CH_3CH(CH_3)CH_2CH_2CH_3$	−153.7	60.3	0.6532^{20}	1.3714
C_6H_{14}	$CH_3CH_2CH(CH_3)CH_2CH_3$	−118	63.3	0.6643^{20}	1.3765
C_6H_{14}	$CH_3CH(CH_3)—CH(CH_3)CH_3$	−128.8	58	0.6616^{20}	1.3750
C_6H_{14}	$CH_3—C(CH_3)_2—CH_2CH_3$	−98	49.7	0.6492^{20}	1.3688

[a] Unless otherwise indicated, all boiling points given in this book are at 1 atm or 760 torr.
[b] The superscript indicates the temperature at which the density was measured.
[c] The index of refraction is a measure of the ability of the alkane to bend (refract) light rays. The values reported are for light of the D line of the sodium spectrum (n_D).

The large numbers in Table 4.3 are based on calculations that must be done with a computer. Similar calculations, which take into account stereoisomers (Chapter 5) as well as constitutional isomers, indicate that an alkane with the formula $C_{167}H_{336}$ would, in theory, have more possible isomers than there are particles in the observed universe!

TABLE 4.3 Number of Alkane Isomers

Molecular Formula	Possible Number of Constitutional Isomers
C_4H_{10}	2
C_5H_{12}	3
C_6H_{14}	5
C_7H_{16}	9
C_8H_{18}	18
C_9H_{20}	35
$C_{10}H_{22}$	75
$C_{15}H_{32}$	4,347
$C_{20}H_{42}$	366,319
$C_{30}H_{62}$	4,111,846,763
$C_{40}H_{82}$	62,481,801,147,341

4.3 IUPAC Nomenclature of Alkanes, Alkyl Halides, and Alcohols

The development of a formal system for naming organic compounds did not come about until near the end of the nineteenth century. Before that time many organic compounds had already been discovered. The names given these compounds sometimes reflected a source of the compound. Acetic acid, for example, can be obtained from vinegar; it takes its name from the Latin word for vinegar, *acetum.* Formic acid can be obtained from

CAS No. Ingredient
7732-18-5 .... Water
Unknown .... Acrylic Polymer
111-77-3 ..2-(2-Methoxyethoxy)-ethanol
13463-67-7 .... Titanium Dioxide
25265-77-4 ..Trimethylpentanediol Isobutyrate
108419-35-8 .... Oxo-Tridecyl Acetate

The Chemical Abstracts Service assigns a CAS Registry Number to every compound. CAS numbers make it easy to find information about a compound in the chemical literature. The CAS numbers for ingredients in a can of latex paint are shown here.

some ants; it is named from the Latin word for ants, *formicae*. Ethanol (or ethyl alcohol) was at one time called grain alcohol because it was obtained by the fermentation of grains.

These older names for organic compounds are now called "common" or "trivial" names. Many of these names are still widely used by chemists and biochemists and in commerce. (Many are even written into laws.) For this reason it is still necessary to learn the common names for some of the common compounds. We shall point out these common names as we go along, and we shall use them occasionally. Most of the time, however, the names that we shall use will be those called IUPAC names.

The formal system of nomenclature used today is one proposed by the International Union of Pure and Applied Chemistry (IUPAC). This system was first developed in 1892 and has been revised at irregular intervals to keep it up to date. The latest revision was in 1993. Underlying the IUPAC system of nomenclature for organic compounds is a fundamental principle: ***Each different compound should have an unambiguous name.*** Thus, through a systematic set of rules, the IUPAC system provides different names for the more than 16 million known organic compounds, and names can be devised for any one of millions of other compounds yet to be synthesized. In addition, the IUPAC system is simple enough to allow any chemist familiar with the rules (or with the rules at hand) to write the name for any compound that might be encountered. In the same way, one is also able to derive the structure of a given compound from its IUPAC name.*

IUPAC Nomenclature System

The **IUPAC system** for naming alkanes is not difficult to learn, and the principles involved are used in naming compounds in other families as well. For these reasons we begin our study of the IUPAC system with the rules for naming alkanes and then study the rules for alkyl halides and alcohols.

The names for several of the unbranched alkanes are listed in Table 4.4. The ending for all of the names of alkanes is *-ane*. The stems of the names of most of the alkanes

TABLE 4.4 The Unbranched Alkanes

Name	Number of Carbon Atoms	Structure	Name	Number of Carbon Atoms	Structure
Methane	1	CH_4	Heptadecane	17	$CH_3(CH_2)_{15}CH_3$
Ethane	2	CH_3CH_3	Octadecane	18	$CH_3(CH_2)_{16}CH_3$
Propane	3	$CH_3CH_2CH_3$	Nonadecane	19	$CH_3(CH_2)_{17}CH_3$
Butane	4	$CH_3(CH_2)_2CH_3$	Eicosane	20	$CH_3(CH_2)_{18}CH_3$
Pentane	5	$CH_3(CH_2)_3CH_3$	Heneicosane	21	$CH_3(CH_2)_{19}CH_3$
Hexane	6	$CH_3(CH_2)_4CH_3$	Docosane	22	$CH_3(CH_2)_{20}CH_3$
Heptane	7	$CH_3(CH_2)_5CH_3$	Tricosane	23	$CH_3(CH_2)_{21}CH_3$
Octane	8	$CH_3(CH_2)_6CH_3$	Triacontane	30	$CH_3(CH_2)_{28}CH_3$
Nonane	9	$CH_3(CH_2)_7CH_3$	Hentriacontane	31	$CH_3(CH_2)_{29}CH_3$
Decane	10	$CH_3(CH_2)_8CH_3$	Tetracontane	40	$CH_3(CH_2)_{38}CH_3$
Undecane	11	$CH_3(CH_2)_9CH_3$	Pentacontane	50	$CH_3(CH_2)_{48}CH_3$
Dodecane	12	$CH_3(CH_2)_{10}CH_3$	Hexacontane	60	$CH_3(CH_2)_{58}CH_3$
Tridecane	13	$CH_3(CH_2)_{11}CH_3$	Heptacontane	70	$CH_3(CH_2)_{68}CH_3$
Tetradecane	14	$CH_3(CH_2)_{12}CH_3$	Octacontane	80	$CH_3(CH_2)_{78}CH_3$
Pentadecane	15	$CH_3(CH_2)_{13}CH_3$	Nonacontane	90	$CH_3(CH_2)_{88}CH_3$
Hexadecane	16	$CH_3(CH_2)_{14}CH_3$	Hectane	100	$CH_3(CH_2)_{98}CH_3$

*The IUPAC rules, as updated in 1993, are published in Panico, R.; Powell, W. H.; Richter, K.-C. *A Guide to IUPAC Nomenclature of Organic Compounds. Recommendations 1993;* Blackwell Scientific Publication: London, 1993. The IUPAC rules for nomenclature can also be found through links at the IUPAC website (www.iupac.org).

(above C_4) are of Greek and Latin origin. Learning the stems is like learning to count in organic chemistry. Thus, one, two, three, four, and five become meth-, eth-, prop-, but-, and pent-.

4.3A Nomenclature of Unbranched Alkyl Groups

If we remove one hydrogen atom from an alkane, we obtain what is called an **alkyl group.** These alkyl groups have names that end in **-yl.** When the alkane is **unbranched,** and the hydrogen atom that is removed is a **terminal** hydrogen atom, the names are straightforward:

CD Tutorial
Alkyl groups

ALKANE		ALKYL GROUP	ABBREVIATION
CH_3—H **Methane**	becomes	CH_3— **Methyl**	Me—
CH_3CH_2—H **Ethane**	becomes	CH_3CH_2— **Ethyl**	Et—
$CH_3CH_2CH_2$—H **Propane**	becomes	$CH_3CH_2CH_2$— **Propyl**	Pr—
$CH_3CH_2CH_2CH_2$—H **Butane**	becomes	$CH_3CH_2CH_2CH_2$— **Butyl**	Bu—

4.3B Nomenclature of Branched-Chain Alkanes

Branched-chain alkanes are named according to the following rules:

1. **Locate the longest continuous chain of carbon atoms; this chain determines the parent name for the alkane.** We designate the following compound, for example, as a *hexane* because the longest continuous chain contains six carbon atoms:

$$\begin{array}{l} CH_3CH_2CH_2CH_2\underset{\displaystyle |\atop \displaystyle CH_3}{C}HCH_3 \end{array}$$

The longest continuous chain may not always be obvious from the way the formula is written. Notice, for example, that the following alkane is designated as a *heptane* because the longest chain contains seven carbon atoms:

Practice drawing bond-line formulas for the compounds shown in Section 4.3.

$$CH_3CH_2CH_2CH_2\underset{\displaystyle |\atop \displaystyle \underset{\displaystyle |\atop \displaystyle CH_3}{CH_2}}{C}HCH_3$$

2. **Number the longest chain beginning with the end of the chain nearer the substituent.** Applying this rule, we number the two alkanes that we illustrated previously in the following way:

$$\overset{6}{C}H_3\overset{5}{C}H_2\overset{4}{C}H_2\overset{3}{C}H_2\overset{2}{C}H\overset{1}{C}H_3 \text{ with } CH_3 \text{ (Substituent) on C2} \qquad \overset{7}{C}H_3\overset{6}{C}H_2\overset{5}{C}H_2\overset{4}{C}H_2\overset{3}{C}HCH_3 \text{ (Substituent) with } {}^{2}CH_2{}^{1}CH_3 \text{ on C3}$$

Substituent

Substituent

3. **Use the numbers obtained by application of rule 2 to designate the location of the substituent group.** The parent name is placed last, and the substituent group, preceded by the number designating its location on the chain, is placed first. Numbers are separated from words by a hyphen. Our two examples are 2-methylhexane and 3-methylheptane, respectively:

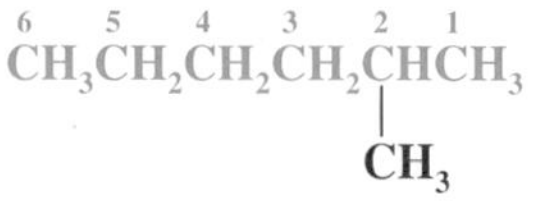

$$\overset{6}{C}H_3\overset{5}{C}H_2\overset{4}{C}H_2\overset{3}{C}H_2\overset{2}{C}H(CH_3)\overset{1}{C}H_3$$

2-Methylhexane

$$\overset{7}{C}H_3\overset{6}{C}H_2\overset{5}{C}H_2\overset{4}{C}H_2\overset{3}{C}H(\overset{2}{C}H_2\overset{1}{C}H_3)CH_3$$

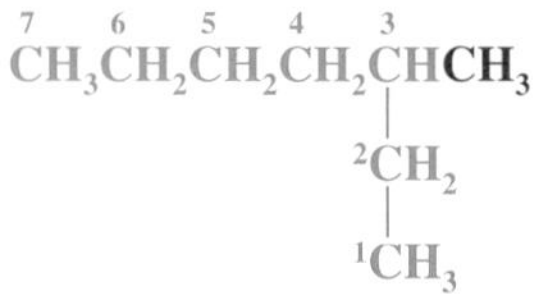

3-Methylheptane

4. **When two or more substituents are present, give each substituent a number corresponding to its location on the longest chain.** For example, we designate the following compound as 4-ethyl-2-methylhexane:

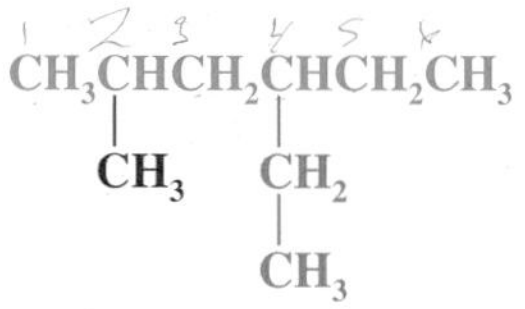

$$CH_3CH(CH_3)CH_2CH(CH_2CH_3)CH_2CH_3$$

4-Ethyl-2-methylhexane

The substituent groups should be listed *alphabetically* (i.e., ethyl before methyl).* In deciding on alphabetical order, disregard multiplying prefixes such as "di" and "tri."

5. **When two substituents are present on the same carbon atom, use that number twice:**

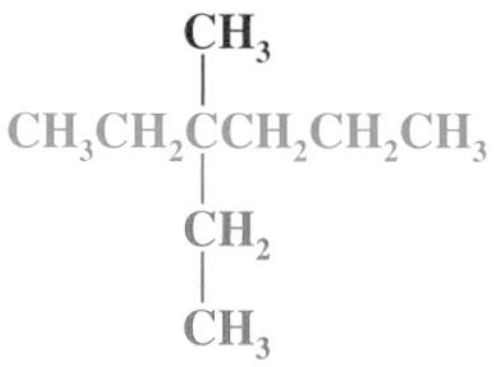

$$CH_3CH_2C(CH_3)(CH_2CH_3)CH_2CH_2CH_3$$

3-Ethyl-3-methylhexane

6. **When two or more substituents are identical, indicate this by the use of the prefixes di-, tri-, tetra-,** and so on. Then make certain that each and every substituent has a number. Commas are used to separate numbers from each other:

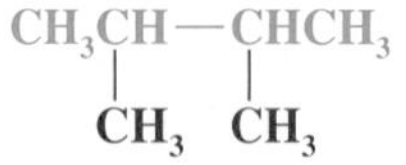

$$CH_3CH(CH_3){-}CH(CH_3)CH_3$$

2,3-Dimethylbutane

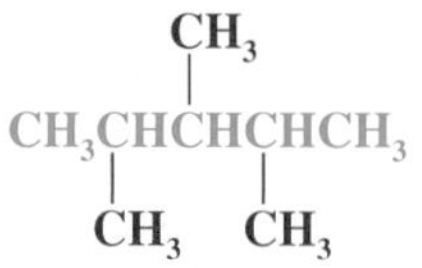

$$CH_3CH(CH_3)CH(CH_3)CH(CH_3)CH_3$$

2,3,4-Trimethylpentane

$$CH_3C(CH_3)_2CH_2C(CH_3)_2CH_3$$

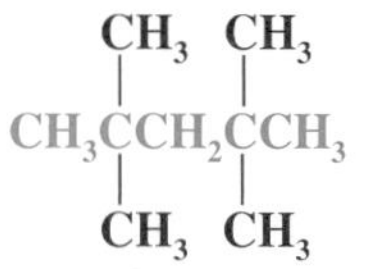

2,2,4,4-Tetramethylpentane

*Some handbooks also list the groups in order of increasing size or complexity (i.e., methyl before ethyl). An alphabetical listing, however, is now by far the most widely used system.

Application of these six rules allows us to name most of the alkanes that we shall encounter. Two other rules, however, may be required occasionally:

7. When two chains of equal length compete for selection as the parent chain, choose the chain with the greater number of substituents:

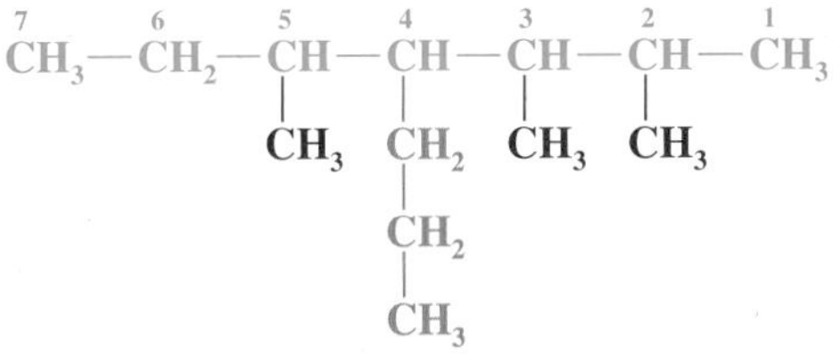

$$\overset{7}{CH_3}-\overset{6}{CH_2}-\overset{5}{CH}(CH_3)-\overset{4}{CH}(CH_2CH_2CH_3)-\overset{3}{CH}(CH_3)-\overset{2}{CH}(CH_3)-\overset{1}{CH_3}$$

2,3,5-Trimethyl-4-propylheptane
(four substituents)

8. When branching first occurs at an equal distance from either end of the longest chain, choose the name that gives the lower number at the first point of difference:

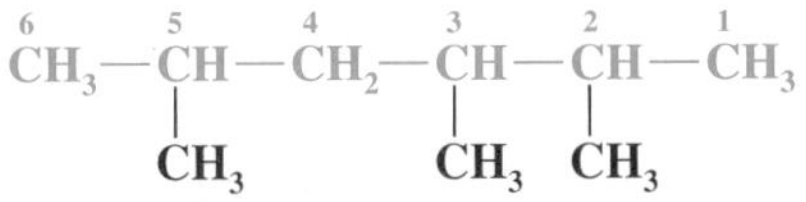

$$\overset{6}{CH_3}-\overset{5}{CH}(CH_3)-\overset{4}{CH_2}-\overset{3}{CH}(CH_3)-\overset{2}{CH}(CH_3)-\overset{1}{CH_3}$$

2,3,5-Trimethylhexane
(*not* 2,4,5-trimethylhexane)

4.3C Nomenclature of Branched Alkyl Groups

In Section 4.3A you learned the names for the unbranched alkyl groups such as methyl, ethyl, propyl, and butyl, groups derived by removing a terminal hydrogen from an alkane. For alkanes with more than two carbon atoms, more than one derived group is possible. Two groups can be derived from propane, for example; the **propyl group** is derived by removal of a terminal hydrogen, and the **1-methylethyl** or **isopropyl group** is derived by removal of a hydrogen from the central carbon:

Three-Carbon Groups

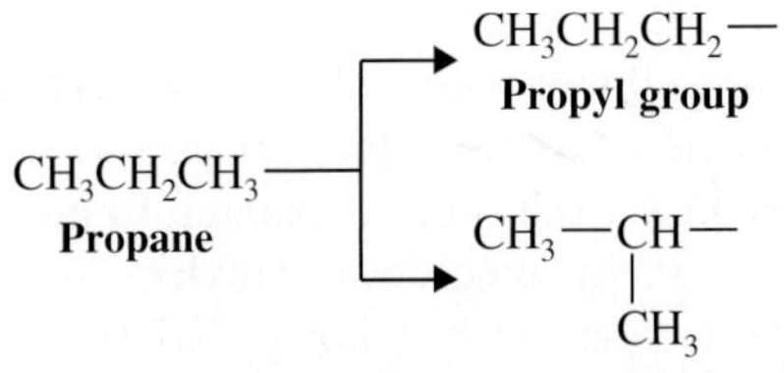

$CH_3CH_2CH_3$ **Propane** → $CH_3CH_2CH_2-$ **Propyl group**

$CH_3CH_2CH_3$ **Propane** → $CH_3-CH(CH_3)-$ **1-Methylethyl or isopropyl group**

1-Methylethyl is the systematic name for this group; isopropyl is a common name. Systematic nomenclature for alkyl groups is similar to that for branched-chain alkanes, with the provision that *numbering always begins at the point where the group is attached to the main chain.* There are four C_4 groups. Two are derived from butane and two are derived from isobutane.*

*Isobutane is a common name for 2-methylpropane that is approved by the IUPAC.

Four-Carbon Groups

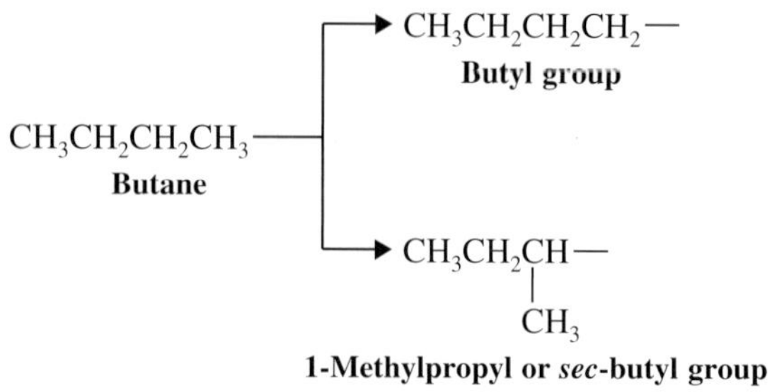

$CH_3CH_2CH_2CH_3$
Butane

→ $CH_3CH_2CH_2CH_2-$
Butyl group

→ $CH_3CH_2CH(CH_3)-$
1-Methylpropyl or *sec*-butyl group

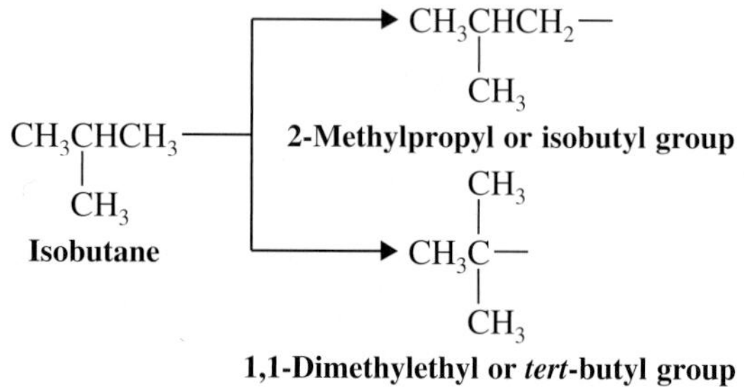

$CH_3CH(CH_3)CH_3$
Isobutane

→ $CH_3CH(CH_3)CH_2-$
2-Methylpropyl or isobutyl group

→ $CH_3C(CH_3)_2-$
1,1-Dimethylethyl or *tert*-butyl group

The following examples show how the names of these groups are employed:

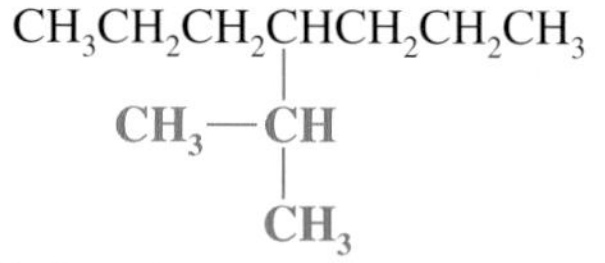

$CH_3CH_2CH_2CH[CH(CH_3)_2]CH_2CH_2CH_3$
4-(1-Methylethyl)heptane or 4-isopropylheptane

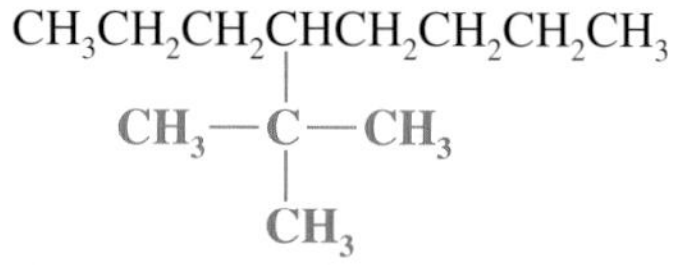

$CH_3CH_2CH_2CH[C(CH_3)_3]CH_2CH_2CH_2CH_3$
4-(1,1-Dimethylethyl)octane or 4-*tert*-butyloctane

The common names **isopropyl, isobutyl, *sec*-butyl,** and ***tert*-butyl** are approved by the IUPAC for the unsubstituted groups, and they are still very frequently used. You should learn these groups so well that you can recognize them any way that they are written. In deciding on alphabetical order for these groups you should disregard structure-defining prefixes that are written in italics and separated from the name by a hyphen. Thus *tert*-butyl precedes ethyl, but ethyl precedes isobutyl.*

There is one five-carbon group with an IUPAC approved common name that you should also know: the 2,2-dimethylpropyl group, commonly called the **neopentyl group:**

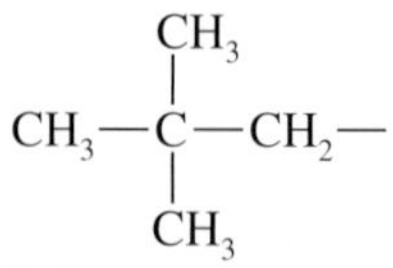

$CH_3-C(CH_3)_2-CH_2-$
2,2-Dimethylpropyl or neopentyl group

*The abbreviations *i*, *s*, and *t* are sometimes used for iso-, *sec*-, and *tert*-, respectively, although they are not recommended.

PROBLEM 4.2

(a) In addition to the 2,2-dimethylpropyl (or neopentyl) group just given, there are seven other five-carbon groups. Write their structures and give each structure its systematic name. **(b)** Provide IUPAC names for the nine C_7H_{16} isomers that you gave as answers to Problem 4.1.

4.3D Classification of Hydrogen Atoms

The hydrogen atoms of an alkane are classified on the basis of the carbon atom to which they are attached. A hydrogen atom attached to a primary carbon atom is a primary (1°) hydrogen atom, and so forth. The following compound, 2-methylbutane, has primary, secondary (2°), and tertiary (3°) hydrogen atoms:

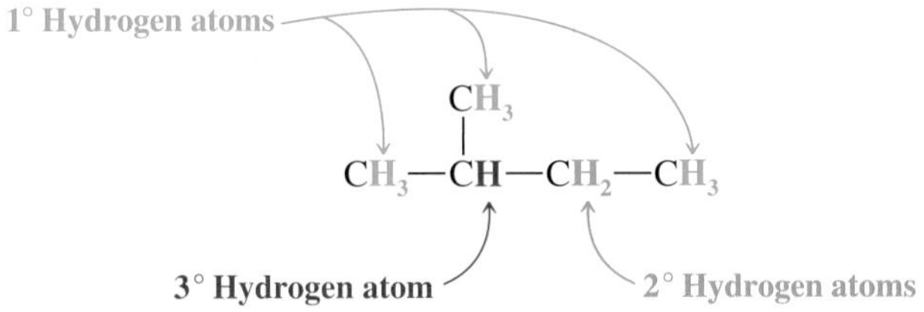

1° Hydrogen atoms

$$CH_3-CH(CH_3)-CH_2-CH_3$$

3° Hydrogen atom   2° Hydrogen atoms

On the other hand, 2,2-dimethylpropane, a compound that is often called **neopentane,** has only primary hydrogen atoms:

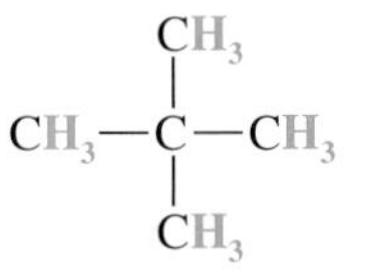

$$CH_3-C(CH_3)_2-CH_3$$

2,2-Dimethylpropane
(neopentane)

4.3E Nomenclature of Alkyl Halides

Alkanes bearing halogen substituents are named in the IUPAC substitutive system as haloalkanes:

CH_3CH_2Cl	$CH_3CH_2CH_2F$	$CH_3CHBrCH_3$
Chloroethane	**1-Fluoropropane**	**2-Bromopropane**

When the parent chain has both a halo and an alkyl substituent attached to it, number the chain from the end nearer the first substituent, regardless of whether it is halo or alkyl. If two substituents are of equal distance from the end of the chain, then number the chain from the end nearer the substituent that has alphabetical precedence:

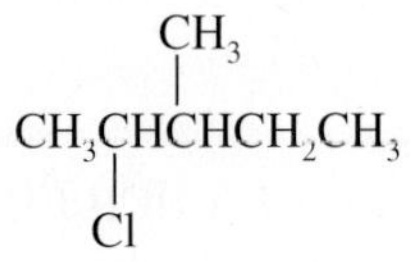

$$CH_3CH(Cl)CH(CH_3)CH_2CH_3 \qquad CH_3CH(Cl)CH_2CH(CH_3)CH_3$$

2-Chloro-3-methylpentane 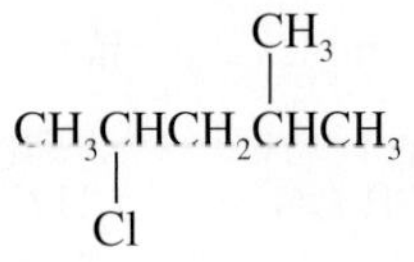   **2-Chloro-4-methylpentane**

Common names for many simple haloalkanes are still widely used, however. In this common nomenclature system, called *functional class nomenclature,* haloalkanes are named as alkyl halides. (The following names are also accepted by the IUPAC.)

CH_3CH_2Cl	$CH_3CH(Br)CH_3$	$(CH_3)_3CBr$	$CH_3CH(CH_3)CH_2Cl$	$CH_3C(CH_3)_2CH_2Br$
Ethyl chloride	**Isopropyl bromide**	***tert*-Butyl bromide**	**Isobutyl chloride**	**Neopentyl bromide**

PROBLEM 4.3 Give IUPAC substitutive names for all of the isomers of **(a)** C_4H_9Cl and **(b)** $C_5H_{11}Br$.

4.3F Nomenclature of Alcohols

In what is called IUPAC **substitutive nomenclature,** a name may have as many as four features: **locants, prefixes, parent compound,** and **suffixes.** Consider the following compound as an illustration without, for the moment, being concerned as to how the name arises:

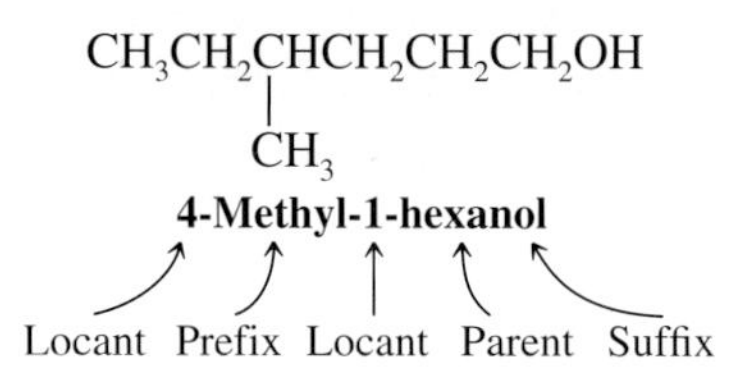

The *locant* **4-** tells that the substituent **methyl** group, named as a *prefix,* is attached to the *parent compound* at C4. The parent compound contains six carbon atoms and no multiple bonds, hence the parent name **hexane,** and it is an alcohol; therefore it has the *suffix* **-ol.** The locant **1-** tells that C1 bears the hydroxyl group. **In general, numbering of the chain always begins at the end nearer the group named as a suffix.**

The locant for a suffix (whether it is for an alcohol or another functional group) may be placed before the parent name, as in the above example or, according to a 1993 IUPAC revision of the rules, immediately before the suffix. Both methods are IUPAC approved. Therefore, the above compound could also be named **4-methylhexan-1-ol.**

The following procedure should be followed in giving alcohols IUPAC substitutive names:

1. Select the longest continuous carbon chain *to which the hydroxyl is directly attached.* Change the name of the alkane corresponding to this chain by dropping the final *e* and adding the suffix *-ol.*

2. Number the longest continuous carbon chain so as to give the carbon atom bearing the hydroxyl group the lower number. Indicate the position of the hydroxyl group by using this number as a locant; indicate the positions of other substituents (as prefixes) by using the numbers corresponding to their positions along the carbon chain as locants.

The following examples show how these rules are applied:

$\overset{3}{C}H_3\overset{2}{C}H_2\overset{1}{C}H_2OH$	$\overset{1}{C}H_3\overset{2}{C}H(OH)\overset{3}{C}H_2\overset{4}{C}H_3$	$\overset{5}{C}H_3\overset{4}{C}H(CH_3)\overset{3}{C}H_2\overset{2}{C}H_2\overset{1}{C}H_2OH$
1-Propanol	**2-Butanol**	**4-Methyl-1-pentanol or 4-methylpentan-1-ol (*not* 2-methyl-5-pentanol)**

$\overset{3}{Cl}CH_2\overset{2}{C}H_2\overset{1}{C}H_2OH$

3-Chloro-1-propanol
or 3-chloropropan-1-ol

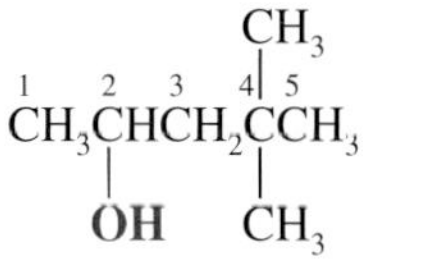

$\overset{1}{C}H_3\overset{2}{C}H(OH)\overset{3}{C}H_2\overset{4}{C}(CH_3)_2\overset{5}{C}H_3$

4,4-Dimethyl-2-pentanol
or 4,4-dimethylpentan-2-ol

PROBLEM 4.4

Give IUPAC substitutive names for all of the isomeric alcohols with the formulas **(a)** $C_4H_{10}O$ and **(b)** $C_5H_{12}O$.

Simple alcohols are often called by *common* functional class names that are also approved by the IUPAC. We have seen several examples already (Section 2.7). In addition to *methyl alcohol, ethyl alcohol,* and *isopropyl alcohol,* there are several others, including the following:

$CH_3CH_2CH_2OH$ — **Propyl alcohol**

$CH_3CH_2CH_2CH_2OH$ — **Butyl alcohol**

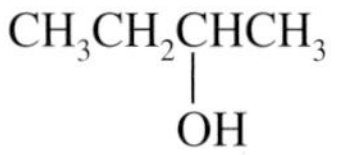

$CH_3CH_2CH(OH)CH_3$ — ***sec*-Butyl alcohol**

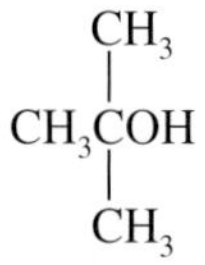

$(CH_3)_3COH$ — ***tert*-Butyl alcohol**

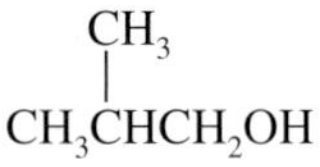

$CH_3CH(CH_3)CH_2OH$ — **Isobutyl alcohol**

$CH_3C(CH_3)_2CH_2OH$ — **Neopentyl alcohol**

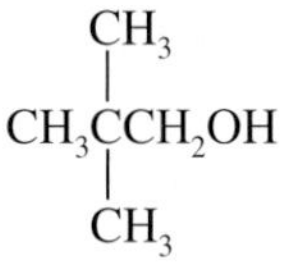

Alcohols containing two hydroxyl groups are commonly called glycols. In the IUPAC substitutive system they are named as **diols:**

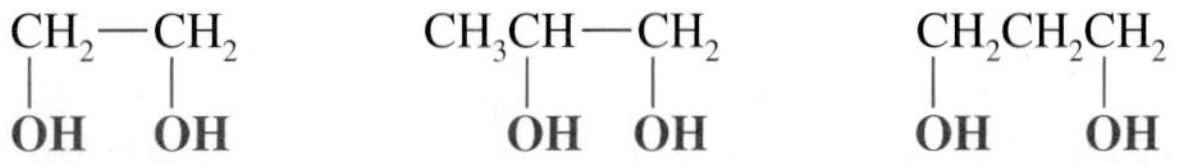

	$CH_2(OH)—CH_2(OH)$	$CH_3CH(OH)—CH_2(OH)$	$CH_2(OH)CH_2CH_2(OH)$
Common	Ethylene glycol	Propylene glycol	Trimethylene glycol
Substitutive	**1,2-Ethanediol or ethane-1,2-diol**	**1,2-Propanediol or propane-1,2-diol**	**1,3-Propanediol or propane-1,3-diol**

4.4 Nomenclature of Cycloalkanes

4.4A Monocyclic Compounds

Cycloalkanes with only one ring are named by attaching the prefix cyclo- to the names of the alkanes possessing the same number of carbon atoms. For example,

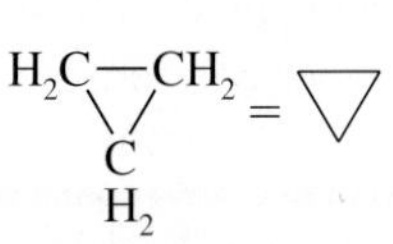

$H_2C—CH_2—CH_2$ (ring) = ▽ — **Cyclopropane**

$H_2C—CH_2—CH_2—CH_2—CH_2$ (ring) = ⬠ — **Cyclopentane**

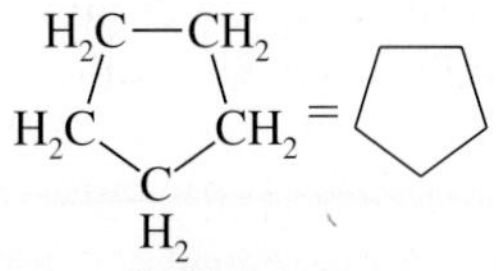

Naming substituted cycloalkanes is straightforward: We name them as *alkylcycloalkanes, halocycloalkanes, alkylcycloalkanols,* and so on. If only one substituent is present, it is not necessary to designate its position. When two substituents are present, we number the ring *beginning with the substituent* first in the alphabet and number in the direction that gives the next substituent the lower number possible. When three or more substituents are present, we begin at the substituent that leads to the lowest set of locants:

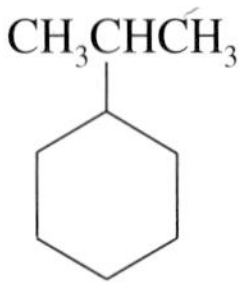

CH_3CHCH_3 — **Isopropylcyclohexane**

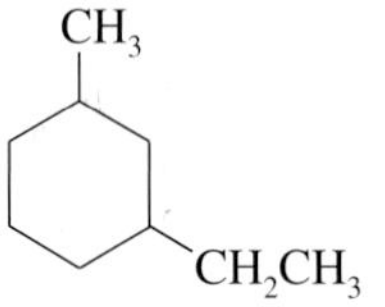

CH_3, CH_2CH_3 — **1-Ethyl-3-methylcyclohexane** (*not* 1-ethyl-5-methylcyclohexane)

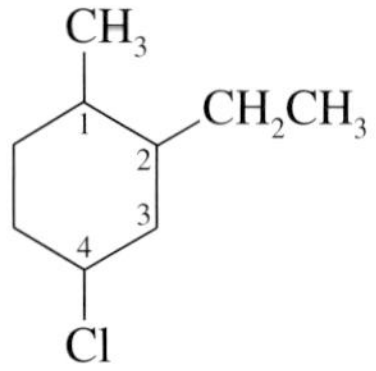

CH_3, CH_2CH_3, Cl — **4-Chloro-2-ethyl-1-methylcyclohexane** (*not* 1-chloro-3-ethyl-4-methylcyclohexane)

Cl — **Chlorocyclopentane**

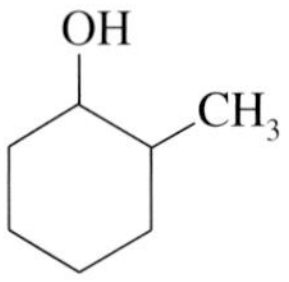

OH, CH_3 — **2-Methylcyclohexanol**

When a single ring system is attached to a single chain with a greater number of carbon atoms, or when more than one ring system is attached to a single chain, then it is appropriate to name the compounds as *cycloalkylalkanes*. For example,

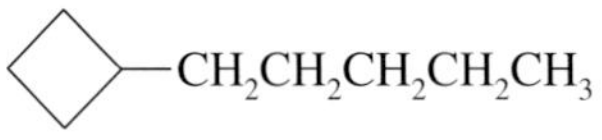

$-CH_2CH_2CH_2CH_2CH_3$ — **1-Cyclobutylpentane**

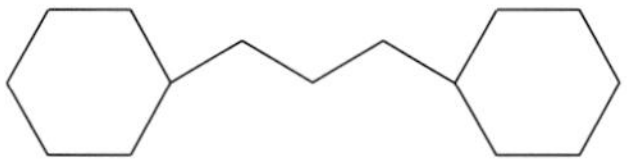

1,3-Dicyclohexylpropane

PROBLEM 4.5 Give names for the following substituted alkanes:

(a) $(CH_3)_3C$, $(CH_3)_2CH$ (cyclopentane)

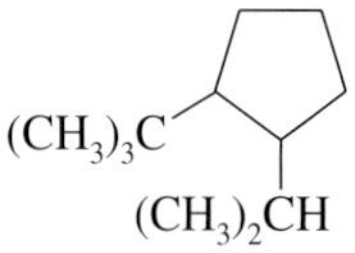

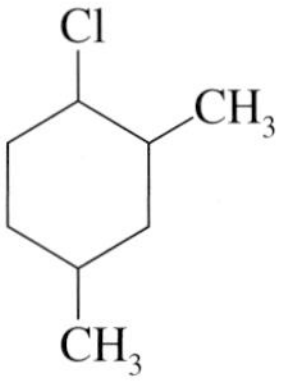

(b) $(CH_3)_2CHCH_2$, H_3C (cyclohexane)

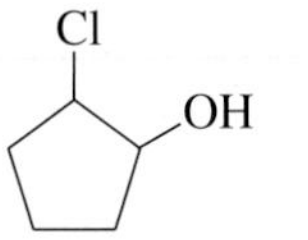

(c) $CH_3(CH_2)_2CH_2-$ (cyclohexane)

(d) Cl, CH_3, CH_3 (cyclohexane)

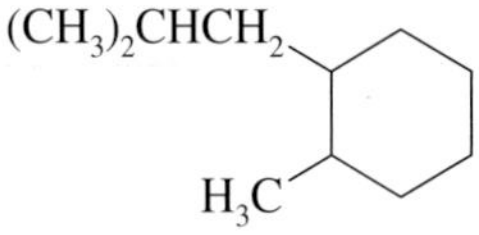

(e) Cl, OH (cyclopentane)

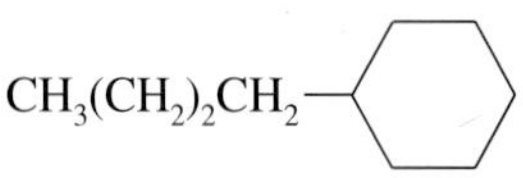

(f) OH, $C(CH_3)_3$ (cyclohexane)

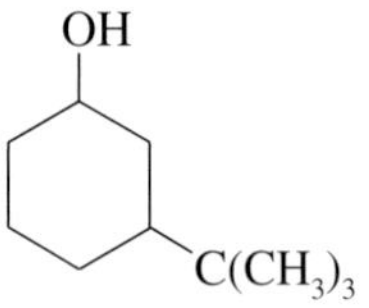

4.4B Bicyclic Compounds

We name compounds containing two fused or bridged rings as **bicycloalkanes,** and we use the name of the alkane corresponding to the total number of carbon atoms in the rings as the parent name. The following compound, for example, contains seven carbon atoms and is, therefore, a bicycloheptane. The carbon atoms common to both rings are called bridgeheads, and each bond, or each chain of atoms, connecting the bridgehead atoms is called a bridge:

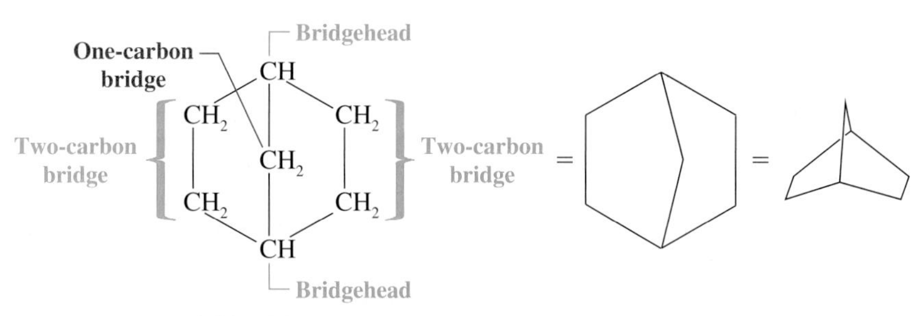

A bicycloheptane

Explore the structures of these bicyclic compounds by building hand-held molecular models.

Then we interpose in the name an expression in brackets that denotes the number of carbon atoms in each bridge (in order of decreasing length). For example,

Bicyclo[2.2.1]heptane (also called *norbornane*)

Bicyclo[1.1.0]butane

If substituents are present, we number the bridged ring system beginning at one bridgehead, proceeding first along the longest bridge to the other bridgehead, then along the next longest bridge back to the first bridgehead. The shortest bridge is numbered last:

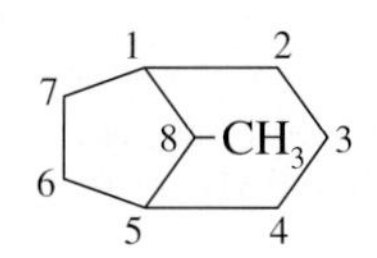

8-Methylbicyclo[3.2.1]octane

8-Methylbicyclo[4.3.0]nonane

PROBLEM 4.6

Give names for each of the following bicyclic alkanes:

(a)

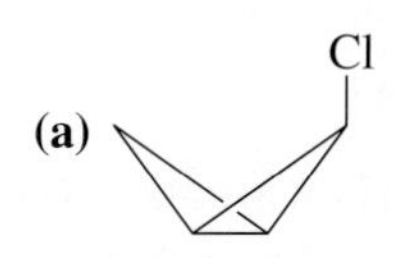

(b)

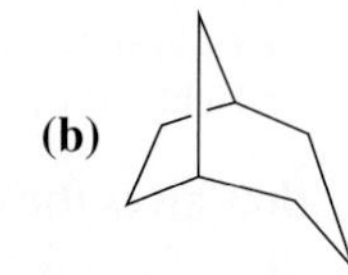

(c)

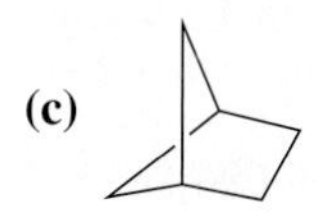

(d) **(e)**

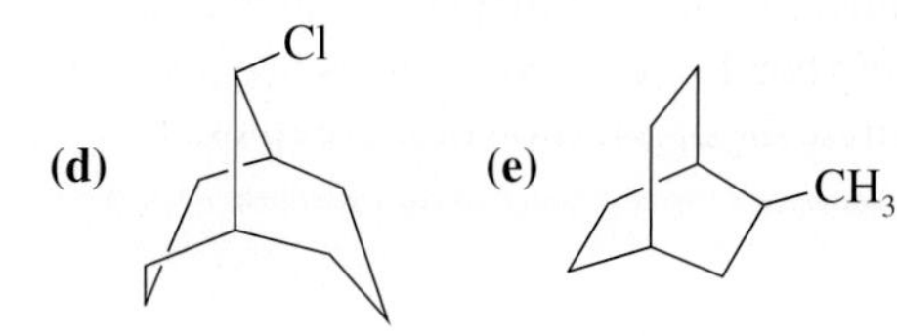

(f) Write the structure of a bicyclic compound that is an isomer of bicyclo[2.2.0]hexane and give its name.

4.5 Nomenclature of Alkenes and Cycloalkenes

Many older names for alkenes are still in common use. Propene is often called propylene, and 2-methylpropene frequently bears the name isobutylene:

	$CH_2{=}CH_2$	$CH_3CH{=}CH_2$	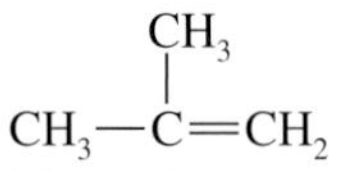 $CH_3{-}C(CH_3){=}CH_2$
IUPAC:	**Ethene**	**Propene**	**2-Methylpropene**
Common:	**Ethylene**	**Propylene**	**Isobutylene**

The IUPAC rules for naming alkenes are similar in many respects to those for naming alkanes:

1. **Determine the parent name by selecting the longest chain that contains the double bond and change the ending of the name of the alkane of identical length from -ane to -ene.** Thus, if the longest chain contains five carbon atoms, the parent name for the alkene is *pentene;* if it contains six carbon atoms, the parent name is *hexene,* and so on.

2. **Number the chain so as to include both carbon atoms of the double bond, and begin numbering at the end of the chain nearer the double bond. Designate the location of the double bond by using the number of the first atom of the double bond as a prefix. The locant for the alkene suffix may precede the parent name or be placed immediately before the suffix.** We will show examples of both styles:

$\overset{1}{C}H_2{=}\overset{2}{C}H\overset{3}{C}H_2\overset{4}{C}H_3$
1-Butene
(*not* 3-butene)

$CH_3CH{=}CHCH_2CH_2CH_3$
2-Hexene
(*not* 4-hexene)

3. **Indicate the locations of the substituent groups by the numbers of the carbon atoms to which they are attached:**

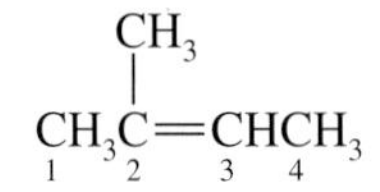

$\underset{1}{C}H_3\underset{2}{C}(CH_3){=}\underset{3}{C}H\underset{4}{C}H_3$
2-Methyl-2-butene
(*not* 3-methyl-2-butene)

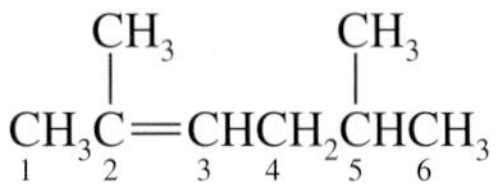

$\underset{1}{C}H_3\underset{2}{C}(CH_3){=}\underset{3}{C}H\underset{4}{C}H_2\underset{5}{C}H(CH_3)\underset{6}{C}H_3$
2,5-Dimethyl-2-hexene
(*not* 2,5-dimethyl-4-hexene)

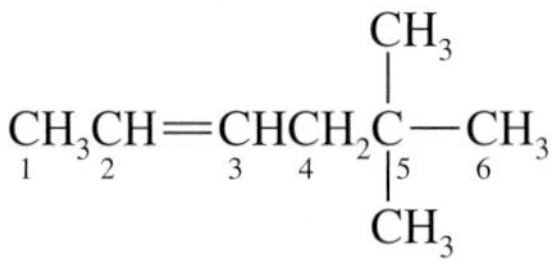

$\underset{1}{C}H_3\underset{2}{C}H{=}\underset{3}{C}H\underset{4}{C}H_2\underset{5}{C}(CH_3)_2{-}\underset{6}{C}H_3$
5,5-Dimethyl-2-hexene
or 5,5-dimethylhex-2-ene

$\overset{4}{C}H_3\overset{3}{C}H{=}\overset{2}{C}H\overset{1}{C}H_2Cl$
1-Chloro-2-butene
or 1-chlorobut-2-ene

4. **Number substituted cycloalkenes in the way that gives the carbon atoms of the double bond the 1 and 2 positions and that also gives the substituent groups the lower numbers at the first point of difference.** With substituted cycloalkenes it is not necessary to specify the position of the double bond since it will always begin with C1 and C2. The two examples shown here illustrate the application of these rules:

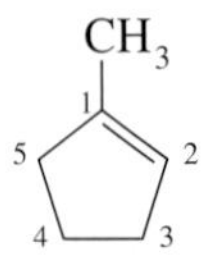

(structure: cyclopentene ring numbered 1–5 with CH_3 at C1)
1-Methylcyclopentene
(*not* 2-methylcyclopentene)

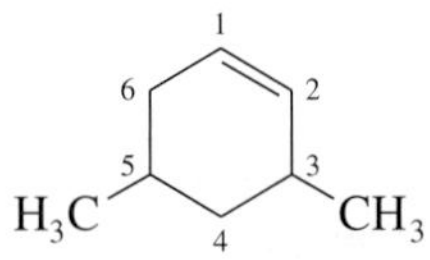

(structure: cyclohexene ring numbered 1–6 with H_3C at C5 and CH_3 at C3)
3,5-Dimethylcyclohexene
(*not* 4,6-dimethylcyclohexene)

5. Name compounds containing a double bond and an alcohol group as alkenols (or cycloalkenols) and give the alcohol carbon the lower number:

$$\overset{5}{CH_3}\overset{4}{\underset{|}{\overset{CH_3}{C}}}=\overset{3}{C}H\overset{2}{C}H(OH)\overset{1}{C}H_3$$

4-Methyl-3-penten-2-ol
or 4-methylpent-3-en-2-ol

2-Methyl-2-cyclohexen-1-ol
or 2-methylcyclohex-2-en-1-ol

6. Two frequently encountered alkenyl groups are the *vinyl group* and the *allyl group:*

$CH_2{=}CH{-}$ — **The vinyl group**

$CH_2{=}CHCH_2{-}$ — **The allyl group**

Using substitutive nomenclature, these groups are called *ethenyl* and *prop-2-en-1-yl,* respectively. The following examples illustrate how these names are employed:

Bromoethene or **vinyl bromide** **(common)**

Ethenylcyclopropane or **vinylcyclopropane**

3-Chloropropene or **allyl chloride** **(common)**

3-(Prop-2-en-1-yl)cyclohexanol or **3-allylcyclohexanol**

7. If two identical groups are on the same side of the double bond, the compound can be designated *cis;* if they are on opposite sides it can be designated *trans:*

***cis*-1,2-Dichloroethene**

***trans*-1,2-Dichloroethene**

In Section 7.2 we shall see another method for designating the geometry of the double bond.

PROBLEM 4.7

Give IUPAC names for the following alkenes:

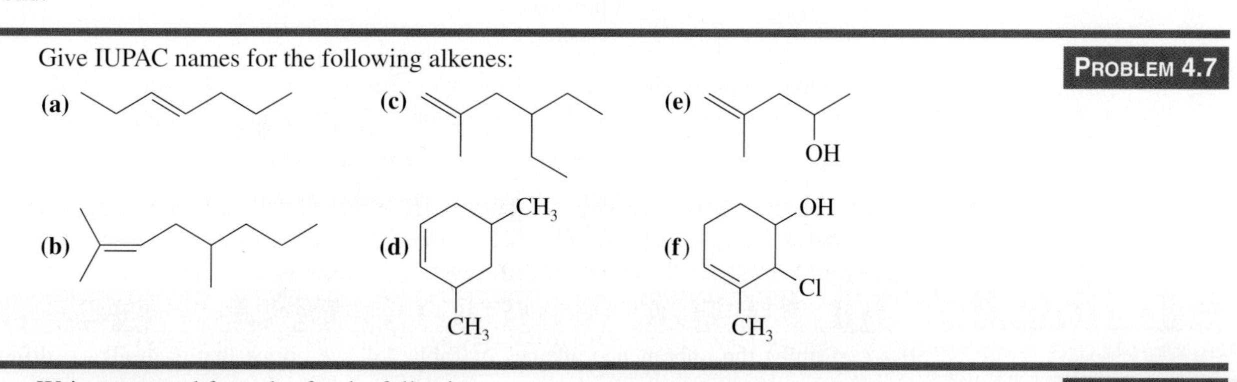

PROBLEM 4.8

Write structural formulas for the following:

(a) *cis*-3-Octene
(b) *trans*-2-Hexene
(c) 2,4-Dimethyl-2-pentene
(d) *trans*-1-Chlorobut-2-ene
(e) 4,5-Dibromo-1-pentene
(f) 1-Bromo-2-methyl-1-(prop-2-en-1-yl)cyclopentane
(g) 3,4-Dimethylcyclopentene
(h) Vinylcyclopentane
(i) 1,2-Dichlorocyclohexene
(j) *trans*-1,4-Dichloro-2-pentene

4.6 Nomenclature of Alkynes

Alkynes are named in much the same way as alkenes. Unbranched alkynes, for example, are named by replacing the **-ane** of the name of the corresponding alkane with the ending **-yne.** The chain is numbered to give the carbon atoms of the triple bond the lower possible numbers. The lower number of the two carbon atoms of the triple bond is used to designate the location of the triple bond. The IUPAC names of three unbranched alkynes are shown here:

$$H{-}C{\equiv}C{-}H \qquad CH_3CH_2C{\equiv}CCH_3 \qquad H{-}C{\equiv}CCH_2CH{=}CH_2$$

Ethyne or acetylene* — **2-Pentyne** — **1-Penten-4-yne† or pent-1-en-4-yne**

The locations of substituent groups of branched alkynes and substituted alkynes are also indicated with numbers. An —OH group has priority over the triple bond when numbering the chain of an alkynol:

$$\overset{3}{Cl{-}CH_2}\overset{2}{C}{\equiv}\overset{1}{CH} \qquad \overset{4}{CH_3}\overset{3}{C}{\equiv}\overset{2}{C}\overset{1}{CH_2}Cl \qquad \overset{4}{HC}{\equiv}\overset{3}{C}\overset{2}{CH_2}\overset{1}{CH_2}OH$$

3-Chloropropyne — **1-Chloro-2-butyne or 1-chlorobut-2-yne** — **3-Butyn-1-ol or but-3-yn-1-ol**

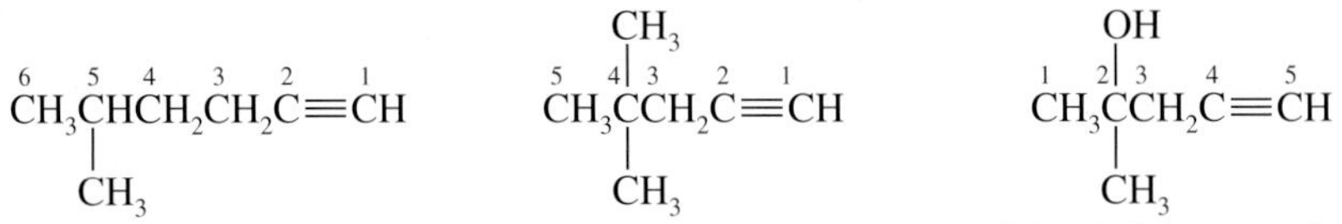

$$\overset{6}{CH_3}\overset{5}{CH}(CH_3)\overset{4}{CH_2}\overset{3}{CH_2}\overset{2}{C}{\equiv}\overset{1}{CH} \qquad \overset{5}{CH_3}\overset{4}{C}(CH_3)_2\overset{3}{CH_2}\overset{2}{C}{\equiv}\overset{1}{CH} \qquad \overset{1}{CH_3}\overset{2}{C}(OH)(CH_3)\overset{3}{CH_2}\overset{4}{C}{\equiv}\overset{5}{CH}$$

5-Methyl-1-hexyne or 5-methylhex-1-yne — **4,4-Dimethyl-1-pentyne or 4,4-dimethylpent-1-yne** — **2-Methyl-4-pentyn-2-ol or 2-methylpent-4-yn-2-ol**

Problem 4.9 Give the structures and IUPAC names for all the alkynes with the formula C_6H_{10}.

Monosubstituted acetylenes or 1-alkynes are called **terminal alkynes,** and the hydrogen attached to the carbon of the triple bond is called the acetylenic hydrogen:

$$R{-}C{\equiv}C{-}H$$

A terminal alkyne (H: **Acetylenic hydrogen**)

The anion obtained when the acetylenic hydrogen is removed is known as an *alkynide ion* or an acetylide ion. As we shall see in Section 4.18C, these ions are useful in synthesis:

$$R{-}C{\equiv}C{:}^{-} \qquad CH_3C{\equiv}C{:}^{-}$$

An alkynide ion (an acetylide ion) — **The propynide ion**

4.7 Physical Properties of Alkanes and Cycloalkanes

If we examine the unbranched alkanes in Table 4.4, we notice that each alkane differs from the preceding alkane by one $-CH_2-$ group. Butane, for example, is $CH_3(CH_2)_2CH_3$ and pentane is $CH_3(CH_2)_3CH_3$. A series of compounds like this, where each member differs from the next member by a constant unit, is called a **homologous series.** Members of a homologous series are called **homologs.**

*The name acetylene is retained by the IUPAC system for the compound HC≡CH and is used frequently.
†Where there is a choice, the double bond is given the lower number.

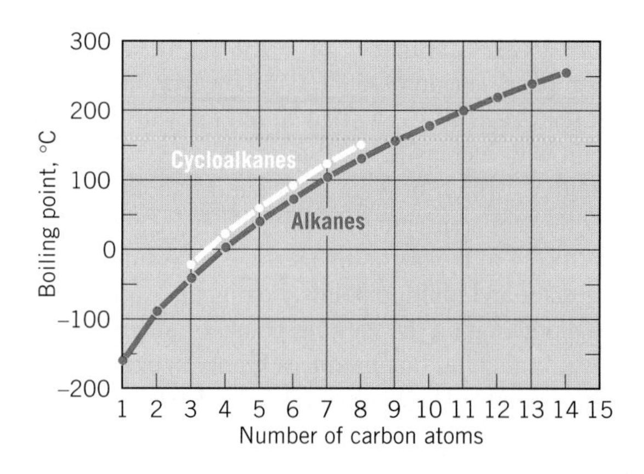

FIGURE 4.3 Boiling points of unbranched alkanes (in red) and cycloalkanes (in white).

At room temperature (25°C) and 1 atm pressure the first four members of the homologous series of unbranched alkanes are gases (Fig. 4.3), the C_5–C_{17} unbranched alkanes (pentane to heptadecane) are liquids, and the unbranched alkanes with 18 and more carbon atoms are solids.

Boiling Points The boiling points of the unbranched alkanes show a regular increase with increasing molecular weight (Fig. 4.3). Branching of the alkane chain, however, lowers the boiling point. As examples, consider the C_6H_{14} isomers in Table 4.2. Hexane boils at 68.7°C and 2-methylpentane and 3-methylpentane, each having one branch, boil lower, at 60.3 and 63.3°C, respectively. 2,3-Dimethylbutane and 2,2-dimethylbutane, each with two branches, boil lower still, at 58 and 49.7°C, respectively.

Part of the explanation for these effects lies in the van der Waals forces that we studied in Section 2.14D. With unbranched alkanes, as molecular weight increases, so too do molecular size and, even more importantly, molecular surface area. With increasing surface area, the van der Waals forces between molecules increase; therefore, more energy (a higher temperature) is required to separate molecules from one another and produce boiling. Chain branching, on the other hand, makes a molecule more compact, reducing its surface area and with it the strength of the van der Waals forces operating between it and adjacent molecules; this has the effect of lowering the boiling point.

CD Tutorial
Chain branching and intermolecular forces

Melting Points The unbranched alkanes do not show the same smooth increase in melting points with increasing molecular weight (blue line in Fig. 4.4) that they show in their boiling points. There is an alternation as one progresses from an unbranched alkane with an even number of carbon atoms to the next one with an odd number of carbon atoms. For example, propane (mp −188°C) melts lower than ethane (mp −183°C) and also

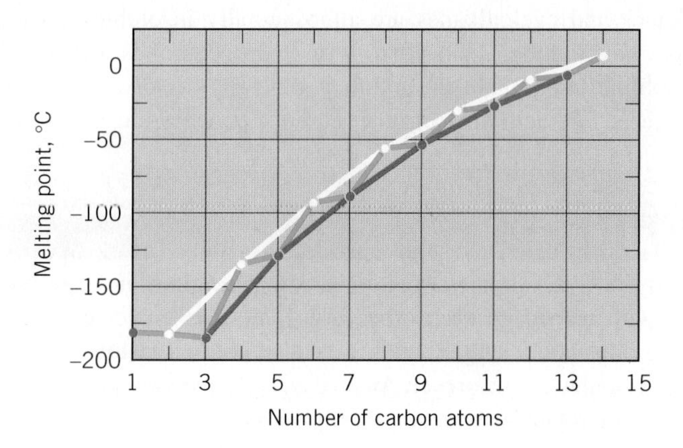

FIGURE 4.4 Melting points of unbranched alkanes.

lower than methane (mp −182°C). Butane (mp −138°C) melts 50°C higher than propane and only 8°C lower than pentane (mp −130°C). If, however, the even- and odd-numbered alkanes are plotted on *separate* curves (white and red lines in Fig. 4.4), there *is* a smooth increase in melting point with increasing molecular weight.

X-ray diffraction studies, which provide information about molecular structure, have revealed the reason for this apparent anomaly. Alkane chains with an even number of carbon atoms pack more closely in the crystalline state. As a result, attractive forces between individual chains are greater and melting points are higher.

The effect of chain branching on the melting points of alkanes is more difficult to predict. Generally, however, branching that produces highly symmetrical structures results in abnormally high melting points. The compound 2,2,3,3-tetramethylbutane, for example, melts at 100.7°C. Its boiling point is only six degrees higher, 106.3°C:

$$CH_3-C(CH_3)_2-C(CH_3)_2-CH_3$$

2,2,3,3-Tetramethylbutane

Cycloalkanes also have much higher melting points than their open-chain counterparts (Table 4.5).

TABLE 4.5 Physical Constants of Cycloalkanes

Number of Carbon Atoms	Name	bp (°C) (1 atm)	mp (°C)	Density, d^{20} (g mL^{-1})	Refractive Index (n_D^{20})
3	Cyclopropane	−33	−126.6	—	—
4	Cyclobutane	13	−90	—	1.4260
5	Cyclopentane	49	−94	0.751	1.4064
6	Cyclohexane	81	6.5	0.779	1.4266
7	Cycloheptane	118.5	−12	0.811	1.4449
8	Cyclooctane	149	13.5	0.834	—

Density As a class, the alkanes and cycloalkanes are the least dense of all groups of organic compounds. All alkanes and cycloalkanes have densities considerably less than 1.00 g mL^{-1} (the density of water at 4°C). As a result, petroleum (a mixture of hydrocarbons rich in alkanes) floats on water.

Solubility Alkanes and cycloalkanes are almost totally insoluble in water because of their very low polarity and their inability to form hydrogen bonds. Liquid alkanes and cycloalkanes are soluble in one another, and they generally dissolve in solvents of low polarity. Good solvents for them are benzene, carbon tetrachloride, chloroform, and other hydrocarbons.

4.8 Sigma Bonds and Bond Rotation

Conformational analysis

Groups bonded by only a sigma (σ) bond (i.e., by a single bond) can undergo rotation about that bond with respect to each other. The temporary molecular shapes that result from rotation of groups about single bonds are called **conformations** of a molecule. Each possible structure is called a **conformer.** An analysis of the energy changes associated with a molecule undergoing rotation about single bonds is called **conformational analysis.**

When we do conformational analysis, we will find that certain types of structural formulas are especially convenient to use. One of these types is called a **Newman projection formula** and another type is a **sawhorse formula.** Sawhorse formulas are much like other three-dimensional formulas we have used so far. In conformational analyses, we will make substantial use of Newman projections.

Learn to draw Newman projections and sawhorse formulas. Build hand-held molecular models and compare them with your drawings.

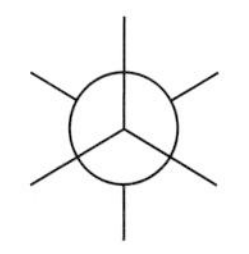

Newman projection formula

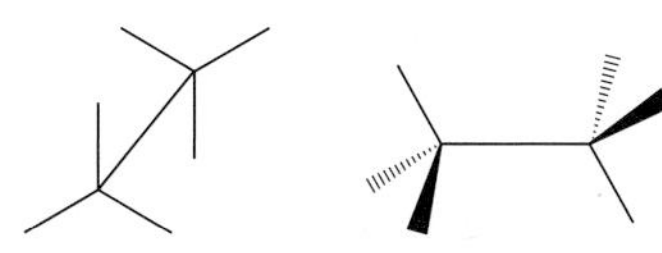

Sawhorse formula

Melvin S. Newman

To write a Newman projection formula, we imagine ourselves taking a view from one atom (usually a carbon) directly along a selected bond axis to the next atom (also usually a carbon atom). The front carbon and its other bonds are represented as Y and those of the back carbon as ⌾. In Figs. 4.5*a*,*b* we show ball-and-stick models and a Newman projection formula for the staggered conformation of ethane. The **staggered conformation** of a molecule is that conformation where the **dihedral angle** between the bonds at each of the carbon–carbon bonds is 180° and where atoms or groups bonded to carbons at each end of a carbon–carbon bond are as far apart as possible. The 180° dihedral angle in the staggered conformation of ethane is indicated in Fig. 4.5*b*.

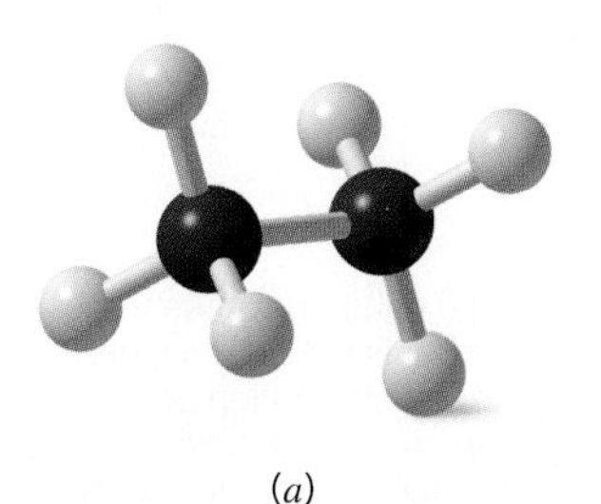

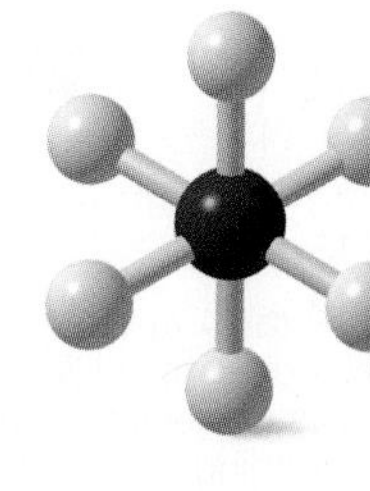

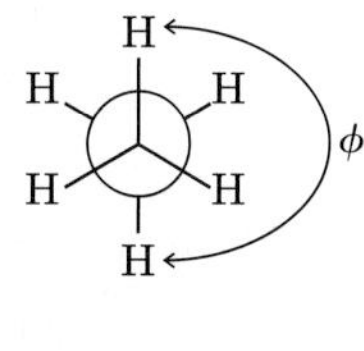

The dihedral angle (φ) between these hydrogens is 180°.

Chem3D Model

(*a*) (*b*)

FIGURE 4.5 *(a)* The staggered conformation of ethane. *(b)* The Newman projection formula for the staggered conformation.

The eclipsed conformation of ethane is shown in Fig. 4.6 using ball-and-stick models and a Newman projection. In an **eclipsed conformation** the atoms bonded to carbons at each end of a carbon–carbon bond are directly opposed to one another. The dihedral angle between them is 0°.

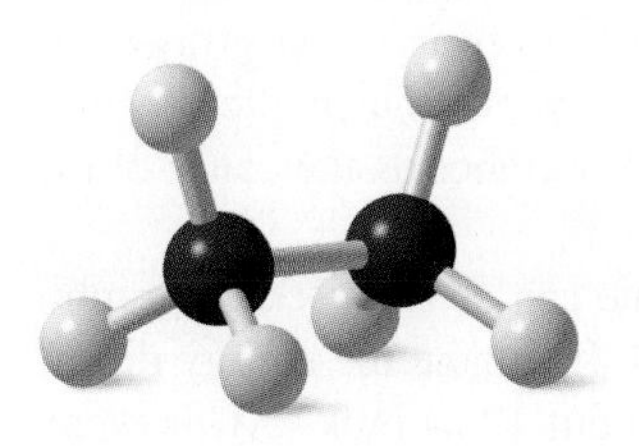

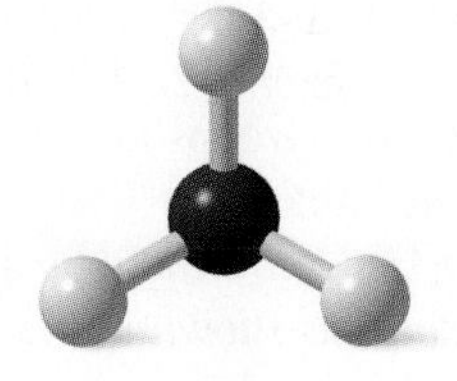

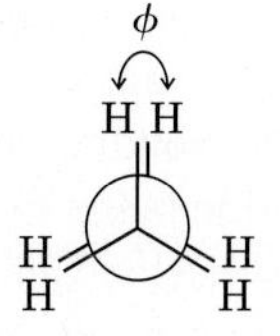

The dihedral angle (φ) between these hydrogens is 0°.

Chem3D Model

(*a*) (*b*)

FIGURE 4.6 *(a)* The eclipsed conformation of ethane. *(b)* The Newman projection formula for the eclipsed conformation.

CD Tutorial

Conformational analysis of ethane

Now let us consider a conformational analysis of ethane. Clearly, infinitesimally small changes in the dihedral angle between C—H bonds at each end of ethane could lead to an infinite number of conformations, including, of course, the staggered and eclipsed conformations. These different conformations are not all of equal stability, however, and it is known that the staggered conformation of ethane is the most stable conformation (i.e., it is the conformation of lowest potential energy). The fundamental reason for this has recently come to light.

Quantum mechanical calculations by L. Goodman and V. T. Pophristic (Rutgers University) have shown that the greater stability of the staggered conformation in ethane over the eclipsed conformation is mainly due to favorable overlap between sigma (σ) bonding orbitals from the C—H bonds at one carbon and unfilled antibonding sigma (σ^*) orbitals at the adjacent carbon. In ethane's staggered conformation, electrons from a given bonding C—H orbital on one carbon can be shared with an unfilled σ^* orbital at the adjacent carbon. This phenomenon of electron delocalization (via orbital overlap) from a filled bonding orbital to an adjacent unfilled orbital is called **hyperconjugation** (and we shall see in later chapters that it is a general stabilizing effect). Figure 4.7*a* shows the favorable overlap of σ and σ^* in ethane by color coding of the orbital phases.

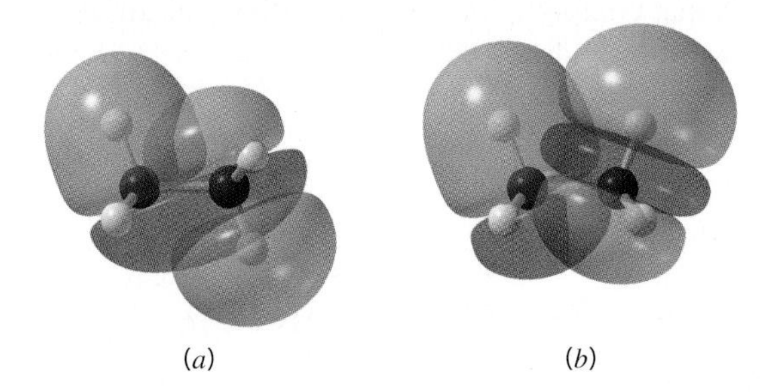

FIGURE 4.7 The bonding C—H orbital is the orbital where one lobe of a single phase (represented by one color) envelops the C—H atoms. The adjacent unfilled antibonding orbital is the orbital where the phase changes between the carbon and its hydrogen atom. The staggered conformation of ethane *(a)* has greater overlap between the bonding C—H orbital and the adjacent antibonding orbital than in the eclipsed conformation *(b)*. The orbital overlap shown in *(a)* leads to the lower potential energy of the staggered conformation of ethane.

If we now consider the eclipsed conformation of ethane (Fig. 4.7*b*), where the C—H bonds at each carbon are directly opposed to each other, we see that the bonding σ C—H orbital at one carbon does not overlap to as great an extent with the adjacent antibonding orbital as in the staggered conformation. The possibility for hyperconjugation is diminished, and therefore the potential energy of this conformation is higher.

The σ–σ^* interactions in ethane are present in more complicated molecules as well. However, where atoms and groups larger than hydrogen are involved in a conformational analysis, such as our example in Section 4.9, it is likely that repulsion of the electron clouds involved in the bonding of those groups increases in importance as the cause of the staggered conformation being most stable.

The energy difference between the conformations of ethane can be represented graphically in a **potential energy diagram,** as shown in Figure 4.8. In ethane the energy difference between the staggered and eclipsed conformations is about 12 kJ mol^{-1}. This small barrier to rotation is called the **torsional barrier** of the single bond. Because of this barrier, some molecules will wag back and forth with their atoms in staggered or nearly staggered conformations, while others with slightly more energy will rotate through an eclipsed conformation to another staggered conformation. At any given moment, unless

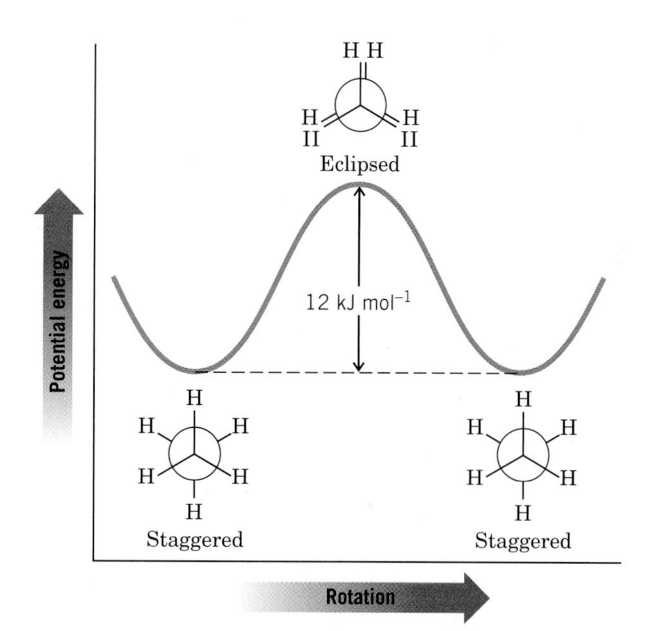

FIGURE 4.8 Potential energy changes that accompany rotation of groups about the carbon–carbon bond of ethane.

the temperature is extremely low (−250°C), most ethane molecules will have enough energy to undergo bond rotation from one conformation to another.

What does all this mean about ethane? We can answer this question in two different ways. If we consider a single molecule of ethane, we can say, for example, that it will spend most of its time in the lowest energy, staggered conformation, or in a conformation very close to being staggered. Many times every second, however, it will acquire enough energy through collisions with other molecules to surmount the torsional barrier and it will rotate through an eclipsed conformation. If we speak in terms of a large number of ethane molecules (a more realistic situation), we can say that at any given moment most of the molecules will be in staggered or nearly staggered conformations.

If we consider substituted ethanes such as GCH_2CH_2G (where G is a group or atom other than hydrogen), the barriers to rotation are somewhat larger, but they are still far too small to allow isolation of the different staggered conformations. The factors involved in this rotational barrier are together called **torsional strain** and include the orbital considerations discussed above as well as repulsive interactions called **steric hindrance** between electron clouds of bonded groups. In the next section we consider a conformational analysis of butane, where groups larger than hydrogen are involved in the analysis.

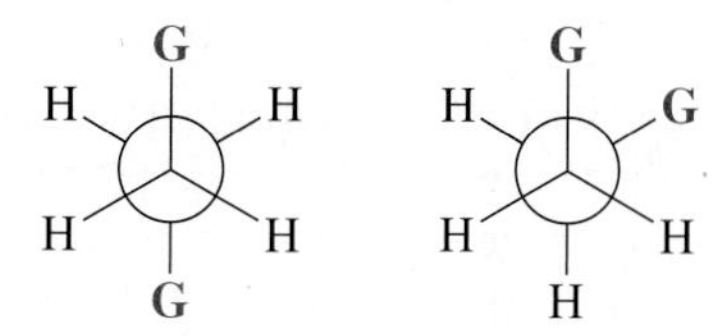

Conformers like these cannot be isolated except at extremely low temperatures.

J. H. van't Hoff.

The idea that certain conformations of molecules are favored originates from the work of van't Hoff. He was also winner of the first Nobel Prize in Chemistry (1901) for his work in chemical kinetics.

CD Tutorial

Conformational analysis of propane

4.9 Conformational Analysis of Butane

If we consider rotation about the C2—C3 bond of butane, we find that there are six important conformers, shown as I–VI below:

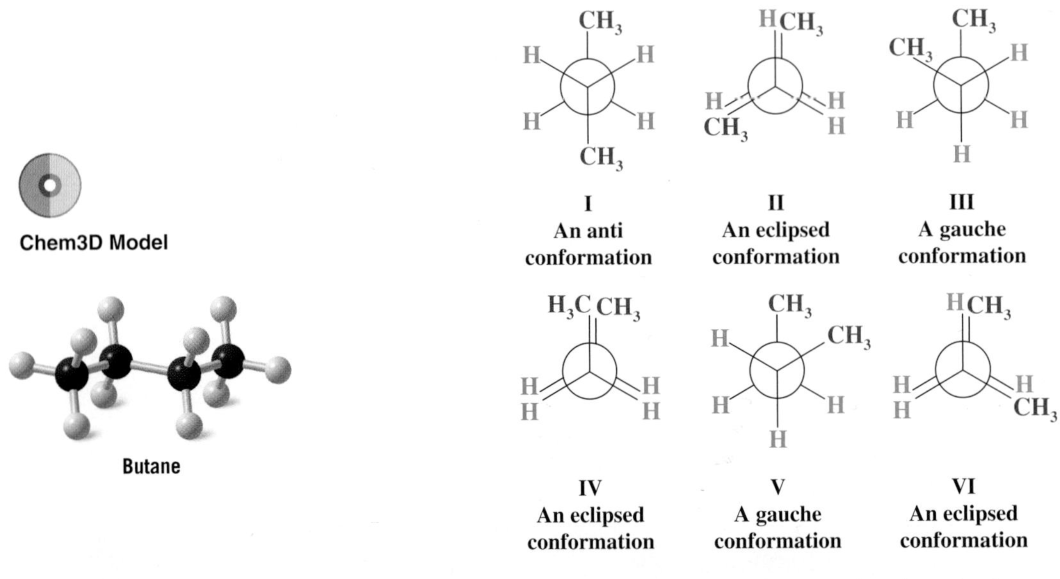

STUDY TIP!

You should build a molecular model of butane and examine its various conformations as we discuss their relative potential energies.

The **anti conformation (I)** does not have torsional strain from steric hindrance because the groups are staggered and the methyl groups are far apart. The anti conformation is the most stable. The methyl groups in the **gauche conformations III** and **V** are close enough to each other that the van der Waals forces between them are *repulsive;* the electron clouds of the two groups are so close that they repel each other. This repulsion causes the gauche conformations to have approximately 3.8 kJ mol^{-1} more energy than the anti conformation.

The eclipsed conformations **(II, IV,** and **VI)** represent energy maxima in the potential energy diagram (Fig. 4.9). Eclipsed conformations **II** and **VI** have van der Waals repulsions arising from the eclipsed methyl groups and hydrogen atoms. Eclipsed conforma-

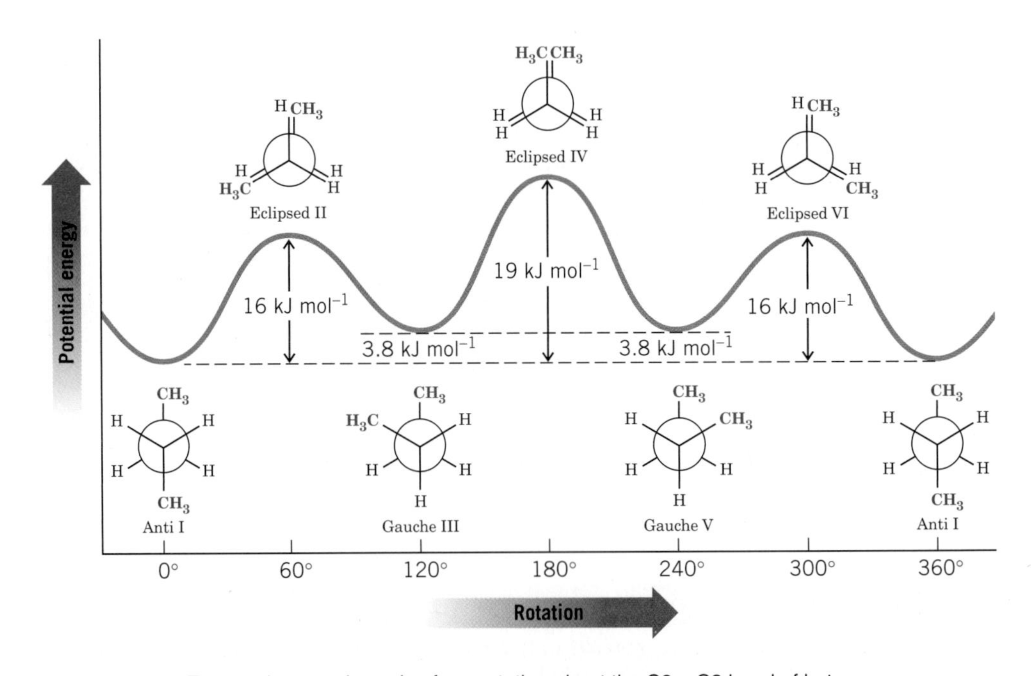

FIGURE 4.9 Energy changes that arise from rotation about the C2—C3 bond of butane.

tion **IV** has the greatest energy of all because of the added large van der Waals repulsive force between the eclipsed methyl groups as compared to **II** and **VI**.

Although the barriers to rotation in a butane molecule are larger than those of an ethane molecule (Section 4.8), they are still far too small to permit isolation of the gauche and anti conformations at normal temperatures. Only at extremely low temperatures would the molecules have insufficient energies to surmount these barriers.

We saw earlier that van der Waals forces can be *attractive.* Here, however, we find that they can also be *repulsive,* leading to steric hindrance. Whether van der Waals interactions lead to attraction or to repulsion depends on the distance that separates the two groups. As two nonpolar groups are brought closer and closer together, the first effect is one in which a momentarily unsymmetrical distribution of electrons in one group induces an opposite polarity in the other. The opposite charges induced in those portions of the two groups that are in closest proximity lead to attraction between them. This attraction increases to a maximum as the internuclear distance of the two groups decreases. The internuclear distance at which the attractive force is at a maximum is equal to the sum of what are called the *van der Waals radii* of the two groups. The van der Waals radius of a group is, in effect, a measure of its size. If the two groups are brought still closer—closer than the sum of their van der Waals radii—the interaction between them becomes repulsive. Their electron clouds begin to penetrate each other, and strong electron–electron interactions begin to occur.

Gauche conformers **III** and **V** of butane are examples of stereoisomers. **Stereoisomers** have the same molecular formula and connectivity but different arrangements of atoms in three-dimensional space. Because conformers **III** and **V** are stereoisomers that are interconvertible with one another by bond rotations, we call them **conformational stereoisomers.**

Conformational analysis is but one of the ways in which we will consider the three-dimensional shapes and stereochemistry of molecules. We shall see that there are other types of stereoisomers that cannot be interconverted simply by rotations about single bonds. Among these are cis–trans cycloalkane isomers (Section 4.14) and others that we shall consider in Chapter 5.

PROBLEM 4.10

Sketch a curve similar to that in Fig. 4.9 showing in general terms the energy changes that arise from rotation about the C2—C3 bond of 2-methylbutane. You need not concern yourself with the actual numerical values of the energy changes, but you should label all maxima and minima with the appropriate conformations.

4.10 The Relative Stabilities of Cycloalkanes: Ring Strain

Cycloalkanes do not all have the same relative stability. Data from heats of combustion (Section 4.10A) show that cyclohexane is the most stable cycloalkane and cyclopropane and cyclobutane are much less stable. The relative instability of cyclopropane and cyclobutane is a direct consequence of their cyclic structures, and for this reason their molecules are said to possess **ring strain.** To see how this can be demonstrated experimentally, we need to examine the relative heats of combustion of cycloalkanes.

4.10A Heats of Combustion

The **heat of combustion** of a compound is the enthalpy change for the complete oxidation of the compound.

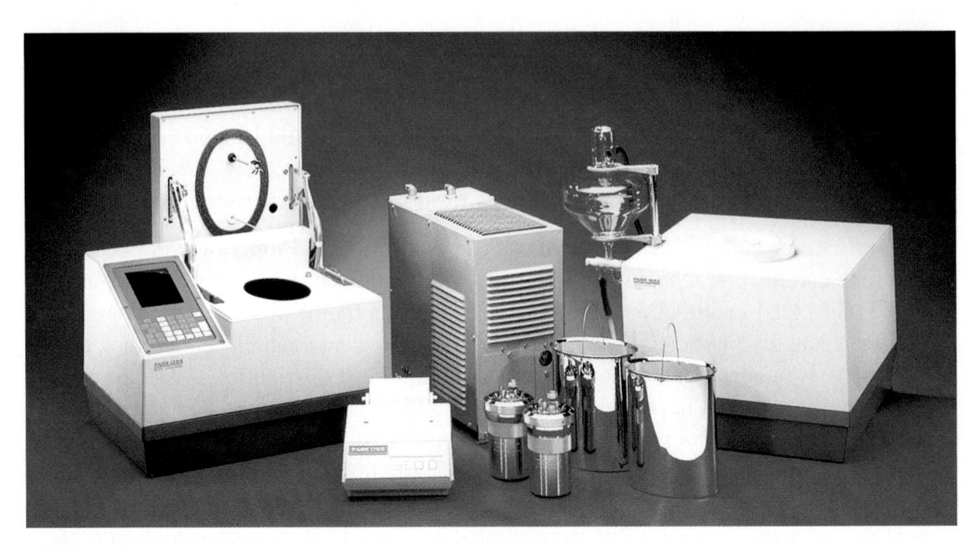

A bomb calorimeter apparatus, such as shown here, is used to precisely measure the amount of heat released upon combustion of a known amount of sample inside a sturdy metal 'bomb' (two are shown at center, front) that has been pressurized with oxygen gas.

For a hydrocarbon complete oxidation means converting it to carbon dioxide and water. This can be accomplished experimentally, and the amount of heat evolved can be accurately measured in a device called a calorimeter. For methane, for example, the heat of combustion is -803 kJ mol^{-1}:

$$CH_4 + 2\,O_2 \longrightarrow CO_2 + 2\,H_2O \qquad \Delta H° = -803 \text{ kJ mol}^{-1}$$

For isomeric hydrocarbons, complete combustion of 1 mol of each will require the same amount of oxygen and will yield the same number of moles of carbon dioxide and water. We can, therefore, use heats of combustion to measure the relative stabilities of the isomers.

Consider, as an example, the combustion of butane and isobutane:

$$\underset{(C_4H_{10})}{CH_3CH_2CH_2CH_3} + 6\tfrac{1}{2}\,O_2 \longrightarrow 4\,CO_2 + 5\,H_2O \qquad \Delta H° = -2877 \text{ kJ mol}^{-1}$$

$$\underset{(C_4H_{10})}{CH_3CH(CH_3)CH_3} + 6\tfrac{1}{2}\,O_2 \longrightarrow 4\,CO_2 + 5\,H_2O \qquad \Delta H° = -2868 \text{ kJ mol}^{-1}$$

Since butane liberates more heat on combustion than isobutane, it must contain relatively more potential energy. Isobutane, therefore, must be *more stable.* Figure 4.10 illustrates this comparison.

4.10B Heats of Combustion of Cycloalkanes

The cycloalkanes constitute a *homologous series;* each member of the series differs from the one immediately preceding it by the constant amount of one —CH_2— group. Thus, the general equation for combustion of a cycloalkane can be formulated as follows:

$$(CH_2)_n + \tfrac{3}{2}n\,O_2 \longrightarrow n\,CO_2 + n\,H_2O + \text{heat}$$

Because the cycloalkanes are not isomeric, their heats of combustion cannot be compared directly. However, we can calculate the amount of heat evolved *per CH_2 group.* On this basis, the stabilities of the cycloalkanes become directly comparable. The results of such an investigation are given in Table 4.6.

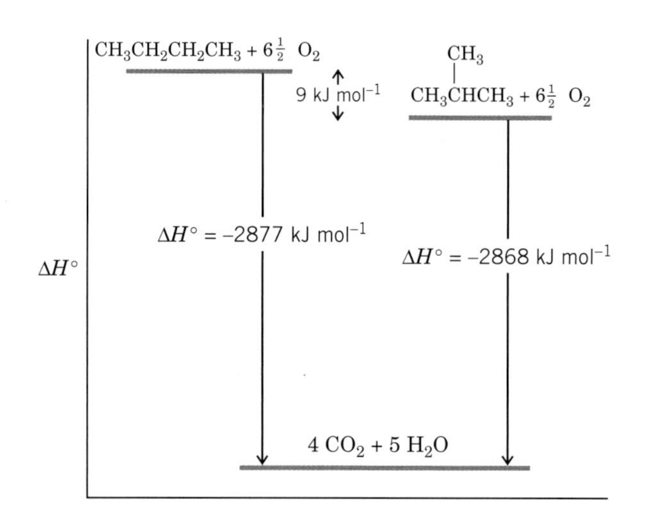

FIGURE 4.10 Heats of combustion show that isobutane is more stable than butane by 9 kJ mol^{-1}.

Several observations emerge from a consideration of these results:

1. Cyclohexane has the lowest heat of combustion per CH_2 group (658.7 kJ mol^{-1}). This amount does not differ from that of unbranched alkanes, which, having no ring, can have no ring strain. We can assume, therefore, that cyclohexane has no ring strain and that it can serve as our standard for comparison with other cycloalkanes. We can calculate ring strain for the other cycloalkanes (Table 4.6) by multiplying 658.7 kJ mol^{-1} by n and then subtracting the result from the heat of combustion of the cycloalkane.

2. The combustion of cyclopropane evolves the greatest amount of heat per CH_2 group. Therefore, molecules of cyclopropane must have the greatest ring strain (115 kJ mol^{-1}, Table 4.6). Since cyclopropane molecules evolve the greatest amount of heat energy per CH_2 group on combustion, they must contain the greatest amount of potential energy per CH_2 group. Thus what we call ring strain is a form of potential energy that the cyclic molecule contains. The more ring strain a molecule possesses, the more potential energy it has and the less stable it is compared to its ring homologs.

3. The combustion of cyclobutane evolves the second largest amount of heat per CH_2 group and, therefore, cyclobutane has the second largest amount of ring strain (109 kJ mol^{-1}).

TABLE 4.6 Heats of Combustion and Ring Strain of Cycloalkanes

Cycloalkane $(CH_2)_n$	n	Heat of Combustion (kJ mol^{-1})	Heat of Combustion per CH_2 Group (kJ mol^{-1})	Ring Strain (kJ mol^{-1})
Cyclopropane	3	2091	697.0	115
Cyclobutane	4	2744	686.0	109
Cyclopentane	5	3320	664.0	27
Cyclohexane	6	3952	658.7	0
Cycloheptane	7	4637	662.4	27
Cyclooctane	8	5310	663.8	42
Cyclononane	9	5981	664.6	54
Cyclodecane	10	6636	663.6	50
Cyclopentadecane	15	9885	659.0	6
Unbranched alkane	—	—	658.6	—

4. While other cycloalkanes possess ring strain to varying degrees, the relative amounts are not large. Cyclopentane and cycloheptane have about the same modest amount of ring strain. Rings of 8, 9, and 10 members have slightly larger amounts of ring strain and then the amount falls off. A 15-membered ring has only a very slight amount of ring strain.

4.11 The Origin of Ring Strain in Cyclopropane and Cyclobutane: Angle Strain and Torsional Strain

The carbon atoms of alkanes are sp^3 hybridized. The normal tetrahedral bond angle of an sp^3-hybridized atom is 109.5°. In cyclopropane (a molecule with the shape of a regular triangle), the internal angles must be 60° and therefore they must depart from this ideal value by a very large amount—by 49.5°:

This compression of the internal bond angle causes what chemists call **angle strain.** Angle strain exists in a cyclopropane ring because the sp^3 orbitals of the carbon atoms cannot overlap as effectively (Fig. 4.11*a*) as they do in alkanes (where perfect end-on overlap is possible). The carbon–carbon bonds of cyclopropane are often described as being "bent." Orbital overlap is less effective. (The orbitals used for these bonds are not purely sp^3; they contain more p character). The carbon–carbon bonds of cyclopropane are weaker, and as a result the molecule has greater potential energy.

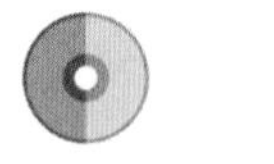

Chem3D Model

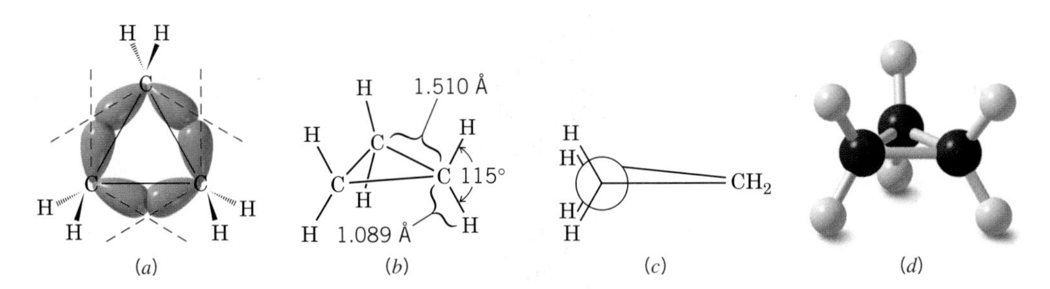

FIGURE 4.11 *(a)* Orbital overlap in the carbon–carbon bonds of cyclopropane cannot occur perfectly end-on. This leads to weaker "bent" bonds and to angle strain. *(b)* Bond distances and angles in cyclopropane. *(c)* A Newman projection formula as viewed along one carbon–carbon bond shows the eclipsed hydrogens. (Viewing along either of the other two bonds would show the same picture.) *(d)* Ball-and-stick model of cyclopropane.

While angle strain accounts for most of the ring strain in cyclopropane, it does not account for it all. Because the ring is (of necessity) planar, the C—H bonds of the ring are all *eclipsed* (Figs. 4.11*b*,*c*), and the molecule has torsional strain as well.

Cyclobutane also has considerable angle strain. The internal angles are 88°—a departure of more than 21° from the normal tetrahedral bond angle. The cyclobutane ring is not planar but is slightly "folded" (Fig. 4.12*a*). If the cyclobutane ring were planar, the angle strain would be somewhat less (the internal angles would be 90° instead of 88°), but torsional strain would be considerably larger because all eight C—H bonds would be eclipsed. By folding or bending slightly the cyclobutane ring relieves more of its torsional strain than it gains in the slight increase in its angle strain.

(a) (b)

Chem3D Model

FIGURE 4.12 *(a)* The "folded" or "bent" conformation of cyclobutane. *(b)* The "bent" or "envelope" form of cyclopentane. In this structure the front carbon atom is bent upward. In actuality, the molecule is flexible and shifts conformations constantly.

4.11A Cyclopentane

The internal angles of a regular pentagon are 108°, a value very close to the normal tetrahedral bond angles of 109.5°. Therefore, if cyclopentane molecules were planar, they would have very little angle strain. Planarity, however, would introduce considerable torsional strain because all 10 C—H bonds would be eclipsed. Consequently, like cyclobutane, cyclopentane assumes a slightly bent conformation in which one or two of the atoms of the ring are out of the plane of the others (Fig. 4.12*b*). This relieves some of the torsional strain. Slight twisting of carbon–carbon bonds can occur with little change in energy and causes the out-of-plane atoms to move into plane and causes others to move out. Therefore, the molecule is flexible and shifts rapidly from one conformation to another. With little torsional strain and angle strain, cyclopentane is almost as stable as cyclohexane.

An understanding of this and subsequent discussions of conformational analysis can be aided immeasurably through the use of a molecular model. We suggest you "follow along" with models as you read Sections 4.12–4.14.

4.12 Conformations of Cyclohexane

There is considerable evidence that the most stable conformation of the cyclohexane ring is the "chair" conformation illustrated in Fig. 4.13. In this nonplanar structure the carbon–carbon bond angles are all 109.5° and are thereby free of angle strain. The chair conformation is free of torsional strain as well. When viewed along any carbon–carbon bond (viewing the structure from an end, Fig. 4.14), the bonds are seen to be perfectly staggered. Moreover, the hydrogen atoms at opposite corners of the cyclohexane ring are maximally separated.

CD Tutorial

Chair conformation of cyclohexane

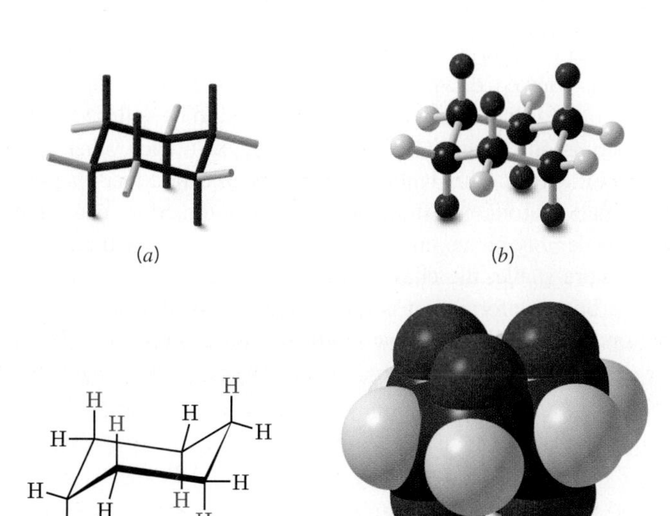

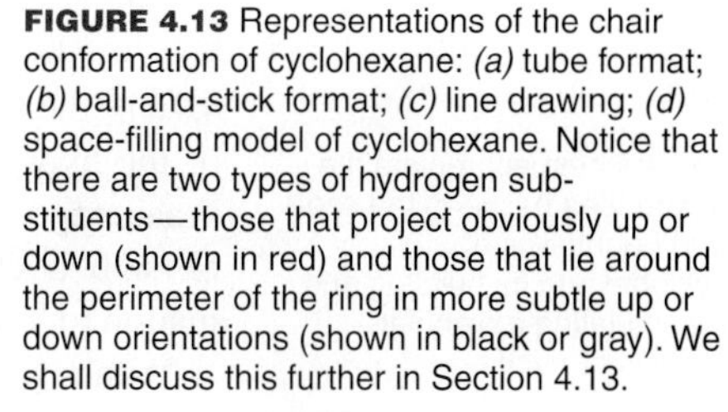

FIGURE 4.13 Representations of the chair conformation of cyclohexane: *(a)* tube format; *(b)* ball-and-stick format; *(c)* line drawing; *(d)* space-filling model of cyclohexane. Notice that there are two types of hydrogen substituents—those that project obviously up or down (shown in red) and those that lie around the perimeter of the ring in more subtle up or down orientations (shown in black or gray). We shall discuss this further in Section 4.13.

Chem3D Model

FIGURE 4.14 *(a)* A Newman projection of the chair conformation of cyclohexane. (Comparisons with an actual molecular model will make this formulation clearer and will show that similar staggered arrangements are seen when other carbon–carbon bonds are chosen for sighting.) *(b)* Illustration of large separation between hydrogen atoms at opposite corners of the ring (designated C1 and C4) when the ring is in the chair conformation.

By partial rotations about the carbon–carbon single bonds of the ring, the chair conformation can assume another shape called the "boat" conformation (Fig. 4.15). The boat conformation is like the chair conformation in that it is also free of angle strain.

FIGURE 4.15 *(a)* The boat conformation of cyclohexane is formed by "flipping" one end of the chair form up (or down). This flip requires only rotations about carbon–carbon single bonds. *(b)* Ball-and-stick model of the boat conformation. *(c)* A space-filling model of the boat conformation.

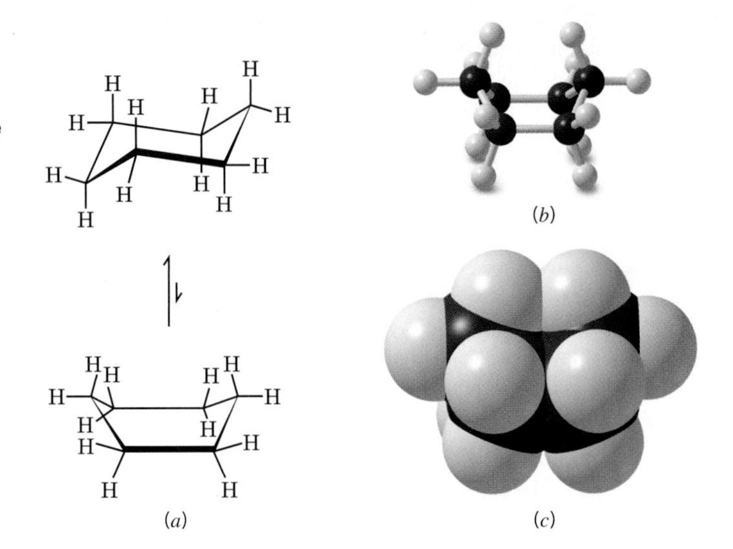

Chem3D Model

The boat conformation, however, is not free of torsional strain. When a model of the boat conformation is viewed down carbon–carbon bond axes along either side (Fig. 4.16*a*), the C—H bonds at those carbon atoms are found to be eclipsed. Additionally, two of the hydrogen atoms on C1 and C4 are close enough to each other to cause van der Waals repulsion (Fig. 4.16*b*). This latter effect has been called the "flagpole" interaction of the boat conformation. Torsional strain and flagpole interactions cause the boat conformation to have considerably higher energy than the chair conformation.

You will best appreciate the differences between the chair and boat forms of cyclohexane by building and manipulating molecular models of each.

Although it is more stable, the chair conformation is much more rigid than the boat conformation. The boat conformation is quite flexible. By flexing to a new form—the twist conformation (Fig. 4.17)—the boat conformation can relieve some of its torsional

FIGURE 4.16 *(a)* Illustration of the eclipsed conformation of the boat conformation of cyclohexane. *(b)* Flagpole interaction of the C1 and C4 hydrogen atoms of the boat conformation. The C1–C4 flagpole interaction is also readily apparent in Fig. 14.15*(c)*.

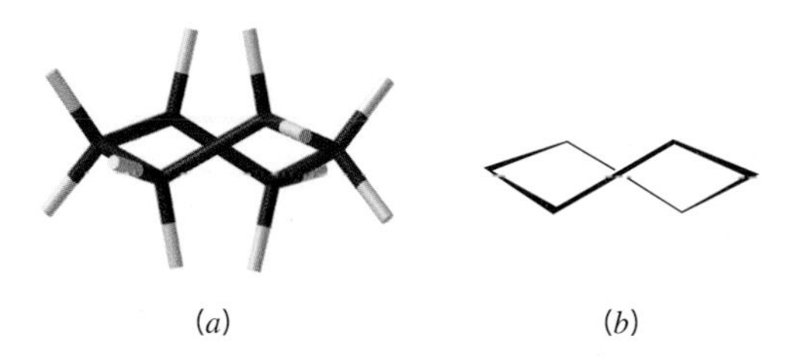

FIGURE 4.17 *(a)* Carbon skeleton and *(b)* line drawing of the twist conformation of cyclohexane.

Chem3D Model

strain and, at the same time, reduce the flagpole interactions. Thus, the twist conformation has a lower energy than the boat conformation. *The stability gained by flexing is insufficient, however, to cause the twist conformation of cyclohexane to be more stable than the chair conformation.* The chair conformation is estimated to be lower in energy than the twist conformation by approximately 21 kJ mol^{-1}.

The energy barriers between the chair, boat, and twist conformations of cyclohexane are low enough (Fig. 4.18) to make separation of the conformers impossible at room temperature. At room temperature the thermal energies of the molecules are great enough to cause approximately 1 million interconversions to occur each second and, ***because of the greater stability of the chair, more than 99% of the molecules are estimated to be in a chair conformation at any given moment.***

CD Tutorial
Interconversion of cyclohexane conformers

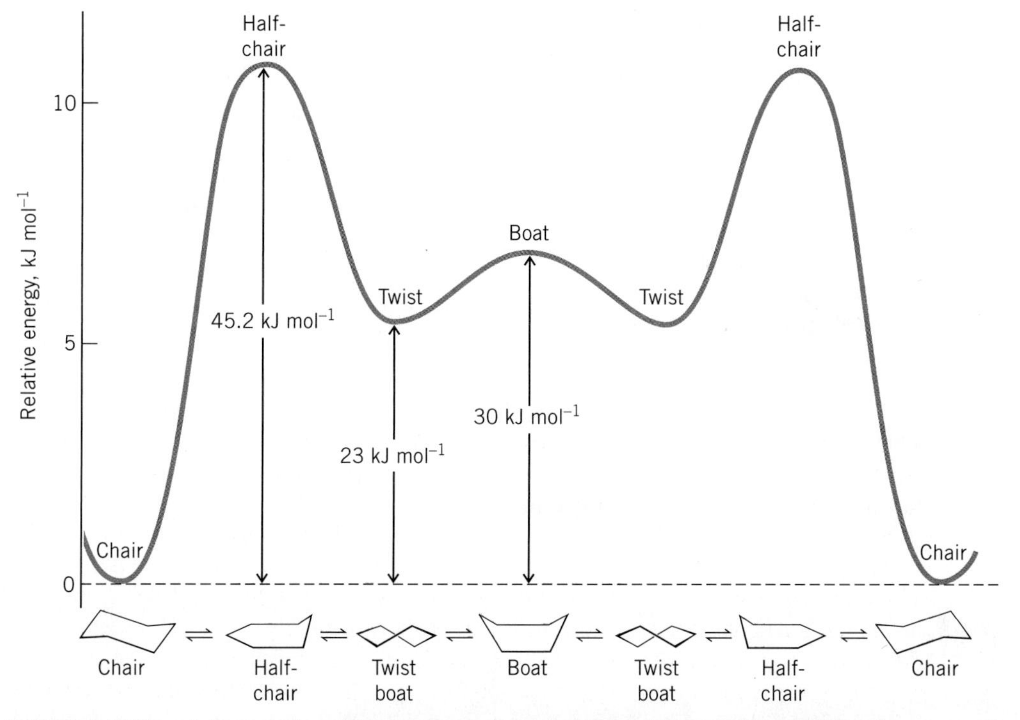

FIGURE 4.18 The relative energies of the various conformations of cyclohexane. The positions of maximum energy are conformations called half-chair conformations, in which the carbon atoms of one end of the ring have become coplanar.

4.12A Conformations of Higher Cycloalkanes

Cycloheptane, cyclooctane, and cyclononane and other higher cycloalkanes also exist in nonplanar conformations. The small instabilities of these higher cycloalkanes (Table 4.6) appear to be caused primarily by torsional strain and van der Waals repulsions between

Derek H. R. Barton (1918–1998, formerly Distinguished Professor of Chemistry at Texas A&M University) and Odd Hassel (1897–1981, formerly Chair of Physical Chemistry of Oslo University) shared the Nobel Prize in 1969 "for developing and applying the principles of conformation in chemistry." Their work led to fundamental understanding of not only the conformations of cyclohexane rings but also the structures of steroids (Section 23.4) and other compounds containing cyclohexane rings.

Derek Barton

Odd Hassel

hydrogen atoms across rings, called *transannular strain.* The nonplanar conformations of these rings, however, are essentially free of angle strain.

X-ray crystallographic studies of cyclodecane reveal that the most stable conformation has carbon–carbon–carbon bond angles of 117°. This indicates some angle strain. The wide bond angles apparently allow the molecules to expand and thereby minimize unfavorable repulsions between hydrogen atoms across the ring.

There is very little free space in the center of a cycloalkane unless the ring is quite large. Calculations indicate that cyclooctadecane, for example, is the smallest ring through which a $—CH_2CH_2CH_2—$ chain can be threaded. Molecules have been synthesized, however, that have large rings threaded on chains and that have large rings which are interlocked like links in a chain. These latter molecules are called **catenanes:**

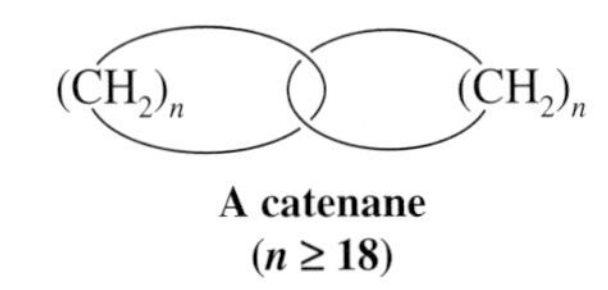

A catenane
($n \geq 18$)

In 1994 J. F. Stoddart and co-workers, then at the University of Birmingham (England), achieved a remarkable synthesis of a catenane containing a linear array of five interlocked rings. Because the rings are interlocked in the same way as those of the olympic symbol, they named the compound **olympiadane.**

The Chemistry of...

Nanoscale Motors and Molecular Switches

Molecular rings that interlock with one another and others that are linear molecules threaded through rings are proving to have fascinating potential for the creation of molecular switches and motors. Molecules consisting of interlocking rings, like a chain, are called **catenanes.** The first catenanes were synthesized in the 1960s and have come to include examples such as olympiadane, mentioned above. Further research by Stoddart (now at UCLA) and collaborators on interlocking molecules has recently led to examples such as the catenane molecular switch shown here in (i). In an application that could be useful in design of binary logic circuits, one ring of this molecule can be made to circumrotate in controlled fashion about the other, such that it switches between two defined states. As a demonstration of its potential for application in electron-

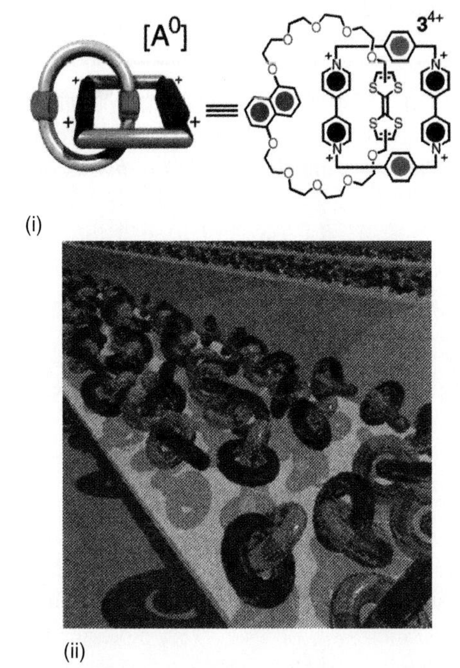

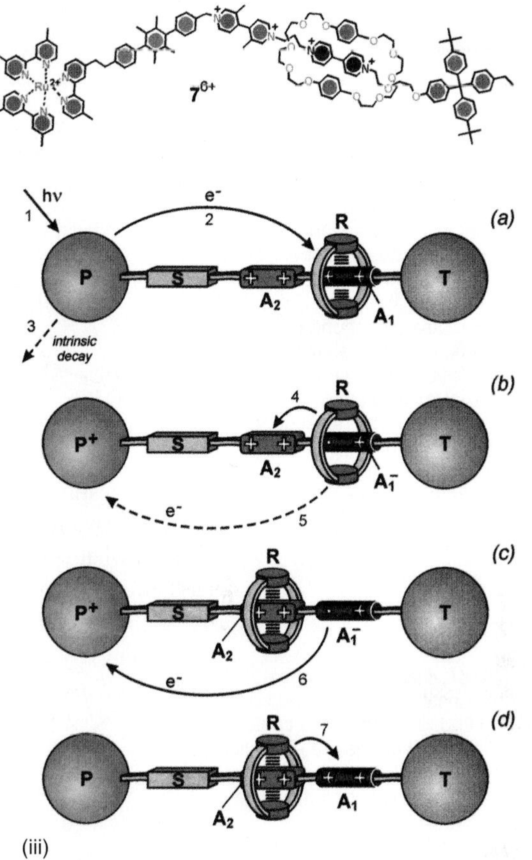

ics fabrication, a monolayer of these molecules has been "tiled" on a surface (ii) and shown to have characteristics like a conventional magnetic memory bit.

Molecules where a linear molecule is threaded through a ring are called **rotaxanes.** One captivating example of a rotaxane system is the one shown here in (iii), under development by V. Balzani (University of Bologna) and collaborators. By conversion of light energy to mechanical energy at the molecular level, this rotaxane behaves like a "four-stroke" shuttle engine. In step (a) light excitation of an electron in the **P** group leads to transfer of the electron to the initially +2 $\mathbf{A_1}$ group, at which point $\mathbf{A_1}$ is reduced to the +1 state. Ring **R,** which was attracted to $\mathbf{A_1}$ when it was in the +2 state, now slides over to $\mathbf{A_2}$ in step (b), which remains +2. Back transfer of the electron from $\mathbf{A_1}$ to $\mathbf{P^+}$ in step (c) restores the +2 state of $\mathbf{A_1}$, causing ring **R** to return to its original location in step (d). Modifications envisioned for this system include attaching binding sites to **R** such that some other molecular species could be transported from one location to another as **R** slides along the linear molecule, or linking **R** by a springlike tether to one end of the "piston rod" such that additional potential and mechanical energy can be incorporated in the system.

4.13 Substituted Cyclohexanes: Axial and Equatorial Hydrogen Atoms

The six-membered ring is the most common ring found among nature's organic molecules. For this reason, we shall give it special attention. We have already seen that the chair conformation of cyclohexane is the most stable one and that it is the predominant conformation of the molecules in a sample of cyclohexane. With this fact in mind, we are in a position to undertake a limited analysis of the conformations of substituted cyclohexanes.

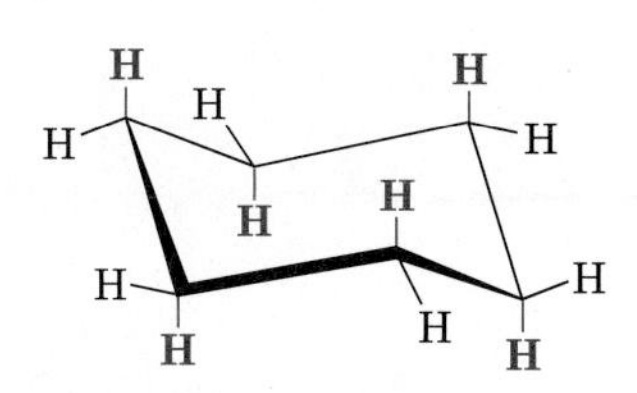

FIGURE 4.19 The chair conformation of cyclohexane. The axial hydrogen atoms are shown in color.

If we look carefully at the chair conformation of cyclohexane (Fig. 4.19), we can see that there are only two different kinds of hydrogen atoms. One hydrogen atom attached to each of the six carbon atoms lies around the perimeter of the ring of carbon atoms. These hydrogen atoms, by analogy with the equator of Earth, are called **equatorial** hydrogen atoms. Six other hydrogen atoms, one on each carbon, are oriented in a direction that is generally perpendicular to the average plane of the ring. These hydrogen atoms, again by

You should now learn how to draw the important chair conformation. Notice (Fig. 4.20*a*) the sets of colored lines that define the parallel relationship between the equatorial bonds and specific bonds of the ring. Notice, too (Fig. 4.20*b*), that when drawn this way, the axial bonds are all vertical, and when the vertex of the ring points up, the axial bond is up; when the vertex is down, the axial bond is down. Compare your drawings of chair conformations with actual models.

analogy with Earth, are called **axial** hydrogen atoms. There are three axial hydrogen atoms on each face of the cyclohexane ring, and their orientation (up or down) alternates from one carbon atom to the next.

We saw in Section 4.12 (and Fig. 4.18) that at room temperature the cyclohexane ring rapidly flips back and forth between two *equivalent* chair conformations. An important thing to notice now is that ***when the ring flips, all of the bonds that were axial become equatorial and vice versa:***

Axial
5 6 1 4 3 2
Axial
ring flip
4 5 6 3 2 1
Equatorial
Equatorial

The question one might next ask is: What is the most stable conformation of a cyclohexane derivative *in which one hydrogen atom has been replaced by an alkyl substituent?* That is, what is the most stable conformation of a monosubstituted cyclohexane? We can answer this question by considering methylcyclohexane as an example.

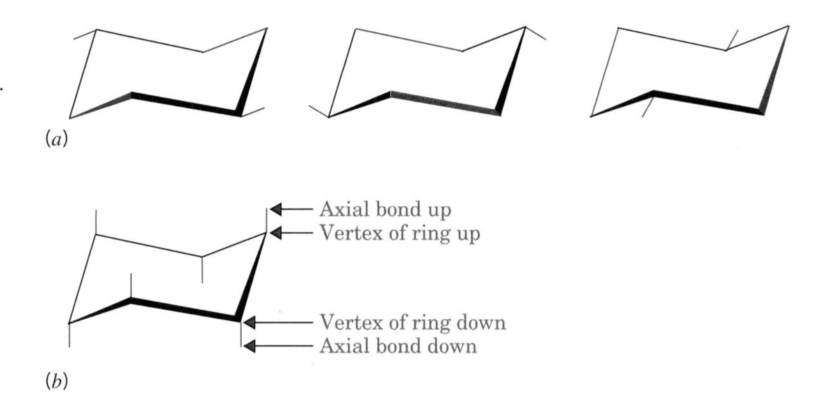

FIGURE 4.20 *(a)* Sets of parallel lines that constitute the ring and equatorial C—H bonds of the chair conformation. *(b)* The axial bonds are all vertical. When the vertex of the ring points up, the axial bond is up and vice versa.

Methylcyclohexane has two possible chair conformations (Fig. 4.21), and these are interconvertible through the partial rotations that constitute a ring flip. In one conformation (Fig. 4.21*a*) the methyl group (with yellow hydrogens) occupies an *axial* position, and in the other the methyl group occupies an *equatorial* position. Studies indicate that the conformation with the methyl group equatorial is more stable than the conformation with the methyl group axial by about 7.6 kJ mol^{-1}. Thus, in the equilibrium mixture, the conformation with the methyl group in the equatorial position is the predominant one. Calculations show that it constitutes about 95% of the equilibrium mixture (Table 4.7).

The greater stability of methylcyclohexane with an equatorial methyl group can be understood through an inspection of the two forms as they are shown in Figs. 4.21*a–c*.

Studies done with models of the two conformations show that when the methyl group is axial, it is so close to the two axial hydrogen atoms on the same side of the molecule (attached to C3 and C5 atoms) that the van der Waals forces between them are repulsive. This type of steric strain, because it arises from an interaction between axial groups on carbon atoms 1 and 3 (or 5), is called a **1,3-diaxial interaction.** Similar studies with other substituents indicate that ***there is generally less repulsive interaction when the groups are equatorial rather than axial.***

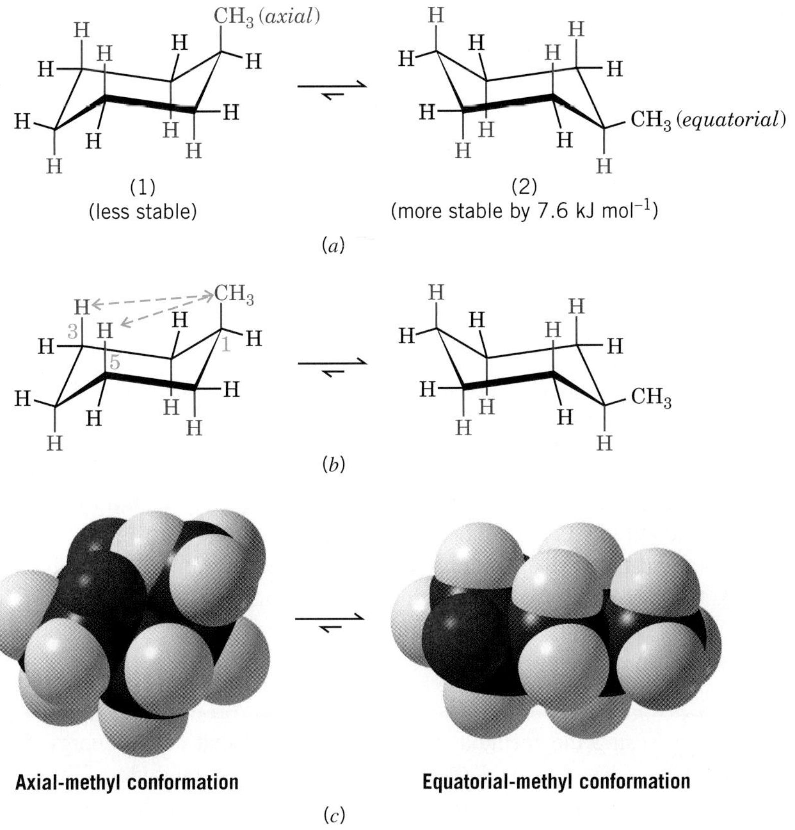

(1)
(less stable)

(2)
(more stable by 7.6 kJ mol^{-1})

(*a*)

(*b*)

Axial-methyl conformation

Equatorial-methyl conformation

(*c*)

FIGURE 4.21 *(a)* The conformations of methylcyclohexane with the methyl group axial (1) and equatorial (2). *(b)* **1,3-Diaxial interactions** between the two axial hydrogen atoms and the axial methyl group in the axial conformation of methylcyclohexane are shown with dashed arrows. Less crowding occurs in the equatorial conformation. *(c)* Space-filling molecular models for the axial–methyl and equatorial–methyl conformers of methylcyclohexane. In the axial–methyl conformer the methyl group (shown with yellow hydrogen atoms) is crowded by the 1,3-diaxial hydrogen atoms (red), as compared to the equatorial–methyl conformer, which has no 1,3-diaxial interactions with the methyl group.

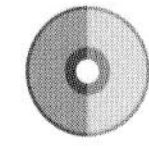

Chem3D Models

The strain caused by a 1,3-diaxial interaction in methylcyclohexane is the same as the strain caused by the close proximity of the hydrogen atoms of methyl groups in the gauche form of butane (Section 4.9A). Recall that the interaction in *gauche*-butane (called, for convenience, a *gauche interaction*) causes *gauche*-butane to be less stable than *anti*-butane by 3.8 kJ mol^{-1}. The following Newman projections will help you to see that the two steric interactions are the same. In the second projec-

TABLE 4.7 Relationship between Free-Energy Difference and Isomer Percentages for Isomers at Equilibrium at 25°C

Free-Energy Difference, $\Delta G°$ (kJ mol^{-1})	More Stable Isomer (%)	Less Stable Isomer (%)
0	50	50
1.7	67	33
2.7	75	25
3.4	80	20
4.0	83	17
5.9	91	9
7.5	95	5
11	99	1
17	99.9	0.1
23	99.99	0.01

tion we view axial methylcyclohexane along the C1—C2 bond and see that what we call a 1,3-diaxial interaction is simply a gauche interaction between the hydrogen atoms of the methyl group and the hydrogen atom at C3:

***gauche*-Butane**
(3.8 kJ mol^{-1} steric strain)

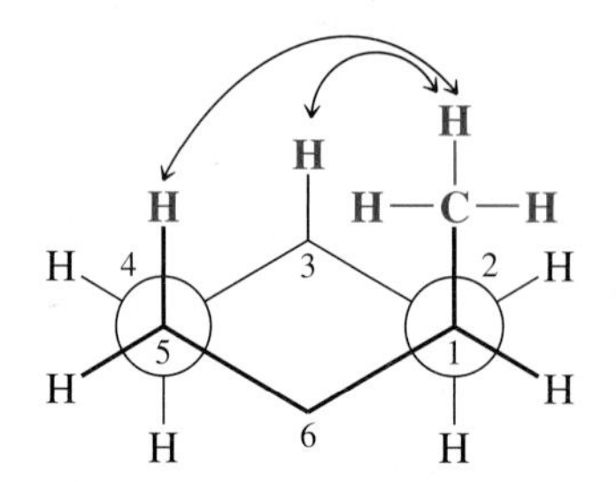

Axial methylcyclohexane
(two gauche interactions =
7.6 kJ mol^{-1} steric strain)

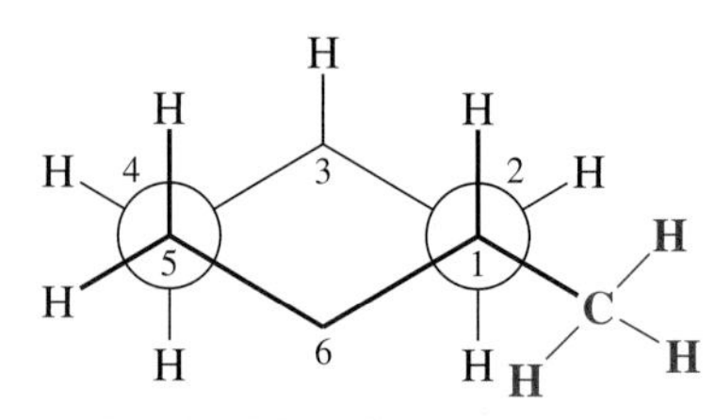

Equatorial methylcyclohexane

Viewing methylcyclohexane along the C1—C6 bond (do this with a model) shows that it has a second identical gauche interaction between the hydrogen atoms of the methyl group and the hydrogen atom at C5. The methyl group of axial methylcyclohexane, therefore, has two gauche interactions and, consequently, it has 7.6 kJ mol^{-1} of strain. The methyl group of equatorial methylcyclohexane does not have a gauche interaction because it is anti to C3 and C5.

PROBLEM 4.11 Show by a calculation (using the formula $\Delta G^\circ = -RT \ln k_{eq}$) that a free-energy difference of 7.6 kJ mol^{-1} between the axial and equatorial forms of methylcyclohexane at 25°C (with the equatorial form being more stable) does correlate with an equilibrium mixture in which the concentration of the equatorial form is approximately 95%.

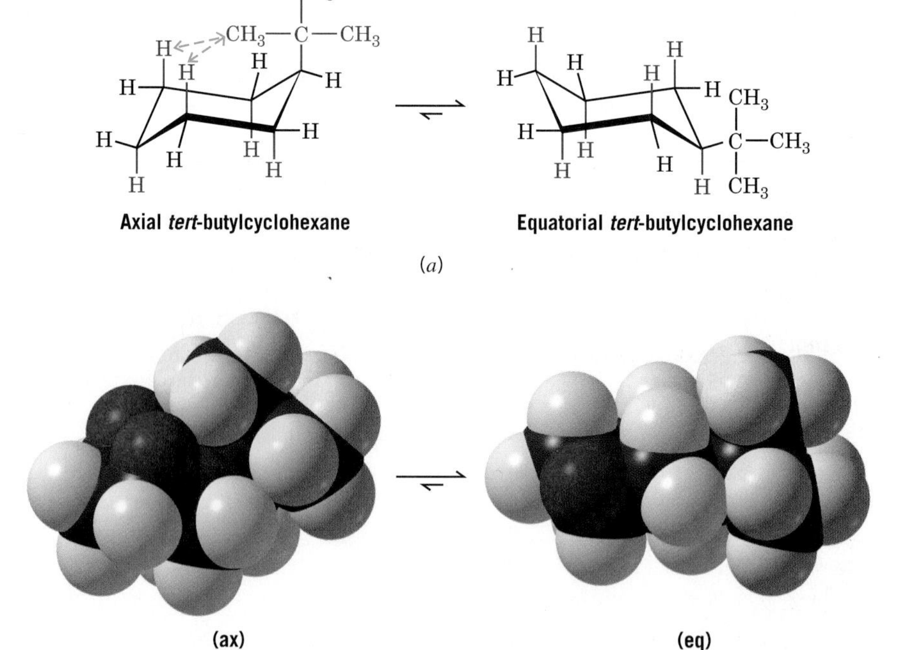

FIGURE 4.22 *(a)* Diaxial interactions with the large *tert*-butyl group axial cause the conformation with the *tert*-butyl group equatorial to be the predominant one to the extent of 99.99%. *(b)* Space-filling molecular models of *tert*-butylcyclohexane in the axial (ax) and equatorial (eq) conformations, highlighting the position of the 1,3-hydrogens (red) and the *tert*-butyl group (shown with yellow hydrogen atoms).

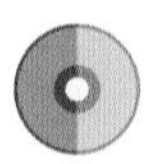

Chem3D Models

In cyclohexane derivatives with larger alkyl substituents, the strain caused by 1,3-diaxial interactions is even more pronounced. The conformation of *tert*-butylcyclohexane with the *tert*-butyl group equatorial is estimated to be approximately 21 kJ mol^{-1} more stable than the axial form (Fig. 4.22, page 170). This large energy difference between the two conformations means that, at room temperature, 99.99% of the molecules of *tert*-butylcyclohexane have the *tert*-butyl group in the equatorial position. (The molecule is not conformationally "locked," however; it still flips from one chair conformation to the other.)

4.14 Disubstituted Cycloalkanes: Cis–Trans Isomerism

The presence of two substituents on the ring of a molecule of any cycloalkane allows for the possibility of **cis–trans isomerism.** Let us consider 1,2-dimethylcyclopentane as an example. To simplify our presentation of the cyclopentane ring, we will draw it as though it were flat and nearly perpendicular to the page. We know, of course, that a cyclopentane ring is not flat in reality but is slightly bent and that at any given moment the various bent conformations are rapidly interconverted. This fact and our simplification of the drawing will not interfere with our analysis of cis–trans isomerism because neither changes the cis–trans orientation of groups bonded to the ring.

In the first structure shown in Figure 4.23 the methyl groups are on the same side of the ring, that is, they are cis. In the second structure the methyl groups are on the opposite side of the ring; they are trans.

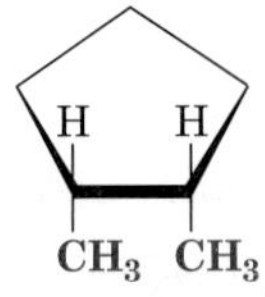

cis-1,2-Dimethylcyclopentane
bp 99.5°C

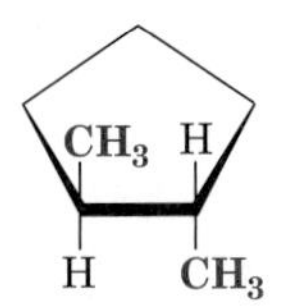

trans-1,2-Dimethylcyclopentane
bp 91.9°C

FIGURE 4.23 The *cis*- and *trans*-1,2-dimethylcyclopentanes.

The *cis*- and *trans*-1,2-dimethylcyclopentanes are stereoisomers: They differ from each other only in the arrangement of the atoms in space. In contrast to conformational stereoisomers, the cis and trans forms cannot be interconverted without breaking carbon–carbon bonds. As a result, the cis and trans forms can be separated, placed in separate bottles, and kept indefinitely.

1,3-Dimethylcyclopentanes show cis–trans isomerism as well:

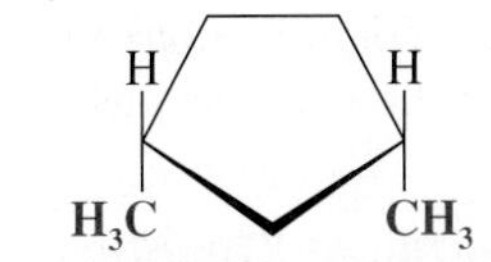

cis-1,3-Dimethylcyclopentane

trans-1,3-Dimethylcyclopentane

The physical properties of cis–trans isomers are different; they have different melting points, boiling points, and so on. Table 4.8 lists these physical constants of the dimethylcyclohexanes.

Problem 4.12

Write structures for the cis and trans isomers of **(a)** 1,2-dichlorocyclopropane and **(b)** 1,3-dibromocyclobutane.

The cyclohexane ring is, of course, not planar either. To simplify visualization of cis–trans relationships in cyclohexane, we will draw the ring as though it were flat and

TABLE 4.8 Physical Constants of Cis- and Trans-Disubstituted Cyclohexane Derivatives

Substituents	Isomer	mp (°C)	bp (°C)[a]
1,2-Dimethyl-	cis	−50.1	130.04^{760}
1,2-Dimethyl-	trans	−89.4	123.7^{760}
1,3-Dimethyl-	cis	−75.6	120.1^{760}
1,3-Dimethyl-	trans	−90.1	123.5^{760}
1,2-Dichloro-	cis	−6	93.5^{22}
1,2-Dichloro-	trans	−7	74.7^{16}

[a] The pressures (in units of torr) at which the boiling points were measured are given as superscripts.

nearly perpendicular to the page, as we did for cyclopentane. Planar representations of the 1,2-, 1,3-, and 1,4-dimethylcyclohexane isomers follow:

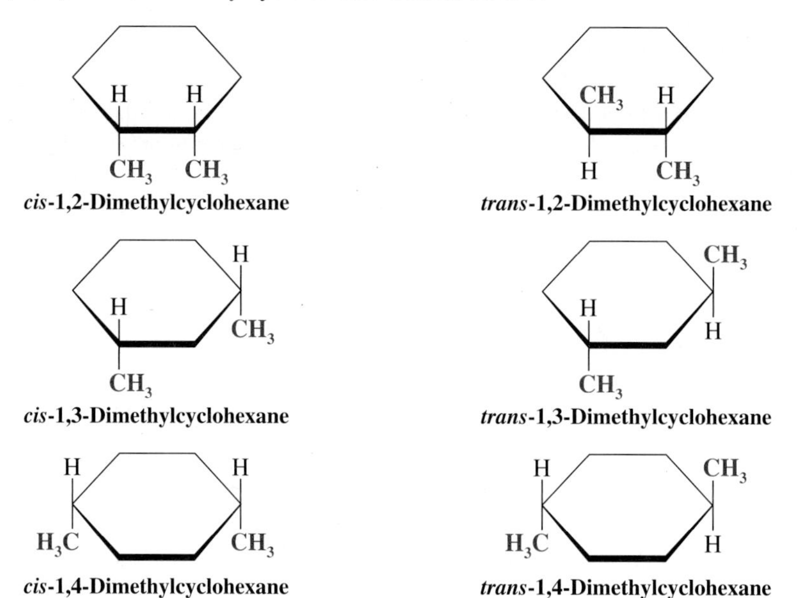

It is important to remember that these drawings are helpful for visualizing cis–trans relationships, but they do not represent the full three-dimensional structure of substituted cyclohexanes accurately. We can use these simplified drawings whenever we do not need to consider aspects of structure relating to the various conformations they can adopt. We shall consider their conformational structures in the following sections.

4.14A Cis–Trans Isomerism and Conformational Structures

If we consider the *actual* conformations of these isomers, the structures are somewhat more complex. Beginning with *trans*-1,4-dimethylcyclohexane, because it is easiest to visualize, we find there are two possible chair conformations (Fig. 4.24). In one conformation both methyl groups are axial; in the other both are equatorial. The diequatorial conformation is, as we would expect it to be, the more stable conformation, and it represents the structure of at least 99% of the molecules at equilibrium.

That the diaxial form of *trans*-1,4-dimethylcyclohexane is a trans isomer is easy to see; the two methyl groups are clearly on opposite sides of the ring. The trans relationship of the methyl groups in the diequatorial form is not as obvious, however. The trans relationship of the methyl groups becomes more apparent if we imagine ourselves "flattening" the molecule by turning one end up and the other down.

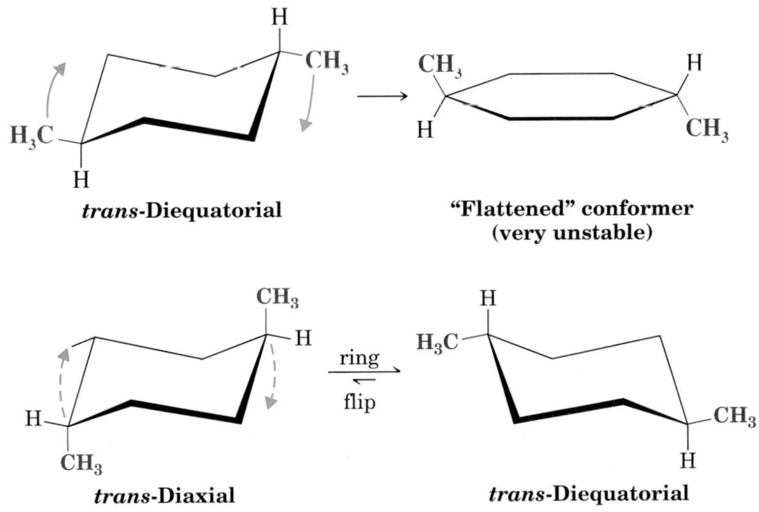

trans-Diequatorial

"Flattened" conformer (very unstable)

trans-Diaxial

ring flip

trans-Diequatorial

FIGURE 4.24 The two chair conformations of *trans*-1,4-dimethylcyclohexane: *trans*-diequatorial and *trans*-diaxial. The "flattened" conformer (very unstable) is shown only to emphasize the trans relationship of the methyl groups. (*Note:* All other C—H bonds have been omitted for clarity.)

A second *and general* way to recognize a trans-disubstituted cyclohexane is to notice that one group is attached by the *upper* bond (of the two to its carbon) and one by the *lower* bond:

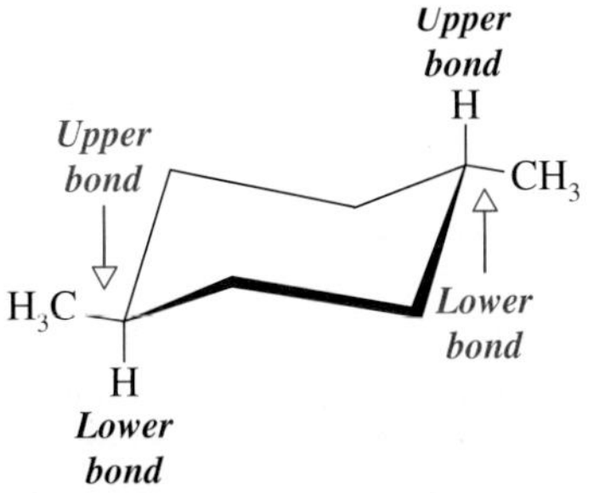

trans-1,4-Dimethylcyclohexane

In a cis-disubstituted cyclohexane both groups are attached by an upper bond or both by a lower bond. For example,

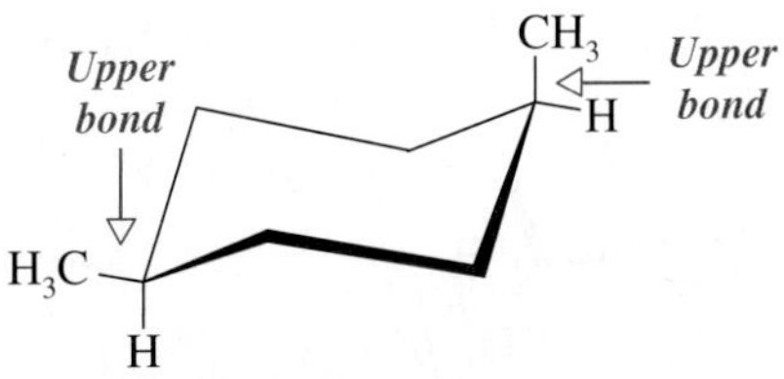

cis-1,4-Dimethylcyclohexane

cis-1,4-Dimethylcyclohexane actually exists in two *equivalent* chair conformations (Fig. 4.25). This cis relationship of the methyl groups, however, precludes the possibility

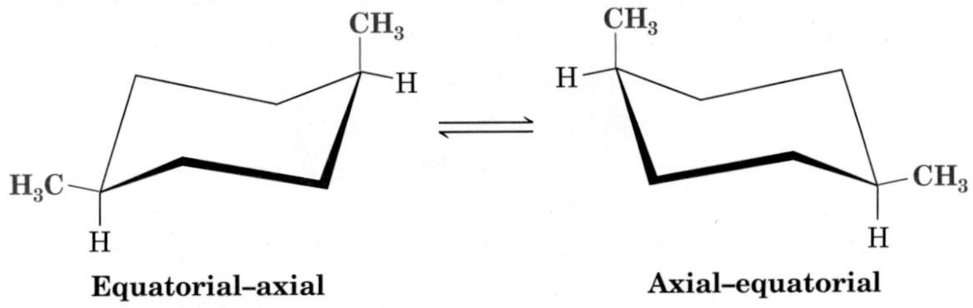

Equatorial–axial

Axial–equatorial

FIGURE 4.25 Equivalent conformations of *cis*-1,4-dimethylcyclohexane.

of a structure with both groups in an equatorial position. One group is axial in either conformation.

SOLVED PROBLEM

Consider each of the following conformational structures and tell whether each is cis or trans:

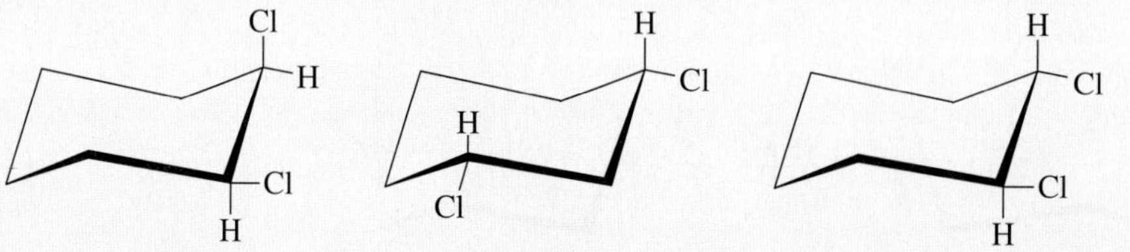

(a) (b) (c)

ANSWER: **(a)** Each chlorine is attached by the upper bond at its carbon; therefore, both chlorine atoms are on the same side of the molecule and this is a cis isomer. This is a *cis*-1,2-dichlorocyclohexane. **(b)** Here both chlorine atoms are attached by a lower bond; therefore, in this example, too, both chlorine atoms are on the same side of the molecule and this, too, is a cis isomer. It is *cis*-1,3-dichlorocyclohexane. **(c)** Here one chlorine atom is attached by a lower bond and one by an upper bond. The two chlorine atoms, therefore, are on opposite sides of the molecule, and this is a trans isomer. It is *trans*-1,2-dichlorocyclohexane.

PROBLEM 4.13 **(a)** Write structural formulas for the two chair conformations of *cis*-1-isopropyl-4-methylcyclohexane. **(b)** Are these two conformations equivalent? **(c)** If not, which would be more stable? **(d)** Which would be the preferred conformation at equilibrium?

trans-1,3-Dimethylcyclohexane is like the cis-1,4 compound in that no chair conformation is possible with both methyl groups in the favored equatorial position. The following two conformations are of equal energy and are equally populated at equilibrium:

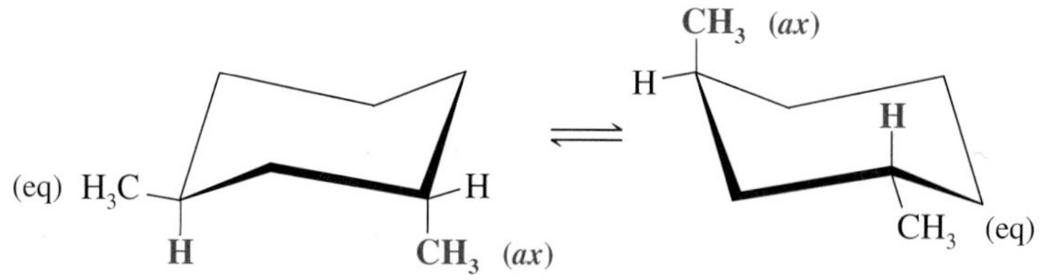

trans-1,3-Dimethylcyclohexane

If, however, we consider some other trans-1,3-disubstituted cyclohexane in which one alkyl group is larger than the other, the conformation of lower energy is the one having the larger group in the equatorial position. For example, the more stable conformation of *trans*-1-*tert*-butyl-3-methylcyclohexane, shown here, has the large *tert*-butyl group occupying the equatorial position:

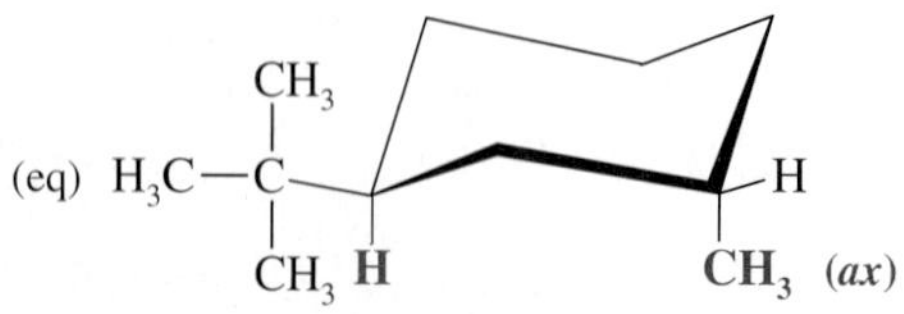

PROBLEM 4.14

(a) Write the two conformations of *cis*-1,2-dimethylcyclohexane. **(b)** Would these two conformations have equal potential energy? **(c)** What about the two conformations of *cis*-1-*tert*-butyl-2-methylcyclohexane? **(d)** Would the two conformations of *trans*-1,2-dimethylcyclohexane have the same potential energy?

4.15 Bicyclic and Polycyclic Alkanes

Many of the molecules that we encounter in our study of organic chemistry contain more than one ring (Section 4.4B). One of the most important bicyclic systems is bicyclo[4.4.0]decane, a compound that is usually called by its common name, *decalin:*

Decalin (bicyclo[4.4.0]decane)
(carbon atoms 1 and 6 are bridgehead carbon atoms)

Decalin shows cis–trans isomerism:

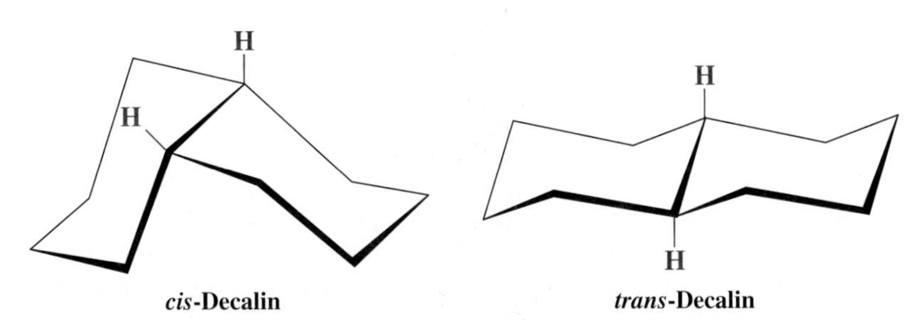

cis-**Decalin** *trans*-**Decalin**

Chemical Abstracts Service (CAS) determines the number of rings by the formula $S - A + 1 = N$, where S is the number of single bonds in the ring system, A is the number of atoms in the ring system, and N is the calculated number of rings (see Problem 4.34).

In *cis*-decalin the two hydrogen atoms attached to the bridgehead atoms lie on the same side of the ring; in *trans*-decalin they are on opposite sides. We often indicate this by writing their structures in the following way:

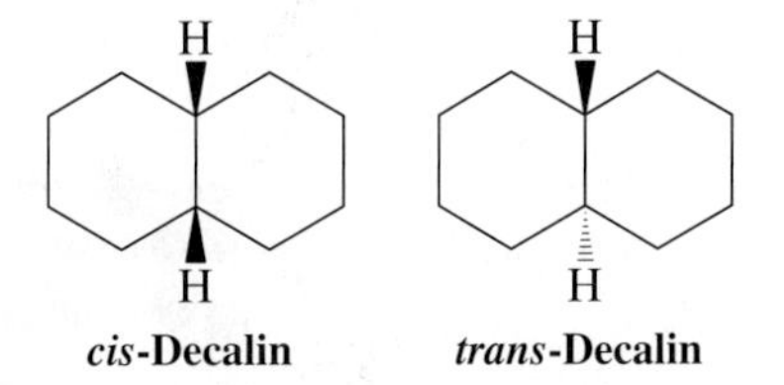

cis-**Decalin** *trans*-**Decalin**

Simple rotations of groups about carbon–carbon bonds do not interconvert *cis*- and *trans*-decalins. In this respect they resemble the isomeric cis- and trans-disubstituted cyclohexanes. (We can, in fact, regard them as being cis- and trans-1,2-disubstituted cyclohexanes in which the 1,2-substituents are the two ends of a four-carbon bridge, that is, $—CH_2CH_2CH_2CH_2—$).

The *cis*- and *trans*-decalins can be separated. *cis*-Decalin boils at 195°C (at 760 torr) and *trans*-decalin boils at 185.5°C (at 760 torr).

Adamantane (see below) is a tricyclic system that contains a three-dimensional array of cyclohexane rings, all of which are in the chair form. Extending the structure of adamantane in three dimensions gives the structure of diamond. The great hardness of diamond results from the fact that the entire diamond crystal is actually one very large molecule—a molecule that is held together by millions of strong covalent bonds:*

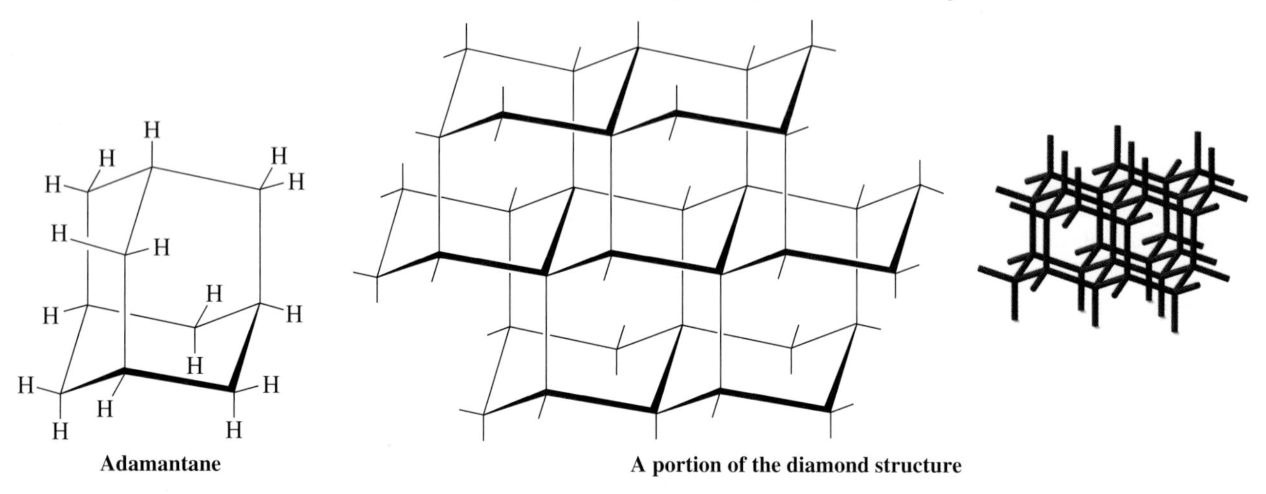

Adamantane

A portion of the diamond structure

One goal of research in recent years has been the synthesis of unusual, and sometimes highly strained, cyclic hydrocarbons. Among those that have been prepared are the compounds that follow:

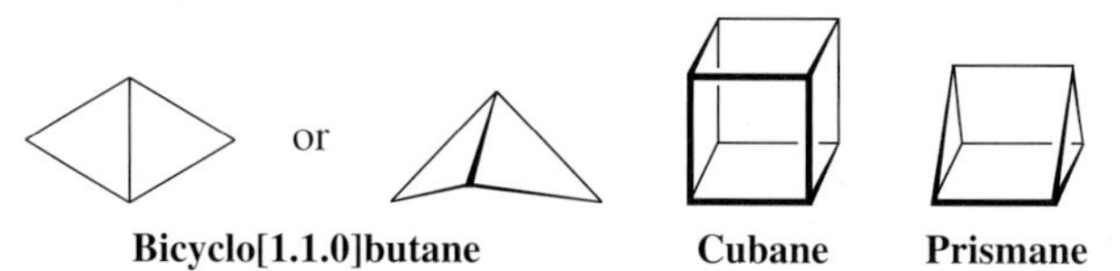

Bicyclo[1.1.0]butane Cubane Prismane

Larger molecules with beautiful symmetry have also been synthesized. One example is dodecahedrane, synthesized in 1982 by Leo A. Paquette and co-workers at Ohio State University. Another is buckminsterfullerene, a compound discovered in 1985 by H. W. Kroto (University of Sussex), R. F. Curl (Rice University), and R. E. Smalley (Rice University). Buckminsterfullerene is named after the famous architect and inventor of the geodesic dome, Buckminster Fuller. The Nobel Prize was awarded to Curl, Kroto, and Smalley in 1996 for their discovery of buckminsterfullerene. Buckminsterfullerene has since been identified in comets and asteroids and synthesized by several methods. Research in compounds related to buckminsterfullerene has expanded tremendously in recent years as potential applications for "buckyballs" from medicine to superconductors have been envisioned (see Section 14.8C).

Chem3D Models

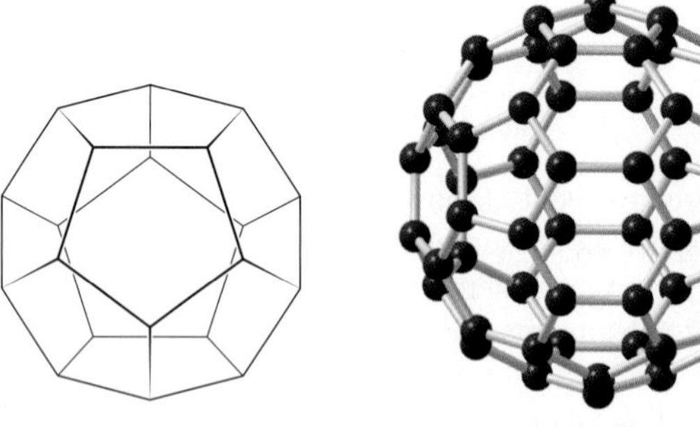

Dodecahedrane Buckminsterfullerene

*There are other allotropic forms of carbon, including graphite and Wurzite carbon [with a structure related to Wurzite (ZnS)].

The Chemistry of...

Pheromones: Communication by Means of Chemicals

Many animals communicate with other members of their species using a language based not on sounds or even visual signals but on the odors of chemicals called **pheromones** that these animals release. For insects, this appears to be the chief method of communication. Although pheromones are secreted by insects in extremely small amounts, they can cause profound and varied biological effects. Insects use some pheromones in courtship as sex attractants. Others use pheromones as warning substances, and still others secrete chemicals called "aggregation compounds" to cause members of their species to congregate. Often these pheromones are relatively simple compounds, and some are hydrocarbons. For example, a species of cockroach uses undecane as an aggregation pheromone:

$$CH_3(CH_2)_9CH_3$$

Undecane
(cockroach aggregation pheromone)

$$(CH_3)_2CH(CH_2)_{14}CH_3$$

2-Methylheptadecane
(sex attractant of female tiger moth)

When a female tiger moth wants to mate, she secretes 2-methylheptadecane, a perfume that the male tiger moth apparently finds irresistible.

The sex attractant of the common housefly *(Musca domestica)* is a 23-carbon alkene with a cis double bond between atoms 9 and 10 called muscalure:

$CH_3(CH_2)_7$ $(CH_2)_{12}CH_3$
C=C
H H

Muscalure
(sex attractant of common housefly)

Many insect sex attractants have been synthesized and are used to lure insects into traps as a means of insect control, a much more environmentally sensitive method than the use of insecticides.

Research suggests there are roles for pheromones in the lives of humans as well. For example, studies have shown that the phenomenon of menstrual synchronization among women who live or work with each other is likely caused by pheromones. Olfactory sensitivity to musk, which includes steroids such as androsterone, large cyclic ketones, and lactones (cyclic esters), also varies cyclically in women, differs between the sexes, and may influence our behavior. Some of these compounds are used in perfumes, including civetone, a natural product isolated from glands of the civet cat, and pentalide, a synthetic musk:

Androsterone **Civetone** **Pentalide**

4.16 Chemical Reactions of Alkanes

Alkanes, as a class, are characterized by a general inertness to many chemical reagents. Carbon–carbon and carbon–hydrogen bonds are quite strong; they do not break unless alkanes are heated to very high temperatures. Because carbon and hydrogen atoms have nearly the same electronegativity, the carbon–hydrogen bonds of alkanes are only slightly polarized. As a consequence, they are generally unaffected by most bases. Molecules of alkanes have no unshared electrons to offer as sites for attack by acids. This low reactivity of alkanes toward many reagents accounts for the fact that alkanes were originally called *paraffins* (Latin: *parum affinis,* little affinity).

The term paraffin, however, is probably not an appropriate one. We all know that alkanes react vigorously with oxygen when an appropriate mixture is ignited. This combustion occurs in the cylinders of automobiles and in oil furnaces, for example. When heated, alkanes also react with chlorine and bromine, and they react explosively with fluorine. We shall study these reactions in Chapter 10.

4.17 Synthesis of Alkanes and Cycloalkanes

Mixtures of alkanes as they are obtained from petroleum are suitable as fuels. However, in our laboratory work we often have the need for a pure sample of a particular alkane. For these purposes, the chemical preparation—or synthesis—of that particular alkane is often the most reliable way of obtaining it. The preparative method that we choose should be one that will lead to the desired product alone or, at least, to products that can be easily and effectively separated.

Several such methods are available, and three are outlined here. In subsequent chapters we shall encounter others.

4.17A Hydrogenation of Alkenes and Alkynes

Alkenes and alkynes react with hydrogen in the presence of metal catalysts such as nickel, palladium, and platinum to produce alkanes. The general reaction is one in which the atoms of the hydrogen molecule add to each atom of the carbon–carbon double or triple bond of the alkene or alkyne. This converts the alkene or alkyne to an alkane:

General Reaction

$$\text{C}{=}\text{C} + \text{H}{-}\text{H} \xrightarrow[\text{solvent, pressure}]{\text{Pt, Pd, or Ni}} {-}\text{C(H)}{-}\text{C(H)}{-}$$

Alkene **Alkane**

$$\text{C}{\equiv}\text{C} + 2\,\text{H}_2 \xrightarrow[\text{solvent, pressure}]{\text{Pt}} \text{H}{-}\text{C}{-}\text{H} / \text{H}{-}\text{C}{-}\text{H}$$

Alkyne **Alkane**

The reaction is usually carried out by dissolving the alkene or alkyne in a solvent such as ethyl alcohol (C_2H_5OH), adding the metal catalyst, and then exposing the mixture to hydrogen gas under pressure in a special apparatus. One molar equivalent of hydrogen is required to reduce an alkene to an alkane. Two molar equivalents are required to reduce an alkyne. (We shall discuss the mechanism of this reaction—called **hydrogenation**—in Chapter 7.)

Specific Examples

$$\underset{\textbf{Propene}}{CH_3CH{=}CH_2} + H{-}H \xrightarrow[C_2H_5OH\ (25°C,\ 50\ atm)]{Ni} \underset{\textbf{Propane}}{CH_3CH(H){-}CH_2(H)}$$

$$\underset{\textbf{2-Methylpropene}}{CH_3{-}C(CH_3){=}CH_2} + H_2 \xrightarrow[C_2H_5OH\ (25°C,\ 50\ atm)]{Ni} \underset{\textbf{Isobutane}}{CH_3{-}C(CH_3)(H){-}CH_2(H)}$$

$$\textbf{Cyclohexene} + H_2 \xrightarrow[C_2H_5OH\ (25°C,\ 1\ atm)]{Pd} \textbf{Cyclohexane}$$

$$\textbf{Cyclononyn-6-one} + 2\,H_2 \xrightarrow[\text{ethyl acetate}]{Pd} \textbf{Cyclononanone}$$

PROBLEM 4.15

Show the reactions involved for hydrogenation of all the alkenes and alkynes that would yield 2-methylbutane.

4.17B Reduction of Alkyl Halides

Most alkyl halides react with zinc and aqueous acid to produce an alkane. The general reaction is as follows:

General Reaction

$$\text{R—X} + \text{Zn} + \text{HX} \longrightarrow \text{R—H} + \text{ZnX}_2$$

or*

$$\text{R—X} \xrightarrow[(-\text{ZnX}_2)]{\text{Zn, HX}} \text{R—H}$$

Specific Examples

$$2\ \text{CH}_3\text{CH}_2\underset{\substack{|\\ \text{Br}}}{\text{CH}}\text{CH}_3 \xrightarrow[\text{Zn}]{\text{HBr}} 2\ \text{CH}_3\text{CH}_2\underset{\substack{|\\ \text{H}}}{\text{CH}}\text{CH}_3 + \text{ZnBr}_2$$

sec-Butyl bromide (2-bromobutane) → **Butane**

$$2\ \text{CH}_3\overset{\substack{\text{CH}_3\\ |}}{\text{C}}\text{HCH}_2\text{CH}_2\text{—Br} \xrightarrow[\text{Zn}]{\text{HBr}} 2\ \text{CH}_3\overset{\substack{\text{CH}_3\\ |}}{\text{C}}\text{HCH}_2\text{CH}_2\text{—H} + \text{ZnBr}_2$$

Isopentyl bromide (1-bromo-3-methylbutane) → **Isopentane (2-methylbutane)**

In these reactions zinc atoms transfer electrons to the carbon atom of the alkyl halide. Therefore, the reaction is a **reduction of the alkyl halide.** Zinc is a good reducing agent because it has two electrons in an orbital far from the nucleus which are readily donated to an electron acceptor. The mechanism for the reaction is complex because the reaction takes place in a separate phase at or near the surface of the zinc metal. It is possible that an alkylzinc halide forms first and then reacts with the acid to produce the alkane:

$$\text{Zn:} + \overset{\delta+}{\text{R}}\text{—}\overset{\delta-}{\ddot{\underset{..}{\text{X}}}}\text{:} \longrightarrow \left[\text{R}\overset{-}{\text{:}}\ \text{Zn}^{2+}\text{:}\ddot{\underset{..}{\text{X}}}\text{:}^{-}\right] \xrightarrow{\text{HX}} \text{R—H} + \text{Zn}^{2+} + 2\ \text{:}\ddot{\underset{..}{\text{X}}}\text{:}^{-}$$

Reducing agent — **Alkylzinc halide** — **Alkane**

PROBLEM 4.16

Your goal is the synthesis of 2,3-dimethylbutane by treating an alkyl halide with zinc and aqueous acid. Show two methods (beginning with different alkyl halides) for doing this.

4.17C Alkylation of Terminal Alkynes

We can replace a hydrogen attached to a triply bonded carbon of a terminal alkyne (called an **acetylenic hydrogen**) by an alkyl group. This type of reaction, called an **alkylation,** is of considerable use in synthesis. The acetylenic hydrogen is weakly acidic, as described

*This illustrates the way organic chemists often write abbreviated equations for chemical reactions. The organic reactant is shown on the left and the organic product on the right. The reagents necessary to bring about the transformation are written over (or under) the arrow. The equations are often left unbalanced, and sometimes byproducts (in this case, ZnX_2) are either omitted or are placed under the arrow in parentheses with a minus sign, for example, $(-\ ZnX_2)$.

in Section 3.14, and can be removed with a strong base such as sodium amide. Once the terminal hydrogen is removed, the alkyne carbon is an anion (called an alkynide anion), and it can be treated with an appropriate alkyl halide. The following provides an overview of the sequence:

$$\underset{\textbf{An alkyne}}{R{-}C{\equiv}C{-}H} \xrightarrow[(-NH_3)]{NaNH_2} \underset{\textbf{An alkynide anion}}{R{-}C{\equiv}C{:}^-\,Na^+} \xrightarrow[(-NaX)]{R'{-}X} R{-}C{\equiv}C{-}R'$$

Sodium amide (NaNH₂); **R′ must be methyl or 1° and unbranched at the second carbon** (R′—X)

A specific example is the synthesis of propyne from acetylene (ethyne) and bromomethane:

$$\underset{\substack{\textbf{Ethyne}\\\textbf{(acetylene)}}}{H{-}C{\equiv}C{-}H} \xrightarrow[(-NH_3)]{NaNH_2} \underset{\substack{\textbf{Ethynide anion}\\\textbf{(acetylide anion)}}}{H{-}C{\equiv}C{:}^-\,Na^+} \xrightarrow[(-NaX)]{CH_3{-}Br} \underset{\substack{\textbf{84\%}\\\textbf{Propyne}}}{H{-}C{\equiv}C{-}CH_3}$$

An important point to note is that the alkyl halide used with the alkynide anion must be methyl or primary and also unbranched at its second (beta) carbon. Alkyl halides that are secondary or tertiary or are primary with branching at the beta carbon react to give other products, predominantly by a mechanism called elimination (which we discuss in detail in Chapter 7).

Formation of a new carbon–carbon bond by alkylation of an alkynide anion is in itself an important transformation, but the alkyne triple bond also can be used for further reactions. For example, hydrogenation of a newly synthesized alkyne constitutes the formal synthesis of an alkane. A possible synthesis of 2-methylpentane from 3-methyl-1-butyne and bromomethane would be the following:

$$CH_3\overset{\overset{\large CH_3}{|}}{C}HC{\equiv}CH \xrightarrow[(-NH_3)]{NaNH_2} CH_3\overset{\overset{\large CH_3}{|}}{C}HC{\equiv}C{:}^-Na^+ \xrightarrow[(-NaBr)]{CH_3Br} CH_3\overset{\overset{\large CH_3}{|}}{C}HC{\equiv}C{-}CH_3 \xrightarrow{\text{excess } H_2,\ Pt,\ \text{pressure}} CH_3\overset{\overset{\large CH_3}{|}}{C}HCH_2CH_2CH_3$$

Note that it would not work to use propyne and 2-bromopropane for the alkylation step of this synthesis because the alkyl halide would have been secondary and elimination would predominate.

We shall see in Chapters 7 and 8 that an alkyne triple bond can lead to many, many other functional groups. Those reactions, together with alkylation of terminal alkynes, make a compound with the carbon–carbon triple bond an extremely versatile intermediate for synthesis.

4.18 Some General Principles of Structure and Reactivity: A Look toward Synthesis

You should pay attention to the bookkeeping of valence electrons and formal charges in the reaction shown in Fig. 4.26, just as with every other reaction you study in organic chemistry.

The alkylation of alkynide anions illustrates several essential aspects of structure and reactivity that we have discussed so far. First, preparation of the alkynide anion involves simple Brønsted–Lowry acid–base chemistry. As you have seen, the hydrogen of a terminal alkyne is weakly acidic ($pK_a \sim 25$), and with a strong base such as sodium amide it can be removed. The reason for this acidity was explained in Section 3.7A. Once formed, the alkynide anion is a Lewis base (Section 3.2B) with which the alkyl halide reacts as an electron pair acceptor (a Lewis acid). The alkynide anion can thus be called a *nucleophile* (Sections 3.3 and 6.4) because of the negative charge concentrated at its terminal carbon—it is a reagent that seeks positive charge. Conversely, the alkyl halide can be called an *elec-*

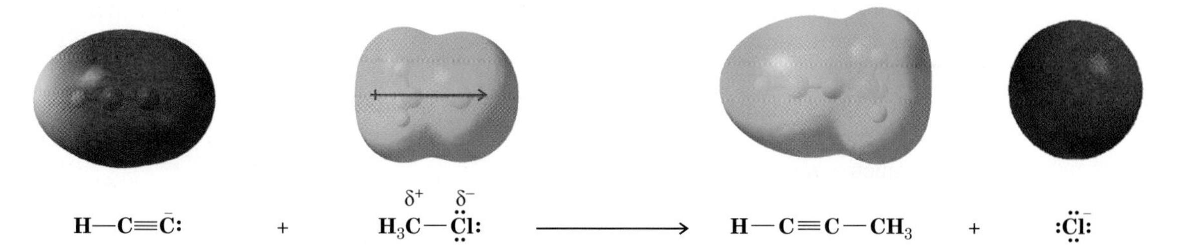

FIGURE 4.26 The reaction of ethynide (acetylide) anion and chloromethane. Electrostatic potential maps illustrate the complementary nucleophilic and electrophilic character of the alkynide anion and the alkyl halide. The dipole moment of chloromethane is shown by the red arrow.

Animated Graphics

trophile (Sections 3.3 and 8.1) because of the partial positive charge at the carbon bearing the halogen—it is a reagent that seeks negative charge. Polarity in the alkyl halide is the direct result of the difference in electronegativity between the halogen atom and carbon atom.

The electrostatic potential maps for ethynide (acetylide) anion and chloromethane in Fig. 4.26 illustrate the complementary nucleophilic and electrophilic character of a typical alkynide anion and alkyl halide. The ethynide anion has strong localization of negative charge at its terminal carbon, indicated by red in the electrostatic potential map. Conversely, chloromethane has partial positive charge at the carbon bonded to the electronegative chlorine atom. (The dipole moment for chloromethane is aligned directly along the carbon–chlorine bond.) Thus, acting as a Lewis base, the alkynide anion is attracted to the partially positive carbon of the alkyl halide. Assuming a collision between the two occurs with the proper orientation and sufficient kinetic energy, as the alkynide anion brings two electrons to the alkyl halide to form a new bond, it will displace the halogen from the alkyl halide. The halogen leaves as an anion with the pair of electrons that formerly bonded it to the carbon. The details of this type of mechanism will be discussed fully in Chapter 6. For now, however, suffice it to say that many of the reactions you will study in organic chemistry involve acid–base transformations (both Brønsted–Lowry and Lewis) and interaction of reagents with complementary charges.

4.19 An Introduction to Organic Synthesis

Organic synthesis is the process of building organic molecules from simpler precursors. Syntheses of organic compounds are carried out for many reasons. Chemists who develop new drugs carry out organic syntheses in order to discover molecules with structural attributes that enhance certain medicinal effects or reduce undesired side effects. Crixivan, whose structure is shown below, was designed by small-scale synthesis in a research laboratory and then quickly moved to large-scale synthesis after its approval as a drug. In other situations, a particular compound may need to be synthesized in order to test some hypothesis about a reaction mechanism or about how a certain organism metabolizes a compound. In cases like these we often will need to synthesize a particular compound "labeled" at a certain position (e.g., with deuterium, tritium, or an isotope of carbon).

N N OH H C_6H_5 HO H N H NH O HN O $C(CH_3)_3$

Crixivan (an HIV protease inhibitor)

A very simple organic synthesis may involve only one chemical reaction. Others may require from several to 20 or more steps. A landmark example of organic synthesis is that of vitamin B_{12}, announced in 1972 by R. B. Woodward (Harvard) and A. Eschenmoser (Swiss Federal Institute of Technology). Their synthesis of Vitamin B_{12} took 11 years, required more than 90 steps, and involved the work of nearly 100 people. We will work with much simpler examples, however.

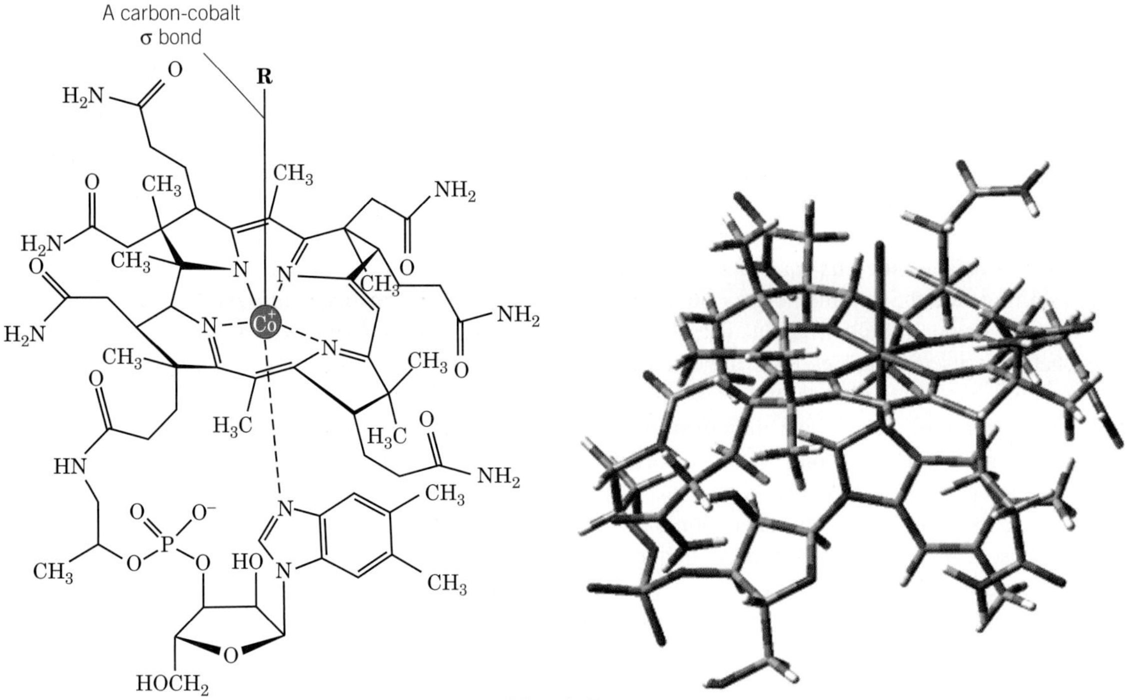

A carbon-cobalt σ bond

Vitamin B_{12}

An organic synthesis typically involves two types of transformations: reactions that convert functional groups from one to another and reactions that create new carbon–carbon bonds. You have studied examples of both types of reactions already—hydrogenation transforms the carbon–carbon double- or triple-bond functional groups in alkenes and alkynes to single bonds (actually removing a functional group in this case), and alkylation of alkynide anions forms carbon–carbon bonds. Ultimately, at the heart of organic synthesis is the orchestration of functional group interconversions and carbon–carbon bond-forming steps. Many methods are available to accomplish both of these things.

4.19A Retrosynthetic Analysis—Planning an Organic Synthesis

Retrosynthetic analysis

Sometimes it is possible to visualize from the start all the steps necessary to synthesize a desired (target) molecule from obvious precursors. Often, however, the sequence of transformations that would lead to the desired compound is too complex for us to "see" a path from the beginning to the end. In this case, since we know where we want to finish (the target molecule) but not where to start, we envision the sequence of steps that is required in a backward fashion, one step at a time. We begin by identifying immediate precursors that could be caused to react together to make the target molecule. Once these have been chosen, they in turn become new, intermediate target molecules, and we identify the next set of precursors that could be caused to react to form them, and so on, and so on. This process is repeated until we have worked backward to compounds that are sufficiently simple that they are readily available in a typical laboratory:

Target molecule ⟹ 1st precursor ⟹ 2nd precursor ⟹ ⟹ Starting compound

The process we have just described is called **retrosynthetic analysis.** The open arrow used in the example above is a **retrosynthetic arrow,** a symbol that relates the target molecule to its most immediate precursors; it signifies a **retro** or **backward** step. Although organic chemists have used retrosynthetic analysis intuitively for many years, E. J. Corey (1990 Chemistry Nobel Prize winner) originated the term retrosynthetic analysis and was the first person to state its principles formally. Once retrosynthetic analysis has been completed, to actually carry out the synthesis we conduct the sequence of reactions from the beginning, starting with the simplest precursors and working step by step until the target molecule is achieved.

E. J. Corey

You may want to examine the book: Corey, E. J.; Cheng, X.-M. *The Logic of Chemical Synthesis;* Wiley: New York, 1989. His studies, dating from the 1960s, have made the designing of complex organic syntheses systematic enough to be aided by computers.

When doing retrosynthetic analysis it is necessary to generate as many possible precursors, and hence different synthetic routes, as possible (Fig. 4.27). We evaluate all the possible advantages and disadvantages of each path and in so doing determine the most efficient route for synthesis. The prediction of which route is most feasible is usually based on specific restrictions or limitations of reactions in the sequence, the availability of materials, or other factors. We shall see an example of this in Section 4.19B. In actuality more than one route may work well. In other cases it may be necessary to try several approaches in the laboratory in order to find the most efficient or successful route.

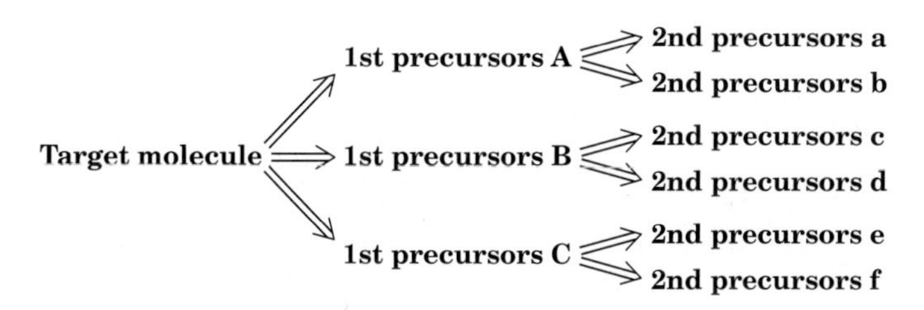

FIGURE 4.27 Retrosynthetic analysis often discloses several routes from the target molecule back to varied precursors.

4.19B Identifying Precursors

How do we go about identifying the immediate retrosynthetic precursors to a given compound? In the case of functional groups we need to have a toolbox of reactions from which to choose those we know can convert one given functional group into another. You will develop such a toolbox of reactions as you proceed through your study of organic chemistry. Similarly, with regard to making carbon–carbon bonds in synthesis, you will develop a repertoire of reactions for that purpose. In order to choose the appropriate reaction for either purpose, you will inevitably consider basic principles of structure and reactivity.

Over time you will develop a toolbox of reactions for two major categories of synthetic operations: carbon–carbon bond formation and functional group interconversion.

As we stated in Sections 3.2C and 4.19, many organic reactions depend on the interaction of molecules that have complementary full or partial charges. One very important aspect of retrosynthetic analysis is being able to identify those atoms in a target molecule that could have had complementary (opposite) charges in synthetic precursors. Consider, for example, the synthesis of 1-cyclohexyl-1-butyne. On the basis of reactions learned in this chapter, you might envision an alkynide anion and an alkyl halide as precursors having complementary polarities that when allowed to react together would lead to this molecule:

Retrosynthetic Analysis

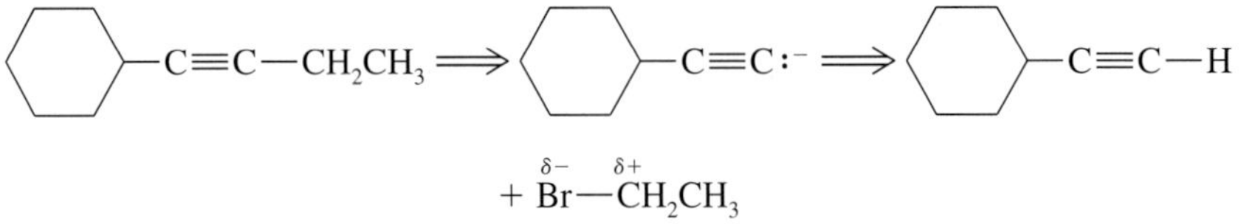

$$C_6H_{11}{-}C{\equiv}C{-}CH_2CH_3 \Longrightarrow C_6H_{11}{-}C{\equiv}C{:}^- \Longrightarrow C_6H_{11}{-}C{\equiv}C{-}H$$

$$+\ \overset{\delta-}{Br}{-}\overset{\delta+}{C}H_2CH_3$$

Synthesis

$$C_6H_{11}{-}C{\equiv}C{-}H \xrightarrow[(-NH_3)]{NaNH_2} C_6H_{11}{-}C{\equiv}C{:}^-Na^+ \xrightarrow[(-NaBr)]{BrCH_2CH_3} C_6H_{11}{-}C{\equiv}C{-}CH_2CH_3$$

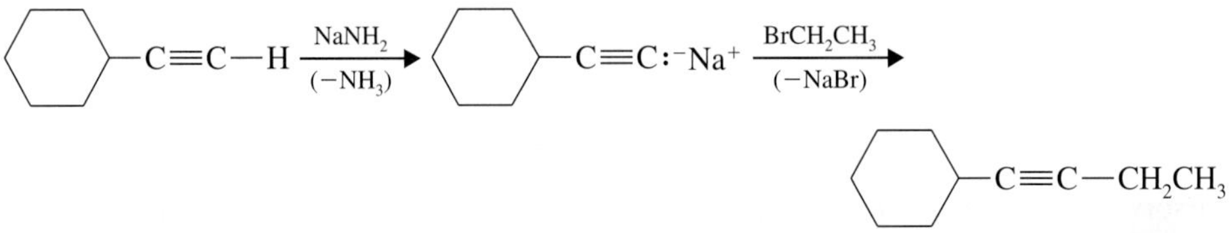

Sometimes, however, it will not at first be obvious where the retrosynthetic bond disconnections are in a target molecule that would lead to oppositely charged or complementary precursors. The synthesis of an alkane would be such an example. An alkane does not contain carbon atoms that could directly have had opposite charges in precursor molecules. However, if one supposes that certain carbon–carbon single bonds in the alkane could have arisen by hydrogenation of a corresponding alkyne (a functional group interconversion), then, in turn, two atoms of the alkyne could have been joined from precursor molecules that had complementary charges (i.e., an alkynide anion and an alkyl halide). Consider the following retrosynthetic analysis for 2-methylhexane:

Retrosynthetic Analysis

$$CH_3CH_2CH_2CH_2CH(CH_3)CH_3 \text{ (2-Methylhexane)} \Longrightarrow HC{\equiv}CCH_2CH_2CH(CH_3)CH_3 \Longrightarrow HC{\equiv}C{:}^- + X{-}CH_2CH_2CH(CH_3)CH_3$$

$$CH_3CH_2CH_2CH_2CH(CH_3)CH_3 \Longrightarrow CH_3C{\equiv}CCH_2CH(CH_3)CH_3$$

$$CH_3C{\equiv}CCH_2CH(CH_3)CH_3 \not\Longrightarrow CH_3C{\equiv}C{:}^- + X{-}CH_2CH(CH_3)CH_3$$

(1°, but branched at second carbon)

$$CH_3C{\equiv}CCH_2CH(CH_3)CH_3 \Longrightarrow CH_3{-}X + {}^-{:}C{\equiv}CCH_2CH(CH_3)CH_3$$

$$CH_3CH_2CH_2CH_2CH(CH_3)CH_3 \Longrightarrow CH_3CH_2C{\equiv}CCH(CH_3)CH_3$$

$$CH_3CH_2C{\equiv}CCH(CH_3)CH_3 \not\Longrightarrow CH_3CH_2C{\equiv}C{:}^- + X{-}CH(CH_3)CH_3$$

(a 2° alkyl halide)

$$CH_3CH_2C{\equiv}CCH(CH_3)CH_3 \Longrightarrow CH_3CH_2{-}X + {}^-{:}C{\equiv}CCH(CH_3)CH_3$$

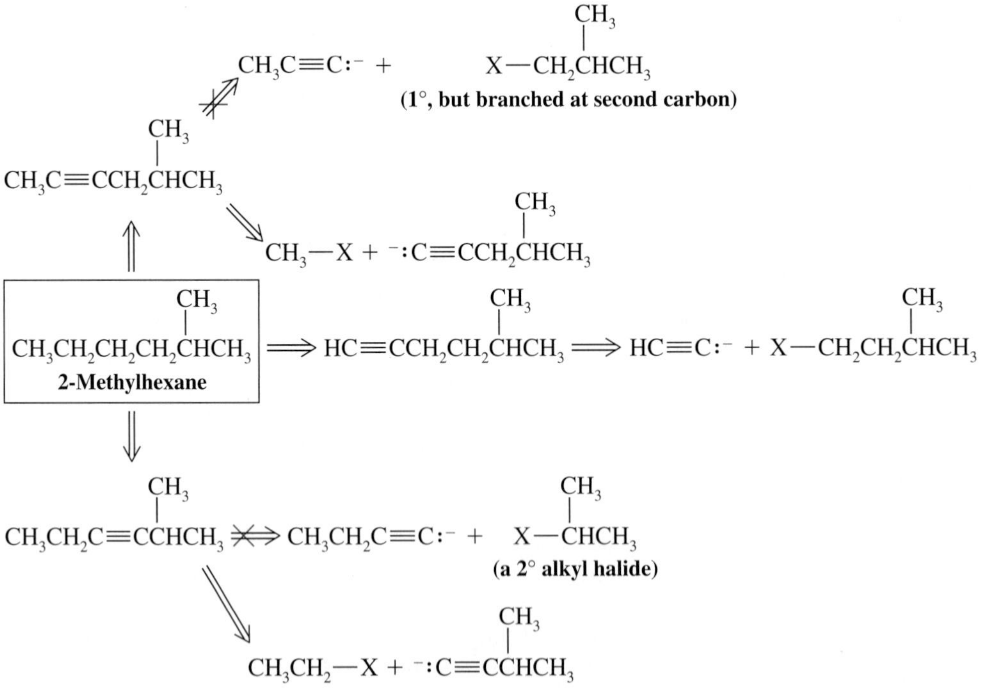

The Chemistry of...

From the Inorganic to the Organic

In 1862, Friedrich Wöhler discovered calcium carbide (CaC_2) by heating carbon with an alloy of zinc and calcium. He then synthesized acetylene by allowing the calcium carbide to react with water:

$$C \xrightarrow{\text{zinc–calcium alloy, heat}} CaC_2 \xrightarrow{2\ H_2O} HC{\equiv}CH + Ca(OH)_2$$

Acetylene produced this way burned in lamps of some lighthouses and in old-time miners' headlamps. From the standpoint of organic synthesis, it is theoretically possible to synthesize *anything* using reactions of alkynes to form carbon–carbon bonds and to prepare other functional groups. Thus, Wöhler's discovery of calcium carbide gives us a link from inorganic materials to all of organic synthesis.

As indicated in the retrosynthetic analysis above, we must bear in mind the limitations that exist for the reactions that would be applied in the synthetic (forward) direction. In the example above, two of the pathways have to be discarded because they involve the use of a 2° alkyl halide or a primary halide branched at the second (beta) carbon (Section 4.17C).

PROBLEM 4.17

Referring to the retrosynthetic analysis for 2-methylhexane in this section, write reactions for those synthesis routes that are feasible.

PROBLEM 4.18

(a) Devise retrosynthetic schemes for all conceivable alkynide anion alkylation syntheses of the insect pheromones undecane and 2-methylheptadecane (see "The Chemistry of . . . Pheromones" box in this chapter). **(b)** Write reactions for two feasible syntheses of each pheromone.

4.19C Raison d'Etre

Solving synthetic puzzles by application of retrosynthetic analysis is one of the joys of learning organic chemistry. As you might imagine, there is skill and some artistry involved. Over the years many chemists have set their minds to organic synthesis, and because of this we have all prospered from the fruits of their endeavors.

Key Terms and Concepts

Alkanes	**Sections 2.2A, 4.1, 4.2, 4.3, 4.7, 4.16**
Alkenes	**Section 4.1**
Alkylation	**Section 4.18C**
Alkynes	**Section 4.1**
Angle strain	**Section 4.11**
Anti conformation	**Section 4.9**
1,3-Diaxial interaction	**Section 4.13**
Axial position	**Section 4.13**
Cis–trans isomerism	**Sections 1.13B, 4.14**
Conformation	**Section 4.8**
Conformational analysis	**Sections 4.8, 4.9, 4.12, 4.13**
Conformational stereoisomers	**Section 4.9**

(continued)

Conformations of cyclohexane	**Sections 4.12, 4.13**
Conformers	**Section 4.8**
Constitutional isomers	**Sections 1.3A, 4.2**
Cycloalkanes	**Sections 4.1, 4.4, 4.7, 4.10, 4.16**
Dihedral angle	**Section 4.8**
Eclipsed conformation	**Section 4.8**
Equatorial position	**Section 4.13**
Gauche conformation	**Section 4.9**
Hydrogenation	**Section 4.18A**
Hyperconjugation	**Section 4.8**
IUPAC system	**Section 4.3**
Newman projection formula	**Section 4.8**
Potential energy diagram	**Section 4.8**
Reduction of alkyl halides	**Section 4.18B**
Retrosynthetic analysis	**Section 4.19A**
Retrosynthetic arrows	**Section 4.19A**
Ring strain	**Sections 4.10, 4.11**
Sawhorse formula	**Section 4.8**
Staggered conformation	**Section 4.8**
Stereoisomers	**Section 4.9**
Steric hindrance	**Section 4.8**
Torsional strain	**Sections 4.8, 4.9**
van der Waals forces	**Sections 2.14D, 4.11**

Additional Problems

4.19 Write a structural formula for each of the following compounds:

(a) 1,4-Dichloropentane
(b) *sec*-Butyl bromide
(c) 4-Isopropylheptane
(d) 2,2,3-Trimethylpentane
(e) 3-Ethyl-2-methylhexane
(f) 1,1-Dichlorocyclopentane
(g) *cis*-1,2-Dimethylcyclopropane
(h) *trans*-1,2-Dimethylcyclopropane
(i) 4-Methyl-2-pentanol
(j) *trans*-4-Isobutylcyclohexanol
(k) 1,4-Dicyclopropylhexane
(l) Neopentyl alcohol
(m) Bicyclo[2.2.2]octane
(n) Bicyclo[3.1.1]heptane
(o) Cyclopentylcyclopentane

4.20 Give systematic IUPAC names for each of the following (problem continues on p. 188):

(a) $CH_3CH_2C(CH_3)_2CH(CH_2CH_3)CH_3$
(b) $CH_3CH_2C(CH_3)_2CH_2OH$

(c)

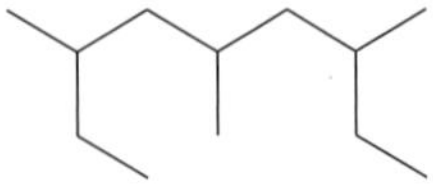

(d)

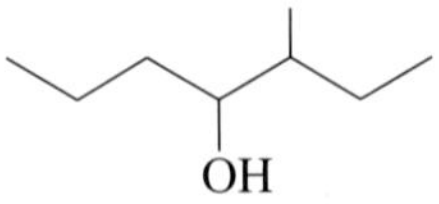

OH

Note: Problems marked with an asterisk are "challenge problems."

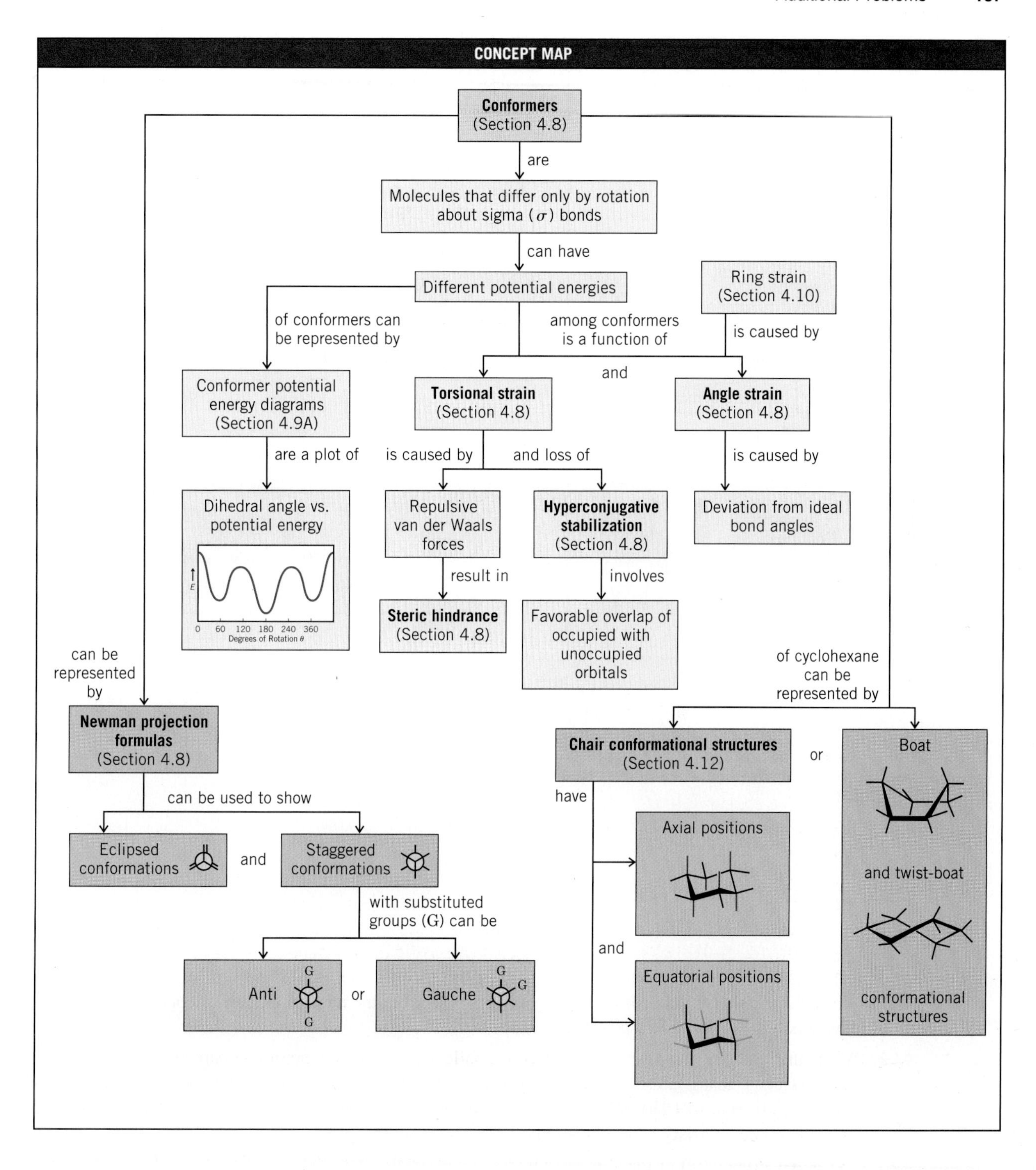
CONCEPT MAP
Conformers
(Section 4.8)
are
Molecules that differ only by rotation about sigma (σ) bonds
can have
Different potential energies
of conformers can be represented by
among conformers is a function of
Ring strain
(Section 4.10)
is caused by
Conformer potential energy diagrams
(Section 4.9A)
Torsional strain
(Section 4.8)
and
Angle strain
(Section 4.8)
are a plot of
is caused by
and loss of
is caused by
Dihedral angle vs. potential energy
E
0 60 120 180 240 360
Degrees of Rotation θ
Repulsive van der Waals forces
Hyperconjugative stabilization
(Section 4.8)
Deviation from ideal bond angles
result in
involves
Steric hindrance
(Section 4.8)
Favorable overlap of occupied with unoccupied orbitals
can be represented by
of cyclohexane can be represented by
Newman projection formulas
(Section 4.8)
Chair conformational structures
(Section 4.12)
or
Boat
can be used to show
have
Axial positions
and twist-boat
Eclipsed conformations
and
Staggered conformations
with substituted groups (G) can be
and
Equatorial positions
conformational structures
Anti
G
G
or
Gauche
G
G

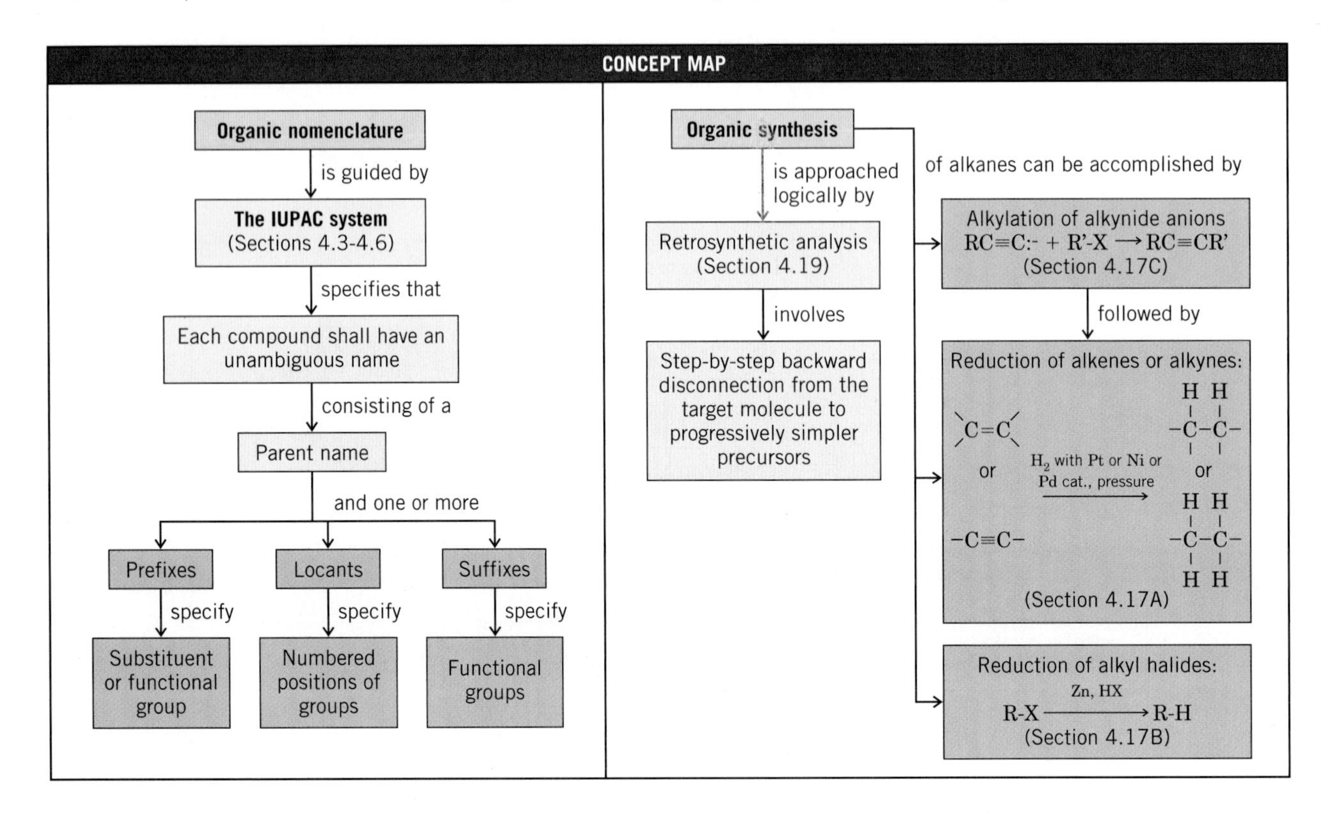

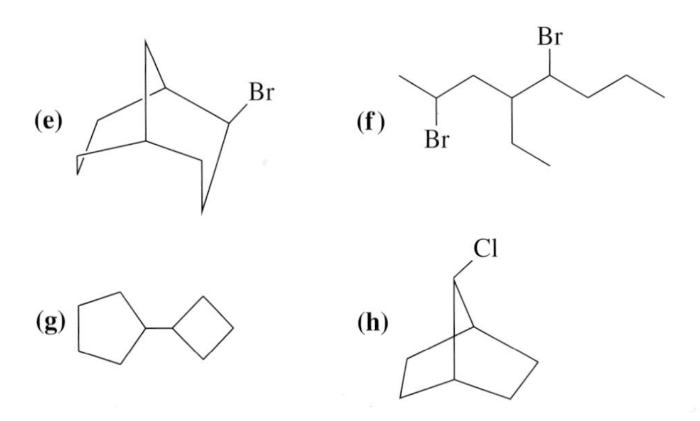

4.21 The name *sec*-butyl alcohol defines a specific structure but the name *sec*-pentyl alcohol is ambiguous. Explain.

4.22 Write the structure and give the IUPAC systematic name of an alkane or cycloalkane with the formulas **(a)** C_8H_{18}, that has only primary hydrogen atoms, **(b)** C_6H_{12}, that has only secondary hydrogen atoms, **(c)** C_6H_{12}, that has only primary and secondary hydrogen atoms, and **(d)** C_8H_{14}, that has 12 secondary and 2 tertiary hydrogen atoms.

4.23 Write the structure(s) of the simplest alkane(s), i.e., one(s) with the fewest number of carbon atoms, wherein each possesses primary, secondary, tertiary, and quaternary carbon atoms. (A quaternary carbon is one that is bonded to four other carbon atoms.) Assign an IUPAC name to each structure.

4.24 **(a)** Three different alkenes yield 2-methylbutane when they are hydrogenated in the presence of a metal catalyst. Give their structural formulas and write equations for the reactions involved. **(b)** One of these alkene isomers has characteristic absorptions at approximately 998 and 914 cm^{-1} in its IR spectrum. Which one is it?

4.25 An alkane with the formula C_6H_{14} can be synthesized by treating (in separate reactions) five different alkyl chlorides ($C_6H_{13}Cl$) with zinc and aqueous acid. Give the structure of the alkane and the structures of the alkyl chlorides.

4.26 An alkane with the formula C_6H_{14} can be prepared by reduction (with Zn and HCl) of only two alkyl chlorides ($C_6H_{13}Cl$) and by the hydrogenation of only two alkenes (C_6H_{12}). Write the structure of this alkane, give its IUPAC name, and show the reactions.

4.27 Four different cycloalkenes will all yield methylcyclopentane when subjected to catalytic hydrogenation. What are their structures? Show the reactions.

4.28 A spiro ring junction is one where two rings originate from a single carbon atom. Alkanes containing such a ring junction are called spiranes.
(a) For the case of bicyclic spiranes of formula C_7H_{12}, write structures for all possibilities where all carbons are incorporated into rings.
(b) Write structures for other bicyclic structures that fit this formula.

4.29 The heats of combustion of three pentane (C_5H_{12}) isomers are $CH_3(CH_2)_3CH_3$, 3536 kJ mol^{-1}; $CH_3CH(CH_3)CH_2CH_3$, 3529 kJ mol^{-1}; and $(CH_3)_3CCH_3$, 3515 kJ mol^{-1}. Which isomer is most stable? Construct a diagram such as that in Fig. 4.10 showing the relative potential energies of the three compounds.

4.30 Tell what is meant by a homologous series and illustrate your answer by writing structures for a homologous series of alkyl halides.

4.31 Write the structures of two chair conformations of 1-*tert*-butyl-1-methylcyclohexane. Which conformation is more stable? Explain your answer.

4.32 Ignoring compounds with double bonds, write structural formulas and give names for all of the isomers with the formula C_5H_{10}.

4.33 Write structures for the following bicyclic alkanes:
(a) Bicyclo[1.1.0]butane
(b) Bicyclo[2.1.0]pentane
(c) 2-Chlorobicyclo[3.2.0]heptane
(d) 7-Methylbicyclo[2.2.1]heptane

4.34 Use the $S - A + 1 = N$ method (study tip, Section 4.15) to determine the number of rings in cubane.

4.35 Rank the following compounds in order of **(a)** increasing heat of combustion and **(b)** increasing stability.

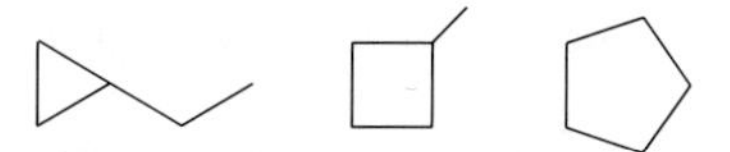

4.36 Sketch curves similar to the one given in Fig. 4.9 showing the energy changes that arise from rotation about the C2—C3 bond of **(a)** 2,3-dimethylbutane and **(b)** 2,2,3,3-tetramethylbutane. You need not concern yourself with actual numerical values of the energy changes, but you should label all maxima and minima with the appropriate conformations.

4.37 Without referring to tables, decide which member of each of the following pairs would have the higher boiling point. Explain your answers.
(a) Pentane or 2-methylbutane
(b) Heptane or pentane
(c) Propane or 2-chloropropane
(d) Butane or 1-propanol
(e) Butane or CH_3COCH_3

4.38 One compound whose molecular formula is C_4H_6 is a bicyclic compound. Another compound with the same formula has an infrared absorption at roughly 2250 cm^{-1} (the bicyclic compound does not). Draw structures for each of these two compounds and explain how the IR absorption allows them to be differentiated.

4.39 Which compound would you expect to be the more stable: *cis*-1,2-dimethylcyclopropane or *trans*-1,2-dimethylcyclopropane? Explain your answer.

4.40 Consider that cyclobutane exhibits a puckered geometry. Judge the relative stabilities of the 1,2-disubstituted cyclobutanes and of the 1,3-disubstituted cyclobutanes. (You may find it helpful to build hand-held molecular models of representative compounds.)

4.41 Which member of each of the following pairs would you expect to have the larger heat of combustion? **(a)** *cis*- or *trans*-1,2-dimethylcyclohexane, **(b)** *cis*- or *trans*-1,3-dimethylcyclohexane, **(c)** *cis*- or *trans*-1,4-dimethylcyclohexane. Explain your answers.

4.42 Write the two chair conformations of each of the following and in each part designate which conformation would be the more stable: **(a)** *cis*-1-*tert*-butyl-3-methylcyclohexane, **(b)** *trans*-1-*tert*-butyl-3-methylcyclohexane, **(c)** *trans*-1-*tert*-butyl-4-methylcyclohexane, **(d)** *cis*-1-*tert*-butyl-4-methylcyclohexane.

4.43 Provide an explanation for the fact that *all-trans*-1,2,3,4,5,6-hexaisopropylcyclohexane is a stable molecule in which all isopropyl groups are axial. (You may find it helpful to build a hand-held molecular model.)

4.44 *trans*-1,3-Dibromocyclobutane has a measurable dipole moment. Explain how this proves that the cyclobutane ring is not planar.

4.45 Specify the missing compounds and/or reagents in each of the following syntheses (more than one step may be necessary in some cases):

(a) *trans*-5-Methyl-2-hexene $\xrightarrow{?}$ 2-methylhexane

(b) 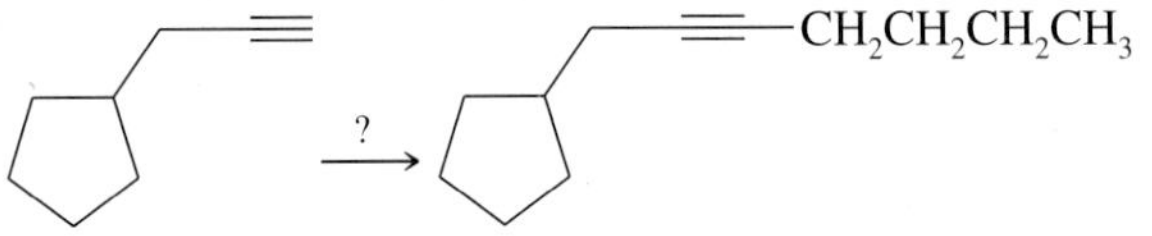 (cyclopentyl-CH_2-C≡CH) $\xrightarrow{?}$ (cyclopentyl-CH_2-C≡C-$CH_2CH_2CH_2CH_3$)

(c) $CH_3CH_2CH_2Br \xrightarrow{?}$ (2-methylhexane skeletal structure)

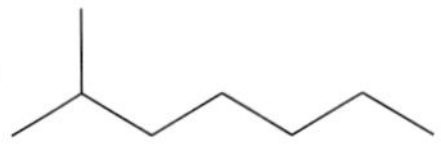

(d) 4-Bromo-3,4-diethylheptane $\xrightarrow{\text{Zn, HBr}}$?

(e) 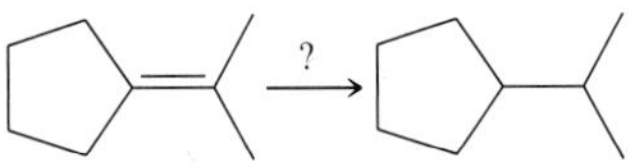 (isopropylidenecyclopentane) $\xrightarrow{?}$ (isopropylcyclopentane)

(f) ? $\xrightarrow{\text{Zn, HX}}$ 2,2-dimethylpropane

(g) Chemical reactions rarely yield products in such initially pure form that no trace can be found of the starting materials used to make them. What evidence in an IR spectrum of each of the crude (unpurified) products from the above reactions would indicate the presence of one of the organic reactants used to synthesize each target molecule? That is, predict one or two key IR absorptions for the reactants that would distinguish it/them from IR absorptions predicted for the product.

4.46 Why can the use of high temperatures in the catalytic hydrogenation of alkenes to alkanes be self-defeating?

***4.47** When 1,2-dimethylcyclohexene (below) is allowed to react with hydrogen in the presence of a platinum catalyst, the product of the reaction is a cycloalkane that has a melting point of $-50°C$ and a boiling point of 130°C (at 760 torr). **(a)** What is the structure of the product of this reaction? **(b)** Consult an appropriate table and tell which stereoisomer it is. **(c)** What does this experiment suggest about the mode of addition of hydrogen to the double bond?

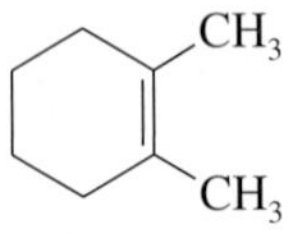

CH_3

CH_3

1,2-Dimethylcyclohexene

***4.48** When cyclohexene is dissolved in an appropriate solvent and allowed to react with chlorine, the product of the reaction, $C_6H_{10}Cl_2$, has a melting point of $-7°C$ and a boiling point (at 16 torr) of 74°C. **(a)** Which stereoisomer is this? **(b)** What does this experiment suggest about the mode of addition of chlorine to the double bond?

4.49 Consider the cis and trans isomers of 1,3-di-*tert*-butylcyclohexane (build molecular models). What unusual feature accounts for the fact that one of these isomers apparently exists in a twist-boat conformation rather than a chair conformation?

***4.50** Using the rules found in this chapter, give systematic names for the following or indicate that more rules need to be provided:

(a) 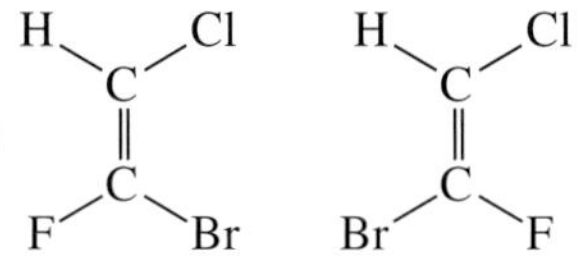

(b) 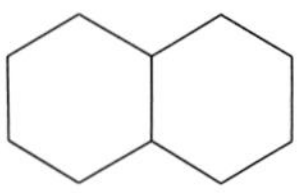

(c) 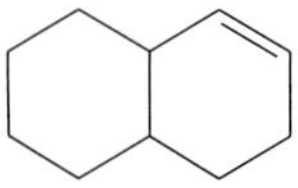

(d) 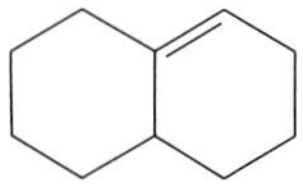

***4.51** This is the predominant conformation for D-glucose:

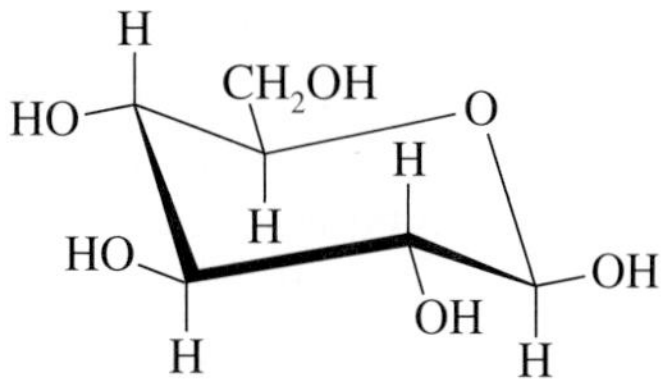

Why is it not surprising that D-glucose is the most commonly found sugar in nature? (*Hint:* Look up structures for sugars such as D-galactose and D-mannose, and compare these with D-glucose.)

***4.52** Using Newman projections, depict the relative positions of the substituents on the bridgehead atoms of *cis*- and *trans*-decalin. Which of these isomers would be expected to be more stable, and why?

***4.53** Starting with any two compounds containing no more than four carbon atoms, write equations for a synthesis of dodecane, $CH_3(CH_2)_{10}CH_3$.

4.54 Open the energy-minimized 3D Molecular Models on the CD for *trans*-1-*tert*-butyl-3-methylcyclohexane and *trans*-1,3-di-*tert*-butylcyclohexane. What conformations of cyclohexane do the rings in these two compounds resemble most closely? How can you account for the difference in ring conformations between them?

Chem3D Model

4.55 Open the 3D Molecular Models on the CD for cyclopentane and vitamin B_{12}. Compare cyclopentane with the nitrogen-containing five-membered rings in vitamin B_{12}. Is the conformation of cyclopentane represented in the specified rings of vitamin B_{12}? What factor(s) account for any differences you observe?

Chem3D Model

4.56 Open the 3D Molecular Model on the CD for buckminsterfullerene. What molecule has its type of ring represented 16 times in the surface of buckminsterfullerene?

Chem3D Model

LEARNING GROUP PROBLEMS

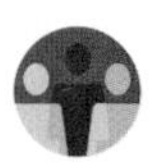

Consider the following compound:

CH_3

$(CH_2)_4-C-$

C_2H_5

CH_3

1. Draw a bond-line formula for this compound (neglecting cis–trans isomerism in the ring).
2. Develop all reasonable retrosynthetic analyses for this compound that, at some point, involve carbon–carbon bond formation by alkylation of an alkynide ion.
3. Write reactions, including reagents and conditions, for syntheses of this compound that correspond to the retrosynthetic analyses you developed above.
4. Write an IUPAC name for the target molecule and for all uncharged synthetic intermediates in your syntheses.
5. Infrared spectroscopy could be used to show the presence of certain impurities in your final product that would result from leftover intermediates in your syntheses. Which of your synthetic intermediates would show IR absorptions that are distinct from those in the final product, and in what regions of the IR spectrum would these absorptions occur?
6. Pick a compound from your synthesis that contains a cyclohexane ring.
   (a) Tell which isomer (cis or trans) you would expect to be the more stable form of this compound.
   (b) Draw a chair conformational structure for this compound in its lowest energy conformation.
   (c) Draw a chair conformational structure for a higher energy conformation of this compound.
   (d) Draw chair conformational structures for two forms of the cis–trans isomer you did *not* use in answering parts (a)–(c) above.
7. Choose a specific carbon–carbon bond in an *acyclic* intermediate from one of your syntheses and draw a Newman projection showing the most stable conformation along this bond.
8. Draw a three-dimensional structure for either the cis or trans form of the target molecule. Use dashes and wedges where appropriate in the alkyl side chain and use a chair conformational structure for the ring. (*Hint:* Draw the structure so that the carbon chain of the most complicated substituent on the cyclohexane ring and the ring carbon where it is attached are all in the plane of the paper. In general, for three-dimensional structures choose an orientation that allows as many carbon atoms as possible to be in the plane of the paper.)

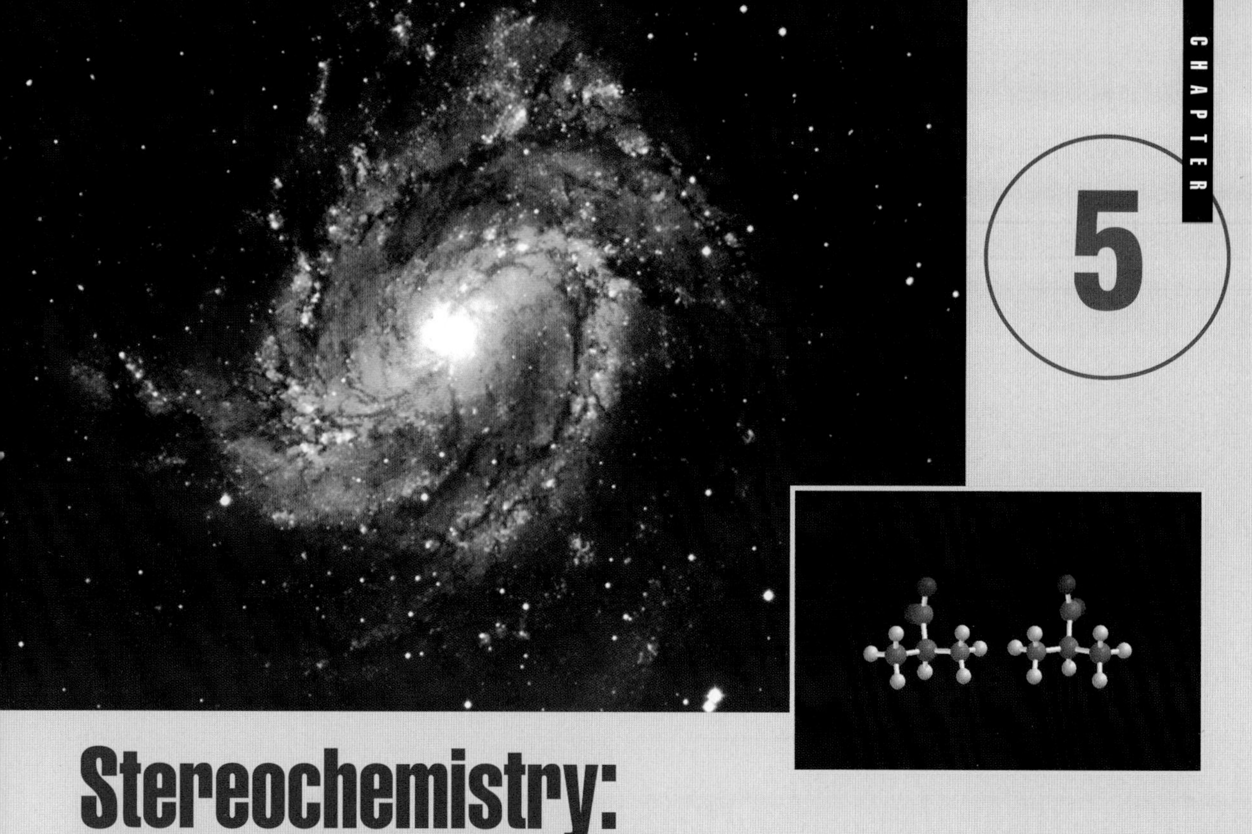

CHAPTER 5

Stereochemistry: Chiral Molecules

The Handedness of Life

Molecules of the amino acids that comprise our proteins have the property of being non-superposable on their mirror image. Because of this, they are said to be chiral, or to possess "handedness." Although both mirror image forms are theoretically possible, such as those for the amino acid alanine above, life on Earth involves amino acids that are mainly of the mirror image form said to be "left-handed" (designated L). "The reason that most amino acids are of the left-handed form is not known, however. In the absence of an influence that possesses handedness such as a living system, chemical reactions produce an equal mixture of both mirror image forms. Since almost all theories about the origin of life presume that amino acids and other molecules central to life were present before self-replicating organisms came into being, it was assumed that they were present in equal mirror image forms in the primordial soup. But could the mirror image forms of these molecules actually have been present in unequal amounts before life began, leading to some sort of preference as life evolved? A meteorite discovered in 1970, known as the Murchison meteorite, fueled speculation about this topic. Analysis of the meteorite showed that amino acids and other complex molecules associated with life were present, proving that molecules required for life could arise outside the confines of Earth. Even more interesting, recent experiments have shown that a 7–9% excess of four L-amino acids is present in the Murchison meteorite. The origin of this unequal distribution is uncertain, but some scientists speculate that electromagnetic radiation emitted in a corkscrew fashion from the poles of spinning neutron stars could lead to a bias of one mirror image isomer over another when molecules form in interstellar space.

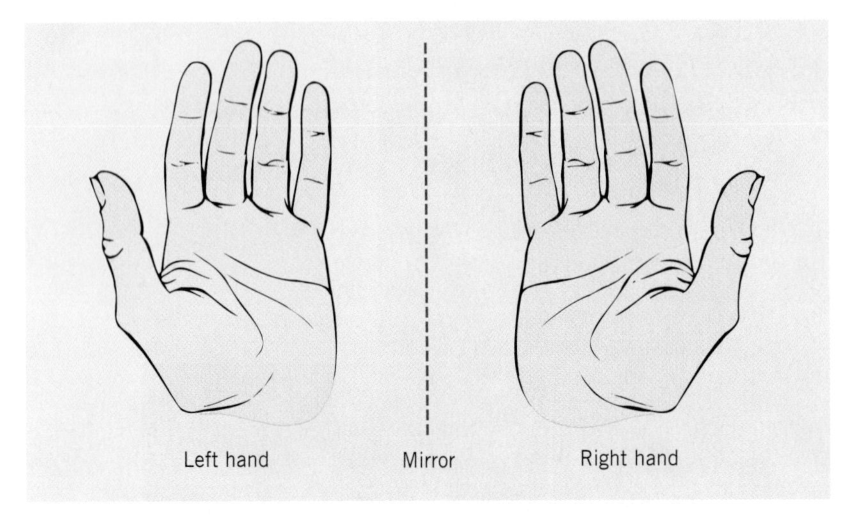

FIGURE 5.1 The mirror image of a left hand is a right hand.

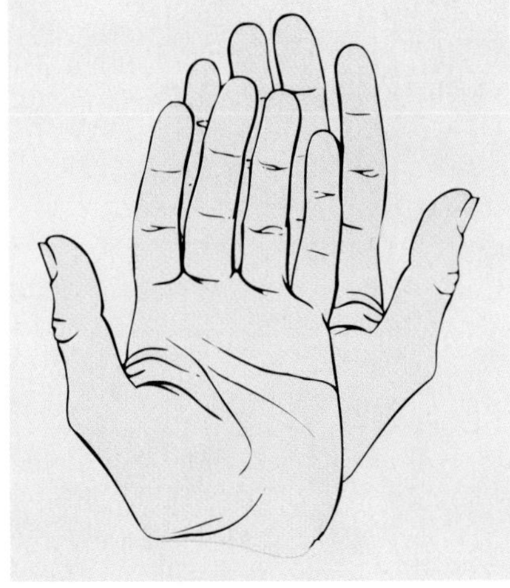

FIGURE 5.2 Left and right hands are not superposable.

5.1 The Biological Significance of Chirality

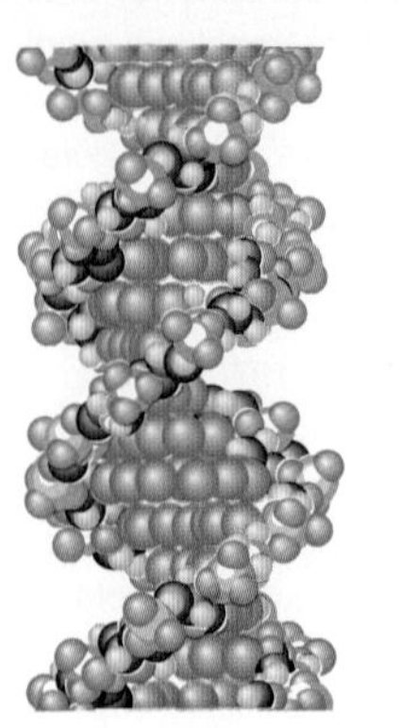

Bindweed (top photo) *(Convolvulus sepium)* winds in a right-handed fashion, like the right-handed helix of DNA.

Chirality is a phenomenon that pervades the universe. A **chiral** object is an object that possesses the property of "handedness" (from the Greek word *cheir,* meaning hand). A chiral object, such as each of our hands, is one that cannot be placed on its mirror image so that all parts coincide. In other words, a chiral object is *not* **superposable** on its mirror image. When you view your left hand in a mirror, for example, the mirror image of your left hand is a right hand (Fig. 5.1). Yet your left and right hands are *not identical because they are not superposable,** even though they are mirror images (Fig. 5.2).

The human body is structurally chiral, with the heart lying to the left of center and the liver to the right. Helical seashells are chiral and most are spiral, such as a right-handed screw. Many plants show chirality in the way they wind around supporting structures. Honeysuckle winds as a left-handed helix; bindweed winds in a right-handed way. DNA is a chiral molecule. The double helical form of DNA turns in a right-handed way.

Chirality in molecules, however, involves more than the fact that some molecules adopt left- or right-handed conformations. As we shall see in this chapter, it is the nature of groups bonded at specific atoms that can bestow chirality upon a molecule. Indeed, all but one of the 20 amino acids that make up naturally occurring proteins are chiral, and all of these are classified as being left-handed. The molecules of natural sugars are almost all classified as being right-handed. In fact, most of the molecules of life are chiral, and most are found in only one mirror image form.† (The chapter opening essay discussed the possible natural origin of this preference.)

Chirality has tremendous importance in our daily lives. Most pharmaceuticals are chiral. Usually only one mirror image form of a drug provides the desired effect. The other

---

*To be superposable is different than to be super*im*posable. Any two objects can be superimposed simply by putting one object on top of the other, whether or not the objects are the same. To *superpose* two objects (as in the property of superposition) means, on the other hand, that **all parts of each object must coincide.** The condition of superposability must be met for two things to be **identical.**

†For interesting reading, see Hegstrum, R. A.; Kondepudi, D. K. The Handedness of the Universe. *Sci. Am.* **1990,** *262*(1), 98–105, and Horgan, J. The Sinister Cosmos. *Sci. Am.* **1997,** *276*(5), 18–19.

mirror image form is often inactive or, at best, less active. In some cases the other mirror image form of a drug actually has severe side effects or toxicity (see Section 5.4 regarding thalidomide). Our senses of taste and smell also depend on chirality. As we shall see, one mirror image form of a chiral molecule may have a certain odor or taste, while its mirror image smells and tastes completely different. The food we eat is largely made of molecules of one mirror image form. If we were to eat food that was somehow made of molecules with the unnatural mirror image form, we would likely starve because the enzymes in our bodies are chiral and preferentially react with the natural mirror image form of their substrates.

Let us now consider what causes some molecules to be chiral. To begin, we will return to aspects of isomerism.

5.2 Isomerism: Constitutional Isomers and Stereoisomers

Isomers are different compounds that have the same molecular formula. In our study thus far, much of our attention has been directed toward isomers we have called constitutional isomers. **Constitutional isomers** have the same molecular formula but different connectivity, meaning that their atoms are connected in a different order. Examples of constitutional isomers are the following:

MOLECULAR FORMULA	CONSTITUTIONAL ISOMERS		
C_4H_{10}	$CH_3CH_2CH_2CH_3$ **Butane**	and	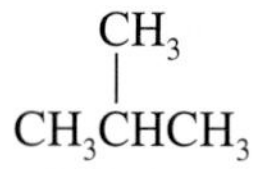 $CH_3CH(CH_3)CH_3$ **Isobutane**
C_3H_7Cl	$CH_3CH_2CH_2Cl$ **1-Chloropropane**	and	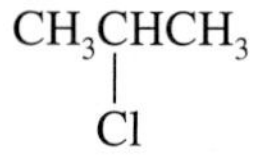 $CH_3CH(Cl)CH_3$ **2-Chloropropane**
C_2H_6O	CH_3CH_2OH **Ethanol**	and	CH_3OCH_3 **Dimethyl ether**

Stereoisomers are not constitutional isomers. Stereoisomers have their atoms connected in the same sequence (the same constitution), but they differ in the arrangement of their atoms in space. We have already seen examples of some types of stereoisomers. The cis and trans forms of alkenes are stereoisomers (Section 1.13B), as are the cis and trans forms of substituted cyclic molecules (Section 4.14).

Stereoisomers can be subdivided into two general categories: **enantiomers** and **diastereomers. Enantiomers are stereoisomers whose molecules are nonsuperposable mirror images of each other. Diastereomers are stereoisomers whose molecules are *not* mirror images of each other.** The alkene isomers *cis*- and *trans*-1,2-dichloroethene, shown here, are stereoisomers that are diastereomers:

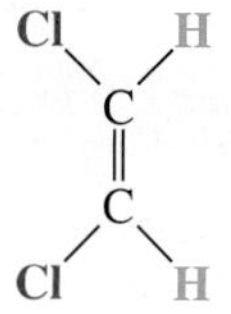

Cl(H)C=C(Cl)H — ***cis*-1,2-Dichloroethene** ($C_2H_2Cl_2$)

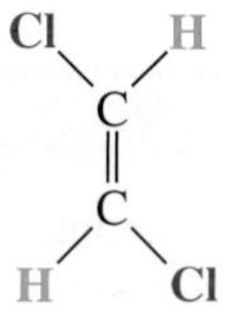

Cl(H)C=C(H)Cl — ***trans*-1,2-Dichloroethene** ($C_2H_2Cl_2$)

By examining the structural formulas for *cis*- and *trans*-1,2-dichloroethene, we see that they have the same molecular formula ($C_2H_2Cl_2$) and the same connectivity (both compounds have two central carbon atoms joined by a double bond, and both compounds have one chlorine and one hydrogen atom attached to each carbon atom). But, their atoms have a different arrangement in space that is not interconvertible from one to another (due to the large barrier to rotation of the carbon–carbon double bond), making them stereoisomers. Furthermore, they are stereoisomers that are not mirror images of each other; therefore they are diastereomers and not enantiomers.

Cis and trans isomers of cycloalkanes furnish us with another example of stereoisomers that are diastereomers. Consider the following two compounds:

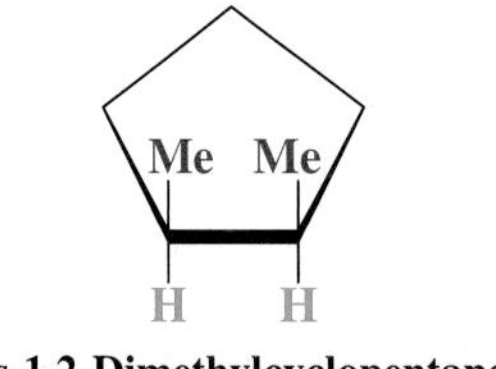

cis-1,2-Dimethylcyclopentane
(C_7H_{14})

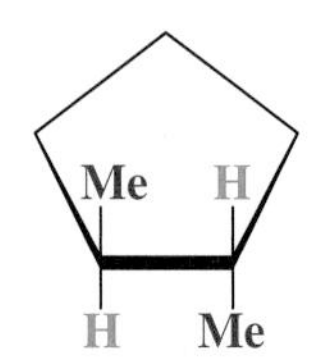

trans-1,2-Dimethylcyclopentane
(C_7H_{14})

These two compounds have the same molecular formula (C_7H_{14}), the same sequence of connections for their atoms, but different arrangements of their atoms in space. In one compound both methyl groups are bonded toward the same face of the ring, while in the other compound the two methyl groups are bonded toward opposite faces of the ring. Furthermore, the positions of the methyl groups cannot be interconverted by conformational changes. Therefore, these compounds are stereoisomers, and because they are stereoisomers that are not mirror images of each other, they can be further classified as diastereomers.

In Section 5.12 we shall study other molecules that can exist as diastereomers but are not cis and trans isomers of each other. First, however, we need to consider enantiomers further.

A tool for subdividing types of isomers

SUBDIVISION OF ISOMERS

ISOMERS
(Different compounds with same molecular formula)

Constitutional isomers
(Isomers whose atoms have a different connectivity)

Stereoisomers
(Isomers that have the same connectivity but that differ in the arrangement of their atoms in space)

Enantiomers
(Stereoisomers that are nonsuperposable mirror images of each other)

Diastereomers
(Stereoisomers that are not mirror images of each other)

5.3 Enantiomers and Chiral Molecules

Enantiomers occur only with compounds whose molecules are chiral. A **chiral molecule is defined as one that is *not* superposable on its mirror image.** Alkene stereoisomers (as those above) are *not* chiral, whereas the *trans*-1,2-dimethylcyclopentane isomers are chiral. A chiral molecule and its mirror image are called a **pair of enantiomers.** The relationship between them is defined as **enantiomeric.** Molecules (and objects) that *are* superposable on their mirror image are **achiral** (meaning not chiral). Most socks, for example, are achiral, whereas shoes are chiral. Many familiar objects are chiral, while other objects can be shown to be chiral only by applying the universal test for chirality—the nonsuperposability of the object and its mirror image.

PROBLEM 5.1

Classify each of the following objects as to whether it is chiral or achiral:

(a) A screwdriver **(d)** A tennis shoe **(g)** A car
(b) A baseball bat **(e)** An ear **(h)** A hammer
(c) A golf club **(f)** A woodscrew

The chirality of molecules can be demonstrated with relatively simple compounds. Consider, for example, 2-butanol:

$$\underset{\displaystyle \text{OH}}{\text{CH}_3\underset{|}{\text{C}}\text{HCH}_2\text{CH}_3}$$

2-Butanol

Until now, we have presented the formula for 2-butanol as though it represented only one compound and we have not mentioned that molecules of 2-butanol are chiral. Because they are, there are actually two different 2-butanols and these two 2-butanols are enantiomers. We can understand this if we examine the drawings and models in Fig. 5.3.

If model **I** is held before a mirror, model **II** is seen in the mirror and vice versa. Models **I** and **II** are not superposable on each other; therefore, they represent different, but isomeric, molecules. ***Because models I and II are nonsuperposable mirror images of each other, the molecules that they represent are enantiomers.***

FIGURE 5.3 *(a)* Three-dimensional drawings of the 2-butanol enantiomers I and II. *(b)* Models of the 2-butanol enantiomers. *(c)* An unsuccessful attempt to superpose models of I and II.

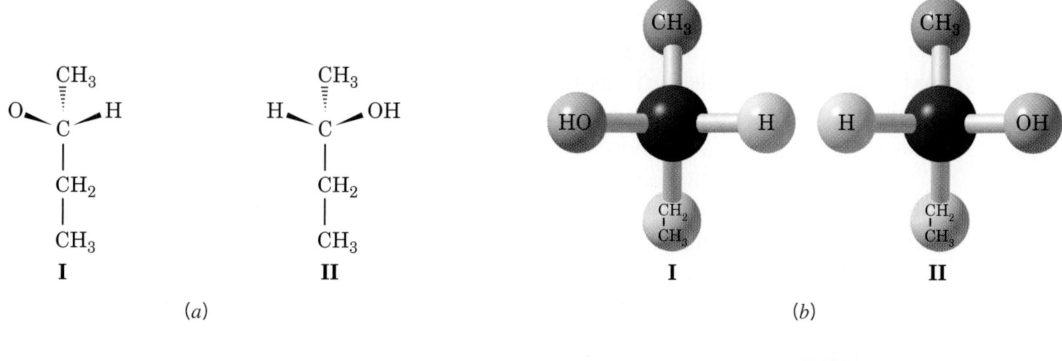

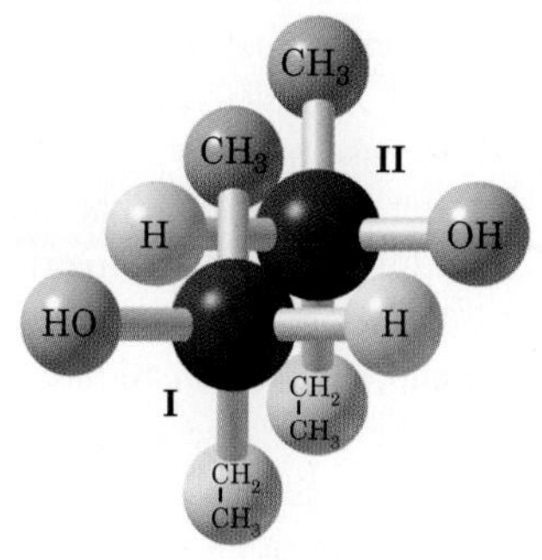

PROBLEM 5.2

Construct hand-held models of the 2-butanols represented in Fig. 5.3 and demonstrate for yourself that they are not mutually superposable. **(a)** Make similar models of 2-bromopropane. Are they superposable? **(b)** Is a molecule of 2-bromopropane chiral? **(c)** Would you expect to find enantiomeric forms of 2-bromopropane?

How do we know when to expect the possibility of enantiomers? One way (but not the only way) is to recognize that **a pair of enantiomers is always possible for molecules that contain one tetrahedral atom with four different groups attached to it.*** In 2-butanol (Fig. 5.4) this atom is C2. The four different groups that are attached to C2 are a hydroxyl group, a hydrogen atom, a methyl group, and an ethyl group.

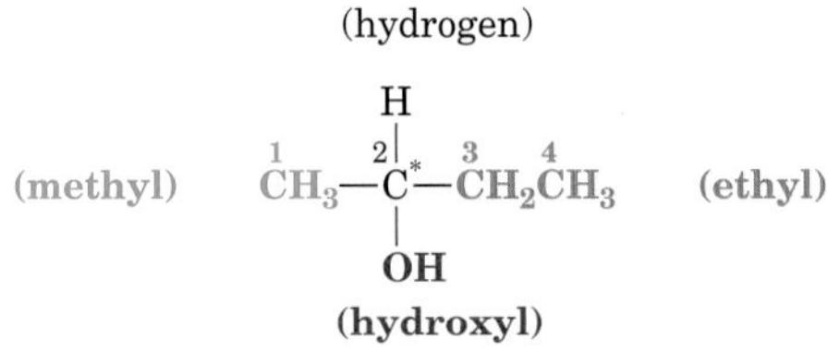

FIGURE 5.4 The tetrahedral carbon atom of 2-butanol that bears four different groups. [By convention, carbons at a tetrahedral stereogenic center are often designated with an asterisk (*).]

(hydrogen)

H

(methyl) $\overset{1}{C}H_3$—$\overset{2}{C}^*$—$\overset{3}{C}H_2\overset{4}{C}H_3$ (ethyl)

OH

(hydroxyl)

Interchanging two groups of a model or three-dimensional formula is a useful test for determining whether structures of two chiral molecules are the same or different.

An important property of enantiomers such as these is that *interchanging any two groups at the tetrahedral atom that bears four different groups converts one enantiomer into the other.* In Fig. 5.3*b* it is easy to see that interchanging the hydroxyl group and the hydrogen atom converts one enantiomer into the other. You should now convince yourself with models that interchanging any other two groups has the same result.

Because interchanging two groups at C2 converts one stereoisomer into another, C2 is an example of what is called a **stereogenic carbon.** A stereogenic carbon is defined as **a carbon atom bearing groups of such nature that an interchange of any two groups will produce a stereoisomer.** Carbon-2 of butanol is an example of a **tetrahedral stereogenic carbon.** Not all stereogenic carbons are tetrahedral, however. The carbon atoms of *cis*- and *trans*-1,2-dichloroethene (Section 5.2) are examples of **trigonal planar stereogenic carbons** because an interchange of groups at either atom also produces a stereoisomer (a diastereomer). In general, any location where an interchange of groups leads to a stereoisomer is called a **stereogenic center.**

When we discuss interchanging groups like this, we must take care to notice that what we are describing is *something we do to a molecular model* or *something we do on paper.* An interchange of groups in a real molecule, if it can be done, requires breaking covalent bonds, and this is something that requires a large input of energy. This means that enantiomers such as the 2-butanol enantiomers ***do not interconvert*** spontaneously.

At one time, *tetrahedral atoms* with four different groups were called *chiral atoms* or *asymmetric atoms.* Then, in 1984, K. Mislow (of Princeton University) and J. Siegel (now at the University of California, San Diego) pointed out that the use of such terms as this had represented a source of conceptual confusion in stereochemistry that had existed from the time of van't Hoff (Section 5.5). Chirality is a geometric property that pervades and affects all parts of a chiral molecule. All of the atoms of 2-butanol, for example, are in a chiral environment and, therefore, all are said to be *chirotopic.* When we consider an atom such

*We shall see later that enantiomers are also possible for molecules that contain more than one tetrahedral atom with four different groups attached to it, but some of these molecules (Section 5.12A) do not exist as enantiomers.

as C2 of 2-butanol in the way that we describe here, however, we are considering it as a *stereogenic carbon* and, therefore, we should designate it as such, and not as a "chiral atom." Further consideration of these issues is beyond our scope here, but those interested may wish to read the original paper; see Mislow, K.; Siegel, J. *J. Am. Chem. Soc.* **1984,** *106,* 3319–3328.

Working with models is a helpful study technique whenever three-dimensional aspects of chemistry are involved.

Figure 5.5 demonstrates the validity of the generalization that enantiomeric compounds necessarily result whenever a molecule contains a single tetrahedral stereogenic carbon.

PROBLEM 5.3

Demonstrate the validity of what we have represented in Fig. 5.5 by constructing models. Demonstrate for yourself that **III** and **IV** are related as an object and its mirror image *and that they are not superposable* (i.e., that **III** and **IV** are chiral molecules and are enantiomers). **(a)** Now take **IV** and exchange the positions of any two groups. What is the new relationship between the molecules? **(b)** Now take either model and exchange the positions of any two groups. What is the relationship between the molecules now?

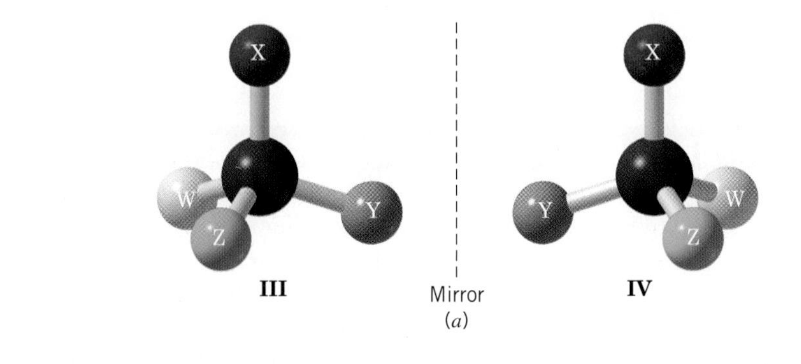

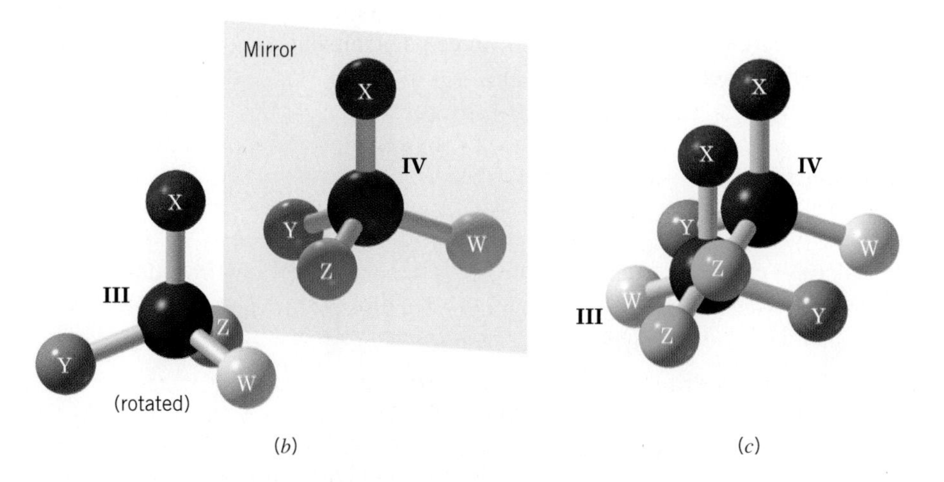

FIGURE 5.5 A demonstration of chirality of a generalized molecule containing one tetrahedral stereogenic carbon. *(a)* The four different groups around the carbon atom in III and IV are arbitrary. *(b)* III is rotated and placed in front of a mirror. III and IV are found to be related as an object and its mirror image. *(c)* III and IV are not superposable; therefore, the molecules that they represent are chiral and are enantiomers.

CD Tutorial

Tetrahedral stereogenic centers

If all of the tetrahedral atoms in a molecule have two or more groups attached that *are the same,* the molecule does not have a stereogenic carbon. The molecule is superposable on its mirror image and is **achiral.** An example of a molecule of this type is 2-propanol; carbon atoms 1 and 3 bear three identical hydrogen atoms and the central atom bears two identical methyl groups. If we write three-dimensional formulas for 2-propanol, we find (Fig. 5.6, on p. 200) that one structure can be superposed on its mirror image.

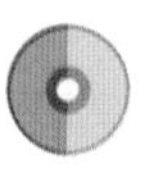

CD Tutorial

Two identical groups

FIGURE 5.6 *(a)* 2-Propanol (V) and its mirror image (VI). *(b)* When either one is rotated, the two structures are superposable and so do not represent enantiomers. They represent two molecules of the same compound. 2-Propanol does not have a stereogenic carbon.

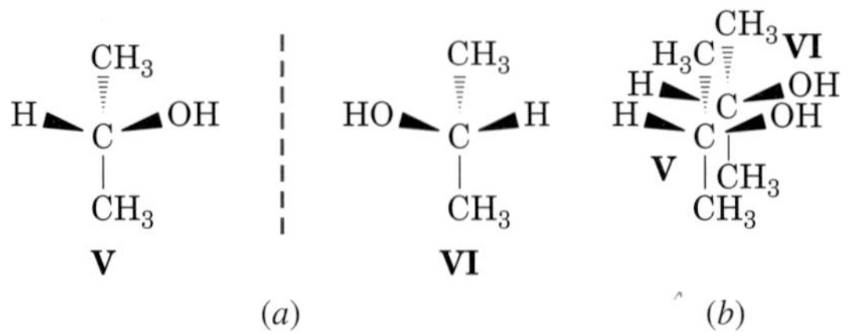

Thus, we would not predict the existence of enantiomeric forms of 2-propanol, and experimentally only one form of 2-propanol has ever been found.

PROBLEM 5.4 Some of the molecules listed here have a stereogenic carbon; some do not. Write three-dimensional formulas for both enantiomers of those molecules that do have a stereogenic carbon.

(a) 2-Fluoropropane
(b) 2-Methylbutane
(c) 2-Chlorobutane
(d) 2-Methyl-1-butanol
(e) 2-Bromopentane
(f) 3-Methylpentane
(g) 3-Methylhexane
(h) 1-Chloro-2-methylbutane

5.4 More About the Biological Importance of Chirality

The origin of biological properties relating to chirality is often likened to the specificity of our hands for their respective gloves; the binding specificity for a chiral molecule (like a hand) at a chiral receptor site (a glove) is only favorable in one way. If either the molecule or the biological receptor site had the wrong handedness, the natural physiological response (e.g., neural impulse, reaction catalysis) would not occur. A diagram showing how only one amino acid in a pair of enantiomers can interact in an optimal way with a hypothetical binding site (e.g., in an enzyme) is shown in Fig. 5.7. Because of the tetrahedral stereogenic carbon of the amino acid, three-point binding can occur with proper alignment for only one of the two enantiomers.

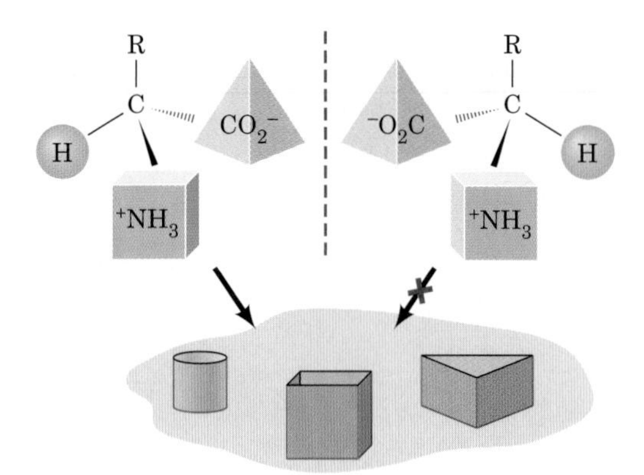

FIGURE 5.7 Only one of the two amino acid enantiomers shown can achieve three-point binding with the hypothetical binding site (e.g., in an enzyme).

Chiral molecules can show their different handedness in many ways, including the way they affect human beings. One enantiomeric form of a compound called limonene (Section 23.3) is primarily responsible for the odor of oranges and the other enantiomer for the odor of lemons.

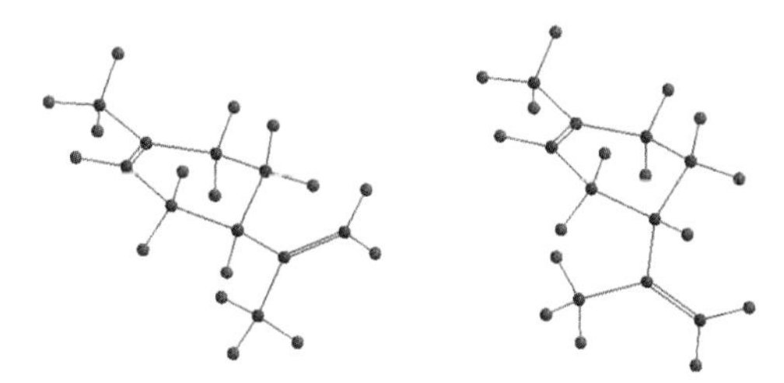

Enantiomeric forms of limonene

One enantiomer of a compound called carvone (Problem 5.14) is the essence of caraway, and the other is the essence of spearmint.

The activity of drugs containing stereogenic carbons can similarly vary between enantiomers, sometimes with serious or even tragic consequences. For several years before 1963 the drug thalidomide was used to alleviate the symptoms of morning sickness in pregnant women. In 1963 it was discovered that thalidomide was the cause of horrible birth defects in many children born subsequent to the use of the drug.

Thalidomide (Thalomid®)

Even later, evidence began to appear indicating that whereas one of the thalidomide enantiomers (the right-handed molecule) has the intended effect of curing morning sickness, the other enantiomer, which was also present in the drug (in an equal amount), may be the cause of the birth defects. The evidence regarding the effects of the two enantiomers is complicated by the fact that, under physiological conditions, the two enantiomers are interconverted. Now, however, thalidomide is approved under highly strict regulations for treatment of some forms of cancer and a serious complication associated with leprosy. Its potential for use against other conditions including AIDS and rheumatoid arthritis is also under investigation. We shall consider other aspects of chiral drugs in Section 5.11.

PROBLEM 5.5

Which atom is the stereogenic center of **(a)** limonene and **(b)** of thalidomide? Draw bond-line formulas for the limonene and thalidomide enantiomers, showing the stereogenic center in each using wedge–dashed wedge notation (Section 1.17E).

5.5 Historical Origin of Stereochemistry

In 1877, Hermann Kolbe (of the University of Leipzig), one of the most eminent organic chemists of the time, wrote the following:

> Not long ago, I expressed the view that the lack of general education and of thorough training in chemistry was one of the causes of the deterioration of chemical research in Germany. . . . Will anyone to whom my worries seem exaggerated please read, if he can, a recent memoir by a Herr van't Hoff on "The Arrangements of Atoms in Space," a document crammed to the hilt with the outpourings of a

> childish fantasy. . . . This Dr. J. H. van't Hoff, employed by the Veterinary College at Utrecht, has, so it seems, no taste for accurate chemical research. He finds it more convenient to mount his Pegasus (evidently taken from the stables of the Veterinary College) and to announce how, on his bold flight to Mount Parnassus, he saw the atoms arranged in space.

Kolbe, nearing the end of his career, was reacting to a publication of a 22-year-old Dutch scientist. This publication had appeared earlier, in September 1874, and in it, van't Hoff had argued that the spatial arrangement of four groups around a central carbon atom is tetrahedral. A young French scientist, J. A. Le Bel, independently put forth the same idea in a publication in November 1874. Within 10 years after Kolbe's comments, however, abundant evidence had accumulated that substantiated the "childish fantasy" of van't Hoff. Later in his career (in 1901), and for other work, van't Hoff was named the first recipient of the Nobel Prize in Chemistry.

The 1901 Nobel Prize in chemistry was awarded to J. H. van't Hoff.

Together, the publications of van't Hoff and Le Bel marked an important turn in a field of study that is concerned with the structures of molecules in three dimensions: **stereochemistry.** Stereochemistry, as we shall see in Section 5.16, had been founded earlier by Louis Pasteur.

It was reasoning based on many observations such as those we presented earlier in this chapter that led van't Hoff and Le Bel to the conclusion that the spatial orientation of groups around carbon atoms is tetrahedral when a carbon atom is bonded to four other atoms. The following information was available to van't Hoff and Le Bel:

1. Only one compound with the general formula CH_3X is ever found.
2. Only one compound with the formula CH_2X_2 or CH_2XY is ever found.
3. Two enantiomeric compounds with the formula CHXYZ are found.

By working Problem 5.6 you can see more about the reasoning of van't Hoff and Le Bel.

PROBLEM 5.6

Show how a square planar structure for carbon compounds can be eliminated from consideration by considering CH_2Cl_2 and CH_2BrCl as examples of disubstituted methanes. **(a)** How many isomers would be possible in each instance if the carbon had a square planar structure? **(b)** How many isomers are possible in each instance if the carbon is tetrahedral? Consider CHBrClF as an example of a trisubstituted methane. **(c)** How many isomers would be possible if the carbon atom were square planar? **(d)** How many isomers are possible for CHBrClF if carbon is tetrahedral?

5.6 Tests for Chirality: Planes of Symmetry

The ultimate way to test for molecular chirality is to construct models of the molecule and its mirror image and then determine whether they are superposable. If the two models are superposable, the molecule that they represent is achiral. If the models are not superposable, then the molecules that they represent are chiral. We can apply this test with actual models, as we have just described, or we can apply it by drawing three-dimensional structures and attempting to superpose them in our minds.

There are other aids, however, that will assist us in recognizing chiral molecules. We have mentioned one already: **the presence of a *single* tetrahedral stereogenic carbon.** The other aids are based on the absence in the molecule of certain symmetry elements. A molecule **will not be chiral,** for example, if it possesses **a plane of symmetry.**

A plane of symmetry (also called a **mirror plane**) is defined as *an imaginary plane that bisects a molecule in such a way that the two halves of the molecule are mirror im-*

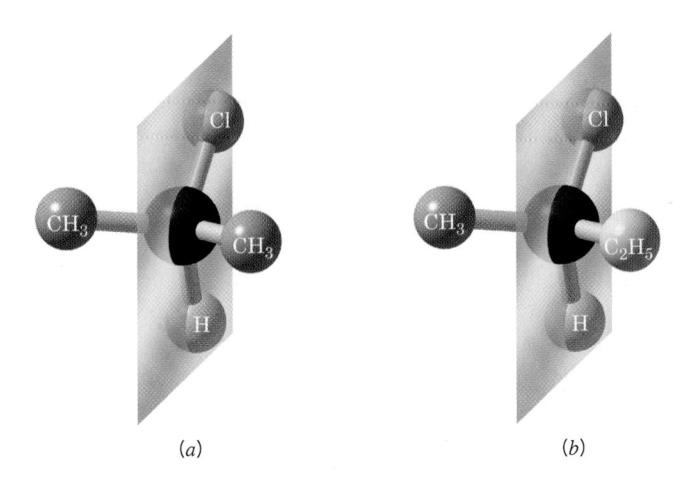

FIGURE 5.8 *(a)* 2-Chloropropane has a plane of symmetry and is achiral. *(b)* 2-Chlorobutane does not possess a plane of symmetry and is chiral.

ages of each other. The plane may pass through atoms, between atoms, or both. For example, 2-chloropropane has a plane of symmetry (Fig. 5.8*a*), whereas 2-chlorobutane does not (Fig. 5.8*b*). **All molecules with a plane of symmetry are achiral.**

PROBLEM 5.7

Which of the objects listed in Problem 5.1 possess a plane of symmetry and are, therefore, achiral?

PROBLEM 5.8

Write three-dimensional formulas and designate a plane of symmetry for all of the achiral molecules in Problem 5.4. (In order to be able to designate a plane of symmetry you may need to write the molecule in an appropriate conformation. This is permissible with all of these molecules because they have only single bonds and groups joined by single bonds are capable of essentially free rotation at room temperature. We discuss this matter further in Section 5.12.)

5.7 Nomenclature of Enantiomers: The *R,S*-System

The two enantiomers of 2-butanol are the following:

I	II
CH_3	CH_3
HO–C–H	H–C–OH
CH_2	CH_2
CH_3	CH_3

If we name these two enantiomers using only the IUPAC system of nomenclature that we have learned so far, both enantiomers will have the same name: 2-butanol (or *sec*-butyl alcohol) (Section 4.3F). This is undesirable because *each compound must have its own distinct name.* Moreover, the name that is given a compound should allow a chemist who is familiar with the rules of nomenclature to write the structure of the compound from its name alone. Given the name 2-butanol, a chemist could write either structure **I** or structure **II**.

Three chemists, R. S. Cahn (England), C. K. Ingold (England), and V. Prelog (Switzerland), devised a system of nomenclature that, when added to the IUPAC system, solves both of these problems. This system, called the *R,S*-system or the Cahn–Ingold–Prelog system, is now widely used and is part of the IUPAC rules.

Cahn (from left), Ingold, and Prelog, shown here at a 1966 meeting, developed the sequence rules for designating the configuration of tetrahedral stereogenic carbon atoms.

According to this system, one enantiomer of 2-butanol should be designated (*R*)-2-butanol and the other enantiomer should be designated (*S*)-2-butanol. [(*R*) and (*S*) are from the Latin words *rectus* and *sinister,* meaning right and left, respectively.] These molecules are said to have opposite **configurations** at C2.

The (*R*) and (*S*) configurations are assigned on the basis of the following procedure.

1. Each of the four groups attached to the stereogenic carbon is assigned a **priority** or **preference** *a*, *b*, *c*, or *d*. Priority is first assigned on the basis of the **atomic number** of the atom that is directly attached to the stereogenic carbon. The group with the lowest atomic number is given the lowest priority, *d*; the group with next higher atomic number is given the next higher priority, *c*; and so on. (In the case of isotopes, the isotope of greatest atomic mass has highest priority.)

We can illustrate the application of the rule with the 2-butanol enantiomer, **I**:

CH_3 (*b* or *c*)
(*a*) HO — C — H (*d*)
CH_2 (*b* or *c*)
CH_3

I

Oxygen has the highest atomic number of the four atoms attached to the stereogenic carbon and is assigned the highest priority, *a*. Hydrogen has the lowest atomic number and is assigned the lowest priority, *d*. A priority cannot be assigned for the methyl group and the ethyl group by this approach because the atom that is directly attached to the stereogenic carbon is a carbon atom in both groups.

2. When a priority cannot be assigned on the basis of the atomic number of the atoms that are directly attached to the stereogenic carbon, then the next set of atoms in the unassigned groups is examined. This process is continued until a decision can be made. *We assign a priority at the first point of difference.**

When we examine the methyl group of enantiomer **I**, we find that the next set of atoms consists of three hydrogen atoms **(H, H, H).** In the ethyl group of **I** the next set of atoms consists of one carbon atom and two hydrogen atoms **(C, H, H).** Carbon has a higher atomic number than hydrogen, so we assign the ethyl group the higher priority, *b*, and the methyl group the lower priority, *c* **(C, H, H) > (H, H, H):**

H—C(H)—H (*c*) **(H, H, H)**
(*a*) HO — C — H (*d*)
H—C—H (*b*) **(C, H, H)**
H—C(H)—H

3. We now rotate the formula (or model) so that the group with lowest priority (*d*) is directed away from us:

*The rules for a branched chain require that we follow the chain with the highest priority atoms.

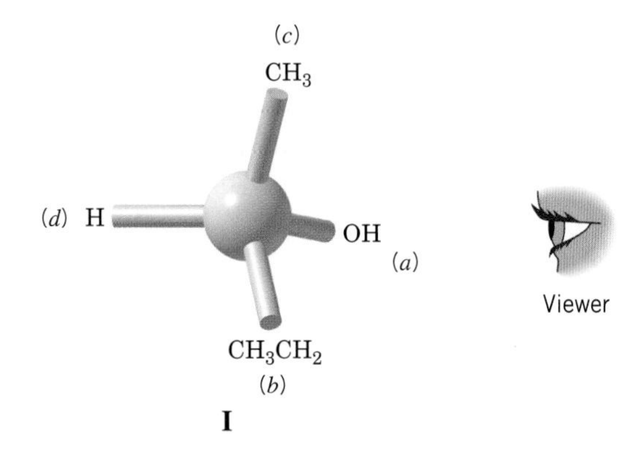

Then we trace a path from *a* to *b* to *c*. If, as we do this, the direction of our finger (or pencil) is *clockwise,* the enantiomer is designated (*R*). If the direction is *counterclockwise,* the enantiomer is designated (*S*). On this basis the 2-butanol enantiomer **I** is (*R*)-2-butanol:

CD Tutorial
(*R,S*) nomenclature example

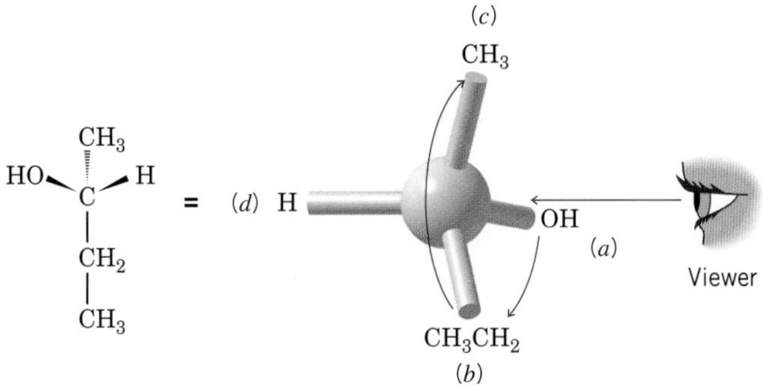

CH_3
HO—C—H
CH_2
CH_3
= (d) H —C— OH (a), (c) CH_3, (b) CH_3CH_2 Viewer

Arrows are clockwise

(*R*)-2-Butanol

Write the enantiomeric forms of bromochlorofluoromethane and assign each enantiomer its correct (*R*) or (*S*) designation. **PROBLEM 5.9**

Give (*R*) and (*S*) designations for each pair of enantiomers given as answers to Problem 5.4. **PROBLEM 5.10**

The first three rules of the Cahn–Ingold–Prelog system allow us to make an (*R*) or (*S*) designation for most compounds containing single bonds. For compounds containing multiple bonds one other rule is necessary:

4. Groups containing double or triple bonds are assigned priorities as if both atoms were duplicated or triplicated, that is,

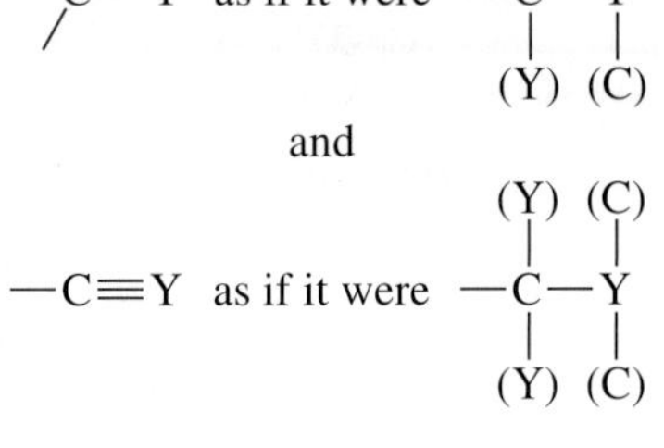

>C=Y as if it were —C(—)(Y)—Y(C)

(i.e., —C—Y with (Y) attached to C and (C) attached to Y)

and

—C≡Y as if it were —C((Y)(Y))—Y((C)(C))

(i.e., —C—Y with two (Y) attached to C and two (C) attached to Y) where the symbols in parentheses are duplicate or triplicate representations of the atoms at the other end of the multiple bond.

Thus, the vinyl group, $—CH{=}CH_2$, is of higher priority than the isopropyl group, $—CH(CH_3)_2$. That is,

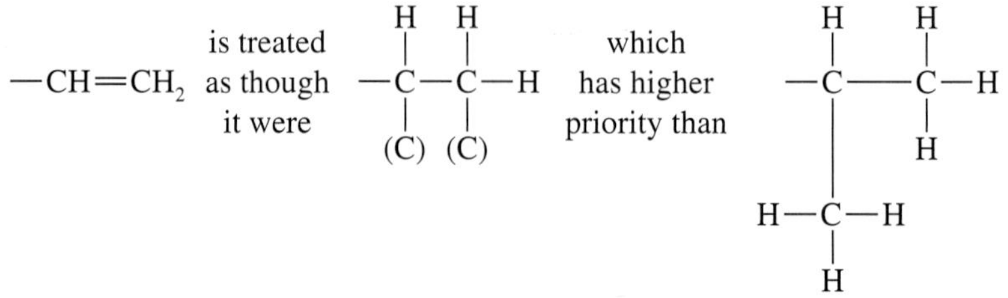

$$—CH{=}CH_2 \text{ is treated as though it were } \begin{array}{c} \text{H}\ \ \text{H} \\ | \ \ \ | \\ —\text{C}—\text{C}—\text{H} \\ | \ \ \ | \\ (\text{C})\ (\text{C}) \end{array} \text{ which has higher priority than } \begin{array}{c} \text{H} \quad\ \text{H} \\ | \quad\ \ | \\ —\text{C}——\text{C}—\text{H} \\ | \quad\ \ | \\ \text{H—C—H} \quad \text{H} \\ | \\ \text{H} \end{array}$$

because at the second set of atoms out, the vinyl group (see the following structure) is **C, H, H**, whereas the isopropyl group along either branch is **H, H, H**. (At the first set of atoms both groups are the same: **C, C, H**.)

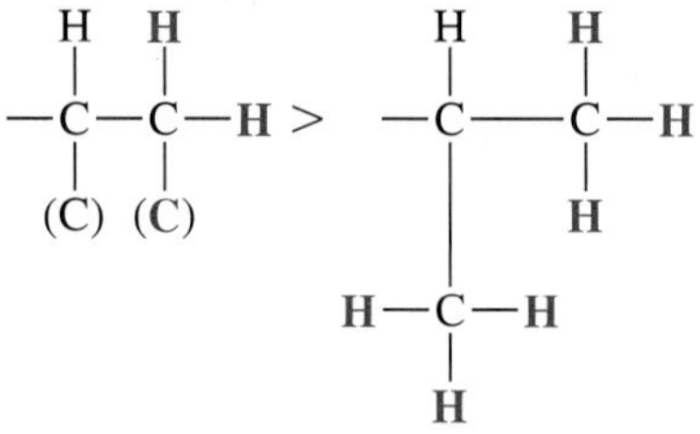

$$\begin{array}{c} \text{H}\ \ \text{H} \\ | \ \ \ | \\ —\text{C}—\text{C}—\text{H} \\ | \ \ \ | \\ (\text{C})\ (\text{C}) \end{array} > \begin{array}{c} \text{H} \quad\ \text{H} \\ | \quad\ \ | \\ —\text{C}——\text{C}—\text{H} \\ | \quad\ \ | \\ \text{H—C—H} \quad \text{H} \\ | \\ \text{H} \end{array}$$

C, H, H > **H, H, H**

Vinyl group **Isopropyl group**

Other rules exist for more complicated structures, but we shall not study them here.*

PROBLEM 5.11 List the substituents in each of the following sets in order of priority, from highest to lowest:

(a) $—Cl, —OH, —SH, —H$

(b) $—CH_3, —CH_2Br, —CH_2Cl, —CH_2OH$

(c) $—H, —OH, —CHO, —CH_3$

(d) $—CH(CH_3)_2, —C(CH_3)_3, —H, —CH{=}CH_2$

(e) $—H, —N(CH_3)_2, —OCH_3, —CH_3$

PROBLEM 5.12 Assign (*R*) or (*S*) designations to each of the following compounds:

(a) Central C bearing CH_3 (hashed, up), $CH_2{=}CH$ (wedge, left), Cl (wedge, right), C_2H_5 (down)

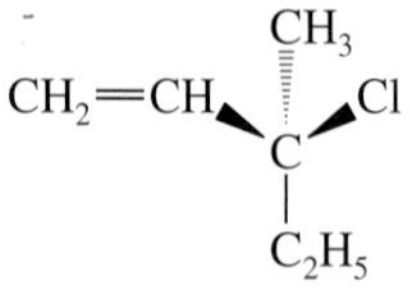

(b) Central C bearing H (hashed, up), $CH_2{=}CH$ (wedge, left), OH (wedge, right), $C(CH_3)_3$ (down)

(c) Central C bearing CH_3 (hashed, up), $H—C{\equiv}C$ (wedge, left), $C(CH_3)_3$ (wedge, right), H (down)

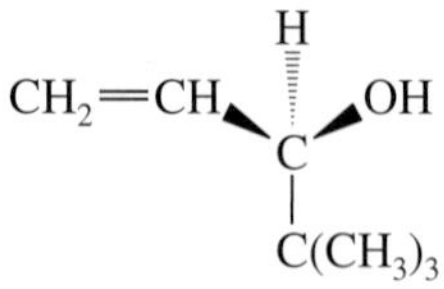

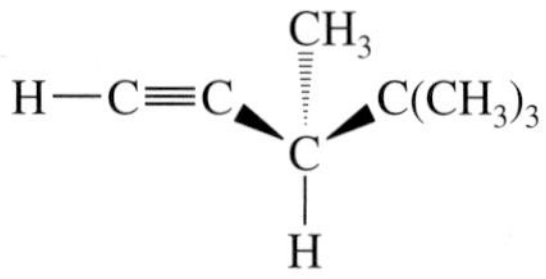

*Further information can be found in the Chemical Abstracts Service *Index Guide*.

SOLVED PROBLEM

Consider the following pair of structures and tell whether they represent enantiomers or two molecules of the same compound in different orientations:

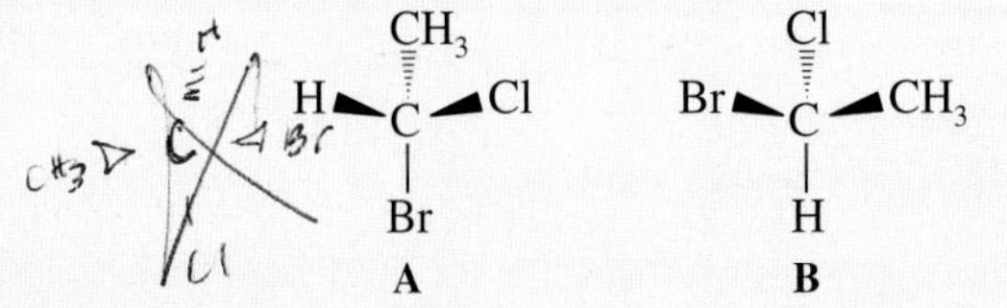

ANSWER: One way to approach this kind of problem is to take one structure and, in your mind, hold it by one group. Then rotate the other groups until at least one group is in the same place as it is in the other structure. (Until you can do this easily in your mind, practice with models.) By a series of rotations like this you will be able to convert the structure you are manipulating into one that is either identical with or the mirror image of the other. For example, take **B**, hold it by the Cl atom and then rotate the other groups about the C*—Cl bond until the bromine is at the bottom (as it is in **A**). Then hold it by the Br and rotate the other groups about the C*—Br bond. This will make **B** identical with **A**:

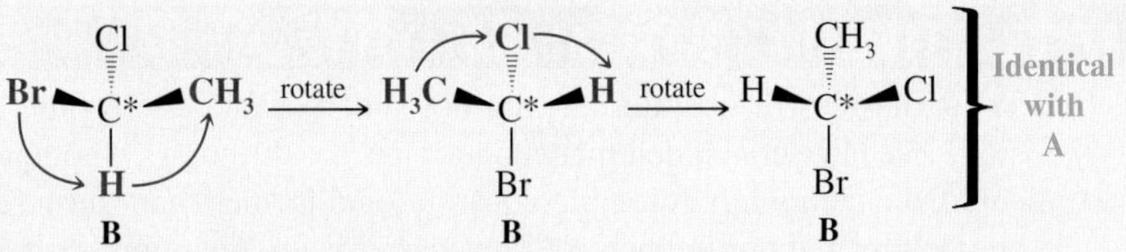

Another approach is to recognize that exchanging two groups at the stereogenic carbon *inverts the configuration of* that carbon atom and converts a structure *with only one stereogenic carbon* into its enantiomer; a second exchange re-creates the original molecule. So we proceed this way, keeping track of how many exchanges are required to convert **B** into **A**. In this instance we find that two exchanges are required, and, again, we conclude that **A** and **B** are the same:

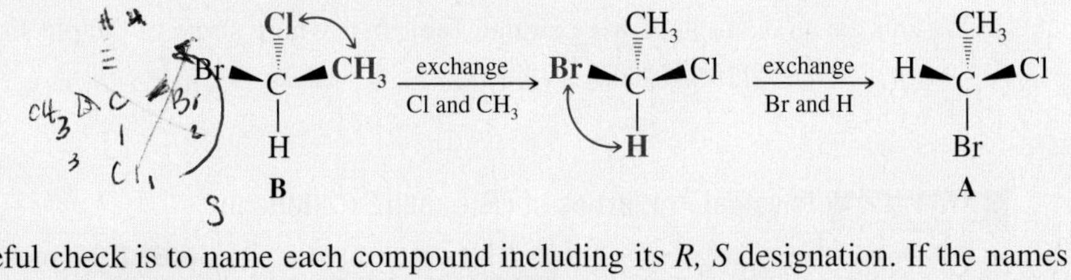

A useful check is to name each compound including its *R, S* designation. If the names are the same, then the structures are the same. In this instance both structures are (*R*)-1-bromo-1-chloroethane.

Another method for assigning *R* and *S* configurations using one's hands as chiral templates has been described (Huheey, J. E. *J. Chem. Educ.* **1986,** *63,* 598–600). Groups at a stereogenic carbon are correlated from lowest to highest priority with one's wrist, thumb, index finger, and second finger, respectively. With the ring and little finger closed against the palm and viewing one's hand with the wrist away, if the correlation between the stereogenic carbon is with the left hand, the configuration is *S*, and if with the right hand, *R*.

PROBLEM 5.13 Tell whether the two structures in each pair represent enantiomers or two molecules of the same compound in different orientations:

(a) Br–C(H)(F)–Cl and Br–C(H)(Cl)–F

(b) F–C(H)(CH$_3$)–Cl and H–C(F)(Cl)–CH$_3$

(c) H–C(CH$_3$)(OH)–CH$_2$–CH$_3$ and HO–C(H)(CH$_2$—CH$_3$)–CH$_3$

5.8 Properties of Enantiomers: Optical Activity

The molecules of enantiomers are not superposable one on the other and, on this basis alone, we have concluded that enantiomers are different compounds. How are they different? Do enantiomers resemble constitutional isomers and diastereomers in having different melting and boiling points? The answer is *no.* Enantiomers have *identical* melting and boiling points. Do enantiomers have different indexes of refraction, different solubilities in common solvents, different infrared spectra, and different rates of reaction with achiral reagents? The answer to each of these questions is also *no.*

Many of these properties (e.g., boiling points, melting points, and solubilities) are dependent on the magnitude of the intermolecular forces operating between the molecules (Section 2.14), and for molecules that are mirror images of each other these forces will be identical.

We can see an example if we examine Table 5.1, where some of the physical properties of the 2-butanol enantiomers are listed.

TABLE 5.1 Physical Properties of (*R*)- and (*S*)-2-Butanol

Physical Property	(*R*)-2-Butanol	(*S*)-2-Butanol
Boiling point (1 atm)	99.5°C	99.5°C
Density (g mL^{-1} at 20°C)	0.808	0.808
Index of refraction (20°C)	1.397	1.397

Enantiomers show different behavior only when they interact with other chiral substances. Enantiomers show different rates of reaction toward other chiral molecules—that is, toward reagents that consist of a single enantiomer or an excess of a single enantiomer. Enantiomers also show different solubilities in solvents that consist of a single enantiomer or an excess of a single enantiomer.

One easily observable way in which enantiomers differ is in *their behavior toward plane-polarized light.* When a beam of plane-polarized light passes through an enantiomer, the plane of polarization **rotates.** Moreover, separate enantiomers rotate the plane of plane-polarized light equal amounts *but in opposite directions.* Because of their effect on plane-polarized light, separate enantiomers are said to be **optically active compounds.**

In order to understand this behavior of enantiomers, we need to understand the nature of plane-polarized light. We also need to understand how an instrument called a **polarimeter** operates.

5.8A Plane-Polarized Light

Light is an electromagnetic phenomenon. A beam of light consists of two mutually perpendicular oscillating fields: an oscillating electric field and an oscillating magnetic field (Fig. 5.9).

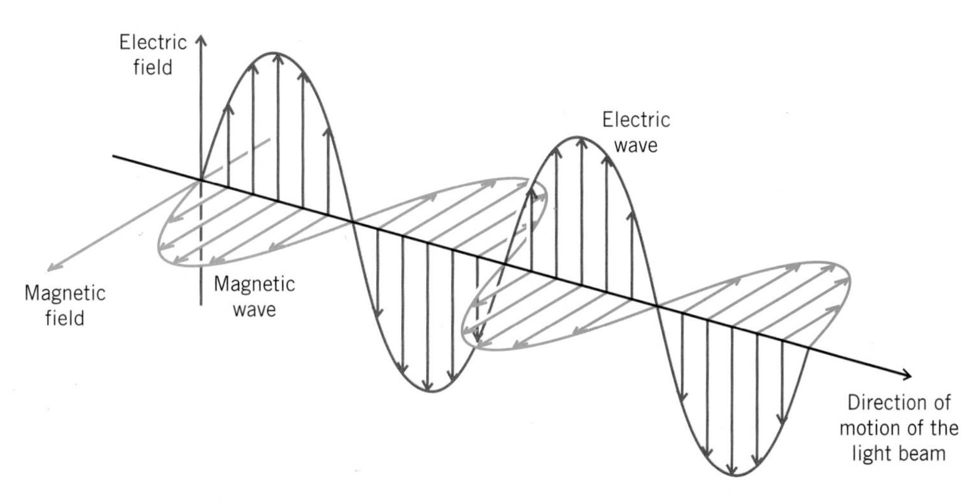

FIGURE 5.9 The oscillating electric and magnetic fields of a beam of ordinary light in one plane. The waves depicted here occur in all possible planes in ordinary light.

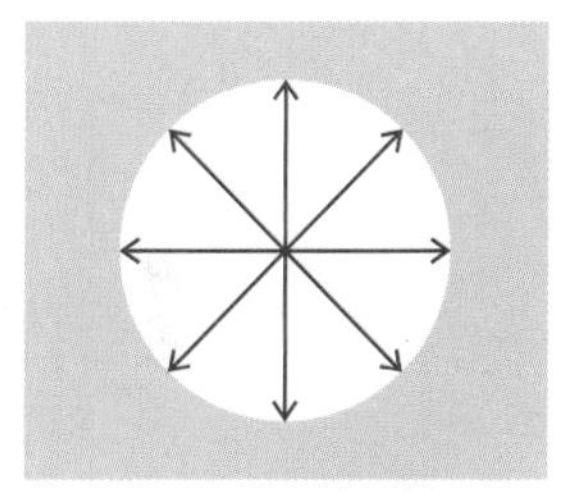

FIGURE 5.10 Oscillation of the electric field of ordinary light occurs in all possible planes perpendicular to the direction of propagation.

If we were to view a beam of ordinary light from one end, and if we could actually see the planes in which the electrical oscillations were occurring, we would find that oscillations of the electric field were occurring in all possible planes perpendicular to the direction of propagation (Fig. 5.10). (The same would be true of the magnetic field.)

When ordinary light is passed through a polarizer, the polarizer interacts with the electric field so that the electric field of the light that emerges from the polarizer (and the magnetic field perpendicular to it) is oscillating only in one plane. Such light is called plane-polarized light (Fig. 5.11).

5.8B The Polarimeter

The device that is used for measuring the effect of optically active compounds on plane-polarized light is a polarimeter. A sketch of a polarimeter is shown in Fig. 5.12. The principal working parts of a polarimeter are (1) a light source (usually a sodium lamp), (2) a polarizer, (3) a tube for holding the optically active substance (or solution) in the light beam, (4) an analyzer, and (5) a scale for measuring the number of degrees that the plane of polarized light has been rotated.

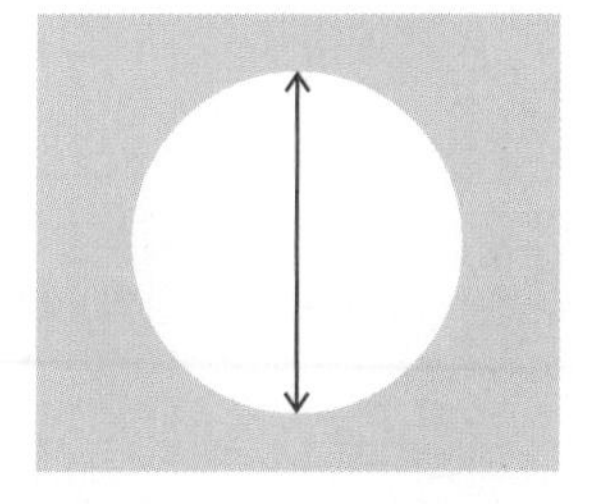

FIGURE 5.11 The plane of oscillation of the electric field of plane-polarized light. In this example the plane of polarization is vertical.

The analyzer of a polarimeter (Fig. 5.12) is nothing more than another polarizer. If the tube of the polarimeter is empty or if an optically *inactive* substance is present, the axes of the plane-polarized light and the analyzer will be exactly parallel when the instrument

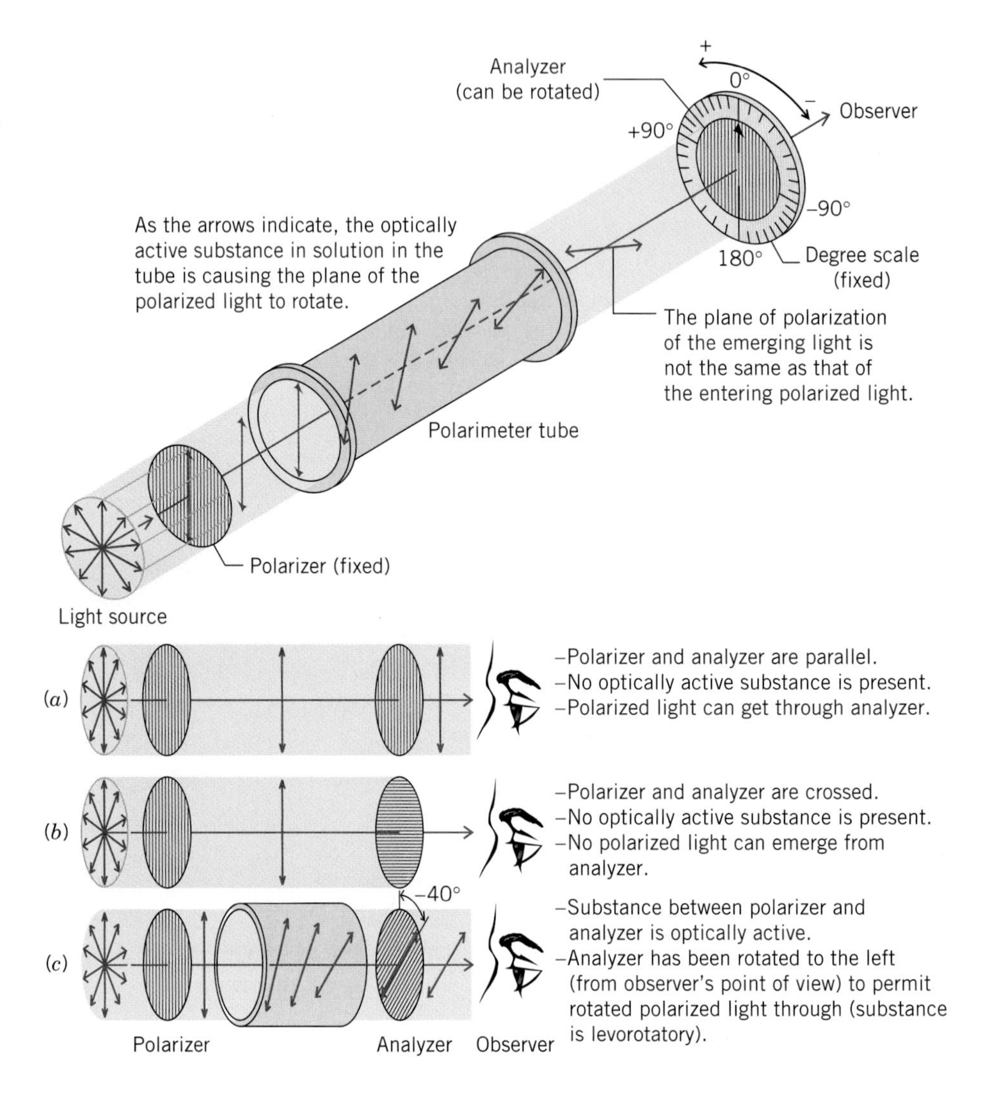

Analyzer
(can be rotated)
+
0°
–
Observer
+90°
–90°
180°
Degree scale
(fixed)
As the arrows indicate, the optically active substance in solution in the tube is causing the plane of the polarized light to rotate.
The plane of polarization of the emerging light is not the same as that of the entering polarized light.
Polarimeter tube
Polarizer (fixed)
Light source
(a)
–Polarizer and analyzer are parallel.
–No optically active substance is present.
–Polarized light can get through analyzer.
(b)
–Polarizer and analyzer are crossed.
–No optically active substance is present.
–No polarized light can emerge from analyzer.
–40°
(c)
–Substance between polarizer and analyzer is optically active.
–Analyzer has been rotated to the left (from observer's point of view) to permit rotated polarized light through (substance is levorotatory).
Polarizer
Analyzer
Observer

FIGURE 5.12 The principal working parts of a polarimeter and the measurement of optical rotation. (From Holum, J.R. *Organic Chemistry: A Brief Course;* Wiley: New York, 1975; p 316.)

reads 0°, and the observer will detect the maximum amount of light passing through. If, by contrast, the tube contains an optically active substance, a solution of one enantiomer, for example, the plane of polarization of the light will be rotated as it passes through the tube. In order to detect the maximum brightness of light, the observer will have to rotate the axis of the analyzer in either a clockwise or counterclockwise direction. If the analyzer is rotated in a clockwise direction, the rotation, α (measured in degrees), is said to be positive (+). If the rotation is counterclockwise, the rotation is said to be negative (−). A substance that rotates plane-polarized light in the clockwise direction is also said to be **dextrorotatory,** and one that rotates plane-polarized light in a counterclockwise direction is said to be **levorotatory** (Latin: *dexter,* right, and *laevus,* left).

5.8C Specific Rotation

The number of degrees that the plane of polarization is rotated as the light passes through a solution of an enantiomer depends on the number of chiral molecules that it encounters. This, of course, depends on the length of the tube and the concentration of the enantiomer. In order to place measured rotations on a standard basis, chemists calculate a quantity called the **specific rotation,** $[\alpha]$, by the following equation:

$$[\alpha] = \frac{\alpha}{c \cdot l}$$

where $[\alpha]$ = the specific rotation

α = the observed rotation

c = the concentration of the solution in grams per milliliter of solution (or density in g mL^{-1} for neat liquids)

l = the length of the tube in decimeters (1 dm = 10 cm)

The specific rotation also depends on the temperature and the wavelength of light that is employed. Specific rotations are reported so as to incorporate these quantities as well. A specific rotation might be given as follows:

$$[\alpha]_{D}^{25} = +3.12°$$

This means that the D line of a sodium lamp (λ = 589.6 nm) was used for the light, that a temperature of 25°C was maintained, and that a sample containing 1.00 g mL^{-1} of the optically active substance, in a 1-dm tube, produced a rotation of 3.12° in a clockwise direction.*

The specific rotations of (*R*)-2-butanol and (*S*)-2-butanol are given here:

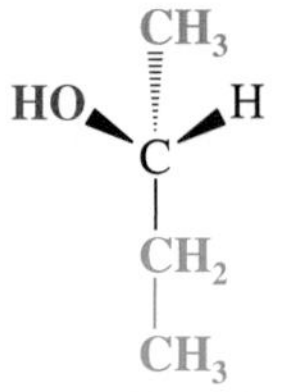

(*R*)-2-Butanol
$[\alpha]_{D}^{25} = -13.52°$

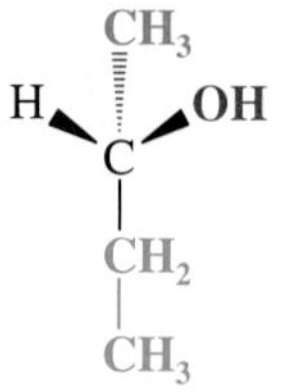

(*S*)-2-Butanol
$[\alpha]_{D}^{25} = +13.52°$

The direction of rotation of plane-polarized light is often incorporated into the names of optically active compounds. The following two sets of enantiomers show how this is done:

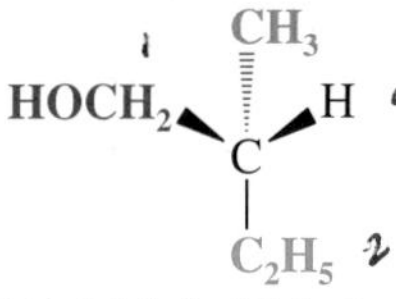

(*R*)-(+)-2-Methyl-1-butanol
$[\alpha]_{D}^{25} = +5.756°$

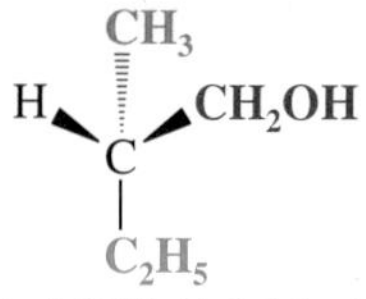

(*S*)-(−)-2-Methyl-1-butanol
$[\alpha]_{D}^{25} = -5.756°$

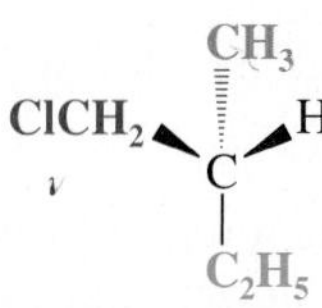

(*R*)-(−)-1-Chloro-2-methylbutane
$[\alpha]_{D}^{25} = -1.64°$

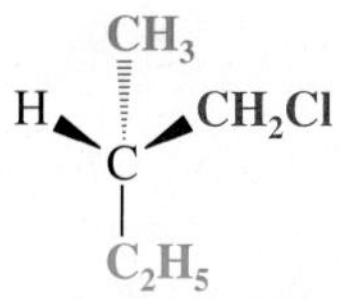

(*S*)-(+)-1-Chloro-2-methylbutane
$[\alpha]_{D}^{25} = +1.64°$

The previous compounds also illustrate an important principle: ***No obvious correlation exists between the configurations of enantiomers and the direction [(+) or (−)] in which they rotate plane-polarized light.***

*The magnitude of rotation is dependent on the solvent when solutions are measured. This is the reason the solvent is specified when a rotation is reported in the chemical literature.

(*R*)-(+)-2-Methyl-1-butanol and (*R*)-(−)-1-chloro-2-methylbutane have the same *configuration;* that is, they have the same general arrangement of their atoms in space. They have, however, an opposite effect on the direction of rotation of the plane of plane-polarized light:

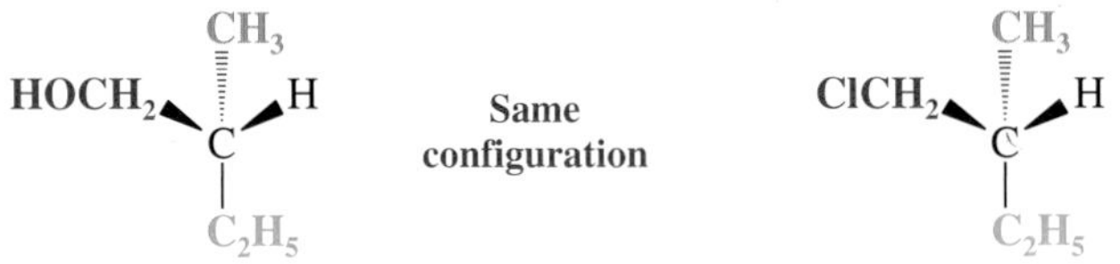

CH_3, $HOCH_2$, H, C, C_2H_5 — (*R*)-(+)-2-Methyl-1-butanol

Same configuration

CH_3, $ClCH_2$, H, C, C_2H_5 — (*R*)-(−)-1-Chloro-2-methylbutane

These same compounds also illustrate a second important principle: ***No necessary correlation exists between the (R) and (S) designation and the direction of rotation of plane-polarized light.*** (*R*)-2-Methyl-1-butanol is dextrorotatory (+), and (*R*)-1-chloro-2-methylbutane is levorotatory (−).

A method based on the measurement of optical rotation at many different wavelengths, called optical rotatory dispersion, has been used to correlate configurations of chiral molecules. A discussion of the technique of optical rotatory dispersion, however, is beyond the scope of this text.

PROBLEM 5.14 Shown below is the configuration of (+)-carvone. (+)-Carvone is the principal component of caraway seed oil and is responsible for its characteristic odor. (−)-Carvone, its enantiomer, is the main component of spearmint oil and gives it its characteristic odor. The fact that the carvone enantiomers do not smell the same suggests that the receptor sites in the nose for these compounds are chiral, and only the correct enantiomer will fit its particular site (just as a hand requires a glove of the correct chirality for a proper fit). Give the correct (*R*) and (*S*) designations for (+)- and (−)-carvone.

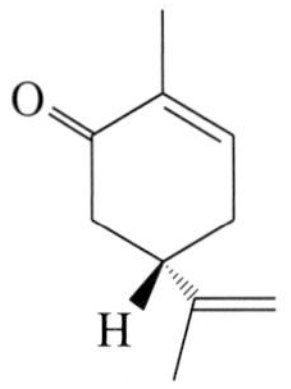

O, H

(+)-Carvone

5.9 The Origin of Optical Activity

Optical activity is measured by the degree of rotation of plane-polarized light upon passage through a chiral medium. The theoretical explanation for the origin of optical activity requires consideration of *circularly* polarized light, however, and its interaction with chiral molecules. While it is not possible to provide a full theoretical explanation for the origin of optical activity here, suffice it to say the following. A beam of plane-polarized light (Fig. 5.13*a*) can be described in terms of circularly polarized light. A beam of circularly polarized light rotating in one direction is shown in Fig. 5.13*b*. The vector sum of *two* counterrotating in-phase circularly polarized beams is a beam of plane-polarized light (Fig. 5.13*c*). The optical activity of chiral molecules results from the fact that the two *counterrotating circularly polarized beams travel with different velocities through the chiral medium.* As the difference between the two circularly polarized beams propagates through the sample, their vector sum describes a plane that is progressively rotated (Fig.

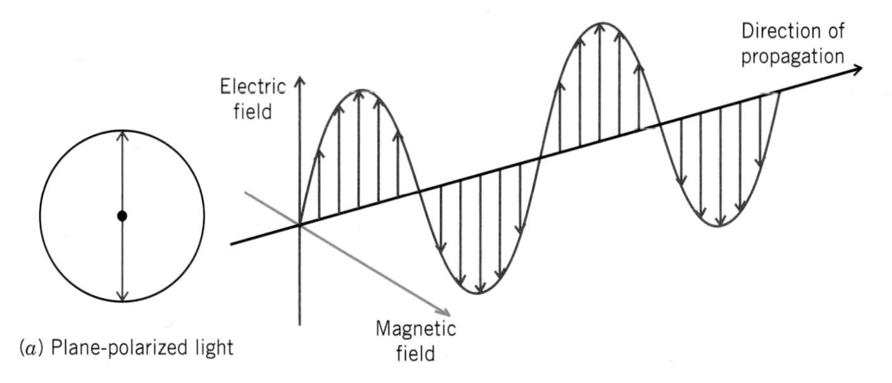

(*a*) Plane-polarized light

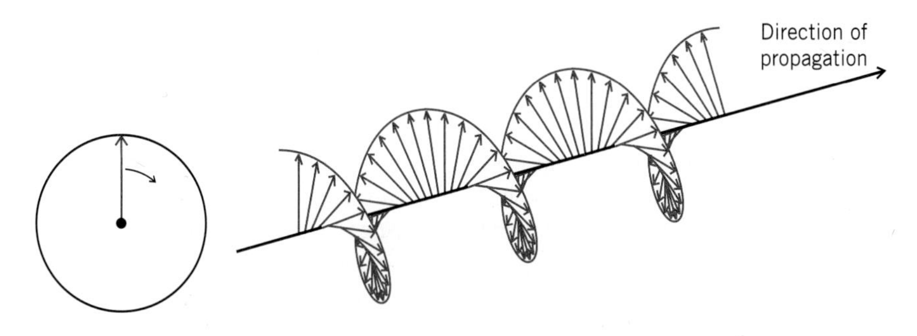

(*b*) Circularly-polarized light

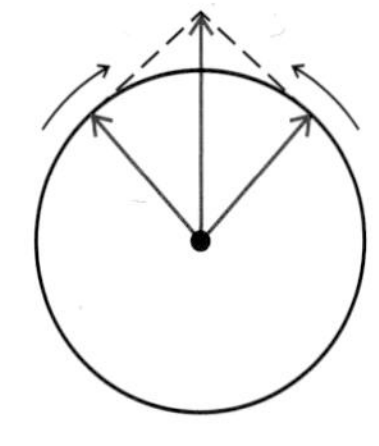

(*c*) Two circularly-polarized beams counter-rotating at the same velocity (in phase), and their vector sum. The net result is like (*a*) above.

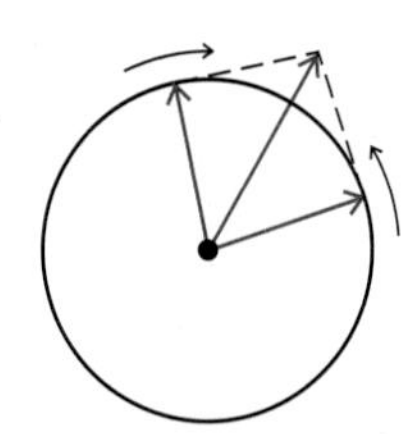

(*d*) Two circularly-polarized beams counter-rotating at the different velocities, such as after interaction with a chiral molecule, and their vector sum. The net result is like (*b*) above.

FIGURE 5.13 *(a)* Plane-polarized light. *(b)* Circularly polarized light. *(c)* Two circularly polarized beams counterrotating at the same velocity (in phase) and their vector sum. The net result is like *(a)*. *(d)* Two circularly polarized beams counterrotating at the different velocities, such as after interaction with a chiral molecule, and their vector sum. The net result is like *(b)*.

5.13*d*). What we measure when light emerges from the sample is the net rotation of the plane-polarized light caused by differences in velocity of the circularly polarized beam components. The origin of the differing velocities has ultimately to do with interactions between electrons in the chiral molecule and light.

Molecules that are not chiral cause no difference in velocity of the two circularly polarized beams; hence there is no rotation of the plane of polarized light described by their vector sum. Achiral molecules, therefore, are not optically active.

5.9A Racemic Forms

A sample that consists exclusively or predominantly of one enantiomer causes a net rotation of plane-polarized light. Figure 5.14*a* depicts a plane of polarized light as it encounters a molecule of (*R*)-2-butanol, causing the plane of polarization to rotate slightly in one

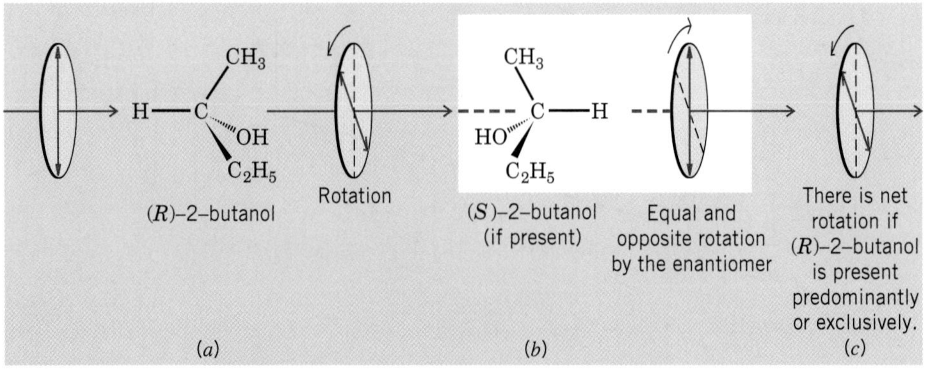

FIGURE 5.14 *(a)* A beam of plane-polarized light encounters a molecule of (*R*)-2-butanol, a chiral molecule. This encounter produces a slight rotation of the plane of polarization. *(b)* Exact cancellation of this rotation occurs if a molecule of (*S*)-2-butanol is encountered. *(c)* Net rotation of the plane of polarization occurs if (*R*)-2-butanol is present predominantly or exclusively.

direction. (For the remaining purposes of our discussion we shall limit our description of polarized light to the resultant plane, neglecting consideration of the circularly polarized components from which plane-polarized light arises.) Each additional molecule of (*R*)-2-butanol that the beam encounters would cause further rotation in the same direction. If, on the other hand, the mixture contained molecules of (*S*)-2-butanol, each molecule of this enantiomer would cause the plane of polarization to rotate in the opposite direction (Fig. 5.14*b*). If the (*R*) and (*S*) enantiomers were present in equal amounts, there would be no net rotation of the plane of polarized light.

An equimolar mixture of two enantiomers such as the example above is called a **racemic mixture** (or **racemate** or **racemic form**). A racemic mixture causes no net rotation of plane-polarized light. In a racemic mixture the effect of each molecule of one enantiomer on the circularly polarized beam cancels the effect of molecules of the other enantiomer, resulting in no net optical activity.

The racemic form of a sample is often designated as being (±). A racemic mixture of (*R*)-(−)-2-butanol and (*S*)-(+)-2-butanol might be indicated as

$$(\pm)\text{-2-butanol} \qquad \text{or as} \qquad (\pm)\text{-}CH_3CH_2CHOHCH_3$$

5.9B Racemic Forms and Enantiomeric Excess

A sample of an optically active substance that consists of a single enantiomer is said to be **enantiomerically pure** or to have an **enantiomeric excess** of 100%. An enantiomerically pure sample of (*S*)-(+)-2-butanol shows a specific rotation of +13.52° ($[\alpha]_D^{25} = +13.52°$). On the other hand, a sample of (*S*)-(+)-2-butanol that contains less than an equimolar amount of (*R*)-(−)-2-butanol will show a specific rotation that is less than +13.52° but greater than 0°. Such a sample is said to have an *enantiomeric excess* less than 100%. The **enantiomeric excess (ee)** is defined as follows:

$$\%\ \text{Enantiomeric excess} = \frac{\text{moles of one enantiomer} - \text{moles of other enantiomer}}{\text{total moles of both enantiomers}} \times 100$$

The enantiomeric excess can be calculated from optical rotations:

$$\%\ \text{Enantiomeric excess}^* = \frac{\text{observed specific rotation}}{\text{specific rotation of the pure enantiomer}} \times 100$$

*This calculation should be applied to a single enantiomer or to mixtures of enantiomers only. It should not be applied to mixtures in which some other compound is present.

Let us suppose, for example, that a mixture of the 2-butanol enantiomers showed a specific rotation of $+6.76°$. We would then say that the enantiomeric excess of the (*S*)-(+)-2-butanol is 50%:

$$\text{Enantiomeric excess} = \frac{+6.76°}{+13.52°} \times 100 = 50\%$$

When we say that the enantiomeric excess of this mixture is 50%, we mean that 50% of the mixture consists of the (+) enantiomer (the excess) and the other 50% consists of the racemic form. Since for the 50% that is racemic the optical rotations cancel one another out, only the 50% of the mixture that consists of the (+) enantiomer contributes to the observed optical rotation. The observed rotation is, therefore, 50% (or one-half) of what it would have been if the mixture had consisted only of the (+) enantiomer.

SOLVED PROBLEM

What is the actual stereoisomeric composition of the mixture referred to above?

ANSWER: Of the total mixture, 50% consists of the racemic form, which contains equal numbers of the two enantiomers. Therefore, half of this 50%, or 25%, is the (−) enantiomer and 25% is the (+) enantiomer. The other 50% of the mixture (the excess) is also the (+) enantiomer. Consequently, the mixture is 75% (+) enantiomer and 25% (−) enantiomer.

PROBLEM 5.15

A sample of 2-methyl-1-butanol (see Section 5.8C) has a specific rotation, $[\alpha]_D^{25}$, equal to $+1.151°$. **(a)** What is the percent enantiomeric excess of the sample? **(b)** Which enantiomer is in excess, the (*R*) or the (*S*)?

5.10 The Synthesis of Chiral Molecules

5.10A Racemic Forms

Reactions carried out with achiral reactants can often lead to *chiral* products. In the absence of any chiral influence from a catalyst, reagent, or solvent, the outcome of such a reaction is a racemic mixture. In other words, the chiral product is obtained as a 50:50 mixture of enantiomers.

An example is the synthesis of 2-butanol by the nickel-catalyzed hydrogenation of butanone. In this reaction the hydrogen molecule adds across the carbon–oxygen double bond in much the same way that it adds to a carbon–carbon double bond.

$$CH_3CH_2\underset{\underset{O}{\|}}{C}CH_3 + H{-}H \xrightarrow{Ni} (\pm)\text{-}CH_3CH_2\underset{\underset{OH}{|}}{\overset{*}{C}}HCH_3$$

Butanone (achiral molecules) — **Hydrogen (achiral molecules)** — **(±)-2-Butanol [chiral molecules but 50:50 mixture (*R*) and (*S*)]**

Figure 5.15 illustrates the stereochemical aspects of this reaction. Because butanone is achiral, there is no difference in presentation of either face of the molecule to the surface of the metal catalyst. The two faces of the trigonal planar carbonyl group interact with the metal surface with equal probability. Transfer of the hydrogen atoms from the metal to the

FIGURE 5.15 The reaction of butanone with hydrogen in the presence of a nickel catalyst. The reaction rate by path *(a)* is equal to that by path *(b)*. (*R*)-(−)-2-Butanol and (*S*)-(+)-2-butanol are produced in equal amounts, as a racemate.

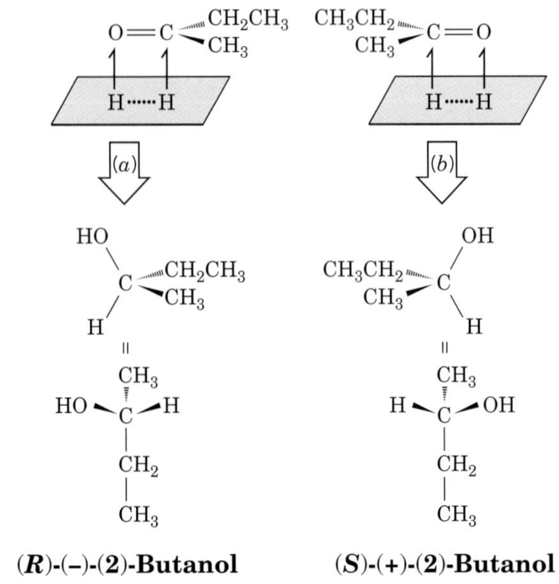

(*R*)-(−)-(2)-Butanol **(*S*)-(+)-(2)-Butanol**

carbonyl group produces a new stereogenic carbon atom at the alcohol functional group. Since there has been no chiral influence in the reaction pathway, the product is obtained as a racemic mixture of the two enantiomers, (*R*)-(−)-2-butanol and (*S*)-(+)-2-butanol.

We shall see that when reactions like this are carried out in the presence of a chiral influence, such as an enzyme or chiral catalyst, the result is not a racemic mixture.

5.10B Stereoselective Syntheses

Stereoselective reactions are reactions that lead to a preponderance of one stereoisomer over other stereoisomers that could possibly be formed. If a reaction produces one enantiomer in preponderance over its mirror image, the reaction is said to be **enantioselective.** If a reaction leads to predominantly one diastereomer over others that are possible, the reaction is said to be **diastereoselective.** For a reaction to be either enantioselective or diastereoselective, a chiral reagent, catalyst, or solvent must assert an influence on the course of the reaction.

In nature, where most reactions are stereoselective, the chiral influences come from protein molecules called **enzymes.** Enzymes are biological catalysts of extraordinary efficiency. They not only have the ability to cause reactions to take place much more rapidly than they would otherwise, they also have the ability to assert a *dramatic chiral influence* on a reaction. Enzymes do this because they, too, are chiral, and they posess an active site where the reactant molecules are momentarily bound while the reaction takes place. The active site is chiral, and only one enantiomer of a chiral reactant fits it properly and is able to undergo the reaction.

Many enzymes have found use in the organic chemistry laboratory, where organic chemists take advantage of their properties to bring about stereoselective reactions. One of these is an enzyme called **lipase.** Lipase catalyzes a reaction called **hydrolysis,** whereby an ester (Section 2.11C) reacts with a molecule of water to produce a carboxylic acid and an alcohol.

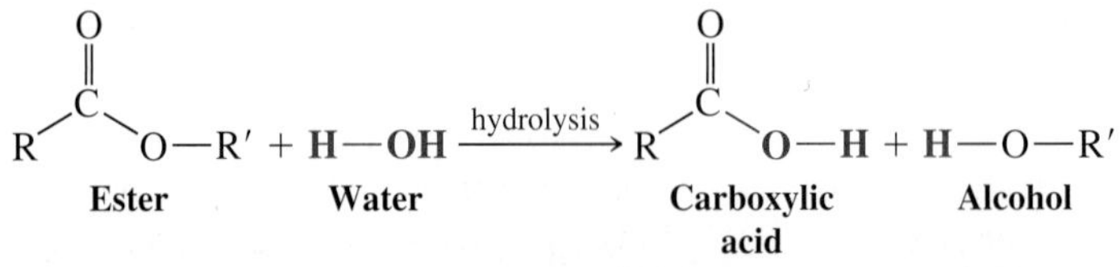

$$RC(=O)O{-}R' + H{-}OH \xrightarrow{\text{hydrolysis}} RC(=O)O{-}H + H{-}O{-}R'$$

Ester **Water** **Carboxylic acid** **Alcohol**

If the starting ester is chiral and present as a mixture of its enantiomers, the lipase enzyme reacts selectively with one enantiomer to release the corresponding chiral carboxylic acid and an alcohol, while the other ester enantiomer remains unchanged or reacts much more slowly. The result is a mixture that consists predominantly of one stereoisomer of the reactant and one stereoisomer of the product, which can usually be separated easily on the basis of their different physical properties. Such a process is called a **kinetic resolution,** where the rate of a reaction with one enantiomer is different than with the other, leading to a preponderance of one product stereoisomer. We shall say more about the resolution of enantiomers in Section 5.16. The following hydrolysis is an example of a kinetic resolution using lipase:

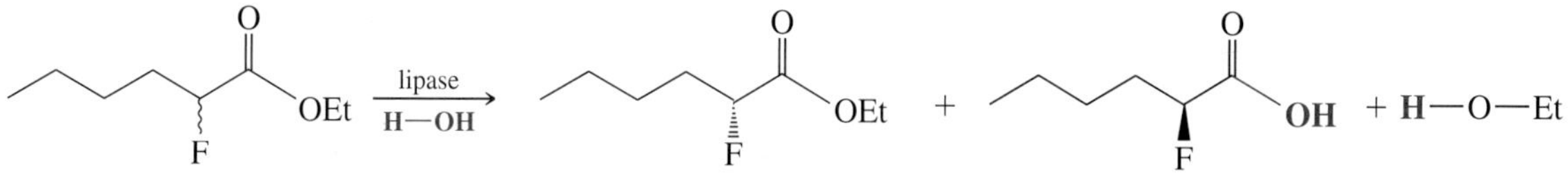

lipase, H—OH → + H—O—Et

Ethyl (±)-2-fluorohexanoate [an ester that is a racemate of (*R*) and (*S*) forms]

Ethyl (*R*)-(+)-2-fluorohexanoate (>99% enantiomeric excess)

(*S*)-(−)-2-Fluorohexanoic acid (>69% enantiomeric excess)

Other enzymes called hydrogenases have been used to effect enantioselective versions of carbonyl reductions like that in Section 5.10A. We shall have more to say about the stereoselectivity of enzymes in Chapter 12.

5.11 Chiral Drugs

The U.S. Food and Drug Administration and the pharmaceutical industry are very interested in the production of chiral drugs, that is, drugs that contain a single enantiomer rather than a racemate. The antihypertensive drug **methyldopa** (Aldomet), for example, owes its effect exclusively to the (*S*) isomer. In the case of **penicillamine,** the (*S*) isomer is a highly potent therapeutic agent for primary chronic arthritis, while the (*R*) isomer has no therapeutic action and is highly toxic. The anti-inflammatory agent **ibuprofen** (Advil, Motrin, Nuprin) is marketed as a racemate even though only the (*S*) enantiomer is the active agent. The (*R*) isomer of ibuprofen has no anti-inflammatory action and is slowly converted to the (*S*) isomer in the body. A formulation of ibuprofen based on solely the (*S*) isomer, however, would be more effective than the racemate.

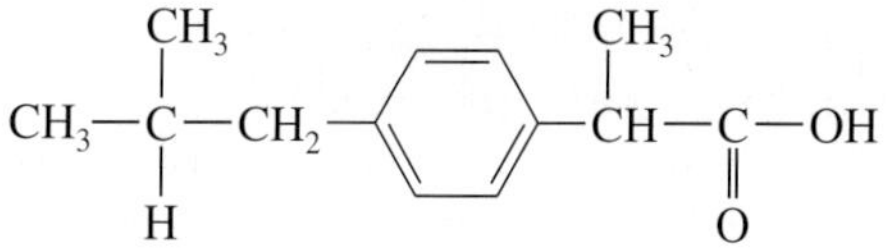

Ibuprofen

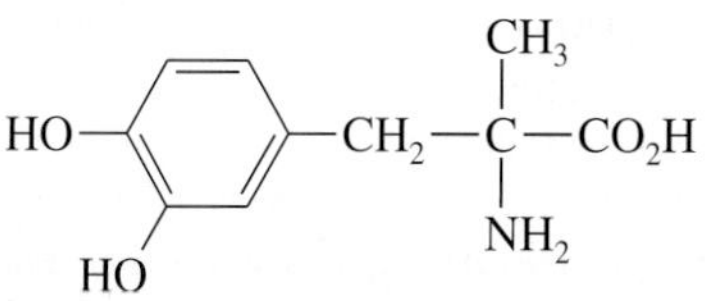

Methyldopa

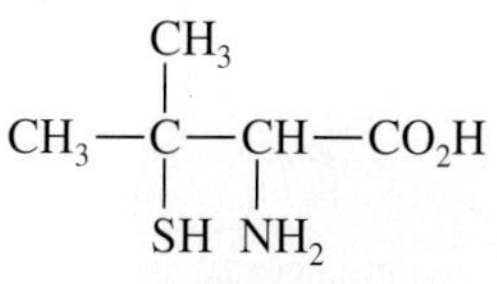

Penicillamine

Problem 5.16

Write three-dimensional formulas for the (*S*) isomers of **(a)** methyldopa, **(b)** penicillamine, and **(c)** ibuprofen.

There are many other examples of drugs like these, including drugs where the enantiomers have distinctly different effects. The preparation of enantiomerically pure drugs, therefore, is one factor that makes stereoselective synthesis (Section 5.10B) and the resolution of racemic drugs (separation into pure enantiomers, Section 5.16) major areas of research today.

Underscoring the importance of stereoselective synthesis is the fact that the 2001 Nobel Prize in Chemistry was given to researchers who developed reaction catalysts that are now widely used in industry and academia. William Knowles (Monsanto Company, retired) and Ryoji Noyori (Nagoya University) were awarded half of the prize for their development of reagents used for catalytic stereoselective hydrogenation reactions. The other half of the prize was awarded to Barry Sharpless (Scripps Research Institute) for development of catalytic stereoselective oxidation reactions (see Chapter 11). An important example resulting from the work of Noyori and based on earlier work by Knowles is a synthesis of the anti-inflammatory agent **naproxen,** involving a stereoselective catalytic hydrogenation reaction:

CH_2 ... COOH + H_2 $\xrightarrow[\text{MeOH}]{(S)\text{-BINAP-Ru}(OCOCH_3)_2\ (0.5\ \text{mol}\ \%)}$ CH_3 H ... COOH

H_3CO H_3CO

(*S*)-Naproxen
(an anti-inflammatory agent)
(92% yield, 97% ee)

The hydrogenation catalyst in this reaction is an organometallic complex formed from ruthenium and a chiral organic ligand called (*S*)-BINAP. The reaction itself is truly remarkable because it proceeds with excellent enantiomeric excess (97%) and in very high yield (92%). We will have more to say about BINAP ligands and the origin of their chirality in Section 5.18.

The Chemistry of...

Selective Binding of Drug Enantiomers to Left- and Right-Hand Coiled DNA

Would you like left- or right-handed DNA with your drug? That's a question that can now be answered due to the recent discovery that each enantiomer of the drug daunorubicin selectively binds DNA coiled with opposite handedness. (+)-Daunorubicin binds selectively to DNA coiled in the typical right-handed conformation (B-DNA). (−)-Daunorubicin binds selectively to DNA coiled in the left-handed conformation (Z-DNA). Furthermore, daunorubicin is capable of inducing conformational changes in DNA from one coiling direction to the other, depending on which coiling form is favored when a given daunorubicin enantiomer binds to the DNA. It has long been known that DNA adopts a number of secondary and tertiary structures, and it is presumed that some of these conformations are involved in turning on or off transcription of a given section of DNA. The discovery of specific interactions between each daunorubicin enantiomer and the left- and right-hand coil forms of DNA will likely assist in design and discovery of new drugs with anticancer or other activities.

Enantiomeric forms of daunorubicin bind with DNA and cause it to coil with opposite handedness.

5.12 Molecules with More than One Stereogenic Center

So far all of the chiral molecules that we have considered have contained only one stereogenic carbon. Many organic molecules, especially those important in biology, contain more than one stereogenic center. Cholesterol (Section 23.4B), for example, contains eight stereogenic centers. (Can you locate them?) We can begin, however, with simpler molecules. Let us consider 2,3-dibromopentane shown here—a structure that has two stereogenic centers:

$$CH_3\overset{*}{C}H(Br)\overset{*}{C}H(Br)CH_2CH_3$$

2,3-Dibromopentane

Cholesterol, having eight stereogenic centers, hypothetically could exist in 2^8 (256) stereoisomeric forms, yet biosynthesis via enzymes produces only *one* stereoisomer.

A useful rule gives the maximum number of stereoisomers: In compounds whose stereoisomerism is due to tetrahedral stereogenic centers, ***the total number of stereoisomers will not exceed 2^n, where n is equal to the number of tetrahedral stereogenic centers.*** For 2,3-dibromopentane we should not expect more than four stereoisomers ($2^2 = 4$).

Our next task is to write three-dimensional formulas for the stereoisomers of the compound. We begin by writing a three-dimensional formula for one stereoisomer and then by writing the formula for its mirror image:

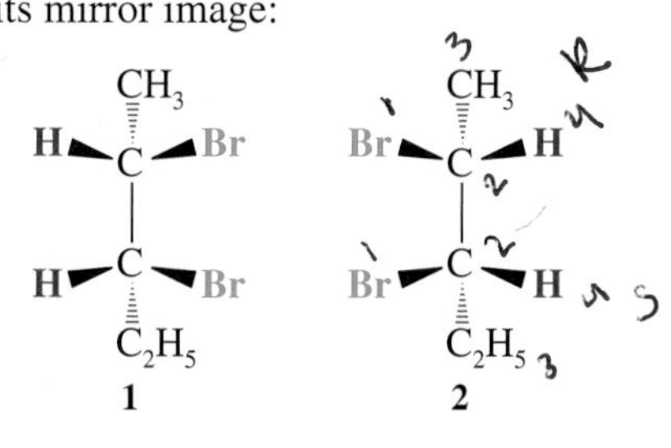

Useful conventions when writing three-dimensional formulas

It is helpful to follow certain conventions when we write these three-dimensional formulas. For example, we usually write our structures in eclipsed conformations. When we do this, we do not mean to imply that eclipsed conformations are the most stable ones—they most certainly are not. We write eclipsed conformations because, as we shall see later, they make it easy for us to recognize planes of symmetry when they are present. We also write the longest carbon chain in a generally vertical orientation on the page; this makes the structures that we write directly comparable. As we do these things, however, *we must remember that molecules can rotate in their entirety* and that *at normal temperatures rotations about all single bonds are also possible.* If rotations of the structure itself or rotations of groups joined by single bonds make one structure superposable with another, then *the structures do not represent different compounds;* instead, they represent different orientations or different conformations of two molecules of the same compound.

Since structures **1** and **2** are not superposable, they represent different compounds. Since structures **1** and **2** differ *only* in the arrangement of their atoms in space, they represent stereoisomers. Structures **1** and **2** are also mirror images of each other; thus **1** and **2** represent enantiomers.

Structures **1** and **2** are not the only possible structures, however. We find that we can write a structure **3** that is different from either **1** or **2**, and we can write a structure **4** that is a nonsuperposable mirror image of structure **3**.

CH$_3$ / Br—C—H / H—C—Br / C$_2$H$_5$ — **3**

CH$_3$ / H—C—Br / Br—C—H / C$_2$H$_5$ — **4**

Structures **3** and **4** correspond to another pair of enantiomers. Structures **1–4** are all different, so there are, in total, four stereoisomers of 2,3-dibromopentane. Essentially what we have done above is to write all the possible structures that result by successively interchanging two groups at all stereogenic carbons. At this point you should convince yourself that there are no other stereoisomers by writing other structural formulas. You will find that rotation about the single bonds (or of the entire structure) of any other arrangement of the atoms will cause the structure to become superposable with one of the structures that we have written here. Better yet, using different colored balls, make molecular models as you work this out.

The compounds represented by structures **1–4** are all optically active compounds. Any one of them, if placed separately in a polarimeter, would show optical activity.

The compounds represented by structures **1** and **2** are enantiomers. The compounds represented by structures **3** and **4** are also enantiomers. But what is the isomeric relation between the compounds represented by **1** and **3**?

We can answer this question by observing that **1** and **3** *are stereoisomers* and that they *are not mirror images of each other.* They are, therefore, *diastereomers.* ***Diastereomers have different physical properties***—different melting points and boiling points, different solubilities, and so forth. In this respect these diastereomers are just like diastereomeric alkenes such as *cis*- and *trans*-2-butene:

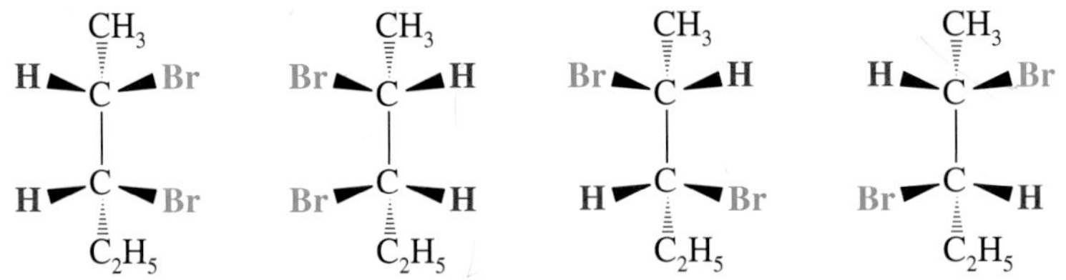

1 **2** **3** **4**

PROBLEM 5.17

(a) If **3** and **4** are enantiomers, what are **1** and **4**? **(b)** What are **2** and **3**, and **2** and **4**? **(c)** Would you expect **1** and **3** to have the same melting point? **(d)** The same boiling point? **(e)** The same vapor pressure?

5.12A Meso Compounds

A structure with two stereogenic carbons does not always have four possible stereoisomers. Sometimes there are only *three.* This happens because some molecules are *achiral even though they contain stereogenic carbons.*

To understand this, let us write stereochemical formulas for 2,3-dibromobutane shown here:

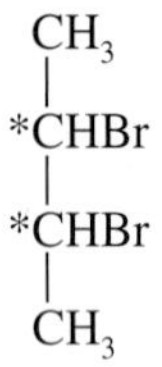

CH_3
|
*CHBr
|
*CHBr
|
CH_3

2,3-Dibromobutane

We begin in the same way as we did before. We write the formula for one stereoisomer and for its mirror image:

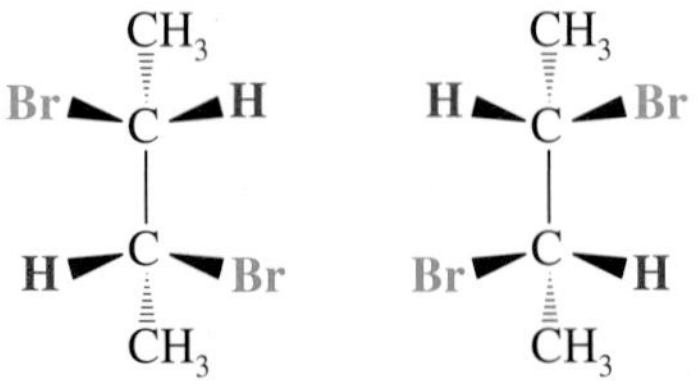

A **B**

Structures **A** and **B** are nonsuperposable and represent a pair of enantiomers.

When we write structure **C** (see below) and its mirror image **D**, however, the situation is different. *The two structures are superposable.* This means that **C** and **D** do not represent a pair of enantiomers. Formulas **C** and **D** represent two different orientations of the same compound:

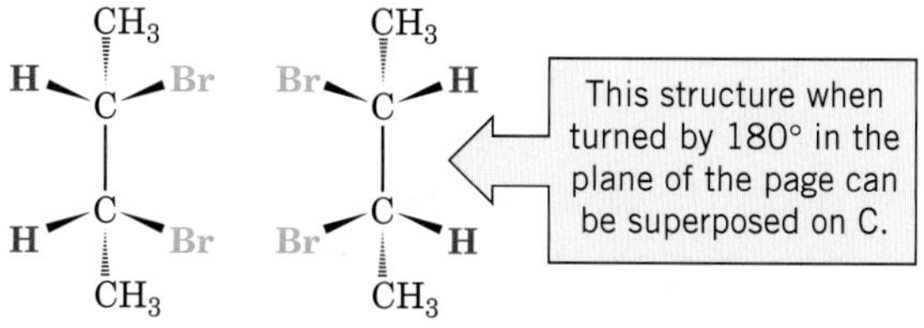

This structure when turned by 180° in the plane of the page can be superposed on C.

C **D**

The molecule represented by structure **C** (or **D**) is not chiral even though it contains tetrahedral atoms with four different attached groups. Such molecules are called **meso compounds.** Meso compounds, *because they are achiral,* are optically inactive.

The ultimate test for molecular chirality is to construct a model (or write the structure) of the molecule and then test whether or not the model (or structure) is superposable on its mirror image. If it is, the molecule is achiral: If it *is not,* the molecule is chiral.

We have already carried out this test with structure **C** and found that it is achiral. We can also demonstrate that **C** is achiral in another way. Figure 5.16 shows that structure **C** *has a plane of symmetry* (Section 5.6).

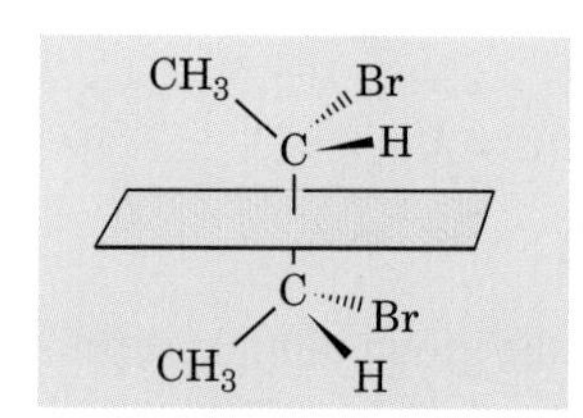

FIGURE 5.16 The plane of symmetry of *meso*-2,3-dibromobutane. This plane divides the molecule into halves that are mirror images of each other.

The following two problems relate to compounds **A–D** in the preceding paragraphs.

PROBLEM 5.18

Which of the following would be optically active?

(a) A pure sample of **A**
(b) A pure sample of **B**
(c) A pure sample of **C**
(d) An equimolar mixture of **A** and **B**

PROBLEM 5.19

The following are formulas for three compounds, written in noneclipsed conformations. In each instance tell which compound (**A**, **B**, or **C**) each formula represents.

(a) **(b)** 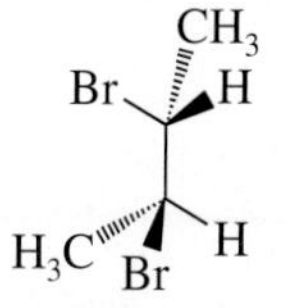 **(c)**

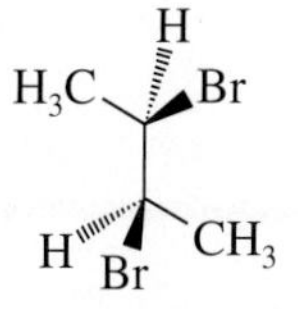

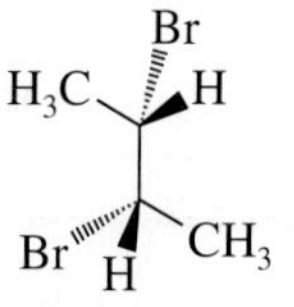

PROBLEM 5.20

Write three-dimensional formulas for all of the stereoisomers of each of the following compounds. Label pairs of enantiomers and label meso compounds.

(a) $CH_3CHClCHClCH_3$
(b) $CH_3CHOHCH_2CHOHCH_3$
(c) $CH_2ClCHFCHFCH_2Cl$
(d) $CH_3CHOHCH_2CHClCH_3$
(e) $CH_3CHBrCHFCH_3$

5.12B Naming Compounds with More than One Stereogenic Carbon

If a compound has more than one tetrahedral stereogenic carbon, we analyze each center separately and decide whether it is (*R*) or (*S*). Then, using numbers, we tell which designation refers to which carbon atom.

Consider the stereoisomer **A** of 2,3-dibromobutane:

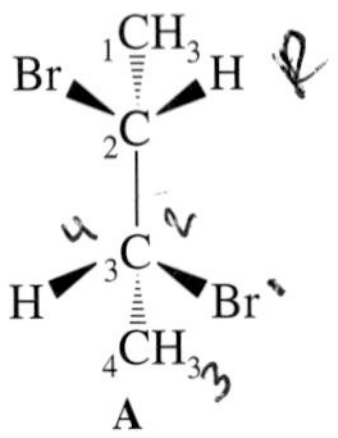

A

2,3-Dibromobutane

When this formula is rotated so that the group of lowest priority attached to C2 is directed away from the viewer, it resembles the following:

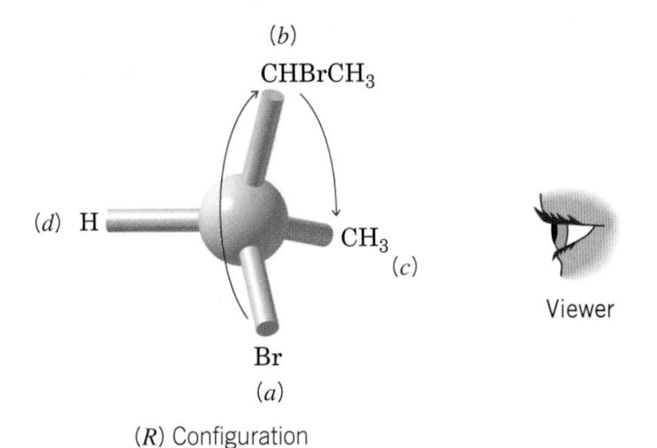

(*R*) Configuration

The order of progression from the group of highest priority to that of next highest priority (from —Br, to —$CHBrCH_3$, to —CH_3) is clockwise. So C2 has the (*R*) configuration.

When we repeat this procedure with C3, we find that C3 also has the (*R*) configuration:

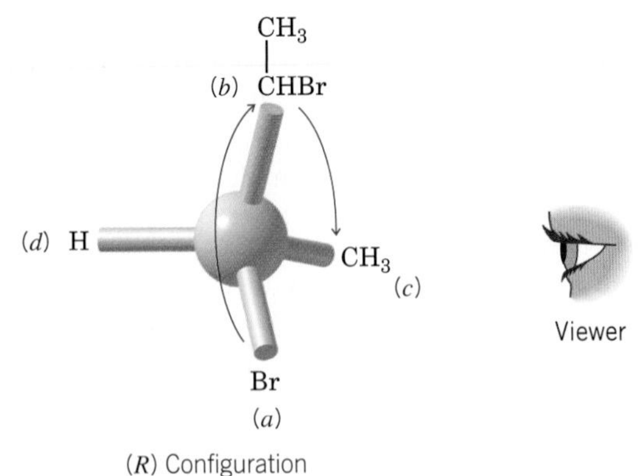

(*R*) Configuration

Compound **A**, therefore, is (2*R*,3*R*)-2,3-dibromobutane.

PROBLEM 5.21 Give names that include (*R*) and (*S*) designations for compounds **B** and **C** in Section 5.12A.

PROBLEM 5.22

Give names that include (*R*) and (*S*) designations for your answers to Problem 5.20.

PROBLEM 5.23

Chloramphenicol (below) is a potent antibiotic, isolated from *Streptomyces venezuelae,* that is particularly effective against typhoid fever. It was the first naturally occurring substance shown to contain a nitro (—NO_2) group attached to an aromatic ring. Both stereogenic carbons in chloramphenicol are known to have the (*R*) configuration. Identify the two stereogenic carbons and write a three-dimensional formula for chloramphenicol.

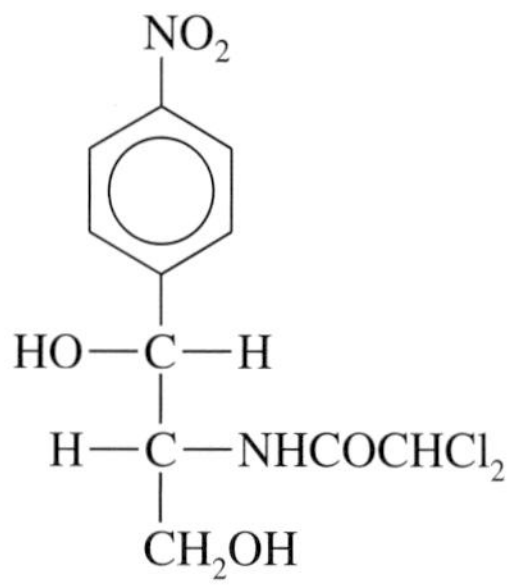

NO_2 (on benzene ring, para)

HO—C—H

H—C—$NHCOCHCl_2$

CH_2OH

Chloramphenicol

5.13 Fischer Projection Formulas

In writing structures for chiral molecules so far, we have used only three-dimensional formulas, and we shall continue to do so until we study carbohydrates in Chapter 22. The reason: Three-dimensional formulas are unambiguous and can be manipulated on paper in any way that we wish, as long as we do not break bonds. Their use, moreover, teaches us to see molecules (in our mind's eye) in three dimensions, and this ability will serve us well.

Chemists sometimes represent structures for chiral molecules with *two-dimensional formulas* called **Fischer projection formulas.** These two-dimensional formulas are especially useful for compounds with several stereogenic carbons because they save space and are easy to write. They are widely used to depict acyclic forms of simple carbohydrates. (See the Learning Group Problem, Part 2). Their use, however, requires a rigid adherence to certain conventions. **Used carelessly, these projection formulas can easily lead to incorrect conclusions.**

The Fischer projection formula for (2*R*,3*R*)-2,3-dibromobutane is written as follows:

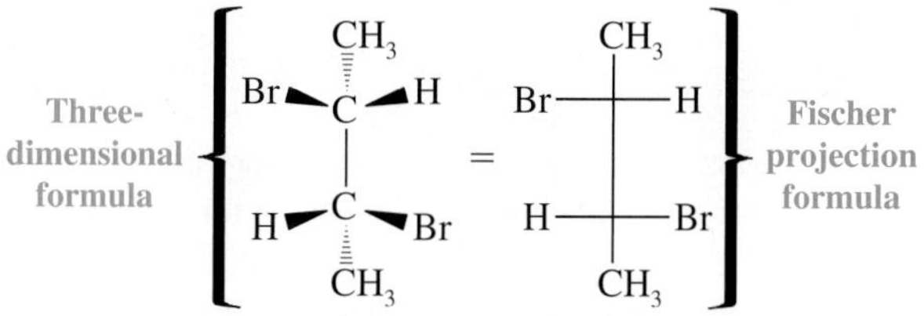

Three-dimensional formula: CH_3 / Br–C–H / H–C–Br / CH_3 (**A**) = Fischer projection formula: CH_3 / Br–+–H / H–+–Br / CH_3 (**A**)

By convention, Fischer projections are written with the main carbon chain extending from top to bottom and with all groups eclipsed. ***Vertical lines represent bonds that project behind the plane of the paper (or that lie in it). Horizontal lines represent bonds that project out of the plane of the paper.*** The intersection of vertical and horizontal lines represents a carbon atom, usually one that is a stereogenic center. By **not** writing the carbon at the intersections in a Fischer projection, we know that we can interpret the formula as indicating the three-dimensional aspects of the molecule. If the carbons were shown (as in Problem 5.23), the formula would not be a Fischer projection and we could not ascertain the stereochemistry of the molecule.

In using Fischer projections to test the superposability for two structures, we are permitted to rotate them in the plane of the paper by 180° *but by no other angle.* We must always keep them in the plane of the paper, and ***we are not allowed to flip them over.***

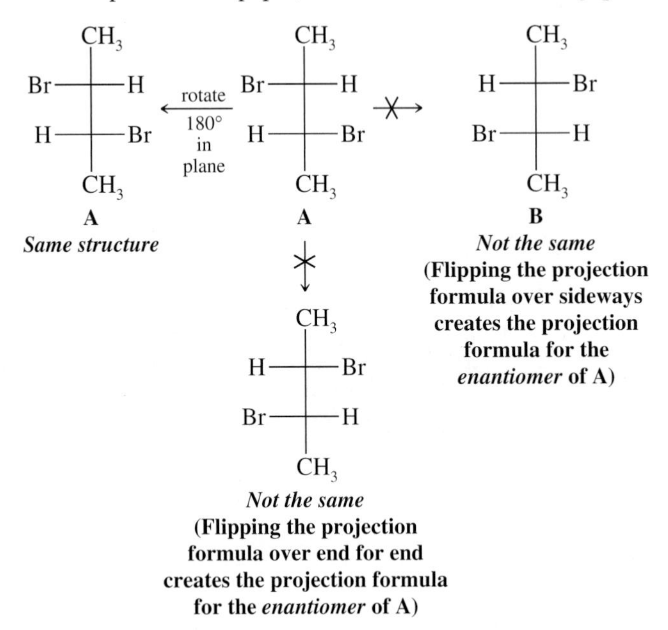

Build hand-held models of **A** and **B** and relate them to the Fischer projections shown here.

Your instructor will advise you about the use you are to make of Fischer projections.

5.14 Stereoisomerism of Cyclic Compounds

Cyclopentane derivatives offer a convenient starting point for a discussion of the stereoisomerism of cyclic compounds. For example, 1,2-dimethylcyclopentane has two stereogenic centers and exists in three stereoisomeric forms **5**, **6**, and **7**:

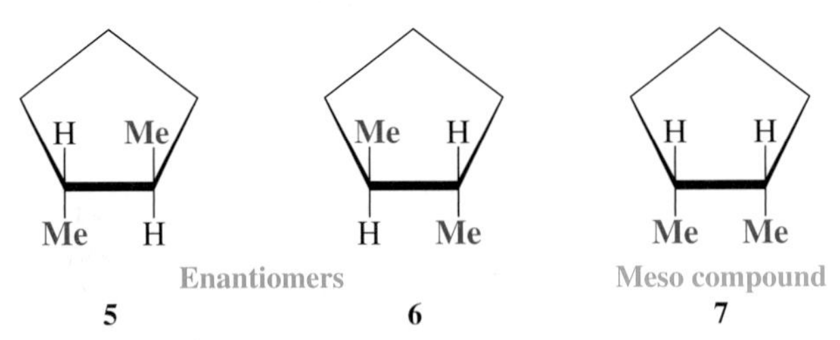

The trans compound exists as a pair of enantiomers **5** and **6**. *cis*-1,2-Dimethylcyclopentane (**7**) is a meso compound. It has a plane of symmetry that is perpendicular to the plane of the ring:

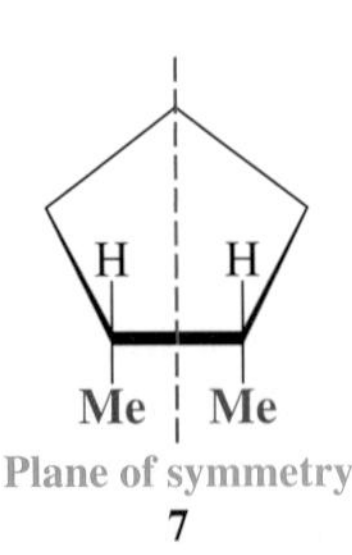

PROBLEM 5.24 **(a)** Is *trans*-1,2-dimethylcyclopentane (**5**) superposable on its mirror image (i.e., on compound **6**)? **(b)** Is *cis*-1,2-dimethylcyclopentane (**7**) superposable on its mirror image? **(c)** Is *cis*-1,2-dimethylcyclopentane a chiral molecule? **(d)** Would *cis*-1,2-dimethylcyclopentane show optical activity? **(e)** What is the stereoisomeric relationship between **5** and **7**? **(f)** Between **6** and **7**?

PROBLEM 5.25 Write structural formulas for all of the stereoisomers of 1,3-dimethylcyclopentane. Label pairs of enantiomers and meso compounds if they exist.

5.14A Cyclohexane Derivatives

Build hand-held molecular models of the 1,4-, 1,3-, and 1,2-dimethylcyclohexane isomers discussed here and examine their stereochemical properties. Experiment with flipping the chairs and also switching between cis and trans isomers.

1,4-Dimethylcyclohexanes If we examine a formula of 1,4-dimethylcyclohexane, we find that it does not contain any tetrahedral atoms with four different groups. However, we learned in Section 4.12 that 1,4-dimethylcyclohexane exists as cis–trans isomers. The cis and trans forms (Fig. 5.17) are *diastereomers.* Neither compound is chiral and, therefore, neither is optically active. Notice that both the cis and trans forms of 1,4-dimethylcyclohexane have a plane of symmetry.

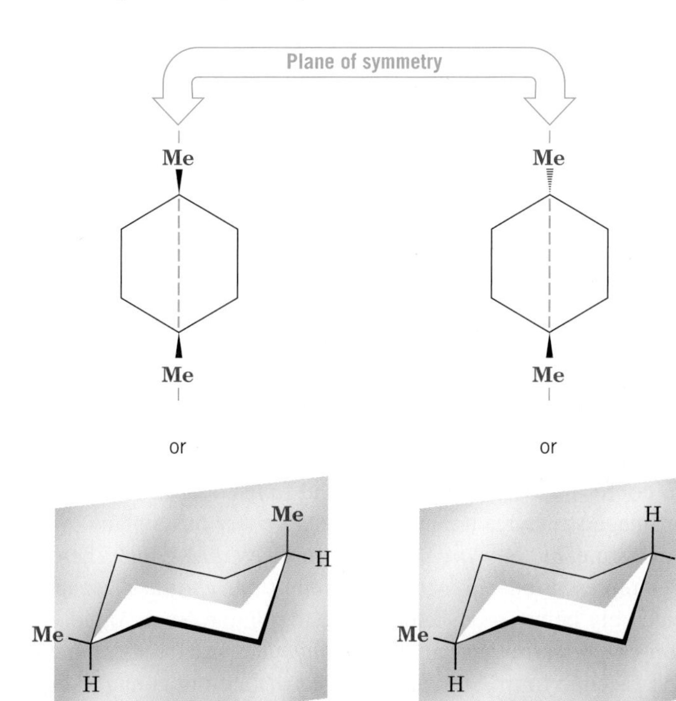

FIGURE 5.17 The cis and trans forms of 1,4-dimethylcyclohexane are diastereomers of each other. Both compounds are achiral.

1,3-Dimethylcyclohexanes 1,3-Dimethylcyclohexane has two stereogenic carbons; we can, therefore, expect as many as four stereoisomers ($2^2 = 4$). In reality there are only three. *cis*-1,3-Dimethylcyclohexane has a plane of symmetry (Fig. 5.18) and is achiral. *trans*-1,3-Dimethylcyclohexane does not have a plane of symmetry and exists as a pair of enantiomers (Fig. 5.19). You may want to make models of the *trans*-1,3-dimethylcyclohexane enantiomers. Having done so, convince yourself that they cannot be superposed as they stand and that they cannot be superposed after one enantiomer has undergone a ring flip.

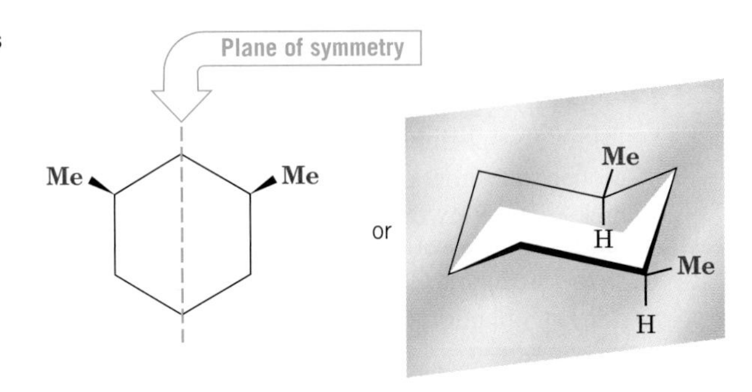

FIGURE 5.18 *cis*-1,3-Dimethylcyclohexane has a plane of symmetry and is therefore achiral.

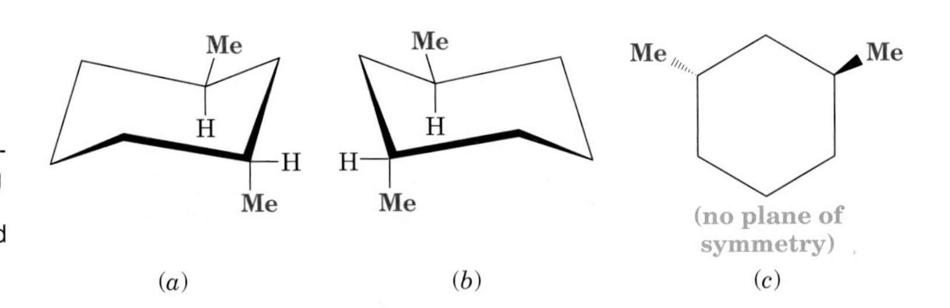

FIGURE 5.19 *trans*-1,3-Dimethylcyclohexane does not have a plane of symmetry and exists as a pair of enantiomers. The two structures (*a* and *b*) shown here are not superposable as they stand, and flipping the ring of either structure does not make it superposable on the other. (*c*) A simplified representation of (*b*).

1,2-Dimethylcyclohexanes 1,2-Dimethylcyclohexane also has two stereogenic carbons, and again we might expect as many as four stereoisomers. *There are four;* however, we find that we can *isolate* only three stereoisomers. *trans*-1,2-Dimethylcyclohexane (Fig. 5.20) exists as a pair of enantiomers. Its molecules do not have a plane of symmetry.

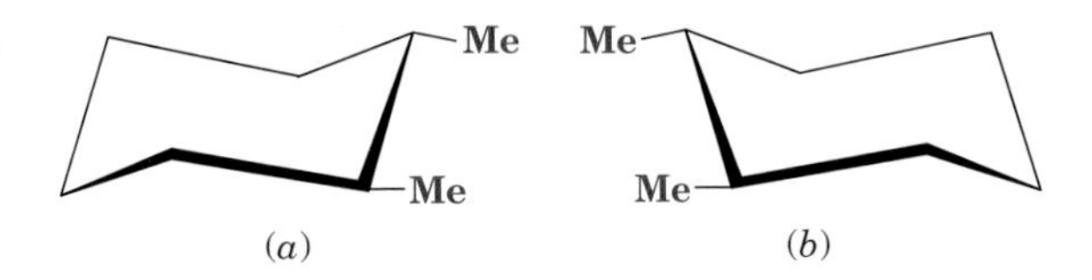

FIGURE 5.20 *trans*-1,2-Dimethylcyclohexane has no plane of symmetry and exists as a pair of enantiomers (*a* and *b*). [Notice that we have written the most stable conformations for (*a*) and (*b*). A ring flip of either (*a*) or (*b*) would cause both methyl groups to become axial.]

With *cis*-1,2-dimethylcyclohexane, the situation is somewhat more complex. If we consider the two conformational structures (*c*) and (*d*) shown in Fig. 5.21, we find that these two mirror image structures are not identical. Neither has a plane of symmetry and each is a chiral molecule, *but they are interconvertible by a ring flip.* Therefore, although the two structures represent enantiomers, *they cannot be separated* because at temperatures even considerably below room temperature they interconvert rapidly. They simply represent *different conformations of the same compound.* In effect, (*c*) and (*d*) comprise an interconverting racemic form. Structures (*c*) and (*d*) are not configurational stereoisomers; they are *conformational stereoisomers* (see Section 4.9). This means that at normal temperatures there are only three *isolable stereoisomers* of 1,2-dimethylcyclohexane.

You should prove this to yourself with models.

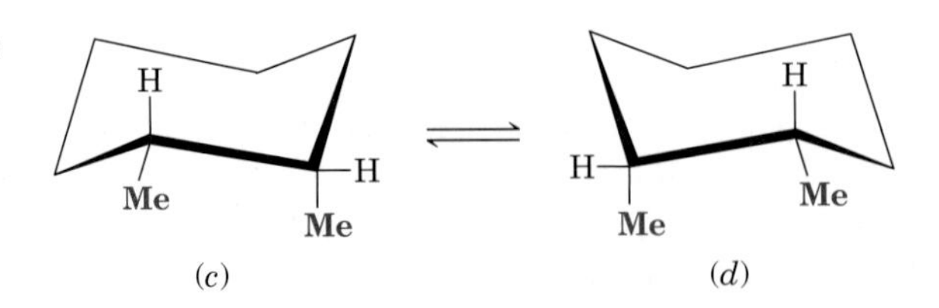

FIGURE 5.21 *cis*-1,2-Dimethylcyclohexane exists as two rapidly interconverting chair conformations (*c*) and (*d*).

PROBLEM 5.26

Write formulas for all of the isomers of each of the following. Designate pairs of enantiomers and achiral compounds where they exist.

(a) 1-Bromo-2-chlorocyclohexane
(b) 1-Bromo-3-chlorocyclohexane
(c) 1-Bromo-4-chlorocyclohexane

PROBLEM 5.27

Give the *R, S* designation for each compound given as an answer to Problem 5.26.

5.15 Relating Configurations through Reactions in Which No Bonds to the Stereogenic Carbon Are Broken

If a reaction takes place in a way so that no bonds to the stereogenic carbon are broken, the product will of necessity have the same general configuration of groups around the stereogenic carbon as the reactant. Such a reaction is said **to proceed with retention of configuration.** Consider as an example the reaction that takes place when (*S*)-(−)-2-methyl-1-butanol is heated with concentrated hydrochloric acid:

Same configuration

CH_3, H, C, CH_2—OH + H—Cl, CH_2, CH_3 $\xrightarrow{\text{heat}}$ CH_3, H, C, CH_2—Cl + H—OH, CH_2, CH_3

(*S*)-(−)-2-Methyl-1-butanol
$[\alpha]_D^{25} = -5.756°$

(*S*)-(+)-1-Chloro-2-methylbutane
$[\alpha]_D^{25} = +1.64°$

We do not need to know now exactly how this reaction takes place to see that the reaction must involve breaking the CH_2—OH bond of the alcohol because the —OH group is replaced by a —Cl. There is no reason to assume that any other bonds are broken. (We shall study how this reaction takes place in Section 11.13.) Since no bonds to the stereogenic carbon are broken, the reaction must take place with retention of configuration, and the product of the reaction *must have the same configuration of groups around the stereogenic carbon that the reactant had.* By saying that the two compounds have the same configuration, we simply mean that comparable or identical groups in the two compounds occupy the same relative positions in space around the stereogenic carbon. (In this instance the —CH_2OH group and the —CH_2Cl are comparable, and they occupy the same relative position in both compounds; all the other groups are identical and they occupy the same positions.)

Notice that in this example while the *R, S* designation *does not change* [both reactant and product are (*S*)], the direction of optical rotation *does change* [the reactant is (−) and the product is (+)]. Neither occurrence is a necessity when a reaction proceeds with retention of configuration. In the next section we shall see examples of reactions in which configurations are retained and where the direction of optical rotation does not change. The following reaction is an example of a reaction that proceeds with retention of configuration but involves a change in the *R, S* designation:

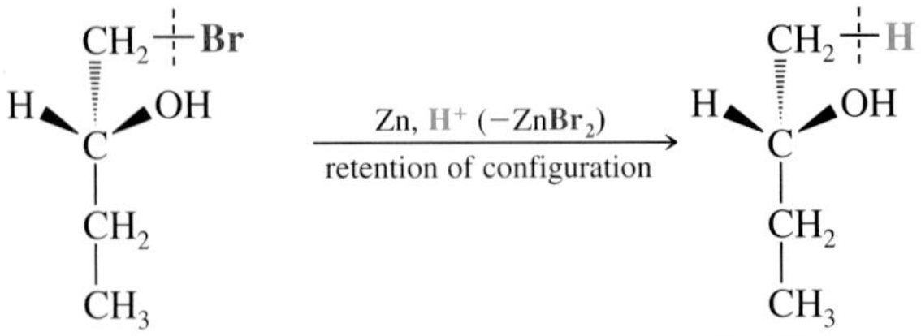

CH_2—Br, H, OH, C, CH_2, CH_3 —— Zn, H^+ ($-ZnBr_2$) / retention of configuration ——> CH_2—H, H, OH, C, CH_2, CH_3

(*R*)-1-Bromo-2-butanol **(*S*)-2-Butanol**

In this example the *R, S* designation changes because the $—CH_2Br$ group of the reactant changes to a $—CH_3$ group in the product ($—CH_2Br$ has a higher priority than $—CH_2CH_3$, and $—CH_3$ has a lower priority than $—CH_2CH_3$).

5.15A Relative and Absolute Configurations

Reactions in which no bonds to the stereogenic carbon are broken are useful in relating configurations of chiral molecules. That is, they allow us to demonstrate that certain compounds have the same **relative configuration.** In each of the examples that we have just cited, the products of the reactions have the same *relative configurations* as the reactants.

Before 1951 only relative configurations of chiral molecules were known. No one prior to that time had been able to demonstrate with certainty what the actual spatial arrangement of groups was in any chiral molecule. To say this another way, no one had been able to determine the **absolute configuration** of an optically active compound.

Configurations of chiral molecules were related to each other *through reactions of known stereochemistry.* Attempts were also made to relate all configurations to a single compound that had been chosen arbitrarily to be the standard. This standard compound was glyceraldehyde:

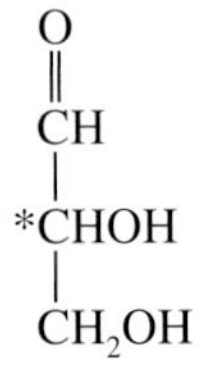

O
‖
CH
|
*CHOH
|
CH_2OH

Glyceraldehyde

Glyceraldehyde molecules have one tetrahedral stereogenic carbon; therefore, glyceraldehyde exists as a pair of enantiomers:

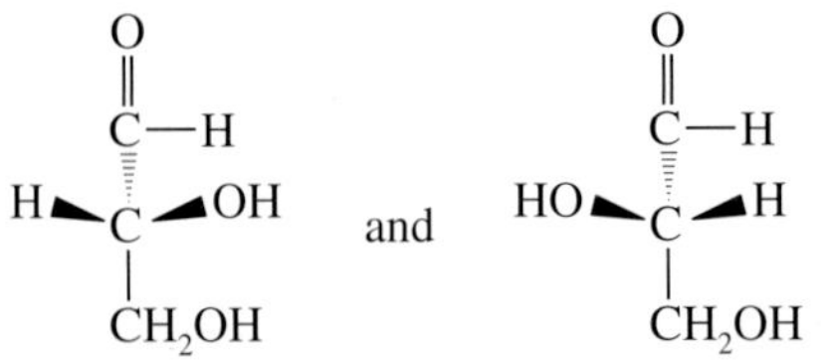

O=C—H, H, C, OH, CH_2OH and O=C—H, HO, C, H, CH_2OH

(*R*)-Glyceraldehyde **(*S*)-Glyceraldehyde**

In the older system for designating configurations (*R*)-glyceraldehyde was called D-glyceraldehyde and (*S*)-glyceraldehyde was called L-glyceraldehyde. This system of nomenclature is still widely used in biochemistry. See Section 22.2B.

One glyceraldehyde enantiomer is dextrorotatory (+) and the other, of course, is levorotatory (−). Before 1951 no one could be sure, however, which configuration belonged to which enantiomer. Chemists decided arbitrarily to assign the (*R*) configuration to the (+)-enantiomer. Then, configurations of other molecules were related to one glyceraldehyde enantiomer or the other through reactions of known stereochemistry.

For example, the configuration of (−)-lactic acid can be related to (+)-glyceraldehyde through the following sequence of reactions in which no bond to the stereogenic carbon is broken:

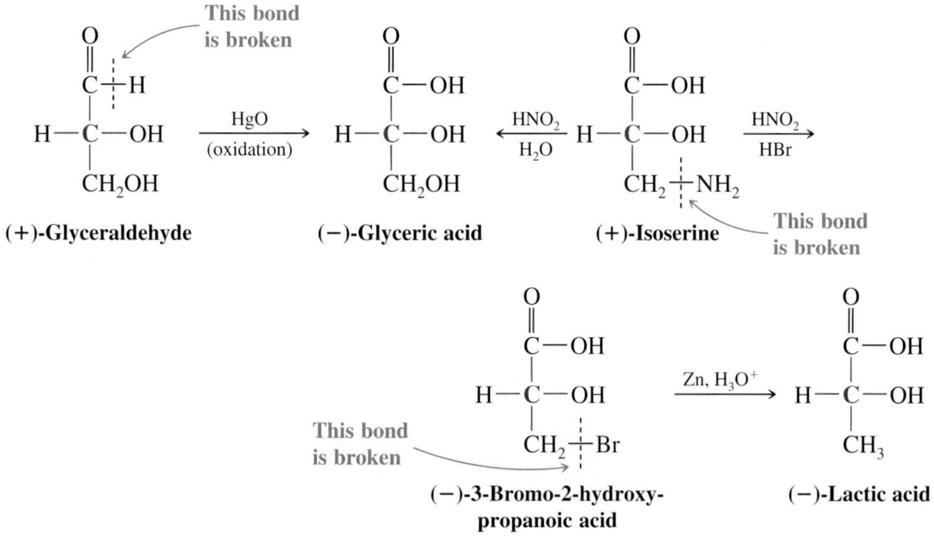

This bond is broken

$$\text{(+)-Glyceraldehyde (HC(=O)–CH(OH)–CH}_2\text{OH)} \xrightarrow[\text{(oxidation)}]{\text{HgO}} \text{(−)-Glyceric acid (HOC(=O)–CH(OH)–CH}_2\text{OH)} \xleftarrow[\text{H}_2\text{O}]{\text{HNO}_2} \text{(+)-Isoserine (HOC(=O)–CH(OH)–CH}_2\text{NH}_2\text{)} \xrightarrow[\text{HBr}]{\text{HNO}_2}$$

This bond is broken

$$\text{(−)-3-Bromo-2-hydroxypropanoic acid (HOC(=O)–CH(OH)–CH}_2\text{Br)} \xrightarrow{\text{Zn, H}_3\text{O}^+} \text{(−)-Lactic acid (HOC(=O)–CH(OH)–CH}_3\text{)}$$

This bond is broken

(+)-Glyceraldehyde **(−)-Glyceric acid** **(+)-Isoserine** **(−)-3-Bromo-2-hydroxy-propanoic acid** **(−)-Lactic acid**

The stereochemistry of all of these reactions is known. Because none of the bonds to the stereogenic carbon (shown in red) has been broken during the sequence, its original configuration is retained. If the assumption is made that the configuration of (+)-glyceraldehyde is

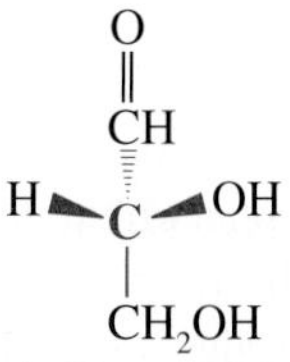

$$\text{O=CH–C(H)(OH)–CH}_2\text{OH}$$

(*R*)-(+)-Glyceraldehyde

then the configuration of (−)-lactic acid is

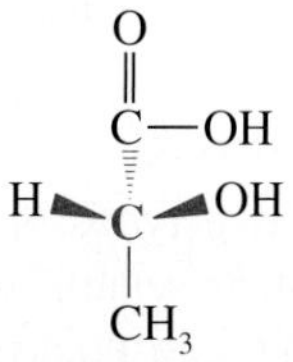

$$\text{HO–C(=O)–C(H)(OH)–CH}_3$$

(*R*)-(−)-Lactic acid

PROBLEM 5.28

Write three-dimensional formulas for the starting compound, the product, and all of the intermediates in a synthesis similar to the one just given that relates the configuration of (−)-glyceraldehyde with (+)-lactic acid. Label each compound with its proper (*R*) or (*S*) and (+) or (−) designation.

The configuration of (−)-glyceraldehyde was also related through reactions of known stereochemistry to (+) or tartaric acid:

$$\begin{array}{c} CO_2H \\ H \blacktriangleright C \blacktriangleleft OH \\ | \\ HO \blacktriangleright C \blacktriangleleft H \\ CO_2H \end{array}$$

(+)-Tartaric acid

In 1951 J. M. Bijvoet, the director of the van't Hoff Laboratory of the University of Utrecht in the Netherlands, using a special technique of X-ray diffraction, was able to show conclusively that (+)-tartaric acid had the absolute configuration shown above. This meant that the original arbitrary assignment of the configurations of (+)- and (−)-glyceraldehyde was also correct. It also meant that the configurations of all of the compounds that had been related to one glyceraldehyde enantiomer or the other were now known with certainty and were now **absolute configurations.**

PROBLEM 5.29 How would you synthesize (*R*)-1-deuterio-2-methylbutane? [*Hint:* Consider one of the enantiomers of 1-chloro-2-methylbutane in Section 5.8C as a starting compound.]

5.16 Separation of Enantiomers: Resolution

So far we have left unanswered an important question about optically active compounds and racemic forms: How are enantiomers separated? Enantiomers have identical solubilities in ordinary solvents, and they have identical boiling points. Consequently, the conventional methods for separating organic compounds, such as crystallization and distillation, fail when applied to a racemic form.

5.16A Pasteur's Method for Separating Enantiomers

It was, in fact, Louis Pasteur's separation of a racemic form of a salt of tartaric acid in 1848 that led to the discovery of the phenomenon called enantiomerism. Pasteur, consequently, is often considered to be the founder of the field of stereochemistry.

(+)-Tartaric acid is one of the byproducts of wine making (nature usually only synthesizes one enantiomer of a chiral molecule). Pasteur had obtained a sample of racemic tartaric acid from the owner of a chemical plant. In the course of his investigation Pasteur began examining the crystal structure of the sodium ammonium salt of racemic tartaric acid. He noticed that two types of crystals were present. One was identical with crystals of the sodium ammonium salt of (+)-tartaric acid that had been discovered earlier and had been shown to be dextrorotatory. Crystals of the other type were *non*superposable mirror images of the first kind. The two types of crystals were actually chiral. Using tweezers and a magnifying glass, Pasteur separated the two kinds of crystals, dissolved them in water, and placed the solutions in a polarimeter. The solution of crystals of the first type was dextrorotatory, and the crystals themselves proved to be identical with the sodium ammonium salt of (+)-tartaric acid that was already known. The solution of crystals of the second type was levorotatory; it rotated plane-polarized light in the opposite direction and by an equal amount. The crystals of the second type were of the sodium ammonium salt of (−)-tartaric acid. The chirality of the crystals themselves disappeared, of course, as the crystals dissolved into their solutions, *but the optical activity* remained. Pasteur reasoned, therefore, that the molecules themselves must be chiral.

Pasteur's discovery of enantiomerism and his demonstration that the optical activity of the two forms of tartaric acid was a property of the molecules themselves led, in 1874, to the proposal of the tetrahedral structure of carbon by van't Hoff and Le Bel.

Unfortunately, few organic compounds give chiral crystals as do the (+)- and (−)-tartaric acid salts. Few organic compounds crystallize into separate crystals (containing separate enantiomers) that are visibly chiral like the crystals of the sodium ammonium salt of tartaric acid. Pasteur's method, therefore, is not generally applicable to the separation of enantiomers.

5.16B Current Methods for Resolution of Enantiomers

One of the most useful procedures for separating enantiomers is based on allowing a racemic form to react with a single enantiomer of some other compound. This changes *a racemic form into a mixture of diastereomers,* and **diastereomers, because they have different melting points, different boiling points, and different solubilities, can be separated by conventional means.** Diastereomeric recrystallization is one such process. We shall see how this is done in Section 20.3E. Another method is **resolution** by enzymes, whereby an enzyme selectively converts one enantiomer in a racemic mixture to another compound, after which the unreacted enantiomer and the new compound are separated. The reaction by lipase in Section 5.10B is an example of this type of resolution. Chromatography using chiral media is also widely used to resolve enantiomers. This approach is applied in high-performance liquid chromatography (HPLC) as well as in other forms of chromatography. Diastereomeric interactions between molecules of the racemic mixture and the chiral chromatography medium cause enantiomers of the racemate to move through the chromatography apparatus at different rates. The enantiomers are then collected separately as they elute from the chromatography device. (See "*The Chemistry of*. . . HPLC Resolution of Enantiomers," Section 20.3.)

5.17 Compounds with Stereogenic Centers Other than Carbon

Any tetrahedral atom with four different groups attached to it is a stereogenic center. Shown here are general formulas of compounds whose molecules contain stereogenic centers other than carbon. Silicon and germanium are in the same group of the periodic table as carbon. They form tetrahedral compounds as carbon does. When four different groups are situated around the central atom in silicon, germanium, and nitrogen compounds, the molecules are chiral and the enantiomers can, in principle, be separated. Sulfoxides, like certain examples of other functional groups where one of the four groups is a nonbonding electron pair, are also chiral. This is not the case for amines, however (Section 20.2B):

R_1, R_2, R_3, R_4 around Si; R_1, R_2, R_3, R_4 around Ge; R_1, R_2, R_3, R_4 around N^+ X^-; R_1, R_2 on S=O (sulfoxide)

5.18 Chiral Molecules That Do Not Possess a Tetrahedral Atom with Four Different Groups

A molecule is chiral if it is not superposable on its mirror image. The presence of a tetrahedral atom with four different groups is only one type of stereogenic center, however. While most of the chiral molecules we shall encounter have tetrahedral stereogenic centers, there are other structural attributes that can confer chirality on a molecule. For example, there are compounds that have such large rotational barriers between conformers that individual conformational isomers can be separated and purified, and some of these conformational isomers are stereoisomers.

Conformational isomers that are stable, isolable compounds are called **atropisomers.** The existence of chiral atropisomers has been exploited to great effect in the development of chiral catalysts for stereoselective reactions. An example is BINAP, shown below in its enantiomeric forms:

$\ddot{P}(Ph)_2$ $\ddot{P}(Ph)_2$ $(Ph)_2\ddot{P}$ $(Ph)_2\ddot{P}$

***(S)*-BINAP** ***(R)*-BINAP**

The origin of chirality in BINAP is the restricted rotation about the bond between the two nearly perpendicular naphthalene rings. This torsional barrier leads to two resolvable enantiomeric conformers, (*S*)- and (*R*)-BINAP. When each enantiomer is used as a ligand for metals such as ruthenium and rhodium (bound by unshared electron pairs on the phosphorus atoms), chiral organometallic complexes result that are capable of catalyzing stereoselective hydrogenation and other important industrial reactions. The significance of chiral ligands is highlighted by the industrial synthesis each year of approximately 3500 *tons* of (−)-menthol using an isomerization reaction involving a rhodium (*S*)-BINAP catalyst.

Allenes are compounds that also exhibit stereoisomerism. Allenes are molecules that contain the following double-bond sequence:

C=C=C

The planes of the π bonds of allenes are perpendicular to each other:

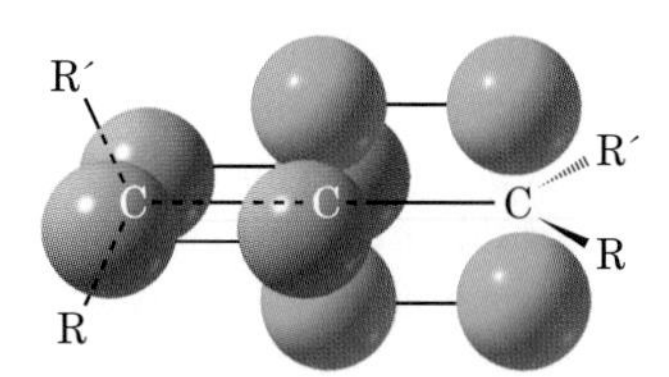

This geometry of the π bonds causes the groups attached to the end carbon atoms to lie in perpendicular planes, and, because of this, allenes with different substituents on the end carbon atoms are chiral (Fig. 5.22). (Allenes do not show cis–trans isomerism.)

H, Cl C=C=C H, Cl | H, Cl C=C=C H, Cl

Mirror

FIGURE 5.22 Enantiomeric forms of 1,3-dichloroallene. These two molecules are nonsuperposable mirror images of each other and are therefore chiral. They do not possess a tetrahedral atom with four different groups, however.

Key Terms and Concepts

Achiral molecule	Sections 5.3, 5.12
Atropisomers	Section 5.18
Chirality	Sections 5.1, 5.4, 5.6
Chiral molecule	Section 5.3
Configuration	Sections 5.7, 5.15
Constitutional isomers	Sections 1.3A, 4.2, 5.2
Diastereomers	Section 5.2
Diastereoselective reaction	Section 5.10A
Enantiomers	Sections 5.3, 5.7, 5.8, 5.16
Enantioselective reaction	Section 5.10A
Fischer projection formulas	Section 5.13
Isomers	Sections 1.13B, 5.2
Kinetic resolution	Section 5.10A
Meso compound	Section 5.12A
Plane of symmetry	Sections 5.6, 5.12A
Racemic form (racemate or racemic mixture)	Sections 5.9A, B
Resolution	Section 5.16
Stereochemistry	Sections 5.2, 5.5
Stereogenic carbon	Section 5.3
Stereogenic center	Sections 5.3, 5.12, 5.17
Stereoisomers	Sections 5.2, 5.14
Stereoselective reaction	Section 5.10A
Superposable	Section 5.1

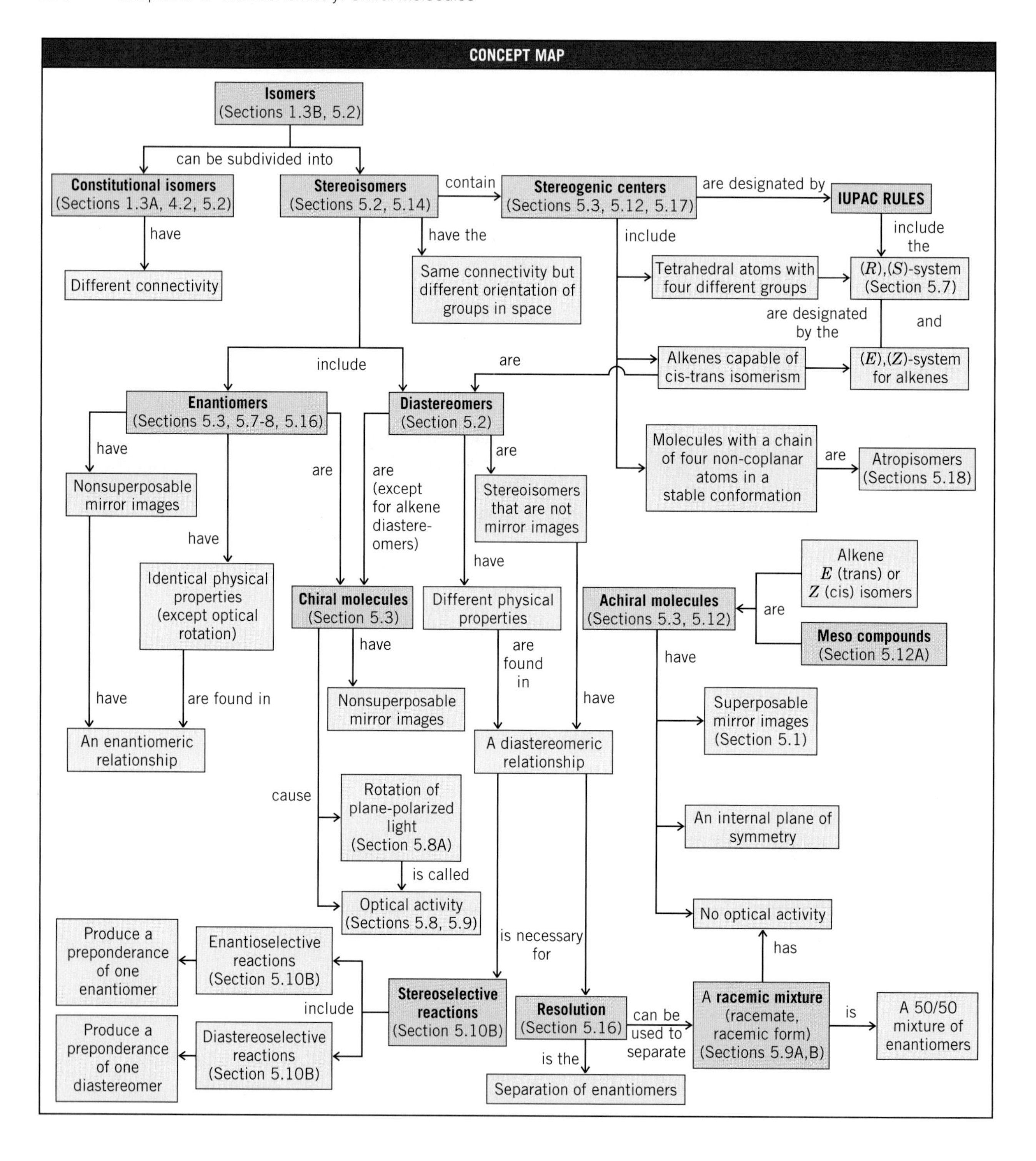
CONCEPT MAP
Isomers (Sections 1.3B, 5.2)
can be subdivided into
Constitutional isomers (Sections 1.3A, 4.2, 5.2)
have
Different connectivity
Stereoisomers (Sections 5.2, 5.14)
contain
Stereogenic centers (Sections 5.3, 5.12, 5.17)
are designated by
IUPAC RULES
include the
have the
Same connectivity but different orientation of groups in space
include
Tetrahedral atoms with four different groups
(R),(S)-system (Section 5.7)
are designated by the
and
Alkenes capable of cis-trans isomerism
(E),(Z)-system for alkenes
include
are
Enantiomers (Sections 5.3, 5.7-8, 5.16)
Diastereomers (Section 5.2)
have
Nonsuperposable mirror images
are
are (except for alkene diastere-omers)
are
Stereoisomers that are not mirror images
Molecules with a chain of four non-coplanar atoms in a stable conformation
are
Atropisomers (Sections 5.18)
have
Identical physical properties (except optical rotation)
Chiral molecules (Section 5.3)
have
Different physical properties
Achiral molecules (Sections 5.3, 5.12)
are
Alkene E (trans) or Z (cis) isomers
Meso compounds (Section 5.12A)
have
Nonsuperposable mirror images
are found in
have
have
are found in
An enantiomeric relationship
A diastereomeric relationship
Superposable mirror images (Section 5.1)
cause
Rotation of plane-polarized light (Section 5.8A)
An internal plane of symmetry
is called
Optical activity (Sections 5.8, 5.9)
No optical activity
is necessary for
has
Produce a preponderance of one enantiomer
Enantioselective reactions (Section 5.10B)
include
Stereoselective reactions (Section 5.10B)
Resolution (Section 5.16)
can be used to separate
A racemic mixture (racemate, racemic form) (Sections 5.9A,B)
is
A 50/50 mixture of enantiomers
Produce a preponderance of one diastereomer
Diastereoselective reactions (Section 5.10B)
is the
Separation of enantiomers

ADDITIONAL PROBLEMS

The CD accompanying this book includes a set of computer molecular model stereochemistry exercises that are keyed to the text.

5.30 Which of the following are chiral and, therefore, capable of existing as enantiomers?

(a) 1,3-Dichlorobutane
(b) 1,2-Dibromopropane
(c) 1,5-Dichloropentane
(d) 3-Ethylpentane
(e) 2-Bromobicyclo[1.1.0]butane
(f) 2-Fluorobicyclo[2.2.2]octane
(g) 2-Chlorobicyclo[2.1.1]hexane
(h) 5-Chlorobicyclo[2.1.1]hexane

5.31 **(a)** How many carbon atoms does an alkane (not a cycloalkane) need before it is capable of existing in enantiomeric forms? **(b)** Give correct names for two sets of enantiomers with this minimum number of carbon atoms.

5.32 **(a)** Write the structure of 2,2-dichlorobicyclo[2.2.1]heptane. **(b)** How many stereogenic carbon atoms does it contain? **(c)** How many stereoisomers are predicted by the 2^n rule? **(d)** Only one pair of enantiomers is possible for 2,2-dichlorobicyclo[2.2.1]heptane. Explain.

5.33 Shown below are Newman projection formulas for (*R,R*)-, (*S,S*)-, and (*R,S*)-2,3-dichlorobutane. **(a)** Which is which? **(b)** Which formula is a meso compound?

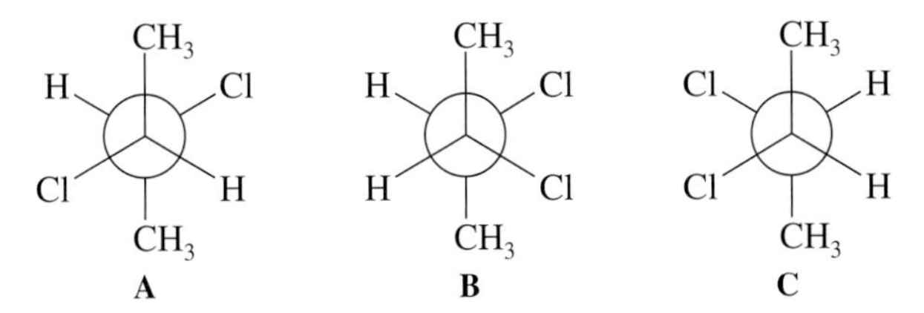

5.34 Write appropriate structural formulas for **(a)** a cyclic molecule that is a constitutional isomer of cyclohexane, **(b)** molecules with the formula C_6H_{12} that contain one ring and that are enantiomers of each other, **(c)** molecules with the formula C_6H_{12} that contain one ring and that are diastereomers of each other, **(d)** molecules with the formula C_6H_{12} that contain no ring and that are enantiomers of each other, and **(e)** molecules with the formula C_6H_{12} that contain no ring and that are diastereomers of each other.

5.35 Consider the following pairs of structures. Identify the relationship between them by describing them as representing enantiomers, diastereomers, constitutional isomers, or two molecules of the same compound. Use hand-held molecular models to check your answers.

(a) H–C(CH$_3$)(Br)–F and F–C(CH$_3$)(H)–Br

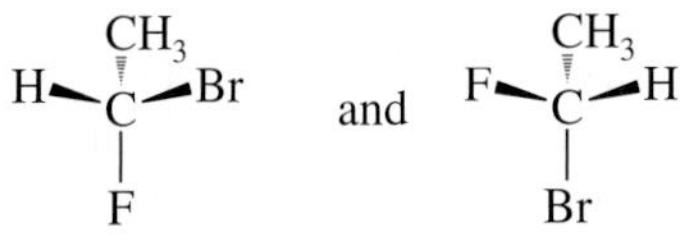

(b) H–C(CH$_3$)(Br)–F and Br–C(CH$_3$)(H)–F

(c) H–C(CH$_3$)(Br)–C(H)(CH$_3$)(F) and Br–C(CH$_3$)(H)–C(F)(H)(CH$_3$)

(d) H–C(CH$_3$)(Br)–C(H)(CH$_3$)(F) and Br–C(H)(CH$_3$)–C(F)(H)(CH$_3$)

(e) H–C(CH$_3$)(Br)–C(H)(F)(CH$_3$) and H–C(F)(CH$_3$)–C(H)(CH$_3$)(Br)

(f) H–C(CH$_3$)(Br)–CH$_3$ and H–C(CH$_3$)(H)–CH$_2$Br

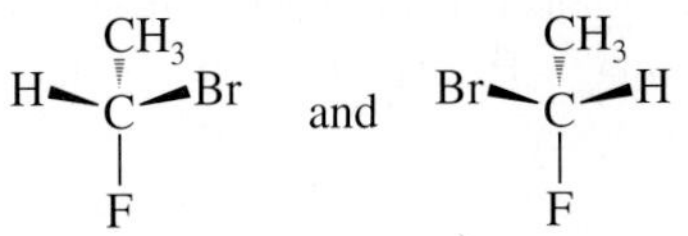

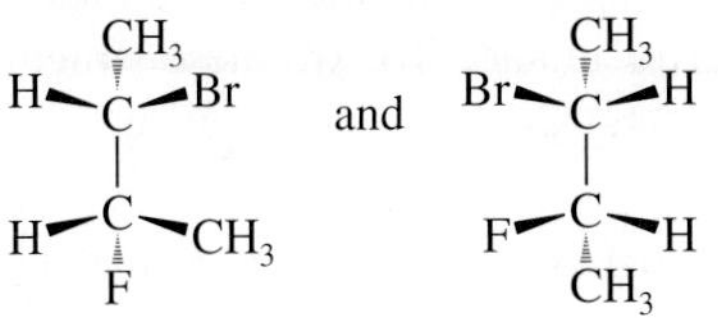

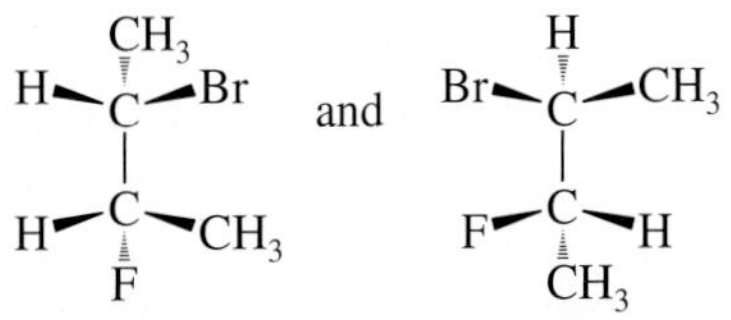

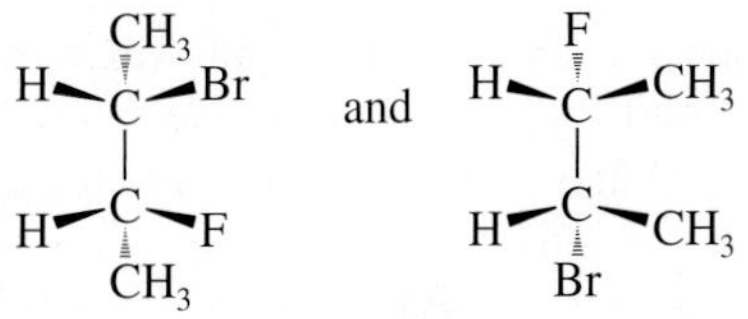

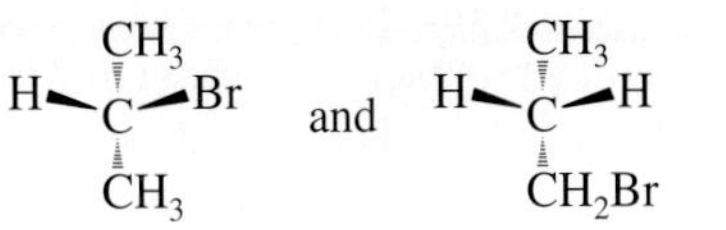

**Note:* Problems marked with an asterisk are "challenge problems."

(g) 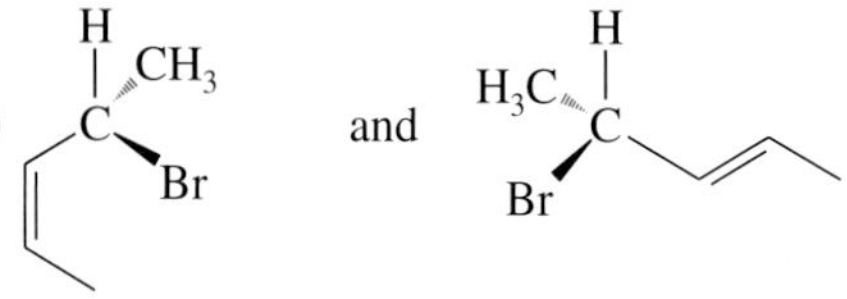 and

(h) 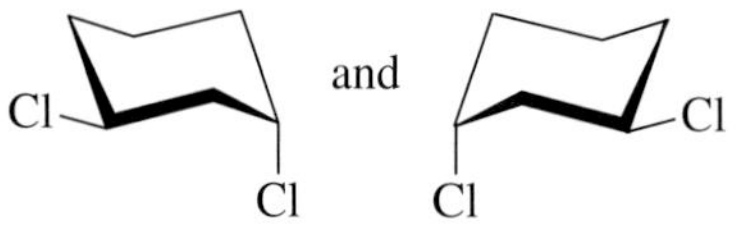 and

(i) 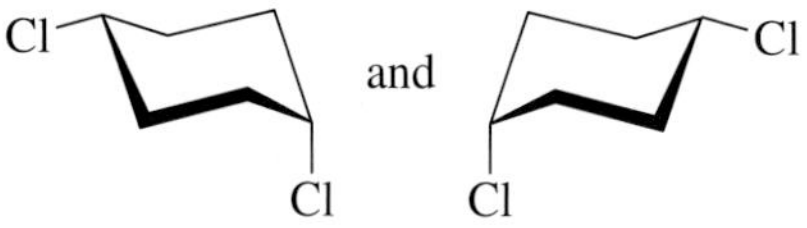 and

(j) 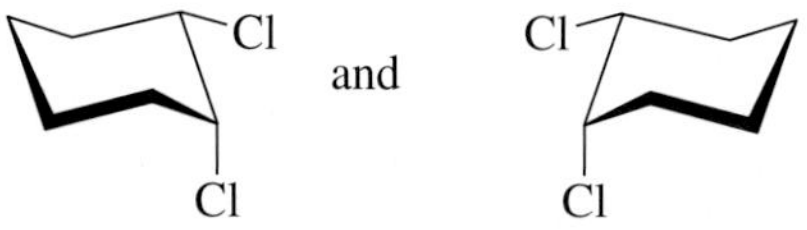 and

(k) 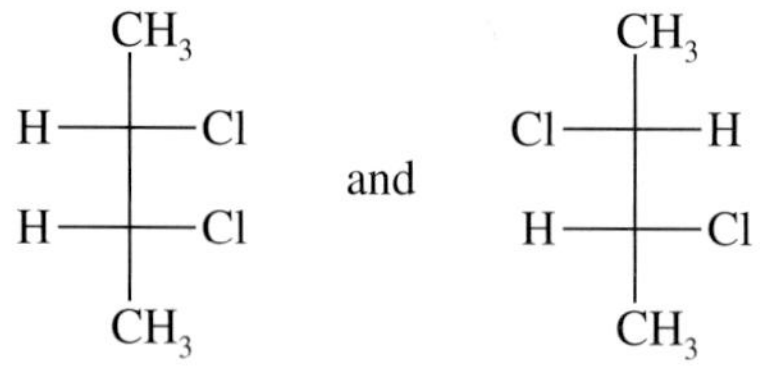 and

(l) 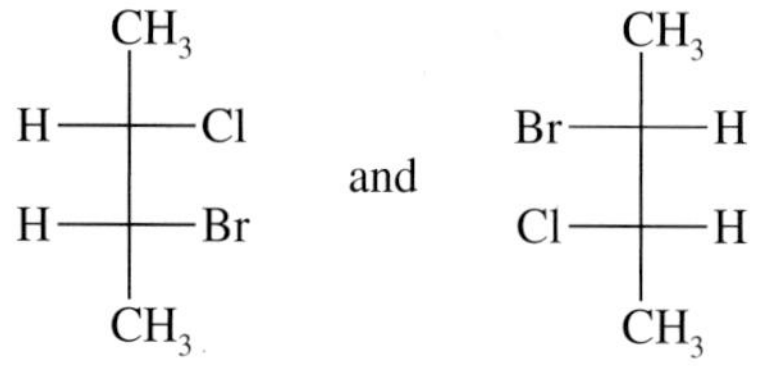 and

(m) 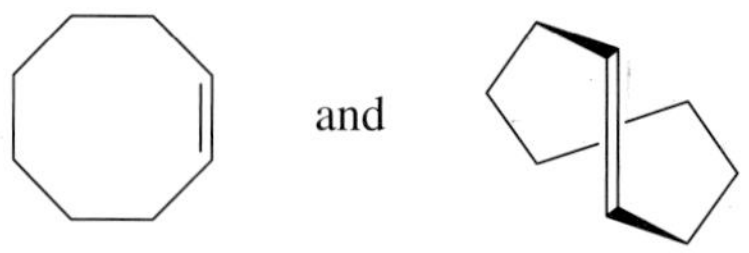 and

(n) 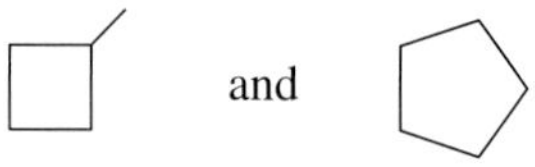 and

(o)  and

(p)  and

(q) 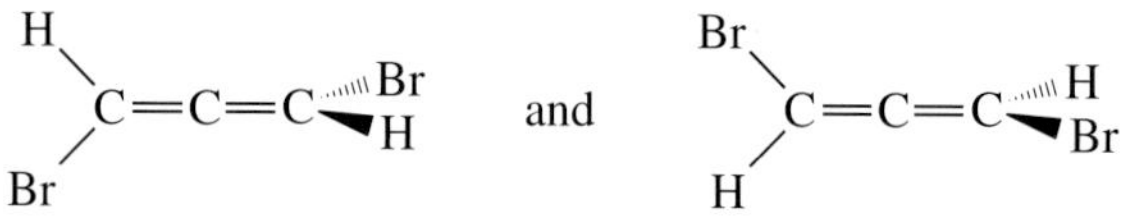 and

5.36 Discuss the anticipated stereochemistry of each of the following compounds.
(a) ClCH=C=C=CHCl
(b) CH_2=C=C=CHCl
(c) ClCH=C=C=CCl_2

5.37 There are four dimethylcyclopropane isomers. **(a)** Write three-dimensional formulas for these isomers. **(b)** Which of these isomers are chiral? **(c)** If a mixture consisting of 1 mol of each of these isomers were subjected to simple gas chromatography, how many fractions would be obtained and which compounds would each fraction contain? **(d)** How many of these fractions would be optically active?

5.38 (Use models to solve this problem.) **(a)** Write a conformational structure for the most stable conformation of *trans*-1,2-diethylcyclohexane and write its mirror image. **(b)** Are these two molecules superposable? **(c)** Are they interconvertible through a ring "flip"? **(d)** Repeat the process in part (a) with *cis*-1,2-diethylcyclohexane. **(e)** Are these structures superposable? **(f)** Are they interconvertible?

5.39 (Use models to solve this problem.) **(a)** Write a conformational structure for the most stable conformation of *trans*-1,4-diethylcyclohexane and for its mirror image. **(b)** Are these structures superposable? **(c)** Do they represent enantiomers? **(d)** Does *trans*-1,4-diethylcyclohexane have a stereoisomer, and if so, what is it? **(e)** Is this stereoisomer chiral?

5.40 (Use models to solve this problem.) Write conformational structures for all of the stereoisomers of 1,3-diethylcyclohexane. Label pairs of enantiomers and meso compounds if they exist.

***5.41** Tartaric acid [$HO_2CCH(OH)CH(OH)CO_2H$] was an important compound in the history of stereochemistry. Two naturally occurring forms of tartaric acid are optically inactive. One form has a melting point of 206°C, the other a melting point of 140°C. The inactive tartaric acid with a melting point of 206°C can be separated into two optically active forms of tartaric acid with the same melting point (170°C). One optically active tartaric acid has $[\alpha]_D^{25} = +12°$, and the other, $[\alpha]_D^{25} = -12°$. All attempts to separate the other inactive tartaric acid (melting point 140°C) into optically active compounds fail. **(a)** Write the three-dimensional structure of the tartaric acid with melting point 140°C. **(b)** What are possible structures for the optically active tartaric acids with melting points of 170°C? **(c)** Can you be sure which tartaric acid in (b) has a positive rotation and which has a negative rotation? **(d)** What is the nature of the form of tartaric acid with a melting point of 206°C?

***5.42** **(a)** An aqueous solution of pure stereoisomer X of concentration 0.10 g mL^{-1} had observed rotation $-30°$ in a 1.0-dm tube at 589.6 nm (the sodium D line) and 25°C. What do you calculate its $[\alpha]_D$ to be at this temperature?

(b) Under identical conditions but with concentration 0.050 g mL^{-1}, a solution of X had observed rotation $+165°$. Rationalize how this could be and recalculate $[\alpha]_D$ for stereoisomer X.

(c) If the optical rotation of a substance studied at only one concentration is 0°, can it definitely be concluded to be achiral? Racemic?

***5.43** If a sample of a pure substance that has two or more stereogenic carbons has an observed rotation of 0°, it could be a racemate. Could it possibly be a pure stereoisomer? Could it possibly be a pure enantiomer?

***5.44** Unknown Y has a molecular formula of $C_3H_6O_2$. It contains one functional group that absorbs infrared radiation in the 3200–3550-cm^{-1} region (when studied as a pure liquid; i.e., "neat"), and it has no absorption in the 1620–1780-cm^{-1} region. No carbon atom in the structure of Y has more than one oxygen atom bonded to it, and Y can exist in two (and only two) stereoisomeric forms. What are the structures of these forms of Y?

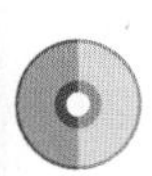

5.45 Complete the stereochemistry exercises on the CD that use computer molecular models.

CD Stereochemistry Exercises

Learning Group Problems

1. Streptomycin is an antibiotic that is especially useful against penicillin-resistant bacteria. The structure of streptomycin is shown in Section 22.17. **(a)** Identify all of the stereogenic carbons in the structure of streptomycin. **(b)** Assign the appropriate (*R*) or (*S*) designation for the configuration of each stereogenic carbon in streptomycin.
2. D-Galactitol is one of the toxic compounds produced by the disease galactosemia. Accumulation of high levels of D-galactitol causes the formation of cataracts. A Fischer projection for D-galactitol is shown here.

$$
\begin{array}{c}
CH_2OH \\
H-\!\!\!\!+\!\!\!\!-OH \\
HO-\!\!\!\!+\!\!\!\!-H \\
HO-\!\!\!\!+\!\!\!\!-H \\
H-\!\!\!\!+\!\!\!\!-OH \\
CH_2OH
\end{array}
$$

(a) Draw a three-dimensional structure for D-galactitol.
(b) Draw the mirror image of D-galactitol and write its Fischer projection formula.
(c) What is the stereochemical relationship between D-galactitol and its mirror image?

3. Cortisone is a natural steroid that can be isolated from the adrenal cortex. It has anti-inflammatory properties and is used to treat a variety of disorders (e.g., as a topical application for common skin diseases). The structure of cortisone is shown in Section 23.4D. **(a)** Identify all of the stereogenic carbons in cortisone. **(b)** Assign the appropriate (*R*) or (*S*) designation for the configuration of each stereogenic carbon in cortisone.

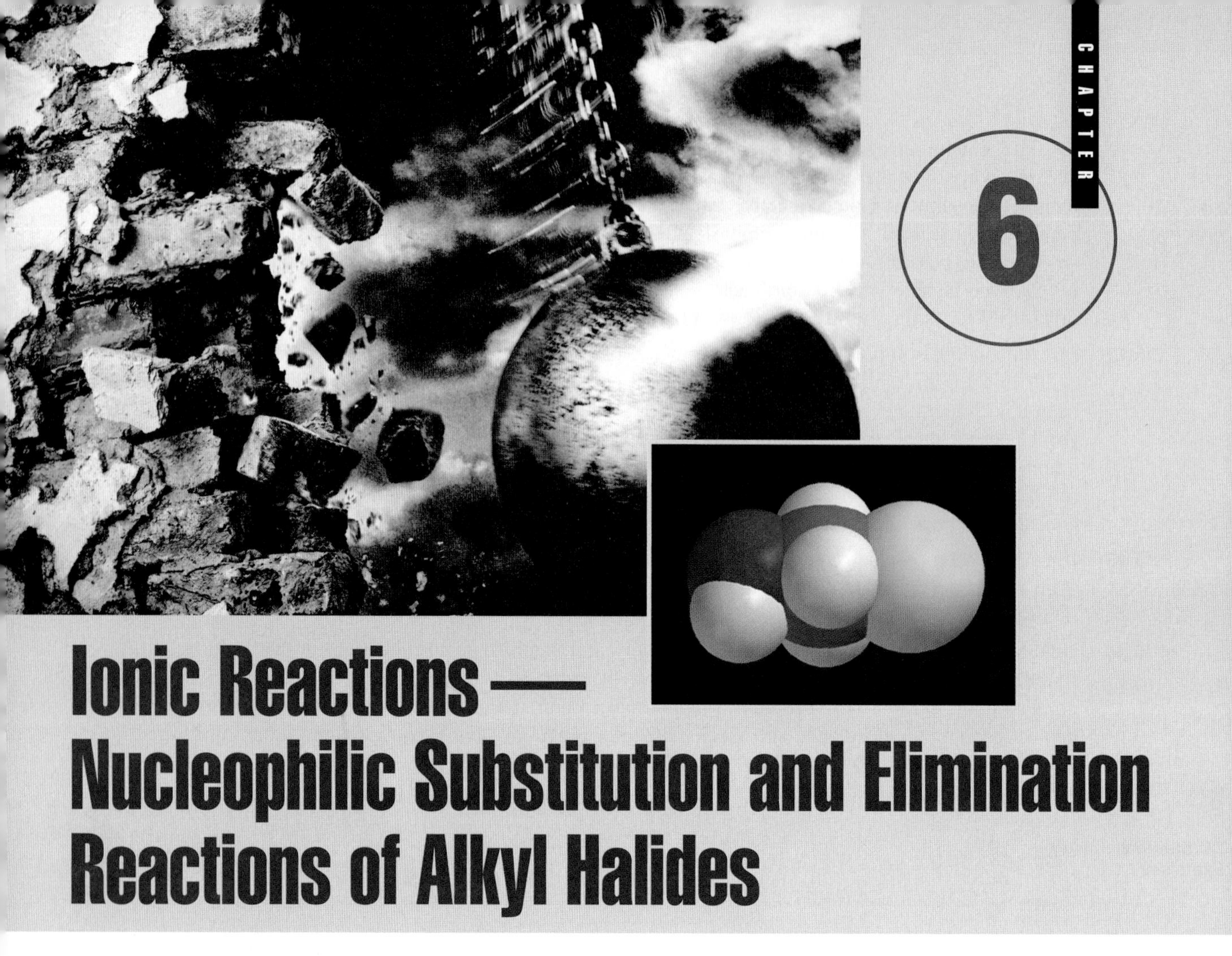

Ionic Reactions—Nucleophilic Substitution and Elimination Reactions of Alkyl Halides

Breaking Bacterial Cell Walls with Organic Chemistry

Enzymes, nature's quintessential chemists, catalyze most reactions of life. Enzymes catalyze metabolic reactions, the flow of genetic information, and the synthesis of molecules that provide biological structure. They also help defend us against infections and disease. Although mechanisms have been elucidated for the action of many enzymes, those for which mechanistic secrets have been unlocked constitute only a fraction of all of the enzymes involved in life's processes. It is widely accepted, however, that all reactions catalyzed by enzymes occur on the basis of rational chemical reactivity. The mechanisms utilized by enzymes are essentially those that we learn about in organic chemistry. One such example involves lysozyme. Lysozyme is an enzyme in nasal mucus that fights infection by degrading bacterial cell walls. Lysozyme employs a mechanism that generates an unstable, positively charged carbon intermediate (called a carbocation) within the molecular architecture of the bacterial cell wall. Lysozyme elegantly stabilizes this carbocation by providing a nearby negatively charged site from its own structure. This facilitates cleavage of the cell wall, yet does not involve bonding of lysozyme itself with the carbocation intermediate in the cell wall. Carbocation intermediates are central to a number of organic reaction types. One of these is called unimolecular nucleophilic substitution (S_N1), and it is one of the important reactions whose mechanism you will study in this chapter. (We shall revisit the mechanism of lysozyme in detail in Section 24.10)

6.1 INTRODUCTION

The halogen atom of an alkyl halide is attached to an sp^3-hybridized carbon. The arrangement of groups around the carbon atom, therefore, is generally tetrahedral. Because halogen atoms are more electronegative than carbon, the carbon–halogen bond of alkyl halides is *polarized;* the carbon atom bears a partial positive charge, the halogen atom a partial negative charge:

$$\overset{\delta+}{C} \rightarrow \overset{\delta-}{X}$$

The size of the halogen atom increases as we go down the periodic table: fluorine atoms are the smallest and iodine atoms the largest. Consequently, the carbon–halogen *bond length increases* and carbon–halogen *bond strength decreases* as we go down the periodic table (Table 6.1). Maps of electrostatic potential at the van der Waals surface for the four methyl halides, with ball-and-stick models inside, illustrate the trend in polarity, C—X bond length, and halogen atom size as one progresses from fluorine to iodine substitution. Fluoromethane is highly polar and has the shortest C—X bond length and the strongest C—X bond. Iodomethane is much less polar and has the longest C—X bond length and the weakest C—X bond.

TABLE 6.1 Carbon–Halogen Bond Lengths and Bond Strengths

	CH_3F	CH_3Cl	CH_3Br	CH_3I
C—X Bond length (Å)	1.39	1.78	1.93	2.14
C—X Bond strength (kJ mol^{-1})	472	350	293	239

In the laboratory and in industry, alkyl halides are used as solvents for relatively nonpolar compounds, and they are used as the starting materials for the synthesis of many compounds. As we shall learn in this chapter, the halogen atom of an alkyl halide can be easily replaced by other groups, and the presence of a halogen atom on a carbon chain also affords us the possibility of introducing a multiple bond.

Compounds in which a halogen atom is bonded to an sp^2-hybridized carbon are called **vinylic halides** or **phenyl halides.** The compound $CH_2{=}CHCl$ has the common name **vinyl chloride,** and the group $CH_2{=}CH—$ is commonly called the **vinyl group.** *Vinylic halide,* therefore, is a general term that refers to a compound in which a halogen is attached to a carbon atom that is also forming a double bond to another carbon atom. *Phenyl halides* are compounds in which a halogen is attached to a benzene ring (Section

2.5B). Phenyl halides belong to a larger group of compounds that we shall study later, called **aryl halides.**

A vinylic halide **A phenyl halide or aryl halide**

Together with alkyl halides, these compounds comprise a larger group of compounds known simply as **organic halides** or **organohalogen compounds.** The chemistry of vinylic and aryl halides is, as we shall also learn later, quite different from alkyl halides, and it is on alkyl halides that we shall focus most of our attention in this chapter.

6.1A Physical Properties of Organic Halides

Most alkyl and aryl halides have very low solubilities in water, but as we might expect, they are miscible with each other and with other relatively nonpolar solvents. Dichloromethane (CH_2Cl_2, also called *methylene chloride*), trichloromethane ($CHCl_3$, also called *chloroform*), and tetrachloromethane (CCl_4, also called *carbon tetrachloride*) are often used as solvents for nonpolar and moderately polar compounds. Many chloroalkanes, including $CHCl_3$ and CCl_4, have a cumulative toxicity and are carcinogenic, however, and should therefore be used only in fume hoods and with great care.

Iodomethane (bp 42°C) is the only monohalomethane that is a liquid at room temperature and 1 atm pressure. Bromoethane (bp 38°C) and iodoethane (bp 72°C) are both liquids, but chloroethane (bp 13°C) is a gas. The chloro-, bromo-, and iodopropanes are all liquids. In general, higher chloro-, bromo-, and iodoalkanes are all liquids and tend to have boiling points near those of alkanes of similar molecular weights.

Polyfluoroalkanes, however, tend to have unusually low boiling points (Section 2.14D). Hexafluoroethane boils at −79°C, even though its molecular weight (MW = 138) is near that of decane (MW = 144; bp 174°C).

Table 6.2 lists the physical properties of some common organic halides.

TABLE 6.2 Organic Halides

	Fluoride		Chloride		Bromide		Iodide[a]	
Group	bp (°C)	Density (g mL^{-1})	bp (°C)	Density (g mL^{-1})	bp (°C)	Density (g mL^{-1})	bp (°C)	Density (g mL^{-1})
Methyl	−78.4	0.84^{-60}	−23.8	0.92^{20}	3.6	1.73^{0}	42.5	2.28^{20}
Ethyl	−37.7	0.72^{20}	13.1	0.91^{15}	38.4	1.46^{20}	72	1.95^{20}
Propyl	−2.5	0.78^{-3}	46.6	0.89^{20}	70.8	1.35^{20}	102	1.74^{20}
Isopropyl	−9.4	0.72^{20}	34	0.86^{20}	59.4	1.31^{20}	89.4	1.70^{20}
Butyl	32	0.78^{20}	78.4	0.89^{20}	101	1.27^{20}	130	1.61^{20}
sec-Butyl	—	—	68	0.87^{20}	91.2	1.26^{20}	120	1.60^{20}
Isobutyl	—	—	69	0.87^{20}	91	1.26^{20}	119	1.60^{20}
tert-Butyl	12	0.75^{12}	51	0.84^{20}	73.3	1.22^{20}	100 dec	1.57^{0}
Pentyl	62	0.79^{20}	108.2	0.88^{20}	129.6	1.22^{20}	155^{740}	1.52^{20}
Neopentyl	—	—	84.4	0.87^{20}	105	1.20^{20}	127 dec	1.53^{13}
$CH_2{=}CH-$	−72	0.68^{26}	−13.9	0.91^{20}	16	1.52^{14}	56	2.04^{20}
$CH_2{=}CHCH_2-$	−3	—	45	0.94^{20}	70	1.40^{20}	102–103	1.84^{22}
C_6H_5-	85	1.02^{20}	132	1.10^{20}	155	1.52^{20}	189	1.82^{20}
$C_6H_5CH_2-$	140	1.02^{25}	179	1.10^{25}	201	1.44^{22}	93^{10}	1.73^{25}

[a]Decomposes is abbreviated dec.

6.2 Nucleophilic Substitution Reactions

There are many reactions of the general type shown here:

$$\underset{\textbf{Nucleophile}}{\text{Nu}:^-} + \underset{\substack{\textbf{Alkyl}\\\textbf{halide}\\\textbf{(substrate)}}}{\text{R}-\ddot{\underset{..}{\text{X}}}:} \longrightarrow \underset{\textbf{Product}}{\text{R}-\text{Nu}} + \underset{\textbf{Halide ion}}{:\ddot{\underset{..}{\text{X}}}:^-}$$

Following are some examples:

In Section 6.14 we shall see examples of biological nucleophilic substitution.

$$\text{H}\ddot{\underset{..}{\text{O}}}:^- + \text{CH}_3-\ddot{\underset{..}{\text{I}}}: \longrightarrow \text{CH}_3-\ddot{\underset{..}{\text{O}}}\text{H} + :\ddot{\underset{..}{\text{I}}}:^-$$

$$\text{CH}_3\ddot{\underset{..}{\text{O}}}:^- + \text{CH}_3\text{CH}_2-\ddot{\underset{..}{\text{Br}}}: \longrightarrow \text{CH}_3\text{CH}_2-\ddot{\underset{..}{\text{O}}}\text{CH}_3 + :\ddot{\underset{..}{\text{Br}}}:^-$$

$$:\ddot{\underset{..}{\text{I}}}:^- + \text{CH}_3\text{CH}_2\text{CH}_2-\ddot{\underset{..}{\text{Cl}}}: \longrightarrow \text{CH}_3\text{CH}_2\text{CH}_2-\ddot{\underset{..}{\text{I}}}: + :\ddot{\underset{..}{\text{Cl}}}:^-$$

In this type of reaction a **nucleophile,** ***a species with an unshared electron pair,*** reacts with an alkyl halide (called the **substrate**) by replacing the halogen substituent. A *substitution reaction* takes place and the halogen substituent, called the leaving group, departs as a halide ion. Because the substitution reaction is initiated by a nucleophile, it is called a **nucleophilic substitution reaction.**

In nucleophilic substitution reactions the carbon–halogen bond of the substrate undergoes *heterolysis,* and the unshared electron pair of the nucleophile is used to form a new bond to the carbon atom:

In color-coded reactions of this chapter, we will use red to indicate a nucleophile and blue to indicate a leaving group.

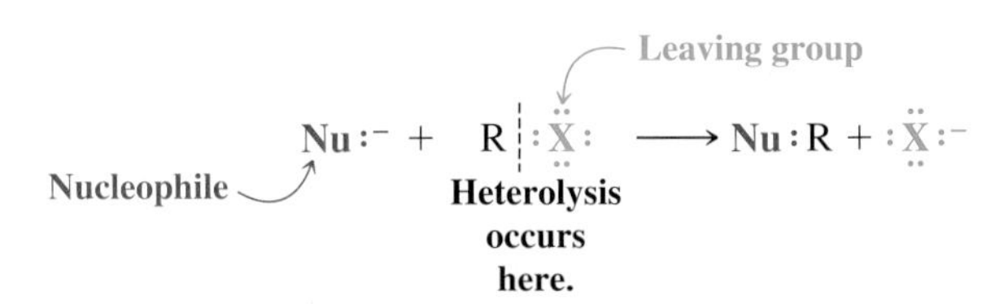

One of the questions we shall want to address later in this chapter is this: When does the carbon–halogen bond break? Does it break at the same time that the new bond between the nucleophile and the carbon forms?

$$\text{Nu}:^- + \text{R}:\ddot{\underset{..}{\text{X}}}: \longrightarrow \overset{\delta-}{\text{Nu}}\text{---R---}\overset{\delta-}{\ddot{\underset{..}{\text{X}}}}: \longrightarrow \text{Nu}:\text{R} + :\ddot{\underset{..}{\text{X}}}:^-$$

Or does the carbon–halogen bond break first?

$$\text{R}:\ddot{\underset{..}{\text{X}}}: \longrightarrow \text{R}^+ + :\ddot{\underset{..}{\text{X}}}:^-$$

And then

$$\text{Nu}:^- + \text{R}^+ \longrightarrow \text{Nu}:\text{R}$$

We shall find that the answer depends primarily on the structure of the alkyl halide.

6.3 Nucleophiles

A nucleophile is a reagent that seeks a positive center. (The word nucleophile comes from *nucleus,* the positive part of an atom, plus *-phile* from the Greek word *philos,* meaning to love.) When a nucleophile reacts with an alkyl halide, the positive center that the nucleophile seeks is the carbon atom that bears the halogen atom. This carbon atom carries a partial positive charge because the electronegative halogen pulls the electrons of the carbon–halogen bond in its direction (see Section 2.4):

You may wish to review Section 3.2C, "Opposite Charges Attract."

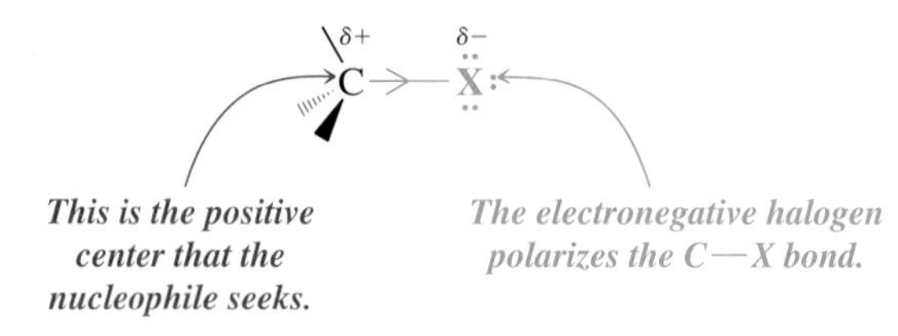

A nucleophile is any negative ion or any neutral molecule that has at least one unshared electron pair. (Later we shall see that pi bonds can be nucleophiles as well.) For example, both hydroxide ions and water molecules can act as nucleophiles by reacting with alkyl halides to produce alcohols.

General Reaction for Nucleophilic Substitution of an Alkyl Halide by Hydroxide Ion

$$\underset{\text{Nucleophile}}{H-\ddot{O}:^-} + \underset{\text{Alkyl halide}}{R-\ddot{X}:} \longrightarrow \underset{\text{Alcohol}}{H-\ddot{O}-R} + \underset{\text{Leaving group}}{:\ddot{X}:^-}$$

General Reaction for Nucleophilic Substitution of an Alkyl Halide by Water

$$\underset{\text{Nucleophile}}{H-\ddot{O}(H):} + \underset{\text{Alkyl halide}}{R-\ddot{X}:} \longrightarrow \underset{\text{Alkyloxonium ion}}{H-\ddot{O}^{+}(H)-R} + :\ddot{X}:^-$$

$$\xrightarrow{H_2\ddot{O}} H-\ddot{O}-R + H_3\ddot{O}^+ + :\ddot{X}:^-$$

A deprotonation step is always required to complete the reaction when the nucleophile was a neutral atom that bore a proton.

In the second reaction, note that the first product is an alkyloxonium ion, $R-\ddot{O}^{+}(H)-H$, which then loses a proton to a water molecule to form an alcohol.

Problem 6.1

Write the following as *net ionic reactions* and designate the nucleophile, substrate, and leaving group in each reaction:

(a) $CH_3I + CH_3CH_2ONa \longrightarrow CH_3OCH_2CH_3 + NaI$

(b) $NaI + CH_3CH_2Br \longrightarrow CH_3CH_2I + NaBr$

(c) $2\,CH_3OH + (CH_3)_3CCl \longrightarrow (CH_3)_3COCH_3 + CH_3OH_2^+ + Cl^-$

(d) $CH_3CH_2CH_2Br + NaCN \longrightarrow CH_3CH_2CH_2CN + NaBr$

(e) $C_6H_5CH_2Br + 2\,NH_3 \longrightarrow C_6H_5CH_2NH_2 + NH_4Br$

6.4 Leaving Groups

To act as the substrate in a nucleophilic substitution reaction, a molecule must have a good **leaving group.** A good leaving group is a substituent that can leave as a relatively stable, weakly basic molecule or anion. The halogen atom of an alkyl halide is a good leaving group because once departed it is a weak base and stable anion. (Alkyl halides are not the only class of compounds that can act as substrates in nucleophilic substitution reactions. We shall see later that other compounds can also react in the same way).

In the following examples we show nucleophilic substitution reactions first with a charged nucleophile and then with a neutral nucleophile. We use L to represent a generic leaving group. Note that the overall charge is balanced in the case of each equation:

$$Nu:^- + R-L \longrightarrow R-Nu + :L^-$$

or

$$\text{Nu}: + \text{R—L} \longrightarrow \text{R—Nu}^+ + :\text{L}^-$$

Specific Examples

$$\text{H}\ddot{\text{O}}:^- + \text{CH}_3\text{—}\ddot{\text{Cl}}: \longrightarrow \text{CH}_3\text{—}\ddot{\text{O}}\text{H} + :\ddot{\text{Cl}}:^-$$

$$\text{H}_3\text{N}: + \text{CH}_3\text{—}\ddot{\text{Br}}: \longrightarrow \text{CH}_3\text{—NH}_3^+ + :\ddot{\text{Br}}:^-$$

Later we shall also see reactions where the substrate bears a formal positive charge and a reaction like the following takes place:

$$\text{Nu}: + \text{R—L}^+ \longrightarrow \text{R—Nu}^+ + :\text{L}$$

In this case, when the leaving group departs with an electron pair, its formal charge goes to zero.

Specific Example

Note that the net charge is the same on each side of a properly written chemical equation.

$$\text{CH}_3\text{—}\underset{\text{H}}{\underset{|}{\ddot{\text{O}}}}: + \text{CH}_3\text{—}\underset{\text{H}}{\underset{|}{\ddot{\text{O}}^+}}\text{—H} \longrightarrow \text{CH}_3\text{—}\underset{\text{H}}{\underset{|}{\ddot{\text{O}}^+}}\text{—CH}_3 + :\underset{\text{H}}{\underset{|}{\ddot{\text{O}}}}\text{—H}$$

Nucleophilic substitution reactions are more understandable and useful if we know something about their mechanisms. How does the nucleophile replace the leaving group? Does the reaction take place in one step or is more than one step involved? If more than one step is involved, what kinds of intermediates are formed? Which steps are fast and which are slow? In order to answer these questions, we need to know something about the rates of chemical reactions.

6.5 Kinetics of a Nucleophilic Substitution Reaction: An S_N2 Reaction

To understand how the rate of a reaction might be measured, let us consider an actual example: the reaction that takes place between chloromethane and hydroxide ion in aqueous solution:

$$\text{CH}_3\text{—Cl} + \text{OH}^- \xrightarrow[\text{H}_2\text{O}]{60°\text{C}} \text{CH}_3\text{—OH} + \text{Cl}^-$$

Although chloromethane is not highly soluble in water, it is soluble enough to carry out our kinetic study in an aqueous solution of sodium hydroxide. Because reaction rates are known to be temperature dependent (Section 6.8), we carry out the reaction at a specific temperature.

The rate of the reaction can be determined experimentally by measuring the rate at which chloromethane or hydroxide ion *disappears* from the solution or the rate at which methanol or chloride ion *appears* in the solution. We can make any of these measurements by withdrawing a small sample from the reaction mixture soon after the reaction begins and analyzing it for the concentrations of CH_3Cl or OH^- and CH_3OH or Cl^-. We are interested in what are called *initial rates,* because as time passes the concentrations of the reactants change. Since we also know the initial concentrations of reactants (because we measured them when we made up the solution), it will be easy to calculate the rate at which the reactants are disappearing from the solution or the products are appearing in the solution.

We perform several such experiments keeping the temperature the same but varying the initial concentrations of the reactants. The results that we might get are shown in Table 6.3.

Notice that the experiments show that the rate depends on the concentration of chloromethane *and* on the concentration of hydroxide ion. When we doubled the concentration of chloromethane in experiment 2, the rate *doubled.* When we doubled the concen-

TABLE 6.3 Rate Study of Reaction of CH_3Cl with OH^- at 60°C

Experiment Number	Initial $[CH_3Cl]$	Initial $[OH^-]$	Initial Rate (mol L^{-1} s^{-1})
1	0.0010	1.0	4.9×10^{-7}
2	0.0020	1.0	9.8×10^{-7}
3	0.0010	2.0	9.8×10^{-7}
4	0.0020	2.0	19.6×10^{-7}

tration of hydroxide ion in experiment 3, the rate *doubled.* When we doubled both concentrations in experiment 4, the rate increased by a factor of *four.*

We can express these results as a proportionality,

$$\text{Rate} \propto [CH_3Cl][OH^-]$$

and this proportionality can be expressed as an equation through the introduction of a proportionality constant (k) called the rate constant:

$$\text{Rate} = k[CH_3Cl][OH^-]$$

For this reaction at this temperature we find that $k = 4.9 \times 10^{-4}$ L mol^{-1} s^{-1}. (Verify this for yourself by doing the calculation.)

This reaction is said to be **second order overall.*** It is reasonable to conclude, therefore, that *for the reaction to take place a hydroxide ion and a chloromethane molecule must collide.* We also say that the reaction is **bimolecular.** (By *bimolecular* we mean that two species are involved in the step whose rate is being measured.) We call this kind of reaction an **S_N2 reaction,** meaning **substitution, nucleophilic, bimolecular.**

6.6 A Mechanism for the S_N2 Reaction

A schematic representation of orbitals involved in an S_N2 reaction—based on ideas proposed by Edward D. Hughes and Sir Christopher Ingold in 1937—is outlined below.

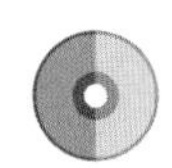

CD Tutorial
Orbitals in S_N2 reactions

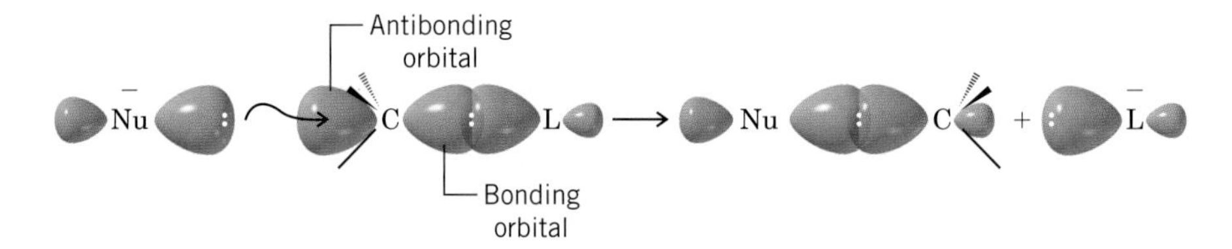

According to this mechanism, the nucleophile approaches the carbon bearing the leaving group from the **back side,** that is, from the side directly opposite the leaving group. The orbital that contains the electron pair of the nucleophile (its highest occupied molecular orbital, or HOMO) begins to overlap with an empty orbital (its lowest unoccupied molecular orbital, or LUMO) of the carbon atom bearing the leaving group. As the reaction progresses, the bond between the nucleophile and the carbon atom strengthens, and the bond between the carbon atom and the leaving group weakens. As this happens, the carbon atom has its configuration turned inside out, it becomes *inverted,*† and the leaving group

*In general, the overall order of a reaction is equal to the sum of the exponents a and b in the rate equation Rate = $k[A]^a[B]^b$. If in some other reaction, for example, we found that Rate = $k[A]^2[B]$, then we would say that the reaction is second order with respect to [A], first order with respect to [B], and third order overall.

†Considerable evidence had appeared in the years prior to Hughes and Ingold's 1937 publication indicating that in reactions like this an inversion of configuration of the carbon bearing the leaving group takes place. The first observation of such an inversion was made by the Latvian chemist Paul Walden in 1896, and such inversions are called **Walden inversions** in his honor. We shall study this aspect of the S_N2 reaction further in Section 6.8.

is pushed away. The formation of the bond between the nucleophile and the carbon atom provides most of the energy necessary to break the bond between the carbon atom and the leaving group. We can represent this mechanism with chloromethane and hydroxide ion as shown in the box "A Mechanism for the S_N2 Reaction" below.

The Hughes–Ingold mechanism for the S_N2 reaction involves only one step. There are no intermediates. The reaction proceeds through the formation of an unstable arrangement of atoms called the **transition state.**

The transition state is a fleeting arrangement of the atoms in which the nucleophile and the leaving group are both partially bonded to the carbon atom undergoing substitution. Because the transition state involves both the nucleophile (e.g., a hydroxide ion) and the substrate (e.g., a molecule of chloromethane), this mechanism accounts for the second-order reaction kinetics that we observe. (Because bond formation and bond breaking occur simultaneously in a single transition state, the S_N2 reaction is an example of what is called a *concerted reaction.*)

CD Tutorial

S_N2 mechanism

The transition state has an extremely brief existence. It lasts only as long as the time required for one molecular vibration, about 10^{-12} s. The structure and energy of the transition state are highly important aspects of any chemical reaction. We shall, therefore, examine this subject further in Section 6.7.

A Mechanism for the S_N2 Reaction

Reaction:

$$HO^- + CH_3Cl \longrightarrow CH_3OH + Cl^-$$

Mechanism:

$$H-\ddot{O}:^- + CH_3\overset{\delta+}{C}-\overset{\delta-}{Cl}: \longrightarrow \left[H-\overset{\delta-}{O}\cdots C\cdots \overset{\delta-}{Cl} \right]^{\ddagger} \longrightarrow H-\ddot{O}-CH_3 + :\ddot{Cl}:^-$$

The negative hydroxide ion brings a pair of electrons to the partially positive carbon from the back side with respect to the leaving group. The chlorine begins to move away with the pair of electrons that bonded it to the carbon.

Transition state

In the transition state, a bond between oxygen and carbon is partially formed and the bond between carbon and chlorine is partially broken. The configuration of the carbon atom begins to invert.

Now the bond between the oxygen and carbon has formed and the chloride ion has departed. The configuration of the carbon has inverted.

6.7 Transition State Theory: Free-Energy Diagrams

A reaction that proceeds with a negative free-energy change is said to be **exergonic;** one that proceeds with a positive free-energy change is said to be **endergonic.** The reaction between chloromethane and hydroxide ion in aqueous solution is highly exergonic;

at 60°C (333 K), $\Delta G^\circ = -100 \text{ kJ mol}^{-1}$. (The reaction is also exothermic, $\Delta H^\circ = -75 \text{ kJ mol}^{-1}$.)

$$\text{CH}_3\text{—Cl} + \text{OH}^- \longrightarrow \text{CH}_3\text{—OH} + \text{Cl}^- \qquad \Delta G^\circ = -100 \text{ kJ mol}^{-1}$$

The equilibrium constant for the reaction is extremely large, as we show by the following calculation:

$$\Delta G^\circ = -RT \ln K_{eq}$$

$$\ln K_{eq} = \frac{-\Delta G^\circ}{RT}$$

$$\ln K_{eq} = \frac{-(-100 \text{ kJ mol}^{-1})}{0.00831 \text{ kJ K}^{-1} \text{ mol}^{-1} \times 333 \text{ K}}$$

$$\ln K_{eq} = 36.1$$

$$K_{eq} = 5.0 \times 10^{15}$$

An equilibrium constant as large as this means that the reaction goes to completion.

Because the free-energy change is negative, we can say that in energy terms the reaction goes **downhill.** The products of the reaction are at a lower level of free energy than the reactants.

CD Tutorial

S_N2 collision

However, considerable experimental evidence exists showing that **if covalent bonds are broken in a reaction, the reactants must go up an energy hill first,** before they can go downhill. This will be true even if the reaction is exergonic.

We can represent this graphically by plotting the free energy of the reacting particles against the reaction coordinate. Such a graph is given in Fig. 6.1. We have chosen as our example a generalized S_N2 reaction.

The reaction coordinate is a quantity that measures the progress of the reaction. It represents the changes in bond orders and bond distances that must take place as the reactants are converted to products. In this instance the R–L distance could be used as the reaction coordinate because as the reaction progresses the R–L distance becomes longer.

In our illustration (Fig. 6.1), we can see that an **energy barrier** exists between the reactants and products. The height of this barrier (in kilojoules per mole) above the level of reactants is called the **free energy of activation, $\Delta G^\ddagger$.**

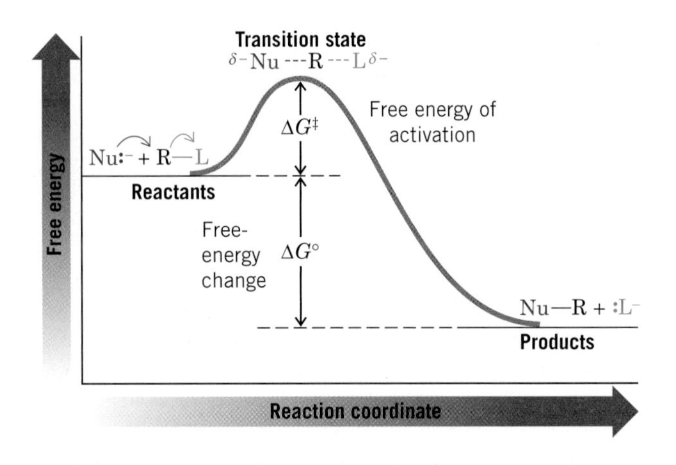

FIGURE 6.1 A free-energy diagram for a hypothetical S_N2 reaction that takes place with a negative ΔG°.

The top of the energy hill corresponds to the transition state. *The difference in free energy between the reactants and the transition state is the free energy of activation, $\Delta G^\ddagger$. The difference in free energy between the reactants and products is the free-energy change for the reaction, ΔG°.* For our example in Fig. 6.1, the free-energy level of the products is lower than that of the reactants. In terms of our analogy, we can say that the reactants in one energy valley must surmount an energy hill (the transition state) in order to reach the lower energy valley of the products.

If a reaction in which covalent bonds are broken proceeds with a positive free-energy change (Fig. 6.2), there will still be a free energy of activation. That is, if the products have greater free energy than reactants, the free energy of activation will be even higher. ($\Delta G^{\ddagger}$ will be larger than ΔG°.) In other words, in the **uphill** (endergonic) reaction an even larger energy hill lies between the reactants in one valley and the products in a higher one.

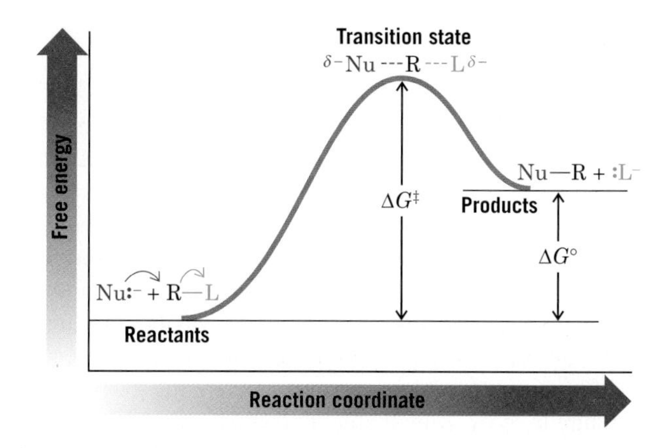

FIGURE 6.2 A free-energy diagram for a hypothetical reaction with a positive free-energy change.

Just as the overall free-energy change for a reaction contains enthalpy and entropy components (Section 3.9),

$$\Delta G^{\circ} = \Delta H^{\circ} - T\,\Delta S^{\circ}$$

the free energy of activation has similar components:

$$\Delta G^{\ddagger} = \Delta H^{\ddagger} - T\,\Delta S^{\ddagger}$$

The enthalpy of activation ($\Delta H^{\ddagger}$) is the difference in bond energies between the reactants and the transition state. It is, in effect, the energy necessary to bring the reactants close together and to bring about the partial breaking of bonds that must happen in the transition state. Some of this energy may be furnished by the bonds that are partially formed. The entropy of activation ($\Delta S^{\ddagger}$) is the difference in entropy between the reactants and the transition state. Most reactions require the reactants to come together with a particular orientation. (Consider, e.g., the specific orientation required in the S_N2 reaction.) This requirement for a particular orientation means that the transition state must be more ordered than the reactants and that $\Delta S^{\ddagger}$ will be negative. The more highly ordered the transition state, the more negative $\Delta S^{\ddagger}$ will be. When a three-dimensional plot of free energy versus the reaction coordinate is made, the transition state is found to resemble a mountain pass or *col* (Fig. 6.3) rather than the top of an energy hill as we have shown in Figs. 6.1 and 6.2. (A plot such as that seen in Figs. 6.1 and 6.2 is simply a two-dimensional slice through the three-dimensional energy surface for the reaction.) That is, the reactants and products appear to be separated by an energy barrier resembling a mountain range. While an infinite number of possible routes lead from reactants to products, the transition state lies at the top of the route that requires the lowest energy climb. Whether the pass is a wide or narrow one depends on $\Delta S^{\ddagger}$. A wide pass means that there is a relatively large number of orientations of reactants that allow a reaction to take place. A narrow pass means just the opposite.

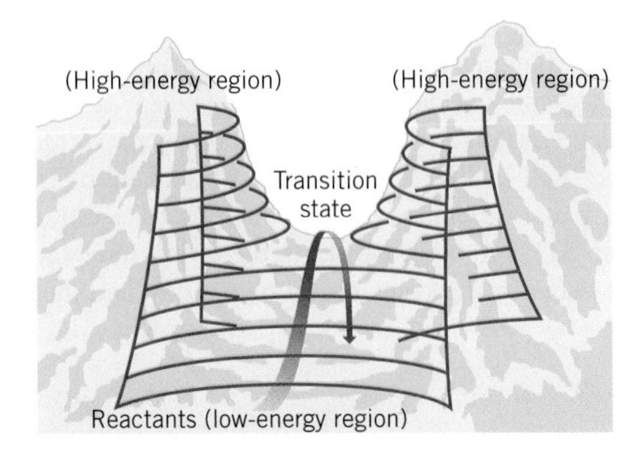

FIGURE 6.3 Mountain pass or col analogy for the transition state. (Adapted with permission from Leffler, J. E.; Grunwald, E. *Rates and Equilibria of Organic Reactions;* Wiley: New York, 1963; p 65.)

The existence of an activation energy ($\Delta G^{\ddagger}$) explains why most chemical reactions occur much more rapidly at higher temperatures. *For many reactions taking place near room temperature, a 10°C increase in temperature will cause the reaction rate to double.*

This dramatic increase in reaction rate results from a large increase in the number of collisions between reactants that together have sufficient energy to surmount the barrier at the higher temperature. The kinetic energies of molecules at a given temperature are not all the same. Figure 6.4 shows the distribution of energies brought to collisions at two temperatures (that do not differ greatly), labeled T_1 and T_2. Because of the way energies are distributed at different temperatures (as indicated by the shapes of the curves), increasing the temperature by only a small amount causes a large increase in the number of collisions with larger energies. In Fig. 6.4 we have designated an arbitrary minimum free energy of activation as being required to bring about a reaction between colliding molecules. The number of collisions having sufficient energy to allow reaction to take place at a given temperature is proportional to the area under the portion of the curve that represents free energies greater than or equal to $\Delta G^{\ddagger}$. At the lower temperature (T_1) this number is relatively small. At the higher temperature (T_2), however, the number of collisions that take place with enough energy to react is very much larger. Consequently, a modest temperature increase produces a large increase in the number of collisions with energy sufficient to lead to a reaction.

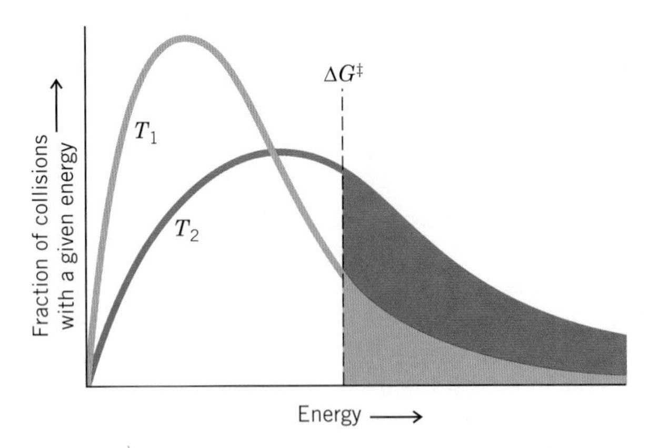

FIGURE 6.4 The distribution of energies at two different temperatures, T_1 and T_2 ($T_2 > T_1$). The number of collisions with energies greater than the free energy of activation is indicated by the appropriately shaded area under each curve.

There is also an important relationship between the rate of a reaction and the magnitude of the free energy of activation. The relationship between the rate constant (k) and $\Delta G^{\ddagger}$ is an *exponential one:*

$$k = k_0 e^{-\Delta G^{\ddagger}/RT}$$

In this equation, $e = 2.718$, the base of natural logarithms, and k_0 is the absolute rate constant, which equals the rate at which all transition states proceed to products. At 25°C, $k_0 = 6.2 \times 10^{12}\ s^{-1}$. Because of this exponential relationship, **a reaction with a lower free energy of activation will occur very much faster than a reaction with a higher one.**

Generally speaking, if a reaction has a $\Delta G^{\ddagger}$ less than 84 kJ mol^{-1}, it will take place readily at room temperature or below. If $\Delta G^{\ddagger}$ is greater than 84 kJ mol^{-1}, heating will be required to cause the reaction to occur at a reasonable rate.

A free-energy diagram for the reaction of chloromethane with hydroxide ion is shown in Fig. 6.5. At 60°C, $\Delta G^{\ddagger} = 103$ kJ mol^{-1}, which means that at this temperature the reaction reaches completion in a matter of a few hours.

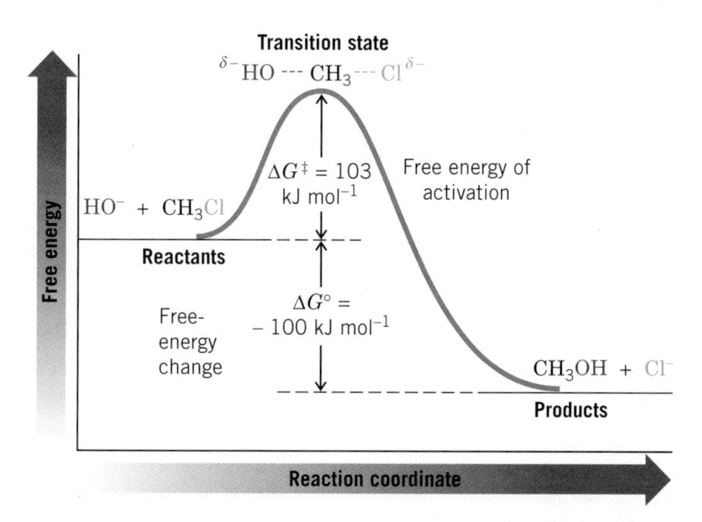

FIGURE 6.5 A free-energy diagram for the reaction of chloromethane with hydroxide ion at 60°C.

6.8 The Stereochemistry of S_N2 Reactions

As we learned earlier (Section 6.6), in an S_N2 reaction **the nucleophile approaches from the back side, that is, from the side directly opposite the leaving group.** This mode of attack (see below) causes a **change in the configuration** of the carbon atom that is the target of nucleophilic attack. (The **configuration** of an atom *is the particular arrangement of groups around that atom in space;* Section 5.7.) As the displacement takes place, the configuration of the carbon atom undergoing substitution **inverts**—it is turned inside out in much the same way that an umbrella is turned inside out, or inverts, when caught in a strong wind:

S_N2 stereochemistry

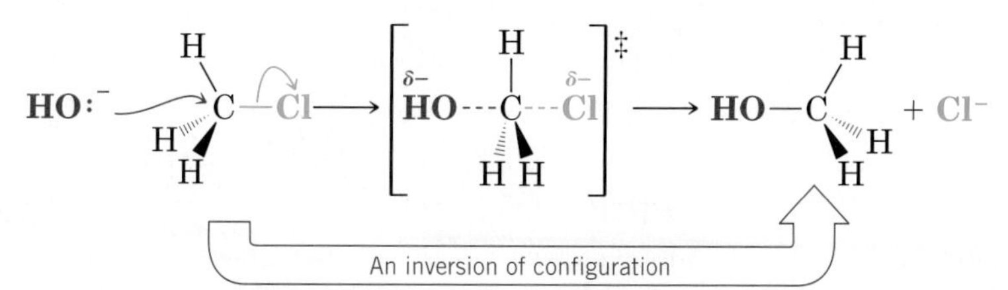

Transition state for an S_N2 reaction.

With a molecule such as chloromethane, however, there is no way to prove that attack by the nucleophile has involved inversion of configuration of the carbon atom because one form of methyl chloride is identical to its inverted form. With a molecule containing stereogenic centers such as *cis*-1-chloro-3-methylcyclopentane, however, we can observe the results of an inversion of configuration by the change in stereochemistry that occurs. When *cis*-1-chloro-3-methylcyclopentane reacts with hydroxide ion in an S_N2 reaction, the product is *trans*-3-methylcyclopentanol. *The hydroxide ion ends up being bonded on the opposite side of the ring from the chlorine it replaces:*

An inversion of configuration

H_3C ... Cl: ... H + :ÖH⁻ $\xrightarrow{S_N2}$ H_3C ... H ... :OH + :Cl:⁻

cis-1-Chloro-3-methylcyclopentane **trans-3-Methylcyclopentanol**

Presumably, the transition state for this reaction is like that shown here.

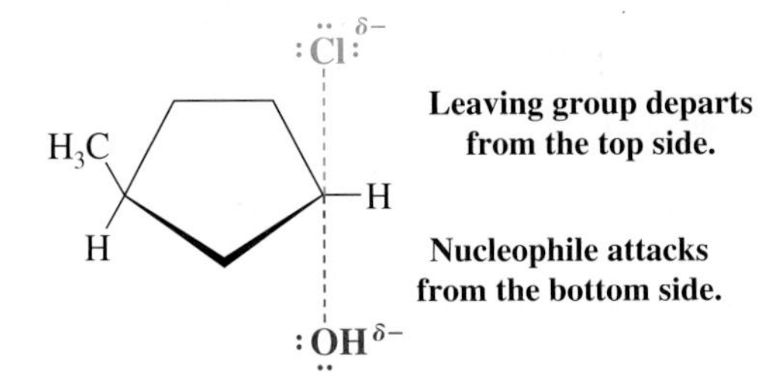

:Cl: δ−
H_3C
H
H
:OH δ−
Leaving group departs from the top side.
Nucleophile attacks from the bottom side.

PROBLEM 6.2

Using chair conformational structures (Sect. 4.12), show the nucleophilic substitution reaction that would take place when *trans*-1-bromo-4-*tert*-butylcyclohexane reacts with iodide ion. (Show the most stable conformation of the reactant and the product.)

CD Tutorial
S_N2 inversion

We can also observe inversion of configuration when an S_N2 reaction occurs at a stereogenic carbon in an acyclic molecule. Indeed, we find that **S_N2 reactions always occur with inversion of configuration.** The reaction of (*R*)-(−)-2-bromooctane with sodium hydroxide provides an example. We can determine whether or not inversion of configuration occurs in this reaction because the configurations and optical rotations for both enantiomers of 2-bromooctane and the expected product, 2-octanol, are known.

C_6H_{13}, H, Br, C, CH_3	C_6H_{13}, Br, H, C, CH_3	C_6H_{13}, H, OH, C, CH_3	C_6H_{13}, HO, H, C, CH_3
(*R*)-(−)-2-Bromooctane $[\alpha]_D^{25} = -34.25°$	**(*S*)-(+)-2-Bromooctane** $[\alpha]_D^{25} = +34.25°$	**(*R*)-(−)-2-Octanol** $[\alpha]_D^{25} = -9.90°$	**(*S*)-(+)-2-Octanol** $[\alpha]_D^{25} = +9.90°$

When the reaction is carried out, we find that enantiomerically pure (*R*)-(−)-2-bromooctane ($[\alpha]_D^{25} = -34.25°$) has been converted to enantiomerically pure (*S*)-(+)-2-octanol ($[\alpha]_D^{25} = +9.90°$).

The Stereochemistry of an S_N2 Reaction

This reaction is S_N2 and takes place with *complete inversion of configuration:*

$$HO{:}^- + (R)\text{-}CH_3CH(Br)C_6H_{13} \longrightarrow \left[\overset{\delta-}{HO}\cdots C \cdots \overset{\delta-}{Br} \right]^{\ddagger} \longrightarrow HO\text{—}C(CH_3)(H)(C_6H_{13}) + Br^-$$

An inversion of configuration

(*R*)-(−)-2-Bromooctane
$[\alpha]_D^{25} = -34.25°$
Enantiomeric purity = 100%

(*S*)-(+)-2-Octanol
$[\alpha]_D^{25} = +9.90°$
Enantiomeric purity = 100%

PROBLEM 6.3

S_N2 reactions that involve breaking a bond to a stereogenic carbon can be used to relate configurations of molecules because the *stereochemistry* of the reaction is known.

(a) Illustrate how this is true by assigning configurations to the 2-chlorobutane enantiomers based on the following data. [The configuration of (−)-2-butanol is given in Section 5.8C.]

$$(+)\text{-2-Chlorobutane} \xrightarrow[S_N2]{OH^-} (-)\text{-2-Butanol}$$

$[\alpha]_D^{25} = +36.00°$ **Enantiomerically pure** — **$[\alpha]_D^{25} = -13.52°$** **Enantiomerically pure**

(b) When optically pure (+)-2-chlorobutane is allowed to react with potassium iodide in acetone in an S_N2 reaction, the 2-iodobutane that is produced has a minus rotation. What is the configuration of (−)-2-iodobutane? Of (+)-2-iodobutane?

6.9 The Reaction of *tert*-Butyl Chloride with Hydroxide Ion: An S_N1 Reaction

When *tert*-butyl chloride reacts with sodium hydroxide in a mixture of water and acetone, the kinetic results are quite different than for the reaction of chloromethane with hydroxide. The rate of formation of *tert*-butyl alcohol is dependent on the concentration of *tert*-butyl chloride, but it is *independent of the concentration of hydroxide ion.* Doubling the *tert*-butyl chloride concentration *doubles* the rate of the reaction, but changing the hydroxide ion concentration (within limits) has no appreciable effect. *tert*-Butyl chloride reacts by substitution at virtually the same rate in pure water (where the hydroxide ion is $10^{-7}M$) as it does in 0.05*M* aqueous sodium hydroxide (where the hydroxide ion concentration is 500,000 times larger). (We shall see in Section 6.10 that the important nucleophile in this reaction is a molecule of water.)

Thus the rate equation for this substitution reaction is first order with respect to *tert*-butyl chloride and *first order overall:*

$$(CH_3)_3C\text{—}Cl + OH^- \xrightarrow[H_2O]{\text{acetone}} (CH_3)_3C\text{—}OH + Cl^-$$

$$\text{Rate} \propto [(CH_3)_3CCl]$$

$$\text{Rate} = k[(CH_3)_3CCl]$$

We can conclude, therefore, that hydroxide ions do not participate in the transition state of the step that controls the rate of the reaction and that only molecules of *tert*-butyl chloride are involved. This reaction is said to be **unimolecular** (first order) in the rate-determining step. We call this type of reaction an S_N1 reaction **(substitution, nucleophilic, unimolecular).**

How can we explain an S_N1 reaction in terms of a mechanism? To do so, we shall need to consider the possibility that the mechanism involves more than one step. But what kind of kinetic results should we expect from a multistep reaction? Let us consider this point further.

6.9 Multistep Reactions and the Rate-Determining Step

If a reaction takes place in a series of steps, and if one step is intrinsically slower than all the others, then the rate of the overall reaction will be essentially the same as the rate of this slow step. This slow step, consequently, is called the **rate-limiting step** or the **rate-determining step.**

Consider a multistep reaction such as the following:

$$\text{Reactant} \xrightarrow[\text{(slow)}]{k_1} \text{intermediate 1} \xrightarrow[\text{(fast)}]{k_2} \text{intermediate 2} \xrightarrow[\text{(fast)}]{k_3} \text{product}$$

Step 1 *Step 2* *Step 3*

When we say that the first step in this example is intrinsically slow, we mean that the rate constant for step 1 is very much smaller than the rate constant for step 2 or for step 3. That is, $k_1 \ll k_2$ or k_3. When we say that steps 2 and 3 are *fast,* we mean that because their rate constants are larger, they could (in theory) take place rapidly if the concentrations of the two intermediates ever became high. In actuality, the concentrations of the intermediates are always very small because of the slowness of step 1.

As an analogy, imagine an hourglass modified in the way shown in Fig. 6.6. The opening between the top chamber and the one just below is considerably smaller than the other two. The overall rate at which sand falls from the top to the bottom of the hourglass is limited by the rate at which sand passes through this small orifice. This step, in the passage of sand, is analogous to the rate-determining step of the multistep reaction.

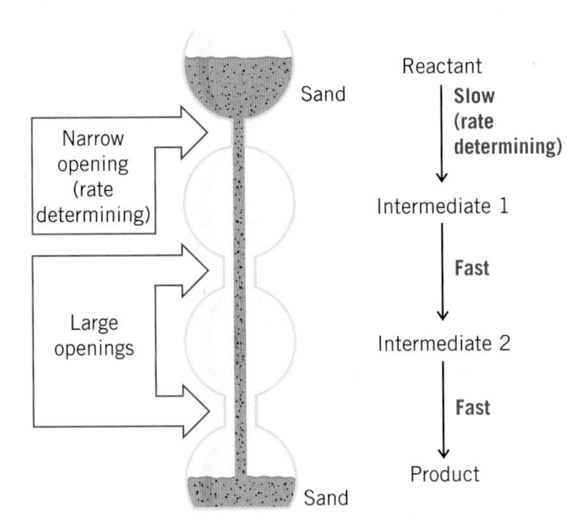

FIGURE 6.6 A modified hourglass that serves as an analogy for a multistep reaction. The overall rate is limited by the rate of the slow step.

6.10 A Mechanism for the S_N1 Reaction

The mechanism for the reaction of *tert*-butyl chloride with water (Section 6.9) apparently involves three steps. See the box "A Mechanism for the S_N1 Reaction" on p. 253, with a schematic free-energy diagram highlighted for each step. Two distinct **intermediates** are formed. The first step is the slow step—it is the rate-determining step. In it a molecule of *tert*-butyl chloride ionizes and becomes a *tert*-butyl cation and a chloride ion. In the transition state for this step the carbon–chlorine bond of *tert*-butyl chloride is largely broken and ions are beginning to develop:

$$(CH_3)_3C^{\delta+}\text{---}Cl^{\delta-}$$

The solvent (water) stabilizes these developing ions by solvation. Carbocation formation, in general, takes place slowly because it is usually a highly endothermic process and is uphill in terms of free energy.

A Mechanism for the S_N1 Reaction

Reaction:

$$(CH_3)_3CCl + 2\,H_2O \longrightarrow (CH_3)_3COH + H_3O^+ + Cl^-$$

Mechanism:

Step 1

$$H_3C\text{—}C(CH_3)_2\text{—}\ddot{\underset{..}{Cl}}: \underset{H_2O}{\overset{slow}{\rightleftharpoons}} H_3C\text{—}C^+(CH_3)_2 + :\ddot{\underset{..}{Cl}}:^-$$

Aided by the polar solvent a chlorine departs with the electron pair that bonded it to the carbon.

This slow step produces the relatively stable 3° carbocation and a chloride ion. Although not shown here, the ions are solvated (and stabilized) by water molecules.

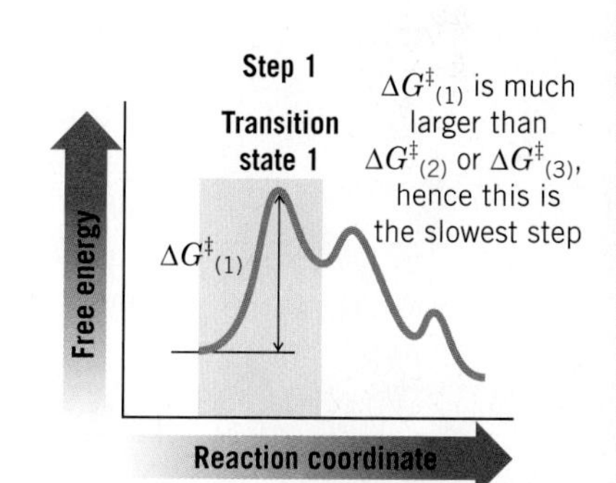

Step 2

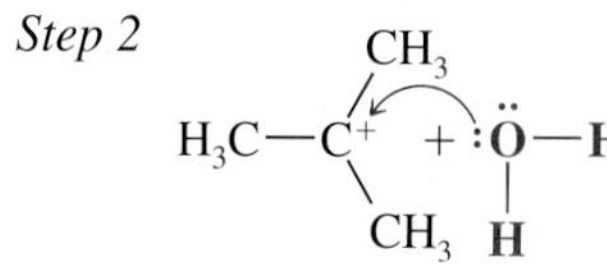

$\overset{fast}{\rightleftharpoons}$

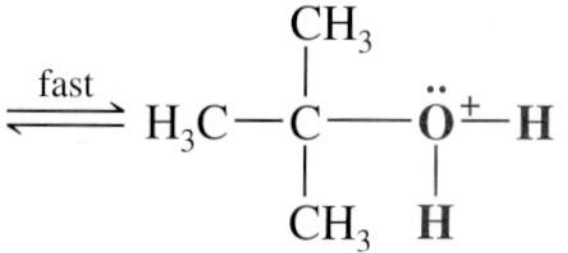

A water molecule acting as a Lewis base donates an electron pair to the carbocation (a Lewis acid). This gives the cationic carbon eight electrons.

The product is a *tert*-butyloxonium ion (or protonated *tert*-butyl alcohol).

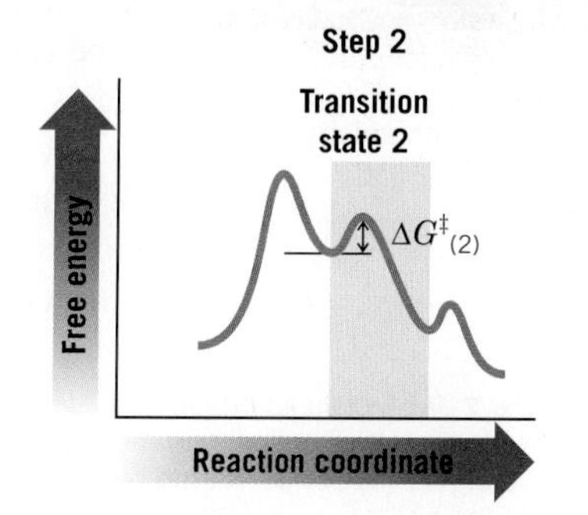

Step 3

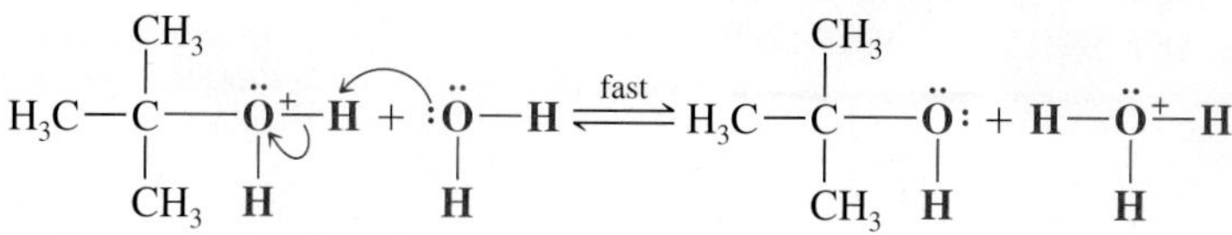

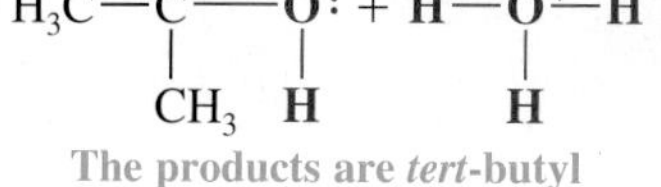

A water molecule acting as a Brønsted base accepts a proton from the *tert*-butyloxonium ion.

The products are *tert*-butyl alcohol and a hydronium ion.

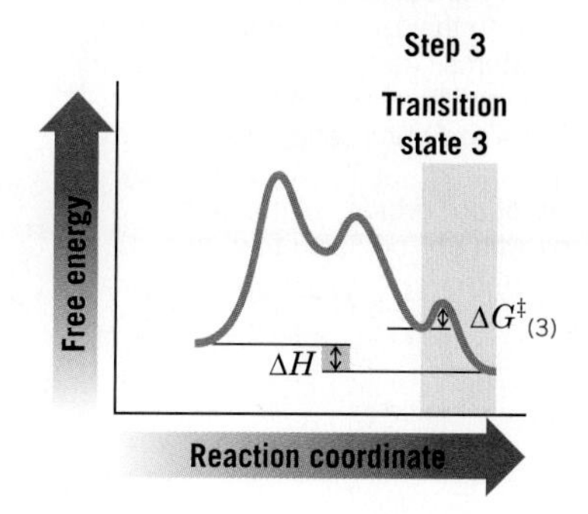

In the second step the intermediate *tert*-butyl cation reacts rapidly with water to produce a *tert*-butyloxonium ion (another intermediate); this, in the third step, rapidly transfers a proton to a molecule of water producing *tert*-butyl alcohol.

The first step requires heterolytic cleavage of the carbon–chlorine bond. Because no other bonds are formed in this step, it should be highly endothermic and it should have a high free energy of activation, as we see in the free-energy diagram. That it takes place at all is largely because of the ionizing ability of the solvent, water. Experiments indicate that in the gas phase (i.e., in the absence of a solvent), the free energy of activation is about 630 kJ mol^{-1}! In aqueous solution, however, the free energy of activation is much lower—about 84 kJ mol^{-1}. Water molecules surround and stabilize the cation and anion that are produced (cf. Section 2.14E).

CD Tutorial
S_N1 mechanism

Even though the *tert*-butyl cation produced in step 1 is stabilized by solvation, it is still a highly reactive species. Almost immediately after it is formed, it reacts with one of the surrounding water molecules to form the *tert*-butyloxonium ion, $(CH_3)_3COH_2^+$. (It may also occasionally react with a hydroxide ion, but water molecules are far more plentiful.)

6.11 Carbocations

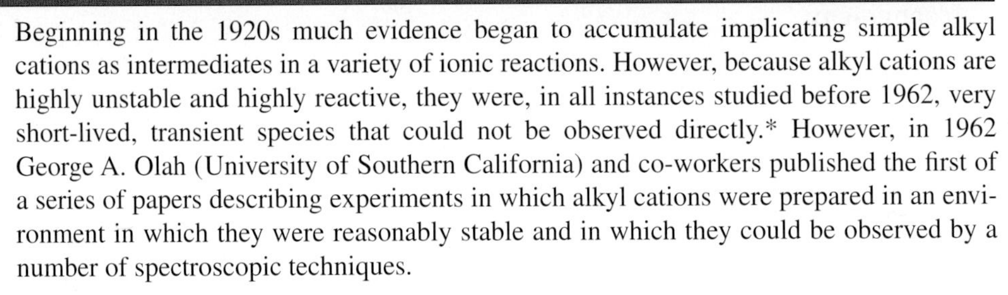

Olah was awarded the 1994 Nobel Prize in chemistry.

Beginning in the 1920s much evidence began to accumulate implicating simple alkyl cations as intermediates in a variety of ionic reactions. However, because alkyl cations are highly unstable and highly reactive, they were, in all instances studied before 1962, very short-lived, transient species that could not be observed directly.* However, in 1962 George A. Olah (University of Southern California) and co-workers published the first of a series of papers describing experiments in which alkyl cations were prepared in an environment in which they were reasonably stable and in which they could be observed by a number of spectroscopic techniques.

6.11A The Structure of Carbocations

Considerable experimental evidence indicates that the structure of **carbocations** is **trigonal planar** like that of BF_3 (Section 1.16D). Just as the trigonal planar structure of BF_3 can be accounted for on the basis of sp^2 hybridization so, too (Fig. 6.7), can the trigonal planar structure of carbocations.

The central carbon atom in a carbocation is electron deficient; it has only six electrons in its outermost energy level. In our model (Fig. 6.7) these six electrons are used to form sigma covalent bonds to hydrogen atoms (or to alkyl groups). The *p* orbital contains no electrons.

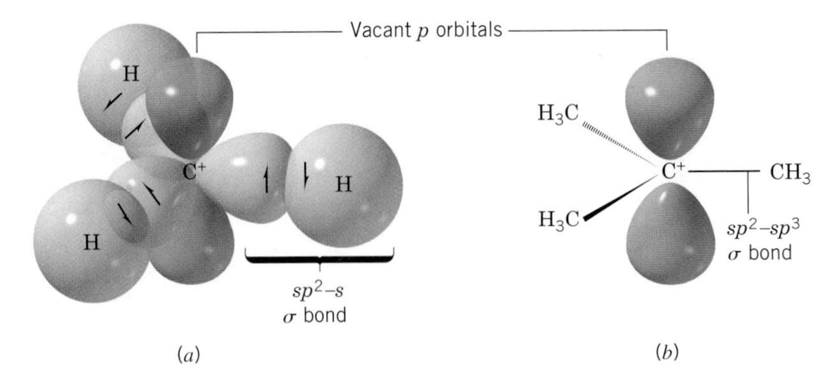

FIGURE 6.7 *(a)* A stylized orbital structure of the methyl cation. The bonds are sigma (σ) bonds formed by overlap of the carbon atom's three sp^2 orbitals with the 1*s* orbitals of the hydrogen atoms. The *p* orbital is vacant. *(b)* A dashed line–wedge representation of the *tert*-butyl cation. The bonds between carbon atoms are formed by overlap of sp^3 orbitals of the methyl groups with sp^2 orbitals of the central carbon atom.

*As we shall learn later, carbocations bearing aromatic groups can be much more stable; one of these had been studied as early as 1901.

6.11B The Relative Stabilities of Carbocations

Knowledge of carbocation structure is an important tool for understanding a variety of reaction processes.

A large body of experimental evidence indicates that the relative stabilities of carbocations are related to the number of alkyl groups attached to the positively charged trivalent carbon atom. Tertiary carbocations are the most stable, and the methyl cation is the least stable. The overall order of stability is as follows:

$$\mathrm{R_3C^+} > \mathrm{R_2CH^+} > \mathrm{RCH_2^+} > \mathrm{H_3C^+}$$

3° (most stable) > 2° > 1° > Methyl (least stable)

Relative carbocation stability

This order of stability of carbocations can be explained on the basis of **hyperconjugation.** Hyperconjugation involves electron delocalization (via partial orbital overlap) from a filled bonding orbital to an adjacent unfilled orbital (Section 4.7). In the case of a carbocation, the unfilled orbital is the vacant p orbital of the carbocation, and the filled orbitals are C—H or C—C sigma bonds at the carbons *adjacent* to the p orbital of the carbocation. Sharing of electron density from adjacent C—H or C—C sigma bonds with the carbocation p orbital delocalizes the positive charge. Any time a charge can be dispersed or delocalized, a system will be stabilized. Figure 6.8 shows a stylized representation of hyperconjugation between a sigma bonding orbital and an adjacent carbocation p orbital.

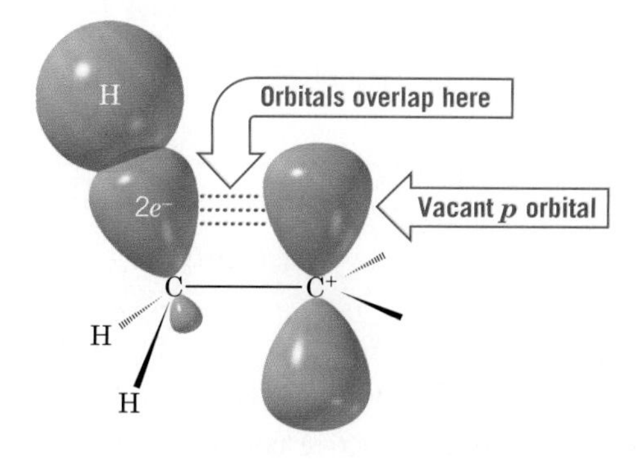

FIGURE 6.8 *How a methyl group helps stabilize the positive charge of a carbocation.* Electron density from one of the carbon–hydrogen sigma bonds of the methyl group flows into the vacant p orbital of the carbocation because the orbitals can partly overlap. Shifting electron density in this way makes the sp^2-hybridized carbon of the carbocation somewhat less positive, and the hydrogens of the methyl group assume some of the positive charge. Delocalization (dispersal) of the charge in this way leads to greater stability. This interaction of a bond orbital with a p orbital is called hyperconjugation.

Tertiary carbocations have three carbons with C—H bonds (or, depending on the specific example, C—C bonds instead of C—H) adjacent to the carbocation that can overlap partially with the vacant p orbital. Secondary carbocations have only two adjacent carbons with C—H or C—C bonds to overlap with the carbocation; hence, the possibility for hyperconjugation is less and the secondary carbocation is less stable. Primary carbocations have only one adjacent carbon from which to derive hyperconjugative stabilization, and so they are even less stable. A methyl carbocation has no possibility for hyperconjugation, and it is the least stable of all in this series. The following are specific examples:

$$(^{\delta+}\mathrm{CH_3})_3\mathrm{C}^{\delta+} \text{ is more stable than } (^{\delta+}\mathrm{CH_3})_2\mathrm{C}^{\delta+}\mathrm{H} \text{ is more stable than } {}^{\delta+}\mathrm{CH_3}\mathrm{C}^{\delta+}\mathrm{H_2} \text{ is more stable than } \mathrm{H_3C^+}$$

***tert*-Butyl cation (3°) (most stable)** — **Isopropyl cation (2°)** — **Ethyl cation (1°)** — **Methyl cation (least stable)**

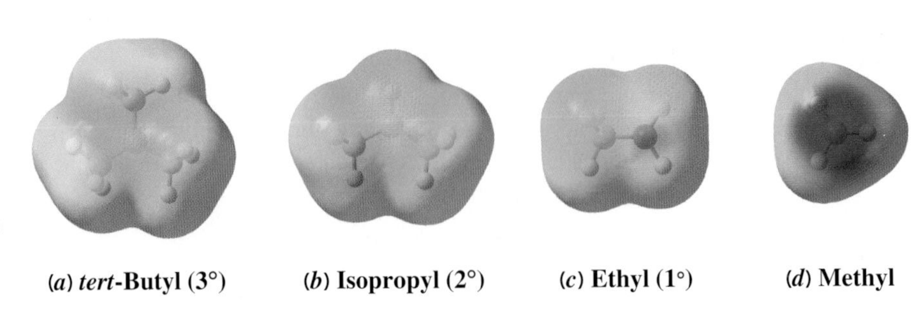

FIGURE 6.9 Maps of electrostatic potential for *(a) tert*-butyl (3°), *(b)* isopropyl (2°), *(c)* ethyl (1°), and *(d)* methyl carbocations show the trend from greater to lesser delocalization (stabilization) of the positive charge in these structures. Less blue color indicates greater delocalization of the positive charge. (The structures are mapped on the same scale of electrostatic potential to allow direct comparison.)

Thus, the relative stability of carbocations is 3° > 2° > 1° > methyl. This trend is also readily seen in electrostatic potential maps for these carbocations (Fig. 6.9).

6.12 The Stereochemistry of S_N1 Reactions

Because the carbocation formed in the first step of an S_N1 reaction has a trigonal planar structure (Section 6.11A), when it reacts with a nucleophile, it may do so from either the front side or the back side (see below). With the *tert*-butyl cation this makes no difference because the same product is formed by either mode of attack. (Convince yourself of this result by examining models.)

$H_2\overset{+}{O}$–$C(CH_3)_3$ ⟵ back-side attack ⟵ $H_2\ddot{O}$: → $(CH_3)_3C^+$ ← :$\ddot{O}H_2$ ⟶ front-side attack ⟶ $(CH_3)_3C$–$\overset{+}{O}H_2$

Same product

With some cations, however, different products arise from the two reaction possibilities. We shall study this point next.

6.12A Reactions That Involve Racemization

A reaction that transforms an optically active compound into a racemic form is said to proceed with **racemization.** If the original compound loses all of its optical activity in the course of the reaction, chemists describe the reaction as having taken place with *complete* racemization. If the original compound loses only part of its optical activity, as would be the case if an enantiomer were only partially converted to a racemic form, then chemists describe this as proceeding with *partial* racemization.

Racemization takes place *whenever the reaction causes chiral molecules to be converted to an achiral intermediate.*

CD Tutorial

S_N1 stereochemistry

Examples of this type of reaction are S_N1 reactions in which the leaving group departs from a stereogenic carbon. These reactions almost always result in extensive and sometimes complete racemization. For example, heating optically active (*S*)-3-bromo-3-methylhexane with aqueous acetone results in the formation of 3-methyl-3-hexanol as a racemic form:

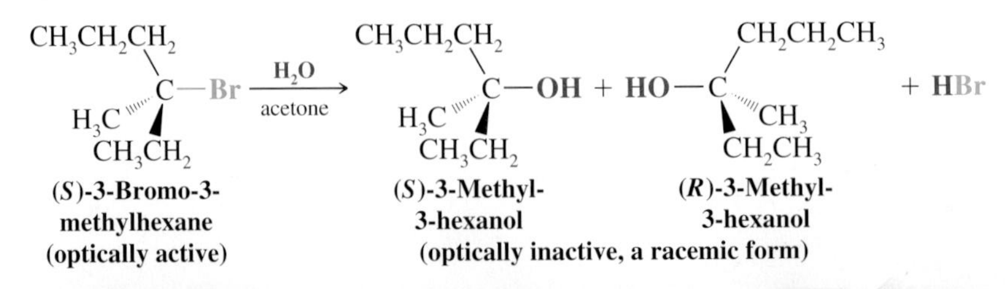

The reason: The S_N1 reaction proceeds through the formation of an intermediate carbocation and the carbocation, because of its trigonal planar configuration, *is achiral.* It reacts with water at equal rates from either side to form the enantiomers of 3-methyl-3-hexanol in equal amounts.

The Stereochemistry of an S_N1 Reaction

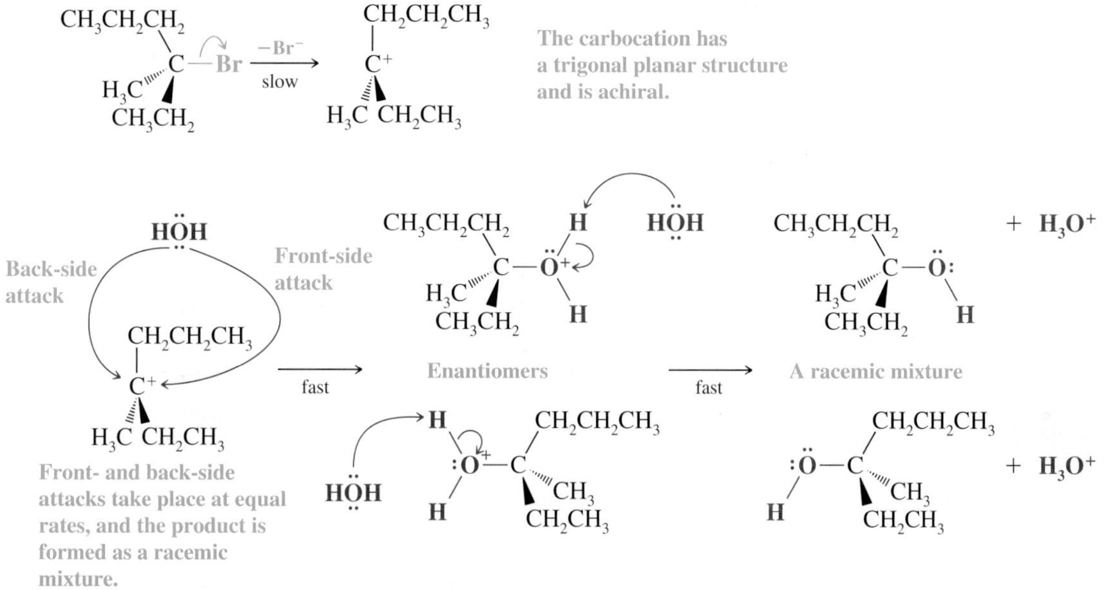

$CH_3CH_2CH_2$–C(H_3C)(CH_3CH_2)–Br $\xrightarrow[\text{slow}]{-Br^-}$ C^+($CH_2CH_2CH_3$)(H_3C)(CH_2CH_3)

The carbocation has a trigonal planar structure and is achiral.

Back-side attack / Front-side attack ($H\ddot{O}H$)

Front- and back-side attacks take place at equal rates, and the product is formed as a racemic mixture.

$\xrightarrow{\text{fast}}$ Enantiomers $\xrightarrow{\text{fast}}$ A racemic mixture $+ H_3O^+$

The S_N1 reaction of (*S*)-3-bromo-3-methylhexane proceeds with racemization because the intermediate carbocation is achiral and attack by the nucleophile can occur from either side.

PROBLEM 6.4

Keeping in mind that carbocations have a trigonal planar structure, **(a)** write a structure for the carbocation intermediate and **(b)** write structures for the alcohol (or alcohols) that you would expect from the following reaction:

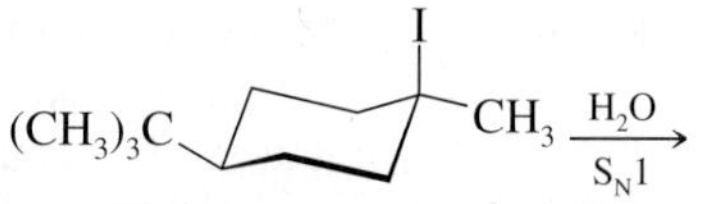

$(CH_3)_3C$–[cyclohexane]–(I)(CH_3) $\xrightarrow[S_N1]{H_2O}$

6.12B Solvolysis

The S_N1 reaction of an alkyl halide with water is an example of **solvolysis.** A solvolysis reaction is a nucleophilic substitution in which *the nucleophile is a molecule of the solvent* (*solvent + lysis:* cleavage by the solvent). Since the solvent in this instance is water, we could also call the reaction a **hydrolysis.** If the reaction had taken place in methanol, we would call it a **methanolysis.**

Examples of Solvolysis

$$(CH_3)_3C\text{—}Br + H_2O \longrightarrow (CH_3)_3C\text{—}OH + HBr$$

$$(CH_3)_3C\text{—}Cl + CH_3OH \longrightarrow (CH_3)_3C\text{—}OCH_3 + HCl$$

$$(CH_3)_3C\text{—}Cl + HCOH \longrightarrow (CH_3)_3C\text{—}OCH + HCl$$

(with C=O in HCOH and OCH)

In the last example the solvent is formic acid (HCO_2H) and the following steps take place:

Step 1 $(CH_3)_3C{-}\ddot{\underset{..}{Cl}}{:} \xrightarrow{\text{slow}} (CH_3)_3C^+ + {:}\ddot{\underset{..}{Cl}}{:}^-$

Step 2

$$(CH_3)_3C^+ + H{-}\ddot{O}{-}\overset{\overset{\ddot{O}}{\|}}{C}H \xrightarrow{\text{fast}} H{-}\ddot{O}{-}\overset{\overset{\overset{+}{:O}{-}C(CH_3)_3}{\|}}{C}H \longleftrightarrow H{-}\overset{+}{\ddot{O}}{=}\overset{\overset{:\ddot{O}{-}C(CH_3)_3}{|}}{C}H$$

Step 3

$$:\ddot{Cl}:^- \;\; H{-}\overset{+}{O}{=}\overset{\overset{:\ddot{O}{-}C(CH_3)_3}{|}}{C}H \xrightarrow{\text{fast}} \ddot{O}{=}\overset{\overset{:\ddot{O}{-}C(CH_3)_3}{|}}{C}H \equiv H\overset{\overset{\ddot{O}}{\|}}{C}{-}\ddot{O}{-}C(CH_3)_3$$

$$+ H{-}\ddot{Cl}:$$

These reactions all involve the initial formation of a carbocation and the subsequent reaction of that cation with a molecule of the solvent. Solvolysis reactions are, therefore, S_N1 reactions.

PROBLEM 6.5 What product(s) would you expect from the methanolysis of the cyclohexane derivative given as the reactant in Problem 6.4?

6.13 Factors Affecting the Rates of S_N1 and S_N2 Reactions

Now that we have an understanding of the mechanisms of S_N2 and S_N1 reactions, our next task is to explain why chloromethane reacts by an S_N2 mechanism and *tert*-butyl chloride by an S_N1 mechanism. We would also like to be able to predict which pathway—S_N1 or S_N2—would be followed by the reaction of any alkyl halide with any nucleophile under varying conditions.

The answer to this kind of question is to be found in the *relative rates of the reactions that occur.* If a given alkyl halide and nucleophile react *rapidly* by an S_N2 mechanism but *slowly* by an S_N1 mechanism under a given set of conditions, then an S_N2 pathway will be followed by most of the molecules. On the other hand, another alkyl halide and another nucleophile may react very slowly (or not at all) by an S_N2 pathway. If they react rapidly by an S_N1 mechanism, then the reactants will follow an S_N1 pathway.

Experiments have shown that a number of factors affect the relative rates of S_N1 and S_N2 reactions. The most important factors are:

1. the structure of the substrate,
2. the concentration and reactivity of the nucleophile (for bimolecular reactions only),
3. the effect of the solvent, and
4. the nature of the leaving group.

6.13A The Effect of the Structure of the Substrate

S_N2 order of reactivity

S_N2 Reactions Simple alkyl halides show the following general order of reactivity in S_N2 reactions:

Methyl > primary > secondary >> (tertiary—unreactive)

Methyl halides react most rapidly and tertiary halides react so slowly as to be unreactive by the S_N2 mechanism. Table 6.4 gives the relative rates of typical S_N2 reactions.

TABLE 6.4 Relative Rates of Reactions of Alkyl Halides in S_N2 Reactions

Substituent	Compound	Approximate Relative Rate
Methyl	CH_3X	30
1°	CH_3CH_2X	1
2°	$(CH_3)_2CHX$	0.03
Neopentyl	$(CH_3)_3CCH_2X$	0.00001
3°	$(CH_3)_3CX$	~0

Neopentyl halides, even though they are primary halides, are very unreactive:

$$CH_3-\overset{\displaystyle CH_3}{\underset{\displaystyle CH_3}{\overset{|}{\underset{|}{C}}}}-CH_2-X$$

A neopentyl halide

The important factor behind this order of reactivity is a **steric effect.** A steric effect is an effect on relative rates caused by the space-filling properties of those parts of a molecule attached at or near the reacting site. One kind of steric effect—the kind that is important here—is called **steric hindrance.** By this we mean that *the spatial arrangement of the atoms or groups at or near the reacting site of a molecule hinders or retards a reaction.*

For particles (molecules and ions) to react, their reactive centers must be able to come within bonding distance of each other. Although most molecules are reasonably flexible, very large and bulky groups can often hinder the formation of the required transition state. In some cases they can prevent its formation altogether.

CD Tutorial

S_N2 steric hindrance

An S_N2 reaction requires an approach by the nucleophile to a distance within the bonding range of the carbon atom bearing the leaving group. Because of this, bulky substituents on *or near* that carbon atom have a dramatic inhibiting effect (Fig. 6.10). They cause the free energy of the required transition state to be increased and, consequently, they increase the free energy of activation for the reaction. Of the simple alkyl halides, methyl halides react most rapidly in S_N2 reactions because only three small hydrogen

The steric effects in these structures can best be appreciated by building models.

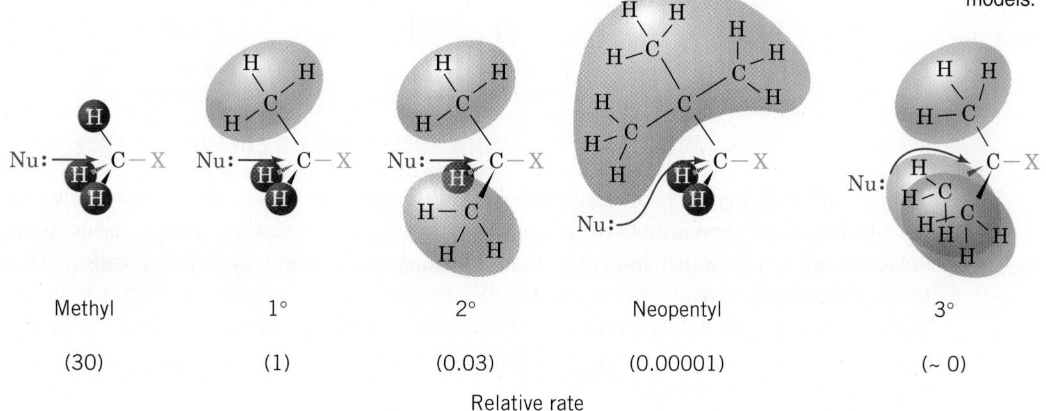

FIGURE 6.10 Steric effects in the S_N2 reaction.

atoms interfere with the approaching nucleophile. Neopentyl and tertiary halides are the least reactive because bulky groups present a strong hindrance to the approaching nucleophile. (Tertiary substrates, for all practical purposes, do not react by an S_N2 mechanism.)

S_N1 Reactions ***The primary factor that determines the reactivity of organic substrates in an S_N1 reaction is the relative stability of the carbocation that is formed.***

S_N1 reactivity and carbocation stability

Except for those reactions that take place in strong acids, which we shall study later, the only organic compounds that undergo reaction by an S_N1 path at a reasonable rate are *those that are capable of forming relatively stable carbocations.* Of the simple alkyl halides that we have studied so far, this means (for all practical purposes) that only tertiary halides react by an S_N1 mechanism. (Later we shall see that certain organic halides, called *allylic halides* and *benzylic halides,* can also react by an S_N1 mechanism because they can form relatively stable carbocations; see Sections 13.4 and 15.15.)

Tertiary carbocations are stabilized because sigma bonds at three adjacent carbons contribute electron density to the carbocation *p* orbital by hyperconjugation (Section 6.11B). Secondary and primary carbocations have less stabilization by hyperconjugation. A methyl carbocation has no stabilization. Formation of a relatively stable carbocation is important in an S_N1 reaction because it means that the free energy of activation for the slow step of the reaction (e.g., $R{-}L \rightarrow R^+ + L^-$) will be low enough for the reaction to take place at a reasonable rate.

The Hammond–Leffler Postulate If you review the free-energy diagrams that accompany the mechanism for the S_N1 reaction of *tert*-butyl chloride and water (Section 6.9), you will see that step 1, the ionization of the leaving group to form the carbocation, is *uphill in terms of free energy* (ΔG° for this step is positive). It is also uphill in terms of enthalpy (ΔH° is also positive), and, therefore, this step is *endothermic.* According to a postulate made by G. S. Hammond (then at California Institute of Technology) and J. E. Leffler (Florida State University), **the transition state for a step that is uphill in energy should show a strong resemblance to the product of that step.** Since the product of this step (actually an intermediate in the overall reaction) is a carbocation, any factor that stabilizes it—such as dispersal of the positive charge by electron-releasing groups—should also stabilize the transition state in which the positive charge is developing.

Ionization of the Leaving Group

$$(CH_3)_3C{-}Cl \xrightarrow{H_2O} \left[(CH_3)_3\overset{\delta+}{C}\cdots\overset{\delta-}{Cl} \right]^{\ddagger} \xrightarrow{H_2O} (CH_3)_3C^{+} + Cl^{-}$$

Reactant — **Transition state** *Resembles product of step because ΔG° is positive* — **Product of step** *Stabilized by three electron-releasing groups*

For a methyl, primary, or secondary halide to react by an S_N1 mechanism, it would have to ionize to form a methyl, primary, or secondary carbocation. These carbocations, however, are much higher in energy than a tertiary carbocation, and the transition states leading to these carbocations are even higher in energy. The activation energy for an S_N1 reaction of a simple methyl, primary, or secondary halide, consequently, is so large (the reaction is so slow) that, for all practical purposes, an S_N1 reaction does not compete with the corresponding S_N2 reaction.

The **Hammond–Leffler postulate** is quite general and can be better understood through consideration of Fig. 6.11. One way that the postulate can be stated is to say that

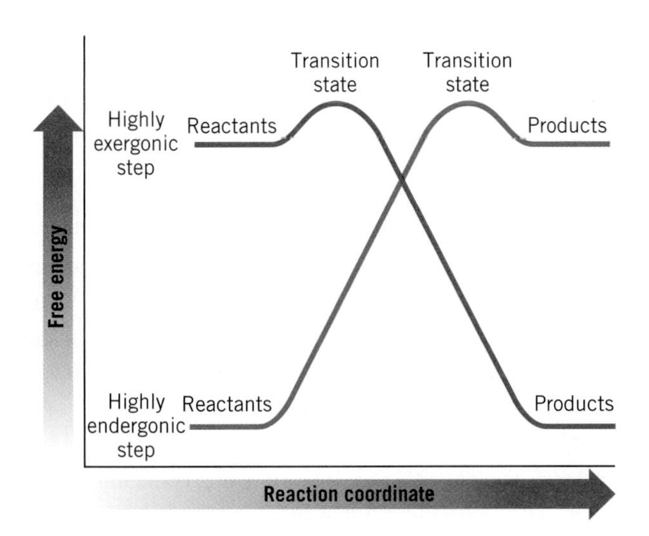

FIGURE 6.11 The transition state for a highly exergonic step (red curve) lies close to and resembles the reactants. The transition state for an endergonic step (blue curve) lies close to and resembles the products of a reaction. (Adapted from Pryor, W. A. *Free Radicals* McGraw-Hill; New York, 1966; p 156. Reprinted by permission.)

the structure of a transition state resembles the stable species that is nearest it in free energy. For example, in a highly **endergonic** step (blue curve) the transition state lies close to the products in free energy, and we assume, therefore, that **it resembles the products of that step in structure.** Conversely, in a highly exergonic step (red curve) the transition state lies close to the reactants in free energy, and we assume **it resembles the reactants in structure** as well. The great value of the Hammond–Leffler postulate is that it gives us an intuitive way of visualizing those important, but fleeting, species that we call transition states. We shall make use of it in many future discussions.

PROBLEM 6.6

The relative rates of ethanolysis of four primary alkyl halides are as follows: CH_3CH_2Br, 1.0; $CH_3CH_2CH_2Br$, 0.28; $(CH_3)_2CHCH_2Br$, 0.030; $(CH_3)_3CCH_2Br$, 0.00000042.

(a) Is each of these reactions likely to be S_N1 or S_N2?

(b) Provide an explanation for the relative reactivities that are observed.

6.13B The Effect of the Concentration and Strength of the Nucleophile

Since the nucleophile does not participate in the rate-determining step of an S_N1 reaction, the rates of S_N1 reactions are unaffected by either the concentration or the identity of the nucleophile. The rates of S_N2 reactions, however, depend on *both* the concentration *and* the identity of the attacking nucleophile. We saw in Section 6.5 how increasing the concentration of the nucleophile increases the rate of an S_N2 reaction. We can now examine how the rate of an S_N2 reaction depends on the identity of the nucleophile.

The relative strength of a nucleophile is measured in terms of the relative rate of its S_N2 reaction with a given substrate. A good nucleophile is one that reacts rapidly in an S_N2 reaction with a given substrate. A poor nucleophile is one that reacts slowly in an S_N2 reaction with the same substrate under comparable reaction conditions. (As mentioned above, we cannot compare nucleophilicities with regard to S_N1 reactions because the nucleophile does not participate in the rate-determining step of an S_N1 reaction.)

Methoxide anion, for example, is a good nucleophile for a substitution reaction with iodomethane. It reacts rapidly by an S_N2 mechanism to form dimethyl ether:

$$CH_3O^- + CH_3I \xrightarrow{\text{rapid}} CH_3OCH_3 + I^-$$

Methanol, on the other hand, is a poor nucleophile for reaction with iodomethane. Under comparable conditions it reacts very slowly. It is not a sufficiently powerful Lewis base (i.e., nucleophile) to cause displacement of the iodide leaving group at a significant rate:

$$CH_3OH + CH_3I \xrightarrow{\text{very slow}} CH_3\overset{+}{O}(H)CH_3 + I^-$$

The relative strengths of nucleophiles can be correlated with two structural features:

Relative nucleophile strength

1. **A negatively charged nucleophile is always a more reactive nucleophile than its conjugate acid.** Thus HO^- is a better nucleophile than H_2O and RO^- is better than ROH.
2. **In a group of nucleophiles in which the nucleophilic atom is the same, nucleophilicities parallel basicities.** Oxygen compounds, for example, show the following order of reactivity:

$$RO^- > HO^- \gg RCO_2^- > ROH > H_2O$$

This is also their order of basicity. An alkoxide ion (RO^-) is a slightly stronger base than a hydroxide ion (HO^-), a hydroxide ion is a much stronger base than a carboxylate ion (RCO_2^-), and so on.

Nucleophilicity versus Basicity While nucleophilicity and basicity are related, they are not measured in the same way. Basicity, as expressed by pK_a, is measured *by the position of an equilibrium* involving an electron pair donor (base), a proton, the conjugate acid, and the conjugate base. Nucleophilicity is measured *by relative rates of reaction,* by how rapidly an electron pair donor reacts at an atom (usually carbon) bearing a leaving group. For example, the hydroxide ion (OH^-) is a stronger base than a cyanide ion (CN^-); at equilibrium it has the greater affinity for a proton (the pK_a of H_2O is ~16, while the pK_a of HCN is ~10). Nevertheless, cyanide ion is a stronger nucleophile; it reacts more rapidly with a carbon bearing a leaving group than does hydroxide ion.

6.13C Solvent Effects on S_N2 Reactions: Polar Protic and Aprotic Solvents

A molecule of a solvent such as water or an alcohol—called a **protic solvent** (Section 3.11)—has a hydrogen atom attached to a strongly electronegative element (oxygen). Molecules of protic solvents can, therefore, form hydrogen bonds to nucleophiles in the following way.

Molecules of the protic solvent, water, solvate a halide ion by forming hydrogen bonds to it.

The effect of this hydrogen bonding interaction is to encumber the nucleophile and hinder its reaction in a substitution reaction. For a strongly solvated nucleophile to react, it must shed some of its solvent molecules so that it can approach the carbon of the substrate that bears the leaving group.

Hydrogen bonds to a small nucleophilic atom are stronger than to larger nucleophilic atoms. This trend is borne out among elements in the *same group (column) of the periodic table.* For example, fluoride anion is more strongly solvated than the other halides because it is the smallest halide anion and its charge is the most concentrated. Hence, in a protic solvent fluoride is not as effective a nucleophile as the other halide anions. Iodide is the largest halide anion and it is the most weakly solvated in a protic solvent; hence, it is the strongest nucleophile among the halide anions. In general, the trend in *nucleophilicity* among the halide anions in a protic solvent is as follows:

Halide Nucleophilicity in Protic Solvents

$$I^- > Br^- > Cl^- > F^-$$

The same effect holds true when we compare sulfur nucleophiles with oxygen nucleophiles. Sulfur atoms are larger than oxygen atoms and hence they are not solvated as strongly in a protic solvent. Thus, thiols (R—SH) are stronger nucleophiles than alcohols, and RS^- anions are better nucleophiles than RO^- anions.

The greater reactivity of nucleophiles with large nucleophilic atoms is not entirely related to solvation. Larger atoms are more **polarizable** (their electron clouds are more easily distorted); therefore, a larger nucleophilic atom can donate a greater degree of electron density to the substrate than a smaller nucleophile whose electrons are more tightly held.

The relative nucleophilicities of some common nucleophiles in protic solvents are as follows:

Relative Nucleophilicity in Protic Solvents

$$SH^- > CN^- > I^- > OH^- > N_3^- > Br^- > CH_3CO_2^- > Cl^- > F^- > H_2O$$

Polar Aprotic Solvents ***Aprotic solvents** are those solvents whose molecules do not have a hydrogen atom that is attached to an atom of an electronegative element.* A number of **polar aprotic solvents** have come into wide use by chemists *because they are especially useful in S_N2 reactions.* Several examples are the following:

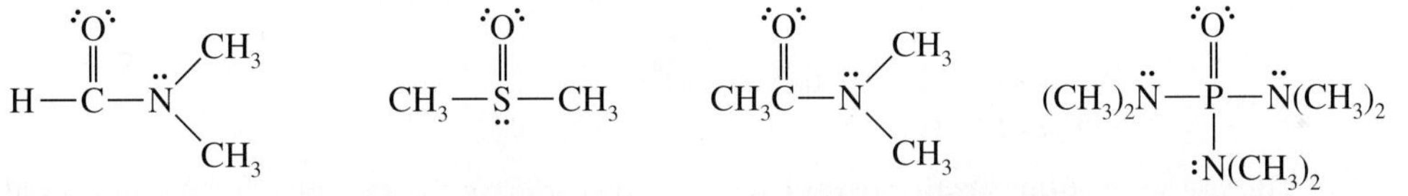

$H-C(=O)-N(CH_3)_2$ — ***N,N*-Dimethylformamide (DMF)**

$CH_3-S(=O)-CH_3$ — **Dimethyl sulfoxide (DMSO)**

$CH_3C(=O)-N(CH_3)_2$ — **Dimethylacetamide (DMA)**

$(CH_3)_2N-P(=O)(N(CH_3)_2)-N(CH_3)_2$ — **Hexamethylphosphoramide (HMPA)**

All of these solvents (DMF, DMSO, DMA, and HMPA) dissolve ionic compounds, and they solvate cations very well. They do so in the same way that protic solvents solvate cations: by orienting their negative ends around the cation and by donating unshared electron pairs to vacant orbitals of the cation:

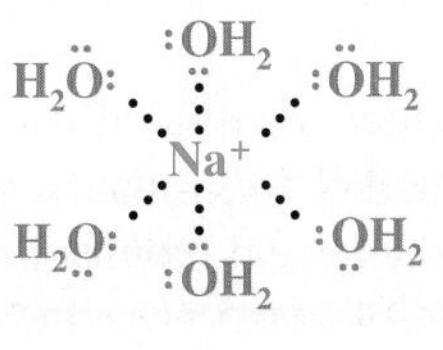

Na^+ surrounded by six $:OH_2$ ($H_2O:$) molecules

A sodium ion solvated by molecules of the protic solvent water

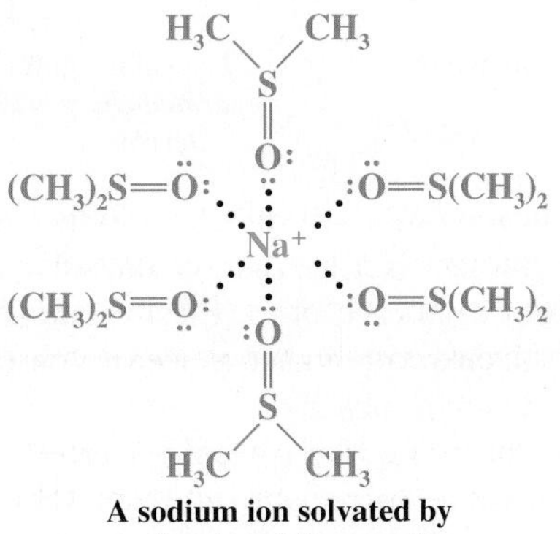

Na^+ surrounded by six $:O{=}S(CH_3)_2$ molecules

A sodium ion solvated by molecules of the aprotic solvent DMSO

However, because they cannot form hydrogen bonds and because their positive centers are well shielded by steric effects from any interaction with anions, *aprotic solvents do not solvate anions to any appreciable extent.* In these solvents anions are unencumbered by a layer of solvent molecules and they are therefore poorly stabilized by solvation. These "naked" anions are highly reactive both *as bases and nucleophiles.* In DMSO, for example, the relative order of reactivity of halide ions is opposite to that in protic solvents, and it follows the same trend as their relative basicity:

Polar aprotic solvents and S_N2 rates

Halide Nucleophilicity in Aprotic Solvents

$$F^- > Cl^- > Br^- > I^-$$

The rates of S_N2 reactions generally are vastly increased when they are carried out in polar aprotic solvents. The increase in rate can be as large as a millionfold.

PROBLEM 6.7 Classify the following solvents as being protic or aprotic: formic acid, HCO_2H; acetone, CH_3COCH_3; acetonitrile, $CH_3C{\equiv}N$; formamide, $HCONH_2$; sulfur dioxide, SO_2; ammonia, NH_3; trimethylamine, $N(CH_3)_3$; ethylene glycol, $HOCH_2CH_2OH$.

PROBLEM 6.8 Would you expect the reaction of propyl bromide with sodium cyanide (NaCN), that is,

$$CH_3CH_2CH_2Br + NaCN \longrightarrow CH_3CH_2CH_2CN + NaBr$$

to occur faster in DMF or in ethanol? Explain your answer.

PROBLEM 6.9 Which would you expect to be the stronger nucleophile in a protic solvent: **(a)** $CH_3CO_2^-$ or CH_3O^- **(b)** H_2O or H_2S? **(c)** $(CH_3)_3P$ or $(CH_3)_3N$?

6.13D Solvent Effects on S_N1 Reactions: The Ionizing Ability of the Solvent

Polar protic solvents and S_N1 rates

The use of a **polar protic solvent** will greatly increase the rate of ionization of an alkyl halide *in any S_N1 reaction* because of its ability to solvate cations *and* anions so effectively. It does this because solvation stabilizes the transition state leading to the intermediate carbocation and halide ion more than it does the reactants; thus the free energy of activation is lower. The transition state for this endothermic step is one in which separated charges are developing, and thus it resembles the ions that are ultimately produced:

$$(CH_3)_3C{-}Cl \longrightarrow \left[(CH_3)_3\overset{\delta+}{C}\cdots\overset{\delta-}{Cl}\right]^{\ddagger} \longrightarrow (CH_3)_3C^+ + Cl^-$$

Reactant **Transition state** *Separated charges are developing.* **Products**

A rough indication of a solvent's polarity is a quantity called the **dielectric constant.** The dielectric constant is a measure of the solvent's ability to insulate opposite charges (or separate ions) from each other. Electrostatic attractions and repulsions between ions are smaller in solvents with higher dielectric constants. Table 6.5 gives the dielectric constants of some common solvents.

Water is the most effective solvent for promoting ionization, but most organic compounds do not dissolve appreciably in water. They usually dissolve, however, in alcohols, and quite often mixed solvents are used. Methanol–water and ethanol–water are common mixed solvents for nucleophilic substitution reactions.

TABLE 6.5 Dielectric Constants of Common Solvents

	Solvent	Formula	Dielectric Constant
↑ Increasing solvent polarity	Water	H_2O	80
	Formic acid	HCO_2H	59
	Dimethyl sulfoxide (DMSO)	CH_3SOCH_3	49
	N,N-Dimethylformamide (DMF)	$HCON(CH_3)_2$	37
	Acetonitrile	$CH_3C{\equiv}N$	36
	Methanol	CH_3OH	33
	Hexamethylphosphoramide (HMPA)	$[(CH_3)_2N]_3P{=}O$	30
	Ethanol	CH_3CH_2OH	24
	Acetone	CH_3COCH_3	21
	Acetic acid	CH_3CO_2H	6

PROBLEM 6.10

When *tert*-butyl bromide undergoes solvolysis in a mixture of methanol and water, the rate of solvolysis (measured by the rate at which bromide ions form in the mixture) *increases* when the percentage of water in the mixture is increased. **(a)** Explain this occurrence. **(b)** Provide an explanation for the observation that the rate of the S_N2 reaction of ethyl chloride with potassium iodide in methanol and water *decreases* when the percentage of water in the mixture is increased.

6.13E The Nature of the Leaving Group

Leaving groups depart with the electron pair that was used to bond them to the substrate. The best leaving groups are those that become either a relatively stable anion or a neutral molecule when they depart. First, let us consider leaving groups that become anions when they separate from the substrate. Because weak bases stabilize a negative charge effectively, leaving groups that become weak bases are good leaving groups. In general, the best leaving groups are those that can be classified as weak bases after they depart.

Good leaving groups are weak bases.

The reason that stabilization of the negative charge is important can be understood by considering the structure of the transition states. In either an S_N1 or S_N2 reaction the leaving group begins to acquire a negative charge as the transition state is reached:

S_N1 Reaction (rate-limiting step)

$$\text{C—X} \longrightarrow \left[\overset{\delta+}{\text{C}}\text{---}\overset{\delta-}{\text{X}}\right]^{\ddagger} \longrightarrow \text{C}^{+} + \text{X}^{-}$$

Transition state

S_N2 Reaction

$$\text{Nu:}^{-} + \text{C—X} \longrightarrow \left[\overset{\delta-}{\text{Nu}}\text{---C---}\overset{\delta-}{\text{X}}\right]^{\ddagger} \longrightarrow \text{Nu—C} + \text{X}^{-}$$

Transition state

Stabilization of this developing negative charge at the leaving group stabilizes the transition state (lowers its free energy); this lowers the free energy of activation and thereby increases the rate of the reaction. Of the halogens, an iodide ion is the best leaving group and a fluoride ion is the poorest:

$$I^- > Br^- > Cl^- \gg F^-$$

The order is the opposite of the basicity:

$$F^- \gg Cl^- > Br^- > I^-$$

Other weak bases that are good leaving groups that we shall study later are alkanesulfonate ions, alkyl sulfate ions, and the *p*-toluenesulfonate ion:

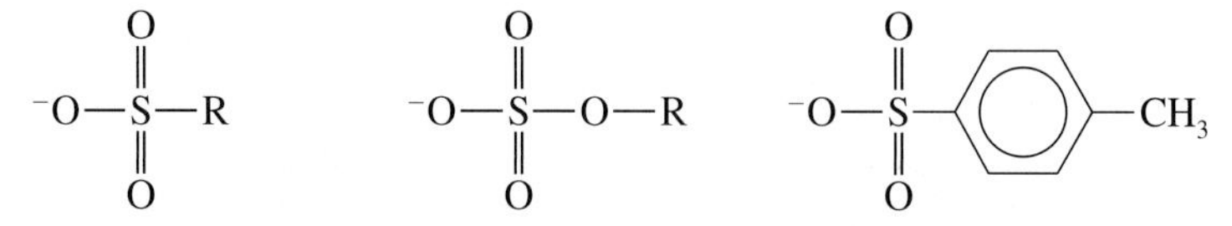

$^-O-S(=O)_2-R$ **An alkanesulfonate ion**

$^-O-S(=O)_2-O-R$ **An alkyl sulfate ion**

$^-O-S(=O)_2-C_6H_4-CH_3$ ***p*-Toluenesulfonate ion**

These anions are all the conjugate bases of very strong acids.

The trifluoromethanesulfonate ion ($CF_3SO_3^-$, commonly called the **triflate ion**) is one of the best leaving groups known to chemists. It is the anion of CF_3SO_3H, an exceedingly strong acid ($pK_a \sim -5$ to -6):

$$CF_3SO_3^-$$

Triflate ion
(a "super" leaving group)

Strongly basic ions rarely act as leaving groups. The hydroxide ion, for example, is a strong base and thus reactions like the following do not take place:

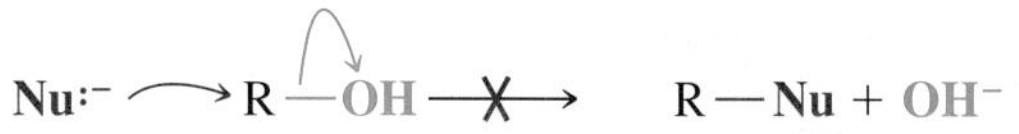

$$Nu{:}^- \; + \; R-OH \not\longrightarrow R-Nu + OH^-$$

This reaction does not take place because the leaving group is a strongly basic hydroxide ion.

However, when an alcohol is dissolved in a strong acid, it can undergo substitution by a nucleophile. Because the acid protonates the —OH group of the alcohol, the leaving group no longer needs to be a hydroxide ion; it is now a molecule of water—a much weaker base than a hydroxide ion and a good leaving group:

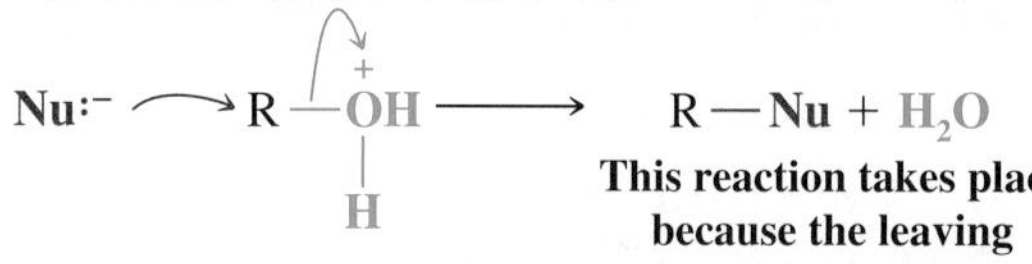

$$Nu{:}^- \; + \; R-\overset{+}{O}H_2 \longrightarrow R-Nu + H_2O$$

This reaction takes place because the leaving group is a weak base.

PROBLEM 6.11

List the following compounds in order of decreasing reactivity toward CH_3O^- in an S_N2 reaction carried out in CH_3OH: CH_3F, CH_3Cl, CH_3Br, CH_3I, $CH_3OSO_2CF_3$, $^{14}CH_3OH$.

Very powerful bases such as hydride ions ($H{:}^-$) and alkanide ions ($R{:}^-$) virtually never act as leaving groups. Therefore, **reactions such as the following are not feasible:**

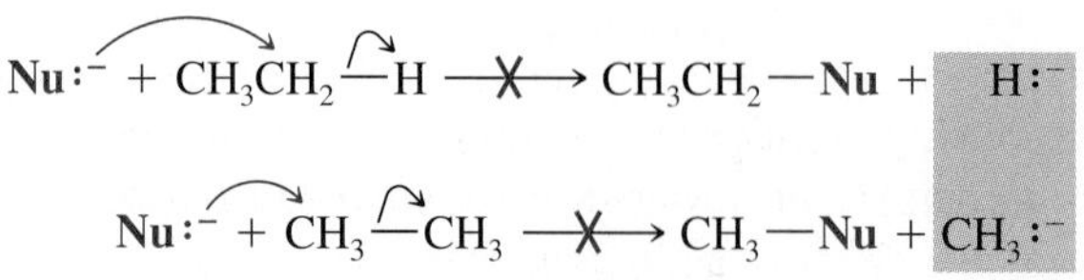

$$Nu{:}^- + CH_3CH_2-H \not\longrightarrow CH_3CH_2-Nu + H{:}^-$$

$$Nu{:}^- + CH_3-CH_3 \not\longrightarrow CH_3-Nu + CH_3{:}^-$$

These are not leaving groups.

6.13F Summary: S_N1 versus S_N2

S_N1 versus S_N2

Reactions of alkyl halides by an S_N1 mechanism are favored by the use of substrates that can form relatively stable carbocations, by the use of weak nucleophiles, and by the use of highly ionizing solvents. S_N1 mechanisms, therefore, are important in solvolysis reactions of tertiary halides, especially when the solvent is highly polar. In a solvolysis the nucleophile is weak because it is a neutral molecule (of the solvent) rather than an anion.

If we want to favor the reaction of an alkyl halide by an S_N2 mechanism, we should use a relatively unhindered alkyl halide, a strong nucleophile, a polar aprotic solvent, and a high concentration of the nucleophile. For substrates, the order of reactivity in S_N2 reactions is

$$\underset{\textbf{Methyl}}{CH_3{-}X} > \underset{\textbf{1°}}{R{-}CH_2{-}X} > \underset{\textbf{2°}}{R{-}\overset{\displaystyle R}{\overset{|}{C}}H{-}X}$$

Tertiary halides do not react by an S_N2 mechanism.

The effect of the leaving group is the same in both S_N1 and S_N2 reactions: alkyl iodides react fastest; fluorides react slowest. (Because alkyl fluorides react so slowly, they are seldom used in nucleophilic substitution reactions.)

$$R{-}I > R{-}Br > R{-}Cl \qquad \mathbf{S_N1 \ \text{or} \ S_N2}$$

These factors are summarized in Table 6.6.

TABLE 6.6 Factors Favoring S_N1 versus S_N2 Reactions

Factor	S_N1	S_N2
Substrate	3° (requires formation of a relatively stable carbocation)	Methyl > 1° > 2° (requires unhindered substrate)
Nucleophile	Weak Lewis base, neutral molecule, nucleophile may be the solvent (solvolysis)	Strong Lewis base, rate favored by high concentration of nucleophile
Solvent	Polar protic (e.g., alcohols, water)	Polar aprotic (e.g., DMF, DMSO)
Leaving group	I > Br > Cl > F for both S_N1 and S_N2 (the weaker the base after the group departs, the better the leaving group)	

6.14 Organic Synthesis: Functional Group Transformations Using S_N2 Reactions

Methods for functional group preparation

In Chapter 4 we had an introduction to the synthesis of organic molecules and retrosynthetic analysis. Now that we have studied nucleophilic substitution reactions, these reactions give us new tools to add to our toolbox.

S_N2 reactions are highly useful in organic synthesis because they enable us to convert one functional group into another—a process that is called a **functional group transformation** or a **functional group interconversion.** With the S_N2 reactions shown in Fig. 6.12, the functional group of a methyl, primary, or secondary alkyl halide can be transformed into that of an alcohol, ether, thiol, thioether, nitrile, ester, and so on. (*Note:* The use of the prefix *thio-* in a name means that a sulfur atom has replaced an oxygen atom in the compound.) In Section 4.18C we saw how carbon–carbon bonds can be formed by alkylation of alkynide anions. This was an S_N2 reaction, too.

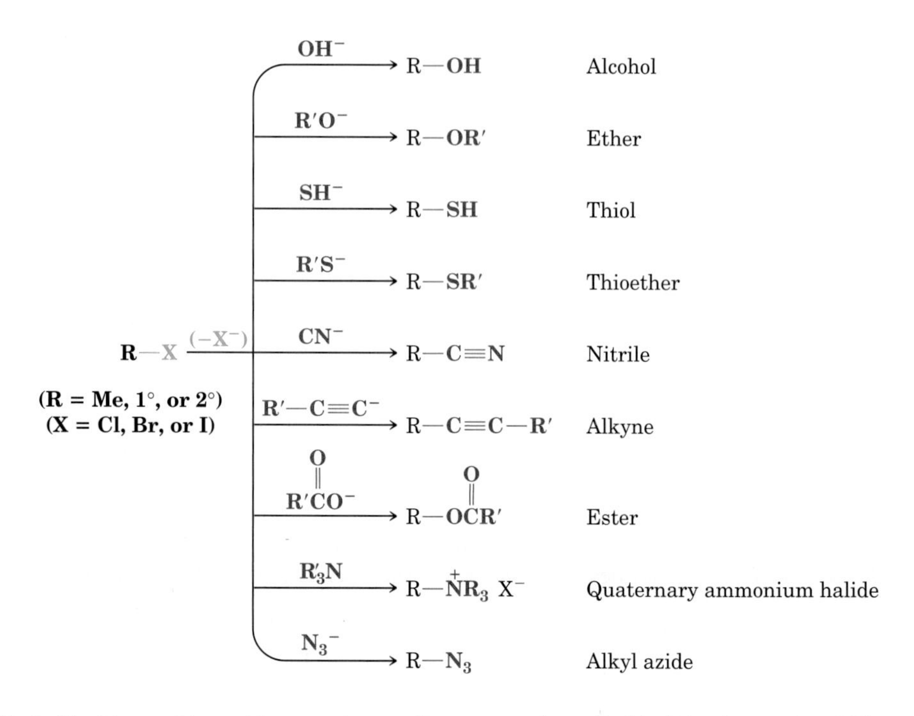

FIGURE 6.12 Functional group interconversions of methyl, primary, and secondary alkyl halides using S_N2 reactions.

Alkyl chlorides and bromides are also easily converted to alkyl iodides by nucleophilic substitution reactions.

$$\begin{matrix} R{-}Cl \\ \text{or} \\ R{-}Br \end{matrix} \xrightarrow{I^-} R{-}I\ (+\ Cl^-\ \text{or}\ Br^-)$$

The Chemistry of...

Biological Methylation: A Biological Nucleophilic Substitution Reaction

The cells of living organisms synthesize many of the compounds they need from smaller molecules. Often these biosyntheses resemble the syntheses organic chemists carry out in their laboratories. Let us examine one example now.

Many reactions taking place in the cells of plants and animals involve the transfer of a methyl group from an amino acid called methionine to some other compound. That this transfer takes place can be demonstrated experimentally by feeding a plant or animal methionine containing an isotopically labeled carbon atom (e.g., ^{13}C or ^{14}C) in its methyl group. Later, other compounds containing the "labeled" methyl group can be isolated from the organism. Some of the compounds that get their methyl groups from methionine are the following. The isotopically labeled carbon atom is shown in green.

$$^{-}O_2C\underset{\underset{NH_3^+}{|}}{C}HCH_2CH_2SCH_3$$

Methionine

Nicotine **Adrenaline** **Choline**

$$CH_3{-}\overset{\overset{CH_3}{|}}{\underset{\underset{CH_3}{|}}{N^+}}{-}CH_2CH_2OH$$

Choline is important in the transmission of nerve impulses, adrenaline causes blood pressure to increase, and nicotine is the compound contained in tobacco that makes smoking tobacco addictive. (In large doses nicotine is poisonous.)

The transfer of the methyl group from methionine to these other compounds does not take place directly. The actual methylating agent is not methionine; it is *S*-adenosylmethionine,* a compound that results when methionine reacts with adenosine triphosphate (ATP):

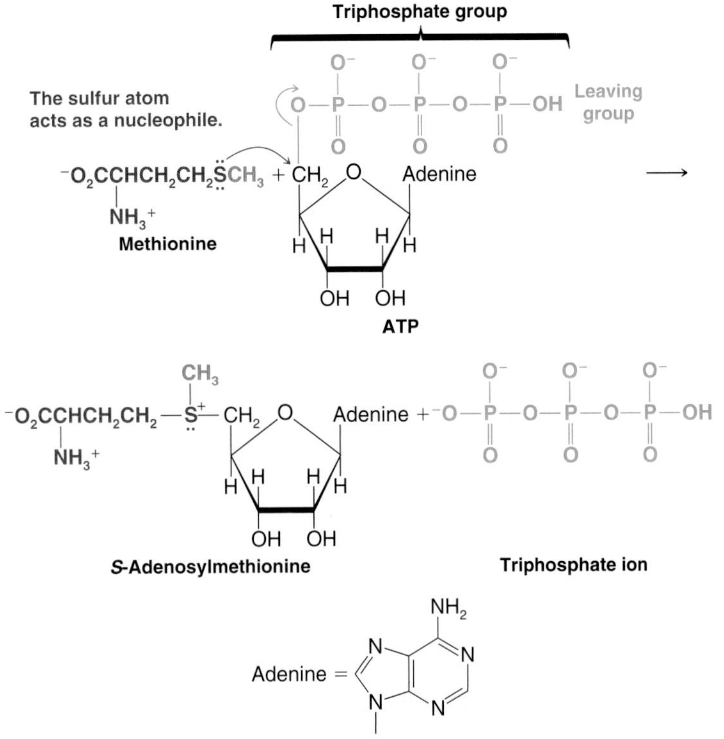

Triphosphate group

The sulfur atom acts as a nucleophile.

Leaving group

$^{-}O_2CCH(NH_3^+)CH_2CH_2\ddot{S}CH_3 + CH_2(\text{ribose–Adenine})–O–P(O)(O^-)–O–P(O)(O^-)–O–P(O)(O^-)–OH \longrightarrow$

Methionine **ATP**

$^{-}O_2CCH(NH_3^+)CH_2CH_2–\overset{CH_3}{S^+}–CH_2(\text{ribose–Adenine}) + {}^{-}O–P(O)(O^-)–O–P(O)(O^-)–O–P(O)(O^-)–OH$

***S*-Adenosylmethionine** **Triphosphate ion**

Adenine = (purine ring with NH_2, attached through N)

This reaction is a nucleophilic substitution reaction. The nucleophilic atom is the sulfur atom of methionine. The leaving group is the weakly basic triphosphate group of ATP. The product, *S*-adenosylmethionine, contains a methylsulfonium group, $CH_3–\overset{|}{S^+}–$.

S-Adenosylmethionine then acts as the substrate for other nucleophilic substitution reactions. In the biosynthesis of choline, for example, it transfers its methyl group to a nucleophilic nitrogen atom of 2-(*N,N*-dimethylamino)ethanol:

$CH_3–\ddot{N}(CH_3)–CH_2CH_2OH + {}^{-}O_2CCH(NH_3^+)CH_2CH_2–\overset{CH_3}{S^+}–CH_2(\text{ribose–Adenine}) \longrightarrow$

2-(*N, N*-Dimethylamino)ethanol

$CH_3–\overset{CH_3}{N^+}(CH_3)–CH_2CH_2OH + {}^{-}O_2CCH(NH_3^+)CH_2CH_2–\ddot{S}–CH_2(\text{ribose–Adenine})$

Choline

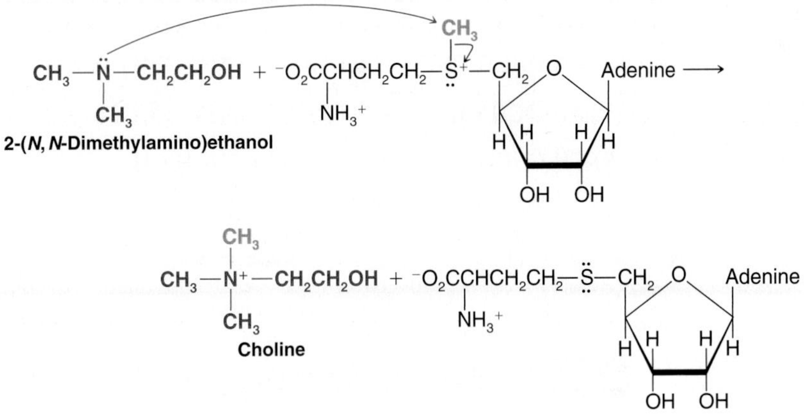

*The prefix *S* is a locant meaning "on the sulfur atom" and should not be confused with the (*S*) used to define absolute configuration. Another example of this kind of locant is *N*, meaning "on the nitrogen atom."

These reactions appear complicated only because the structures of the nucleophiles and substrates are complex. Yet conceptually they are simple, and they illustrate many of the principles we have encountered thus far in Chapter 6. In them we see how nature makes use of the high nucleophilicity of sulfur atoms. We also see how a weakly basic group (e.g., the triphosphate group of ATP) functions as a leaving group. In the reaction of 2-(*N,N*-dimethylamino)ethanol we see that the more basic $(CH_3)_2N—$ group acts as the nucleophile rather than the less basic —OH group. And when a nucleophile attacks *S*-adenosylmethionine, we see that the attack takes place at the less hindered $CH_3—$ group rather than at one of the more hindered $—CH_2—$ groups.

STUDY PROBLEM

(a) What is the leaving group when 2-(*N,N*-dimethylamino)ethanol reacts with *S*-adenosylmethionine? **(b)** What would the leaving group have to be if methionine itself were to react with 2-(*N,N*-dimethylamino)ethanol? **(c)** Of what special significance is this difference?

One other aspect of the S_N2 reaction that is of great importance is **stereochemistry** (Section 6.8). S_N2 reactions always occur **with inversion of configuration** at the atom that bears the leaving group. This means that when we use S_N2 reactions in syntheses we can be sure of the configuration of our product if we know the configuration of our reactant. For example, suppose we need a sample of the following nitrile with the (*S*) configuration:

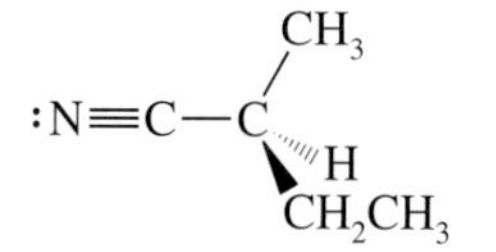

$$:N{\equiv}C{-}C(CH_3)(H)(CH_2CH_3)$$

(*S*)-2-Methylbutanenitrile

If we have available (*R*)-2-bromobutane, we can carry out the following synthesis:

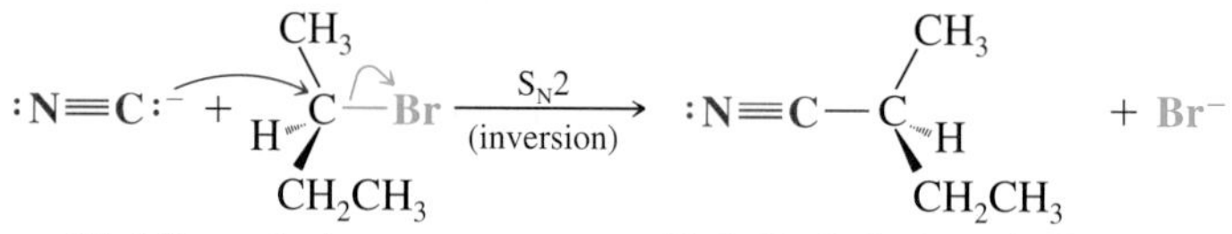

$$:N{\equiv}C:^- + H{-}C(CH_3)(CH_2CH_3){-}Br \xrightarrow[\text{(inversion)}]{S_N2} :N{\equiv}C{-}C(CH_3)(H)(CH_2CH_3) + Br^-$$

(*R*)-2-Bromobutane **(*S*)-2-Methylbutanenitrile**

PROBLEM 6.12

Starting with (*S*)-2-bromobutane, outline syntheses of each of the following compounds:

(a) (*R*)-$CH_3CH(OCH_2CH_3)CH_2CH_3$

(b) (*R*)-$CH_3CH(OC(=O)CH_3)CH_2CH_3$

(c) (*R*)-$CH_3CH(SH)CH_2CH_3$

(d) (*R*)-$CH_3CH(SCH_3)CH_2CH_3$

6.14A The Unreactivity of Vinylic and Phenyl Halides

As we learned in Section 6.1, compounds that have a halogen atom attached to one carbon atom of a double bond are called **vinylic halides;** those that have a halogen atom attached to a benzene ring are called **phenyl halides:**

$$\begin{array}{cc} \text{>C=C<X} & \text{C}_6\text{H}_5\text{—X} \\ \textbf{A vinylic halide} & \textbf{A phenyl halide} \end{array}$$

Vinylic and phenyl halides are generally unreactive in S_N1 or S_N2 reactions. They are unreactive in S_N1 reactions because vinylic and phenyl cations are highly unstable and do not form readily. They are unreactive in S_N2 reactions because the carbon–halogen bond of a vinylic or phenyl halide is stronger than that of an alkyl halide (we shall see why later), and the electrons of the double bond or benzene ring repel the approach of a nucleophile from the back side.

6.15 Elimination Reactions of Alkyl Halides

Elimination reactions of alkyl halides are important reactions that compete with substitution reactions. In an **elimination reaction** the fragments of some molecule (YZ) are removed (eliminated) from adjacent atoms of the reactant. This elimination leads to the introduction of a multiple bond:

$$-\underset{\text{Y}}{\overset{|}{\text{C}}}-\underset{\text{Z}}{\overset{|}{\text{C}}}- \xrightarrow[(-\text{YZ})]{\text{elimination}} \text{>C=C<}$$

6.15A Dehydrohalogenation

A widely used method for synthesizing alkenes is the elimination of HX from adjacent atoms of an alkyl halide. Heating the alkyl halide with a strong base causes the reaction to take place. The following are two examples:

$$CH_3\underset{\text{Br}}{CH}CH_3 \xrightarrow[C_2H_5OH,\ 55°C]{C_2H_5ONa} \underset{(79\%)}{CH_2{=}CH{-}CH_3} + NaBr + C_2H_5OH$$

$$CH_3-\underset{CH_3}{\overset{CH_3}{C}}-Br \xrightarrow[C_2H_5OH,\ 55°C]{C_2H_5ONa} \underset{(91\%)}{CH_3\overset{CH_3}{C}{=}CH_2} + NaBr + C_2H_5OH$$

Reactions like these are not limited to the elimination of hydrogen bromide. Chloroalkanes also undergo the elimination of hydrogen chloride, iodoalkanes undergo the elimination of hydrogen iodide, and, in all cases, alkenes are produced. When the elements of a hydrogen halide are eliminated from a haloalkane in this way, the reaction is often called **dehydrohalogenation:**

$$-\overset{\text{H}}{\underset{|}{\text{C}^{\beta}}}-\underset{\text{X}}{\overset{|}{\text{C}^{\alpha}}}- + \underset{\textbf{A base}}{:\text{B}^-} \longrightarrow \text{>C=C<} + \text{H:B} + :\text{X}^-$$

Dehydrohalogenation

In these eliminations, as in S_N1 and S_N2 reactions, there is a leaving group and an attacking Lewis base that possesses an electron pair.

Chemists often call the carbon atom that bears the substituent (e.g., the halogen atom in the previous reaction) the **alpha (α) carbon atom** and any carbon atom adjacent to it a **beta (β) carbon atom.** A hydrogen atom attached to the β carbon atom is called a **β hydrogen atom.** Since the hydrogen atom that is eliminated in dehydrohalogenation is from the β carbon atom, these reactions are often called **β eliminations.** They are also often referred to as **1,2 eliminations.**

We shall have more to say about dehydrohalogenation in Chapter 7, but we can examine several important aspects here.

6.15B Bases Used in Dehydrohalogenation

Various strong bases have been used for dehydrohalogenations. Potassium hydroxide dissolved in ethanol is a reagent sometimes used, but the sodium salts of alcohols, such as sodium ethoxide, often offer distinct advantages.

The sodium salt of an alcohol (a sodium alkoxide) can be prepared by treating an alcohol with sodium metal:

$$2\,R{-}\ddot{\underset{..}{O}}H + 2\,Na \longrightarrow 2\,R{-}\ddot{\underset{..}{O}}{:}^{-}\,Na^{+} + H_2$$

Alcohol **Sodium alkoxide**

This reaction is an **oxidation–reduction reaction.** Metallic sodium, an alkali metal, reacts with hydrogen atoms that are bonded to oxygen atoms to generate hydrogen gas, sodium cations, and the hydroxide anion. The reaction is vigorous and at times explosive.

$$2\,H\ddot{\underset{..}{O}}H + 2\,Na \longrightarrow 2\,H\ddot{\underset{..}{O}}{:}^{-}Na^{+} + H_2$$

Sodium hydroxide

Sodium alkoxides can also be prepared by allowing an alcohol to react with sodium hydride (NaH). The hydride ion ($H{:}^{-}$) is a very strong base. (The pK_a of H_2 is 36.)

$$R{-}\ddot{\underset{..}{O}}{-}H + Na^{+}{:}H^{-} \longrightarrow R{-}\ddot{\underset{..}{O}}{:}^{-}Na^{+} + H{-}H$$

Sodium (and potassium) alkoxides are usually prepared by using an excess of the alcohol, and the excess alcohol becomes the solvent for the reaction. Sodium ethoxide is frequently prepared in this way.

$$2\,CH_3CH_2\ddot{\underset{..}{O}}H + 2\,Na \longrightarrow 2\,CH_3CH_2\ddot{\underset{..}{O}}{:}^{-}Na^{+} + H_2$$

Ethanol (excess) **Sodium ethoxide dissolved in excess ethanol**

Potassium *tert*-butoxide is another highly effective dehydrohalogenating reagent.

$$2\,CH_3\overset{CH_3}{\underset{CH_3}{C}}{-}\ddot{\underset{..}{O}}H + 2\,K \longrightarrow 2\,CH_3\overset{CH_3}{\underset{CH_3}{C}}{-}\ddot{\underset{..}{O}}{:}^{-}K^{+} + H_2$$

***tert*-Butyl alcohol (excess)** **Potassium *tert*-butoxide**

6.15C Mechanisms of Dehydrohalogenations

Elimination reactions occur by a variety of mechanisms. With alkyl halides, two mechanisms are especially important because they are closely related to the S_N2 and S_N1 reactions that we have just studied. One mechanism, called the **E2 reaction,** is bimolecular in the rate-determining step; the other mechanism is the **E1 reaction,** which is unimolecular in the rate-determining step.

6.16 The E2 Reaction

When isopropyl bromide is heated with sodium ethoxide in ethanol to form propene, the reaction rate depends on the concentration of isopropyl bromide and the concentration of ethoxide ion. The rate equation is first order in each reactant and second order overall:

$$\text{Rate} \propto [CH_3CHBrCH_3][C_2H_5O^-]$$
$$\text{Rate} = k[CH_3CHBrCH_3][C_2H_5O^-]$$

From this we infer that the transition state for the rate-determining step must involve both the alkyl halide and the alkoxide ion. The reaction must be bimolecular. Considerable experimental evidence indicates that the reaction takes place in the following way:

A Mechanism for the E2 Reaction

Reaction:

$$C_2H_5O^- + CH_3CHBrCH_3 \longrightarrow CH_2{=}CHCH_3 + C_2H_5OH + Br^-$$

CD Tutorial

E2 elimination mechanism

Mechanism:

$C_2H_5{-}\ddot{O}{:}^-$ H, CH_3, H, C—C (β α), H H, Br: $\longrightarrow$ $\left[C_2H_5{-}\overset{\delta-}{\ddot{O}}{\cdots}H \;\; CH_3 \;\; H \;\; C{\cdots}C \; (\beta\ \alpha) \;\; H\ H \;\; Br{:}^{\delta-} \right]^{\ddagger}$ $\longrightarrow$

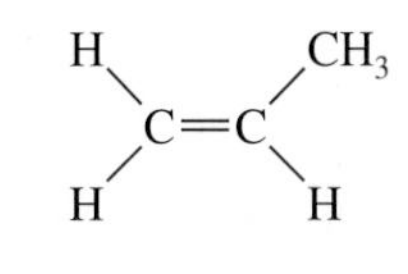

$+$

$C_2H_5{-}\ddot{O}H + {:}\ddot{Br}{:}^-$

Transition state

The basic ethoxide ion begins to remove a proton from the β carbon using its electron pair to form a bond to it. At the same time, the electron pair of the β C—H bond begins to move in to become the π bond of a double bond, and the bromine begins to depart with the electrons that bonded it to the α carbon

Partial bonds in the transition state extend from the oxygen atom that is removing the β hydrogen, through the carbon skeleton of the developing double bond, to the departing leaving group. The flow of electron density is from the base toward the leaving group as an electron pair fills the π bonding orbital of the alkene.

At completion of the reaction, the double bond is fully formed and the alkene has a trigonal planar geometry at each carbon atom. The other products are a molecule of ethanol and a bromide ion.

When we study the E2 reaction further in Section 7.6C, we shall find that the orientations of the hydrogen atom being removed and the leaving group are not arbitrary and that the orientation where they are all in the same plane, as shown above, is required.

6.17 The E1 Reaction

Eliminations may take a different pathway from that given in Section 6.16. Treating *tert*-butyl chloride with 80% aqueous ethanol at 25°C, for example, gives *substitution products* in 83% yield and an elimination product (2-methylpropene) in 17% yield:

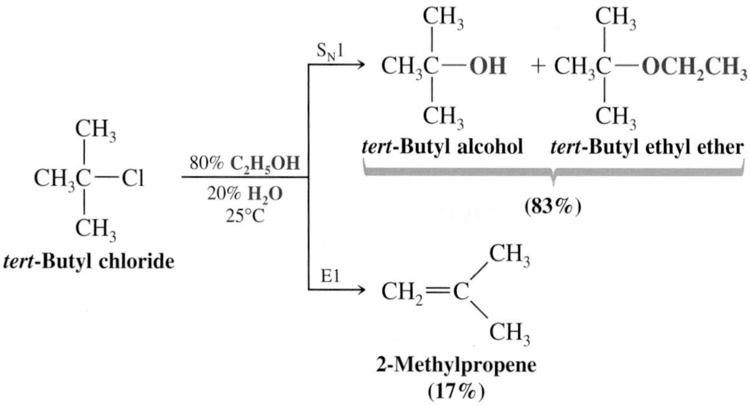

$$(CH_3)_3C{-}Cl \xrightarrow[20\%\ H_2O,\ 25°C]{80\%\ C_2H_5OH} \begin{cases} \xrightarrow{S_N1} (CH_3)_3C{-}OH + (CH_3)_3C{-}OCH_2CH_3 \\ \xrightarrow{E1} CH_2{=}C(CH_3)_2 \end{cases}$$

***tert*-Butyl chloride** → ***tert*-Butyl alcohol** + ***tert*-Butyl ethyl ether** **(83%)**; **2-Methylpropene (17%)**

The initial step for both reactions is the formation of a *tert*-butyl cation. This is also the rate-determining step for both reactions; thus both reactions are unimolecular:

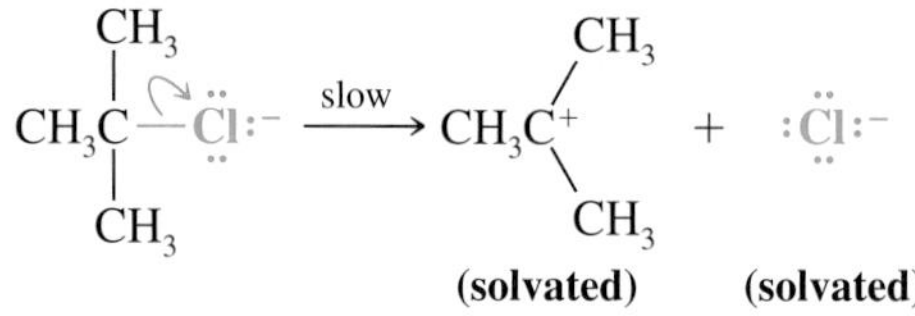

$$CH_3C(CH_3)_2{-}\ddot{Cl}{:} \xrightarrow{slow} CH_3C^+(CH_3)_2 + {:}\ddot{Cl}{:}^-$$

(solvated) **(solvated)**

Whether substitution or elimination takes place depends on the next step (the fast step). If a solvent molecule reacts as a nucleophile at the positive carbon atom of the *tert*-butyl cation, the product is *tert*-butyl alcohol or *tert*-butyl ethyl ether and the reaction is S_N1:

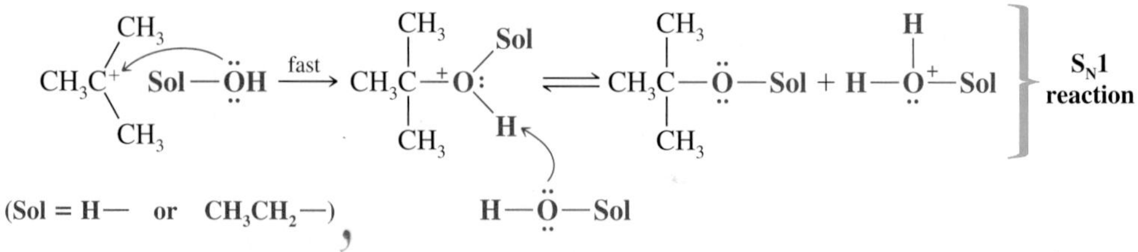

$$CH_3C^+(CH_3)_2 + Sol{-}\ddot{O}H \xrightarrow{fast} CH_3C(CH_3)_2{-}\overset{+}{O}(H){-}Sol \xrightleftharpoons{H{-}\ddot{O}{-}Sol} CH_3C(CH_3)_2{-}\ddot{O}{-}Sol + H{-}\overset{+}{O}(H){-}Sol$$

S_N1 reaction

$(Sol = H{-}$ or $CH_3CH_2{-})$,

If, however, a solvent molecule acts as a base and removes one of the β hydrogen atoms as a proton, the product is 2-methylpropene and the reaction is E1.

E1 reactions almost always accompany S_N1 reactions.

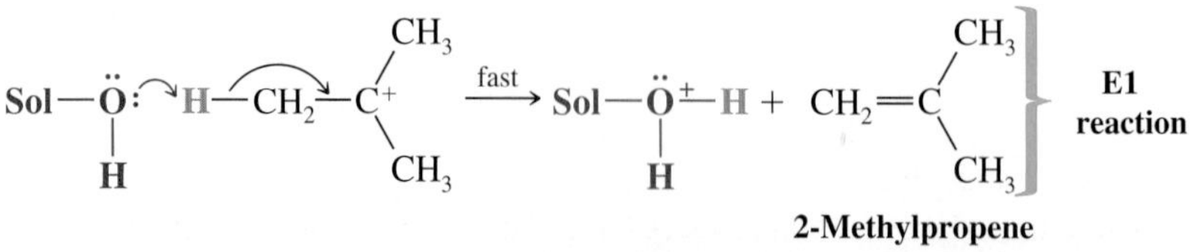

$$Sol{-}\ddot{O}(H){:} + H{-}CH_2{-}C^+(CH_3)_2 \xrightarrow{fast} Sol{-}\overset{+}{O}(H){-}H + CH_2{=}C(CH_3)_2$$

2-Methylpropene — **E1 reaction**

A Mechanism for the E1 Reaction

Reaction:

$$(CH_3)_3CCl + \mathbf{H_2O} \longrightarrow CH_2{=}C(CH_3)_2 + \mathbf{H_3O^+} + Cl^-$$

Mechanism:

Step 1

$$H_3C{-}C(CH_3)_2{-}\ddot{\underset{..}{Cl}}: \xrightarrow[H_2O]{\text{slow}} H_3C{-}\overset{+}{C}(CH_3)_2 + :\ddot{\underset{..}{Cl}}:^-$$

Aided by the polar solvent, a chlorine departs with the electron pair that bonded it to the carbon.

This slow step produces the relatively stable 3° carbocation and a chloride ion. The ions are solvated (and stabilized) by surrounding water molecules.

Step 2

$$H{-}\ddot{O}{:}(H) + H{-}\underset{\beta}{C}H_2{-}\underset{\alpha}{\overset{+}{C}}(CH_3)_2 \longrightarrow H{-}\overset{+}{\ddot{O}}(H){-}H + H_2C{=}C(CH_3)_2$$

A molecule of water removes one of the hydrogens from the β carbon of the carbocation. These hydrogens are acidic due to the adjacent positive charge. At the same time an electron pair moves in to form a double bond between the α and β carbon atoms.

This step produces the alkene and a hydronium ion.

6.18 Substitution versus Elimination

All nucleophiles are potential bases and all bases are potential nucleophiles. This is because the reactive part of both nucleophiles and bases is an unshared electron pair. It should not be surprising, then, that nucleophilic substitution reactions and elimination reactions often compete with each other.

This section draws together the various factors that influence the competition between substitution and elimination.

6.18A S_N2 versus E2

S_N2 and E2 reactions are both favored by a high concentration of a strong nucleophile or base. When the nucleophile (base) attacks a β hydrogen atom, elimination occurs. When the nucleophile attacks the carbon atom bearing the leaving group, substitution results:

$$\mathrm{Nu:^-} + \mathrm{H{-}\overset{|}{C}{-}\overset{|}{C}{-}X} \xrightarrow[\text{elimination E2}]{(a)} \mathrm{C{=}C} + \mathrm{Nu{-}H} + \mathrm{:X^-}$$

$$\mathrm{Nu:^-} + \mathrm{H{-}\overset{|}{C}{-}\overset{|}{C}{-}X} \xrightarrow[\text{substitution } S_N2]{(b)} \mathrm{H{-}\overset{|}{C}{-}\overset{|}{C}{-}Nu} + \mathrm{X:^-}$$

Consider the following examples with small (unhindered) nucleophiles and alkyl halides of different classes.

Primary Substrate When the substrate is a *primary* halide and the base is unhindered, like ethoxide ion, substitution is highly favored because the base can easily approach the carbon bearing the leaving group:

$$CH_3CH_2O^-Na^+ + CH_3CH_2Br \xrightarrow[55°C\ (-NaBr)]{C_2H_5OH} \underset{\substack{S_N2\\(90\%)}}{CH_3CH_2OCH_2CH_3} + \underset{\substack{E2\\(10\%)}}{CH_2{=}CH_2}$$

Secondary Substrate With *secondary* halides, however, a strong base favors elimination because steric hindrance in the substrate makes substitution more difficult:

$$CH_3CH_2O^-Na^+ + CH_3\underset{Br}{\underset{|}{C}}HCH_3 \xrightarrow[55°C\ (-NaBr)]{C_2H_5OH} \underset{\substack{S_N2\\(21\%)}}{CH_3\underset{OCH_2CH_3}{\underset{|}{C}}HCH_3} + \underset{\substack{E2\\(79\%)}}{CH_2{=}CHCH_3}$$

Tertiary Substrate With *tertiary* halides, steric hindrance in the substrate is severe and an S_N2 reaction cannot take place. Elimination is highly favored, especially when the reaction is carried out at higher temperatures. Any substitution that occurs must take place through an S_N1 mechanism:

$$CH_3CH_2O^-Na^+ + CH_3\underset{Br}{\underset{|}{\overset{CH_3}{\overset{|}{C}}}}CH_3 \xrightarrow[55°C\ (-NaBr)]{C_2H_5OH} \underset{\substack{S_N1\\(9\%)}}{CH_3\underset{OCH_2CH_3}{\underset{|}{\overset{CH_3}{\overset{|}{C}}}}CH_3} + \underset{\substack{\text{Mainly E2}\\(91\%)}}{CH_2{=}\overset{CH_3}{\overset{|}{C}}CH_3}$$

$$CH_3CH_2O^-Na^+ + CH_3\underset{Br}{\underset{|}{\overset{CH_3}{\overset{|}{C}}}}CH_3 \xrightarrow[55°C\ (-NaBr)]{C_2H_5OH} \underset{\substack{E2 + E1\\(100\%)}}{CH_2{=}\overset{CH_3}{\overset{|}{C}}CH_3 + CH_3CH_2OH}$$

Temperature Increasing the reaction temperature favors elimination (E1 and E2) over substitution. Elimination reactions have greater free energies of activation than substitution reactions because more bonding changes occur during elimination. When higher temperature is used, the proportion of molecules able to surmount the energy of activation barrier for elimination increases more than the proportion of molecules able to undergo substitution, although the rate of both substitution and elimination will be increased.

Furthermore, elimination reactions are entropically favored over substitution because the products of an elimination reaction are greater in number than the reactants. Additionally, because temperature is the coefficient of the entropy term in the Gibbs free-energy equation $\Delta G° = \Delta H° - T\,\Delta S°$, an increase in temperature further enhances the entropy effect.

Size of the Base/Nucleophile Increasing the reaction temperature is one way of favorably influencing an elimination reaction of an alkyl halide. Another way is to use a **strong sterically hindered base** such as the *tert*-butoxide ion. The bulky methyl groups of the *tert*-butoxide ion inhibit its reacting by substitution, allowing elimination reactions to take precedence. We can see an example of this effect in the following two reactions. The relatively unhindered methoxide ion reacts with octadecyl bromide primarily by *substitution;* the bulky *tert*-butoxide ion gives mainly *elimination.*

Unhindered (small) Base/Nucleophile

$$CH_3O^- + CH_3(CH_2)_{15}CH_2CH_2{-}Br \xrightarrow[65°C]{CH_3OH} \underset{\substack{\textbf{E2} \\ \textbf{(1\%)}}}{CH_3(CH_2)_{15}CH{=}CH_2} + \underset{\substack{\textbf{S}_\textbf{N}\textbf{2} \\ \textbf{(99\%)}}}{CH_3(CH_2)_{15}CH_2CH_2OCH_3}$$

Hindered Base/Nucleophile

$$CH_3{-}\overset{\overset{\large CH_3}{|}}{\underset{\underset{\large CH_3}{|}}{C}}{-}O^- + CH_3(CH_2)_{15}CH_2CH_2{-}Br \xrightarrow[40°C]{(CH_3)_3COH}$$

$$\underset{\substack{\textbf{E2} \\ \textbf{(85\%)}}}{CH_3(CH_2)_{15}CH{=}CH_2} + \underset{\substack{\textbf{S}_\textbf{N}\textbf{2} \\ \textbf{(15\%)}}}{CH_3(CH_2)_{15}CH_2CH_2{-}O{-}\overset{\overset{\large CH_3}{|}}{\underset{\underset{\large CH_3}{|}}{C}}{-}CH_3}$$

Basicity and Polarizability Another factor that affects the relative rates of E2 and S_N2 reactions is the relative basicity and polarizability of the base/nucleophile. Use of a strong, slightly polarizable base such as amide ion (NH_2^-) or alkoxide ion (especially a hindered one) tends to increase the likelihood of elimination (E2). Use of a weakly basic ion such as a chloride ion (Cl^-) or an acetate ion ($CH_3CO_2^-$) or a weakly basic and highly polarizable one such as Br^-, I^-, or RS^- increases the likelihood of substitution (S_N2). Acetate ion, for example, reacts with isopropyl bromide almost exclusively by the S_N2 path:

$$CH_3\overset{\overset{\large O}{\|}}{C}{-}O^- + CH_3\overset{\overset{\large CH_3}{|}}{CH}{-}Br \xrightarrow[\substack{S_N2 \\ (\sim 100\%)}]{} CH_3\overset{\overset{\large O}{\|}}{C}{-}O{-}\overset{\overset{\large CH_3}{|}}{C}HCH_3 + Br^-$$

The more strongly basic ethoxide ion (Section 6.15B) reacts with the same compound mainly by an E2 mechanism.

6.18B Tertiary Halides: S_N1 versus E1

Because E1 and S_N1 reactions proceed through the formation of a common intermediate, the two types respond in similar ways to factors affecting reactivities. E1 reactions are favored with substrates that can form stable carbocations (i.e., tertiary halides); they are also favored by the use of poor nucleophiles (weak bases) and they are generally favored by the use of polar solvents.

It is usually difficult to influence the relative partition between S_N1 and E1 products because the free energy of activation for either reaction proceeding from the carbocation (loss of a proton or combination with a molecule of the solvent) is very small.

In most unimolecular reactions the S_N1 reaction is favored over the E1 reaction, especially at lower temperatures. *In general, however, substitution reactions of tertiary halides do not find wide use as synthetic methods. Such halides undergo eliminations much too easily.*

Increasing the temperature of the reaction favors reaction by the E1 mechanism at the expense of the S_N1 mechanism. ***If the elimination product is desired, however, it is more convenient to add a strong base and force an E2 reaction to take place instead.***

6.19 Overall Summary

The most important reaction pathways for the substitution and elimination reactions of simple alkyl halides can be summarized in the way shown in Table 6.7. Let us examine several sample exercises that will illustrate how the information in Table 6.7 can be used.

Overall summary

TABLE 6.7 Overall Summary of S_N1, S_N2, E1, and E2 Reactions

CH_3X	RCH_2X	$RCHX$ (with R above C)	R_3CX (R—C(R)(R)—X)
Methyl	1°	2°	3°
Bimolecular Reactions Only			S_N1/E1 or E2
Gives S_N2 reactions	Gives mainly S_N2 except with a hindered strong base [e.g., $(CH_3)_3CO^-$] and then gives mainly E2	Gives mainly S_N2 with weak bases (e.g., I^-, CN^-, RCO_2^-) and mainly E2 with strong bases (e.g., RO^-)	No S_N2 reaction. In solvolysis gives S_N1/E1, and at lower temperatures S_N1 is favored. When a strong base (e.g., RO^-) is used, E2 predominates.

SOLVED PROBLEM

Give the product (or products) that you would expect to be formed in each of the following reactions. In each case give the mechanism (S_N1, S_N2, E1, or E2) by which the product is formed and predict the relative amount of each (i.e., would the product be the only product, the major product, or a minor product?).

(a) $CH_3CH_2CH_2Br + CH_3O^- \xrightarrow[CH_3OH]{50°C}$

(b) $CH_3CH_2CH_2Br + (CH_3)_3CO^- \xrightarrow[(CH_3)_3COH]{50°C}$

(c) $(CH_3)(CH_3CH_2)(H)C\text{—}Br + HS^- \xrightarrow[CH_3OH]{50°C}$ (chiral carbon bearing CH_3, CH_3CH_2, H, Br)

(d) $(CH_3CH_2)_3CBr + OH^- \xrightarrow[CH_3OH]{50°C}$

(e) $(CH_3CH_2)_3CBr \xrightarrow[CH_3OH]{25°C}$

ANSWER:

(a) The substrate is a 1° halide. The base/nucleophile is CH_3O^-, a strong base (but not a hindered one) and a good nucleophile. According to Table 6.7, we should expect an S_N2 reaction mainly, and the major product should be $CH_3CH_2CH_2OCH_3$. A minor product might be $CH_3CH{=}CH_2$ by an E2 pathway.

(b) Again the substrate is a 1° halide, but the base/nucleophile, $(CH_3)_3CO^-$, is a strong hindered base. We should expect, therefore, the major product to be $CH_3CH{=}CH_2$ by an E2 pathway and a minor product to be $CH_3CH_2CH_2OC(CH_3)_3$ by an S_N2 pathway.

(c) The reactant is (*S*)-2-bromobutane, a 2° halide and one in which the leaving group is attached to a stereogenic center. The base/nucleophile is HS^-, a strong nucleophile but a weak base. We should expect mainly an S_N2 reaction, causing an inversion of configuration at the stereogenic center and producing the (*R*) stereoisomer:

HS—C(CH_3)(CH_2CH_3)(H)

(d) The base/nucleophile is OH^-, a strong base and a strong nucleophile. However, the substrate is a 3° halide; therefore, we should not expect an S_N2 reaction. The major product should be $CH_3CH{=}C(CH_2CH_3)_2$ via an E2 reaction. At this higher temperature and in the presence of a strong base, we should not expect an appreciable amount of the S_N1 product, $CH_3OC(CH_2CH_3)_3$.

(e) This is solvolysis; the only base/nucleophile is the solvent, CH_3OH, which is a weak base (therefore, no E2 reaction) and a poor nucleophile. The substrate is tertiary (therefore, no S_N2 reaction). At this lower temperature we should expect mainly an S_N1 pathway leading to $CH_3OC(CH_2CH_3)_3$. A minor product, by an E1 pathway, would be $CH_3CH{=}C(CH_2CH_3)_2$.

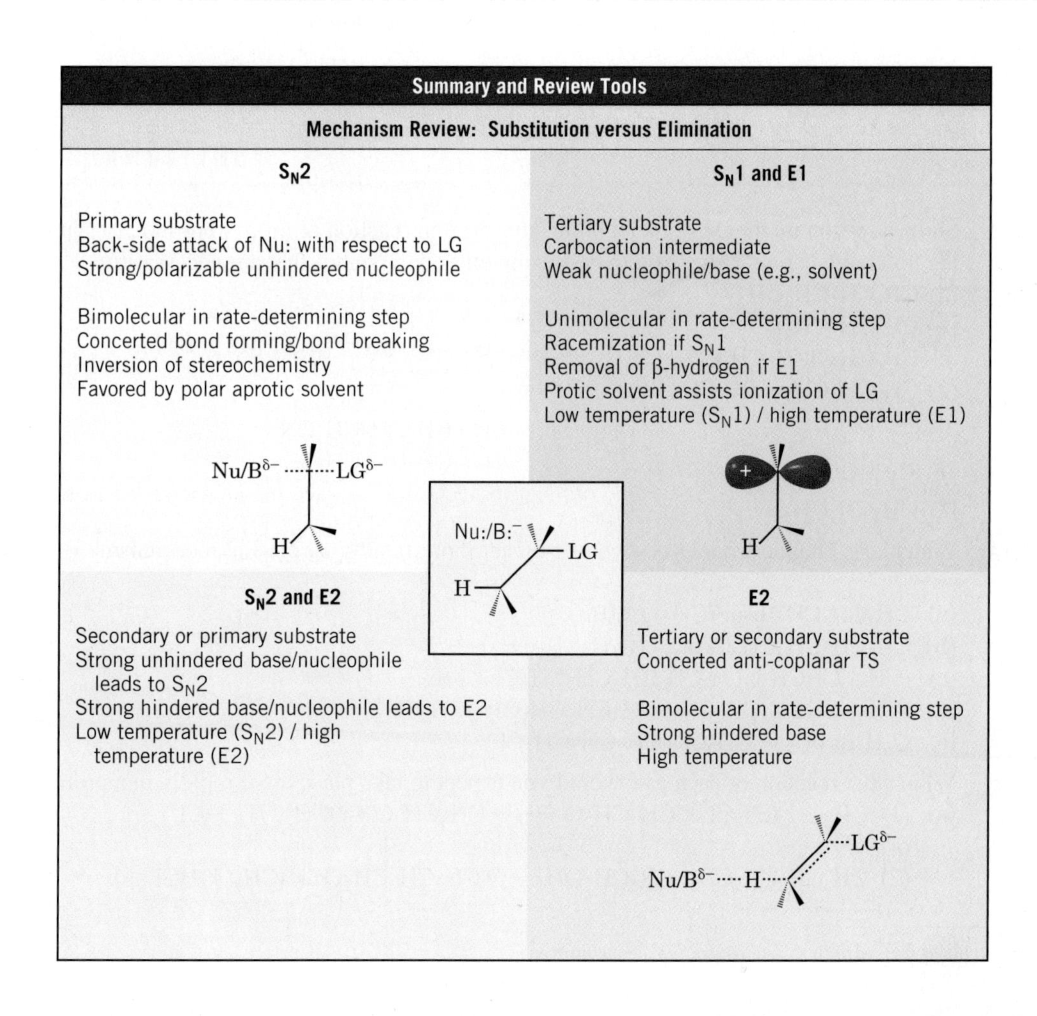

Summary and Review Tools
Mechanism Review: Substitution versus Elimination
S_N2
Primary substrate
Back-side attack of Nu: with respect to LG
Strong/polarizable unhindered nucleophile
Bimolecular in rate-determining step
Concerted bond forming/bond breaking
Inversion of stereochemistry
Favored by polar aprotic solvent
S_N1 and E1
Tertiary substrate
Carbocation intermediate
Weak nucleophile/base (e.g., solvent)
Unimolecular in rate-determining step
Racemization if S_N1
Removal of β-hydrogen if E1
Protic solvent assists ionization of LG
Low temperature (S_N1) / high temperature (E1)
Nu/B$^{\delta-}$ LG$^{\delta-}$
H
Nu:/B:$^-$
LG
H
+
H
S_N2 and E2
Secondary or primary substrate
Strong unhindered base/nucleophile leads to S_N2
Strong hindered base/nucleophile leads to E2
Low temperature (S_N2) / high temperature (E2)
E2
Tertiary or secondary substrate
Concerted anti-coplanar TS
Bimolecular in rate-determining step
Strong hindered base
High temperature
LG$^{\delta-}$
Nu/B$^{\delta-}$ H

Key Terms and Concepts

Aprotic solvent	Section 6.13C
Carbocation	Section 6.11
E1 reaction	Sections 6.15C, 6.17, 6.18B
E2 reaction	Sections 6.15C, 6.16, 6.18A
Elimination reaction	Section 6.15
Inversion of configuration	Section 6.8
Leaving group	Sections 6.4, 6.13E
Nucleophile	Sections 6.2, 6.3, 6.13B
Nucleophilic substitution reaction	Section 6.2
Polar aprotic solvent	Section 6.13C
Polar protic solvent	Sections 6.13C, 6.13D
Protic solvent	Section 6.13C
Racemization	Section 6.12A
S_N1 reaction	Sections 6.9, 6.10, 6.12, 6.13, 6.18B
S_N2 reaction	Sections 6.5–6.8, 6.13, 6.18A
Solvent effects	Sections 6.13C, 6.13D
Solvolysis	Section 6.12B
Steric effect	Section 6.13A
Steric hindrance	Section 6.13A
Substrate	Section 6.2
Transition state	Section 6.6

Additional Problems

6.13 Show how you might use a nucleophilic substitution reaction of propyl bromide to synthesize each of the following compounds. (You may use any other compounds that are necessary.)

(a) $CH_3CH_2CH_2OH$

(b) $CH_3CH_2CH_2I$

(c) $CH_3CH_2OCH_2CH_2CH_3$

(d) $CH_3CH_2CH_2{-}S{-}CH_3$

(e) $CH_3\overset{O}{\overset{\|}{C}}OCH_2CH_2CH_3$

(f) $CH_3CH_2CH_2N_3$

(g) $CH_3{-}\overset{+}{N}(CH_3)_2{-}CH_2CH_2CH_3\ Br^-$ (N bearing two CH_3 groups, drawn above and below)

(h) $CH_3CH_2CH_2CN$

(i) $CH_3CH_2CH_2SH$

6.14 Which alkyl halide would you expect to react more rapidly by an S_N2 mechanism? Explain your answer.

(a) $CH_3CH_2CH_2Br$ or $(CH_3)_2CHBr$

(b) $CH_3CH_2CH_2CH_2Cl$ or $CH_3CH_2CH_2CH_2I$

(c) $(CH_3)_2CHCH_2Cl$ or $CH_3CH_2CH_2CH_2Cl$

(d) $(CH_3)_2CHCH_2CH_2Cl$ or $CH_3CH_2CH(CH_3)CH_2Cl$

(e) C_6H_5Br or $CH_3CH_2CH_2CH_2CH_2CH_2Cl$

6.15 Which S_N2 reaction of each pair would you expect to take place more rapidly in a protic solvent?

(a) (1) $CH_3CH_2CH_2Cl + CH_3CH_2O^- \longrightarrow CH_3CH_2CH_2OCH_2CH_3 + Cl^-$

or

(2) $CH_3CH_2CH_2Cl + CH_3CH_2OH \longrightarrow CH_3CH_2CH_2OCH_2CH_3 + HCl$

Note: Problems marked with an asterisk are "challenge problems."

(b) (1) $CH_3CH_2CH_2Cl + CH_3CH_2O^- \longrightarrow CH_3CH_2CH_2OCH_2CH_3 + Cl^-$

or

(2) $CH_3CH_2CH_2Cl + CH_3CH_2S^- \longrightarrow CH_3CH_2CH_2SCH_2CH_3 + Cl^-$

(c) (1) $CH_3CH_2CH_2Br + (C_6H_5)_3N \longrightarrow CH_3CH_2CH_2N(C_6H_5)_3^+ + Br^-$

or

(2) $CH_3CH_2CH_2Br + (C_6H_5)_3P \longrightarrow CH_3CH_2CH_2P(C_6H_5)_3^+ + Br^-$

(d) (1) $CH_3CH_2CH_2Br\ (1.0M) + CH_3O^-\ (1.0M) \longrightarrow CH_3CH_2CH_2OCH_3 + Br^-$

or

(2) $CH_3CH_2CH_2Br\ (1.0M) + CH_3O^-\ (2.0M) \longrightarrow CH_3CH_2CH_2OCH_3 + Br^-$

6.16 Which S_N1 reaction of each pair would you expect to take place more rapidly? Explain your answer.

(a) (1) $(CH_3)_3CCl + H_2O \longrightarrow (CH_3)_3COH + HCl$

or

(2) $(CH_3)_3CBr + H_2O \longrightarrow (CH_3)_3COH + HBr$

(b) (1) $(CH_3)_3CCl + H_2O \longrightarrow (CH_3)_3COH + HCl$

or

(2) $(CH_3)_3CCl + CH_3OH \longrightarrow (CH_3)_3COCH_3 + HCl$

(c) (1) $(CH_3)_3CCl\ (1.0M) + CH_3CH_2O^-\ (1.0M) \xrightarrow{EtOH} (CH_3)_3COCH_2CH_3 + Cl^-$

or

(2) $(CH_3)_3CCl\ (2.0M) + CH_3CH_2O^-\ (1.0M) \xrightarrow{EtOH} (CH_3)_3COCH_2CH_3 + Cl^-$

(d) (1) $(CH_3)_3CCl\ (1.0M) + CH_3CH_2O^-\ (1.0M) \xrightarrow{EtOH} (CH_3)_3COCH_2CH_3 + Cl^-$

or

(2) $(CH_3)_3CCl\ (1.0M) + CH_3CH_2O^-\ (2.0M) \xrightarrow{EtOH} (CH_3)_3COCH_2CH_3 + Cl^-$

(e) (1) $(CH_3)_3CCl + H_2O \longrightarrow (CH_3)_3COH + HCl$

or

(2) $C_6H_5Cl + H_2O \longrightarrow C_6H_5OH + HCl$

6.17 With methyl, ethyl, or cyclopentyl halides as your organic starting materials and using any needed solvents or inorganic reagents, outline syntheses of each of the following. More than one step may be necessary and you need not repeat steps carried out in earlier parts of this problem.

(a) CH_3I
(b) CH_3CH_2I
(c) CH_3OH
(d) CH_3CH_2OH
(e) CH_3SH
(f) CH_3CH_2SH
(g) CH_3CN
(h) CH_3CH_2CN
(i) CH_3OCH_3
(j) $CH_3OCH_2CH_3$
(k) Cyclopentene

6.18 Listed below are several hypothetical nucleophilic substitution reactions. None is synthetically useful because the product indicated is not formed at an appreciable rate. In each case provide an explanation for the failure of the reaction to take place as indicated.

(a) $CH_3CH_2CH_3 + OH^- \not\longrightarrow CH_3CH_2OH + CH_3^-$

(b) $CH_3CH_2CH_3 + OH^- \not\longrightarrow CH_3CH_2CH_2OH + H^-$

(c) Cyclobutane $+ OH^- \not\longrightarrow {}^-CH_2CH_2CH_2CH_2OH$

(d) $CH_3CH_2-\overset{CH_3}{\underset{CH_3}{\overset{|}{\underset{|}{C}}}}-Br + CN^- \not\longrightarrow CH_3CH_2-\overset{CH_3}{\underset{CH_3}{\overset{|}{\underset{|}{C}}}}-CN + Br^-$

(e) $NH_3 + CH_3OCH_3 \not\longrightarrow CH_3NH_2 + CH_3OH$

(f) $NH_3 + CH_3OH_2^+ \not\longrightarrow CH_3NH_3^+ + H_2O$

6.19 You have the task of preparing styrene ($C_6H_5CH{=}CH_2$) by dehydrohalogenation of either 1-bromo-2-phenylethane or 1-bromo-1-phenylethane using KOH in ethanol. Which halide would you choose as your starting material to give the better yield of the alkene? Explain your answer.

6.20 Your task is to prepare isopropyl methyl ether, $CH_3OCH(CH_3)_2$, by one of the following reactions. Which reaction would give the better yield? Explain your choice.

(1) $CH_3ONa + (CH_3)_2CHI \longrightarrow CH_3OCH(CH_3)_2$

(2) $(CH_3)_2CHONa + CH_3I \longrightarrow CH_3OCH(CH_3)_2$

6.21 Which product (or products) would you expect to obtain from each of the following reactions? In each part give the mechanism (S_N1, S_N2, E1, or E2) by which each product is formed and predict the relative amount of each product (i.e., would the product be the only product, the major product, a minor product, etc.?).

(a) $CH_3CH_2CH_2CH_2CH_2Br + CH_3CH_2O^- \xrightarrow[CH_3CH_2OH]{50°C}$

(b) $CH_3CH_2CH_2CH_2CH_2Br + (CH_3)_3CO^- \xrightarrow[(CH_3)_3COH]{50°C}$

(c) $(CH_3)_3CBr + CH_3O^- \xrightarrow[CH_3OH]{50°C}$

(d) $(CH_3)_3CBr + (CH_3)_3CO^- \xrightarrow[(CH_3)_3COH]{50°C}$

(e) 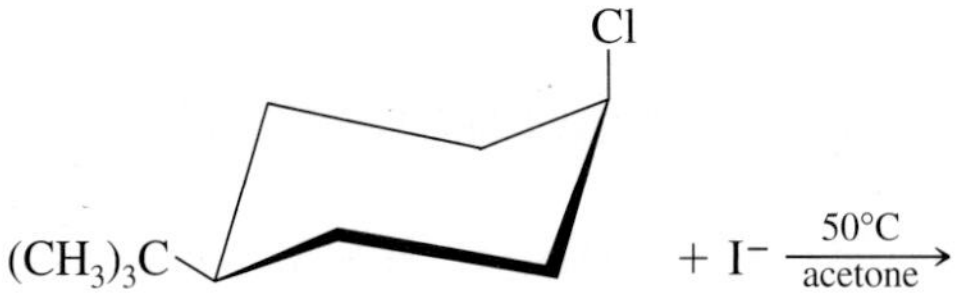 $(CH_3)_3C$– (cyclohexane chair with axial Cl) $+ I^- \xrightarrow[acetone]{50°C}$

(f) 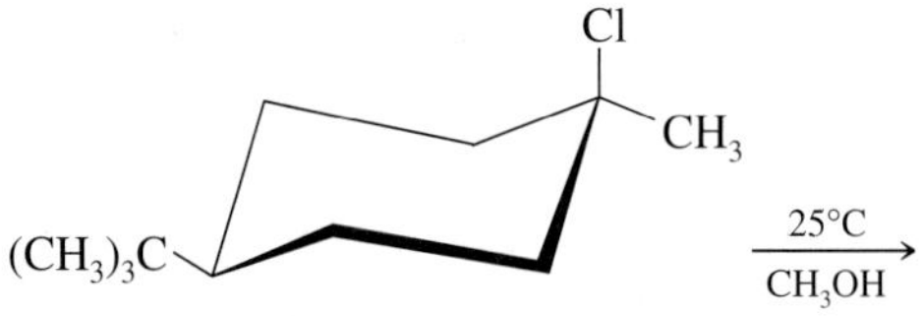 $(CH_3)_3C$– (cyclohexane chair with axial Cl and equatorial CH_3) $\xrightarrow[CH_3OH]{25°C}$

(g) 3-Chloropentane $+ CH_3O^- \xrightarrow[CH_3OH]{50°C}$

(h) 3-Chloropentane $+ CH_3CO_2^- \xrightarrow[CH_3CO_2H]{50°C}$

(i) HO^- + (*R*)-2-bromobutane $\xrightarrow{25°C}$

(j) (*S*)-3-Bromo-3-methylhexane $\xrightarrow[CH_3OH]{25°C}$

(k) (*S*)-2-Bromooctane $+ I^- \xrightarrow[CH_3OH]{50°C}$

6.22 Write conformational structures for the substitution products of the following deuterium-labeled compounds:

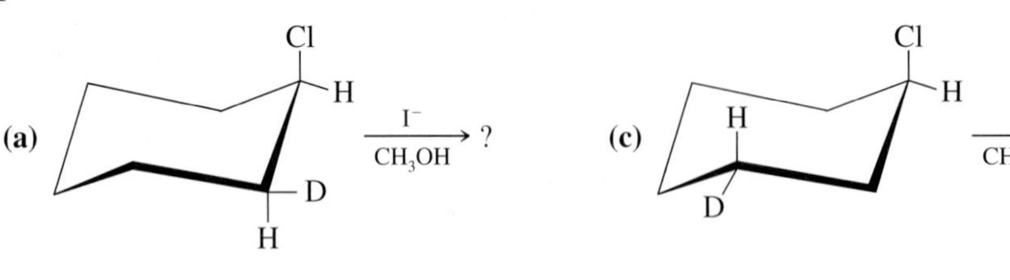

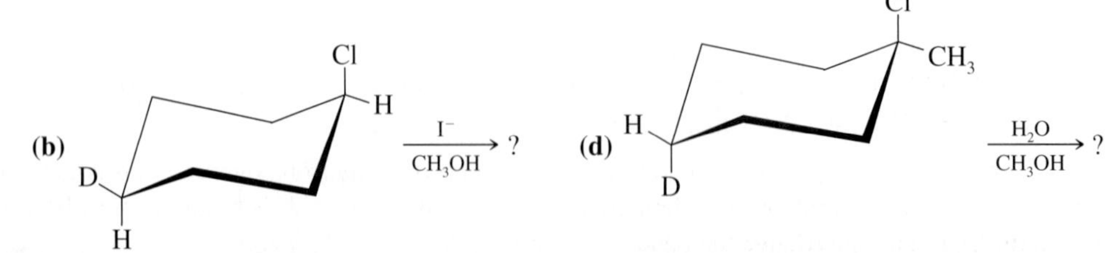

6.23 Although ethyl bromide and isobutyl bromide are both primary halides, ethyl bromide undergoes S_N2 reactions more than 10 times faster than isobutyl bromide does. When each compound is treated with a strong base/nucleophile ($CH_3CH_2O^-$), isobutyl bromide gives a greater yield of elimination products than substitution products, whereas with ethyl bromide this behavior is reversed. What factor accounts for these results?

6.24 Consider the reaction of I^- with CH_3CH_2Cl.
(a) Would you expect the reaction to be S_N1 or S_N2? The rate constant for the reaction at 60°C is 5×10^{-5} L mol^{-1} s^{-1}.
(b) What is the reaction rate if $[I^-] = 0.1$ mol L^{-1} and $[CH_3CH_2Cl] = 0.1$ mol L^{-1}?
(c) If $[I^-] = 0.1$ mol L^{-1} and $[CH_3CH_2Cl] = 0.2$ mol L^{-1}?
(d) If $[I^-] = 0.2$ mol L^{-1} and $[CH_3CH_2Cl] = 0.1$ mol L^{-1}?
(e) If $[I^-] = 0.2$ mol L^{-1} and $[CH_3CH_2Cl] = 0.2$ mol L^{-1}?

6.25 Which reagent in each pair listed here would be the more reactive nucleophile in a protic solvent?
(a) CH_3NH^- or CH_3NH_2
(b) CH_3O^- or $CH_3C(=O)O^-$
(c) CH_3SH or CH_3OH
(d) $(C_6H_5)_3N$ or $(C_6H_5)_3P$
(e) H_2O or H_3O^+
(f) NH_3 or NH_4^+
(g) H_2S or HS^-
(h) $CH_3C(=O)O^-$ or OH^-

6.26 Write mechanisms that account for the products of the following reactions:

(a) 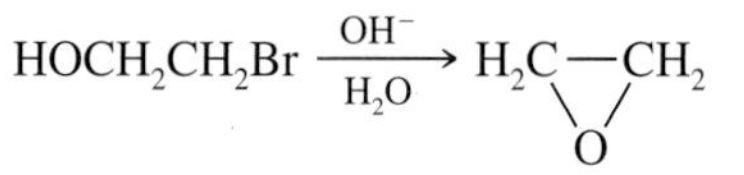 $HOCH_2CH_2Br \xrightarrow[H_2O]{OH^-} H_2C{-}CH_2$ (epoxide: H_2C and CH_2 bridged by O)

(b) 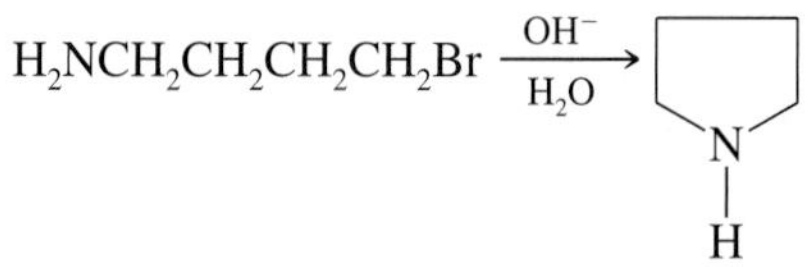 $H_2NCH_2CH_2CH_2CH_2Br \xrightarrow[H_2O]{OH^-}$ pyrrolidine (five-membered ring with N–H)

6.27 Many S_N2 reactions of alkyl chlorides and alkyl bromides are catalyzed by the addition of sodium or potassium iodide. For example, the hydrolysis of methyl bromide takes place much faster in the presence of sodium iodide. Explain.

6.28 Explain the following observations: When *tert*-butyl bromide is treated with sodium methoxide in a mixture of methanol and water, the rate of formation of *tert*-butyl alcohol and *tert*-butyl methyl ether does not change appreciably as the concentration of sodium methoxide is increased. However, increasing the concentration of sodium methoxide causes a marked increase in the rate at which *tert*-butyl bromide disappears from the mixture.

6.29 **(a)** Consider the general problem of converting a tertiary alkyl halide to an alkene, for example, the conversion of *tert*-butyl chloride to 2-methylpropene. What experimental conditions would you choose to ensure that elimination is favored over substitution?
(b) Consider the opposite problem, that of carrying out a substitution reaction on a tertiary alkyl halide. Use as your example the conversion of *tert*-butyl chloride to *tert*-butyl ethyl ether. What experimental conditions would you employ to ensure the highest possible yield of the ether?

6.30 1-Bromobicyclo[2.2.1]heptane is extremely unreactive in either S_N2 or S_N1 reactions. Provide explanations for this behavior.

6.31 When ethyl bromide reacts with potassium cyanide in methanol, the major product is CH_3CH_2CN. Some CH_3CH_2NC is formed as well, however. Write Lewis structures for the cyanide ion and for both products and provide a mechanistic explanation of the course of the reaction.

6.32 Starting with an appropriate alkyl halide and using any other needed reagents, outline syntheses of each of the following. When alternative possibilities exist for a synthesis, you should be careful to choose the one that gives the better yield.

(a) Butyl *sec*-butyl ether
(b) $CH_3CH_2SC(CH_3)_3$
(c) Methyl neopentyl ether
(d) Methyl phenyl ether
(e) $C_6H_5CH_2CN$
(f) $CH_3CO_2CH_2C_6H_5$
(g) (*S*)-2-Pentanol
(h) (*R*)-2-Iodo-4-methylpentane
(i) $(CH_3)_3CCH{=}CH_2$
(j) *cis*-4-Isopropylcyclohexanol
(k) (*R*)-$CH_3CH(CN)CH_2CH_3$
(l) *trans*-1-Iodo-4-methylcyclohexane

6.33 Give structures for the products of each of the following reactions:

(a) [cyclopentane with H and Br on one carbon, F and H on another] + NaI (1 mol) $\xrightarrow{\text{acetone}}$ C_5H_8FI + NaBr

(b) 1,4-Dichlorohexane (1 mol) + NaI (1 mol) $\xrightarrow{\text{acetone}}$ $C_6H_{12}ClI$ + NaCl

(c) 1,2-Dibromoethane (1 mol) + $NaSCH_2CH_2SNa \longrightarrow C_4H_8S_2 + 2\ NaBr$

(d) 4-Chloro-1-butanol + NaH $\xrightarrow[\text{Et}_2\text{O}]{(-\text{H}_2)}$ C_4H_8ClONa $\xrightarrow[\text{Et}_2\text{O}]{\text{heat}}$ C_4H_8O + NaCl

(e) Propyne + $NaNH_2$ $\xrightarrow[\text{liq. NH}_3]{(-\text{NH}_3)}$ C_3H_3Na $\xrightarrow{\text{CH}_3\text{I}}$ C_4H_6 + NaI

6.34 When *tert*-butyl bromide undergoes S_N1 hydrolysis, adding a "common ion" (e.g., NaBr) to the aqueous solution has no effect on the rate. On the other hand, when $(C_6H_5)_2CHBr$ undergoes S_N1 hydrolysis, adding NaBr retards the reaction. Given that the $(C_6H_5)_2CH^+$ cation is known to be much more stable than the $(CH_3)_3C^+$ cation (and we shall see why in Section 15.12A), provide an explanation for the different behavior of the two compounds.

6.35 When the alkyl bromides (listed here) were subjected to hydrolysis in a mixture of ethanol and water (80% C_2H_5OH/20% H_2O) at 55°C, the rates of the reaction showed the following order:

$$(CH_3)_3CBr > CH_3Br > CH_3CH_2Br > (CH_3)_2CHBr$$

Provide an explanation for this order of reactivity.

6.36 The reaction of 1° alkyl halides with nitrite salts produces both RNO_2 and RONO. Account for this behavior.

6.37 What would be the effect of increasing solvent polarity on the rate of each of the following nucleophilic substitution reactions?

(a) $\text{Nu:} + \text{R—L} \longrightarrow \text{R—Nu}^+ + \text{:L}^-$
(b) $\text{R—L}^+ \longrightarrow \text{R}^+ + \text{:L}$

6.38 Competition experiments are those in which two reactants at the same concentration (or one reactant with two reactive sites) compete for a reagent. Predict the major product resulting from each of the following competition experiments:

(a) $\text{Cl—CH}_2\text{—C(CH}_3)_2\text{—CH}_2\text{—CH}_2\text{—Cl} + \text{I}^- \xrightarrow{\text{DMF}}$

(b) $\text{Cl—C(CH}_3)_2\text{—CH}_2\text{—CH}_2\text{—Cl} + \text{H}_2\text{O} \xrightarrow{\text{acetone}}$

6.39 In contrast to S_N2 reactions, S_N1 reactions show relatively little nucleophile selectivity. That is, when more than one nucleophile is present in the reaction medium, S_N1 reactions show only a slight tendency to discriminate between weak nucleophiles and strong nucleophiles, whereas S_N2 reactions show a marked tendency to discriminate.

(a) Provide an explanation for this behavior.
(b) Show how your answer accounts for the fact that $CH_3CH_2CH_2CH_2Cl$ reacts with 0.01 *M* NaCN

in ethanol to yield primarily $CH_3CH_2CH_2CH_2CN$, whereas under the same conditions $(CH_3)_3CCl$ reacts to give primarily $(CH_3)_3COCH_2CH_3$.

***6.40** In the gas phase, the homolytic bond dissociation energy (Section 10.2A) for the carbon–chlorine bond of *tert*-butyl chloride is $+349\ \text{kJ mol}^{-1}$, the ionization potential for a *tert*-butyl radical is $+715\ \text{kJ mol}^{-1}$, and the electron affinity of chlorine is $-330\ \text{kJ mol}^{-1}$. Using these data, calculate the enthalpy change for the gas-phase ionization of *tert*-butyl chloride to a *tert*-butyl cation and a chloride ion (this is the heterolytic bond dissociation energy of the carbon–chlorine bond).

***6.41** The reaction of chloroethane with water *in the gas phase* to produce ethanol and hydrogen chloride has a $\Delta H° = +26.6\ \text{kJ mol}^{-1}$ and a $\Delta S° = +4.81\ \text{J K}^{-1}\ \text{mol}^{-1}$ at 25°C.
(a) Which of these terms, if either, favors the reaction going to completion?
(b) Calculate $\Delta G°$ for the reaction. What can you now say about whether the reaction will proceed to completion?
(c) Calculate the equilibrium constant for the reaction.
(d) In aqueous solution the equilibrium constant is very much larger than the one you just calculated. How can you account for this fact?

***6.42** When (*S*)-2-bromopropanoic acid [(*S*)-$CH_3CHBrCO_2H$] reacts with concentrated sodium hydroxide, the product formed (after acidification) is (*R*)-2-hydroxypropanoic acid [(*R*)-$CH_3CHOHCO_2H$, commonly known as (*R*)-lactic acid]. This is, of course, the normal stereochemical result for an S_N2 reaction. However, when the same reaction is carried out with a low concentration of hydroxide ion in the presence of Ag_2O (where Ag^+ acts as a Lewis acid), it takes place with overall *retention of configuration* to produce (*S*)-2-hydroxypropanoic acid. The mechanism of this reaction involves a phenomenon called *neighboring-group participation.* Write a detailed mechanism for this reaction that accounts for the net retention of configuration when Ag^+ and a low concentration of hydroxide are used.

***6.43** The phenomenon of configuration inversion in a chemical reaction was discovered in 1896 by Paul von Walden (Section 6.6). Walden's proof of configuration inversion was based on the following cycle:

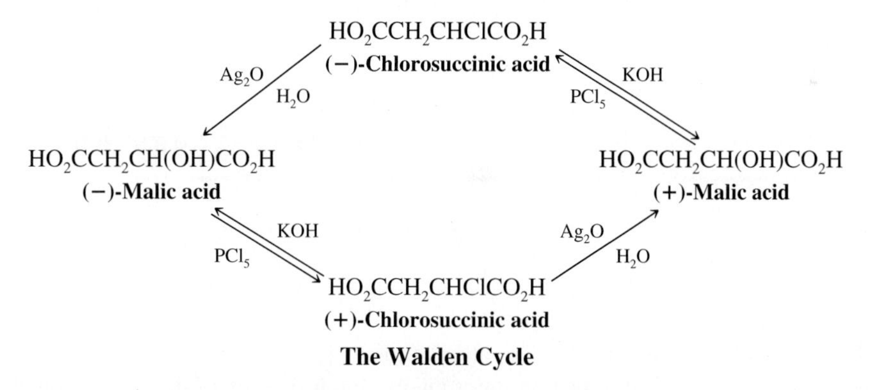

The Walden Cycle

(a) Basing your answer on the preceding problem, which reactions of the Walden cycle are likely to take place with overall inversion of configuration and which are likely to occur with overall retention of configuration?
(b) Malic acid with a negative optical rotation is now known to have the (*S*) configuration. What are the configurations of the other compounds in the Walden cycle?
(c) Walden also found that when (+)-malic acid is treated with thionyl chloride (rather than PCl_5), the product of the reaction is (+)-chlorosuccinic acid. How can you explain this result?
(d) Assuming that the reaction of (−)-malic acid and thionyl chloride has the same stereochemistry, outline a Walden cycle based on the use of thionyl chloride instead of PCl_5.

***6.44** (*R*)-(3-Chloro-2-methylpropyl) methyl ether (**A**) on reaction with azide ion (N_3^-) in aqueous ethanol gives (*S*)-(3-azido-2-methylpropyl) methyl ether (**B**). Compound **A** has the structure $ClCH_2CH(CH_3)CH_2OCH_3$.
(a) Draw wedge–dashed wedge–line formulas of both **A** and **B**.
(b) Is there a change of configuration during this reaction?

***6.45** Predict the structure of the product of this reaction:

$$\text{(cyclohexane with HS, H / H, Cl)} \xrightarrow[\text{aqueous EtOH}]{\text{NaOH in}} C_6H_{10}S$$

The product has no infrared absorption in the 1620 to 1680 cm^{-1} region.

***6.46** *cis*-4-Bromocyclohexanol $\xrightarrow[t\text{-BuOH}]{t\text{-BuO}^- \text{ in}}$ racemic $C_6H_{10}O$ (compound **C**)

Compound **C** has infrared absorption in the 1620–1680-cm^{-1} and in the 3590–3650-cm^{-1} regions. Draw and label the (*R*) and (*S*) enantiomers of product **C**.

6.47 1-Bromo[2.2.1]bicycloheptane is unreactive toward both S_N2 and S_N1 reactions. Open the CD molecular model titled "1-Bromo[2.2.1]bicycloheptane" and examine the structure. What barriers are there to substitution of 1-bromo[2.2.1]bicycloheptane by both S_N2 and S_N1 reaction mechanisms?

Chem3D Model

6.48 The Concept Unit titled "S_N2 Orbitals" on the CD explains the importance of the HOMO (highest occupied molecular orbital) and LUMO (lowest unoccupied molecular orbital) in S_N2 reactions. Open the CD molecular model titled "1-Bromo[2.2.1]bicycloheptane LUMO" for the LUMO molecular orbital of this compound. Where is the lobe of the LUMO with which the HOMO of a nucleophile would interact in an S_N2 reaction?

Chem3D Model

6.49 In the previous problem and the associated Concept Unit on the CD, you considered the role of HOMO and LUMO orbitals in an S_N2 reaction.

(a) What is the LUMO in an S_N1 reaction and in what reactant and species is it found?

(b) Open the molecular model on the CD titled "Isopropyl Methyl Ether Carbocation LUMO." Identify the lobe of the LUMO in this carbocation model with which a nucleophile would interact.

(c) Open the CD model titled "Isopropyl Methyl Ether Carbocation HOMO." Why is there a large orbital lobe between the oxygen and the carbon of the carbocation?

Chem3D Model

Learning Group Problems

1. Consider the solvolysis reaction of (1*S*,2*R*)-1-bromo-1,2-dimethylcyclohexane in 80% H_2O/20% CH_3CH_2OH at room temperature.
   (a) Write the structure of all chemically reasonable products from this reaction and predict which would be the major one.
   (b) Write a detailed mechanism for formation of the major product.
   (c) Write the structure of all transition states involved in formation of the major product.
2. Consider the following sequence of reactions, taken from the early steps in a synthesis of ω-fluorooleic acid, a toxic natural compound from an African shrub. (ω-Fluorooleic acid, also called "ratsbane," has been used to kill rats and also as an arrow poison in tribal warfare. Two more steps beyond those below are required to complete its synthesis.)

   (i) 1-Bromo-8-fluorooctane + sodium acetylide (the sodium salt of ethyne) $\longrightarrow$ compound **A** ($C_{10}H_{17}F$)
   (ii) Compound **A** + $NaNH_2 \longrightarrow$ compound **B** ($C_{10}H_{16}FNa$)
   (iii) Compound **B** + I—$(CH_2)_7$—Cl $\longrightarrow$ compound **C** ($C_{17}H_{30}ClF$)
   (iv) Compound **C** + NaCN $\longrightarrow$ compound **D** ($C_{18}H_{30}NF$)

   (a) Elucidate the structure of compounds **A, B, C,** and **D** above.
   (b) Write the mechanism for each of the reactions above.
   (c) Write the structure of the transition state for each reaction.

CHAPTER 7

Alkenes and Alkynes I: Properties and Synthesis. Elimination Reactions of Alkyl Halides

Cell Membrane Fluidity

Cell membranes must remain fluid in order to function properly as a selective barrier between the cell's internal and external environments. One factor that influences cell membrane fluidity is the ratio of saturated to unsaturated fatty acids in the membrane. Cold-blooded animals like fish adjust the fluidity of their cell membranes to the temperature of their environment by altering the ratio of unsaturated to saturated fatty acids. Reindeer, because they live in a cold environment, have a higher proportion of unsaturated fatty acids in the cell membranes of their legs than in the core of their body, thus keeping the cell membranes in their extremities fluid. In these examples we see the important relationship between molecular structure, physical properties, and, ultimately, biological function. We shall now consider further the structures of saturated and unsaturated fatty acids.

Fatty acids are carboxylic acids with long hydrocarbon tails extending from a carboxylic acid group. Most fatty acids in higher animals have 16 or 18 carbons. A saturated fatty acid is one that contains no double bonds in its hydrocarbon tail (it is "saturated" with hydrogen). An unsaturated fatty acid contains at least one double bond. But what difference does the presence of double bonds make to membrane fluidity? Consider the difference between butter and olive oil. Butter largely contains saturated fatty acids. The hydrocarbon tails of saturated fatty acids in butter align in a sufficiently regular and ordered fashion that a solid results at room temperature. Olive oil, on the other hand, contains a large percentage of unsaturated fatty acids. The alkene double bonds in the fatty acids of olive oil, which are mostly in the cis configuration, introduce a kink in the hydrocarbon chain of the fatty acid. The result is that unsaturated fatty acids in olive oil cannot easily form an orderly array and thus have lower melting points than saturated fatty acids with the same number of carbons. Since butter is richer in saturated fatty acids than olive oil, it is a solid. The fluidity of cell membranes is similarly affected. The higher the proportion of saturated fatty acids, the less fluid is the membrane, and vice versa.

7.1 Introduction

Alkenes are hydrocarbons whose molecules contain the carbon–carbon double bond. An old name for this family of compounds that is still often used is the name *olefins.* Ethene (ethylene), the simplest olefin (alkene), was called olefiant gas (Latin: *oleum,* oil + *facere,* to make) because gaseous ethene (C_2H_4) reacts with chlorine to form $C_2H_4Cl_2$, a liquid (oil).

Hydrocarbons whose molecules contain the carbon–carbon triple bond are called alkynes. The common name for this family is *acetylenes,* after the first member, HC≡CH:

$$\begin{array}{ccc} \mathrm{H_2C{=}CH_2} & \mathrm{H_2C{=}CHCH_3} & \mathrm{H{-}C{\equiv}C{-}H} \\ \textbf{Ethene} & \textbf{Propene} & \textbf{Ethyne} \end{array}$$

7.1A Physical Properties of Alkenes and Alkynes

Alkenes and alkynes have physical properties similar to those of corresponding alkanes. Alkenes and alkynes up to four carbons (except 2-butyne) are gases at room temperature. Being relatively nonpolar themselves, alkenes and alkynes dissolve in nonpolar solvents or in solvents of low polarity. Alkenes and alkynes are only *very slightly soluble* in water (with alkynes being slightly more soluble than alkenes). The densities of alkenes and alkynes are lower than that of water.

7.2 The (*E*)–(*Z*) System for Designating Alkene Diastereomers

In Section 4.5 we learned to use the terms cis and trans to designate the stereochemistry of alkene diastereomers. These terms are unambiguous, however, only when applied to disubstituted alkenes. If the alkene is trisubstituted or tetrasubstituted, the terms cis and trans are either ambiguous or do not apply at all. Consider the following alkene as an example:

$$\mathrm{(Br)(H)C{=}C(Cl)(F)}\quad \textbf{A}$$

It is impossible to decide whether **A** is cis or trans since no two groups are the same.

A system that works in all cases is based on the priorities of groups in the Cahn–Ingold–Prelog convention (Section 5.6). This system, called the **(*E*)–(*Z*) system** applies to alkene diastereomers of all types. In the (*E*)–(*Z*) system, we examine the two groups attached to one carbon atom of the double bond and decide which has higher priority. Then we repeat that operation at the other carbon atom:

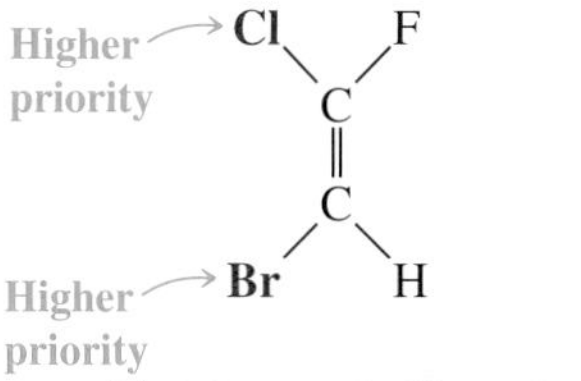

Higher priority → Cl, F; Higher priority → Br, H

(*Z*)-2-Bromo-1-chloro-1-fluoroethene

F, Cl ← Higher priority; Higher priority → Br, H

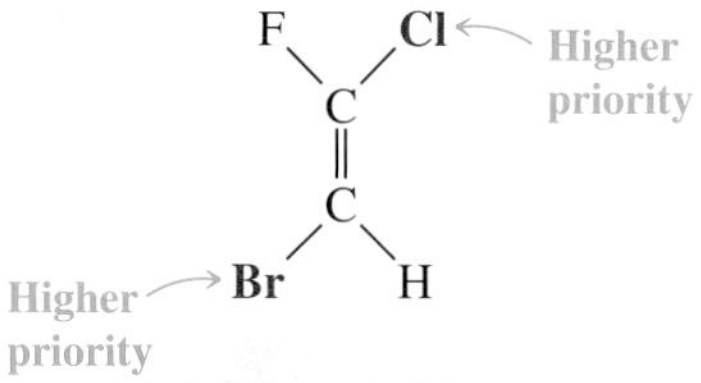

(*E*)-2-Bromo-1-chloro-1-fluoroethene

Cl > F

Br > H

We take the group of higher priority on one carbon atom and compare it with the group of higher priority on the other carbon atom. If the two groups of higher priority are on the same side of the double bond, the alkene is designated (*Z*) (from the German word *zusammen,* meaning together). If the two groups of higher priority are on opposite sides of the double bond, the alkene is designated (*E*) (from the German word *entgegen,* meaning opposite). The following examples illustrate this:

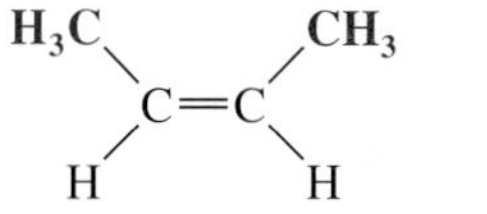

(*Z*)-2-Butene or (*Z*)-but-2-ene (*cis*-2-butene)

$CH_3 > H$

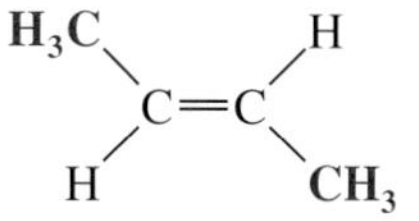

(*E*)-2-Butene or (*E*)-but-2-ene (*trans*-2-butene)

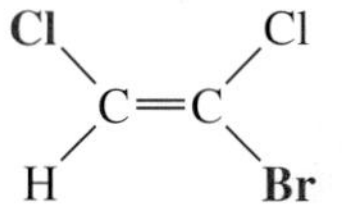

(*E*)-1-Bromo-1,2-dichloroethene

Cl > H
Br > Cl

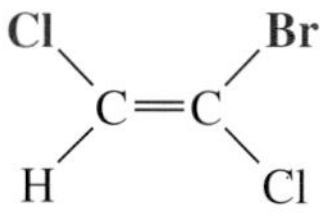

(*Z*)-1-Bromo-1,2-dichloroethene

PROBLEM 7.1

Using the (*E*)–(*Z*) designation [and in parts (e) and (f) the (*R*)–(*S*) designation as well] give IUPAC names for each of the following:

(a) 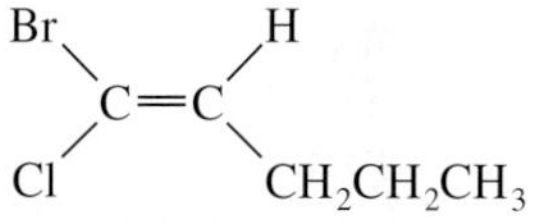 Br(Cl)C=CH($CH_2CH_2CH_3$)

(b) Cl(I)C=C(Br)(CH_2CH_3)

(c) H_3C(H)C=C($CH_2CH(CH_3)_2$)(CH_3)

(d) Cl(I)C=C(CH_3)(CH_2CH_3)

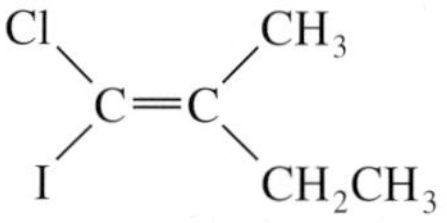

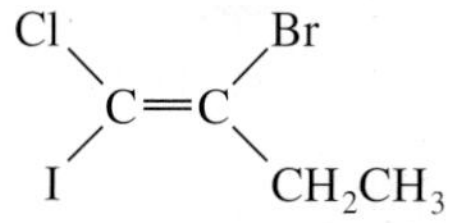

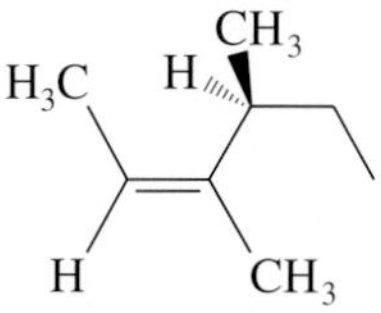

(e)

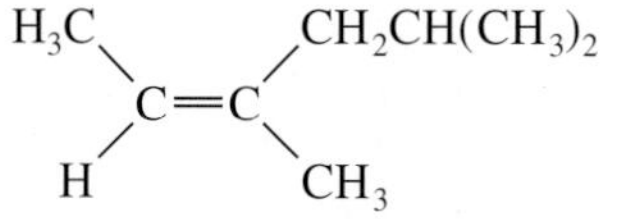

(f)

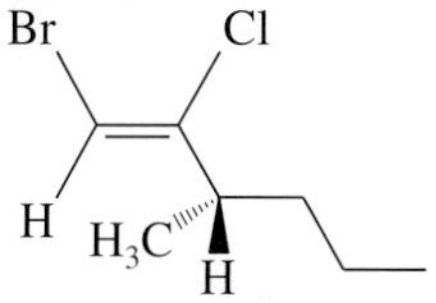

7.3 Relative Stabilities of Alkenes

Cis and trans isomers of alkenes do not have the same stability. Strain caused by crowding of two alkyl groups on the same side of a double bond makes cis isomers generally less stable than trans isomers (Fig. 7.1). This effect can be measured quantitatively by comparing thermodynamic data from experiments involving alkenes with related structures, as we shall see below.

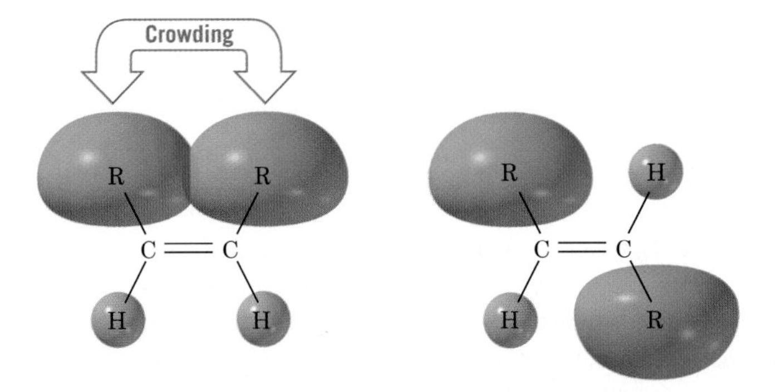

FIGURE 7.1 *cis*- and *trans*-Alkene isomers. The cis isomer is less stable due to greater strain from crowding by the adjacent alkyl groups.

7.3A Heat of Reaction

The addition of hydrogen to an alkene (Sections 4.18 and 7.12) is an exothermic reaction; the enthalpy change involved is called the **heat of reaction** or, in this specific case, the **heat of hydrogenation.**

$$\text{>C=C<} + \text{H—H} \xrightarrow{\text{Pt}} \text{—}\underset{\text{H}}{\overset{|}{\text{C}}}\text{—}\underset{\text{H}}{\overset{|}{\text{C}}}\text{—} \qquad \Delta H° \cong -120 \text{ kJ mol}^{-1}$$

We can gain a quantitative measure of relative alkene stabilities by comparing the heats of hydrogenation for a family of alkenes that all become the same alkane product upon hydrogenation. The results of such an experiment involving platinum-catalyzed hydrogenation of three butene isomers are shown in Fig. 7.2. All three isomers yield the same prod-

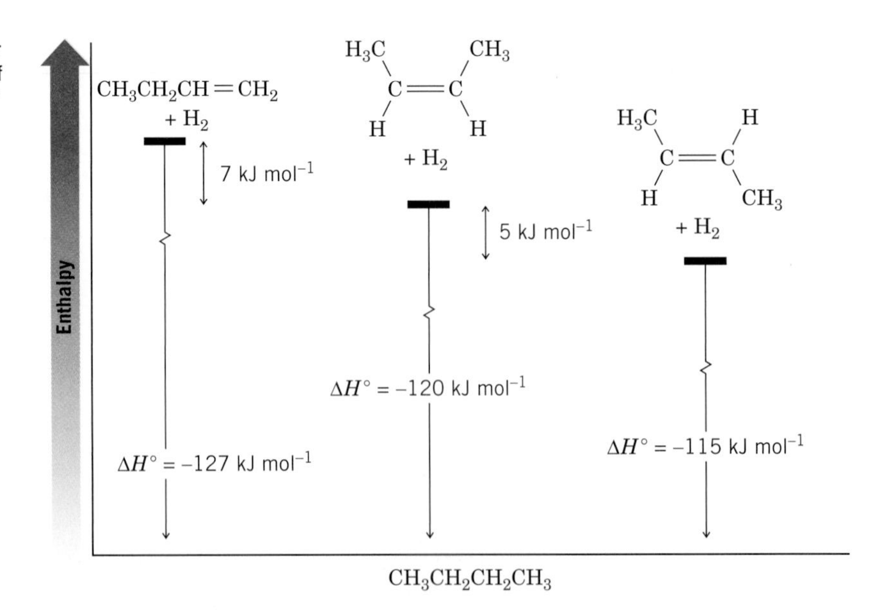

FIGURE 7.2 An energy diagram for platinum-catalyzed hydrogenation of the three butene isomers. The order of stability based on the differences in their heats of hydrogenation is *trans*-2-butene > *cis*-2-butene > 1-butene.

uct—butane—but the heat of reaction is different in each case. Upon conversion to butane, 1-butene liberates the most heat (127 kJ mol^{-1}), followed by *cis*-2-butene (120 kJ mol^{-1}), with *trans*-2-butene producing the least heat (115 kJ mol^{-1}). These data indicate that the trans isomer is more stable than the cis isomer, since less energy is released when the trans isomer is converted to butane. Furthermore, it shows that the terminal alkene, 1-butene, is less stable than either of the disubstituted alkenes, since its reaction is the most exothermic. Of course, alkenes that do not yield the same hydrogenation products cannot be compared on the basis of their respective heats of hydrogenation. In such cases it is necessary to compare other thermochemical data, such as heats of combustion, although we will not go into analyses of that type here.

7.3B Overall Relative Stabilities of Alkenes

Studies of numerous alkenes reveal a pattern of stabilities that is related to the number of alkyl groups attached to the carbon atoms of the double bond. **The greater the number of attached alkyl groups (i.e., the more highly substituted the carbon atoms of the double bond), the greater is the alkene's stability.** This order of stabilities can be given in general terms as follows:*

Relative alkene stability

Relative Stabilities of Alkenes

$R_2C{=}CR_2$ > $R_2C{=}CRH$ > $R_2C{=}CH_2$ > $RHC{=}CHR$ (trans) > $RHC{=}CHR$ (cis) > $RHC{=}CH_2$ > $H_2C{=}CH_2$

Tetrasubstituted **Trisubstituted** ⟵ **Disubstituted** ⟶ **Monosubstituted** **Unsubstituted**

PROBLEM 7.2

Heats of hydrogenation of three alkenes are as follows:

2-methyl-1-butene (− 119 kJ mol^{-1})
3-methyl-1-butene (− 127 kJ mol^{-1})
2-methyl-2-butene (− 113 kJ mol^{-1})

(a) Write the structure of each alkene and classify it as to whether its doubly bonded atoms are monosubstituted, disubstituted, trisubstituted, or tetrasubstituted. **(b)** Write the structure of the product formed when each alkene is hydrogenated. **(c)** Can heats of hydrogenation be used to relate the relative stabilities of these three alkenes? **(d)** If so, what is the predicted order of stability? If not, why not? **(e)** What other alkene isomers are possible for these alkenes? Write their structures. **(f)** What are the relative stabilities among just these isomers?

PROBLEM 7.3

Predict the more stable alkene of each pair: **(a)** 2-Methyl-2-pentene or 2,3-dimethyl-2-butene; **(b)** *cis*-3-Hexene or *trans*-3-hexene; **(c)** 1-Hexene or *cis*-3-hexene; **(d)** *trans*-2-Hexene or 2-methyl-2-pentene.

PROBLEM 7.4

Reconsider the pairs of alkenes given in Problem 7.3. Explain how IR spectroscopy can be used to differentiate between the members of each pair.

---

*This order of stabilities may seem contradictory when compared with the explanation given for the relative stabilities of cis and trans isomers. Although a detailed explanation of the trend given here is beyond our scope, the relative stabilities of substituted alkenes can be rationalized. Part of the explanation can be given in terms of the electron-releasing effect of alkyl groups (Section 6.11B), an effect that satisfies the electron-withdrawing properties of the sp^2-hybridized carbon atoms of the double bond.

7.4 Cycloalkenes

The rings of cycloalkenes containing five carbon atoms or fewer exist only in the cis form (Fig. 7.3). The introduction of a trans double bond into rings this small would, if it were possible, introduce greater strain than the bonds of the ring atoms could accommodate. (Verify this with hand-held molecular models.) *trans*-Cyclohexene might resemble the structure shown in Fig. 7.4. There is evidence that it can be formed as a very reactive short-lived intermediate in some chemical reactions, but it is not isolable as a stable molecule.

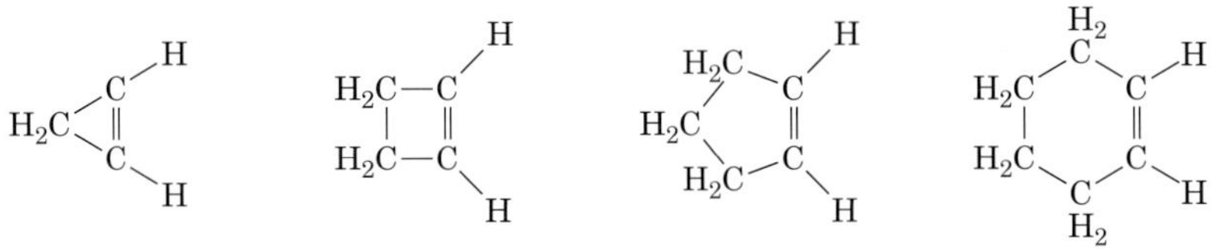

Cyclopropene Cyclobutene Cyclopentene Cyclohexene

FIGURE 7.3 *cis*-Cycloalkenes.

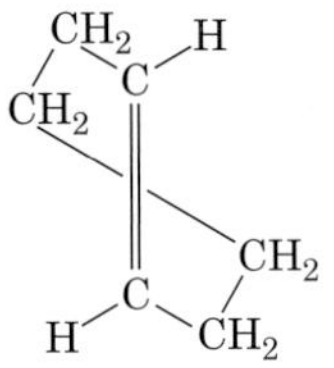

FIGURE 7.4 Hypothetical *trans*-cyclohexene. This molecule is apparently too highly strained to exist at room temperature.

trans-Cycloheptene has been observed spectroscopically, but it is a substance with a very short lifetime and has not been isolated.

trans-Cyclooctene (Fig. 7.5) has been isolated, however. Here the ring is large enough to accommodate the geometry required by a trans double bond and still be stable at room temperature. *trans*-Cyclooctene is chiral and exists as a pair of enantiomers. You may wish to verify this using hand-held models.

Exploring all of these cycloalkenes with hand-held molecular models, including both enantiomers of *trans*-cyclooctene, will help illustrate their structural differences.

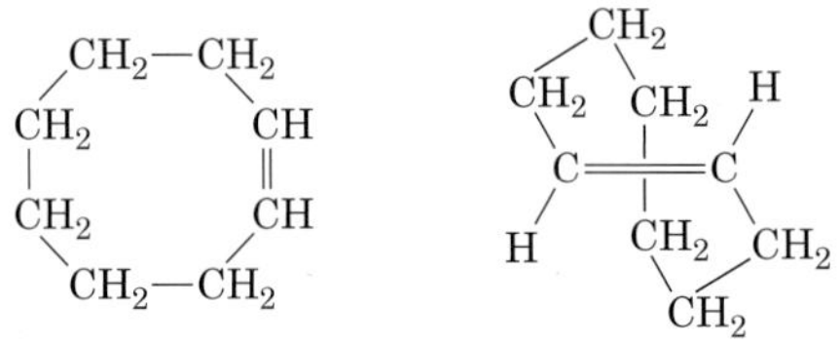

cis-Cyclooctene *trans*-Cyclooctene

FIGURE 7.5 The *cis* and *trans* forms of cyclooctene.

7.5 Synthesis of Alkenes via Elimination Reactions

Some alkene syntheses

Elimination reactions are the most important means for synthesizing alkenes. In this chapter we shall study two methods for alkene synthesis based on elimination reactions: dehydrohalogenation of alkyl halides and dehydration of alcohols.

Dehydrohalogenation of Alkyl Halides (Sections 6.15, 6.16, and 7.6)

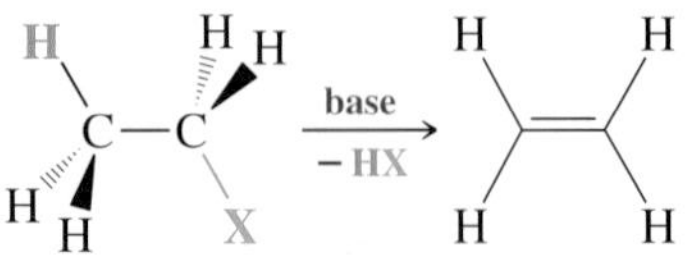

$$\xrightarrow[-\text{HX}]{\text{base}}$$

Dehydration of Alcohols (Sections 7.7 and 7.8)

$$\text{H}_2\text{C(H)}-\text{C(H}_2\text{)OH} \xrightarrow[-\text{HOH}]{\text{H}^+,\ \text{heat}} \text{H}_2\text{C}=\text{CH}_2$$

7.6 Dehydrohalogenation of Alkyl Halides

The best reaction conditions to use when synthesizing an alkene by **dehydrohalogenation** are those that promote an E2 mechanism.

$$\text{B:}^- + \text{H}-\overset{\beta}{\text{C}}-\overset{\alpha}{\text{C}}-\text{X} \xrightarrow{\text{E2}} \text{C}=\text{C} + \text{B:H} + \text{:X}^-$$

CD Tutorial

E2 elimination mechanism

Reaction conditions that favor elimination by an E1 mechanism should be avoided, in general, because the results can be too variable. The carbocation intermediate that accompanies an E1 reaction can undergo rearrangement of the carbon skeleton, as we shall see in Section 7.8, and it can also undergo substitution by an S_N1 mechanism, which competes strongly with formation of products by an E1 path.

If possible, a secondary or tertiary alkyl halide should be used to favor the E2 mechanism. When a synthesis must begin with a primary alkyl halide, then a bulky base should be used. A high concentration of a strong, relatively nonpolarizable base, such as an alkoxide anion should be employed to minimize competition from E1 and S_N1 mechanisms. Typical bases for dehydrohalogenation are sodium ethoxide in ethanol and, for a bulky base, potassium *tert*-butoxide in *tert*-butyl alcohol. As these examples show, the alkoxide base is typically dissolved in its corresponding alcohol. Potassium hydroxide in ethanol is also used sometimes; in this reagent the reactive bases probably include both ethoxide and hydroxide anions present in equilibrium. Finally, elevated temperature should be used because heat favors elimination in general.

Reaction conditions for E2 elimination

7.6A Zaitsev's Rule: Formation of the Most Substituted Alkene Is Favored with a Small Base

We showed examples in Sections 6.15–6.17 of dehydrohalogenations where only a single elimination product was possible. For example:

$$CH_3CH(Br)CH_3 \xrightarrow[\substack{C_2H_5OH \\ 55°C}]{C_2H_5O^-Na^+} CH_2{=}CHCH_3 \quad \textbf{(79\%)}$$

$$CH_3C(CH_3)(Br)CH_3 \xrightarrow[\substack{C_2H_5OH \\ 55°C}]{C_2H_5O^-Na^+} CH_2{=}C(CH_3){-}CH_3 \quad \textbf{(100\%)}$$

$$CH_3(CH_2)_{15}CH_2CH_2Br \xrightarrow[\substack{(CH_3)_3COH \\ 40°C}]{(CH_3)_3CO^-K^+} CH_3(CH_2)_{15}CH{=}CH_2 \quad \textbf{(85\%)}$$

Dehydrohalogenation of many alkyl halides, however, yields more than one product. For example, dehydrohalogenation of 2-bromo-2-methylbutane can yield two products: 2-

methyl-2-butene and 2-methyl-1-butene, as shown here by pathways (a) and (b), respectively:

$$\text{B:}^- + CH_3CH_2C(CH_3)_2Br \xrightarrow{(a)} CH_3CH{=}C(CH_3)_2 + H{-}B + :\ddot{Br}:^-$$

$$\text{B:}^- + CH_3CH_2C(CH_3)_2Br \xrightarrow{(b)} CH_3CH_2C(CH_3){=}CH_2 + H{-}B + :\ddot{Br}:^-$$

2-Bromo-2-methylbutane → **2-Methyl-2-butene** (a); **2-Methyl-1-butene** (b)

If we use a small base such as ethoxide ion or hydroxide ion, the major product of the reaction will be the **more stable alkene.** The more stable alkene, as we know from Section 7.3, has the more highly substituted double bond:

$$CH_3CH_2O^- + CH_3CH_2C(CH_3)(Br){-}CH_3 \xrightarrow[CH_3CH_2OH]{70°C} CH_3CH{=}C(CH_3)_2 + CH_3CH_2C(CH_3){=}CH_2$$

2-Methyl-2-butene (69%) (more stable) | **2-Methyl-1-butene (31%) (less stable)**

2-Methyl-2-butene is a trisubstituted alkene (three methyl groups are attached to carbon atoms of the double bond), whereas 2-methyl-1-butene is only disubstituted. 2-Methyl-2-butene is the major product.

The reason for this behavior is related to the double-bond character that develops in the transition state (cf. Section 6.16) for each reaction:

$$C_2H_5O^- + \text{H}{-}C{-}C{-}\text{Br} \longrightarrow \left[C_2H_5\overset{\delta-}{O}\cdots H \cdots C \overset{\cdots}{=} C \cdots Br^{\delta-} \right]^{\ddagger} \longrightarrow C_2H_5OH + C{=}C + Br^-$$

Transition state for an E2 reaction
The carbon–carbon bond has some of the character of a double bond.

The transition state for the reaction leading to 2-methyl-2-butene (Fig. 7.6) has the developing character of the double bond in a trisubstituted alkene. The transition state for the reaction leading to 2-methyl-1-butene has the developing character of a double bond in a disubstituted alkene. Because the transition state leading to 2-methyl-2-butene resembles a more stable alkene, this transition state is more stable (recall the Hammond–Leffler Postulate, Fig. 6.11). Because this transition state is more stable (occurs at lower free energy), the free energy of activation for this reaction is lower and 2-methyl-2-butene is formed faster. This explains why 2-methyl-2-butene is the major product. In general, the preferential formation of one product because the free energy of activation leading to its formation is lower than that for another product, and therefore the rate of its formation faster, is called **kinetic control** of product formation. (See also Section 13.10A.)

The Zaitsev product is that which is the most stable product.

Whenever an elimination occurs to give the most stable, most highly substituted alkene, chemists say that the elimination follows the **Zaitsev rule,** named for the nineteenth-century Russian chemist A. N. Zaitsev (1841–1910) who formulated it. (Zaitsev's name is also transliterated as Zaitzev, Saytzeff, Saytseff, or Saytzev.)

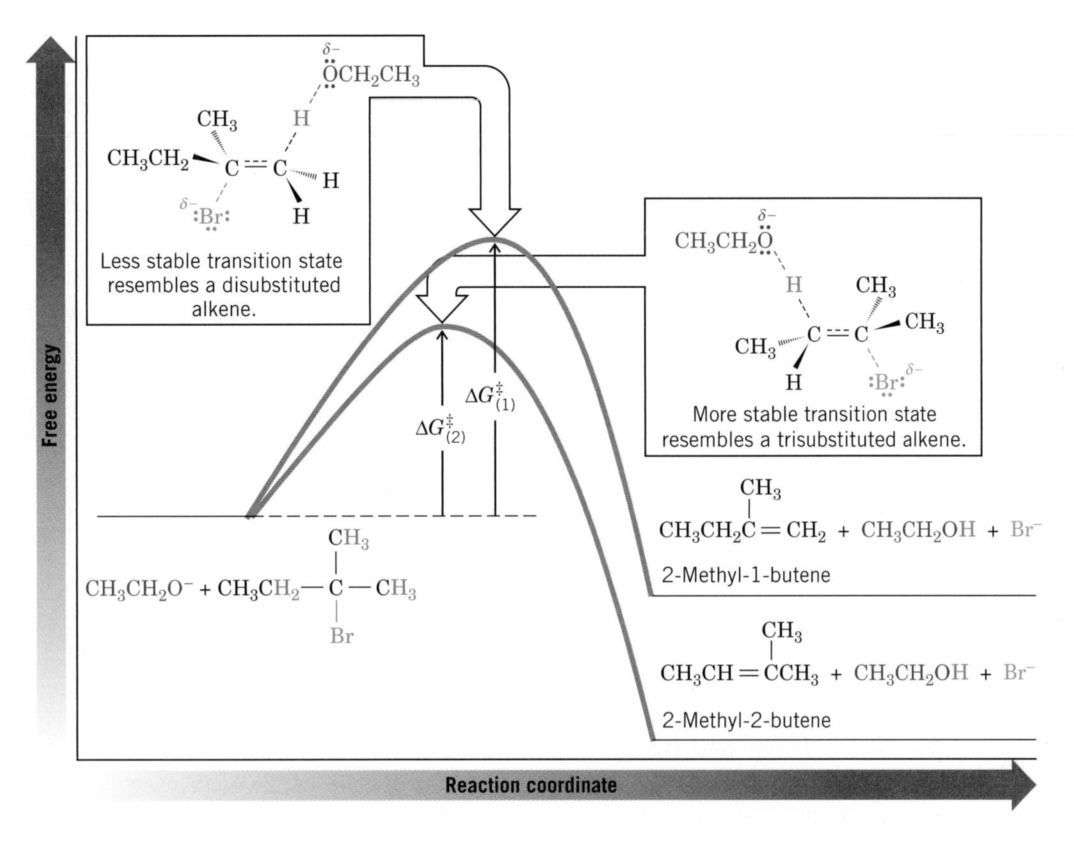

FIGURE 7.6 Reaction (2) leading to the more stable alkene occurs faster than reaction (1) leading to the less stable alkene; $\Delta G^{\ddagger}_{(2)}$ is less than $\Delta G^{\ddagger}_{(1)}$.

PROBLEM 7.5

List the alkenes that would be formed when each of the following alkyl halides is subjected to dehydrohalogenation with potassium ethoxide in ethanol and use Zaitsev's rule to predict the major product of each reaction: **(a)** 2-bromo-3-methylbutane and **(b)** 2-bromo-2,3-dimethylbutane.

7.6B Formation of the Least Substituted Alkene Using a Bulky Base

Carrying out dehydrohalogenations with a bulky base such as potassium *tert*-butoxide in *tert*-butyl alcohol favors the formation of the **less substituted alkene:**

$$(CH_3)_3C{-}O^- + CH_3CH_2{-}C(CH_3)_2{-}Br \xrightarrow[(CH_3)_3COH]{75^\circ C} CH_3CH{=}C(CH_3)_2 + CH_3CH_2C({=}CH_2)CH_3$$

2-Methyl-2-butene (27.5%) (more substituted) | **2-Methyl-1-butene (72.5%) (less substituted)**

Synthesizing the less substituted alkene

The reasons for this behavior are related in part to the steric bulk of the base and to the fact that in *tert*-butyl alcohol the base is associated with solvent molecules and thus made even larger. The large *tert*-butoxide ion appears to have difficulty removing one of the internal (2°) hydrogen atoms because of greater crowding at that site in the transition state. It removes one of the more exposed (1°) hydrogen atoms of the methyl group instead.

When an elimination yields the less substituted alkene, we say that it follows the **Hofmann rule** (see Section 20.13A).

7.6C The Stereochemistry of E2 Reactions: The Orientation of Groups in the Transition State

Stereochemistry of the E2 transition state

Considerable experimental evidence indicates that the five atoms involved in the transition state of an E2 reaction (including the base) must lie in the same plane. The requirement for coplanarity of the H—C—C—L unit arises from a need for proper overlap of orbitals in the developing π bond of the alkene that is being formed. There are two ways that this can happen:

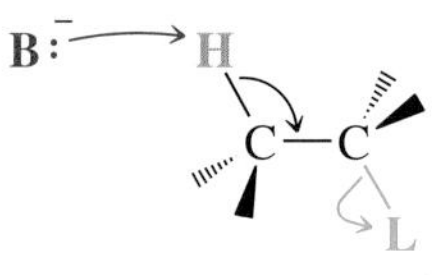

Anti coplanar transition state (preferred)

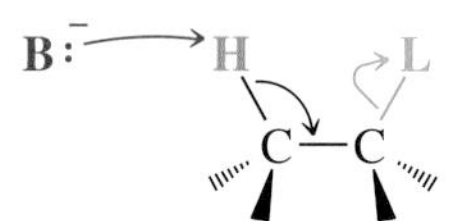

Syn coplanar transition state (only with certain rigid molecules)

Evidence also indicates that of these two arrangements for the transition state, the arrangement called the **anti coplanar** conformation is the preferred one. The **syn coplanar** transition state occurs only with rigid molecules that are unable to assume the anti arrangement. The reason: The anti coplanar transition state is staggered (and therefore of lower energy), while the syn coplanar transition state is eclipsed. Problem 7.6 will help to illustrate this difference.

PROBLEM 7.6 Consider a simple molecule such as ethyl bromide and show with Newman projection formulas how the anti coplanar transition state would be favored over the syn coplanar one.

Part of the evidence for the preferred anti coplanar arrangement of groups comes from experiments done with cyclic molecules. Two groups axially oriented on adjacent carbons in a chair conformation of cyclohexane are anti coplanar. If one of these groups is a hydrogen and the other a leaving group, the geometric requirements for an anti coplanar E2 transition state are met. Neither an axial–equatorial nor an equatorial–equatorial orientation of the groups allows formation of an anti coplanar transition state. (Note that there are no syn coplanar groups in a chair conformation, either.)

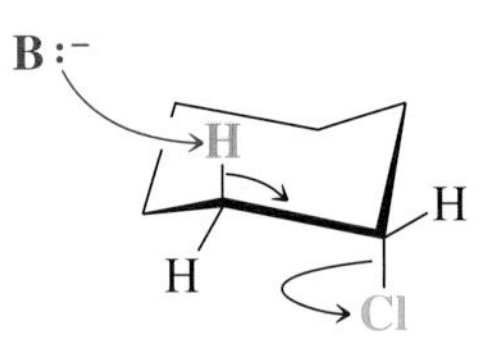

Here the β-hydrogen and the chlorine are both axial. This allows an anti coplanar transition state.

A Newman projection formula shows that the β hydrogen and the chlorine are anti coplanar when they are both axial.

As examples, let us consider the different behavior in E2 reactions shown by two compounds containing cyclohexane rings, neomenthyl chloride and menthyl chloride:

Neomenthyl chloride **Menthyl chloride**

In the more stable conformation of neomenthyl chloride (see the following mechanism), the alkyl groups are both equatorial and the chlorine is axial. There are also axial hydrogen atoms on both C1 and C3. The base can attack either of these hydrogen atoms and achieve an anti coplanar transition state for an E2 reaction. Products corresponding to each of these transition states (2-menthene and 1-menthene) are formed rapidly. In accordance with Zaitsev's rule, 1-menthene (with the more highly substituted double bond) is the major product.

Examine the conformations of neomenthyl chloride using hand-held models.

A Mechanism for the Reaction

E2 Elimination Where There Are Two Axial Cyclohexane β Hydrogens

(a) Et—Ö:⁻ (b) Et—Ö:⁻

Neomenthyl chloride

Both green hydrogens are anti to the chlorine in this the more stable conformation. Elimination by path (a) leads to 1-menthene; by path (b) to 2-menthene.

(a) → **1-Menthene (78%)** **(more stable alkene)**

(b) → **2-Menthene (22%)** **(less stable alkene)**

On the other hand, the more stable conformation of menthyl chloride has all three groups (including the chlorine) equatorial. For the chlorine to become axial, menthyl chloride has to assume a conformation in which the large isopropyl group and the methyl group are also axial. This conformation is of much higher energy, and the free energy of activation for the reaction is large because it includes the energy necessary for the conformational change. Consequently, menthyl chloride undergoes an E2 reaction very slowly, and the product is entirely 2-menthene (contrary to Zaitsev's rule). This product (or any resulting from an elimination to yield the less substituted alkene) is sometimes called the *Hofmann product* (Sections 7.6B and 20.13A).

A Mechanism for the Reaction

E2 Elimination Where the Only Eligible Axial Cyclohexane β Hydrogen Is from a Less Stable Conformer

Menthyl chloride
(more stable conformation)

Elimination is not possible for this conformation because no hydrogen is anti to the leaving group.

Menthyl chloride
(less stable conformation)

Elimination is possible from this conformation because the green hydrogen is anti to the chlorine.

2-Menthene (100%)

PROBLEM 7.7 When *cis*-1-bromo-4-*tert*-butylcyclohexane is treated with sodium ethoxide in ethanol, it reacts rapidly; the product is 4-*tert*-butylcyclohexene. Under the same conditions, *trans*-1-bromo-4-*tert*-butylcyclohexane reacts very slowly. Write conformational structures and explain the difference in reactivity of these cis–trans isomers.

PROBLEM 7.8 **(a)** When *cis*-1-bromo-2-methylcyclohexane undergoes an E2 reaction, two products (cycloalkenes) are formed. What are these two cycloalkenes, and which would you expect to be the major product? Write conformational structures showing how each is formed.
(b) When *trans*-1-bromo-2-methylcyclohexane reacts in an E2 reaction, only one cycloalkene is formed. What is this product? Write conformational structures showing why it is the only product.

7.7 ACID-CATALYZED DEHYDRATION OF ALCOHOLS

Heating most alcohols with a strong acid causes them to lose a molecule of water (to **dehydrate**) and form an alkene:

$$-\underset{H}{\overset{|}{\underset{|}{C}}}-\underset{OH}{\overset{|}{\underset{|}{C}}}- \xrightarrow[\text{heat}]{\text{HA}} \;\rangle C{=}C\langle \; + H_2O$$

The reaction is an **elimination** and is favored at higher temperatures (Section 6.18). The most commonly used acids in the laboratory are Brønsted acids—proton donors such as sulfuric acid and phosphoric acid. Lewis acids such as alumina (Al_2O_3) are often used in industrial, gas-phase dehydrations.

Dehydration reactions of alcohols show several important characteristics:

1. **The temperature and concentration of acid required to dehydrate an alcohol depend on the structure of the alcohol substrate:**

(a) Primary alcohols are the most difficult to dehydrate. Dehydration of ethanol, for example, requires concentrated sulfuric acid and a temperature of 180°C:

$$H-\underset{H}{\overset{H}{\underset{|}{\overset{|}{C}}}}-\underset{OH}{\overset{H}{\underset{|}{\overset{|}{C}}}}-H \xrightarrow[\substack{H_2SO_4\\180°C}]{\text{concd}} \underset{H}{\overset{H}{\rangle}}C{=}C\underset{H}{\overset{H}{\langle}} + H_2O$$

Ethanol
(a 1° alcohol)

(b) Secondary alcohols usually dehydrate under milder conditions. Cyclohexanol, for example, dehydrates in 85% phosphoric acid at 165–170°C:

$$\text{Cyclohexanol} \xrightarrow[165-170°C]{85\%\ H_3PO_4} \text{Cyclohexene} + H_2O$$

Cyclohexanol **Cyclohexene (80%)**

(c) Tertiary alcohols are usually so easily dehydrated that extremely mild conditions can be used. *tert*-Butyl alcohol, for example, dehydrates in 20% aqueous sulfuric acid at a temperature of 85°C:

$$(CH_3)_3C\text{—}OH \xrightarrow[85°C]{20\%\ H_2SO_4} CH_2{=}C(CH_3)_2 + H_2O$$

tert-Butyl alcohol → **2-Methylpropene (84%)**

Thus, overall, the relative ease with which alcohols undergo dehydration is in the following order:

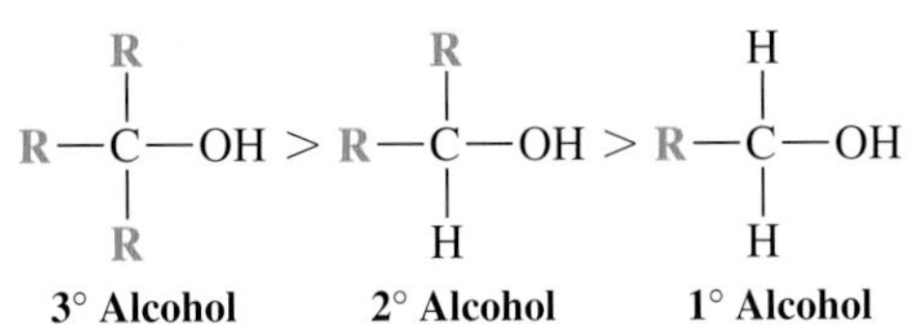

Relative ease of alcohol dehydration

This behavior, as we shall see in Section 7.7B, is related to the relative stabilities of carbocations.

2. **Some primary and secondary alcohols also undergo rearrangements of their carbon skeletons during dehydration.** Such a rearrangement occurs in the dehydration of 3,3-dimethyl-2-butanol:

$$CH_3\text{—}C(CH_3)_2\text{—}CH(OH)\text{—}CH_3 \xrightarrow[80°C]{85\%\ H_3PO_4} CH_3\text{—}C(CH_3){=}C(CH_3)\text{—}CH_3 + CH_2{=}C(CH_3)\text{—}CH(CH_3)CH_3$$

3,3-Dimethyl-2-butanol → **2,3-Dimethyl-2-butene (80%)** + **2,3-Dimethyl-1-butene (20%)**

Notice that the carbon skeleton of the reactant is

$$C\text{—}C(C)(C)\text{—}C\text{—}C \quad \text{while that of the products is} \quad C\text{—}C(C)\text{—}C(C)\text{—}C$$

We shall see in Section 7.8 that this reaction involves the migration of a methyl group from one carbon to the next so as to form a more stable carbocation.

7.7A Mechanism for Dehydration of Secondary and Tertiary Alcohols: An E1 Reaction

Explanations for these observations can be based on a stepwise mechanism originally proposed by F. Whitmore (of Pennsylvania State University). The mechanism is *an E1 reaction in which the substrate is a protonated alcohol (or an alkyloxonium ion,* see Section 6.13E). Consider the dehydration of *tert*-butyl alcohol as an example:

Step 1

$$(CH_3)_3C\text{—}\ddot{O}\text{—}H + H\text{—}\overset{+}{O}H_2 \rightleftharpoons (CH_3)_3C\text{—}\overset{+}{O}H_2 + H\text{—}\ddot{O}\text{—}H$$

Protonated alcohol or alkyloxonium ion

In this step, an acid–base reaction, a proton is rapidly transferred from the acid to one of the unshared electron pairs of the alcohol. In dilute sulfuric acid the acid is a hydronium ion; in concentrated sulfuric acid the initial proton donor is sulfuric acid itself. This step is characteristic of all reactions of an alcohol with a strong acid.

The presence of the positive charge on the oxygen of the protonated alcohol weakens all bonds to oxygen, including the carbon–oxygen bond, and in step 2 the carbon–oxygen bond breaks. The leaving group is a molecule of water:

Step 2 $(CH_3)_3C{-}\ddot{O}^{+}H_2 \rightleftharpoons (CH_3)_3C^{+} + H{-}\ddot{\underset{..}{O}}{-}H$

A carbocation

The carbon–oxygen bond breaks **heterolytically.** The bonding electrons depart with the water molecule and leave behind a carbocation. The carbocation is, of course, highly reactive because the central carbon atom has only six electrons in its valence level, not eight.

Finally, in step 3, the carbocation transfers a proton to a molecule of water. The result is the formation of a hydronium ion and an alkene:

Step 3 $H_3C{-}C^{+}(CH_3)_2 + H{-}\ddot{O}{-}H \rightleftharpoons CH_2{=}C(CH_3)_2 + H{-}\ddot{O}^{+}H_2$

2-Methylpropene

In step 3, also an acid–base reaction, any one of the nine protons available at the three methyl groups can be transferred to a molecule of water. The electron pair left behind when a proton is removed becomes the second bond of the double bond of the alkene. Notice that this step restores an octet of electrons to the central carbon atom.

PROBLEM 7.9

Dehydration of 2-propanol occurs in 75% H_2SO_4 at 100°C. **(a)** Using curved arrows, write all steps in a mechanism for the dehydration. **(b)** Explain the essential role performed in alcohol dehydrations by the acid catalyst. (*Hint:* Consider what would have to happen if no acid were present.)

7.7B Carbocation Stability and the Transition State

We saw in Section 6.11 that the order of stability of carbocations is tertiary > secondary > primary > methyl:

$$R_3C^{+} > R_2HC^{+} > RH_2C^{+} > H_3C^{+}$$

3° > 2° > 1° > **Methyl**

In the dehydration of secondary and tertiary alcohols the slowest step is formation of the carbocation (step 2 below). The first and third steps involve simple acid–base proton transfers, which occur very rapidly. The second step involves loss of the protonated hydroxyl as a leaving group, a highly endergonic process (Section 6.7), and hence it is the rate-determining step.

A Mechanism for the Reaction

Acid-Catalyzed Dehydration of Secondary or Tertiary Alcohols: An E1 Reaction

Step 1

2° or 3° Alcohol (R′ may be H) + Strong acid (typically sulfuric or phosphoric acid) ⇌ (fast) Protonated alcohol + Conjugate base

The alcohol accepts a proton from the acid in a fast step.

Step 2

Protonated alcohol ⇌ (slow, rate determining) carbocation + H_2O

The protonated alcohol loses a molecule of water to become a carbocation. This step is slow and rate determining.

Step 3

Carbocation + :A⁻ ⇌ (fast) Alkene + H—A

The carbocation loses a proton to a base. In this step, the base may be another molecule of the alcohol, water, or the conjugate base of the acid. The proton transfer results in the formation of the alkene. Note that the overall role of the acid is catalytic (it is used in the reaction and regenerated).

Because step 2 is the rate-determining step, it is this step that determines the overall reactivity of alcohols toward dehydration. With that in mind, we can now understand why tertiary alcohols are the most easily dehydrated. The formation of a tertiary carbocation is easiest because the free energy of activation for step 2 of a reaction leading to a tertiary carbocation is lowest (see Fig. 7.7). Secondary alcohols are not so easily dehydrated because the free energy of activation for their dehydration is higher—a secondary carbocation is less stable. The free energy of activation for dehydration of primary alcohols via a carbocation is so high that they undergo dehydration by another mechanism (Section 7.7C).

The reactions by which carbocations are formed from protonated alcohols are all highly *endergonic*. According to the Hammond–Leffler postulate (Section 6.13A), there should be a strong resemblance between the transition state and the product (here, a carbocation intermediate) in each case. Of the three, ***the transition state that leads to the tertiary carbocation is lowest in free energy because it resembles the most stable product, that is, the carbocation.*** By contrast, the transition state that leads to the primary carbocation occurs at highest free energy because it resembles the least stable product. In each instance, moreover, the same factor stabilizes the transition state that stabilizes the carbocation itself: **delocalization of the charge.** We can understand this if we examine the process by which the transition state is formed:

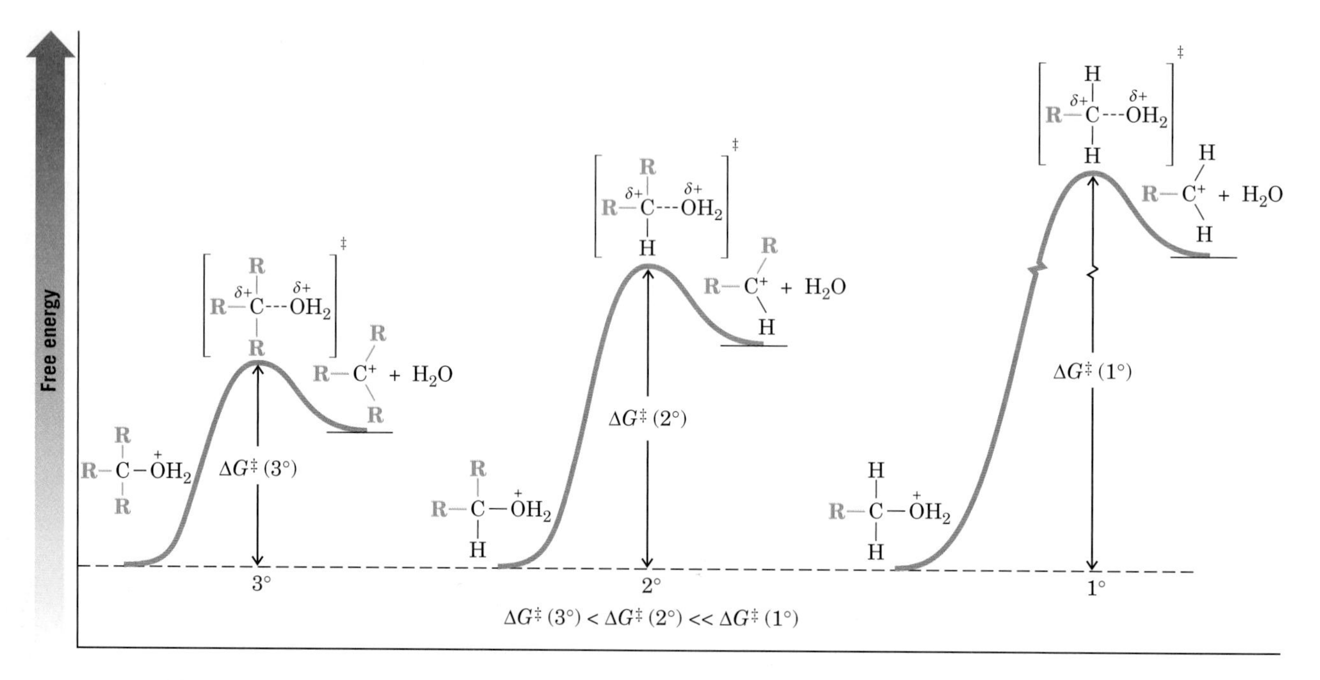

FIGURE 7.7 Free-energy diagrams for the formation of carbocations from protonated tertiary, secondary, and primary alcohols. The relative free energies of activation are tertiary < secondary ≪ primary.

$$-\overset{|}{\underset{|}{C}}-\overset{+}{O}(H)-H \rightleftharpoons \left[-\overset{|}{\underset{|}{C}}\overset{\delta+}{\cdots}\overset{\delta+}{O}(H)-H\right]^{\ddagger} \rightleftharpoons -C^{+} + :\ddot{O}(H)-H$$

Protonated alcohol **Transition state** **Carbocation**

The oxygen atom of the protonated alcohol bears a full positive charge. As the transition state develops, this oxygen atom begins to separate from the carbon atom to which it is attached. The carbon atom, because it is losing the electrons that bonded it to the oxygen atom, begins to develop a partial positive charge. This developing positive charge ***is most effectively delocalized in the transition state leading to a tertiary carbocation because of the presence of three alkyl groups that contribute electron density by hyperconjugation (Section 6.10A) to the developing carbocation.*** The positive charge is less effectively delocalized in the transition state leading to a secondary carbocation (*two* electron-releasing groups) and is least effectively delocalized in the transition state leading to a primary carbocation (*one* electron-releasing group). For this reason the dehydration of a primary alcohol proceeds through a different mechanism—an E2 mechanism:

Transition state leading to 3° carbocation (most stable) **Transition state leading to 2° carbocation** **Transition state leading to 1° carbocation (least stable)**

7.7C A Mechanism for Dehydration of Primary Alcohols: An E2 Reaction

The dehydration of primary alcohols is believed to proceed through an E2 mechanism because the primary carbocation required for dehydration by an E1 mechanism would be too unstable. The first step in dehydration of a primary alcohol is protonation, just as in the E1 mechanism. Then, with the protonated hydroxyl as a good leaving group, a Lewis base in the reaction mixture removes a β hydrogen simultaneously with formation of the alkene double bond and departure of the protonated hydroxyl group (water).

A Mechanism for the Reaction

Dehydration of a Primary Alcohol: An E2 Reaction

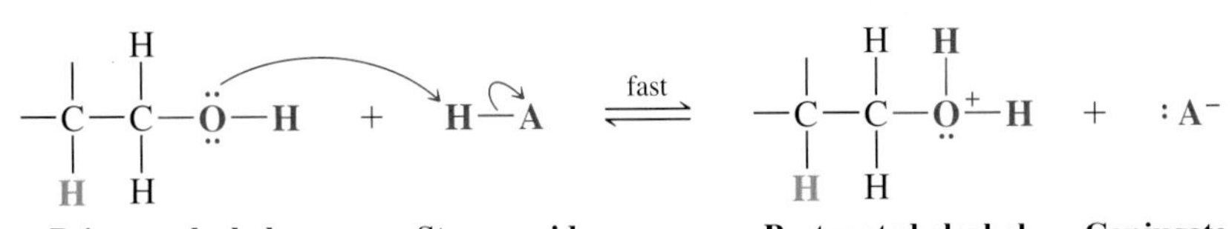

The alcohol accepts a proton from the acid in a fast step.

$$:A^- + \text{H—}\overset{|}{\underset{|}{C}}\text{—}\overset{|}{\underset{|}{C}}\text{—}\overset{+}{\ddot{O}}\text{—H} \xrightarrow[\text{(rate determining)}]{\text{slow}} \text{C=C} + \text{H—A} + :\ddot{O}\text{—H}$$

Alkene

A base removes a hydrogen from the β carbon as the double bond forms and the protonated hydroxyl group departs. (The base may be another molecule of the alcohol or the conjugate base of the acid.)

7.8 Carbocation Stability and the Occurrence of Molecular Rearrangements

With an understanding of carbocation stability and its effect on transition states, we can now proceed to explain the rearrangements of carbon skeletons that occur in some alcohol dehydrations.

7.8A Rearrangements During Dehydration of Secondary Alcohols

Consider again the rearrangement that occurs when 3,3-dimethyl-2-butanol is dehydrated:

$$CH_3\text{—}C(CH_3)_2\text{—}CH(OH)\text{—}CH_3 \xrightarrow[\text{heat}]{85\%\ H_3PO_4} CH_3\text{—}C(CH_3)\text{=}C(CH_3)\text{—}CH_3 + CH_2\text{=}C(CH_3)\text{—}CH(CH_3)\text{—}CH_3$$

3,3-Dimethyl-2-butanol **2,3-Dimethyl-2-butene (major product)** **2,3-Dimethyl-1-butene (minor product)**

The first step of this dehydration is the formation of the protonated alcohol in the usual way:

Step 1

$$CH_3\text{—}C(CH_3)_2\text{—}CH(:\ddot{O}\text{—H})\text{—}CH_3 + H\text{—}\overset{+}{\ddot{O}}H_2 \rightleftharpoons CH_3\text{—}C(CH_3)_2\text{—}CH(:\overset{+}{O}H_2)\text{—}CH_3 + H_2\ddot{O}:$$

Protonated alcohol

In the second step the protonated alcohol loses water and a secondary carbocation forms:

Step 2

$$CH_3-\underset{CH_3}{\overset{CH_3}{C}}-\underset{:\overset{+}{O}H_2}{CH}-CH_3 \rightleftharpoons CH_3-\underset{CH_3}{\overset{CH_3}{C}}-\overset{+}{C}H-CH_3 + H_2\ddot{O}:$$

A 2° carbocation

Now the rearrangement occurs. ***The less stable, secondary carbocation rearranges to a more stable tertiary carbocation:***

Step 3

$$CH_3-\underset{CH_3}{\overset{CH_3}{C}}-\overset{+}{C}HCH_3 \longrightarrow \left[CH_3-\underset{CH_3}{\overset{\delta+\quad CH_3 \quad\delta+}{C}}-CHCH_3\right]^{\ddagger} \longrightarrow CH_3-\underset{CH_3}{\overset{+}{C}}-\overset{CH_3}{C}HCH_3$$

2° Carbocation (less stable) — **Transition state** — **3° Carbocation (more stable)**

The rearrangement occurs through the migration of an alkyl group (methyl) from the carbon atom adjacent to the one with the positive charge. The methyl group migrates **with its pair of electrons,** that is, as a methyl anion, $^{-}{:}CH_3$ (called a **methanide** ion). After the migration is complete, the carbon atom that the methyl anion left has become a carbocation, and the positive charge on the carbon atom to which it migrated has been neutralized. Because a group migrates from one carbon to the next, this kind of rearrangement is often called a **1,2 shift.**

In the transition state the shifting methyl is partially bonded to both carbon atoms by the pair of electrons with which it migrates. It never leaves the carbon skeleton.

The final step of the reaction is the removal of a proton from the new carbocation (by a Lewis base in the reaction mixture) and the formation of an alkene. This step, however, can occur in two ways:

Step 4

$$A:^{-}\ \text{(a) or (b)}\qquad H-CH_2-\underset{CH_3}{\overset{+}{C}}-\underset{CH_3}{\overset{H}{C}}-CH_3 \xrightarrow{(a)} CH_2{=}\underset{CH_3}{C}-\underset{CH_3}{CH}-CH_3 \qquad \text{Less stable alkene (minor product)}$$

$$\xrightarrow{(b)} CH_3-\underset{CH_3}{C}{=}\underset{CH_3}{C}-CH_3 \qquad \text{More stable alkene (major product)} \qquad + HA$$

Alcohol dehydration follows Zaitsev's rule.

The more favored route is dictated by the type of alkene being formed. Path (b) leads to the highly stable tetrasubstituted alkene, and this is the path followed by most of the carbocations. Path (a), on the other hand, leads to a less stable, disubstituted alkene and produces the minor product of the reaction. ***The formation of the more stable alkene is the general rule (Zaitsev's rule) in the acid-catalyzed dehydration reactions of alcohols.***

Studies of thousands of reactions involving carbocations show that rearrangements like those just described are general phenomena. ***They occur almost invariably when the migration of an alkanide ion or hydride ion can lead to a more stable carbocation.*** The following are examples:

$$CH_3-\overset{CH_3}{\underset{CH_3}{C}}-\overset{+}{C}H-CH_3 \xrightarrow[\text{migration}]{\text{methanide}} CH_3-\overset{+}{\underset{CH_3}{C}}-\overset{CH_3}{CH}-CH_3$$

2° Carbocation **3° Carbocation**

$$CH_3-\overset{H}{\underset{CH_3}{C}}-\overset{+}{C}H-CH_3 \xrightarrow[\text{migration}]{\text{hydride}} CH_3-\overset{+}{\underset{CH_3}{C}}-\overset{H}{CH}-CH_3$$

2° Carbocation **3° Carbocation**

We shall see biological examples of methanide and hydride migrations in "The Chemistry of . . . Cholesterol Biosynthesis" (Chapter 8).

Rearrangements of carbocations can also lead to a change in ring size, as the following example shows:

OH, CH_3, $CH-CH_3$ $\xrightarrow[(-H_2O)]{\text{HA, heat}}$ CH_3, $\overset{+}{C}H-CH_3$ ⟶

2° Carbocation

CH_3, $CHCH_3$ ≡ CH_3, CH_3, H, :A$^-$ $\xrightarrow{-HA}$ CH_3, CH_3

3° Carbocation

This process is especially favorable if a relief in ring strain occurs.

PROBLEM 7.10

Acid-catalyzed dehydration of neopentyl alcohol, $(CH_3)_3CCH_2OH$, yields 2-methyl-2-butene as the major product. Outline a mechanism showing all steps in its formation.

PROBLEM 7.11

Acid-catalyzed dehydration of either 2-methyl-1-butanol or 3-methyl-1-butanol gives 2-methyl-2-butene as the major product. Write plausible mechanisms that explain these results.

PROBLEM 7.12

When the compound called *isoborneol* is heated with 50% sulfuric acid, the product of the reaction is the compound called camphene and not bornylene, as one might expect. Using models to assist you, write a step-by-step mechanism showing how camphene is formed:

HO $\xrightarrow[\text{heat}]{H_3O^+}$ CH_2 not

Isoborneol **Camphene** **Bornylene**

7.8B Rearrangement after Dehydration of a Primary Alcohol

Rearrangements also accompany the dehydration of primary alcohols. Since a primary carbocation is unlikely to be formed during dehydration of a primary alcohol, the alkene that is produced initially from a primary alcohol arises by an E2 mechanism, as described in Section 7.7C. However, an alkene can accept a proton to *generate* a carbocation in a process that is essentially the reverse of the *deprotonation* step in the E1 mechanism for dehydration of an alcohol (Section 7.7A). When a terminal alkene does this by using its π electrons to bond a proton at the terminal carbon, a carbocation forms at the second carbon of the chain.* This carbocation, since it is internal to the chain, will be secondary or tertiary, depending on the specific substrate. Various processes that you have already learned can now occur from this carbocation: (1) a different β hydrogen may be removed, leading to a more stable alkene than the initially formed terminal alkene; (2) a hydride or alkanide rearrangement may occur leading to a yet more stable carbocation (e.g., moving from a 2° to a 3° carbocation), after which the elimination may be completed; or (3) a nucleophile may attack any of these carbocations to form a substitution product. Under the high-temperature conditions for alcohol dehydration the principal products will be alkenes rather than substitution products.

A Mechanism for the Reaction

Formation of a Rearranged Alkene During Dehydration of a Primary Alcohol

Primary alcohol (R may be H) $\rightleftharpoons$ (E2) The initial alkene + $H_2\ddot{O}$ + H—A

The primary alcohol initially undergoes acid-catalyzed dehydration by an E2 mechanism (Section 7.7C).

(protonation)

The π electrons of the initial alkene can then be used to form a bond with a proton at the terminal carbon, forming a secondary or tertiary carbocation.

(deprotonation) Final alkene + H—A

A different β hydrogen can be removed from the carbocation, so as to form a more highly substituted alkene than the initial alkene. This deprotonation step is the same as the usual completion of an E1 elimination. (This carbocation could experience other fates, such as further rearrangement before elimination or substitution by an S_N1 process.)

*The carbocation could also form directly from the primary alcohol by a hydride shift from its β carbon to the terminal carbon as the protonated hydroxyl group departs:

$\longrightarrow$ + A—H + H_2O

7.9 Synthesis of Alkynes by Elimination Reactions

Alkynes can be synthesized from alkenes via compounds called vicinal dihalides. A vicinal dihalide (abbreviated ***vic*-dihalide**) is a compound bearing the halogens on adjacent carbons (vicinal derives from the Latin *vicinus,* meaning adjacent). A vicinal dibromide, for example, can be synthesized by addition of bromine to an alkene (Section 8.6). The *vic*-dibromide can then be subjected to a double dehydrohalogenation reaction with a strong base to yield an alkyne.

$$\text{RCH{=}CHR} + \mathbf{Br_2} \longrightarrow \underset{\textbf{A } \textit{vic}\textbf{-dibromide}}{\text{R—CH(Br)—CH(Br)—R}} \xrightarrow{2\ \text{NaNH}_2} \text{R—C}{\equiv}\text{C—R} + 2\text{NH}_3 + 2\ \text{Na}\mathbf{Br}$$

The dehydrohalogenations occur in two steps, the first yielding a bromoalkene, and the second, the alkyne.

A Mechanism for the Reaction

Dehydrohalogenation of *vic*-Dibromides to form Alkynes

Reaction:

$$\text{R—CH(Br)—CH(Br)—R} + 2\ \text{NH}_2^- \longrightarrow \text{R—C}{\equiv}\text{C—R} + 2\ \text{NH}_3 + 2\ \text{Br}^-$$

Mechanism:

Step 1

$$\underset{\textbf{Amide ion}}{\text{H—}\ddot{\text{N}}{:}^-\text{—H}} + \underset{\textit{vic}\textbf{-Dibromide}}{\text{R—CH(Br)—CH(Br)—R}} \longrightarrow \underset{\textbf{Bromoalkene}}{\text{R(H)C{=}C(Br)R}} + \underset{\textbf{Ammonia}}{\text{H—}\ddot{\text{N}}\text{—H(H)}} + \underset{\textbf{Bromide ion}}{:\ddot{\text{Br}}:^-}$$

The strongly basic amide ion brings about an E2 reaction.

Step 2

$$\underset{\textbf{Bromoalkene}}{\text{R(Br)C{=}C(H)R}} + \underset{\textbf{Amide ion}}{:\ddot{\text{N}}^-\text{—H(H)}} \longrightarrow \underset{\textbf{Alkyne}}{\text{R—C}{\equiv}\text{C—R}} + \underset{\textbf{Ammonia}}{\text{H—}\ddot{\text{N}}\text{—H(H)}} + \underset{\textbf{Bromide ion}}{:\ddot{\text{Br}}:^-}$$

A second E2 reaction produces the alkyne.

Depending on the conditions, these two dehydrohalogenations may be carried out as separate reactions, or they may be carried out consecutively in a single mixture. Sodium amide, a strong base, is capable of effecting both dehydrohalogenations in a single reaction mixture. (At least two molar equivalents of sodium amide per mole of the dihalide must be used. If the product is to be a terminal alkyne, then three molar equivalents must be used because the terminal alkyne is deprotonated by sodium amide as it is formed in the mixture, consuming the sodium amide otherwise needed for the remaining dehydrohalogenation steps. See Section 7.10.) Dehydrohalogenations with sodium amide are usually carried out in liquid ammonia or in an inert medium such as mineral oil.

The following example illustrates this method:

$$CH_3CH_2CH{=}CH_2 \xrightarrow[CCl_4]{Br_2} CH_3CH_2CHBrCH_2Br \xrightarrow[\text{mineral oil, } 110\text{–}160^\circ C]{NaNH_2}$$

$$\left[\begin{array}{c} CH_3CH_2CH{=}CHBr \\ + \\ CH_3CH_2CBr{=}CH_2 \end{array}\right] \xrightarrow[\text{mineral oil, } 110\text{–}160^\circ C]{NaNH_2} [CH_3CH_2C{\equiv}CH] \xrightarrow{NaNH_2}$$

$$CH_3CH_2C{\equiv}C{:}^- \, Na^+ \xrightarrow{NH_4Cl} CH_3CH_2C{\equiv}CH + NH_3 + NaCl$$

Geminal dihalides can also be converted to alkynes by dehydrohalogenation. A geminal dihalide (abbreviated ***gem*-dihalide**) has two halogen atoms bonded to the same carbon (from the Latin *geminus,* twins). Ketones can be converted to *gem*-dichlorides through their reactions with phosphorus pentachloride, and these products can be used to synthesize alkynes. Ketones can be converted to *gem*-dichlorides through their reaction with phosphorus pentachloride, and these products can also be used to synthesize alkynes.

$$C_6H_{11}COCH_3 \xrightarrow[0^\circ C \; (-POCl_3)]{PCl_5} C_6H_{11}CCl_2CH_3 \xrightarrow[(2)\ HA]{(1)\ 3\ NaNH_2,\ \text{mineral oil, heat}} C_6H_{11}C{\equiv}CH$$

Cyclohexyl methyl ketone — **A *gem*-dichloride (70–80%)** — **Cyclohexylacetylene (46%)**

PROBLEM 7.13

Outline all steps in a synthesis of propyne from each of the following:

(a) CH_3COCH_3 **(c)** $CH_3CHBrCH_2Br$

(b) $CH_3CH_2CHBr_2$ **(d)** $CH_3CH{=}CH_2$

7.10 The Acidity of Terminal Alkynes

The hydrogen bonded to the carbon of a terminal alkyne is considerably more acidic than those bonded to carbons of an alkene or alkane (see Section 3.7). The pK_a values for ethyne, ethene, and ethane illustrate this point:

$$H{-}C{\equiv}C{-}H \qquad H_2C{=}CH_2 \qquad H_3C{-}CH_3$$

$$pK_a = 25 \qquad pK_a = 44 \qquad pK_a = 50$$

The order of basicity of their anions is opposite that of their relative acidity:

Relative Basicity

$$CH_3CH_2{:}^- > CH_2{=}CH{:}^- > HC{\equiv}C{:}^-$$

If we include in our comparison hydrogen compounds of other first-row elements of the periodic table, we can write the following orders of relative acidities and basicities:

Relative Acidity

$$H{-}\ddot{\underset{..}{O}}H > H{-}\ddot{\underset{..}{O}}R > H{-}C{\equiv}CR > H{-}\ddot{N}H_2 > H{-}CH{=}CH_2 > H{-}CH_2CH_3$$

pK_a 15.7 16–17 25 38 44 50

Relative Basicity

$$^-{:}\ddot{\underset{..}{O}}H < {}^-{:}\ddot{\underset{..}{O}}R < {}^-{:}C{\equiv}CR < {}^-{:}\ddot{N}H_2 < {}^-{:}CH{=}CH_2 < {}^-{:}CH_2CH_3$$

We see from the order just given that while terminal alkynes are more acidic than ammonia, they are less acidic than alcohols and are less acidic than water.

The arguments just made apply only to acid–base reactions that take place in solution. In the gas phase, acidities and basicities are very much different. For example, in the gas phase the hydroxide ion is a stronger base than the acetylide ion. The explanation for this shows us again the important roles solvents play in reactions that involve ions (see Section 6.13). In solution, smaller ions (e.g., hydroxide ions) are more effectively solvated than larger ones (e.g., ethynide ions). Because they are more effectively solvated, smaller ions are more stable and are therefore less basic. In the gas phase, large ions are stabilized by polarization of their bonding electrons, and the bigger a group is, the more polarizable it will be. Thus, in the gas phase larger ions are less basic.

PROBLEM 7.14

Predict the products of the following acid–base reactions. If the equilibrium would not result in the formation of appreciable amounts of products, you should so indicate. In each case label the stronger acid, the stronger base, the weaker acid, and the weaker base:

(a) $CH_3CH{=}CH_2 + NaNH_2 \longrightarrow$
(b) $CH_3C{\equiv}CH + NaNH_2 \longrightarrow$
(c) $CH_3CH_2CH_3 + NaNH_2 \longrightarrow$
(d) $CH_3C{\equiv}C{:}^- + CH_3CH_2OH \longrightarrow$
(e) $CH_3C{\equiv}C{:}^- + NH_4Cl \longrightarrow$

7.11 Replacement of the Acetylenic Hydrogen Atom of Terminal Alkynes

Sodium ethynide and other sodium alkynides can be prepared by treating terminal alkynes with sodium amide in liquid ammonia:

$$H{-}C{\equiv}C{-}H + NaNH_2 \xrightarrow{\text{liq. } NH_3} H{-}C{\equiv}C{:}^-Na^+ + NH_3$$

$$CH_3C{\equiv}C{-}H + NaNH_2 \xrightarrow{\text{liq. } NH_3} CH_3C{\equiv}C{:}^-Na^+ + NH_3$$

These are acid–base reactions. The amide ion, by virtue of its being the anion of a very weak acid, ammonia (pK_a = 38), is able to remove the acetylenic protons of terminal alkynes (pK_a = 25). These reactions, for all practical purposes, go to completion.

As we saw in Section 4.18C, sodium alkynides are useful intermediates for the synthesis of other alkynes. These syntheses can be accomplished by treating the sodium alkynide with a primary alkyl halide:

$$\underset{\textbf{Sodium alkynide}}{R{-}C{\equiv}C{:}^-\ Na^+} + \underset{\textbf{Primary alkyl halide}}{R'CH_2{-}Br} \longrightarrow \underset{\textbf{Mono- or disubstituted acetylene}}{R{-}C{\equiv}C{-}CH_2R'} + NaBr$$

(R or R′ or both may be hydrogen.)

The following example illustrates this synthesis of higher alkyne homologues:

$$CH_3CH_2C{\equiv}C{:}^- \; Na^+ + CH_3CH_2{-}Br \xrightarrow[6\text{ h}]{\text{liq. }NH_3} CH_3CH_2C{\equiv}CCH_2CH_3 + NaBr$$

3-Hexyne
(75%)

In this and other examples we have seen that the alkynide ion acts as a nucleophile and displaces a halide ion from the primary alkyl halide. We now recognize this as **an S_N2 reaction** (Section 6.5):

$$RC{\equiv}C{:}^- \; Na^+ \;\; + \;\; R'CH_2{-}\ddot{B}r{:} \xrightarrow[S_N2]{\text{nucleophilic substitution}} RC{\equiv}C{-}CH_2R' + NaBr$$

Sodium alkynide — **1° Alkyl halide**

The unshared electron pair of the alkynide ion attacks the back side of the carbon atom that bears the halogen atom and forms a bond to it. The halogen atom departs as a halide ion.

This synthesis fails when secondary or tertiary halides are used because the alkynide ion acts as a base rather than as a nucleophile, and the major result is an **E2 elimination** (Section 6.16). The products of the elimination are an alkene and the alkyne from which the sodium alkynide was originally formed:

$$RC{\equiv}C{:}^- \;\; + \;\; R'CH_2{-}CH(R''){-}Br \xrightarrow{E2} RC{\equiv}CH + R'CH{=}CHR'' + Br^-$$

2° Alkyl halide

PROBLEM 7.15

Your goal is to synthesize 4,4-dimethyl-2-pentyne. You have a choice of beginning with any of the following reagents:

$CH_3C{\equiv}CH$ $\quad$ $(CH_3)_3C{-}Br$ $\quad$ $(CH_3)_3C{-}C{\equiv}CH$ $\quad$ CH_3I

Assume that you also have available sodium amide and liquid ammonia. Outline the best synthesis of the required compound.

7.12 Hydrogenation of Alkenes

Alkenes react with hydrogen in the presence of a variety of metal catalysts to add one hydrogen atom to each carbon atom of the double bond (Sections 4.17, 5.10, 5.17). Hydrogenation reactions that involve finely divided *insoluble* platinum, palladium, or nickel catalysts (Section 4.17) are said to proceed by **heterogeneous catalysis** because the substrate is soluble in the reaction mixture but the catalyst is not. Hydrogenation reactions where the catalyst is *soluble* in the reaction mixture involve **homogeneous catalysis.** Typical homogeneous hydrogenation catalysts include rhodium and ruthenium complexes that bear various phosphorus and other ligands. One of the most well-known homogeneous hydrogenation catalysts is Wilkinson's catalyst, tris(triphenylphosphine)-rhodium chloride, $Rh[(C_6H_5)_3P]_3Cl$. The following are some examples of hydrogenation reactions under hetereogeneous and homogeneous catalysis:

$$CH_2{=}CH_2 + H_2 \xrightarrow[\text{or Pt}\ 25°C]{\text{Ni,Pd,}} CH_3{-}CH_3$$

$$CH_3CH{=}CH_2 + H_2 \xrightarrow[\text{or Pt}\ 25°C]{\text{Ni,Pd,}} CH_3CH_2{-}CH_3$$

$$CH_3CH_2CH_2CH_2CH{=}CH_2 + H_2 \xrightarrow{Rh[(C_6H_5)_3P]_3Cl} CH_3CH_2CH_2CH_2CH_2CH_3$$

The type of reaction that takes place in these examples is an **addition reaction.** The product that results from the addition of hydrogen to an alkene is an alkane. Alkanes have only single bonds and contain the maximum number of hydrogen atoms that a hydrocarbon can possess. For this reason, alkanes are said to be **saturated compounds.** Alkenes, because they contain a double bond and possess fewer than the maximum number of hydrogen atoms, are capable of adding hydrogen and are said to be **unsaturated.** The process of adding hydrogen to an alkene is sometimes described as being one of **reduction.** Most often, however, the term used to describe the addition of hydrogen is **catalytic hydrogenation.** Now let us see what the mechanism is for heterogeneous catalytic hydrogenation. (The mechanism of homogeneous catalysis is discussed in Special Topic G.)

The Chemistry of...

Hydrogenation in the Food Industry

The food industry makes use of catalytic hydrogenation to convert liquid vegetable oils to semisolid fats in making margarine and solid cooking fats. Examine the labels of many prepared foods and you will find that they contain "partially hydrogenated vegetable oils." There are several reasons why foods contain these oils, but one is that partially hydrogenated vegetable oils have a longer shelf life.

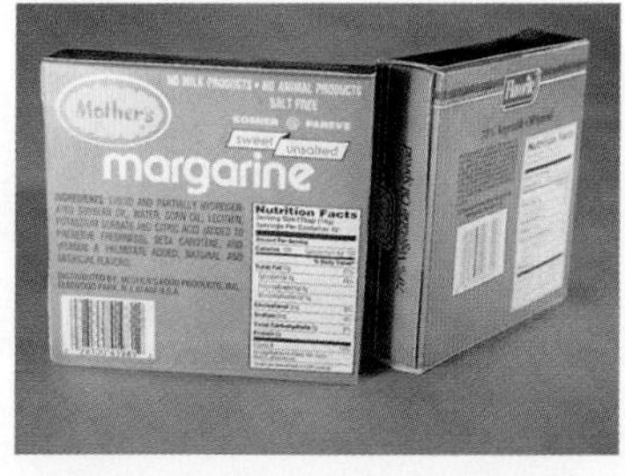

A product used in baking that contains oils and mono- and diacylglycerols that are partially hydrogenated.

Fats and oils (Section 23.2) are glyceryl esters of carboxylic acids with long carbon chains, called "fatty acids." Fatty acids are saturated (no double bonds), monounsaturated (one double bond), or polyunsaturated (more than one double bond). Oils typically contain a higher proportion of fatty acids with one or more double bonds than fats do. Hydrogenation of an oil converts some of its double bonds to single bonds, and this conversion has the effect of producing a fat with the consistency of margarine or a semisolid cooking fat.

Our bodies are incapable of making polyunsaturated fats, and therefore, such fats must be present in our diets in moderate amounts in order to maintain health. Saturated fats can be made in the cells of our body from other food sources, for example, from carbohydrates (i.e., from sugars and starches). For this reason saturated fats in our diet are not absolutely necessary, and, indeed, too much saturated fat has been implicated in the development of cardiovascular disease.

One potential problem that arises from using catalytic hydrogenation to produce partially hydrogenated vegetable oils is that the catalysts used for hydrogenation cause isomerization of some of the double bonds of the fatty acids (some of those that do not absorb hydrogen). In most natural fats and oils, the double bonds of the fatty acids have the cis configuration. The catalysts used for hydrogenation convert some of these cis double bonds to the unnatural trans configuration. The health effects of trans fatty acids are still under study, but experiments thus far indicate that they cause an increase in serum levels of cholesterol and triacylglycerols, which in turn increases the risk of cardiovascular disease.

7.13 Hydrogenation: The Function of the Catalyst

Hydrogenation of an alkene is an exothermic reaction ($\Delta H° \cong -120$ kJ mol^{-1}):

$$R{-}CH{=}CH{-}R + H_2 \xrightarrow{\text{hydrogenation}} R{-}CH_2{-}CH_2{-}R + \text{heat}$$

Although the process is exothermic, there is usually a high free energy of activation for uncatalyzed alkene hydrogenation, and therefore, the uncatalyzed reaction does not take place at room temperature. However, hydrogenation will take place readily at room temperature in the presence of a catalyst because the catalyst provides a new pathway for the reaction that involves lower free energy of activation (Fig. 7.8).

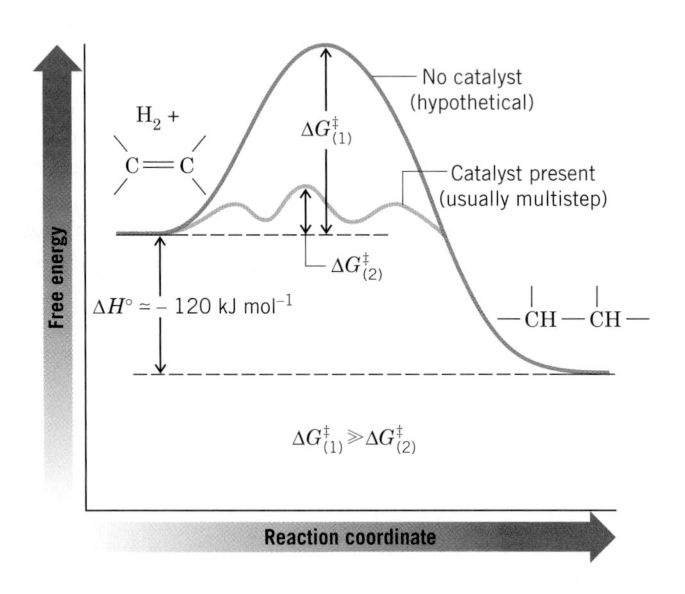

FIGURE 7.8 Free-energy diagram for the hydrogenation of an alkene in the presence of a catalyst and the hypothetical reaction in the absence of a catalyst. The free energy of activation for the uncatalyzed reaction ($\Delta G^{\ddagger}_{(1)}$) is very much larger than the largest free energy of activation for the catalyzed reaction ($\Delta G^{\ddagger}_{(2)}$) (the uncatalyzed hydrogenation reaction does not occur).

CD Tutorial
Hydrogenation

Heterogeneous hydrogenation catalysts typically involve finely divided platinum, palladium, nickel, or rhodium deposited on the surface of powdered carbon (charcoal). Hydrogen gas introduced into the atmosphere of the reaction vessel adsorbs to the metal by a chemical reaction where unpaired electrons on the surface of the metal *pair* with the electrons of hydrogen (Fig. 7.9*a*) and bind the hydrogen to the surface. The collision of an alkene with the surface bearing adsorbed hydrogen causes adsorption of the alkene as well (Fig. 7.9*b*). A stepwise transfer of hydrogen atoms takes place, and this produces an alkane before the organic molecule leaves the catalyst surface (Figs. 7.9*c*,*d*). As a conse-

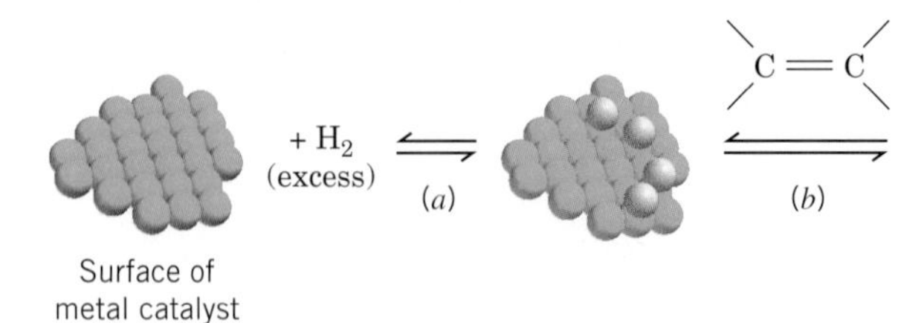

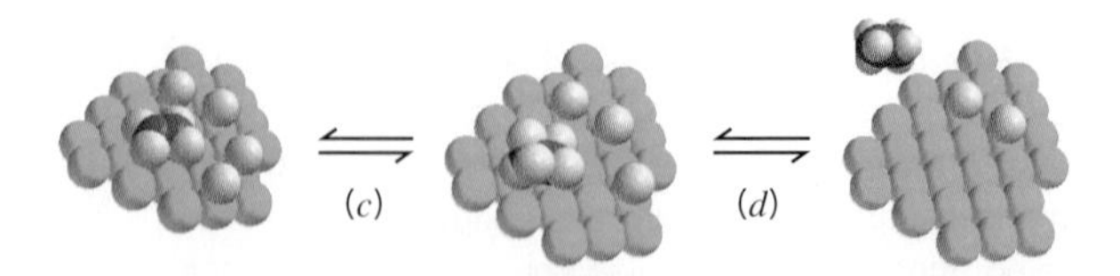

FIGURE 7.9 The mechanism for the hydrogenation of an alkene as catalyzed by finely divided platinum metal: (*a*) hydrogen adsorption; (*b*) adsorption of the alkene; (*c*,*d*), stepwise transfer of both hydrogen atoms to the same face of the alkene (*syn* addition).

quence, *both hydrogen atoms usually add from the same side of the molecule.* This mode of addition is called a **syn** addition (Section 7.13A):

Catalytic hydrogenation is a syn addition.

7.13A Syn and Anti Additions

An addition that places the parts of the adding reagent on the same side (or face) of the reactant is called **syn addition.** We have just seen that the platinum-catalyzed addition of hydrogen (X = Y = H) is a syn addition:

A syn addition

The opposite of a syn addition is an **anti addition.** An anti addition places the parts of the adding reagent on opposite faces of the reactant.

An anti addition

In Chapter 8 we shall study a number of important syn and anti additions.

The Chemistry of...

Homogeneous Asymmetric Catalytic Hydrogenation: Examples Involving L-DOPA, (*S*)-Naproxen, and Aspartame

Development by Geoffrey Wilkinson of a soluble catalyst for hydrogenation [tris(triphenylphosphine)rhodium chloride, Section 7.12] led to Wilkinson's earning a share of the 1973 Nobel Prize in Chemistry. His initial discovery, while at Imperial College, University of London, inspired many other researchers to create novel catalysts based on the Wilkinson catalyst. Some of these researchers were themselves recognized by the 2001 Nobel Prize in Chemistry, 50% of which was awarded to William S. Knowles (Monsanto Corporation, retired) and Ryoji Noyori (Nagoya University). (The other half of the 2001 prize was awarded to K. B. Sharpless for asymmetric oxidation reactions. See Chapter 8.) Knowles, Noyori, and others developed chiral catalysts for homogeneous hydrogenation that have proven extraordinarily useful for enantioselective syntheses ranging from small laboratory-scale reactions to industrial- (ton-) scale reactions. An important example is the method developed by Knowles and co-workers at Monsanto Corporation for synthesis of L-DOPA, a compound used in the treatment of Parkinson's disease:

Asymmetric Synthesis of L-DOPA

H_2 (100%), $[(Rh(R,R)\text{-DIPAMP})COD]^+BF_4^-$ (cat.)

(100% yield, 95% ee)

H_3O^+

L-DOPA

(*continued on page 304*)

(*R*,*R*)-DIPAMP
(Chiral ligand for rhodium)

COD =
1,5-Cyclooctadiene

Another example is synthesis of the over-the-counter analgesic naproxen using a BINAP rhodium catalyst developed by Noyori (Sections 5.10 and 5.17).

Asymmetric Synthesis of (S)-Naproxen

(*S*)-BINAP-Ru($OCOCH_3$)$_2$ (0.5 mol%), MeOH

(*S*)-Naproxen
(an anti-inflammatory agent)
(92% yield, 97% ee)

(*S*)-BINAP **(*R*)-BINAP**
(*S*)-BINAP and (*R*)-BINAP are chiral atropisomers
(see Section 5.18).

Catalysts like these are important for asymmetric chemical synthesis of amino acids (Section 24.3E), as well. A final example is the synthesis of (*S*)-phenylalanine methyl ester, a compound used in the synthesis of the artificial sweetener aspartame. This preparation employs yet a different chiral ligand for the rhodium catalyst.

Asymmetric Synthesis of Aspartame

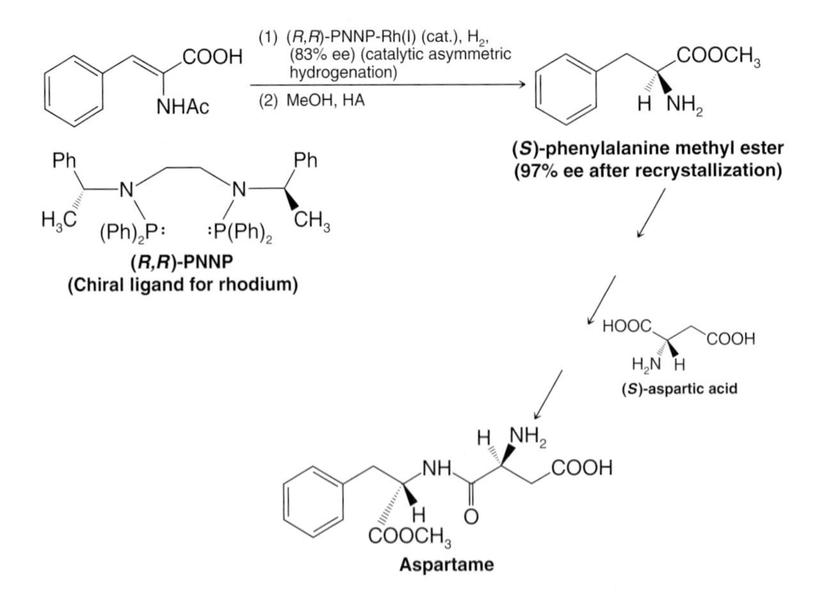

The mechanism of homogeneous catalytic hydrogenation involves reactions characteristic of transition metal organometallic compounds. A general scheme for hydrogenation using Wilkinson's catalyst is shown here. Structural details of the mechanism are provided in Special Topic G.

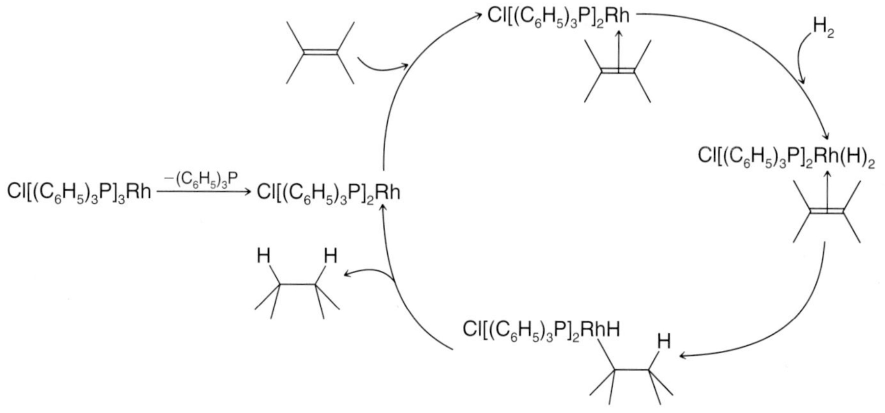

$$Cl[(C_6H_5)_3P]_3Rh \xrightarrow{-(C_6H_5)_3P} Cl[(C_6H_5)_3P]_2Rh$$

$Cl[(C_6H_5)_3P]_2Rh$ + alkene → $Cl[(C_6H_5)_3P]_2Rh$(alkene) $\xrightarrow{H_2}$ $Cl[(C_6H_5)_3P]_2Rh(H)_2$(alkene) → $Cl[(C_6H_5)_3P]_2RhH$(alkyl) → $Cl[(C_6H_5)_3P]_2Rh$ + alkane (H, H)

7.14 Hydrogenation of Alkynes

Depending on the conditions and the catalyst employed, one or two molar equivalents of hydrogen will add to a carbon–carbon triple bond. When a platinum catalyst is used, the alkyne generally reacts with two molar equivalents of hydrogen to give an alkane:

$$CH_3C{\equiv}CCH_3 \xrightarrow[H_2]{Pt} [CH_3CH{=}CHCH_3] \xrightarrow[H_2]{Pt} CH_3CH_2CH_2CH_3$$

However, hydrogenation of an alkyne to an alkene can be accomplished through the use of special catalysts or reagents. Moreover, these special methods allow the preparation of either (E)- or (Z)-alkenes from disubstituted alkynes.

7.14A Syn Addition of Hydrogen: Synthesis of *cis*-Alkenes

A heterogeneous catalyst that permits hydrogenation of an alkyne to an alkene is the nickel boride compound called P-2 catalyst. The P-2 catalyst can be prepared by the reduction of nickel acetate with sodium borohydride:

$$Ni\left(OCCH_3\right)_2 \xrightarrow[C_2H_5OH]{NaBH_4} Ni_2B$$

(the acetate carbon bears C=O)

P-2

Hydrogenation of alkynes in the presence of P-2 catalyst causes **syn addition of hydrogen** to take place, and the alkene that is formed from an alkyne with an internal triple bond has the (Z) or cis configuration. The hydrogenation of 3-hexyne illustrates this method. The reaction takes place on the surface of the catalyst (Section 7.13), accounting for the syn addition:

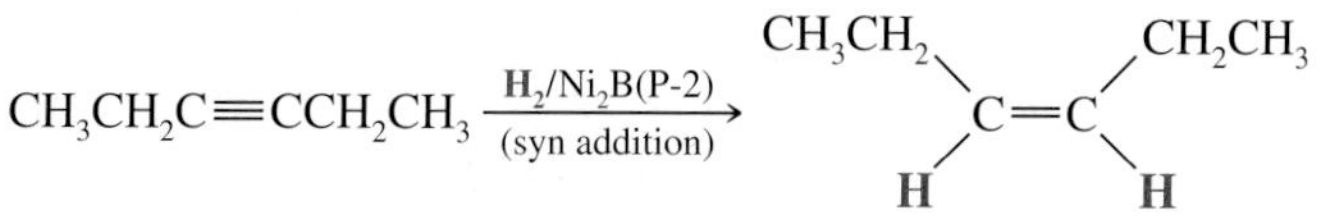

$$CH_3CH_2C{\equiv}CCH_2CH_3 \xrightarrow[\text{(syn addition)}]{H_2/Ni_2B(P\text{-}2)} \underset{H}{\overset{CH_3CH_2}{}}C{=}C\underset{H}{\overset{CH_2CH_3}{}}$$

3-Hexyne → **(Z)-3-Hexene (*cis*-3-hexene) (97%)**

Other specially conditioned catalysts can be used to prepare *cis*-alkenes from disubstituted alkynes. Metallic palladium deposited on calcium carbonate can be used in this way after it has been conditioned with lead acetate and quinoline (Section 20.1B). This special catalyst is known as Lindlar's catalyst:

$$R{-}C{\equiv}C{-}R \xrightarrow[\text{quinoline (syn addition)}]{H_2,\ Pd/CaCO_3\ \text{(Lindlar's catalyst)}} \underset{H\quad\quad H}{\overset{R\quad\quad R}{C{=}C}}$$

7.14B Anti Addition of Hydrogen: Synthesis of *trans*-Alkenes

An **anti addition** of hydrogen atoms to the triple bond occurs when alkynes are reduced with lithium or sodium metal in ammonia or ethylamine at low temperatures. This reaction, called a **dissolving metal reduction,** takes place in solution and produces an (*E*)- or *trans*-alkene. The mechanism involves radicals, which are molecules that have unpaired electrons (see Chapter 10):

$$CH_3(CH_2)_2{-}C{\equiv}C{-}(CH_2)_2CH_3 \xrightarrow[(2)\ NH_4Cl]{(1)\ Li,\ C_2H_5NH_2,\ -78°C} \underset{H\quad\quad (CH_2)_2CH_3}{\overset{CH_3(CH_2)_2\quad\quad H}{C{=}C}}$$

4-Octyne

(*E*)-4-Octene
(*trans*-4-octene)
(52%)

A Mechanism for the Reaction

The Dissolving Metal Reduction of an Alkyne

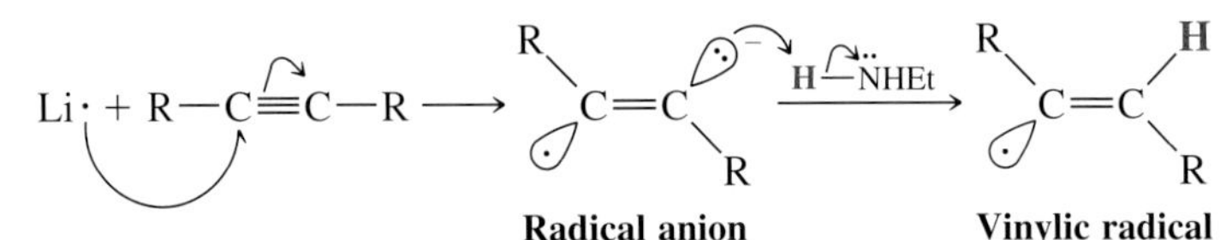

A lithium atom donates an electron to the π bond of the alkyne. An electron pair shifts to one carbon as the hybridization states change to sp^2.

The radical anion acts as a base and removes a proton from a molecule of the ethylamine.

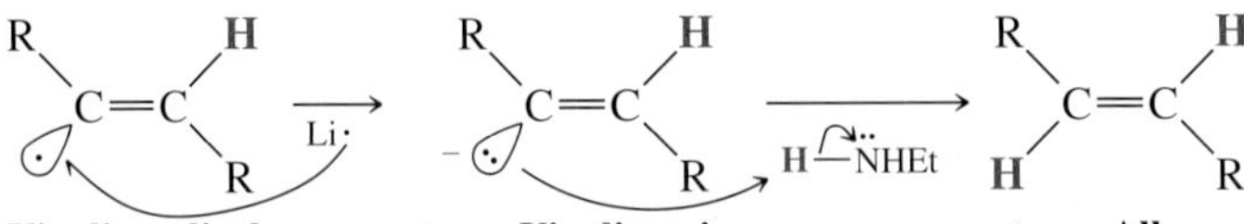

A second lithium atom donates an electron to the vinylic radical.

The anion acts as a base and removes a proton from a second molecule of ethylamine.

The mechanism for this reduction, shown in the preceding box, involves successive electron transfers from lithium (or sodium) atoms and proton transfers from amines (or ammonia). In the first step, a lithium atom transfers an electron to the alkyne to produce an intermediate that bears a negative charge and has an unpaired electron, called a **radical anion.** In the second step, an amine transfers a proton to produce a **vinylic radical.** Then, transfer of another electron gives a **vinylic anion.** It is this step that determines the stereochemistry of the reaction. The *trans*-vinylic anion is formed preferentially because it is more stable; the bulky alkyl groups are farther apart. Protonation of the *trans*-vinylic anion leads to the *trans*-alkene.

7.15 Structural Information from Molecular Formulas and the Index of Hydrogen Deficiency

7.15A Unsaturated and Cyclic Compounds

Alkenes whose molecules contain only one double bond have the general formula C_nH_{2n}. They are isomeric with cycloalkanes. For example, 1-hexene and cyclohexane have the same molecular formula (C_6H_{12}) and they are constitutional isomers:

$CH_2{=}CHCH_2CH_2CH_2CH_3$

1-Hexene **(C_6H_{12})**

Cyclohexane **(C_6H_{12})**

Alkynes and alkenes with two double bonds (alkadienes) have the general formula C_nH_{2n-2}. Hydrocarbons with one triple bond and one double bond (alkenynes) and alkenes with three double bonds (alkatrienes) have the general formula C_nH_{2n-4}, and so forth:

$CH_2{=}CH{-}CH{=}CH_2$

1,3-Butadiene **(C_4H_6)**

$CH_2{=}CH{-}CH{=}CH{-}CH{=}CH_2$

1,3,5-Hexatriene **(C_6H_8)**

A chemist working with an unknown hydrocarbon can obtain considerable information about its structure from its molecular formula and its **index of hydrogen deficiency.** The molecular formula for a compound can be determined in several ways, including by mass spectrometry (Section 9.15), and from this its index of hydrogen deficiency can be calculated. The index of hydrogen deficiency is defined as the number of *pairs* of hydrogen atoms that must be subtracted from the molecular formula of the corresponding alkane to give the molecular formula of the compound under consideration.*

For example, both cyclohexane and 1-hexene have an index of hydrogen deficiency equal to 1 (meaning one ***pair*** of hydrogen atoms). The corresponding alkane (i.e., the alkane with the same number of carbon atoms) is hexane:

C_6H_{14} = formula of corresponding alkane (hexane)
C_6H_{12} = formula of compound (1-hexene or cyclohexane)
H_2 = difference = 1 pair of hydrogen atoms

Index of hydrogen deficiency = 1

The index of hydrogen deficiency of ethyne (acetylene) or of 1,3-butadiene equals 2; the index of hydrogen deficiency of 1,3,5-hexatriene equals 3.

*Some organic chemists refer to the index of hydrogen deficiency as the "degree of unsaturation" or "the number of double-bond equivalencies."

Determination of the number of rings present in a given compound is easily done experimentally. Molecules with double bonds and triple bonds add hydrogen readily in the presence of a hydrogenation catalyst. **Each double bond consumes one molar equivalent of hydrogen; each triple bond consumes two. Rings are not affected by hydrogenation at room temperature, but each ring equates with one unit of hydrogen deficiency.** Hydrogenation, therefore, allows us to distinguish between rings on the one hand and double or triple bonds on the other. Consider as an example two compounds with the molecular formula C_6H_{12}: 1-hexene and cyclohexane. 1-Hexene reacts with one molar equivalent of hydrogen to yield hexane; under the same conditions cyclohexane does not react:

$$CH_2{=}CH(CH_2)_3CH_3 + H_2 \xrightarrow[25°C]{Pt} CH_3(CH_2)_4CH_3$$

$$\text{(cyclohexane)} + H_2 \xrightarrow[25°C]{Pt} \text{no reaction}$$

Or consider another example. Cyclohexene and 1,3-hexadiene have the same molecular formula (C_6H_{10}). Both compounds react with hydrogen in the presence of a catalyst, but cyclohexene, because it has a ring and only one double bond, reacts with only one molar equivalent. 1,3-Hexadiene adds two molar equivalents:

$$\text{Cyclohexene} + H_2 \xrightarrow[25°C]{Pt} \text{(cyclohexane)}$$

Cyclohexene

$$CH_2{=}CHCH{=}CHCH_2CH_3 + 2\ H_2 \xrightarrow[25°C]{Pt} CH_3(CH_2)_4CH_3$$

1,3-Hexadiene

PROBLEM 7.16 **(a)** What is the index of hydrogen deficiency of 2-hexene? **(b)** Of methylcyclopentane? **(c)** Does the index of hydrogen deficiency reveal anything about the location of the double bond in the chain? **(d)** About the size of the ring? **(e)** What is the index of hydrogen deficiency of 2-hexyne? **(f)** In general terms, what structural possibilities exist for a compound with the molecular formula $C_{10}H_{16}$?

PROBLEM 7.17 Zingiberene, a fragrant compound isolated from ginger, has the molecular formula $C_{15}H_{24}$ and is known not to contain any triple bonds. **(a)** What is the index of hydrogen deficiency of zingiberene? **(b)** When zingiberene is subjected to catalytic hydrogenation using an excess of hydrogen, 1 mol of zingiberene absorbs 3 mol of hydrogen and produces a compound with the formula $C_{15}H_{30}$. How many double bonds does a molecule of zingiberene have? **(c)** How many rings?

7.15B Compounds Containing Halogens, Oxygen, or Nitrogen

Calculating the index of hydrogen deficiency (IHD) for compounds other than hydrocarbons is relatively easy.

For compounds containing halogen atoms, we simply count the halogen atoms as though they were hydrogen atoms. Consider a compound with the formula $C_4H_6Cl_2$. To calculate the IHD, we change the two chlorine atoms to hydrogen atoms, considering the formula as though it were C_4H_8. This formula has two hydrogen atoms fewer than the formula for a saturated alkane (C_4H_{10}), and this tells us that the compound has an IHD = 1. It could, therefore, have either one ring or one double bond. [We can tell which it has from a hydrogenation experiment: If the compound adds one molar equivalent of hydrogen (H_2) on catalytic hydrogenation at room temperature, then it must have a double bond; if it does not add hydrogen, then it must have a ring.]

For compounds containing oxygen, we simply ignore the oxygen atoms and calculate the IHD from the remainder of the formula. Consider as an example a compound with the formula C_4H_8O. For the purposes of our calculation we consider the compound to be simply C_4H_8 and we calculate an IHD = 1. Again, this means that the compound contains either a ring or a double bond. Some structural possibilities for this compound are shown next. Notice that the double bond may be present as a carbon–oxygen double bond:

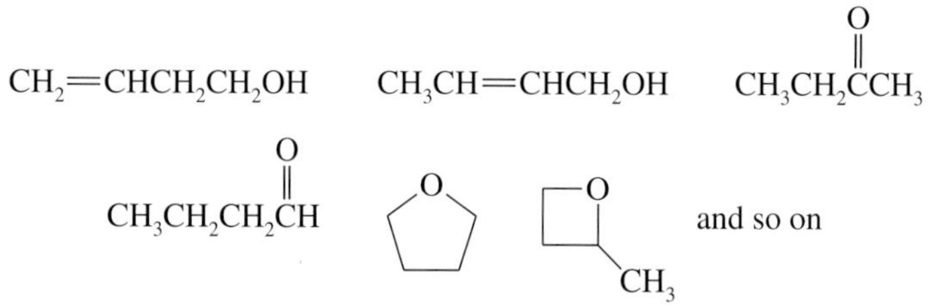

$CH_2{=}CHCH_2CH_2OH$ $\quad$ $CH_3CH{=}CHCH_2OH$ $\quad$ $CH_3CH_2C(=O)CH_3$

$CH_3CH_2CH_2CH(=O)$ $\quad$ [tetrahydrofuran ring] $\quad$ [2-methyloxetane ring, CH_3] $\quad$ and so on

For compounds containing nitrogen atoms we subtract one hydrogen for each nitrogen atom, and then we ignore the nitrogen atoms. For example, we treat a compound with the formula C_4H_9N as though it were C_4H_8, and again we get an IHD = 1. Some structural possibilities are the following:

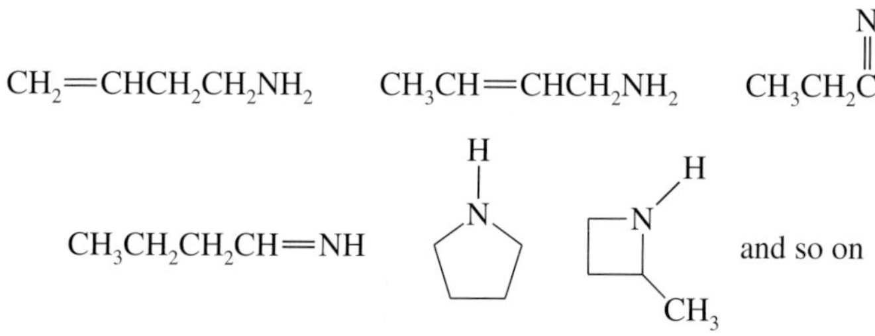

$CH_2{=}CHCH_2CH_2NH_2$ $\quad$ $CH_3CH{=}CHCH_2NH_2$ $\quad$ $CH_3CH_2C(=NH)CH_3$

$CH_3CH_2CH_2CH{=}NH$ $\quad$ [pyrrolidine ring, N–H] $\quad$ [2-methylazetidine ring, N–H, CH_3] $\quad$ and so on

Summary and Review Tools

Mechanism Review: E2 and E1 Elimination

E2 via small base

- Strong **unhindered base**, e.g., CH_3CH_2ONa (EtONa), HO^-
- Predominant formation of **most-substituted alkene** (Zaitsev product)
- Anti coplanar transition state
- Bimolecular in the rate-determining step

E2 via bulky base

- Strong **hindered** base, e.g., $(CH_3)_3COK$
- Predominant formation of **least-substituted alkene** (Hofmann product)
- Anti coplanar transition state
- Bimolecular in the rate-determining step

E1 (including Alcohol Dehydration)

- Absence of strong base (solvent is often the base)
- Alcohols require **strong acid catalyst**
- Carbocation formation is unimolecular rate-determining step
- Carbocation **may rearrange**
- Predominant formation of most substituted alkene (Zaitsev product)
- Leaving group for alcohols is $—OH_2^+$
- 1° Alcohols react by E2-type mechanism

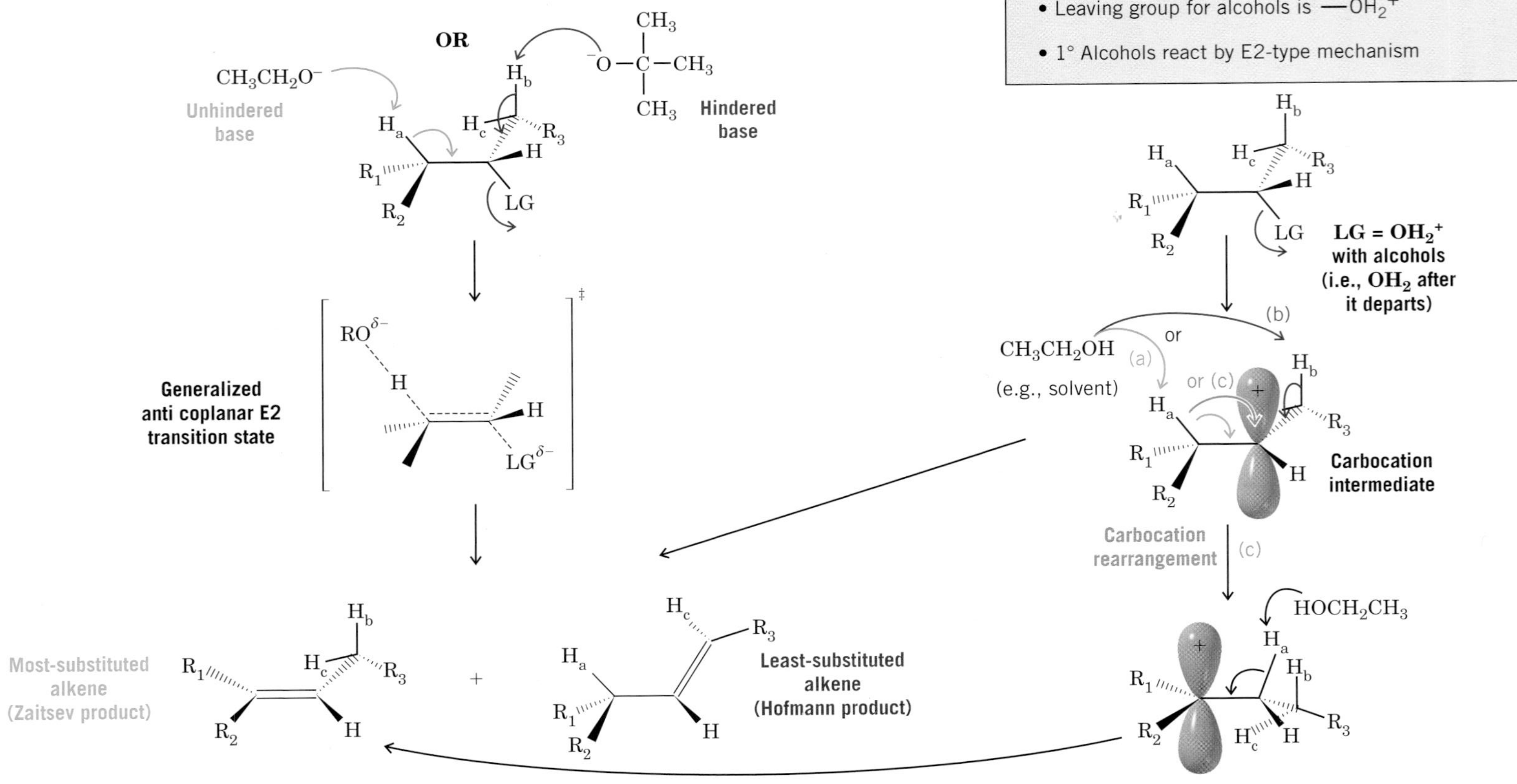

Summary of Methods for the Preparation of Alkenes and Alkynes

In this chapter we described four general methods for the preparation of alkenes.

1. Dehydrohalogenation of alkyl halides (Section 7.6):

General Reaction

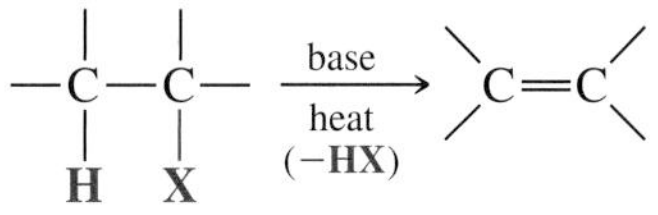

$$-\underset{\substack{|\\H}}{\overset{|}{C}}-\underset{\substack{|\\X}}{\overset{|}{C}}- \xrightarrow[\text{heat}\;(-HX)]{\text{base}} \;\rangle C{=}C\langle$$

Specific Examples

$$\underset{\substack{|\\Br}}{CH_3CH_2CHCH_3} \xrightarrow[C_2H_5OH]{C_2H_5ONa} \underset{\textbf{(cis and trans, 81\%)}}{CH_3CH{=}CHCH_3} + \underset{\textbf{(19\%)}}{CH_3CH_2CH{=}CH_2}$$

$$\underset{\substack{|\\Br}}{CH_3CH_2CHCH_3} \xrightarrow[\substack{70°C\\(CH_3)_3COH}]{(CH_3)_3COK} \underset{\substack{\textbf{Disubstituted alkenes}\\\textbf{(cis and trans, 47\%)}}}{CH_3CH{=}CHCH_3} + \underset{\substack{\textbf{Monosubstituted alkene}\\\textbf{(53\%)}}}{CH_3CH_2CH{=}CH_2}$$

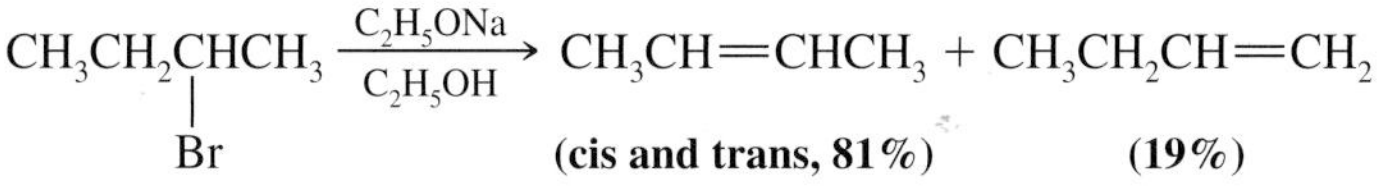

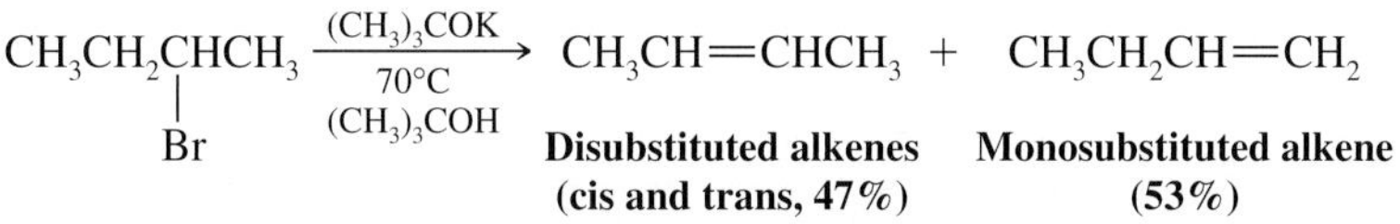

2. Dehydration of alcohols (Sections 7.7 and 7.8):

General Reaction

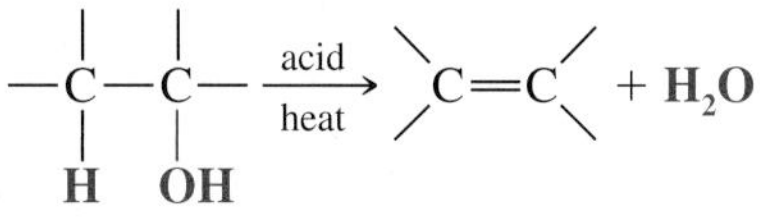

$$-\underset{\substack{|\\H}}{\overset{|}{C}}-\underset{\substack{|\\OH}}{\overset{|}{C}}- \xrightarrow[\text{heat}]{\text{acid}} \;\rangle C{=}C\langle \; + H_2O$$

Specific Examples

$$CH_3CH_2OH \xrightarrow[180°C]{\text{concd } H_2SO_4} CH_2{=}CH_2 + H_2O$$

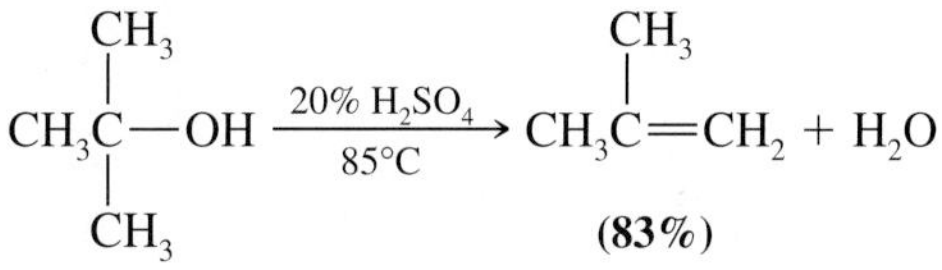

$$CH_3\underset{\substack{|\\CH_3}}{\overset{\substack{CH_3\\|}}{C}}{-}OH \xrightarrow[85°C]{20\%\ H_2SO_4} \underset{\textbf{(83\%)}}{CH_3\overset{\substack{CH_3\\|}}{C}{=}CH_2} + H_2O$$

3. Hydrogenation of alkynes (Section 7.14):

General Reaction

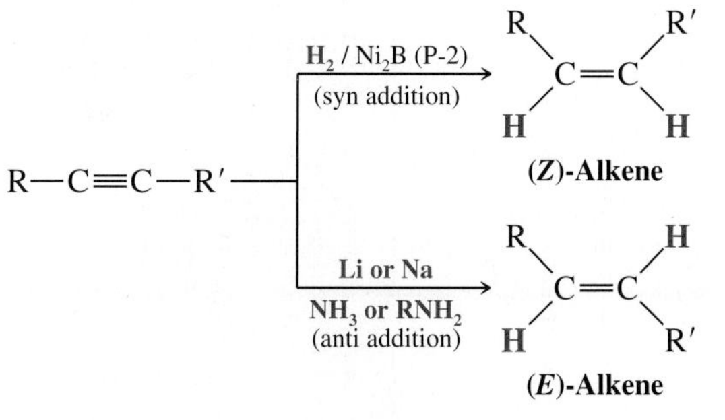

$$R{-}C{\equiv}C{-}R' \xrightarrow[\text{(syn addition)}]{H_2\ /\ Ni_2B\ (P\text{-}2)} \underset{\textbf{(Z)-Alkene}}{\overset{R\qquad R'}{C{=}C}\atop{H\qquad H}}$$

$$R{-}C{\equiv}C{-}R' \xrightarrow[\substack{NH_3 \text{ or } RNH_2\\\text{(anti addition)}}]{\text{Li or Na}} \underset{\textbf{(E)-Alkene}}{\overset{R\qquad H}{C{=}C}\atop{H\qquad R'}}$$

In subsequent chapters we shall see a number of other methods for alkene synthesis.

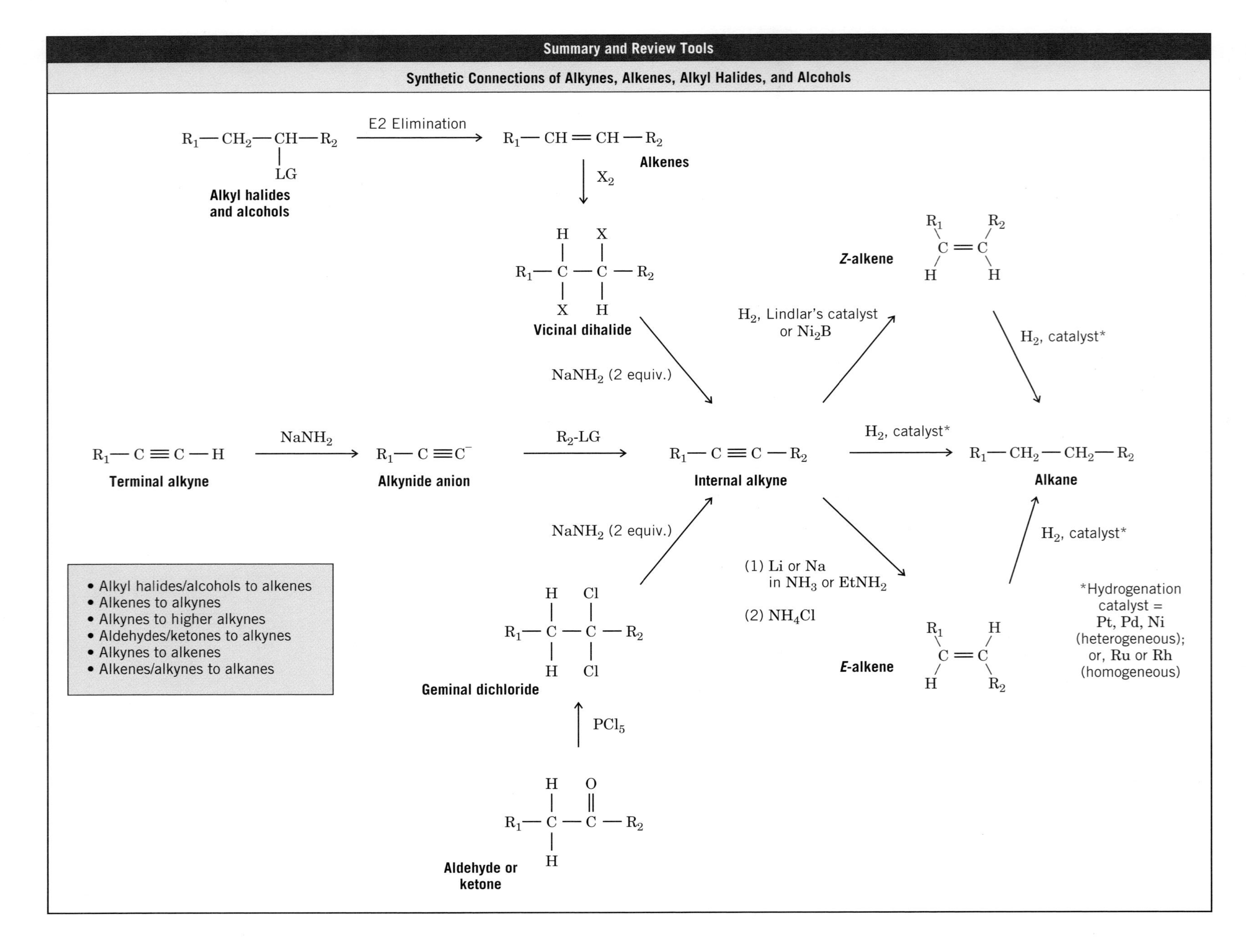
Summary and Review Tools
Synthetic Connections of Alkynes, Alkenes, Alkyl Halides, and Alcohols
R1—CH2—CH—R2
LG
Alkyl halides and alcohols
E2 Elimination
R1—CH=CH—R2
Alkenes
X2
Vicinal dihalide
NaNH2 (2 equiv.)
Z-alkene
H2, Lindlar's catalyst or Ni2B
H2, catalyst*
R1—C≡C—H
Terminal alkyne
NaNH2
R1—C≡C⁻
Alkynide anion
R2-LG
R1—C≡C—R2
Internal alkyne
H2, catalyst*
R1—CH2—CH2—R2
Alkane
NaNH2 (2 equiv.)
(1) Li or Na in NH3 or EtNH2
(2) NH4Cl
H2, catalyst*
• Alkyl halides/alcohols to alkenes
• Alkenes to alkynes
• Alkynes to higher alkynes
• Aldehydes/ketones to alkynes
• Alkynes to alkenes
• Alkenes/alkynes to alkanes
Geminal dichloride
PCl5
E-alkene
*Hydrogenation catalyst = Pt, Pd, Ni (heterogeneous); or, Ru or Rh (homogeneous)
Aldehyde or ketone

Key Terms and Concepts

1,2 Shifts and molecular rearrangements	Section 7.8A
Dehydration	Section 7.7
Dehydrohalogenation	Section 7.6
(*E*)–(*Z*) System	Section 7.2
gem-Dihalide	Section 7.9
Hetereogeneous catalysis	Section 7.12
Hofmann rule	Section 7.6B
Homogeneous catalysis	Section 7.12
Hydride, methanide, and alkanide rearrangements	Section 7.8A
Hydrogenation	Sections 7.3A, 7.12–7.14
Index of hydrogen deficiency	Section 7.15
Overall relative stabilities of alkenes	Section 7.3B
Saturated and unsaturated compounds	Section 7.12
Syn and anti additions	Section 7.13A
Syn and anti coplanar	Section 7.6C
vic-Dihalide	Section 7.9
Zaitsev's rule	Sections 7.6A,B, 7.8

Additional Problems

7.18 Each of the following names is incorrect. Give the correct name and explain your reasoning.

(a) *trans*-3-Pentene
(b) 1,1-Dimethylethene
(c) 2-Methylcyclohexene
(d) 4-Methylcyclobutene
(e) (*Z*)-3-Chloro-2-butene
(f) 5,6-Dichlorocyclohexene

7.19 Write a structural formula for each of the following:

(a) 3-Methylcyclobutene
(b) 1-Methylcyclopentene
(c) 2,3-Dimethyl-2-pentene
(d) (*Z*)-3-Hexene
(e) (*E*)-2-Pentene
(f) 3,3,3-Tribromopropene
(g) (*Z*,4*R*)-4-Methyl-2-hexene
(h) (*E*,4*S*)-4-Chloro-2-pentene
(i) (*Z*)-1-Cyclopropyl-1-pentene
(j) 5-Cyclobutyl-1-pentene
(k) (*R*)-4-Chloro-2-pentyne
(l) (*E*)-4-Methylhex-4-en-1-yne

7.20 Write three-dimensional formulas for and give names using (*R*)–(*S*) and (*E*)–(*Z*) designations for the isomers of:

(a) 4-Bromo-2-hexene
(b) 3-Chloro-1,4-hexadiene
(c) 2,4-Dichloro-2-pentene
(d) 2-Bromo-4-chlorohex-2-en-5-yne

7.21 Give the IUPAC names for each of the following:

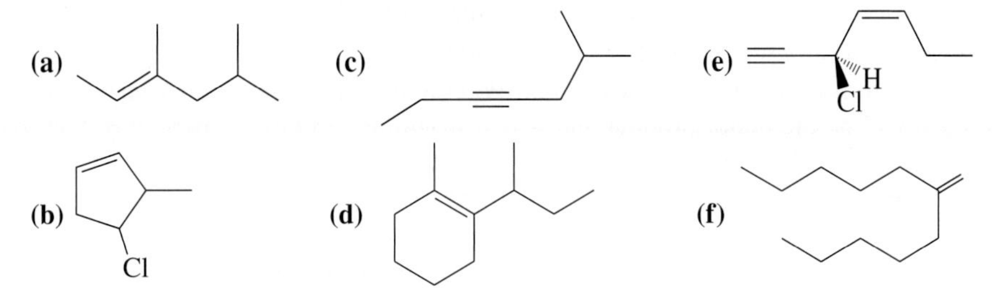

Note: Problems marked with an asterisk are "challenge problems."

7.22 Outline a synthesis of propene from each of the following:

(a) Propyl chloride
(b) Isopropyl chloride
(c) Propyl alcohol
(d) Isopropyl alcohol
(e) 1,2-Dibromopropane
(f) Propyne

7.23 Outline a synthesis of cyclopentene from each of the following:

(a) Bromocyclopentane **(b)** 1,2-Dichlorocyclopentane **(c)** Cyclopentanol

7.24 Starting with ethyne, outline syntheses of each of the following. You may use any other needed reagents, and you need not show the synthesis of compounds prepared in earlier parts of this problem.

(a) Propyne
(b) 1-Butyne
(c) 2-Butyne
(d) *cis*-2-Butene
(e) *trans*-2-Butene
(f) 1-Pentyne
(g) 2-Hexyne
(h) (*Z*)-2-Hexene
(i) (*E*)-2-Hexene
(j) 3-Hexyne
(k) $CH_3CH_2C{\equiv}CD$
(i) (*E*)-2-Hexene
(l) 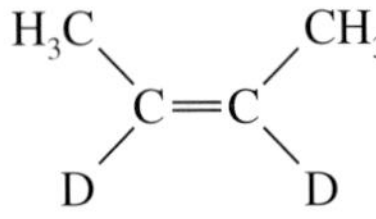 H_3C, CH_3, C=C, D, D

7.25 Starting with 1-methylcyclohexene and using any other needed reagents, outline a synthesis of the following deuterium-labeled compound:

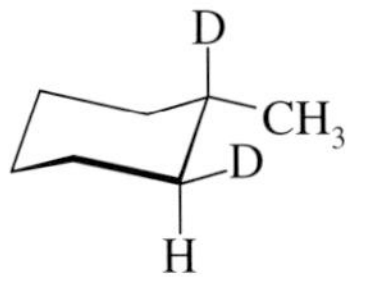

D, CH_3, D, H

7.26 When *trans*-2-methylcyclohexanol (see the following reaction) is subjected to acid-catalyzed dehydration, the major product is 1-methylcyclohexene:

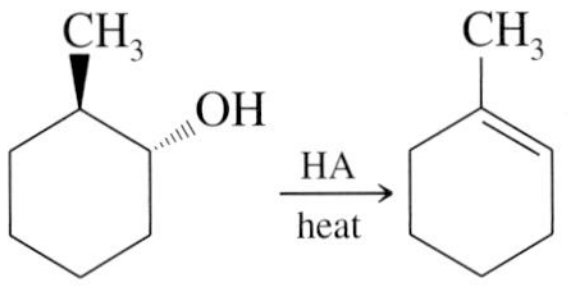

CH_3, OH $\xrightarrow[\text{heat}]{\text{HA}}$ CH_3

However, when *trans*-1-bromo-2-methylcyclohexane is subjected to dehydrohalogenation, the major product is 3-methylcyclohexene:

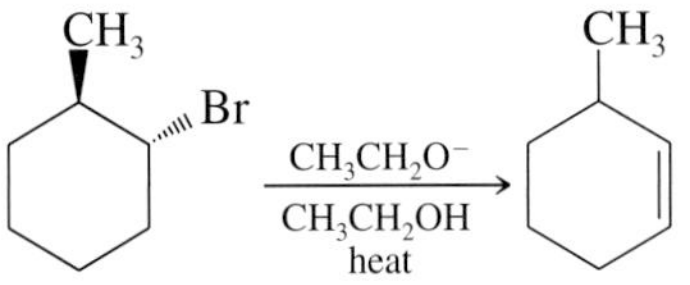

CH_3, Br $\xrightarrow[CH_3CH_2OH \text{ heat}]{CH_3CH_2O^-}$ CH_3

Account for the different products of these two reactions.

7.27 Outline a synthesis of phenylethyne from each of the following:

(a) 1,1-Dibromo-1-phenylethane
(b) 1,1-Dibromo-2-phenylethane
(c) Phenylethene (styrene)
(d) Acetophenone ($C_6H_5COCH_3$)

7.28 Without consulting tables, arrange the following compounds in order of decreasing acidity:

Pentane 1-Pentene 1-Pentyne 1-Pentanol

7.29 Write structural formulas for all the products that would be obtained when each of the following alkyl halides is heated with sodium ethoxide in ethanol. When more than one product results, you should indicate which would be the major product and which would be the minor product(s). You may neglect cis–trans isomerism of the products when answering this question.

(a) 2-Bromo-3-methylbutane
(b) 3-Bromo-2,2-dimethylbutane
(c) 3-Bromo-3-methylpentane
(d) 1-Bromo-1-methylcyclohexane
(e) *cis*-1-Bromo-2-ethylcyclohexane
(f) *trans*-1-Bromo-2-ethylcyclohexane

7.30 Write structural formulas for all the products that would be obtained when each of the following alkyl halides is heated with potassium *tert*-butoxide in *tert*-butyl alcohol. When more than one product results, you should indicate which would be the major product and which would be the minor product(s). You may neglect cis–trans isomerism of the products when answering this question.

(a) 2-Bromo-3-methylbutane
(b) 4-Bromo-2,2-dimethylbutane
(c) 3-Bromo-3-methylpentane
(d) 4-Bromo-2,2-dimethylpentane
(e) 1-Bromo-1-methylcyclohexane

7.31 Starting with an appropriate alkyl halide and base, outline syntheses that would yield each of the following alkenes as the major (or only) product:

(a) 1-Pentene
(b) 3-Methyl-1-butene
(c) 2,3-Dimethyl-1-butene
(d) 4-Methylcyclohexene
(e) 1-Methylcyclopentene

7.32 Arrange the following alcohols in order of their reactivity toward acid-catalyzed dehydration (with the most reactive first):

1-Pentanol 2-Methyl-2-butanol 3-Methyl-2-butanol

7.33 Give the products that would be formed when each of the following alcohols is subjected to acid-catalyzed dehydration. If more than one product would be formed, designate the alkene that would be the major product. (Neglect cis–trans isomerism.)

(a) 2-Methyl-2-propanol
(b) 3-Methyl-2-butanol
(c) 3-Methyl-3-pentanol
(d) 2,2-Dimethyl-1-propanol
(e) 1,4-Dimethylcyclohexanol

7.34 1-Bromobicyclo[2.2.1]heptane does not undergo elimination (below) when heated with a base. Explain this failure to react. (Construction of molecular models may help.)

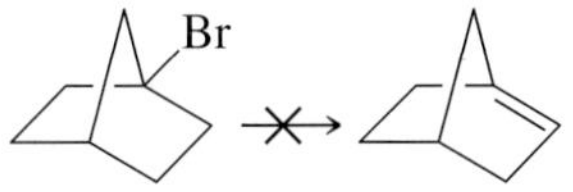

Br

7.35 When the deuterium-labeled compound shown below is subjected to dehydrohalogenation using sodium ethoxide in ethanol, the only alkene product is 3-methylcyclohexene. (The product contains no deuterium.) Provide an explanation for this result.

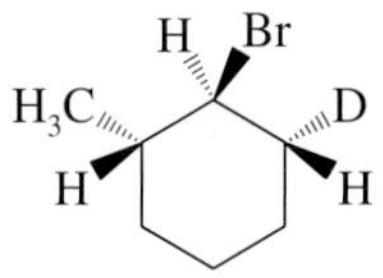

H Br
H_3C D
H H

7.36 Provide mechanistic explanations for each of the following occurrences:

(a) Acid-catalyzed dehydration of 1-butanol produces *trans*-2-butene as the major product.

(b) Acid-catalyzed dehydration of 2,2-dimethylcyclohexanol produces 1,2-dimethylcyclohexene as the major product.

(c) Treating 3-iodo-2,2-dimethylbutane with silver nitrate in ethanol produces 2,3-dimethyl-2-butene as the major product.

(d) Dehydrohalogenation of (1*S*,2*R*)-1-bromo-1,2-diphenylpropane produces only (*E*)-1,2-diphenylpropene.

7.37 Caryophyllene, a compound found in oil of cloves, has the molecular formula $C_{15}H_{24}$ and has no triple bonds. Reaction of caryophyllene with an excess of hydrogen in the presence of a platinum catalyst produces a compound with the formula $C_{15}H_{28}$. How many **(a)** double bonds, and **(b)** rings does a molecule of caryophyllene have?

7.38 Squalene, an important intermediate in the biosynthesis of steroids, has the molecular formula $C_{30}H_{50}$ and has no triple bonds.

(a) What is the index of hydrogen deficiency of squalene?

(b) Squalene undergoes catalytic hydrogenation to yield a compound with the molecular formula $C_{30}H_{62}$. How many double bonds does a molecule of squalene have?

(c) How many rings?

7.39 Consider the interconversion of *cis*-2-butene and *trans*-2-butene.
(a) What is the value of $\Delta H°$ for the reaction *cis*-2-butene $\rightarrow$ *trans*-2-butene?
(b) Assume $\Delta H° \cong \Delta G°$. What minimum value of $\Delta G^{\ddagger}$ would you expect for this reaction?
(c) Sketch a free-energy diagram for the reaction and label $\Delta G°$ and $\Delta G^{\ddagger}$.

7.40 Propose structures for compounds **E–H**. Compound **E** has the molecular formula C_5H_8 and is optically active. On catalytic hydrogenation **E** yields **F**. Compound **F** has the molecular formula C_5H_{10}, is optically inactive, and cannot be resolved into separate enantiomers. Compound **G** has the molecular formula C_6H_{10} and is optically active. Compound **G** contains no triple bonds. On catalytic hydrogenation **G** yields **H**. Compound **H** has the molecular formula C_6H_{14}, is optically inactive, and cannot be resolved into separate enantiomers.

7.41 Compounds **I** and **J** both have the molecular formula C_7H_{14}. Compounds **I** and **J** are both optically active and both rotate plane-polarized light in the same direction. On catalytic hydrogenation **I** and **J** yield the same compound **K** (C_7H_{16}). Compound **K** is optically active. Propose possible structures for **I**, **J**, and **K**.

7.42 Compounds **L** and **M** have the molecular formula C_7H_{14}. Compounds **L** and **M** are optically inactive, are nonresolvable, and are diastereomers of each other. Catalytic hydrogenation of either **L** or **M** yields **N**. Compound **N** is optically inactive but can be resolved into separate enantiomers. Propose possible structures for **L**, **M**, and **N**.

7.43 **(a)** Partial dehydrohalogenation of either (1*R*,2*R*)-1,2-dibromo-1,2-diphenylethane or (1*S*,2*S*)-1,2-dibromo-1,2-diphenylethane enantiomers (or a racemate of the two) produces (*Z*)-1-bromo-1,2-diphenylethene as the product, whereas **(b)** partial dehydrohalogenation of (1*R*,2*S*)-1,2-dibromo-1,2-diphenylethane (the meso compound) gives only (*E*)-1-bromo-1,2-diphenylethene. **(c)** Treating (1*R*,2*S*)-1,2-dibromo-1,2-diphenylethane with sodium iodide in acetone produces only (*E*)-1,2-diphenylethene. Explain these results.

***7.44** **(a)** Using reactions studied in this chapter, show steps by which this alkyne could be converted to the seven-membered ring homologue of the product obtained in Problem 7.36(b).
(b) Could the homologous products obtained in these two cases be relied upon to show infrared absorption in the 1620–1680-cm^{-1} region?

CH_3
—C≡CH

***7.45** Predict the structures of compounds **A**, **B**, and **C**:
A is an unbranched C_6 alkyne that is also a primary alcohol.
B is obtained from **A** by use of hydrogen and nickel boride catalyst or dissolving metal reduction.
C is formed from **B** on treatment with aqueous acid at room temperature. This compound **C** has no infrared absorption in either the 1620–1680-cm^{-1} or the 3590–3650-cm^{-1} region. It has an index of hydrogen deficiency of 1 and has one stereogenic center but forms as the racemate.

***7.46** What is the index of hydrogen deficiency for **(a)** $C_7H_{10}O_2$ and **(b)** $C_5H_4N_4$?

Learning Group Problems

1. Write the structure of the product(s) obtained from the reaction of 2-chloro-2,3-dimethylbutane when it reacts with **(a)** sodium ethoxide (NaOEt) in ethanol (EtOH) at 80°C or (in a separate reaction) **(b)** potassium *tert*-butoxide [$KOC(CH_3)_3$] in *tert*-butyl alcohol [$HOC(CH_3)_3$] at 80°C. If more than one product is formed, indicate which one would be expected to be the major product. Provide a detailed mechanism for formation of the major product from each reaction, including transition states.

2. Explain using mechanistic arguments why the reaction of 2-bromo-1,2-diphenylpropane with sodium ethoxide (NaOEt) in ethanol (EtOH) at 80°C produces mainly (*E*)-1,2-diphenylpropene (little of the *Z* diastereomer is formed).
3. Write the structure of the product(s) formed when 1-methylcyclohexanol reacts with 85% H_3PO_4 at 150°C. Write a detailed mechanism for the reaction.
4. Consider the reaction of 1-cyclobutylethanol (1-hydroxyethylcyclobutane) with concentrated H_2SO_4 at 120°C. Write structures of all reasonable organic products. Assuming that methylcyclopentene is one product, write a mechanism that accounts for its formation. Write mechanisms that account for formation of all other products as well.

CHAPTER 8

Alkenes and Alkynes II: Addition Reactions

The Sea: A Treasury of Biologically Active Natural Products

The world's oceans are a vast storehouse of dissolved halide ions. The concentration of halides in the ocean is approximately 0.5*M* in chloride, 1m*M* in bromide, and 1μ*M* in iodide ions. Perhaps it is not surprising, then, that marine organisms have incorporated halogen atoms into the structures of many of their metabolites. Among these are such intriguing polyhalogenated compounds as halomon, dactylyne (shown above), tetrachloromertensene, (3*E*)-laureatin, and (3*R*)- and (3*S*)-cyclocymopol (see page 329). Just the sheer number of halogen atoms in these metabolites is cause for wonder. For the organisms that make them, some of these molecules are part of defense mechanisms that serve to promote the species' survival by deterring predators or inhibiting the growth of competing organisms. For humans, the vast resource of marine natural products shows ever-greater potential as a source of new therapeutic agents. Halomon, for example, is in preclinical evaluation as a cytotoxic agent against certain tumor cell types, dactylyne is an inhibitor of pentobarbital metabolism, and the cyclocymopol enantiomers show agonistic or antagonistic effects on the human progesterone receptor, depending on which enantiomer is used.

The biosynthesis of certain halogenated marine natural products is intriguing. Some of their halogens appear to have been introduced as *electrophiles* rather than as Lewis bases or nucleophiles, which is their character when they are solutes in seawater. But how do marine organisms transform nucleophilic halide anions into *electrophilic* species for incorporation into their metabolites? It happens that many marine organisms have enzymes called haloperoxidases that convert nucleophilic iodide, bromide, or chloride anions into electrophilic species that react like I^+, Br^+, or Cl^+. In the biosynthetic schemes proposed for some halogenated natural products, positive halogen intermediates are attacked by electrons from the π bond of an alkene or alkyne in what is called an addition reaction.

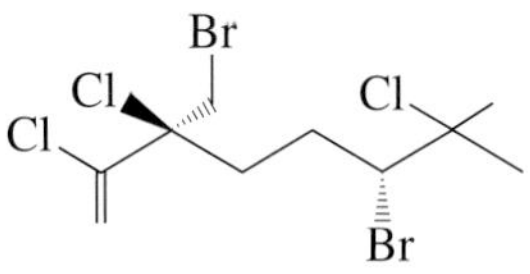

Halomon

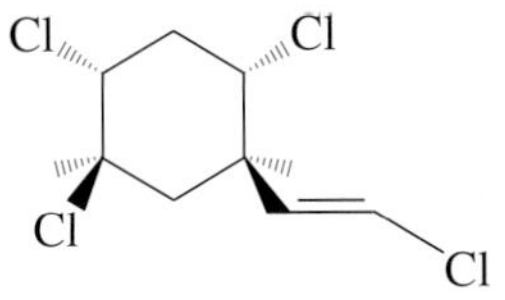

Tetrachloromertensene

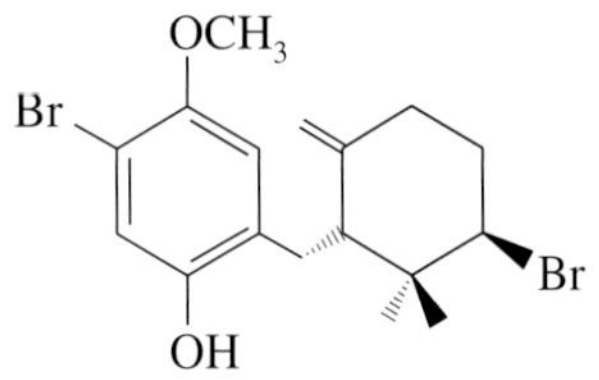

Cyclocymopol monomethyl ether

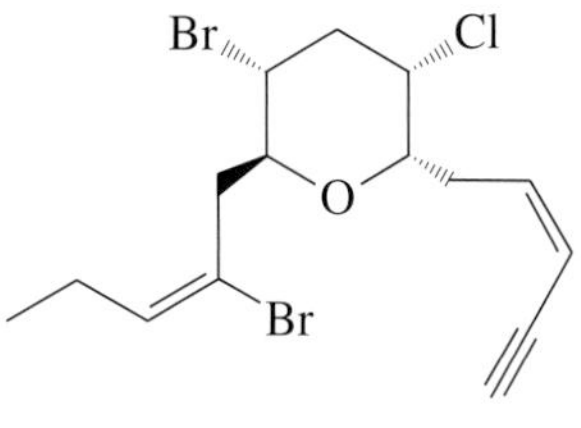

Dactylyne

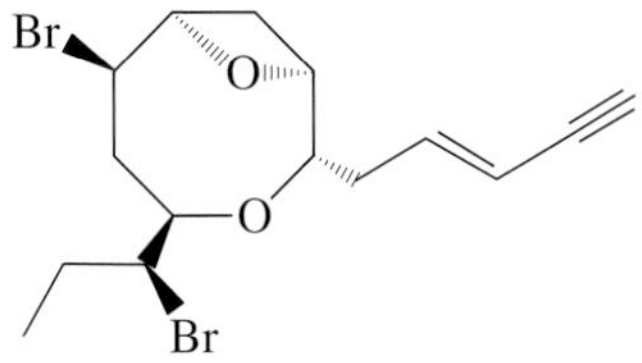

(3*E*)-Laureatin

Although you may not have considered halogens as electrophilic reagents before, in this chapter we shall see how molecular halogens often react this way with alkenes and alkynes. In addition, a Learning Group Problem for this chapter asks you to propose a scheme for biosynthesis of the marine natural product kumepaloxane by electrophilic halogen addition. Kumepaloxane is a fish antifeedant synthesized by the Guam bubble snail *Haminoea cymbalum,* presumably as a defense mechanism for the snail. In later chapters we shall see other examples of truly remarkable marine natural products, such as brevetoxin B, associated with deadly "red tides," and eleutherobin, a promising anti-cancer agent.

8.1 Introduction: Additions to Alkenes

A characteristic reaction of compounds with a carbon–carbon double bond is an **addition**—a reaction of the general type shown below.

$$\mathrm{C{=}C} + \mathrm{A{-}B} \xrightarrow{\text{addition}} \mathrm{A{-}\overset{|}{\underset{|}{C}}{-}\overset{|}{\underset{|}{C}}{-}B}$$

We saw in Section 7.12 that alkenes undergo the addition of hydrogen. In this chapter we shall study other examples of additions to the double bonds of alkenes. We begin with the additions of hydrogen halides, sulfuric acid, water (in the presence of an acid catalyst), and halogens. Later we shall study some specialized reagents that also add to alkenes.

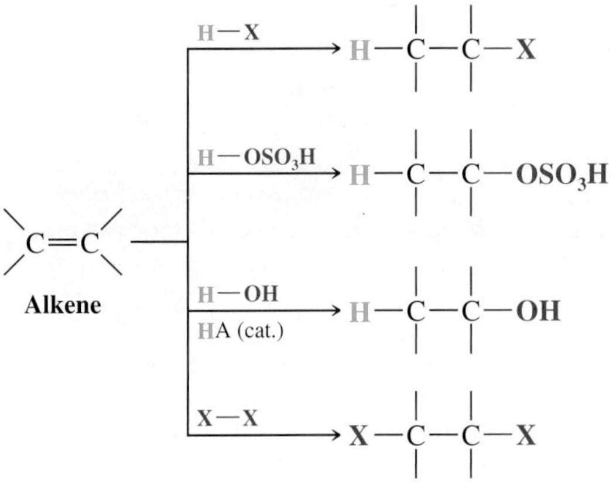

Alkene $\mathrm{C{=}C}$

$\xrightarrow{\mathrm{H{-}X}}$ $\mathrm{H{-}\overset{|}{\underset{|}{C}}{-}\overset{|}{\underset{|}{C}}{-}X}$ **Alkyl halide** (Sections 8.2, 8.3, and 10.9)

$\xrightarrow{\mathrm{H{-}OSO_3H}}$ $\mathrm{H{-}\overset{|}{\underset{|}{C}}{-}\overset{|}{\underset{|}{C}}{-}OSO_3H}$ **Alkyl hydrogen sulfate** (Section 8.4)

$\xrightarrow[\text{HA (cat.)}]{\mathrm{H{-}OH}}$ $\mathrm{H{-}\overset{|}{\underset{|}{C}}{-}\overset{|}{\underset{|}{C}}{-}OH}$ **Alcohol** (Section 8.5)

$\xrightarrow{\mathrm{X{-}X}}$ $\mathrm{X{-}\overset{|}{\underset{|}{C}}{-}\overset{|}{\underset{|}{C}}{-}X}$ **Dihaloalkane** (Sections 8.12, 8.13)

Two characteristics of the double bond help us understand why these addition reactions occur:

1. An addition reaction results in the conversion of one π bond and one σ bond (Sections 1.12 and 1.13) into two σ bonds. The result of this change is usually energetically favorable. The heat evolved in making two σ bonds exceeds that needed to break one σ bond and one π bond (because π bonds are weaker), and, therefore, addition reactions are usually exothermic:

$$\text{>C=C<} + \text{X—Y} \longrightarrow \text{—}\overset{|}{\underset{X}{\text{C}}}\text{—}\overset{|}{\underset{Y}{\text{C}}}\text{—}$$

π bond $\quad$ σ bond $\longrightarrow$ 2σ bonds

Bonds broken $\quad$ **Bonds formed**

2. The electrons of the π bond are exposed. Because the π bond results from overlapping p orbitals, the π electrons lie above and below the plane of the double bond:

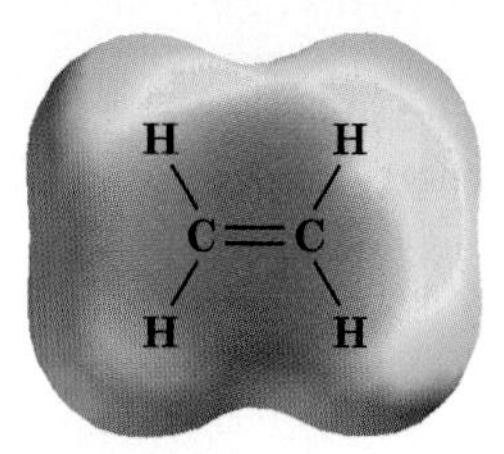

An electrostatic potential map for ethene shows the higher density of negative charge in the region of the π bond.

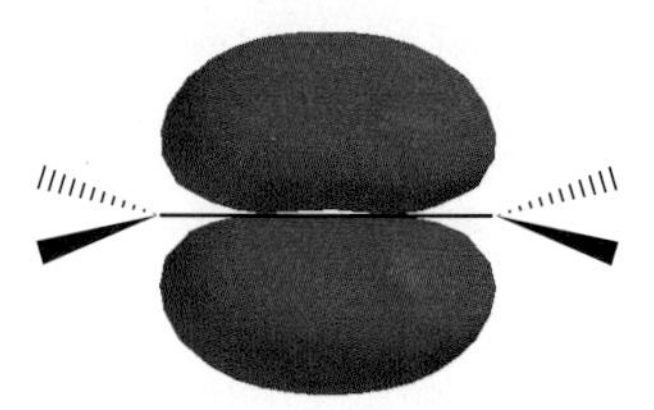

The electron pair of the π bond is distributed throughout both lobes of the π molecular orbital.

The π bond is particularly susceptible to electron-seeking reagents. Such reagents are said to be **electrophilic** (electron seeking) and are called **electrophiles.** Electrophiles include positive reagents such as protons (H^+), neutral reagents such as bromine (because it can be polarized so that one end is positive), and the Lewis acids such as BH_3, BF_3, and $AlCl_3$. Metal ions that contain vacant orbitals—the silver ion (Ag^+), the mercuric ion (Hg^{2+}), and the platinum ion (Pt^{2+}), for example—also act as electrophiles.

Hydrogen halides, for example, react with alkenes by donating a proton to the π bond. The proton uses the two electrons of the π bond to form a σ bond to one of the carbon atoms. This leaves a vacant p orbital and a + charge on the other carbon. The overall result is the formation of a carbocation and a halide ion from the alkene and HX:

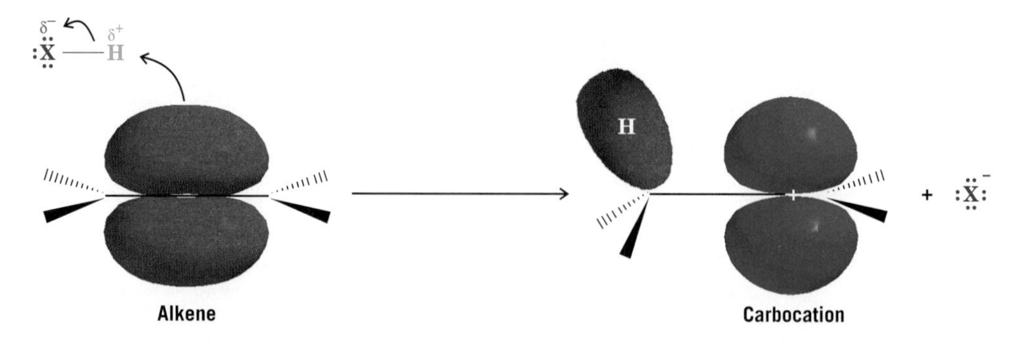

Being highly reactive, the carbocation may then combine with the halide ion by accepting one of its electron pairs:

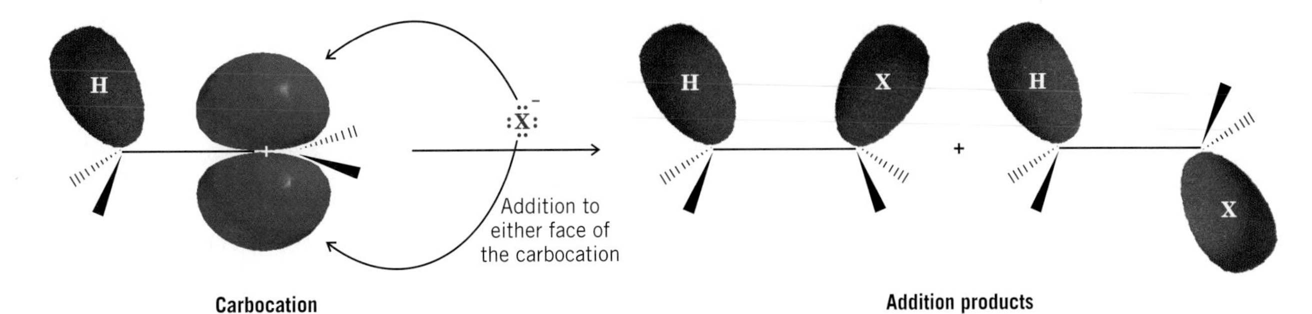

Electrophiles Are Lewis Acids Electrophiles are molecules or ions that can accept an electron pair. Nucleophiles are molecules or ions that can furnish an electron pair (i.e., Lewis bases). Any reaction of an electrophile also involves a nucleophile. In the protonation of an alkene the electrophile is the proton donated by an acid; the nucleophile is the alkene:

Essential properties of electrophiles

$$:\ddot{X}-H + \rangle C=C\langle \longrightarrow -\overset{H}{\overset{|}{C}}-C^{+}\langle + :\ddot{X}:^{-}$$

Electrophile **Nucleophile**

In the next step, the reaction of the carbocation with a halide ion, the carbocation is the electrophile and the halide ion is the nucleophile:

$$-\overset{H}{\overset{|}{C}}-C^{+}\langle + :\ddot{X}:^{-} \longrightarrow -\overset{H}{\overset{|}{C}}-\overset{:\ddot{X}:}{\overset{|}{C}}-$$

Electrophile **Nucleophile**

8.2 Addition of Hydrogen Halides to Alkenes: Markovnikov's Rule

Hydrogen halides (HI, HBr, HCl, and HF) add to the double bond of alkenes:

$$\rangle C=C\langle + HX \longrightarrow -\underset{H}{\underset{|}{C}}-\underset{X}{\underset{|}{C}}-$$

These additions are sometimes carried out by dissolving the hydrogen halide in a solvent, such as acetic acid or CH_2Cl_2, or by bubbling the gaseous hydrogen halide directly into the alkene and using the alkene itself as the solvent. HF is prepared as polyhydrogen fluoride in pyridine. The order of reactivity of the hydrogen halides is HI > HBr > HCl > HF, and unless the alkene is highly substituted, HCl reacts so slowly that the reaction is not one that is useful as a preparative method. HBr adds readily, but as we shall learn in Section 10.9, unless precautions are taken, the reaction may follow an alternate course. However, adding silica gel or alumina to the mixture of the alkene and HCl or HBr in CH_2Cl_2 increases the rate of addition dramatically and makes the reaction an easy one to carry out.*

The addition of HX to an unsymmetrical alkene could conceivably occur in two ways. In practice, however, one product usually predominates. The addition of HBr to propene,

*See Kropp, P. J.; Daus, K. A.; Crawford, S. D.; Tubergen, M. W.; Kepler, K. D.; Craig, S. L.; Wilson, V. P. *J. Am. Chem. Soc.* **1990,** *112,* 7433–7434.

for example, could conceivably lead to either 1-bromopropane or 2-bromopropane. The main product, however, is 2-bromopropane:

$$CH_2{=}CHCH_3 + HBr \longrightarrow \underset{\textbf{2-Bromopropane}}{CH_3\underset{|}{\overset{}{C}}HCH_3 \text{ (Br)}} \qquad (\textit{little } BrCH_2CH_2CH_3)$$

2-Bromopropane **1-Bromopropane**

When 2-methylpropene reacts with HBr, the main product is *tert*-butyl bromide, not isobutyl bromide:

$$(H_3C)_2C{=}CH_2 + HBr \longrightarrow CH_3{-}C(CH_3)(Br){-}CH_3 \quad \left(\textit{little} \; CH_3{-}CH(CH_3){-}CH_2{-}Br\right)$$

2-Methylpropene (isobutylene) — ***tert*-Butyl bromide** — **Isobutyl bromide**

Consideration of many examples like this led the Russian chemist Vladimir Markovnikov in 1870 to formulate what is now known as **Markovnikov's rule.** One way to state this rule is to say that ***in the addition of HX to an alkene, the hydrogen atom adds to the carbon atom of the double bond that already has the greater number of hydrogen atoms.**** The addition of HBr to propene is an illustration:

Carbon atom with the greater number of hydrogen atoms → $CH_2{=}CHCH_3$ (H adds to CH_2, Br adds to CH) $\longrightarrow CH_2(H){-}CH(Br)CH_3$

Markovnikov addition product

Reactions that illustrate Markovnikov's rule are said to be *Markovnikov additions.*

A mechanism for addition of a hydrogen halide to an alkene involves the following two steps:

A Mechanism for the Reaction

Addition of a Hydrogen Halide to an Alkene

Step 1

$$\text{>C=C<} + H{-}\ddot{\underset{..}{X}}{:} \xrightarrow{\text{slow}} {}^{+}C{-}C(H){-} + {:}\ddot{\underset{..}{X}}{:}^{-}$$

The π electrons of the alkene form a bond with a proton from HX to form a carbocation and a halide ion.

Step 2

$${:}\ddot{\underset{..}{X}}{:}^{-} + {}^{+}C{-}C(H){-} \xrightarrow{\text{fast}} {-}C(X){-}C(H){-}$$

The halide ion reacts with the carbocation by donating an electron pair; the result is an alkyl halide.

*In his original publication, Markovnikov described the rule in terms of the point of attachment of the halogen atom, stating that "if an unsymmetrical alkene combines with a hydrogen halide, the halide ion adds to the carbon atom with the fewer hydrogen atoms."

The important step—because it is the **rate-determining step**—is step 1. In step 1 the alkene donates a pair of electrons to the proton of the hydrogen halide and forms a carbocation. This step (Fig. 8.1) is highly endergonic and has a high free energy of activation. Consequently, it takes place slowly. In step 2 the highly reactive carbocation stabilizes itself by combining with a halide ion. This exergonic step has a very low free energy of activation and takes place very rapidly.

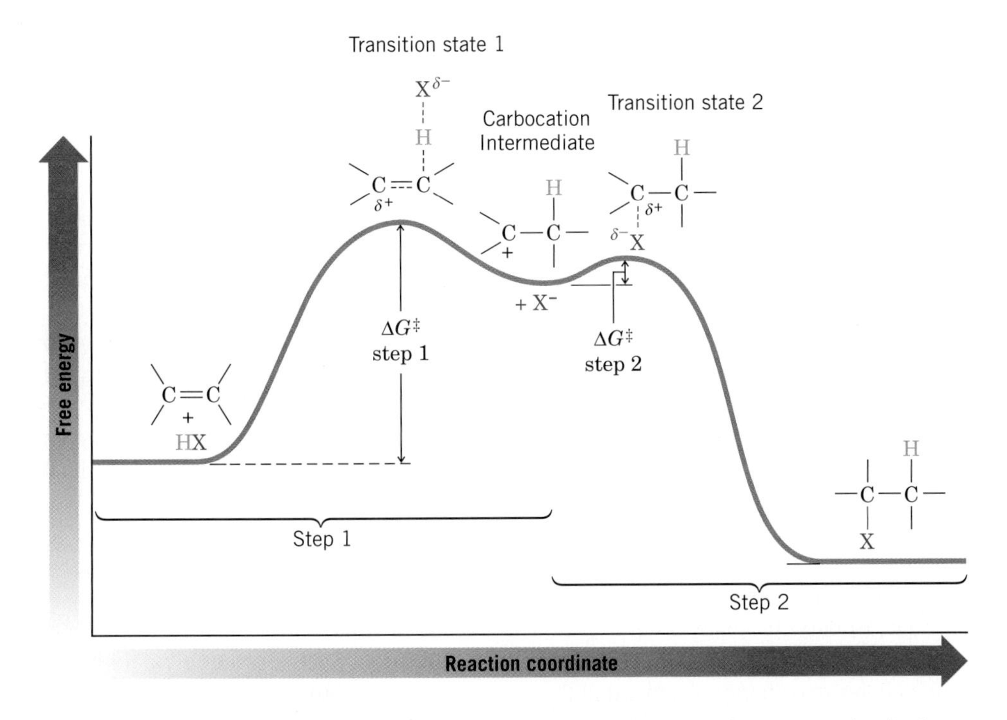

FIGURE 8.1 Free-energy diagram for the addition of HX to an alkene. The free energy of activation for step 1 is much larger than that for step 2.

8.2A Theoretical Explanation of Markovnikov's Rule

If the alkene that undergoes addition of a hydrogen halide is an unsymmetrical alkene such as propene, then step 1 could conceivably lead to two different carbocations:

$$X{-}H + CH_3CH{=}CH_2 \longrightarrow CH_3CH(H){-}\overset{+}{C}H_2 + X^-$$

1° Carbocation (less stable)

$$CH_3CH{=}CH_2 + H{-}X \longrightarrow CH_3\overset{+}{C}H{-}CH_2{-}H + X^-$$

2° Carbocation (more stable)

These two carbocations are not of equal stability, however. The secondary carbocation is *more stable,* and it is the greater stability of the secondary carbocation that accounts for the correct prediction of the overall addition by Markovnikov's rule. In the addition of HBr to propene, for example, the reaction takes the following course:

$$CH_3CH{=}CH_2 \xrightarrow[\text{slow}]{HBr} \begin{cases} \not\rightarrow CH_3CH_2\overset{+}{C}H_2 \ (1°) \xrightarrow{Br^-} CH_3CH_2CH_2Br \quad \textbf{1-Bromopropane } (\textit{little} \textbf{ formed}) \\ \rightarrow CH_3\overset{+}{C}HCH_3 \ (2°) \xrightarrow[\text{fast}]{Br^-} CH_3CH(Br)CH_3 \quad \textbf{2-Bromopropane (main product)} \end{cases}$$

Step 1 —— **Step 2**

The chief product of the reaction is 2-bromopropane because the more stable secondary carbocation is formed preferentially in the first step.

The more stable carbocation predominates because it is formed faster. We can understand why this is true if we examine the free-energy diagrams in Fig. 8.2.

The reaction (Fig. 8.2) leading to the secondary carbocation (and ultimately to 2-bromopropane) has the lower free energy of activation. That is reasonable because its transition state resembles the more stable carbocation. The reaction leading to the primary carbocation (and ultimately to 1-bromopropane) has a higher free energy of activation because its transition state resembles a less stable primary carbocation. This second reaction is much slower and does not compete with the first reaction.

The reaction of HBr with 2-methylpropene produces only *tert*-butyl bromide and for the same reason. Here, in the first step (i.e., the attachment of the proton) the choice is even more pronounced—between a tertiary carbocation and a primary carbocation.

Thus, isobutyl bromide is *not* obtained as a product of the reaction because its formation would require the formation of a primary carbocation. Such a reaction would have a much higher free energy of activation than that leading to a tertiary carbocation.

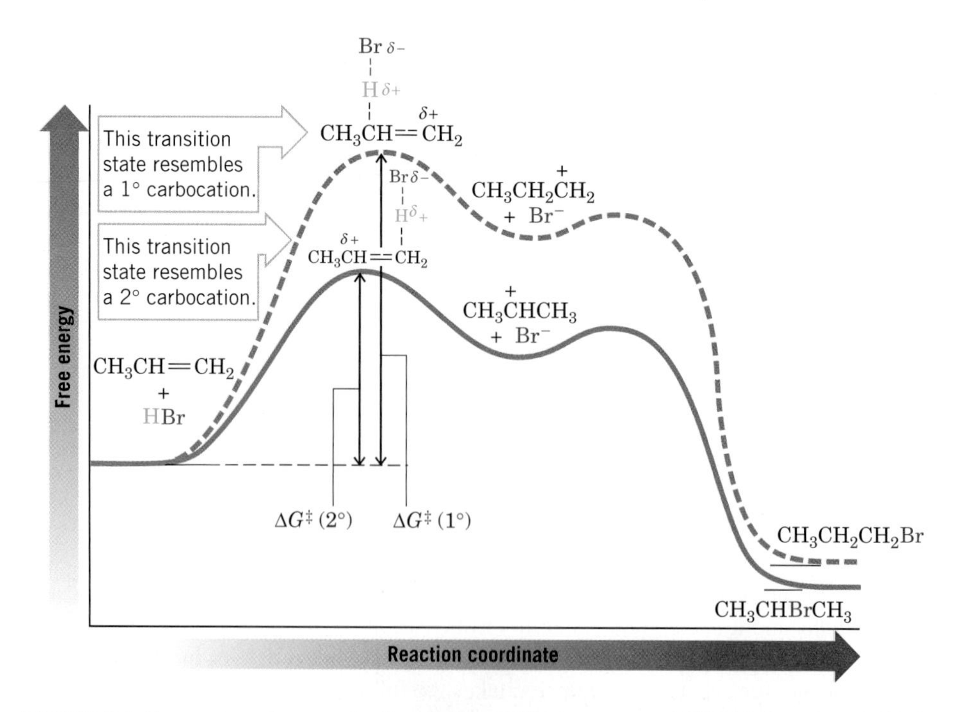

FIGURE 8.2 Free-energy diagrams for the addition of HBr to propene. $\Delta G^{\ddagger}(2°)$ is less than $\Delta G^{\ddagger}(1°)$.

A Mechanism for the Reaction

Addition of HBr to 2-Methylpropene

This reaction takes place:

$$(CH_3)_2C{=}CH_2 + H{-}\ddot{Br}: \longrightarrow (CH_3)_2\overset{+}{C}{-}CH_2{-}H + :\ddot{Br}:^- \longrightarrow (CH_3)_3C{-}\ddot{Br}:$$

3° Carbocation (more stable carbocation) — *tert*-Butyl bromide (Major product)

This reaction *does not* occur to any appreciable extent:

$$H_3C{-}\underset{}{C}(CH_3){=}CH_2 + H{-}\ddot{Br}: \not\longrightarrow H_3C{-}C(CH_3)(H){-}\overset{+}{C}H_2 + :\ddot{Br}:^- \not\longrightarrow H_3C{-}CH(CH_3){-}CH_2{-}\ddot{Br}:$$

1° Carbocation (less stable carbocation) — Isobutyl bromide (Little formed)

Because carbocations are formed in the addition of HX to an alkene, rearrangements invariably occur when the carbocation initially formed can rearrange to a more stable one (see Section 7.8 and Problem 8.3).

8.2B Modern Statement of Markovnikov's Rule

With this understanding of the mechanism for the ionic addition of hydrogen halides to alkenes behind us, we are now in a position to give the following **modern statement of Markovnikov's rule.** ***In the ionic addition of an unsymmetrical reagent to a double bond, the positive portion of the adding reagent attaches itself to a carbon atom of the double bond so as to yield the more stable carbocation as an intermediate.*** Because this is the step that occurs first (before the addition of the nucleophilic portion of the adding reagent), it is the step that determines the overall orientation of the reaction.

Markovnikov's rule

Notice that this formulation of Markovnikov's rule allows us to predict the outcome of the addition of a reagent such as ICl. Because of the greater electronegativity of chlorine, the positive portion of this molecule is iodine. The addition of ICl to 2-methylpropene takes place in the following way and produces 2-chloro-1-iodo-2-methylpropane:

$$(H_3C)_2C{=}CH_2 + :\overset{\delta+}{\ddot{I}}{-}\overset{\delta-}{\ddot{Cl}}: \longrightarrow (H_3C)_2\overset{+}{C}{-}CH_2{-}\ddot{I}: + :\ddot{Cl}:^- \longrightarrow CH_3{-}C(CH_3)(\ddot{Cl}:){-}CH_2{-}\ddot{I}:$$

2-Methylpropene — 2-Chloro-1-iodo-2-methylpropane

PROBLEM 8.1

Give the structure and name of the product that would be obtained from the ionic addition of IBr to propene.

PROBLEM 8.2

Outline mechanisms for the ionic additions of **(a)** HBr to 2-methyl-1-butene, **(b)** ICl to 2-methyl-2-butene, and **(c)** HI to 1-methylcyclopentene.

PROBLEM 8.3

Provide mechanistic explanations for the following observations:
(a) The addition of hydrogen chloride to 3-methyl-1-butene produces two products: 2-chloro-3-methylbutane and 2-chloro-2-methylbutane.
(b) The addition of hydrogen chloride to 3,3-dimethyl-1-butene produces two products: 3-chloro-2,2-dimethylbutane and 2-chloro-2,3-dimethylbutane.
(The first explanations for the course of both of these reactions were provided by F. Whitmore; see Section 7.7A.)

8.2C Regioselective Reactions

Regioselectivity

Chemists describe reactions like the Markovnikov additions of hydrogen halides to alkenes as being **regioselective**. *Regio* comes from the Latin word *regionem* meaning direction. ***When a reaction that can potentially yield two or more constitutional isomers actually produces only one (or a predominance of one), the reaction is said to be regioselective.*** The addition of HX to an unsymmetrical alkene such as propene could conceivably yield two constitutional isomers, for example. However, as we have seen, the reaction yields only one, and therefore it is regioselective.

8.2D An Exception to Markovnikov's Rule

In Section 10.9 we shall study an exception to Markovnikov's rule. This exception concerns the addition of HBr to alkenes ***when the addition is carried out in the presence of peroxides*** (i.e., compounds with the general formula ROOR). When alkenes are treated with HBr in the presence of peroxides, an **anti-Markovnikov addition** occurs in the sense that the hydrogen atom becomes attached to the carbon atom with the fewer hydrogen atoms. With propene, for example, the addition takes place as follows:

$$CH_3CH{=}CH_2 + HBr \xrightarrow{ROOR} CH_3CH_2CH_2Br$$

In Section 10.9 we shall find that this addition occurs by *a radical mechanism,* and not by the ionic mechanism given at the beginning of Section 8.2. This anti-Markovnikov addition occurs ***only when HBr is used in the presence of peroxides*** and does not occur significantly with HF, HCl, and HI even when peroxides are present.

8.3 Stereochemistry of the Ionic Addition to an Alkene

Consider the following addition of HX to 1-butene and notice that the reaction leads to the formation of a product, 2-halobutane, that contains a stereogenic center:

$$CH_3CH_2CH{=}CH_2 + HX \longrightarrow CH_3CH_2\overset{*}{C}HCH_3 \text{ (with X on the starred carbon)}$$

The product, therefore, can exist as a pair of enantiomers. The question now arises as to how these enantiomers are formed. Is one enantiomer formed in greater amount than the other? The answer is *no*; the carbocation that is formed in the first step of the addition (see the following scheme) is trigonal planar and ***is achiral*** (a model will show that it has a plane of symmetry). When the halide ion reacts with this achiral carbocation in the second step, ***reaction is equally likely at either face.*** The reactions leading to the two enantiomers occur at the same rate, and the enantiomers, therefore, are produced in equal amounts ***as a racemic form.***

The Stereochemistry of the Reaction

Ionic Addition to an Alkene

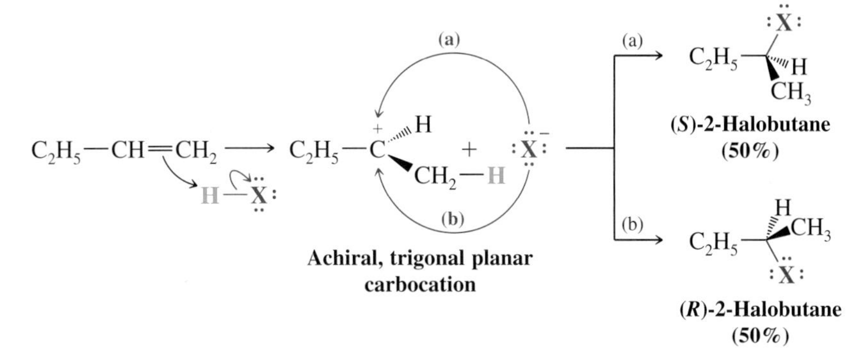

$C_2H_5—CH=CH_2 + H—\ddot{X}: \longrightarrow C_2H_5—\overset{+}{C}(H)CH_2—H + :\ddot{X}:^-$

Achiral, trigonal planar carbocation

(a) → (*S*)-2-Halobutane (50%)

(b) → (*R*)-2-Halobutane (50%)

1-Butene donates a pair of electrons to the proton of HX to form an achiral carbocation.

The carbocation reacts with the halide ion at equal rates by path (a) or (b) to form the enantiomers as a racemate.

8.4 Addition of Sulfuric Acid to Alkenes

When alkenes are treated with **cold** concentrated sulfuric acid, *they dissolve* because they react by addition to form alkyl hydrogen sulfates. The mechanism is similar to that for the addition of HX. In the first step of this reaction the alkene donates a pair of electrons to a proton from sulfuric acid to form a carbocation; in the second step the carbocation reacts with a hydrogen sulfate ion to form an alkyl hydrogen sulfate:

$$>C=C< + H—O—SO_2—O—H \rightleftharpoons —\overset{+}{C}—\overset{H}{C}— + {}^-O—SO_2—O—H \rightleftharpoons —\overset{HO_3SO}{C}—\overset{H}{C}—$$

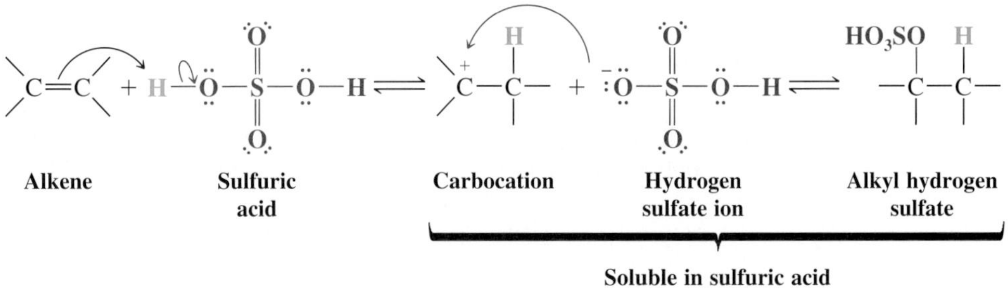

Alkene — Sulfuric acid — Carbocation — Hydrogen sulfate ion — Alkyl hydrogen sulfate

Soluble in sulfuric acid

The addition of sulfuric acid is also regioselective, and it follows Markovnikov's rule. Propene, for example, reacts to yield isopropyl hydrogen sulfate rather than propyl hydrogen sulfate:

$$H_3C(H)C=CH_2 + H—\ddot{O}SO_3H \longrightarrow H_3C(H)\overset{+}{C}—CH_2—H + {}^-:\ddot{O}SO_3H \longrightarrow H_3C—\underset{:\ddot{O}SO_3H}{\overset{H}{C}}—CH_3$$

2° Carbocation (more stable carbocation)

Isopropyl hydrogen sulfate

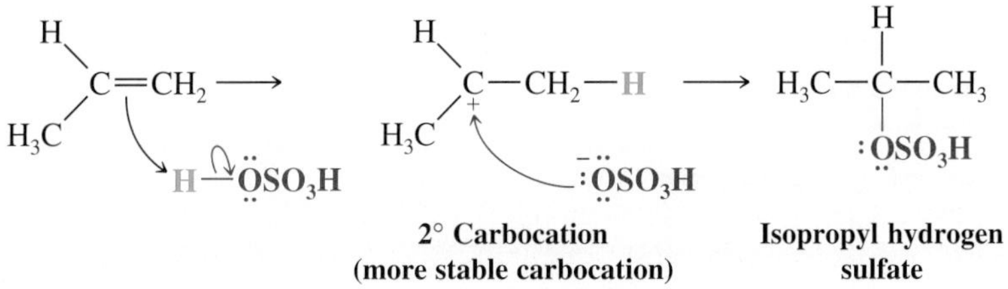

8.4A Alcohols from Alkyl Hydrogen Sulfates

Alkyl hydrogen sulfates can be easily hydrolyzed to alcohols by **heating** them with water. The overall result of the addition of sulfuric acid to an alkene followed by hydrolysis is the Markovnikov addition of H— and —OH:

$$CH_3CH{=}CH_2 \xrightarrow[H_2SO_4]{\text{cold}} CH_3\underset{\substack{|\\ OSO_3H}}{C}HCH_3 \xrightarrow{H_2O,\ \text{heat}} CH_3\underset{\substack{|\\ OH}}{C}HCH_3 + H_2SO_4$$

PROBLEM 8.4 In one industrial synthesis of ethanol, ethene is first dissolved in 95% sulfuric acid. In a second step water is added and the mixture is heated. Outline the reactions involved.

8.5 Addition of Water to Alkenes: Acid-Catalyzed Hydration

The acid-catalyzed addition of water to the double bond of an alkene (**hydration** of an alkene) is a method for the preparation of low-molecular-weight alcohols that has its greatest utility in large-scale industrial processes. The acids most commonly used to catalyze the hydration of alkenes are dilute aqueous solutions of sulfuric acid and phosphoric acid. These reactions, too, are usually regioselective, and the addition of water to the double bond follows Markovnikov's rule. In general, the reaction takes the form that follows:

$$\rangle C{=}C\langle + HOH \xrightarrow{H_3O^+} -\underset{\substack{|\\ H}}{\overset{|}{C}}-\underset{\substack{|\\ OH}}{\overset{|}{C}}-$$

An example is the hydration of 2-methylpropene:

$$(CH_3)_2C{=}CH_2 + HOH \xrightarrow[25°C]{H_3O^+} CH_3-\underset{\substack{|\\ OH}}{\overset{\substack{CH_3\\ |}}{C}}-CH_2-H$$

2-Methylpropene (isobutylene) → ***tert*-Butyl alcohol**

Because the reactions follow Markovnikov's rule, acid-catalyzed hydrations of alkenes do not yield primary alcohols except in the special case of the hydration of ethene:

$$CH_2{=}CH_2 + HOH \xrightarrow[300°C]{H_3PO_4} CH_3CH_2OH$$

The mechanism for the hydration of an alkene is simply the reverse of the mechanism for the dehydration of an alcohol. We can illustrate this by giving the mechanism for the **hydration** of 2-methylpropene and by comparing it with the mechanism for the **dehydration** of *tert*-butyl alcohol given in Section 7.7A.

A Mechanism for the Reaction

Acid-Catalyzed Hydration of an Alkene

Step 1

$$(H_3C)(CH_3)C{=}CH_2 + H-\overset{\substack{H\\ |}}{\underset{+}{\ddot{O}}}-H \underset{}{\overset{\text{slow}}{\rightleftharpoons}} H_3C-\underset{\substack{|\\ CH_3}}{\overset{H_2C-H}{C^+}} + :\overset{\substack{H\\ |}}{\ddot{O}}-H$$

The alkene donates an electron pair to a proton to form the more stable 3° carbocation.

Step 2

$$\text{H}_3\text{C}-\overset{\text{CH}_3}{\text{C}^+}-\text{CH}_3 + :\ddot{\text{O}}(\text{H})-\text{H} \xrightleftharpoons{\text{fast}} \text{H}_3\text{C}-\underset{\text{CH}_3}{\overset{\text{CH}_3}{\text{C}}}-\overset{\text{H}}{\underset{+}{\ddot{\text{O}}}}-\text{H}$$

The carbocation reacts with a molecule of water to form a protonated alcohol.

Step 3

$$\text{H}_3\text{C}-\underset{\text{CH}_3}{\overset{\text{CH}_3}{\text{C}}}-\overset{\text{H}}{\overset{+}{\text{O}}}-\text{H} + :\ddot{\text{O}}(\text{H})-\text{H} \xrightleftharpoons{\text{fast}} \text{H}_3\text{C}-\underset{\text{CH}_3}{\overset{\text{CH}_3}{\text{C}}}-\ddot{\text{O}}-\text{H} + \text{H}-\overset{\text{H}}{\underset{+}{\ddot{\text{O}}}}-\text{H}$$

A transfer of a proton to a molecule of water leads to the product.

The rate-determining step in the *hydration* mechanism is step 1: the formation of the carbocation. It is this step, too, that accounts for the Markovnikov addition of water to the double bond. The reaction produces *tert*-butyl alcohol because step 1 leads to the formation of the more stable *tert*-butyl cation (3°) rather than the much less stable isobutyl (1°) cation:

$$\underset{\text{H}_3\text{C}\quad\text{CH}_3}{\overset{\text{CH}_2}{\text{C}}} + \text{H}-\overset{\text{H}}{\underset{+}{\ddot{\text{O}}}}-\text{H} \xrightleftharpoons[\text{slow}]{\text{very}} \text{CH}_3-\underset{\text{CH}_3}{\overset{\text{CH}_2^{\,+}}{\text{C}}}-\text{H} + :\ddot{\text{O}}(\text{H})-\text{H}$$

For all practical purposes this reaction does not take place because it produces a 1° carbocation.

The reactions whereby *alkenes are hydrated or alcohols are dehydrated* are reactions in which the ultimate product is governed by the position of an equilibrium. Therefore, in the *dehydration of an alcohol* it is best to use a concentrated acid so that the concentration of water is low. (The water can be removed as it is formed, and it helps to use a high temperature.) In the *hydration of an alkene* it is best to use dilute acid so that the concentration of water is high. (It also usually helps to use a lower temperature.)

PROBLEM 8.5

(a) Show all steps in the acid-catalyzed hydration of cyclohexene to produce cyclohexanol and give the general conditions that you would use to ensure a good yield of the product. **(b)** Give the general conditions that you would use to carry out the reverse reaction, the dehydration of cyclohexanol to produce cyclohexene. **(c)** What product would you expect to obtain from the acid-catalyzed hydration of 1-methylcyclohexene? Explain your answer.

One complication associated with alkene hydrations is the occurrence of **rearrangements.** Because the reaction involves the formation of a carbocation in the first step, the carbocation formed initially invariably rearranges to a more stable one if such a rearrangement is possible. An illustration is the formation of 2,3-dimethyl-2-butanol as the major product when 3,3-dimethyl-1-butene is hydrated:

$$\text{CH}_3-\underset{\text{CH}_3}{\overset{\text{CH}_3}{\text{C}}}-\text{CH}=\text{CH}_2 \xrightarrow[\text{H}_2\text{O}]{\text{H}_2\text{SO}_4} \text{CH}_3-\underset{\text{CH}_3}{\overset{\text{OH}}{\text{C}}}-\underset{\text{CH}_3}{\text{CH}}-\text{CH}_3$$

3,3-Dimethyl-1-butene → **2,3-Dimethyl-2-butanol (major product)**

PROBLEM 8.6 Outline all steps in a mechanism showing how 2,3-dimethyl-2-butanol is formed in the acid-catalyzed hydration of 3,3-dimethyl-1-butene.

PROBLEM 8.7 The following order of reactivity is observed when the following alkenes are subjected to acid-catalyzed hydration:

$$(CH_3)_2C{=}CH_2 > CH_3CH{=}CH_2 > CH_2{=}CH_2$$

Explain this order of reactivity.

PROBLEM 8.8 When 2-methylpropene (isobutylene) is dissolved in methanol containing a strong acid, a reaction takes place to produce *tert*-butyl methyl ether, $CH_3OC(CH_3)_3$. Write a mechanism that accounts for this.

8.6 Alcohols from Alkenes through Oxymercuration–Demercuration: Markovnikov Addition

Mercury compounds are extremely hazardous. Before you carry out a reaction involving mercury or its compounds, you should familiarize yourself with current procedures for its use and containment. There are no satisfactory methods for disposal of mercury.

A useful laboratory procedure for synthesizing alcohols from alkenes that avoids rearrangement is a two-step method called **oxymercuration–demercuration.**

Alkenes react with mercuric acetate in a mixture of tetrahydrofuran (THF) and water to produce (hydroxyalkyl)mercury compounds. These (hydroxyalkyl)mercury compounds can be reduced to alcohols with sodium borohydride:

Step 1: Oxymercuration

$$\text{>C=C<} + H_2O + Hg\left(\overset{O}{\overset{\|}{O}}CCH_3\right)_2 \xrightarrow{THF} -\underset{HO}{\overset{|}{C}}-\underset{Hg-OCCH_3}{\overset{|}{C}}- + CH_3\overset{O}{\overset{\|}{C}}OH$$

Step 2: Demercuration

$$-\underset{HO}{\overset{|}{C}}-\underset{Hg-O\overset{O}{\overset{\|}{C}}CH_3}{\overset{|}{C}}- + OH^- + NaBH_4 \longrightarrow -\underset{HO}{\overset{|}{C}}-\underset{H}{\overset{|}{C}}- + Hg + CH_3\overset{O}{\overset{\|}{C}}O^-$$

In the first step, **oxymercuration,** water and mercuric acetate add to the double bond; in the second step, **demercuration,** sodium borohydride reduces the acetoxymercury group and replaces it with hydrogen. (The acetate group is often abbreviated —OAc.)

Both steps can be carried out in the same vessel, and both reactions take place very rapidly at room temperature or below. The first step—oxymercuration—usually goes to completion within a period of 20 s–10 min. The second step—demercuration—normally requires less than an hour. The overall reaction gives alcohols in very high yields, usually greater than 90%.

Regioselectivity of oxymercuration–demercuration

Oxymercuration–demercuration is also highly regioselective. The net orientation of the addition of the elements of water, H— and —OH, ***is in accordance with Markovnikov's rule.*** The H— becomes attached to the carbon atom of the double bond with the greater number of hydrogen atoms:

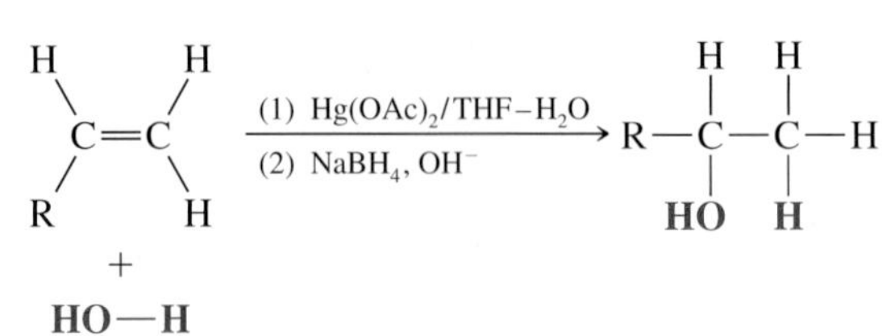

The following are specific examples:

$$\underset{\textbf{1-Pentene}}{CH_3(CH_2)_2CH{=}CH_2} \xrightarrow[\substack{THF\text{–}H_2O \\ (15\ s)}]{Hg(OAc)_2} CH_3(CH_2)_2\underset{OH}{CH}{-}\underset{HgOAc}{CH_2} \xrightarrow[\substack{OH^- \\ (1\ h)}]{NaBH_4} \underset{\substack{\textbf{2-Pentanol} \\ \textbf{(93\%)}}}{CH_3(CH_2)_2\underset{OH}{C}HCH_3} + Hg$$

$$\textbf{1-Methylcyclopentene} \xrightarrow[\substack{THF\text{–}H_2O \\ (20\ s)}]{Hg(OAc)_2} \text{(1-methyl-1-hydroxy-2-(HgOAc)cyclopentane: } H_3C,\ OH,\ HgOAc,\ H) \xrightarrow[\substack{OH^- \\ (6\ min)}]{NaBH_4} \textbf{1-Methylcyclopentanol} + Hg$$

Rearrangements of the carbon skeleton seldom occur in oxymercuration–demercuration. The oxymercuration–demercuration of 3,3-dimethyl-1-butene is a striking example illustrating this feature. It is in direct contrast to the hydration of 3,3-dimethyl-1-butene we studied previously (Section 8.5).

Oxymercuration–demercuration is not prone to hydride or alkanide rearrangements.

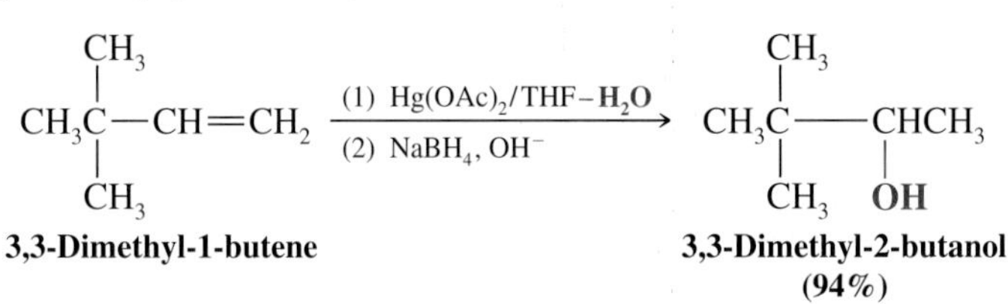

Analysis of the mixture of products by gas chromatography failed to reveal the presence of any 2,3-dimethyl-2-butanol. The acid-catalyzed hydration of 3,3-dimethyl-1-butene, by contrast, gives 2,3-dimethyl-2-butanol as the major product.

A mechanism that accounts for the orientation of addition in the oxymercuration stage, and one that also explains the general lack of accompanying rearrangements, is shown below. Central to this mechanism is an electrophilic attack by the mercury species, $\overset{+}{H}gOAc$, at the less substituted carbon of the double bond (i.e., at the carbon atom that bears the greater number of hydrogen atoms), and the formation of a bridged intermediate. We illustrate the mechanism using 3,3-dimethyl-1-butene as the example:

A Mechanism for the Reaction

Oxymercuration

Step 1

$$Hg(OAc)_2 \rightleftharpoons \overset{+}{H}gOAc + OAc^-$$

Mercuric acetate dissociates to form a $\overset{+}{H}gOAc$ cation and an acetate anion.

Step 2

$$\underset{\textbf{3,3-Dimethyl-1-butene}}{H_3C{-}\overset{CH_3}{\underset{CH_3}{C}}{-}CH{=}CH_2} + \overset{+}{H}gOAc \longrightarrow \underset{\textbf{Mercury-bridged carbocation}}{H_3C{-}\overset{CH_3}{\underset{CH_3}{C}}{-}\overset{\delta+}{CH}{-}\underset{\underset{\delta+}{HgOAc}}{CH_2}}$$

The alkene donates a pair of electrons to the electrophilic $HgOAc^+$ cation to form a mercury-bridged carbocation. In this carbocation, the positive charge is shared between the 2° (more substituted) carbon atom and the mercury atom. The charge on the carbon atom is large enough to account for the Markovnikov orientation of the addition, but not large enough for a rearrangement to occur.

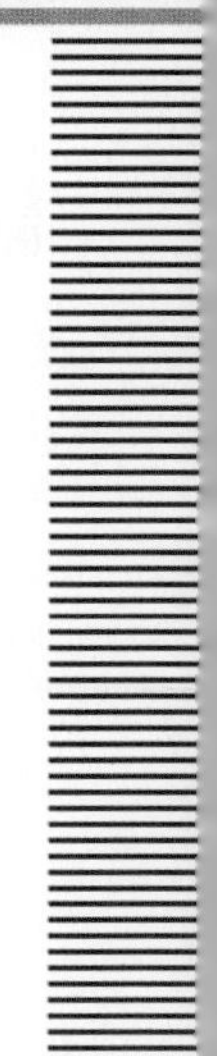

Step 3

$$H_3C-\overset{CH_3}{\underset{CH_3}{C}}-\overset{\delta+}{CH}-\underset{HgOAc}{CH_2} \xrightarrow{:\ddot{O}H_2} H_3C-\overset{H_3C\ \ ^{+}:OH_2}{\underset{H_3C}{C}}-\!\!-CH-\underset{HgOAc}{CH_2}$$

A water molecule attacks the carbon of the bridged mercurinium ion that is better able to bear the partial positive charge.

Step 4

$$H_3C-\overset{H_3C\ \ ^{+}:OH_2}{\underset{H_3C}{C}}-CH-\underset{HgOAc}{CH_2} \xrightarrow{:\ddot{O}H_2} H_3C-\overset{H_3C\ \ :\ddot{O}H}{\underset{H_3C}{C}}-CH-\underset{HgOAc}{CH_2} + H-\overset{+}{\ddot{O}}-H \ (H)$$

(Hydroxyalkyl)mercury compound

An acid–base reaction transfers a proton to another water molecule (or to an acetate ion). This step produces the (hydroxyalkyl)mercury compound.

Calculations indicate that mercury-bridged carbocations (termed mercurinium ions) such as those formed in this reaction retain much of the positive charge on the mercury moiety. Only a small portion of the positive charge resides on the more substituted carbon atom. The charge is large enough to account for the observed Markovnikov addition, but it is too small to allow the usual rapid carbon skeleton rearrangements that take place with more fully developed carbocations.

Although attack by water on the bridged mercurinium ion leads to anti addition of the hydroxyl and mercury groups, the reaction that replaces mercury with hydrogen is not stereocontrolled (it likely involves radicals; see Chapter 10). This step scrambles the overall stereochemistry of oxymercuration–demercuration, and the net result is a mixture of syn and anti addition of —H and —OH to the alkene. As already noted, oxymercuration–demercuration takes place with anti-Markovnikov regiochemistry, however.

PROBLEM 8.9 Starting with an appropriate alkene, show all steps in the synthesis of each of the following alcohols by oxymercuration–demercuration:
(a) *tert*-Butyl alcohol **(b)** Isopropyl alcohol **(c)** 2-Methyl-2-butanol

PROBLEM 8.10 When an alkene is treated with mercuric trifluoroacetate, $Hg(O_2CCF_3)_2$, in THF containing an alcohol, ROH, the product is an (alkoxyalkyl)mercury compound. Treating this product with $NaBH_4/OH^-$ results in the formation of an ether. The overall process is called *solvomercuration–demercuration:*

$$\text{>C=C<} \xrightarrow[\text{solvomercuration}]{Hg(O_2CCF_3)_2/\text{THF–ROH}} -\overset{RO}{C}-\underset{HgO_2CCF_3}{C}- \xrightarrow[\text{demercuration}]{NaBH_4,\ OH^-} -\overset{RO}{C}-\underset{H}{C}-$$

Alkene **(Alkoxyalkyl)mercuric trifluoroacetate** **Ether**

(a) Outline a likely mechanism for the solvomercuration step of this ether synthesis. **(b)** Show how you would use solvomercuration–demercuration to prepare *tert*-butyl methyl ether. **(c)** Why would one use $Hg(O_2CCF_3)_2$ instead of $Hg(OAc)_2$?

8.7 Alcohols from Alkenes through Hydroboration–Oxidation: Anti-Markovnikov Syn Hydration

Hydration of a double bond in an **anti-Markovnikov** sense can be achieved through the use of diborane (B_2H_6) or a solution of borane in tetrahydrofuran (BH_3:THF). The addition of water is indirect, and two reactions are involved. The first is the addition of a boron atom and hydrogen atom to the double bond, called **hydroboration;** the second is **oxidation** and hydrolysis of the alkylborane intermediate to an alcohol and boric acid. The anti-Markovnikov regiochemistry of the addition is illustrated by the hydroboration–oxidation of propene:

$$\underset{\textbf{Propene}}{3\ H_3CCH{=}CH_2} \xrightarrow[\text{hydroboration}]{BH_3:THF} \underset{\textbf{Tripropylborane}}{(CH_3CH_2CH_2)_3B} \xrightarrow[\text{oxidation}]{H_2O_2/OH^-} \underset{\textbf{1-Propanol}}{3\ CH_3CH_2CH_2OH}$$

Hydroboration–oxidation takes place with **syn** stereochemistry, as well as anti-Markovnikov regiochemistry, as can be seen in the following example with 1-methylcyclopentene:

$$\text{1-methylcyclopentene} \xrightarrow[(2)\ H_2O_2,\ HO^-]{(1)\ BH_3:THF} \text{trans-2-methylcyclopentanol (H and OH syn; } CH_3 \text{ and OH trans)}$$

In the following sections we shall consider details of the mechanism that lead to the anti-Markovnikov regiochemistry and syn stereochemistry of hydroboration–oxidation.

8.8 Hydroboration: Synthesis of Alkylboranes

Hydroboration of an alkene is the starting point for a number of useful synthetic procedures, including the anti-Markovnikov syn hydration procedure we have just mentioned. Hydroboration was discovered by Herbert C. Brown (Purdue University), and it can be represented in its simplest terms as follows:

$$\underset{\textbf{Alkene}}{>C{=}C<} + \underset{\textbf{Boron hydride}}{H{-}B<} \xrightarrow{\text{hydroboration}} \underset{\textbf{Alkylborane}}{-\overset{|}{\underset{H}{C}}-\overset{|}{\underset{B<}{C}}-}$$

Brown's discovery of hydroboration led to his being named a co-winner of the 1979 Nobel Prize in Chemistry.

Hydroboration can be accomplished with diborane (B_2H_6), which is a gaseous dimer of borane (BH_3), or more conveniently with a reagent prepared by dissolving diborane in THF. When diborane is introduced to THF, it reacts to form a Lewis acid–base complex of borane (the Lewis acid) and THF. The complex is represented as BH_3:THF.

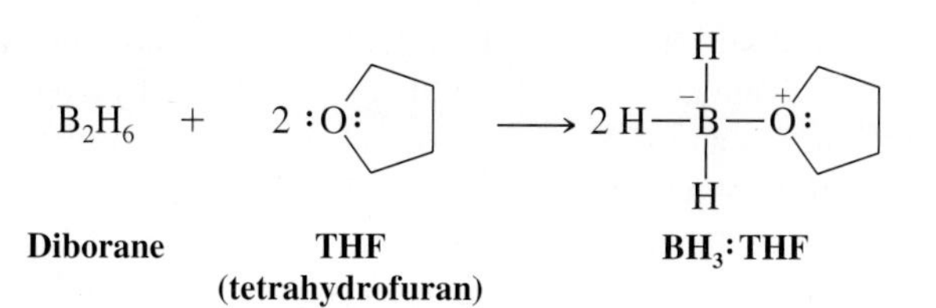

Solutions containing the BH_3:THF complex can be obtained commercially. Hydroboration reactions are usually carried out in ethers: either in diethyl ether

$(CH_3CH_2)_2O$, or in some higher molecular weight ether such as "diglyme" [$(CH_3OCH_2CH_2)_2O$, *di*ethylene *gly*col di*me*thyl ether]. Great care must be used in handling diborane and alkylboranes because they ignite spontaneously in air (with a green flame). The solution of BH_3:THF must be used in an inert atmosphere (e.g., argon or nitrogen) and with care.

8.8A Mechanism of Hydroboration

When a terminal alkene such as propene is treated with a solution containing BH_3:THF, the boron hydride adds successively to the double bonds of three molecules of the alkene to form a trialkylborane:

More substituted carbon — Less substituted carbon

$$H_3CCH{=}CH_2 + H{-}BH_2 \longrightarrow H_3CCH(H){-}CH_2(BH_2) \xrightarrow[\text{(2nd equiv.)}]{H_3CCH{=}CH_2} (H_3CCH(H){-}CH_2)_2BH \xrightarrow[\text{(3rd equiv.)}]{H_3CCH{=}CH_2} (H_3CCH(H){-}CH_2)_3B$$

Tripropylborane

Regiochemistry of hydroboration

In each addition step ***the boron atom becomes attached to the less substituted carbon atom of the double bond,*** and a hydrogen atom is transferred from the boron atom to the other carbon atom of the double bond. Thus, hydroboration is regioselective and it is **anti-Markovnikov** (the hydrogen atom becomes attached to the carbon atom with fewer hydrogen atoms).

Other examples that illustrate this tendency for the boron atom to become attached to the less substituted carbon atom are shown here. The percentages designate where the boron atom becomes attached.

$CH_3CH_2C(CH_3){=}CH_2$: 1% (C-2), **99%** (CH$_2$, **Less substituted**)

$CH_3C(CH_3){=}CHCH_3$: 2% (C-2), **98%** (CHCH$_3$, **Less substituted**)

This observed attachment of boron to the less substituted carbon atom of the double bond seems to result in part from **steric factors**—the bulky boron-containing group can approach the less substituted carbon atom more easily.

A mechanism that has been proposed for the addition of BH_3 to the double bond begins with a donation of π electrons from the double bond to the vacant *p* orbital of BH_3 (see mechanism below). In the next step this complex becomes the addition product by passing through a four-atom transition state in which the boron atom is partially bonded to the less substituted carbon atom of the double bond and one hydrogen atom is partially bonded to the other carbon atom. As this transition state is approached, electrons shift in the direction of the boron atom and away from the more substituted carbon atom of the double bond. This makes the more substituted carbon atom develop a partial positive charge, *and because it bears an electron-releasing alkyl group, it is better able to accommodate this positive charge.* Thus, both *electronic* and *steric factors* account for the anti-Markovnikov orientation of the addition.

Hydrolysis of the Borate Ester

$$R{-}\ddot{O}{:}_2B{-}\ddot{O}{-}R + {:}\ddot{O}H^- \longrightarrow (R{-}\ddot{O})_2\overset{-}{B}(OR)(\ddot{O}H) \longrightarrow (R{-}\ddot{O})_2B{-}\ddot{O}H + {}^-{:}\ddot{O}{-}R \longrightarrow (R{-}\ddot{O})_2B{-}\ddot{O}{:}^- + {:}\ddot{O}(H){-}R$$

Trialkyl borate ester ... **Alcohol**

Hydroxide anion attacks the boron atom of the borate ester.

An alkoxide anion departs from the borate anion, reducing the formal change on boron to zero.

Proton transfer completes the formation of one alcohol molecule. The sequence repeats until all three alkoxy groups are released as alcohols and inorganic borate remains.

8.9A Regiochemistry and Stereochemistry of Alkylborane Oxidation and Hydrolysis

Because hydroboration reactions are regioselective, the net result of hydroboration–oxidation is anti-Markovnikov addition of water. As a consequence, hydroboration–oxidation gives us a method for the preparation of alcohols that cannot normally be obtained through the acid-catalyzed hydration of alkenes or by oxymercuration–demercuration. For example, the acid-catalyzed hydration (or oxymercuration–demercuration) of 1-hexene yields 2-hexanol:

Regiochemistry of hydroboration–oxidation

$$CH_3CH_2CH_2CH_2CH{=}CH_2 \xrightarrow{H_3O^+,\ H_2O} CH_3CH_2CH_2CH_2\underset{\displaystyle OH}{\underset{|}{C}H}{-}CH_3$$

1-Hexene **2-Hexanol**

Hydroboration–oxidation of 1-hexene, by contrast, yields 1-hexanol:

$$CH_3CH_2CH_2CH_2CH{=}CH_2 \xrightarrow[(2)\ H_2O_2,\ HO^-]{(1)\ BH_3{:}THF} CH_3CH_2CH_2CH_2CH_2CH_2OH$$

1-Hexene **1-Hexanol (90%)**

$$CH_3{-}\overset{\displaystyle CH_3}{\overset{|}{C}}{=}CH{-}CH_3 \xrightarrow[(2)\ H_2O_2,\ HO^-]{(1)\ BH_3{:}THF} CH_3{-}\overset{\displaystyle CH_3}{\overset{|}{\underset{\displaystyle H}{\underset{|}{C}}}}{-}\underset{\displaystyle OH}{\underset{|}{C}H}{-}CH_3$$

2-Methyl-2-butene **3-Methyl-2-butanol**

Because the oxidation step in the hydroboration–oxidation synthesis of alcohols takes place with retention of configuration, **the hydroxyl group replaces the boron atom where it stands in the alkylboron compound.** The net result of the two steps (hydroboration and oxidation) is the syn addition of —H and —OH. We can review the anti-Markovnikov and syn aspects of hydroboration–oxidation by considering the hydration of 1-methylcyclopentene, as shown in Figure 8.3.

Overall stereochemistry of hydroboration–oxidation

PROBLEM 8.13

Starting with the appropriate alkene, show how you could use hydroboration–oxidation to prepare each of the alcohols:

(a) 1-Pentanol
(b) 2-Methyl-1-pentanol
(c) 3-Methyl-2-pentanol
(d) 2-Methyl-3-pentanol
(e) *trans*-2-Methylcyclobutanol

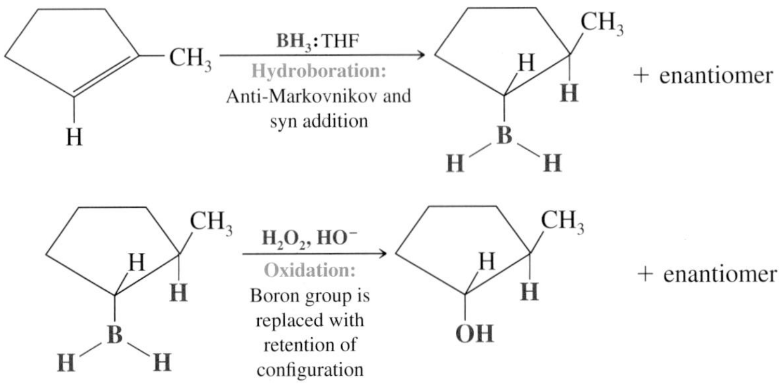

FIGURE 8.3 The hydroboration–oxidation of 1-methylcyclopentene. The first reaction is a syn addition of borane. (In this illustration we have shown the boron and hydrogen entering from the bottom side of 1-methylcyclopentene. The reaction also takes place from the top side at an equal rate to produce the enantiomer.) In the second reaction the boron atom is replaced by a hydroxyl group with retention of configuration. The product is a compound (*trans*-2-methylcyclopentanol), and the overall result is the syn addition of —H and —OH.

8.10 Summary of Alkene Hydration Methods

The three methods we have studied for alcohol synthesis by addition reactions to alkenes have different regiochemical and stereochemical characteristics. Acid-catalyzed hydration of alkenes takes place with Markovnikov regiochemistry but may lead to a mixture of constitutional isomers if the carbocation intermediate in the reaction undergoes rearrangement to a more stable carbocation. Oxymercuration–demercuration occurs with Markovnikov regiochemistry and results in hydration of alkenes without complication from carbocation rearrangement. It is often the preferred choice over acid-catalyzed hydration for Markovnikov addition. The overall stereochemistry of addition in acid-catalyzed hydration and oxymercuration–demercuration is not controlled—they both result in a mixture of cis and trans addition products. Hydroboration–oxidation results in anti-Markovnikov and syn hydration of an alkene. The complementary regiochemical and stereochemical aspects of these methods provide useful alternatives when we desire to synthesize a specific alcohol by hydration of an alkene.

8.11 Protonolysis of Alkylboranes

Heating an alkylborane with acetic acid causes cleavage of the carbon–boron bond and replacement with hydrogen:

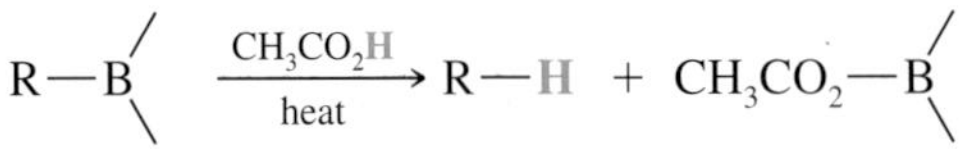

$$\mathrm{R{-}B\langle} \xrightarrow[\text{heat}]{\mathrm{CH_3CO_2H}} \mathrm{R{-}H} + \mathrm{CH_3CO_2{-}B\langle}$$

Alkylborane **Alkane**

This reaction also takes place with retention of configuration; therefore, hydrogen replaces boron **where it stands** in the alkylborane. The stereochemistry of this reaction, therefore, is **syn,** like that of the oxidation of alkylboranes. Hydroboration followed by protonolysis of the resulting alkylborane can be used as an alternative method for hydrogenation of alkenes, although catalytic hydrogenation (Section 7.12) is the more common procedure. Reaction of alkylboranes with deuterated or tritiated acetic acid also provides a very useful way to introduce these isotopes into a compound in a specific way.

PROBLEM 8.14

Starting with any needed alkene (or cycloalkene) and assuming you have deuterioacetic acid (CH_3CO_2D) available, outline syntheses of the following deuterium-labeled compounds.

(a) $(CH_3)_2CHCH_2CH_2D$ **(b)** $(CH_3)_2CHCHDCH_3$ **(c)** 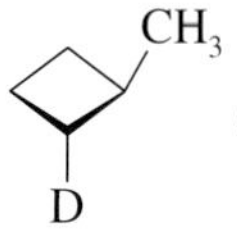 (cyclobutane with CH_3 and D) (+ enantiomer)

(d) Assuming you also have available BD_3:THF and CH_3CO_2T, can you suggest a synthesis of the following?

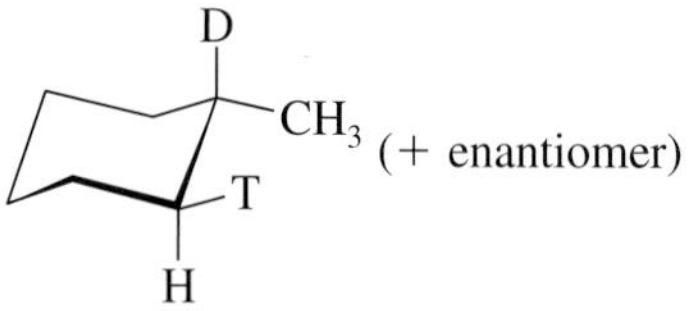

(cyclohexane bearing D, CH_3, T, H) (+ enantiomer)

8.12 Addition of Bromine and Chlorine to Alkenes

Alkenes react rapidly with chlorine and bromine in nonnucleophilic solvents to form vicinal dihalides. The following are specific examples of this reaction:

$$CH_3CH{=}CHCH_3 + Cl_2 \xrightarrow{-9°C} CH_3CHCl{-}CHClCH_3 \quad \textbf{(100\%)}$$

$$CH_3CH_2CH{=}CH_2 + Cl_2 \xrightarrow{-9°C} CH_3CH_2CHCl{-}CH_2Cl \quad \textbf{(97\%)}$$

$$\text{Cyclohexene} + Br_2 \xrightarrow[CCl_4]{-5°C} \text{trans-1,2-dibromocyclohexane} + \text{enantiomer} \quad (95\%)$$

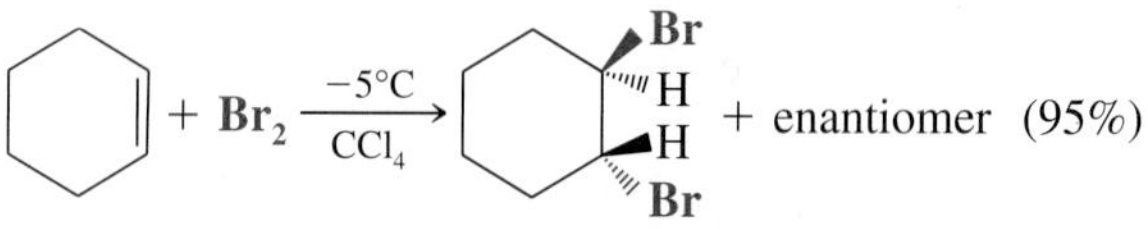

***trans*-1,2-Dibromocyclohexane (as a racemic form)**

When bromine is used for this reaction, it can serve as a test for the presence of carbon–carbon multiple bonds. If we add bromine to an alkene (or alkyne, see Section 8.18), the red-brown color of the bromine disappears almost instantly as long as the alkene (or alkyne) is present in excess:

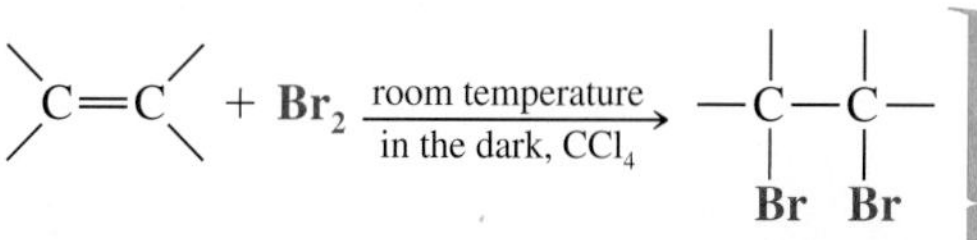

$$\text{C=C} + Br_2 \xrightarrow[\text{in the dark, } CCl_4]{\text{room temperature}} {-}\text{CBr}{-}\text{CBr}{-}$$

An alkene (colorless) → ***vic*-Dibromide (a colorless compound)**

Rapid decolorization of Br_2/CCl_4 is a test for alkenes and alkynes.

This behavior contrasts markedly with that of **alkanes.** Alkanes do not react appreciably with bromine or chlorine at room temperature and in the absence of light. When alkanes *do* react under those conditions, however, it is by substitution rather than addition and by a mechanism involving radicals that we shall discuss in Chapter 10:

$$\underset{\textbf{Alkane (colorless)}}{R{-}H} + \underset{\textbf{Bromine (red brown)}}{Br_2} \xrightarrow[\text{in the dark, } CCl_4]{\text{room temperature}} \text{no appreciable reaction}$$

8.12A Mechanism of Halogen Addition

One mechanism that has been proposed for halogen addition is an **ionic mechanism.*** In the first step the exposed electrons of the π bond of the alkene attack the halogen in the following way.

As the π electrons of the alkene approach the bromine molecule, the electrons of the bromine–bromine bond drift in the direction of the bromine atom more distant from the approaching alkene. The bromine molecule becomes *polarized* as a result. The more distant bromine develops a partial negative charge; the nearer bromine becomes partially positive. Polarization weakens the bromine–bromine bond, causing it to *break heterolytically.* A bromide ion departs, and a **bromonium ion** forms, similar to the way a bridged mercurinium ion forms during oxymercuration. In the bromonium ion a positively charged bromine atom is bonded to two carbon atoms by *two pairs of electrons:* one pair from the π bond of the alkene, the other pair from the bromine atom (one of its unshared pairs). In this way, all atoms of the bromonium ion have an octet of electrons, though one large, polarizable bromine atom still carries a formal positive charge.

In the second step, one of the bromide ions produced in step 1 attacks the back side of one of the carbon atoms of the bromonium ion. The nucleophilic attack results in the formation of a *vic*-dibromide by opening the three-membered ring.

A Mechanism for the Reaction

Addition of Bromine to an Alkene

Step 1

δ+ :Br: δ− :Br:

Bromonium ion **Bromide ion**

As a bromine molecule approaches an alkene, the electron density of the alkene π bond repels electron density in the closer bromine, polarizing the bromine molecule and making the closer bromine atom electrophilic. The alkene donates a pair of electrons to the closer bromine, causing displacement of the distant bromine atom. As this occurs, the newly bonded bromine atom, due to its size and polarizability, donates an electron pair to the carbon that would otherwise be a carbocation, thereby stabilizing the positive charge by delocalization. The result is a bromonium ion intermediate.

Step 2

Bromonium ion **Bromide ion** ***vic*-Dibromide**

A bromide anion attacks at the back side of one carbon (or the other) of the bromonium ion in an S_N2 reaction, causing the ring to open and resulting in the formation of a *vic*-dibromide.

This ring opening (see the preceding scheme) is an S_N2 reaction. The bromide ion, acting as a nucleophile, uses a pair of electrons to form a bond to one carbon atom of the bromonium ion while the positive bromine of the bromonium ion acts as a leaving group.

*There is evidence that in the absence of oxygen some reactions between alkenes and chlorine proceed through a radical mechanism. We shall not discuss that mechanism here, however.

The mechanisms for addition of Cl_2 and I_2 to alkenes are similar to that for Br_2, involving formation and ring opening of their respective halonium ions.

As with bridged mercurinium ions, the bromonium ion does not necessarily have symmetrical charge distribution at its two carbon atoms. If one carbon of the bromonium ion is more highly substituted than the other, and therefore able to stabilize positive charge better, it may bear a greater fraction of positive charge than the other carbon (i.e., the positively charged bromine draws electron density from the two carbon atoms of the ring, but not equally if they are of different substitution). Consequently, the more positively charged carbon may be attacked by the reaction nucleophile more often than the other carbon. However, in reactions with symmetrical reagents (e.g., Br_2, Cl_2, and I_2) there is no observed difference. (We shall discuss this point further in Section 8.14, where we will study a reaction where we can discern regioselectivity of attack by the nucleophile on a halonium ion. See also "The Chemistry of. . ." box within Section 8.14.)

8.13 Stereochemistry of the Addition of Halogens to Alkenes

The addition of bromine to cyclopentene provides additional evidence for bromonium ion intermediates in bromine additions. When cyclopentene reacts with bromine in carbon tetrachloride, **anti addition** occurs, and the products of the reaction are *trans*-1,2-dibromocyclopentane enantiomers (as a racemate):

$\xrightarrow[CCl_4]{Br_2}$ Br H / H Br + enantiomer

***trans*-1,2-Dibromocyclopentane**

This anti addition of bromine to cyclopentene can be explained by a mechanism that involves the formation of a bromonium ion in the first step. In the second step, a bromide ion attacks a carbon atom of the ring from the side opposite that of the bromonium ion. The reaction is an S_N2 reaction. Nucleophilic attack by the bromide ion causes ***inversion of the configuration of the carbon being attacked*** (Section 6.8). This inversion of configuration at one carbon atom of the ring leads to the formation of one *trans*-1,2-dibromocyclopentane enantiomer. The other enantiomer results from attack of the bromide ion at the other carbon of the bromonium ion:

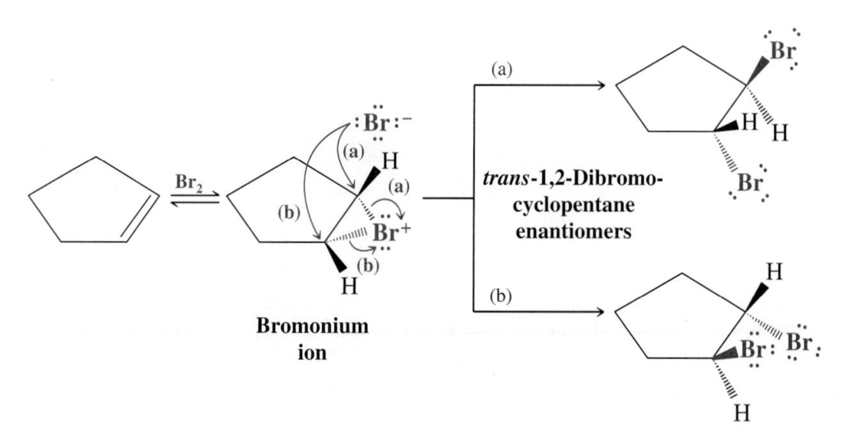

Exploring this reaction with molecular models helps to illustrate the stereochemical outcome.

When cyclohexene undergoes addition of bromine, the product is a racemate of the *trans*-1,2-dibromocyclohexane enantiomers (Section 8.12). In this instance, too, anti addition results from the intermediate formation of a bromonium ion followed by S_N2 attack by a bromide ion. The reaction shown here illustrates the formation of one enantiomer. (The other enantiomer is formed when the bromide ion attacks the other carbon of the bromonium ion or if the bromonium ion forms on the other face of the ring.)

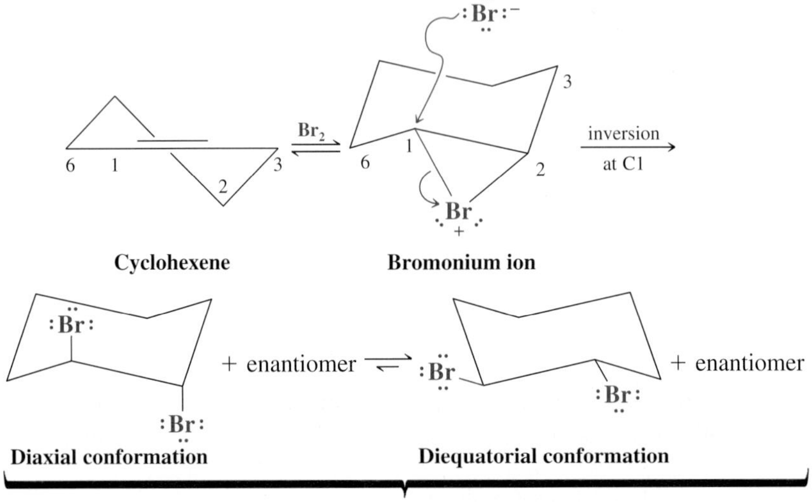

:Br:⁻

Cyclohexene ⇌ (Br_2) Bromonium ion → (inversion at C1)

Diaxial conformation + enantiomer ⇌ Diequatorial conformation + enantiomer

***trans*-1,2-Dibromocyclohexane**

Notice that the initial product of the reaction is the *diaxial conformer.* This rapidly converts to the diequatorial form, and when equilibrium is reached the diequatorial form predominates. We saw earlier (Section 7.6C) that when cyclohexane derivatives undergo elimination, the required conformation is the diaxial one. Here we find that when cyclohexene undergoes addition (the opposite of elimination), the initial product is also diaxial.

8.13A Stereospecific Reactions

Stereospecificity

The anti addition of a halogen to an alkene provides us with an example of what is called a **stereospecific reaction.**

A reaction is ***stereospecific*** when ***a particular stereoisomeric form of the starting material reacts in such a way that it gives a specific stereoisomeric form of the product.*** It does this because the reaction mechanism requires that the configurations of the atoms involved change in a characteristic way.

Consider the reactions of *cis*- and *trans*-2-butene with bromine shown below. When *trans*-2-butene adds bromine, the product is the meso compound, (2*R*,3*S*)-2,3-dibromobutane. When *cis*-2-butene adds bromine, the product is a *racemic form* of (2*R*,3*R*)-2,3-dibromobutane and (2*S*,3*S*)-2,3-dibromobutane:

Reaction 1

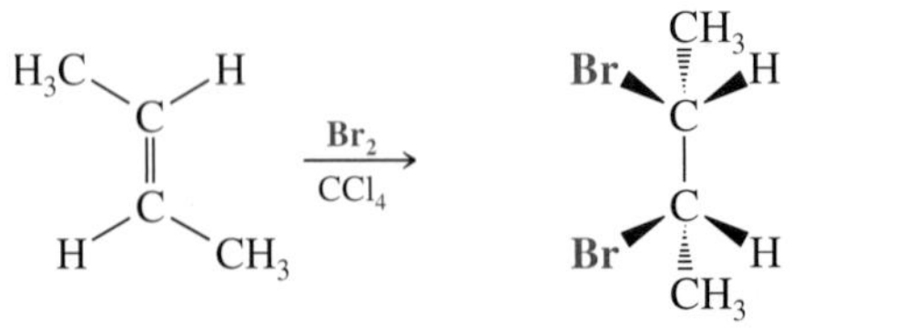

***trans*-2-Butene** →(Br_2, CCl_4) **(2*R*,3*S*)-2,3-Dibromobutane (a meso compound)**

Reaction 2

cis-2-Butene $\xrightarrow[CCl_4]{Br_2}$ (2*R*,3*R*) + (2*S*,3*S*)

The reactants *cis*-2-butene and *trans*-2-butene are stereoisomers; they are *diastereomers.* The product of reaction 1, (2*R*,3*S*)-2,3-dibromobutane, is a meso compound, and it is a stereoisomer of both of the products of reaction 2 (the enantiomeric 2,3-dibromobutanes). Thus, by definition, both reactions are stereospecific. One stereoisomeric form of the reactant (e.g., *trans*-2-butene) gives one product (the meso compound), whereas the other stereoisomeric form of the reactant (*cis*-2-butene) gives a stereoisomerically different product (the enantiomers).

We can better understand the results of these two reactions if we examine their mechanisms.

The first mechanism in the following box shows how *cis*-2-butene adds bromine to yield intermediate bromonium ions that are achiral. (The bromonium ion has a plane of symmetry. Can you find it?) These bromonium ions can then react with bromide ions by either path (a) or path (b). Reaction by path (a) yields one 2,3-dibromobutane enantiomer; reaction by path (b) yields the other enantiomer. The reaction occurs at the same rate by either path; therefore, the two enantiomers are produced in equal amounts (as a racemic form).

The second mechanism in the box shows how *trans*-2-butene reacts at the bottom face to yield an intermediate bromonium ion that is chiral. (Reaction at the other face would produce the enantiomeric bromonium ion.) Reaction of this chiral bromonium ion (or its enantiomer) with a bromide ion either by path (a) or by path (b) yields the same compound, the *meso*-2,3-dibromobutane.

The Stereochemistry of the Reaction

Addition of Bromine to *cis*- and *trans*-2-Butene

cis-2-Butene reacts with bromine to yield the enantiomeric 2,3-dibromobutanes by the following mechanism:

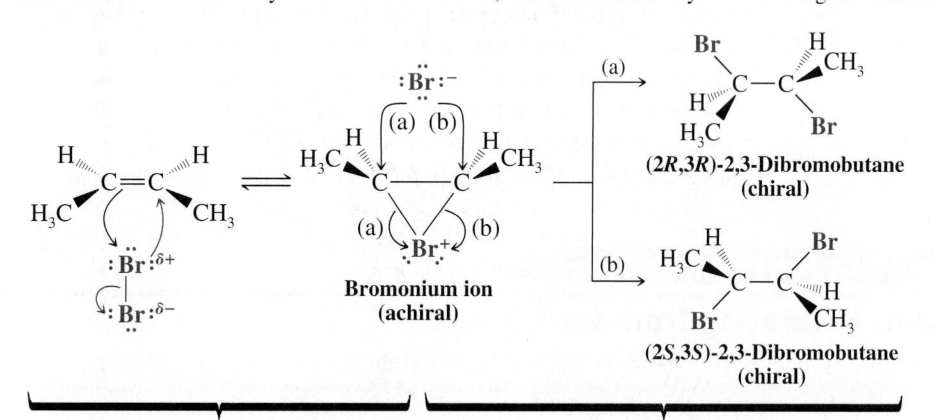
Bromonium ion
(achiral)
(2R,3R)-2,3-Dibromobutane
(chiral)
(2S,3S)-2,3-Dibromobutane
(chiral)
cis-2-Butene reacts with bromine to yield an achiral bromonium ion and a bromide ion. [Reaction at the other face of the alkene (top) would yield the same bromonium ion.]
The bromonium ion reacts with the bromide ions at equal rates by paths (a) and (b) to yield the two enantiomers in equal amounts (i.e., as the racemic form).

trans-2-Butene reacts with bromine to yield *meso*-2,3-dibromobutane.

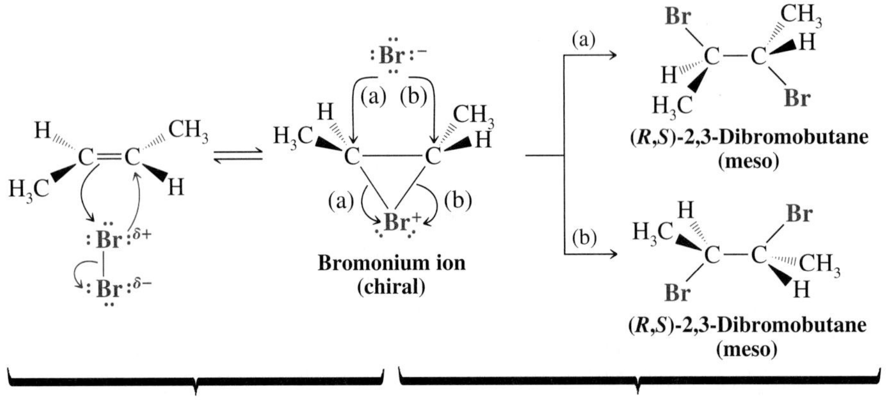

Bromonium ion (chiral)

(*R*,*S*)-2,3-Dibromobutane (meso)

(*R*,*S*)-2,3-Dibromobutane (meso)

trans-2-Butene reacts with bromine to yield chiral bromonium ions and bromide ions. [Reaction at the other face (top) would yield the enantiomer of the bromonium ion as shown here.]

When the bromonium ions react by either path (a) or path (b), they yield the *same* achiral meso compound. [Reaction of the enantiomer of the intermediate bromonium ion would produce the same result.]

PROBLEM 8.15 In Section 8.13 you studied a mechanism for the formation of one enantiomer of *trans*-1,2-dibromocyclopentane when bromine adds to cyclopentene. Now write a mechanism showing how the other enantiomer forms.

8.14 Halohydrin Formation

If the halogenation of an alkene is carried out in aqueous solution (rather than in carbon tetrachloride), the major product of the overall reaction is not a *vic*-dihalide; instead, it is a **halo alcohol** called a **halohydrin.** In this case, molecules of the solvent become reactants, too:

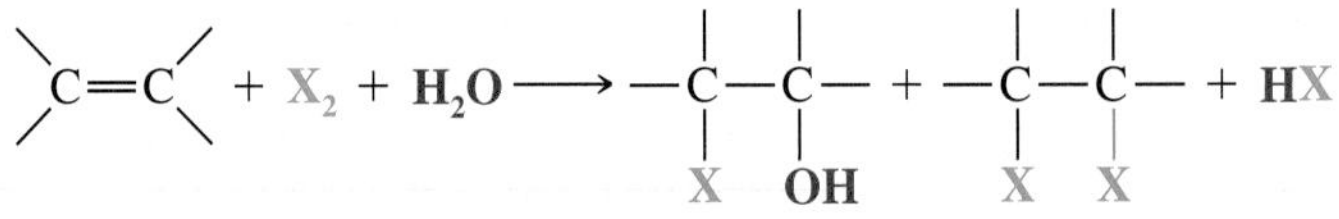

$$\mathrm{C{=}C} + X_2 + H_2O \longrightarrow \mathrm{-C(X)-C(OH)-} + \mathrm{-C(X)-C(X)-} + HX$$

X = Cl or Br — **Halohydrin (major)** — ***vic*-Dihalide (minor)**

Halohydrin formation can be explained by the following mechanism.

A Mechanism for the Reaction

Halohydrin Formation from an Alkene

Step 1

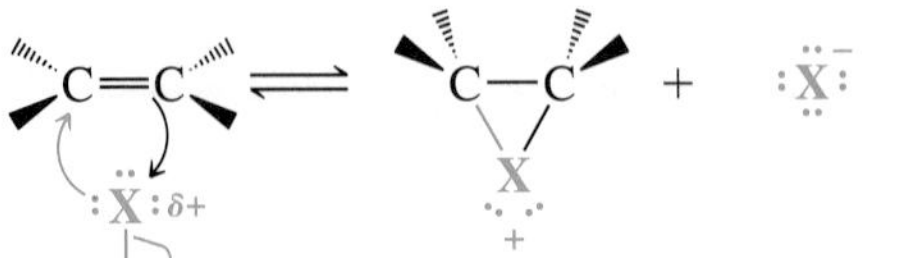

Halonium ion **Halide ion**

This step is the same as for halogen addition to an alkene (see Section 8.12A).

Steps 2 and 3

Halonium ion $\longrightarrow$ **Protonated halohydrin** $\longrightarrow$ **Halohydrin** + H_3O^+

Here, however, a water molecule acts as the nucleophile and attacks a carbon of the ring, causing the formation of a protonated halohydrin.

The protonated halohydrin loses a proton (it is transferred to a molecule of water). This step produces the halohydrin and hydronium ion.

The first step is the same as that for halogen addition. In the second step, however, the two mechanisms differ. In halohydrin formation, water acts as the nucleophile and attacks one carbon atom of the halonium ion. The three-membered ring opens, and a protonated halohydrin is produced. Loss of a proton then leads to the formation of the halohydrin itself.

The Chemistry of...

Regiospecificity in Unsymmetrically Substituted Bromonium Ions: Bromonium Ions of Ethene, Propene, and 2-Methylpropene

When a nucleophile reacts with a bromonium ion, the addition takes place with Markovnikov regiochemistry. In the formation of bromohydrins, for example, bromine bonds at the least substituted carbon, and the hydroxyl group (from nucleophilic attack by water) bonds at the more substituted carbon (i.e., the carbon that accommodated more of the positive charge in the bromonium ion). A closer look at the bromonium ions from ethene, propene, and 2-methylpropene with respect to charge distribution, bond lengths, and the shape of their lowest unoccupied molecular orbital (LUMO) shows that these factors support the observed result of Markovnikov addition.

The relative distributions of electron densities in the bromonium ions of ethene, propene, and 2-methylpropene are shown in the calculated electrostatic potential maps below (Fig. 8.A). Just as with electrostatic potential maps that we have used earlier, red indicates relatively negative areas and blue indicates relatively positive (or less negative) areas. In the three structures, the same absolute color scale is used to map charge distribution for all of them, so that they can be directly compared with one another. The salient feature apparent in these charge density maps

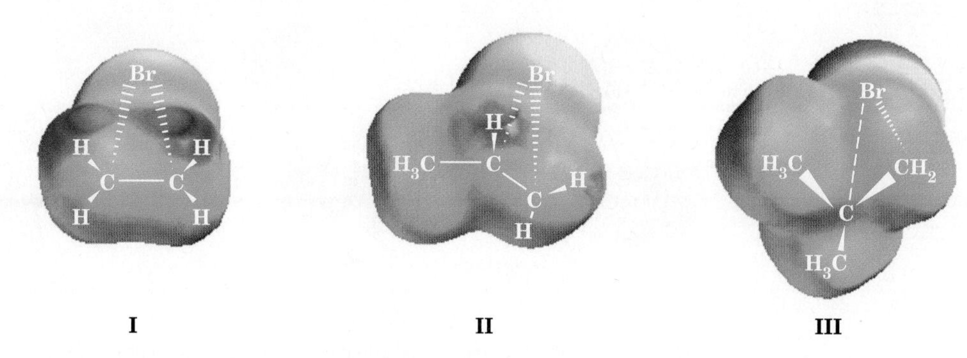

FIGURE 8.A As alkyl substitution increases, carbon is able to accommodate greater positive charge and bromine contributes less of its electron density.

is that as alkyl substitution increases in bromonium ions, the carbon having greater substitution requires less stabilization by contribution of electron density from bromine. In the bromonium ion of ethene **(I)**, for example, the bromine atom contributes substantial electron density. This is indicated in the electrostatic potential map by more green and yellow (i.e., less red) near the bromine. At the other extreme, in the bromonium ion from 2-methylpropene **(III)**, the tertiary carbon can accommodate substantial positive charge, and hence most of the positive charge is localized there (as indicated by deep blue at the tertiary carbon in the electrostatic potential map). The bromine retains the bulk of its electron density (as indicated by the mapping of red color near the bromine). This structure shows that the bromonium ion from 2-methylpropene has essentially the charge distribution of a tertiary carbocation at its carbon atoms. The bromonium ion from propene **(II)**, which has a secondary carbon, utilizes some electron density from the bromine but not as much as in ethene **(I)**, as indicated by the moderate extent of yellow near the bromine in **II**. Overall, when a nucleophile reacts with either of the substituted bromonium ions **II** or **III**, it does so at the carbon of each that bears the greater positive charge, in accord with Markovnikov regiochemistry. That is, when water (a nucleophile) reacts with **II**, the product is the halohydrin $CH_3CHOHCH_2Br$, and when water reacts with **III**, the product is the halohydrin $(CH_3)_2COHCH_2Br$. (See page 354.)

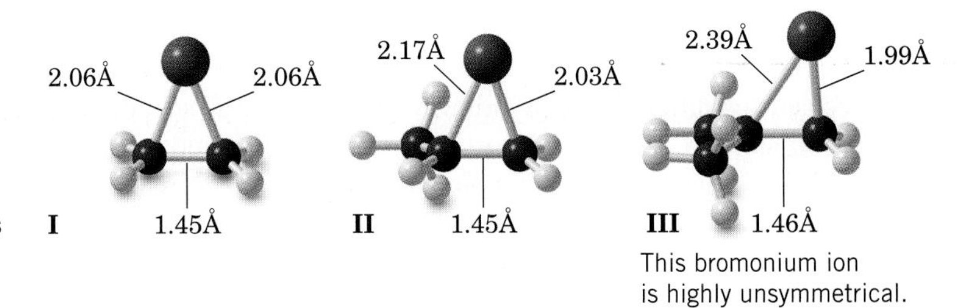

FIGURE 8.B The carbon–bromine bond length (shown in angstroms) at the central carbon increases as less electron density from the bromine is needed to stabilize the positive charge. A lesser electron density contribution from bromine is needed because additional alkyl groups help stabilize the charge.

The C—Br bond lengths in these three bromonium ions tell a similar story (Fig. 8.B). In the bromonium ion of ethene **(I)** the C—Br bond lengths are identical, of course, due to symmetry in the ion, and they measure 2.06 Å. In the bromonium ion of propene **(II)** the C—Br bond involving the secondary carbon is 2.17 Å, whereas with the primary carbon it is 2.03 Å. The longer bond length to the secondary carbon is consistent with the lesser contribution of electron density from the bromine to the secondary carbon, because the secondary carbon can accommodate the charge better than the primary carbon. In the bromonium ion of 2-methylpropene **(III)**, the C—Br bond involving the tertiary carbon is very long, measuring 2.39 Å. On the other hand, the C—Br bond involving the primary carbon is 1.99 Å. This indicates that significantly less contribution of electron density from the bromine is necessary to stabilize the charge at the tertiary carbon (it becomes essentially like a tertiary carbocation), while the bond at the primary carbon is like that expected for a typical alkyl bromide.

Lastly, calculations that show the LUMOs of the three bromonium ions are consistent with the regiochemistry of their reactivity as well. When considering reactions that involve nucleophiles, the LUMO of the electrophile represents the orbital to which electron density from the nucleophile will be contributed when it forms a bond. For the three bromonium ions in Fig. 8.C, the complete shape of the LUMO is shown for each one (the different colors simply indicate the mathematical phase of the orbital lobes). The lobes of the LUMO on which we should focus our at-

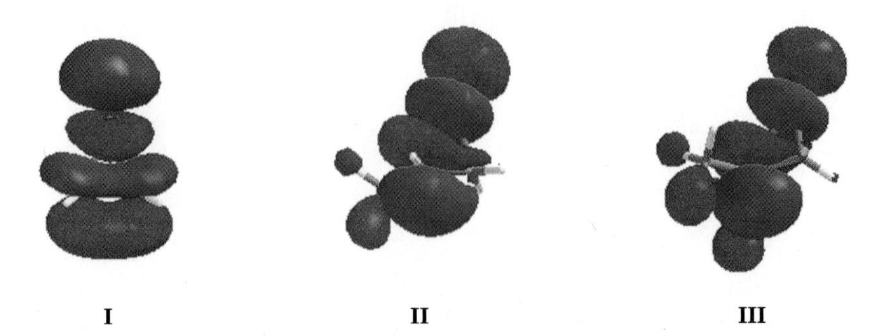

FIGURE 8.C With increasing alkyl substitution of the bromonium ion, the lobe of the LUMO where electron density from the nucleophile will be contributed shifts more and more to the more substituted carbon.

tention, however, are those opposite the three-membered ring portion of the bromonium ion in each ("underneath" the carbons of the ion). The LUMO of the ethene bromonium ion **(I)** shows, due to symmetry, equal distribution of the LUMO lobe near the two carbons where the nucleophile could attack. The nucleophile bonds equally well, of course, at either carbon of the ethene bromonium ion. In the propene bromonium ion **(II)**, the corresponding LUMO lobe has more of its volume associated with the more substituted carbon, indicating that electron density from the nucleophile will be best accommodated here. This is in accord with the observed Markovnikov addition. Finally, the bromonium ion of 2-methylpropene **(III)** has nearly all of the volume from this lobe of the LUMO associated with the tertiary carbon and virtually none associated with the primary carbon. This, too, is consistent with observed attack by the nucleophile at the tertiary carbon.

Water, because of its unshared electron pairs, acts as a nucleophile in this and in many other reactions. In this instance water molecules far outnumber halide ions because water is the solvent for the reactants. This accounts for the halohydrin being the major product.

PROBLEM 8.16

Outline a mechanism that accounts for the formation of *trans*-2-bromocyclopentanol and its enantiomer when cyclopentene is treated with an aqueous solution of bromine:

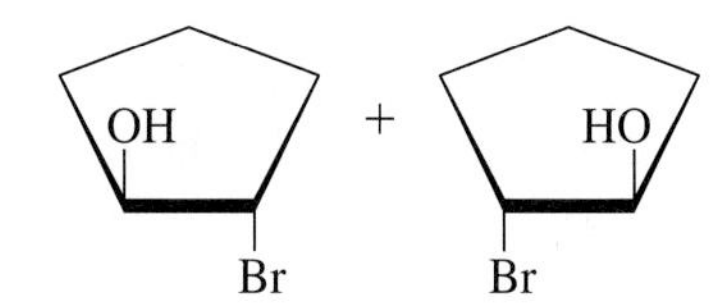

***trans*-2-Bromocyclopentanol enantiomers**

If the alkene is unsymmetrical, the halogen ends up on the carbon atom with the greater number of hydrogen atoms. Bonding in the intermediate bromonium ion is *unsymmetrical* (see the preceding boxed section). The more highly substituted carbon atom bears the greater positive charge because it resembles the more stable carbocation. Consequently, water attacks this carbon atom preferentially. The greater positive charge on the tertiary carbon permits a pathway with a lower free energy of activation even though attack at the primary carbon atom is less hindered:

$$(H_3C)_2C{=}CH_2 \xrightleftharpoons{Br_2} CH_3-\overset{CH_3}{\overset{|}{C}}{}^{\delta+}-CH_2\ (\text{bridged by } Br^{\delta+}) \xrightarrow{:\ddot{O}H_2} CH_3-\underset{CH_3}{\underset{|}{\overset{H-\overset{+}{\ddot{O}}-H}{\overset{|}{C}}}}-CH_2Br \xrightarrow{:\ddot{O}H_2} CH_3-\underset{CH_3}{\underset{|}{\overset{OH}{\overset{|}{C}}}}-CH_2Br + H_3O^+$$

(73%)

PROBLEM 8.17

When ethene gas is passed into an aqueous solution containing bromine and sodium chloride, the products of the reaction are $BrCH_2CH_2Br$, $BrCH_2CH_2OH$, *and* $BrCH_2CH_2Cl$. Write mechanisms showing how each product is formed.

8.15 Divalent Carbon Compounds: Carbenes

There is a group of compounds in which carbon forms only *two bonds.* These neutral divalent carbon compounds are called **carbenes.** Most carbenes are highly unstable compounds that are capable of only fleeting existence. Soon after carbenes are formed, they usually react with another molecule. The reactions of carbenes are especially interesting because, in many instances, the reactions show a remarkable degree of stereospecificity.

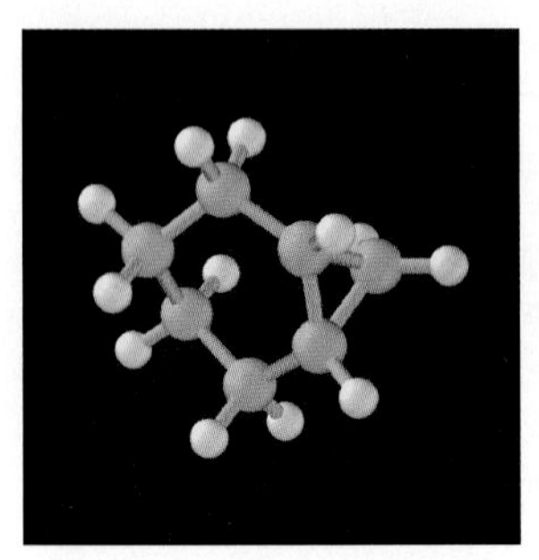
Bicyclo[4.1.0]heptane

The reactions of carbenes are also of great synthetic use in the preparation of compounds that have three-membered rings, for example, bicyclo[4.1.0]heptane, shown at left.

8.15A Structure and Reactions of Methylene

The simplest carbene is the compound called methylene ($:CH_2$). Methylene can be prepared by the decomposition of diazomethane (CH_2N_2), a very poisonous yellow gas. This decomposition can be accomplished by heating diazomethane (thermolysis) or by irradiating it with light of a wavelength that it can absorb (photolysis):

$$:\overset{-}{C}H_2-\overset{+}{N}\equiv N: \xrightarrow[\text{or light}]{\text{heat}} :CH_2 + :N\equiv N:$$

Diazomethane **Methylene** **Nitrogen**

The structure of diazomethane is actually a resonance hybrid of the three structures

$$:\overset{-}{C}H_2-\overset{+}{N}\equiv N: \longleftrightarrow CH_2=\overset{+}{N}=\overset{-}{\ddot{N}}: \longleftrightarrow :\overset{-}{C}H_2-\ddot{N}=\overset{+}{N}:$$

I **II** **III**

We have chosen resonance structure **I** to illustrate the decomposition of diazomethane because with **I** it is readily apparent that heterolytic cleavage of the carbon–nitrogen bond results in the formation of methylene and molecular nitrogen.

Methylene reacts with alkenes by adding to the double bond to form cyclopropanes:

$$>C=C< + :CH_2 \longrightarrow \text{cyclopropane } (>C-C< \text{ bridged by } CH_2)$$

Alkene **Methylene** **Cyclopropane**

8.15B Reactions of Other Carbenes: Dihalocarbenes

Dihalocarbenes are also frequently employed in the synthesis of cyclopropane derivatives from alkenes. Most reactions of dihalocarbenes are stereospecific:

$$\text{trans-RCH=CHR} + :CCl_2 \longrightarrow \text{trans-1,1-dichloro-2,3-dialkylcyclopropane}$$

+ enantiomer

The addition of $:CX_2$ is stereospecific. If the R groups of the alkene are trans, they will be trans in the product. (If the R groups were initially cis, they would be cis in the product.)

Dichlorocarbene can be synthesized by the ***α elimination*** of the elements of hydrogen chloride from chloroform. (The hydrogen of chloroform is mildly acidic ($pK_a \approx 24$) due to the inductive effect of the chlorine atoms.) This reaction resembles the β-elimination reactions by which alkenes are synthesized from alkyl halides (Section 6.15):

$$R-\ddot{\underset{..}{O}}:^-K^+ + H:CCl_3 \rightleftharpoons R-\ddot{\underset{..}{O}}:H + ^-:CCl_3 + K^+ \xrightarrow[\text{slow}]{} :CCl_2 + :\ddot{\underset{..}{Cl}}:^-$$

Dichlorocarbene

Compounds *with a β hydrogen* react by β elimination preferentially. Compounds with no β hydrogen but with an α hydrogen (such as chloroform) react by α elimination.

A variety of cyclopropane derivatives have been prepared by generating dichlorocarbene in the presence of alkenes. Cyclohexene, for example, reacts with dichlorocarbene generated by treating chloroform with potassium *tert*-butoxide to give a bicyclic product:

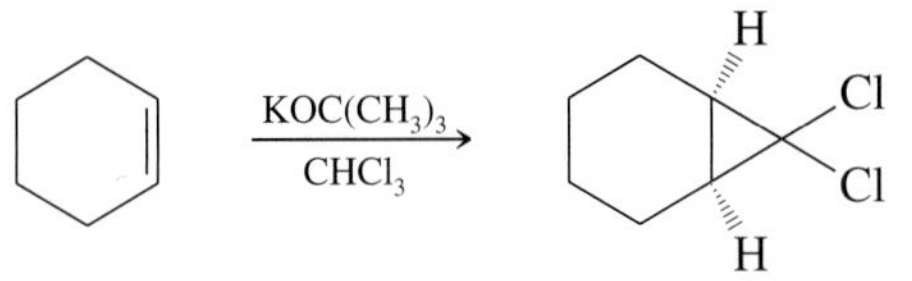

$$\text{Cyclohexene} \xrightarrow[CHCl_3]{KOC(CH_3)_3} \text{7,7-Dichlorobicyclo[4.1.0]heptane}$$

7,7-Dichlorobicyclo[4.1.0]heptane
(59%)

8.15C Carbenoids: The Simmons–Smith Cyclopropane Synthesis

A useful cyclopropane synthesis has been developed by H. E. Simmons and R. D. Smith of the DuPont Company. In this synthesis diiodomethane and a zinc–copper couple are stirred together with an alkene. The diiodomethane and zinc react to produce a carbene-like species called a *carbenoid:*

$$CH_2I_2 + Zn(Cu) \longrightarrow ICH_2ZnI$$

A carbenoid

The carbenoid then brings about the stereospecific addition of a CH_2 group directly to the double bond.

Problem 8.18

What products would you expect from each of the following reactions?

(a) *trans*-2-Butene $\xrightarrow[CHCl_3]{KOC(CH_3)_3}$

(b) Cyclopentene $\xrightarrow[CHBr_3]{KOC(CH_3)_3}$

(c) *cis*-2-Butene $\xrightarrow[\text{diethyl ether}]{CH_2I_2/Zn(Cu)}$

Problem 8.19

Starting with cyclohexene and using any other needed reagents, outline a synthesis of 7,7-dibromobicyclo[4.1.0]heptane.

Problem 8.20

Treating cyclohexene with 1,1-diiodoethane and a zinc–copper couple leads to two isomeric products. What are their structures?

8.16 Oxidations of Alkenes: Syn 1,2-Dihydroxylation

Alkenes undergo a number of reactions in which the carbon–carbon double bond is oxidized. One important reaction is the 1,2-dihydroxylation of alkenes. Osmium tetroxide, for example, is widely used to synthesize **1,2-diols** (sometimes called **glycols**). Potassium permanganate can also be used, although because it is a stronger oxidizing agent it is prone to cleave the diol through further oxidation (Section 8.17).

$$CH_3CH{=}CH_2 \xrightarrow[\text{(2) } Na_2SO_3/H_2O \text{ or } NaHSO_3/H_2O]{\text{(1) } OsO_4\text{, pyridine}} CH_3CH(OH){-}CH_2(OH)$$

Propene → **1,2-Propanediol (propylene glycol)**

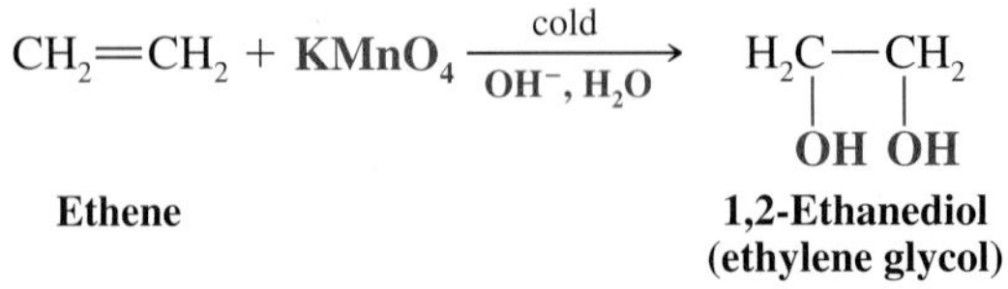

$$CH_2{=}CH_2 + KMnO_4 \xrightarrow[OH^-,\ H_2O]{cold} \underset{OH\ \ OH}{H_2C{-}CH_2}$$

Ethene **1,2-Ethanediol (ethylene glycol)**

8.16A Mechanisms for Syn Dihydroxylation of Alkenes

The mechanisms for the formation of 1,2-diols by osmium tetroxide and potassium permanganate involve cyclic intermediates that result in **syn addition** of the oxygen atoms, as shown below. After formation of the cyclic intermediate with osmium or manganese, cleavage at the oxygen–metal bonds takes place without altering the stereochemistry of the two new C—O bonds.

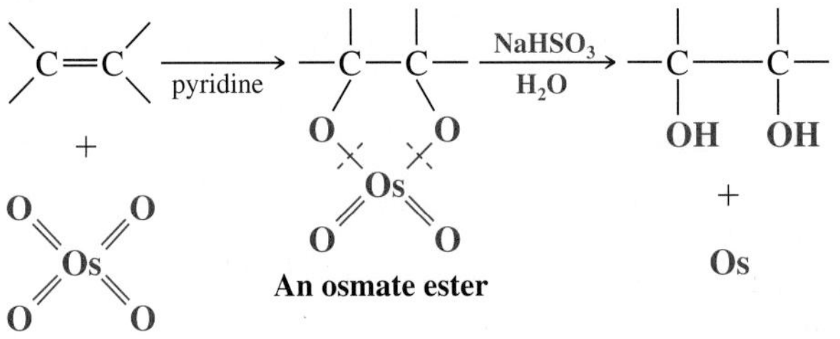

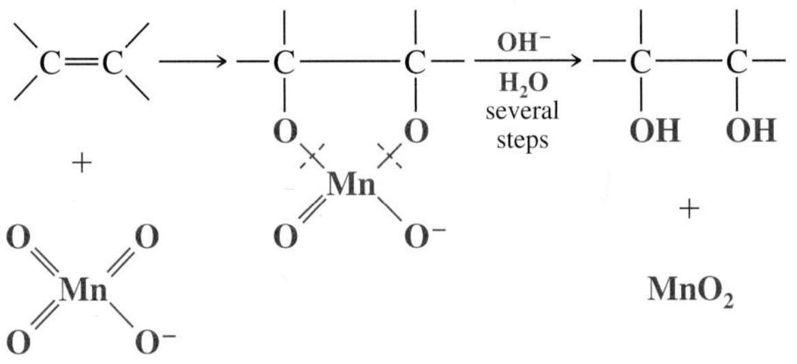

C=C + OsO_4 $\xrightarrow{\text{pyridine}}$ —C—C— (O, O, Os, O, O) **An osmate ester** $\xrightarrow[H_2O]{NaHSO_3}$ —C—C— (OH OH) + Os

C=C + MnO_4^- (O, O, Mn, O, O^-) ⟶ —C—C— (O, O, Mn, O, O^-) $\xrightarrow[H_2O,\ \text{several steps}]{OH^-}$ —C—C— (OH OH) + MnO_2

The syn stereochemistry of these dihydroxylations can readily be observed by the reaction of cyclopentene with osmium tetroxide or potassium permanganate. The product in either case is *cis*-1,2-cyclopentanediol.

Cyclopentene + OsO_4 $\xrightarrow[\text{pyridine}]{25°C}$ (H, H, O, O, Os, O, O) $\xrightarrow[H_2O]{NaHSO_3}$ (H, H, OH, OH)

***cis*-1,2-Cyclopentanediol (a meso compound)**

Cyclopentene + MnO_4^- $\xrightarrow{\text{cold}}$ [(H, H, O, O, Mn, O, O^-)] $\xrightarrow[OH^-]{H_2O}$ (H, H, OH, OH)

***cis*-1,2-Cyclopentanediol (a meso compound)**

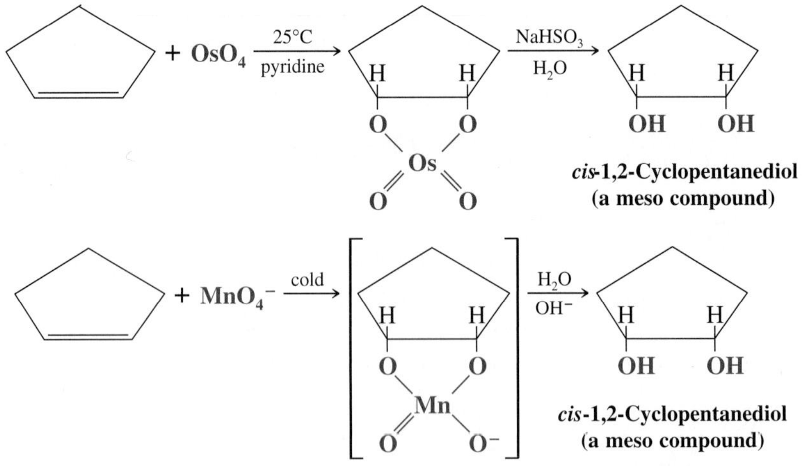

Of the two reagents mentioned for syn dihydroxylation, osmium tetroxide is the more widely applied in organic synthesis. However, osmium tetroxide is highly toxic, volatile, and very expensive. For these reasons, methods have been developed that permit OsO_4 to be used *catalytically* in conjunction with a co-oxidant.* A very small molar percentage of OsO_4 is placed in the reaction mixture to do the dihydroxylation step, while a stoichiometric amount of co-oxidant reoxidizes the OsO_4 as it is used in each cycle, allowing oxidation of the alkene to continue until all has been converted to the diol. *N*-Methylmorpholine *N*-oxide (NMO) is one of the most commonly used co-oxidants with catalytic OsO_4. The method was discovered at Upjohn Corporation in the context of reactions for synthesis of a prostaglandin† (Section 23.5):

Catalytic OsO_4 1,2-Dihydroxylation

OsO_4 (0.2%), NMO, 25°C

NMO (stoichiometric co-oxidant for catalytic dihydroxylation)

>95% Yield (used in synthesis of a prostaglandin)

PROBLEM 8.21

Starting with an alkene, outline syntheses of each of the following:

(a) (b) (plus enantiomer) (c) (plus enantiomer)

PROBLEM 8.22

Explain the following facts:

(a) Treating (*Z*)-2-butene with OsO_4 in pyridine and then $NaHSO_3$ in water gives a diol that is optically inactive and cannot be resolved.

(b) Treating (*E*)-2-butene with OsO_4 and then $NaHSO_3$ gives a diol that is optically inactive but can be resolved into enantiomers.

Sharpless shared the 2001 Nobel Prize in Chemistry for his development of asymmetric oxidation methods.

The Chemistry of...

Catalytic Asymmetric Dihydroxylation

Methods for catalytic *asymmetric* syn dihydroxylation have been developed that significantly extend the synthetic utility of dihydroxylation. K. B. Sharpless (The Scripps Research Institute) and co-workers discovered that addition of a chiral amine to the oxidizing mixture leads to enantioselective catalytic syn dihydroxylation. Asymmetric dihydroxylation has become an important and widely used tool in the synthesis of complex organic molecules. In recognition of this and other advances in asymmetric oxidation procedures developed by his group (Section 11.17), Sharpless was awarded half of the 2001 Nobel Prize in Chemistry. (The other half of the 2001 prize was awarded to W. Knowles and R. Noyori for their development of catalytic asymmetric reduction reactions; see Section 7.13.) The following reaction, involved in an enantiose-

*See Nelson, D. W., et al. *J. Am. Chem. Soc.* **1997,** *119,* 1840–1858, and Corey, E. J., et al. *J. Am. Chem. Soc.* **1996,** *118,* 319–329.

†Van Rheenan, V.; Kelley, R. C.; Cha, D. Y. *Tetrahedron Lett.* **1976,** 25, 1973.

lective synthesis of the side chain of the anticancer drug Paclitaxel (Taxol), serves to illustrate Sharpless's catalytic asymmetric dihydroxylation. The example utilizes a catalytic amount of $K_2OsO_2(OH)_4$, an OsO_4 equivalent, a chiral amine ligand to induce enantioselectivity, and NMO as the stoichiometric co-oxidant. The product is obtained in 99% enantiomeric excess(ee):

Asymmetric Catalytic OsO_4 1,2-Dihydroxylation*

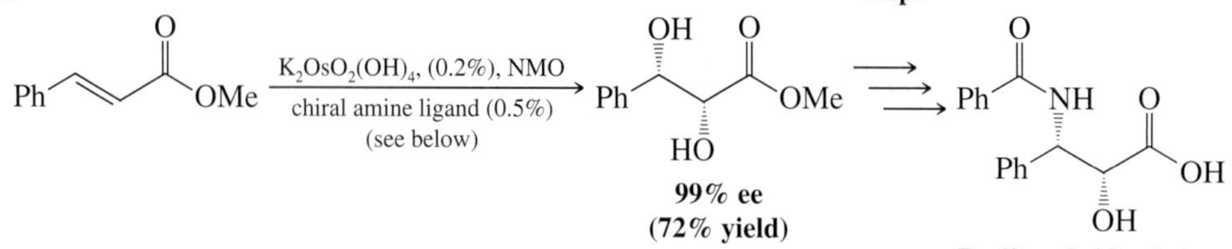

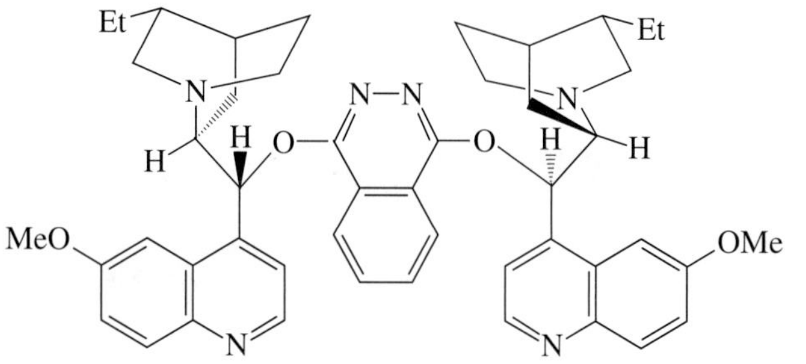

$K_2OsO_2(OH)_4$, (0.2%), NMO; chiral amine ligand (0.5%) (see below)

99% ee (72% yield)

Multiple steps

Paclitaxel side chain

A chiral amine ligand used in catalytic asymmetric dihydroxylation

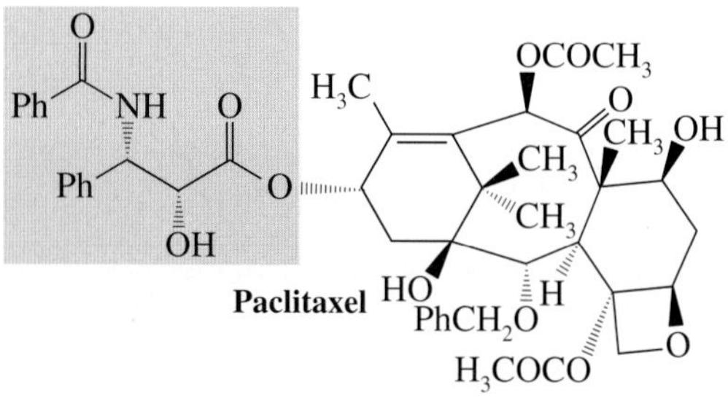

Paclitaxel

*Wang, Z.-M.; Kolb, H. C.; Sharpless, K. B. *J. Org. Chem.* **1994,** *59,* 5104.

8.17 Oxidative Cleavage of Alkenes

Alkenes can be **oxidatively cleaved** using potassium permanganate or ozone (as well as by other reagents). Potassium permanganate ($KMnO_4$) is used when strong oxidation is needed. Ozone (O_3) is used when mild oxidation is desired. (Alkynes and aromatic rings are also oxidized by $KMnO_4$ and O_3; see Sections 8.19 and 15.13.)

8.17A Cleavage with Hot Basic Potassium Permanganate

Treatment with hot basic potassium permanganate oxidatively cleaves the double bond of an alkene. Cleavage is believed to occur via intermediate formation of a 1,2-diol. (Undesired cleavage of this type detracts from the use of $KMnO_4$ as a reagent for 1,2-diol synthesis.) Alkenes with monosubstituted carbon atoms are oxidatively cleaved to salts of carboxylic acids. Disubstituted alkene carbons are oxidatively cleaved to ketones. Unsubstituted alkene carbons are oxidized to carbon dioxide. The following examples illustrate the results of potassium permanganate cleavage of alkenes with different substitution patterns. In the case where the product is a carboxylate salt, an acidification step is required to obtain the carboxylic acid.

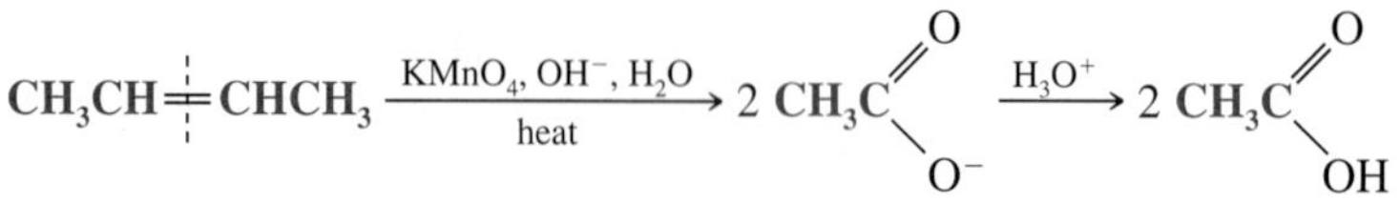

$$CH_3CH{=}CHCH_3 \xrightarrow[\text{heat}]{KMnO_4,\ OH^-,\ H_2O} 2\ CH_3CO_2^- \xrightarrow{H_3O^+} 2\ CH_3CO_2H$$

(cis or trans) — Acetate ion — Acetic acid

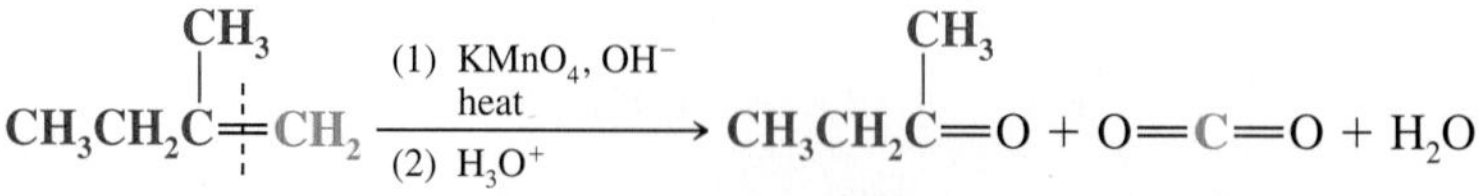

$$CH_3CH_2C(CH_3){=}CH_2 \xrightarrow[(2)\ H_3O^+]{(1)\ KMnO_4,\ OH^-,\ \text{heat}} CH_3CH_2C(CH_3){=}O + O{=}C{=}O + H_2O$$

One of the uses of potassium permanganate, other than for desired oxidative cleavage, is as a chemical test for the presence of unsaturation in an unknown compound. Solutions of potassium permanganate are purple. If an alkene is present (or an alkyne, Section 8.20), the purple color is discharged and a brown precipitate of manganese dioxide (MnO_2) forms as the oxidation takes place.

The oxidative cleavage of alkenes has also been used to establish the location of the double bond in an alkene chain or ring. The reasoning process requires us to think backward much as we do with retrosynthetic analysis. Here we are required to work backward from the products to the reactant that might have led to those products. We can see how this might be done with the following examples:

SOLVED PROBLEM

An unknown alkene with the formula C_8H_{16} was found, on oxidation with hot basic permanganate, to yield a three-carbon carboxylic acid (propanoic acid) and a five-carbon carboxylic acid (pentanoic acid). What was the structure of this alkene?

$$C_8H_{16} \xrightarrow[(2)\ H_3O^+]{(1)\ KMnO_4,\ H_2O,\ OH^-,\ heat} \underset{\textbf{Propanoic acid}}{CH_3CH_2\overset{O}{\overset{\|}{C}}-OH} + \underset{\textbf{Pentanoic acid}}{HO-\overset{O}{\overset{\|}{C}}CH_2CH_2CH_2CH_3}$$

ANSWER: Oxidative cleavage must have occurred as follows, and the unknown alkene must have been *cis-* or *trans*-3-octene.

Cleavage occurs here

$$\underset{\substack{\textbf{Unknown alkene}\\ \textbf{(either } \textbf{\textit{cis-}} \textbf{ or } \textbf{\textit{trans}}\textbf{-3-octene)}}}{CH_3CH_2CH{=}CHCH_2CH_2CH_2CH_3} \xrightarrow[(2)\ H_3O^+]{(1)\ KMnO_4,\ H_2O,\ OH^-,\ heat} CH_3CH_2\overset{O}{\overset{\|}{C}}-OH + HO-\overset{O}{\overset{\|}{C}}CH_2CH_2CH_2CH_3$$

SOLVED PROBLEM

An unknown alkene with the formula C_7H_{12} undergoes oxidation by hot basic $KMnO_4$ to yield, after acidification, *only the following product:*

$$C_7H_{12} \xrightarrow[(2)\ H_3O^+]{(1)\ KMnO_4,\ H_2O,\ OH^-,\ heat} CH_3\overset{O}{\overset{\|}{C}}CH_2CH_2CH_2CH_2\overset{O}{\overset{\|}{C}}-OH$$

ANSWER: Since the product contains the same number of carbon atoms as the reactant, the only reasonable explanation is that the reactant has a double bond contained in a ring. Oxidative cleavage of the double bond opens the ring:

$$\underset{\substack{\textbf{Unknown alkene}\\ \textbf{(1-methylcyclohexene)}}}{\text{(1-methylcyclohexene, }CH_3\text{)}} \xrightarrow[(2)\ H_3O^+]{(1)\ KMnO_4,\ H_2O,\ OH^-,\ heat} CH_3\overset{O}{\overset{\|}{C}}CH_2CH_2CH_2CH_2\overset{O}{\overset{\|}{C}}-OH$$

8.17B Ozonolysis of Alkenes

Treatment of an alkene with ozone followed by workup with zinc and acetic acid [or dimethyl sulfide $(CH_3)_2S$] results in oxidative cleavage. This process is called **ozonolysis,** and it results in milder oxidative cleavage of alkenes than potassium permanganate. Unsubstituted alkene carbons are oxidized to formaldehyde, rather than to carbon dioxide as with $KMnO_4$. Monosubstituted alkene carbons are oxidized to aldehydes, rather than to carboxylate salts as with basic $KMnO_4$. Disubstituted alkene carbons are oxidized to ketones, the same as with $KMnO_4$. The process can be generalized for mono- and disubstituted alkenes in the following fashion:

$$RR'C{=}CR''H \xrightarrow[\text{(2) Zn/HOAc}]{\text{(1) } O_3,\ CH_2Cl_2,\ -78°C} RR'C{=}O + O{=}CR''H$$

Notice that a —H attached to the double bond is not oxidized to —OH as it was with permanganate oxidations. Besides use as a synthetic tool for preparation of compounds with carbonyl functional groups, ozonolysis has also been used to determine the location of double bonds in alkenes by structural analysis of the cleavage products. The following examples illustrate specifically the various alkene cleavage patterns that result from ozonolysis:

$$\underset{\textbf{2-Methyl-2-butene}}{CH_3C(CH_3){=}CHCH_3} \xrightarrow[\text{(2) Zn/HOAc}]{\text{(1) } O_3,\ CH_2Cl_2,\ -78°C} \underset{\textbf{Acetone}}{CH_3C(CH_3){=}O} + \underset{\textbf{Acetaldehyde}}{CH_3CH{=}O}$$

$$\underset{\textbf{3-Methyl-1-butene}}{CH_3CH(CH_3){-}CH{=}CH_2} \xrightarrow[\text{(2) Zn/HOAc}]{\text{(1) } O_3,\ CH_2Cl_2,\ -78°C} \underset{\textbf{Isobutyraldehyde}}{CH_3CH(CH_3){-}CH{=}O} + \underset{\textbf{Formaldehyde}}{HCH{=}O}$$

The mechanism of ozone addition to alkenes begins with formation of unstable compounds called *initial ozonides* (sometimes called molozonides). The process occurs vigorously and leads to spontaneous (and sometimes noisy) rearrangement to compounds known as **ozonides.** The rearrangement is believed to occur with dissociation of the initial ozonide into reactive fragments that recombine to yield the ozonide.

A Mechanism for the Reaction

Ozonide Formation from an Alkene

Initial ozonide

Ozone adds to the alkene to form an initial ozonide.

The initial ozonide fragments.

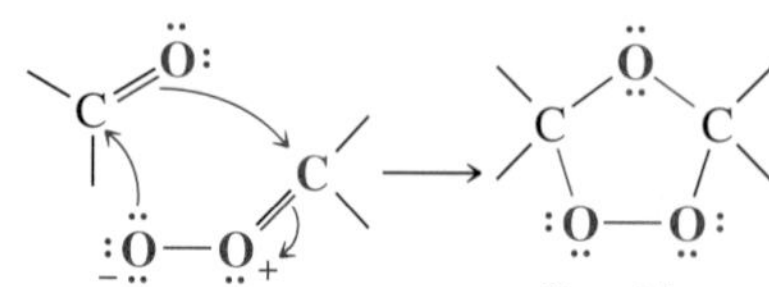

Ozonide

The fragments recombine to form the ozonide.

Ozonides are very unstable compounds, and low-molecular-weight ozonides often explode violently. Because of this property, they are not usually isolated but are reduced directly to the carbonyl products by treatment with zinc and acetic acid (HOAc) or dimethyl sulfide:

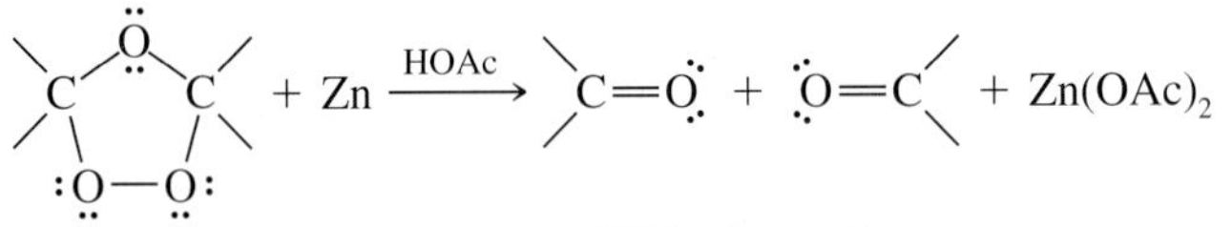

$$\text{Ozonide} + Zn \xrightarrow{HOAc} \text{C=O} + \text{O=C} + Zn(OAc)_2$$

Ozonide **Aldehydes and/or ketones**

PROBLEM 8.23

Write the structures of the alkenes that would produce the following products when treated with ozone and then with zinc and acetic acid:

(a) CH_3COCH_3 and $CH_3CH(CH_3)CHO$

(b) CH_3CH_2CHO only (2 mol is produced from 1 mol of alkene)

(c) 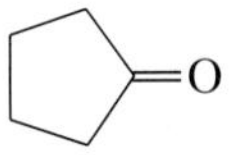 Cyclopentanone (cyclopentane=O) and HCHO

8.18 ADDITION OF BROMINE AND CHLORINE TO ALKYNES

Alkynes show the same kind of reactions toward chlorine and bromine that alkenes do: ***They react by addition.*** However, with alkynes the addition may occur once or twice, depending on the number of molar equivalents of halogen we employ:

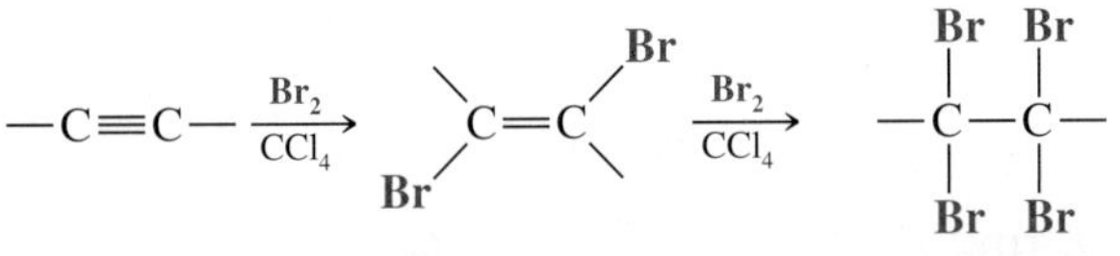

$$-C\equiv C- \xrightarrow[CCl_4]{Br_2} \underset{\text{Dibromoalkene}}{(Br)C{=}C(Br)} \xrightarrow[CCl_4]{Br_2} \underset{\text{Tetrabromoalkane}}{-CBr_2-CBr_2-}$$

$$-C\equiv C- \xrightarrow[CCl_4]{Cl_2} \underset{\text{Dichloroalkene}}{(Cl)C{=}C(Cl)} \xrightarrow[CCl_4]{Cl_2} \underset{\text{Tetrachloroalkane}}{-CCl_2-CCl_2-}$$

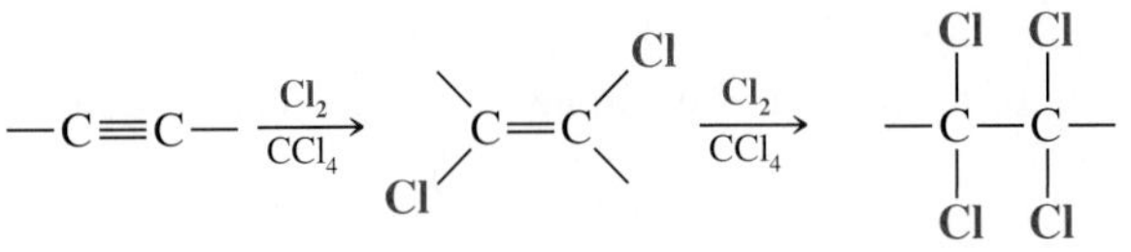

It is usually possible to prepare a dihaloalkene by simply adding one molar equivalent of the halogen:

$$CH_3CH_2CH_2CH_2C\equiv CCH_2OH \xrightarrow[CCl_4,\ 0°C]{Br_2\ (1\ mol)} CH_3CH_2CH_2CH_2CBr{=}CBrCH_2OH$$

Most additions of chlorine and bromine to alkynes are anti additions and yield trans-dihaloalkenes. Addition of bromine to acetylenedicarboxylic acid, for example, gives the trans isomer in 70% yield:

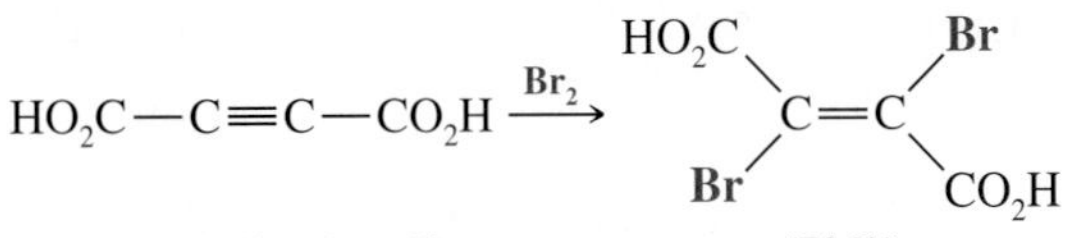

$$\underset{\text{Acetylenedicarboxylic acid}}{HO_2C-C\equiv C-CO_2H} \xrightarrow{Br_2} \underset{(70\%)}{(HO_2C)(Br)C{=}C(Br)(CO_2H)}$$

PROBLEM 8.24 Alkenes are more reactive than alkynes toward addition of electrophilic reagents (i.e., Br_2, Cl_2, or HCl). Yet when alkynes are treated with one molar equivalent of these same electrophilic reagents, it is easy to stop the addition at the "alkene stage." This appears to be a paradox and yet it is not. Explain.

8.19 ADDITION OF HYDROGEN HALIDES TO ALKYNES

Alkynes react with hydrogen chloride and hydrogen bromide to form haloalkenes or geminal dihalides depending, once again, on whether one or two molar equivalents of the hydrogen halide are used. **Both additions are regioselective and follow Markovnikov's rule:**

$$-C{\equiv}C- \xrightarrow{HX} \underset{\textbf{Haloalkene}}{(H)C{=}C(X)} \xrightarrow{HX} \underset{\textbf{\textit{gem}-Dihalide}}{-CH_2-CX_2-}$$

The hydrogen atom of the hydrogen halide becomes attached to the carbon atom that has the greater number of hydrogen atoms. 1-Hexyne, for example, reacts slowly with one molar equivalent of hydrogen bromide to yield 2-bromo-1-hexene and with two molar equivalents to yield 2,2-dibromohexane:

$$C_4H_9C{\equiv}CH \xrightarrow{HBr} \underset{\textbf{2-Bromo-1-hexene}}{C_4H_9-C(Br){=}CH_2} \xrightarrow{HBr} \underset{\textbf{2,2-Dibromohexane}}{C_4H_9-CBr_2-CH_3}$$

The addition of HBr to an alkyne can be facilitated by using acetyl bromide (CH_3COBr) and alumina instead of aqueous HBr. Acetyl bromide acts as an HBr precursor by reacting with the alumina to generate HBr. The alumina by its presence (Section 8.2) increases the rate of reaction. For example, 1-heptyne can be converted to 2-bromo-1-heptene in good yield using this method:

$$C_5H_{11}C{\equiv}CH \xrightarrow[CH_3COBr/\text{alumina},\ CH_2Cl_2]{\text{“HBr”}} \underset{\textbf{(82\%)}}{(Br)(C_5H_{11})C{=}CH_2}$$

Anti-Markovnikov addition of hydrogen bromide to alkynes occurs when peroxides are present in the reaction mixture. These reactions take place through a free-radical mechanism (Section 10.9):

$$CH_3CH_2CH_2CH_2C{\equiv}CH \xrightarrow[\text{peroxides}]{\textbf{HBr}} \underset{\textbf{(74\%)}}{CH_3CH_2CH_2CH_2CH{=}CHBr}$$

The addition of water to alkynes, a method for synthesizing ketones, is discussed in Section 16.5.

8.20 Oxidative Cleavage of Alkynes

Treating alkynes with ozone or with basic potassium permanganate leads to cleavage at the carbon–carbon triple bond. The products are carboxylic acids:

$$R{-}C{\equiv}C{-}R' \xrightarrow[(2)\ HOAc]{(1)\ O_3} RCO_2H + R'CO_2H$$

or

$$R{-}C{\equiv}C{-}R' \xrightarrow[(2)\ H_3O^+]{(1)\ KMnO_4,\ OH^-} RCO_2H + R'CO_2H$$

Problem 8.25

Alkynes **A** and **B** have the molecular formula C_8H_{14}. Treating either compound **A** or **B** with excess hydrogen in the presence of a metal catalyst leads to the formation of octane. Similar treatment of a compound **C** (C_8H_{12}) leads to a product with the formula C_8H_{16}. Treating alkyne **A** with ozone and then acetic acid furnishes a single product, $CH_3CH_2CH_2CO_2H$. Treating alkyne **C** with ozone and then water produces a single product, $HO_2C(CH_2)_6CO_2H$. Compound **B** has an absorption in its IR spectrum at ~3300 cm^{-1}. What are compounds **A**, **B**, and **C**?

8.21 Synthetic Strategies Revisited

In planning a synthesis we often have to consider four interrelated aspects:

1. Construction of the carbon skeleton
2. Functional group interconversions
3. Control of regiochemistry
4. Control of stereochemistry

You have had some experience in the first two aspects of synthetic strategy in earlier sections. In Section 4.19 you were introduced to the ideas of *retrosynthetic analysis* and how this kind of thinking could be applied to the construction of carbon skeletons of alkanes and cycloalkanes. In Section 6.14 you learned the meaning of a *functional group interconversion* and how nucleophilic substitution reactions could be used for this purpose. In other sections, perhaps without realizing it, you have begun adding to your basic store of methods for construction of carbon skeletons and for making functional group interconversions. This is the time to begin keeping a card file for all the reactions that you have learned, noting especially their applications to synthesis. This file will become your **Tool Kit for Organic Synthesis.**

Now is the time to look at some new examples and to see how we integrate all four aspects of synthesis into our planning.

Consider a problem in which we are asked to outline a synthesis of 2-bromobutane from compounds of two carbon atoms or fewer. This synthesis, as we shall see, involves construction of the carbon skeleton, functional group interconversion, and control of regiochemistry.

We begin by thinking backward. One way to make 2-bromobutane is by addition of hydrogen bromide to 1-butene. The regiochemistry of this functional group interconversion must be Markovnikov addition:

Retrosynthetic Analysis

$$CH_3CH_2CH(Br)CH_3 \Rightarrow CH_3CH_2CH{=}CH_2 + H{-}Br$$

Markovnikov addition

Synthesis

$$CH_3CH_2CH{=}CH_2 + HBr \xrightarrow[\text{peroxides}]{\text{no}} CH_3CH_2CH(Br)CH_3$$

Remember: The open arrow is a symbol used to show a retrosynthetic process that relates the target molecule to its precursors:

$$\text{Target molecule} \Longrightarrow \text{precursors}$$

Next we try to think of a way to synthesize 1-butene, keeping in mind that we have to construct the carbon skeleton from compounds with two carbon atoms or fewer:

Retrosynthetic Analysis

$$CH_3CH_2CH{=}CH_2 \Longrightarrow CH_3CH_2C{\equiv}CH + H_2$$
$$CH_3CH_2C{\equiv}CH \Longrightarrow CH_3CH_2Br + NaC{\equiv}CH$$
$$NaC{\equiv}CH \Longrightarrow HC{\equiv}CH + NaNH_2$$

Synthesis

$$HC{\equiv}C{-}H + Na^+ \ {}^-NH_2 \xrightarrow[-33°C]{\text{liq. } NH_3} HC{\equiv}C{:}^- Na^+$$

$$CH_3CH_2{-}Br + Na^+ \ {}^-{:}C{\equiv}CH \xrightarrow[-33°C]{\text{liq. } NH_3} CH_3CH_2C{\equiv}CH$$

$$CH_3CH_2C{\equiv}CH + H_2 \xrightarrow{Ni_2B\ (P\text{-}2)} CH_3CH_2CH{=}CH_2$$

One approach to retrosynthetic analysis is to consider a retrosynthetic step as a "disconnection" of one of the bonds (Section 4.20A).* For example, an important step in the synthesis that we have just given is the one in which a new carbon–carbon bond is formed. Retrosynthetically, it can be shown in the following way:

$$CH_3CH_2{-}C{\equiv}CH \Longrightarrow CH_3\overset{+}{C}H_2 + {}^-{:}C{\equiv}CH$$

The fragments of this disconnection are an ethyl cation and an ethynide anion. These fragments are called **synthons.** Seeing these synthons may help us to reason that we could, in theory, synthesize a molecule of 1-butyne by combining an ethyl cation with an ethynide anion. We know, however, that bottles of carbocations and carbanions are not to be found on our laboratory shelves. What we need are the **synthetic equivalents** of these synthons. The synthetic equivalent of an ethynide ion is sodium ethynide, because sodium ethynide contains an ethynide ion (and a sodium cation). The synthetic equivalent of an ethyl cation is ethyl bromide. To understand how this is true, we reason as follows: If ethyl bromide were to react by an S_N1 reaction, it would produce an ethyl cation and a bromide ion. However, we know that, being a primary halide, ethyl bromide is unlikely to react by an S_N1 reaction. Ethyl bromide, however, will react readily with a strong nucleophile such as sodium ethynide by an S_N2 reaction, and when it reacts, the product that is obtained is the same as the product that would have been obtained from the reaction of an

---

*For an excellent detailed treatment of this approach you may want to read the following: Warren, S. *Organic Synthesis, The Disconnection Approach;* Wiley: New York, 1982, and Warren, S. *Workbook for Organic Synthesis, The Disconnection Approach;* Wiley: New York, 1982.

ethyl cation with sodium ethynide. Thus, ethyl bromide, in this reaction, functions as the synthetic equivalent of an ethyl cation.

2-Bromobutane could also be synthesized from compounds of two carbons or fewer by a route in which (*E*)- or (*Z*)-2-butene is an intermediate. You may wish to work out the details of that synthesis for yourself.

Consider another example, a synthesis that requires stereochemical control: the synthesis of the enantiomeric 2,3-butanediols, (2*R*,3*R*)-2,3-butanediol and (2*S*,3*S*)-2,3-butanediol, from compounds of two carbon atoms or fewer.

Here (recall Problem 8.22) we see that a possible final step to the enantiomers is syn hydroxylation of *trans*-2-butene:

Retrosynthetic Analysis

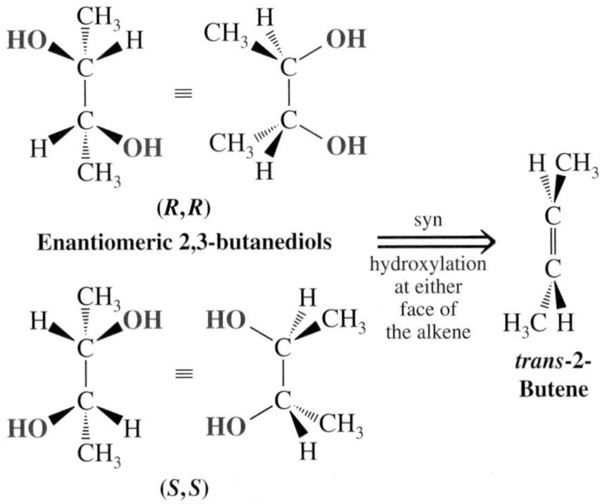

(***R***,***R***)

Enantiomeric 2,3-butanediols

(***S***,***S***)

syn hydroxylation at either face of the alkene

***trans*-2-Butene**

Synthesis

***trans*-2-Butene**

(1) OsO_4
(2) $NaHSO_3$, H_2O

(***R***,***R***) (***S***,***S***)

Enantiomeric 2,3-butanediols

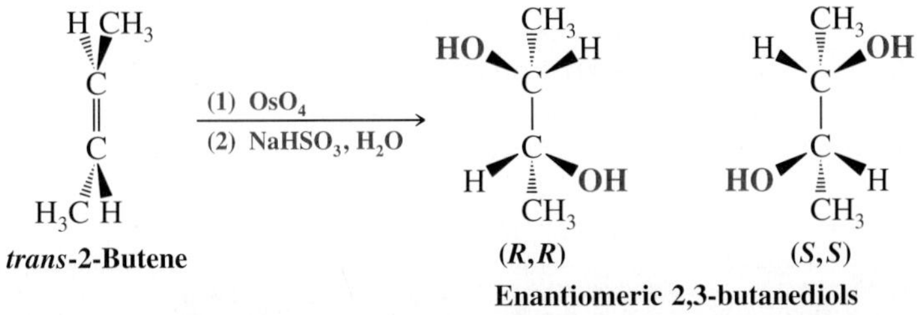

This reaction is stereospecific and produces the desired enantiomeric 2,3-butanediols as a racemic form.

Next, a synthesis of *trans*-2-butene can be accomplished by treating 2-butyne with lithium in liquid ammonia. The anti addition of hydrogen by this reaction gives us the trans product we need. This reaction is an example of a **stereoselective reaction**. A stereoselective reaction is one in which the reactant is not necessarily chiral (as in the case of an alkyne) but in which the reaction produces predominantly or exclusively one stereoisomeric form of the product (or a certain subset of stereoisomers from among all those that are possible). Note the difference between stereoselective and stereospecific. A stereospecific reaction is one that produces predominantly or exclusively one stereoisomer of the product when a specific stereoisomeric form of the reactant is used. (All stereospecific reactions are stereoselective, but the reverse is not necessarily true.)

Retrosynthetic Analysis

trans-2-Butene $\xrightarrow[\text{addition}]{\text{anti}}$ 2-Butyne + H_2

$$\underset{\textit{trans}\text{-2-Butene}}{(H)(CH_3)C{=}C(H_3C)(H)} \stackrel{\text{anti addition}}{\Longrightarrow} \underset{\text{2-Butyne}}{CH_3-C{\equiv}C-CH_3} + H_2$$

Synthesis

$$\underset{\text{2-Butyne}}{CH_3-C{\equiv}C-CH_3} \xrightarrow[\text{(2) } NH_4Cl \text{ anti addition of } H_2]{\text{(1) Li, } EtNH_2} \underset{\textit{trans}\text{-2-Butene}}{(H)(CH_3)C{=}C(H_3C)(H)}$$

Finally, we can synthesize 2-butyne from propyne by first converting it to sodium propynide and then alkylating sodium propynide with methyl iodide:

Retrosynthetic Analysis

$$CH_3-C{\equiv}C \nmid CH_3 \Longrightarrow CH_3-C{\equiv}C{:}^-Na^+ + CH_3-I$$

$$CH_3-C{\equiv}C{:}^-Na^+ \Longrightarrow CH_3-C{\equiv}C-H + NaNH_2$$

Synthesis

$$CH_3-C{\equiv}C-H \xrightarrow[\text{(2) } CH_3I]{\text{(1) } NaNH_2\text{/liq. } NH_3} CH_3-C{\equiv}C-CH_3$$

Finally, we can synthesize propyne from ethyne:

Retrosynthetic Analysis

$$H-C{\equiv}C \nmid CH_3 \Longrightarrow H-C{\equiv}C{:}^-Na^+ + CH_3-I$$

Synthesis

$$H-C{\equiv}C-H \xrightarrow[\text{(2) } CH_3I]{\text{(1) } NaNH_2\text{/liq. } NH_3} CH_3-C{\equiv}C-H$$

The Chemistry of. . .

Cholesterol Biosynthesis: Elegant and Familiar Reactions in Nature

Cholesterol is the biochemical precursor of cortisone, estradiol, and testosterone. In fact, it is the parent of all of the steroid hormones and bile acids in the body (Section 23.4). The biosynthesis of cholesterol involves some of the most beautiful and marvelous metabolic transformations known in biochemistry. It is a prime example of the synthetic prowess of enzymes, and it is a biochemical performance that should not be missed during the study of organic chemistry.

Our study will join the biosynthesis of cholesterol at the point of its last acyclic precursor, squalene. Squalene consists of a linear polyalkene chain having 30 carbons. From squalene, the first cyclic precursor called lanosterol is created by a remarkable set of enzyme-catalyzed addition reactions and rearrangements that create four fused rings and seven stereogenic centers. In theory, 2^7 (or 128) stereoisomers are possible for a structure having the constitution of lanosterol because lanosterol has seven stereogenic centers, yet only a single stereoisomer is produced by this enzymatic reaction. We shall now examine these reactions in detail. Even though the molecules involved are complex, you will find that the chemical principles behind their biosynthesis are readily recognizable because they are among those you have already learned in your study of organic chemistry:

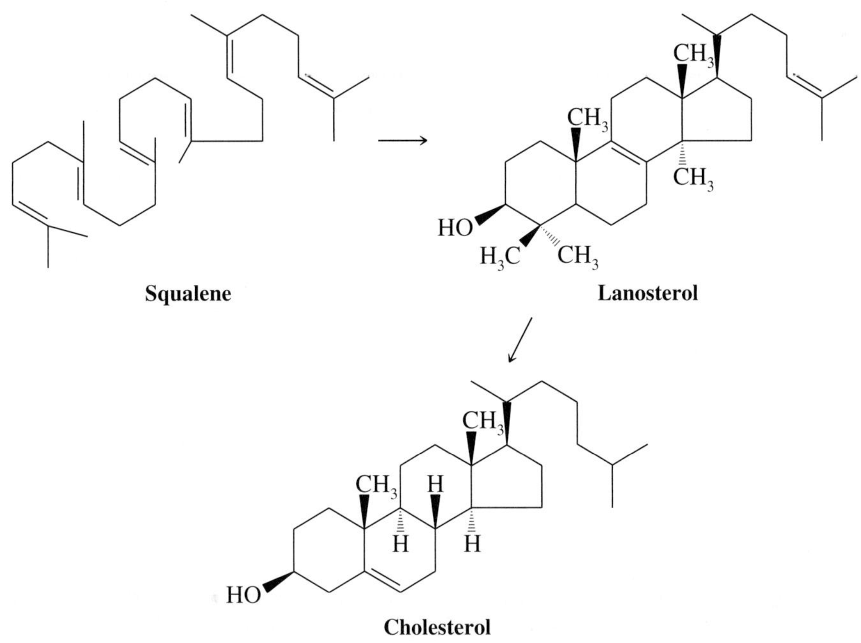

Squalene → Lanosterol → Cholesterol

Polyalkene Cyclization of Squalene to Lanosterol

The remarkable sequence of transformations from squalene to lanosterol begins by enzymatic oxidation of the 2,3 double bond of squalene to form (3*S*)-2,3-oxidosqualene [also called squalene 2,3-epoxide (an epoxide is a three-membered ring cyclic ether, see Section 11.18)]. At this point, a cascade of alkene addition reactions begin that in a three-dimensional sense proceeds through a so-called chair–boat–chair conformation (see the following scheme). Protonation of (3*S*)-2,3-oxidosqualene by squalene oxidocyclase gives the oxygen a formal positive charge and converts it to a good leaving group. The protonated epoxide makes the tertiary carbon (C2) electron deficient (resembling a tertiary carbocation), and C2 serves as the electrophile for an addition reaction with the double bond between C6 and C7 in the squalene chain. As this alkene attacks the tertiary carbon of the epoxide, another tertiary carbocation begins to develop at C6. This is attacked by the next double bond, and so on for two more alkene additions (see below), until the exocyclic tertiary carbocation shown below results. This intermediate is called the protosteryl cation:

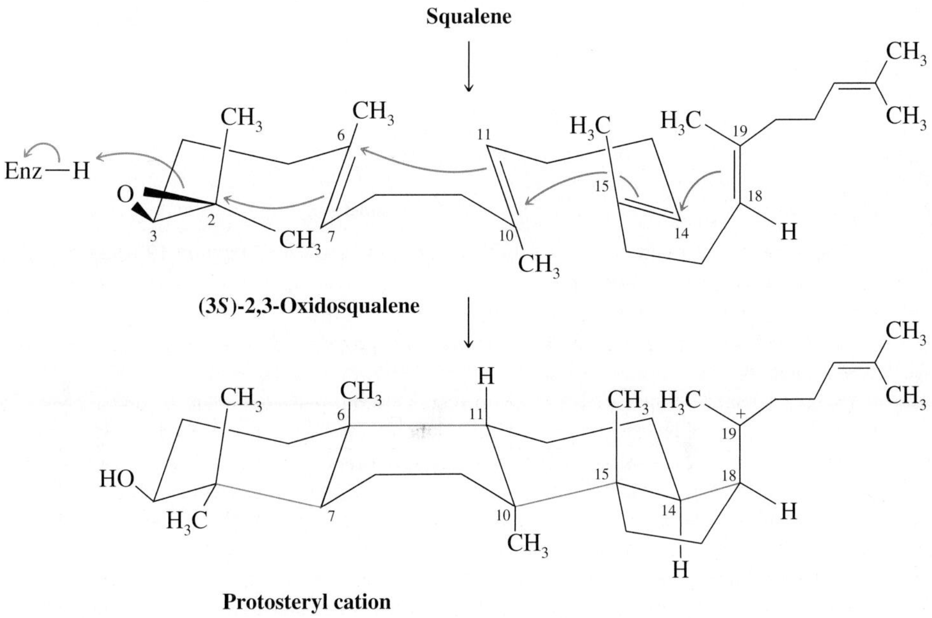

Squalene

(3*S*)-2,3-Oxidosqualene

Protosteryl cation

STUDY PROBLEM

All of the alkene additions except one in the cyclization leading to lanosterol were formally Markovnikov in character. Which alkene addition step during cyclization of 2,3-oxidosqualene occurred in an anti-Markovnikov sense? (Identify the alkene and the electrophilic carbon according to their numbers along the 2,3-oxidosqualene chain.) What final ring structure would have resulted, instead of the protosteryl cation, from the cyclization of 2,3-oxidosqualene if the anti-Markovnikov addition step had occurred in a Markovnikov fashion instead? Use conformational structures or an ordinary bond-line formula to show its structure.

An Elimination Reaction Involving a Sequence of 1,2-Methanide and 1,2-Hydride Rearrangements

The next aspect of lanosterol biosynthesis (still under the guidance of squalene oxidocyclase) is as remarkable as the first part, yet it remains consistent with principles of chemical logic. The transformations involved are a series of migrations (carbocation rearrangements) followed by removal of a proton to form an alkene. The process begins with a 1,2-hydride migration from C17 to C18 (using the standard numbering system for steroids; see Section 23.4), leading to development of positive charge at C17. The developing positive charge at C17 facilitates another hydride migration from C13 to C17, which is accompanied by methanide migrations from C14 to C13 and C8 to C14. Finally, enzymatic removal of a proton from C9 forms the C8–C9 double bond leading to lanosterol, the stable product of the squalene oxidocyclase reaction. These steps are shown below:

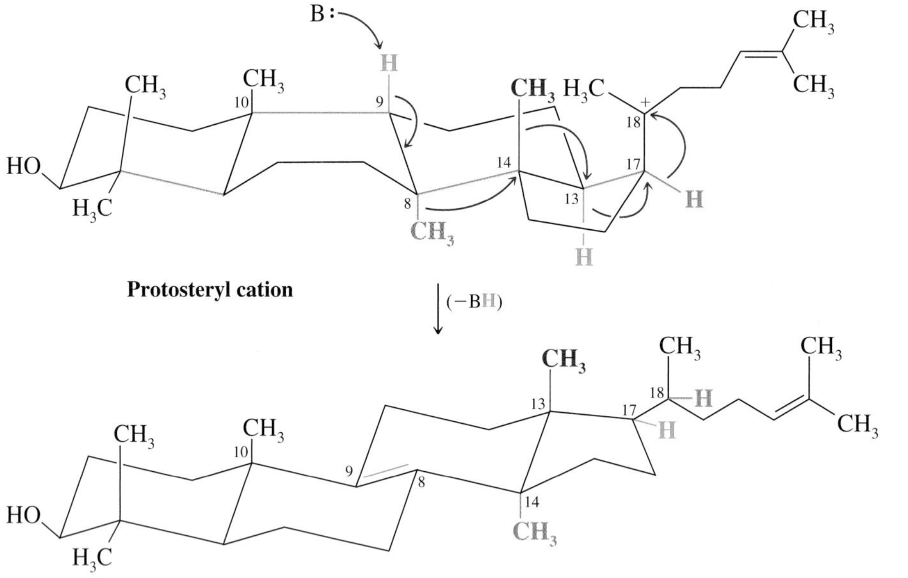

Protosteryl cation

$\downarrow$ (−BH)

Lanosterol

The remaining steps to cholesterol involve loss of three carbons through 19 oxidation–reduction steps, but we will not discuss them here, however:

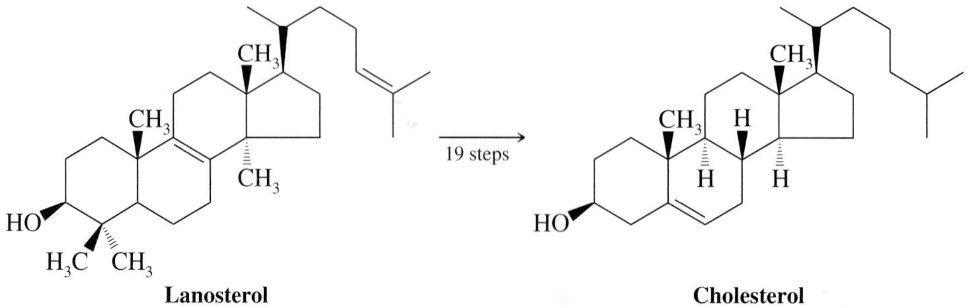

Lanosterol $\xrightarrow{\text{19 steps}}$ Cholesterol

These elegant transformations illustrate the beauty of organic chemistry in nature. As you can see, biosynthetic reactions occur on the basis of the same fundamental principles and reaction pathways that you have been studying in classic organic chemistry. In biological chemistry there are acid–base reactlons, nucleophiles, electrophiles, leaving groups, rearrangements, and so on. In Section 6.14, "The Chemistry of . . . Biological Methylation," we saw an example of biosynthetic nucleophilic substitution involving biological amines and S-adenosylmethionine. Here, in the biosynthesis of cholesterol, we have seen nature's use of carbocations, electrophiles, alkene addition reactions, and skeletal rearrangements. All of these processes are analogous to reactions you have studied recently.

SOLVED PROBLEM

ILLUSTRATING A STEREOSPECIFIC MULTISTEP SYNTHESIS Starting with compounds of two carbon atoms or fewer, outline a stereospecific synthesis of *meso*-3,4-dibromohexane.

ANSWER: We begin by working backward from the product. The addition of bromine to an alkene is stereospecifically anti. Therefore, adding bromine to *trans*-3-hexene will give *meso*-3,4-dibromohexane:

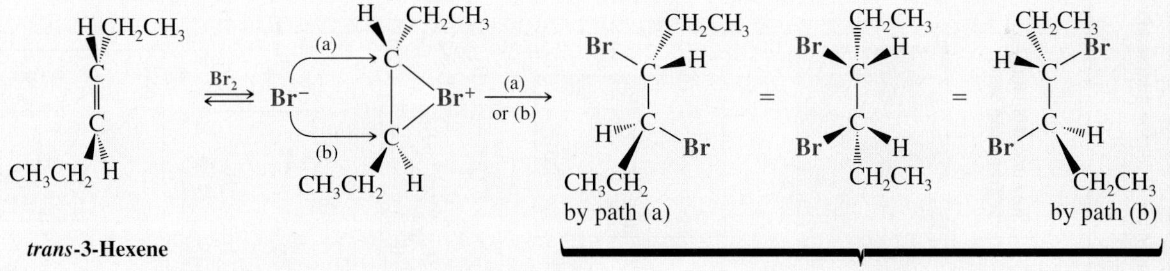

trans-3-Hexene $\xrightleftharpoons{Br_2}$ Br^- + bromonium ion $\xrightarrow[\text{or (b)}]{\text{(a)}}$ by path (a) = = by path (b)

***meso*-3,4-Dibromohexane**

We can make *trans*-3-hexene in a stereoselective way from 3-hexyne by reducing it with lithium in liquid ammonia (Section 7.14B). Again the addition is anti:

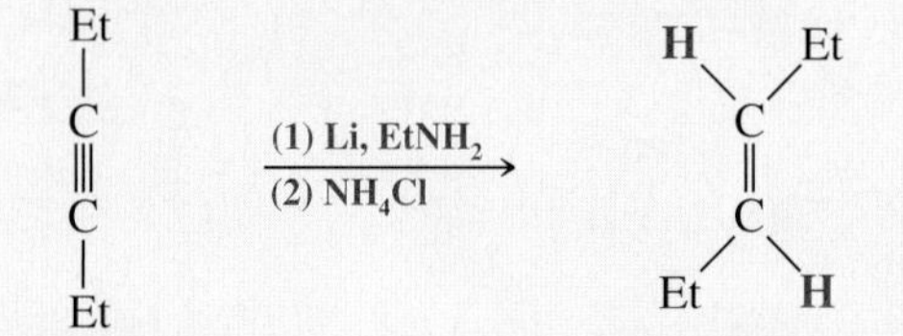

$$\text{Et–C}\equiv\text{C–Et} \xrightarrow[\text{(2) } NH_4Cl]{\text{(1) Li, } EtNH_2} \text{trans-EtCH=CHEt}$$

3-Hexyne **trans-3-Hexene**

3-Hexyne can be made from acetylene and ethyl bromide by successive alkylations using sodium amide as a base:

$$H{-}C{\equiv}C{-}H \xrightarrow[\text{(2) } CH_3CH_2Br]{\text{(1) } NaNH_2\text{, liq. } NH_3} CH_3CH_2C{\equiv}CH \xrightarrow[\text{(2) } CH_3CH_2Br]{\text{(1) } NaNH_2\text{, liq. } NH_3} CH_3CH_2C{\equiv}CCH_2CH_3$$

PROBLEM 8.26

How would you modify the procedure given in the solved problem so as to synthesize a racemic form of (3*R*,4*R*)- and (3*S*,4*S*)-3,4-dibromohexane?

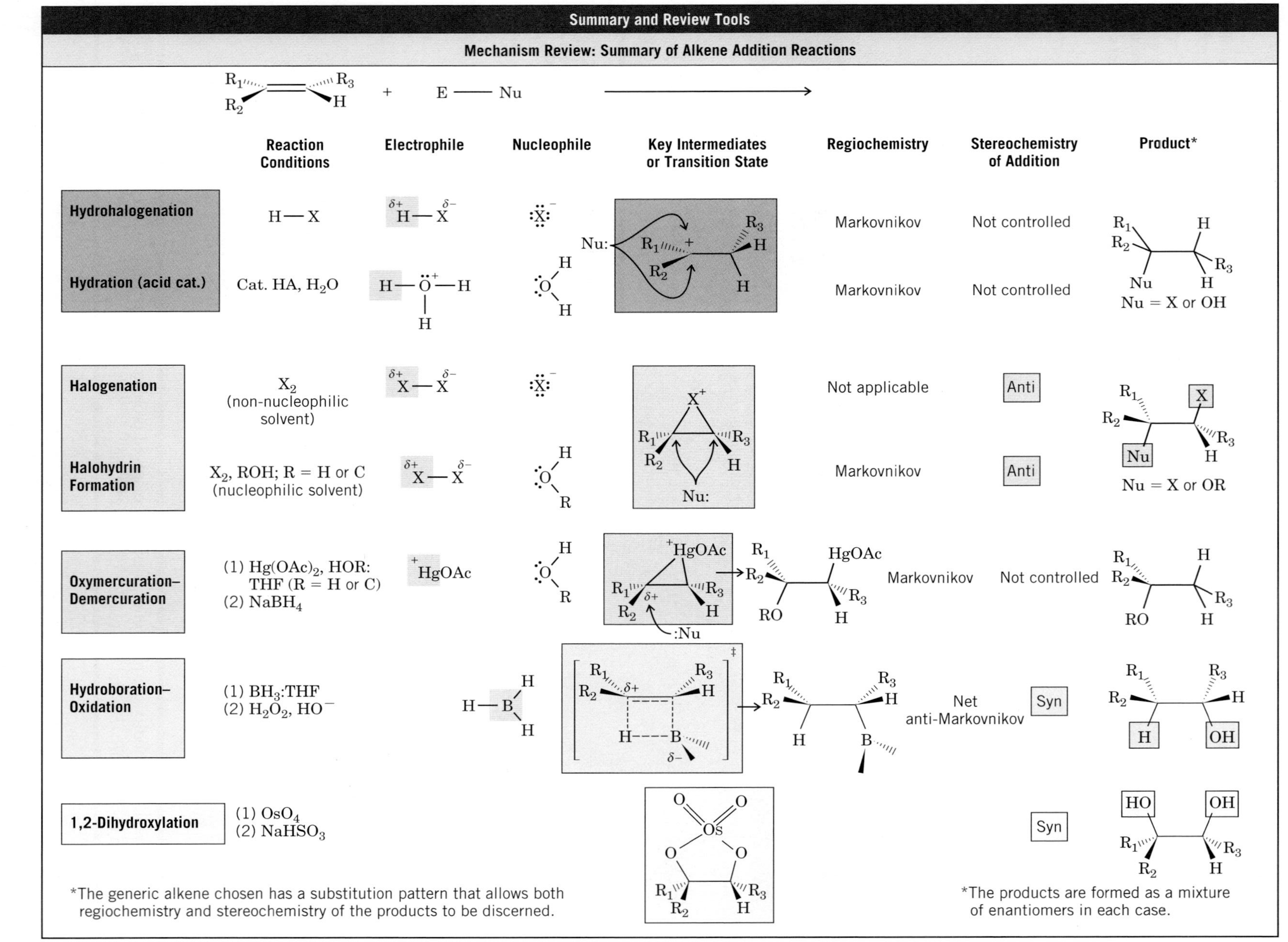

Summary and Review Tools

Mechanism Review: Summary of Alkene Addition Reactions

Alkene (R_1, R_2 / R_3, H) + E — Nu →

	Reaction Conditions	Electrophile	Nucleophile	Key Intermediates or Transition State	Regiochemistry	Stereochemistry of Addition	Product*
Hydrohalogenation	H — X	$H^{\delta+}$ — $X^{\delta-}$	$:\ddot{X}:^-$		Markovnikov	Not controlled	
Hydration (acid cat.)	Cat. HA, H_2O	H — $\ddot{O}^+H_2$	$H_2\ddot{O}:$		Markovnikov	Not controlled	Nu = X or OH
Halogenation	X_2 (non-nucleophilic solvent)	$X^{\delta+}$ — $X^{\delta-}$	$:\ddot{X}:^-$		Not applicable	Anti	
Halohydrin Formation	X_2, ROH; R = H or C (nucleophilic solvent)	$X^{\delta+}$ — $X^{\delta-}$	$H\ddot{O}R$		Markovnikov	Anti	Nu = X or OR
Oxymercuration–Demercuration	(1) $Hg(OAc)_2$, HOR:THF (R = H or C) (2) $NaBH_4$	^+HgOAc	$H\ddot{O}R$		Markovnikov	Not controlled	
Hydroboration–Oxidation	(1) BH_3:THF (2) H_2O_2, HO^-		H—BH_2		Net anti-Markovnikov	Syn	
1,2-Dihydroxylation	(1) OsO_4 (2) $NaHSO_3$					Syn	

*The generic alkene chosen has a substitution pattern that allows both regiochemistry and stereochemistry of the products to be discerned.

*The products are formed as a mixture of enantiomers in each case.

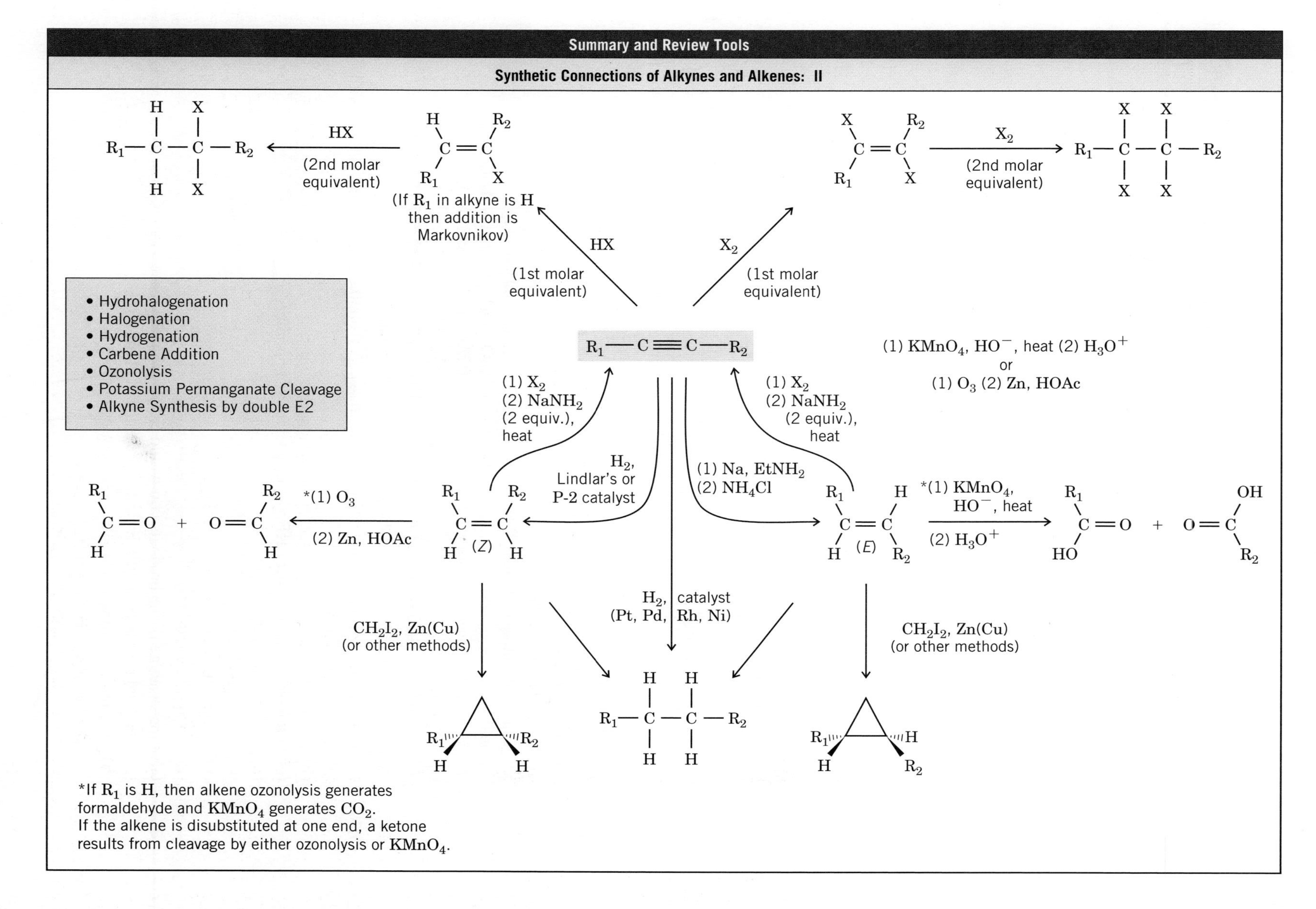
Summary and Review Tools
Synthetic Connections of Alkynes and Alkenes: II
• Hydrohalogenation
• Halogenation
• Hydrogenation
• Carbene Addition
• Ozonolysis
• Potassium Permanganate Cleavage
• Alkyne Synthesis by double E2
HX
(2nd molar equivalent)
(If R_1 in alkyne is H then addition is Markovnikov)
HX
(1st molar equivalent)
X_2
(1st molar equivalent)
X_2
(2nd molar equivalent)
$R_1—C≡C—R_2$
(1) $KMnO_4$, HO^-, heat (2) H_3O^+
or
(1) O_3 (2) Zn, HOAc
(1) X_2
(2) $NaNH_2$
(2 equiv.),
heat
H_2,
Lindlar's or
P-2 catalyst
(1) Na, $EtNH_2$
(2) NH_4Cl
*(1) O_3
(2) Zn, HOAc
(Z)
(E)
*(1) $KMnO_4$, HO^-, heat
(2) H_3O^+
H_2, catalyst
(Pt, Pd, Rh, Ni)
CH_2I_2, Zn(Cu)
(or other methods)
*If R_1 is H, then alkene ozonolysis generates formaldehyde and $KMnO_4$ generates CO_2.
If the alkene is disubstituted at one end, a ketone results from cleavage by either ozonolysis or $KMnO_4$.

Key Terms and Concepts

Addition reaction	Sections 8.1–8.9, 8.12, 8.13
anti-Markovnikov addition	Sections 8.2D, 8.19
Bromonium ion	Section 8.12
Carbenes	Section 8.15
Electrophile	Section 8.1
Halohydrin	Section 8.14
Hydration	Section 8.5
Markovnikov's rule	Sections 8.2, 8.19
Oxidative cleavage	Sections 8.17, 8.20
Oxymercuration–demercuration	Section 8.6
Ozonolysis	Sections 8.17A, 8.20
Regioselective reaction	Sections 8.2C, 8.19
Stereoselective reaction	Section 8.21
Stereospecific reaction	Section 8.13A
Syn dihydroxylation	Section 8.16A

Additional Problems

8.27 Write structural formulas for the products that form when 1-butene reacts with each of the following reagents:

(a) HI
(b) H_2, Pt
(c) Dilute H_2SO_4, warm
(d) Cold concentrated H_2SO_4
(e) Cold concentrated H_2SO_4, then H_2O and heat
(f) HBr in the presence of alumina
(g) Br_2 in CCl_4
(h) Br_2 in CCl_4, then KI in acetone
(i) Br_2 in H_2O
(j) HCl in the presence of alumina
(k) Cold dilute $KMnO_4$, OH^-
(l) O_3, then Zn, HOAc
(m) OsO_4, then $NaHSO_3/H_2O$
(n) $KMnO_4$, OH^-, heat, then H_3O^+
(o) $Hg(OAc)_2$ in THF and H_2O, then $NaBH_4$, OH^-
(p) BH_3:THF, then H_2O_2, OH^-

8.28 Repeat Problem 8.27 using cyclohexene instead of 1-butene.

8.29 Give the structure of the products that you would expect from the reaction of 1-butyne with:

(a) One molar equivalent of Br_2
(b) One molar equivalent of HBr in the presence of alumina
(c) Two molar equivalents of HBr in the presence of alumina
(d) H_2 (in excess)/Pt
(e) H_2, Ni_2B (P-2)
(f) $NaNH_2$ in liquid NH_3, then CH_3I
(g) $NaNH_2$ in liquid NH_3, then $(CH_3)_3CBr$

8.30 Give the structure of the products you would expect from the reaction (if any) of 2-butyne with:

(a) One molar equivalent of HBr in the presence of alumina
(b) Two molar equivalents of HBr in the presence of alumina
(c) One molar equivalent of Br_2
(d) Two molar equivalents of Br_2

**Note:* Problems marked with an asterisk are "challenge problems."

(e) H_2, Ni_2B (P-2)
(f) One molar equivalent of HCl in the presence of alumina
(g) Li/liquid NH_3
(h) H_2 (in excess), Pt
(i) Two molar equivalents of H_2, Pt
(j) $KMnO_4$, OH^-, then H_3O^+
(k) O_3, then HOAc
(l) $NaNH_2$, liquid NH_3

8.31 Show how 1-butyne could be synthesized from each of the following:
(a) 1-Butene
(b) 1-Chlorobutane
(c) 1-Chloro-1-butene
(d) 1,1-Dichlorobutane
(e) Ethyne and ethyl bromide

8.32 Starting with 2-methylpropene (isobutylene) and using any other needed reagents, outline a synthesis of each of the following:
(a) $(CH_3)_3COH$
(b) $(CH_3)_3CCl$
(c) $(CH_3)_3CBr$
(d) $(CH_3)_3CF$
(e) $(CH_3)_2C(OH)CH_2Cl$
(f) $(CH_3)_2CHCH_2OH$

8.33 Write a three-dimensional formula for the product formed when 1-methylcyclohexene is treated with each of the following reagents. In each case, designate the location of deuterium or tritium atoms.
(a) (1) BH_3:THF, (2) CH_3CO_2T
(b) (1) BD_3:THF, (2) CH_3CO_2D
(c) (1) BD_3:THF, (2) NaOH, H_2O_2, H_2O

8.34 Myrcene, a fragrant compound found in bayberry wax, has the formula $C_{10}H_{16}$ and is known not to contain any triple bonds. **(a)** What is the index of hydrogen deficiency of myrcene? When treated with excess hydrogen and a platinum catalyst, myrcene is converted to a compound **(A)** with the formula $C_{10}H_{22}$. **(b)** How many rings does myrcene contain? **(c)** How many double bonds? Compound **A** can be identified as 2,6-dimethyloctane. Ozonolysis of myrcene followed by treatment with zinc and acetic acid yields 2 mol of formaldehyde (HCHO), 1 mol of acetone (CH_3COCH_3), and a third compound **(B)** with the formula $C_5H_6O_3$. **(d)** What is the structure of myrcene? **(e)** Of compound **B**?

8.35 When propene is treated with hydrogen chloride in ethanol, one of the products of the reaction is ethyl isopropyl ether. Write a plausible mechanism that accounts for the formation of this product.

8.36 When, in separate reactions, 2-methylpropene, propene, and ethene are allowed to react with HI under the same conditions (i.e., identical concentration and temperature), 2-methylpropene is found to react fastest and ethene slowest. Provide an explanation for these relative rates.

8.37 Farnesene (below) is a compound found in the waxy coating of apples. Give the structure and IUPAC name of the product formed when farnesene is allowed to react with excess hydrogen in the presence of a platinum catalyst.

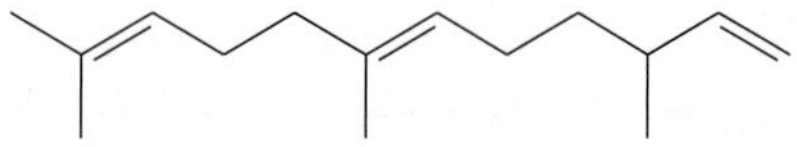

Farnesene

8.38 Write structural formulas for the products that would be formed when geranial, a component of lemongrass oil, is treated with ozone and then with zinc and acetic acid.

H
O

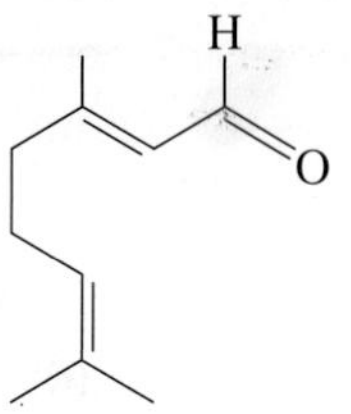

Geranial

8.39 Limonene is a compound found in orange oil and lemon oil. When limonene is treated with excess hydrogen and a platinum catalyst, the product of the reaction is 1-isopropyl-4-methylcyclohexane. When limonene is treated with ozone and then with zinc and acetic acid, the products of the reaction are HCHO and the following compound. Write a structural formula for limonene.

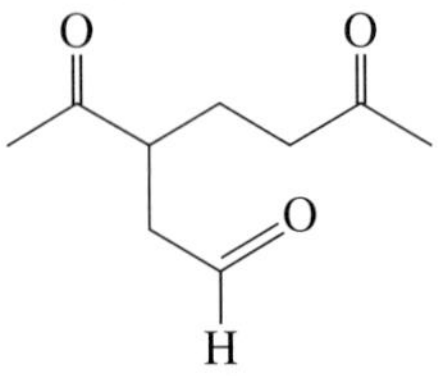

O O O H

8.40 When 2,2-diphenyl-1-ethanol is treated with aqueous HI, the main product of the reaction is 1-iodo-1,1-diphenylethane. Propose a plausible mechanism that accounts for this product.

8.41 When 3,3-dimethyl-2-butanol is treated with concentrated HI, a rearrangement takes place. Which alkyl iodide would you expect from the reaction? (Show the mechanism by which it is formed.)

8.42 Pheromones (Section 4.16) are substances secreted by animals (especially insects) that produce a specific behavioral response in other members of the same species. Pheromones are effective at very low concentrations and include sex attractants, warning substances, and "aggregation" compounds. The sex attractant pheromone of the codling moth has the molecular formula $C_{13}H_{24}O$. On catalytic hydrogenation this pheromone absorbs two molar equivalents of hydrogen and is converted to 3-ethyl-7-methyl-1-decanol. On treatment with ozone followed by treatment with zinc and acetic acid, the pheromone produces $CH_3CH_2CH_2COCH_3$, $CH_3CH_2COCH_2CH_2CHO$, and $OHCCH_2OH$.
(a) Neglecting the stereochemistry of the double bonds, write a general structure for this pheromone.
(b) The double bonds are known (on the basis of other evidence) to be (2*Z*,6*E*). Write a stereochemical formula for the codling moth sex attractant.

8.43 The sex attractant of the common housefly *(Musca domestica)* is a compound called muscalure. Muscalure is (*Z*)-$CH_3(CH_2)_{12}CH{=}CH(CH_2)_7CH_3$. Starting with ethyne and any other needed reagents, outline a possible synthesis of muscalure.

***8.44** Starting with ethyne and 1-bromopentane as your only organic reagents (except for solvents) and using any needed inorganic compounds, outline a synthesis of the following compound:

Br Br

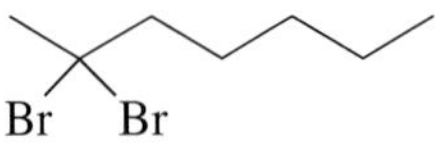

8.45 Shown below is the final step in a synthesis of an important perfume constituent, *cis*-jasmone. Which reagents would you choose to carry out this last step?

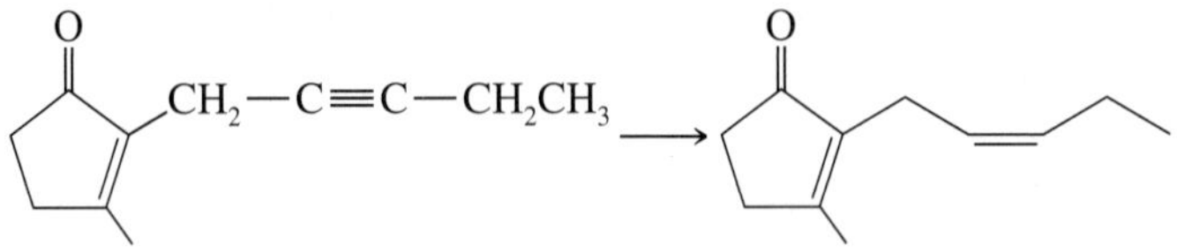

O $CH_2{-}C{\equiv}C{-}CH_2CH_3$ ⟶ O

***cis*-Jasmone**

8.46 Write stereochemical formulas for all of the products that you would expect from each of the following reactions. (You may find models helpful.)

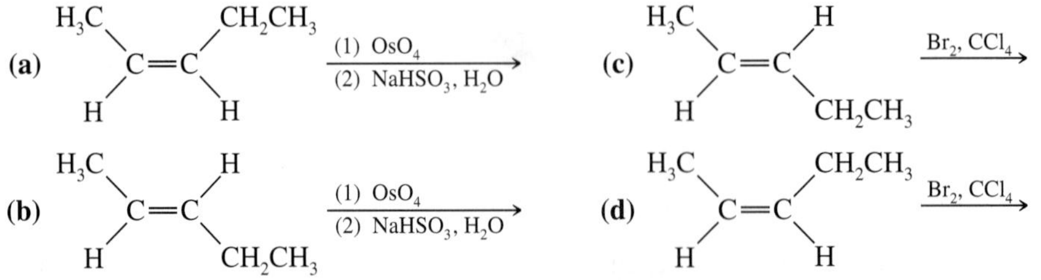

(a) $H_3C(H)C{=}C(H)CH_2CH_3$ $\xrightarrow[(2)\ NaHSO_3,\ H_2O]{(1)\ OsO_4}$

(b) $H_3C(H)C{=}C(CH_2CH_3)H$ $\xrightarrow[(2)\ NaHSO_3,\ H_2O]{(1)\ OsO_4}$

(c) $H_3C(H)C{=}C(H)CH_2CH_3$ $\xrightarrow{Br_2,\ CCl_4}$

(d) $H_3C(H)C{=}C(CH_2CH_3)H$ $\xrightarrow{Br_2,\ CCl_4}$

8.47 Give (*R*)–(*S*) designations for each different compound given as an answer to Problem 8.46.

8.48 When cyclohexene is allowed to react with bromine in an aqueous solution of sodium chloride, the products are *trans*-1,2-dibromocyclohexane, *trans*-2-bromocyclohexanol, and *trans*-1-bromo-2-chlorocyclohexane. Write a plausible mechanism that explains the formation of this last product.

8.49 Predict features of their IR spectra that you could use to distinguish between the members of the following pairs of compounds:
(a) Pentane and 1-pentyne
(b) Pentane and 1-pentene
(c) 1-Pentene and 1-pentyne
(d) Pentane and 1-bromopentane
(e) 2-Pentyne and 1-pentyne
(f) 1-Pentene and 1-pentanol
(g) Pentane and 1-pentanol
(h) 1-Bromo-2-pentene and 1-bromopentane
(i) 1-Pentanol and 2-penten-1-ol

8.50 The double bond of tetrachloroethene is undetectable in the bromine/carbon tetrachloride test for unsaturation. Give a plausible explanation for this behavior.

***8.51** Three compounds **A**, **B**, and **C** all have the formula C_6H_{10}. All three compounds rapidly decolorize bromine in CCl_4; all three are soluble in cold concentrated sulfuric acid. Compound **A** has an absorption in its IR spectrum at about 3300 cm^{-1}, but compounds **B** and **C** do not. Compounds **A** and **B** both yield hexane when they are treated with excess hydrogen in the presence of a platinum catalyst. Under these conditions **C** absorbs only one molar equivalent of hydrogen and gives a product with the formula C_6H_{12}. When **A** is oxidized with hot basic potassium permanganate and the resulting solution acidified, the only organic product that can be isolated is $CH_3(CH_2)_3CO_2H$. Similar oxidation of **B** gives only $CH_3CH_2CO_2H$, and similar treatment of **C** gives only $HO_2C(CH_2)_4CO_2H$. What are structures for **A**, **B**, and **C**?

8.52 Ricinoleic acid, a compound that can be isolated from castor oil, has the structure $CH_3(CH_2)_5CHOHCH_2CH{=}CH(CH_2)_7CO_2H$.
(a) How many stereoisomers of this structure are possible?
(b) Write these structures.

8.53 There are two dicarboxylic acids with the general formula $HO_2CCH{=}CHCO_2H$. One dicarboxylic acid is called maleic acid; the other is called fumaric acid. In 1880, Kekulé found that on treatment with cold dilute $KMnO_4$ maleic acid yields *meso*-tartaric acid and fumaric acid yields (±)-tartaric acid. Show how this information allows one to write stereochemical formulas for maleic acid and fumaric acid.

8.54 Use your answers to the preceding problem to predict the stereochemical outcome of the addition of bromine to maleic acid and to fumaric acid. **(a)** Which dicarboxylic acid would add bromine to yield a meso compound? **(b)** Which would yield a racemic form?

8.55 Alkyl halides add to alkenes in the presence of $AlCl_3$; yields are the highest when tertiary halides are used. Predict the outcome of the reaction of *tert*-pentyl chloride with propene and specify the mechanistic steps.

8.56 Explain the stereochemical results observed in this catalytic hydrogenation. (You may find it helpful to build hand-held molecular models.)

CH_2 CH_3 H $\xrightarrow[\text{Pt catalyst, HOAc}]{H_2}$ CH_3 H CH_3 H (**I** (70%)) + H CH_3 CH_3 H (**II** (30%))

8.57 The reaction of bromine with cyclohexene involves anti addition, which generates, initially, the diaxial conformation of the addition product that then undergoes a ring flip to the diequatorial conformation of *trans*-1,2-dibromocyclohexane.

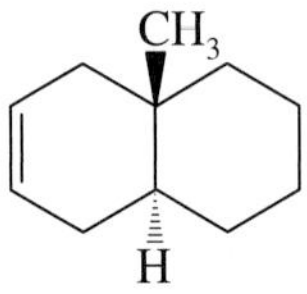

CH_3

H

I

However, when the unsaturated bicyclic compound **I** is the alkene, the addition product is exclusively in a stable diaxial conformation. Account for this. (You may find it helpful to build hand-held molecular models.)

8.58 Under acidic conditions, 2-methylpropene forms a carbocation that then attacks a second molecule of 2-methylpropene to form two eight-carbon alkenes. Propose an explanation for the fact that with a terminal double bond **I** is formed in considerably greater amount than **II**, which has an interior double bond:

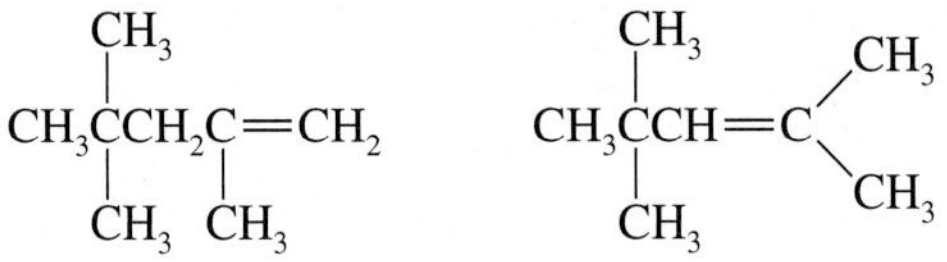

$CH_3C(CH_3)_2CH_2C(CH_3){=}CH_2$ **I**

$CH_3C(CH_3)_2CH{=}C(CH_3)_2$ **II**

8.59 Internal alkynes can be isomerized to terminal alkynes on treatment with $NaNH_2$. The process is much less successful when NaOH is used. Why is there this difference?

8.60 Terminal alkynes, RC≡CH, are not reduced by Na in liquid NH_3, but reduction to terminal alkenes occurs if $(NH_4)_2SO_4$ is included in the reaction mixture. Explain.

***8.61** An optically active compound **A** (assume that it is dextrorotatory) has the molecular formula $C_7H_{11}Br$. **A** reacts with hydrogen bromide, in the absence of peroxides, to yield isomeric products, **B** and **C**, with the molecular formula $C_7H_{12}Br_2$. Compound **B** is optically active; **C** is not. Treating **B** with 1 mol of potassium *tert*-butoxide yields (+)-**A**. Treating **C** with 1 mol of potassium *tert*-butoxide yields (±)-**A**. Treating **A** with potassium *tert*-butoxide yields **D** (C_7H_{10}). Subjecting 1 mol of **D** to ozonolysis followed by treatment with zinc and acetic acid yields 2 mol of formaldehyde and 1 mol of 1,3-cyclopentanedione.

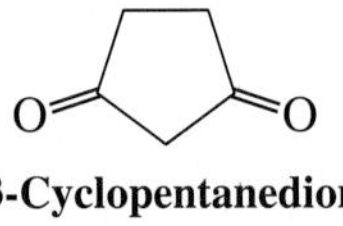

O O

1,3-Cyclopentanedione

Propose stereochemical formulas for **A**, **B**, **C**, and **D** and outline the reactions involved in these transformations.

8.62 A naturally occurring antibiotic called mycomycin has the structure shown here. Mycomycin is optically active. Explain this by writing structures for the enantiomeric forms of mycomycin.

$$HC{\equiv}C{-}C{\equiv}C{-}CH{=}C{=}CH{-}(CH{=}CH)_2CH_2CO_2H$$

Mycomycin

8.63 An optically active compound **D** has the molecular formula C_6H_{10} and shows a peak at about 3300 cm^{-1} in its IR spectrum. On catalytic hydrogenation **D** yields **E** (C_6H_{14}). Compound **E** is optically inactive and cannot be resolved. Propose structures for **D** and **E**.

***8.64** **(a)** By analogy with the mechanism of bromine addition to alkenes, draw the likely three-dimensional structures of **A**, **B**, and **C**.

Reaction of cyclopentene with bromine in water gives **A**.

Reaction of **A** with aqueous NaOH (1 equivalent, cold) gives **B**, C_5H_8O (no 3590–3650 cm^{-1} infrared absorption). (See the squalene cyclization discussion for a hint.)

Heating of **B** in methanol containing a catalytic amount of strong acid gives **C**, $C_6H_{12}O_2$, which does show 3590–3650 cm^{-1} infrared absorption.

(b) Specify the *R* or *S* configuration of the stereogenic centers in your predicted structures for **C**. Would it be formed as a single stereoisomer or as a racemate?

(c) How could you experimentally confirm your predictions about the stereochemistry of **C**?

*8.65 Triethylamine, $(C_2H_5)_3N$, like all amines, has a nitrogen atom with an unshared pair of electrons. Dichlorocarbene also has an unshared pair of electrons. Both can be represented as shown below. Draw the structures of compounds **D**, **E**, and **F**.

$$(C_2H_5)_3N\!: + \,:\!CCl_2 \longrightarrow \textbf{D} \text{ (an unstable adduct)}$$

$$\textbf{D} \longrightarrow \textbf{E} + C_2H_4 \text{ (by an intramolecular E2 reaction)}$$

$\textbf{E} \xrightarrow{H_2O} \textbf{F}$ (Water effects a replacement that is the reverse of that used to make *gem*-dichlorides.)

8.66 In Chapter 3 we first mentioned the importance of the interaction of a HOMO (highest occupied molecular orbital) of one molecule with the LUMO (lowest unoccupied molecular orbital) of another when two molecules react with each other (see "The Chemistry of. . . " box, Section 3.2). These ideas carry forth into our understanding of addition reactions between alkenes and electrophiles. Open the molecular models on the CD for ethene and BH_3 and view the HOMO and LUMO for each reactant. Which reactant is likely to have its HOMO orbital involved in the hydroboration of ethene? Which molecule's LUMO will be involved? As you view the models, can you envision favorable overlap of these orbitals as the reaction occurs?

8.67 Hydroboration reactions are frequently done using BH_3:THF as a complex in solution. BH_3 in pure form is a gas, and in the absence of other Lewis bases it exists as the dimer, diborane, B_2H_6. Open the molecular model on the CD for the BH_3:THF complex and display its LUMO orbital. Does the LUMO have lobes suitably disposed to allow the BH_3 portion of the BH_3:THF complex to interact with other Lewis bases, e.g., an alkene π bond in the course of a hydroboration reaction? Such an interaction is required at the beginning of a hydroboration reaction with BH_3:THF because the reaction begins with a complex between BH_3 and the alkene π bond, which then changes to the four-atom transition state of the addition as the reaction proceeds.

8.68 Open the molecular model for diborane (B_2H_6) on the CD and examine its HOMO and LUMO. Is the LUMO of B_2H_6 readily accessible to the HOMO of an alkene or other Lewis base? How does the orientation of the diborane LUMO compare with that of its HOMO?

1. **(a)** Synthesize (3*S*,4*R*)-3,4-dibromo-1-cyclohexylpentane (and its enantiomer, since a racemic mixture will be formed) from ethyne, 1-chloro-2-cyclohexylethane, bromomethane, and any other reagents necessary. (Use ethyne, 1-chloro-2-cyclohexylethane, and bromomethane as the sole sources of carbon atoms.) Start the problem by showing a retrosynthetic analysis. In the process, decide which atoms of the target molecule will come from which atoms of the starting reagents. Also, bear in mind how the stereospecificity of the reactions you employ can be used to achieve the required stereochemical form of the final product.

   (b) Explain why a racemic mixture of products results from this synthesis.

   (c) How could the synthesis be modified to produce a racemic mixture of the (3*R*,4*R*) and (3*S*,4*S*) isomers instead?

2. Write a reasonable and detailed mechanism for the following transformation:

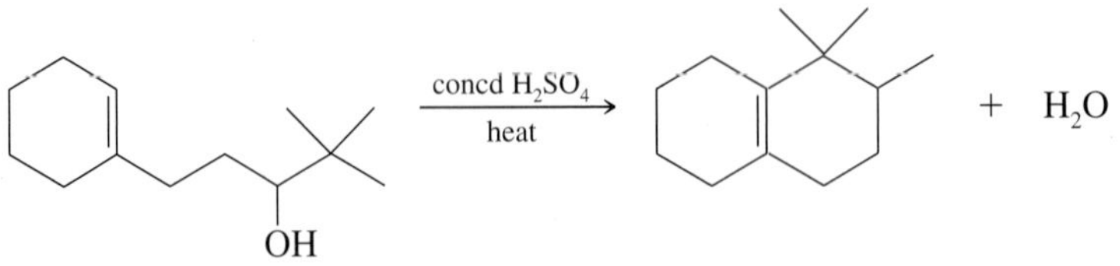

concd H_2SO_4, heat; + H_2O

OH

3. Deduce the structures of compounds **A–D**. Draw structures that show stereochemistry where appropriate:

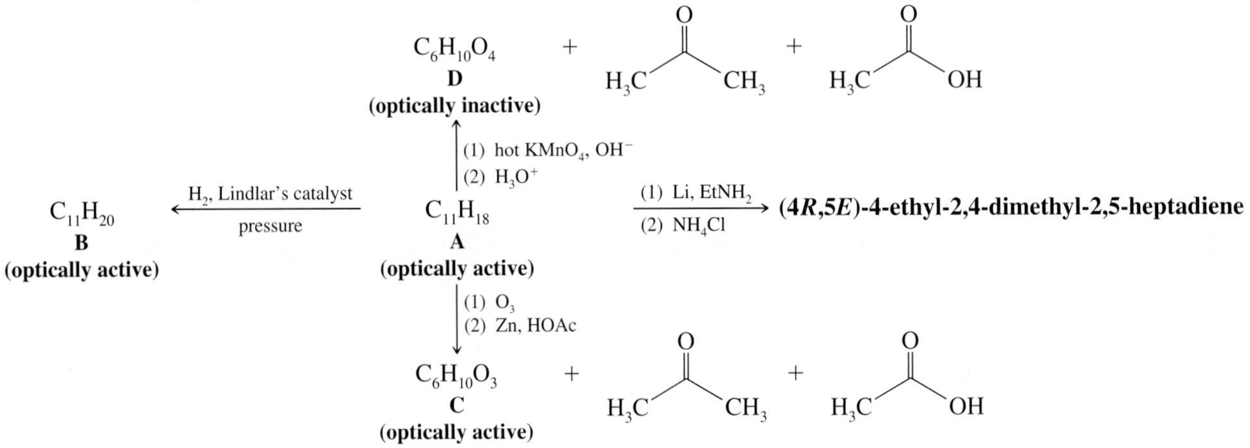

$C_6H_{10}O_4$ **D** **(optically inactive)** + H_3C–CO–CH_3 + H_3C–CO–OH

(1) hot $KMnO_4$, OH^-
(2) H_3O^+

$C_{11}H_{20}$ **B** **(optically active)** ← H_2, Lindlar's catalyst, pressure — $C_{11}H_{18}$ **A** **(optically active)** — (1) Li, $EtNH_2$ (2) NH_4Cl → **(4*R*,5*E*)-4-ethyl-2,4-dimethyl-2,5-heptadiene**

(1) O_3
(2) Zn, HOAc

$C_6H_{10}O_3$ **C** **(optically active)** + H_3C–CO–CH_3 + H_3C–CO–OH

4. The Guam bubble snail *(Haminoea cymbalum)* contains kumepaloxane (shown below), a chemical signal agent discharged when this mollusk is disturbed by predatory carnivorous fish. The biosynthesis of bromoethers like kumepaloxane is thought to occur via the enzymatic intermediacy of a "Br^+" agent. Draw the structure of a possible biosynthetic precursor (*hint:* an alkene alcohol) to kumepaloxane and write a plausible and detailed mechanism by which it could be converted to kumepaloxane using Br^+ and some generic proton acceptor Y^-.

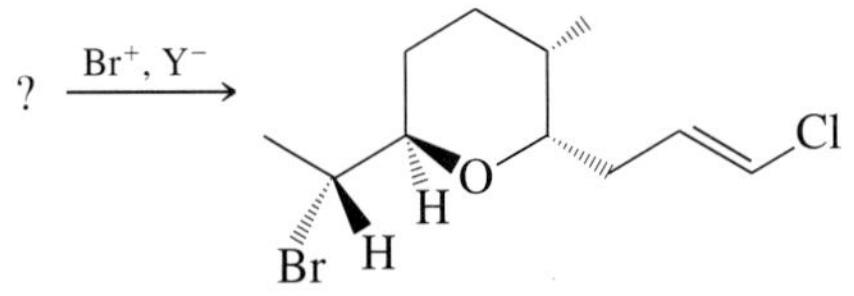

? $\xrightarrow{Br^+,\ Y^-}$

Br H H O Cl

Kumepaloxane

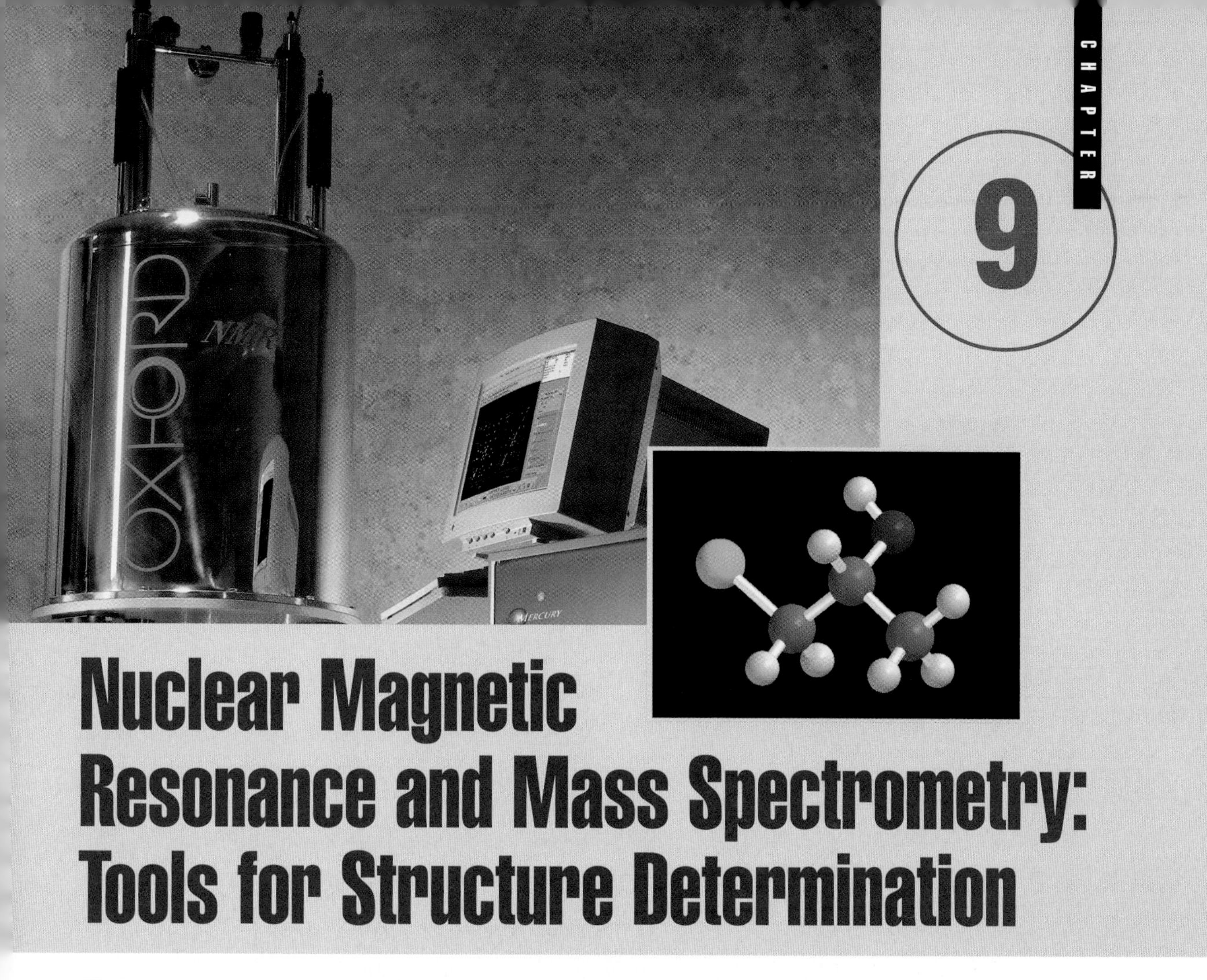

CHAPTER 9

Nuclear Magnetic Resonance and Mass Spectrometry: Tools for Structure Determination

A Thermos of Liquid Helium

Strong magnets immersed in a bath of liquid helium are used in some of the most important instruments in chemistry and medicine. The high field magnets used in these instruments are superconducting magnets, which means that the magnet coil in the superconducting magnet conducts electricity with essentially zero resistance. Instruments that use superconducting magnets include Fourier transform nuclear magnetic resonance (FT NMR) spectrometers, certain types of mass spectrometers, and magnetic resonance imaging (MRI) machines. The NMR spectra in this book were taken on a Fourier transform instrument with a superconducting magnet. Once electricity is applied to a superconducting magnet coil and the circuit is closed, the current in this loop can, theoretically, flow forever. For the coil to be superconducting, however, it must be kept very, very cold—as cold as liquid helium. Helium boils at 4.3 K (4.3 degrees above absolute zero). If the temperature of the coil ever warms above liquid helium temperature, resistance in the wire builds up and heat is generated, the liquid helium begins to rapidly boil away, and the magnetic field is quenched.

To keep the superconducting coil bathed in liquid helium, the entire magnet is housed in what is a called a Dewar chamber (named after James Dewar, a Scottish chemist who invented the original vacuum-jacketed vessel). The Dewar chamber is essentially a giant thermos bottle. In a superconducting magnet, the innermost chamber of the Dewar holds liquid helium. In this chamber rests the superconducting magnet coil. Surrounding the inner compartment of helium is a concentric chamber with liquid nitrogen (which has a boiling point of 77.4 K), and surrounding that is a vacuum jacket. Ordinary thermos bottles do not use liquid helium or nitrogen, of course, but they do rely on the insulating properties of a vacuum. Superconducting magnets, on the other hand, rely on their operators to feed them liquid nitrogen and liquid helium periodically.

The superconducting magnet of many research-grade FT NMR spectrometers has a magnetic field strength of 14 tesla (approximately 140,000 times as strong as Earth's magnetic field). In FT NMR spectrometers, the bigger the magnet, the better the instrument, and the more it costs. A stronger magnet means greater sample sensitivity and less complicated signals. In mass spectrometers, most common instruments use regular electromagnets, permanent magnets, or devices called quadrupoles; ICR (ion cyclotron resonance) mass spectrometers, however, use superconducting magnets. The principles of ion cyclotron resonance cause ICR mass spectrometers to be able to achieve extremely high mass resolution. Magnetic resonance imaging machines, which were developed from principles used in NMR spectrometers, typically use superconducting magnets with a field strength of 1–2 tesla (approximately 10,000–20,000 times Earth's magnetic field). One comforting fact about MRI instruments and their superconducting magnets is that the sample (human subject) is *not* placed in a chamber at 4.3 degrees above absolute zero. The subject, instead, rests comfortably at ambient temperature. Just as in FT NMR spectrometers, the part of the instrument where the magnetic field is focused is at approximately room temperature (except for special experiments in NMR).

9.1 Introduction

The structure of a molecule determines its physical properties, reactivity, and biological activity. Throughout this book we emphasize molecular structure as a basis for understanding mechanisms, predicting physical properties, and organizing our knowledge of reactivity. But how are molecular structures determined? One way is through the use of spectroscopic methods. Although there are other approaches to determining the structure of a molecule, such as confirmation by independent synthesis or correlation with known materials, spectroscopic tools are usually most expedient.

Spectroscopy is the study of the interaction of energy with matter. When energy is applied to matter, it can be absorbed, be emitted, cause a chemical change, or be transmitted. In this chapter we shall see how detailed information about molecular structure can be obtained by interpreting results from the interaction of energy with molecules. In our study of nuclear magnetic resonance (NMR) spectroscopy we shall focus our attention on energy absorption by molecules that have been placed in a strong magnetic field. When we study mass spectrometry (MS), we shall learn how molecular structure can be probed by bombarding molecules with a beam of high-energy electrons. These two techniques (NMR and MS) are a powerful combination for elucidating the structures of organic molecules. Together with infrared (IR) spectroscopy (Section 2.16), these methods comprise the typical array of spectroscopic tools used by organic chemists. Later, we shall briefly discuss how gas chromatography (GC) is linked with mass spectrometry in GC/MS instruments to obtain mass spectrometric data from individual components of a mixture.

9.2 The Electromagnetic Spectrum

Electromagnetic radiation is one type of energy. The names of most forms of electromagnetic energy have become familiar terms. The *X-rays* used in medicine, the *light* that we see, the *ultraviolet* (UV) rays that produce sunburns, and the *radio* and *radar* waves used in communication are all different forms of the same phenomenon: electromagnetic radiation.

According to quantum mechanics, electromagnetic radiation has a dual and seemingly contradictory nature. Electromagnetic radiation has the properties of both a wave and a particle. Electromagnetic radiation can be described as a wave occurring simultaneously in electrical and magnetic fields. It can also be described as if it consisted of particles called quanta or photons. Different experiments disclose these two different aspects of electromagnetic radiation. They are not seen together in the same experiment.

A wave is usually described in terms of its **wavelength** (λ) or its **frequency** (ν). A simple wave is shown in Fig. 9.1. The distance between consecutive crests (or troughs) is the wavelength. The number of full cycles of the wave that pass a given point each second, as the wave moves through space, is called the *frequency* and is measured in cycles per second (cps), or **hertz (Hz).***

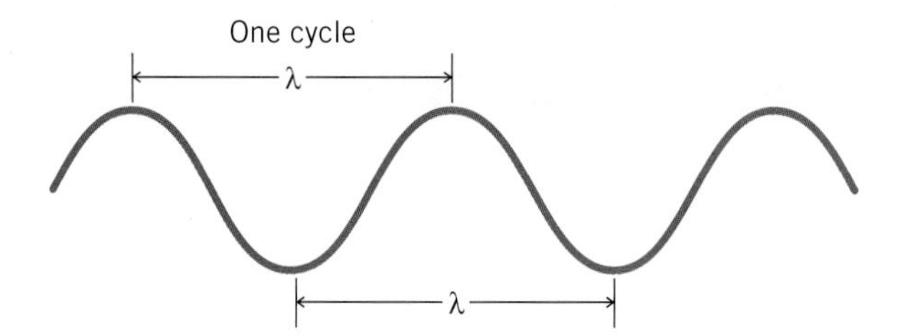

FIGURE 9.1 A simple wave and its wavelength, λ.

All electromagnetic radiation travels through a vacuum at the same velocity. This velocity (c), called the velocity of light, is 2.99792458×10^8 m s^{-1} and $c = \lambda\nu$. The wavelengths of electromagnetic radiation are expressed either in meters (m), millimeters (1 mm = 10^{-3} m), micrometers (1 μm = 10^{-6} m), or nanometers (1 nm = 10^{-9} m). [An older term for micrometer is *micron* (abbreviated μ) and an older term for nanometer is *millimicron.*]

The energy of a quantum of electromagnetic energy is directly related to its frequency:

$$E = h\nu$$

where h = Planck's constant, 6.63×10^{-34} J s
ν = frequency (Hz)

This means that the higher the frequency of radiation, the greater is its energy. X-rays, for example, are much more energetic than rays of visible light. The frequencies of X-rays are on the order of 10^{19} Hz, while those of visible light are on the order of 10^{15} Hz.

Since $\nu = c/\lambda$, the energy of electromagnetic radiation is inversely proportional to its wavelength:

$$E = \frac{hc}{\lambda}$$

where c = velocity of light

Thus, per quantum, electromagnetic radiation of long wavelength has low energy, whereas that of short wavelength has high energy. X-rays have wavelengths on the order of 0.1 nm, whereas visible light has wavelengths between 400 and 750 nm.†

---

*The term hertz (after the German physicist H. R. Hertz), abbreviated Hz, is used in place of the older term *cycles per second* (cps). Frequency of electromagnetic radiation is also sometimes expressed in *wavenumbers,* that is, the number of waves per centimeter.

†A convenient formula that relates wavelength (in nm) to the energy of electromagnetic radiation is the following:

$$E \text{ (in kJ mol}^{-1}) = \frac{1.20 \times 10^{-9} \text{ kJ mol}^{-1}}{\text{wavelength in nanometers}}$$

It may be helpful to point out, too, that for visible light, wavelengths (and thus frequencies) are related to what we perceive as colors. The light that we call red light has a wavelength of approximately 750 nm. The light we call violet light has a wavelength of approximately 400 nm. All of the other colors of the visible spectrum (the rainbow) lie in between these wavelengths.

The different regions of the electromagnetic spectrum are shown in Fig. 9.2. Nearly every portion of the electromagnetic spectrum from the region of X-rays to that of microwaves and radio waves has been used in elucidating structures of atoms and molecules. Although techniques differ according to the portion of the electromagnetic spectrum in which we are working, there is a consistency and unity of basic principles.

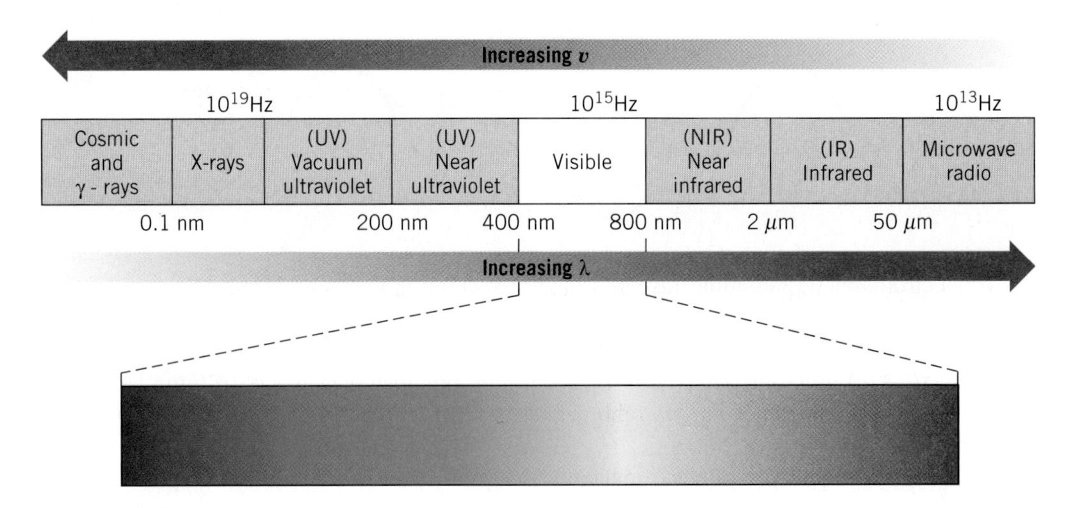

FIGURE 9.2 The electromagnetic spectrum.

In Chapter 2 we saw how the infrared region of the electromagnetic spectrum could be used to probe the stretching and bending frequencies of covalent bonds and hence indicate what functional groups are present in a molecule. We begin this chapter by considering nuclear magnetic resonance, a form of spectroscopy where absorption of radio frequency energy by hydrogen and carbon atoms in a magnetic field provides information about their molecular environment. Later, we shall discuss mass spectrometry, which involves the collision of a beam of electrons with a sample of an organic molecule to generate ions and fragments whose mass can be detected and related to the original structure. In Chapter 13 we shall see how spectroscopy using visible and ultraviolet light can provide structural information.

9.3 Nuclear Magnetic Resonance Spectroscopy

The 1952 Nobel Prize in Physics was awarded to Felix Bloch (Stanford) and Edward M. Purcell (Harvard) for their discoveries relating to nuclear magnetic resonance.

The nuclei of certain elements and isotopes behave as though they were magnets spinning about an axis. The nuclei of ordinary hydrogen (^{1}H) and carbon-13 (^{13}C) nuclei have this property. When one places a compound containing ^{1}H or ^{13}C atoms in a very strong magnetic field and simultaneously irradiates it with electromagnetic energy, the nuclei of the compound may absorb energy through a process called magnetic resonance.* This absorption of energy is quantized and produces a characteristic spectrum for the compound.

*Magnetic resonance is an entirely different phenomenon from the resonance theory of chemical structures that we have discussed in earlier chapters.

Absorption of energy does not occur unless the strength of the magnetic field and the frequency of electromagnetic radiation are at specific values.

Instruments known as NMR spectrometers allow chemists to measure the absorption of energy by ^{1}H or ^{13}C nuclei and by the nuclei of other elements that we shall discuss in Section 9.4. These instruments use very powerful magnets and irradiate the sample with electromagnetic radiation in the radio frequency (RF) region. Two types of NMR spectrometers are used, each based on a different design: continuous-wave (CW) and Fourier transform (FT) spectrometers.

9.3A Continuous-wave NMR Spectrometers

Continuous-wave (CW) NMR spectrometers irradiate a sample with electromagnetic energy of a constant frequency while the magnetic field strength is varied or swept (Fig. 9.3). When the magnetic field reaches the correct strength, the nuclei absorb energy and resonance occurs. This absorption causes a tiny electrical current to flow in a receiver coil surrounding the sample. The instrument then amplifies this current and displays it as a signal (a peak or series of peaks) along an axis calibrated in frequency units (hertz or ppm, see Section 9.6). The result is an NMR spectrum.

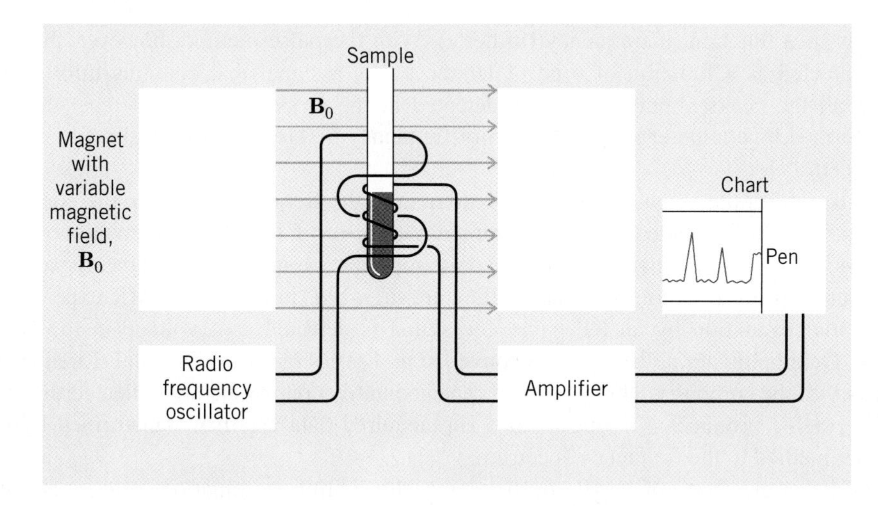

FIGURE 9.3 Diagram of a sweep (CW) NMR spectrometer.

9.3B Fourier Transform NMR Spectrometers

Fourier transform NMR instruments use superconducting magnets that have a much higher magnetic field strength than CW instruments (see the chapter opening vignette). They also employ computers that allow signal averaging followed by a mathematical calculation known as a Fourier transform. In practice, an FT instrument accumulates a number of data scans and averages them so as to cancel out random electronic noise, thus enhancing the actual NMR signals. Fourier transform instruments (Fig. 9.4) achieve very high resolution and much greater sensitivity than CW NMR instruments. Instead of slowly sweeping the magnetic field while irradiating the sample with electromagnetic energy in the RF region, as in the CW spectrometer, the FT instrument irradiates the sample with a short pulse of RF radiation (for $\sim 10^{-5}$ s). This RF pulse excites all the nuclei at once, as opposed to each nucleus being individually excited as with the sweep method. The data obtained from the pulse excitation method, however, are different from those obtained by the sweep method. One difference is that the sweep method takes at least 2–5 min to give a complete spectrum, whereas the pulse method can produce a spectrum in as little as a few seconds. Another difference is that the sweep method provides a spectrum

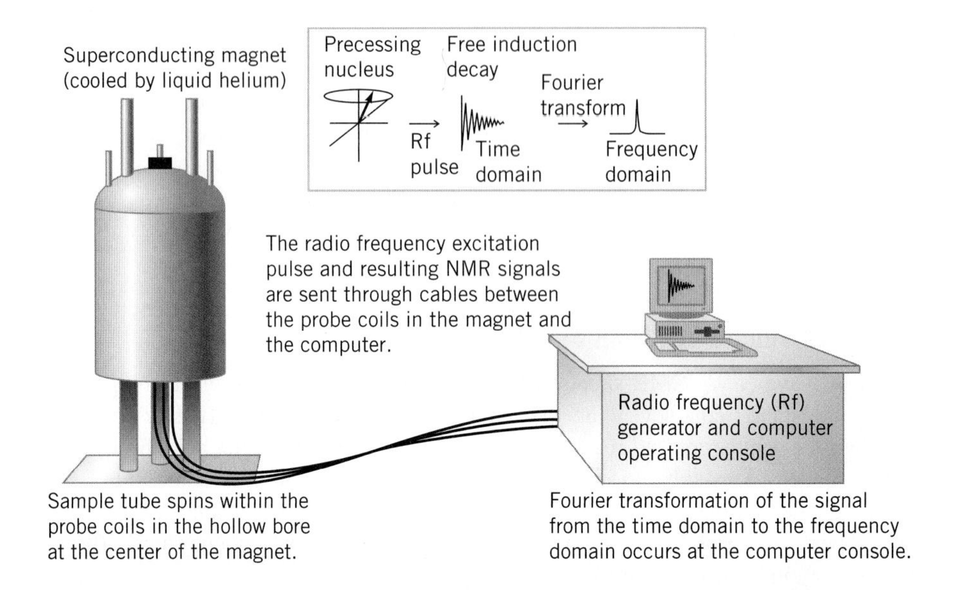
Superconducting magnet (cooled by liquid helium)
Precessing nucleus
Free induction decay
Fourier transform
Rf pulse
Time domain
Frequency domain
The radio frequency excitation pulse and resulting NMR signals are sent through cables between the probe coils in the magnet and the computer.
Radio frequency (Rf) generator and computer operating console
Sample tube spins within the probe coils in the hollow bore at the center of the magnet.
Fourier transformation of the signal from the time domain to the frequency domain occurs at the computer console.

FIGURE 9.4 Diagram of a Fourier transform NMR spectrometer.

directly as a function of frequency (in hertz). With the pulse method, however, the data are collected as a function of time. After the pulse, a signal that contains information about all the peaks simultaneously is detected in the probe. This signal must then be transformed by computer into a function of frequency before the individual peaks can be identified.

To transform the signal from the time domain to the frequency domain, the computer carries out what is called a Fourier transformation (or FT). The mathematics of this process need not be of concern to us; the only important point is that the data are sampled and stored as discrete points—that is, they are *digitized.* In a pulsed NMR experiment, then, after excitation by an RF pulse, the signal is detected as a voltage in the NMR probe. On amplification, the signal is converted to a series of data points and stored in the memory of the computer. This process is repeated until enough data are collected that signal averaging produces a strong signal. The acquired data are then transformed by the Fourier method to the frequency spectrum.

Tools for interpreting NMR spectra

We begin our study of NMR spectroscopy with a brief examination of the main features of spectra arising from hydrogen nuclei. These spectra are often called *proton magnetic resonance (PMR) spectra* or **^{1}H NMR spectra.** The features we shall explore are **chemical shift**, peak area (**integration**), and **signal splitting**. After this brief overview we shall examine these and other aspects of NMR spectroscopy in more detail.

9.3C Chemical Shift: Peak Position in an NMR Spectrum

If hydrogen nuclei were stripped of their electrons and were completely isolated from other nuclei, all hydrogen nuclei (protons) would absorb energy at the same magnetic field strength for a given frequency of electromagnetic radiation. Fortunately this is not the case. In a given molecule some hydrogen nuclei are in regions of greater electron density than others, and as a result, the nuclei (protons) absorb energy at *slightly different* magnetic field strengths. The signals for these protons consequently occur at different positions in the NMR spectrum; they are said to have different **chemical shifts.** The actual field strength at which absorption occurs (the chemical shift) is highly dependent on the magnetic environment of each proton. This magnetic environment depends on two factors: the magnetic fields generated by circulating electrons and the magnetic fields that result from other nearby protons (or other magnetic nuclei).

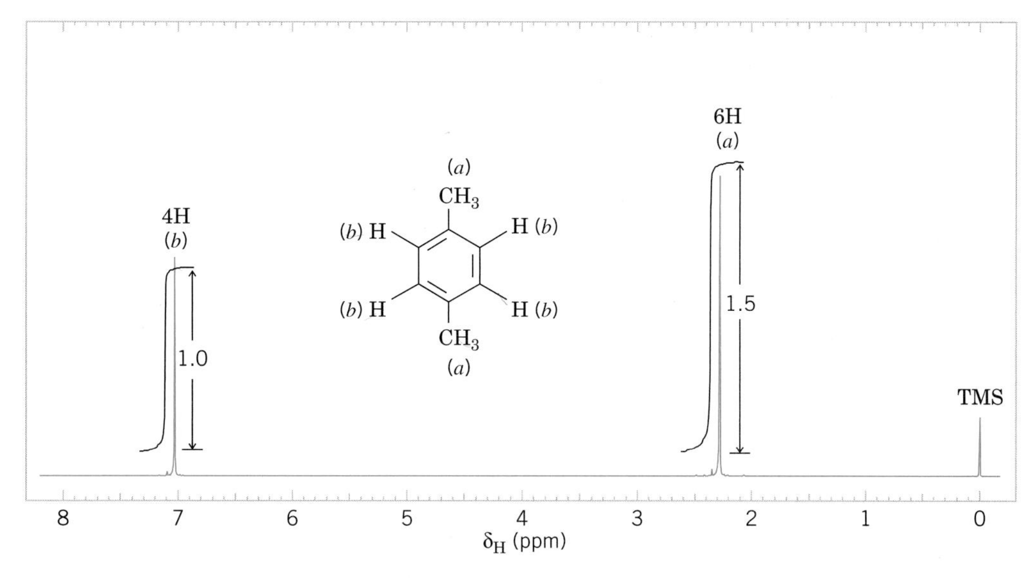

FIGURE 9.5 The 300-MHz ^{1}H NMR spectrum of 1,4-dimethylbenzene.

Figure 9.5 shows the ^{1}H NMR spectrum of 1,4-dimethylbenzene (a compound that is also called *p*-xylene).

Chemical shifts are measured along the bottom of the spectrum on a delta (δ) scale (in units of parts per million). We shall have more to say about these units later; for the moment, we need only point out that the externally applied magnetic field strength increases from left to right. A signal that occurs at a chemical shift of δ 7 occurs at a lower external magnetic field strength than one that occurs at δ 2. Signals on the left of the spectrum are said to occur **downfield** and those on the right are said to be **upfield**.

The spectrum in Fig. 9.5 shows a small signal at δ 0. This signal arises from a compound called tetramethylsilane (TMS) that has been added to the sample to allow calibration of the chemical shift scale.

The first feature we want to notice is **the relation between the number of signals in the spectrum and the number of different types of hydrogen atoms in the compound.**

Because of its symmetry, 1,4-dimethylbenzene has only *two* different types of hydrogen atoms, and it gives only *two* signals in its NMR spectrum. The two types of hydrogen atoms of 1,4-dimethylbenzene are the hydrogen atoms of the methyl groups and the hydrogen atoms of the benzene ring. The six methyl hydrogens of 1,4-dimethylbenzene *are all equivalent,* and they are in a different environment from the four *equivalent* hydrogens of the ring. The six methyl hydrogens give rise to a signal that occurs at δ 2.30. The four hydrogen atoms of the benzene ring give a signal at δ 7.05.

9.3D Integration of Peak Areas. The Integral Curve

Next we want to examine the relative magnitudes of the signals, for these are often helpful in assigning signals to particular groups of hydrogen atoms. What is important here is not necessarily the height of each peak, *but the area underneath it.* These areas, when accurately measured, are in the same ratio as the number of hydrogen atoms causing each signal. We can see in Fig. 9.5 that the area under the signal for the methyl hydrogen atoms of 1,4-dimethylbenzene (6H) is larger than that for the ring hydrogen atoms (4H). Spectrometers measure these areas automatically and plot curves, called integral curves, over each signal. The heights of the integral curves are proportional to the areas under the signals. In this instance, the ratio of heights is 1.5 : 1 or 6 : 4.

9.3E Signal Splitting

A third feature of ^1H NMR spectra that provides us with information about the structure of a compound can be illustrated if we examine the spectrum for 1,1,2-trichloroethane (Fig. 9.6).

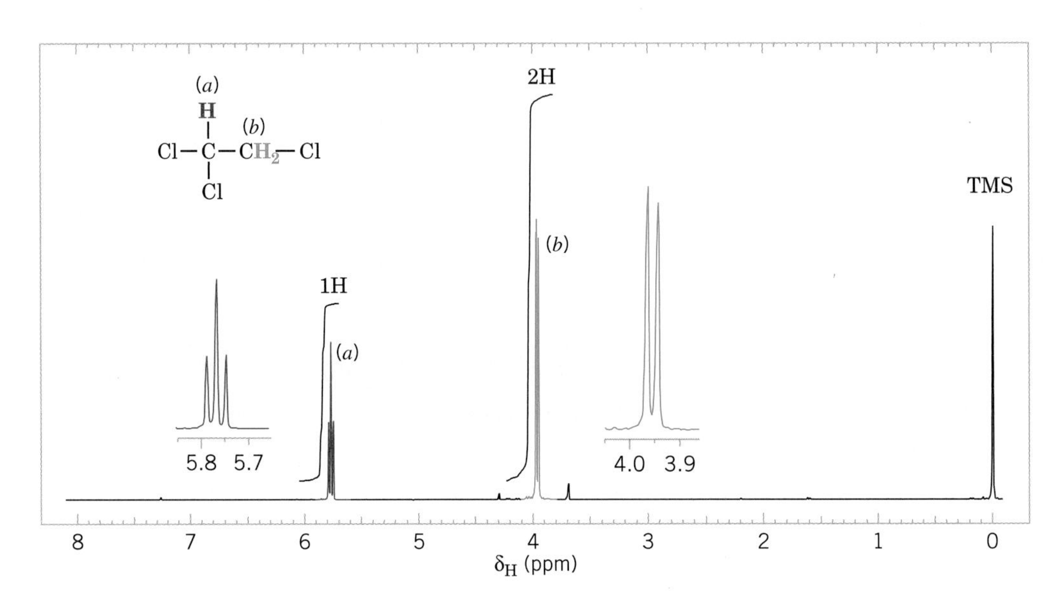

FIGURE 9.6 The 300-MHz ^1H NMR spectrum of 1,1,2-trichloroethane. Expansions of the signals are shown in the offset plots.

In Fig 9.6 we have an example of **signal splitting**. Signal splitting is a phenomenon that arises from magnetic influences of hydrogens on atoms adjacent to those bearing the hydrogen atoms causing the signal being considered. The signal (*b*) from the two equivalent hydrogen atoms of the $—CH_2Cl$ group is split into two peaks (a **doublet**) by the magnetic influence of the hydrogen of the $—CHCl_2$ group. Conversely, the signal (*a*) from the hydrogen of the $—CHCl_2$ group is split into three peaks (a **triplet**) by the magnetic influences of the two equivalent hydrogens of the $—CH_2Cl$ group. You will notice that we are careful not to use the word "peak" when we mean "signal." Due to signal splitting, a given signal may be comprised of several peaks. Hence we make a distinction in terminology.

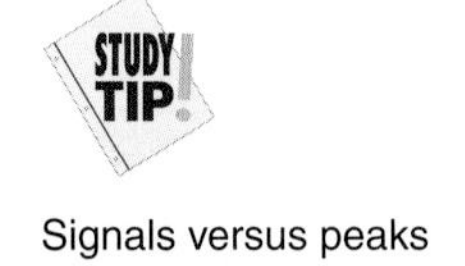

Signals versus peaks

At this point signal splitting may seem like an unnecessary complication. As we gain experience in interpreting ^1H NMR spectra, we shall find that because signal splitting occurs in a predictable way, it often provides us with important information about the structure of the compound.

Now that we have had an introduction to the important features of ^1H NMR spectra, we are in a position to consider them in greater detail.

9.4 Nuclear Spin: The Origin of the Signal

We are already familiar with the concept of electron spin and with the fact that the spins of electrons confer on them the spin quantum states of $+\frac{1}{2}$ or $-\frac{1}{2}$. Electron spin is the basis for the Pauli exclusion principle (Section 1.10); it allows us to understand how two electrons with paired spins may occupy the same atomic or molecular orbital.

The nuclei of certain isotopes also spin, and therefore these nuclei possess spin quantum numbers, I. The nucleus of ordinary hydrogen, ^1H (i.e., a proton), is like the electron;

its spin quantum number I is $\frac{1}{2}$ and it can assume either of two spin states: $+\frac{1}{2}$ or $-\frac{1}{2}$. These correspond to the magnetic moments allowed for $I = \frac{1}{2}$, $m = +\frac{1}{2}$ or $m = -\frac{1}{2}$. Other nuclei with spin quantum numbers $I = \frac{1}{2}$ are ^{13}C, ^{19}F, and ^{31}P. Some nuclei, such as ^{12}C, ^{16}O, and ^{32}S, have no spin ($I = 0$), and these nuclei do not give an NMR spectrum. Other nuclei have spin quantum numbers greater than $\frac{1}{2}$. In our treatment here, however, we shall be concerned primarily with the spectra that arise from protons and from ^{13}C, both of which have $I = \frac{1}{2}$. We shall begin with proton spectra.

Since the proton is electrically charged, the spinning proton generates a tiny magnetic moment—one that coincides with the axis of spin (Fig. 9.7). This tiny magnetic moment confers on the spinning proton the properties of a tiny bar magnet.

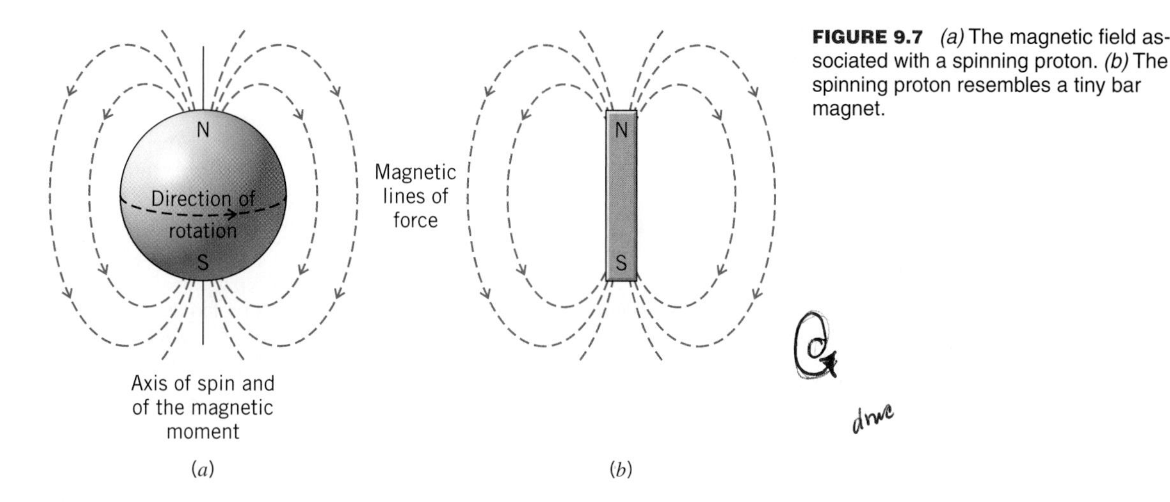

FIGURE 9.7 *(a)* The magnetic field associated with a spinning proton. *(b)* The spinning proton resembles a tiny bar magnet.

In the absence of a magnetic field (Fig. 9.8*a*), the magnetic moments of the protons of a given sample are randomly oriented. When a compound containing hydrogen (and thus protons) is placed in an applied external magnetic field, however, the protons may assume one of two possible orientations with respect to the external magnetic field. The magnetic moment of the proton may be aligned "with" the external field or "against" it (Fig. 9.8*b*). These alignments correspond to the two spin states mentioned earlier.

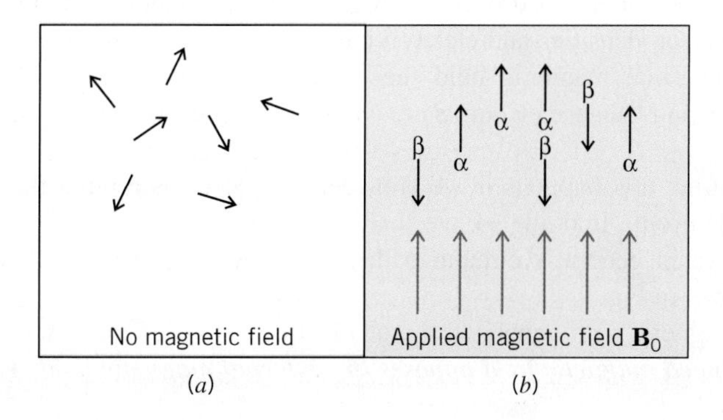

FIGURE 9.8 *(a)* In the absence of a magnetic field the magnetic moments of protons (represented by arrows) are randomly oriented. *(b)* When an external magnetic field (B_0) is applied, the protons orient themselves. Some are aligned with the applied field (α spin state) and some against it (β spin state). The difference in the number of protons aligned with and against the applied field is very small, but observable with an NMR spectrometer.

As we might expect, the two alignments of the proton in an external field are not of equal energy. When the proton is aligned with the magnetic field, its energy is lower than when it is aligned against the magnetic field.

Energy is required to "flip" the proton from its lower energy state (with the field) to its higher energy state (against the field). In an NMR spectrometer this energy is supplied by electromagnetic radiation in the RF region. When this energy absorption occurs, the nuclei are said to be *in resonance* with the electromagnetic radiation. The energy required is proportional to the strength of the magnetic field (Fig. 9.9). One can show by relatively simple calculations that, in a magnetic field of approximately 7.04 tesla, for example, electromagnetic radiation of 300×10^6 cps (300 MHz) supplies the correct amount of energy for protons.* The proton NMR spectra given in this chapter are 300-MHz spectra.

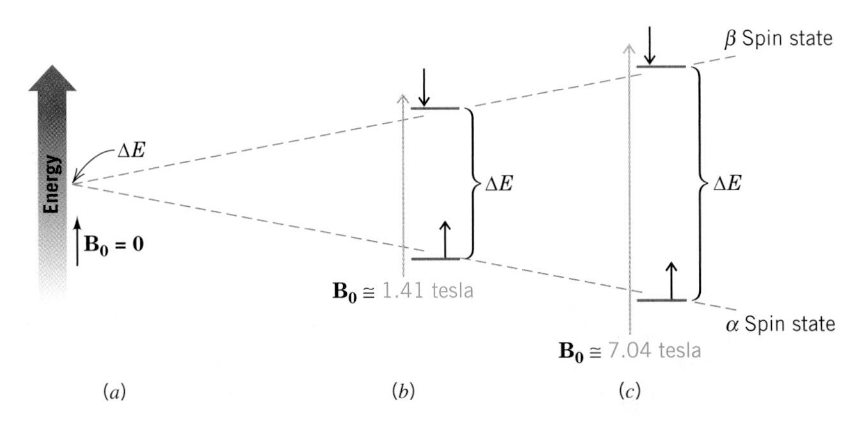

FIGURE 9.9 The energy difference between the two spin states of a proton depends on the strength of the applied external magnetic field, B_0. *(a)* If there is no applied field (B = 0), there is no energy difference between the two states. *(b)* If $B_0 \cong 1.41$ tesla, the energy difference corresponds to that of electromagnetic radiation of 60×10^6 Hz (60 MHz). *(c)* In a magnetic field of approximately 7.04 tesla, the energy difference corresponds to electromagnetic radiation of 300×10^6 Hz (300 MHz). Instruments are available that operate at these and even higher frequencies (as high as 800 MHz–1 GHz).

9.5 Shielding and Deshielding of Protons

All protons do not absorb energy at the same frequency in a given external magnetic field. The two spectra that we examined earlier demonstrate this for us. The aromatic protons of 1,4-dimethylbenzene absorb at lower frequency (δ 7.05); the various alkyl protons of 1,4-dimethylbenzene and 1,1,2-trichloroethane all absorb at higher frequency.

The general position of a signal in a NMR spectrum—that is, the frequency of radiation required to bring about absorption of energy at a given magnetic field strength—can be related to electron densities and electron circulations in the compounds. Under the influence of an external magnetic field the electrons move in certain preferred paths. Because they do, and because electrons are charged particles, they generate tiny magnetic fields.

We can see how this happens if we consider the electrons around the proton in a σ bond of a C—H group. In doing so, we shall oversimplify the situation by assuming that σ electrons move in generally circular paths. The magnetic field generated by these σ electrons is shown in Fig. 9.10.

The small magnetic field generated by the electrons is called **an induced field.** ***At the proton, the induced magnetic field opposes the external magnetic field.*** This means that

*The relationship between the frequency of the radiation (ν) and the strength of the magnetic field ($\mathbf{B}_0$) is

$$\nu = \frac{\gamma \mathbf{B}_0}{2\pi}$$

where γ is the magnetogyric (or gyromagnetic) ratio. For a proton, $\gamma = 26.753$ rad s^{-1} tesla^{-1}.

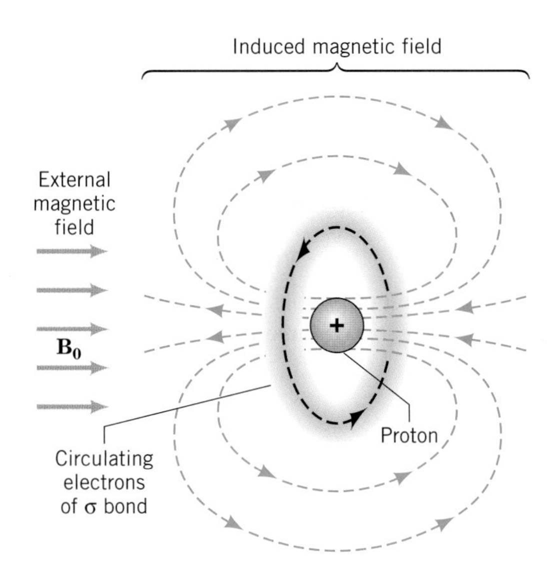

FIGURE 9.10 Circulations of the electrons of a C—H bond under the influence of an external magnetic field. The electron circulations generate a small magnetic field (an induced field) that shields the proton from the external field.

the actual magnetic field sensed by the proton is slightly less than the external field. The electrons are said *to shield* the proton.

A proton strongly shielded by electrons does not, of course, absorb at the same frequency as a proton that is less shielded by electrons. A shielded proton will absorb *at higher frequency (upfield)* (Fig. 9.11).

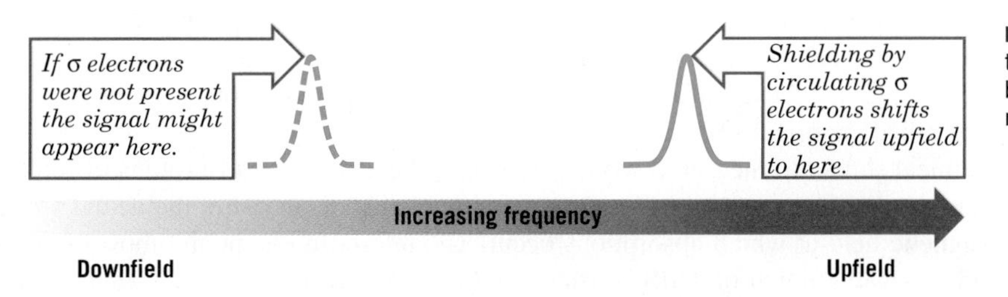

FIGURE 9.11 Shielding by σ electrons causes ¹H NMR absorptions to be shifted to higher frequency external magnetic field strengths.

The extent to which a proton is shielded by the circulation of σ electrons depends on the relative electron density around the proton. This electron density depends largely on the presence or absence of electronegative groups. Electronegative groups withdraw electron density from the C—H bond, particularly if they are attached to the same carbon. We can see an example of this effect in the spectrum of 1,1,2-trichloroethane (Fig. 9.6). The proton of C1 absorbs at a lower magnetic field strength (δ 5.77) than the protons of C2 (δ 3.95). Carbon 1 bears two highly electronegative chloro groups, whereas C2 bears only one. The protons of C2, consequently, are more effectively shielded because the σ electron density around them is greater.

The circulations of delocalized π electrons generate magnetic fields that can either **shield** or **deshield** nearby protons. Whether shielding or deshielding occurs depends on the location of the proton in the *induced* field. The aromatic protons of benzene derivatives (Section 14.7B) are *deshielded* because their locations are such that the induced magnetic field reinforces the applied magnetic field.

Because of this deshielding effect, the absorption of energy by aromatic protons occurs downfield at relatively low magnetic field strength. The protons of benzene itself absorb at δ 7.27. The aromatic protons of 1,4-dimethylbenzene (Fig. 9.5) absorb at δ 7.05.

Magnetic fields created by circulating π electrons *shield* the protons of ethyne (and other terminal alkynes), causing them to absorb at higher magnetic field strengths than we might otherwise expect. If we were to consider *only* the relative electronegativities of carbon in its three hybridization states (Section 3.7A), we might expect the following order of protons attached to each type of carbon:

$$\text{(low field strength)} \qquad sp < sp^2 < sp^3 \qquad \text{(high field strength)}$$

In fact, protons of terminal alkynes absorb between δ 2.0 and δ 3.0, and the order is

$$\text{(low field strength)} \qquad sp^2 < sp < sp^3 \qquad \text{(high field strength)}$$

This upfield shift of the absorption of protons of terminal alkynes is a result of shielding produced by the circulating π electrons of the triple bond. The origin of this shielding is illustrated in Fig. 9.12.

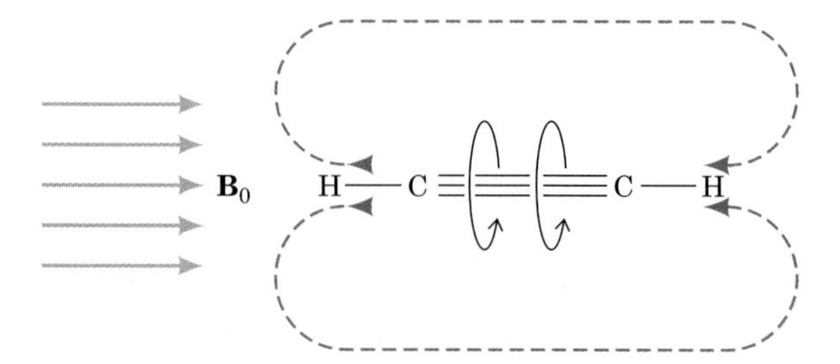

FIGURE 9.12 The shielding of protons of ethyne by π-electron circulations. Shielding causes protons attached to the *sp* carbons to absorb further upfield than vinylic protons.

9.6 The Chemical Shift

We see now that shielding and deshielding effects cause the absorptions of protons to be shifted from the position at which a bare proton would absorb (i.e., a proton stripped of its electrons). Since these shifts result from the circulation of electrons in *chemical* bonds, they are called **chemical shifts.**

Chemical shifts are measured with reference to the absorption of protons of reference compounds. A reference is used because it is impractical to measure the actual value of the magnetic field at which absorptions occur. The reference compound most often used is TMS. A small amount of TMS is usually added to the sample, and its signal is used to establish the zero point on the delta (δ) scale.

$$Si(CH_3)_4$$

Tetramethylsilane
(TMS)

Tetramethylsilane was chosen as a reference compound for several reasons. It has 12 equivalent hydrogen atoms, and, therefore, a very small amount of TMS gives a relatively large signal. Because the hydrogen atoms are all equivalent, they give a *single signal.* Since silicon is less electronegative than carbon, the protons of TMS are in regions of high electron density. They are, as a result, highly shielded, and the signal from TMS occurs in a region of the spectrum where few other hydrogen atoms absorb. Thus, their signal seldom interferes with the signals from other hydrogen atoms. Tetramethylsilane, like an alkane, is relatively inert. It is also volatile, having a boiling point of 27°C. After the spectrum has been determined, the TMS can be removed from the sample easily by evaporation.

The chemical shift of a proton, when expressed in hertz, is proportional to the strength of the external magnetic field. Since spectrometers with different magnetic field strengths are commonly used, it is desirable to express chemical shifts in a form that is independent of the strength of the external field. This can be done easily by dividing the chemical shift by the frequency of the spectrometer, with both numerator and denominator of the fraction expressed in frequency units (hertz). Since chemical shifts are always very small (typically <5000 Hz) compared with the total field strength (commonly the equivalent of 60, 300, or 600 *million* hertz), it is convenient to express these fractions in units of *parts per million* (ppm). This is the origin of the delta scale for the expression of chemical shifts relative to TMS:

$$\delta = \frac{(\text{observed shift from TMS in hertz}) \times 10^6}{(\text{operating frequency of the instrument in hertz})}$$

For example, the chemical shift for benzene protons is 2181 Hz when the instrument is operating at 300 MHz. Therefore,

$$\delta = \frac{2181 \text{ Hz} \times 10^6}{300 \times 10^6 \text{ Hz}} = 7.27$$

The chemical shift of benzene protons in a 60-MHz instrument is 436 Hz:

$$\delta = \frac{436 \text{ Hz} \times 10^6}{60 \times 10^6 \text{ Hz}} = 7.27$$

Thus, the chemical shift expressed in ppm is the same whether measured with an instrument operating at 300 or 60 MHz (or any other field strength).

Figure 9.13 gives the *approximate* values of proton chemical shifts for some common hydrogen-containing groups.

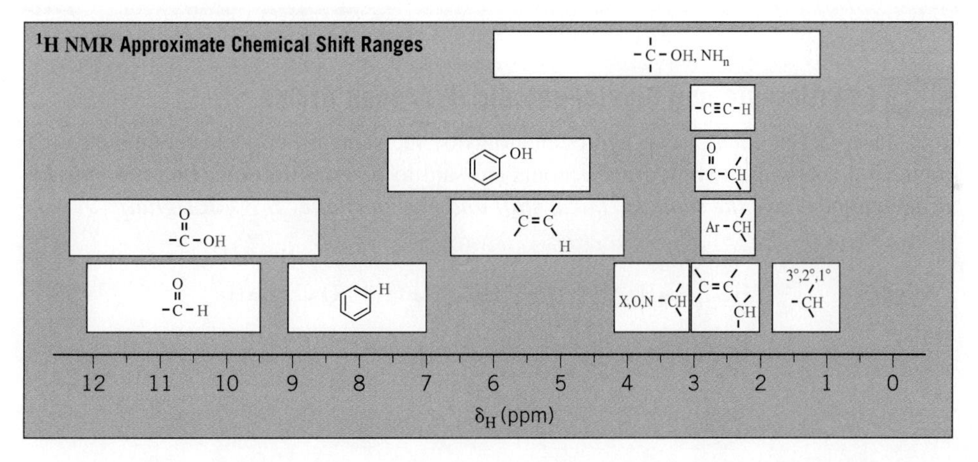

FIGURE 9.13 Approximate proton chemical shifts.

9.7 Chemical Shift Equivalent and Nonequivalent Protons

Two or more protons that are in identical environments have the same chemical shift and, therefore, give only one ^{1}H NMR signal. How do we know when protons are in the same environment? For most compounds, protons that are in the same environment are also equivalent in chemical reactions. That is, **chemically equivalent** protons are **chemical shift equivalent** in ^{1}H NMR spectra.

9.7A Homotopic Hydrogen Atoms

A simple way to decide whether two or more protons in a given compound are chemical shift equivalent is to replace each hydrogen in turn by some other group. The group may be real or imaginary. If, in making these replacements, you get the same compound, then the hydrogens being replaced are said to be **chemically equivalent** or **homotopic,** and **homotopic atoms (or groups) are chemical shift equivalent.**

Consider 2-methylpropene as an example:

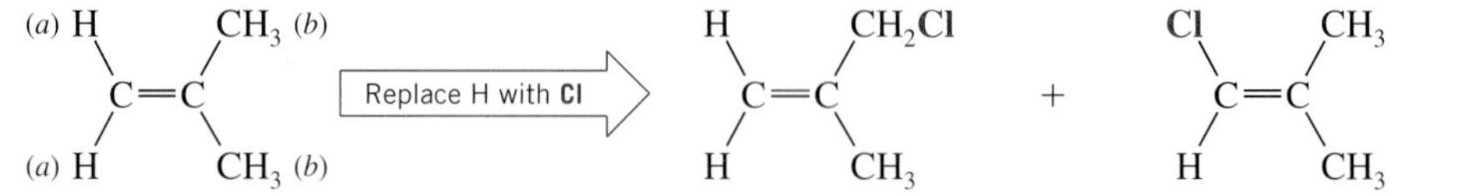

(a) H, (a) H, CH_3 (b), CH_3 (b) — C=C —[Replace H with Cl]→ H, H, CH_2Cl, CH_3 C=C + Cl, H, CH_3, CH_3 C=C

3-Chloro-2-methylpropene **1-Chloro-2-methylpropene**

Here there are two sets of homotopic hydrogens. The six methyl hydrogens (*b*) form one set; replacing any one of them with chlorine, for example, leads to the same compound, 3-chloro-2-methylpropene. The two vinyl hydrogens (*a*) form another set; replacing either of these leads to 1-chloro-2-methylpropene. 2-Methylpropene, therefore, gives two ^1H NMR signals.

PROBLEM 9.1 Carry out a similar analysis for 1,4-dimethylbenzene (Section 9.3C) and show that it has two sets of chemical shift equivalent protons, explaining why its spectrum (Fig. 9.5) has only two signals.

PROBLEM 9.2 How many signals would each compound give in its ^1H NMR spectrum?

(a) Ethane
(b) Propane
(c) *tert*-Butyl methyl ether
(d) 2,3-Dimethyl-2-butene
(e) (*Z*)-2-Butene
(f) (*E*)-2-Butene

9.7B Enantiotopic and Diastereotopic Hydrogen Atoms

If replacement of each of two hydrogen atoms by the same group yields compounds that are enantiomers, the two hydrogen atoms are said to be **enantiotopic.** ***Enantiotopic hydrogen atoms have the same chemical shift and give only one ^1H NMR signal:****

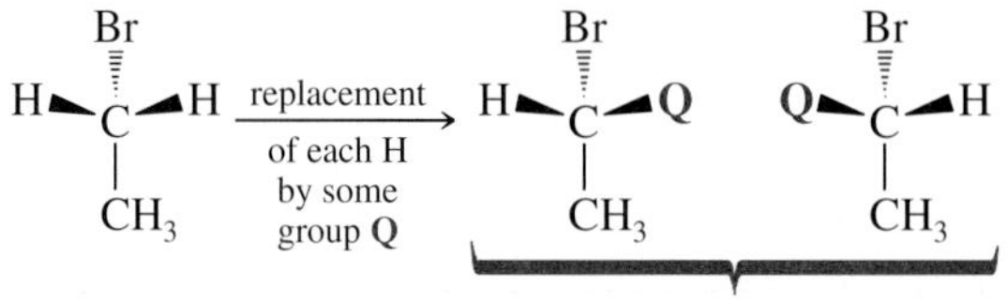

Br, H, H, CH_3 on C —replacement of each H by some group Q→ Br, H, Q, CH_3 on C and Br, Q, H, CH_3 on C

Enantiomers

The two hydrogen atoms of the $—CH_2Br$ group of ethyl bromide are enantiotopic. Ethyl bromide, then, gives two signals in its ^1H NMR spectrum. The three equivalent protons of the $—CH_3$ group give one signal; the two enantiotopic protons of the $—CH_2Br$ group give the other signal. [The ^1H NMR spectrum of ethyl bromide, as we shall see, actually consists of seven peaks (three in one signal, four in the other). This is a result of signal splitting, which is explained in Section 9.8.]

If replacement of each of two hydrogen atoms by a group, **Q**, gives compounds that are diastereomers, the two hydrogens are said to be **diastereotopic.** Except for accidental coincidence, ***diastereotopic protons do not have the same chemical shift and give rise to different ^1H NMR signals.***

---

*Enantiotopic hydrogen atoms may not have the same chemical shift if the compound is dissolved in a chiral solvent. However, most ^1H NMR spectra are determined using achiral solvents, and in this situation enantiotopic protons have the same chemical shift.

The two protons of the $=CH_2$ group of chloroethene are diastereotopic:

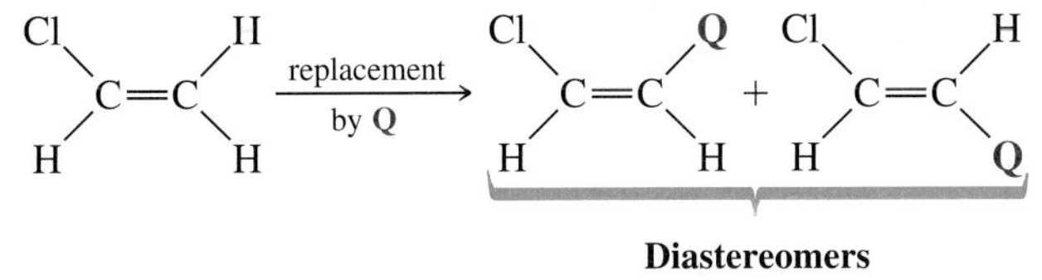

replacement by Q

Diastereomers

Chloroethene, then, should give signals from three nonequivalent protons: one for the proton of the $ClCH=$ group, and one for each of the diastereotopic protons of the $=CH_2$ group.

The two methylene ($-CH_2-$) protons of 2-butanol are also diastereotopic. We can illustrate this with one enantiomer of 2-butanol in the following way:

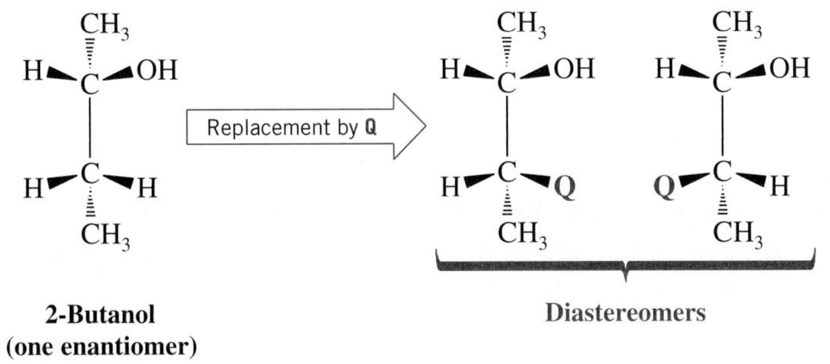

Replacement by **Q**

2-Butanol (one enantiomer)

Diastereomers

These two protons have different chemical shifts and give two signals in the ^1H NMR spectrum. The two signals may be close enough to overlap, however.

PROBLEM 9.3

(a) Show that replacing each of the two methylene protons of the other 2-butanol enantiomer by **Q** also leads to a pair of diastereomers.
(b) How many chemically different kinds of protons are there in 2-butanol?
(c) How many ^1H NMR signals would you expect to find in the spectrum of 2-butanol?

PROBLEM 9.4

How many ^1H NMR signals would you expect from each of the following compounds?

(a) $CH_3CH_2CH_2CH_3$
(b) CH_3CH_2OH
(c) $CH_3CH{=}CH_2$
(d) *trans*-2-Butene
(e) 1,2-Dibromopropane
(f) 1,1-Dimethylcyclopropane
(g) *trans*-1,2-Dimethylcyclopropane
(h) *cis*-1,2-Dimethylcyclopropane
(i) 1-Pentene
(j) 1-Chloro-2-propanol

9.8 Signal Splitting: Spin–Spin Coupling

Signal splitting is caused by the effect of magnetic fields of protons on nearby atoms. We have seen an example of signal splitting in the spectrum of 1,1,2-trichloroethane (Fig. 9.6). The signal from the two equivalent protons of the $-CH_2Cl$ group of 1,1,2-trichloroethane is split into two peaks by the single proton of the $CHCl_2-$ group. The signal from the proton of the $CHCl_2-$ group is split into three peaks by the two protons of the $-CH_2Cl$ group. This is further illustrated in Fig. 9.14.

Signal splitting arises from a phenomenon known as **spin–spin coupling,** which we shall soon examine. Spin–spin coupling effects are transferred primarily through the bonding electrons and ***are not usually observed if the coupled protons are separated by***

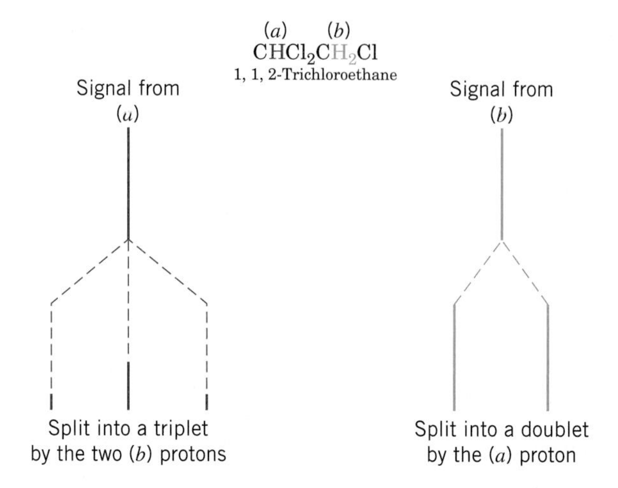

FIGURE 9.14 Signal splitting in 1,1,2-trichloroethane.

more than three σ bonds.* Thus, we observe signal splitting from the protons of *adjacent σ-bonded* atoms as in 1,1,2-trichloroethane (Fig. 9.6). However, we would not observe splitting of either signal of *tert*-butyl methyl ether (see following structure) because the protons labeled (*b*) are separated from those labeled (*a*) by more than three σ bonds. Both signals from *tert*-butyl methyl ether are singlets:

$$\begin{array}{c} \scriptstyle(a) \\ CH_3 \\ \scriptstyle(a) \quad | \quad\quad\quad\quad \scriptstyle(b) \\ CH_3—C—O—CH_3 \\ | \\ CH_3 \\ \scriptstyle(a) \end{array}$$

***tert*-Butyl methyl ether**
(no signal splitting)

Signal splitting is not observed for protons that are chemically equivalent (homotopic) or enantiotopic. That is, signal splitting does not occur between protons that have ***exactly the same chemical shift.*** Thus, we would not expect, and do not find, signal splitting in the signal from the six equivalent hydrogen atoms of ethane:

CH_3CH_3 (no signal splitting)

Nor do we find signal splitting occurring from enantiotopic protons of methoxyacetonitrile (Fig. 9.15):

> There is a subtle distinction between *spin–spin coupling* and signal splitting. Spin–spin coupling often occurs between sets of protons that have the same chemical shift (and this coupling can be detected by methods that we shall not go into here). However, spin–spin coupling *leads to signal splitting only when the sets of protons have different chemical shifts.*

Let us now explain how signal splitting arises from coupled sets of protons that are not chemical shift equivalent.

We have seen that protons can be aligned in only two ways in an external magnetic field: with the field or against it. Therefore, the magnetic moment of a proton on an adjacent atom may affect the magnetic field at the proton whose signal we are observing in

*Long-range coupling can be observed over more than three bond lengths in some conformationally inflexible molecules and systems where π bonds are involved.

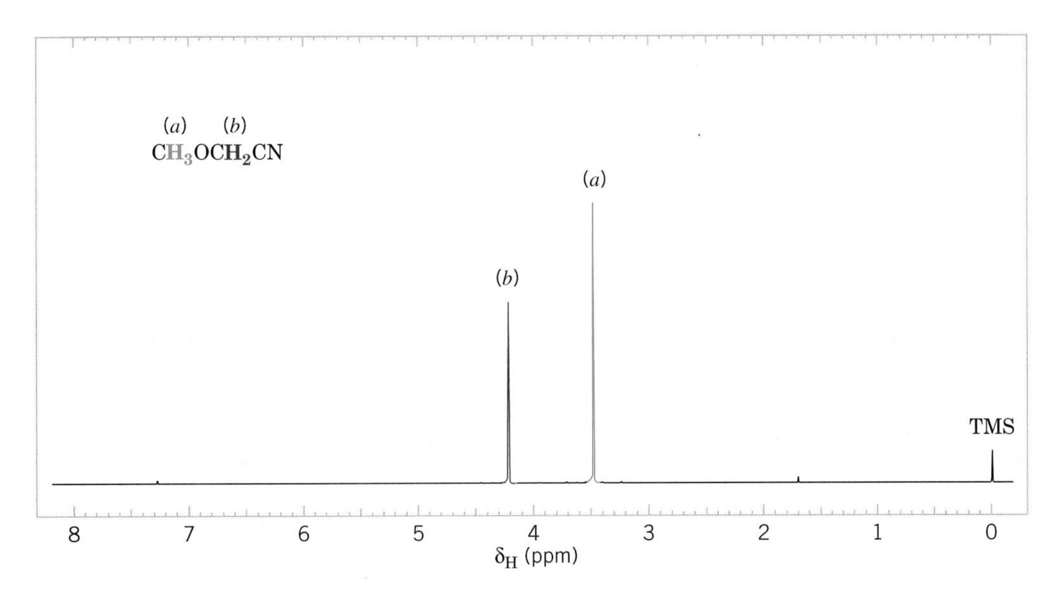

FIGURE 9.15 The 300-MHz ^{1}H NMR spectrum of methoxyacetonitrile. The signal of the enantiotopic protons *(b)* is not split.

only one of two ways. The occurrence of these two slightly different effects causes the appearance of two smaller peaks, one somewhat upfield from where the signal would have occurred and the other peak somewhat downfield.

Figure 9.16 shows how two possible orientations of a neighboring proton, H_b, split the signal of the proton H_a. (H_b and H_a are not equivalent.)

The separation of these peaks in frequency units is called the **coupling constant** and is abbreviated J_{ab}. Coupling constants are generally reported in hertz. Because coupling is

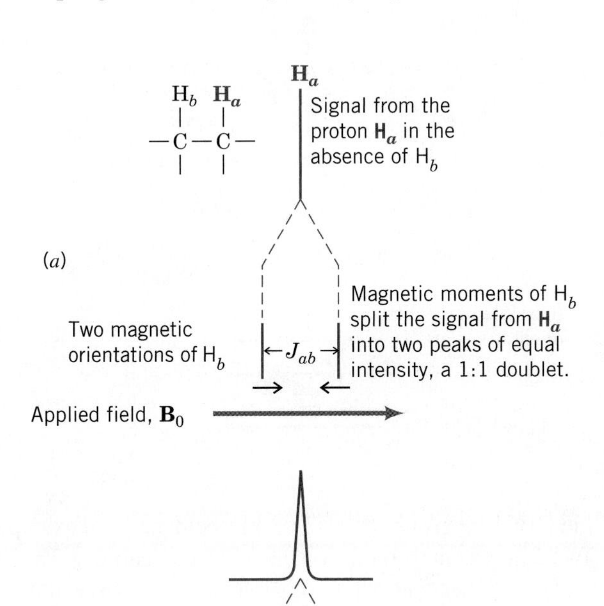

FIGURE 9.16 Signal splitting arising from spin–spin coupling with one nonequivalent proton of a neighboring hydrogen atom. A theoretical analysis is shown in *(a)* and the actual appearance of the spectrum in *(b)*. The distance between the centers of the peaks of the doublet is called the coupling constant, J_{ab}. The term J_{ab} is expressed in hertz. The magnitudes of coupling constants are *not* dependent on the magnitude of the applied field and their values (in hertz) are the same, regardless of the operating frequency of the spectrometer.

caused entirely by internal forces, the magnitudes of coupling constants *are not* dependent on the magnitude of the applied magnetic field. Coupling constants measured (in hertz) on an instrument operating at 60 MHz will be the same as those measured on an instrument operating at 300 MHz or any other magnetic field strength.

When we determine ^1H NMR spectra we are, of course, observing effects produced by billions of molecules. Since the difference in energy between the two possible orientations of H_b protons is very small, the two orientations will be present in roughly (but not exactly) equal numbers. The signal that we observe from H_a is, therefore, split into two peaks of roughly equal intensity, a 1 : 1 ***doublet.***

PROBLEM 9.5 Sketch the ^1H NMR spectrum of $CHBr_2CHCl_2$. Which signal would you expect to occur at lower magnetic field strength: that of the proton of the $CHBr_2$— group or of the —$CHCl_2$ group? Why?

Two equivalent protons on an adjacent carbon atom (or on adjacent equivalent carbon atoms) split the signal from an absorbing proton into a 1 : 2 : 1 ***triplet.*** Figure 9.17 illustrates how this pattern occurs.

In compounds of either type shown in Figure 9.17, both protons may be aligned with the applied field. This orientation causes a peak to appear at a lower applied field strength than would occur in the absence of the two hydrogen atoms H_b. Conversely, both protons may be aligned against the applied field. This orientation of H_b protons causes a peak to appear at higher applied field strengths than would occur in their absence. Finally, there are two ways in which the two protons may be aligned in which one opposes the applied field and one reinforces it. These arrangements do not displace the signal. Since the probability of this last arrangement is twice that of either of the other two, the center peak of the triplet is twice as intense.

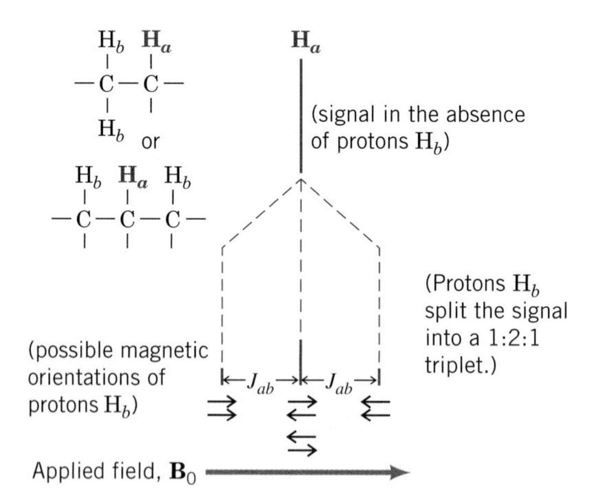

FIGURE 9.17 Two equivalent protons (H_b) on an adjacent carbon atom split the signal from H_a into a 1:2:1 triplet.

The proton of the $CHCl_2$— group of 1,1,2-trichloroethane is an example of a proton of the type having two equivalent protons on an adjacent carbon. The signal from the $CHCl_2$— group (Fig. 9.6) appears as a 1 : 2 : 1 triplet, and, as we would expect, the signal of the —CH_2Cl group of 1,1,2-trichloroethane is split into a 1 : 1 doublet by the proton of the $CHCl_2$— group.

The spectrum of 1,1,2,3,3-pentachloropropane (Fig. 9.18) is similar to that of 1,1,2-trichloroethane in that it also consists of a 1 : 2 : 1 triplet and a 1 : 1 doublet. The two hy-

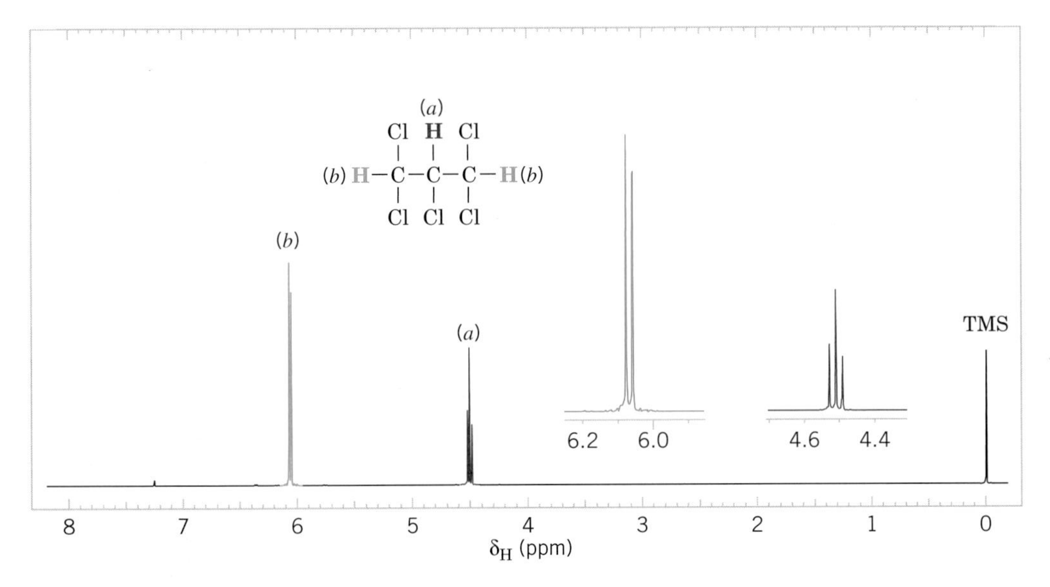

FIGURE 9.18 The 300-MHz ^{1}H NMR spectrum of 1,1,2,3,3-pentachloropropane. Expansions of the signals are shown in the offset plots.

PROBLEM 9.6 The relative positions of the doublet and triplet of 1,1,2-trichloroethane (Fig. 9.6) and 1,1,2,3,3-pentachloropropane (Fig. 9.18) are reversed. Explain this.

drogen atoms H_b of 1,1,2,3,3-pentachloropropane are equivalent even though they are on separate carbon atoms.

Three equivalent protons (H_b) on a neighboring carbon split the signal from the H_a proton into a 1 : 3 : 3 : 1 ***quartet.*** This pattern is shown in Fig. 9.19.

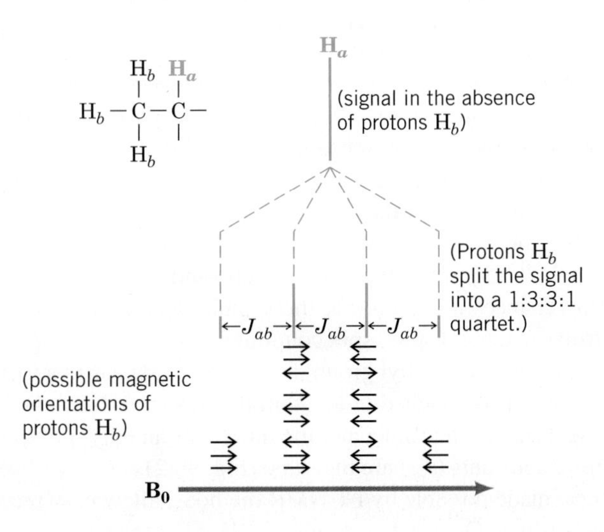

FIGURE 9.19 Three equivalent protons (H_b) on an adjacent carbon split the signal from H_a into a 1:3:3:1 quartet.

The signal from two equivalent protons of the $—CH_2Br$ group of ethyl bromide (Fig. 9.20) appears as a 1 : 3 : 3 : 1 quartet because of this type of signal splitting. The three equivalent protons of the $CH_3—$ group are split into a 1 : 2 : 1 triplet by the two protons of the $—CH_2Br$ group.

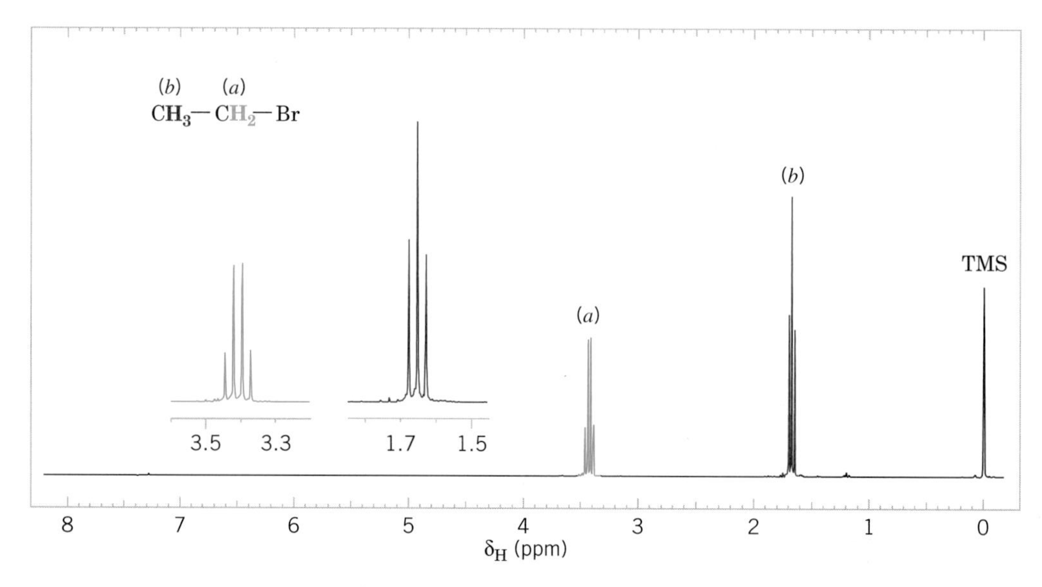

FIGURE 9.20 The 300-MHz ^{1}H NMR spectrum of ethyl bromide. Expansions of the signals are shown in the offset plots.

The kind of analysis that we have just given can be extended to compounds with even larger numbers of equivalent protons on adjacent atoms. These analyses show that ***if there are n equivalent protons on adjacent atoms, these will split a signal into n + 1 peaks.*** (We may not always see all of these peaks in actual spectra, however, because some of them may be very small.)

PROBLEM 9.7 Sketch the ^{1}H NMR spectrum you would expect for the compound $(Cl_2CH)_3CH$, showing the splitting patterns and relative position of each signal.

PROBLEM 9.8 Propose a structure for each of the compounds whose spectra are shown in Fig. 9.21, and account for the splitting pattern of each signal.

Reciprocity of coupling constants

The splitting patterns shown in Fig. 9.21 are fairly easy to recognize because in each compound there are only two sets of nonequivalent hydrogen atoms. One feature present in all spectra, however, helps us recognize splitting patterns in more complicated spectra: the **reciprocity of coupling constants.**

The separation of the peaks in hertz gives us the value of the coupling constants. Therefore, if we look for doublets, triplets, quartets, and so on, that have *the same coupling constants,* the chances are good that these multiplets are related to each other because they arise from reciprocal spin–spin couplings.

The two sets of protons of an ethyl group, for example, appear as a triplet and a quartet as long as the ethyl group is attached to an atom that does not bear any hydrogen atoms. The spacings of the peaks of the triplet and the quartet of an ethyl group will be the same because the coupling constants (J_{ab}) are the same (Fig. 9.22).

Other techniques, made possible by FT NMR methods, allow easy recognition of complex splitting relationships. One such technique is ^{1}H–^{1}H **co**rrelation **s**pectroscop**y**, called COSY (Section 9.11).

Proton NMR spectra have other features, however, that are not at all helpful when we try to determine the structure of a compound. For example:

1. Signals may overlap. This happens when the chemical shifts of the signals are very nearly the same. In the 60-MHz spectrum of ethyl chloroacetate (Fig. 9.23, top) we see

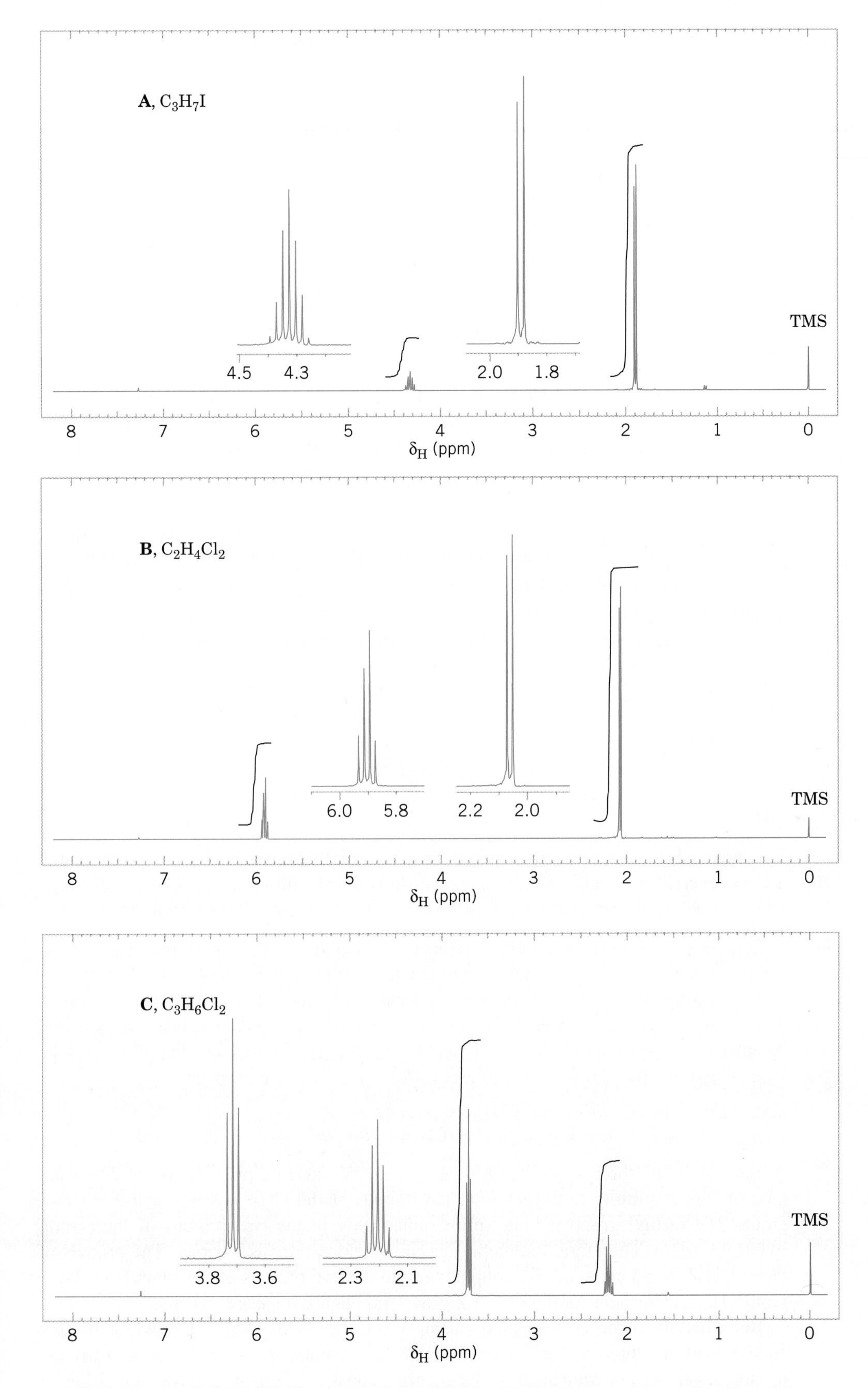

FIGURE 9.21 The 300-MHz ^1H NMR spectra for Problem 9.8. Expansions are shown in the offset plots.

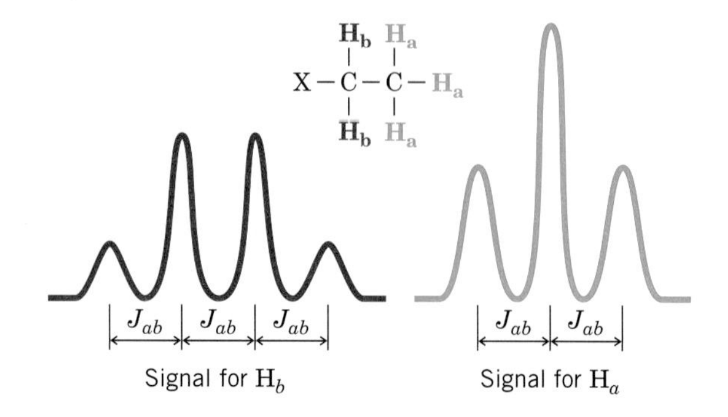

FIGURE 9.22 A theoretical splitting pattern for an ethyl group. For an actual example, see the spectrum of ethyl bromide (Fig. 9.20).

that the singlet of the $—CH_2Cl$ group falls directly on top of one of the outermost peaks of the ethyl quartet. Using NMR spectrometers with higher magnetic field strength (corresponding to ^1H resonance frequencies of 300, 500, or 600 MHz) often allows separation of signals that would overlap at lower magnetic field strengths (for an example, see Fig. 9.23, bottom).

2. Spin–spin couplings between the protons of nonadjacent atoms may occur. This long-range coupling happens frequently in compounds when π-bonded atoms intervene between the atoms bearing the coupled protons and in some cyclic molecules that are rigid.

3. The splitting patterns of aromatic groups can be difficult to analyze. A monosubstituted benzene ring (a phenyl group) has three different kinds of protons:

[Structure: benzene ring with X at top; H_a, H_a ortho; H_b, H_b meta; H_c para]

The chemical shifts of these protons may be so similar that the phenyl group gives a signal that resembles a singlet. Or the chemical shifts may be different, and because of long-range couplings, the phenyl group signal appears as a very complicated multiplet.

In all of the ^1H NMR spectra that we have considered so far, we have restricted our attention to signal splittings arising from interactions of only two sets of equivalent protons on adjacent atoms. What kinds of patterns should we expect from compounds in which more than two sets of equivalent protons are interacting? We cannot answer this question completely because of limitations of space, but we can give an example that illustrates the kind of analysis that is involved. Let us consider a 1-substituted propane:

$$\overset{(a)}{CH_3}—\overset{(b)}{CH_2}—\overset{(c)}{CH_2}—Z$$

Here, there are three sets of equivalent protons. We have no problem in deciding what kind of signal splitting to expect from the protons of the $CH_3—$ group or the $—CH_2Z$ group. The methyl group is spin–spin coupled only to the two protons of the central $—CH_2—$ group. Therefore, the methyl group should appear as a triplet. The protons of the $—CH_2Z$ group are similarly coupled only to the two protons of the central $—CH_2—$ group. Thus, the protons of the $—CH_2Z$ group should also appear as a triplet.

But what about the protons of the central $—CH_2—$ group (*b*)? They are spin–spin coupled with the three protons at (*a*) and with two protons at (*c*). The protons at (*a*) and (*c*), moreover, are not equivalent. If the coupling constants J_{ab} and J_{bc} have quite different values, then the protons at (*b*) could be split into a quartet by the three protons at (*a*) and each line of the quartet could be split into a triplet by the two protons at (*c*) (Fig. 9.24).

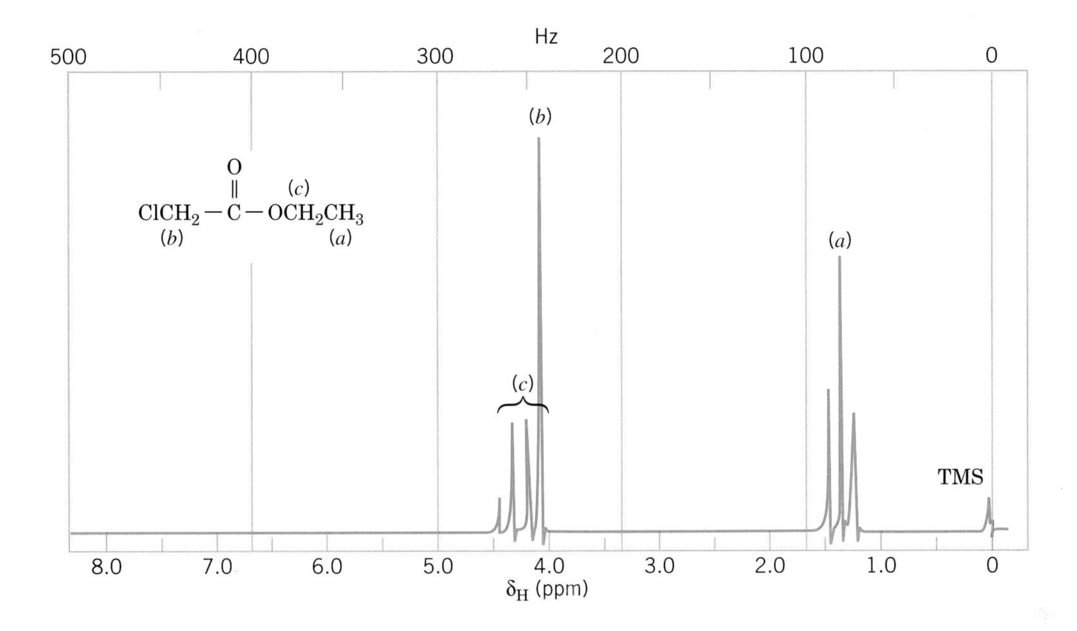

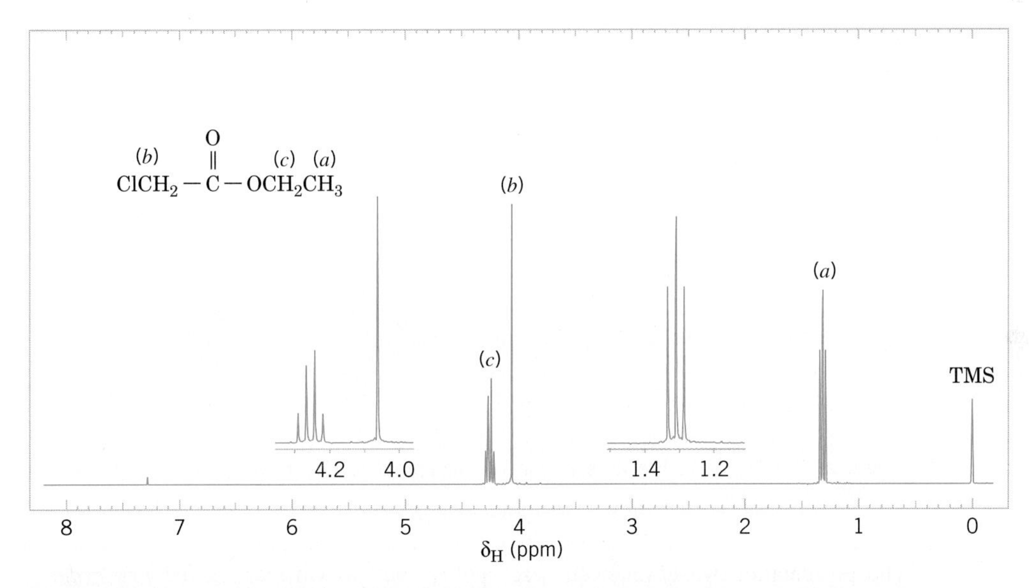

FIGURE 9.23 *Top:* The 60-MHz ^{1}H NMR spectrum of ethyl chloroacetate. Note the overlapping signals at δ 4. *Bottom:* The 300-MHz ^{1}H NMR spectrum of ethyl chloroacetate, showing resolution at higher magnetic field strength of the signals that overlapped at 60 MHz. Expansions of the signals are shown in the offset plots.

It is unlikely, however, that we would observe as many as 12 peaks in an actual spectrum because the coupling constants are such that peaks usually fall on top of peaks. The ^{1}H NMR spectrum of 1-nitropropane (Fig. 9.25) is typical of 1-substituted propane compounds. We see that the (*b*) protons are split into six major peaks, each of which shows a slight sign of further splitting.

PROBLEM 9.9

Carry out an analysis like that shown in Fig. 9.24 and show how many peaks the signal from (*b*) would be split into if $J_{ab} = 2J_{bc}$ and if $J_{ab} = J_{bc}$. (*Hint:* In both cases peaks will fall on top of peaks so that the total number of peaks in the signal is fewer than 12.)

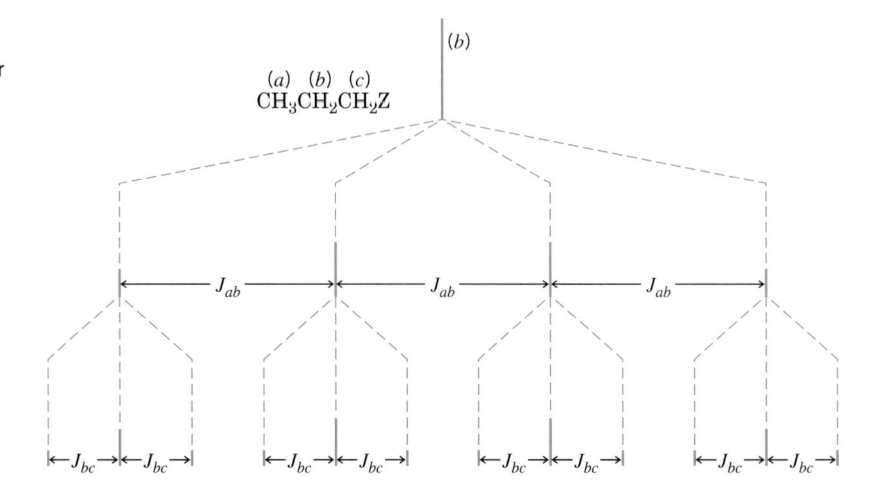

FIGURE 9.24 The splitting pattern that would occur for the *(b)* protons of $CH_3CH_2CH_2Z$ if J_{ab} is much larger than J_{bc}. Here $J_{ab} = 3J_{bc}$.

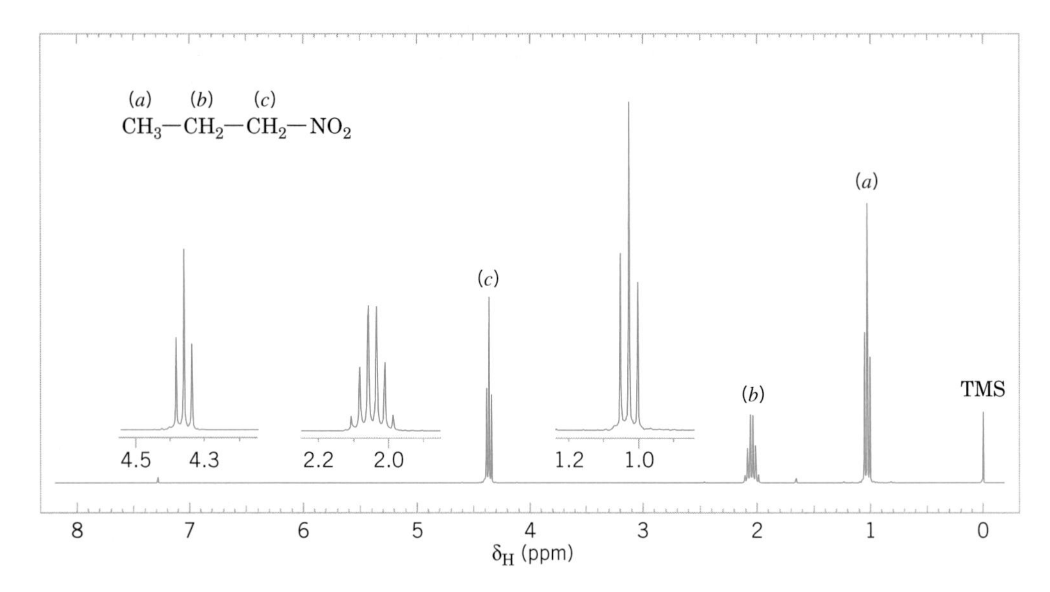

FIGURE 9.25 The 300-MHz ^{1}H NMR spectrum of 1-nitropropane. Expansions of the signals are shown in the offset plots.

The presentation we have given here applies only to what are called *first-order spectra.* In first-order spectra, the distance in hertz ($\Delta\nu$) that separates the coupled signals is very much larger than the coupling constant, *J*. That is, $\Delta\nu \gg J$. In *second-order spectra* (which we have not discussed), $\Delta\nu$ approaches *J* in magnitude and the situation becomes much more complex. The number of peaks increases and the intensities are not those that might be expected from first-order considerations.

9.9 Proton NMR Spectra and Rate Processes

J. D. Roberts (Emeritus Professor, California Institute of Technology), a pioneer in the application of NMR spectroscopy to problems of organic chemistry, has compared the NMR spectrometer to a camera with a relatively slow shutter speed. Just as a camera with a slow shutter speed blurs photographs of objects that are moving rapidly, the NMR spectrometer blurs its picture of molecular processes that are occurring rapidly.

What are some of the rapid processes that occur in organic molecules?

At temperatures near room temperature, groups connected by carbon–carbon single bonds rotate very rapidly (unless rotation is prevented by some structural constraint, e.g., a rigid ring system). Because of this, when we determine spectra of compounds with single bonds that allow rotation, the spectra that we obtain often reflect the individual hydrogen atoms in their average environment—that is, in an environment that is an average of all the environments that the protons have as a result of the group rotations.

To see an example of this effect, let us consider the spectrum of ethyl bromide again. The most stable conformation of ethyl bromide is the one in which the groups are perfectly staggered. In this staggered conformation one hydrogen of the methyl group (in red in the following structure) is in a different environment from that of the other two methyl hydrogen atoms. If the NMR spectrometer were to detect this specific conformation of ethyl bromide, it would show the protons of the methyl group at *a different chemical shift.* We know, however, that in the spectrum of ethyl bromide (Fig. 9.20), the three protons of the methyl group give *a single signal* (a signal that is split into a triplet by spin–spin coupling with the two protons of the adjacent carbon):

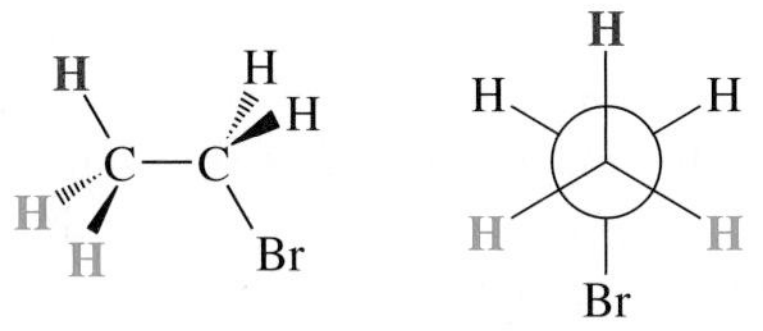

The methyl protons of ethyl bromide give a single signal because at room temperature the groups connected by the carbon–carbon single bond rotate approximately 1 million times each second. The "shutter speed" of the NMR spectrometer is too slow to "photograph" this rapid rotation; instead, it photographs the methyl hydrogen atoms in their average environments, and in this sense, it gives us a blurred picture of the methyl group.

Rotations about single bonds slow down as the temperature of the compound is lowered. Sometimes, this slowing of rotations allows us to "see" the different conformations of a molecule when we determine the spectrum at a sufficiently low temperature.

An example of this phenomenon, and one that also shows the usefulness of deuterium labeling, can be seen in the low-temperature ^1H NMR spectra of cyclohexane and of undecadeuteriocyclohexane. (These experiments originated with F. A. L. Anet, Emeritus Professor, University of California, Los Angeles, another pioneer in the applications of NMR spectroscopy to organic chemistry, especially to conformational analysis.)

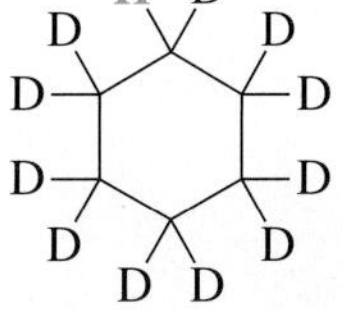

Undecadeuteriocyclohexane

At room temperature, ordinary cyclohexane gives one signal because interconversions between the various chair forms occur very rapidly. At low temperatures, however, ordinary cyclohexane gives a very complex ^1H NMR spectrum. At low temperatures interconversions are slow; the chemical shifts of the axial and equatorial protons are resolved, and complex spin–spin couplings occur.

At $-100°C$, however, undecadeuteriocyclohexane gives only two signals of equal intensity. These signals correspond to the axial and equatorial hydrogen atoms of the following two chair conformations. Interconversions between these conformations occur at this low temperature, but they happen slowly enough for the NMR spectrometer to detect the individual conformations. [The nucleus of a deuterium atom (a deuteron) has a much smaller magnetic moment than a proton, and signals from deuteron absorption do not occur in ^{1}H NMR spectra.]

PROBLEM 9.10 How many signals would you expect to obtain in the ^{1}H NMR spectrum of undecadeuteriocyclohexane at room temperature?

Another example of a rapidly occurring process can be seen in ^{1}H NMR spectra of ethanol. The ^{1}H NMR spectrum of ordinary ethanol shows the hydroxyl proton as a singlet and the protons of the $—CH_2—$ group as a quartet (Fig. 9.26). In ordinary ethanol we observe *no signal splitting arising from coupling between the hydroxyl proton and the protons of the* $—CH_2—$ *group even though they are on adjacent atoms.*

If we were to examine a ^{1}H NMR spectrum of *very pure* ethanol, however, we would find that the signal from the hydroxyl proton was split into a triplet and that the signal from the protons of the $—CH_2—$ group was split into a multiplet of eight peaks. Clearly, in very pure ethanol the spin of the proton of the hydroxyl group is coupled with the spins of the protons of the $—CH_2—$ groups.

Whether coupling occurs between the hydroxyl protons and the methylene protons depends on the length of time the proton spends on a particular ethanol molecule. Protons attached to electronegative atoms with lone pairs such as oxygen (or nitrogen) can undergo rapid **chemical exchange.** That is, they can be transferred rapidly from one molecule to another. The chemical exchange in very pure ethanol is slow and, as a consequence, we see the signal splitting of and by the hydroxyl proton in the spectrum. In ordinary ethanol, acidic and basic impurities catalyze the chemical exchange; the exchange occurs so rapidly that the hydroxyl proton gives an unsplit signal and that of the methylene protons is split only by coupling with the protons of the methyl group. We say, then, that rapid exchange causes **spin decoupling.**

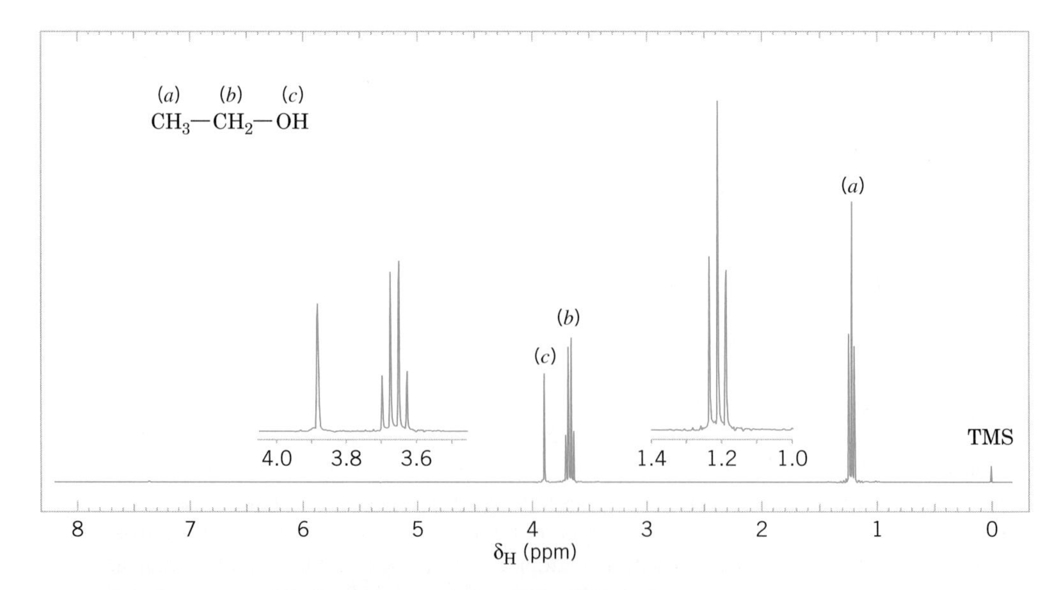

FIGURE 9.26 The 300-MHz ^{1}H NMR spectrum of ordinary ethanol. There is no signal splitting by the hydroxyl proton due to rapid chemical exchange. Expansions of the signals are shown in the offset plots.

Spin decoupling is found in the ^{1}H NMR spectra of alcohols, amines, and carboxylic acids, and the signals of OH and NH protons are normally unsplit.

Protons that undergo rapid chemical exchange (i.e., those attached to oxygen or nitrogen) can be easily detected by placing the compound in D_2O. The protons are rapidly replaced by deuterons, and the proton signal disappears from the spectrum.

Deuterium exchange

9.10 Carbon-13 NMR Spectroscopy

9.10A Interpretation of ^{13}C NMR Spectra

We begin our study of **^{13}C NMR spectroscopy** with a brief examination of some special features of spectra arising from carbon-13 nuclei. These spectra are often called carbon magnetic resonance (CMR) spectra or ^{13}C NMR spectra. Although ^{13}C accounts for only 1.1% of naturally occurring carbon, the fact that ^{13}C can produce an NMR signal has profound importance for the analysis of organic compounds. In some important ways ^{13}C spectra are usually less complex and easier to interpret than ^{1}H NMR spectra. The major isotope of carbon, on the other hand, carbon-12 (^{12}C), with natural abundance of about 99%, has no net magnetic spin and therefore cannot produce NMR signals.

9.10B One Peak for Each Unique Carbon Atom

One aspect of ^{13}C NMR spectra that greatly simplifies the interpretation process is that **each unique carbon atom in an ordinary organic molecule produces only one ^{13}C NMR peak.** There is no carbon–carbon coupling that causes splitting of signals into multiple peaks. Recall that in ^{1}H NMR spectra, hydrogen nuclei that are near each other (within a few bonds) couple with each other and cause the signal for each hydrogen to become a multiplet of peaks. This does not occur for adjacent carbons because only one carbon atom of every 100 carbon atoms is a ^{13}C nucleus (1.1% natural abundance). Therefore, the probability of there being two ^{13}C atoms adjacent to each other in a molecule is only about 1 in 10,000 ($1.1\% \times 1.1\%$), essentially eliminating the possibility of two neighboring carbon atoms splitting each other's signal into a multiplet of peaks. The low natural abundance of ^{13}C nuclei and its inherently low sensitivity also have another effect: Carbon-13 NMR spectra can be obtained only on pulse FT NMR spectrometers, where signal averaging is possible.

Whereas carbon–carbon signal splitting does not occur in ^{13}C NMR spectra, hydrogen atoms attached to carbon can split ^{13}C NMR signals into multiple peaks. However, it is useful to simplify the appearance of ^{13}C NMR spectra by initially eliminating signal splitting for $^{1}H-^{13}C$ coupling. This can be done by choosing instrumental parameters that decouple the proton–carbon interactions, and such a spectrum is said to be **broadband (BB) proton decoupled**. Thus, in a broadband proton-decoupled ^{13}C NMR spectrum, each carbon atom in a unique environment gives a signal consisting of only one peak. Most ^{13}C NMR spectra are obtained in the simplified broadband decoupled mode first and then in modes that provide information from the $^{1}H-^{13}C$ couplings (Sections 9.10D and 9.10E).

9.10C ^{13}C Chemical Shifts

As we found with ^{1}H spectra, the chemical shift of a given nucleus depends on the relative electron density around that atom. Decreased electron density around an atom **deshields** the atom from the magnetic field and causes its signal to occur further **downfield** (higher ppm, to the left) in the NMR spectrum. Relatively higher electron density around an atom **shields** the atom from the magnetic field and causes the signal to occur **upfield** (lower ppm, to the right) in the NMR spectrum. For example, carbon atoms that are attached only to other carbon and hydrogen atoms are relatively shielded from the magnetic field by the density of electrons around them, and, as a consequence, carbon atoms of this type produce peaks that are upfield in ^{13}C NMR spectra. On the other hand, carbon atoms bearing electronegative groups are deshielded from the magnetic field by the electron-withdrawing effects of these groups and, therefore, produce peaks that are downfield in the NMR spectrum. Electronegative groups such as halogens, hydroxyl groups, and other electron-withdrawing functional groups deshield the carbons to which they are attached, causing their ^{13}C NMR peaks to occur further downfield than those of unsubstituted carbon atoms. Reference tables of approximate chemical shift ranges for carbons bearing different substituents are available. Figure 9.27 is an example. [The reference standard assigned as zero ppm in ^{13}C NMR spectra is also tetramethylsilane (TMS), $(CH_3)_4Si$.]

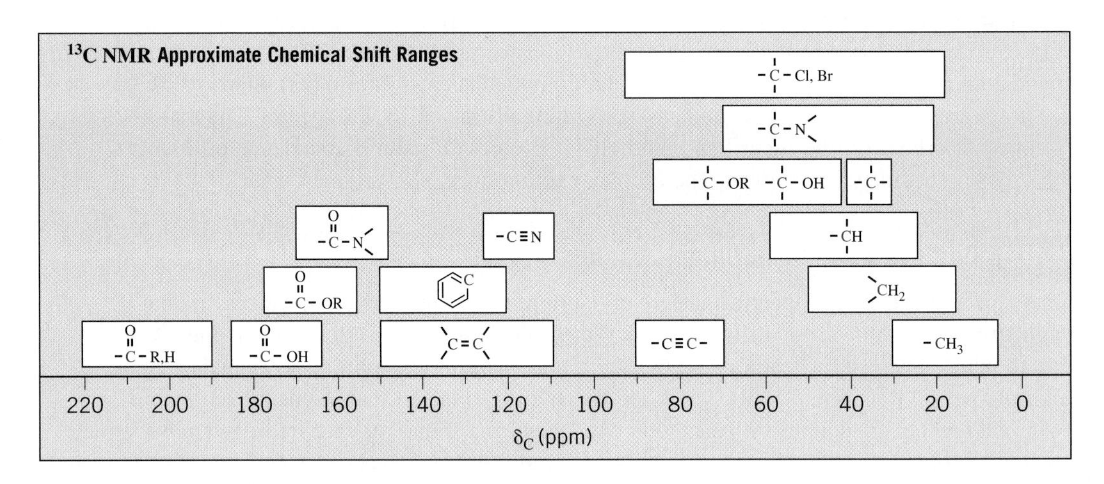

FIGURE 9.27 Approximate ^{13}C chemical shifts.

As a first example of the interpretation of ^{13}C NMR spectra, let us consider the ^{13}C spectrum of 1-chloro-2-propanol (Fig. 9.28*a*):

$$\overset{(a)}{\mathrm{Cl{-}CH_2}}{-}\overset{(b)}{\underset{\mathrm{OH}}{\mathrm{CH}}}{-}\overset{(c)}{\mathrm{CH_3}}$$

1-Chloro-2-propanol

This compound contains three unique carbons and therefore produces only three peaks in its broadband decoupled ^{13}C NMR spectrum: approximately at δ 20, δ 51, and δ 67. Figure 9.28 also shows a close grouping of three peaks at δ 77. These peaks come from the deuteriochloroform ($CDCl_3$) used as a solvent for the sample. Many ^{13}C NMR spectra contain these peaks. Although not of concern to us here, the signal for the single carbon of deuteriochloroform is split into three peaks by an effect of the attached deuterium. The $CDCl_3$ peaks should be disregarded when interpreting ^{13}C spectra.

As we can see, the chemical shifts of the three peaks from 1-chloro-2-propanol are well separated from one another. This separation results from differences in shielding by circulating electrons in the local environment of each carbon. Remember: The lower the electron density in the vicinity of a given carbon, the less that carbon will be shielded, and the more downfield will be the signal for that carbon. The oxygen of the hydroxyl group is the most electronegative atom; it withdraws electrons most effectively. Therefore, the carbon bearing the —OH group is the most *deshielded* carbon, and so this carbon gives the signal that is the furthest downfield, at δ 67. Chlorine is less electronegative than oxygen, causing the peak for the carbon to which it is attached to occur more

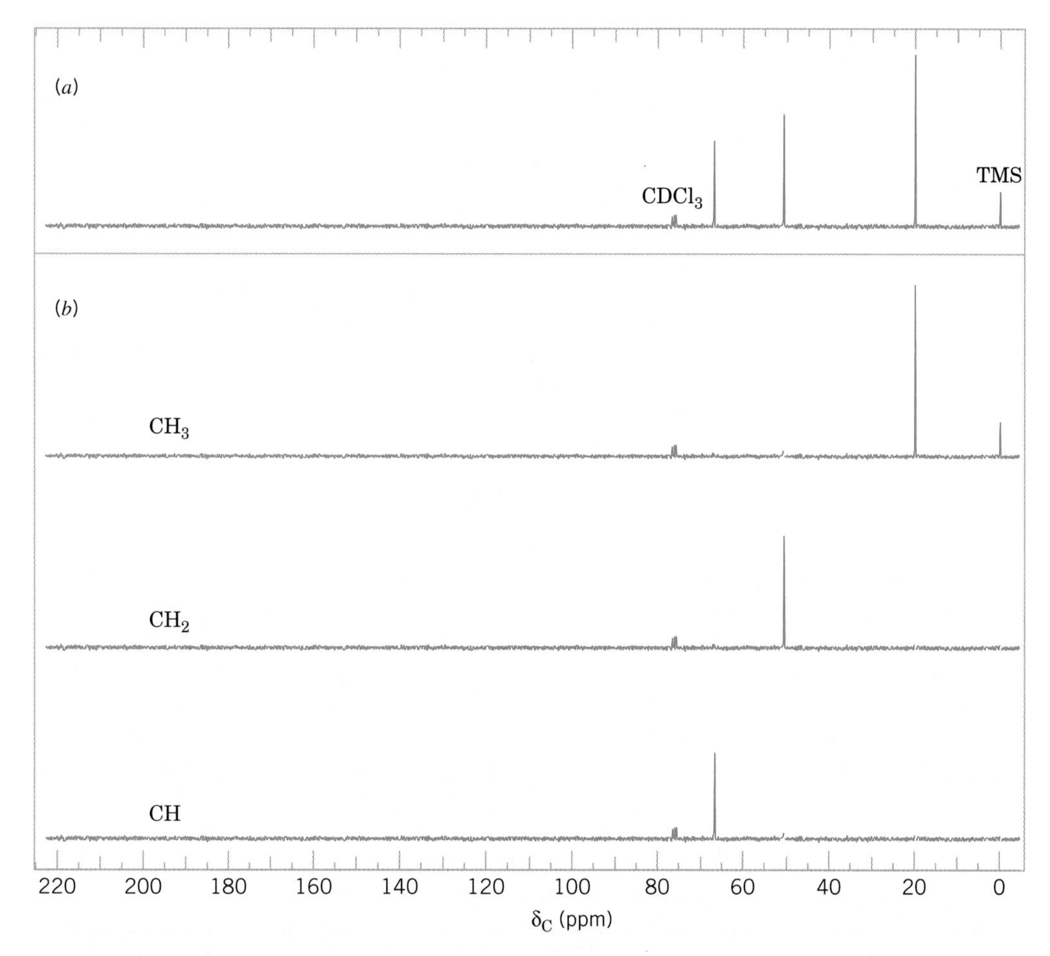

FIGURE 9.28 *(a)* The broadband proton-decoupled ^{13}C NMR spectrum of 1-chloro-2-propanol. *(b)* These three spectra show the DEPT ^{13}C NMR data from 1-chloro-2-propanol (see Section 9.10E). (This will be the only full display of a DEPT spectrum in the text. Other ^{13}C NMR figures will show the full broadband proton-decoupled spectrum but with information from the DEPT ^{13}C NMR spectra indicated near each peak as C, CH, CH_2, or CH_3.)

upfield, at δ 51. The methyl group carbon has no electronegative groups directly attached to it, so it occurs the furthest upfield, at δ 20. Using tables of typical chemical shift values (such as Figure 9.27), one can usually assign ^{13}C NMR signals to each carbon in a molecule, on the basis of the groups attached to each carbon.

9.10D Off-Resonance Decoupled Spectra

At times, more information than a predicted chemical shift is needed to assign an NMR peak to a specific carbon atom of a molecule. Fortunately, NMR spectrometers can differentiate among carbon atoms on the basis of the number of hydrogen atoms that are attached to each carbon. Several methods to accomplish this are available. One method not widely used anymore is called **off-resonance decoupling**. In an off-resonance decoupled ^{13}C NMR spectrum, each carbon signal is split into a multiplet of peaks, depending on how many hydrogens are attached to that carbon. An $n + 1$ rule applies, where n is the number of hydrogens on the carbon in question. Thus, a carbon with no hydrogens produces a singlet ($n = 0$), a carbon with one hydrogen produces a doublet (two peaks), a carbon with two hydrogens produces a triplet (three peaks), and a methyl group carbon produces a quartet (four peaks). Interpretation of off-resonance decoupled ^{13}C spectra, however, is often complicated by overlapping peaks from the multiplets.

9.10E DEPT ^{13}C Spectra

A **DEPT ^{13}C NMR spectrum** is a very simple type of carbon spectrum to interpret. It includes the information from a broadband decoupled ^{13}C NMR spectrum as well as information about the number of hydrogen atoms bonded to the carbon(s) producing each signal. A DEPT (**d**istortionless **e**nhanced **p**olarization **t**ransfer) spectrum is actually produced using data from several spectra, with the final presentation giving the information about the hydrogen substitution at each carbon. Signals from CH_3, CH_2, and CH carbons can be displayed separately (Fig. 9.28*b*), along with a broadband decoupled spectrum showing all signals (Fig. 9.28*a*). Together, these spectra uniquely identify each type of carbon. **From this point forward in this text, rather than reproducing the entire family of four sorted DEPT spectra for all subsequent spectra, we will show the ^{13}C peaks labeled according to the information gained from the DEPT spectra for the compound being considered.**

As a second example of interpreting ^{13}C NMR spectra, let us look at the spectrum of methyl methacrylate (Fig. 9.29). (This compound is the monomeric starting material for the commercial polymers Lucite and Plexiglas, see Special Topic A.) The five carbons of methyl methacrylate represent carbon types from several chemical shift regions of ^{13}C spectra. Furthermore, because there is no symmetry in the structure of methyl methacrylate, all of its carbon atoms are chemically unique and so produce five distinct carbon NMR signals. Making use of our table of approximate ^{13}C chemical shifts (Figure 9.27), we can readily deduce that the peak at δ 167.3 is due to the ester carbonyl carbon, the peak at δ 51.5 is for the methyl carbon attached to the ester oxygen, the peak at δ 18.3 is for the methyl attached to C2, and the peaks at δ 136.9 and δ 124.7 are for the alkene carbons. Additionally, employing the information from the DEPT ^{13}C spectra, we can unambiguously assign signals to the alkene carbons. The DEPT spectra tell us definitively that the peak at δ 124.7 has two attached hydrogens, and so it is due to C3, the terminal alkene carbon of methyl methacrylate. The alkene carbon with no attached hydrogens is then, of course, C2.

PROBLEM 9.11 Compounds **A**, **B**, and **C** are isomers with the formula $C_5H_{11}Br$. Their broadband proton-decoupled ^{13}C NMR spectra are given in Fig. 9.30. Information from the DEPT ^{13}C NMR spectra is given near each peak. Give structures for **A**, **B**, and **C**.

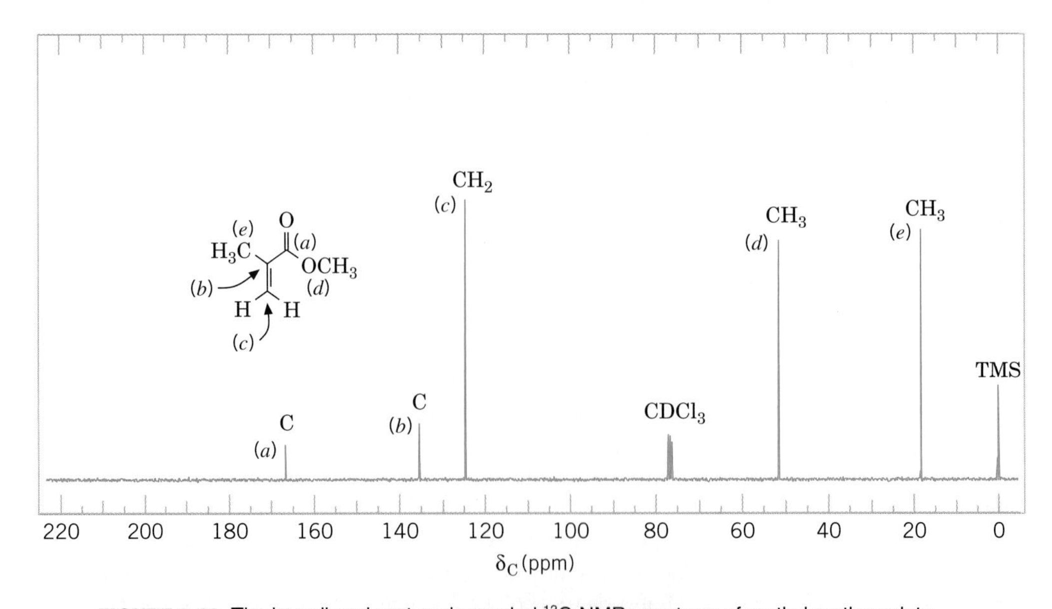

FIGURE 9.29 The broadband proton-decoupled ^{13}C NMR spectrum of methyl methacrylate. Information from the DEPT ^{13}C NMR spectra is given above the peaks.

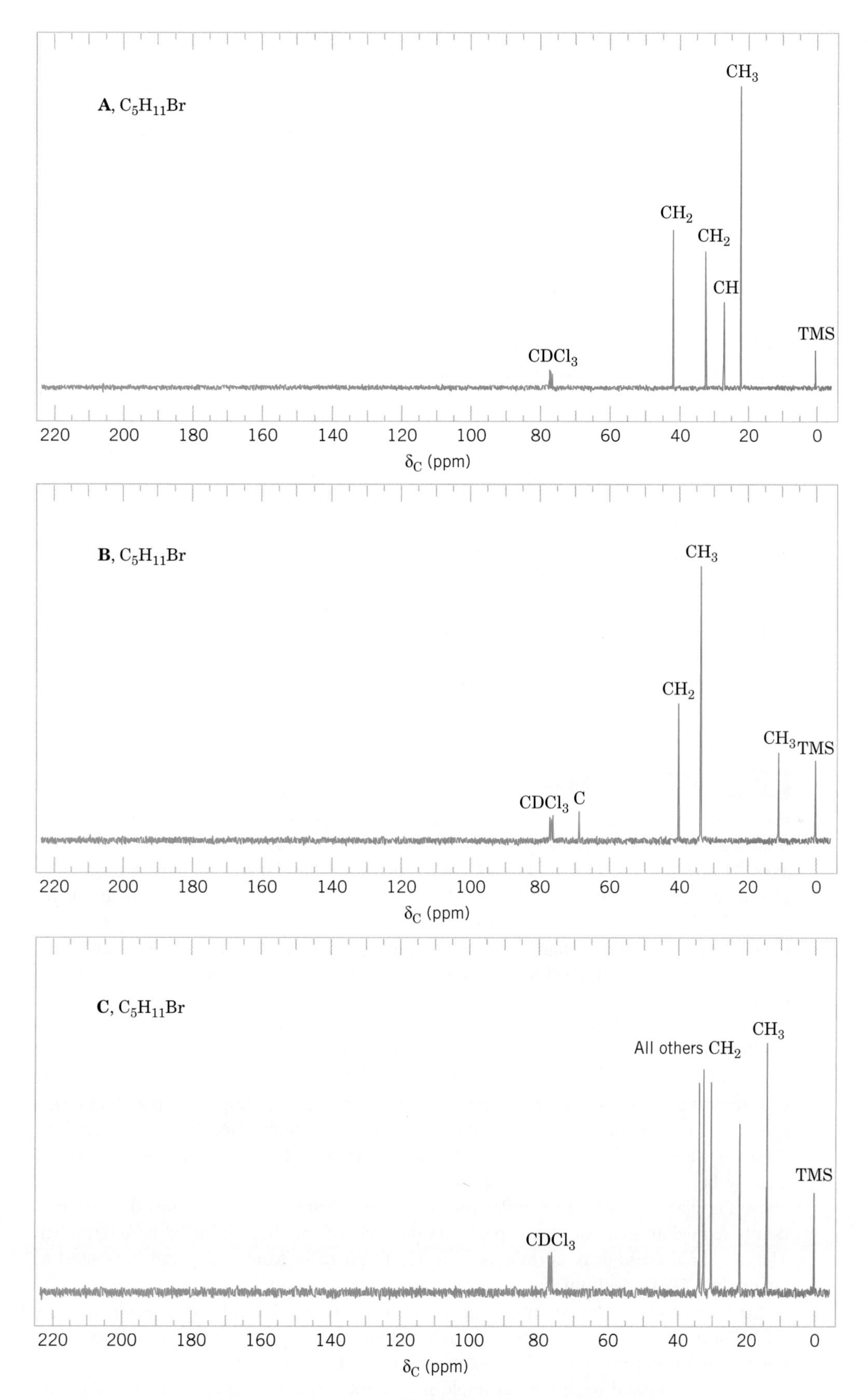

FIGURE 9.30 The broadband proton-decoupled ^{13}C NMR spectra of compounds A, B, and C, Problem 9.11. Information from the DEPT ^{13}C NMR spectra is given above the peaks.

9.11 Two-Dimensional (2D) NMR Techniques

Many NMR techniques are now available that greatly simplify the interpretation of NMR spectra. Chemists can now readily glean information about spin–spin coupling and the exact *connectivity* of atoms in molecules through techniques called **multidimensional NMR spectroscopy.** These techniques require an NMR spectrometer of the pulse (Fourier transform) type. The most common multidimensional techniques utilize two-dimensional NMR (2D NMR) and go by acronyms such as COSY, HETCOR, and a variety of others. [Even three-dimensional techniques (and beyond) are possible, although computational requirements can limit their feasibility.] The two-dimensional sense of 2D NMR spectra does not refer to the way they appear on paper but instead reflects the fact that the data are accumulated using two radio frequency pulses with a varying time delay between them. Sophisticated application of other instrumental parameters is involved as well. Discussion of these parameters and the physics behind multidimensional NMR is beyond the scope of this text. The result, however, is an NMR spectrum with the usual one-dimensional spectrum along the horizontal and vertical axes and a set of correlation peaks that appear in the $x-y$ field of the graph.

When 2D NMR is applied to ^1H NMR it is called **^1H–^1H correlation spectroscopy** (or **COSY** for short). COSY spectra are exceptionally useful for deducing proton–proton coupling relationships. Two-dimensional NMR spectra can also be obtained that indicate coupling between hydrogens and the *carbons* to which they are attached. In this case it is called **heteronuclear correlation spectroscopy** (**HETCOR**, or **C–H HETCOR**). When ambiguities are present in the one-dimensional ^1H and ^{13}C NMR spectra, a HETCOR spectrum can be very useful for assigning precisely which hydrogens and carbons are producing their respective peaks.

9.11A COSY Cross-Peak Correlations

Figure 9.31 shows the COSY spectrum for 1-chloro-2-propanol. In a COSY spectrum the ordinary one-dimensional ^1H spectrum is shown along both the horizontal and the vertical axes. Meanwhile, the $x-y$ field of a COSY spectrum is similar to a topographic map and can be thought of as looking down on the contour lines of a map of a mountain range. Along the diagonal of the COSY spectrum is a view that corresponds to looking down on the ordinary one-dimensional spectrum of 1-chloro-2-propanol as though each peak were a mountain. The one-dimensional counterpart of a given peak on the diagonal lies directly below that peak on each axis. The peaks on the diagonal provide no new information relative to that obtained from the one-dimensional spectrum along each axis.

The important and new information from the COSY spectrum, however, comes from the correlation peaks ("mountains") that appear off the diagonal (called "cross peaks"). If one starts at a given cross peak and imagines two perpendicular lines (i.e., parallel to each spectrum axis) leading back to the diagonal, the peaks intersected on the diagonal by these lines are coupled to each other. Hence, the peaks on the one-dimensional spectrum directly below the coupled diagonal peaks are coupled to each other. The cross peaks above the diagonal are mirror reflections of those below the diagonal; thus the information is redundant and only cross peaks on one side of the diagonal need be interpreted. The $x-y$ field cross-peak correlations are the result of instrumental parameters used to obtain the COSY spectrum.

Let's trace the coupling relationships in 1-chloro-2-propanol made evident in its COSY spectrum (Fig. 9.31). (Even though coupling relationships from the ordinary one-dimensional spectrum for 1-chloro-2-propanol are fairly readily interpreted, this compound makes a good beginning example for interpretation of COSY spectra.) First, one chooses a starting point in the COSY spectrum from which to begin tracing the coupling

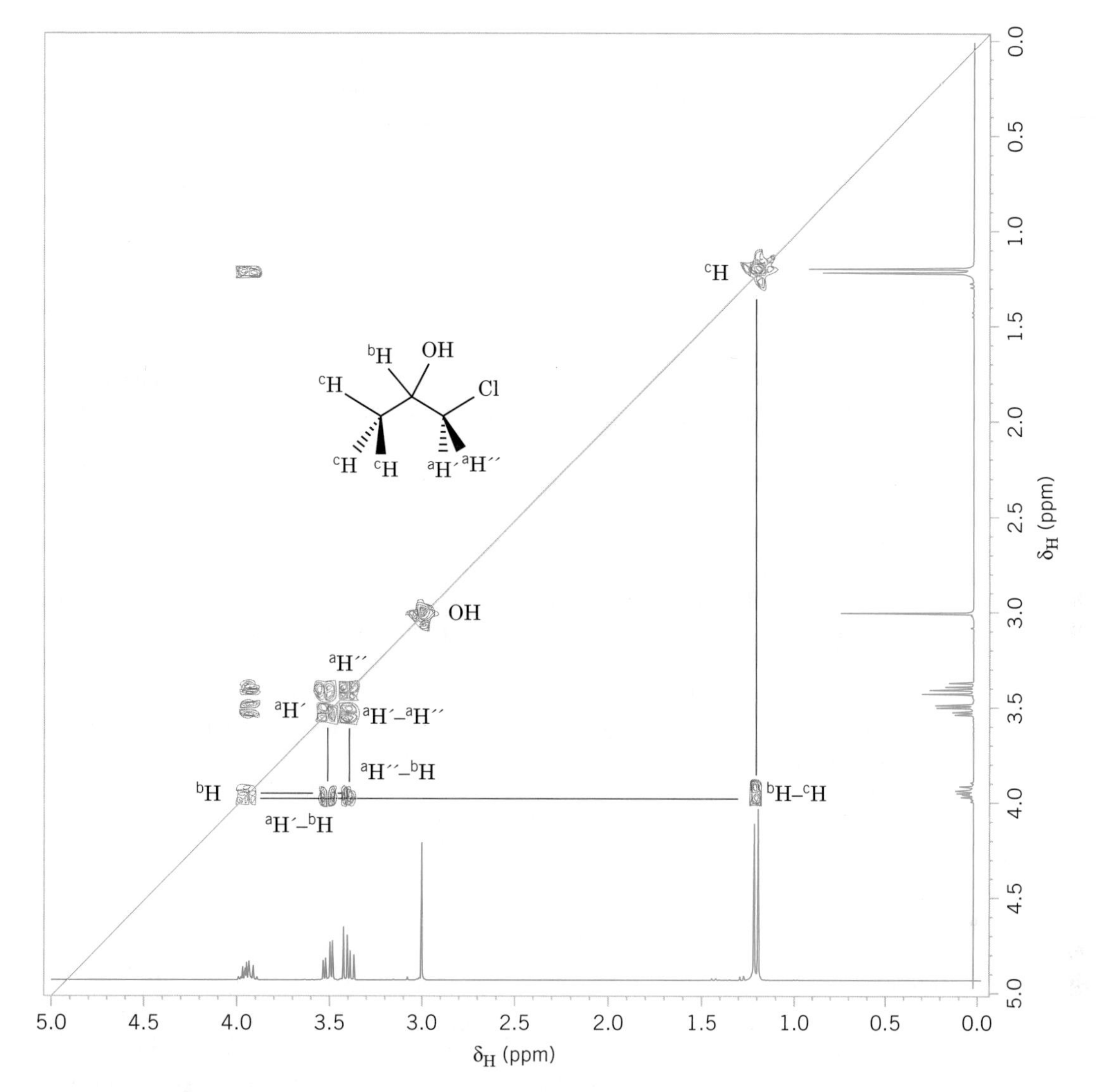

FIGURE 9.31 COSY spectrum of 1-chloro-2-propanol.

relationships. A peak whose assignment is relatively apparent in the one-dimensional spectrum is a good point of reference. For this compound, the doublet from the methyl group at 1.2 ppm is quite obvious and readily assigned. If we find the peak on the diagonal that corresponds to the methyl doublet (labeled ^{c}H in Fig. 9.31 and directly above the one-dimensional methyl doublet on both axes), an imaginary line can be drawn parallel to the vertical axis that intersects a correlation peak (labeled ^{b}H–^{c}H) in the x–y field off the diagonal. From here a perpendicular imaginary line can be drawn back to its intersection with the diagonal peaks. At its intersection we see that this diagonal peak is directly above the one-dimensional spectrum peak at δ 3.9. Thus, the methyl hydrogens at δ 1.2 are coupled to the hydrogen whose signal appears at δ 3.9. It is now clear that the peaks at δ 3.9 are due to the hydrogen on the alcohol carbon in 1-chloro-2-propanol (^{b}H on C2).

Returning to the peak on the diagonal above δ 3.9, we can trace a line back parallel to the horizontal axis that intersects a pair of cross peaks between δ 3.4 and δ 3.5. Moving back up to the diagonal from each of these cross peaks (^{a}H′–^{b}H and ^{a}H″–^{b}H) indicates that the hydrogen whose signal appears at δ 3.9 is coupled to the hydrogens whose signals appear at δ 3.4 and δ 3.5. The hydrogens at δ 3.4 and δ 3.5 are therefore the two hydrogens on the carbon that bears the chlorine (^{a}H′ and ^{a}H″). One can even see that ^{a}H′ and ^{a}H″ couple with each other by the cross peak they have in common between them right next to their diagonal peaks. Thus, from the COSY spectrum we can quickly see which hydrogens are coupled to each other. Furthermore, from the reference starting point, we can "walk around" a molecule, tracing the neighboring coupling relationships along the molecule's carbon skeleton as we go through the COSY spectrum.

9.11B HETCOR Cross-Peak Correlations

In a HETCOR spectrum a ^{13}C spectrum is presented along one axis and a ^{1}H spectrum is shown along the other. Cross peaks relating the two types of spectra to each other are found in the *x–y* field. Specifically, the cross peaks in a HETCOR spectrum indicate which hydrogens are attached to which carbons in a molecule, or vice versa. These cross-peak correlations are the result of instrumental parameters used to obtain the HETCOR spectrum. There is no diagonal spectrum in the *x–y* field like that found in the COSY. If imaginary lines are drawn from a given cross peak in the *x–y* field to each respective axis, the cross peak indicates that the hydrogen giving rise to the ^{1}H NMR signal on one axis is coupled (and attached) to the carbon that gives rise to the corresponding ^{13}C NMR signal on the other axis. Therefore, it is readily apparent which hydrogens are attached to which carbons.

Let us take a look at the HETCOR spectrum for 1-chloro-2-propanol (Fig. 9.32). Having interpreted the COSY spectrum already, we know precisely which hydrogens of 1-chloro-2-propanol produce each signal in the ^{1}H spectrum. If an imaginary line is taken from the methyl doublet of the proton spectrum at 1.2 ppm (vertical axis) out to the correlation peak in the *x–y* field and then dropped down to the ^{13}C spectrum axis (horizontal

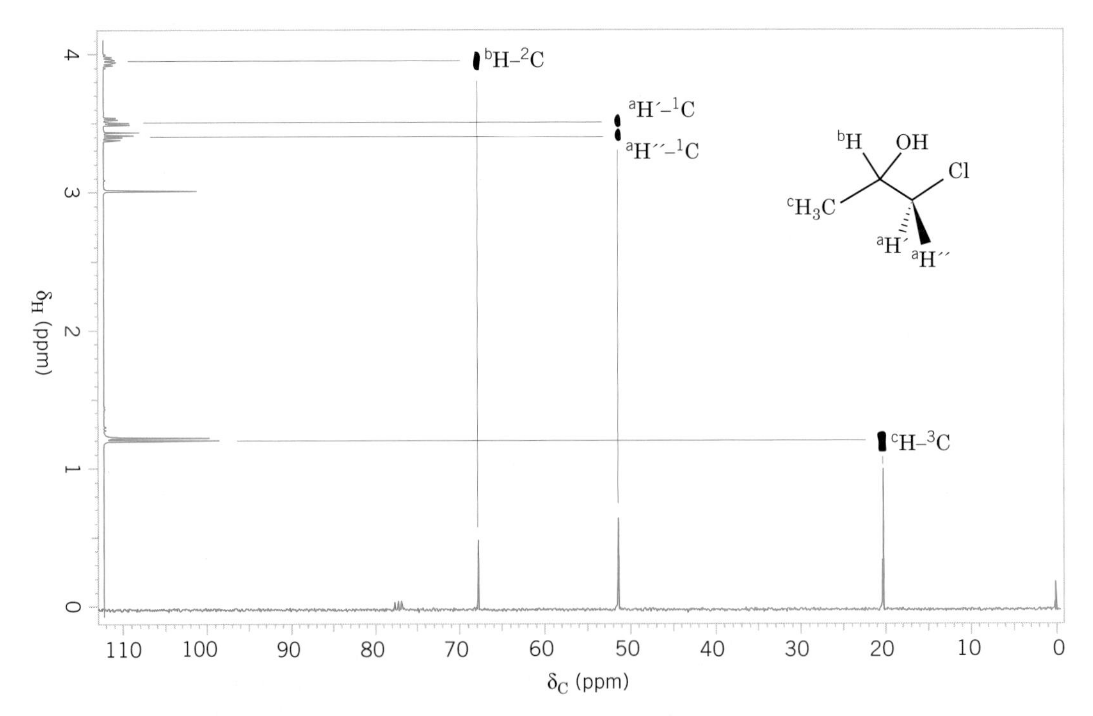

FIGURE 9.32 ^{1}H–^{13}C HETCOR NMR spectrum of 1-chloro-2-propanol. The ^{1}H NMR spectrum is shown in blue and the ^{13}C NMR spectrum is shown in green. Correlations of the ^{1}H–^{13}C cross peaks with the one-dimensional spectra are indicated by red lines.

axis), it is apparent that the ^{13}C peak at 20 ppm is produced by the methyl carbon of 1-chloro-2-propanol (C3). Having assigned the ^{1}H NMR peak at 3.9 ppm to the hydrogen on the alcohol carbon of the molecule (C2), tracing out to the correlation peak and down to the ^{13}C spectrum indicates that the ^{13}C NMR signal at 67 ppm arises from the alcohol carbon (C2). Finally, from the ^{1}H NMR peaks at 3.4–3.5 ppm for the two hydrogens on the carbon bearing the chlorine, our interpretation leads us out to the cross peak and down to the ^{13}C peak at 51 ppm (C1).

The Chemistry of. . .

Magnetic Resonance Imaging in Medicine

An important application of ^{1}H NMR spectroscopy in medicine today is a technique called **magnetic resonance imaging**, or **MRI**. One great advantage of MRI is that, unlike X-rays, it does not use dangerous ionizing radiation, and it does not require the injection of potentially harmful chemicals in order to produce contrasts in the image. In MRI, a portion of the patient's body is placed in a powerful magnetic field and irradiated with RF energy.

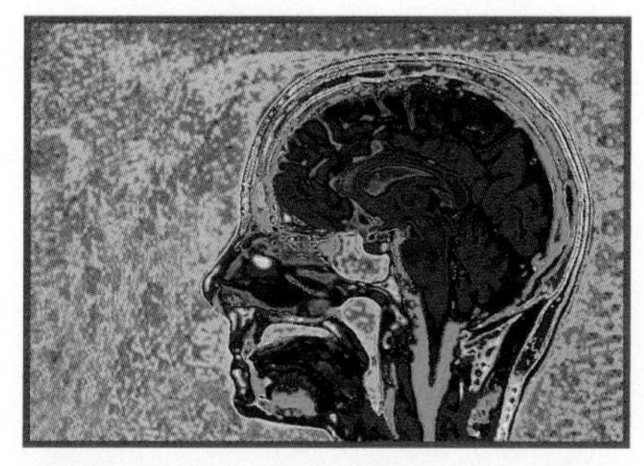

An image obtained by magnetic resonance imaging.

A typical MRI image is shown in the figure above. The instruments used in producing images like this one use the pulse method (Section 9.3) to excite the protons in the tissue under observation and use a Fourier transformation to translate the information into an image. The brightness of various regions of the image is related to two things.

The first factor is the number of protons in the tissue at that particular place. The second factor arises from what are called the **relaxation times** of the protons. When protons are excited to a higher energy state by the pulse of RF energy, they absorb energy. They must lose this energy to return to the lower energy spin state before they can be excited again by a second pulse. The process by which the nuclei lose this energy is called **relaxation,** and the time it takes to occur is the relaxation time.

There are two basic modes of relaxation available to protons. In one, called *spin–lattice relaxation,* the extra energy is transferred to neighboring molecules in the surroundings (or lattice). The time required for this to happen is called T_1 and is characteristic of the time required for the spin system to return to thermal equilibrium with its surroundings. In solids, T_1 can be hours long. For protons in pure liquid water, T_1 is only a few seconds. In the other type of relaxation, called *spin–spin relaxation,* the extra energy is dissipated by being transferred to nuclei of nearby atoms. The time required for this is called T_2. In liquids the magnitude of T_2 is approximately equal to T_1. In solids, however, the T_1 is very much longer.

Various techniques based on the time between pulses of RF radiation have been developed to utilize the differences in relaxation times in order to produce contrasts between different regions in soft tissues. The soft tissue contrast is inherently higher than that produced with X-ray techniques. Magnetic resonance imaging is being used to great effect in locating tumors, lesions, and edemas. Improvements in this technique are occurring rapidly, and the method is not restricted to observation of proton signals.

One important area of medical research is based on the observation of signals from ^{31}P. Compounds that contain phosphorus as phosphate esters (Section 11.12) such as adenosine triphosphate (ATP) and adenosine diphosphate (ADP) are involved in most metabolic processes. By using techniques based on NMR, researchers now have a noninvasive way to follow cellular metabolism.

Thus, by a combination of COSY and HETCOR spectra, all ^{13}C and ^{1}H peaks can be unambiguously assigned to their respective carbon and hydrogen atoms in 1-chloro-2-propanol. (In this simple example using 1-chloro-2-propanol, we could have arrived at complete assignment of these spectra without COSY and HETCOR data. For many compounds, however, the assignments are quite difficult to make without the aid of these 2D NMR techniques.)

9.12 An Introduction to Mass Spectrometry

A mass spectrometer produces a spectrum of masses based on the structure of a molecule. A mass spectrum is not a spectrum with respect to electromagnetic radiation, as in the case of IR (Section 2.16) and NMR (Sections 9.3–9.11). Instead, it is a spectrum or plot of the distribution of ion masses corresponding to the formula weight of a molecule, fragments derived from the molecule, or both. A typical mass spectrum is shown in Figure 9.33. Peaks along the *x* axis of a mass spectrum correspond to the distribution of masses produced from molecules of the compound when it is in the mass spectrometer. The height of each peak as measured against the *y* axis indicates the relative amount of each ion produced from the molecule. Together, the mass and abundance of each ion can tell us a great deal about the structure of a molecule. In the next section we shall see how these molecular ions and fragment ions come about in a conventional electron impact (EI) mass spectrometer.

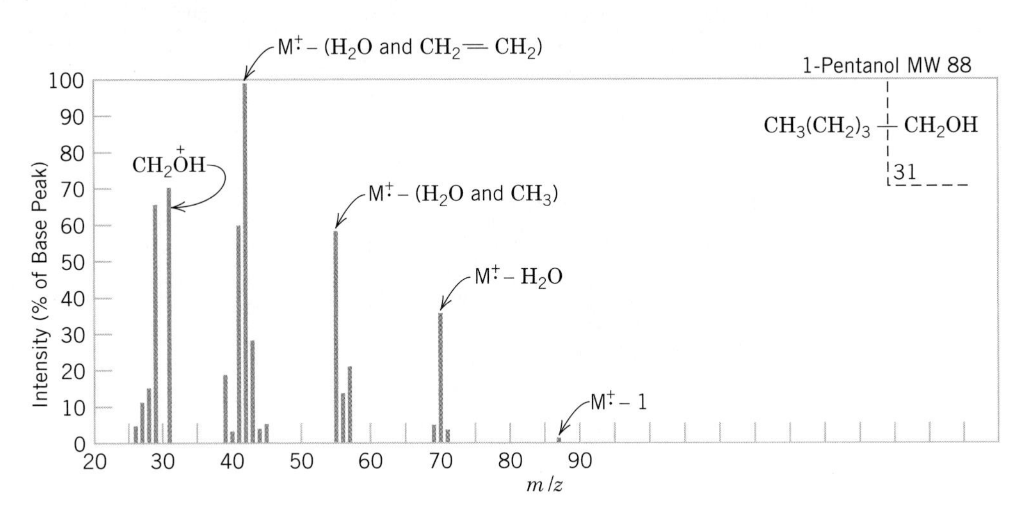

FIGURE 9.33 Mass spectrum of 1-pentanol. (Adapted from Silverstein, R. M.; Webster F. X. *Spectrometric Identification of Organic Compounds,* 6th ed.; Wiley: New York, 1998; p. 19.)

9.13 The Mass Spectrometer

9.13A Ionization

In an EI mass spectrometer (Fig. 9.34) molecules in the gaseous state under low pressure are bombarded with a beam of high-energy electrons. The energy of the beam of electrons is usually 70 eV (electron volts), and one of the things this bombardment can do is dislodge one of the electrons of the molecule and produce a positively charged ion called the **molecular ion:**

$$\underset{\textbf{Molecule}}{M} + \underset{\textbf{High-energy electron}}{e^-} \longrightarrow \underset{\textbf{Molecular ion}}{M^{+\cdot}} + 2\,e^-$$

The molecular ion is not only a cation, but because it contains an odd number of electrons, it also is a radical. Radicals contain unpaired electrons (Section 3.1A). Thus it be-

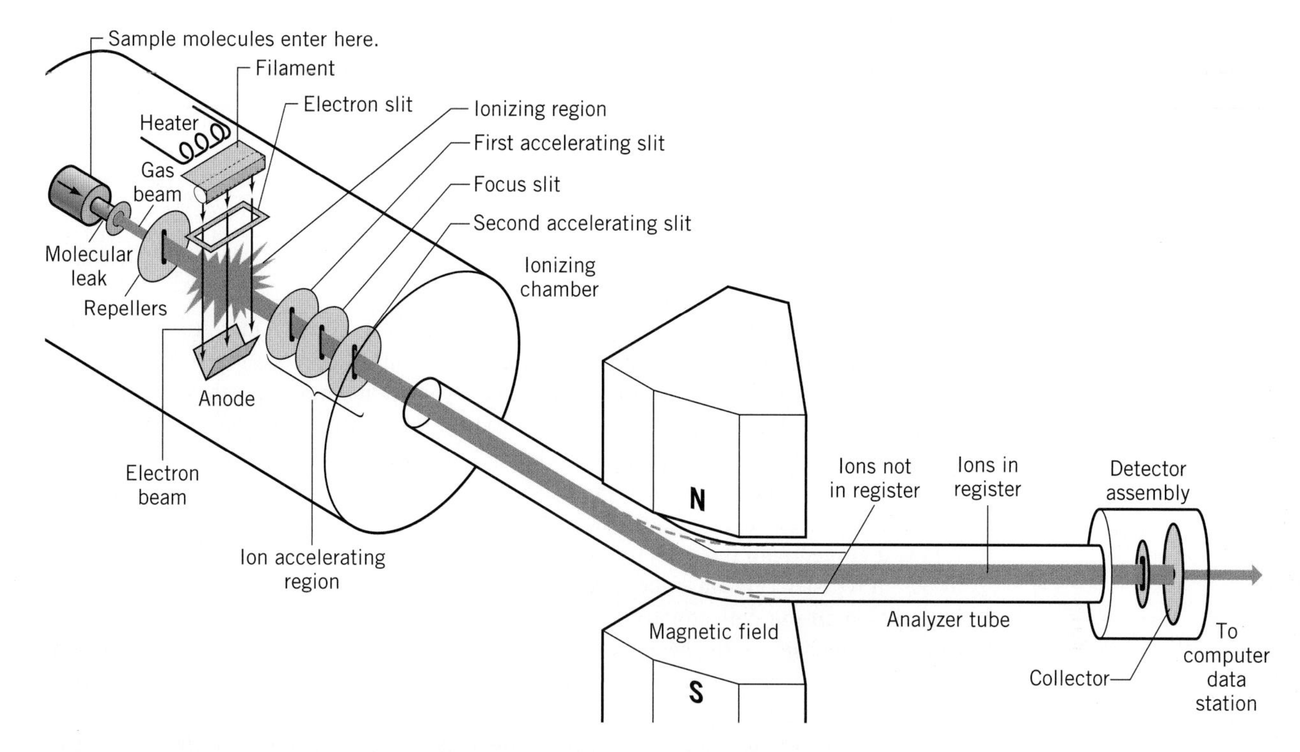

FIGURE 9.34 Mass spectrometer. Schematic diagram of CEG model 21-103. The magnetic field that brings ions of varying mass-to-charge ratios (*m/z*) into register is perpendicular to the page. (From Holum, J. R. *Organic Chemistry: A Brief Course;* Wiley: New York, 1975. Used with permission.)

longs to a general group of ions called *radical cations.* If, for example, the molecule under bombardment is a molecule of ammonia, the following reaction will take place.

$$\mathrm{H{:}\ddot{N}{:}H\ (with\ H\ below)} + e^- \longrightarrow \left[\mathrm{H{:}\dot{N}{:}H\ (with\ H\ below)}\right]^{+} + 2\,e^-$$

Molecular ion, M$^{+\cdot}$
(a radical cation)

9.13B Fragmentation

An electron beam with an energy of 70 eV ($\sim 6.7 \times 10^3$ kJ mol^{-1}) not only dislodges electrons from molecules, producing molecular ions, it also imparts to the molecular ions considerable surplus energy. Not all molecular ions have the same amount of surplus energy, but for most, the surplus is far in excess of that required to break covalent bonds (200–400 kJ mol^{-1}). Soon after they are formed, therefore, most molecular ions literally fly apart—they undergo ***fragmentation.*** Fragmentation can take place in a variety of ways depending on the nature of the particular molecular ion, and as we shall see later (Section 9.16), the way a molecular ion fragments can give us highly useful information about the structure of a complex molecule. Even with a relatively simple molecule such as ammonia, however, fragmentation can produce several new cations. The molecular ion can eject a hydrogen atom by homolytic bond cleavage and produce the cation NH_2^+:

$$\mathrm{H{:}\overset{\cdot +}{N}{:}H\ (with\ H\ below)} \longrightarrow \mathrm{H{:}\overset{+}{N}{:}\ (with\ H\ below)} + \mathrm{H}\cdot$$

Recall that we use single-barbed arrows to show the movement of single electrons, as in the case of these homolytic bond cleavages and other processes involving radicals (see Section 3.1A).

This $\overset{+}{N}H_2$ cation can then lose another hydrogen atom to produce $\overset{+}{N}H\cdot$, which can lead, in turn, to $\overset{+}{N}$.

$$H:\overset{+}{N}:H \longrightarrow H:\overset{+}{N}: + H\cdot$$

$$H:\overset{+}{N}: \longrightarrow :\overset{+}{N}: + H\cdot$$

9.13C Ion Sorting

The mass spectrometer then *sorts* the cations on the basis of their mass-to-charge ratio, or *m/z*. Since for all practical purposes the charge on all of the ions is +1, this amounts to sorting them on the basis of their mass. The conventional mass spectrometer does this by accelerating the ions through a series of slits and then sending the ion beam into a curved tube (see Fig. 9.34 again). This curved tube passes through a variable magnetic field and the magnetic field exerts an influence on the moving ions. Depending on its strength at a given moment, the magnetic field will cause ions with a particular *m/z* to follow a curved path that exactly matches the curvature of the tube. These ions are said to be "in register." Because they are in register, these ions pass through another slit and impinge on an ion collector where the intensity of the ion beam is measured electronically. The intensity of the beam is simply a measure of the relative abundance of the ions with a particular *m/z*. Some mass spectrometers are so sensitive that they can detect the arrival of a *single ion.*

The actual sorting of ions takes place in the magnetic field, and this sorting takes place because laws of physics govern the paths followed by charged particles when they move through magnetic fields. Generally speaking, a magnetic field such as this will cause ions moving through it to move in a path that represents part of a circle. The radius of curvature of this circular path is related to the *m/z* of the ions, to the strength of the magnetic field ($\mathbf{B}_0$, in tesla), and to the accelerating voltage. If we keep the accelerating voltage constant and progressively increase the magnetic field, ions whose *m/z* values are progressively larger will travel in a circular path that exactly matches that of the curved tube. Hence, by steadily increasing B_0, ions with progressively increasing *m/z* will be brought into register and so will be detected at the ion collector. Since, as we said earlier, the charge on nearly all of the ions is unity, this means that *ions of progressively increasing mass arrive at the collector and are detected.*

What we have described is called "magnetic focusing" (or "magnetic scanning"), and all of this is done automatically by the mass spectrometer. The spectrometer displays the results by plotting a series of peaks of varying intensity in which each peak corresponds to ions of a particular *m/z*. This display (see Fig. 9.33) is one form of a *mass spectrum.*

A variety of other methods are used for **ion sorting** in mass spectrometers, including quadrupole mass filtering, ion trapping, and time-of-flight mass analyzers. In a quadrupole mass analyzer, ions are filtered by varying the electrical signal in four parallel charged rods. At any given instant, only ions of certain mass-to-charge ratio are able to travel through the quadrupole region to the detector. Other ions collide with the rods and are neutralized. By varying the electrical state of the four rods in pairs, a range of masses can be scanned. Ion trap mass analyzers involve a ring electrode charged with a varying radio frequency voltage. Ions enter the cavity enclosed by the ring, and those of appropriate mass take up a stable orbit. As the voltage state of the ring varies, so does the mass for which a stable orbit is possible. Varying the ring voltage allows a range of masses to be scannned by progressively trapping and releasing them to the detector. In time-of-flight mass analyzers, ions are accelerated into a tube that is free of electrical fields. The ions drift toward the detector, and the time it takes to traverse the tube is correlated with their respective masses.

To summarize: An EI mass spectrometer bombards organic molecules with a beam of high energy electrons, causing them to ionize and fragment. It then separates the resulting mixture of ions on the basis of their *m/z* values and records the relative abundance of each ionic fragment. It displays this result as a plot of ion abundance versus *m/z*.

9.14 THE MASS SPECTRUM

Mass spectra are usually published as bar graphs or in tabular form, as illustrated in Fig. 9.35 for the mass spectrum of ammonia. In either presentation, the most intense peak—called the **base peak**—is arbitrarily assigned an intensity of 100%. The intensities of all other peaks are given proportionate values, as percentages of the base peak.

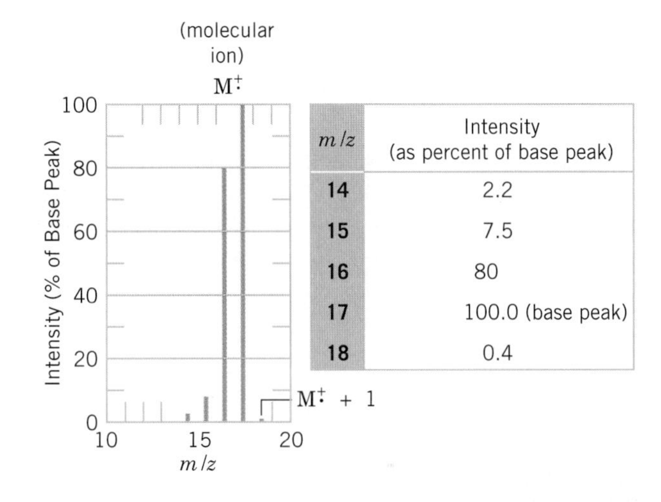

m/z	Intensity (as percent of base peak)
14	2.2
15	7.5
16	80
17	100.0 (base peak)
18	0.4

FIGURE 9.35 The mass spectrum of NH_3 presented as a bar graph and in tabular form.

The masses of the ions given in a mass spectrum are those that we would calculate for the ion by assigning to the constituent atoms *masses rounded off to the nearest whole number*. For the commonly encountered atoms the nearest whole-number masses are

$$H = 1 \qquad O = 16$$
$$C = 12 \qquad F = 19$$
$$N = 14$$

In the mass spectrum of ammonia we see major peaks at *m/z* 14, 15, 16, and 17. These correspond to the molecular ion and to the fragments we saw earlier.

$$NH_3 \xrightarrow{-e^-} [NH_3]^{+\cdot} \xrightarrow{-H\cdot} [NH_2]^+ \xrightarrow{-H\cdot} [NH]^{+\cdot} \xrightarrow{-H\cdot} [N]^+$$

$$m/z = \mathbf{17} \;(\textbf{molecular ion}) \qquad \mathbf{16} \qquad \mathbf{15} \qquad \mathbf{14}$$

By convention, we express

$$H{:}\overset{\cdot +}{\underset{\cdot\cdot}{N}}{:}H \text{ (with H below N) as } [NH_3]^{+\cdot}$$

$$H{:}\overset{+}{\underset{\cdot\cdot}{N}}{:} \text{ (with H below N) as } [NH_2]^+$$

$$H{:}\overset{+}{\underset{\cdot}{N}}{:} \text{ as } [NH]^{+\cdot}$$

$${:}\overset{+}{\underset{\cdot}{N}}{:} \text{ as } [N]^+$$

In the case of ammonia, the base peak is the peak arising from the molecular ion. This is not always the case, however; in many of the spectra that we shall see later, the base peak (the most intense peak) will be at an *m/z* value different from that of the molecular ion. This happens because in many instances the molecular ion fragments so rapidly that some other ion at a smaller *m/z* value produces the most intense peak. In a few cases the molecular ion peak is extremely small, and sometimes it is absent altogether.

One other feature in the spectrum of ammonia requires explanation: the small peak that occurs at *m/z* 18. In the bar graph we have labeled this peak $M^{+\cdot} + 1$ to indicate that it is one mass unit greater than the molecular ion. The $M^{+\cdot} + 1$ peak appears in the spectrum because most elements (e.g., nitrogen and hydrogen) have more than one naturally occurring isotope (Table 9.1). Isotopes differ in mass, of course, because they differ in the number of neutrons the atoms of a given element have. Although most of the NH_3 molecules in a sample of ammonia are composed of $^{14}N^{1}H_3$, a small but detectable fraction of molecules is composed of $^{15}N^{1}H_3$. (A very tiny fraction of molecules is also composed of $^{14}N^{1}H_2{}^{2}H$.) Therefore, these molecules ($^{15}N^{1}H_3$ or $^{14}N^{1}H_2{}^{2}H$) produce molecular ions at *m/z* 18, that is, at $M^{+\cdot} + 1$.

TABLE 9.1 Principal Stable Isotopes of Common Elements[a]

Element	Most Common Isotope		Natural Abundance of Other Isotopes (Based on 100 Atoms of Most Common Isotope)			
Carbon	^{12}C	100	^{13}C	1.11		
Hydrogen	^{1}H	100	^{2}H	0.016		
Nitrogen	^{14}N	100	^{15}N	0.38		
Oxygen	^{16}O	100	^{17}O	0.04	^{18}O	0.20
Fluorine	^{19}F	100				
Silicon	^{28}Si	100	^{29}Si	5.10	^{30}Si	3.35
Phosphorus	^{31}P	100				
Sulfur	^{32}S	100	^{33}S	0.78	^{34}S	4.40
Chlorine	^{35}Cl	100	^{37}Cl	32.5		
Bromine	^{79}Br	100	^{81}Br	98.0		
Iodine	^{127}I	100				

[a]Data obtained from Silverstein, R. M.; Webster, F. X. *Spectrometric Identification of Organic Compounds,* 6th ed.; Wiley: New York, 1998; p 7.

The spectrum of ammonia begins to show us with a simple example how the masses (or *m/z* values) of individual ions can give us information about the composition of the ions and how this information can allow us to arrive at possible structures for a compound. Problems 9.12–9.14 will allow us further practice with this technique.

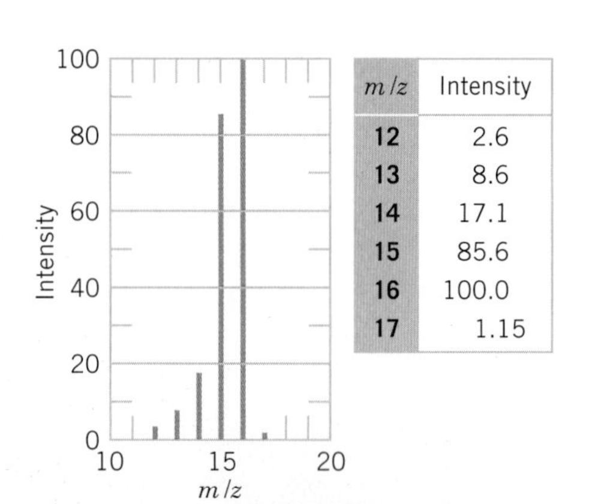

m/z	Intensity
12	2.6
13	8.6
14	17.1
15	85.6
16	100.0
17	1.15

FIGURE 9.36 Mass spectrum for Problem 9.12.

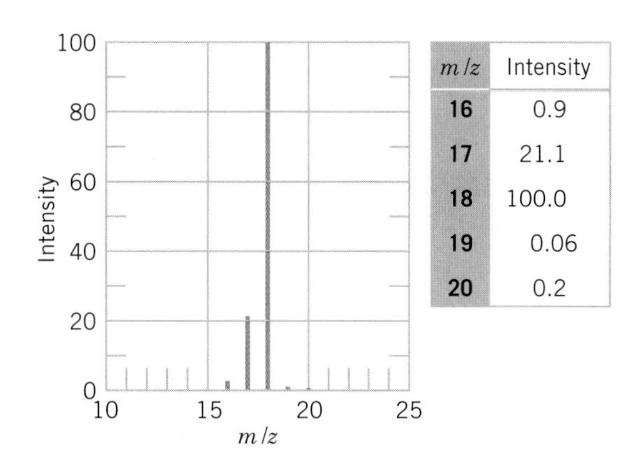

m/z	Intensity
16	0.9
17	21.1
18	100.0
19	0.06
20	0.2

FIGURE 9.37 Mass spectrum for Problem 9.13.

PROBLEM 9.12

Propose a structure for the compound whose mass spectrum is given in Fig. 9.36 and make a reasonable assignment for each peak.

PROBLEM 9.13

Propose a structure for the compound whose mass spectrum is given in Fig. 9.37 and make a reasonable assignment for each peak.

PROBLEM 9.14

The compound whose mass spectrum is given in Fig. 9.38 contains three elements, one of which is fluorine. Propose a structure for the compound and make a reasonable assignment for each peak.

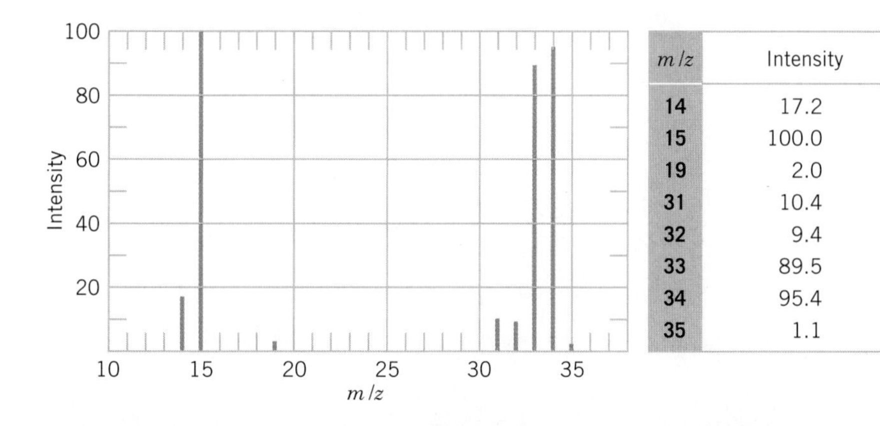

m/z	Intensity
14	17.2
15	100.0
19	2.0
31	10.4
32	9.4
33	89.5
34	95.4
35	1.1

FIGURE 9.38 Mass spectrum for Problem 9.14.

9.15 Determination of Molecular Formulas and Molecular Weights

9.15A The Molecular Ion and Isotopic Peaks

Look at Table 9.1 for a moment. Notice that most of the common elements found in organic compounds have naturally occurring *heavier* isotopes. For three of the elements—carbon, hydrogen, and nitrogen—the principal heavier isotope is one mass unit greater than the most common isotope. The presence of these elements in a compound gives rise to a small isotopic peak with mass one unit greater than the molecular ion—at $M^{\ddagger} + 1$. For four of the elements—oxygen, sulfur, chlorine, and bromine—the principal heavier isotope is two mass units greater than the most common isotope. The presence of these elements in a compound gives rise to an isotopic peak at $M^{\ddagger} + 2$:

$M^{\ddagger}$ + 1 Elements: C, H, N

$M^{\ddagger}$ + 2 Elements: O, S, Br, Cl

Isotopic peaks give us one method for determining molecular formulas. To understand how this can be done, let us begin by noticing that the isotope abundances in Table 9.1 are based on 100 atoms of the normal isotope. Now let us suppose, as an example, that we have 100 molecules of methane (CH_4). On the average there will be 1.11 molecules that contain a ^{13}C atom and 4 × 0.016 molecules that contain a ^{2}H atom. Altogether, then, these heavier isotopes should contribute an $M^{+\cdot} + 1$ peak whose intensity is about 1.17% of the intensity of the peak for the molecular ion:

$$1.11 + 4(0.016) \simeq 1.17\%$$

This correlates well with the observed intensity of the $M^{+\cdot} + 1$ peak in the actual spectrum of methane given in Fig. 9.36.

For molecules with a modest number of atoms we can determine molecular formulas in the following way. If the $M^{+\cdot}$ peak is not the base peak, the first thing we do with the mass spectrum of an unknown compound is to recalculate the intensities of the $M^{+\cdot} + 1$ and $M^{+\cdot} + 2$ peaks to express them as percentages of the intensity of the $M^{+\cdot}$ peak. Consider, for example, the mass spectrum given in Fig. 9.39. The $M^{+\cdot}$ peak at *m/z* 72 is not the base peak. Therefore, we need to recalculate the intensities of the peaks in our spectrum at *m/z* 72, 73, and 74 as percentages of the peak at *m/z* 72. We do this by dividing each intensity by the intensity of the $M^{+\cdot}$ peak, which is 73%, and multiplying by 100. These results are shown here and in the second column of Fig. 9.39.

m/z	Intensity (% of $M^{+\cdot}$)
72	73.0/73 × 100 = 100
73	3.3/73 × 100 = 4.5
74	0.2/73 × 100 = 0.3

Then we use the following guides to determine the molecular formula:

FIGURE 9.39 Mass spectrum of an unknown compound.

m/z	Intensity (as percent of base peak)	*m/z*		Intensity (as percent of $M^{+\cdot}$)
27	59.0	72	$M^{+\cdot}$	100.0
28	15.0	73	$M^{+\cdot} + 1$	4.5
29	54.0	74	$M^{+\cdot} + 2$	0.3
39	23.0			
41	60.0			
42	12.0			
43	79.0			
44	100.0 (base)			
72	73.0 $M^{+\cdot}$			
73	3.3			
74	0.2			

Recalculated to base on $M^{+\cdot}$

Tools for determining a molecular formula by MS

1. **Is $M^{+\cdot}$ odd or even? According to the nitrogen rule, if it is even, then the compound must contain an even number of nitrogen atoms (zero is an even number).** For our unknown, $M^{+\cdot}$ is even. The compound must have an even number of nitrogen atoms.

2. **The relative abundance of the $M^{+\cdot} + 1$ peak indicates the number of carbon atoms. Number of C atoms = relative abundance of $(M^{+\cdot} + 1)/1.1$.** For our unknown (Fig. 9.39),

$$\text{Number of C atoms} = \frac{4.5}{1.1} \cong 4$$

(This formula works because ^{13}C is the most important contributor to the $M^{+\cdot} + 1$ peak and the approximate natural abundance of ^{13}C is 1.1%.)

3. **The relative abundance of the $M^{+\cdot} + 2$ peak indicates the presence (or absence) of S (4.4%), Cl (33%), or Br (98%)** (see Table 9.1). For our unknown, $M^{+\cdot} + 2 = 0.3\%$; thus, we can assume that S, Cl, and Br are absent.

4. **The molecular formula can now be established by determining the number of hydrogen atoms and adding the appropriate number of oxygen atoms, if necessary.**

For our unknown the $M^{+\cdot}$ peak at m/z 72 gives us the molecular weight. It also tells us (since it is even) that nitrogen is absent because a compound with four carbons (as established above) and two nitrogens (to get an even molecular weight) would have a molecular weight (76) greater than that of our compound.

For a molecule composed of C and H only,

$$H = 72 - (4 \times 12) = 24$$

but C_4H_{24} is impossible.

For a molecule composed of C, H, and one O,

$$H = 72 - (4 \times 12) - 16 = 8$$

and thus our unknown has the molecular formula C_4H_8O.

PROBLEM 9.15

(a) Write structural formulas for at least 14 stable compounds that have the formula C_4H_8O.
(b) The IR spectrum of the unknown compound shows a strong peak near 1730 cm^{-1}. Which structures now remain as possible formulas for the compound? (We continue with this compound in Problem 9.25.)

PROBLEM 9.16

Determine the molecular formula for a compound that gives the following mass spectral data:

m/z	Intensity (as % of base peak)
86 $M^{+\cdot}$	10.00
87	0.56
88	0.04

PROBLEM 9.17

(a) What approximate intensities would you expect for the $M^{+\cdot}$ and $M^{+\cdot} + 2$ peaks of CH_3Cl?
(b) For the $M^{+\cdot}$ and $M^{+\cdot} + 2$ peaks of CH_3Br?
(c) An organic compound gives an $M^{+\cdot}$ peak at m/z 122 and a peak of nearly equal intensity at m/z 124. What is a likely molecular formula for the compound?

PROBLEM 9.18 Use the mass spectral data below to determine the molecular formula for the compound:

m/z	Intensity (as % of base peak)
14	8.0
15	38.6
18	16.3
28	39.7
29	23.4
42	46.6
43	10.7
44	100.0 (base)
73	86.1 $M^{+\cdot}$
74	3.2
75	0.2

PROBLEM 9.19 **(a)** Determine the molecular formula of the compound whose mass spectrum is given in the following tabulation:

m/z	Intensity (as % of base peak)
27	34
39	11
41	22
43	100 (base)
63	26
65	8
78	24 $M^{+\cdot}$
79	0.8
80	8

(b) The ^1H NMR spectrum of this compound consists only of a large doublet and a small septet. What is the structure of the compound?

As the number of atoms in a molecule increases, calculations like this become more and more complex and time consuming. Fortunately, however, these calculations can be done readily with computers, and tables are now available that give relative values for the $M^{+\cdot} + 1$ and $M^{+\cdot} + 2$ peaks for all combinations of common elements with molecular formulas up to mass 500. Part of the data obtained from one of these tables is given in Table 9.2. Use Table 9.2 to check the results of our example (Fig. 9.39) and your answer to Problem 9.18.

9.15B High-Resolution Mass Spectrometry

All of the spectra that we have described so far have been determined on what are called "low-resolution" mass spectrometers. These spectrometers, as we noted earlier, measure *m/z* values to the nearest whole-number mass unit. Many laboratories are equipped with this type of mass spectrometer.

Some laboratories, however, are equipped with the more expensive "high-resolution" mass spectrometers. These spectrometers can measure *m/z* values to three or four decimal

TABLE 9.2 **Relative Intensities of $M^{+\cdot} + 1$ and $M^{+\cdot} + 2$ Peaks for Various Combinations of C, H, N, and O for Masses 72 and 73**

		Percentage of $M^{+\cdot}$ Intensity				Percentage of $M^{+\cdot}$ Intensity	
$M^{+\cdot}$	Formula	$M^{+\cdot} + 1$	$M^{+\cdot} + 2$	$M^{+\cdot}$	Formula	$M^{+\cdot} + 1$	$M^{+\cdot} + 2$
72	CH_2N_3O	2.30	0.22	73	CHN_2O_2	1.94	0.41
	CH_4N_4	2.67	0.03		CH_3N_3O	2.31	0.22
	$C_2H_2NO_2$	2.65	0.42		CH_5N_4	2.69	0.03
	$C_2H_4N_2O$	3.03	0.23		C_2HO_3	2.30	0.62
	$C_2H_6N_3$	3.40	0.04		$C_2H_3NO_2$	2.67	0.42
	$C_3H_4O_2$	3.38	0.44		$C_2H_5N_2O$	3.04	0.23
	C_3H_6NO	3.76	0.25		$C_2H_7N_3$	3.42	0.04
	$C_3H_8N_2$	4.13	0.07		$C_3H_5O_2$	3.40	0.44
	C_4H_8O	4.49	0.28		C_3H_7NO	3.77	0.25
	$C_4H_{10}N$	4.86	0.09		$C_3H_9N_2$	4.15	0.07
	C_5H_{12}	5.60	0.13		C_4H_9O	4.51	0.28
					$C_4H_{11}N$	4.88	0.10
					C_6H	6.50	0.18

[a]Data from Beynon, J. H. *Mass Spectrometry and Its Applications to Organic Chemistry;* Elsevier: Amsterdam, 1960; p 489.

places and thus provide an extremely accurate method for determining molecular weights. And because molecular weights can be measured so accurately, these spectrometers also allow us to determine molecular formulas.

The determination of a molecular formula by an accurate measurement of a molecular weight is possible because the actual masses of atomic particles (nuclides) are not integers (see Table 9.3). Consider, as examples, the three molecules O_2, N_2H_4, and CH_3OH. The actual atomic masses of the molecules are all different (though nominally they all have atomic mass of 32):

$$O_2 = 2(15.9949) = 31.9898$$

$$N_2H_4 = 2(14.0031) + 4(1.00783) = 32.0375$$

$$CH_4O = 12.00000 + 4(1.00783) + 15.9949 = 32.0262$$

High-resolution mass spectrometers are available that are capable of measuring mass with an accuracy of 1 part in 40,000 or better. Thus, such a spectrometer can easily distinguish among these three molecules and, in effect, tell us the molecular formula.

The ability of high-resolution instruments to measure exact masses has been put to great use in the analysis of biomolecules such as proteins and nucleic acids. For example,

TABLE 9.3 **Exact Masses of Nuclides**

Isotope	Mass	Isotope	Mass
^{1}H	1.00783	^{19}F	18.9984
^{2}H	2.01410	^{32}S	31.9721
^{12}C	12.00000 (std)	^{33}S	32.9715
^{13}C	13.00336	^{34}S	33.9679
^{14}N	14.0031	^{35}Cl	34.9689
^{15}N	15.0001	^{37}Cl	36.9659
^{16}O	15.9949	^{79}Br	78.9183
^{17}O	16.9991	^{81}Br	80.9163
^{18}O	17.9992	^{127}I	126.9045

one method that has been used to determine the amino acid sequence in oligopeptides is to measure the exact mass of fragments derived from an original oligopeptide, where the mixture of fragments includes oligopeptides differing in length by one amino acid residue. The exact mass difference between each fragment uniquely indicates the amino acid residue that occupies that position in the intact oligopeptide (see Section 24.5F). Another application of exact mass determinations is the identification of peptides in mixtures by comparison of mass spectral data with a database of exact masses for known peptides. This technique has become increasingly important in the field of proteomics (Section 24.14).

9.16 Fragmentation

In EI mass spectrometry the molecular ion is a highly energetic species, and in the case of a complex molecule a great many things can happen to it. The molecular ion can break apart in a variety of ways, the fragments that are produced can undergo further fragmentation, and so on. In a certain sense, EI mass spectroscopy is a "brute-force" technique. Striking an organic molecule with 70-eV electrons is a little like firing a howitzer at a house made of matchsticks. That fragmentation takes place in any sort of predictable way is truly remarkable—and yet it does. Many of the same factors that govern ordinary chemical reactions seem to apply to fragmentation processes, and many of the principles that we have learned about the relative stabilities of carbocations, radicals, and molecules will help us to make some sense out of what takes place. And as we learn something about what kinds of fragmentations to expect, we shall be much better able to use mass spectra as aids in determining the structures of organic molecules.

We cannot, of course, in the limited space that we have here, look at these processes in great detail, but we can examine some of the more important ones.

As we begin, keep two important principles in mind. (1) The reactions that take place in a mass spectrometer are usually *unimolecular*—that is, they involve only a *single* molecular fragment. This is true because the pressure in a mass spectrometer is kept so low ($\sim 10^{-6}$ torr) that reactions requiring bimolecular collisions usually do not occur. (There are specialized techniques where bimolecular collisions are involved, but we will not discuss them here.) (2) The relative ion abundances, as measured by peak intensities, are extremely important. We shall see that the appearance of certain prominent peaks in the spectrum gives us important information about the structures of the fragments produced and about their original locations in the molecule.

9.16A Fragmentation by Cleavage at a Single Bond

One important type of fragmentation is the simple cleavage at a single bond. With a radical cation this cleavage can take place in at least two ways; each way produces a *cation* and a *radical*. Only the cations are detected in a positive ion mass spectrometer. (The radicals, because they are not charged, are not deflected by the magnetic field and, therefore, are not detected.) With the molecular ion obtained from propane, for example, two possible modes of cleavage are

$$CH_3CH_2CH_3 \xrightarrow[-e^-]{} CH_3CH_2 \overset{+}{\vdots} CH_3 \longrightarrow \underset{m/z\ 29}{CH_3CH_2^+} + \cdot CH_3$$

or

$$CH_3CH_2 \overset{+}{\vdots} CH_3 \longrightarrow CH_3CH_2\cdot + \underset{m/z\ 15}{{}^+CH_3}$$

These two modes of cleavage do not take place at equal rates, however. Although the relative abundance of cations produced by such a cleavage is influenced both by the stability of the carbocation and by the stability of the radical, *the carbocation's stability is more important.** In the spectrum of propane the peak at *m/z* 29 ($CH_3CH_2^+$) is the most intense peak; the peak at *m/z* 15 (CH_3^+) has an intensity of only 5.6%. This reflects the greater stability of $CH_3CH_2^+$ when compared to CH_3^+.

9.16B Fragmentation Equations

Before we go further, we need to examine some of the conventions that are used in writing equations for fragmentation reactions. In the two equations for cleavage at the single bond of propane that we have just written, we have localized the odd electron and the charge on one of the carbon–carbon sigma bonds of the molecular ion. When we write structures this way, the choice of just where to localize the odd electron and the charge is sometimes arbitrary. When possible, however, we write the structure showing the molecular ion that would result from the removal of one of the most loosely held electrons of the original molecule. Just which electrons these are can usually be estimated from ionization potentials (Table 9.4). [The ionization potential of a molecule is the amount of energy (in electron volts) required to remove an electron from the molecule.]

TABLE 9.4 Ionization Potentials of Selected Molecules

Compound	Ionization Potential (eV)
$CH_3(CH_2)_3NH_2$	8.7
C_6H_6 (benzene)	9.2
C_2H_4	10.5
CH_3OH	10.8
C_2H_6	11.5
CH_4	12.7

As we might expect, ionization potentials indicate that the nonbonding electrons of nitrogen and oxygen and the π electrons of alkenes and aromatic molecules are held more loosely than the electrons of carbon–carbon and carbon–hydrogen sigma bonds. Therefore, the convention of localizing the odd electron and charge is especially applicable when the molecule contains an oxygen, a nitrogen, a double bond, or an aromatic ring. If the molecule contains only carbon–carbon and carbon–hydrogen sigma bonds and if it contains a great many of these, then the choice of where to localize the odd electron and the charge is so arbitrary as to be impractical. In these instances we usually resort to another convention. We write the formula for the radical cation in brackets and place the odd electron and charge outside. Using this convention, we would write the two fragmentation reactions of propane in the following way:

$$[CH_3CH_2CH_3]^{\ddagger} \longrightarrow CH_3CH_2^+ + \cdot CH_3$$

$$[CH_3CH_2CH_3]^{\ddagger} \longrightarrow CH_3CH_2\cdot + {}^+CH_3$$

PROBLEM 9.20

The most intense peak in the mass spectrum of 2,2-dimethylbutane occurs at *m/z* 57. **(a)** What carbocation does this peak represent? **(b)** Using the convention that we have just described, write an equation that shows how this carbocation arises from the molecular ion.

*This can be demonstrated through thermochemical calculations that we cannot go into here. The interested student is referred to McLafferty, F. W. *Interpretation of Mass Spectra,* 2nd ed.; Benjamin: Reading, MA, 1973; pp. 41, 210–211.

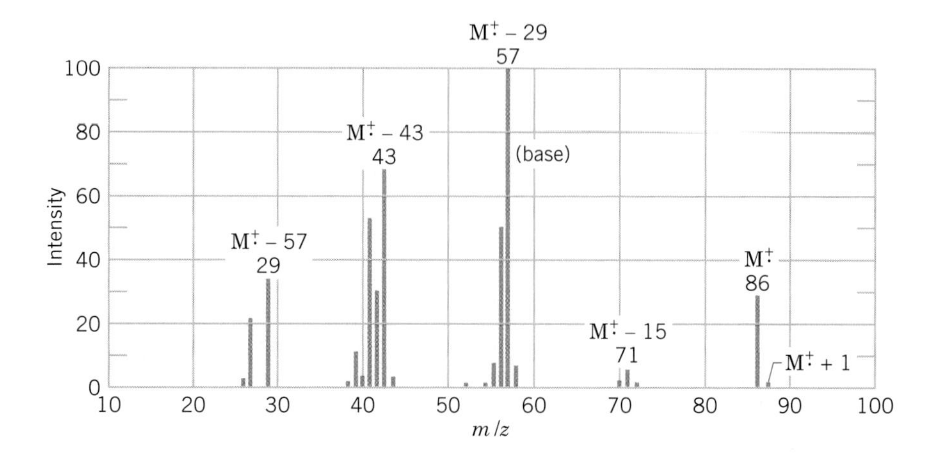

FIGURE 9.40 Mass spectrum of hexane.

Figure 9.40 shows us the kind of fragmentation a longer chain alkane can undergo. The example here is hexane, and we see a reasonably abundant molecular ion at *m/z* 86 accompanied by a small $M^{\ddagger} + 1$ peak. There is also a smaller peak at *m/z* 71 ($M^{\ddagger} - 15$) corresponding to the loss of $\cdot CH_3$, and the base peak is at *m/z* 57 ($M^{\ddagger} - 29$) corresponding to the loss of $\cdot CH_2CH_3$. The other prominent peaks are at *m/z* 43 ($M^{\ddagger} - 43$) and *m/z* 29 ($M^{\ddagger} - 57$), corresponding to the loss of $\cdot CH_2CH_2CH_3$ and $\cdot CH_2CH_2CH_2CH_3$, respectively. The important fragmentations are just the ones we would expect:

$$[CH_3CH_2CH_2CH_2CH_2CH_3]^{\ddagger} \longrightarrow CH_3CH_2CH_2CH_2CH_2^+ + \cdot CH_3 \quad (m/z\ 71)$$

$$\longrightarrow CH_3CH_2CH_2CH_2^+ + \cdot CH_2CH_3 \quad (m/z\ 57)$$

$$\longrightarrow CH_3CH_2CH_2^+ + \cdot CH_2CH_2CH_3 \quad (m/z\ 43)$$

$$\longrightarrow CH_3CH_2^+ + \cdot CH_2CH_2CH_2CH_3 \quad (m/z\ 29)$$

Chain branching increases the likelihood of cleavage at a branch point because a more stable carbocation can result. When we compare the mass spectrum of 2-methylbutane (Fig. 9.41) with the spectrum of hexane, we see a much more intense peak at $M^{\ddagger} - 15$.

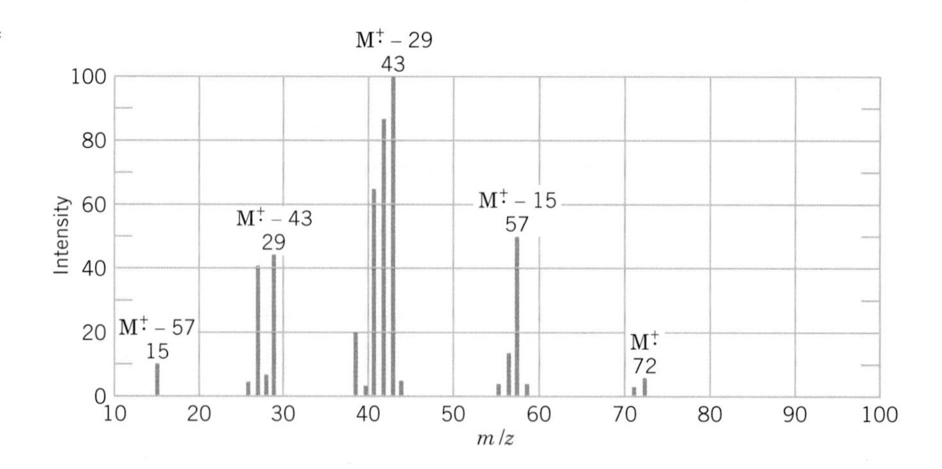

FIGURE 9.41 Mass spectrum of 2-methylbutane.

Loss of a methyl radical from the molecular ion of 2-methylbutane can give a secondary carbocation:

$$\left[\begin{array}{c} \quad CH_3 \\ \quad | \\ CH_3CHCH_2CH_3 \end{array}\right]^{+\cdot} \longrightarrow CH_3\overset{+}{C}HCH_2CH_3 + \cdot CH_3$$

$$\underset{M^{+\cdot}}{m/z\ 72} \qquad\qquad \underset{M^{+\cdot} - 15}{m/z\ 57}$$

whereas with hexane loss of a methyl radical can yield only a primary carbocation.

With neopentane (Fig. 9.42), this effect is even more dramatic. Loss of a methyl radical by the molecular ion produces a *tertiary* carbocation, and this reaction takes place so readily that virtually none of the molecular ions survive long enough to be detected:

$$\left[\begin{array}{c} CH_3 \\ | \\ CH_3{-}C{-}CH_3 \\ | \\ CH_3 \end{array}\right]^{+\cdot} \longrightarrow CH_3{-}\overset{+}{C}(CH_3)_2 + \cdot CH_3$$

$$\underset{M^{+\cdot}}{m/z\ 72} \qquad\qquad \underset{M^{+\cdot} - 15}{m/z\ 57}$$

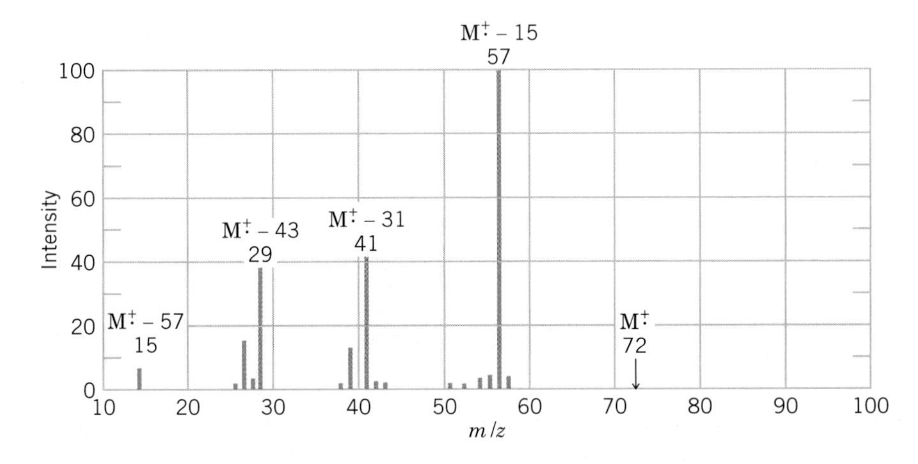

FIGURE 9.42 Mass spectrum of neopentane.

PROBLEM 9.21

In contrast to 2-methylbutane and neopentane, the mass spectrum of 3-methylpentane (not given) has a peak of very low intensity at $M^{+\cdot} - 15$. It has a peak of very high intensity at $M^{+\cdot} - 29$, however. Explain.

Carbocations stabilized by resonance are usually also prominent in mass spectra. Several ways that resonance-stabilized carbocations can be produced are outlined in the following list. These examples begin by illustrating the likely sites for initial ionization (π and nonbonding electrons), as well.

1. Alkenes ionize and frequently undergo fragmentations that yield resonance-stabilized allylic cations:

$$CH_2{=}CH{-}CH_2{-}R \xrightarrow[-e^-]{\text{ionization}} CH_2\overset{+\cdot}{-}CH{-}CH_2{:}R \xrightarrow{\text{fragmentation}} {}^{+}CH_2{-}CH{=}CH_2 + \cdot R$$

$$m/z\ 41 \qquad \updownarrow$$

$$CH_2{=}CH{-}\overset{+}{C}H_2$$

2. Carbon–carbon bonds next to an atom with an unshared electron pair usually break readily because the resulting carbocation is resonance stabilized:

$$R{-}\ddot{Z}{-}CH_2{-}CH_3 \xrightarrow[-e^-]{\text{ionization}} R{-}\overset{+\cdot}{Z}{-}CH_2{:}CH_3 \xrightarrow{\text{fragmentation}} R{-}\overset{+}{Z}{=}CH_2 + \cdot CH_3 \longleftrightarrow R{-}\ddot{Z}{-}\overset{+}{C}H_2$$

where Z = N, O, or S; R may also be H.

3. Carbon–carbon bonds next to the carbonyl group of an aldehyde or ketone break readily because resonance-stabilized ions called acylium ions are produced:

$$RR'C{=}\ddot{O}{:} \xrightarrow[-e^-]{\text{ionization}} RR'C{=}\overset{\cdot+}{O}{:} \xrightarrow{\text{fragmentation}} R'{-}C{\equiv}\overset{+}{O}{:} + R\cdot \longleftrightarrow R'{-}\overset{+}{C}{=}\ddot{O}{:}$$

Acylium ion

or

$$RR'C{=}\overset{\cdot+}{O}{:} \xrightarrow{\text{fragmentation}} R{-}C{\equiv}\overset{+}{O}{:} + R'\cdot \longleftrightarrow R{-}\overset{+}{C}{=}\ddot{O}{:}$$

Acylium ion

4. Alkyl-substituted benzenes ionize by loss of a π electron and undergo loss of a hydrogen atom or methyl group to yield the relatively stable tropylium ion (see Section 14.7C). This fragmentation gives a prominent peak (sometimes the base peak) at m/z 91:

Toluene $\xrightarrow[-e^-]{\text{ionization}}$ CH_3-substituted benzene radical cation (+·) $\xrightarrow[-H\cdot]{\text{fragmentation and rearrangement}}$ (m/z 91) ≡ **Tropylium ion**

Dimethylbenzene radical cation (+·) $\xrightarrow[-CH_3\cdot]{\text{fragmentation and rearrangement}}$ (m/z 91)

5. Monosubstituted benzenes with other than alkyl groups also ionize by loss of a π electron and then lose their substituent to yield a phenyl cation with m/z 77:

$C_6H_5{-}Y \xrightarrow[-e^-]{\text{ionization}}$ $[C_6H_5{-}Y]^{+\cdot} \xrightarrow[-Y\cdot]{\text{fragmentation}}$ $C_6H_5^+$ (m/z 77)

Y = halogen, $-NO_2$, $-\overset{O}{\overset{\|}{C}}R$, $-R$, and so on

PROBLEM 9.22 The mass spectrum of 4-methyl-1-hexene (not given) shows intense peaks at m/z 57 and m/z 41. What fragmentation reactions account for these peaks?

Problem 9.23

Explain the following observations that can be made about the mass spectra of alcohols:
(a) The molecular ion peak of a primary or secondary alcohol is very small; with a tertiary alcohol it is usually undetectable.
(b) Primary alcohols show a prominent peak at *m/z* 31.
(c) Secondary alcohols usually give prominent peaks at *m/z* 45, 59, 73, and so on.
(d) Tertiary alcohols have prominent peaks at *m/z* 59, 73, 87, and so on.

Problem 9.24

The mass spectra of butyl isopropyl ether and butyl propyl ether are given in Figs. 9.43 and 9.44. **(a)** Which spectrum represents which ether? **(b)** Explain your choice.

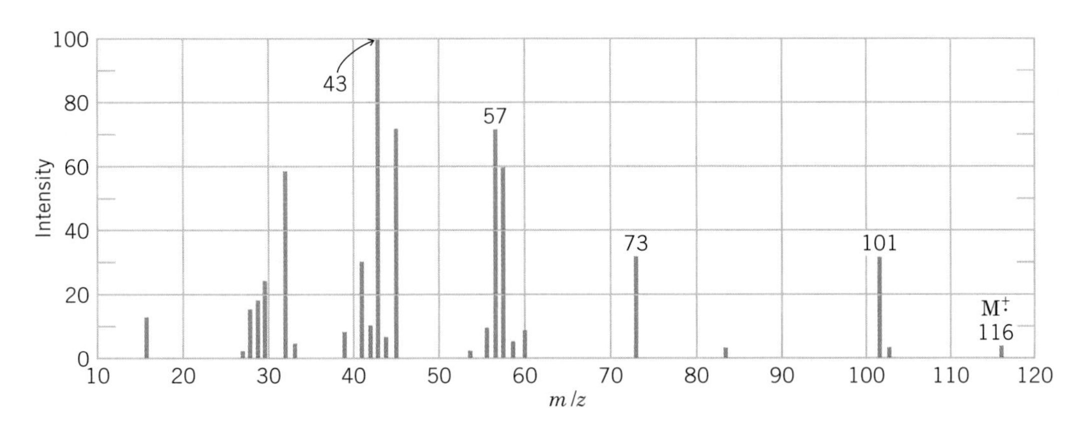

FIGURE 9.43 Mass spectrum for Problem 9.24.

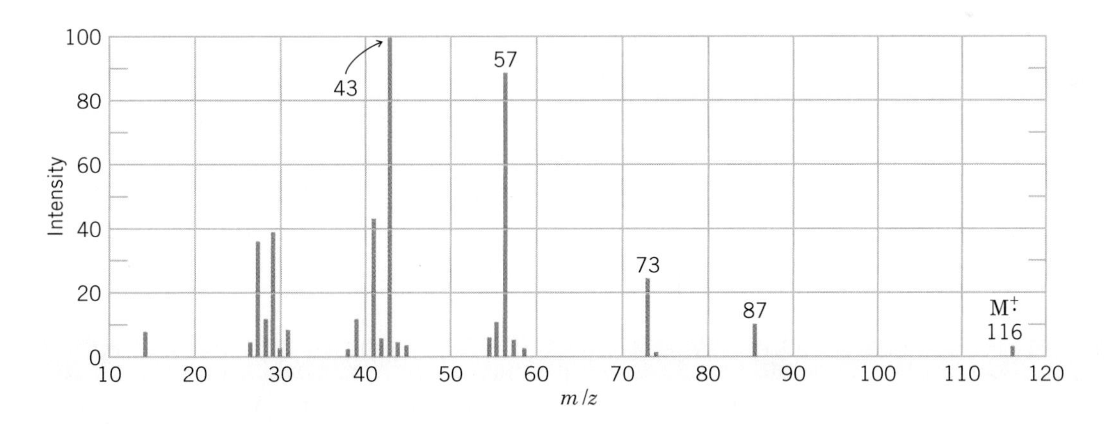

FIGURE 9.44 Mass spectrum for Problem 9.24.

9.16C Fragmentation by Cleavage of Two Bonds

Many peaks in mass spectra can be explained by fragmentation reactions that involve the breaking of two covalent bonds. When a radical cation undergoes this type of fragmentation, the products are *a new radical cation* and *a neutral molecule*. Some important examples, starting from the initial radical cation, are the following:

1. Alcohols frequently show a prominent peak at $M^{\dagger} - 18$. This corresponds to the loss of a molecule of water:

$$\text{R—CH(H)—CH}_2\text{—}\overset{+}{\text{O}}\text{H} \longrightarrow \text{R—CH}\overset{\cdot +}{=}\text{CH}_2 + \text{H—}\ddot{\text{O}}\text{—H}$$

M‡ **M‡ − 18**

which can also be written as

$$[\text{R—CH}_2\text{—CH}_2\text{—OH}]^{\cdot +} \longrightarrow [\text{R—CH=CH}_2]^{\cdot +} + \text{H}_2\text{O}$$

M‡ **M‡ − 18**

2. Cycloalkenes can undergo a retro-Diels–Alder reaction (Section 13.11) that produces an alkene and an alkadienyl radical cation:

$$[\text{cyclohexene}]^{\cdot +} \longrightarrow [\text{butadiene}]^{\cdot +} + \text{CH}_2\text{=CH}_2$$

which can also be written as

$$[\text{cyclohexene}]^{\cdot +} \longrightarrow [\text{butadiene}]^{\cdot +} + \|$$

3. Carbonyl compounds with a hydrogen on their γ carbon undergo a fragmentation called the *McLafferty rearrangement.*

$$\text{Y—C(=}\overset{+}{\text{O}}\text{)—CH}_2\text{—CH}_2\text{—CH(H)—R} \longrightarrow \text{Y—C(—}\overset{+}{\ddot{\text{O}}}\text{—H)=CH}_2 + \text{R—CH=CH}_2$$

where Y may be R, H, OR, OH, and so on.

In addition to these reactions, we frequently find peaks in mass spectra that result from the elimination of other small stable neutral molecules, for example, H_2, NH_3, CO, HCN, H_2S, alcohols, and alkenes.

9.17 GC/MS Analysis

Gas chromatography is often coupled with mass spectrometry in a technique called **GC/MS analysis**. The gas chromatograph separates components of a mixture, while the mass spectrometer then gives structural information about each one (Fig. 9.45). GC/MS can also provide quantitative data when standards of known concentration are used with the unknown.

In GC analysis, a minute amount of a mixture to be analyzed, typically 0.001 mL (1.0 μL) or less of a dilute solution containing the sample, is injected by syringe into a heated port of the gas chromatograph. The sample is vaporized in the injector port and swept by a flow of inert gas into a capillary column. The capillary column is a thin tube usually 10–30 meters long and 0.1–0.5 mm in diameter. It is contained in a chamber (the "oven") whose temperature can be varied according to the volatility of the samples being

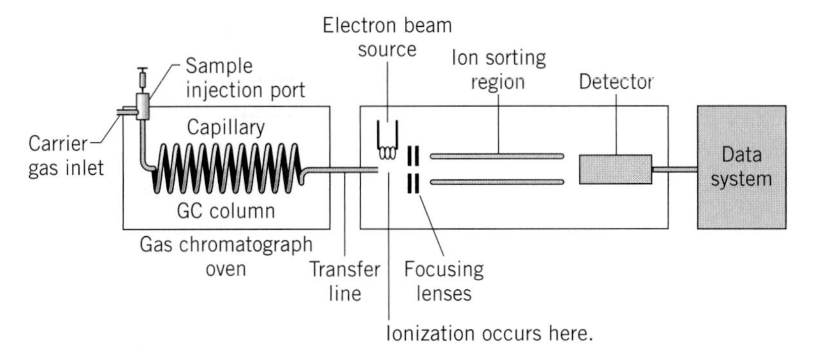

FIGURE 9.45 Schematic of a typical capillary gas chromatograph/mass spectrometer (GC/MS).

analyzed. The inside of the capillary column is typically coated with a "stationary phase" of low polarity (essentially a high-boiling and very viscous liquid that is often a nonpolar silicon-based polymer). As molecules of the mixture are swept by the inert gas through the column, they travel at different rates according to their boiling points and the degree of affinity for the stationary phase. Materials with higher boiling points or stronger affinity for the stationary phase take longer to pass through the column. Low-boiling and nonpolar materials pass through very quickly. The length of time each component takes to travel through the column is called the retention time. Retention times typically range from 1 to about 30 minutes, depending on the sample and the specific column used.

As each component of the mixture exits the GC column it travels into a mass spectrometer. Here, molecules of the sample are bombarded by electrons; ions and fragments of the molecule are formed, and a mass spectrum results similar to those we have studied earlier in this chapter. The important thing, however, is that mass spectra are obtained for *each* component of the original mixture that is separated. This ability of GC/MS to separate mixtures and give information about the structure of each component makes it a virtually indispensable tool in analytical, forensic, and organic synthesis laboratories.

9.18 Mass Spectrometry of Biomolecules

Advances in mass spectrometry have made it a tool of exceptional power for analysis of large biomolecules. Although the EI mass spectrometry techniques described earlier are invaluable for analysis of volatile compounds with molecular weights up to about 1000 daltons (atomic mass units), it is now possible to use other "soft" ionization techniques for analysis of nonvolatile compounds such as proteins, nucleic acids, and other biologically relevant compounds with molecular weights up to and in excess of 100,000 daltons. Techniques such as electrospray ionization (ESI), ion spray, matrix-assisted laser desorption ionization (MALDI), and fast atom bombardment (FAB) mass spectrometry have become virtually indispensable for some analyses of protein molecular weight, enzyme–substrate complexes, antibody–antigen binding, drug–receptor interactions, and DNA oligonucleotide sequence determination. The sorting methods (Section 9.13C) for ions generated by these techniques include quadrupole mass filters, ion traps, time-of-flight (TOF), and Fourier transform–ion cyclotron resonance (FT ICR, or FT MS). We discuss ESI and MALDI mass spectrometry, as well as applications of mass spectrometry to protein sequencing and analysis, in Sections 24.5F, 24.13B, and 24.14.

A classic and highly useful reference on MS, NMR, and IR methods is Silverstein, R. M.; Webster, F. X. *Spectrometric Identification of Organic Compounds,* 6th ed.; Wiley: New York, 1998.

Key Terms and Concepts

Base peak	Section 9.14
Broadband (BB) proton decoupling	Section 9.10B
Carbon-13 NMR spectroscopy	Section 9.10A
Chemical shift	Sections 9.3C, 9.6, 9.10C
COSY	Section 9.11
Coupling constant ($J_{a,b}$)	Section 9.8
DEPT ^{13}C spectrum	Section 9.10E
Diastereotopic hydrogen atoms	Section 9.7B
Enantiotopic hydrogen atoms	Section 9.7B
Fragmentation	Sections 9.13B, 9.16
Frequency	Section 9.2
GC/MS analysis	Section 9.17
Hertz	Section 9.2
HETCOR	Section 9.11
Homotopic (chemically equivalent) hydrogen atoms	Section 9.7A
Integration	Section 9.3B
Ionization	Section 9.13A
Ion sorting	Section 9.13C
Magnetic resonance imaging (MRI)	Section 9.11B
Molecular ion	Section 9.13
Off-resonance decoupling	Section 9.10D
Shielding/deshielding of protons	Section 9.5
Signal splitting	Sections 9.3E, 9.8
Spectroscopy	Section 9.1
Spin–spin splitting	Section 9.8
Upfield/downfield	Section 9.3C
Wavelength	Section 9.2

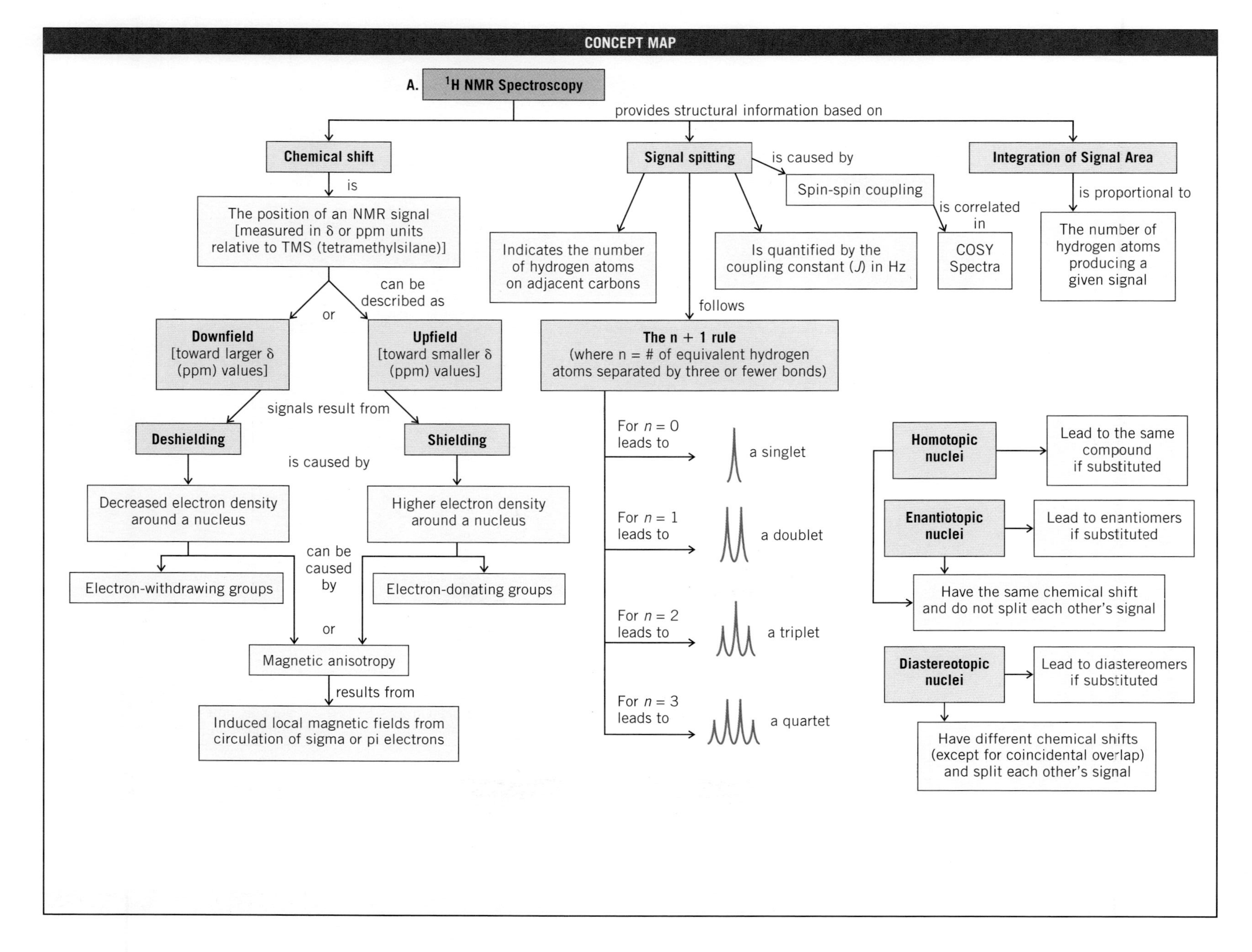
CONCEPT MAP
A. 1H NMR Spectroscopy
provides structural information based on
Chemical shift
is
The position of an NMR signal [measured in δ or ppm units relative to TMS (tetramethylsilane)]
can be described as
or
Downfield [toward larger δ (ppm) values]
Upfield [toward smaller δ (ppm) values]
signals result from
Deshielding
Shielding
is caused by
Decreased electron density around a nucleus
Higher electron density around a nucleus
can be caused by
or
Electron-withdrawing groups
Electron-donating groups
Magnetic anisotropy
results from
Induced local magnetic fields from circulation of sigma or pi electrons
Signal spitting
is caused by
Spin-spin coupling
is correlated in
COSY Spectra
Indicates the number of hydrogen atoms on adjacent carbons
Is quantified by the coupling constant (J) in Hz
follows
The n + 1 rule
(where n = # of equivalent hydrogen atoms separated by three or fewer bonds)
For n = 0 leads to
a singlet
For n = 1 leads to
a doublet
For n = 2 leads to
a triplet
For n = 3 leads to
a quartet
Integration of Signal Area
is proportional to
The number of hydrogen atoms producing a given signal
Homotopic nuclei
Lead to the same compound if substituted
Enantiotopic nuclei
Lead to enantiomers if substituted
Have the same chemical shift and do not split each other's signal
Diastereotopic nuclei
Lead to diastereomers if substituted
Have different chemical shifts (except for coincidental overlap) and split each other's signal

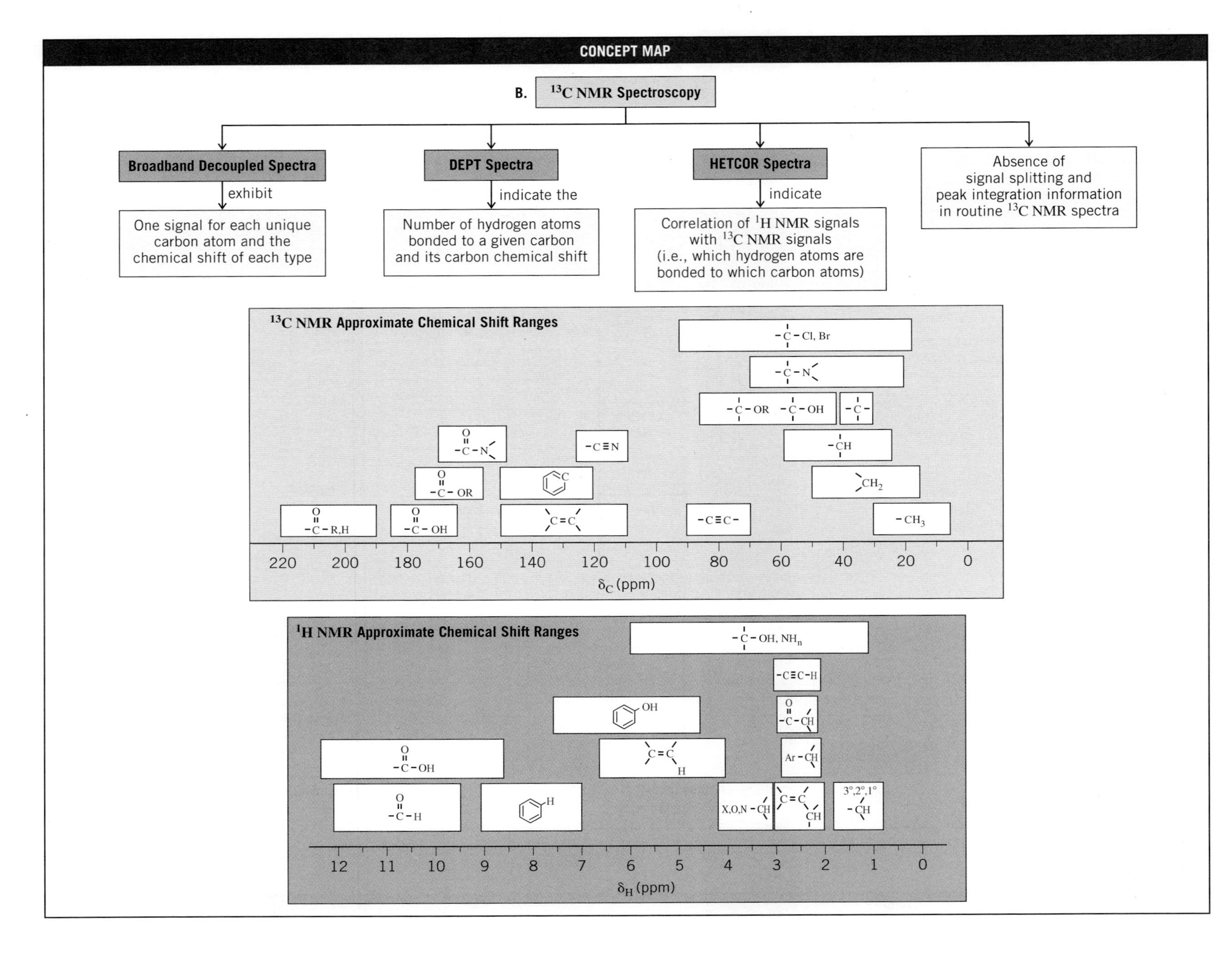
CONCEPT MAP
B. ^{13}C NMR Spectroscopy
Broadband Decoupled Spectra
exhibit
One signal for each unique carbon atom and the chemical shift of each type
DEPT Spectra
indicate the
Number of hydrogen atoms bonded to a given carbon and its carbon chemical shift
HETCOR Spectra
indicate
Correlation of ^{1}H NMR signals with ^{13}C NMR signals (i.e., which hydrogen atoms are bonded to which carbon atoms)
Absence of signal splitting and peak integration information in routine ^{13}C NMR spectra
^{13}C NMR Approximate Chemical Shift Ranges
220 200 180 160 140 120 100 80 60 40 20 0
δ_C (ppm)
^{1}H NMR Approximate Chemical Shift Ranges
12 11 10 9 8 7 6 5 4 3 2 1 0
δ_H (ppm)

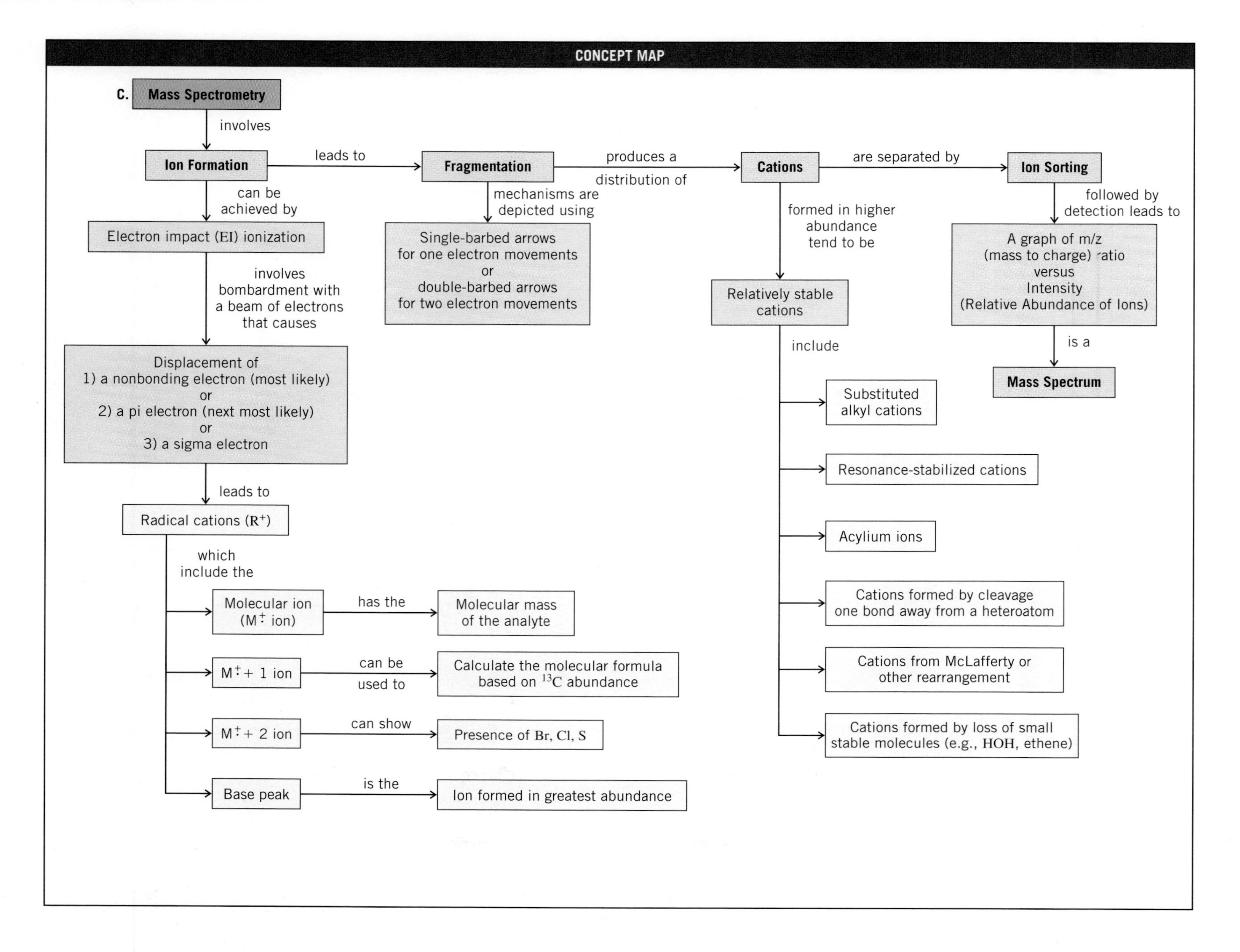

CONCEPT MAP
C.
Mass Spectrometry
involves
Ion Formation
leads to
Fragmentation
produces a distribution of
Cations
are separated by
Ion Sorting
can be achieved by
Electron impact (EI) ionization
involves bombardment with a beam of electrons that causes
Displacement of 1) a nonbonding electron (most likely) or 2) a pi electron (next most likely) or 3) a sigma electron
leads to
Radical cations (R⁺)
which include the
Molecular ion (M⁺ ion)
has the
Molecular mass of the analyte
M⁺ + 1 ion
can be used to
Calculate the molecular formula based on ¹³C abundance
M⁺ + 2 ion
can show
Presence of Br, Cl, S
Base peak
is the
Ion formed in greatest abundance
mechanisms are depicted using
Single-barbed arrows for one electron movements or double-barbed arrows for two electron movements
formed in higher abundance tend to be
Relatively stable cations
include
Substituted alkyl cations
Resonance-stabilized cations
Acylium ions
Cations formed by cleavage one bond away from a heteroatom
Cations from McLafferty or other rearrangement
Cations formed by loss of small stable molecules (e.g., HOH, ethene)
followed by detection leads to
A graph of m/z (mass to charge) ratio versus Intensity (Relative Abundance of Ions)
is a
Mass Spectrum

ADDITIONAL PROBLEMS

9.25 Reconsider Problem 9.15 and the spectrum given in Fig. 9.39. Important clues to the structure of this compound are the peaks at *m/z* 44 (the base peak) and *m/z* 29. Propose a structure for the compound and write fragmentation equations showing how these peaks arise.

9.26 The homologous series of primary amines, $CH_3(CH_2)_nNH_2$, from CH_3NH_2 to $CH_3(CH_2)_{13}NH_2$ all have their base (largest) peak at *m/z* 30. What ion does this peak represent, and how is it formed?

9.27 The mass spectrum of compound **A** is given in Fig. 9.46. The ^1H NMR spectrum of **A** consists of two singlets with area ratios of 9:2. The larger singlet is at δ 1.2, the smaller one at δ 1.3. Propose a structure for compound **A**.

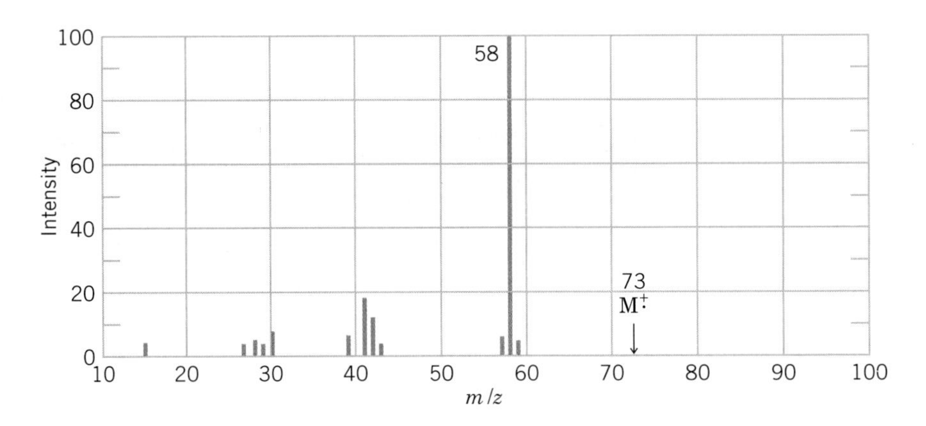

FIGURE 9.46 Mass spectrum of compound A (Problem 9.27).

9.28 The mass spectrum of compound **B** is given in Fig. 9.47. The IR spectrum of **B** shows a broad peak between 3200 and 3550 cm^{-1}. The ^1H NMR spectrum of **B** shows the following peaks: a triplet at δ 0.9, a singlet at δ 1.1, and a quartet at δ 1.6. The area ratio of these peaks is 3:7:2, respectively. Propose a structure for **B**.

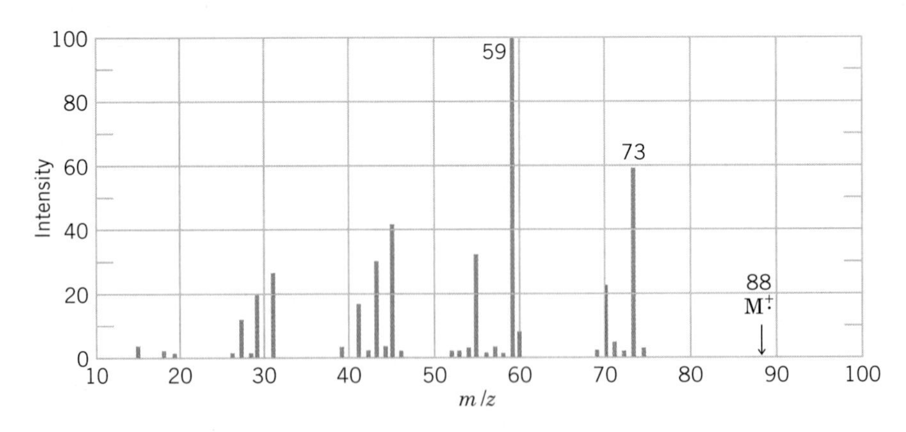

FIGURE 9.47 Mass spectrum of compound B (Problem 9.28).

Note: Problems marked with an asterisk are "challenge problems."

9.29 Listed here are ^{1}H NMR absorption peaks for several compounds. Propose a structure that is consistent with each set of data. (In some cases characteristic IR absorptions are given as well.)

(a) $C_4II_{10}O$	**^{1}H NMR spectrum**		
	singlet,	δ 1.28 (9H)	
	singlet,	δ 1.35 (1H)	
(b) C_3H_7Br	**^{1}H NMR spectrum**		
	doublet,	δ 1.71 (6H)	
	septet,	δ 4.32 (1H)	
(c) C_4H_8O	**^{1}H NMR spectrum**		**IR spectrum**
	triplet,	δ 1.05 (3H)	strong peak
	singlet,	δ 2.13 (3H)	near 1720 cm^{-1}
	quartet,	δ 2.47 (2H)	
(d) C_7H_8O	**^{1}H NMR spectrum**		**IR spectrum**
	singlet,	δ 2.43 (1H)	broad peak in
	singlet,	δ 4.58 (2H)	3200–3550 cm^{-1}
	multiplet,	δ 7.28 (5H)	region
(e) C_4H_9Cl	**^{1}H NMR spectrum**		
	doublet,	δ 1.04 (6H)	
	multiplet,	δ 1.95 (1H)	
	doublet,	δ 3.35 (2H)	
(f) $C_{15}H_{14}O$	**^{1}H NMR spectrum**		**IR spectrum**
	singlet,	δ 2.20 (3H)	strong peak
	singlet,	δ 5.08 (1H)	near 1720 cm^{-1}
	multiplet,	δ 7.25 (10H)	
(g) $C_4H_7BrO_2$	**^{1}H NMR spectrum**		**IR spectrum**
	triplet,	δ 1.08 (3H)	broad peak in
	multiplet,	δ 2.07 (2H)	2500–3000 cm^{-1}
	triplet,	δ 4.23 (1H)	region and a peak
	singlet,	δ 10.97 (1H)	at 1715 cm^{-1}
(h) C_8H_{10}	**^{1}H NMR spectrum**		
	triplet,	δ 1.25 (3H)	
	quartet,	δ 2.68 (2H)	
	multiplet,	δ 7.23 (5H)	
(i) $C_4H_8O_3$	**^{1}H NMR spectrum**		**IR spectrum**
	triplet,	δ 1.27 (3H)	broad peak in
	quartet,	δ 3.66 (2H)	2500–3000 cm^{-1}
	singlet	δ 4.13 (2H)	region and a peak
	singlet,	δ 10.95 (1H)	at 1715 cm^{-1}
(j) $C_3H_7NO_2$	**^{1}H NMR spectrum**		
	doublet,	δ 1.55 (6H)	
	septet,	δ 4.67 (1H)	
(k) $C_4H_{10}O_2$	**^{1}H NMR spectrum**		
	singlet,	δ 3.25 (6H)	
	singlet,	δ 3.45 (4H)	
(l) $C_5H_{10}O$	**^{1}H NMR spectrum**		**IR spectrum**
	doublet,	δ 1.10 (6H)	strong peak
	singlet,	δ 2.10 (3H)	near 1720 cm^{-1}
	septet,	δ 2.50 (1H)	
(m) C_8H_9Br	**^{1}H NMR spectrum**		
	doublet,	δ 2.0 (3H)	
	quartet,	δ 5.15 (1H)	
	multiplet,	δ 7.35 (5H)	

9.30 The IR spectrum of compound **E** (C_8H_6) is shown in Fig. 9.48. Compound **E** reacts with bromine in carbon tetrachloride and has an IR absorption band at about 3300 cm^{-1}. What is the structure of **E**?

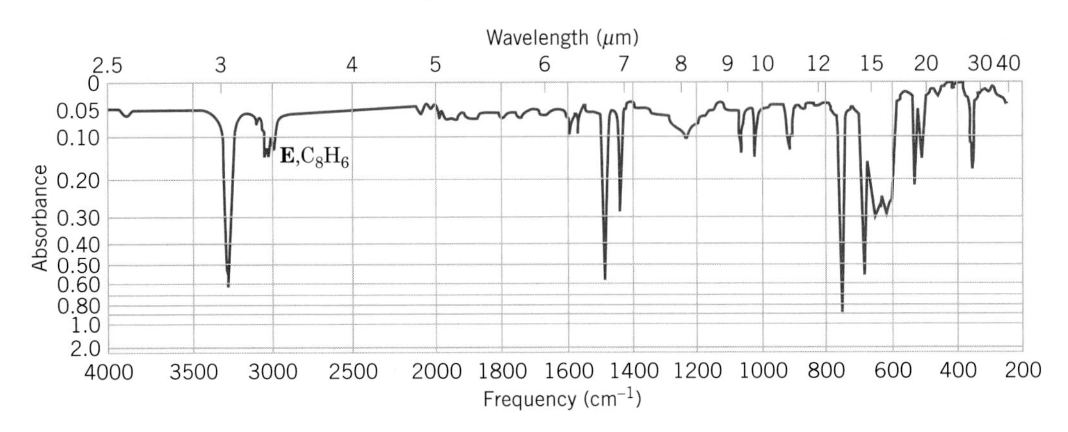

FIGURE 9.48 The IR spectrum of compound E, Problem 9.30. (Spectrum courtesy of Sadtler Research Laboratories, Inc., Philadelphia.)

9.31 Propose structures for the compounds **G** and **H** whose ^1H NMR spectra are shown in Figs. 9.49 and 9.50.

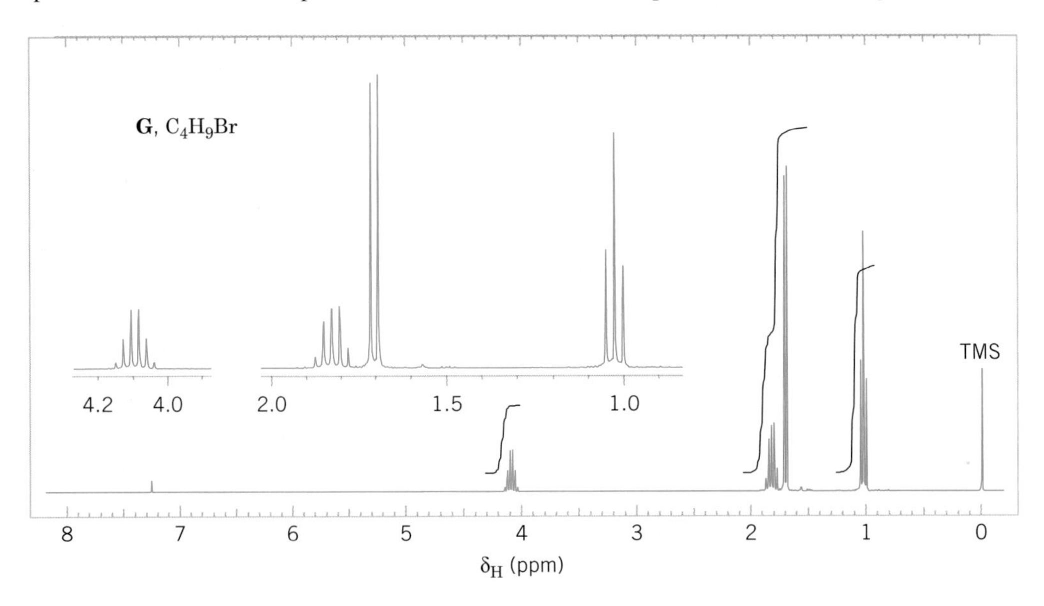

FIGURE 9.49 The 300-MHz ^1H NMR spectrum of compound G, Problem 9.31. Expansions of the signals are shown in the offset plots.

9.32 A two-carbon compound (**J**) contains only carbon, hydrogen, and chlorine. Its IR spectrum is relatively simple and shows the following absorption peaks: 3125 cm^{-1} (multiplet), 1625 cm^{-1} (multiplet), 1280 cm^{-1} (multiplet), 820 cm^{-1} (singlet), 695 cm^{-1} (singlet). The ^1H NMR spectrum of **J** consists of a singlet at δ 6.3. Using Table 2.4, make as many IR assignments as you can and propose a structure for compound **J**.

9.33 When dissolved in $CDCl_3$, a compound (**K**) with the molecular formula $C_4H_8O_2$ gives a ^1H NMR spectrum that consists of a doublet at δ 1.35, a singlet at δ 2.15, a broad singlet at δ 3.75 (1H), and a quartet at δ 4.25 (1H). When dissolved in D_2O, the compound gives a similar ^1H NMR spectrum, with the exception that the signal at δ 3.75 has disappeared. The IR spectrum of the compound shows a strong absorption peak near 1720 cm^{-1}.

(a) Propose a structure for compound **K.**

(b) Explain why the NMR signal at δ 3.75 disappears when D_2O is used as the solvent.

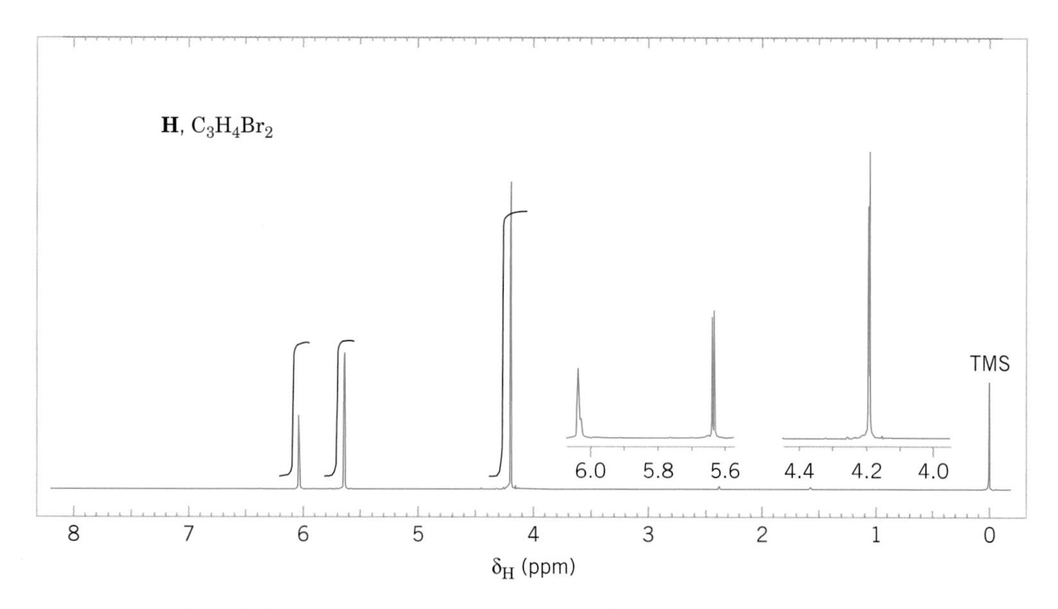

FIGURE 9.50 The 300-MHz ^1H NMR spectrum of compound H, Problem 9.31. Expansions of the signals are shown in the offset plots.

***9.34** Assume that in a certain ^1H NMR spectrum you find two peaks of roughly equal intensity. You are not certain whether these two peaks are *singlets* arising from uncoupled protons at different chemical shifts or are two peaks of a *doublet* that arises from protons coupling with a single adjacent proton. What simple experiment would you perform to distinguish between these two possibilities?

9.35 Compound **O** (C_6H_8) reacts with two molar equivalents of hydrogen in the presence of a catalyst to produce **P** (C_6H_{12}). The proton-decoupled ^{13}C spectrum of **O** consists of two singlets, one at δ 26.0 and one at δ 124.5. In the DEPT ^{13}C spectra of **O** the signal at δ 26.0 appears as a CH_2 group and the one at δ 124.5 appears as a CH group. Propose structures for **O** and **P**.

***9.36** Compound **Q** has the molecular formula C_7H_8. On catalytic hydrogenation **Q** is converted to **R** (C_7H_{12}). The broadband proton-decoupled ^{13}C spectrum of **Q** is given in Fig. 9.51. Propose structures for **Q** and **R**.

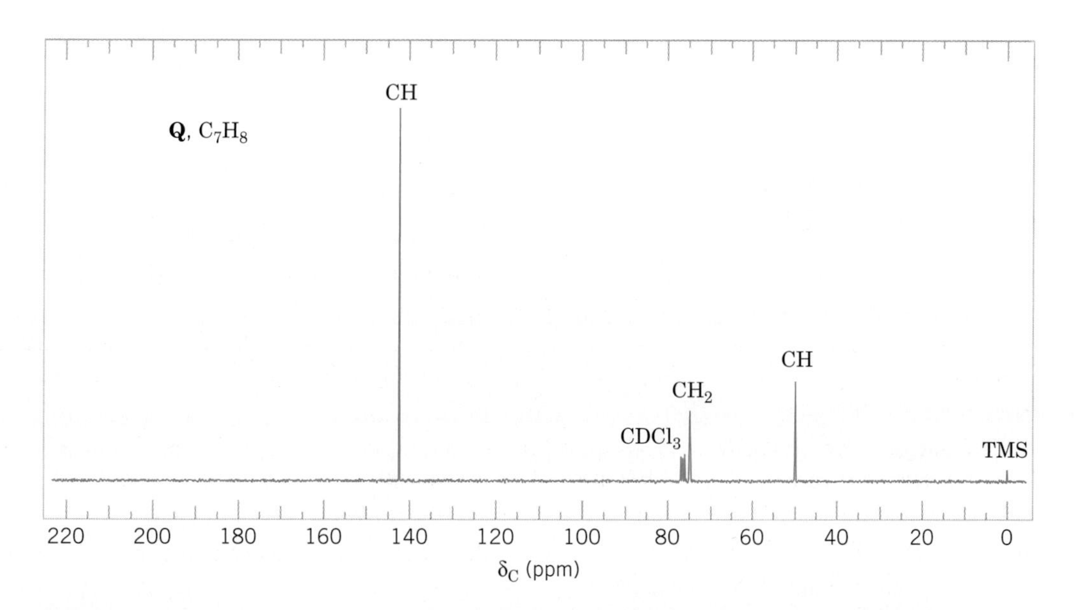

FIGURE 9.51 The broadband proton-decoupled ^{13}C NMR spectrum of compound Q, Problem 9.36. Information from the DEPT ^{13}C NMR spectra is given above each peak.

***9.37** Compound **S** (C_8H_{16}) decolorizes a solution of bromine in carbon tetrachloride. The broadband proton-decoupled ^{13}C spectrum of **S** is given in Fig. 9.52. Propose a structure for **S**.

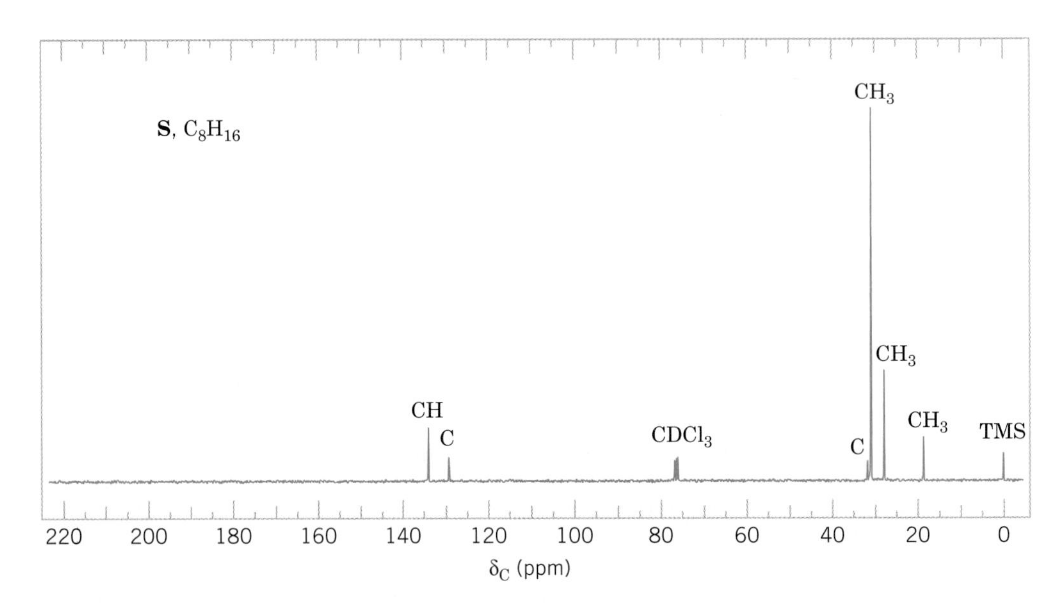

FIGURE 9.52 The broadband proton-decoupled ^{13}C NMR spectrum of compound S, Problem 9.37. Information from the DEPT ^{13}C NMR spectra is given above each peak.

9.38 Compound **T** (C_5H_8O) has a strong IR absorption band at 1745 cm^{-1}. The broadband proton-decoupled ^{13}C spectrum of **T** is given in Fig. 9.53. Propose a structure for **T**.

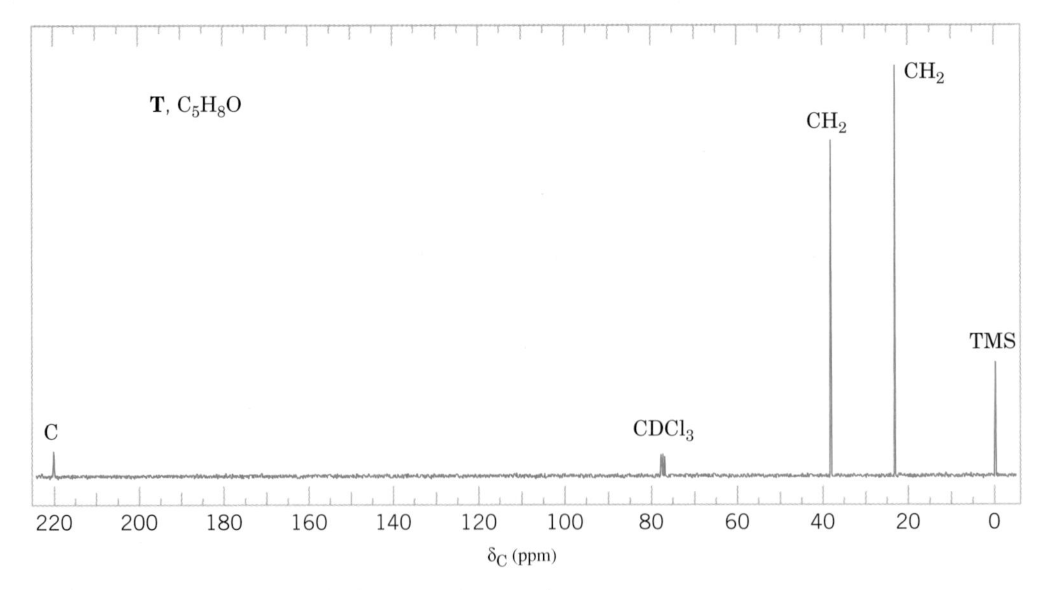

FIGURE 9.53 The broadband proton-decoupled ^{13}C NMR spectrum of compound T, Problem 9.38. Information from the DEPT ^{13}C NMR spectra is given near each peak.

9.39 The ^1H NMR examination of a solution of 1,3-dimethylcyclopentadiene in concentrated sulfuric acid shows three peaks with relative areas of 6:4:1. What is the explanation for the appearance of the spectrum?

9.40 Acetic acid has a mass spectrum showing a molecular ion peak at m/z 60. Other unbranched monocarboxylic acids with four or more carbon atoms also have a peak, frequently prominent, at m/z 60. Show how this can occur.

9.41 The ^{1}H NMR peak for the hydroxyl proton of alcohols can be found anywhere from δ 0.5 to δ 5.4. Explain this variability.

9.42 The ^{1}H NMR study of DMF (*N,N*-dimethylformamide) results in different spectra according to the temperature of the sample. At room temperature, two signals are observed for the protons of the two methyl groups. On the other hand, at elevated temperatures (>130°C) a singlet is observed that integrates for six hydrogens. Explain these differences.

9.43 The mass spectra of many benzene derivatives show a peak at m/z 51. What could account for this fragment?

***9.44**

OH OH $\xrightarrow{\text{60\% HBr}}$ **X**

The infrared spectrum of product **X** has no absorption at 3590–3650 cm^{-1} but does absorb at about 1370 and 1380 cm^{-1}, which is characteristic of *gem*-dimethyl groups. The mass spectrum of **X** contains these peaks:

A cluster at m/z 270, 272, and 274 with relative peak intensities of about 1 : 2 : 1.

Another cluster at m/z 191 and 193 with about equal peak intensities.

The NMR spectra show:

(a) In the ^{1}H spectrum, singlets at δ 1.70 and 1.80, with relative areas of 3 : 1.

(b) The ^{13}C spectrum has peaks at δ 32 (CH_3), 40 (CH_2), and 54 (quaternary C).

What is the structure of **X**? Suggest a mechanism by which it is formed.

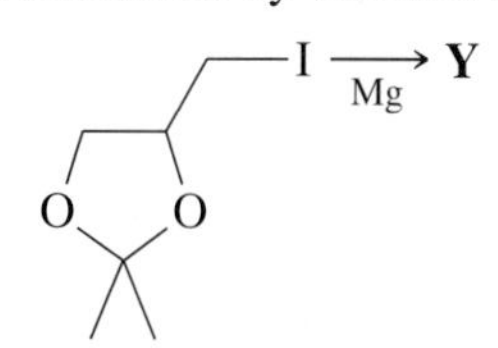

(This reaction is, incidentally, not a typical reaction of magnesium with alkyl halides.)

The infrared spectrum of product **Y** provides no easily interpretable data.

The mass spectrum of **Y** has its heaviest (and most intense) peak at m/z 215. By analogy with similar materials, this is known to be an $M^{\ddagger}-CH_3$ peak.

The ^{1}H NMR spectrum of **Y** is essentially like that of the starting material except that a doublet at δ 3.4 (representing 2 Hs) is replaced by a multiplet at δ 1.64 (representing 4 Hs).

What is the structure of **Y**?

LEARNING GROUP PROBLEMS

1. Given the following information, elucidate the structures of compounds **A** and **B**. Both compounds are soluble in dilute aqueous HCl, and both have the same molecular formula. The mass spectrum of **A** has $M^{\ddagger}$ 149 (intensity 37.1% of base peak) and $M^{\ddagger}+1$ 150 (intensity 4.2% of base peak). Other spectroscopic data for **A** and **B** are given below. Justify the structures you propose by assigning specific aspects of the data to the structures. Make sketches of the NMR spectra.

a. The IR spectrum for compound **A** shows two bands in the 3300–3500 cm^{-1} region. The broadband proton-decoupled ^{13}C NMR spectrum displayed the following signals (information from the DEPT ^{13}C spectra is given in parentheses with the ^{13}C chemical shifts):

^{13}C NMR: δ 140 (C), 127 (C), 125 (CH), 118 (CH), 24 (CH_2), 13 (CH_3)

b. The IR spectrum for compound **B** shows no bands in the 3300–3500 cm^{-1} region. The broadband proton-decoupled ^{13}C NMR spectrum displayed the following signals (information from the DEPT ^{13}C spectra is given in parentheses with the ^{13}C chemical shifts):

^{13}C NMR: δ 147 (C), 129 (CH), 115 (CH), 111 (CH), 44 (CH_2), 13 (CH_3)

2. Two compounds with the molecular formula $C_5H_{10}O$ have the following ^1H and ^{13}C NMR data. Both compounds have a strong IR absorption band in the 1710–1740 cm^{-1} region. Elucidate the structure of these two compounds and interpret the spectra. Make a sketch of each NMR spectrum.

a. ^1H NMR: δ 2.55 (septet, 1H), 2.10 (singlet, 3H), 1.05 (doublet, 6H);
^{13}C NMR: δ 212.6, 41.5, 27.2, 17.8.

b. ^1H NMR: δ 2.38 (triplet, 2H), 2.10 (singlet, 3H), 1.57 (sextet, 2H), 0.88 (triplet, 3H);
^{13}C NMR: δ 209.0, 45.5, 29.5, 17.0, 13.2.

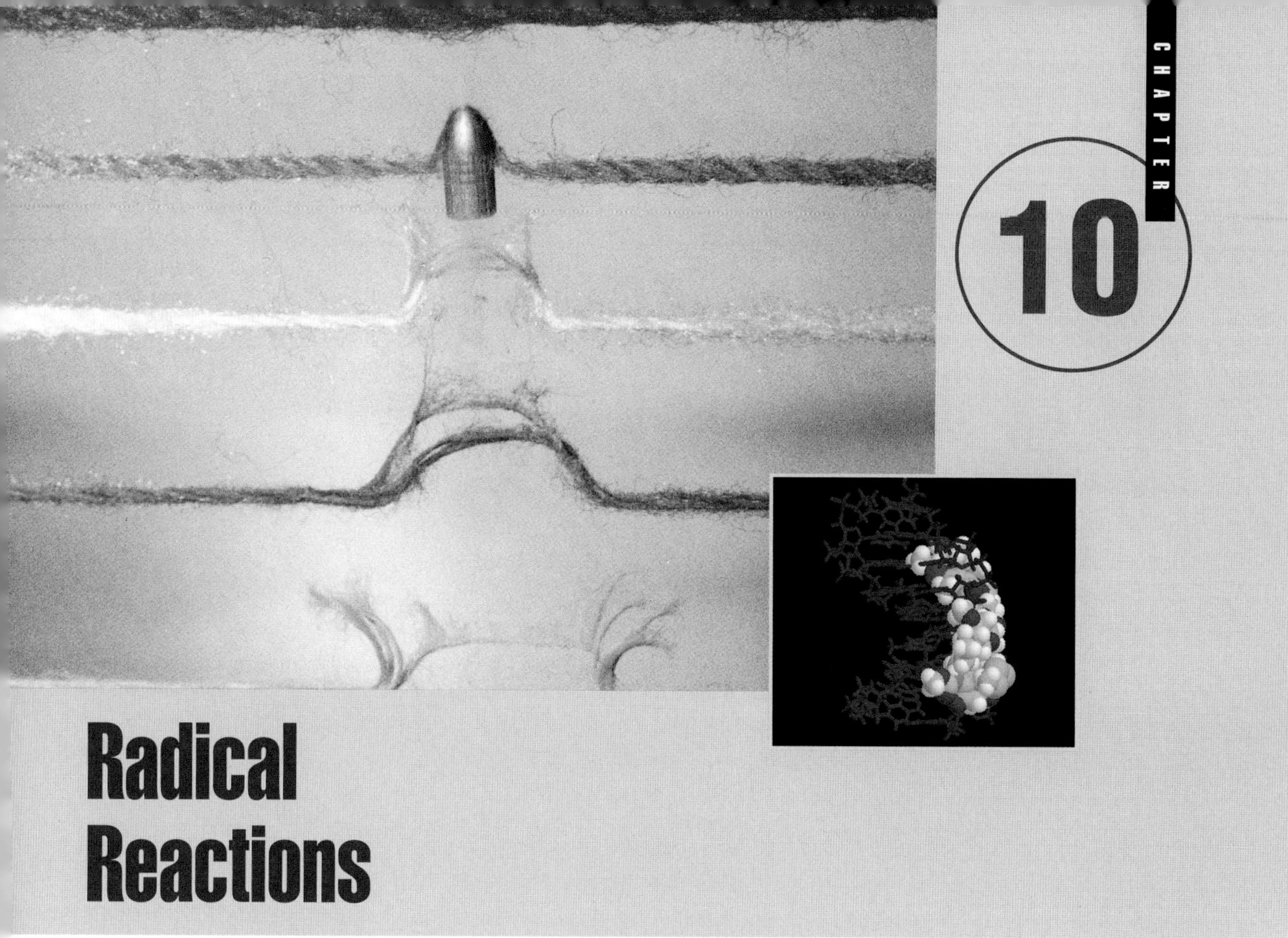

CHAPTER 10

Radical Reactions

Calicheamicin γ_1^{I}: A Radical Device for Slicing the Backbone of DNA

The beautiful architecture of calicheamicin γ_1^{I} conceals a lethal reactivity. Calicheamicin γ_1^{I} binds to the minor groove of DNA where its unusual enediyne (pronounced ēn dī īn) moiety reacts to form a highly effective device for slicing the backbone of DNA. A model of calicheamicin bound to DNA is shown above, and its structural formula is shown on the next page. Calicheamicin γ_1^{I} and its analogues are of great clinical interest because they are extraordinarily deadly for cancer cells, having been shown to initiate apoptosis (programmed cell death). Indeed, research on calicheamicin has since led to development of the drug Mylotarg, now used to treat some cases of acute mylogenous leukemia. Mylotarg carries two calicheamicin "warheads" on an antibody that delivers it specifically to the cancerous cells. In nature, bacteria called *Micromonospora echinospora* synthesize calicheamicin γ_1^{I} as part of their normal metabolism, presumably as a chemical defense against other organisms. The laboratory synthesis of of this complex molecule by the research group of K. C. Nicolaou (Scripps Research Institute, University of California, San Diego), on the other hand, represents a *tour de force* achievement in synthetic organic chemistry. Synthesis of calicheamicin and analogues, as well as investigations by many other researchers, has led to fascinating insights about its mechanism of action and biological properties.

The DNA-slicing property of calicheamicin γ_1^{I} arises because it acts as a molecular machine for producing carbon radicals. A carbon radical is a highly reactive and unstable intermediate that has an unpaired electron. Once formed, a carbon radical can become a stable molecule again by removing a proton and one electron (i.e., a hydrogen atom) from another molecule. In this way, its unpaired electron becomes part of a bonding electron

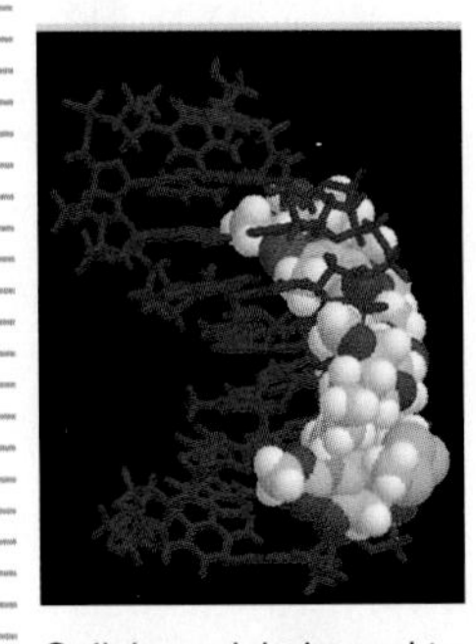

Calicheamicin bound to DNA.

Calicheamicin γ_1^I

pair. (Other paths to achieve this are possible, too). The molecule that lost the hydrogen atom, however, becomes a new reactive radical intermediate. When the radical weaponry of each calicheamicin γ_1^I is activated, it removes a hydrogen atom from the backbone of DNA. This leaves the DNA molecule as an unstable radical intermediate which, in turn, results in double-strand cleavage of the DNA and cell death.

(1) nucleophilic attack
(2) conjugate addition

Bergman cyclo-aromatization

Calicheamicin γ_1^I

DNA

DNA diradical

O_2

DNA double strand cleavage

In Problem 10.29 and in "The Chemistry of . . . Calicheamicin γ_1^I Activation for Cleavage of DNA" box in Chapter 17 we shall revisit calicheamicin γ_1^I to consider the reactions that remodel its structure into a machine for producing radicals.

10.1 INTRODUCTION

So far almost all of the reactions whose mechanisms we have studied have been **ionic reactions.** Ionic reactions are those in which covalent bonds break **heterolytically** and in which ions are involved as reactants, intermediates, or products.

Another broad category of reactions has mechanisms that involve **homolysis** of covalent bonds with the production of intermediates possessing unpaired electrons called **radicals** (or **free radicals**):

$$A:B \xrightarrow{\text{homolysis}} A\cdot + \cdot B$$

Radicals

This simple example illustrates the way we use **single-barbed** curved arrows to show the movement of **a single electron** (not of an electron pair as we have done earlier). In this instance, each group, A and B, comes away with one of the electrons of the covalent bond that joined them.

10.1A Production of Radicals

Energy must be supplied to cause homolysis of covalent bonds (Section 10.2), and this is usually done in two ways: by heating or by irradiation with light. For example, compounds with an oxygen–oxygen single bond, called **peroxides,** undergo homolysis readily when heated, because the oxygen–oxygen bond is weak. The products are two radicals, called alkoxyl radicals:

$$R{-}\ddot{\underset{..}{O}}{:}\ddot{\underset{..}{O}}{-}R \xrightarrow{\text{heat}} 2\,R{-}\ddot{\underset{..}{O}}\cdot$$

Dialkyl peroxide **Alkoxyl radicals**

Halogen molecules (X_2) also contain a relatively weak bond. As we shall soon see, halogens undergo homolysis readily when heated or when irradiated with light of a wavelength that can be absorbed by the halogen molecule:

$$:\ddot{\underset{..}{X}}:\ddot{\underset{..}{X}}: \xrightarrow[\text{heat or light}]{\text{homolysis}} 2\,:\ddot{\underset{..}{X}}\cdot$$

The products of this homolysis are halogen atoms, and because halogen atoms contain an unpaired electron, they are radicals.

10.1B Reactions of Radicals

Almost all small radicals are short-lived, highly reactive species. When they collide with other molecules, they tend to react in a way that leads to pairing of their unpaired electron. One way they can do this is by abstracting an atom from another molecule. For example, a halogen atom may abstract a hydrogen atom from an alkane. This **hydrogen abstraction** gives the halogen atom an electron (from the hydrogen atom) to pair with its unpaired electron. Notice, however, that the other product of this abstraction *is another radical,* in this case, an alkyl radical, R·:

General Reaction

$$:\ddot{\underset{..}{X}}\cdot + H:R \longrightarrow :\ddot{\underset{..}{X}}:H + R\cdot$$

Alkane **Alkyl radical**

The Chemistry of...

Radicals in Biology, Medicine, and Industry

Radical reactions are of vital importance in biology and medicine. Radical reactions are ubiquitous in living things, because radicals are produced in the normal course of metabolism. Radicals are all *around* us, too, because molecular oxygen ($\cdot\ddot{\underset{..}{O}}-\ddot{\underset{..}{O}}\cdot$) is itself a diradical (Section 10.11A). Another radical, one that was never suspected to be very important in normal cell function, is nitric oxide ($\cdot\ddot{N}=\ddot{O}:$). It has been shown, however, that nitric oxide plays a remarkable number of important roles in living systems. Although in its free form nitric oxide is a relatively unstable and potentially toxic gas, in biological systems it is involved in blood pressure regulation and blood clotting, neurotransmission, and the immune response against tumor cells. Nitric oxide has taken the spotlight as one of nature's most surprising yet ubiquitous chemical messengers (Section 10.11B).

Because radicals are highly reactive, however, they are also capable of randomly damaging all components of the body. Accordingly, they are believed to be important in the "aging process" in the sense that radicals are involved in the development of the chronic diseases that are life limiting. For example, there is growing evidence that radical reactions are important in the development of cancers and in the development of atherosclerosis. A naturally occurring radical called superoxide ($O_2^{-}\cdot$) paradoxically is associated with both the immune response against pathogens and at the same time the development of certain diseases. An enzyme called superoxide dismutase regulates the level of superoxide in the body (Section 10.11A). Radicals in cigarette smoke have been implicated in inactivation of an antiprotease in the lungs, an inactivation that leads to the development of emphysema. We also saw in the chapter opening vignette how calicheamicin, a natural product from bacteria, has clinical potential for fighting tumor cells by cleaving their DNA by a radical reaction.

Radical reactions are important in many industrial processes, as well. We shall learn in Section 10.10 how radical reactions are used to produce a whole class of useful "plastics" or *polymers* such as polyethylene, Teflon, and polystyrene. (Additional information is provided in Special Topic A, which follows this chapter.) Radical reactions are also central to the "cracking" process by which gasoline and other fuels are made from petroleum. Moreover, the combustion process by which these fuels are converted to energy involves radical reactions (Section 10.11C).

Specific Example

$$:\ddot{\underset{..}{Cl}}\cdot + H:CH_3 \longrightarrow :\ddot{\underset{..}{Cl}}:H + CH_3\cdot$$

Methane   **Methyl radical**

This behavior is characteristic of radical reactions. Consider another example, one that shows another way in which radicals can react: They can combine with a compound containing a multiple bond to produce a new, larger radical. (We shall study reactions of this type in Section 10.10.)

$$R\cdot + \rangle C{=}C\langle \longrightarrow -\overset{R}{\overset{|}{\underset{|}{C}}}-\dot{C}\langle$$

Alkene   **New radical**

10.2 Homolytic Bond Dissociation Energies

When atoms combine to form molecules, energy is released as covalent bonds form. The molecules of the products have lower enthalpy than the separate atoms. When hydrogen atoms combine to form hydrogen molecules, for example, the reaction is *exothermic;* it evolves 436 kJ of heat for every mole of hydrogen that is produced. Similarly, when chlorine atoms combine to form chlorine molecules, the reaction evolves 243 kJ mol^{-1} of chlorine produced:

$$H\cdot + H\cdot \longrightarrow H{-}H \qquad \Delta H^\circ = -436 \text{ kJ mol}^{-1}$$

$$Cl\cdot + Cl\cdot \longrightarrow Cl{-}Cl \qquad \Delta H^\circ = -243 \text{ kJ mol}^{-1}$$

Bond formation is an exothermic process.

To break covalent bonds, energy must be supplied. Reactions in which only bond breaking occurs are always endothermic. The energy required to break the covalent bonds of hydrogen or chlorine homolytically is exactly equal to that evolved when the separate atoms combine to form molecules. In the bond cleavage reaction, however, ΔH° is positive:

$$H{-}H \longrightarrow H\cdot + H\cdot \qquad \Delta H^\circ = +436 \text{ kJ mol}^{-1}$$

$$Cl{-}Cl \longrightarrow Cl\cdot + Cl\cdot \qquad \Delta H^\circ = +243 \text{ kJ mol}^{-1}$$

The energies required to break covalent bonds homolytically have been determined experimentally for many types of covalent bonds. These energies are called **homolytic bond dissociation energies,** and they are usually abbreviated by the symbol DH°. The homolytic bond dissociation energies of hydrogen and chlorine, for example, might be written in the following way:

$$\underset{(DH^\circ = 436 \text{ kJ mol}^{-1})}{H{-}H} \qquad \underset{(DH^\circ = 243 \text{ kJ mol}^{-1})}{Cl{-}Cl}$$

The homolytic bond dissociation energies of a variety of covalent bonds are listed in Table 10.1.

10.2A Homolytic Bond Dissociation Energies and Heats of Reaction

Bond dissociation energies have, as we shall see, a variety of uses. They can be used, for example, to calculate the enthalpy change (ΔH°) for a reaction. To make such a calculation (see following reaction), we must remember that **for bond breaking ΔH° is positive and for bond formation ΔH° is negative.** Let us consider, for example, the reaction of hydrogen and chlorine to produce 2 mol of hydrogen chloride. From Table 10.1 we get the following values of DH°:

$$\underset{(DH^\circ = 436 \text{ kJ mol}^{-1})}{H{-}H} + \underset{(DH^\circ = 243 \text{ kJ mol}^{-1})}{Cl{-}Cl} \longrightarrow \underset{(DH^\circ = 432 \text{ kJ mol}^{-1}) \times 2}{2\,H{-}Cl}$$

+679 kJ is required to cleave 1 mol of H_2 bonds and 1 mol of Cl_2 bonds.

−864 kJ is evolved in formation of the bonds in 2 mol of HCl.

Overall, the reaction of 1 mol of H_2 and 1 mol of Cl_2 to form 2 mol of HCl is exothermic:

$$\Delta H^\circ = (-864 \text{ kJ} + 679 \text{ kJ}) = -185 \text{ kJ} \quad \text{for 2 mol HCl produced}$$

For the purpose of our calculation, we have assumed a particular pathway, which amounts to

$$H{-}H \longrightarrow 2\,H\cdot$$

and

$$Cl{-}Cl \longrightarrow 2\,Cl\cdot$$

then

$$2\,H\cdot + 2\,Cl\cdot \longrightarrow 2\,H{-}Cl$$

This is not the way the reaction actually occurs. Nevertheless, the heat of reaction, ΔH°, is a thermodynamic quantity that is dependent *only* on the initial and final states of the reacting molecules. Here, ΔH° is independent of the path followed (Hess's law), and, for this reason, our calculation is valid.

TABLE 10.1 Single-Bond Homolytic Dissociation Energies $DH°$ at 25°C[a]

$$A{:}B \longrightarrow A\cdot + B\cdot$$

Bond Broken (shown in red)	kJ mol^{-1}	Bond Broken (shown in red)	kJ mol^{-1}
H—H	436	$(CH_3)_2CH$—Br	298
D—D	443	$(CH_3)_2CH$—I	222
F—F	159	$(CH_3)_2CH$—OH	402
Cl—Cl	243	$(CH_3)_2CH$—OCH_3	359
Br—Br	193	$(CH_3)_2CHCH_2$—H	422
I—I	151	$(CH_3)_3C$—H	400
H—F	570	$(CH_3)_3C$—Cl	349
H—Cl	432	$(CH_3)_3C$—Br	292
H—Br	366	$(CH_3)_3C$—I	227
H—I	298	$(CH_3)_3C$—OH	400
CH_3—H	440	$(CH_3)_3C$—OCH_3	348
CH_3—F	461	$C_6H_5CH_2$—H	375
CH_3—Cl	352	$CH_2{=}CHCH_2$—H	369
CH_3—Br	293	$CH_2{=}CH$—H	465
CH_3—I	240	C_6H_5—H	474
CH_3—OH	387	$HC{\equiv}C$—H	547
CH_3—OCH_3	348	CH_3—CH_3	378
CH_3CH_2—H	421	CH_3CH_2—CH_3	371
CH_3CH_2—F	444	$CH_3CH_2CH_2$—CH_3	374
CH_3CH_2—Cl	353	CH_3CH_2—CH_2CH_3	343
CH_3CH_2—Br	295	$(CH_3)_2CH$—CH_3	371
CH_3CH_2—I	233	$(CH_3)_3C$—CH_3	363
CH_3CH_2—OH	393	HO—H	499
CH_3CH_2—OCH_3	352	HOO—H	356
$CH_3CH_2CH_2$—H	423	HO—OH	214
$CH_3CH_2CH_2$—F	444	$(CH_3)_3CO$—$OC(CH_3)_3$	157
$CH_3CH_2CH_2$—Cl	354	$C_6H_5C(=O)O$—$OC(=O)C_6H_5$	139
$CH_3CH_2CH_2$—Br	294	CH_3CH_2O—OCH_3	184
$CH_3CH_2CH_2$—I	176	CH_3CH_2O—H	431
$CH_3CH_2CH_2$—OH	395	$CH_3C(=O)$—H	364
$CH_3CH_2CH_2$—OCH_3	355		
$(CH_3)_2CH$—H	413		
$(CH_3)_2CH$—F	439		
$(CH_3)_2CH$—Cl	355		

[a]Data compiled from the *National Institute of Standards (NIST) Standard Reference Database Number 69*, July 2001 Release, accessed via *NIST Chemistry WebBook* (http://webbook.nist.gov/chemistry/) and the *CRC Handbook of Chemistry and Physics*, 3rd Electronic Edition (updated from content in the 81st print edition), accessed via Knovel Engineering and Scientific Online References (http://www.knovel.com). $DH°$ values were obtained directly or calculated from heat of formation (H_f) data using the equation $DH°[A—B] = H_f[A\cdot] + H_f[B\cdot] - H_f[A—B]$.

PROBLEM 10.1

Calculate the heat of reaction, $\Delta H°$, for the following reactions:

(a) $H_2 + F_2 \longrightarrow 2\ HF$
(b) $CH_4 + F_2 \longrightarrow CH_3F + HF$
(c) $CH_4 + Cl_2 \longrightarrow CH_3Cl + HCl$
(d) $CH_4 + Br_2 \longrightarrow CH_3Br + HBr$
(e) $CH_4 + I_2 \longrightarrow CH_3I + HI$
(f) $CH_3CH_3 + Cl_2 \longrightarrow CH_3CH_2Cl + HCl$
(g) $CH_3CH_2CH_3 + Cl_2 \longrightarrow CH_3CHClCH_3 + HCl$
(h) $(CH_3)_3CH + Cl_2 \longrightarrow (CH_3)_3CCl + HCl$

10.2B Homolytic Bond Dissociation Energies and the Relative Stabilities of Radicals

Homolytic bond dissociation energies also provide us with a convenient way to estimate the relative stabilities of radicals. If we examine the data given in Table 10.1, we find the following values of $DH°$ for the primary and secondary C—H bonds of propane:

$$CH_3CH_2CH_2—H \quad (DH° = 423\ kJ\ mol^{-1}) \qquad (CH_3)_2CH—H \quad (DH° = 413\ kJ\ mol^{-1})$$

This means that for the reaction in which the designated C—H bonds are broken homolytically, the values of $\Delta H°$ are those given here.

$$CH_3CH_2CH_2—H \longrightarrow CH_3CH_2CH_2\cdot + H\cdot \qquad \Delta H° = +423\ kJ\ mol^{-1}$$

Propyl radical (a 1° radical)

$$CH_3CH(H)CH_3 \longrightarrow CH_3\dot{C}HCH_3 + H\cdot \qquad \Delta H° = +413\ kJ\ mol^{-1}$$

Isopropyl radical (a 2° radical)

These reactions resemble each other in two respects: They both begin with the same alkane (propane), and they both produce an alkyl radical and a hydrogen atom. They differ, however, in the amount of energy required and in the type of carbon radical produced. These two differences are related to each other.

Alkyl radicals are classified as being 1°, 2°, or 3° on the basis of the carbon atom that has the unpaired electron. More energy must be supplied to produce a primary alkyl radical (the propyl radical) from propane than is required to produce a secondary carbon radical (the isopropyl radical) from the same compound. This must mean that the primary radical has absorbed more energy and thus has greater *potential energy.* Because the relative stability of a chemical species is inversely related to its potential energy, the secondary radical must be the *more stable* radical (Fig. 10.1*a*). In fact, the secondary isopropyl radical is more stable than the primary propyl radical by 10 kJ mol^{-1}.

We can use the data in Table 10.1 to make a similar comparison of the *tert*-butyl radical (a 3° radical) and the isobutyl radical (a 1° radical) relative to isobutane:

$$CH_3—C(CH_3)(H)—CH_2—H \longrightarrow CH_3\dot{C}(CH_3)CH_3 + H\cdot \qquad \Delta H° = +400\ kJ\ mol^{-1}$$

***tert*-Butyl radical (a 3° radical)**

$$CH_3—C(CH_3)(H)—CH_2—H \longrightarrow CH_3C(CH_3)(H)CH_2\cdot + H\cdot \qquad \Delta H° = +422\ kJ\ mol^{-1}$$

Isobutyl radical (a 1° radical)

Here we find (Fig. 10.1*b*) that the difference in stability of the two radicals is even larger. The tertiary radical is more stable than the primary radical by 22 kJ mol^{-1}.

Relative radical stability

The kind of pattern that we find in these examples is found with alkyl radicals generally; overall, their relative stabilities are the following:

FIGURE 10.1 *(a)* Comparison of the potential energies of the propyl radical (+H·) and the isopropyl radical (+H·) relative to propane. The isopropyl radical—a 2° radical—is more stable than the 1° radical by 10 kJ mol^{-1}. *(b)* Comparison of the potential energies of the *tert*-butyl radical (+H·) and the isobutyl radical (+H·) relative to isobutane. The 3° radical is more stable than the 1° radical by 22 kJ mol^{-1}.

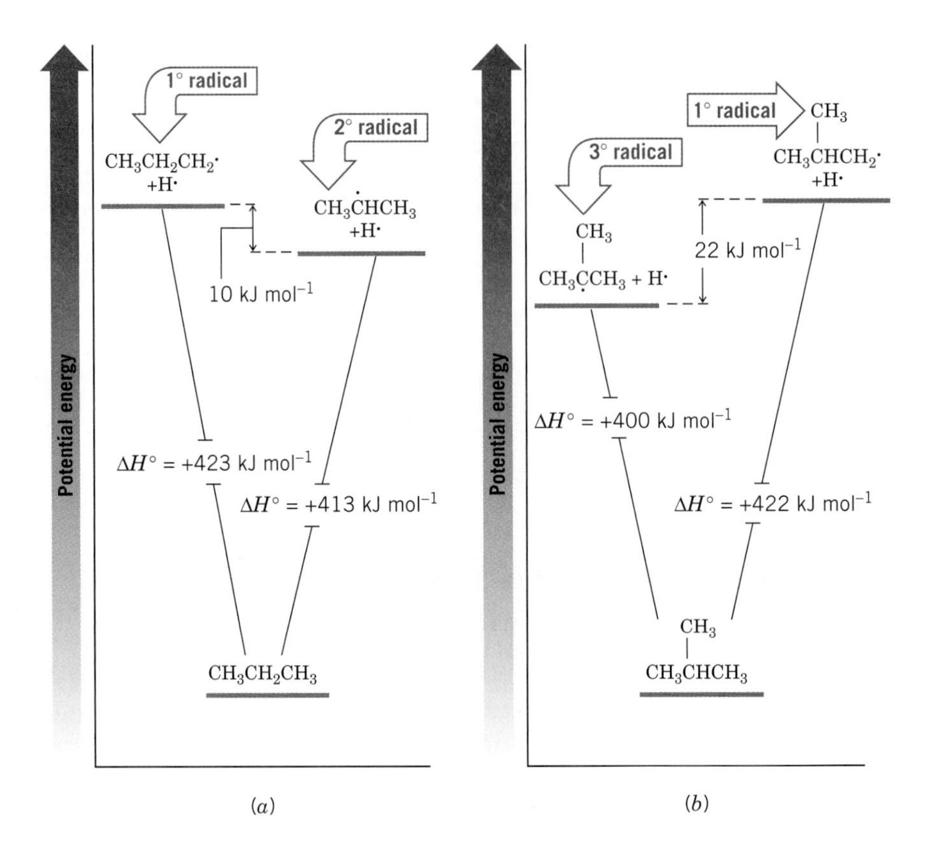

Tertiary > **Secondary** > **Primary** > **Methyl**

$$\mathrm{C_3C\cdot} > \mathrm{C_2HC\cdot} > \mathrm{CH_2C\cdot} > \mathrm{H_3C\cdot}$$

The order of stability of alkyl radicals is the same as for carbocations (Section 6.11B), and the reasons are similar. Although alkyl radicals are uncharged, the carbon that bears the odd electron is *electron deficient.* Therefore, alkyl groups attached to this carbon provide a stabilizing effect through hyperconjugation, and the more alkyl groups bonded to it, the more stable the radical is.

PROBLEM 10.2 List the following radicals in order of decreasing stability:

$CH_3\cdot$ $\qquad$ $(CH_3)_2CHCH_2\cdot$ $\qquad$ $CH_3CH_2C(CH_3)_2\cdot$ $\qquad$ $CH_3CH_2CH(CH_3)\cdot$

10.3 The Reactions of Alkanes with Halogens

Methane, ethane, and other alkanes react with the first three members of the halogen family: fluorine, chlorine, and bromine. Alkanes do not react appreciably with iodine. With methane the reaction produces a mixture of halomethanes and a hydrogen halide:

$$\underset{\substack{\textbf{Methane}}}{\mathrm{H{-}\overset{\displaystyle H}{\underset{\displaystyle H}{C}}{-}H}} + \underset{\textbf{Halogen}}{X_2} \xrightarrow[\text{light}]{\text{heat or}} \underset{\textbf{Halomethane}}{\mathrm{H{-}\overset{\displaystyle H}{\underset{\displaystyle H}{C}}{-}X}} + \underset{\textbf{Dihalomethane}}{\mathrm{H{-}\overset{\displaystyle X}{\underset{\displaystyle H}{C}}{-}X}} + \underset{\textbf{Trihalomethane}}{\mathrm{H{-}\overset{\displaystyle X}{\underset{\displaystyle X}{C}}{-}X}} + \underset{\textbf{Tetrahalomethane}}{\mathrm{X{-}\overset{\displaystyle X}{\underset{\displaystyle X}{C}}{-}X}} + \underset{\textbf{Hydrogen halide}}{\mathrm{H{-}X}}$$

X = F, Cl, or Br

(The sum of the number of moles of each halogenated methane produced equals the number of moles of methane that reacted.)

The reaction of an alkane with a halogen is a **substitution reaction,** called **halogenation.** The general reaction to produce a monohaloalkane can be written as follows:

$$\mathrm{R{-}H} + X_2 \longrightarrow \mathrm{R{-}X} + \mathrm{HX}$$

In these reactions a halogen atom replaces one or more of the hydrogen atoms of the alkane.

10.3A Multiple Substitution Reactions versus Selectivity

One complicating characteristic of alkane halogenations is that multiple substitution reactions almost always occur. As we saw at the beginning of this section, the halogenation of methane produces a mixture of monohalomethane, dihalomethane, trihalomethane, and tetrahalomethane. This happens because all hydrogen atoms attached to carbon are capable of reacting with fluorine, chlorine, or bromine.

Let us consider the reaction that takes place between chlorine and methane as an example. If we mix methane and chlorine (both substances are gases at room temperature) and then either heat the mixture or irradiate it with light, a reaction begins to occur vigorously. At the outset, the only compounds that are present in the mixture are chlorine and methane, and the only reaction that can take place is one that produces chloromethane and hydrogen chloride:

$$\mathrm{H{-}\overset{\displaystyle H}{\underset{\displaystyle H}{C}}{-}H} + Cl_2 \longrightarrow \mathrm{H{-}\overset{\displaystyle H}{\underset{\displaystyle H}{C}}{-}Cl} + \mathrm{H{-}Cl}$$

As the reaction progresses, however, the concentration of chloromethane in the mixture increases, and a second substitution reaction begins to occur. Chloromethane reacts with chlorine to produce dichloromethane:

$$\mathrm{H{-}\overset{\displaystyle H}{\underset{\displaystyle H}{C}}{-}Cl} + Cl_2 \longrightarrow \mathrm{H{-}\overset{\displaystyle Cl}{\underset{\displaystyle H}{C}}{-}Cl} + \mathrm{H{-}Cl}$$

The dichloromethane produced can then react to form trichloromethane, and trichloromethane, as it accumulates in the mixture, can react with chlorine to produce tetrachloromethane. Each time a substitution of —Cl for —H takes place, a molecule of H—Cl is produced.

SOLVED PROBLEM

If the goal of a synthesis is to prepare chloromethane (CH_3Cl), its formation can be maximized and the formation of CH_2Cl_2, $CHCl_3$, and CCl_4 minimized by using a large excess of methane in the reaction mixture. Explain why this is possible.

ANSWER: The use of a large excess of methane maximizes the probability that chlorine will attack methane molecules because the concentration of methane in the mixture will always be relatively large. It also minimizes the probability that chlorine will attack molecules of CH_3Cl, CH_2Cl_2, and $CHCl_3$, because their concentrations will always be relatively small. After the reaction is over, the unreacted excess methane can be recovered and recycled.

Chlorination of most higher alkanes gives a mixture of isomeric monochloro products as well as more highly halogenated compounds. Chlorine is relatively ***unselective;*** it does not discriminate greatly among the different types of hydrogen atoms (primary, secondary, and tertiary) in an alkane. An example is the light-promoted chlorination of isobutane:

$$CH_3CH(CH_3)CH_3 \xrightarrow[\text{light}]{Cl_2} CH_3CH(CH_3)CH_2Cl + CH_3C(CH_3)(Cl)CH_3 + \text{polychlorinated products} + HCl$$

Isobutane → **Isobutyl chloride (48%)** + ***tert*-Butyl chloride (29%)** + **(23%)**

STUDY TIP!

Chlorination is unselective.

Because alkane chlorinations usually yield a complex mixture of products, they are not generally useful as synthetic methods when our goal is the preparation of a specific alkyl chloride. An exception is the halogenation of an alkane (or cycloalkane) whose hydrogen atoms *are all equivalent.* [Equivalent hydrogen atoms are defined as those which on replacement by some other group (e.g., chlorine) yield the same compound.] Neopentane, for example, can form only one monohalogenation product, and the use of a large excess of neopentane minimizes polychlorination:

$$CH_3-C(CH_3)_2-CH_3 + Cl_2 \xrightarrow[\text{or light}]{\text{heat}} CH_3-C(CH_3)_2-CH_2Cl + HCl$$

Neopentane (excess) → **Neopentyl chloride**

Bromine is generally less reactive toward alkanes than chlorine, and bromine is ***more selective*** in the site of attack when it does react. We shall examine this topic further in Section 10.6A.

10.4 Chlorination of Methane: Mechanism of Reaction

The *halogenation* reactions of alkanes take place by a radical mechanism. Let us begin our study of them by examining a simple example of an alkane halogenation—the reaction of methane with chlorine that takes place in the gas phase:

$$CH_4 + Cl_2 \longrightarrow CH_3Cl + HCl\ (+\ CH_2Cl_2,\ CHCl_3,\ \text{and}\ CCl_4)$$

Several important experimental observations can be made about this reaction:

1. **The reaction is promoted by heat or light.** At room temperature methane and chlorine do not react at a perceptible rate as long as the mixture is kept away from light. Methane and chlorine do react, however, at room temperature if the gaseous reaction mix-

ture is irradiated with UV light at a wavelength absorbed by Cl_2, and they react in the dark if the gaseous mixture is heated to temperatures greater than 100°C.

2. **The light-promoted reaction is highly efficient.** A relatively small number of light photons permits the formation of relatively large amounts of chlorinated product.

A mechanism that is consistent with these observations has several steps, shown below. The first step involves the fragmentation of a chlorine molecule, by heat or light, into two chlorine atoms. The second step involves hydrogen abstraction by a chlorine atom.

A Mechanism for the Reaction

Radical Chlorination of Methane

Reaction:

$$CH_4 + Cl_2 \xrightarrow[\text{or light}]{\text{heat}} CH_3Cl + HCl$$

Mechanism:

Step 1 $:\ddot{Cl}:\ddot{Cl}: \xrightarrow[\text{or light}]{\text{heat}} :\ddot{Cl}\cdot + \cdot\ddot{Cl}:$

Under the influence of heat or light a molecule of chlorine dissociates; each atom takes one of the bonding electrons.

This step produces two highly reactive chlorine atoms.

Remember: These conventions are used in illustrating reaction mechanisms in this text.
1. Arrows ↷ or ↷ always show the direction of movement of electrons.
2. Single-barbed arrows ↷ show the attack (or movement) of an unpaired electron.
3. Double-barbed arrows ↷ show the attack (or movement) of an electron pair.

Step 2 $:\ddot{Cl}\cdot + H:CH_3 \longrightarrow :\ddot{Cl}:H + \cdot CH_3$

A chlorine atom abstracts a hydrogen atom from a methane molecule.

This step produces a molecule of hydrogen chloride and a methyl radical.

Step 3 $H_3C\cdot + :\ddot{Cl}:\ddot{Cl}: \longrightarrow H_3C:\ddot{Cl}: + \cdot\ddot{Cl}:$

A methyl radical abstracts a chlorine atom from a chlorine molecule.

This step produces a molecule of methyl chloride and a chlorine atom. The chlorine atom can now cause a repetition of step 2.

In step 3 the highly reactive methyl radical reacts with a chlorine molecule by abstracting a chlorine atom. This results in the formation of a molecule of chloromethane (one of the ultimate products of the reaction) and a *chlorine atom.* The latter product is particularly significant, for the chlorine atom formed in step 3 can attack another methane molecule and cause a repetition of step 2. Then, step 3 is repeated, and so forth, for hundreds or thousands of times. (With each repetition of step 3 a molecule of chloromethane is produced.) This type of sequential, stepwise mechanism, in which each step generates the reactive intermediate that causes the next cycle of the reaction to occur, is called a **chain reaction.**

Step 1 is called the **chain-initiating step.** In the chain-initiating step ***radicals are created.*** Steps 2 and 3 are called **chain-propagating steps.** In chain-propagating steps ***one radical generates another.***

Chain Initiation

$$\textit{Step 1}\quad Cl_2 \xrightarrow[\text{or light}]{\text{heat}} 2\,Cl\cdot$$

Chain Propagation

$$\textit{Step 2}\quad CH_4 + Cl\cdot \longrightarrow CH_3\cdot + H{-}Cl$$

$$\textit{Step 3}\quad CH_3\cdot + Cl_2 \longrightarrow CH_3Cl + Cl\cdot$$

The chain nature of the reaction accounts for the observation that the light-promoted reaction is highly efficient. The presence of a relatively few atoms of chlorine at any given moment is all that is needed to cause the formation of many thousands of molecules of chloromethane.

What causes the chains to terminate? Why does one photon of light not promote the chlorination of all of the methane molecules present? We know that this does not happen because we find that, at low temperatures, continuous irradiation is required or the reaction slows and stops. The answer to these questions is the existence of ***chain-terminating steps:*** steps that occur infrequently but occur often enough to *use up one or both of the reactive intermediates.* The continuous replacement of intermediates used up by chain-terminating steps requires continuous irradiation. Plausible chain-terminating steps are as follows.

Chain Termination

$$H_3C\cdot + \cdot\ddot{Cl}: \longrightarrow H_3C:\ddot{Cl}:$$

$$H_3C\cdot + \cdot CH_3 \longrightarrow H_3C:CH_3$$

$$:\ddot{Cl}\cdot + \cdot\ddot{Cl}: \longrightarrow :\ddot{Cl}:\ddot{Cl}:$$

This last step probably occurs least frequently. The two chlorine atoms are highly energetic; as a result, the simple diatomic chlorine molecule that is formed must dissipate its excess energy rapidly by colliding with some other molecule or the walls of the container. Otherwise it simply flies apart again. By contrast, chloromethane and ethane, formed in the other two chain-terminating steps, can dissipate their excess energy through vibrations of their C—H bonds.

Our radical mechanism also explains how the reaction of methane with chlorine produces the more highly halogenated products, CH_2Cl_2, $CHCl_3$, and CCl_4 (as well as additional HCl). As the reaction progresses, chloromethane (CH_3Cl) accumulates in the mixture and its hydrogen atoms, too, are susceptible to abstraction by chlorine. Thus chloromethyl radicals are produced that lead to dichloromethane (CH_2Cl_2).

$$\text{Step 2a} \quad Cl\cdot + H{:}\overset{Cl}{\underset{H}{C}}{-}H \longrightarrow H{:}Cl + \cdot\overset{Cl}{\underset{H}{C}}{-}H$$

$$\text{Step 3a} \quad H{-}\overset{Cl}{\underset{H}{C}}\cdot + Cl{:}Cl \longrightarrow H{-}\overset{Cl}{\underset{H}{C}}{:}Cl + Cl\cdot$$

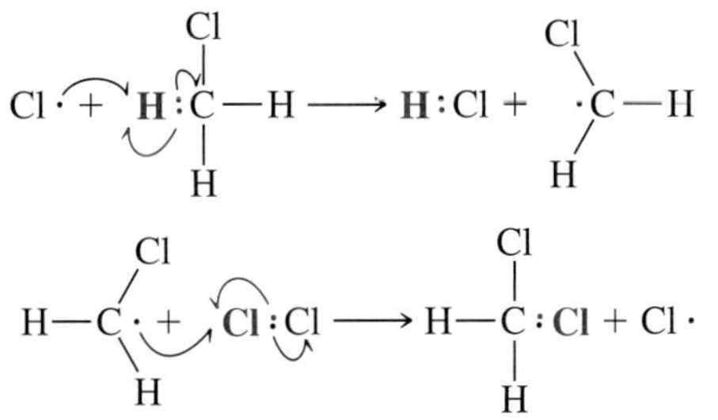

Then step 2a is repeated, then step 3a is repeated, and so on. Each repetition of step 2a yields a molecule of HCl, and each repetition of step 3a yields a molecule of CH_2Cl_2.

PROBLEM 10.3

Suggest a method for separating the mixture of CH_4, CH_3Cl, CH_2Cl_2, $CHCl_3$, and CCl_4 that is formed when methane is chlorinated. (You may want to consult a handbook.) What analytical method could be used to separate this mixture and give structural information about each component? How would the molecular ion peaks in their respective mass spectra differ on the basis of the number of chlorines (remember that chlorine has two predominant isotopes, ^{35}Cl and ^{37}Cl)?

PROBLEM 10.4

When methane is chlorinated, among the products found are traces of chloroethane. How is it formed? Of what significance is its formation?

PROBLEM 10.5

If our goal is to synthesize CCl_4 in maximum yield, this can be accomplished by using a large excess of chlorine. Explain.

10.5 Chlorination of Methane: Energy Changes

We saw in Section 10.2A that we can calculate the overall heat of reaction from bond dissociation energies. We can also calculate the heat of reaction for each individual step of a mechanism:

Chain Initiation

Step 1 $Cl{-}Cl \longrightarrow 2\,Cl\cdot$ $\quad \Delta H° = +243\ \text{kJ mol}^{-1}$

($DH°$ = 243)

Chain Propagation

Step 2 $CH_3{-}H + Cl\cdot \longrightarrow CH_3\cdot + H{-}Cl$ $\quad \Delta H° = +8\ \text{kJ mol}^{-1}$

($DH°$ = 440) **($DH°$ = 432)**

Step 3 $CH_3\cdot + Cl{-}Cl \longrightarrow CH_3{-}Cl + Cl\cdot$ $\quad \Delta H° = -109\ \text{kJ mol}^{-1}$

($DH°$ = 243) **($DH°$ = 352)**

Chain Termination

$CH_3\cdot + Cl\cdot \longrightarrow CH_3{-}Cl$ $\quad \Delta H° = -352\ \text{kJ mol}^{-1}$

($DH°$ = 352)

$CH_3\cdot + \cdot CH_3 \longrightarrow CH_3{-}CH_3$ $\quad \Delta H° = -378\ \text{kJ mol}^{-1}$

($DH°$ = 378)

$Cl\cdot + Cl\cdot \longrightarrow Cl{-}Cl$ $\quad \Delta H° = -243\ \text{kJ mol}^{-1}$

($DH°$ = 243)

In the chain-initiating step only one bond is broken—the bond between two chlorine atoms—and no bonds are formed. The heat of reaction for this step is simply the bond dissociation energy for a chlorine molecule, and it is highly endothermic.

In the chain-terminating steps bonds are formed, but no bonds are broken. As a result, all of the chain-terminating steps are highly exothermic.

Each of the chain-propagating steps, on the other hand, requires the breaking of one bond and the formation of another. The value of $\Delta H°$ for each of these steps is the difference between the bond dissociation energy of the bond that is broken and the bond dissociation energy for the bond that is formed. The first chain-propagating step is slightly endothermic ($\Delta H° = +8$ kJ mol^{-1}), but the second is exothermic by a large amount ($\Delta H° = -109$ kJ mol^{-1}).

PROBLEM 10.6 Assuming the same mechanism applies, calculate $\Delta H°$ for the chain-initiating, chain-propagating, and chain-terminating steps involved in the fluorination of methane.

Calculating overall $\Delta H°$ for a chain reaction

The addition of the chain-propagating steps yields the overall equation for the chlorination of methane:

$$Cl\cdot + CH_3{-}H \longrightarrow CH_3\cdot + H{-}Cl \qquad \Delta H° = +8 \text{ kJ mol}^{-1}$$
$$CH_3\cdot + Cl{-}Cl \longrightarrow CH_3{-}Cl + Cl\cdot \qquad \Delta H° = -109 \text{ kJ mol}^{-1}$$
$$CH_3{-}H + Cl{-}Cl \longrightarrow CH_3{-}Cl + H{-}Cl \qquad \Delta H° = -101 \text{ kJ mol}^{-1}$$

and the addition of the values of $\Delta H°$ for the individual chain-propagating steps yields the overall value of $\Delta H°$ for the reaction.

PROBLEM 10.7 Show how you can use the chain-propagating steps (see Problem 10.6) to calculate the overall value of $\Delta H°$ for the fluorination of methane.

10.5A The Overall Free-Energy Change

For many reactions the entropy change is so small that the term $T\,\Delta S°$ in the expression

$$\Delta G° = \Delta H° - T\,\Delta S°$$

is almost zero, and $\Delta G°$ is approximately equal to $\Delta H°$. This happens when the reaction is one in which the relative order of reactants and products is about the same. Recall (Section 3.9) that entropy measures the relative disorder or randomness of a system. For a chemical system the relative disorder of the molecules can be related to the number of *degrees of freedom* available to the molecules and their constituent atoms. Degrees of freedom are associated with ways in which *movement or changes in relative position can occur.* Molecules have three sorts of degrees of freedom: translational degrees of freedom associated with movements of the whole molecule through space, rotational degrees of freedom associated with the tumbling motions of the molecule, and vibrational degrees of freedom associated with the stretching and bending motion of atoms about the bonds that connect them (Fig. 10.2). If the atoms of the products of a reaction have more degrees of freedom available than they did as reactants, the entropy change ($\Delta S°$) for the reaction will be positive. If, on the other hand, the atoms of the products are more constrained (have fewer degrees of freedom) than the reactants, a negative $\Delta S°$ results.

Consider the reaction of methane with chlorine:

$$CH_4 + Cl_2 \longrightarrow CH_3Cl + HCl$$

Here, 2 mol of the products are formed from the same number of moles of the reactants. Thus the number of translational degrees of freedom available to products and reactants is the same. Furthermore, CH_3Cl is a tetrahedral molecule like CH_4, and HCl is a diatomic

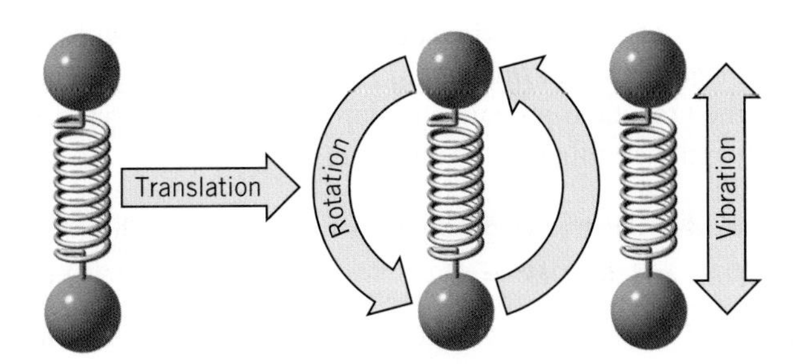

FIGURE 10.2 Translational, rotational, and vibrational degrees of freedom for a simple diatomic molecule.

molecule like Cl_2. This means that vibrational and rotational degrees of freedom available to products and reactants should also be approximately the same. The actual entropy change for this reaction is quite small, $\Delta S° = +2.8 \text{ J K}^{-1} \text{ mol}^{-1}$. Therefore, at room temperature (298 K) the $T\,\Delta S°$ term is 0.8 kJ mol^{-1}, and thus the enthalpy change for the reaction and the free-energy change are almost equal: $\Delta H° = -101 \text{ kJ mol}^{-1}$ and $\Delta G° = -102 \text{ kJ mol}^{-1}$.

In situations like this one it is often convenient to make predictions about whether a reaction will proceed to completion on the basis of $\Delta H°$ rather than $\Delta G°$ since $\Delta H°$ values are readily obtained from bond dissociation energies.

10.5B Activation Energies

For many reactions that we shall study in which entropy changes are small, it is also often convenient to base our estimates of reaction rates simply on **energies of activation, E_{act},** rather than on free energies of activation, $\Delta G^{\ddagger}$. Without going into detail, suffice it to say that these two quantities are closely related and that **both measure the difference in energy between the reactants and the transition state.** Therefore, a low energy of activation means a reaction will take place rapidly; a high energy of activation means that a reaction will take place slowly.

Having seen earlier in this section how to calculate $\Delta H°$ for each step in the chlorination of methane, let us consider the energy of activation for each step. These values are as follows:

Chain Initiation

Step 1 $Cl_2 \longrightarrow 2\,Cl\cdot$ $\qquad E_{act} = +243 \text{ kJ mol}^{-1}$

Chain Propagation

Step 2 $Cl\cdot + CH_4 \longrightarrow HCl + CH_3\cdot$ $\qquad E_{act} = +16 \text{ kJ mol}^{-1}$

Step 3 $CH_3\cdot + Cl_2 \longrightarrow CH_3Cl + Cl\cdot$ $\qquad E_{act} = \sim 8 \text{ kJ mol}^{-1}$

How does one know what the energy of activation for a reaction will be? Could we, for example, have predicted from bond dissociation energies that the energy of activation for the reaction $Cl\cdot + CH_4 \longrightarrow HCl + CH_3\cdot$ would be precisely 16 kJ mol^{-1}? The answer is *no*. The energy of activation must be determined from other experimental data. It cannot be directly measured—it is calculated. Certain principles have been established, however, that enable one to arrive at estimates of energies of activation:

1. **Any reaction in which *bonds are broken* will have an energy of activation greater than zero. This will be true even if a stronger bond is formed and the reaction is exothermic. The reason: Bond formation and bond breaking do not occur simultaneously in the transition state. Bond formation lags behind, and its energy is not all available for bond breaking.**

2. **Activation energies of *endothermic reactions that involve both bond formation and bond rupture will be greater than the heat of reaction, ΔH°*.** Two examples illustrate this principle, namely, the first chain-propagating step in the chlorination of methane and the corresponding step in the bromination of methane:

$$\underset{}{Cl\cdot} + \underset{(DH^\circ = 440)}{CH_3{-}H} \longrightarrow \underset{(DH^\circ = 432)}{H{-}Cl} + CH_3\cdot \qquad \Delta H^\circ = +8 \text{ kJ mol}^{-1} \quad E_{act} = +16 \text{ kJ mol}^{-1}$$

$$Br\cdot + \underset{(DH^\circ = 440)}{CH_3{-}H} \longrightarrow \underset{(DH^\circ = 366)}{H{-}Br} + CH_3\cdot \qquad \Delta H^\circ = +74 \text{ kJ mol}^{-1} \quad E_{act} = +78 \text{ kJ mol}^{-1}$$

In both of these reactions the energy released in bond formation is less than that required for bond rupture; both reactions are, therefore, endothermic. We can easily see why the energy of activation for each reaction is greater than the heat of reaction by looking at the potential energy diagrams in Fig. 10.3. In each case the path from reactants to products is from a lower energy plateau to a higher one. In each case the intervening energy hill is higher still, and since the energy of activation is the vertical (energy) distance between the plateau of reactants and the top of this hill, the energy of activation exceeds the heat of reaction.

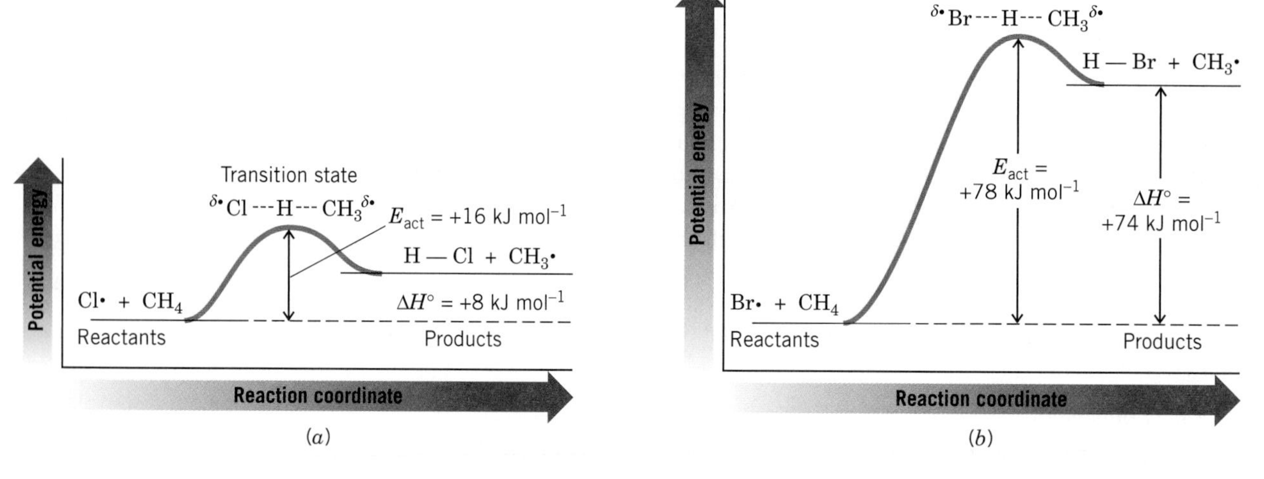

FIGURE 10.3 Potential energy diagrams for *(a)* the reaction of a chlorine atom with methane and *(b)* the reaction of a bromine atom with methane.

3. **The energy of activation of a gas-phase reaction where bonds are broken homolytically but no bonds are formed is equal to ΔH°.*** An example of this type of reaction is the chain-initiating step in the chlorination of methane—the dissociation of chlorine molecules into chlorine atoms:

$$\underset{(DH^\circ = 243)}{Cl{-}Cl} \longrightarrow 2\,Cl\cdot \qquad \Delta H^\circ = +243 \text{ kJ mol}^{-1} \quad E_{act} = +243 \text{ kJ mol}^{-1}$$

The potential energy diagram for this reaction is shown in Fig. 10.4.

*This rule applies only to radical reactions taking place in the gas phase. It does not apply to reactions taking place in solution, especially where ions are involved, because solvation energies are also important.

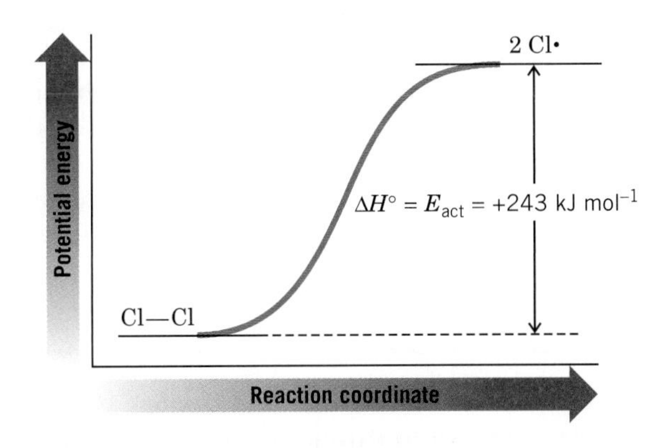

FIGURE 10.4 Potential energy diagram for the dissociation of a chlorine molecule into chlorine atoms.

4. **The energy of activation for a gas-phase reaction in which small radicals combine to form molecules is usually zero.** In reactions of this type the problem of nonsimultaneous bond formation and bond rupture does not exist; only one process occurs: that of bond formation. All of the chain-terminating steps in the chlorination of methane fall into this category. An example is the combination of two methyl radicals to form a molecule of ethane:

$$2\,CH_3\cdot \longrightarrow \underset{(DH^\circ = 378)}{CH_3{-}CH_3} \qquad \Delta H^\circ = -378 \text{ kJ mol}^{-1}$$
$$E_{act} = 0$$

Figure 10.5 illustrates the potential energy changes that occur in this reaction.

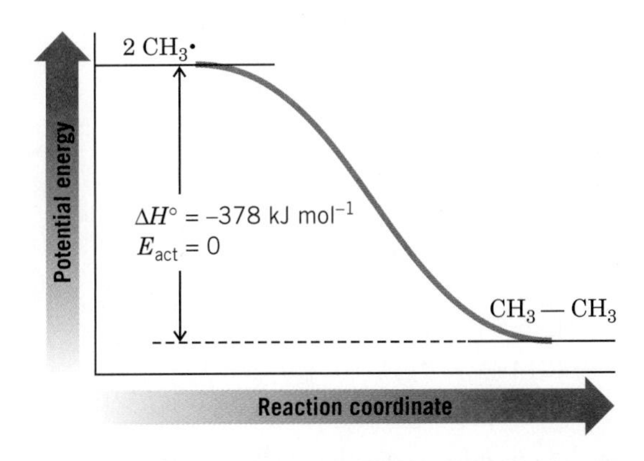

FIGURE 10.5 Potential energy diagram for the combination of two methyl radicals to form a molecule of ethane.

PROBLEM 10.8

When pentane is heated to a very high temperature, radical reactions take place that produce (among other products) methane, ethane, propane, and butane. This type of change is called **thermal cracking.** Among the reactions that take place are the following:

(1) $CH_3CH_2CH_2CH_2CH_3 \longrightarrow CH_3\cdot + CH_3CH_2CH_2CH_2\cdot$
(2) $CH_3CH_2CH_2CH_2CH_3 \longrightarrow CH_3CH_2\cdot + CH_3CH_2CH_2\cdot$
(3) $CH_3\cdot + CH_3\cdot \longrightarrow CH_3CH_3$
(4) $CH_3\cdot + CH_3CH_2CH_2CH_2CH_3 \longrightarrow CH_4 + CH_3CH_2CH_2CH_2CH_2\cdot$
(5) $CH_3\cdot + CH_3CH_2\cdot \longrightarrow CH_3CH_2CH_3$
(6) $CH_3CH_2\cdot + CH_3CH_2\cdot \longrightarrow CH_3CH_2CH_2CH_3$

(a) For which of these reactions would you expect E_{act} to equal zero? **(b)** To be greater than zero? **(c)** To equal ΔH°?

PROBLEM 10.9 The energy of activation for the first chain-propagating step in the fluorination of methane (cf. Problem 10.6) is known to be $+5.0$ kJ mol^{-1}. The energy of activation for the second chain-propagating step is known to be very small. Assuming that it is $+1.0$ kcal mol^{-1}, then **(a)** and **(b)** sketch potential energy diagrams for these two chain-propagating steps. Sketch a potential energy diagram for **(c)** the chain-initiating step in the fluorination of methane and **(d)** the chain-terminating step that produces CH_3F. **(e)** Sketch a potential energy diagram for the following reaction:

$$CH_3\cdot + HF \longrightarrow CH_4 + F\cdot$$

10.5C Reaction of Methane with Other Halogens

The *reactivity* of one substance toward another is measured by the *rate* at which the two substances react. A reagent that reacts very rapidly with a particular substance is said to be highly reactive toward that substance. One that reacts slowly or not at all under the same experimental conditions (e.g., concentration, pressure, and temperature) is said to have a low relative reactivity or to be unreactive. The reactions of the halogens (fluorine, chlorine, bromine, and iodine) with methane show a wide spread of relative reactivities. Fluorine is most reactive—so reactive, in fact, that without special precautions mixtures of fluorine and methane explode. Chlorine is the next most reactive. However, the chlorination of methane is easily controlled by the judicious control of heat and light. Bromine is much less reactive toward methane than chlorine, and iodine is so unreactive that for all practical purposes we can say that no reaction takes place.

If the mechanisms for fluorination, bromination, and iodination of methane are the same as for its chlorination, we can explain the wide variation in reactivity of the halogens by a careful examination of $\Delta H°$ and E_{act} for each step.

FLUORINATION

	$\Delta H°$ (kJ mol^{-1})	E_{act} (kJ mol^{-1})
Chain Initiation		
$F_2 \longrightarrow 2\,F\cdot$	$+159$	$+159$
Chain Propagation		
$F\cdot + CH_4 \longrightarrow HF + CH_3\cdot$	-130	$+5.0$
$CH_3\cdot + F_2 \longrightarrow CH_3F + F\cdot$	-302	Small
Overall $\Delta H° =$	-432	

The chain-initiating step in fluorination is highly endothermic and thus has a high energy of activation.

If we did not know otherwise, we might carelessly conclude from the energy of activation of the chain-initiating step alone that fluorine would be quite unreactive toward methane. (If we then proceeded to try the reaction, as a result of this careless assessment, the results would be literally disastrous.) We know, however, that the chain-initiating step occurs only infrequently relative to the chain-propagating steps. One initiating step is able to produce thousands of fluorination reactions. As a result, the high activation energy for this step is not an impediment to the reaction.

Chain-propagating steps, by contrast, cannot afford to have high energies of activation. If they do, the highly reactive intermediates are consumed by chain-terminating steps before the chains progress very far. Both of the chain-propagating steps in fluorination have very small energies of activation. This allows a relatively large fraction of energetically favorable collisions even at room temperature. Moreover, the overall heat of reaction, $\Delta H°$, is very large. This means that as the reaction occurs, a large quantity of heat is evolved. This heat may accumulate in the mixture faster than it dissipates to the surroundings, causing the temperature to rise and with it a rapid increase in the frequency of additional chain-initiating steps that would generate additional chains. These two factors, the low energy of activation for the chain-propagating steps and the large overall heat of reaction, account for the high reactivity of fluorine toward methane. (Fluorination reactions can be controlled. This is usually accomplished by diluting both the hydrocarbon and the fluorine with an inert gas such as helium before bringing them together. The reaction is also carried out in a reactor packed with copper shot. The copper, by absorbing the heat produced, moderates the reaction.)

CHLORINATION

	$\Delta H°$ (kJ mol^{-1})	E_{act} (kJ mol^{-1})
Chain Initiation		
$Cl_2 \longrightarrow 2\,Cl\cdot$	+243	+243
Chain Propagation		
$Cl\cdot + CH_4 \longrightarrow HCl + CH_3\cdot$	+8	+16
$CH_3\cdot + Cl_2 \longrightarrow CH_3Cl + Cl\cdot$	−109	Small
Overall $\Delta H°$ =	−101	

The higher energy of activation of the first chain-propagating step (the hydrogen abstraction step) in the chlorination of methane (+16 kJ mol^{-1}), versus the lower energy of activation (+5.0 kJ mol^{-1}) in the fluorination, partly explains the lower reactivity of chlorine. The greater energy required to break the chlorine–chlorine bond in the initiating step (243 kJ mol^{-1} for Cl_2 versus 159 kJ mol^{-1} for F_2) has some effect, too. However, the much greater overall heat of reaction in fluorination probably plays the greatest role in accounting for the much greater reactivity of fluorine.

BROMINATION

	$\Delta H°$ (kJ mol^{-1})	E_{act} (kJ mol^{-1})
Chain Initiation		
$Br_2 \longrightarrow 2\,Br\cdot$	+193	+193
Chain Propagation		
$Br\cdot + CH_4 \longrightarrow HBr + CH_3\cdot$	+74	+78
$CH_3\cdot + Br_2 \longrightarrow CH_3Br + Br\cdot$	−100	Small
Overall $\Delta H°$ =	−26	

In contrast to chlorination, the hydrogen atom abstraction step in bromination has a very high energy of activation (E_{act} = 78 kJ mol^{-1}). This means that only a very tiny fraction of all of the collisions between bromine atoms and methane molecules will be energetically effective even at a temperature of 300°C. Bromine, as a result, is much less reactive toward methane than chlorine, even though the net reaction is slightly exothermic.

IODINATION

	$\Delta H°$ (kJ mol^{-1})	E_{act} (kJ mol^{-1})
Chain Initiation		
$I_2 \longrightarrow 2\,I\cdot$	+151	+151
Chain Propagation		
$I\cdot + CH_4 \longrightarrow HI + CH_3\cdot$	+142	+140
$CH_3\cdot + I_2 \longrightarrow CH_3I + I\cdot$	−89	Small
Overall $\Delta H°$ =	+53	

The thermodynamic quantities for iodination of methane make it clear that the chain-initiating step is not responsible for the observed order of reactivities: $F_2 > Cl_2 > Br_2 > I_2$. The iodine–iodine bond is even weaker than the fluorine–fluorine bond. On this basis alone, one would predict iodine to be the most reactive of the halogens. This clearly is not the case. Once again, it is the hydrogen atom–abstraction step that correlates with the experimentally determined order of reactivities. The energy of activation of this step in the iodine reaction (140 kJ mol^{-1}) is so large that only two collisions out of every 10^{12} have sufficient energy to produce reactions at 300°C. As a result, iodination is not a feasible reaction experimentally.

Before we leave this topic, one further point needs to be made. We have given explanations of the relative reactivities of the halogens toward methane that have been based on energy considerations alone. This has been possible *only because the reactions are quite similar and thus have similar entropy changes.* Had the reactions been of different types, this kind of analysis would not have been proper and might have given incorrect explanations.

10.6 Halogenation of Higher Alkanes

Higher alkanes react with halogens by the same kind of chain mechanism as those that we have just seen. Ethane, for example, reacts with chlorine to produce chloroethane (ethyl chloride). The mechanism is as follows:

A Mechanism for the Reaction

Radical Halogenation of Ethane

Chain Initiation

$$\textit{Step 1}\quad Cl_2 \xrightarrow[\text{or heat}]{\text{light}} 2\,Cl\cdot$$

Chain Propagation

$$\textit{Step 2}\quad CH_3CH_2{:}H + \cdot Cl \longrightarrow CH_3CH_2\cdot + H{:}Cl$$

$$\textit{Step 3}\quad CH_3CH_2\cdot + Cl{:}Cl \longrightarrow CH_3CH_2{:}Cl + Cl\cdot$$

Chain propagation continues with steps 2, 3, 2, 3, and so on.

Chain Termination

$$CH_3CH_2\cdot + \cdot Cl \longrightarrow CH_3CH_2{:}Cl$$

$$CH_3CH_2\cdot + \cdot CH_2CH_3 \longrightarrow CH_3CH_2{:}CH_2CH_3$$

$$Cl\cdot + \cdot Cl \longrightarrow Cl{:}Cl$$

Problem 10.10 The energy of activation for the hydrogen atom abstraction step in the chlorination of ethane is 4.2 kJ mol^{-1}. **(a)** Use the homolytic bond dissociation energies in Table 10.1 to calculate $\Delta H°$ for this step. **(b)** Sketch a potential energy diagram for the hydrogen atom abstraction step in the chlorination of ethane similar to that for the chlorination of methane shown in Fig. 10.3*a*. **(c)** When an equimolar mixture of methane and ethane is chlorinated, the reaction yields far more chloroethane than chloromethane (~400 molecules of chloroethane for every molecule of chloromethane). Explain this greater yield of chloroethane.

PROBLEM 10.11

When ethane is chlorinated, 1,1-dichloroethane and 1,2-dichloroethane, as well as more highly chlorinated ethanes, are formed in the mixture (see Section 10.3A). Write chain mechanisms accounting for the formation of 1,1-dichloroethane and 1,2-dichloroethane.

Chlorination of most alkanes whose molecules contain more than two carbon atoms gives a mixture of isomeric monochloro products (as well as more highly chlorinated compounds). Several examples follow. The percentages given are based on the total amount of monochloro products formed in each reaction.

$$CH_3CH_2CH_3 \xrightarrow[\text{light, 25°C}]{Cl_2} CH_3CH_2CH_2Cl + CH_3CHClCH_3$$

Propane → **1-Chloropropane (45%)** + **2-Chloropropane (55%)**

$$(CH_3)_2CHCH_3 \xrightarrow[\text{light, 25°C}]{Cl_2} CH_3CH(CH_3)CH_2Cl + CH_3CCl(CH_3)CH_3$$

2-Methylpropane → **1-Chloro-2-methylpropane (63%)** + **2-Chloro-2-methylpropane (37%)**

$$CH_3CH(CH_3)CH_2CH_3 \xrightarrow[300°C]{Cl_2} ClCH_2CH(CH_3)CH_2CH_3 + CH_3CCl(CH_3)CH_2CH_3 + CH_3CH(CH_3)CHClCH_3 + CH_3CH(CH_3)CH_2CH_2Cl$$

2-Methylbutane → **1-Chloro-2-methylbutane (30%)** + **2-Chloro-2-methylbutane (22%)** + **2-Chloro-3-methylbutane (33%)** + **1-Chloro-3-methylbutane (15%)**

The ratios of products that we obtain from chlorination reactions of higher alkanes are not identical with what we would expect if all the hydrogen atoms of the alkane were equally reactive. We find that there is a correlation between reactivity of different hydrogen atoms and the type of hydrogen atom (1°, 2°, or 3°) being replaced. The tertiary hydrogen atoms of an alkane are most reactive, secondary hydrogen atoms are next most reactive, and primary hydrogen atoms are the least reactive (see Problem 10.12).

PROBLEM 10.12

(a) What percentages of 1-chloropropane and 2-chloropropane would you expect to obtain from the chlorination of propane if 1° and 2° hydrogen atoms were equally reactive? **(b)** What percentages of 1-chloro-2-methylpropane and 2-chloro-2-methylpropane would you expect from the chlorination of 2-methylpropane if the 1° and 3° hydrogen atoms were equally reactive? **(c)** Compare these calculated answers with the results actually obtained (above) and justify the assertion that the order of reactivity of the hydrogen atoms is 3° > 2° > 1°.

We can account for the relative reactivities of the primary, secondary, and tertiary hydrogen atoms in a chlorination reaction on the basis of the homolytic bond dissociation energies we saw earlier (Table 10.1). Of the three types, breaking a tertiary C—H bond requires the least energy, and breaking a primary C—H bond requires the most. Since the step in which the C—H bond is broken (i.e., the hydrogen atom abstraction step) determines the location or orientation of the chlorination, we would expect the E_{act} for abstracting a tertiary hydrogen atom to be least and the E_{act} for abstracting a primary hydrogen atom to be greatest. Thus tertiary hydrogen atoms should be most reactive, secondary hydrogen atoms should be the next most reactive, and primary hydrogen atoms should be the least reactive.

The differences in the rates with which primary, secondary, and tertiary hydrogen atoms are replaced by chlorine are not large, however, Chlorine, as a result, does not discriminate among the different types of hydrogen atoms in a way that makes chlorination of higher alkanes a generally useful laboratory synthesis. (Alkane chlorinations do find use in some industrial processes, especially in those instances where mixtures of alkyl chlorides can be used.)

PROBLEM 10.13 Chlorination reactions of certain alkanes can be used for laboratory preparations. Examples are the preparation of chlorocyclopropane from cyclopropane and chlorocyclobutane from cyclobutane. What structural feature of these molecules makes this possible?

PROBLEM 10.14 Each of the following alkanes reacts with chlorine to give a single monochloro substitution product. On the basis of this information, deduce the structure of each alkane.
(a) C_5H_{10} **(b)** C_8H_{18} **(c)** C_5H_{12}

10.6A Selectivity of Bromine

STUDY TIP! Bromination is selective.

Bromine is less reactive toward alkanes in general than chlorine, but bromine is more *selective* in the site of attack when it does react. Bromine shows a much greater ability to discriminate among the different types of hydrogen atoms. The reaction of 2-methylpropane and bromine, for example, gives almost exclusive replacement of the tertiary hydrogen atom:

$$CH_3-\underset{H}{\overset{CH_3}{C}}-CH_3 \xrightarrow[\text{light, 127°C}]{Br_2} \underset{(>99\%)}{CH_3-\underset{Br}{\overset{CH_3}{C}}-CH_3} + \underset{(\text{trace})}{CH_3-\underset{H}{\overset{CH_3}{C}}-CH_2Br}$$

A very different result is obtained when 2-methylpropane reacts with chlorine:

$$CH_3\overset{CH_3}{C}HCH_3 \xrightarrow[25°C]{Cl_2,\ h\nu} \underset{(37\%)}{CH_3\underset{Cl}{\overset{CH_3}{C}}CH_3} + \underset{(63\%)}{CH_3\overset{CH_3}{C}HCH_2Cl}$$

Fluorine, being much more reactive than chlorine, *is even less selective than chlorine.* Because the energy of activation for the abstraction of any type of hydrogen by a fluorine atom is low, there is very little difference in the rate at which a 1°, 2°, or 3° hydrogen reacts with fluorine. Reactions of alkanes with fluorine give (almost) the distribution of products that we would expect if all of the hydrogens of the alkane were equally reactive.

PROBLEM 10.15

Why is temperature an important variable to consider when using isomer distribution to evaluate the reactivities of the hydrogens of an alkane toward radical chlorination?

10.7 The Geometry of Alkyl Radicals

Experimental evidence indicates that the geometric structure of most alkyl radicals is trigonal planar at the carbon having the unpaired electron. This structure can be accommodated by an sp^2-hybridized central carbon. In an alkyl radical, the p orbital contains the unpaired electron (Fig. 10.6).

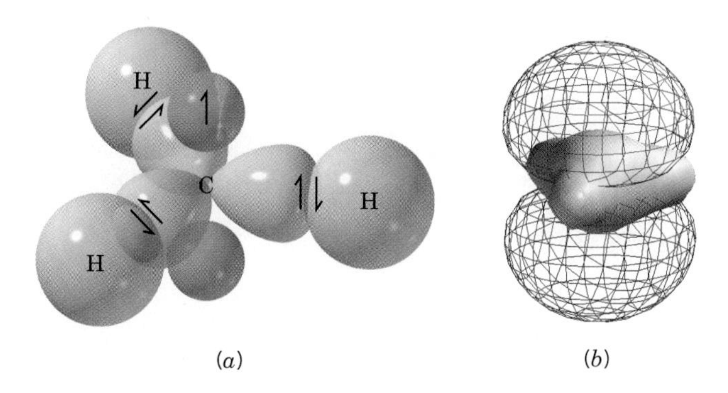

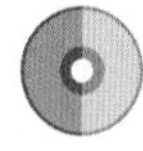

FIGURE 10.6 *(a)* Drawing of a methyl radical showing the sp^2-hybridized carbon atom at the center, the unpaired electron in the half-filled p orbital, and the three pairs of electrons involved in covalent bonding. The unpaired electron could be shown in either lobe. *(b)* Calculated structure for the methyl radical showing the highest occupied molecular orbital, where the unpaired electron resides, in red and blue. The region of bonding electron density around the carbons and hydrogens is in gray.

10.8 Reactions that Generate Tetrahedral Stereogenic Carbons

When achiral molecules react to produce a compound with a single tetrahedral stereogenic carbon, the product will be obtained as a racemic form. This will always be true in the absence of any chiral influence on the reaction such as an enzyme or the use of a chiral solvent.

Let us examine a reaction that illustrates this principle, the radical chlorination of pentane:

$$CH_3CH_2CH_2CH_2CH_3 \xrightarrow[\text{(achiral)}]{Cl_2} CH_3CH_2CH_2CH_2CH_2Cl + CH_3CH_2CH_2\overset{*}{C}HClCH_3$$

Pentane (achiral) — **1-Chloropentane (achiral)** — **(±)-2-Chloropentane (a racemic form)**

$$+ CH_3CH_2CHClCH_2CH_3$$

3-Chloropentane (achiral)

The reaction will lead to the products shown here, as well as more highly chlorinated products. (We can use an excess of pentane to minimize multiple chlorinations.) Neither 1-chloropentane nor 3-chloropentane contains a stereogenic carbon, but 2-chloropentane does, and it is *obtained as a racemic form.* If we examine the mechanism we shall see why.

A Mechanism for the Reaction

The Stereochemistry of Chlorination at C2 of Pentane

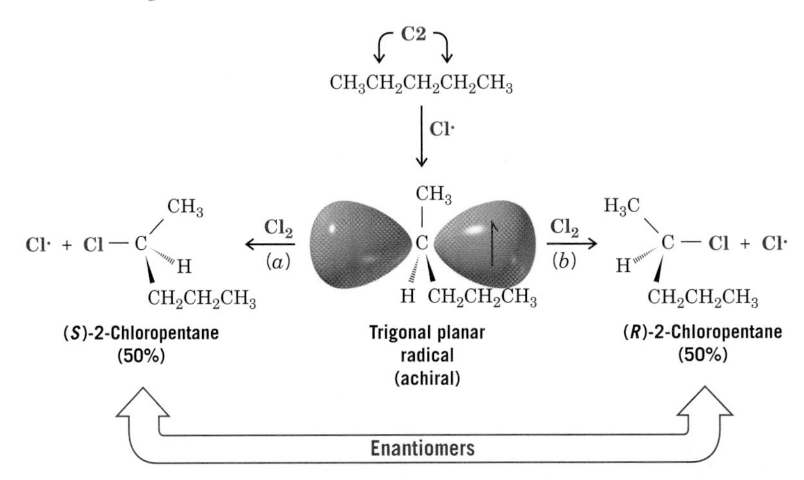

Abstraction of a hydrogen atom from C2 produces a trigonal planar radical that is achiral. This radical then reacts with chlorine at either face [by path (a) or path (b)]. Because the radical is achiral, the probability of reaction by either path is the same; therefore, the two enantiomers are produced in equal amounts, and a racemic form of 2-chloropentane results.

10.8A Generation of a Second Stereogenic Carbon in a Radical Halogenation

Let us now examine what happens when a chiral molecule (containing one stereogenic carbon) reacts so as to yield a product with a second stereogenic carbon. As an example consider what happens when (*S*)-2-chloropentane undergoes chlorination at C3 (other products are formed, of course, by chlorination at other carbon atoms). The results of chlorination at C3 are shown in the box at the top of page 471.

The products of the reactions are (2*S*,3*S*)-2,3-dichloropentane and (2*S*,3*R*)-2,3-dichloropentane. These two compounds are **diastereomers.** (They are stereoisomers but they are not mirror images of each other.) The two diastereomers are *not* produced in equal amounts. Because the intermediate radical itself is chiral, reactions at the two faces are not equally likely. The radical reacts with chlorine to a greater extent at one face than the other (although we cannot easily predict which). That is, the presence of a stereogenic carbon in the radical (at C2) influences the reaction that introduces the new stereogenic carbon (at C3).

Both of the 2,3-dichloropentane diastereomers are chiral and, therefore, each exhibits optical activity. Moreover, because the two compounds are *diastereomers,* they have different physical properties (e.g., different melting points and boiling points) and are separable by conventional means (by gas chromatography or by careful fractional distillation).

Problem 10.16

Consider the chlorination of (*S*)-2-chloropentane at C4. **(a)** Write stereochemical structures for the products that would be obtained and give each its proper (*R*)–(*S*) designation. **(b)** What is the stereoisomeric relationship between these products? **(c)** Are both products chiral? **(d)** Are both optically active? **(e)** Could the products be separated by conventional means? **(f)** What other dichloropentanes would be obtained by chlorination of (*S*)-2-chloropentane? **(g)** Which of these are optically active?

A Mechanism for the Reaction

The Stereochemistry of Chlorination at C3 of (S)-2-Chloropentane

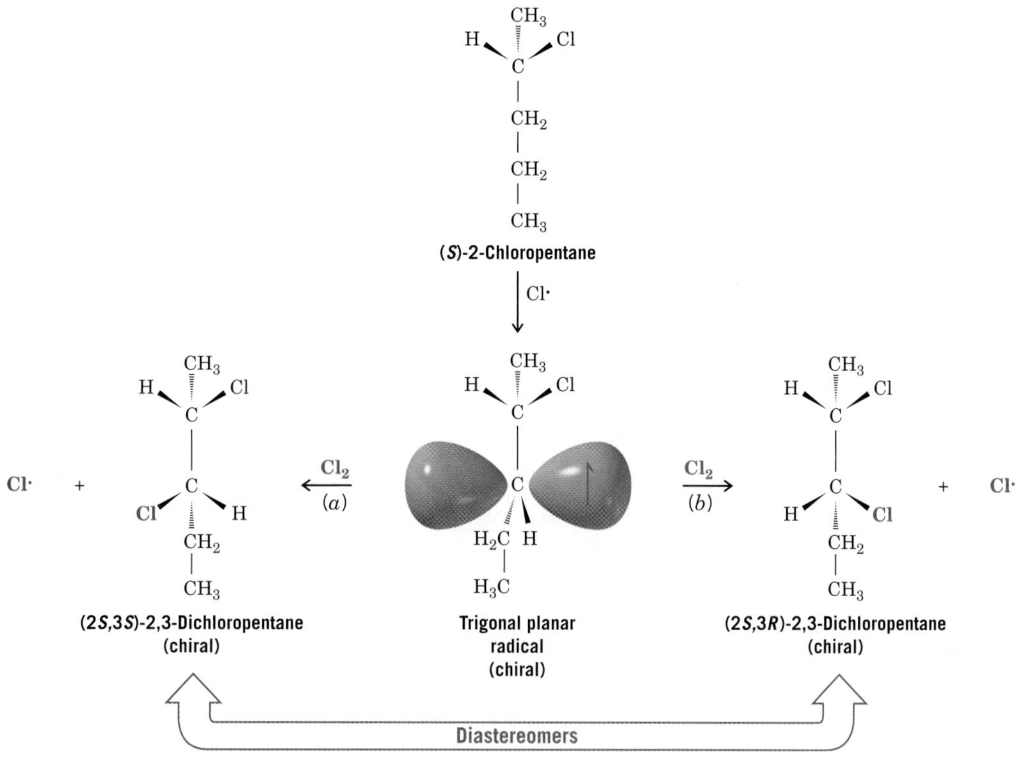

(S)-2-Chloropentane

Cl·

Cl· + (2S,3S)-2,3-Dichloropentane (chiral) $\xleftarrow[(a)]{Cl_2}$ Trigonal planar radical (chiral) $\xrightarrow[(b)]{Cl_2}$ (2S,3R)-2,3-Dichloropentane (chiral) + Cl·

Diastereomers

Abstraction of a hydrogen atom from C3 of (S)-2-chloropentane produces a radical that is chiral (it contains a stereogenic carbon at C2). This chiral radical can then react with chlorine at one face [path (a)] to produce (2S,3S)-2,3-dichloropentane and at the other face [path (b)] to yield (2S,3R)-2,3-dichloropentane. These two compounds are diastereomers, and they are not produced in equal amounts. Each product is chiral, and each alone would be optically active.

PROBLEM 10.17

Consider the bromination of butane using sufficient bromine to cause dibromination. After the reaction is over, you isolate all of the dibromo isomers by gas chromatography or by fractional distillation. **(a)** How many fractions would you obtain? **(b)** What compound (or compounds) would the individual fractions contain? **(c)** Which, if any, of the fractions would show optical activity? **(d)** Knowing that the bromine isotopes ^{79}Br and ^{81}Br are almost equally abundant in nature, what mass-to-charge peaks would predominate in the mass spectra of these dibromo isomers?

PROBLEM 10.18

The chlorination of 2-methylbutane yields 1-chloro-2-methylbutane, 2-chloro-2-methylbutane, 2-chloro-3-methylbutane, and 1-chloro-3-methylbutane. **(a)** Assuming that these compounds were separated after the reaction by fractional distillation, tell whether any fractions would show optical activity. **(b)** Would any of these fractions be resolvable into enantiomers? **(c)** How would the ^{1}H NMR spectra of these compounds differ at the position where the chlorine is bonded? Could each fraction from the distillation be identified on the basis of ^{1}H NMR spectroscopy?

10.9 Radical Addition to Alkenes: The Anti-Markovnikov Addition of Hydrogen Bromide

Before 1933, the orientation of the addition of hydrogen bromide to alkenes was the subject of much confusion. At times addition occurred in accordance with Markovnikov's rule; at other times it occurred in just the opposite manner. Many instances were reported where, under what seemed to be the same experimental conditions, Markovnikov additions were obtained in one laboratory and anti-Markovnikov additions in another. At times even the same chemist would obtain different results using the same conditions but on different occasions.

The mystery was solved in 1933 by the research of M. S. Kharasch and F. R. Mayo (of the University of Chicago). The explanatory factor turned out to be organic peroxides present in the alkenes—peroxides that were formed by the action of atmospheric oxygen on the alkenes (Section 10.11D). Kharasch and Mayo found that when alkenes that contained peroxides or hydroperoxides reacted with hydrogen bromide, anti-Markovnikov addition of hydrogen bromide occurred:

$$\text{R—}\ddot{\underset{..}{\text{O}}}\text{—}\ddot{\underset{..}{\text{O}}}\text{—R} \qquad \text{R—}\ddot{\underset{..}{\text{O}}}\text{—}\ddot{\underset{..}{\text{O}}}\text{—H}$$

An organic peroxide **An organic hydroperoxide**

Under these conditions, for example, propene yields 1-bromopropane. In the absence of peroxides, or in the presence of compounds that would "trap" radicals, normal Markovnikov addition occurs.

$$CH_3CH{=}CH_2 + HBr \xrightarrow{\text{ROOR}} CH_3CH_2CH_2Br \quad \text{Anti-Markovnikov addition}$$

1-Bromopropane

$$CH_3CH{=}CH_2 + HBr \xrightarrow[\text{peroxides}]{\text{no}} CH_3\underset{\text{Br}}{\underset{|}{C}}HCH_3 \quad \text{Markovnikov addition}$$

2-Bromopropane

Hydrogen fluoride, hydrogen chloride, and hydrogen iodide *do not* give anti-Markovnikov addition even when peroxides are present.

According to Kharasch and Mayo, the mechanism for **anti-Markovnikov addition of hydrogen bromide** is a **radical chain reaction** initiated by peroxides.

A Mechanism for the Reaction

Anti-Markovnikov Addition

Chain Initiation

Step 1
$$\text{R—}\ddot{\underset{..}{\text{O}}}\text{:}\ddot{\underset{..}{\text{O}}}\text{—R} \xrightarrow{\text{heat}} 2\ \text{R—}\ddot{\underset{..}{\text{O}}}\cdot$$

Heat brings about homolytic cleavage of the weak oxygen–oxygen bond.

Step 2
$$\text{R—}\ddot{\underset{..}{\text{O}}}\cdot + \text{H}:\ddot{\underset{..}{\text{Br}}}: \longrightarrow \text{R—}\ddot{\underset{..}{\text{O}}}:\text{H} + :\ddot{\underset{..}{\text{Br}}}\cdot$$

The alkoxyl radical abstracts a hydrogen atom from HBr, producing a bromine atom.

Step 3 $:\ddot{Br}\cdot + H_2C{=}CH{-}CH_3 \longrightarrow :\ddot{Br}:CH_2{-}\dot{C}H{-}CH_3$

2° Radical

A bromine atom adds to the double bond to produce the more stable 2° radical.

Step 4 $:\ddot{Br}{-}CH_2{-}\dot{C}H{-}CH_3 + H:\ddot{Br}: \longrightarrow :\ddot{Br}{-}CH_2{-}CH(H){-}CH_3 + \cdot\ddot{Br}:$

1-Bromopropane

The 2° radical abstracts a hydrogen atom from HBr. This leads to the product and regenerates a bromine atom. Then repetitions of steps 3 and 4 lead to a chain reaction.

Step 1 is the simple homolytic cleavage of the peroxide molecule to produce two alkoxyl radicals. The oxygen–oxygen bond of peroxides is weak, and such reactions are known to occur readily:

$$R{-}\ddot{O}{:}\ddot{O}{-}R \longrightarrow 2\,R{-}\ddot{O}\cdot \qquad \Delta H^\circ \cong +150\ \text{kJ mol}^{-1}$$

Peroxide **Alkoxyl radical**

Step 2 of the mechanism, abstraction of a hydrogen atom by the radical, is exothermic and has a low energy of activation:

$$R{-}\ddot{O}\cdot + H{:}\ddot{Br}{:} \longrightarrow R{-}\ddot{O}{:}H + {:}\ddot{Br}\cdot \qquad \Delta H^\circ \cong -96\ \text{kJ mol}^{-1}$$

E_{act} is low

Step 3 of the mechanism determines the final orientation of bromine in the product. It occurs as it does because a *more stable secondary radical* is produced and because *attack at the primary carbon atom is less hindered.* Had the bromine attacked propene at the secondary carbon atom, a less stable, primary radical would have been the result,

$$Br\cdot + CH_2{=}CHCH_3 \not\longrightarrow \cdot CH_2CH(Br)CH_3$$

1° Radical (less stable)

and attack at the secondary carbon atom would have been more hindered.

Step 4 of the mechanism is simply the abstraction of a hydrogen atom from hydrogen bromide by the radical produced in step 3. This hydrogen atom abstraction produces a bromine atom that can bring about step 3 again; then step 4 occurs again—a chain reaction.

10.9A Summary of Markovnikov versus Anti-Markovnikov Addition of HBr to Alkenes

We can now see the contrast between the two ways that HBr can add to an alkene. In the *absence* of peroxides, the reagent that attacks the double bond first is a proton. Because a proton is small, steric effects are unimportant. It attaches itself to a carbon atom by an ionic mechanism so as to form the more stable carbocation. The result is Markovnikov addition. Polar, protic solvents favor this process.

A tip for alkyl halide synthesis

Ionic Addition

$$CH_3CH{=}CH_2 \xrightarrow{H-Br} CH_3\overset{+}{C}HCH_3 \xrightarrow{Br^-} CH_3\underset{\displaystyle Br}{\underset{|}{C}}HCH_3$$

More stable carbocation **Markovnikov product**

In the *presence* of peroxides, the reagent that attacks the double bond first is the larger bromine atom. It attaches itself to the less hindered carbon atom by a radical mechanism, so as to form the more stable radical intermediate. The result is anti-Markovnikov addition. Nonpolar solvents are preferable for reactions involving radicals.

Radical Addition

$$CH_3CH{=}CH_2 \xrightarrow{Br\cdot} CH_3\dot{C}HCH_2Br \xrightarrow{HBr} CH_3CH_2CH_2Br + Br\cdot$$

More stable radical **Anti-Markovnikov product**

10.10 Radical Polymerization of Alkenes: Chain-Growth Polymers

Polymers are substances that consist of very large molecules called **macromolecules** that are made up of many repeating subunits. The molecular subunits that are used to synthesize polymers are called **monomers,** and the reactions by which monomers are joined together are called **polymerizations.** Many polymerizations can be initiated by radicals.

Ethylene (ethene), for example, is the monomer that is used to synthesize the familiar polymer called *polyethylene.*

Monomeric units

$$m\ CH_2{=}CH_2 \xrightarrow{\text{polymerization}} -CH_2CH_2 \left(CH_2CH_2 \right)_n CH_2CH_2-$$

Ethylene *monomer* **Polyethylene** *polymer*

(m and n are large numbers)

Because polymers such as polyethylene are made by addition reactions, they are often called **chain-growth polymers** or **addition polymers.** Let us now examine in some detail how polyethylene is made.

Ethylene polymerizes by a radical mechanism when it is heated at a pressure of 1000 atm with a small amount of an organic peroxide (called a diacyl peroxide).

A Mechanism for the Reaction

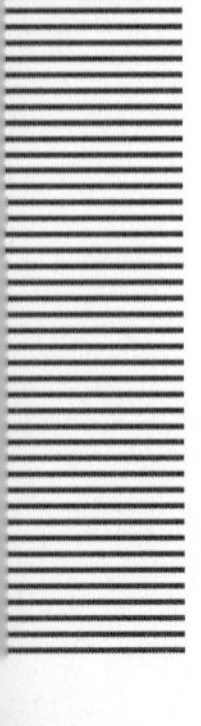

Radical Polymerization of Ethene

Chain Initiation

Step 1 $$R-\overset{O}{\overset{\|}{C}}-O{:}O-\overset{O}{\overset{\|}{C}}-R \longrightarrow 2\ R{:}\overset{O}{\overset{\|}{C}}-O\cdot \longrightarrow 2\ CO_2 + 2\ R\cdot$$

Diacyl peroxide

Step 2 $$R\cdot + CH_2{=}CH_2 \longrightarrow R{:}CH_2-CH_2\cdot$$

The diacyl peroxide dissociates and releases carbon dioxide gas. Alkyl radicals are produced, which in turn initiate chains.

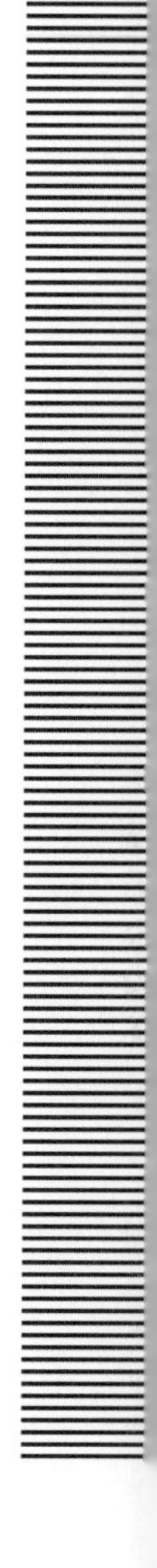

Chain Propagation

$$\text{Step 3}\quad R{-}CH_2CH_2\cdot + n\,CH_2{=}CH_2 \longrightarrow R{+}CH_2CH_2{+}_n CH_2CH_2\cdot$$

Chains propagate by adding successive ethylene units, until their growth is stopped by combination or disproportionation.

Chain Termination

$$\text{Step 4}\quad 2\,R{+}CH_2CH_2{+}_n CH_2CH_2\cdot \xrightarrow{\text{combination}} [R{+}CH_2CH_2{+}_n CH_2CH_2{+}_2$$

$$\xrightarrow{\text{disproportionation}} R{+}CH_2CH_2{+}_n CH{=}CH_2 + R{+}CH_2CH_2{+}_n CH_2CH_3$$

The radical at the end of the growing polymer chain can also abstract a hydrogen atom from itself by what is called "back biting." This leads to chain branching.

Chain Branching

$$R{-}CH_2\overset{H}{C}H(CH_2CH_2)_n CH_2\dot{C}H_2 \longrightarrow RCH_2\dot{C}H{+}CH_2CH_2{+}_n CH_2CH_2{-}H$$

$$\big\downarrow\ CH_2{=}CH_2$$

$$RCH_2\underset{\underset{\dot{C}H_2}{|}}{\underset{CH_2}{|}}{CH}{+}CH_2CH_2{+}_n CH_2CH_3$$

$$\big\downarrow\ \text{etc.}$$

The polyethylene produced by radical polymerization is not generally useful unless it has a molecular weight of nearly 1,000,000. Very high molecular weight polyethylene can be obtained by using a low concentration of the initiator. This initiates the growth of only a few chains and ensures that each chain will have a large excess of the monomer available. More initiator may be added as chains terminate during the polymerization, and, in this way, new chains are begun.

Polyethylene has been produced commercially since 1943. It is used in manufacturing flexible bottles, films, sheets, and insulation for electric wires. Polyethylene produced by radical polymerization has a softening point of about 110°C.

Polyethylene can be produced in a different way using catalysts called **Ziegler–Natta catalysts** that are organometallic complexes of transition metals. In this process no radicals are produced, no back biting occurs, and, consequently, there is no chain branching. The polyethylene that is produced is of higher density, has a higher melting point, and has greater strength. (Ziegler–Natta catalysts are discussed in greater detail in Special Topic A.)

Another familiar polymer is *polystyrene*. The monomer used in making polystyrene is phenylethene, a compound commonly known as *styrene*.

$$m\ CH_2{=}CH(C_6H_5) \xrightarrow{\text{polymerization}} {-}CH_2CH(C_6H_5){+}CH_2CH(C_6H_5){+}_n CH_2CH(C_6H_5){-}$$

Styrene **Polystyrene**

Table 10.2 lists several other common chain-growth polymers. Further information on each is provided in Special Topic A.

TABLE 10.2 Other Common Chain-Growth Polymers

Monomer	Polymer	Names
$CH_2{=}CHCH_3$	$\text{+}CH_2{-}CH(CH_3)\text{+}_n$	Polypropylene
$CH_2{=}CHCl$	$\text{+}CH_2{-}CH(Cl)\text{+}_n$	Poly(vinyl chloride), PVC
$CH_2{=}CHCN$	$\text{+}CH_2{-}CH(CN)\text{+}_n$	Polyacrylonitrile, Orlon
$CF_2{=}CF_2$	$\text{+}CF_2{-}CF_2\text{+}_n$	Polytetrafluoroethene, Teflon
$CH_2{=}C(CH_3)CO_2CH_3$	$\text{+}CH_2{-}C(CH_3)(CO_2CH_3)\text{+}_n$	Poly(methyl methacrylate), Lucite, Plexiglas, Perspex

10.11 Other Important Radical Reactions

Radical mechanisms are important in understanding many other organic reactions. We shall see other examples in later chapters, but let us examine a few important radicals and radical reactions here: oxygen and superoxide, the combustion of alkanes, autoxidation, antioxidants, and some reactions of chlorofluoromethanes that have threatened the protective layer of ozone in the stratosphere.

10.11A Molecular Oxygen and Superoxide

One of the most important radicals (and one that we encounter every moment of our lives) is molecular oxygen. Molecular oxygen in the ground state is a diradical with one unpaired electron on each oxygen. As a radical, oxygen can abstract hydrogen atoms just like other radicals we have seen. This is one way oxygen is involved in combustion reactions (Section 10.11C) and autoxidation (Section 10.11D). In biological systems, oxygen is an electron acceptor. When molecular oxygen accepts one electron, it becomes a radical anion called superoxide ($O_2^{\cdot-}$). Superoxide is involved in both positive and negative physiological roles: The immune system uses superoxide in its defense against pathogens, yet superoxide is also suspected of being involved in degenerative disease processes associated with aging and oxidative damage to healthy cells. The enzyme superoxide dismutase regulates the level of superoxide by catalyzing conversion of superoxide to hydrogen peroxide and molecular oxygen. Hydrogen peroxide, however, is also harmful because it can produce hydroxyl ($HO\cdot$) radicals. The enzyme catalase helps to prevent release of hydroxyl radicals by converting hydrogen peroxide to water and oxygen:

$$2\,O_2^{\cdot -} + 2\,H^+ \xrightarrow{\text{superoxide dismutase}} H_2O_2 + O_2$$

$$2\,H_2O_2 \xrightarrow{\text{catalase}} 2\,H_2O + O_2$$

10.11B Nitric Oxide

Nitric oxide, synthesized in the body from the amino acid arginine, serves as a chemical messenger in a variety of biological processes, including blood pressure regulation and the immune response (see "The Chemistry of . . . Radicals in Biology, Medicine, and Industry" in Section 10.1B). Its role in relaxation of smooth muscle in vascular tissues is shown in Fig. 10.7.

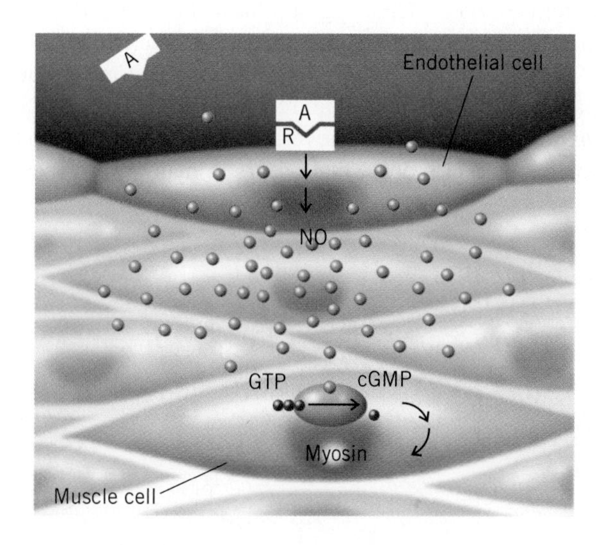

FIGURE 10.7 The neurotransmitter acetylcholine (A) binds to a specific endothelial cell receptor (R), activating it to begin synthesis of nitric oxide (NO). NO diffuses to muscle cells and binds to the enzyme guanylyl cyclase, signaling it to convert guanosine triphosphate (GTP) to cyclic guanosine monophosphate (cGMP). In turn, cGMP signals smooth muscle relaxation. The drug Viagra prolongs the effect of cGMP on certain smooth muscle tissue by inhibiting cGMP degradation.

The 1998 Nobel Prize in Physiology or Medicine was awarded to R. F. Furchgott, L. J. Ignarro, and F. Murad for their discovery that NO is an important signaling molecule.

10.11C Combustion of Alkanes

When alkanes react with oxygen (e.g., in oil and gas furnaces and in internal combustion engines) a complex series of reactions takes place, ultimately converting the alkane to carbon dioxide and water (Section 4.10A). Although our understanding of the detailed mechanism of combustion is incomplete, we do know that the important reactions occur by radical chain mechanisms with chain-initiating and chain-propagating steps such as the following reactions:

$$RH + O_2 \longrightarrow R\cdot + \cdot OOH \qquad \textbf{Initiating}$$

$$\left.\begin{aligned} R\cdot + O_2 &\longrightarrow R{-}OO\cdot \\ R{-}OO\cdot + R{-}H &\longrightarrow R{-}OOH + R\cdot \end{aligned}\right\} \textbf{Propagating}$$

One product of the second chain-propagating step is R—OOH, called an alkyl hydroperoxide. The oxygen–oxygen bond of an alkyl hydroperoxide is quite weak, and it can break and produce radicals that can initiate other chains:

$$RO{-}OH \longrightarrow RO\cdot + \cdot OH$$

10.11D Autoxidation

Linoleic acid is an example of a *polyunsaturated fatty acid,* the kind of polyunsaturated acid that occurs as an ester in **polyunsaturated fats** (Section 7.12, "Hydrogenation in the Food Industry," and Chapter 23). By polyunsaturated, we mean that the compound contains two or more double bonds:

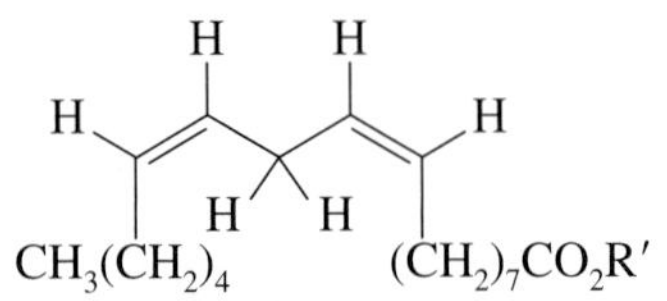

Linoleic acid (as an ester)

Polyunsaturated fats occur widely in the fats and oils that are components of our diets. They are also widespread in the tissues of the body where they perform numerous vital functions.

The hydrogen atoms of the $—CH_2—$ group located between the two double bonds of linoleic ester (Lin—H) are especially susceptible to abstraction by radicals (we shall see why in Chapter 13). Abstraction of one of these hydrogen atoms produces a new radical (Lin·) that can react with oxygen in a chain reaction that belongs to a general type of reaction called **autoxidation** (Fig. 10.8). The result of autoxidation is the formation of a hydroperoxide. Autoxidation is a process that occurs in many substances; for example, au-

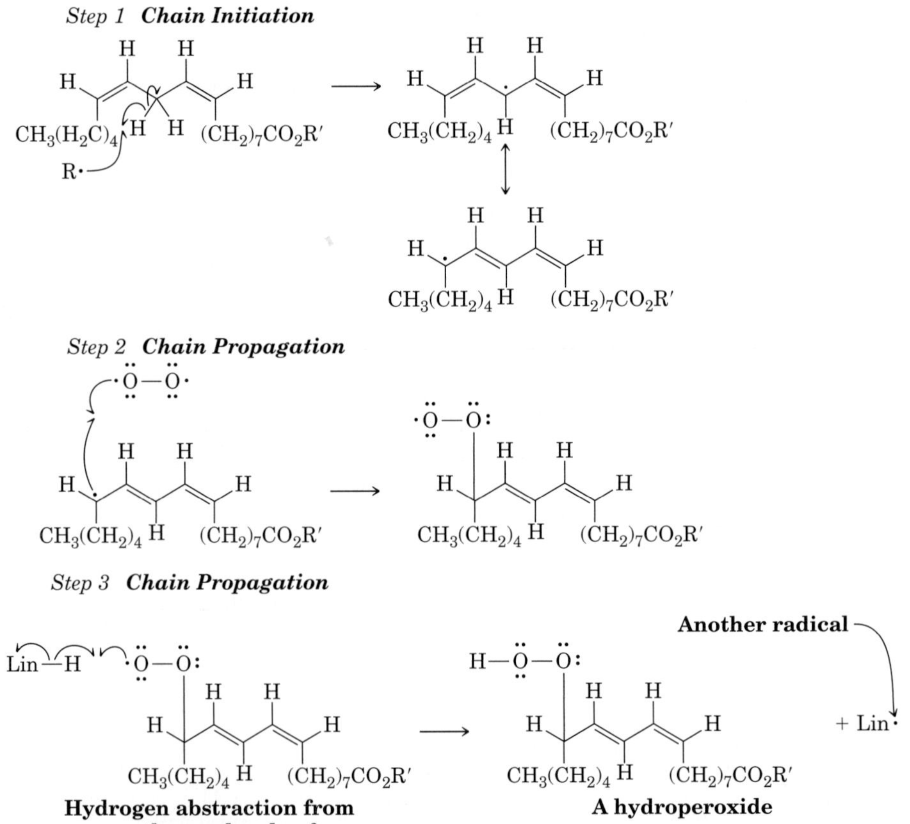

Step 1 ***Chain Initiation***

Step 2 ***Chain Propagation***

Step 3 ***Chain Propagation***

Hydrogen abstraction from another molecule of the linoleic ester

A hydroperoxide

Another radical

FIGURE 10.8 Autoxidation of a linoleic acid ester. In step 1 the reaction is initiated by the attack of a radical on one of the hydrogen atoms of the $—CH_2—$ group between the two double bonds; this hydrogen abstraction produces a radical that is a resonance hybrid. In step 2 this radical reacts with oxygen in the first of two chain-propagating steps to produce an oxygen-containing radical, which in step 3 can abstract a hydrogen from another molecule of the linoleic ester (Lin—H). The result of this second chain-propagating step is the formation of a hydroperoxide and a radical (Lin·) that can bring about a repetition of step 2.

toxidation is responsible for the development of the rancidity that occurs when fats and oils spoil and for the spontaneous combustion of oily rags left open to the air. Autoxidation also occurs in the body, and here it may cause irreversible damage.

10.11E Antioxidants

Autoxidation is inhibited when compounds called antioxidants are present that can rapidly "trap" peroxyl radicals by reacting with them to give stabilized radicals that do not continue the chain.

Vitamin E (α-tocopherol) is capable of acting as a radical trap, and one of the important roles that vitamin E plays in the body may be in inhibiting radical reactions that could cause cell damage. Vitamin C is also an antioxidant, although recent work indicates that supplements over 500 mg per day may have prooxidant effects. Compounds such as BHT are added to foods to prevent autoxidation. BHT is also known to trap radicals.

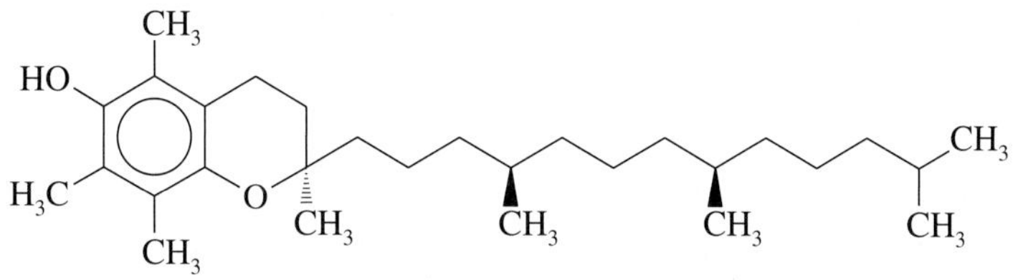

Vitamin E
(α-tocopherol)

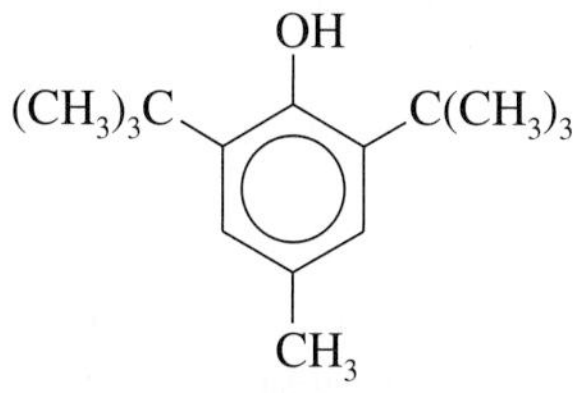

BHT
(butylated hydroxytoluene)

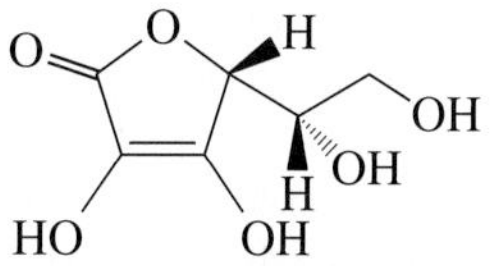

Vitamin C

10.11F Ozone Depletion and Chlorofluorocarbons (CFCs)

In the stratosphere at altitudes of about 25 km, very high-energy (very short wavelength) UV light converts diatomic oxygen (O_2) into ozone (O_3). The reactions that take place may be represented as follows:

$$\textit{Step 1}\quad O_2 + h\nu \longrightarrow O + O$$

$$\textit{Step 2}\quad O + O_2 + M \longrightarrow O_3 + M + \text{heat}$$

where M is some other particle that can absorb some of the energy released in the second step.

The ozone produced in step 2 can also interact with high-energy UV light in the following way:

$$\textit{Step 3}\quad O_3 + h\nu \longrightarrow O_2 + O + \text{heat}$$

The oxygen atom formed in step 3 can cause a repetition of step 2, and so forth. The net result of these steps is to convert highly energetic UV light into heat. This is important because the existence of this cycle shields Earth from radiation that is destructive to living organisms. This shield makes life possible on Earth's surface. Even a relatively small increase in high-energy UV radiation at Earth's surface would cause a large increase in the incidence of skin cancers.

Production of chlorofluoromethanes (and of chlorofluoroethanes) called chlorofluorocarbons (CFCs) or *freons* began in 1930. These compounds have been used as refrigerants, solvents, and propellants in aerosol cans. Typical freons are trichlorofluoromethane, $CFCl_3$ (called Freon-11), and dichlorodifluoromethane, CF_2Cl_2 (called Freon-12).

By 1974 world freon production was about 2 billion pounds annually. Most freon, even that used in refrigeration, eventually makes its way into the atmosphere where it diffuses unchanged into the stratosphere. In June 1974 F. S. Rowland and M. J. Molina published an article indicating, for the first time, that in the stratosphere freon is able to initiate radical chain reactions that can upset the natural ozone balance. The 1995 Nobel Prize in Chemistry was awarded to P. J. Crutzen, M. J. Molina, and F. S. Rowland for their combined work in this area. The reactions that take place are the following. (Freon-12 is used as an example.)

Chain Initiation

$$\textit{Step 1} \quad CF_2Cl_2 + h\nu \longrightarrow CF_2Cl\cdot + Cl\cdot$$

Chain Propagation

$$\textit{Step 2} \quad Cl\cdot + O_3 \longrightarrow ClO\cdot + O_2$$

$$\textit{Step 3} \quad ClO\cdot + O \longrightarrow O_2 + Cl\cdot$$

In the chain-initiating step, UV light causes homolytic cleavage of one C—Cl bond of the freon. The chlorine atom thus produced is the real villain; it can set off a chain reaction that destroys thousands of molecules of ozone before it diffuses out of the stratosphere or reacts with some other substance.

In 1975 a study by the National Academy of Sciences supported the predictions of Rowland and Molina, and since January 1978 the use of freons in aerosol cans in the United States has been banned.

In 1985 a hole was discovered in the ozone layer above Antarctica. Studies done since then strongly suggest that chlorine atom destruction of the ozone is a factor in the formation of the hole. This ozone hole has continued to grow in size, and such a hole has also been discovered in the Arctic ozone layer. Should the ozone layer be depleted, more of the sun's damaging rays would penetrate to the surface of Earth.

Recognizing the global nature of the problem, the "Montreal Protocol" was initiated in 1987. This treaty required the signing nations to reduce their production and consumption of chlorofluorocarbons. Accordingly, the industrialized nations of the world ceased production of chlorofluorocarbons as of January 1, 1996, and over 120 nations have now signed the "Montreal Protocol." Increased worldwide understanding of stratospheric ozone depletion, in general, has accelerated the phasing out of chlorofluorocarbons.

Summary and Review Tools

Concept Map

Mechanism Review of Radical Reactions

Radical Halogenation of Alkanes

$$X-X \xrightarrow{\text{heat or light}} X\cdot + X\cdot \quad \text{Chain initiation}$$

If X = Br, hydrogen abstraction is selective.
If X = Cl, hydrogen abstraction is not selective.

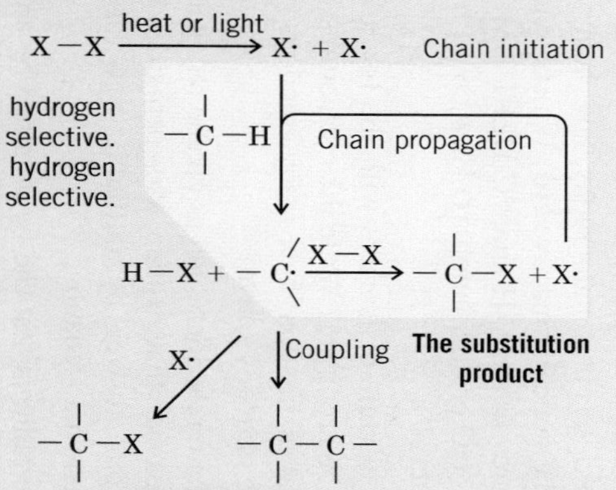

$$-\overset{|}{\underset{|}{C}}-H \longrightarrow H-X + -\dot{C}\langle \xrightarrow{X-X} -\overset{|}{\underset{|}{C}}-X + X\cdot \quad \text{Chain propagation}$$

The substitution product

X· → $-\overset{|}{\underset{|}{C}}-X$; Coupling → $-\overset{|}{\underset{|}{C}}-\overset{|}{\underset{|}{C}}-$

Some possible chain-terminating steps

Anti-Markovnikov Addition of HBr to Alkenes

$$RO-OR \xrightarrow{\text{heat}} RO\cdot + RO\cdot$$
$$RO\cdot + H-Br \longrightarrow ROH + Br\cdot$$
Chain initiation

Addition of the bromine radical to the alkene occurs so as to form the more stable carbon radical intermediate (The alkene reactant shown is meant to indicate any alkene where a difference exists in the extent of alkyl substitution at the initial alkene carbons.)

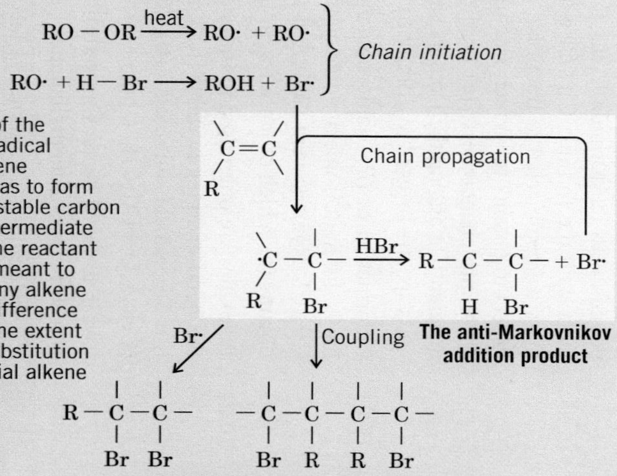

$$R(C=C) \longrightarrow \cdot\underset{R}{C}-\underset{Br}{C}- \xrightarrow{HBr} R-\underset{H}{C}-\underset{Br}{C}- + Br\cdot \quad \text{Chain propagation}$$

The anti-Markovnikov addition product

Br· → $R-\underset{Br}{C}-\underset{Br}{C}-$; Coupling → $-\underset{Br}{C}-\underset{R}{C}-\underset{R}{C}-\underset{Br}{C}-$

Some possible chain-terminating steps

Radical Polymerization

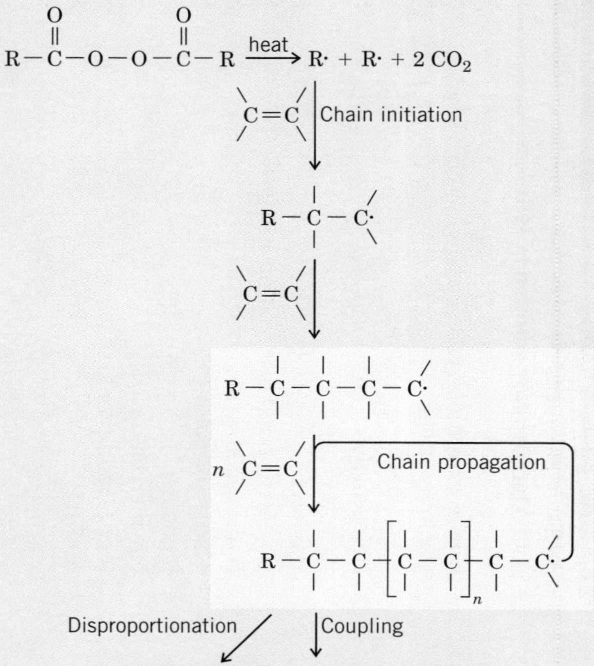

$$R-\overset{O}{\overset{\|}{C}}-O-O-\overset{O}{\overset{\|}{C}}-R \xrightarrow{\text{heat}} R\cdot + R\cdot + 2\,CO_2$$

$$C=C \longrightarrow R-\overset{|}{\underset{|}{C}}-\dot{C}\langle \quad \text{Chain initiation}$$

$$C=C \longrightarrow R-\overset{|}{\underset{|}{C}}-\overset{|}{\underset{|}{C}}-\overset{|}{\underset{|}{C}}-\dot{C}\langle$$

$$n\ C=C \longrightarrow R-\overset{|}{\underset{|}{C}}-\overset{|}{\underset{|}{C}}-\left[\overset{|}{\underset{|}{C}}-\overset{|}{\underset{|}{C}}\right]_n-\overset{|}{\underset{|}{C}}-\dot{C}\langle \quad \text{Chain propagation}$$

Disproportionation; Coupling

Possible chain-terminating steps

Key Terms and Concepts

Addition polymers	Section 10.10
Anti-Markovnikov addition of HBr	Section 10.9
Autoxidation	Section 10.11D
Chain-growth polymers	Section 10.10
Chain reactions	Sections 10.4, 10.5, 10.6, 10.11
Halogenation	Sections 10.3, 10.4, 10.5, 10.6, 10.8
Homolysis	Section 10.1
Homolytic bond dissociation energy ($DH°$)	Section 10.2
Hydrogen abstraction	Section 10.1B
Macromolecules	Section 10.10
Monomers	Section 10.10
Polymerizations	Section 10.10
Radical addition to alkenes	Section 10.9
Radicals	Section 10.1

Additional Problems

10.19 The radical reaction of propane with chlorine yields (in addition to more highly halogenated compounds) 1-chloropropane and 2-chloropropane. Write chain-initiating and chain-propagating steps showing how each compound is formed.

***10.20** In addition to more highly chlorinated products, chlorination of butane yields a mixture of compounds with the formula C_4H_9Cl. **(a)** Taking stereochemistry into account, how many different isomers with the formula C_4H_9Cl would you expect to be produced? **(b)** If the mixture of C_4H_9Cl isomers were subjected to fractional distillation or gas chromatography, how many fractions would you expect to obtain? **(c)** Which fractions would be optically inactive? **(d)** Which would you be able to resolve into enantiomers? **(e)** Predict the features in their ^{1}H and ^{13}C DEPT NMR spectra that would differentiate among the isomers separated by GC or distillation. **(f)** How could fragmentation in their mass spectra be used to differentiate the isomers?

10.21 Chlorination of (R)-2-chlorobutane yields a mixture of isomers with the formula $C_4H_8Cl_2$. **(a)** How many different isomers would you expect to be produced? Write their structures. **(b)** If the mixture of $C_4H_8Cl_2$ isomers were subjected to fractional distillation, how many fractions would you expect to obtain? **(c)** Which of these fractions would be optically active?

10.22 Peroxides are often used to initiate radical chain reactions such as alkane halogenations. **(a)** Examine the bond energies in Table 10.1 and give reasons that explain why peroxides are especially effective as radical initiators. **(b)** Illustrate your answer by outlining how di-*tert*-butyl peroxide, $(CH_3)_3CO{-}OC(CH_3)_3$, might initiate an alkane halogenation.

10.23 List in order of decreasing stability all of the radicals that can be obtained by abstraction of a hydrogen atom from 2-methylbutane.

10.24 Shown below is an alternative mechanism for the chlorination of methane:

(1) $Cl_2 \longrightarrow 2\,Cl\cdot$

(2) $Cl\cdot + CH_4 \longrightarrow CH_3Cl + H\cdot$

(3) $H\cdot + Cl_2 \longrightarrow HCl + Cl\cdot$

Calculate $\Delta H°$ for each step of this mechanism and then explain whether this mechanism is likely to compete with the one discussed in Sections 10.4 and 10.5.

Note: Problems marked with an asterisk are "challenge problems."

10.25 Starting with the compound or compounds indicated in each part and using any other needed reagents, outline syntheses of each of the following compounds. (You need not repeat steps carried out in earlier parts of this problem.)

(a) Iodoethane from ethane
(b) Diethyl ether from ethane
(c) Cyclopentene from cyclopentane
(d) 2-Bromo-3-methylbutane from 2-methylbutane
(e) 2-Butyne from methane and acetylene
(f) 2-Butanol from ethane and acetylene
(g) Ethyl azide ($CH_3CH_2N_3$) from ethane

10.26 Consider the relative rates of chlorination at the various positions in 1-fluorobutane:

$$\underset{\mathbf{1.0}}{H_3C}-\underset{\mathbf{3.7}}{CH_2}-\underset{\mathbf{1.7}}{CH_2}-\underset{\mathbf{0.9}}{CH_2}-F$$

Explain this order of reactivity.

10.27 The relative stability of a series of primary, secondary, and tertiary alkyl radicals can be compared using $R-CH_3$ carbon–carbon bond dissociation energies instead of R—H bond dissociation energies (the method used Section 10.2B). Bond dissociation energies (*DH*) needed to make such a comparison for various $R-CH_3$ species can be calculated from values for the heat of formation (H_f) of radicals $R\cdot$, $CH_3\cdot$, and the molecule $R-CH_3$ using the following equation: $DH[R-R'] = H_f[R\cdot] + H_f[CH_3\cdot] - H_f[R-CH_3]$. Using the data below, calculate the $R-CH_3$ bond dissociation energies for the examples given, and from your results compare the relative stabilities of the respective primary, secondary, and tertiary radicals in this series.

CHEMICAL SPECIES	H_f (HEAT OF FORMATION, kJ/mol)
$CH_3CH_2CH_2CH_2-CH_3$	−146.8
$CH_3CH_2CH(CH_3)-CH_3$	−153.7
$(CH_3)_3C-CH_3$	−167.9
$CH_3CH_2CH_2CH_2\cdot$	80.9
$CH_3CH_2CH(CH_3)\cdot$	69
$(CH_3)_3C\cdot$	48
$CH_3\cdot$	147

10.28 In general, the bond dissociation energy of a bond A—B in a molecule can be calculated using the heats of formation for the molecule and the repective radicals $A\cdot$ and $B\cdot$ using the general equation $DH[A-B] = H_f[A\cdot] + H_f[B\cdot] - H_f[A-B]$. Use the heat of formation data below to calculate C—H bond dissociation energies for chloromethane, dichloromethane, and trichloromethane.

CHEMICAL SPECIES	H_f (HEAT OF FORMATION, kJ/mol)
CH_2Cl-H	−83.7
$CHCl_2-H$	−95.7
CCl_3-H	−103.2
$CH_2Cl\cdot$	117.3
$CHCl_2\cdot$	89
$CCl_3\cdot$	71.1
$H\cdot$	218

10.29 Draw mechanism arrows to show electron movements in the Bergman cycloaromatization reaction that leads to the diradical believed responsible for the DNA-cleaving action of the antitumor agent calicheamicin (see the chapter opening vignette).

***10.30** In the radical chlorination of 2,2-dimethylhexane, chlorine substitution occurs much more rapidly at C5 than it does at a typical secondary carbon (e.g., C2 in butane). Review the discussion on radical polymerization and then suggest an explanation for the enhanced rate of substitution at C5 in 2,2-dimethylhexane.

***10.31** Hydrogen peroxide and ferrous sulfate react to produce hydroxyl radical (HO·), as reported in 1894 by English chemist H. J. H. Fenton. When *tert*-butyl alcohol is treated with HO· generated this way, it affords a crystalline reaction product **X**, mp 92°, which has these spectral properties:
MS: heaviest mass peak is at *m/z* 131
IR: 3620, 3350 (broad), 2980, 2940, 1385, 1370 cm^{-1}
^{1}H NMR: sharp singlets at δ 1.22, 1.58, and 2.95 (6 : 2 : 1 area ratio)
^{13}C NMR: δ 28 (CH_3), 35 (CH_2), 68 (C)
Draw the structure of **X** and write a mechanism for its formation.

***10.32** Molecular orbital calculations can be used to model the location of electron density from unpaired electrons in a radical. Open the molecular models on the CD for the methyl, ethyl, and *tert*-butyl radicals. The gray wire mesh surfaces in these models represent volumes enclosing electron density from unpaired electrons. What do you notice about the distribution of unpaired electron density in the ethyl radical and *tert*-butyl radical, as compared to the methyl radical? What bearing does this have on the relative stabilities of the radicals in this series?

***10.33** If one were to try to draw the simplest Lewis structure for molecular oxygen, the result might be the following (:Ö=Ö:). However, it is known from the properties of molecular oxygen and experiments that O_2 contains two unpaired electrons, and therefore, the Lewis structure above is incorrect. To understand the structure of O_2, it is necessary to employ a molecular orbital representation. To do so, we will need to recall (1) the shapes of bonding and antibonding σ and π molecular orbitals, (2) that each orbital can contain a maximum of two electrons, (3) that molecular oxygen has 16 electrons in total, and (4) that the two unpaired electrons in oxygen occupy separate degenerate (equal-energy) orbitals. Now, open the molecular model on the CD for oxygen and examine its molecular orbitals in sequence from the HOMO-7 orbital to the LUMO. [HOMO-7 means the seventh orbital in energy below the highest occupied molecular orbital (HOMO), HOMO-6 means the sixth below the HOMO, and so forth.] Orbitals HOMO-7 through HOMO-4 represent the $\sigma 1s$, $\sigma 1s^*$, $\sigma 2s$, and $\sigma 2s^*$ orbitals, respectively, each containing a pair of electrons.
(a) What type of orbital is represented by HOMO-3 and HOMO-2? (*Hint:* What types of orbitals are possible for second-row elements like oxygen, and which orbitals have already been used?)
(b) What type of orbital is HOMO-1? [*Hint:* The $\sigma 2s$ and $\sigma 2s^*$ orbitals are already filled, as are the HOMO-3 and HOMO-2 orbitals identified in part (b). What bonding orbital remains?]
(c) The orbitals designated HOMO and LUMO in O_2 have the same energy (they are degenerate), and each contains one of the unpaired electrons of the oxygen molecule. What type of orbital are these?

Learning Group Problems

1. **(a)** Draw structures for all organic products that would result when an *excess* of *cis*-1,3-dimethylcyclohexane reacts with Br_2 in the presence of heat and light. Use three-dimensional formulas to show stereochemistry.
(b) Draw structures for all organic products that would result when an *excess* of *cis*-1,3-dimethylcyclohexane reacts with Cl_2 in the presence of heat and light. Use three-dimensional formulas to show stereochemistry.
(c) As an alternative, use *cis*-1,2-dimethylcyclohexane to answer parts **(a)** and **(b)** above.
2. **(a)** Propose a synthesis of 2-methoxypropene starting with propane and methane as the sole source for carbon atoms. You may use any other reagents necessary. Devise a retrosynthetic analysis first.
(b) 2-Methoxypropene will form a polymer when treated with a radical initiator. Write the structure of this polymer and a mechanism for the polymerization reaction assuming a radical mechanism initiated by a diacyl peroxide.

SPECIAL TOPIC

A

Chain-Growth Polymers

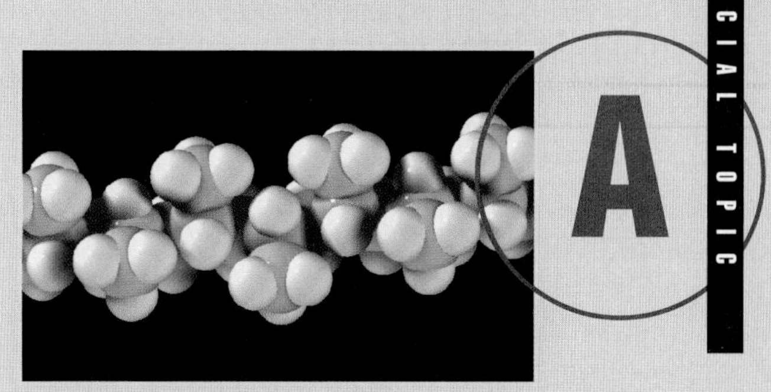

Polypropylene (syndiotactic)

The names *Orlon, Plexiglas, Lucite, polyethylene,* and *Teflon* are now familiar to most of us. These "plastics" or polymers are used in the construction of many objects around us—from the clothing we wear to portions of the houses we live in. Yet all of these compounds were unknown 70 years ago. The development of the processes by which synthetic polymers are made, more than any other single factor, was responsible for the remarkable growth of the chemical industry in the twentieth century.

At the same time, some scientists are now expressing concern about the reliance we have placed on these synthetic materials. Because they are the products of laboratory and industrial processes rather than processes that occur in nature, nature often has no way of disposing of many of them. Although progress has been made in the development of "biodegradable plastics" in recent years, many materials are still used that are not biodegradable. Although most of these objects are combustible, incineration is not always a feasible method of disposal because of attendant air pollution.

Not all polymers are synthetic. Many naturally occurring compounds are polymers as well. Silk and wool are polymers that we call proteins. The starches of our diet are polymers and so is the cellulose of cotton and wood.

Polymers are compounds that consist of very large molecules made up of many repeating subunits. The molecular subunits that are used to synthesize polymers are called *monomers,* and the reactions by which monomers are joined together are called polymerization reactions.

Propylene (propene), for example, can be polymerized to form *polypropylene.* This polymerization occurs by a chain reaction, and, as a consequence, polymers such as polypropylene are called ***chain-growth*** or ***addition polymers:***

$$\underset{\textbf{Propylene}}{CH_2{=}\underset{\displaystyle CH_3}{\underset{|}{CH}}} \xrightarrow{\text{polymerization}} \underset{\textbf{Polypropylene}}{-CH_2\underset{\displaystyle CH_3}{\underset{|}{CH}}\!-\!\left(CH_2\underset{\displaystyle CH_3}{\underset{|}{CH}}\right)_n\!\!-\!CH_2\underset{\displaystyle CH_3}{\underset{|}{CH}}-}$$

As we saw in Section 10.10, alkenes are convenient starting materials for the preparation of chain-growth polymers. The addition reactions occur through radical, cationic, or anionic mechanisms depending on how they are initiated. The following examples illustrate these mechanisms. All of these reactions are chain reactions:

Radical Polymerization

$$R\cdot \;+\; \diagdown C{=}C \diagup \longrightarrow R{:}\overset{|}{\underset{|}{C}}-\overset{\diagup}{\underset{\diagdown}{C}}\cdot \xrightarrow{\;\diagdown C=C \diagup\;} R-\overset{|}{\underset{|}{C}}-\overset{|}{\underset{|}{C}}-\overset{|}{\underset{|}{C}}-\overset{\diagup}{\underset{\diagdown}{C}}\cdot \xrightarrow{\;\diagdown C=C \diagup\;} \text{etc.}$$

Cationic Polymerization

$$\mathrm{R^+} + \mathrm{C{=}C} \longrightarrow \mathrm{R{-}C{-}C^+} \xrightarrow{\mathrm{C{=}C}} \mathrm{R{-}C{-}C{-}C{-}C^+} \xrightarrow{\mathrm{C{=}C}} \text{etc.}$$

Anionic Polymerization

$$\mathrm{Z{:}^-} + \mathrm{C{=}C} \longrightarrow \mathrm{Z{-}C{-}C{:}^-} \xrightarrow{\mathrm{C{=}C}} \mathrm{Z{-}C{-}C{-}C{-}C{:}^-} \xrightarrow{\mathrm{C{=}C}} \text{etc.}$$

Radical polymerization of chloroethene (vinyl chloride) produces a polymer called poly(vinyl chloride), also known as **PVC**:

$$n\ \underset{\text{Vinyl chloride}}{\mathrm{CH_2{=}CH(Cl)}} \longrightarrow \underset{\text{Poly(vinyl chloride) (PVC)}}{\left(\mathrm{CH_2{-}CH(Cl)}\right)_n}$$

This reaction produces a polymer that has a molecular weight of about 1,500,000 and that is a hard, brittle, and rigid material. In this form it is often used to make pipes, rods, and compact discs. Poly(vinyl chloride) can be softened by mixing it with esters (called plasticizers). The softer material is used for making "vinyl leather," plastic raincoats, shower curtains, and garden hoses.

Exposure to vinyl chloride has been linked to the development of a rare cancer of the liver called angiocarcinoma. This link was first noted in 1974 and 1975 among workers in vinyl chloride factories. Since that time, standards have been set to limit workers' exposure to less than one part per million average over an 8-h day. The U.S. Food and Drug Administration (FDA) has banned the use of PVC in packaging materials for food. [There is evidence that poly(vinyl chloride) contains traces of vinyl chloride.]

Acrylonitrile ($CH_2{=}CHCN$) polymerizes to form polyacrylonitrile or Orlon. The initiator for the polymerization is a mixture of ferrous sulfate and hydrogen peroxide. These two compounds react to produce hydroxyl radicals ($\cdot OH$), which act as chain initiators.

$$n\ \underset{\text{Acrylonitrile}}{\mathrm{CH_2{=}CH(CN)}} \xrightarrow[\mathrm{H{-}O{-}O{-}H}]{\mathrm{FeSO_4}} \underset{\text{Polyacrylonitrile (Orlon)}}{\left(\mathrm{CH_2{-}CH(CN)}\right)_n}$$

Polyacrylonitrile decomposes before it melts, so melt spinning cannot be used for the production of fibers. Polyacrylonitrile, however, is soluble in *N,N*-dimethylformamide, and these solutions can be used to spin fibers. Fibers produced in this way are used in making carpets and clothing.

Teflon is made by polymerizing tetrafluoroethene in aqueous suspension:

$$n\ \mathrm{CF_2{=}CF_2} \xrightarrow[\mathrm{H_2O_2},\ \mathrm{H_2O}]{\mathrm{Fe^{2+}}} \left(\mathrm{CF_2{-}CF_2}\right)_n$$

The reaction is highly exothermic, and water helps dissipate the heat that is produced. Teflon has a melting point (327°C) that is unusually high for an addition polymer. It is also highly resistant to chemical attack (due to the strength of the C—F bonds) and has a low coefficient of friction. Because of these properties, Teflon is used in greaseless bearings, in liners for pots and pans, and in many special situations that require a substance that is highly resistant to corrosive chemicals.

Vinyl alcohol is an unstable compound that rearranges spontaneously to acetaldehyde (see Section 17.2):

$$\underset{\textbf{Vinyl alcohol}}{CH_2{=}CH(OH)} \rightleftharpoons \underset{\textbf{Acetaldehyde}}{CH_3{-}CH{=}O}$$

Consequently, the water-soluble polymer, poly(vinyl alcohol), cannot be made directly. It can be made, however, by an indirect method that begins with the polymerization of vinyl acetate to poly(vinyl acetate). This is then hydrolyzed to poly(vinyl alcohol). Hydrolysis is rarely carried to completion, however, because the presence of a few ester groups helps confer water solubility on the product. The ester groups apparently help keep the polymer chains apart, and this permits hydration of the hydroxyl groups. Poly(vinyl alcohol) in which 10% of the ester groups remain dissolves readily in water. Poly(vinyl alcohol) is used to manufacture water-soluble films and adhesives. Poly(vinyl acetate) is used as an emulsion in water-base paints.

$$\underset{\textbf{Vinyl acetate}}{n\,CH_2{=}CH{-}O{-}C({=}O){-}CH_3} \longrightarrow \underset{\textbf{Poly(vinyl acetate)}}{\left(CH_2{-}CH(O{-}C({=}O){-}CH_3)\right)_n} \xrightarrow[H_2O]{OH^-} \underset{\textbf{Poly(vinyl alcohol)}}{\left(CH_2{-}CH(OH)\right)_n}$$

A polymer with excellent optical properties can be made by the radical polymerization of methyl methacrylate. Poly(methyl methacrylate) is marketed under the names Lucite, Plexiglas, and Perspex:

$$\underset{\textbf{Methyl methacrylate}}{n\,CH_2{=}C(CH_3){-}C({=}O){-}OCH_3} \longrightarrow \underset{\textbf{Poly(methyl methacrylate)}}{\left(CH_2{-}C(CH_3)(C({=}O){-}OCH_3)\right)_n}$$

A mixture of vinyl chloride and vinylidene chloride (1,1-dichloroethene) polymerizes to form what is known as a *copolymer*. The familiar *Saran Wrap* used in food packaging is made by polymerizing a mixture in which the vinylidene chloride predominates:

$$\underset{\substack{\textbf{Vinylidene chloride}\\\textbf{(excess)}}}{CH_2{=}CCl_2} + \underset{\textbf{Vinyl chloride}}{CH_2{=}CHCl} \xrightarrow{R\cdot} \underset{\textbf{Saran Wrap}}{\left[\left(CH_2{-}CCl_2\right)\left(CH_2{-}CHCl\right)\right]}$$

The subunits do not necessarily alternate regularly along the polymer chain.

PROBLEM A.1

Can you suggest an explanation that accounts for the fact that the radical polymerization of styrene ($C_6H_5CH{=}CH_2$) to produce polystyrene occurs in a head-to-tail fashion,

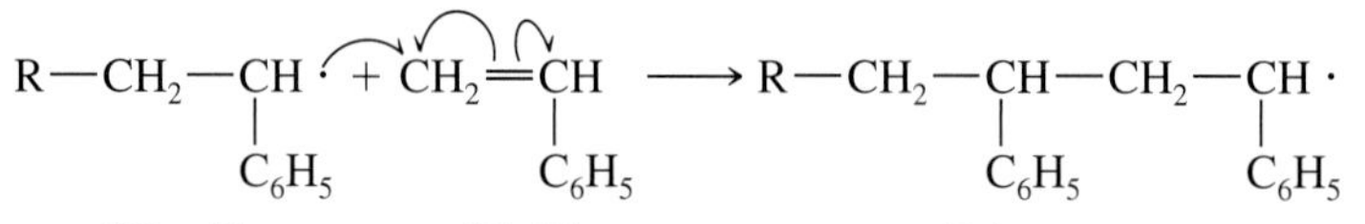

$$R{-}CH_2{-}\underset{\underset{C_6H_5}{|}}{CH}\cdot + CH_2{=}\underset{\underset{C_6H_5}{|}}{CH} \longrightarrow R{-}CH_2{-}\underset{\underset{C_6H_5}{|}}{CH}{-}CH_2{-}\underset{\underset{C_6H_5}{|}}{CH}\cdot$$

"Head" **"Tail"** **Polystyrene**

rather than the head-to-head manner shown here?

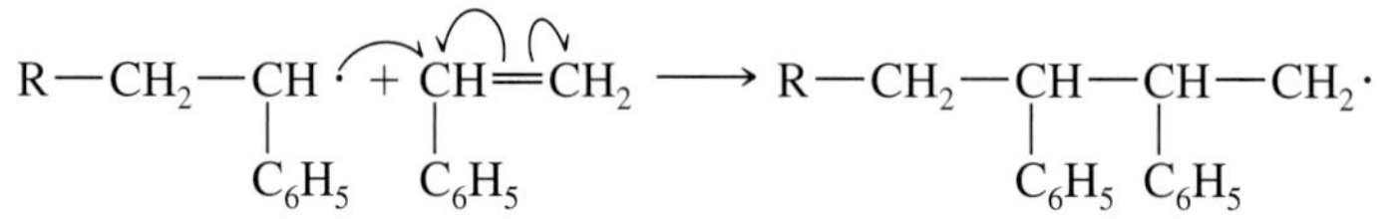

$$R{-}CH_2{-}\underset{\underset{C_6H_5}{|}}{CH}\cdot + \underset{\underset{C_6H_5}{|}}{CH}{=}CH_2 \longrightarrow R{-}CH_2{-}\underset{\underset{C_6H_5}{|}}{CH}{-}\underset{\underset{C_6H_5}{|}}{CH}{-}CH_2\cdot$$

"Head" **"Head"**

PROBLEM A.2

Outline a general method for the synthesis of each of the following polymers by radical polymerization. Show the monomers that you would use.

(a) 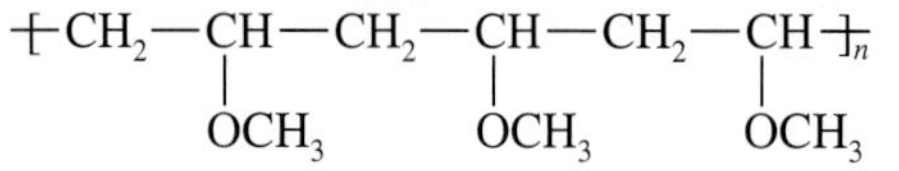 $\left[CH_2{-}\underset{\underset{OCH_3}{|}}{CH}{-}CH_2{-}\underset{\underset{OCH_3}{|}}{CH}{-}CH_2{-}\underset{\underset{OCH_3}{|}}{CH}\right]_n$

(b) $\left[CH_2{-}CCl_2{-}CH_2{-}CCl_2{-}CH_2{-}CCl_2\right]_n$

Alkenes also polymerize when they are treated with strong acids. The growing chains in acid-catalyzed polymerizations are *cations* rather than radicals. The following reactions illustrate the cationic polymerization of isobutylene:

Step 1 $H{-}\ddot{O}{:} + BF_3 \rightleftharpoons H{-}\overset{+}{\ddot{O}}{-}\overset{-}{B}F_3$ (with H on O)

Step 2 $H{-}\overset{+}{\ddot{O}}{-}\overset{-}{B}F_3 + CH_2{=}C(CH_3)_2 \longrightarrow CH_3{-}\overset{+}{C}(CH_3)_2$ (H on O)

Step 3 $CH_3{-}\overset{+}{C}(CH_3)_2 + CH_2{=}C(CH_3)_2 \longrightarrow CH_3{-}C(CH_3)_2{-}CH_2{-}\overset{+}{C}(CH_3)_2$

Step 4 $CH_3{-}C(CH_3)_2{-}CH_2{-}\overset{+}{C}(CH_3)_2 \xrightarrow{CH_2=C(CH_3)_2} CH_3{-}C(CH_3)_2{-}CH_2{-}C(CH_3)_2{-}CH_2{-}\overset{+}{C}(CH_3)_2 \xrightarrow{\text{etc.}}$

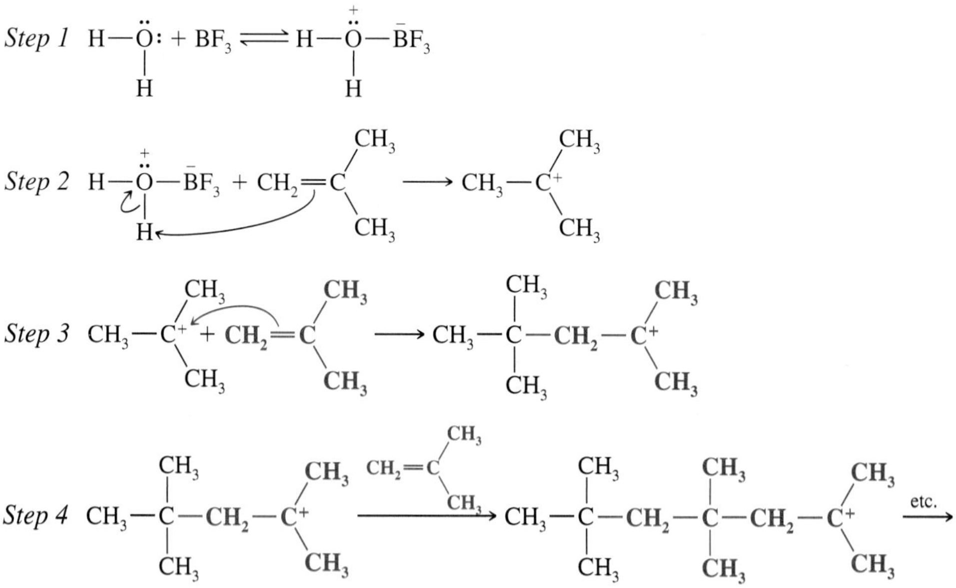

The catalysts used for cationic polymerizations are usually Lewis acids that contain a small amount of water. The polymerization of isobutylene illustrates how the catalyst (BF_3 and H_2O) functions to produce growing cationic chains.

PROBLEM A.3

Alkenes such as ethene, vinyl chloride, and acrylonitrile do not undergo cationic polymerization very readily. On the other hand, isobutylene undergoes cationic polymerization rapidly. Provide an explanation for this behavior.

Alkenes containing electron-withdrawing groups polymerize in the presence of strong bases. Acrylonitrile, for example, polymerizes when it is treated with sodium amide ($NaNH_2$) in liquid ammonia. The growing chains in this polymerization are anions:

$$H_2\ddot{N}:^- + CH_2{=}CH(CN) \xrightarrow{NH_3} H_2N{-}CH_2{-}CH(CN):^-$$

$$H_2N{-}CH_2{-}CH(CN):^- \xrightarrow{CH_2=CHCN} H_2N{-}CH_2{-}CH(CN){-}CH_2{-}CH(CN):^- \xrightarrow{etc.}$$

Anionic polymerization of acrylonitrile is less important in commercial production than the radical process we illustrated earlier.

PROBLEM A.4

The remarkable adhesive called "superglue" is a result of anionic polymerization. Superglue is a solution containing purified methyl α-cyanoacrylate:

$$CH_2{=}C(CN)(CO_2CH_3)$$

Methyl α-cyanoacrylate

Methyl cyanoacrylate can be polymerized by anions such as hydroxide ion, but it is even polymerized by traces of water found on the surfaces of the two objects that are being glued together. (These two objects, unfortunately, have often been two fingers of the person doing the gluing.) Show how methyl α-cyanoacrylate would undergo anionic polymerization.

A.1 Stereochemistry of Chain-Growth Polymerization

Head-to-tail polymerization of propylene produces a polymer in which every other carbon atom is a stereogenic center. Many of the physical properties of the polypropylene produced in this way depend on the stereochemistry of these stereogenic carbons:

$$CH_2{=}CH(CH_3) \xrightarrow[\text{(head to tail)}]{\text{polymerization}} {-}CH_2\overset{*}{C}H(CH_3)CH_2\overset{*}{C}H(CH_3)CH_2\overset{*}{C}H(CH_3)CH_2\overset{*}{C}H(CH_3){-}$$

There are three general arrangements of the methyl groups and hydrogen atoms along the chain. These arrangements are described as being *atactic, syndiotactic,* and *isotactic.*

If the stereochemistry at the stereogenic carbons is random (Fig. A.1), the polymer is said to be atactic (*a*, without + Greek: *taktikos,* order).

In atactic polypropylene the methyl groups are randomly disposed on either side of the stretched carbon chain. If we were to arbitrarily designate one end of the chain as having higher preference than the other, we could give (*R*)–(*S*) designations (Section 5.7) to the stereogenic carbons. In atactic polypropylene the sequence of (*R*)–(*S*) designations along the chain is random.

Polypropylene produced by radical polymerization at high pressures is atactic. Because the polymer is atactic, it is noncrystalline, has a low softening point, and has poor mechanical properties.

A second possible arrangement of the groups along the carbon chain is that of *syndiotactic* polypropylene. In syndiotactic polypropylene the methyl groups alternate regularly

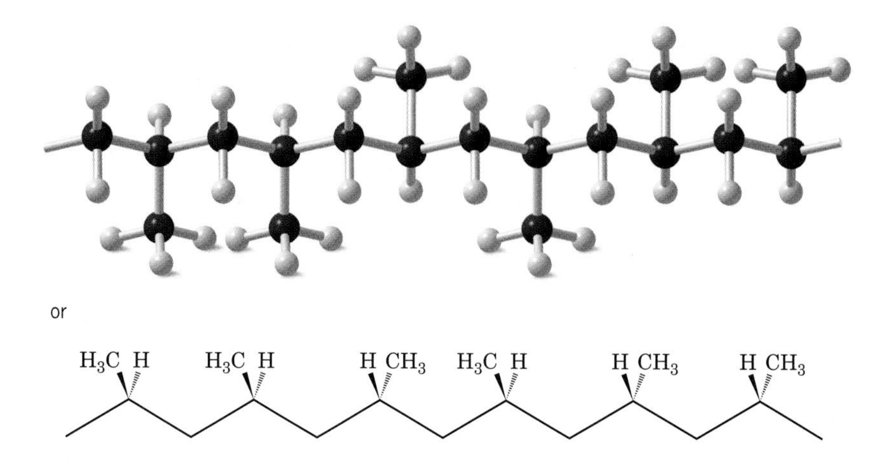

FIGURE A.1 Atactic polypropylene. (In this illustration a "stretched" carbon chain is used for clarity.)

from one side of the stretched chain to the other (Fig. A.2). If we were to arbitrarily designate one end of the chain of syndiotactic polypropylene as having higher preference, the configuration of the stereogenic carbons would alternate, (*R*), (*S*), (*R*), (*S*), (*R*), (*S*), (*R*), (*S*), and so on.

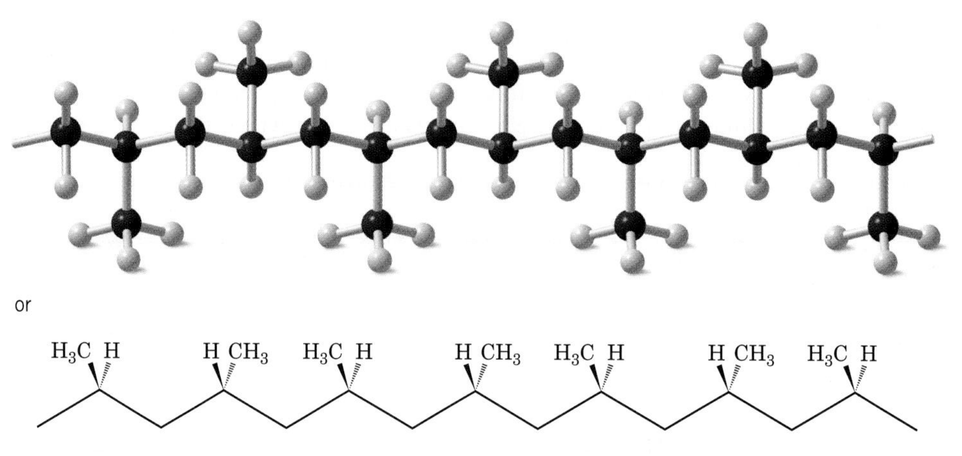

FIGURE A.2 Syndiotactic polypropylene.

The third possible arrangement of stereogenic carbons is the *isotactic* arrangement shown in Fig. A.3. In the isotactic arrangement all of the methyl groups are on the same side of the stretched chain. The configurations of the stereogenic carbons are either all (*R*) or all (*S*) depending on which end of the chain is assigned higher preference.

The names isotactic and syndiotactic come from the Greek term *taktikos* (order) plus *iso* (same) and *syndyo* (two together).

Ziegler and Natta were awarded the Nobel Prize in Chemistry for their discoveries in 1963.

Before 1953 isotactic and syndiotactic addition polymers were unknown. In that year, however, a German chemist, Karl Ziegler, and an Italian chemist, Giulio Natta, announced independently the discovery of catalysts that permit stereochemical control of polymerization reactions. The Ziegler–Natta catalysts, as they are now called, are prepared from transition metal halides and a reducing agent. The catalysts most commonly used are prepared from titanium tetrachloride ($TiCl_4$) and a trialkylaluminum (R_3Al).

Ziegler–Natta catalysts are generally employed as suspended solids, and polymerization probably occurs at metal atoms on the surfaces of the particles. The mechanism for

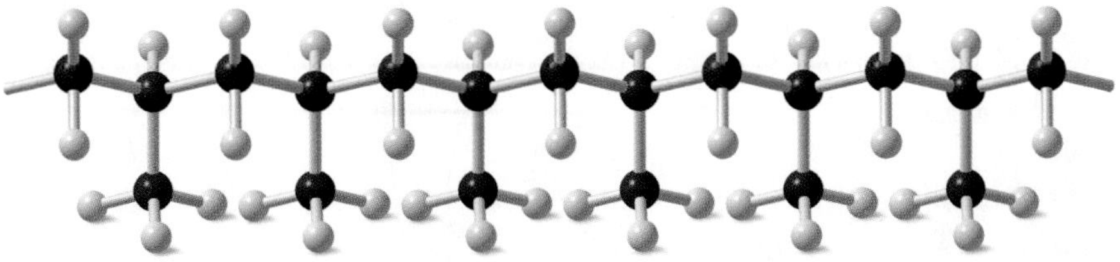

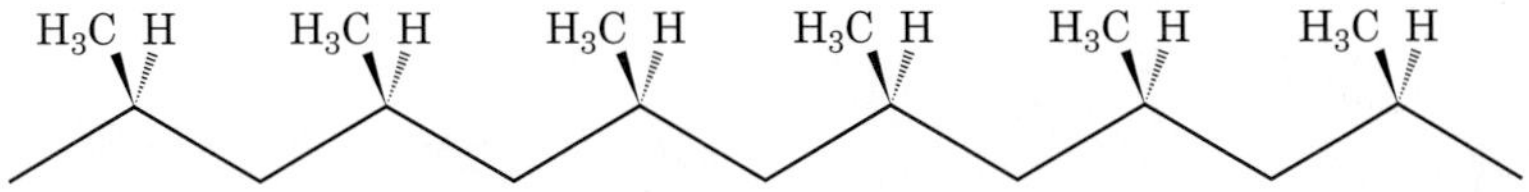

FIGURE A.3 Isotactic polypropylene.

or

H_3C H H_3C H H_3C H H_3C H H_3C H H_3C H

the polymerization is an ionic mechanism, but its details are not fully understood. There is evidence that polymerization occurs through an insertion of the alkene monomer between the metal and the growing polymer chain.

Both syndiotactic and isotactic polypropylene have been made using Ziegler–Natta catalysts. The polymerizations occur at much lower pressures, and the polymers that are produced are much higher melting than atactic polypropylene. Isotactic polypropylene, for example, melts at 175°C. Isotactic and syndiotactic polymers are also much more crystalline than atactic polymers. The regular arrangement of groups along the chains allows them to fit together better in a crystal structure.

Atactic, syndiotactic, and isotactic forms of poly(methyl methacrylate) are known. The atactic form is a noncrystalline glass. The crystalline syndiotactic and isotactic forms melt at 160 and 200°C, respectively.

PROBLEM A.5

(a) Write structural formulas for portions of the chain of the atactic, syndiotactic, and isotactic forms of polystyrene (see Problem A.1). **(b)** If solutions were made of each of these forms of polystyrene, which solutions would you expect to show optical activity?

CHAPTER

11

Alcohols and Ethers

Molecular Hosts

The cell membrane establishes critical concentration gradients between the interior and exterior of cells, like the robotic box stacker above alters the "concentration" of boxes between one place and another. An intracellular-to-extracellular difference in sodium and potassium ion concentrations, for example, is essential to the function of nerves, transport of important nutrients into the cell, and maintenance of proper cell volume.* There is a family of antibiotics whose effectiveness results from disrupting this crucial ion gradient. These antibiotics are called ionophores. Monensin is one such ionophore antibiotic.

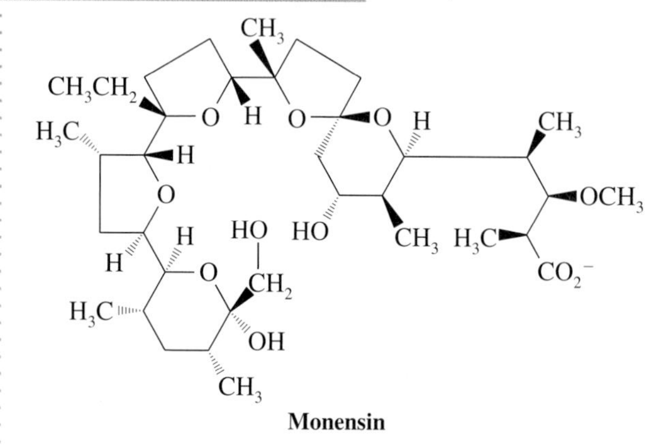

Monensin

Monensin is called a *carrier* ionophore because it is a compound that binds with sodium ions and carries them across the cell membrane. (Other ionophore antibiotics such as gramicidin and valinomycin are *channel-forming* ionophores because they open pores that extend through the membrane.)

The ion-transporting ability of monensin results principally from its many ether functional groups, and, as such, it is an example of a polyether antibiotic. The oxygen atoms of these molecules bind with metal cations by Lewis acid–base interactions. Each monensin molecule forms an

*Discovery and characterization of the actual molecular pump that establishes the sodium and potassium concentration gradient (Na^+, K^+-ATPase) earned Jens Skou (Aarhus University, Denmark) one half of the 1997 Nobel Prize in Chemistry. The other half went to Paul D. Boyer (UCLA) and John E. Walker (Cambridge) for elucidating the enzymatic mechanism of ATP synthesis.

octahedral complex with a sodium ion (this complex is shown as an inset by the photo). The complex is a hydrophobic "host" for the cation that allows it to be carried as a "guest" of monensin from one side of the nonpolar cell membrane to the other. This transport process destroys the critical sodium concentration gradient needed for cell function.

Compounds called crown ethers are molecular "hosts" that are also polyether ionophores. Though not used as antibiotics, crown ethers are useful for conducting reactions with ionic reagents in nonpolar solvents. The 1987 Nobel Prize in Chemistry was awarded to Charles J. Pedersen, Donald J. Cram, and Jean-Marie Lehn for their work on crown ethers and related compounds, research that marked the beginning of a field of study called host–guest chemistry. We shall consider crown ethers further in Section 11.16.

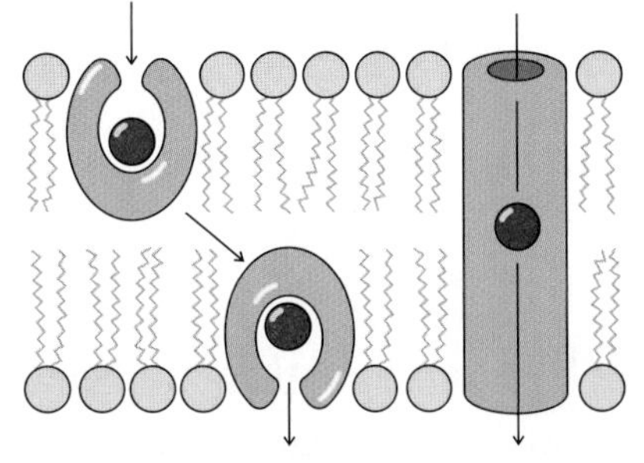

Carrier (left) and channel-forming modes of transport ionophores.

11.1 Structure and Nomenclature

Alcohols are compounds whose molecules have a hydroxyl group attached to a *saturated* carbon atom.* The saturated carbon atom may be that of a simple alkyl group, as in the following examples:

CH_3OH — **Methanol (methyl alcohol)**

CH_3CH_2OH — **Ethanol (ethyl alcohol)** *a 1° alcohol*

$CH_3CH(OH)CH_3$ — **2-Propanol (or propan-2-ol, or isopropyl alcohol)** *a 2° alcohol*

$CH_3C(CH_3)(OH)CH_3$ — **2-Methyl-2-propanol (or 2-methylpropan-2-ol, or *tert*-butyl alcohol)** *a 3° alcohol*

The alcohol carbon atom may also be a saturated carbon atom of an alkenyl or alkynyl group, or the carbon atom may be a saturated carbon atom that is attached to a benzene ring:

$C_6H_5-CH_2OH$ — **Benzyl alcohol** *a benzylic alcohol*

$CH_2{=}CHCH_2OH$ — **2-Propenol (or prop-2-en-1-ol, or allyl alcohol)** *an allylic alcohol*

$H-C{\equiv}CCH_2OH$ — **2-Propynol (or prop-2-yn-1-ol, or propargyl alcohol)**

Compounds that have a hydroxyl group attached directly to a benzene ring are called *phenols.* (Phenols are discussed in detail in Chapter 21.)

C_6H_5-OH — **Phenol**

$H_3C-C_6H_4-OH$ — ***p*-Methylphenol** *a substituted phenol*

$Ar-OH$ — **General formula for a phenol**

*Compounds in which a hydroxyl group is attached to an unsaturated carbon atom of a double bond (i.e., $C{=}C-OH$) are called enols (see Section 17.2).

Ethers differ from alcohols in that the oxygen atom of an ether is bonded to two carbon atoms. The hydrocarbon groups may be alkyl, alkenyl, vinyl, alkynyl, or aryl. Several examples are shown here:

$CH_3CH_2{-}O{-}CH_2CH_3$ **Diethyl ether**

$CH_2{=}CHCH_2{-}O{-}CH_3$ **Allyl methyl ether**

$(CH_3)_3COCH_3$ ***tert*-Butyl methyl ether**

$CH_2{=}CH{-}O{-}CH{=}CH_2$ **Divinyl ether**

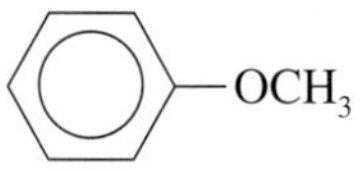

$C_6H_5{-}OCH_3$ **Methyl phenyl ether**

11.1A Nomenclature of Alcohols

We studied the IUPAC system of nomenclature for alcohols in Section 4.3F. As a review consider the following example.

SOLVED PROBLEM

Give IUPAC substitutive names for the following alcohols:

(a) 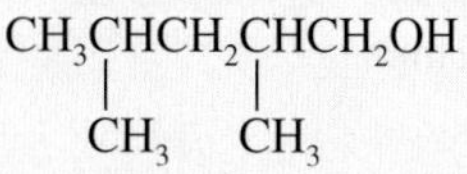 $CH_3CH(CH_3)CH_2CH(CH_3)CH_2OH$

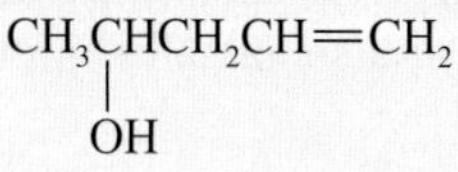

(b) 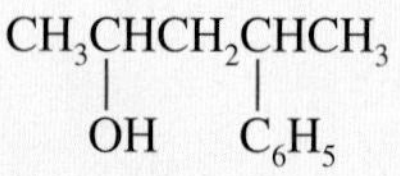 $CH_3CH(OH)CH_2CH(C_6H_5)CH_3$

(c) $CH_3CH(OH)CH_2CH{=}CH_2$

ANSWER: The longest chain *to which the hydroxyl group is attached* gives us the *base name.* The ending is **-ol**. We then number *the longest chain from the end that gives the carbon bearing the hydroxyl group the lower number.* Thus, the names, in both of the accepted IUPAC formats, are

(a) 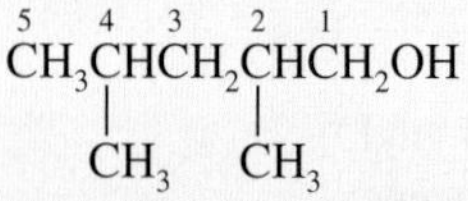 $\overset{5}{C}H_3\overset{4}{C}H(CH_3)\overset{3}{C}H_2\overset{2}{C}H(CH_3)\overset{1}{C}H_2OH$

2,4-Dimethyl-1-pentanol
(or 2,4-dimethylpentan-1-ol)

(b) $\overset{1}{C}H_3\overset{2}{C}H(OH)\overset{3}{C}H_2\overset{4}{C}H(C_6H_5)\overset{5}{C}H_3$

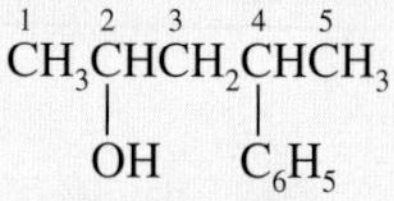

4-Phenyl-2-pentanol
(or 4-phenylpentan-2-ol)

(c) $\overset{1}{C}H_3\overset{2}{C}H(OH)\overset{3}{C}H_2\overset{4}{C}H{=}\overset{5}{C}H_2$

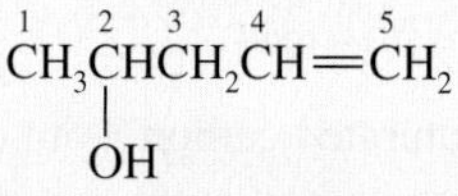

4-Penten-2-ol
(or pent-4-en-2-ol)

The hydroxyl group [see example (c) above] has precedence over double bonds and triple bonds in deciding which functional group to name as the suffix.

In common functional class nomenclature (Section 2.7) alcohols are called **alkyl alcohols** such as methyl alcohol, ethyl alcohol, and so on.

PROBLEM 11.1 What is wrong with thc usc of such names as "isopropanol" and "*tert*-butanol"?

11.1B Nomenclature of Ethers

Simple ethers are frequently given common functional class names. One simply lists (in alphabetical order) both groups that are attached to the oxygen atom and adds the word *ether:*

$CH_3OCH_2CH_3$ — **Ethyl methyl ether**

$CH_3CH_2OCH_2CH_3$ — **Diethyl ether**

$C_6H_5OC(CH_3)_2-CH_3$ — ***tert*-Butyl phenyl ether**

IUPAC substitutive names should be used for complicated ethers, however, and for compounds with more than one ether linkage. In this IUPAC style, ethers are named as alkoxyalkanes, alkoxyalkenes, and alkoxyarenes. The RO— group is an **alkoxy** group.

$CH_3CH(OCH_3)CH_2CH_2CH_3$ — **2-Methoxypentane**

$CH_3CH_2O-C_6H_4-CH_3$ — **1-Ethoxy-4-methylbenzene**

$CH_3OCH_2CH_2OCH_3$ — **1,2-Dimethoxyethane**

Cyclic ethers can be named in several ways. One simple way is to use **replacement nomenclature,** in which we relate the cyclic ether to the corresponding hydrocarbon ring system and use the prefix **oxa-** to indicate that an oxygen atom replaces a CH_2 group. In another system, a cyclic three-membered ether is named **oxirane** and a four-membered ether is called **oxetane.** Several simple cyclic ethers also have common names; in the examples below, these common names are given in parentheses. Tetrahydrofuran (THF) and 1,4-dioxane are useful solvents:

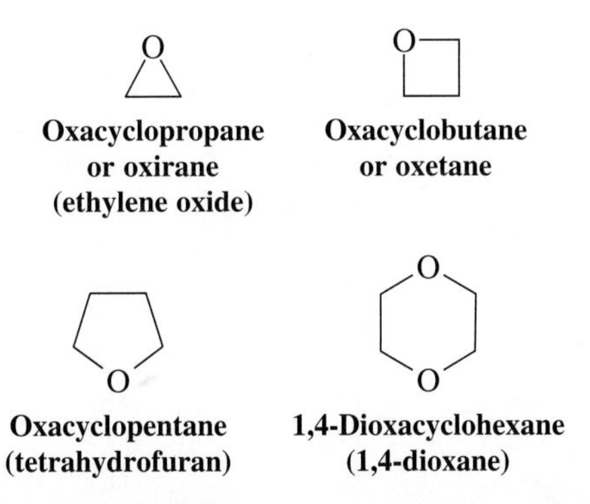

PROBLEM 11.2

Give bond-line formulas and appropriate names for all of the alcohols and ethers with the formulas **(a)** C_3H_8O and **(b)** $C_4H_{10}O$.

11.2 Physical Properties of Alcohols and Ethers

The physical properties of a number of alcohols and ethers are given in Tables 11.1 and 11.2.

Ethers have boiling points that are roughly comparable with those of hydrocarbons of the same molecular weight (MW). For example, the boiling point of diethyl ether (MW = 74) is 34.6°C; that of pentane (MW = 72) is 36°C. Alcohols, on the other hand, have much higher boiling points than comparable ethers or hydrocarbons. The boiling point of

TABLE 11.1 Physical Properties of Ethers

Name	Formula	mp (°C)	bp (°C)	Density d_4^{20} (g mL^{-1})
Dimethyl ether	CH_3OCH_3	−138	−24.9	0.661
Ethyl methyl ether	$CH_3OCH_2CH_3$		10.8	0.697
Diethyl ether	$CH_3CH_2OCH_2CH_3$	−116	34.6	0.714
Dipropyl ether	$(CH_3CH_2CH_2)_2O$	−122	90.5	0.736
Diisopropyl ether	$(CH_3)_2CHOCH(CH_3)_2$	−86	68	0.725
Dibutyl ether	$(CH_3CH_2CH_2CH_2)_2O$	−97.9	141	0.769
1,2-Dimethoxyethane	$CH_3OCH_2CH_2OCH_3$	−68	83	0.863
Tetrahydrofuran	(five-membered ring with O)	−108	65.4	0.888
1,4-Dioxane	(six-membered ring with two O)	11	101	1.033
Anisole (methoxybenzene)	C_6H_5–OCH_3	−37.3	158.3	0.994

TABLE 11.2 Physical Properties of Alcohols

Compound	Name	mp (°C)	bp (°C) (1 atm)	Density d_4^{20} (g mL^{-1})	Water Solubility (g/100 mL^{-1} H_2O)
Monohydroxy Alcohols					
CH_3OH	Methanol	−97	64.7	0.792	∞
CH_3CH_2OH	Ethanol	−117	78.3	0.789	∞
$CH_3CH_2CH_2OH$	Propyl alcohol	−126	97.2	0.804	∞
$CH_3CH(OH)CH_3$	Isopropyl alcohol	−88	82.3	0.786	∞
$CH_3CH_2CH_2CH_2OH$	Butyl alcohol	−90	117.7	0.810	8.3
$CH_3CH(CH_3)CH_2OH$	Isobutyl alcohol	−108	108.0	0.802	10.0
$CH_3CH_2CH(OH)CH_3$	*sec*-Butyl alcohol	−114	99.5	0.808	26.0
$(CH_3)_3COH$	*tert*-Butyl alcohol	25	82.5	0.789	∞
$CH_3(CH_2)_3CH_2OH$	Pentyl alcohol	−78.5	138.0	0.817	2.4
$CH_3(CH_2)_4CH_2OH$	Hexyl alcohol	−52	156.5	0.819	0.6
$CH_3(CH_2)_5CH_2OH$	Heptyl alcohol	−34	176	0.822	0.2
$CH_3(CH_2)_6CH_2OH$	Octyl alcohol	−15	195	0.825	0.05
$CH_3(CH_2)_7CH_2OH$	Nonyl alcohol	−5.5	212	0.827	
$CH_3(CH_2)_8CH_2OH$	Decyl alcohol	6	228	0.829	
$CH_2{=}CHCH_2OH$	Allyl alcohol	−129	97	0.855	∞
(cyclopentyl)–OH	Cyclopentanol	−19	140	0.949	
(cyclohexyl)–OH	Cyclohexanol	24	161.5	0.962	3.6
$C_6H_5CH_2OH$	Benzyl alcohol	−15	205	1.046	4
Diols and Triols					
CH_2OHCH_2OH	Ethylene glycol	−12.6	197	1.113	∞
$CH_3CHOHCH_2OH$	Propylene glycol	−59	187	1.040	∞
$CH_2OHCH_2CH_2OH$	Trimethylene glycol	−30	215	1.060	∞
$CH_2OHCHOHCH_2OH$	Glycerol	18	290	1.261	∞

butyl alcohol (MW = 74) is 117.7°C. We learned the reason for this behavior in Section 2.14C; the molecules of alcohols can associate with each other through **hydrogen bonding,** whereas those of ethers and hydrocarbons cannot:

CH_3 H—Ö: O:---H—Ö: CH_3 H CH_3

Hydrogen bonding between molecules of methanol

Ethers, however, *are* able to form hydrogen bonds with compounds such as water. Ethers, therefore, have solubilities in water that are similar to those of alcohols of the same molecular weight and that are very different from those of hydrocarbons.

Diethyl ether and 1-butanol, for example, have the same solubility in water, approximately 8 g per 100 mL at room temperature. Pentane, by contrast, is virtually insoluble in water.

Methanol, ethanol, both propyl alcohols, and *tert*-butyl alcohol are completely miscible with water (Table 11.2). The remaining butyl alcohols have solubilities in water between 8.3 and 26.0 g per 100 mL. The solubility of alcohols in water gradually decreases as the hydrocarbon portion of the molecule lengthens; long-chain alcohols are more "alkanelike" and are, therefore, less like water.

PROBLEM 11.3

1,2-Propanediol and 1,3-propanediol (propylene glycol and trimethylene glycol, respectively; see Table 11.2) have higher boiling points than any of the butyl alcohols, even though all of the compounds have roughly the same molecular weight. How can you explain this observation?

11.3 Important Alcohols and Ethers

11.3A Methanol

At one time, most methanol was produced by the destructive distillation of wood (i.e., heating wood to a high temperature in the absence of air). It was because of this method of preparation that methanol came to be called "wood alcohol." Today, most methanol is prepared by the catalytic hydrogenation of carbon monoxide. This reaction takes place under high pressure and at a temperature of 300–400°C:

$$CO + 2\,H_2 \xrightarrow[\substack{200\text{--}300\ \text{atm} \\ ZnO\text{–}Cr_2O_3}]{300\text{–}400°C} CH_3OH$$

Methanol is highly toxic. Ingestion of even small quantities of methanol can cause blindness; large quantities cause death. Methanol poisoning can also occur by inhalation of the vapors or by prolonged exposure to the skin.

11.3B Ethanol

Ethanol can be made by the fermentation of sugars, and it is the alcohol of all alcoholic beverages. The synthesis of ethanol in the form of wine by the fermentation of the sugars of fruit juices was probably our first accomplishment in the field of organic synthesis. Sugars from a wide variety of sources can be used in the preparation of alcoholic beverages. Often, these sugars are from grains, and it is this derivation that accounts for ethanol having the synonym "grain alcohol."

Grapes fermenting in a stone vat.

Fermentation is usually carried out by adding yeast to a mixture of sugars and water. Yeast contains enzymes that promote a long series of reactions that ultimately convert a simple sugar ($C_6H_{12}O_6$) to ethanol and carbon dioxide:

$$C_6H_{12}O_6 \xrightarrow{\text{yeast}} 2\ CH_3CH_2OH + 2\ CO_2$$

(~95% yield)

Fermentation alone does not produce beverages with an ethanol content greater than 12–15% because the enzymes of the yeast are deactivated at higher concentrations. To produce beverages of higher alcohol content, the aqueous solution must be distilled. Brandy, whiskey, and vodka are produced in this way. The "proof" of an alcoholic beverage is simply twice the percentage of ethanol (by volume). One hundred proof whiskey is 50% ethanol. The flavors of the various distilled liquors result from other organic compounds that distill with the alcohol and water.

Distillation of a solution of ethanol and water does not yield ethanol more concentrated than 95%. A mixture of 95% ethanol and 5% water boils at a lower temperature (78.15°C) than either pure ethanol (bp 78.3°C) or pure water (bp 100°C). Such a mixture is an example of an **azeotrope.** (Azeotropes can also have boiling points that are higher than that of either of the pure components.) Pure ethanol can be prepared by adding benzene to the mixture of 95% ethanol and water and then distilling this solution. Benzene forms a different azeotrope with ethanol and water that is 7.5% water. This azeotrope boils at 64.9°C and allows removal of the water (along with some ethanol). Eventually pure ethanol distills over. Pure ethanol is called **absolute alcohol.**

Ethanol is quite inexpensive, but when it is used for beverages, it is highly taxed. (The tax is greater than $20 per gallon in most states.) Federal law requires that some ethanol used for scientific and industrial purposes be adulterated or "denatured" to make it toxic or unpalatable. Various denaturants are used, including methanol.

Ethanol is an important industrial chemical. Most ethanol for industrial purposes is produced by the acid-catalyzed hydration of ethene:

$$CH_2{=}CH_2 + H_2O \xrightarrow[\text{acid}]{} CH_3CH_2OH$$

Ethanol is a *hypnotic* (sleep producer). It depresses activity in the upper brain even though it gives the illusion of being a stimulant. Ethanol is also toxic, but it is much less toxic than methanol. In rats the lethal dose of ethanol is 13.7 g kg^{-1} of body weight. Abuse of ethanol is a major drug problem in most countries.

11.3C Ethylene Glycol

Ethylene glycol ($HOCH_2CH_2OH$) has a low molecular weight and a high boiling point and is miscible with water. These properties make ethylene glycol a very good automobile antifreeze. Unfortunately, ethylene glycol is toxic. Much ethylene glycol is still sold as antifreeze under a variety of trade names. The use of propylene glycol as a low-toxicity, environmentally friendly alternative is increasing, however.

11.3D Diethyl Ether

Diethyl ether is a very low boiling, highly flammable liquid. Care should always be taken when diethyl ether is used in the laboratory, because open flames or sparks from light switches can cause explosive combustion of mixtures of diethyl ether and air.

Most ethers react slowly with oxygen by a radical process called **autoxidation** (see Section 10.11D) to form hydroperoxides and peroxides:

Step 1 $\cdot\ddot{O}{-}\ddot{O}\cdot + {-}\overset{H}{\underset{|}{C}}{-}OR' \longrightarrow \cdot\ddot{O}{-}\ddot{O}{-}H + {-}\dot{C}(OR')\backslash$

Step 2 ${-}\dot{C}(OR')\backslash + O_2 \longrightarrow {-}\overset{OO\cdot}{\underset{|}{C}}{-}OR'$

Step 3a ${-}\overset{OO\cdot}{\underset{|}{C}}{-}OR' + {-}\overset{H}{\underset{|}{C}}{-}OR' \longrightarrow {-}\overset{OOH}{\underset{|}{C}}{-}OR' + {-}\dot{C}(OR')\backslash$

A hydroperoxide

or

Step 3b ${-}\overset{OO\cdot}{\underset{|}{C}}{-}OR' + {-}\dot{C}(OR')\backslash \longrightarrow R'O{-}\overset{|}{\underset{|}{C}}{-}OO{-}\overset{|}{\underset{|}{C}}{-}OR'$

A peroxide

These hydroperoxides and peroxides, which often accumulate in ethers that have been stored for months or longer in contact with air (the air in the top of the bottle is enough), are dangerously explosive. They often detonate without warning when ether solutions are distilled to near dryness. Since ethers are used frequently in extractions, one should take care to test for and decompose any peroxides present in the ether before a distillation is carried out. (Consult a laboratory manual for instructions.)

Diethyl ether was first employed as a surgical anesthetic by C. W. Long of Jefferson, Georgia, in 1842. Long's use of diethyl ether was not published, but shortly thereafter, diethyl ether was introduced into surgical use at the Massachusetts General Hospital in Boston by J. C. Warren.

The most popular modern anesthetic is halothane ($CF_3CHBrCl$). Unlike diethyl ether, halothane is not flammable.

11.4 Synthesis of Alcohols from Alkenes

We have already studied the acid-catalyzed **hydration of alkenes, oxymercuration–demercuration,** and **hydroboration–oxidation** as methods for the synthesis of alcohols from alkenes (see Sections 8.5, 8.6, and 8.7, respectively). Below, we briefly summarize these methods.

1. Acid-Catalyzed Hydration of Alkenes Alkenes add water in the presence of an acid catalyst to yield alcohols (Section 8.5). The addition takes place with **Markovnikov regioselectivity.** The reaction is reversible, and the mechanism for the acid-catalyzed hydration of an alkene is simply the reverse of that for the dehydration of an alcohol (Section 7.7).

$$\underset{\textbf{Alkene}}{>C{=}C<} + HA \rightleftharpoons {-}\underset{H}{\underset{|}{\overset{|}{C}}}{-}\overset{+}{C}< + A^- \underset{-H_2O}{\overset{+H_2O}{\rightleftharpoons}} {-}\underset{H}{\underset{|}{\overset{|}{C}}}{-}\underset{H\overset{+}{O}H}{\underset{|}{\overset{|}{C}}}{-} + A^- \rightleftharpoons \underset{\textbf{Alcohol}}{{-}\underset{H}{\underset{|}{\overset{|}{C}}}{-}\underset{:\ddot{O}H}{\underset{|}{\overset{|}{C}}}{-}} + HA$$

Acid-catalyzed hydration of alkenes has limited synthetic utility, however, because the carbocation intermediate may rearrange if a more stable carbocation is possible by hydride or alkanide migration. Thus, a mixture of isomeric alcohol products may result.

2. **Oxymercuration–Demercuration** Alkenes react with mercuric acetate in a mixture of water and tetrahydrofuran (THF) to produce (hydroxyalkyl)mercury compounds. These can be reduced to alcohols with sodium borohydride and water (Section 8.6).

Mercury compounds are hazardous. Before you carry out a reaction involving mercury or its compounds, you should familiarize yourself with current procedures for its use and disposal.

Oxymercuration

$$\mathrm{>C{=}C<} + H_2O + Hg(OCOCH_3)_2 \xrightarrow{\text{THF}} -\underset{HO}{\overset{|}{C}}-\underset{Hg-OCOCH_3}{\overset{|}{C}}- + CH_3COOH$$

Demercuration

$$-\underset{HO}{\overset{|}{C}}-\underset{Hg-OCOCH_3}{\overset{|}{C}}- + OH^- + NaBH_4 \longrightarrow -\underset{HO}{\overset{|}{C}}-\underset{H}{\overset{|}{C}}- + Hg + CH_3CO_2^-$$

Regioselectivity of oxymercuration–demercuration

In the oxymercuration step, water and mercuric acetate add to the double bond; in the demercuration step, sodium borohydride reduces the acetoxymercury group and replaces it with hydrogen. The net addition of H— and —OH takes place with **Markovnikov regioselectivity** and **generally takes place without the complication of rearrangements,** as sometimes occurs with acid-catalyzed hydration of alkenes. The overall alkene hydration is not stereoselective because even though the oxymercuration step occurs with anti addition, the demercuration step is not stereoselective (radicals are thought to be involved), and hence a mixture of syn and anti products results.

3. **Hydroboration–Oxidation** An alkene reacts with BH_3:THF or diborane to produce an alkylborane. Oxidation and hydrolysis of the alkylborane with hydrogen peroxide and base yields an alcohol (Section 8.7).

Hydroboration

1-Methylcyclopentene $\xrightarrow[\text{Anti-Markovnikov and syn addition}]{BH_3\text{:THF}}$ trans-2-methylcyclopentyl borane (BH_2, with H and CH_3) + enantiomer + dialkyl- and trialkylborane

Oxidation

Alkylborane $\xrightarrow[\text{Boron group is replaced with retention of configuration}]{H_2O_2,\ HO^-}$ trans-2-methylcyclopentanol (OH, with H and CH_3) + enantiomer

In the first step, boron and hydrogen undergo syn addition to the alkene; in the second step, treatment with hydrogen peroxide and base replaces the boron with —OH with retention of configuration. The net addition of —H and —OH occurs with **anti-Markovnikov regioselectivity** and **syn stereoselectivity.** Hydroboration–oxidation, therefore, serves as a useful regiochemical complement to oxymercuration–demercuration.

Problem 11.4

Write the products you would expect when each of the following alkenes is subjected to (i) acid-catalyzed hydration, (ii) hydration by oxymercuration–demercuration, and (iii) hydration by hydroboration–oxidation:

(a) Ethene **(c)** 2-Methylpropene
(b) Propene **(d)** 2-Methyl-1-butene

Problem 11.5

Treating 3,3-dimethyl-1-butene with dilute sulfuric acid is largely unsuccessful as a method for preparing 3,3-dimethyl-2-butanol because an isomeric compound is the major product. **(a)** What is the isomeric product, and how is it formed? (Writed a detailed mechanism.) **(b)** What reaction conditions would you use to obtain 3,3-dimethyl-2-butanol from 3,3-dimethyl-1-butene?

11.5 Reactions of Alcohols

The reactions of alcohols have mainly to do with the reactivity of their hydroxyl oxygen atom as a weak base and their hydroxyl proton as a weak acid and with reactions that convert the hydroxyl group to a leaving group so as to allow substitution or elimination reactions. Our understanding of the reactions of alcohols will be aided by an initial examination of the electron distribution in the alcohol functional group and of how this distribution affects its reactivity. The oxygen atom of an alcohol polarizes both the C—O bond and the O—H bond of an alcohol:

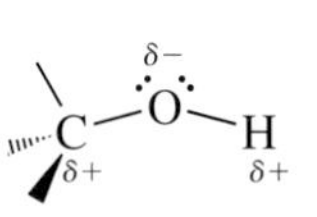
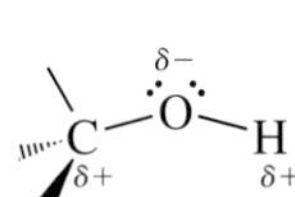

The functional group of an alcohol

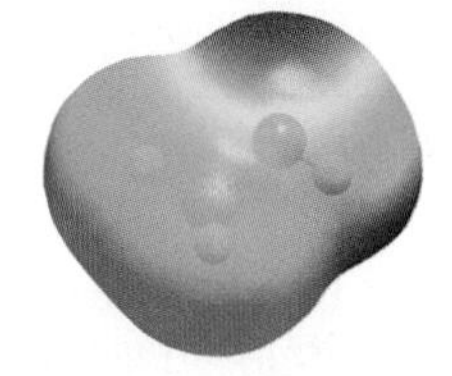

An electrostatic potential map for methanol

Polarization of the O—H bond makes the hydrogen partially positive and explains why alcohols are weak acids (Section 11.6). Polarization of the C—O bond makes the carbon atom partially positive, and if it were not for the fact that OH^- is a strong base and, therefore, a very poor leaving group, this carbon would be susceptible to nucleophilic attack.

The electron pairs on the oxygen atom make it both ***basic*** and ***nucleophilic.*** In the presence of strong acids, alcohols act as bases and accept protons in the following way:

$$-\overset{|}{\underset{|}{C}}-\ddot{\underset{..}{O}}-H + H-A \rightleftharpoons -\overset{|}{\underset{|}{C}}-\overset{H}{\overset{|}{\underset{..}{O}}}\!^{+}-H + A^-$$

Alcohol **Strong acid** **Protonated alcohol**

Protonation of the alcohol converts a poor leaving group (OH^-) into a good one (H_2O). It also makes the carbon atom even more positive (because $—OH_2^+$ is more electron withdrawing than —OH) and, therefore, even more susceptible to nucleophilic attack. Now substitution reactions become possible (S_N2 or S_N1, depending on the class of alcohol, Section 11.8):

$$Nu:^- + -\overset{|}{\underset{|}{C}}-\overset{H}{\overset{|}{\underset{..}{O}}}\!^{+}-H \xrightarrow{S_N2} Nu-\overset{|}{\underset{|}{C}}- + :\overset{H}{\overset{|}{\underset{..}{O}}}-H$$

Protonated alcohol

Because alcohols are nucleophiles, they, too, can react with protonated alcohols. This, as we shall see in Section 11.11A, is an important step in one synthesis of ethers:

$$R-\ddot{O}(H): + -C-\overset{+}{O}H_2 \xrightarrow{S_N2} R-\overset{+}{O}(H)-C- + :\ddot{O}H_2$$

Protonated ether

At a high enough temperature and in the absence of a good nucleophile, protonated alcohols are capable of undergoing E1 or E2 reactions. This is what happens in alcohol dehydrations (Section 7.7).

Alcohols also react with PBr_3 and $SOCl_2$ to yield alkyl bromides and alkyl chlorides. These reactions, as we shall see in Section 11.9, are initiated by the alcohol using its unshared electron pairs to act as a nucleophile.

11.6 Alcohols as Acids

As we might expect, alcohols have acidities similar to that of water. Methanol is a slightly stronger acid than water ($pK_a = 15.7$) but most alcohols are somewhat weaker acids. Values of pK_a for several alcohols are listed in Table 11.3.

TABLE 11.3
pK_a Values for Some Weak Acids

Acid	pK_a
CH_3OH	15.5
H_2O	15.74
CH_3CH_2OH	15.9
$(CH_3)_3COH$	18.0

$$R-\ddot{O}-H + H-\ddot{O}-H \rightleftharpoons R-\ddot{O}:^- + H-\overset{+}{O}(H)-H$$

Alcohol **Alkoxide ion**

(If R is bulky, there is less stabilization of the alkoxide by solvation, and consequently the equilibrium lies even further toward the alcohol.)

It is noteworthy that sterically hindered alcohols such as *tert*-butyl alcohol are even less acidic, and hence their conjugate bases more basic, than unhindered alcohols such as ethanol or methanol. One reason has to do with the effect of solvation. With an unhindered alcohol, water molecules can easily surround, solvate, and hence stabilize the alkoxide anion that would form by loss of the alcohol proton to a base. As a consequence of this stabilization, formation of the alcohol's conjugate base is easier, and therefore its acidity is increased. If the R— group of the alcohol is bulky, solvation of the alkoxide anion is hindered. Stabilization of the conjugate base is not as effective, and consequently the hindered alcohol is a weaker acid. Another reason that hindered alcohols are less acidic has to do with the inductive electron-donating effect of alkyl groups. The alkyl groups of a hindered alcohol donate electron density, making formation of an alkoxide anion more difficult than with a less hindered alcohol.

Remember: Any factor that stabilizes the conjugate base of an acid increases its acidity.

All alcohols, however, are much stronger acids than terminal alkynes, and they are very much stronger acids than hydrogen, ammonia, and alkanes (see Table 3.1).

Relative Acidity

$$H_2O > ROH > RC{\equiv}CH > H_2 > NH_3 > RH$$

Sodium and potassium alkoxides can be prepared by treating alcohols with sodium or potassium metal or with the metal hydride (Section 6.15B). Because most alcohols are weaker acids than water, most alkoxide ions are stronger bases than the hydroxide ion.

Relative Basicity

$$R^- > NH_2^- > H^- > RC{\equiv}C^- > RO^- > HO^-$$

PROBLEM 11.6

Write equations for the acid–base reactions that would occur (if any) if ethanol were added to solutions of each of the following compounds. In each reaction, label the stronger acid, the stronger base, and so forth: **(a)** sodium amide, **(b)** sodium ethynide, and **(c)** sodium acetate (consult Table 3.1).

Sodium and potassium alkoxides are often used as bases in organic syntheses (Section 6.15B). We use alkoxides, such as ethoxide and *tert*-butoxide, when we carry out reactions that require stronger bases than hydroxide ion but do not require exceptionally powerful bases, such as the amide ion or the anion of an alkane. We also use alkoxide ions when (for reasons of solubility) we need to carry out a reaction in an alcohol solvent rather than in water.

11.7 Conversion of Alcohols into Alkyl Halides

In this and several following sections we will be concerned with reactions that involve substitution of the alcohol hydroxyl group. Because a hydroxyl group is such a poor leaving group (it would depart as hydroxide), a common theme of these reactions will be conversion of the hydroxyl to a group that can depart as a weak base. These processes begin by reaction of the alcohol oxygen as a base or nucleophile, after which the modified oxygen group undergoes substitution. First, we shall consider reactions that convert alcohols to alkyl halides.

Alcohols react with a variety of reagents to yield alkyl halides. The most commonly used reagents are hydrogen halides (HCl, HBr, and HI), phosphorus tribromide (PBr_3), and thionyl chloride ($SOCl_2$). Examples of the use of these reagents are the following. All of these reactions result in cleavage of the C—O bond of the alcohol. In each case, the hydroxyl group is first converted to a suitable leaving group. We will see how this is accomplished when we study each type of reaction.

$$(CH_3)_3C\text{—OH} + \mathbf{HCl}_{(concd)} \xrightarrow{25°C} (CH_3)_3C\text{—}\mathbf{Cl} + H_2O \quad \mathbf{(94\%)}$$

$$CH_3CH_2CH_2CH_2OH + \mathbf{HBr}_{(concd)} \xrightarrow[\text{reflux}]{} CH_3CH_2CH_2CH_2\mathbf{Br} \quad \mathbf{(95\%)}$$

$$3\ (CH_3)_2CHCH_2OH + \mathbf{PBr_3} \xrightarrow[4\text{ h}]{-10\text{ to }0°C} 3\ (CH_3)_2CHCH_2\mathbf{Br} + H_3PO_3 \quad \mathbf{(55\text{–}60\%)}$$

$$m\text{-}CH_3OC_6H_4CH_2OH + \mathbf{SOCl_2} \xrightarrow[\text{pyridine (an organic base)}]{} m\text{-}CH_3OC_6H_4CH_2\mathbf{Cl}\ \mathbf{(91\%)} + SO_2 + HCl\ \textbf{(forms a salt with pyridine)}$$

11.8 Alkyl Halides from the Reaction of Alcohols with Hydrogen Halides

When alcohols react with a hydrogen halide, a substitution takes place producing an alkyl halide and water:

$$R\text{—}OH + HX \longrightarrow R\text{—}X + H_2O$$

The order of reactivity of the hydrogen halides is HI > HBr > HCl (HF is generally unreactive), and the order of reactivity of alcohols is 3° > 2° > 1° < methyl.

The reaction is *acid catalyzed.* Alcohols react with the strongly acidic hydrogen halides HCl, HBr, and HI, but they do not react with nonacidic NaCl, NaBr, or NaI. Primary and secondary alcohols can be converted to alkyl chlorides and bromides by allowing them to react with a mixture of a sodium halide and sulfuric acid:

$$\text{ROH} + \text{NaX} \xrightarrow{\text{H}_2\text{SO}_4} \text{RX} + \text{NaHSO}_4 + \text{H}_2\text{O}$$

11.8A Mechanisms of the Reactions of Alcohols with HX

Secondary, tertiary, allylic, and benzylic alcohols appear to react by a mechanism that involves the formation of a carbocation—one that we studied first in Section 3.13 and that you should now recognize *as an S_N1 reaction with the protonated alcohol acting as the substrate.* We again illustrate this mechanism with the reaction of *tert*-butyl alcohol and aqueous hydrochloric acid (H_3O^+, Cl^-).

The first two steps are the same as in the mechanism for the dehydration of an alcohol (Section 7.7). The alcohol accepts a proton and then the protonated hydroxyl group departs as a leaving group to form a carbocation and water:

Step 1 $$(\text{CH}_3)_3\text{C}{-}\ddot{\text{O}}{-}\text{H} + \text{H}{-}\overset{+}{\ddot{\text{O}}}(\text{H}){-}\text{H} \overset{\text{fast}}{\rightleftharpoons} (\text{CH}_3)_3\text{C}{-}\overset{+}{\ddot{\text{O}}}(\text{H}){-}\text{H} + :\ddot{\text{O}}(\text{H}){-}\text{H}$$

Step 2 $$(\text{CH}_3)_3\text{C}{-}\overset{+}{\ddot{\text{O}}}(\text{H}){-}\text{H} \overset{\text{slow}}{\rightleftharpoons} (\text{CH}_3)_3\text{C}^+ + :\ddot{\text{O}}(\text{H}){-}\text{H}$$

In step 3 the mechanisms for the dehydration of an alcohol and the formation of an alkyl halide differ. In dehydration reactions the carbocation loses a proton in an E1 reaction to form an alkene. In the formation of an alkyl halide, the carbocation reacts with a nucleophile (a halide ion) in an S_N1 reaction.

Step 3 $$(\text{CH}_3)_3\text{C}^+ + :\ddot{\text{Cl}}:^- \overset{\text{fast}}{\rightleftharpoons} (\text{CH}_3)_3\text{C}{-}\ddot{\text{Cl}}:$$

How can we account for the different course of these two reactions?

When we dehydrate alcohols, we usually carry out the reaction in concentrated sulfuric acid. The only nucleophiles present in this reaction mixture are water and hydrogen sulfate (HSO_4^-) ions. Both are poor nucleophiles and both are usually present in low concentrations. Under these conditions, the highly reactive carbocation stabilizes itself by losing a proton and becoming an alkene. The net result is *an E1 reaction.*

In the reverse reaction, that is, the hydration of an alkene (Section 8.5), the carbocation *does* react with a nucleophile. It reacts with water. Alkene hydrations are carried out in dilute sulfuric acid, where the water concentration is high. In some instances, too, carbocations may react with HSO_4^- ions or with sulfuric acid, itself. When they do, they form alkyl hydrogen sulfates ($R{-}OSO_2OH$).

When we convert an alcohol to an alkyl halide, we carry out the reaction in the presence of acid and *in the presence of halide ions.* Halide ions are good nucleophiles (much

stronger nucleophiles than water), and since halide ions are present in high concentration, most of the carbocations stabilize themselves by accepting the electron pair of a halide ion. The overall result is an S_N1 reaction.

These two reactions, dehydration and the formation of an alkyl halide, also furnish us another example of the competition between nucleophilic substitution and elimination (see Section 6.18). Very often, in conversions of alcohols to alkyl halides, we find that the reaction is accompanied by the formation of some alkene (i.e., by elimination). The free energies of activation for these two reactions of carbocations are not very different from one another. Thus, not all of the carbocations react with nucleophiles; some stabilize themselves by losing protons.

Not all acid-catalyzed conversions of alcohols to alkyl halides proceed through the formation of carbocations. Primary alcohols and methanol apparently react through a mechanism that we recognize as *an S_N2 type.* In these reactions the function of the acid is to produce *a protonated alcohol.* The halide ion then displaces a molecule of water (a good leaving group) from carbon; this produces an alkyl halide:

$$:\ddot{X}:^- + R{-}CH_2{-}\overset{+}{\ddot{O}}H{-}H \longrightarrow :\ddot{X}{-}CH_2{-}R + :\ddot{O}H{-}H$$

(Protonated 1° alcohol or methanol) **(A good leaving group)**

Although halide ions (particularly iodide and bromide ions) are strong nucleophiles, they are not strong enough to carry out substitution reactions with alcohols themselves. That is, reactions like the following do not occur because the leaving group would have to be a strongly basic hydroxide ion:

The reverse reaction, that is, the reaction of an alkyl halide with hydroxide ion, does occur and is a method for the synthesis of alcohols. We saw this reaction in Chapter 6.

$$:\ddot{Br}:^- + {-}\overset{|}{\underset{|}{C}}{-}\ddot{O}H \not\longrightarrow :\ddot{Br}{-}\overset{|}{\underset{|}{C}}{-} + {}^-:\ddot{O}H$$

We can see now why the reactions of alcohols with hydrogen halides are acid promoted. With tertiary and secondary alcohols, the function of the acid is to help produce a carbocation. With methanol and primary alcohols, the function of the acid is to produce a substrate in which the leaving group is a weakly basic water molecule rather than a strongly basic hydroxide ion.

As we might expect, many reactions of alcohols with hydrogen halides, particularly those in which carbocations are formed, *are accompanied by rearrangements.*

Because the chloride ion is a weaker nucleophile than bromide or iodide ions, hydrogen chloride does not react with primary or secondary alcohols unless zinc chloride or some similar Lewis acid is added to the reaction mixture as well. Zinc chloride, a good Lewis acid, forms a complex with the alcohol through association with an unshared pair of electrons on the oxygen atom. This provides a better leaving group for the reaction than H_2O:

$$R{-}\ddot{O}H: + ZnCl_2 \rightleftharpoons R{-}\overset{+}{\ddot{O}}H{-}\overset{-}{Zn}Cl_2$$

$$:\ddot{Cl}:^- + R{-}\overset{+}{\ddot{O}}H{-}\overset{-}{Zn}Cl_2 \longrightarrow :\ddot{Cl}{-}R + [Zn(OH)Cl_2]^-$$

$$[Zn(OH)Cl_2]^- + H^+ \rightleftharpoons ZnCl_2 + H_2O$$

PROBLEM 11.7

(**a**) What factor explains the observation that tertiary alcohols react with HX faster than secondary alcohols? (**b**) What factor explains the observation that methanol reacts with HX faster than a primary alcohol?

PROBLEM 11.8

Treating 3-methyl-2-butanol (see following reaction) with HBr yields 2-bromo-2-methylbutane as the sole product. Outline a mechanism for the reaction.

$$CH_3CH(OH)CH(CH_3)CH_3 \xrightarrow{HBr} CH_3C(Br)(CH_3)CH_2CH_3$$

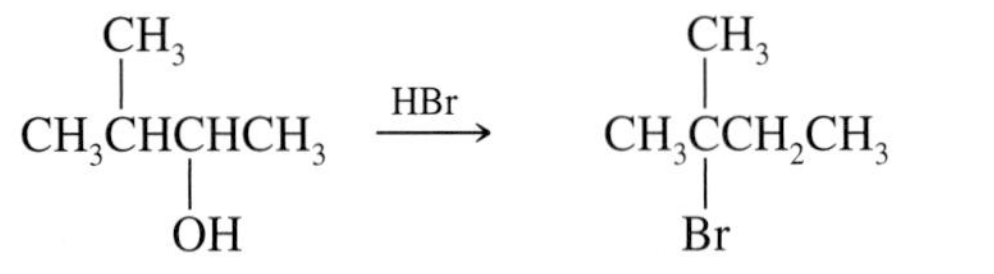

3-Methyl-2-butanol → **2-Bromo-2-methylbutane**

11.9 Alkyl Halides from the Reaction of Alcohols with PBr_3 or $SOCl_2$

Primary and secondary alcohols react with phosphorus tribromide to yield alkyl bromides.

$$3\ R{-}OH + PBr_3 \longrightarrow 3\ R{-}Br + H_3PO_3$$

(1° or 2°)

PBr_3: A reagent for synthesizing 1° and 2° alkyl bromides

Unlike the reaction of an alcohol with HBr, the reaction of an alcohol with PBr_3 does not involve the formation of a carbocation and *usually occurs without rearrangement* of the carbon skeleton (especially if the temperature is kept below 0°C). For this reason phosphorus tribromide is often preferred as a reagent for the transformation of an alcohol to the corresponding alkyl bromide.

The mechanism for the reaction involves attack of the alcohol group on the phosphorus atom, displacing a bromide ion and forming a protonated alkyl dibromophosphite (see following reaction):

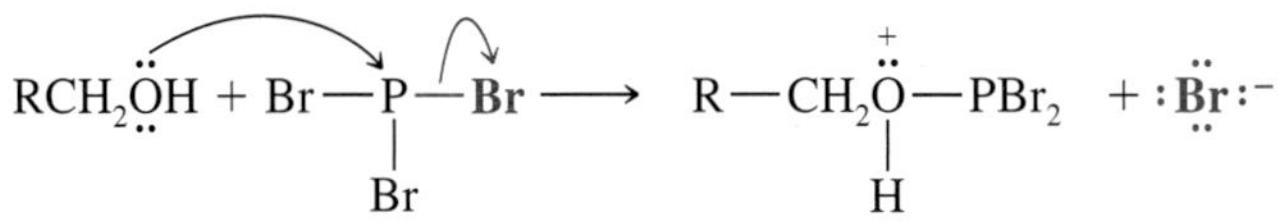

$$RCH_2\ddot{O}H + Br{-}P(Br){-}Br \longrightarrow R{-}CH_2\overset{+}{\ddot{O}}(H){-}PBr_2 + :\ddot{Br}:^-$$

Protonated alkyl dibromophosphite

In a second step a bromide ion acts as a nucleophile to displace $HOPBr_2$, a good leaving group due to the electronegative atoms bonded to the phosphorus:

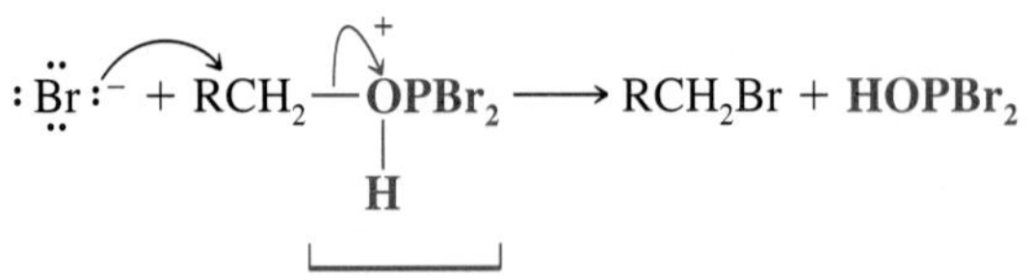

$$:\ddot{Br}:^- + RCH_2{-}\overset{+}{\ddot{O}}(H)PBr_2 \longrightarrow RCH_2Br + \mathbf{HOPBr_2}$$

A good leaving group

$HOPBr_2$ can react with 2 more moles of alcohol, so the net result is conversion of 3 mol of alcohol to alkyl bromide by 1 mol of phosphorus tribromide.

Thionyl chloride ($SOCl_2$) converts primary and secondary alcohols to alkyl chlorides (usually without rearrangement):

$$R{-}OH + SOCl_2 \xrightarrow{\text{reflux}} R{-}Cl + SO_2 + HCl$$

(1° or 2°)

$SOCl_2$: A reagent for synthesizing 1° and 2° alkyl chlorides

Often a tertiary amine is added to the mixture to promote the reaction by reacting with the HCl:

$$R_3N: + HCl \longrightarrow R_3NH^+ + Cl^-$$

The reaction mechanism involves initial formation of the alkyl chlorosulfite:

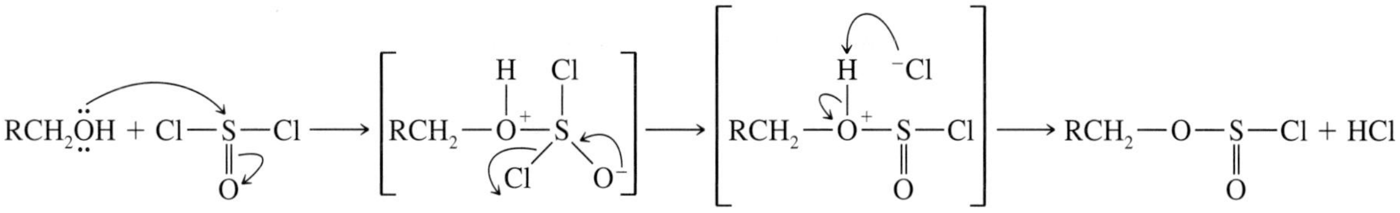

$$RCH_2\ddot{O}H + Cl{-}S(=O){-}Cl \longrightarrow \left[RCH_2{-}\overset{H}{O^+}{-}S(Cl)(Cl)(O^-)\right] \longrightarrow \left[RCH_2{-}\overset{H}{O^+}{-}S(=O){-}Cl\ \ {}^-Cl\right] \longrightarrow RCH_2{-}O{-}S(=O){-}Cl + HCl$$

Alkyl chlorosulfite

Then a chloride ion (from $R_3N + HCl \longrightarrow R_3NH^+ + Cl^-$) can bring about an S_N2 displacement of a very good leaving group, $ClSO_2^-$, which, by decomposing (to the gas SO_2 and Cl^- ion), helps drive the reaction to completion:

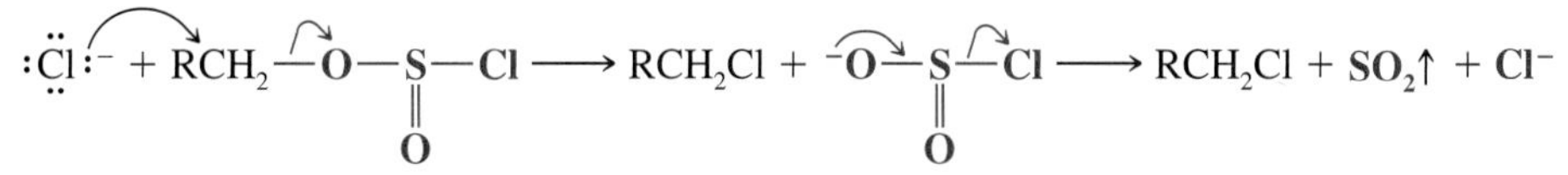

$$:\ddot{Cl}:^- + RCH_2{-}O{-}S(=O){-}Cl \longrightarrow RCH_2Cl + {}^-O{-}S(=O){-}Cl \longrightarrow RCH_2Cl + SO_2\uparrow + Cl^-$$

SOLVED PROBLEM

Starting with alcohols, outline a synthesis of each of the following: **(a)** benzyl bromide, **(b)** cyclohexyl chloride, and **(c)** butyl bromide.

POSSIBLE ANSWERS:

(a) $C_6H_5CH_2OH \xrightarrow{PBr_3} C_6H_5CH_2Br$

(b) 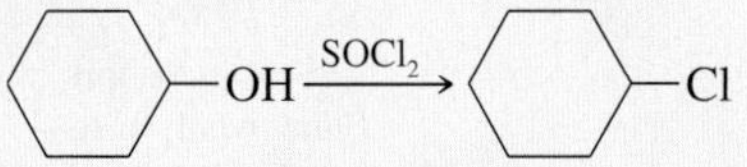 cyclohexyl–OH $\xrightarrow{SOCl_2}$ cyclohexyl–Cl

(c) $CH_3CH_2CH_2CH_2OH \xrightarrow{PBr_3} CH_3CH_2CH_2CH_2Br$

11.10 Tosylates, Mesylates, and Triflates: Leaving Group Derivatives of Alcohols

The hydroxyl group of an alcohol can be converted to a good leaving group by conversion to a **sulfonate ester** derivative. The most common sulfonate esters used for this purpose are methanesulfonate esters ("mesylates"), *p*-toluenesulfonate esters ("tosylates"), and trifluoromethanesulfonates ("triflates").

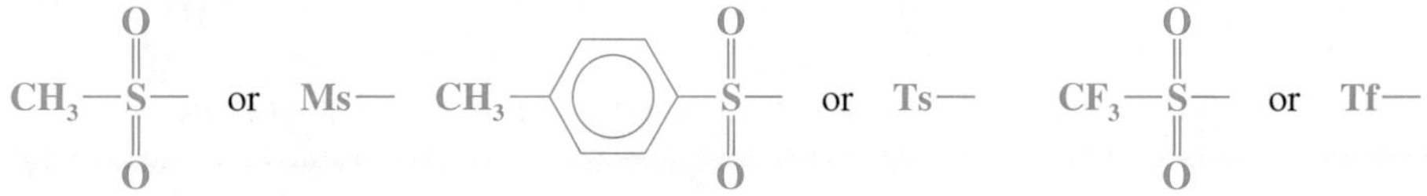

$CH_3{-}S(=O)_2{-}$ or Ms—

The mesyl group

$CH_3{-}C_6H_4{-}S(=O)_2{-}$ or Ts—

The tosyl group

$CF_3{-}S(=O)_2{-}$ or Tf—

The trifyl group

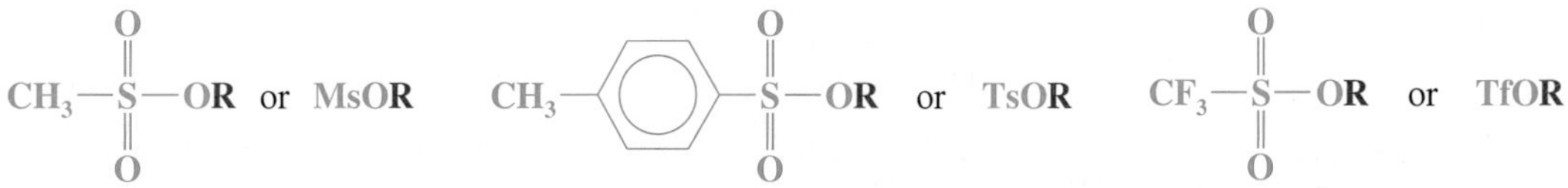

$CH_3{-}S(=O)_2{-}OR$ or MsOR

An alkyl mesylate

$CH_3{-}C_6H_4{-}S(=O)_2{-}OR$ or TsOR

An alkyl tosylate

$CF_3{-}S(=O)_2{-}OR$ or TfOR

An alkyl triflate

A method for making an alcohol hydroxyl group into a leaving group

The desired sulfonate ester is usually prepared by reaction of the alcohol with the appropriate sulfonyl chloride, that is, methanesulfonyl chloride ("mesyl" chloride) for a mesylate, *p*-toluenesulfonyl chloride ("tosyl" chloride) for a tosylate, or trifluoromethanesulfonyl chloride [or trifluoromethanesulfonic anhydride ("triflic" anhydride)] for a triflate. Ethanol, for example, reacts with methanesulfonyl chloride to form ethyl methanesulfonate and with *p*-toluenesulfonyl chloride to form ethyl *p*-toluenesulfonate:

$$CH_3S(=O)_2{-}Cl + H{-}OCH_2CH_3 \xrightarrow[(-HCl)]{\text{base}} CH_3S(=O)_2{-}OCH_2CH_3$$

Methanesulfonyl chloride **Ethanol** **Ethyl methanesulfonate (ethyl mesylate)**

$$CH_3{-}C_6H_4{-}S(=O)_2{-}Cl + H{-}OCH_2CH_3 \xrightarrow[(-HCl)]{\text{base}} CH_3{-}C_6H_4{-}S(=O)_2{-}OCH_2CH_3$$

***p*-Toluenesulfonyl chloride** **Ethanol** **Ethyl *p*-toluenesulfonate (ethyl tosylate)**

Mesylates, tosylates, and triflates, because they are good leaving groups, are frequently used as substrates for nucleophilic substitution reactions. They are good leaving groups because the sulfonate anions they become when they depart are very weak bases:

$$Nu\!:^- + RCH_2{-}O{-}S(=O)_2{-}R' \longrightarrow Nu{-}CH_2R + {}^-O{-}S(=O)_2{-}R'$$

Alkyl sulfonate (tosylate, mesylate, etc.) **Sulfonate ion (very weak base— a good leaving group)**

To carry out a nucleophilic substitution on an alcohol, we first convert the alcohol to an alkyl sulfonate and then, in a second reaction, allow it to react with a nucleophile. It is important to note that formation of the sulfonate ester does not affect the stereochemistry of the alcohol carbon, because C—O bond is not involved in this step. Thus, if the alcohol carbon is a stereogenic center, no change in configuration occurs upon making the sulfonate ester—the reaction proceeds with **retention of configuration.** Upon reaction of the sulfonate ester with a nucleophile, the usual parameters of nucleophilic substitution reactions become involved. If the mechanism is S_N2, as shown in the second reaction of the following example, **inversion of configuration** takes place at the carbon that originally bore the alcohol hydroxyl group:

Step 1 $$(R)(R')(H)C{-}O{-}H + Cl{-}Ts \xrightarrow[-HCl]{\text{retention}} (R)(R')(H)C{-}O{-}Ts$$

Step 2 $$Nu\!:^- + (R)(R')(H)C{-}O{-}Ts \xrightarrow[S_N2]{\text{inversion}} Nu{-}C(R)(R')(H) + {}^-O{-}Ts$$

The fact that the C—O bond of the alcohol does not break during formation of the sulfonate ester is accounted for by the following mechanism. Methanesulfonyl chloride is used in the example:

A Mechanism for the Reaction

Conversion of an Alcohol into an Alkyl Methanesulfonate

$$\text{Me—SO}_2\text{—Cl} + \text{H—}\ddot{\text{O}}\text{—R} \longrightarrow \left[\text{Me—S(O}^-\text{)(=O)(Cl)—}\ddot{\text{O}}^+\text{(R)(H)}\right] \longrightarrow$$

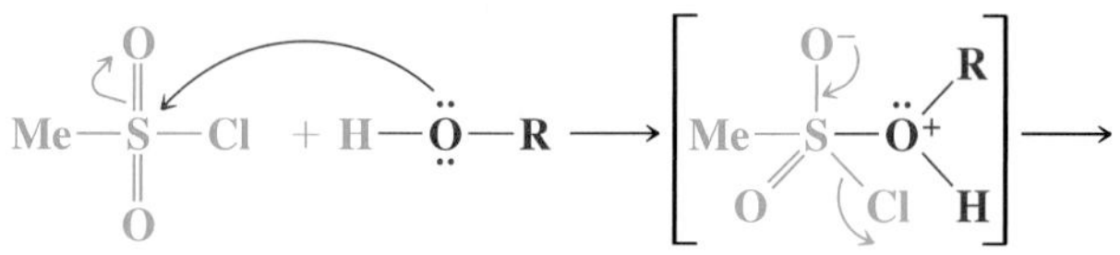

Methanesulfonyl chloride **Alcohol**

The alcohol oxygen attacks the sulfur atom of the sulfonyl chloride.

The intermediate loses a chloride ion.

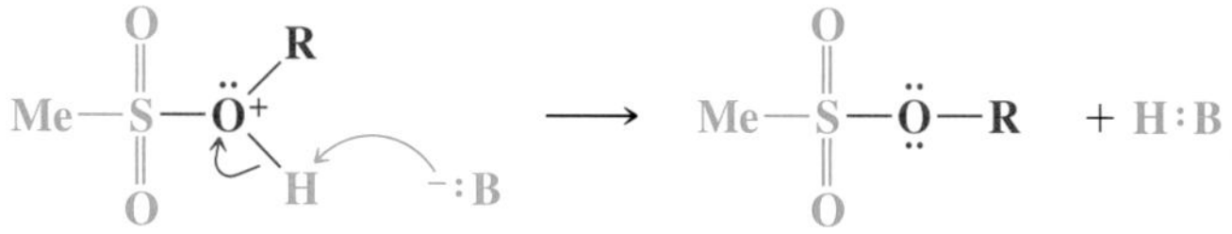

$$\text{Me—SO}_2\text{—}\ddot{\text{O}}^+\text{(R)—H} \quad {}^-\text{:B} \longrightarrow \text{Me—SO}_2\text{—}\ddot{\text{O}}\text{—R} + \text{H:B}$$

(a base) **Alkyl methanesulfonate**

Loss of a proton leads to the product.

The trifluoromethanesulfonate (triflate) anion is one of the best of all known leaving groups. Alkyl triflates react extremely rapidly in nucleophilic substitution reactions. The triflate anion is such a good leaving group that even vinyl triflates undergo S_N1 reactions to yield vinylic cations:

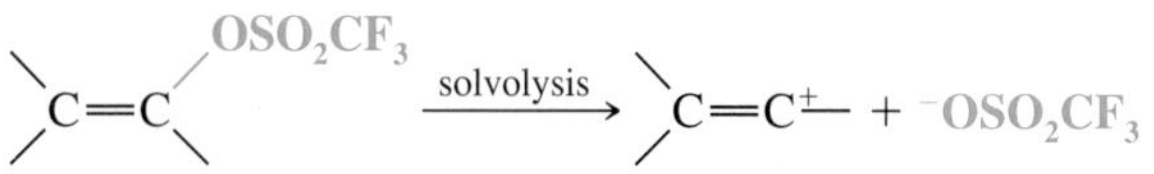

$$\text{>C=C(OSO}_2\text{CF}_3\text{)—} \xrightarrow{\text{solvolysis}} \text{>C=C}^+\text{—} + {}^-\text{OSO}_2\text{CF}_3$$

Vinylic triflate **Vinylic cation** **Triflate ion**

Sulfonyl chlorides needed to make common sulfonate esters can usually be purchased. If necessary, however, a sulfonyl choride can be prepared by treating a sulfonic acid with phosphorus pentachloride, as shown here. (We shall study syntheses of sulfonic acids in Chapter 15.)

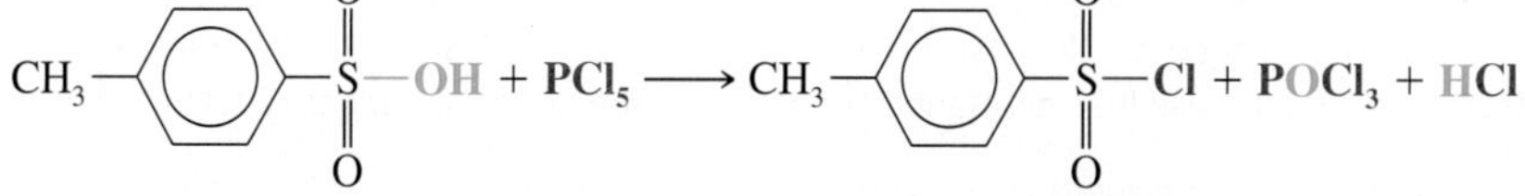

$$CH_3\text{—C}_6\text{H}_4\text{—SO}_2\text{—OH} + PCl_5 \longrightarrow CH_3\text{—C}_6\text{H}_4\text{—SO}_2\text{—Cl} + POCl_3 + HCl$$

***p*-Toluenesulfonic acid** ***p*-Toluenesulfonyl chloride (tosyl chloride)**

PROBLEM 11.9

Show the configurations of products formed when **(a)** (*R*)-2-butanol is converted to a tosylate and **(b)** this tosylate reacts with hydroxide ion by an S_N2 reaction. **(c)** Converting *cis*-4-methylcyclohexanol to a tosylate and then allowing the tosylate to react with LiCl (in an appropriate solvent) yields *trans*-1-chloro-4-methylcyclohexane. Outline the stereochemistry of these steps.

PROBLEM 11.10

Starting with the appropriate sulfonic acid and PCl_5 or with the appropriate sulfonyl chloride, show how you would prepare **(a)** methyl *p*-toluenesulfonate, **(b)** isobutyl methanesulfonate, and **(c)** *tert*-butyl methanesulfonate.

PROBLEM 11.11 Suggest an experiment using an isotopically labeled alcohol that would prove that the formation of an alkyl sulfonate does not cause cleavage at the C—O bond of the alcohol.

The Chemistry of...

Alkyl Phosphates

Alcohols react with phosphoric acid to yield alkyl phosphates:

$$\underset{\text{Phosphoric acid}}{ROH + HO{-}\overset{O}{\overset{\|}{P}}{-}OH \atop \quad\quad\ |\ OH} \xrightarrow[(-H_2O)]{} \underset{\text{Alkyl dihydrogen phosphate}}{RO{-}P(=O)(OH){-}OH} \xrightarrow[(-H_2O)]{ROH} \underset{\text{Dialkyl hydrogen phosphate}}{RO{-}P(=O)(OR){-}OH} \xrightarrow[(-H_2O)]{ROH} \underset{\text{Trialkyl phosphate}}{RO{-}P(=O)(OR){-}OR}$$

Esters of phosphoric acids are important in biochemical reactions. Especially important are triphosphate esters. Although hydrolysis of the ester group or of one of the anhydride linkages of an alkyl triphosphate is exothermic, these reactions occur very slowly in aqueous solutions. Near pH 7, these triphosphates exist as negatively charged ions and hence are much less susceptible to nucleophilic attack. Alkyl triphosphates are, consequently, relatively stable compounds in the aqueous medium of a living cell.

$$R{-}O{-}P(=O)(O^-){-}O{-}P(=O)(O^-){-}O{-}P(=O)(O^-){-}O^- \xrightarrow[\text{slow}]{H_2O} \begin{cases} ROH + HO{-}P(=O)(O^-){-}O{-}P(=O)(O^-){-}O{-}P(=O)(O^-){-}O^- \\ RO{-}P(=O)(O^-){-}O^- + HO{-}P(=O)(O^-){-}O{-}P(=O)(O^-){-}O^- \\ RO{-}P(=O)(O^-){-}O{-}P(=O)(O^-){-}O^- + HO{-}P(=O)(O^-){-}O^- \end{cases}$$

Phosphate ester linkage

Phosphoric anhydride linkages

Enzymes, on the other hand, are able to catalyze reactions of these triphosphates in which the energy made available when their anhydride linkages break helps the cell make other chemical bonds. We have more to say about this in Chapter 22 when we discuss the important triphosphate called adenosine triphosphate (or ATP).

11.11 Synthesis of Ethers

11.11A Ethers by Intermolecular Dehydration of Alcohols

Alcohols can dehydrate to form alkenes. We studied this in Sections 7.7 and 7.8. Primary alcohols can also dehydrate to form ethers:

$$R{-}OH + HO{-}R \xrightarrow[(-H_2O)]{HA} R{-}O{-}R$$

Dehydration to an ether usually takes place at a lower temperature than dehydration to the alkene, and dehydration to the ether can be aided by distilling the ether as it is formed. Diethyl ether is made commercially by dehydration of ethanol. Diethyl ether is the predominant product at 140°C; ethene is the major product at 180°C:

$$CH_3CH_2OH \xrightarrow[180^\circ C]{H_2SO_4} CH_2{=}CH_2 \quad \textbf{Ethene}$$

$$CH_3CH_2OH \xrightarrow[140^\circ C]{H_2SO_4} CH_3CH_2OCH_2CH_3 \quad \textbf{Diethyl ether}$$

The formation of the ether occurs by an S_N2 mechanism with one molecule of the alcohol acting as the nucleophile and another protonated molecule of the alcohol acting as the substrate (see Section 11.5).

A Mechanism for the Reaction

Intermolecular Dehydration of Alcohols to Form an Ether

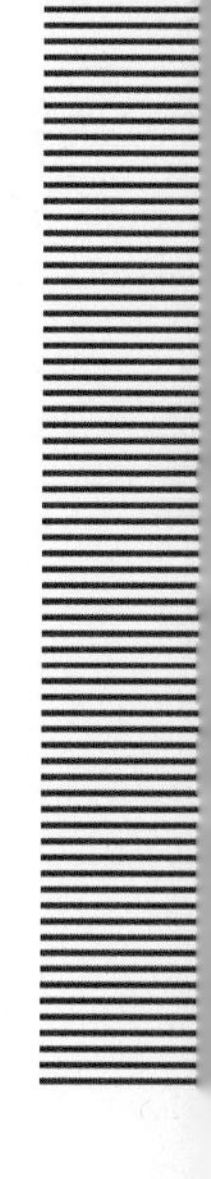

Step 1 $CH_3CH_2{-}\ddot{O}{-}H + H{-}OSO_3H \rightleftharpoons CH_3CH_2{-}\overset{H}{\overset{|}{\ddot{O}^+}}{-}H + {}^-OSO_3H$

This is an acid–base reaction in which the alcohol accepts a proton from the sulfuric acid.

Step 2 $CH_3CH_2{-}\ddot{O}{-}H + CH_3CH_2{-}\overset{H}{\overset{|}{\ddot{O}^+}}{-}H \rightleftharpoons CH_3CH_2{-}\overset{H}{\overset{|}{\ddot{O}^+}}{-}CH_2CH_3 + :\overset{H}{\overset{|}{\ddot{O}}}{-}H$

Another molecule of the alcohol acts as a nucleophile and attacks the protonated alcohol in an S_N2 reaction.

Step 3 $CH_3CH_2{-}\overset{H}{\overset{|}{\ddot{O}^+}}{-}CH_2CH_3 + :\overset{H}{\overset{|}{\ddot{O}}}{-}H \rightleftharpoons CH_3CH_2{-}\ddot{O}{-}CH_2CH_3 + H{-}\overset{H}{\overset{|}{\ddot{O}^+}}{-}H$

Another acid–base reaction converts the protonated ether to an ether by transferring a proton to a molecule of water (or to another molecule of the alcohol).

This method of preparing ethers is of limited usefulness, however. Attempts to synthesize ethers with secondary alkyl groups by intermolecular dehydration of secondary alcohols are usually unsuccessful because alkenes form too easily. Attempts to make ethers with tertiary alkyl groups lead exclusively to the alkenes. Finally, this method is not useful for the preparation of unsymmetrical ethers from primary alcohols because the reaction leads to a mixture of products:

$$\underbrace{ROH + \mathbf{R'OH}}_{1^\circ \text{ alcohols}} \xrightarrow{H_2SO_4} \begin{matrix} ROR \\ + \\ \mathbf{ROR'} + H_2O \\ + \\ \mathbf{R'OR'} \end{matrix}$$

Problem 11.12 An exception to what we have just said has to do with syntheses of unsymmetrical ethers in which one alkyl group is a *tert*-butyl group and the other group is primary. This synthesis can be accomplished by adding *tert*-butyl alcohol to a mixture of the primary alcohol and H_2SO_4 at room temperature. Give a likely mechanism for this reaction and explain why it is successful.

11.11B The Williamson Synthesis of Ethers

Alexander William Williamson was an English chemist who lived between 1824 and 1904. His method is especially useful for synthesis of unsymmetrical ethers.

An important route to unsymmetrical ethers is a nucleophilic substitution reaction known as the **Williamson synthesis.** This synthesis consists of an S_N2 reaction of a sodium alkoxide with an alkyl halide, alkyl sulfonate, or alkyl sulfate.

A Mechanism for the Reaction

The Williamson Ether Synthesis

$$R{-}\ddot{\underset{..}{O}}{:}^{-}\ Na^{+} + R'{-}L \longrightarrow R{-}\ddot{\underset{..}{O}}{-}R' + Na^{+}\ {:}L^{-}$$

Sodium (or potassium) alkoxide; Alkyl halide, alkyl sulfonate, or dialkyl sulfate; Ether

The alkoxide ion reacts with the substrate in an S_N2 reaction, with the resulting formation of an ether. The substrate must be unhindered and bear a good leaving group. Typical substrates are 1° or 2° alkyl halides, alkyl sulfonates, and dialkyl sulfates, that is,

$$—L = —\ddot{\underset{..}{Br}}{:},\ —\ddot{\underset{..}{I}}{:},\ —OSO_2R'',\ \text{or}\ —OSO_2OR''$$

The following reaction is a specific example of the Williamson synthesis. The sodium alkoxide can be prepared by allowing an alcohol to react with NaH:

$$CH_3CH_2CH_2OH + NaH \longrightarrow CH_3CH_2CH_2\ddot{\underset{..}{O}}{:}^{-}Na^{+} + H{-}H$$

Propyl alcohol; Sodium propoxide

$$\downarrow CH_3CH_2I$$

$$CH_3CH_2OCH_2CH_2CH_3 + Na^{+}I^{-}$$

Ethyl propyl ether (70%)

Conditions that favor a Williamson ether synthesis

The usual limitations of S_N2 reactions apply here. Best results are obtained when the alkyl halide, sulfonate, or sulfate is primary (or methyl). If the substrate is tertiary, elimination is the exclusive result. Substitution is also favored over elimination at lower temperatures.

Problem 11.13 **(a)** Outline two methods for preparing isopropyl methyl ether by a Williamson synthesis.
(b) One method gives a much better yield of the ether than the other. Explain which is the better method and why.

PROBLEM 11.14 The two syntheses of 2-ethoxy-1-phenylpropane shown here give products with opposite optical rotations:

$$\underset{[\alpha] = +33.0°}{\underset{\text{OH}}{C_6H_5CH_2\underset{|}{C}HCH_3}} \xrightarrow{K} \begin{matrix}\text{potassium}\\ \text{alkoxide}\\ + H_2\end{matrix} \xrightarrow[(-\text{KBr})]{C_2H_5Br} \underset{[\alpha] = +23.5°}{\underset{OC_2H_5}{C_6H_5CH_2\underset{|}{C}HCH_3}}$$

$$\downarrow \text{TsCl/base (Ts} = p\text{-toluenesulfonyl, Section 11.10)}$$

$$\underset{\text{OTs}}{C_6H_5CH_2\underset{|}{C}HCH_3} \xrightarrow[K_2CO_3]{C_2H_5OH} \underset{[\alpha] = -19.9°}{\underset{OC_2H_5}{C_6H_5CH_2\underset{|}{C}HCH_3}} + \text{KOTs}$$

How can you explain this result?

PROBLEM 11.15 Write a mechanism that explains the formation of tetrahydrofuran (THF) from the reaction of 4-chloro-1-butanol and aqueous sodium hydroxide.

PROBLEM 11.16 Epoxides can be synthesized by treating halohydrins with aqueous base. For example, treating $ClCH_2CH_2OH$ with aqueous sodium hydroxide yields ethylene oxide. **(a)** Propose a mechanism for this reaction. **(b)** *trans*-2-Chlorocyclohexanol reacts readily with sodium hydroxide to yield cyclohexene oxide. *cis*-2-Chlorocyclohexanol does not undergo this reaction, however. How can you account for this difference?

11.11C Synthesis of Ethers by Alkoxymercuration–Demercuration

Alkoxymercuration–demercuration is another method for synthesizing ethers. The reaction of an alkene with an alcohol in the presence of a mercury salt such as mercuric acetate or trifluoroacetate leads to an alkoxymercury intermediate, which upon reaction with sodium borohydride yields an ether. When the alcohol reactant is also the solvent, the method is called solvomercuration–demercuration. This method directly parallels hydration by oxymercuration–demercuration (Section 8.6):

$$\text{Cyclohexene} \xrightarrow[\text{(2) NaBH}_4\text{, HO}^-]{\text{(1) Hg(O}_2\text{CCF}_3)_2\text{, HOCH(CH}_3)_2} \text{C}_6\text{H}_{11}\text{—OCH(CH}_3)_2$$

(98% Yield)

11.11D *tert*-Butyl Ethers by Alkylation of Alcohols: Protecting Groups

Primary alcohols can be converted to *tert*-butyl ethers by dissolving them in a strong acid such as sulfuric acid and then adding isobutylene to the mixture. (This procedure minimizes dimerization and polymerization of the isobutylene.)

$$RCH_2OH + CH_2{=}\underset{CH_3}{\underset{|}{C}}CH_3 \xrightarrow{H_2SO_4} RCH_2O{-}\overset{CH_3}{\underset{CH_3}{\overset{|}{\underset{|}{C}}}}CH_3 \quad \left.\right\} \textbf{\textit{tert}-Butyl protecting group}$$

This method can be used to "protect" the hydroxyl group of a primary alcohol while another reaction is carried out on some other part of the molecule. The protecting *tert*-butyl group can be removed easily by treating the ether with dilute aqueous acid.

Suppose, for example, we wanted to prepare 4-pentyn-1-ol from 3-bromo-1-propanol and sodium acetylide. If we allow them to react directly, the strongly basic sodium acetylide will react first with the hydroxyl group:

$$\underset{\textbf{3-Bromo-1-propanol}}{HOCH_2CH_2CH_2Br} + NaC{\equiv}CH \longrightarrow NaOCH_2CH_2CH_2Br + HC{\equiv}CH$$

However, if we protect the —OH group first, the synthesis becomes feasible:

$$HOCH_2CH_2CH_2Br \xrightarrow[(2)\ CH_2{=}C(CH_3)_2]{(1)\ H_2SO_4} (CH_3)_3COCH_2CH_2CH_2Br \xrightarrow{NaC{\equiv}CH}$$

$$(CH_3)_3COCH_2CH_2CH_2C{\equiv}CH \xrightarrow{H_3O^+/H_2O} \underset{\textbf{4-Pentyn-1-ol}}{HOCH_2CH_2CH_2C{\equiv}CH} + (CH_3)_3COH$$

PROBLEM 11.17

(a) The mechanism for the formation of the *tert*-butyl ether from a primary alcohol and isobutylene is similar to that discussed in Problem 11.12. Propose such a mechanism. **(b)** What factor makes it possible to remove the protecting *tert*-butyl group so easily? (Other ethers require much more forcing conditions for their cleavage, as we shall see in Section 11.12.) **(c)** Propose a mechanism for the removal of the protecting *tert*-butyl group.

11.11E Silyl Ether Protecting Groups

TBDMS ethers

A hydroxyl group can also be protected by converting it to a silyl ether group. One of the most common is the *tert*-butyldimethylsilyl ether group [*tert*-butyl$(CH_3)_2Si$—O—R, or TBDMS—O—R], although triethylsilyl, triisopropylsilyl, *tert*-butyldiphenylsilyl, and others can be used. The *tert*-butyldimethylsilyl ether is stable over a pH range of roughly 4–12. A TBDMS group can be added by allowing the alcohol to react with *tert*-butylchlorodimethylsilane in the presence of an aromatic amine (a base) such as imidazole or pyridine:

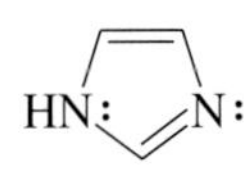

Imidazole

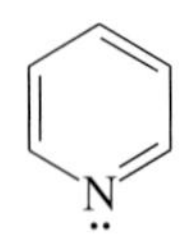

Pyridine

$$R{-}O{-}H + \underset{\substack{\textbf{\textit{tert}-Butylchlorodimethylsilane} \\ \textbf{(TBDMSCl)}}}{Cl{-}Si(CH_3)_2{-}C(CH_3)_3} \xrightarrow[\substack{DMF \\ (-HCl)}]{imidazole} \underset{\textbf{(R—O—TBDMS)}}{R{-}O{-}Si(CH_3)_2{-}C(CH_3)_3}$$

The TBDMS group can be removed by treatment with fluoride ion (tetrabutylammonium fluoride or aqueous HF is frequently used):

$$\underset{\textbf{(R—O—TBDMS)}}{R{-}O{-}Si(CH_3)_2{-}C(CH_3)_3} \xrightarrow[THF]{Bu_4N^+F^-} R{-}O{-}H + F{-}Si(CH_3)_2{-}C(CH_3)_3$$

Converting an alcohol to a silyl ether also makes it much more volatile. This increased volatility makes the alcohol (as a silyl ether) much more amenable to analysis by gas chromatography. Trimethylsilyl ethers are often used for this purpose. (The trimethylsilyl ether group is too labile to use as a protecting group in most reactions, however.)

11.12 Reactions of Ethers

Dialkyl ethers react with very few reagents other than acids. The only reactive sites that molecules of a dialkyl ether present to another reactive substance are the C—H bonds of the alkyl groups and the $-\ddot{O}-$ group of the ether linkage. Ethers resist attack by nucleophiles (why?) and by bases. This lack of reactivity coupled with the ability of ethers to solvate cations (by donating an electron pair from their oxygen atom) makes ethers especially useful as solvents for many reactions.

Ethers are like alkanes in that they undergo halogenation reactions (Chapter 10), but these reactions are of little synthetic importance. They also undergo slow autoxidation to form explosive peroxides (see Section 11.3D).

The oxygen of the ether linkage makes ethers basic. Ethers can react with proton donors to form **oxonium salts:**

$$CH_3CH_2\ddot{O}CH_2CH_3 + \mathbf{HBr} \rightleftharpoons CH_3CH_2-\overset{+}{\ddot{O}}(H)-CH_2CH_3\ \mathbf{Br^-}$$

An oxonium salt

Heating dialkyl ethers with very strong acids (HI, HBr, and H_2SO_4) causes them to undergo reactions in which the carbon–oxygen bond breaks. Diethyl ether, for example, reacts with hot concentrated hydrobromic acid to give two molecular equivalents of ethyl bromide:

$$CH_3CH_2OCH_2CH_3 + \mathbf{2\ HBr} \longrightarrow 2\ CH_3CH_2\mathbf{Br} + \mathbf{H_2O}$$

Cleavage of an ether

The mechanism for this reaction begins with formation of an oxonium cation. Then, an S_N2 reaction with a bromide ion acting as the nucleophile produces ethanol and ethyl bromide. Excess HBr reacts with the ethanol produced to form the second molar equivalent of ethyl bromide.

A Mechanism for the Reaction

Ether Cleavage by Strong Acids

Step 1 $$CH_3CH_2\ddot{O}CH_2CH_3 + H-\ddot{Br}: \rightleftharpoons CH_3CH_2\overset{+}{\ddot{O}}(H)-CH_2CH_3 + :\ddot{Br}:^- \longrightarrow CH_3CH_2\ddot{O}(H): + CH_3CH_2Br$$

Ethanol **Ethyl bromide**

In step 2, the ethanol (just formed) reacts with HBr (present in excess) to form a second molar equivalent of ethyl bromide.

Step 2 $$CH_3CH_2\ddot{O}H + H-\ddot{Br}: \rightleftharpoons :\ddot{Br}:^- + CH_3CH_2-\overset{+}{\ddot{O}}(H)-H \longrightarrow CH_3CH_2-\ddot{Br}: + :\ddot{O}(H)-H$$

Problem 11.18

When an ether is treated with *cold* concentrated HI, cleavage occurs as follows:

$$R-O-R + HI \longrightarrow ROH + RI$$

When mixed ethers are used, the alcohol and alkyl iodide that form depend on the nature of the alkyl groups. Explain the following observations: **(a)** When (R)-2-methoxybutane reacts, the products are methyl iodide and (R)-2-butanol. **(b)** When *tert*-butyl methyl ether reacts, the products are methanol and *tert*-butyl iodide.

11.13 Epoxides

Epoxides are cyclic ethers with three-membered rings. In IUPAC nomenclature epoxides are called **oxiranes.** The simplest epoxide has the common name ethylene oxide:

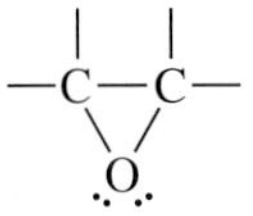

An epoxide

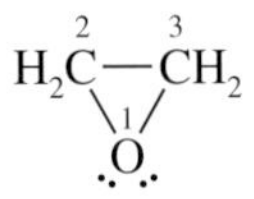

IUPAC name: oxirane
Common name: ethylene oxide

The most widely used method for synthesizing epoxides is the reaction of an alkene with an organic **peroxy acid** (sometimes called simply a **peracid**), a process that is called **epoxidation:**

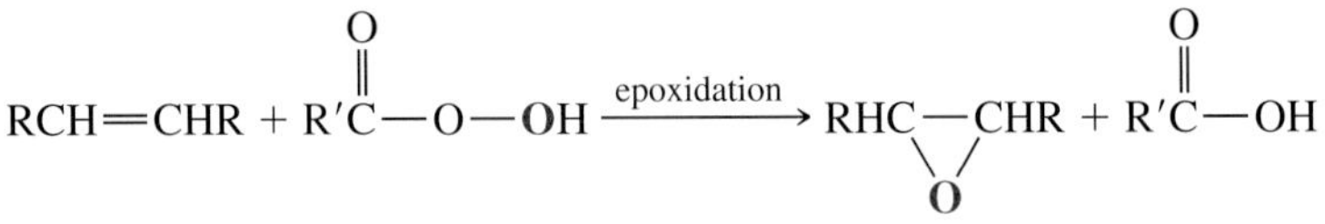

$$RCH{=}CHR + R'\overset{O}{\overset{\|}{C}}{-}O{-}OH \xrightarrow{\text{epoxidation}} RHC{-}CHR\ (\text{O bridge}) + R'\overset{O}{\overset{\|}{C}}{-}OH$$

An alkene **A peroxy acid** **An epoxide (or oxirane)**

In this reaction the peroxy acid transfers an oxygen atom to the alkene. The following mechanism has been proposed.

A Mechanism for the Reaction

Alkene Epoxidation

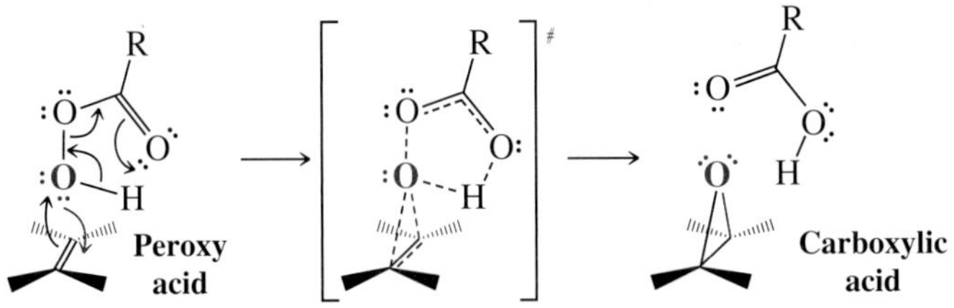

Alkene **Peroxy acid** → **Concerted transition state** → **Epoxide** **Carboxylic acid**

The peroxy acid transfers an oxygen atom to the alkene in a cyclic, single-step mechanism. The result is the syn addition of the oxygen to the alkene, with the formation of an epoxide and a carboxylic acid.

The addition of oxygen to the double bond in an epoxidation reaction is, of necessity, a **syn** addition. In order to form a three-membered ring, the oxygen atom must add to both carbon atoms of the double bond at the same face.

Some peroxy acids used in the past for preparing epoxides are unstable and, therefore, possibly unsafe. Because of its stability, the peroxy acid magnesium monoperoxyphthalate (MMPP) offers a good and safe alternative:

Magnesium monoperoxyphthalate (MMPP)

Cyclohexene, for example, reacts with magnesium monoperoxyphthalate in ethanol to give 1,2-epoxycyclohexane:

$$\xrightarrow[CH_3CH_2OH]{MMPP}$$

1,2-Epoxycyclohexane (cyclohexene oxide) (85%)

The Chemistry of...

The Sharpless Asymmetric Epoxidation

In 1980, K. B. Sharpless (then at the Massachusetts Institute of Technology, presently at The Scripps Research Institute) and co-workers reported a method that has since become one of the most valuable tools for chiral synthesis. The Sharpless asymmetric epoxidation is a method for converting allylic alcohols (Section 11.1) to chiral epoxy alcohols with very high enantioselectivity (i.e., with preference for one enantiomer rather than formation of a racemic mixture). In recognition of this and other work in asymmetric oxidation methods (see Section 8.16A), Sharpless received half of the 2001 Nobel Prize in Chemistry (the other half was awarded to W. S. Knowles and R. Noyori; see Section 7.13). The Sharpless asymmetric epoxidation involves treating the allylic alcohol with *tert*-butyl hydroperoxide, titanium(IV) tetraisopropoxide [Ti(O—*i*Pr)$_4$], and a specific stereoisomer of a tartrate ester. (The tartrate stereoisomer that is chosen depends on the specific enantiomer of the epoxide desired). The following is an example:

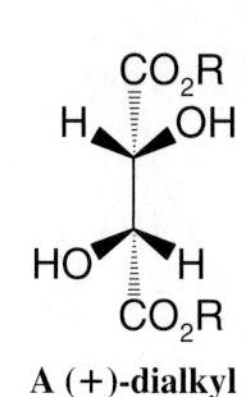

A (+)-dialkyl tartrate ester

$$\xrightarrow[CH_2Cl_2,\ -20°C,\ (+)\text{-diethyl tartrate}]{tert\text{-BuOOH, Ti(O—}i\text{Pr)}_4}$$

Geraniol → **77% yield (95% enantiomeric excess)**

The oxygen that is transferred to the allylic alcohol to form the epoxide is derived from *tert*-butyl hydroperoxide. The enantioselectivity of the reaction results from a titanium complex among the reagents that includes the enantiomerically pure tartrate ester as one of the ligands. The choice of whether to use the (+)- or (−)-tartrate ester for stereochemical control depends on which enantiomer of the epoxide is desired. [The (+)- and (−)-tartrates are either diethyl or diisopropyl esters.] The stereochemical preferences of the reaction have been well studied, such that it is possible to prepare either enantiomer of a chiral epoxide in high enantiomeric excess, simply by choosing the appropriate (+)- or (−)-tartrate stereoisomer as the chiral ligand:

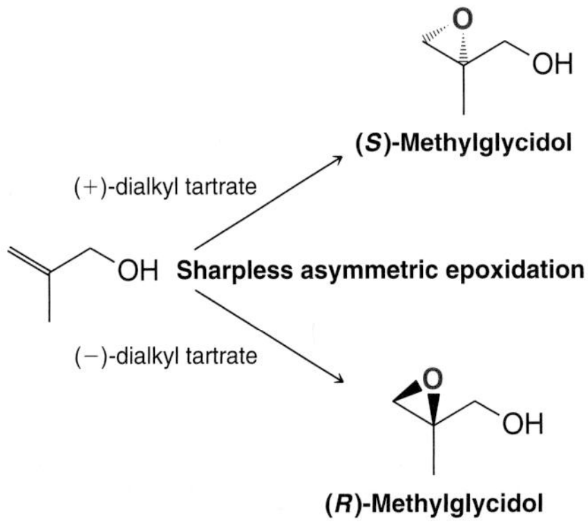

(+)-dialkyl tartrate

OH Sharpless asymmetric epoxidation

(−)-dialkyl tartrate

(S)-Methylglycidol

(R)-Methylglycidol

Compounds of this general structure are extremely useful and versatile synthons because combined in one molecule are an epoxide functional group (a highly reactive electrophilic site), an alcohol functional group (a potentially nucleophilic site), and at least one stereogenic center that is present in high enantiomeric purity. The synthetic utility of chiral epoxy alcohol synthons produced by the Sharpless asymmetric epoxidation has been demonstrated over and over again in enantioselective syntheses of many important compounds. Some examples include the synthesis of the polyether antibiotic X-206 by E. J. Corey (Harvard), the J. T. Baker commercial synthesis of the gypsy moth pheromone (7*R*,8*S*)-disparlure, and synthesis by K. C. Nicolaou (University of California San Diego and Scripps Research Institute) of zaragozic acid A (which is also called squalestatin S1 and has been shown to lower serum cholesterol levels in test animals by inhibition of squalene biosynthesis; see "The Chemistry of Cholesterol Biosynthesis," Chapter 8).

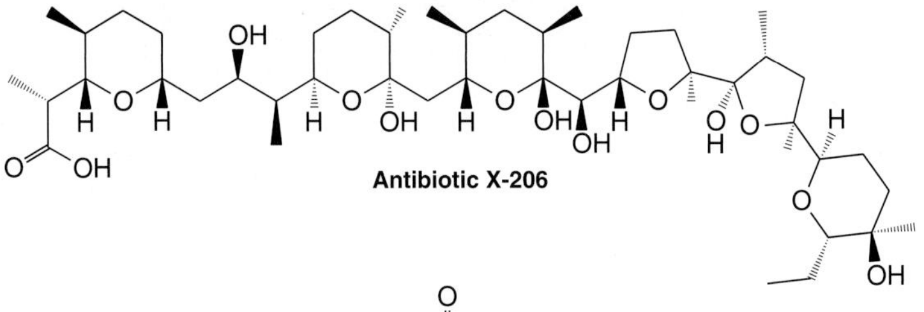

Antibiotic X-206

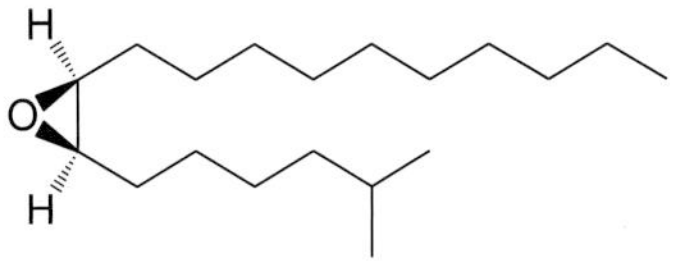

(7R,8S)-Disparlure

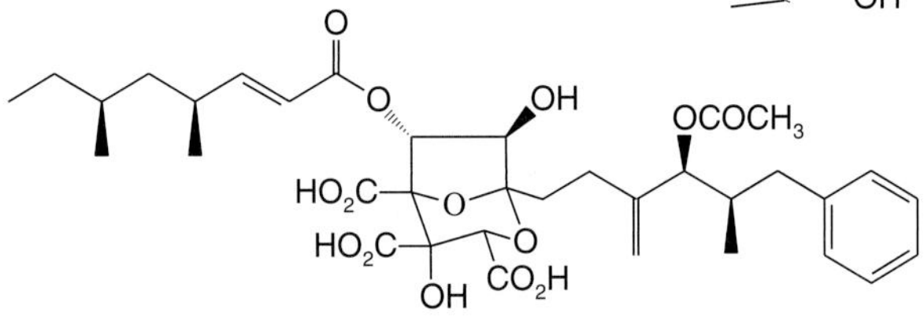

Zaragozic acid A (squalestatin S1)

The reaction of alkenes with peroxy acids takes place in a stereospecific way. *cis*-2-Butene, for example, yields only *cis*-2,3-dimethyloxirane, and *trans*-2-butene yields only the racemic *trans*-2,3-dimethyloxiranes:

$$\text{cis-2-Butene} + \text{RCOOH} \longrightarrow \text{cis-2,3-Dimethyloxirane}$$

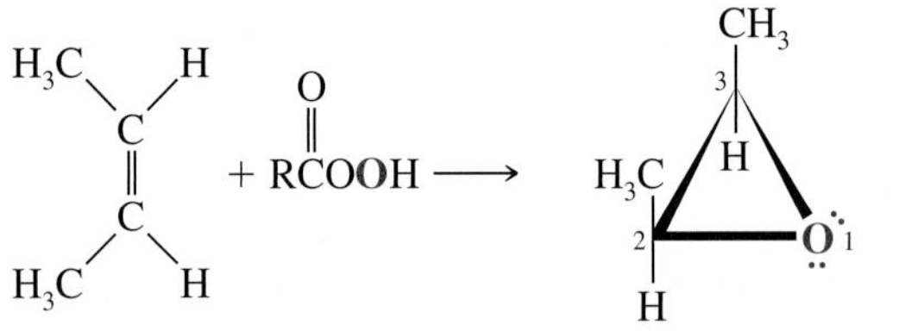

***cis*-2-Butene**

***cis*-2,3-Dimethyloxirane (a meso compound)**

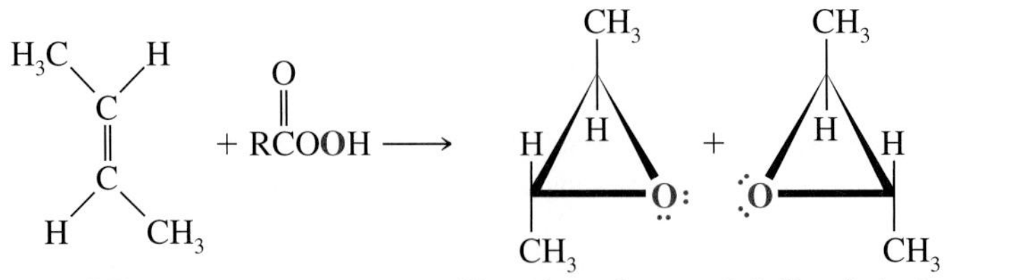

$$\text{trans-2-Butene} + \text{RCOOH} \longrightarrow \text{Enantiomeric trans-2,3-dimethyloxiranes}$$

***trans*-2-Butene**

Enantiomeric *trans*-2,3-dimethyloxiranes

In Special Topic C (Section C.3) we present a method for synthesizing epoxides from aldehydes and ketones.

11.14 Reactions of Epoxides

The highly strained three-membered ring in molecules of epoxides makes them much more reactive toward nucleophilic substitution than other ethers.

Acid catalysis assists epoxide ring opening by providing a better leaving group (an alcohol) at the carbon atom undergoing nucleophilic attack. This catalysis is especially important if the nucleophile is a weak nucleophile such as water or an alcohol. An example is the acid-catalyzed hydrolysis of an epoxide.

A Mechanism for the Reaction

Acid-Catalyzed Ring Opening of an Epoxide

Epoxide + $H-\ddot{O}^{+}-H$ (H) ⟶ Protonated epoxide + $:\ddot{O}-H$ (H)

Epoxide — Protonated epoxide

The acid reacts with the epoxide to produce a protonated epoxide.

Protonated epoxide + Weak nucleophile ($:\ddot{O}-H$, H) ⟶ Protonated 1,2-diol ⟶ (with $:\ddot{O}-H$, H) 1,2-Diol + $H-\ddot{O}^{+}-H$ (H)

Protonated epoxide — Weak nucleophile — Protonated 1,2-diol — 1,2-Diol

The protonated epoxide reacts with the weak nucleophile (water) to form a protonated 1,2-diol, which then transfers a proton to a molecule of water to form the 1,2-diol and a hydronium ion.

Epoxides can also undergo base-catalyzed ring opening. Such reactions do not occur with other ethers, but they are possible with epoxides (because of ring strain), provided that the attacking nucleophile is also a strong base such as an alkoxide ion or hydroxide ion.

A Mechanism for the Reaction

Base-Catalyzed Ring Opening of an Epoxide

$R-\ddot{O}:^{-}$ + Epoxide ⟶ $R\ddot{O}-C-C-\ddot{O}:^{-}$ ⟶ ($H-\ddot{O}R$) $R\ddot{O}-C-C-\ddot{O}H$ + $R\ddot{O}:^{-}$

Strong nucleophile — Epoxide — An alkoxide ion

A strong nucleophile such as an alkoxide ion or a hydroxide ion is able to open the strained epoxide ring in a direct S_N2 reaction.

Regioselectivity in the opening of epoxides

If the epoxide is unsymmetrical, in **base-catalyzed ring opening,** attack by the alkoxide ion occurs primarily ***at the less substituted carbon atom.*** For example, methyloxirane reacts with an alkoxide ion mainly at its primary carbon atom:

1° Carbon atom is less hindered.

$$CH_3CH_2\ddot{O}:^- + H_2C\overset{O}{-}CHCH_3 \longrightarrow CH_3CH_2OCH_2\underset{O^-}{CH}CH_3 \xrightarrow{CH_3CH_2OH}$$

Methyloxirane

$$\longrightarrow CH_3CH_2OCH_2\underset{OH}{CH}CH_3 + CH_3CH_2O^-$$

1-Ethoxy-2-propanol

This is just what we should expect: The reaction is, after all, an S_N2 reaction, and, as we learned earlier (Section 6.13A), primary substrates react more rapidly in S_N2 reactions because they are less sterically hindered.

In the **acid-catalyzed ring opening** of an unsymmetrical epoxide the nucleophile attacks primarily ***at the more substituted carbon atom.*** For example,

$$CH_3OH + CH_3-\overset{CH_3}{C}\overset{O}{-}CH_2 \xrightarrow{HA} CH_3-\overset{CH_3}{\underset{OCH_3}{C}}-CH_2OH$$

The reason: Bonding in the protonated epoxide (see following reaction) is unsymmetrical, with the more highly substituted carbon atom bearing a considerable positive charge; the reaction is S_N1 like. The nucleophile, therefore, attacks this carbon atom even though it is more highly substituted:

This carbon resembles a 3° carbocation.

$$CH_3\ddot{O}H + CH_3-\overset{CH_3}{\overset{\delta+}{C}}\overset{O^{\delta+}H}{-}CH_2 \longrightarrow CH_3-\overset{CH_3}{\underset{^+OCH_3\ H}{C}}-CH_2OH$$

Protonated epoxide

The more highly substituted carbon atom bears a greater positive charge because it resembles a more stable tertiary carbocation. [Notice how this reaction (and its explanation) resembles that given for halohydrin formation from unsymmetrical alkenes in Section 8.14 and attack on mercurinium ions.]

PROBLEM 11.19

Propose structures for each of the following products derived from oxirane (ethylene oxide):

(a) Oxirane $\xrightarrow[CH_3OH]{HA}$ $C_3H_8O_2$ (an industrial solvent called Methyl Cellosolve)

(b) Oxirane $\xrightarrow[CH_3CH_2OH]{HA}$ $C_4H_{10}O_2$ (Ethyl Cellosolve)

(c) Oxirane $\xrightarrow[H_2O]{KI}$ C_2H_5IO

(d) Oxirane $\xrightarrow{NH_3}$ C_2H_7NO

(e) Oxirane $\xrightarrow[CH_3OH]{CH_3ONa}$ $C_3H_8O_2$

PROBLEM 11.20

Treating 2,2-dimethyloxirane, $H_2C\!-\!C(CH_3)_2$ (epoxide, bridged by O), with sodium methoxide in methanol gives primarily 1-methoxy-2-methyl-2-propanol. What factor accounts for this result?

PROBLEM 11.21

When sodium ethoxide reacts with 1-(chloromethyl)oxirane, labeled with ^{14}C as shown by the asterisk in **I**, the major product is an epoxide bearing the label as in **II**. Provide an explanation for this reaction:

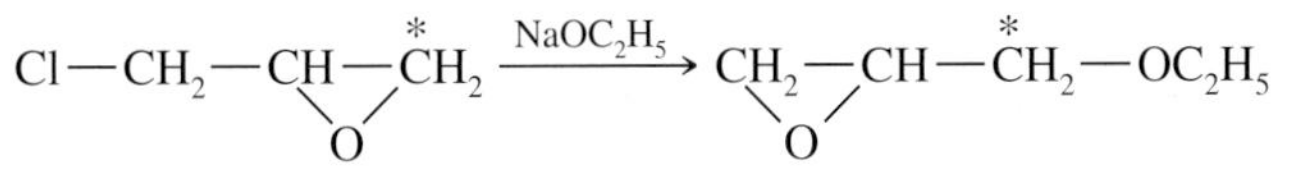

$$Cl-CH_2-CH-\overset{*}{C}H_2 \xrightarrow{NaOC_2H_5} CH_2-CH-\overset{*}{C}H_2-OC_2H_5$$

(in both structures, O bridges the two ring carbons)

I
1-(Chloromethyl)oxirane
(epichlorohydrin)

II

The Chemistry of...

Epoxides, Carcinogens, and Biological Oxidation

Certain molecules from the environment become carcinogenic by "activation" through metabolic processes that are normally involved in preparing them for excretion. This is the case with two of the most carcinogenic compounds known: dibenzo[*a,l*]pyrene, a polycyclic aromatic hydrocarbon, and aflatoxin B_1, a fungal metabolite. During the course of oxidative processing in the liver and intestines, these molecules undergo epoxidation by enzymes called P450 cytochromes. Their epoxide products, as you might expect, are exceptionally reactive electrophiles, and it is precisely because of this that they are carcinogenic. The dibenzo[*a,l*]pyrene and aflatoxin B_1 epoxides undergo very facile nucleophilic substitution reactions with DNA. Nucleophilic sites on DNA react to open the epoxide ring, causing alkylation of the DNA by formation of a covalent bond with the carcinogen. Modification of the DNA in this way causes onset of the disease state.

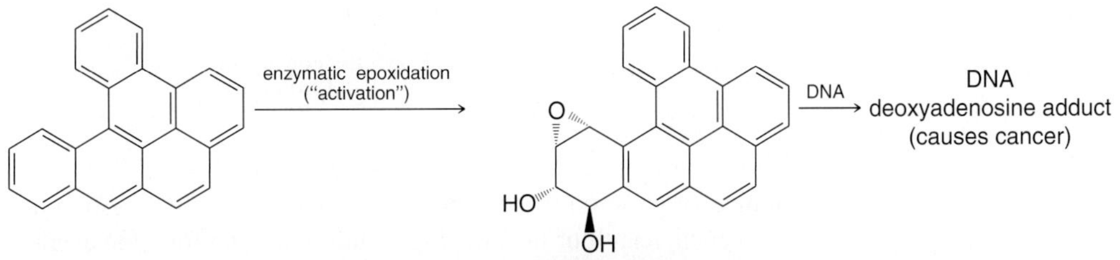

enzymatic epoxidation ("activation") →

DNA → DNA deoxyadenosine adduct (causes cancer)

HO OH O

Dibenzo[*a,l*]pyrene

Dibenzo[*a,l*]pyrene-11,12-diol-13,14-epoxide

The normal pathway toward excretion of foreign molecules like dibenzo[*a,l*]pyrene and aflatoxin B_1, however, also involves nucleophilic substitution reactions of their epoxides. One pathway involves opening of the epoxide ring by nucleophilic substitution with glutathione. Glutathione is a relatively polar molecule that has a strongly nucleophilic sulfhydryl group. After reaction of the sulfhydryl group with the epoxide, the newly formed covalent derivative, because it is substantially more polar than the original epoxide, is readily excreted through aqueous pathways.

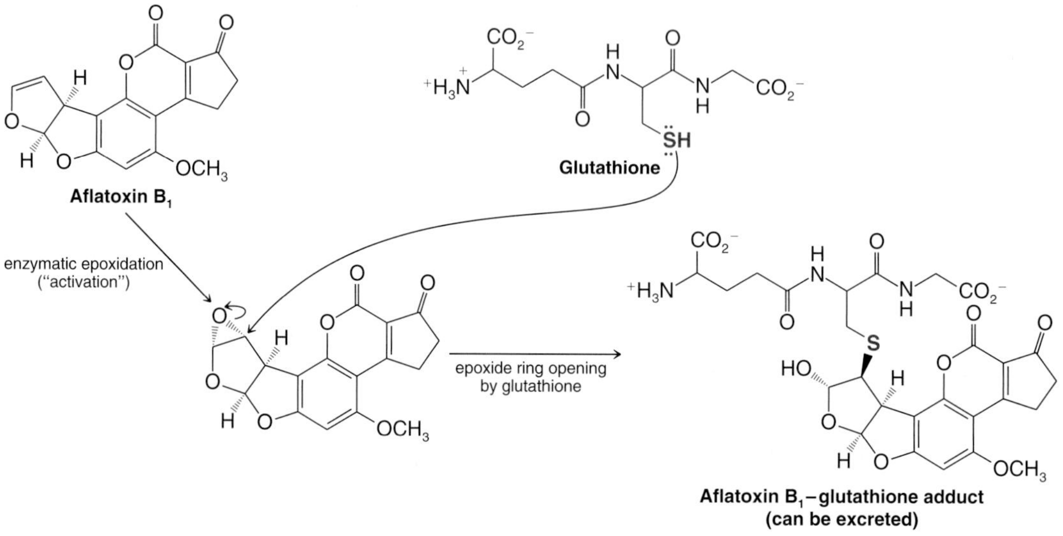

Aflatoxin B_1

Glutathione

enzymatic epoxidation ("activation")

epoxide ring opening by glutathione

Aflatoxin B_1–glutathione adduct (can be excreted)

11.14A Polyether Formation

Treating ethylene oxide with sodium methoxide (in the presence of a small amount of methanol) can result in the formation of a **polyether:**

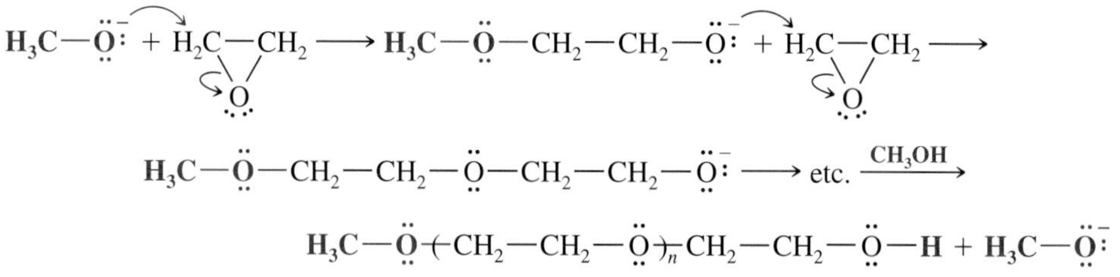

$$H_3C{-}\ddot{O}{:}^- + H_2C{-}CH_2 \text{ (epoxide, } \ddot{O}\text{)} \longrightarrow H_3C{-}\ddot{O}{-}CH_2{-}CH_2{-}\ddot{O}{:}^- + H_2C{-}CH_2 \text{ (epoxide, } \ddot{O}\text{)} \longrightarrow$$

$$H_3C{-}\ddot{O}{-}CH_2{-}CH_2{-}\ddot{O}{-}CH_2{-}CH_2{-}\ddot{O}{:}^- \longrightarrow \text{etc.} \xrightarrow{CH_3OH}$$

$$H_3C{-}\ddot{O}{+}CH_2{-}CH_2{-}\ddot{O}{+}_{n}CH_2{-}CH_2{-}\ddot{O}{-}H + H_3C{-}\ddot{O}{:}^-$$

Poly(ethylene glycol)
(a polyether)

This is an example of **anionic polymerization** (Special Topic A). The polymer chains continue to grow until methanol protonates the alkoxide group at the end of the chain. The average length of the growing chains and, therefore, the average molecular weight of the polymer can be controlled by the amount of methanol present. The physical properties of the polymer depend on its average molecular weight.

Polyethers have high water solubilities because of their ability to form multiple hydrogen bonds to water molecules. Marketed commercially as **carbowaxes,** these polymers have a variety of uses, ranging from use in gas chromatography columns to applications in cosmetics.

11.15 Anti 1,2-Dihydroxylation of Alkenes via Epoxides

Epoxidation of cyclopentene produces 1,2-epoxycyclopentane:

(1)

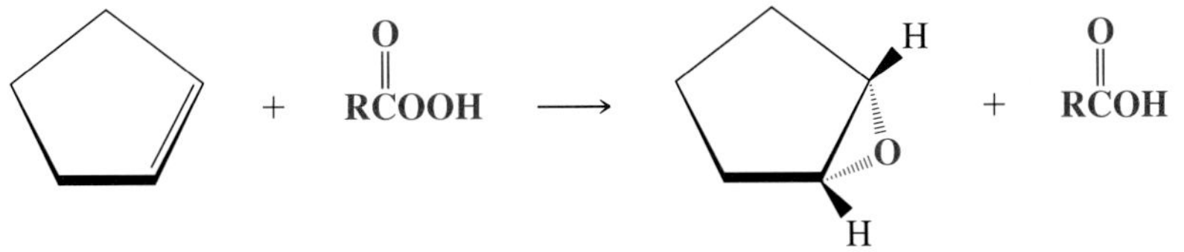

Cyclopentene **1,2-Epoxycyclopentane**

Acid-catalyzed hydrolysis of 1,2-epoxycyclopentane yields a trans diol, *trans*-1,2-cyclopentanediol. Water acting as a nucleophile attacks the protonated epoxide from the side opposite the epoxide group. The carbon atom being attacked undergoes an inversion of configuration. We show here only one carbon atom being attacked. Attack at the other carbon atom of this symmetrical system is equally likely and produces the enantiomeric form of *trans*-1,2-cyclopentanediol:

A synthetic method for anti 1,2-dihydroxylation

(2)

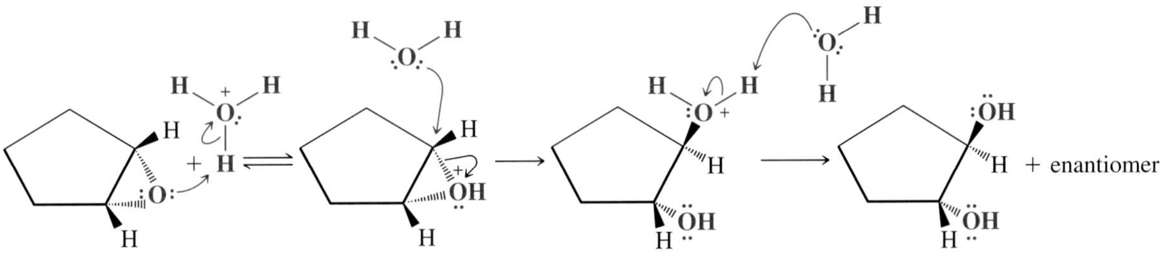

***trans*-1,2-Cyclopentanediol**

Epoxidation followed by acid-catalyzed hydrolysis gives us, therefore, a method for **anti 1,2-dihydroxylation** of a double bond (as opposed to syn 1,2-dihydroxylation, Section 8.16). The stereochemistry of this technique parallels closely the stereochemistry of the bromination of cyclopentene given earlier (Section 8.13).

PROBLEM 11.22

Outline a mechanism similar to the one just given that shows how the enantiomeric form of *trans*-1,2-cyclopentanediol is produced.

SOLVED PROBLEM

In Section 11.13 we showed the epoxidation of *cis*-2-butene to yield *cis*-2,3-dimethyloxirane and epoxidation of *trans*-2-butene to yield *trans*-2,3-dimethyloxirane. **(a)** Now consider acid-catalyzed hydrolysis of these two epoxides and show what product or products would result from each. **(b)** Are these reactions stereospecific?

ANSWER: The meso compound, *cis*-2,3-dimethyloxirane (Fig. 11.1), yields on hydrolysis (2*R*,3*R*)-2,3-butanediol and (2*S*,3*S*)-2,3-butanediol. These products are enantiomers. Since the attack by water at either carbon [path (a) or path (b) in Fig. 11.1] occurs at the same rate, the product is obtained in a racemic form.

When either of the *trans*-2,3-dimethyloxirane enantiomers undergoes acid-catalyzed hydrolysis, the only product that is obtained is the meso compound, (2*R*,3*S*)-2,3-butanediol. The hydrolysis of one enantiomer is shown in Fig. 11.2. (You might construct a similar diagram showing the hydrolysis of the other enantiomer to convince yourself that it, too, yields the same product.)

Since both steps in this method for the conversion of an alkene to a 1,2-diol (glycol) are stereospecific (i.e., both the epoxidation step and the acid-catalyzed hydrolysis), the net result is a stereospecific anti 1,2-dihydroxylation of the double bond (Fig. 11.3).

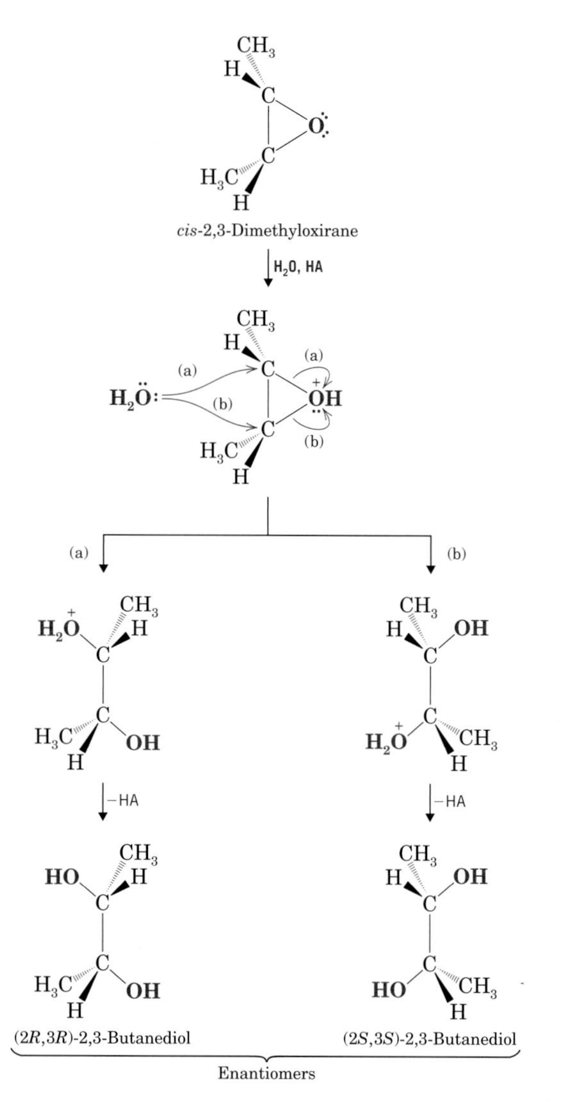

FIGURE 11.1 Acid-catalyzed hydrolysis of ***cis*-2,3-dimethyloxirane yields** (2*R*,3*R*)-2,3-butanediol by path *(a)* and (2*S*,3*S*)-2,3-butanediol by path *(b)*. (Use models to convince yourself.)

One *trans*-2,3-dimethyloxirane enantiomer

H_2O, HA

(a)

(b)

−HA

These molecules are identical: they both represent the meso compound (2*R*,3*S*)-2,3-butanediol.

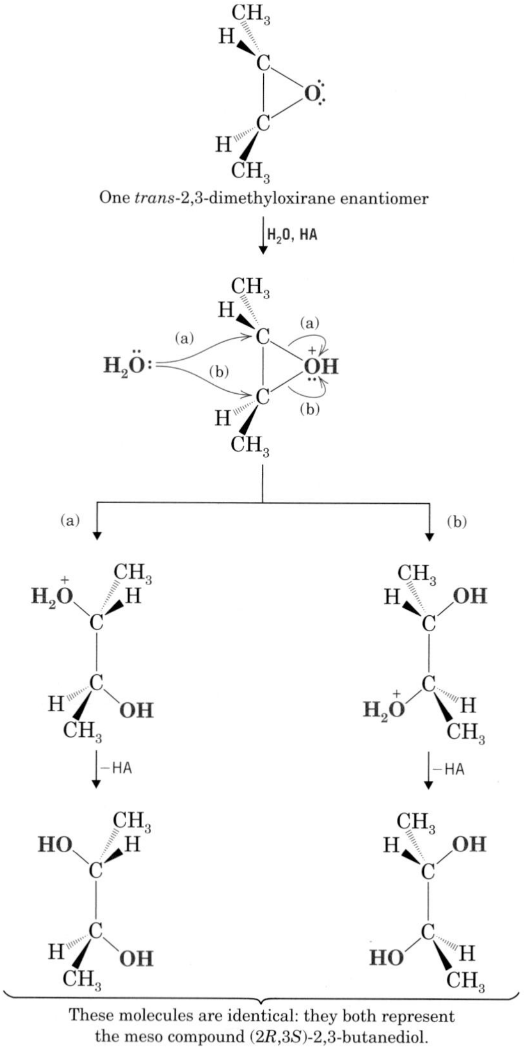

FIGURE 11.2 The acid-catalyzed hydrolysis of one *trans*-2,3-dimethyloxirane enantiomer produces the meso compound, (2*R*,3*S*)-2,3-butanediol, by path *(a)* or path *(b)*. Hydrolysis of the other enantiomer (or the racemic modification) would yield the same product. (You should use models to convince yourself that the two structures given for the products do represent the same compound.)

cis-2-Butene → (1) $RCOOH$ (2) HA, H_2O (anti 1,2-dihydroxylation) → Enantiomeric 2,3-butanediols

trans-2-Butene → (1) $RCOOH$ (2) HA, H_2O (anti 1,2-dihydroxylation) → *meso*-2,3-Butanediol

FIGURE 11.3 The overall result of epoxidation followed by acid-catalyzed hydrolysis is a stereospecific anti 1,2-dihydroxylation of the double bond. *cis*-2-Butene yields the enantiomeric 2,3-butanediols; *trans*-2-butene yields the meso compound.

The Chemistry of...

Environmentally Friendly Alkene Oxidation Methods

The effort to develop synthetic methods that are environmentally friendly is a very active area of chemistry research. The push to devise "green chemistry" procedures includes not only replacing the use of potentially hazardous or toxic reagents with ones that are more friendly to the environment but also developing catalytic procedures that use smaller quantities of potentially harmful reagents when other alternatives are not available. The catalytic syn 1,2-dihydroxylation methods that we described in Section 8.16 (including the Sharpless asymmetric dihydroxylation procedure) are environmentally friendly modifications of the original procedures because they require only a small amount of OsO_4 or other heavy metal oxidant.

Nature has provided hints for ways to carry out environmentally sound oxidations as well. The enzyme methane monooxygenase (MMO) uses iron to catalyze hydrogen peroxide oxidation of small hydrocarbons, yielding alcohols or epoxides, and this example has inspired development of new laboratory methods for alkene oxidation. A 1,2-dihydroxylation procedure developed by L. Que (University of Minnesota) yields a mixture of 1,2-diols and epoxides upon action of an iron catalyst and hydrogen peroxide on an alkene. (The ratio of diol to epoxide formed depends on the reaction conditions, and in the case of dihydroxylation, the procedure shows some enantioselectivity.) Another green reaction is the epoxidation method developed by E. Jacobsen (Harvard University). Jacobsen's procedure

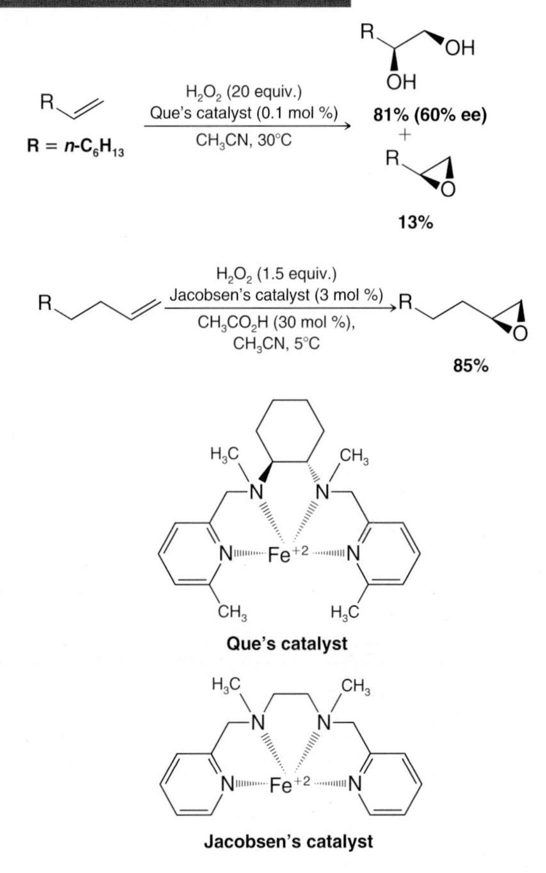

uses hydrogen peroxide and a similar iron catalyst to epoxidize alkenes (without complication of diol formation). Que's and Jacobsen's methods are environmentally friendly because their procedures employ catalysts containing a nontoxic metal, and a cheap, relatively safe oxidizing reagent is used that is converted to water in the course of the reaction.

The quest for more methods in green chemistry, with benign reagents and byproducts, catalytic cycles, and high yields, will no doubt drive further research by present and future chemists. In coming chapters we will see more examples of green chemistry in use or under development.

11.16 Crown Ethers: Nucleophilic Substitution Reactions in Relatively Nonpolar Aprotic Solvents by Phase-Transfer Catalysis

When we studied the effect of the solvent on nucleophilic substitution reactions in Section 6.13C, we found that S_N2 reactions take place much more rapidly in polar aprotic solvents such as dimethyl sulfoxide and *N,N*-dimethylformamide. The reason: ***In these polar aprotic solvents the nucleophile is only very slightly solvated and is, consequently, highly reactive.***

This increased reactivity of nucleophiles is a distinct advantage. Reactions that might have taken many hours or days are often over in a matter of minutes. There are, unfortunately, certain disadvantages that accompany the use of solvents such as DMSO and DMF. These solvents have very high boiling points, and as a result they are often difficult to remove after the reaction is over. Purification of these solvents is also time consuming, and they are expensive. At high temperatures certain of these polar aprotic solvents decompose.

In some ways the ideal solvent for an S_N2 reaction would be a *nonpolar* aprotic solvent such as a hydrocarbon or a relatively nonpolar chlorinated hydrocarbon. They have low boiling points, they are inexpensive, and they are relatively stable. The development of a procedure called **phase-transfer catalysis** has made the use of nonpolar solvents possible in reactions involving polar reagents.

With phase-transfer catalysis, we usually use two immiscible solutions that are in contact—often an aqueous phase containing an ionic reactant and an organic phase (benzene, $CHCl_3$, etc.) containing the organic substrate. Normally the reaction of two substances in separate phases like this is inhibited because of the inability of the reagents to come together. Adding a phase-transfer catalyst solves this problem by transferring the ionic reactant into the organic phase. And, again, because the reaction medium is aprotic, an S_N2 reaction occurs rapidly.

An example of phase-transfer catalysis is outlined in Fig. 11.4. The phase-transfer catalyst (Q^+X^-) is usually a quaternary ammonium halide ($R_4N^+X^-$) such as tetrabutylammonium halide, $(CH_3CH_2CH_2CH_2)_4N^+X^-$. The phase-transfer catalyst causes the transfer of the nucleophile (e.g., CN^-) as an ion pair [Q^+CN^-] into the organic phase. This transfer apparently takes place because the cation (Q^+) of the ion pair, with its four alkyl groups, resembles a hydrocarbon in spite of its positive charge. It is said to be **lipophilic**—it prefers a nonpolar environment to an aqueous one. In the organic phase the nucleophile of the ion pair (CN^-) reacts with the organic substrate RX. The cation (Q^+) [and anion (X^-)] then migrates back into the aqueous phase to complete the cycle. This process continues until all of the nucleophile or the organic substrate has reacted.

An example of a nucleophilic substitution reaction carried out with phase-transfer catalysis is the reaction of 1-chlorooctane (in decane) and sodium cyanide (in water). The reaction (at 105°C) is complete in less than 2 h and gives a 95% yield of the substitution product:

$$CH_3(CH_2)_6CH_2Cl \text{ (in decane)} \xrightarrow[\text{aqueous NaCN, 105°C}]{R_4N^+Br^-} \underset{\textbf{(95\%)}}{CH_3(CH_2)_6CH_2CN}$$

Many other nucleophilic substitution reactions have been carried out in a similar way.

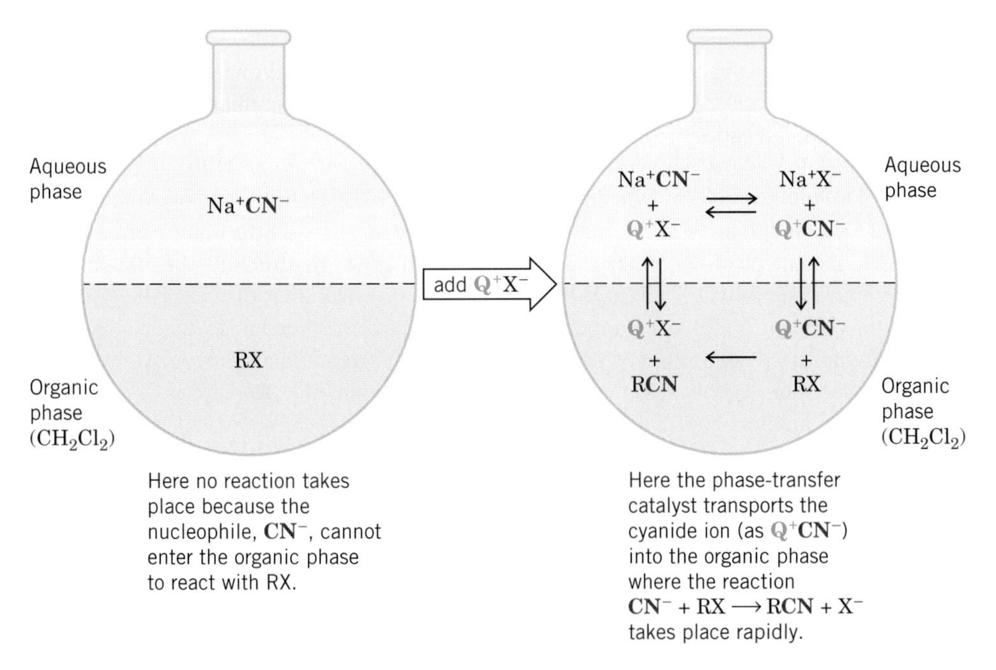

FIGURE 11.4 Phase-transfer catalysis of the S_N2 reaction between sodium cyanide and an alkyl halide.

Phase-transfer catalysis, however, is not limited to nucleophilic substitutions. Many other types of reactions are also amenable to phase-transfer catalysis. Oxidations of alkenes dissolved in benzene can be accomplished in excellent yield using potassium permanganate (in water) when a quaternary ammonium salt is present:

$$CH_3(CH_2)_5CH{=}CH_2 \text{ (in benzene)} \xrightarrow[\text{aqueous } KMnO_4,\ 35°C]{R_4N^+X^-} \underset{(99\%)}{CH_3(CH_2)_5CO_2H} + HCO_2H$$

Potassium permanganate can be transferred to benzene by quaternary ammonium salts for the purpose of chemical tests as well. The resulting "purple benzene" can be used as a test reagent for unsaturated compounds. As an unsaturated compound is added to the benzene solution of $KMnO_4$, the purple color disappears and the solution becomes brown (because of the presence of MnO_2), indicating a positive test for a double or triple bond.

PROBLEM 11.23

Outline a scheme such as the one shown in Fig. 11.4 showing how the reaction of $CH_3(CH_2)_6CH_2Cl$ with cyanide ion (just shown) takes place by phase-transfer catalysis. Be sure to indicate which ions are present in the organic phase, which are in the aqueous phase, and which pass from one phase to the other.

11.16A Crown Ethers

Compounds called **crown ethers** are also phase-transfer catalysts and are able to transport ionic compounds into an organic phase. Crown ethers are cyclic polymers of ethylene glycol such as 18-crown-6:

K^+

18-Crown-6

Crown ethers are named as *x*-crown-*y* where *x* is the total number of atoms in the ring and *y* is the number of oxygen atoms. The relationship between the crown ether and the ion that it transports is called a **host–guest relationship.** The crown ether acts as the **host,** and the coordinated cation is the **guest.**

The Nobel Prize in Chemistry in 1987 was awarded to Charles J. Pedersen (retired from the DuPont company), Donald J. Cram (University of California, Los Angeles, deceased 2001), and Jean-Marie Lehn (Louis Pasteur University, Strasbourg, France) for their development of crown ethers and other molecules "with structure specific interactions of high selectivity." Their contributions to our understanding of what is now called "molecular recognition" have implications for how enzymes recognize their substrates, how hormones cause their effects, how antibodies recognize antigens, how neurotransmitters propagate their signals, and many other aspects of biochemistry.

When crown ethers coordinate with a metal cation, they thereby convert the metal ion into a species with a hydrocarbon-like exterior. The crown ether 18-crown-6, for example, coordinates very effectively with potassium ions because the cavity size is correct and because the six oxygen atoms are ideally situated to donate their electron pairs to the central ion. This is also how the antibiotic monensin facilitates transport of sodium ions across the lipid bilayer of the cell membrane (see the chapter opening vignette and Section 11.16B).

Crown ethers render many salts soluble in nonpolar solvents. Salts such as KF, KCN, and CH_3CO_2K, for example, can be transferred into aprotic solvents by using catalytic amounts of 18-crown-6. In the organic phase the relatively unsolvated anions of these salts can carry out a nucleophilic substitution reaction on an organic substrate:

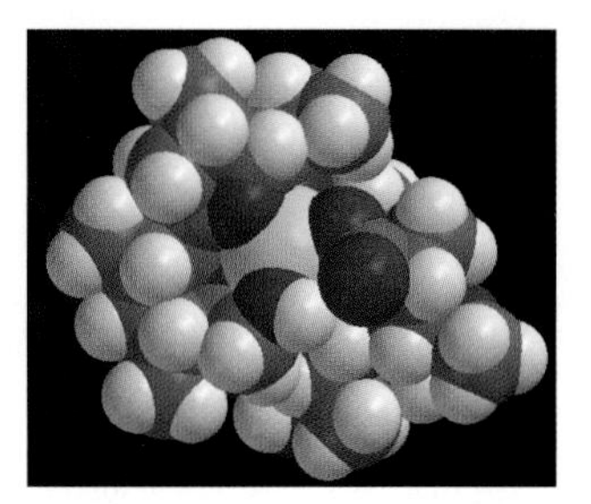

The ionophore antibiotic monensin complexed with a sodium cation.

$$K^+CN^- + RCH_2X \xrightarrow[\text{benzene}]{\text{18-crown-6}} RCH_2CN + K^+X^-$$

$$C_6H_5CH_2Cl + K^+F^- \xrightarrow[\text{acetonitrile}]{\text{18-crown-6}} C_6H_5CH_2F + K^+Cl^-$$

(100%)

Crown ethers can also be used as phase-transfer catalysts for many other types of reactions. The following reaction is one example of the use of a crown ether in an oxidation:

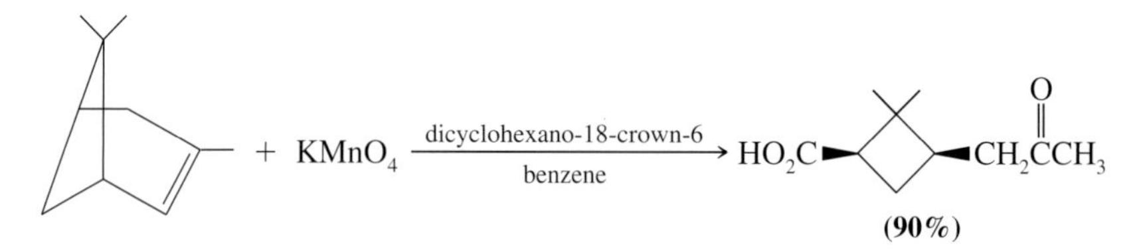

Dicyclohexano-18-crown-6 has the following structure:

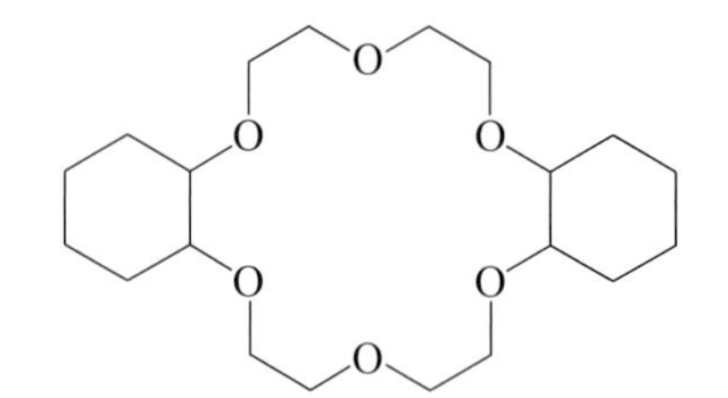

Dicyclohexano-18-crown-6

PROBLEM 11.24 Write structures for **(a)** 15-crown-5 and **(b)** 12-crown-4.

11.16B Transport Antibiotics and Crown Ethers

There are several antibiotics called ionophores (see the chapter opening vignette), most notably *nonactin* and *valinomycin,* that coordinate with metal cations in a manner similar to that of crown ethers. Normally, cells must maintain a gradient between the concentrations of sodium and potassium ions inside and outside the cell wall. Potassium ions are "pumped" in; sodium ions are pumped out. The cell membrane, in its interior, is like a hydrocarbon, because it consists in this region primarily of the hydrocarbon portions of lipids (Chapter 23). The transport of hydrated sodium and potassium ions through the cell membrane is slow, and this transport requires an expenditure of energy by the cell. Nonactin, for example, upsets the concentration gradient of these ions by coordinating more strongly with potassium ions than with sodium ions. Because the potassium ions are bound in the interior of the nonactin, this host–guest complex becomes hydrocarbon-like on its surface and passes readily through the interior of the membrane. The cell membrane thereby becomes permeable to potassium ions, and the essential concentration gradient is destroyed.

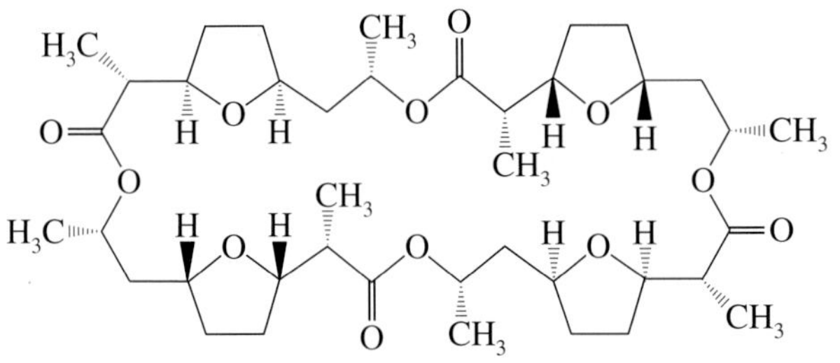

Nonactin

11.17 Summary of Reactions of Alkenes, Alcohols, and Ethers

We have studied reactions in this chapter and in Chapter 8 that can be extremely useful in designing syntheses. Most of these reactions involving alcohols and ethers are summarized in Fig. 11.5. We can use alcohols to make alkyl halides, sulfonate esters, ethers, and alkenes. We can oxidize alkenes to make epoxides, diols, aldehydes, ketones, and carboxylic acids (depending on the specific alkene and conditions). We can use alkenes to make alkanes, alcohols, and alkyl halides. If we have a terminal alkyne, such as could be made from an appropriate vicinal dihalide, we can use the alkynide anion derived from it to form carbon–carbon bonds by nucleophilic substitution. All together, we have a repertoire of reactions that can be used to directly or indirectly interconvert almost all of the functional groups we have studied so far. In Section 11.17A we summarize some reactions of alkenes.

Some tools for synthesis

11.17A Alkenes in Synthesis

Alkenes are an entry point to virtually all of the other functional groups that we have studied. For this reason, and because many of the reactions afford us some degree of control over the regiochemical and/or stereochemical form of the products, alkenes are versatile intermediates for synthesis. For example, if we want to **hydrate a double bond in a Markovnikov orientation,** we have two methods for doing so: (1) *oxymercuration–demercuration* (Section 8.6), and (2) *acid-catalyzed hydration* (Section 8.5). Of these methods oxymercuration–demercuration is the most useful in the laboratory because it is easy to carry out and *is not accompanied by rearrangements.*

If we want to **hydrate a double bond in an anti-Markovnikov orientation,** we can use *hydroboration–oxidation* (Section 8.7). With hydroboration–oxidation we can also achieve a *syn addition of the H— and —OH groups.* Remember, too, **the boron group**

of an organoborane can be replaced by hydrogen, deuterium, or tritium (Section 8.11), and that hydroboration, itself, involves a *syn addition of H— and —B—*.

If we want to **add HX to a double bond in a Markovnikov sense** (Section 8.2), we treat the alkene with HF, HCl, HBr, or HI.

If we want to **add HBr in an anti-Markovnikov orientation** (Section 10.9), we treat the alkene with HBr *and a peroxide.* (The other hydrogen halides do not undergo anti-Markovnikov addition when peroxides are present.)

We can **add bromine or chlorine to a double bond** (Section 8.12), and the addition is an *anti addition* (Section 8.13). We can also **add X— and —OH** to a double bond (i.e., synthesize a halohydrin) by carrying out the bromination or chlorination in water (Section 8.14). This addition, too, is an *anti addition.*

If we want to carry out a **syn 1,2-dihydroxylation of a double bond,** we can use either $KMnO_4$ in cold, dilute, and basic solution or OsO_4 followed by $NaHSO_3$ (Section 8.16). Of these two methods, the latter is preferable because of the tendency of $KMnO_4$ to overoxidize the alkene and cause cleavage at the double bond.

Anti 1,2-dihydroxylation of a double bond can be achieved by converting the alkene to an *epoxide* and then carrying out an acid-catalyzed hydrolysis (Section 11.15).

Equations for most of these reactions are given in the Synthetic Connections reviews for Chapters 7 and 8 and this chapter.

Key Terms and Concepts

Crown ethers	Section 11.16A
1,2-Dihydroxylation of alkenes	Sections 8.16, 11.15
Host–guest relationship	Section 11.16A
Hydration of alkenes	Sections 8.5–8.7, 11.4
Hydroboration–oxidation	Sections 8.7, 11.4
Oxiranes (epoxides)	Sections 11.13, 11.14
Oxonium salts	Section 11.12
Oxymercuration–demercuration	Sections 8.6, 11.4
Phase-transfer catalysis	Section 11.16
Protecting groups	Sections 11.11C, 11.11D
Sulfonate esters	Section 11.10
Williamson synthesis	Section 11.11B

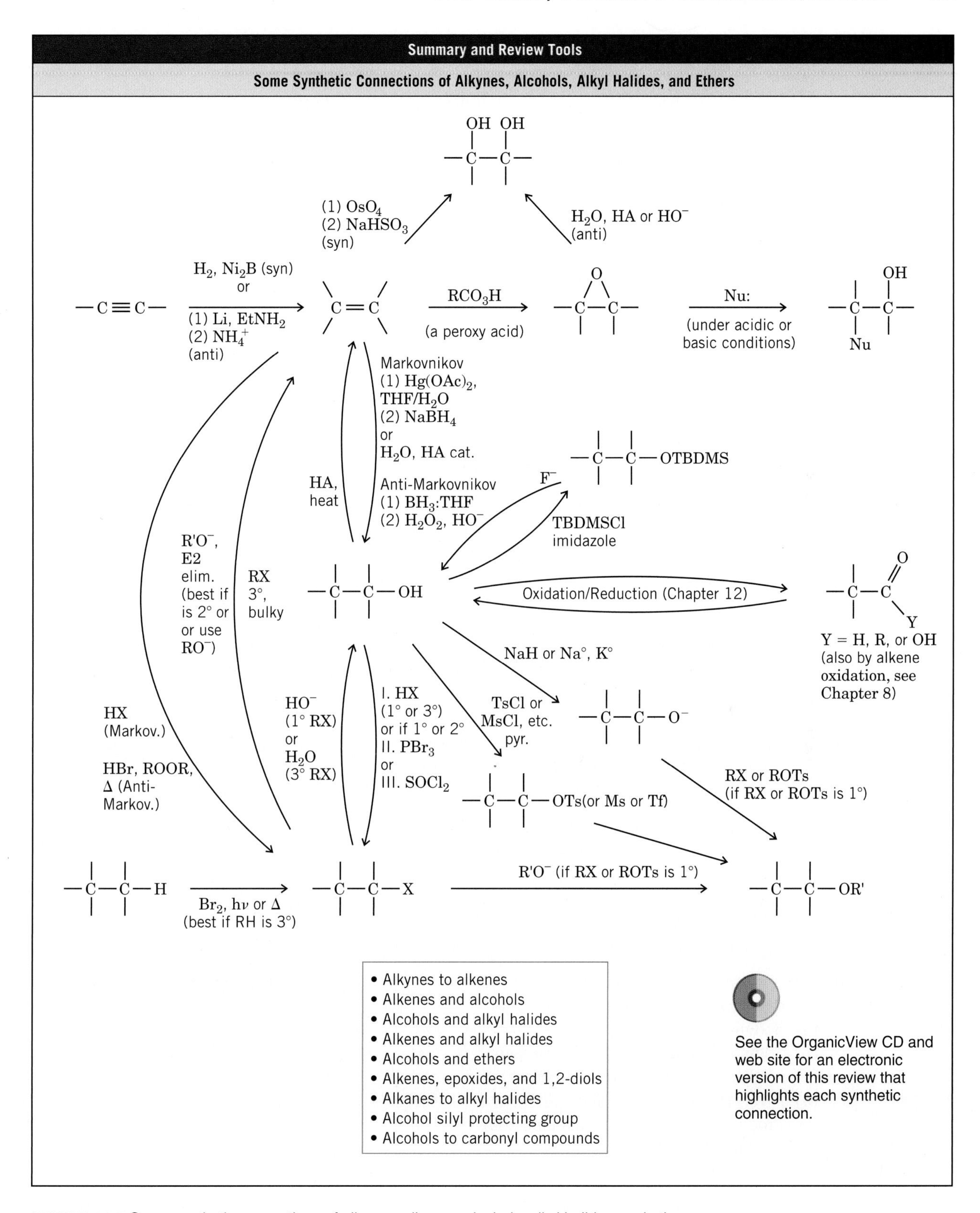

FIGURE 11.5 Some synthetic connections of alkynes, alkenes, alcohols, alkyl halides, and ethers.

Additional Problems

11.25 Give an IUPAC substitutive name for each of the following alcohols:

(a) $(CH_3)_3CCH_2CH_2OH$

(b) $CH_2{=}CHCH_2CH(CH_3)OH$

(c) $HOCH_2CH(CH_3)CH_2CH_2OH$

(d) $C_6H_5CH_2CH_2OH$

(e) OH, CH_3

(f)

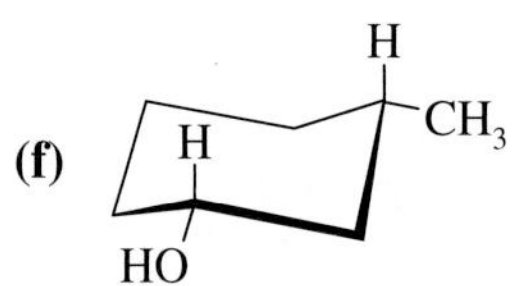

11.26 Write structural formulas for each of the following:

(a) (*Z*)-But-2-en-1-ol
(b) (*R*)-Butane-1,2,4-triol
(c) (1*R*,2*R*)-Cyclopentane-1,2-diol
(d) 1-Ethylcyclobutanol
(e) 2-Chlorohex-3-yn-1-ol
(f) Tetrahydrofuran
(g) 2-Ethoxypentane
(h) Ethyl phenyl ether
(i) Diisopropyl ether
(j) 2-Ethoxyethanol

11.27 Starting with each of the following, outline a practical synthesis of 1-butanol:

(a) 1-Butene
(b) 1-Chlorobutane
(c) 2-Chlorobutane
(d) 1-Butyne

11.28 Show how you might prepare 2-bromobutane from

(a) 2-Butanol, $CH_3CH_2CHOHCH_3$
(b) 1-Butanol, $CH_3CH_2CH_2CH_2OH$
(c) 1-Butene
(d) 1-Butyne

11.29 Show how you might carry out the following transformations:

(a) Cyclohexanol ⟶ chlorocyclohexane
(b) Cyclohexene ⟶ chlorocyclohexane
(c) 1-Methylcyclohexene ⟶ 1-bromo-1-methylcyclohexane
(d) 1-Methylcyclohexene ⟶ *trans*-2-methylcyclohexanol
(e) 1-Bromo-1-methylcyclohexane ⟶ cyclohexylmethanol

11.30 Give the structures and acceptable names for the compounds that would be formed when 1-butanol is treated with each of the following reagents:

(a) Sodium hydride
(b) Sodium hydride, then 1-bromopropane
(c) Methanesulfonyl chloride and base
(d) *p*-Toluenesulfonyl chloride
(e) Product of (c), then sodium methoxide
(f) Product of (d), then KI

Note: Problems marked with an asterisk are "challenge problems."

(g) Phosphorus trichloride
(h) Thionyl chloride
(i) Sulfuric acid at 140°C
(j) Refluxing concentrated HBr
(k) *tert*-Butylchlorodimethylsilane
(l) Product of (k), then fluoride ion

11.31 Give the structures and names for the compounds that would be formed when 2-butanol is treated with each of the reagents in Problem 11.30.

11.32 What compounds would you expect to be formed when each of the following ethers is refluxed with excess concentrated hydrobromic acid?

(a) Ethyl methyl ether
(b) *tert*-Butyl ethyl ether
(c) Tetrahydrofuran
(d) 1,4-Dioxane

11.33 Write a mechanism that accounts for the following reaction:

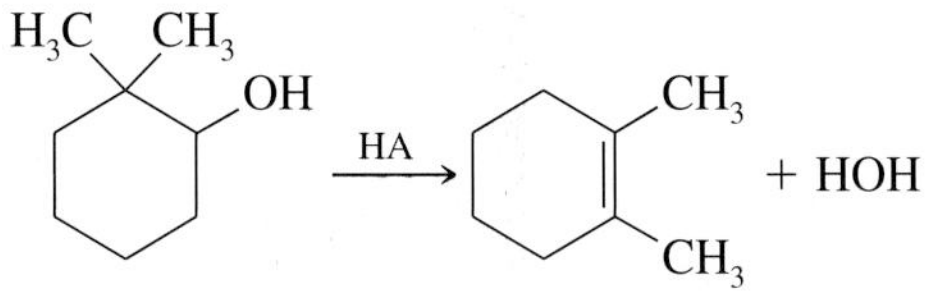

$$\text{2,2-dimethylcyclohexanol } (H_3C, CH_3, OH) \xrightarrow{HA} \text{1,2-dimethylcyclohexene } (CH_3, CH_3) + HOH$$

11.34 Show how you would utilize the hydroboration–oxidation procedure to prepare each of the following alcohols:

(a) 3,3-Dimethyl-1-butanol
(b) 1-Hexanol
(c) 2-Phenylethanol
(d) *trans*-2-Methylcyclopentanol

11.35 Starting with isobutane, show how each of the following could be synthesized. (You need not repeat the synthesis of a compound prepared in an earlier part of this problem.)

(a) *tert*-Butyl bromide
(b) 2-Methylpropene
(c) Isobutyl bromide
(d) Isobutyl iodide
(e) Isobutyl alcohol (two ways)
(f) *tert*-Butyl bromide
(g) Isobutyl methyl ether
(h) $CH_3CH(CH_3)CH_2OC(=O)CH_3$
(i) $CH_3CH(CH_3)CH_2CN$
(j) $CH_3CH(CH_3)CH_2SCH_3$ (two ways)
(k) $CH_3C(CH_3)(Br)CH_2CBr_3$

11.36 Vicinal halo alcohols (halohydrins) can be synthesized by treating epoxides with HX. **(a)** Show how you would use this method to synthesize 2-chlorocyclopentanol from cyclopentene. **(b)** Would you expect the product to be *cis*-2-chlorocyclopentanol or *trans*-2-chlorocyclopentanol; that is, would you expect a net syn addition or a net anti addition of —Cl and —OH? Explain.

11.37 Outlined below is a synthesis of the gypsy moth sex attractant **E** (a type of pheromone, see Section 4.16). Give the structures of **E** and the intermediates **A–D** in the synthesis.

$$HC{\equiv}CNa \xrightarrow[\text{liq. }NH_3]{\text{1-bromo-5-methylhexane}} \mathbf{A}\ (C_9H_{16}) \xrightarrow[\text{liq. }NH_3]{NaNH_2} \mathbf{B}\ (C_9H_{15}Na)$$

$$\xrightarrow{\text{1-bromodecane}} \mathbf{C}\ (C_{19}H_{36}) \xrightarrow[Ni_2B\ (P\text{-}2)]{H_2} \mathbf{D}\ (C_{19}H_{38}) \xrightarrow{C_6H_5CO_3H} \mathbf{E}\ (C_{19}H_{38}O)$$

11.38 Starting with 2-methylpropene (isobutylene) and using any other needed reagents, outline a synthesis of each of the following:

(a) $(CH_3)_2CHCH_2OH$
(b) $(CH_3)_2CHCH_2T$
(c) $(CH_3)_2CDCH_2T$
(d) $(CH_3)_2CHCH_2OCH_2CH_3$

11.39 Show you would use oxymercuration–demercuration to prepare each of the following alcohols from the appropriate alkene:

(a) 2-Pentanol
(b) 1-Cyclopentylethanol
(c) 3-Methyl-3-pentanol
(d) 1-Ethylcyclopentanol

11.40 Occasionally, alcohols (especially secondary and tertiary ones) are named as derivatives of methanol, the core structures, $-\overset{|}{\underset{|}{C}}-OH$, being designated by "carbinol." Assign both derived and IUPAC names to each of the following structures:

(a) $(CH_3CH_2)_3COH$

(b) 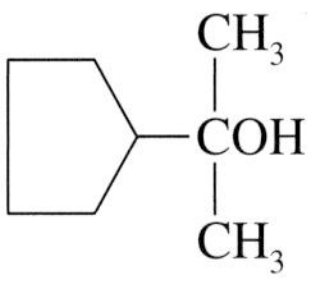 cyclopentyl–C(CH₃)₂OH [cyclopentane ring bonded to a carbon bearing CH₃ above, CH₃ below, and OH: $C_5H_9C(CH_3)_2OH$]

(c) $C_6H_5\underset{\underset{OH}{|}}{C}HC_6H_5$

11.41 Compounds of the type $R\overset{\overset{OH}{|}}{\underset{\underset{X}{|}}{C}}R'$ are unstable and cannot be isolated. Propose a mechanistic explanation for why this is so.

11.42 While simple alcohols yield alkenes on reaction with dehydrating acids, diols form carbonyl compounds. Rationalize mechanistically the outcome of the following reaction:

$$(CH_3)_2\underset{\underset{HO}{|}}{C}-\underset{\underset{OH}{|}}{C}(CH_3)_2 \xrightarrow{HA} (CH_3)_3C\overset{\overset{O}{\|}}{C}CH_3$$

11.43 When the bicyclic alkene **I**, a *trans*-decalin derivative, reacts with a peroxy acid, **II** is the major product. What factor favors the formation of **II** in preference to **III**? (You may find it helpful to build a hand-held molecular model.)

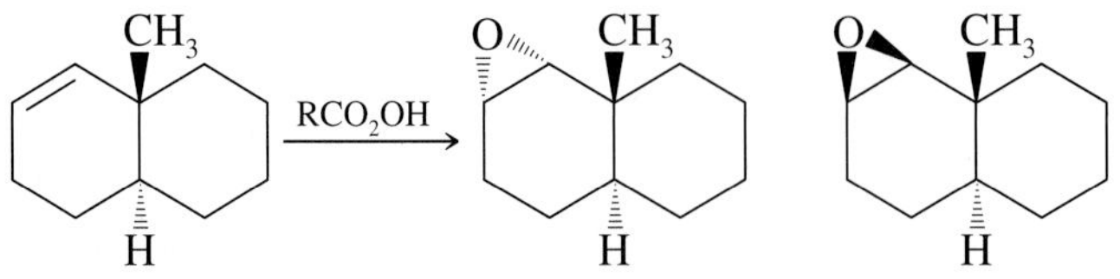

$$\mathbf{I} \xrightarrow{RCO_2OH} \mathbf{II}$$

(Structures: **I**, trans-decalin alkene with angular CH_3 and H; **II**, epoxide with O hashed (trans to CH_3); **III**, epoxide with O wedged (cis to CH_3).)

I **II** **III**

11.44 Both glycerol and epichlorohydrin, $ClCH_2\overset{O}{CH-CH_2}$ (epoxide), are synthesized from propene. Write equations for their preparation.

11.45 In comparing ethylene glycol ($HOCH_2CH_2OH$) to butane, explain why the gauche conformer of ethylene glycol would be expected to contribute more to the complete ensemble of ethylene glycol conformers than would the gauche conformer of butane to its ensemble of conformers.

***11.46** Give stereochemical formulas for each product **A–L** and answer the questions given in parts (b) and (g):

(a) 1-Methylcyclobutene $\xrightarrow[\text{(2) } H_2O_2,\ OH^-]{\text{(1) } BH_3\text{:THF}}$ **A** ($C_5H_{10}O$) $\xrightarrow[OH^-]{TsCl}$ **B** ($C_{12}H_{16}SO_3$) $\xrightarrow{OH^-}$ **C** ($C_5H_{10}O$)

(b) What is the stereoisomeric relationship between **A** and **C**?

(c) **B** ($C_{12}H_{16}SO_3$) $\xrightarrow{I^-}$ **D** (C_5H_9I)

(d) *trans*-4-Methylcyclohexanol $\xrightarrow[OH^-]{MsCl}$ **E** ($C_8H_{16}SO_3$) $\xrightarrow{HC\equiv CNa}$ **F** (C_9H_{14})

(e) (*R*)-2-Butanol $\xrightarrow{NaH}$ [**H** (C_4H_9ONa)] $\xrightarrow{CH_3I}$ **J** ($C_5H_{12}O$)

(f) (*R*)-2-Butanol $\xrightarrow{MsCl}$ **K** ($C_5H_{12}SO_3$) $\xrightarrow{CH_3ONa}$ **L** ($C_5H_{12}O$)

(g) What is the stereoisomeric relationship between **J** and **L**?

***11.47** When the 3-bromo-2-butanol with the stereochemical structure **A** is treated with concentrated HBr, it yields *meso*-2,3-dibromobutane; a similar reaction of the 3-bromo-2-butanol **B** yields (±)-2,3-dibromobutane. This classic experiment performed in 1939 by S. Winstein and H. J. Lucas was the starting point for a series of investigations of what are called *neighboring group effects.* Propose mechanisms that will account for the stereochemistry of these reactions.

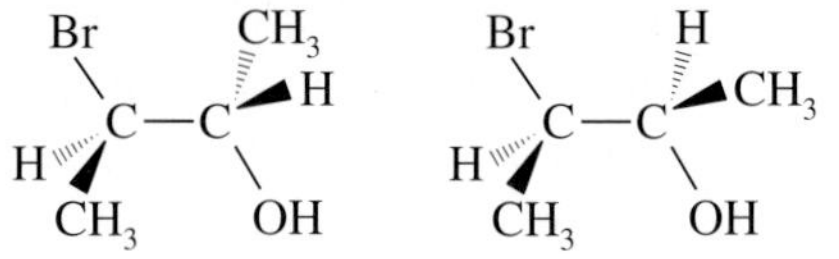

Br, H, CH_3 — C—C — CH_3, H, OH
A

Br, H, CH_3 — C—C — H, CH_3, OH
B

***11.48** Reaction of an alcohol with thionyl chloride in the presence of a tertiary amine (e.g., pyridine) affords replacement of the OH group by Cl *with inversion of configuration* (Section 11.19). However, if the amine is omitted, the result is usually replacement with retention of configuration. The same chlorosulfite intermediate is involved in both cases. Suggest a mechanism by which this intermediate can give the chloro product without inversion.

11.49** Draw the stereoisomers that are possible for the compound 1,2,3-cyclopentanetriol. Label their stereogenic carbons and say which are enantiomers and which are diastereomers. [Some of the isomers contain a "pseudoasymmetric center," one that has two possible configurations, each affording a different stereoisomer, each of which is identical to its mirror image. Such stereoisomers can only be distinguished by the order of attachment of ***R versus ***S*** groups at the pseudoasymmetric center. Of these the ***R*** group is given higher priority than the ***S***, and this permits assignment of configuration as ***r*** or ***s***, lowercase letters being used to designate the pseudoasymmetry.]

***11.50** Dimethyldioxirane (DMDO), whose structure is shown below, is another reagent commonly used for alkene epoxidation. Write a mechanism for the epoxidation of (*Z*)-2-butene by DMDO, including a possible transition state structure. What is the byproduct of a DMDO epoxidation?

Dimethyldioxirane (DMDO)

*11.51  Two configurations can actually be envisioned for the transition state in the DMDO epoxidation of (*Z*)-2-butene, based on analogy with geometric possibilities fitting within the general outline for the transition state in a peroxycarboxylic acid epoxidation of (*Z*)-2-butene. Draw these geometries for the DMDO epoxidation of (*Z*)-2-butene. Then, open the molecular models on the CD for these two possible transition state geometries in the DMDO epoxidation of (*Z*)-2-butene and speculate as to which transition state would be lower in energy.

LEARNING GROUP PROBLEMS

1. Devise two syntheses for *meso*-2,3-butanediol starting with acetylene (ethyne) and methane. Your two pathways should take different approaches during the course of the reactions for controlling the origin of the stereochemistry required in the product.
2. **(a)** Write as many chemically reasonable syntheses as you can think of for ethyl 2-methylpropyl ether (ethyl isobutyl ether). Be sure that at some point in one or more of your syntheses you utilize the following reagents (not all in the same synthesis, however): PBr_3, $SOCl_2$, *p*-toluenesulfonyl chloride (tosyl chloride), NaH, ethanol, 2-methyl-1-propanol (isobutyl alcohol), concentrated H_2SO_4, $Hg(OAc)_2$, ethene (ethylene).
   (b) Evaluate the relative merits of your syntheses on the basis of selectivity and efficiency. [Decide which ones could be argued to be the "best" syntheses and which might be "poorer" syntheses.]
3. Synthesize the compound shown below from methylcyclopentane and 2-methylpropane using those compounds as the source of the carbon atoms and any other reagents necessary. Synthetic tools you may need to use could include Markovnikov or anti-Markovnikov hydration, Markovnikov or anti-Markovnikov hydrobromination, radical halogenation, elimination, and nucleophilic substitution reactions.

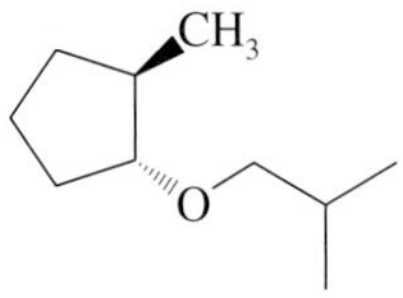

CHAPTER 12

Alcohols from Carbonyl Compounds. Oxidation–Reduction and Organometallic Compounds

The Two Aspects of the Coenzyme NADH

The role of many of the vitamins in our diet is to become coenzymes for enzymatic reactions. Coenzymes are molecules that are part of the organic machinery used by some enzymes to catalyze reactions. The vitamins niacin (nicotinic acid) and its amide niacinamide (nicotinamide, shown by the photo of soybeans above) are precursors to the coenzyme nicotinamide adenine dinucleotide. Soybeans are one dietary source of niacin:

Niacin (Nicotinic acid) **NAD^+** 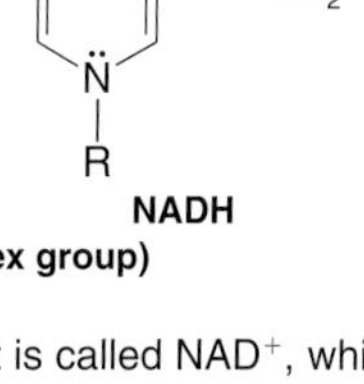 **NADH**
(R is a complex group)

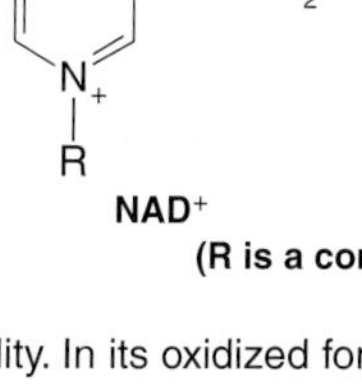

This coenzyme has a split personality. In its oxidized form it is called NAD^+, while in its reduced form it is known as NADH. In glycolysis, the citric acid cycle, and many other biochemical pathways, NAD^+ serves as an oxidizing agent. On the other hand, in the electron transport chain and other metabolic processes, its alter ego NADH is a reducing agent that acts as an electron donor and frequently as a biochemical source of hydride ("H⁻"). We shall learn in this chapter about laboratory reagents that are used to carry out oxidation and reduction reactions like those of NAD^+ and NADH. In "The Chemistry of. . . Alcohol Dehydrogenase" we look more closely at how nature uses NADH and NAD^+ to interconvert ethanol and acetaldehyde (ethanal) by this enzyme.

12.1 Introduction

Carbonyl compounds are a broad group of compounds that includes aldehydes, ketones, carboxylic acids, and esters:

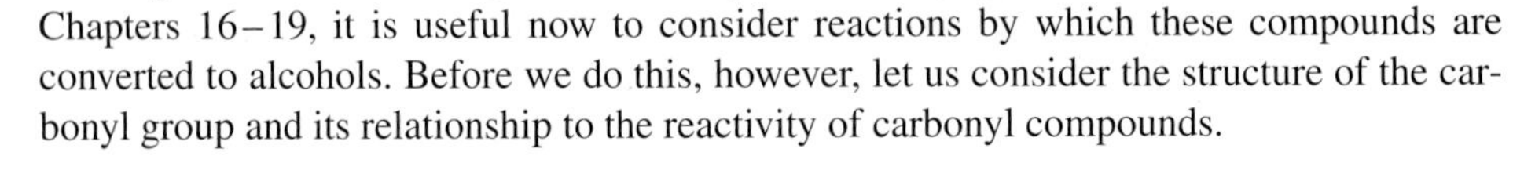

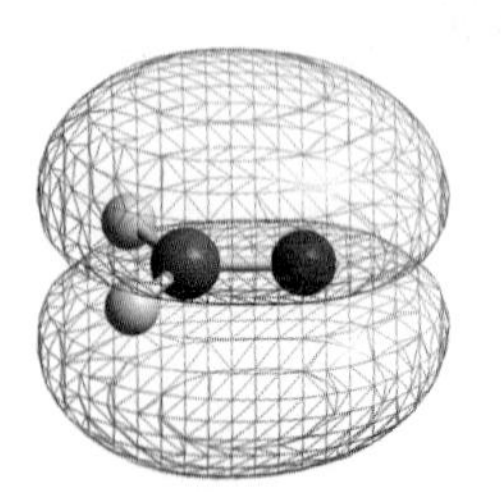

The π bonding molecular orbital of formaldehyde (HCHO). The electron pair of the π bond occupies both lobes.

Carbonyl group

Although we shall not study the chemistry of these compounds in detail until we reach Chapters 16–19, it is useful now to consider reactions by which these compounds are converted to alcohols. Before we do this, however, let us consider the structure of the carbonyl group and its relationship to the reactivity of carbonyl compounds.

12.1A Structure of the Carbonyl Group

The carbonyl carbon atom is sp^2 hybridized; thus it and the three atoms attached to it lie in the same plane. The bond angles between the three attached atoms are what we would expect of a trigonal planar structure; they are approximately 120°:

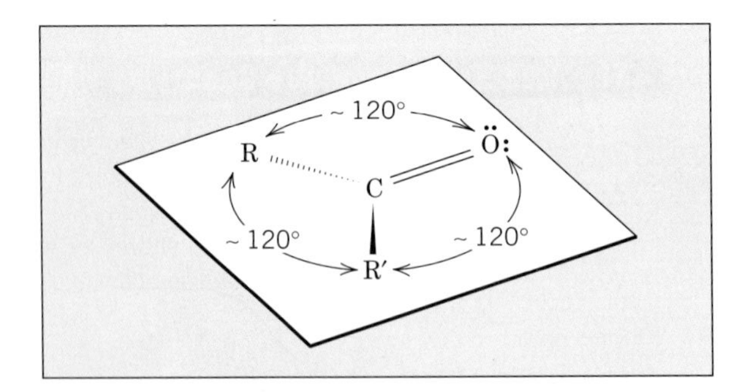

The carbon–oxygen double bond consists of two electrons in a σ bond and two electrons in a π bond. The π bond is formed by overlap of the carbon p orbital with a p orbital from the oxygen atom. The electron pair in the π bond occupies both lobes (above and below the plane of the σ bonds).

The more electronegative oxygen atom strongly attracts the electrons of both the σ bond and the π bond, causing the carbonyl group to be highly polarized; the carbon atom bears a substantial positive charge and the oxygen atom bears a substantial negative charge. Polarization of the π bond can be represented by the following resonance structures for the carbonyl group (see also Section 3.10):

$$\gt C{=}\ddot{O}: \longleftrightarrow \gt \overset{+}{C}{-}\ddot{\underset{..}{O}}:^{-} \qquad \text{or} \qquad \gt \overset{\delta+}{C}{=}\overset{\delta-}{O}$$

Resonance structures for the carbonyl group **Hybrid**

Evidence for the polarity of the carbon–oxygen bond can be found in the rather large dipole moments associated with carbonyl compounds.

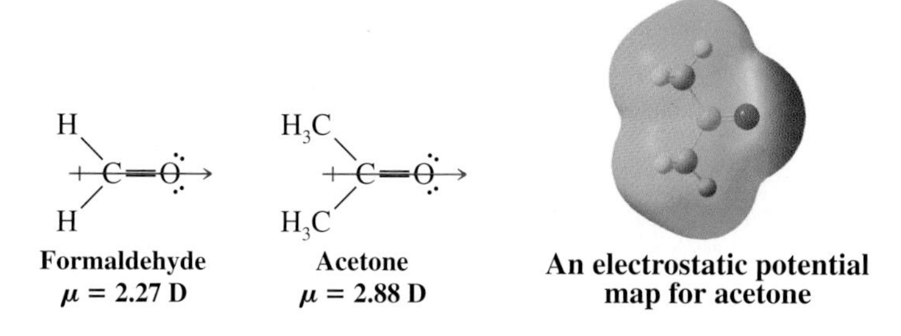

Formaldehyde $\mu = 2.27$ **D** **Acetone** $\mu = 2.88$ **D** **An electrostatic potential map for acetone**

12.1B Reactions of Carbonyl Compounds with Nucleophiles

From a synthetic point of view, one of the most important reactions of carbonyl compounds is one in which the carbonyl compound undergoes **nucleophilic addition.** The carbonyl group is susceptible to nucleophilic attack because, as we have just seen, the carbonyl carbon bears a partial positive charge. When the nucleophile adds to the carbonyl group, it uses its electron pair to form a bond to the carbonyl carbon atom. The carbonyl carbon can accept this electron pair because one pair of electrons of the carbon–oxygen double bond can shift out to the oxygen:

$$Nu:^{-} \; + \; \gt \overset{\delta+}{C}{=}\overset{\delta-}{\ddot{O}} \longrightarrow Nu{-}\overset{|}{\underset{|}{C}}{-}\ddot{\underset{..}{O}}:^{-}$$

Nucleophilic addition to a carbonyl

As the reaction takes place, the carbon atom undergoes a change in its geometry and its hybridization state. It goes from a trigonal planar geometry and sp^2 hybridization to a tetrahedral geometry and sp^3 hybridization.

Two important nucleophiles that add to carbonyl compounds are **hydride ions** from compounds such as $NaBH_4$ or $LiAlH_4$ (Section 12.3) and **carbanions** from compounds such as RLi or RMgX (Section 12.7C).

Another related set of reactions are reactions in which alcohols and carbonyl compounds are **oxidized** and **reduced** (Sections 12.2–12.4). For example, primary alcohols can be oxidized to aldehydes, and aldehydes can be reduced to alcohols:

$$R{-}\overset{H}{\underset{H}{C}}{-}\ddot{\underset{..}{O}}{-}H \underset{\text{reduction [H]}}{\overset{\text{[O] oxidation}}{\rightleftharpoons}} \overset{R}{\underset{H}{\gt}}C{=}\ddot{O}$$

A primary alcohol **An aldehyde**

Let us begin by examining some general principles that apply to the oxidation and reduction of organic compounds.

12.2 Oxidation–Reduction Reactions in Organic Chemistry

***Reduction** of an organic molecule usually corresponds to increasing its hydrogen content or to decreasing its oxygen content.* For example, converting a carboxylic acid to an aldehyde is a reduction because the oxygen content is decreased:

Oxygen content decreases

$$\underset{\text{Carboxylic acid}}{\text{RC(=O)OH}} \xrightarrow[\text{reduction}]{[H]} \underset{\text{Aldehyde}}{\text{RC(=O)H}}$$

Converting an aldehyde to an alcohol is a reduction:

Hydrogen content increases

$$\text{RC(=O)H} \xrightarrow[\text{reduction}]{[H]} RCH_2OH$$

Converting an alcohol to an alkane is also a reduction:

Oxygen content decreases

$$RCH_2OH \xrightarrow[\text{reduction}]{[H]} RCH_3$$

In these examples we have used the symbol [H] to indicate that a reduction of the organic compound has taken place. We do this when we want to write a general equation without specifying what the reducing agent is.

The opposite of reduction is **oxidation.** ***Thus, increasing the oxygen content of an organic molecule or decreasing its hydrogen content is an oxidation*** of the organic molecule. The reverse of each reaction that we have just given is an oxidation of the organic molecule, and we can summarize these oxidation–reduction reactions as shown below. We use the symbol [O] to indicate in a general way that the organic molecule has been oxidized.

Note the general interpretation of oxidation–reduction regarding organic compounds.

$$\underset{\text{Lowest oxidation state}}{RCH_3} \underset{[H]}{\overset{[O]}{\rightleftharpoons}} RCH_2OH \underset{[H]}{\overset{[O]}{\rightleftharpoons}} \text{RC(=O)H} \underset{[H]}{\overset{[O]}{\rightleftharpoons}} \underset{\text{Highest oxidation state}}{\text{RC(=O)OH}}$$

Oxidation of an organic compound may be more broadly defined as a reaction that increases its content of any element more electronegative than carbon. For example, replacing hydrogen atoms by chlorine atoms is an oxidation:

$$Ar{-}CH_3 \underset{[H]}{\overset{[O]}{\rightleftharpoons}} Ar{-}CH_2Cl \underset{[H]}{\overset{[O]}{\rightleftharpoons}} Ar{-}CHCl_2 \underset{[H]}{\overset{[O]}{\rightleftharpoons}} Ar{-}CCl_3$$

Of course, when an organic compound is reduced, something else—the **reducing agent**—must be oxidized. And when an organic compound is oxidized, something else—the **oxidizing agent**—is reduced. These oxidizing and reducing agents are often inorganic compounds, and in the next two sections we shall see what some of them are.

PROBLEM 12.1

One method for assigning an oxidation state to a carbon atom of an organic compound is to base that assignment on the groups attached to the carbon; a bond to hydrogen (or anything less electronegative than carbon) makes it −1, a bond to oxygen, nitrogen, or halogen (or to anything more electronegative than carbon) makes it +1, and a bond to another carbon makes it 0. Thus the carbon of methane is assigned an oxidation state of −4, and that of carbon dioxide, +4.

(a) Use this method to assign oxidation states to the carbon atoms of methanol (CH_3OH), formic acid $\left(\mathrm{HC(=O)OH}\right)$, and formaldehyde $\left(\mathrm{HC(=O)H}\right)$.

(b) Arrange the compounds methane, carbon dioxide, methanol, formic acid, and formaldehyde in order of increasing oxidation state.

(c) What change in oxidation state accompanies the following reaction: methanol ⟶ formaldehyde?

(d) Is this an oxidation or a reduction?

(e) When H_2CrO_4 acts as an oxidizing agent in this reaction, the chromium of H_2CrO_4 becomes Cr^{3+}. What change in oxidation state does chromium undergo?

STUDY TIP!

A method for balancing organic oxidation–reduction reactions is described in the Study Guide that accompanies this text.

PROBLEM 12.2

(a) Use the method described in the preceding problem to assign oxidation states to each carbon of ethanol and to each carbon of acetaldehyde. **(b)** What do these numbers reveal about the site of oxidation when ethanol is oxidized to acetaldehyde? **(c)** Repeat this procedure for the oxidation of acetaldehyde to acetic acid.

PROBLEM 12.3

(a) Although we have described the hydrogenation of an alkene as an addition reaction, organic chemists often refer to it as a "reduction." Refer to the method described in Problem 12.1 and explain. **(b)** Make similar comments about this reaction:

$$\mathrm{CH_3{-}C({=}O){-}H} + \mathrm{H_2} \xrightarrow{\mathrm{Ni}} \mathrm{CH_3CH_2OH}$$

12.3 Alcohols by Reduction of Carbonyl Compounds

Primary and secondary alcohols can be synthesized by the reduction of a variety of compounds that contain the carbonyl group. Several general examples are shown here:

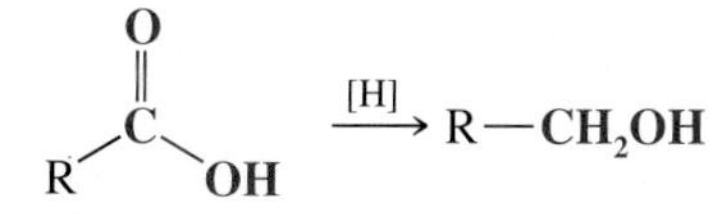

$$\underset{\textbf{Carboxylic acid}}{\mathrm{R{-}C({=}O){-}OH}} \xrightarrow{[\mathrm{H}]} \underset{\textbf{1° Alcohol}}{\mathrm{R{-}CH_2OH}}$$

$$\underset{\textbf{Ester}}{\mathrm{R{-}C({=}O){-}OR'}} \xrightarrow{[\mathrm{H}]} \underset{\textbf{1° Alcohol}}{\mathrm{R{-}CH_2OH}}\ (+\ \mathrm{R'OH})$$

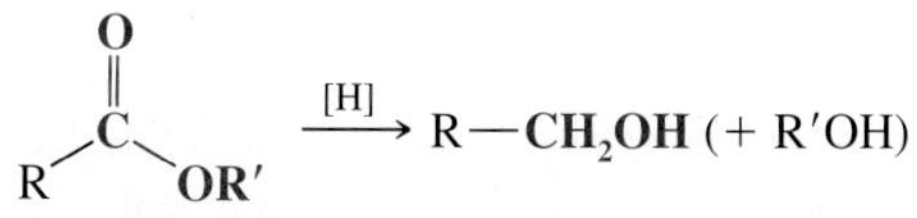

$$\underset{\textbf{Aldehyde}}{\mathrm{R{-}C({=}O){-}H}} \xrightarrow{[\mathrm{H}]} \underset{\textbf{1° Alcohol}}{\mathrm{R{-}CH_2OH}}$$

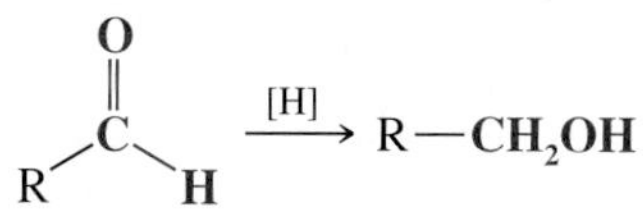

$$\underset{\textbf{Ketone}}{R\overset{O}{\overset{\|}{C}}R'} \xrightarrow{[H]} \underset{\textbf{2° Alcohol}}{R-\underset{OH}{\underset{|}{CH}}-R'}$$

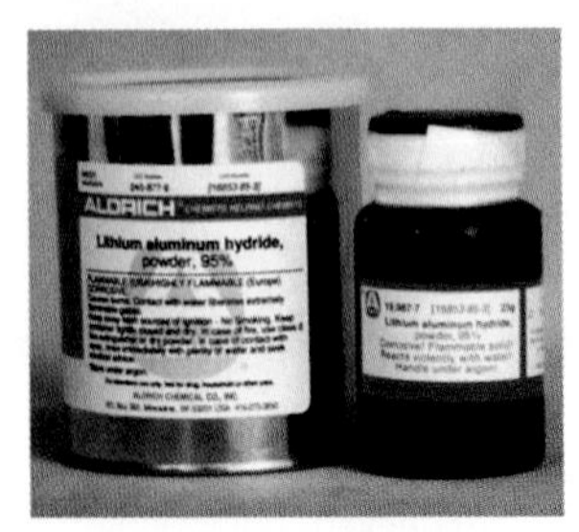

Unless special precautions are taken, lithium aluminum hydride reductions can be very dangerous. You should consult an appropriate laboratory manual before attempting such a reduction, and the reaction should be carried out on a small scale.

Reductions of carboxylic acids are the most difficult, but they can be accomplished with the powerful reducing agent **lithium aluminum hydride** ($LiAlH_4$, abbreviated LAH). It reduces carboxylic acids to primary alcohols in excellent yields:

$$4\,RCO_2H + \underset{\textbf{Lithium aluminum hydride}}{3\,LiAlH_4} \xrightarrow{Et_2O} [(RCH_2O)_4Al]Li + 4\,H_2 + 2\,LiAlO_2$$

$$\xrightarrow{H_2O/H_2SO_4} 4\,RCH_2OH + Al_2(SO_4)_3 + Li_2SO_4$$

An example is the lithium aluminum hydride reduction of 2,2-dimethylpropanoic acid:

$$\underset{\textbf{2,2-Dimethylpropanoic acid}}{CH_3-\overset{CH_3}{\underset{CH_3}{C}}-CO_2H} \xrightarrow[(2)\ H_2O/H_2SO_4]{(1)\ LiAlH_4/Et_2O} \underset{\textbf{Neopentyl alcohol (92\%)}}{CH_3-\overset{CH_3}{\underset{CH_3}{C}}-CH_2OH}$$

Esters can be reduced by high-pressure hydrogenation (a reaction preferred for industrial processes and often referred to as "hydrogenolysis" because a carbon–oxygen bond is cleaved in the process) or through the use of lithium aluminum hydride:

$$R\overset{O}{\overset{\|}{C}}-OR' + H_2 \xrightarrow[175°C,\ 5000\ psi]{CuO\cdot CuCr_2O_4} RCH_2OH + R'OH$$

$$R\overset{O}{\overset{\|}{C}}-OR' \xrightarrow[(2)\ H_2O/H_2SO_4]{(1)\ LiAlH_4/Et_2O} RCH_2OH + R'OH$$

The latter method is the one most commonly used now in small-scale laboratory syntheses.

Aldehydes and ketones can also be reduced to alcohols by hydrogen and a metal catalyst, by sodium in alcohol, and by lithium aluminum hydride. The reducing agent most often used, however, is sodium borohydride ($NaBH_4$):

$$4\,R\overset{O}{\overset{\|}{C}}H + NaBH_4 + 3\,H_2O \longrightarrow 4\,RCH_2OH + NaH_2BO_3$$

$$\underset{\textbf{Butanal}}{CH_3CH_2CH_2\overset{O}{\overset{\|}{C}}H} \xrightarrow[H_2O]{NaBH_4} \underset{\textbf{1-Butanol (85\%)}}{CH_3CH_2CH_2CH_2OH}$$

$$\underset{\textbf{Butanone}}{CH_3CH_2\underset{O}{\underset{\|}{C}}CH_3} \xrightarrow[H_2O]{NaBH_4} \underset{\textbf{2-Butanol (87\%)}}{CH_3CH_2\underset{OH}{\underset{|}{C}H}CH_3}$$

The key step in the reduction of a carbonyl compound by either lithium aluminum hydride or sodium borohydride is the transfer of a **hydride ion** from the metal to the carbonyl carbon. In this transfer the hydride ion acts as a *nucleophile.* The mechanism for the reduction of a ketone by sodium borohydride is illustrated here.

A Mechanism for the Reaction

Reduction of Aldehydes and Ketones by Hydride Transfer

$$H_3\bar{B}{-}H + \underset{R'}{\overset{R}{\diagdown}}\overset{\delta+}{C}{=}\overset{\delta-}{\ddot{O}}: \longrightarrow H{-}\underset{R'}{\overset{R}{C}}{-}\ddot{O}:^- \xrightarrow{H{-}OH} H{-}\underset{R'}{\overset{R}{C}}{-}\ddot{O}{-}H$$

Hydride transfer **Alkoxide ion** **Alcohol**

These steps are repeated until all hydrogen atoms attached to boron have been transferred.

Sodium borohydride is a less powerful reducing agent than lithium aluminum hydride. Lithium aluminum hydride reduces acids, esters, aldehydes, and ketones, but sodium borohydride reduces only aldehydes and ketones:

Reduced by $LiAlH_4$

Reduced by $NaBH_4$

$$RCO^-(=O) < RCOR'(=O) < RCR'(=O) < RCH(=O)$$

Ease of reduction

Lithium aluminum hydride reacts violently with water, and therefore reductions with lithium aluminum hydride must be carried out in anhydrous solutions, usually in anhydrous ether. (Ethyl acetate is added cautiously after the reaction is over to decompose excess $LiAlH_4$; then water is added to decompose the aluminum complex.) Sodium borohydride reductions, by contrast, can be carried out in water or alcohol solutions.

PROBLEM 12.4

Which reducing agent, $LiAlH_4$ or $NaBH_4$, would you use to carry out the following transformations?

(a) $CH_3{-}C_6H_{10}{-}\overset{O}{\overset{\|}{C}}OH \longrightarrow CH_3{-}C_6H_{10}{-}CH_2OH$

(b) $CH_3\overset{O}{\overset{\|}{C}}{-}C_6H_{10}{-}\overset{O}{\overset{\|}{C}}OH \longrightarrow CH_3\overset{OH}{\overset{|}{C}}H{-}C_6H_{10}{-}CH_2OH$

(c) $H\overset{O}{\overset{\|}{C}}{-}C_6H_{10}{-}\overset{O}{\overset{\|}{C}}OCH_3 \longrightarrow HOCH_2{-}C_6H_{10}{-}\overset{O}{\overset{\|}{C}}OCH_3$

The Chemistry of...

Alcohol Dehydrogenase

When the enzyme alcohol dehydrogenase converts acetaldehyde to ethanol, NADH acts as a reducing agent by transferring a hydride from C4 of the nicotinamide ring to the carbonyl group of acetaldehyde. The nitrogen of the nicotinamide ring facilitates this process by contributing its nonbonding electron pair to the ring, which together with loss of the hydride converts the ring to the energetically more stable ring found in NAD^+ (we shall see why it is more stable in Chapter 14). The ethoxide anion resulting from hydride transfer to acetaldehyde is then protonated by the enzyme to form ethanol.

Although the carbonyl carbon of acetaldehyde that accepts the hydride is inherently electrophilic because of its electronegative oxygen, the enzyme enhances this property by providing a zinc ion as a Lewis acid to coordinate with the carbonyl oxygen. The Lewis acid stabilizes the negative charge that develops on the oxygen in the transition state. The role of the enzyme's protein scaffold, then, is to hold the zinc ion, coenzyme, and substrate in the three-dimensional array required to lower the energy of the transition state. The reaction is entirely reversible, of course, and when the relative concentration of ethanol is high, alcohol dehydrogenase carries out the oxidation of ethanol by removal of a hydride. This role of alcohol dehydrogenase is important in detoxification. In "The Chemistry of... Stereoselective Reductions of Carbonyl Groups" we discuss the stereochemical aspect of alcohol dehydrogenase reactions.

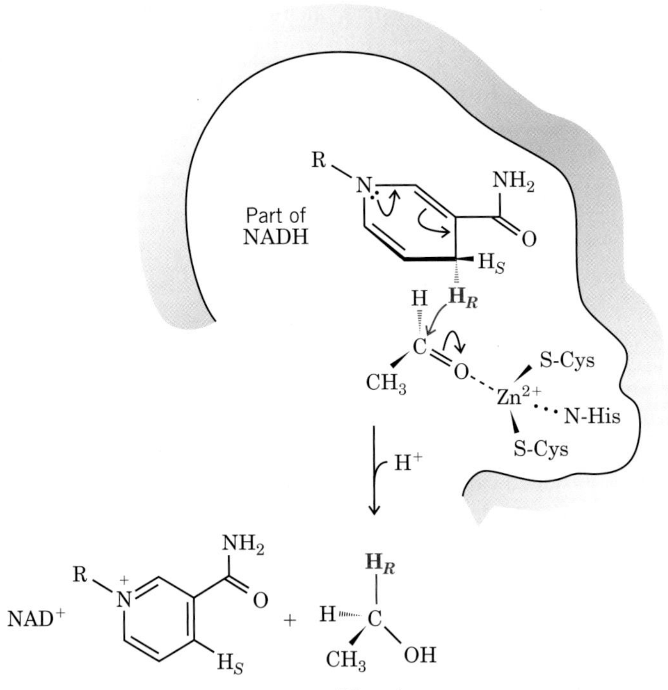

The Chemistry of...

Stereoselective Reductions of Carbonyl Groups

Enantioselectivity

The possibility of **stereoselective** reduction of a carbonyl group is an important consideration in many syntheses. Depending on the structure about the carbonyl group that is being reduced, the tetrahedral carbon that is formed by transfer of a hydride could be a new stereogenic center. Achiral reagents, like $NaBH_4$ and $LiAlH_4$, react with equal rates at either face of an

Thermophilic bacteria, growing in hot springs like these at Yellowstone National Park, produce heat-stable enzymes called extremozymes that have proven useful for a variety of chemical processes.

achiral trigonal planar substrate, leading to a racemic form of the product. But enzymes, for example, are chiral, and reactions involving a chiral reactant typically lead to a predominance of one enantiomeric form of a chiral product. Such a reaction is said to be **enantioselective.** Thus, when enzymes like alcohol dehydrogenase reduce carbonyl groups using the coenzyme NADH (see the chapter opening vignette), they discriminate between the two faces of the trigonal planar carbonyl substrate, such that a predominance of one of the two possible stereoisomeric forms of the tetrahedral product results. (If the original reactant was chiral, then formation of the new stereogenic center may result in preferential formation of one *diastereomer* of the product, in which case the reaction is said to be **diastereoselective.**)

The two faces of a trigonal planar center are designated *re* and *si*, according to the direction of Cahn–Ingold–Prelog priorities (Section 5.7) for the groups bonded at the trigonal center when viewed from one face or the other (*re* is clockwise, *si* is counterclockwise):

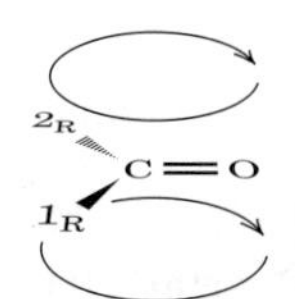

***re* face (when looking at this face, there is a clockwise sequence of priorities)**

***si* face (when looking at this face, there is a counterclockwise sequence of priorities)**

The *re* and *si* faces of a carbonyl group (where O > ^{1}R > ^{2}R in terms of Cahn-Ingold-Prelog priorities)

The preference of many NADH-dependent enzymes for either the *re* or *si* face of their respective substrates is known. This knowledge has allowed some of these enzymes to become exceptionally useful stereoselective reagents for synthesis. One of the most widely used is yeast alcohol dehydrogenase. Others that have become important are enzymes from thermophilic bacteria (bacteria that grow at elevated temperatures). Use of heat-stable enzymes (called **extremozymes**) allows reactions to be completed faster due to the rate-enhancing factor of elevated temperature (over 100°C in some cases), although greater enantioselectivity is achieved at lower temperatures.

Thermoanaerobium brockii

96% enantiomeric excess (85% yield)

A number of chemical reagents that are chiral have also been developed for the purpose of stereoselective reduction of carbonyl groups. Most of them are derivatives of standard aluminum or boron hydride reducing agents that involve one or more chiral organic ligands. (*S*)-Alpine-Borane and (*R*)-Alpine-Borane, for example, are reagents derived from diborane (B_2H_6) and either (−)-α-pinene or (+)-α-pinene (enantiomeric natural hydrocarbons), respectively. Reagents derived from $LiAlH_4$ and chiral amines have also been developed. The extent of stereoselectivity achieved either by enzymatic reduction or reduction by a chiral reducing agent depends on the specific structure of the substrate. Often it is necessary to test several reaction conditions in order to achieve optimal stereoselectivity.

(*S*)-(−)Alpine-Borane

97% enantiomeric excess (60–65% yield)

Prochirality

A second aspect of the stereochemistry of NADH reactions results from NADH having two hydrogens at C4, either of which could, in principle, be transferred as a hydride in a reduction process. For a given enzymatic reaction, however, only one specific hydride from C4 in NADH is transferred. Just which hydride is transferred depends on the specific enzyme involved, and we designate it by a useful extension of stereochemical nomenclature. The hydrogens at C4 of NADH are said to be **prochiral.** We designate one **pro-*R*,** and the other **pro-*S*,** depending on whether the configuration would be *R* or *S* when (in our imagination) each is replaced by a group of higher priority than hydrogen. If this exercise produces the *R* configuration, the hydrogen "replaced" is pro-*R*, and if it produces the *S* configuration it is pro-*S*. In general, a **prochiral center** is one for which addition of a group to a trigonal planar atom (as in reduction of a ketone) or replacement of one of two identical groups at a tetrahedral atom leads to a new stereogenic center.

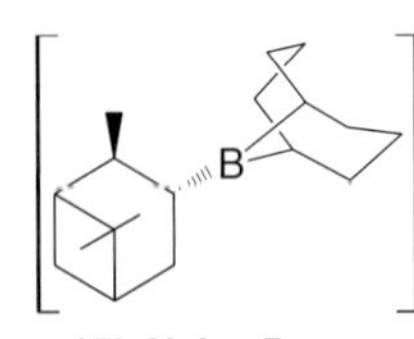

(*R*)-Alpine-Borane

Nicotinamide ring of NADH, showing the pro-*R* and pro-*S* hydrogens

12.4 Oxidation of Alcohols

12.4A Oxidation of Primary Alcohols to Aldehydes: $RCH_2OH \longrightarrow RCHO$

Primary alcohols can be oxidized to aldehydes and carboxylic acids:

$$\underset{\textbf{1° Alcohol}}{R{-}CH_2OH} \xrightarrow{[O]} \underset{\textbf{Aldehyde}}{RC(=O)H} \xrightarrow{[O]} \underset{\textbf{Carboxylic acid}}{RC(=O)OH}$$

The oxidation of aldehydes to carboxylic acids in aqueous solutions is easier than oxidation of primary alcohols to aldehydes; thus it is difficult to stop the oxidation at the aldehyde stage. [Notice that dehydrogenation of an organic compound corresponds to oxidation, whereas hydrogenation (see Problem 12.3) corresponds to reduction.] Therefore, in most laboratory preparations we must rely on special conditions to prepare aldehydes from primary alcohols. A variety of reagents is available, and discussing all of them here is beyond our scope. An excellent reagent for this purpose is **pyridinium chlorochromate** (abbreviated PCC), the compound formed when CrO_3 is dissolved in hydrochloric acid and then treated with pyridine:

$$CrO_3 + HCl + \underset{\substack{\textbf{Pyridine}\\ (\mathbf{C_5H_5N})}}{C_5H_5N:} \longrightarrow \underset{\substack{\textbf{Pyridinium chlorochromate}\\ \textbf{(PCC)}}}{C_5H_5\overset{+}{N}{-}H \quad CrO_3Cl^-}$$

PCC, when dissolved in CH_2Cl_2, will oxidize a primary alcohol to an aldehyde and stop at that stage:

$$\underset{\textbf{2-Ethyl-2-methyl-1-butanol}}{(C_2H_5)_2C(CH_3){-}CH_2OH} + PCC \xrightarrow[25°C]{CH_2Cl_2} \underset{\textbf{2-Ethyl-2-methylbutanal}}{(C_2H_5)_2C(CH_3){-}CH{=}O}$$

Pyridinium chlorochromate also does not attack double bonds.

One reason for the success of oxidation with pyridinium chlorochromate is that the oxidation can be carried out in a solvent such as CH_2Cl_2, in which PCC is soluble. Aldehydes themselves are not nearly so easily oxidized as are the *aldehyde hydrates, $RCH(OH)_2$*, that form (Section 16.7) when aldehydes are dissolved in water, the usual medium for oxidation by chromium compounds:

$$RCHO + H_2O \rightleftharpoons RCH(OH)_2$$

We explain this further in Section 12.4D.

12.4B Oxidation of Primary Alcohols to Carboxylic Acids: $RCH_2OH \longrightarrow RCO_2H$

Primary alcohols can be oxidized to **carboxylic acids** by potassium permanganate. The reaction is usually carried out in basic aqueous solution from which MnO_2 precipitates as the oxidation takes place. After the oxidation is complete, filtration allows removal of the MnO_2 and acidification of the filtrate gives the carboxylic acid:

$$R{-}CH_2OH + KMnO_4 \xrightarrow[\substack{H_2O \\ \text{heat}}]{OH^-} RCO_2^-K^+ + MnO_2$$

$$\downarrow H_3O^+$$

$$RCO_2H$$

12.4C Oxidation of Secondary Alcohols to Ketones: $RCH(OH)R' \longrightarrow RC(=O)R'$

Secondary alcohols can be oxidized to ketones. The reaction usually stops at the ketone stage because further oxidation requires the breaking of a carbon–carbon bond:

$$R{-}CH(OH){-}R' \xrightarrow{[O]} RC(=O)R'$$

2° Alcohol **Ketone**

Various oxidizing agents based on Cr(VI) have been used to oxidize secondary alcohols to ketones. The most commonly used reagent is chromic acid (H_2CrO_4). Chromic acid is usually prepared by adding Cr(VI) oxide (CrO_3) or sodium dichromate ($Na_2Cr_2O_7$) to aqueous sulfuric acid. Oxidations of secondary alcohols are generally carried out in acetone or acetic acid solutions. The balanced equation is shown here:

It is the color change from orange to green that accompanies this change in oxidation state that allows chromic acid to be used as a test for primary and secondary alcohols (Section 12.4E).

$$3\ RR'CHOH + 2\,H_2CrO_4 + 6\,H^+ \longrightarrow 3\ RR'C{=}O + 2\,Cr^{3+} + 8\,H_2O$$

As chromic acid oxidizes the alcohol to the ketone, chromium is reduced from the +6 oxidation state (H_2CrO_4) to the +3 oxidation state (Cr^{3+}). Chromic acid oxidations of secondary alcohols generally give ketones in excellent yields if the temperature is controlled. A specific example is the oxidation of cyclooctanol to cyclooctanone:

$$\text{Cyclooctanol} \xrightarrow[\substack{\text{acetone} \\ 35°C}]{H_2CrO_4} \text{Cyclooctanone (92–96\%)}$$

Cyclooctanol **Cyclooctanone (92–96%)**

The use of CrO_3 in aqueous acetone is usually called the **Jones oxidation** (or oxidation by the Jones reagent). This procedure rarely affects double bonds present in the molecule.

12.4D Mechanism of Chromate Oxidations

The mechanism of chromic acid oxidations of alcohols has been investigated thoroughly. It is interesting because it shows how changes in oxidation states occur in a reaction between an organic and an inorganic compound. The first step is the formation of a chromate ester of the alcohol. Here we show this step using a 2° alcohol.

A Mechanism for the Reaction

Chromate Oxidations: Formation of the Chromate Ester

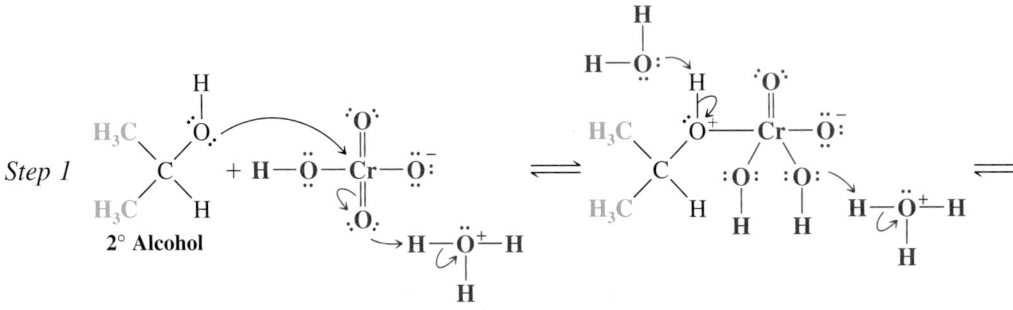

Step 1

2° Alcohol

The alcohol donates an electron pair to the chromium atom, as an oxygen accepts a proton.

One oxygen loses a proton; another oxygen accepts a proton.

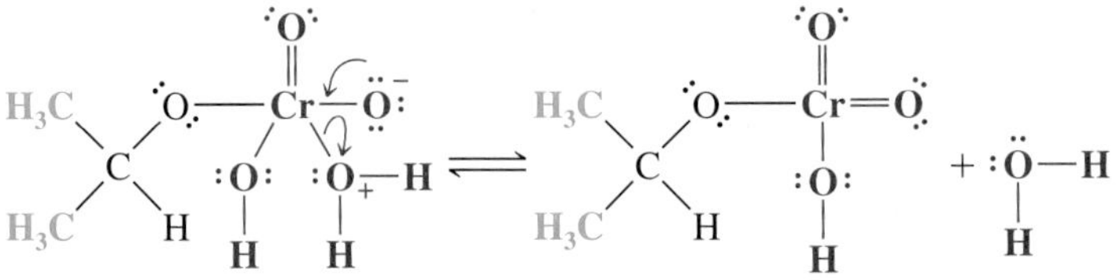

Chromate ester

A molecule of water departs as a leaving group as a chromium–oxygen double bond forms.

The chromate ester is unstable and is not isolated. It transfers a proton to a base (usually water) and simultaneously eliminates an $HCrO_3^-$ ion.

A Mechanism for the Reaction

Chromate Oxidations: The Oxidation Step

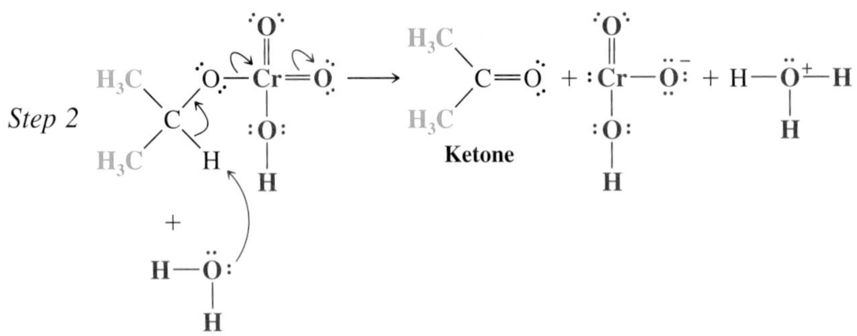

Step 2

Ketone

The chromium atom departs with a pair of electrons that formerly belonged to the alcohol; the alcohol is thereby oxidized and the chromium reduced.

The overall result of the second step is the reduction of $HCrO_4^-$ to $HCrO_3^-$, a two-electron ($2\ e^-$) change in the oxidation state of chromium, from Cr(VI) to Cr(IV). At the same time the alcohol undergoes a $2\ e^-$ oxidation to the ketone.

The remaining steps of the mechanism are complicated and we need not give them in detail. Suffice it to say that further oxidations (and disproportionations) take place, ultimately converting Cr(IV) compounds to Cr^{3+} ions.

The requirement for the formation of a chromate ester in step 1 of the mechanism helps us understand why 1° alcohols are easily oxidized beyond the aldehyde stage in aqueous solutions (and, therefore, why oxidation with PCC in CH_2Cl_2 stops at the aldehyde stage). The aldehyde initially formed from the 1° alcohol (produced by a mechanism similar to the one we have just given) reacts with water to form an aldehyde hydrate. The aldehyde hydrate can then react with $HCrO_4^-$ (and H^+) to form a chromate ester, and this can then be oxidized to the carboxylic acid. In the absence of water (i.e., using PCC in CH_2Cl_2), the aldehyde hydrate does not form; therefore, further oxidation does not take place.

$HCrO_4^-$, H^+

Aldehyde hydrate

$+ HCrO_3^- + H_3O^+$

Carboxylic acid

The elimination that takes place in step 2 of the preceding mechanism helps us to understand why 3° alcohols do not generally react in chromate oxidations. Although 3° alcohols have no difficulty in forming chromate esters, the ester that is formed does not bear a hydrogen that can be eliminated, and therefore no oxidation takes place.

$+ H_2O$

3° Alcohol

This chromate ester cannot undergo elimination of H_2CrO_3.

12.4E A Chemical Test for Primary and Secondary Alcohols

The relative ease of oxidation of primary and secondary alcohols compared with the difficulty of oxidizing tertiary alcohols forms the basis for a convenient chemical test. Primary and secondary alcohols are rapidly oxidized by a solution of CrO_3 in aqueous sulfuric acid. Chromic oxide (CrO_3) dissolves in aqueous sulfuric acid to give a clear orange solution containing $Cr_2O_7^{2-}$ ions. A positive test is indicated when this clear orange solution becomes opaque and takes on a greenish cast within 2 s:

This color change, associated with the reduction of $Cr_2O_7^{2-}$ to Cr^{3+}, formed the basis for "Breathalyzer tubes," once widely used to detect intoxicated motorists. In the Breathalyzer the dichromate salt is coated on granules of silica gel. (IR spectroscopy has now largely replaced the chromium colorimetric method for breath analysis.)

$$\underbrace{\begin{array}{c} RCH_2OH \\ \text{or} \\ RCHOH \\ | \\ R \end{array} + CrO_3/\text{aqueous } H_2SO_4}_{\textbf{Clear orange solution}} \longrightarrow \underbrace{Cr^{3+} \text{ and oxidation products}}_{\textbf{Greenish opaque solution}}$$

Not only will this test distinguish primary and secondary alcohols from tertiary alcohols, it will distinguish primary and secondary alcohols from most other compounds except aldehydes.

PROBLEM 12.5 Show how each of the following transformations could be accomplished:

(a) cyclopentyl–$CH_2OH \xrightarrow{?}$ cyclopentyl–CH(=O)

(b) cyclopentyl–$CH_2OH \xrightarrow{?}$ cyclopentyl–C(=O)OH

(c) cyclopentanol (OH) $\xrightarrow{?}$ cyclopentanone (O)

(d) cyclopentene $\xrightarrow{?}$ HC(=O)$CH_2CH_2CH_2$C(=O)H

12.4F Spectroscopic Evidence for Alcohols

Alcohols give rise to O—H stretching absorptions from 3200 to 3600 cm^{-1} in infrared spectra. The alcohol hydroxyl hydrogen typically produces a broad ^1H NMR signal of variable chemical shift which can be eliminated by exchange with deuterium from D_2O (see Table 9.1). The ^{13}C NMR spectrum of an alcohol shows a signal between δ 50 and δ 90 for the alcohol carbon (see Table 9.2). Hydrogen atoms on the carbon of a primary or secondary alcohol produce a signal in the ^1H NMR spectrum between δ 3.3 and δ 4.0 (see Table 9.1) that integrates for 2 and 1 hydrogens, respectively.

12.5 Organometallic Compounds

Compounds that contain carbon–metal bonds are called **organometallic compounds.** The nature of the carbon–metal bond varies widely, ranging from bonds that are essentially ionic to those that are primarily covalent. Whereas the structure of the organic portion of the organometallic compound has some effect on the nature of the carbon–metal bond, the identity of the metal itself is of far greater importance. Carbon–sodium and carbon–potassium bonds are largely ionic in character; carbon–lead, carbon–tin, carbon–thallium, and carbon–mercury bonds are essentially covalent. Carbon–lithium and carbon–magnesium bonds lie between these extremes.

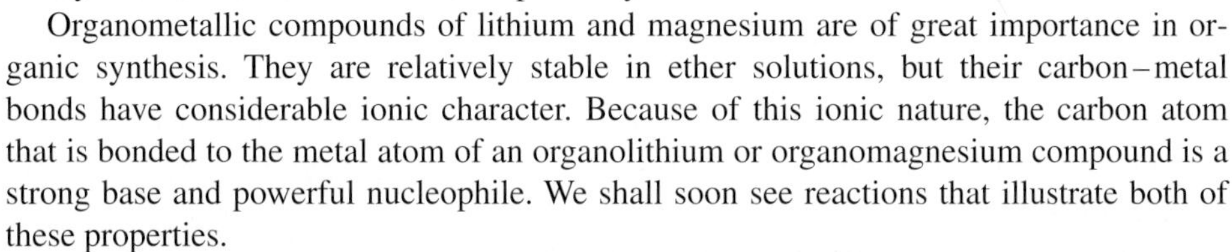

The reactivity of organometallic compounds increases with the percent ionic character of the carbon–metal bond. Alkylsodium and alkylpotassium compounds are highly reactive and are among the most powerful of bases. They react explosively with water and burst into flame when exposed to air. Organomercury and organolead compounds are much less reactive; they are often volatile and are stable in air. They are all poisonous. They are generally soluble in nonpolar solvents. Tetraethyllead, for example, was once used as an "antiknock" compound in gasoline, but because of the lead pollution it contributed to the environment it has been replaced by other antiknock agents. *tert*-Butyl methyl ether is an antiknock additive presently in use.

A number of organometallic reagents are very useful for carbon–carbon bond forming reactions (see Sections 12.8, 12.9, and Special Topic I).

Organometallic compounds of lithium and magnesium are of great importance in organic synthesis. They are relatively stable in ether solutions, but their carbon–metal bonds have considerable ionic character. Because of this ionic nature, the carbon atom that is bonded to the metal atom of an organolithium or organomagnesium compound is a strong base and powerful nucleophile. We shall soon see reactions that illustrate both of these properties.

12.6 Preparation of Organolithium and Organomagnesium Compounds

12.6A Organolithium Compounds

Organolithium compounds are often prepared by the reduction of organic halides with lithium metal. These reductions are usually carried out in ether solvents, and since organolithium compounds are strong bases, care must be taken to exclude moisture. (Why?) The ethers most commonly used as solvents are diethyl ether and tetrahydrofuran. (Tetrahydrofuran is a cyclic ether.)

$$CH_3CH_2\ddot{O}CH_2CH_3$$

Diethyl ether (Et_2O)

Tetrahydrofuran (THF)

For example, butyl bromide reacts with lithium metal in diethyl ether to give a solution of butyllithium:

$$\underset{\textbf{Butyl bromide}}{CH_3CH_2CH_2CH_2Br} + 2\,Li \xrightarrow[Et_2O]{-10°C} \underset{\substack{\textbf{Butyllithium}\\ \textbf{(80–90\%)}}}{CH_3CH_2CH_2CH_2Li} + LiBr$$

Other organolithium compounds, such as methyllithium, ethyllithium, and phenyllithium, can be prepared in the same general way:

$$\underset{\textbf{(or Ar—X)}}{R{-}X} + 2\,Li \xrightarrow{Et_2O} \underset{\textbf{(or ArLi)}}{RLi} + LiX$$

The order of reactivity of halides is RI > RBr > RCl. (Alkyl and aryl fluorides are seldom used in the preparation of organolithium compounds.)

Most organolithium compounds slowly attack ethers by bringing about an elimination reaction:

$$\overset{\delta-}{R}:\overset{\delta+}{Li} + H{-}CH_2{-}CH_2{-}OCH_2CH_3 \longrightarrow RH + CH_2{=}CH_2 + \overset{+}{Li}\overset{-}{O}CH_2CH_3$$

For this reason, ether solutions of organolithium reagents are not usually stored but are used immediately after preparation. Organolithium compounds are much more stable in hydrocarbon solvents. Several alkyl- and aryllithium reagents are commercially available in hexane and other hydrocarbon solvents.

12.6B Grignard Reagents

Victor Grignard

Organomagnesium halides were discovered by the French chemist Victor Grignard in 1900. Grignard received the Nobel Prize for his discovery in 1912, and organomagnesium halides are now called **Grignard reagents** in his honor. Grignard reagents have great use in organic synthesis.

Grignard reagents are usually prepared by the reaction of an organic halide and magnesium metal (turnings) in an ether solvent:

$$\left.\begin{array}{l} RX + Mg \xrightarrow{Et_2O} RMgX \\ ArX + Mg \xrightarrow{Et_2O} ArMgX \end{array}\right\} \textbf{Grignard reagents}$$

The order of reactivity of halides with magnesium is also RI > RBr > RCl. Very few organomagnesium fluorides have been prepared. Aryl Grignard reagents are more easily prepared from aryl bromides and aryl iodides than from aryl chlorides, which react very sluggishly.

Grignard reagents are seldom isolated but are used for further reactions in ether solution. The ether solutions can be analyzed for the content of the Grignard reagent, however, and the yields of Grignard reagents are almost always very high (85–95%). Two examples are shown here:

$$CH_3I + Mg \xrightarrow[35°C]{Et_2O} CH_3MgI$$

Methylmagnesium iodide (95%)

$$C_6H_5Br + Mg \xrightarrow[35°C]{Et_2O} C_6H_5MgBr$$

Phenylmagnesium bromide (95%)

The actual structures of Grignard reagents are more complex than the general formula RMgX indicates. Experiments done with radioactive magnesium have established that, for most Grignard reagents, there is an equilibrium between an alkylmagnesium halide and a dialkylmagnesium.

$$2\ RMgX \rightleftharpoons R_2Mg + MgX_2$$

Alkylmagnesium halide — **Dialkylmagnesium**

For convenience in this text, however, we shall write the formula for the Grignard reagent as though it were simply RMgX.

A Grignard reagent forms a complex with its ether solvent; the structure of the complex can be represented as follows:

$$\begin{array}{c} R\diagdown\ \ddot{O}\ \diagup R \\ R-Mg-X \\ R\diagup\ \ddot{O}\ \diagdown R \end{array}$$

Complex formation with molecules of ether is an important factor in the formation and stability of Grignard reagents. Organomagnesium compounds can be prepared in nonethereal solvents, but the preparations are more difficult.

The mechanism by which Grignard reagents form is complicated and has been a matter of debate.* There seems to be general agreement that radicals are involved and that a mechanism similar to the following is likely:

$$R{-}X + :Mg \longrightarrow R\cdot + \cdot MgX$$
$$R\cdot + \cdot MgX \longrightarrow RMgX$$

12.7 Reactions of Organolithium and Organomagnesium Compounds

12.7A Reactions with Compounds Containing Acidic Hydrogen Atoms

Grignard reagents and organolithium compounds are very strong bases. They react with any compound that has a hydrogen atom attached to an electronegative atom such as oxygen, nitrogen, or sulfur. We can understand how these reactions occur if we represent the Grignard reagent and organolithium compounds in the following ways:

$$\overset{\delta-}{R}:\overset{\delta+}{MgX} \quad \text{and} \quad \overset{\delta-}{R}:\overset{\delta+}{Li}$$

When we do this, we can see that the reactions of Grignard reagents with water and alcohols are nothing more than acid–base reactions; they lead to the formation of the weaker conjugate acid and weaker conjugate base. The Grignard reagent behaves as if it contained the anion of an alkane, *as if it contained a carbanion:*

$$\overset{\delta-}{R}:\overset{\delta+}{MgX} + H:\ddot{O}H \longrightarrow R:H + H\ddot{O}:^- + Mg^{2+} + X^-$$

Grignard reagent (stronger base) | **Water (stronger acid, pK_a 15.7)** | **Alkane (weaker acid, pK_a 40–50)** | **Hydroxide ion (weaker base)**

$$\overset{\delta-}{R}:\overset{\delta+}{MgX} + H:\ddot{O}R \longrightarrow R:H + R\ddot{O}:^- + Mg^{2+} + X^-$$

Grignard reagent (stronger base) | **Alcohol (stronger acid, pK_a 15–18)** | **Alkane (weaker acid, pK_a 40–50)** | **Alkoxide ion (weaker base)**

---

*Those interested may want to read the following articles: Garst, J. L.; Swift, B. L. *J. Am. Chem. Soc.* **1989,** *111,* 241–250; Walborsky, H. M. *Acc. Chem Res.* **1990,** *23,* 286–293; and Garst, J. L. *Acc. Chem. Res.* **1991,** *24,* 95–97.

PROBLEM 12.6 Write equations similar to the above for the reactions that take place when phenyllithium is treated with **(a)** water and **(b)** ethanol. Designate the stronger and weaker acids and stronger and weaker bases.

PROBLEM 12.7 Assuming you have bromobenzene (C_6H_5Br), magnesium, dry (anhydrous) ether, and deuterium oxide (D_2O) available, show how you might synthesize the following deuterium-labeled compound.

C_6H_5—D

Grignard reagents and organolithium compounds remove protons that are much less acidic than those of water and alcohols. They react with the terminal hydrogen atoms of 1-alkynes, for example, and this is a useful method for the preparation of alkynylmagnesium halides and alkynyllithiums. These reactions are also acid–base reactions.

$$R{-}C{\equiv}C{-}H + \overset{\delta-}{R'}{:}\overset{\delta+}{MgX} \longrightarrow R{-}C{\equiv}\overset{\delta-}{C}{:}\overset{\delta+}{MgX} + R'{:}H$$

Terminal alkyne (stronger acid, pK_a ~25) — **Grignard reagent (stronger base)** — **Alkynylmagnesium halide (weaker base)** — **Alkane (weaker acid, pK_a 40–50)**

$$R{-}C{\equiv}C{-}H + \overset{\delta-}{R'}{:}\overset{\delta+}{Li} \longrightarrow R{-}C{\equiv}\overset{\delta-}{C}{:}\overset{\delta+}{Li} + R'{:}H$$

Terminal alkyne (stronger acid) — **Alkyllithium (stronger base)** — **Alkynyllithium (weaker base)** — **Alkane (weaker acid)**

The fact that these reactions go to completion is not surprising when we recall that alkanes have pK_a values of 40–50, whereas those of terminal alkynes are ~25 (Table 3.1).

Not only are Grignard reagents strong bases, they are also *powerful nucleophiles.* Reactions in which Grignard reagents act as nucleophiles are by far the most important. At this point, let us consider general examples that illustrate the ability of a Grignard reagent to act as a nucleophile by attacking saturated and unsaturated carbon atoms.

12.7B Reactions of Grignard Reagents with Oxiranes (Epoxides)

Grignard reagents carry out nucleophilic attack at a saturated carbon when they react with oxiranes. These reactions take the general form shown below and give us a convenient synthesis of primary alcohols.

The nucleophilic alkyl group of the Grignard reagent attacks the partially positive carbon of the oxirane ring. Because it is highly strained, the ring opens, and the reaction leads to the salt of a primary alcohol. Subsequent acidification produces the alcohol. (Compare this reaction with the base-catalyzed ring opening we studied in Section 11.14).

$$\overset{\delta-}{R}{:}\overset{\delta+}{MgX} + \overset{\delta+}{H_2C}{-}\overset{\delta+}{CH_2}\,(\overset{\delta-}{O}) \longrightarrow R{-}CH_2CH_2{-}\ddot{O}{:}^{-}\,Mg^{2+}X^{-} \xrightarrow{H_3O^+} R{-}CH_2CH_2\ddot{O}H$$

Oxirane — **A primary alcohol**

Specific Example

$$C_6H_5MgBr + H_2C\overset{O}{\frown}CH_2 \xrightarrow[Et_2O]{} C_6H_5CH_2CH_2OMgBr \xrightarrow{H_3O^+} C_6H_5CH_2CH_2OH$$

Grignard reagents react primarily at the less-substituted ring carbon atom of substituted oxiranes.

Specific Example

$$C_6H_5MgBr + H_2C\overset{O}{\frown}CH{-}CH_3 \xrightarrow[Et_2O]{} C_6H_5CH_2\underset{OMgBr}{CHCH_3} \xrightarrow{H_3O^+} C_6H_5CH_2\underset{OH}{CHCH_3}$$

12.7C Reactions of Grignard Reagents with Carbonyl Compounds

From a synthetic point of view, the most important reactions of Grignard reagents and organolithium compounds are those in which these reagents act as nucleophiles and attack an unsaturated carbon—*especially the carbon of a carbonyl group.*

We saw in Section 12.1B that carbonyl compounds are highly susceptible to nucleophilic attack. Grignard reagents react with carbonyl compounds (aldehydes and ketones) in the following way.

A Mechanism for the Reaction

The Grignard Reaction

Reaction:

$$RMgX + \;>C{=}O \xrightarrow[(2)\ H_3O^+X^-]{(1)\ ether^*} R{-}\overset{|}{\underset{|}{C}}{-}O{-}H + MgX_2$$

Mechanism:

Step 1

$$\overset{\delta-}{R}{:}\overset{\delta+}{MgX} + \;>\overset{\delta+}{C}{=}\overset{\delta-}{\ddot{O}}{:} \longrightarrow R{-}\overset{|}{\underset{|}{C}}{-}\ddot{O}{:}^-\ Mg^{2+}\ X^-$$

Grignard reagent **Carbonyl compound** **Halomagnesium alkoxide**

The strongly nucleophilic Grignard reagent uses its electron pair to form a bond to the carbon atom. One electron pair of the carbonyl group shifts out to the oxygen. This reaction is a nucleophilic addition to the carbonyl group, and it results in the formation of an alkoxide ion associated with Mg^{2+} and X^-.

Step 2

$$R{-}\overset{|}{\underset{|}{C}}{-}\ddot{O}{:}^-\ Mg^{2+}\ X^- + H{-}\underset{H}{\overset{+}{\ddot{O}}}{-}H + X^- \longrightarrow R{-}\overset{|}{\underset{|}{C}}{-}\ddot{O}{-}H + {:}\underset{H}{\ddot{O}}{-}H + MgX_2$$

Halomagnesium alkoxide **Alcohol**

In the second step, the addition of aqueous HX causes protonation of the alkoxide ion; this leads to the formation of the alcohol and MgX_2.

*By writing "(1) ether" over the arrow and "(2) H_3O^+ X^-" under the arrow, we mean that in the first step the Grignard reagent and the carbonyl compound are allowed to react in an ether solvent. Then in a second step, after the reaction of the Grignard reagent and the carbonyl compound is over, we add aqueous acid (e.g., dilute HX) to convert the salt of the alcohol (ROMgX) to the alcohol itself. If the alcohol is tertiary, it will be susceptible to acid-catalyzed dehydration. In this case, a solution of NH_4Cl in water is often used because it is acidic enough to convert ROMgX to ROH without causing dehydration.

12.8 Alcohols from Grignard Reagents

Grignard additions to carbonyl compounds are especially useful because they can be used to prepare primary, secondary, or tertiary alcohols:

1. Grignard Reagents React with Formaldehyde to Give a Primary Alcohol

$$\overset{\delta-}{R}{:}\overset{\delta+}{MgX} + H_2C{=}\ddot{O}{:} \longrightarrow R{-}CH_2{-}\ddot{O}{:}MgX \xrightarrow{H_3O^+} R{-}CH_2{-}\ddot{O}H$$

Formaldehyde — 1° Alcohol

2. Grignard Reagents React with All Other Aldehydes to Give Secondary Alcohols

$$\overset{\delta-}{R}{:}\overset{\delta+}{MgX} + R'(H)C{=}\ddot{O}{:} \longrightarrow R{-}C(R')(H){-}\ddot{O}{:}MgX \xrightarrow{H_3O^+} R{-}C(R')(H){-}\ddot{O}H$$

Higher aldehyde — 2° Alcohol

3. Grignard Reagents React with Ketones to Give Tertiary Alcohols

$$\overset{\delta-}{R}{:}\overset{\delta+}{MgX} + R'(R'')C{=}\ddot{O}{:} \longrightarrow R{-}C(R')(R''){-}\ddot{O}{:}MgX \xrightarrow[H_2O]{NH_4Cl} R{-}C(R')(R''){-}\ddot{O}H$$

Ketone — 3° Alcohol

4. Esters React with Two Molar Equivalents of a Grignard Reagent to Form Tertiary Alcohols When a Grignard reagent adds to the carbonyl group of an ester, the initial product is unstable and loses a magnesium alkoxide to form a ketone. Ketones, however, are more reactive toward Grignard reagents than esters. Therefore, as soon as a molecule of the ketone is formed in the mixture, it reacts with a second molecule of the Grignard reagent. After hydrolysis, **the product is a tertiary alcohol with two identical alkyl groups,** groups that correspond to the alkyl portion of the Grignard reagent:

$$R{:}MgX + R'(R''\ddot{O})C{=}\ddot{O}{:} \longrightarrow \left[R{-}C(R')(\ddot{O}{-}R''){-}\ddot{O}{-}MgX\right] \xrightarrow[\text{spontaneously}]{-R''OMgX}$$

Ester — Initial product (unstable)

$$\left[R'(R)C{=}\ddot{O}{:}\right] \xrightarrow{RMgX} R{-}C(R')(R){-}\ddot{O}MgX \xrightarrow[H_2O]{NH_4Cl} R{-}C(R')(R){-}OH$$

Ketone — Salt of an alcohol (not isolated) — 3° Alcohol

Specific examples of these reactions are shown here.

Grignard Reagent	Carbonyl Reactant				Final Product

Reaction with Formaldehyde

$$C_6H_5MgBr + \underset{\text{H}}{\overset{\text{H}}{>}}C{=}O \xrightarrow{Et_2O} C_6H_5CH_2{-}OMgBr \xrightarrow{H_3O^+} C_6H_5CH_2OH$$

Phenylmagnesium bromide — **Formaldehyde** — **Benzyl alcohol (90%)**

Reaction with a Higher Aldehyde

$$CH_3CH_2MgBr + \underset{\text{H}}{\overset{CH_3}{>}}C{=}O \xrightarrow{Et_2O} CH_3CH_2\underset{H}{\overset{CH_3}{C}}{-}OMgBr \xrightarrow{H_3O^+} CH_3CH_2\underset{OH}{CH}CH_3$$

Ethylmagnesium bromide — **Acetaldehyde** — **2-Butanol (80%)**

Reaction with a Ketone

$$CH_3CH_2CH_2CH_2MgBr + \underset{CH_3}{\overset{CH_3}{>}}C{=}O \xrightarrow{Et_2O} CH_3CH_2CH_2CH_2\underset{CH_3}{\overset{CH_3}{C}}{-}OMgBr \xrightarrow[H_2O]{NH_4Cl} CH_3CH_2CH_2CH_2\underset{OH}{\overset{CH_3}{C}}{-}CH_3$$

Butylmagnesium bromide — **Acetone** — **2-Methyl-2-hexanol (92%)**

Reaction with an Ester

$$CH_3CH_2MgBr + \underset{C_2H_5O}{\overset{CH_3}{>}}C{=}O \xrightarrow{Et_2O} \left[CH_3CH_2{-}\underset{OC_2H_5}{\overset{CH_3}{C}}{-}OMgBr\right] \xrightarrow{-C_2H_5OMgBr}$$

Ethylmagnesium bromide — **Ethyl acetate**

$$\left[\underset{CH_3CH_2}{\overset{CH_3}{>}}C{=}O\right] \xrightarrow{CH_3CH_2MgBr} CH_3CH_2\underset{OMgBr}{\overset{CH_3}{C}}{-}CH_2CH_3 \xrightarrow[H_2O]{NH_4Cl} CH_3CH_2\underset{OH}{\overset{CH_3}{C}}CH_2CH_3$$

3-Methyl-3-pentanol (67%)

PROBLEM 12.8

Phenylmagnesium bromide reacts with benzoyl chloride, $C_6H_5\overset{O}{\overset{\|}{C}}Cl$, to form triphenylmethanol, $(C_6H_5)_3COH$. This reaction is typical of the reaction of Grignard reagents with acyl chlorides, and the mechanism is similar to that for the reaction of a Grignard reagent with an ester just shown. Show the steps that lead to the formation of triphenylmethanol.

12.8A Planning a Grignard Synthesis

By using Grignard syntheses skillfully, we can synthesize almost any alcohol we wish. In planning a Grignard synthesis we must simply choose the correct Grignard reagent and the correct aldehyde, ketone, ester, or epoxide. We do this by examining the alcohol we wish to prepare and by paying special attention to the groups attached to the carbon atom bearing the —OH group. Many times there may be more than one way of carrying out the synthesis. In these cases our final choice will probably be dictated by the availability of starting compounds. Let us consider an example.

Suppose we want to prepare 3-phenyl-3-pentanol. We examine its structure and we see that the groups attached to the carbon atom bearing the —OH are a *phenyl group* and *two ethyl groups:*

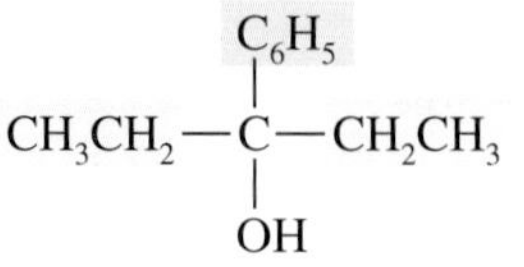

$$CH_3CH_2-\overset{C_6H_5}{\underset{OH}{C}}-CH_2CH_3$$

3-Phenyl-3-pentanol

This means that we can synthesize this compound in several different ways:

1. We can use a ketone with two ethyl groups (3-pentanone) and allow it to react with phenylmagnesium bromide:

Retrosynthetic Analysis

$$CH_3CH_2-\overset{C_6H_5}{\underset{OH}{C}}-CH_2CH_3 \Longrightarrow CH_3CH_2-\underset{O}{\overset{\|}{C}}-CH_2CH_3 + C_6H_5MgBr$$

Synthesis

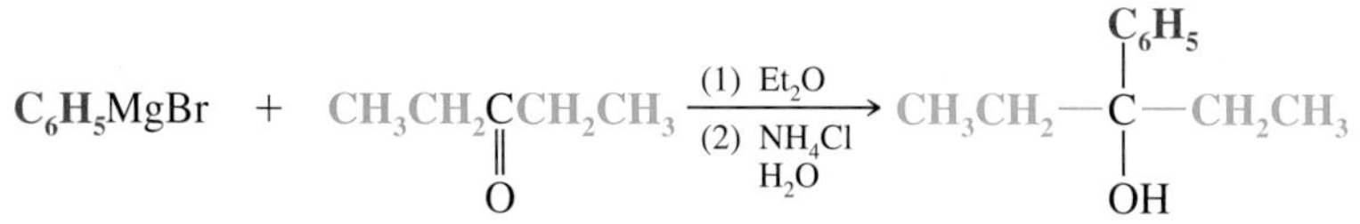

$$\underset{\textbf{Phenylmagnesium bromide}}{C_6H_5MgBr} + \underset{\textbf{3-Pentanone}}{CH_3CH_2\underset{O}{\overset{\|}{C}}CH_2CH_3} \xrightarrow[(2)\ NH_4Cl,\ H_2O]{(1)\ Et_2O} \underset{\textbf{3-Phenyl-3-pentanol}}{CH_3CH_2-\overset{C_6H_5}{\underset{OH}{C}}-CH_2CH_3}$$

2. We can use a ketone containing an ethyl group and a phenyl group (ethyl phenyl ketone) and allow it to react with ethylmagnesium bromide:

Retrosynthetic Analysis

$$CH_3CH_2-\overset{C_6H_5}{\underset{OH}{C}}-CH_2CH_3 \Longrightarrow CH_3CH_2-C(=O)C_6H_5 + CH_3CH_2MgBr$$

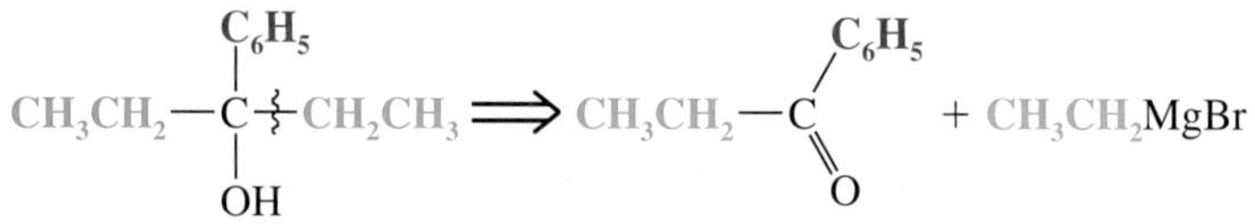

Synthesis

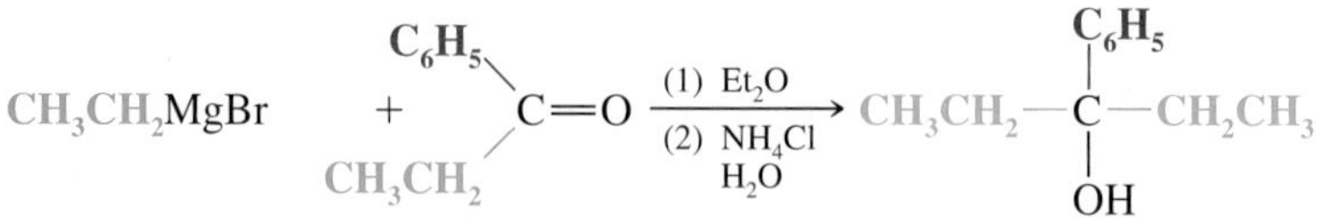

$$\underset{\textbf{Ethylmagnesium bromide}}{CH_3CH_2MgBr} + \underset{\textbf{Ethyl phenyl ketone}}{(C_6H_5)(CH_3CH_2)C{=}O} \xrightarrow[(2)\ NH_4Cl,\ H_2O]{(1)\ Et_2O} \underset{\textbf{3-Phenyl-3-pentanol}}{CH_3CH_2-\overset{C_6H_5}{\underset{OH}{C}}-CH_2CH_3}$$

3. We can use an ester of benzoic acid and allow it to react with two molar equivalents of ethylmagnesium bromide:

Retrosynthetic Analysis

$$CH_3CH_2-\overset{C_6H_5}{\underset{OH}{C}}-CH_2CH_3 \Longrightarrow C_6H_5C(=O)OCH_3 + 2\ CH_3CH_2MgBr$$

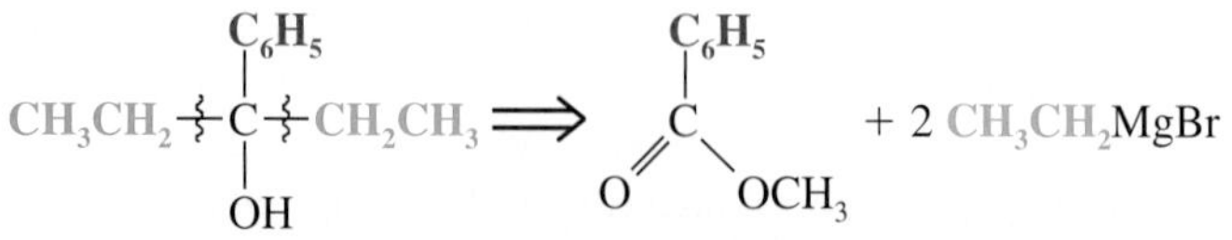

Synthesis

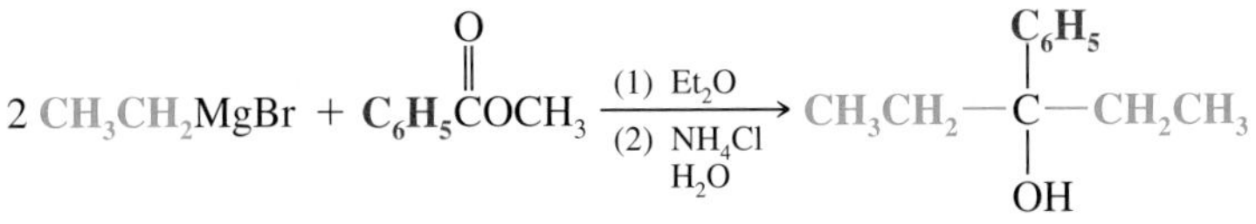

$$2\,CH_3CH_2MgBr + C_6H_5\overset{\overset{\displaystyle O}{\|}}{C}OCH_3 \xrightarrow[(2)\ NH_4Cl,\ H_2O]{(1)\ Et_2O} CH_3CH_2-\underset{\underset{\displaystyle OH}{|}}{\overset{\overset{\displaystyle C_6H_5}{|}}{C}}-CH_2CH_3$$

Ethylmagnesium bromide + **Methyl benzoate** → **3-Phenyl-3-pentanol**

All of these methods will be likely to give us our desired compound in yields greater than 80%.

SOLVED PROBLEM

ILLUSTRATING A MULTISTEP SYNTHESIS Using an alcohol of no more than four carbon atoms as your only organic starting material, outline a synthesis of **A**:

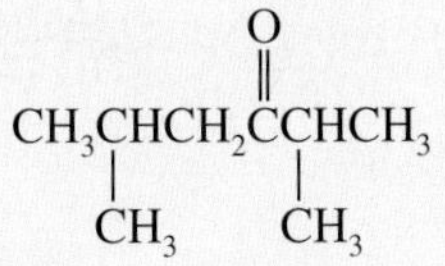

$$\underset{\underset{\displaystyle CH_3}{|}}{CH_3CH}CH_2\overset{\overset{\displaystyle O}{\|}}{C}\underset{\underset{\displaystyle CH_3}{|}}{CH}CH_3$$

A

ANSWER: We can construct the carbon skeleton from two four-carbon compounds using a Grignard reaction. Then oxidation of the alcohol produced will yield the desired ketone.

Retrosynthetic Analysis

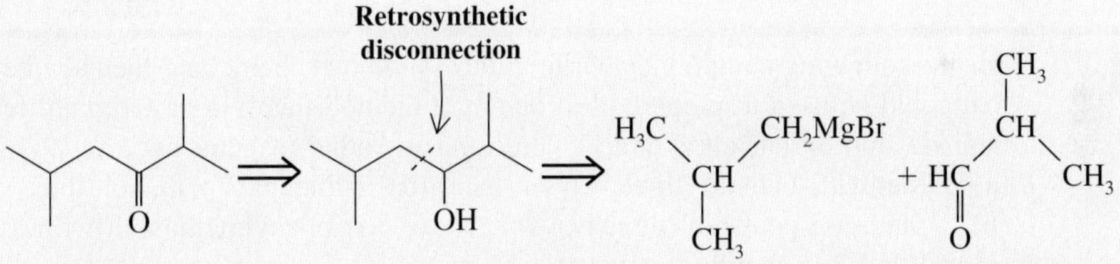

Retrosynthetic disconnection

A (ketone, skeletal) ⟹ (alcohol, OH, skeletal) ⟹ $H_3C\underset{\underset{\displaystyle CH_3}{|}}{CH}CH_2MgBr$ (**B**) + $H\overset{\overset{\displaystyle O}{\|}}{C}\underset{}{CH}(CH_3)CH_3$ (**C**)

Synthesis

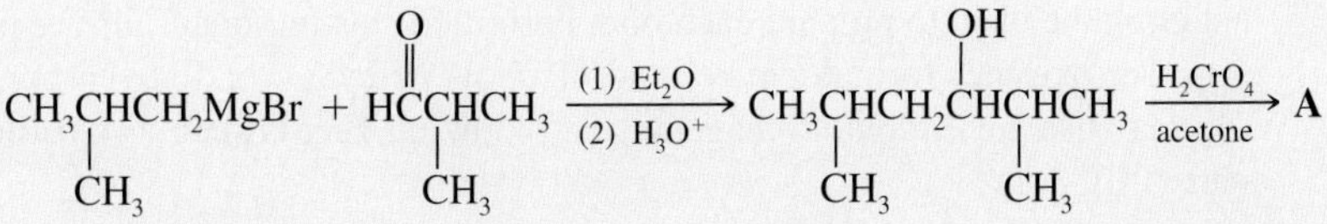

$$\underset{\mathbf{B}}{\underset{\underset{\displaystyle CH_3}{|}}{CH_3CH}CH_2MgBr} + \underset{\mathbf{C}}{H\overset{\overset{\displaystyle O}{\|}}{C}\underset{\underset{\displaystyle CH_3}{|}}{CH}CH_3} \xrightarrow[(2)\ H_3O^+]{(1)\ Et_2O} \underset{\underset{\displaystyle CH_3}{|}}{CH_3CH}CH_2\overset{\overset{\displaystyle OH}{|}}{CH}\underset{\underset{\displaystyle CH_3}{|}}{CH}CH_3 \xrightarrow[\text{acetone}]{H_2CrO_4} \mathbf{A}$$

We can synthesize the Grignard reagent (**B**) and the aldehyde (**C**) from isobutyl alcohol:

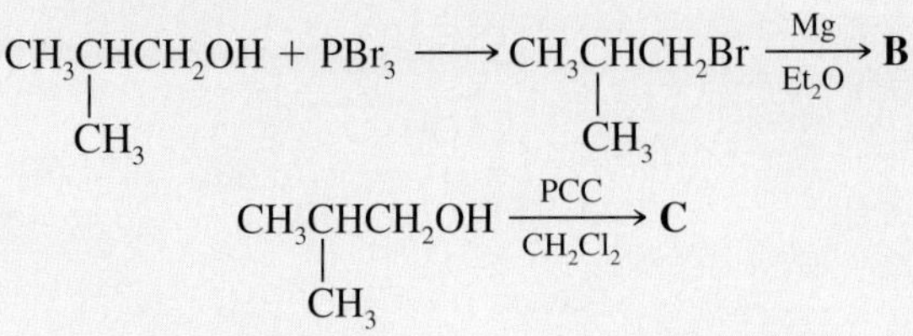

$$\underset{\underset{\displaystyle CH_3}{|}}{CH_3CH}CH_2OH + PBr_3 \longrightarrow \underset{\underset{\displaystyle CH_3}{|}}{CH_3CH}CH_2Br \xrightarrow[Et_2O]{Mg} \mathbf{B}$$

$$\underset{\underset{\displaystyle CH_3}{|}}{CH_3CH}CH_2OH \xrightarrow[CH_2Cl_2]{PCC} \mathbf{C}$$

SOLVED PROBLEM

ILLUSTRATING A MULTISTEP SYNTHESIS Starting with bromobenzene and any other needed reagents, outline a synthesis of the following aldehyde:

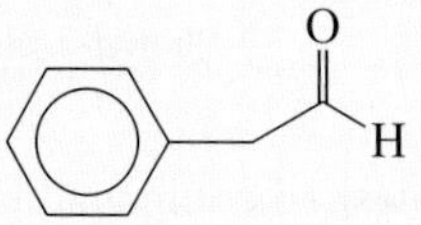

$$C_6H_5CH_2CHO$$

ANSWER: Working backward, we remember that we can synthesize the aldehyde from the corresponding alcohol by oxidation with PCC (Section 12.4A). The alcohol can be made by treating phenylmagnesium bromide with oxirane. [Adding oxirane to a Grignard reagent is a very useful method for adding a $—CH_2CH_2OH$ unit to an organic group (Section 12.7B).] Phenylmagnesium bromide can be made in the usual way, by treating bromobenzene with magnesium in an ether solvent.

Retrosynthetic Analysis

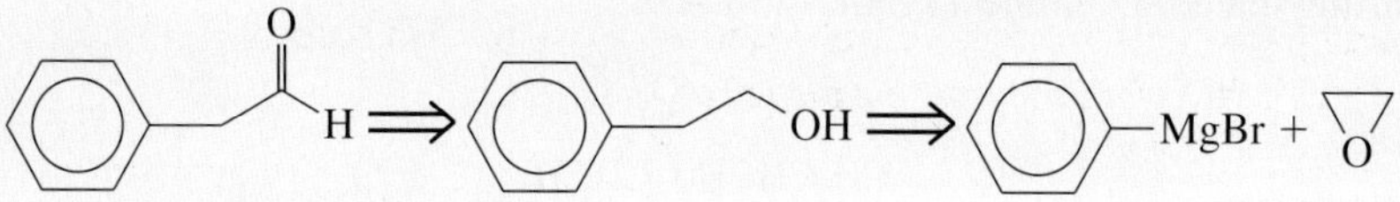

$$C_6H_5CH_2CHO \Longrightarrow C_6H_5CH_2CH_2OH \Longrightarrow C_6H_5MgBr + \text{oxirane}$$

Synthesis

$$C_6H_5Br \xrightarrow[Et_2O]{Mg} C_6H_5MgBr \xrightarrow[(2)\ H_3O^+]{(1)\ \text{oxirane}} C_6H_5CH_2CH_2OH \xrightarrow[CH_2Cl_2]{PCC} C_6H_5CH_2CHO$$

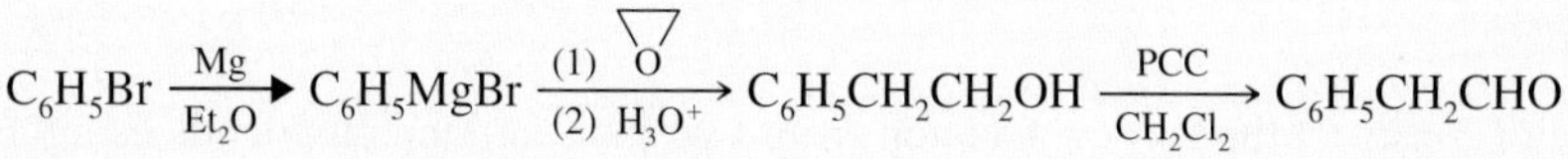

PROBLEM 12.9 For the following compounds, write a retrosynthetic scheme and then synthetic reactions that could be used to prepare each one by a method involving a Grignard reaction. Start with any appropriate alkyl or aryl halide and any other reagents necessary.

(a) 2-Methyl-2-butanol (three ways)
(b) 3-Methyl-3-pentanol (three ways)
(c) 3-Ethyl-2-pentanol (two ways)
(d) 2-Phenyl-2-pentanol (three ways)
(e) Triphenylmethanol (two ways)

PROBLEM 12.10 For the following compounds, write a retrosynthetic scheme and then synthetic reactions that could be used to prepare each one. Permitted starting materials are phenylmagnesium bromide, oxirane, formaldehyde, and alcohols or esters of four carbon atoms or fewer. You may use any inorganic reagents and oxidizing agents such as pyridinium chlorochromate (PCC).

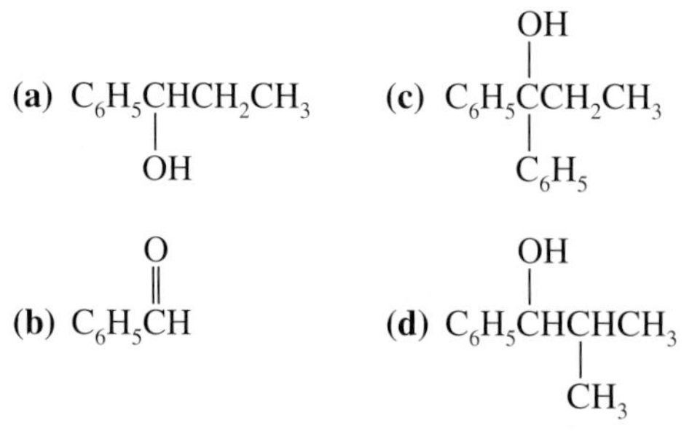

(a) $C_6H_5CH(OH)CH_2CH_3$

(b) C_6H_5CHO

(c) $C_6H_5C(OH)(C_6H_5)CH_2CH_3$

(d) $C_6H_5CH(OH)CH(CH_3)CH_3$

12.8B Restrictions on the Use of Grignard Reagents

Although the Grignard synthesis is one of the most versatile of all general synthetic procedures, it is not without its limitations. Most of these limitations arise from the very feature of the Grignard reagent that makes it so useful, its *extraordinary reactivity as a nucleophile and a base.*

The Grignard reagent is a very powerful base; in effect it contains a carbanion. Thus, it is not possible to prepare a Grignard reagent from an organic group that contains an *acidic hydrogen;* by an acidic hydrogen, we mean any hydrogen more acidic than the hydrogen atoms of an alkane or alkene. We cannot, for example, prepare a Grignard reagent from a compound containing an —OH group, an —NH— group, an —SH group, a $—CO_2H$ group, or an $—SO_3H$ group. If we were to attempt to prepare a Grignard reagent from an organic halide containing any of these groups, the formation of the Grignard reagent would simply fail to take place. (Even if a Grignard reagent were to form, it would immediately react with the acidic group.)

Since Grignard reagents are powerful nucleophiles, we cannot prepare a Grignard reagent from any organic halide that contains a carbonyl, epoxy, nitro, or cyano (—CN) group. If we were to attempt to carry out this kind of reaction, any Grignard reagent that formed would only react with the unreacted starting material:

$$—OH, —NH_2, —NHR, —CO_2H, —SO_3H, —SH, —C{\equiv}C—H$$

$$—\overset{O}{\overset{\|}{C}}H, —\overset{O}{\overset{\|}{C}}R, —\overset{O}{\overset{\|}{C}}OR, —\overset{O}{\overset{\|}{C}}NH_2, —NO_2, —C{\equiv}N, —\overset{|}{C}\underset{O}{\diagdown\diagup}\overset{|}{C}—$$

Grignard reagents containing these groups cannot be prepared.

A protecting group can sometimes be used to mask the reactivity of an incompatible group (see Sections 11.11C, 11.11D, and 12.10).

This means that when we prepare Grignard reagents, we are effectively limited to alkyl halides or to analogous organic halides containing carbon–carbon double bonds, internal triple bonds, ether linkages, and $—NR_2$ groups.

Grignard reactions are so sensitive to acidic compounds that when we prepare a Grignard reagent we must take special care to exclude moisture from our apparatus, and we must use an anhydrous ether as our solvent.

As we saw earlier, acetylenic hydrogens are acidic enough to react with Grignard reagents. This is a limitation that we can use, however. We can make acetylenic Grignard reagents by allowing terminal alkynes to react with alkyl Grignard reagents (cf. Section 12.7A). We can then use these acetylenic Grignard reagents to carry out other syntheses. For example,

$$C_6H_5C{\equiv}CH + C_2H_5MgBr \longrightarrow C_6H_5C{\equiv}CMgBr + C_2H_6\uparrow$$

$$C_6H_5C{\equiv}CMgBr + C_2H_5\overset{O}{\overset{\|}{C}}H \xrightarrow[(2)\ H_3O^+]{} C_6H_5C{\equiv}C—\underset{OH}{\underset{|}{C}}HC_2H_5$$

(52%)

When we plan Grignard syntheses, we must also take care not to plan a reaction in which a Grignard reagent is treated with an aldehyde, ketone, epoxide, or ester that contains an acidic group (other than when we deliberately let it react with a terminal alkyne). If we were to do this, the Grignard reagent would simply react as a base with the acidic hydrogen rather than reacting at the carbonyl or epoxide carbon as a nucleophile. If we were to treat 4-hydroxy-2-butanone with methylmagnesium bromide, for example, the reaction that would take place first is

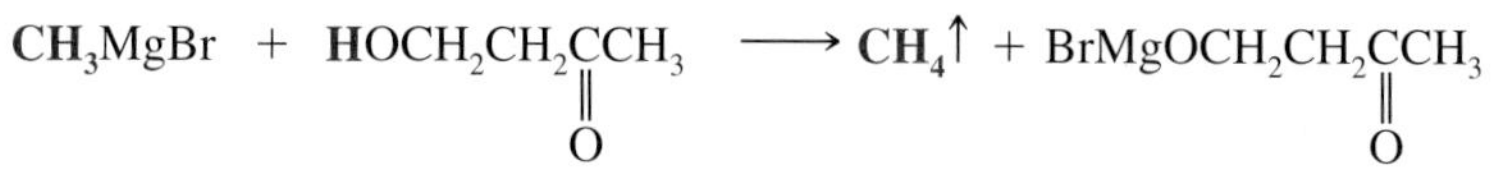

$$CH_3MgBr + HOCH_2CH_2C(=O)CH_3 \longrightarrow CH_4\uparrow + BrMgOCH_2CH_2C(=O)CH_3$$

4-Hydroxy-2-butanone

rather than

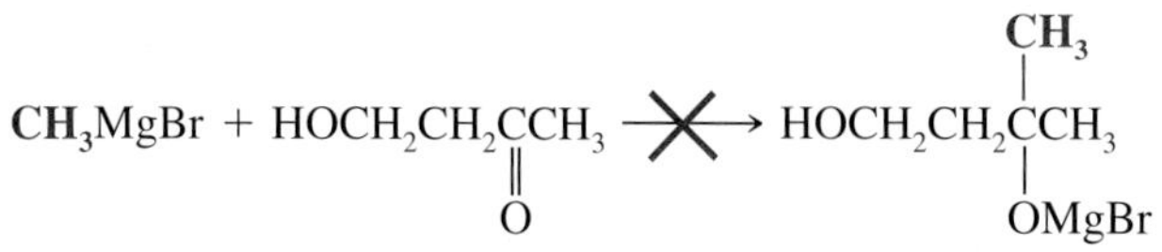

$$CH_3MgBr + HOCH_2CH_2C(=O)CH_3 \not\longrightarrow HOCH_2CH_2C(CH_3)(OMgBr)CH_3$$

If we were prepared to waste one molar equivalent of the Grignard reagent, we can treat 4-hydroxy-2-butanone with two molar equivalents of the Grignard reagent and thereby get addition to the carbonyl group:

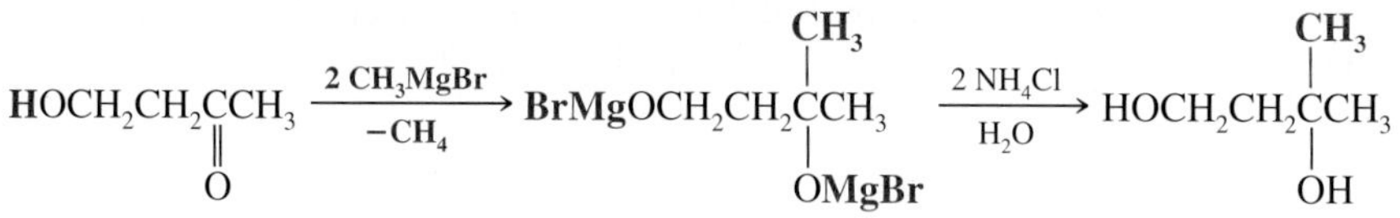

$$HOCH_2CH_2C(=O)CH_3 \xrightarrow[-CH_4]{2\ CH_3MgBr} BrMgOCH_2CH_2C(CH_3)(OMgBr)CH_3 \xrightarrow[H_2O]{2\ NH_4Cl} HOCH_2CH_2C(CH_3)(OH)CH_3$$

This technique is sometimes employed in small-scale reactions when the Grignard reagent is inexpensive and the other reagent is expensive.

12.8C The Use of Lithium Reagents

Organolithium reagents (RLi) react with carbonyl compounds in the same way as Grignard reagents and thus provide an alternative method for preparing alcohols.

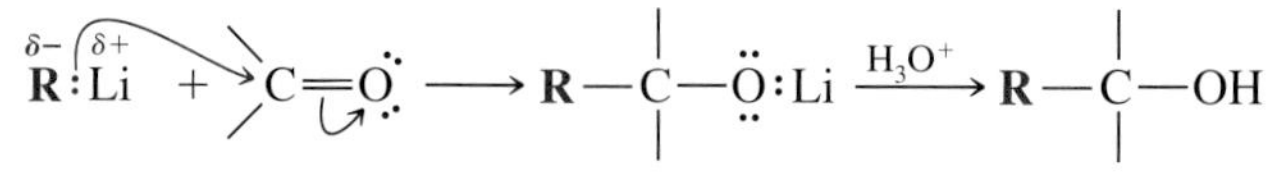

$$\overset{\delta-}{R}:\overset{\delta+}{Li} + \rangle C{=}\ddot{O}: \longrightarrow R{-}\overset{|}{\underset{|}{C}}{-}\ddot{O}:Li \xrightarrow{H_3O^+} R{-}\overset{|}{\underset{|}{C}}{-}OH$$

Organolithium reagent — **Aldehyde or ketone** — **Lithium alkoxide** — **Alcohol**

Organolithium reagents have the advantage of being somewhat more reactive than Grignard reagents.

12.8D The Use of Sodium Alkynides

Sodium alkynides also react with aldehydes and ketones to yield alcohols. An example is the following:

$$CH_3C{\equiv}CH \xrightarrow[-NH_3]{NaNH_2} CH_3C{\equiv}CNa$$

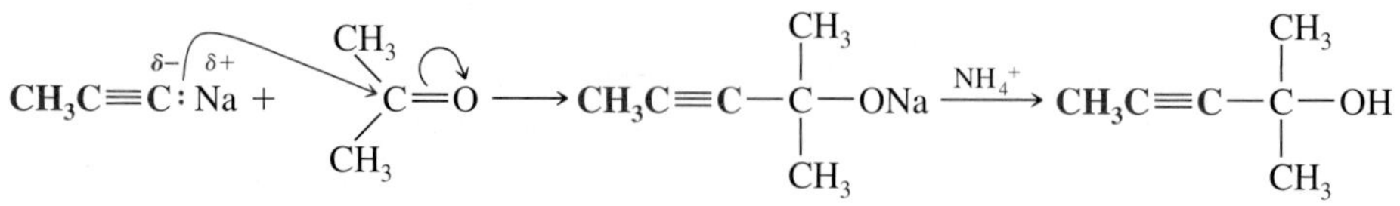

$$CH_3C{\equiv}\overset{\delta-}{C}:\overset{\delta+}{Na} + (CH_3)_2C{=}O \longrightarrow CH_3C{\equiv}C{-}C(CH_3)_2{-}ONa \xrightarrow{NH_4^+} CH_3C{\equiv}C{-}C(CH_3)_2{-}OH$$

SOLVED PROBLEM

ILLUSTRATING MULTISTEP SYNTHESES For the following compounds, write a retrosynthetic scheme and then synthetic reactions that could be used to prepare each one. Use hydrocarbons, organic halides, alcohols, aldehydes, ketones, or esters containing six carbon atoms or fewer and any other needed reagents.

(a) 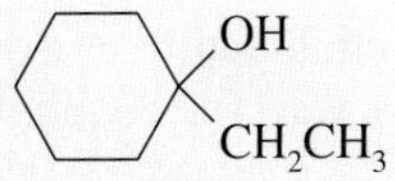 1-ethylcyclohexanol (cyclohexane ring bearing OH and CH_2CH_3) (b) 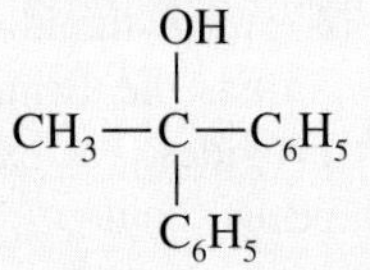 $CH_3-C(OH)(C_6H_5)-C_6H_5$ (c) 1-ethynylcyclopentanol (cyclopentane ring bearing OH and $C{\equiv}CH$)

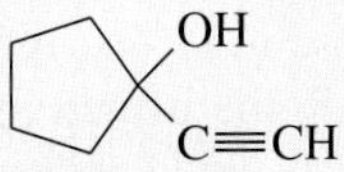

ANSWERS:

(a) ***Retrosynthetic Analysis***

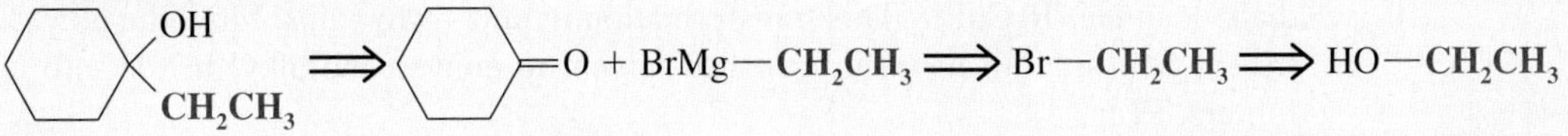

$$\text{1-ethylcyclohexanol (OH, } CH_2CH_3) \Rightarrow \text{cyclohexanone}{=}O + BrMg-CH_2CH_3 \Rightarrow Br-CH_2CH_3 \Rightarrow HO-CH_2CH_3$$

Synthesis

$$CH_3CH_2OH \xrightarrow{PBr_3} CH_3CH_2Br \xrightarrow[Et_2O]{Mg} CH_3CH_2MgBr \xrightarrow[(2)\ NH_4Cl,\ H_2O]{(1)\ \text{cyclohexanone}} \text{1-ethylcyclohexanol (HO, } CH_2CH_3)$$

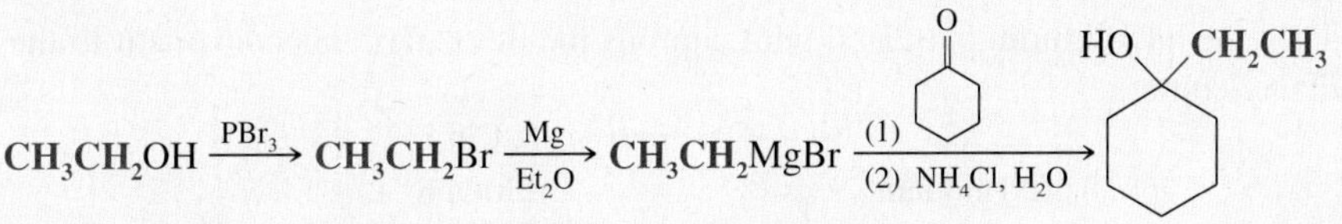

(b) ***Retrosynthetic Analysis***

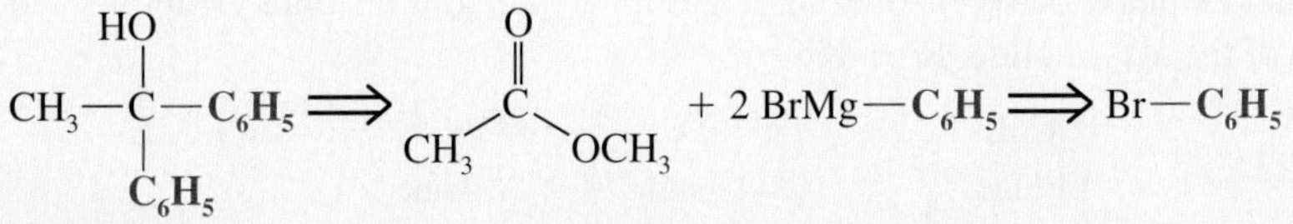

$$CH_3-C(HO)(C_6H_5)-C_6H_5 \Rightarrow CH_3C(=O)OCH_3 + 2\ BrMg-C_6H_5 \Rightarrow Br-C_6H_5$$

Synthesis

$$C_6H_5Br \xrightarrow[Et_2O]{Mg} C_6H_5MgBr \xrightarrow[(2)\ NH_4Cl,\ H_2O]{(1)\ CH_3\overset{O}{\overset{\|}{C}}OCH_3} CH_3-C(OH)(C_6H_5)-C_6H_5$$

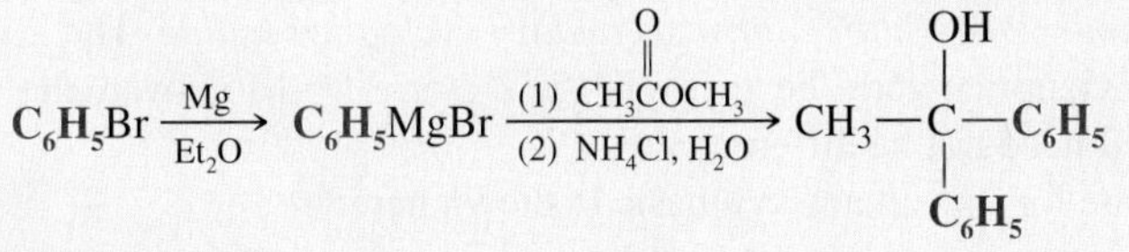

(c) ***Retrosynthetic Analysis***

$$\text{1-ethynylcyclopentanol (HO, } C{\equiv}CH) \Rightarrow \text{cyclopentanone} + HC{\equiv}C{:}^{-}Na^{+} \Rightarrow HC{\equiv}CH$$

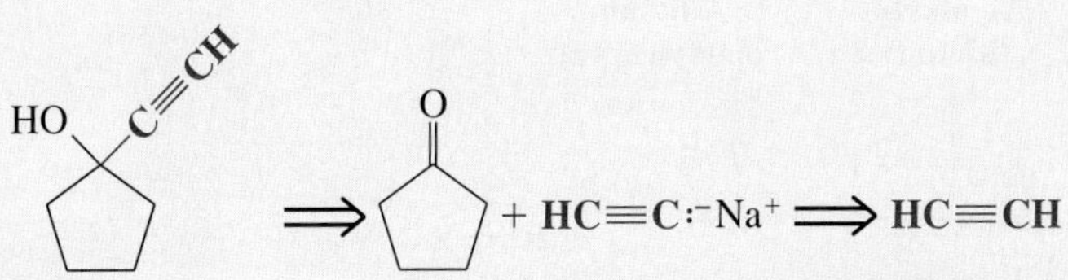

Synthesis

$$HC{\equiv}CH \xrightarrow{NaNH_2} HC{\equiv}CNa \xrightarrow[(2)\ NH_4Cl,\ H_2O]{(1)\ \text{cyclopentanone}} \text{1-ethynylcyclopentanol (HO, } C{\equiv}CH)$$

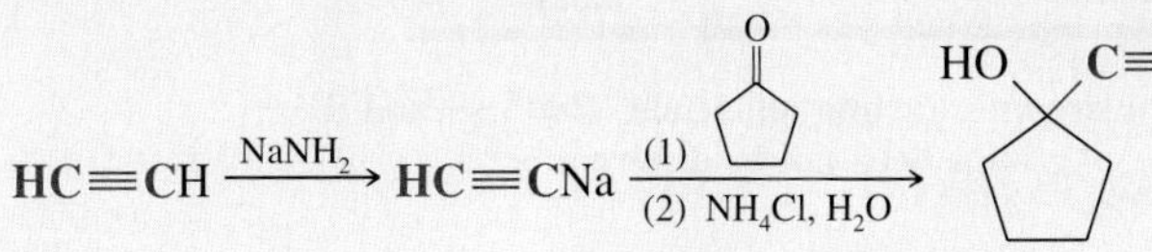

12.9 Lithium Dialkylcuprates: The Corey–Posner, Whitesides–House Synthesis

Corey was awarded the Nobel Prize in Chemistry in 1990 for finding new ways of synthesizing organic compounds, which, in the words of the Nobel committee, "have contributed to the high standards of living and health enjoyed . . . in the Western world."

A highly versatile method for the synthesis of alkanes and other hydrocarbons from organic halides has been developed by E. J. Corey (Harvard University) and G. H. Posner (Johns Hopkins University) and by G. M. Whitesides (Harvard University) and H. O. House (Georgia Institute of Technology). Although it does not create a new functional group for use in further reactions, as does the Grignard reaction and others discussed in Section 12.8, the overall synthesis provides, for example, a way for coupling the alkyl groups of two alkyl halides to produce an alkane:

$$\text{R—X} + \mathbf{R'—X} \xrightarrow[\text{steps}\ (-2\,\text{X})]{\text{several}} \text{R—}\mathbf{R'}$$

In order to accomplish this coupling, we must transform one alkyl halide into a lithium dialkylcuprate (R_2CuLi). This transformation requires two steps. First, the alkyl halide is treated with lithium metal in an ether solvent to convert the alkyl halide into an alkyllithium, RLi:

$$\text{R—X} + 2\,\text{Li} \xrightarrow{\text{diethyl ether}} \underset{\textbf{Alkyllithium}}{\text{RLi}} + \text{LiX}$$

Lithium dialkylcuprates were first synthesized by Henry Gilman (of Iowa State University) and are often called Gilman reagents.

Then the alkyllithium is treated with cuprous iodide (CuI). This converts it to the lithium dialkylcuprate:

$$\underset{\textbf{Alkyllithium}}{2\,\text{RLi}} + \text{CuI} \longrightarrow \underset{\textbf{Lithium dialkylcuprate}}{\text{R}_2\text{CuLi}} + \text{LiI}$$

When the lithium dialkylcuprate is treated with the second alkyl halide (R′—X), coupling takes place between one alkyl group of the lithium dialkylcuprate and the alkyl group of the alkyl halide, R′—X:

$$\underset{\textbf{Lithium dialkylcuprate}}{\text{R}_2\text{CuLi}} + \underset{\textbf{Alkyl halide}}{\mathbf{R'—X}} \longrightarrow \underset{\textbf{Alkane}}{\text{R—}\mathbf{R'}} + \text{RCu} + \text{LiX}$$

For the last step to give a good yield of the alkane, the alkyl halide R′—X must be a methyl halide, a primary alkyl halide, or a secondary cycloalkyl halide. The alkyl groups of the lithium dialkylcuprate may be methyl, 1°, 2°, or 3°.* Moreover, the two alkyl groups being coupled need not be different.

The overall scheme for this alkane synthesis is shown here:

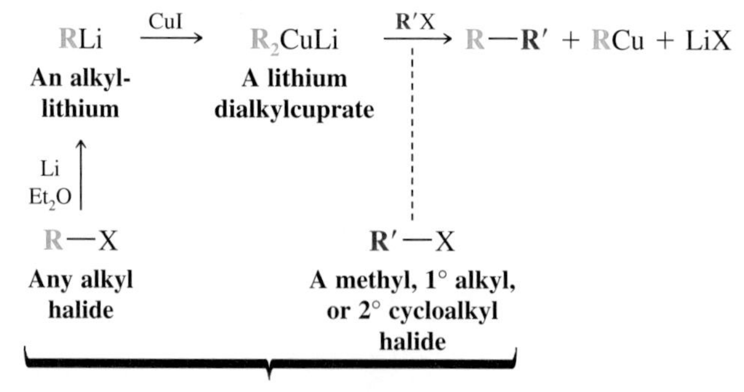

*Special techniques, which we shall not discuss here, are required when R is tertiary. For an excellent review of these reactions, see Posner, G. H. Substitution Reactions Using Organocopper Reagents. In *Organic Reactions,* Vol. 22; Wiley: New York, 1975; pp 253–400.

Consider as examples the synthesis of hexane from methyl iodide and pentyl iodide and the synthesis of nonane from butyl bromide and pentyl bromide:

$$CH_3I \xrightarrow[Et_2O]{Li} CH_3Li \xrightarrow{CuI} (CH_3)_2CuLi \xrightarrow{CH_3CH_2CH_2CH_2CH_2I} CH_3{-}CH_2CH_2CH_2CH_2CH_3$$

Hexane
(98%)

$$CH_3CH_2CH_2CH_2Br \xrightarrow[Et_2O]{Li} CH_3CH_2CH_2CH_2Li \xrightarrow{CuI}$$

$$(CH_3CH_2CH_2CH_2)_2CuLi \xrightarrow{CH_3CH_2CH_2CH_2CH_2Br} CH_3CH_2CH_2CH_2{-}CH_2CH_2CH_2CH_2CH_3$$

Nonane
(98%)

Lithium dialkylcuprates couple with other organic groups. Coupling reactions of lithium dimethylcuprate with two cycloalkyl halides are shown here:

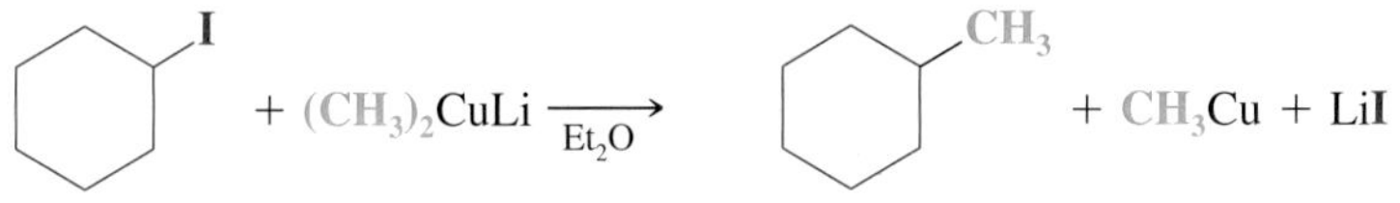

$$\text{cyclohexyl-I} + (CH_3)_2CuLi \xrightarrow[Et_2O]{} \text{cyclohexyl-}CH_3 + CH_3Cu + LiI$$

Methylcyclohexane
(75%)

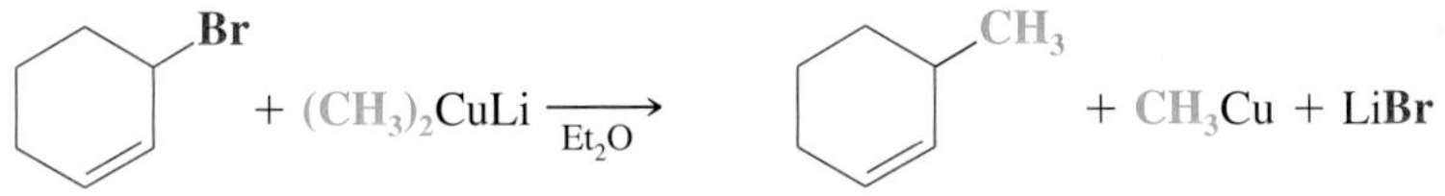

$$\text{3-bromocyclohexene} + (CH_3)_2CuLi \xrightarrow[Et_2O]{} \text{3-methylcyclohexene} + CH_3Cu + LiBr$$

3-Methylcyclohexene
(75%)

Lithium dialkylcuprates also couple with phenyl and vinyl halides. An example with a phenyl halide is the following synthesis of butylbenzene:

$$(CH_3CH_2CH_2CH_2)_2CuLi + I{-}C_6H_5 \xrightarrow[Et_2O]{} CH_3CH_2CH_2CH_2{-}C_6H_5$$

Butylbenzene
(75%)

The following scheme summarizes the coupling reactions of lithium dialkylcuprates:

R_2CuLi —
- $\xrightarrow{CH_3X \text{ or } R'CH_2X}$ $R{-}CH_3$ or $R'CH_2{-}R$
- $\xrightarrow{\text{cyclohexyl-X}}$ cyclohexyl–R
- $\xrightarrow{\text{phenyl-X}}$ phenyl–R
- $\xrightarrow{\text{vinyl-X}}$ vinyl–R

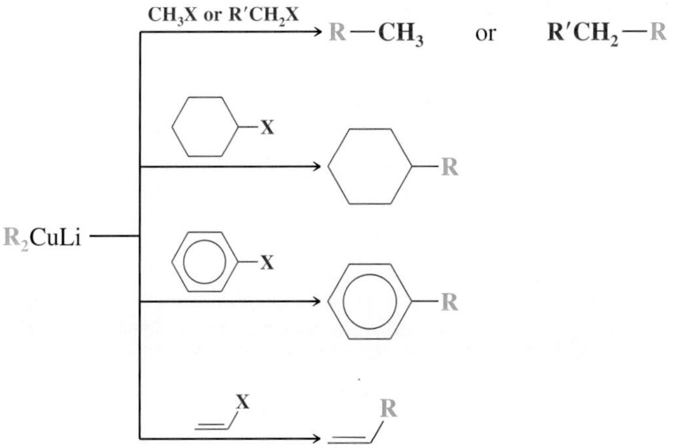

The mechanism of the Corey–Posner, Whitesides–House synthesis is beyond our scope but is of a type discussed in Special Topic G.

12.10 Protecting Groups

A **protecting group** can be used in some cases where a reactant contains a group that is incompatible with the reaction conditions necessary for a given transformation. For example, if it is necessary to prepare a Grignard reagent from an alkyl halide that already contains an alcohol hydroxyl group, the Grignard reagent can still be prepared if the alcohol is first protected by conversion to a functional group that is stable in the presence of a Grignard reagent, for example, a *tert*-butyldimethylsilyl ether (Section 11.11D). The Grignard reaction can be conducted, and then the original alcohol group can be liberated by cleavage of the silyl ether with fluoride ion (see Problem 12.26). This same strategy can be used when an organolithium reagent or alkynide anion must be prepared in the presence of an incompatible group. In later chapters we will encounter strategies that can be used to protect other functional groups during various reactions (Section 16.7C).

SUMMARY OF REACTIONS

Summaries of reactions discussed in this chapter are shown below. Detailed conditions for the reactions that are summarized can be found in the chapter section where each is discussed.

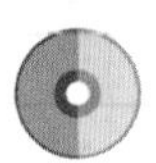

Synthetic Connections of Alcohols and Carbonyl Compounds

1. Carbonyl Reduction Reactions

- Aldehydes to primary alcohols
- Ketones to secondary alcohols
- Esters to alcohols
- Carboxylic acids to primary alcohols

	Substrate		Reducing agent: $NaBH_4$	Reducing agent: $LiAlH_4$ (LAH)
Aldehydes	R—C(=O)—H	$\xrightarrow{[H]}$	R—CH(OH)—H	R—CH(OH)—H
Ketones	R—C(=O)—R'	$\xrightarrow{[H]}$	R—CH(OH)—R'	R—CH(OH)—R'
Esters	R—C(=O)—OR'	$\xrightarrow{[H]}$	—	R—CH_2—OH + HO—R'
Carboxylic acids	R—C(=O)—OH	$\xrightarrow{[H]}$	—	R—CH_2—OH

(Hydrogen atoms in blue are added during the reaction workup by water or aqueous acid.)

2. Alcohol Oxidation Reactions

- Primary alcohols to aldehydes
- Primary alcohols to carboxylic acids
- Secondary alcohols to ketones

	Substrate		Oxidizing agent [O]: PCC	Oxidizing agent [O]: H_2CrO_4	Oxidizing agent [O]: $KMnO_4$
Primary alcohols	R—CH(OH)—H	$\xrightarrow{[O]}$	R—C(=O)—H	R—C(=O)—OH	R—C(=O)—OH
Secondary alcohols	R—CH(OH)—R'	$\xrightarrow{[O]}$	R—C(=O)—R'	R—C(=O)—R'	R—C(=O)—R'
Tertiary alcohols	R—C(OH)(R')—R''	$\xrightarrow{[O]}$	—	—	—

$KMnO_4$ in basic solution followed by acidification, either **A** or **C** produces the meso form of 1,3-cyclopentanedicarboxylic acid (see the following structure). Give structural formulas for **A–C.**

HO_2C CO_2H

1,3-Cyclopentanedicarboxylic acid

9. Starting with propyne and using any other required reagents, show how you would synthesize each of the following compounds. You need not repeat steps carried out in earlier parts of this problem.

(a) 2-Butyne
(b) *cis*-2-Butene
(c) *trans*-2-Butene
(d) 1-Butene
(e) 1,3-Butadiene
(f) 1-Bromobutane
(g) 2-Bromobutane (as a racemic form)
(h) (2*R*,3*S*)-2,3-Dibromobutane
(i) (2*R*,3*R*)- and (2*S*,3*S*)-2,3-Dibromobutane (as a racemic form)
(j) *meso*-2,3-Butanediol
(k) (*Z*)-2-Bromo-2-butene

10. Bromination of 2-methylbutane yields predominantly one product with the formula $C_5H_{11}Br$. What is this product? Show how you could use this compound to synthesize each of the following. (You need not repeat steps carried out in earlier parts.)

(a) 2-Methyl-2-butene
(b) 2-Methyl-2-butanol
(c) 3-Methyl-2-butanol
(d) 3-Methyl-1-butyne
(e) 1-Bromo-3-methylbutane
(f) 2-Chloro-3-methylbutane
(g) 2-Chloro-2-methylbutane
(h) 1-Iodo-3-methylbutane
(i) 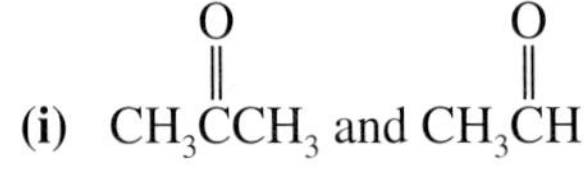

(j) 

11. An alkane **(A)** with the formula C_6H_{14} reacts with chlorine to yield three compounds with the formula $C_6H_{13}Cl$: **B, C,** and **D.** Of these only **C** and **D** undergo dehydrohalogenation with sodium ethoxide in ethanol to produce an alkene. Moreover, **C** and **D** yield the same alkene **E** (C_6H_{12}). Hydrogenation of **E** produces **A.** Treating **E** with HCl produces a compound **(F)** that is an isomer of **B, C,** and **D.** Treating **F** with Zn and acetic acid gives a compound **(G)** that is isomeric with **A.** Propose structures for **A–G.**

12. Compound **A** (C_4H_6) reacts with hydrogen and a platinum catalyst to yield butane. Compound **A** reacts with Br_2 in CCl_4 and aqueous $KMnO_4$. The IR spectrum of **A** does not have an absorption in the 2200–2300-cm^{-1} region. On treatment with hydrogen and Ni_2B (P-2 catalyst), **A** is converted to **B** (C_4H_8). When **B** is treated with OsO_4 and then with $NaHSO_3$, **B** is converted to C ($C_4H_{10}O_2$). Compound **C** cannot be resolved. Provide structures for **A–C.**

13. Dehalogenation of *meso*-2,3-dibromobutane occurs when it is treated with potassium iodide in ethanol. The product is *trans*-2-butene. Similar dehalogenation of either of the enantiomeric forms of 2,3-dibromobutane produces *cis*-2-butene. Give a mechanistic explanation of these results.

14. Dehydrohalogenation of *meso*-1,2-dibromo-1,2-diphenylethane by the action of sodium ethoxide in ethanol yields (*E*)-1-bromo-1,2-diphenylethene. Similar dehydrohalogenation of either of the enantiomeric forms of 1,2-dibromo-1,2-diphenylethane yields (*Z*)-1-bromo-1,2-diphenylethene. Provide an explanation for the results.

15. Give conformational structures for the major product formed when 1-*tert*-butylcyclohexene reacts with each of the following reagents. If the product would be obtained as a racemic form, you should so indicate.

(a) Br_2, CCl_4
(b) OsO_4, then aqueous $NaHSO_3$
(c) $C_6H_5CO_3H$, then H_3O^+, H_2O
(d) BH_3:THF, then H_2O_2, OH^-

(e) $Hg(OAc)_2$ in $THF-H_2O$, then $NaBH_4$, OH^-
(f) Br_2, H_2O
(g) ICl
(h) O_3, then Zn, HOAc (conformational structure not required)
(i) D_2, Pt
(j) BD_3:THF, then CH_3CO_2T

16. Give structures for **A–C**.

$$\underset{\underset{\text{Br}}{|}}{\overset{\overset{\text{CH}_3}{|}}{\text{CH}_3\text{CCH}_2\text{CH}_2\text{CH}_3}} \xrightarrow{\text{EtO}^-/\text{EtOH}} \textbf{A}\ (\text{C}_6\text{H}_{12})\ \text{major product} \xrightarrow{\text{BH}_3\text{:THF}} \textbf{B}\ (\text{C}_6\text{H}_{13})_2\text{BH} \xrightarrow{\text{H}_2\text{O}_2,\ \text{OH}^-} \textbf{C}\ (\text{C}_6\text{H}_{14}\text{O})$$

17. (*R*)-3-Methyl-1-pentene is treated separately with the following reagents and the products in each case are separated by fractional distillation. Write appropriate formulas for all of the components of each fraction and tell whether each fraction would be optically active.

(a) Br_2, CCl_4
(b) H_2, Pt
(c) OsO_4, then $NaHSO_3$
(d) BH_3:THF, then H_2O_2, OH^-
(e) $Hg(OAc)_2$, $THF-H_2O$, then $NaBH_4$, OH^-
(f) Magnesium perphthalate, then H_3O^+, H_2O

18. Compound **A** ($C_8H_{15}Cl$) exists as a racemic form. Compound **A** does not react with either Br_2/CCl_4 or dilute aqueous $KMnO_4$. When **A** is treated with zinc and acetic acid and the mixture is separated by gas chromatography, two fractions, **B** and **C**, are obtained. The components of both fractions have the formula C_8H_{16}. Fraction **B** consists of a racemic form and can be resolved. Fraction **C** cannot be resolved. Treating **A** with sodium ethoxide in ethanol converts **A** into **D** (C_8H_{14}). Hydrogenation of **D** using a platinum catalyst yields **C.** Ozonolysis of **D** followed by treatment with zinc and acetic acid yields

$$\text{CH}_3\overset{\overset{\text{O}}{\|}}{\text{C}}\text{CH}_2\text{CH}_2\text{CH}_2\text{CH}_2\overset{\overset{\text{O}}{\|}}{\text{C}}\text{CH}_3$$

Propose structures for **A, B, C,** and **D**, including, where appropriate, their stereochemistry.

19. Elucidate the structure of the compound that gives the following spectroscopic data. Assign the data to specific aspects of your proposed structure.

MS (*m*/*z*); 120, 105 (base peak), 77

^{1}H NMR (δ): 7.2–7.6 (m, 5H), 2.95 (septet, 1H), 1.29 (d, 6H)

20. Compound **X** ($C_5H_{10}O$) shows a strong IR absorption band near 1710 cm^{-1}. The broadband proton-decoupled ^{13}C NMR spectrum of **X** is shown in Fig. 1 (page 584). Propose a structure for **X.**

21. There are nine stereoisomers of 1,2,3,4,5,6-hexachlorocyclohexane. Seven of these isomers are meso compounds and two are a pair of enantiomers.

(a) Write structures for all of these stereoisomers, labeling meso forms and the pair of enantiomers.
(b) One of these stereoisomers undergoes E2 reactions much more slowly than any of the others. Which isomer is this and why does it react so slowly in an E2 reaction?

22. In addition to more highly fluorinated products, fluorination of 2-methylbutane yields a mixture of compounds with the formula $C_5H_{11}F$.

(a) How many different isomers with the formula $C_5H_{11}F$ would you expect to be produced, taking stereochemistry into account?
(b) If the mixture of $C_5H_{11}F$ isomers were subjected to fractional distillation, how many fractions would you expect to obtain?
(c) Which fractions would be optically inactive?
(d) Which would you be able to resolve into enantiomers?

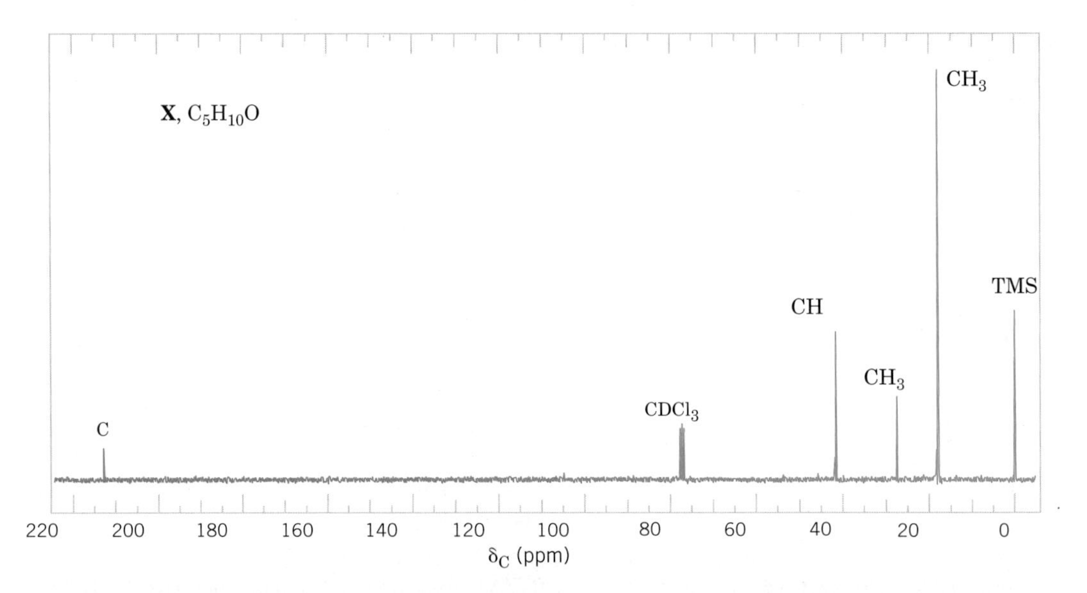

FIGURE 1 The broadband proton-decoupled ^{13}C NMR spectrum of compound X (Problem 20). Information from the DEPT ^{13}C NMR spectra is given near each peak.

23. Fluorination of (*R*)-2-fluorobutane yields a mixture of isomers with the formula $C_4H_8F_2$.
(a) How many different isomers would you expect to be produced? Write their structures.
(b) If the mixture of $C_4H_8F_2$ isomers were subjected to fractional distillation, how many fractions would you expect to obtain?
(c) Which of these fractions would be optically active?

24. There are two optically inactive (and nonresolvable) forms of 1,3-di-*sec*-butylcyclohexane. Write their structures.

25. When the following deuterium-labeled isomer undergoes elimination, the reaction yields *trans*-2-butene and *cis*-2-butene-2-*d* (as well as some 1-butene-3-*d*).

CH_3; H, C, D; H, C, Br; CH_3 $\xrightarrow[C_2H_5OH]{^-OC_2H_5}$ H, H_3C C=C CH_3, H + D, H_3C C=C H, CH_3

trans*-2-Butene** ***cis*-2-Butene-2-*d

(+ $CH_3CHDCH{=}CH_2$)

These compounds are not produced:

H, H_3C C=C H, CH_3 or H_3C, D C=C H, CH_3

cis*-2-Butene** ***trans*-2-Butene-2-*d

How can you explain these results?

CHAPTER

13

Conjugated Unsaturated Systems

Molecules with the Nobel Prize in Their Synthetic Lineage

Many organic molecules that have been among the great targets for synthesis by organic chemists have in their synthetic lineage a common reaction. This reaction is deceptively simple by appearances, yet it is extraordinarily powerful in what it can accomplish. From acyclic precursors it can form a six-membered ring, with as many as four new stereogenic centers created in a single stereospecific step. It also produces a double bond that can be used to introduce other functionality. The reaction is the Diels–Alder reaction, and we shall study it later in this chapter. Otto Diels and Kurt Alder won the Nobel Prize in Chemistry in 1950 for developing this reaction.

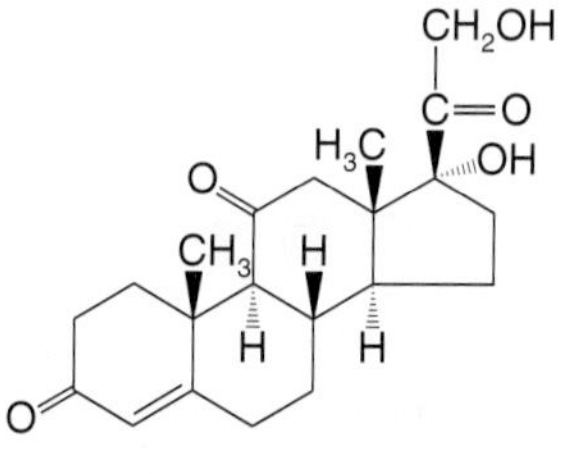

Cortisone

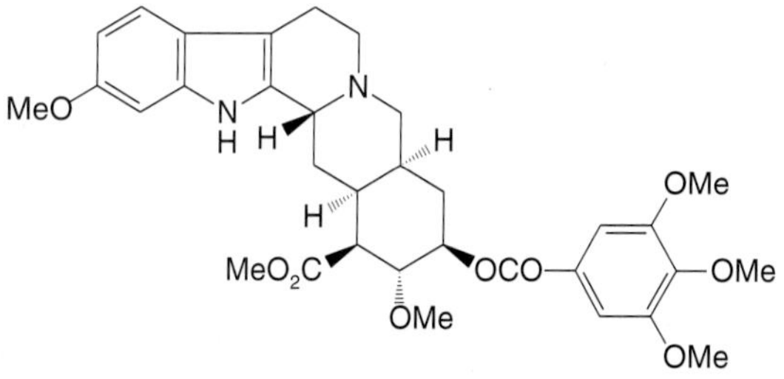

Reserpine

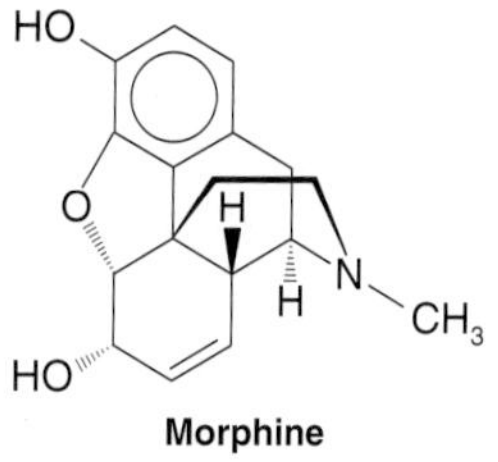

Morphine

Molecules that have been synthesized using the Diels–Alder reaction (and the chemists who led the work) include morphine (on the previous page, and by the chapter-opening photo), the hypnotic sedative used after many surgical procedures (M. Gates); reserpine (also on page 585), a clinically used antihypertensive agent (R. B. Woodward); cholesterol, precursor of all steroids in the body, and cortisone (also on page 585), the anti-inflammatory agent (both by R. B. Woodward); prostaglandins $F_{2\alpha}$ and E_2 (Section 13.11D), members of a family of hormones that mediate blood pressure, smooth muscle contraction, and inflammation (E. J. Corey); vitamin B_{12} (Section 4.20), used in the production of blood and nerve cells (A. Eschenmoser and R. B. Woodward); and Taxol® (chemical name paclitaxel, Section 13.11), a potent cancer chemotherapy agent (K. C. Nicolaou). This list alone is a veritable litany of monumental synthetic accomplishments, yet there are many other molecules that have also succumbed to synthesis using the Diels–Alder reaction. It could be said that all of these molecules have a certain sense of "Nobel-ity" in their heritage.

13.1 Introduction

In our study of the reactions of alkenes in Chapter 8 we saw how important the π bond is in understanding the chemistry of unsaturated compounds. In this chapter we study a special group of unsaturated compounds, and again we shall find that the π bond is the important part of the molecule. Here we shall examine ***species that have a p orbital on an atom adjacent to a double bond.*** The p orbital may be one that contains a single electron as in the allyl radical ($CH_2{=}CHCH_2\cdot$) (Section 13.2), it may be a vacant p orbital as in the allyl cation ($CH_2{=}CHCH_2^+$) (Section 13.4), or it may be the p orbital of another double bond as in 1,3-butadiene ($CH_2{=}CH{-}CH{=}CH_2$) (Section 13.7). We shall see that having a p orbital on an atom adjacent to a double bond allows the formation of an extended π bond—one that encompasses more than two nuclei.

Systems that have a p orbital on an atom adjacent to a double bond—molecules with delocalized π bonds—are called **conjugated unsaturated systems.** This general phenomenon is called **conjugation.** As we shall see, conjugation gives these systems special properties. We shall find, for example, that conjugated radicals, ions, or molecules are more stable than nonconjugated ones. We shall demonstrate this with the allyl radical, the allyl cation, and 1,3-butadiene. We shall see that conjugated molecules absorb energy in the ultraviolet and visible regions of the electromagnetic spectrum (Section 13.9), and this can be used for UV–Vis spectroscopy. Conjugation also allows molecules to undergo unusual reactions, and we shall study these, too, including an important reaction for forming rings called the Diels–Alder reaction (Section 13.11).

13.2 Allylic Substitution and the Allyl Radical

When propene reacts with bromine or chlorine at low temperatures, the reaction that takes place is the usual addition of halogen to the double bond:

$$CH_2{=}CH{-}CH_3 + X_2 \xrightarrow[\substack{CCl_4 \\ \text{(addition reaction)}}]{\text{low temperature}} \underset{\displaystyle X}{\underset{|}{CH_2}}{-}\underset{\displaystyle X}{\underset{|}{CH}}{-}CH_3$$

However, when propene reacts with chlorine or bromine at very high temperatures or under conditions in which the concentration of the halogen is very small, the reaction that occurs is a **substitution.** These two examples illustrate how we can often change the course of an organic reaction simply by changing the conditions. (They also illustrate the need for specifying the conditions of a reaction carefully when we report experimental results.)

$$\underset{\textbf{Propene}}{CH_2{=}CH{-}CH_3} + X_2 \xrightarrow[\text{or low concentration of } X_2 \text{ (substitution reaction)}]{\text{high temperature}} CH_2{=}CH{-}CH_2X + HX$$

In this substitution a halogen atom replaces one of the hydrogen atoms of the methyl group of propene. These hydrogen atoms are called the **allylic hydrogen atoms,** and the substitution reaction is known as an **allylic substitution:**

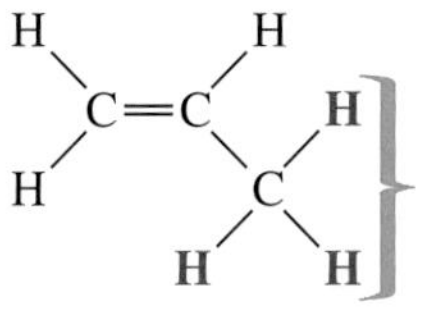

Allylic hydrogen atoms

Allylic hydrogen atom and *allylic substitution* are general terms as well. The hydrogen atoms of any saturated carbon atom adjacent to a double bond, that is,

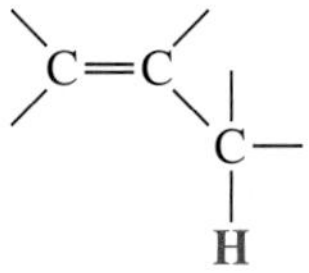

are called allylic hydrogen atoms, and any reaction in which an allylic hydrogen atom is replaced is called an allylic substitution.

13.2A Allylic Chlorination (High Temperature)

Propene undergoes allylic chlorination when propene and chlorine react in the gas phase at 400°C. This method for synthesizing allyl chloride is called the "Shell process."

$$CH_2{=}CH{-}CH_3 + Cl_2 \xrightarrow[\text{gas phase}]{400^\circ C} \underset{\substack{\textbf{3-Chloropropene} \\ \textbf{(allyl chloride)}}}{CH_2{=}CH{-}CH_2Cl} + HCl$$

The mechanism for allylic substitution is the same as the chain mechanism for alkane halogenations that we saw in Chapter 10. In the chain-initiating step, the chlorine molecule dissociates into chlorine atoms.

Chain-Initiating Step

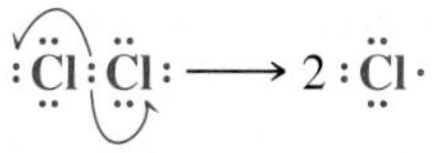

$$:\ddot{Cl}:\ddot{Cl}: \longrightarrow 2\,:\ddot{Cl}\cdot$$

In the first chain-propagating step the chlorine atom abstracts one of the allylic hydrogen atoms.

First Chain-Propagating Step

$$H_2C{=}CH{-}CH_2{-}H + \cdot\ddot{Cl}: \longrightarrow H_2C{=}CH{-}\dot{C}H_2 + H:\ddot{Cl}:$$

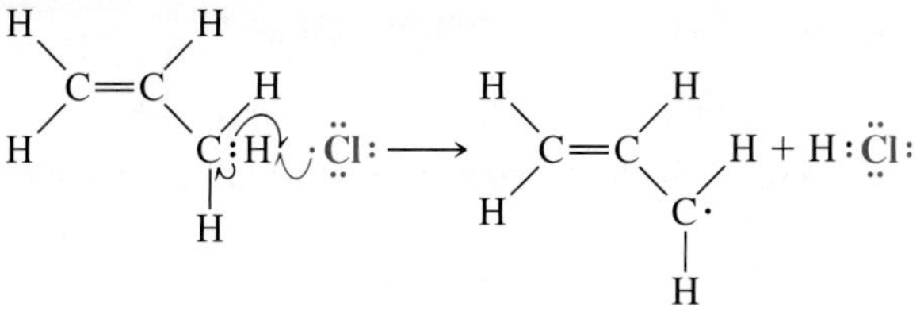

Allyl radical

The radical that is produced in this step is called an ***allyl radical.*** A radical of the general type $\diagdown C{=}C\diagup$ with an allylic $\dot{C}$— is called an *allylic* radical.

In the second chain-propagating step the allyl radical reacts with a molecule of chlorine.

Second Chain-Propagating Step

$$H_2C{=}CH{-}CH_2\cdot + :\ddot{Cl}:\ddot{Cl}: \longrightarrow H_2C{=}CH{-}CH_2:\ddot{Cl}: + :\ddot{Cl}\cdot$$

Allyl chloride

This step results in the formation of a molecule of allyl chloride and a chlorine atom. The chlorine atom then brings about a repetition of the first chain-propagating step. The chain reaction continues until the usual chain-terminating steps consume the radicals.

The reason for substitution at the allylic hydrogen atoms of propene will be more understandable if we examine the bond dissociation energy of an allylic carbon–hydrogen bond and compare it with the bond dissociation energies of other carbon–hydrogen bonds (see Table 10.1):

$$\underset{\textbf{Propene}}{CH_2{=}CHCH_2{-}H} \longrightarrow \underset{\textbf{Allyl radical}}{CH_2{=}CHCH_2\cdot} + H\cdot \qquad DH^\circ = 369\ \text{kJ mol}^{-1}$$

$$\underset{\textbf{Isobutane}}{(CH_3)_3C{-}H} \longrightarrow \underset{\textbf{3° Radical}}{(CH_3)_3C\cdot} + H\cdot \qquad DH^\circ = 400\ \text{kJ mol}^{-1}$$

$$\underset{\textbf{Propane}}{(CH_3)_2CH{-}H} \longrightarrow \underset{\textbf{2° Radical}}{(CH_3)_2CH\cdot} + H\cdot \qquad DH^\circ = 413\ \text{kJ mol}^{-1}$$

$$\underset{\textbf{Propane}}{CH_3CH_2CH_2{-}H} \longrightarrow \underset{\textbf{1° Radical}}{CH_3CH_2CH_2\cdot} + H\cdot \qquad DH^\circ = 423\ \text{kJ mol}^{-1}$$

$$\underset{\textbf{Ethene}}{CH_2{=}CH{-}H} \longrightarrow \underset{\textbf{Vinyl radical}}{CH_2{=}CH\cdot} + H\cdot \qquad DH^\circ = 465\ \text{kJ mol}^{-1}$$

We see that an allylic carbon–hydrogen bond of propene is broken with greater ease than even the tertiary carbon–hydrogen bond of isobutane and with far greater ease than a vinylic carbon–hydrogen bond:

$$CH_2{=}CH{-}CH_2{-}H + \cdot\ddot{X}: \longrightarrow \underset{\textbf{Allyl radical}}{CH_2{=}CH{-}CH_2\cdot} + HX \qquad E_{act}\ \textbf{is low.}$$

$$:\ddot{X}\cdot + H{-}CH{=}CH{-}CH_3 \longrightarrow \underset{\textbf{Vinylic radical}}{\cdot CH{=}CH{-}CH_3} + HX \qquad E_{act}\ \textbf{is high.}$$

The ease with which an allylic carbon–hydrogen bond is broken means that relative to primary, secondary, tertiary, and vinylic free radicals the allyl radical is the *most stable* (Fig. 13.1):

Relative stability: allylic or allyl > 3° > 2° > 1° > vinyl or vinylic.

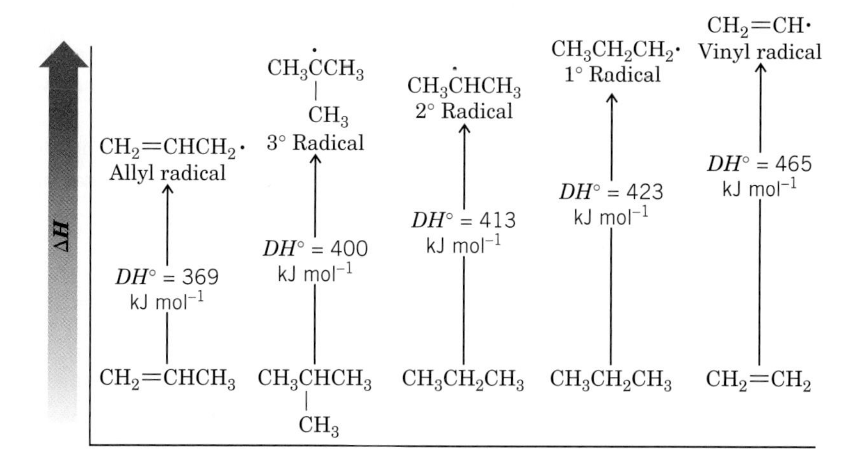

FIGURE 13.1 The relative stability of the allyl radical compared to 1°, 2°, 3°, and vinyl radicals. (The stabilities of the radicals are relative to the hydrocarbon from which each was formed, and the overall order of stability is allyl > 3° > 2° > 1° > vinyl.)

13.2B Allylic Bromination with *N*-Bromosuccinimide (Low Concentration of Br_2)

Propene undergoes allylic bromination when it is treated with *N*-bromosuccinimide (NBS) in CCl_4 in the presence of peroxides or light:

$$CH_2{=}CH{-}CH_3 + \text{NBS} \xrightarrow[CCl_4]{\text{light or ROOR}} CH_2{=}CH{-}CH_2Br + \text{succinimide}$$

N-Bromosuccinimide (NBS) **3-Bromopropene (allyl bromide)** **Succinimide**

The reaction is initiated by the formation of a small amount of Br· (possibly formed by dissociation of the N—Br bond of the NBS). The main propagation steps for this reaction are the same as for allylic chlorination (Section 13.2A):

$$CH_2{=}CH{-}CH_2{-}H + \cdot Br \longrightarrow CH_2{=}CH{-}CH_2\cdot + HBr$$

$$CH_2{=}CH{-}CH_2\cdot + Br{-}Br \longrightarrow CH_2{=}CH{-}CH_2Br + \cdot Br$$

N-Bromosuccinimide is nearly insoluble in CCl_4 and provides a constant but very low concentration of bromine in the reaction mixture. It does this by reacting very rapidly with the HBr formed in the substitution reaction. Each molecule of HBr is replaced by one molecule of Br_2:

$$\text{NBS (N—Br)} + HBr \longrightarrow \text{succinimide (N—H)} + Br_2$$

Under these conditions, that is, *in a nonpolar solvent and with a very low concentration of bromine,* very little bromine adds to the double bond; it reacts by substitution and replaces an allylic hydrogen atom instead.

OPTIONAL MATERIAL

Why, we might ask, does a low concentration of bromine favor allylic substitution over addition? To understand this, we must recall the mechanism for addition and notice that in the first step only one atom of the bromine molecule becomes attached to the alkene *in a reversible process*. The other atom (now a bromide anion) becomes attached in the second step:

$$\text{Br—Br} + \;\text{>C=C<}\; \rightleftharpoons \; \overset{+}{\text{Br}}\text{(bridged)}\!\!<\!\!\begin{matrix}\text{—C—}\\|\\\text{—C—}\end{matrix} + \text{Br}^- \longrightarrow \begin{matrix}\text{—C—Br}\\|\\\text{Br—C—}\end{matrix}$$

With a low concentration of bromine initially, the concentration of the bromonium ion and bromide anion after the first step will also be low. Consequently, the probability of a bromide anion finding a bromonium ion in its vicinity for the second step is also low, and hence the overall rate of addition is slow and allylic substitution competes successfully.

The use of a nonpolar solvent also slows addition. Since there are no polar molecules to solvate (and thus stabilize) the bromide ion formed in the first step, the bromide ion uses a bromine molecule as a substitute:

$$2\,\text{Br}_2 + \;\text{>C=C<}\; \underset{\text{solvent}}{\overset{\text{nonpolar}}{\rightleftharpoons}} \; \overset{+}{\text{Br}}\text{(bridged)}\!\!<\!\!\begin{matrix}\text{—C—}\\|\\\text{—C—}\end{matrix} + \text{Br}_3^-$$

This means that in a nonpolar solvent the rate equation is second order with respect to bromine,

$$\text{Rate} = k\left[\text{>C=C<}\right][\text{Br}_2]^2$$

and that the low bromine concentration has an even more pronounced effect in slowing the rate of addition.

Understanding why a high temperature favors allylic substitution over addition requires a consideration of the effect of entropy changes on equilibria (Section 3.9). The addition reaction, because it combines two molecules into one, has a substantial negative entropy change. At low temperatures, the $T\,\Delta S°$ term in $\Delta G° = \Delta H° - T\,\Delta S°$ is not large enough to offset the favorable $\Delta H°$ term. But as the temperature is increased, the $T\,\Delta S°$ term becomes more significant, $\Delta G°$ becomes more positive, and the equilibrium becomes more unfavorable.

13.3 THE STABILITY OF THE ALLYL RADICAL

An explanation of the stability of the allyl radical can be approached in two ways: in terms of molecular orbital theory and in terms of resonance theory (Section 1.8). As we shall see soon, both approaches give us equivalent descriptions of the allyl radical. The molecular orbital approach is easier to visualize, so we shall begin with it. (As preparation for this section, it would be a good idea to review the molecular orbital theory given in Sections 1.11 and 1.13.)

13.3A Molecular Orbital Description of the Allyl Radical

As an allylic hydrogen atom is abstracted from propene (see the following diagram), the sp^3-hybridized carbon atom of the methyl group changes its hybridization state to sp^2 (see Section 10.7). The p orbital of this new sp^2-hybridized carbon atom overlaps with the p orbital of the central carbon atom. Thus, in the allyl radical three p orbitals overlap to form a set of π molecular orbitals that encompass all three carbon atoms. The new p orbital of the allyl radical is said to be *conjugated* with those of the double bond, and the allyl radical is said to be a *conjugated unsaturated system.*

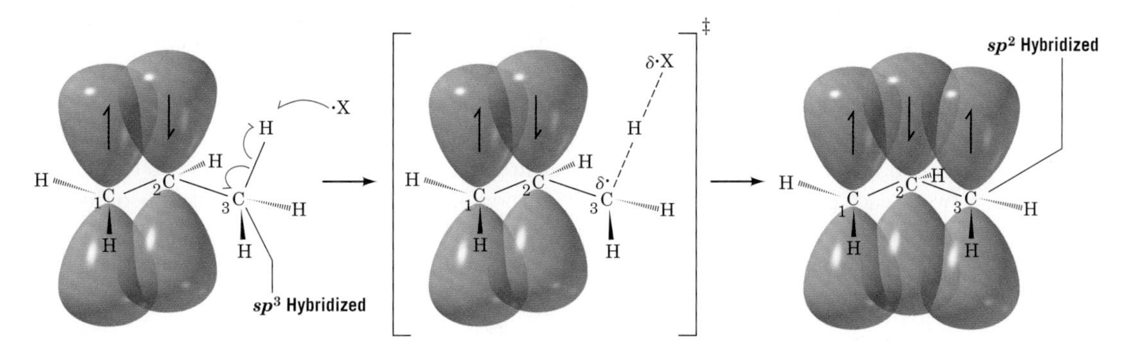

The unpaired electron of the allyl radical and the two electrons of the π bond are **delocalized** over all three carbon atoms. This delocalization of the unpaired electron accounts for the greater stability of the allyl radical when compared to primary, secondary, and tertiary radicals. Although some delocalization occurs in primary, secondary, and tertiary radicals, delocalization is not as effective because it occurs only through hyperconjugation (Section 6.11B) with σ bonds.

These orbitals are stylized as spheres for simplicity of presentation.

The diagram in Figure 13.2 illustrates how the three p orbitals of the allyl radical combine to form three π molecular orbitals. (*Remember:* The number of molecular orbitals that results always equals the number of atomic orbitals that combine; see Section 1.11.)

Chem 3D Models

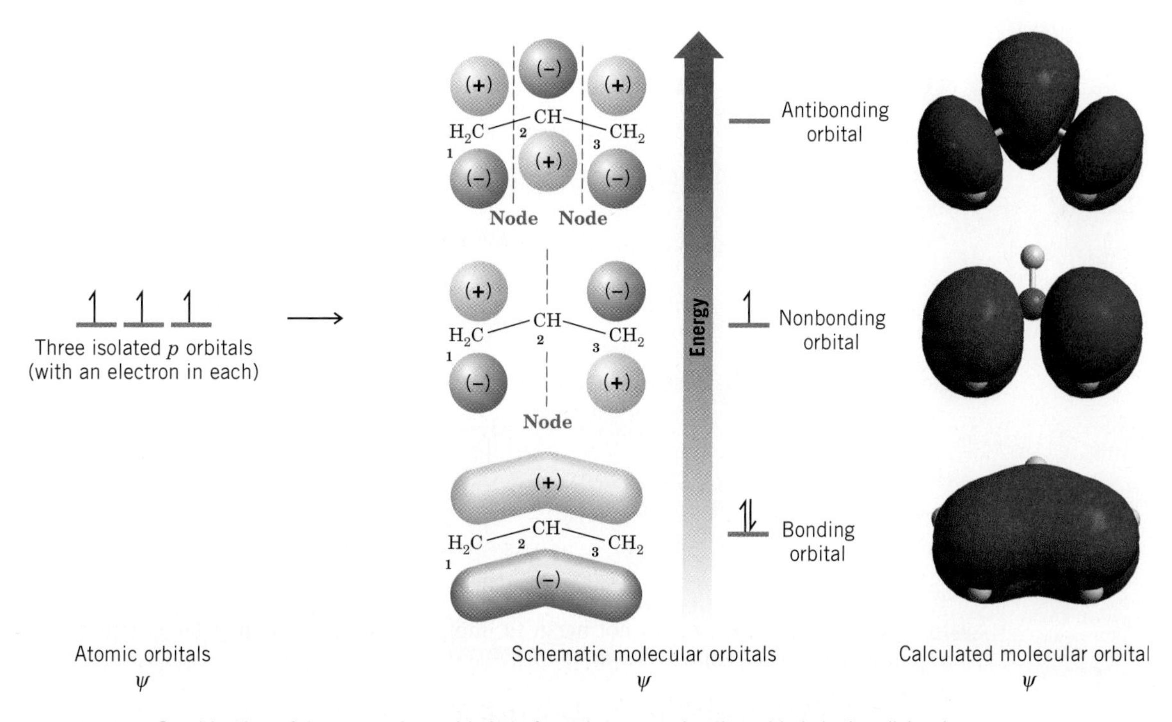

FIGURE 13.2 Combination of three atomic p orbitals to form three π molecular orbitals in the allyl radical. The bonding π molecular orbital is formed by the combination of the three p orbitals with lobes of the same sign overlapping above and below the plane of the atoms. The nonbonding π molecular orbital has a node at C2. The antibonding π molecular orbital has two nodes: between C1 and C2 and between C2 and C3. The shapes of molecular orbitals for the allyl radical calculated using quantum mechanical principles are shown alongside the schematic orbitals.

The bonding π molecular orbital is of lowest energy; it encompasses all three carbon atoms and is occupied by two spin-paired electrons. This bonding π orbital is the result of having p orbitals with lobes of the same sign overlap between adjacent carbon atoms. This type of overlap, as we recall, increases the π-electron density in the regions between the atoms where it is needed for bonding. The nonbonding π orbital is occupied by one unpaired electron, and it has a node at the central carbon atom. This node means that the unpaired electron is located in the vicinity of carbon atoms 1 and 3 only. The antibonding π molecular orbital results when orbital lobes of opposite sign overlap between adjacent carbon atoms: Such overlap means that in the antibonding π orbital there is a node between each pair of carbon atoms. This antibonding orbital of the allyl radical is of highest energy and is empty in the ground state of the radical.

We can illustrate the picture of the allyl radical given by molecular orbital theory with the following structure:

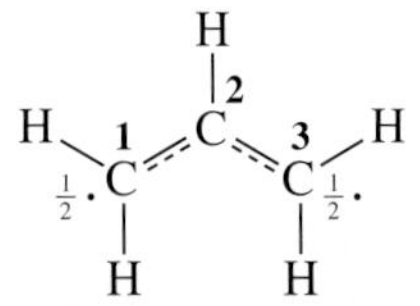

We indicate with dashed lines that both carbon–carbon bonds are partial double bonds. This accommodates one of the things that molecular orbital theory tells us: *that there is a π bond encompassing all three atoms.* We also place the symbol $\frac{1}{2}\cdot$ beside the C1 and C3 atoms. This denotes a second thing molecular orbital theory tells us: *that the unpaired electron spends its time in the vicinity of C1 and C3.* Finally, implicit in the molecular orbital picture of the allyl radical is this: The two ends of the allyl radical are *equivalent.* This aspect of the molecular orbital description is also implicit in the formula just given.

13.3B Resonance Description of the Allyl Radical

In section 13.2A we wrote the structure of the allyl radical as **A**:

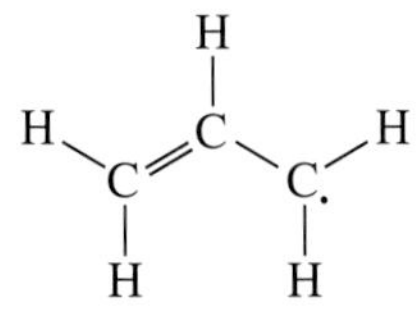

A

However, we might just as well have written the equivalent structure, **B**:

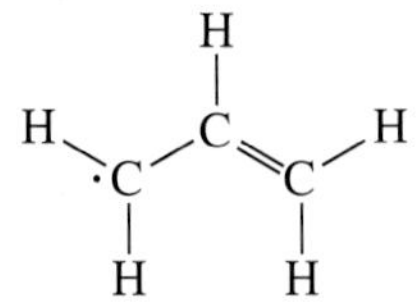

B

In writing structure **B**, we do not mean to imply that we have simply taken structure **A** and turned it over. What we have done is move the electrons in the following way:

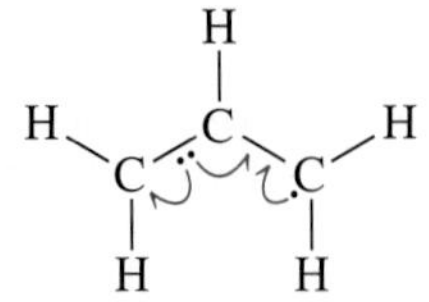

We have not moved the atomic nuclei themselves.

Resonance theory (Section 1.8) tells us that whenever we can write two structures for a chemical entity ***that differ only in the positions of the electrons,*** the entity cannot be represented by either structure alone but is a *hybrid* of both. We can represent the hybrid in two ways. We can write both structures **A** and **B** and connect them with a double-headed arrow, the special arrow we use to indicate they are resonance structures:

$$\underset{\mathbf{A}}{\overset{1}{C}H_2{=}\overset{2}{C}H{-}\overset{3}{C}H_2\cdot} \longleftrightarrow \underset{\mathbf{B}}{\cdot\overset{1}{C}H_2{-}\overset{2}{C}H{=}\overset{3}{C}H_2}$$

Or we can write a single structure, **C**, that blends the features of both resonance structures:

$$\underset{\mathbf{C}}{\tfrac{1}{2}\cdot\overset{1}{C}H_2{\overset{\cdots}{-}}\overset{2}{C}H{\overset{\cdots}{-}}\overset{3}{C}H_2\tfrac{1}{2}\cdot}$$

We see, then, that resonance theory gives us exactly the same picture of the allyl radical that we obtained from molecular orbital theory. Structure **C** describes the carbon–carbon bonds of the allyl radical as partial double bonds. The resonance structures **A** and **B** also tell us that the unpaired electron is associated only with the C1 and C3 atoms. We indicate this in structure **C** by placing a $\frac{1}{2}\cdot$ beside C1 and C3.* Because resonance structures **A** and **B** are equivalent, *C1 and C3 are also equivalent.*

Another rule in resonance theory is that *whenever equivalent resonance structures* can be written for a chemical species, *the chemical species is much more stable than any resonance structure (when taken alone) would indicate.* If we were to examine either **A** or **B** alone, we might decide that it resembled a primary radical. Thus, we might estimate the stability of the allyl radical as approximately that of a primary radical. In doing so, we would greatly underestimate the stability of the allyl radical. Resonance theory tells us, however, that since **A** and **B** are *equivalent resonance structures,* the allyl radical should be much more stable than either, that is, much more stable than a primary radical. This correlates with what experiments have shown to be true: **The allyl radical is even more stable than a tertiary radical.**

Relative stability of the allyl radical

PROBLEM 13.1

(a) What product(s) would you expect to obtain if propene labeled with ^{14}C at C1 were subjected to allylic chlorination or bromination?

$$^{14}CH_2{=}CHCH_3 + X_2 \xrightarrow[\text{low concentration of } X_2]{\text{high temperature or}} ?$$

(b) Explain your answer by writing a mechanism and drawing all relevant resonance structures.

(c) If more than one product would be obtained, what relative proportions would you expect?

---

*A resonance structure such as the one shown below would indicate that an unpaired electron is associated with C2. This structure is not a proper resonance structure because resonance theory dictates that *all resonance structures must have the same number of unpaired electrons* (see Section 13.5).

$$\cdot CH_2{-}\dot{C}H{-}CH_2\cdot$$

(an incorrect resonance structure)

13.4 The Allyl Cation

The allyl cation ($CH_2{=}CHCH_2^+$) is an unusually stable carbocation (although we cannot go into the experimental evidence here). It is even more stable than a secondary carbocation and is almost as stable as a tertiary carbocation. In general terms, the relative order of stabilities of carbocations is that given here.

Relative stability of the allyl cation

Relative Order of Carbocation Stability

$$-\overset{|}{C}{=}\overset{|}{C}-\underset{|}{\overset{+}{C}}-\underset{|}{\overset{|}{C}}- > C-\underset{C}{\overset{C}{C^+}} > CH_2{=}CHCH_2^+ > C-\underset{H}{\overset{C}{C^+}} > C-\underset{H}{\overset{H}{C^+}} > CH_2{=}CH^+$$

Substituted allylic > **3°** > **Allyl** > **2°** > **1°** > **Vinyl**

As we might expect, the unusual stability of the allyl cation and other allylic cations can also be accounted for in terms of molecular orbital or resonance theory.

The molecular orbital description of the allyl cation is shown in Fig. 13.3.

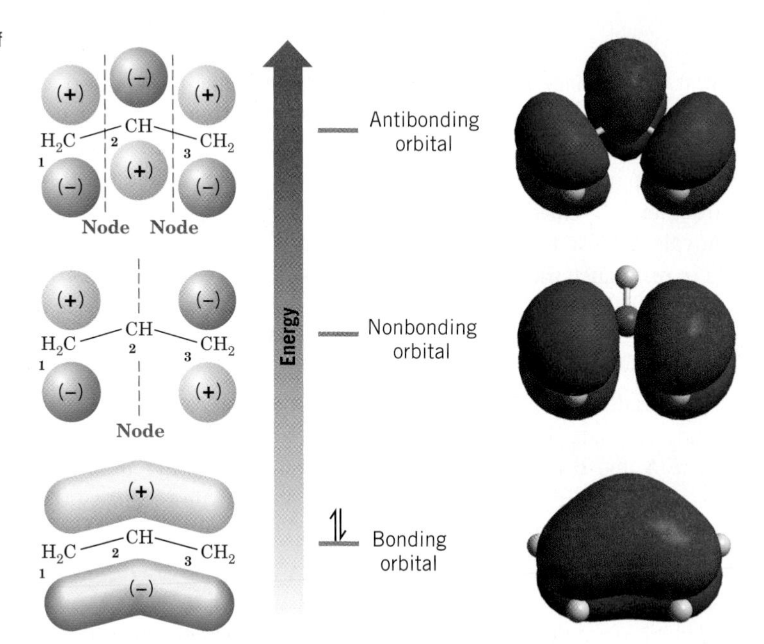

FIGURE 13.3 The π molecular orbitals of the allyl cation. The allyl cation, like the allyl radical (Fig. 13.2), is a conjugated unsaturated system. The shapes of molecular orbitals for the allyl cation calculated using quantum mechanical principles are shown alongside the schematic orbitals.

Chem 3D Models

The bonding π molecular orbital of the allyl cation, like that of the allyl radical (Fig. 13.2), contains two spin-paired electrons. The nonbonding π molecular orbital of the allyl cation, however, is empty. Since an allyl cation is what we would get if we removed an electron from an allyl radical, we can say, in effect, that we remove the electron from the nonbonding molecular orbital:

$$CH_2{=}CHCH_2\cdot \xrightarrow{-e^-} CH_2{=}CHCH_2^+$$

Removal of an electron from a nonbonding orbital (see Fig. 13.2) is known to require less energy than removal of an electron from a bonding orbital. In addition, the positive charge that forms on the allyl cation is *effectively delocalized* between C1 and C3. Thus, in molecular orbital theory these two factors, the ease of removal of a nonbonding electron and the delocalization of charge, account for the unusual stability of the allyl cation.

Resonance theory depicts the allyl cation as a hybrid of structures **D** and **E** represented here:

$$\underset{\mathbf{D}}{H_2\overset{1}{C}=\overset{2}{C}H-\overset{3}{C}H_2^{+}} \longleftrightarrow \underset{\mathbf{E}}{{}^{+}H_2\overset{1}{C}-\overset{2}{C}H=\overset{3}{C}H_2}$$

Because **D** and **E** are *equivalent* resonance structures, resonance theory predicts that the allyl cation should be unusually stable. Since the positive charge is located on C3 in **D** and on C1 in **E**, resonance theory also tells us that the positive charge should be delocalized over both carbon atoms. Carbon atom **2** carries none of the positive charge. The hybrid structure **F** includes charge and bond features of both **D** and **E**:

$$\underset{\mathbf{F}}{\overset{\frac{1}{2}+}{H_2\overset{1}{C}}\text{---}\overset{2}{C}H\text{---}\overset{\frac{1}{2}+}{\overset{3}{C}H_2}}$$

PROBLEM 13.2

(a) Write structures corresponding to **D**, **E**, and **F** for the carbocation shown:

$$CH_3-\overset{+}{C}H-CH=CH_2$$

(b) This carbocation appears to be even more stable than a tertiary carbocation; how can you explain this?

(c) What product(s) would you expect to be formed if this carbocation reacted with a chloride ion?

13.5 Summary of Rules for Resonance

We have already used resonance theory in earlier chapters, and we have been using it extensively in this chapter because we are describing radicals and ions with delocalized electrons (and charges) in π bonds. Resonance theory is especially useful with systems like this, and we shall use it again and again in the chapters that follow. In Section 1.8 we had an introduction to resonance theory and an initial presentation of some rules for writing resonance structures. It should now be helpful, in light of our previous discussions of relative carbocation stability, radicals, and our growing understanding of conjugated systems to review and expand upon those rules as well as those for the ways in which we estimate the relative contribution a given structure will make to the overall hybrid.

Resonance is an important tool we use frequently when discussing structure and reactivity.

13.5A Rules for Writing Resonance Structures

1. **Resonance structures exist only on paper** Although they have no real existence of their own, **resonance structures** are useful because they allow us to describe molecules, radicals, and ions for which a single Lewis structure is inadequate. We write two or more Lewis structures, calling them resonance structures or resonance contributors. We connect these structures by double-headed arrows ($\longleftrightarrow$), and we say that the hybrid of all of them represents the real molecule, radical, or ion.

2. In writing resonance structures, we are only allowed to move electrons The positions of the nuclei of the atoms must remain the same in all of the structures. Structure **3** is not a resonance structure for the allylic cation, for example, because in order to form it we would have to move a hydrogen atom and this is not permitted:

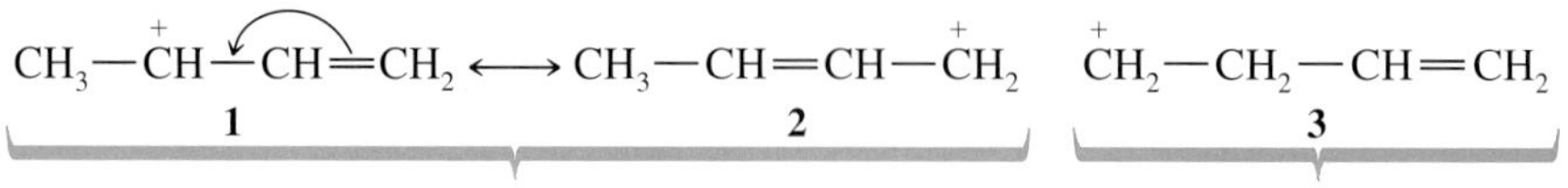

$$\underset{\mathbf{1}}{CH_3-\overset{+}{C}H-CH=CH_2} \longleftrightarrow \underset{\mathbf{2}}{CH_3-CH=CH-\overset{+}{C}H_2} \qquad \underset{\mathbf{3}}{\overset{+}{C}H_2-CH_2-CH=CH_2}$$

These are resonance structures for the allylic cation formed when 1,3-butadiene accepts a proton.

This is not a proper resonance structure for the allylic cation because a hydrogen atom has been moved.

Generally speaking, when we move electrons we move only those of π bonds (as in the example above) and those of lone pairs.

3. All of the structures must be proper Lewis structures We should not write structures in which carbon has five bonds, for example:

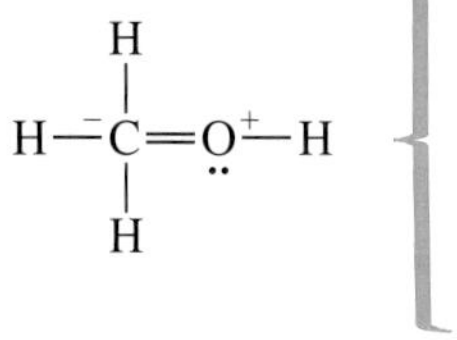

$$H-\underset{H}{\overset{H}{\overset{|}{\underset{|}{C}}}}=\ddot{O}^{+}-H$$

This is not a proper resonance structure for methanol because carbon has five bonds. Elements of the first major row of the periodic table cannot have more than eight electrons in their valence shell.

4. All resonance structures must have the same number of unpaired electrons The following structure is not a resonance structure for the allyl radical because it contains three unpaired electrons and the allyl radical contains only one:

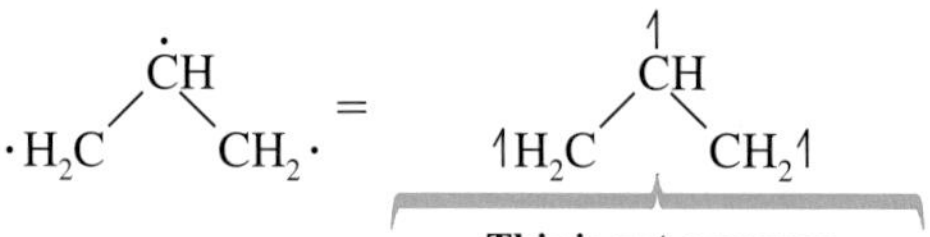

$$\cdot H_2C-\dot{C}H-CH_2\cdot \quad = \quad \uparrow H_2C-\overset{\uparrow}{C}H-CH_2\uparrow$$

This is not a proper resonance structure for the allyl radical because it does not contain the same number of unpaired electrons as $CH_2=CHCH_2\cdot$.

5. All atoms that are a part of the delocalized π-electron system must lie in a plane or be nearly planar For example, 2,3-di-*tert*-butyl-1,3-butadiene behaves like a *nonconjugated* diene because the large *tert*-butyl groups twist the structure and prevent the double bonds from lying in the same plane. Because they are not in the same plane, the *p* orbitals at C2 and C3 do not overlap and delocalization (and therefore resonance) is prevented:

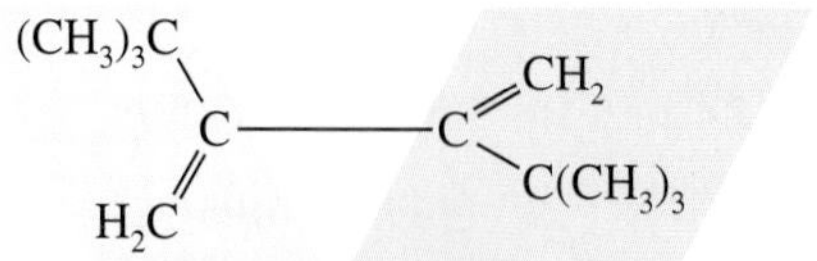

$$H_2C=C(C(CH_3)_3)-C(C(CH_3)_3)=CH_2$$

2,3-Di-*tert*-butyl-1,3-butadiene

6. The energy of the actual molecule is lower than the energy that might be estimated for any contributing structure The actual allyl cation, for example, is more stable than either resonance structure **4** or **5** taken separately would indicate. Structures **4**

and **5** resemble primary carbocations and yet the allyl cation is more stable (has lower energy) than a secondary carbocation. Chemists often call this kind of stabilization *resonance stabilization:*

$$\underset{\mathbf{4}}{CH_2{=}CH{-}\overset{+}{C}H_2} \longleftrightarrow \underset{\mathbf{5}}{\overset{+}{C}H_2{-}CH{=}CH_2}$$

In Chapter 14 we shall find that benzene is highly resonance stabilized because it is a hybrid of the two equivalent forms that follow:

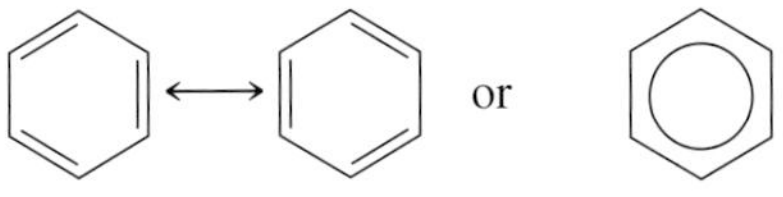

Resonance structures for benzene or **Representation of hybrid**

7. **Equivalent resonance structures make equal contributions to the hybrid, and a system described by them has a large resonance stabilization** Structures **4** and **5** make equal contributions to the allylic cation because they are equivalent. They also make a large stabilizing contribution and account for allylic cations being unusually stable. The same can be said about the contributions made by the equivalent structures **A** and **B** (Section 13.3B) for the allyl radical and by the equivalent structures for benzene.

8. **The more stable a structure is (when taken by itself), the greater is its contribution to the hybrid** Structures that are not equivalent do not make equal contributions. For example, the following cation is a hybrid of structures **6** and **7**. Structure **6** makes a greater contribution than **7** because structure **6** is a more stable tertiary carbocation while structure **7** is a primary cation:

$$\overset{a}{C}H_3{-}\underset{\delta+}{\overset{CH_3}{\overset{|b}{C}}}\text{---}\overset{c}{C}H\text{---}\underset{\delta+}{\overset{d}{C}H_2} = \left[\underset{\mathbf{6}}{CH_3{-}\underset{+}{\overset{CH_3}{\overset{|}{C}}}{-}CH{=}CH_2} \longleftrightarrow \underset{\mathbf{7}}{CH_3{-}\overset{CH_3}{\overset{|}{C}}{=}CH{-}\overset{}{\underset{+}{C}H_2}}\right]$$

That **6** makes a larger contribution means that the partial positive charge on carbon *b* of the hybrid will be larger than the partial positive charge on carbon *d*. It also means that the bond between carbon atoms *c* and *d* will be more like a double bond than the bond between carbon atoms *b* and *c*.

13.5B Estimating the Relative Stability of Resonance Structures

The following rules will help us in making decisions about the relative stabilities of resonance structures.

a. **The more covalent bonds a structure has, the more stable it is** This is exactly what we would expect because we know that forming a covalent bond lowers the energy of atoms. This means that of the following structures for 1,3-butadiene, **8** is by far the most stable and makes by far the largest contribution because it contains one more bond. (It is also more stable for the reason given under rule **c**.)

$$\underset{\mathbf{8}}{CH_2{=}CH{-}CH{=}CH_2} \longleftrightarrow \underset{\mathbf{9}}{\overset{+}{C}H_2{-}CH{=}CH{-}\overset{..}{C}H_2^{-}} \longleftrightarrow \underset{\mathbf{10}}{\overset{..}{C}H_2^{-}{-}CH{=}CH{-}\overset{+}{C}H_2}$$

This structure is the most stable because it contains more covalent bonds.

b. Structures in which all of the atoms have a complete valence shell of electrons (i.e., the noble gas structure) are especially stable and make large contributions to the hybrid Again, this is what we would expect from what we know about bonding. This means, for example, that **12** makes a larger stabilizing contribution to the cation below than **11** because all of the atoms of **12** have a complete valence shell. (Notice too that **12** has more covalent bonds than **11**; see rule **a.**)

$$\overset{+}{C}H_2-\ddot{\underset{..}{O}}-CH_3 \longleftrightarrow CH_2=\overset{+}{\underset{..}{O}}-CH_3$$

11   **12**

Here this carbon atom has only six electrons. (11)

Here the carbon atom has eight electrons. (12)

c. Charge separation decreases stability Separating opposite charges requires energy. Therefore, structures in which opposite charges are separated have greater energy (lower stability) than those that have no charge separation. This means that of the following two structures for vinyl chloride, structure **13** makes a larger contribution because it does not have separated charges. (This does not mean that structure **14** does not contribute to the hybrid; it just means that the contribution made by **14** is smaller.)

$$CH_2=CH-\ddot{\underset{..}{Cl}}: \longleftrightarrow \overset{-}{:}CH_2-CH=\underset{..}{\overset{..}{Cl}}:^+$$

13   **14**

PROBLEM 13.3

Give the important resonance structures for each of the following:

(a) $CH_2=\underset{|}{\overset{CH_3}{C}}-CH_2\cdot$

(b) $CH_2=CH-\underset{+}{C}H-CH=CH_2$

(c) 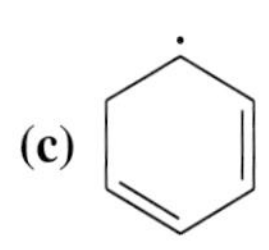

(d) (cyclohexadienyl cation, + on ring carbon)

(e) $CH_3CH=CH-CH=\overset{+}{\underset{..}{O}}H$

(f) $CH_2=CH-Br$

(g) (benzene ring)$-CH_2^+$

(h) $^-:CH_2-\overset{O}{\overset{\|}{C}}-CH_3$

(i) $CH_3-S-CH_2^+$

(j) CH_3-NO_2

PROBLEM 13.4

From each set of resonance structures that follow, designate the one that would contribute most to the hybrid and explain your choice:

(a) $CH_3CH_2\overset{CH_3}{\overset{|}{C}}=CH-CH_2^+ \longleftrightarrow CH_3CH_2\overset{CH_3}{\overset{|}{\underset{+}{C}}}-CH=CH_2$

(b) (1-cyclopentenyl)$\overset{+}{C}H_2$ $\longleftrightarrow$ (methylenecyclopentane with + on ring carbon)

(c) $\overset{+}{C}H_2-\ddot{N}(CH_3)_2 \longleftrightarrow CH_2=\overset{+}{N}(CH_3)_2$

(d) $CH_3-C(=\ddot{O}:)-\ddot{O}-H \longleftrightarrow CH_3-C(-\ddot{O}:^-)=\ddot{O}^+-H$

(e) $\dot{C}H_2CH{=}CHCH{=}CH_2 \longleftrightarrow CH_2{=}CH\dot{C}HCH{=}CH_2 \longleftrightarrow CH_2{=}CHCH{=}CH\dot{C}H_2$

(f) $:NH_2-C{\equiv}N: \longleftrightarrow \overset{+}{N}H_2{=}C{=}\ddot{N}:^-$

PROBLEM 13.5

The following keto and enol forms differ in the positions for their electrons, but they are not resonance structures. Explain why they are not.

$$H_2C{=}CH-\ddot{O}-H \qquad H_3C-CH{=}\ddot{O}:$$

Enol form **Keto form**

13.6 ALKADIENES AND POLYUNSATURATED HYDROCARBONS

Many hydrocarbons are known whose molecules contain more than one double or triple bond. A hydrocarbon whose molecules contain two double bonds is called an **alkadiene;** one whose molecules contain three double bonds is called an **alkatriene,** and so on. Colloquially, these compounds are often referred to simply as "dienes" or "trienes." A hydrocarbon with two triple bonds is called an **alkadiyne,** and a hydrocarbon with a double and triple bond is called an **alkenyne.**

The following examples of polyunsaturated hydrocarbons illustrate how specific compounds are named. Recall from IUPAC rules (Sections 4.5 and 4.6) that the numerical locants for double and triple bonds can be placed at the beginning of the name or immediately preceding the respective suffix. We will provide examples of both styles.

$\overset{1}{C}H_2{=}\overset{2}{C}{=}\overset{3}{C}H_2$
1,2-Propadiene (allene, or propa-1,2-diene)

$\overset{1}{C}H_2{=}\overset{2}{C}H-\overset{3}{C}H{=}\overset{4}{C}H_2$
1,3-Butadiene (buta-1,3-diene)

(3*Z*)-Penta-1,3-diene (*cis*-penta-1,3-diene)

(2*E*,4*E*)-2,4-Hexadiene (*trans,trans*-2,4-hexadiene)

(2*Z*,4*E*)-Hexa-2,4-diene (*cis,trans*-hexa-2,4-diene)

$H\overset{5}{C}{\equiv}\overset{4}{C}-\overset{3}{C}H_2\overset{2}{C}H{=}\overset{1}{C}H_2$
Pent-1-en-4-yne

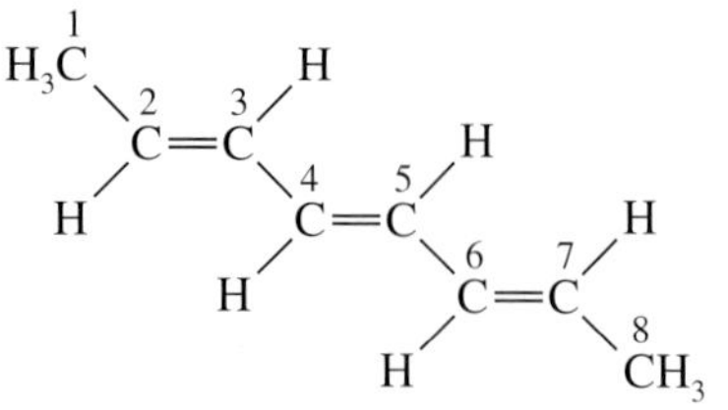

(2*E*,4*E*,6*E*)-Octa-2,4,6-triene
(*trans,trans,trans*-octa-2,4,6-triene)

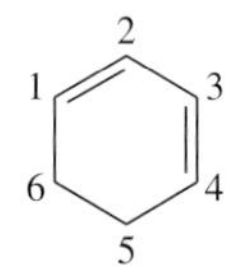

1,3-Cyclohexadiene 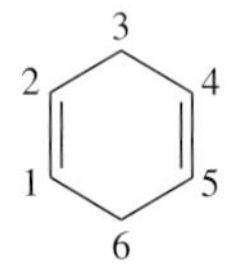 **1,4-Cyclohexadiene**

The multiple bonds of polyunsaturated compounds are classified as being **cumulated, conjugated,** or **isolated.** The double bonds of allene (1,2-propadiene) are said to be cumulated because one carbon (the central carbon) participates in two double bonds. Hydrocarbons whose molecules have cumulated double bonds are called **cumulenes.** The name **allene** (Section 5.18) is also used as a class name for molecules with two cumulated double bonds:

$CH_2{=}C{=}CH_2$ **Allene**

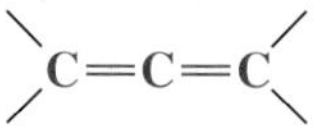

$>C{=}C{=}C<$ **A cumulated diene**

An example of a conjugated diene is 1,3-butadiene. In conjugated polyenes the double and single bonds *alternate* along the chain:

$CH_2{=}CH{-}CH{=}CH_2$ **1,3-Butadiene**

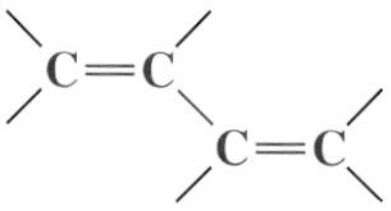

$>C{=}C{-}C{=}C<$ **A conjugated diene**

(2*E*,4*E*,6*E*)-Octa-2,4,6-triene is an example of a conjugated alkatriene.

If one or more saturated carbon atoms intervene between the double bonds of an alkadiene, the double bonds are said to be *isolated.* An example of an isolated diene is 1,4-pentadiene:

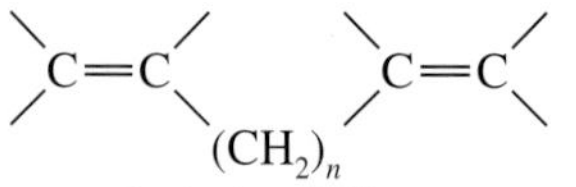

$>C{=}C{-}(CH_2)_n{-}C{=}C<$ **An isolated diene** **($n \neq 0$)**

$CH_2{=}CH{-}CH_2{-}CH{=}CH_2$ **1,4-Pentadiene**

PROBLEM 13.6

(a) Which other compounds in Section 13.6 are conjugated dienes?
(b) Which other compounds are isolated dienes?
(c) Which compound is an isolated enyne?

In Chapter 5 we saw that appropriately substituted cumulated dienes (allenes) give rise to chiral molecules. Cumulated dienes have had some commercial importance, and cumulated double bonds are occasionally found in naturally occurring molecules. In general, cumulated dienes are less stable than isolated dienes.

The double bonds of isolated dienes behave just as their name suggests—as isolated "enes." They undergo all of the reactions of alkenes, and except for the fact that they are capable of reacting twice, their behavior is not unusual. Conjugated dienes are far more interesting because we find that their double bonds interact with each other. This interaction leads to unexpected properties and reactions. We shall therefore consider the chemistry of conjugated dienes in detail.

13.7 1,3-Butadiene: Electron Delocalization

13.7A Bond Lengths of 1,3-Butadiene

The carbon–carbon bond lengths of 1,3-butadiene have been determined and are shown here:

$$\overset{1}{C}H_2{=}\overset{2}{C}H{-}\overset{3}{C}H{=}\overset{4}{C}H_2$$

1.34 Å   1.47 Å   1.34 Å

The C1—C2 bond and the C3—C4 bond are (within experimental error) the same length as the carbon–carbon double bond of ethene. The central bond of 1,3-butadiene (1.47 Å), however, is considerably shorter than the single bond of ethane (1.54 Å).

This should not be surprising. All of the carbon atoms of 1,3-butadiene are sp^2 hybridized and, as a result, the central bond of butadiene results from overlapping sp^2 orbitals. And, as we know, a sigma bond that is sp^3–sp^3 is *longer.* There is, in fact, a steady decrease in bond length of carbon–carbon single bonds as the hybridization state of the bonded atoms changes from sp^3 to sp (Table 13.1).

TABLE 13.1 Carbon–Carbon Single-Bond Lengths and Hybridization State

Compound	Hybridization State	Bond Length (Å)
$H_3C{-}CH_3$	sp^3-sp^3	1.54
$CH_2{=}CH{-}CH_3$	sp^2-sp^3	1.50
$CH_2{=}CH{-}CH{=}CH_2$	sp^2-sp^2	1.47
$HC{\equiv}C{-}CH_3$	$sp-sp^3$	1.46
$HC{\equiv}C{-}CH{=}CH_2$	$sp-sp^2$	1.43
$HC{\equiv}C{-}C{\equiv}CH$	$sp-sp$	1.37

13.7B Conformations of 1,3-Butadiene

There are two possible planar conformations of 1,3-butadiene: the s-cis and the s-trans conformations.

2 3 — rotate about C2—C3 → 2 3

s-cis Conformation of 1,3-butadiene   **s-trans Conformation of 1,3-butadiene**

These are not true cis and trans forms since the s-cis and s-trans conformations of 1,3-butadiene can be interconverted through rotation about the single bond (hence the prefix s). The s-trans conformation is the predominant one at room temperature. We shall see that the s-cis conformation of 1,3-butadiene and other 1,3-conjugated alkenes is necessary for the Diels–Alder reaction (Section 13.11).

13.7C Molecular Orbitals of 1,3-Butadiene

The central carbon atoms of 1,3-butadiene (Fig. 13.4) are close enough for overlap to occur between the *p* orbitals of C2 and C3. This overlap is not as great as that between the orbitals of C1 and C2 (or those of C3 and C4). The C2–C3 orbital overlap, however, gives the central bond partial double-bond character and allows the four π electrons of 1,3-butadiene to be delocalized over all four atoms.

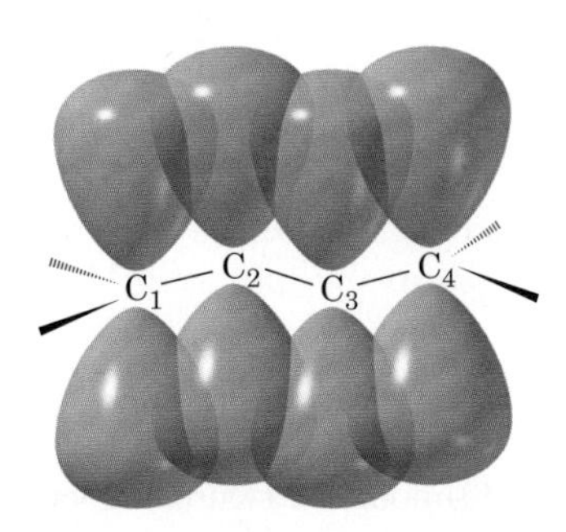

FIGURE 13.4 The *p* orbitals of 1,3-butadiene, stylized as spheres. (See Figure 13.5 for the shapes of calculated molecular orbitals for 1,3-butadiene.)

Figure 13.5 shows how the four *p* orbitals of 1,3-butadiene combine to form a set of four π molecular orbitals.

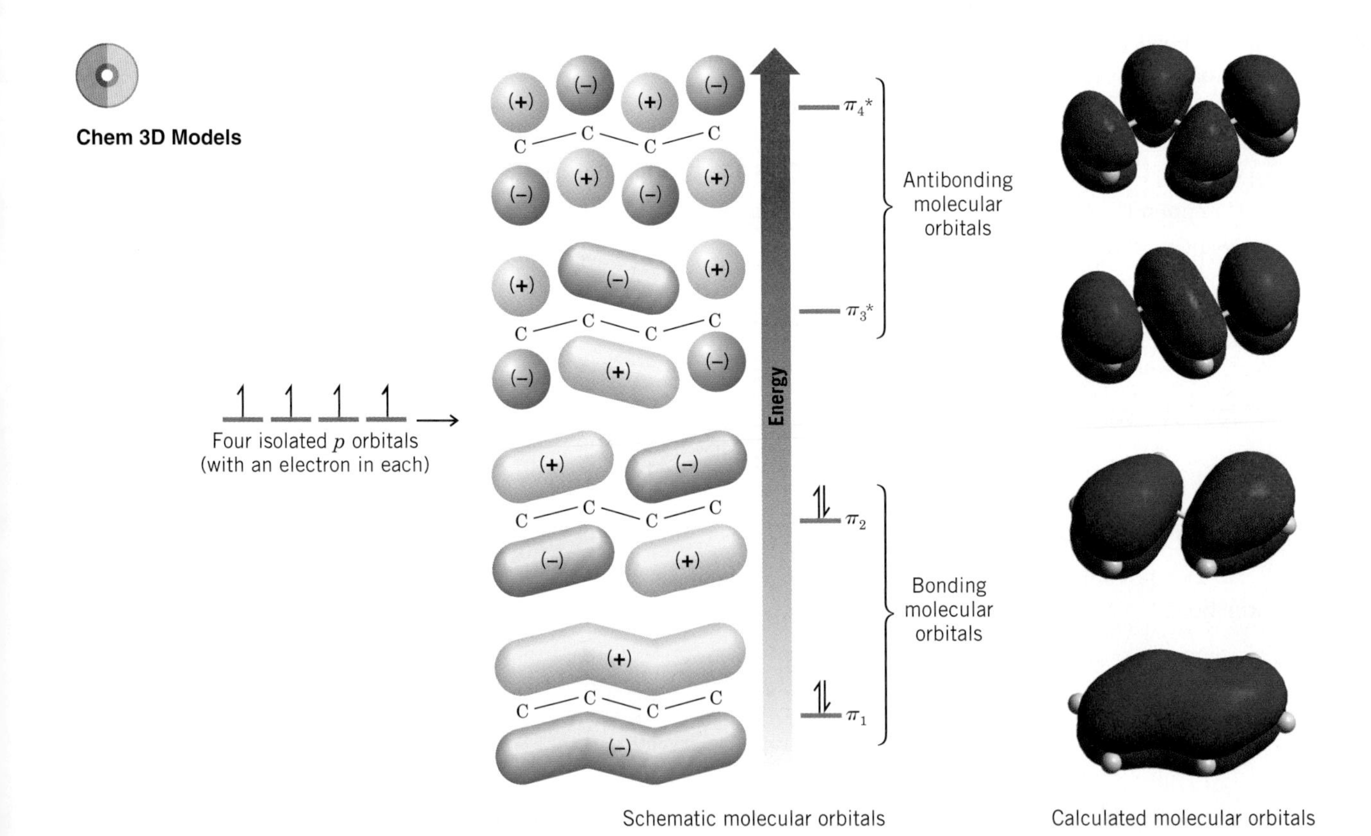

FIGURE 13.5 Formation of the π molecular orbitals of 1,3-butadiene from four isolated *p* orbitals. The shapes of molecular orbitals for 1,3-butadiene calculated using quantum mechanical principles are shown alongside the schematic orbitals.

Two of the π molecular orbitals of 1,3-butadiene are bonding molecular orbitals. In the ground state these orbitals hold the four π electrons with two spin-paired electrons in each. The other two π molecular orbitals are antibonding molecular orbitals. In the ground state these orbitals are unoccupied. An electron can be excited from the highest occupied molecular orbital (HOMO) to the lowest unoccupied molecular orbital (LUMO) when 1,3-butadiene absorbs light with a wavelength of 217 nm. (We shall study the absorption of light by unsaturated molecules in Section 13.9.)

The delocalized bonding that we have just described for 1,3-butadiene is characteristic of all conjugated polyenes.

13.8 The Stability of Conjugated Dienes

Conjugated alkadienes are thermodynamically more stable than isomeric isolated alkadienes. Two examples of this extra stability of conjugated dienes can be seen in an analysis of the heats of hydrogenation given in Table 13.2.

TABLE 13.2 Heats of Hydrogenation of Alkenes and Alkadienes

Compound	H_2 (mol)	$\Delta H°$ (kJ mol^{-1})
1-Butene	1	−127
1-Pentene	1	−126
trans-2-Pentene	1	−115
1,3-Butadiene	2	−239
trans-1,3-Pentadiene	2	−226
1,4-Pentadiene	2	−254
1,5-Hexadiene	2	−253

In itself, 1,3-butadiene cannot be compared directly with an isolated diene of the same chain length. However, a comparison can be made between the heat of hydrogenation of 1,3-butadiene and that obtained when two molar equivalents of 1-butene are hydrogenated:

$$\Delta H° \text{ (kJ mol}^{-1}\text{)}$$

$$\underset{\textbf{1-Butene}}{2\ CH_2{=}CHCH_2CH_3} + 2\ H_2 \longrightarrow 2\ CH_3CH_2CH_2CH_3 \qquad 2 \times (-127) = -254$$

$$\underset{\textbf{1,3-Butadiene}}{CH_2{=}CHCH{=}CH_2} + 2\ H_2 \longrightarrow CH_3CH_2CH_2CH_3 \qquad = -239$$

$$\text{Difference} \qquad 15 \text{ kJ mol}^{-1}$$

Because 1-butene has the same kind of monosubstituted double bond as either of those in 1,3-butadiene, we might expect hydrogenation of 1,3-butadiene to liberate the same amount of heat (254 kJ mol^{-1}) as two molar equivalents of 1-butene. We find, however, that 1,3-butadiene liberates only 239 kJ mol^{-1}, 15 kJ mol^{-1} *less* than expected. We conclude, therefore, that conjugation imparts some extra stability to the conjugated system (Fig. 13.6).

An assessment of the stabilization that conjugation provides *trans*-1,3-pentadiene can be made by comparing the heat of hydrogenation of *trans*-1,3-pentadiene to the sum of the heats of hydrogenation of 1-pentene and *trans*-2-pentene. This way we are comparing double bonds of comparable types:

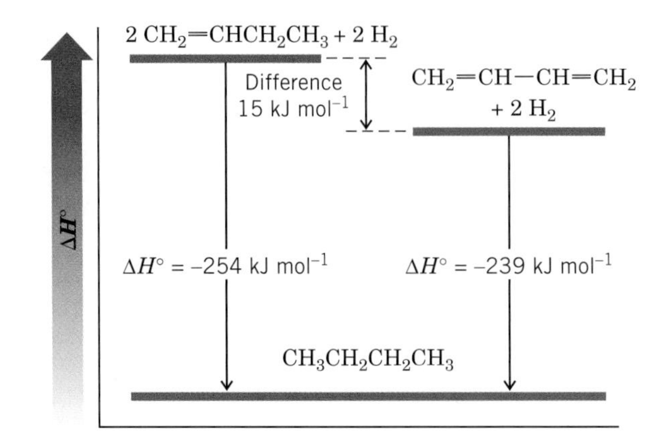

FIGURE 13.6 Heats of hydrogenation of 2 mol of 1-butene and 1 mol of 1,3-butadiene.

$CH_2{=}CHCH_2CH_2CH_3$ **1-Pentene** $\Delta H° = -126$ kJ mol^{-1}

$CH_3CH_2(H)C{=}C(H)CH_3$ ***trans*-2-Pentene** $\Delta H° = -115$ kJ mol^{-1}

Sum $= -241$ kJ mol^{-1}

$CH_2{=}CH(H)C{=}C(H)CH_3$ ***trans*-1,3-Pentadiene** $\Delta H° = -226$ kJ mol^{-1}

Difference $=$ 15 kJ mol^{-1}

We see from these calculations that conjugation affords *trans*-1,3-pentadiene an extra stability of 15 kJ mol^{-1}, a value that is equivalent, to two significant figures, to the one we obtained for 1,3-butadiene (15 kJ mol^{-1}).

When calculations like these are carried out for other conjugated dienes, similar results are obtained; *conjugated dienes are found to be more stable than isolated dienes.* The question, then, is this: What is the source of the extra stability associated with conjugated dienes? There are two factors that contribute. The extra stability of conjugated dienes arises in part from the stronger central bond that they contain and, in part, from the additional delocalization of the π electrons that occurs in conjugated dienes.

13.9 Ultraviolet–Visible Spectroscopy

The extra stability of conjugated dienes when compared to corresponding unconjugated dienes can also be seen in data from **ultraviolet–visible (UV–Vis) spectroscopy.** When electromagnetic radiation in the UV and visible regions passes through a compound containing multiple bonds, a portion of the radiation is usually absorbed by the compound. Just how much radiation is absorbed depends on the wavelength of the radiation and the structure of the compound. (You may wish to review Section 9.2 regarding properties of electromagnetic radiation.) The absorption of radiation in UV–Vis spectroscopy is caused by the subtraction of energy from the radiation beam when electrons in orbitals of lower energy are excited into orbitals of higher energy. In Section 13.9B we shall return to discuss specifically how data from UV–Vis spectroscopy demonstrate the additional stability of conjugated dienes. First, we briefly look at how data from a UV–Vis spectrophotometer are obtained.

13.9A UV–Vis Spectrophotometers

A UV–Vis spectrophotometer measures the amount of light absorbed at each wavelength of the UV and visible regions of the electromagnetic spectrum. UV and visible radiation are of higher energy (shorter wavelength) than infrared radiation (used in IR spectroscopy) and radio frequency radiation (used in NMR) but not as energetic as X-radiation (Fig. 13.7).

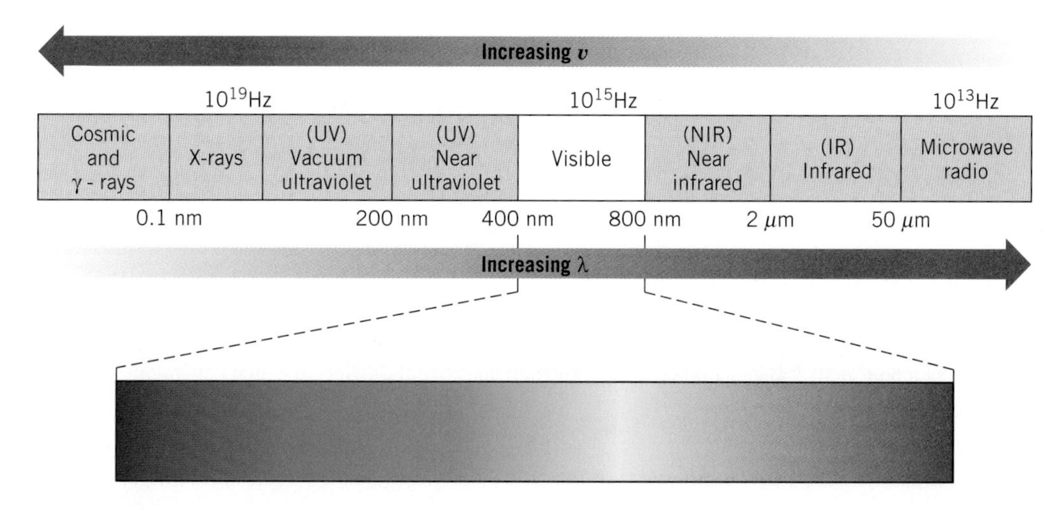

FIGURE 13.7 The electromagnetic spectrum.

In a standard UV–Vis spectrophotometer a beam of light is split; one half of the beam (the sample beam) is directed through a transparent cell containing a solution of the compound being analyzed and one half (the reference beam) is directed through an identical cell that does not contain the compound but contains the solvent. Solvents are chosen to be transparent in the region of the spectrum being used for analysis. The instrument is designed so that it can make a comparison of the intensities of the two beams as it scans over the desired region of wavelengths. If the compound absorbs light at a particular wavelength, the intensity of the sample beam (I_S) will be less than that of the reference beam (I_R). The instrument indicates this by producing a graph—a plot of the wavelength of the entire region versus the absorbance (A) of light at each wavelength. [The absorbance at a particular wavelength is defined by the equation $A_\lambda = \log(I_R/I_S)$.] Such a graph is called an **absorption spectrum.** (In diode-array UV–Vis spectrophotometers the absorption of all wavelengths of light in the region of analysis is measured simultaneously by an array of photodiodes. The absorption of the solvent is measured over all wavelengths of interest first, and then the absorption of the sample is recorded over the same range. Data from the solvent are electronically subtracted from the data for the sample. The difference is then displayed as the absorption spectrum for the sample.)

A typical UV absorption spectrum, that of 2,5-dimethyl-2,4-hexadiene, is given in Fig. 13.8. It shows a broad absorption band in the region between 210 and 260 nm. The absorption is at a maximum at 242.5 nm. It is this wavelength that is usually reported in the chemical literature as λ_{max}.

In addition to reporting the wavelength of maximum absorption (λ_{max}), chemists often report another quantity that indicates the strength or the *intensity* of the absorption, called the **molar absorptivity, ε**. (In older literature, the molar absorptivity, ε, is often referred to as the molar extinction coefficient.)

The molar absorptivity is simply the proportionality constant that relates the observed absorbance (A) at a particular wavelength (λ) to the molar concentration (C) of the sample

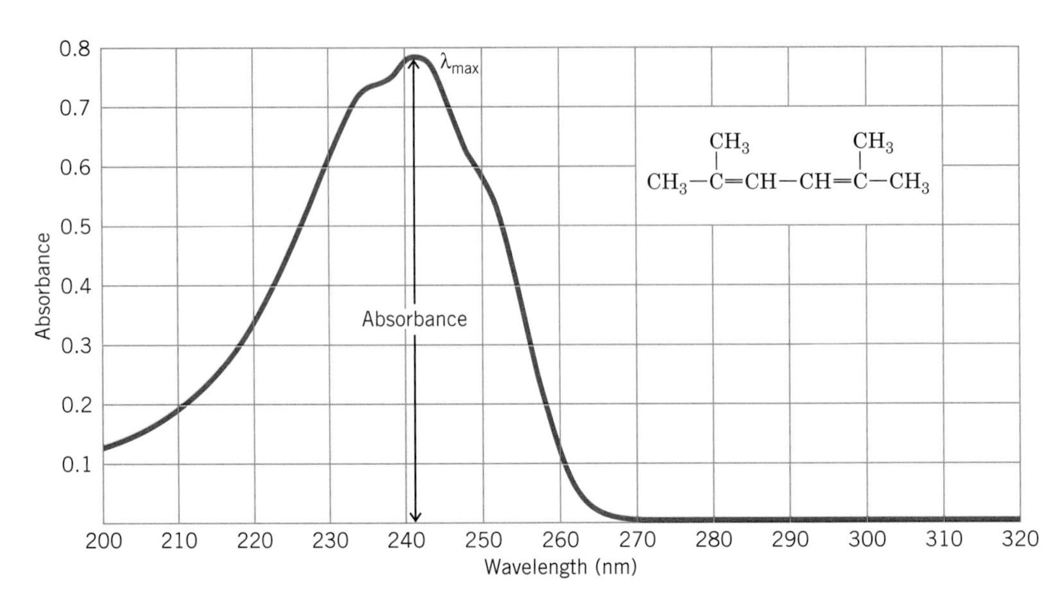

FIGURE 13.8 The UV absorption spectrum of 2,5-dimethyl-2,4-hexadiene in methanol at a concentration of 5.95×10^{-5} *M* in a 1.00-cm cell. (Spectrum courtesy of Sadtler Research Laboratories, Inc., Philadelphia.)

and the length (l) (in centimeters) of the path of the light beam through the sample cell:

$$A = \varepsilon \times C \times l \quad \text{or} \quad \varepsilon = \frac{A}{C \times l}$$

For 2,5-dimethyl-2,4-hexadiene dissolved in methanol the molar absorptivity at the wavelength of maximum absorbance (242.5 nm) is 13,100 M^{-1} cm^{-1}. In the chemical literature this would be reported as

2,5-Dimethyl-2,4-hexadiene, $\lambda_{max}^{methanol}$ 242.5 nm (ε = 13,100)

13.9B Absorption Maxima for Nonconjugated and Conjugated Dienes

As we noted earlier, when compounds absorb light in the UV and visible regions, electrons are excited from lower electronic energy levels to higher ones. For this reason, visible and UV spectra are often called **electronic spectra.** The absorption spectrum of 2,5-dimethyl-2,4-hexadiene is a typical electronic spectrum because the absorption band (or peak) is very broad. Most absorption bands in the visible and UV region are broad because each electronic energy level has associated with it vibrational and rotational levels. Thus, electron transitions may occur from any of several vibrational and rotational states of one electronic level to any of several vibrational and rotational states of a higher level.

Alkenes and nonconjugated dienes usually have absorption maxima below 200 nm. Ethene, for example, gives an absorption maximum at 171 nm; 1,4-pentadiene gives an absorption maximum at 178 nm. These absorptions occur at wavelengths that are out of the range of operation of most ultraviolet–visible spectrometers because they occur where the oxygen in air also absorbs. Special air-free techniques must be employed in measuring them.

UV–Vis spectroscopic evidence for conjugated π-electron systems

Compounds whose molecules contain *conjugated* multiple bonds have absorption maxima at wavelengths longer than 200 nm. For example, 1,3-butadiene absorbs at 217 nm. This longer wavelength absorption by conjugated dienes is a direct consequence of conjugation.

We can understand how conjugation of multiple bonds brings about absorption of light at longer wavelengths if we examine Fig. 13.9.

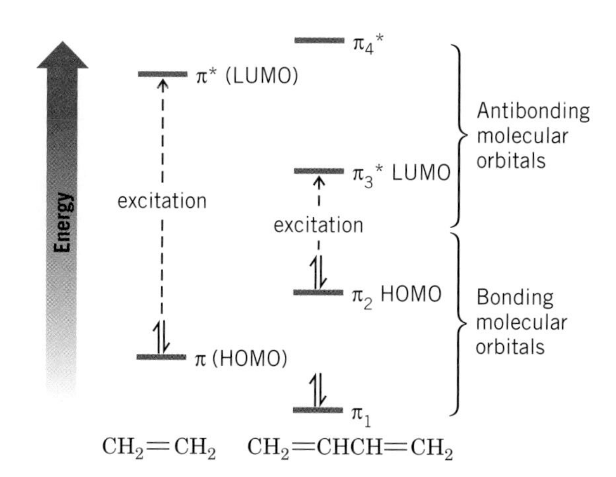

FIGURE 13.9 The relative energies of the π molecular orbitals of ethene and 1,3-butadiene (Section 13.7).

When a molecule absorbs light at its longest wavelength, an electron is excited from its **highest occupied molecular orbital (HOMO)** to the **lowest unoccupied molecular orbital (LUMO).** For most alkenes and alkadienes the HOMO is a bonding π orbital and the LUMO is an antibonding π^* orbital. The wavelength of the absorption maximum is determined by the difference in energy between these two levels. The energy gap between the HOMO and LUMO of ethene is greater than that between the corresponding orbitals of 1,3-butadiene. Thus, the $\pi \longrightarrow \pi^*$ electron excitation of ethene requires absorption of light of greater energy (shorter wavelength) than the corresponding $\pi_2 \longrightarrow \pi_3{}^*$ excitation in 1,3-butadiene. The energy difference between the HOMOs and the LUMOs of the two compounds is reflected in their absorption spectra. Ethene has its λ_{max} at 171 nm; 1,3-butadiene has a λ_{max} at 217 nm.

The narrower gap between the HOMO and the LUMO in 1,3-butadiene results from the conjugation of the double bonds. Molecular orbital calculations indicate that a much larger gap should occur in isolated alkadienes. This is borne out experimentally. Isolated alkadienes give absorption spectra similar to those of alkenes. Their λ_{max} are at shorter wavelengths, usually below 200 nm. As we mentioned, 1,4-pentadiene has its λ_{max} at 178 nm.

Conjugated alkatrienes absorb at longer wavelengths than conjugated alkadienes, and this too can be accounted for in molecular orbital calculations. The energy gap between the HOMO and the LUMO of an alkatriene is even smaller than that of an alkadiene. In fact, there is a general rule that states that ***the greater the number of conjugated multiple bonds a compound contains, the longer will be the wavelength at which the compound absorbs light.***

Polyenes with eight or more conjugated double bonds absorb light in the visible region of the spectrum. For example, β-carotene, a precursor of vitamin A and a compound that imparts its orange color to carrots, has 11 conjugated double bonds; β-carotene has an absorption maximum at 497 nm, well into the visible region. Light of 497 nm has a blue-green color; this is the light that is absorbed by β-carotene. We perceive the complementary color of blue green, which is red orange.

β-Carotene

The Chemistry of...

The Photochemistry of Vision

The chemical changes that occur when light impinges on the retina of the eye involve several of the phenomena that we have studied. Central to an understanding of the visual process at the molecular level are two phenomena in particular: the absorption of light by conjugated polyenes and the interconversion of cis–trans isomers.

The retina of the human eye contains two types of receptor cells. Because of their shapes, these cells have been named *rods* and *cones.* Rods are located primarily at the periphery of the retina and are responsible for vision in dim light. Rods, however, are color blind and "see" only in shades of gray. Cones are found mainly in the center of the retina and are responsible for vision in bright light. Cones also possess the pigments that are responsible for color vision.

Some animals do not possess both rods and cones. The retinas of pigeons contain only cones. Thus, while pigeons have color vision, they see only in the bright light of day. The retinas of owls, on the other hand, have only rods; owls see very well in dim light but are color blind.

When light strikes rod cells, it is absorbed by a compound called rhodopsin. This initiates a series of chemical events that ultimately result in the transmission of a nerve impulse to the brain. Rhodopsin was discovered in 1877 by the German physiologist Franz Boll, who noticed that the initial red-purple color of a pigment in the retina of frogs was "bleached" by the action of light. The bleaching process led first to a yellow retina and then to a colorless one. In 1878 this purple pigment was isolated and named rhodopsin by Willy Kuhne, a German scientist. Later, studies by UV–Vis spectroscopy showed that the absorption curve of rhodopsin and the sensitivity curve for human rod vision coincided, providing strong evidence that rhodopsin was indeed the light-sensitive material in rod vision (Fig. 13.A)

Early understanding of the chemical nature of rhodopsin and the conformational changes that occur when rhodopsin absorbs light resulted largely from the work of George Wald and co-workers at Harvard University. In 1952, Wald and one of his students, Ruth Hubbard, showed that the chromophore (light-absorbing group) of rhodopsin is a polyunsaturated aldehyde called 11-*cis*-retinal. Rhodopsin is the product of covalent attachment of the aldehyde group of 11-*cis*-retinal to an amino group of the protein opsin (Fig.13.B), through formation of an imine group (or Schiff's base, Section 16.8) and loss of a water molecule. The conjugated polyunsaturated chain of 11-*cis*-retinal gives rhodopsin the ability to absorb light over a broad range of the visible spectrum (with an absorption maximum at 498 nm, corresponding to its purple color).

When rhodopsin absorbs a photon of light, the 11-*cis*-retinal chromophore isomerizes to the all-trans form, causing the cyclohexene ring of the chromophore to swing into a different orientation (Fig. 13.B). The first photoproduct is an intermediate called bathorhodopsin, which through a series of steps becomes metarhodopsin II. It is believed that repositioning of the retinal cyclohexene ring, through the 11-cis to all-*trans* isomerization, causes further conformational changes in the protein that ultimately initiate a cascade of enzymatic reactions and transmission of a neural signal to the brain. In 2001, an X-ray crystal structure of rhodopsin was obtained by David Teller and coworkers at the University of Washington that provided the first high-resolution three-dimensional model of this complex macromolecule (Figure 13.C).

Together, all-*trans*-retinal and opsin have an absorbance maximum at 387 nm and, thus, are yellow. Bleaching to the colorless form is the result of enzymatic reduction of all-*trans*-retinal to all-*trans*-retinol, vitamin A (after release from the protein). The reduction converts the aldehyde group of retinal to the primary

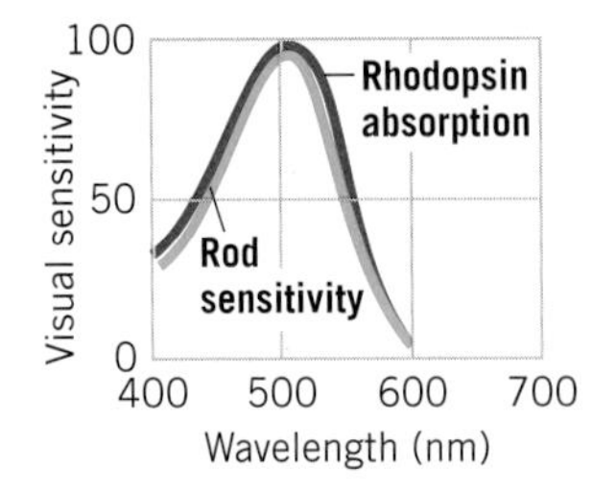

FIGURE 13.A A comparison of the visible absorption spectrum of rhodopsin and the sensitivity curve for rod vision. (Adapted from Hecht, S; Shlaer, S.; Pirenne, M. H. *J. Gen. Physiol.* **1942,** *25,* 819–840.)

$\xrightarrow[\text{enzyme}]{[H]}$

all-*trans*-Retinal

all-*trans*-Vitamin A

FIGURE 13.B The formation of rhodopsin from 11-*cis*-retinal and opsin. The important chemical steps of the visual process. Absorption of a photon of light by the 11-*cis*-retinal portion of rhodopsin generates a nerve impulse as a result of an isomerization that leads, through a series of steps, to metarhodopsin II. Then hydrolysis of metarhodopsin II produces all-*trans*-retinal and opsin.

alcohol function of vitamin A. The reverse reaction (oxidation) is important in biochemical utilization of vitamin A for the production of retinal needed in the eyes and for other functions.

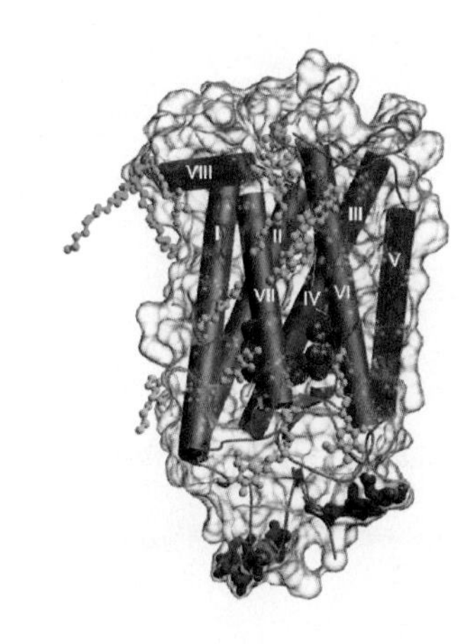

FIGURE 13.C A three-dimensional model of rhodopsin based on X-ray crystallographic data. The retinal chromophore is shown in red embedded within the larger structure of the protein. A transparent envelope around the protein represents the overall molecular surface. Protein helices and other substructures and groups are highlighted by various display formats. (See Teller, D. C.; Okada, T.; Behnke, C. A.; Palczewski, K.: Stenkamp, R. E. *Biochemistry* **2001,** *40,* 7761–7772. Reprinted with permission.)

Lycopene, a compound partly responsible for the red color of tomatoes, also has 11 conjugated double bonds. Lycopene has an absorption maximum at 505 nm, and it absorbs there intensely. (Approximately 0.02 g of lycopene can be isolated from 1 kg of fresh, ripe tomatoes.)

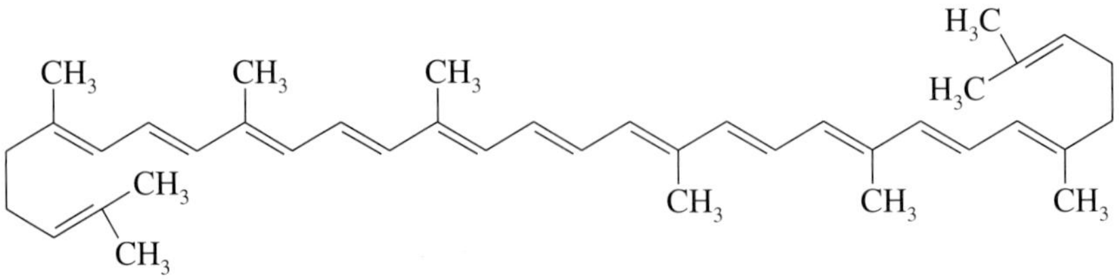

Lycopene

Table 13.3 gives the values of λ_{max} for a number of unsaturated compounds.

TABLE 13.3 Long-Wavelength Absorption Maxima of Unsaturated Hydrocarbons

Compound	Structure	λ_{max} (nm)	ε_{max} (M^{-1} cm^{-1})
Ethene	$CH_2{=}CH_2$	171	15,530
trans-3-Hexene	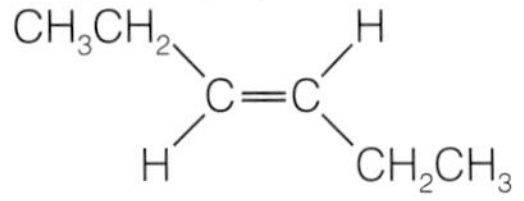 CH_3CH_2 H C=C H CH_2CH_3	184	10,000
Cyclohexene	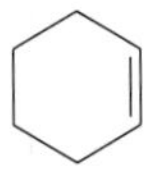	182	7,600
1-Octene	$CH_3(CH_2)_5CH{=}CH_2$	177	12,600
1-Octyne	$CH_3(CH_2)_5C{\equiv}CH$	185	2,000
1,3-Butadiene	$CH_2{=}CHCH{=}CH_2$	217	21,000
cis-1,3-Pentadiene	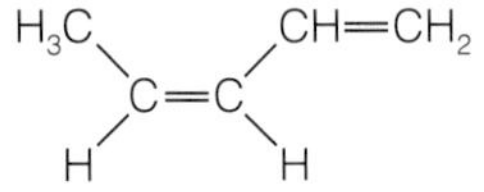 H_3C $CH{=}CH_2$ C=C H H	223	22,600
trans-1,3-Pentadiene	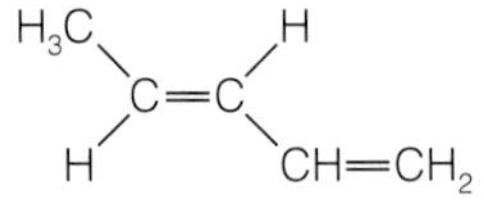 H_3C H C=C H $CH{=}CH_2$	223.5	23,000
But-1-en-3-yne	$CH_2{=}CHC{\equiv}CH$	228	7,800
1,4-Pentadiene	$CH_2{=}CHCH_2CH{=}CH_2$	178	17,000
1,3-Cyclopentadiene		239	3,400
1,3-Cyclohexadiene		256	8,000
trans-1,3,5-Hexatriene	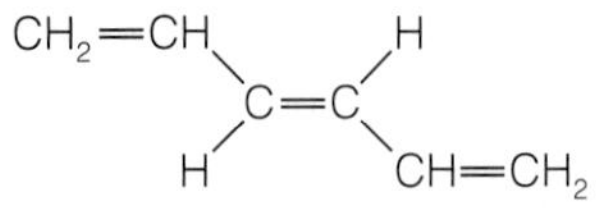 $CH_2{=}CH$ H C=C H $CH{=}CH_2$	274	50,000

Compounds with carbon–oxygen double bonds also absorb light in the UV region. Acetone, for example, has a broad absorption peak at 280 nm that corresponds to the excitation of an electron from one of the unshared pairs (a nonbonding or "*n*" electron) to the π^* orbital of the carbon–oxygen double bond:

$$\underset{\substack{\textbf{Acetone}\\ \lambda_{max} = 280\text{ nm}\\ \varepsilon_{max} = 15}}{(H_3C)_2C{=}\ddot{\underset{\cdot\cdot}{O}}} \xrightarrow{n \longrightarrow \pi^*} (H_3C)_2C\dot{=}\underset{\cdot\cdot}{O}\cdot$$

Compounds in which the carbon–oxygen double bond is conjugated with a carbon–carbon double bond have absorption maxima corresponding to $n \longrightarrow \pi^*$ excitations and $\pi \longrightarrow \pi^*$ excitations. The $n \longrightarrow \pi^*$ absorption maxima occur at longer wavelengths but are much weaker (i.e., have smaller molar absorptivities):

$$CH_2{=}CH{-}C({=}O){-}CH_3$$

$$n \longrightarrow \pi^*\ \lambda_{max} = 324\text{ nm},\ \varepsilon_{max} = 24$$
$$\pi \longrightarrow \pi^*\ \lambda_{max} = 219\text{ nm},\ \varepsilon_{max} = 3600$$

13.9C Analytical Uses of UV–Vis Spectroscopy

UV–Vis spectroscopy can be used in the structure elucidation of organic molecules to indicate whether conjugation is present in a given sample. Although conjugation in a molecule may be indicated by data from IR, NMR, or mass spectrometry, UV–Vis analysis can provide corroborating information.

A more widespread use of UV–Vis spectroscopy, however, has to do with determining the concentration of an unknown sample. As mentioned in Section 13.9A, the relationship $A = \varepsilon Cl$ indicates that the amount of absorption by a sample at a certain wavelength is dependent on its concentration. This relationship is usually linear over a range of concentrations suitable for analysis. To determine the unknown concentration of a sample, a graph of absorbance versus concentration is made for a set of standards of known concentrations. The wavelength used for analysis is usually the λ_{max} of the sample. The concentration of the sample is obtained by measuring its absorbance and determining the corresponding value of concentration from the graph of known concentrations. Quantitative analysis using UV–Vis spectroscopy is routinely used in biochemical studies to measure the rates of enzymatic reactions. The concentration of a species involved in the reaction (as related to its UV–Vis absorbance) is plotted versus time to determine the rate of reaction. UV–Vis spectroscopy is also used in environmental chemistry to determine the concentration of various metal ions (sometimes involving absorption spectra for organic complexes with the metal) and as a detection method in high-performance liquid chromatography (HPLC).

PROBLEM 13.7

Two compounds, **A** and **B**, have the same molecular formula, C_6H_8. Both **A** and **B** react with two molar equivalents of hydrogen in the presence of platinum to yield cyclohexane. Compound **A** shows three signals in its broadband decoupled ^{13}C NMR spectrum. Compound **B** shows only two ^{13}C NMR signals. Compound **A** shows an absorption maximum at 256 nm, whereas **B** shows no absorption maximum at wavelengths longer than 200 nm. What are the structures of **A** and **B**?

PROBLEM 13.8

Three compounds, **D**, **E**, and **F**, have the same molecular formula, C_5H_6. In the presence of a platinum catalyst, all three compounds absorb three molar equivalents of hydrogen and yield pentane. Compounds **E** and **F** have an IR absorption at roughly 3300 cm^{-1}; compound **D** has no IR absorption in that region. Compounds **D** and **E** show an absorption maximum in their UV–Vis spectra near 230 nm. Compound **F** shows no absorption maximum beyond 200 nm. Propose structures for **D**, **E**, and **F**.

13.10 Electrophilic Attack on Conjugated Dienes: 1,4 Addition

Not only are conjugated dienes somewhat more stable than nonconjugated dienes, they also display special behavior when they react with electrophilic reagents. For example, 1,3-butadiene reacts with one molar equivalent of hydrogen chloride to produce two products, 3-chloro-1-butene and 1-chloro-2-butene:

$$CH_2{=}CH{-}CH{=}CH_2 \xrightarrow[25°C]{HCl} CH_3{-}\underset{\displaystyle Cl}{\underset{|}{CH}}{-}CH{=}CH_2 + CH_3{-}CH{=}CH{-}CH_2Cl$$

1,3-Butadiene **3-Chloro-1-butene (78%)** **1-Chloro-2-butene (22%)**

If only the first product (3-chloro-1-butene) were formed, we would not be particularly surprised. We would conclude that hydrogen chloride had added to one double bond of 1,3-butadiene in the usual way:

$$\overset{1}{C}H_2{=}\overset{2}{C}H{-}\overset{3}{C}H{=}\overset{4}{C}H_2 + H{-}Cl \xrightarrow{\text{1,2 addition}} \underset{\displaystyle H}{\underset{|}{CH_2}}{-}\underset{\displaystyle Cl}{\underset{|}{CH}}{-}CH{=}CH_2$$

3-Chloro-1-butene

It is the second product, 1-chloro-2-butene, that is unusual. Its double bond is between the central atoms, and the elements of hydrogen chloride have added to the C1 and C4 atoms:

$$\overset{1}{C}H_2{=}\overset{2}{C}H{-}\overset{3}{C}H{=}\overset{4}{C}H_2 + H{-}Cl \xrightarrow{\text{1,4 addition}} \underset{\displaystyle H}{\underset{|}{CH_2}}{-}CH{=}CH{-}\underset{\displaystyle Cl}{\underset{|}{CH_2}}$$

1-Chloro-2-butene

This unusual behavior of 1,3-butadiene can be attributed directly to the stability and the delocalized nature of an allylic cation (Section 13.4). In order to see this, consider a mechanism for the addition of hydrogen chloride:

Step 1

$$:\ddot{Cl}{-}H + CH_2{=}CH{-}CH{=}CH_2 \longrightarrow CH_3{-}\overset{+}{C}H{-}CH{=}CH_2 \longleftrightarrow CH_3{-}CH{=}CH{-}\overset{+}{C}H_2 + :\ddot{Cl}:^-$$

An allylic cation equivalent to

$$CH_3{-}\underset{\delta+}{CH}{\overset{...}{=}}CH{\overset{...}{=}}\underset{\delta+}{CH_2}$$

Step 2

$$CH_3\overset{\delta+}{C}H{\overset{...}{=}}CH{\overset{...}{=}}\overset{\delta+}{C}H_2 + :\ddot{Cl}:^- \longrightarrow$$

(a) $\longrightarrow CH_3\underset{\displaystyle Cl}{\underset{|}{CH}}{-}CH{=}CH_2$ **1,2 Addition**

(b) $\longrightarrow CH_3CH{=}CHCH_2Cl$ **1,4 Addition**

In step 1 a proton adds to one of the terminal carbon atoms of 1,3-butadiene to form, as usual, the more stable carbocation, in this case a resonance-stabilized allylic cation. Addition to one of the inner carbon atoms would have produced a much less stable primary cation, one that could not be stabilized by resonance:

$$CH_2{=}CH{-}CH{=}CH_2 + H{-}Cl \not\longrightarrow {}^{+}CH_2{-}CH_2{-}CH{=}CH_2 + Cl^-$$

A 1° carbocation

In step 2 a chloride ion forms a bond to one of the carbon atoms of the allylic cation that bears a partial positive charge. Reaction at one carbon atom results in the 1,2-addition product; reaction at the other gives the 1,4-addition product.

PROBLEM 13.9

(a) What products would you expect to obtain if hydrogen chloride were allowed to react with 2,4-hexadiene, $CH_3CH{=}CHCH{=}CHCH_3$? **(b)** With 1,3-pentadiene, $CH_2{=}CHCH{=}CHCH_3$? (Neglect cis–trans isomerism.)

1,3-Butadiene shows 1,4-addition reactions with electrophilic reagents other than hydrogen chloride. Two examples are shown here, the addition of hydrogen bromide (in the absence of peroxides) and the addition of bromine:

$$CH_2{=}CHCH{=}CH_2 \xrightarrow[40^\circ C]{HBr} \underset{(20\%)}{CH_3CHBrCH{=}CH_2} + \underset{(80\%)}{CH_3CH{=}CHCH_2Br}$$

$$CH_2{=}CHCH{=}CH_2 \xrightarrow[-15^\circ C]{Br_2} \underset{(54\%)}{CH_2BrCHBrCH{=}CH_2} + \underset{(46\%)}{CH_2BrCH{=}CHCH_2Br}$$

Reactions of this type are quite general with other conjugated dienes. Conjugated trienes often show 1,6 addition. An example is the 1,6 addition of bromine to 1,3,5-cyclooctatriene:

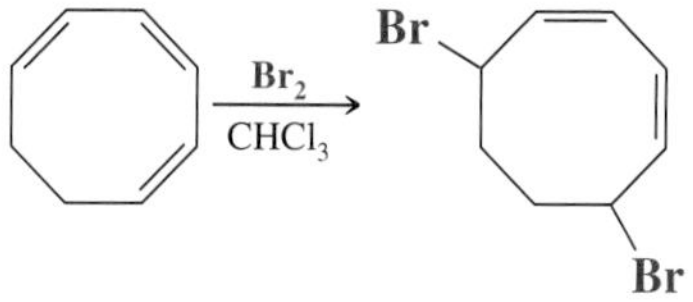

1,3,5-Cyclooctatriene $\xrightarrow[CHCl_3]{Br_2}$ 5,8-dibromo-1,3-cyclooctadiene (Br, Br) (>68%)

13.10A Kinetic Control versus Thermodynamic Control of a Chemical Reaction

The addition of hydrogen bromide to 1,3-butadiene is interesting in another respect. The relative amounts of 1,2- and 1,4-addition products that we obtain are dependent on the temperature at which we carry out the reaction.

When 1,3-butadiene and hydrogen bromide react at a low temperature (−80°C) in the absence of peroxides, the major reaction is 1,2 addition; we obtain about 80% of the 1,2 product and only about 20% of the 1,4 product. At a higher temperature (40°C) the result is reversed. The major reaction is 1,4 addition; we obtain about 80% of the 1,4 product and only about 20% of the 1,2 product.

When the mixture formed at the lower temperature is brought to the higher temperature, moreover, the relative amounts of the two products change. This new reaction mixture eventually contains the same proportion of products given by the reaction carried out at the higher temperature:

$$CH_2{=}CHCH{=}CH_2 + HBr \xrightarrow{-80^\circ C} \underset{(80\%)}{CH_3CH(Br)CH{=}CH_2} + \underset{(20\%)}{CH_3CH{=}CHCH_2Br}$$

$$\downarrow 40^\circ C$$

$$CH_2{=}CHCH{=}CH_2 + HBr \xrightarrow{40^\circ C} \underset{(20\%)}{CH_3CH(Br)CH{=}CH_2} + \underset{(80\%)}{CH_3CH{=}CHCH_2Br}$$

It can also be shown that at the higher temperature and in the presence of hydrogen bromide, the 1,2-addition product rearranges to the 1,4 product and that an equilibrium exists between them:

$$\underset{\textbf{1,2-Addition product}}{CH_3\underset{\textbf{Br}}{\underset{|}{C}}HCH{=}CH_2} \xrightleftharpoons{40°C,\ HBr} \underset{\textbf{1,4-Addition product}}{CH_3CH{=}CHCH_2\textbf{Br}}$$

Because this equilibrium favors the 1,4-addition product, *that product must be more stable.*

The reactions of hydrogen bromide with 1,3-butadiene serve as a striking illustration of the way that the outcome of a chemical reaction can be determined, in one instance, by relative rates of competing reactions and, in another, by the relative stabilities of the final products. At the lower temperature, the relative amounts of the products of the addition are determined by the relative rates at which the two additions occur; 1,2 addition occurs faster so the 1,2-addition product is the major product. At the higher temperature, the relative amounts of the products are determined by the position of an equilibrium. The 1,4-addition product is the more stable, so it is the major product.

This behavior of 1,3-butadiene and hydrogen bromide can be more fully understood if we examine the diagram shown in Fig. 13.10.

The step that determines the overall outcome of the reaction is the step in which the hybrid allylic cation combines with a bromide ion, that is,

$$CH_2{=}CH{-}CH{=}CH_2 \xrightarrow{HBr} H_3C{-}\overset{\delta+}{C}H{\cdots}CH{\cdots}\overset{\delta+}{C}H_2 \begin{cases} \xrightarrow{Br^-} \underset{\textbf{1,2 Product}}{CH_3{-}\underset{\textbf{Br}}{\underset{|}{C}H}{-}CH{=}CH_2} \\ \xrightarrow{Br^-} \underset{\textbf{1,4 Product}}{CH_3{-}CH{=}CH{-}CH_2\textbf{Br}} \end{cases}$$

This step determines the regioselectivity of the reaction.

We see in Fig. 13.10 that, for this step, the free energy of activation leading to the 1,2-addition product is less than the free energy of activation leading to the 1,4-addition product, even though the 1,4 product is more stable. At low temperatures, a larger fraction of collisions between the intermediate ions has enough energy to cross the lower barrier (leading to the 1,2 product), and only a very small fraction of collisions has enough energy to cross the higher barrier (leading to the 1,4 product). In either case (and this is the *key point*), whichever barrier is crossed at low temperature (i.e., −80°C in this example), product formation is essentially *irreversible* because there is not enough energy available to lift either product out of its deep potential energy valley. Since 1,2 addition occurs faster, the 1,2 product predominates and the reaction is said to be under **kinetic control** or **rate control.**

At higher temperatures, the intermediate ions have sufficient energy to cross both barriers with relative ease. More importantly, however, *at higher temperatures both reactions are reversible.* Sufficient energy is also available to take the products back over their energy barriers to the intermediate level of allylic cations and bromide ions. The 1,2 product is still formed faster, but, being less stable than the 1,4 product, it also reverts to the allylic cation faster. Under these conditions, that is, at higher temperatures, the relative proportions of the products *do not reflect* the relative heights of the energy barriers leading from allylic cation to products. Instead, *they reflect the relative stabilities of the products themselves.* Since the 1,4 product is more stable, it is formed at the expense of the 1,2

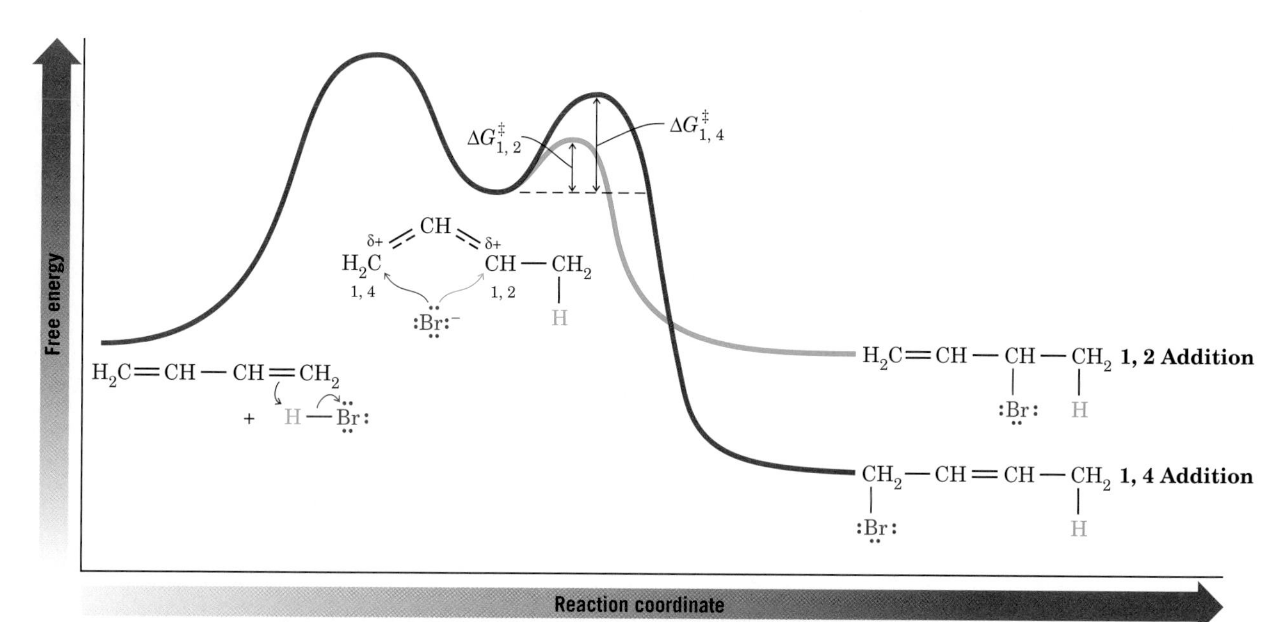

FIGURE 13.10 A schematic free-energy versus reaction coordinate diagram for the 1,2 and 1,4 addition of HBr to 1,3-butadiene. An allylic carbocation is common to both pathways. The energy barrier for attack of bromide ion on the allylic cation to form the 1,2-addition product is less than that to form the 1,4-addition product. The 1,2-addition product is kinetically favored. The 1,4-addition product is more stable, and so it is the thermodynamically favored product.

product because the overall change from 1,2 product to 1,4 product is energetically favored. Such a reaction is said to be under **thermodynamic control** or **equilibrium control.**

Before we leave this subject, one final point should be made. This example clearly demonstrates that predictions of relative reaction rates made on the basis of product stabilities alone can be wrong. This is not always the case, however. For many reactions in which a common intermediate leads to two or more products, the most stable product is formed fastest.

PROBLEM 13.10

(a) Can you suggest a possible explanation for the fact that the 1,2-addition reaction of 1,3-butadiene and hydrogen bromide occurs faster than 1,4 addition? (*Hint:* Consider the relative contributions that the two forms $CH_3\overset{+}{C}HCH{=}CH_2$ and $CH_3CH{=}CH\overset{+}{C}H_2$ make to the resonance hybrid of the allylic cation.)
(b) How can you account for the fact that the 1,4-addition product is more stable?

13.11 The Diels–Alder Reaction: A 1,4-Cycloaddition Reaction of Dienes

In 1928 two German chemists, Otto Diels and Kurt Alder, developed a 1,4-cycloaddition reaction of dienes that has since come to bear their names. The reaction proved to be one of such great versatility and synthetic utility that Diels and Alder were awarded the Nobel Prize in Chemistry in 1950.

An example of the Diels–Alder reaction is the reaction that takes place when 1,3-butadiene and maleic anhydride are heated together at 100°C. The product is obtained in quantitative yield:

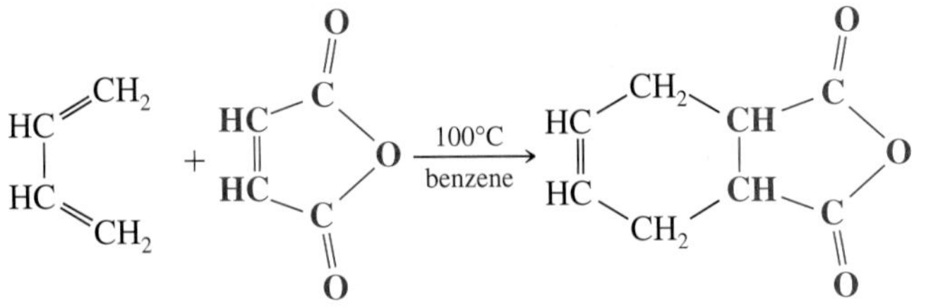

1,3-Butadiene (diene) + **Maleic anhydride (dienophile)** $\xrightarrow[\text{benzene}]{100°C}$ **Adduct (100%)**

This reaction can be more simply written as

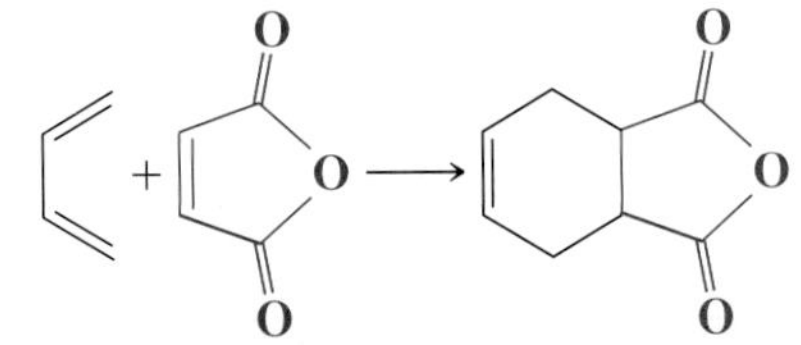

Diels–Alder Reaction

In general terms, the reaction is one between a conjugated **diene** (a 4π-electron system) and a compound containing a double bond (a 2π-electron system) called a **dienophile** (diene + Greek: *philein,* to love). The product of a Diels–Alder reaction is often called an **adduct.** In the Diels–Alder reaction, two new σ bonds are formed at the expense of two π bonds of the diene and dienophile. The adduct contains a new six-membered ring with a double bond. Since σ bonds are usually stronger than π bonds, formation of the adduct is usually favored energetically, *but most Diels–Alder reactions are reversible.*

The Diels–Alder reaction is a very useful synthetic tool for preparing cyclohexene rings.

We can account for all of the bond changes in a Diels–Alder reaction by using curved arrows in the following way:

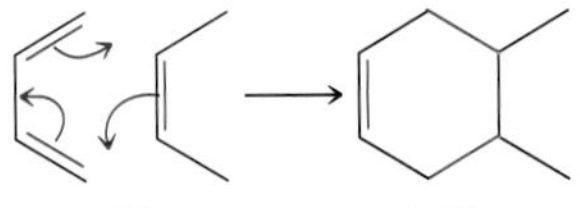

Diene **Dienophile** → **Adduct**

The simplest example of a Diels–Alder reaction is the one that takes place between 1,3-butadiene and ethene. This reaction, however, takes place much more slowly than the reaction of butadiene with maleic anhydride and also must be carried out under pressure:

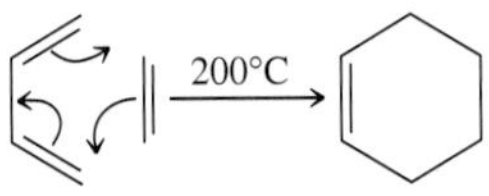

$\xrightarrow{200°C}$ **(20%)**

Another example is the preparation of an intermediate in the synthesis of the anticancer drug Taxol® (paclitaxel) by K. C. Nicolaou (Scripps Research Institute and the University of California, San Diego):

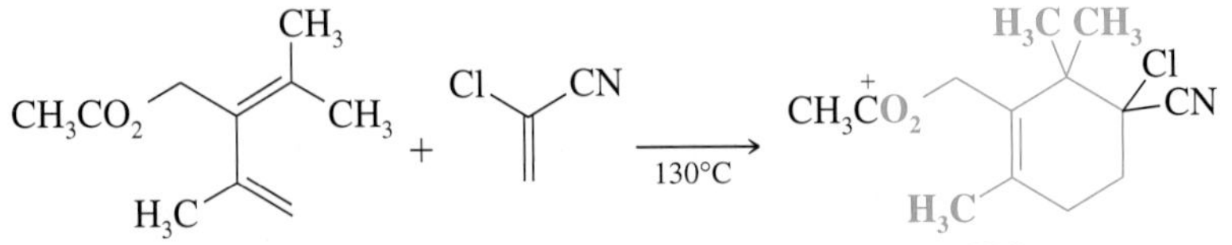

CH_3CO_2 … CH_3, CH_3, H_3C + Cl, CN $\xrightarrow{130°C}$ CH_3CO_2 … H_3C CH_3, Cl, CN, H_3C

85%
(Used in a synthesis of Taxol®)

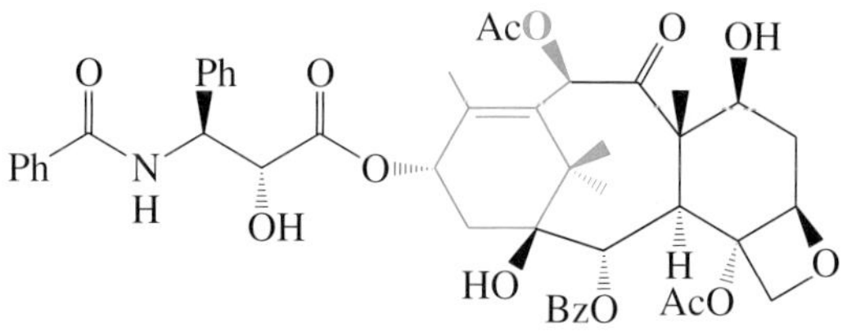

Taxol®

13.11A Factors Favoring the Diels–Alder Reaction

Alder originally stated that the Diels–Alder reaction is favored by the presence of electron-withdrawing groups in the dienophile and by electron-releasing groups in the diene. Maleic anhydride, a very potent dienophile, has two electron-withdrawing carbonyl groups on carbon atoms adjacent to the double bond.

The helpful effect of electron-releasing groups in the diene can also be demonstrated; 2,3-dimethyl-1,3-butadiene, for example, is nearly five times as reactive in Diels–Alder reactions as is 1,3-butadiene. When 2,3-dimethyl-1,3-butadiene reacts with propenal (acrolein) at only 30°C, the adduct is obtained in quantitative yield:

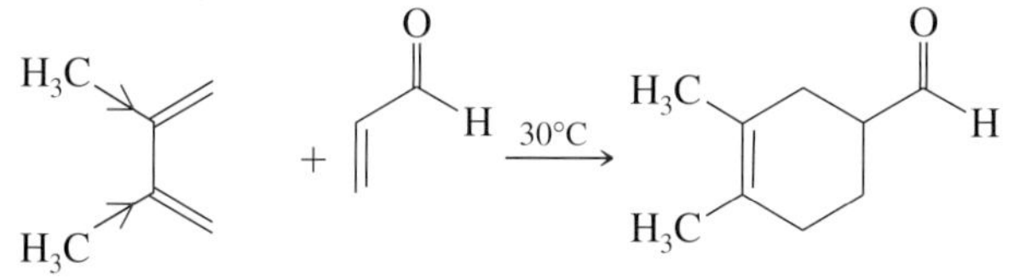

2,3-Dimethyl-1,3-butadiene + **Propenal** $\xrightarrow{30°C}$ **(100%)**

Research (by C. K. Bradsher of Duke University) has shown that the locations of electron-withdrawing and electron-releasing groups in the dienophile and diene can be reversed without reducing the yields of the adducts. Dienes with electron-withdrawing groups have been found to react readily with dienophiles containing electron-releasing groups.

Besides the use of dienes and dienophiles that have complementary electron-releasing and electron-donating properties, other factors found to enhance the rate of Diels–Alder reactions include high temperature and high pressure. Another widely used method is the use of Lewis acid catalysts. The following reaction is one of many examples where Diels–Alder adducts form readily at ambient temperature in the presence of a Lewis acid catalyst. (In Section 13.11D we see how Lewis acids can be used with chiral ligands to induce asymmetry in the reaction products.)

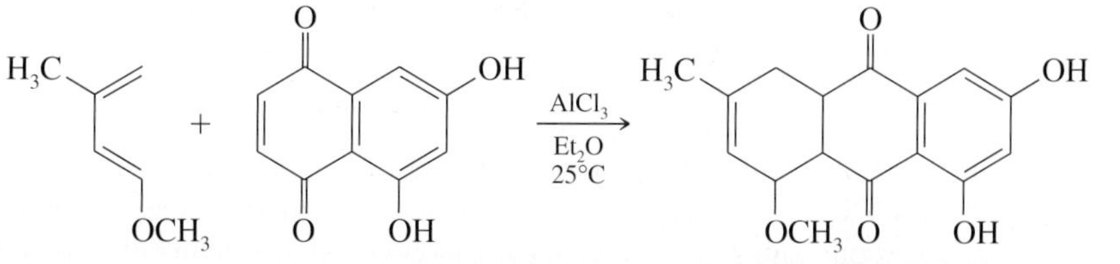

$\xrightarrow[Et_2O,\ 25°C]{AlCl_3}$ **80%**

13.11B Stereochemistry of the Diels–Alder Reaction

Now let us consider some stereochemical aspects of the Diels–Alder reaction. The following factors are among the reasons why Diels–Alder reactions are so extraordinarily useful in synthesis.

1. **The Diels–Alder reaction is highly stereospecific: The reaction is a syn addition, and the configuration of the dienophile is *retained* in the product.** Two examples that illustrate this aspect of the reaction are shown here:

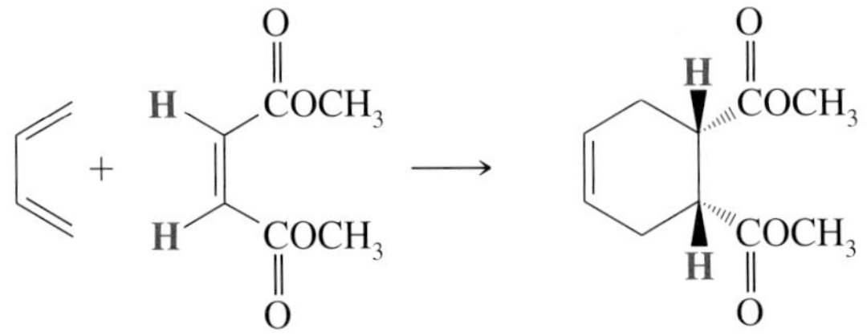

Dimethyl maleate (a *cis*-dienophile) → **Dimethyl cyclohex-4-ene-*cis*-1,2-dicarboxylate**

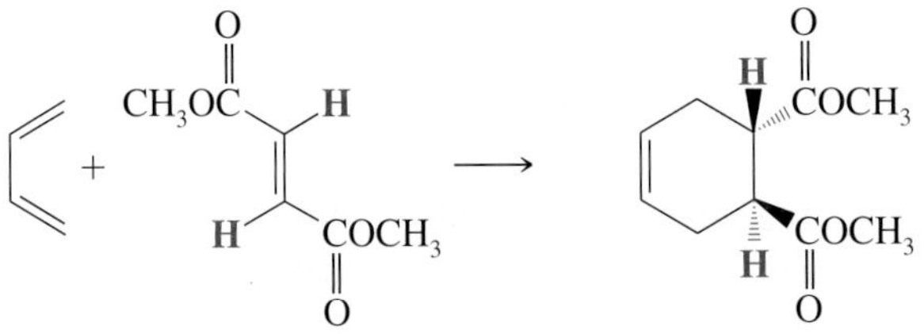

Dimethyl fumarate (a *trans*-dienophile) → **Dimethyl cyclohex-4-ene-*trans*-1,2-dicarboxylate**

In the first example, a dienophile with cis ester groups reacts with 1,3-butadiene to give an adduct with cis ester groups. In the second example just the reverse is true. A *trans*-dienophile gives a trans adduct.

2. **The diene, of necessity, reacts in the s-cis conformation rather than the s-trans:**

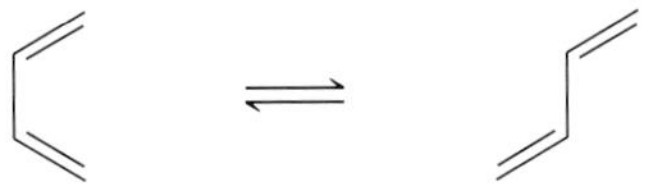

s-cis Conformation ⇌ **s-trans Conformation**

Reaction in the s-trans conformation would, if it occurred, produce a six-membered ring with a highly strained trans double bond. This course of the Diels–Alder reaction has never been observed.

Use hand-held molecular models to investigate the strained nature of hypothetical *trans*-cyclohexene

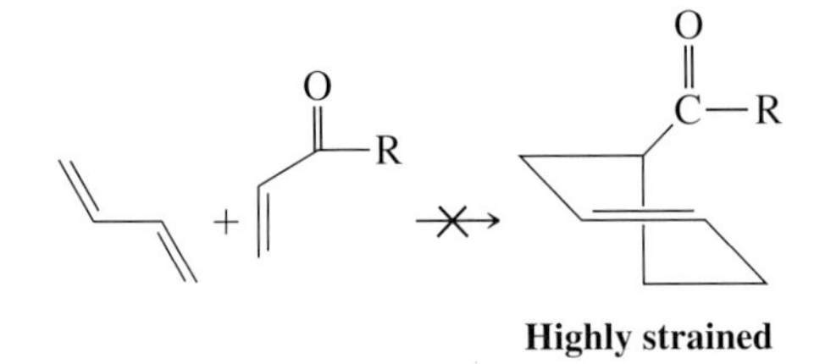

Cyclic dienes in which the double bonds are held in the s-cis conformation are usually highly reactive in the Diels–Alder reaction. Cyclopentadiene, for example, reacts with maleic anhydride at room temperature to give the following adduct in quantitative yield:

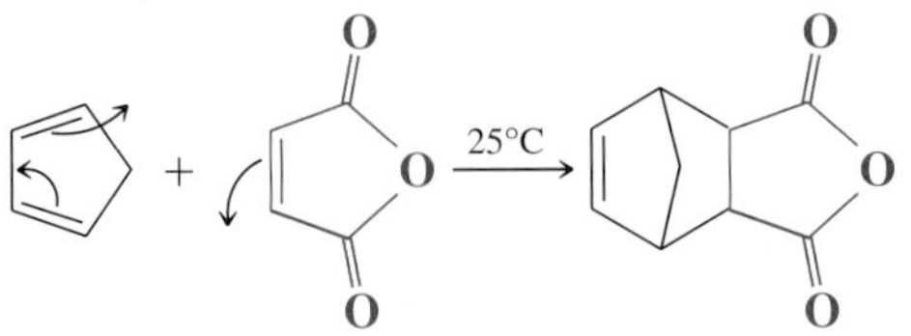

25°C

Cyclopentadiene is so reactive that on standing at room temperature it slowly undergoes a Diels–Alder reaction with itself:

25°C

"Dicyclopentadiene"

The reaction is reversible, however. When "dicyclopentadiene" is distilled, it dissociates (is "cracked") into two molar equivalents of cyclopentadiene.

The reactions of cyclopentadiene illustrate a third stereochemical characteristic of the Diels–Alder reaction.

3. **The Diels–Alder reaction occurs primarily in an endo rather than an exo fashion when the reaction is kinetically controlled** (see Problem 13.28). **Endo** and **exo** are terms used to designate the stereochemistry of bridged rings such as bicyclo[2.2.1]heptane. The point of reference is the longest bridge. A group that is anti to the longest bridge (the two-carbon bridge) is said to be exo; if it is on the same side, it is endo:

Exo/endo Diels–Alder reaction

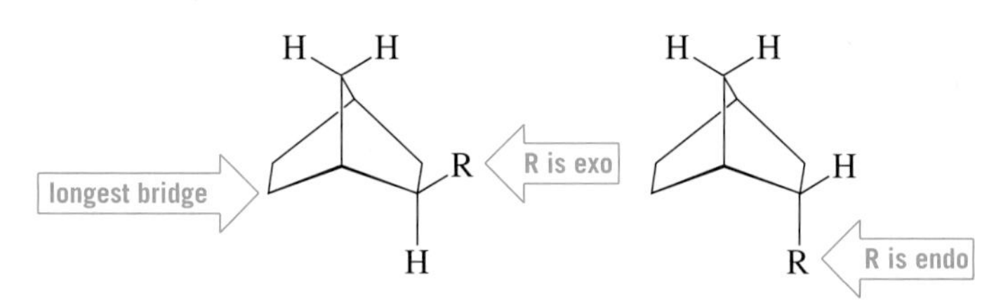

In general, the exo substituent is always on the side anti to the *longer* bridge of a bicyclic structure (exo, outside; endo, inside). For example,

exo endo exo endo exo endo

13.11C Molecular Orbital Considerations That Favor an Endo Transition State

In the Diels–Alder reaction of cyclopentadiene with maleic anhydride the major product is the one in which the anhydride linkage, $-\underset{\underset{\displaystyle \mathrm{O}}{\|}}{\mathrm{C}}-\mathrm{O}-\underset{\underset{\displaystyle \mathrm{O}}{\|}}{\mathrm{C}}-$, has assumed the endo configuration. This favored endo stereochemistry seems to arise from favorable interactions between the π electrons of the developing double bond in the diene and the π electrons of unsaturated groups of the dienophile. In Fig. 13.11 we can see that when the two molecules approach each other in the endo orientation, as shown, orbitals in the LUMO of maleic anhydride and the HOMO of cyclopentadiene can interact at the carbons where the new σ bonds will form (the interaction of these orbitals is indicated by purple in Fig. 13.11*b*). We can also see that this same orientation of approach (endo) has overlap between the LUMO lobes at the carbonyl groups of maleic anhydride and the HOMO lobes in cyclopentadiene above them (the interaction of these orbitals is indicated by green). This so-called secondary orbital interaction is also favorable, and it leads to a preference for endo approach of the dienophile, such that the unsaturated groups of the dienophile are tucked in and under the diene, rather than out and away in the exo orientation.

The transition state for the endo product is thus of lower energy because of the favorable orbital interactions described above, and therefore the endo form is the kinetic (and major) product of this Diels–Alder reaction. The exo form is the thermodynamic product because steric interactions are fewer in the exo adduct than in the endo adduct (Fig. 13.12). Thus, the exo adduct is more stable overall, but it is not the major product because it is formed more slowly.

As we have seen, the Diels–Alder reaction is stereospecific because (1) the configuration of the dienophile is retained in the product and (2) endo addition is favored under kinetic control of the reaction. Even though this results in formation of predominantly one

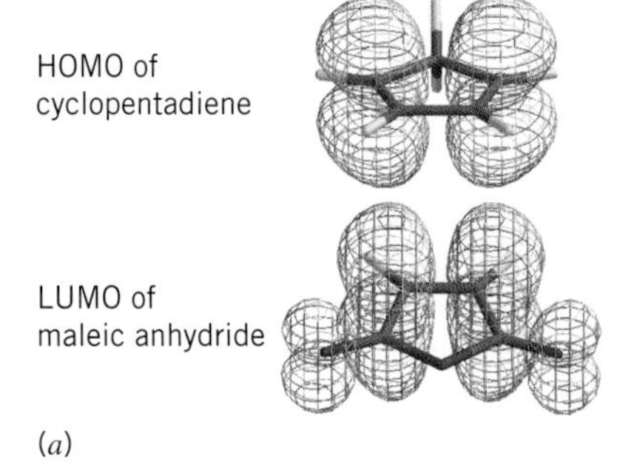

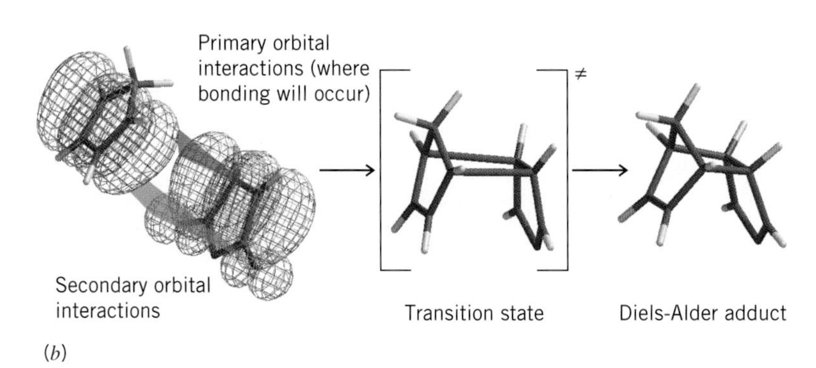

FIGURE 13.11 Diels–Alder reaction of cyclopentadiene and maleic anhydride. *(a)* When the highest occupied molecular orbital (HOMO) of the diene (cyclopentadiene) interacts with the lowest unoccupied molecular orbital (LUMO) of the dienophile (maleic anhydride), favorable secondary orbital interactions occur involving orbitals of the dienophile. *(b)* This interaction is indicated by the purple plane. Favorable overlap of secondary orbitals (indicated by the green plane) leads to a preference for the endo transition state shown above.

CD Tutorial:
Diels–Alder reaction

Animated Graphic
Cyclopentadiene, maleic anhydride Diels–Alder reaction

stereoisomeric form (endo with retention of the original dienophile configuration), the product is nevertheless formed as a racemic mixture. The reason for this is that either face of the diene can interact with the dienophile. When the dienophile bonds with one face of the diene, the product is formed as one enantiomer, and when the dienophile bonds at the other face of the diene, the product is the other enantiomer. In the absence of chiral influences, both faces of the diene are equally likely to be attacked.

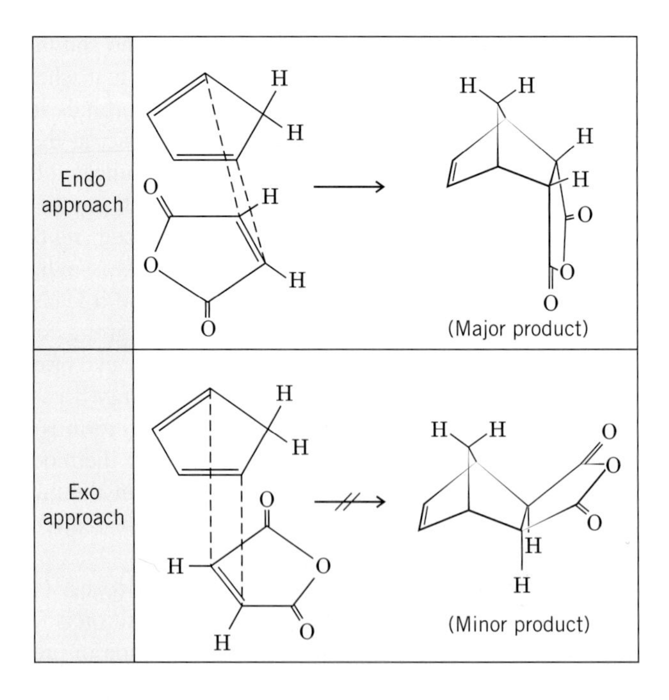

FIGURE 13.12 Endo and exo product formation in the Diels–Alder reaction of cyclopentadiene and maleic anhydride.

PROBLEM 13.11

The dimerization of cyclopentadiene also occurs in an endo way. **(a)** Show how this happens. **(b)** Which π electrons interact? **(c)** What is the three-dimensional structure of the product?

PROBLEM 13.12

What products would you expect from the following reactions?

(a) H_3C, H_3C (2,3-dimethyl-1,3-butadiene) + methyl vinyl ketone (CH_3) ⟶

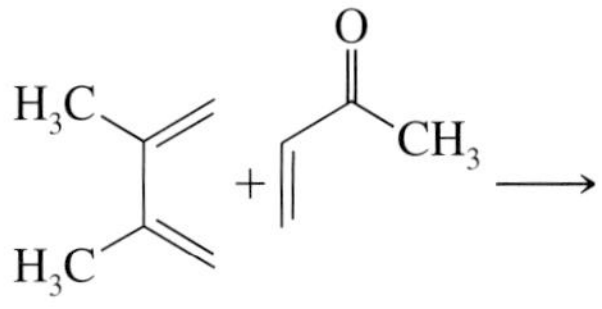

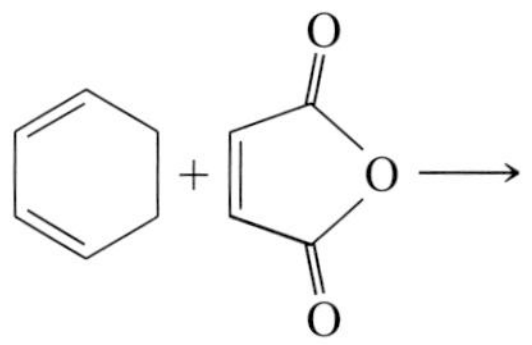

(b) H_3C, H_3C (2,3-dimethyl-1,3-butadiene) + dimethyl maleate (OCH_3, OCH_3) ⟶

(c) 1,3-cyclohexadiene + maleic anhydride ⟶

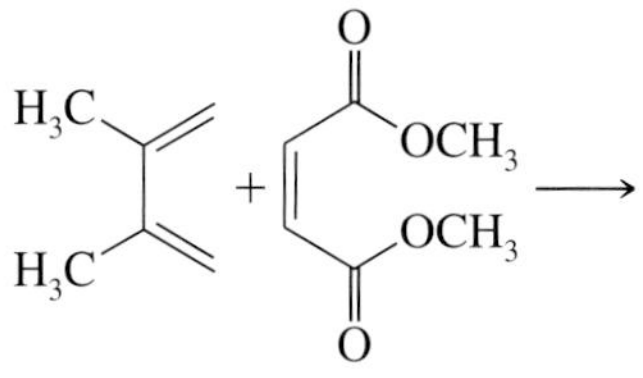

PROBLEM 13.13

Which diene and dienophile would you employ to synthesize the following compound?

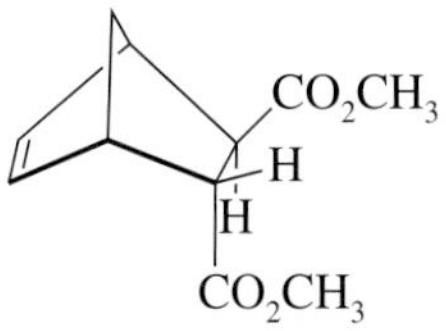

(bicyclic product with CO_2CH_3, H, H, CO_2CH_3)

PROBLEM 13.14

Diels–Alder reactions also take place with triple-bonded (acetylenic) dienophiles. Which diene and which dienophile would you use to prepare the following?

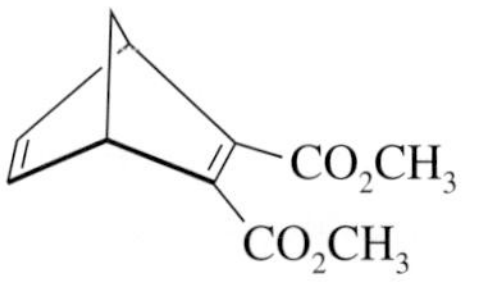

(bicyclic product with CO_2CH_3, CO_2CH_3)

PROBLEM 13.15

1,3-Butadiene and the dienophile shown below were used by A. Eschenmoser in his synthesis of vitamin B_{12} with R. B. Woodward. Draw the structure of the enantiomeric Diels–Alder adducts that would form in this reaction and the two transition states that lead to them.

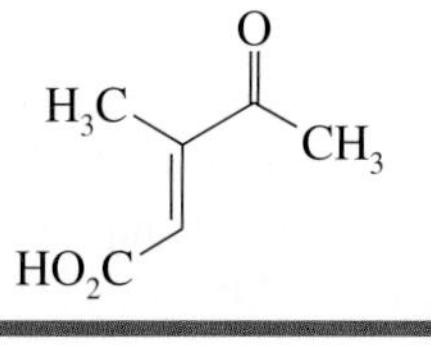

(dienophile with H_3C, CH_3, O, HO_2C)

13.11D Asymmetric Diels–Alder Reactions

Several methods have been developed for inducing enantioselectivity in Diels–Alder reactions. One method involves the use of chiral auxiliaries. A **chiral auxiliary** is a group, present in one enantiomeric form only, that is appended by a functional group to the diene or dienophile to provide a chiral influence on the course of the reaction. After the reaction is over and the influence of the chiral auxiliary is no longer needed, it is removed by an appropriate reaction.

A more elegant approach, because it does not require separate reactions to append or remove the chiral auxiliary, is the use of a chiral Lewis acid catalyst. A powerful example is that of the reaction used by E. J. Corey and co-workers in their refinement of his original synthesis of prostaglandins $F_{2\alpha}$ and E_2:

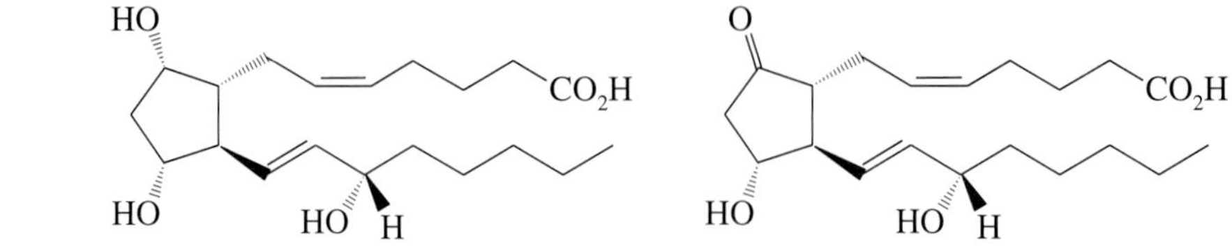

HO, HO, HO H, CO_2H — **$PGF_{2\alpha}$**; O, HO, HO H, CO_2H — **PGE_2**

Prostaglandins

Here, the chiral Lewis acid catalyst not only caused extraordinarily effective enantioselective product formation but also was recovered and used again in future reactions:

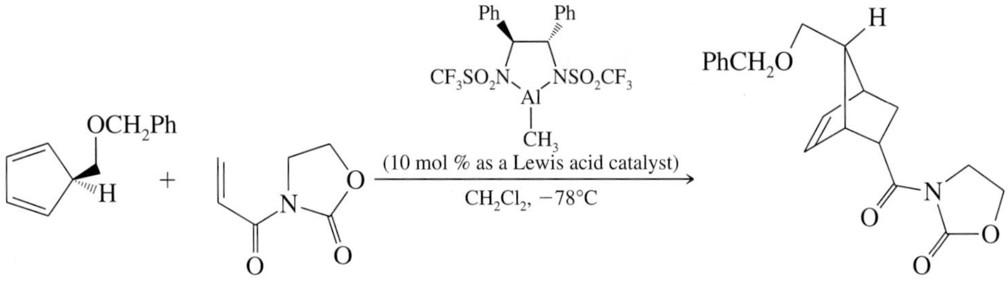

OCH_2Ph, H + (N-acryloyl oxazolidinone) —[Ph, Ph, CF_3SO_2N, NSO_2CF_3, Al, CH_3 (10 mol % as a Lewis acid catalyst); CH_2Cl_2, −78°C]→ $PhCH_2O$, H, O, N, O, O

In this example, the transition state involving the chiral catalyst strongly favors approach of the dienophile from the face of the diene opposite the ether functional group.

Another method for inducing enantioselectivity is the **"chiron approach."** In this method, a stereogenic center that is ultimately part of the target molecule is included from the start in a single enantiomeric form of one of the Diels–Alder reactants. The chiral influence of the stereogenic center in the "chiron" leads to enantioselective interaction of the diene and dienophile. Many examples of this approach can be found in the chemical literature.

13.11E Intramolecular Diels–Alder Reactions

An intriguing version of the Diels–Alder reaction is one where the diene and dienophile are both part of the same molecule. Such a reaction is called an intramolecular Diels–Alder reaction. Reactants of this type have been used in many syntheses of complex molecules where the structure of the desired adduct lent itself to this strategy. An example is the following reaction used by K. C. Nicolaou (Scripps Research Institute and the University of California, San Diego) and co-workers in their synthesis of an intermediate toward the endiandric acids A–D:

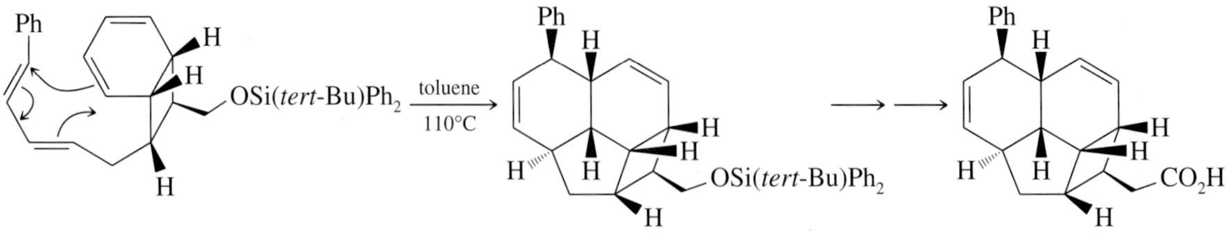

Ph, H, H, H, OSi(*tert*-Bu)Ph_2 —[toluene, 110°C]→ Ph, H, H, H, H, H, H, OSi(*tert*-Bu)Ph_2 — **100%** → → Ph, H, H, H, H, H, H, CO_2H — **Endiandric acid A**

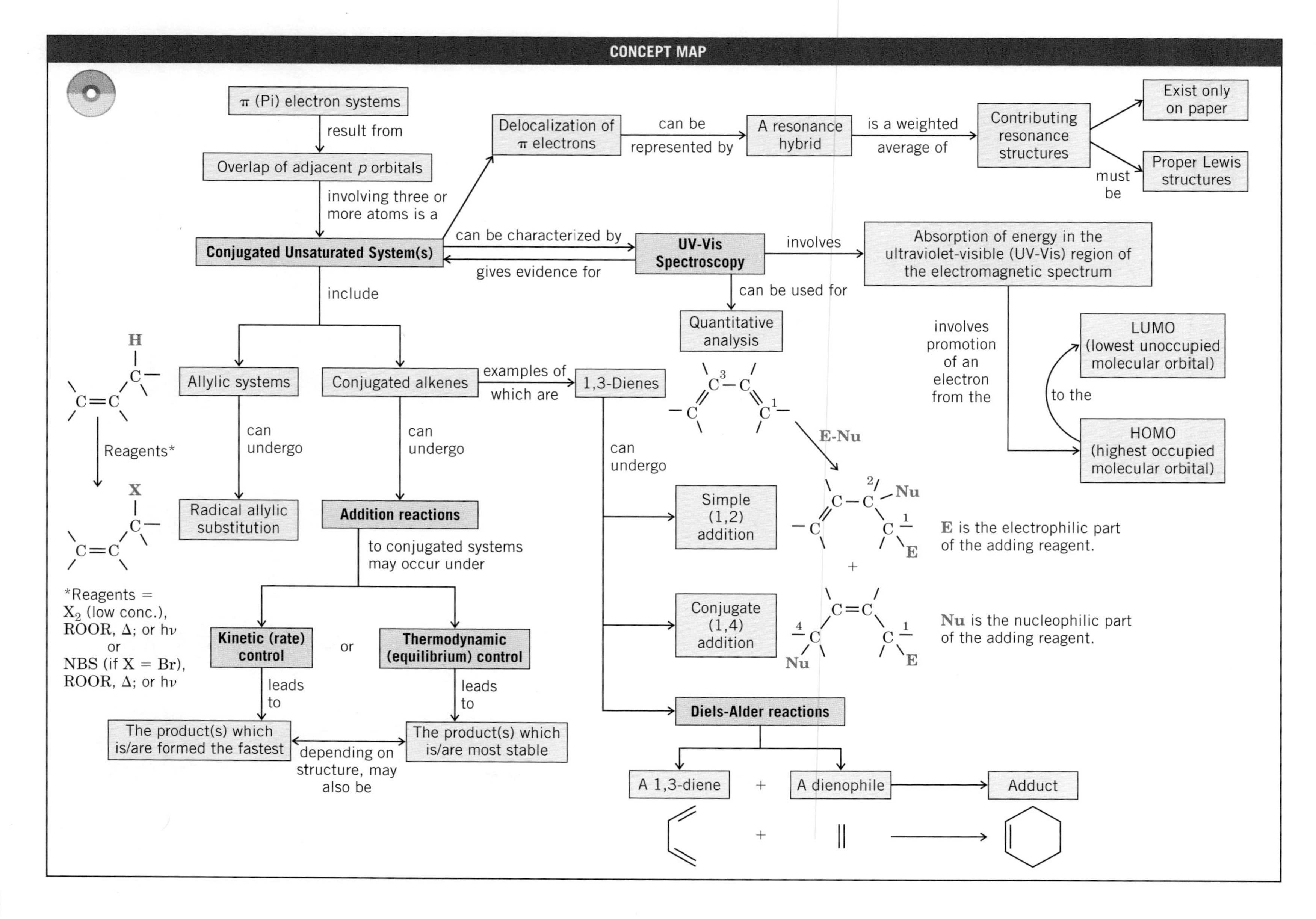
CONCEPT MAP
π (Pi) electron systems
result from
Overlap of adjacent p orbitals
involving three or more atoms is a
Conjugated Unsaturated System(s)
Delocalization of π electrons
can be represented by
A resonance hybrid
is a weighted average of
Contributing resonance structures
Exist only on paper
must be
Proper Lewis structures
can be characterized by
gives evidence for
UV-Vis Spectroscopy
involves
Absorption of energy in the ultraviolet-visible (UV-Vis) region of the electromagnetic spectrum
involves promotion of an electron from the
HOMO (highest occupied molecular orbital)
to the
LUMO (lowest unoccupied molecular orbital)
can be used for
Quantitative analysis
include
Allylic systems
can undergo
Radical allylic substitution
Reagents*
*Reagents = X2 (low conc.), ROOR, Δ; or hν or NBS (if X = Br), ROOR, Δ; or hν
Conjugated alkenes
examples of which are
1,3-Dienes
can undergo
Addition reactions
to conjugated systems may occur under
Kinetic (rate) control
or
Thermodynamic (equilibrium) control
leads to
The product(s) which is/are formed the fastest
depending on structure, may also be
The product(s) which is/are most stable
Simple (1,2) addition
Conjugate (1,4) addition
E-Nu
E is the electrophilic part of the adding reagent.
Nu is the nucleophilic part of the adding reagent.
Diels-Alder reactions
A 1,3-diene
A dienophile
Adduct

Key Terms and Concepts

Adduct	Section 13.11
Allyl cation	Section 13.4
Allyl radical	Sections 13.2A, 13.3
Allylic substitution	Section 13.2
Chiral auxiliary	Section 13.11D
Chiron	Section 13.11D
Conjugation	Section 13.1
Diene	Section 13.11
Dienophile	Section 13.11
Endo and exo groups	Sections 13.11B, 13.11C
HOMO	Section 13.9B
Kinetic (rate) control	Section 13.10A
LUMO	Section 13.9B
Resonance structures (contributors)	Sections 13.3B, 13.5
Thermodynamic (equilibrium) control	Section 13.10A
Ultraviolet–visible (UV–Vis) spectroscopy	Section 13.9

Additional Problems

13.16 Outline a synthesis of 1,3-butadiene starting from

(a) 1,4-Dibromobutane
(b) $HOCH_2(CH_2)_2CH_2OH$
(c) $CH_2{=}CHCH_2CH_2OH$
(d) $CH_2{=}CHCH_2CH_2Cl$
(e) $CH_2{=}CHCHClCH_3$
(f) $CH_2{=}CHCH(OH)CH_3$
(g) $HC{\equiv}CCH{=}CH_2$

13.17 What product would you expect from the following reaction?

$$(CH_3)_2\underset{\displaystyle Cl}{\underset{|}{C}}{-}\underset{\displaystyle Cl}{\underset{|}{C}}(CH_3)_2 + 2\ KOH \xrightarrow[\text{heat}]{\text{ethanol}}$$

13.18 What products would you expect from the reaction of 1 mol of 1,3-butadiene and each of the following reagents? (If no reaction would occur, you should indicate that as well.)

(a) 1 mol of Cl_2
(b) 2 mol of Cl_2
(c) 2 mol of Br_2
(d) 2 mol of H_2, Ni
(e) 1 mol of Cl_2 in H_2O
(f) Hot $KMnO_4$ (excess)
(g) H_2O, cat. H_2SO_4

13.19 Show how you might carry out each of the following transformations. (In some transformations several steps may be necessary.)

(a) 1-Butene $\longrightarrow$ 1,3-butadiene
(b) 1-Pentene $\longrightarrow$ 1,3-pentadiene
(c) $CH_3CH_2CH_2CH_2OH \longrightarrow CH_2BrCH{=}CHCH_2Br$
(d) $CH_3CH{=}CHCH_3 \longrightarrow CH_3CH{=}CHCH_2Br$
(e)

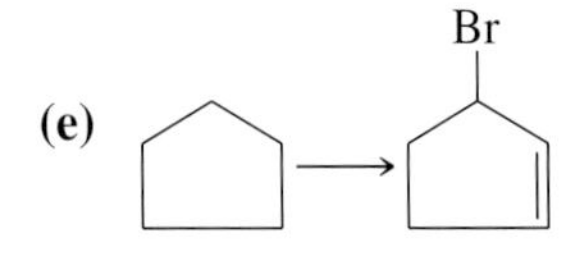

Note: Problems marked with an asterisk are "challenge problems."

(f) 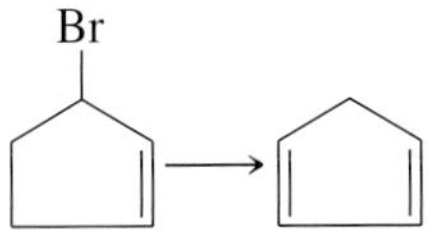 [structure: 3-bromocyclopentene → cyclopentadiene]

13.20 Conjugated dienes react with free radicals by both 1,2 and 1,4 addition. Write a detailed mechanism to account for this fact using the peroxide-promoted addition of one molar equivalent of HBr to 1,3-butadiene as an illustration.

13.21 UV–Vis, IR, NMR, and mass spectrometry are spectroscopic tools we use to obtain structural information about compounds. For each pair of compounds below, describe at least one aspect from each of two spectroscopic methods (UV–Vis, IR, NMR, or mass spectrometry) that would distinguish one compound in a pair from the other.

(a) 1,3-Butadiene and 1-butyne
(b) 1,3-Butadiene and butane
(c) Butane and $CH_2{=}CHCH_2CH_2OH$
(d) 1,3-Butadiene and $CH_2{=}CHCH_2CH_2Br$
(e) $CH_2BrCH{=}CHCH_2Br$ and $CH_3CBr{=}CBrCH_3$

13.22 **(a)** The hydrogen atoms attached to C3 of 1,4-pentadiene are unusually susceptible to abstraction by radicals. How can you account for this? **(b)** Can you provide an explanation for the fact that the protons attached to C3 of 1,4-pentadiene are more acidic than the methyl hydrogen atoms of propene?

13.23 When 2-methyl-1,3-butadiene (isoprene) undergoes a 1,4 addition of hydrogen chloride, the major product that is formed is 1-chloro-3-methyl-2-butene. Little or no 1-chloro-2-methyl-2-butene is formed. How can you explain this?

13.24 Which diene and dienophile would you employ in a synthesis of each of the following?

(a) 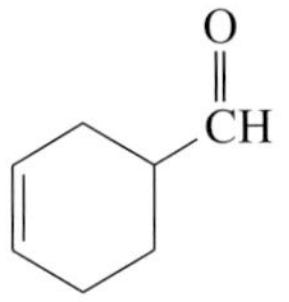 [structure: cyclohexene ring bearing a CHO group (C=O, CH)]

(d) [structure: bicyclic anhydride fused to cyclohexene ring, with CH_3, H, H, C=O, O, C=O labels]

(b) [structure: 4,5-dimethylcyclohexene ring (H_3C, H_3C) with two cis $COCH_3$ ester groups (C=O)]

(e) [structure: cyclohexene ring bearing CN]

(c) [structure: 4,5-dimethylcyclohexene ring (H_3C, H_3C) with two trans $COCH_3$ ester groups (C=O)]

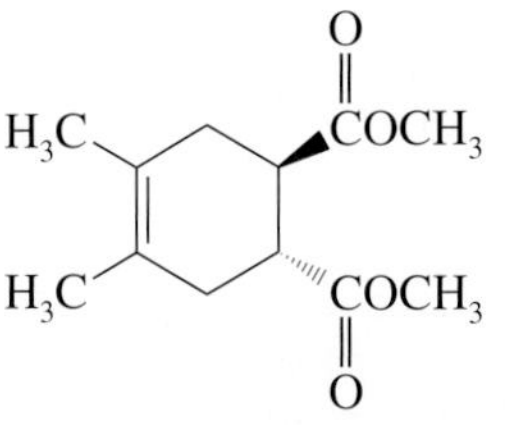

(f) 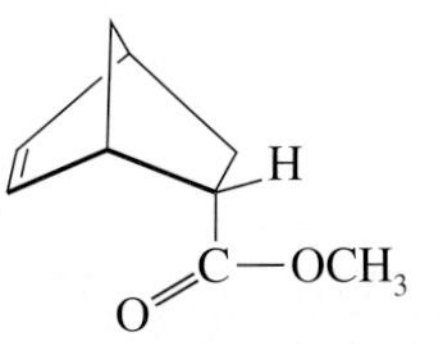 [structure: bicyclic norbornene with H and $C{-}OCH_3$ (C=O) ester group]

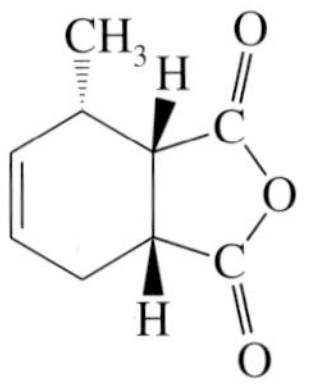

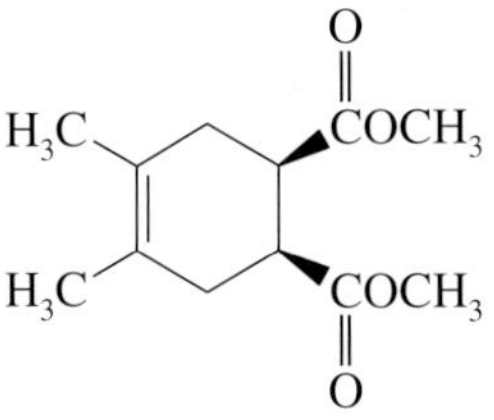

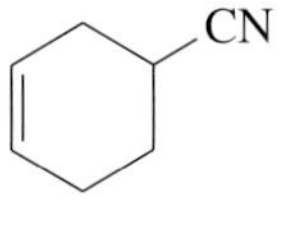

13.25 Account for the fact that neither of the following compounds undergoes a Diels–Alder reaction with maleic anhydride:

$HC{\equiv}C{-}C{\equiv}CH$ or [structure: 3-methylenecyclohexene, ring $={CH_2}$]

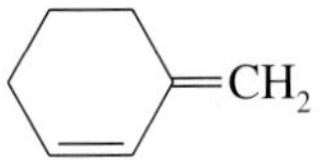

13.26 Acetylenic compounds may be used as dienophiles in the Diels–Alder reaction (see Problem 13.14). Write structures for the adducts that you expect from the reaction of 1,3-butadiene with

(a) $CH_3O\overset{\overset{O}{\|}}{C}C{\equiv}C\overset{\overset{O}{\|}}{C}OCH_3$ (dimethyl acetylenedicarboxylate)

(b) $CF_3C{\equiv}CCF_3$ (hexafluoro-2-butyne)

13.27 Cyclopentadiene undergoes a Diels–Alder reaction with ethene at 160–180°C. Write the structure of the product of this reaction.

13.28 When furan and maleimide undergo a Diels–Alder reaction at 25°C, the major product is the endo adduct **G**. When the reaction is carried out at 90°C, however, the major product is the exo isomer **H**. The endo adduct isomerizes to the exo adduct when it is heated to 90°C. Propose an explanation that will account for these results.

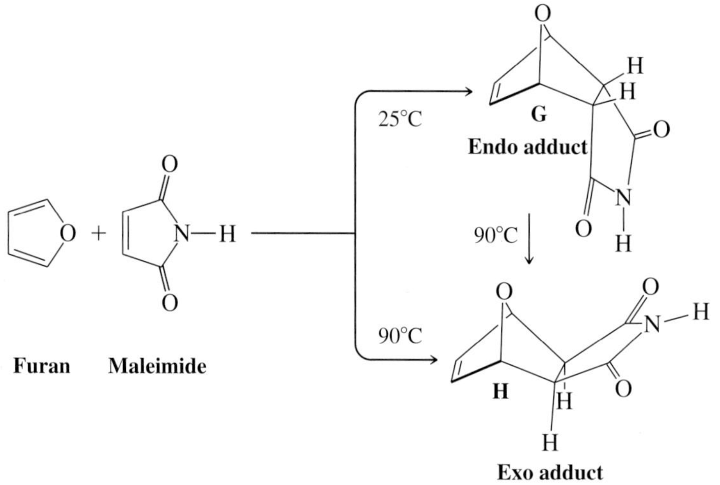

25°C → **G** **Endo adduct**

90°C ↓

90°C → **Exo adduct**

Furan **Maleimide**

13.29 Two controversial "hard" insecticides are aldrin and dieldrin (see the following). [The Environmental Protection Agency (EPA) halted the use of these insecticides because of possible harmful side effects and because they are not biodegradable.] The commercial synthesis of aldrin begins with hexachloro-cyclopentadiene and norbornadiene. Dieldrin is synthesized from aldrin. Show how these syntheses might be carried out.

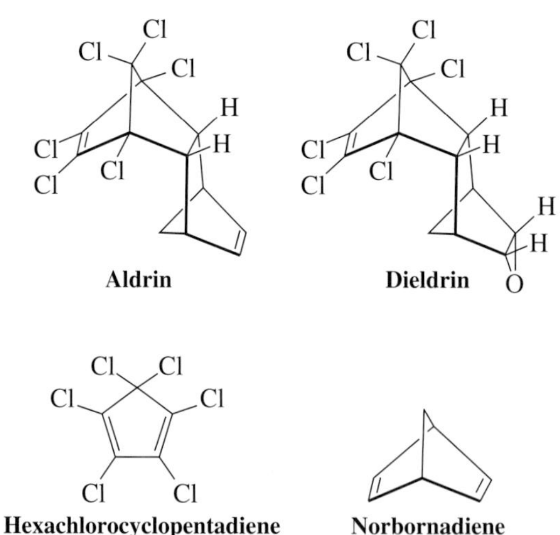

13.30 **(a)** Norbornadiene for the aldrin synthesis (Problem 13.29) can be prepared from cyclopentadiene and acetylene. Show the reaction involved. **(b)** It can also be prepared by allowing cyclopentadiene to react with vinyl chloride and treating the product with a base. Outline this synthesis.

13.31 Two other hard insecticides (see Problem 13.29) are chlordan and heptachlor. Their commercial syntheses begin with cyclopentadiene and hexachlorocyclopentadiene. Show how these syntheses might be carried out.

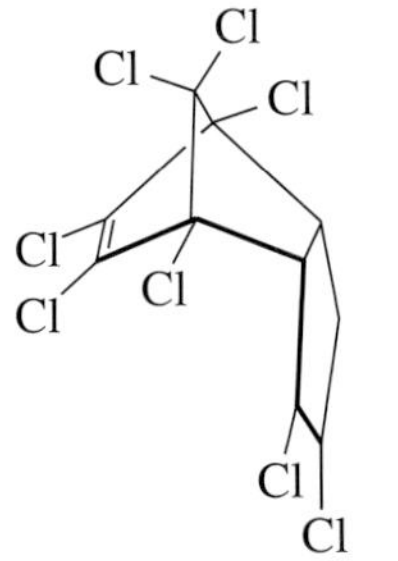

Chlordan 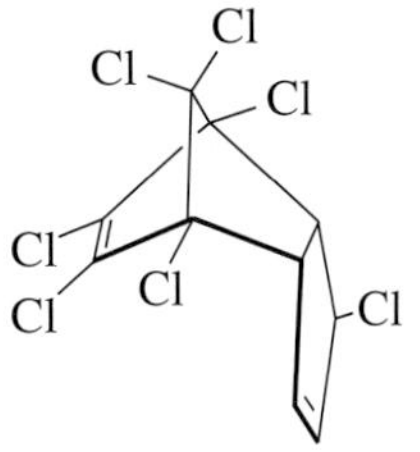 **Heptachlor**

13.32 Isodrin, an isomer of aldrin, is obtained when cyclopentadiene reacts with the hexachloronorbornadiene, shown here. Propose a structure for isodrin.

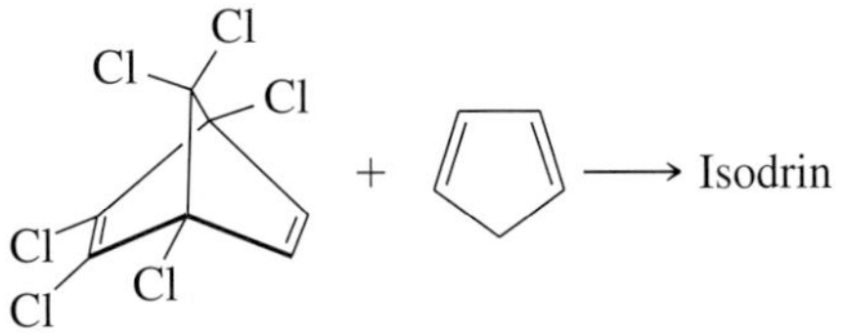

$+ \longrightarrow$ Isodrin

13.33 When $CH_3CH{=}CHCH_2OH$ is treated with concentrated HCl, two products are produced, $CH_3CH{=}CHCH_2Cl$ and $CH_3CHClCH{=}CH_2$. Outline a mechanism that explains this.

13.34 When a solution of 1,3-butadiene in CH_3OH is treated with chlorine, the products are $ClCH_2CH{=}CHCH_2OCH_3$ (30%) and $ClCH_2CH(OCH_3)CH{=}CH_2$ (70%). Write a mechanism that accounts for their formation.

13.35 Dehydrohalogenation of 1,2-dihalides (with the elimination of two molar equivalents of HX) normally leads to an alkyne rather than to a conjugated diene. However, when 1,2-dibromocyclohexane is dehydrohalogenated, 1,3-cyclohexadiene is produced in good yield. What factor accounts for this?

13.36 When 1-pentene reacts with *N*-bromosuccinimide, two products with the formula C_5H_9Br are obtained. What are these products and how are they formed?

13.37 Treating either 1-chloro-3-methyl-2-butene or 3-chloro-3-methyl-1-butene with Ag_2O in water gives (in addition to AgCl) the same mixture of alcohols: $(CH_3)_2C{=}CHCH_2OH$ (15%) and $(CH_3)_2C(OH)CH{=}CH_2$ (85%).

(a) Write a mechanism that accounts for the formation of these products.
(b) What might explain the relative proportions of the two alkenes that are formed?

13.38 The heat of hydrogenation of allene is 298 kJ mol^{-1}, whereas that of propyne is 290 kJ mol^{-1}. **(a)** Which compound is more stable? **(b)** Treating allene with a strong base causes it to isomerize to propyne. Explain.

13.39 Mixing furan (Problem 13.28) with maleic anhydride in diethyl ether yields a crystalline solid with a melting point of 125°C. When melting of this compound takes place, however, one can notice that the melt evolves a gas. If the melt is allowed to resolidify, one finds that it no longer melts at 125°C but instead it melts at 56°C. Consult an appropriate chemistry handbook and provide an explanation for what is taking place.

13.40 Give the structures of the products that would be formed when 1,3-butadiene reacts with each of the following:

(a) (*E*)-$CH_3CH{=}CHCO_2CH_3$
(b) (*Z*)-$CH_3CH{=}CHCO_2CH_3$
(c) (*E*)-$CH_3CH{=}CHCN$
(d) (*Z*)-$CH_3CH{=}CHCN$

13.41 Although both 1-bromobutane and 4-bromo-1-butene are primary halides, the latter undergoes elimination more rapidly. How can this behavior be explained?

13.42 Why does the molecule shown below, although a conjugated diene, fail to undergo a Diels–Alder reaction?

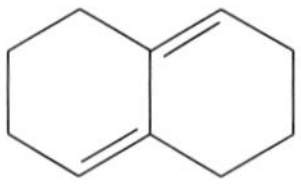

13.43 Draw the structure of the product from the following reaction (formed during a synthesis of one of the endiandric acids by K. C. Nicolaou):

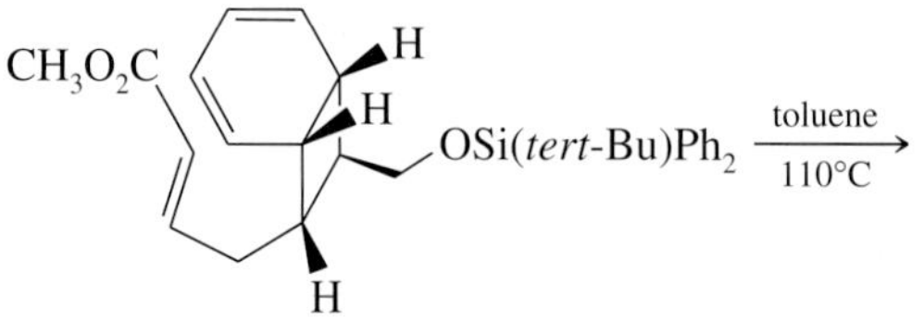

CH_3O_2C ... OSi(*tert*-Bu)Ph_2 $\xrightarrow[110°C]{\text{toluene}}$

***13.44** When tetraphenylcyclopentadienone (**A**) is heated with maleic anhydride (**B**), the deep purple color of **A** disappears, carbon monoxide is evolved, and a final product **C** is formed. The reaction proceeds via a Diels–Alder intermediate. Compound **C** has ^{1}H NMR signals at δ 3.7, 7.1, 7.3, and 7.4 (area ratios 1 : 2 : 4 : 4).

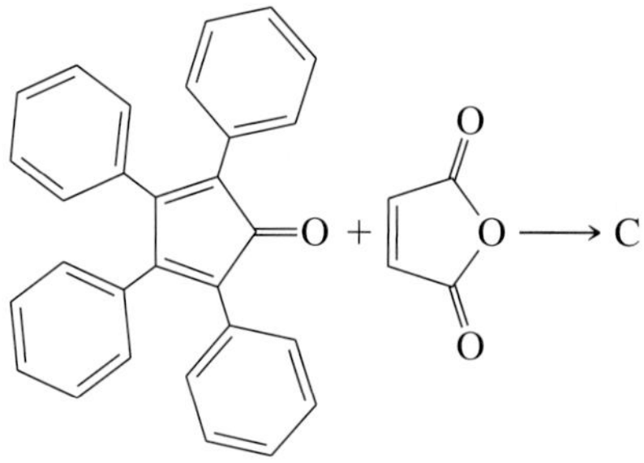

A + **B** ⟶ C

If **C** is reacted with one molar equivalent of bromine, two molar equivalents of HBr are produced by oxidative removal of two hydrogen atoms, affording product **D**. **D** has ^{1}H NMR signals at δ 7.2, 7.3, and 7.5 (area ratio 1 : 2 : 2) only.

What are the structures of **C** and **D**?

***13.45** **(a)** In a study of cyclopentadiene, predominantly 1,2 addition of BrCl occurred when BrCl was used in the form of its pyridine addition product, *N*-bromopyridinium chloride (shown below). The addition was Markovnikov and analogous to the stereochemistry of bromine addition to simple alkenes. Draw the structure of the product.

(b) When free BrCl itself was used (i.e., not as the pyridine addition complex), *cis*-1,4 addition was predominant. Draw the structure of this product.

(c) The two addition products above can be distinguished by the nature of the ^{1}H NMR spectra of their methylene groups, which are:

For isomer **1**:

H_a of the CH_2: δ 2.57 (double triplet, J values of 2.5 and 16 Hz)
H_b of the CH_2: δ 3.14 (double triplet, J values of 6.6 and 16 Hz)

For isomer **2**:

H_a of the CH_2: δ 2.76 (broad doublet, J value 18 Hz)
H_b of the CH_2: δ 3.35 (double doublet, J values of 5.5 and 18 Hz)

Which isomer is which?

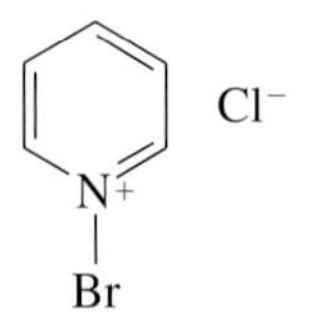

N-Bromopyridinium chloride

13.46 Draw all of the contributing resonance structures and the resonance hybrid for the carbocation that would result from ionization of bromine from 5-bromo-1,3-pentadiene. Open the CD molecular model depicting a map of electrostatic potential for the pentadienyl carbocation. Based on the model, which is the most important contributing resonance structure for this cation? Is this consistent with what you would have predicted based on your knowledge of relative carbocation stabilities? Why or why not?

LEARNING GROUP PROBLEMS

1. Elucidate the structures of compounds **A** through **I** in the following "road map" problem. Specify any missing reagents.

A (C_5H_8) + **B** (C_9H_{10}) ⟶ **C** $\xrightarrow[\text{(or } RCO_3H)]{\text{MMPP}}$ [epoxide with CH_3, CH_3, C_6H_5] + [epoxide with CH_3, CH_3, C_6H_5]

A ↓ Br_2, warm (1 molar equiv.) **F** ↓ CH_3ONa (2 molar equiv.) **G** ↓ HBr (no ROOR) **H** ↓ $KOC(CH_3)_3$, heat **I** ($C_7H_{14}O_2$) $\xrightarrow{\text{reagents?}}$ [structure with OH, OCH_3, OCH_3]

D (C_9H_{12}) ↑ NBS, ROOR, heat **E** ↑ NaOEt, heat **B**

2. **(a)** Write reactions to show how you could convert 2-methyl-2-butene into 2-methyl-1,3-butadiene.
(b) Write reactions to show how you could convert ethylbenzene into the following compound:

[structure: C_6H_5–CH=CH–CN]

(c) Write structures for the various Diels–Alder adduct(s) that could result on reaction of 2-methyl-1,3-butadiene with the compound shown in part (b).

Aromatic Compounds

Green Chemistry

The twenty-first century brings an urgent need for chemists to develop environmentally friendly "green" methods. We have already discussed several green methods over the course of our study (e.g., "The Chemistry of Environmentally Friendly Alkene Oxidation Methods" in Chapter 11.) Environmentally benign chemistry is especially important in the chemical industry, where the worldwide synthesis of a compound may involve billions of pounds of chemicals a year. We can think of many ways to reduce the impact on the environment of a large-scale process: We can run the reaction in safer aqueous-based systems instead of potentially hazardous organic solvents, we can carry out the reaction at ambient temperature instead of with applied heat, we can recycle materials, and we can use pathways that do not use or generate toxic substances. All of these may lessen the impact of the process on the environment in terms of pollution or consumption of resources. Consider two new possibilities that replace benzene, a known carcinogen, with a safer alternative. (A molecular model of benzene is shown by the photo of recycling materials above.)

The chemical industry consumes styrene, the monomer used to form polystyrene (and a component of other polymers as well) in huge quantities each year. The current industrial method for making it converts benzene into styrene in two steps: a Friedel–Crafts alkylation (Section 15.6) followed by a dehydrogenation. O. L. Chapman of the University of California, Los Angeles, has developed a new styrene synthesis that is benign by design. Chapman's method uses a single step to convert mixed xylenes (compounds that are not carcinogenic) into styrene. This new method has the potential to eliminate the use of billions of pounds of benzene each year.

Another opportunity to eliminate industrial reliance on benzene would come from the development of an alternative method for the production of adipic acid. Industry requires adipic acid in large quantities—almost 2 billion pounds a year—for the synthesis of nylon. Currently adipic acid syntheses start with benzene. J. W. Frost of Michigan State University, however, is investigating the use of genetically engineered microbes to synthesize adipic acid. This method, in addition to eliminating the need for benzene, also eliminates production of nitrous oxide, an undesired by-product of the benzene process. Nitrous oxide contributes to the greenhouse effect as well as destruction of the ozone layer.

These examples illustrate the opportunities that await students of today as they become the well-trained chemists of the future who will tackle challenges like these.

14.1 Introduction

The study of the class of compounds that organic chemists call aromatic compounds (Section 2.2D) began with the discovery in 1825 of a new hydrocarbon by the English chemist Michael Faraday (Royal Institution). Faraday called this new hydrocarbon "bicarburet of hydrogen"; we now call it benzene. Faraday isolated benzene from a compressed illuminating gas that had been made by pyrolyzing whale oil.

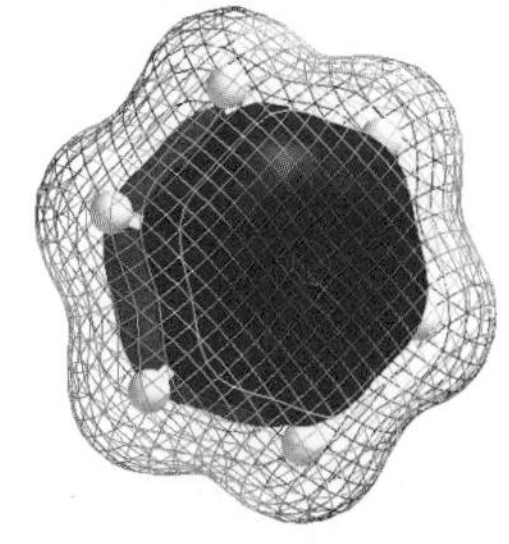

One of the π molecular orbitals of benzene, seen through a mesh representation of its electrostatic potential at its van der Waals surface.

In 1834 the German chemist Eilhardt Mitscherlich (University of Berlin) synthesized benzene by heating benzoic acid with calcium oxide. Using vapor density measurements, Mitscherlich further showed that benzene has the molecular formula C_6H_6:

$$\underset{\textbf{Benzoic acid}}{C_6H_5CO_2H} + CaO \xrightarrow{\text{heat}} \underset{\textbf{Benzene}}{C_6H_6} + CaCO_3$$

The molecular formula itself was surprising. Benzene has *only as many hydrogen atoms as it has carbon atoms.* Most compounds that were known then had a far greater proportion of hydrogen atoms, usually twice as many. Benzene with the formula of C_6H_6 (or C_nH_{2n-6}) should be a highly unsaturated compound, because it has an index of hydrogen deficiency equal to 4. Eventually, chemists began to recognize that benzene was a member of a new class of organic compounds with unusual and interesting properties. As we shall see in Section 14.3, benzene does not show the behavior expected of a highly unsaturated compound.

During the latter part of the nineteenth century the Kekulé–Couper–Butlerov theory of valence was systematically applied to all known organic compounds. One result of this effort was the placing of organic compounds in either of two broad categories; compounds were classified as being either **aliphatic** or **aromatic.** To be classified as aliphatic meant then that the chemical behavior of a compound was "fatlike." (Now it means that the compound reacts like an alkane, an alkene, an alkyne, or one of their derivatives.) To be classified as aromatic meant then that the compound had a low hydrogen–carbon ratio and that it was "fragrant." Most of the early aromatic compounds were obtained from balsams, resins, or essential oils. Included among these were benzaldehyde (from oil of bitter almonds), benzoic acid and benzyl alcohol (from gum benzoin), and toluene (from tolu balsam).

Kekulé was the first to recognize that these early aromatic compounds all contain a six-carbon unit and that they retain this six-carbon unit through most chemical transformations and degradations. Benzene was eventually recognized as being the parent compound of this new series.

Since this new group of compounds proved to be distinctive in ways that are far more important than their odors, the term *"aromatic"* began to take on a purely chemical connotation. We shall see in this chapter that the meaning of aromatic has evolved as chemists have learned more about the reactions and properties of aromatic compounds.

14.2 Nomenclature of Benzene Derivatives

Two systems are used in naming monosubstituted benzenes. In certain compounds, *benzene* is the parent name and the substituent is simply indicated by a prefix. We have, for example,

F Cl Br NO_2

Fluorobenzene **Chlorobenzene** **Bromobenzene** **Nitrobenzene**

For other compounds, the substituent and the benzene ring taken together may form a new parent name. Methylbenzene is usually called *toluene,* hydroxybenzene is almost always called *phenol,* and aminobenzene is almost always called *aniline.* These and other examples are indicated here:

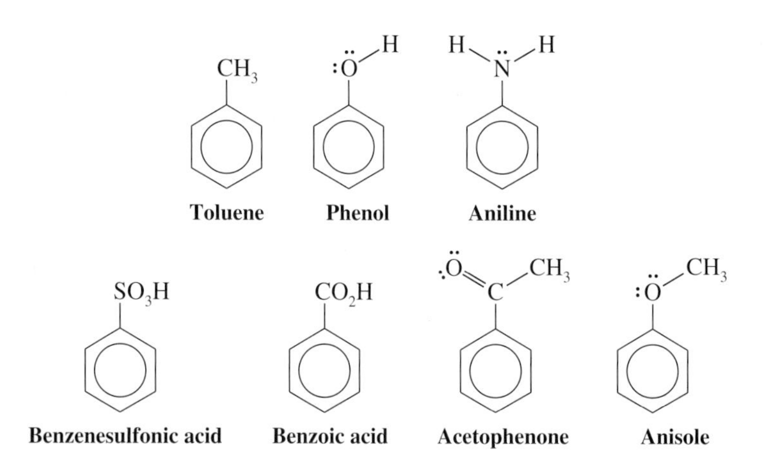

When two substituents are present, their relative positions are indicated by the prefixes ***ortho, meta,*** and ***para*** (abbreviated ***o-, m-,*** and ***p-***) or by the use of numbers. For the dibromobenzenes we have

Numbers can be used for two or more substituents, but ortho, meta, and para must never be used for more than two.

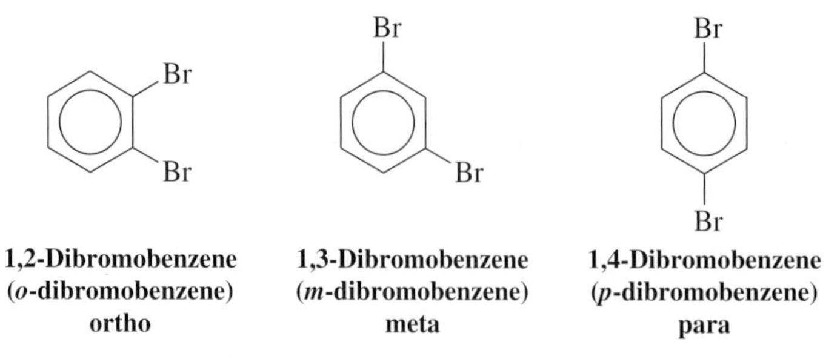

1,2-Dibromobenzene (*o*-dibromobenzene) ortho **1,3-Dibromobenzene (*m*-dibromobenzene) meta** **1,4-Dibromobenzene (*p*-dibromobenzene) para**

and for the nitrobenzoic acids

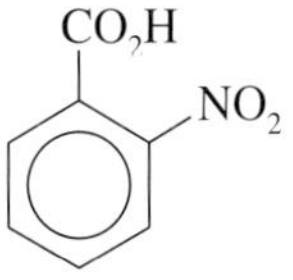

2-Nitrobenzoic acid (*o*-nitrobenzoic acid) **3-Nitrobenzoic acid (*m*-nitrobenzoic acid)** **4-Nitrobenzoic acid (*p*-nitrobenzoic acid)**

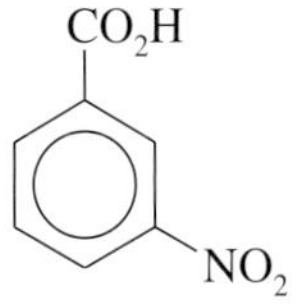

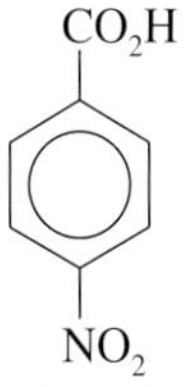

The dimethylbenzenes are often called *xylenes:*

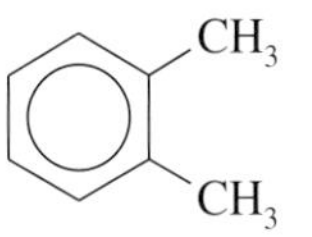

1,2-Dimethylbenzene (*o*-xylene) **1,3-Dimethylbenzene (*m*-xylene)** **1,4-Dimethylbenzene (*p*-xylene)**

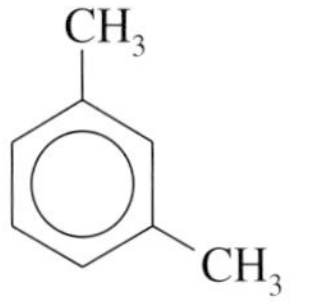

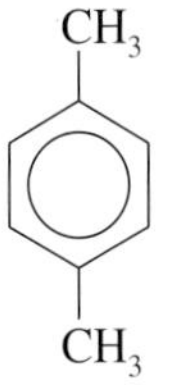

If more than two groups are present on the benzene ring, their positions must be indicated by the use of *numbers.* As examples, consider the following two compounds:

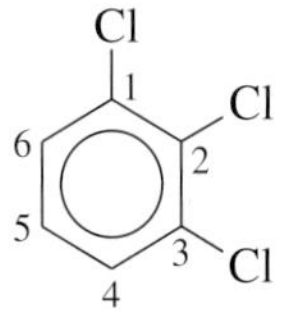

1,2,3-Trichlorobenzene 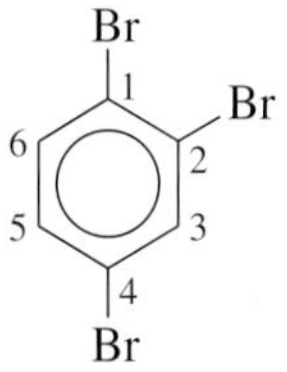 **1,2,4-Tribromobenzene (*not* 1,3,4-tribromobenzene)**

We notice, too, that the benzene ring is numbered so as to give ***the lowest possible numbers to the substituents.***

When more than two substituents are present and the substituents are different, they are listed in alphabetical order.

When a substituent is one that when taken together with the benzene ring gives a new base name, that substituent is assumed to be in position 1 and the new parent name is used:

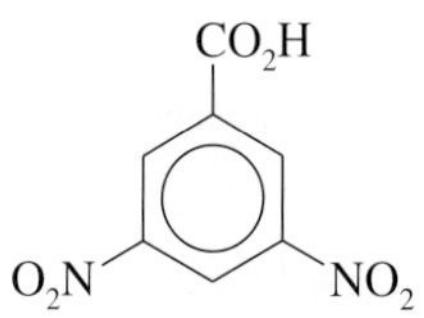

3,5-Dinitrobenzoic acid 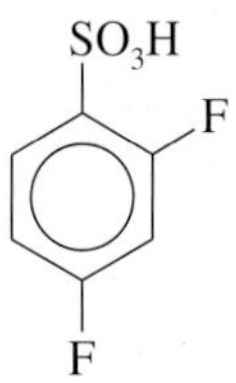 **2,4-Difluorobenzenesulfonic acid**

When the C_6H_5— group is named as a substituent, it is called a **phenyl** group. A hydrocarbon composed of one saturated chain and one benzene ring is usually named as a derivative of the larger structural unit. However, if the chain is unsaturated, the compound may be named as a derivative of that chain, regardless of ring size. The following are examples:

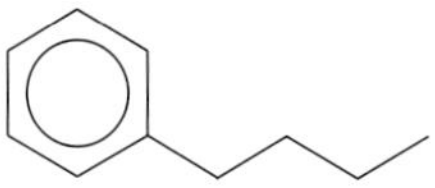

Butylbenzene 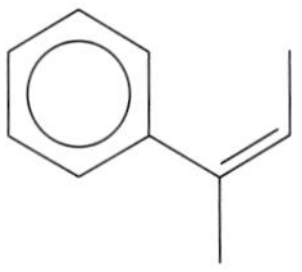 **(Z)-2-Phenyl-2-butene**

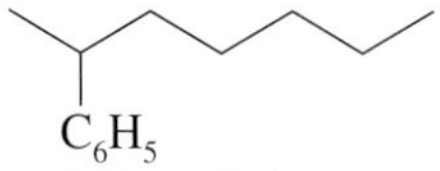

C_6H_5
2-Phenylheptane

The phenyl group is often abbreviated as C_6H_5—, Ph—, or ϕ—.

The name **benzyl** is an alternative name for the phenylmethyl group. It is sometimes abbreviated Bz:

Note the abbreviations for common aromatic groups.

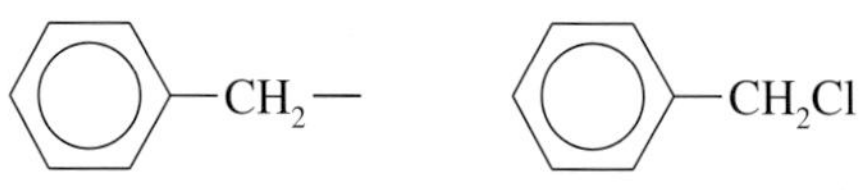

The benzyl group (the phenylmethyl group) **Benzyl chloride (phenylmethyl chloride or BzCl)**

14.3 Reactions of Benzene

In the mid-nineteenth century, benzene presented chemists with a real puzzle. They knew from its formula (Section 14.1) that benzene was highly unsaturated, and they expected it to react accordingly. They expected it to react like an alkene by decolorizing bromine in carbon tetrachloride through *addition of bromine.* They expected that it would change the color of aqueous potassium permanganate by being *oxidized,* that it would *add hydrogen* rapidly in the presence of a metal catalyst, and that it would *add water* in the presence of strong acids.

Benzene does none of these. When benzene is treated with bromine in carbon tetrachloride in the dark or with aqueous potassium permanganate or with dilute acids, none of the expected reactions occurs. Benzene does add hydrogen in the presence of finely divided nickel, but only at high temperatures and under high pressures:

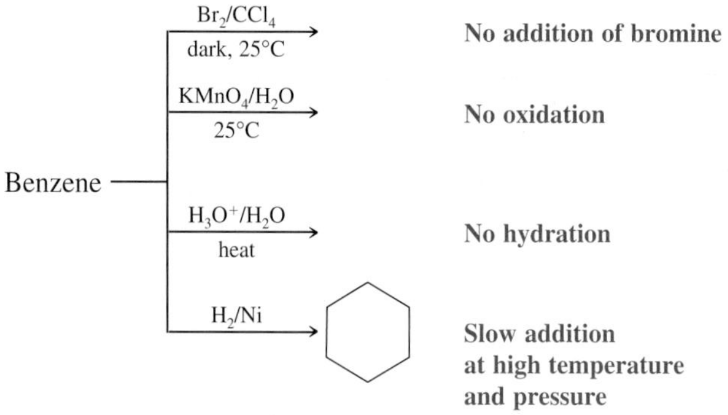

Benzene
- $\xrightarrow[\text{dark, 25°C}]{Br_2/CCl_4}$ **No addition of bromine**
- $\xrightarrow[\text{25°C}]{KMnO_4/H_2O}$ **No oxidation**
- $\xrightarrow[\text{heat}]{H_3O^+/H_2O}$ **No hydration**
- $\xrightarrow{H_2/Ni}$ (cyclohexane) **Slow addition at high temperature and pressure**

Benzene *does* react with bromine but only in the presence of a Lewis acid catalyst such as ferric bromide. Most surprisingly, however, it reacts not by addition but by *substitution*—**benzene substitution.**

Substitution

$$C_6H_6 + Br_2 \xrightarrow{FeBr_3} C_6H_5Br + HBr \qquad \text{Observed}$$

Addition

$$C_6H_6 + Br_2 \not\rightarrow C_6H_6Br_2 + C_6H_6Br_4 + C_6H_6Br_6 \quad \textbf{Not observed}$$

When benzene reacts with bromine, *only one monobromobenzene* is formed. That is, only one compound with the formula C_6H_5Br is found among the products. Similarly, when benzene is chlorinated, *only one monochlorobenzene* results.

Two possible explanations can be given for these observations. The first is that only one of the six hydrogen atoms in benzene is reactive toward these reagents. The second is that all six hydrogen atoms in benzene are equivalent, and replacing any one of them with a substituent results in the same product. As we shall see, the second explanation is correct.

PROBLEM 14.1

Listed below are four compounds that have the molecular formula C_6H_6. Which of these compounds would yield only one monosubstitution product, if, for example, one hydrogen were replaced by bromine?

(a) $CH_3C{\equiv}C{-}C{\equiv}CCH_3$ **(b)** 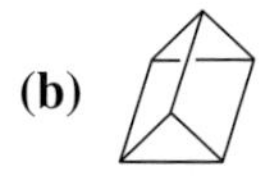 **(c)** 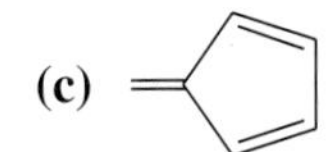 **(d)** 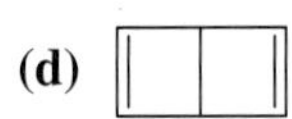

14.4 The Kekulé Structure for Benzene

In 1865, August Kekulé, the originator of the structural theory (Section 1.3), proposed the first definite structure for benzene,* a structure that is still used today (although as we shall soon see, we give it a meaning different from the meaning Kekulé gave it). Kekulé suggested that the carbon atoms of benzene are in a ring, that they are bonded to each other by alternating single and double bonds, and that one hydrogen atom is attached to each carbon atom. This structure satisfied the requirements of the structural theory that carbon atoms form four bonds and that all the hydrogen atoms of benzene are equivalent:

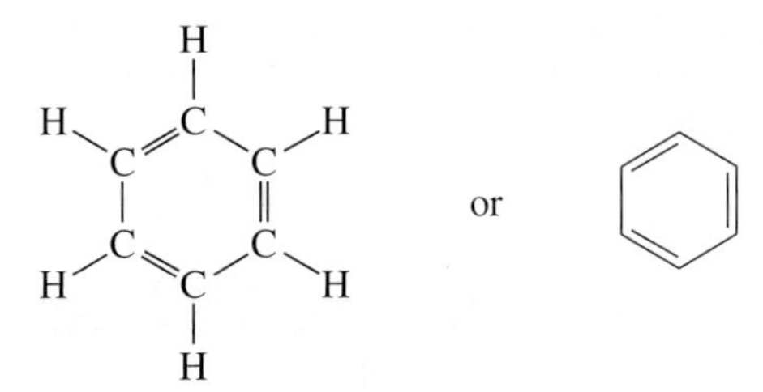

The Kekulé formula for benzene

A problem soon arose with the Kekulé structure, however. The Kekulé structure predicts that there should be two different 1,2-dibromobenzenes. In one of these hypothetical compounds (below), the carbon atoms that bear the bromines are separated by a single bond, and in the other they are separated by a double bond. *Only one 1,2-dibromobenzene has ever been found, however.*

Br Br and Br Br

*In 1861 the Austrian chemist Johann Josef Loschmidt represented the benzene ring with a circle, but he made no attempt to indicate how the carbon atoms were actually arranged in the ring.

To accommodate this objection, Kekulé proposed that the two forms of benzene (and of benzene derivatives) are in a state of equilibrium and that this equilibrium is so rapidly established that it prevents isolation of the separate compounds. Thus, the two 1,2-dibromobenzenes would also be rapidly equilibrated, and this would explain why chemists had not been able to isolate the two forms:

We now know that this proposal was incorrect and that *no such equilibrium exists.* Nonetheless, the Kekulé formulation of benzene's structure was an important step forward and, for very practical reasons, it is still used today. We understand its meaning differently, however.

The tendency of benzene to react by substitution rather than addition gave rise to another concept of aromaticity. For a compound to be called aromatic meant, experimentally, that it gave substitution reactions rather than addition reactions even though it was highly unsaturated.

Before 1900, chemists assumed that the ring of alternating single and double bonds was the structural feature that gave rise to the aromatic properties. Since benzene and benzene derivatives (i.e., compounds with six-membered rings) were the only aromatic compounds known, chemists naturally sought other examples. The compound cyclooctatetraene seemed to be a likely candidate:

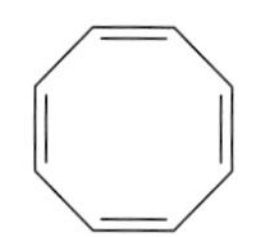

Cyclooctatetraene

In 1911, Richard Willstätter succeeded in synthesizing cyclooctatetraene. Willstätter found, however, that it is not at all like benzene. Cyclooctatetraene reacts with bromine by addition, it adds hydrogen readily, it is oxidized by solutions of potassium permanganate, and thus it is clearly *not aromatic.* While these findings must have been a keen disappointment to Willstätter, they were very significant for what they did not prove. Chemists, as a result, had to look deeper to discover the origin of benzene's aromaticity.

14.5 The Stability of Benzene

We have seen that benzene shows unusual behavior by undergoing substitution reactions when, on the basis of its Kekulé structure, we should expect it to undergo addition. Benzene is unusual in another sense: It is *more stable* than the Kekulé structure suggests. To see how, consider the following thermochemical results.

Cyclohexene, a six-membered ring containing one double bond, can be hydrogenated easily to cyclohexane. When the $\Delta H°$ for this reaction is measured, it is found to be -120 kJ mol^{-1}, very much like that of any similarly substituted alkene:

$$\text{Cyclohexene} + H_2 \xrightarrow{Pt} \text{Cyclohexane} \qquad \Delta H° = -120 \text{ kJ mol}^{-1}$$

We would expect that hydrogenation of 1,3-cyclohexadiene would liberate roughly twice as much heat and thus have a $\Delta H°$ equal to about -240 kJ mol^{-1}. When this experiment is done, the result is $\Delta H° = -232$ kJ mol^{-1}. This result is quite close to what we calculated, and the difference can be explained by taking into account the fact that compounds containing conjugated double bonds are usually somewhat more stable than those that contain isolated double bonds (Section 13.8):

$$\text{1,3-Cyclohexadiene} + 2\,H_2 \xrightarrow{Pt} \text{Cyclohexane}$$

Calculated
$\Delta H° = 2 \times (-120) = -240$ kJ mol^{-1}
Observed
$\Delta H° = -232$ kJ mol^{-1}

If we extend this kind of thinking, and if benzene is simply 1,3,5-cyclohexatriene, we would predict benzene to liberate approximately 360 kJ mol^{-1} [$3 \times (-120)$] when it is hydrogenated. When the experiment is actually done, the result is surprisingly different. The reaction is exothermic, but only by 208 kJ mol^{-1}:

$$\text{Benzene} + 3\,H_2 \xrightarrow{Pt} \text{Cyclohexane}$$

Calculated
$\Delta H° = 3 \times (-120) = -360$ kJ mol^{-1}
Observed $\Delta H° = -208$ kJ mol^{-1}
Difference $= 152$ kJ mol^{-1}

When these results are represented as in Fig. 14.1, it becomes clear that benzene is much more stable than we calculated it to be. Indeed, it is more stable than the hypothetical 1,3,5-cyclohexatriene by 152 kJ mol^{-1}. This difference between the amount of heat actually released and that calculated on the basis of the Kekulé structure is now called the **resonance energy** of the compound.

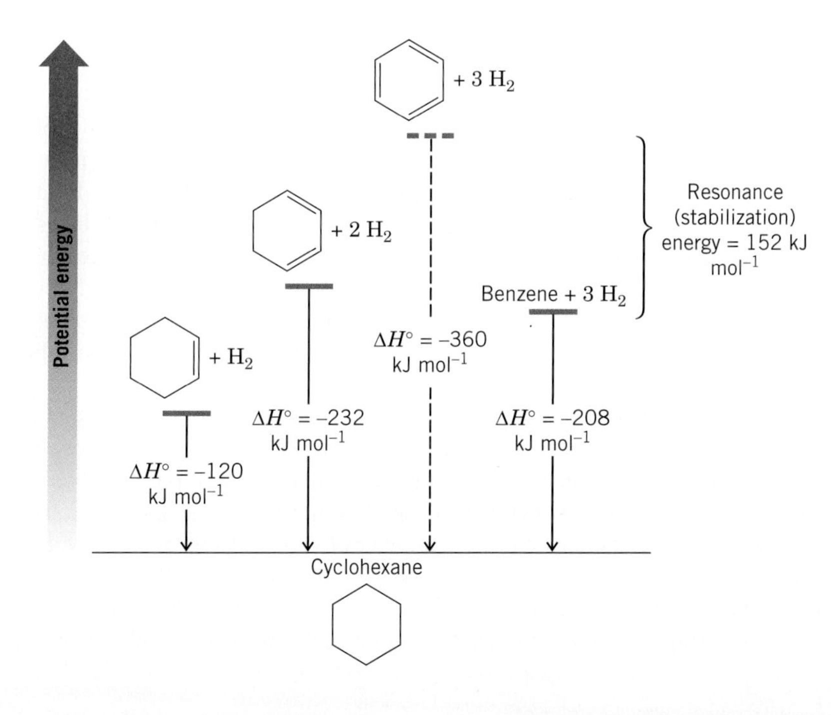

FIGURE 14.1 Relative stabilities of cyclohexene, 1,3-cyclohexadiene, 1,3,5-cyclohexatriene (hypothetical), and benzene.

14.6 Modern Theories of the Structure of Benzene

It was not until the development of quantum mechanics in the 1920s that the unusual behavior and stability of benzene began to be understood. Quantum mechanics, as we have seen, produced two ways of viewing bonds in molecules: resonance theory and molecular orbital theory. We now look at both of these as they apply to benzene.

14.6A The Resonance Explanation of the Structure of Benzene

A basic postulate of resonance theory (Sections 1.8 and 13.5) is that whenever two or more Lewis structures can be written for a molecule that *differ only in the positions of their electrons,* none of the structures will be in complete accord with the compound's chemical and physical properties. If we recognize this, we can now understand the true nature of the two Kekulé structures (**I** and **II**) for benzene. The two Kekulé structures differ only in the positions of their electrons. Structures **I** and **II**, then, do not represent two separate molecules in equilibrium as Kekulé had proposed. Instead, they are the closest we can get to a structure for benzene within the limitations of its molecular formula, the classic rules of valence, and the fact that the six hydrogen atoms are chemically equivalent. The problem with the Kekulé structures is that they are Lewis structures, and Lewis structures portray electrons in localized distributions. (With benzene, as we shall see, the electrons are delocalized.) Resonance theory, fortunately, does not stop with telling us when to expect this kind of trouble; it also gives us a way out. Resonance theory tells us to use structures **I** and **II** as resonance contributors to a picture of the real molecule of benzene. As such, **I** and **II** should be connected with a double-headed arrow and not with two separate ones (because we must reserve the symbol of two separate arrows for chemical equilibria). Resonance contributors, we emphasize again, are not in equilibrium. They are not structures of real molecules. They are the closest we can get if we are bound by simple rules of valence, but they are very useful in helping us visualize the actual molecule as a hybrid:

CD Tutorial
Benzene resonance

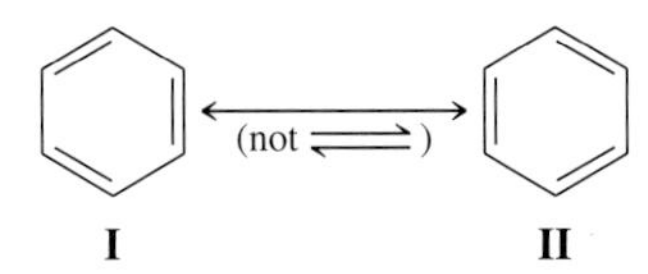

Look at the structures carefully. All of the single bonds in structure **I** are double bonds in structure **II**. If we blend **I** and **II**, that is, if we fashion a hybrid of them, then the carbon–carbon bonds in benzene are neither single bonds nor double bonds. Rather, they have a bond order between that of a single bond and that of a double bond. This is exactly what we find experimentally. Spectroscopic measurements show that the molecule of benzene is planar and that all of its carbon–carbon bonds are of equal length. Moreover, the carbon–carbon bond lengths in benzene (Fig. 14.2) are 1.39 Å, a value in between that for a carbon–carbon single bond between sp^2-hybridized atoms (1.47 Å) (see Table 13.1) and that for a carbon–carbon double bond (1.33Å).

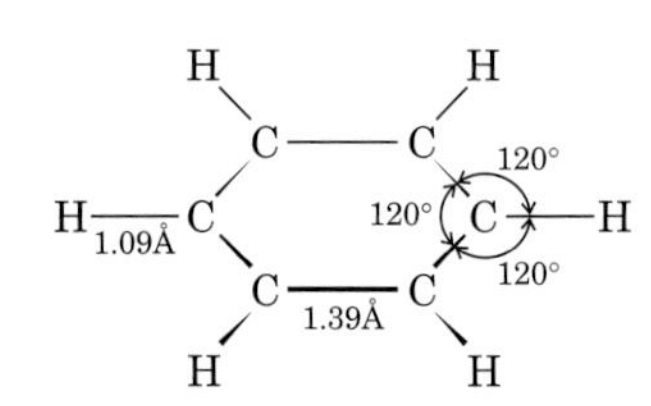

FIGURE 14.2 Bond lengths and angles in benzene. (Only the σ bonds are shown.)

The hybrid structure is represented by inscribing a circle in the hexagon, and it is this new formula (**III**) that is most often used for benzene today. There are times, however, when an accounting of the electrons must be made, and for these purposes we may use one or the other of the Kekulé structures. We do this simply because the electron count in a Kekulé structure is obvious, whereas the number of electrons represented by a circle or portion of a circle is ambiguous. With benzene the circle represents the six electrons that are delocalized about the six carbon atoms of the benzene ring. With other systems, however, a circle in a ring may represent numbers of delocalized electrons other than six:

III

PROBLEM 14.2 If benzene were 1,3,5-cyclohexatriene, the carbon–carbon bonds would be alternately long and short as indicated in the following structures. However, to consider the structures here as resonance contributors (or to connect them by a double-headed arrow) violates a basic principle of resonance theory. Explain.

Resonance theory (Section 13.5) also tells us that whenever equivalent resonance structures can be drawn for a molecule, the molecule (or hybrid) is much more stable than any of the resonance structures would be individually if they could exist. In this way resonance theory accounts for the much greater stability of benzene when compared to the hypothetical 1,3,5-cyclohexatriene. For this reason the extra stability associated with benzene is called its *resonance energy.*

14.6B The Molecular Orbital Explanation of the Structure of Benzene

The fact that the bond angles of the carbon atoms in the benzene ring are all 120° strongly suggests that the carbon atoms are sp^2 hybridized. If we accept this suggestion and construct a planar six-membered ring from sp^2 carbon atoms, a representation like that shown in Fig. 14.3*a* emerges. In this model, each carbon is sp^2 hybridized and has a *p* orbital available for overlap with *p* orbitals of its neighboring carbons. If we consider favorable overlap of these *p* orbitals all around the ring, the result is the model shown in Fig. 14.3*b*.

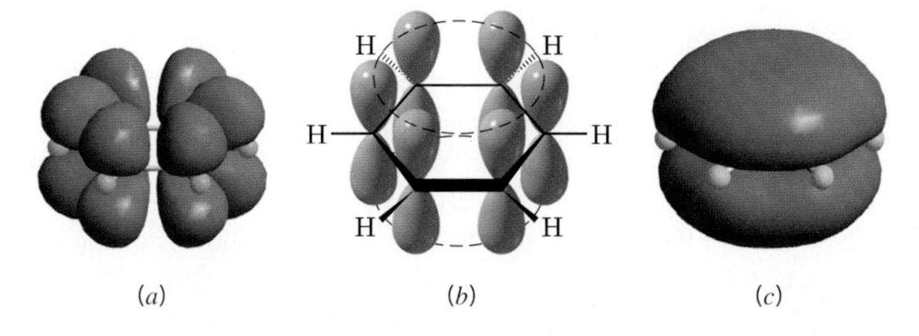

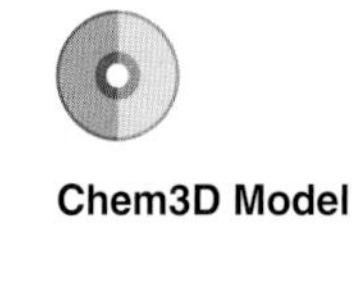

FIGURE 14.3 (*a*) Six sp^2-hybridized carbon atoms joined in a ring (each carbon also bears a hydrogen atom). Each carbon has a *p* orbital with lobes above and below the plane of the ring. (*b*) A stylized depiction of the *p* orbitals in (a). (*c*) Overlap of the *p* orbitals around the ring results in a molecular orbital encompassing the top and bottom faces of the ring. (Differences in the mathematical phase of the orbital lobes are not shown in these representations.)

As we recall from the principles of quantum mechanics (Section 1.9), the number of molecular orbitals in a molecule is the same as the number of atomic orbitals combined to form the molecule, and each orbital can accommodate a maximum of two electrons if their spins are opposed. If we consider only the *p* atomic orbitals contributed by the carbon atoms of benzene, there should be six π molecular orbitals. These orbitals are shown in Fig. 14.4.

CD Tutorial
Benzene molecular orbitals

The electronic structure of the ground state of benzene is obtained by adding the six π electrons to the π molecular orbitals shown in Fig. 14.4, starting with the orbitals of lowest energy. The lowest energy π molecular orbital in benzene has overlap of *p* orbitals with the same mathematical phase sign all around the top and bottom faces of the ring. In this orbital there are no nodal planes (changes in orbital phase sign) perpendicular to the atoms of the ring. The orbitals of next higher energy each have one nodal plane. (In general, each set of higher energy π molecular orbitals has an additional nodal plane.) Each of these orbitals is filled with a pair of electons, as well. These orbitals are of equal en-

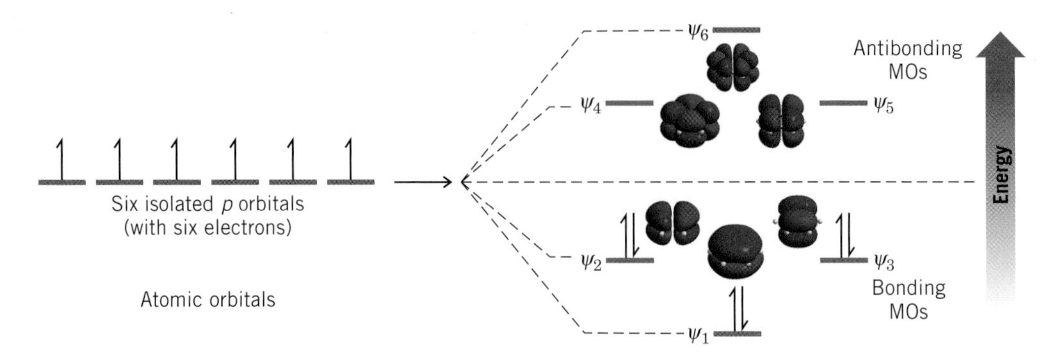

FIGURE 14.4 How six *p* atomic orbitals (one from each carbon of the benzene ring) combine to form six π molecular orbitals. Three of the molecular orbitals have energies lower than that of an isolated *p* orbital; these are the bonding molecular orbitals. Three of the molecular orbitals have energies higher than that of an isolated *p* orbital; these are the antibonding molecular orbitals. Orbitals ψ_2 and ψ_3 have the same energy and are said to be degenerate; the same is true of orbitals ψ_4 and ψ_5.

ergy (degenerate) because they both have one nodal plane. Together, these three orbitals comprise the bonding π molecular orbitals of benzene. The next higher energy set of π molecular orbitals each has two nodal planes, and the highest energy π molecular orbital of benzene has three nodal planes. These three orbitals are the antibonding π molecular orbitals of benzene, and they are unoccupied in the ground state. Benzene is said to have a closed bonding shell of delocalized π electrons because all of its bonding orbitals are filled with electrons that have their spins paired, and no electrons are found in antibonding orbitals. This closed bonding shell accounts, in part, for the stability of benzene.

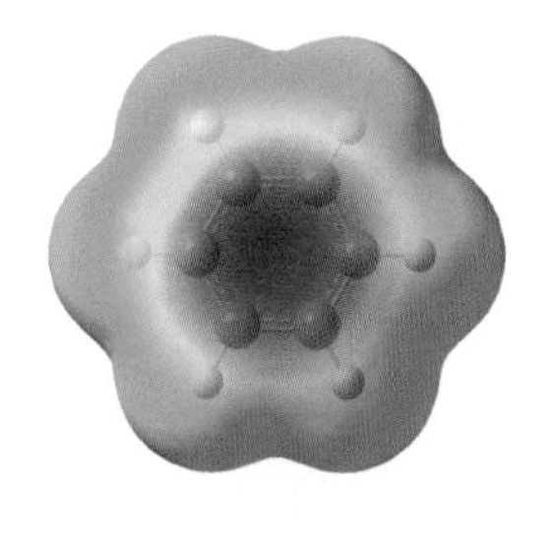

FIGURE 14.5 Electrostatic potential map of benzene.

Having considered the molecular orbitals of benzene, it is now useful to view an electrostatic potential map of the van der Waals surface for benzene, also calculated from quantum mechanical principles (Fig. 14.5). We can see that this representation is consistent with our understanding that the π electrons of benzene are not localized but are evenly distributed around the top face and bottom face (not shown) of the carbon ring in benzene.

It is interesting to note the recent discovery that crystalline benzene involves perpendicular interactions between benzene rings, so that the relatively positive periphery of one molecule associates with the relatively negative faces of the benzene molecules aligned above and below it.

14.7 Hückel's Rule: The $4n + 2$ π Electron Rule

In 1931 the German physicist Erich Hückel carried out a series of mathematical calculations based on the kind of theory that we have just described. **Hückel's rule** is concerned with compounds containing **one planar ring in which each atom has a *p* orbital** as in benzene. His calculations show that planar monocyclic rings containing $4n + 2$ π electrons, where $n = 0, 1, 2, 3, \ldots$, and so on (i.e., rings containing 2, 6, 10, 14, . . . , etc., π electrons), have closed shells of delocalized electrons like benzene and should have substantial resonance energies. In other words, **planar monocyclic rings with 2, 6, 10, 14, . . . , delocalized electrons should be aromatic.**

There is a simple way to make a diagram of the relative energies that would be obtained from Hückel's calculations for the π molecular orbitals of conjugated monocyclic systems. To so so, we inscribe a regular polygon corresponding to the ring of the compound inside a circle, *placing a corner of the polygon at the bottom.* At the points where vertices of the polygon touch the circle, we draw short horizontal lines outside that indicate the relative energies of the π molecular orbitals of the system. A dashed horizontal line placed halfway up the circle divides the bonding from the antibonding orbitals. Nonbonding orbitals fall in the line. Based on the number of π electrons in the molecule, we then place electron arrows on the lines representing the respective orbitals, beginning at the lowest energy level and working upward. In doing so, we fill degenerate orbitals each with one electron first and then add to each unpaired electron another with opposite spin if it is available. Applying this method to benzene, for example (Fig. 14.6), furnishes the same energy levels that we saw earlier in Fig. 14.4, energy levels that were based on quantum mechanical calculations.

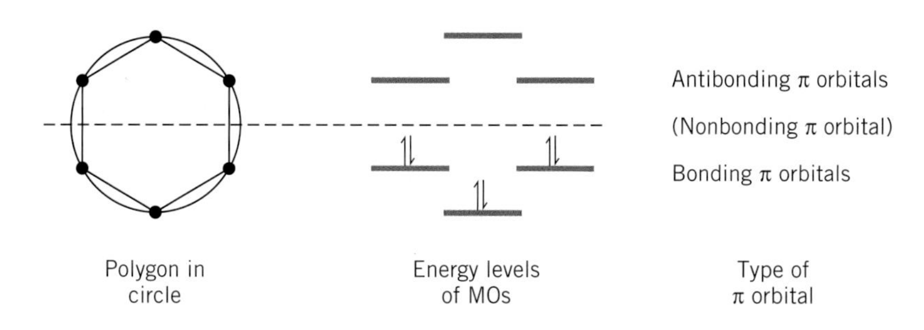

FIGURE 14.6 The polygon-and-circle method for deriving the relative energies of the π molecular orbitals of benzene. A horizontal line halfway up the circle divides the bonding orbitals from the antibonding orbitals. If an orbital falls on this line, it is a nonbonding orbital. This method was developed by C. A. Coulson (of Oxford University).

We can now understand why cyclooctatetraene is not aromatic. Cyclooctatetraene has a total of eight π electrons. Eight is not a Hückel number; it is a *4n number,* not a *4n + 2 number.* Using the polygon-and-circle method (Fig. 14.7), we find that cyclooctatetraene, if it were planar, *would not* have a closed shell of π electrons like benzene; it would have an unpaired electron in each of two nonbonding orbitals. Molecules with unpaired electrons (radicals) are *not* unusually stable; they are typically highly reactive and unstable. A planar form of cyclooctatetraene, therefore, should not be at all like benzene and should not be aromatic.

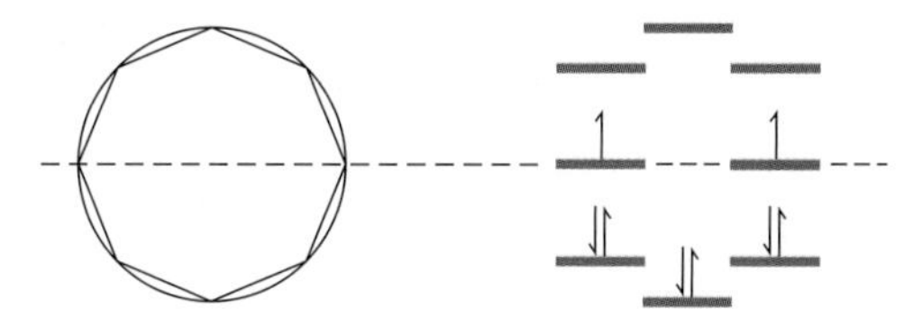

FIGURE 14.7 The π molecular orbitals that cyclooctatetraene would have if it were planar. Notice that, unlike benzene, this molecule is predicted to have two nonbonding orbitals, and because it has eight π electrons, it would have an unpaired electron in each of the two nonbonding orbitals (Hund's rule, Section 1.10). Such a system would not be expected to be aromatic.

Because cyclooctatetraene does not gain stability by becoming planar, it assumes the tub shape shown below. (In Section 14.7D we shall see that cyclooctatetraene would actually lose stability by becoming planar.)

The bonds of cyclooctatetraene are known to be alternately long and short; X-ray studies indicate that they are 1.48 and 1.34 Å, respectively.

14.7A The Annulenes

The name **annulene** has been proposed as a general name for monocyclic compounds that can be represented by structures having alternating single and double bonds. The ring size of an annulene is indicated by a number in brackets. Thus, benzene is [6]annulene and cyclooctatetraene is [8]annulene. Hückel's rule predicts that annulenes will be aromatic, provided their molecules have $4n + 2$ π electrons and have a planar carbon skeleton:

These names are often used for conjugated rings of 10 or more carbon atoms, but they are seldom used for benzene and cyclooctatetraene.

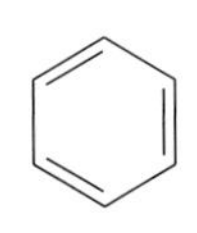
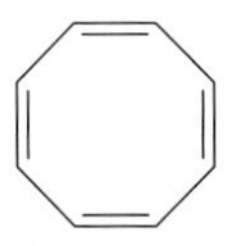

Benzene ([6]annulene) **Cyclooctatetraene ([8]annulene)**

Before 1960 the only annulenes that were available to test Hückel's predictions were benzene and cyclooctatetraene. During the 1960s, and largely as a result of research by F. Sondheimer, a number of large-ring annulenes were synthesized, and the predictions of Hückel's rule were verified.

Consider the [14], [16], [18], [20], [22], and [24]annulenes as examples. Of these, *as Hückel's rule predicts,* the [14], [18], and [22]annulenes ($4n + 2$ when $n = 3, 4, 5$, respectively) have been found to be aromatic. The [16]annulene and the [24]annulene are not aromatic. They are $4n$ compounds, not $4n + 2$ compounds:

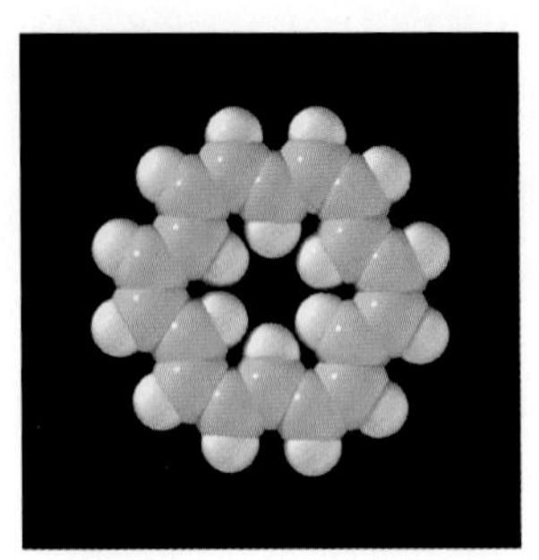

[18]Annulene

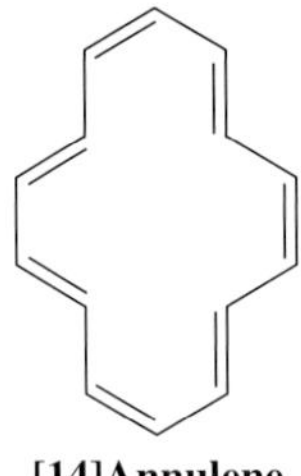
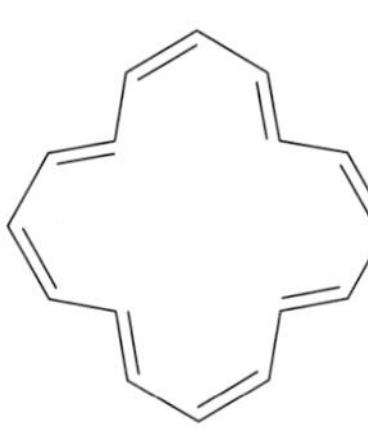
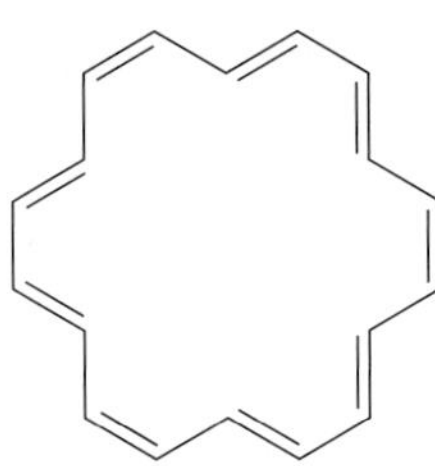

[14]Annulene (aromatic) **[16]Annulene (*not* aromatic)** **[18]Annulene (aromatic)**

Examples of [10] and [12]annulenes have also been synthesized and none is aromatic. We would not expect [12]annulenes to be aromatic since they have 12 π electrons and do not obey Hückel's rule. The following [10]annulenes would be expected to be aromatic on the basis of electron count, but their rings are not planar.

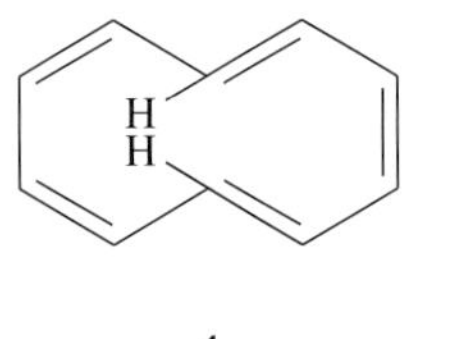

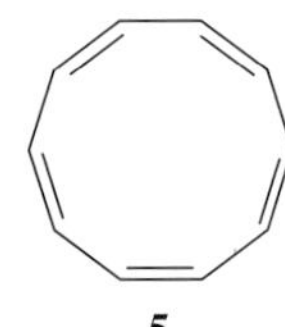
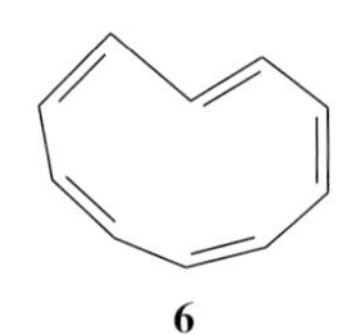

4 **5** **6**

[10]Annulenes
None is aromatic because none is planar.

The [10]annulene (**4**) has two trans double bonds. Its bond angles are approximately 120°; therefore, it has no appreciable angle strain. The carbon atoms of its ring, however, are prevented from becoming coplanar because the two hydrogen atoms in the center of the ring interfere with each other. Because the ring is not planar, the *p* orbitals of the carbon atoms are not parallel and, therefore, cannot overlap effectively around the ring to form the π molecular orbitals of an aromatic system.

The [10]annulene with all cis double bonds (**5**) would, if it were planar, have considerable angle strain because the internal bond angles would be 144°. Consequently, any stability this isomer gained by becoming planar in order to become aromatic would be more than offset by the destabilizing effect of the increased angle strain. A similar problem of a large angle strain associated with a planar form prevents molecules of the [10]annulene isomer with one trans double bond (**6**) from being aromatic.

After many unsuccessful attempts over many years, in 1965 [4]annulene (or cyclobutadiene) was synthesized by R. Pettit and co-workers at the University of Texas, Austin.

Cyclobutadiene is a $4n$ molecule, not a $4n + 2$ molecule, and, as we would expect, it is a highly unstable compound and *it is not aromatic:*

Cyclobutadiene or [4]annulene (*not* aromatic)

PROBLEM 14.3

Use the polygon-and-circle method to outline the π molecular orbitals of cyclobutadiene and explain why, on this basis, you would not expect it to be aromatic.

14.7B NMR Spectroscopy: Evidence for Electron Delocalization in Aromatic Compounds

The ^{1}H NMR spectrum of benzene consists of a single unsplit signal at δ 7.27. That only a single unsplit signal is observed is further proof that all of the hydrogens of benzene are equivalent. That the signal occurs at such a low magnetic field strength, is, as we shall see, compelling evidence for the assertion that the π electrons of benzene are delocalized.

We learned in Section 9.5 that circulations of σ electrons of C—H bonds cause the protons of alkanes to be *shielded* from the applied magnetic field of an NMR spectrometer and, consequently, these protons absorb at higher magnetic field strengths. We shall now explain the low field strength of absorption of benzene protons on the basis of *deshielding caused by circulation of the π electrons of benzene,* and this explanation, as you will see, requires that the π electrons be delocalized.

When benzene molecules are placed in the powerful magnetic field of the NMR spectrometer, electrons circulate in the direction shown in Fig. 14.8; by doing so, they generate a **ring current.** (If you have studied physics, you will understand why the electrons circulate in this way.) This π-electron circulation creates an induced magnetic field that, *at the position of the protons of benzene, reinforces the applied magnetic field,* and this reinforcement causes the protons to be strongly *deshielded.* By "deshielded" we mean that the protons sense the sum of the two fields, and, therefore, the applied magnetic field strength does not have to be as high as might have been required in the absence of the induced field. This strong deshielding, which we attribute to a ring current created by the

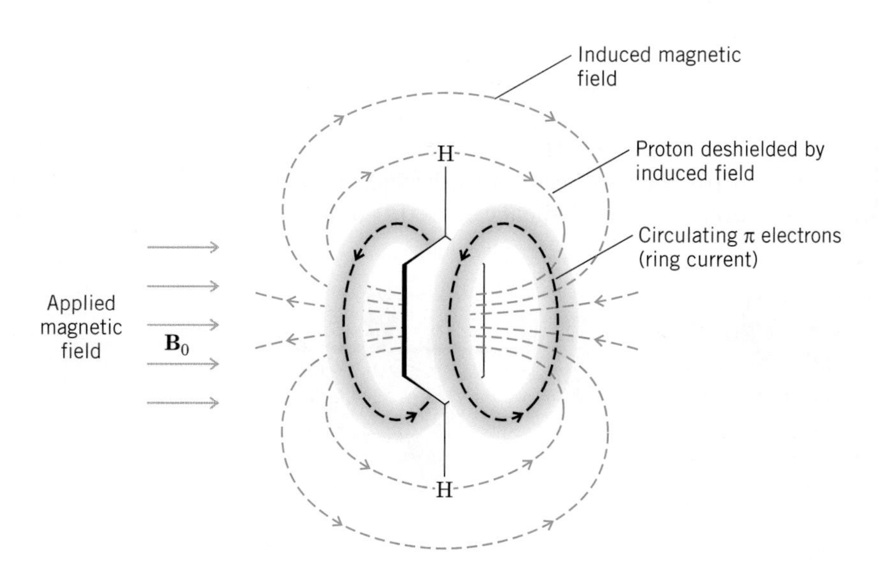

FIGURE 14.8 The induced magnetic field of the π electrons of benzene deshields the benzene protons. Deshielding occurs because at the location of the protons the induced field is in the same direction as the applied field.

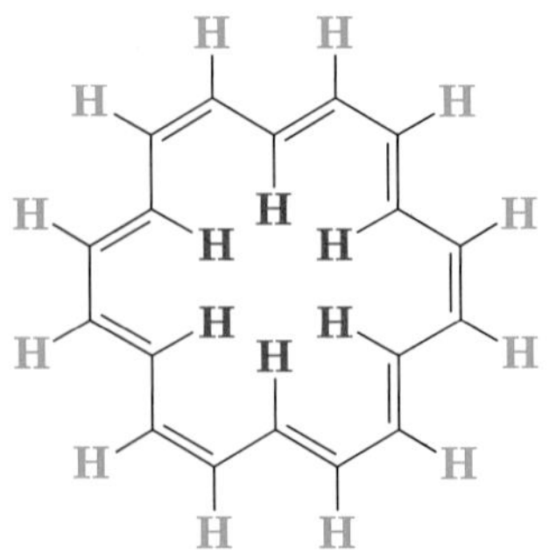

FIGURE 14.9 [18]Annulene. The internal protons (red) are highly shielded and absorb at δ −3.0. The external protons (blue) are highly deshielded and absorb at δ 9.3.

delocalized π electrons, explains why aromatic protons absorb at very low magnetic field strengths.

The deshielding of external aromatic protons that results from the ring current is one of the best pieces of physical evidence that we have for π-electron delocalization in aromatic rings. In fact, low-field-strength proton absorption is often used as a criterion for aromaticity in newly synthesized conjugated cyclic compounds.

Not all aromatic protons absorb at low magnetic field strengths, however. The internal protons of large-ring aromatic compounds that have hydrogens in the center of the ring (in the π-electron cavity) absorb at unusually high magnetic field strengths because they are highly shielded by the opposing induced magnetic field in the center of the ring (see Fig. 14.8). An example is [18]annulene (Fig. 14.9). The internal protons of [18]annulene absorb far upfield at δ −3.0, above the signal for tetramethylsilane (TMS); the external protons, on the other hand, absorb far downfield at δ 9.3. Considering that [18]annulene has $4n + 2$ π electrons, this evidence provides strong support for π-electron delocalization as a criterion for aromaticity and for the predictive power of Hückel's rule.

14.7C Aromatic Ions

In addition to the neutral molecules that we have discussed so far, there are a number of monocyclic species that bear either a positive or a negative charge. Some of these ions show unexpected stabilities that suggest that they, too, are **aromatic.** Hückel's rule is helpful in accounting for the properties of these ions as well. We shall consider two examples: the cyclopentadienyl anion and the cycloheptatrienyl cation.

Cyclopentadiene is not aromatic; however, it is unusually acidic for a hydrocarbon. (The pK_a for cyclopentadiene is 16 and, by contrast, the pK_a for cycloheptatriene is 36.) Because of its acidity, cyclopentadiene can be converted to its anion by treatment with moderately strong bases. The cyclopentadienyl anion, moreover, is unusually stable, and NMR spectroscopy shows that all five hydrogen atoms in the cyclopentadienyl anion are equivalent and absorb downfield.

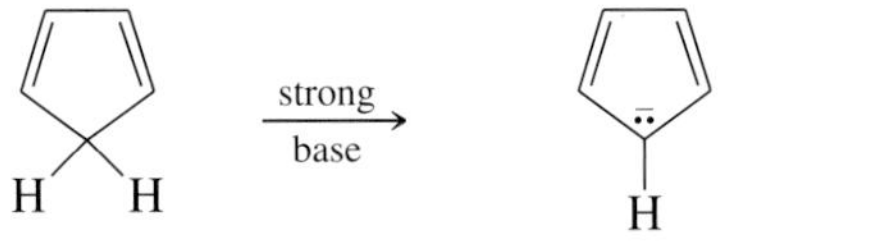

strong base

Cyclopentadiene **Cyclopentadienyl anion**

The orbital structure of cyclopentadiene (Fig. 14.10) shows why cyclopentadiene, itself, is not aromatic. Not only does it not have the proper number of π electrons, but the π electrons cannot be delocalized about the entire ring because of the intervening sp^3-hybridized $—CH_2—$ group with no available p orbital.

On the other hand, if the $—CH_2—$ carbon atom becomes sp^2 hybridized after it loses a proton (Fig. 14.10), the two electrons left behind can occupy the new p orbital that is produced. Moreover, this new p orbital can overlap with the p orbitals on either side of it

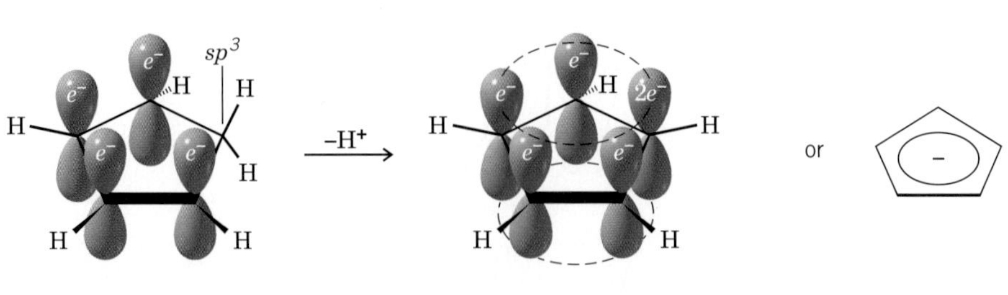

FIGURE 14.10 The stylized p orbitals of cyclopentadiene and of the cyclopentadienyl anion.

and give rise to a ring with *six* delocalized π electrons. Because the electrons are delocalized, all of the hydrogen atoms are equivalent, and this agrees with what NMR spectroscopy tells us. A calculated electrostatic potential map for cyclopentadienyl anion (Fig. 14.11) also shows the symmetrical distribution of negative charge within the ring, and the overall symmetry of the ring structure.

Six is, of course, a Hückel number ($4n + 2$, where $n = 1$), and the cyclopentadienyl anion is, in fact, an **aromatic anion.** The unusual acidity of cyclopentadiene is a result of the unusual stability of its anion.

Cycloheptatriene (Fig. 14.12) (a compound with the common name tropylidene) has six π electrons. However, the six π electrons of cycloheptatriene cannot be fully delocalized because of the presence of the $-CH_2-$ group, a group that does not have an available p orbital (Fig. 14.12).

When cycloheptatriene is treated with a reagent that can abstract a hydride ion, it is converted to the cycloheptatrienyl (or tropylium) cation. The loss of a hydride ion from cycloheptatriene occurs with unexpected ease, and the cycloheptatrienyl cation is found to be unusually stable. The NMR spectrum of the cycloheptatrienyl cation indicates that all seven hydrogen atoms are equivalent. If we look closely at Fig. 14.12, we see how we can account for these observations.

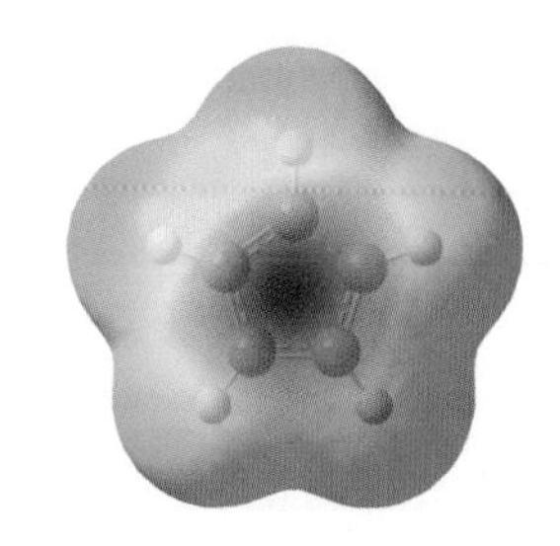
FIGURE 14.11 An electrostatic potential map of the cyclopentadienyl anion. The ion is negatively charged overall, of course, but regions with greatest negative potential are shown in red, and regions with least negative potential are in blue. The concentration of negative potential in the center of the top face and bottom face (not shown) indicates that the extra electron of the ion is involved in the aromatic π-electron system.

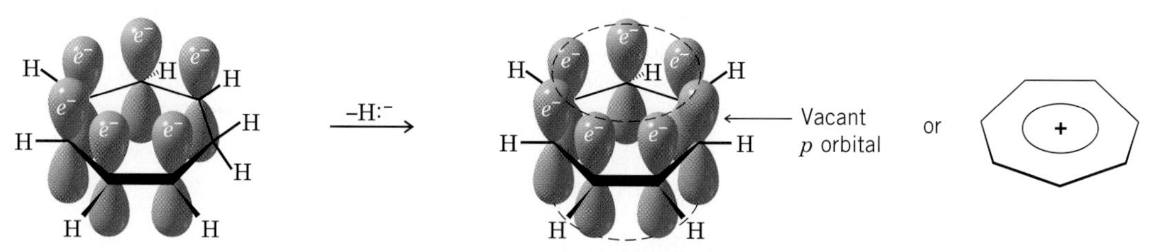

FIGURE 14.12 The stylized p orbitals of cycloheptatriene and of the cycloheptatrienyl (tropylium) cation.

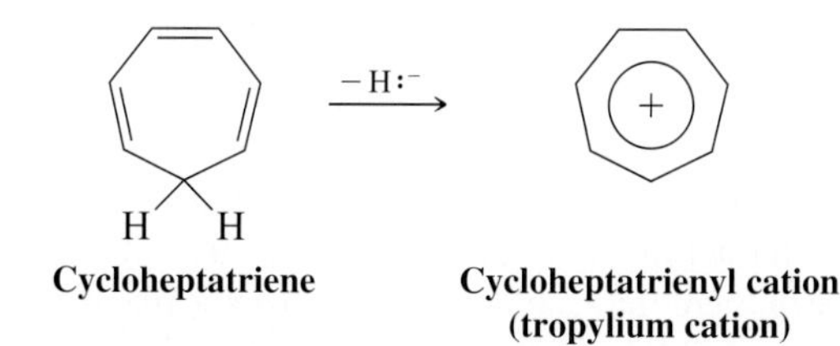

As a hydride ion is removed from the $-CH_2-$ group of cycloheptatriene, a vacant p orbital is created, and the carbon atom becomes sp^2 hybridized. The cation that results has seven overlapping p orbitals containing *six* delocalized π electrons. The cycloheptatrienyl cation is, therefore, an aromatic cation, and all of its hydrogen atoms should be equivalent; again, this is exactly what we find experimentally.

The calculated electrostatic potential map for cycloheptatrienyl (tropylium) cation (Fig. 14.13) also shows the symmetry of this ion. Electrostatic potential from the π electrons involved in the aromatic system is indicated by the relatively red color that is evenly distributed around the top face (and bottom face, though not shown) of the carbon framework. The entire ion is positive, of course, and the region of greatest positive potential is indicated by blue around the periphery of the ion.

FIGURE 14.13 An electrostatic potential map of the tropylium cation. The ion is positive overall, of course, but a region of relatively greater negative electrostatic potential can clearly be seen around the top face (and bottom face, though not shown) of the ring where electrons are involved in the π system of the aromatic ring.

PROBLEM 14.4 **(a)** Outline the π molecular orbitals of the cyclopentadienyl system by inscribing a regular pentagon in a circle and explain on this basis why the cyclopentadienyl anion is aromatic. **(b)** What electron distribution would you expect for the cyclopentadienyl cation? **(c)** Would you expect it to be aromatic? Explain your answer. **(d)** Would you expect the cyclopentadienyl cation to be aromatic on the basis of Hückel's rule?

PROBLEM 14.5 **(a)** Use the polygon-and-circle method to sketch the relative energies of the π molecular orbitals of the cycloheptatrienyl cation and explain why it is aromatic. **(b)** Would you expect the cycloheptatrienyl anion to be aromatic on the basis of the electron distribution in its π molecular orbitals? Explain. **(c)** Would you expect the cycloheptatrienyl anion to be aromatic on the basis of Hückel's rule?

PROBLEM 14.6 1,3,5-Cycloheptatriene (shown on page 637) is even less acidic than 1,3,5-heptatriene. Explain how this experimental observation might help to confirm your answer to parts (b) and (c) of the previous problem.

PROBLEM 14.7 When 1,3,5-cycloheptatriene reacts with one molar equivalent of bromine in CCl_4 at 0°C, it undergoes 1,6 addition. **(a)** Write the structure of this product. **(b)** On heating, this 1,6-addition product loses HBr readily to form a compound with the molecular formula C_7H_7Br, called *tropylium bromide*. Tropylium bromide is insoluble in nonpolar solvents but is soluble in water; it is unexpectedly high melting (mp 203°C), and when treated with silver nitrate, an aqueous solution of tropylium bromide gives a precipitate of AgBr. What do these experimental results suggest about the bonding in tropylium bromide?

14.7D Aromatic, Antiaromatic, and Nonaromatic Compounds

What do we mean when we say that a compound is aromatic? We mean that its π electrons are *delocalized* over the entire ring and that it is *stabilized* by the π-electron delocalization.

As we have seen, the best way to determine whether the π electrons of a cyclic system are delocalized is through the use of NMR spectroscopy. It provides direct physical evidence of whether or not the π electrons are delocalized.

But what do we mean by saying that a compound is stabilized by π-electron delocalization? We have an idea of what this means from our comparison of the heat of hydrogenation of benzene and that calculated for the hypothetical 1,3,5-cyclohexatriene. We saw that benzene—in which the π electrons are delocalized—is much more stable than 1,3,5-cyclohexatriene (a model in which the π electrons are not delocalized). We call the energy difference between them the resonance energy (delocalization energy) or stabilization energy.

In order to make similar comparisons for other aromatic compounds, we need to choose proper models. But what should these models be?

One proposal is that we should compare the π-electron energy of the cyclic system with that of the corresponding open-chain compound. This approach is particularly useful because it furnishes us with models not only for annulenes but for aromatic cations and anions as well. (Corrections need to be made, of course, when the cyclic system is strained.)

When we use this approach, we take as our model a linear chain of sp^2-hybridized atoms that carries the same number of π electrons as our cyclic compound. Then we imagine ourselves removing two hydrogen atoms from the ends of this chain and joining the ends to form a ring. If the ring has *lower* π-electron energy than the open chain, then the ring is *aromatic*. If the ring and chain have *the same* π-electron energy, then the ring is *nonaromatic*. If the ring has *greater* π-electron energy than the open chain, then the ring is *antiaromatic*.

The actual calculations and experiments used in determining π-electron energies are beyond our scope, but we can study four examples that illustrate how this approach has been used.

Cyclobutadiene For cyclobutadiene we consider the change in π-electron energy for the following *hypothetical* transformation:

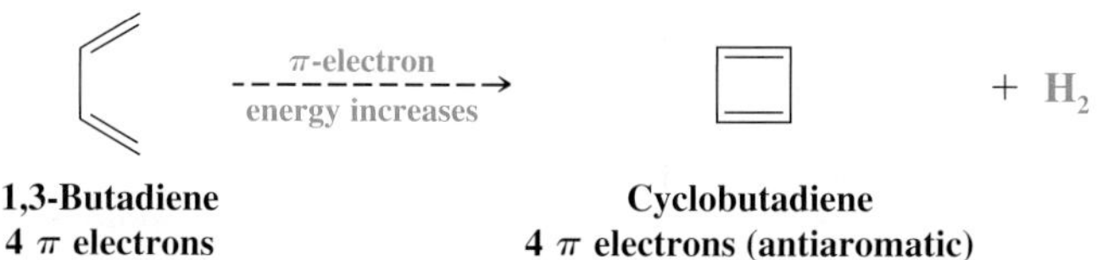

Calculations indicate and experiments appear to confirm that the π-electron energy of cyclobutadiene is higher than that of its open-chain counterpart. Thus cyclobutadiene is classified as antiaromatic.

Benzene Here our comparison is based on the following hypothetical transformation:

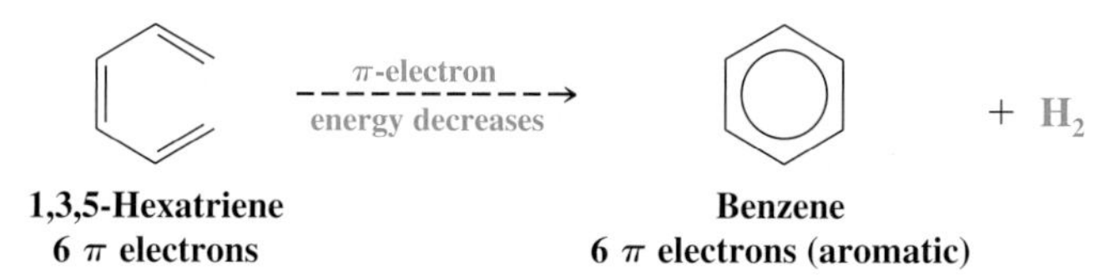

Calculations indicate and experiments confirm that benzene has a much lower π-electron energy than 1,3,5-hexatriene. Benzene is classified as being aromatic on the basis of this comparison as well.

Cyclopentadienyl Anion Here we use a linear anion for our hypothetical transformation:

$HC{:}^-$ $\xrightarrow{\pi\text{-electron energy decreases}}$ + H_2

6 π electrons **Cyclopentadienyl anion 6 π electrons (aromatic)**

Both calculations and experiments confirm that the cyclic anion has a lower π-electron energy than its open-chain counterpart. Therefore the cyclopentadienyl anion is classified as aromatic.

Cyclooctatetraene For cyclooctatetraene we consider the following hypothetical transformation:

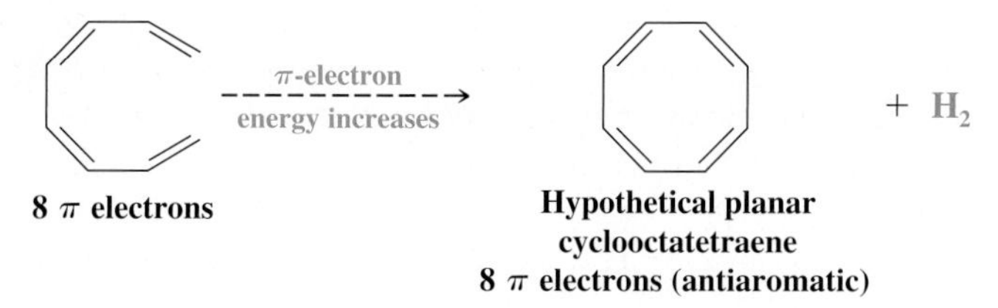

Here calculations and experiments indicate that a planar cyclooctatetraene would have higher π-electron energy than the open-chain octatetraene. Therefore, a planar form of cyclooctatetraene would, if it existed, be *antiaromatic*. As we saw earlier, cyclooctatetraene is not planar and behaves like a simple cyclic polyene.

PROBLEM 14.8 Calculations indicate that the π-electron energy decreases for the hypothetical transformation from the allyl cation to the cyclopropenyl cation below.

$$CH_2{=}CH{-}CH_2^+ \dashrightarrow \text{(cyclopropenyl cation)}^+ + H_2$$

What does this indicate about the possible aromaticity of the cyclopropenyl cation? (See Problem 14.10 for more on this cation.)

PROBLEM 14.9 The cyclopentadienyl cation is apparently *antiaromatic*. Explain what this means in terms of the π-electron energies of a cyclic and an open-chain compound.

PROBLEM 14.10 In 1967 R. Breslow (of Columbia University) and co-workers showed that adding $SbCl_5$ to a solution of 3-chlorocyclopropene in CH_2Cl_2 caused the precipitation of a white solid with the composition $C_3H_3^+SbCl_6^-$. NMR spectroscopy of a solution of this salt showed that all of its hydrogen atoms were equivalent. What new aromatic ion had Breslow and co-workers prepared?

14.8 Other Aromatic Compounds

14.8A Benzenoid Aromatic Compounds

In addition to those that we have seen so far, there are many other examples of aromatic compounds. Representatives of one broad class of **benzenoid aromatic compounds,** called **polycyclic benzenoid aromatic hydrocarbons,** are illustrated in Fig. 14.14. All of these consist of molecules having two or more benzene rings *fused* together. A close look at one, naphthalene, will illustrate what we mean by this.

According to resonance theory, a molecule of naphthalene can be considered to be a hybrid of three Kekulé structures. One of these Kekulé structures, the most important one,

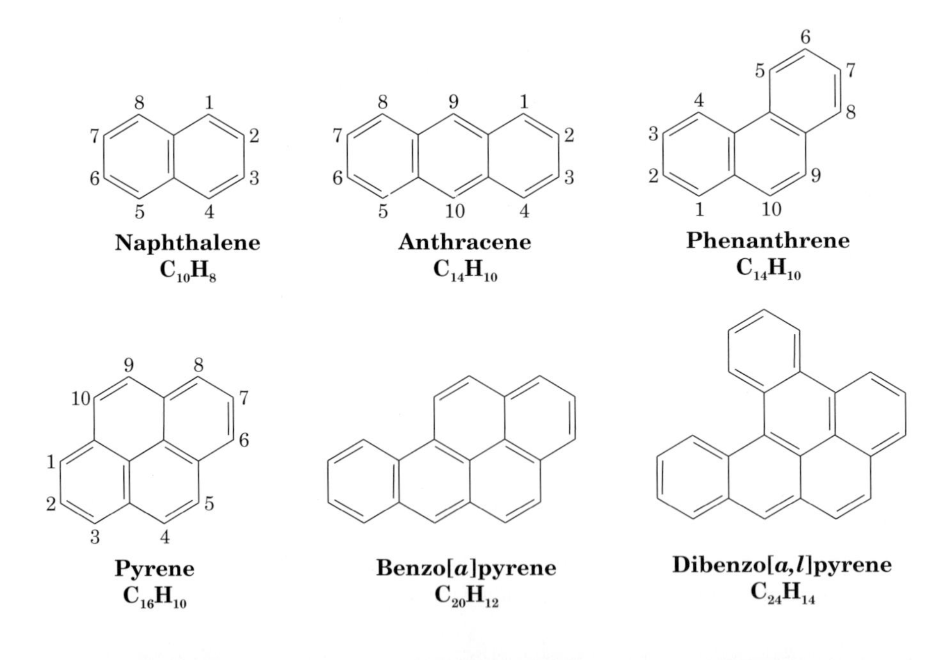

FIGURE 14.14 Benzenoid aromatic hydrocarbons. Some polycyclic aromatic hydrocarbons (PAHs), such as dibenzo[*a,l*]pyrene, are carcinogenic. (See "The Chemistry of . . ." in Section 11.14.)

is shown in Fig. 14.15. There are two carbon atoms in naphthalene (C4*a* and C8*a*) that are common to both rings. These two atoms are said to be at the points of *ring fusion.* They direct all of their bonds toward other carbon atoms and do not bear hydrogen atoms.

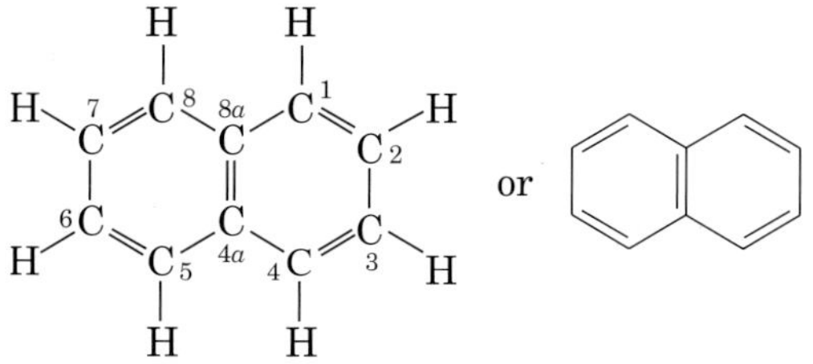

FIGURE 14.15 One Kekulé structure for naphthalene.

Molecular orbital calculations for naphthalene begin with the model shown in Fig. 14.16. The *p* orbitals overlap around the periphery of both rings and across the points of ring fusion.

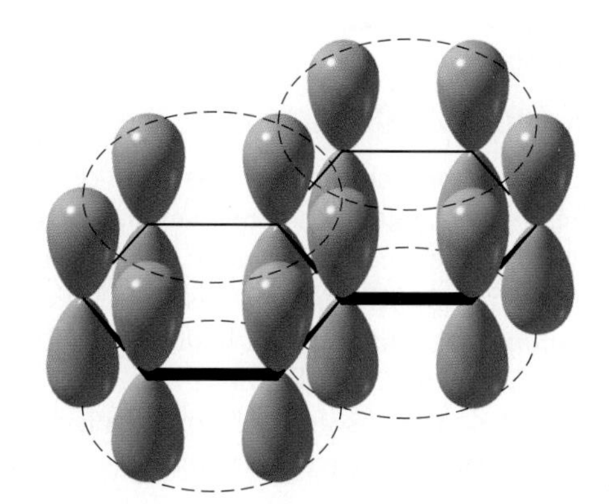

FIGURE 14.16 The stylized *p* orbitals of naphthalene.

When molecular orbital calculations are carried out for naphthalene using the model shown in Fig. 14.16, the results of the calculations correlate well with our experimental knowledge of naphthalene. The calculations indicate that delocalization of the 10 π electrons over the two rings produces a structure with considerably lower energy than that calculated for any individual Kekulé structure. Naphthalene, consequently, has a substantial resonance energy. Based on what we know about benzene, moreover, naphthalene's tendency to react by substitution rather than addition and to show other properties associated with aromatic compounds is understandable.

Anthracene and phenanthrene are isomers. In anthracene the three rings are fused in a linear way, and in phenanthrene they are fused so as to produce an angular molecule. Both of these molecules also show large resonance energies and chemical properties typical of aromatic compounds.

Pyrene is also aromatic. Pyrene itself has been known for a long time; a pyrene derivative, however, has been the object of research that shows another interesting application of Hückel's rule.

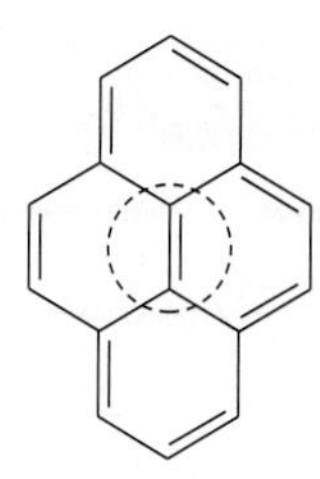

FIGURE 14.17 One Kekulé structure for pyrene. The internal double bond is enclosed in a dotted circle for emphasis.

To understand this particular research, we need to pay special attention to the Kekulé structure for pyrene (Fig. 14.17). The total number of π electrons in pyrene is 16 (8 double bonds = 16 π electrons). Sixteen is a non-Hückel number, but Hückel's rule is intended to be applied only to monocyclic compounds and pyrene is clearly tetracyclic. If we disregard the internal double bond of pyrene, however, and look only at the periphery, we see that the periphery is a planar ring with 14 π electrons. The periphery is, in fact, very much like that of [14]annulene. Fourteen *is* a Hückel number ($4n + 2$, where $n = 3$), and one might then predict that the periphery of pyrene would be aromatic by itself, in the absence of the internal double bond.

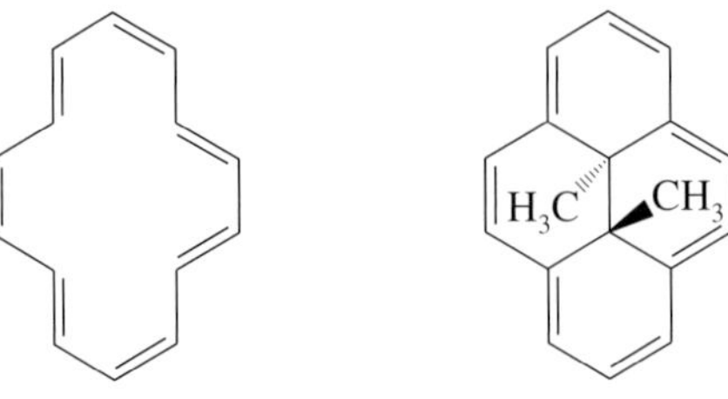

This prediction was confirmed when V. Boekelheide (University of Oregon) synthesized *trans*-15,16-dimethyldihydropyrene and showed that it is aromatic.

PROBLEM 14.11 In addition to a signal downfield, the ^{1}H NMR spectrum of *trans*-15,16-dimethyldihydropyrene has a signal at δ −4.2. Account for the presence of this high field signal.

14.8B Nonbenzenoid Aromatic Compounds

Naphthalene, phenanthrene, and anthracene are examples of *benzenoid* aromatic compounds. On the other hand, the cyclopentadienyl anion, the cycloheptatrienyl cation, *trans*-15,16-dimethyldihydropyrene, and the aromatic annulenes (except for [6]annulene) are classified as **nonbenzenoid aromatic compounds.**

Another example of a *nonbenzenoid* aromatic hydrocarbon is the compound azulene. Azulene has a resonance energy of 205 kJ mol^{-1}. There is substantial separation of charge between the rings in azulene, as is indicated by the electrostatic potential map for azulene shown in Fig. 14.18. Factors related to aromaticity account for this property of azulene (see Problem 14.12).

FIGURE 14.18 A calculated electrostatic potential map for azulene. (Red areas are more negative and blue areas are less negative.)

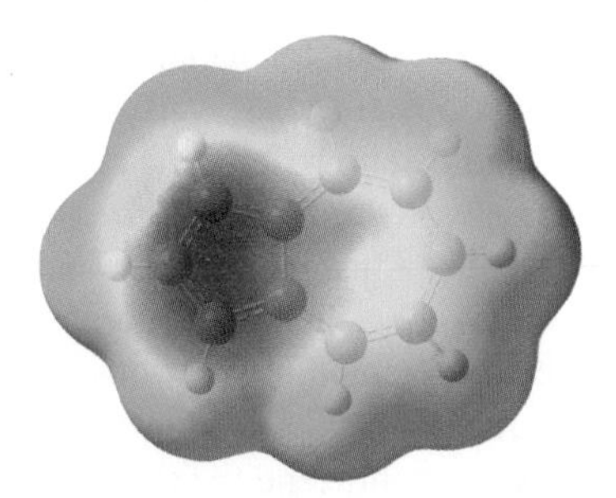

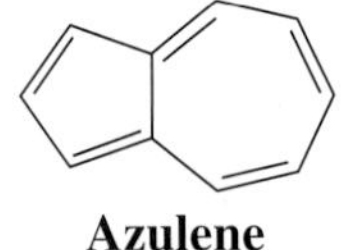

Azulene

PROBLEM 14.12 Azulene has an appreciable dipole moment. Write resonance structures for azulene that explain this dipole moment and that help explain its aromaticity.

14.8C Fullerenes

In 1990 W. Krätschmer (Max Planck Institute, Heidelberg), D. Huffman (University of Arizona), and their co-workers described the first practical synthesis of C_{60}, a molecule shaped like a soccer ball and called buckminsterfullerene. Formed by the resistive heating of graphite in an inert atmosphere, C_{60} is a member of an exciting new group of aromatic compounds called **fullerenes.** Fullerenes are cagelike molecules with the geometry of a

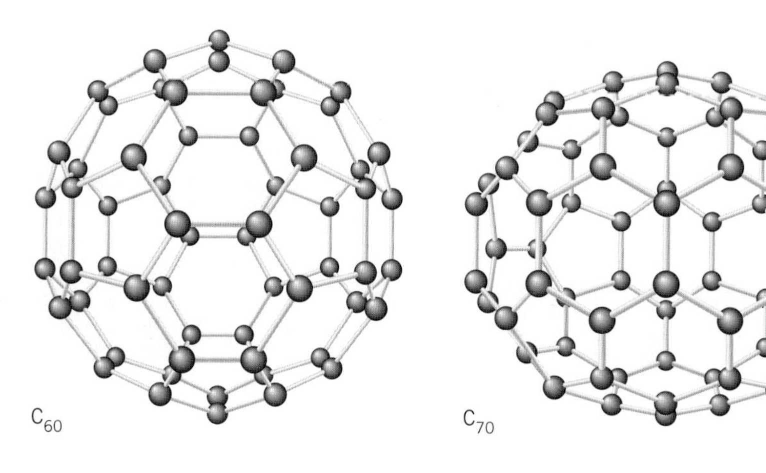

FIGURE 14.19 The structures of C_{60} and C_{70}. (Adapted from Diederich, F.; Whetten, R. L. *Acc. Chem. Res.* **1992,** *25,* 119–126.)

truncated icosahedron or geodesic dome, named after the architect Buckminster Fuller, renowned for his development of structures with geodesic domes. The structure of C_{60} and its existence had been established five years earlier, by H. W. Kroto (University of Sussex), R. E. Smalley and R. F. Curl (Rice University), and their co-workers. Kroto, Curl, and Smalley had found both C_{60} and C_{70} (Fig. 14.19) as highly stable components of a mixture of carbon clusters formed by laser-vaporizing graphite. Since 1990 chemists have synthesized many other higher and lower fullerenes and have begun exploring their interesting chemistry.

The Chemistry of...

Nanotubes

Nanotubes are a relatively new class of carbon-based materials related to buckminsterfullerenes. A **nanotube** is a structure that looks as though it were formed by rolling a sheet of graphitelike carbon (a flat network of fused benzene rings resembling chicken wire) into the shape of a tube and capping each end with half of a buckyball. Nanotubes are very tough—about 100 times as strong as steel. Besides their potential as strengtheners for new composite materials, some nanotubes have been shown to act as electrical conductors or semiconductors depending on their precise form. They are also being used as probe tips for analysis of DNA and proteins by atomic force microscopy (AFM). Many other applications have been envisioned for them as well, including use as molecular-size test tubes or capsules for drug delivery.

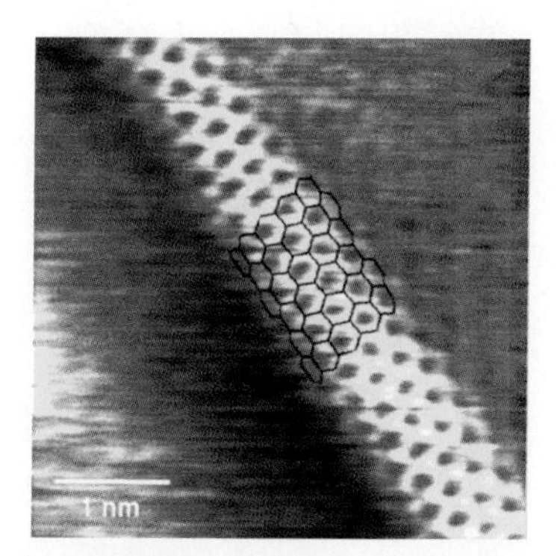

The wall of a nanotube is comprised of a network of benzene rings, highlighted in black on this scanning tunneling microscopy (STM) image. Image courtesy of C. M. Lieber (Harvard University).

Like a geodesic dome, a fullerene is composed of a network of pentagons and hexagons. To close into a spheroid, a fullerene must have exactly 12 five-membered faces, but the number of six-membered faces can vary widely. The structure of C_{60} has 20 hexagonal faces; C_{70} has 25. Each carbon of a fullerene is sp^2 hybridized and forms σ bonds to three other carbon atoms. The remaining electron at each carbon is delocalized into a system of molecular orbitals that gives the whole molecule aromatic character.

The Nobel Prize in Chemistry was awarded in 1996 to Professors Curl, Kroto, and Smalley for their discovery of fullerenes.

Robert F. Curl

Harold W. Kroto

Richard E. Smalley

The chemistry of fullerenes is proving to be even more fascinating than their synthesis. Fullerenes have a high electron affinity and readily accept electrons from alkali metals to produce a new metallic phase—a "buckide" salt. One such salt, K_3C_{60}, is a stable metallic crystal consisting of a face-centered-cubic structure of "buckyballs" with a potassium atom in between; it becomes a superconductor when cooled below 18 K. Fullerenes have even been synthesized that have metal atoms in the interior of the carbon atom cage.

14.9 Heterocyclic Aromatic Compounds

Almost all of the cyclic molecules that we have discussed so far have had rings composed solely of carbon atoms. However, in molecules of many cyclic compounds an element other than carbon is present in the ring. These compounds are called **heterocyclic compounds.** Heterocyclic molecules are quite commonly encountered in nature. For this reason, and because the structures of some of these molecules are closely related to the compounds that we discussed earlier, we shall now describe a few examples.

Heterocyclic compounds containing nitrogen, oxygen, or sulfur are by far the most common. Four important examples are given here in their Kekulé forms. *These four compounds are all aromatic:*

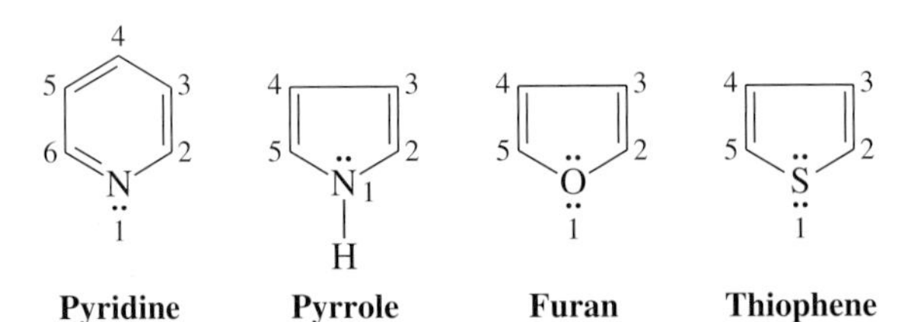

If we examine these structures, we shall see that pyridine is electronically related to benzene, and that pyrrole, furan, and thiophene are related to the cyclopentadienyl anion.

The nitrogen atoms in molecules of both pyridine and pyrrole are sp^2 hybridized. In pyridine (Fig. 14.20) the sp^2-hybridized nitrogen donates one bonding electron to the π system. This electron, together with one from each of the five carbon atoms, gives pyridine a sextet of electrons like benzene. The two unshared electrons of the nitrogen of pyridine are in an sp^2 orbital that lies in the same plane as the atoms of the ring. This sp^2 orbital does not overlap with the p orbitals of the ring (it is, therefore, said to be *orthogonal* to the p orbitals). The unshared pair on nitrogen is not a part of the π system, and these electrons confer on pyridine the properties of a weak base.

In pyrrole (Fig. 14.21) the electrons are arranged differently. Because only four π electrons are contributed by the carbon atoms of the pyrrole ring, the sp^2-hybridized nitrogen

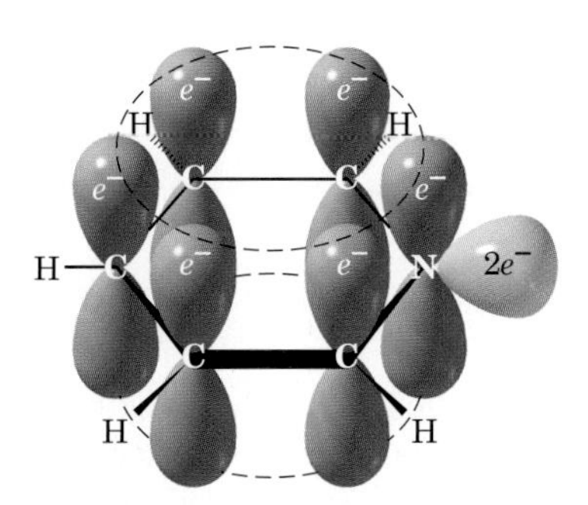

FIGURE 14.20 The stylized *p* orbital structure of pyridine.

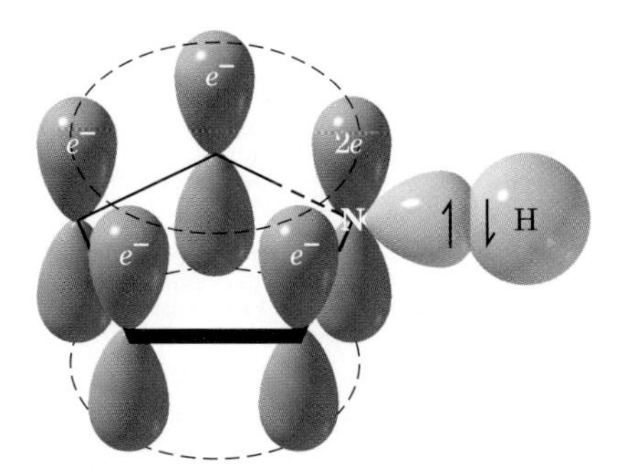

FIGURE 14.21 The stylized *p* orbital structure of pyrrole. (Compare with the orbital structure of the cyclopentadienyl anion in Fig. 14.10.)

must contribute two electrons to give an aromatic sextet. Because these electrons are a part of the aromatic sextet, they are not available for donation to a proton. Thus, in aqueous solution, pyrrole is not appreciably basic.

Furan and thiophene are structurally quite similar to pyrrole. The oxygen atom in furan and the sulfur atom in thiophene are sp^2 hybridized. In both compounds the *p* orbital of the heteroatom donates two electrons to the π system. The oxygen and sulfur atoms of furan and thiophene carry an unshared pair of electrons in an sp^2 orbital (Fig. 14.22) that is orthogonal to the π system.

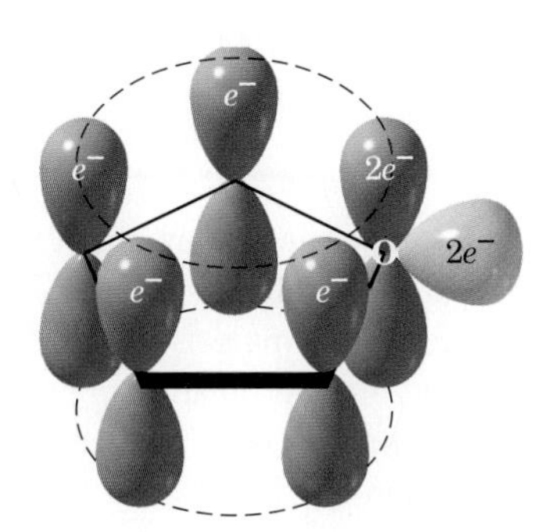

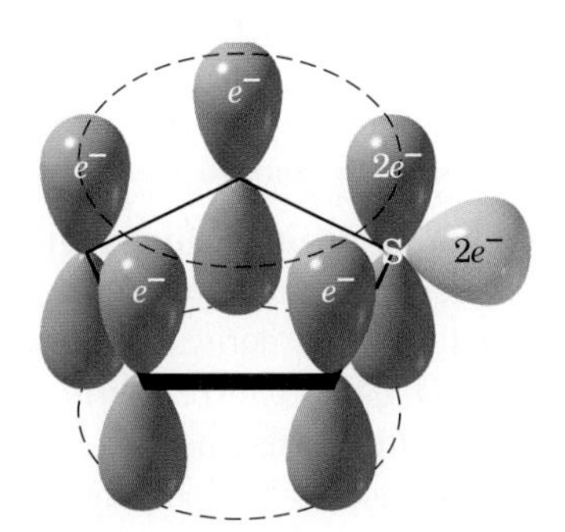

FIGURE 14.22 The stylized *p* orbital structures of furan and thiophene.

14.10 Aromatic Compounds in Biochemistry

Compounds with aromatic rings occupy numerous and important positions in reactions that occur in living systems. It would be impossible to describe them all in this chapter. We shall, however, point out a few examples now and we shall see others later.

Two amino acids necessary for protein synthesis contain the benzene ring:

CO_2^- H NH_3^+
Phenylalanine

HO CO_2^- H NH_3^+
Tyrosine

A third aromatic amino acid, tryptophan, contains a benzene ring fused to a pyrrole ring. (This aromatic ring system is called an indole system, see Section 20.1B.)

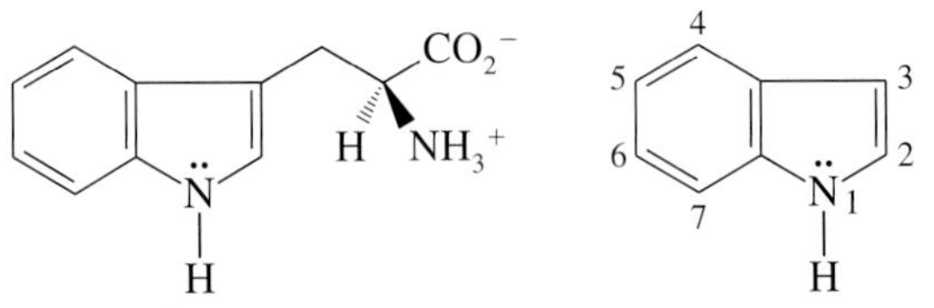

Tryptophan Indole

It appears that humans, because of the course of evolution, do not have the biochemical ability to synthesize the benzene ring. As a result, phenylalanine and tryptophan derivatives are essential in the human diet. Because tyrosine can be synthesized from phenylalanine in a reaction catalyzed by an enzyme known as *phenylalanine hydroxylase,* it is not essential in the diet as long as phenylalanine is present.

Heterocyclic aromatic compounds are also present in many biochemical systems. Derivatives of purine and pyrimidine are essential parts of DNA and RNA:

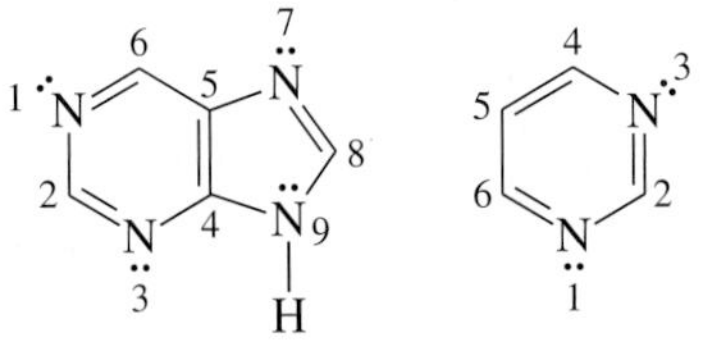

Purine Pyrimidine

DNA is the molecule responsible for the storage of genetic information, and RNA is prominently involved in the synthesis of enzymes and other proteins (Chapter 25).

PROBLEM 14.13 (**a**) The —SH group is sometimes called the *mercapto group.* 6-Mercaptopurine is used in the treatment of acute leukemia. Write its structure. (**b**) Allopurinol, a compound used to treat gout, is 6-hydroxypurine. Write its structure.

Nicotinamide adenine dinucleotide, one of the most important coenzymes (Section 24.9) in biological oxidations and reductions, includes both a pyridine derivative (nicotinamide) and a purine derivative (adenine) in its structure. Its formula is shown in Fig. 14.23 as NAD^+, the oxidized form that contains the pyridinium aromatic ring. The reduced form of the coenzyme is NADH, in which the pyridine ring is no longer aromatic due to presence of an additional hydrogen and two electrons in the ring.

A key role of NAD^+ in metabolism is to serve as a coenzyme for glyceraldehyde-3-phosphate dehydrogenase (GAPDH) in glycolysis, the pathway by which glucose is bro-

FIGURE 14.23 Nicotinamide adenine dinucleotide (NAD^+).

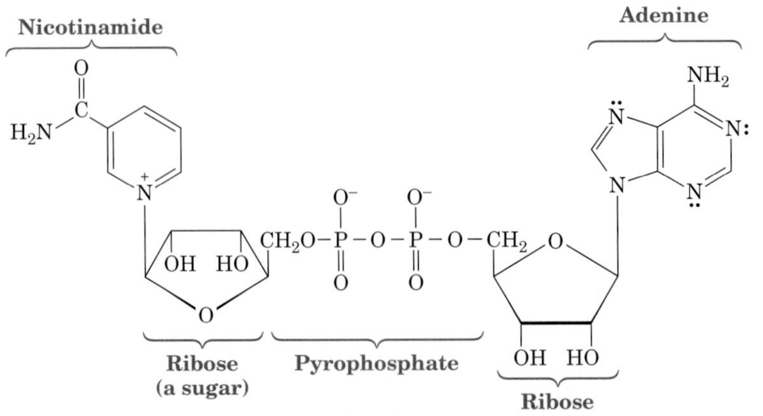

Nicotinamide Adenine Ribose (a sugar) Pyrophosphate Ribose

ken down for energy production. In the reaction catalyzed by GAPDH (Fig. 14.24), the aldehyde group of glyceraldehyde-3-phosphate (GAP) is oxidized to a carboxyl group (incorporated as a phosphoric anhydride) in 1,3-bisphosphoglycerate (1,3-BPG). Concurrently, the aromatic pyridinium ring of NAD^+ is reduced to its higher energy from, NADH. One of the ways the chemical energy stored in the nonaromatic ring of NADH is used is in the mitochondria for the production of ATP, where cytochrome electron transport and oxidative phosphorylation take place. There, release of chemical energy from NADH by oxidation to the more stable aromatic form NAD^+ (and a proton) is coupled with the pumping of protons across the inner mitochondrial membrane. An electrochemical gradient is created across the mitochondrial membrane, which drives the synthesis of ATP by the enzyme ATP synthase.

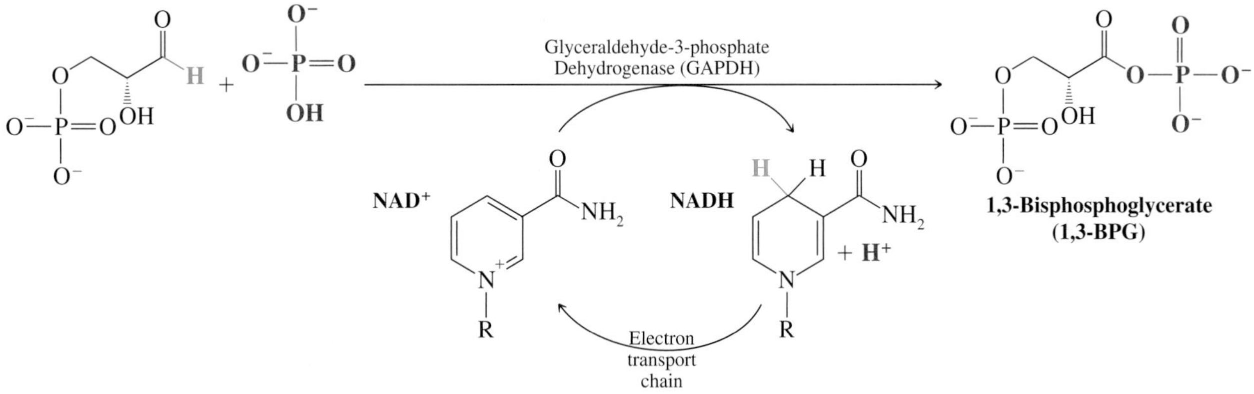

Glyceraldehyde-3-phosphate Dehydrogenase (GAPDH)

NAD^+ NADH + H^+

Electron transport chain

1,3-Bisphosphoglycerate (1,3-BPG)

FIGURE 14.24 NAD^+, as the coenzyme in glyceraldehyde-3-phosphate dehydrogenase (GAPDH), is used to oxidize glyceraldehyde-3-phosphate (GAP) to 1,3-bisphosphoglycerate during the degradation of glucose in glycolysis. One of the ways that NADH can be reoxidized to NAD^+ is by the electron transport chain in mitochondria, where, under aerobic conditions, rearomatization of NADH helps to drive ATP synthesis.

The chemical energy stored in NADH is used to bring about many other essential biochemical reactions, as well. NADH is part of an enzyme called lactate dehydrogenase that reduces the ketone group of pyruvic acid to the alcohol group of lactic acid. Here, the nonaromatic ring of NADH is converted to the aromatic ring of NAD^+. This process is important in muscles operating under oxygen-depleted conditions (anaerobic metabolism), where reduction of pyruvic acid to lactic acid by NADH serves to regenerate NAD^+ that is needed to continue glycolytic synthesis of ATP:

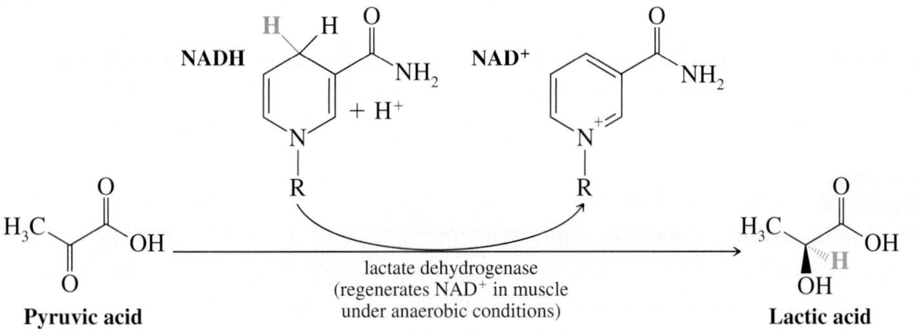

NADH + H^+ NAD^+

Pyruvic acid — lactate dehydrogenase (regenerates NAD^+ in muscle under anaerobic conditions) → **Lactic acid**

Additional information about NADH and NAD^+ was given in the Chapter 12 opening vignette, "Two Aspects of the Coenzyme NADH."

Yeasts growing under anaerobic conditions (fermentation) also have a pathway for regenerating NAD^+ from NADH. Under oxygen-deprived conditions, yeasts convert pyruvic acid to acetaldehyde by decarboxylation (CO_2 is released, Section 18.11); then NADH in alcohol dehydrogenase reduces acetaldehyde to ethanol. As in oxygen-starved muscles, this pathway occurs for the purpose of regenerating NAD^+ needed to continue glycolytic ATP synthesis:

$$\text{Pyruvic acid} \xrightarrow[\text{pyruvate decarboxylase (with thiamine as a coenzyme, Section 18.11)}]{CO_2} \text{Acetaldehyde} \xrightarrow[\text{alcohol dehydrogenase (regenerates } NAD^+ \text{ in yeast under anaerobic conditions)}]{NADH \rightarrow NAD^+} \text{Ethanol}$$

Although many aromatic compounds are essential to life, others are hazardous. Many are quite toxic, and several benzenoid compounds, including benzene itself, are **carcinogenic.** Two other examples are benzo[*a*]pyrene and 7-methylbenz[*a*]anthracene:

Benzo[*a*]pyrene **7-Methylbenz[*a*]anthracene**

The mechanism for the carcinogenic effects of compounds like benzo[*a*]pyrene was discussed in "The Chemistry of . . . Epoxides, Carcinogens, and Biological Oxidation" in Section 11.14.

The hydrocarbon benzo[*a*]pyrene has been found in cigarette smoke and in the exhaust from automobiles. It is also formed in the incomplete combustion of any fossil fuel. It is found on charcoal-broiled steaks and exudes from asphalt streets on a hot summer day. Benzo[*a*]pyrene is so carcinogenic that one can induce skin cancers in mice with almost total certainty simply by shaving an area of the body of the mouse and applying a coating of benzo[*a*]pyrene.

14.11 Spectroscopy of Aromatic Compounds

14.11A ^{1}H NMR Spectra

As we learned in Chapter 9, the hydrogens of benzene derivatives absorb downfield in the region between δ 6.0 and δ 9.5. In Section 14.7B we found that absorption takes place far downfield because a ring current generated in the benzene ring creates a magnetic field, called "the induced field," which reinforces the applied magnetic field at the position of the protons of the ring. This reinforcement causes the protons of benzene to be highly deshielded.

We also learned in Section 14.7B that internal hydrogens of large-ring aromatic compounds such as [18]annulene, because of their position, are highly shielded by this induced field. They therefore absorb at unusually high field strength, often at negative delta values.

14.11B ^{13}C NMR Spectra

The carbon atoms of benzene rings generally absorb in the δ 100–170 region of ^{13}C NMR spectra. Figure 14.25 gives the broadband proton-decoupled ^{13}C NMR spectrum of 4-*N*,*N*-diethylaminobenzaldehyde and permits an exercise in making ^{13}C assignments of a compound with both aromatic and aliphatic carbon atoms.

The DEPT spectra (not given to save space) show that the signal at δ 45 arises from a CH_2 group and the one at δ 13 arises from a CH_3 group. This allows us to assign these two signals immediately to the two carbons of the equivalent ethyl groups.

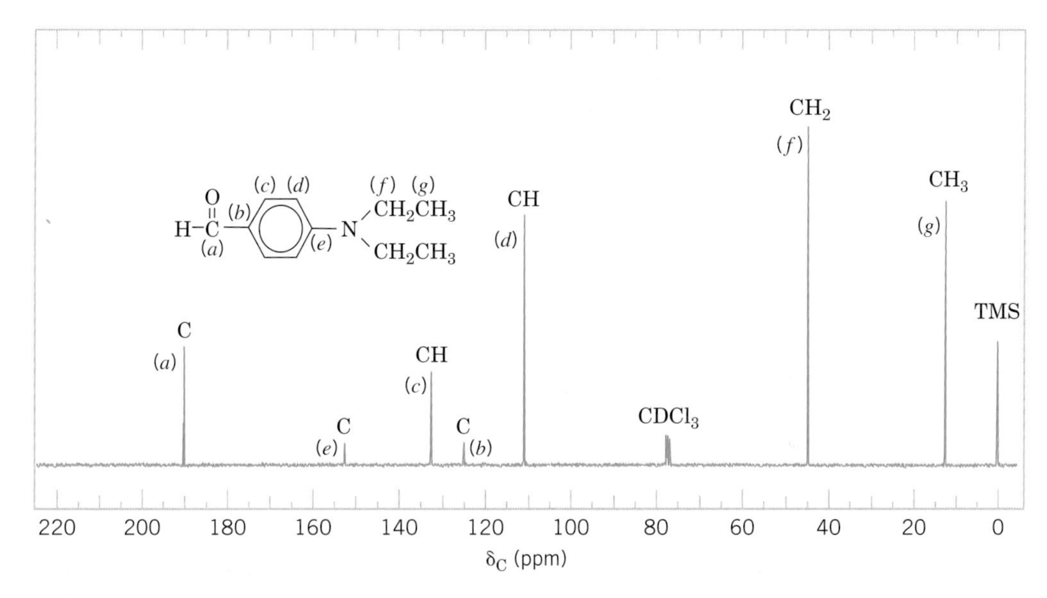

FIGURE 14.25 The broadband proton-decoupled ^{13}C NMR spectrum of 4-*N*,*N*-diethylaminobenzaldehyde. DEPT information and carbon assignments are shown by each peak.

The signals at δ 126 and δ 153 appear in the DEPT spectra as carbon atoms that do not bear hydrogen atoms and are assigned to carbons (*b*) and (*e*) (see Fig. 14.25). The greater electronegativity of nitrogen (when compared to carbon) causes the signal from (*e*) to be further downfield (at δ 153). The signal at δ 190 appears as a CH group in the DEPT spectra and arises from the carbon of the aldehyde group. Its chemical shift is the most downfield of all the peaks because of the great electronegativity of its oxygen and because the second resonance structure below contributes to the hybrid. Both factors cause the electron density at this carbon to be very low, and, therefore, this carbon is strongly deshielded.

Resonance contributors for an aldehyde group

This leaves the signals at δ 112 and δ 133 and the two sets of carbon atoms of the benzene ring labeled (*c*) and (*d*) to be accounted for. Both signals are indicated as CH groups in the DEPT spectra. But which signal belongs to which set of carbon atoms? Here we find another interesting application of resonance theory.

If we write resonance structures **A–D** involving the unshared electron pair of the amino group, we see that contributions made by **B** and **D** increase the electron density at the set of carbon atoms labeled (*d*):

On the other hand, writing structures **E–H** involving the aldehyde group shows us that contributions made by **F** and **H** decrease the electron density at the set of carbon atoms labeled (*c*):

(*d*) (*c*) **E** ⟷ **F** ⟷ **G** ⟷ **H** (*d*) + (*c*)

(Other resonance structures are possible but are not pertinent to the argument here.)

Increasing the electron density at a carbon should increase its shielding and should shift its signal upfield. Therefore, we assign the signal at δ 112 to the set of carbon atoms labeled (*d*). Conversely, decreasing the electron density at a carbon should shift its signal downfield, so we assign the signal at δ 133 to the set labeled (*c*).

Carbon-13 spectroscopy can be especially useful in recognizing a compound with a high degree of symmetry. The following solved problem illustrates one such application.

SOLVED PROBLEM

The broadband proton-decoupled ^{13}C spectrum given in Fig. 14.26 is of a tribromobenzene ($C_6H_3Br_3$). Which tribromobenzene is it?

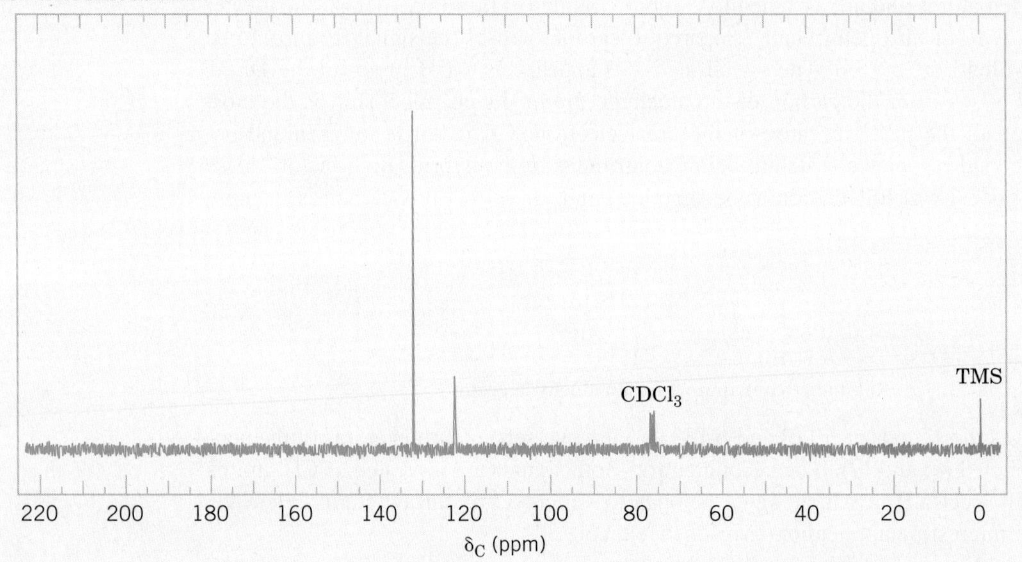

FIGURE 14.26 The broadband proton-decoupled ^{13}C NMR spectrum of a tribromobenzene.

ANSWER: There are three possible tribromobenzenes:

Br (*a*) Br (*b*) Br (*b*) (*c*) (*c*) (*d*) — **1,2,3-Tribromobenzene**

Br (*a*) Br (*b*) (*f*) (*e*) (*c*) (*d*) Br — **1,2,4-Tribromobenzene**

Br (*a*) (*b*) (*b*) (*a*) (*a*) Br Br (*b*) — **1,3,5-Tribromobenzene**

Our spectrum (Fig. 14.26) consists of only two signals, indicating that only two different types of carbon atoms are present in the compound. Only 1,3,5-tribromobenzene has a degree of symmetry such that it would give only two signals, and, therefore, it is the correct answer. 1,2,3-Tribromobenzene would give four ^{13}C signals and 1,2,4-tribromobenzene would give six.

PROBLEM 14.14 Explain how ^{13}C spectroscopy could be used to distinguish the *ortho-*, *meta-*, and *para-*dibromobenzene isomers one from another.

14.11C Infrared Spectra of Substituted Benzenes

Benzene derivatives give characteristic C—H stretching peaks near 3030 cm^{-1} (Table 2.7). Stretching motions of the benzene ring can give as many as four bands in the 1450–1600-cm^{-1} region, with two peaks near 1500 and 1600 cm^{-1} being stronger.

Absorption peaks in the 680–860-cm^{-1} region from out-of-plane C—H bending can often (but not always) be used to characterize the substitution patterns of benzene compounds (Table 14.1). **Monosubstituted benzenes** give two very strong peaks, between 690 and 710 cm^{-1} and between 730 and 770 cm^{-1}.

Ortho-disubstituted benzenes show a strong absorption peak between 735 and 770 cm^{-1} that arises from bending motions of the C—H bonds. **Meta-disubstituted benzenes** show two peaks: one strong peak between 680 and 725 cm^{-1} and one very strong peak between 750 and 810 cm^{-1}. **Para-disubstituted benzenes** give a single very strong absorption between 800 and 860 cm^{-1}.

TABLE 14.1 Infrared Absorptions in the 680–860-cm^{-1} Region[a]

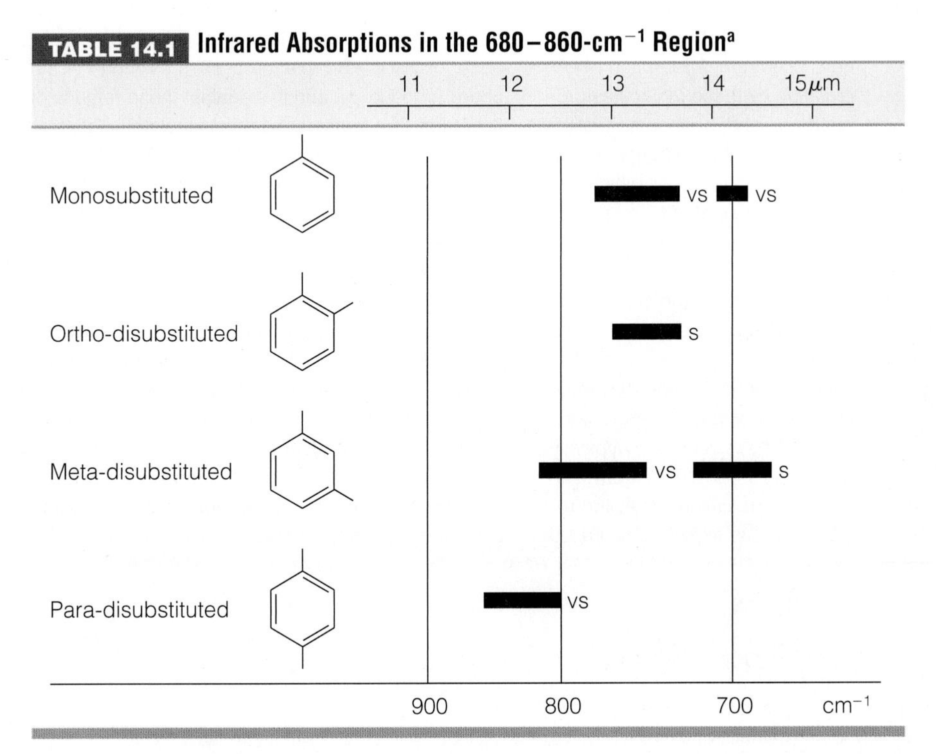

[a]s, strong; vs, very strong.

PROBLEM 14.15 Four benzenoid compounds, all with the formula C_7H_7Br, gave the following IR peaks in the 680–860-cm^{-1} region:

A, 740 cm^{-1} (strong)
B, 800 cm^{-1} (very strong)
C, 680 cm^{-1} (strong) and 760 cm^{-1} (very strong)
D, 693 cm^{-1} (very strong) and 765 cm^{-1} (very strong)

Propose structures for **A, B, C,** and **D.**

14.11D Ultraviolet–Visible Spectra of Aromatic Compounds

The conjugated π electrons of a benzene ring give characteristic ultraviolet absorptions that indicate the presence of a benzene ring in an unknown compound. One absorption band of moderate intensity occurs near 205 nm and another, less intense band appears in the 250–275-nm range. Conjugation outside the benzene ring leads to absorptions at other wavelengths.

The Chemistry of...

Sunscreens (Catching the Sun's Rays and What Happens to Them)

The use of sunscreens in recent years has increased due to heightened concern over the risk of skin cancer and other conditions caused by exposure to UV radiation. In DNA, for example, UV radiation can cause adjacent thymine bases to form mutagenic dimers. Sunscreens afford protection from UV radiation because they contain aromatic molecules that absorb energy in the UV region of the electromagnetic spectrum. Absorption of UV radiation by these molecules promotes π and nonbonding electrons to higher energy levels (Section 13.9B), after which the energy is dissipated by relaxation through molecular vibration. In essence, the UV radiation is converted to heat (IR radiation).

A UV-A and UV-B sunscreen product whose active ingredients are octyl 4-methoxycinnamate and 2-hydroxy-4-methoxybenzophenone (Oxybenzone).

Sunscreens are classified according to the portion of the UV spectrum where their maximum absorption occurs. Three regions of the UV spectrum are typically discussed. The region from 320 to 400 nm is called UV-A, the region from 280 to 320 nm is called UV-B, and the region from 100 to 280 nm is called UV-C. The UV-C region is potentially the most dangerous because it encompasses the shortest UV wavelengths and is therefore of the highest energy. However, ozone and other components in Earth's atmosphere absorb UV-C wavelengths, and thus we are protected from radiation in this part of the spectrum so long as Earth's atmosphere is not compromised further by ozone-depleting pollutants. Most of the UV-A and some of the UV-B radiation passes through the atmosphere to reach us, and it is against these regions of the spectrum that sunscreens are formulated. Tanning and sunburn are caused by UV-B radiation. Risk of skin cancer is primarily associated with UV-B radiation, although some UV-A wavelengths may be important as well.

The specific range of protection provided by a sunscreen depends on the structure of its UV-absorbing groups. Most sunscreens have structures derived from the following parent compounds: *p*-aminobenzoic acid (PABA), cinnamic acid (3-phenylpropenoic acid), benzophenone (diphenyl ketone), and salicylic acid (*o*-hydroxybenzoic acid). The structures and λ_{max} for a few of the most common sunscreen agents are given below. The common theme among them is an aromatic core in conjugation with other functional groups.

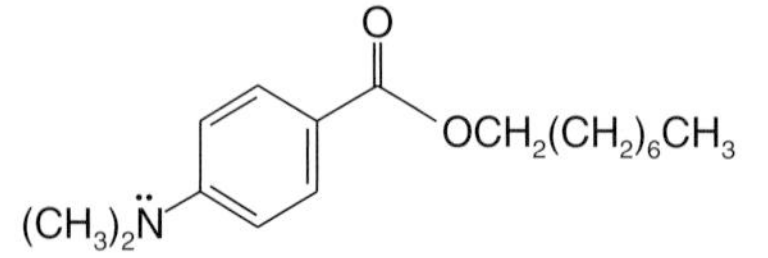

Octyl 4-*N,N*-dimethylaminobenzoate (Padimate O) λ_{max} 310 nm

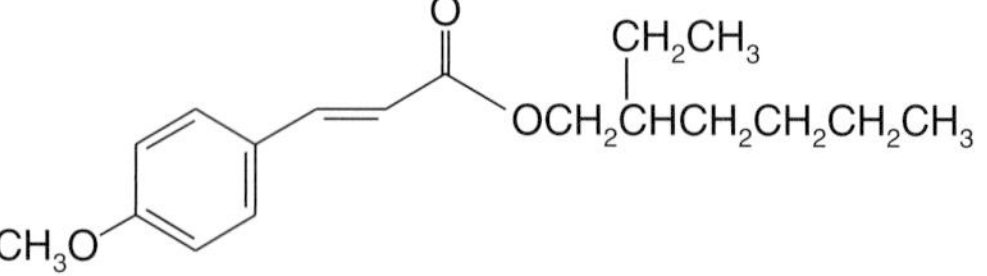

2-Ethylhexyl 4-methoxycinnamate λ_{max} 310 nm (Parsol MCX)

2-Hydroxy-4-methoxybenzophenone (Oxybenzone) λ_{max} 288 and 325 nm

Homomenthyl salicylate (Homosalate) λ_{max} 309 nm

2-Ethylhexyl 2-cyano-3,3-diphenylacrylate (Octocrylene) λ_{max} 310 nm

14.11E Mass Spectra of Aromatic Compounds

The major ion in the mass spectrum of an alkyl-substituted benzene is often *m/z* 91 ($C_6H_5CH_2^+$), resulting from cleavage between the first and second carbons of the alkyl chain attached to the ring. The ion presumably originates as a benzylic cation that rearranges to a tropylium cation ($C_7H_7^+$, Section 14.7C). Another ion frequently seen in mass spectra of monoalkylbenzene compounds is *m/z* 77, corresponding to $C_6H_5^+$.

Key Terms and Concepts

Aliphatic compounds	**Section 14.1**
Annulenes	**Section 14.7A**
Aromatic compounds	**Sections 14.1, 14.7D**
Aromatic ions	**Section 14.7C**
Benzene substitution	**Section 14.3**
Benzenoid aromatic compounds	**Sections 14.7D, 14.8A**
Fullerenes and nanotubes	**Section 14.8C**
Heterocyclic aromatic compounds	**Section 14.9**
Hückel's rule	**Section 14.7**
Nonbenzenoid aromatic compounds	**Sections 14.7D, 14.8B**
Resonance energy	**Section 14.5**

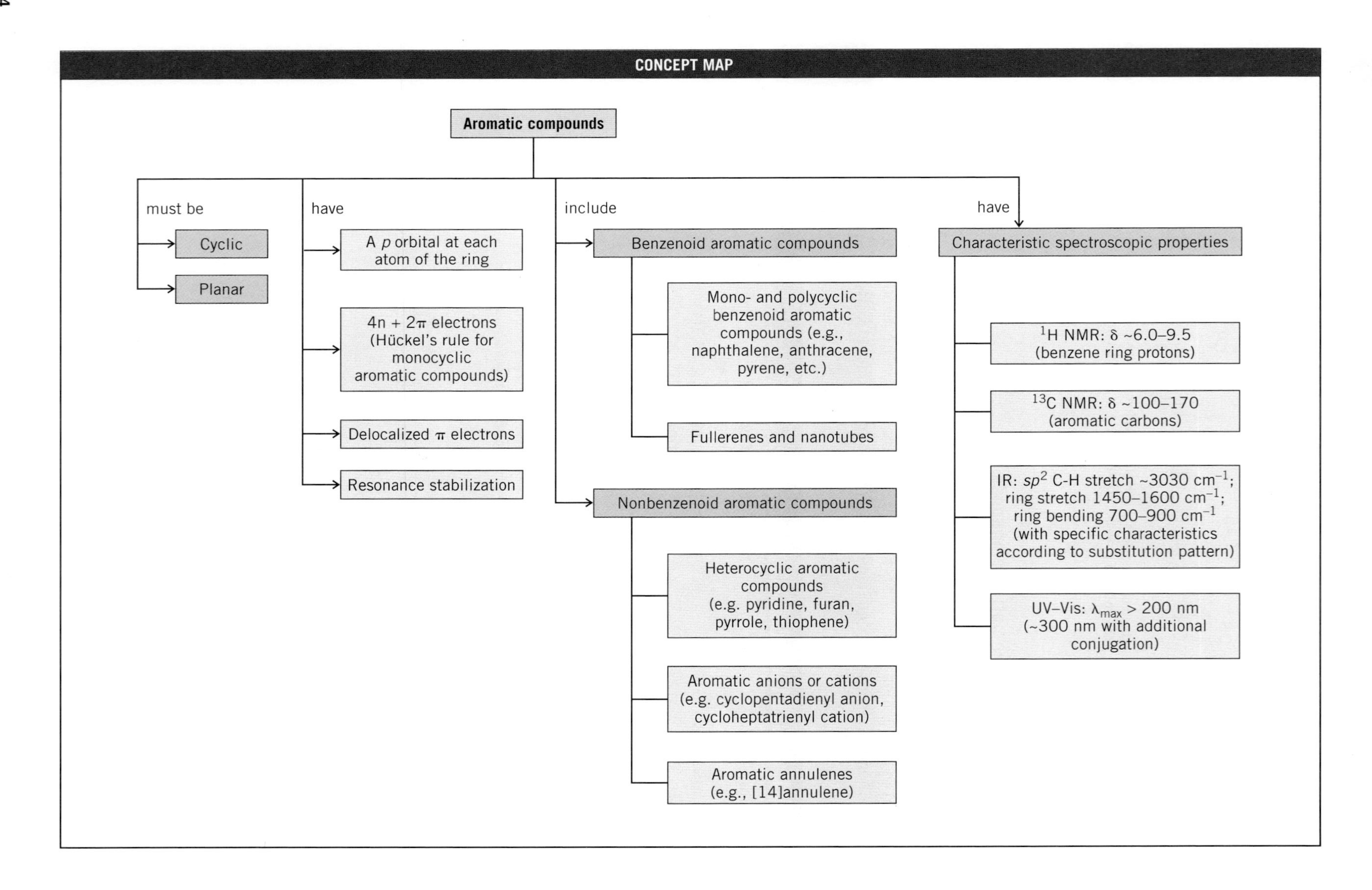
CONCEPT MAP
Aromatic compounds
must be
Cyclic
Planar
have
A p orbital at each atom of the ring
4n + 2π electrons (Hückel's rule for monocyclic aromatic compounds)
Delocalized π electrons
Resonance stabilization
include
Benzenoid aromatic compounds
Mono- and polycyclic benzenoid aromatic compounds (e.g., naphthalene, anthracene, pyrene, etc.)
Fullerenes and nanotubes
Nonbenzenoid aromatic compounds
Heterocyclic aromatic compounds (e.g. pyridine, furan, pyrrole, thiophene)
Aromatic anions or cations (e.g. cyclopentadienyl anion, cycloheptatrienyl cation)
Aromatic annulenes (e.g., [14]annulene)
have
Characteristic spectroscopic properties
^{1}H NMR: δ ~6.0–9.5 (benzene ring protons)
^{13}C NMR: δ ~100–170 (aromatic carbons)
IR: sp^2 C-H stretch ~3030 cm^{-1}; ring stretch 1450–1600 cm^{-1}; ring bending 700–900 cm^{-1} (with specific characteristics according to substitution pattern)
UV–Vis: λ_{max} > 200 nm (~300 nm with additional conjugation)

Additional Problems

14.16 Write structural formulas for each of the following:

(a) 3-Nitrobenzoic acid
(b) *p*-Bromotoluene
(c) *o*-Dibromobenzene
(d) *m*-Dinitrobenzene
(e) 3,5-Dinitrophenol
(f) *p*-Nitrobenzoic acid
(g) 3-Chloro-1-ethoxybenzene
(h) *p*-Chlorobenzenesulfonic acid
(i) Methyl *p*-toluenesulfonate
(j) Benzyl bromide
(k) *p*-Nitroaniline
(l) *o*-Xylene
(m) *tert*-Butylbenzene
(n) *p*-Cresol
(o) *p*-Bromoacetophenone
(p) 3-Phenylcyclohexanol
(q) 2-Methyl-3-phenyl-1-butanol
(r) *o*-Chloroanisole

14.17 Write structural formulas and give acceptable names for all representatives of the following:

(a) Tribromobenzenes
(b) Dichlorophenols
(c) Nitroanilines
(d) Methylbenzenesulfonic acids
(e) Isomers of $C_6H_5—C_4H_9$

14.18 Although Hückel's rule (Section 14.7) strictly applies only to monocyclic compounds, it does appear to have application to certain bicyclic compounds, provided the important resonance structures involve only the perimeter double bonds, as in naphthalene below.

Both naphthalene (Section 14.8A) and azulene (Section 14.8B) have 10 π electrons and are aromatic. Pentalene (below) is apparently antiaromatic and is unstable even at $-100°C$. Heptalene has been made but it adds bromine, it reacts with acids, and it is not planar. Is Hückel's rule applicable to these compounds? If so, explain their lack of aromaticity.

Pentalene **Heptalene**

14.19 **(a)** In 1960 T. Katz (Columbia University) showed that cyclooctatetraene adds two electrons when treated with potassium metal and forms a stable, planar dianion, $C_8H_8^{2-}$ (as the dipotassium salt):

$$\xrightarrow[\text{THF}]{2\text{ K}} 2\text{ K}^+\ C_8H_8^{2-}$$

Use the molecular orbital diagram given in Fig. 14.7 and explain this result.

(b) In 1964 Katz also showed that removing two protons from the compound below (using butyllithium as the base) leads to the formation of a stable dianion with the formula $C_8H_6^{2-}$ (as the dilithium salt).

$$+\ 2\ C_4H_9Li \longrightarrow 2\ Li^+\ C_8H_6^{2-} + 2\ C_4H_{10}$$

Consider your answer to Problem 14.18 and provide an explanation for what is happening here.

Note: Problems marked with an asterisk are "challenge problems."

14.20 Although none of the [10]annulenes given in Section 14.7A is aromatic, the following 10 π-electron system is aromatic:

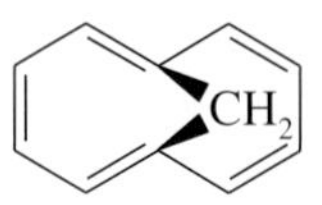

What factor makes this possible?

14.21 Cycloheptatrienone (**I**) is very stable. Cyclopentadienone (**II**) by contrast is quite unstable and rapidly undergoes a Diels–Alder reaction with itself.
(**a**) Propose an explanation for the different stabilities of these two compounds.
(**b**) Write the structure of the Diels–Alder adduct of cyclopentadienone.

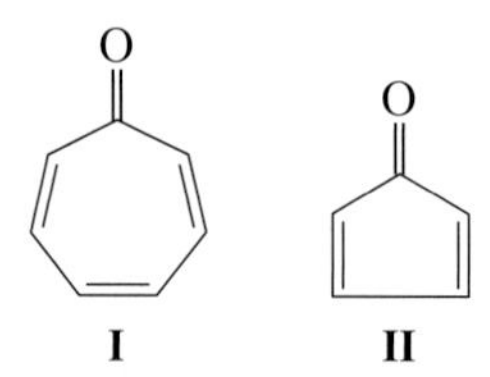

14.22 5-Chloro-1,3-cyclopentadiene (below) undergoes S_N1 solvolysis in the presence of silver ion extremely slowly even though the chlorine is doubly allylic and allylic halides normally ionize readily (Section 15.15). Provide an explanation for this behavior.

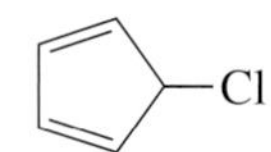

14.23 Explain the following: (**a**) The cycloheptatrienyl anion is antiaromatic, whereas the cyclononatetraenyl anion is planar (in spite of the angle strain involved) and appears to be aromatic. (**b**) Although [16]annulene is not aromatic, it adds two electrons readily to form an aromatic dianion.

14.24 Furan possesses less aromatic character than benzene as measured by their resonance energies (96 kJ mol^{-1} for furan; 151 kJ mol^{-1} for benzene). What chemical evidence have we studied earlier that shows that furan is less aromatic than benzene?

14.25 Assign structures to each of the compounds **A, B,** and **C** whose ^1H NMR spectra are shown in Fig. 14.27.

14.26 The ^1H NMR spectrum of cyclooctatetraene consists of a single line located at δ 5.78. What does the location of this signal suggest about electron delocalization in cyclooctatetraene?

14.27 Give a structure for compound **F** that is consistent with the ^1H NMR and IR spectra in Fig. 14.28.

14.28 A compound (**L**) with the molecular formula C_9H_{10} reacts with bromine in carbon tetrachloride and gives an IR absorption spectrum that includes the following absorption peaks: 3035 cm^{-1}(m), 3020 cm^{-1}(m), 2925 cm^{-1}(m), 2853 cm^{-1}(w), 1640 cm^{-1}(m), 990 cm^{-1}(s), 915 cm^{-1}(s), 740 cm^{-1}(s), 695 cm^{-1}(s). The ^1H NMR spectrum of **L** consists of:

Doublet δ 3.1 (2H)	Multiplet δ 5.1	Multiplet δ 7.1 (5H)
Multiplet δ 4.8	Multiplet δ 5.8	

The UV spectrum shows a maximum at 255 nm. Propose a structure for compound **L** and make assignments for each of the IR peaks.

14.29 Compound **M** has the molecular formula C_9H_{12}. The ^1H NMR spectrum of **M** is given in Fig. 14.29 and the IR spectrum in Fig. 14.30. Propose a structure for **M**.

14.30 A compound (**N**) with the molecular formula $C_9H_{10}O$ reacts with osmium tetroxide. The ^1H NMR spectrum of **N** is shown in Fig. 14.31 and the IR spectrum of **N** is shown in Fig. 14.32. Propose a structure for **N**.

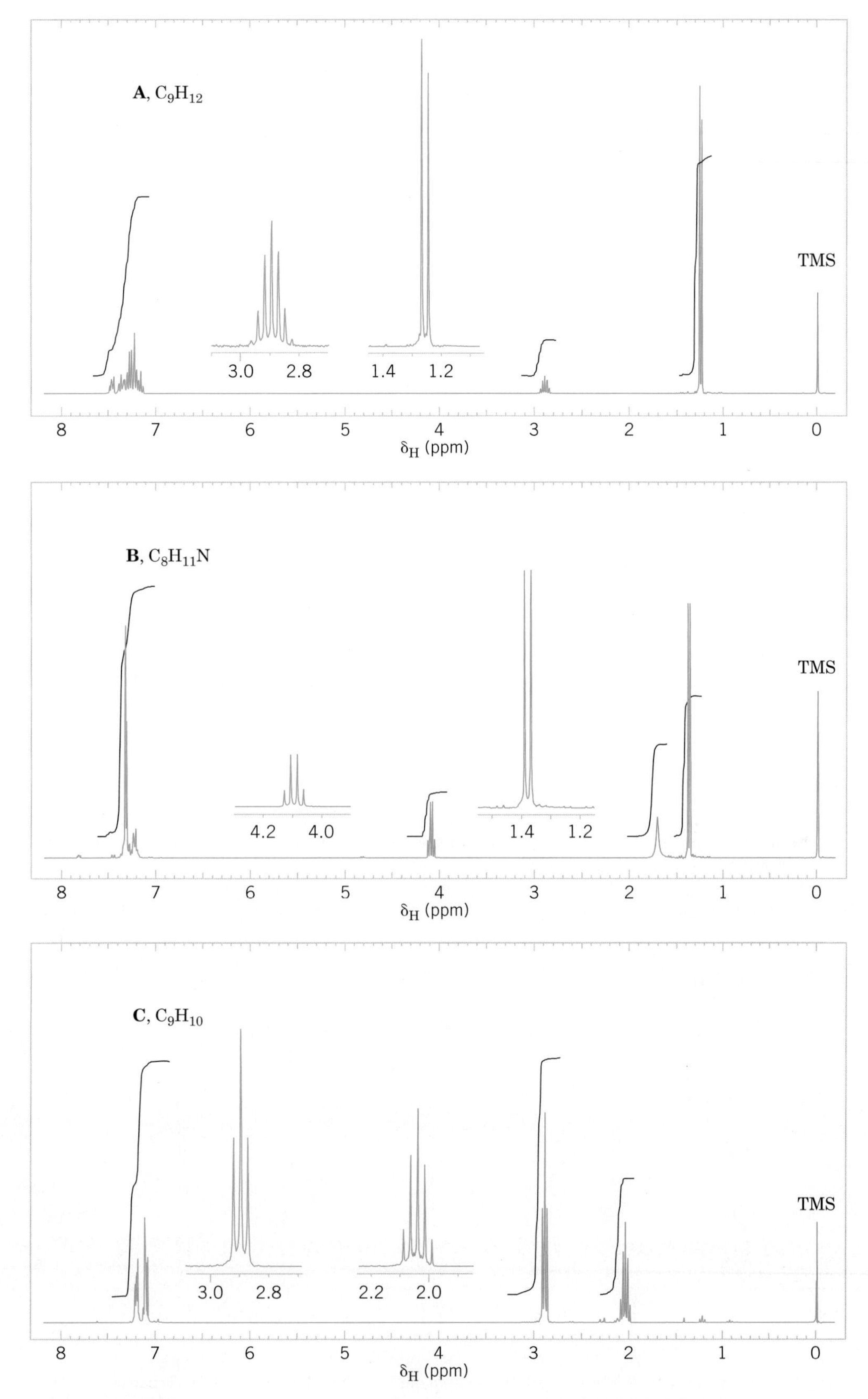

FIGURE 14.27 The 300-MHz ^1H NMR spectra for Problem 14.25. Expansions of the signals are shown in the offset plots.

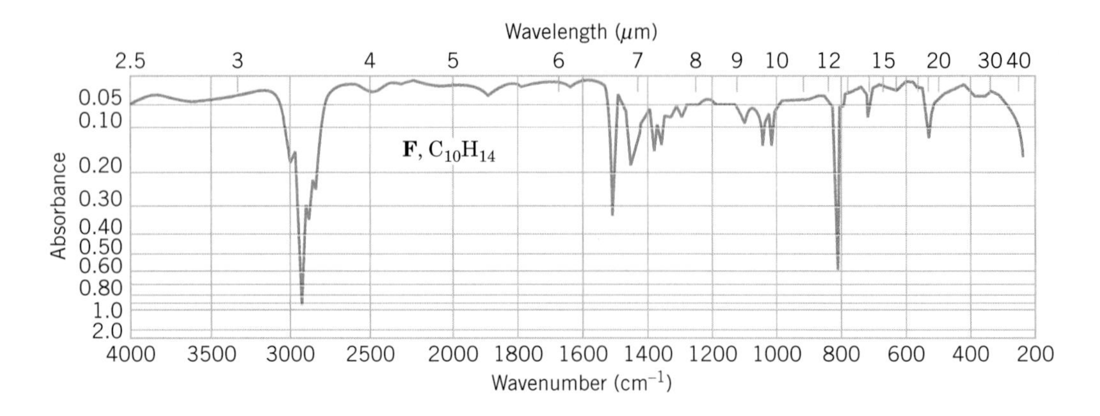

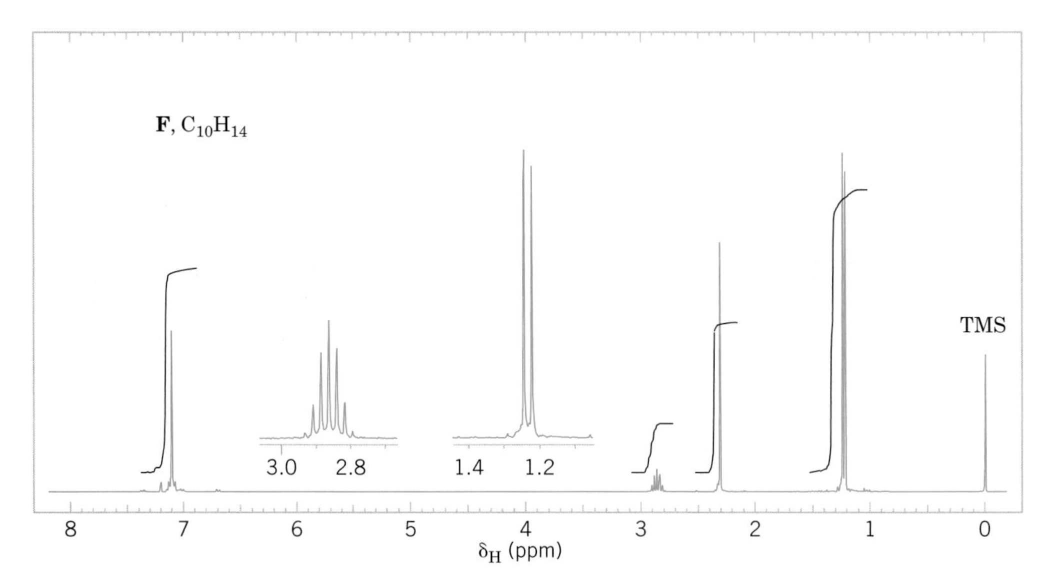

FIGURE 14.28 The 300-MHz ^1H NMR and IR spectra of compound F, Problem 14.27. Expansions of the signals are shown in the offset plots.

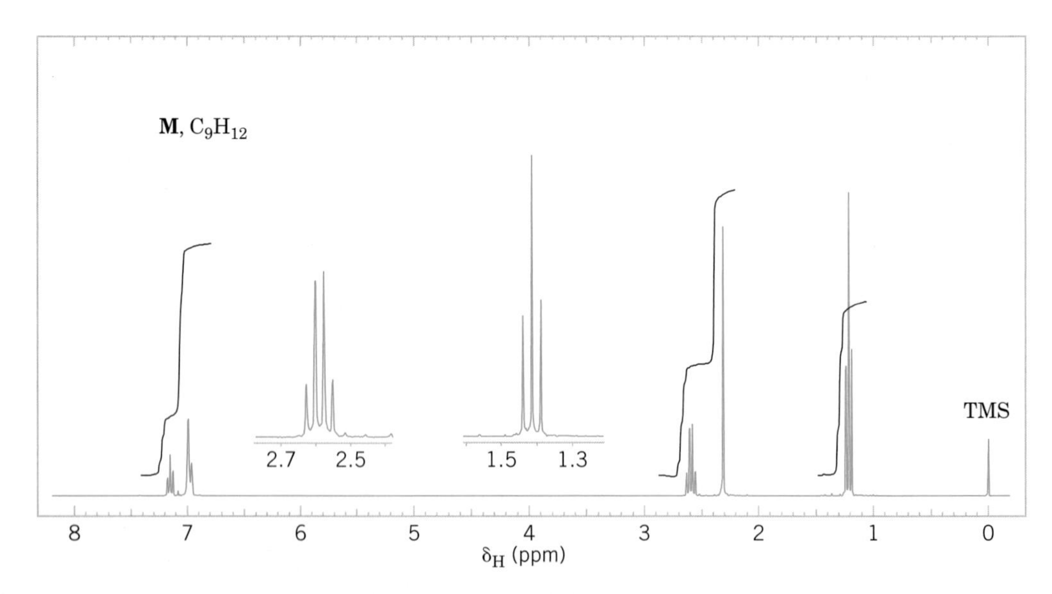

FIGURE 14.29 The 300-MHz ^1H NMR spectrum of compound M, Problem 14.29. Expansions of the signals are shown in the offset plots.

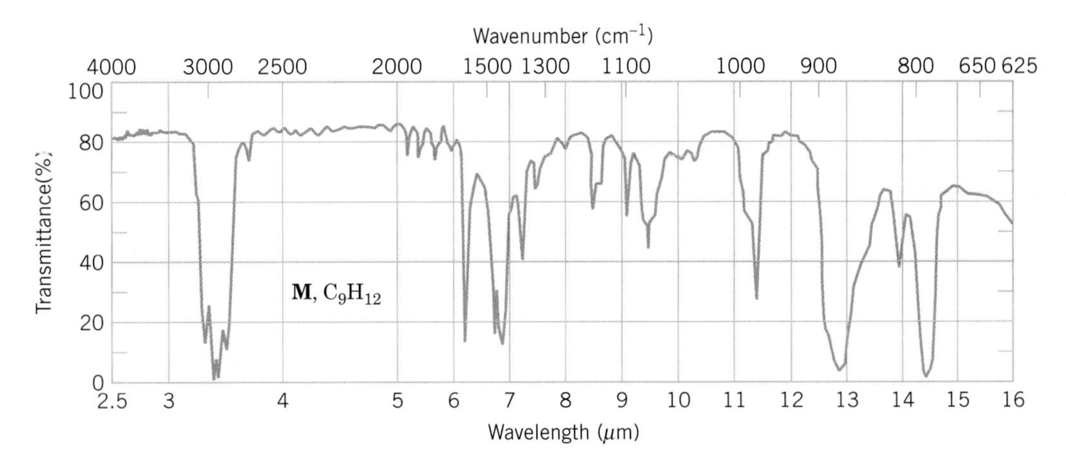

FIGURE 14.30 The IR spectrum of compound M, Problem 14.29. (Spectrum courtesy of Aldrich Chemical Co., Milwaukee, WI.)

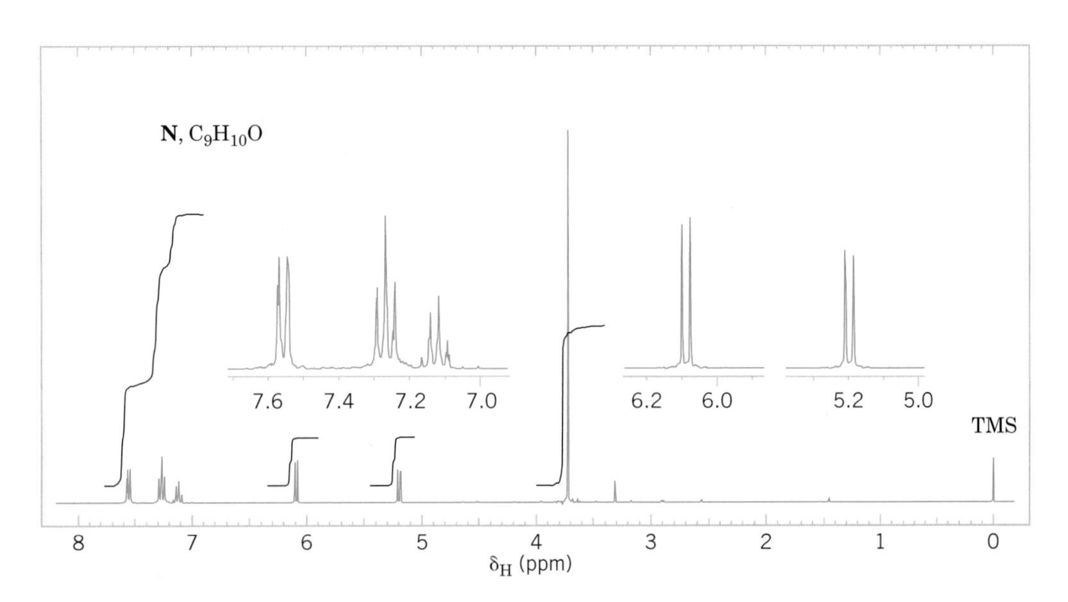

FIGURE 14.31 The 300-MHz ^{1}H NMR spectrum of compound N, Problem 14.30. Expansions of the signals are shown in the offset plots.

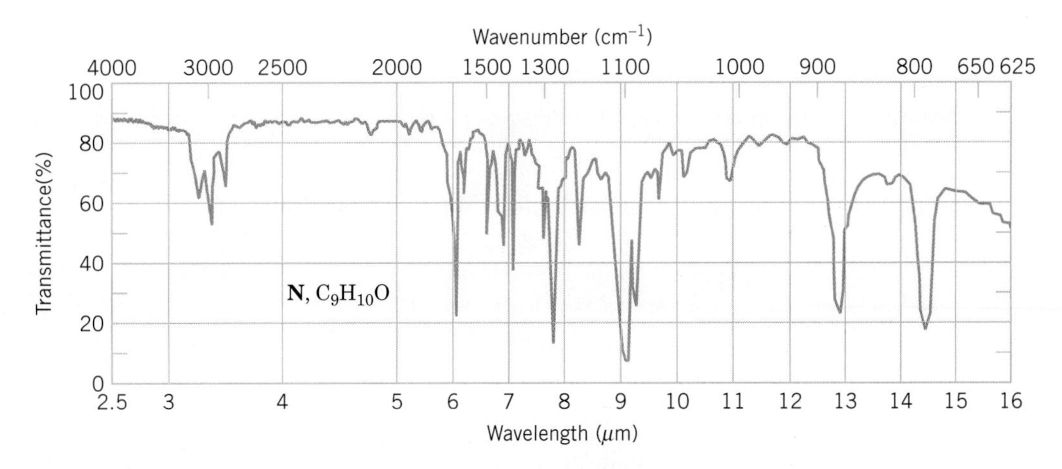

FIGURE 14.32 The IR spectrum of compound N, Problem 14.30. (Spectrum courtesy of Aldrich Chemical Co., Milwaukee, WI.)

14.31 The IR and ^{1}H NMR spectra for compound **X** (C_8H_{10}) are given in Fig. 14.33. Propose a structure for compound **X**.

14.32 The IR and ^{1}H NMR spectra of compound **Y** ($C_9H_{12}O$) are given in Fig. 14.34. Propose a structure for **Y**.

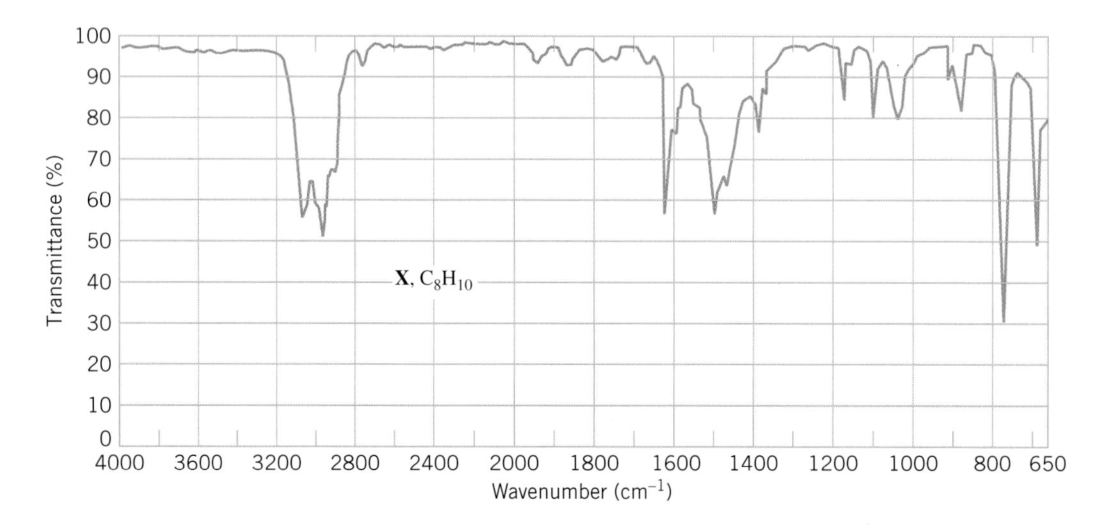

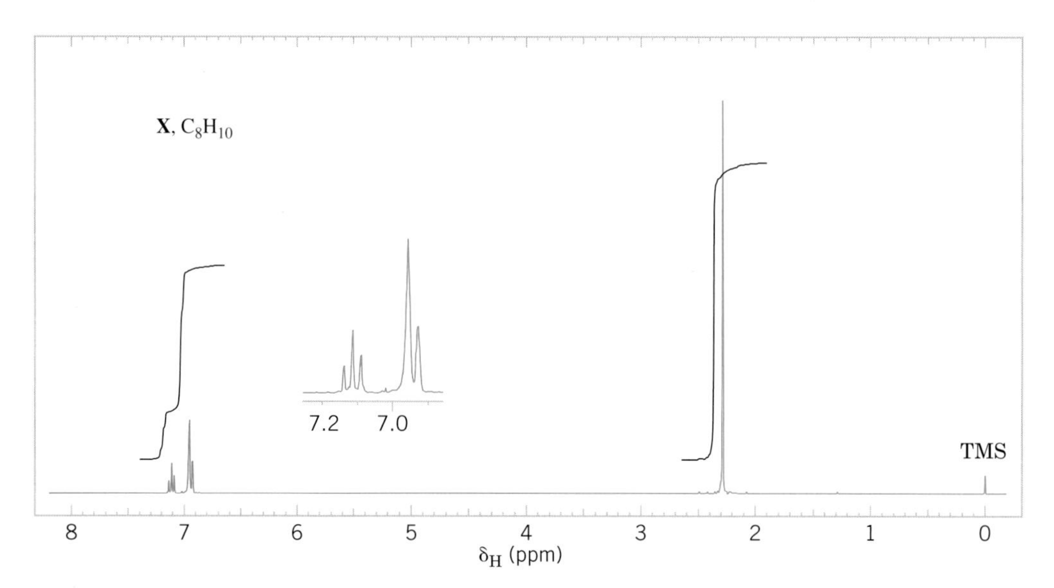

FIGURE 14.33 The IR and 300-MHz ^{1}H NMR spectra of compound X, Problem 14.31. Expansions of the signals are shown in the offset plots.

14.33 **(a)** How many peaks would you expect to find in the ^{1}H NMR spectrum of caffeine?

Caffeine

(b) What characteristic peaks would you expect to find in the IR spectrum of caffeine?

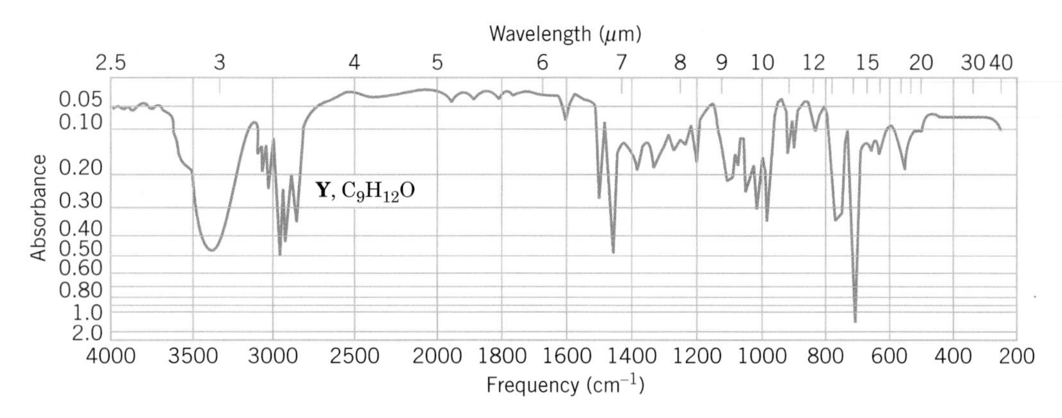

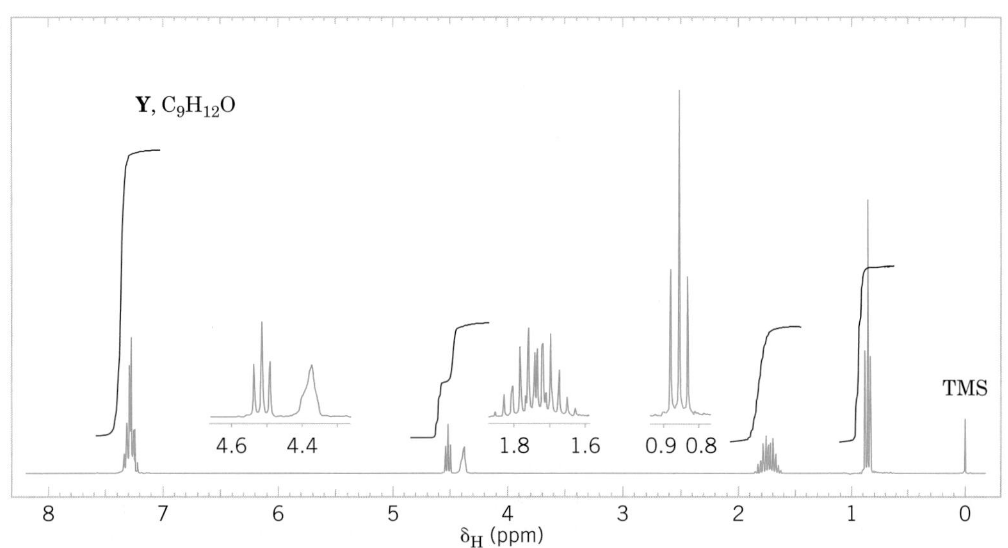

FIGURE 14.34 The IR and 300-MHz ¹H NMR spectra of compound Y, Problem 14.32. Expansions of the signals are shown in the offset plots.

***14.34** Given the following information, predict the appearance of the ¹H NMR spectrum arising from the vinyl hydrogen atoms of *p*-chlorostyrene.

(*c*) H, H (*b*), C=C, H (*a*), Cl

Deshielding by the induced magnetic field of the ring is greatest at proton (*c*) (δ 6.7) and is least at proton (*b*) (δ 5.3). The chemical shift of (*a*) is about δ 5.7. The coupling constants have the following approximate magnitudes: $J_{ac} \cong 18$ Hz, $J_{bc} \cong 11$ Hz, and $J_{ab} \cong 2$ Hz. (These coupling constants are typical of those given by vinylic systems: Coupling constants for trans hydrogen atoms are larger than those for cis hydrogen atoms, and coupling constants for geminal vinylic hydrogen atoms are very small.)

***14.35** Consider these reactions:

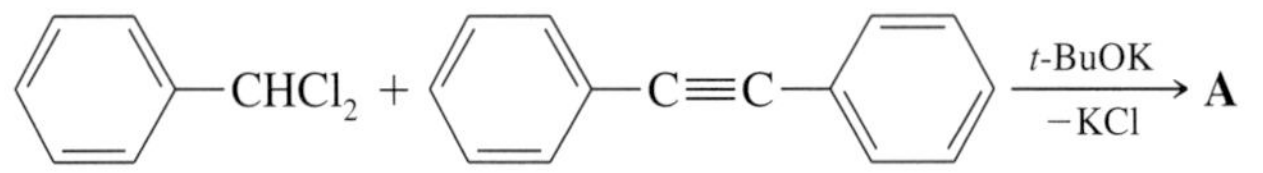

$$\mathbf{A} \xrightarrow[-t\text{-BuOH}]{\text{HBr}} \mathbf{B}$$

The intermediate **A** is a covalently bonded compound that has typical ^{1}H NMR peaks for aromatic ring hydrogens and only one additional peak at δ 1.21, with an area ratio of 5:3, respectively. Final product **B** is ionic and has only aromatic hydrogen peaks.

What are the structures of **A** and **B**?

***14.36** The final product of this sequence, **D**, is an orange, crystalline solid melting at 174°C and having molecular weight 186:

$$\text{Cyclopentadiene} + \text{Na} \longrightarrow \mathbf{C} + H_2$$
$$2\,\mathbf{C} + FeCl_2 \longrightarrow \mathbf{D} + 2\,NaCl$$

In its ^{1}H and ^{13}C NMR spectra, product **D** shows only one kind of hydrogen and only one kind of carbon, respectively.

Draw the structure of **C** and make a structural suggestion as to how the high degree of symmetry of **D** can be explained. (**D** belongs to a group of compounds named after something you might get at a deli for lunch.)

***14.37** Compound **E** has the spectral features given below. What is its structure?

MS (*m/z*): heaviest peak is at 202

IR (cm^{-1}): 3030–3080, 2150 (very weak), 1600, 1490, 760, and 690

^{1}H NMR (δ): narrow multiplet centered at 7.34

UV (nm): 287 (ϵ = 25,000), 305 (ϵ = 36,000), and 326 (ϵ = 33,000)

14.38 Draw all of the π molecular orbitals for (3*E*)-1,3,5-hexatriene, order them from lowest to highest in energy, and indicate the number of electrons that would be found in each in the ground state for the molecule. After doing so, open the computer molecular model for (3*E*)-1,3,5-hexatriene and display the calculated molecular orbitals. How well does the appearance and sequence of the orbitals you drew (e.g., number of nodes, overall symmetry of each, etc.) compare with the orbitals in the calculated model? Are the same orbitals populated with electrons in your analysis as in the calculated model?

Learning Group Problems

1. Write mechanism arrows for the following step in the chemical synthesis by A. Robertson and R. Robinson (*J. Chem. Soc.* **1928,** 1455–1472) of callistephin chloride, a red flower pigment from the purple-red aster. Explain why this transformation is a reasonable process.

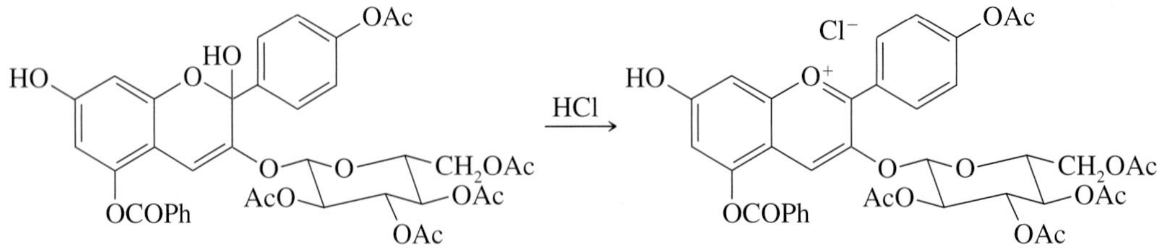

Callistephin chloride

2. The following reaction sequence was used by E. J. Corey (*J. Am. Chem. Soc.* **1969,** *91,* 5675–5677) at the beginning of a synthesis of prostaglandin $F_{2\alpha}$ and prostaglandin E_2. Explain what is involved in this reaction and why it is a reasonable process.

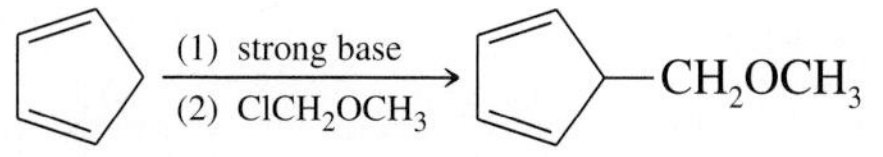

$$\text{cyclopentadiene} \xrightarrow[\text{(2) } ClCH_2OCH_3]{\text{(1) strong base}} \text{cyclopentadiene}-CH_2OCH_3$$

3. The ^1H NMR signals for the aromatic hydrogens of methyl *p*-hydroxybenzoate appear as two doublets at approximately 7.05 and 8.04 ppm (δ). Assign these two doublets to the respective hydrogens that produce each signal. Justify your assignments using arguments of relative electron density based on contributing resonance structures.

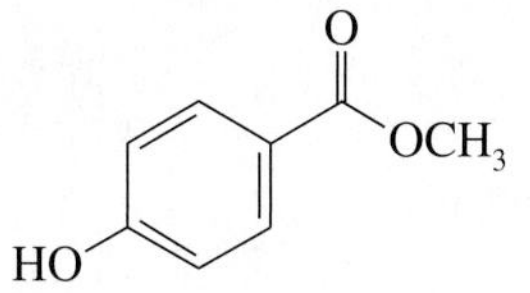

O
OCH$_3$
HO

4. Draw the structure of adenine, a heterocyclic aromatic compound incorporated in the structure of DNA. Identify the nonbonding electron pairs that are *not* part of the aromatic system in the rings of adenine. Which nitrogen atoms in the rings would you expect to be more basic and which should be less basic?

5. Draw structures of the nicotinamide ring in NADH and NAD^+. In the transformation of NADH to NAD^+, in what form must a hydrogen be transferred in order to produce the aromatic pyridinium ion in NAD^+?

CHAPTER 15

Reactions of Aromatic Compounds

Biosynthesis of Thyroxine: Aromatic Substitution Involving Iodine

Thyroxine (see the model above) is one of the key hormones involved in regulating metabolic rate. It causes an increase in the metabolism of carbohydrates, proteins, and lipids, as well as a general increase in oxygen consumption by most tissues. Low levels of thyroxine (hypothyroidism) can lead to obesity and lethargy. High levels of thyroxine (hyperthyroidism) can cause the opposite effects. The thyroid gland makes thyroxine from iodine and tyrosine, two essential components of our diet. Most of us obtain iodine from iodized salt, although foods with ingredients derived from kelp (see the photo above) are a significant source of iodine in the diet of some cultures. An abnormal level of thyroid hormones is a relatively common malady, however. Chronically low thyroxine levels (if untreated) can lead to a condition called goiter, in which the thyroid gland is enlarged. Fortunately, low levels of thyroxine are easily corrected by hormone supplements.

HO–(I,I-phenyl)–O–(I,I-phenyl)–CH_2–C(NH_3^+)(H)–CO_2^-

Thyroxine

Thyroxine biosynthesis, as noted above, requires tyrosine and iodine. The thyroid gland stores iodine and tyrosine in the form of a protein (a polymer of amino acids) called thyroglobulin. Each molecule of thyroglobulin contains 140 tyrosine units (as well as many other amino acids). Iodine is incorporated into approximately 20% of the tyrosine units in thyroglobulin. To introduce iodine into thyroglobulin, an enzyme called iodoperoxidase converts *nucleophilic* iodide anions from our diet (e.g., from iodized table salt) into an *electrophilic* iodine species. This electrophilic form of iodine reacts with tyrosine in thyroglobulin by a mechanism called *electrophilic aromatic substitution*—a topic central to this chapter. After we study electrophilic aromatic substitution, we shall return to thyroxine in "The Chemistry of. . . Iodine Incorporation in Thyroxine Biosynthesis" later in this chapter.

15.1 Electrophilic Aromatic Substitution Reactions

Aromatic hydrocarbons are known generally as **arenes.** An **aryl group** is one derived from an arene by removal of a hydrogen atom, and its symbol is Ar—. Thus, arenes are designated ArH just as alkanes are designated RH.

The most characteristic reactions of benzenoid arenes are the substitution reactions that occur when they react with electrophilic reagents. These reactions are of the general type shown below.

$$\text{Ar—H} + \text{E—A} \longrightarrow \text{Ar—E} + \text{H—A}$$

or (benzene) + E—A ⟶ (benzene)—E + H—A

The electrophiles are either a positive ion (E^+) or some other electron-deficient species with a large partial positive charge. As we shall learn in Section 15.3, for example, benzene can be brominated when it reacts with bromine in the presence of $FeBr_3$. Bromine and $FeBr_3$ react to produce positive bromine ions, Br^+. These positive bromine ions act as electrophiles and attack the benzene ring, replacing one of the hydrogen atoms in a reaction that is called an **electrophilic aromatic substitution** (EAS).

Electrophilic aromatic substitutions allow the direct introduction of a wide variety of groups onto an aromatic ring, and because of this, they provide synthetic routes to many important compounds. The five electrophilic aromatic substitutions that we shall study in this chapter are outlined in Fig. 15.1. In Sections 15.3–15.7 we shall learn what the electrophile is in each instance.

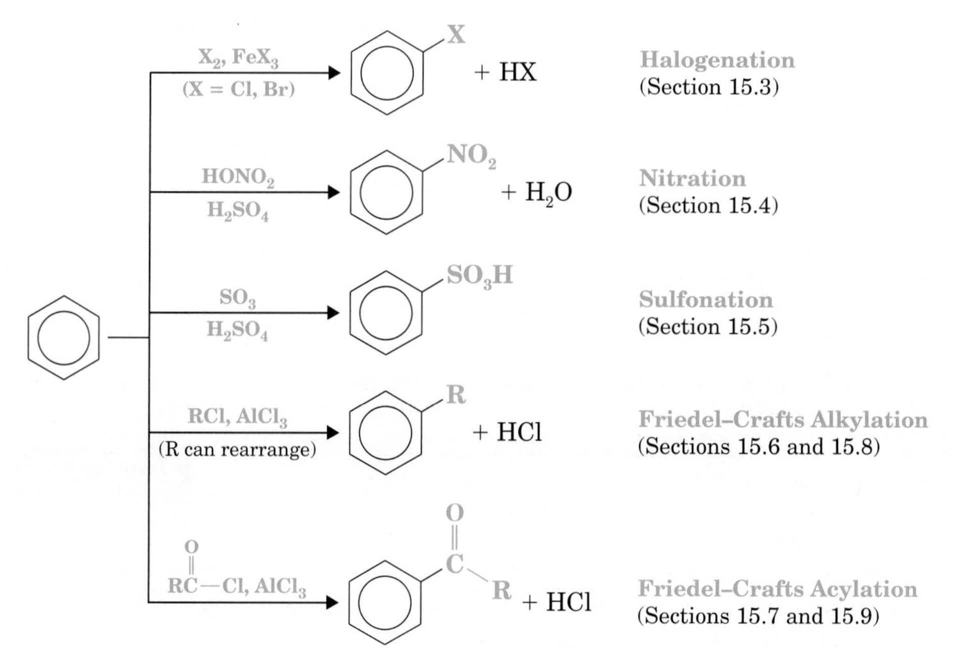

FIGURE 15.1 Electrophilic aromatic substitution reactions.

15.2 A General Mechanism for Electrophilic Aromatic Substitution: Arenium Ions

Benzene is susceptible to electrophilic attack primarily because of its exposed π electrons. In this respect benzene resembles an alkene, for in the reaction of an alkene with an electrophile the site of attack is the exposed π bond.

We saw in Chapter 14, however, that benzene differs from an alkene in a very significant way. Benzene's closed shell of six π electrons gives it a special stability. So although benzene is susceptible to electrophilic attack, it undergoes *substitution reactions* rather than *addition reactions*. Substitution reactions allow the aromatic sextet of π electrons to be regenerated after attack by the electrophile has occurred. We can see how this happens if we examine a general mechanism for electrophilic aromatic substitution.

STUDY TIP!

Resonance structures (like those used here for the arenium ion) will be important for our study of electrophilic aromatic substitution.

A considerable body of experimental evidence indicates that electrophiles attack the π system of benzene to form a ***nonaromatic cyclohexadienyl carbocation*** known as an **arenium ion.** In showing this step, it is convenient to use Kekulé structures, because these make it much easier to keep track of the π electrons:

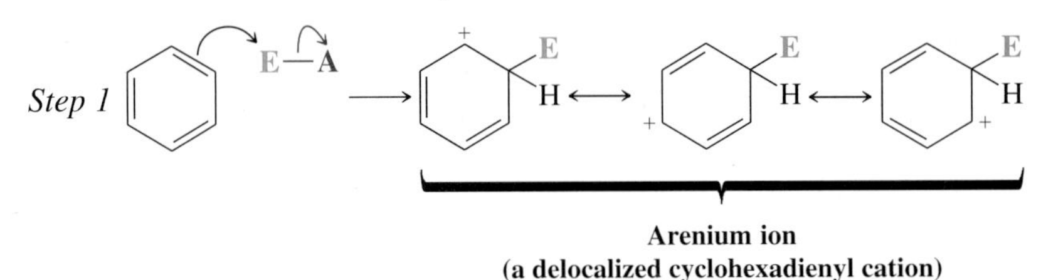

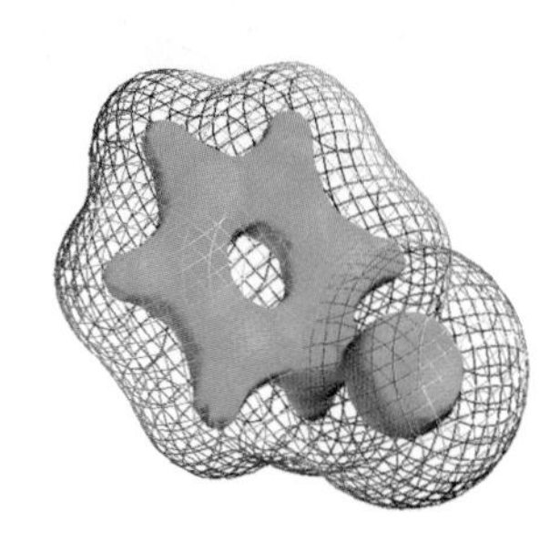

FIGURE 15.2 A calculated structure for the arenium ion intermediate formed by electrophilic addition of bromine to benzene (Section 15.3). The electrostatic potential map for the principal location of bonding electrons (indicated by the solid surface) shows that positive charge (blue) resides primarily at the ortho and para carbons relative to the carbon where the electrophile has bonded. This distribution of charge is consistent with the resonance model for an arenium ion. (The van der Waals surface is indicated by the wire mesh.)

In step 1 the electrophile takes two electrons of the six-electron π system to form a σ bond to one carbon atom of the benzene ring. Formation of this bond interrupts the cyclic system of π electrons, because in the formation of the arenium ion the carbon that forms a bond to the electrophile becomes sp^3 hybridized and, therefore, no longer has an available p orbital. Now only five carbon atoms of the ring are still sp^2 hybridized and still have p orbitals. The four π electrons of the arenium ion are delocalized through these five p orbitals. A calculated electrostatic potential map for the arenium ion formed by electrophilic addition of bromine to benzene indicates that positive charge is distributed in the arenium ion ring (Fig. 15.2), just as was shown in the contributing resonance structures.

In step 2 a proton is removed from the carbon atom of the arenium ion that bears the electrophile. The two electrons that bonded this proton to carbon become a part of the π system. The carbon atom that bears the electrophile becomes sp^2 hybridized again, and a benzene derivative with six fully delocalized π electrons is formed. We can represent step 2 with any one of the resonance structures for the arenium ion:

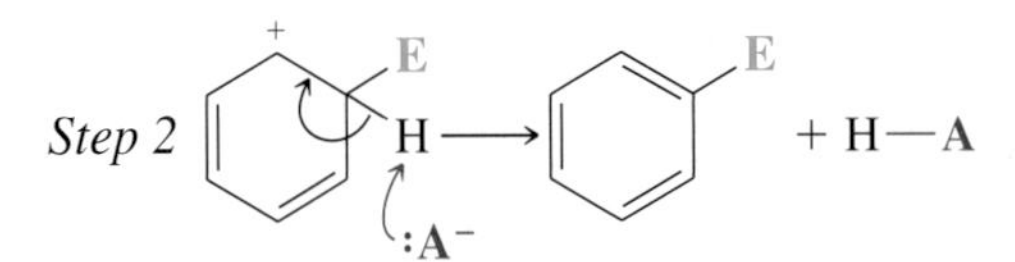

CD Animated Graphics

(The proton is removed by any of the bases present, for example, by the anion derived from the electrophile.)

PROBLEM 15.1

Show how loss of a proton can be represented using each of the three resonance structures for the arenium ion and show how each representation leads to the formation of a benzene ring with three alternating double bonds (i.e., six fully delocalized π electrons).

Kekulé structures are more appropriate for writing mechanisms such as electrophilic aromatic substitution because they permit the use of resonance theory, which, as we shall soon see, is invaluable as an aid to our understanding. If, for brevity, however, we wish to show the mechanism using the modern formula for benzene we can do it in the following way. We draw the arenium ion as a delocalized cyclohexadienyl cation:

In our color scheme for chemical formulas blue generally indicates groups that are electrophilic or have electron-withdrawing character. Red indicates groups that are or become Lewis bases, or have electron-donating character.

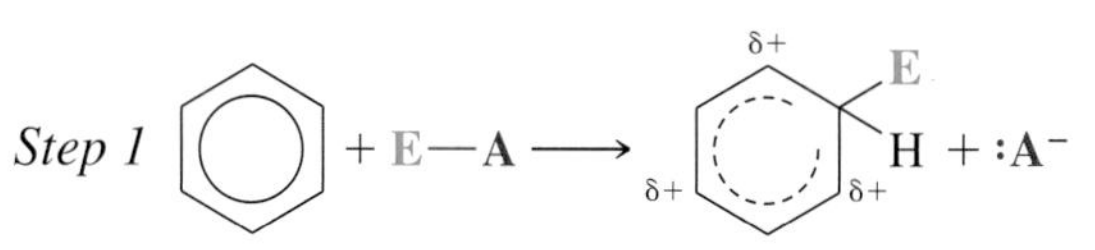

Step 2 (arenium ion, δ+ δ+ δ+, E, H, :A⁻) ⟶ (benzene–E) + H—A

There is firm experimental evidence that the arenium ion is a true *intermediate* in electrophilic substitution reactions. It is not a transition state. This means that in a free-energy diagram (Fig. 15.3) the arenium ion lies in an energy valley between two transition states.

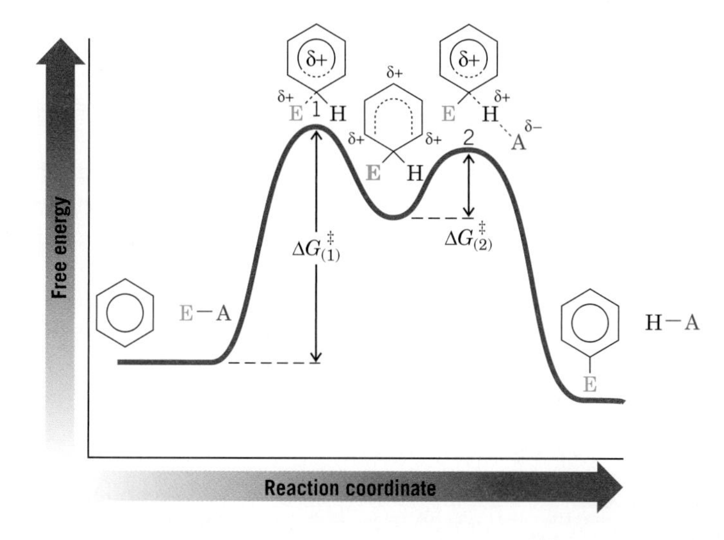

FIGURE 15.3 The free-energy diagram for an electrophilic aromatic substitution reaction. The arenium ion is a true intermediate lying between transition states 1 and 2. In transition state 1 the bond between the electrophile and one carbon atom of the benzene ring is only partially formed. In transition state 2 the bond between the same benzene carbon atom and its hydrogen atom is partially broken. The bond between the hydrogen atom and the conjugate base is partially formed.

The free energy of activation, $\Delta G^{\ddagger}_{(1)}$, for the reaction leading from benzene and the electrophile, E^+, to the arenium ion has been shown to be much greater than the free energy of activation, $\Delta G^{\ddagger}_{(2)}$, leading from the arenium ion to the final product. This is consistent with what we would expect. The reaction leading from benzene and an electrophile to the arenium ion is highly endothermic, because the benzene ring loses its resonance energy. The reaction leading from the arenium ion to the substituted benzene, by contrast, is highly exothermic because in it the benzene ring regains its resonance energy.

Of the following two steps, step 1—the formation of the arenium ion—is usually the rate-determining step in electrophilic aromatic substitution:

Step 1 (benzene) + E—A ⟶ (arenium ion, δ+ δ+ δ+, H E) + :A⁻ **Slow, rate determining**

Step 2 (arenium ion, δ+ δ+ δ+, H E, :A⁻) ⟶ (benzene–E) + H—A **Fast**

Step 2, the removal of a proton, occurs rapidly relative to step 1 and has no effect on the overall rate of reaction.

15.3 Halogenation of Benzene

Benzene does not react with bromine or chlorine unless a Lewis acid is present in the mixture. (As a consequence, benzene does not decolorize a solution of bromine in carbon tetrachloride.) When Lewis acids are present, however, benzene reacts readily with bromine or chlorine, and the reactions give bromobenzene and chlorobenzene, respectively, in good yields:

$$C_6H_6 + Cl_2 \xrightarrow[25^\circ C]{FeCl_3} C_6H_5Cl + HCl$$

Chlorobenzene (90%)

$$C_6H_6 + Br_2 \xrightarrow[heat]{FeBr_3} C_6H_5Br + HBr$$

Bromobenzene (75%)

The Lewis acids most commonly used to effect chlorination and bromination reactions are $FeCl_3$, $FeBr_3$, and $AlCl_3$, all in the anhydrous form. Ferric chloride and ferric bromide are usually generated in the reaction mixture by adding iron to it. The iron then reacts with halogen to produce the ferric halide:

$$2\ Fe + 3\ X_2 \longrightarrow 2\ FeX_3$$

The mechanism for electrophilic aromatic bromination is as follows:

A Mechanism for the Reaction

Electrophillic Aromatic Bromination

Step 1 $:\ddot{Br}-\ddot{Br}: + FeBr_3 \longrightarrow :\ddot{Br}-\overset{+}{\ddot{Br}}-\overset{-}{FeBr_3} \longrightarrow :\ddot{Br}^+ + :\ddot{Br}-\overset{-}{FeBr_3}$

Bromine combines with $FeBr_3$ to form a complex that dissociates to form a positive bromine ion and $FeBr_4^-$.

Step 2 $C_6H_6 + \overset{+}{\ddot{Br}}: \xrightarrow{slow}$ [arenium ion resonance structures] $\longleftrightarrow$ $\longleftrightarrow$

Arenium ion

The positive bromine ion attacks benzene to form an arenium ion.

An electrostatic potential map for this arenium ion is shown in Figure 15.2.

Step 3 [arenium ion] $+ :\ddot{Br}-\overset{-}{FeBr_3} \longrightarrow C_6H_5\ddot{Br}: + H-\ddot{Br}: + FeBr_3$

A proton is removed from the arenium ion to become bromobenzene.

The function of the Lewis acid can be seen in step 1. The ferric bromide reacts with bromine to produce a positive bromine ion, Br^+ (and $FeBr_4^-$). In step 2 this Br^+ ion attacks the benzene ring to produce an arenium ion. Then, finally in step 3 a proton is removed from the arenium ion by $FeBr_4^-$. This results in the formation of bromobenzene and hydrogen bromide, the products of the reaction. At the same time this step regenerates the catalyst, $FeBr_3$.

The mechanism of the chlorination of benzene in the presence of ferric chloride is analogous to the one for bromination. Ferric chloride serves the same purpose in aromatic chlorinations as ferric bromide does in aromatic brominations. It assists in the generation and transfer of a positive halogen ion.

Fluorine reacts so rapidly with benzene that aromatic fluorination requires special conditions and special types of apparatus. Even then, it is difficult to limit the reaction to monofluorination. Fluorobenzene can be made, however, by an indirect method that we shall see in Section 20.8D.

Iodine, on the other hand, is so unreactive that a special technique has to be used to effect direct iodination; the reaction has to be carried out in the presence of an oxidizing agent such as nitric acid:

An electrophilic iodination reaction in thyroxine biosynthesis was mentioned in the opening vignette for this chapter. It is discussed further in "The Chemistry of..." box after Section 15.11F.

$$C_6H_6 + I_2 \xrightarrow{HNO_3} C_6H_5I \quad (86\%)$$

15.4 Nitration of Benzene

Benzene reacts slowly with hot concentrated nitric acid to yield nitrobenzene. The reaction is much faster if it is carried out by heating benzene with a mixture of concentrated nitric acid and concentrated sulfuric acid.

$$C_6H_6 + HNO_3 + H_2SO_4 \xrightarrow{50-55^\circ C} C_6H_5NO_2 \; (85\%) + H_3O^+ + HSO_4^-$$

Concentrated sulfuric acid increases the rate of the reaction by increasing the concentration of the electrophile, the nitronium ion (NO_2^+), as shown in the first two steps of the following mechanism.

A Mechanism for the Reaction

Nitration of Benzene

Step 1 $$HO_3SO{-}H + H{-}\ddot{O}{-}N^+(=O)(O^-) \rightleftharpoons H{-}O^+(H){-}N^+(=O)(O^-) + HSO_4^-$$

(H_2SO_4)

In this step nitric acid accepts a proton from the stronger acid, sulfuric acid.

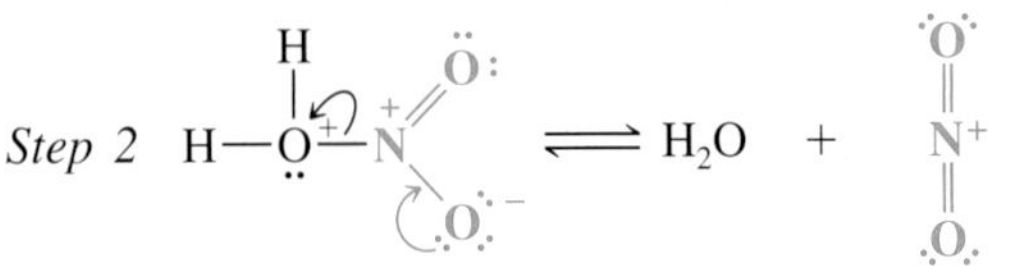

Step 2 $H-\ddot{O}^{+}(H)-NO_2 \rightleftharpoons H_2O + O{=}N^{+}{=}O$

Nitronium ion

Now that it is protonated, nitric acid can dissociate to form a nitronium ion.

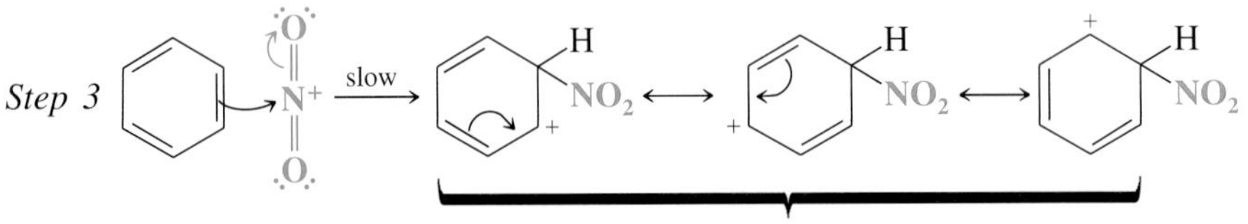

Step 3 benzene + NO_2^+ $\xrightarrow{\text{slow}}$ arenium ion (three resonance structures with H and NO_2)

Arenium ion

The nitronium ion is the actual electrophile in nitration; it reacts with benzene to form a resonance-stabilized arenium ion.

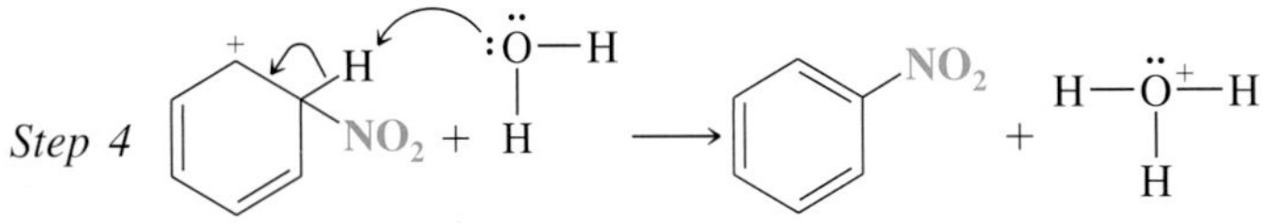

Step 4 arenium ion + $H-\ddot{O}-H$ $\longrightarrow$ nitrobenzene + $H-\ddot{O}^{+}(H)-H$

The arenium ion then loses a proton to a Lewis base and becomes nitrobenzene.

PROBLEM 15.2 Given that the pK_a of H_2SO_4 is -9 and that of HNO_3 is -1.4, explain why nitration occurs more rapidly in a mixture of concentrated nitric and sulfuric acids rather than in concentrated nitric acid alone.

15.5 Sulfonation of Benzene

Benzene reacts with fuming sulfuric acid at room temperature to produce benzenesulfonic acid. Fuming sulfuric acid is sulfuric acid that contains added sulfur trioxide (SO_3). Sulfonation also takes place in concentrated sulfuric acid alone, but more slowly:

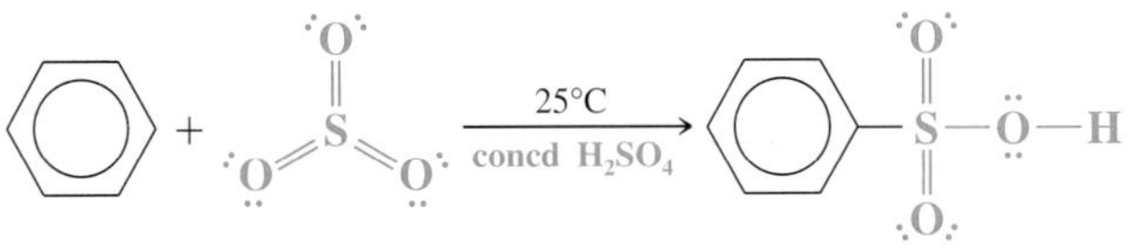

$$C_6H_6 + SO_3 \xrightarrow[\text{concd } H_2SO_4]{25°C} C_6H_5SO_3H$$

Sulfur trioxide **Benzenesulfonic acid (56%)**

In either reaction the electrophile appears to be sulfur trioxide. In concentrated sulfuric acid, sulfur trioxide is produced in an equilibrium in which H_2SO_4 acts as both an acid and a base (see step 1 of the following mechanism).

A Mechanisms for the Reaction

Sulfonation of Benzene

Step 1 $2\,H_2SO_4 \rightleftharpoons SO_3 + H_3O^+ + HSO_4^-$

This equilibrium produces SO_3 in concentrated H_2SO_4.

Step 2 Benzene + SO_3 $\xrightleftharpoons{\text{slow}}$ arenium ion ⟷ other resonance structures

SO_3 is the actual electrophile that reacts with benzene to form an arenium ion.

Step 3 HSO_4^- + arenium ion $\xrightleftharpoons{\text{fast}}$ benzenesulfonate ion + H_2SO_4

A proton is removed from the arenium ion to form the benzenesulfonate ion.

Step 4 benzenesulfonate ion + H_3O^+ $\xrightleftharpoons{\text{fast}}$ benzenesulfonic acid + H_2O

The benzenesulfonate ion accepts a proton to become benzenesulfonic acid.

All of the steps are equilibria, including step 1, in which sulfur trioxide is formed from sulfuric acid. This means that the overall reaction is an equilibrium as well. In concentrated sulfuric acid, the overall equilibrium is the sum of steps 1–4:

$$C_6H_6 + H_2SO_4 \rightleftharpoons C_6H_5SO_3H + H_2O$$

In fuming sulfuric acid, step 1 is unimportant because the dissolved sulfur trioxide reacts directly.

Because all of the steps are equilibria, the position of equilibrium can be influenced by the conditions we employ. If we want to sulfonate benzene, we use concentrated sulfuric acid or—better yet—fuming sulfuric acid. Under these conditions the position of equilibrium lies appreciably to the right, and we obtain benzenesulfonic acid in good yield.

On the other hand, we may want to remove a sulfonic acid group from a benzene ring. To do this, we employ dilute sulfuric acid and usually pass steam through the mixture. Under these conditions—with a high concentration of water—the equilibrium lies appreciably to the left and desulfonation occurs. The equilibrium is shifted even further to the left with volatile aromatic compounds because the aromatic compound distills with the steam.

Sulfonation–desulfonation is a useful tool in syntheses involving electrophilic aromatic substitution.

We shall see later that sulfonation and desulfonation reactions are often used in synthetic work. We may, for example, introduce a sulfonic acid group into a benzene ring to influence the course of some further reaction. Later, we may remove the sulfonic acid group by desulfonation.

15.6 Friedel–Crafts Alkylation

In 1877 a French chemist, Charles Friedel, and his American collaborator, James M. Crafts, discovered new methods for the preparation of alkylbenzenes (ArR) and acylbenzenes (ArCOR). These reactions are now called the Friedel–Crafts alkylation and acylation reactions. We shall study the Friedel–Crafts alkylation reaction here and take up the Friedel–Crafts acylation reaction in Section 15.7.

Charles Friedel

James Mason Crafts

A general equation for a Friedel–Crafts alkylation reaction is the following:

$$C_6H_6 + R{-}X \xrightarrow{AlCl_3} C_6H_5{-}R + HX$$

The mechanism for the reaction (shown in the following steps, with isopropyl chloride as R—X) starts with the formation of a carbocation (step 1). The carbocation then acts as an electrophile (step 2) and attacks the benzene ring to form an arenium ion. The arenium ion (step 3) then loses a proton to generate isopropylbenzene.

A Mechanism for the Reaction

Friedel–Crafts Alkylation

Step 1

$$(H_3C)_2CH{-}\ddot{Cl}: + AlCl_3 \rightleftharpoons$$

This is a Lewis acid–base reaction (see Section 3.2B).

$$(H_3C)_2CH{-}\overset{+}{Cl}{-}\overset{-}{Al}Cl_3 \rightleftharpoons (H_3C)_2\overset{+}{C}H + :\ddot{Cl}{-}\overset{-}{Al}Cl_3$$

The complex dissociates to form a carbocation and $AlCl_4^-$.

Step 2

$$C_6H_6 + {}^{+}CH(CH_3)_2 \rightleftharpoons \text{arenium ion } (C_6H_6^{+}{-}CH(CH_3)_2) \longleftrightarrow \text{other resonance structures}$$

The carbocation, acting as an electrophile, reacts with benzene to produce an arenium ion.

Step 3

$$\text{arenium ion} + :\ddot{Cl}{-}\overset{-}{Al}Cl_3 \rightleftharpoons C_6H_5{-}CH(CH_3)_2 + HCl + AlCl_3$$

A proton is removed from the arenium ion to form isopropylbenzene. This step also regenerates the $AlCl_3$ and liberates HCl.

When R—X is a primary halide, a simple carbocation probably does not form. Instead, the aluminum chloride forms a complex with the alkyl halide, and this complex acts as the electrophile. The complex is one in which the carbon–halogen bond is nearly broken—and one in which the carbon atom has a considerable positive charge:

$$\overset{\delta+}{RCH_2}\text{----}\ddot{\underset{..}{Cl}}\!:\!\overset{\delta-}{AlCl_3}$$

Even though this complex is not a simple carbocation, it acts as if it were and it transfers a positive alkyl group to the aromatic ring. As we shall see in Section 15.8, these complexes react so much like carbocations that they also undergo typical carbocation rearrangements.

Friedel–Crafts alkylations are not restricted to the use of alkyl halides and aluminum chloride. Many other pairs of reagents that form carbocations (or species like carbocations) may be used as well. These possibilities include the use of a mixture of an alkene and an acid:

$$C_6H_6 + CH_3CH{=}CH_2 \xrightarrow[HF]{0°C} C_6H_5CH(CH_3)_2$$

Propene → **Isopropylbenzene (cumene) (84%)**

$$C_6H_6 + \text{cyclohexene} \xrightarrow[HF]{0°C} \text{cyclohexylbenzene}$$

Cyclohexene → **Cyclohexylbenzene (62%)**

A mixture of an alcohol and an acid may also be used:

$$C_6H_6 + HO\text{–}C_6H_{11} \xrightarrow[BF_3]{60°C} C_6H_5\text{–}C_6H_{11} + H_2O$$

Cyclohexanol → **Cyclohexylbenzene (56%)**

There are several important limitations of the Friedel–Crafts reaction. These are discussed in Section 15.8.

PROBLEM 15.3

Outline all steps in a reasonable mechanism for the formation of isopropylbenzene from propene and benzene in liquid HF (just shown). Your mechanism must account for the product being isopropylbenzene, not propylbenzene.

15.7 Friedel–Crafts Acylation

The $RC(=O)\text{—}$ group is called an **acyl group,** and a reaction whereby an acyl group is introduced into a compound is called an **acylation** reaction. Two common acyl groups are the acetyl group and the benzoyl group. (The benzoyl group is not to be confused with the benzyl group, $\text{—}CH_2C_6H_5$, see Section 14.2.)

$$CH_3C(=O)\text{—} \qquad C_6H_5C(=O)\text{—}$$

Acetyl group (ethanoyl group) **Benzoyl group**

The Friedel–Crafts acylation reaction is an effective means of introducing an acyl group into an aromatic ring. The reaction is often carried out by treating the aromatic compound with an acyl chloride. Unless the aromatic compound is one that is highly reactive, the reaction requires the addition of at least one equivalent of a Lewis acid (such as $AlCl_3$) as well. The product of the reaction is an aryl ketone:

$$C_6H_6 + CH_3C(=O){-}Cl \xrightarrow[\text{excess benzene, } 80^\circ C]{AlCl_3} C_6H_5C(=O)CH_3 + HCl$$

Acetyl chloride → Acetophenone (methyl phenyl ketone) (97%)

Acyl chlorides, also called **acid chlorides,** are easily prepared (Section 18.5) by treating carboxylic acids with thionyl chloride ($SOCl_2$) or phosphorus pentachloride (PCl_5):

$$CH_3C(=O)OH + SOCl_2 \xrightarrow{80^\circ C} CH_3C(=O)Cl + SO_2 + HCl$$

Acetic acid + Thionyl chloride → Acetyl chloride (80–90%)

$$C_6H_5C(=O)OH + PCl_5 \longrightarrow C_6H_5C(=O)Cl + POCl_3 + HCl$$

Benzoic acid + Phosphorus pentachloride → Benzoyl chloride

Friedel–Crafts acylations can also be carried out using carboxylic acid anhydrides. For example,

$$C_6H_6 + (CH_3C(=O))_2O \xrightarrow[\text{excess benzene, } 80^\circ C]{AlCl_3} C_6H_5C(=O)CH_3 + CH_3C(=O)OH$$

Acetic anhydride (a carboxylic acid anhydride) → Acetophenone (82–85%)

In most Friedel–Crafts acylations the electrophile appears to be an **acylium ion** formed from an acyl halide in the following way:

Step 1 $$R{-}C(=\ddot{O}){-}\ddot{Cl}: + AlCl_3 \rightleftharpoons R{-}C(=\ddot{O}){-}\overset{+}{\ddot{Cl}}{:}\overset{-}{Al}Cl_3$$

Step 2 $$R{-}C(=\ddot{O}){-}\overset{+}{\ddot{Cl}}{:}\overset{-}{Al}Cl_3 \rightleftharpoons R{-}\overset{+}{C}{=}\ddot{O}: \longleftrightarrow R{-}C{\equiv}\overset{+}{O}: + \overset{-}{Al}Cl_4$$

An acylium ion (a resonance hybrid)

PROBLEM 15.4

Show how an acylium ion could be formed from acetic anhydride in the presence of $AlCl_3$.

The remaining steps in the Friedel–Crafts acylation of benzene are the following:

A Mechanism for the Reaction

Friedel–Crafts Acylation

Step 3 — other resonance structures

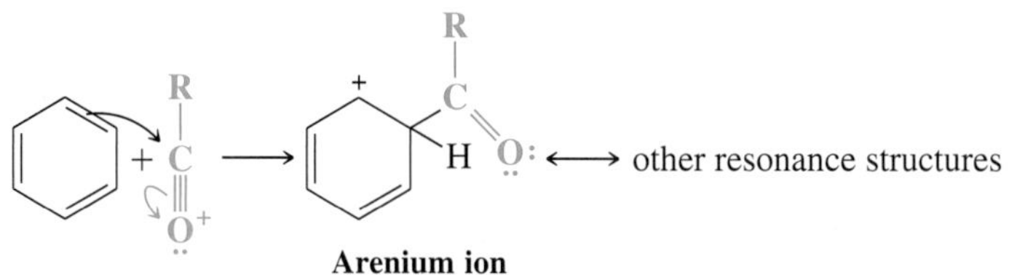

Arenium ion

The acylium ion, acting as an electrophile, reacts with benzene to form the arenium ion.

Step 4 — $\longrightarrow$ + HCl + $AlCl_3$

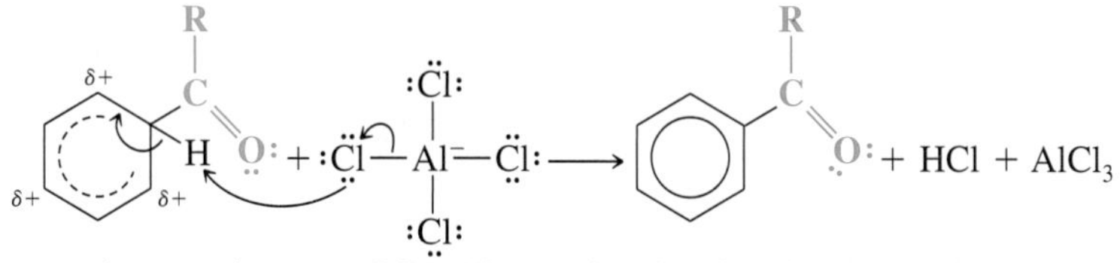

A proton is removed from the arenium ion, forming the aryl ketone.

Step 5 — + $AlCl_3$ $\rightleftharpoons$ $:AlCl_3$

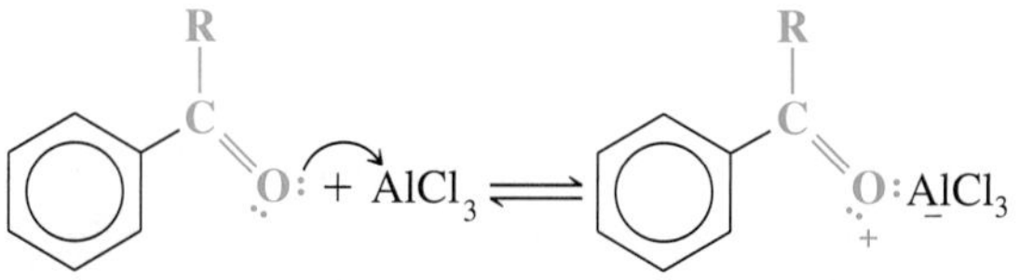

The ketone, acting as a Lewis base, reacts with aluminum chloride (a Lewis acid) to form a complex.

Step 6

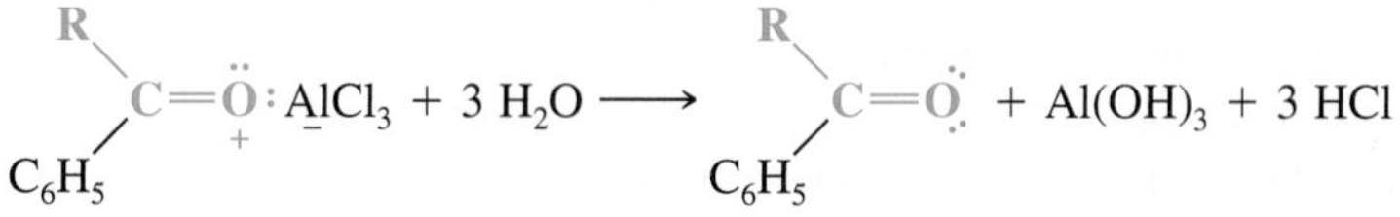

$$\mathrm{R(C_6H_5)C{=}\overset{+}{O}{:}\overset{-}{Al}Cl_3 + 3\,H_2O \longrightarrow R(C_6H_5)C{=}O + Al(OH)_3 + 3\,HCl}$$

Treating the complex with water liberates the ketone and hydrolyzes the Lewis acid.

Several important synthetic applications of the Friedel–Crafts reaction are given in Section 15.9.

15.8 Limitations of Friedel–Crafts Reactions

Several restrictions limit the usefulness of Friedel–Crafts reactions:

1. **When the carbocation formed from an alkyl halide, alkene, or alcohol can rearrange to a more stable carbocation, it usually does so, and the major product obtained from the reaction is usually the one from the more stable carbocation.**

When benzene is alkylated with butyl bromide, for example, some of the developing butyl cations rearrange by a hydride shift—some developing 1° carbocations (see following reactions) become more stable 2° carbocations. Then benzene reacts with both kinds of carbocations to form both butylbenzene and *sec*-butylbenzene:

$$CH_3CH_2CH_2CH_2Br \xrightarrow{AlCl_3} CH_3CH_2\underset{H}{CH}\overset{\delta+}{CH_2}\cdots\overset{\delta-}{Br\bar{A}lCl_3} \xrightarrow[(-BrAlCl_3^-)]{} CH_3CH_2\overset{+}{C}HCH_3$$

Benzene, $(-AlCl_3)$, $(-HBr)$ → $CH_3CH_2CH_2CH_2$–benzene

Benzene, $(-H^+)$ → $CH_3CH_2CHCH_3$–benzene

Butylbenzene
(32–36% of mixture)

***sec*-Butylbenzene**
(64–68% of mixture)

2. **Friedel–Crafts reactions usually give poor yields when powerful electron-withdrawing groups** (Section 15.11) **are present on the aromatic ring or when the ring bears an $-NH_2$, $-NHR$, or $-NR_2$ group.** This applies to both alkylations and acylations.

NO_2, $\overset{+}{N}(CH_3)_3$, $COOH$, COR, CF_3, SO_3H, NH_2

These usually give poor yields in Friedel–Crafts reactions.

We shall learn in Section 15.10 that groups present on an aromatic ring can have a large effect on the reactivity of the ring toward electrophilic aromatic substitution. Electron-withdrawing groups make the ring less reactive by making it electron deficient. Any substituent more electron withdrawing (or deactivating) than a halogen, that is, **any meta-directing group** (Section 15.11C), **makes an aromatic ring too electron deficient to undergo a Friedel–Crafts reaction.** The amino groups, $-NH_2$, $-NHR$, and $-NR_2$, are changed into powerful electron-withdrawing groups by the Lewis acids used to catalyze Friedel–Crafts reactions. For example,

$$C_6H_5\ddot{N}H_2 + AlCl_3 \longrightarrow C_6H_5\overset{+}{N}H_2-\bar{A}lCl_3$$

Does not undergo a Friedel–Crafts reaction.

3. **Aryl and vinylic halides cannot be used as the halide component because they do not form carbocations readily** (see Section 6.14A):

$$C_6H_6 \xrightarrow{C_6H_5Cl,\ AlCl_3} \text{no Friedel–Crafts reaction}$$

$$C_6H_6 \xrightarrow{\text{Cl}-\text{C}{=}\text{C},\ AlCl_3} \text{no Friedel–Crafts reaction}$$

4. **Polyalkylations often occur.** Alkyl groups are electron-releasing groups, and once one is introduced into the benzene ring, it activates the ring toward further substitution (see Section 15.10):

$$C_6H_6 + (H_3C)_2CHOH \xrightarrow[60°C]{BF_3} C_6H_5CH(CH_3)_2 + p\text{-}C_6H_4[CH(CH_3)_2]_2$$

Isopropylbenzene (24%)   ***p*-Diisopropylbenzene (14%)**

Polyacylations are not a problem in Friedel–Crafts acylations, however. The acyl group (RCO—) by itself is an electron-withdrawing group, and when it forms a complex with $AlCl_3$ in the last step of the reaction (Section 15.7), it is made even more electron withdrawing. This strongly inhibits further substitution and makes monoacylation easy.

PROBLEM 15.5

When benzene reacts with neopentyl chloride, $(CH_3)_3CCH_2Cl$, in the presence of aluminum chloride, the major product is 2-methyl-2-phenylbutane, not neopentylbenzene. Explain this result.

PROBLEM 15.6

When benzene reacts with propyl alcohol in the presence of boron trifluoride, both propylbenzene and isopropylbenzene are obtained as products. Write a mechanism that accounts for this.

15.9 SYNTHETIC APPLICATIONS OF FRIEDEL–CRAFTS ACYLATIONS: THE CLEMMENSEN REDUCTION

Rearrangements of the carbon chain do not occur in Friedel–Crafts acylations. The acylium ion, because it is stabilized by resonance, is more stable than most other carbocations. Thus, there is no driving force for a rearrangement. Because rearrangements do not occur, Friedel–Crafts acylations followed by reduction of the carbonyl group to a CH_2 group often give us much better routes to unbranched alkylbenzenes than do Friedel–Crafts alkylations.

As an example, let us consider the problem of synthesizing propylbenzene. If we attempt this synthesis through a Friedel–Crafts alkylation, a rearrangement occurs and the major product is isopropylbenzene (see also Problem 15.6):

$$C_6H_6 + CH_3CH_2CH_2Br \xrightarrow{AlCl_3} C_6H_5CH(CH_3)_2 + C_6H_5CH_2CH_2CH_3 + HBr$$

Isopropylbenzene (major product)   **Propylbenzene (minor product)**

By contrast, the Friedel–Crafts acylation of benzene with propanoyl chloride produces a ketone with an unrearranged carbon chain in excellent yield:

$$C_6H_6 + CH_3CH_2\overset{O}{\overset{\|}{C}}Cl \xrightarrow{AlCl_3} C_6H_5\overset{O}{\overset{\|}{C}}CH_2CH_3 + HCl$$

Propanoyl chloride → **Ethyl phenyl ketone (90%)**

Friedel–Crafts *acylation* followed by *ketone reduction* is the synthetic equivalent of Friedel–Crafts *alkylation.*

This ketone can then be reduced to propylbenzene by several methods. One general method—called the **Clemmensen reduction**—consists of refluxing the ketone with hydrochloric acid containing amalgamated zinc. [*Caution:* As we shall discuss later (Section 20.5B), zinc and hydrochloric acid will also reduce nitro groups to amino groups.]

$$C_6H_5\overset{O}{\overset{\|}{C}}CH_2CH_3 \xrightarrow[\text{HCl, reflux}]{Zn(Hg)} C_6H_5CH_2CH_2CH_3$$

Ethyl phenyl ketone → **Propylbenzene (80%)**

$$Ar\overset{O}{\overset{\|}{C}}R \xrightarrow[\text{HCl, reflux}]{Zn(Hg)} ArCH_2R$$

When cyclic anhydrides are used as one component, the Friedel–Crafts acylation provides a means of adding a new ring to an aromatic compound. One illustration is shown here. Note that only the ketone is reduced in the Clemmensen reduction step. The carboxylic acid is unaffected:

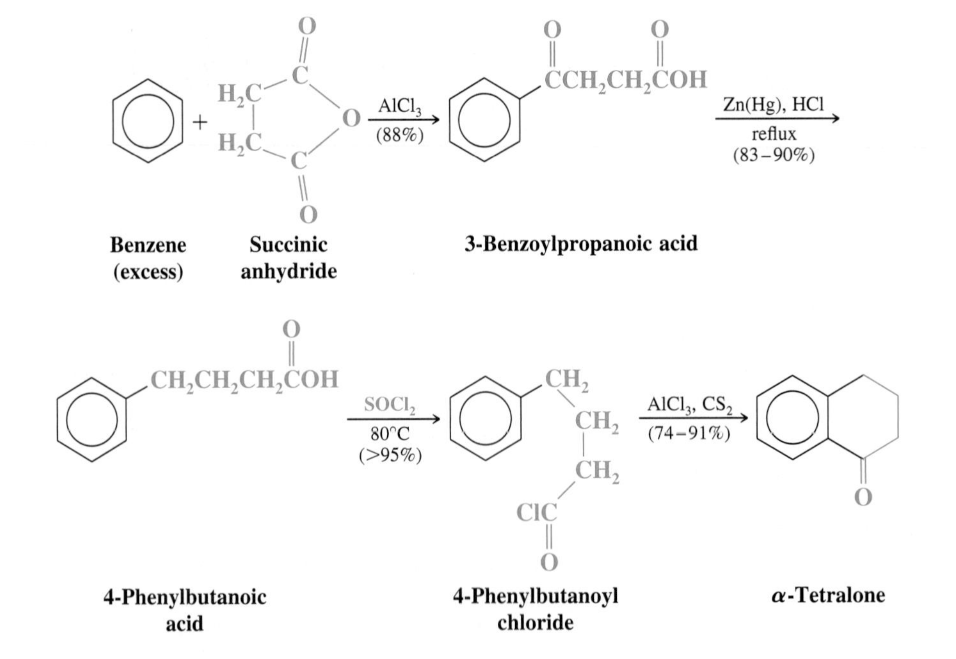

PROBLEM 15.7

Starting with benzene and the appropriate acyl chloride or acid anhydride, outline a synthesis of each of the following:

(a) Butylbenzene
(b) $(CH_3)_2CHCH_2CH_2C_6H_5$
(c) Benzophenone ($C_6H_5COC_6H_5$)
(d) 9,10-Dihydroanthracene

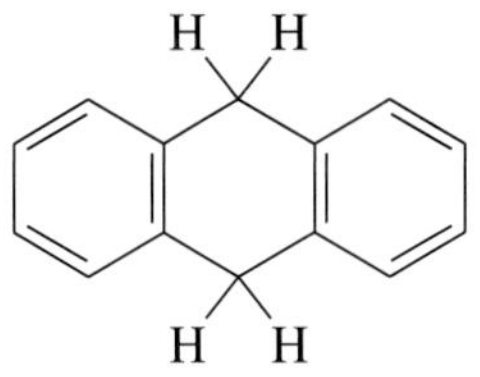

9,10-Dihydroanthracene

15.10 Effect of Substituents on Reactivity and Orientation

When substituted benzenes undergo electrophilic attack, groups already on the ring affect both the rate of the reaction and the site of attack. We say, therefore, that substituent groups affect both **reactivity** and **orientation** in electrophilic aromatic substitutions.

We can divide substituent groups into two classes according to their influence on the reactivity of the ring. Those that cause the ring to be more reactive than benzene itself we call **activating groups.** Those that cause the ring to be less reactive than benzene we call **deactivating groups.**

We also find that we can divide substituent groups into two classes according to the way they influence the orientation of attack by the incoming electrophile. Substituents in one class tend to bring about electrophilic substitution primarily at the positions ***ortho*** and ***para*** to themselves. We call these groups **ortho–para directors** because they tend to *direct* the incoming group into the ortho and para positions. Substituents in the second category tend to direct the incoming electrophile to the ***meta*** position. We call these groups **meta directors.**

Several examples will illustrate more clearly what we mean by these terms.

15.10A Activating Groups: Ortho–Para Directors

The methyl group is an **activating** group and an **ortho–para director.** Toluene reacts considerably faster than benzene in all electrophilic substitutions:

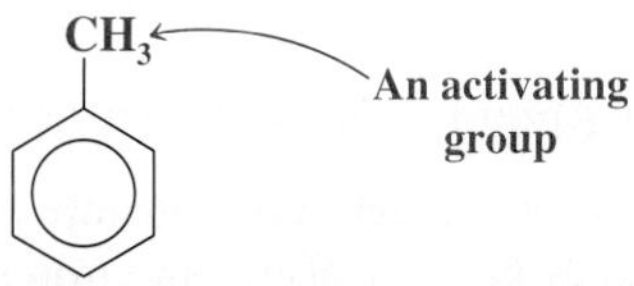

CH_3 — An activating group

More reactive than benzene toward electrophilic substitution

We observe the greater reactivity of toluene in several ways. We find, for example, that with toluene, milder conditions—lower temperatures and lower concentrations of the electrophile—can be used in electrophilic substitutions than with benzene. We also find that under the same conditions toluene reacts faster than benzene. In nitration, for example, toluene reacts 25 times as fast as benzene.

We find, moreover, that when toluene undergoes electrophilic substitution, most of the substitution takes place at its ortho and para positions. When we nitrate toluene with nitric and sulfuric acids, we get mononitrotoluenes in the following relative proportions:

Ortho | Para | Meta

CH_3-benzene $\xrightarrow[H_2SO_4]{HNO_3}$ o-Nitrotoluene (59%) + p-Nitrotoluene (37%) + m-Nitrotoluene (4%)

Of the mononitrotoluenes obtained from the reaction, 96% (59% + 37%) have the nitro group in an ortho or para position. Only 4% have the nitro group in a meta position.

PROBLEM 15.8 Explain how the percentages just given show that the methyl group exerts an ortho–para directive effect by considering the percentages that would be obtained if the methyl group had no effect on the orientation of the incoming electrophile.

Predominant substitution of toluene at the ortho and para positions is not restricted to nitration reactions. The same behavior is observed in halogenation, sulfonation, and so forth.

All alkyl groups are activating groups, and they are all also ortho–para directors. The methoxyl group, CH_3O—, and the acetamido group, CH_3CONH—, are strong activating groups, and both are ortho–para directors.

The hydroxyl group and the amino group are very powerful activating groups and are also powerful ortho–para directors. Phenol and aniline react with bromine in water (no catalyst is required) to produce products in which both of the ortho positions and the para position are substituted. These tribromo products are obtained in nearly quantitative yield:

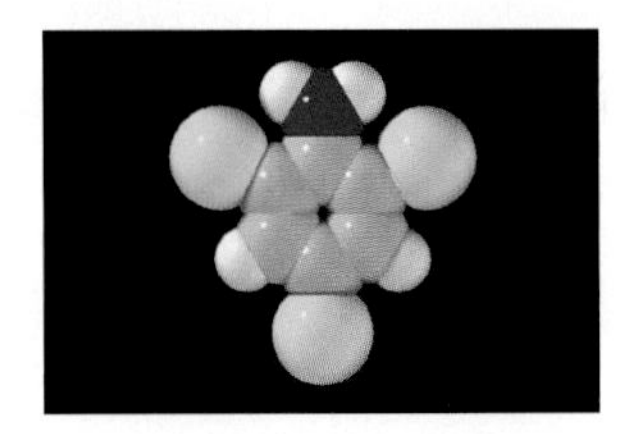

2,4,6-Tribromoaniline

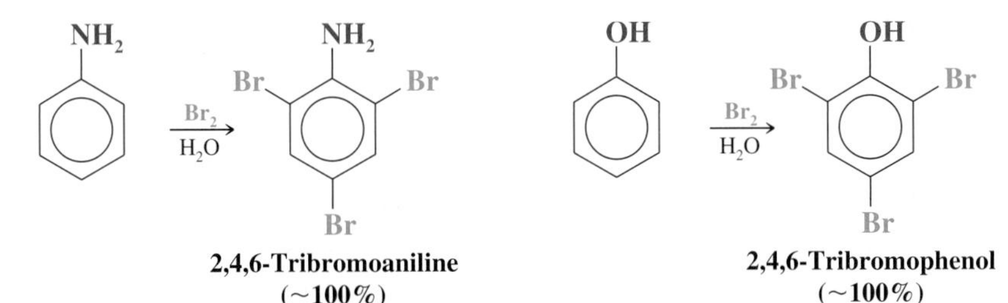

15.10B Deactivating Groups: Meta Directors

The nitro group is a very strong **deactivating group.** Nitrobenzene undergoes nitration at a rate only 10^{-4} times that of benzene. The nitro group is a meta director. When nitrobenzene is nitrated with nitric and sulfuric acids, 93% of the substitution occurs at the meta position:

Nitrobenzene $\xrightarrow[H_2SO_4]{HNO_3}$ o-dinitrobenzene (6%) + p-dinitrobenzene (1%) + m-dinitrobenzene (93%)

The carboxyl group (—CO_2H), the sulfo group (—SO_3H), and the trifluoromethyl group (—CF_3) are also deactivating groups; they are also meta directors.

15.10C Halo Substituents: Deactivating Ortho–Para Directors

The chloro and bromo groups are weak deactivating groups. Chlorobenzene and bromobenzene undergo nitration at rates that are, respectively, 33 and 30 times slower than for benzene. The chloro and bromo groups are ortho–para directors, however. The relative percentages of monosubstituted products that are obtained when chlorobenzene is chlorinated, brominated, nitrated, and sulfonated are shown in Table 15.1.

TABLE 15.1 Electrophilic Substitutions of Chlorobenzene

Reaction	Ortho Product (%)	Para Product (%)	Total Ortho and Para (%)	Meta Product (%)
Chlorination	39	55	94	6
Bromination	11	87	98	2
Nitration	30	70	100	
Sulfonation		100	100	

Similar results are obtained from electrophilic substitutions of bromobenzene.

15.10D Classification of Substituents

Studies like the ones that we have presented in this section have been done for a number of other substituted benzenes. The effects of these substituents on reactivity and orientation are included in Table 15.2.

TABLE 15.2 Effect of Substituents on Electrophilic Aromatic Substitution

Ortho–Para Directors	Meta Directors
Strongly Activating	**Moderately Deactivating**
$-\ddot{N}H_2$, $-\ddot{N}HR$, $-\ddot{N}R_2$	$-C{\equiv}N$
$-\ddot{O}H$, $-\ddot{O}:^-$	$-SO_3H$
Moderately Activating	$-CO_2H$, $-CO_2R$
$-\ddot{N}HCOCH_3$, $-\ddot{N}HCOR$	$-CHO$, $-COR$
$-\ddot{O}CH_3$, $-\ddot{O}R$	**Strongly Deactivating**
Weakly Activating	$-NO_2$
$-CH_3$, $-C_2H_5$, $-R$	$-NR_3^+$
$-C_6H_5$	$-CF_3$, $-CCl_3$
Weakly Deactivating	
$-\ddot{F}:$, $-\ddot{Cl}:$, $-\ddot{Br}:$, $-\ddot{I}:$	

PROBLEM 15.9

Use Table 15.2 to predict the major products formed when:
(a) Toluene is sulfonated. **(c)** Nitrobenzene is brominated.
(b) Benzoic acid is nitrated. **(d)** Phenol is subjected to Friedel–Crafts acetylation.
If the major products would be a mixture of ortho and para isomers, you should so state.

15.11 Theory of Substituent Effects on Electrophilic Aromatic Substitution

15.11A Reactivity: The Effect of Electron-Releasing and Electron-Withdrawing Groups

We have now seen that certain groups *activate* the benzene ring toward electrophilic substitution, whereas other groups *deactivate* the ring. When we say that a group activates the ring, what we mean, of course, is that the group increases the relative rate of the reaction. We mean that an aromatic compound with an activating group reacts faster in electrophilic substitutions than benzene. When we say that a group deactivates the ring, we mean that an aromatic compound with a deactivating group reacts slower than benzene.

We have also seen that we can account for relative reaction rates by examining the transition state for the rate-determining steps. We know that any factor that increases the energy of the transition state relative to that of the reactants decreases the relative rate of the reaction. It does this because it increases the free energy of activation of the reaction. In the same way, any factor that decreases the energy of the transition state relative to that of the reactants lowers the free energy of activation and increases the relative rate of the reaction.

The rate-determining step in electrophilic substitutions of substituted benzenes is the step that results in the formation of the arenium ion. We can write the formula for a substituted benzene in a generalized way if we use the letter **Q** to represent any ring substituent including hydrogen. (If **Q** is hydrogen, the compound is benzene itself.) We can also write the structure for the arenium ion in the way shown here. By this formula we mean that **Q** can be in any position—ortho, meta, or para—relative to the electrophile, **E.** Using these conventions, then, we are able to write the rate-determining step for electrophilic aromatic substitution in the following general way:

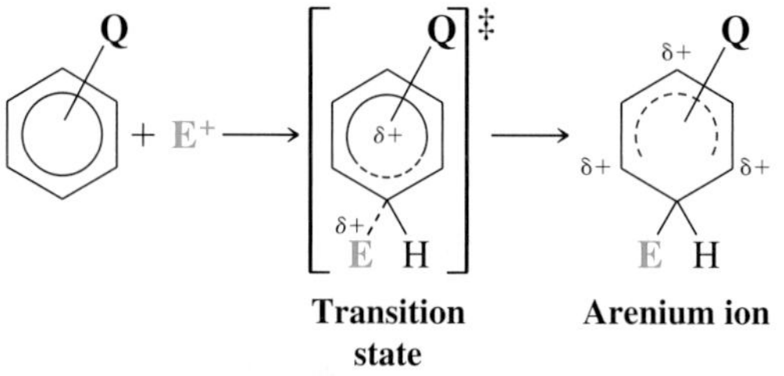

Transition state **Arenium ion**

When we examine this step for a large number of reactions, we find that the relative rates of the reactions depend on whether **Q withdraws** or **releases** electrons. If **Q** is an electron-releasing group (relative to hydrogen), the reaction occurs faster than the corresponding reaction of benzene. If Q is an electron-withdrawing group, the reaction is slower than that of benzene:

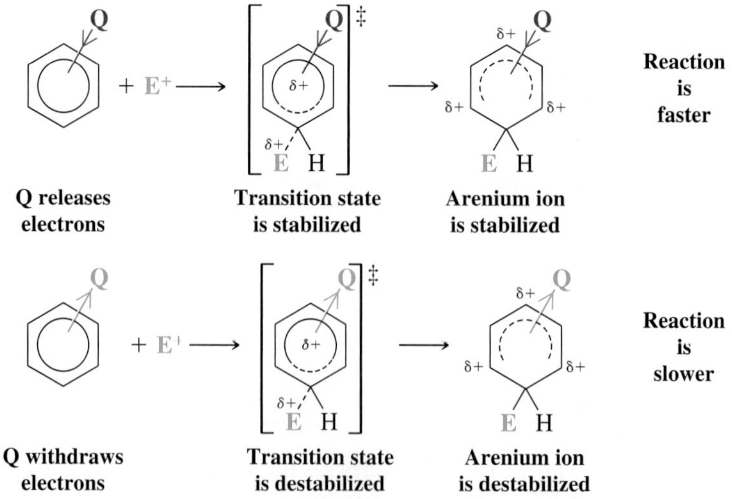

Q releases electrons **Transition state is stabilized** **Arenium ion is stabilized** **Reaction is faster**

Q withdraws electrons **Transition state is destabilized** **Arenium ion is destabilized** **Reaction is slower**

It appears, then, that the substituent (**Q**) must affect the stability of the transition state relative to that of the reactants. Electron-releasing groups apparently make the transition state more stable, whereas electron-withdrawing groups make it less stable. That this is so is entirely reasonable, because the transition state resembles the arenium ion, and the arenium ion is a delocalized *carbocation.*

This effect illustrates another application of the Hammond–Leffler postulate (Section 6.13A). The arenium ion is a high-energy intermediate, and the step that leads to it is a *highly endothermic step.* Thus, according to the Hammond–Leffler postulate, there should be a strong resemblance between the arenium ion itself and the transition state leading to it.

Since the arenium ion is postively charged, we would expect an electron-releasing group to stabilize it *and the transition state leading to the arenium ion,* for the transition state is a developing delocalized carbocation. We can make the same kind of arguments about the effect of electron-withdrawing groups. An electron-withdrawing group should make the arenium ion *less stable,* and in a corresponding way it should make the transition state leading to the arenium ion *less stable.*

Figure 15.4 shows how the electron-withdrawing and electron-releasing abilities of substituents affect the relative free energies of activation of electrophilic aromatic substitution reactions.

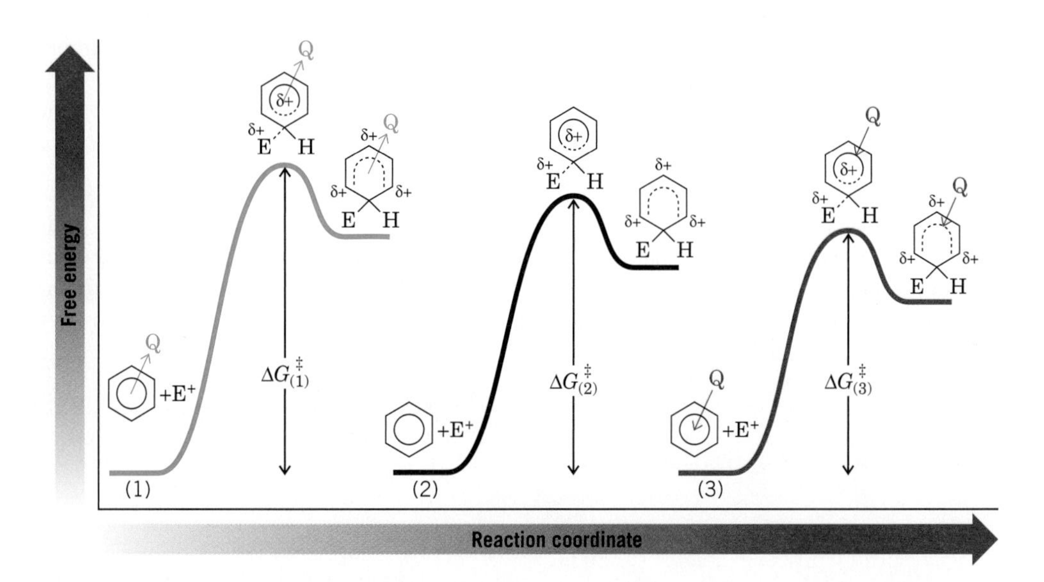

FIGURE 15.4 A comparison of free-energy profiles for arenium ion formation in a ring with an electron-withdrawing substituent (⇸ Q), no substituent, and an electon-donating substituent (⇷ Q). In (1) (blue energy profile), the electron-withdrawing group Q raises the transition state energy. The energy of activation barrier is the highest, and therefore the reaction is the slowest. Reaction (2), with no substituent, serves as a reference for comparison. In (3) (red energy profile), an electron-donating group Q stabilizes the transition state. The energy of activation barrier is lowest, and therefore the reaction is the fastest.

Calculated electrostatic potential maps for two arenium ions comparing the charge-stabilizing effect of an electron-donating methyl group with the charge-destabilizing effect of an electron-withdrawing trifluoromethyl group are shown in Fig. 15.5. The arenium ion at the left (Fig. 15.5*a*) is that from electrophilic addition of bromine to methylbenzene (toluene) at the para position. The arenium ion at the right (Fig. 15.5*b*) is that from electrophilic addition of bromine to trifluoromethylbenzene at the meta position. Notice that the atoms of the ring in Fig. 15.5*a* have much less blue color associated with them, showing that they are much less positive and that the ring is stabilized.

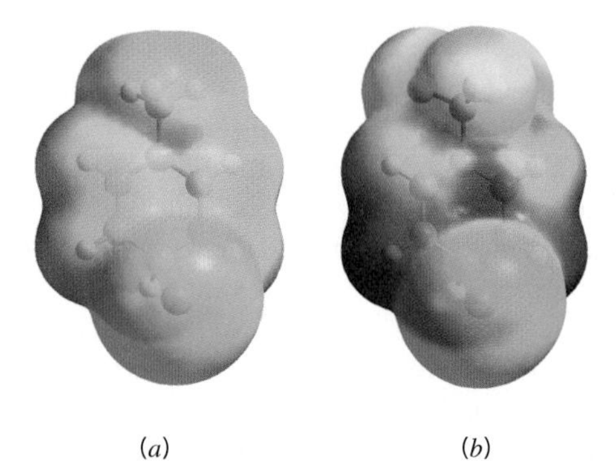

FIGURE 15.5 Calculated electrostatic potential maps for the arenium ions from electrophilic addition of bromine to *(a)* methylbenzene (toluene) and *(b)* trifluoromethylbenzene. The positive charge in the arenium ion ring of methylbenzene *(a)* is delocalized by the electron-releasing ability of the methyl group, whereas the positive charge in the arenium ion of trifluoromethylbenzene *(b)* is enhanced by the electron-withdrawing effect of the trifluoromethyl group. (The electrostatic potential maps for the two structures use the same color scale with respect to potential so that they can be directly compared.)

CD Animated Graphics

15.11B Inductive and Resonance Effects: Theory of Orientation

We can account for the electron-withdrawing and electron-releasing properties of groups on the basis of two factors: *inductive effects* and *resonance effects.* We shall also see that these two factors determine orientation in aromatic substitution reactions.

The **inductive effect** of a substituent **Q** arises from the electrostatic interaction of the polarized bond to **Q** with the developing positive charge in the ring as it is attacked by an electrophile. If, for example, **Q** is a more electronegative atom (or group) than carbon, then the ring will be at the positive end of the dipole:

$$\overset{\delta-}{\text{Q}} \leftarrow \overset{\delta+}{\text{C}_6\text{H}_5} \qquad \text{(e.g., Q = F, Cl, or Br)}$$

Attack by an electrophile will be retarded because this will lead to an additional full positive charge on the ring. The halogens are all more electronegative than carbon and exert an electron-withdrawing inductive effect. Other groups have an electron-withdrawing inductive effect because the atom directly attached to the ring bears a full or partial positive charge. Examples are the following:

$$\leftarrow \overset{+}{\text{N}}\text{R}_3 \quad (\text{R = alkyl or H}) \qquad \leftarrow \text{C}^{\delta+}(\rightarrow \text{X}^{\delta-})_3 \qquad \leftarrow \text{N}^+(=\text{O})\text{O}^- \qquad \leftarrow \text{S}^{+}(\text{O}^-)(=\text{O})\text{—OH}$$

$$\leftarrow \text{C}(=\ddot{\text{O}})\text{—G} \longleftrightarrow \leftarrow \text{C}^+(\text{—}\ddot{\text{O}}\text{:}^-)\text{—G} \qquad (\text{G = H, R, OH, or OR})$$

Electron-withdrawing groups with a full or partial charge on the atom attached to the ring

The **resonance effect** of a substituent **Q** refers to the possibility that the presence of **Q** may increase or decrease the resonance stabilization of the intermediate arenium ion. The **Q** substituent may, for example, cause one of the three contributors to the resonance hybrid for the arenium ion to be better or worse than the case when **Q** is hydrogen. Moreover, when **Q** is an atom bearing one or more nonbonding electron pairs, it may lend extra stability to the arenium ion by providing a *fourth* resonance contributor in which the positive charge resides on **Q**:

$$\text{(E,H)C}_6\text{H}_4^{+}\text{—}\ddot{\text{Q}} \longleftrightarrow \text{(E,H)C}_6\text{H}_4\text{=Q}^+$$

This electron-donating resonance effect applies with decreasing strength in the following order:

Most electron donating $-\ddot{N}H_2$, $-\ddot{N}R_2$ > $-\ddot{O}H$, $-\ddot{O}R$ > $-\ddot{X}:$ Least electron donating

This is also the order of the activating ability of these groups. Amino groups are highly activating, hydroxyl and alkoxyl groups are somewhat less activating, and halogen substituents are weakly deactivating. When X = F, this order can be related to the electronegativity of the atoms with the nonbonding pair. The more electronegative the atom is, the less able it is to accept the positive charge (fluorine is the most electronegative, nitrogen the least). When X = Cl, Br, or I, the relatively poor electron-donating ability of the halogens by resonance is understandable on a different basis. These atoms (Cl, Br, and I) are all larger than carbon, and, therefore, the orbitals that contain the nonbonding pairs are further from the nucleus and do not overlap well with the $2p$ orbital of carbon. (This is a general phenomenon: Resonance effects are not transmitted well between atoms of different rows in the periodic table.)

15.11C Meta-Directing Groups

All meta-directing groups have either a partial positive charge or a full positive charge on the atom directly attached to the ring. As a typical example let us consider the trifluoromethyl group.

STUDY TIP! Properties of meta-directing groups

The trifluoromethyl group, because of the three highly electronegative fluorine atoms, is strongly electron withdrawing. It is a strong deactivating group and a powerful meta director in electrophilic aromatic substitution reactions. We can account for both of these characteristics of the trifluoromethyl group in the following way.

The trifluoromethyl group affects the rate of reaction by causing the transition state leading to the arenium ion to be highly unstable. It does this by withdrawing electrons from the developing carbocation, thus increasing the positive charge on the ring:

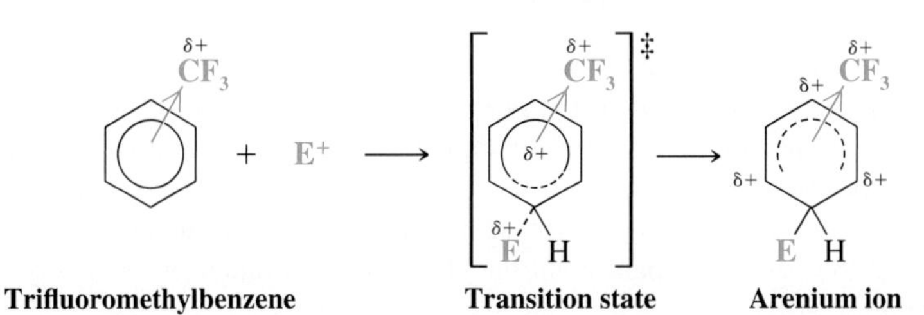

Trifluoromethylbenzene **Transition state** **Arenium ion**

We can understand how the trifluoromethyl group affects *orientation* in electrophilic aromatic substitution if we examine the resonance structures for the arenium ion that would be formed when an electrophile attacks the ortho, meta, and para positions of trifluoromethylbenzene.

Ortho Attack

CF3 E+ ⟶ CF3 E H + ⟷ CF3 E H + ⟷ CF3 + E H

Highly unstable contributor

Meta Attack

Para Attack

Highly unstable contributor

We see in the resonance structures for the arenium ion arising from ortho and para attack that *one contributing structure is highly unstable relative to all the others because the positive charge is located on the ring carbon that bears the electron-withdrawing group.* We see *no* such highly unstable resonance structure in the arenium ion arising from meta attack. This means that the arenium ion formed by meta attack should be the most stable of the three. By the usual reasoning we would also expect the transition state leading to the meta-substituted arenium ion to be the most stable and, therefore, that meta attack would be favored. This is exactly what we find experimentally. The trifluoromethyl group is a powerful meta director:

$$C_6H_5CF_3 + HNO_3 \xrightarrow{H_2SO_4} m\text{-}O_2NC_6H_4CF_3$$

Trifluoromethylbenzene **(~100%)**

Bear in mind, however, that meta substitution is favored only in the sense that *it is the least unfavorable of three unfavorable pathways.* The free energy of activation for substitution at the meta position of trifluoromethylbenzene is less than that for attack at an ortho or para position, but it is still far greater than that for an attack on benzene. Substitution occurs at the meta position of trifluoromethylbenzene faster than substitution takes place at the ortho and para positions, but it occurs much more slowly than it does with benzene.

STUDY TIP!

Other examples of meta-directing groups

The nitro group, the carboxyl group, and other meta-directing groups are all powerful electron-withdrawing groups and act in a similar way.

15.11D Ortho–Para-Directing Groups

Except for the alkyl and phenyl substituents, all of the ortho–para-directing groups in Table 15.2 are of the following general type:

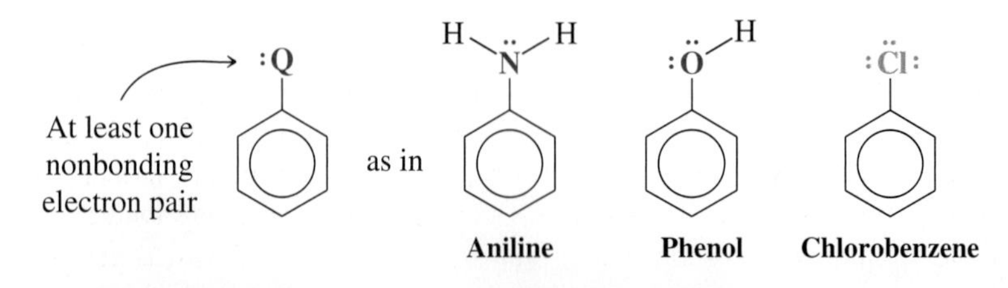

This structural feature—an unshared electron pair on the atom adjacent to the ring—determines the orientation and influences reactivity in electrophilic substitution reactions.

All of these ortho–para directors have at least one pair of nonbonding electrons on the atom adjacent to the benzene ring.

The *directive effect* of these groups with an unshared pair is predominantly caused by an electron-releasing resonance effect. The resonance effect, moreover, operates primarily in the arenium ion and, consequently, in the transition state leading to it.

Except for the halogens, the primary effect on reactivity of these groups is also caused by an electron-releasing resonance effect. And, again, this effect operates primarily in the transition state leading to the arenium ion.

In order to understand these resonance effects, let us begin by recalling the effect of the amino group on electrophilic aromatic substitution reactions. The amino group is not only a powerful activating group, it is also a powerful ortho–para director. We saw earlier (Section 15.10A) that aniline reacts with bromine in aqueous solution at room temperature and in the absence of a catalyst to yield a product in which both ortho positions and the para position are substituted.

The inductive effect of the amino group makes it slightly electron withdrawing. Nitrogen, as we know, is more electronegative than carbon. The difference between the electronegativities of nitrogen and carbon in aniline is not large, however, because the carbon of the benzene ring is sp^2 hybridized and so it is somewhat more electronegative than it would be if it were sp^3 hybridized.

The resonance effect of the amino group is far more important than its inductive effect in electrophilic aromatic substitution, and this resonance effect makes the amino group electron releasing. We can understand this effect if we write the resonance structures for the arenium ions that would arise from ortho, meta, and para attack on aniline:

Ortho Attack

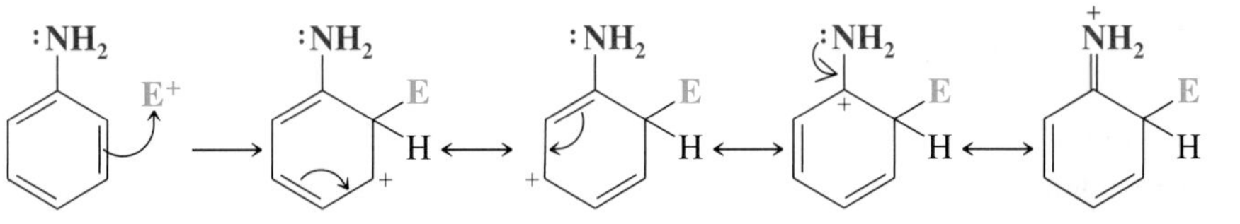

Relatively stable contributor

Meta Attack

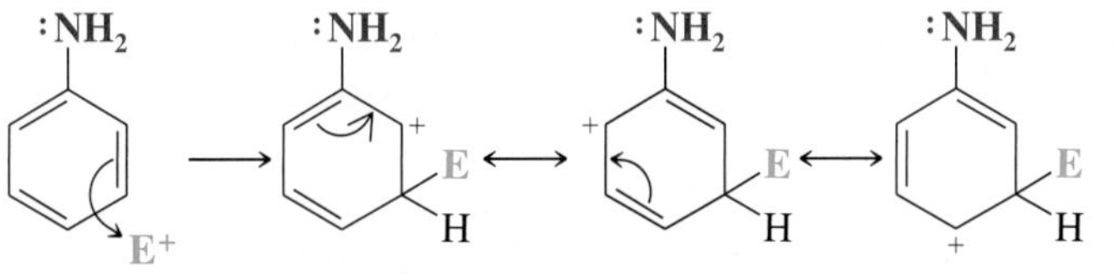

Para Attack

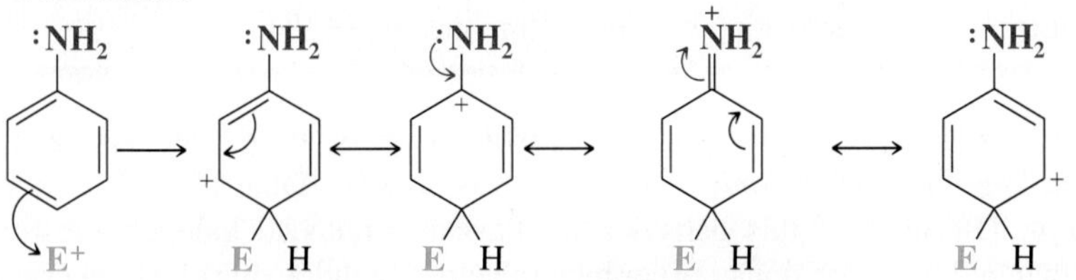

Relatively stable contributor

We see that four reasonable resonance structures can be written for the arenium ions resulting from ortho and para attack, whereas only three can be written for the arenium ion that results from meta attack. This, in itself, suggests that the ortho- and para-substituted arenium ions should be more stable. Of greater importance, however, are the relatively stable structures that contribute to the hybrid for the ortho- and para-substituted arenium ions. In these structures, nonbonding pairs of electrons from nitrogen form an extra bond to the carbon of the ring. This extra bond—and the fact that every atom in each of these structures has a complete outer octet of electrons—makes these structures the most stable of all of the contributors. Because these structures are unusually stable, they make a large—*and stabilizing*—contribution to the hybrid. This means, of course, that the ortho- and para-substituted arenium ions themselves are considerably more stable than the arenium ion that results from the meta attack. The transition states leading to the ortho- and para-substituted arenium ions occur at unusually low free energies. As a result, electrophiles react at the ortho and para positions very rapidly.

PROBLEM 15.10 Use resonance theory to explain why the hydroxyl group of phenol is an activating group and an ortho–para director. Illustrate your explanation by showing the arenium ions formed when phenol reacts with a Br^+ ion at the ortho, meta, and para positions.

PROBLEM 15.11 Phenol reacts with acetic anhydride in the presence of sodium acetate to produce the ester phenyl acetate:

$$\text{C}_6\text{H}_5\text{—OH} \xrightarrow[\text{CH}_3\text{CO}_2\text{Na}]{(\text{CH}_3\text{CO})_2\text{O}} \text{C}_6\text{H}_5\text{—O—}\overset{\text{O}}{\overset{\|}{\text{C}}}\text{—CH}_3$$

Phenol **Phenyl acetate**

The $CH_3COO—$ group of phenyl acetate, like the —OH group of phenol (Problem 15.10), is an ortho–para director.

(a) What structural feature of the $CH_3COO—$ group explains this?

(b) Phenyl acetate, although undergoing reaction at the ortho and para positions, is less reactive toward electrophilic aromatic substitution than phenol. Use resonance theory to explain why this is so.

(c) Aniline is often so highly reactive toward electrophilic substitution that undesirable reactions take place (see Section 15.14A). One way to avoid these undesirable reactions is to convert aniline to acetanilide (below) by treating aniline with acetyl chloride or acetic anhydride:

$$\text{C}_6\text{H}_5\text{—NH}_2 \xrightarrow{(\text{CH}_3\text{CO})_2\text{O}} \text{C}_6\text{H}_5\text{—NH—}\overset{\text{O}}{\overset{\|}{\text{C}}}\text{—CH}_3$$

Aniline **Acetanilide**

What kind of directive effect would you expect the acetamido group ($CH_3CONH—$) to have?

(d) Explain why it is much less activating than the amino group, $—NH_2$.

The directive and reactivity effects of halo substituents may, at first, seem to be contradictory. *The halo groups are the only ortho–para directors* (in Table 15.2) *that are deactivating groups.* [Because of this behavior we have color coded halogen substituents green rather than red (electron donating) or blue (electron withdrawing).] All other deactivating groups are meta directors. We can readily account for the behavior of halo substituents, however, if we assume that their electron-withdrawing inductive effect influences reactivity and their electron-donating resonance effect governs orientation.

Let us apply these assumptions specifically to chlorobenzene. The chlorine atom is highly electronegative. Thus, we would expect a chlorine atom to withdraw electrons from the benzene ring and thereby deactivate it:

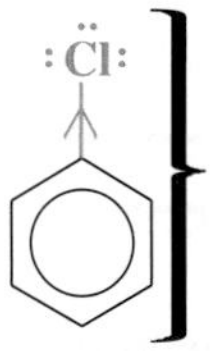

Inductive effect of chlorine atom deactivates ring.

On the other hand, when electrophilic attack does take place, the chlorine atom stabilizes the arenium ions resulting from ortho and para attack relative to that from meta attack. The chlorine atom does this in the same way as amino groups and hydroxyl groups do—***by donating an unshared pair of electrons.*** These electrons give rise to relatively stable resonance structures contributing to the hybrids for the ortho- and para-substituted arenium ions (Section 15.11D).

Ortho Attack

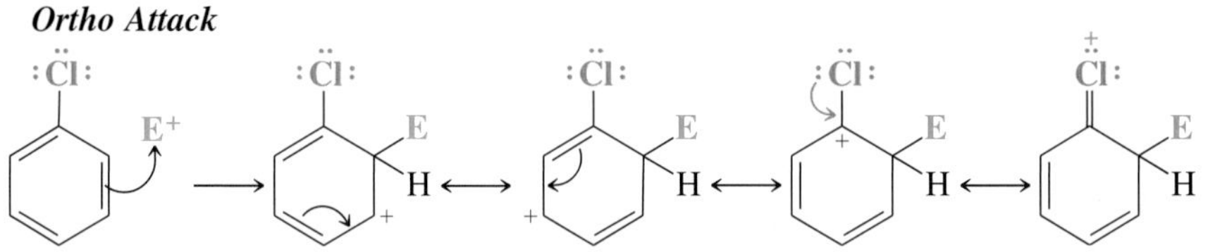

Relatively stable contributor

Meta attack

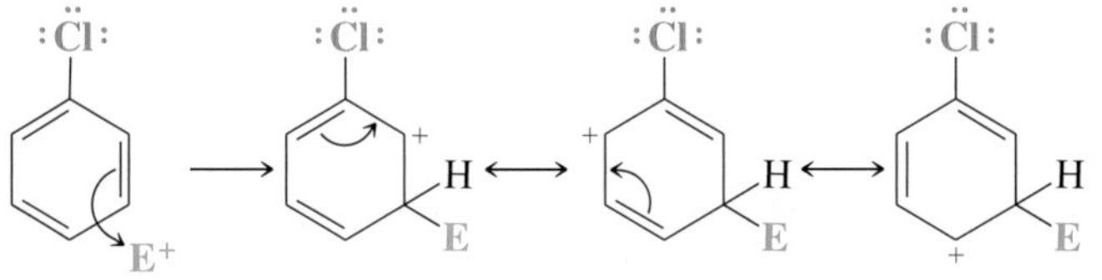

Para Attack

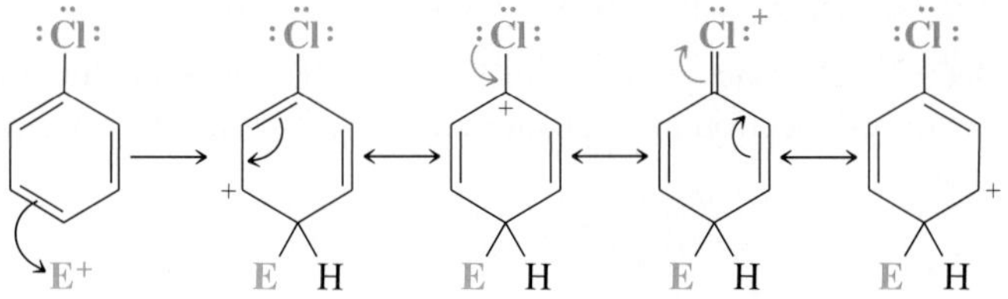

Relatively stable contributor

What we have said about chlorobenzene is, of course, true of bromobenzene.

We can summarize the inductive and resonance effects of halo substituents in the following way. Through their electron-withdrawing inductive effect, halo groups make the ring more electron deficient than that of benzene. This causes the free energy of activation for any electrophilic aromatic substitution reaction to be greater than that for benzene, and, therefore, halo groups are deactivating. Through their electron-donating resonance effect, however, halo substituents cause the free energies of activation leading to ortho and para substitution to be lower than the free energy of activation leading to meta substitution. This makes halo substituents ortho–para directors.

You may have noticed an apparent contradiction between the rationale offered for the unusual effects of the halogens and that offered earlier for amino or hydroxyl groups. That is, oxygen is *more* electronegative than chlorine or bromine (and especially iodine). Yet the hydroxyl group is an activating group, whereas halogens are deactivating groups. An explanation for this can be obtained if we consider the relative stabilizing contributions made to the transition state leading to the arenium ion by resonance structures involving a group —Q̈ (—Q̈ = —N̈H$_2$, —Ö—H, —F̈:, —C̈l:, —B̈r:, —Ï:) that is directly attached to the benzene ring in which Q donates an electron pair. If —Q̈ is —ÖH or —N̈H$_2$, these resonance structures arise because of the overlap of a $2p$ orbital of carbon with that of oxygen or nitrogen. Such overlap is favorable because the atoms are almost the same size. With chlorine, however, donation of an electron pair to the benzene ring requires overlap of a carbon $2p$ orbital with a chlorine $3p$ orbital. Such overlap is less effective; the chlorine atom is much larger and its $3p$ orbital is much further from its nucleus. With bromine and iodine, overlap is even less effective. Justification for this explanation can be found in the observation that fluorobenzene (Q = —F̈:) is the most reactive halobenzene in spite of the high electronegativity of fluorine and the fact that —F̈: is the most powerful ortho–para director of the halogens. With fluorine, donation of an electron pair arises from overlap of a $2p$ orbital of fluorine with a $2p$ orbital of carbon (as with —N̈H$_2$ and —ÖH). This overlap is effective because the orbitals of =C< and —F̈: are of the same relative size.

PROBLEM 15.12 Chloroethene adds hydrogen chloride more slowly than ethene, and the product is 1,1-dichloroethane. How can you explain this using resonance and inductive effects?

$$Cl{-}CH{=}CH_2 \xrightarrow{HCl} Cl{-}\underset{\displaystyle Cl}{\underset{|}{CH}}{-}\underset{\displaystyle H}{\underset{|}{CH_2}}$$

15.11E Ortho–Para Direction and Reactivity of Alkylbenzenes

Alkyl groups are better electron-releasing groups than hydrogen. Because of this, they can activate a benzene ring toward electrophilic substitution by stabilizing the transition state leading to the arenium ion:

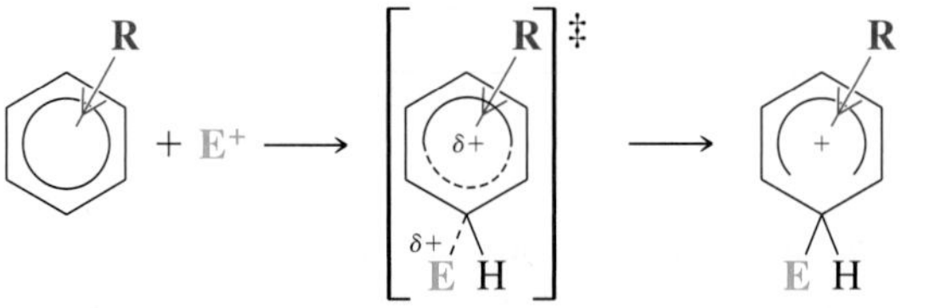

Transition state is stabilized. **Arenium ion is stabilized.**

For an alkylbenzene the free energy of activation of the step leading to the arenium ion (just shown) is lower than that for benzene, and alkylbenzenes react faster.

Alkyl groups are ortho–para directors. We can also account for this property of alkyl groups on the basis of their ability to release electrons—an effect that is particularly important when the alkyl group is attached directly to a carbon that bears a positive charge. (Recall the ability of alkyl groups to stabilize carbocations that we discussed in Section 6.11 and in Fig. 6.9.)

If, for example, we write resonance structures for the arenium ions formed when toluene undergoes electrophilic substitution, we get the results shown below (after the "Chemistry of . . ." box):

The Chemistry of...

Iodine Incorporation in Thyroxine Biosynthesis

The biosynthesis of thyroxine involves introduction of iodine atoms into tyrosine units of thyroglobulin (see the opening vignette for this chapter). This process occurs by a biochemical version of electrophilic aromatic substitution. An iodoperoxidase enzyme catalyzes the reaction between iodide anions and hydrogen peroxide to generate an electrophilic form of iodine (presumably a species like I—OH). Nucleophilic attack by the aromatic ring of tyrosine on the electrophilic iodine leads to incorporation of iodine at the 3 and 5 positions of the tyrosine rings in thyroglobulin. These are the positions ortho to the phenol hydroxyl group, precisely where we would expect electrophilic aromatic substitution to occur in tyrosine. (Substitution para to the hydroxyl cannot occur in tyrosine because that position is blocked, and substitution ortho to the alkyl group is less favored than ortho to the hydroxyl.) Electrophilic iodine is also involved in the coupling of two tyrosine units necessary to complete biosynthesis of thyroxine.

Electrophilic aromatic substitution also plays a role in the 1927 laboratory synthesis of thyroxine by C. Harington and G. Barger. Their synthesis helped prove the structure of this important hormone by comparison of the synthetic material with natural thyroxine. Harington and Barger used electrophilic aromatic substitution to introduce the iodine atoms at the ortho positions in the phenol ring of thyroxine. They used a different reaction, however, to introduce the iodine atoms in the other ring of thyroxine (*nucleophilic* aromatic substitution—a reaction we shall study in Chapter 21.)

Tyr

Tyr

iodoperoxidase

$I^- + H_2O_2$

$I^- + H_2O_2$

I^-

protein hydrolysis

Thyroxine

Thyroglobin (Two tyrosine groups are shown. The remainder of the thyroglobulin protein is indicated by the shaded area.)

The biosynthesis of thyroxine in the thyroid gland through the iodination, rearrangement, and hydrolysis (proteolysis) of thyroglobin Tyr residues. The relatively scarce I^- is actively sequestered by the thyroid gland.

Ortho Attack

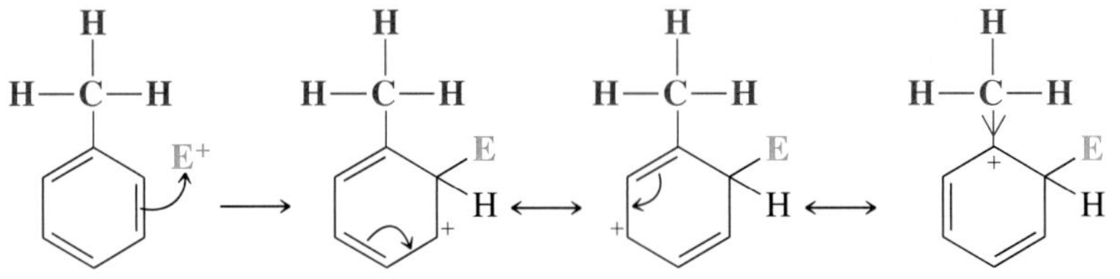

Relatively stable contributor

Meta Attack

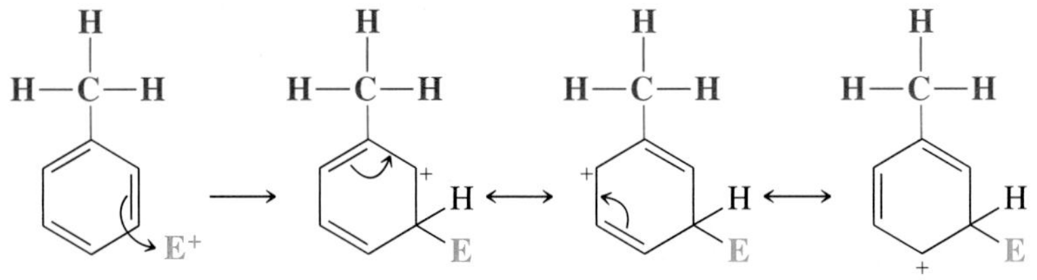

Para Attack

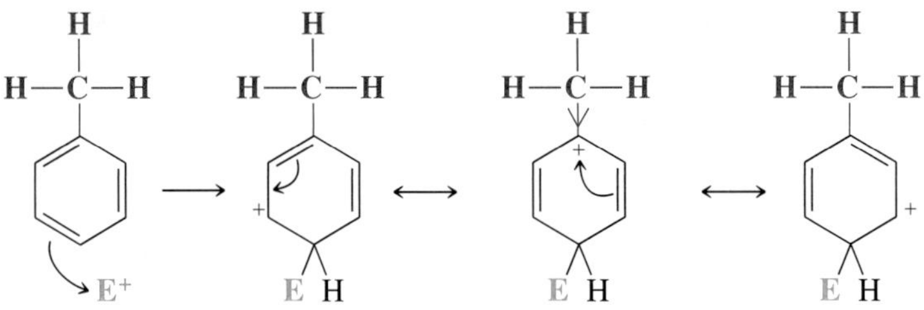

Relatively stable contributor

In ortho attack and para attack we find that we can write resonance structures in which the methyl group is directly attached to a positively charged carbon of the ring. These structures are more *stable* relative to any of the others because in them the stabilizing influence of the methyl group (by electron release) is most effective. These structures, therefore, make a large (stabilizing) contribution to the overall hybrid for ortho- and para-substituted arenium ions. No such relatively stable structure contributes to the hybrid for the meta-substituted arenium ion, and as a result it is less stable than the ortho- or para-substituted arenium ion. Since the ortho- and para-substituted arenium ions are more stable, the transition states leading to them occur at lower energy and ortho and para substitutions take place most rapidly.

PROBLEM 15.13 Write resonance structures for the ortho and para arenium ions formed when ethylbenzene reacts with a Br^+ ion (as formed from $Br_2/FeBr_3$).

PROBLEM 15.14 When biphenyl ($C_6H_5—C_6H_5$) undergoes nitration, it reacts more rapidly than benzene, and the major products are 1-nitro-2-phenylbenzene and 1-nitro-4-phenylbenzene. Explain these results.

15.11F Summary of Substituent Effects on Orientation and Reactivity

With a theoretical understanding now in hand of substituent effects on orientation and reactivity, we refer you to Table 15.2 for a summary of specific groups and their effects.

15.12 Reactions of the Side Chain of Alkylbenzenes

Hydrocarbons that consist of both aliphatic and aromatic groups are also known as **arenes.** Toluene, ethylbenzene, and isopropylbenzene are **alkylbenzenes:**

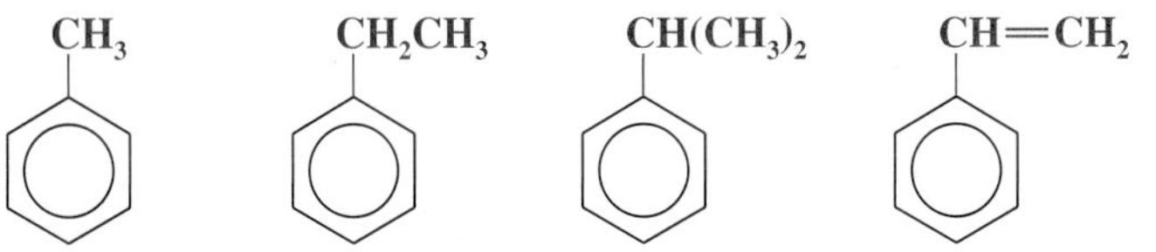

CH_3 / CH_2CH_3 / $CH(CH_3)_2$ / $CH{=}CH_2$

Methylbenzene (toluene) **Ethylbenzene** **Isopropylbenzene (cumene)** **Phenylethene (styrene or vinylbenzene)**

Phenylethene, usually called styrene, is an example of an **alkenylbenzene.** The aliphatic portion of these compounds is commonly called the **side chain.**

15.12A Benzylic Radicals and Cations

Hydrogen abstraction from the methyl group of methylbenzene (toluene) produces a radical called the **benzyl radical:**

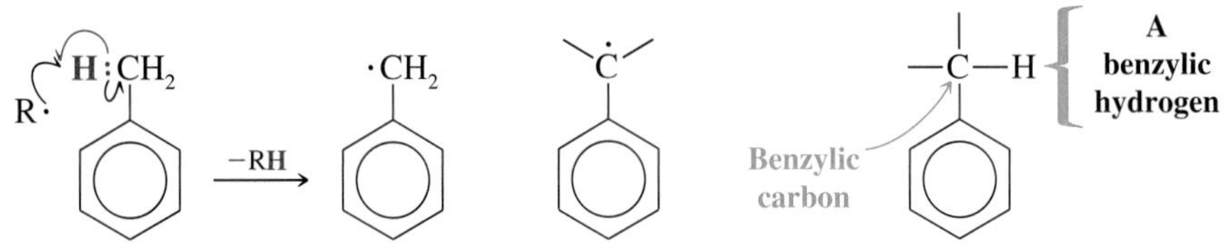

R· H:CH_2 —(−RH)→ ·CH_2

Methylbenzene (toluene) **The benzyl radical** A benzylic radical

Benzylic carbon —C—H **A benzylic hydrogen**

The name benzyl radical is used as a specific name for the radical produced in this reaction. The general name **benzylic radical** applies to all radicals that have an unpaired electron on the side-chain carbon atom that is directly attached to the benzene ring. The hydrogen atoms of the carbon atom directly attached to the benzene ring are called **benzylic hydrogen atoms.**

Departure of a leaving group (LG) from a benzylic position produces a **benzylic cation:**

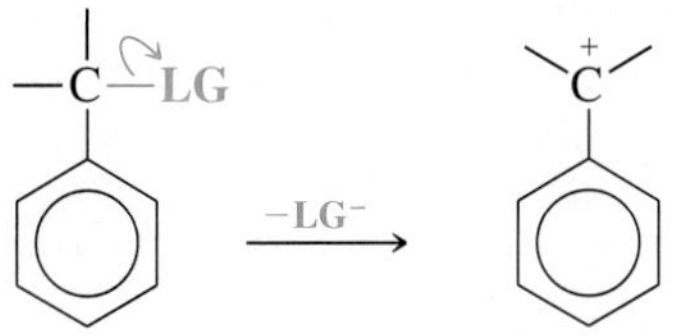

—C—LG —($-LG^-$)→ C^+

A benzylic cation

Benzylic radicals and benzylic cations are *conjugated unsaturated systems* and *both are unusually stable.* They have approximately the same stabilities as allylic radicals and cations. This exceptional stability of benzylic radicals and cations can be explained by resonance theory. In the case of each entity, resonance structures can be written that place either the unpaired electron (in the case of the radical) or the positive charge (in the case of the cation) on an ortho or para carbon of the ring (see the following structures). Thus resonance delocalizes the unpaired electron or the charge, and this delocalization causes the radical or cation to be highly stabilized.

The Chemistry of...

Industrial Styrene Synthesis

Styrene is one of the most important industrial chemicals—more than 11 billion pounds is produced each year. The starting material for a major commercial synthesis of styrene is ethylbenzene, produced by Friedel–Crafts alkylation of benzene:

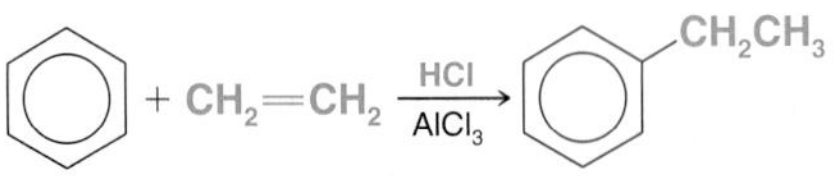

$$C_6H_6 + CH_2{=}CH_2 \xrightarrow[AlCl_3]{HCl} C_6H_5CH_2CH_3$$

Ethylbenzene

Ethylbenzene is then dehydrogenated in the presence of a catalyst (zinc oxide or chromium oxide) to produce styrene. (Another method for the synthesis of styrene was discussed in the opening vignette for Chapter 14.)

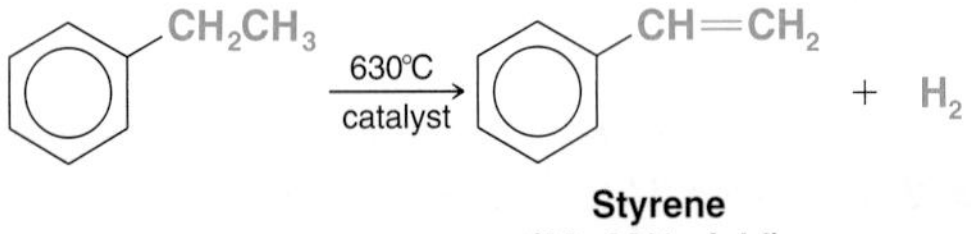

$$C_6H_5CH_2CH_3 \xrightarrow[\text{catalyst}]{630°C} C_6H_5CH{=}CH_2 + H_2$$

Styrene (90–92% yield)

Most styrene is polymerized (Special Topic A) to the familiar plastic, polystyrene:

$$C_6H_5CH{=}CH_2 \xrightarrow{\text{catalyst}} -CH_2CH(C_6H_5)-(CH_2CH(C_6H_5))_n-CH_2CH(C_6H_5)-$$

Polystyrene

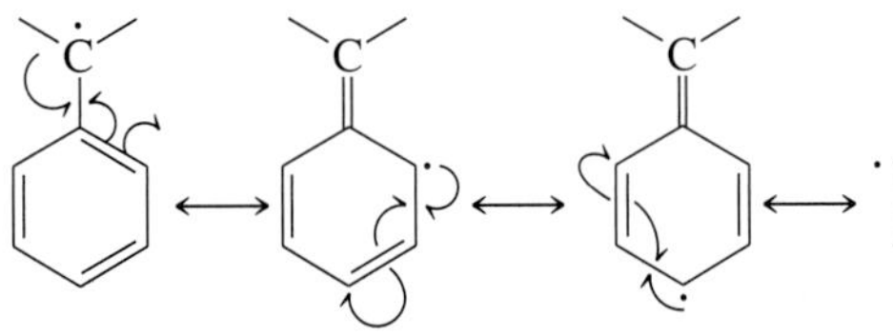

Benzylic radicals are stabilized by resonance.

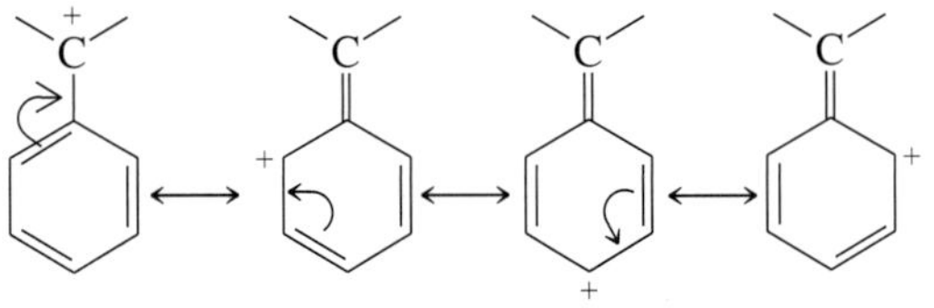

Benzylic cations are stabilized by resonance.

Calculated structures for the benzyl radical and benzyl cation are presented in Fig. 15.6. These structures show the presence at their ortho and para carbons of unpaired electron density in the radical and positive charge in the cation, consistent with the resonance structures above.

15.12B Halogenation of the Side Chain: Benzylic Radicals

We have seen that bromine and chlorine replace hydrogen atoms on the ring of toluene when the reaction takes place in the presence of a Lewis acid. In ring halogenations the electrophiles are *positive* chlorine or bromine ions or they are Lewis acid complexes that have positive halogens. These positive electrophiles attack the π electrons of the benzene ring, and aromatic substitution takes place.

Chlorine and bromine can also be made to replace hydrogens of the methyl group of toluene. Side-chain, or *benzylic,* halogenation takes place when the reaction is carried out ***in the absence of Lewis acids*** **and under conditions that favor the formation of radicals.** When toluene reacts with *N*-bromosuccinimide (NBS) in the presence of light, for

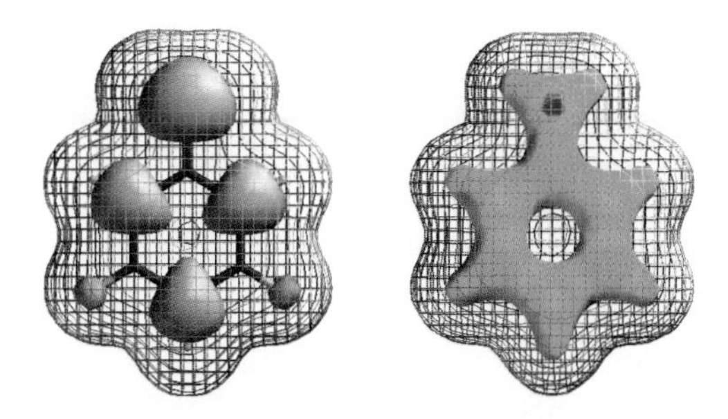

FIGURE 15.6 The gray lobes in the calculated structure for the benzyl radical *(left)* show the location of density from the unpaired electron. This model indicates that the unpaired electron resides primarily at the benzylic, ortho, and para carbons, which is consistent with the resonance model for the benzylic radical discussed earlier. The calculated electrostatic potential map for the bonding electrons in the benzyl cation *(right)* indicates that positive charge (blue regions) resides primarily at the benzylic, ortho, and para carbons, which is consistent with the resonance model for the benzylic cation. The van der Waals surface of both structures is represented by the wire mesh.

example, the major product is benzyl bromide. *N*-Bromosuccinimide furnishes a low concentration of Br_2, and the reaction is analogous to that for allylic bromination that we studied in Section 13.2B.

$$C_6H_5{-}CH_3 + \text{NBS} \xrightarrow[CCl_4]{\text{light}} C_6H_5{-}CH_2Br + \text{succinimide (N—H)}$$

NBS

Benzyl bromide
(α-bromotoluene)
(64%)

Side-chain chlorination of toluene takes place in the gas phase at 400–600°C or in the presence of UV light. When an excess of chlorine is used, multiple chlorinations of the side chain occur:

$$C_6H_5CH_3 \xrightarrow[\text{heat or light}]{Cl_2} C_6H_5CH_2Cl \xrightarrow[\text{heat or light}]{Cl_2} C_6H_5CHCl_2 \xrightarrow[\text{heat or light}]{Cl_2} C_6H_5CCl_3$$

Benzyl chloride **Dichloromethylbenzene** **Trichloromethylbenzene**

These halogenations take place through the same radical mechanism we saw for alkanes in Section 10.4. The halogens dissociate to produce halogen atoms and then the halogen atoms initiate chain reactions by abstracting hydrogens of the methyl group.

A Mechanism for the Reaction

Benzylic Halogenation

Chain Initiation

Step 1: $X{-}X \xrightarrow[\text{or light}]{\text{peroxides, heat}} 2\,X\cdot$

Peroxides, heat, or light cause halogen molecules to cleave into radicals.

Chain Propagation

Step 2:
$$C_6H_5{-}\underset{H}{\overset{H}{C}}{-}H + X\cdot \longrightarrow C_6H_5{-}\dot{C}H_2 + H{-}X$$

Benzyl radical

A halogen radical abstracts a benzylic hydrogen atom, forming a benzylic radical and a molecule of the hydrogen halide.

Step 3:
$$C_6H_5{-}\dot{C}H_2 + X{-}X \longrightarrow C_6H_5{-}\underset{H}{\overset{H}{C}}{-}X + X\cdot$$

Benzyl radical — **Benzyl halide**

The benzylic radical reacts with a halogen molecule to form the benzylic halide product and a halogen radical that propagates the chain.

Chain Termination

Step 4:
$$C_6H_5CH_2\cdot + \cdot X \longrightarrow C_6H_5CH_2{-}X$$
and
$$C_6H_5CH_2\cdot + \cdot CH_2C_6H_5 \longrightarrow C_6H_5CH_2{-}CH_2C_6H_5$$

Various radical coupling reactions terminate the chain.

Benzylic halogenations are similar to allylic halogenations (Section 13.2) in that they involve the formation of *unusually stable radicals* (Section 15.12A). Benzylic and allylic radicals are even more stable than tertiary radicals.

The greater stability of benzylic radicals accounts for the fact that when ethylbenzene is halogenated, the major product is the 1-halo-1-phenylethane. The benzylic radical is formed much faster than the 1° radical:

$$C_6H_5{-}CH_2CH_3 \xrightarrow[(-HX)]{X\cdot} \begin{cases} \xrightarrow{\text{fast}} C_6H_5{-}\dot{C}HCH_3 \xrightarrow{X_2} C_6H_5{-}CHXCH_3 + X\cdot \\ \xrightarrow{\text{slow}} C_6H_5{-}CH_2CH_2\cdot \xrightarrow{X_2} C_6H_5{-}CH_2CH_2X + X\cdot \end{cases}$$

Benzylic radical (more stable) → **1-Halo-1-phenylethane (major product)**

1° Radical (less stable) → **1-Halo-2-phenylethane (minor product)**

PROBLEM 15.15

When propylbenzene reacts with chlorine in the presence of UV radiation, the major product is 1-chloro-1-phenylpropane. Both 2-chloro-1-phenylpropane and 3-chloro-1-phenylpropane are minor products. Write the structure of the radical leading to each product and account for the fact that 1-chloro-1-phenylpropane is the major product.

SOLVED PROBLEM

ILLUSTRATING A MULTISTEP SYNTHESIS

Show how phenylacetylene ($C_6H_5C{\equiv}CH$) could be synthesized from ethylbenzene (phenylethane). Begin by writing a retrosynthetic analysis, and then write reactions needed for the synthesis.

ANSWER: Working backward, that is, using *retrosynthetic analysis,* we find that we can easily envision two syntheses of phenylacetylene. We can make phenylacetylene by dehydrohalogenation of 1,1-dibromo-1-phenylethane, which could have been prepared by allowing ethylbenzene (phenylethane) to react with 2 mol of NBS. Alternatively, we can prepare phenylacetylene from 1,2-dibromo-1-phenylethane, which could be prepared from styrene (phenylethene). Styrene can be made from 1-bromo-1-phenylethane, which can be made from ethylbenzene.

$$C_6H_5C{\equiv}CH \Rightarrow C_6H_5CBr_2CH_3 \Rightarrow C_6H_5CH_2CH_3$$

$$C_6H_5C{\equiv}CH \Rightarrow C_6H_5CHBrCH_2Br \Rightarrow C_6H_5CH{=}CH_2 \Rightarrow C_6H_5CHBrCH_3$$

Following are the synthetic reactions we need for the two retrosynthetic analyses above:

$$C_6H_5CH_2CH_3 \xrightarrow[CCl_4]{\text{NBS, light}} C_6H_5CBr_2CH_3$$

$$C_6H_5CBr_2CH_3 \xrightarrow[(2)\ H_3O^+]{(1)\ NaNH_2\text{, mineral oil, heat}} C_6H_5C{\equiv}CH$$

or

$$C_6H_5CH_2CH_3 \xrightarrow[CCl_4]{\text{NBS, light}} C_6H_5CHBrCH_3 \xrightarrow{\text{KOH, heat}} C_6H_5CH{=}CH_2 \xrightarrow{Br_2,\ CCl_4} C_6H_5CHBrCH_2Br$$

$$C_6H_5CHBrCH_2Br \xrightarrow[(2)\ H_3O^+]{(1)\ NaNH_2\text{, mineral oil, heat}} C_6H_5C{\equiv}CH$$

PROBLEM 15.16

Show how the following compounds could be synthesized from phenylacetylene ($C_6H_5C{\equiv}CH$): **(a)** 1-phenylpropyne, **(b)** 1-phenyl-1-butyne, **(c)** (*Z*)-1-phenylpropene, and **(d)** (*E*)-1-phenylpropene. Begin each synthesis by writing a retrosynthetic analysis.

15.13 Alkenylbenzenes

15.13A Stability of Conjugated Alkenylbenzenes

Alkenylbenzenes that have their side-chain double bond conjugated with the benzene ring are more stable than those that do not:

[Structure: phenyl–C=C–C (conjugated)] more stable than [Structure: phenyl–C–C=C (nonconjugated)]

Conjugated system **Nonconjugated system**

Part of the evidence for this comes from acid-catalyzed alcohol dehydrations, which are known to yield the most stable alkene (Section 7.8). For example, dehydration of an alcohol such as the one that follows yields exclusively the conjugated system:

$$\text{C}_6\text{H}_5\text{—CH—CH(OH)—CH—} \xrightarrow[(-H_2O)]{\text{HA, heat}} \text{C}_6\text{H}_5\text{—C{=}C—C(H)<}$$

Because conjugation always lowers the energy of an unsaturated system by allowing the π electrons to be delocalized, this behavior is just what we would expect.

15.13B Additions to the Double Bond of Alkenylbenzenes

In the presence of peroxides, hydrogen bromide adds to the double bond of 1-phenylpropene to give 2-bromo-1-phenylpropane as the major product:

$$C_6H_5\text{—CH{=}CHCH}_3 \xrightarrow[\text{peroxides}]{\text{HBr}} C_6H_5\text{—CH}_2\text{CH(Br)CH}_3$$

1-Phenylpropene **2-Bromo-1-phenylpropane**

In the absence of peroxides, HBr adds in just the opposite way:

$$C_6H_5\text{—CH{=}CHCH}_3 \xrightarrow[\text{(no peroxides)}]{\text{HBr}} C_6H_5\text{—CH(Br)CH}_2\text{CH}_3$$

1-Phenylpropene **1-Bromo-1-phenylpropane**

The addition of hydrogen bromide to 1-phenylpropene proceeds through a benzylic radical in the presence of peroxides and through a benzylic cation in their absence (see Problem 15.17 and Section 10.9).

PROBLEM 15.17 Write mechanisms for the reactions whereby HBr adds to 1-phenylpropene **(a)** in the presence of peroxides and **(b)** in the absence of peroxides. In each case account for the regiochemistry of the addition (i.e., explain why the major product is 2-bromo-1-phenylpropane when peroxides are present and why it is 1-bromo-1-phenylpropane when peroxides are absent).

PROBLEM 15.18 **(a)** What would you expect to be the major product when 1-phenylpropene reacts with HCl? **(b)** When it is subjected to oxymercuration–demercuration?

15.13C Oxidation of the Side Chain

Strong oxidizing agents oxidize toluene to benzoic acid. The oxidation can be carried out by the action of hot alkaline potassium permanganate. This method gives benzoic acid in almost quantitative yield:

$$C_6H_5\text{—CH}_3 \xrightarrow[\text{(2) } H_3O^+]{\text{(1) KMnO}_4\text{, OH}^-\text{, heat}} C_6H_5\text{—COH (C=O)}$$

Benzoic acid
(~100%)

An important characteristic of side-chain oxidations is that oxidation takes place initially at the benzylic carbon; **alkylbenzenes with alkyl groups longer than methyl are ultimately degraded to benzoic acids:**

$$C_6H_5CH_2CH_2CH_2R \xrightarrow[\text{(2) } H_3O^+]{\text{(1) } KMnO_4,\ OH^-,\ \text{heat}} C_6H_5C(=O){-}OH$$

An alkylbenzene → **Benzoic acid**

Side-chain oxidations are similar to benzylic halogenations, because in the first step the oxidizing agent abstracts a benzylic hydrogen. Once oxidation is begun at the benzylic carbon, it continues at that site. Ultimately, the oxidizing agent oxidizes the benzylic carbon to a carboxyl group, and, in the process, it cleaves off the remaining carbon atoms of the side chain. (*tert*-Butylbenzene is resistant to side-chain oxidation. Why?)

Side-chain oxidation is not restricted to alkyl groups. **Alkenyl, alkynyl, and acyl groups are oxidized by hot alkaline potassium permanganate in the same way:**

$$\left.\begin{array}{c} C_6H_5CH{=}CHCH_3 \\ \text{or} \\ C_6H_5C{\equiv}CCH_3 \\ \text{or} \\ C_6H_5C(=O)CH_2CH_3 \end{array}\right\} \xrightarrow[\text{(2) } H_3O^+]{\text{(1) } KMnO_4,\ OH^-,\ \text{heat}} C_6H_5C(=O)OH$$

15.13D Oxidation of the Benzene Ring

The benzene ring of an alkylbenzene can be converted to a carboxyl group by ozonolysis, followed by treatment with hydrogen peroxide:

$$R{-}C_6H_5 \xrightarrow[\text{(2) } H_2O_2]{\text{(1) } O_3,\ CH_3CO_2H} R{-}C(=O)OH$$

15.14 Synthetic Applications

The substitution reactions of aromatic rings and the reactions of the side chains of alkyl- and alkenylbenzenes, when taken together, offer us a powerful set of reactions for organic synthesis. By using these reactions skillfully, we shall be able to synthesize a large number of benzene derivatives.

Part of the skill in planning a synthesis is in deciding the order in which reactions should be carried out. Let us suppose, for example, that we want to synthesize *o*-bromonitrobenzene. We can see very quickly that we should introduce the bromine into the ring first because it is an ortho–para director:

$$C_6H_6 \xrightarrow[FeBr_3]{Br_2} C_6H_5Br \xrightarrow[H_2SO_4]{HNO_3} o\text{-}BrC_6H_4NO_2 + p\text{-}BrC_6H_4NO_2$$

***o*-Bromonitrobenzene** + ***p*-Bromonitrobenzene**

The ortho and para compounds that we get as products can be separated by various methods because they have different physical properties. However, had we introduced the nitro group first, we would have obtained *m*-bromonitrobenzene as the major product.

Other examples in which choosing the proper order for the reactions is important are the syntheses of the *ortho*-, *meta*-, and *para*-nitrobenzoic acids. We can synthesize the *ortho*- and *para*-nitrobenzoic acids from toluene by nitrating it, separating the *ortho*- and *para*-nitrotoluenes, and then oxidizing the methyl groups to carboxyl groups:

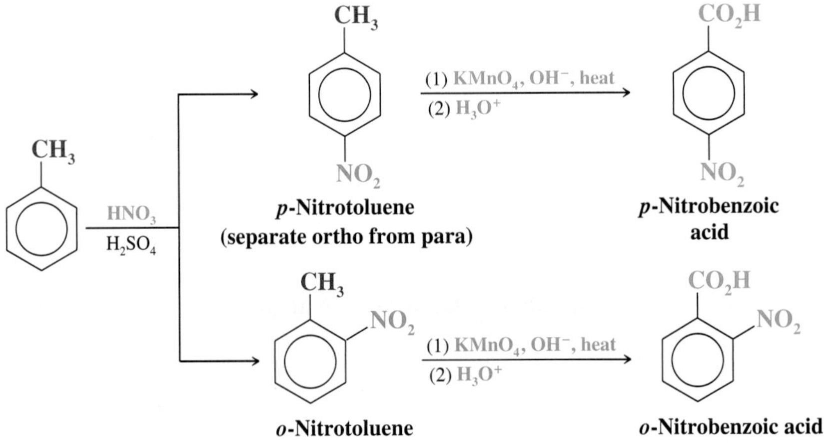

CH_3 CH_3 CO_2H

HNO_3 / H_2SO_4 → (1) $KMnO_4$, OH^-, heat / (2) H_3O^+ →

NO_2 NO_2

***p*-Nitrotoluene**
(separate ortho from para)

***p*-Nitrobenzoic acid**

CH_3 NO_2 (1) $KMnO_4$, OH^-, heat / (2) H_3O^+ → CO_2H NO_2

***o*-Nitrotoluene** ***o*-Nitrobenzoic acid**

We can synthesize *m*-nitrobenzoic acid by reversing the order of the reactions:

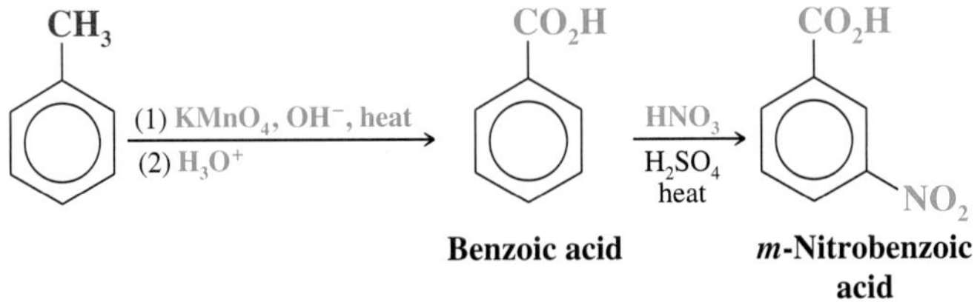

CH_3 (1) $KMnO_4$, OH^-, heat / (2) H_3O^+ → CO_2H HNO_3 / H_2SO_4 heat → CO_2H NO_2

Benzoic acid ***m*-Nitrobenzoic acid**

SOLVED PROBLEM

Starting with toluene, outline a synthesis of **(a)** 1-bromo-2-trichloromethylbenzene, **(b)** 1-bromo-3-trichloromethylbenzene, and **(c)** 1-bromo-4-trichloromethylbenzene.

ANSWER: Compounds (a) and (c) can be obtained by ring bromination of toluene followed by chlorination of the side chain using three molar equivalents of chlorine:

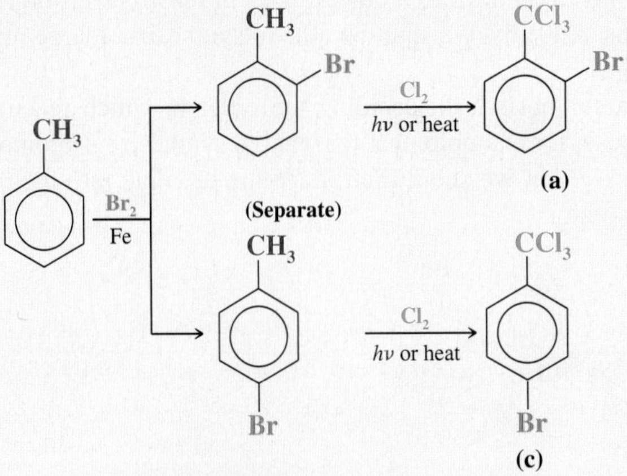

CH_3 Br_2 / Fe → CH_3 Br Cl_2 / *hv* or heat → CCl_3 Br

(a)

(Separate)

CH_3 Br Cl_2 / *hv* or heat → CCl_3 Br

(c)

To make compound (b), we reverse the order of the reactions. By converting the side chain to a $—CCl_3$ group first, we create a meta director, which causes the bromine to enter the desired position:

$$C_6H_5CH_3 \xrightarrow[hv \text{ or heat}]{Cl_2} C_6H_5CCl_3 \xrightarrow[Fe]{Br_2} m\text{-}BrC_6H_4CCl_3 \ \textbf{(b)}$$

PROBLEM 15.19

Suppose you needed to synthesize *m*-chloroethylbenzene from benzene.

$$C_6H_6 \xrightarrow{?} m\text{-}ClC_6H_4CH_2CH_3$$

You could begin by chlorinating benzene and then follow with a Friedel–Crafts alkylation using CH_3CH_2Cl and $AlCl_3$, or you could begin with a Friedel–Crafts alkylation followed by chlorination. Neither method will give the desired product, however.

(a) Why will neither method give the desired product?

(b) There is a three-step method that will work if the steps are done in the right order. What is this method?

15.14A Use of Protecting and Blocking Groups

Very powerful activating groups such as amino groups and hydroxyl groups cause the benzene ring to be so reactive that undesirable reactions may take place. Some reagents used for electrophilic substitution reactions, such as nitric acid, are also strong *oxidizing agents.* (Both electrophiles and oxidizing agents seek electrons.) Thus, amino groups and hydroxyl groups not only activate the ring toward electrophilic substitution but also activate it toward oxidation. Nitration of aniline, for example, results in considerable destruction of the benzene ring because it is oxidized by the nitric acid. Direct nitration of aniline, consequently, is not a satisfactory method for the preparation of *o*- and *p*-nitroaniline.

Treating aniline with acetyl chloride, CH_3COCl, or acetic anhydride, $(CH_3CO)_2O$, converts aniline to acetanilide. The amino group is converted to an acetamido group ($—NHCOCH_3$), a group that is only moderately activating and one that does not make the ring highly susceptible to oxidation (see Problem 15.11). With acetanilide, direct nitration becomes possible:

Aniline (NH_2) → (CH_3COCl, base) → Acetanilide ($NHCOCH_3$) → (HNO_3, H_2SO_4) → **p-Nitroacetanilide (90%)** + **o-Nitroacetanilide (trace)**

p-Nitroacetanilide → (1) H_2O, H_2SO_4, heat (2) OH^- → **p-Nitroaniline** + $CH_3CO_2^-$

This step removes the $CH_3C(=O)—$ group and replaces it with an —H.

Nitration of acetanilide gives *p*-nitroacetanilide in excellent yield with only a trace of the ortho isomer. Acidic hydrolysis of *p*-nitroacetanilide (Section 18.8F) removes the acetyl group and gives *p*-nitroaniline, also in good yield.

Suppose, however, that we need *o*-nitroaniline. The synthesis that we just outlined would obviously not be a satisfactory method, for only a trace of *o*-nitroacetanilide is obtained in the nitration reaction. (The acetamido group is purely a para director in many reactions. Bromination of acetanilide, for example, gives *p*-bromoacetanilide almost exclusively.)

We can synthesize *o*-nitroaniline, however, through the reactions that follow:

Acetanilide ($NHCOCH_3$) → (concd H_2SO_4) → ($NHCOCH_3$, SO_3H) → (HNO_3) → ($NHCOCH_3$, NO_2, SO_3H) → ((1) H_2O, H_2SO_4, heat (2) OH^-) → **o-Nitroaniline (56%)**

Here we see how a sulfonic acid group can be used as a "blocking group." We can remove the sulfonic acid group by desulfonation at a later stage. In this example, the reagent used for desulfonation (dilute H_2SO_4) also conveniently removes the acetyl group that we employed to "protect" the benzene ring from oxidation by nitric acid.

15.14B Orientation in Disubstituted Benzenes

When two different groups are present on a benzene ring, ***the more powerful activating group*** (Table 15.2) ***generally determines the outcome of the reaction.*** Let us consider, as an example, the orientation of electrophilic substitution of *p*-methylacetanilide. The acetamido group is a much stronger activating group than the methyl group. The following example shows that the acetamido group determines the outcome of the reaction. Substitution occurs primarily at the position ortho to the acetamido group:

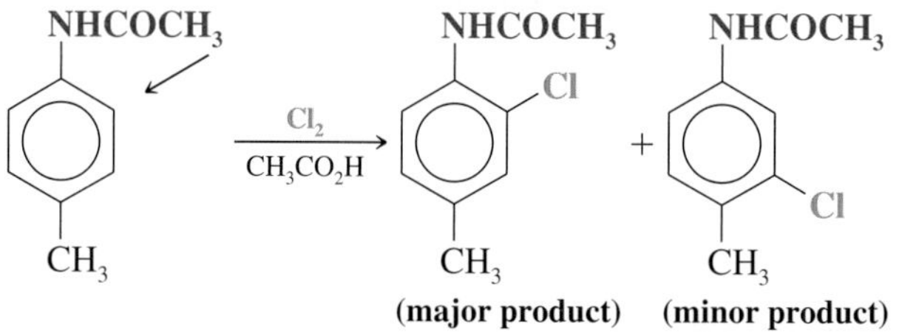

$NHCOCH_3$ / CH_3 $\xrightarrow[CH_3CO_2H]{Cl_2}$ $NHCOCH_3$ / Cl / CH_3 (major product) + $NHCOCH_3$ / Cl / CH_3 (minor product)

Because all ortho–para-directing groups are more activating than meta directors, ***the ortho–para director determines the orientation of the incoming group.***

Steric effects are also important in aromatic substitutions. ***Substitution does not occur to an appreciable extent between meta substituents if another position is open.*** A good example of this effect can be seen in the nitration of *m*-bromochlorobenzene:

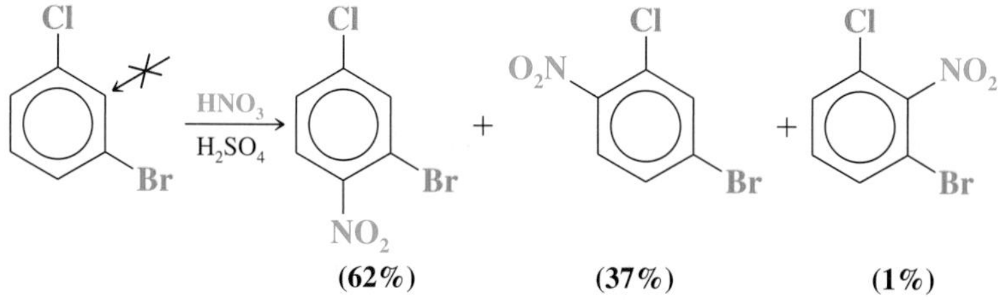

Cl / Br $\xrightarrow[H_2SO_4]{HNO_3}$ Cl / Br / NO_2 (62%) + Cl / O_2N / Br (37%) + Cl / NO_2 / Br (1%)

Only 1% of the mononitro product has the nitro group between the bromine and chlorine.

PROBLEM 15.20

Predict the major product (or products) that would be obtained when each of the following compounds is nitrated:

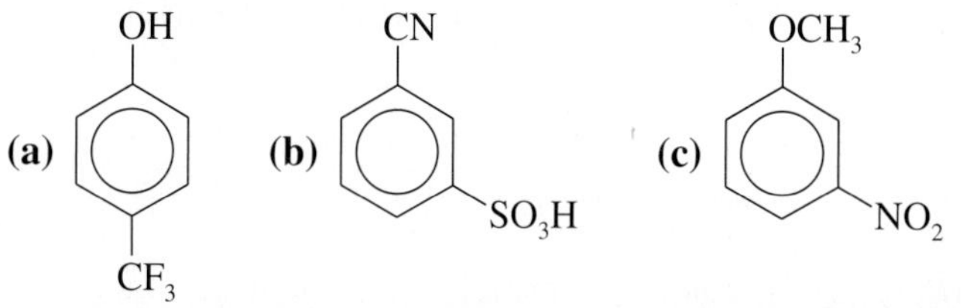

(a) OH / CF_3 (b) CN / SO_3H (c) OCH_3 / NO_2

15.15 Allylic and Benzylic Halides in Nucleophilic Substitution Reactions

Allylic and benzylic halides can be classified in the same way that we have classified other organic halides:

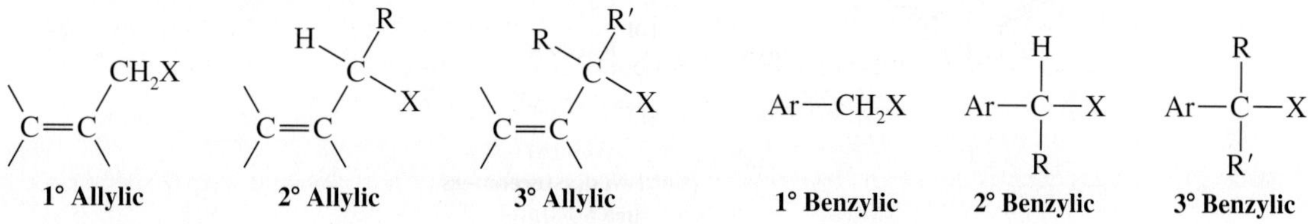

$C{=}C{-}CH_2X$ (1° Allylic); $C{=}C{-}CH(R)X$ (2° Allylic); $C{=}C{-}C(R)(R')X$ (3° Allylic); $Ar{-}CH_2X$ (1° Benzylic); $Ar{-}CH(R){-}X$ (2° Benzylic); $Ar{-}C(R)(R'){-}X$ (3° Benzylic)

All of these compounds undergo nucleophilic substitution reactions. As with other tertiary halides (Section 6.13), the steric hindrance associated with having three bulky groups on the carbon bearing the halogen prevents tertiary allylic and tertiary benzylic halides from reacting by an S_N2 mechanism. They react with nucleophiles only by an S_N1 mechanism.

TABLE 15.3 A Summary of Alkyl, Allylic, and Benzylic Halides in S_N Reactions

These halides give mainly S_N2 reactions.	These halides give mainly S_N1 reactions.
CH_3—X R—CH_2—X R—CH(R′)—X	R′—C(R)(R″)—X
These halides may give either S_N1 or S_N2 reactions.	
Ar—CH_2—X Ar—CH(R)—X	Ar—C(R)(R′)—X
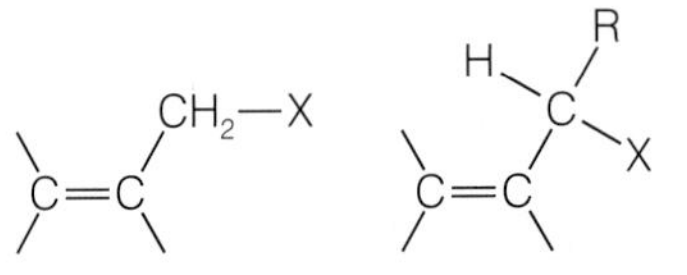 C=C—CH_2—X C=C—C(H)(R)—X	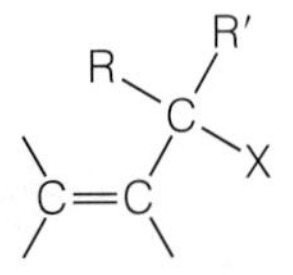 C=C—C(R)(R′)—X

Primary and secondary allylic and benzylic halides can react either by an S_N2 mechanism or by an S_N1 mechanism in ordinary nonacidic solvents. We would expect these halides to react by an S_N2 mechanism because they are structurally similar to primary and secondary alkyl halides. (Having only one or two groups attached to the carbon bearing the halogen does not prevent S_N2 attack.) But primary and secondary allylic and benzylic halides can also react by an S_N1 mechanism because they can form relatively stable carbocations, and in this regard they differ from primary and secondary alkyl halides.*

Overall we can summarize the effect of structure on the reactivity of alkyl, allylic, and benzylic halides in the ways shown in Table 15.3.

SOLVED PROBLEM

When either enantiomer of 3-chloro-1-butene [(*R*) or (*S*)] is subjected to hydrolysis, the products of the reaction are optically inactive. Explain these results.

ANSWER: The solvolysis reaction is S_N1. The intermediate allylic cation is achiral and therefore reacts with water to give the enantiomeric 3-buten-2-ols in equal amounts and to give some of the achiral 2-buten-1-ol:

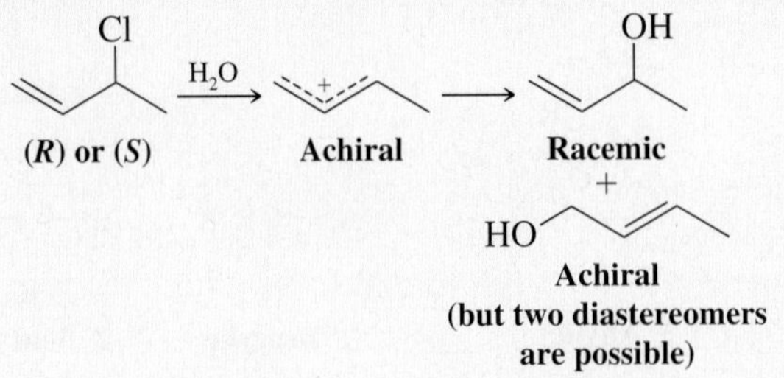

Cl $\xrightarrow{H_2O}$ + ⟶ OH

(*R*) or (*S*) **Achiral** **Racemic**

\+

HO

Achiral
(but two diastereomers are possible)

*There is some dispute as to whether 2° alkyl halides react by an S_N1 mechanism to any appreciable extent in ordinary nonacidic solvents such as mixtures of water and alcohol or acetone, but it is clear that reaction by an S_N2 mechanism is, for all practical purposes, the more important pathway.

PROBLEM 15.21

Account for the following observations: **(a)** When 1-chloro-2-butene is allowed to react with a relatively concentrated solution of sodium ethoxide in ethanol, the reaction rate depends on the concentration of the allylic halide and on the concentration of ethoxide ion. The product of the reaction is almost exclusively $CH_3CH{=}CHCH_2OCH_2CH_3$. **(b)** When 1-chloro-2-butene is allowed to react with very dilute solutions of sodium ethoxide in ethanol (or with ethanol alone), the reaction rate is independent of the concentration of ethoxide ion; it depends only on the concentration of the allylic halide. Under these conditions the reaction produces a mixture of $CH_3CH{=}CHCH_2OCH_2CH_3$ and $CH_3CH(OCH_2CH_3)CH{=}CH_2$. **(c)** In the presence of traces of water 1-chloro-2-butene is slowly converted to a mixture of 1-chloro-2-butene and 3-chloro-1-butene.

PROBLEM 15.22

1-Chloro-3-methyl-2-butene undergoes hydrolysis in a mixture of water and dioxane at a rate that is more than a thousand times that of 1-chloro-2-butene. **(a)** What factor accounts for the difference in reactivity? **(b)** What products would you expect to obtain? [Dioxane is a cyclic ether (below) that is miscible with water in all proportions and is a useful cosolvent for conducting reactions like these. Dioxane is carcinogenic (i.e., cancer causing), however, and like most ethers, it tends to form peroxides.]

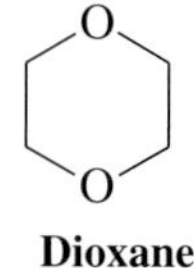

Dioxane

PROBLEM 15.23

Primary halides of the type $ROCH_2X$ apparently undergo S_N1-type reactions, whereas most primary halides do not. Can you propose a resonance explanation for the ability of halides of the type $ROCH_2X$ to undergo S_N1 reactions?

PROBLEM 15.24

The following chlorides undergo solvolysis in ethanol at the relative rates given in parentheses. How can you explain these results?

$C_6H_5CH_2Cl$	$C_6H_5CHClCH_3$	$(C_6H_5)_2CHCl$	$(C_6H_5)_3CCl$
(0.08)	**(1)**	**(300)**	**(3×10^6)**

15.16 Reduction of Aromatic Compounds

Hydrogenation of benzene under pressure using a metal catalyst such as nickel results in the addition of three molar equivalents of hydrogen and the formation of cyclohexane (Section 14.3). The intermediate cyclohexadienes and cyclohexene cannot be isolated because these undergo catalytic hydrogenation faster than benzene does.

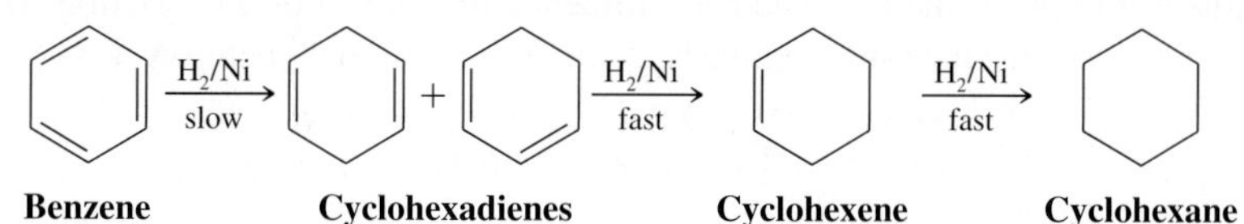

Benzene **Cyclohexadienes** **Cyclohexene** **Cyclohexane**

15.16A The Birch Reduction

Benzene can be reduced to 1,4-cyclohexadiene by treating it with an alkali metal (sodium, lithium, or potassium) in a mixture of liquid ammonia and an alcohol:

$$\text{Benzene} \xrightarrow[\text{NH}_3,\ \text{EtOH}]{\text{Na}} \text{1,4-Cyclohexadiene}$$

Benzene **1,4-Cyclohexadiene**

This is another dissolving metal reduction, and the mechanism for it resembles the mechanism for the reduction of alkynes that we studied in Section 7.14B. A sequence of electron transfers from the alkali metal and proton transfers from the alcohol takes place. Formation of a 1,4-cyclohexadiene in a reaction of this type is quite general, but the reason for its formation in preference to the more stable conjugated 1,3-cyclohexadiene is not understood.

A Mechanism for the Reaction

Birch Reduction

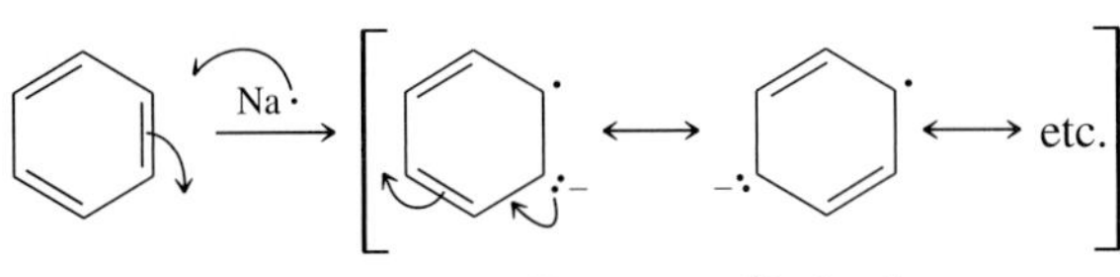

Benzene **Benzene radical anion**

The first electron transfer produces a delocalized benzene radical anion.

EtOH ↓

H H etc.

Cyclohexadienyl radical

Protonation produces a cyclohexadienyl radical (also a delocalized species).

Na· ↓

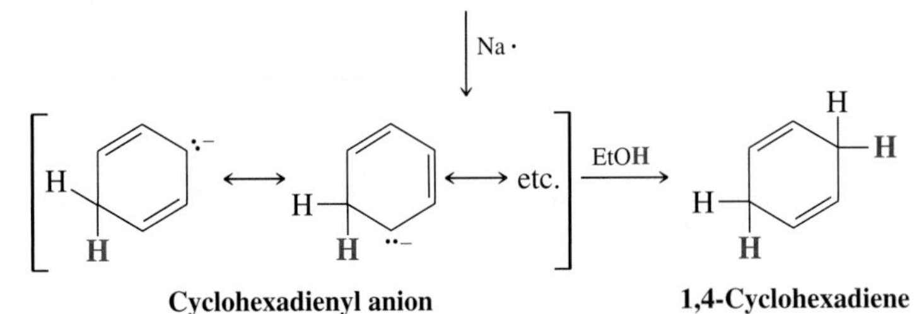

Cyclohexadienyl anion **1,4-Cyclohexadiene**

Transfer of another electron leads to the formation of a delocalized cyclohexadienyl anion, and protonation of this produces the 1,4-cyclohexadiene.

Dissolving metal reductions of this type were developed by the Australian chemist A. J. Birch and have come to be known as **Birch reductions.**

Substituent groups on the benzene ring influence the course of the reaction. Birch reduction of methoxybenzene (anisole) leads to the formation of 1-methoxy-1,4-cyclohexadiene, a compound that can be hydrolyzed by dilute acid to 2-cyclohexenone. This method provides a useful synthesis of 2-cyclohexenones:

Methoxybenzene (anisole) —[Li / liq. NH_3, EtOH]→ 1-Methoxy-1,4-cyclohexadiene (84%) —[H_3O^+ / H_2O]→ 2-Cyclohexenone

PROBLEM 15.25

Birch reduction of toluene leads to a product with the molecular formula C_7H_{10}. On ozonolysis followed by reduction with zinc and acetic acid, the product is transformed into CH_3COCH_2CHO and $OHCCH_2CHO$. What is the structure of the Birch reduction product?

KEY TERMS AND CONCEPTS

Activating group	**Sections 15.10, 15.10A, 15.11A**
Allylic carbocation	**Sections 13.10, 15.15**
Arenium ion	**Section 15.2**
Benzylic cation	**Sections 15.12A, 15.15**
Benzylic radical	**Sections 15.12A, 15.12B**
Deactivating group	**Sections 15.10, 15.10B, 15.10C, 15.11A**
Electrophilic aromatic substitution	**Sections 15.1, 15.2**
Meta director	**Sections 15.10, 15.10B, 15.11C**
Ortho–para director	**Sections 15.10, 15.10A, 15.10C, 15.11D, 15.11E**
Protecting and blocking groups	**Section 15.14A**

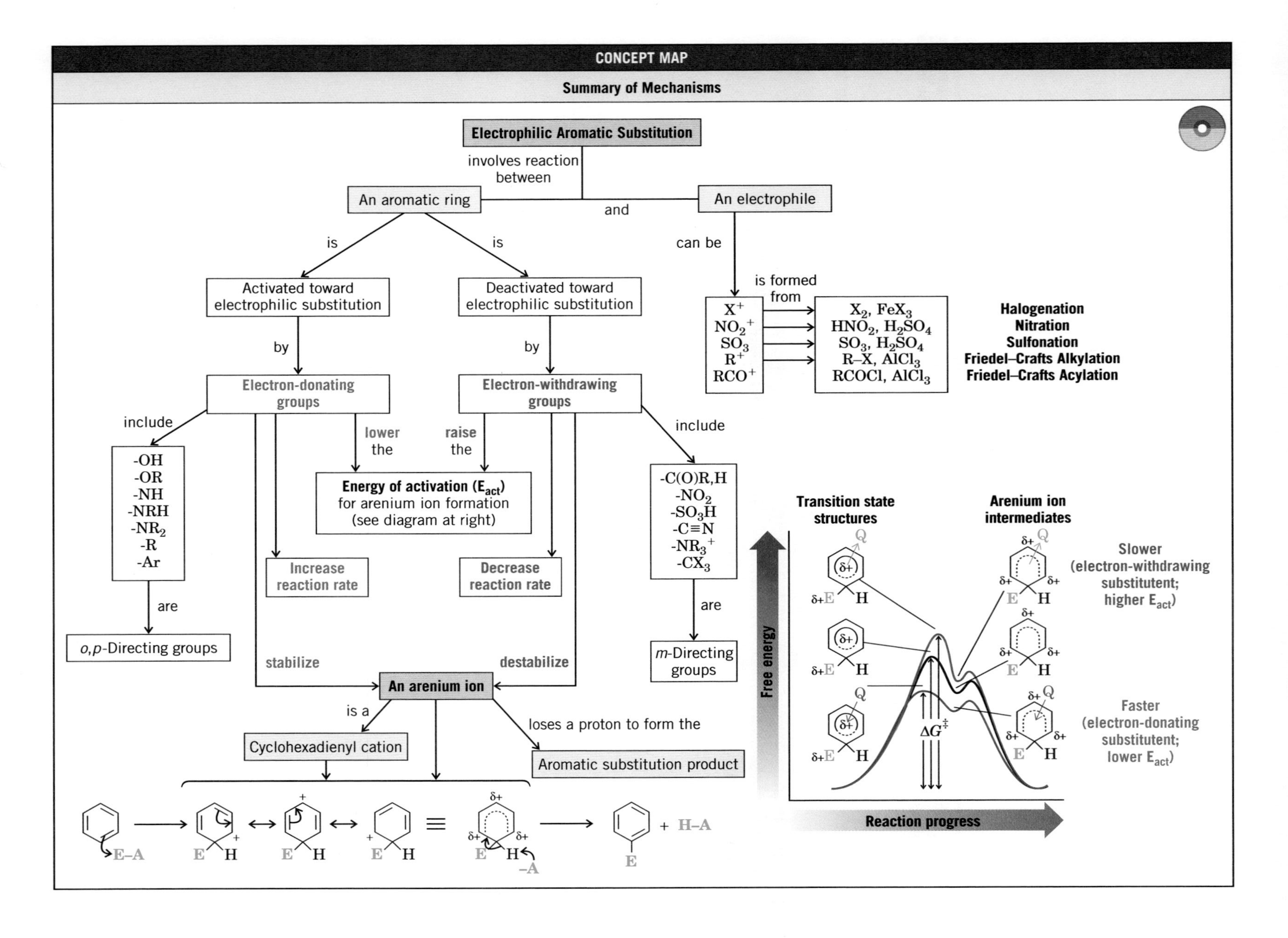

CONCEPT MAP
Summary of Mechanisms
Electrophilic Aromatic Substitution
involves reaction between
An aromatic ring
and
An electrophile
is
is
can be
Activated toward electrophilic substitution
Deactivated toward electrophilic substitution
is formed from
X+ NO2+ SO3 R+ RCO+
X2, FeX3
HNO2, H2SO4
SO3, H2SO4
R–X, AlCl3
RCOCl, AlCl3
Halogenation
Nitration
Sulfonation
Friedel–Crafts Alkylation
Friedel–Crafts Acylation
by
by
Electron-donating groups
Electron-withdrawing groups
include
include
-OH -OR -NH -NRH -NR2 -R -Ar
lower the
raise the
Energy of activation (Eact) for arenium ion formation (see diagram at right)
-C(O)R,H -NO2 -SO3H -C≡N -NR3+ -CX3
Increase reaction rate
Decrease reaction rate
are
are
o,p-Directing groups
m-Directing groups
stabilize
destabilize
An arenium ion
is a
Cyclohexadienyl cation
loses a proton to form the
Aromatic substitution product
E–A
E H
δ+
–A
+ H–A
Transition state structures
Arenium ion intermediates
Q
δ+E
Slower (electron-withdrawing substituent; higher Eact)
Faster (electron-donating substituent; lower Eact)
ΔG‡
Free energy
Reaction progress

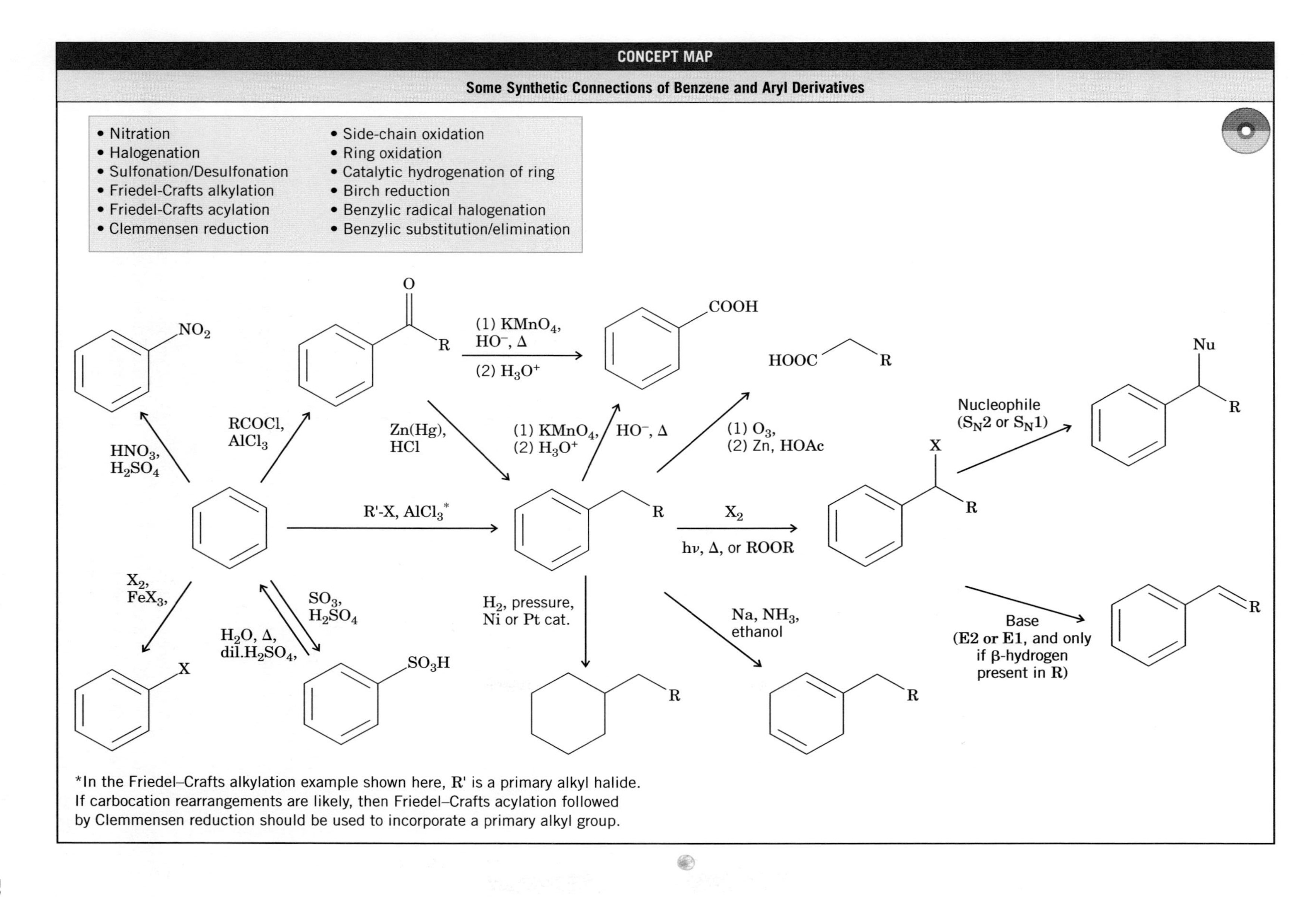
CONCEPT MAP
Some Synthetic Connections of Benzene and Aryl Derivatives
• Nitration
• Halogenation
• Sulfonation/Desulfonation
• Friedel-Crafts alkylation
• Friedel-Crafts acylation
• Clemmensen reduction
• Side-chain oxidation
• Ring oxidation
• Catalytic hydrogenation of ring
• Birch reduction
• Benzylic radical halogenation
• Benzylic substitution/elimination
NO_2
HNO_3, H_2SO_4
RCOCl, $AlCl_3$
O
R
(1) $KMnO_4$, HO^-, Δ
(2) H_3O^+
COOH
Zn(Hg), HCl
(1) $KMnO_4$, (2) H_3O^+
HO^-, Δ
HOOC
R
(1) O_3, (2) Zn, HOAc
R'-X, $AlCl_3$*
R
X_2
hν, Δ, or ROOR
X
R
Nucleophile (S_N2 or S_N1)
Nu
R
X_2, FeX_3,
X
SO_3, H_2SO_4
H_2O, Δ, dil.H_2SO_4,
SO_3H
H_2, pressure, Ni or Pt cat.
R
Na, NH_3, ethanol
R
Base (E2 or E1, and only if β-hydrogen present in R)
R
*In the Friedel–Crafts alkylation example shown here, R' is a primary alkyl halide. If carbocation rearrangements are likely, then Friedel–Crafts acylation followed by Clemmensen reduction should be used to incorporate a primary alkyl group.

Additional Problems

15.26 Give the major product (or products) that would be obtained when each of the following compounds is subjected to ring chlorination with Cl_2 and $FeCl_3$:

(a) Ethylbenzene
(b) Anisole ($C_6H_5OCH_3$)
(c) Fluorobenzene
(d) Benzoic acid
(e) Nitrobenzene
(f) Chlorobenzene
(g) Biphenyl ($C_6H_5—C_6H_5$)
(h) Ethyl phenyl ether

15.27 Predict the major product (or products) formed when each of the following compounds is subjected to ring nitration:

(a) Acetanilide ($C_6H_5NHCOCH_3$)
(b) Phenyl acetate ($CH_3CO_2C_6H_5$)
(c) 4-Chlorobenzoic acid
(d) 3-Chlorobenzoic acid
(e) $C_6H_5COC_6H_5$

15.28 Give the structures of the major products of the following reactions:

(a) Styrene + HCl $\longrightarrow$
(b) 2-Bromo-1-phenylpropane + C_2H_5ONa $\longrightarrow$
(c) $C_6H_5CH_2CHOHCH_2CH_3 \xrightarrow{\text{HA, heat}}$
(d) Product of (c) + HBr $\xrightarrow{\text{peroxides}}$
(e) Product of (c) + $H_2O \xrightarrow[\text{heat}]{\text{HA}}$
(f) Product of (c) + H_2 (1 molar equivalent) $\xrightarrow[25°C]{\text{Pt}}$
(g) Product of (f) $\xrightarrow[\text{(2) } H_3O^+]{\text{(1) } KMnO_4,\ OH^-,\ \text{heat}}$

15.29 Starting with benzene, outline a synthesis of each of the following:

(a) Isopropylbenzene
(b) *tert*-Butylbenzene
(c) Propylbenzene
(d) Butylbenzene
(e) 1-*tert*-Butyl-4-chlorobenzene
(f) 1-Phenylcyclopentene
(g) *trans*-2-Phenylcyclopentanol
(h) *m*-Dinitrobenzene
(i) *m*-Bromonitrobenzene
(j) *p*-Bromonitrobenzene
(k) *p*-Chlorobenzenesulfonic acid
(l) *o*-Chloronitrobenzene
(m) *m*-Nitrobenzenesulfonic acid

15.30 Starting with styrene, outline a synthesis of each of the following:

(a) $C_6H_5CHClCH_2Cl$
(b) $C_6H_5CH_2CH_3$
(c) $C_6H_5CHOHCH_2OH$
(d) $C_6H_5CO_2H$
(e) $C_6H_5CHOHCH_3$
(f) $C_6H_5CHBrCH_3$
(g) $C_6H_5CH_2CH_2OH$
(h) $C_6H_5CH_2CH_2D$
(i) $C_6H_5CH_2CH_2Br$
(j) $C_6H_5CH_2CH_2I$
(k) $C_6H_5CH_2CH_2CN$
(l) $C_6H_5CHDCH_2D$
(m) Cyclohexylbenzene
(n) $C_6H_5CH_2CH_2OCH_3$

15.31 Starting with toluene, outline a synthesis of each of the following:

(a) *m*-Chlorobenzoic acid
(b) *p*-Methylacetophenone
(c) 2-Bromo-4-nitrotoluene
(d) *p*-Bromobenzoic acid
(e) 1-Chloro-3-trichloromethylbenzene
(f) *p*-Isopropyltoluene (*p*-cymene)
(g) 1-Cyclohexyl-4-methylbenzene
(h) 2,4,6-Trinitrotoluene (TNT)
(i) 4-Chloro-2-nitrobenzoic acid
(j) 1-Butyl-4-methylbenzene

15.32 Starting with aniline, outline a synthesis of each of the following:

Note: Problems marked with an asterisk are "challenge problems."

(a) *p*-Bromoaniline
(b) *o*-Bromoaniline
(c) 2-Bromo-4-nitroaniline
(d) 4-Bromo-2-nitroaniline
(e) 2,4,6-Tribromoaniline

15.33 Both of the following syntheses will fail. Explain what is wrong with each one.

(a) Benzene $\xrightarrow[(3)\ Zn(Hg),\ HCl]{(1)\ HNO_3/H_2SO_4\quad (2)\ CH_3COCl/AlCl_3}$ 1-ethyl-3-nitrobenzene (NO_2, CH_2CH_3 meta)

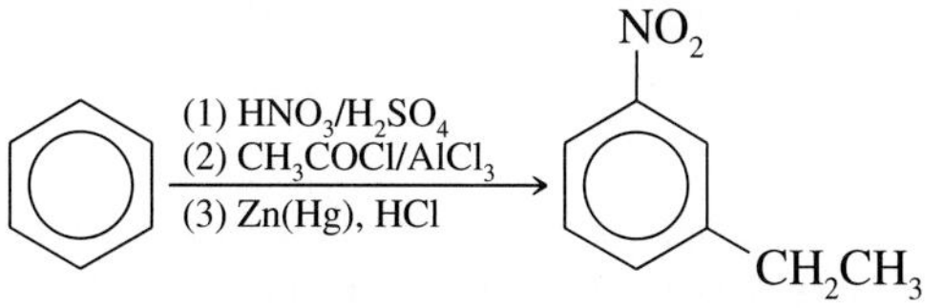

(b) Ethylbenzene (CH_2CH_3) $\xrightarrow[(3)\ Br_2,\ FeBr_3]{(1)\ NBS,\ CCl_4,\ light\quad (2)\ NaOEt,\ EtOH,\ heat}$ 4-bromostyrene ($CH{=}CH_2$, Br para)

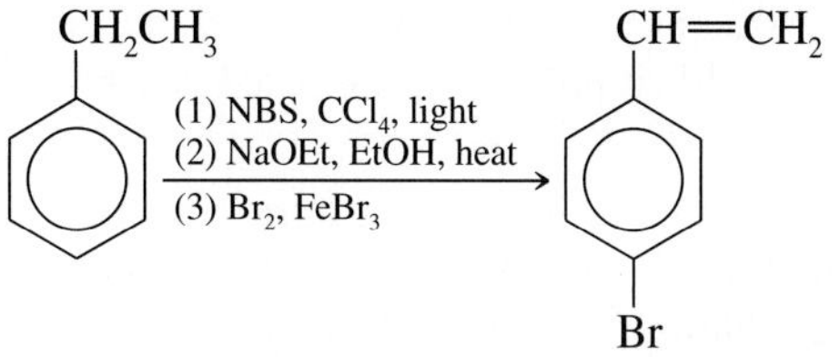

15.34 One ring of phenyl benzoate undergoes electrophilic aromatic substitution much more readily than the other. **(a)** Which one is it? **(b)** Explain your answer.

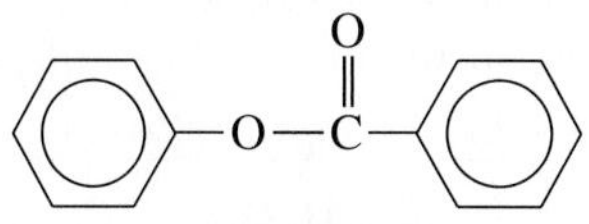

$C_6H_5{-}O{-}C(=O){-}C_6H_5$

Phenyl benzoate

15.35 What monobromination product (or products) would you expect to obtain when the following compounds undergo ring bromination with Br_2 and $FeBr_3$?

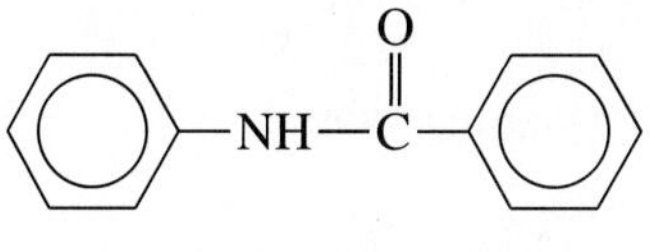

(a) $C_6H_5{-}CH_2{-}C(=O){-}C_6H_5$

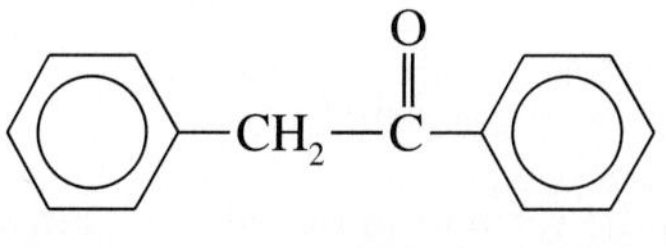

(b) $C_6H_5{-}NH{-}C(=O){-}C_6H_5$

(c) $C_6H_5{-}C(=O){-}C(=O){-}C_6H_5$

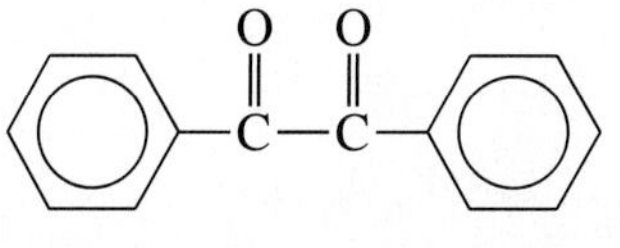

15.36 Many polycyclic aromatic compounds have been synthesized by a cyclization reaction known as the **Bradsher reaction** or **aromatic cyclodehydration.** This method can be illustrated by the following synthesis of 9-methylphenanthrene:

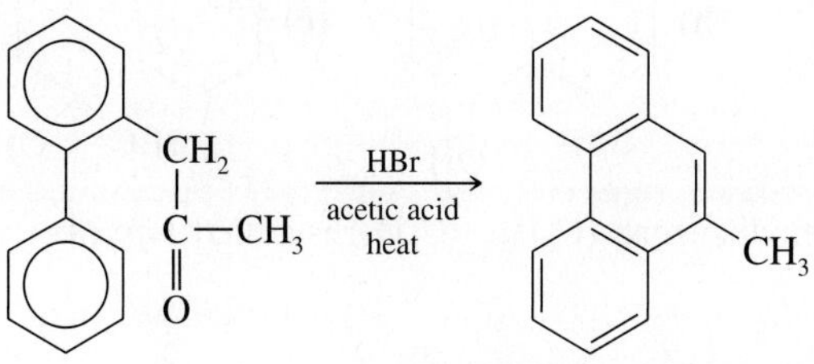

2-Biphenylyl–$CH_2{-}C(=O){-}CH_3$ $\xrightarrow[acetic\ acid,\ heat]{HBr}$ 9-methylphenanthrene (CH_3)

9-Methylphenanthrene

An arenium ion is an intermediate in this reaction, and the last step involves the dehydration of an alcohol. Propose a plausible mechanism for this example of the Bradsher reaction.

15.37 Propose structures for compounds **G–I:**

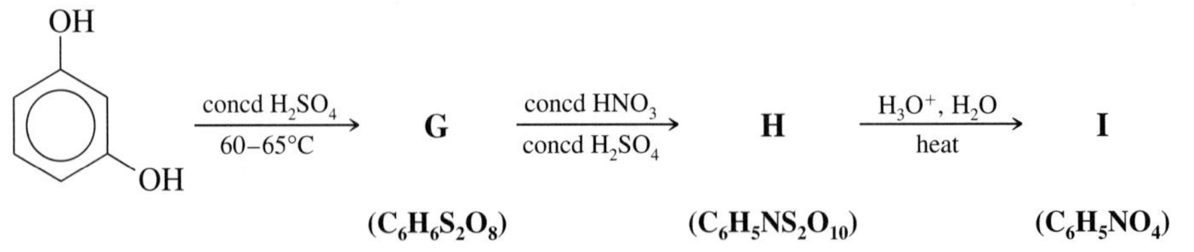

1,3-Benzenediol (resorcinol) $\xrightarrow[\text{60–65°C}]{\text{concd } H_2SO_4}$ **G** $(C_6H_6S_2O_8)$ $\xrightarrow[\text{concd } H_2SO_4]{\text{concd } HNO_3}$ **H** $(C_6H_5NS_2O_{10})$ $\xrightarrow[\text{heat}]{H_3O^+,\ H_2O}$ **I** $(C_6H_5NO_4)$

15.38 2,6-Dichlorophenol has been isolated from the females of two species of ticks (*Amblyomma americanum* and *A. maculatum*), where it apparently serves as a sex attractant. Each female tick yields about 5 ng of 2,6-dichlorophenol. Assume that you need larger quantities than this and outline a synthesis of 2,6-dichlorophenol from phenol. (*Hint:* When phenol is sulfonated at 100°C, the product is chiefly *p*-hydroxybenzenesulfonic acid.)

15.39 The addition of a hydrogen halide (hydrogen bromide or hydrogen chloride) to 1-phenyl-1,3-butadiene produces (only) 1-phenyl-3-halo-1-butene. **(a)** Write a mechanism that accounts for the formation of this product. **(b)** Is this 1,4 addition or 1,2 addition to the butadiene system? **(c)** Is the product of the reaction consistent with the formation of the most stable intermediate carbocation? **(d)** Does the reaction appear to be under kinetic control or equilibrium control? Explain.

15.40 2-Methylnaphthalene can be synthesized from toluene through the following sequence of reactions. Write the structure of each intermediate.

Toluene + succinic anhydride $\xrightarrow{AlCl_3}$ **A** $(C_{11}H_{12}O_3)$ $\xrightarrow[HCl]{Zn(Hg)}$ **B** $(C_{11}H_{14}O_2)$

$\xrightarrow{SOCl_2}$ **C** $(C_{11}H_{13}ClO)$ $\xrightarrow{AlCl_3}$ **D** $(C_{11}H_{12}O)$ $\xrightarrow{NaBH_4}$ **E** $(C_{11}H_{14}O)$

$\xrightarrow[\text{heat}]{H_2SO_4}$ **F** $(C_{11}H_{12})$ $\xrightarrow[CCl_4,\ \text{light}]{\text{NBS}}$ **G** $(C_{11}H_{12}Br)$ $\xrightarrow[\text{EtOH, heat}]{\text{NaOEt}}$ 2-methylnaphthalene

15.41 Ring nitration of a dimethylbenzene (a xylene) results in the formation of only one nitrodimethylbenzene. What is the structure of the dimethylbenzene?

15.42 Write mechanisms that account for the products of the following reactions:

(a) 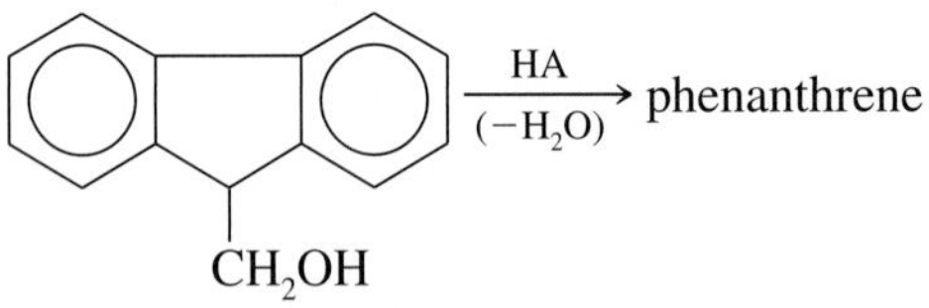 9-Fluorenylmethanol (CH_2OH) $\xrightarrow[(-H_2O)]{HA}$ phenanthrene

(b) 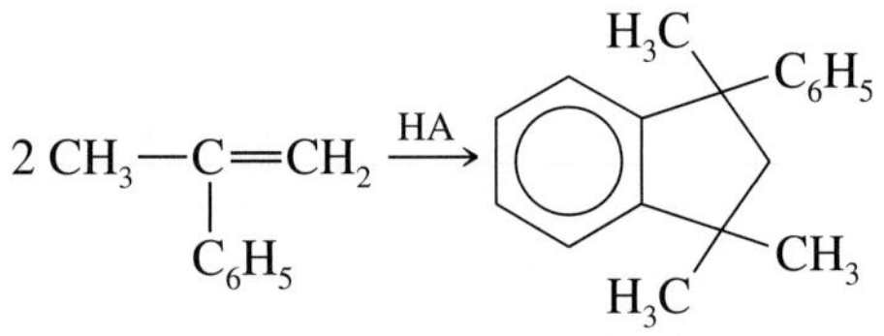 $2\ CH_3{-}C(C_6H_5){=}CH_2 \xrightarrow{HA}$ indane bearing H_3C, C_6H_5, CH_3, H_3C substituents

15.43 Show how you might synthesize each of the following starting with α-tetralone (Section 15.9):

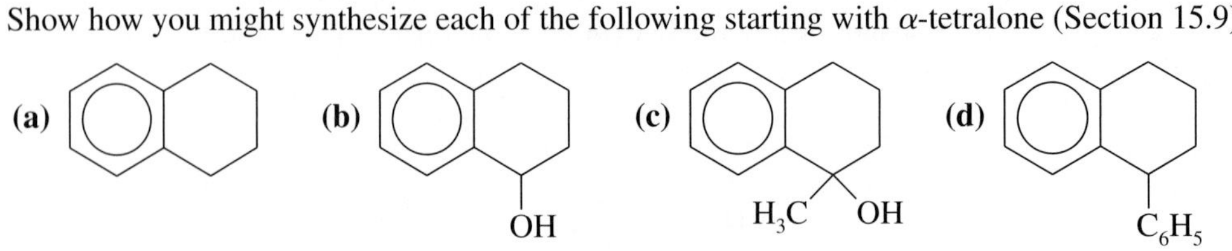

(a) tetralin **(b)** 1-tetralol (OH) **(c)** 1-methyl-1-tetralol (H_3C, OH) **(d)** 1-phenyltetralin (C_6H_5)

15.44 The compound phenylbenzene ($C_6H_5{-}C_6H_5$) is called *biphenyl,* and the rings are numbered in the following manner:

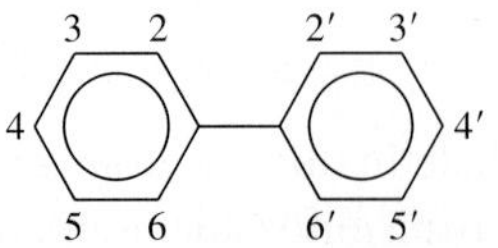

Biphenyl: positions 2, 3, 4, 5, 6 on one ring and 2′, 3′, 4′, 5′, 6′ on the other.

Use models to answer the following questions about substituted biphenyls. **(a)** When certain large groups occupy three or four of the *ortho* positions (e.g., 2, 6, 2′, and 6′), the substituted biphenyl may exist in enantiomeric forms. An example of a biphenyl that exists in enantiomeric forms is the compound in which the following substituents are present: 2-NO_2, 6-CO_2H, 2′-NO_2, 6′-CO_2H. What factors account for this? **(b)** Would you expect a biphenyl with 2-Br, 6-CO_2H, 2′-CO_2H, 6′-H to exist in enantiomeric forms? **(c)** The biphenyl with 2-NO_2, 6-NO_2, 2′-CO_2H, 6′-Br cannot be resolved into enantiomeric forms. Explain.

15.45 Give structures (including stereochemistry where appropriate) for compounds **A–G:**

(a) Benzene + $CH_3CH_2\overset{O}{\overset{\|}{C}}Cl \xrightarrow{AlCl_3}$ **A** $\xrightarrow[0°C]{PCl_5}$ **B** ($C_9H_{10}Cl_2$) $\xrightarrow[\text{mineral oil, heat}]{2\ NaNH_2}$

C (C_9H_8) $\xrightarrow{H_2, Ni_2B\ (P\text{-}2)}$ **D** (C_9H_{10})

Hint: The ^1H NMR spectrum of compound **C** consists of a multiplet at δ 7.20 (5H) and a singlet at δ 2.0 (3H).

(b) **C** $\xrightarrow[(2)\ H_2O]{(1)\ Li,\ liq.\ NH_3}$ **E** (C_9H_{10})

(c) **D** $\xrightarrow[2\text{–}5°C]{Br_2,\ CCl_4}$ **F** + enantiomer (major products)

(d) **E** $\xrightarrow[2\text{–}5°C]{Br_2,\ CCl_4}$ **G** + enantiomer (major products)

15.46 Treating cyclohexene with acetyl chloride and $AlCl_3$ leads to the formation of a product with the molecular formula $C_8H_{13}ClO$. Treating this product with a base leads to the formation of 1-acetylcyclohexene. Propose mechanisms for both steps of this sequence of reactions.

15.47 The *tert*-butyl group can be used as a blocking group in certain syntheses of aromatic compounds. **(a)** How would you introduce a *tert*-butyl group? **(b)** How would you remove it? **(c)** What advantage might a *tert*-butyl group have over a —SO_3H group as a blocking group?

15.48 When toluene is sulfonated (concentrated H_2SO_4) at room temperature, predominantly (about 95% of the total) ortho and para substitution occurs. If elevated temperatures (150–200°C) and longer reaction times are employed, meta (chiefly) and para substitution account for some 95% of the products. Account for these differences. (*Hint: m*-Toluenesulfonic acid is the most stable isomer.)

15.49 A C—D bond is harder to break than a C—H bond, and, consequently, reactions in which C—D bonds are broken proceed more slowly than reactions in which C—H bonds are broken. What mechanistic information comes from the observation that perdeuterated benzene, C_6D_6, is nitrated at the same rate as normal benzene, C_6H_6?

15.50 Show how you might synthesize each of the following compounds starting with either benzyl bromide or allyl bromide:

(a) $C_6H_5CH_2CN$ **(c)** $C_6H_5CH_2O_2CCH_3$ **(e)** $CH_2{=}CHCH_2N_3$
(b) $C_6H_5CH_2OCH_3$ **(d)** $C_6H_5CH_2I$ **(f)** $CH_2{=}CHCH_2OCH_2CH(CH_3)_2$

15.51 Provide structures for compounds **A, B,** and **C**:

Benzene $\xrightarrow[liq.\ NH_3,\ EtOH]{Na}$ **A** (C_6H_8) $\xrightarrow[CCl_4]{NBS}$ **B** (C_6H_7Br) $\xrightarrow{(CH_3)_2CuLi}$ **C** (C_7H_{10})

15.52 Heating 1,1,1-triphenylmethanol with ethanol containing a trace of a strong acid causes the formation of 1-ethoxy-1,1,1-triphenylmethane. Write a plausible mechanism that accounts for the formation of this product.

15.53 **(a)** Which of the following halides would you expect to be most reactive in an S_N2 reaction? **(b)** In an S_N1 reaction? Explain your answers.

$CH_3CH_2CH{=}CHCH_2Br$ $CH_3CH{=}CHCHBrCH_3$ $CH_2{=}CHCBr(CH_3)_2$

15.54 Acetanilide was subjected to the following sequence of reactions: (1) concd H_2SO_4; (2) HNO_3, heat; (3) H_2O, H_2SO_4, heat, then OH^-. The ^{13}C NMR spectrum of the final product gives six signals. Write the structure of the final product.

***15.55** The lignins are macromolecules that are major components of the many types of wood, where they bind cellulose fibers together in these natural composites. The lignins are built up out of a variety of small molecules (most having phenylpropane skeletons). These precursor molecules are covalently connected in varying ways, and this gives the lignins great complexity. To explain the formation of compound **B** below as one of many products obtained when lignins are ozonized, lignin model compound **A** was treated as shown. Use the following information to determine the structure of **B**.

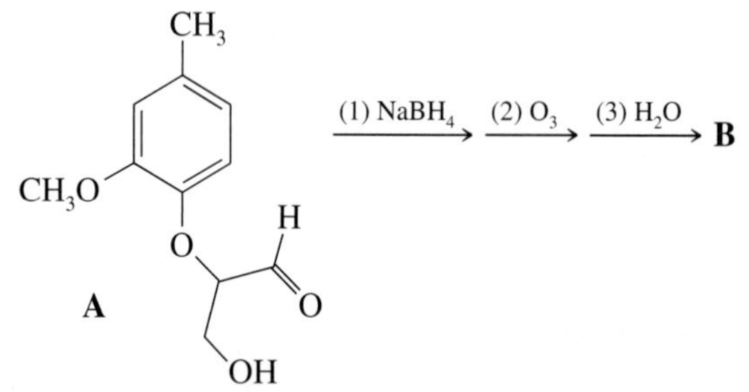

CH_3, CH_3O, O, H, O, OH — **A** $\xrightarrow{(1)\ NaBH_4}\xrightarrow{(2)\ O_3}\xrightarrow{(3)\ H_2O}$ **B**

To make **B** volatile enough for GC/MS (gas chromatography–mass spectrometry, Section 9.17), it was first converted to its tris(*O*-trimethylsilyl) derivative, which had $M^{+\cdot}$ 308 *m/z*. ["Tris" means that three of the indicated complex groups named (e.g., trimethylsilyl groups here) are present. The capital, italicized *O* means these are attached to oxygen atoms of the parent compound, taking the place of hydrogen atoms. Similarly, the prefix "bis" indicates the presence of two complex groups subsequently named, and "tetrakis" (used in the problem below), means four.] The IR spectrum of **B** had a broad absorption at 3400 cm^{-1}, and its ^1H NMR spectrum showed a single multiplet at δ 3.6. What is the structure of **B**?

***15.56** When compound **C**, which is often used to model a more frequently occurring unit in lignins, was ozonized, product **D** was obtained. In a variety of ways it has been established that the stereochemistry of the three-carbon side chain of such lignin units remains largely if not completely unchanged during oxidations like this.

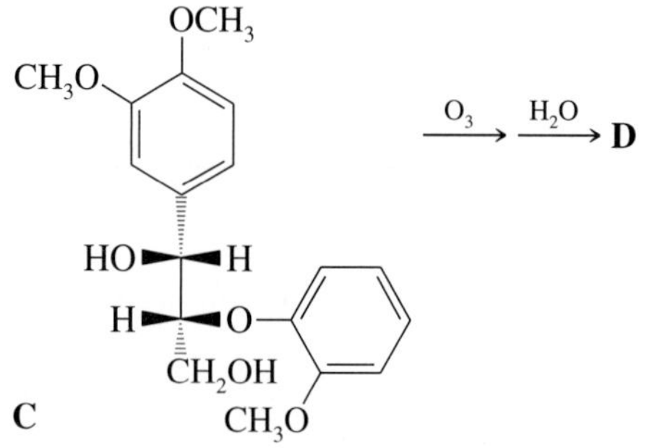

OCH_3, CH_3O, HO, H, H, O, CH_2OH, CH_3O — **C** $\xrightarrow{O_3}\xrightarrow{H_2O}$ **D**

For GC/MS, **D** was converted to its tetrakis(*O*-trimethylsilyl) derivative, which had $M^{+\cdot}$ 424 *m/z*. The IR spectrum of **D** has bands at 3000 cm^{-1} (broad, strong) and 1710 cm^{-1} (strong). Its ^1H NMR spectrum had peaks at δ 3.7 (multiplet, 3H) and δ 4.2 (doublet, 1H) after treatment with D_2O. Its DEPT ^{13}C NMR spectra had peaks at δ 64 (CH_2), δ 75 (CH), δ 82 (CH), and δ 177 (C). What is the structure of **D**, including its stereochemistry?

LEARNING GROUP PROBLEMS

1. The structure of thyroxine, a thyroid hormone that helps to regulate metabolic rate, was determined in part by comparison with a synthetic compound believed to have the same structure as natural thyroxine. The final step in the laboratory synthesis of thyroxine by Harington and Barger, shown below, involves an electrophilic aromatic substitution. Draw a detailed mechanism for this step and explain why the iodine substitutions occur ortho to the phenolic hydroxyl and not ortho to the oxygen of the aryl ether. [One reason iodine is required in our diet (e.g., in iodized salt), of course, is for the biosynthesis of thyroxine.]

I_2, HO^- → **Thyroxine**

2. Synthesize 2-chloro-4-nitrobenzoic acid from toluene and any other reagents necessary. Begin by writing a retrosynthetic analysis.
3. Deduce the structures of compounds **E–L** in the roadmap below.

E ($C_8H_{13}Br$) —HBr (no peroxides)→ **F** (meso) + **G** (racemate) ($C_8H_{14}Br_2$)

E —*tert*-BuOK, *tert*-BuOH, heat→ **H** (C_8H_{12})

H —Br_2, warm→ dibromide

H —EtO_2C–CH=CH–CO_2Et→ **I** —(1) O_3 (2) $NaHSO_3$→ diketone diester

H —maleic anhydride→ **J** ($C_{12}H_{14}O_3$)

J —toluene, $AlCl_3$→ **K** ($C_{19}H_{22}O_3$) —(1) Zn(Hg), HCl, reflux (2) $SOCl_2$ (3) $AlCl_3$→ **L** —Zn(Hg), HCl, reflux→ product (CH_3)

CHAPTER 16

Aldehydes and Ketones I. Nucleophilic Addition to the Carbonyl Group

A Very Versatile Vitamin, Pyridoxine (Vitamin B_6)

Pyridoxal phosphate (PLP) is at the heart of chemistry conducted by a number of enzymes. Many of us know the coenzyme pyridoxal phosphate by the closely related vitamin from which it is derived in our diet—pyridoxine, or vitamin B_6. Wheat is a good dietary source of vitamin B_6. Although pyridoxal phosphate (shown by the photo above) is a member of the aldehyde family, when it is involved in biological chemistry, it often contains a closely related functional group that has a carbon–nitrogen double bond, called an imine. We will study aldehydes, imines, and related groups over the course of this chapter.

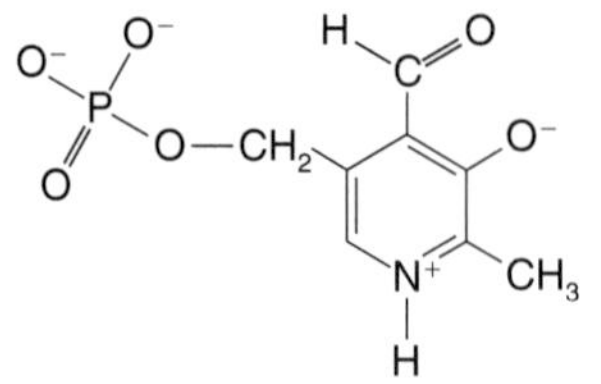

Pyridoxal phosphate 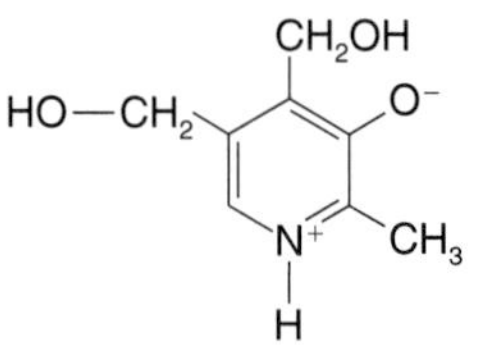 **Pyridoxine**

Some enzymatic reactions that involve PLP include *transaminations,* which convert amino acids to ketones for use in the citric acid cycle and other pathways; *decarboxylation* of amino acids for biosynthesis of neurotransmitters such as histamine, dopamine, and serotonin; and *inversion* of amino acid stereogenic centers, such as required for the biosynthesis of cell walls in bacteria.

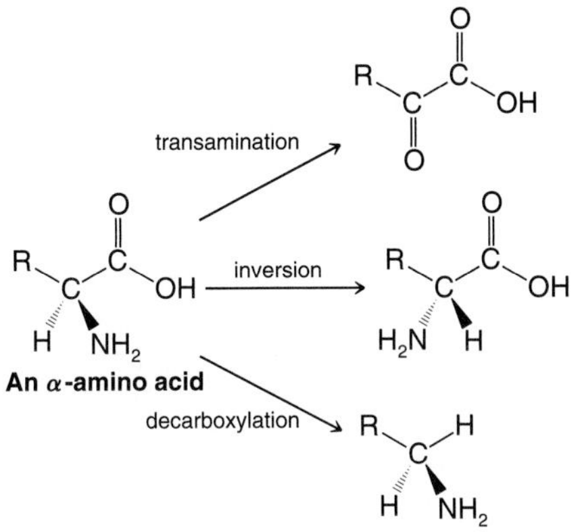

transamination

inversion

decarboxylation

An α-amino acid

In all of these reactions, as well as a number of other types, the essential role of PLP is to stabilize a carbanion intermediate by acting as a sink for electron density. Aspects of these transformations are described further in "The Chemistry of Pyridoxal Phosphate" later in this chapter (see Section 16.8). All of the reactions of PLP are wonderful examples of how biological processes exemplify organic chemistry in action.

16.1 Introduction

All aldehydes have a carbonyl group $\mathrm{>C{=}O}$, bonded on one side to a carbon and on the other side to a hydrogen. In ketones, the carbonyl group is situated between two carbon atoms.

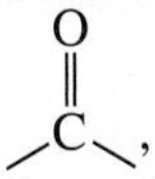

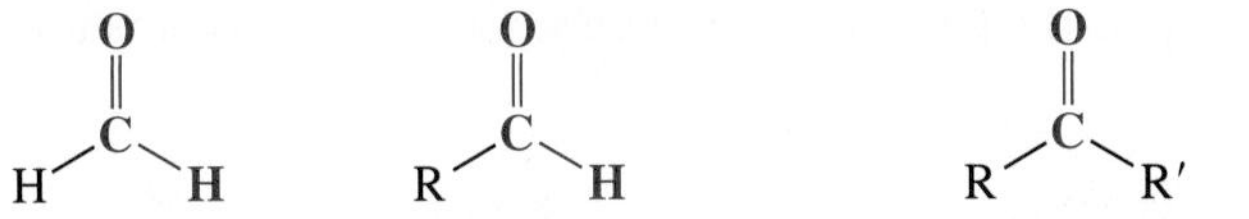

Formaldehyde	RCHO **General formulas for an aldehyde**	RCOR′ **General formulas for a ketone**

Although earlier chapters have given us some insight into the chemistry of carbonyl compounds, we shall now consider their chemistry in detail. The reason: The chemistry of the carbonyl group is central to the chemistry of most of the chapters that follow.

In this chapter our attention is focused on the preparation of aldehydes and ketones, on their physical properties, and especially on *the nucleophilic addition reactions that take place at their carbonyl groups.* In Chapter 17 we shall study the chemistry of aldehydes and ketones *that results from the acidity of the hydrogen atoms on the carbon atoms adjacent to their carbonyl groups.*

16.2 Nomenclature of Aldehydes and Ketones

In the IUPAC system aliphatic aldehydes are named *substitutively* by replacing the final **-e** of the name of the corresponding alkane with **-al.** Since the aldehyde group must be at an end of the chain of carbon atoms, there is no need to indicate its position. When other substituents are present, however, the carbonyl group carbon is assigned position 1. Many aldehydes also have common names; these are given here in parentheses. These common names are derived from the common names for the corresponding carboxylic acids (Section 18.2A), and some of them are retained by the IUPAC as acceptable names.

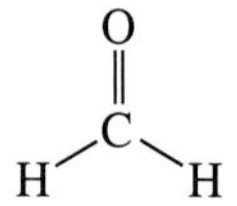

Methanal (formaldehyde) 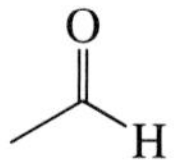 **Ethanal (acetaldehyde)** 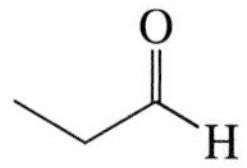 **Propanal (propionaldehyde)**

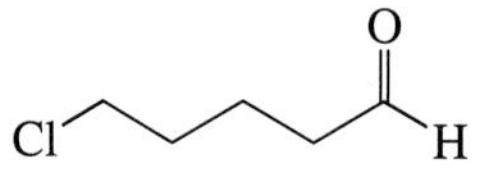

5-Chloropentanal 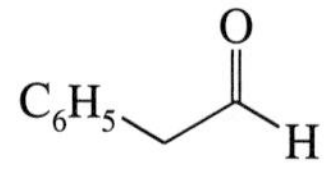 **Phenylethanal (phenylacetaldehyde)**

Aldehydes in which the —CHO group is attached to a ring system are named substitutively by adding the suffix *carbaldehyde.* Several examples follow:

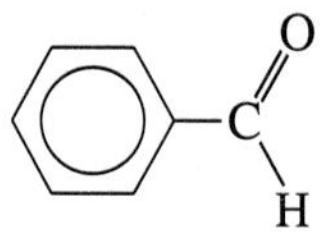

Benzenecarbaldehyde (benzaldehyde) 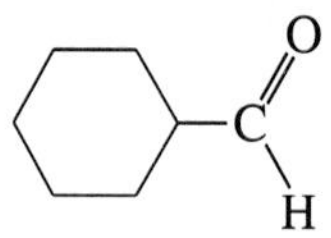 **Cyclohexanecarbaldehyde** 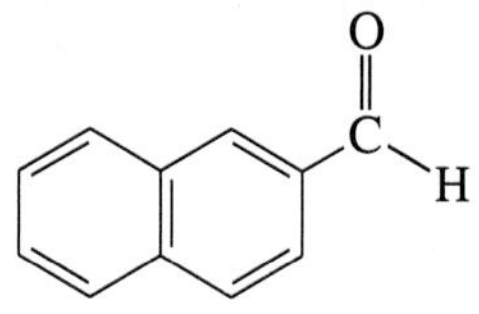 **2-Naphthalenecarbaldehyde**

The common name *benzaldehyde* is far more frequently used than benzenecarbaldehyde for C_6H_5CHO, and it is the name we shall use in this text.

Aliphatic ketones are named substitutively by replacing the final **-e** of the name of the corresponding alkane with **-one.** The chain is then numbered in the way that gives the carbonyl carbon atom the lower possible number, and this number is used to designate its position.

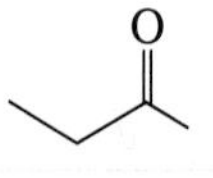

Butanone (ethyl methyl ketone) 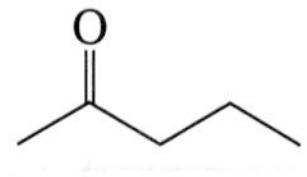 **2-Pentanone (methyl propyl ketone)** 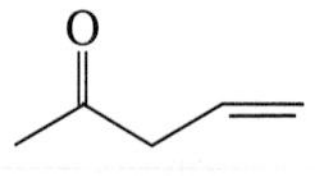 **Pent-4-en-2-one (*not* 1-penten-4-one) (allyl methyl ketone)**

Common functional group names for ketones (in parentheses above) are obtained simply by separately naming the two groups attached to the carbonyl group and adding the word **ketone** as a separate word.

Some ketones have common names that are retained in the IUPAC system:

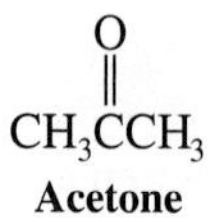

CH_3COCH_3 **Acetone (propanone)** $C_6H_5COCH_3$ **Acetophenone (1-phenylethanone or methyl phenyl ketone)** 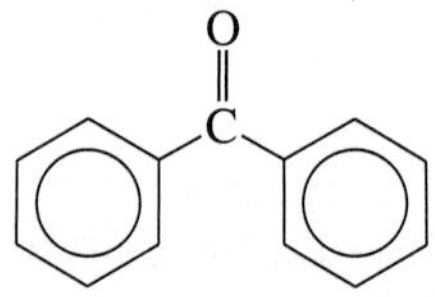 **Benzophenone (diphenylmethanone or diphenyl ketone)**

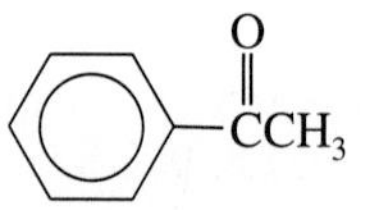

When it is necessary to name the $-\overset{\text{O}}{\overset{\|}{\text{C}}}\text{H}$ group as a prefix, it is the **methanoyl** or **formyl group.** The $\text{CH}_3\overset{\text{O}}{\overset{\|}{\text{C}}}-$ group is called the **ethanoyl** or **acetyl group** (often abbreviated as Ac). When $\text{R}\overset{\text{O}}{\overset{\|}{\text{C}}}-$ groups are named as substituents, they are called **alkanoyl** or **acyl groups.**

2-Methanoylbenzoic acid
(*o*-formylbenzoic acid)

4-Ethanoylbenzenesulfonic acid
(*p*-acetylbenzenesulfonic acid)

PROBLEM 16.1

(a) Give IUPAC substitutive names for the seven isomeric aldehydes and ketones with the formula $C_5H_{10}O$. **(b)** Give structures and names (common or IUPAC substitutive names) for all the aldehydes and ketones that contain a benzene ring and have the formula C_8H_8O.

16.3 Physical Properties

The carbonyl group is a polar group; therefore, aldehydes and ketones have higher boiling points than hydrocarbons of the same molecular weight. However, since aldehydes and ketones cannot have strong hydrogen bonds *between their molecules,* they have lower boiling points than the corresponding alcohols. The following compounds that have similar molecular weights exemplify this trend:

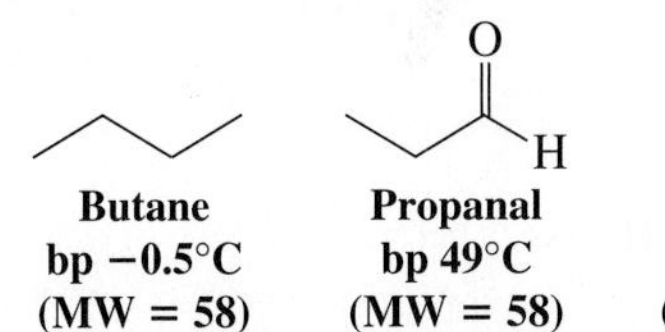

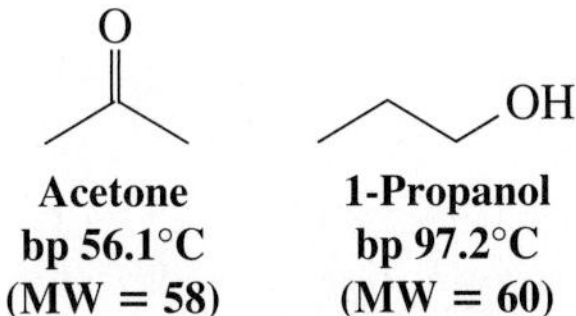

Butane	Propanal	Acetone	1-Propanol
bp −0.5°C	bp 49°C	bp 56.1°C	bp 97.2°C
(MW = 58)	(MW = 58)	(MW = 58)	(MW = 60)

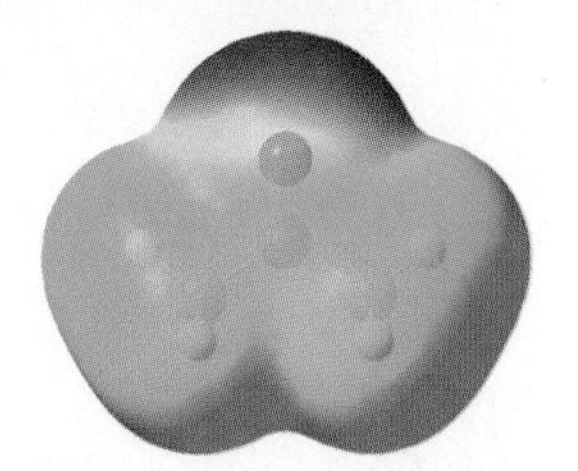

A map of electrostatic potential for acetone shows the polarity of the carbonyl C═O bond.

PROBLEM 16.2

Which compound in each of the following pairs has the higher boiling point? (Answer this problem without consulting tables.)

(a) Pentanal or 1-pentanol
(b) 2-Pentanone or 2-pentanol
(c) Pentane or pentanal
(d) Acetophenone or 2-phenylethanol
(e) Benzaldehyde or benzyl alcohol

The carbonyl oxygen atom allows molecules of aldehydes and ketones to form strong hydrogen bonds to molecules of water. As a result, low-molecular-weight aldehydes and ketones show appreciable solubilities in water. Acetone and acetaldehyde are soluble in water in all proportions.

Hydrogen bonding (shown in red) between water molecules and acetone

TABLE 16.1 Physical Properties of Aldehydes and Ketones

Formula	Name	mp (°C)	bp (°C)	Solubility in Water
HCHO	Formaldehyde	−92	−21	Very soluble
CH_3CHO	Acetaldehyde	−125	21	∞
CH_3CH_2CHO	Propanal	−81	49	Very soluble
$CH_3(CH_2)_2CHO$	Butanal	−99	76	Soluble
$CH_3(CH_2)_3CHO$	Pentanal	−91.5	102	Slightly soluble
$CH_3(CH_2)_4CHO$	Hexanal	−51	131	Slightly soluble
C_6H_5CHO	Benzaldehyde	−26	178	Slightly soluble
$C_6H_5CH_2CHO$	Phenylacetaldehyde	33	193	Slightly soluble
CH_3COCH_3	Acetone	−95	56.1	∞
$CH_3COCH_2CH_3$	Butanone	−86	79.6	Very soluble
$CH_3COCH_2CH_2CH_3$	2-Pentanone	−78	102	Soluble
$CH_3CH_2COCH_2CH_3$	3-Pentanone	−39	102	Soluble
$C_6H_5COCH_3$	Acetophenone	21	202	Insoluble
$C_6H_5COC_6H_5$	Benzophenone	48	306	Insoluble

Table 16.1 lists the physical properties of a number of common aldehydes and ketones. Some aromatic aldehydes obtained from natural sources have very pleasant fragrances. Some of these are the following:

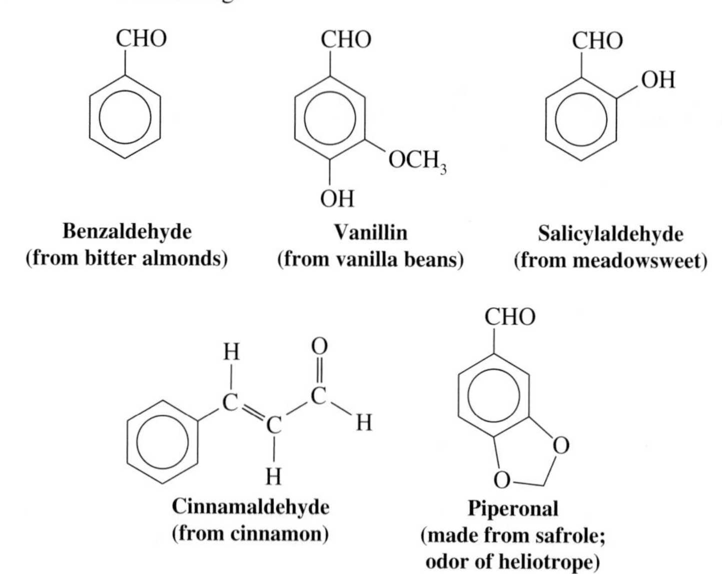

Benzaldehyde (from bitter almonds) **Vanillin (from vanilla beans)** **Salicylaldehyde (from meadowsweet)**

Cinnamaldehyde (from cinnamon) **Piperonal (made from safrole; odor of heliotrope)**

16.4 Synthesis of Aldehydes

16.4A Aldehydes by Oxidation of 1° Alcohols

We learned in Section 12.4A that the oxidation state of aldehydes lies between that of 1° alcohols and carboxylic acids and that aldehydes can be prepared from 1° alcohols by oxidation with pyridinium chlorochromate (PCC):

$$\mathrm{R{-}CH_2OH} \underset{[H]}{\overset{[O]}{\rightleftharpoons}} \mathrm{R{-}\overset{O}{\overset{\|}{C}}{-}H} \underset{[H]}{\overset{[O]}{\rightleftharpoons}} \mathrm{R{-}\overset{O}{\overset{\|}{C}}{-}OH}$$

1° Alcohol **Aldehyde** **Carboxylic acid**

$$\mathrm{R{-}CH_2OH} \xrightarrow[\mathrm{CH_2Cl_2}]{\mathrm{PCC}\ (\mathrm{C_5H_5NH^+CrO_3Cl^-})} \mathrm{R{-}\overset{O}{\overset{\|}{C}}{-}H}$$

1° Alcohol **Aldehyde**

An example of this synthesis of aldehydes is the oxidation of 1-heptanol to heptanal:

$$\mathrm{CH_3(CH_2)_5CH_2OH} \xrightarrow[\mathrm{CH_2Cl_2}]{\mathrm{PCC}\ (\mathrm{C_5H_5NH^+CrO_3Cl^-})} \mathrm{CH_3(CH_2)_5CHO}$$

1-Heptanol **Heptanal (93%)**

16.4B Aldehydes by Reduction of Acyl Chlorides, Esters, and Nitriles

Theoretically, it ought to be possible to prepare aldehydes by reduction of carboxylic acids. In practice, this is not possible, because the reagent normally used to reduce a carboxylic acid directly is lithium aluminum hydride ($LiAlH_4$ or LAH), and when any carboxylic acid is treated with LAH, it is reduced all the way to the 1° alcohol. This happens because LAH is a very powerful reducing agent and aldehydes are very easily reduced. Any aldehyde that might be formed in the reaction mixture is immediately reduced by LAH to the 1° alcohol. (It does not help to use a stoichiometric amount of LAH, because as soon as the first few molecules of aldehyde are formed in the mixture, there will still be much unreacted LAH present and it will reduce the aldehyde.)

$$\mathrm{R{-}\overset{O}{\overset{\|}{C}}{-}OH} \xrightarrow{\mathrm{LiAlH_4}} \left[\mathrm{R{-}\overset{O}{\overset{\|}{C}}{-}H}\right] \xrightarrow{\mathrm{LiAlH_4}} \mathrm{R{-}CH_2OH}$$

Carboxylic acid **Aldehyde** **1° Alcohol**

The secret to success here is not to use a carboxylic acid itself, but to use a derivative of a carboxylic acid that is more easily reduced and an aluminum hydride derivative that is less reactive than LAH. We shall study derivatives of carboxylic acids in detail in Chapter 18, but suffice it to say here that acyl chlorides (RCOCl), esters (RCO_2R'), and nitriles (RCN) are all easily prepared from carboxylic acids, and they all are more easily reduced. (Acyl chlorides, esters, and nitriles all also have the same oxidation state as carboxylic acids. Convince yourself of this by applying the principles that you learned in Problem 12.1.) Two derivatives of aluminum hydride that are less reactive than LAH (in part because they are much more sterically hindered and, therefore, have difficulty in transferring hydride ions) are lithium tri-*tert*-butoxyaluminum hydride and diisobutylaluminum hydride (DIBAL-H):

$$\mathrm{Li^+}\left[\mathrm{H{-}Al(O{-}C(CH_3)_3)_3}\right]^-$$

Lithium tri-*tert*-butoxyaluminum hydride

$$\mathrm{H{-}Al(CH_2CH(CH_3)_2)_2}$$

Diisobutylaluminum hydride (abbreviated *i*-Bu_2AlH or DIBAL-H)

The following scheme summarizes how these reagents are used to synthesize aldehydes from acid derivatives:

$$\underset{\textbf{Acyl chloride}}{\text{RCOCl}} \xrightarrow[\text{(2) } H_2O]{\text{(1) LiAlH(O-}t\text{-Bu)}_3,\ -78°C} \underset{\textbf{Aldehyde}}{\text{RCHO}}$$

$$\underset{\textbf{Ester}}{\text{RCOOR}'} \xrightarrow[\text{(2) } H_2O]{\text{(1) DIBAL-H, hexane, } -78°C} \underset{\textbf{Aldehyde}}{\text{RCHO}}$$

$$\underset{\textbf{Nitrile}}{\text{R—C}\equiv\text{N}} \xrightarrow[\text{(2) } H_2O]{\text{(1) DIBAL-H, hexane}} \underset{\textbf{Aldehyde}}{\text{RCHO}}$$

We now examine each of these aldehyde syntheses in more detail.

Aldehydes from Acyl Chlorides: RCOCl ⟶ RCHO Acyl chlorides can be reduced to aldehydes by treating them with lithium tri-*tert*-butoxyaluminum hydride, $LiAlH[OC(CH_3)_3]_3$, at −78°C. (Carboxylic acids can be converted to acyl chlorides by using $SOCl_2$; see Section 15.7.)

$$\text{RCOOH} \xrightarrow{SOCl_2} \text{RCOCl} \xrightarrow[\text{(2) } H_2O]{\text{(1) LiAlH(O-}t\text{-Bu)}_3,\ Et_2O,\ -78°C} \text{RCHO}$$

The following is a specific example:

$$\underset{\textbf{3-Methoxy-4-methylbenzoyl chloride}}{\text{3-CH}_3\text{O-4-CH}_3\text{C}_6\text{H}_3\text{COCl}} \xrightarrow[\text{(2) } H_2O]{\text{(1) LiAlH(O-}t\text{-Bu)}_3,\ Et_2O,\ -78°C} \underset{\textbf{3-Methoxy-4-methylbenzaldehyde}}{\text{3-CH}_3\text{O-4-CH}_3\text{C}_6\text{H}_3\text{CHO}}$$

Mechanistically, the reduction is brought about by the transfer of a hydride ion from the aluminum atom to the carbonyl carbon of the acyl chloride (see Section 12.3). Subsequent hydrolysis frees the aldehyde.

A Mechanism for the Reaction

Reduction of an Acyl Chloride to an Aldehyde

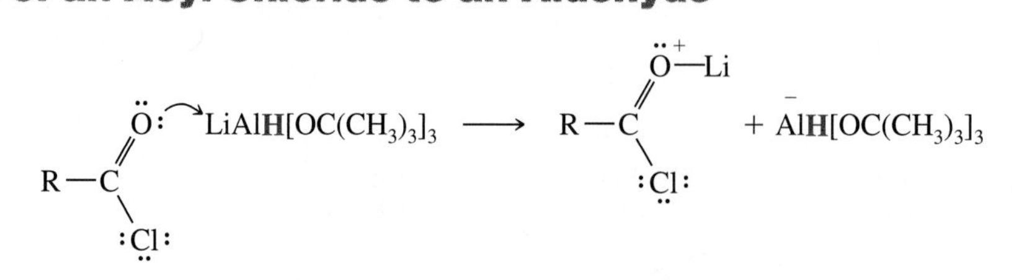

Transfer of a hydride ion to the carbonyl carbon brings about the reduction.

Acting as a Lewis acid, the aluminum atom accepts an electron pair from oxygen.

This intermediate loses a chloride ion as an electron pair from the oxygen assists.

The addition of water causes hydrolysis of this aluminum complex to take place, producing the aldehyde. (Several steps are involved.)

Aldehydes from Esters and Nitriles: $RCO_2R' \longrightarrow RCHO$ and $RC{\equiv}N \longrightarrow RCHO$ Both esters and nitriles can be reduced to aldehydes by use of DIBAL-H. Carefully controlled amounts of the reagent must be used to avoid overreduction, and the ester reduction must be carried out at low temperatures. Both reductions result in the formation of a relatively stable intermediate by the addition of a hydride ion to the carbonyl carbon of the ester or to the carbon of the $—C{\equiv}N$ group of the nitrile. Hydrolysis of the intermediate liberates the aldehyde. Schematically, the reactions can be viewed in the following way.

A Mechanism for the Reaction

Reduction of an Ester to an Aldehyde

The aluminum atom accepts an electron pair from the carbonyl oxygen atom in a Lewis acid–base reaction.

Transfer of a hydride ion to the carbonyl carbon brings about its reduction.

This intermediate loses an alkoxide ion as an electron pair from the oxygen assists.

The addition of water causes hydrolysis of this aluminum complex to take place, producing the aldehyde. (Several steps are involved.)

A Mechanism for the Reaction

Reduction of a Nitrile to an Aldehyde

$$R-C\equiv N{:} \;\; Al(i\text{-}Bu)_2H \longrightarrow \left[R-C\equiv \overset{+}{N}-\overset{-}{Al}(i\text{-}Bu)_2H\right] \longrightarrow$$

The aluminum atom accepts an electron pair from the nitrile in a Lewis acid–base reaction.

Transfer of a hydride ion to the nitrile carbon brings about its reduction.

$$\left[R-\underset{H}{C}{=}\ddot{N}-Al(i\text{-}Bu)_2\right] \xrightarrow{H_2O} R-\underset{H}{C}{=}\ddot{O}{:}$$

The addition of water causes hydrolysis of this aluminum complex to take place, producing the aldehyde. (Several steps are involved.)

The following specific examples illustrate these syntheses:

$$CH_3(CH_2)_{10}\overset{O}{\overset{\|}{C}}OEt \xrightarrow[\text{hexane, } -78°C]{(i\text{-}Bu)_2AlH} CH_3(CH_2)_{10}\underset{OEt}{\overset{OAl(i\text{-}Bu)_2}{CH}} \xrightarrow{H_2O} CH_3(CH_2)_{10}\overset{O}{\overset{\|}{C}}H \;\; \mathbf{(88\%)}$$

$$CH_3CH{=}CHCH_2CH_2CH_2C\equiv N \xrightarrow[\text{hexane}]{(i\text{-}Bu)_2AlH} CH_3CH{=}CHCH_2CH_2CH_2\overset{NAl(i\text{-}Bu)_2}{\overset{\|}{C}H} \xrightarrow{H_2O} CH_3CH{=}CHCH_2CH_2CH_2\overset{O}{\overset{\|}{C}}H$$

PROBLEM 16.3

Show how you would synthesize propanal from each of the following: **(a)** 1-propanol and **(b)** propanoic acid ($CH_3CH_2CO_2H$).

16.5 Synthesis of Ketones

16.5A Ketones from Alkenes, Arenes, and 2° Alcohols

We have seen three laboratory methods for the preparation of ketones in earlier chapters:

1. Ketones (and aldehydes) by ozonolysis of alkenes (discussed in Section 8.17B).

$$\underset{R'}{\overset{R}{}}C{=}C\underset{H}{\overset{R''}{}} \xrightarrow[(2)\ Zn,\ HOAc]{(1)\ O_3} \underset{\textbf{Ketone}}{\underset{R'}{\overset{R}{}}C{=}O} + \underset{\textbf{Aldehyde}}{O{=}C\underset{H}{\overset{R''}{}}}$$

2. Ketones from arenes by Friedel–Crafts acylations (discussed in Section 15.7):

$$ArH + R-\overset{O}{\overset{\|}{C}}-Cl \xrightarrow{AlCl_3} \underset{\textbf{An alkyl aryl ketone}}{Ar-\overset{O}{\overset{\|}{C}}-R} + HCl$$

Alternatively,

$$\text{ArH} + \text{Ar}-\overset{\overset{\displaystyle O}{\|}}{C}-\text{Cl} \xrightarrow{AlCl_3} \underset{\textbf{A diaryl ketone}}{\text{Ar}-\overset{\overset{\displaystyle O}{\|}}{C}-\text{Ar}} + \text{HCl}$$

3. Ketones from secondary alcohols by oxidation (discussed in Section 12.4):

$$\text{R}-\overset{\overset{\displaystyle OH}{|}}{CH}-\text{R}' \xrightarrow{H_2CrO_4} \text{R}-\overset{\overset{\displaystyle O}{\|}}{C}-\text{R}'$$

16.5B Ketones from Alkynes

Alkynes add water readily when the reaction is catalyzed by strong acids and mercuric ions (Hg^{2+}). Aqueous solutions of sulfuric acid and mercuric sulfate are often used for this purpose. The vinylic alcohol that is initially produced is usually unstable, and it rearranges rapidly to a ketone (or in the case of ethyne, to ethanal). The rearrangement involves the loss of a proton from the hydroxyl group, the addition of a proton to the vicinal carbon atom, and the relocation of the double bond:

$$-C{\equiv}C- + H-OH \xrightarrow[H_2SO_4]{HgSO_4} \left[\underset{\textbf{A vinylic alcohol (unstable)}}{\overset{H}{\ }C{=}C\,OH}\right] \longrightarrow \underset{\textbf{Ketone}}{-\overset{H}{\underset{H}{C}}-\overset{}{C}{=}O}$$

This kind of rearrangement, known as a **tautomerization,** is acid catalyzed and occurs in the following way:

$$H-\overset{+}{\underset{H}{O}}-H + \underset{\textbf{Vinylic alcohol}}{C{=}C-\ddot{O}-H} \longrightarrow -\overset{H}{\underset{H}{C}}-C{=}\overset{+}{O}-H \xrightarrow{H_2\ddot{O}:} \underset{\textbf{Ketone}}{-\overset{H}{\underset{H}{C}}-C{=}\ddot{O}:} + H_3O^+$$

The vinylic alcohol accepts a proton at one carbon atom of the double bond to yield a cationic intermediate, which then loses a proton from the oxygen atom to produce a ketone.

Vinylic alcohols are often called **enols** (after *-en*, the ending for alkenes, plus *-ol*, the ending for alcohols). The product of the rearrangement is usually a ketone, and these rearrangements are known as **keto–enol tautomerizations:**

$$\underset{\textbf{Enol form}}{C{=}C-\ddot{O}-H} \overset{HA}{\rightleftharpoons} \underset{\textbf{Keto form}}{-\underset{H}{C}-C{=}\ddot{O}:}$$

We examine this phenomenon in greater detail in Section 17.2.

The addition of water to alkynes follows Markovnikov's rule—the hydrogen atom becomes attached to the carbon atom with the greater number of hydrogen atoms. Therefore, when terminal alkynes other than ethyne (acetylene) are hydrated, ketones, rather than aldehydes, are the products. Also note that the method is not useful for internal alkynes because a mixture results.

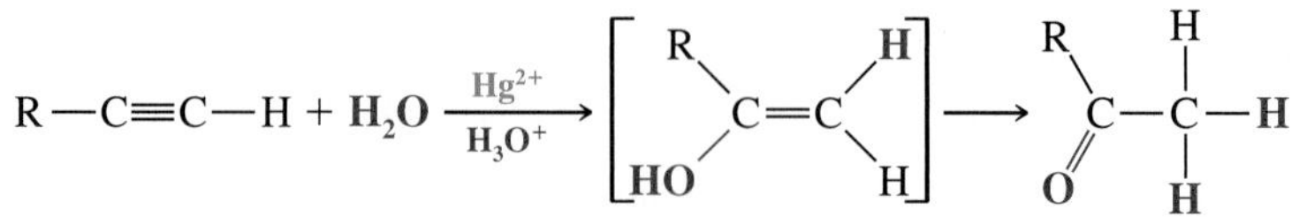

$$R{-}C{\equiv}C{-}H + H_2O \xrightarrow[H_3O^+]{Hg^{2+}} \left[(R)(HO)C{=}CH_2 \right] \longrightarrow R{-}C({=}O){-}CH_3$$

A ketone

Two examples of this ketone synthesis are listed here:

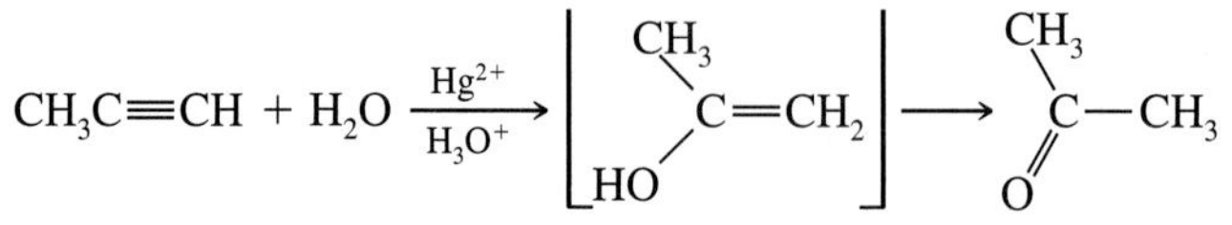

$$CH_3C{\equiv}CH + H_2O \xrightarrow[H_3O^+]{Hg^{2+}} \left[(CH_3)(HO)C{=}CH_2 \right] \longrightarrow CH_3{-}C({=}O){-}CH_3$$

Acetone

$$CH_3CH_2CH_2CH_2C{\equiv}CH + H_2O \xrightarrow[H_2SO_4]{HgSO_4} CH_3CH_2CH_2CH_2CCH_3 \ (C{=}O)$$

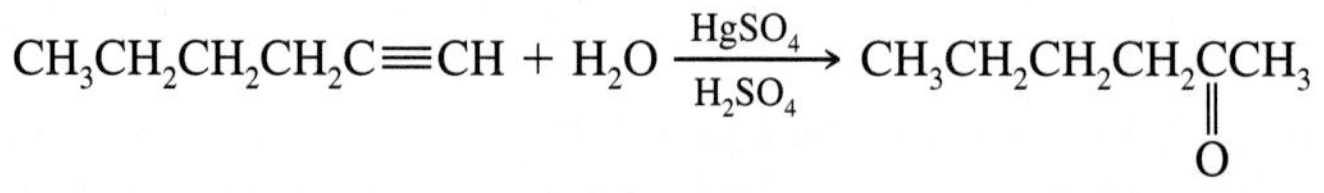

(80%)

When ethyne itself undergoes addition of water, the product is an aldehyde:

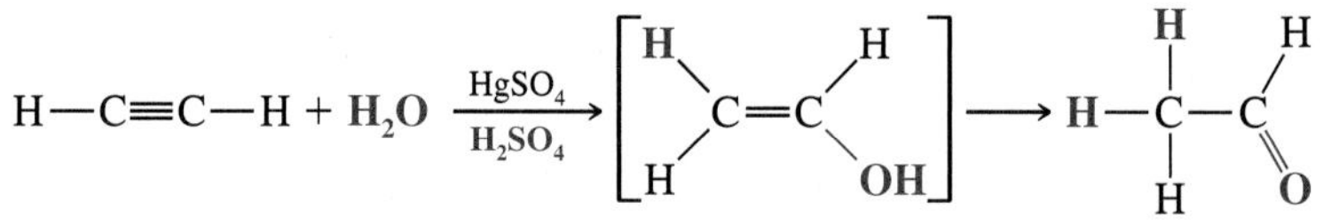

$$H{-}C{\equiv}C{-}H + H_2O \xrightarrow[H_2SO_4]{HgSO_4} \left[(H)(H)C{=}C(H)(OH) \right] \longrightarrow H{-}CH_2{-}C({=}O){-}H$$

Ethyne **Ethanal (acetaldehyde)**

This method has been important in the commercial production of ethanal.

Two other laboratory methods for the preparation of ketones are based on the use of organometallic compounds as discussed next.

16.5C Ketones from Lithium Dialkylcuprates

When an ether solution of a lithium dialkylcuprate is treated with an acyl chloride at −78°C, the product is a ketone. This ketone synthesis is a variation of the Corey–Posner, Whitesides–House alkane synthesis (Section 12.9).

General Reaction

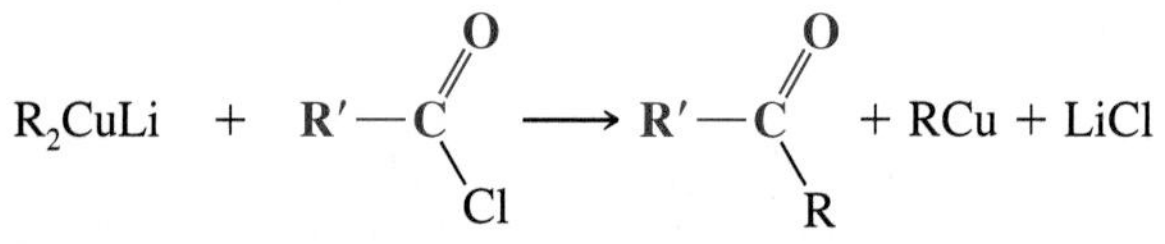

$$R_2CuLi + R'{-}C({=}O){-}Cl \longrightarrow R'{-}C({=}O){-}R + RCu + LiCl$$

Lithium dialkylcuprate **Acyl chloride** **Ketone**

Specific Example

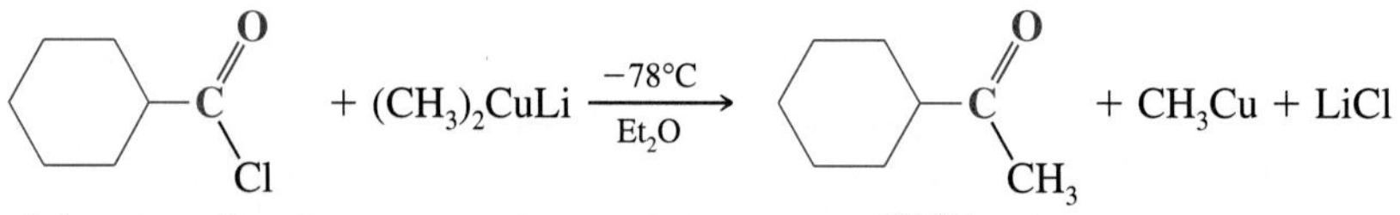

$$C_6H_{11}{-}C({=}O){-}Cl + (CH_3)_2CuLi \xrightarrow[Et_2O]{-78°C} C_6H_{11}{-}C({=}O){-}CH_3 + CH_3Cu + LiCl$$

Cyclohexanecarbonyl chloride

(81%) 1-Cyclohexylethanone (Cyclohexyl methyl ketone)

16.5D Ketones from Nitriles

Treating a nitrile (R—C≡N) with either a Grignard reagent or an organolithium reagent followed by hydrolysis yields a ketone.

General Reactions

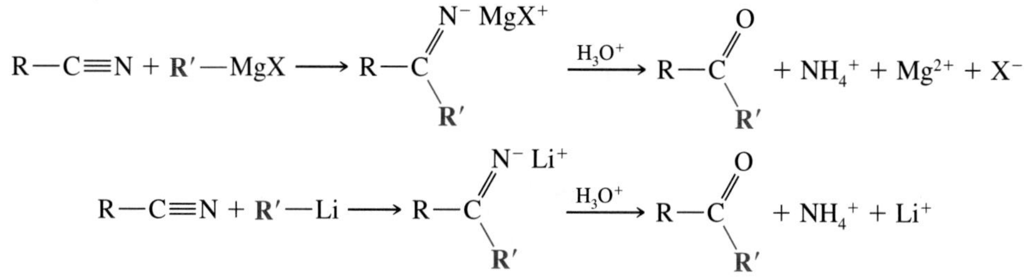

$$R{-}C{\equiv}N + R'{-}MgX \longrightarrow R{-}C(=N^- MgX^+){-}R' \xrightarrow{H_3O^+} R{-}C(=O){-}R' + NH_4^+ + Mg^{2+} + X^-$$

$$R{-}C{\equiv}N + R'{-}Li \longrightarrow R{-}C(=N^- Li^+){-}R' \xrightarrow{H_3O^+} R{-}C(=O){-}R' + NH_4^+ + Li^+$$

The mechanism for the acidic hydrolysis step is the reverse of one that we shall study for imine formation in Section 16.8.

Specific Examples

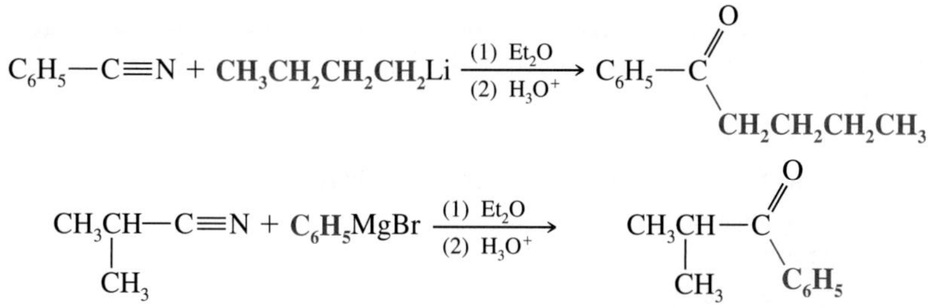

$$C_6H_5{-}C{\equiv}N + CH_3CH_2CH_2CH_2Li \xrightarrow[(2)\ H_3O^+]{(1)\ Et_2O} C_6H_5{-}C(=O){-}CH_2CH_2CH_2CH_3$$

$$CH_3CH(CH_3){-}C{\equiv}N + C_6H_5MgBr \xrightarrow[(2)\ H_3O^+]{(1)\ Et_2O} CH_3CH(CH_3){-}C(=O){-}C_6H_5$$

2-Cyanopropane → **2-Methyl-1-phenylpropanone (isopropyl phenyl ketone)**

Even though a nitrile has a triple bond, addition of the Grignard or lithium reagent takes place only once. The reason: If addition took place twice, this would place a double negative charge on the nitrogen.

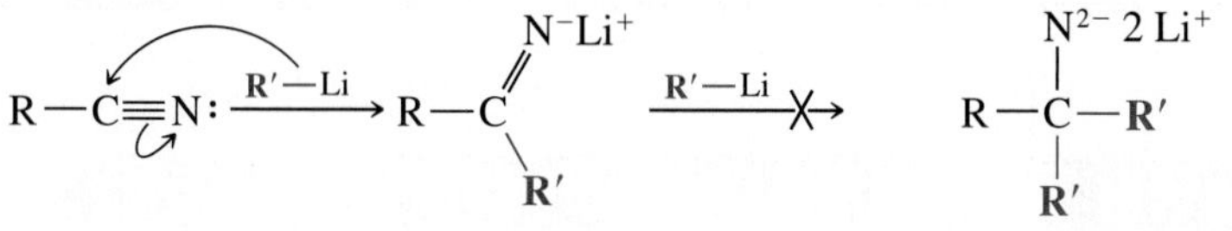

$$R{-}C{\equiv}N{:} \xrightarrow{R'{-}Li} R{-}C(=N^-Li^+){-}R' \xrightarrow{R'{-}Li}\!\!\!\!\!\!\times\ R{-}C(N^{2-}\ 2\,Li^+)(R'){-}R'$$

(The dianion does not form.)

SOLVED PROBLEM

ILLUSTRATING A MULTISTEP SYNTHESIS

With 1-butanol as your only organic starting compound, devise a synthesis of 5-nonanone. Begin by writing a retrosynthetic analysis.

ANSWER Retrosynthetic disconnection of 5-nonanone (see the following page) suggests butylmagnesium bromide and pentane-nitrile as immediate precursors. Butylmagnesium bromide can, in turn, be synthesized from 1-bromobutane. Pentanenitrile can also be synthesized from 1-bromobutane, via S_N2 reaction of 1-bromobutane with cyanide. To begin the synthesis, 1-bromobutane can be prepared from 1-butanol by reaction with phosphorus tribromide.

Retrosynthetic Analysis

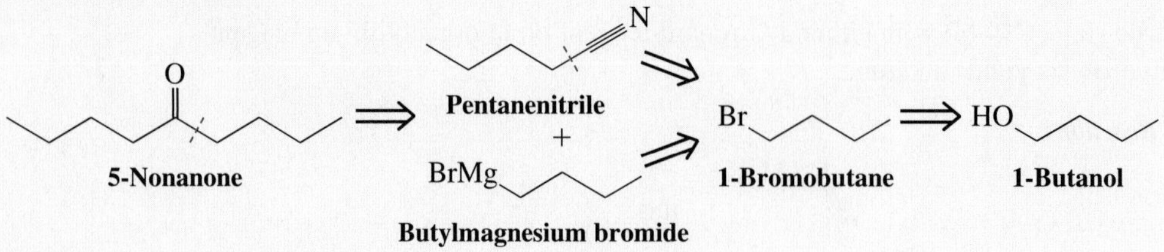

5-Nonanone ⟹ Pentanenitrile + Butylmagnesium bromide ⟹ 1-Bromobutane ⟹ 1-Butanol

Synthesis

1-Butanol $\xrightarrow{PBr_3}$ 1-Bromobutane; NaCN → pentanenitrile; Mg → BrMg-butyl; (1) Add RMgBr to RCN (2) H_3O^+ → 5-nonanone

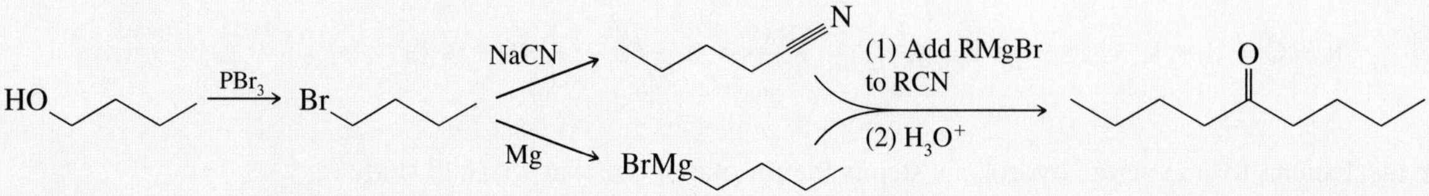

PROBLEM 16.4 Which reagents would you use to carry out each of the following reactions?

(a) Benzene ⟶ bromobenzene ⟶ phenylmagnesium bromide ⟶ benzyl alcohol ⟶ benzaldehyde
(b) Toluene ⟶ benzoic acid ⟶ benzoyl chloride ⟶ benzaldehyde
(c) Ethyl bromide ⟶ 1-butyne ⟶ butanone
(d) 2-Butyne ⟶ butanone
(e) 1-Phenylethanol ⟶ acetophenone
(f) Benzene ⟶ acetophenone
(g) Benzoyl chloride ⟶ acetophenone
(h) Benzoic acid ⟶ acetophenone
(i) Benzyl bromide ⟶ $C_6H_5CH_2CN$ ⟶ 1-phenyl-2-butanone
(j) $C_6H_5CH_2CN$ ⟶ 2-phenylethanal
(k) $CH_3(CH_2)_4CO_2CH_3$ ⟶ hexanal

16.6 Nucleophilic Addition to the Carbon–Oxygen Double Bond

The most characteristic reaction of aldehydes and ketones is *nucleophilic addition* to the carbon–oxygen double bond.

General Reaction

$$\text{R(H)C{=}O} + \text{H{-}Nu} \rightleftharpoons \text{R{-}C(Nu)(H){-}OH}$$

Specific Examples

$$\text{H}_3\text{C(H)C{=}O} + \text{H{-}OCH}_2\text{CH}_3 \rightleftharpoons \text{CH}_3\text{{-}C(OCH}_2\text{CH}_3\text{)(H){-}OH}$$

A hemiacetal (see Section 16.7)

$$(H_3C)_2C{=}O + H{-}CN \rightleftharpoons CH_3{-}C(CN)(OH){-}CH_3$$

A cyanohydrin (see Section 16.9)

Aldehydes and ketones are especially susceptible to nucleophilic addition because of the structural features that we discussed in Section 12.1 and which are shown below.

Aldehyde or ketone (R or R′ may be H) — **The nucleophile may attack from above or below.**

The trigonal planar arrangement of groups around the carbonyl carbon atom means that the carbonyl carbon atom is relatively open to attack from above or below the plane of the carbonyl group (see above). The positive charge on the carbonyl carbon atom means that it is especially susceptible to attack by a nucleophile. The negative charge on the carbonyl oxygen atom means that nucleophilic addition is susceptible to acid catalysis. Nucleophilic addition to the carbon–oxygen double bond occurs, therefore, in either of two general ways.

1. When the reagent is a strong nucleophile (Nu:⁻), addition usually takes place in the following way, converting the trigonal planar aldehyde or ketone into a tetrahedral product.

A Mechanism for the Reaction

Addition of a Strong Nucleophile to an Aldehyde or Ketone

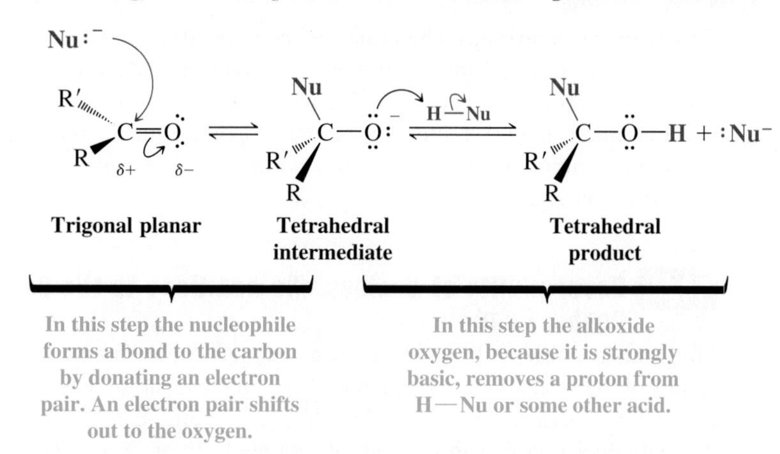
Nu:⁻
R′
R
C=O
δ+
δ−
Nu
R′
R
C—O:⁻
H—Nu
Nu
R′
R
C—O—H + :Nu⁻
Trigonal planar
Tetrahedral intermediate
Tetrahedral product
In this step the nucleophile forms a bond to the carbon by donating an electron pair. An electron pair shifts out to the oxygen.
In this step the alkoxide oxygen, because it is strongly basic, removes a proton from H—Nu or some other acid.

In this type of addition the nucleophile uses its electron pair to form a bond to the carbonyl carbon atom. As this happens the electron pair of the carbon–oxygen π bond shifts out to the electronegative carbonyl oxygen atom and the hybridization state of both the carbon and the oxygen changes from sp^2 to sp^3. ***The important aspect of this step is the ability of the carbonyl oxygen atom to accommodate the electron pair of the carbon–oxygen double bond.***

In the second step the oxygen atom accepts a proton. This happens because the oxygen atom is now much more basic; it carries a full negative charge as an alkoxide anion.

2. A second general mechanism that operates in nucleophilic additions to carbon–oxygen double bonds is an acid-catalyzed mechanism.

A Mechanism for the Reaction

Acid-Catalyzed Nucleophilic Addition to an Aldehyde or Ketone

Step 1

$$\underset{R}{\overset{R'}{>}}\overset{\delta+}{C}{=}\overset{\delta-}{\ddot{O}}: + H{-}A \rightleftharpoons \left[\underset{R}{\overset{R'}{>}}C{=}\overset{+}{\ddot{O}}H \longleftrightarrow \underset{R}{\overset{R'}{>}}\overset{+}{C}{-}\ddot{O}H\right] + A:^-$$

(or a Lewis acid)

In this step an electron pair of the carbonyl oxygen accepts a proton from the acid (or associates with a Lewis acid), producing an oxonium cation. The carbon of the oxonium cation is more susceptible to nucleophilic attack than the carbonyl of the starting ketone.

Step 2

$$\underset{R}{\overset{R'}{>}}C{=}\overset{+}{\ddot{O}}H \;\; :Nu{-}H \rightleftharpoons R'R\,\overset{\overset{H}{|}\;^{+}Nu}{C}{-}\ddot{O}{-}H \;\; :A^- \rightleftharpoons R'R\,\overset{Nu:}{C}{-}\ddot{O}{-}H + H{-}A$$

In the first of these two steps, the oxonium cation accepts the electron pair of the nucleophile. In the second step, a base removes a proton from the positively charged atom, regenerating the acid.

Any compound containing a positively charged oxygen atom that forms three covalent bonds is an *oxonium cation.*

This mechanism operates when carbonyl compounds are treated with *strong acids* in the presence of *weak nucleophiles.* In the first step the acid donates a proton to an electron pair of the carbonyl oxygen atom. The resulting protonated carbonyl compound, an **oxonium cation,** is highly reactive toward nucleophilic attack at the carbonyl carbon atom because the carbonyl carbon atom carries more positive charge than it does in the unprotonated compound.

16.6A Reversibility of Nucleophilic Additions to the Carbon–Oxygen Double Bond

Many nucleophilic additions to carbon–oxygen double bonds are reversible; the overall results of these reactions depend, therefore, on the position of an equilibrium. This behavior contrasts markedly with most electrophilic additions to carbon–carbon double bonds and with nucleophilic substitutions at saturated carbon atoms. The latter reactions are essentially irreversible, and overall results are a function of relative reaction rates.

16.6B Relative Reactivity: Aldehydes versus Ketones

In general, **aldehydes are more reactive in nucleophilic additions than are ketones.** Both steric and electronic factors favor aldehydes: With one group being the small hydrogen atom, the central carbon of the tetrahedral product formed from an aldehyde is less crowded and the product is more stable. Formation of the product, therefore, is favored at equilibrium. With ketones, the two alkyl substituents at the carbonyl carbon cause greater steric crowding in the tetrahedral product and make it less stable. Therefore, a smaller concentration of the product is present at equilibrium.

Because alkyl groups are electron releasing, **aldehydes are more reactive on electronic grounds as well.** Aldehydes have only one electron-releasing group to partially neutralize, and thereby stabilize, the positive charge at their carbonyl carbon atom. Ketones have two electron-releasing groups and are stabilized more. Greater stabilization of the ketone (the reactant) relative to its product means that the equilibrium constant for the formation of the tetrahedral product from a ketone is smaller and the reaction is less favorable:

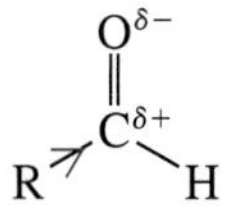

$O^{\delta-}$ / $C^{\delta+}$ / R, H

Aldehyde

Carbonyl carbon is more positive.

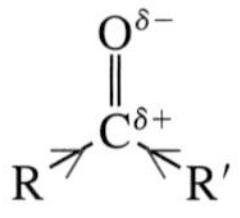

$O^{\delta-}$ / $C^{\delta+}$ / R, R′

Ketone

Carbonyl carbon is less positive.

In this regard, electron-withdrawing substituents (e.g., $—CF_3$ or $—CCl_3$ groups) cause the carbonyl carbon to be more positive (and the starting compound to become less stable) and cause the addition reaction to be more favorable.

16.6C Subsequent Reactions of Addition Products

Nucleophilic addition to a carbon–oxygen double bond may lead to a product that is stable under the reaction conditions that we employ. If this is the case we are then able to isolate products with the following general structure:

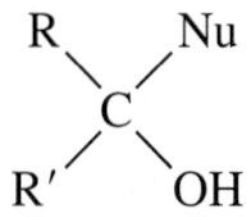

R, Nu, C, R′, OH

In other reactions the product formed initially may be unstable and may spontaneously undergo subsequent reactions. Even if the initial addition product is stable, however, we may deliberately bring about a subsequent reaction by changing the reaction conditions. When we begin our study of specific reactions, we shall see that one common subsequent reaction is an *elimination reaction,* especially *dehydration.*

PROBLEM 16.5

The reaction of an aldehyde or ketone with a Grignard reagent (Section 12.8) is a nucleophilic addition to the carbon–oxygen double bond. **(a)** What is the nucleophile? **(b)** The magnesium portion of the Grignard reagent plays an important part in this reaction. What is its function? **(c)** What product is formed initially? **(d)** What product forms when water is added?

PROBLEM 16.6

The reactions of aldehydes and ketones with $LiAlH_4$ and $NaBH_4$ (Section 12.3) are nucleophilic additions to the carbonyl group. What is the nucleophile in these reactions?

16.7 The Addition of Alcohols: Hemiacetals and Acetals

16.7A Hemiacetals

Dissolving an aldehyde or ketone in an alcohol causes the slow establishment of an equilibrium between these two compounds and a new compound called a **hemiacetal.** The hemiacetal results by nucleophilic addition of the alcohol oxygen to the carbonyl carbon of the aldehyde or ketone.

A Mechanism for the Reaction

Hemiacetal Formation

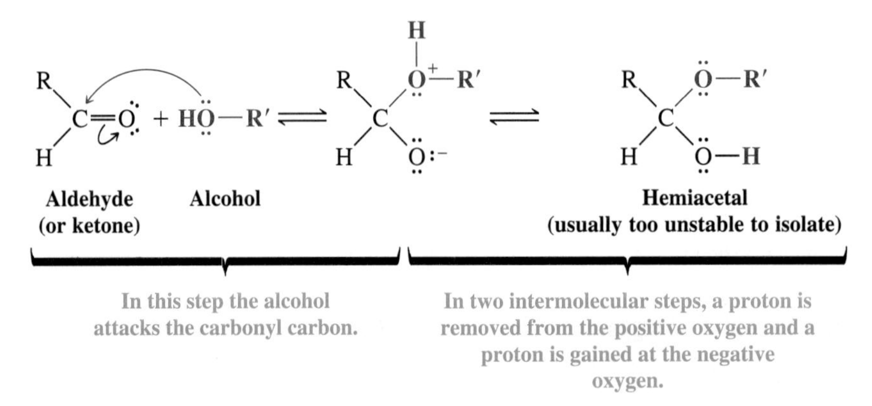

Aldehyde (or ketone)
Alcohol
Hemiacetal (usually too unstable to isolate)
In this step the alcohol attacks the carbonyl carbon.
In two intermolecular steps, a proton is removed from the positive oxygen and a proton is gained at the negative oxygen.

The essential structural features of a hemiacetal are an —OH and an —OR group attached to the same carbon atom.

Most open-chain hemiacetals are not sufficiently stable to allow their isolation. Cyclic hemiacetals with five- or six-membered rings, however, are usually much more stable:

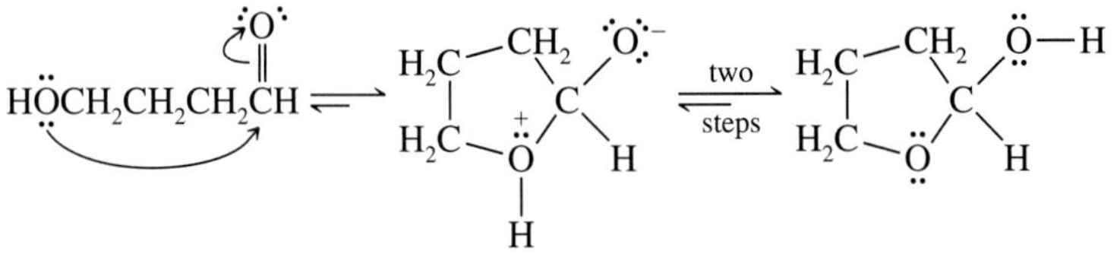

Most simple sugars (Chapter 22) exist primarily in a cyclic hemiacetal form. Glucose is an example:

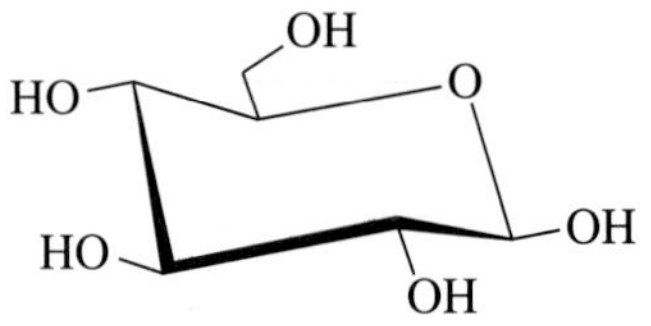

(+)-Glucose
(a cyclic hemiacetal)

Ketones undergo similar reactions when they are dissolved in an alcohol. The products (which are also unstable as open-chain compounds) are called **hemiacetals.**

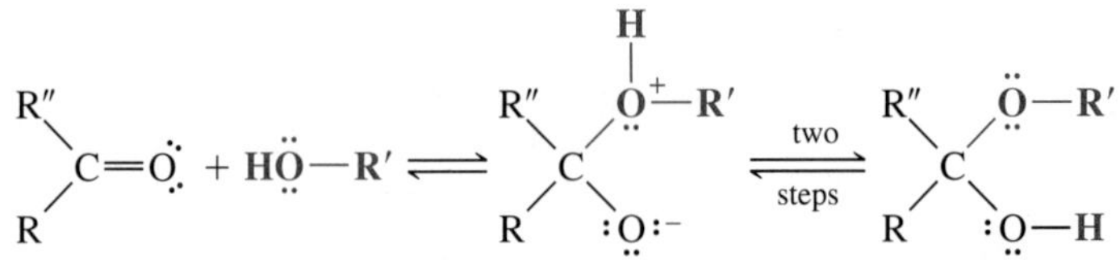

Ketone **Hemiacetal**

The formation of hemiacetals is catalyzed by acids and bases.

A Mechanisms for the Reaction

Acid-Catalyzed Hemiacetal Formation

$$\mathrm{R''R}\overset{\delta+}{\mathrm{C}}{=}\overset{\delta-}{\ddot{\mathrm{O}}}: + \mathrm{H}{-}\overset{+}{\ddot{\mathrm{O}}}(\mathrm{H}){-}\mathrm{R'} \rightleftharpoons \mathrm{R''RC}{=}\overset{+}{\ddot{\mathrm{O}}}{-}\mathrm{H} + :\ddot{\mathrm{O}}(\mathrm{H}){-}\mathrm{R'} \rightleftharpoons$$

(R″ may be H)

Protonation of the aldehyde or ketone oxygen atom makes the carbonyl carbon more susceptible to nucleophilic attack. [The protonated alcohol results from reaction of the alcohol (present in excess) with the acid catalyst, e.g., HCl.]

An alcohol molecule adds to the carbon of the oxonium cation.

$$\mathrm{R''}{-}\mathrm{C}(\mathrm{R})(\overset{+}{\ddot{\mathrm{O}}}\mathrm{H}{-}\mathrm{R'}){-}\ddot{\mathrm{O}}{-}\mathrm{H} + :\ddot{\mathrm{O}}(\mathrm{H}){-}\mathrm{R'} \rightleftharpoons \mathrm{R''}{-}\mathrm{C}(\mathrm{R})(:\ddot{\mathrm{O}}{-}\mathrm{R'}){-}\ddot{\mathrm{O}}{-}\mathrm{H} + \mathrm{H}{-}\overset{+}{\ddot{\mathrm{O}}}(\mathrm{H}){-}\mathrm{R'}$$

The transfer of a proton from the positive oxygen to another molecule of the alcohol leads to the hemiacetal.

A Mechanism for the Reaction

Base-Catalyzed Hemiacetal Formation

$$\mathrm{R''R}\overset{\delta+}{\mathrm{C}}{=}\overset{\delta-}{\ddot{\mathrm{O}}}: + {}^{-}:\ddot{\mathrm{O}}{-}\mathrm{R'} \rightleftharpoons \mathrm{R''}{-}\mathrm{C}(\mathrm{R})(\mathrm{R'}{-}\ddot{\mathrm{O}}:){-}\ddot{\mathrm{O}}:^{-} \xrightleftharpoons{\mathrm{H}{-}\mathrm{OR'}} \mathrm{R''}{-}\mathrm{C}(\mathrm{R})(\mathrm{R'}{-}\ddot{\mathrm{O}}:){-}\ddot{\mathrm{O}}{-}\mathrm{H} + {}^{-}:\ddot{\mathrm{O}}{-}\mathrm{R'}$$

(R″ may be H)

An alkoxide anion acting as a nucleophile attacks the carbonyl carbon atom. An electron pair shifts onto the oxygen atom, producing a new alkoxide anion.

The alkoxide anion abstracts a proton from an alcohol molecule to produce the hemiacetal and regenerates an alkoxide anion.

Aldehyde Hydrates: *Gem*-Diols Dissolving an aldehyde such as acetaldehyde in water causes the establishment of an equilibrium between the aldehyde and its **hydrate.** This hydrate is in actuality a 1,1-diol, called a *gem*-diol:

$$\mathrm{H_3C(H)C{=}O} + \mathrm{H_2O} \rightleftharpoons \mathrm{H_3C(H)C(O{-}H)_2}$$

Acetaldehyde — **Hydrate (a *gem*-diol)**

The *gem*-diol results from a nucleophilic addition of water to the carbonyl group of the aldehyde.

A Mechanism for the Reaction

Hydrate Formation

$$H_3C(H)C{=}\ddot{O}: + :\ddot{O}H_2 \rightleftharpoons H_3C(H)C(\overset{+}{\ddot{O}}H_2)(:\ddot{O}:^-) \rightleftharpoons H_3C(H)C(:\ddot{O}H)(:\ddot{O}H)$$

In this step water attacks the carbonyl carbon atom.

In two intermolecular steps a proton is lost from the positive oxygen atom and a proton is gained at the negative oxygen atom.

The equilibrium for the addition of water to most ketones is unfavorable, whereas some aldehydes (e.g., formaldehyde) exist primarily as the *gem*-diol in aqueous solution.

It is not possible to isolate most *gem*-diols from the aqueous solutions in which they are formed. Evaporation of the water, for example, simply displaces the overall equilibrium to the right and the *gem*-diol (or hydrate) reverts to the carbonyl compound:

$$R(H)C(OH)_2 \xrightarrow{\text{distillation}} R(H)C{=}O + H_2O$$

Compounds with strong electron-withdrawing groups attached to the carbonyl group can form stable *gem*-diols. An example is the compound called chloral hydrate:

$$Cl_3C(H)C(OH)_2$$

Chloral hydrate

PROBLEM 16.7 Dissolving formaldehyde in water leads to a solution containing primarily the *gem*-diol $CH_2(OH)_2$. Show the steps in its formation from formaldehyde.

PROBLEM 16.8 When acetone is dissolved in water containing ^{18}O instead of ordinary ^{16}O (i.e., $H_2{}^{18}O$ instead of $H_2{}^{16}O$), the acetone soon begins to acquire ^{18}O and becomes $CH_3\overset{^{18}O}{\overset{\|}{C}}CH_3$. The formation of this oxygen-labeled acetone is catalyzed by traces of strong acids and by strong bases (e.g., OH^-). Show the steps that explain both the acid-catalyzed reaction and the base-catalyzed reaction.

16.7B Acetals

If we take an alcohol solution of an aldehyde (or ketone) and pass into it a small amount of gaseous HCl, the hemiacetal forms, and then a second reaction takes place. The hemiacetal reacts with a second molar equivalent of the alcohol to produce an **acetal.** An acetal has two —OR groups attached to the same carbon atom:

$$R{-}\underset{R''}{\overset{OH}{C}}{-}OR' \xrightarrow[R'{-}OH]{HCl_{(g)}} R{-}\underset{R''}{\overset{OR'}{C}}{-}OR' + H_2O$$

Hemiacetal (R″ may be H) **An acetal (R″ may be H)**

PROBLEM 16.9 Shown below is the structural formula for sucrose (table sugar). Sucrose has two acetal groupings. Identify these.

The mechanism for acetal formation involves acid-catalyzed formation of the hemiacetal, then an acid-catalyzed elimination of water, followed by a second *addition* of the alcohol and loss of a proton.

A Mechanism for the Reaction

Acid-Catalyzed Acetal Formation

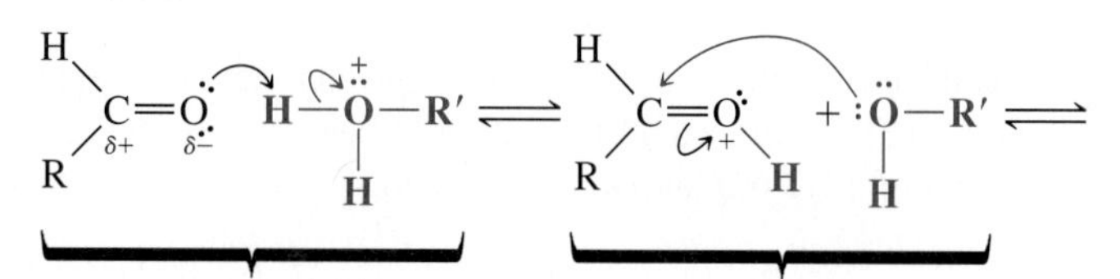

Proton transfer to the carbonyl oxygen

Nucleophilic addition of the first alcohol molecule

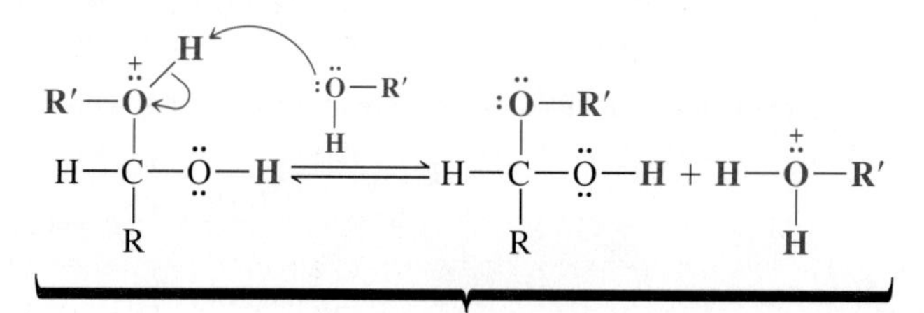

Proton removal from the positive oxygen results in formation of a hemiacetal.

Protonation of the hydroxyl group leads to elimination of water and formation of a highly reactive oxonium cation.

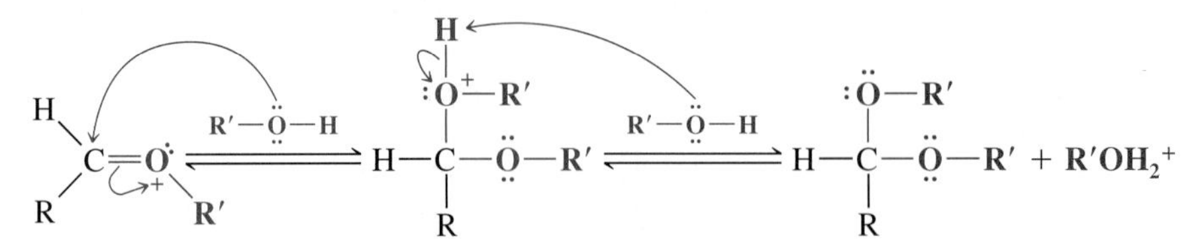

Attack on the carbon of the oxonium ion by a second molecule of the alcohol, followed by removal of a proton, leads to the acetal.

PROBLEM 16.10 Write a detailed mechanism for the formation of an acetal from benzaldehyde and methanol in the presence of an acid catalyst.

Equilibrium conditions govern the formation and hydrolysis of hemiacetals and acetals.

All steps in the formation of an acetal from an aldehyde are reversible. If we dissolve an aldehyde in a large excess of an anhydrous alcohol and add a small amount of an anhydrous acid (e.g., gaseous HCl or concentrated H_2SO_4), the equilibrium will strongly favor the formation of an acetal. After the equilibrium is established, we can isolate the acetal by neutralizing the acid and evaporating the excess alcohol.

If we then place the acetal in water and add a small amount of acid, all of the steps reverse. Under these conditions (an excess of water), the equilibrium favors the formation of the aldehyde. The acetal undergoes ***hydrolysis:***

$$\text{RCH(OR')}_2 \text{ (Acetal)} + H_2O \underset{\text{(several steps)}}{\overset{H_3O^+}{\rightleftharpoons}} R{-}\overset{O}{\overset{\|}{C}}{-}H \text{ (Aldehyde)} + 2\,R'OH$$

Acetal formation is not favored when ketones are treated with simple alcohols and gaseous HCl. Cyclic acetal formation *is* favored, however, when a ketone is treated with an excess of a 1,2-diol and a trace of acid:

$$R'RC{=}O \text{ (Ketone)} + \begin{matrix} HOCH_2 \\ | \\ HOCH_2 \end{matrix} \text{ (excess)} \overset{H_3O^+}{\rightleftharpoons} R'RC\begin{matrix} O{-}CH_2 \\ \quad | \\ O{-}CH_2 \end{matrix} \text{ (Cyclic acetal)} + H_2O$$

This reaction, too, can be reversed by treating the acetal with aqueous acid:

$$R'RC\begin{matrix} O{-}CH_2 \\ \quad | \\ O{-}CH_2 \end{matrix} + H_2O \overset{H_3O^+}{\rightleftharpoons} R'RC{=}O + \begin{matrix} CH_2OH \\ | \\ CH_2OH \end{matrix}$$

PROBLEM 16.11 Outline all steps in the mechanism for the formation of a cyclic acetal from acetone and ethylene glycol in the presence of gaseous HCl.

16.7C Acetals as Protecting Groups

Although acetals are hydrolyzed to aldehydes and ketones in aqueous acid, ***they are stable in basic solutions:***

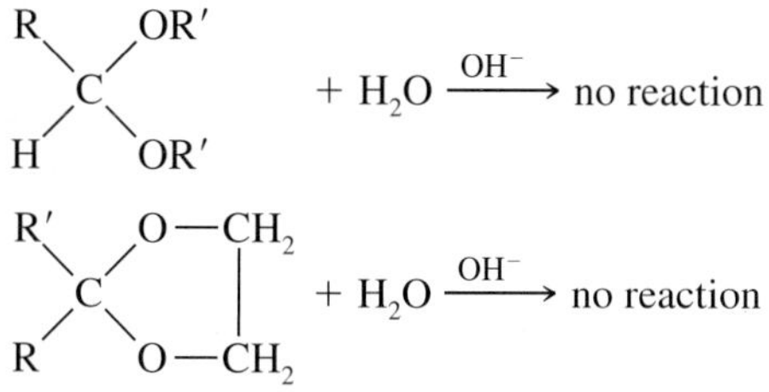

$$\text{R(H)C(OR')}_2 + H_2O \xrightarrow{OH^-} \text{no reaction}$$

$$\text{R}_2\text{C(OCH}_2\text{CH}_2\text{O)} + H_2O \xrightarrow{OH^-} \text{no reaction}$$

Because of this property, acetals give us a convenient method for **protecting aldehyde and ketone groups from undesired reactions in basic solutions.** (Acetals are really *gem*-diethers, and, like ethers, they are relatively unreactive toward bases.) We can convert an aldehyde or ketone to an acetal, carry out a reaction on some other part of the molecule, and then hydrolyze the acetal with aqueous acid.

Protecting groups are strategic tools for synthesis. See Sections 11.5C, 11.5D, and 12.10 also.

As an example, let us consider the problem of converting

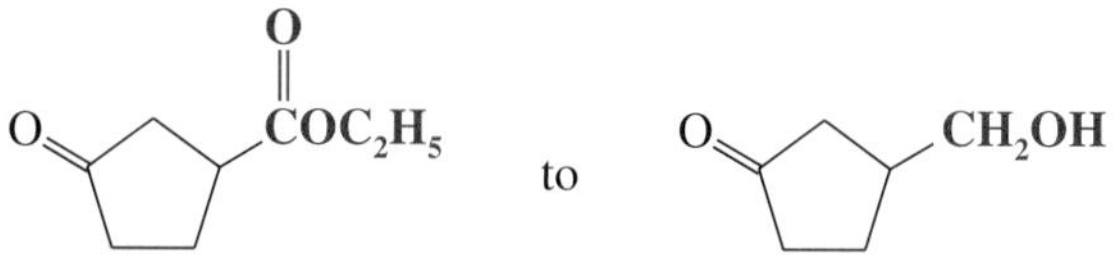

A (3-oxocyclopentyl–COC_2H_5 ester) to B (3-oxocyclopentyl–CH_2OH)

Keto groups are more easily reduced than ester groups. Any reducing agent (e.g., $LiAlH_4$ or H_2/Ni) that can reduce the ester group of **A** reduces the keto group as well. But if we "protect" the keto group by converting it to a cyclic acetal (the ester group does not react), we can reduce the ester group in basic solution without affecting the cyclic acetal. After we finish the ester reduction, we can hydrolyze the cyclic acetal and obtain our desired product, **B:**

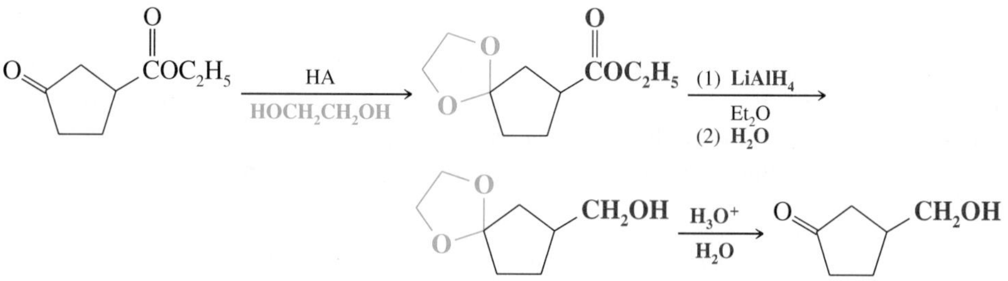

A $\xrightarrow[HOCH_2CH_2OH]{HA}$ cyclic acetal–COC_2H_5 $\xrightarrow[(2)\ H_2O]{(1)\ LiAlH_4,\ Et_2O}$ cyclic acetal–CH_2OH $\xrightarrow[H_2O]{H_3O^+}$ 3-oxocyclopentyl–CH_2OH

PROBLEM 16.12

What product would be obtained if **A** were treated with lithium aluminum hydride without first converting it to a cyclic acetal?

PROBLEM 16.13

(a) Show how you might use a cyclic acetal in carrying out the following transformation:

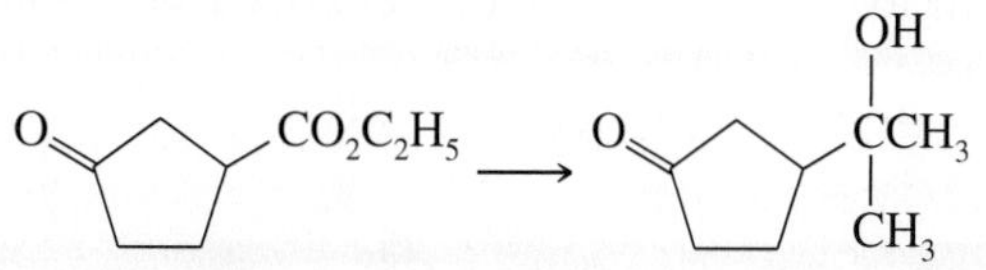

A (3-oxocyclopentyl–$CO_2C_2H_5$) $\longrightarrow$ C (3-oxocyclopentyl–$C(OH)(CH_3)_2$)

(b) Why would a direct addition of methylmagnesium bromide to **A** fail to give **C**?

PROBLEM 16.14

Dihydropyran reacts readily with an alcohol in the presence of a trace of anhydrous HCl or H_2SO_4 to form a tetrahydropyranyl (THP) ether:

$$\text{Dihydropyran} + ROH \xrightarrow{HA} \text{Tetrahydropyranyl ether}$$

Dihydropyran **Tetrahydropyranyl ether**

(a) Write a plausible mechanism for this reaction. **(b)** Tetrahydropyranyl ethers are stable in aqueous base but hydrolyze rapidly in aqueous acid to yield the original alcohol and another compound. Explain. (What is the other compound?) **(c)** The tetrahydropyranyl group can be used as a protecting group for alcohols and phenols. Show how you might use it in a synthesis of 5-methyl-1,5-hexanediol starting with 4-chloro-1-butanol.

16.7D Thioacetals

Aldehydes and ketones react with thiols to form *thioacetals:*

$$RHC{=}O + 2\ CH_3CH_2SH \xrightarrow{HA} RHC(S{-}CH_2CH_3)_2 + H_2O$$

Thioacetal

$$RR'C{=}O + HSCH_2CH_2SH \xrightarrow{BF_3} RR'C(S{-}CH_2)(S{-}CH_2) + H_2O$$

Cyclic thioacetal

A synthesis tool for reducing the carbonyl group of aldehydes and ketones to $-CH_2-$ groups.

Thioacetals are important in organic synthesis because they react with Raney nickel to yield hydrocarbons. Raney nickel is a special nickel catalyst that contains absorbed hydrogen. These reactions (i.e., thioacetal formation and subsequent "desulfurization") give us an additional method for converting carbonyl groups of aldehydes and ketones to $-CH_2-$ groups:

$$RR'C(S{-}CH_2)(S{-}CH_2) \xrightarrow[(H_2)]{\text{Raney Ni}} RR'CH_2 + H{-}CH_2CH_2{-}H + NiS$$

The other method we have studied is the **Clemmensen reduction** (Section 15.9).

PROBLEM 16.15 Show how you might use thioacetal formation and Raney nickel desulfurization to convert: **(a)** cyclohexanone to cyclohexane and **(b)** benzaldehyde to toluene.

16.8 The Addition of Primary and Secondary Amines

Aldehydes and ketones react with primary amines to form **imines** and with secondary amines to form **enamines.** Imines have a carbon–nitrogen double bond. Enamines have an amino group joined to a carbon–carbon double bond (they are alk*ene*-*amines*).

Imine (R, R, N–R′)

Enamine (R, R, R, N–R″, R′)

R′, R″ = C;
R = C or H

16.8A Imines

A general equation for the formation of an imine from a primary amine and an aldehyde or ketone is shown here. Imine formation is acid catalyzed, and the product can form as a mixture of (*E*) and (*Z*) isomers:

$$\text{C=O} + H_2N{-}R \xrightleftharpoons{H_3O^+} \text{C=N}{-}R + H_2O$$

Aldehyde or ketone **1° Amine** **Imine [(*E*) and (*Z*) isomers]**

Imine formation generally takes place fastest between pH 4 and 5 and is slow at very low or very high pH. We can understand why an acid catalyst is necessary if we consider the mechanism that has been proposed for imine formation. The important step is the step in which the protonated aminoalcohol loses a molecule of water to become an iminium ion. By protonating the alcohol group, the acid converts a poor leaving group (an —OH group) into a good one (an $-OH_2^+$ group).

A Mechanism for the Reaction

Imine Formation

$$\text{C=O} + H_2N{-}R \rightleftharpoons \text{C}(NH_2^+{-}R)(O^-) \rightleftharpoons \text{C}(NH{-}R)(OH) \xrightleftharpoons{H_3O^+}$$

Aldehyde or ketone **1° Amine** **Dipolar intermediate** **Aminoalcohol**

The amine adds to the carbonyl group to form a dipolar tetrahedral intermediate.

Intermolecular proton transfer from nitrogen to oxygen produces an aminoalcohol.

$$\text{C}(NH{-}R)(OH_2^+) \xrightleftharpoons{-H_2O} \text{C=N}^+(H)(R) \overset{:OH_2}{\rightleftharpoons} \text{C=N}{-}R + H_3O^+$$

Protonated aminoalcohol **Iminium ion** **Imine [(*E*) and (*Z*) isomers]**

Protonation of the oxygen produces a good leaving group. Loss of a molecule of water yields an iminium ion.

Transfer of a proton to water produces the imine and regenerates the catalytic hydronium ion.

The chapter opening vignette discussed pyridoxal phosphate, an important coenzyme that reacts via the imine functional group. See also "The Chemistry of Pyridoxal Phosphate" in Section 16.8B.

The reaction proceeds more slowly if the hydronium ion concentration is too high, because protonation of the amine itself takes place to a considerable extent; this has the effect of decreasing the concentration of the nucleophile needed in the first step. If the concentration of the hydronium ion is too low, the reaction becomes slower because the concentration of the protonated aminoalcohol becomes lower. A pH between pH 4 and pH 5 is an effective compromise.

Imine formation occurs in many biochemical reactions because enzymes often use an $—NH_2$ group to react with an aldehyde or ketone. Formation of an imine linkage is important in one step of the reactions that take place during the visual process (see "The Photochemistry of Vision," Section 13.9).

Imines are also formed as intermediates in a useful laboratory synthesis of amines that we shall study in Section 20.5.

16.8B Oximes, Hydrazones, and Semicarbazones

Compounds such as hydroxylamine (NH_2OH), hydrazine (NH_2NH_2), and substituted hydrazines such as phenylhydrazine ($C_6H_5NHNH_2$), 2,4-dinitrophenylhydrazine, and semicarbazide ($H_2NCONHNH_2$) form C═N derivatives of aldehydes and ketones. These derivatives are called oximes, hydrazones, phenylhydrazones, 2,4-dinitrophenylhydrazones, and semicarbazones, respectively. The mechanisms by which these C═N derivatives form are similar to the mechanism for imine formation from a primary amine. As with imines, the formation of (*E*) and (*Z*) isomers is possible. Table 16.2 shows general examples of these reactions.

Oxime, semicarbazone, and the various **hydrazone** derivatives of aldehydes and ketones are sometimes used to identify unknown aldehydes and ketones. These derivatives are usually relatively insoluble solids that have sharp, characteristic melting points. The melting point of the derivative of an unknown compound can be compared with the melting point for the same derivative of a known compound or with data found in a reference table, and on this basis one can propose an identity for the unknown compound. Most laboratory textbooks for organic chemistry include extensive tables of derivative melting points. The method of comparing melting points is only useful, however, for compounds that have derivative melting points previously reported in the literature. Spectroscopic methods (especially IR, NMR, and mass spectrometry) are more generally applicable to identification of unknown compounds (Section 16.14).

16.8C Enamines

Aldehydes and ketones react with secondary amines to form enamines. The following is a general equation for enamine formation:

$$\text{H}-\overset{|}{\underset{|}{\text{C}}}-\text{C}(=\ddot{\text{O}}:)- \;+\; \text{H}-\ddot{\text{N}}(\text{R})-\text{R} \;\underset{}{\overset{\text{cat. HA}}{\rightleftharpoons}}\; \text{C}{=}\text{C}(-\ddot{\text{N}}\text{R}_2) \;+\; \text{H}_2\text{O}$$

Secondary amine **Enamine**

A mechanism for the reaction is given in the box on page 742. Note the difference between the previously described mechanism for imine formation and this mechanism for

The Chemistry of...

Pyridoxal Phosphate

Pyridoxal phosphate (PLP) is a biological molecule involved in many important reactions of α-amino acids (see the chapter opening vignette). Depending on the substrate and specific enzyme, PLP can catalyze reactions that result in interconversion of amino and ketone groups, decarboxylation, stereochemical inversion, elimination, and replacement. In all of the reactions catalyzed by PLP its role is to serve as a reversible acceptor of an electron pair. As a specific example of this function, let us consider a racemization reaction involving PLP. Bacteria require (*R*)-(−)-alanine (also called D-alanine, Section 24.2) for construction of cell walls. The more abundant form of alanine, however, is the enantiomer (*S*)-(+)-alanine (or L-alanine). To obtain the necessary (*R*) enantiomer for cell wall synthesis, bacteria use an enzyme called alanine racemase to catalyze formation of a racemic mixture of alanine, thereby obtaining a supply of (*R*)-(−)-alanine.

The first stage in enzymatic reactions involving PLP is exchange of the imine linkage that tethers PLP to its enzyme for an imine linkage between PLP and the amino substrate. Here, this process is shown with (*S*)-(+)-alanine:

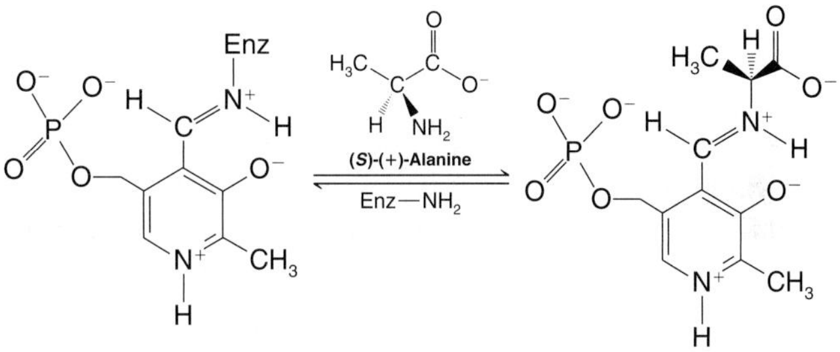

(*S*)-(+)-Alanine

Enz—NH_2

The next stage involves removal of a proton from the α carbon of alanine by a basic group within the enzyme. Formation of this anion is a feasible process because the positively charged pyridine ring in PLP serves as an electron pair acceptor (see Problem 16.B2). Formation of the conjugated carbanion intermediate results in the loss of chirality at the α carbon of alanine. Subsequent reprotonation of the trigonal planar α-carbon intermediate (on either face) results in the racemic form of alanine. Release of the racemized amino acid takes place with exchange of the imine linkage back to the enzyme amino group (a hydrolysis step that is the reverse of the first stage shown above).

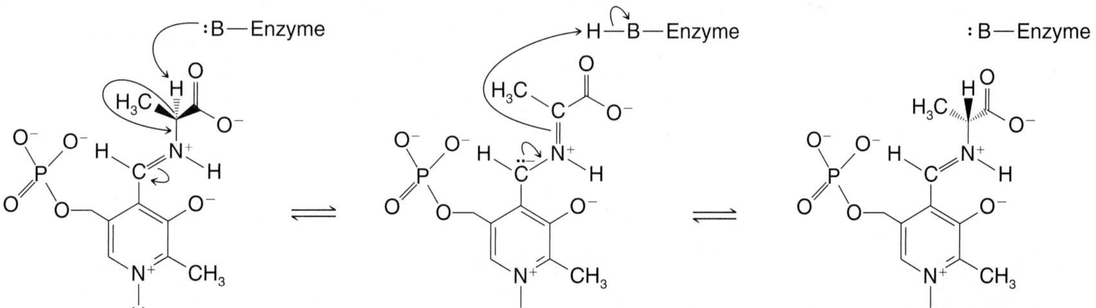

:B—Enzyme

H—B—Enzyme

:B—Enzyme

Write a mechanism showing how the imine of PLP with an enzyme can exchange for an imine of PLP with a substrate such as alanine.

PROBLEM 16.B1

Write resonance structures that show how the α-carbanion intermediate in the racemization of alanine is stabilized by conjugation with the pyridine ring.

PROBLEM 16.B2

A Mechanism for the Reaction

Enamine Formation

Step 1

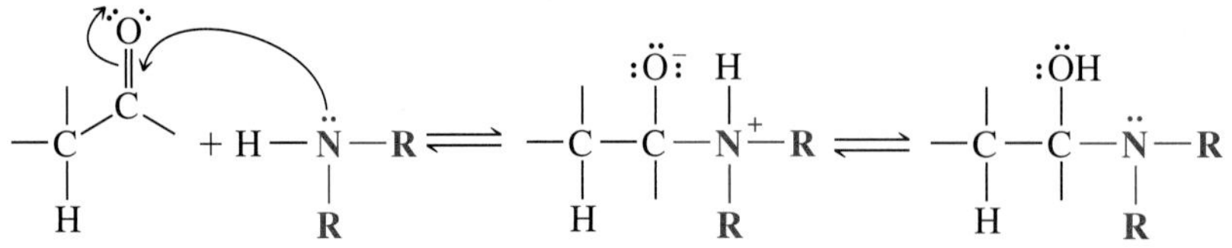

Aldehyde or ketone + Secondary amine ⇌ ⇌ Aminoalcohol intermediate

The amine adds to the ketone or aldehyde carbonyl to form a tetrahedral adduct. Intermolecular proton transfer leads to the aminoalcohol intermediate.

Step 2

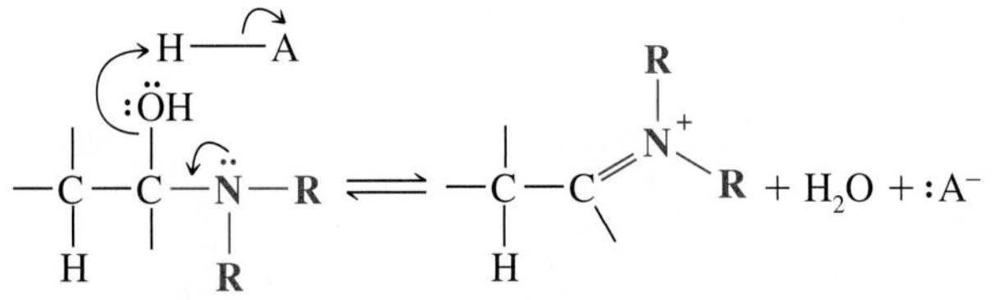

Aminoalcohol intermediate ⇌ Iminium ion intermediate + H_2O + $:A^-$

The aminoalcohol intermediate is protonated by the catalytic acid and a water molecule departs. Contribution of an unshared electron pair from the nitrogen atom leads to an iminium cation intermediate.

Step 3

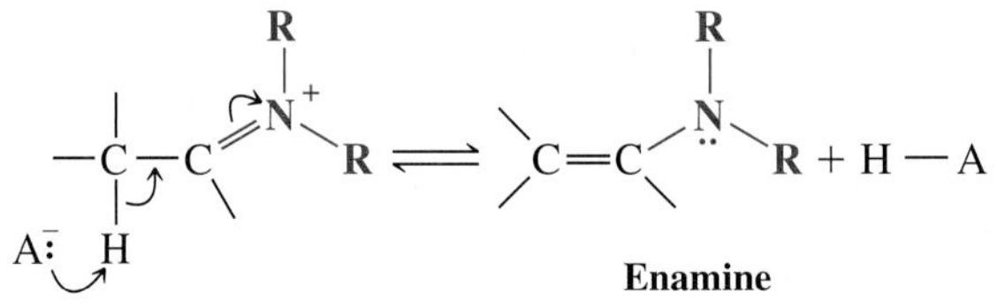

⇌ Enamine + H—A

A proton is removed from the carbon adjacent to the iminium group. Proton removal occurs from the carbon because there is no proton to remove from the nitrogen of the iminium cation (as there would have been if a primary amine had been used). This step forms the enamine, neutralizes the formal charge, and regenerates the catalytic acid. (If there had been a proton to remove from the nitrogen of the iminium cation, the final product would have been an imine.)

enamine formation. In enamine formation, which involves a secondary amine, there is no proton for removal from the nitrogen in the iminium cation intermediate. Hence, a neutral imine cannot be formed. A proton is removed from a carbon adjacent to the former carbonyl group instead, resulting in an enamine. We shall see in Chapter 19 that enamines are very useful for carbon–carbon bond formation (Section 19.11).

Tertiary amines do not form stable addition products with aldehydes and ketones because, upon forming the tetrahedral intermediate, the resulting formal positive charge cannot be neutralized by loss of a proton.

TABLE 16.2 Reactions of Aldehydes and Ketones with Derivatives of Ammonia

1. Imine formation—reaction with a primary amine

$$\text{>C=O} + \text{H}_2\ddot{\text{N}}\text{—R} \longrightarrow \text{>C=}\ddot{\text{N}}\text{—R} + \text{H}_2\text{O}$$

Aldehyde or ketone **A 1° amine** **An imine [(*E*) and (*Z*) isomers)]**

2. Oxime formation—reaction with hydroxylamine

$$\text{>C=O} + \text{H}_2\text{N—OH} \longrightarrow \text{>C=}\ddot{\text{N}}\text{—OH} + \text{H}_2\text{O}$$

Aldehyde or ketone **Hydroxylamine** **An oxime [(*E*) and (*Z*) isomers)]**

3. Hydrazone and substituted hydrazone formation—reactions with hydrazine, phenylhydrazine, 2,4-dinitrophenylhydrazine, and semicarbazide [each derivative can form as an (*E*) or (*Z*) isomer]

$$\text{>C=O} + \text{H}_2\text{NNH}_2 \longrightarrow \text{>C=}\ddot{\text{N}}\text{—NH}_2 + \text{H}_2\text{O}$$

Aldehyde or ketone **Hydrazine** **A hydrazone**

$$\text{>C=O} + \text{H}_2\text{NNHC}_6\text{H}_5 \longrightarrow \text{>C=NNHC}_6\text{H}_5 + \text{H}_2\text{O}$$

Phenylhydrazine **A phenylhydrazone**

$$\text{>C=O} + \text{H}_2\text{NNH—C}_6\text{H}_3(\text{NO}_2)_2 \longrightarrow \text{>C=NNH—C}_6\text{H}_3(\text{NO}_2)_2 + \text{H}_2\text{O}$$

2,4-Dinitrophenylhydrazine **A 2,4-dinitrophenylhydrazone**

$$\text{>C=O} + \text{H}_2\text{NNHCNH}_2(\text{C=O}) \longrightarrow \text{>C=NNHCNH}_2(\text{C=O}) + \text{H}_2\text{O}$$

Semicarbazide **A semicarbazone**

4. Enamine formation—reaction with a secondary amine

$$\text{H—C—C(=}\ddot{\text{O}}\text{:)—} + \text{H—}\ddot{\text{N}}\text{R—R} \overset{\text{cat. HA}}{\rightleftharpoons} \text{>C=C(—}\ddot{\text{N}}\text{R}_2\text{)—} + \text{HOH}$$

Secondary amine **Enamine**

16.9 The Addition of Hydrogen Cyanide

Hydrogen cyanide adds to the carbonyl groups of aldehydes and most ketones to form compounds called **cyanohydrins.** (Ketones in which the carbonyl group is highly hindered do not undergo this reaction.)

$$RCHO + HCN \rightleftharpoons RCH(OH)CN$$

$$RCOR' + HCN \rightleftharpoons RR'C(OH)CN$$

Cyanohydrins

Cyanohydrins form fastest under conditions where cyanide anions are present to act as the nucleophile. Use of potassium cyanide, or any base that can generate cyanide anions from HCN, increases the reaction rate as compared to the use of HCN alone. The addition of hydrogen cyanide itself to a carbonyl group is slow because the weak acidity of HCN (pK_a ~9) provides only a small concentration of the nucleophilic cyanide anion. The following is a mechanism for formation of a cyanohydrin.

A Mechanism for the Reaction

Cyanohydrin Formation

$$\text{C=O} + {}^{-}:C\equiv N: \xrightleftharpoons{\text{slow}} \text{C(O}^{-}\text{)(C}\equiv\text{N:)} \xrightleftharpoons[-H-C\equiv N:]{+H-C\equiv N:} \text{C(OH)(C}\equiv\text{N:)} + {}^{-}:C\equiv N:$$

Great care must be taken when working with hydrogen cyanide due to its high toxicity and volatility. Reactions involving HCN must be conducted in an efficient fume hood.

Cyanohydrins are useful intermediates in organic synthesis. Depending on the conditions used, acidic hydrolysis converts cyanohydrins to α-hydroxy acids or to α,β-unsaturated acids. (The mechanism for this hydrolysis is discussed in Section 18.8H.) The preparation of α-hydroxy acids from cyanohydrins is part of the Kiliani–Fischer synthesis of simple sugars (Section 22.9A):

$$CH_3CH_2COCH_3 \xrightarrow{HCN} CH_3CH_2C(OH)(CH_3)CN \xrightarrow[\text{heat}]{HCl,\ H_2O} CH_3CH_2C(OH)(CH_3)COOH$$

α-Hydroxy acid

$$CH_3CH_2C(OH)(CH_3)CN \xrightarrow[\text{heat}]{95\%\ H_2SO_4} CH_3CH{=}C(CH_3)COOH$$

α,β-Unsaturated acid

Reduction of a cyanohydrin with lithium aluminum hydride gives a β-aminoalcohol:

$$\text{cyclohexanone} \xrightarrow{HCN} \text{1-hydroxycyclohexane-CN} \xrightarrow[(2)\ H_2O]{(1)\ LiAlH_4} \text{1-hydroxycyclohexane-}CH_2NH_2$$

PROBLEM 16.16

(a) Show how you might prepare lactic acid ($CH_3CHOHCO_2H$) from acetaldehyde through a cyanohydrin intermediate. **(b)** What stereoisomeric form of lactic acid would you expect to obtain?

16.10 The Addition of Ylides: The Wittig Reaction

Aldehydes and ketones react with phosphorus ylides to yield *alkenes* and triphenylphosphine oxide. (An **ylide** is a neutral molecule having a negative carbon adjacent to a positive heteroatom.) Phosphorus ylides are also called phosphoranes:

$$RR'C{=}O + (C_6H_5)_3\overset{+}{P}{-}\overset{-}{\ddot{C}}R''R''' \longrightarrow RR'C{=}CR''R''' + O{=}P(C_6H_5)_3$$

Aldehyde or ketone | **Phosphorus ylide (or phosphorane)** | **Alkene [(*E*) and (*Z*) isomers)]** | **Triphenylphosphine oxide**

This reaction, known as the **Wittig reaction,** has proved to be a valuable method for synthesizing alkenes. The Wittig reaction is applicable to a wide variety of compounds, and although a mixture of (*E*) and (*Z*) isomers may result, the Wittig reaction offers a great advantage over most other alkene syntheses in that *no ambiguity exists as to the location of the double bond in the product.* (This is in contrast to E1 eliminations, which may yield multiple alkene products by rearrangement to more stable carbocation intermediates, and both E1 and E2 elimination reactions, which may produce multiple products when different β hydrogens are available for removal.)

Phosphorus ylides are easily prepared from triphenylphosphine and primary or secondary alkyl halides. Their preparation involves two reactions:

General Reaction

Reaction 1

$$(C_6H_5)_3P{:} + R''R'''CH{-}X \longrightarrow (C_6H_5)_3\overset{+}{P}{-}CHR''R''' \; X^-$$

Triphenylphosphine | **An alkyltriphenylphosphonium halide**

Reaction 2

$$(C_6H_5)_3\overset{+}{P}{-}CR''R'''{-}H \;\; {:}\overset{-}{B} \longrightarrow (C_6H_5)_3\overset{+}{P}{-}CR''R'''{:}^- + H{:}B$$

A phosphorus ylide

Specific Example

Reaction 1

$$(C_6H_5)_3P{:} + CH_3Br \xrightarrow{C_6H_6} (C_6H_5)_3\overset{+}{P}{-}CH_3 \; Br^-$$

Methyltriphenylphosphonium bromide (89%)

Reaction 2

$$(C_6H_5)_3\overset{+}{P}{-}CH_3 \; Br^- + C_6H_5Li \longrightarrow (C_6H_5)_3\overset{+}{P}{-}CH_2{:}^- + C_6H_6 + LiBr$$

The first reaction is a nucleophilic substitution reaction. Triphenylphosphine is an excellent nucleophile and a weak base. It reacts readily with 1° and 2° alkyl halides by an S_N2 mechanism to displace a halide ion from the alkyl halide to give an alkyltriphenylphosphonium salt. The second reaction is an acid–base reaction. A strong base (usually an alkyllithium or phenyllithium) removes a proton from the carbon that is attached to phosphorus to give the ylide.

Phosphorus ylides can be represented as a hybrid of the two resonance structures shown here. Quantum mechanical calculations indicate that the contribution made by the first structure is relatively unimportant.

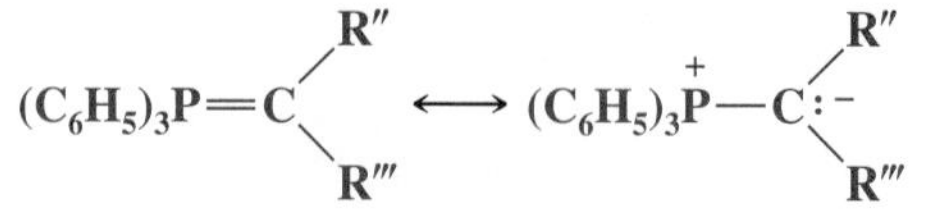

$$(C_6H_5)_3P{=}C(R'')(R''') \longleftrightarrow (C_6H_5)_3\overset{+}{P}{-}\ddot{C}^{-}(R'')(R''')$$

The mechanism of the Wittig reaction has been the subject of considerable study. An early mechanistic proposal suggested that the ylide, acting as a carbanion, attacks the carbonyl carbon of the aldehyde or ketone to form an unstable intermediate with separated charges called a **betaine.** In the next step, the betaine is envisioned as becoming an unstable four-membered cyclic system called an **oxaphosphetane,** which then spontaneously loses triphenylphosphine oxide to become an alkene. However, studies by E. Vedejs (of the University of Wisconsin) and others suggest that the betaine is not an intermediate and that the oxaphosphetane is formed directly by a cycloaddition reaction. The driving force for the Wittig reaction is the formation of the very strong ($DH° = 540$ kJ mol^{-1}) phosphorus–oxygen bond in triphenylphosphine oxide.

A Mechanism for the Reaction

The Wittig Reaction

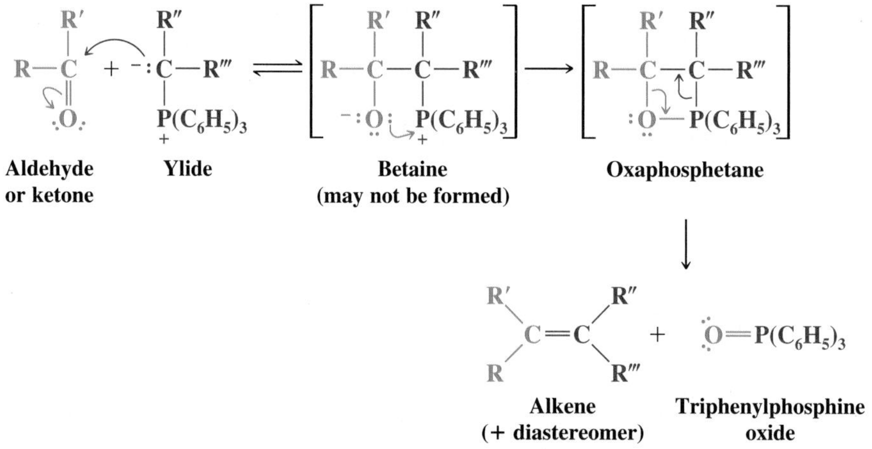

$$R{-}C(R'){=}\ddot{O}: + {}^{-}{:}C(R'')(R'''){-}\overset{+}{P}(C_6H_5)_3 \rightleftharpoons \left[R{-}C(R')({-}{:}\ddot{O}{:}^{-}){-}C(R'')(R'''){-}\overset{+}{P}(C_6H_5)_3\right] \longrightarrow \left[\text{oxaphosphetane}\right]$$

Aldehyde or ketone **Ylide** **Betaine (may not be formed)** **Oxaphosphetane**

$$\longrightarrow R(R')C{=}C(R'')(R''') + \ddot{O}{=}P(C_6H_5)_3$$

Alkene (+ diastereomer) **Triphenylphosphine oxide**

Specific Example

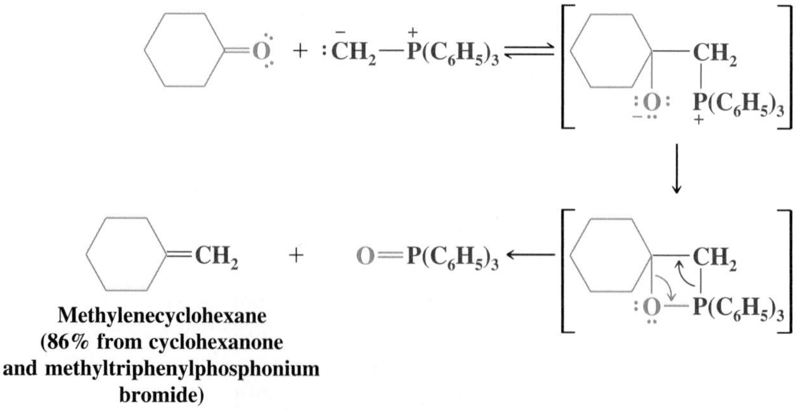

$$\text{cyclohexanone} + {:}\overset{-}{C}H_2{-}\overset{+}{P}(C_6H_5)_3 \rightleftharpoons \left[\text{betaine}\right] \longrightarrow \left[\text{oxaphosphetane}\right] \longrightarrow \text{methylenecyclohexane} + O{=}P(C_6H_5)_3$$

Methylenecyclohexane (86% from cyclohexanone and methyltriphenylphosphonium bromide)

The elimination of triphenylphosphine oxide from the betaine (if, indeed, it forms) may occur in two separate steps, as we have just shown, or both steps may occur simultaneously.

While Wittig syntheses may appear to be complicated, in actual practice they are easy to carry out. Most of the steps can be carried out in the same reaction vessel, and the entire synthesis can be accomplished in a matter of hours.

The overall result of a Wittig synthesis is

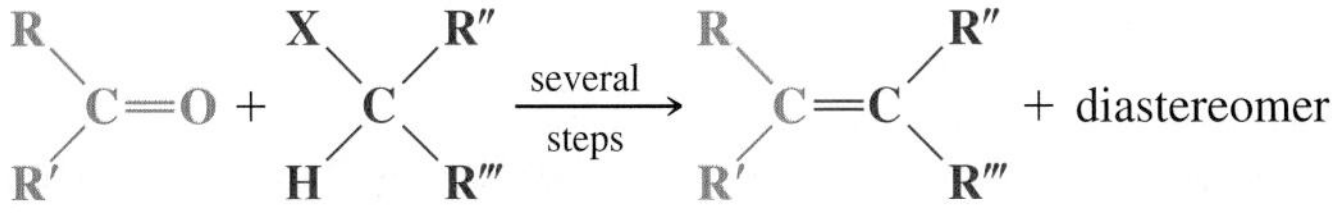

$$\underset{R'}{\overset{R}{}}\!\!>C{=}O + \underset{H}{\overset{X}{}}\!\!>C<\!\!\underset{R'''}{\overset{R''}{}} \xrightarrow[\text{steps}]{\text{several}} \underset{R'}{\overset{R}{}}\!\!>C{=}C<\!\!\underset{R'''}{\overset{R''}{}} + \text{diastereomer}$$

Planning a Wittig synthesis begins with recognizing in the desired alkene what can be the aldehyde or ketone component and what can be the halide component. Any or all of the R groups may be hydrogen, although yields are generally better when at least one group is hydrogen. The halide component must be a primary, secondary, or methyl halide.

SOLVED PROBLEM

Synthesize 2-methyl-1-phenylprop-1-ene using a Wittig reaction. Begin by writing a retrosynthetic analysis.

ANSWER: We examine the structure of the compound, paying attention to the groups on each side of the double bond:

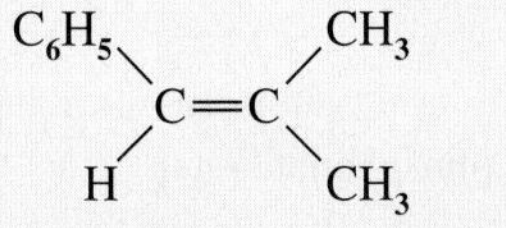

$$\underset{H}{\overset{C_6H_5}{}}\!\!>C{=}C<\!\!\underset{CH_3}{\overset{CH_3}{}}$$

2-Methyl-1-phenylprop-1-ene

We see that two retrosynthetic analyses are possible.

Retrosynthetic Analysis

(a)

$$\underset{H}{\overset{C_6H_5}{}}\!\!>C{=}C<\!\!\underset{CH_3}{\overset{CH_3}{}} \Longrightarrow \underset{H}{\overset{C_6H_5}{}}\!\!>C{=}O + {:}\overset{CH_3}{\underset{P(C_6H_5)_3}{C}}{-}CH_3 \Longrightarrow H{-}\overset{CH_3}{\underset{X}{C}}{-}CH_3 + O{=}P(C_6H_5)_3$$

(b)

$$\underset{H}{\overset{C_6H_5}{}}\!\!>C{=}C<\!\!\underset{CH_3}{\overset{CH_3}{}} \Longrightarrow H{-}\overset{C_6H_5}{\underset{^{+}P(C_6H_5)_3}{C}}{:}^{-} + O{=}C<\!\!\underset{CH_3}{\overset{CH_3}{}}$$

$$\Downarrow$$

$$H{-}\overset{C_6H_5}{\underset{X}{C}}{-}H + O{=}P(C_6H_5)_3$$

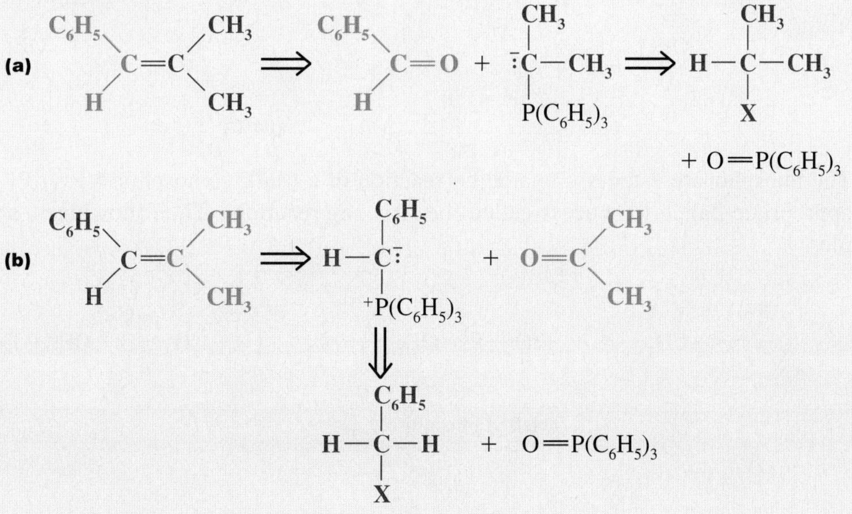

Synthesis

Following retrosynthetic analysis (a), we begin by making the ylide from a 2-halopropane and then allow the ylide to react with benzaldehyde:

(a) $(CH_3)_2CHBr + (C_6H_5)_3P \longrightarrow (CH_3)_2CH-\overset{+}{P}(C_6H_5)_3\ Br^- \xrightarrow{RLi}$

$$(CH_3)_2\ddot{C}-\overset{+}{P}(C_6H_5)_3 \xrightarrow{C_6H_5CHO} (CH_3)_2C{=}CHC_6H_5 + (C_6H_5)_3P{=}O$$

Following retrosynthetic analysis (b), we make the ylide from a benzyl halide and allow it to react with acetone:

(b) $C_6H_5CH_2Br + (C_6H_5)_3P \longrightarrow C_6H_5CH_2-\overset{+}{P}(C_6H_5)_3\ Br^- \xrightarrow{RLi}$

$$C_6H_5\ddot{C}H-\overset{+}{P}(C_6H_5)_3 \xrightarrow{(CH_3)_2C=O} C_6H_5CH{=}C(CH_3)_2 + (C_6H_5)_3P{=}O$$

A widely used variation of the Wittig reaction is the **Horner–Wadsworth–Emmons** modification. The Horner–Wadsworth–Emmons reaction involves use of a phosphonate ester instead of a triphenylphosphonium salt. The major product is usually the (*E*)-alkene isomer. Some bases that are typically used to form the phosphonate ester carbanion include sodium hydride, potassium *tert*-butoxide, and butyllithium. The following reaction sequence is an example:

Step 1

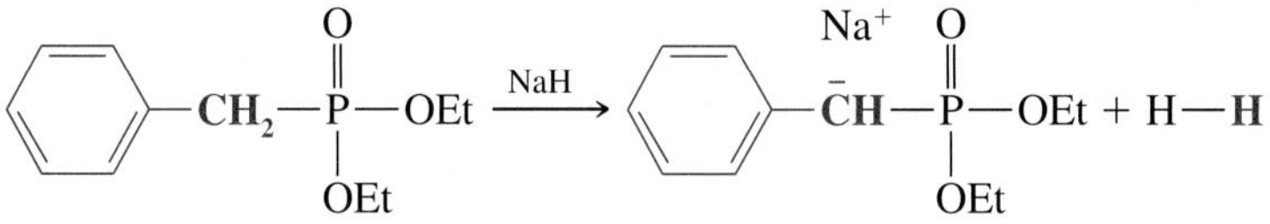

$$C_6H_5-CH_2-\overset{O}{\overset{\|}{P}}(OEt)-OEt \xrightarrow{NaH} C_6H_5-\overset{-}{C}H-\overset{O}{\overset{\|}{P}}(OEt)-OEt\ Na^+ + H-H$$

A phosphonate ester

Step 2

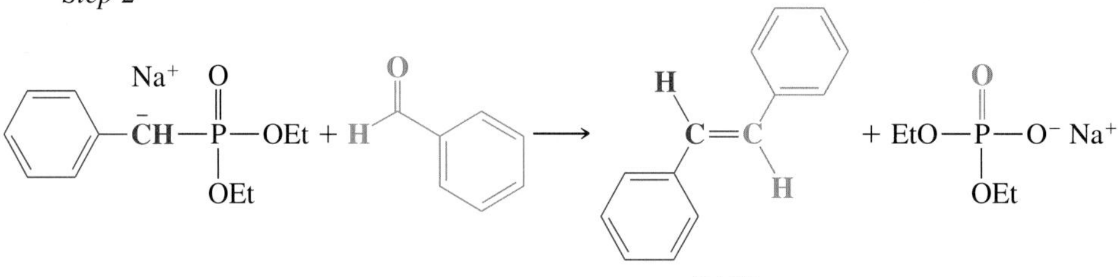

$$C_6H_5-\overset{-}{C}H-\overset{O}{\overset{\|}{P}}(OEt)-OEt\ Na^+ + C_6H_5CHO \longrightarrow (E)\text{-}C_6H_5CH{=}CHC_6H_5 + EtO-\overset{O}{\overset{\|}{P}}(OEt)-O^-\ Na^+$$

(84%)

The phosphonate ester is prepared by reaction of a trialkyl phosphite $[(RO)_3P]$ with an appropriate halide (a process called the Arbuzov reaction). The following is an example:

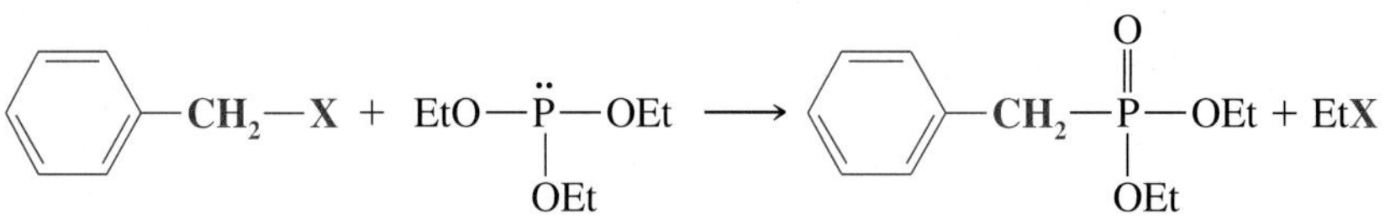

$$C_6H_5-CH_2-X + EtO-\ddot{P}(OEt)-OEt \longrightarrow C_6H_5-CH_2-\overset{O}{\overset{\|}{P}}(OEt)-OEt + EtX$$

Triethyl phosphite

PROBLEM 16.17 In addition to triphenylphosphine, assume that you have available as starting materials any necessary aldehydes, ketones, and organic halides. Show how you might synthesize each of the following alkenes using the Wittig reaction:

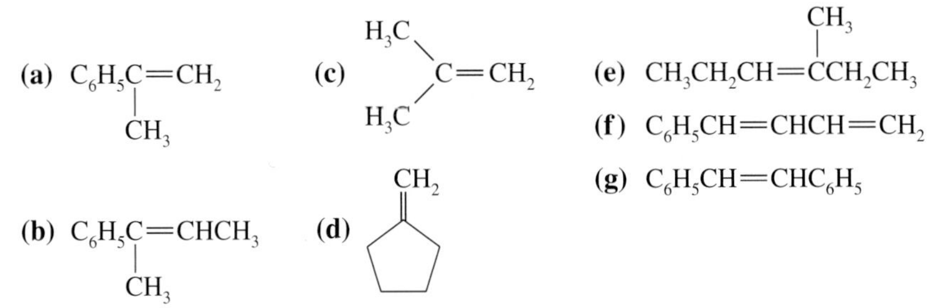

(a) $C_6H_5C(CH_3){=}CH_2$

(b) $C_6H_5C(CH_3){=}CHCH_3$

(c) $(H_3C)_2C{=}CH_2$

(d) methylenecyclopentane ($CH_2{=}$cyclopentane)

(e) $CH_3CH_2CH{=}C(CH_3)CH_2CH_3$

(f) $C_6H_5CH{=}CHCH{=}CH_2$

(g) $C_6H_5CH{=}CHC_6H_5$

PROBLEM 16.18

Triphenylphosphine can be used to convert epoxides to alkenes, for example,

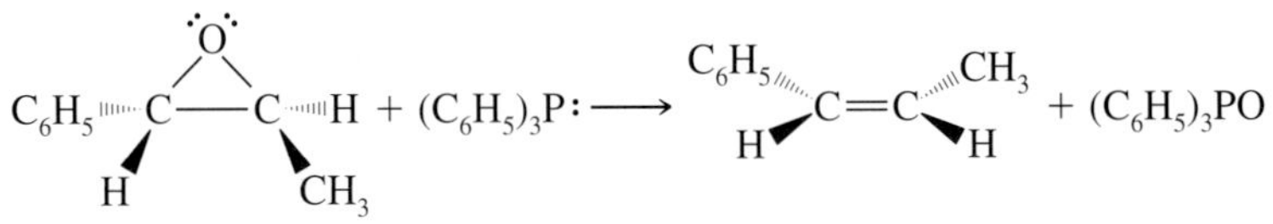

$$C_6H_5(H)C\overset{\ddot{O}}{\frown}C(H)CH_3 + (C_6H_5)_3P\!: \longrightarrow (C_6H_5)(H)C{=}C(CH_3)(H) + (C_6H_5)_3PO$$

Propose a likely mechanism for this reaction.

16.11 The Addition of Organometallic Reagents: The Reformatsky Reaction

In Section 12.8 we studied the addition of Grignard reagents, organolithium compounds, and sodium alkynides to aldehydes and ketones. These reactions, as we saw then, can be used to produce a wide variety of alcohols:

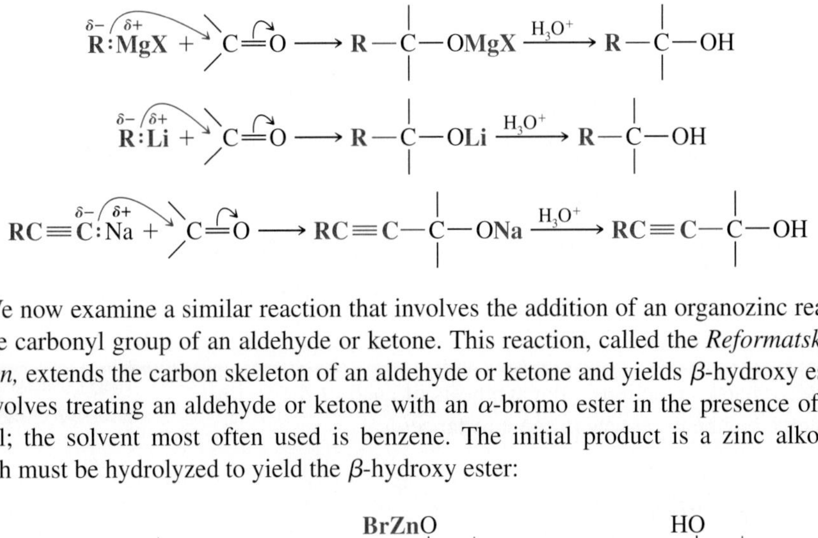

$$\overset{\delta-}{R}\!:\!\overset{\delta+}{MgX} + \;\rangle C{=}O \longrightarrow R{-}\overset{|}{\underset{|}{C}}{-}OMgX \xrightarrow{H_3O^+} R{-}\overset{|}{\underset{|}{C}}{-}OH$$

$$\overset{\delta-}{R}\!:\!\overset{\delta+}{Li} + \;\rangle C{=}O \longrightarrow R{-}\overset{|}{\underset{|}{C}}{-}OLi \xrightarrow{H_3O^+} R{-}\overset{|}{\underset{|}{C}}{-}OH$$

$$RC{\equiv}\overset{\delta-}{C}\!:\!\overset{\delta+}{Na} + \;\rangle C{=}O \longrightarrow RC{\equiv}C{-}\overset{|}{\underset{|}{C}}{-}ONa \xrightarrow{H_3O^+} RC{\equiv}C{-}\overset{|}{\underset{|}{C}}{-}OH$$

We now examine a similar reaction that involves the addition of an organozinc reagent to the carbonyl group of an aldehyde or ketone. This reaction, called the *Reformatsky reaction,* extends the carbon skeleton of an aldehyde or ketone and yields β-hydroxy esters. It involves treating an aldehyde or ketone with an α-bromo ester in the presence of zinc metal; the solvent most often used is benzene. The initial product is a zinc alkoxide, which must be hydrolyzed to yield the β-hydroxy ester:

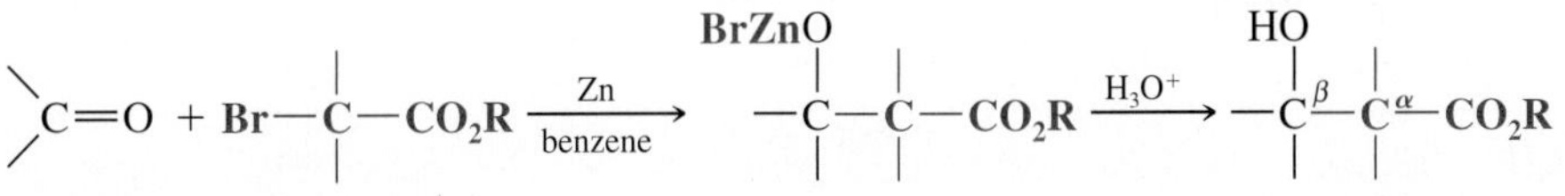

$$\underset{\textbf{Aldehyde or ketone}}{\rangle C{=}O} + \underset{\alpha\textbf{-Bromo ester}}{Br{-}\overset{|}{\underset{|}{C}}{-}CO_2R} \xrightarrow[\text{benzene}]{Zn} {-}\overset{BrZnO}{\overset{|}{\underset{|}{C}}}{-}\overset{|}{\underset{|}{C}}{-}CO_2R \xrightarrow{H_3O^+} \underset{\beta\textbf{-Hydroxy ester}}{{-}\overset{HO}{\overset{|}{\underset{|}{C}}}\overset{\beta}{-}\overset{|}{\underset{|}{C}}\overset{\alpha}{-}CO_2R}$$

The intermediate in the reaction appears to be an organozinc reagent that adds to the carbonyl group in a manner analogous to that of a Grignard reagent.

A Mechanisms for the Reaction

Reformatsky Reaction

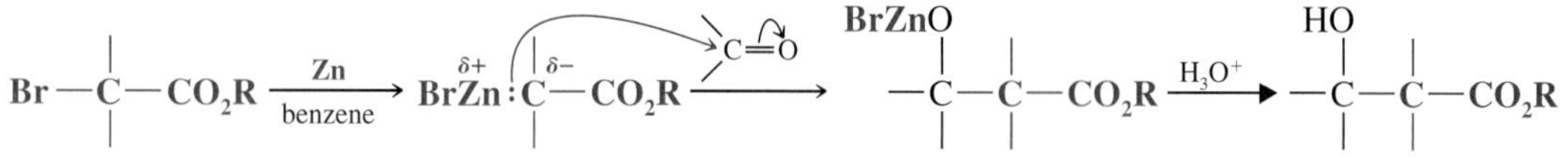

$$Br-\overset{|}{\underset{|}{C}}-CO_2R \xrightarrow[\text{benzene}]{Zn} \overset{\delta+}{BrZn}:\overset{\delta-}{\underset{|}{\overset{|}{C}}}-CO_2R \xrightarrow{>C=O} -\overset{BrZnO}{\underset{|}{\overset{|}{C}}}-\underset{|}{\overset{|}{C}}-CO_2R \xrightarrow{H_3O^+} -\overset{HO}{\underset{|}{\overset{|}{C}}}-\underset{|}{\overset{|}{C}}-CO_2R$$

Because the organozinc reagent is less reactive than a Grignard reagent, it does not add to the ester group. The β-hydroxy esters produced in the Reformatsky reaction are easily dehydrated to α,β-unsaturated esters, because dehydration yields a system in which the carbon–carbon double bond is conjugated with the carbon–oxygen double bond of the ester:

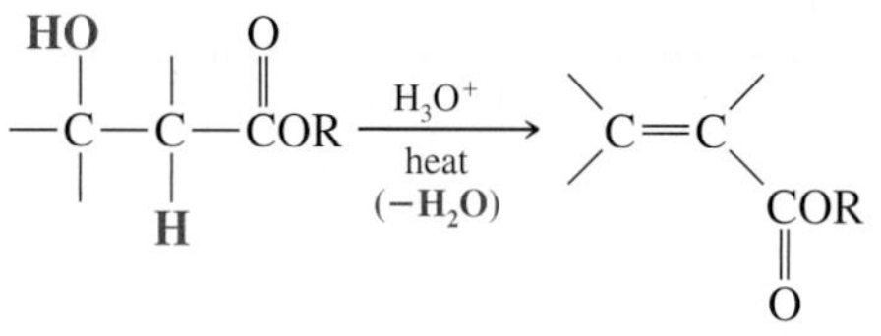

$$-\overset{HO}{\underset{|}{\overset{|}{C}}}-\underset{H}{\overset{|}{C}}-\overset{O}{\overset{\|}{C}}OR \xrightarrow[\text{heat}\ (-H_2O)]{H_3O^+} >C=C<_{\underset{O}{\overset{|}{\underset{\|}{C}}}OR}$$

β-Hydroxy ester **α,β-Unsaturated ester**

Examples of the Reformatsky reaction are the following (where Et = CH_3CH_2—):

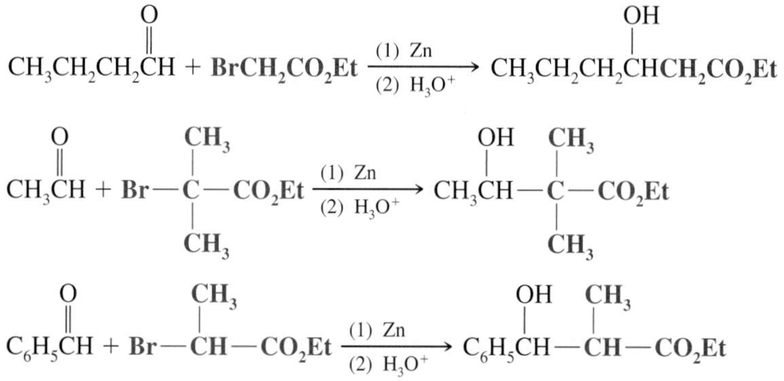

$$CH_3CH_2CH_2\overset{O}{\overset{\|}{C}}H + BrCH_2CO_2Et \xrightarrow[(2)\ H_3O^+]{(1)\ Zn} CH_3CH_2CH_2\overset{OH}{\overset{|}{C}}HCH_2CO_2Et$$

$$CH_3\overset{O}{\overset{\|}{C}}H + Br-\overset{CH_3}{\underset{CH_3}{\overset{|}{\underset{|}{C}}}}-CO_2Et \xrightarrow[(2)\ H_3O^+]{(1)\ Zn} CH_3\overset{OH}{\overset{|}{C}}H-\overset{CH_3}{\underset{CH_3}{\overset{|}{\underset{|}{C}}}}-CO_2Et$$

$$C_6H_5\overset{O}{\overset{\|}{C}}H + Br-\overset{CH_3}{\overset{|}{C}}H-CO_2Et \xrightarrow[(2)\ H_3O^+]{(1)\ Zn} C_6H_5\overset{OH}{\overset{|}{C}}H-\overset{CH_3}{\overset{|}{C}}H-CO_2Et$$

PROBLEM 16.19

Show how you would use a Reformatsky reaction in the synthesis of each of the following compounds. (Additional steps may be necessary in some instances.)

(a)  $(CH_3)_2\overset{OH}{\overset{|}{C}}CH_2CO_2CH_2CH_3$

(b) 1-hydroxycyclohexyl—$\underset{CH_3}{\underset{|}{C}}HCO_2CH_2CH_3$ (cyclohexane ring bearing OH and $-CHCO_2CH_2CH_3$ with CH_3)

(c) $CH_3CH_2CH_2CH_2CO_2CH_2CH_3$

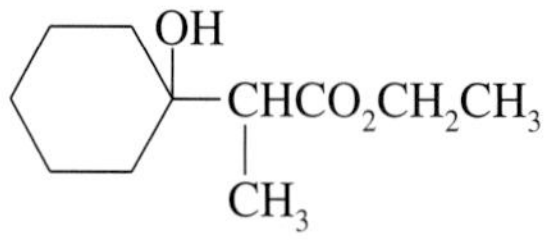

We shall see other methods for synthesizing β-hydroxy and α,β-unsaturated carbonyl compounds in Chapters 17 and 19.

16.12 Oxidation of Aldehydes and Ketones

Aldehydes are much more easily oxidized than ketones. Aldehydes are readily oxidized by strong oxidizing agents such as potassium permanganate, and they are also oxidized by such mild oxidizing agents as silver oxide:

$$\text{RCH}(=\text{O}) \xrightarrow{\text{KMnO}_4,\ \text{OH}^-} \text{RCO}^-(=\text{O}) \xrightarrow{\text{H}_3\text{O}^+} \text{RCOH}(=\text{O})$$

$$\text{RCH}(=\text{O}) \xrightarrow{\text{Ag}_2\text{O},\ \text{OH}^-} \text{RCO}^-(=\text{O}) \xrightarrow{\text{H}_3\text{O}^+} \text{RCOH}(=\text{O})$$

Notice that in these oxidations aldehydes lose the hydrogen that is attached to the carbonyl carbon atom. Because ketones lack this hydrogen, they are more resistant to oxidation.

16.12A The Baeyer–Villiger Oxidation of Aldehydes and Ketones

The Baeyer–Villiger oxidation is a useful method for conversion of aldehydes and ketones to esters by the insertion of an oxygen atom from a peroxycarboxylic acid (RCO_3H). For example, treating acetophenone with a peroxycarboxylic acid converts it to the ester phenyl acetate:

$$\underset{\textbf{Acetophenone}}{\text{C}_6\text{H}_5-\text{C}(=\text{O})-\text{CH}_3} \xrightarrow{\text{RCOOH}\ (\text{RC}(=\text{O})\text{OOH})} \underset{\textbf{Phenyl acetate}}{\text{C}_6\text{H}_5-\text{O}-\text{C}(=\text{O})-\text{CH}_3}$$

The Baeyer–Villiger oxidation is also widely used for synthesizing lactones (cyclic esters) from cyclic ketones. A common reagent used to carry out the Baeyer–Villiger oxidation is *m*-chloroperoxybenzoic acid (MCPBA). Certain other peroxycarboxylic acids can be used as well, and we will discuss shortly a "green" variation of the Baeyer–Villiger oxidation by a peroxycarboxylic acid. The following is a mechanism proposed for Baeyer-Villiger oxidation by a peroxycarboxylic acid.

A Mechanism for the Reaction

Baeyer–Villiger Oxidation

$$\text{CH}_3\text{C}(=\ddot{\text{O}}\text{:})\text{C}_6\text{H}_5 + \text{H}-\text{A} \rightleftharpoons \text{CH}_3\text{C}(=\overset{+}{\text{O}}\text{H})\text{C}_6\text{H}_5 + \text{H}-\ddot{\text{O}}-\ddot{\text{O}}-\text{C}(=\text{O})-\text{R} \rightleftharpoons \text{CH}_3-\text{C}(\text{OH})(\text{C}_6\text{H}_5)-\overset{+}{\text{O}}(\text{H})-\ddot{\text{O}}-\text{C}(=\text{O})-\text{R} \rightleftharpoons$$

The carbonyl reactant removes a proton from an acid.

The peroxy acid attacks the protonated carbonyl reactant.

A proton is removed from the oxonium ion. (:A⁻)

The peroxy acid carbonyl group in this intermediate is protonated, preparing the RCO_2H portion to be a leaving group.

The phenyl group migrates with an electron pair to the adjacent oxygen, simultaneous with departure of RCO_2H as a leaving group.

A proton is removed, resulting in the ester product.

The products of this reaction show that a phenyl group has a greater tendency to migrate than a methyl group. Had this not been the case, the product would have been $C_6H_5COOCH_3$ and not $CH_3COOC_6H_5$. This tendency of a group to migrate is called its **migratory aptitude.** Studies of the Baeyer–Villiger oxidation and other reactions have shown that the migratory aptitude of groups is H > phenyl > 3° alkyl > 2° alkyl > 1° alkyl > methyl. In all cases, this order is for groups migrating with their electron pairs, that is, as anions.

A green version of the Baeyer–Villiger oxidation has been reported by A. Corma (University of Valencia, Italy) and colleagues. Their method uses a tin–zeolite catalyst with hydrogen peroxide (H_2O_2) as the oxygen atom donor. The following reaction is an example:

+ HOOH —Sn–zeolite catalyst (1.5 wt %), 30°C, 2 h→ + HOH

95%

The method is environmentally friendly because oxidation with hydrogen peroxide produces only water as the byproduct, whereas oxidation with a peroxycarboxylic acid generates a carboxylic acid byproduct that must be separated from the desired product and either recycled or discarded. The mechanism proposed for the tin–zeolite-catalyzed reaction involves polarization of the carbonyl group through a Lewis acid–base interaction of the carbonyl oxygen of the substrate with tin in the matrix of the catalyst, thus enhancing the electrophilicity of the carbonyl carbon. Attack of hydrogen peroxide at the carbonyl carbon and loss of a proton from the attacking oxygen results in a tetrahedral intermediate called the Criegee adduct (as is also formed in the traditional Baeyer–Villiger oxidation). A carbon of the substrate migrates to the proximate oxygen as a molecule of water departs upon protonation of the distal oxygen and cleavage of the oxygen–oxygen bond. The tin–zeolite catalyst developed by Corma is highly selective and holds great promise as a more environmentally benign version of the Baeyer–Villiger oxidation.

PROBLEM 16.20 When benzaldehyde reacts with a peroxy acid, the product is benzoic acid. The mechanism for this reaction is analogous to the one just given for the oxidation of acetophenone, and the outcome illustrates the greater migratory aptitude of a hydrogen atom compared to phenyl. Outline all the steps involved.

PROBLEM 16.21 Give the structure of the product that would result from a Baeyer–Villiger oxidation of cyclopentanone.

PROBLEM 16.22 What would be the major product formed in the Baeyer–Villiger oxidation of 3-methyl-2-butanone?

16.13 Chemical Analyses for Aldehydes and Ketones

16.13A Derivatives of Aldehydes and Ketones

Aldehydes and ketones can be differentiated from noncarbonyl compounds through their reactions with derivatives of ammonia (Section 16.8). Semicarbazide, 2,4-dinitrophenylhydrazine, and hydroxylamine react with aldehydes and ketones to form precipitates. Semicarbazones and oximes are usually colorless, whereas 2,4-dinitrophenylhydrazones are usually orange. The melting points of these derivatives can also be used in identifying specific aldehydes and ketones.

16.13B Tollens' Test (Silver Mirror Test)

The ease with which aldehydes undergo oxidation provides a useful test that differentiates aldehydes from most ketones. Mixing aqueous silver nitrate with aqueous ammonia produces a solution known as Tollens' reagent. The reagent contains the diamminosilver(I) ion, $Ag(NH_3)_2^+$. Although this ion is a very weak oxidizing agent, it oxidizes aldehydes to carboxylate anions. As it does this, silver is reduced from the +1 oxidation state [of $Ag(NH_3)_2^+$] to metallic silver. If the rate of reaction is slow and the walls of the vessel are clean, metallic silver deposits on the walls of the test tube as a mirror; if not, it deposits as a gray-to-black precipitate. Tollens' reagent gives a negative result with all ketones except α-hydroxy ketones:

$$\underset{\textbf{Aldehyde}}{R-\overset{O}{\overset{\|}{C}}-H} \xrightarrow[H_2O]{Ag(NH_3)_2^+} R-\overset{O}{\overset{\|}{C}}-O^- + \underset{\textbf{Silver mirror}}{Ag\downarrow}$$

$$\underset{\alpha\textbf{-Hydroxy ketone}}{R-\overset{O}{\overset{\|}{C}}-\overset{OH}{\overset{|}{C}H}-R'} \xrightarrow[H_2O]{Ag(NH_3)_2^+} R-\overset{O}{\overset{\|}{C}}-\overset{O}{\overset{\|}{C}}-R' + \underset{\textbf{Silver mirror}}{Ag\downarrow}$$

$$\underset{\textbf{Ketone}}{R-\overset{O}{\overset{\|}{C}}-R'} \xrightarrow[H_2O]{Ag(NH_3)_2^+} \text{no reaction}$$

16.14 Spectroscopic Properties of Aldehydes and Ketones

16.14A IR Spectra of Aldehydes and Ketones

Carbonyl groups of aldehydes and ketones give rise to very strong C═O stretching absorption bands in the 1665–1780-cm^{-1} region. The exact location of the absorption (Table 16.3) depends on the structure of the aldehyde or ketone and is one of the most useful and characteristic absorptions in the IR spectrum. Saturated acyclic aldehydes typically absorb near 1730 cm^{-1}; similar ketones absorb near 1715 cm^{-1}.

TABLE 16.3 IR Carbonyl Stretching Bands of Aldehydes and Ketones

C═O Stretching Frequencies			
Compound	Range (cm^{-1})	Compound	Range (cm^{-1})
R—CHO	1720–1740	RCOR	1705–1720
Ar—CHO	1695–1715	ArCOR	1680–1700
>C═C(—)CHO	1680–1690	>C═C(—)COR	1665–1680
		Cyclohexanone	1715
		Cyclopentanone	1751
		Cyclobutanone	1785

Conjugation of the carbonyl group with a double bond or a benzene ring shifts the C═O *absorption to lower frequencies by about 40 cm^{-1}.* This shift to lower frequencies occurs because the double bond of a conjugated compound has more single-bond character (see the resonance structures below), and single bonds are easier to stretch than double bonds.

$$\text{>C=C-C(=O)-} \longleftrightarrow \text{>}\overset{+}{\text{C}}\text{-C=C-O}^{-}$$

Single bond

The location of the carbonyl absorption of cyclic ketones depends on the size of the ring (compare the cyclic compounds in Table 16.3). *As the ring grows smaller, the* C═O *stretching peak is shifted to higher frequencies.*

Vibrations of the C—H bond of the CHO group of aldehydes also give two weak bands in the 2700–2775- and 2820–2900-cm^{-1} regions that are easily identified.

Figure 16.1 shows the IR spectrum of phenylethanal.

16.14B NMR Spectra of Aldehydes and Ketones

^{13}C NMR Spectra The carbonyl carbon atom of an aldehyde or ketone gives characteristic NMR signals in the δ 180–220 region of ^{13}C spectra. Since almost no other signals occur in this region, *the presence of a signal in this region (near δ 200) strongly suggests the presence of a carbonyl group.*

^{1}H NMR Spectra An aldehyde proton gives a signal far downfield in ^{1}H NMR spectra in a region (δ 9–12) where almost no other protons absorb; therefore, it is easily identified.

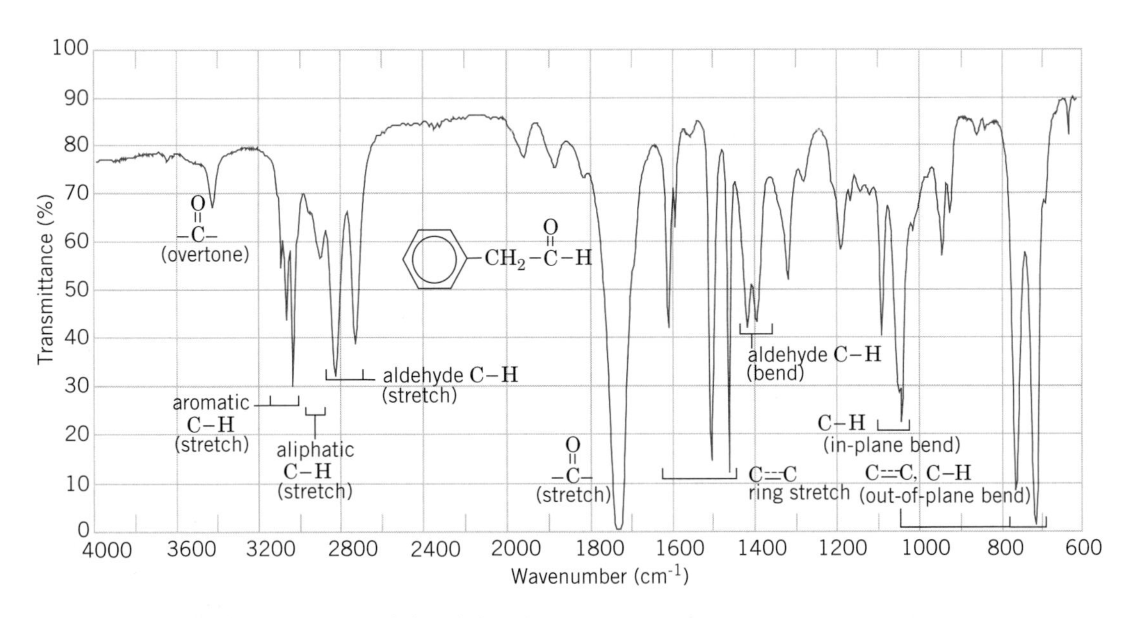

FIGURE 16.1 The infrared spectrum of phenylethanal.

The aldehyde proton of an aliphatic aldehyde shows spin–spin coupling with protons on the adjacent α carbon, and the splitting pattern reveals the degree of substitution of the α carbon. For example, in acetaldehyde (CH_3CHO) the aldehyde proton signal is split into a quartet by the three methyl protons, and the methyl proton signal is split into a doublet by the aldehyde proton. The coupling constant is about 3 Hz.

Protons on the α carbon are deshielded by the carbonyl group, and their signals generally appear in the δ 2.0–2.3 region. Methyl ketones show a characteristic (3H) singlet near δ 2.1.

Figures 16.2 and 16.3 show annotated ^{1}H and ^{13}C spectra of phenylethanal.

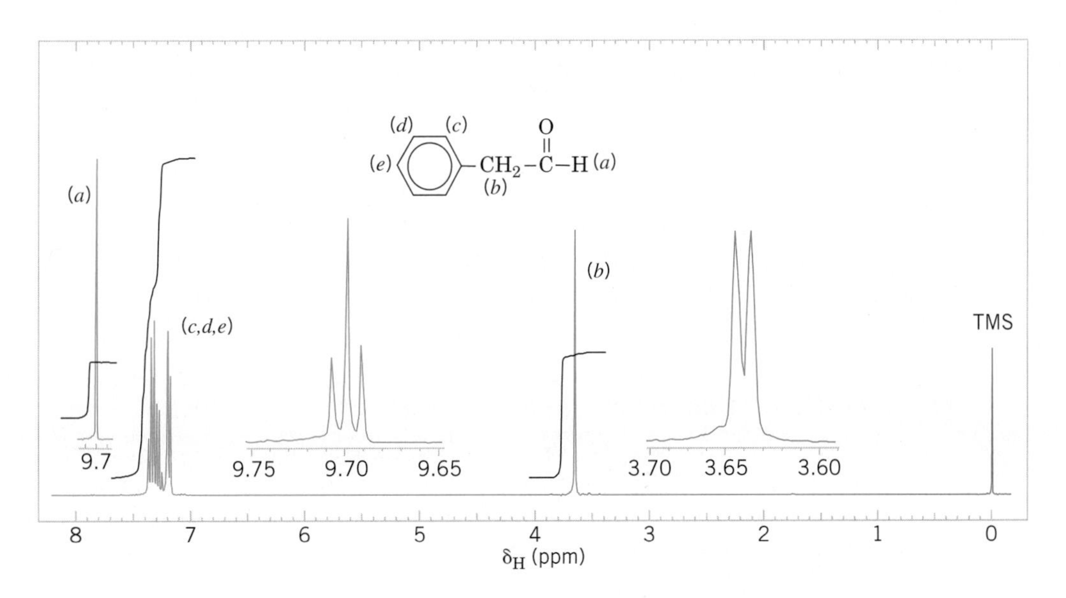

FIGURE 16.2 The 300-MHz ^{1}H NMR spectrum of phenylethanal. The small coupling between the aldehyde and methylene protons (2.6 Hz) is shown in the expanded offset plots.

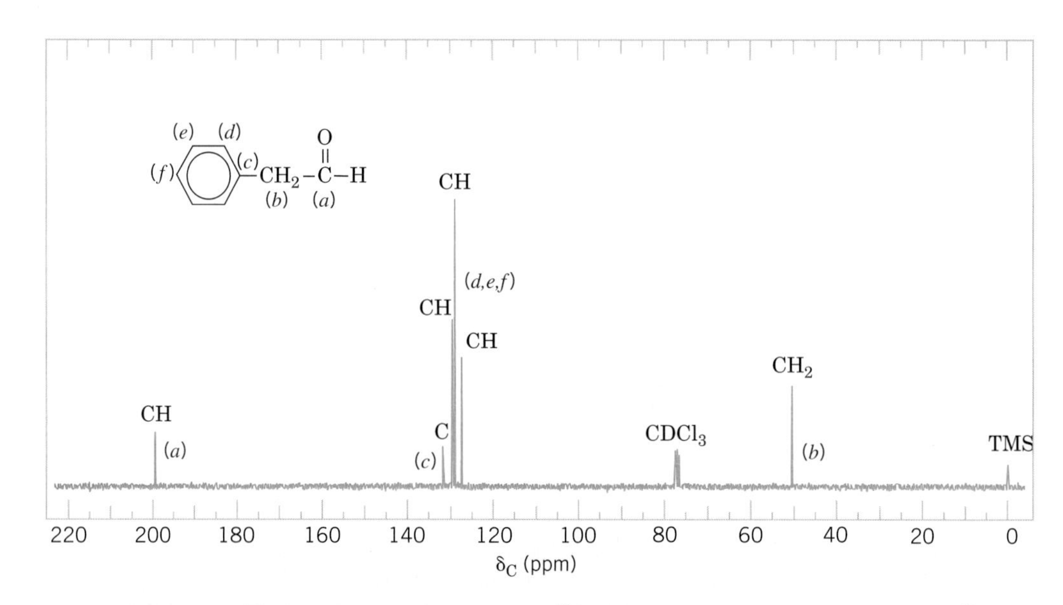

FIGURE 16.3 The broadband proton-decoupled ^{13}C NMR spectrum of phenylethanal. DEPT ^{13}C NMR information and carbon assignments are shown near each peak.

16.14C Mass Spectra of Aldehydes and Ketones

The mass spectra of ketones usually show a peak corresponding to the molecular ion. Aldehydes typically produce a prominent $M^{\dot{+}} - 1$ peak in their mass spectra from cleavage of the aldehyde hydrogen. Ketones usually undergo cleavage on either side of the carbonyl group to produce acylium ions, $RC{\equiv}\ddot{O}^{+}$, where R can be the alkyl group from either side of the ketone carbonyl. Cleavage via the McLafferty rearrangement (Section 9.16C) is also possible in many aldehydes and ketones.

16.14D UV Spectra

The carbonyl groups of saturated aldehydes and ketones give a weak absorption band in the UV region between 270 and 300 nm. This band is shifted to longer wavelengths (300–350 nm) when the carbonyl group is conjugated with a double bond.

Summary of Mechanisms

Acetals, Imines, and Enamines: Common Mechanistic Themes in their Acid-catalyzed Formation from Aldehydes and Ketones

Many steps are nearly the same in acid-catalyzed reactions of aldehydes and ketones with alcohols and amines. Compare the mechanisms vertically to see the similarities and differences. Note differences in completion of the mechanism for each type of product.

I. Hemiacetal and acetal formation: reaction with alcohols

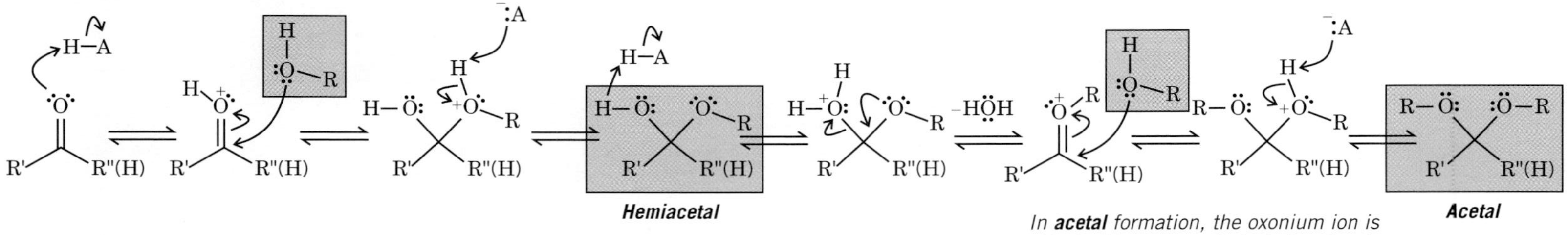

Hemiacetal

Acetal

*In **acetal** formation, the oxonium ion is attacked by a second alcohol molecule.*

II. Imine formation: reaction with primary amines

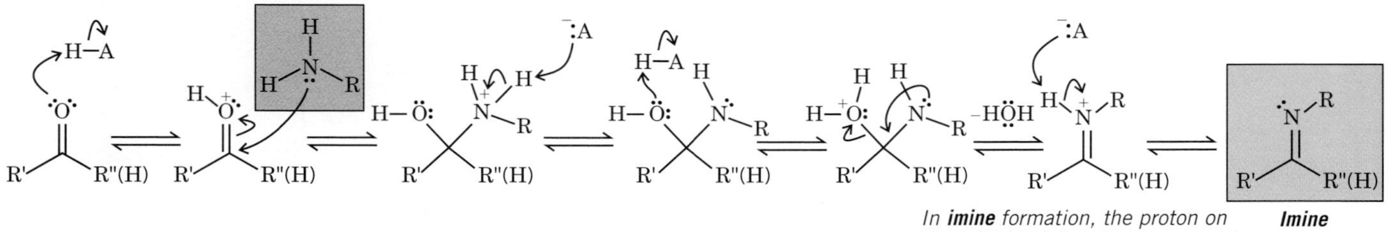

Imine

*In **imine** formation, the proton on the initial iminium ion is removed, leading to the stable imine product.*

III. Enamine formation: reaction with secondary amines

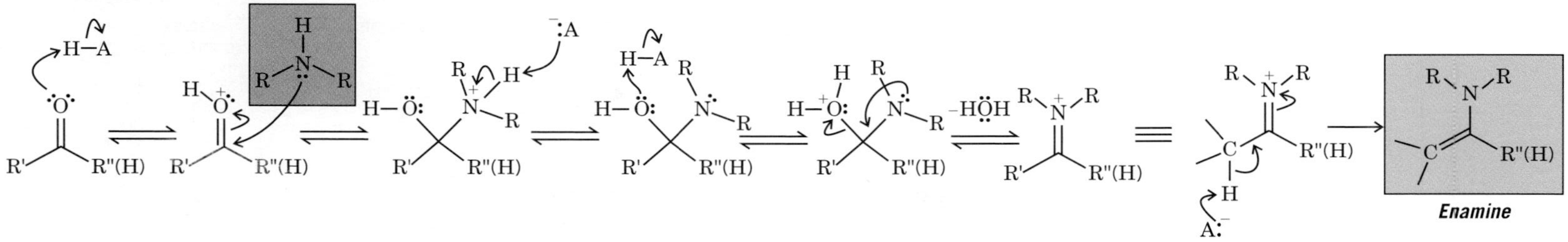

Enamine

*In **enamine** formation, a proton is removed from a carbon adjacent to the iminium carbon (because no proton is available for removal from the nitrogen).*

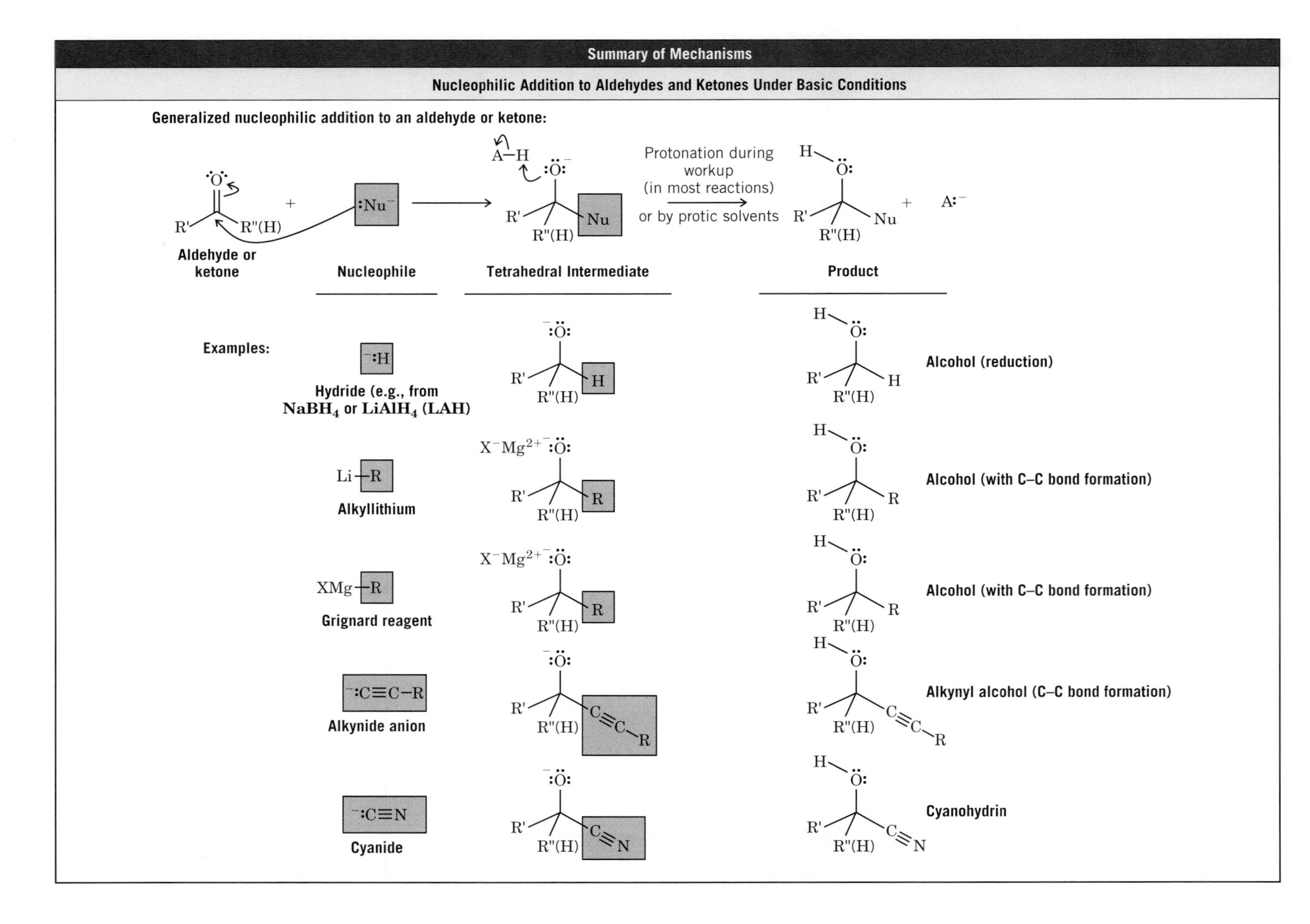
Summary of Mechanisms
Nucleophilic Addition to Aldehydes and Ketones Under Basic Conditions
Generalized nucleophilic addition to an aldehyde or ketone:
Aldehyde or ketone
Nucleophile
Tetrahedral Intermediate
Protonation during workup (in most reactions) or by protic solvents
Product
Examples:
Hydride (e.g., from NaBH4 or LiAlH4 (LAH)
Alkyllithium
Grignard reagent
Alkynide anion
Cyanide
Alcohol (reduction)
Alcohol (with C–C bond formation)
Alcohol (with C–C bond formation)
Alkynyl alcohol (C–C bond formation)
Cyanohydrin

Summary of Mechanisms

Nucleophilic Addition to Aldehydes and Ketones Under Basic Conditions

Generalized nucleophilic addition to an aldehyde or ketone:

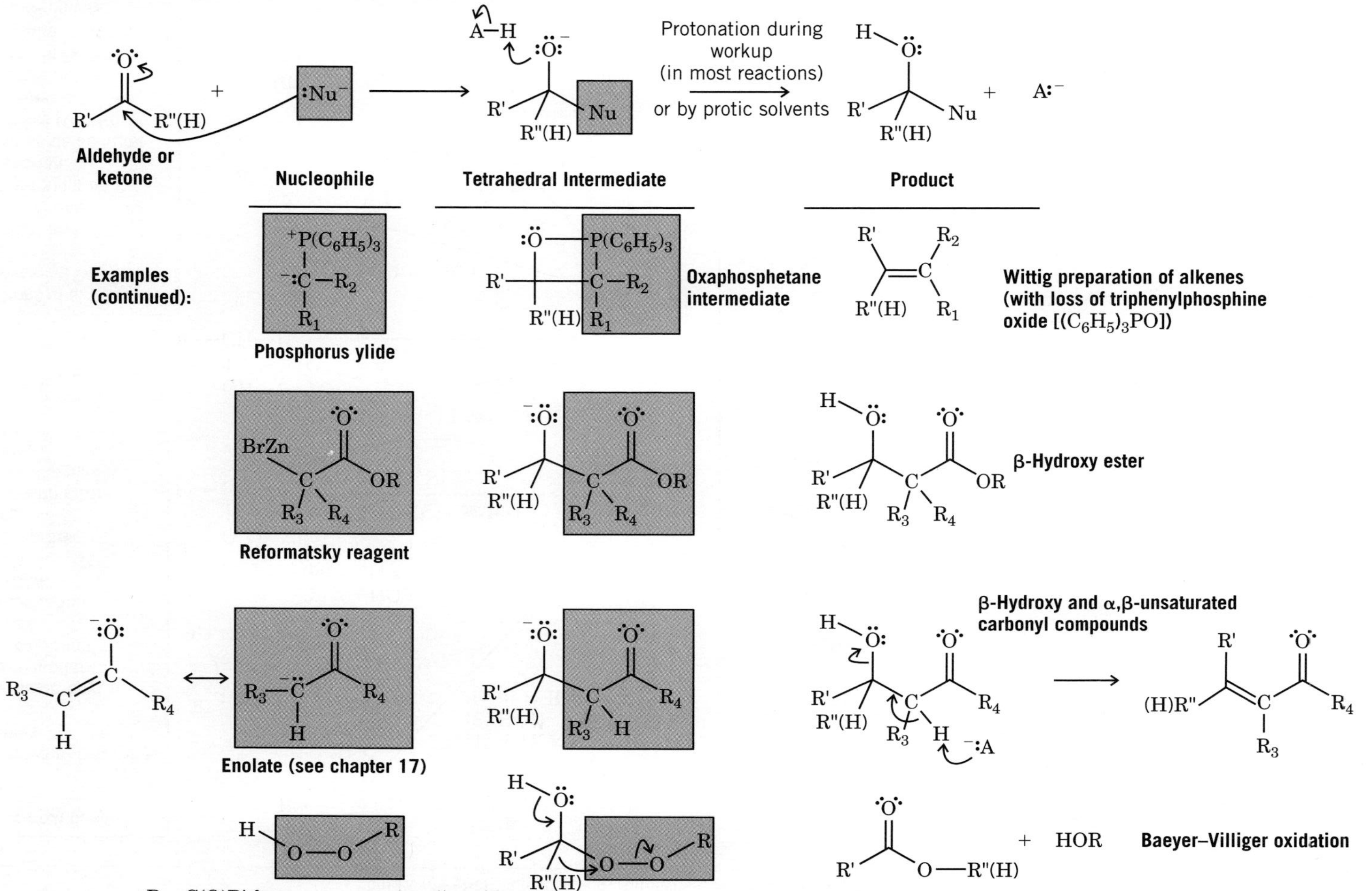

Aldehyde or ketone + :Nu⁻ ⟶ tetrahedral alkoxide (A–H) — Protonation during workup (in most reactions) or by protic solvents ⟶ R'C(OH)(R''(H))Nu + A:⁻

Aldehyde or ketone

Nucleophile	Tetrahedral Intermediate	Product
Phosphorus ylide: $^{+}P(C_6H_5)_3$, $^{-}$:C–R_2, R_1	Oxaphosphetane intermediate	Wittig preparation of alkenes (with loss of triphenylphosphine oxide [$(C_6H_5)_3PO$])
Reformatsky reagent: BrZn–C(R_3)(R_4)–C(=O)OR	Alkoxide	β-Hydroxy ester
Enolate (see chapter 17): R_3–C⁻(H)–C(=O)R_4 ⟷ R_3–CH=C(O⁻)R_4	Alkoxide	β-Hydroxy and α,β-unsaturated carbonyl compounds
H–O–O–R	Tetrahedral intermediate	R'C(=O)O–R''(H) + HOR — Baeyer–Villiger oxidation

Examples (continued):

R = C(O)R' from a peroxycarboxylic acid, or R = H, from H_2O_2 if Sn–zeolite method

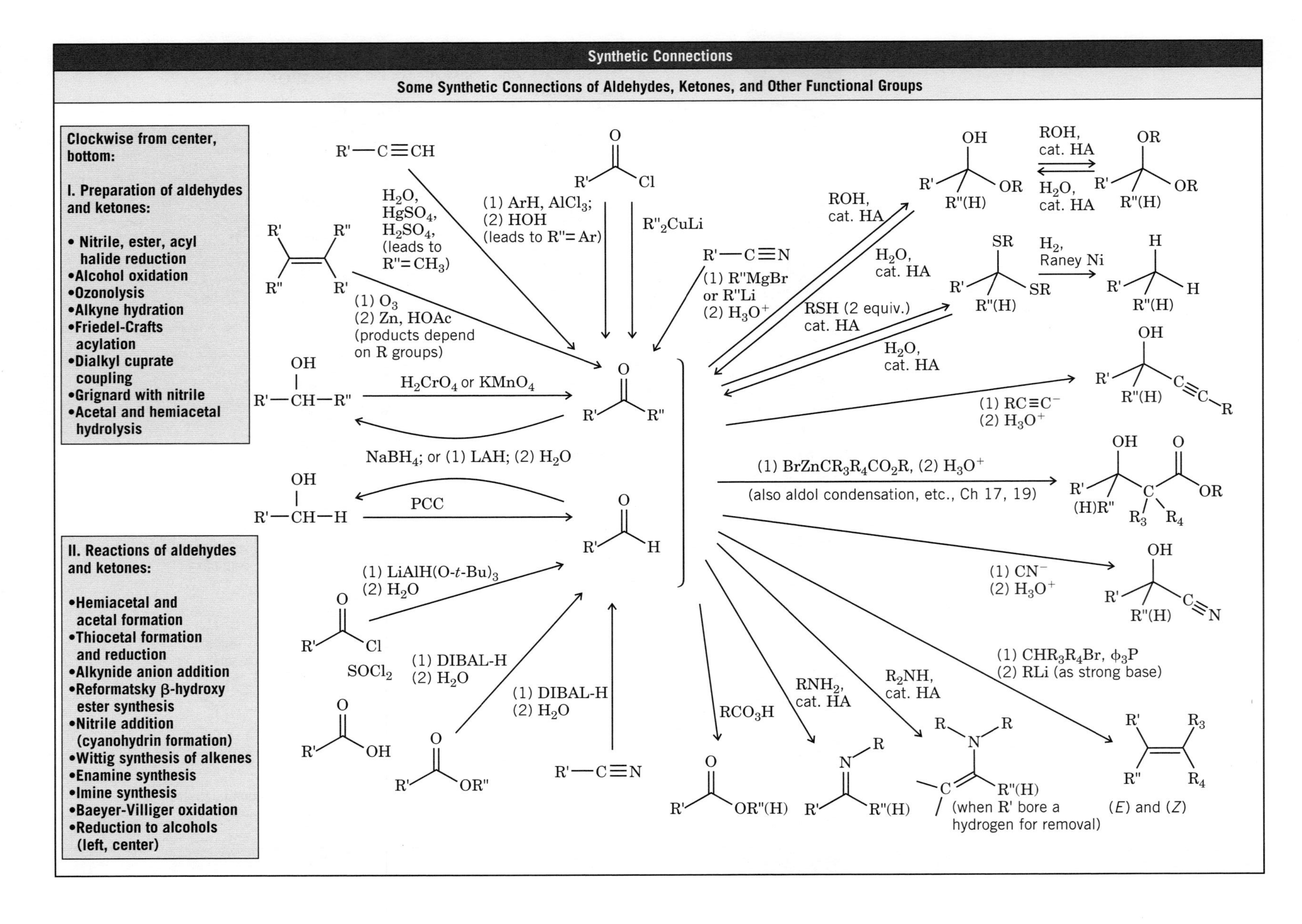

Synthetic Connections
Some Synthetic Connections of Aldehydes, Ketones, and Other Functional Groups
Clockwise from center, bottom:
I. Preparation of aldehydes and ketones:
• Nitrile, ester, acyl halide reduction
•Alcohol oxidation
•Ozonolysis
•Alkyne hydration
•Friedel-Crafts acylation
•Dialkyl cuprate coupling
•Grignard with nitrile
•Acetal and hemiacetal hydrolysis
II. Reactions of aldehydes and ketones:
•Hemiacetal and acetal formation
•Thiocetal formation and reduction
•Alkynide anion addition
•Reformatsky β-hydroxy ester synthesis
•Nitrile addition (cyanohydrin formation)
•Wittig synthesis of alkenes
•Enamine synthesis
•Imine synthesis
•Baeyer-Villiger oxidation
•Reduction to alcohols (left, center)
R'—C≡CH
H_2O, $HgSO_4$, H_2SO_4, (leads to R"= CH_3)
(1) ArH, $AlCl_3$; (2) HOH (leads to R"= Ar)
R"$_2$CuLi
R'—C≡N
(1) R"MgBr or R"Li (2) H_3O^+
(1) O_3 (2) Zn, HOAc (products depend on R groups)
H_2CrO_4 or $KMnO_4$
$NaBH_4$; or (1) LAH; (2) H_2O
PCC
(1) LiAlH(O-t-Bu)$_3$ (2) H_2O
$SOCl_2$
(1) DIBAL-H (2) H_2O
(1) DIBAL-H (2) H_2O
R'—C≡N
ROH, cat. HA
H_2O, cat. HA
ROH, cat. HA
H_2O, cat. HA
RSH (2 equiv.) cat. HA
H_2O, cat. HA
H_2, Raney Ni
(1) RC≡C$^-$ (2) H_3O^+
(1) $BrZnCR_3R_4CO_2R$, (2) H_3O^+
(also aldol condensation, etc., Ch 17, 19)
(1) CN^- (2) H_3O^+
(1) CHR_3R_4Br, ϕ_3P (2) RLi (as strong base)
(E) and (Z)
R_2NH, cat. HA
(when R' bore a hydrogen for removal)
RNH_2, cat. HA
RCO_3H
OR"(H)

Key Terms and Concepts

Acetals	Section 16.7B
Cyanohydrins	Section 16.9
Enamines	Section 16.8
Enols	Section 16.5B
Hemiacetals	Section 16.7A
Hydrazones, semicarbazones, and oximes	Section 16.8B
Imines	Section 16.8
Keto–enol tautomerization	Section 16.5B
Nucleophilic addition to the carbonyl carbon	Section 16.6
Tautomerization	Section 16.5B
Ylides	Section 16.10

Additional Problems

16.23 Give a structural formula and another acceptable name for each of the following compounds:

(a) Formaldehyde
(b) Acetaldehyde
(c) Phenylacetaldehyde
(d) Acetone
(e) Ethyl methyl ketone
(f) Acetophenone
(g) Benzophenone
(h) Salicylaldehyde
(i) Vanillin
(j) Diethyl ketone
(k) Ethyl isopropyl ketone
(l) Diisopropyl ketone
(m) Dibutyl ketone
(n) Dipropyl ketone
(o) Cinnamaldehyde

16.24 Write structural formulas for the products formed when propanal reacts with each of the following reagents:

(a) $NaBH_4$ in aqueous NaOH
(b) C_6H_5MgBr, then H_3O^+
(c) $LiAlH_4$, then H_2O
(d) Ag_2O, OH^-
(e) $(C_6H_5)_3P{=}CH_2$
(f) H_2 and Pt
(g) $HOCH_2CH_2OH$ and HA
(h) $CH_3\ddot{C}^{-}H{-}\overset{+}{P}(C_6H_5)_3$
(i) (1) $BrCH_2CO_2C_2H_5$, Zn; (2) H_3O^+
(j) $Ag(NH_3)_2{}^+$
(k) Hydroxylamine
(l) Semicarbazide
(m) Phenylhydrazine
(n) Cold dilute $KMnO_4$
(o) $HSCH_2CH_2SH$, HA
(p) $HSCH_2CH_2SH$, HA, then Raney nickel

16.25 Give structural formulas for the products formed (if any) from the reaction of acetone with each reagent in Problem 16.24.

16.26 What products would be obtained from each of the following reactions of acetophenone?

(a) Acetophenone + $HNO_3 \xrightarrow[H_2SO_4]{}$
(b) Acetophenone + $C_6H_5NHNH_2 \longrightarrow$
(c) Acetophenone + $^{-}{:}CH_2{-}\overset{+}{P}(C_6H_5)_3 \longrightarrow$
(d) Acetophenone + $NaBH_4 \xrightarrow[OH^-]{H_2O}$
(e) Acetophenone + $C_6H_5MgBr \xrightarrow[(2)\ NH_4Cl]{}$

Note: Problems marked with an asterisk are "challenge problems."

16.27 **(a)** Give three methods for synthesizing phenyl propyl ketone from benzene and any other needed reagents.
(b) Give three methods for transforming phenyl propyl ketone into butylbenzene.

16.28 Show how you would convert benzaldehyde into each of the following. You may use any other needed reagents, and more than one step may be required.

(a) Benzyl alcohol
(b) Benzoic acid
(c) Benzoyl chloride
(d) Benzophenone
(e) Acetophenone
(f) 1-Phenylethanol
(g) 3-Methyl-1-phenyl-1-butanol
(h) Benzyl bromide
(i) Toluene
(j) $C_6H_5CH(OCH_3)_2$
(k) $C_6H_5CH^{18}O$
(l) C_6H_5CHDOH
(m) $C_6H_5CH(OH)CN$
(n) $C_6H_5CH{=}NOH$
(o) $C_6H_5CH{=}NNHC_6H_5$
(p) $C_6H_5CH{=}NNHCONH_2$
(q) $C_6H_5CH{=}CHCH{=}CH_2$

16.29 Show how ethyl phenyl ketone ($C_6H_5COCH_2CH_3$) could be synthesized from each of the following:

(a) Benzene
(b) Benzoyl chloride
(c) Benzonitrile, C_6H_5CN
(d) Benzaldehyde

16.30 Show how benzaldehyde could be synthesized from each of the following:

(a) Benzyl alcohol
(b) Benzoic acid
(c) Phenylethyne
(d) Phenylethene (styrene)
(e) $C_6H_5CO_2CH_3$
(f) $C_6H_5C{\equiv}N$

16.31 Give structures for compounds **A–E**.

$$\text{Cyclohexanol} \xrightarrow[\text{acetone}]{H_2CrO_4} \mathbf{A}\ (C_6H_{10}O) \xrightarrow[(2)\ H_3O^+]{(1)\ CH_3MgI} \mathbf{B}\ (C_7H_{14}O) \xrightarrow[\text{heat}]{HA}$$

$$\mathbf{C}\ (C_7H_{12}) \xrightarrow[(2)\ \text{Zn, HOAc}]{(1)\ O_3} \mathbf{D}\ (C_7H_{12}O_2) \xrightarrow[(2)\ H_3O^+]{(1)\ Ag_2O,\ OH^-} \mathbf{E}\ (C_7H_{12}O_3)$$

16.32 The following reaction sequence shows how the carbon chain of an aldehyde may be lengthened by two carbon atoms. What are the intermediates **K–M**?

$$\text{Ethanal} \xrightarrow[(2)\ H_3O^+]{(1)\ BrCH_2CO_2Et,\ Zn} \mathbf{K}\ (C_6H_{12}O_3) \xrightarrow{\text{HA, heat}}$$

$$\mathbf{L}\ (C_6H_{10}O_2) \xrightarrow{H_2,\ Pt} \mathbf{M}\ (C_6H_{12}O_2) \xrightarrow[(2)\ H_2O]{(1)\ \text{DIBAL-H}} \text{butanal}$$

Hint: The ^{13}C NMR spectrum of **L** consists of signals at δ 166.7, δ 144.5, δ 122.8, δ 60.2, δ 17.9, and δ 14.3.

16.33 Warming piperonal (Section 16.3) with dilute aqueous HCl converts it to a compound with the formula $C_7H_6O_3$. What is this compound, and what type of reaction is involved?

16.34 Starting with benzyl bromide, show how you would synthesize each of the following:

(a) $C_6H_5CH_2CHOHCH_3$
(b) $C_6H_5CH_2CH_2CHO$
(c) $C_6H_5CH{=}CH{-}CH{=}CHC_6H_5$
(d) $C_6H_5CH_2COCH_2CH_3$

16.35 Compounds **A** and **D** do not give positive Tollens' tests; however, compound **C** does. Give structures for **A–D**.

$$\text{4-Bromobutanal} \xrightarrow{HOCH_2CH_2OH,\ HA} \mathbf{A}\ (C_6H_{11}O_2Br) \xrightarrow{Mg,\ Et_2O}$$

$$[\mathbf{B}\ (C_6H_{11}MgO_2Br)] \xrightarrow[(2)\ H_3O^+,\ H_2O]{(1)\ CH_3CHO} \mathbf{C}\ (C_6H_{12}O_2) \xrightarrow[HA]{CH_3OH} \mathbf{D}\ (C_7H_{14}O_2)$$

16.36 In Chapter 4 we discussed pheromones and several compounds used in the manufacture of perfumes. **(a)** Civetone, shown below, is a natural compound produced by the civet cat and used in some perfumes. Write an IUPAC name for civetone. **(b)** Pentalide is a synthetic musk that is also used in some perfumes. Suggest a method for preparing pentalide from an appropriate cyclic ketone and give an IUPAC name to this precursor ketone.

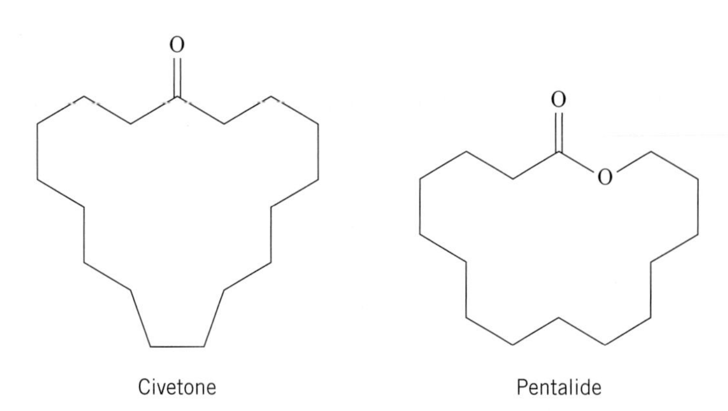

Civetone Pentalide

16.37 Dianeackerone is a volatile natural product isolated from secretory glands of the adult African dwarf crocodile. The compound is believed to be a pheromone associated with nesting and mating. Dianeackerone is named after Diane Ackerman, an author in the field of natural history and champion of the importance of preserving biodiversity. The IUPAC name of dianeackerone is 3,7-diethyl-9-phenylnonan-2-one, and it is found as both the (3*S*, 7*S*) and (3*S*, 7*R*) stereoisomers. Draw structures for both stereoisomers of dianeackerone.

16.38 Provide the missing reagents and intermediate in the following synthesis:

$$HO-C_6H_4-CH_2OH \xrightarrow{(a)} CH_3O-C_6H_4-CH_2OH \xrightarrow{(b)} ? \xrightarrow{(c)}$$

$$CH_3O-C_6H_4-CH(OH)CH(CH_3)CO_2Et \xrightarrow{(d)} CH_3O-C_6H_4-CH(OH)CH(CH_3)CH_2OH$$

16.39 Outlined here is a synthesis of glyceraldehyde (Section 5.15A). What are the intermediates **A–C** and what stereoisomeric form of glyceraldehyde would you expect to obtain?

$$CH_2{=}CHCH_2OH \xrightarrow[CH_2Cl_2]{PCC} \mathbf{A}\ (C_3H_4O) \xrightarrow{CH_3OH,\ HA}$$

$$\mathbf{B}\ (C_5H_{10}O_2) \xrightarrow[\text{cold, dilute}]{KMnO_4, OH^-} \mathbf{C}\ (C_5H_{12}O_4) \xrightarrow[H_2O]{H_3O^+} \text{glyceraldehyde}$$

16.40 Consider the reduction of (*R*)-3-phenyl-2-pentanone by sodium borohydride. After the reduction is complete, the mixture is separated by chromatography into two fractions. These fractions contain isomeric compounds, and each isomer is optically active. What are these two isomers and what is the stereoisomeric relationship between them?

16.41 The structure of the sex pheromone (attractant) of the female tsetse fly has been confirmed by the following synthesis. Compound **C** appears to be identical to the natural pheromone in all respects (including the response of the male tsetse fly). Provide structures for **A, B,** and **C.**

$$BrCH_2(CH_2)_7CH_2Br \xrightarrow[(2)\ 2\ RLi]{(1)\ 2\ (C_6H_5)_3P} \mathbf{A}\ (C_{45}H_{46}P_2) \xrightarrow{2\ CH_3(CH_2)_{11}\overset{O}{\overset{\|}{C}}CH_3} \mathbf{B}\ (C_{37}H_{72}) \xrightarrow{H_2,\ Pt} \mathbf{C}\ (C_{37}H_{76})$$

16.42 Outline simple chemical tests that would distinguish between each of the following:
(a) Benzaldehyde and benzyl alcohol
(b) Hexanal and 2-hexanone

(c) 2-Hexanone and hexane
(d) 2-Hexanol and 2-hexanone
(e) $C_6H_5CH{=}CHCOC_6H_5$ and $C_6H_5COC_6H_5$
(f) Pentanal and diethyl ether

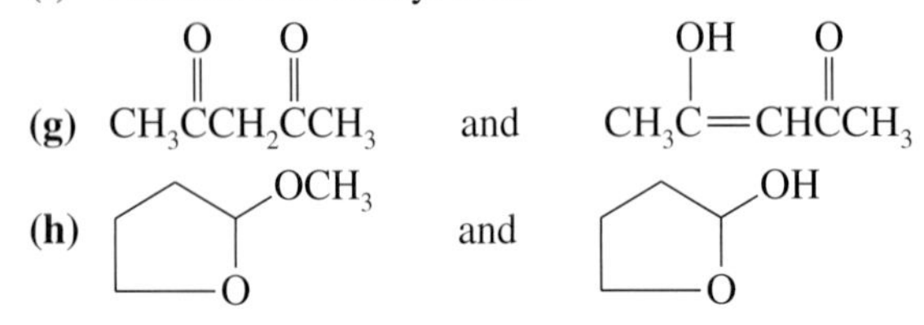

16.43 Compounds **W** and **X** are isomers; they have the molecular formula C_9H_8O. The IR spectrum of each compound shows a strong absorption band near 1715 cm^{-1}. Oxidation of either compound with hot, basic potassium permanganate followed by acidification yields phthalic acid. The ^1H NMR spectrum of **W** shows a multiplet at δ 7.3 and a singlet at δ 3.4. The ^1H NMR spectrum of **X** shows a multiplet at δ 7.5, a triplet at δ 3.1, and a triplet at δ 2.5. Propose structures for **W** and **X**.

CO_2H
CO_2H
Phthalic acid

16.44 Compounds **Y** and **Z** are isomers with the molecular formula $C_{10}H_{12}O$. The IR spectrum of each compound shows a strong absorption band near 1710 cm^{-1}. The ^1H NMR spectra of **Y** and **Z** are given in Figs. 16.4 and 16.5. Propose structures for **Y** and **Z.**

16.45 Compound **A** ($C_9H_{18}O$) forms a phenylhydrazone, but it gives a negative Tollens' test. The IR spectrum of **A** has a strong band near 1710 cm^{-1}. The broadband proton-decoupled ^{13}C NMR spectrum of **A** is given in Fig. 16.6. Propose a structure for **A.**

16.46 Compound **B** ($C_8H_{12}O_2$) shows a strong carbonyl absorption in its IR spectrum. The broadband proton-decoupled ^{13}C NMR spectrum of **B** is given in Fig. 16.7. Propose a structure for **B.**

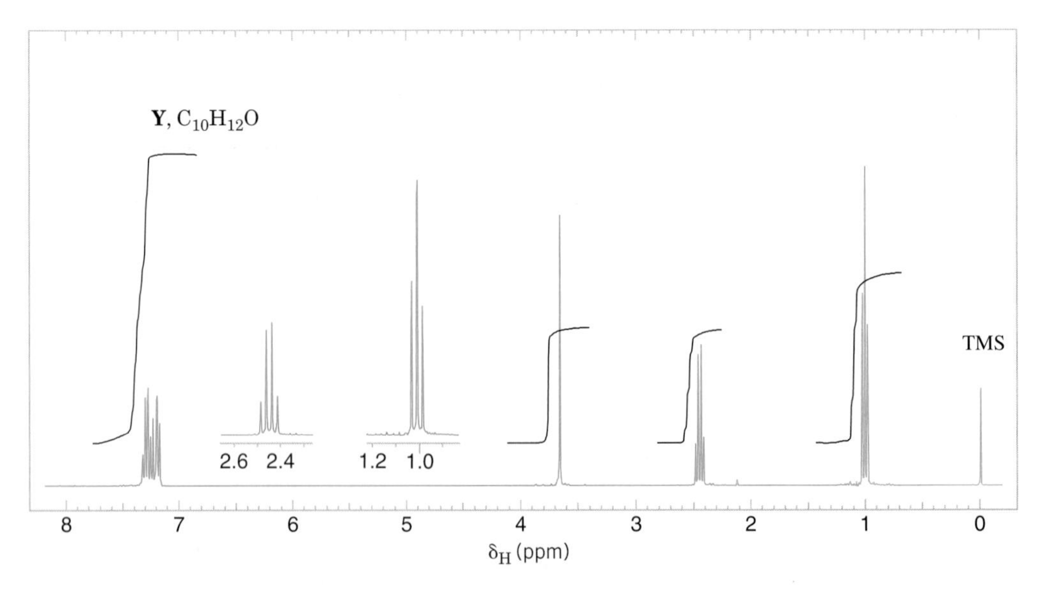

FIGURE 16.4 The 300-MHz ^1H NMR spectrum of compound Y, Problem 16.44. Expansions of the signals are shown in the offset plots.

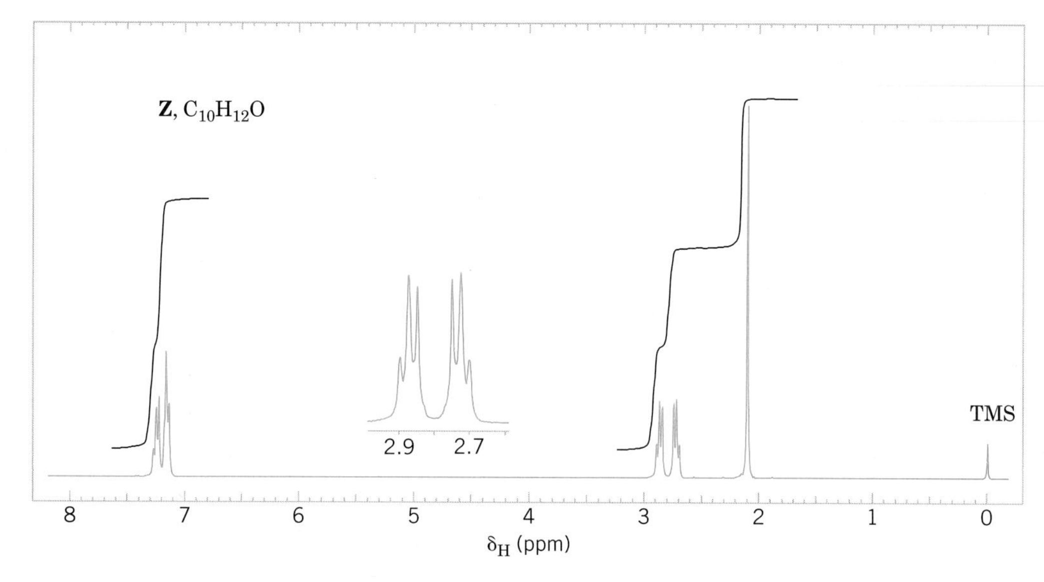

FIGURE 16.5 The 300-MHz ^{1}H NMR spectrum of compound Z, Problem 16.44. Expansions of the signals are shown in the offset plots.

16.47 When semicarbazide ($H_2NNHCONH_2$) reacts with a ketone (or an aldehyde) to form a semicarbazone (Section 16.8A), only one nitrogen atom of semicarbazide acts as a nucleophile and attacks the carbonyl carbon atom of the ketone. The product of the reaction, consequently, is $R_2C{=}NNHCONH_2$ rather than $R_2C{=}NCONHNH_2$. What factor accounts for the fact that two nitrogen atoms of semicarbazide are relatively nonnucleophilic?

16.48 Dutch elm disease is caused by a fungus transmitted to elm trees by the elm bark beetle. The female beetle, when she has located an attractive elm tree, releases several pheromones, including multistriatin, below. These pheromones attract male beetles, which bring with them the deadly fungus.

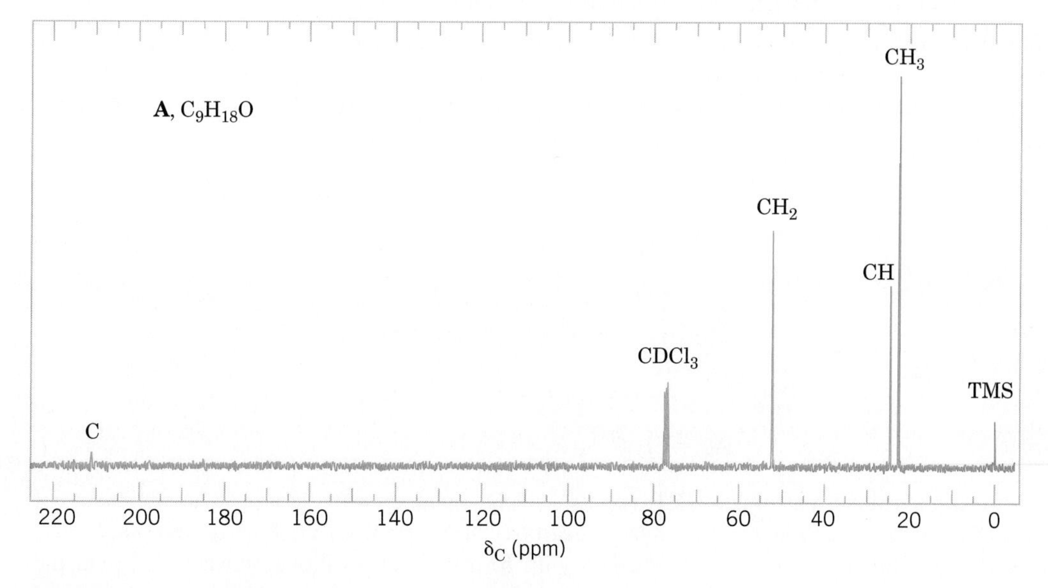

FIGURE 16.6 The broadband proton-decoupled ^{13}C NMR spectrum of compound A, Problem 16.45. Information from the DEPT ^{13}C NMR spectra is given above the peaks.

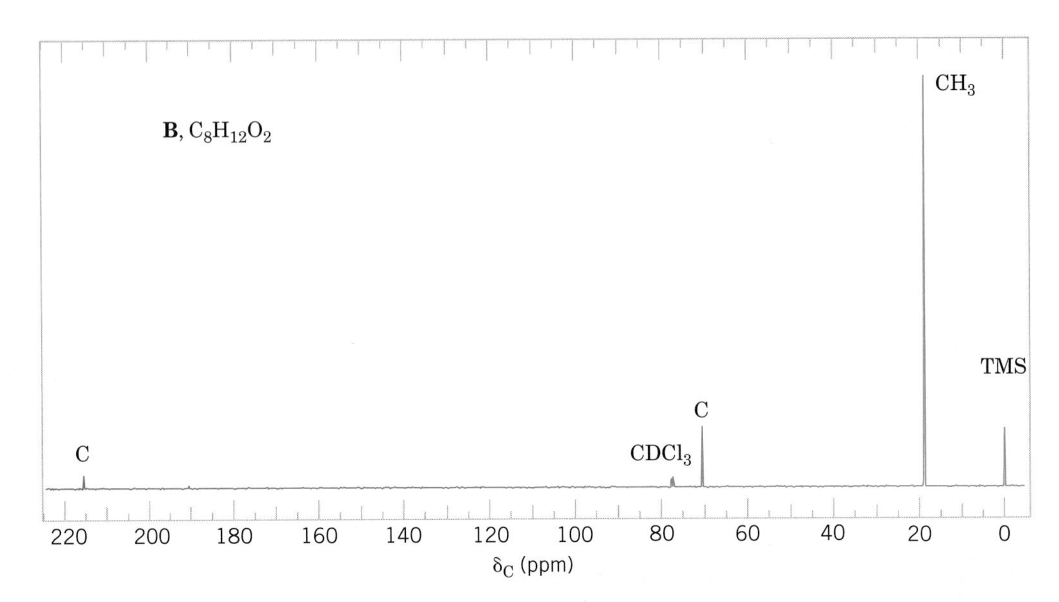

FIGURE 16.7 The broadband proton-decoupled ^{13}C NMR spectrum of compound B, Problem 16.46. Information from the DEPT ^{13}C NMR spectra is given above the peaks.

Multistriatin

Treating multistriatin with dilute aqueous acid at room temperature leads to the formation of a product, $C_{10}H_{20}O_3$, which shows a strong infrared peak near 1715 cm^{-1}. Propose a structure for this product.

16.49 An optically active compound, $C_6H_{12}O$, gives a positive test with 2,4-dinitrophenylhydrazine but a negative test with Tollens' reagent. What is the structure of the compound?

16.50 The following structure is an intermediate in a synthesis of prostaglandins $F_{2\alpha}$ and E_2 by E. J. Corey (Harvard University). A Horner–Wadsworth–Emmons reaction was used to form the (*E*)-alkene. Write structures for the phosphonate ester and carbonyl reactant that were used in this process. (*Note:* The carbonyl component of the reaction included the cyclopentyl group.)

AcO H

Ac = $CH_3\overset{O}{\overset{\|}{C}}$

***16.51** The coenzyme pyridoxal phosphate (PLP, see the chapter opening vignette and "The Chemistry of Pyridoxal Phosphate," Section 16.8) catalyzes decarboxylation reactions of some α-amino acids. As with all PLP reactions, the process commences with formation of an imine between PLP and the amino acid substrate. In the mechanism, an intermediate corresponding to an anion at the amino acid α carbon is produced, which is stablized by the PLP group as a temporary electron pair acceptor. Once the carboxylic acid group is lost, it is replaced by a proton from some site within the enzyme (in general, from a group "H–B–Enzyme"). Given this information, propose a detailed mechanism for decarboxylation of an α-amino acid catalyzed by PLP.

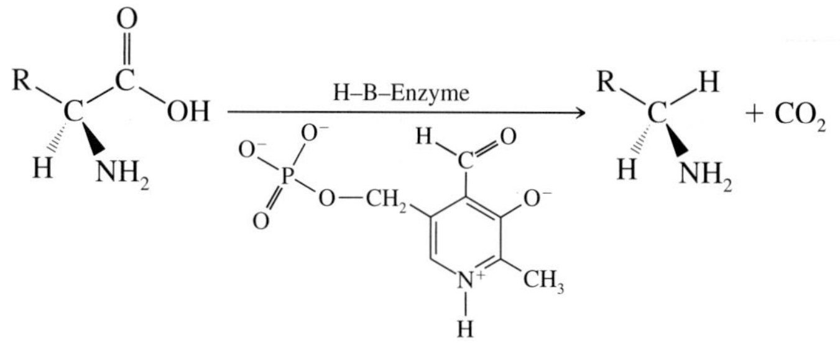

Pyridoxal phosphate

***16.52** **(a)** What would be the frequencies of the two absorption bands expected to be most prominent in the infrared spectrum of 4-hydroxycycloheptanone (**A**)?

(b) In reality, the lower frequency band of these two is very weak. Draw the structure of an isomer that would exist in equilibrium with **A** and that explains this observation.

***16.53** One of the important reactions of benzylic alcohols, ethers, and esters is the ease of cleavage of the benzyl–oxygen bond during hydrogenation. This is another example of "hydrogenolysis," the cleavage of a bond by hydrogen. It is facilitated by the presence of acid. Hydrogenolysis can also occur with strained-ring compounds.

On hydrogenation of compound **B** (see below) using Raney nickel catalyst in a dilute solution of hydrogen chloride in dioxane and water, most products have a 3,4-dimethoxyphenyl group attached to a side chain. Among these, an interesting product is **C**, whose formation illustrates not only hydrogenolysis but also the migratory aptitude of phenyl groups. For product **C,** these are key spectral data:

MS (*m/z*): 196.1084 ($M^{+\cdot}$, at high resolution), 178
IR (cm^{-1}): 3400 (broad), 3050, 2850 (CH_3–O stretch)
^{1}H NMR (δ, in $CDCl_3$): 1.21 (d, 3H, $J = 7$ Hz), 2.25 (s, 1H), 2.83 (m, 1H), 3.58 (d, 2H, $J = 7$ Hz), 3.82 (s, 6H), 6.70 (s, 3H).

What is the structure of compound **C**?

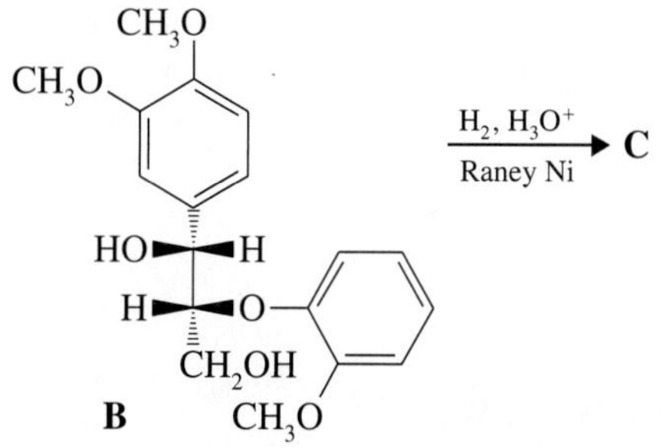

B

LEARNING GROUP PROBLEMS

A synthesis of ascorbic acid (vitamin C, **1**) starting from D-(+)-galactose (**2**) is shown below (Haworth, W. N., et al., *J. Chem. Soc.* **1933,** 1419–1423). Consider the following questions about the design and reactions used in this synthesis:

(a) Why did Haworth and co-workers introduce the acetal functional groups in **3**?

(b) Write a mechanism for the formation of one of the acetals.

(c) Write a mechanism for the hydrolysis of one of the acetals (**4** to **5**). Assume that water was present in the reaction mixture.

(d) In the reaction from **5** to **6** you can assume that there was acid (e.g., HCl) present with the sodium amalgam. What reaction occurred here and from what functional group did that reaction actually proceed?

(e) Write a mechanism for the formation of a phenylhydrazone from the aldehyde carbonyl of **7**. [Do not be concerned about the phenylhydrazone group at C2. We shall study the formation of bishydrazones of this type (called an osazone) in Chapter 22.]

(f) What reaction was used to add the carbon atom that ultimately became the lactone carbonyl carbon in ascorbic acid (**1**)?

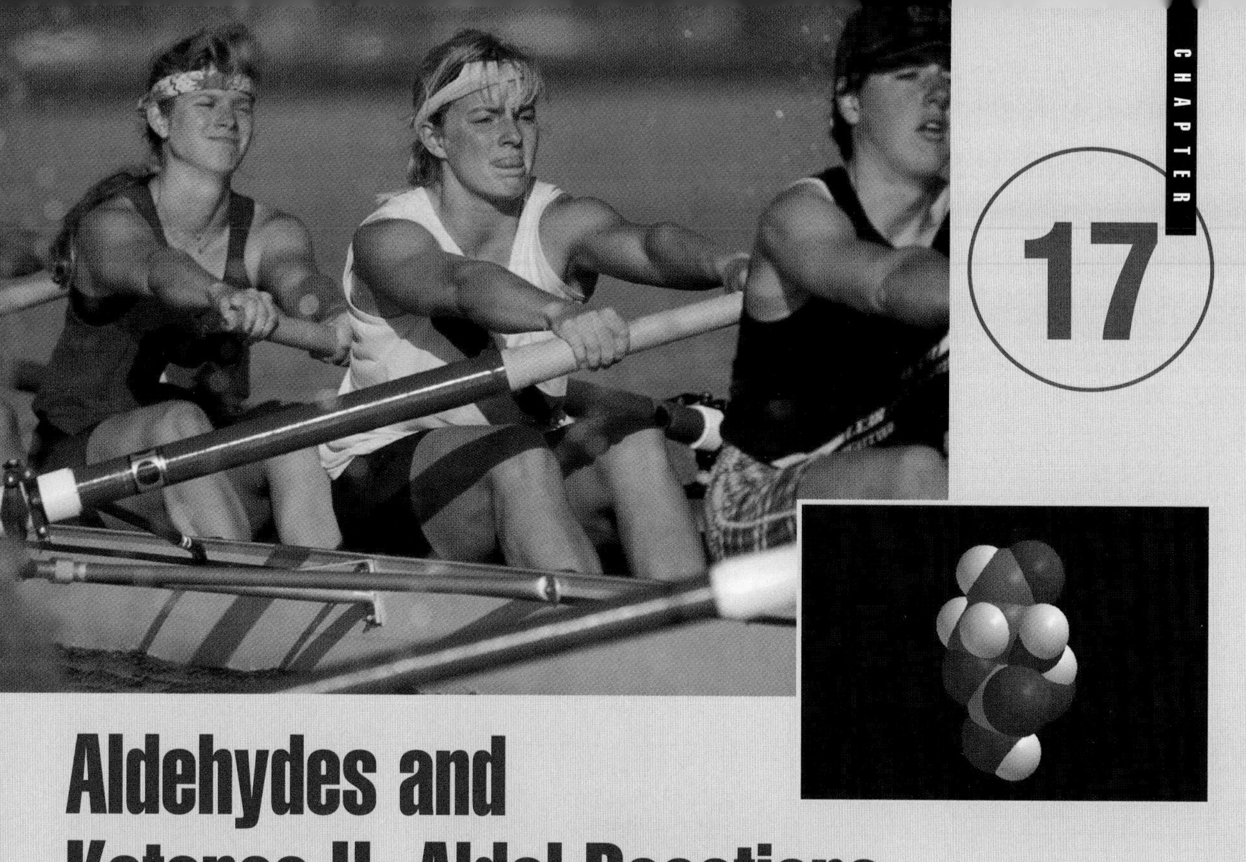

CHAPTER 17

Aldehydes and Ketones II. Aldol Reactions

TIM (Triose Phosphate Isomerase) Recycles Carbon via an Enol

An enol is a vinylic alcohol, or an alk**ene**–alcoh**ol**. Enols are central to life as well as to reactions we shall study in this chapter. For example, an enol intermediate plays a key role in glycolysis, a pathway used by all living things for production of energy through the breakdown of glucose. Were it not for the intermediacy of an enol, the net yield of ATP from glycolysis alone would be nil.

$$\mathrm{>C{=}C<^{OH}}$$

An enol

In the first stage of glycolysis, a C_6 molecule of glucose is divided into two different C_3 molecules [dihydroxyacetone phosphate (DHAP) and glyceraldehyde-3-phosphate (GAP), shown by the photo above]. This process *consumes* energy in the form of two ATP molecules. In the second stage of glycolysis, metabolism of one of the C_3 intermediates (GAP) causes the *formation* of two ATP molecules. Thus, to this point the energy yield of glycolysis is zero. However, an enzyme called triose phosphate isomerase (TIM, or TPI) recycles the unused C_3 intermediate (DHAP) formed from glucose so that a second passage through stage II of glycolysis is possible. Metabolism of the second C_3 unit produces two more ATP molecules, resulting in an overall yield by glycolysis of two ATP from one glucose molecule. An outline of this sequence is shown in the following scheme:

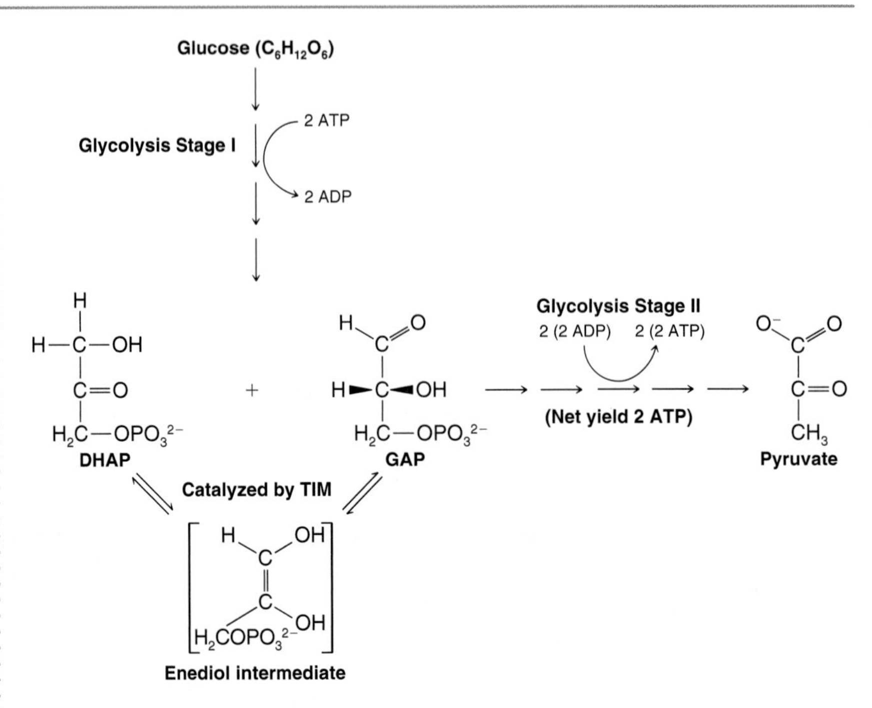

We shall use enols and enolates (the conjugate base of an enol) in reactions to make carbon–carbon bonds, one example of which is known as an aldol reaction. Interestingly, the direct precursor of DHAP and GAP in glycolysis is a type of molecule called an aldol (an aldehyde or ketone with a β-hydroxyl group). This precursor is cleaved to DHAP and GAP by an enzyme called aldolase. We shall consider the mechanism of this biological retro-aldol reaction later in this chapter.

17.1 The Acidity of the α Hydrogens of Carbonyl Compounds: Enolate Anions

In Chapter 16, we found that one important characteristic of aldehydes and ketones is their ability to undergo nucleophilic addition at their carbonyl groups:

$$\mathrm{>C{=}O + H{-}Nu \longrightarrow >C(OH)(Nu)}$$

Nucleophilic addition

A second important characteristic of carbonyl compounds is the unusual acidity of hydrogen atoms on carbon atoms adjacent to the carbonyl group. (These hydrogen atoms are usually called the ***α* hydrogens,** and the carbon to which they are attached is called the ***α* carbon.**)

$$\mathrm{R{-}\overset{\overset{\displaystyle :\ddot{O}:}{\|}}{C}{-}\underset{H}{\overset{|}{C^{\alpha}}}{-}\underset{H}{\overset{|}{C^{\beta}}}{-}}$$

α Hydrogens are unusually acidic (pK_a = 19–20).

***β* Hydrogens are not acidic (pK_a = 40–50).**

When we say that the α hydrogens are acidic, *we mean that they are unusually acidic for hydrogen atoms attached to carbon.* The pK_a values for the α hydrogens of most simple aldehydes or ketones are of the order of 19–20 ($K_a = 10^{-19}$–10^{-20}). This means that they are more acidic than hydrogen atoms of ethyne, $pK_a = 25$ ($K_a = 10^{-25}$), and are far more acidic than the hydrogens of ethene ($pK_a = 44$) or of ethane ($pK_a = 50$).

The reasons for the unusual acidity of the α hydrogens of carbonyl compounds are straightforward: The carbonyl group is strongly electron withdrawing (Section 3.10), and when a carbonyl compound loses an α proton, the anion that is produced is *stabilized by resonance. The negative charge of the anion is delocalized:*

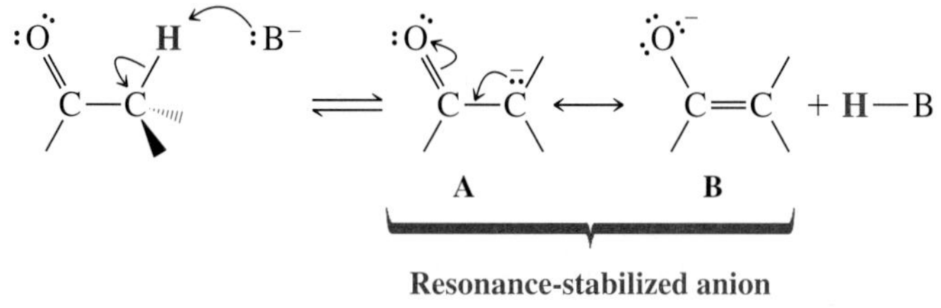

A

B

Resonance-stabilized anion

We see from this reaction that two resonance structures, **A** and **B**, can be written for the anion. In structure **A** the negative charge is on carbon, and in structure **B** the negative charge is on oxygen. Both structures contribute to the hybrid. Although structure **A** is favored by the strength of its carbon–oxygen π bond relative to the weaker carbon–carbon π bond of **B**, structure **B** makes a greater contribution to the hybrid because oxygen, being highly electronegative, is better able to accommodate the negative charge. We can depict the hybrid in the following way:

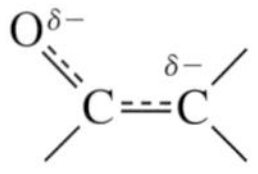

When this resonance-stabilized anion accepts a proton, it can do so in either of two ways: It can accept the proton at carbon to form the original carbonyl compound in what is called the **keto form** or it may accept the proton at oxygen to form an **enol** (alk**ene** alcoh**ol**):

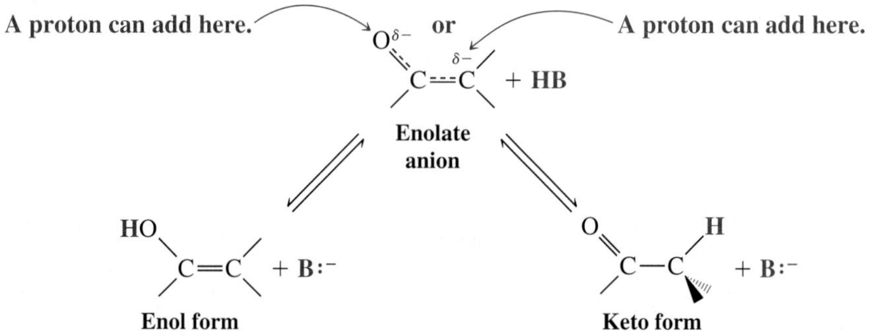

A proton can add here. or **A proton can add here.**

+ HB

Enolate anion

+ B:⁻

Enol form

+ B:⁻

Keto form

Both of these reactions are reversible. Because of its relation to the enol, the resonance-stabilized anion is called an **enolate anion.**

A calculated electrostatic potential map for the enolate anion of acetone is shown here. The map indicates approximately the outermost extent of electron density (the van der Waals surface) of the acetone enolate anion. Red color near the oxygen is consistent with oxygen being better able to stabilize the excess negative charge of the anion. Yellow at the carbon where the α hydrogen was removed indicates that some of the excess negative charge is localized there as well. These implications are parallel with the conclusions above about charge distribution in the hybrid based on resonance and electronegativity effects.

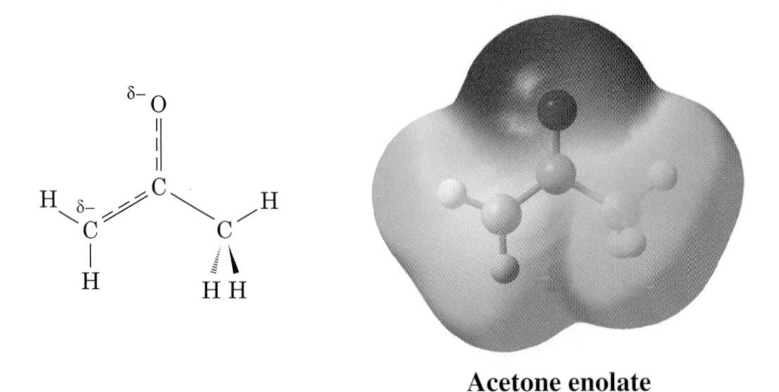

Acetone enolate

17.2 Keto and Enol Tautomers

The keto and enol forms of carbonyl compounds are constitutional isomers, but of a special type. Because they are easily interconverted in the presence of traces of acids and bases, chemists use a special term to describe this type of constitutional isomerism. Interconvertible keto and enol forms are said to be **tautomers,** and their interconversion is called **tautomerization.**

Under most circumstances, we encounter keto–enol tautomers in a state of equilibrium. (The surfaces of ordinary laboratory glassware are able to catalyze the interconversion and establish the equilibrium.) For simple monocarbonyl compounds such as acetone and acetaldehyde, the amount of the enol form present at equilibrium is *very small.* In acetone it is much less than 1%; in acetaldehyde the enol concentration is too small to be detected. The greater stability of the following keto forms of monocarbonyl compounds can be related to the greater strength of the carbon–oxygen π bond compared to the carbon–carbon π bond (~364 versus ~250 kJ mol^{-1}):

Note that keto–enol tautomers are not resonance structures. They are constitutional isomers in equilibrium (generally favoring the keto form).

	Keto Form		*Enol Form*
Acetaldehyde	$CH_3C(=O)H$	⇌	$CH_2{=}C(OH)H$
	(~100%)		**(extremely small)**
Acetone	$CH_3C(=O)CH_3$	⇌	$CH_2{=}C(OH)CH_3$
	(>99%)		**(1.5 × 10^{-4}%)**
Cyclohexanone	cyclohexanone (C=O)	⇌	cyclohex-1-en-1-ol (OH)
	(98.8%)		**(1.2%)**

In compounds whose molecules have two carbonyl groups separated by one —CH_2— group (called β-dicarbonyl compounds), the amount of enol present at equilibrium is far higher. For example, pentane-2,4-dione exists in the enol form to an extent of 76%:

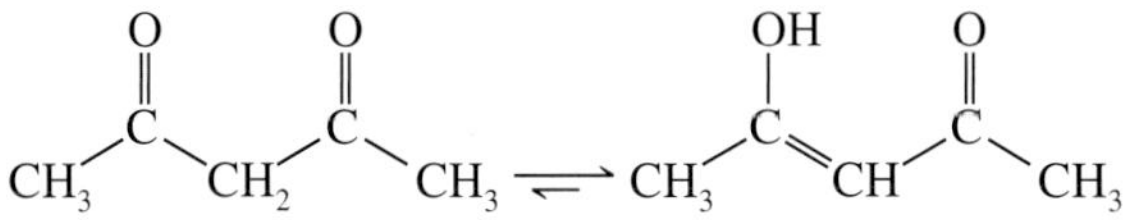

$$CH_3COCH_2COCH_3 \rightleftharpoons CH_3C(OH){=}CHCOCH_3$$

Pentane-2,4-dione (24%) **Enol form** (76%)

The greater stability of the enol form of β-dicarbonyl compounds can be attributed to stability gained through resonance stabilization of the conjugated double bonds and (in a cyclic form) through hydrogen bonding:

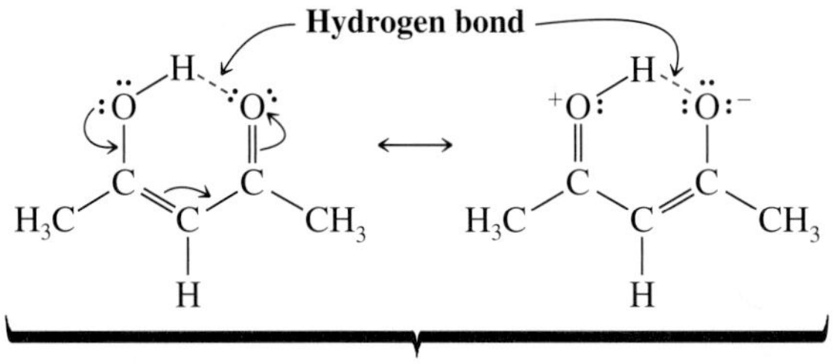

Hydrogen bond

Resonance stabilization of the enol form

PROBLEM 17.1

For all practical purposes, the compound cyclohexa-2,4-dien-1-one exists totally in its enol form. Write the structure of cyclohexa-2,4-dien-1-one and of its enol form. What special factor accounts for the stability of the enol form?

17.3 Reactions via Enols and Enolate Anions

17.3A Racemization

When a solution of (+)-2-methyl-1-phenylbutan-1-one (see the following reaction) in aqueous ethanol is treated with either acids or bases, the solution gradually loses its optical activity. After a time, isolation of the ketone shows that it has been racemized:

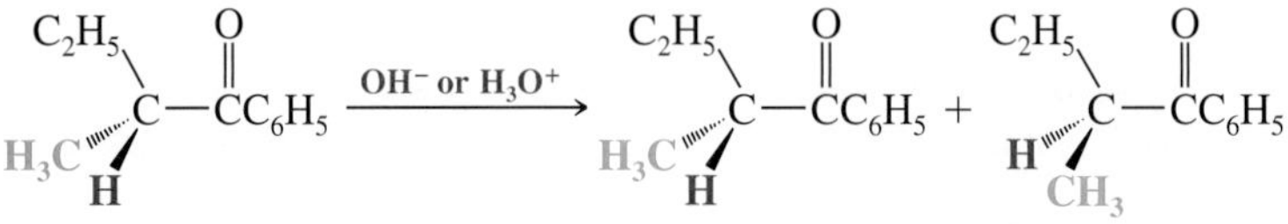

$$\xrightarrow{OH^- \text{ or } H_3O^+}$$

(R)-(+)-2-Methyl-1-phenylbutan-1-one → **(±)-2-Methyl-1-phenylbutan-1-one (racemic form)**

Racemization takes place in the presence of acids or bases because the ketone slowly but reversibly changes to its enol *and the enol is achiral.* When the enol reverts to the keto form, it produces equal amounts of the two enantiomers:

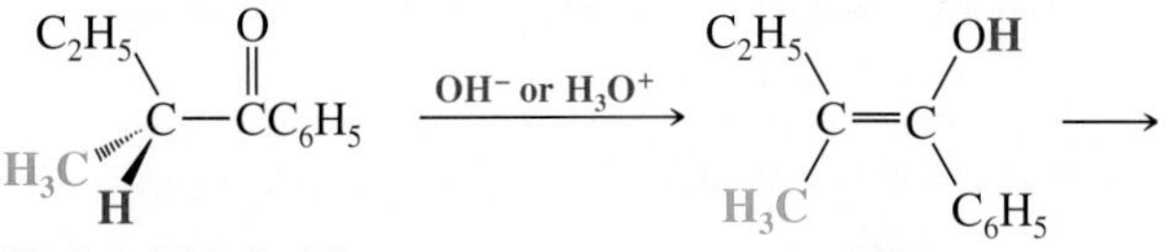

$$\xrightarrow{OH^- \text{ or } H_3O^+}$$

(R)-(+)-2-Methyl-1-phenylbutan-1-one (chiral) → **Enol (achiral)** → (+)- and (−)-2-methyl-1-phenylbutan-1-one (in the racemic form shown above)

A base catalyzes the formation of an enol through the intermediate formation of an enolate anion.

A Mechanism for the Reaction

Base-Catalyzed Enolization

H—OH

:Ö: :Ö:⁻   :Ö—H

C—C ⇌ C=C ⇌ C=C + :ÖH⁻

HÖ:⁻   H

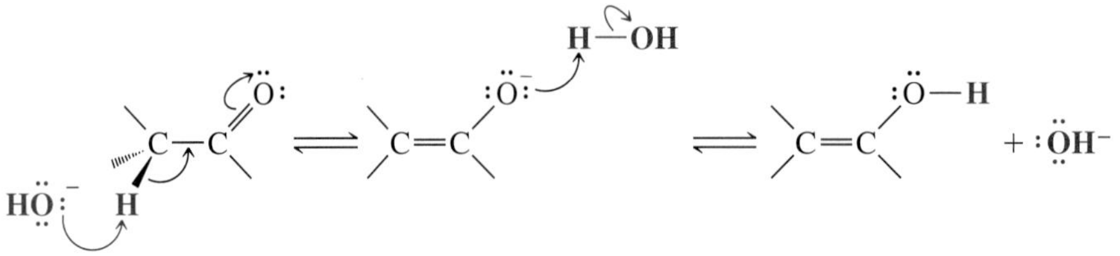

Enolate anion (achiral)

Enol (achiral)

An acid can catalyze enolization in the following way.

A Mechanism for the Reaction

Acid-Catalyzed Enolization

Ö: H—Ö⁺—H   Ö⁺—H   :Ö—H

C—C + H ⇌ C—C ⇌ C=C + H—Ö⁺—H

H   H + :Ö—H   H

H

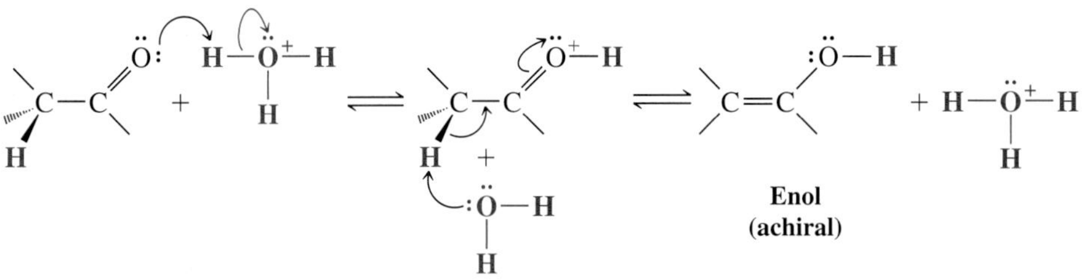

Enol (achiral)

In acyclic ketones, the enol or enolate anion formed can be (*E*) or (*Z*). Protonation on one face of the (*E*) isomer and protonation on the same face of the (*Z*) isomer produces enantiomers.

PROBLEM 17.2 Would optically active ketones such as the following undergo acid- or base-catalyzed racemization? Explain your answer.

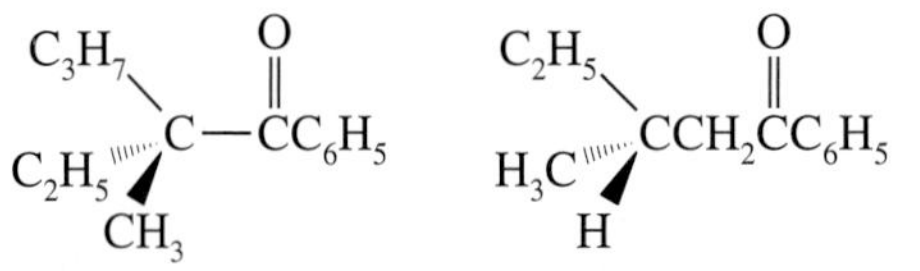

C_3H_7, C_2H_5, CH_3 on C—$\overset{O}{\overset{\|}{C}}C_6H_5$ C_2H_5, H_3C, H on C—$CH_2\overset{O}{\overset{\|}{C}}C_6H_5$

PROBLEM 17.3 When 2-methyl-1-phenylbutan-1-one is treated with either OD^- or D_3O^+ in the presence of D_2O, the ketone undergoes hydrogen–deuterium exchange and produces this compound:

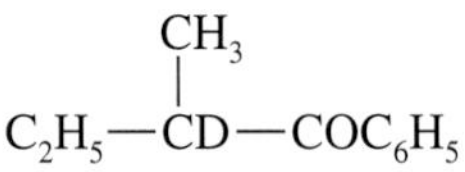

$$C_2H_5{-}\overset{CH_3}{\overset{|}{C}}D{-}COC_6H_5$$

Write mechanisms that account for this behavior.

Diastereomers that differ in configuration at only one stereogenic center are sometimes called **epimers.** Keto–enol tautomerization can sometimes be used to convert a less stable epimer to a more stable one. This equilibration process is an example of **epimerization.** An example is the epimerization of *cis*-decalone to *trans*-decalone:

cis-Decalone ⇌ *trans*-Decalone

PROBLEM 17.4

Write a mechanism using sodium ethoxide in ethanol for the epimerization of *cis*-decalone to *trans*-decalone. Draw chair conformational structures that show why *trans*-decalone is more stable than *cis*-decalone. You may find it helpful to also examine hand-held molecular models of *cis*- and *trans*-decalone.

17.3B Halogenation of Ketones

Ketones that have an α hydrogen react readily with halogens by substitution. The rates of these halogenation reactions *increase when acids or bases are added, and substitution takes place almost exclusively at the α carbon:*

$$\text{—}\overset{\text{H}}{\underset{|}{\overset{|}{\text{C}}}}\text{—}\overset{\text{O}}{\overset{\|}{\text{C}}}\text{—} + X_2 \xrightarrow[\text{or base}]{\text{acid}} \text{—}\overset{\text{X}}{\underset{|}{\overset{|}{\text{C}}}}\text{—}\overset{\text{O}}{\overset{\|}{\text{C}}}\text{—} + HX$$

This behavior of ketones can be accounted for in terms of two related properties that we have already encountered: the acidity of the α hydrogens of ketones and the tendency of ketones to form enols.

Base-Promoted Halogenation In the presence of bases, halogenation takes place through the slow formation of an enolate anion or an enol followed by a rapid reaction of the enolate anion or enol with halogen.

A Mechanism for the Reaction

Base-Promoted Halogenation of Aldehydes and Ketones

Step 1 $\text{B:}^- + \text{—}\overset{\text{H}}{\overset{|}{\text{C}}}\text{—C(=O)—} \xrightleftharpoons{\text{slow}} \text{B:H} + \text{C}{\overset{\delta-}{\cdots}}\text{C}{\cdots}\text{O}^{\delta-} \xrightleftharpoons{\text{fast}} \text{C=C—OH} + \text{B:}^-$

Enolate anion — Enol

Step 2 $\text{C=C—O}^- \longleftrightarrow \text{—}\overset{..}{\text{C}}^-\text{—C(=O)—}$

$\text{X—X} + \text{—}\overset{..}{\text{C}}^-\text{—C(=O)—} \xrightarrow{\text{fast}} \text{—}\overset{\text{X}}{\overset{|}{\text{C}}}\text{—C(=O)—} + X^-$

Enolate anion

As we shall see in Section 17.3C, multiple halogenations can occur.

Acid-Catalyzed Halogenation In the presence of acids, halogenation takes place through the slow formation of an enol followed by rapid reaction of the enol with the halogen.

A Mechanism for the Reaction

Acid-Catalyzed Halogenation of Aldehydes and Ketones

Step 1 $\text{H}-\overset{|}{\underset{|}{\text{C}}}-\text{C}(=\ddot{\text{O}}\text{:}) + \text{H:B} \underset{}{\overset{\text{fast}}{\rightleftharpoons}} \text{H}-\overset{|}{\underset{|}{\text{C}}}-\text{C}(=\overset{+}{\text{O}}\text{H}) \xrightleftharpoons{\text{slow}} \text{C}=\text{C}-\ddot{\text{O}}\text{H} + \text{H:B}$ (:B⁻ removes the α hydrogen)

Enol

Step 2 $\text{X}-\text{X} + \text{C}=\text{C}-\ddot{\text{O}}-\text{H} \overset{\text{fast}}{\rightleftharpoons} \text{X}-\overset{|}{\underset{|}{\text{C}}}-\text{C}(=\overset{+}{\text{O}}-\text{H}) + \text{X}^-$

Step 3 $\text{X}-\overset{|}{\underset{|}{\text{C}}}-\text{C}(=\overset{+}{\text{O}}-\text{H}) + \text{X}^- \overset{\text{fast}}{\rightleftharpoons} \text{X}-\overset{|}{\underset{|}{\text{C}}}-\text{C}(=\ddot{\text{O}}\text{:}) + \text{HX}$

Part of the evidence that supports these mechanisms comes from studies of reaction kinetics. Both base-promoted and acid-catalyzed halogenations of ketones *show initial rates that are independent of the halogen concentration.* The mechanisms that we have written are in accord with this observation: In both instances the slow step of the mechanism occurs before the intervention of the halogen. (The initial rates are also independent of the nature of the halogen; see Problem 17.6.)

PROBLEM 17.5 Why do we say that the halogenation of ketones in a base is "base promoted" rather than "base catalyzed"?

PROBLEM 17.6 Additional evidence for the halogenation mechanisms that we just presented comes from the following facts: **(a)** Optically active 2-methyl-l-phenylbutan-1-one undergoes acid-catalyzed racemization at a rate exactly equivalent to the rate at which it undergoes acid-catalyzed halogenation. **(b)** 2-Methyl-1-phenylbutan-1-one undergoes acid-catalyzed iodination at the same rate that it undergoes acid-catalyzed bromination. **(c)** 2-Methyl-1-phenylbutan-1-one undergoes base-catalyzed hydrogen–deuterium exchange at the same rate that it undergoes base-promoted halogenation. Explain how each of these observations supports the mechanisms that we have presented.

17.3C The Haloform Reaction

When methyl ketones react with halogens in the presence of base, multiple halogenations always occur at the carbon of the methyl group. Multiple halogenations occur because introduction of the first halogen (owing to its electronegativity) makes the remaining α hydrogens on the methyl carbon more acidic:

$$C_6H_5-\overset{O}{\overset{\|}{C}}-\underset{H}{\overset{H}{C}}-H + 3\,X_2 + 3\,OH^- \xrightarrow[\text{base}]{} C_6H_5-\overset{O}{\overset{\|}{C}}-\underset{X}{\overset{X}{C}}-X + 3\,X^- + 3\,H_2O$$

A Mechanism for the Reaction

Halogenation Step of the Haloform Reaction

$$C_6H_5-\overset{O}{\overset{\|}{C}}-\underset{H}{\overset{H}{C}}-H + :\bar{B} \rightleftharpoons C_6H_5-\overset{O}{\overset{\|}{C}}-\underset{H}{\overset{..}{C}}-H \xrightarrow{X-X} C_6H_5-\overset{O}{\overset{\|}{C}}-\underset{H}{\overset{X}{C}}-H + X^-$$

$$\updownarrow$$

$$^-:O\text{(}C_6H_5\text{)}C=CH_2$$

Enolate anion

$$C_6H_5-\overset{O}{\overset{\|}{C}}-\underset{H}{\overset{X}{C}}-H + :\bar{B} \rightleftharpoons C_6H_5-\overset{O}{\overset{\|}{C}}-\underset{H}{\overset{X}{C}}:^- \xrightarrow{X-X} C_6H_5-\overset{O}{\overset{\|}{C}}-\underset{H}{\overset{X}{C}}-X + X^-$$

Acidity is increased by the electron-withdrawing halogen atom.

$$\updownarrow$$

$$^-:O\text{(}C_6H_5\text{)}C=C\text{(}H\text{)}X$$

Enolate anion

$$\xrightarrow{:\bar{B}\text{ then }X_2} C_6H_5-\overset{O}{\overset{\|}{C}}-\underset{X}{\overset{X}{C}}-X$$

When methyl ketones react with halogens in aqueous sodium hydroxide (i.e., in *hypohalite solutions**), an additional reaction takes place. Hydroxide ion attacks the carbonyl carbon atom of the trihalo ketone and causes a cleavage at the carbon–carbon bond between the carbonyl group and the trihalomethyl group, a moderately good leaving group. This cleavage ultimately produces a carboxylate anion and a *haloform* (i.e., $CHCl_3$, $CHBr_3$, or CHI_3). The initial step is a nucleophilic attack by hydroxide ion on the carbonyl carbon atom. In the next step, carbon–carbon bond cleavage occurs and the trihalomethyl anion ($:CX_3^-$) departs. This is one of the rare instances in which a carbanion acts as a leaving group. This step can occur because the trihalomethyl anion is unusually stable; its negative charge is dispersed by the three electronegative halogen atoms (when X = Cl, the conjugate acid, $CHCl_3$, has $pK_a = 13.6$). In the last step, a proton transfer takes place between the carboxylic acid and the trihalomethyl anion.

*Dissolving a halogen in aqueous sodium hydroxide produces a solution containing sodium hypohalite (NaOX) because of the following equilibrium:

$$X_2 + 2\,NaOH \rightleftharpoons NaOX + NaX + H_2O$$

You may be familiar with bleach, which is aqueous NaOCl.

A Mechanism for the Reaction

Cleavage Step of the Haloform Reaction

$$C_6H_5-C(=O)-CX_3 + {}^-OH \rightleftharpoons C_6H_5-C(O^-)(OH)-CX_3$$

$$C_6H_5-C(O^-)(OH)-CX_3 \rightleftharpoons C_6H_5-C(=O)-O-H + {}^-:CX_3$$

$$\updownarrow$$

$$C_6H_5-C(=O)-O^- + H-CX_3$$

Carboxylate anion **Haloform**

The **haloform reaction** is of synthetic utility as a means of converting methyl ketones to carboxylic acids. When the haloform reaction is used in synthesis, chlorine and bromine are most commonly used as the halogen component. Chloroform ($CHCl_3$) and bromoform ($CHBr_3$) are both liquids which are immiscible with water and are easily separated from the aqueous solution containing the carboxylate anion. When iodine is the halogen component, the bright yellow solid iodoform (CHI_3) results. This version is the basis of a laboratory classification test for methyl ketones and methyl secondary alcohols (which are oxidized to methyl ketones first under the reaction conditions):

$$-C(=O)-CH_3 + 3\,I_2 + 3\,OH^- \longrightarrow -C(=O)-CI_3 + 3\,I^- + 3\,H_2O$$

$$-C(=O)-CI_3 + OH^- \longrightarrow -C(=O)-O^- + CHI_3\downarrow$$

Yellow precipitate

When water is chlorinated to purify it for public consumption, chloroform is produced from organic impurities in the water via the haloform reaction. (Many of these organic impurities are naturally occurring, such as humic substances.) The presence of chloroform in public water is of concern for water treatment plants and environmental officers, because chloroform is carcinogenic. Thus, the technology that solves one problem creates another. It is worth recalling, however, that before chlorination of water was introduced, thousands of people died in epidemics of diseases such as cholera and dysentery.

17.4 The Aldol Reaction: The Addition of Enolate Anions to Aldehydes and Ketones

When acetaldehyde reacts with dilute sodium hydroxide at room temperature (or below), a dimerization takes place producing 3-hydroxybutanal. Since 3-hydroxybutanal is both an **ald**ehyde and an alcoh**ol**, it has been given the common name **"aldol,"** and reactions of this general type have come to be known as **aldol additions** (or **aldol reactions**):

$$2\ CH_3\overset{O}{\overset{\|}{C}}H \xrightarrow[5°C]{10\%\ NaOH,\ H_2O} CH_3\overset{OH}{\overset{|}{C}}HCH_2\overset{O}{\overset{\|}{C}}H$$

3-Hydroxybutanal
("aldol")
(50%)

STUDY TIP!

Note the two key aspects of carbonyl chemistry: α-hydrogen acidity and susceptibility to nucleophilic attack.

The mechanism for the aldol addition illustrates two important characteristics of carbonyl compounds: the acidity of their α hydrogens and the tendency of their carbonyl groups to undergo nucleophilic addition.

A Mechanism for the Reaction

The Aldol Addition

Step 1

$$H\!-\!\ddot{O}:^- + H\!-\!CH_2\!-\!\overset{:O:}{\overset{\|}{C}}\!-\!H \rightleftharpoons \left[^-:CH_2\!-\!\overset{:O:}{\overset{\|}{C}}\!-\!H \longleftrightarrow CH_2\!=\!\overset{:\ddot{O}:^-}{\overset{|}{C}}\!-\!H \right] + H\!-\!\ddot{O}\!-\!H$$

Enolate anion

In this step the base (a hydroxide ion) removes a proton from the α carbon of one molecule of acetaldehyde to give a resonance-stabilized enolate anion.

Step 2

$$CH_3\!-\!\overset{:O:}{\overset{\|}{C}}\!-\!H + ^-:CH_2\!-\!\overset{:O:}{\overset{\|}{C}}\!-\!H \rightleftharpoons CH_3\!-\!\overset{:\ddot{O}:^-}{\overset{|}{C}}H\!-\!CH_2\!-\!\overset{:O:}{\overset{\|}{C}}\!-\!H$$

An alkoxide anion

$$\updownarrow$$

$$CH_2\!=\!\overset{:\ddot{O}:^-}{\overset{|}{C}}\!-\!H$$

The enolate anion then acts as a nucleophile and attacks the carbonyl carbon of a second molecule of acetaldehyde, producing an alkoxide anion.

Step 3

$$CH_3\!-\!\overset{:\ddot{O}:^-}{\overset{|}{C}}H\!-\!CH_2\!-\!\overset{:O:}{\overset{\|}{C}}\!-\!H + H\!-\!\ddot{O}\!-\!H \longrightarrow CH_3\!-\!\overset{:\ddot{O}-H}{\overset{|}{C}}H\!-\!CH_2\!-\!\overset{:O:}{\overset{\|}{C}}\!-\!H + ^-:\ddot{O}\!-\!H$$

Stronger base **Aldol** **Weaker base**

The alkoxide anion now removes a proton from a molecule of water to form the aldol.

17.4A Dehydration of the Aldol Addition Product

If the basic mixture containing the aldol (in the previous example) is heated, dehydration takes place and 2-butenal (crotonaldehyde) is formed. Dehydration occurs readily because of the acidity of the remaining α hydrogens (even though the leaving group is a hydroxide ion) *and because the product is stabilized by having conjugated double bonds.*

A Mechanism for the Reaction

Dehydration of the Aldol Addition Product

The α hydrogens are acidic.

The double bonds of the alkene and carbonyl groups are conjugated.

$$CH_3{-}CH(OH){-}CH(H){-}C(=O){-}H \xrightarrow{H{-}\ddot{O}:^-} CH_3{-}CH{=}CH{-}C(=O){-}H + H{-}\ddot{O}{-}H + H{-}\ddot{O}:^-$$

2-Butenal (crotonaldehyde)

In some aldol reactions, dehydration occurs so readily that we cannot isolate the product in the aldol form; we obtain the derived *enal* (alk*ene* *al*dehyde) instead. An **aldol condensation** occurs instead of an aldol *addition.* A condensation reaction is one in which molecules are joined through the intermolecular elimination of a small molecule such as water or an alcohol:

$$2\ RCH_2CH(=O) \xrightarrow{\text{base}} \left[RCH_2CH(OH)CH(R)CH(=O)\right] \xrightarrow{-H_2O} RCH_2CH{=}C(R){-}CH(=O)$$

Addition product — **Not isolated**

Condensation product — **An enal (an α,β-unsaturated aldehyde)**

17.4B Synthetic Applications

The aldol reaction is a general reaction of aldehydes that possess an α hydrogen. Propanal, for example, reacts with aqueous sodium hydroxide to give 3-hydroxy-2-methylpentanal:

$$2\ CH_3CH_2CH(=O) \xrightarrow[0-10°C]{OH^-} CH_3CH_2CH(OH)CH(CH_3)CH(=O)$$

Propanal → **3-Hydroxy-2-methylpentanal (55–60%)**

PROBLEM 17.7 **(a)** Show all steps in the aldol addition that occur when propanal is treated with base. **(b)** How can you account for the fact that the product of the aldol addition is 3-hydroxy-2-methylpentanal and not 4-hydroxyhexanal? **(c)** What products would be formed if the reaction mixture were heated?

The aldol reaction is important in organic synthesis because it gives us a method for linking two smaller molecules by introducing a carbon–carbon bond between them. Because aldol products contain two functional groups, —OH and —CHO, we can use them to carry out a number of subsequent reactions. Examples are the following:

$$2\ RCH_2\overset{O}{\overset{\|}{C}}H \xrightarrow[H_2O]{OH^-} RCH_2\overset{OH}{\overset{|}{C}}H\underset{R}{\underset{|}{C}}H\overset{O}{\overset{\|}{C}}H \xrightarrow{NaBH_4} RCH_2\overset{OH}{\overset{|}{C}}H\underset{R}{\underset{|}{C}}HCH_2OH$$

Aldehyde — **An aldol** — **A 1,3-diol**

The aldol reaction: a tool for synthesis. See also the Synthetic Connections review at the end of the chapter.

An aldol $\xrightarrow{HA,\ -H_2O}$

$$RCH_2CH_2\underset{R}{\underset{|}{C}}HCH_2OH \xleftarrow[\text{high pressure}]{H_2/Ni} RCH_2CH{=}\underset{R}{\underset{|}{C}}\overset{O}{\overset{\|}{C}}H \xrightarrow{LiAlH_4^*} RCH_2CH{=}\underset{R}{\underset{|}{C}}CH_2OH$$

A saturated alcohol — **An α,β-unsaturated aldehyde** — **An allylic alcohol**

$\downarrow H_2, Pd\text{–}C$

$$RCH_2CH_2\underset{R}{\underset{|}{C}}H\overset{O}{\overset{\|}{C}}H$$

An aldehyde

PROBLEM 17.8

One industrial process for the synthesis of 1-butanol begins with acetaldehyde. Show how this synthesis might be carried out.

PROBLEM 17.9

Show how each of the following products could be synthesized from butanal:
(a) 2-Ethyl-3-hydroxyhexanal **(c)** 2-Ethylhexan-1-ol
(b) 2-Ethylhex-2-en-1-ol **(d)** 2-Ethylhexane-1,3-diol (the insect repellent "6–12")

Ketones also undergo base-catalyzed aldol additions, but for them the equilibrium is unfavorable. This complication can be overcome, however, by carrying out the reaction in a special apparatus that allows the product to be removed from contact with the base as it is formed. This removal of product displaces the equilibrium to the right and permits successful aldol additions with many ketones. Acetone, for example, reacts as follows:

$$2\ CH_3\overset{O}{\overset{\|}{C}}CH_3 \overset{OH^-}{\rightleftharpoons} CH_3\underset{CH_3}{\underset{|}{\overset{OH}{\overset{|}{C}}}}CH_2\overset{O}{\overset{\|}{C}}CH_3$$

(80%)

17.4C The Reversibility of Aldol Additions

The aldol addition is reversible. If, for example, the aldol addition product obtained from acetone (see above) is heated with a strong base, it reverts to an equilibrium mixture that consists largely (~95%) of acetone. This type of reaction is called a *retro-aldol* reaction:

$$CH_3\underset{CH_3}{\underset{|}{\overset{OH}{\overset{|}{C}}}}{-}CH_2\overset{O}{\overset{\|}{C}}CH_3 \underset{H_2O}{\overset{OH^-}{\rightleftharpoons}} CH_3\underset{CH_3}{\underset{|}{\overset{O^-}{\overset{|}{C}}}}{-}CH_2\overset{O}{\overset{\|}{C}}CH_3 \rightleftharpoons CH_3\underset{CH_3}{\underset{|}{\overset{O}{\overset{\|}{C}}}} + {}^-{:}CH_2\overset{O}{\overset{\|}{C}}CH_3 \underset{OH^-}{\overset{H_2O}{\rightleftharpoons}} 2\ CH_3\overset{O}{\overset{\|}{C}}CH_3$$

(5%) **(95%)**

*$LiAlH_4$ reduces the carbonyl group of α,β-unsaturated aldehydes and ketones cleanly. $NaBH_4$ often reduces the carbon–carbon double bond as well.

The Chemistry of...

A Retro-Aldol Reaction in Glycolysis—Dividing Assets to Double the ATP Yield

Glycolysis is a fundamental pathway for production of ATP in living systems. The pathway begins with glucose and ends with two molecules of pyruvate and a net yield of two ATP molecules. Aldolase, an enzyme in glycolysis, plays a key role by dividing the six-carbon compound fructose-1,6-diphosphate (derived from glucose) into two compounds that each have three carbons, glyceraldehyde-3-phosphate (GAP) and 1,3-dihydroxyacetone phosphate (DHAP). This process is essential because it provides two three-carbon units for the final stage of glycolysis, wherein the net yield of two ATP molecules per glucose is realized. (See the chapter opening vignette. Two ATP molecules are consumed to form fructose-1,6-diphosphate, and only two are generated per pyruvate. Thus, two passages through the second stage of glycolysis are necessary to obtain a net yield of two ATP molecules per glucose.)

The cleavage reaction catalyzed by aldolase is a net retro-aldol reaction. Details of the mechanism are shown here, beginning at the left with fructose-1,6-diphosphate.

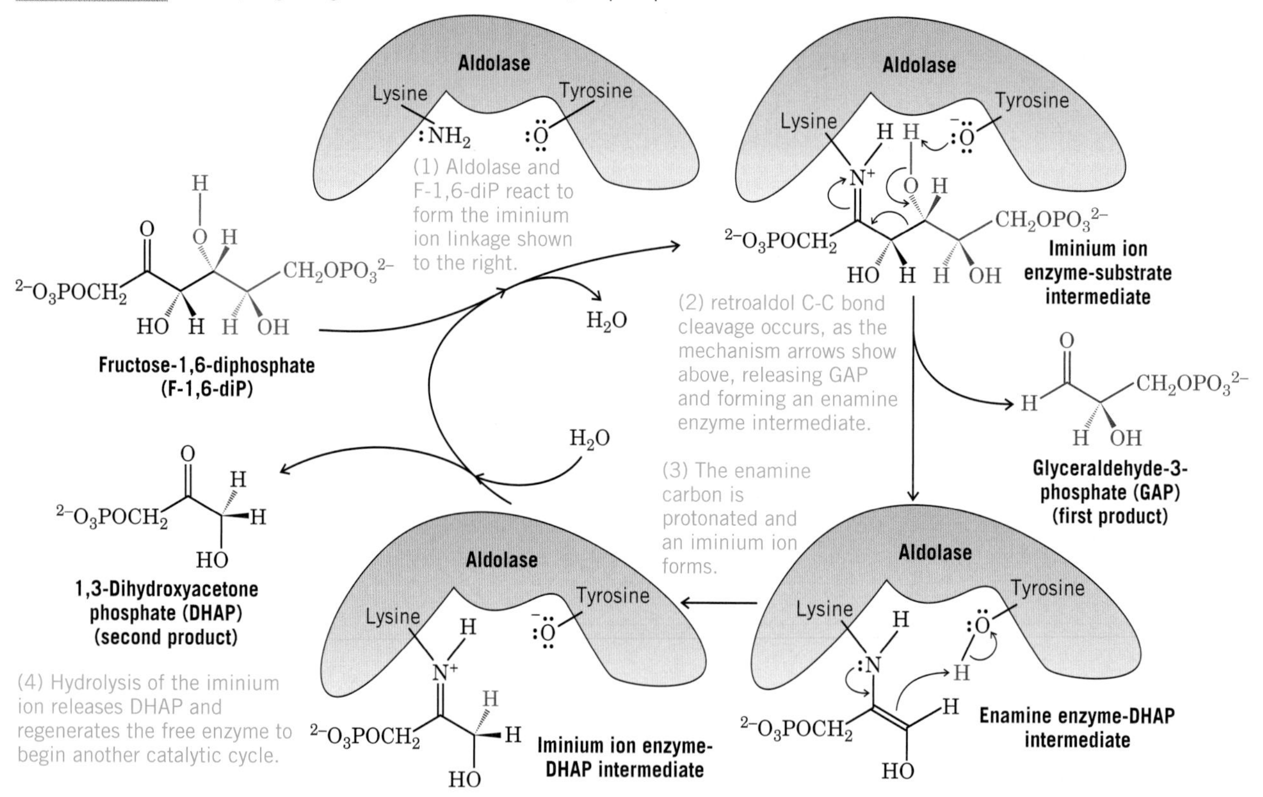

Aldolase
Lysine
Tyrosine
(1) Aldolase and F-1,6-diP react to form the iminium ion linkage shown to the right.
Fructose-1,6-diphosphate (F-1,6-diP)
H_2O
Iminium ion enzyme-substrate intermediate
(2) retroaldol C-C bond cleavage occurs, as the mechanism arrows show above, releasing GAP and forming an enamine enzyme intermediate.
Glyceraldehyde-3-phosphate (GAP) (first product)
(3) The enamine carbon is protonated and an iminium ion forms.
Enamine enzyme-DHAP intermediate
Iminium ion enzyme-DHAP intermediate
1,3-Dihydroxyacetone phosphate (DHAP) (second product)
(4) Hydrolysis of the iminium ion releases DHAP and regenerates the free enzyme to begin another catalytic cycle.

Two key intermediates in the aldolase mechanism involve functional groups that we have studied recently (Chapter 16)—an imine (protonated in the form of an iminium cation) and an enamine. In the mechanism of aldolase, an iminium cation acts as a sink for electron density during C—C bond cleavage (step 2), much like a carbonyl group does in a typical retro-aldol reaction. In this step the iminium cation is converted to an enamine, corresponding to the enolate or enol that is formed when a carbonyl group accepts electron density during C—C bond cleavage in an ordinary retro-aldol reaction. The enamine intermediate is then a source of an electron pair used to bond with a proton taken from the tyrosine hydroxyl at the aldolase active site (step 3). Lastly, the resulting iminium group undergoes hydrolysis (step 4), freeing aldolase for another catalytic cycle and releasing DHAP, the second product of the retro-aldol reaction. Then, by the process described in the chapter opening vignette, DHAP undergoes isomerization to GAP for processing to pyruvate and synthesis of two more ATP molecules.

As we have seen with aldolase and from our earlier discussion of vitamin B_6 (see the Chapter 16 opening vignette), imine and enamine functional groups have widespread roles in biological chemistry. Yet the functions of imines and enamines in biology are just as we would predict based on their native chemical reactivity.

17.4D Acid-Catalyzed Aldol Condensations

Aldol condensations can also be brought about with acid catalysis. Treating acetone with hydrogen chloride, for example, leads to the formation of 4-methylpent-3-en-2-one, the aldol condensation product. In general, acid-catalyzed aldol reactions lead to dehydration of the initially formed aldol addition product.

A Mechanism for the Reaction

The Acid-Catalyzed Aldol Reaction

Reaction:

$$2\,H_3C{-}\overset{O}{\overset{\|}{C}}{-}CH_3 \xrightarrow{HCl} H_3C{-}\overset{O}{\overset{\|}{C}}{-}CH{=}\overset{CH_3}{\overset{|}{C}}{-}CH_3 + H_2O$$

4-Methylpent-3-en-2-one

Mechanism:

$$H_3C{-}\overset{O}{\overset{\|}{C}}{-}CH_3 + H{-}Cl \rightleftharpoons H_3C{-}\overset{\overset{+}{O}{-}H}{\overset{\|}{C}}{-}CH_2{-}H + Cl^- \rightleftharpoons H_3C{-}\overset{O{-}H}{\overset{|}{C}}{=}CH_2 + H{-}Cl$$

The mechanism begins with the acid-catalyzed formation of the enol.

$$H_3C{-}\overset{O{-}H}{\overset{|}{C}}{=}CH_2 + \underset{CH_3}{\overset{CH_3}{C}}{=}\overset{+}{O}{-}H \rightleftharpoons H_3C{-}\overset{\overset{+}{O}{-}H}{\overset{\|}{C}}{-}CH_2{-}\underset{CH_3}{\overset{CH_3}{\overset{|}{\underset{|}{C}}}}{-}O{-}H$$

Then the enol adds to the protonated carbonyl group of another molecule of acetone.

$$H_3C{-}\overset{\overset{+}{O}{-}H}{\overset{\|}{C}}{-}CH_2{-}\underset{CH_3}{\overset{CH_3}{\overset{|}{\underset{|}{C}}}}{-}O{-}H \rightleftharpoons H_3C{-}\overset{O}{\overset{\|}{C}}{-}\underset{H}{\underset{|}{CH}}{-}\underset{CH_3}{\overset{CH_3}{\overset{|}{\underset{|}{C}}}}{-}\underset{H}{\underset{|}{\overset{+}{O}}}{-}H \rightleftharpoons \quad (:Cl:^-)$$

$$H_3C{-}\overset{O}{\overset{\|}{C}}{-}CH{=}\overset{CH_3}{\overset{|}{C}}{-}CH_3 + H{-}Cl + H{-}\underset{H}{\underset{|}{O}}$$

Finally, proton transfers and dehydration lead to the product.

PROBLEM 17.10 The acid-catalyzed aldol condensation of acetone (just shown) also produces some 2,6-dimethylhepta-2,5-dien-4-one. Give a mechanism that explains the formation of this product.

PROBLEM 17.11 Heating acetone with sulfuric acid leads to the formation of mesitylene (1,3,5-trimethylbenzene). Propose a mechanism for this reaction.

17.5 Crossed Aldol Reactions

An aldol reaction that starts with two different carbonyl compounds is called a **crossed aldol reaction.** Crossed aldol reactions using aqueous sodium hydroxide solutions are of little synthetic importance if both reactants have α hydrogens, because these reactions give a complex mixture of products. If, for example, we were to carry out a crossed aldol addition using acetaldehyde and propanal, we would obtain at least four products:

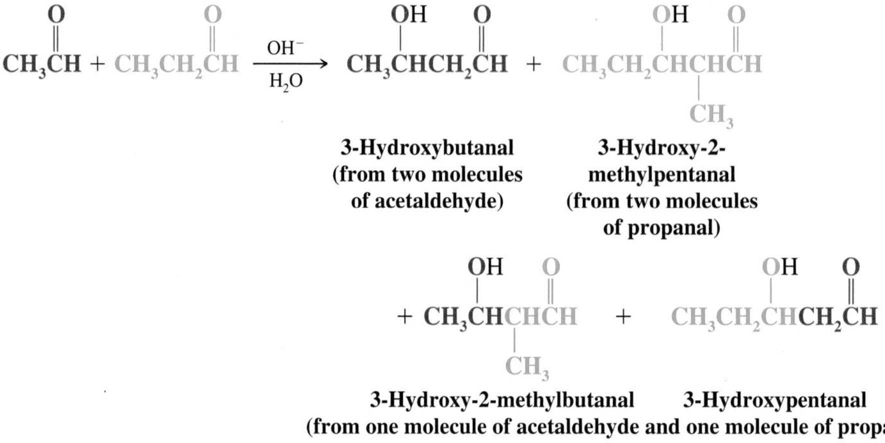

$$CH_3\overset{O}{\overset{\|}{C}}H + CH_3CH_2\overset{O}{\overset{\|}{C}}H \xrightarrow[H_2O]{OH^-} CH_3\overset{OH}{\overset{|}{C}}HCH_2\overset{O}{\overset{\|}{C}}H + CH_3CH_2\overset{OH}{\overset{|}{C}}H\underset{CH_3}{\underset{|}{C}}H\overset{O}{\overset{\|}{C}}H$$

3-Hydroxybutanal (from two molecules of acetaldehyde) — **3-Hydroxy-2-methylpentanal (from two molecules of propanal)**

$$+ CH_3\overset{OH}{\overset{|}{C}}H\underset{CH_3}{\underset{|}{C}}H\overset{O}{\overset{\|}{C}}H + CH_3CH_2\overset{OH}{\overset{|}{C}}HCH_2\overset{O}{\overset{\|}{C}}H$$

3-Hydroxy-2-methylbutanal — **3-Hydroxypentanal**
(from one molecule of acetaldehyde and one molecule of propanal)

SOLVED PROBLEM

Show how each of the four products just given is formed in the crossed aldol addition between acetaldehyde and propanal.

ANSWER In the basic aqueous solution, four organic entities will initially be present: molecules of acetaldehyde, molecules of propanal, enolate anions derived from acetaldehyde, and enolate anions derived from propanal.

We have already seen (Section 17.4) how a molecule of acetaldehyde can react with its enolate anion to form 3-hydroxybutanal (aldol):

Reaction 1

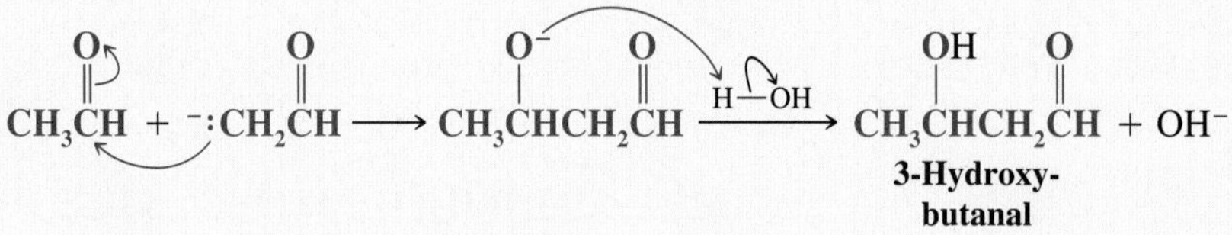

$$CH_3\overset{O}{\overset{\|}{C}}H + {}^{-}\!:CH_2\overset{O}{\overset{\|}{C}}H \longrightarrow CH_3\overset{O^-}{\overset{|}{C}}HCH_2\overset{O}{\overset{\|}{C}}H \xrightarrow{H-OH} CH_3\overset{OH}{\overset{|}{C}}HCH_2\overset{O}{\overset{\|}{C}}H + OH^-$$

3-Hydroxybutanal

We have also seen (Problem 17.7) how propanal can react with its enolate anion to form 3-hydroxy-2-methylpentanal:

Reaction 2

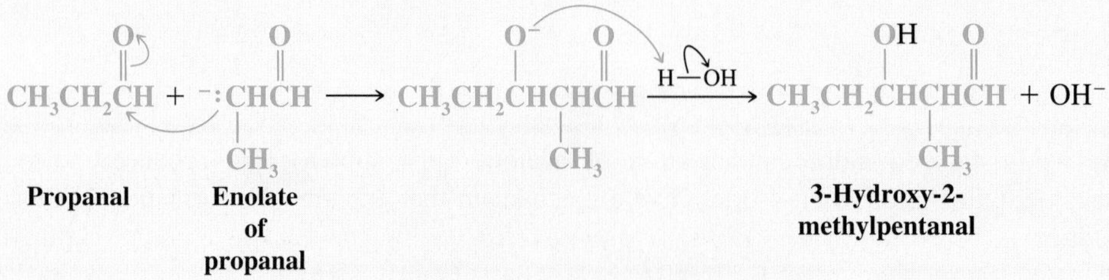

$$CH_3CH_2\overset{O}{\overset{\|}{C}}H + {}^{-}\!:\underset{CH_3}{\underset{|}{C}}H\overset{O}{\overset{\|}{C}}H \longrightarrow CH_3CH_2\overset{O^-}{\overset{|}{C}}H\underset{CH_3}{\underset{|}{C}}H\overset{O}{\overset{\|}{C}}H \xrightarrow{H-OH} CH_3CH_2\overset{OH}{\overset{|}{C}}H\underset{CH_3}{\underset{|}{C}}H\overset{O}{\overset{\|}{C}}H + OH^-$$

Propanal — **Enolate of propanal** — **3-Hydroxy-2-methylpentanal**

Acetaldehyde can also react with the enolate of propanal. This reaction leads to the third product, 3-hydroxy-2-methylbutanal:

Reaction 3

$$CH_3\overset{O}{\overset{\|}{C}}H + {}^{-}{:}\underset{CH_3}{\underset{|}{C}}H\overset{O}{\overset{\|}{C}}H \longrightarrow CH_3\overset{O^-}{\overset{|}{C}}H\underset{CH_3}{\underset{|}{C}}H\overset{O}{\overset{\|}{C}}H \xrightarrow{H-OH} CH_3\overset{OH}{\overset{|}{C}}H\underset{CH_3}{\underset{|}{C}}H\overset{O}{\overset{\|}{C}}H + OH^-$$

Acetaldehyde — Enolate of propanal — 3-Hydroxy-2-methylbutanal

And finally, propanal can react with the enolate of acetaldehyde. This reaction accounts for the fourth product:

Reaction 4

$$CH_3CH_2\overset{O}{\overset{\|}{C}}H + {}^{-}{:}CH_2\overset{O}{\overset{\|}{C}}H \longrightarrow CH_3CH_2\overset{O^-}{\overset{|}{C}}HCH_2\overset{O}{\overset{\|}{C}}H \xrightarrow{H-OH} CH_3CH_2\overset{OH}{\overset{|}{C}}HCH_2\overset{O}{\overset{\|}{C}}H + OH^-$$

Propanal — Enolate of acetaldehyde — 3-Hydroxypentanal

17.5A Practical Crossed Aldol Reactions

Crossed aldol reactions are practical, using bases such as NaOH, when one reactant does not have an α hydrogen and so cannot undergo self-condensation because it cannot form an enolate anion. We can avoid other side reactions by placing this component in a base and then slowly adding the reactant with an α hydrogen to the mixture. Under these conditions, the concentration of the reactant with an α hydrogen is always low and much of the reactant is present as an enolate anion. The main reaction that takes place is one between this enolate anion and the component that has no α hydrogen. The examples listed in Table 17.1 illustrate this technique. In Section 17.7 we study another method for crossed aldol reactions.

As the examples in Table 17.1 also show, the crossed aldol reaction is often accompanied by dehydration. Whether dehydration occurs can, at times, be determined by our choice of reaction conditions, but ***dehydration is especially easy when it leads to an extended conjugated system.***

PROBLEM 17.12

Outlined below is a synthesis of a compound used in perfumes, called lily aldehyde. Provide all of the missing structures.

$$p\text{-}tert\text{-Butylbenzyl alcohol} \xrightarrow[CH_2Cl_2]{PCC} C_{11}H_{14}O \xrightarrow[OH^-]{\text{propanal}}$$

$$C_{14}H_{18}O \xrightarrow{H_2,\ Pd\text{–}C} \text{lily aldehyde } (C_{14}H_{20}O)$$

PROBLEM 17.13

Show how you could use a crossed aldol reaction to synthesize cinnamaldehyde ($C_6H_5CH{=}CHCHO$). Write a detailed mechanism for the reaction.

TABLE 17.1 Crossed Aldol Reactions

This Reactant with No α Hydrogen Is Placed in Base		This Reactant with an α Hydrogen Is Added Slowly		Product
$C_6H_5CH{=}O$ Benzaldehyde	+	$CH_3CH_2CH{=}O$ Propanal	$\xrightarrow[10°C]{OH^-}$	$C_6H_5CH{=}C(CH_3)CH{=}O$ 2-Methyl-3-phenyl-2-propenal (α-methylcinnamaldehyde) (68%)
$C_6H_5CH{=}O$ Benzaldehyde	+	$C_6H_5CH_2CH{=}O$ Phenylacetaldehyde	$\xrightarrow[20°C]{OH^-}$	$C_6H_5CH{=}C(C_6H_5)CH{=}O$ 2,3-Diphenyl-2-propenal
$HCH{=}O$ Formaldehyde	+	$CH_3CH(CH_3)CH{=}O$ 2-Methylpropanal	$\xrightarrow[40°C]{\text{dilute } Na_2CO_3}$	$CH_3C(CH_3)(CH_2OH)CH{=}O$ 3-Hydroxy-2,2-dimethylpropanal (>64%)

PROBLEM 17.14 When excess formaldehyde in basic solution is treated with acetaldehyde, the following reaction takes place:

$$3\ \overset{O}{\overset{\|}{HCH}} + CH_3\overset{O}{\overset{\|}{CH}} \xrightarrow[40°C]{\text{dilute } Na_2CO_3} HOCH_2{-}\underset{CH_2OH}{\overset{CH_2OH}{C}}{-}CHO$$

(82%)

Write a mechanism that accounts for the formation of the product.

17.5B Claisen–Schmidt Reactions

When ketones are used as one component, the crossed aldol reactions are called **Claisen–Schmidt reactions,** after the German chemists J. G. Schmidt (who discovered the reaction in 1880) and Ludwig Claisen (who developed it between 1881 and 1889). These reactions are practical when bases such as sodium hydroxide are used because under these conditions ketones do not self-condense appreciably. (The equilibrium is unfavorable; see Section 17.4C.)

Two examples of Claisen–Schmidt reactions are the following:

$$C_6H_5\overset{O}{\overset{\|}{C}}H + CH_3\overset{O}{\overset{\|}{C}}CH_3 \xrightarrow[100°C]{OH^-} C_6H_5CH{=}CH\overset{O}{\overset{\|}{C}}CH_3$$

4-Phenylbut-3-en-2-one (benzalacetone) (70%)

$$C_6H_5\overset{O}{\overset{\|}{C}}H + CH_3\overset{O}{\overset{\|}{C}}C_6H_5 \xrightarrow[20°C]{OH^-} C_6H_5CH{=}CH\overset{O}{\overset{\|}{C}}C_6H_5$$

1,3-Diphenylprop-2-en-1-one (benzalacetophenone) (85%)

A Mechanism for the Reaction

The Claisen–Schmidt Reaction

Step 1

$$H{-}\ddot{O}{:}^- + H{-}CH_2{-}\overset{O}{\overset{\|}{C}}{-}CH_3 \rightleftharpoons \left[{}^-{:}CH_2{-}\overset{O}{\overset{\|}{C}}{-}CH_3 \longleftrightarrow CH_2{=}\overset{{:}\ddot{O}{:}^-}{\overset{|}{C}}{-}CH_3 \right] + H{-}\ddot{O}{:}{-}H$$

Enolate anion

In this step the base (a hydroxide ion) removes a proton from the α carbon of one molecule of the ketone to give a resonance-stabilized enolate anion.

Step 2

$$C_6H_5{-}\overset{O}{\overset{\|}{C}}{-}H + {}^-{:}CH_2{-}\overset{O}{\overset{\|}{C}}{-}CH_3 \rightleftharpoons C_6H_5{-}\overset{{:}\ddot{O}{:}^-}{\overset{|}{C}H}{-}CH_2{-}\overset{O}{\overset{\|}{C}}{-}CH_3$$

An alkoxide anion

$$\updownarrow$$

$$CH_2{=}\overset{{:}\ddot{O}{:}^-}{\overset{|}{C}}{-}CH_3$$

The enolate anion then acts as a nucleophile—as a carbanion—and attacks the carbonyl carbon of a molecule of aldehyde, producing an alkoxide anion.

Step 3

$$C_6H_5{-}\overset{{:}\ddot{O}{:}^-}{\overset{|}{C}H}{-}CH_2{-}\overset{O}{\overset{\|}{C}}{-}CH_3 + H{-}\ddot{O}{-}H \longrightarrow C_6H_5{-}\overset{{:}\ddot{O}{-}H}{\overset{|}{C}H}{-}CH_2{-}\overset{O}{\overset{\|}{C}}{-}CH_3 + {}^-{:}\ddot{O}{-}H$$

The alkoxide anion now removes a proton from a molecule of water.

Step 4

$$C_6H_5{-}\overset{{:}\ddot{O}{-}H}{\overset{|}{C}H}{-}\underset{H}{\underset{|}{C}H}{-}\overset{O}{\overset{\|}{C}}{-}CH_3 \;(+\; H{-}\ddot{O}{:}^-) \longrightarrow C_6H_5{-}\overset{{:}\ddot{O}{-}H}{\overset{|}{C}H}{-}CH{=}\overset{{:}\ddot{O}{:}^-}{\overset{|}{C}}{-}CH_3 \longrightarrow$$

$$C_6H_5{-}CH{=}CH{-}\overset{O}{\overset{\|}{C}}{-}CH_3 + H{-}\ddot{O}{:}{-}H + H{-}\ddot{O}{:}^-$$

4-Phenylbut-3-en-2-one
(benzalacetone)

Dehydration produces the conjugated product.

In the Claisen–Schmidt reactions given above dehydration occurs readily because the double bond that forms is conjugated both with the carbonyl group and with the benzene ring. The conjugated system is thereby extended.

An important step in a commercial synthesis of vitamin A makes use of a Claisen–Schmidt reaction between geranial and acetone:

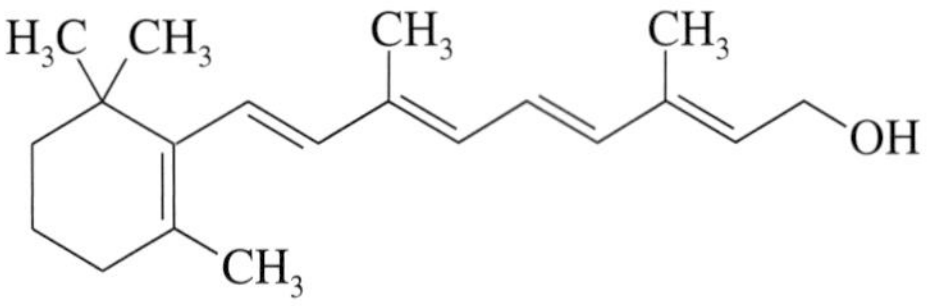

Vitamin A

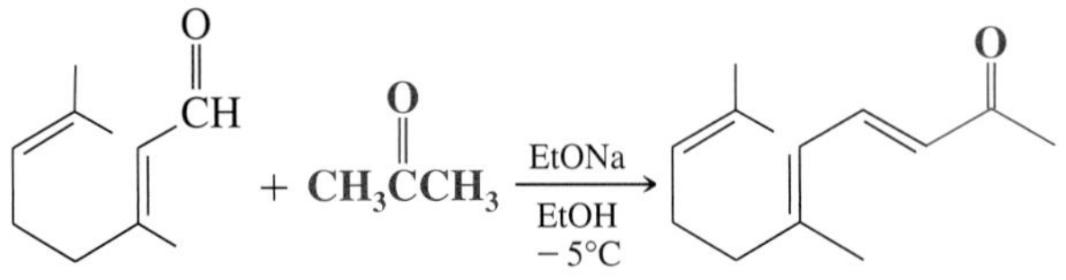

$$\text{Geranial} + CH_3\overset{O}{\overset{\|}{C}}CH_3 \xrightarrow[\text{EtOH},\ -5°C]{\text{EtONa}} \text{Pseudoionone (49\%)}$$

Geranial is a naturally occurring aldehyde that can be obtained from lemongrass oil. Its α hydrogen is *vinylic* and, therefore, not appreciably acidic. Notice, in this reaction, too, dehydration occurs readily because dehydration extends the conjugated system.

PROBLEM 17.15

When pseudoionone is treated with BF_3 in acetic acid, ring closure takes place and α- and β-ionone are produced. This is the next step in the vitamin A synthesis.

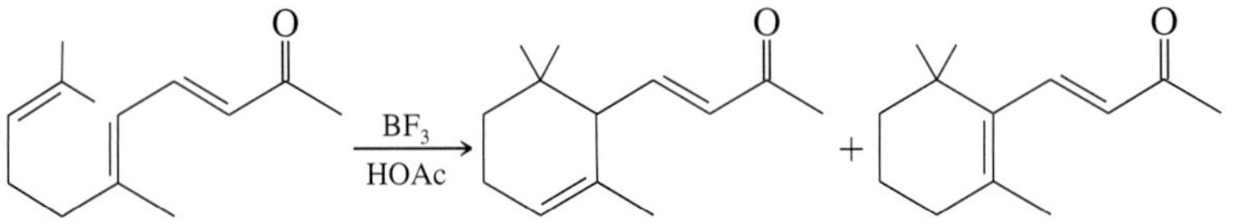

$$\text{Pseudoionone} \xrightarrow[\text{HOAc}]{BF_3} \alpha\text{-Ionone} + \beta\text{-Ionone}$$

(a) Write mechanisms that explain the formation of α- and β-ionone.
(b) β-Ionone is the major product. How can you explain this?
(c) Which ionone would you expect to absorb at longer wavelengths in the UV–Visible region? Why?

17.5C Condensations with Nitroalkanes

The α hydrogens of nitroalkanes are appreciably acidic ($pK_a = 10$), much more acidic than those of aldehydes and ketones. The acidity of these hydrogen atoms, like the α hydrogens of aldehydes and ketones, can be explained by the powerful electron-withdrawing effect of the nitro group and by resonance stabilization of the anion that is produced:

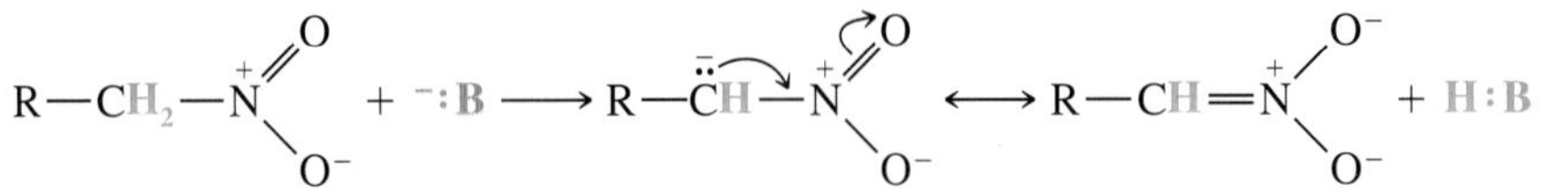

$$R{-}CH_2{-}\overset{+}{N}(=O){-}O^- + {}^-{:}B \longrightarrow R{-}\ddot{C}H^-{-}\overset{+}{N}(=O){-}O^- \longleftrightarrow R{-}CH{=}\overset{+}{N}(O^-){-}O^- + H{:}B$$

Resonance-stabilized anion

Nitroalkanes that have α hydrogens undergo base-catalyzed condensations with aldehydes and ketones that resemble aldol condensations. An example is the condensation of benzaldehyde with nitromethane:

$$C_6H_5\overset{\overset{\displaystyle O}{\|}}{C}H + CH_3NO_2 \xrightarrow{OH^-} C_6H_5CH{=}CHNO_2$$

This condensation is especially useful because the nitro group of the product can be easily reduced to an amino group. One technique that brings about this transformation uses hydrogen and a nickel catalyst. This combination not only reduces the nitro group but also reduces the double bond:

$$C_6H_5CH{=}CHNO_2 \xrightarrow{H_2,\ Ni} C_6H_5CH_2CH_2NH_2$$

PROBLEM 17.16

Assuming that you have available the required aldehydes, ketones, and nitroalkanes, show how you would synthesize each of the following compounds and write a detailed mechanism for each reaction:

(a) $C_6H_5CH{=}\underset{\underset{\displaystyle CH_3}{|}}{C}NO_2$ **(b)** $HOCH_2CH_2NO_2$

17.5D Condensations with Nitriles

The α hydrogens of nitriles are also appreciably acidic, but less so than those of aldehydes and ketones. The acidity constant for acetonitrile (CH_3CN) is about 10^{-25} ($pK_a \cong 25$). Other nitriles with α hydrogens show comparable acidities, and consequently these nitriles undergo condensations of the aldol type. An example is the condensation of benzaldehyde with phenylacetonitrile:

$$C_6H_5\overset{\overset{\displaystyle O}{\|}}{C}H + C_6H_5CH_2CN \xrightarrow[EtOH]{EtO^-} C_6H_5CH{=}\underset{\underset{\displaystyle C_6H_5}{|}}{C}{-}CN$$

PROBLEM 17.17

(a) Write resonance structures for the anion of acetonitrile that account for its being much more acidic than ethane. **(b)** Give a step-by-step mechanism for the condensation of benzaldehyde with acetonitrile.

17.6 Cyclizations via Aldol Condensations

The aldol condensation also offers a convenient way to synthesize molecules with five- and six-membered rings (and sometimes even larger rings). This can be done by an intramolecular aldol condensation using a dialdehyde, a keto aldehyde, or a diketone as the substrate. For example, the following keto aldehyde cyclizes to yield 1-cyclopentenyl methyl ketone:

$$CH_3\overset{\overset{\displaystyle O}{\|}}{C}CH_2CH_2CH_2CH_2\overset{\overset{\displaystyle O}{\|}}{C}H \xrightarrow{OH^-} \text{(1-cyclopentenyl)}{-}\overset{\overset{\displaystyle O}{\|}}{C}CH_3$$

(73%)

This reaction almost certainly involves the formation of at least three different enolates. However, it is the enolate from the ketone side of the molecule that adds to the aldehyde group leading to the product.

A Mechanism for the Reaction

The Aldol Cyclization

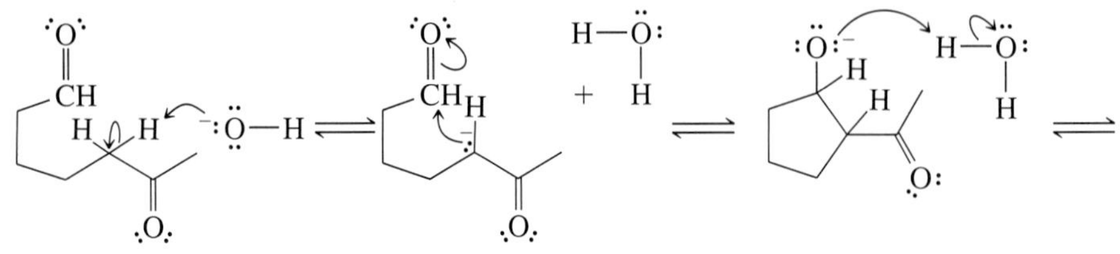

Other enolate anions

This enolate leads to the main product via an intramolecular aldol reaction.

The alkoxide anion removes a proton from water.

Base-promoted dehydration leads to a product with conjugated double bonds.

Selectivity in aldol cyclizations is influenced by carbonyl type and ring size.

The reason the aldehyde group undergoes addition preferentially may arise from the greater reactivity of aldehydes toward nucleophilic addition generally. The carbonyl carbon atom of a ketone is less positive (and therefore less reactive toward a nucleophile) because it bears two electron-releasing alkyl groups; it is also more sterically hindered:

Ketones are less reactive toward nucleophiles.

Aldehydes are more reactive toward nucleophiles.

In reactions of this type, five-membered rings form far more readily than seven-membered rings and six-membered rings are more favorable than four- or eight-membered rings when possible.

Problem 17.18 Assuming that dehydration occurs in all instances, write the structures of the two other products that might have resulted from the aldol cyclization just given. (One of these products will have a five-membered ring and the other will have a seven-membered ring.)

Problem 17.19 What starting compound would you use in an aldol cyclization to prepare each of the following?

(a) (b) (c)

PROBLEM 17.20

What experimental conditions would favor the cyclization process in the intramolecular aldol reaction over intermolecular condensation?

17.7 Lithium Enolates

The extent to which an enolate anion forms depends on the strength of the base used. If the base employed is a weaker base than the enolate anion, then the equilibrium lies to the left. This is the case, for example, when a ketone is treated with an aqueous solution containing sodium hydroxide:

$$CH_3COCH_3 + Na^+OH^- \rightleftharpoons CH_3C(O^{\delta-}Na^+){=}CH_2^{\delta-} + H_2O$$

Weaker acid (pK_a = 20) — **Weaker base** — **Stronger base** — **Stronger acid (pK_a = 16)**

On the other hand, if a very strong base is employed, the equilibrium lies far to the right. One very useful strong base for converting ketones to enolates is **lithium diisopropylamide,** $(i\text{-}C_3H_7)_2N^-Li^+$:

$$CH_3COCH_3 + (i\text{-}C_3H_7)_2N^-Li^+ \longrightarrow CH_3C(O^{\delta-}Li^+){=}CH_2^{\delta-} + (i\text{-}C_3H_7)_2NH$$

Stronger acid (pK_a = 20) — **Stronger base** — **Weaker base** — **Weaker acid (pK_a = 38)**

Lithium diisopropylamide (abbreviated **LDA**) can be prepared by dissolving diisopropylamine in a solvent such as diethyl ether or THF and treating it with an alkyllithium:

$$(i\text{-}C_3H_7)_2NH + \overset{\delta-}{C_4H_9}\overset{\delta+}{Li} \xrightarrow{THF} (i\text{-}C_3H_7)_2N^-Li^+ + C_4H_{10}$$

Stronger acid (pK_a = 38) — **Stronger base** — **Weaker base** — **Weaker acid (pK_a = 50)**

17.7A Regioselective Formation of Enolate Anions

An unsymmetrical ketone such as 2-methylcyclohexanone can form two possible enolates. Just which enolate is formed predominantly depends on the base used and on the conditions employed. The enolate *with the more highly substituted double bond is the* ***thermodynamically more stable enolate*** in the same way that an alkene with the more highly substituted double bond is the more stable alkene (Section 7.3). This enolate, called the **thermodynamic enolate,** is formed predominantly under conditions that permit the establishment of an equilibrium. This will generally be the case if the enolate is produced using a relatively weak base in a protic solvent:

B:⁻ or

weak base in a protic solvent

Kinetic (less stable) enolate ⇌ **2-Methylcyclohexanone** ⇌ **Thermodynamic (more stable) enolate**

This enolate is more stable because the double bond is more highly substituted. It is the predominant enolate at equilibrium.

On the other hand, *the enolate with the less substituted double bond is usually formed faster,* because removal of the hydrogen necessary to produce this enolate is less sterically hindered. This enolate, called the **kinetic enolate,** is formed predominantly when the reaction is kinetically controlled (or rate controlled).

The kinetically favored enolate can be formed cleanly through the use of lithium diisopropylamide (LDA). This strong, sterically hindered base rapidly removes the proton from the less substituted α carbon of the ketone. The following example, using 2-methylcyclohexanone, is an illustration. The solvent for the reaction is 1,2-dimethoxyethane ($CH_3OCH_2CH_2OCH_3$), abbreviated **DME.** The LDA removes the hydrogen from the $-CH_2-$ α carbon more rapidly because it is less hindered (and because there are twice as many hydrogens there to react):

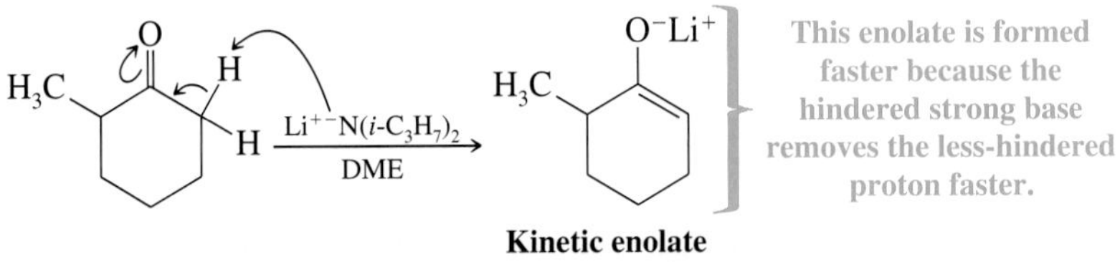

$\xrightarrow[\text{DME}]{Li^{+}\,^{-}N(i\text{-}C_3H_7)_2}$

Kinetic enolate

This enolate is formed faster because the hindered strong base removes the less-hindered proton faster.

17.7B Lithium Enolates in Directed Aldol Reactions

Lithium enolates are a tool for crossed aldol syntheses.

One of the most effective and versatile ways to bring about a crossed aldol reaction is to use a lithium enolate obtained from a ketone as one component and an aldehyde or ketone as the other. An example of what is called a **directed aldol reaction** is shown in Fig. 17.1.

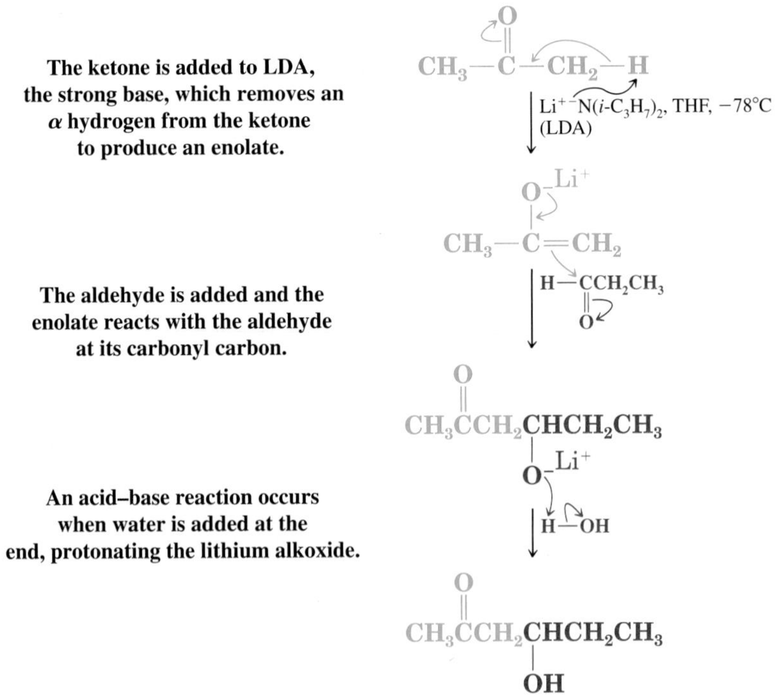

FIGURE 17.1 A directed aldol synthesis using a lithium enolate.

The ketone is added to LDA, the strong base, which removes an α hydrogen from the ketone to produce an enolate.

$CH_3-\overset{O}{\overset{\|}{C}}-CH_2-H$

$\downarrow$ $Li^{+}\,^{-}N(i\text{-}C_3H_7)_2$, THF, $-78°C$ (LDA)

$CH_3-\overset{O^{-}Li^{+}}{\overset{|}{C}}=CH_2$

The aldehyde is added and the enolate reacts with the aldehyde at its carbonyl carbon.

$\downarrow$ $H-\overset{O}{\overset{\|}{C}}CH_2CH_3$

$CH_3\overset{O}{\overset{\|}{C}}CH_2\underset{O^{-}Li^{+}}{\underset{|}{C}H}CH_2CH_3$

An acid–base reaction occurs when water is added at the end, protonating the lithium alkoxide.

$\downarrow$ $H-OH$

$CH_3\overset{O}{\overset{\|}{C}}CH_2\underset{OH}{\underset{|}{C}H}CH_2CH_3$

Regioselectivity can be achieved when unsymmetrical ketones are used in directed aldol reactions by generating the kinetic enolate using lithium diisopropylamide. This ensures production of the enolate in which the proton has been removed from the less-substituted α carbon. The following is an example:

An Aldol Reaction via the Kinetic Enolate (Using LDA)

$$CH_3CH_2\overset{O}{\overset{\|}{C}}CH_3 \xrightarrow[-78°C]{LDA,\ THF} CH_3CH_2\overset{O^-Li^+}{\overset{|}{C}}{=}CH_2 \xrightarrow{CH_3\overset{O}{\overset{\|}{C}}H}$$

$$CH_3CH_2\overset{O}{\overset{\|}{C}}CH_2\overset{O^-Li^+}{\overset{|}{C}}HCH_3 \xrightarrow{H_2O} CH_3CH_2\overset{O}{\overset{\|}{C}}CH_2\overset{OH}{\overset{|}{C}}HCH_3$$

(75%)

A single crossed aldol product results.

If the aldol (Claisen–Schmidt) reaction had been carried out in the classic way (Section 17.5B) using hydroxide ion as the base, then at least two products would have been formed in significant amounts. Both the kinetic and thermodynamic enolates would have been formed from the ketone, and each of these would have added to the carbonyl carbon of the aldehyde:

An Aldol Reaction That Produces a Mixture via Both Kinetic and Thermodynamic Enolates (Using a Weaker Base under Protic Conditions)

$$CH_3CH_2{-}\overset{O}{\overset{\|}{C}}{-}CH_3 \underset{\text{protic solvent}}{\overset{^-OH}{\rightleftharpoons}} \underset{\textbf{Kinetic enolate}}{CH_3CH_2{-}\overset{O^-}{\overset{|}{C}}{=}CH_2} + \underset{\textbf{Thermodynamic enolate}}{CH_3CH{=}\overset{O^-}{\overset{|}{C}}{-}CH_3}$$

Kinetic enolate $\xrightarrow{CH_3\overset{O}{\overset{\|}{C}}H}$ $CH_3CH_2\overset{O}{\overset{\|}{C}}CH_2\overset{O^-}{\overset{|}{C}}HCH_3$ $\xrightarrow{H_2O}$ $CH_3CH_2\overset{O}{\overset{\|}{C}}CH_2\overset{OH}{\overset{|}{C}}HCH_3$

Thermodynamic enolate $\xrightarrow{CH_3\overset{O}{\overset{\|}{C}}H}$ $CH_3\underset{CHCH_3(O^-)}{\underset{|}{C}}H\overset{O}{\overset{\|}{C}}CH_3$ $\xrightarrow{H_2O}$ $CH_3\underset{CHCH_3(OH)}{\underset{|}{C}}H\overset{O}{\overset{\|}{C}}CH_3$

A mixture of crossed aldol products results.

PROBLEM 17.21

Starting with ketones and aldehydes of your choice, outline a directed aldol synthesis of each of the following using lithium enolates:

(a) 2-methyl-5-(1-hydroxyethyl)cyclopentanone: H_3C–(cyclopentanone ring, C=O)–$\overset{OH}{\overset{|}{C}}HCH_3$

(b) $CH_3CH_2\overset{O}{\overset{\|}{C}}CH_2\overset{OH}{\overset{|}{C}}HC_6H_5$

(c) $CH_3\underset{CH_3}{\underset{|}{C}}H\overset{O}{\overset{\|}{C}}CH_2\overset{OH}{\overset{|}{C}}HCH_2CH_3$

(d) $CH_3CH{=}CH\overset{O}{\overset{\|}{C}}CH_2\overset{OH}{\overset{|}{C}}HCH_3$

PROBLEM 17.22

The compounds called α-bisabolanone and ocimenone have both been synthesized by directed aldol syntheses. In both syntheses one starting compound was $(CH_3)_2C{=}CHCOCH_3$. Choose other appropriate starting compounds and outline syntheses of **(a)** α-bisabolanone and **(b)** ocimenone.

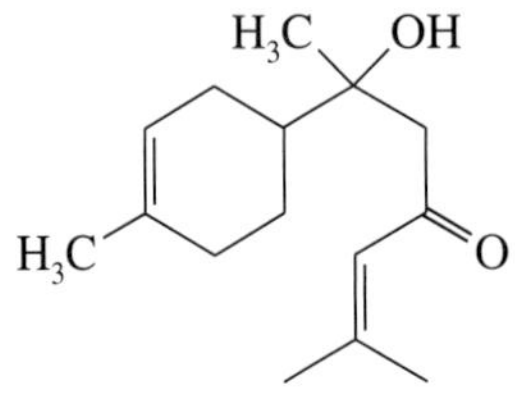

α-Bisabolanone **Ocimenone**

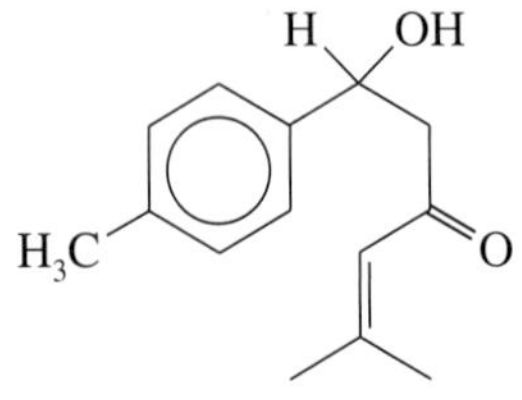

Alkylation of lithium enolates is a tool for synthesis.

17.7C Direct Alkylation of Ketones via Lithium Enolates

The formation of lithium enolates using lithium diisopropylamide furnishes a useful way of alkylating ketones in a regioselective way. For example, the lithium enolate formed from 2-methylcyclohexanone (Section 17.7A) can be methylated or benzylated by allowing it to react with methyl iodide or benzyl bromide, respectively:

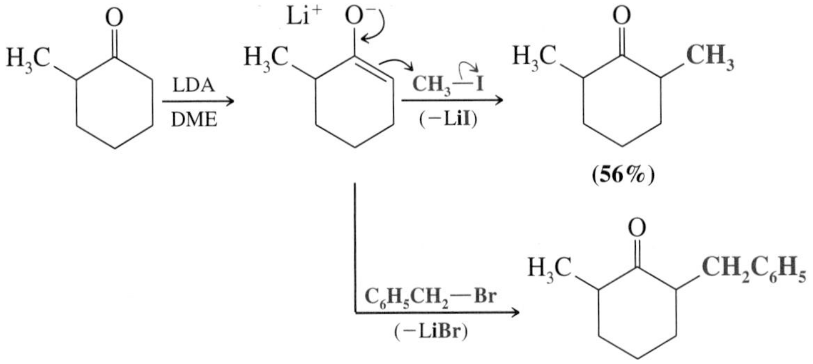

LDA / DME; $CH_3{-}I$ $(-LiI)$ → **(56%)**; $C_6H_5CH_2{-}Br$ $(-LiBr)$ → **(42–45%)**

Proper choice of the alkylating agent is key to successful lithium enolate alkylation.

Alkylation reactions like these have an important limitation. Because the reactions are S_N2 reactions and because enolate anions are strong bases, *successful alkylations occur only when primary alkyl, primary benzylic, and primary allylic halides are used.* With secondary and tertiary halides, elimination becomes the main course of the reaction.

PROBLEM 17.23

(a) Write a reaction involving a lithium enolate for introduction of the methyl group in the following compound (an intermediate in a synthesis by E. J. Corey of cafestol, an anti-inflammatory agent found in coffee beans):

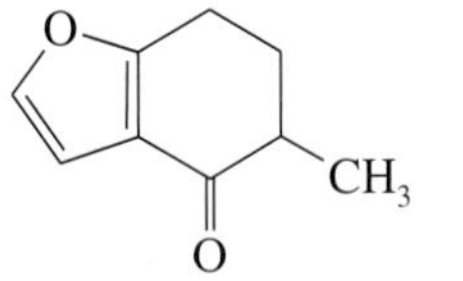

(b) Dienolates can be formed from β-keto esters using two equivalents of LDA. The dienolate can then be alkylated selectively at the more basic of the two enolate carbons. Write a reaction for synthesis of the following compound using a dienolate and the appropriate alkyl halide:

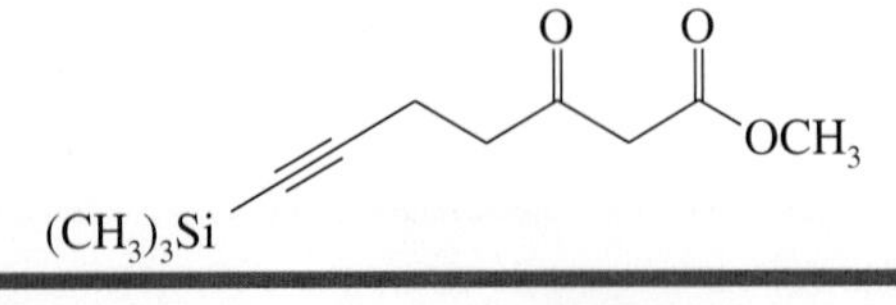

The Chemistry of...

Silyl Enol Ethers

Because enolate anions have a partial negative charge on an oxygen atom, they can react in nucleophilic substitution reactions as if they were **alkoxide anions.** Because they have a partial negative charge on a carbon atom, they can also react as **carbanions.** Nucleophiles like this, ***those that are capable of reacting at two sites,*** are called **ambident nucleophiles:**

This site reacts as an alkoxide anion.

$CH_3C(=O^{\delta-})CH_2^{\delta-}$

This site reacts as a carbanion.

Just how an enolate anion reacts depends, in part, on the substrate with which it reacts. ***Chlorotrialkylsilanes tend to react almost exclusively at the oxygen atom of an enolate.*** Reagents used include chlorotrimethylsilane, *tert*-butylchlorodimethylsilane (TBDMSCl), and *tert*-butylchlorodiphenylsilane (TBDPSCl).

$$CH_3C(O^{\delta-})CH_2^{\delta-} + (CH_3)_3Si{-}Cl \xrightarrow{THF} CH_3C(OSi(CH_3)_3){=}CH_2 + Cl^-$$

Trimethylsilyl enol ether (85%)

This reaction, called **silylation** (see Section 11.11D), is a nucleophilic substitution at the silicon atom by the oxygen atom of the enolate, and it takes place as it does because the oxygen–silicon bond that forms in the trimethylsilyl enol ether is very strong (much stronger than a carbon–silicon bond). This factor makes formation of the trimethylsilyl enol ether highly exothermic, and, consequently, the free energy of activation for reaction at the oxygen atom is lower than that for reaction at the α carbon:

2-Methylcyclohexanone $\xrightarrow[DME]{Li^+\,{}^-N(i\text{-}C_3H_7)_2}$ **Kinetic enolate** (O^-Li^+) $\xrightarrow{(CH_3)_3Si{-}Cl}$ trimethylsilyl enol ether ($OSi(CH_3)_3$) **(99%)**

The example just given shows how the enolate anion can be "trapped" by converting it to the trimethylsilyl enol ether. This procedure is especially useful because the trimethylsilyl enol ether can be purified, if necessary, and then converted back to an enolate. One way of achieving this conversion is by treating the trimethylsilyl enol ether with an aprotic solution containing fluoride ions:

Trimethylsilyl enol ether ($O{-}Si(CH_3)_3$) $\xrightarrow{F^-}$ **Kinetic enolate** (O^-) $+ (CH_3)_3Si{-}F$

This reaction is a nucleophilic substitution at the silicon atom brought about by a fluoride ion. Fluoride ions have an extremely high affinity for silicon atoms because Si—F bonds are very strong.

Another way to convert a trimethylsilyl enol ether back to an enolate is to treat it with methyllithium:

Trimethylsilyl enol ether ($O{-}Si(CH_3)_3$) $+ CH_3{-}Li \longrightarrow$ enolate (O^-Li^+) $+ (CH_3)_3Si{-}CH_3$

PROBLEM 17.B1 Treating the trimethylsilyl enol ether derived from cyclohexanone with benzaldehyde and tetrabutylammonium fluoride, $(C_4H_9)_4N^+F^-$ (abbreviated TBAF), gave the following product. Outline the steps that occur in this reaction.

1-(Trimethylsilyloxy)cyclohexene ($O{-}Si(CH_3)_3$) $+ C_6H_5CHO \xrightarrow[(2)\ H_2O]{(1)\ TBAF}$ 2-($CH(OH)C_6H_5$)cyclohexanone

17.8 α Selenation: A Synthesis of α,β-Unsaturated Carbonyl Compounds

Lithium enolates react with benzeneselenenyl bromide (C_6H_5SeBr) (or with C_6H_5SeCl) to yield products containing a C_6H_5Se— group at the α position:

$$C_6H_5COCH_2CH_3 + Li^+ \ ^-:\ddot{N}(i\text{-}C_3H_7)_2 \xrightarrow{[-HN(i\text{-}C_3H_7)_2]} C_6H_5C(O^-)=CHCH_3$$

Lithium enolate

$$C_6H_5C(O^-)=CHCH_3 + C_6H_5Se{-}Br \longrightarrow Br^- + C_6H_5COCH(SeC_6H_5)CH_3$$

This step is a nucleophilic substitution at the selenium atom.

An α-benzene-selenenyl ketone

Treating the α-benzeneselenenyl ketone with hydrogen peroxide at room temperature converts it to an α,β-unsaturated ketone:

$$C_6H_5COCH(SeC_6H_5)CH_3 \xrightarrow[25°C]{H_2O_2} C_6H_5COCH{=}CH_2 + C_6H_5SeOH$$

(84–89% overall from the ketone)

These are very mild conditions for the introduction of a double bond (room temperature and a neutral solution), and this is one reason why this method is a valuable one.

Mechanistically, two steps are involved in the conversion of the α-benzeneselenenyl ketone to the α,β-unsaturated ketone. The first step is an oxidation brought about by the H_2O_2. The second step is a spontaneous intramolecular elimination in which the negatively charged oxygen atom attached to the selenium atom acts as a base:

$$C_6H_5COCH(SeC_6H_5)CH_3 \xrightarrow[\text{(oxidation)}]{H_2O_2} C_6H_5COCH(CH_2{-}H)\overset{+}{Se}(C_6H_5){-}\ddot{O}:^-$$

$$\xrightarrow[(-C_6H_5SeOH)]{\text{intramolecular elimination}} C_6H_5COCH{=}CH_2$$

When we study the Cope elimination in Section 20.13B, we shall find another example of this kind of intramolecular elimination.

PROBLEM 17.24

Starting with 2-methylcyclohexanone, show how you would use α selenation in a synthesis of the following compound:

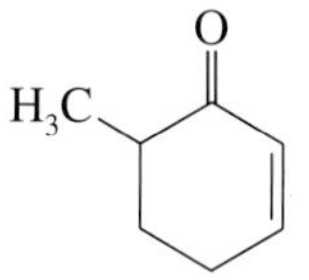

17.9 ADDITIONS TO α,β-UNSATURATED ALDEHYDES AND KETONES

When α,β-unsaturated aldehydes and ketones react with nucleophilic reagents, they may do so in two ways. They may react by a **simple addition,** that is, one in which the nucleophile adds across the double bond of the carbonyl group; or they may react by a **conjugate addition.** These two processes resemble the 1,2- and the 1,4-addition reactions of conjugated dienes (Section 13.10):

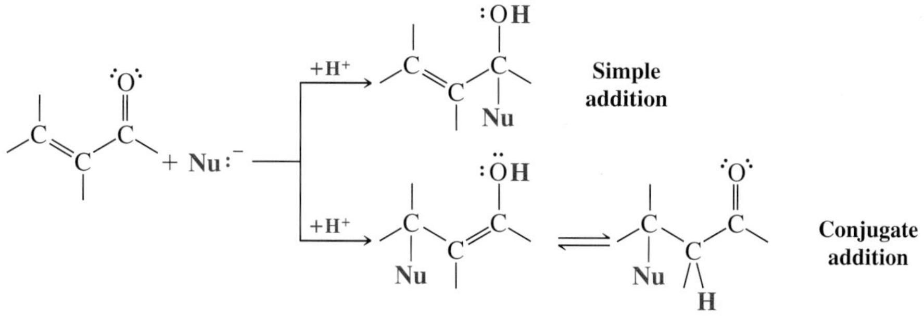

Simple addition

Conjugate addition

Enol form

Keto form

In many instances both modes of addition occur in the same mixture. As an example, let us consider the Grignard reaction shown here:

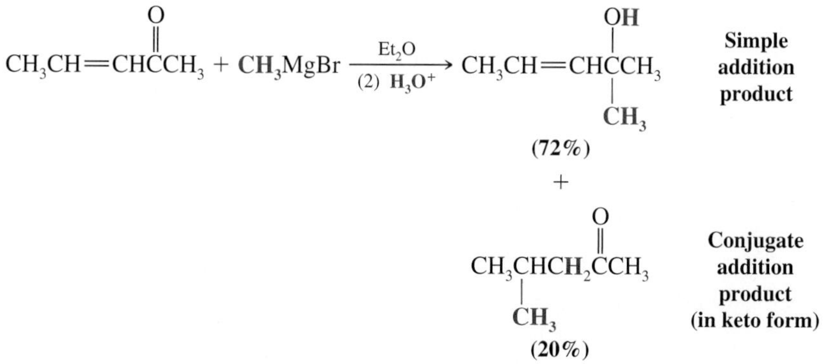

$$CH_3CH{=}CHCOCH_3 + CH_3MgBr \xrightarrow[(2)\ H_3O^+]{Et_2O} CH_3CH{=}CHC(OH)(CH_3)CH_3 \quad (72\%) \text{ Simple addition product}$$

$$+ \; CH_3CH(CH_3)CH_2COCH_3 \quad (20\%) \text{ Conjugate addition product (in keto form)}$$

In this example we see that simple addition is favored, and this is generally the case with strong nucleophiles. Conjugate addition is favored when weaker nucleophiles are employed.

STUDY TIP! Note the influence of nucleophile strength on conjugate versus simple addition.

If we examine the resonance structures that contribute to the overall hybrid for an α,β-unsaturated aldehyde or ketone (see structures **A–C**), we shall be in a better position to understand these reactions:

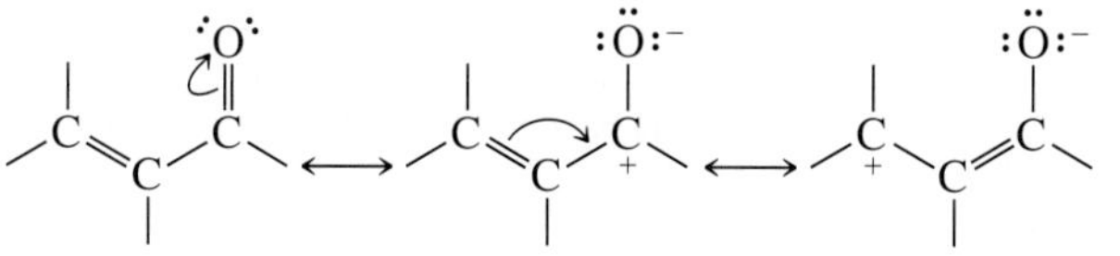

A **B** **C**

Although structures **B** and **C** involve separated charges, they make a significant contribution to the hybrid because, in each, the negative charge is carried by electronegative oxygen. Structures **B** and **C** also indicate that *both the carbonyl carbon and the β carbon should bear a partial positive charge.* They indicate that we should represent the hybrid in the following way:

$$\overset{|}{\underset{\delta+}{C}}\!\!=\!\!\underset{|}{C}\!\!=\!\!\underset{\delta+}{\overset{O^{\delta-}}{\overset{\|}{C}}}$$

This structure tells us that we should expect a nucleophilic reagent to attack either the carbonyl carbon or the β carbon.

Almost every nucleophilic reagent that adds at the carbonyl carbon of a simple aldehyde or ketone is capable of adding at the β carbon of an α,β-unsaturated carbonyl compound. In many instances when weaker nucleophiles are used, conjugate addition is the major reaction path. Consider the following addition of hydrogen cyanide:

$$C_6H_5CH{=}CH\overset{O}{\overset{\|}{C}}C_6H_5 + CN^- \xrightarrow[CH_3CO_2H]{C_2H_5OH} C_6H_5\underset{CN}{\underset{|}{C}}H{-}CH_2\overset{O}{\overset{\|}{C}}C_6H_5$$

(95%)

A Mechanism for the Reaction

The Conjugate Addition of HCN

$$C_6H_5CH{=}CH{-}\overset{O}{\overset{\|}{C}}C_6H_5 \longrightarrow \left[C_6H_5\underset{CN}{\underset{|}{C}}H{-}CH{=}\overset{O^-}{\overset{|}{C}}C_6H_5 \longleftrightarrow C_6H_5\underset{CN}{\underset{|}{C}}H{-}\ddot{C}H{-}\overset{O}{\overset{\|}{C}}C_6H_5 \right]$$

$:CN^-$

Enolate anion intermediate

Then, the enolate intermediate accepts a proton in either of two ways:

$$C_6H_5\underset{CN}{\underset{|}{C}}H{-}\overset{\delta-}{C}H{=\!=}\overset{O^{\delta-}}{\overset{\|}{C}}C_6H_5 \xrightarrow{H^+} C_6H_5\underset{CN}{\underset{|}{C}}H{-}CH{=}\overset{OH}{\overset{|}{C}}C_6H_5 \quad \textbf{Enol form}$$

$$\rightleftharpoons$$

$$C_6H_5\underset{CN}{\underset{|}{C}}H{-}CH_2{-}\overset{O}{\overset{\|}{C}}C_6H_5 \quad \textbf{Keto form}$$

Another example of this type of addition is the following:

$$CH_3\overset{CH_3}{\overset{|}{C}}{=}CH\overset{O}{\overset{\|}{C}}CH_3 + CH_3NH_2 \xrightarrow[H_2O]{} CH_3\underset{CH_3NH}{\underset{|}{\overset{CH_3}{\overset{|}{C}}}}{-}CH_2\overset{O}{\overset{\|}{C}}CH_3$$

(75%)

A Mechanism for the Reaction

The Conjugate Addition of an Amine

$$H_3C-\overset{\delta+}{C}(CH_3)=CH-\overset{\delta+}{C}(=\ddot{O}^{\delta-})-CH_3 + CH_3\ddot{N}H_2 \rightleftharpoons H_3C-C(CH_3)(N^+H_2CH_3)-CH=C(-\ddot{O}:^-)-CH_3 \rightleftharpoons$$

The nucleophile attacks the partially positive β carbon.

In two separate steps, a proton is lost from the nitrogen atom and a proton is gained at the oxygen.

$$H_3C-C(CH_3)(\ddot{N}HCH_3)-CH=C(\ddot{O}-H)-CH_3 \rightleftharpoons H_3C-C(CH_3)(\ddot{N}HCH_3)-CH_2-C(=O)-CH_3$$

Enol form Keto form

17.9A Conjugate Addition of Organocopper Reagents

Organocopper reagents, either RCu or R_2CuLi, add to α,β-unsaturated carbonyl compounds, and **they add almost exclusively in the conjugate manner:**

$$CH_3CH=CH-\overset{O}{\overset{\|}{C}}-CH_3 \xrightarrow[\text{(2) } H_2O]{\text{(1) } CH_3Cu} CH_3\underset{CH_3}{\underset{|}{C}}HCH_2\overset{O}{\overset{\|}{C}}CH_3 \quad (85\%)$$

4-Methylcyclohex-2-enone $\xrightarrow[\text{(2) } H_2O]{\text{(1) } (CH_3)_2CuLi}$ trans-3,4-dimethylcyclohexanone (98%) + cis-3,4-dimethylcyclohexanone (2%)

With an alkyl-substituted cyclic α,β-unsaturated ketone, as the example just cited shows, lithium dialkylcuprates add predominantly in the less hindered way to give the product with the alkyl groups trans to each other.

We shall see examples of biochemically relevant conjugate additions in "The Chemistry of... Calicheamicin $\gamma_1{}^I$ Activation for Cleavage of DNA" (see the next section) and in "The Chemistry of... A Suicide Enzyme Substrate" (Chapter 19).

17.9B Michael Additions

Conjugate additions of enolate anions to α,β-unsaturated carbonyl compounds are known generally as Michael additions (after their discovery, in 1887, by Arthur Michael, of Tufts University and later of Harvard University). An example is the addition of cyclohexanone to $C_6H_5CH{=}CHCOC_6H_5$:

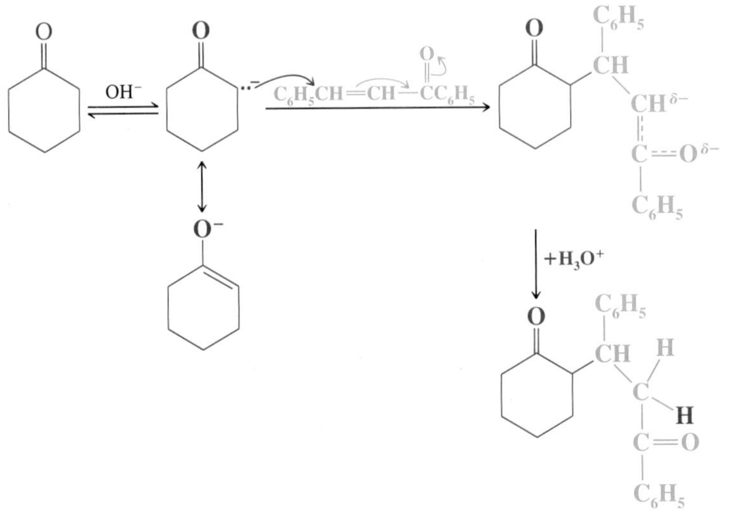

The sequence that follows illustrates how a conjugate aldol addition (Michael addition) followed by a simple aldol condensation may be used to build one ring onto another. This procedure is known as the *Robinson annulation* (ring-forming) reaction (after the English chemist, Sir Robert Robinson, who won the Nobel Prize in Chemistry in 1947 for his research on naturally occurring compounds):

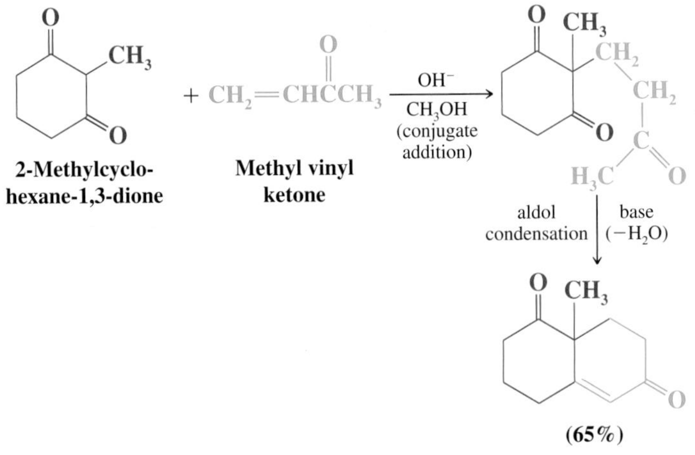

2-Methylcyclohexane-1,3-dione + $CH_2{=}CHCCH_3$ **Methyl vinyl ketone** $\xrightarrow[\text{(conjugate addition)}]{OH^-,\ CH_3OH}$ → aldol condensation, base $(-H_2O)$ → **(65%)**

PROBLEM 17.25 **(a)** Propose step-by-step mechanisms for both transformations of the Robinson annulation sequence just shown. **(b)** Would you expect 2-methylcyclohexane-1,3-dione to be more or less acidic than cyclohexanone? Explain your answer.

PROBLEM 17.26 What product would you expect to obtain from the base-catalyzed Michael reaction of **(a)** of 1,3-diphenylprop-2-en-1-one (Section 17.5B) and acetophenone and **(b)** of 1,3-diphenylprop-2-en-1-one and cyclopentadiene? Show all steps in each mechanism.

Problem 17.27

When acrolein (propenal) reacts with hydrazine, the product is a dihydropyrazole:

$$CH_2{=}CHCHO + H_2N{-}NH_2 \longrightarrow$$ (A dihydropyrazole)

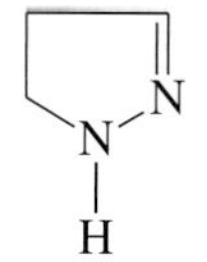

Acrolein **Hydrazine** **A dihydropyrazole**

Suggest a mechanism that explains this reaction.

We shall study further examples of the Michael addition in Chapter 19.

The Chemistry of...

Calicheamicin γ_1^{I} Activation for Cleavage of DNA

In the Chapter 10 opening vignette we described a potent antitumor antibiotic called calicheamicin γ_1^{I}. Now that we have considered conjugate addition reactions, it is time to revisit this fascinating molecule. The molecular machinery of calicheamicin γ_1^{I} for destroying DNA is unleashed by attack of a nucleophile on the trisulfide linkage shown in the accompanying scheme. The sulfur anion that initially was a leaving group from the trisulfide immediately becomes a nucleophile that attacks the bridgehead alkene carbon. This alkene carbon is electrophilic because it is conjugated with the adjoining carbonyl group. Attack by the sulfur nucleophile on the alkene carbon is a *conjugate addition.* Now that the bridgehead carbon is tetrahedral, the geometry of the bicyclic structure favors conversion of the enediyne to a 1,4-benzenoid diradical by a reaction called the Bergman cycloaromatization (after R. G. Bergman of the University of California, Berkeley). Once the calicheamicin diradical is formed it can pluck two hydrogen atoms from the DNA backbone, converting the DNA to a reactive diradical and ultimately resulting in DNA cleavage and the death of the cell.

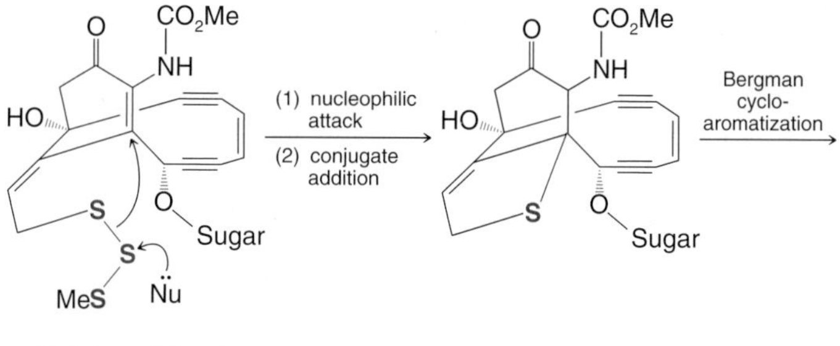

O, CO_2Me, NH, HO, S, O, Sugar, S, MeS, Nu

Calicheamicin γ_1^{I}

(1) nucleophilic attack
(2) conjugate addition

Bergman cyclo-aromatization

Hydrogen abstraction from DNA backbone

Radicals in DNA backbone $\xrightarrow{O_2}$ Oxidative double strand cleavage of DNA

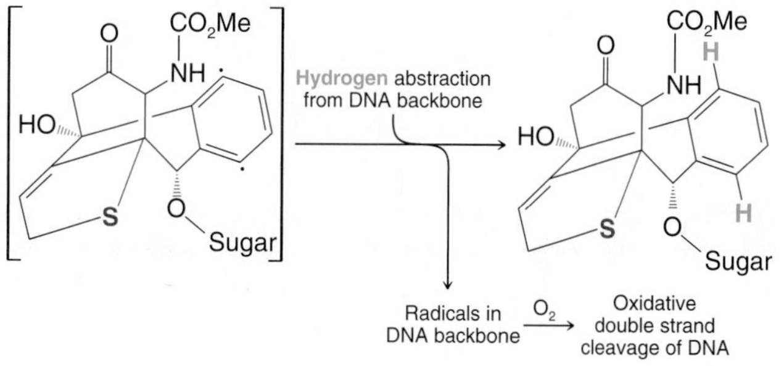

Summary of Mechanisms

Enolates: Formation and Reaction with Electrophiles by Substitution or Addition

General Reaction

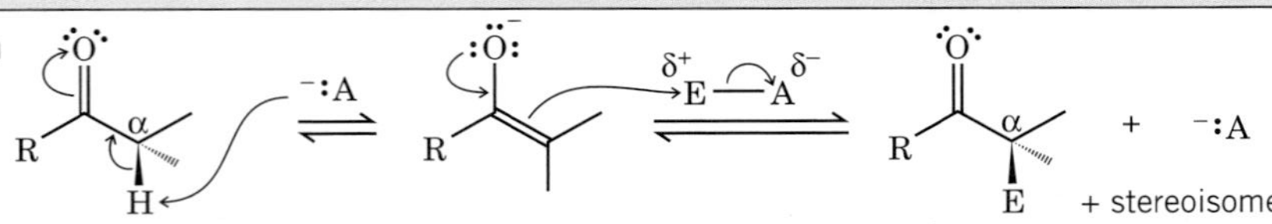

+ stereoisomer (if α–carbon is, and/or if E contains, a stereogenic center)

Some groups that increase α-hydrogen acidity	Typical bases ($^-$:A) and solvents for enolate formation	Possible electrophiles (E-A)	Product(s)
Carbonyl	I. HO^- in H_2O or ROH; or RO^- in ROH; Useful for aldol condensations (especially those involving thermodynamically-favored enolates and equilibrium product control)	H—A **Deprotonation-protonation** (may lead to racemization or epimerization)	Substitution of enolate α-hydrogen by H, X, or R
Nitrile (cyano group)	II. LDA (lithium diisopropylamide) in THF or DME; Useful, in general, for forming enolates in aprotic solvents (especially kinetically-favored enolates and for directed aldol and direct alkylation)	:X—X: **Halogenation**	
		R′—X: **Alkylation**	

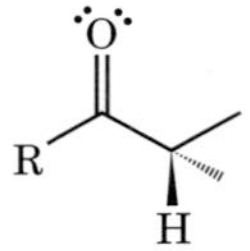

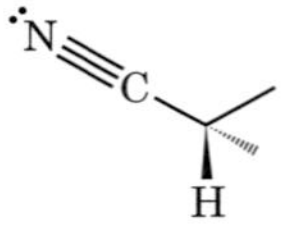

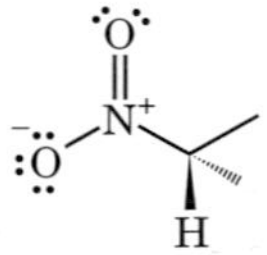

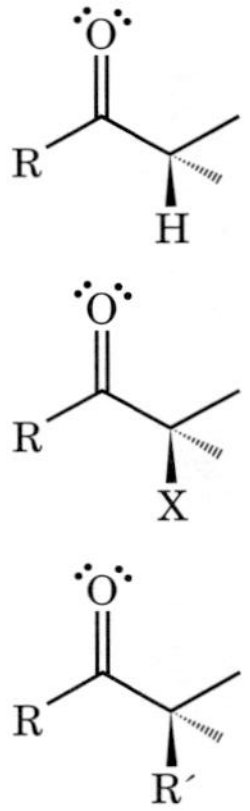

Nitro

(and in general, other groups that can stabilize an α-carbanion)

Enolate addition to a carbonyl electrophile:

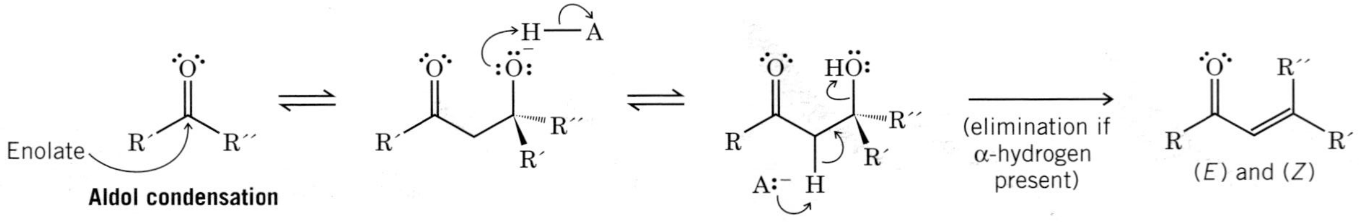

Aldol condensation — (elimination if α-hydrogen present) — (*E*) and (*Z*)

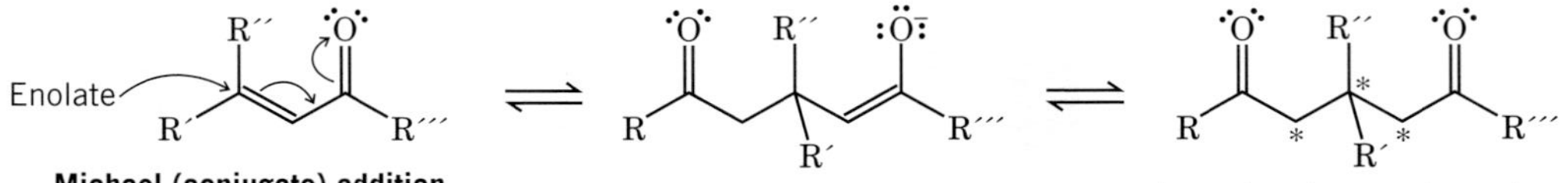

Michael (conjugate) addition — * may be stereogenic centers

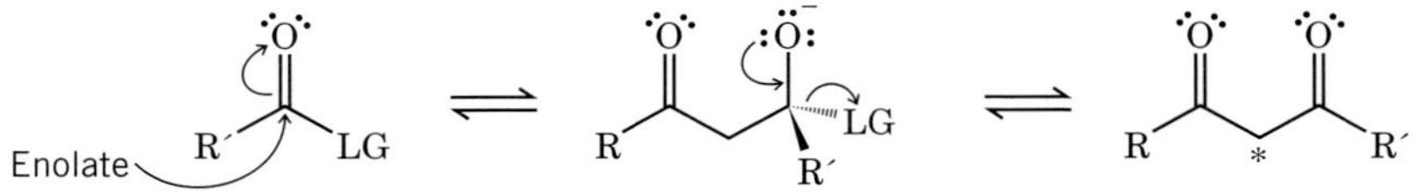

Acyl substitution [addition-elimination (see Ch 18)]

Synthetic Connections

Some Synthetic Connections Involving Enolates

- Keto-enol tautomerism
- Enolate formation
- Racemization/epimerization
- Halogenation
- Alkylation
- α-Selenation/selenoxide elimination

Enolates provide a variety of ways to functionalize the α-carbon of a carbonyl compound. Most importantly, enolates provide ways to form new carbon–carbon bonds. Some of these synthetic connections are shown below. Previously studied reactions of carbonyl, alcohol, and alkene functional groups (e.g., reduction, oxidation, addition, substitution, etc.) lead to or from some of the pathways shown below.

- Aldol condensation
- Michael addition
- Conjugate addition of R_2CuLi
- Addition of Grignard and RLi
- Conjugate addition of HCN
- Conjugate addition of R_2NH

Reactions of X as LG, etc.

X_2, base

HA or $A:^-$

(1) LDA (2) R′X

(1) LDA (2) C_6H_5SeBr

R′ = CHR′R″

SeC_6H_5

H_2O_2

(1) LDA (2) R′COR″ (3) H_3O^+

(1) LDA, HCH_2COR'''

Strong base (e.g., LDA)

HA

HNR_2

HCN

R'''_2CuLi

(1) R‴MgBr or R‴Li (2) H_3O^+

+1,4-addition

Enolate resonance contributors

Key Terms and Concepts

Aldol	Section 17.4
Aldol reactions (additions and condensations)	Sections 17.4, 17.4A, 17.5, and 17.6
α Carbon	Section 17.1
α Hydrogens	Sections 17.1, 17.5C, 17.5D
Ambident nucleophile	Section 17.7C
Conjugate addition (Michael addition)	Section 17.9
Crossed aldol reaction	Section 17.5
Directed aldol reaction	Section 17.7B
Enolate anion	Sections 17.1, 17.3, 17.4, 17.7
Epimers, epimerization	Section 17.3A
Haloform reaction	Section 17.3C
Keto and enol forms	Sections 17.1, 17.2, 17.3
Kinetic and thermodynamic enolate anions	Section 17.7A
Lithium diisopropylamide (LDA)	Section 17.7
Silylation	Sections 11.15, 17.7C
Simple addition	Section 17.9
Tautomers, tautomerization	Section 17.2

Additional Problems

17.28 Give structural formulas for the products of the reaction (if one occurs) when propanal is treated with each of the following reagents:

(a) OH^-, H_2O
(b) C_6H_5CHO, OH^-
(c) HCN
(d) $NaBH_4$
(e) $HOCH_2CH_2OH$, *p*-TsOH
(f) Ag_2O, OH^-, then H_3O^+
(g) CH_3MgI, then H_3O^+
(h) $Ag(NH_3)_2{}^+OH^-$, then H_3O^+
(i) NH_2OH
(j) $C_6H_5\bar{C}H—\overset{+}{P}(C_6H_5)_3$
(k) C_6H_5Li, then H_3O^+
(l) $HC{\equiv}CNa$, then H_3O^+
(m) $HSCH_2CH_2SH$, HA, then Raney Ni, H_2
(n) $CH_3CH_2CHBrCO_2Et$ and Zn, then H_3O^+

17.29 Give structural formulas for the products of the reaction (if one occurs) when acetone is treated with each reagent of the preceding problem.

17.30 What products would form when 4-methylbenzaldehyde reacts with each of the following?

(a) CH_3CHO, OH^-
(b) $CH_3C{\equiv}CNa$, then H_3O^+
(c) CH_3CH_2MgBr, then H_3O^+
(d) Cold dilute $KMnO_4$, OH^-, then H_3O^+
(e) Hot $KMnO_4$, OH^-, then H_3O^+
(f) $^-{:}CH_2—\overset{+}{P}(C_6H_5)_3$
(g) $CH_3COC_6H_5$, OH^-
(h) $BrCH_2CO_2Et$ and Zn, then H_3O^+

17.31 Show how each of the following transformations could be accomplished. You may use any other required reagents.

(a) $CH_3COC(CH_3)_3 \longrightarrow C_6H_5CH{=}CHCOC(CH_3)_3$

(b) $C_6H_5CHO \longrightarrow C_6H_5CH{=}$(cyclopentanone, at C-2)

(c) $C_6H_5CHO \longrightarrow C_6H_5CH_2CH(CH_3)NH_2$

Note: Problems marked with an asterisk are "challenge problems."

(d) 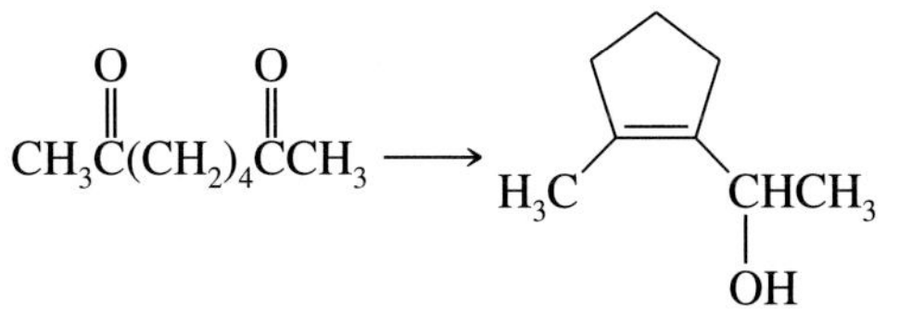 $CH_3\overset{O}{\overset{\|}{C}}(CH_2)_4\overset{O}{\overset{\|}{C}}CH_3 \longrightarrow$ (cyclopentene bearing H_3C and $\underset{OH}{\underset{|}{CHCH_3}}$)

(e) 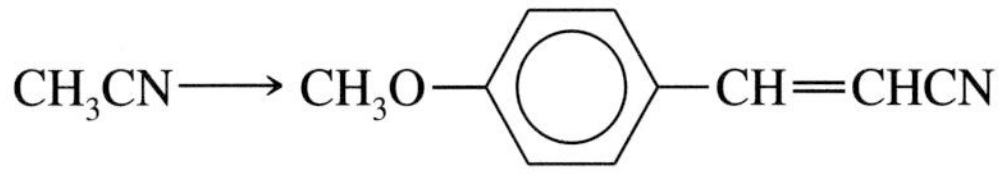 $CH_3CN \longrightarrow CH_3O-C_6H_4-CH{=}CHCN$

(f) $CH_3CH_2CH_2CH_2\overset{O}{\overset{\|}{C}}H \longrightarrow CH_3(CH_2)_3CH{=}\overset{CH_2OH}{\overset{|}{C}}(CH_2)_2CH_3$

(g) 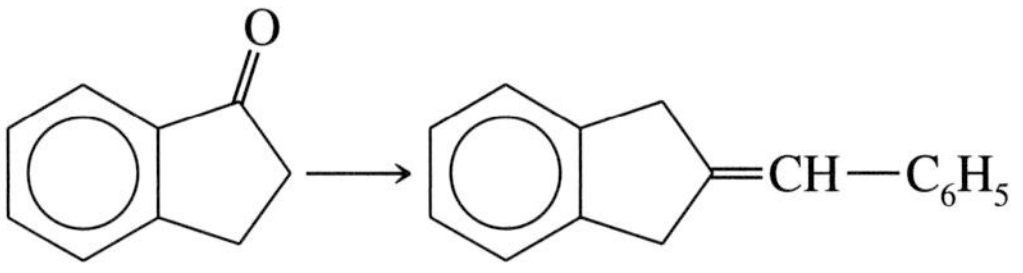 (1-indanone) $\longrightarrow$ (indane)$=CH-C_6H_5$

17.32 The following reaction illustrates the Robinson annulation reaction (Section 17.9B). Give mechanisms for the steps that occur.

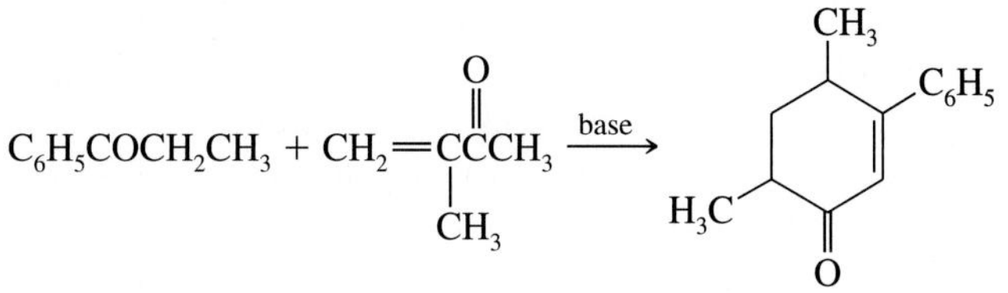

$$C_6H_5COCH_2CH_3 + CH_2{=}\underset{CH_3}{\underset{|}{C}}\overset{O}{\overset{\|}{C}}CH_3 \xrightarrow{\text{base}}$$ (cyclohexenone bearing CH_3, C_6H_5, and H_3C substituents)

17.33 Write structural formulas for **A**, **B**, and **C**.

$$HC{\equiv}CH \xrightarrow[\substack{(2)\ CH_3COCH_3 \\ (3)\ NH_4Cl/H_2O}]{(1)\ NaNH_2,\ \text{liq. } NH_3} \mathbf{A}\ (C_5H_8O) \xrightarrow[H_2O]{Hg^{2+},\ H_3O^+} \mathbf{B}\ (C_5H_{10}O_2) \xrightarrow{C_6H_5CHO,\ OH^-} \mathbf{C}\ (C_{12}H_{14}O_2)$$

17.34 The hydrogen atoms of the γ carbon of crotonaldehyde are appreciably acidic ($pK_a \cong 20$).

$$\overset{\gamma}{C}H_3\overset{\beta}{C}H{=}\overset{\alpha}{C}HCHO$$
Crotonaldehyde

(a) Write resonance structures that will explain this fact.
(b) Write a mechanism that accounts for the following reaction:

$$C_6H_5CH{=}CHCHO + CH_3CH{=}CHCHO \xrightarrow[EtOH]{\text{base}} C_6H_5(CH{=}CH)_3CHO$$
(87%)

17.35 What reagents would you use to bring about each step of the following syntheses?

(a) 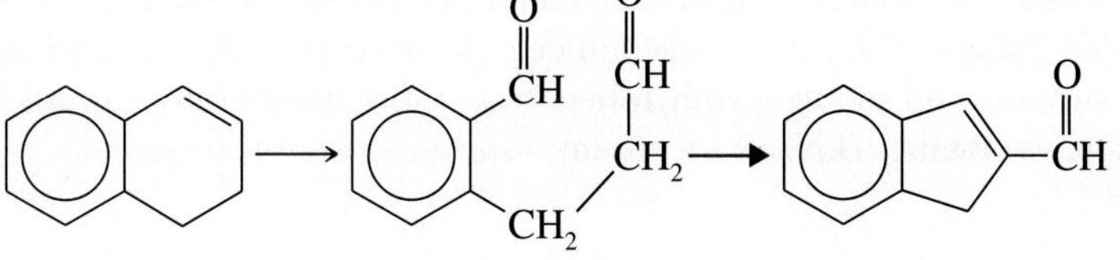 (1,2-dihydronaphthalene) $\longrightarrow$ (C_6H_4 bearing $\overset{O}{\overset{\|}{C}}H$ and $CH_2CH_2\overset{O}{\overset{\|}{C}}H$) $\longrightarrow$ (indene-2-carbaldehyde, $-\overset{O}{\overset{\|}{C}}H$)

(b) 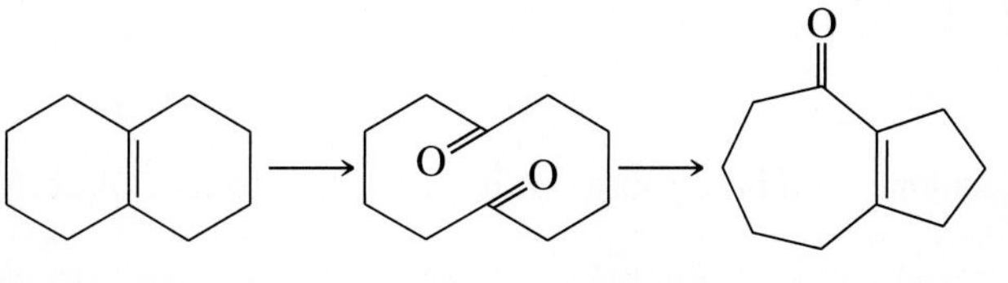 (bicyclic octahydronaphthalene) $\longrightarrow$ (cyclodecane-1,6-dione, O, O) $\longrightarrow$ (bicyclic enone, O)

(c) 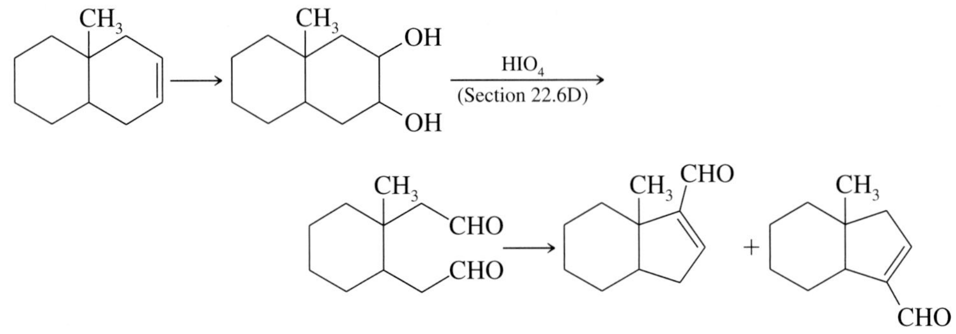 [decalin structures with CH_3] $\longrightarrow$ [diol: CH_3, OH, OH] $\xrightarrow[\text{(Section 22.6D)}]{HIO_4}$

[CH_3, CHO, CHO] $\longrightarrow$ [CH_3, CHO] + [CH_3, CHO]

(d) 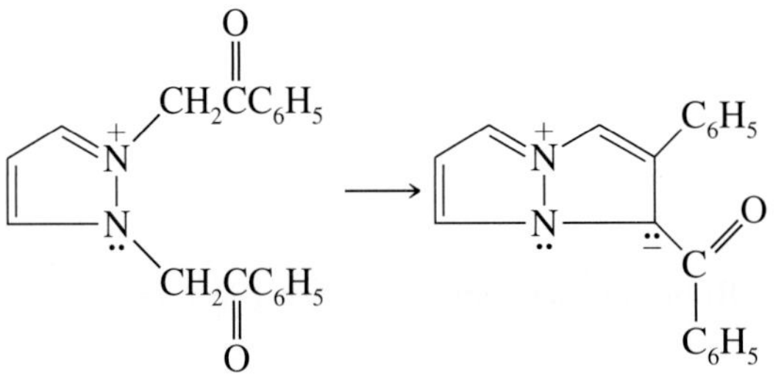 [pyrazolium with $CH_2\overset{O}{\overset{\|}{C}}C_6H_5$ substituents] $\longrightarrow$ [bicyclic product with C_6H_5, O, C_6H_5]

17.36 **(a)** Infrared spectroscopy provides an easy method for deciding whether the product obtained from the addition of a Grignard reagent to an α,β-unsaturated ketone is the simple addition product or the conjugate addition product. Explain. (What peak or peaks would you look for?)
(b) How might you follow the rate of the following reaction using UV spectroscopy?

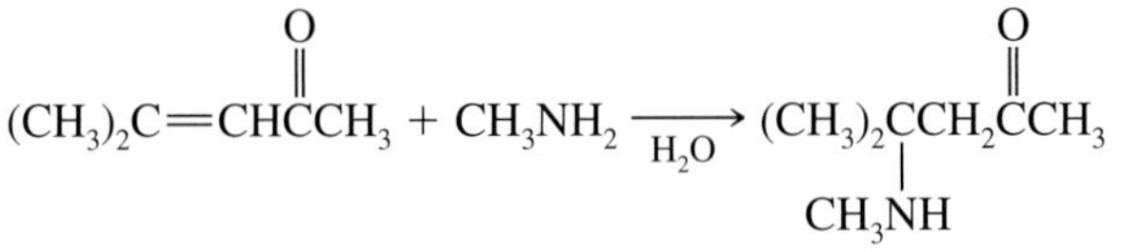

$$(CH_3)_2C{=}CH\overset{O}{\overset{\|}{C}}CH_3 + CH_3NH_2 \xrightarrow[H_2O]{} (CH_3)_2\underset{CH_3NH}{\underset{|}{C}}CH_2\overset{O}{\overset{\|}{C}}CH_3$$

17.37 **(a)** A compound **U** ($C_9H_{10}O$) gives a negative iodoform test. The IR spectrum of **U** shows a strong absorption peak at 1690 cm^{-1}. The ^1H NMR spectrum of **U** gives the following:

Triplet δ 1.2 (3H)
Quartet δ 3.0 (2H)
Multiplet δ 7.7 (5H)

What is the structure of **U**?

(b) A compound **V** is an isomer of **U**. Compound **V** gives a positive iodoform test; its IR spectrum shows a strong peak at 1705 cm^{-1}. The ^1H NMR spectrum of **V** gives the following:

Singlet δ 2.0 (3H)
Singlet δ 3.5 (2H)
Multiplet δ 7.1 (5H)

What is the structure of **V**?

17.38 Compound **A** has the molecular formula $C_6H_{12}O_3$ and shows a strong IR absorption peak at 1710 cm^{-1}. When treated with iodine in aqueous sodium hydroxide, **A** gives a yellow precipitate. When **A** is treated with Tollens' reagent, no reaction occurs; however, if **A** is treated first with water containing a drop of sulfuric acid and then with Tollens' reagent, a silver mirror forms in the test tube. Compound **A** shows the following ^1H NMR spectrum:

Singlet δ 2.1
Doublet δ 2.6
Singlet δ 3.2 (6H)
Triplet δ 4.7

Write a structure for **A**.

17.39 Treating a solution of *cis*-1-decalone with base causes an isomerization to take place. When the system reaches equilibrium, the solution is found to contain about 95% *trans*-1-decalone and about 5% *cis*-1-decalone. Explain this isomerization.

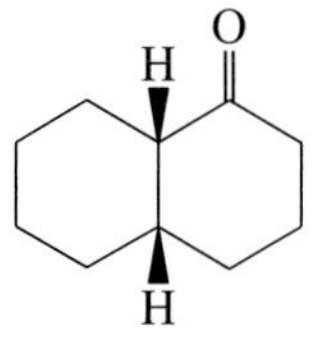

***cis*-1-Decalone**

17.40 The Wittig reaction (Section 16.10) can be used in the synthesis of aldehydes, for example,

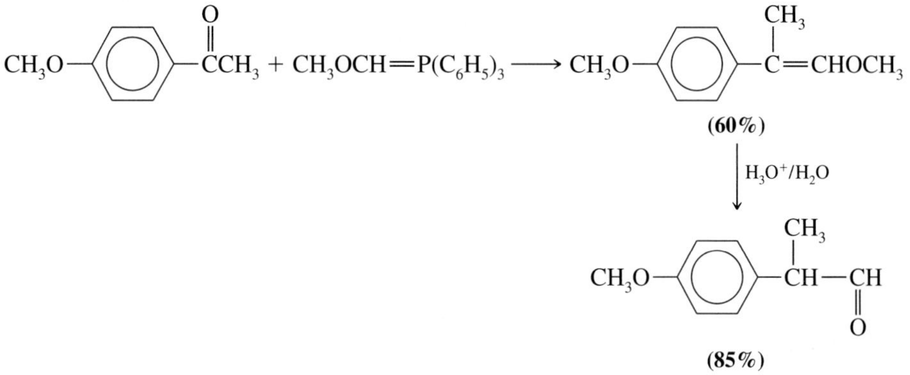

$$CH_3O-C_6H_4-\overset{O}{\overset{\|}{C}}CH_3 + CH_3OCH{=}P(C_6H_5)_3 \longrightarrow CH_3O-C_6H_4-\overset{CH_3}{\overset{|}{C}}{=}CHOCH_3 \quad \textbf{(60\%)}$$

$$\xrightarrow{H_3O^+/H_2O} CH_3O-C_6H_4-\overset{CH_3}{\overset{|}{C}}H-\underset{\underset{O}{\|}}{C}H \quad \textbf{(85\%)}$$

(a) How would you prepare $CH_3OCH{=}P(C_6H_5)_3$?
(b) Show with a mechanism how the second reaction produces an aldehyde.
(c) How would you use this method to prepare cyclohexyl–CHO from cyclohexanone?

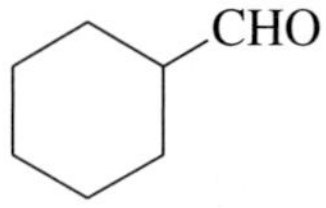

17.41 Aldehydes that have no α hydrogen undergo an intermolecular oxidation–reduction called the **Cannizzaro reaction** when they are treated with concentrated base. An example is the following reaction of benzaldehyde:

$$2\,C_6H_5{-}CHO \xrightarrow[H_2O]{OH^-} C_6H_5{-}CH_2OH + C_6H_5{-}CO_2^-$$

(a) When the reaction is carried out in D_2O, the benzyl alcohol that is isolated contains no deuterium bound to carbon. It is $C_6H_5CH_2OD$. What does this suggest about the mechanism for the reaction?
(b) When $(CH_3)_2CHCHO$ and $Ba(OH)_2/H_2O$ are heated in a sealed tube, the reaction produces only $(CH_3)_2CHCH_2OH$ and $[(CH_3)_2CHCO_2]_2Ba$. Provide an explanation for the formation of these products instead of those expected from an aldol reaction.

17.42 When the aldol reaction of acetaldehyde is carried out in D_2O, no deuterium is found in the methyl group of unreacted aldehyde. However, in the aldol reaction of acetone, deuterium is incorporated in the methyl group of the unreacted acetone. Explain this difference in behavior.

17.43 Shown below is a synthesis of the elm bark beetle pheromone, multistriatin (see Problem 16.48). Give structures for compounds **A**, **B**, **C**, and **D**.

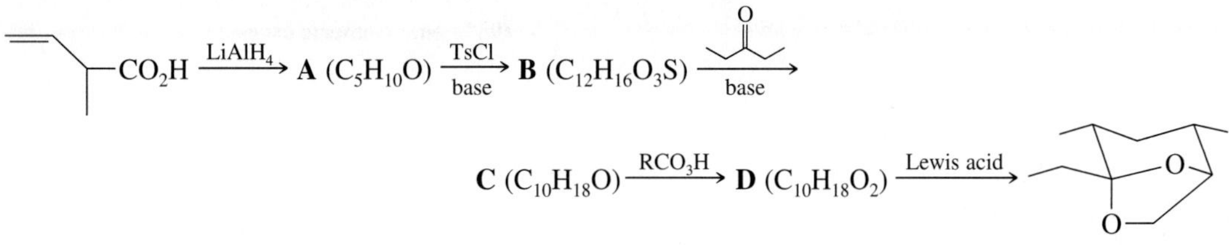

$$\text{(alkenyl)}{-}CO_2H \xrightarrow{LiAlH_4} \textbf{A}\ (C_5H_{10}O) \xrightarrow[\text{base}]{TsCl} \textbf{B}\ (C_{12}H_{16}O_3S) \xrightarrow[\text{base}]{\text{(3-pentanone)}}$$

$$\textbf{C}\ (C_{10}H_{18}O) \xrightarrow{RCO_3H} \textbf{D}\ (C_{10}H_{18}O_2) \xrightarrow{\text{Lewis acid}} \text{Multistriatin}$$

Multistriatin

17.44 Allowing acetone to react with 2 molar equivalents of benzaldehyde in the presence of KOH in ethanol leads to the formation of compound **X**. The ^{13}C NMR spectrum of **X** is given in Fig. 17.2. Propose a structure for compound **X**.

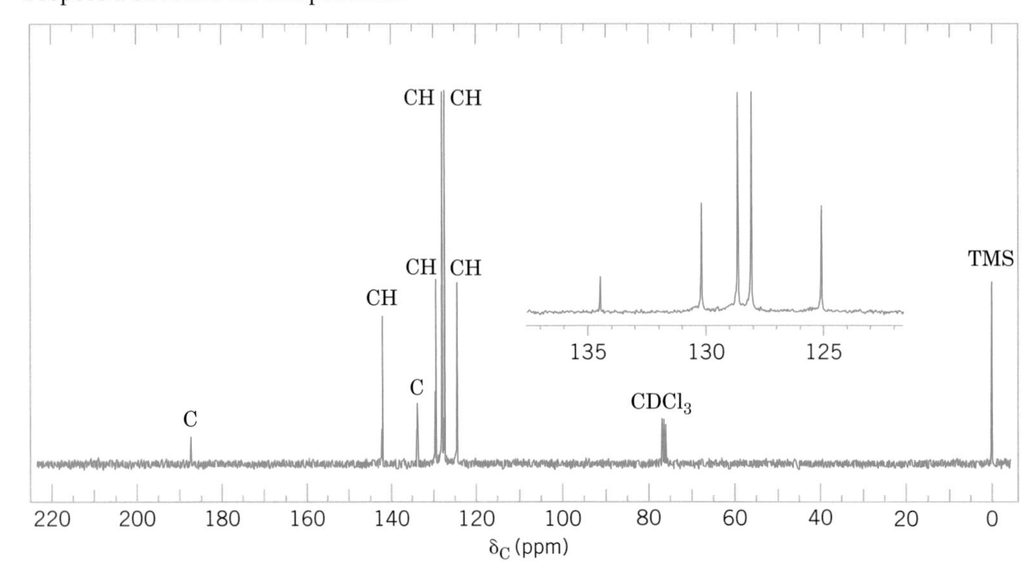

FIGURE 17.2 The broadband proton-decoupled ^{13}C NMR spectrum of compound X, Problem 17.44. Information from the DEPT ^{13}C NMR spectra is given above the peaks.

***17.45** The following is an example of a reaction sequence developed by Derin C. D'Amico and Michael E. Jung (UCLA) that results in enantiospecific formation of an aldol addition product by nonaldol reactions. The sequence includes a Horner–Wadsworth–Emmons reaction (Section 16.10), a Sharpless asymmetric epoxidation (Section 11.13), and a novel rearrangement that ultimately leads to the aldol-type product. Propose a mechanism for rearrangement of the epoxy alcohol under the conditions shown to form the aldol product. [*Hint:* The rearrangement can also be accomplished by preparing a trialkylsilyl ether from the epoxy alcohol in a separate reaction first and then treating the resulting silyl ether with a Lewis acid catalyst (e.g., BF_3).]

(1) $CH_3CHCO_2CH_3$ with $PO(OCH_3)_2$, NaH
(2) DIBAL-H

tert-BuOOH, Ti(O-*i*-Pr)$_4$, D-(−)-diisopropyl tartrate

94%
(95% enantiomeric excess)

TBDMSOTf (*tert*-butyldimethylsilyl triflate), 1.3 equiv.
diisopropylethylamine, 1.35 equiv., molecular sieves, −42°C

TBDMSO, O, H, OH, CH$_3$

LEARNING GROUP PROBLEMS

Steroids are an extremely important class of natural and pharmaceutical compounds. Synthetic efforts directed toward steroids have been underway for many years and continue to be an area of important research. The syntheses of cholesterol and cortisone by R. B. Woodward (Harvard University, recipient of the Nobel Prize in Chemistry for 1965) and co-workers represent paramount accomplishments in steroid synthesis. The following are selected reactions from Woodward's synthesis of cholesterol. This synthesis is rich with examples of carbonyl chemistry and other reactions we have studied.

(a) Name the type of reaction involved from **2** to **3**. Classify each reactant according to its role in the reaction.

(b) Write a mechanism for the reaction that occurs from **3** to **4**. The reaction can occur under either acidic or basic conditions.

(c) The reaction from **5** to **6** converts an enol ether to an enone by hydrolysis and dehydration. Write a mechanism for this process.

(d) Write a mechanism for the reaction from **7** to **8** (for clarification, other ways to write "EtO_2CH" are HCO_2Et, ethyl formate, and ethyl methanoate). Comment on why **8** exists in the enol form.

(e) What is the name of the reaction from **8** to **9**? Write a reaction mechanism for this reaction. [EVK (ethyl vinyl ketone) is a common abbreviation for pent-4-en-3-one.]

(f) What is the name of the type of reaction from **9** to **10**? Write a mechanism for this reaction.

(g) Note that you have studied the reactions that occur from **10** to **11** and from **11** to **12**. What functional group is formed in **12**?

(h) Write a mechanism for step (1) of the reaction between **14** and **15/16.** (The initial product of this step has a nitrile group where the carboxyl group is in **15** and **16**. In Chapter 18 we shall learn how to convert a nitrile to a carboxylic acid.) Comment on why a mixture of **15** and **16** is formed.

(i) Write a mechanism that explains the combination of steps between **17** and **19**.

(j) Name the type of reaction that occurs from **20** to **21** and write a mechanism.

(k) What reaction occurs in the step from **24** to **25**? Explain why a mixture of configurations results at the alcohol carbon.

(l) Write a mechanism for the reaction that converts the ketone shown immediately before **27** to **27** itself. (The abbreviation pyr stands for pyridine, a base.)

Cholesterol (1)

2 + (butadiene) —100°C→ **3** —(1) NaOH, (2) HCl⇌ **4** —$LiAlH_4$→ **5** (next page)

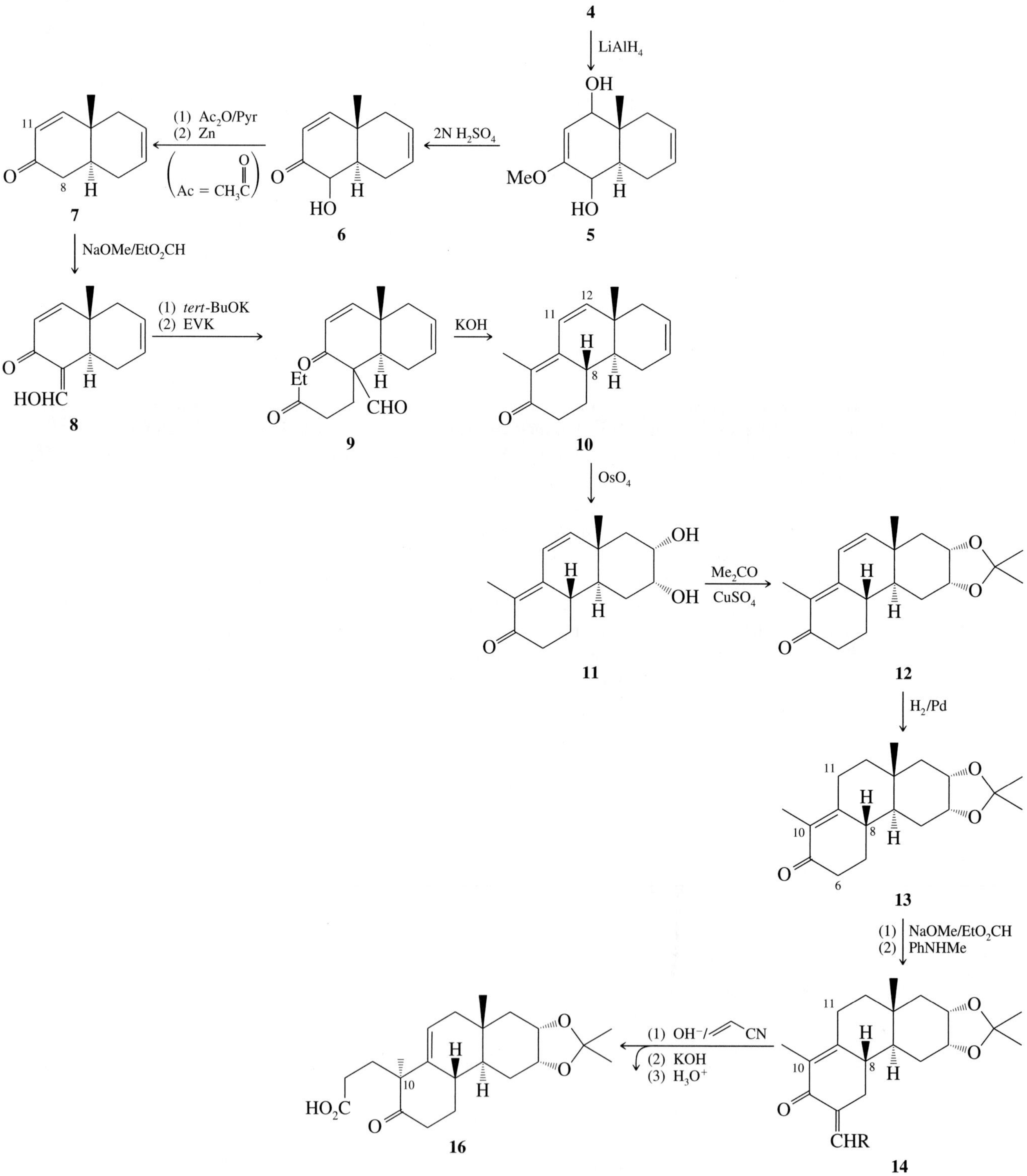

4

$LiAlH_4$

5

$2N\ H_2SO_4$

6

(1) Ac_2O/Pyr
(2) Zn
(Ac = $CH_3C(=O)$)

7

$NaOMe/EtO_2CH$

8

(1) *tert*-BuOK
(2) EVK

9

KOH

10

OsO_4

11

Me_2CO
$CuSO_4$

12

H_2/Pd

13

(1) $NaOMe/EtO_2CH$
(2) PhNHMe

14

(1) OH^-/$CH_2=CHCN$
(2) KOH
(3) H_3O^+

16

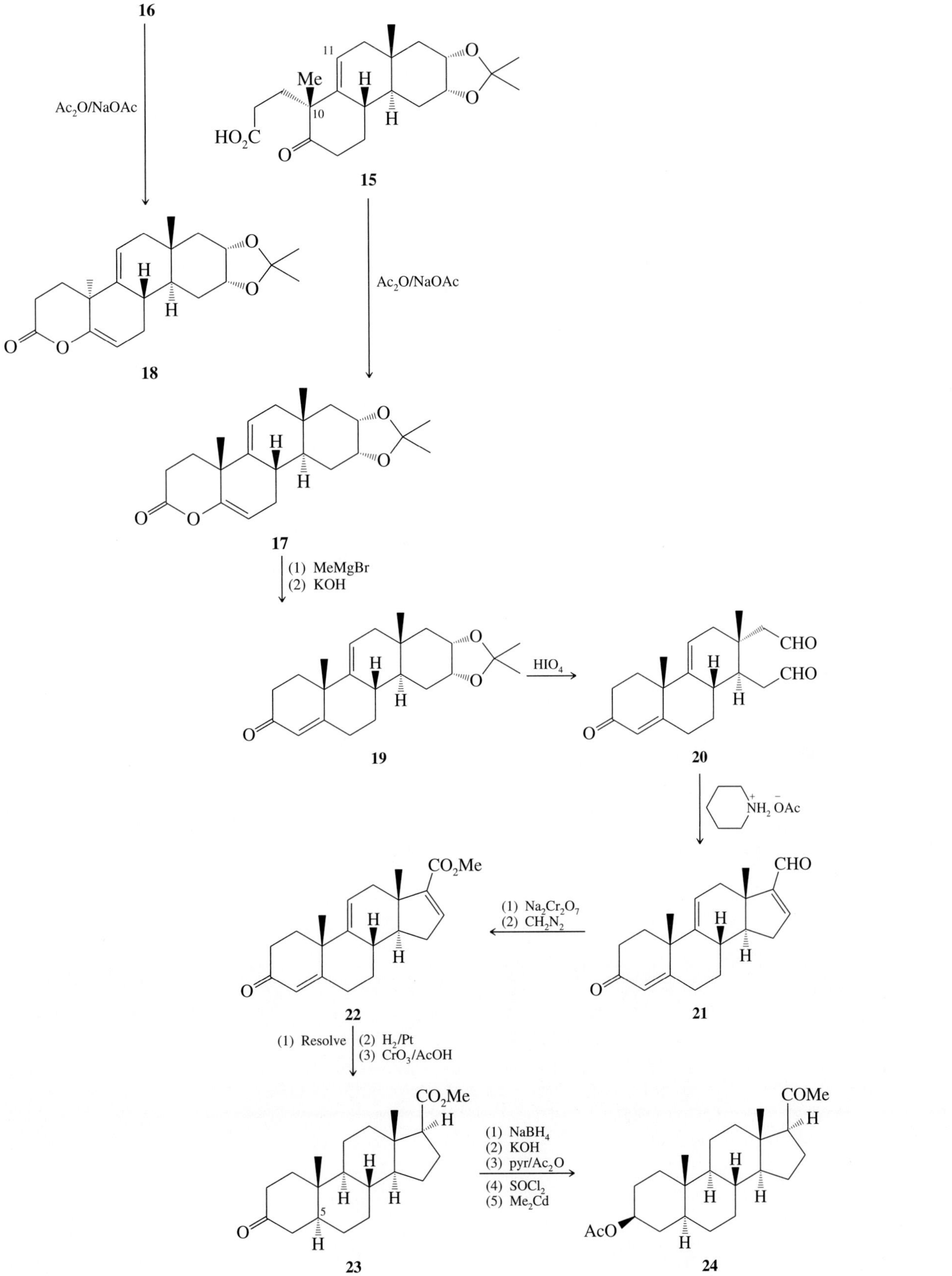

16

$Ac_2O/NaOAc$

11

Me

H

10

HO_2C

O

O

O

H

15

$Ac_2O/NaOAc$

H

O

O

H

O

O

18

H

O

O

H

O

O

17

(1) MeMgBr
(2) KOH

H

O

O

H

O

19

HIO_4

CHO

H

CHO

H

O

20

$\overset{+}{N}H_2 \overset{-}{O}Ac$

CO_2Me

H

H

O

22

(1) $Na_2Cr_2O_7$
(2) CH_2N_2

CHO

H

H

O

21

(1) Resolve (2) H_2/Pt
(3) $CrO_3/AcOH$

CO_2Me

H

H

H

H

5

O

H

23

(1) $NaBH_4$
(2) KOH
(3) pyr/Ac_2O
(4) $SOCl_2$
(5) Me_2Cd

COMe

H

H

H

H

AcO

H

24

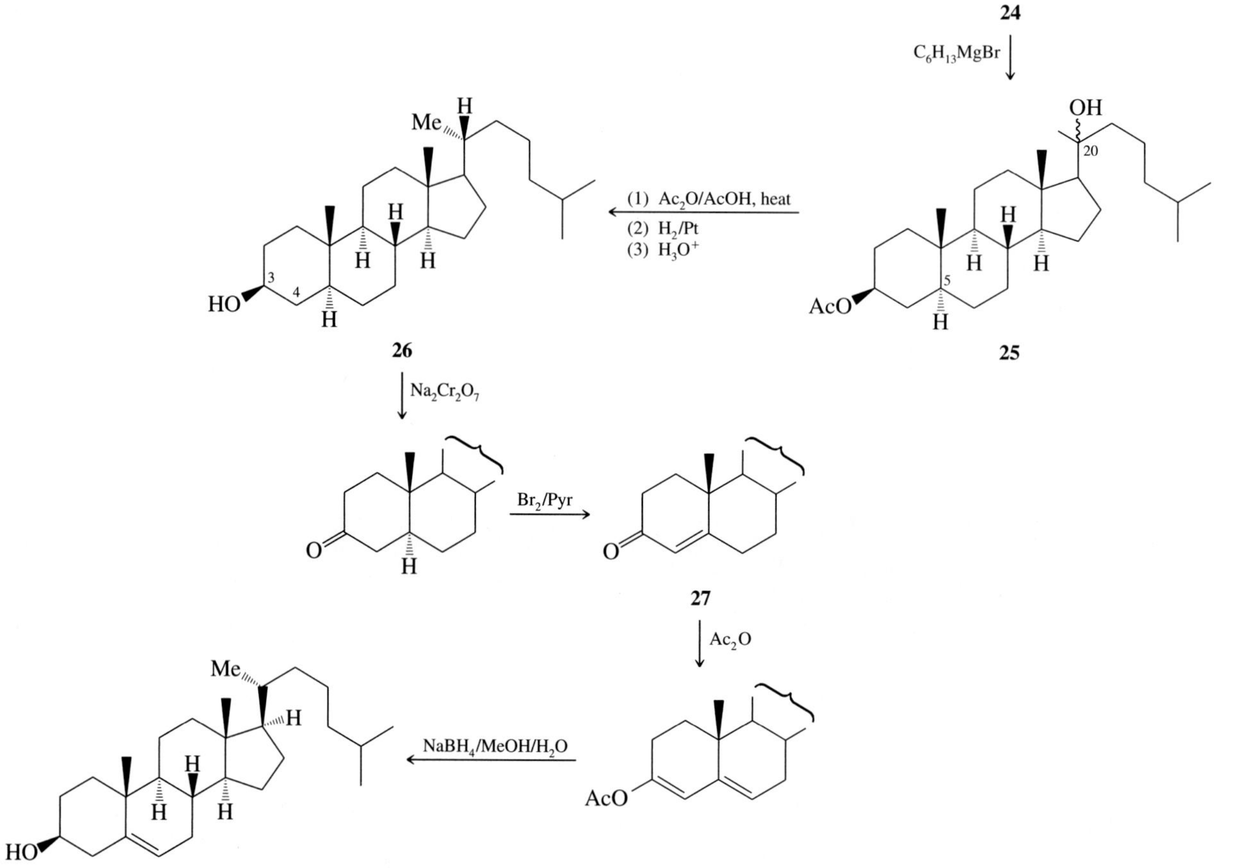

24

$C_6H_{13}MgBr$

25

(1) $Ac_2O/AcOH$, heat
(2) H_2/Pt
(3) H_3O^+

26

$Na_2Cr_2O_7$

Br_2/Pyr

27

Ac_2O

$NaBH_4/MeOH/H_2O$

Cholesterol (1)

Carboxylic Acids and Their Derivatives. Nucleophilic Addition–Elimination at the Acyl Carbon

A Common Bond

Polyesters, nylon, and many biological molecules share a common aspect of bond formation during their synthesis. This process is called acyl transfer, and it involves creation of a bond by nucleophilic addition and elimination at a carbonyl group. Acyl transfer reactions occur every moment of every day in our bodies as we biosynthesize proteins, fats, precursors to steroids, and other molecules and as we degrade food molecules to provide energy and biosynthetic raw materials. Acyl transfer reactions are used virtually nonstop in industry as well. Approximately 3 billion pounds of nylon and 4 billion pounds of polyester fibers are made by acyl transfer reactions every year. The photo above shows the process of nylon fiber production and the inset molecular graphic is a portion of a nylon 6,6 polymer.

The functional groups of acyl transfer reactions all relate to carboxylic acids. They include acyl chlorides, anhydrides, esters, amides, thioesters, carboxylic acids themselves, and others that we shall study in this chapter. In Special Topic B we shall see how acyl transfer reactions are used to synthesize polymers such as nylon and Mylar. In Special Topic D we shall consider the biosynthesis of fatty acids and other biological molecules by acyl transfer reactions. Although many functional groups participate in acyl transfer reactions, their reactions are all readily understandable because of the common mechanistic theme that unites them—nucleophilic addition–elimination at an acyl carbon.

18.1 Introduction

The carboxyl group, $-C(=O)OH$ (abbreviated $—CO_2H$ or $—COOH$), is one of the most widely occurring functional groups in chemistry and biochemistry. Not only are carboxylic acids themselves important, but the carboxyl group is the parent group of a large family of related compounds called **acyl compounds** or **carboxylic acid derivatives,** shown in Table 18.1.

TABLE 18.1 Carboxylic Acid Derivatives

Structure	Name	Structure	Name
$RC(=O)Cl$	Acyl (or acid) chloride	$RC(=O)NH_2$	Amide
$RC(=O)OC(=O)R'$	Acid anhydride	$RC(=O)NHR'$	Amide
$RC(=O)OR'$	Ester	$RC(=O)NR'R''$	Amide
$R—C{\equiv}N$	Nitrile		

18.2 Nomenclature and Physical Properties

18.2A Carboxylic Acids

The International Union of Pure and Applied Chemistry (IUPAC) systematic or substitutive names for carboxylic acids are obtained by dropping the final *-e* of the name of the alkane corresponding to the longest chain in the acid and by adding *-oic acid.* The carboxyl carbon atom is assigned number 1. The examples listed here illustrate how this is done:

$$\overset{6}{C}H_3\overset{5}{C}H_2\overset{4}{C}H(CH_3)\overset{3}{C}H_2\overset{2}{C}H_2\overset{O}{\overset{\|}{C}}OH$$

4-Methylhexanoic acid

$$\overset{6}{C}H_3\overset{5}{C}H{=}\overset{4}{C}H\overset{3}{C}H_2\overset{2}{C}H_2\overset{O}{\overset{\|}{C}}OH$$

4-Hexenoic acid
(or hex-4-enoic acid)

Many carboxylic acids have common names that are derived from Latin or Greek words that indicate one of their natural sources (Table 18.2). Methanoic acid is called formic acid (Latin: *formica,* ant). Ethanoic acid is called acetic acid (Latin: *acetum,* vinegar). Butanoic acid is one compound responsible for the odor of rancid butter, so its common name is butyric acid (Latin: *butyrum,* butter). Pentanoic acid, as a result of its occurrence in valerian, a perennial herb, is named valeric acid. Hexanoic acid is one compound associated with the odor of goats, hence its common name, caproic acid (Latin: *caper,* goat). Octadecanoic acid takes its common name, stearic acid, from the Greek word *stear,* for tallow.

TABLE 18.2 Carboxylic Acids

Structure	Systematic Name	Common Name	mp (°C)	bp (°C)	Water Solubility (g 100 mL^{-1} H_2O), 25°C	pK_a
HCO_2H	Methanoic acid	Formic acid	8	100.5	∞	3.75
CH_3CO_2H	Ethanoic acid	Acetic acid	16.6	118	∞	4.76
$CH_3CH_2CO_2H$	Propanoic acid	Propionic acid	−21	141	∞	4.87
$CH_3(CH_2)_2CO_2H$	Butanoic acid	Butyric acid	−6	164	∞	4.81
$CH_3(CH_2)_3CO_2H$	Pentanoic acid	Valeric acid	−34	187	4.97	4.82
$CH_3(CH_2)_4CO_2H$	Hexanoic acid	Caproic acid	−3	205	1.08	4.84
$CH_3(CH_2)_6CO_2H$	Octanoic acid	Caprylic acid	16	239	0.07	4.89
$CH_3(CH_2)_8CO_2H$	Decanoic acid	Capric acid	31	269	0.015	4.84
$CH_3(CH_2)_{10}CO_2H$	Dodecanoic acid	Lauric acid	44	179^{18}	0.006	5.30
$CH_3(CH_2)_{12}CO_2H$	Tetradecanoic acid	Myristic acid	59	200^{20}	0.002	
$CH_3(CH_2)_{14}CO_2H$	Hexadecanoic acid	Palmitic acid	63	219^{17}	0.0007	6.46
$CH_3(CH_2)_{16}CO_2H$	Octadecanoic acid	Stearic acid	70	383	0.0003	
CH_2ClCO_2H	Chloroethanoic acid	Chloroacetic acid	63	189	Very soluble	2.86
$CHCl_2CO_2H$	Dichloroethanoic acid	Dichloroacetic acid	10.8	192	Very soluble	1.48
CCl_3CO_2H	Trichloroethanoic acid	Trichloroacetic acid	56.3	198	Very soluble	0.70
$CH_3CHClCO_2H$	2-Chloropropanoic acid	α-Chloropropionic acid		186	Soluble	2.83
$CH_2ClCH_2CO_2H$	3-Chloropropanoic acid	β-Chloropropionic acid	61	204	Soluble	3.98
$C_6H_5CO_2H$	Benzoic acid	Benzoic acid	122	250	0.34	4.19
p-$CH_3C_6H_4CO_2H$	4-Methylbenzoic acid	*p*-Toluic acid	180	275	0.03	4.36
p-$ClC_6H_4CO_2H$	4-Chlorobenzoic acid	*p*-Chlorobenzoic acid	242		0.009	3.98
p-$NO_2C_6H_4CO_2H$	4-Nitrobenzoic acid	*p*-Nitrobenzoic acid	242		0.03	3.41
CO_2H	1-Naphthoic acid	α-Naphthoic acid	160	300	Insoluble	3.70
CO_2H	2-Naphthoic acid	β-Naphthoic acid	185	>300	Insoluble	4.17

Valerian is a source of valeric acid.

Most of these common names have been with us for a long time and some are likely to remain in common usage for even longer, so it is helpful to be familiar with them. In this text we shall always refer to methanoic acid and ethanoic acid as formic acid and acetic acid, respectively. However, in almost all other instances we shall use IUPAC systematic or substitutive names.

Carboxylic acids are polar substances. Their molecules can form strong hydrogen bonds with each other and with water. As a result, carboxylic acids generally have high boiling points, and low-molecular-weight carboxylic acids show appreciable solubility in water. The first four carboxylic acids (Table 18.2) are miscible with water in all proportions. As the length of the carbon chain increases, water solubility declines.

18.2B Carboxylate Salts

Salts of carboxylic acids are named as *-ates;* in both common and systematic names, *-ate* replaces *-ic acid.* Thus, CH_3CO_2Na is sodium acetate or sodium ethanoate.

Sodium and potassium salts of most carboxylic acids are readily soluble in water. This is true even of the long-chain carboxylic acids. Sodium or potassium salts of long-chain carboxylic acids are the major ingredients of soap (see Section 23.2C).

PROBLEM 18.1 Give an IUPAC systematic name for each of the following:

(a) $CH_3CH_2CHCO_2H$ (with CH_3 on the CH carbon)

(b) $CH_3CH{=}CHCH_2CO_2H$

(c) $BrCH_2CH_2CH_2CO_2Na$

(d) $C_6H_5CH_2CH_2CH_2CH_2CO_2H$

(e) $CH_3CH{=}CCH_2CO_2H$ (with CH_3 on the C carbon)

PROBLEM 18.2 Experiments show that the molecular weight of acetic acid in the vapor state (just above its boiling point) is approximately 120. Explain the discrepancy between this experimental value and the true value of approximately 60.

18.2C Acidity of Carboxylic Acids

Solubility tests such as these are rapid and useful ways of classifying unknown compounds.

Most unsubstituted carboxylic acids have K_a values in the range of 10^{-4}–10^{-5} (pK_a = 4–5) as seen in Table 18.2. The pK_a of water is about 16, and the apparent pK_a of H_2CO_3 is about 7. These relative acidities mean that carboxylic acids react readily with aqueous solutions of sodium hydroxide and sodium bicarbonate to form soluble sodium salts. We can use solubility tests, therefore, to distinguish water-insoluble carboxylic acids from water-insoluble phenols (Chapter 21) and alcohols. Water-insoluble carboxylic acids dissolve in either aqueous sodium hydroxide or aqueous sodium bicarbonate:

$$C_6H_5COOH + NaOH \xrightarrow{H_2O} C_6H_5CO^-Na^+ + H_2O$$

Benzoic acid (water insoluble) *Stronger acid* + *Stronger base* → **Sodium benzoate (water soluble)** *Weaker base* + *Weaker acid*

$$C_6H_5COOH + NaHCO_3 \xrightarrow{H_2O} C_6H_5CO^-Na^+ + \underbrace{CO_2\uparrow + H_2O}_{H_2CO_3}$$

(water insoluble) *Stronger acid* + *Stronger base* → **(water soluble)** *Weaker base* + **H_2CO_3** *Weaker acid*

Water-insoluble phenols (Section 21.5) dissolve in aqueous sodium hydroxide but (except for some nitrophenols) do not dissolve in aqueous sodium bicarbonate. Water-insoluble alcohols do not dissolve in either aqueous sodium hydroxide or sodium bicarbonate.

We see in Table 18.2 that carboxylic acids having electron-withdrawing groups are stronger than unsubstituted acids. The chloroacetic acids, for example, show the following order of acidities:

$$\mathrm{Cl_3C{-}CO_2H > Cl_2CH{-}CO_2H > ClCH_2{-}CO_2H > H_3C{-}CO_2H}$$

	$\mathrm{Cl_3CCO_2H}$	$\mathrm{Cl_2CHCO_2H}$	$\mathrm{ClCH_2CO_2H}$	$\mathrm{CH_3CO_2H}$
pK_a	0.70	1.48	2.86	4.76

As we saw in Sections 3.5B and 3.11, this acid-strengthening effect of electron-withdrawing groups arises from a combination of inductive effects and entropy effects. We can visualize inductive charge delocalization when we compare the electrostatic potential maps for carboxylate anions of acetic acid and trichloroacetetic acid in Fig. 18.1. The maps show more negative charge localized near the acetate carboxyl group than the trichloroacetate carboxyl group. Delocalization of the negative charge in trichloroacetate by the electron-withdrawing effect of its three chlorine atoms contributes to its being a stronger acid than acetic acid. (In general, the less localized the charge, the more stable the conjugate base and the stronger the acid.)

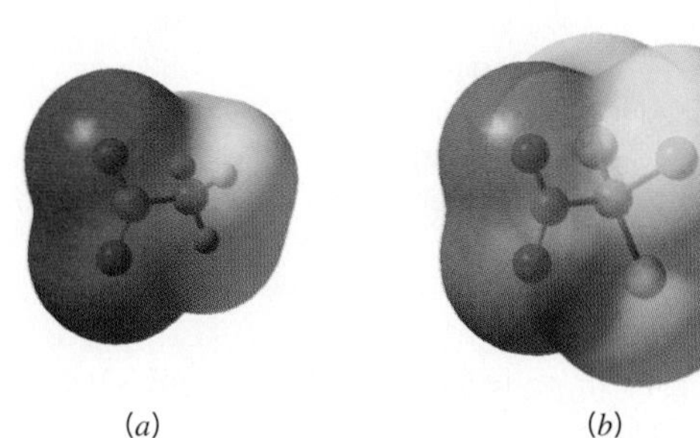

FIGURE 18.1 Electrostatic potential maps for the carboxylate anions of *(a)* acetic acid and *(b)* trichloroacetic acid. There is greater delocalization of negative charge in trichloroacetate than acetate due to the inductive electron-withdrawing effect of the three chlorine atoms in trichloroacetate.

Since inductive effects are not transmitted very effectively through covalent bonds, the acid-strengthening effect decreases as the distance between the electron-withdrawing group and the carboxyl group increases. Of the chlorobutanoic acids that follow, the strongest acid is 2-chlorobutanoic acid:

$$\mathrm{CH_3{-}CH_2{-}CHCl{-}C({=}O){-}OH} \qquad \mathrm{CH_3{-}CHCl{-}CH_2{-}C({=}O){-}OH} \qquad \mathrm{ClCH_2{-}CH_2{-}CH_2{-}C({=}O){-}OH}$$

2-Chlorobutanoic acid	**3-Chlorobutanoic acid**	**4-Chlorobutanoic acid**
($pK_a = 2.85$)	($pK_a = 4.05$)	($pK_a = 4.50$)

PROBLEM 18.3

Which acid of each pair shown here would you expect to be stronger?

(a) CH_3CO_2H or CH_2FCO_2H
(b) CH_2FCO_2H or CH_2ClCO_2H
(c) CH_2ClCO_2H or CH_2BrCO_2H
(d) $CH_2FCH_2CH_2CO_2H$ or $CH_3CHFCH_2CO_2H$
(e) $CH_3CH_2CHFCO_2H$ or $CH_3CHFCH_2CO_2H$
(f) 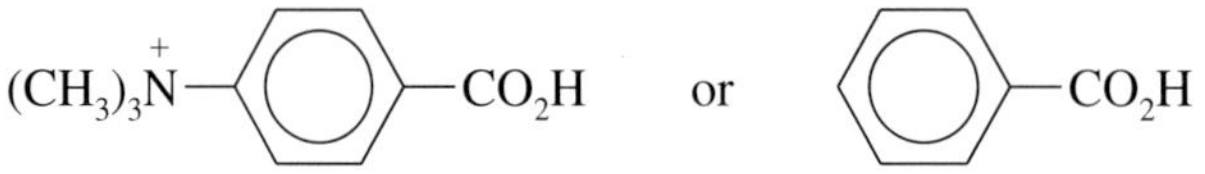 $(CH_3)_3\overset{+}{N}-C_6H_4-CO_2H$ (para) or $C_6H_5-CO_2H$
(g) 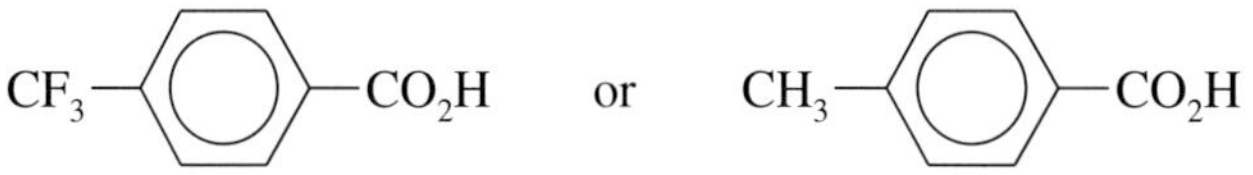 $CF_3-C_6H_4-CO_2H$ (para) or $CH_3-C_6H_4-CO_2H$ (para)

18.2D Dicarboxylic Acids

Dicarboxylic acids are named as **alkanedioic acids** in the IUPAC systematic or substitutive system. Most simple dicarboxylic acids have common names (Table 18.3), and these are the names that we shall use.

Succinic and fumaric acids are key metabolites in the citric acid pathway. Adipic acid is used in the synthesis of nylon. The isomers of phthalic acid are used in making polyesters. See Special Topic B for further information on polymers.

TABLE 18.3 Dicarboxylic Acids

Structure	Common Name	mp (°C)	pK_a (at 25°C) pK_1	pK_2
HO_2C-CO_2H	Oxalic acid	189 dec	1.2	4.2
$HO_2CCH_2CO_2H$	Malonic acid	136	2.9	5.7
$HO_2C(CH_2)_2CO_2H$	Succinic acid	187	4.2	5.6
$HO_2C(CH_2)_3CO_2H$	Glutaric acid	98	4.3	5.4
$HO_2C(CH_2)_4CO_2H$	Adipic acid	153	4.4	5.6
cis-$HO_2C-CH=CH-CO_2H$	Maleic acid	131	1.9	6.1
trans-$HO_2C-CH=CH-CO_2H$	Fumaric acid	287	3.0	4.4
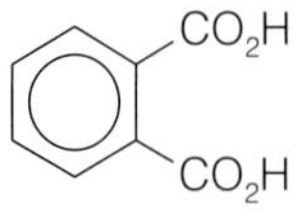 $C_6H_4(CO_2H)_2$ (1,2-)	Phthalic acid	206–208 dec	2.9	5.4
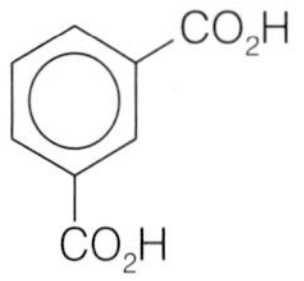 $C_6H_4(CO_2H)_2$ (1,3-)	Isophthalic acid	345–348	3.5	4.6
 $C_6H_4(CO_2H)_2$ (1,4-)	Terephthalic acid	Sublimes	3.5	4.8

PROBLEM 18.4

Suggest an explanation for the following facts: **(a)** The pK_1 for all of the dicarboxylic acids in Table 18.3 is smaller than the pK_a for monocarboxylic acids with the same number of carbon atoms. **(b)** The difference between pK_1 and pK_2 for dicarboxylic acids of type $HO_2C(CH_2)_nCO_2H$ decreases as n increases.

18.2E Esters

The names of esters are derived from the names of the alcohol (with the ending **-yl**) and the acid (with the ending **-ate** or **-oate**). The portion of the name derived from the alcohol comes first:

$$\mathrm{CH_3\overset{O}{\overset{\|}{C}}{-}OCH_2CH_3}$$

Ethyl acetate **or** **ethyl** ethanoate

$$\mathrm{CH_3CH_2\overset{O}{\overset{\|}{C}}{-}O\underset{CH_3}{\overset{CH_3}{C}}{-}CH_3}$$

***tert*-Butyl** propanoate

$$\mathrm{CH_3\overset{O}{\overset{\|}{C}}OCH{=}CH_2}$$

Vinyl acetate **or** **ethenyl** ethanoate

$$\mathrm{Cl{-}C_6H_4{-}\overset{O}{\overset{\|}{C}}OCH_3}$$

Methyl ***p*-chlorobenzoate**

$$\mathrm{CH_3CH_2O\overset{O}{\overset{\|}{C}}CH_2\overset{O}{\overset{\|}{C}}OCH_2CH_3}$$

Diethyl malonate

Esters are polar compounds, but, lacking a hydrogen attached to oxygen, their molecules cannot form strong hydrogen bonds to each other. As a result, esters have boiling points that are lower than those of acids and alcohols of comparable molecular weight. The boiling points (Table 18.4) of esters are about the same as those of comparable aldehydes and ketones.

TABLE 18.4 Carboxylic Esters

Name	Structure	mp (°C)	bp (°C)	Solubility in Water (g 100 mL^{-1} at 20°C)
Methyl formate	HCO_2CH_3	−99	31.5	Very soluble
Ethyl formate	$HCO_2CH_2CH_3$	−79	54	Soluble
Methyl acetate	$CH_3CO_2CH_3$	−99	57	24.4
Ethyl acetate	$CH_3CO_2CH_2CH_3$	−82	77	7.39 (25°C)
Propyl acetate	$CH_3CO_2CH_2CH_2CH_3$	−93	102	1.89
Butyl acetate	$CH_3CO_2CH_2(CH_2)_2CH_3$	−74	125	1.0 (22°C)
Ethyl propanoate	$CH_3CH_2CO_2CH_2CH_3$	−73	99	1.75
Ethyl butanoate	$CH_3(CH_2)_2CO_2CH_2CH_3$	−93	120	0.51
Ethyl pentanoate	$CH_3(CH_2)_3CO_2CH_2CH_3$	−91	145	0.22
Ethyl hexanoate	$CH_3(CH_2)_4CO_2CH_2CH_3$	−68	168	0.063
Methyl benzoate	$C_6H_5CO_2CH_3$	−12	199	0.15
Ethyl benzoate	$C_6H_5CO_2CH_2CH_3$	−35	213	0.08
Phenyl acetate	$CH_3CO_2C_6H_5$		196	Slightly soluble
Methyl salicylate	*o*-$HOC_6H_4CO_2CH_3$	−9	223	0.74 (30°C)

Unlike the low-molecular-weight acids, esters usually have pleasant odors, some resembling those of fruits, and these are used in the manufacture of synthetic flavors:

$$\mathrm{CH_3\overset{O}{\overset{\|}{C}}OCH_2CH_2\underset{CH_3}{\underset{|}{C}}HCH_3}$$

Isopentyl acetate **(used in synthetic banana flavor)**

$$\mathrm{CH_3CH_2CH_2CH_2\overset{O}{\overset{\|}{C}}{-}OCH_2CH_2\underset{CH_3}{\underset{|}{C}}HCH_3}$$

Isopentyl pentanoate **(used in synthetic apple flavor)**

18.2F Carboxylic Anhydrides

Most anhydrides are named by dropping the word **acid** from the name of the carboxylic acid and then adding the word **anhydride:**

Acetic anhydride (ethanoic anhydride) mp −73°C — $(CH_3CO)_2O$

Succinic anhydride mp 121°C

Phthalic anhydride mp 131°C

Maleic anhydride mp 53°C

18.2G Acyl Chlorides

Acyl chlorides are also called **acid chlorides.** They are named by dropping **-ic acid** from the name of the acid and then adding **-yl chloride.** Examples are

CH_3COCl — **Acetyl chloride (ethanoyl chloride)** mp −112°C; bp 51°C

CH_3CH_2COCl — **Propanoyl chloride** mp −94°C; bp 80°C

C_6H_5COCl — **Benzoyl chloride** mp −1°C; bp 197°C

Acyl chlorides and carboxylic anhydrides have boiling points in the same range as esters of comparable molecular weight.

18.2H Amides

Amides that have no substituent on nitrogen are named by dropping **-ic acid** from the common name of the acid (or *-oic acid* from the substitutive name) and then adding **-amide.** Alkyl groups on the nitrogen atom of amides are named as substituents, and the named substituent is prefaced by *N*- or *N,N*-. Examples are

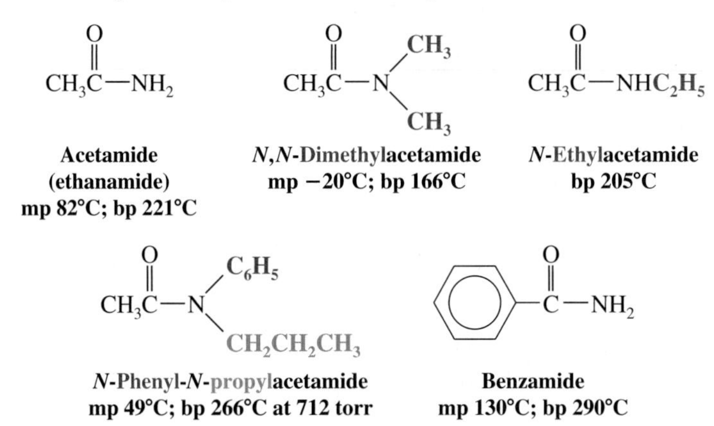

Amides with nitrogen atoms bearing one or two hydrogen atoms are able to form strong hydrogen bonds to each other and, consequently, such amides have high melting points and boiling points. On the other hand, molecules of *N,N*-disubstituted amides cannot form strong hydrogen bonds to each other, and they have lower melting points and boiling points. The melting and boiling data given above illustrate this trend.

Hydrogen bonding between amide groups plays a key role in the way proteins and peptides fold to achieve their overall shape (Chapter 24). Proteins and peptides (short proteins) are polymers of amino acids joined by amide groups. One feature common to the structure of many proteins is the β sheet, shown below:

Hydrogen bonding (red dots) between amide molecules

Hydrogen bonding between amide groups of peptide chains. This interaction between chains (called a β sheet) is important to the structure of many proteins.

18.2I Nitriles

Carboxylic acids can be converted to nitriles and vice versa. In IUPAC substitutive nomenclature, acyclic nitriles are named by adding the suffix *-nitrile* to the name of the corresponding hydrocarbon. The carbon atom of the $-C{\equiv}N$ group is assigned number 1. Additional examples of nitriles were presented in Section 2.12 with other functional groups of organic molecules. The name acetonitrile is an acceptable common name for CH_3CN and acrylonitrile is an acceptable common name for $CH_2{=}CHCN$:

$$\overset{2}{C}H_3-\overset{1}{C}{\equiv}N{:} \qquad \overset{3}{C}H_2{=}\overset{2}{C}H-\overset{1}{C}{\equiv}N{:}$$

Ethanenitrile (acetonitrile) **Propenenitrile (acrylonitrile)**

PROBLEM 18.5

Write structural formulas for the following:

(a) Methyl propanoate
(b) Ethyl *p*-nitrobenzoate
(c) Dimethyl malonate
(d) *N,N*-Dimethylbenzamide
(e) Pentanenitrile
(f) Dimethyl phthalate
(g) Dipropyl maleate
(h) *N,N*-Dimethylformamide
(i) 2-Bromopropanoyl bromide
(j) Diethyl succinate

18.2J Spectroscopic Properties of Acyl Compounds

IR Spectra Infrared spectroscopy is of considerable importance in identifying carboxylic acids and their derivatives. The C=O stretching band is one of the most prominent in their IR spectra since it is always a strong band. The C=O stretching band occurs at different frequencies for acids, esters, and amides, and its precise location is often helpful in structure determination. Figure 18.2 gives the location of this band for most acyl compounds. Notice that conjugation shifts the location of the C=O absorption to lower frequencies.

Infrared spectroscopy is a highly useful tool for classifying acyl compounds.

The hydroxyl groups of carboxylic acids also give rise to a broad peak in the 2500–3100-cm^{-1} region arising from O—H stretching vibrations. The N—H stretching vibrations of amides absorb between 3140 and 3500 cm^{-1}. Presence or absence of an O—H or N—H absorption can be an important clue as to which carbonyl functional group is present in an unknown compound.

Functional Group	Approximate Frequency Range (cm^{-1})	1840 1820 1800 1780 1760 1740 1720 1700 1680 1660 1640 1620 1600
Acid chloride	1815–1785 1800–1770 (conj.)	*
Acid anhydride	1820–1750 1775–1720 (conj.)	(Two C=O absorptions)
Ester/Lactone	1750–1735 1730–1715 (conj.)	Also C–O (1300–1000); no O–H absorption
Carboxylic acid	~1760 or 1720–1705 1710–1680 (conj.)	(monomer) (dimer) Also C–O (1315–1280) and O–H (~3300, broad)
Aldehyde	1740–1720 1710–1685 (conj.)	Also C–H (2830–2695)
Ketone	1720–1710 1685–1665 (conj.)	
Amide/lactam	1650–1640	(solid) (solution)
Carboxylate salt	1650–1550	(Two C=O absorptions)

*Orange bars represent absorption ranges for conjugated species.

FIGURE 18.2 Approximate carbonyl IR absorption frequencies Frequency ranges based on Silverstein, R. M.; Webster, F. X. *Spectrometric Identification of Organic Compounds,* 6th ed.; Wiley: New York, 1998.

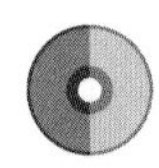

See IR Tutor on the CD for interpreted IR spectra and animated vibrations of heptanoic acid, ethyl acetate, butyric anhydride, and octanenitrile.

Figure 18.3 shows an annotated spectrum of propanoic acid. Nitriles show an intense and characteristic infrared absorption band near 2250 cm^{-1} that arises from stretching of the carbon–nitrogen triple bond.

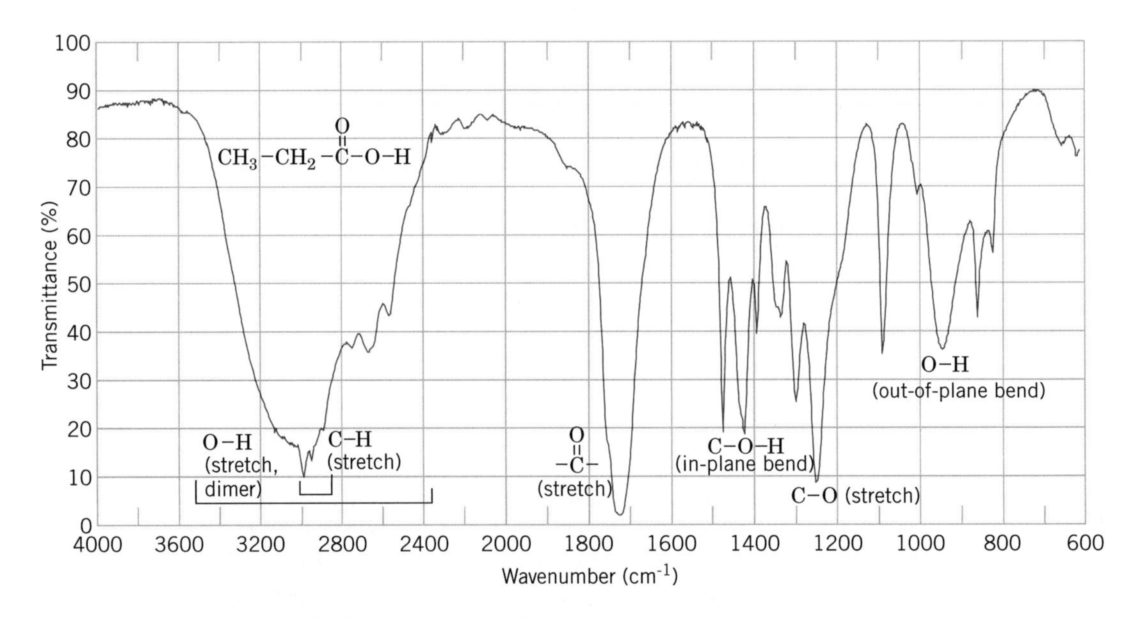

FIGURE 18.3 The infrared spectrum of propanoic acid.

^{1}H NMR Spectra The acidic protons of carboxylic acids are highly deshielded and absorb far downfield in the δ 10–12 region. The protons of the α carbon of carboxylic acids absorb in the δ 2.0–2.5 region. Figure 18.4 gives an annotated ^{1}H NMR spectrum of an ester, methyl propanoate; it shows the normal splitting pattern (quartet and triplet) of an ethyl group, and, as we would expect, it shows an unsplit methyl group.

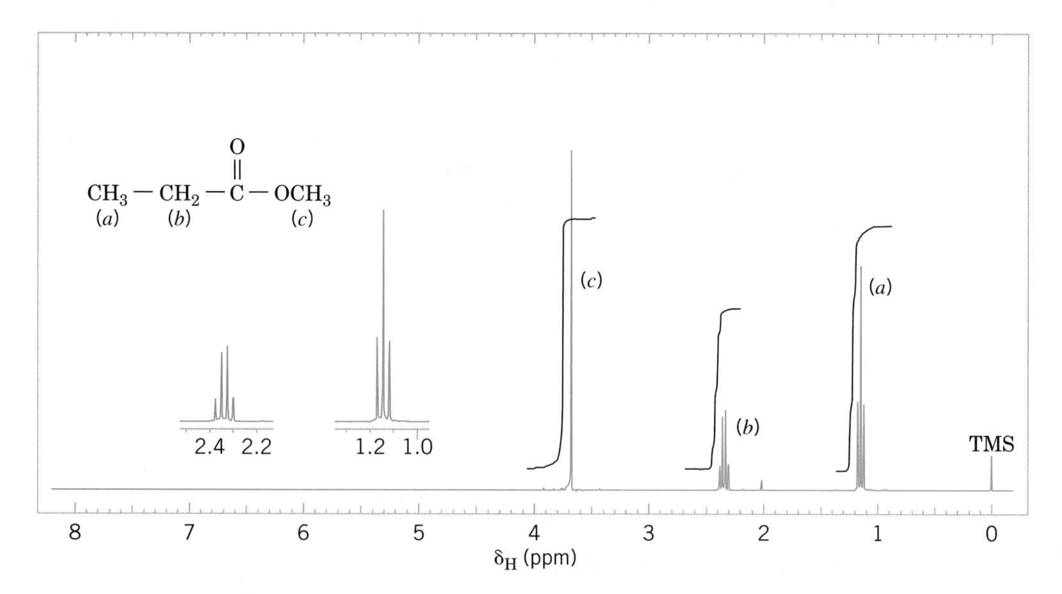

FIGURE 18.4 The 300-MHz ^{1}H NMR spectrum of methyl propanoate. Expansions of the signals are shown in the offset plots.

^{13}C NMR Spectra The carbonyl carbon of carboxylic acids and their derivatives occurs far downfield in the δ 160–180 region (see the following examples). The nitrile carbon is not shifted so far downfield and absorbs in the δ 115–120 region. Notice, however, that the downfield shift for these derivatives is not as much as for aldehydes and ketones (δ 180–220):

$H_3C{-}C(=O){-}OH$	$H_3C{-}C(=O){-}OCH_2CH_3$	$H_3C{-}C(=O){-}Cl$	$H_3C{-}C(=O){-}NH_2$	$H_3C{-}C{\equiv}N$
δ 177.2	δ 170.7	δ 170.3	δ 172.6	δ 117.4

^{13}C NMR chemical shifts for the carbonyl or nitrile carbon atom

The carbon atoms of the alkyl groups of carboxylic acids and their derivatives have ^{13}C chemical shifts much further upfield. The chemical shifts for each carbon of pentanoic acid are as follows:

$H_3C{-}CH_2{-}CH_2{-}CH_2{-}C(=O){-}OH$

δ 13.5 22.0 27.0 34.1 179.7

^{13}C NMR chemical shifts

18.3 Preparation of Carboxylic Acids

Most of the methods for the preparation of carboxylic acids have been presented previously:

1. **By oxidation of alkenes.** We learned in Section 8.17 that alkenes can be oxidized to carboxylic acids with hot alkaline $KMnO_4$:

$$RCH{=}CHR' \xrightarrow[\text{heat; (2) } H_3O^+]{\text{(1) } KMnO_4,\ OH^-} RCO_2H + R'CO_2H$$

Alternatively, ozonides (Section 8.17A) can be subjected to an oxidative workup that yields carboxylic acids:

$$RCH{=}CHR' \xrightarrow[\text{(2) } H_2O_2]{\text{(1) } O_3} RCO_2H + R'CO_2H$$

2. **By oxidation of aldehydes and primary alcohols.** Aldehydes can be oxidized to carboxylic acids with mild oxidizing agents such as $Ag(NH_3)_2{}^+OH^-$ (Section 16.12). Primary alcohols can be oxidized with $KMnO_4$. Aldehydes and primary alcohols are oxidized to carboxylic acids with chromic acid (H_2CrO_4) in aqueous acetone (the Jones oxidation; Section 12.4C).

$$R{-}CHO \xrightarrow[\text{(2) } H_3O^+]{\text{(1) } Ag_2O \text{ or } Ag(NH_3)_2{}^+OH^-} RCO_2H$$

$$RCH_2OH \xrightarrow[\text{heat; (2) } H_3O^+]{\text{(1) } KMnO_4,\ OH^-} RCO_2H$$

$$R{-}CHO \text{ or } RCH_2OH \xrightarrow{H_2CrO_4} RCO_2H$$

3. **By oxidation of alkylbenzenes.** Primary and secondary alkyl groups (but not 3° groups) directly attached to a benzene ring are oxidized by $KMnO_4$ to a $-CO_2H$ group (Section 15.13C):

$$C_6H_5{-}CH_3 \xrightarrow[\text{heat; (2) } H_3O^+]{\text{(1) } KMnO_4,\ OH^-} C_6H_5{-}CO_2H$$

4. **By oxidation of the benzene ring.** The benzene ring of an alkylbenzene can be converted to a carboxyl group by ozonolysis, followed by treatment with hydrogen peroxide (Section 15.13D):

$$R{-}C_6H_5 \xrightarrow[\text{(2) } H_2O_2]{\text{(1) } O_3,\ CH_3CO_2H} R{-}\overset{O}{\overset{\|}{C}}OH$$

5. **By oxidation of methyl ketones.** Methyl ketones can be converted to carboxylic acids via the haloform reaction (Section 17.3C):

$$Ar{-}\overset{O}{\overset{\|}{C}}{-}CH_3 \xrightarrow[\text{(2) } H_3O^+]{\text{(1) } X_2/NaOH} Ar{-}\overset{O}{\overset{\|}{C}}{-}OH + CHX_3$$

6. **By hydrolysis of cyanohydrins and other nitriles.** We saw, in Section 16.9, that aldehydes and ketones can be converted to **cyanohydrins** and that these can be hydrolyzed to α-hydroxy acids. In the hydrolysis the $-CN$ group is converted to a $-CO_2H$ group. The mechanism of nitrile hydrolysis is discussed in Section 18.8H:

$$RR'C{=}O + HCN \rightleftharpoons RR'C(OH)CN \xrightarrow[H_2O]{HA} R{-}\underset{R'}{\underset{|}{\overset{OH}{\overset{|}{C}}}}{-}CO_2H$$

Nitriles can also be prepared by nucleophilic substitution reactions of alkyl halides with sodium cyanide. Hydrolysis of the nitrile yields a carboxylic acid *with a chain one carbon atom longer* than the original alkyl halide:

General Reaction

$$R{-}CH_2X + CN^- \longrightarrow RCH_2CN \xrightarrow[H_2O,\ heat]{HA} RCH_2CO_2H + NH_4^+$$

$$RCH_2CN \xrightarrow[H_2O,\ heat]{OH^-} RCH_2CO_2^- + NH_3$$

Specific Examples

$$HOCH_2CH_2Cl \xrightarrow[(80\%)]{NaCN} HOCH_2CH_2CN \xrightarrow[(2)\ H_3O^+\ (75\text{–}80\%)]{(1)\ OH^-,\ H_2O} HOCH_2CH_2CO_2H$$

3-Hydroxypropanenitrile → **3-Hydroxypropanoic acid**

$$BrCH_2CH_2CH_2Br \xrightarrow[(77\text{–}86\%)]{NaCN} NCCH_2CH_2CH_2CN \xrightarrow[(77\text{–}86\%)]{H_3O^+} HO_2CCH_2CH_2CH_2CO_2H$$

Pentanenitrile → **Glutaric acid**

This synthetic method is generally limited to the use of *primary alkyl halides.* The cyanide ion is a relatively strong base, and the use of a secondary or tertiary alkyl halide leads primarily to an alkene (through elimination) rather than to a nitrile (through substitution). Aryl halides (except for those with ortho and para nitro groups) do not react with sodium cyanide.

7. **By carbonation of Grignard reagents.** Grignard reagents react with carbon dioxide to yield magnesium carboxylates. Acidification produces carboxylic acids:

$$R{-}X + Mg \xrightarrow[Et_2O]{} RMgX \xrightarrow{CO_2} RCO_2MgX \xrightarrow{H_3O^+} RCO_2H$$

or

$$Ar{-}Br + Mg \xrightarrow[Et_2O]{} ArMgBr \xrightarrow{CO_2} ArCO_2MgBr \xrightarrow{H_3O^+} ArCO_2H$$

This synthesis of carboxylic acids is applicable to primary, secondary, tertiary, allyl, benzyl, and aryl halides, provided they have no groups incompatible with a Grignard reaction (see Section 12.8B):

$$(CH_3)_3C{-}Cl \xrightarrow[Et_2O]{Mg} (CH_3)_3CMgCl \xrightarrow[(2)\ H_3O^+]{(1)\ CO_2} (CH_3)_3CCO_2H$$

***tert*-Butyl chloride** → **2,2-Dimethylpropanoic acid (79–80% overall)**

$$CH_3CH_2CH_2CH_2Cl \qquad CH_3CH_2CH_2CH_2MgCl \xrightarrow[(2)\ H_3O^+]{(1)\ CO_2} CH_3CH_2CH_2CH_2CO_2H$$

Butyl chloride → **Pentanoic acid (80% overall)**

$$C_6H_5{-}Br \xrightarrow[Et_2O]{Mg} C_6H_5{-}MgBr \xrightarrow[(2)\ H_3O^+]{(1)\ CO_2} C_6H_5{-}CO_2H$$

Benzoic acid (85%)

PROBLEM 18.6 Show how each of the following compounds could be converted to benzoic acid:
(a) Ethylbenzene **(c)** Acetophenone **(e)** Benzyl alcohol
(b) Bromobenzene **(d)** Phenylethene (styrene) **(f)** Benzaldehyde

PROBLEM 18.7 Show how you would prepare each of the following carboxylic acids through a Grignard synthesis:
(a) Phenylacetic acid **(c)** 3-Butenoic acid **(e)** Hexanoic acid
(b) 2,2-Dimethylpentanoic acid **(d)** 4-Methylbenzoic acid

PROBLEM 18.8 **(a)** Which of the carboxylic acids in Problem 18.7 could be prepared by a nitrile synthesis as well? **(b)** Which synthesis, Grignard or nitrile, would you choose to prepare $HOCH_2CH_2CH_2CH_2CO_2H$ from $HOCH_2CH_2CH_2CH_2Br$? Why?

18.4 NUCLEOPHILIC ADDITION–ELIMINATION AT THE ACYL CARBON

In our study of carbonyl compounds in Chapter 17, we saw that a characteristic reaction of aldehydes and ketones is one of *nucleophilic addition* to the carbon–oxygen double bond:

$$R'(R)C{=}\ddot{O}: + Nu:^- \rightleftharpoons Nu{-}C(R')(R){-}\ddot{O}:^- \xrightleftharpoons{H-Nu} Nu{-}C(R')(R){-}\ddot{O}{-}H + Nu:^-$$

Aldehyde or ketone

Nucleophilic addition

As we study carboxylic acids and their derivatives, we shall find that their reactions are characterized by **nucleophilic addition–elimination** at their acyl (carbonyl) carbon atoms. Key to this mechanism is formation of a **tetrahedral intermediate** that returns to a carbonyl group after the elimination. We shall encounter many reactions of this general type, as shown in the following box.

A Mechanism for the Reaction

Acyl Transfer by Nucleophilic Addition–Elimination

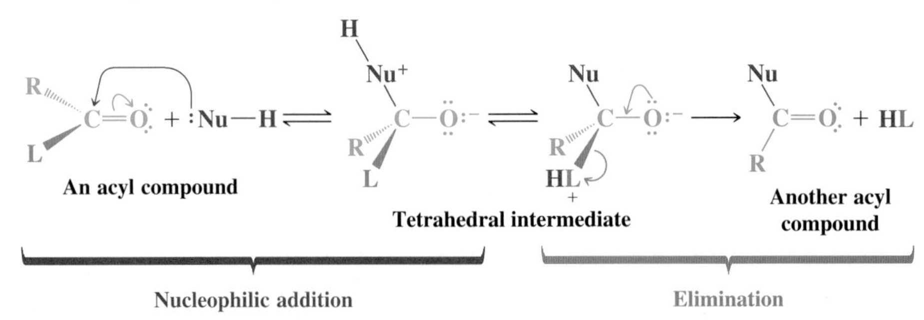

Many reactions like this occur in living organisms, and biochemists call them **acyl transfer reactions.** Acetyl-coenzyme A, discussed in Special Topic D, often serves as a biochemical acyl transfer agent. Acyl transfer reactions are of tremendous importance in industry as well, as described in the chapter opening essay and Special Topic B.

Although the final results obtained from the reactions of acyl compounds with nucleophiles (substitutions) differ from those obtained from aldehydes and ketones (additions), the two reactions have one characteristic in common: *The initial step in both reactions involves nucleophilic addition at the carbonyl carbon atom.* With both groups of compounds, this initial attack is facilitated by the same factors: the relative steric openness of the carbonyl carbon atom and the ability of the carbonyl oxygen atom to accommodate an electron pair of the carbon–oxygen double bond.

If you bear in mind the general mechanism for acyl transfer, you will see the common theme among reactions in this chapter.

It is after the initial nucleophilic attack that the two reactions differ. The tetrahedral intermediate formed from an aldehyde or ketone usually accepts a proton to form a stable addition product. In contrast, the intermediate formed from an acyl compound usually *eliminates* a leaving group (**L** in the mechanism above); this **elimination** leads to regeneration of the carbon–oxygen double bond and to a *substitution product.* The overall process in the case of **acyl substitution** occurs, therefore, by a **nucleophilic addition–elimination** mechanism.

Acyl compounds react as they do because they all have good, or reasonably good, leaving groups (or they can be protonated to form good leaving groups) attached to the carbonyl carbon atom. An acyl chloride, for example, generally reacts by losing *a chloride ion*—a very weak base and thus a very good leaving group. The reaction of an acyl chloride with water is an example.

Acyl substitution requires a leaving group at the carbonyl carbon.

Specific Example

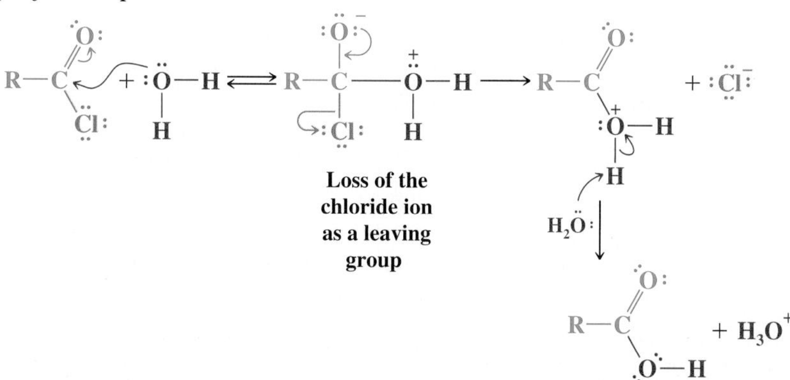

An acid anhydride generally reacts by losing *a carboxylate anion* or a molecule of a *carboxylic acid*—both are weak bases and good leaving groups. We consider the mechanism of acid anhydride hydrolysis in Section 18.6B.

As we shall see later, esters generally undergo nucleophilic addition–elimination by losing a molecule of an *alcohol* (Section 18.7B), acids react by losing a molecule of *water* (Section 18.7A), and amides react by losing a molecule of *ammonia* or of an *amine* (Section 18.8F). All of the molecules lost in these reactions are weak bases and are reasonably good leaving groups.

For an aldehyde or ketone to react by nucleophilic addition–elimination, the tetrahedral intermediate would need to eject a hydride ion (H:$^-$) or an alkanide ion (R:$^-$). Both are *very powerful bases,* and both are therefore *very poor leaving groups:*

R—C(O$^-$)(H)—Nu ⇏ R—C(=O)—Nu + **H:$^-$ Hydride ion**

R—C(O$^-$)(R)—Nu ⇏ R—C(=O)—Nu + **R:$^-$ Alkanide ion**

These reactions rarely occur.

[The haloform reaction (Section 17.3C) is one of the rare instances in which an alkanide anion can act as a leaving group, but then only because the leaving group is a weakly basic trihalomethyl anion.]

18.4A Relative Reactivity of Acyl Compounds

Of the acid derivatives that we study in this chapter, acyl chlorides are the most reactive toward nucleophilic addition–elimination, and amides are the least reactive. In general, the overall order of reactivity is

$$\underset{\text{Acyl chloride}}{R{-}\overset{O}{\overset{\|}{C}}{-}Cl} > \underset{\text{Acid anhydride}}{R{-}\overset{O}{\overset{\|}{C}}{-}O{-}\overset{O}{\overset{\|}{C}}{-}R'} > \underset{\text{Ester}}{R{-}\overset{O}{\overset{\|}{C}}{-}OR'} > \underset{\text{Amide}}{R{-}\overset{O}{\overset{\|}{C}}{-}NH_2}$$

The green groups in the structures above can be related to the green **L** group in the Mechanism for the Reaction box at the beginning of Section 18.4.

The general order of reactivity of acid derivatives can be explained by taking into account the basicity of the leaving groups. When acyl chlorides react, the leaving group is a *chloride ion.* When acid anhydrides react, the leaving group is a carboxylic acid or a carboxylate ion. When esters react, the leaving group is an alcohol, and when amides react, the leaving group is an amine (or ammonia). Of all of these bases, chloride ions are the *weakest bases* and acyl chlorides are the *most reactive* acyl compounds. Amines (or ammonia) are the *strongest bases* and so amides are the *least reactive* acyl compounds.

18.4B Synthesis of Acid Derivatives

Synthesis of acid derivatives by acyl transfer requires that the reactant have a better leaving group at the acyl carbon than the product.

As we begin now to explore the syntheses of carboxylic acid derivatives, we shall find that in many instances one acid derivative can be synthesized through a nucleophilic addition–elimination reaction of another. The order of reactivities that we have presented gives us a clue as to which syntheses are practical and which are not. In general, ***less reactive acyl compounds can be synthesized from more reactive ones, but the reverse is usually difficult and, when possible, requires special reagents.***

18.5 Acyl Chlorides

18.5A Synthesis of Acyl Chlorides

Since acyl chlorides are the most reactive of the acid derivatives, we must use special reagents to prepare them. We use other acid chlorides, *the acid chlorides of inorganic acids:* We use PCl_5 (an acid chloride of phosphoric acid), PCl_3 (an acid chloride of phosphorous acid), and $SOCl_2$ (an acid chloride of sulfurous acid).

All of these reagents react with carboxylic acids to give acyl chlorides in good yield:

General Reactions

$$RCOOH + \underset{\textbf{Thionyl chloride}}{SOCl_2} \longrightarrow R{-}C({=}O){-}Cl + SO_2 + HCl$$

$$3\,RCOOH + \underset{\textbf{Phosphorus trichloride}}{PCl_3} \longrightarrow 3\,RCOCl + H_3PO_3$$

$$RCOOH + \underset{\textbf{Phosphorus pentachloride}}{PCl_5} \longrightarrow RCOCl + POCl_3 + HCl$$

These reactions all involve nucleophilic addition–elimination by a chloride ion on a highly reactive intermediate: a protonated acyl chlorosulfite, a protonated acyl chlorophosphite, or a protonated acyl chlorophosphate. These intermediates contain even better acyl leaving groups than the acyl chloride product. Thionyl chloride, for example, reacts with a carboxylic acid in the following way.

A Mechanism for the Reaction

Synthesis of Acyl Chlorides Using Thionyl Chloride

$$RCOOH + SOCl_2 \longrightarrow RC(=O){-}\overset{+}{O}(H){-}S(Cl)_2{-}O^- \longrightarrow RC(=O){-}\overset{+}{O}(H){-}S(Cl){=}O$$

$$\underset{\textbf{Protonated acyl chlorosulfite}}{RC(=O){-}\overset{+}{O}(H){-}S(Cl){=}O} + Cl^- \longrightarrow R{-}C(O^-)(Cl){-}\overset{+}{O}(H){-}S(Cl){=}O \longrightarrow RCOCl + [HO{-}S(Cl){=}O] \longrightarrow HCl + SO_2$$

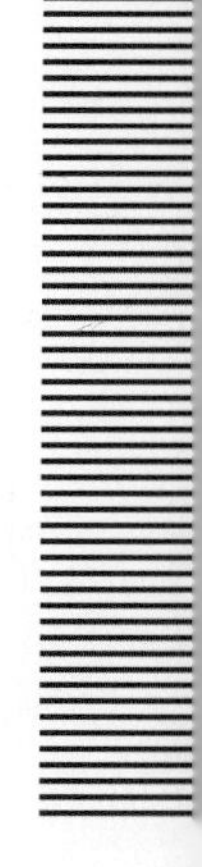

18.5B Reactions of Acyl Chlorides

Because acyl chlorides are the most reactive of the acyl derivatives, they are easily converted to less reactive ones. Many times, therefore, the best synthetic route to an anhydride, an ester, or an amide will involve an initial synthesis of the acyl chloride from the acid and then conversion of the acyl chloride to the desired acid derivative. The scheme given in Fig. 18.5 illustrates how this can be done. We examine these reactions in detail in Sections 18.6–18.8.

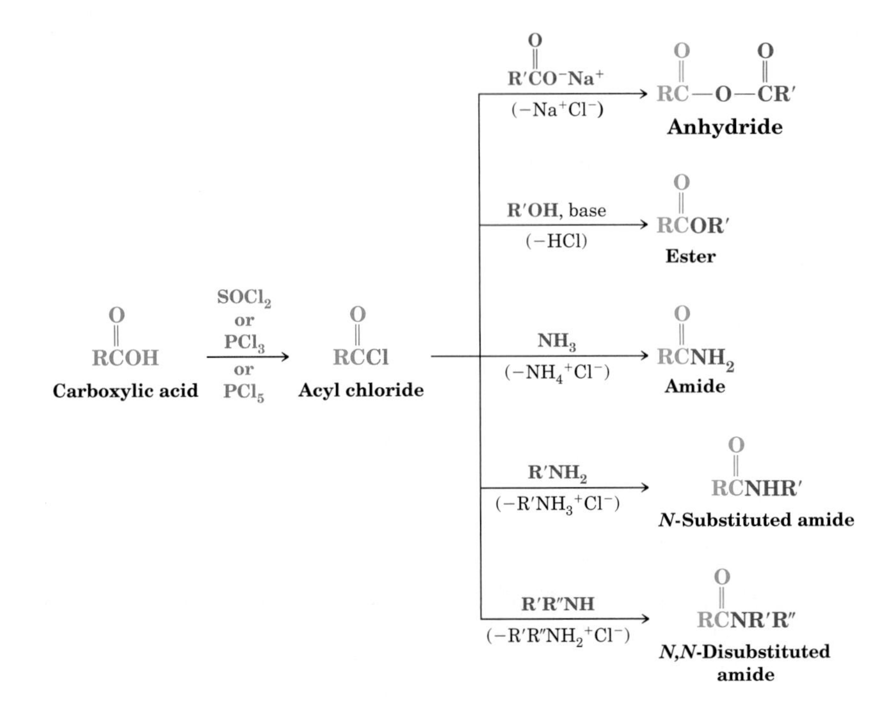

FIGURE 18.5 Preparation of an acyl chloride and reactions of acyl chlorides.

Acyl chlorides also react with water and (even more rapidly) with aqueous base, but these reactions are usually not carried out deliberately because they destroy the useful acyl chloride reactant by regenerating either the carboxylic acid or its salt:

$$\underset{\text{Acyl chloride}}{\text{RCCl}(=\text{O})} \xrightarrow{H_2O} \text{RC}(=\text{O})\text{OH} + \text{HCl}$$

$$\underset{\text{Acyl chloride}}{\text{RCCl}(=\text{O})} \xrightarrow{OH^-/H_2O} \text{RC}(=\text{O})\text{O}^- + \text{Cl}^-$$

18.6 Carboxylic Acid Anhydrides

18.6A Synthesis of Carboxylic Acid Anhydrides

Carboxylic acids react with acyl chlorides in the presence of pyridine to give carboxylic acid anhydrides:

$$\text{RC}(=\text{O})\text{OH} + \text{R}'\text{C}(=\text{O})\text{Cl} + \text{pyridine} \longrightarrow \text{RC}(=\text{O})\text{OC}(=\text{O})\text{R}' + \text{pyridinium (N}^+\text{–H)}\ \text{Cl}^-$$

This method is frequently used in the laboratory for the preparation of anhydrides. The method is quite general and can be used to prepare mixed anhydrides ($R \neq R'$) or symmetric anhydrides ($R = R'$).

Sodium salts of carboxylic acids also react with acyl chlorides to give anhydrides:

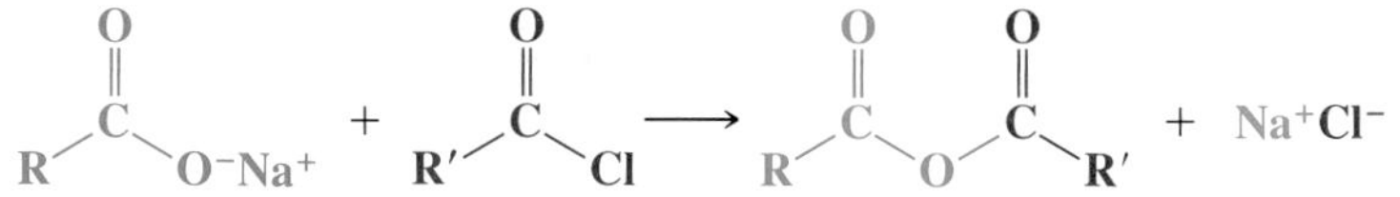

$$RCO^-Na^+ + R'COCl \longrightarrow RCOOCR' + Na^+Cl^-$$

In this reaction a carboxylate anion acts as a nucleophile and brings about a nucleophilic substitution reaction at the acyl carbon of the acyl chloride.

Cyclic anhydrides can sometimes be prepared simply by heating the appropriate dicarboxylic acid. This method succeeds, however, only when anhydride formation leads to a five- or six-membered ring:

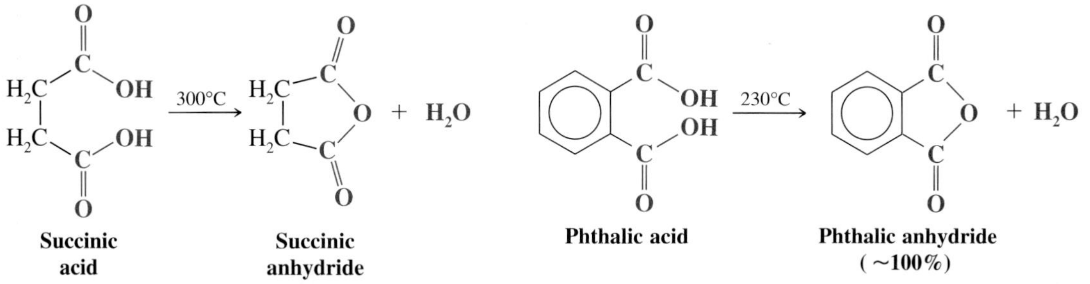

Succinic acid $\xrightarrow{300°C}$ **Succinic anhydride** + H_2O

Phthalic acid $\xrightarrow{230°C}$ **Phthalic anhydride** (~100%) + H_2O

PROBLEM 18.9

When maleic acid is heated to 200°C, it loses water and becomes maleic anhydride. Fumaric acid, a diastereomer of maleic acid, requires a much higher temperature before it dehydrates; when it does, it also yields maleic anhydride. Provide an explanation for these observations.

18.6B Reactions of Carboxylic Acid Anhydrides

Because carboxylic acid anhydrides are highly reactive, they can be used to prepare esters and amides (Fig. 18.6). We study these reactions in detail in Sections 18.7 and 18.8.

Carboxylic acid anhydrides also undergo hydrolysis:

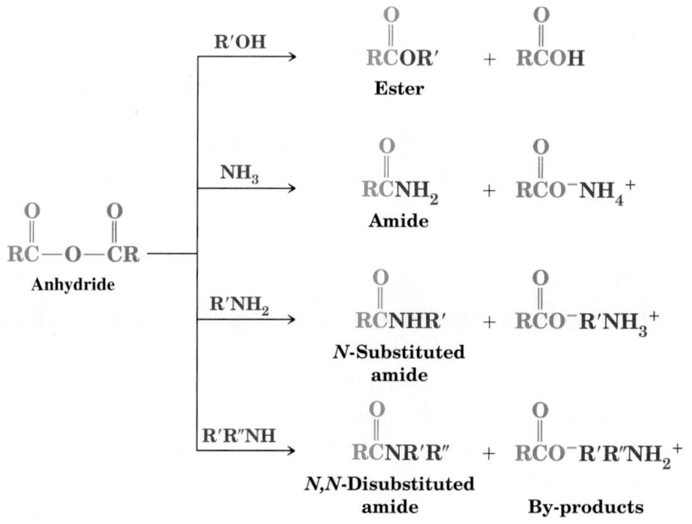

Anhydride RCOOCR:

- $\xrightarrow{R'OH}$ RCOR' (**Ester**) + RCOH
- $\xrightarrow{NH_3}$ $RCNH_2$ (**Amide**) + $RCO^-NH_4^+$
- $\xrightarrow{R'NH_2}$ RCNHR' (***N*-Substituted amide**) + $RCO^-R'NH_3^+$
- $\xrightarrow{R'R''NH}$ RCNR'R'' (***N,N*-Disubstituted amide**) + $RCO^-R'R''NH_2^+$ (**By-products**)

FIGURE 18.6 Reactions of carboxylic acid anhydrides.

$$\underset{\textbf{Anhydride}}{RC(=O){-}O{-}C(=O)R'} \xrightarrow{H_2O} RC(=O)OH + HOC(=O)R'$$

$$\underset{\textbf{Anhydride}}{RC(=O){-}O{-}C(=O)R'} \xrightarrow{OH^-/H_2O} RC(=O)O^- + {}^-OC(=O)R'$$

18.7 Esters

18.7A Synthesis of Esters: Esterification

Carboxylic acids react with alcohols to form esters through a condensation reaction known as **esterification:**

General Reaction

$$RC(=O)OH + R'{-}OH \overset{HA}{\rightleftharpoons} RC(=O)OR' + H_2O$$

Specific Examples

$$\underset{\textbf{Acetic acid}}{CH_3COH} + \underset{\textbf{Ethanol}}{CH_3CH_2OH} \overset{HA}{\rightleftharpoons} \underset{\textbf{Ethyl acetate}}{CH_3COCH_2CH_3} + H_2O$$

$$\underset{\textbf{Benzoic acid}}{C_6H_5COH} + \underset{\textbf{Methanol}}{CH_3OH} \overset{HA}{\rightleftharpoons} \underset{\textbf{Methyl benzoate}}{C_6H_5COCH_3} + H_2O$$

Acid-catalyzed esterifications, such as these examples, are called **Fischer esterifications**. They proceed very slowly in the absence of strong acids, but they reach equilibrium within a matter of a few hours when an acid and an alcohol are refluxed with a small amount of concentrated sulfuric acid or hydrogen chloride. Since the position of equilibrium controls the amount of the ester formed, the use of an excess of either the carboxylic acid or the alcohol increases the yield based on the limiting reagent. Just which component we choose to use in excess will depend on its availability and cost. The yield of an esterification reaction can also be increased by removing water from the reaction mixture as it is formed.

When benzoic acid reacts with methanol that has been labeled with ^{18}O, the labeled oxygen appears in the ester. This result reveals just which bonds break in the esterification:

$$C_6H_5C(=O){\vdots}OH + CH_3{-}^{18}O{\vdots}H \overset{HA}{\rightleftharpoons} C_6H_5C(=O){-}^{18}OCH_3 + H_2O$$

The results of the labeling experiment and the fact that esterifications are acid catalyzed are both consistent with the mechanism that follows. This mechanism is typical of acid-catalyzed nucleophilic addition–elimination reactions at acyl carbon atoms.

A Mechanism for the Reaction

Acid-Catalyzed Esterification

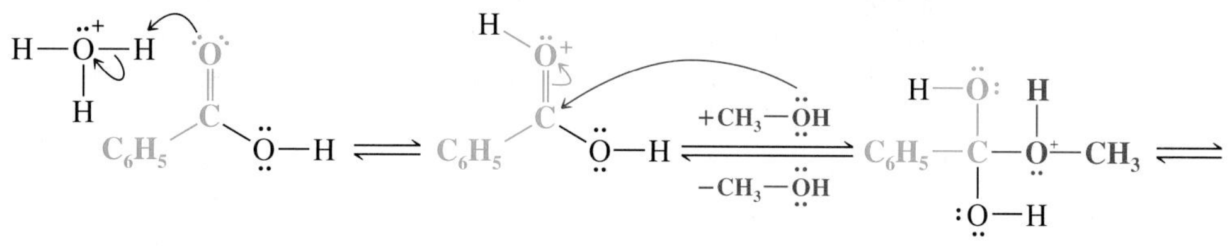

The carboxylic acid accepts a proton from the strong acid catalyst.

The alcohol attacks the protonated carbonyl group to give a tetrahedral intermediate.

A proton is lost at one oxygen atom and gained at another.

$$C_6H_5-C(OH)(\overset{+}{O}H_2)-O-CH_3 \rightleftharpoons C_6H_5-C(=\overset{+}{O}H)-O-CH_3 \underset{+H_3O^+}{\overset{-H_3O^+}{\rightleftharpoons}} C_6H_5-C(=O)-O-CH_3$$

Loss of a molecule of water gives a protonated ester.

Transfer of a proton to a base leads to the ester.

If we follow the forward reactions in this mechanism, we have the mechanism for the *acid-catalyzed esterification of an acid.* If, however, we follow the reverse reactions, we have the mechanism for the *acid-catalyzed hydrolysis of an ester:*

Acid-Catalyzed Ester Hydrolysis

$$R-\overset{O}{\overset{\|}{C}}-OR' + H_2O \underset{}{\overset{H_3O^+}{\rightleftharpoons}} R-\overset{O}{\overset{\|}{C}}-OH + R'-OH$$

Note the mechanism and conditions for acid-catalyzed ester *hydrolysis.*

Which result we obtain will depend on the conditions we choose. If we want to esterify an acid, we use an excess of the alcohol and, if possible, remove the water as it is formed. If we want to hydrolyze an ester, we use a large excess of water; that is, we reflux the ester with dilute aqueous HCl or dilute aqueous H_2SO_4.

PROBLEM 18.10

Where would you expect to find the labeled oxygen if you carried out an acid-catalyzed hydrolysis of methyl benzoate in ^{18}O-labeled water? Write a detailed mechanism to support your answer.

Steric factors strongly affect the rates of acid-catalyzed hydrolyses of esters. Large groups near the reaction site, whether in the alcohol component or the acid component, slow both reactions markedly. Tertiary alcohols, for example, react so slowly in acid-catalyzed esterifications that they usually undergo elimination instead. However, they can be converted to esters safely through the use of acyl chlorides and anhydrides in the ways that follow.

Esters from Acyl Chlorides Esters can also be synthesized by the reaction of acyl chlorides with alcohols. Since acyl chlorides are much more reactive toward nucleophilic addition-elimination than carboxylic acids, the reaction of an acyl chloride and an alcohol occurs rapidly and does not require an acid catalyst. Pyridine is often added to the reaction mixture to react with the HCl that forms. (Pyridine may also react with the acyl chloride to form an acylpyridinium ion, an intermediate that is even more reactive toward the nucleophile than the acyl chloride is.)

General Reaction

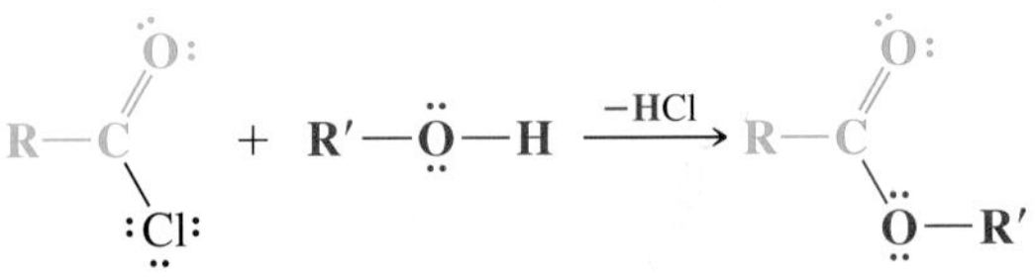

$$R{-}C(=\ddot{O}{:}){-}\ddot{Cl}{:} + R'{-}\ddot{O}{-}H \xrightarrow{-HCl} R{-}C(=\ddot{O}{:}){-}\ddot{O}{-}R'$$

Specific Example

$$C_6H_5C(=O){-}Cl + CH_3CH_2OH + \text{pyridine} \longrightarrow C_6H_5C(=O)OCH_2CH_3 + \text{pyridinium } (N^+{-}H)\ Cl^-$$

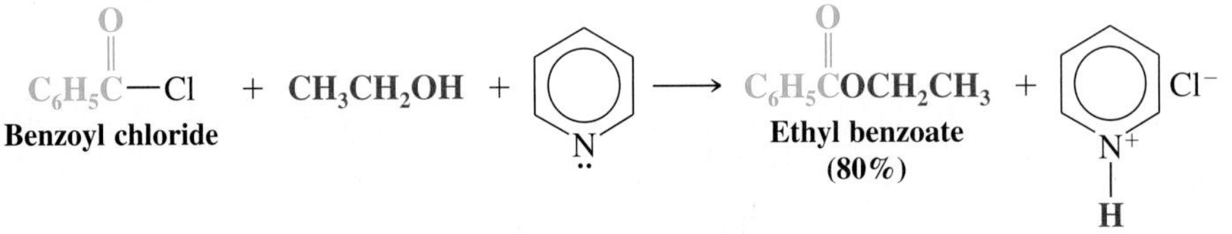

Benzoyl chloride; Ethyl benzoate (80%)

Esters from Carboxylic Acid Anhydrides Carboxylic acid anhydrides also react with alcohols to form esters in the absence of an acid catalyst.

General Reaction

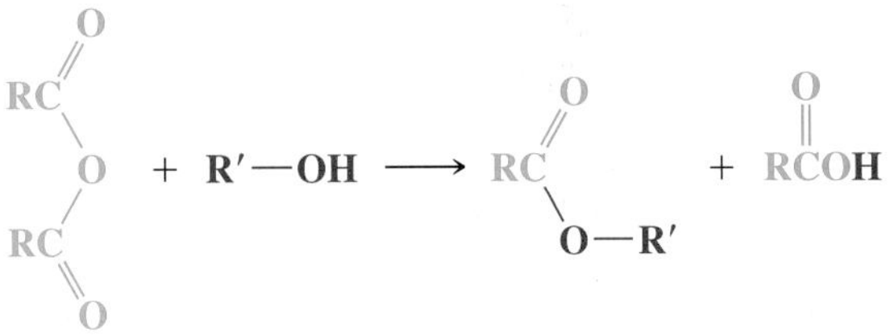

$$(RC(=O))_2O + R'{-}OH \longrightarrow RC(=O){-}O{-}R' + RC(=O)OH$$

Specific Example

$$\left(CH_3C(=O){-}\right)_2O + C_6H_5CH_2OH \longrightarrow CH_3C(=O)OCH_2C_6H_5 + CH_3CO_2H$$

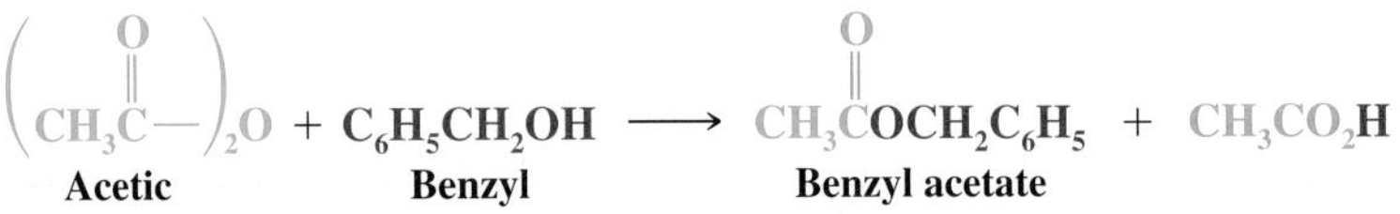

Acetic anhydride; Benzyl alcohol; Benzyl acetate

The reaction of an alcohol with an anhydride or an acyl chloride is often the best method for synthesizing an ester. These reagents avoid the use of a strong acid, as is needed for acid-catalyzed esterification. A strong acid may cause side reactions depending on what other functional groups are present.

Ester synthesis is often accomplished best by the reaction of an alcohol with an acyl chloride or anhydride.

Cyclic anhydrides react with one molar equivalent of an alcohol to form compounds that are *both esters and acids:*

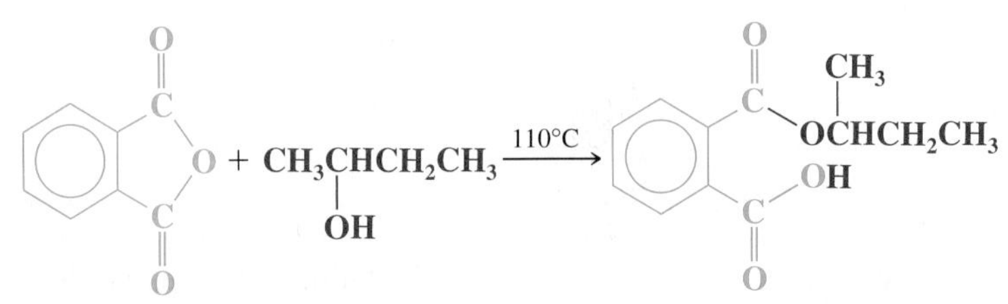

PROBLEM 18.11

Esters can also be synthesized by *transesterification:*

$$\underset{\text{High-boiling ester}}{RC(=O)OR'} + \underset{\text{High-boiling alcohol}}{R''-OH} \xrightleftharpoons{\text{HA, heat}} \underset{\text{Higher boiling ester}}{RC(=O)OR''} + \underset{\text{Lower boiling alcohol}}{R'-OH}$$

In this procedure we shift the equilibrium to the right by allowing the low-boiling alcohol to distill from the reaction mixture. The mechanism for transesterification is similar to that for an acid-catalyzed esterification (or an acid-catalyzed ester hydrolysis). Write a detailed mechanism for the following transesterification:

$$\underset{\text{Methyl acrylate}}{CH_2{=}CHC(=O)OCH_3} + \underset{\text{Butyl alcohol}}{CH_3CH_2CH_2CH_2OH} \xrightleftharpoons{\text{HA}} \underset{\text{Butyl acrylate (94\%)}}{CH_2{=}CHC(=O)OCH_2CH_2CH_2CH_3} + \underset{\text{Methanol}}{CH_3OH}$$

18.7B Base-Promoted Hydrolysis of Esters: Saponification

Esters not only undergo acid hydrolysis, they also undergo *base-promoted hydrolysis.* Base-promoted hydrolysis is sometimes called **saponification,** from the Latin word *sapo,* soap (see Section 23.2C). Refluxing an ester with aqueous sodium hydroxide, for example, produces an alcohol and the sodium salt of the acid:

$$\underset{\text{Ester}}{RC(=O)-OR'} + NaOH \xrightarrow{H_2O} \underset{\text{Sodium carboxylate}}{RC(=O)-O^-Na^+} + \underset{\text{Alcohol}}{R'OH}$$

The carboxylate ion is very unreactive toward nucleophilic substitution because it is negatively charged. Base-promoted hydrolysis of an ester, as a result, is an essentially irreversible reaction.

The mechanism for base-promoted hydrolysis of an ester also involves a nucleophilic addition–elimination at the acyl carbon.

A Mechanism for the Reaction

Base-Promoted Hydrolysis of an Ester

$$RC(=O)-OR' + {}^-:\ddot{O}-H \xrightleftharpoons{\text{slow}} R-C(O^-)(OH)-OR' \rightleftharpoons$$

A hydroxide ion attacks the carbonyl carbon atom.

The tetrahedral intermediate expels an alkoxide ion.

$$RC(=O)-\ddot{O}-H + {}^-:\ddot{O}-R' \longrightarrow RC(=O)-\ddot{O}:^- + H-\ddot{O}-R'$$

Transfer of a proton leads to the products of the reaction.

Evidence for this mechanism comes from studies done with isotopically labeled esters. When ethyl propanoate labeled with ^{18}O in the ether-type oxygen of the ester (below) is subjected to hydrolysis with aqueous NaOH, all of the ^{18}O shows up in the ethanol that is produced. None of the ^{18}O appears in the propanoate ion:

$$CH_3CH_2-\overset{O}{\overset{\|}{C}}-{}^{18}O-CH_2CH_3 + NaOH \xrightarrow{H_2O} CH_3CH_2-\overset{O}{\overset{\|}{C}}-O^-Na^+ + H-{}^{18}O-CH_2CH_3$$

This labeling result is completely consistent with the mechanism given above (outline the steps for yourself and follow the labeled oxygen through to the products). If the hydroxide ion had attacked the alkyl carbon instead of the acyl carbon, the alcohol obtained would not have been labeled. Attack at the alkyl carbon is almost never observed. (For one exception see Problem 18.13.)

$$CH_3CH_2-\overset{O}{\overset{\|}{C}}-{}^{18}O-CH_2CH_3 + {}^-OH \xrightarrow[\times]{H_2O} CH_3CH_2-\overset{O}{\overset{\|}{C}}-{}^{18}O^- + HO-CH_2CH_3$$

$$\updownarrow$$

$$CH_3CH_2-\overset{O^-}{\overset{|}{C}}={}^{18}O$$

These products are not formed.

Although nucleophilic attack at the alkyl carbon seldom occurs with esters of carboxylic acids, it is the preferred mode of attack with esters of sulfonic acids (e.g., tosylates and mesylates, Section 11.10).

$$R-\overset{O}{\overset{\|}{\underset{O}{\underset{\|}{S}}}}-O-C(R')(R'')(H) + {}^-:\ddot{O}-H \xrightarrow{\text{Inversion of configuration}} R-\overset{O}{\overset{\|}{\underset{O}{\underset{\|}{S}}}}-O^- + R'R''HC-OH$$

An alkyl sulfonate

This mechanism is preferred with alkyl sulfonates.

PROBLEM 18.12

(a) Write stereochemical formulas for compounds **A–F**:

1. *cis*-3-Methylcyclopentanol + $C_6H_5SO_2Cl \longrightarrow$ **A** $\xrightarrow[\text{heat}]{OH^-}$ **B** + $C_6H_5SO_3^-$

2. *cis*-3-Methylcyclopentanol + $C_6H_5\overset{O}{\overset{\|}{C}}-Cl \longrightarrow$ **C** $\xrightarrow[\text{reflux}]{OH^-}$ **D** + $C_6H_5CO_2^-$

3. (*R*)-2-Bromooctane + $CH_3CO_2^-Na^+ \longrightarrow$ **E** + NaBr $\xrightarrow[\text{(reflux)}]{OH^-,\ H_2O}$ **F**

4. (*R*)-2-Bromooctane + $OH^- \xrightarrow{\text{acetone}}$ **F** + Br^-

(b) Which of the last two methods, **3** or **4**, would you expect to give a higher yield of **F**? Why?

PROBLEM 18.13

Base-promoted hydrolysis of methyl mesitoate occurs through an attack on the alcohol carbon instead of the acyl carbon:

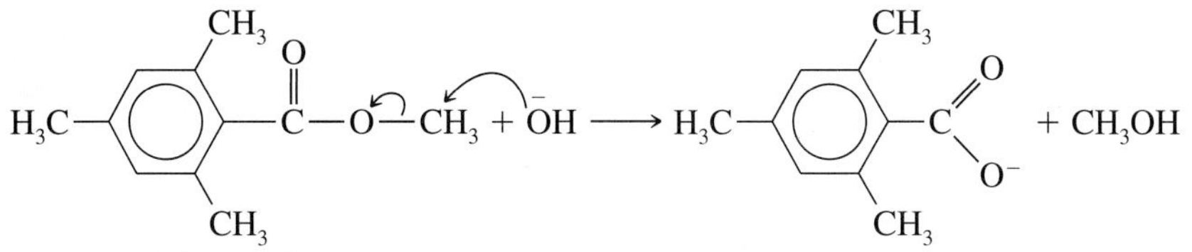

$$H_3C\text{-}C_6H_2(CH_3)_2\text{-}C(=O)\text{-}O\text{-}CH_3 + {}^-OH \longrightarrow H_3C\text{-}C_6H_2(CH_3)_2\text{-}CO_2^- + CH_3OH$$

Methyl mesitoate

(a) Can you suggest a reason that accounts for this unusual behavior? **(b)** Suggest an experiment with labeled compounds that would confirm this mode of attack.

18.7C Lactones

Carboxylic acids whose molecules have a hydroxyl group on a γ or δ carbon undergo an intramolecular esterification to give cyclic esters known as γ- or δ-*lactones.* The reaction is acid catalyzed:

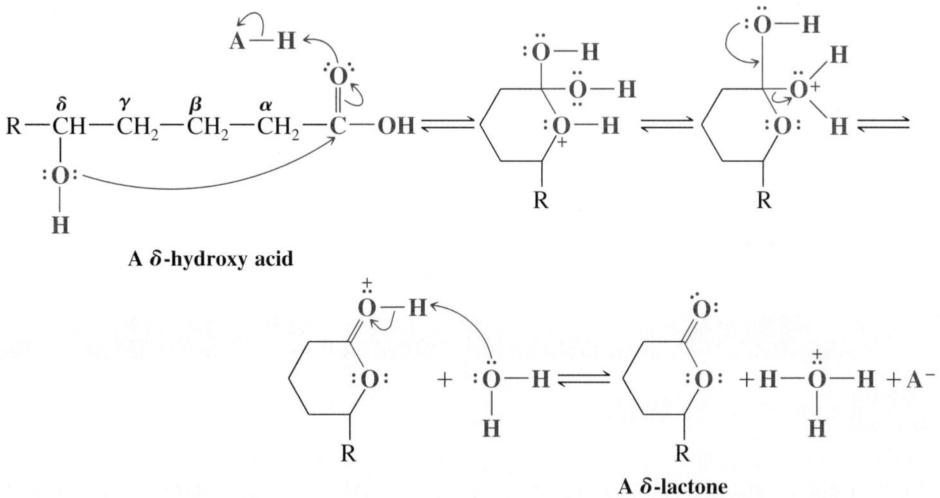

$$\underset{\text{A } \delta\text{-hydroxy acid}}{R\overset{\delta}{C}H(OH)\overset{\gamma}{C}H_2\overset{\beta}{C}H_2\overset{\alpha}{C}H_2COOH} \xrightarrow{A-H} \rightleftharpoons \rightleftharpoons \rightleftharpoons + H_2O \rightleftharpoons \underset{\text{A } \delta\text{-lactone}}{\text{lactone}} + H_3O^+ + A^-$$

A δ-hydroxy acid

A δ-lactone

Lactones are hydrolyzed by aqueous base just as other esters are. Acidification of the sodium salt, however, may lead spontaneously back to the γ- or δ-lactone, particularly if excess acid is used:

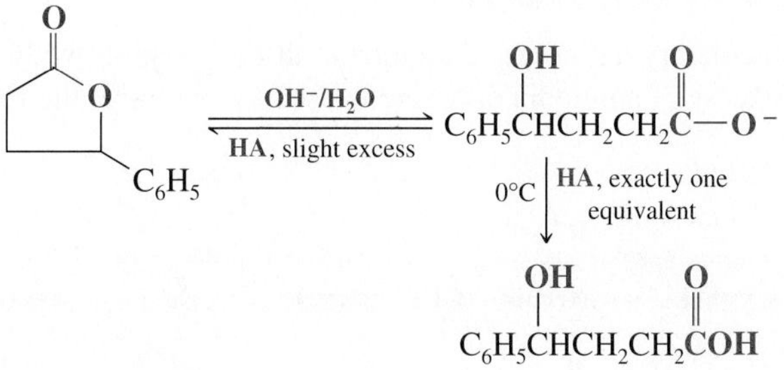

$$\gamma\text{-lactone } (C_6H_5) \underset{\text{HA, slight excess}}{\overset{OH^-/H_2O}{\rightleftharpoons}} C_6H_5CH(OH)CH_2CH_2C(=O)\text{—}O^- \xrightarrow[0^\circ C]{\text{HA, exactly one equivalent}} C_6H_5CH(OH)CH_2CH_2COOH$$

Many lactones occur in nature. Vitamin C (below), for example, is a γ-lactone. Some antibiotics, such as erythromycin and nonactin (Section 11.16B), are lactones with very large rings (called macrocyclic lactones), but most naturally occurring lactones are γ- or δ-lactones; that is, most contain five- or six-membered rings.

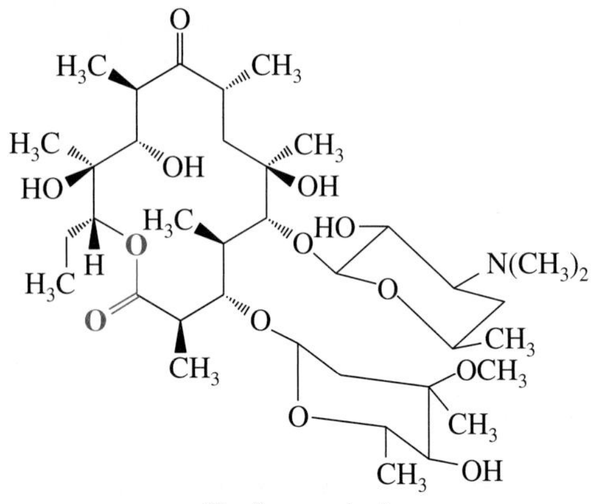

Erythromycin A

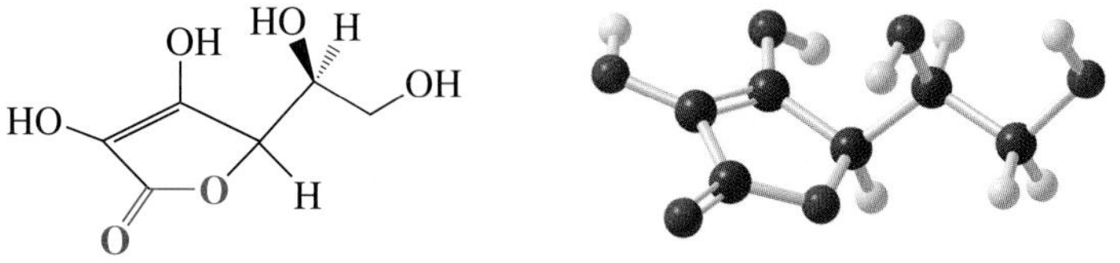

Vitamin C (ascorbic acid)

β-Lactones (lactones with four-membered rings) have been detected as intermediates in some reactions, and several have been isolated. They are highly reactive, however. If one attempts to prepare a β-lactone from a β-hydroxy acid, β elimination usually occurs instead:

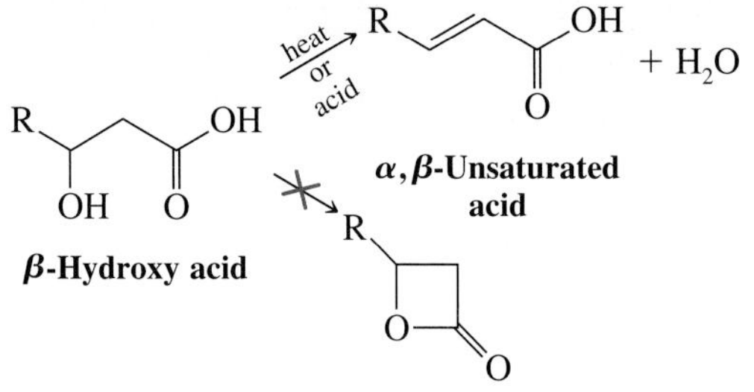

β-Hydroxy acid —(heat or acid)→ **α,β-Unsaturated acid** + H_2O

β-Lactone (does not form)

18.8 Amides

18.8A Synthesis of Amides

Amides can be prepared in a variety of ways, starting with acyl chlorides, acid anhydrides, esters, carboxylic acids, and carboxylate salts. All of these methods involve nucleophilic addition–elimination reactions by ammonia or an amine at an acyl carbon. As we might expect, acid chlorides are the most reactive and carboxylate anions are the least.

18.8B Amides from Acyl Chlorides

Primary amines, secondary amines, and ammonia all react rapidly with acid chlorides to form amides. An excess of ammonia or amine is used to neutralize the HCl that would be formed otherwise:

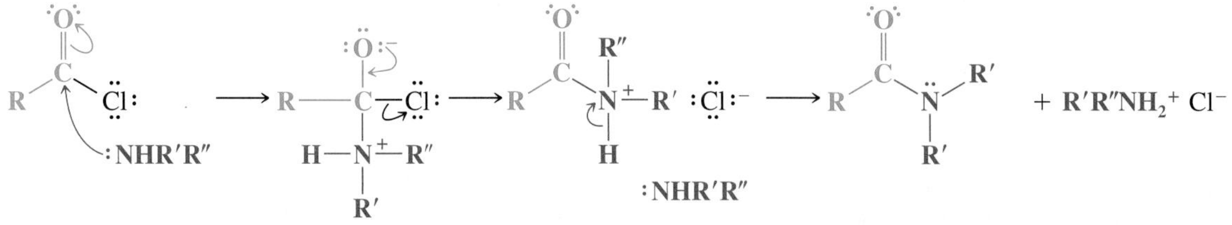

$$RCOCl + :NHR'R'' \longrightarrow \cdots \longrightarrow RCONR'R'' + R'R''NH_2^+\ Cl^-$$

Reactant
Ammonia; R′, R″ = H
1° Amine; R′ = H, R″ = alkyl, aryl
2° Amine; R′, R″ = alkyl, aryl

Product
Unsubstituted amide; R′, R″ = H
***N*-Substituted amide; R′ = H, R″ = alkyl, aryl**
***N,N*-Disubstituted amide; R′,R″ = alkyl, aryl**

Since acyl chlorides are easily prepared from carboxylic acids, this is one of the most widely used laboratory methods for the synthesis of amides. The reaction between the acyl chloride and the amine (or ammonia) usually takes place at room temperature (or below) and produces the amide in high yield.

Acyl chlorides also react with tertiary amines by a nucleophilic addition–elimination reaction. The acylammonium ion that forms, however, is not stable in the presence of water or any hydroxylic solvent:

$$\underset{\textbf{Acyl chloride}}{RCOCl} + \underset{\textbf{3° Amine}}{R_3N:} \longrightarrow \underset{\textbf{Acylammonium chloride}}{RC(=O)\overset{+}{N}R_3\ Cl^-} \xrightarrow{H_2O} RC(=O)OH + H\overset{+}{N}R_3\ Cl^-$$

Acylpyridinium ions are probably involved as intermediates in those reactions of acyl chlorides that are carried out in the presence of pyridine.

18.8C Amides from Carboxylic Anhydrides

Acid anhydrides react with ammonia and with primary and secondary amines and form amides through reactions that are analogous to those of acyl chlorides:

$$(RC(=O)-)_2O + 2\ R'-\ddot{N}H(R'') \longrightarrow RC(=O)-\ddot{N}(R'')-R' + RCO_2^-\ R'R''\overset{+}{N}H_2$$

R′, R″ can be H, alkyl or aryl

Cyclic anhydrides react with ammonia or an amine in the same general way as acyclic anhydrides; however, the reaction yields a product that is both an amide and an ammonium salt. Acidifying the ammonium salt gives a compound that is both an amide and an acid:

$$\underset{\textbf{Phthalic anhydride}}{C_6H_4(CO)_2O} + 2\ NH_3 \xrightarrow[\text{warm}]{H_2O} \underset{\substack{\textbf{Ammonium phthalamate}\\ \textbf{(94\%)}}}{C_6H_4(CONH_2)(CO_2^-\ \overset{+}{N}H_4)} \xrightarrow[(-NH_4^+)]{H_3O^+} \underset{\substack{\textbf{Phthalamic acid}\\ \textbf{(81\%)}}}{C_6H_4(CONH_2)(CO_2H)}$$

Heating the amide acid causes dehydration to occur and gives an *imide.* Imides contain the linkage $-\overset{O}{\overset{\|}{C}}-NH-\overset{O}{\overset{\|}{C}}-$

$$\underset{\textbf{Phthalamic acid}}{C_6H_4(CONH_2)(CO_2H)} \xrightarrow{150-160°C} \underset{\substack{\textbf{Phthalimide}\\ \textbf{(\sim 100\%)}}}{C_6H_4(CO)_2N-H} + H_2O$$

18.8D Amides from Esters

Esters undergo nucleophilic addition–elimination at their acyl carbon atoms when they are treated with ammonia (called *ammonolysis*) or with primary and secondary amines. These reactions take place more slowly than those of acyl chlorides and anhydrides, but they are synthetically useful:

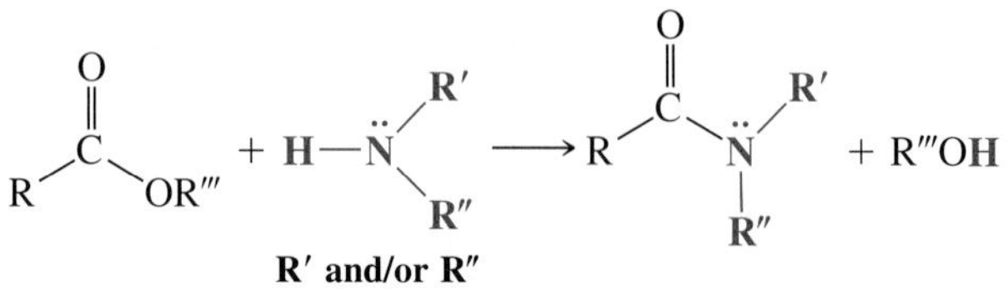

$$RCOOR''' + H{-}\ddot{N}R'R'' \longrightarrow RCO\ddot{N}R'R'' + R'''OH$$

R′ and/or R″ may be H

$$ClCH_2COOC_2H_5 + NH_{3(aq)} \xrightarrow{0-5°C} ClCH_2CONH_2 + C_2H_5OH$$

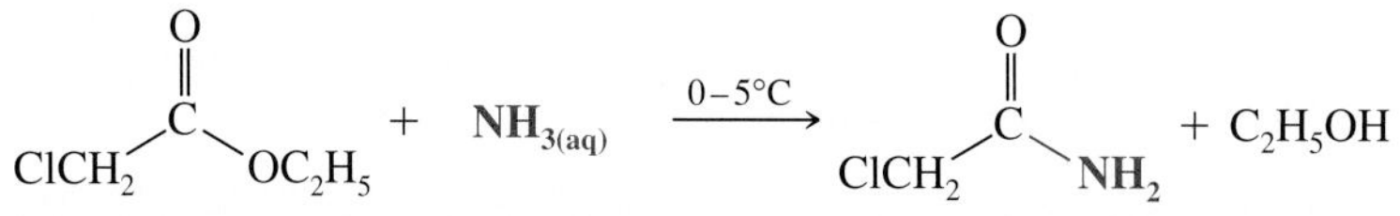

Ethyl chloroacetate **Chloroacetamide (62–87%)**

18.8E Amides from Carboxylic Acids and Ammonium Carboxylates

Carboxylic acids react with aqueous ammonia to form ammonium salts:

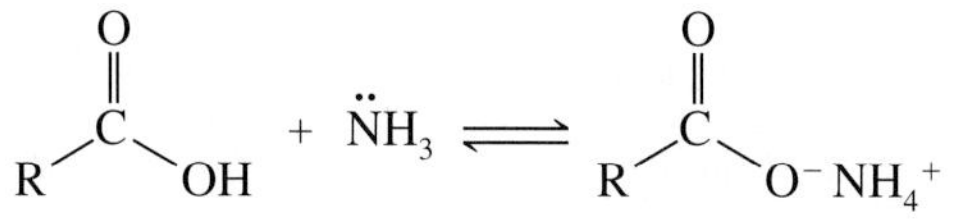

$$RCOOH + \ddot{N}H_3 \rightleftharpoons RCOO^- NH_4^+$$

An ammonium carboxylate

Because of the low reactivity of the carboxylate ion toward nucleophilic addition–elimination, further reaction does not usually take place in aqueous solution. However, if we evaporate the water and subsequently heat the dry salt, dehydration produces an amide:

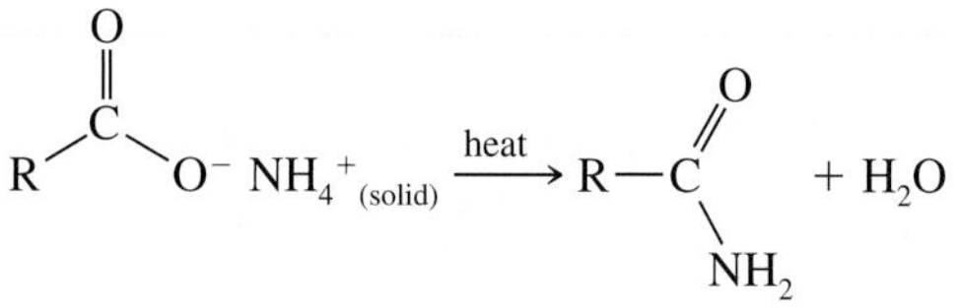

$$RCOO^- NH_4^+{}_{(solid)} \xrightarrow{heat} R{-}C(=O)NH_2 + H_2O$$

This is generally a poor method for preparing amides. A much better method is to convert the acid to an acyl chloride and then treat the acyl chloride with ammonia or an amine (Section 18.8B).

Amides are of great importance in biochemistry. The linkages that join individual amino acids together to form proteins are primarily amide linkages. As a consequence, much research has been done to find convenient and mild ways for amide synthesis. Dialkylcarbodiimides (R—N=C=N—R), such as diisopropylcarbodiimide and dicyclohexylcarbodiimide (DCC) are especially useful reagents for amide synthesis. Dialkylcarbodiimides promote amide formation by reacting with the carboxyl group of an acid and activating it toward nucleophilic addition–elimination.

A Mechanism for the Reaction

DCC-Promoted Amide Synthesis

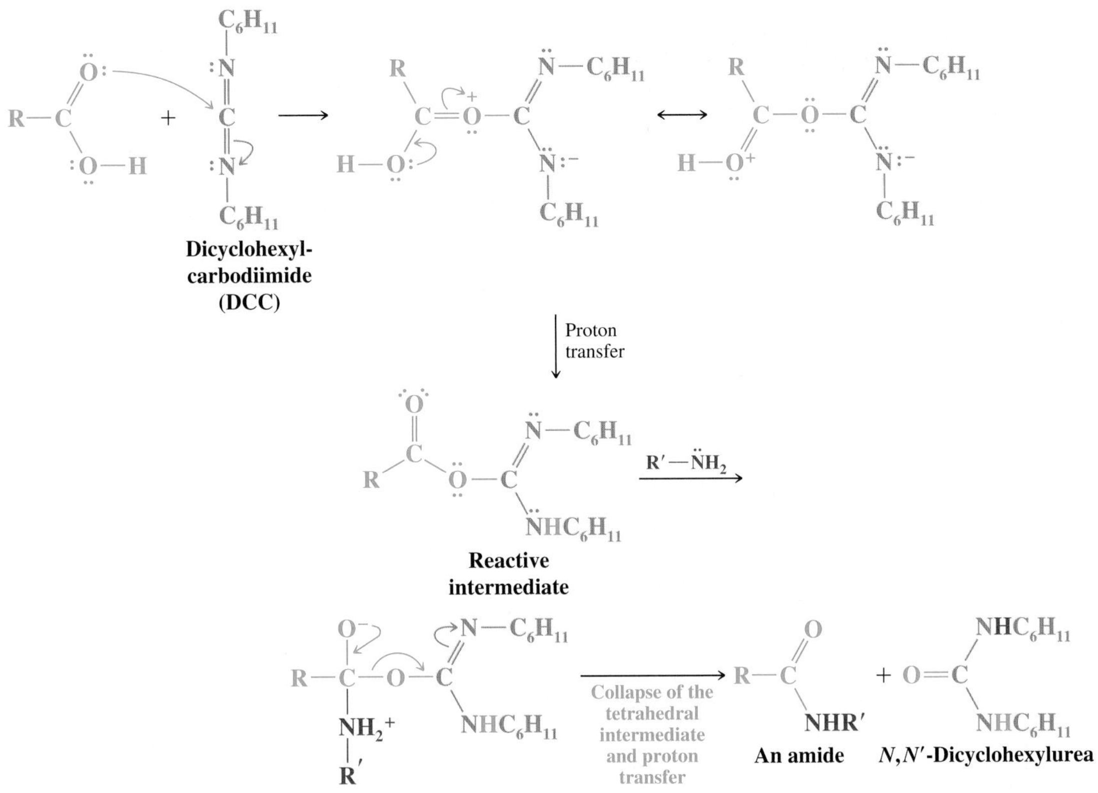

$$R{-}C(=O){-}OH + C_6H_{11}{-}N{=}C{=}N{-}C_6H_{11} \longrightarrow R(HO{-})C{=}\overset{+}{O}{-}C(=N{-}C_6H_{11})(N^{-}{-}C_6H_{11}) \longleftrightarrow R{-}C(={}^{+}OH){-}O{-}C(=N{-}C_6H_{11})(N^{-}{-}C_6H_{11})$$

Dicyclohexyl-carbodiimide (DCC)

Proton transfer

$$R{-}C(=O){-}O{-}C(=N{-}C_6H_{11})(NHC_6H_{11}) \xrightarrow{R'{-}\ddot{N}H_2}$$

Reactive intermediate

$$R{-}C(O^{-})(NH_2^{+}R'){-}O{-}C(=N{-}C_6H_{11})(NHC_6H_{11}) \xrightarrow[\text{Collapse of the tetrahedral intermediate and proton transfer}]{} R{-}C(=O){-}NHR' + O{=}C(NHC_6H_{11})_2$$

An amide $\quad$ *N,N′*-Dicyclohexylurea

The intermediate in this synthesis does not need to be isolated, and both steps take place at room temperature. Amides are produced in very high yield. In Chapter 24 we shall see how diisopropylcarbodiimide is used in an automated synthesis of peptides.

18.8F Hydrolysis of Amides

Amides undergo hydrolysis when they are heated with aqueous acid or aqueous base.

Acidic Hydrolysis

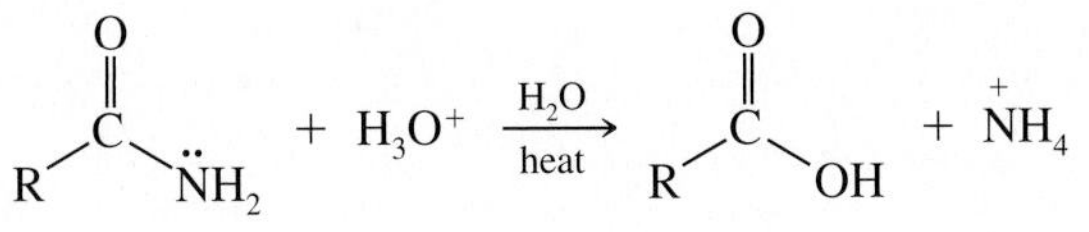

$$R{-}C(=O){-}\ddot{N}H_2 + H_3O^+ \xrightarrow[\text{heat}]{H_2O} R{-}C(=O){-}OH + \overset{+}{N}H_4$$

Basic Hydrolysis

$$R{-}C(=O){-}\ddot{N}H_2 + Na^+OH^- \xrightarrow[\text{heat}]{H_2O} R{-}C(=O){-}O^-Na^+ + \ddot{N}H_3$$

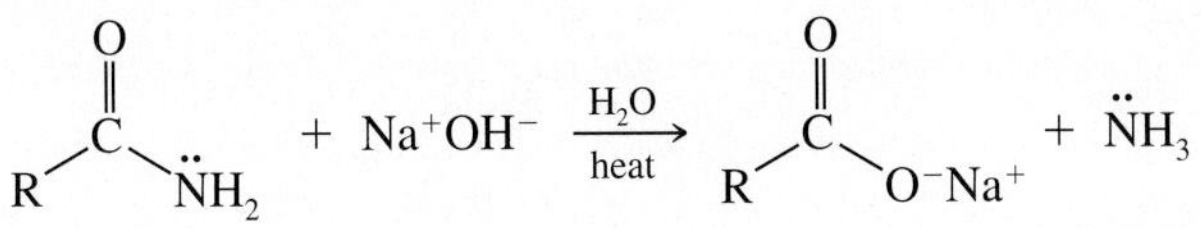

N-Substituted amides and *N,N*-disubstituted amides also undergo hydrolysis in aqueous acid or base. Amide hydrolysis by either method takes place more slowly than the corresponding hydrolysis of an ester. Thus, amide hydrolyses generally require the forcing conditions of heat and strong acid or base.

The mechanism for acid hydrolysis of an amide is similar to that given in Section 18.7A for the acid hydrolysis of an ester. Water acts as a nucleophile and attacks the protonated amide. The leaving group in the acidic hydrolysis of an amide is ammonia (or an amine).

A Mechanism for the Reaction

Acidic Hydrolysis of an Amide

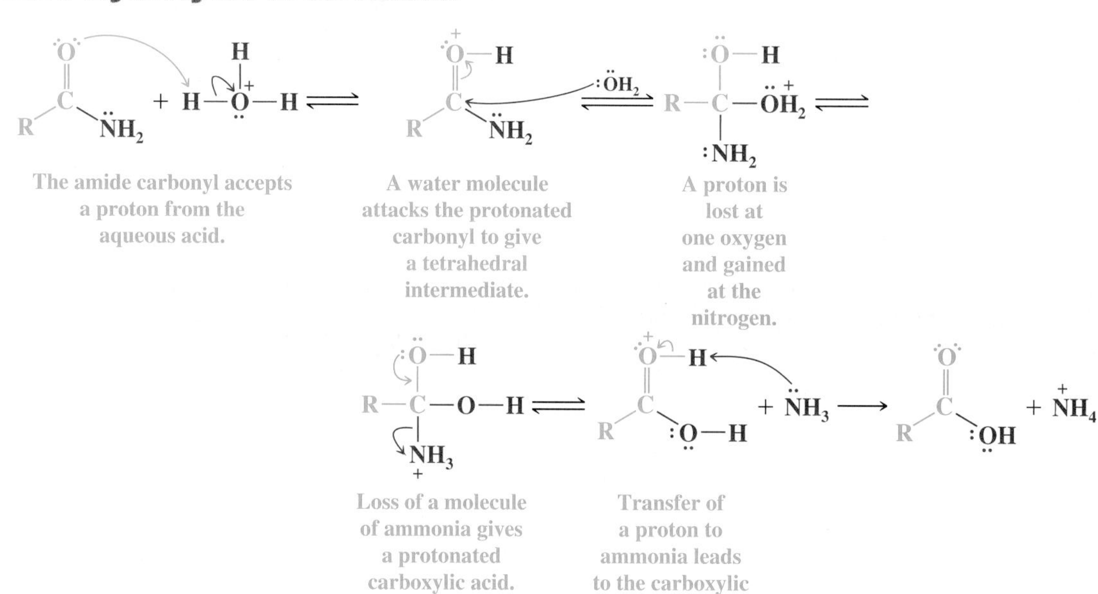

The amide carbonyl accepts a proton from the aqueous acid.
A water molecule attacks the protonated carbonyl to give a tetrahedral intermediate.
A proton is lost at one oxygen and gained at the nitrogen.
Loss of a molecule of ammonia gives a protonated carboxylic acid.
Transfer of a proton to ammonia leads to the carboxylic acid and an ammonium ion.

There is evidence that in basic hydrolyses of amides, hydroxide ions act both as nucleophiles and as bases.

A Mechanism for the Reaction

Basic Hydrolysis of an Amide

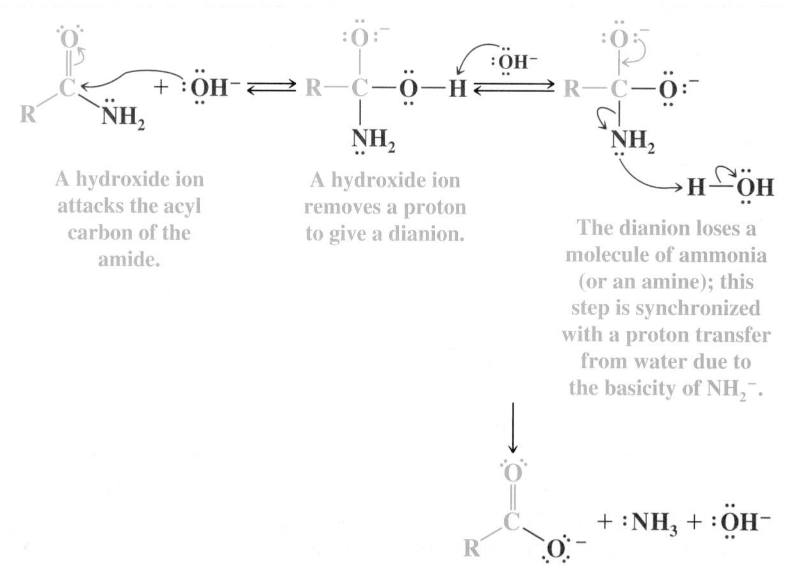

A hydroxide ion attacks the acyl carbon of the amide.
A hydroxide ion removes a proton to give a dianion.
The dianion loses a molecule of ammonia (or an amine); this step is synchronized with a proton transfer from water due to the basicity of NH_2^-.

Hydrolysis of amides by enzymes is central to the digestion of proteins. The mechanism for protein hydrolysis by the enzyme chymotrypsin is presented in Section 24.11.

PROBLEM 18.14

What products would you obtain from acidic and basic hydrolysis of each of the following amides?

(a) *N,N*-Diethylbenzamide

(b) 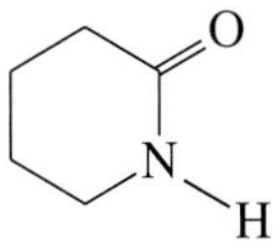 (cyclic amide: six-membered ring with C=O and N–H)

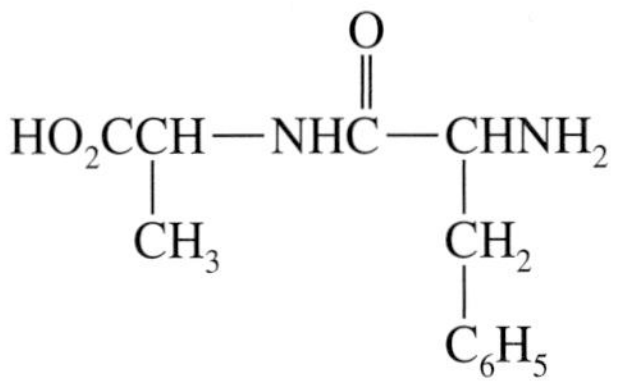

(c) $HO_2CCH(CH_3)\text{—}NHC(=O)\text{—}CHNH_2(CH_2C_6H_5)$ (a dipeptide)

18.8G Nitriles from the Dehydration of Amides

Amides react with P_4O_{10} (a compound that is often called phosphorus pentoxide and written P_2O_5) or with boiling acetic anhydride to form nitriles:

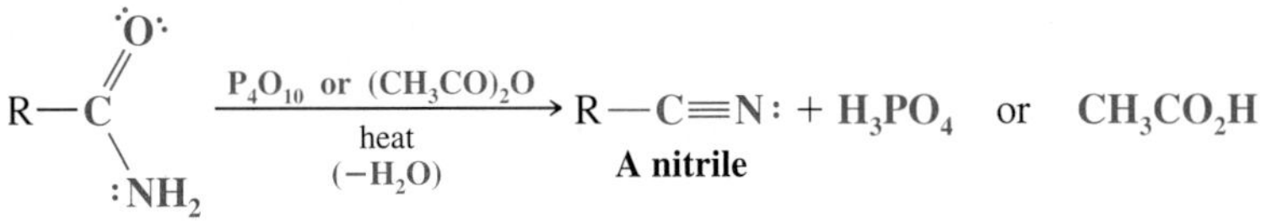

$$R\text{—}C(=\ddot{O}:)\ddot{N}H_2 \xrightarrow[\text{heat} \atop (-H_2O)]{P_4O_{10}\text{ or }(CH_3CO)_2O} \underset{\textbf{A nitrile}}{R\text{—}C{\equiv}N:} + H_3PO_4 \text{ or } CH_3CO_2H$$

This is a useful synthetic method for preparing nitriles that are not available by nucleophilic substitution reactions between alkyl halides and cyanide ion.

PROBLEM 18.15

(a) Show all steps in the synthesis of $(CH_3)_3CCN$ from $(CH_3)_3CCO_2H$.
(b) What product would you expect to obtain if you attempted to synthesize $(CH_3)_3CCN$ using the following method?

$$(CH_3)_3C\text{—}Br + CN^- \longrightarrow$$

18.8H Hydrolysis of Nitriles

Although nitriles do not contain a carbonyl group, they are usually considered to be derivatives of carboxylic acids because complete hydrolysis of a nitrile produces a carboxylic acid or a carboxylate anion (Sections 16.9 and 18.3):

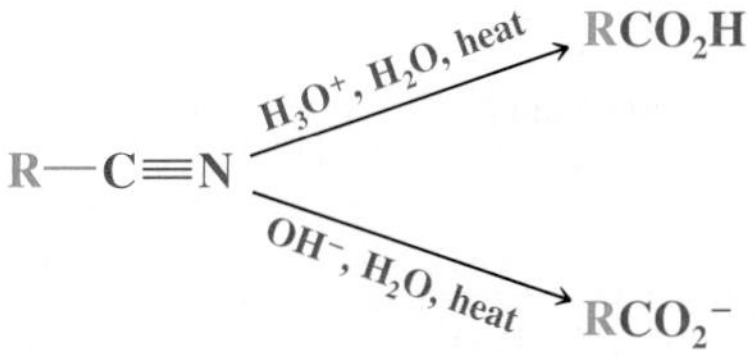

$$R\text{—}C{\equiv}N \xrightarrow{H_3O^+,\ H_2O,\ \text{heat}} RCO_2H$$

$$R\text{—}C{\equiv}N \xrightarrow{OH^-,\ H_2O,\ \text{heat}} RCO_2^-$$

The mechanisms for these hydrolyses are related to those for the acidic and basic hydrolyses of amides. In **acidic hydrolysis** of a nitrile the first step is protonation of the nitrogen atom. This protonation (in the following sequence) enhances polarization of the nitrile group and makes the carbon atom more susceptible to nucleophilic attack by the weak nucleophile, water. The loss of a proton from the oxygen atom then produces a tautomeric form of an amide. Gain of a proton at the nitrogen atom gives a **protonated amide,** and from this point on the steps are the same as those given for the acidic hydrolysis of an amide in Section 18.8F. (In concentrated H_2SO_4 the reaction stops at the protonated amide, and this is a useful way of making amides from nitriles.)

A Mechanism for the Reaction

Acidic Hydrolysis of a Nitrile

$$R{-}C{\equiv}N{:} + H{-}\overset{+}{O}H_2 \rightleftharpoons R{-}C{\equiv}\overset{+}{N}H \longleftrightarrow R{-}\overset{+}{C}{=}NH + {:}\ddot{O}H_2 \overset{\text{slow}}{\rightleftharpoons}$$

Protonated nitrile

$$R{-}C(\overset{+}{O}H_2){=}\ddot{N}H \overset{:\ddot{O}H_2}{\rightleftharpoons} R{-}C(\ddot{O}H){=}NH + H{-}\overset{+}{O}H_2 \rightleftharpoons$$

Amide tautomer

$$R{-}C(\ddot{O}H){=}\overset{+}{N}H_2 \longleftrightarrow R{-}C(={\overset{+}{O}H}){-}\ddot{N}H_2 \xrightarrow[\text{(amide hydrolysis)}]{\text{several steps}} RCO_2H + \overset{+}{N}H_4$$

Protonated amide

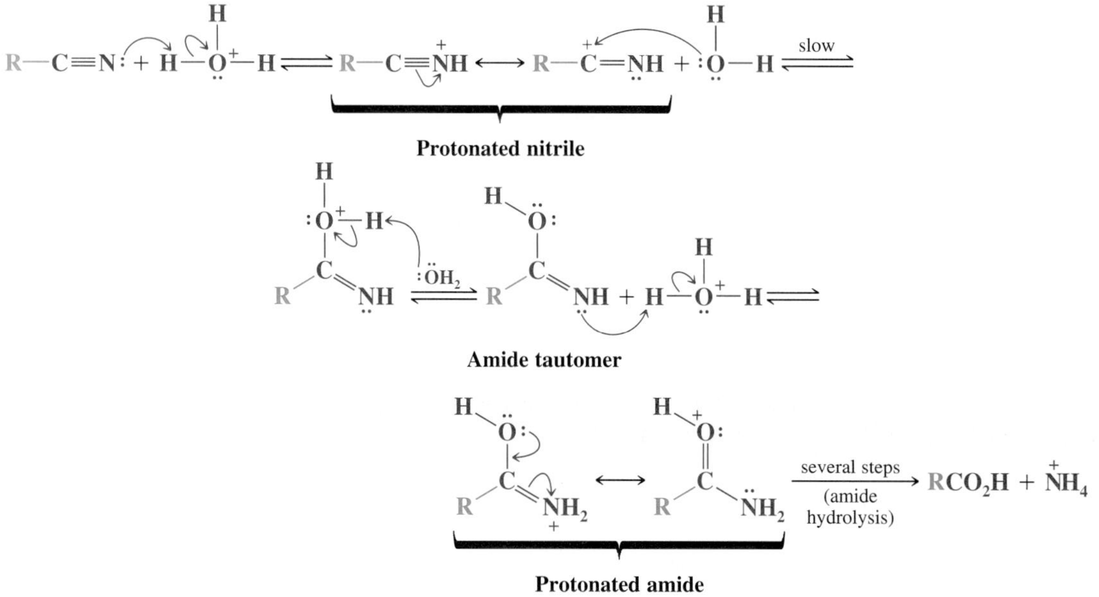

In **basic hydrolysis,** a hydroxide ion attacks the nitrile carbon atom, and subsequent protonation leads to the amide tautomer. Further attack by the hydroxide ion leads to hydrolysis in a manner analogous to that for the basic hydrolysis of an amide (Section 18.8F). (Under the appropriate conditions, amides can be isolated when nitriles are hydrolyzed.)

A Mechanism for the Reaction

Basic Hydrolysis of a Nitrile

$$R{-}C{\equiv}N{:} + {}^{-}{:}\ddot{O}{-}H \rightleftharpoons R{-}C(OH){=}\ddot{N}{:}^{-} \overset{H{-}OH}{\rightleftharpoons} R{-}C(OH){=}\ddot{N}H + {}^{-}{:}\ddot{O}{-}H \rightleftharpoons$$

Amide tautomer

$$R{-}C(OH)_2{-}\ddot{N}H^{-} \underset{^{-}OH}{\overset{H{-}OH}{\rightleftharpoons}} R{-}C(OH)_2{-}\ddot{N}H_2 \underset{H_2O}{\overset{^{-}OH}{\rightleftharpoons}} R{-}C(O^{-})(OH){-}\ddot{N}H_2 \underset{^{-}OH}{\overset{HOH}{\rightleftharpoons}}$$

$$R{-}C(O^{-})_2{-}\ddot{N}H_2 \overset{H{-}OH}{\rightleftharpoons} R{-}C(=O){-}O^{-} + NH_3 + OH^{-}$$

Carboxylate anion

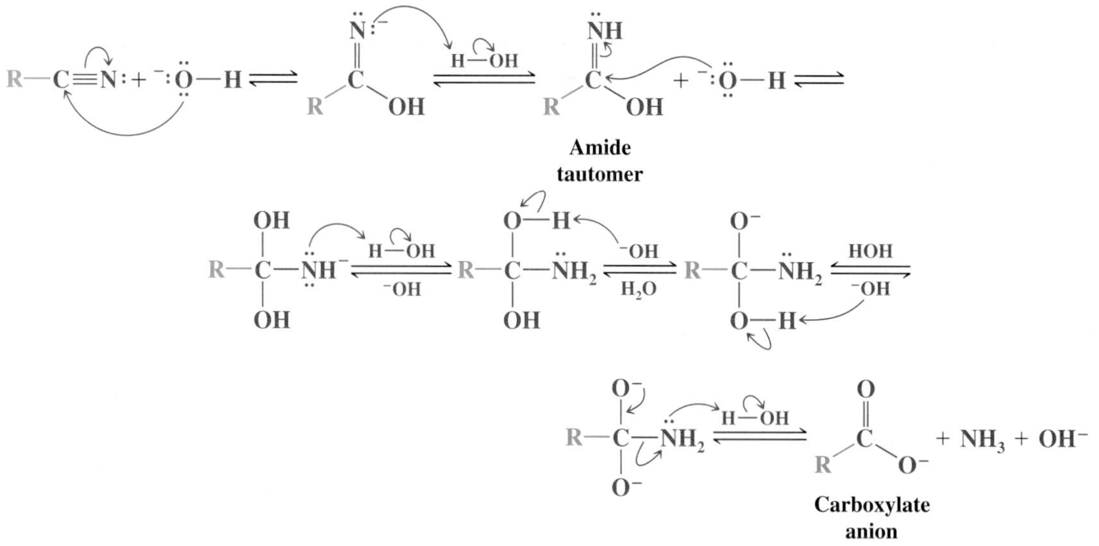

18.8I Lactams

Cyclic amides are called **lactams.** The size of the lactam ring is designated by Greek letters in a way that is analogous to lactone nomenclature (Section 18.7C):

A β-lactam **A γ-lactam** **A δ-lactam**

γ-Lactams and δ-lactams often form spontaneously from γ- and δ-amino acids. β-Lactams, however, are highly reactive; their strained four-membered rings open easily in the presence of nucleophilic reagents.

The Chemistry of...

Penicillins

The penicillin antibiotics (see the following structures) contain a β-lactam ring:

$R = C_6H_5CH_2—$ Penicillin G

$R = C_6H_5CH(NH_2)—$ Ampicillin

$R = C_6H_5OCH_2—$ Penicillin V

The penicillins apparently act by interfering with the synthesis of bacterial cell walls. It is thought that they do this by reacting with an amino group of an essential enzyme of the cell wall biosynthetic pathway. This reaction involves ring opening of the β-lactam and acylation of the enzyme, inactivating it.

Active enzyme **A penicillin** **Inactive enzyme** **Inactive enzyme**

Bacterial resistance to the penicillin antibiotics is a serious problem for the treatment of infections. Bacteria that have developed resistance to penicillin produce an enzyme called penicillinase. Penicillinase hydrolyzes the β-lactam ring of penicillin, resulting in penicilloic acid. Because penicilloic acid cannot act as an acylating agent, it is incapable of blocking bacterial cell wall synthesis by the mechanism shown above.

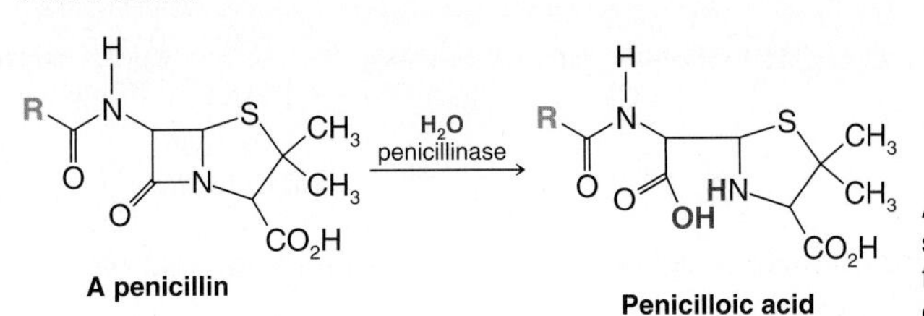

An industrial-scale reactor for preparation of an antibiotic.

18.9 Derivatives of Carbonic Acid

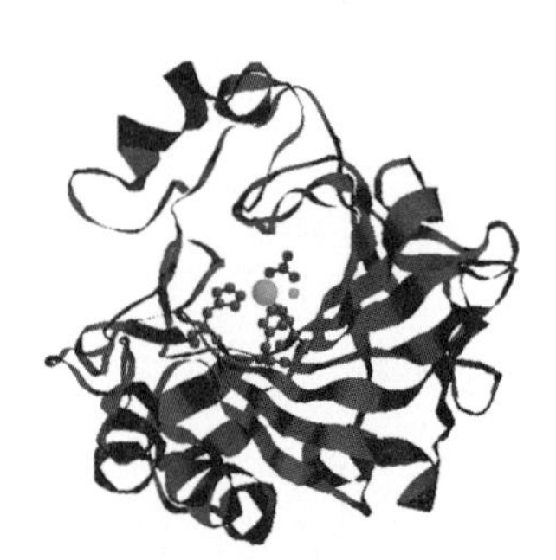

Carbonic anhydrase

The opening vignette for Chapter 3 discussed an enzyme called carbonic anhydrase that interconverts water and carbon dioxide with carbonic acid. A carbonate dianion is shown in red within the structure of carbonic anhydrase above.

Carbonic acid $\left(\mathrm{HO{-}C({=}O){-}OH}\right)$ is an unstable compound that decomposes spontaneously to produce carbon dioxide and water and, therefore, cannot be isolated. However, many acyl chlorides, esters, and amides that are derived from carbonic acid are stable compounds that have important applications.

Carbonyl dichloride (ClCOCl), a highly toxic compound that is also called *phosgene,* can be thought of as the diacyl chloride of carbonic acid. Carbonyl dichloride reacts by nucleophilic addition–elimination with two molar equivalents of an alcohol to yield a **dialkyl carbonate:**

$$\underset{\textbf{Carbonyl dichloride}}{\mathrm{Cl{-}C({=}O){-}Cl}} + 2\,\mathrm{CH_3CH_2OH} \longrightarrow \underset{\textbf{Diethyl carbonate}}{\mathrm{CH_3CH_2O\overset{O}{\overset{\|}{C}}OCH_2CH_3}} + 2\,\mathrm{HCl}$$

A tertiary amine is usually added to the reaction to neutralize the hydrogen chloride that is produced.

Carbonyl dichloride reacts with ammonia to yield **urea** (Section 1.2A):

$$\mathrm{Cl{-}C({=}O){-}Cl} + 4\,\mathrm{NH_3} \longrightarrow \underset{\textbf{Urea}}{\mathrm{H_2N\overset{O}{\overset{\|}{C}}NH_2}} + 2\,\mathrm{NH_4Cl}$$

Urea is the end product of the metabolism of nitrogen-containing compounds in most mammals and is excreted in the urine.

18.9A Alkyl Chloroformates and Carbamates (Urethanes)

Treating carbonyl dichloride with one molar equivalent of an alcohol leads to the formation of an alkyl chloroformate:

$$\mathrm{ROH} + \mathrm{Cl{-}C({=}O){-}Cl} \longrightarrow \underset{\textbf{Alkyl chloroformate}}{\mathrm{RO{-}C({=}O){-}Cl}} + \mathrm{HCl}$$

Specific Example

$$\mathrm{C_6H_5CH_2OH} + \mathrm{Cl{-}C({=}O){-}Cl} \longrightarrow \underset{\textbf{Benzyl chloroformate}}{\mathrm{C_6H_5CH_2O{-}C({=}O){-}Cl}} + \mathrm{HCl}$$

Alkyl chloroformates react with ammonia or amines to yield compounds called *carbamates* or *urethanes:*

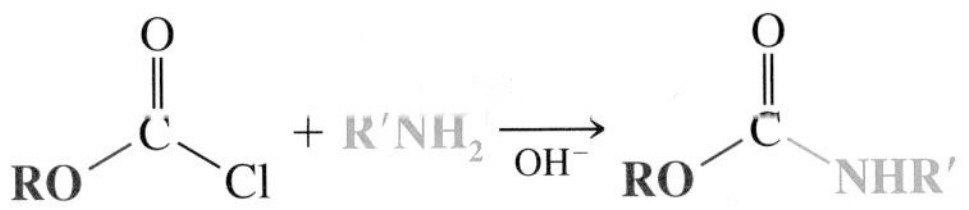

$$RO-\overset{O}{\overset{\|}{C}}-Cl + R'NH_2 \xrightarrow{OH^-} RO-\overset{O}{\overset{\|}{C}}-NHR'$$

A carbamate (or urethane)

Benzyl chloroformate is used to install an amino protecting (blocking) group called the benzyloxycarbonyl (BOC) group. We shall see in Section 24.7 how this protecting group is used in the synthesis of peptides and proteins. One advantage of the BOC group is that it can be removed under mild conditions. Treating the BOC derivative with hydrogen and a catalyst or with cold HBr in acetic acid removes the protecting group:

$$R-NH_2 + C_6H_5CH_2O\overset{O}{\overset{\|}{C}}Cl \xrightarrow{OH^-} R-NH-\overset{O}{\overset{\|}{C}}OCH_2C_6H_5$$

Protected amine

$$R-NH-\overset{O}{\overset{\|}{C}}OCH_2C_6H_5 \xrightarrow{H_2,\ Pd} R-NH_2 + CO_2 + C_6H_5CH_3$$

$$R-NH-\overset{O}{\overset{\|}{C}}OCH_2C_6H_5 \xrightarrow{HBr,\ CH_3CO_2H} R-\overset{+}{N}H_3 + CO_2 + C_6H_5CH_2Br$$

Carbamates can also be synthesized by allowing an alcohol to react with an isocyanate, R—N=C=O. (Carbamates tend to be nicely crystalline solids and are useful derivatives for identifying alcohols.) The reaction is an example of nucleophilic addition to the acyl carbon:

$$ROH + C_6H_5-N=C=O \longrightarrow RO\overset{O}{\overset{\|}{C}}-NH-C_6H_5$$

Phenyl isocyanate

The insecticide called *Sevin* is a carbamate made by allowing 1-naphthol to react with methyl isocyanate:

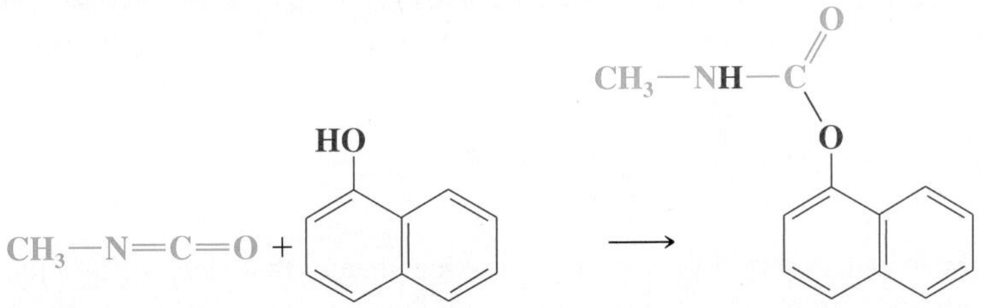

$$CH_3-N=C=O + \text{1-Naphthol (HO on naphthalene)} \longrightarrow CH_3-NH-\overset{O}{\overset{\|}{C}}-O-\text{(1-naphthyl)}$$

Methyl isocyanate **1-Naphthol** **Sevin**

A tragic accident that occurred at Bhopal, India in 1984 was caused by leakage of methyl isocyanate from a manufacturing plant. Methyl isocyanate is a highly toxic gas, and more than 1800 people living near the plant lost their lives.

PROBLEM 18.16

Write structures for the products of the following reactions:

(a) $C_6H_5CH_2OH + C_6H_5N=C=O \longrightarrow$

(b) $ClCOCl$ + excess $CH_3NH_2 \longrightarrow$

(c) Glycine ($H_3\overset{+}{N}CH_2CO_2^-$) + $C_6H_5CH_2OCOCl \xrightarrow{OH^-}$

(d) Product of (c) + H_2, Pd $\longrightarrow$
(e) Product of (c) + cold HBr, CH_3CO_2H $\longrightarrow$
(f) Urea + OH^-, H_2O, heat

Although alkyl chloroformates (ROCOCl), dialkyl carbonates (ROCOOR), and carbamates ($ROCONH_2$, ROCONHR, etc.) are stable, chloroformic acid (HOCOCl), alkyl hydrogen carbonates (ROCOOH), and carbamic acid ($HOCONH_2$) are not. These latter compounds decompose spontaneously to liberate carbon dioxide:

Chloroformic acid HO–C(=O)–Cl $\longrightarrow$ HCl + CO_2
Unstable

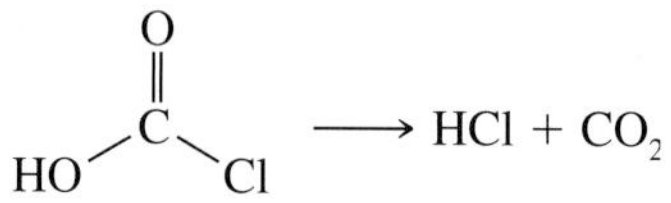

An alkyl hydrogen carbonate RO–C(=O)–OH $\longrightarrow$ ROH + CO_2
Unstable

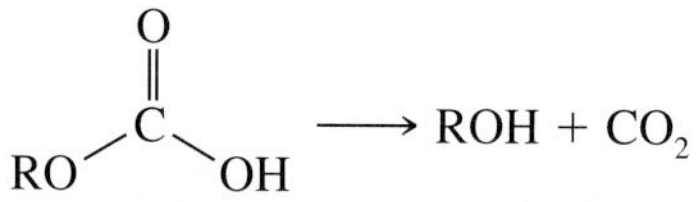

A carbamic acid HO–C(=O)–NH_2 $\longrightarrow$ NH_3 + CO_2
Unstable

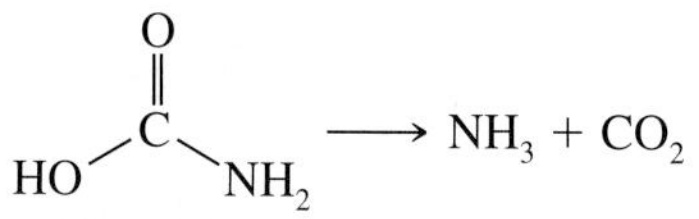

This instability is a characteristic that these compounds share with their functional parent, carbonic acid:

Carbonic acid HO–C(=O)–OH $\longrightarrow$ H_2O + CO_2
Unstable

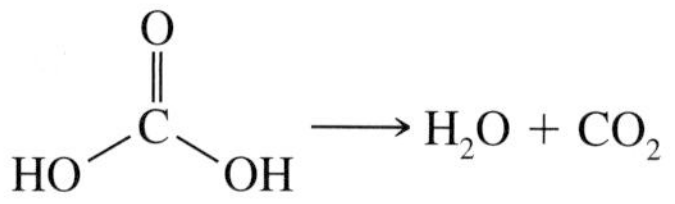

18.10 Decarboxylation of Carboxylic Acids

The reaction whereby a carboxylic acid loses CO_2 is called a **decarboxylation:**

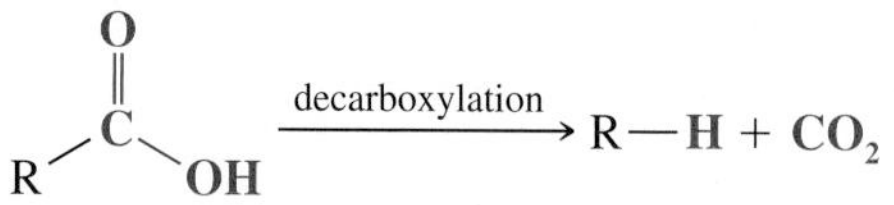

$$\text{R–C(=O)–OH} \xrightarrow{\text{decarboxylation}} \text{R–H} + CO_2$$

Although the unusual stability of carbon dioxide means that decarboxylation of most acids is exothermic, in practice the reaction is not always easy to carry out because the reaction is very slow. Special groups usually have to be present in the molecule for decarboxylation to be rapid enough to be synthetically useful.

Acids whose molecules have a carbonyl group one carbon removed from the carboxylic acid group, called ***β*-keto acids,** decarboxylate readily when they are heated to 100–150°C. (Some *β*-keto acids even decarboxylate slowly at room temperature.)

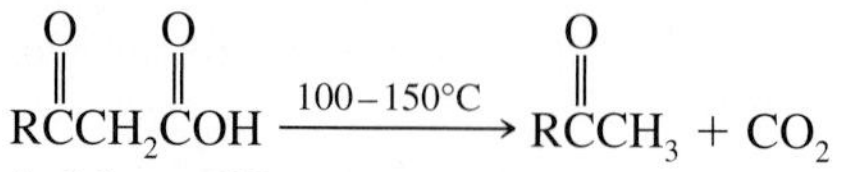

$$\text{RCCH}_2\text{COH} \xrightarrow{100\text{–}150°\text{C}} \text{RCCH}_3 + CO_2$$

(both carbons marked C bear =O)

A *β*-keto acid

There are two reasons for this ease of decarboxylation:

1. When the acid itself decarboxylates, it can do so through a six-membered cyclic transition state:

$$\text{R-CO-CH}_2\text{-COOH} \xrightarrow{-CO_2} \text{R-C(OH)=CH}_2 \rightleftharpoons \text{R-CO-CH}_3$$

β-Keto acid **Enol** **Ketone**

This reaction produces an enol directly and avoids an anionic intermediate. The enol then tautomerizes to a methyl ketone.

2. When the carboxylate anion decarboxylates, it forms a resonance-stabilized enolate anion:

$$\text{R-CO-CH}_2\text{-CO}_2^- \xrightarrow{-CO_2} \text{R-CO-}\ddot{\text{C}}\text{H}_2^- \longleftrightarrow \text{R-C(O}^-\text{)=CH}_2 \xrightarrow{HA} \text{R-CO-CH}_3$$

Acylacetate ion

Resonance-stabilized anion

This anion is much more stable than the anion $RCH_2{:}^-$ that would be produced by decarboxylation of an ordinary carboxylic acid anion.

The Chemistry of...

Thiamine

Thiamine is a vitamin that catalyzes the conversion of pyruvate to acetyl coenzyme A, a step that makes a key link between glycolysis and other metabolic pathways. Pyruvate is the end product of glycolysis (see the Chapter 17 opener), and acetyl coenzyme A is the starting point for other critical biochemical processes, including lipid biosynthesis (see Special Topic D). Pyruvate is converted to acetyl coenzyme A by a decarboxylation reaction catalyzed by the enzyme pyruvate dehydrogenase. Thiamine serves as a coenzyme for pyruvate dehydrogenase. Nucleophilic attack of enzyme-bound thiamine (in an ylide form) on the ketone carbonyl of pyruvate leads to a tetrahedral intermediate (see below), much like other carbonyl addition reactions we have seen. A molecule of CO_2 is lost when the tetrahedral intermediate accepts an electron pair from the pyruvate carboxylate group, releasing CO_2 and leading to a resonance-stabilized carbanion involving the thiamine thiazole ring. Further reaction transforms the two remaining carbons from pyruvate into acetyl coenzyme A and regenerates the thiamine coenzyme for another reaction cycle. An essential role of thiamine, therefore, is to stabilize the carbanion intermediate during this decarboxylation reaction.

Whole-grain breads are a dietary source of thiamine (vitamin B_1).

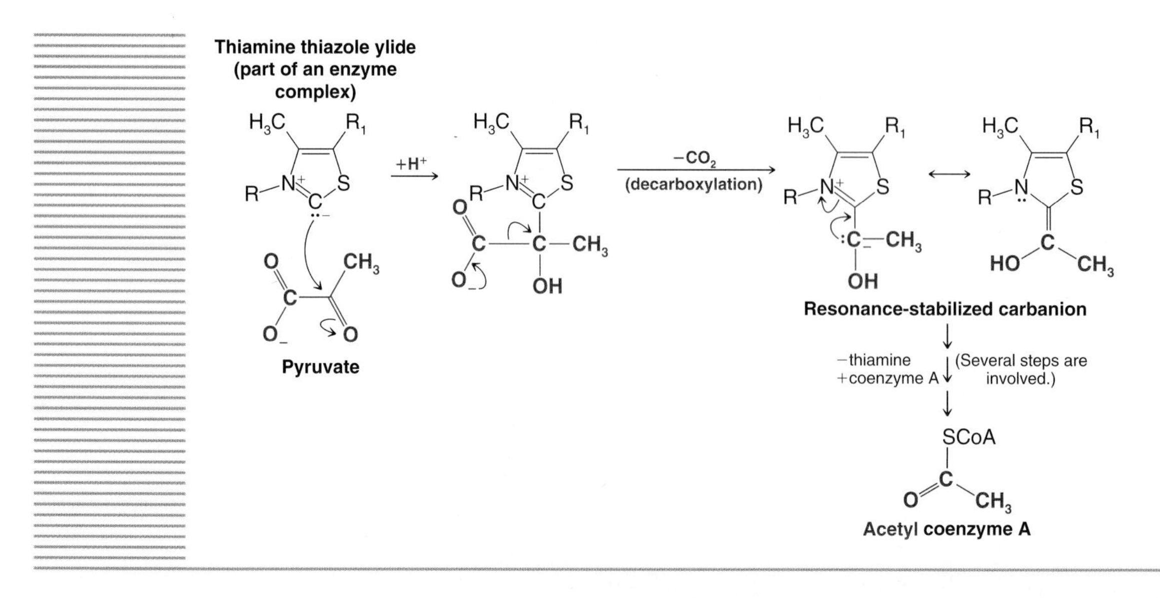

Malonic acids also decarboxylate readily for similar reasons:

$$\text{HO}\overset{\text{O}}{\overset{\|}{\text{C}}}-\underset{\text{R}}{\overset{\text{R}}{\text{C}}}-\overset{\text{O}}{\overset{\|}{\text{C}}}\text{OH} \xrightarrow{100-150°\text{C}} \text{H}-\underset{\text{R}}{\overset{\text{R}}{\text{C}}}-\overset{\text{O}}{\overset{\|}{\text{C}}}\text{OH} + \mathbf{CO_2}$$

A malonic acid

Notice that malonic acids undergo decarboxylation so readily that they do not form cyclic anhydrides (Section 18.6).

We shall see in Chapter 19 how decarboxylation of β-keto acids and malonic acids is synthetically useful.

18.10A Decarboxylation of Carboxyl Radicals

Although the carboxylate ions (RCO_2^-) of simple aliphatic acids do not decarboxylate readily, carboxyl radicals ($RCO_2\cdot$) do. They decarboxylate by losing CO_2 and producing alkyl radicals:

$$RCO_2\cdot \longrightarrow R\cdot + CO_2$$

PROBLEM 18.17

Using decarboxylation reactions, outline a synthesis of each of the following from appropriate starting materials:

(a) 2-Hexanone
(b) 2-Methylbutanoic acid
(c) Cyclohexanone
(d) Pentanoic acid

PROBLEM 18.18

Diacyl peroxides $\left(\text{R}\overset{\text{O}}{\overset{\|}{\text{C}}}-\text{O}-\text{O}-\overset{\text{O}}{\overset{\|}{\text{C}}}\text{R}\right)$ decompose readily when heated.

(a) What factor accounts for this instability?

(b) The decomposition of a diacyl peroxide produces CO_2. How is it formed?

(c) Diacyl peroxides are often used to initiate radical reactions, for example, the polymerization of an alkene:

$$n\,\text{CH}_2{=}\text{CH}_2 \xrightarrow[-\text{CO}_2]{\text{R}\overset{\text{O}}{\overset{\|}{\text{C}}}-\text{O}-\text{O}-\overset{\text{O}}{\overset{\|}{\text{C}}}\text{R}} \text{R}{+}\text{CH}_2\text{CH}_2{+_n}\text{H}$$

Show the steps involved.

18.11 Chemical Tests for Acyl Compounds

Carboxylic acids are weak acids, and their acidity helps us to detect them. Aqueous solutions of water-soluble carboxylic acids give an acid test with blue litmus paper. Water-insoluble carboxylic acids dissolve in aqueous sodium hydroxide and aqueous sodium bicarbonate (see Section 18.2C). The latter reagent helps us distinguish carboxylic acids from most phenols. Except for the di- and trinitrophenols, phenols do not dissolve in aqueous sodium bicarbonate. When carboxylic acids dissolve in aqueous sodium bicarbonate, they also cause the evolution of carbon dioxide.

Acyl chlorides hydrolyze in water, and the resulting chloride ion gives a precipitate when treated with aqueous silver nitrate. Acid anhydrides dissolve when heated briefly with aqueous sodium hydroxide.

Esters and amides hydrolyze slowly when they are refluxed with sodium hydroxide. An ester produces a carboxylate anion and an alcohol; an amide produces a carboxylate anion and an amine or ammonia. The hydrolysis products, the acid and the alcohol or amine, can be isolated and identified. Since base-promoted hydrolysis of an unsubstituted amide produces ammonia, this ammonia can often be detected by holding moist red litmus paper in the vapors above the reaction mixture.

Amides can be distinguished from amines with dilute HCl. Most amines dissolve in dilute HCl, whereas most amides do not (see Problem 18.38).

Summary of the Reactions of Carboxylic Acids and Their Derivatives

The reactions of carboxylic acids and their derivatives are summarized here. Many (but not all) of the reactions in this summary are acyl transfer reactions (they are principally the reactions referenced to Sections 18.5 and beyond). As you use this summary, you will find it helpful to also review Section 18.4, which presents the general nucleophilic addition–elimination mechanism for acyl transfer. It is instructive to relate aspects of the specific acyl transfer reactions below to this general mechanism. In some cases proton transfer steps are also involved, such as to make a leaving group more suitable by prior protonation or to transfer a proton to a stronger base at some point in a reaction, but in all acyl transfers the essential nucleophilic addition–elimination steps are identifiable.

Reactions of Carboxylic Acids

1. As acids (discussed in Sections 3.10 and 18.2C):

$$RCO_2H + NaOH \longrightarrow RCO_2^-Na^+ + H_2O$$

$$RCO_2H + NaHCO_3 \longrightarrow RCO_2^-Na^+ + H_2O + CO_2$$

2. Reduction (discussed in Section 12.3):

$$RCO_2H + LiAlH_4 \xrightarrow[(2)\ H_2O]{(1)\ Et_2O} RCH_2OH$$

3. Conversion to acyl chlorides (discussed in Section 18.5):

$$RCO_2H \xrightarrow{SOCl_2 \text{ or } PCl_5} RCOCl$$

4. Conversion to esters (Fischer esterification) or lactones (discussed in Section 18.7):

$$RCOOH + R'-OH \underset{}{\overset{HA}{\rightleftharpoons}} RCOOR' + H_2O$$

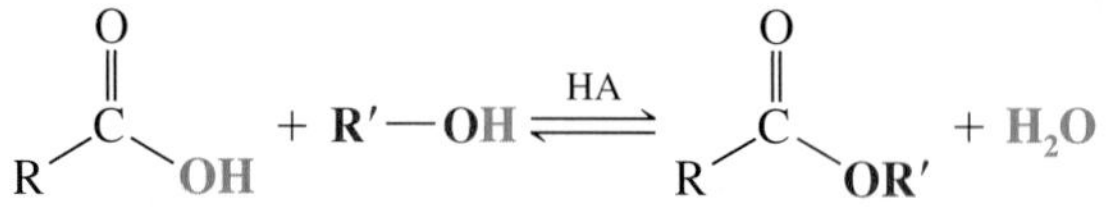

5. Conversion to amides (discussed in Section 18.8):

$$R-COOH + NH_3 \rightleftharpoons R-COO^- NH_4^+ \xrightarrow{heat} R-CONH_2 + H_2O$$

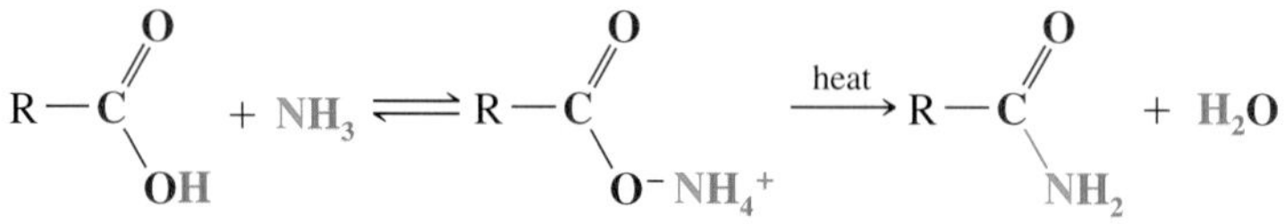

An amide

6. Decarboxylation (discussed in Section 18.10):

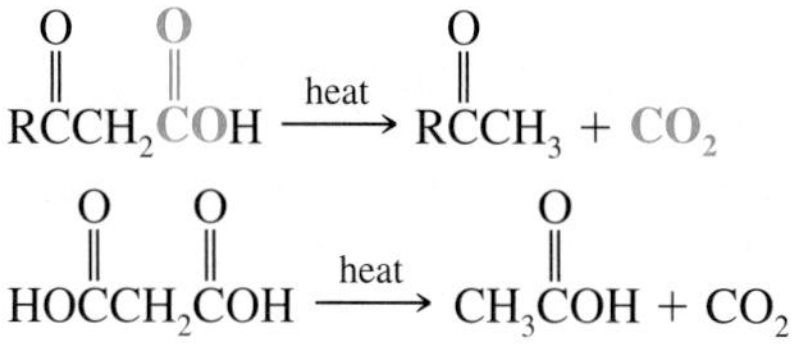

$$RCOCH_2COOH \xrightarrow{heat} RCOCH_3 + CO_2$$

$$HOOCCH_2COOH \xrightarrow{heat} CH_3COOH + CO_2$$

Reactions of Acyl Chlorides

1. Conversion to acids (discussed in Section 18.5B):

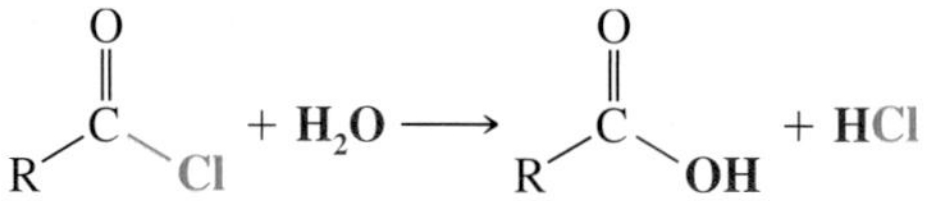

$$RCOCl + H_2O \longrightarrow RCOOH + HCl$$

2. Conversion to anhydrides (discussed in Section 18.6A):

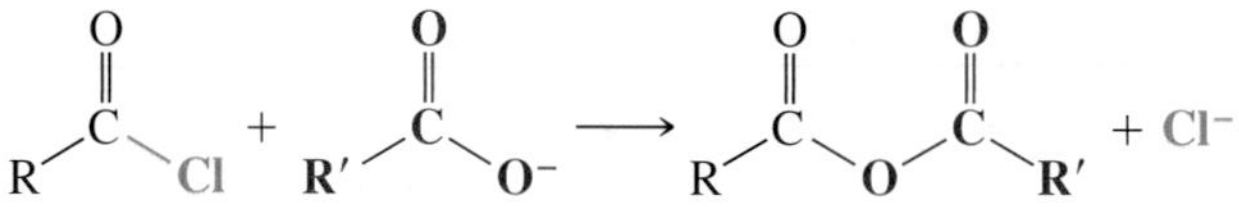

$$RCOCl + R'COO^- \longrightarrow RCOOCOR' + Cl^-$$

3. Conversion to esters (discussed in Section 18.7A):

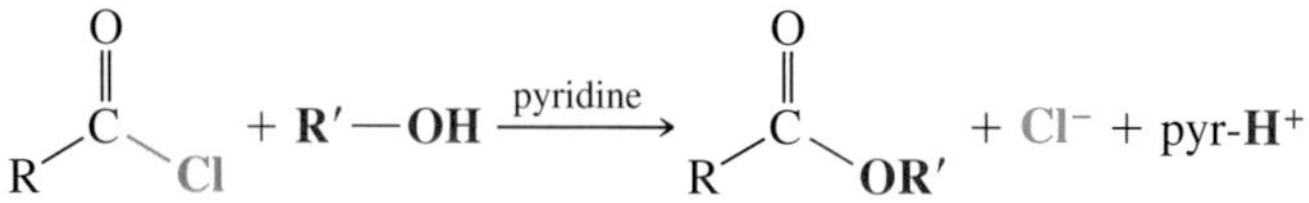

$$RCOCl + R'-OH \xrightarrow{pyridine} RCOOR' + Cl^- + pyr\text{-}H^+$$

4. Conversion to amides (discussed in Section 18.8B):

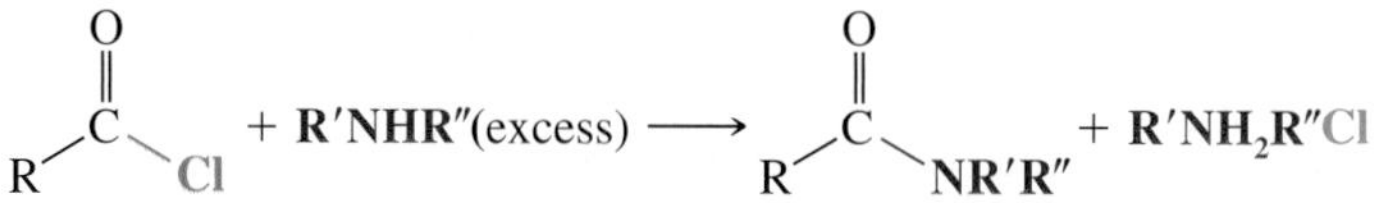

$$RCOCl + R'NHR''(excess) \longrightarrow RCONR'R'' + R'NH_2R''Cl$$

R′ and/or R″ may be H.

5. Conversion to ketones:

$$RCOCl + C_6H_6 \xrightarrow{AlCl_3} C_6H_5COR$$

(Friedel-Crafts acylation, Sections 15.7–15.9)

$$RCOCl + R'_2CuLi \longrightarrow RCOR'$$

(reactions of dialkylcuprates, Section 16.5)

6. Conversion to aldehydes (discussed in Section 16.4):

$$RCOCl + LiAlH[OC(CH_3)_3]_3 \xrightarrow[(2)\ H_2O]{(1)\ Et_2O} RCHO$$

Reactions of Acid Anhydrides

1. Conversion to acids (discussed in Section 18.6B):

$$(RCO)_2O + H_2O \longrightarrow 2\ RCOOH$$

2. Conversion to esters (discussed in Sections 18.6B and 18.7A):

$$(RCO)_2O + R'OH \longrightarrow RCOOR' + RCOOH$$

3. Conversion to amides (discussed in Section 18.8C):

$$(RCO)_2O + H{-}NR'R'' \longrightarrow RCONR'R'' + RCOOH$$

R′ and/or R″ may be H.

4. Conversion to aryl ketones (Friedel-Crafts acylation, Sections 15.7–15.9):

$$(RCO)_2O + C_6H_6 \xrightarrow{AlCl_3} C_6H_5COR + RCOOH$$

Reactions of Esters

1. Hydrolysis (discussed in Section 18.7):

$$RCOOR' + H_2O \underset{}{\overset{HA}{\rightleftharpoons}} RCOOH + R'{-}OH$$

$$RCOOR' + OH^- \xrightarrow[H_2O]{} RCOO^- + R'{-}OH$$

2. Conversion to other esters: transesterification (discussed in Problem 18.11):

$$RCOOR' + R''{-}OH \overset{HA}{\rightleftharpoons} RCOOR'' + R'{-}OH$$

3. Conversion to amides (discussed in Section 18.8D):

$$RCOOR' + HNR''R''' \longrightarrow RCONR''R''' + R'{-}OH$$

R″ and/or R‴ may be H.

4. Reaction with Grignard reagents (discussed in Section 12.8):

$$RCOOR' + 2\,R''MgX \xrightarrow{Et_2O} R{-}C(OMgX)(R''){-}R'' + R'OMgX \xrightarrow{H_3O^+} R{-}C(OH)(R''){-}R''$$

5. Reduction (discussed in Section 12.3):

$$RCO{-}O{-}R' + LiAlH_4 \xrightarrow[(2)\ H_2O]{(1)\ Et_2O} R{-}CH_2OH + R'{-}OH$$

Reactions of Amides

1. Hydrolysis (discussed in Section 18.8F):

$$R-C(=O)-NR'R'' + H_3O^+ \xrightarrow{H_2O} R-C(=O)-OH + R'-\overset{+}{N}H_2R''$$

$$R-C(=O)-NR'R'' + OH^- \xrightarrow{H_2O} R-C(=O)-O^- + R'-NHR''$$

R, R′, and/or R″ may be H.

2. Conversion to nitriles: dehydration (discussed in Section 18.8G):

$$R-C(=O)-NH_2 \xrightarrow[\text{heat}\ (-H_2O)]{P_4O_{10}} R-C\equiv N$$

Reactions of Nitriles

1. Hydrolysis to a carboxylic acid or carboxylate anion (Section 18.8H):

$$R-C\equiv N \xrightarrow{H_3O^+,\ \text{heat}} RCO_2H$$

$$R-C\equiv N \xrightarrow{HO^-,\ H_2O,\ \text{heat}} RCO_2^-$$

2. Reduction to an aldehyde with $(i\text{-Bu})_2AlH$ (DIBAL-H, Section 16.4):

$$R-C\equiv N \xrightarrow[(2)\ H_2O]{(1)\ (i\text{-Bu})_2AlH} R-C(=O)-H$$

3. Conversion to a ketone by a Grignard or organolithium reagent (Section 16.5D):

$$R-C\equiv N + R'-M \xrightarrow[(2)\ H_3O^+]{(1)\ Et_2O} R-C(=O)-R'$$

M = MgBr or Li

Some Synthetic Connections of Carboxylic Acids and Related Functional Groups

Each functional group in the following three-dimensional array (on p. 856) can be synthesized from at least one other functional group that occupies an adjacent or diagonal cor-

Summary and Review Tools

Synthetic Connections of Carboxylic Acids and Related Functional Groups: A 3-D Array of Linked Functional Groups

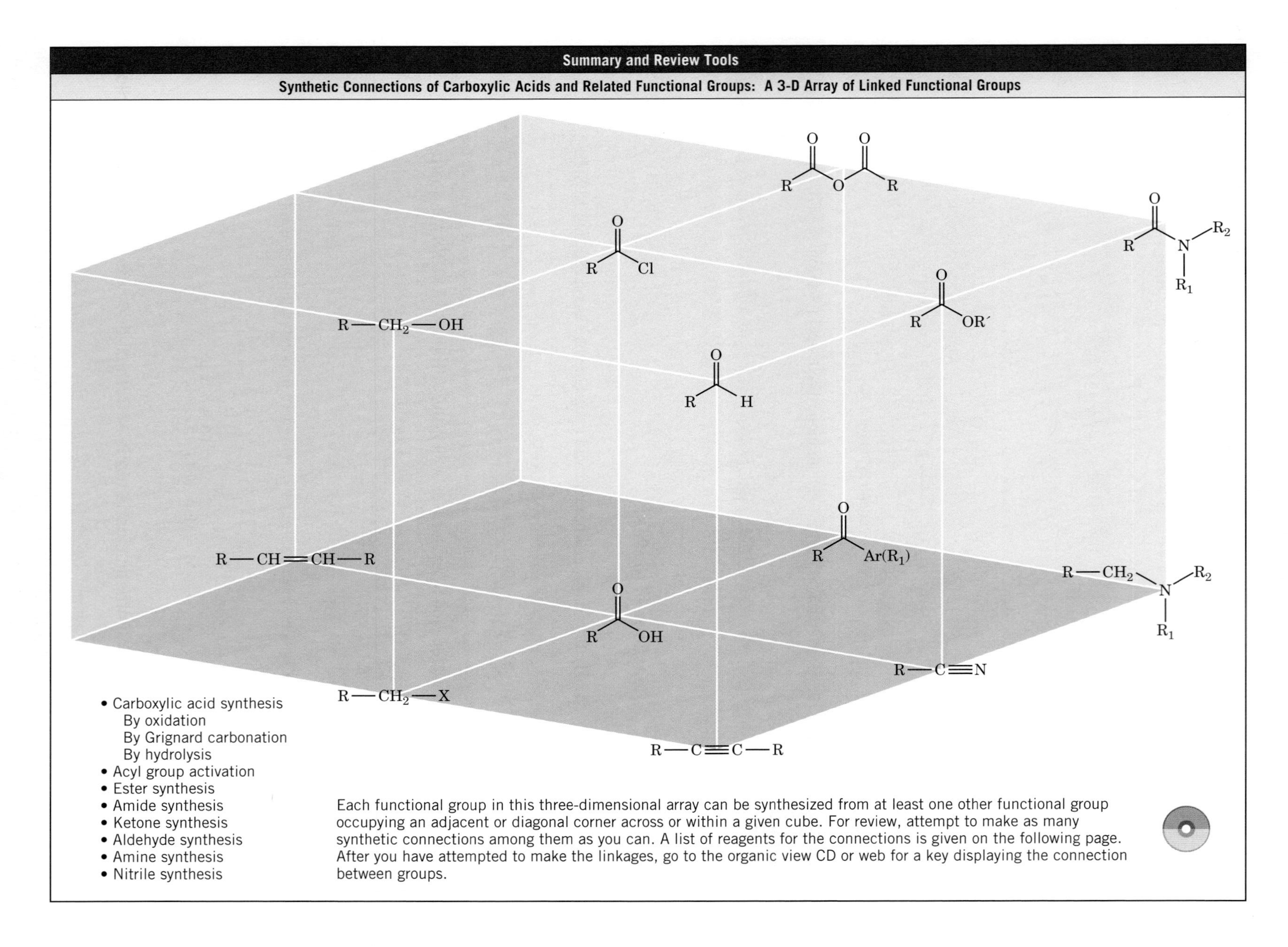

- Carboxylic acid synthesis
  - By oxidation
  - By Grignard carbonation
  - By hydrolysis
- Acyl group activation
- Ester synthesis
- Amide synthesis
- Ketone synthesis
- Aldehyde synthesis
- Amine synthesis
- Nitrile synthesis

Each functional group in this three-dimensional array can be synthesized from at least one other functional group occupying an adjacent or diagonal corner across or within a given cube. For review, attempt to make as many synthetic connections among them as you can. A list of reagents for the connections is given on the following page. After you have attempted to make the linkages, go to the organic view CD or web for a key displaying the connection between groups.

ner within or across a cube. The reagents listed below make viable synthetic links between specific functional groups in this functional group–reaction space. For review, try to make as many synthetic connections as you can between functional groups in the array; then go to the CD for a graphical key to links made by each set of reagents.

Carboxylic Acid Synthesis

- **By oxidation**

i. H_2CrO_4 or (1) $KMnO_4$, OH^-; (2) H_3O^+

ii. (1) Ag_2O; (2) H_3O^+

iii. (1) O_3, HOAc; (2) H_2O_2

iv. (1) O_3; (2) Zn, HOAc

v. (1) $KMnO_4$, OH^-; (2) H_3O^+

- **By Grignard carbonation**

(1) Mg; (2) CO_2; (3) H_3O^+

- **By hydrolysis**

i. H_2O

ii. H_2O

iii. H_3O^+ or (1) HO^-, heat (2) H_3O^+

iv. H_3O^+ or (1) HO^-, heat (2) H_3O^+

v. H_3O^+ or (1) HO^-, heat (2) H_3O^+

Acyl Group Activation

i. $SOCl_2$ or PCl_5

ii. RCO_2^- (react with (RCOCl)

iii. Heat (to form cyclic anhydrides)

Ester Synthesis

i. R′OH, pyridine

ii. R′OH

iii. R′OH, cat. HA

iv. RCO_3H

Amide Synthesis

i. NHR_1R_2 (excess)

ii. NHR_1R_2 (excess)

iii. NHR_1R_2 (heat)

iv. Heat and intramolecular amine (for lactam)

Amine Synthesis

i. (1) $LiAlH_4$; (2) H_2O (where R_1, R_2 = H)

ii. (1) NHR_1R_2; (2) $NaBH_4$

iia. (1) NHR_1R_2; (2) $NaBH_4$ (product has R—CHAr(R_1)— instead of R—CH_2)—

iii. (1) $LiAlH_4$; (2) H_2O

Ketone Synthesis

i. ArH, $AlCl_3$; or R_2CuLi

ii. ArH, $AlCl_3$

iii. (1) Ar(R_1)MgBr; (2) H_3O^+

Aldehyde Synthesis

i. (1) O_3; (2) Zn, HOAc

ii. PCC

iii. (1) (LiAlH[OC(CH_3)$_3$]$_3$; (2) H_2O

iv. (1) (LiAlH[OC(CH_3)$_3$]$_3$; (2) H_2O

v. (1) (i-Bu)$_2$AlH; (2) H_2O

Nitrile Synthesis

i. CN^- (product has R = R—CH_2)

ii. HCN (cyanohydrin; product has R = RCOHAr(R_1)

iii. P_4O_{10}

Key Terms and Concepts

Acyl compounds	**Section 18.1**
Acyl transfer reaction	**Section 18.4**
Carboxylic acid derivatives	**Section 18.1**
Cyanohydrins	**Sections 16.9, 18.3**
Decarboxylation	**Section 18.10**

Esterification	Section 18.7A
Lactams	Section 18.8I
Lactones	Section 18.7C
Nucleophilic addition–elimination	Section 18.4
Saponification	Section 18.7B
Tetrahedral intermediate	Section 18.4
Transesterification	Section 18.7

Additional Problems

18.19 Write a structural formula for each of the following compounds:

(a) Hexanoic acid
(b) Hexanamide
(c) *N*-Ethylhexanamide
(d) *N,N*-Diethylhexanamide
(e) 3-Hexenoic acid
(f) 2-Methyl-4-hexenoic acid
(g) Hexanedioic acid
(h) Phthalic acid
(i) Isophthalic acid
(j) Terephthalic acid
(k) Diethyl oxalate
(l) Diethyl adipate
(m) Isobutyl propanoate
(n) 2-Napththoic acid
(o) Maleic acid
(p) 2-Hydroxybutanedioic acid (malic acid)
(q) Fumaric acid
(r) Succinic acid
(s) Succinimide
(t) Malonic acid
(u) Diethyl malonate

18.20 Give an IUPAC systematic or common name for each of the following compounds:

(a) $C_6H_5CO_2H$
(b) C_6H_5COCl
(c) $C_6H_5CONH_2$
(d) $(C_6H_5CO)_2O$
(e) $C_6H_5CO_2CH_2C_6H_5$
(f) $C_6H_5CO_2C_6H_5$
(g) $CH_3CO_2CH(CH_3)_2$
(h) $CH_3CON(CH_3)_2$
(i) CH_3CN

18.21 Show how *p*-chlorotoluene could be converted to each of the following:

(a) *p*-Chlorobenzoic acid
(b) *p*-Chlorophenylacetic acid
(c) *p*-$ClC_6H_4CH(OH)CO_2H$
(d) *p*-$ClC_6H_4CH{=}CHCO_2H$

18.22 Outline each of the following syntheses:

(a) Succinic acid from 1,4-butanediol
(b) Adipic acid from cyclohexanol

18.23 Show how pentanoic acid can be prepared from each of the following:

(a) 1-Pentanol
(b) 1-Bromobutane (two ways)
(c) 5-Decene
(d) Pentanal

18.24 What major organic product would you expect to obtain when acetyl chloride reacts with each of the following?

(a) H_2O
(b) CH_3CH_2Li (excess)
(c) $CH_3(CH_2)_2CH_2OH$ and pyridine
(d) NH_3 (excess)
(e) $C_6H_5CH_3$ and $AlCl_3$
(f) $LiAlH[OC(CH_3)_3]_3$
(g) $(CH_3)_2CuLi$
(h) $NaOH/H_2O$
(i) CH_3NH_2 (excess)
(j) $C_6H_5NH_2$ (excess)
(k) $(CH_3)_2NH$ (excess)
(l) CH_3CH_2OH and pyridine
(m) $CH_3CO_2^-Na^+$
(n) CH_3CO_2H and pyridine
(o) Phenol and pyridine

Note: Problems marked with an asterisk are "challenge problems."

18.25 What major organic product would you expect to obtain when acetic anhydride reacts with each of the following?

(a) NH_3 (excess) **(c)** $CH_3CH_2CH_2OH$ **(e)** $CH_3CH_2NH_2$ (excess)
(b) H_2O **(d)** $C_6H_6 + AlCl_3$ **(f)** $(CH_3CH_2)_2NH$ (excess)

18.26 What major organic product would you expect to obtain when succinic anhydride reacts with each of the reagents given in Problem 18.25?

18.27 Starting with benzene and succinic anhydride and using any other needed reagents, outline a synthesis of 1-phenylnaphthalene.

18.28 Starting with either *cis*- or *trans*-$HO_2C—CH=CH—CO_2H$ (i.e., either maleic or fumaric acid) and using any other needed compounds, outline syntheses of each of the following:

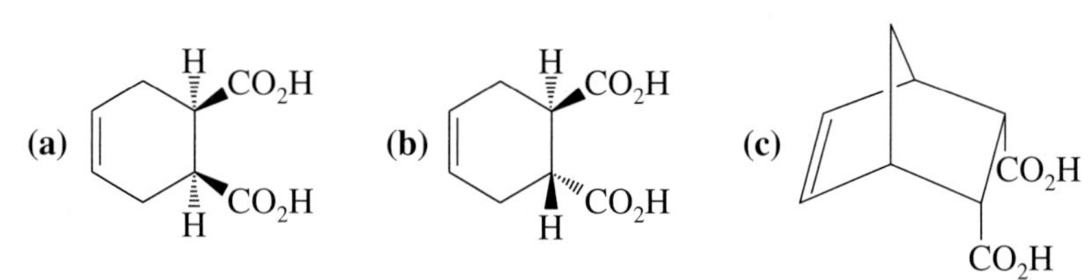

18.29 What products would you expect to obtain when ethyl propanoate reacts with each of the following?

(a) H_3O^+, H_2O **(c)** 1-Octanol, HCl **(e)** $LiAlH_4$, then H_2O
(b) OH^-, H_2O **(d)** CH_3NH_2 **(f)** Excess C_6H_5MgBr, then H_2O, NH_4Cl

18.30 What products would you expect to obtain when propanamide reacts with each of the following?

(a) H_3O^+, H_2O **(b)** OH^-, H_2O **(c)** P_4O_{10} and heat

18.31 Write detailed mechanisms for Problem 18.30 (a) and (b).

18.32 What products would you expect to obtain when each of the following compounds is heated?

(a) 4-Hydroxybutanoic acid
(b) 3-Hydroxybutanoic acid
(c) 2-Hydroxybutanoic acid
(d) Glutaric acid
(e) $CH_3CH(NH_3^+)CH_2CH_2CH_2CO^-$ (with C=O, i.e. $CH_3CHCH_2CH_2CH_2\overset{O}{\overset{\|}{C}}O^-$ bearing NH_3^+ on C-2)
(f) benzene ring bearing $\overset{O}{\overset{\|}{C}}—O^-$ and, ortho to it, $CH_2NH_3^+$

18.33 Give stereochemical formulas for compounds **A–Q**:

(a) (R)-(−)-2-Butanol $\xrightarrow[\text{pyridine}]{p\text{-toluenesulfonyl chloride (TsCl)}}$ **A** $\xrightarrow{CN^-}$ **B** (C_5H_9N) $\xrightarrow[H_2O]{H_2SO_4}$

(+)-**C** $(C_5H_{10}O_2)$ $\xrightarrow[(2)\ H_2O]{(1)\ LiAlH_4}$ (−)-**D** $(C_5H_{12}O)$

(b) (R)-(−)-2-Butanol $\xrightarrow[\text{pyridine}]{PBr_3}$ **E** (C_4H_9Br) $\xrightarrow{CN^-}$ **F** (C_5H_9N) $\xrightarrow[H_2O]{H_2SO_4}$

(−)-**C** $(C_5H_{10}O_2)$ $\xrightarrow[(2)\ H_2O]{(1)\ LiAlH_4}$ (+)-**D** $(C_5H_{12}O)$

(c) **A** $\xrightarrow{CH_3CO_2^-}$ **G** $(C_6H_{12}O_2)$ $\xrightarrow{OH^-}$ (+)-**H** $(C_4H_{10}O)$ + $CH_3CO_2^-$

(d) (−)-**D** $\xrightarrow{PBr_3}$ **J** $(C_5H_{11}Br)$ $\xrightarrow[Et_2O]{Mg}$ **K** $(C_5H_{11}MgBr)$ $\xrightarrow[(2)\ H_3O^+]{(1)\ CO_2}$ **L** $(C_6H_{12}O_2)$

(e) (*R*)-(+)-Glyceraldehyde $\xrightarrow{\text{HCN}}$ **M** ($C_4H_7NO_3$) + **N** ($C_4H_7NO_3$)

Diastereomers, separated by fractional crystallization

(f) **M** $\xrightarrow[\text{H}_2\text{O}]{\text{H}_2\text{SO}_4}$ **P** ($C_4H_8O_5$) $\xrightarrow[\text{HNO}_3]{[\text{O}]}$ *meso*-tartaric acid

(g) **N** $\xrightarrow[\text{H}_2\text{O}]{\text{H}_2\text{SO}_4}$ **Q** ($C_4H_8O_5$) $\xrightarrow[\text{HNO}_3]{[\text{O}]}$ (−)-tartaric acid

18.34 **(a)** (±)-Pantetheine and (±)-pantothenic acid, important intermediates in the synthesis of coenzyme A, were prepared by the following route. Give structures for compounds **A–D:**

$CH_3CH(CH_3)CHO$ + $HCHO$ $\xrightarrow[\text{H}_2\text{O}]{\text{K}_2\text{CO}_3}$ **A** ($C_5H_{10}O_2$) $\xrightarrow{\text{HCN}}$

(±)-**B** ($C_6H_{11}NO_2$) $\xrightarrow{\text{H}_3\text{O}^+}$ [(±)-**C** ($C_6H_{12}O_4$)] $\xrightarrow{-\text{H}_2\text{O}}$

(±)-**D** ($C_6H_{10}O_3$) $\xrightarrow{\text{H}_3\overset{+}{\text{N}}\text{CH}_2\text{CH}_2\text{CO}_2^-}$ $(CH_3)_2C(CH_2OH)—CH(OH)C(=O)—NHCH_2CH_2C(=O)OH$

γ-Lactone from A **(±)-Pantothenic acid**

(±)-**D** $\xrightarrow{\text{H}_2\text{NCH}_2\text{CH}_2\text{C(=O)NHCH}_2\text{CH}_2\text{SH}}$ $(CH_3)_2C(CH_2OH)—CH(OH)—C(=O)—NHCH_2CH_2C(=O)NHCH_2CH_2SH$

(±)-Pantetheine

(b) The γ-lactone, (±)-**D,** can be resolved. If the (−)-γ-lactone is used in the last step, the pantetheine that is obtained is identical with that obtained naturally. The (−)-γ-lactone has the (*R*) configuration. What is the stereochemistry of naturally occurring pantetheine? **(c)** What products would you expect to obtain when (±)-pantetheine is heated with aqueous sodium hydroxide?

18.35 The IR and ^1H NMR spectra of phenacetin ($C_{10}H_{13}NO_2$) are given in Fig. 18.7. Phenacetin is an analgesic and antipyretic compound and was the P of A–P–C tablets (**a**spirin–**p**henacetin–**c**affeine). (Because of its toxicity, phenacetin is no longer used medically.) When phenacetin is heated with aqueous sodium hydroxide, it yields phenetidine ($C_8H_{11}NO$) and sodium acetate. Propose structures for phenacetin and phenetidine.

18.36 Given here are the ^1H NMR spectra and carbonyl IR absorption peaks of five acyl compounds. Propose a structure for each.

	^1H NMR Spectrum		IR Spectrum
(a) $C_8H_{14}O_4$	Triplet	δ 1.2 (6H)	1740 cm^{-1}
	Singlet	δ 2.5 (4H)	
	Quartet	δ 4.1 (4H)	
(b) $C_{11}H_{14}O_2$	Doublet	δ 1.0 (6H)	1720 cm^{-1}
	Multiplet	δ 2.1 (1H)	
	Doublet	δ 4.1 (2H)	
	Multiplet	δ 7.8 (5H)	
(c) $C_{10}H_{12}O_2$	Triplet	δ 1.2 (3H)	1740 cm^{-1}
	Singlet	δ 3.5 (2H)	

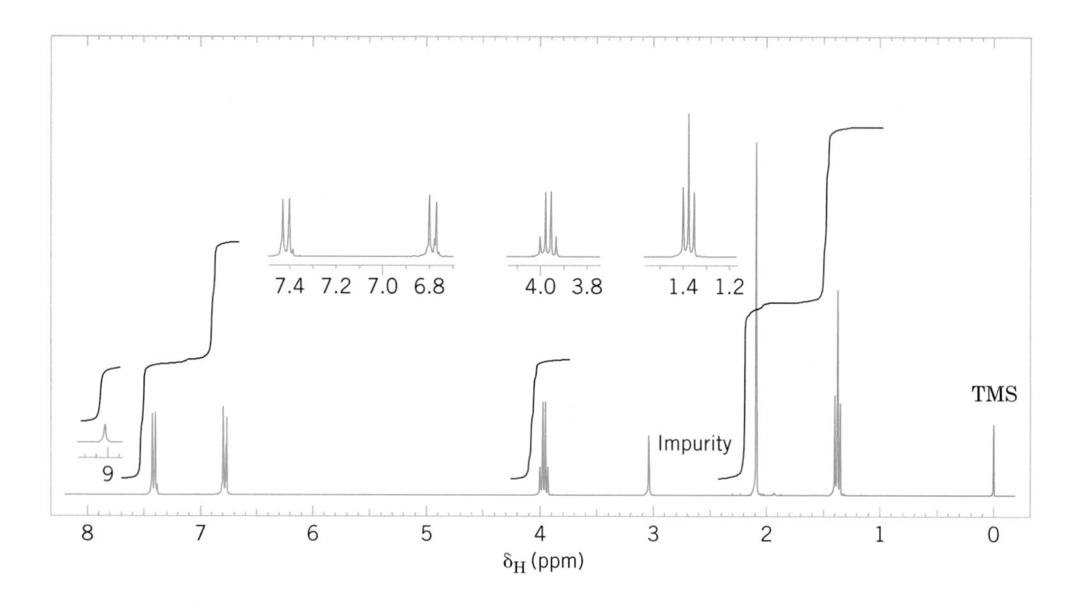

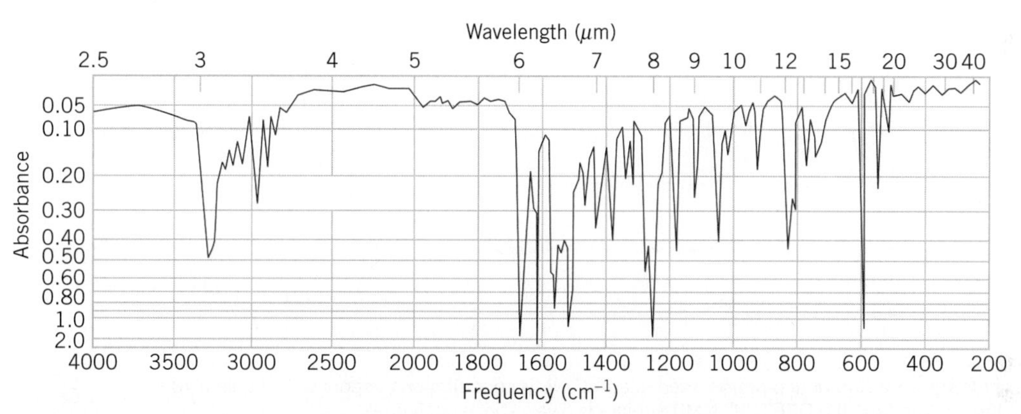

FIGURE 18.7 The 300-MHz ^{1}H NMR and IR spectra of phenacetin. Expansions of the ^{1}H NMR signals are shown in the offset plots. (Infrared spectrum courtesy Sadtler Research Laboratories, Philadelphia.)

	Quartet δ 4.1 (2H)	
	Multiplet δ 7.3 (5H)	
(d) $C_2H_2Cl_2O_2$	^{1}H NMR Spectrum	IR Spectrum
	Singlet δ 6.0	Broad peak 2500–2700 cm^{-1}
	Singlet δ 11.70	1705 cm^{-1}
(e) $C_4H_7ClO_2$	^{1}H NMR Spectrum	IR Spectrum
	Triplet δ 1.3	1745 cm^{-1}
	Singlet δ 4.0	
	Quartet δ 4.2	

18.37 The active ingredient of the insect repellent "Off" is *N,N*-diethyl-*m*-toluamide, *m*-$CH_3C_6H_4CON(CH_2CH_3)_2$. Outline a synthesis of this compound starting with *m*-toluic acid.

18.38 Amides are weaker bases than corresponding amines. For example, most water-insoluble amines (RNH_2) will dissolve in dilute aqueous acids (aqueous HCl, H_2SO_4, etc.) by forming water-soluble alkylammonium salts ($RNH_3^+X^-$). Corresponding amides ($RCONH_2$) *do not dissolve in dilute aqueous acids,* however. Propose an explanation for the much lower basicity of amides when compared to amines.

18.39 While amides are much less basic than amines, they are much stronger acids. Amides have pK_a values in the range 14–16, whereas for amines, $pK_a = 33–35$.

(a) What factor accounts for the much greater acidity of amides?

(b) Imides, that is, compounds with the structure $(\overset{O}{\overset{\|}{R C}})_2NH$, are even stronger acids than amides. For imides, $pK_a = 9–10$, and as a consequence, water-insoluble imides dissolve in aqueous NaOH by forming soluble sodium salts. What extra factor accounts for the greater acidity of imides?

18.40 Compound **X** ($C_7H_{12}O_4$) is insoluble in aqueous sodium bicarbonate. The IR spectrum of **X** has a strong absorption peak near 1740 cm^{-1}, and its broadband proton-decoupled ^{13}C spectrum is given in Fig. 18.8. Propose a structure for **X.**

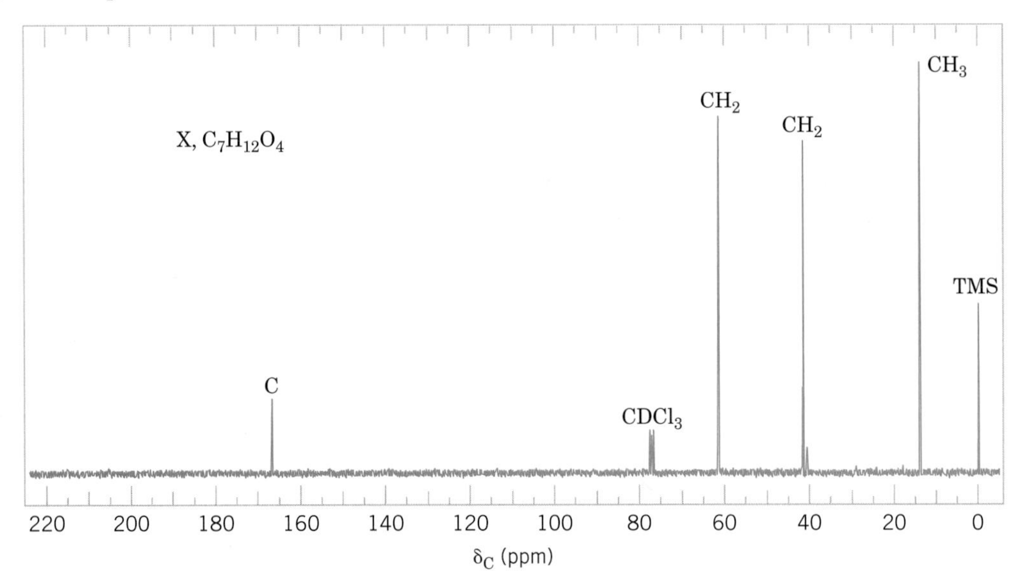

FIGURE 18.8 Broadband proton-decoupled ^{13}C NMR spectrum of compound X, Problem 18.40. Information from the DEPT ^{13}C NMR spectra is given above each peak.

18.41 Alkylthio acetates $\left(CH_3\overset{O}{\overset{\|}{C}}SCH_2CH_2R\right)$ can be prepared by a peroxide-initiated reaction between thiolacetic acid $\left(CH_3\overset{O}{\overset{\|}{C}}SH\right)$ and an alkene ($CH_2{=}CHR$). **(a)** Outline a reasonable mechanism for this reaction. **(b)** Show how you might use this reaction in a synthesis of 3-methyl-2-butanethiol from 2-methyl-2-butene.

18.42 On heating, *cis*-4-hydroxycyclohexanecarboxylic acid forms a lactone but *trans*-4-hydroxycyclohexanecarboxylic acid does not. Explain.

18.43 (*R*)-(+)-Glyceraldehyde can be transformed into (+)-malic acid by the following synthetic route. Give stereochemical structures for the products of each step.

$$(R)\text{-}(+)\text{-Glyceraldehyde} \xrightarrow[\text{oxidation}]{Br_2,\ H_2O} (-)\text{-glyceric acid} \xrightarrow{PBr_3}$$

$$(-)\text{-3-bromo-2-hydroxypropanoic acid} \xrightarrow{NaCN} C_4H_5NO_3 \xrightarrow[\text{heat}]{H_3O^+} (+)\text{-malic acid}$$

18.44 (*R*)-(+)-Glyceraldehyde can also be transformed into (−)-malic acid. This synthesis begins with the conversion of (*R*)-(+)-glyceraldehyde into (−)-tartaric acid, as shown in Problem 18.33, parts (e) and (g). Then (−)-tartaric acid is allowed to react with phosphorus tribromide in order to replace one alcoholic —OH group with —Br. This step takes place with inversion of configuration at the carbon that undergoes attack. Treating the product of this reaction with zinc and acid produces (−)-malic acid. **(a)** Outline all steps in this synthesis by writing stereochemical structures for each intermediate. **(b)** The step in which (−)-tartaric acid is treated with phosphorus tribromide produces only one stereoisomer, even though there are two replaceable —OH groups. How is this possible? **(c)** Suppose that the step in which (−)-tartaric acid is treated with phosphorus tribromide had taken place with "mixed" stereochemistry, that is, with both inversion and retention at the carbon under attack. How many stereoisomers would have been produced? **(d)** What difference would this have made to the overall outcome of the synthesis?

18.45 Cantharidin is a powerful vesicant that can be isolated from dried beetles (*Cantharis vesicatoria,* or "Spanish fly"). Outlined here is the stereospecific synthesis of cantharidin reported by Gilbert Stork of Columbia University in 1953. Supply the missing reagents (a)–(n):

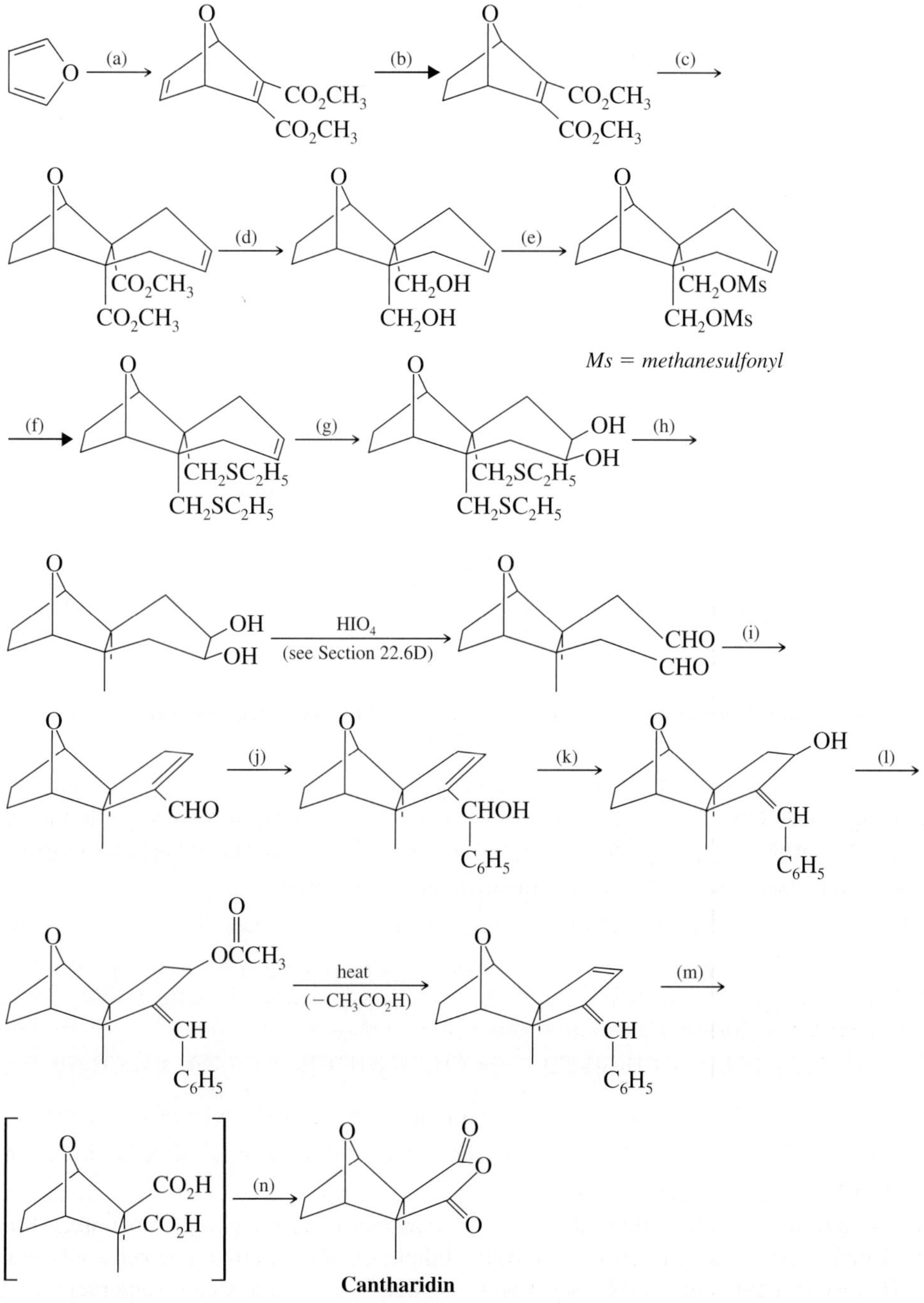

(a) CO_2CH_3 CO_2CH_3 (b) CO_2CH_3 CO_2CH_3 (c)

CO_2CH_3 CO_2CH_3 (d) CH_2OH CH_2OH (e) CH_2OMs CH_2OMs

Ms = methanesulfonyl

(f) $CH_2SC_2H_5$ $CH_2SC_2H_5$ (g) OH OH $CH_2SC_2H_5$ $CH_2SC_2H_5$ (h)

OH OH HIO_4 (see Section 22.6D) CHO CHO (i)

CHO (j) CHOH C_6H_5 (k) OH CH C_6H_5 (l)

$OCCH_3$ CH C_6H_5 heat ($-CH_3CO_2H$) CH C_6H_5 (m)

CO_2H CO_2H (n)

Cantharidin

18.46 Examine the structure of cantharidin (Problem 18.45) carefully and **(a)** suggest a possible two-step synthesis of cantharidin starting with furan (Section 14.9). **(b)** F. von Bruchhausen and H. W. Bersch at the University of Münster attempted this two-step synthesis in 1928, only a few months after Diels and Alder published their first paper describing their new diene addition, and found that the expected addition failed to take place. von Bruchhausen and Bersch also found that although cantharidin is stable at relatively high temperatures, heating cantharidin with a palladium catalyst causes cantharidin to decompose. They identified furan and dimethylmaleic anhydride among the decomposition products. What has happened in the decomposition, and what does this suggest about why the first step of their attempted synthesis failed?

18.47 Compound **Y** ($C_8H_4O_3$) dissolves slowly when warmed with aqueous sodium bicarbonate. The IR spectrum of **Y** has strong peaks at 1779 and at 1854 cm^{-1}. The broadband proton-decoupled ^{13}C spectrum of **Y** is given in Fig. 18.9. Acidification of the bicarbonate solution of **Y** gave compound **Z.** The proton-decoupled ^{13}C NMR spectrum of **Z** showed four signals. When **Y** was warmed in ethanol, a compound **AA** was produced. The ^{13}C NMR spectrum of **AA** displayed 10 signals. Propose structures for **Y, Z,** and **AA.**

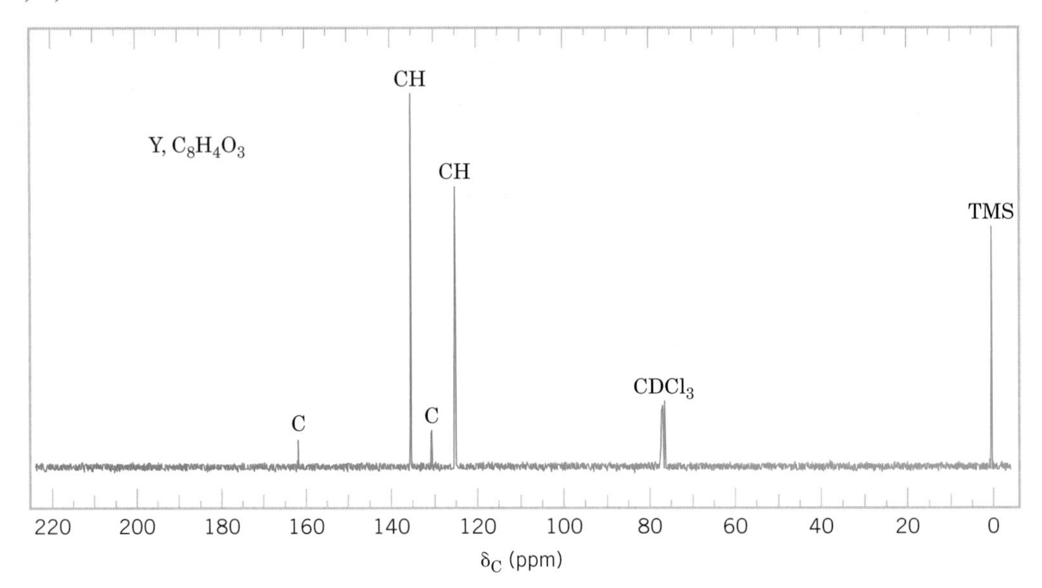

FIGURE 18.9 Broadband proton-decoupled ^{13}C NMR spectrum of compound Y, Problem 18.47. Information from the DEPT ^{13}C NMR spectra is given above each peak.

18.48 Ketene, $H_2C{=}C{=}O$, an important industrial chemical, can be prepared by dehydration of acetic acid at a high temperature or by the pyrolysis of acetone. Predict the products that would be formed when ketene reacts with **(a)** ethanol, **(b)** acetic acid, and **(c)** ethylamine. (*Hint:* Markovnikov addition occurs.)

18.49 Two unsymmetrical anhydrides react with ethylamine as follows:

$$\overset{O}{\overset{\|}{HC}}O\overset{O}{\overset{\|}{C}}CH_3 + CH_3CH_2NH_2 \longrightarrow CH_3CH_2NHCHO + CH_3CO_2^-\,CH_3CH_2NH_3^+$$

$$CF_3\overset{O}{\overset{\|}{C}}O\overset{O}{\overset{\|}{C}}CH_3 + CH_3CH_2NH_2 \longrightarrow CH_3CH_2NH\overset{O}{\overset{\|}{C}}CH_3 + CF_3CO_2^-\,CH_3CH_2NH_3^+$$

Explain the factors that might account for the formation of the products in each reaction.

18.50 Starting with 1-naphthol, suggest an alternative synthesis of the insecticide Sevin to the one given in Section 18.9A.

18.51 Suggest a synthesis of ibuprofen (Section 5.10) from benzene, employing **chloromethylation** as one step. Chloromethylation is a special case of the Friedel–Crafts reaction in which a mixture of HCHO and HCl, in the presence of $ZnCl_2$, introduces a $—CH_2Cl$ group into an aromatic ring.

18.52 An alternative synthesis of ibuprofen is given below. Supply the structural formulas for compounds **A–D:**

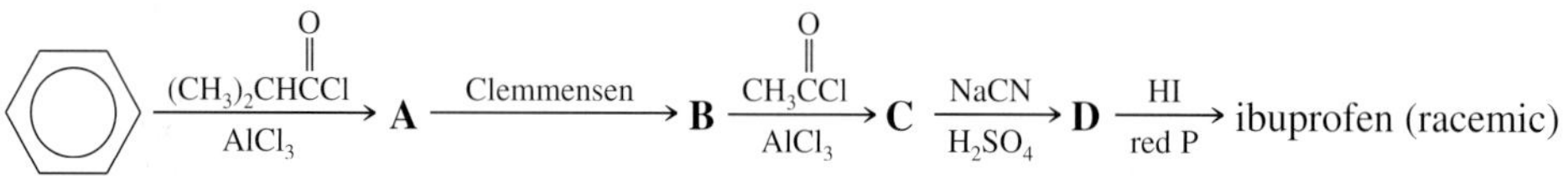

$$\text{C}_6\text{H}_6 \xrightarrow[\text{AlCl}_3]{(\text{CH}_3)_2\text{CHCCl (O)}} \textbf{A} \xrightarrow{\text{Clemmensen}} \textbf{B} \xrightarrow[\text{AlCl}_3]{\text{CH}_3\text{CCl (O)}} \textbf{C} \xrightarrow[\text{H}_2\text{SO}_4]{\text{NaCN}} \textbf{D} \xrightarrow[\text{red P}]{\text{HI}} \text{ibuprofen (racemic)}$$

18.53 As a method for the synthesis of cinnamaldehyde (3-phenyl-2-propenal), a chemist treated 3-phenyl-2-propen-1-ol with $K_2Cr_2O_7$ in sulfuric acid. The product obtained from the reaction gave a signal at δ 164.5 in its ^{13}C NMR spectrum. Alternatively, when the chemist treated 3-phenyl-2-propen-1-ol with PCC in CH_2Cl_2, the ^{13}C NMR spectrum of the product displayed a signal at δ 193.8. (All other signals in the spectra of both compounds appeared at similar chemical shifts.) **(a)** Which reaction produced cinnamaldehyde? **(b)** What was the other product?

***18.54** Two stereoisomers **A** and **C** have the structure

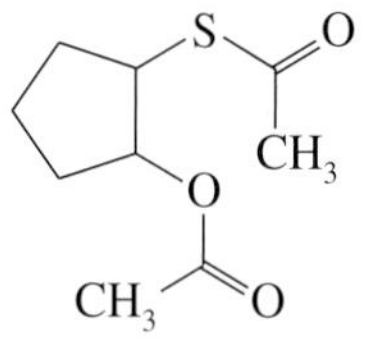

[Structure: cyclopentane bearing S–C(=O)CH₃ and O–C(=O)CH₃ on adjacent carbons]

On base-promoted hydrolysis, isomer **A** yields product **B**. These are some of the spectral data for **B**:

MS (*m/z*): 118 ($M^{\cdot+}$)

IR (cm^{-1}): 3415, 2550

^{1}H NMR (δ): multiplets at 1.51, 1.66, 1.77, 2.65, and 3.55 (area ratios 2:2:2:1:1)

^{13}C NMR (δ): 16(CH_2), 28(CH_2), 30(CH_2), 39(CH), and 77(CH)

Under the same conditions, isomer **C** yields product **D**. These are the spectral data for **D**:

MS (*m/z*): 100 ($M^{\cdot+}$)

IR (cm^{-1}): 3020

^{1}H NMR (δ): multiplets at 1.51, 1.84, and 2.25 (area ratios 1:2:1)

^{13}C NMR (δ): 22(CH_2), 33(CH_2), 35(CH_2)

(a) What are the structures of **A** and **B**?
(b) Of **C** and **D**?
(c) Write mechanisms for the formation of **B** and of **D**.

***18.55** Consider this two-step reaction sequence that proceeds through intermediate **E** (which ordinarily is not isolated) to final product **F**:

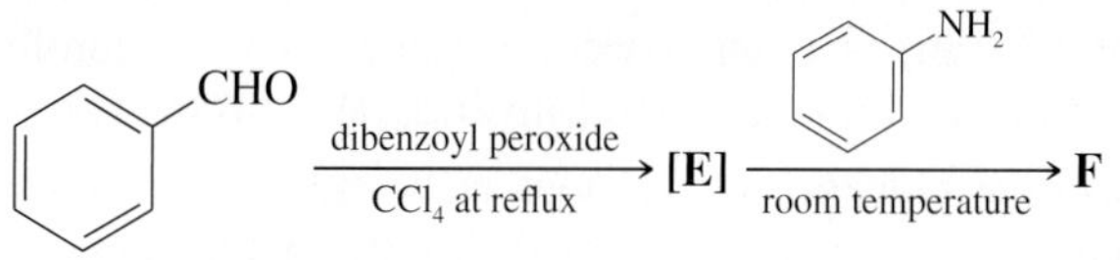

$$\text{C}_6\text{H}_5\text{CHO} \xrightarrow[\text{CCl}_4\text{ at reflux}]{\text{dibenzoyl peroxide}} [\textbf{E}] \xrightarrow[\text{room temperature}]{\text{C}_6\text{H}_5\text{NH}_2} \textbf{F}$$

These are some of the spectral data for **E**:

MS (*m/z*): 105 (*not* $M^{\cdot+}$), 77

IR (cm^{-1}): 3065 (the only band from 2600 to 3600), 1774, 1595, 1485, 775, 685

^{1}H NMR (δ): 7.6 (m, 3H) and 8.1 (m, 2H)

^{13}C NMR (δ): 129(CH), 131(CH), 133(C), 135(CH), and 168(C)

And these are selected data for **F**:

MS (*m/z*): 197 ($M^{\cdot+}$)

IR (cm^{-1}): (in CCl_4): 3200, 3065, 1690, 1530

^1H NMR (δ): 10.0 (s), 7.9 (m, 4H), and 7.3 (m, 6H)

(a) What are the structures of **E** and **F**?

(b) Write a mechanism for the formation of **E** and of **F**.

Learning Group Problems

The Chemical Synthesis of Peptides Carboxylic acids and acyl derivatives of the carboxyl functional group are very important in biochemistry. For example, the carboxylic acid functional group is present in the family of lipids called fatty acids. Lipids called glycerides contain the ester functional group, a derivative of carboxylic acids. Furthermore, the entire class of biopolymers called proteins contain repeating amide functional group linkages. Amides are also derivatives of carboxylic acids. Both laboratory and biochemical syntheses of proteins require reactions that involve substitution at activated acyl carbons.

This Learning Group Problem focuses on the chemical synthesis of small proteins, called peptides. The essence of peptide or protein synthesis is formation of the amide functional group by reaction of an activated carboxylic acid derivative with an amine.

First we shall consider reactions for traditional chemical synthesis of peptides and then we look at reactions used in automated solid-phase peptide synthesis. The method for solid-phase peptide synthesis was invented by R. B. Merrifield (Rockefeller University), for which he earned the 1984 Nobel Prize in Chemistry. Solid-phase peptide synthesis reactions are so reliable that they have been incorporated into machines called peptide synthesizers (Section 24.7D).

1. The first step in peptide synthesis is blocking (protection) of the amine functional group of an amino acid (a compound that contains both amine and carboxylic acid functional groups). Such a reaction is shown in Section 24.7C in the reaction between Ala (alanine) and benzyl chloroformate. The functional group formed in the structure labeled Z-Ala is called a carbamate (or urethane). [Z is a benzyloxycarbonyl group ($C_6H_5CH_2O\overset{O}{\overset{\|}{C}}—$)].
   (a) Write a detailed mechanism for formation of Z-Ala from Ala and benzyl chloroformate in the presence of hydroxide.
   (b) In the reaction of part (a), why does the amino group act as the nucleophile preferentially over the carboxylate anion?
   (c) Another widely used amino protecting group is the 9-fluorenylmethoxycarbonyl (Fmoc) group. Fmoc is the protecting group most often used in automated solid-phase peptide synthesis (see part 4 below). Write a detailed mechanism for formation of an Fmoc-protected amino acid under the conditions given in Section 24.7A.
2. The second step in the reactions of Section 24.7C is the formation of a mixed anhydride. Write a detailed mechanism for the reaction between Z-Ala and ethyl chloroformate ($ClCO_2C_2H_5$) in the presence of triethylamine to form the mixed anhydride. What is the purpose of this step?
3. The third step in the sequence of reactions in Section 24.7C is the one that actually joins the new amino acid (in this case leucine, abbreviated Leu) by another amide functional group. Write a detailed mechanism for this step (from the mixed anhydride of Z-Ala to Z-Ala-Leu). Show how CO_2 and ethanol are formed in the course of this mechanism.
4. A sequence of reactions commonly used for solid-phase peptide synthesis is shown in Section 24.7D.
   (a) Write a detailed mechanism for step 1, in which diisopropylcarbodiimide is used to join the carboxyl group of the first amino acid (in Fmoc-protected form) to a hydroxyl group on the polymer solid support.
   (b) Step 3 of the automated synthesis involves removal of the Fmoc group by reaction with piperidine (a reaction also shown in Section 24.7A). Write a detailed mechanism for this step.

Step-Growth Polymers

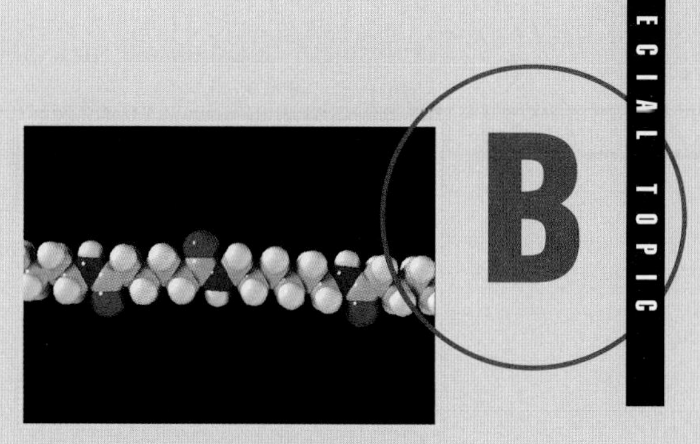

Polypropylene (syndiotactic)

SPECIAL TOPIC B

We saw, in Special Topic A, that large molecules with many repeating subunits—called *polymers*—can be prepared by addition reactions of alkenes. These polymers, we noted, are called *chain-growth polymers* or *addition polymers.*

Another broad group of polymers has been called *condensation polymers* but is now more often called *step-growth polymers.* These polymers, as their older name suggests, are prepared by condensation reactions—reactions in which monomeric subunits are joined through intermolecular eliminations of small molecules such as water or alcohols. Among the most important condensation polymers are *polyamides, polyesters, polyurethanes,* and *formaldehyde resins.*

B.1 Polyamides

Silk and wool are two naturally occurring polymers that humans have used for centuries to fabricate articles of clothing. They are examples of a family of compounds that are called *proteins*—a group of compounds that we shall discuss in detail in Chapter 24. At this point we need only to notice (below) that the repeating subunits of proteins are derived from α-amino acids and that these subunits are joined by amide linkages. Proteins, therefore, are polyamides:

$$H_2N-\underset{\underset{R}{|}}{CH}-\overset{\overset{O}{\|}}{C}-OH$$

An α-amino acid

Amide linkages

$$-NH-\underset{\underset{R}{|}}{CH}-\overset{\overset{O}{\|}}{C}-NH-\underset{\underset{R}{|}}{CH}-\overset{\overset{O}{\|}}{C}-NH-\underset{\underset{R}{|}}{CH}-\overset{\overset{O}{\|}}{C}-NH-\underset{\underset{R}{|}}{CH}-$$

A portion of a polyamide chain as it might occur in a protein.

The search for a synthetic material with properties similar to those of silk led to the discovery of a family of synthetic polyamides called nylons.

One of the most important nylons, called *nylon 6,6,* (shown as model above) can be prepared from the six-carbon dicarboxylic acid, adipic acid, and the six-carbon diamine, hexamethylenediamine (hexane-1,6-diamine). In the commercial process these two compounds are allowed to react in equimolar proportions in order to produce a 1:1 salt,

$$n\,\mathrm{HO\overset{O}{\overset{\|}{C}}\!\left(CH_2\right)_4\!\overset{O}{\overset{\|}{C}}OH} + n\,\mathrm{H_2N\!\left(CH_2\right)_6\!NH_2} \longrightarrow$$

Adipic acid **Hexamethylenediamine**

$$n\left[{}^{-}\mathrm{O\overset{O}{\overset{\|}{C}}\!\left(CH_2\right)_4\!\overset{O}{\overset{\|}{C}}\!-\!O^{-}\quad H_3\overset{+}{N}\!\left(CH_2\right)_6\!\overset{+}{N}H_3}\right] \xrightarrow[\text{(polymerization)}]{\text{heat}}$$

1 : 1 salt (nylon salt)

$${}^{-}\mathrm{O\overset{O}{\overset{\|}{C}}\!\left(CH_2\right)_4\!\overset{O}{\overset{\|}{C}}}\!\left[\mathrm{NH\!\left(CH_2\right)_6\!NH\!-\!\overset{O}{\overset{\|}{C}}\!\left(CH_2\right)_4\!\overset{O}{\overset{\|}{C}}}\right]_{n-1}\!\mathrm{NH\!\left(CH_2\right)_6\!\overset{+}{N}H_3} + (2n-1)\,\mathrm{H_2O}$$

Nylon 6,6
(a polyamide)

Then, heating the 1:1 salt (nylon salt) to a temperature of 270°C at a pressure of 250 psi (pounds per square inch) causes a polymerization to take place. Water molecules are lost as condensation reactions occur between $-\overset{O}{\overset{\|}{C}}-O^{-}$ and $-NH_3^{+}$ groups of the salt to give the polyamide.

The nylon 6,6 produced in this way has a molecular weight of about 10,000, has a melting point of about 250°C, and when molten can be spun into fibers from a melt. The fibers are then stretched to about four times their original length. This orients the linear polyamide molecules so that they are parallel to the fiber axis and allows hydrogen bonds to form between —NH— and C═O groups on adjacent chains. Called "cold drawing," stretching greatly increases the fibers' strength.

Another type of nylon, nylon 6, can be prepared by a ring-opening polymerization of ε-caprolactam:

$$\text{ε-caprolactam} \xrightarrow{H_2O} {}^{-}\mathrm{O\overset{O}{\overset{\|}{C}}\!\left(CH_2\right)_5\!\overset{+}{N}H_3} + \text{ε-caprolactam} \xrightarrow[250^\circ C]{(-H_2O)}$$

ε-Caprolactam
(a cyclic amide)

$$-\mathrm{NH}\!\left[\mathrm{\overset{O}{\overset{\|}{C}}\!\left(CH_2\right)_5\!NH\!-\!\overset{O}{\overset{\|}{C}}\!\left(CH_2\right)_5\!NH}\right]_n\!\mathrm{\overset{O}{\overset{\|}{C}}}-$$

Nylon 6

In this process ε-caprolactam is allowed to react with water, converting some of it to ε-aminocaproic acid. Then, heating this mixture at 250°C drives off water as ε-caprolactam and ε-aminocaproic acid (6-aminohexanoic acid) react to produce the polyamide. Nylon 6 can also be converted into fibers by melt spinning.

PROBLEM B.1

The raw materials for the production of nylon 6,6 can be obtained in several ways, as indicated below. Write structural formulas for each compound whose molecular formula is given in the following syntheses of adipic acid and of hexamethylenediamine:

(a) Benzene $\xrightarrow{H_2,\ \text{cat.}} C_6H_{12} \xrightarrow{O_2,\ \text{cat.}} C_6H_{12}O + C_6H_{10}O \xrightarrow{HNO_3,\ \text{cat.}}$ Adipic acid + N_2O

IR: ~3300 cm^{-1} (broad) **IR: ~1714 cm^{-1}**
(as a mixture)

(b) Adipic acid $\xrightarrow{2\ NH_3}$ a salt $\xrightarrow{\text{heat}}$ $C_6H_{12}N_2O_2$ $\xrightarrow[\text{catalyst}]{350°C}$

$C_6H_8N_2$ $\xrightarrow[\text{catalyst}]{4\ H_2}$ hexamethylenediamine

(c) 1,3-Butadiene $\xrightarrow{Cl_2}$ $C_4H_6Cl_2$ $\xrightarrow{2\ NaCN}$ $C_6H_6N_2$ $\xrightarrow[Ni]{H_2}$

$C_6H_8N_2$ $\xrightarrow[\text{catalyst}]{4\ H_2}$ hexamethylenediamine

(d) Tetrahydrofuran $\xrightarrow{2\ HCl}$ $C_4H_8Cl_2$ $\xrightarrow{2\ NaCN}$

$C_6H_8N_2$ $\xrightarrow[\text{catalyst}]{4\ H_2}$ hexamethylenediamine

The Chemistry of...

A Green Feedstock for Nylon

Billions of pounds of adipic acid are needed per year as feedstock for the synthesis of nylon. Presently, the predominant industrial source of adipic acid is by a synthesis from benzene. Benzene, however, is a known carcinogen, and it is derived from a nonrenewable natural resource, petroleum. In Problem B.1 the industrial synthesis of adipic acid from benzene is outlined. This synthesis, besides beginning with an undesirable starting material, also produces N_2O in the final step. Nitrous oxide is a greenhouse and ozone-destroying gas. The environmental *unfriendliness* of the benzene

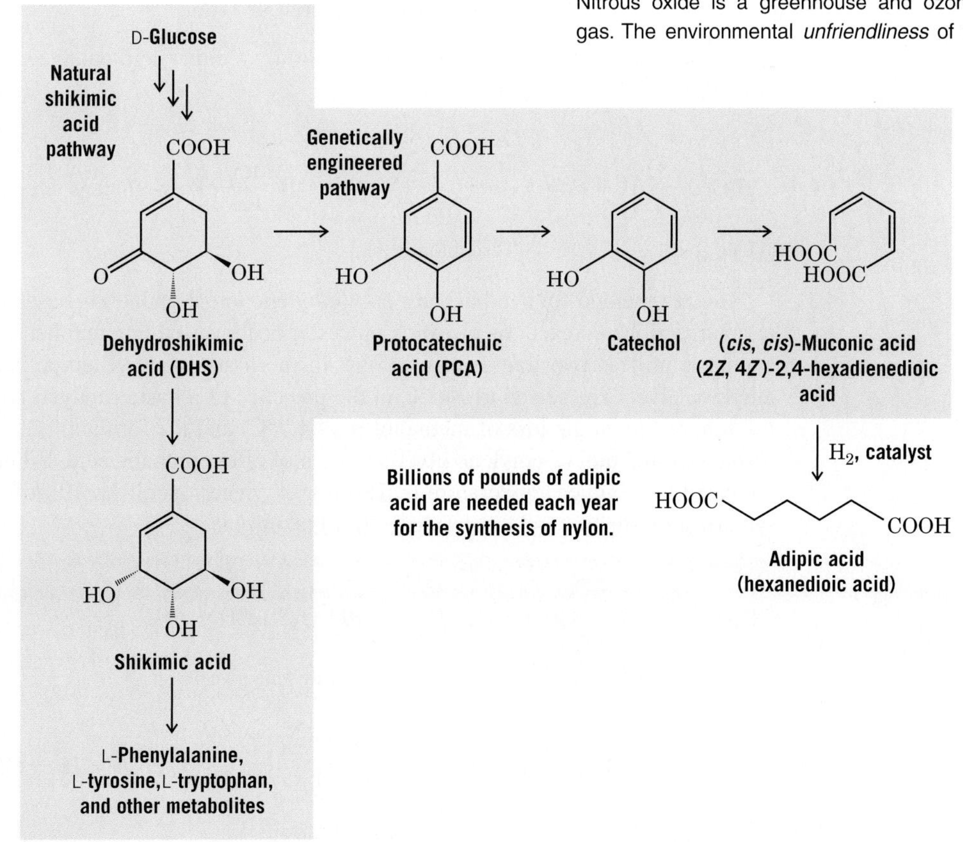

synthesis of adipic acid has caused chemists to look for alternative routes. One method that holds promise is one developed by John W. Frost and Karen Draths of Michigan State University, who used molecular biology and organic chemistry to create an environmentally friendly route to adipic acid.

Frost and Draths' ingenious preparation of adipic acid, shown on the previous page, involves genetically engineered bacteria. All bacteria (as well as plants and some other microorganisms) use a natural metabolic pathway called the shikimic acid pathway to convert glucose to aromatic amino acids and other vital metabolites. Frost and Draths used gene-splicing techniques to create genetically altered *Escherichia coli* bacteria that, instead of producing the normal end products of the shikimic acid pathway, produce *cis,cis*-muconic acid [2*Z*,4*Z*)-2,4-hexadienedioic acid]. *cis,cis-Muconic acid can, in turn, be converted to adipic acid (1,6-hexanedioic acid) by a simple catalytic hydrogenation.*

Although not yet adapted to an industrial scale, the method of Frost and Draths represents the sort of innovation that promises to have a dramatic influence on the environmental friendliness of industry. Their blending of organic chemistry with biotechnology earned them the Presidential Green Chemistry Challenge Award in 1998.*

*For more information see Cann, M. C.; Connelly, M. E. *Real World Cases in Green Chemistry;* American Chemical Society: Washington DC, 2000, and references cited therein.

B.2 Polyesters

One of the most important polyesters is poly(ethylene terephthalate), a polymer that is marketed under the names *Dacron, Terylene,* and *Mylar:*

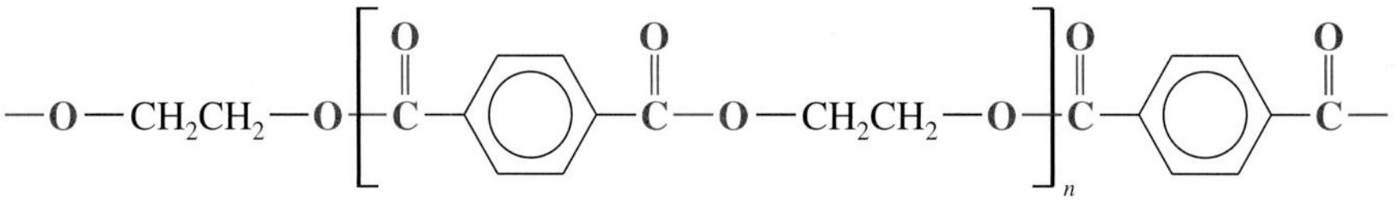

$$-O-CH_2CH_2-O\left[-\overset{O}{\overset{\|}{C}}-C_6H_4-\overset{O}{\overset{\|}{C}}-O-CH_2CH_2-O-\right]_n\overset{O}{\overset{\|}{C}}-C_6H_4-\overset{O}{\overset{\|}{C}}-$$

Poly(ethylene terephthalate) or PET
(Dacron, Terylene, or Mylar)

One can obtain poly(ethylene terephthalate) by a direct acid-catalyzed esterification of ethylene glycol and terephthalic acid:

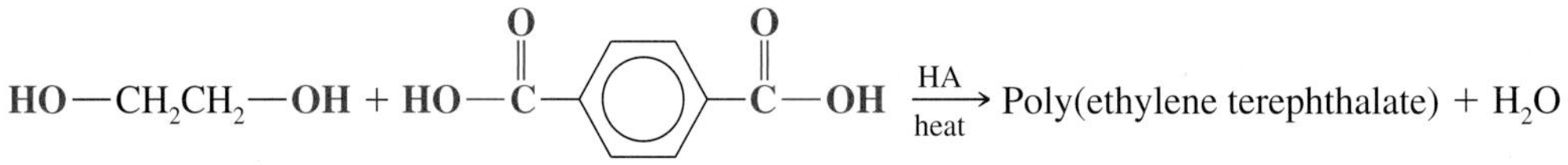

$$HO-CH_2CH_2-OH + HO-\overset{O}{\overset{\|}{C}}-C_6H_4-\overset{O}{\overset{\|}{C}}-OH \xrightarrow[\text{heat}]{HA} \text{Poly(ethylene terephthalate)} + H_2O$$

Ethylene glycol **Terephthalic acid**

Another method for synthesizing poly(ethylene terephthalate) is based on transesterification reactions—reactions in which one ester is converted into another. One commercial synthesis utilizes two transesterifications. In the first, dimethyl terephthalate and excess ethylene glycol are heated to 200°C in the presence of a basic catalyst. Distillation of the mixture results in the loss of methanol (bp 64.7°C) and the formation of a new ester, one formed from 2 mol of ethylene glycol and 1 mol of terephthalic acid. When this new ester is heated to a higher temperature (~280°C), ethylene glycol (bp 198°C) distills and polymerization (the second transesterification) takes place:

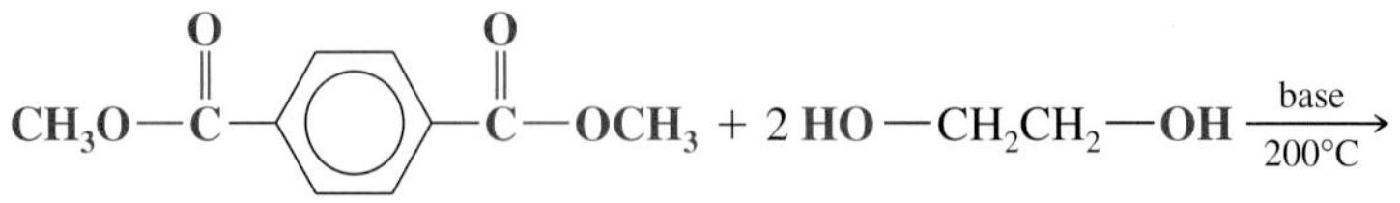

$$CH_3O-\overset{O}{\overset{\|}{C}}-C_6H_4-\overset{O}{\overset{\|}{C}}-OCH_3 + 2\,HO-CH_2CH_2-OH \xrightarrow[200°C]{\text{base}}$$

Dimethyl terephthalate **Ethylene glycol**

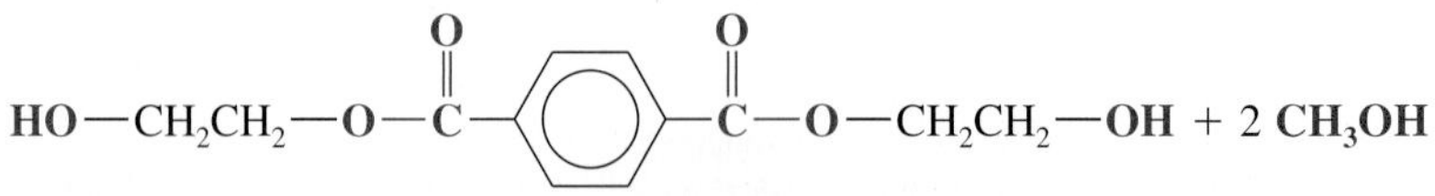

$$HO-CH_2CH_2-O-\overset{O}{\overset{\|}{C}}-C_6H_4-\overset{O}{\overset{\|}{C}}-O-CH_2CH_2-OH + 2\,CH_3OH$$

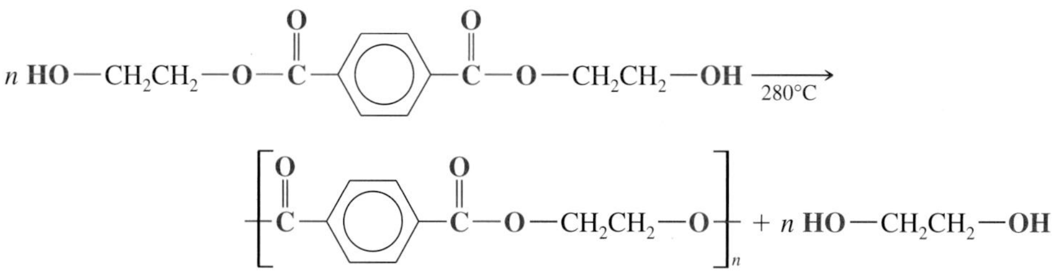

$$n\ HO-CH_2CH_2-O-\overset{O}{\overset{\|}{C}}-C_6H_4-\overset{O}{\overset{\|}{C}}-O-CH_2CH_2-OH \xrightarrow[280°C]{}$$

$$\left[\overset{O}{\overset{\|}{C}}-C_6H_4-\overset{O}{\overset{\|}{C}}-O-CH_2CH_2-O\right]_n + n\ HO-CH_2CH_2-OH$$

Poly(ethylene terephthalate)

The poly(ethylene terephthalate) thus produced melts at about 270°C. It can be melt-spun into fibers to produce Dacron or Terylene; it can also be made into a film, in which form it is marketed as Mylar.

PROBLEM B.2

Transesterifications are catalyzed by either acids or bases. Using the transesterification reaction that takes place when dimethyl terephthalate is heated with ethylene glycol as an example, outline reasonable mechanisms for **(a)** the base-catalyzed reaction and **(b)** the acid-catalyzed reaction.

PROBLEM B.3

Kodel is another polyester that enjoys wide commercial use:

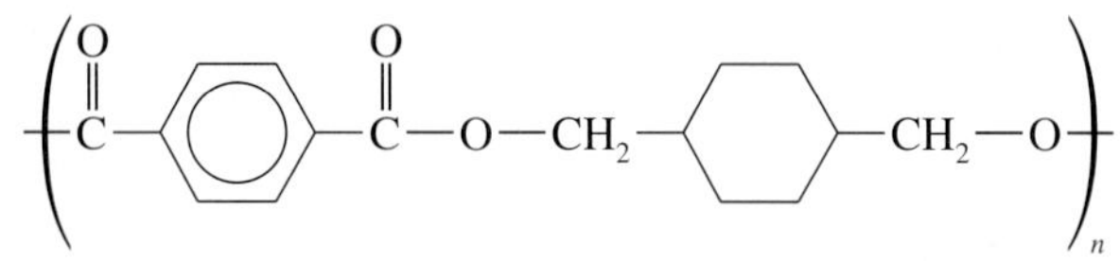

$$\left(\overset{O}{\overset{\|}{C}}-C_6H_4-\overset{O}{\overset{\|}{C}}-O-CH_2-C_6H_{10}-CH_2-O\right)_n$$

Kodel

Kodel is also produced by a transesterification. **(a)** What methyl ester and what alcohol are required for the synthesis of Kodel? **(b)** The alcohol can be prepared from dimethyl terephthalate. How might this be done?

PROBLEM B.4

Heating phthalic anhydride and glycerol together yields a polyester called a glyptal resin. A glyptal resin is especially rigid because the polymer chains are "cross-linked." Write a portion of the structure of a glyptal resin and show how cross-linking occurs.

PROBLEM B.5

Lexan, a high-molecular-weight "polycarbonate," is manufactured by mixing bisphenol A with phosgene in the presence of pyridine. Suggest a structure for Lexan.

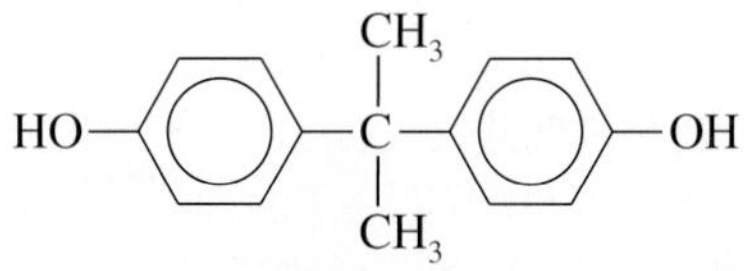

$$HO-C_6H_4-C(CH_3)_2-C_6H_4-OH \qquad Cl-\overset{O}{\overset{\|}{C}}-Cl$$

Bisphenol A 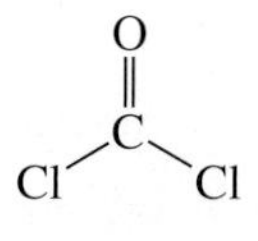 **Phosgene**

PROBLEM B.6

The familiar "epoxy resins" or "epoxy glues" usually consist of two components that are sometimes labeled "resin" and "hardener." The resin is manufactured by allowing bisphenol A (Problem B.5) to react with an excess of epichlorohydrin, $H_2C\overset{O}{—}CHCH_2Cl$, in the presence of a base until a low-molecular-weight polymer is obtained? **(a)** What is a likely structure for this polymer? **(b)** What is the purpose of using an excess of epichlorohydrin? **(c)** The hardener is usually an amine such as $H_2NCH_2CH_2NHCH_2CH_2NH_2$. What reaction takes place when the resin and hardener are mixed?

The Chemistry of...

A PET Green Recycling Method

It is essential that we recycle polymers so as to conserve natural resources and minimize waste. Polymers that cannot be recycled are either incinerated or sent to landfills. In the case of PET [poly(ethylene terephthalate)], recycling of scrap PET for use again in food and beverage containers poses a special challenge because only the highest purity recycled PET is acceptable for food packaging. Fortunately, the DuPont Company has developed a way to depolymerize PET into high-purity monomers that can be recycled for PET synthesis. Called the Petretec process, the DuPont method hinges on a transesterification reaction between methanol and PET:

Scrap PET [poly(ethylene terephthalate)] —(1) Dissolve in DMT with heat; (2) CH_3OH, pressure, heat→ DMT (dimethyl terephthalate) + Ethylene glycol

Scrap PET is first dissolved in liquid dimethyl terephthalate (DMT), one of the monomers used to make PET. This solution is heated, and methanol is added under high pressure. A transesterification reaction occurs, whereby the ethylene glycol units that linked the terephthalate groups in the PET polymer are liberated in exchange for formation of the dimethyl ester of terephthalic acid (DMT). The resulting mixture of DMT, ethylene glycol, and excess methanol can be separated and purified and the DMT and ethylene glycol submitted for fresh polymerization to form new PET. DuPont has a plant capable of recycling 100 million pounds of PET per year by the Petretec method, and higher throughput is possible. The Petretec method therefore has great promise as a green method in polymer chemistry.

*For more information see Cann, M. C.; Connelly, M. E. *Real World Cases in Green Chemistry;* American Chemical Society: Washington, DC, 2000, and references cited therein.

B.3 Polyurethanes

A *urethane* is the product formed when an alcohol reacts with an isocyanate:

$$R{-}OH + O{=}C{=}N{-}R' \longrightarrow R{-}O{-}C({=}O){-}N(H){-}R'$$

Alcohol **An isocyanate** **A urethane (a carbamate)**

The reaction takes place in the following way:

$$R{-}\ddot{O}H + C({=}O){=}N{-}R' \longrightarrow R{-}\overset{+}{O}(H){-}C(O^-){=}N{-}R' \longleftrightarrow R{-}\overset{+}{O}(H){-}C({=}O){-}\overset{-}{N}{-}R' \xrightarrow{H{-}A,\ :A^-} R{-}\ddot{O}{-}C({=}O){-}N(H){-}R'$$

A urethane is also called a *carbamate* because formally it is an ester of an alcohol (ROH) and a carbamic acid ($R'NHCO_2H$).

Polyurethanes are usually made by allowing a *diol* to react with a *diisocyanate.* The diol is typically a polyester with $-CH_2OH$ end groups. The diisocyanate is usually toluene 2,4-diisocyanate:*

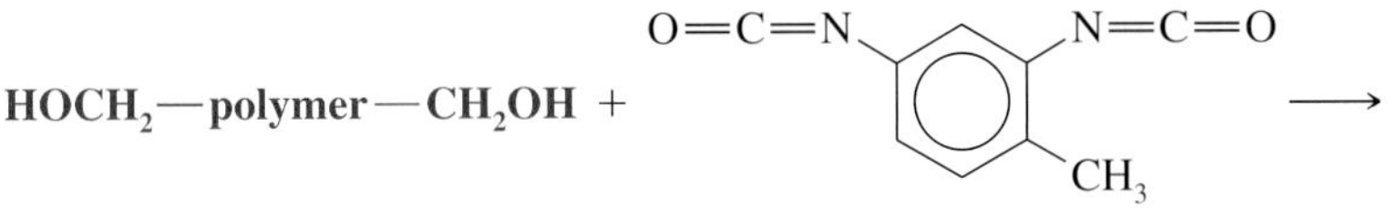

$$HOCH_2-\text{polymer}-CH_2OH + O{=}C{=}N-C_6H_3(CH_3)-N{=}C{=}O \longrightarrow$$

Toluene 2,4-diisocyanate

$$\left[-C_6H_3(CH_3)-NH-\overset{O}{\overset{\|}{C}}-OCH_2-\text{polymer}-CH_2O-\overset{O}{\overset{\|}{C}}-NH- \right]_n$$

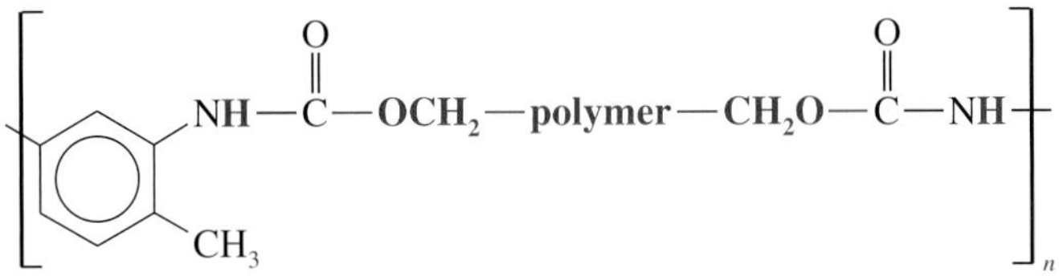

A polyurethane

PROBLEM B.7

A typical polyurethane can be made in the following way. Adipic acid is polymerized with an excess of ethylene glycol. The resulting polyester is then treated with toluene 2,4-diisocyanate. **(a)** Write the structure of the polyurethane. **(b)** Why is an excess of ethylene glycol used in making the polyester?

Polyurethane foams, as used in pillows and paddings, are made by adding small amounts of water to the reaction mixture during the polymerization with the diisocyanate. Some of the isocyanate groups react with water to produce carbon dioxide, and this gas acts as the foaming agent:

$$R-N{=}C{=}O + H_2O \longrightarrow R-NH_2 + CO_2\uparrow$$

B.4 Phenol–Formaldehyde Polymers

One of the first synthetic polymers to be produced was a polymer (or resin) known as *Bakelite.* Bakelite is made by a condensation reaction between phenol and formaldehyde; the reaction can be catalyzed by either acids or bases. The base-catalyzed reaction probably takes place in the general way shown here. Reaction can take place at the ortho and para positions of phenol.

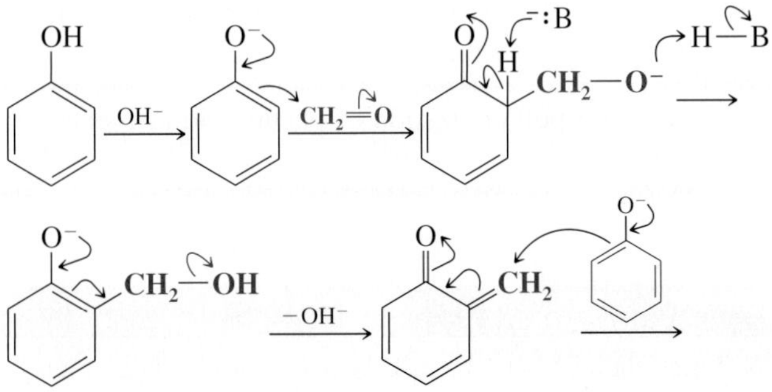

OH (phenol) $\xrightarrow{OH^-}$ O^- (phenoxide) $\xrightarrow{CH_2{=}O}$ (cyclohexadienone with H and CH_2-O^-; :B, H—B) $\longrightarrow$

O^- (with CH_2-OH) $\xrightarrow{-OH^-}$ (quinone methide, $=CH_2$) + O^- (phenoxide) $\longrightarrow$

*Toluene 2,4-diisocyanate is a hazardous chemical that has caused acute respiratory problems among workers synthesizing polyurethanes.

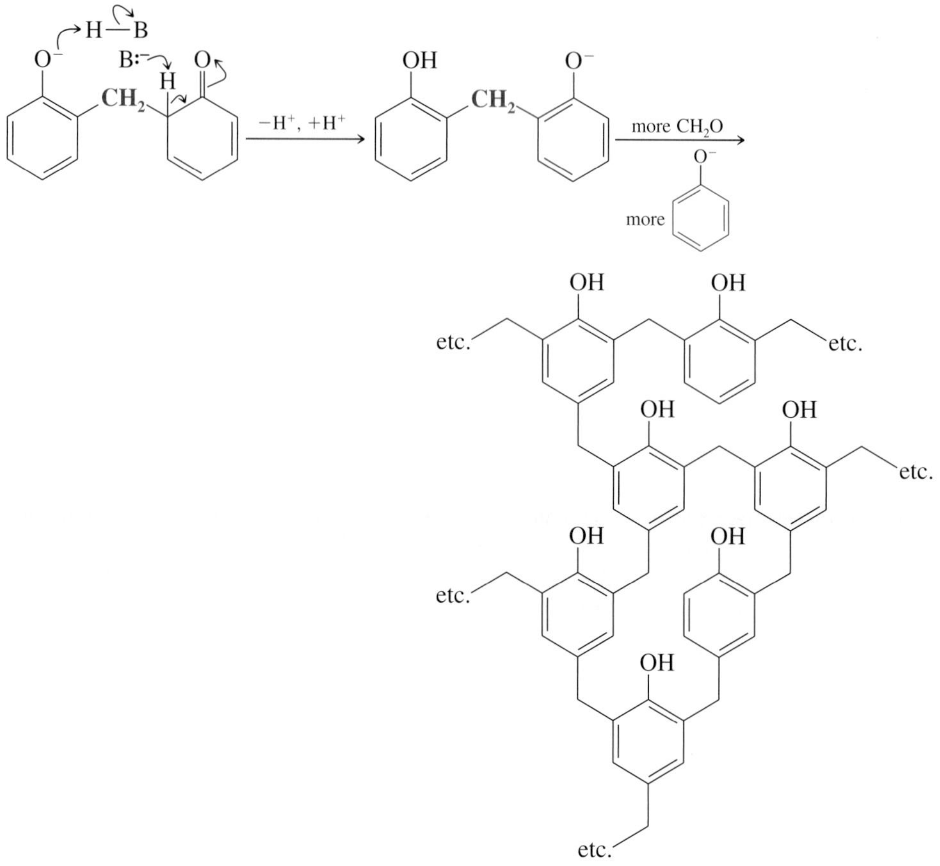

$\xrightarrow{-H^+, +H^+}$ $\xrightarrow{\text{more } CH_2O}$ more

etc. etc. etc. etc. etc.

Bakelite

Generally, the polymerization is carried out in two stages. The first polymerization produces a low-molecular-weight fusible (meltable) polymer called a *resole.* The resole can be molded to the desired shape, and then further polymerization produces a very high molecular weight polymer, which, because it is highly cross-linked, is infusible.

PROBLEM B.8 Using a para-substituted phenol such as *p*-cresol yields a phenol–formaldehyde polymer that is *thermoplastic* rather than *thermosetting.* That is, the polymer remains fusible; it does *not* become impossible to melt. What accounts for this?

PROBLEM B.9 Outline a general mechanism for acid-catalyzed polymerization of phenol and formaldehyde.

B.5 CASCADE POLYMERS

One exciting development in polymer chemistry in the last decade or so has been the synthesis of high-molecular-weight, symmetrical, highly branched, polyfunctional molecules called **cascade polymers.** G. R. Newkome (of the University of South Florida) and D. A. Tomalia (of the Michigan Molecular Institute) have been pioneers in this area of research.

All of the polymers that we have considered so far are inevitably nonhomogeneous. Although they consist of molecules with common repetitive monomeric units, the molecules of the material obtained from the polymerization reactions vary widely in molecular weight (and, therefore, in size). Cascade polymers, by contrast, can be synthesized in ways that yield polymers consisting of molecules of uniform molecular weight and size.

Syntheses of cascade polymers begin with a core building block that can lead to branching in one, two, three, or even four directions. Starting with this core molecule, through repetitive reactions, layers (called **cascade spheres**) are added. Each new sphere increases (usually by three times) the number of branch points from which the next sphere can be constructed. Because of this multiplying effect, very large molecules can be built up very quickly.

Figures B.1 and B.2 show how a four-directional-cascade molecule has been constructed. All of the reactions are closely related to ones that we have studied already. The starting material for construction of the core molecule is a branched tetraol, **1**. In the first step (i), **1** is allowed to react with propenenitrile ($CH_2{=}CHCN$) in a conjugate addition called *cyanoethylation* to produce **2**. Treating **2** with methanol and acid [step (ii)] converts the cyano groups to methyl carboxylate groups. (Instead of hydrolyzing the cyano groups to carboxyl groups and then esterifying them, this process accomplishes the same result in one step.) In step (iii), the ester groups are hydrolyzed, and in step (iv), the carboxyl groups are converted to acyl chloride groups. Compound **5** is the core building block.

The synthesis of the compound used in constructing the next cascade sphere is shown in the sequence **6** → **7** → **8** (cyanoethylation followed by esterification). Treating the core compound, **5**, with an excess of the amine, **8**, produced compound **9** with 12 surface ester groups (called, for convenience, the [12]-ester). The key to this step is the formation

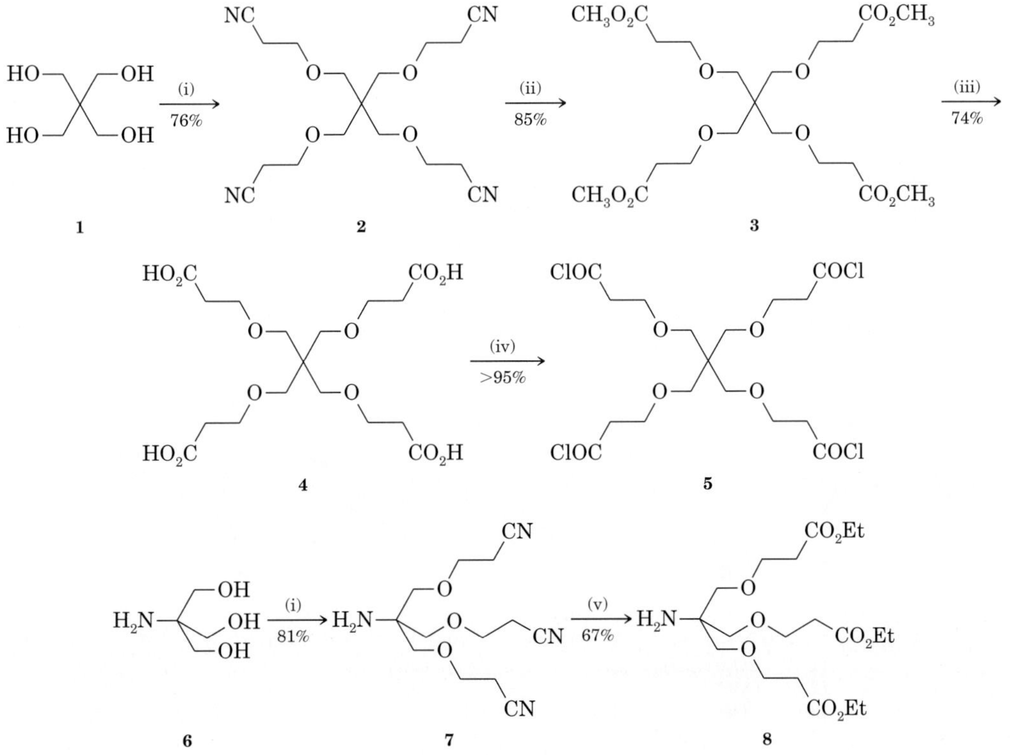

FIGURE B.1 Synthesis of the starting materials for a cascade polymer. Reagents and conditions: (i) $CH_2{=}CHCN$, KOH, *p*-dioxane, 25°C, 24 h; (ii) MeOH, dry HCl, reflux, 2 h; (iii) 3 *N* NaOH, 70°C, 24 h; (iv) $SOCl_2$, CH_2Cl_2, reflux, 1 h; (v) EtOH, dry HCl, reflux, 3 h. (Adapted from Newkome, G. R.; Lin, X. *Macromolecules* **1991,** *24,* 1443–1444.)

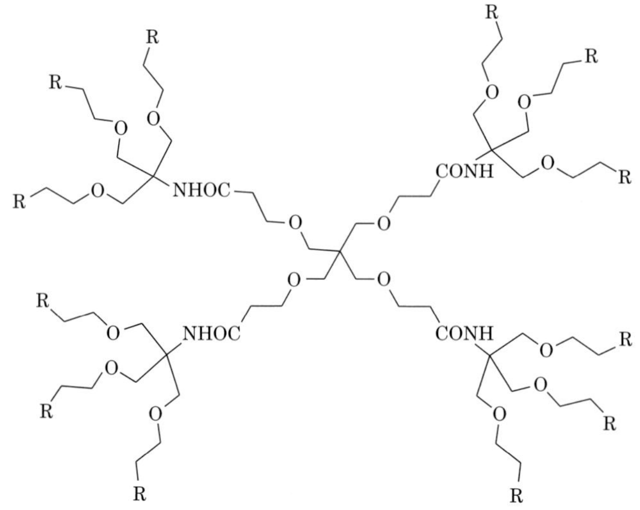

9 R = $CO_2CH_2CH_3$
10 R = COOH

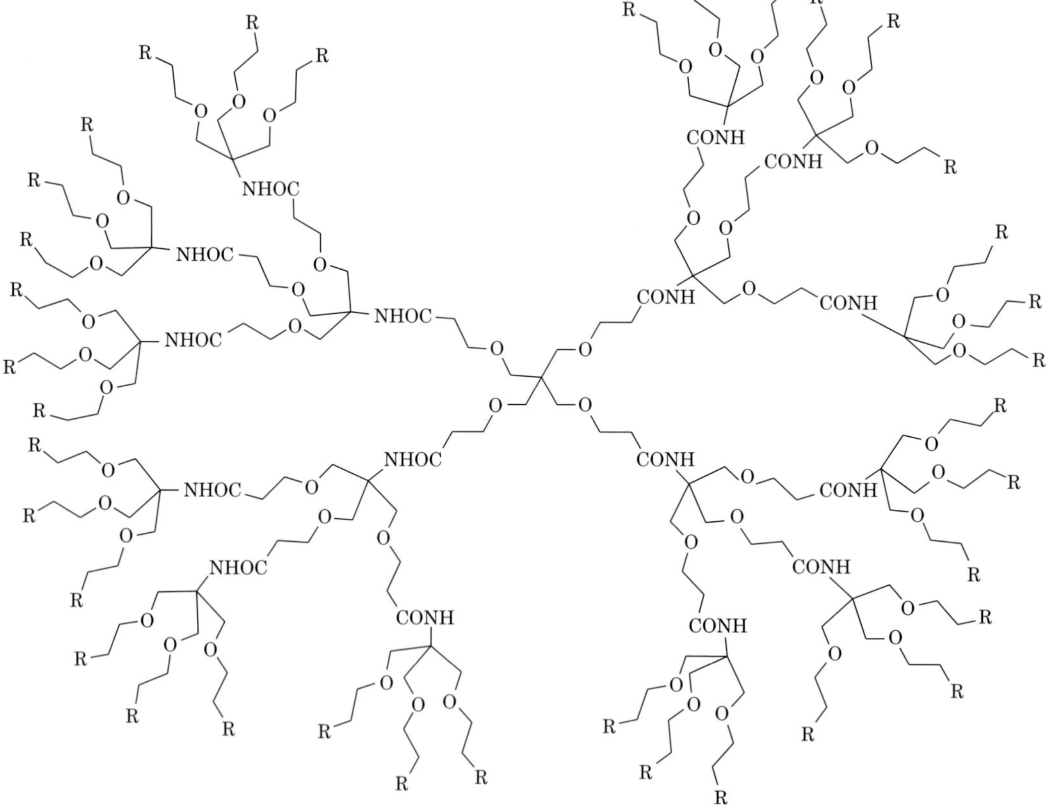

11 R = $CO_2CH_2CH_3$
12 R = COOH
13 R = $CONHC(CH_2OCH_2CH_2CO_2CH_2CH_3)_3$
14 R = $CONHC(CH_2OCH_2CH_2CO_2H)_3$
15 R = $CONHC[CH_2OCH_2CH_2CONHC(CH_2OCH_2CH_2CO_2CH_2CH_3)_3]_3$

FIGURE B.2 Cascade polymers. (Adapted from Newkome, G. R.; Lin, X. *Macromolecules* **1991,** *24,* 1443–1444.)

of amide linkages between **5** and four molecules of **8**. The [12]-ester, **9**, was hydrolyzed to the [12]-acid, **10**. Treating **10** with **8** using dicyclohexylcarbodiimide (Section 18.8E) to promote amide formation led to the [36]-ester, **11**. The [36]-ester, **11**, was then hydrolyzed to the [36]-acid, **12,** which in turn was allowed to react with **8** to produce the next cascade molecule, a [108]-ester, **13**.

Repeating these steps one more time produced the [324]-ester, **15**, a compound with a molecular weight of 60,604! At each step the cascade molecules were isolated, purified, and identified. Because the yields for each step are 40–60%, and because the starting materials are inexpensive, this method offers a reasonable route to large homogeneous spherical polymers.

Synthesis and Reactions of β-Dicarbonyl Compounds: More Chemistry of Enolate Anions

Imposters

There are many important roles for chemical imposters. In biochemistry, molecules that masquerade as natural compounds often cause profound effects by blocking a receptor site or altering the function of an enzyme. An example is 5-fluorouracil (shown above), a clinically used anticancer drug that masquerades as uracil, a natural metabolite needed for DNA synthesis. The enzyme thymidylate synthase mistakes 5-fluorouracil for uracil and acts on it as though it were the natural substrate. Having infiltrated the enzyme, however, the imposter derails the normal mechanism and leaves thymidylate synthase irreversibly damaged, with the effect that DNA synthesis is inhibited. Later, when we consider aspects of this process in "The Chemistry of . . . A Suicide Enzyme Substrate," we shall see that mechanism-based inhibition of thymidylate synthase involves a conjugate addition (like a Michael addition, Sections 17.9B and 19.9), the reaction of an enolate with an imine (like the Mannich reaction, Section 19.10), and an E2-type elimination reaction that is blocked by the presence of a fluorine atom.

Chemical imposters are also useful as so-called synthetic equivalents. A synthetic equivalent is a reagent whose structure, when incorporated into a product, gives the appearance of having come from one type of precursor when as a reactant it actually had a different structural origin. Two examples of thinly veiled synthetic equivalents are anions from ethyl acetoacetate and diethyl malonate, synthetic equivalents of enolate nucleophiles from acetone and acetic acid, respectively. Enamines are another type of synthetic equivalent. Enamines are alkenyl amines which also manifest themselves in reactions as though they are enolates. Yet other synthetic equivalents add to the intrigue by providing umpolung, or "polarity reversal." Carbanions from dithioacetals are examples of synthetic equivalents used for umpolung. A dithioacetal carbanion provides the chemical disguise for a *nucleophilic* (rather than electrophilic) carbonyl carbon atom. We shall unveil the disguises of all these synthetic equivalents in this chapter as well as see how the chemical imposter 5-fluorouracil inhibits thymidylate synthase.

19.1 Introduction

Compounds having two carbonyl groups separated by an intervening carbon atom are called **β-dicarbonyl compounds,** and these compounds are highly versatile reagents for organic synthesis. In this chapter we shall explore some of the methods for preparing β-dicarbonyl compounds and some of their important reactions:

$$-\overset{O}{\overset{\|}{\underset{\beta}{C}}}-\overset{|}{\underset{|\alpha}{C}}-\overset{O}{\overset{\|}{C}}- \qquad R-\overset{O}{\overset{\|}{\underset{\beta}{C}}}-\overset{|}{\underset{|\alpha}{C}}-\overset{O}{\overset{\|}{C}}-OR' \qquad RO-\overset{O}{\overset{\|}{\underset{\beta}{C}}}-\overset{|}{\underset{|\alpha}{C}}-\overset{O}{\overset{\|}{C}}-OR$$

The β-dicarbonyl system **A β-keto ester (Section 19.2)** **A malonic ester (Section 19.4)**

Central to the chemistry of β-dicarbonyl compounds is the acidity of protons located on the carbon between the two carbonyl groups. The pK_a for such a proton is in the range 9–11, acidic enough to be removed easily by an alkoxide base to form an enolate:

$$-\overset{O}{\overset{\|}{C}}-\overset{|}{\underset{\underset{H}{|}}{C}}-\overset{O}{\overset{\|}{C}}- \xrightleftharpoons{^{-}OR} -\overset{O}{\overset{\|}{C}}-\overset{|}{\underset{\cdot\cdot -}{C}}-\overset{O}{\overset{\|}{C}}- + HOR$$

H ← pK_a = 9–11

We can account for the greater acidity of β-dicarbonyl systems, as compared to single carbonyl systems, by delocalization of the negative charge to two oxygen atoms instead of one. We can represent this delocalization by drawing contributing resonance structures for a β-dicarbonyl enolate and its resonance hybrid:

$$\overset{\ddot{O}}{\overset{\|}{C}}\overset{\cdot\cdot -}{C}\overset{\ddot{O}}{\overset{\|}{C}} \longleftrightarrow \overset{\ddot{O}}{\overset{\|}{C}}C{=}\overset{:\ddot{O}:^{-}}{\overset{|}{C}} \longleftrightarrow \overset{:\ddot{O}:^{-}}{\overset{|}{C}}{=}C\overset{\ddot{O}}{\overset{\|}{C}} \equiv \overset{\delta-}{\ddot{O}}\cdots C\cdots \overset{\delta-}{C}\cdots C\cdots\overset{\delta-}{\ddot{O}}$$

Contributing resonance structures **Resonance hybrid**

We can visualize the enhanced charge delocalization of a β-dicarbonyl enolate by examining maps of electrostatic potential for enolates derived from pentane-2,4-dione and acetone. Here we see that the negative charge of the enolate from pentane-2,4-dione enolate is associated substantially with the two oxygen atoms, as compared with the enolate from acetone, where significant negative charge in the enolate remains at the α-carbon atom:

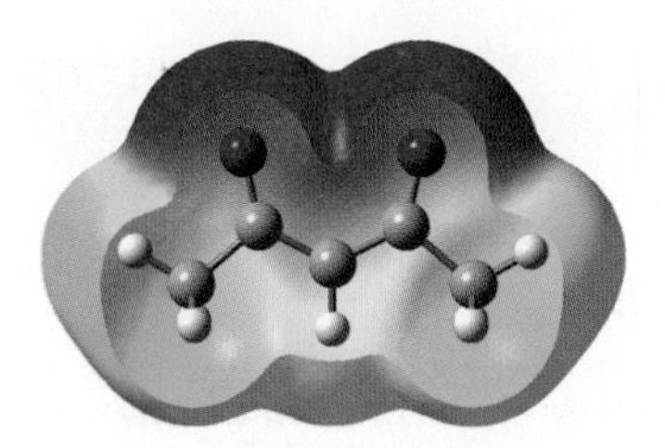

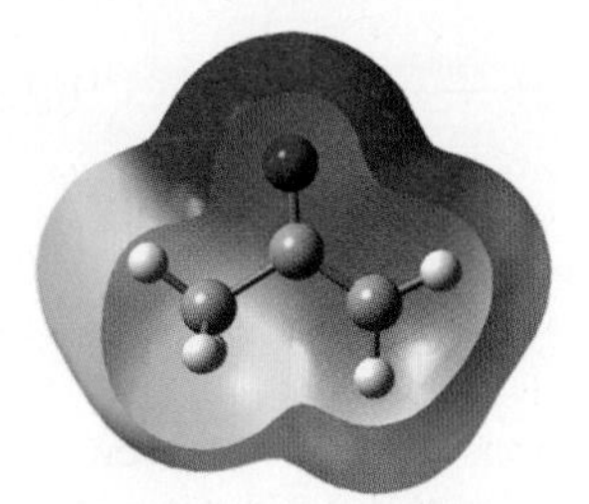

Pentane-2,4-dione enolate **Acetone enolate**

Early in this chapter we shall see how the acidity of these protons allows the synthesis of β-dicarbonyl compounds through reactions that are called *Claisen syntheses* (Section 19.2):

$$\text{H—}\underset{\text{H}}{\overset{|}{\text{C}}}\text{—}\overset{\text{O}}{\overset{\|}{\text{C}}}\text{—OR} \xrightarrow[\text{(2) } H_3O^+]{\text{(1) NaOR, R}'\text{—}\overset{\text{O}}{\overset{\|}{\text{C}}}\text{—OR}} \text{R}'\text{—}\overset{\text{O}}{\overset{\|}{\text{C}}}\text{—}\underset{\text{H}}{\overset{|}{\text{C}}}\text{—}\overset{\text{O}}{\overset{\|}{\text{C}}}\text{—OR} + \text{HOR}$$

Later in the chapter we shall study the *acetoacetic ester synthesis* (Section 19.3) and the *malonic ester synthesis* (Section 19.4):

$$\text{G—}\overset{\text{O}}{\overset{\|}{\text{C}}}\text{—}\underset{\text{H}}{\overset{|}{\text{C}}}\text{—}\overset{\text{O}}{\overset{\|}{\text{C}}}\text{—OR} \xrightarrow{\text{(1) NaOR}} \text{G—}\overset{\text{O}}{\overset{\|}{\text{C}}}\text{—}\underset{\cdot\cdot\,-}{\overset{|}{\text{C}}}\text{—}\overset{\text{O}}{\overset{\|}{\text{C}}}\text{—OR} \xrightarrow[\text{(or acylation)}]{\text{(2) R—X}} \text{G—}\overset{\text{O}}{\overset{\|}{\text{C}}}\text{—}\underset{\text{R}}{\overset{|}{\text{C}}}\text{—}\overset{\text{O}}{\overset{\|}{\text{C}}}\text{—OR}$$

Acetoacetic ester synthesis, G = CH_3
Malonic ester synthesis, G = RO

All of these syntheses involve β-dicarbonyl compounds as nucleophiles in reactions whose mechanisms should be familiar to you: (1) acylation by nucleophilic addition–elimination at a carbonyl group, (2) alkylation by S_N2 substitution, (3) aldol-type condensation, and (4) conjugate addition to an α,β-unsaturated carbonyl compound.

One other feature that appears again and again in the syntheses that we study here is the decarboxylation of a β-keto acid:

$$\text{G—}\overset{\text{O}}{\overset{\|}{\text{C}}}\text{—}\underset{\text{R}}{\overset{|}{\text{C}}}\text{—}\overset{\text{O}}{\overset{\|}{\text{C}}}\text{—OH} \xrightarrow{\text{heat}} [\text{cyclic transition state}] \longrightarrow \text{G—}\overset{\text{O}}{\overset{\|}{\text{C}}}\text{—}\underset{\text{R}}{\overset{|}{\text{C}}}\text{—H} + CO_2$$

(Section 18.10)

We learned in Section 18.10 that these decarboxylations occur under mild conditions, and it is this ease of decarboxylation that makes many of the syntheses in this chapter useful. If **G** in the reaction above is a methyl group, we have a synthesis of *substituted acetones (methyl ketones).* If **G** in the reaction above is the hydroxyl group of a carboxylic acid (which would have been an alkoxyl group of an ester before a hydrolysis step), we have a synthesis of *substituted acetic acids.*

19.2 The Claisen Condensation: The Synthesis of β-Keto Esters

When ethyl acetate reacts with sodium ethoxide, it undergoes a **condensation reaction.** After acidification, the product is a β-keto ester, ethyl acetoacetate (commonly called *acetoacetic ester*):

$$2\,CH_3\overset{O}{\overset{\|}{C}}OC_2H_5 \xrightarrow{NaOC_2H_5} \left[CH_3\overset{O}{\overset{\|}{C}}\underset{\cdot\cdot}{C}H\overset{O}{\overset{\|}{C}}OC_2H_5 \;\; Na^+\right] + C_2H_5OH \xrightarrow{HCl} CH_3\overset{O}{\overset{\|}{C}}CH_2\overset{O}{\overset{\|}{C}}OC_2H_5$$

Sodioacetoacetic ester — **(removed by distillation)** — **Ethyl acetoacetate (acetoacetic ester) (76%)**

Condensations of this type occur with many other esters and are known generally as **Claisen condensations.** Like the aldol condensation (Section 17.4), Claisen condensations involve the α carbon of one molecule and the carbonyl group of another. Ethyl pentanoate, for example, reacts with sodium ethoxide to give the β-keto ester that follows:

$$2\ \text{Ethyl pentanoate} \xrightarrow{NaOCH_2CH_3} \left[Na^+ \text{ enolate of ethyl 2-propyl-3-oxoheptanoate} \right] + C_2H_5OH \xrightarrow{CH_3CO_2H} \text{ethyl 2-propyl-3-oxoheptanoate } (77\%)$$

Ethyl pentanoate **(77%)**

If we look closely at these examples, we can see that, overall, both reactions involve a condensation in which one ester loses an α hydrogen and the other loses an ethoxide ion:

$$R{-}CH_2\overset{O}{\overset{\|}{C}}{-}OC_2H_5 + H{-}\underset{R}{\underset{|}{C}}H\overset{O}{\overset{\|}{C}}{-}OC_2H_5 \xrightarrow[(2)\ H_3O^+]{(1)\ NaOC_2H_5} R{-}CH_2\overset{O}{\overset{\|}{C}}{-}\underset{R}{\underset{|}{C}}H\overset{O}{\overset{\|}{C}}OC_2H_5 + C_2H_5OH$$

(R may also be H) **A β-keto ester**

We can understand how this happens if we examine the reaction mechanism in detail. In doing so, we shall see that the Claisen condensation mechanism is a classic example of nucleophilic addition–elimination at a carbonyl group.

A Mechanism for the Reaction

The Claisen Condensation

Step 1

$$R\overset{\alpha}{C}H(H){-}\overset{O}{\overset{\|}{C}}OC_2H_5 + {:}\ddot{O}C_2H_5^- \rightleftharpoons R\ddot{C}H^-{-}\overset{O}{\overset{\|}{C}}OC_2H_5 \longleftrightarrow RCH{=}\overset{:\ddot{O}:^-}{\overset{|}{C}}OC_2H_5 + C_2H_5OH$$

An alkoxide base removes an α proton from the ester, generating a nucleophilic enolate anion. (The alkoxide base used to form the enolate should have the same alkyl group as the ester, e.g., ethoxide for an ethyl ester; otherwise transesterification may occur.) Although the α protons of an ester are not as acidic as those of aldehydes and ketones, the resulting enolate anion is stabilized by resonance in a similar way.

Step 2

$$RCH_2\overset{O}{\overset{\|}{C}}{-}OC_2H_5 + {^-}{:}\underset{R}{\underset{|}{C}}H{-}\overset{O}{\overset{\|}{C}}OC_2H_5 \rightleftharpoons RCH_2\overset{:\ddot{O}:^-}{\overset{|}{C}}(\ddot{O}C_2H_5){-}\underset{R}{\underset{|}{C}}H{-}\overset{O}{\overset{\|}{C}}OC_2H_5 \rightleftharpoons RCH_2\overset{O}{\overset{\|}{C}}{-}\underset{R}{\underset{|}{C}}H{-}\overset{O}{\overset{\|}{C}}OC_2H_5 + {^-}{:}\ddot{O}C_2H_5$$

Nucleophilic addition **Tetrahedral intermediate and elimination**

The enolate anion attacks the carbonyl carbon of another ester molecule, forming a tetrahedral intermediate. The tetrahedral intermediate expels an alkoxide anion, resulting in substitution of the alkoxide by the group derived from the enolate. The net result is nucleophilic addition–elimination at the ester carbonyl group. *The overall equilibrium for the process is unfavorable thus far, however,* but it is drawn toward the final product by removal of the acidic α hydrogen from the new β-dicarbonyl system.

Step 3

$$\mathrm{RCH_2\overset{O}{\overset{\|}{C}}-\underset{R}{\underset{|}{\overset{H}{\overset{|}{C}}}}-\overset{O}{\overset{\|}{C}}OC_2H_5} + :\ddot{O}C_2H_5^- \rightleftharpoons \mathrm{RCH_2\overset{O}{\overset{\|}{C}}-\underset{R}{\underset{|}{\ddot{C}}}-\overset{O}{\overset{\|}{C}}OC_2H_5} + C_2H_5OH$$

β-Keto ester (pK_a ~ 9; stronger acid)	Ethoxide ion (stronger base)	β-Keto ester anion (weaker base)	Ethanol (pK_a 16; weaker acid)

An alkoxide anion removes an α proton from the newly formed condensation product, resulting in a resonance stabilized β-ketoester anion. This step is highly favorable and draws the overall equilibrium toward product formation. The alcohol byproduct (ethanol in this case) can be distilled from the reaction mixture as it forms, thereby further drawing the equilibrium toward the desired product.

Step 4

$$\mathrm{RCH_2-\overset{\delta-}{\overset{O}{\overset{\vdots}{C}}}\cdots\underset{R}{\underset{|}{\overset{\delta-}{C}}}\cdots\overset{\delta-}{\overset{O}{\overset{\vdots}{C}}}OC_2H_5} \xrightarrow[\text{(rapid)}]{H_3O^+} \mathrm{RCH_2-\overset{O}{\overset{\|}{C}}-\underset{R}{\underset{|}{CH}}-\overset{O}{\overset{\|}{C}}OC_2H_5} \rightleftharpoons \mathrm{RCH_2-\overset{OH}{\overset{|}{C}}=\underset{R}{\underset{|}{C}}-\overset{O}{\overset{\|}{C}}OC_2H_5}$$

Keto form   **Enol form**

Addition of acid quenches the reaction by neutralizing the base and protonating the Claisen condensation product. The β-ketoester product exists as an equilibrium mixture of its keto and enol forms.

When planning a reaction with an ester and an alkoxide anion it is important to use an alkoxide that has the same alkyl group as the alkoxyl group of the ester. We do this to avoid the possibility of transesterification (which occurs with alkoxides by the same mechanism as base-promoted ester hydrolysis; Section 18.7B). Ethyl esters and methyl esters, as it turns out, are the most common ester reactants in these types of syntheses. Therefore, we use sodium ethoxide when ethyl esters are involved and sodium methoxide when methyl esters are involved. (There are some occasions when we shall choose to use other bases, but we shall discuss these later.)

Esters that have only one α hydrogen do not undergo the usual Claisen condensation. An example of an ester that does not react in a normal Claisen condensation is ethyl 2-methylpropanoate:

Only one α hydrogen

$$\mathrm{CH_3\underset{CH_3}{\underset{|}{C}}H\overset{O}{\overset{\|}{C}}OCH_2CH_3}$$

Ethyl 2-methylpropanoate

This ester does not undergo a Claisen condensation.

Inspection of the mechanism just given will make clear why this is so. An ester with only one α hydrogen will not have an acidic hydrogen when step 3 is reached, and step 3 provides the favorable equilibrium that ensures the success of the reaction. (In Section 19.2A we see how esters with only one α hydrogen can be converted to a β-keto ester through the use of very strong bases.)

Problem 19.1

(a) Write a mechanism for all steps of the Claisen condensation that take place when ethyl propanoate reacts with ethoxide ion. **(b)** What products form when the reaction mixture is acidified?

The Dieckmann Condensation When diethyl hexanedioate is heated with sodium ethoxide, subsequent acidification of the reaction mixture gives ethyl 2-oxocyclopentanecarboxylate:

$$C_2H_5O\overset{O}{\overset{\|}{C}}(CH_2)_4\overset{O}{\overset{\|}{C}}OC_2H_5 \xrightarrow[(2)\ H_3O^+]{(1)\ NaOC_2H_5} \text{(2-oxocyclopentyl)}-\overset{O}{\overset{\|}{C}}OC_2H_5$$

Diethyl hexanedioate (diethyl adipate) → **Ethyl 2-oxocyclopentanecarboxylate (74–81%)**

This reaction, called the *Dieckmann condensation,* is an intramolecular Claisen condensation. The α-carbon atom and the ester group for the condensation come from the same molecule. In general, the Dieckmann condensation is useful only for the preparation of five- and six-membered rings. Smaller rings are disfavored due to angle strain. Larger rings are entropically less favorable due to the greater number of conformations available to a longer chain precursor. Intermolecular condensation begins to compete strongly.

A Mechanism for the Reaction

The Dieckmann Condensation

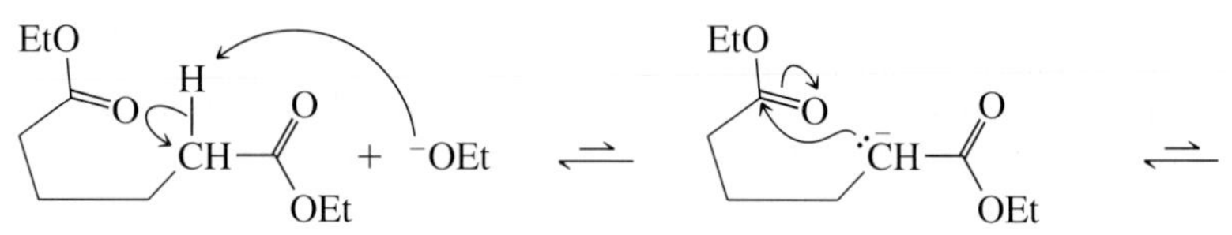

Ethoxide anion removes an α hydrogen.

The enolate anion attacks the carbonyl group at the other end of the chain.

An ethoxide anion is expelled.

The ethoxide anion removes the acidic hydrogen located between two carbonyl groups. This favorable equilibrium drives the reaction.

Addition of aqueous acid rapidly protonates the anion, giving the final product.

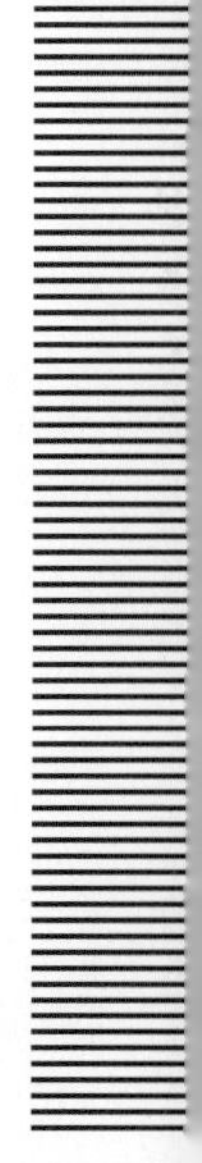

PROBLEM 19.2

(a) What product would you expect from a Dieckmann condensation of diethyl heptanedioate (diethyl pimelate)? **(b)** Can you account for the fact that diethyl pentanedioate (diethyl glutarate) does not undergo a Dieckmann condensation?

19.2A Crossed Claisen Condensations

Crossed Claisen condensations (like crossed aldol condensations) are possible **when one ester component has no α hydrogens** and, therefore, is unable to form an enolate ion and undergo self-condensation. Ethyl benzoate, for example, condenses with ethyl acetate to give ethyl benzoylacetate:

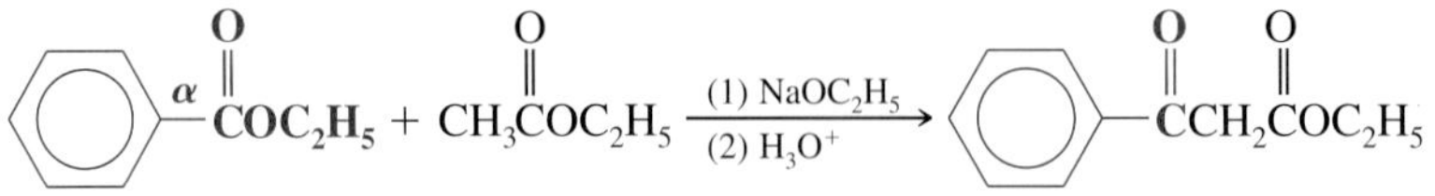

$$C_6H_5COC_2H_5\ (\text{C=O}) + CH_3COC_2H_5\ (\text{C=O}) \xrightarrow[(2)\ H_3O^+]{(1)\ NaOC_2H_5} C_6H_5CCH_2COC_2H_5\ (\text{two C=O})$$

Ethyl benzoate (no α hydrogen) **Ethyl benzoylacetate (60%)**

Ethyl phenylacetate condenses with diethyl carbonate to give diethyl phenylmalonate:

$$C_6H_5CH_2COC_2H_5 + C_2H_5OCOC_2H_5 \xrightarrow[(2)\ H_3O^+]{(1)\ NaOC_2H_5} C_6H_5CH(COC_2H_5)_2$$

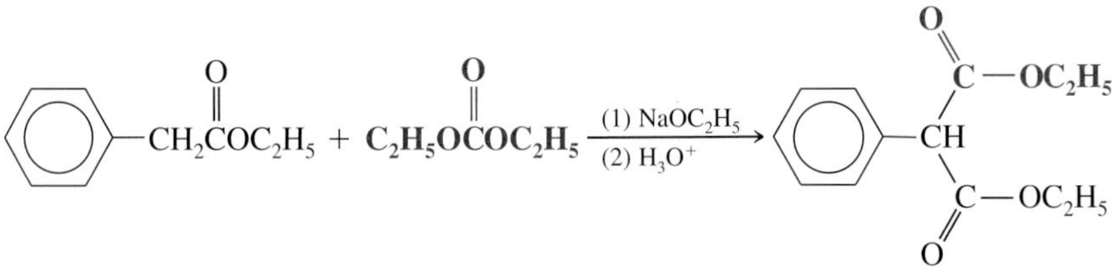

Ethyl phenylacetate **Diethyl carbonate (no α carbon)** **Diethyl phenylmalonate (65%)**

PROBLEM 19.3 Write mechanisms that account for the products that are formed in the two crossed Claisen condensations just illustrated.

PROBLEM 19.4 What products would you expect to obtain from each of the following crossed Claisen condensations?

(a) Ethyl propanoate + diethyl oxalate $\xrightarrow[(2)\ H_3O^+]{(1)\ NaOCH_2CH_3}$

(b) Ethyl acetate + ethyl formate $\xrightarrow[(2)\ H_3O^+]{(1)\ NaOCH_2CH_3}$

As we learned earlier in this section, esters that have only one α hydrogen cannot be converted to β-keto esters by sodium ethoxide. However, they can be converted to β-keto esters by reactions that use very strong bases. The strong base converts the ester to its enolate anion in nearly quantitative yield. This allows us to *acylate* the enolate anion by treating it with an acyl chloride or an ester. An example of this technique that makes use of the very powerful base sodium triphenylmethanide is shown next:

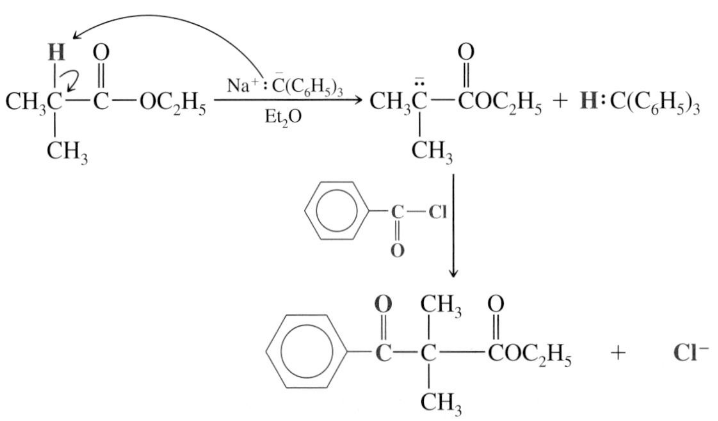

$$(CH_3)_2CH\text{—}COC_2H_5 \xrightarrow[Et_2O]{Na^+\ {}^-{:}C(C_6H_5)_3} (CH_3)_2\ddot{C}^-\text{—}COC_2H_5 + \mathbf{H}{:}C(C_6H_5)_3$$

$$\xrightarrow{C_6H_5COCl} C_6H_5C(=O)\text{—}C(CH_3)_2\text{—}COC_2H_5 + Cl^-$$

Ethyl 2,2-dimethyl-3-oxo-3-phenylpropanoate

19.2B Acylation of Other Carbanions

Enolate anions derived from ketones also react with esters in nucleophilic substitution reactions that resemble Claisen condensations. In the following first example, although two anions are possible from the reaction of the ketone with sodium amide, the major product is derived from the primary carbanion. This is because (a) the primary α hydrogens are slightly more acidic than the secondary α hydrogens and (b) in the presence of the strong base ($NaNH_2$) in an aprotic solvent (Et_2O), the kinetic enolate is formed (see Section 17.7A):

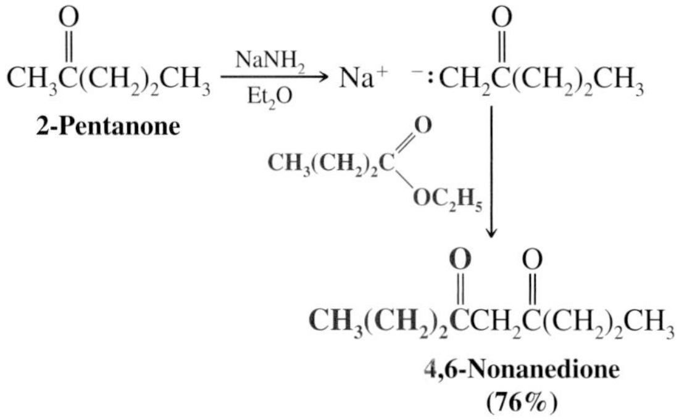

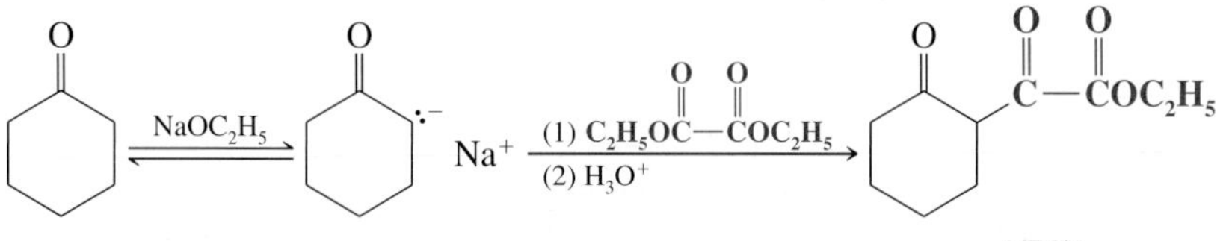

$$CH_3\overset{O}{\overset{\|}{C}}(CH_2)_2CH_3 \xrightarrow[Et_2O]{NaNH_2} Na^+ \; {}^-{:}CH_2\overset{O}{\overset{\|}{C}}(CH_2)_2CH_3$$

2-Pentanone

$$\xrightarrow{CH_3(CH_2)_2C(=O)OC_2H_5} CH_3(CH_2)_2\overset{O}{\overset{\|}{C}}CH_2\overset{O}{\overset{\|}{C}}(CH_2)_2CH_3$$

4,6-Nonanedione (76%)

$$\text{cyclohexanone} \underset{}{\overset{NaOC_2H_5}{\rightleftharpoons}} \text{cyclohexanone enolate } Na^+ \xrightarrow[(2)\ H_3O^+]{(1)\ C_2H_5O\overset{O}{\overset{\|}{C}}-\overset{O}{\overset{\|}{C}}OC_2H_5} \text{2-}(\overset{O}{\overset{\|}{C}}-\overset{O}{\overset{\|}{C}}OC_2H_5)\text{cyclohexanone}$$

(67%)

PROBLEM 19.5

Show how you might synthesize each of the following compounds using, as your starting materials, esters, ketones, acyl halides, and so on:

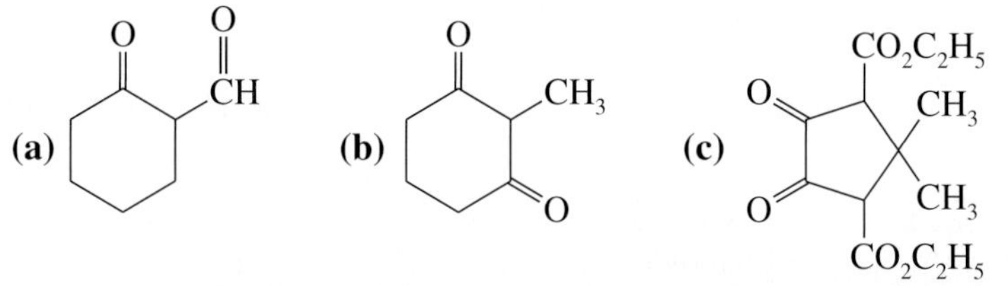

(a) 2-formylcyclohexanone (cyclohexanone with $-\overset{O}{\overset{\|}{C}}H$)
(b) 2-methyl-1,3-cyclohexanedione (ring with two C=O and CH_3)
(c) cyclopentane-1,2-dione bearing $CO_2C_2H_5$ and CH_3 on each of C-4 and C-5 (groups: $CO_2C_2H_5$, CH_3, CH_3, $CO_2C_2H_5$)

PROBLEM 19.6

Keto esters are capable of undergoing cyclization reactions similar to the Dieckmann condensation. Write a mechanism that accounts for the product formed in the following reaction:

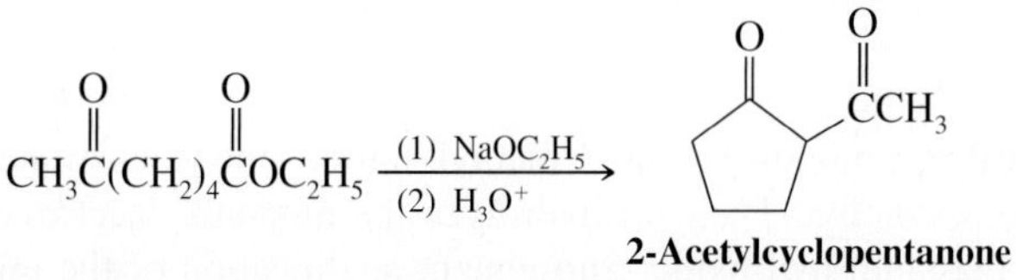

$$CH_3\overset{O}{\overset{\|}{C}}(CH_2)_4\overset{O}{\overset{\|}{C}}OC_2H_5 \xrightarrow[(2)\ H_3O^+]{(1)\ NaOC_2H_5} \text{cyclopentanone-2-}\overset{O}{\overset{\|}{C}}CH_3$$

2-Acetylcyclopentanone

19.3 The Acetoacetic Ester Synthesis: Synthesis of Methyl Ketones (Substituted Acetones)

19.3A Alkylation

As we have seen (Section 19.2), the methylene protons of ethyl acetoacetate (acetoacetic ester) are more acidic than the —OH proton of ethanol because they are located between two carbonyl groups. This acidity means that we can convert ethyl acetoacetate to a highly stabilized enolate anion using sodium ethoxide as a base. We can then carry out an alkylation reaction by treating the nucleophilic enolate anion with an alkyl halide. This process is called an **acetoacetic ester synthesis:**

$$CH_3\overset{O}{\overset{\|}{C}}-CH_2-\overset{O}{\overset{\|}{C}}OC_2H_5 + C_2H_5O^-Na^+ \rightleftharpoons CH_3\overset{O}{\overset{\|}{C}}-\overset{Na^+}{\ddot{C}H}-\overset{O}{\overset{\|}{C}}-OC_2H_5 + C_2H_5OH$$

Acetoacetic ester **Sodium ethoxide** **Sodioacetoacetic ester**

$$\downarrow R-X$$

$$CH_3\overset{O}{\overset{\|}{C}}-\underset{R}{\underset{|}{C}H}-\overset{O}{\overset{\|}{C}}-OC_2H_5 + NaX$$

Monoalkylacetoacetic ester

Since the alkylation in the reaction above is an S_N2 reaction, the best yields are obtained from the use of primary alkyl halides (including primary allylic and benzylic halides) or methyl halides. Secondary halides give lower yields, and tertiary halides give only elimination.

The monoalkylacetoacetic ester shown above still has one appreciably acidic hydrogen, and, if we desire, we can carry out a second alkylation. Because a monoalkylacetoacetic ester is somewhat less acidic than acetoacetic ester itself (why?), it is usually helpful to use a stronger base than ethoxide ion for the second alkylation. Use of potassium *tert*-butoxide is common because it is a stronger base than sodium ethoxide. Potassium *tert*-butoxide, because of its steric bulk, is also not likely to cause transesterification.

$$CH_3\overset{O}{\overset{\|}{C}}-\underset{R}{\underset{|}{C}H}-\overset{O}{\overset{\|}{C}}-OC_2H_5 + (CH_3)_3CO^-K^+ \rightleftharpoons CH_3\overset{O}{\overset{\|}{C}}-\underset{R}{\underset{|}{\overset{K^+}{\ddot{C}}}}-\overset{O}{\overset{\|}{C}}OC_2H_5 + (CH_3)_3COH$$

Monoalkylacetoacetic ester **Potassium *tert*-butoxide**

$$\downarrow R'-X$$

$$CH_3\overset{O}{\overset{\|}{C}}-\underset{R}{\underset{|}{\overset{R'}{\overset{|}{C}}}}-\overset{O}{\overset{\|}{C}}-OC_2H_5 + KX$$

Dialkylacetoacetic ester

To synthesize a monosubstituted methyl ketone (monosubstituted acetone), we carry out only one alkylation. Then we hydrolyze the monoalkylacetoacetic ester using dilute sodium or potassium hydroxide. Subsequent acidification of the mixture gives an alkylacetoacetic acid, and heating this β-keto acid to 100°C brings about decarboxylation (Section 18.10):

$$CH_3\overset{O}{\overset{\|}{C}}-\underset{R}{\underset{|}{C}H}-\overset{O}{\overset{\|}{C}}OC_2H_5 \xrightarrow[\text{heat}]{\text{dilute NaOH}} CH_3\overset{O}{\overset{\|}{C}}-\underset{R}{\underset{|}{C}H}-\overset{O}{\overset{\|}{C}}-O^-Na^+$$

Basic hydrolysis of the ester group

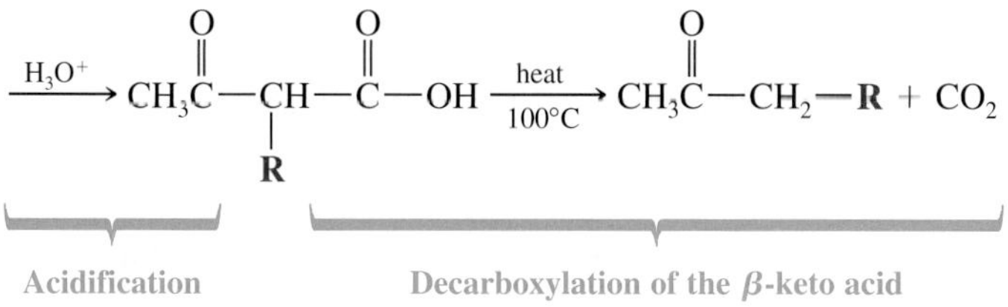

$$\xrightarrow{H_3O^+} CH_3\overset{O}{\overset{\|}{C}}-\underset{R}{\underset{|}{C}H}-\overset{O}{\overset{\|}{C}}-OH \xrightarrow[100°C]{heat} CH_3\overset{O}{\overset{\|}{C}}-CH_2-\mathbf{R} + CO_2$$

Acidification   Decarboxylation of the β-keto acid

A specific example is the following synthesis of 2-heptanone:

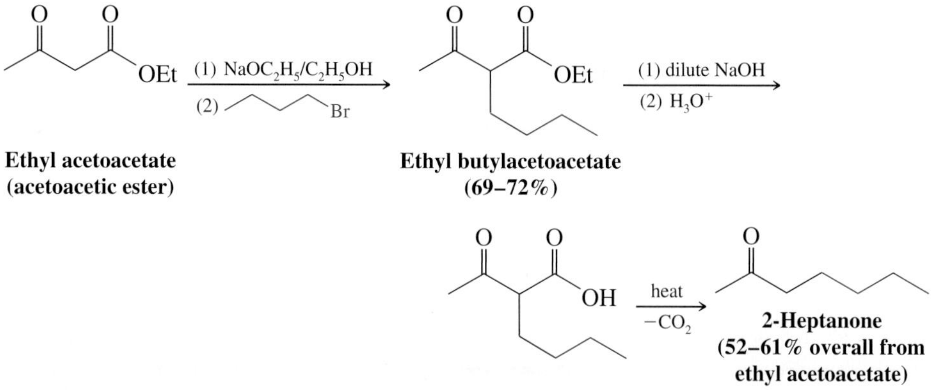

Ethyl acetoacetate (acetoacetic ester) → (1) $NaOC_2H_5/C_2H_5OH$; (2) 1-bromobutane (Br) → Ethyl butylacetoacetate (69–72%) → (1) dilute NaOH; (2) H_3O^+ → butylacetoacetic acid → heat, $-CO_2$ → **2-Heptanone (52–61% overall from ethyl acetoacetate)**

If our goal is the preparation of a disubstituted acetone, we carry out two successive alkylations, we hydrolyze the dialkylacetoacetic ester that is produced, and then we decarboxylate the dialkylacetoacetic acid. An example of this procedure is the synthesis of 3-butyl-2-heptanone.

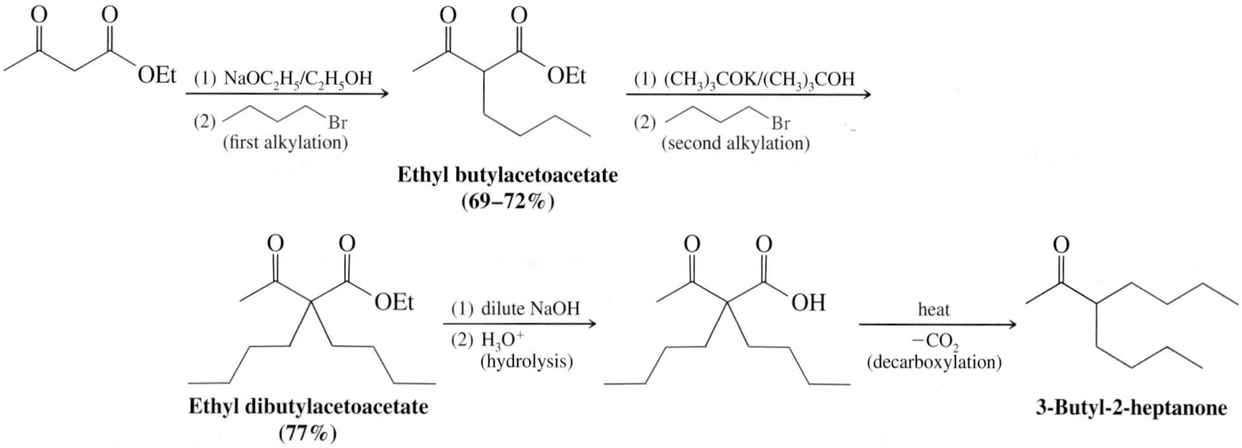

Ethyl acetoacetate → (1) $NaOC_2H_5/C_2H_5OH$; (2) 1-bromobutane (Br) (first alkylation) → **Ethyl butylacetoacetate (69–72%)** → (1) $(CH_3)_3COK/(CH_3)_3COH$; (2) 1-bromobutane (Br) (second alkylation) → **Ethyl dibutylacetoacetate (77%)** → (1) dilute NaOH; (2) H_3O^+ (hydrolysis) → dibutylacetoacetic acid → heat, $-CO_2$ (decarboxylation) → **3-Butyl-2-heptanone**

Although both alkylations in the example just given were carried out with the same alkyl halide, we could have used different alkyl halides if our synthesis had required it.

As we have seen, ethyl acetoacetate is a useful reagent for the preparation of substituted acetones (methyl ketones) of the types shown here:

$$H_3C-\overset{O}{\overset{\|}{C}}-\underset{\mathbf{R}}{\underset{|}{CH_2}} \qquad H_3C-\overset{O}{\overset{\|}{C}}-\underset{\mathbf{R}}{\underset{|}{\overset{\mathbf{R'}}{\overset{|}{CH}}}}$$

A monosubstituted acetone   **A disubstituted acetone**

Ethyl acetoacetate therefore serves as the synthetic equivalent of the enolate from acetone shown below. A **synthetic equivalent** is a reagent whose structure, when incorporated into a product, gives the appearance of having come from one type of precursor when as a reactant it actually had a different structural origin. Although it is possible to form the enolate of acetone, use of ethyl acetoacetate as a synthetic equivalent is often more convenient because its α hydrogens are so much more acidic ($pK_a = 9-11$) than those of acetone itself ($pK_a = 19-20$). If we had wanted to use the acetone enolate directly, we would have had to use a much stronger base and other special conditions (see Section 19.6).

Synthetic equivalents

$$H_3C-\overset{O}{\overset{\|}{C}}-\overset{-}{\underset{\cdot\cdot}{C}H}-\overset{O}{\overset{\|}{C}}-OC_2H_5 \textbf{ is the synthetic equivalent of } H_3C-\overset{O}{\overset{\|}{C}}-\overset{-}{\underset{\cdot\cdot}{C}H_2}$$

Ethyl acetoacetate anion **Acetone enolate**

PROBLEM 19.7 Occasional side products of alkylations of sodioacetoacetic esters are compounds with the following general structure:

$$CH_3\overset{R\ddot{O}:}{\overset{|}{C}}=CH\overset{:\ddot{O}:}{\overset{\|}{C}}OC_2H_5$$

Explain how these are formed.

PROBLEM 19.8 Show how you would use the acetoacetic ester synthesis to prepare **(a)** 2-pentanone, **(b)** 3-propyl-2-hexanone, and **(c)** 4-phenyl-2-butanone.

PROBLEM 19.9 The acetoacetic ester synthesis generally gives best yields when primary halides are used in the alkylation step. Secondary halides give low yields, and tertiary halides give practically no alkylation product at all. **(a)** Explain. **(b)** What products would you expect from the reaction of sodioacetoacetic ester and *tert*-butyl bromide? **(c)** Bromobenzene cannot be used as an arylating agent in an acetoacetic ester synthesis in the manner we have just described. Why not?

PROBLEM 19.10 Since the products obtained from Claisen condensations are β-keto esters, subsequent hydrolysis and decarboxylation of these products give a general method for the synthesis of ketones. Show how you would employ this technique in a synthesis of 4-heptanone.

The acetoacetic ester synthesis can also be carried out using halo esters and halo ketones. The use of an α-halo ester provides a convenient synthesis of γ-keto acids:

$$CH_3\overset{O}{\overset{\|}{C}}-CH_2-\overset{O}{\overset{\|}{C}}-OC_2H_5 \xrightarrow{C_2H_5ONa} CH_3\overset{O}{\overset{\|}{C}}-\overset{Na^+}{\underset{\cdot\cdot}{\overset{-}{C}H}}-\overset{O}{\overset{\|}{C}}-OC_2H_5 \xrightarrow{BrCH_2\overset{O}{\overset{\|}{C}}-OC_2H_5}$$

$$CH_3\overset{O}{\overset{\|}{C}}-\underset{CH_2\underset{\underset{O}{\|}}{C}-OC_2H_5}{\underset{|}{C}H}-\overset{O}{\overset{\|}{C}}-OC_2H_5 \xrightarrow[(2)\ H_3O^+]{(1)\ \text{dilute NaOH}} CH_3\overset{O}{\overset{\|}{C}}-\underset{CH_2\underset{\underset{O}{\|}}{C}-OH}{\underset{|}{C}H}-\overset{O}{\overset{\|}{C}}-OH \xrightarrow[-CO_2]{\text{heat}}$$

$$\underset{\gamma}{CH_3\overset{O}{\overset{\|}{C}}}-\underset{\beta}{CH_2}\underset{\alpha}{CH_2}-\overset{O}{\overset{\|}{C}}-OH$$

4-Oxopentanoic acid

PROBLEM 19.11

In the synthesis of the keto acid just given, the dicarboxylic acid decarboxylates in a specific way; it gives

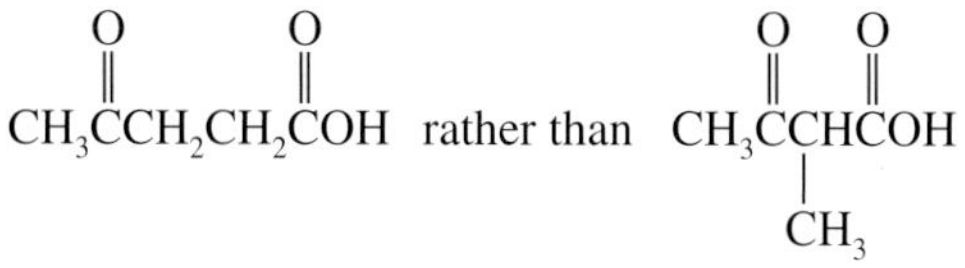

$$CH_3\overset{O}{\overset{\|}{C}}CH_2CH_2\overset{O}{\overset{\|}{C}}OH \text{ rather than } CH_3\overset{O}{\overset{\|}{C}}\underset{CH_3}{\underset{|}{C}H}\overset{O}{\overset{\|}{C}}OH$$

Explain.

The use of an α-halo ketone in an acetoacetic ester synthesis provides a general method for preparing γ-diketones:

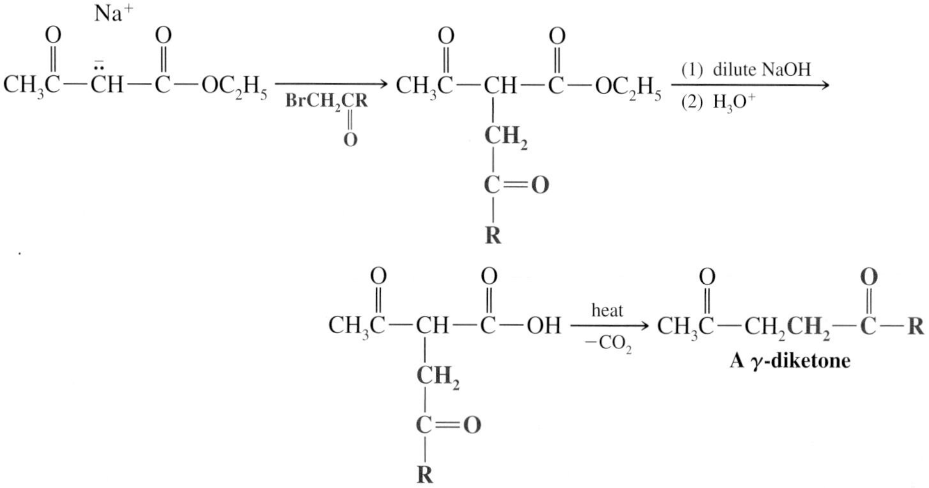

$$CH_3\overset{O}{\overset{\|}{C}}-\overset{Na^+}{\ddot{C}H}-\overset{O}{\overset{\|}{C}}-OC_2H_5 \xrightarrow[\text{BrCH}_2\overset{\|}{\underset{O}{C}}\text{R}]{} CH_3\overset{O}{\overset{\|}{C}}-\underset{\underset{\underset{R}{|}}{\underset{C=O}{|}}}{\underset{CH_2}{\underset{|}{CH}}}-\overset{O}{\overset{\|}{C}}-OC_2H_5 \xrightarrow[\text{(2) } H_3O^+]{\text{(1) dilute NaOH}}$$

$$CH_3\overset{O}{\overset{\|}{C}}-\underset{\underset{\underset{R}{|}}{\underset{C=O}{|}}}{\underset{CH_2}{\underset{|}{CH}}}-\overset{O}{\overset{\|}{C}}-OH \xrightarrow[-CO_2]{\text{heat}} \underset{\textbf{A } \gamma\textbf{-diketone}}{CH_3\overset{O}{\overset{\|}{C}}-CH_2CH_2-\overset{O}{\overset{\|}{C}}-R}$$

PROBLEM 19.12

How would you use the acetoacetic ester synthesis to prepare the following?

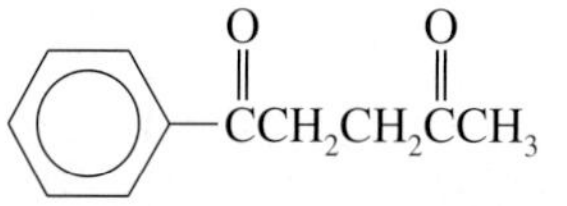

$$C_6H_5-\overset{O}{\overset{\|}{C}}CH_2CH_2\overset{O}{\overset{\|}{C}}CH_3$$

19.3B Acylation

Anions obtained from acetoacetic esters undergo acylation when they are treated with acyl chlorides or acid anhydrides. Because both of these acylating agents react with alcohols, acylation reactions cannot be carried out in ethanol and must be carried out in aprotic solvents such as DMF or DMSO (Section 6.13C). (If the reaction were to be carried out in ethanol, using sodium ethoxide, for example, then the acyl chloride would be rapidly converted to an ethyl ester and the ethoxide ion would be neutralized.) Sodium hydride can be used to generate the enolate anion in an aprotic solvent:

$$CH_3-\overset{O}{\overset{\|}{C}}-CH_2-\overset{O}{\overset{\|}{C}}-OC_2H_5 \xrightarrow[\text{aprotic solvent}\ (-H_2)]{Na^+H^-}$$

$$CH_3-\overset{O}{\overset{\|}{C}}-\ddot{C}H-\overset{O}{\overset{\|}{C}}-OC_2H_5\ Na^+ \xrightarrow[(-NaCl)]{R\overset{O}{\overset{\|}{C}}Cl} CH_3-\overset{O}{\overset{\|}{C}}-\underset{\underset{R}{|}}{\underset{C=O}{\underset{|}{CH}}}-\overset{O}{\overset{\|}{C}}-OC_2H_5 \xrightarrow[(2)\ H_3O^+]{(1)\ \text{dilute NaOH}}$$

$$CH_3-\overset{O}{\overset{\|}{C}}-\underset{\underset{R}{|}}{\underset{C=O}{\underset{|}{CH}}}-\overset{O}{\overset{\|}{C}}-OH \xrightarrow[-CO_2]{\text{heat}} \underset{\textbf{A }\beta\textbf{-diketone}}{CH_3-\overset{O}{\overset{\|}{C}}-CH_2-\overset{O}{\overset{\|}{C}}-R}$$

A method for synthesizing β-dicarbonyl compounds

Acylations of acetoacetic esters followed by hydrolysis and decarboxylation give us another method, in addition to the Claisen condensation, for preparing β-dicarbonyl compounds.

PROBLEM 19.13

How would you use the acetoacetic ester synthesis to prepare the following?

$$C_6H_5-\overset{O}{\overset{\|}{C}}CH_2\overset{O}{\overset{\|}{C}}CH_3$$

19.3C Acetoacetic Ester Dianion: Alkylation at the Terminal Carbon

One further variation of the acetoacetic ester synthesis involves the conversion of an acetoacetic ester to a resonance-stabilized *dianion* by using a very strong base such as potassium amide in liquid ammonia:

$$CH_3-\overset{:\ddot{O}:}{\overset{\|}{C}}-CH_2-\overset{:\ddot{O}:}{\overset{\|}{C}}-OC_2H_5 \xrightarrow[\text{liq. }NH_3]{2\,K^+\,:\ddot{N}H_2^-} \left[{}^-:CH_2-\overset{:\ddot{O}:}{\overset{\|}{C}}-\ddot{C}H-\overset{:\ddot{O}:}{\overset{\|}{C}}-\ddot{O}C_2H_5\right]\ 2\,K^+$$

$$\updownarrow$$

$$\left[CH_2=\overset{:\ddot{O}:^-}{\overset{|}{C}}-CH=\overset{:\ddot{O}:^-}{\overset{|}{C}}-\ddot{O}C_2H_5\right]\ 2\,K^+$$

$$\updownarrow$$

etc.

When this dianion is treated with 1 mol of a primary (or methyl) halide, it undergoes alkylation at its terminal carbon rather than at its interior one. This orientation of the alkylation reaction apparently results from the greater basicity (and thus nucleophilicity) of the terminal carbanion. This carbanion is more basic because it is stabilized by only one adjacent carbonyl group. After monoalkylation has taken place, the anion that remains can be protonated by adding ammonium chloride:

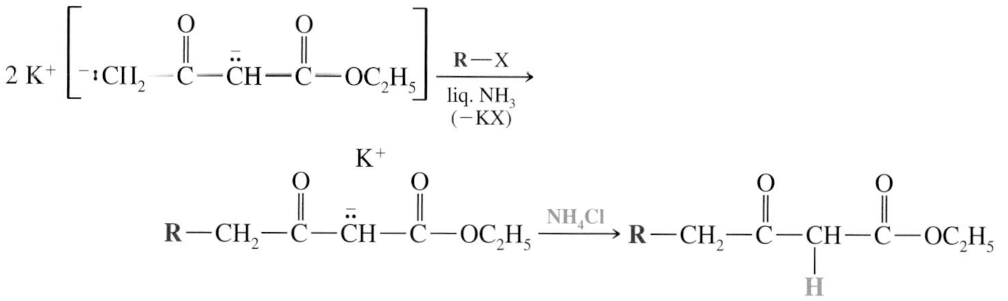

$$2\,K^+\left[{}^-{:}CH_2-\overset{O}{\overset{\|}{C}}-\ddot{C}H-\overset{O}{\overset{\|}{C}}-OC_2H_5\right]\xrightarrow[\substack{\text{liq. }NH_3\\(-KX)}]{R-X}$$

$$R-CH_2-\overset{O}{\overset{\|}{C}}-\overset{K^+}{\ddot{C}H}-\overset{O}{\overset{\|}{C}}-OC_2H_5\xrightarrow{NH_4Cl}R-CH_2-\overset{O}{\overset{\|}{C}}-\underset{H}{\underset{|}{C}H}-\overset{O}{\overset{\|}{C}}-OC_2H_5$$

PROBLEM 19.14 Show how you could use ethyl acetoacetate in a synthesis of

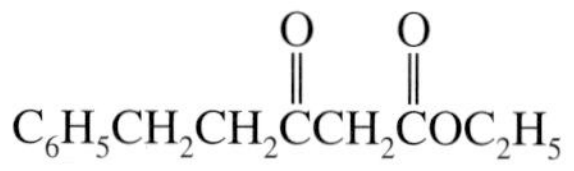

$$C_6H_5CH_2CH_2\overset{O}{\overset{\|}{C}}CH_2\overset{O}{\overset{\|}{C}}OC_2H_5$$

19.4 The Malonic Ester Synthesis: Synthesis of Substituted Acetic Acids

A useful counterpart of the acetoacetic ester synthesis—one that allows the synthesis of *mono-* and *disubstituted acetic acids*—is called the **malonic ester synthesis.** The starting compound is the diester of a β-dicarboxylic acid, called a malonic ester. The most commonly used malonic ester is diethyl malonate.

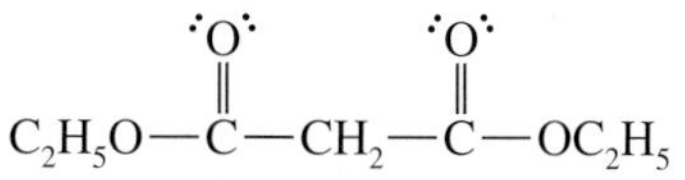

$$C_2H_5O-\overset{\ddot{O}}{\overset{\|}{C}}-CH_2-\overset{\ddot{O}}{\overset{\|}{C}}-OC_2H_5$$

Diethyl malonate
(a β-dicarboxylic acid ester)

We shall see by examining the following mechanism that the malonic ester synthesis resembles the acetoacetic ester synthesis in several respects.

A Mechanism for the Reaction

The Malonic Ester Synthesis of Substituted Acetic Acids

Step 1 Diethyl malonate, the starting compound, forms a relatively stable enolate anion:

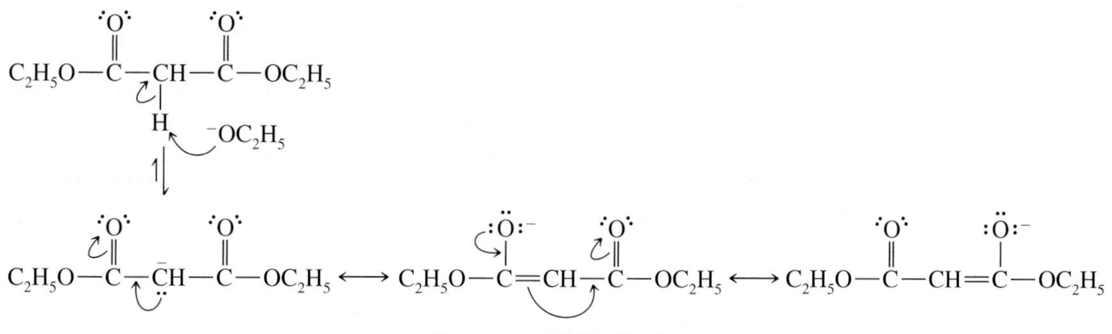

$$C_2H_5O-\overset{\ddot{O}}{\overset{\|}{C}}-\underset{H}{\underset{|}{C}H}-\overset{\ddot{O}}{\overset{\|}{C}}-OC_2H_5 \quad {}^-OC_2H_5$$

$$\Big\updownarrow$$

$$C_2H_5O-\overset{\ddot{O}}{\overset{\|}{C}}-\overset{-}{\ddot{C}H}-\overset{\ddot{O}}{\overset{\|}{C}}-OC_2H_5 \longleftrightarrow C_2H_5O-\overset{:\ddot{O}:^-}{\overset{|}{C}}=CH-\overset{\ddot{O}}{\overset{\|}{C}}-OC_2H_5 \longleftrightarrow C_2H_5O-\overset{\ddot{O}}{\overset{\|}{C}}-CH=\overset{:\ddot{O}:^-}{\overset{|}{C}}-OC_2H_5$$

$$+\ HOC_2H_5$$

Resonance-stabilized anion

Step 2 This enolate anion can be alkylated in an S_N2 reaction,

$$C_2H_5O-\overset{O}{\overset{\|}{C}}-\overset{-}{\ddot{C}}H-\overset{O}{\overset{\|}{C}}-OC_2H_5 + R-X \longrightarrow C_2H_5O-\overset{O}{\overset{\|}{C}}-\underset{R}{\underset{|}{C}H}-\overset{O}{\overset{\|}{C}}-OC_2H_5 + X^-$$

Enolate ion **Monoalkylmalonic ester**

and the product can be alkylated again if our synthesis requires it:

$$C_2H_5O-\overset{O}{\overset{\|}{C}}-\underset{R}{\underset{|}{C}H}-\overset{O}{\overset{\|}{C}}-OC_2H_5 \xrightleftharpoons{(CH_3)_3CO^-} C_2H_5O-\overset{O}{\overset{\|}{C}}-\underset{R}{\underset{|}{\overset{-}{\ddot{C}}}}-\overset{O}{\overset{\|}{C}}-OEt \xrightarrow{R'-X} C_2H_5O-\overset{O}{\overset{\|}{C}}-\underset{R}{\underset{|}{\overset{R'}{\overset{|}{C}}}}-\overset{O}{\overset{\|}{C}}-OC_2H_5$$

Dialkylmalonic ester

Step 3 The mono- or dialkylmalonic ester can then be hydrolyzed to a mono- or dialkylmalonic acid, and substituted malonic acids decarboxylate readily. Decarboxylation gives a mono- or disubstituted acetic acid:

$$C_2H_5O-\overset{O}{\overset{\|}{C}}-\underset{R}{\underset{|}{C}H}-\overset{O}{\overset{\|}{C}}-OC_2H_5 \xrightarrow[(2)\ H_3O^+]{(1)\ HO^-,\ H_2O} HO-\overset{O}{\overset{\|}{C}}-\underset{R}{\underset{|}{C}H}-\overset{O}{\overset{\|}{C}}-OH \xrightarrow{heat}$$

Monoalkylmalonic ester

$$\left[\text{cyclic transition state: } HO-C(=O)-C(R)(H)-C(=O)-O-H\right] \xrightarrow{-CO_2} H_2\underset{R}{\underset{|}{C}}-\overset{O}{\overset{\|}{C}}-OH$$

Monoalkylacetic acid

or after dialkylation,

$$C_2H_5O-\overset{O}{\overset{\|}{C}}-\underset{R}{\underset{|}{\overset{R'}{\overset{|}{C}}}}-\overset{O}{\overset{\|}{C}}-OC_2H_5 \xrightarrow[(2)\ H_3O^+]{(1)\ HO^-,\ H_2O} HO-\overset{O}{\overset{\|}{C}}-\underset{R}{\underset{|}{\overset{R'}{\overset{|}{C}}}}-\overset{O}{\overset{\|}{C}}-OH \xrightarrow{heat}$$

Dialkylmalonic ester

$$\left[\text{cyclic transition state: } HO-C(=O)-C(R)(R')-C(=O)-O-H\right] \xrightarrow{-CO_2} H\underset{R}{\underset{|}{\overset{R'}{\overset{|}{C}}}}-\overset{O}{\overset{\|}{C}}-OH$$

Dialkylacetic acid

Two specific examples of the malonic ester synthesis are the syntheses of hexanoic acid and 2-ethylpentanoic acid that follow.

A Malonic Ester Synthesis of Hexanoic Acid

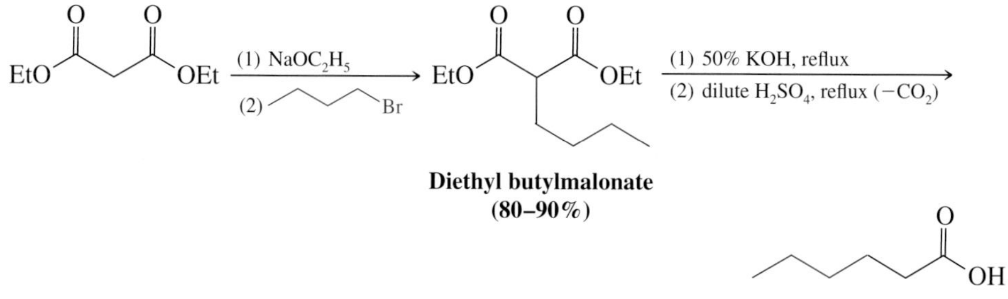

EtO OEt —(1) $NaOC_2H_5$; (2) 1-bromobutane (Br)→ EtO OEt —(1) 50% KOH, reflux; (2) dilute H_2SO_4, reflux ($-CO_2$)→

Diethyl butylmalonate (80–90%)

OH

Hexanoic acid (75%)

A Malonic Ester Synthesis of 2-Ethylpentanoic Acid

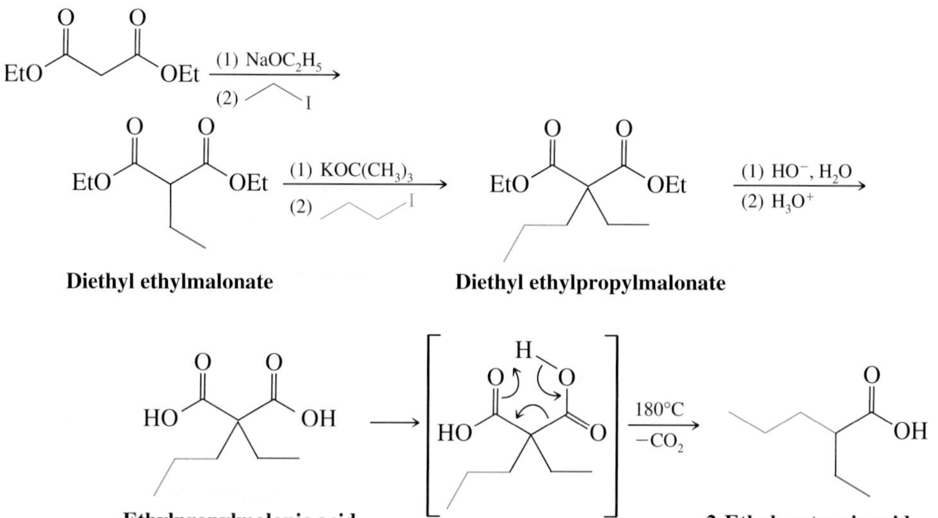

EtO OEt —(1) $NaOC_2H_5$; (2) ethyl iodide (I)→

EtO OEt —(1) $KOC(CH_3)_3$; (2) 1-iodopropane (I)→ EtO OEt —(1) HO^-, H_2O; (2) H_3O^+→

Diethyl ethylmalonate **Diethyl ethylpropylmalonate**

HO OH → [HO, O, H, O, O] —180°C, $-CO_2$→ OH

Ethylpropylmalonic acid **2-Ethylpentanoic acid**

PROBLEM 19.15

Outline all steps in a malonic ester synthesis of each of the following: **(a)** pentanoic acid, **(b)** 2-methylpentanoic acid, and **(c)** 4-methylpentanoic acid.

Two variations of the malonic ester synthesis make use of dihaloalkanes. In the first of these, two molar equivalents of sodiomalonic ester are allowed to react with a dihaloalkane. Two consecutive alkylations occur giving a tetraester; hydrolysis and decarboxylation of the tetraester yields a dicarboxylic acid. An example is the synthesis of glutaric acid:

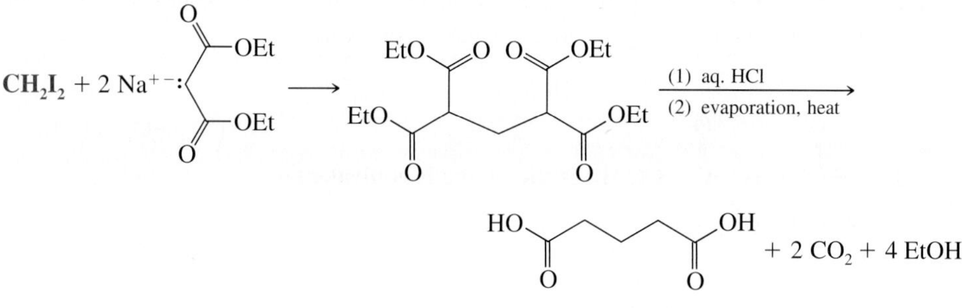

$CH_2I_2 + 2\ Na^{+}\ {}^{-}$:CH(COOEt)$_2$ → (EtO$_2$C)$_2$CH–CH$_2$–CH(CO$_2$Et)$_2$ —(1) aq. HCl; (2) evaporation, heat→

HO$_2$C–CH$_2$CH$_2$CH$_2$–CO$_2$H $+ 2\ CO_2 + 4\ EtOH$

Glutaric acid (80% from tetraester)

In a second variation, one molar equivalent of sodiomalonic ester is allowed to react with one molar equivalent of a dihaloalkane. This reaction gives a haloalkylmalonic ester, which, when treated with sodium ethoxide, undergoes an internal alkylation reaction. This method has been used to prepare three-, four-, five-, and six-membered rings. An example is the synthesis of cyclobutanecarboxylic acid:

$$(EtO_2C)_2CH:^-Na^+ + Br(CH_2)_3Br \xrightarrow{(-NaBr)} (EtO_2C)_2CH(CH_2)_3Br \xrightarrow{C_2H_5O^-Na^+} (EtO_2C)_2\ddot{C}^-(CH_2)_3Br \longrightarrow \text{diethyl cyclobutane-1,1-dicarboxylate}$$

$$\xrightarrow{\text{hydrolysis and decarboxylation}}$$

Cyclobutanecarboxylic acid

As we have seen, the malonic ester synthesis is a useful method for preparing mono- and dialkylacetic acids:

Malonic ester synthesis is a tool for synthesizing substituted acetic acids.

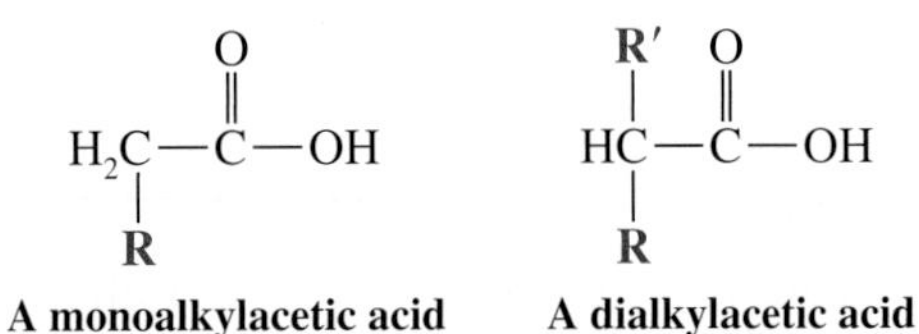

A monoalkylacetic acid **A dialkylacetic acid**

Thus, the malonic ester synthesis provides us with a synthetic equivalent of an ester enolate of acetic acid or acetic acid dianion. Direct formation of such anions is possible (Section 19.6), but it is often more convenient to use diethyl malonate as a synthetic equivalent because its α hydrogens are more easily removed:

$$C_2H_5O-\overset{O}{\overset{\|}{C}}-\overset{-}{\underset{\cdot\cdot}{C}}H-\overset{O}{\overset{\|}{C}}-OC_2H_5 \quad \textbf{is the synthetic equivalent of} \quad {}^-{:}CH_2-\overset{O}{\overset{\|}{C}}-\ddot{\underset{\cdot\cdot}{O}}C_2H_5 \ \textbf{and}\ {}^-{:}CH_2-\overset{O}{\overset{\|}{C}}-\ddot{\underset{\cdot\cdot}{O}}{:}^-$$

Diethyl malonate anion

In Special Topic D we shall see biosynthetic equivalents of these anions.

19.5 Further Reactions of Active Hydrogen Compounds

Because of the acidity of their methylene hydrogens, malonic esters, acetoacetic esters, and similar compounds are often called **active hydrogen compounds** or **active methylene compounds.** Generally speaking, active hydrogen compounds have two electron-withdrawing groups attached to the same carbon atom:

$$Z{-}CH_2{-}Z'$$

Active hydrogen compound
(Z and Z′ are electron-withdrawing groups.)

The electron-withdrawing groups can be a variety of substituents, including

$$-\overset{O}{\overset{\|}{C}}R \quad -\overset{O}{\overset{\|}{C}}H \quad -\overset{O}{\overset{\|}{C}}OR \quad -\overset{O}{\overset{\|}{C}}NR_2 \quad -C{\equiv}N \quad -NO_2$$

$$-\overset{O}{\overset{\|}{S}}-R \quad -\underset{O}{\underset{\|}{\overset{O}{\overset{\|}{S}}}}-R \quad -\underset{O}{\underset{\|}{\overset{O}{\overset{\|}{S}}}}-OR \ \text{ or } \ -\underset{O}{\underset{\|}{\overset{O}{\overset{\|}{S}}}}-NR_2$$

The range of pK_a values for such active methylene compounds is 3–13.

Ethyl cyanoacetate, for example, reacts with a base to yield a resonance-stabilized anion:

$$:N{\equiv}C{-}CH_2{-}\overset{\ddot{O}}{\overset{\|}{C}}OEt \xrightarrow[-H^+]{\text{base}} :N{\equiv}C{-}\overset{..}{C}H^{-}{-}\overset{\ddot{O}}{\overset{\|}{C}}OEt \longleftrightarrow {}^{-}:\ddot{N}{=}C{=}CH{-}\overset{\ddot{O}}{\overset{\|}{C}}OEt \longleftrightarrow :N{\equiv}C{-}CH{=}\overset{:\ddot{O}:^-}{\overset{|}{C}}OEt$$

Ethyl cyanoacetate

Ethyl cyanoacetate anions also undergo alkylations. They can be dialkylated with isopropyl iodide, for example:

$$(CH_3)_2CHI + NCCH_2COOEt \xrightarrow[(2)\ H_3O^+]{(1)\ C_2H_5ONa/C_2H_5OH} NCCH(CH(CH_3)_2)COOEt \xrightarrow[(2)\ (CH_3)_2CHI]{(1)\ C_2H_5ONa/C_2H_5OH} NCC(CH(CH_3)_2)_2COOEt$$

(63%) **(95%)**

Another way of preparing ketones is to use a β-keto sulfoxide as an active hydrogen compound:

$$R\overset{O}{\overset{\|}{C}}{-}CH_2{-}\overset{O}{\overset{\|}{S}}R' \xrightarrow[(2)\ R''X]{(1)\ \text{base}} R\overset{O}{\overset{\|}{C}}{-}\underset{R''}{\underset{|}{C}H}{-}\overset{O}{\overset{\|}{S}}R' \xrightarrow{Al{-}Hg} R\overset{O}{\overset{\|}{C}}{-}CH_2{-}R''$$

A β-keto sulfoxide

The β-keto sulfoxide is first converted to an anion and then the anion is alkylated. Treating the product of these steps with aluminum amalgam (Al–Hg) causes cleavage at the carbon–sulfur bond and gives the ketone in high yield.

PROBLEM 19.16 The antiepileptic drug valproic acid is 2-propylpentanoic acid (administered as the sodium salt). One commercial synthesis of valproic acid begins with ethyl cyanoacetate. The penultimate step of this synthesis involves a decarboxylation, and the last step involves hydrolysis of a nitrile. Outline this synthesis.

19.6 Direct Alkylation of Esters and Nitriles

We have seen in Sections 19.3–19.5 that it is easy to alkylate β-keto esters and other active hydrogen compounds. The hydrogens situated on the carbon atom between the two electron-withdrawing groups are unusually acidic and are easily removed by bases such as ethoxide ion. It is also possible, however, to alkylate esters and nitriles that do not have a β-keto group. To do this, we must use a stronger base, one that will convert the ester or nitrile into its enolate anion rapidly so that all of the ester or nitrile is converted to its enolate before it can undergo Claisen condensation. We must also use a base that is sufficiently bulky not to react at the carbonyl carbon of the ester or at the carbon of the nitrile group. Such a base is lithium diisopropylamide (LDA).

Lithium diisopropylamide is a very strong base because it is the conjugate base of a very weak acid, diisopropylamine ($pK_a = 38$). Lithium diisopropylamide is prepared by treating diisopropylamine with butyllithium. Solvents commonly used for reactions in which LDA is the base are ethers such as tetrahydrofuran (THF) and 1,2-dimethoxyethane (DME). (The use of LDA in other syntheses was described in Section 17.7.)

$$C_4H_9{:}^-Li^+ + [(CH_3)_2CH]_2NH \xrightarrow{THF} [(CH_3)_2CH]_2N^-\,Li^+ + C_4H_{10}$$

Butyllithium | **Diisopropylamine** $pK_a = 38$ | **Lithium diisopropylamide [LDA or $(i\text{-}C_3H_7)_2NLi$]** | **Butane** $pK_a = 50$

Examples of the **direct alkylation** of esters are shown below. In the second example the ester is a lactone (Section 18.7C):

$$CH_3CH_2CH_2{-}\overset{O}{\overset{\|}{C}}{-}OCH_3 \xrightarrow[THF]{LDA} \left[CH_3CH_2\overset{..}{C}H^-{-}\overset{O}{\overset{\|}{C}}{-}OCH_3\ Li^+\right] \xrightarrow{CH_3CH_2{-}I} CH_3CH_2\underset{CH_2CH_3}{\underset{|}{C}H}{-}\overset{O}{\overset{\|}{C}}{-}OCH_3$$

Methyl butanoate → **Methyl 2-ethylbutanoate (96%)**

Butyrolactone $\xrightarrow[THF]{LDA}$ [enolate anion] $\xrightarrow{CH_3{-}I}$ **2-Methylbutyrolactone (88%)**

19.7 Alkylation of 1,3-Dithianes

Two sulfur atoms attached to the same carbon of 1,3-dithiane cause the hydrogen atoms of that carbon to be more acidic ($pK_a = 32$) than those of most alkyl carbon atoms:

1,3-Dithiane
$pK_a = 32$

Sulfur atoms, because they are easily polarized, can aid in stabilizing the negative charge of the anion. Strong bases such as butyllithium are usually used to convert a dithiane to its anion:

$$\text{1,3-dithiane} + \overset{\delta-\ \delta+}{C_4H_9Li} \longrightarrow \text{dithiane anion } Li^+ + C_4H_{10}$$

1,3-Dithianes are thioacetals (see Section 16.7D); they can be prepared by treating an aldehyde with 1,3-propanedithiol in the presence of a trace of acid:

$$RCHO + HSCH_2CH_2CH_2SH \xrightarrow{H_3O^+} \text{A 1,3-dithiane} + H_2O$$

A 1,3-dithiane

Alkylating the 1,3-dithiane with a primary halide by an S_N2 reaction and then hydrolyzing the product (a thioacetal) provides a method for converting an aldehyde to a ketone. Hydrolysis is usually carried out by using $HgCl_2$ either in methanol or in aqueous acetonitrile, CH_3CN:

$$\text{dithiane (R, H)} \xrightarrow[(2)\ R'CH_2X\ (-LiX)]{(1)\ C_4H_9Li\ (-C_4H_{10})} \text{dithiane (R, } CH_2R'\text{)} \xrightarrow[(-HSCH_2CH_2CH_2SH)]{HgCl_2,\ CH_3OH,\ H_2O} R-\overset{O}{\overset{\|}{C}}-CH_2R'$$

Thioacetal **Ketone**

Notice that in these 1,3-dithiane syntheses the usual mode of reaction of an aldehyde is reversed. Normally the carbonyl carbon atom of an aldehyde is partially positive; it is electrophilic and, consequently, reacts with nucleophiles. When the aldehyde is converted to a 1,3-dithiane and treated with butyllithium, this same carbon atom becomes negatively charged and reacts with electrophiles. This reversal of polarity of the carbonyl carbon atom is called **umpolung** (German for **polarity reversal**). The 1,3-dithiane anion therefore becomes a synthetic equivalent of an anionic carbonyl carbon:

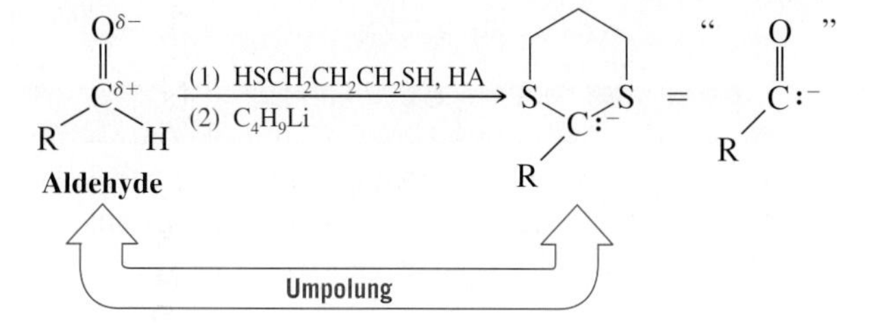

A 1,3-dithiane anion is the synthetic equivalent of an aldehyde carbanion.

The synthetic use of 1,3-dithianes was developed by E. J. Corey and D. Seebach and is often called the *Corey–Seebach* method.

PROBLEM 19.17 (**a**) Which aldehyde would you use to prepare 1,3-dithiane? (**b**) How would you synthesize $C_6H_5CH_2CHO$ using a 1,3-dithiane as an intermediate? (**c**) How would you convert benzaldehyde to acetophenone?

PROBLEM 19.18 The Corey–Seebach method can also be used to synthesize molecules with the structure RCH_2CH_2R'. How might this be done?

PROBLEM 19.19 (**a**) The Corey–Seebach method has been used to prepare the following highly strained molecule called a metaparacyclophane. What are the structures of the intermediates **A–D**?

$$\text{(isophthalaldehyde)} \xrightarrow[\text{acid}]{2\ HSCH_2CH_2CH_2SH} \mathbf{A}\ (C_{14}H_{18}S_4) \xrightarrow[(2)\ BrCH_2-C_6H_4-CH_2Br]{(1)\ 2\ C_4H_9Li}$$

$$\mathbf{B}\ (C_{22}H_{24}S_4) \xrightarrow{\text{hydrolysis}} \mathbf{C}\ (C_{16}H_{12}O_2) \xrightarrow{NaBH_4} \mathbf{D}\ (C_{16}H_{16}O_2) \xrightarrow[(2)\ 2\ KOC(CH_3)_3]{(1)\ 2\ TsCl}$$

A metaparacyclophane

(**b**) What compound would be obtained by treating **B** with excess Raney Ni?

19.8 THE KNOEVENAGEL CONDENSATION

Active hydrogen compounds condense with aldehydes and ketones. Known as **Knoevenagel condensations,** these aldol-like condensations are catalyzed by weak bases such as amines. An example is the following:

$$Cl-C_6H_4-CHO + CH_3\overset{O}{\overset{\|}{C}}CH_2\overset{O}{\overset{\|}{C}}OC_2H_5 \xrightarrow[C_2H_5OH]{(C_2H_5)_2NH}$$

$$\left[Cl-C_6H_4-\overset{OH}{\overset{|}{C}H}-CH(\overset{O}{\overset{\|}{C}}OC_2H_5)(\overset{\|}{\underset{O}{C}}CH_3)\right] \xrightarrow{-H_2O} Cl-C_6H_4-CH{=}C(\overset{O}{\overset{\|}{C}}OC_2H_5)(\underset{O}{\overset{\|}{C}}CH_3)$$

(86%)

19.9 MICHAEL ADDITIONS

Active hydrogen compounds also undergo conjugate additions to α,β-unsaturated carbonyl compounds. These reactions are known as **Michael additions,** a reaction that we studied in Section 17.9B. Nucleophiles such as enolates tend to undergo conjugate addition (Section 17.9).

A Mechanism for the Reaction

Michael Addition of an Active Hydrogen Compound

Overall Reaction:

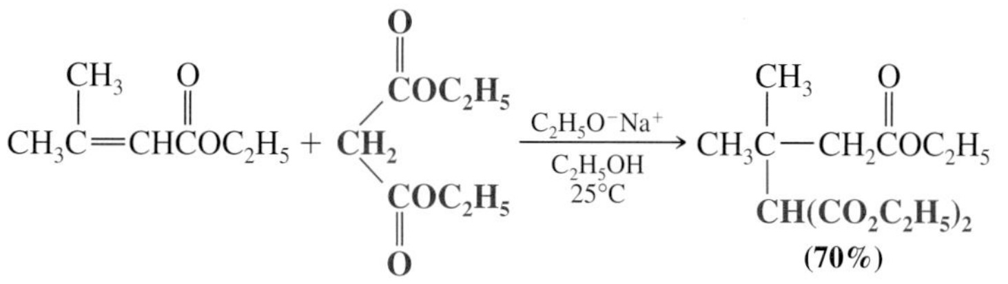

$$CH_3\overset{CH_3}{\overset{|}{C}}{=}CH\overset{O}{\overset{\|}{C}}OC_2H_5 + CH_2(\overset{O}{\overset{\|}{C}}OC_2H_5)_2 \xrightarrow[\substack{C_2H_5OH \\ 25°C}]{C_2H_5O^-Na^+} CH_3\overset{CH_3}{\overset{|}{C}}\underset{CH(CO_2C_2H_5)_2}{}{-}CH_2\overset{O}{\overset{\|}{C}}OC_2H_5$$

(70%)

Mechanism:

Step 1

$$C_2H_5O^- + H{-}CH(\overset{O}{\overset{\|}{C}}OC_2H_5)_2 \rightleftharpoons C_2H_5OH + {}^-{:}CH(\overset{O}{\overset{\|}{C}}OC_2H_5)_2$$

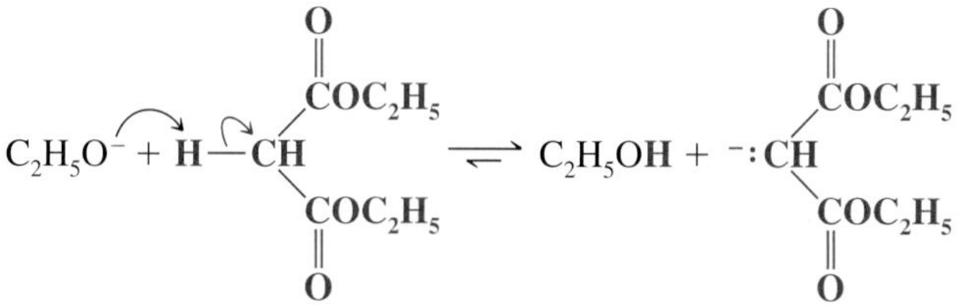

An alkoxide anion removes a proton to form the anion of the active methylene compound.

Step 2

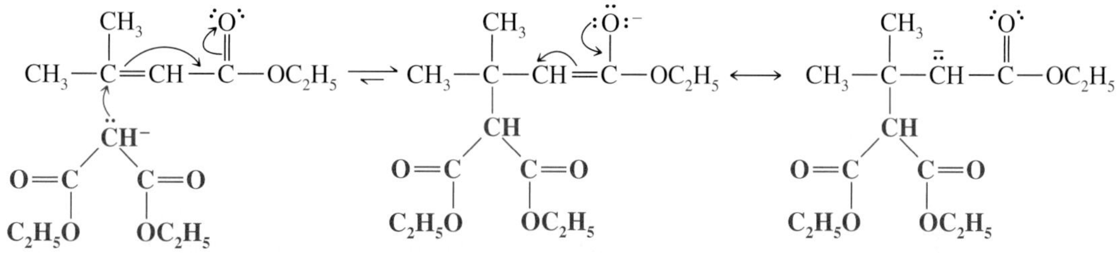

$$CH_3{-}\overset{CH_3}{\overset{|}{C}}{=}CH{-}\overset{:\ddot{O}}{\overset{\|}{C}}{-}OC_2H_5 + {}^-\ddot{C}H(C{=}O)(OC_2H_5)_2 \rightleftharpoons CH_3{-}\overset{CH_3}{\overset{|}{C}}(CH(CO_2C_2H_5)_2){-}CH{=}\overset{:\ddot{O}:^-}{\overset{|}{C}}{-}OC_2H_5 \longleftrightarrow CH_3{-}\overset{CH_3}{\overset{|}{C}}(CH(CO_2C_2H_5)_2){-}\ddot{C}H{-}\overset{:O:}{\overset{\|}{C}}{-}OC_2H_5$$

Conjugate addition of the anion to the α,β-unsaturated ester leads to a new enolate anion.

Step 3

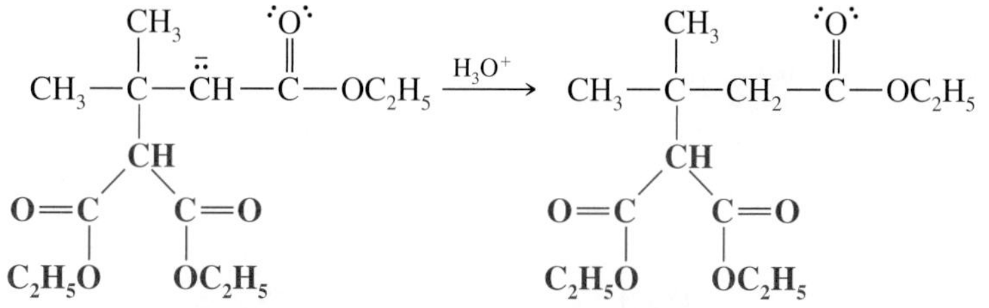

$$CH_3{-}\overset{CH_3}{\overset{|}{C}}(CH(CO_2C_2H_5)_2){-}\ddot{C}H{-}\overset{:O:}{\overset{\|}{C}}{-}OC_2H_5 \xrightarrow{H_3O^+} CH_3{-}\overset{CH_3}{\overset{|}{C}}(CH(CO_2C_2H_5)_2){-}CH_2{-}\overset{:O:}{\overset{\|}{C}}{-}OC_2H_5$$

The enolate anion is protonated by an acid during the workup of the reaction.

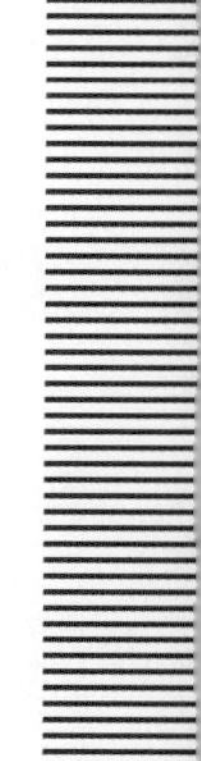

PROBLEM 19.20

How would you prepare $HO\overset{O}{\overset{\|}{C}}CH_2\overset{CH_3}{\underset{CH_3}{C}}CH_2\overset{O}{\overset{\|}{C}}OH$ from the product of the Michael addition given above?

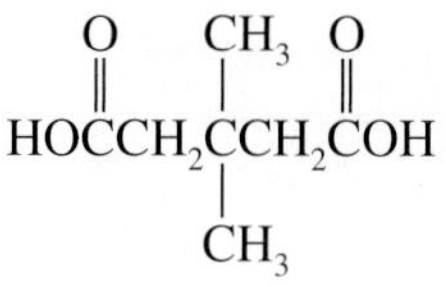

Michael additions take place with a variety of other reagents; these include acetylenic esters and α,β-unsaturated nitriles:

$$H{-}C{\equiv}C{-}\overset{O}{\overset{\|}{C}}{-}OC_2H_5 + CH_3\overset{O}{\overset{\|}{C}}{-}CH_2{-}\overset{O}{\overset{\|}{C}}{-}OC_2H_5 \xrightarrow[C_2H_5OH]{C_2H_5O^-} HC{=}CH{-}\overset{O}{\overset{\|}{C}}{-}OC_2H_5 \text{ (with CH(COCH}_3\text{)(CO}_2C_2H_5\text{) on the first carbon)}$$

$$CH_2{=}CH{-}C{\equiv}N + CH_2(COOC_2H_5)_2 \xrightarrow[C_2H_5OH]{C_2H_5O^-} (C_2H_5OOC)_2CH{-}CH_2{-}CH_2{-}C{\equiv}N$$

19.10 THE MANNICH REACTION

Compounds capable of forming an enol react with imines from formaldehyde and a primary or secondary amine to yield β-aminoalkyl carbonyl compounds called Mannich bases. The following reaction of acetone, formaldehyde, and diethylamine is an example:

$$CH_3{-}\overset{O}{\overset{\|}{C}}{-}CH_3 + H{-}\overset{O}{\overset{\|}{C}}{-}H + (C_2H_5)_2NH \xrightarrow{HCl} CH_3{-}\overset{O}{\overset{\|}{C}}{-}CH_2{-}CH_2{-}N(C_2H_5)_2 + H_2O$$

A Mannich base

The **Mannich reaction** apparently proceeds through a variety of mechanisms depending on the reactants and the conditions that are employed. The mechanism that follows the "Chemistry of . . ." box, on page 909, appears to operate in neutral or acidic media. Note the aspects in common with imine formation and with reactions of enols and carbonyl groups.

A Mechanism for the Reaction

The Mannich Reaction

Step 1 $R_2\ddot{N}H + H_2C{=}\ddot{O} \rightleftharpoons R_2\ddot{N}{-}CH_2{-}\ddot{O}{-}H \overset{HA}{\rightleftharpoons} R_2\ddot{N}{-}CH_2{-}\overset{+}{O}H_2 \overset{-H_2O}{\rightleftharpoons} R_2\overset{+}{N}{=}CH_2$

Iminium cation

Reaction of the secondary amine with the aldehyde forms a hemiaminal.

The hemiaminal loses a molecule of water to form an iminium cation.

Step 2 $CH_3{-}\overset{O}{\overset{\|}{C}}{-}CH_3 \overset{HA}{\rightleftharpoons} CH_3{-}\overset{O-H}{\overset{|}{C}}{=}CH_2 + CH_2{=}\overset{+}{N}R_2 \longrightarrow CH_3{-}\overset{O}{\overset{\|}{C}}{-}CH_2{-}CH_2{-}\ddot{N}R_2 + HA$

Enol **Iminium cation** **Mannich base**

The enol form of the active hydrogen compound reacts with the iminium cation to form a β-aminocarbonyl compound (a Mannich base).

PROBLEM 19.21

Outline reasonable mechanisms that account for the products of the following Mannich reactions:

(a) cyclohexanone + CH_2O + $(CH_3)_2NH$ $\longrightarrow$ cyclohexanone bearing $CH_2N(CH_3)_2$

(b) $C_6H_5\overset{O}{\overset{\|}{C}}CH_3$ + CH_2O + pyrrolidine (N–H) $\longrightarrow$ $C_6H_5\overset{O}{\overset{\|}{C}}CH_2CH_2$—N(pyrrolidine)

(c) 4-methylphenol (OH, CH_3) + 2 CH_2O + 2 $(CH_3)_2NH$ $\longrightarrow$ phenol (OH, CH_3) bearing $(CH_3)_2NCH_2$ and $CH_2N(CH_3)_2$

The Chemistry of...

A Suicide Enzyme Substrate

5-Fluorouracil is a chemical imposter for uracil and a potent clinical anticancer drug. This effect arises because 5-fluorouracil irreversibly destroys the ability of thymidylate synthase (an enzyme) to catalyze a key transformation needed for DNA synthesis. 5-Fluorouracil acts as a mechanism-based inhibitor (or suicide substrate) because it engages thymidylate synthase as though it were the normal substrate but then leads to self-destruction of the enzyme's activity by its own mechanistic pathway. The initial deception is possible because the fluorine atom in the inhibitor occupies roughly the same amount of space as the hydrogen atom does in the natural substrate. Disruption of the enzyme's mechanism occurs because a fluorine atom cannot be removed by a base in the way that is possible for a hydrogen atom to be removed.

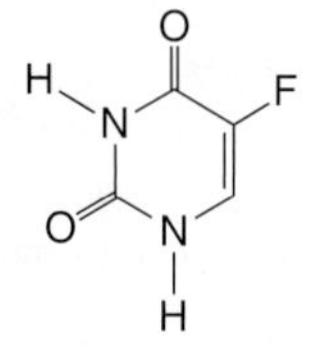

5-Fluorouracil

The mechanism of thymidylate synthase in both its normal mode and when it is about to be blocked by the inhibitor involves attack of an enolate anion on an iminium cation. This process is closely analogous to the Mannich reaction discussed in Section 19.10. The enolate anion in this attack arises by conjugate addition of a thiol group from thymidylate synthase to the α,β-unsaturated carbonyl group of the substrate. This process is analogous to the way an enolate intermediate occurs in a Michael addition. The iminium ion that is attacked in this process derives from the coenzyme N^5,N^{10}-methylenetetrahydrofolate (N^5,N^{10}-methylene-THF). Attack by the enolate in this step forms the bond that covalently links the substrate to the enzyme. It is this bond that cannot be broken when the fluorinated inhibitor is used. The mechanism of inhibition is shown on the following page.

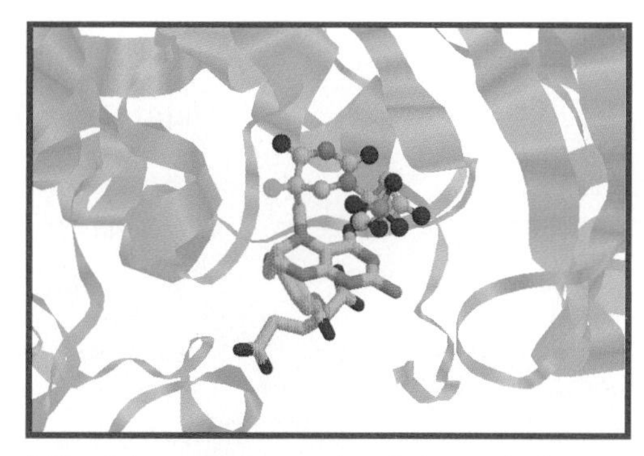

5-Fluorodeoxyuracil monophosphate covalently bound to tetrahydrofolate in thymidylate synthase, blocking the enzyme's catalytic activity.

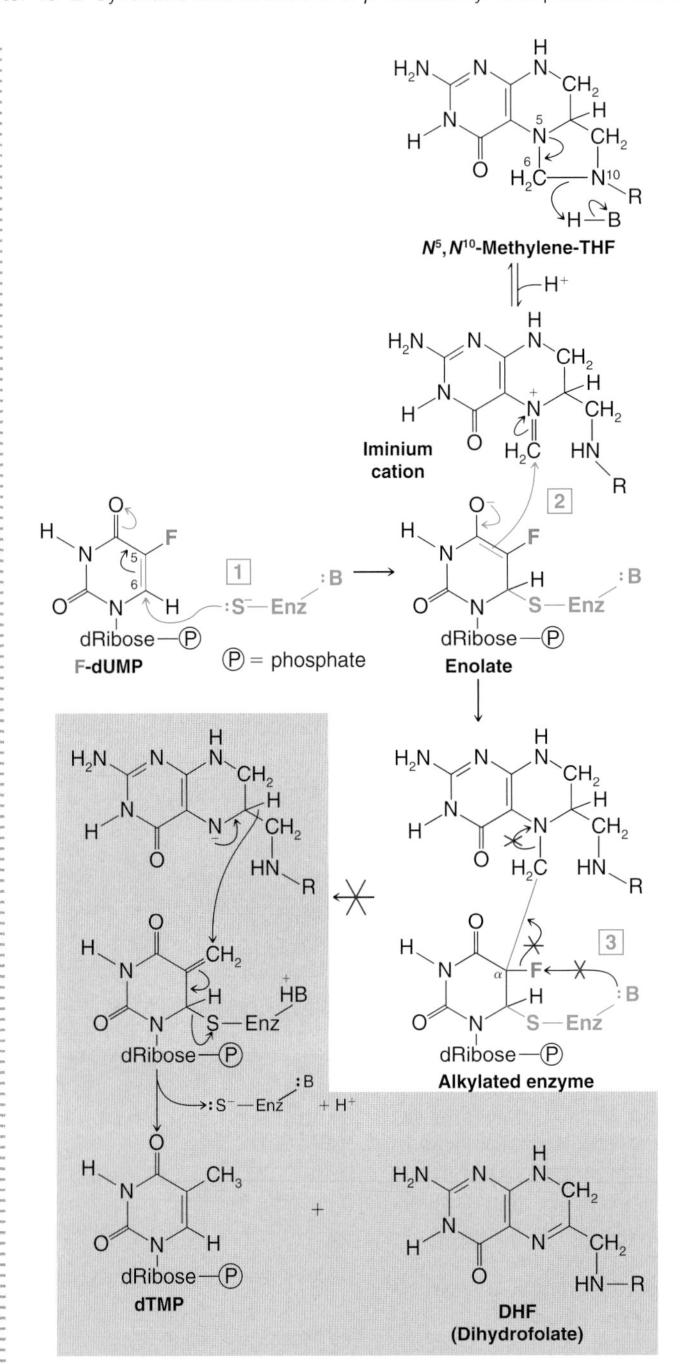

1 Conjugate addition of a thiol group from thymidylate synthase to the β carbon of the α,β-unsaturated carbonyl group in the inhibitor leads to an enolate intermediate.

2 Attack of the enolate anion on the iminium cation of N^5,N^{10}-methylene-THF forms a covalent bond between the inhibitor and the coenzyme.

3 The next step in the normal mechanism would be an elimination reaction involving loss of proton at the carbon α to the substrate's carbonyl group, releasing the tetrahydrofolate coenzyme as a leaving group. In the case of the fluorinated inhibitor, this step is not possible because a fluorine atom takes the place of the hydrogen atom needed for removal in the elimination. The enzyme cannot undergo the elimination reaction necessary to free it from the tetrahydrofolate coenzyme. These blocked steps are marked by cross-outs. Neither can the subsequent hydride transfer occur from the coenzyme to the substrate, which would complete formation of the methyl group and allow release of the product from the enzyme thiol group. These blocked steps are shown in the shaded area. The enzyme's activity is destroyed because it is irreversibly bonded to the inhibitor.

19.11 Synthesis of Enamines: Stork Enamine Reactions

Aldehydes and ketones react with secondary amines to form compounds called **enamines.** The general reaction for enamine formation can be written as follows:

$$\text{—CH—C(=O)— } + \text{HNR}_2 \rightleftharpoons \text{—CH—C(OH)—NR}_2 \rightleftharpoons \text{C=C—NR}_2 + H_2O$$

Aldehyde or ketone 2° Amine Enamine

See Section 16.8C for the mechanism of enamine formation.

Since enamine formation requires the loss of a molecule of water, enamine preparations are usually carried out in a way that allows water to be removed as an azeotrope or by a drying agent. This removal of water drives the reversible reaction to completion. Enamine formation is also catalyzed by the presence of a trace of an acid. The secondary amines most commonly used to prepare enamines are cyclic amines such as pyrrolidine, piperidine, and morpholine:

Pyrrolidine **Piperidine** **Morpholine**

Cyclohexanone, for example, reacts with pyrrolidine in the following way:

p-TsOH, $-H_2O$

***N*-(1-Cyclohexenyl)pyrrolidine (an enamine)**

Enamines are good nucleophiles. Examination of the resonance structures that follow show that we should expect enamines to have both a nucleophilic nitrogen and a *nucleophilic carbon.* A map of electrostatic potential highlights the nucleophilic region of an enamine.

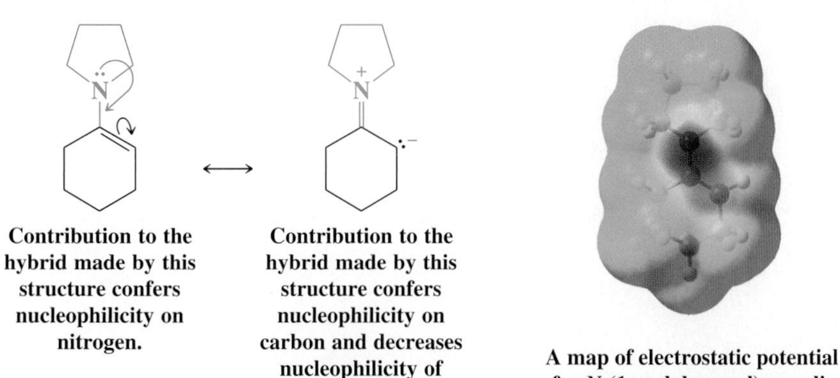

Contribution to the hybrid made by this structure confers nucleophilicity on nitrogen.

Contribution to the hybrid made by this structure confers nucleophilicity on carbon and decreases nucleophilicity of nitrogen.

A map of electrostatic potential for *N*-(1-cyclohexenyl)pyrrolidine shows the distribution of negative charge and nucleophilic region of an enamine.

The nucleophilicity of the carbon of enamines makes them particularly useful reagents in organic synthesis because they can be **acylated, alkylated,** and used in **Michael additions.** Enamines can be used as synthetic equivalents of aldehyde or ketone enolates because the alkene carbon of an enamine reacts the same way as does the α-carbon of an aldehyde or ketone enolate, and after hydrolysis, the products are the same. Development of these techniques originated with the work of Gilbert Stork of Columbia University, and in his honor they have come to be known as **Stork enamine reactions.**

Enamines are the synthetic equivalents of aldehyde and ketone enolates.

When an enamine reacts with an acyl halide or an acid anhydride, the product is the *C*-acylated compound. The iminium ion that forms hydrolyzes when water is added, and the overall reaction provides a synthesis of β-diketones:

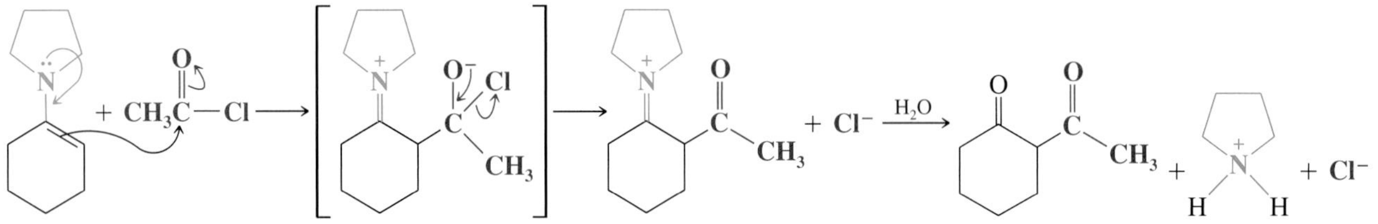

Iminium salt

2-Acetylcyclohexanone (a β-diketone)

Although *N*-acylation may occur in this synthesis, the *N*-acyl product is unstable and can act as an acylating agent itself:

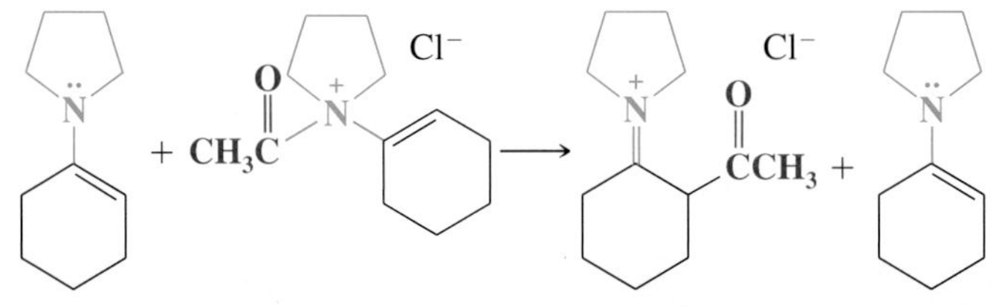

Enamine | *N*-Acylated enamine | *C*-Acylated iminium salt | Enamine

As a consequence, the yields of *C*-acylated products are generally high.

Enamines can be alkylated as well as acylated. Although alkylation may lead to the formation of a considerable amount of *N*-alkylated product, heating the *N*-alkylated product often converts it to a *C*-alkyl compound. This rearrangement is particularly favored when the alkyl halide is an allylic halide, benzylic halide, or α-haloacetic ester:

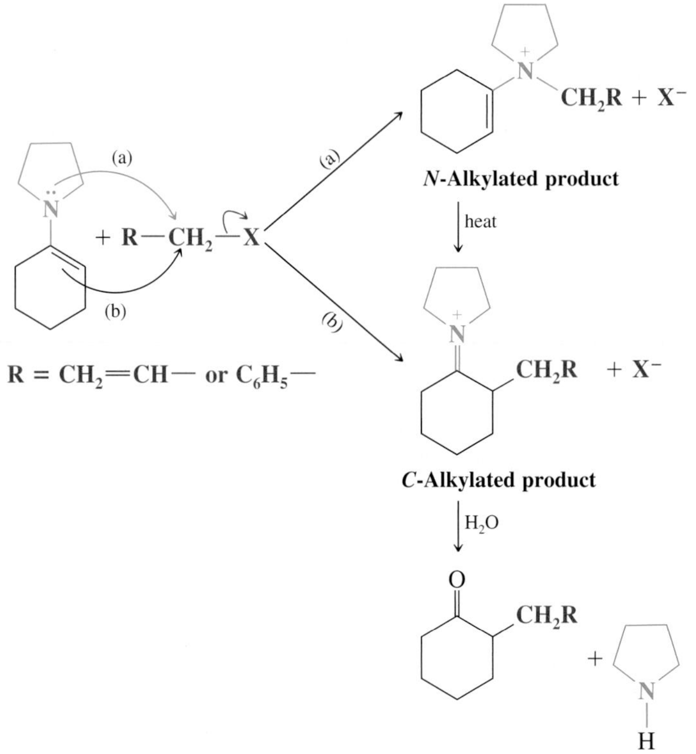

$R = CH_2{=}CH{-}$ or $C_6H_5{-}$

N-Alkylated product

C-Alkylated product

Enamine alkylations are S_N2 reactions; therefore, when we choose our alkylating agents, we are usually restricted to the use of methyl, primary, allylic, and benzylic halides. α-Halo esters can also be used as the alkylating agents, and this reaction provides a convenient synthesis of γ-keto esters:

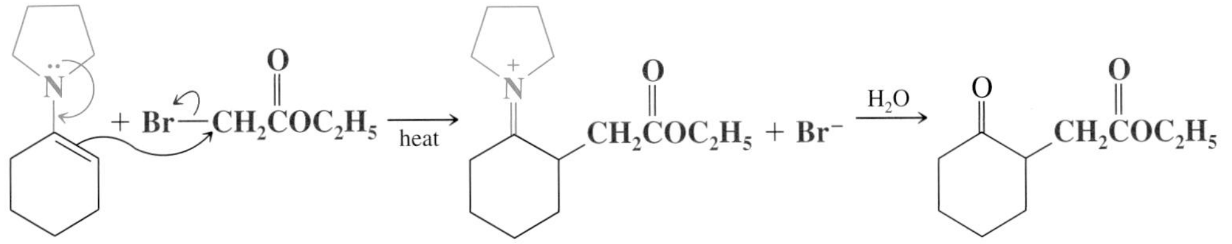

$$\text{enamine} + Br{-}CH_2\overset{O}{\overset{\|}{C}}OC_2H_5 \xrightarrow{\text{heat}} \text{iminium}{-}CH_2\overset{O}{\overset{\|}{C}}OC_2H_5 + Br^- \xrightarrow{H_2O} \text{cyclohexanone}{-}CH_2\overset{O}{\overset{\|}{C}}OC_2H_5$$

A γ-keto ester (75%)

PROBLEM 19,22

Show how you could employ enamines in syntheses of the following compounds:

(a) 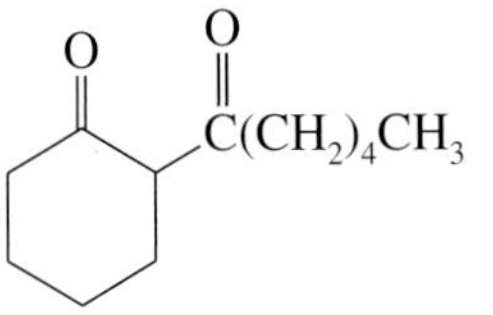 2-[$\overset{O}{\overset{\|}{C}}(CH_2)_4CH_3$]cyclohexanone

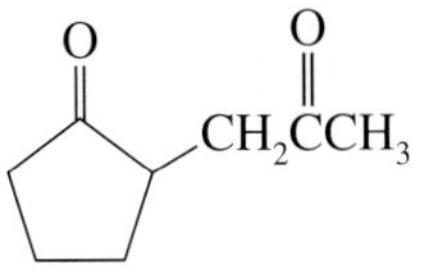

(b) 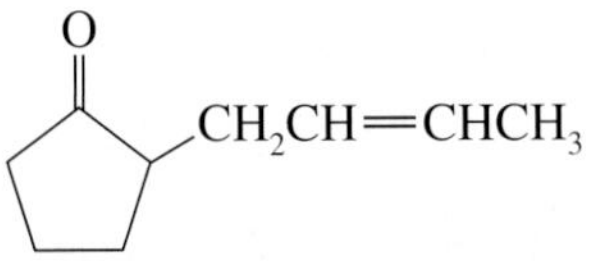 2-[$CH_2CH{=}CHCH_3$]cyclopentanone

(c) 2-[$CH_2\overset{O}{\overset{\|}{C}}CH_3$]cyclopentanone

(d) 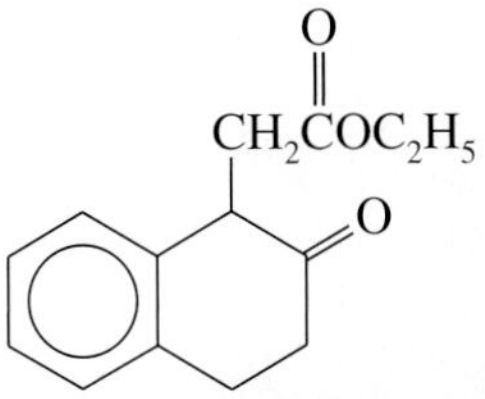 tetralone bearing $CH_2\overset{O}{\overset{\|}{C}}OC_2H_5$

An especially interesting set of enamine alkylations is shown in the following reactions (developed by J. K. Whitesell, North Carolina State University). The enamine (prepared from a single enantiomer of the secondary amine) is chiral. Alkylation from the bottom of the enamine is severely hindered by the methyl group. (Notice that this hindrance exists even if rotation of the groups takes place about the bond connecting the two rings.) Consequently, alkylation takes place much more rapidly from the top side. This reaction yields (after hydrolysis) 2-substituted cyclohexanones consisting almost entirely of a single enantiomer:

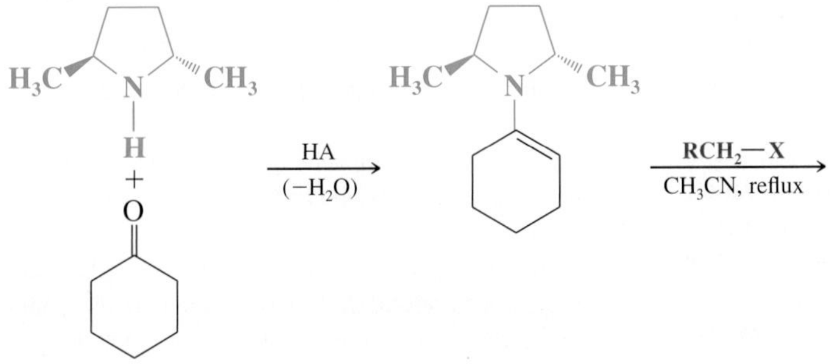

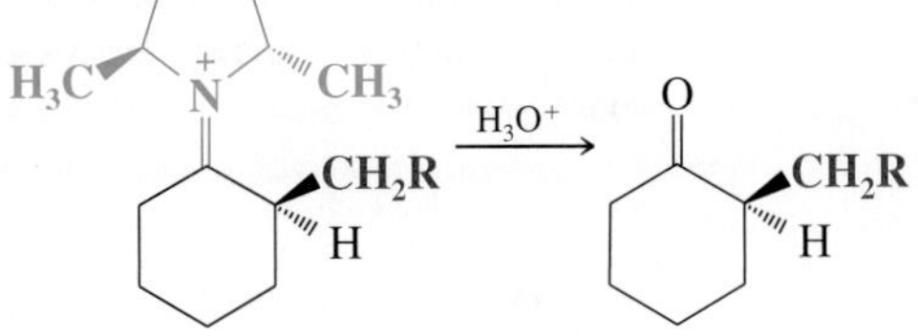

$$\text{2,5-dimethylpyrrolidine } (H_3C, CH_3) + \text{cyclohexanone} \xrightarrow[(-H_2O)]{HA} \text{enamine} \xrightarrow[CH_3CN,\ \text{reflux}]{RCH_2{-}X} \text{iminium } (CH_2R, H) \xrightarrow{H_3O^+} \text{2-}CH_2R\text{-cyclohexanone}$$

R Group	Chemical Yield (%)	Enantiomeric Excess (%)
H—	50	83
CH_3CH_2—	57	93
$CH_2{=}CH$—	80	82

Enamines can also be used in Michael additions. An example is the following:

$$\text{(1-pyrrolidinyl)cyclohexene} + CH_2{=}CHCN \xrightarrow[\text{reflux}]{C_2H_5OH} \text{iminium}{-}CH_2CH_2CN \xrightarrow{H_2O} \text{2-}(CH_2CH_2CN)\text{cyclohexanone}$$

The Chemistry of...

Antibody-Catalyzed Aldol Condensations

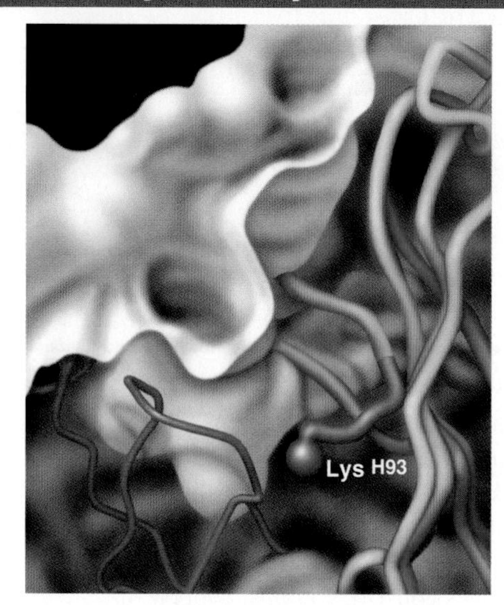

An aldolase catalytic antibody active site.

Enamines are involved in the antibody-catalyzed mechanisms for these two reactions.

In a wonderful confluence of chemistry and biology, chemists have combined aspects of immunology and enzymology to create antibodies that can catalyze chemical reactions (see the Chapter 24 opening vignette). Relevant to our present studies is antibody 38C2 (Ab 38C2), which catalyzes aldol reactions between a variety of aldehyde and ketone substrates. An example of such an antibody-catalyzed reaction is the aldol addition of acetone to (*E*)-3-(4-nitrophenyl)-2-propenal. The product is formed in 67% yield and 99% enantiomeric excess (ee) when antibody 38C2 is present*:

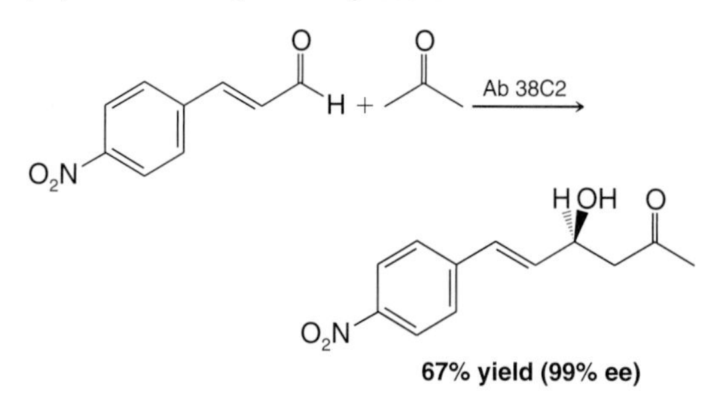

The mechanism of antibody-catalyzed aldol condensations involves imine and enamine intermediates (Section 19.11) formed from an amino group of the antibody and the aldol carbonyl reactant that becomes the nucleophile. After the enamine from the antibody and one aldol reactant attacks the electrophilic aldol component, the resulting imine is hydrolyzed. This step frees the aldol product and releases the antibody amino group for another reaction cycle.

Another remarkable application of antibody catalysis, also with enamines, involves the Robinson annulation. The aldol ring closure and dehydration steps shown here, constituting the second stage of a Robinson annulation, result in greater than 95% enantiomeric excess of the product when catalyzed by Ab 38C2†:

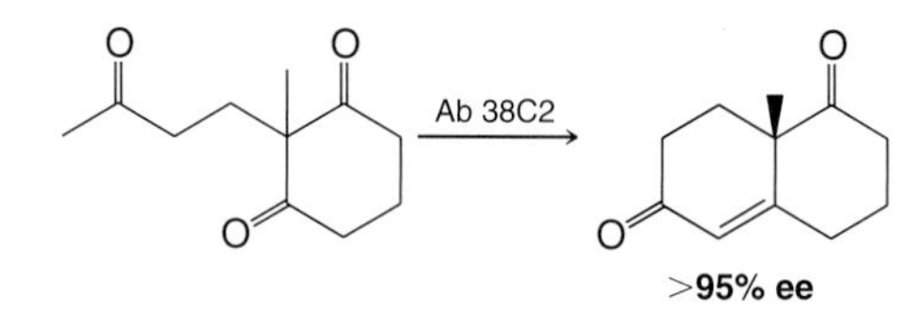

Ab 38C2 also facilitates reaction between 2-methyl-1,3-cyclohexanedione and 3-buten-2-one (methyl vinyl ketone), thereby accomplishing *both* the initial Michael addition and the cyclodehydration steps of a complete Robinson annulation.

Owing to its ability to act on a broad range of aldol substrates, antibody 38C2 is now a commercially available catalytic antibody reagent. As we shall see in Chapter 24, catalytic antibodies are being developed that can catalyze a variety of other reactions as well, including such well-known processes as the Diels–Alder reaction.

*Hoffmann, G.: Zhong, G.; List, B.; Shabat, D.; Anderson, J.; Gramatikova, S.; Lerner, R. A.; Barbas III, C. F. *J. Am. Chem. Soc.* **1998,** *120,* 2768–2779.

†Zhong, G.; Hoffmann, G.; Lerner, R. A.; Danishefsky, S.; Barbas III, C. F. *J. Am. Chem. Soc.* **1997,** *119,* 8131–8132.

19.12 BARBITURATES

In the presence of sodium ethoxide, diethyl malonate reacts with urea to yield a compound called barbituric acid:

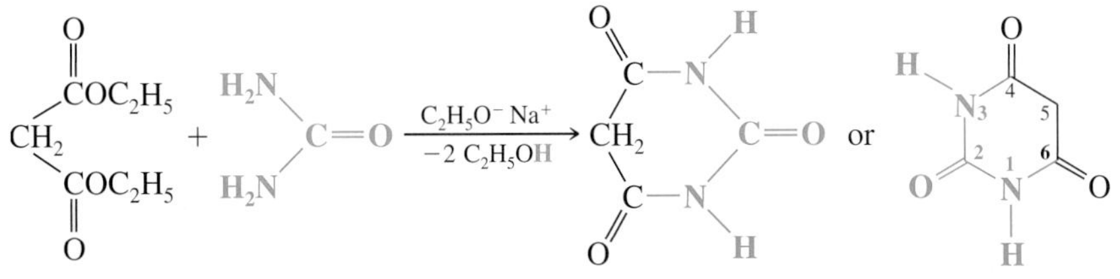

$$CH_2(COOC_2H_5)_2 + (H_2N)_2C{=}O \xrightarrow[-2\,C_2H_5OH]{C_2H_5O^-\,Na^+} \text{Barbituric acid}$$

Barbituric acid

Barbituric acid is a pyrimidine derivative (see Section 20.1), and it exists in several tautomeric forms, including one with an aromatic ring:

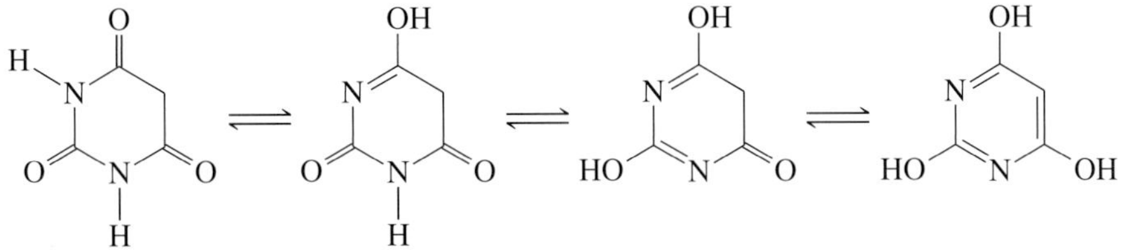

As its name suggests, barbituric acid is a moderately strong acid, stronger even than acetic acid. Its anion is highly resonance stabilized.

Derivatives of barbituric acid are *barbiturates.* Barbiturates have been used in medicine as soporifics (sleep inducers) since 1903. One of the earliest barbiturates introduced into medical use is the compound veronal (5,5-diethylbarbituric acid). Veronal is usually used as its sodium salt. Other barbiturates are seconal and phenobarbital:

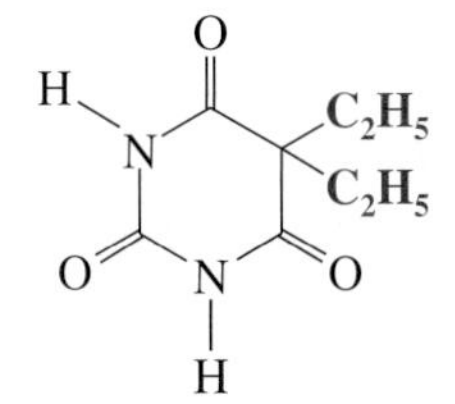

Veronal (5,5-diethylbarbituric acid)

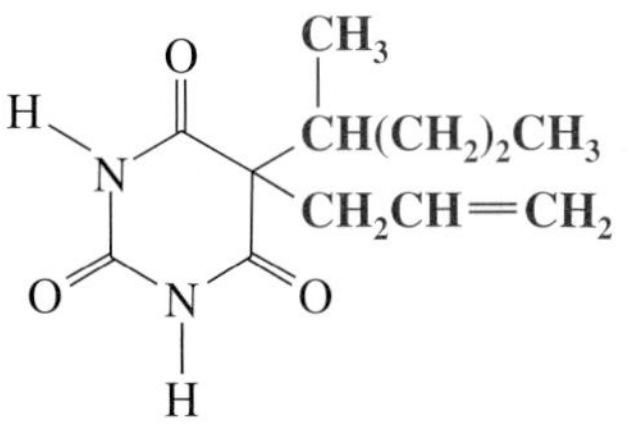

Seconal [5-allyl-5-(1-methylbutyl) barbituric acid]

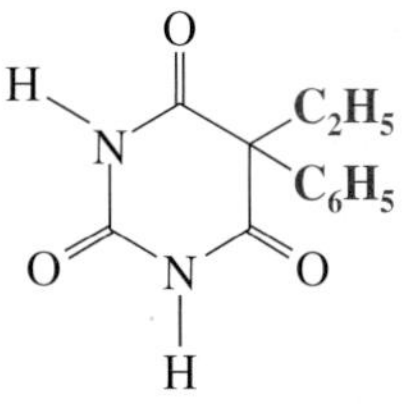

Phenobarbital (5-ethyl-5-phenylbarbituric acid)

Although barbiturates are very effective soporifics, their use is also hazardous. They are addictive, and overdosage, often with fatal results, is common.

PROBLEM 19.23

Outlined here is a synthesis of phenobarbital.
(a) What are compounds **A–F**? **(b)** Propose an alternative synthesis of **E** from diethyl malonate.

$$C_6H_5{-}CH_3 \xrightarrow[CCl_4]{NBS,ROOR} \mathbf{A}\ (C_7H_7Br) \xrightarrow[(2)\ CO_2,\ \text{then}\ H_3O^+]{(1)\ Mg,\ Et_2O} \mathbf{B}\ (C_8H_8O_2) \xrightarrow{SOCl_2}$$

$$\mathbf{C}\ (C_8H_7ClO) \xrightarrow{EtOH} \mathbf{D}\ (C_{10}H_{12}O_2) \xrightarrow[NaOEt]{EtOC(O)OEt} \mathbf{E}\ (C_{13}H_{16}O_4) \xrightarrow[CH_3CH_2Br]{KOC(CH_3)_3}$$

$$\mathbf{F}\ (C_{15}H_{20}O_4) \xrightarrow{H_2NC(O)NH_2,\ NaOEt} \text{phenobarbital}$$

PROBLEM 19.24 Starting with diethyl malonate, urea, and any other required reagents, outline a synthesis of veronal and seconal.

SUMMARY OF IMPORTANT REACTIONS

1. **Claisen condensation** (Section 19.2):

$$2\,R{-}CH_2{-}\overset{O}{\overset{\|}{C}}{-}OEt \xrightarrow[(2)\ H_3O^+]{(1)\ NaOEt} R{-}CH_2{-}\overset{O}{\overset{\|}{C}}{-}\underset{R}{\underset{|}{CH}}{-}\overset{O}{\overset{\|}{C}}{-}OEt$$

2. **Crossed Claisen condensation** (Section 19.2A):

$$R{-}CH_2{-}\overset{O}{\overset{\|}{C}}{-}OEt \xrightarrow[(2)\ H_3O^+]{(1)\ C_6H_5CO_2Et/NaOEt} C_6H_5{-}\overset{O}{\overset{\|}{C}}{-}\underset{R}{\underset{|}{CH}}{-}\overset{O}{\overset{\|}{C}}{-}OEt$$

$$R{-}CH_2{-}\overset{O}{\overset{\|}{C}}{-}OEt \xrightarrow[(2)\ H_3O^+]{(1)\ EtO\overset{O}{\overset{\|}{C}}OEt/NaOEt} R{-}\underset{\underset{\underset{O}{\|}}{C{-}OEt}}{\underset{|}{CH}}{-}\overset{O}{\overset{\|}{C}}{-}OEt$$

$$R{-}CH_2{-}\overset{O}{\overset{\|}{C}}{-}OEt \xrightarrow[(2)\ H_3O^+]{(1)\ HCO_2Et/NaOEt} R{-}\underset{\underset{\underset{O}{\|}}{C{-}H}}{\underset{|}{CH}}{-}\overset{O}{\overset{\|}{C}}{-}OEt$$

$$R{-}CH_2{-}\overset{O}{\overset{\|}{C}}{-}OEt \xrightarrow[(2)\ H_3O^+]{(1)\ EtO_2CCO_2Et/NaOEt} R{-}\underset{\underset{\underset{O\ \ \ O}{\|\ \ \ \ \|}}{C{-}C{-}OEt}}{\underset{|}{CH}}{-}\overset{O}{\overset{\|}{C}}{-}OEt$$

3. **Acetoacetic ester synthesis** (Section 19.3):

$$CH_3-\overset{O}{\overset{\|}{C}}-CH_2-\overset{O}{\overset{\|}{C}}-OEt \xrightarrow[(2)\ RBr]{(1)\ NaOEt} CH_3-\overset{O}{\overset{\|}{C}}-\underset{R}{\underset{|}{C}H}-\overset{O}{\overset{\|}{C}}-OEt \xrightarrow[(2)\ H_3O^+ \ (3)\ heat\ (-CO_2)]{(1)\ OH^-,\ heat} CH_3-\overset{O}{\overset{\|}{C}}-CH_2-R$$

$$CH_3-\overset{O}{\overset{\|}{C}}-\underset{R}{\underset{|}{C}H}-\overset{O}{\overset{\|}{C}}-OEt \xrightarrow[(2)\ R'Br]{(1)\ KOC(CH_3)_3} CH_3-\overset{O}{\overset{\|}{C}}-\underset{R}{\underset{|}{\overset{R'}{\overset{|}{C}}}}-\overset{O}{\overset{\|}{C}}-OEt \xrightarrow[(2)\ H_3O^+ \ (3)\ heat\ (-CO_2)]{(1)\ OH^-,\ heat} CH_3-\overset{O}{\overset{\|}{C}}-\underset{R'}{\underset{|}{C}H}-R$$

4. **Malonic ester synthesis** (Section 19.4):

$$EtO-\overset{O}{\overset{\|}{C}}-CH_2-\overset{O}{\overset{\|}{C}}-OEt \xrightarrow[(2)\ RBr]{(1)\ NaOEt} EtO-\overset{O}{\overset{\|}{C}}-\underset{R}{\underset{|}{C}H}-\overset{O}{\overset{\|}{C}}-OEt \xrightarrow[(2)\ H_3O^+ \ (3)\ heat\ (-CO_2)]{(1)\ OH^-,\ heat} HO-\overset{O}{\overset{\|}{C}}-CH_2-R$$

$$EtO-\overset{O}{\overset{\|}{C}}-\underset{R}{\underset{|}{C}H}-\overset{O}{\overset{\|}{C}}-OEt \xrightarrow[(2)\ R'Br]{(1)\ KOC(CH_3)_3}$$

$$EtO-\overset{O}{\overset{\|}{C}}-\underset{R}{\underset{|}{\overset{R'}{\overset{|}{C}}}}-\overset{O}{\overset{\|}{C}}-OEt \xrightarrow[(2)\ H_3O^+ \ (3)\ heat\ (-CO_2)]{(1)\ HO^-,\ heat} HO-\overset{O}{\overset{\|}{C}}-\underset{R'}{\underset{|}{C}H}-R$$

5. **Direct alkylation of esters** (Section 19.6):

$$R-CH_2-\overset{O}{\overset{\|}{C}}-OEt \xrightarrow[THF]{LDA} R-\overset{Li^+}{\ddot{C}H^-}-\overset{O}{\overset{\|}{C}}-OEt \xrightarrow{R'CH_2-Br} R-\underset{\underset{R'}{|}}{\underset{CH_2}{\underset{|}{C}H}}-\overset{O}{\overset{\|}{C}}-OEt$$

6. **Alkylation of dithianes** (Section 19.7):

$$\mathrm{R{-}\overset{O}{\overset{\|}{C}}{-}H} \xrightarrow[\text{HA}]{\mathrm{HSCH_2CH_2CH_2SH}} \text{(1,3-dithiane, C2: R, H)} \xrightarrow[\text{(2) }\mathrm{R'CH_2X}]{\text{(1) BuLi}} \text{(1,3-dithiane, C2: R, }\mathrm{CH_2R'}\text{)} \xrightarrow[\mathrm{H_2O}]{\mathrm{HgCl_2,\ CH_3OH}} \mathrm{R{-}\overset{O}{\overset{\|}{C}}{-}CH_2R'}$$

7. **Knoevenagel condensation** (Section 19.8):

$$\mathrm{C_6H_5{-}CHO} + \mathrm{CH_2(CO_2R)_2} \xrightarrow[(-\mathrm{H_2O})]{\text{base}} \mathrm{C_6H_5{-}CH{=}C(CO_2R)_2}$$

8. **Michael addition** (Section 19.9):

$$\mathrm{R{-}\overset{R'}{\overset{|}{C}}{=}CH{-}\overset{O}{\overset{\|}{C}}{-}R''} + \mathrm{CH_2(CO_2R)_2} \xrightarrow{\text{base}} \mathrm{R{-}\overset{R'}{\overset{|}{C}}(CH(CO_2R)_2){-}CH_2{-}\overset{O}{\overset{\|}{C}}{-}R''}$$

α,β-Unsaturated carbonyl compound   **A malonic ester (or other active methylene compound)**

9. **Mannich reaction** (Section 19.10):

$$\mathrm{R{-}\overset{O}{\overset{\|}{C}}{-}CH_3} + \mathrm{H{-}\overset{O}{\overset{\|}{C}}{-}H} + \mathrm{H{-}NR'R''} \longrightarrow \mathrm{R{-}\overset{O}{\overset{\|}{C}}{-}CH_2{-}CH_2{-}NR'R''}$$

10. **Stork enamine reaction** (Section 19.11):

$$\mathrm{RCH_2{-}\overset{O}{\overset{\|}{C}}{-}CH_2R} + \mathrm{R'_2NH} \longrightarrow \underset{\textbf{Enamine}}{\mathrm{RCH_2{-}\overset{NR'_2}{\overset{|}{C}}{=}CHR}} \xrightarrow[\text{(2) heat; (3) }\mathrm{H_2O}]{\text{(1) }\mathrm{R''CH_2Br}} \mathrm{RCH_2{-}\overset{O}{\overset{\|}{C}}{-}CHR(CH_2R'')}$$

Summary of Mechanisms

Some Synthetic Connections Involving β-Dicarbonyl Compounds

- Claisen/Dieckmann ester condensation
- Keto-enol tautomerism in β-dicarbonyl compounds
- β-Dicarbonyl enolate resonance contributors
- Acetoacetic ester synthesis
- Malonic ester synthesis
- Direct alkylation/acylation (w/ LDA)
- Enamine formation and alkylation/acylation
- Knoevenagel condensation

- Michael addition (conjugate addition of enolates)
- Conjugate addition of R_2CuLi
- Conjugate addition of HCN
- Conjugate addition of R_2NH
- Mannich reaction
- 1,3-Dithiane formation, alkylation, and hydrolysis

Acetoacetic ester synthesis
G = CH_3, R' (initially) = H

Malonic ester synthesis
G = OR, R' (initially) = H

Key Terms and Concepts

Acetoacetic ester synthesis	Section 19.3
Active hydrogen (methylene) compounds	Section 19.5
Claisen condensation	Section 19.2
Condensation reaction	Section 19.2
β-Dicarbonyl compounds	Section 19.1
Direct alkylation	Section 19.6
Enamines	Section 19.11
Knoevenagel condensation	Section 19.8
Malonic ester synthesis	Section 19.4
Mannich reaction	Section 19.10
Michael addition	Section 19.9
Synthetic equivalent	Sections 8.6, 19.3, 19.4
Umpolung (polarity reversal)	Section 19.7

Additional Problems

19.25 Show all steps in the following syntheses. You may use any other needed reagents but you should begin with the compound given. You need not repeat steps carried out in earlier parts of this exercise.

(a) $CH_3CH_2CH_2\overset{\overset{O}{\|}}{C}OC_2H_5 \longrightarrow CH_3CH_2CH_2\overset{\overset{O}{\|}}{C}\underset{\underset{CH_3}{|}}{\underset{CH_2}{|}}{C}H\overset{\overset{O}{\|}}{C}OC_2H_5$

(b) $CH_3CH_2CH_2\overset{\overset{O}{\|}}{C}OC_2H_5 \longrightarrow CH_3CH_2CH_2\overset{\overset{O}{\|}}{C}CH_2CH_2CH_3$

(c) $C_6H_5CH_2\overset{\overset{O}{\|}}{C}OC_2H_5 \longrightarrow C_6H_5\overset{\overset{CH_3}{|}}{C}HCO_2H$

(d) $CH_3CH_2CH_2\overset{\overset{O}{\|}}{C}OC_2H_5 \longrightarrow CH_3CH_2\underset{\underset{\underset{\underset{O}{\|}}{C}-\underset{\underset{O}{\|}}{C}OC_2H_5}{|}}{C}H\overset{\overset{O}{\|}}{C}OC_2H_5$

(e) $CH_3CH_2CH_2\overset{\overset{O}{\|}}{C}OC_2H_5 \longrightarrow CH_3CH_2CH_2\overset{\overset{O}{\|}}{C}-\overset{\overset{O}{\|}}{C}OC_2H_5$

(f) $C_6H_5CH_2\overset{\overset{O}{\|}}{C}OC_2H_5 \longrightarrow C_6H_5\underset{\underset{\underset{\underset{O}{\|}}{CH}}{|}}{C}H\overset{\overset{O}{\|}}{C}OC_2H_5$

Note: Problems marked with an asterisk are "challenge problems."

(g) 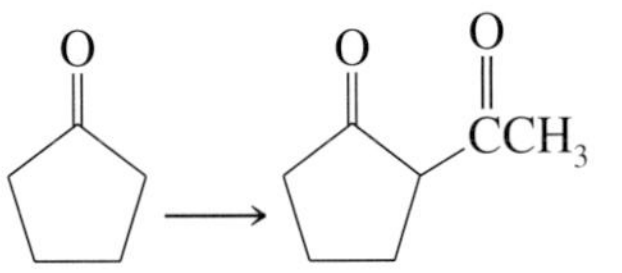 Cyclopentanone $\longrightarrow$ 2-acetylcyclopentanone (CCH_3 with C=O)

(h) 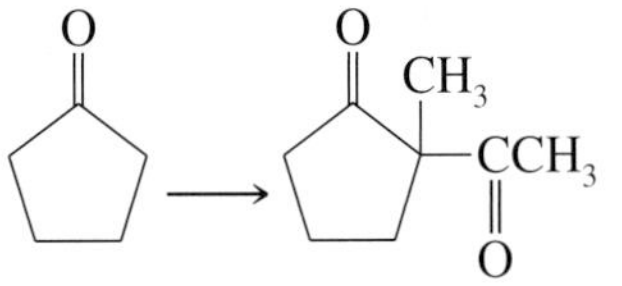 Cyclopentanone $\longrightarrow$ 2-methyl-2-acetylcyclopentanone (CH_3, CCH_3 with C=O)

(i) 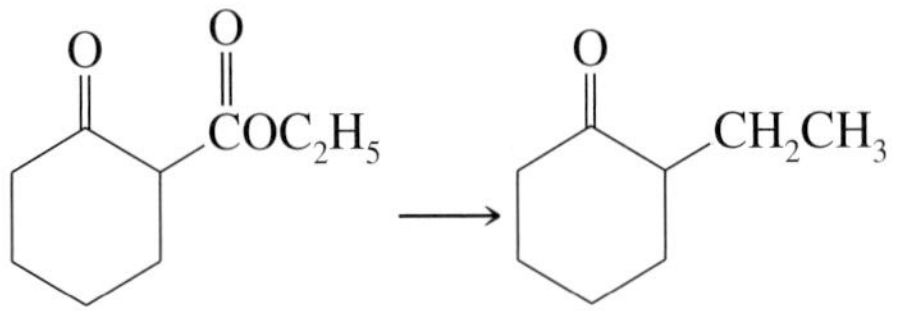 2-(COC_2H_5)cyclohexanone $\longrightarrow$ 2-(CH_2CH_3)cyclohexanone

19.26 Outline syntheses of each of the following from acetoacetic ester and any other required reagents:

(a) *tert*-Butyl methyl ketone

(b) 2-Hexanone

(c) 2,5-Hexanedione

(d) 4-Hydroxypentanoic acid

(e) 2-Ethyl-1,3-butanediol

(f) 1-Phenyl-1,3-butanediol

19.27 Outline syntheses of each of the following from diethyl malonate and any other required reagents:

(a) 2-Methylbutanoic acid

(b) 4-Methyl-1-pentanol

(c) $CH_3CH_2CHCH_2OH$ with CH_2OH on the CH

(d) $HOCH_2CH_2CH_2CH_2OH$

19.28 The synthesis of cyclobutanecarboxylic acid given in Section 19.4 was first carried out by William Perkin, Jr., in 1883, and it represented one of the first syntheses of an organic compound with a ring smaller than six carbon atoms. (There was a general feeling at the time that such compounds would be too unstable to exist.) Earlier in 1883, Perkin reported what he mistakenly believed to be a cyclobutane derivative obtained from the reaction of acetoacetic ester and 1,3-dibromopropane. The reaction that Perkin had expected to take place was the following:

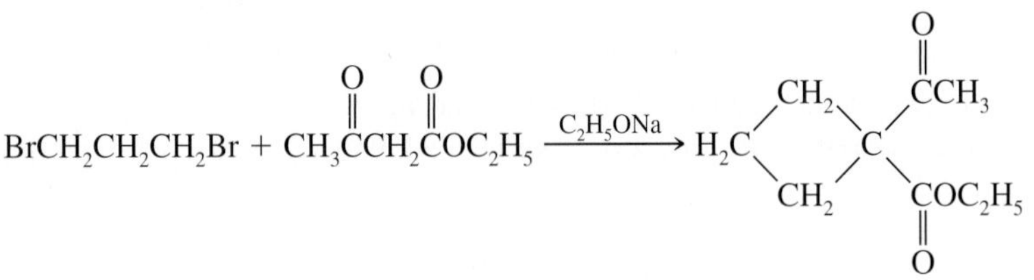

$$BrCH_2CH_2CH_2Br + CH_3\overset{O}{\overset{\|}{C}}CH_2\overset{O}{\overset{\|}{C}}OC_2H_5 \xrightarrow{C_2H_5ONa} H_2C{<}^{CH_2}_{CH_2}{>}C{<}^{\overset{O}{\overset{\|}{C}}CH_3}_{\underset{\|}{\underset{O}{C}}OC_2H_5}$$

The molecular formula for his product agreed with the formulation given in the preceding reaction, and alkaline hydrolysis and acidification gave a nicely crystalline acid (also having the expected molecular formula). The acid, however, was quite stable to heat and resisted decarboxylation. Perkin later found that both the ester and the acid contained six-membered rings (five carbon atoms and one oxygen atom). Recall the charge distribution in the enolate ion obtained from acetoacetic ester and propose structures for Perkin's ester and acid.

19.29 **(a)** In 1884 Perkin achieved a successful synthesis of cyclopropanecarboxylic acid from sodiomalonic ester and 1,2-dibromoethane. Outline the reactions involved in this synthesis.

(b) In 1885 Perkin synthesized five-membered carbocyclic compounds **D** and **E** in the following way:

$$2\ Na^+ :\bar{C}H(CO_2C_2H_5)_2 + BrCH_2CH_2CH_2Br \longrightarrow \mathbf{A}\ (C_{17}H_{28}O_8) \xrightarrow{2\ C_2H_5O^-Na^+} \xrightarrow{Br_2}$$

$$\mathbf{B}\ (C_{17}H_{26}O_8) \xrightarrow[(2)\ H_3O^+]{(1)\ OH^-/H_2O} \mathbf{C}\ (C_9H_{10}O_8) \xrightarrow{heat} \mathbf{D}\ (C_7H_{10}O_4) + \mathbf{E}\ (C_7H_{10}O_4)$$

where **D** and **E** are diastereomers; **D** can be resolved into enantiomeric forms while **E** cannot. What are the structures of **A–E**?

(c) Ten years later Perkin was able to synthesize 1,4-dibromobutane; he later used this compound and diethyl malonate to prepare cyclopentanecarboxylic acid. Show the reactions involved.

19.30 Write mechanisms that account for the products of the following reactions:

(a) $C_6H_5CH{=}CHCOC_2H_5$ (C=O) $+ CH_2(COC_2H_5)_2$ (C=O) $\xrightarrow{NaOCH_2CH_3}$ $C_6H_5CH{-}CH_2COC_2H_5$ (C=O), where the CH bears a $CH(CO_2C_2H_5)_2$ group ($C_2H_5O{-}C(=O){-}CH{-}C(=O){-}OC_2H_5$)

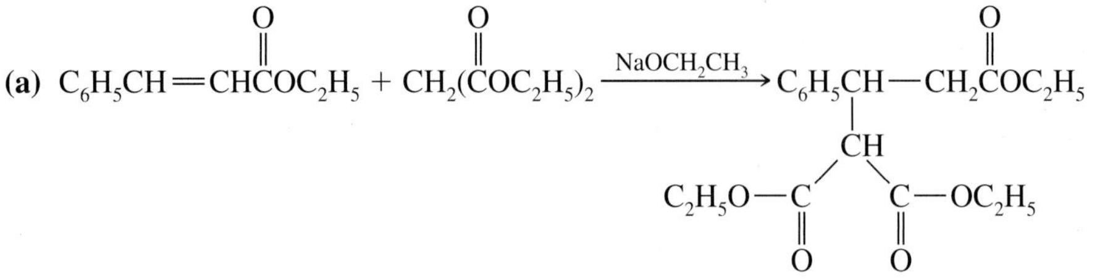

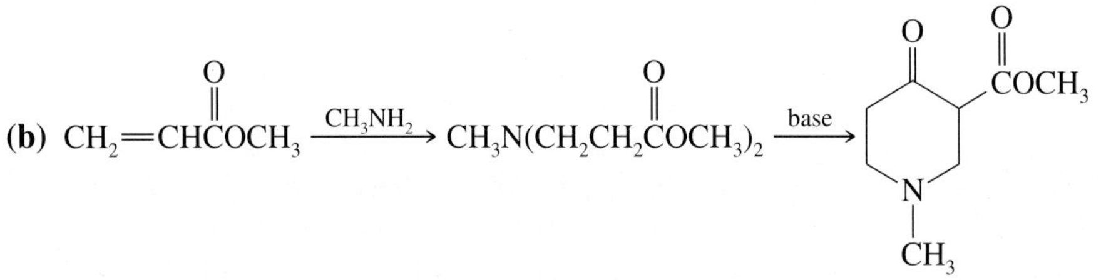

(b) $CH_2{=}CHCOCH_3$ (C=O) $\xrightarrow{CH_3NH_2}$ $CH_3N(CH_2CH_2COCH_3)_2$ $\xrightarrow{base}$ 1-methyl-4-oxopiperidine bearing a $COCH_3$ group at C-3 (N–CH_3)

(c) $CH_3{-}C(CH_3)(CH(CO_2C_2H_5)_2){-}CH_2COC_2H_5$ (C=O) $\xrightarrow[(-C_2H_5OH)]{C_2H_5O^-}$ $CH_3C(CH_3){=}CHCOC_2H_5$ (C=O) $+ {}^{-}{:}CH(CO_2C_2H_5)_2$

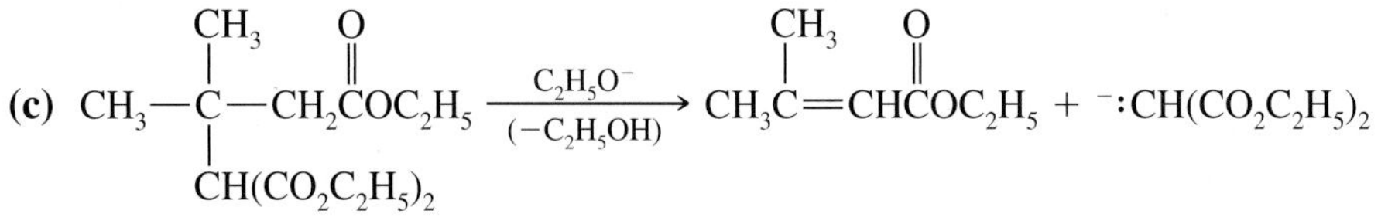

19.31 Knoevenagel condensations in which the active hydrogen compound is a β-keto ester or a β-diketone often yield products that result from one molecule of aldehyde or ketone and two molecules of the active methylene component. For example,

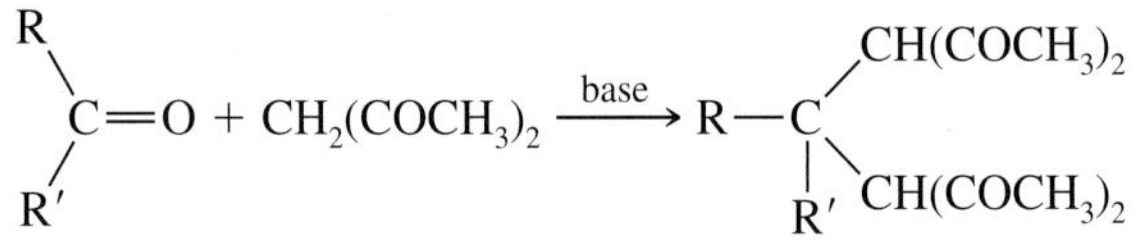

$$RR'C{=}O + CH_2(COCH_3)_2 \xrightarrow{base} R{-}C(R')(CH(COCH_3)_2)_2$$

Suggest a reasonable mechanism that accounts for the formation of these products.

19.32 Thymine is one of the heterocyclic bases found in DNA (see the chapter opening vignette). Starting with ethyl propanoate and using any other needed reagents, show how you might synthesize thymine.

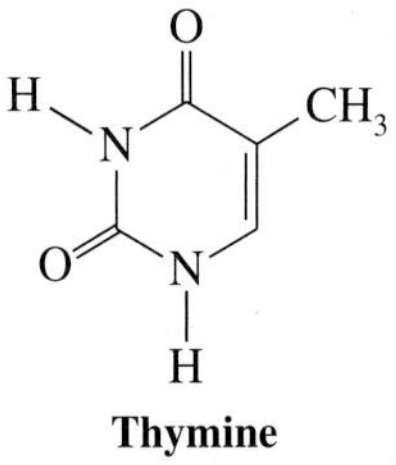

[Structure: thymine ring with N–H, C=O groups, and CH_3]

Thymine

19.33 The mandibular glands of queen bees secrete a fluid that contains a remarkable compound known as "queen substance." When even an exceedingly small amount of the queen substance is transferred to worker bees, it inhibits the development of their ovaries and prevents the workers from bearing new queens. Queen substance, a monocarboxylic acid with the molecular formula $C_{10}H_{16}O_3$, has been synthesized by the following route:

$$\text{Cycloheptanone} \xrightarrow[(2)\ H_3O^+]{(1)\ CH_3MgI} \mathbf{A}\ (C_8H_{16}O) \xrightarrow{HA,\ heat} \mathbf{B}\ (C_8H_{14}) \xrightarrow[(2)\ Zn,\ HOAc]{(1)\ O_3}$$

$$\mathbf{C}\ (C_8H_{14}O_2) \xrightarrow[pyridine]{CH_2(CO_2H)_2} \text{queen substance}\ (C_{10}H_{16}O_3)$$

On catalytic hydrogenation, queen substance yields compound **D**, which, on treatment with iodine in sodium hydroxide and subsequent acidification, yields a dicarboxylic acid **E**; that is,

$$\text{Queen substance} \xrightarrow[\text{Pd}]{H_2} \mathbf{D}\ (C_{10}H_{18}O_3) \xrightarrow[(2)\ H_3O^+]{(1)\ I_2 \text{ in aq. NaOH}} \mathbf{E}\ (C_9H_{16}O_4)$$

Provide structures for the queen substance and compounds **A–E**.

19.34 Linalool, a fragrant compound that can be isolated from a variety of plants, is 3,7-dimethyl-1,6-octadien-3-ol. Linalool is used in making perfumes, and it can be synthesized in the following way:

$$CH_2{=}C(CH_3){-}CH{=}CH_2 + HBr \longrightarrow \mathbf{F}\ (C_5H_9Br) \xrightarrow[\text{ester}]{\text{sodioacetoacetic}}$$

$$\mathbf{G}\ (C_{11}H_{18}O_3) \xrightarrow[(2)\ H_3O^+,\ (3)\ \text{heat}]{(1)\ \text{dilute NaOH}} \mathbf{H}\ (C_8H_{14}O) \xrightarrow[(2)\ H_3O^+]{(1)\ LiC{\equiv}CH}$$

$$\mathbf{I}\ (C_{10}H_{16}O) \xrightarrow[\text{Lindlar's catalyst}]{H_2} \text{linalool}$$

Outline the reactions involved. (*Hint:* Compound **F** is the more stable isomer capable of being produced in the first step.)

19.35 Compound **J**, a compound with two four-membered rings, has been synthesized by the following route. Outline the steps that are involved.

$$NaCH(CO_2C_2H_5)_2 + BrCH_2CH_2CH_2Br \longrightarrow C_{10}H_{17}BrO_4 \xrightarrow{NaOC_2H_5}$$

$$C_{10}H_{16}O_4 \xrightarrow[(2)\ H_2O]{(1)\ LiAlH_4} C_6H_{12}O_2 \xrightarrow{HBr} C_6H_{10}Br_2 \xrightarrow[2\ NaOC_2H_5]{CH_2(CO_2C_2H_5)_2}$$

$$C_{13}H_{20}O_4 \xrightarrow[(2)\ H_3O^+]{(1)\ OH^-,\ H_2O} C_9H_{12}O_4 \xrightarrow{\text{heat}} \mathbf{J}\ (C_8H_{12}O_2) + CO_2$$

19.36 When an aldehyde or a ketone is condensed with ethyl α-chloroacetate in the presence of sodium ethoxide, the product is an α,β-epoxy ester called a *glycidic ester.* The synthesis is called the Darzens condensation.

$$R(R')C{=}O + ClCH_2CO_2C_2H_5 \xrightarrow{C_2H_5ONa} R{-}C(R')\overset{O}{\frown}CHCO_2C_2H_5 + NaCl + C_2H_5OH$$

A glycidic ester

(a) Outline a reasonable mechanism for the Darzens condensation. **(b)** Hydrolysis of the epoxy ester leads to an epoxy acid that, on heating with pyridine, furnishes an aldehyde. What is happening here?

$$R{-}C(R')\overset{O}{\frown}CHCO_2H \xrightarrow[\text{heat}]{C_5H_5N} R{-}CH(R'){-}CH{=}O + CO_2$$

(c) Starting with β-ionone (Problem 17.15), show how you might synthesize the following aldehyde. (This aldehyde is an intermediate in an industrial synthesis of vitamin A.)

[Structure: 2,6,6-trimethylcyclohexenyl ring bearing a side chain –CH=C(CH₃)... terminating in CH=O]

19.37 The *Perkin condensation* is an aldol-type condensation in which an aromatic aldehyde (ArCHO) reacts with a carboxylic acid anhydride, $(RCH_2CO)_2O$, to give an α,β-unsaturated acid ($ArCH{=}CRCO_2H$). The catalyst that is usually employed is the potassium salt of the carboxylic acid (RCH_2CO_2K). **(a)** Outline the Perkin condensation that takes place when benzaldehyde reacts with propanoic anhydride in the presence of potassium propanoate. **(b)** How would you use a Perkin condensation to prepare *p*-chlorocinnamic acid, *p*-$ClC_6H_4CH{=}CHCO_2H$?

19.38 (+)-Fenchone is a terpenoid that can be isolated from fennel oil. (±)-Fenchone has been synthesized through the following route. Supply the missing intermediates and reagents.

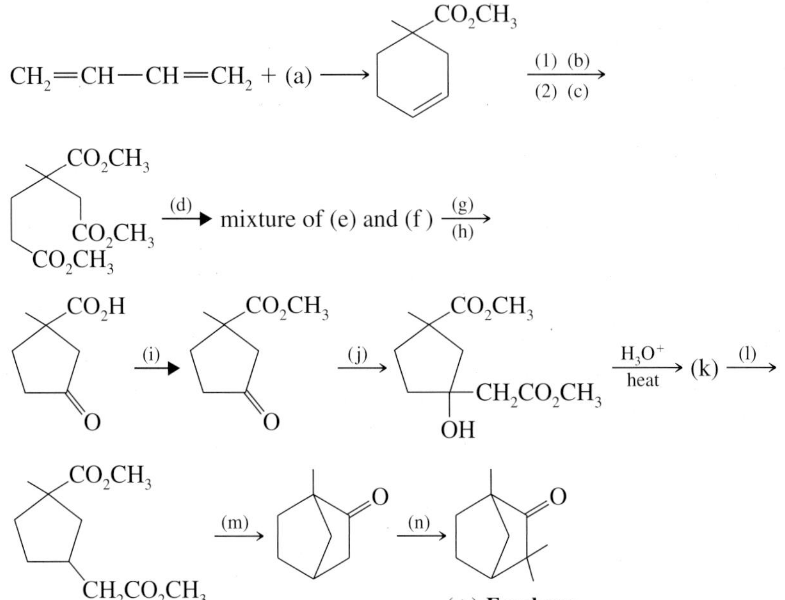

$CH_2{=}CH{-}CH{=}CH_2$ + (a) ⟶ [methyl cyclohexene carboxylate, CO_2CH_3] $\xrightarrow[(2)\ (c)]{(1)\ (b)}$

[CO_2CH_3, CO_2CH_3, CO_2CH_3 triester] $\xrightarrow{(d)}$ mixture of (e) and (f) $\xrightarrow[(h)]{(g)}$

[CO_2H, O] $\xrightarrow{(i)}$ [CO_2CH_3, O] $\xrightarrow{(j)}$ [CO_2CH_3, OH, $CH_2CO_2CH_3$] $\xrightarrow[heat]{H_3O^+}$ (k) $\xrightarrow{(l)}$

[CO_2CH_3, $CH_2CO_2CH_3$] $\xrightarrow{(m)}$ [bicyclic ketone, O] $\xrightarrow{(n)}$ [O]

(±)-Fenchone

19.39 Outline a synthesis of the analgesic Darvon (below) starting with ethyl phenyl ketone.

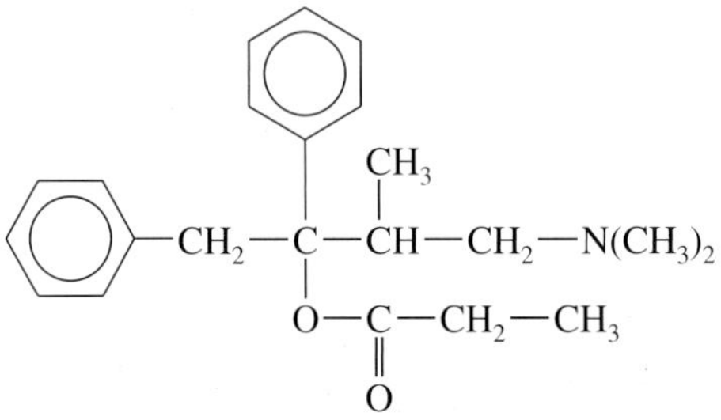

$C_6H_5{-}CH_2{-}C(C_6H_5)(O{-}C({=}O){-}CH_2{-}CH_3){-}CH(CH_3){-}CH_2{-}N(CH_3)_2$

Darvon

19.40 Explain the variation in enol content that is observed for solutions of acetylacetone (pentane-2,4-dione) in the several solvents indicated:

Solvent	% Enol
H_2O	15
CH_3CN	58
C_6H_{14}	92
Gas phase	92

19.41 When a Dieckmann condensation is attempted with diethyl succinate, the product obtained has the molecular formula $C_{12}H_{16}O_6$. What is the structure of this compound?

19.42 Ethyl crotonate, $CH_3CH{=}CHCOOC_2H_5$, reacts with diethyl oxalate, $C_2H_5OOCCOOC_2H_5$, to form a Claisen-type condensation product:

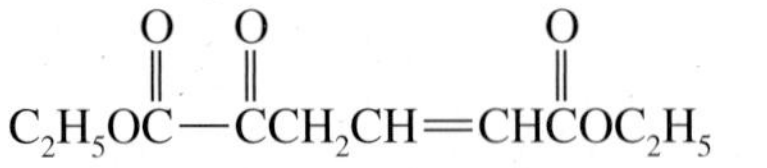

$C_2H_5OC(=O){-}C(=O)CH_2CH{=}CHC(=O)OC_2H_5$

Write a detailed mechanism for the formation of this compound.

19.43 Show how this diketone could be prepared by a condensation reaction:

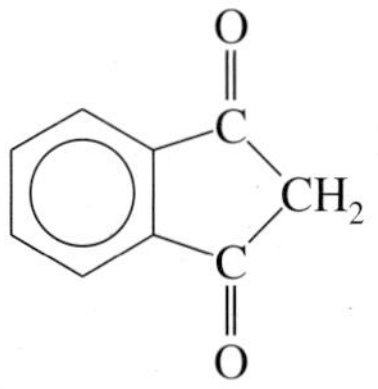

19.44 In contrast to the reaction with dilute alkali (Section 19.3), when concentrated solutions of NaOH are used, acetoacetic esters undergo cleavage as shown below.

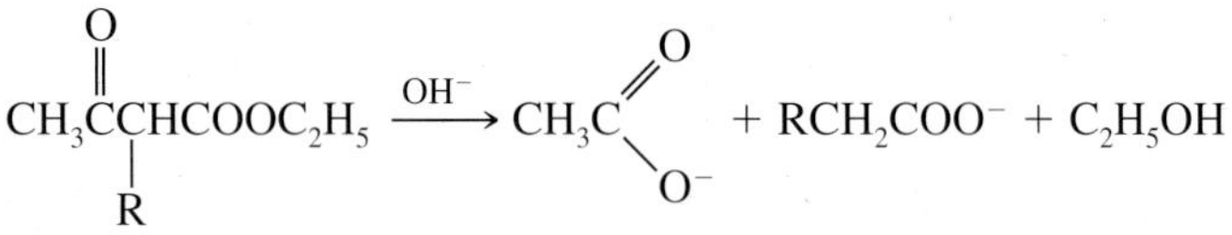

$$CH_3C(=O)CH(R)COOC_2H_5 \xrightarrow{OH^-} CH_3COO^- + RCH_2COO^- + C_2H_5OH$$

Provide a mechanistic explanation for this outcome.

19.45 Show how dimedone can be synthesized from malonic ester and mesityl oxide (4-methyl-3-penten-2-one) under basic conditions.

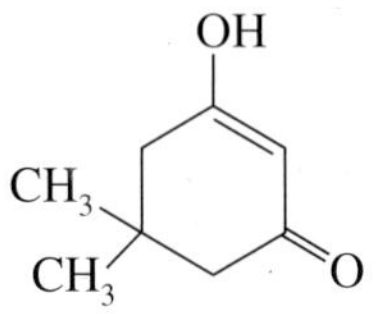

Dimedone

19.46 Show how the Claisen-type product from acetone and ethyl formate can, in effect, trimerize to form 1,3,5-triacetylbenzene.

19.47 In the presence of sodium ethoxide the following transformation occurs. Explain.

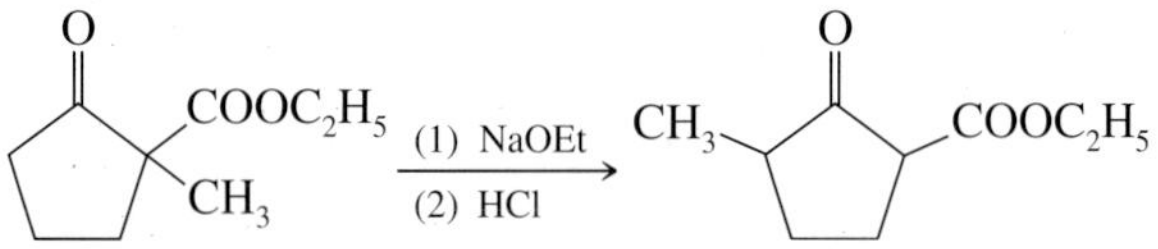

(1) NaOEt
(2) HCl

19.48 Write the mechanistic steps in the cyclization of ethyl phenylacetoacetate (ethyl 3-oxo-4-phenylbutanoate) in concentrated sulfuric acid to form naphthoresorcinol (1,3-naphthalenediol).

19.49 What is the structure of the *cyclic* compound formed in the Michael addition of **1** to **2** in the presence of sodium ethoxide?

$$CH_3CH(COOC_2H_5)C(=O)COOC_2H_5 \qquad CH_2{=}CHC(=O)CH_3$$

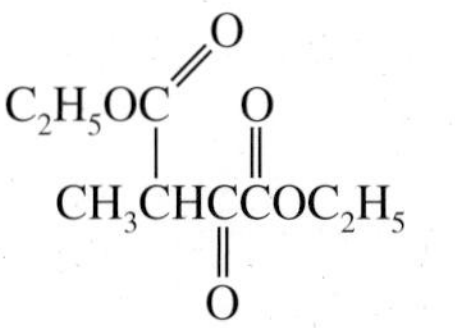

1 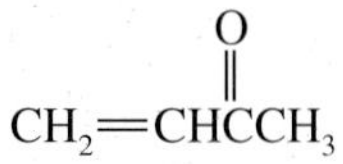 **2**

***19.50** **(a)** Deduce the structure of product **A**, which is highly symmetrical:

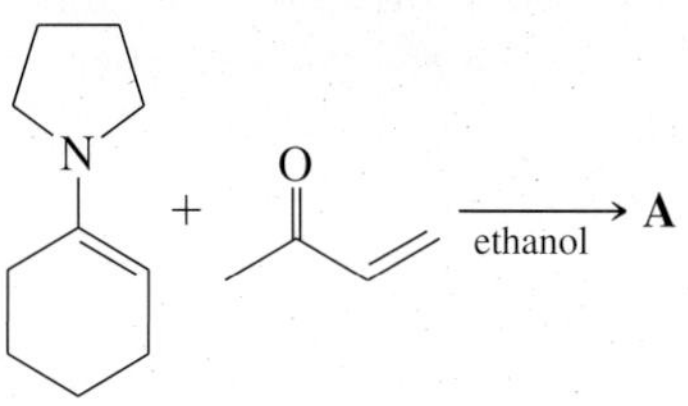

$\xrightarrow{\text{ethanol}}$ **A**

The following are selected spectral data for **A**:

MS (*m/z*): 220 ($M^{\cdot+}$)

IR (cm^{-1}): 2930, 2860, 1715

^{1}H NMR (δ): 1.25 (m), 1.29 (m), 1.76 (m), 1.77 (m), 2.14 (s), and 2.22 (t); (area ratios 2:1:2:1:2:2, respectively)

^{13}C NMR (δ): 23 (CH_2), 26 (CH_2), 27 (CH_2), 29 (C), 39 (CH), 41 (CH_2), 46 (CH_2), 208 (C)

(b) Write a mechanism that explains the formation of **A**.

***19.51** Write the structures of the three products involved in this reaction sequence:

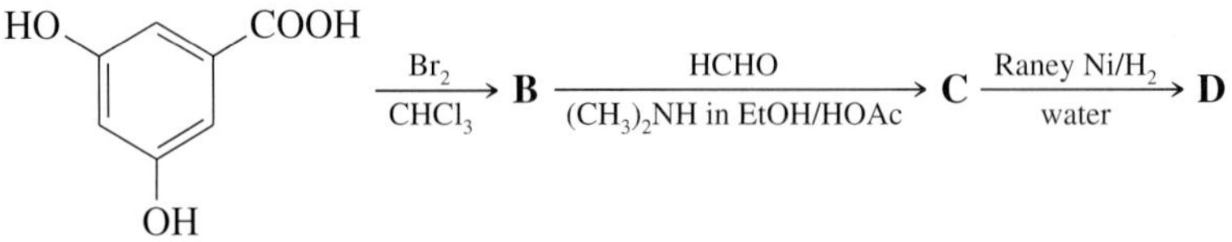

3,5-dihydroxybenzoic acid (HO, COOH, OH on benzene ring) $\xrightarrow[CHCl_3]{Br_2}$ **B** $\xrightarrow[(CH_3)_2NH \text{ in EtOH/HOAc}]{HCHO}$ **C** $\xrightarrow[\text{water}]{\text{Raney Ni/}H_2}$ **D**

Spectral data for **B**:

MS (*m/z*): 314, 312, 310 (relative abundance 1:2:1)

^{1}H NMR (δ): only 6.80 (s) after treatment with D_2O

Data for **C**:

MS (*m/z*): 371, 369, 367 (relative abundance 1:2:1)

^{1}H NMR (δ): 2.48 (s) and 4.99 (s) in area ratio 3:1; broad singlets at 5.5 and 11 disappeared after treatment with D_2O.

Data for **D**:

MS (*m/z*): 369 ($M^{\cdot+} - CH_3$) [when studied as its tris(trimethylsilyl) derivative]

^{1}H NMR (δ): 2.16 (s) and 7.18 (s) in area ratio 3:2; broad singlets at 5.4 and 11 disappeared after treatment with D_2O.

LEARNING GROUP PROBLEMS

β-Carotene, Lycopodine, and Dehydroabeitic Acid

1. β-Carotene is a highly conjugated hydrocarbon with an orange-red color. Its biosynthesis occurs via the isoprene pathway, and it is found in, among other sources, pumpkins. One of the chemical syntheses of β-carotene was accomplished near the turn of the twentieth century by W. Ipatiew (*Ber.* **1901,** *34*, 594–6). The first few steps of this synthesis involve chemistry that should be familiar to you. Write mechanisms for the reactions from all of the steps in this synthesis except between compounds **6** and **7**:

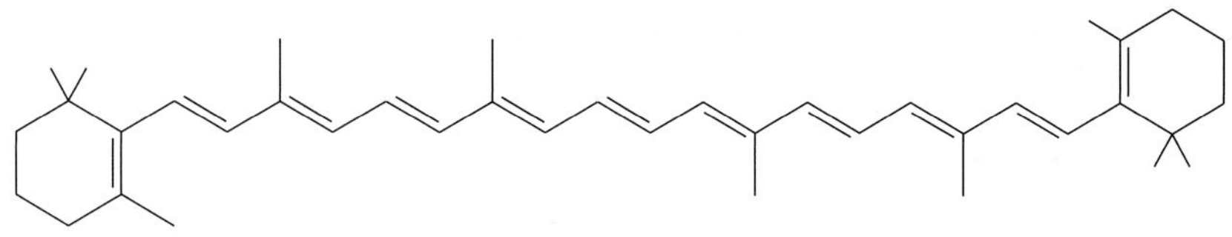

β-carotene

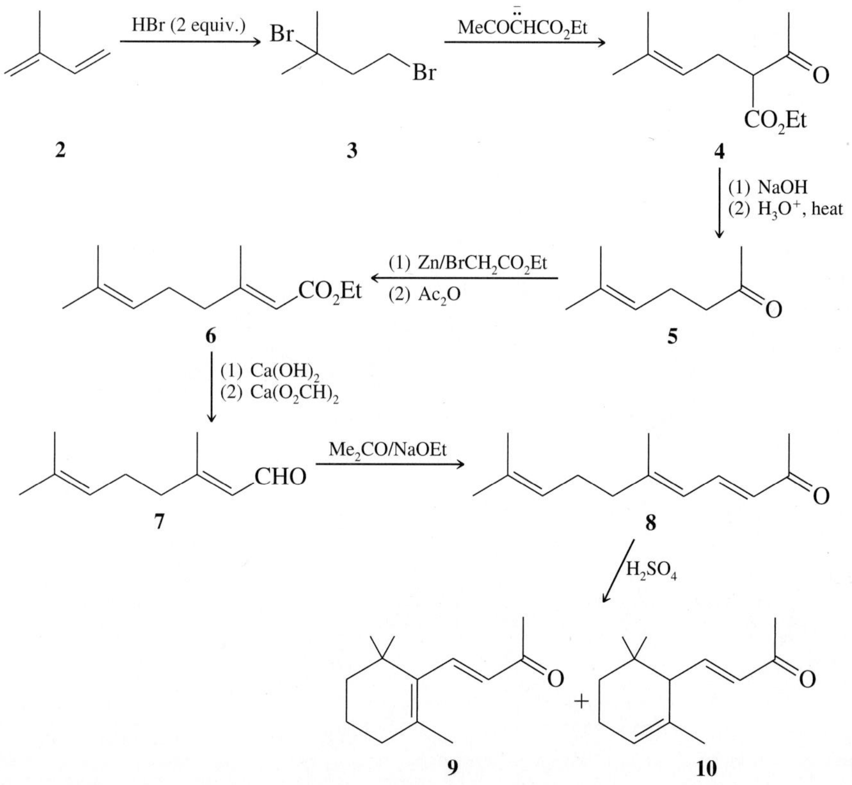

HBr (2 equiv.) Br Br $MeCO\ddot{C}HCO_2Et$ O CO_2Et

2 3 4

(1) NaOH
(2) H_3O^+, heat

CO_2Et (1) $Zn/BrCH_2CO_2Et$ (2) Ac_2O O

6 5

(1) $Ca(OH)_2$
(2) $Ca(O_2CH)_2$

CHO $Me_2CO/NaOEt$ O

7 8

H_2SO_4

O + O

9 10

2. Lycopodine is a naturally occurring amine. As such, it belongs to the family of natural products called alkaloids. Its synthesis (*J. Am. Chem. Soc.* **1968,** *90*, 1647–1648) was accomplished by one of the great synthetic organic chemists of our time, Gilbert Stork (Columbia University). Write a detailed mechanism for all the steps that occur when **2** reacts with ethyl acetoacetate in the presence of ethoxide ion. Note that a necessary part of the mechanism will be a base-catalyzed isomerization (via a conjugated enolate) of the alkene in **2** to form the corresponding α,β-unsaturated ester:

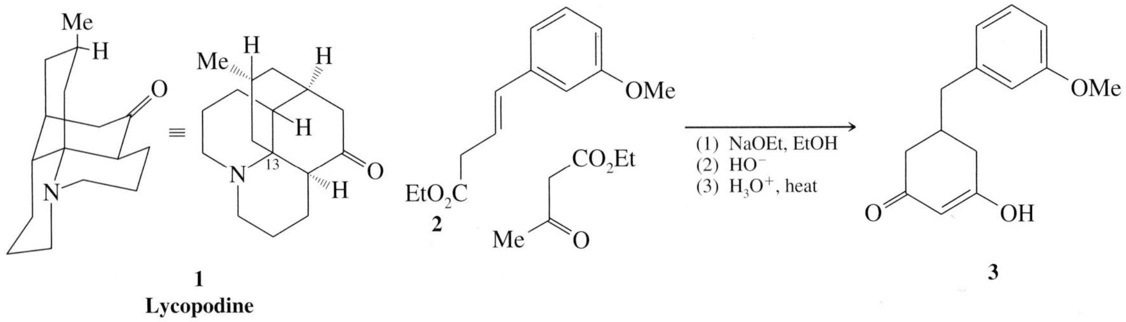

1
Lycopodine

2

(1) NaOEt, EtOH
(2) HO^-
(3) H_3O^+, heat

3

3. Dehydroabietic acid is a natural product isolated from *Pinus palustris.* It is structurally related to abietic acid, which comes from rosin. The synthesis of dehydroabietic acid (*J. Am. Chem. Soc.* **1962,** *84,* 284–292) was also accomplished by Gilbert Stork. In the course of this synthesis, Stork discovered his famous enamine reaction.

(a) Write detailed mechanisms for the reactions from **5** to **7** below.

(b) Write detailed mechanisms for all of the reactions from compound **7** to compound **9** in Stork's synthesis of dehydroabietic acid. Include in your mechanism an account for conversion of one of the reactants to 1-penten-3-one as a preliminary step. What is the name for the process that was accomplished in the synthesis of **8** from **7**?

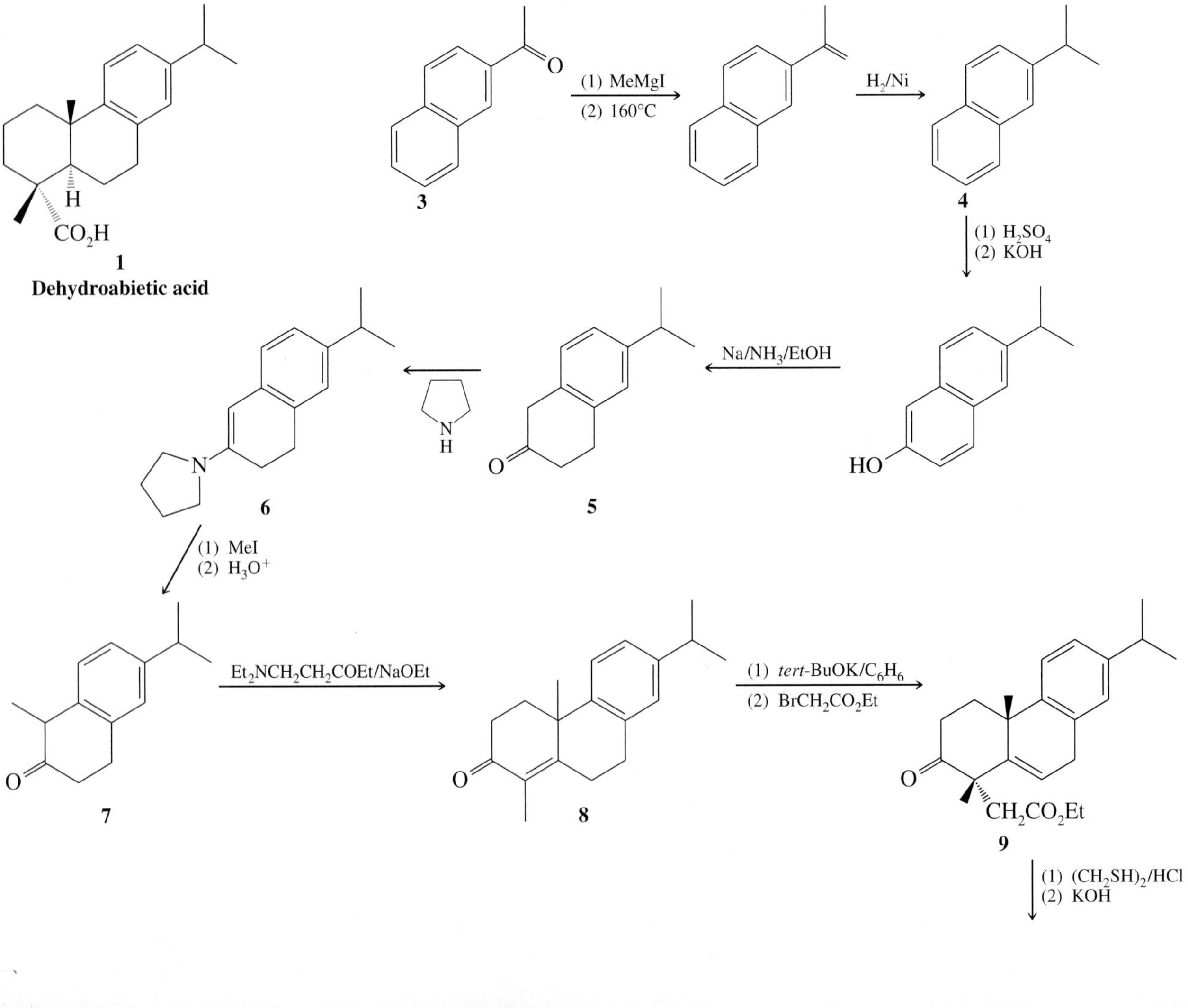

1
Dehydroabietic acid

3 → (1) MeMgI (2) 160°C → H_2/Ni → 4

4 → (1) H_2SO_4 (2) KOH → Na/NH_3/EtOH → 5

5 → pyrrolidine → 6

6 → (1) MeI (2) H_3O^+ → 7

7 → $Et_2NCH_2CH_2COEt$/NaOEt → 8

8 → (1) *tert*-BuOK/C_6H_6 (2) $BrCH_2CO_2Et$ → 9

9 → (1) $(CH_2SH)_2$/HCl (2) KOH →

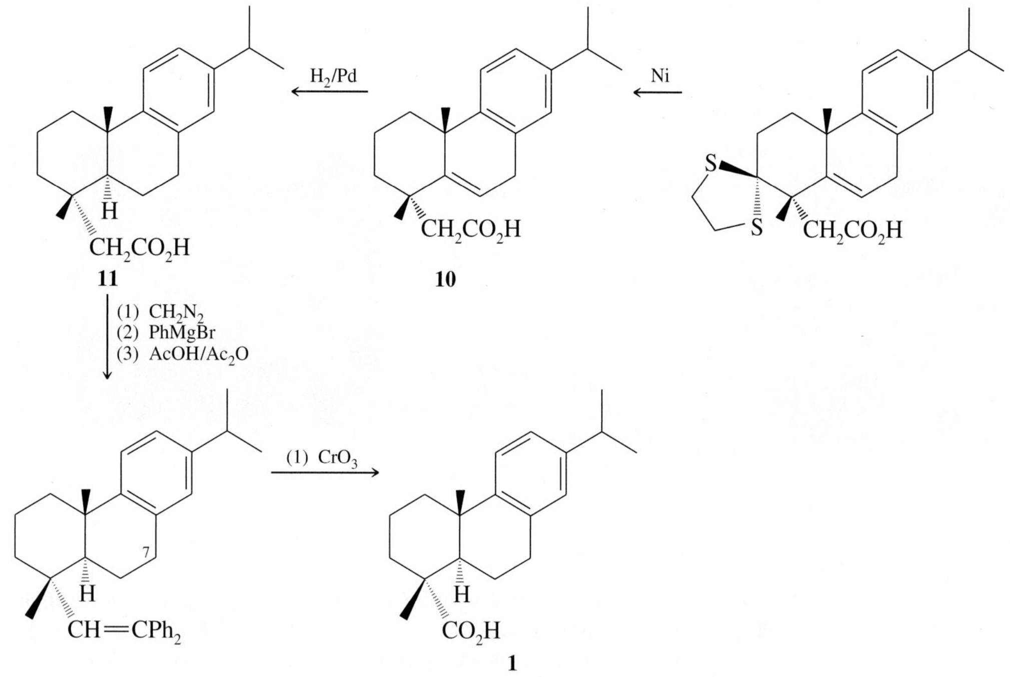

H$_2$/Pd

Ni

CH_2CO_2H

11

10

CH_2CO_2H

CH_2CO_2H

(1) CH_2N_2
(2) PhMgBr
(3) $AcOH/Ac_2O$

7

$CH{=}CPh_2$

(1) CrO_3

CO_2H

1

Structures from Fleming, I. *Selected Organic Syntheses;* Wiley: New York, 1973; p 76.

Thiols, Sulfur Ylides, and Disulfides

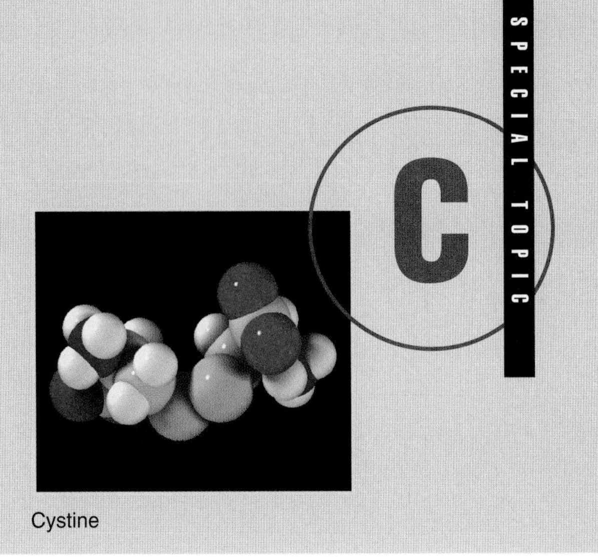

Cystine

Sulfur is directly below oxygen in Group 6 of the periodic table, and, as we might expect, there are sulfur counterparts of the oxygen compounds that we studied in earlier chapters.

Important examples of organosulfur compounds are the following:

$R{-}SH$	$R{-}S{-}R'$	$ArSH$	$R{-}S{-}S{-}R'$	$R{-}\overset{+}{S}(R'){-}R''$
Thiols	**Thioethers**	**Thiophenols**	**Disulfides**	**Trialkylsulfonium ions**
$R{-}S({=}O){-}R'$	$R{-}S({=}O)_2{-}R'$	$R{-}C({=}S){-}R'$	$R{-}S({=}O){-}OH$	$R{-}S({=}O)_2{-}OH$
Sulfoxides	**Sulfones**	**Thioketones**	**Sulfinic acids**	**Sulfonic acids**

The sulfur counterpart of an alcohol is called a *thiol* or a *mercaptan.* The name mercaptan comes from the Latin phrase *mercurium captans,* meaning "capturing mercury." Mercaptans react with mercuric ions and the ions of other heavy metals to form precipitates. The compound $CH_2(SH)CH(SH)CH_2OH$, known as British Anti-Lewisite (BAL), was developed as an antidote for poisonous arsenic compounds used as war gases. British Anti-Lewisite is also an effective antidote for mercury poisoning.

Several simple thiols are shown below:

CH_3CH_2SH	$CH_3CH_2CH_2SH$	$CH_3CH(CH_3)CH_2CH_2SH$	$CH_2{=}CHCH_2SH$
Ethanethiol (added to natural gas)	**1-Propanethiol (found in onions)**	**3-Methyl-1-butanethiol (produced by skunks)**	**2-Propene-1-thiol (found in garlic)**

Compounds of sulfur, in general, and the low-molecular-weight thiols, in particular, are noted for their disagreeable odors. Anyone who has passed anywhere near a general chemistry laboratory when hydrogen sulfide (H_2S) was being used has noticed the strong odor of that substance—the odor of rotten eggs. Another sulfur compound, 3-methyl-1-butanethiol, is one unpleasant constituent of the liquid that skunks use as a defensive weapon. 1-Propanethiol evolves from freshly chopped onions, and allyl mercaptan is one of the compounds responsible for the odor and flavor of garlic.

Aside from their odors, analogous sulfur and oxygen compounds show other chemical differences. These arise largely from the following features of sulfur compounds:

1. The sulfur atom is larger and more polarizable than the oxygen atom. As a result, sulfur compounds are more powerful nucleophiles, and compounds containing —SH groups are stronger acids than their oxygen analogues. The ethanethiolate ion ($CH_3CH_2S{:}^-$), for example, is a much stronger nucleophile when it reacts at carbon atoms than is the ethoxide ion ($CH_3CH_2O{:}^-$). On the other hand, since ethanol is a weaker acid than ethanethiol, the ethoxide ion is the stronger of the two conjugate bases.

2. The bond dissociation energy of the S—H bond of thiols (~365 kJ mol^{-1}) is much less than that of the O—H bond of alcohols (~430 kJ mol^{-1}). The weakness of the S—H bond allows thiols to undergo an oxidative coupling reaction when they react with mild oxidizing agents; the product is a disulfide:

$$\underset{\textbf{A thiol}}{2\,RS{-}H} + H_2O_2 \longrightarrow \underset{\textbf{A disulfide}}{RS{-}SR} + 2\,H_2O$$

Alcohols do not undergo an analogous reaction. When alcohols are treated with oxidizing agents, oxidation takes place at the weaker C—H (~380 kJ mol^{-1}) bond rather than at the stronger O—H bond.

3. Because sulfur atoms are easily polarized, they can stabilize a negative charge on an adjacent atom. This means that hydrogen atoms on carbon atoms that are adjacent to an alkylthio group are more acidic than those adjacent to an alkoxyl group. Thioanisole, for example, reacts with butyllithium in the following way:

$$\underset{\textbf{Thioanisole}}{C_6H_5{-}SCH_3} + C_4H_9{:}^-\,Li^+ \longrightarrow C_6H_5{-}SCH_2{:}^-\,Li^+ + C_4H_{10}$$

Anisole ($CH_3OC_6H_5$) does not undergo an analogous reaction. The $>S{=}O$ group of sulfoxides and the positive sulfur of sulfonium ions are even more effective in delocalizing negative charge on an adjacent atom:

$$\underset{\textbf{Dimethyl sulfoxide}}{CH_3\overset{O}{\overset{\|}{S}}CH_3} \xrightarrow{\overset{+}{Na}{:}\overset{-}{H}} \left[CH_3\overset{O}{\overset{\|}{S}}{-}\ddot{C}H_2^- \longleftrightarrow CH_3\overset{:O:^-}{\overset{|}{S}}{=}CH_2 \right] + H_2$$

$$\underset{\textbf{Trimethylsulfonium bromide}}{CH_3\overset{CH_3}{\overset{|}{\underset{+}{S}}}CH_3\ Br^-} \xrightarrow[(-HBr)]{\text{base}} \underset{\textbf{A sulfur ylide}}{CH_3\overset{CH_3}{\overset{|}{\underset{+}{S}}}{-}\ddot{C}H_2^-}$$

The anions formed in the reactions just given are of synthetic use. They can be used to synthesize epoxides, for example (see Section C.3).

C.1 Preparation of Thiols

Alkyl bromides and iodides react with potassium hydrogen sulfide to form thiols. (Potassium hydrogen sulfide can be generated by passing gaseous H_2S into an alcoholic solution of potassium hydroxide.)

$$\text{R—Br} + \text{KOH} + \mathbf{H_2S} \xrightarrow[\text{heat}]{C_2H_5OH} \text{R—}\mathbf{SH} + \text{KBr} + H_2O$$

(excess)

The thiol that forms is sufficiently acidic to form a thiolate ion in the presence of potassium hydroxide. Thus, if excess H_2S is not employed in the reaction, the major product of the reaction will be a thioether. The thioether results from the following reactions:

$$\text{R—}\mathbf{SH} + \text{KOH} \longrightarrow \text{R—}\ddot{\underset{..}{S}}{:}^- \, K^+ + H_2O$$

$$\text{R—}\ddot{\underset{..}{S}}{:}^- \, K^+ + \text{R—}\ddot{\underset{..}{Br}}{:} \longrightarrow \text{R—}\ddot{\underset{..}{S}}\text{—R} + \text{KBr}$$

Thioether

Alkyl halides also react with thiourea to form (stable) *S*-alkylisothiouronium salts. These can be used to prepare thiols.

$$(H_2N)_2C{=}\ddot{S}{:} + CH_3CH_2\text{—}\ddot{\underset{..}{Br}}{:} \xrightarrow{C_2H_5OH} (H_2N)_2C{=}\overset{+}{\underset{..}{S}}\text{—}CH_2CH_3 \; Br^-$$

Thiourea → ***S*-Ethylisothiouronium bromide (95%)**

$$\Big\downarrow \; OH^-_{aq}, \text{ then } H_3O^+$$

$$(H_2N)_2C{=}O + CH_3CH_2SH$$

Urea **Ethanethiol (90%)**

C.2 Physical Properties of Thiols

Thiols form very weak hydrogen bonds; their hydrogen bonds are not nearly as strong as those of alcohols. Because of this, low-molecular-weight thiols have lower boiling points than the corresponding alcohols. Ethanethiol, for example, boils more than 40°C lower than ethanol (37°C versus 78°C). The relative weakness of hydrogen bonds between molecules of thiols is also evident when we compare the boiling points of ethanethiol and its isomer dimethyl sulfide:

$$\underset{\textbf{bp 37°C}}{CH_3CH_2SH} \qquad \underset{\textbf{bp 38°C}}{CH_3SCH_3}$$

Physical properties of several thiols are given in Table C.1.

TABLE C.1 Physical Properties of Thiols

Compound	Structure	mp (°C)	bp (°C)
Methanethiol	CH_3SH	−123	6
Ethanethiol	CH_3CH_2SH	−144	37
1-Propanethiol	$CH_3CH_2CH_2SH$	−113	67
2-Propanethiol	$(CH_3)_2CHSH$	−131	58
1-Butanethiol	$CH_3(CH_2)_2CH_2SH$	−116	98

C.3 THE ADDITION OF SULFUR YLIDES TO ALDEHYDES AND KETONES

Sulfur ylides also react as nucleophiles at the carbonyl carbon of aldehydes and ketones. The betaine that forms usually decomposes to an *epoxide* rather than to an alkene:

In Section 11.8 we discussed methods for synthesizing epoxides from alkenes.

$$(CH_3)_2\ddot{S}: + CH_3I \longrightarrow (CH_3)_2\overset{+}{S}-CH_3 \;\; I^- \xrightarrow[\text{CH}_3\text{SOCH}_3,\ 0^\circ\text{C}]{\overset{+}{\text{Na}}\ \overset{-}{\ddot{\text{C}}}\text{H}_2\text{SOCH}_3}$$

Trimethylsulfonium iodide

$$\left[(CH_3)_2\overset{+}{S}-\overset{-}{\ddot{C}}H_2 \longleftrightarrow (CH_3)_2S=CH_2\right]$$

Resonance-stabilized sulfur ylide

$$C_6H_5CH=O + :\overset{-}{C}H_2\overset{+}{S}(CH_3)_2 \longrightarrow C_6H_5CH(\ddot{O}:^-)-CH_2-\overset{+}{S}(CH_3)_2 \longrightarrow C_6H_5\underset{\diagdown O \diagup}{CH-CH_2} + :\ddot{S}(CH_3)_2$$

Benzaldehyde

(75%)

PROBLEM C.1

Show how you might use a sulfur ylide to prepare

(a) (cyclohexane spiro-fused epoxide, ring CH_2 bearing H, H)

(b) $(H_3C)_2C$—CH_2 epoxide (2,2-dimethyloxirane)

C.4 THIOLS AND DISULFIDES IN BIOCHEMISTRY

Thiols and disulfides are important compounds in living cells, and in many biochemical oxidation–reduction reactions they are interconverted:

$$2\ RSH \underset{[H]}{\overset{[O]}{\rightleftharpoons}} R-S-S-R$$

Lipoic acid, for example, an important cofactor in biological oxidations, undergoes this oxidation–reduction reaction:

$$\text{(1,2-dithiolane)}-(CH_2)_4CO_2H \underset{[O]}{\overset{[H]}{\rightleftharpoons}} HS\ \ SH\text{-}(CH_2)_4CO_2H$$

Lipoic acid **Dihydrolipoic acid**

The amino acids *cysteine* and *cystine* are interconverted in a similar way:

$$2\ HO_2C\underset{NH_2}{\underset{|}{C}}HCH_2SH \underset{[H]}{\overset{[O]}{\rightleftharpoons}} HO_2C\underset{NH_2}{\underset{|}{C}}HCH_2S-SCH_2\underset{NH_2}{\underset{|}{C}}HCO_2H$$

Cysteine **Cystine**

As we shall see in Chapter 24, the disulfide linkages of cystine units are important in determining the overall shapes of protein molecules.

PROBLEM C.2

Give structures for the products of the following reactions:

(a) Benzyl bromide + thiourea $\longrightarrow$

(b) Product of (a) + OH^-/H_2O, then H_3O^+ $\longrightarrow$

(c) Product of (b) + H_2O_2 $\longrightarrow$

(d) Product of (b) + NaOH $\longrightarrow$

(e) Product of (d) + benzyl bromide $\longrightarrow$

PROBLEM C.3

Allyl disulfide, $CH_2{=}CHCH_2S{-}SCH_2CH{=}CH_2$, is another important component of oil of garlic. Suggest a synthesis of allyl disulfide starting with allyl bromide.

PROBLEM C.4

Starting with allyl alcohol, outline a synthesis of BAL, $HSCH_2CH(SH)CH_2OH$.

PROBLEM C.5

A synthesis of lipoic acid (see structure just given) is outlined here. Supply the missing reagents and intermediates.

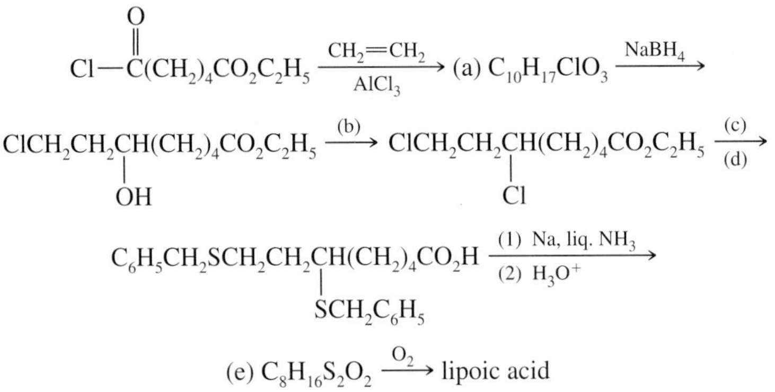

$$Cl{-}\overset{O}{\overset{\|}{C}}(CH_2)_4CO_2C_2H_5 \xrightarrow[AlCl_3]{CH_2=CH_2} \text{(a) } C_{10}H_{17}ClO_3 \xrightarrow{NaBH_4}$$

$$ClCH_2CH_2\underset{OH}{\underset{|}{C}H}(CH_2)_4CO_2C_2H_5 \xrightarrow{(b)} ClCH_2CH_2\underset{Cl}{\underset{|}{C}H}(CH_2)_4CO_2C_2H_5 \xrightarrow[(d)]{(c)}$$

$$C_6H_5CH_2SCH_2CH_2\underset{SCH_2C_6H_5}{\underset{|}{C}H}(CH_2)_4CO_2H \xrightarrow[(2)\ H_3O^+]{(1)\ Na,\ liq.\ NH_3}$$

$$\text{(e) } C_8H_{16}S_2O_2 \xrightarrow{O_2} \text{lipoic acid}$$

Thiol Esters and Lipid Biosynthesis

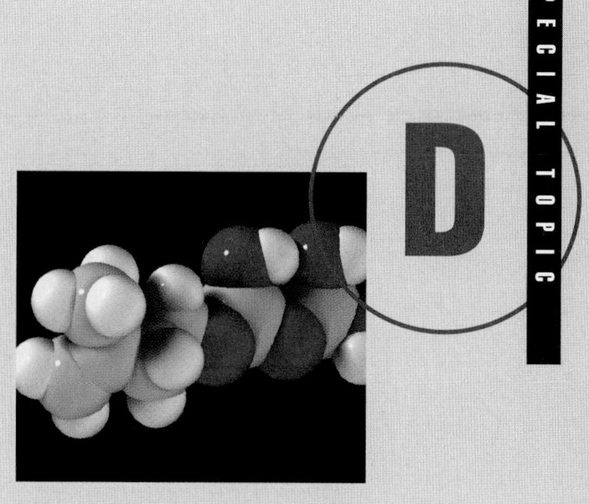

3-Methyl-3-butenyl pyrophosphate

D.1 Thiol Esters

Thiol esters can be prepared by reaction of a thiol with an acyl chloride:

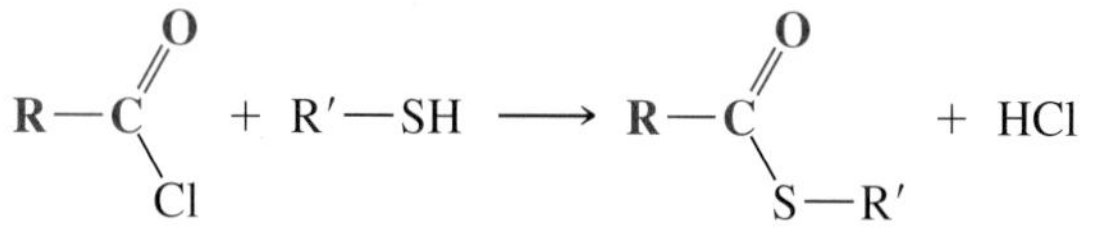

$$R{-}C(=O){-}Cl + R'{-}SH \longrightarrow R{-}C(=O){-}S{-}R' + HCl$$

Thiol ester

$$CH_3C(=O)Cl + CH_3SH \xrightarrow{\text{pyridine}} CH_3C(=O)SCH_3 + C_5H_5N^+{-}H\ Cl^-$$

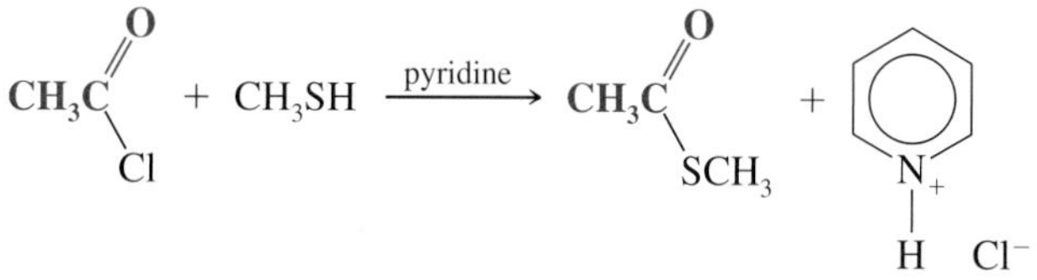

Although thiol esters are not often used in laboratory syntheses, they are of great importance in syntheses that occur within living cells. One of the important thiol esters in biochemistry is "acetyl-coenzyme A":

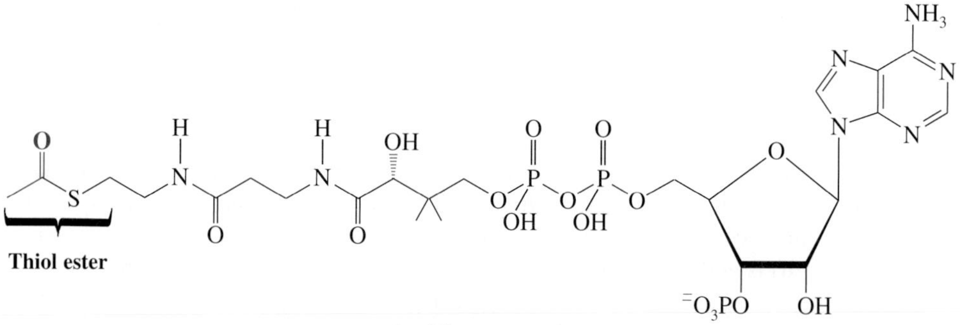

Thiol ester

Acetyl-coenzyme A

The important part of this rather complicated structure is the thiol ester at the beginning of the chain; because of this, acetyl-coenzyme A is usually abbreviated as follows:

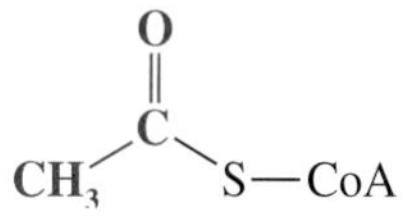

$$CH_3{-}C(=O){-}S{-}CoA$$

and coenzyme A is abbreviated CoA—SH.

In certain biochemical reactions, an *acyl*-coenzyme A operates as an *acylating agent;* it transfers an acyl group to another nucleophile in a reaction that involves a nucleophilic attack at the acyl carbon of the thiol ester. For example,

$$\mathrm{R{-}C({=}O){-}SCoA} + {}^{-}\mathrm{O{-}P({=}O)(OH){-}O^{-}} \underset{}{\overset{\text{enzyme}}{\rightleftharpoons}} \mathrm{R{-}C({=}O){-}O{-}P({=}O)(O^{-}){-}O^{-}} + \mathrm{CoA{-}SH}$$

An acyl phosphate

This reaction is catalyzed by the enzyme *phosphotransacetylase.*

The α hydrogens of the acetyl group of acetyl-coenzyme A are appreciably acidic. Acetyl-coenzyme A, as a result, also functions as a nucleophilic *alkylating agent.* Acetyl-coenzyme A, for example, reacts with oxaloacetate ion to form citrate ion in a reaction that resembles an aldol addition:

This reaction is the entry point for C_2 units into the citric acid metabolic cycle, and it forms the namesake compound of the pathway.

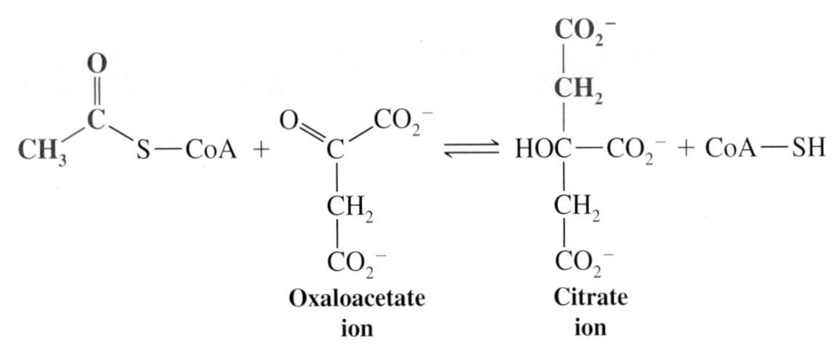

Oxaloacetate ion **Citrate ion**

One might well ask, "Why has nature made such prominent use of thiol esters?" Or, "In contrast to ordinary esters, what advantages do thiol esters offer the cell?" In answering these questions we can consider three factors:

1. Resonance contributions of type **(b)** in the following reaction stabilize an ordinary ester and make the carbonyl group less susceptible to nucleophilic attack:

$$\mathrm{H_3C{-}C({=}\ddot{O}{:}){-}\ddot{O}{-}R} \longleftrightarrow \mathrm{H_3C{-}C({-}\ddot{O}{:}^{-}){=}\overset{+}{O}{-}R}$$

(a) **(b)** **This structure makes an important contribution.**

In contrast, thiol esters are not so effectively stabilized by a similar resonance contribution because structure **(d)** among the following ones requires overlap between the 3*p* orbital of sulfur and a 2*p* orbital of carbon. Since this overlap is not large, resonance stabilization by **(d)** is not so effective. Structure **(e)** does, however, make an important contribution—one that makes the carbonyl group more susceptible to nucleophilic attack.

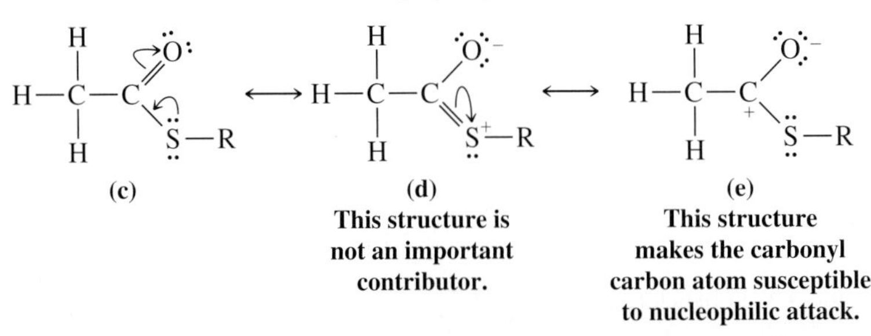

(c) **(d) This structure is not an important contributor.** **(e) This structure makes the carbonyl carbon atom susceptible to nucleophilic attack.**

2. A resonance contribution from the similar structure **(g)** makes the α hydrogens of thiol esters more acidic than those of ordinary esters:

$$B:^- + H-CH_2-C(=O)-S-R \xrightarrow{-BH} H-\ddot{C}H_2^--C(=O)-S-R \longleftrightarrow H_2C{=}C(-O^-)-S-R$$

(f) **(g)**

This structure's contribution stabilizes the anion of a thiol ester.

3. The carbon–sulfur bond of a thiol ester is weaker than the carbon–oxygen bond of an ordinary ester; ^-SR is a better leaving group than ^-OR.

Factors **1** and **3** make thiol esters effective *acylating agents;* factor **2** makes them effective nucleophilic *alkylating* agents. Therefore, we should not be surprised when we encounter reactions similar to the following one:

$$CH_3C(=O)S-R + CH_3C(=O)S-R \longrightarrow CH_3\overset{O}{\overset{\|}{C}}-CH_2-\overset{O}{\overset{\|}{C}}-S-R + HS-R$$

In this reaction, 1 mol of a thiol ester acts as an acylating agent and the other acts as an alkylating agent (see Section D.2).

D.2 Biosynthesis of Fatty Acids

Cell membranes, fats, and oils contain esters of long-chain (mainly C_{14}, C_{16}, and C_{18}) carboxylic acids, called fatty acids. Fatty acids are lipids, a largely hydrophobic family of biomolecules that we shall study in Chapter 23. An example of a fatty acid is hexadecanoic acid, also called palmitic acid:

$$H_3C-(CH_2)_{14}-C(=O)OH$$

Palmitic acid

The fact that most naturally occurring fatty acids are made up of an even number of carbon atoms suggests that they are assembled from two-carbon units. The idea that these might be acetate ($CH_3CO_2^-$) units was put forth as early as 1893. Many years later, when radioactively labeled compounds became available, it became possible to test and confirm this hypothesis.

When an organism is fed acetic acid labeled with carbon-14 at the carboxyl group, the fatty acids that it synthesizes contain the label at alternate carbon atoms beginning with the carboxyl carbon:

$$H_3C-C^*(=O)OH$$

Feeding ^{14}C-carboxyl-labeled acetic acid . . .

$$H_3C-\overset{*}{C}H_2-CH_2-\overset{*}{C}H_2-CH_2-\overset{*}{C}H_2-CH_2-\overset{*}{C}H_2-CH_2-\overset{*}{C}H_2-CH_2-\overset{*}{C}H_2-CH_2-\overset{*}{C}H_2-CH_2-C^*(=O)OH$$

yields palmitic acid labeled at these positions.

Conversely, feeding acetic acid labeled at the methyl carbon yields a fatty acid labeled at the other set of alternate carbon atoms:

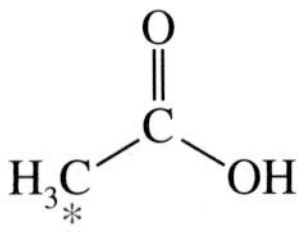

$$H_3\overset{*}{C}-\overset{O}{\overset{\|}{C}}-OH$$

Feeding ¹⁴C-methyl-labeled acetic acid . . .

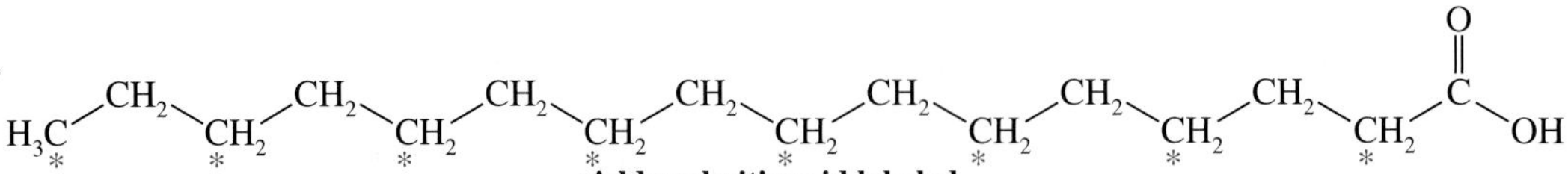

$$H_3\overset{*}{C}-CH_2-\overset{*}{C}H_2-CH_2-\overset{*}{C}H_2-CH_2-\overset{*}{C}H_2-CH_2-\overset{*}{C}H_2-CH_2-\overset{*}{C}H_2-CH_2-\overset{*}{C}H_2-CH_2-\overset{*}{C}H_2-\overset{O}{\overset{\|}{C}}-OH$$

yields palmitic acid labeled at these positions.

The biosynthesis of fatty acids is now known to begin with acetyl-coenzyme A:

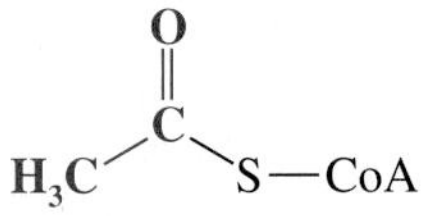

$$H_3C-\overset{O}{\overset{\|}{C}}-S-CoA$$

The following bond-line formula shows the positions of the two-carbon units incorporated into palmitic acid from acetyl-coenzyme A:

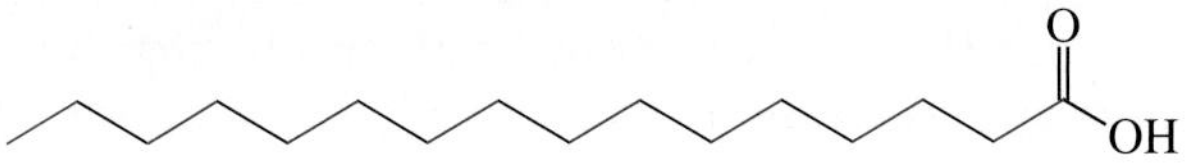

$$CH_3(CH_2)_{14}\overset{O}{\overset{\|}{C}}OH$$

The acetyl portion of acetyl-coenzyme A can be synthesized in the cell from acetic acid; it can also be synthesized from carbohydrates, proteins, and fats:

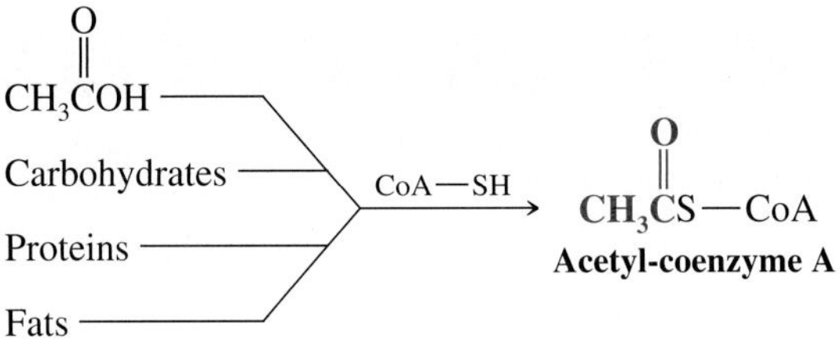

$$\left.\begin{array}{l} CH_3\overset{O}{\overset{\|}{C}}OH \\ \text{Carbohydrates} \\ \text{Proteins} \\ \text{Fats} \end{array}\right\} \xrightarrow{CoA-SH} \underset{\textbf{Acetyl-coenzyme A}}{CH_3\overset{O}{\overset{\|}{C}}S-CoA}$$

Although the methyl group of acetyl-coenzyme A is already activated toward condensation reactions by virtue of its being a part of a thiol ester (Section D.1), nature activates it again by converting it to *malonyl-coenzyme A:*

$$\underset{\textbf{Acetyl-CoA}}{CH_3\overset{O}{\overset{\|}{C}}S-CoA} + CO_2 \xrightleftharpoons{\text{acetyl-CoA carboxylase*}} \underset{\textbf{Malonyl-CoA}}{HO\overset{O}{\overset{\|}{C}}CH_2\overset{O}{\overset{\|}{C}}S-CoA}$$

The next steps in fatty acid synthesis involve the transfer of acyl groups of malonyl-CoA and acetyl-coenzyme A to the thiol group of a coenzyme called *acyl carrier protein* or ACP—SH:

$$\underset{\textbf{Malonyl-CoA}}{HO\overset{O}{\overset{\|}{C}}CH_2\overset{O}{\overset{\|}{C}}S-CoA} + ACP-SH \rightleftharpoons \underset{\textbf{Malonyl-S-ACP}}{HO\overset{O}{\overset{\|}{C}}CH_2\overset{O}{\overset{\|}{C}}S-ACP} + CoA-SH$$

$$\underset{\textbf{Acetyl-CoA}}{CH_3\overset{O}{\overset{\|}{C}}S-CoA} + ACP-SH \rightleftharpoons \underset{\textbf{Acetyl-S-ACP}}{CH_3\overset{O}{\overset{\|}{C}}S-ACP} + CoA-SH$$

---

*This step also requires 1 mol of adenosine triphosphate (Section 22.1B) and an enzyme that transfers the carbon dioxide.

Acetyl-S-ACP and malonyl-S-ACP then condense with each other to form acetoacetyl-S-ACP:

$$\underset{\text{Acetyl-S-ACP}}{CH_3\overset{O}{\overset{\|}{C}}S{-}ACP} + \underset{\text{Malonyl-S-ACP}}{HO\overset{O}{\overset{\|}{C}}CH_2\overset{O}{\overset{\|}{C}}S{-}ACP} \rightleftharpoons \underset{\text{Acetoacetyl-S-ACP}}{CH_3\overset{O}{\overset{\|}{C}}CH_2\overset{O}{\overset{\|}{C}}S{-}ACP} + CO_2 + ACP{-}SH$$

Note the similarity of this reaction to malonic ester syntheses we studied in Chapter 19.

The molecule of CO_2 that is lost in this reaction is the same molecule that was incorporated into malonyl-CoA in the acetyl-CoA carboxylase reaction.

This remarkable reaction bears a strong resemblance to the malonic ester syntheses that we saw earlier (Section 19.4) and deserves special comment. One can imagine, for example, a more economical synthesis of acetoacetyl-S-ACP, that is, a simple condensation of 2 mol of acetyl-S-ACP:

$$CH_3\overset{O}{\overset{\|}{C}}S{-}ACP + CH_3\overset{O}{\overset{\|}{C}}S{-}ACP \rightleftharpoons CH_3\overset{O}{\overset{\|}{C}}CH_2\overset{O}{\overset{\|}{C}}S{-}ACP + ACP{-}SH$$

Studies of this last reaction, however, have revealed that it is highly *endothermic* and that the position of equilibrium lies very far to the left. In contrast, the condensation of acetyl-S-ACP and malonyl-S-ACP is highly exothermic, and the position of equilibrium lies far to the right. The favorable thermodynamics of the condensation utilizing malonyl-S-ACP comes about because *the reaction also produces a highly stable substance: carbon dioxide.* Thus, decarboxylation of the malonyl group provides the condensation with thermodynamic assistance.

The next three steps in fatty acid synthesis transform the acetoacetyl group of acetoacetyl-S-ACP into a butyryl (butanoyl) group. These steps involve (1) reduction of the keto group (utilizing NADPH* as the reducing agent), (2) dehydration of an alcohol, and (3) reduction of a double bond (again utilizing NADPH).

Reduction of the Keto Group

$$\underset{\text{Acetoacetyl-S-ACP}}{CH_3\overset{O}{\overset{\|}{C}}CH_2\overset{O}{\overset{\|}{C}}S{-}ACP} + NADPH + H^+ \rightleftharpoons \underset{\beta\text{-Hydroxybutyryl-S-ACP}}{CH_3\overset{OH}{\overset{|}{C}}HCH_2\overset{O}{\overset{\|}{C}}S{-}ACP} + NADP^+$$

Dehydration of the Alcohol

$$\underset{\beta\text{-Hydroxybutyryl-S-ACP}}{CH_3\overset{OH}{\overset{|}{C}}HCH_2\overset{O}{\overset{\|}{C}}S{-}ACP} \rightleftharpoons \underset{\text{Crotonyl-S-ACP}}{CH_3CH{=}CH\overset{O}{\overset{\|}{C}}S{-}ACP} + H_2O$$

Reduction of the Double Bond

$$\underset{\text{Crotonyl-S-ACP}}{CH_3CH{=}CH\overset{O}{\overset{\|}{C}}S{-}ACP} + NADPH + H^+ \rightleftharpoons \underset{\text{Butyryl-S-ACP}}{CH_3CH_2CH_2\overset{O}{\overset{\|}{C}}S{-}ACP} + NADP^+$$

These steps complete one cycle of the overall fatty acid synthesis. The net result is the conversion of two acetate units into the four-carbon butyrate unit of butyryl-S-ACP. (This conversion requires, of course, the crucial intervention of a molecule of carbon dioxide.) At this point, another cycle begins and the chain is lengthened by two more carbon atoms.

Subsequent turns of the cycle continue to lengthen the chain by two-carbon units until a long-chain fatty acid is produced. The overall equation for the synthesis of palmitic acid, for example, can be written as follows:

---

*NADPH is *nicotinamide adenine dinucleotide phosphate (reduced form),* a coenzyme that is very similar in structure and function to NADH (see Chapter 12 opening vignette and Section 14.10)

$$\mathrm{CH_3\overset{\overset{\displaystyle O}{\|}}{C}S{-}CoA + 7\ HO\overset{\overset{\displaystyle O}{\|}}{C}CH_2\overset{\overset{\displaystyle O}{\|}}{C}S{-}CoA + 14\ NADPH + 14\ H^+ \longrightarrow}$$

$$\mathrm{CH_3(CH_2)_{14}CO_2H + 7\ CO_2 + 8\ CoA{-}SH + 14\ NADP^+ + 6\ H_2O}$$

One of the most remarkable aspects of fatty acid synthesis is that the entire cycle appears to be carried out by a dimeric multifunctional enzyme. The molecular weight of this enzyme, called *fatty acid synthetase,* has been estimated as 2,300,000.* The synthesis begins with a single molecule of acetyl-S-ACP serving as a primer. Then, in the synthesis of palmitic acid, for example, successive condensations of seven molecules of malonyl-S-ACP occur, with each condensation followed by reduction, dehydration, and reduction. All of these steps, which result in the synthesis of a C_{16} chain, take place before the fatty acid is released from the enzyme.

The acyl carrier protein from *Escherichia coli* has been isolated and purified; its molecular weight is approximately 10,000. In animals the carrier is part of the larger multifunctional enzyme. Both types of carrier protein contain a chain of groups called a *phosphopantetheine group* that is identical to that of coenzyme A (Section D.1). In ACP this chain is attached to a protein (rather than to an adenosine phosphate as it is in coenzyme A):

Pantothenic acid **2-Aminoethanethiol**

Protein—O—P(=O)(OH)—O—CH₂C(CH₃)₂CH(OH)C(=O)NHCH₂CH₂C(=O)NHCH₂CH₂SH

Phosphopantetheine group

← 20.2 Å →

The length of the phosphopantetheine group is 20.2 Å, and it has been postulated that it acts to transport the growing acyl chain from one active site of the enzyme to the next (Fig. D.1).

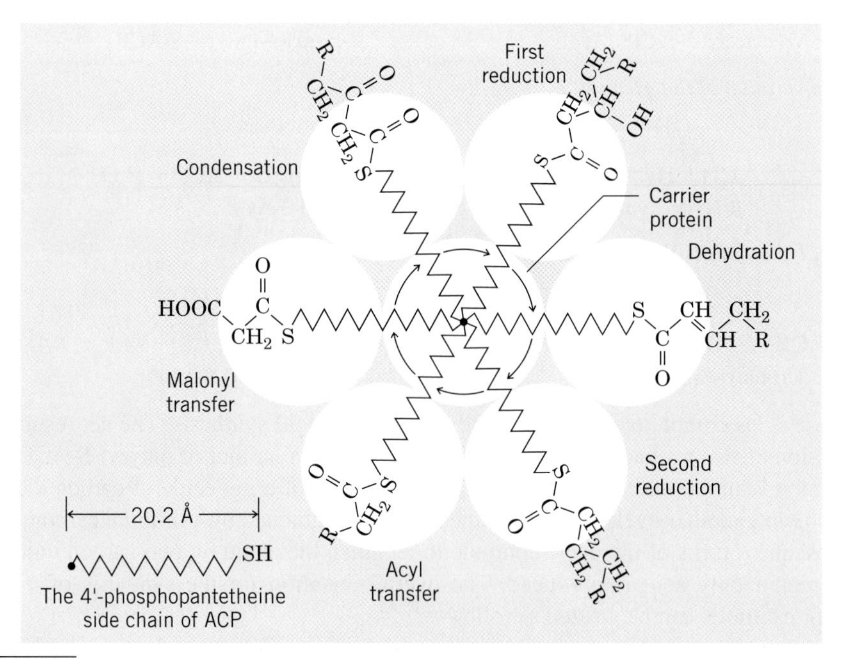

FIGURE D.1 The phosphopantetheine group as a swinging arm in the fatty acid synthetase complex. (Adapted from Lehninger, A. L. *Biochemistry;* Worth: New York, 1970 p 519. Used with permission.)

*As isolated from yeast cells. Fatty acid synthetases from different sources have different molecular weights; that from pigeon liver, for example, has a molecular weight of 450,000.

D.3 Biosynthesis of Isoprenoid Compounds

Isoprenoid compounds are another class of lipid biomolecules. Among them are natural products such as α-terpineol, geraniol, vitamin A, β-carotene, steroids (e.g., cholesterol, cortisone, the estrogens, and testosterone), and many related compounds. We shall study terpenes further in Chapter 23. Now, however, let us consider aspects of their biosynthesis that involve reactions parallel to some that we have recently studied as well as reactions that we have seen in earlier chapters:

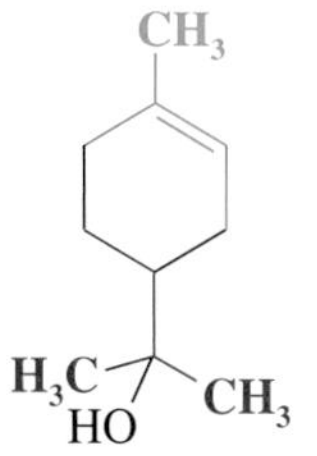

α-Terpineol 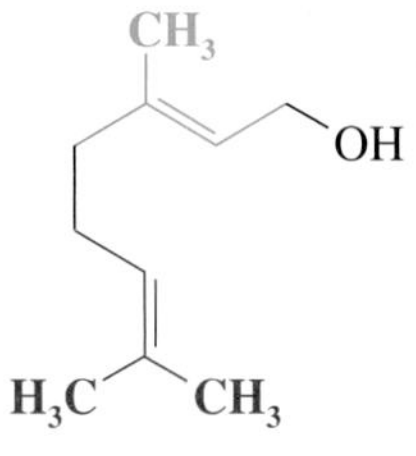 **Geraniol** 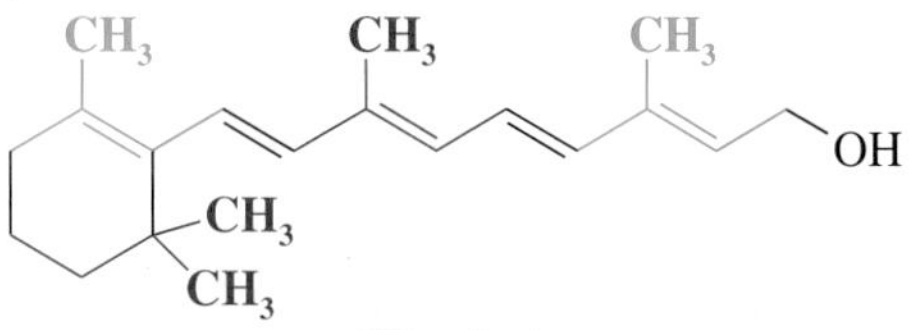 **Vitamin A**

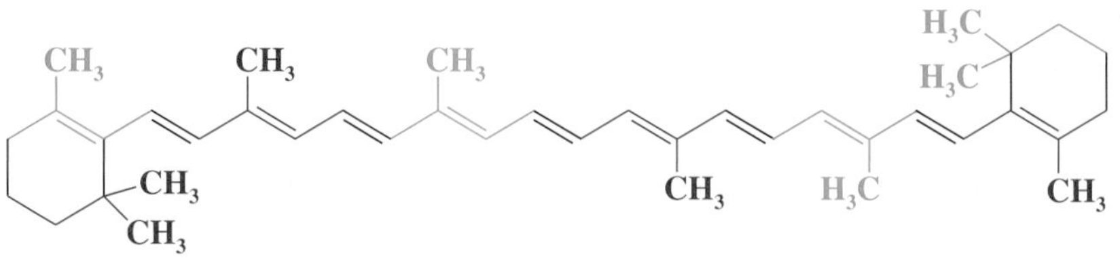

β-Carotene

The basic building block for the synthesis of terpenes and terpenoids is 3-methyl-3-butenyl pyrophosphate. The five carbon atoms of this compound are the source of all the "isoprene units" in isoprenoid compounds. (Isoprene units in the preceding structures are shown in blue and red.)

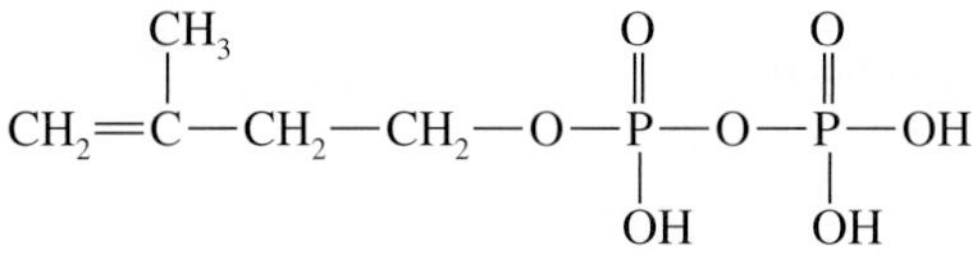

$$CH_2{=}C(CH_3){-}CH_2{-}CH_2{-}O{-}P(=O)(OH){-}O{-}P(=O)(OH){-}OH$$

3-Methyl-3-butenyl pyrophosphate

We consider how 3-methyl-3-butenyl pyrophosphate is biosynthesized in Section D.4. First, however, let us look at the way C_5 isoprene units are joined together. A necessary first step is enzymatic formation of 3-methyl-2-butenyl pyrophosphate from 3-methyl-3-butenyl pyrophosphate. This isomerization establishes an equilibrium that makes both compounds available to the cell:

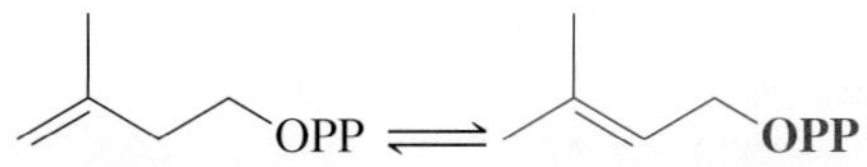

OPP ⇌ OPP **OPP = pyrophosphate**

3-Methyl-3-butenyl pyrophosphate **3-Methyl-2-butenyl pyrophosphate**

The joining of 3-methyl-2-butenyl pyrophosphate and 3-methyl-3-butenyl pyrophosphate involves enzymatic formation of an allylic cation. Here, the pyrophosphate group functions as a natural leaving group. This is one of many instances where nature relies on the pyrophosphate group for biochemical processes. Condensation of the two C_5 units yields a C_{10} compound called geranyl pyrophosphate:

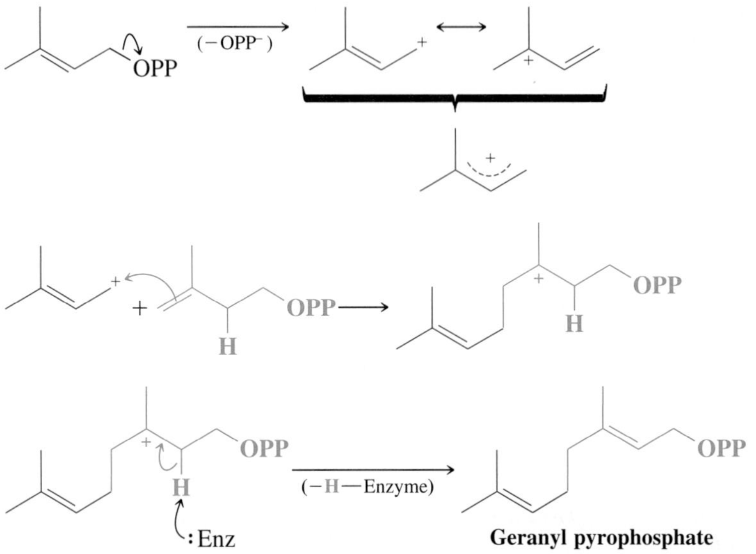

(−OPP⁻)

(−H—Enzyme)

:Enz

Geranyl pyrophosphate

Geranyl pyrophosphate is the precursor of the monoterpenes; hydrolysis of geranyl pyrophosphate, for example, yields geraniol:

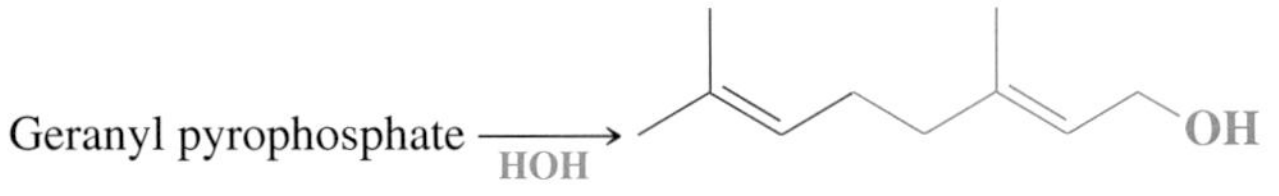

Geranyl pyrophosphate $\xrightarrow{\text{HOH}}$

Geraniol

Geranyl pyrophosphate can also condense with 3-methyl-3-butenyl pyrophosphate to form the C_{15} precursor for sesquiterpenes, farnesyl pyrophosphate:

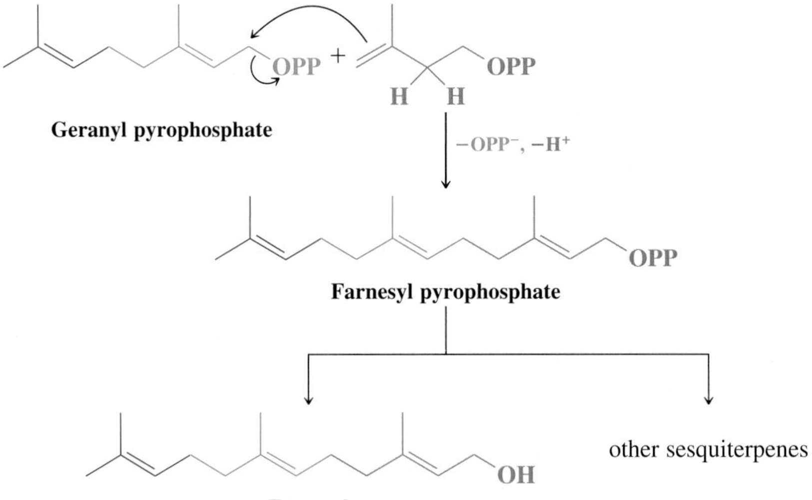

Geranyl pyrophosphate

$-OPP^-, -H^+$

Farnesyl pyrophosphate

Farnesol

other sesquiterpenes

Farnesol is a common component in the essential oils of plants and flowers. It has been isolated from roses, lemon and citronella grasses, and ambrette oil. It has the odor of lily of the valley. Farnesol also functions as a hormone in certain insects and initiates the change from caterpillar to pupa to moth. It is released by female mites as a sex attractant for male mites.

Similar condensation reactions yield the precursors for all of the other terpenes (Fig. D.2). In addition, a tail-to-tail reductive coupling of two molecules of farnesyl pyrophosphate produces squalene, the precursor for the important group of isoprenoids known as *steroids* (see Sections 23.4 and D.4).

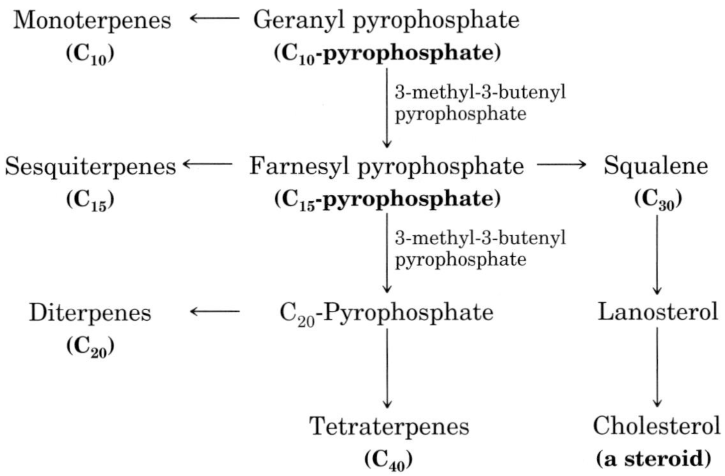

Monoterpenes (C_{10}) ⟵ Geranyl pyrophosphate (C_{10}-pyrophosphate)
↓ 3-methyl-3-butenyl pyrophosphate
Sesquiterpenes (C_{15}) ⟵ Farnesyl pyrophosphate (C_{15}-pyrophosphate) ⟶ Squalene (C_{30})
↓ 3-methyl-3-butenyl pyrophosphate ↓
Diterpenes (C_{20}) ⟵ C_{20}-Pyrophosphate Lanosterol
↓ ↓
Tetraterpenes (C_{40}) Cholesterol (a steroid)

FIGURE D.2 Biosynthetic paths for terpenes and steroids.

PROBLEM D.1

When farnesol is treated with sulfuric acid, it is converted to bisabolene. Outline a possible mechanism for this reaction.

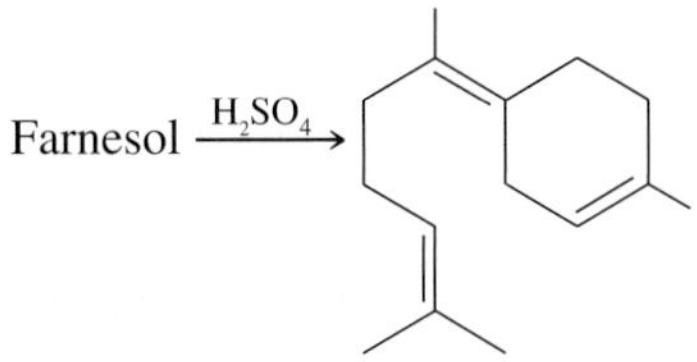

Farnesol $\xrightarrow{H_2SO_4}$ **Bisabolene**

D.4 Biosynthesis of Steroids

We saw in the previous section that the C_5 compound 3-methyl-3-butenyl pyrophosphate is the actual "isoprene unit" that nature uses in constructing terpenoids and carotenoids. We can now extend that biosynthetic pathway in two directions. We can show how 3-methyl-3-butenyl pyrophosphate (like the fatty acids) is ultimately derived from acetate units and how cholesterol, the precursor of most of the important steroids, is synthesized from 3-methyl-3-butenyl pyrophosphate.

Bloch shared the 1964 Nobel Prize in Physiology or Medicine with Feodor Lynen for their work on the biosynthesis of cholesterol and fatty acids.

In the 1940s, Konrad Bloch of Harvard University used labeling experiments to demonstrate that all of the carbon atoms of cholesterol can be derived from acetic acid. Using *methyl-labeled* acetic acid, for example, Bloch found the following label distribution in the cholesterol that was synthesized:

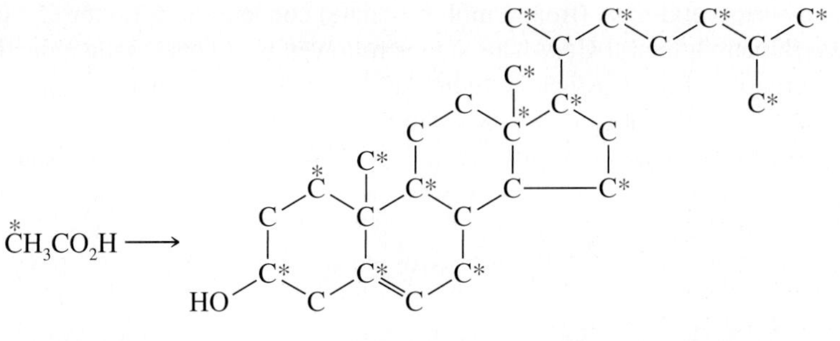

$\overset{*}{C}H_3CO_2H$ ⟶

Bloch also found that feeding *carboxyl-labeled* acetic acid led to incorporation of the label into all of the other carbon atoms of cholesterol (the unstarred carbon atoms of the formula just given).

Subsequent research by a number of investigators has shown that 3-methyl-3-butenyl pyrophosphate is synthesized from acetate units through the following sequence of reactions:

$$\underset{\substack{(C_2)\\ \textbf{Acetyl-CoA}}}{CH_3\overset{O}{\overset{\|}{C}}S{-}CoA} + \underset{\substack{(C_4)\\ \textbf{Acetoacetyl-CoA}}}{CH_3\overset{O}{\overset{\|}{C}}CH_2\overset{O}{\overset{\|}{C}}S{-}CoA} \xrightarrow{CoA{-}SH} \underset{\substack{(C_6)\\ \boldsymbol{\beta}\textbf{-Hydroxy-}\boldsymbol{\beta}\textbf{-methylglutaryl-CoA}}}{HO\overset{O}{\overset{\|}{C}}CH_2{-}C(CH_3)(OH){-}CH_2\overset{O}{\overset{\|}{C}}S{-}CoA}$$

$$\xrightarrow[2\ NADP^+ + CoA{-}SH]{2\ NADPH + 2\ H^+}$$

$$\underset{\substack{(C_6)\\ \textbf{Mevalonic acid}}}{HO\overset{O}{\overset{\|}{C}}CH_2{-}C(H_3C)(OH){-}CH_2CH_2OH}$$

$$\xrightarrow[3\ ADP]{3\ ATP}\ \text{Three successive steps}$$

$$\underset{\substack{(C_6)\\ \textbf{3-Phospho-5-pyrophosphomevalonic acid}}}{H{-}O{-}\overset{O}{\overset{\|}{C}}{-}CH_2{-}C(H_3C)(OPO_3H^-){-}CH_2CH_2O{-}\overset{O}{\overset{\|}{P}}(O^-){-}O{-}\overset{O}{\overset{\|}{P}}(O^-){-}O^-}$$

$$\xrightarrow{} CO_2 + H_2PO_3^-$$

$$\underset{\substack{(C_5)\\ \textbf{3-Methyl-3-butenyl pyrophosphate}}}{(H_2C{=})(H_3C)C{-}CH_2CH_2{-}O{-}\overset{O}{\overset{\|}{P}}(O^-){-}O{-}\overset{O}{\overset{\|}{P}}(O^-){-}O^-}$$

The first step of this synthetic pathway is straightforward. Acetyl-CoA (from 1 mol of acetate) and acetoacetyl-CoA (from 2 mol of acetate) condense to form the C_6 compound, β-hydroxy-β-methylglutaryl-CoA. This step is followed by an enzymatic reduction of the thiol ester group of β-hydroxy-β-methylglutaryl-CoA to the primary alcohol of mevalonic acid. The enzyme that catalyzes this step is called HMG-CoA reductase (HMG is β-hydroxy-β-methylglutaryl), and this step is the rate-limiting step in cholesterol biosynthesis. The key to finding this pathway was the discovery that mevalonic acid was an intermediate and that this C_6 compound could be transformed into the five-carbon 3-methyl-3-butenyl pyrophosphate by successive phosphorylations and decarboxylation.

As we saw earlier (Section D.3), 3-methyl-3-butenyl pyrophosphate isomerizes to produce an equilibrium mixture that contains 3-methyl-2-butenyl pyrophosphate, and these two compounds condense to form geranyl pyrophosphate, a C_{10} compound. Geranyl pyrophosphate subsequently condenses with another mole of 3-methyl-3-butenyl pyrophosphate to form farnesyl pyrophosphate, a C_{15} compound. (Geranyl pyrophosphate and farnesyl pyrophosphate are the precursors of the mono- and sesquiterpenes; see Section D.3.)

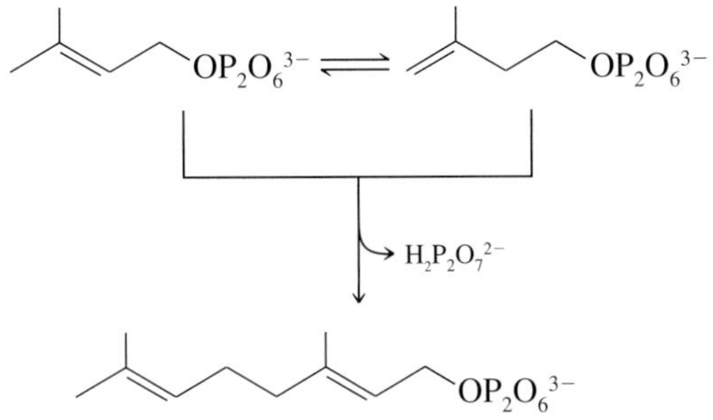

$OP_2O_6^{3-}$ ⇌ $OP_2O_6^{3-}$

→ $H_2P_2O_7^{2-}$

$OP_2O_6^{3-}$

Geranyl pyrophosphate

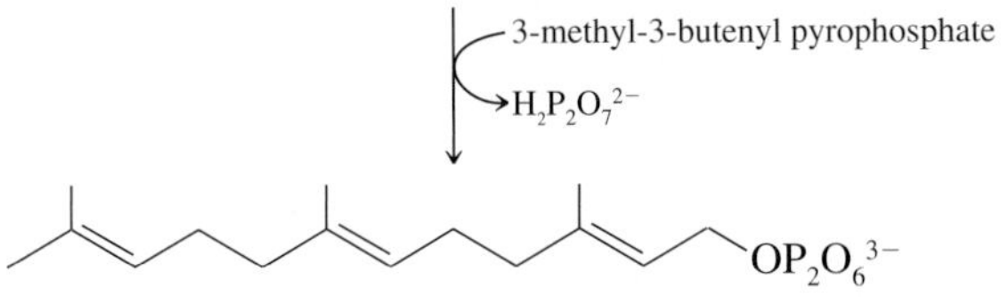

3-methyl-3-butenyl pyrophosphate

→ $H_2P_2O_7^{2-}$

$OP_2O_6^{3-}$

Farnesyl pyrophosphate

Two molecules of farnesyl pyrophosphate then undergo a reductive condensation to produce squalene:

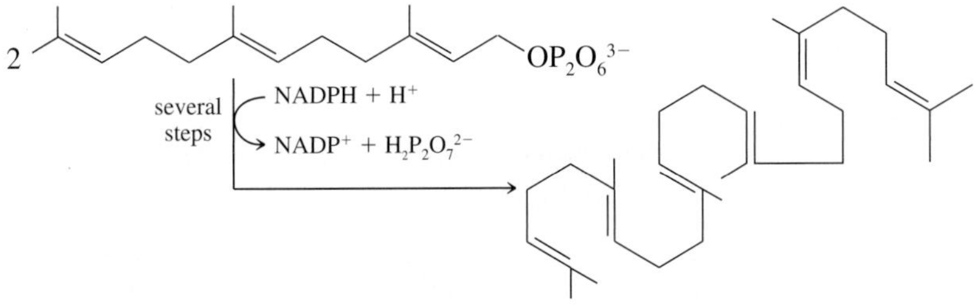

2 $OP_2O_6^{3-}$

several steps: NADPH + H^+ → $NADP^+$ + $H_2P_2O_7^{2-}$

Squalene

Squalene is the direct precursor of cholesterol. Oxidation of squalene yields squalene 2,3-epoxide, which undergoes a remarkable series of ring closures accompanied by concerted methanide and hydride migrations to yield lanosterol. This process was discussed in detail in "The Chemistry of . . . Cholesterol Biosynthesis" in Chapter 8. Lanosterol is then converted to cholesterol through a series of enzyme-catalyzed reactions:

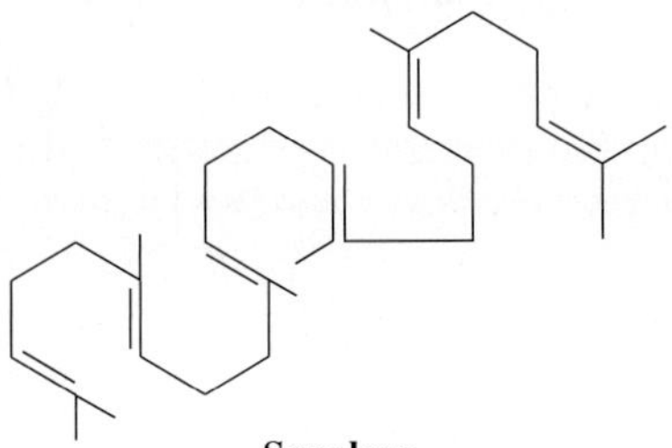

Squalene

↓

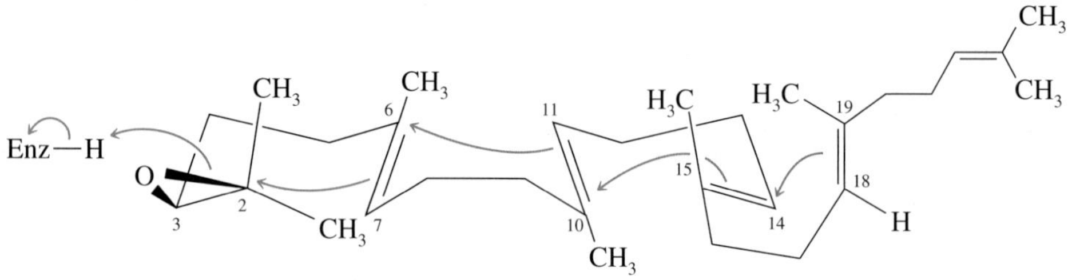

Enz—H, O, CH$_3$, CH$_3$, CH$_3$, CH$_3$, H$_3$C, H$_3$C, CH$_3$, CH$_3$, H, 2, 3, 6, 7, 10, 11, 14, 15, 18, 19

(3*S*)-2,3-Oxidosqualene

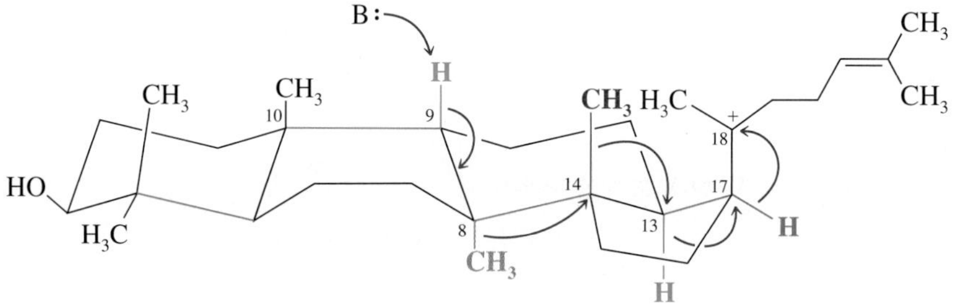

B:, H, HO, H$_3$C, CH$_3$, CH$_3$, CH$_3$, H$_3$C, CH$_3$, CH$_3$, CH$_3$, H, H, 8, 9, 10, 13, 14, 17, 18

Protosteryl cation

(−BH)

HO, H$_3$C, CH$_3$, CH$_3$, CH$_3$, CH$_3$, CH$_3$, CH$_3$, CH$_3$, H, H, 8, 9, 10, 13, 14, 17, 18

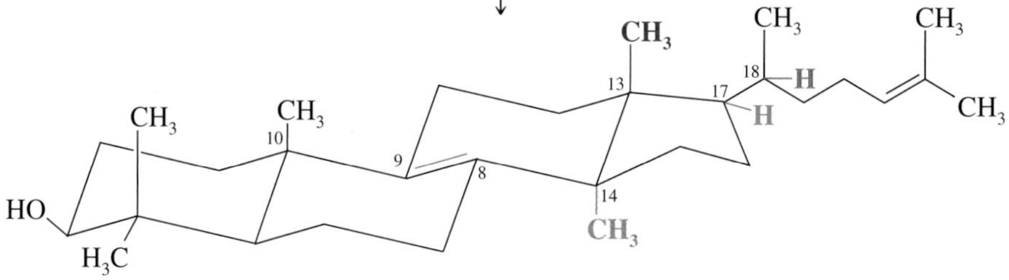

Lanosterol

(a number of enzymatic steps)

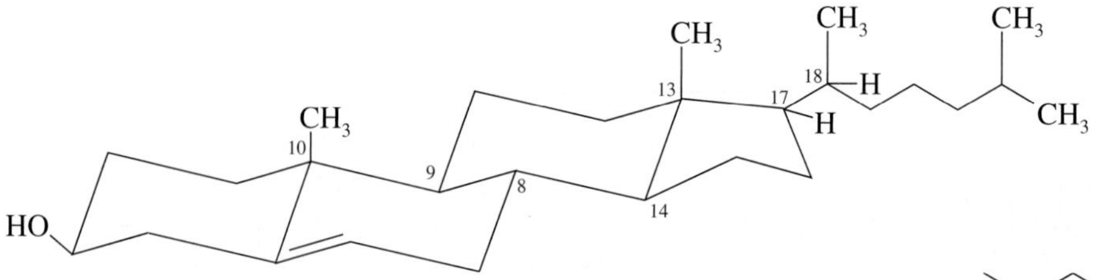

HO, CH$_3$, CH$_3$, CH$_3$, CH$_3$, CH$_3$, H, H, 8, 9, 10, 13, 14, 17, 18

Cholesterol (represented by both structures)

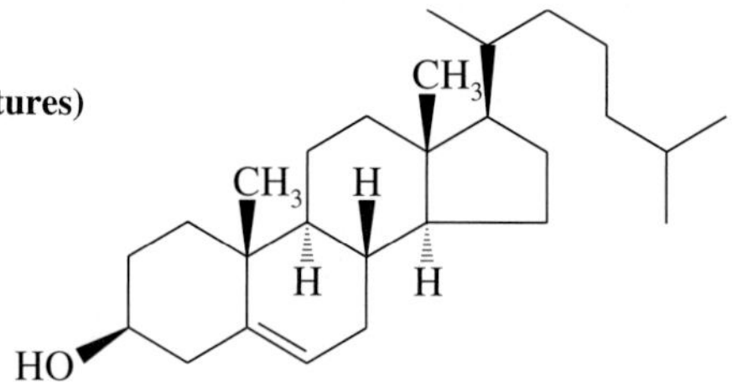

HO, CH$_3$, CH$_3$, H, H, H

Cholesterol

D.5 Cholesterol and Heart Disease

Because cholesterol is the precursor of steroid hormones and is a vital constituent of cell membranes, it is essential to life. On the other hand, deposition of cholesterol in arteries is a cause of heart disease and atherosclerosis, two leading causes of death in humans. For an organism to remain healthy, there has to be an intricate balance between the biosynthesis of cholesterol and its utilization, so that arterial deposition is kept at a minimum.

For some individuals with high blood levels of cholesterol, the remedy is as simple as following a diet low in cholesterol and in fat. For those who suffer from the genetic disease **familial hypercholesterolemia** (FH), other means of blood cholesterol reduction are required. One remedy involves using the drug *lovastatin* (also called *mevinolin*):

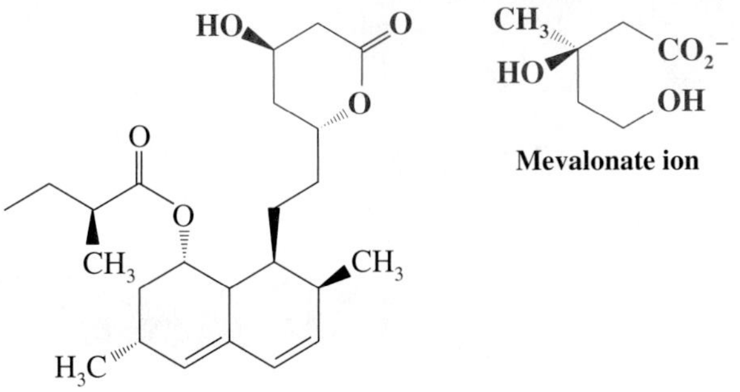

HO, O, O, O, O, CH_3, CH_3, H_3C

Lovastatin

CH_3, HO, CO_2^-, OH

Mevalonate ion

Lovastatin, because part of its structure resembles mevalonate ion, can apparently bind at the active site of HMG-CoA reductase (Section D.4), the enzyme that catalyzes the rate-limiting step in cholesterol biosynthesis. Lovastatin acts as a competitive inhibitor of this enzyme and thereby reduces cholesterol synthesis. Reductions of up to 30% in serum cholesterol levels are possible with lovastatin therapy.

Cholesterol synthesized in the liver either is converted to bile acids that are used in digestion or is esterified for transport by the blood. Cholesterol is transported in the blood, and taken up in cells, in the form of lipoprotein complexes named on the basis of their density. **Low-density lipoproteins (LDLs)** transport cholesterol from the liver to peripheral tissues. **High-density lipoproteins (HDLs)** transport cholesterol back to the liver, where surplus cholesterol is disposed of by the liver as bile acids. High-density lipoproteins have come to be called "good cholesterol" because high levels of HDL may reduce cholesterol deposits in arteries. Because high levels of LDL are associated with the arterial deposition of cholesterol that causes cardiovascular disease, it has come to be called "bad cholesterol."

Bile acids that flow from the liver to the intestines, however, are efficiently recycled to the liver. Recognition of this has led to another method of cholesterol reduction, namely, the ingestion of resins that bind bile acids and thereby prevent their reabsorption in the intestines.

CHAPTER 20

Amines

Neurotoxins and Neurotransmitters

Colombian poison dart frogs are tiny, beautiful, and deadly. They produce a poison called histrionicotoxin, which is an amine that causes paralysis. Death from histrionicotoxin results by suffocation through paralysis of the victim's respiratory muscles. (A molecular model of histrionicotoxin is shown above.) Curare, the Amazonian arrow poison that is a mixture of compounds from a woody vine, contains another paralytic neurotoxin, called *d*-tubocurarine. Histrionicotoxin and *d*-tubocurarine both block the action of acetylcholine, an important neurotransmitter.

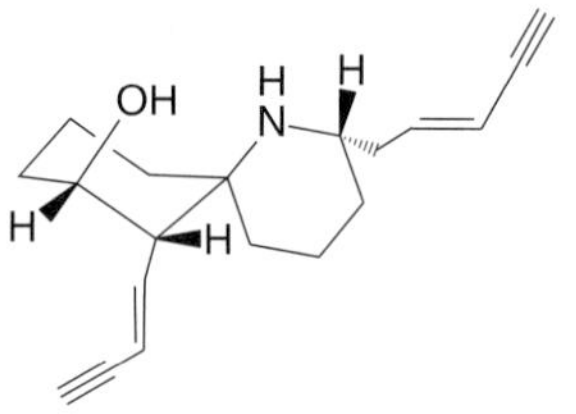

Histrionicotoxin

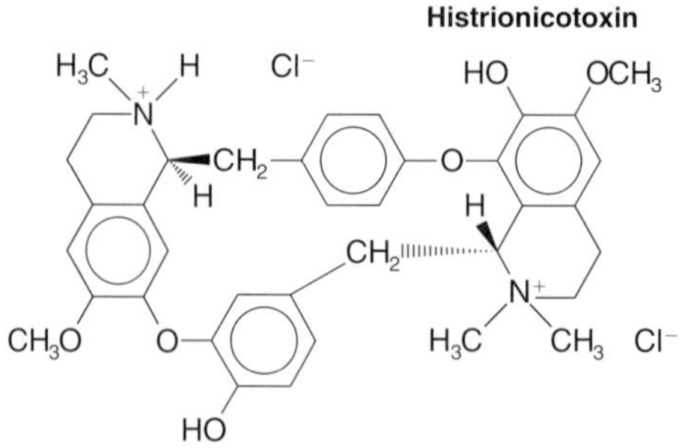

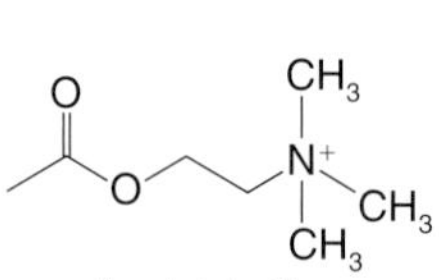

Acetylcholine

***d*-Tubocurarine chloride**

A key aspect of the structure of both *d*-tubocurarine and acetylcholine is that each contains a nitrogen atom (two in the case of *d*-tubocurarine) bearing four groups. This feature gives both molecules a formal positive charge on their nitrogen atoms and places them in a class called quaternary ammonium salts (Section 20.3D). The presence of the quaternary ammonium group is important for binding at the acetylcholine receptor. During normal transmission of a nerve impulse, two acetylcholine molecules bind at the receptor. This causes a conformational change that opens a channel for Na^+ and K^+ cations to diffuse in and out of the cell, respectively, depolarizing the membrane. About 20,000 cations of each type pass through the membrane over 2 ms. When *d*-tubocurarine blocks the binding of acetylcholine by binding at the acetylcholine receptor site, it prevents opening of the ion channel. Inability to depolarize the membrane (initiate a nerve impulse) results in paralysis.

Even though *d*-tubocurarine and histrionicotoxin are deadly poisons, both have been useful in research. For example, experiments in respiratory physiology that require absence of normal breathing patterns have involved curare-induced temporary (and voluntary!) respiratory paralysis of a researcher. While the experiment is underway and until the effects of the curare are reversed, the researcher is kept alive by a hospital respirator. In similar fashion, *d*-tubocurarine, as well as succinylcholine bromide, is used as a muscle relaxant during some surgeries.

Br⁻ H_3C H_3C—N^+ CH_3 O O O O CH_3 N^+ Br^- CH_3 CH_3

Succinylcholine bromide

20.1 NOMENCLATURE

In common nomenclature most primary amines are named as *alkylamines.* In systematic nomenclature (blue names in parentheses below) they are named by adding the suffix *-amine* to the name of the chain or ring system to which the NH_2 group is attached with replacement of the final *-e.* Amines are classified as being **primary (1°), secondary (2°),** or **tertiary (3°)** on the basis of the number of organic groups attached to the nitrogen (Section 2.9).

Primary Amines

CH_3NH_2
Methylamine
(methanamine)

$CH_3CH_2NH_2$
Ethylamine
(ethanamine)

$CH_3CHCH_2NH_2$ with CH_3 on C-2
Isobutylamine
(2-methyl-1-propanamine)

Cyclohexyl—NH_2
Cyclohexylamine
(cyclohexanamine)

Most secondary and tertiary amines are named in the same general way. In common nomenclature we either designate the organic groups individually if they are different or use the prefixes di- or tri- if they are the same. In systematic nomenclature we use the locant *N* to designate substituents attached to a nitrogen atom.

Secondary Amines

$CH_3NHCH_2CH_3$
Ethylmethylamine
(*N*-methylethanamine)

$(CH_3CH_2)_2NH$
Diethylamine
(*N*-ethylethanamine)

Tertiary Amines

$(CH_3CH_2)_3N$
Triethylamine
(*N*,*N*-diethylethanamine)

$CH_3N(CH_2CH_3)CH_2CH_2CH_3$
Ethylmethylpropylamine
(*N*-ethyl-*N*-methyl-1-propanamine)

In the IUPAC system, the substituent $—NH_2$ is called the *amino* group. We often use this system for naming amines containing an OH group or a CO_2H group:

$H_2NCH_2CH_2OH$
2-Aminoethanol

$H_2NCH_2CH_2COH$ (C=O)
3-Aminopropanoic acid

20.1A Arylamines

Some common **arylamines** have the following names:

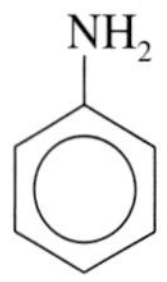

NH_2 on benzene ring	$NHCH_3$ on benzene ring	NH_2 / CH_3 on benzene ring (para)	NH_2 / OCH_3 on benzene ring (para)
Aniline (benzenamine)	***N*-Methylaniline** (*N*-methylbenzenamine)	***p*-Toluidine** (4-methylbenzenamine)	***p*-Anisidine** (4-methoxybenzenamine)

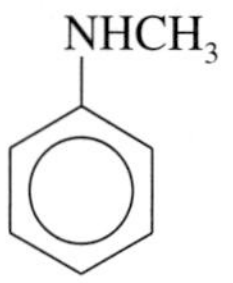

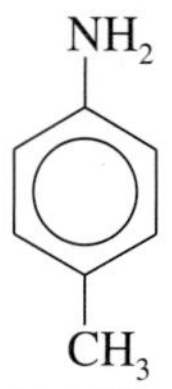

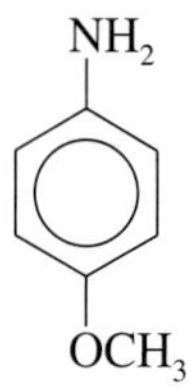

20.1B Heterocyclic Amines

The important **heterocyclic amines** all have common names. In systematic replacement nomenclature the prefixes *aza-*, *diaza-*, and *triaza-* are used to indicate that nitrogen atoms have replaced carbon atoms in the corresponding hydrocarbon. A nitrogen atom in the ring (or the highest atomic weight heteroatom, as in the case of thiazole) is designated position 1 and numbering proceeds to give the lowest overall set of locants to the heteroatoms:

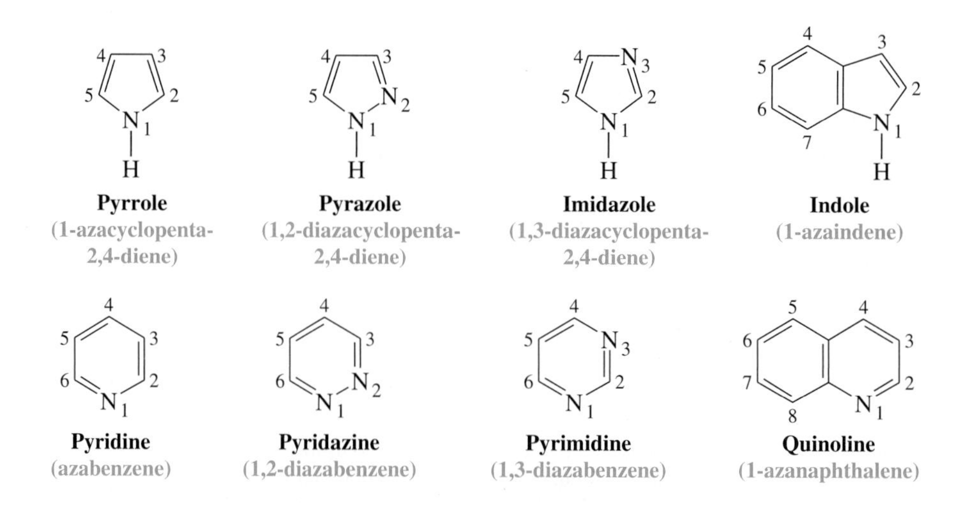

Piperidine (azacyclohexane)

Pyrrolidine (azacyclopentane)

Thiazole (1-thia-3-azacyclopenta-2,4-diene)

Purine

20.2 Physical Properties and Structure of Amines

20.2A Physical Properties

Amines are moderately polar substances; they have boiling points that are higher than those of alkanes but generally lower than those of alcohols of comparable molecular weight. Molecules of primary and secondary amines can form strong hydrogen bonds to each other and to water. Molecules of tertiary amines cannot form hydrogen bonds to each other, but they can form hydrogen bonds to molecules of water or other hydroxylic solvents. As a result, tertiary amines generally boil at lower temperatures than primary and secondary amines of comparable molecular weight, but all low-molecular-weight amines are very water soluble.

Table 20.1 lists the physical properties of some common amines.

TABLE 20.1 Physical Properties of Amines

Name	Structure	mp (°C)	bp (°C)	Water Solubility (25°C) (g 100 mL^{-1})	pK_a (aminium ion)
Primary Amines					
Methylamine	CH_3NH_2	−94	−6	Very soluble	10.64
Ethylamine	$CH_3CH_2NH_2$	−81	17	Very soluble	10.75
Propylamine	$CH_3CH_2CH_2NH_2$	−83	49	Very soluble	10.67
Isopropylamine	$(CH_3)_2CHNH_2$	−101	33	Very soluble	10.73
Butylamine	$CH_3(CH_2)_2CH_2NH_2$	−51	78	Very soluble	10.61
Isobutylamine	$(CH_3)_2CHCH_2NH_2$	−86	68	Very soluble	10.49
sec-Butylamine	$CH_3CH_2CH(CH_3)NH_2$	−104	63	Very soluble	10.56
tert-Butylamine	$(CH_3)_3CNH_2$	−68	45	Very soluble	10.45
Cyclohexylamine	Cyclo-$C_6H_{11}NH_2$	−18	134	Slightly soluble	10.64
Benzylamine	$C_6H_5CH_2NH_2$	10	185	Slightly soluble	9.30
Aniline	$C_6H_5NH_2$	−6	184	3.7	4.58
p-Toluidine	*p*-$CH_3C_6H_4NH_2$	44	200	Slightly soluble	5.08
p-Anisidine	*p*-$CH_3OC_6H_4NH_2$	57	244	Very slightly soluble	5.30
p-Chloroaniline	*p*-$ClC_6H_4NH_2$	73	232	Insoluble	4.00
p-Nitroaniline	*p*-$NO_2C_6H_4NH_2$	148	332	Insoluble	1.00
Secondary Amines					
Dimethylamine	$(CH_3)_2NH$	−92	7	Very soluble	10.72
Diethylamine	$(CH_3CH_2)_2NH$	−48	56	Very soluble	10.98
Dipropylamine	$(CH_3CH_2CH_2)_2NH$	−40	110	Very soluble	10.98
N-Methylaniline	$C_6H_5NHCH_3$	−57	196	Slightly soluble	4.70
Diphenylamine	$(C_6H_5)_2NH$	53	302	Insoluble	0.80
Tertiary Amines					
Trimethylamine	$(CH_3)_3N$	−117	2.9	Very soluble	9.70
Triethylamine	$(CH_3CH_2)_3N$	−115	90	14	10.76
Tripropylamine	$(CH_3CH_2CH_2)_3N$	−93	156	Slightly soluble	10.64
N,*N*-Dimethylaniline	$C_6H_5N(CH_3)_2$	3	194	Slightly soluble	5.06

20.2B Structure of Amines

The nitrogen atom of most amines is like that of ammonia; it is approximately sp^3 hybridized. The three alkyl groups (or hydrogen atoms) occupy corners of a tetrahedron; the sp^3 orbital containing the unshared electron pair is directed toward the other corner. We describe the shape of the amine by the location of the atoms as being **trigonal pyramidal** (Section 1.16). However, if we were to consider the unshared electron pair as being a group we would describe the geometry of the amine as being tetrahedral. The electrostatic potential map for the van der Waals surface of trimethylamine indicates localization of negative charge where the nonbonding electrons are found on the nitrogen:

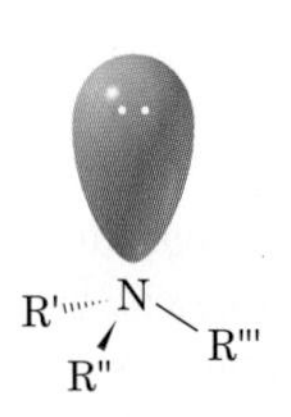

Structure of an amine

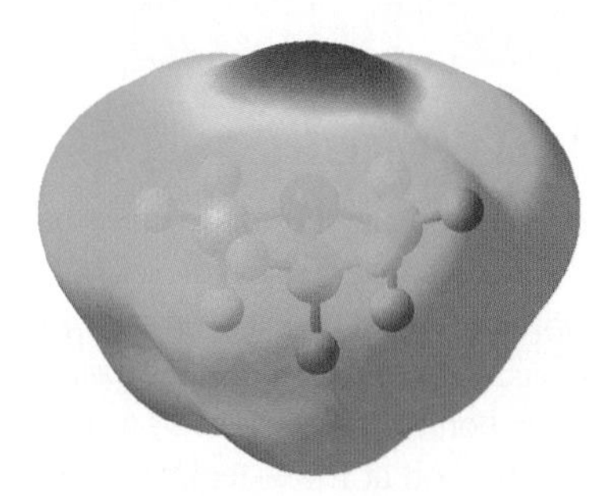

A calculated structure for trimethylamine
The electrostatic potential map shows charge associated with the nitrogen unshared electron pair.

The bond angles are what one would expect of a tetrahedral structure; they are very close to 109.5°. The bond angles for trimethylamine, for example, are 108°.

If the alkyl groups of a tertiary amine are all different, the amine will be chiral. There will be two enantiomeric forms of the tertiary amine, and, theoretically, we ought to be able to resolve (separate) these enantiomers. In practice, however, resolution is usually impossible because the enantiomers interconvert rapidly:

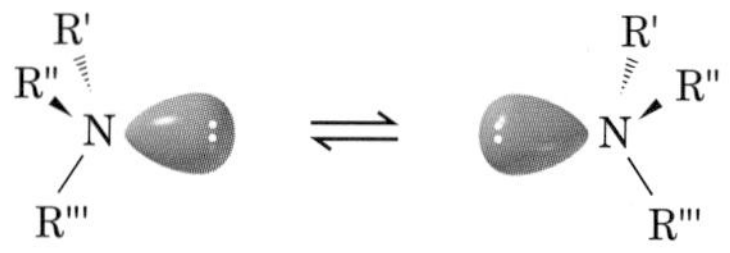

Interconversion of amine enantiomers

This interconversion occurs through what is called a **pyramidal** or **nitrogen inversion.** The barrier to the interconversion is about 25 kJ mol^{-1} for most simple amines, low enough to occur readily at room temperature. In the transition state for the inversion, the nitrogen atom becomes sp^2 hybridized with the unshared electron pair occupying a p orbital.

Ammonium salts cannot undergo inversion because they do not have an unshared pair. Therefore, those quaternary ammonium salts with four different groups are chiral and can be resolved into separate (relatively stable) enantiomers:

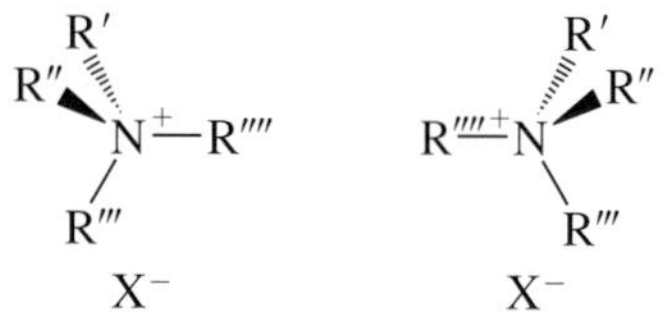

Quaternary ammonium salts such as these can be resolved.

20.3 Basicity of Amines: Amine Salts

Amines are relatively weak bases. They are stronger bases than water but are far weaker bases than hydroxide ions, alkoxide ions, and alkanide anions.

A convenient way to compare the base strengths of amines is to compare the acidity constants (or pK_a values) of their conjugate acids, the corresponding alkylaminium ions (Section 3.5C). The expression for this acidity constant is as follows:

$$R\overset{+}{N}H_3 + H_2O \rightleftharpoons RNH_2 + H_3O^+$$

$$K_a = \frac{[RHN_2][H_3O^+]}{[RNH_3^+]} \qquad pK_a = -\log K_a$$

The equilibrium for an amine that is relatively more basic will lie more toward the left in the above chemical equation than for an amine that is less basic. Consequently, the aminium ion from a more basic amine will have a smaller K_a (larger pK_a) than the aminium ion of a less basic amine.

When we compare aminium ion acidities in terms of this equilibrium, we see that most primary alkylaminium ions (RNH_3^+) are less acidic than ammonium ion (NH_4^+). In other words, primary alkylamines (RHN_2) are more basic than ammonia (NH_3):

	$\ddot{N}H_3$	$CH_3\ddot{N}H_2$	$CH_3CH_2\ddot{N}H_2$	$CH_3CH_2CH_2\ddot{N}H_2$
Conjugate acid pK_a	**9.26**	**10.64**	**10.75**	**10.67**

We can account for this on the basis of the electron-releasing ability of an alkyl group. An alkyl group releases electrons, and it *stabilizes* the alkylaminium ion that results from the acid–base reaction *by dispersing its positive charge.* It stabilizes the alkylaminium ion to a greater extent than it stabilizes the amine:

$$R{\rightarrow}\ddot{N}H_2 + H{-}\ddot{\underset{..}{O}}H \rightleftharpoons R{\rightarrow}\overset{+}{N}H_3 + {}^{-}{:}\ddot{\underset{..}{O}}H$$

By releasing electrons, R→ stabilizes the alkylaminium ion through dispersal of charge.

This explanation is supported by measurements showing that in the *gas phase* the basicities of the following amines increase with increasing methyl substitution:

$$(CH_3)_3N > (CH_3)_2NH > CH_3NH_2 > NH_3$$

This is not the order of basicity of these amines in aqueous solution, however. In aqueous solution (Table 20.1) the order is

$$(CH_3)_2NH > CH_3NH_2 > (CH_3)_3N > NH_3$$

The reason for this apparent anomaly is now known. In aqueous solution the aminium ions formed from secondary and primary amines are stabilized by solvation through hydrogen bonding much more effectively than are the aminium ions formed from tertiary amines. The aminium ion formed from a tertiary amine such as $(CH_3)_3NH^+$ has only one hydrogen to use in hydrogen bonding to water molecules, whereas the aminium ions from secondary and primary amines have two and three hydrogens, respectively. Poorer solvation of the aminium ion formed from a tertiary amine more than counteracts the electron-releasing effect of the three methyl groups and makes the tertiary amine less basic than primary and secondary amines in aqueous solution. The electron-releasing effect does, however, make the tertiary amine more basic than ammonia.

20.3A Basicity of Arylamines

When we examine the pK_a values of the aminium ions of aromatic amines (e.g., aniline and *p*-toluidine) in Table 20.1, we see that they are much weaker bases than the corresponding nonaromatic amine, cyclohexylamine:

	Cyclo-$C_6H_{11}NH_2$	$C_6H_5NH_2$	*p*-$CH_3C_6H_4NH_2$
Conjugate acid pK_a	**10.64**	**4.58**	**5.08**

We can account for this effect, in part, on the basis of resonance contributions to the overall hybrid of an arylamine. For aniline, the following contributors are important:

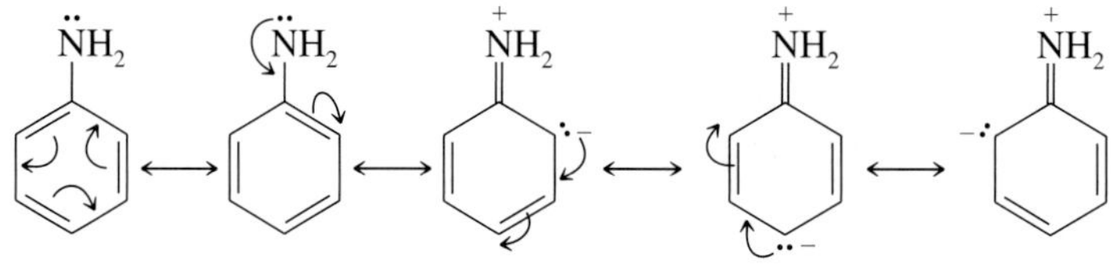

1 ⟷ **2** ⟷ **3** ⟷ **4** ⟷ **5**

Structures **1** and **2** are the Kekulé structures that contribute to any benzene derivative. Structures **3–5,** however, *delocalize* the unshared electron pair of the nitrogen over the ortho and para positions of the ring. This delocalization of the electron pair makes it less available to a proton, and *delocalization of the electron pair stabilizes aniline.*

When aniline accepts a proton it becomes an anilinium ion:

$$C_6H_5\ddot{N}H_2 + H_2O \rightleftharpoons C_6H_5\overset{+}{N}H_3 + \overset{-}{O}H$$

Anilinium ion

Once the electron pair of the nitrogen atom accepts the proton, it is no longer available to participate in resonance, and hence we are only able to write *two* resonance structures for the anilinium ion—the two Kekulé structures:

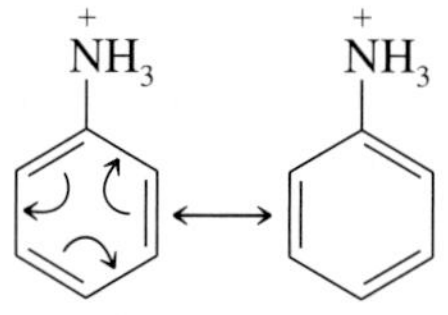

Structures corresponding to **3–5** are not possible for the anilinium ion, and, consequently, although resonance does stabilize the anilinium ion considerably, resonance does not stabilize the anilinium ion to as great an extent as it does aniline itself. This greater stabilization of the reactant (aniline) when compared to that of the product (anilinium ion) means that $\Delta H°$ for the reaction

$$\text{Aniline} + H_2O \longrightarrow \text{anilinium ion} + OH^-$$

will be a larger positive quantity than that for the reaction

$$\text{Cyclohexylamine} + H_2O \longrightarrow \text{cyclohexylaminium ion} + OH^-$$

(See Fig. 20.1.) Aniline, as a result, is the weaker base.

Another important effect in explaining the lower basicity of aromatic amines is the **electron-withdrawing effect of a phenyl group.** Because the carbon atoms of a phenyl group are sp^2 hybridized, they are more electronegative (and therefore more electron withdrawing) than the sp^3-hybridized carbon atoms of alkyl groups. We shall discuss this effect further in Section 21.5A.

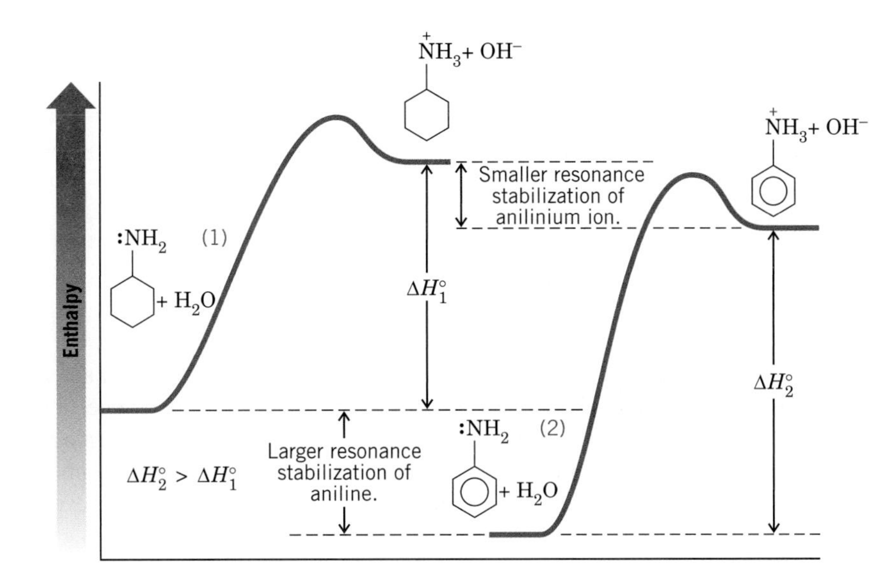

FIGURE 20.1 Enthalpy diagram for (1) the reaction of cyclohexylamine with H_2O and (2) the reaction of aniline with H_2O. (The curves are aligned for comparison only and are not to scale.)

20.3B Basicity of Heterocyclic Amines

Nonaromatic heterocyclic amines have basicities that are approximately the same as those of acyclic amines:

Conjugate acid pK_a	**Piperidine** pK_a = 11.20	**Pyrrolidine** pK_a = 11.11	**Diethylamine** pK_a = 10.98

In aqueous solution, aromatic heterocyclic amines such as pyridine, pyrimidine, and pyrrole are much weaker bases than nonaromatic amines or ammonia. (In the gas phase, however, pyridine and pyrrole are more basic than ammonia, indicating that solvation has a very important effect on their relative basicities; see Section 20.3.)

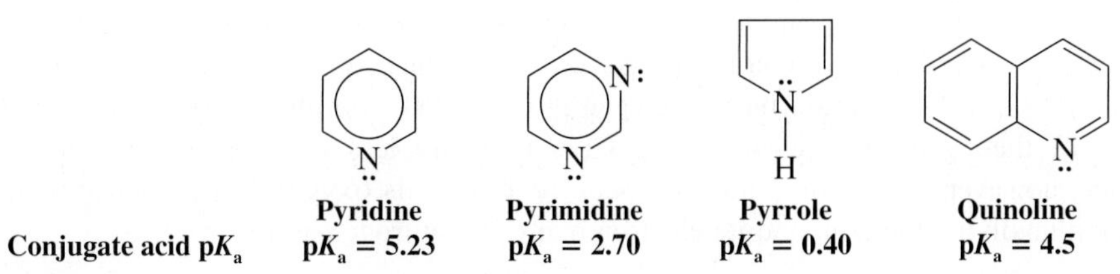

20.3C Amines versus Amides

Although amides are superficially similar to amines, they are far less basic (even less basic than arylamines). The pK_a of the conjugate acid of a typical amide is about zero.

Amides are not basic like amines.

This lower basicity of amides when compared to amines can also be understood in terms of resonance and inductive effects. An amide is stabilized by resonance involving the nonbonding pair of electrons on the nitrogen atom. However, an amide protonated on its nitrogen atom lacks this type of resonance stabilization. This is shown in the following resonance structures:

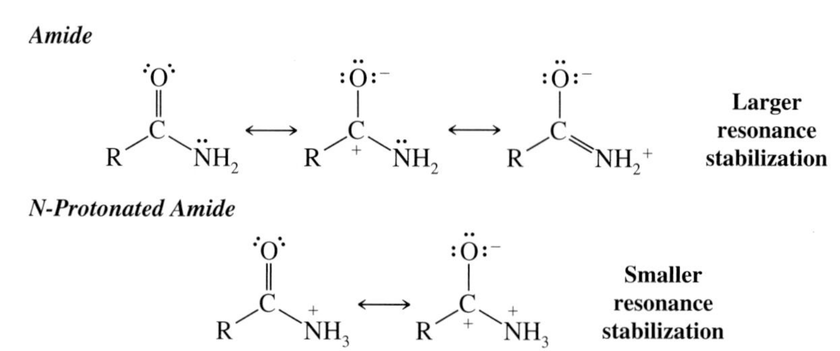

However, a more important factor accounting for amides being weaker bases than amines is the powerful electron-withdrawing effect of the carbonyl group of the amide. This effect is illustrated by the electrostatic potential maps for ethylamine and acetamide shown in Fig. 20.2. Significant negative charge is localized at the position of the nonbonding electron pair in ethylamine (as indicated by the red color). In acetamide, however, less negative charge resides near the nitrogen than in ethylamine.

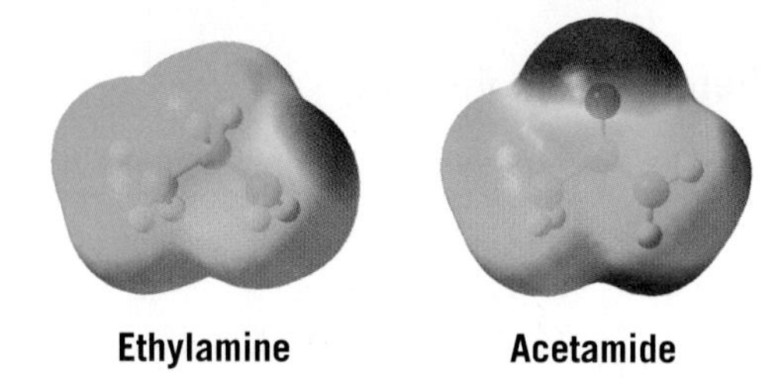

FIGURE 20.2 Calculated electrostatic potential maps (calibrated to the same charge scale) for ethylamine and acetamide. The map for ethylamine shows localization of negative charge at the unshared electron pair of nitrogen. The map for acetamide shows most of the negative charge at its oxygen atom instead of at nitrogen, due to the electron-withdrawing effect of the carbonyl group.

Comparing the following equilibria, the reaction with the amide lies more to the left than the corresponding reaction with an amine. This is consistent with the amine being a stronger base than an amide.

$$R-C(=O)-NH_2 + H_2O \rightleftharpoons R-C(=O)-\overset{+}{N}H_3 + \overset{-}{O}H$$

$$R \rightarrow \ddot{N}H_2 + H_2O \rightleftharpoons R-\overset{+}{N}H_3 + \overset{-}{O}H$$

The nitrogen atoms of amides are so weakly basic that when an amide accepts a proton, it does so on its oxygen atom instead (see the mechanism for hydrolysis of an amide, Section 18.8F). Protonation on the oxygen atom occurs even though oxygen atoms (because of their greater electronegativity) are typically less basic than nitrogen atoms. Notice, however, that if an amide accepts a proton on its oxygen atom, resonance stabilization involving the nonbonding electron pair of the nitrogen atom is possible:

$$R-C(=\overset{+}{O}H)-\ddot{N}H_2 \longleftrightarrow R-\overset{+}{C}(-\ddot{O}H)-\ddot{N}H_2 \longleftrightarrow R-C(-\ddot{O}H)=\overset{+}{N}H_2$$

20.3D Aminium Salts and Quaternary Ammonium Salts

When primary, secondary, and tertiary amines act as bases and react with acids, they form compounds called **aminium salts.** In an aminium salt the positively charged nitrogen atom is attached to at least one hydrogen atom:

$$CH_3CH_2\ddot{N}H_2 + HCl \xrightarrow[H_2O]{} CH_3CH_2\overset{+}{N}H_3\ Cl^-$$

Ethylaminium chloride (an aminium salt)

$$(CH_3CH_2)_2\ddot{N}H + HBr \xrightarrow[H_2O]{} (CH_3CH_2)_2\overset{+}{N}H_2\ Br^-$$

Diethylaminium bromide

$$(CH_3CH_2)_3\ddot{N} + HI \xrightarrow[H_2O]{} (CH_3CH_2)_3\overset{+}{N}H\ I^-$$

Triethylaminium iodide

When the central nitrogen atom of a compound is positively charged *but is not attached to a hydrogen atom,* the compound is called a **quaternary ammonium salt.** For example,

$$CH_3CH_2-\overset{\overset{\displaystyle CH_2CH_3}{|}}{\underset{\underset{\displaystyle CH_2CH_3}{|}}{N^+}}-CH_2CH_3\quad Br^-$$

Tetraethylammonium bromide (a quaternary ammonium salt)

Quaternary ammonium halides—because they do not have an unshared electron pair on the nitrogen atom—cannot act as bases:

$$(CH_3CH_2)_4\overset{+}{N}\ Br^-$$

Tetraethylammonium bromide (does not undergo reaction with acid)

Quaternary ammonium *hydroxides,* however, are strong bases. As solids, or in solution, they consist *entirely* of quaternary ammonium cations (R_4N^+) and hydroxide ions (OH^-); they are, therefore, strong bases—as strong as sodium or potassium hydroxide. Quaternary ammonium hydroxides react with acids to form quaternary ammonium salts:

$$(CH_3)_4\overset{+}{N}\ OH^- + HCl \longrightarrow (CH_3)_4\overset{+}{N}\ Cl^- + H_2O$$

In Section 11.16 we saw how quaternary ammonium salts can be used as phase-transfer catalysts. In Section 20.13A we shall see how they can form alkenes by a reaction called the *Hofmann elimination.*

20.3E Solubility of Amines in Aqueous Acids

Almost all alkylaminium chlorides, bromides, iodides, and sulfates are soluble in water. Thus, primary, secondary, or tertiary amines that are not soluble in water do dissolve in dilute aqueous HCl, HBr, HI, and H_2SO_4. Solubility in dilute acid provides a convenient chemical method for distinguishing amines from nonbasic compounds that are insoluble in water. Solubility in dilute acid also gives us a useful method for separating amines from nonbasic compounds that are insoluble in water. The amine can be extracted into aqueous acid (dilute HCl) and then recovered by making the aqueous solution basic and extracting the amine into ether or CH_2Cl_2.

You may make use of the basicity of amines in your organic chemistry laboratory work for the separation of compounds or for the characterization of unknowns.

$$-\ddot{N}: + H-X \text{ (or } H_2SO_4) \longrightarrow -\overset{+}{N}-H \quad X^- \text{ (or } HSO_4^-)$$

Water-insoluble amine → **Water-soluble aminium salt**

Because amides are far less basic than amines, water-insoluble amides *do not dissolve* in dilute aqueous HCl, HBr, HI, or H_2SO_4:

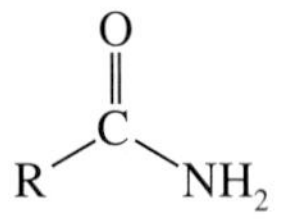

$$\mathrm{R{-}C({=}O){-}NH_2}$$

Water-insoluble amide
(not soluble in aqueous acids)

PROBLEM 20.1 Outline a procedure for separating hexylamine from cyclohexane using dilute HCl, aqueous NaOH, and diethyl ether.

PROBLEM 20.2 Outline a procedure for separating a mixture of benzoic acid, *p*-cresol, aniline, and benzene using acids, bases, and organic solvents.

20.3F Amines as Resolving Agents

Enantiomerically pure amines are often used to resolve racemic forms of acidic compounds by the formation of diastereomeric salts. We can illustrate the principles involved in this procedure by showing how a racemic form of an organic acid might be resolved (separated) into its enantiomers with the single enantiomer of an **amine as a resolving agent** (Fig. 20.3).

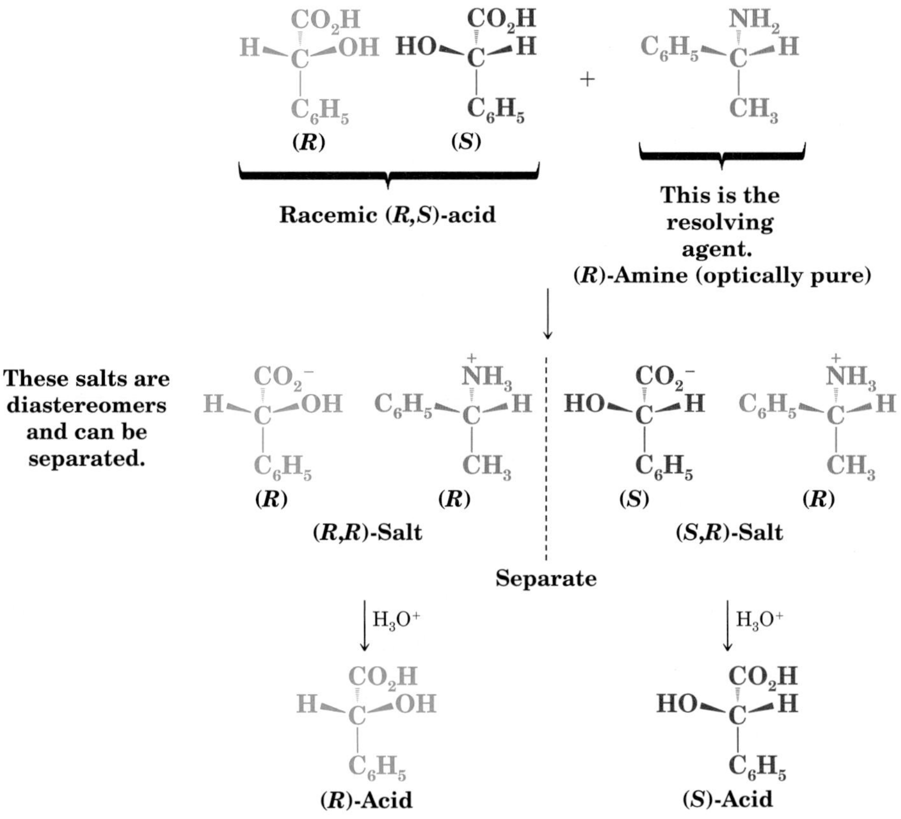

CO_2H, H, OH, C_6H_5 (***R***) — CO_2H, HO, H, C_6H_5 (***S***) — **Racemic (*R*,*S*)-acid**

+ NH_2, C_6H_5, H, CH_3 — **This is the resolving agent. (*R*)-Amine (optically pure)**

These salts are diastereomers and can be separated.

CO_2^-, H, OH, C_6H_5 (***R***) — $\overset{+}{N}H_3$, C_6H_5, H, CH_3 (***R***) — **(*R*,*R*)-Salt**

CO_2^-, HO, H, C_6H_5 (***S***) — $\overset{+}{N}H_3$, C_6H_5, H, CH_3 (***R***) — **(*S*,*R*)-Salt**

Separate

H_3O^+ ↓ CO_2H, H, OH, C_6H_5 — **(*R*)-Acid**

H_3O^+ ↓ CO_2H, HO, H, C_6H_5 — **(*S*)-Acid**

FIGURE 20.3 Resolution of the racemic form of an organic acid by the use of an optically active amine. Acidification of the separated diastereomeric salts causes the enantiomeric acids to precipitate (assuming they are insoluble in water) and leaves the resolving agent in solution as its conjugate acid.

In this procedure the single enantiomer of an amine, (*R*)-1-phenylethylamine, is added to a solution of the racemic form of an acid. The salts that form are *diastereomers.* The stereogenic centers of the acid portion of the salts are enantiomerically related to each other, but the stereogenic centers of the amine portion are not. The diastereomers have different solubilities and can be separated by careful crystallization. The separated salts are then acidified with hydrochloric acid and the enantiomeric acids are obtained from the separate solutions. The amine remains in solution as its hydrochloride salt.

Single enantiomers that are employed as resolving agents are often readily available from natural sources. Because most of the chiral organic molecules that occur in living organisms are synthesized by enzymatically catalyzed reactions, most of them occur as single enantiomers. Naturally occurring optically active amines such as (−)-quinine (Section 20.4), (−)-strychnine, and (−)-brucine are often employed as resolving agents for racemic acids. Acids such as (+)- or (−)-tartaric acid (Section 5.14B) are often used for resolving racemic bases.

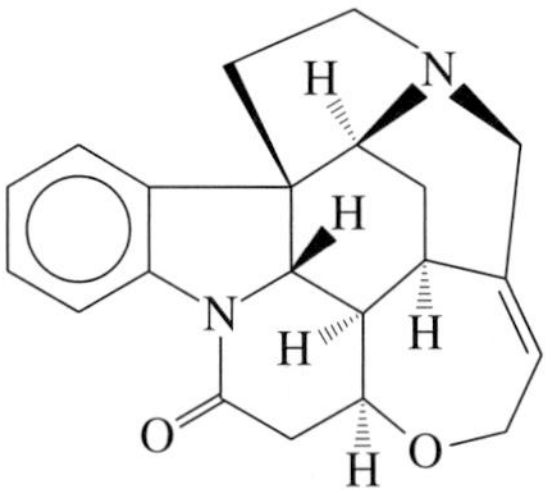

(−)-Strychnine 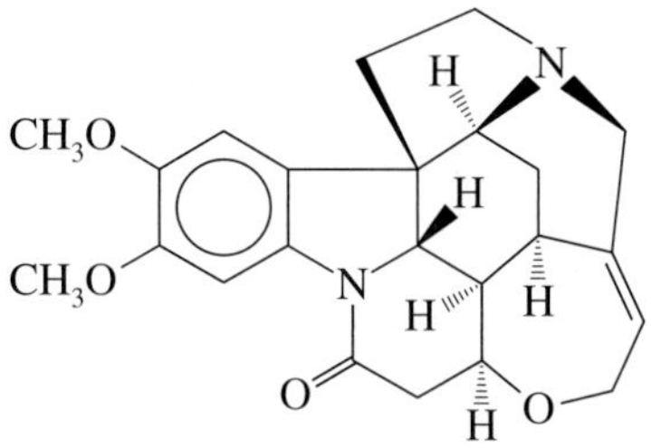 **(−)-Brucine**

The Chemistry of...

HPLC Resolution of Enantiomers

One technique for resolving racemates is based on high-performance liquid chromatography (HPLC) using a **chiral stationary phase** (CSP). This technique, developed by William H. Pirkle of the University of Illinois, has been used to resolve many racemic amines, alcohols, amino acids, and related compounds. We do not have the space here to discuss this technique in detail,* but suffice it to say that a solution of the racemate is passed through a column (called a **Pirkle column**) containing small silica microporous beads. Chemically attached to the surface of the beads is a chiral group such as the one that follows:

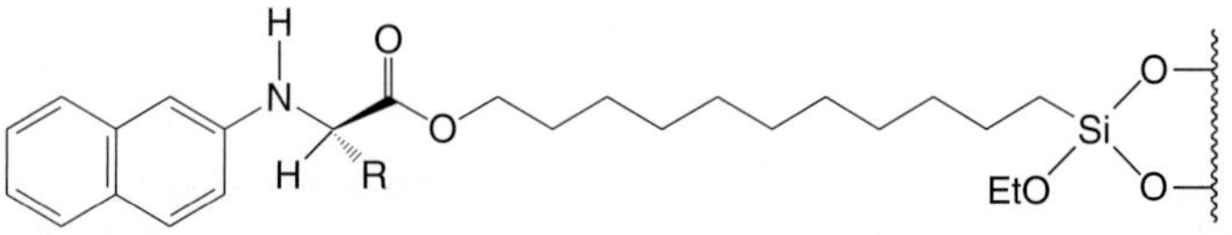

Silica

The compound to be resolved is first converted to a derivative containing a 3,5-dinitrophenyl group. An amine, for example, is converted to a 3,5-dinitrobenzamide:

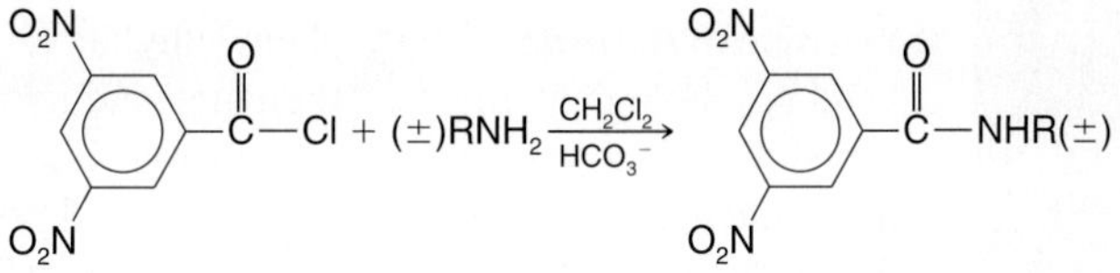

$$\text{3,5-}(O_2N)_2C_6H_3\text{—}\overset{O}{\overset{\|}{C}}\text{—Cl} + (\pm)RNH_2 \xrightarrow[HCO_3^-]{CH_2Cl_2} \text{3,5-}(O_2N)_2C_6H_3\text{—}\overset{O}{\overset{\|}{C}}\text{—NHR}(\pm)$$

*You may want to consult a laboratory manual or read the following article: Pirkle, W. H.; Pochapsky, T. C.; Mahler, G. S.; Corey, D. E.; Reno, D. S.; Alessi, D. M. *J. Org. Chem.* **1986,** *51,* 4991–5000.

An alcohol is converted to a carbamate (Section 18.9A) through a variation of the Curtius rearrangement (Section 20.5E):

$$3,5\text{-}(O_2N)_2C_6H_3\text{—}\overset{O}{\overset{\|}{C}}\text{—}N_3 + (\pm)ROH \xrightarrow[\text{reflux}]{\text{toluene}} 3,5\text{-}(O_2N)_2C_6H_3\text{—}NH\text{—}\overset{O}{\overset{\|}{C}}\text{—}OR(\pm)$$

The stationary phase, because it is chiral, binds one enantiomer much more tightly than the other. This binding increases the retention time of that enantiomer and permits separation. The binding comes partially from hydrogen-bonding interactions between the derivative and the CSP, but highly important is a $\pi-\pi$ interaction between the electron-deficient 3,5-dinitrophenyl ring of the derivative and the electron-rich naphthalene ring of the CSP.

20.4 Some Biologically Important Amines

A large number of medically and biologically important compounds are amines. Listed here are some important examples:

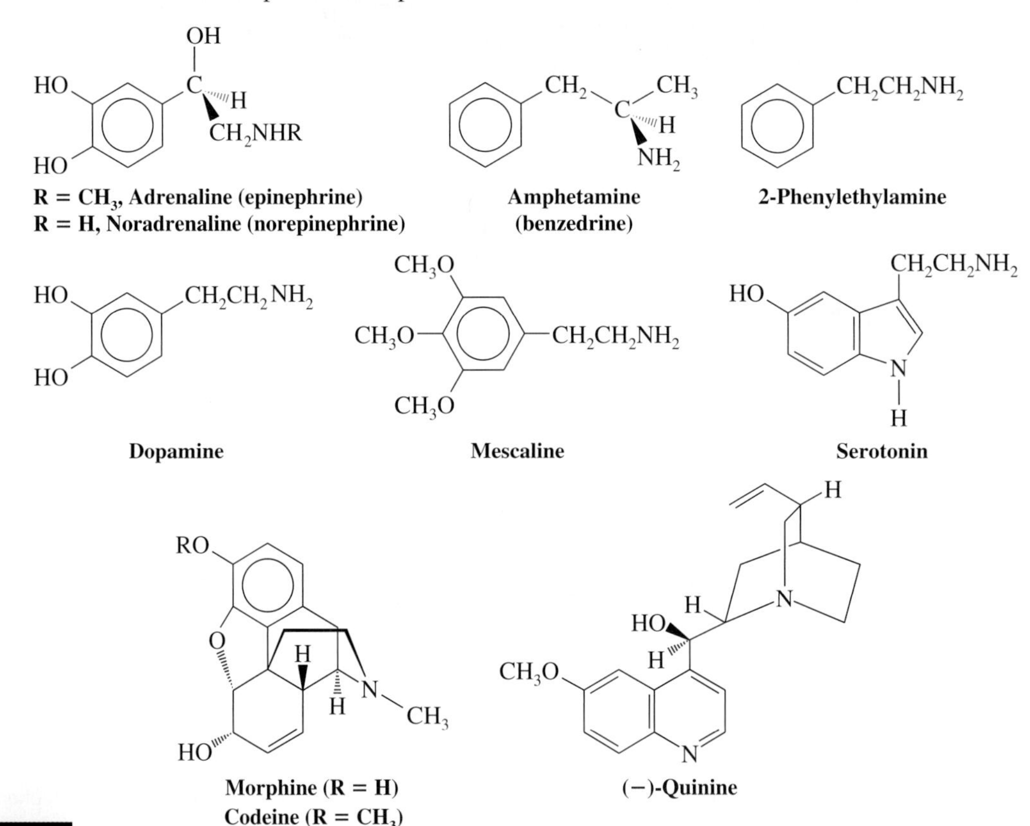

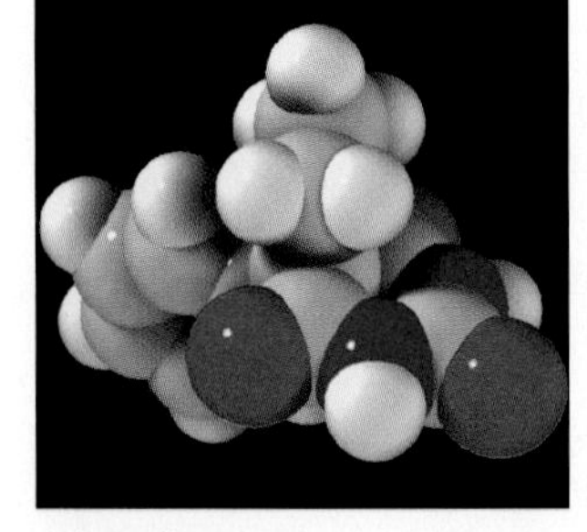
Adrenaline

2-Phenylethylamines Many phenylethylamine compounds have powerful physiological and psychological effects. Adrenaline and noradrenaline are two hormones secreted in the medulla of the adrenal gland. Released into the bloodstream when an animal senses danger, adrenaline causes an increase in blood pressure, a strengthening of the heart rate, and a widening of the passages of the lungs. All of these effects prepare the animal to fight or to flee. Noradrenaline also causes an increase in blood pressure, and it is involved in the transmission of impulses from the end of one nerve fiber to the next. Dopamine and serotonin are important neurotransmitters in the brain. Abnormalities in the level of dopamine in the brain are associated with many psychiatric disorders, including

Parkinson's disease. Dopamine plays a pivotal role in the regulation and control of movement, motivation, and cognition. Serotonin is a compound of particular interest because it appears to be important in maintaining stable mental processes. It has been suggested that the mental disorder schizophrenia may be connected with abnormalities in the metabolism of serotonin.

Amphetamine (a powerful stimulant) and mescaline (a hallucinogen) have structures similar to those of serotonin, adrenaline, and noradrenaline. They are all derivatives of 2-phenylethylamine (see structure given). (In serotonin the nitrogen is connected to the benzene ring to create a five-membered ring.) The structural similarities of these compounds must be related to their physiological and psychological effects because many other compounds with similar properties are also derivatives of 2-phenylethylamine. Examples (not shown) are *N*-methylamphetamine and LSD (lysergic acid diethylamide). Even morphine (see the Chapter 13 opening molecular graphic) and codeine, two powerful analgesics, have a 2-phenylethylamine system as a part of their structures. [Morphine and codeine are examples of compounds called alkaloids (Special Topic E). Try to locate the 2-phenylethylamine system in their structures.]

Vitamins and Antihistamines A number of amines are vitamins. These include nicotinic acid and nicotinamide (the antipellagra factors, see the Chapter 12 opening vignette), pyridoxine (vitamin B_6, see the Chapter 16 opening vignette), and thiamine chloride (vitamin B_1, see "The Chemistry of... Thiamine," Chapter 18). Nicotine is a toxic alkaloid found in tobacco that makes smoking habit forming. Histamine, another toxic amine, is found bound to proteins in nearly all tissues of the body. Release of free histamine causes the symptoms associated with allergic reactions and the common cold. Chlorpheniramine, an "antihistamine," is an ingredient of many over-the-counter cold remedies.

Pyridoxine (vitamin B_6) 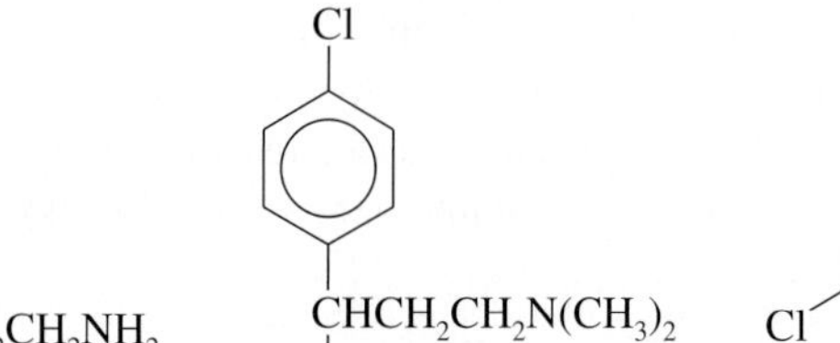 — **Thiamine chloride (vitamin B_1)** 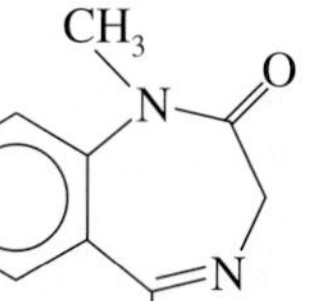 — **Nicotine** — **Nicotinic acid (niacin)**

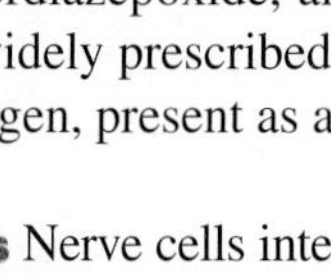

Histamine 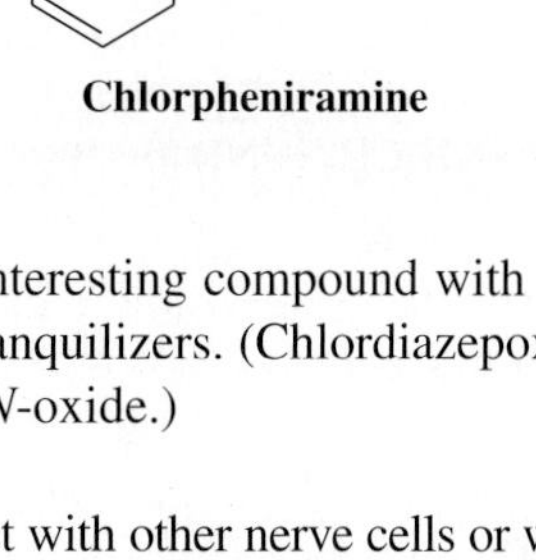 — **Chlorpheniramine** 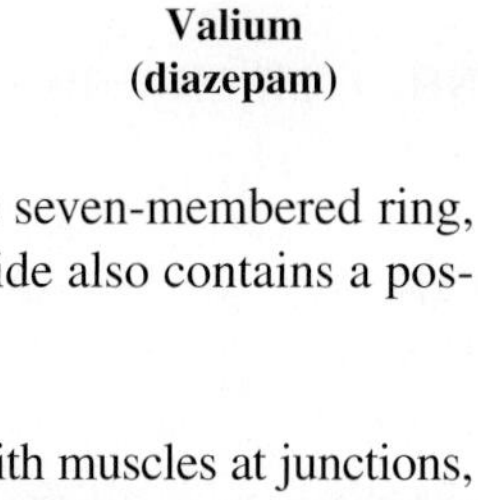 — **Valium (diazepam)**

Tranquilizers Chlordiazepoxide, an interesting compound with a seven-membered ring, is one of the most widely prescribed tranquilizers. (Chlordiazepoxide also contains a positively charged nitrogen, present as an *N*-oxide.)

Neurotransmitters Nerve cells interact with other nerve cells or with muscles at junctions, or gaps, called **synapses.** Nerve impulses are carried across the synaptic gap by chemical compounds called *neurotransmitters*. Acetylcholine (see the following reaction) is an important neurotransmitter at neuromuscular synapses called *cholinergic synapses*. Acetylcholine contains a quaternary ammonium group. Being small and ionic, acetylcholine is highly soluble in water and highly diffusible, qualities that suit its role as a neurotransmitter. Acetylcholine molecules are released by the presynaptic membrane in the neuron in packets of about 10^4 molecules. The packet of molecules then diffuses across the synaptic gap.

$$\underset{\textbf{Acetylcholine}}{CH_3\overset{O}{\overset{\|}{C}}OCH_2CH_2\overset{+}{N}(CH_3)_3} + H_2O \xrightleftharpoons{\text{acetylcholinesterase}} CH_3CO_2H + \underset{\textbf{Choline}}{HOCH_2CH_2\overset{+}{N}(CH_3)_3}$$

Having carried a nerve impulse across the synapse to the muscle where it triggers an electrical response, the acetylcholine molecules must be hydrolyzed (to choline) within a few milliseconds to allow the arrival of the next impulse. This hydrolysis is catalyzed by an enzyme of almost perfect efficiency called *acetylcholinesterase*.

The acetylcholine receptor on the postsynaptic membrane of muscle is the target for some of the most deadly neurotoxins. Some of these were discussed in the chapter opening vignette.

20.5 Preparation of Amines

In this section we will discuss a variety of ways to synthesize amines. Some of these methods will be new to you, while others are methods you have studied earlier in the context of related functional groups and reactions. Later, in Chapter 24, you shall see how some of the methods presented here, as well as some others for asymmetric synthesis, can be used to synthesize α-amino acids, the building blocks of peptides and proteins.

20.5A Through Nucleophilic Substitution Reactions

Alkylation of Ammonia Salts of primary amines can be prepared from ammonia and alkyl halides by nucleophilic substitution reactions. Subsequent treatment of the resulting aminium salts with a base gives primary amines:

$$\ddot{N}H_3 + R{-}X \longrightarrow R{-}\overset{+}{N}H_3\ X^- \xrightarrow{OH^-} RNH_2$$

This method is of very limited synthetic application because multiple alkylations occur. When ethyl bromide reacts with ammonia, for example, the ethylaminium bromide that is produced initially can react with ammonia to liberate ethylamine. Ethylamine can then compete with ammonia and react with ethyl bromide to give diethylaminium bromide. Repetitions of alkylation and proton transfer reactions ultimately produce some tertiary amines and even some quaternary ammonium salts if the alkyl halide is present in excess.

A Mechanism for the Reaction

Alkylation of NH_3

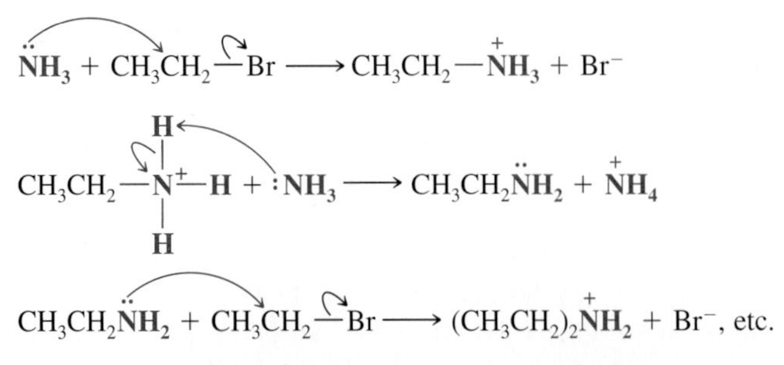

$$\ddot{N}H_3 + CH_3CH_2{-}Br \longrightarrow CH_3CH_2{-}\overset{+}{N}H_3 + Br^-$$

$$CH_3CH_2{-}\overset{+}{N}H_3 + {:}NH_3 \longrightarrow CH_3CH_2\ddot{N}H_2 + \overset{+}{N}H_4$$

$$CH_3CH_2\ddot{N}H_2 + CH_3CH_2{-}Br \longrightarrow (CH_3CH_2)_2\overset{+}{N}H_2 + Br^-,\ \text{etc.}$$

Multiple alkylations can be minimized by using a large excess of ammonia. (Why?) An example of this technique can be seen in the synthesis of alanine from 2-bromopropanoic acid:

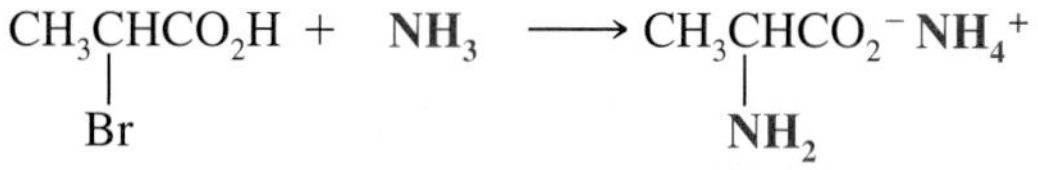

$$\underset{\textbf{(1 mol)}}{CH_3\underset{Br}{C}HCO_2H} + \underset{\textbf{(70 mol)}}{NH_3} \longrightarrow \underset{\textbf{Alanine (65–70\%)}}{CH_3\underset{NH_2}{C}HCO_2^- NH_4^+}$$

Alkylation of Azide Ion and Reduction A much better method for preparing a primary amine from an alkyl halide is first to convert the alkyl halide to an alkyl azide ($R—N_3$) by a nucleophilic substitution reaction:

$$R—X + \underset{\textbf{Azide ion (A good nucleophile)}}{^-N{=}N^+{=}N^-} \xrightarrow[(-X^-)]{S_N2} \underset{\textbf{Alkyl azide}}{R—N{=}N^+{=}N^-} \xrightarrow[\text{or } LiAlH_4]{\text{Na/alcohol}} R\ddot{N}H_2$$

Then the alkyl azide can be reduced to a primary amine with sodium and alcohol or with lithium aluminum hydride. *A word of caution:* Alkyl azides are explosive, and low-molecular-weight alkyl azides should not be isolated but should be kept in solution. Sodium azide is used in automotive airbags.

The Gabriel Synthesis Potassium phthalimide (see the following reaction) can also be used to prepare primary amines by a method known as the *Gabriel synthesis.* This synthesis also avoids the complications of multiple alkylations that occur when alkyl halides are treated with ammonia:

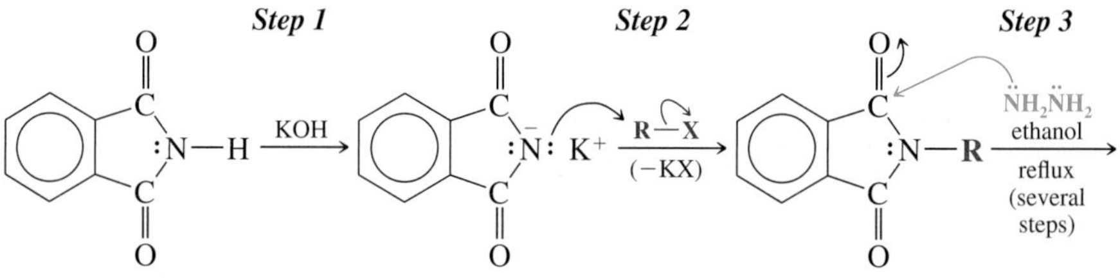

Step 1: **Phthalimide** $\xrightarrow{KOH}$ potassium phthalimide ($:N^-:\ K^+$)

Step 2: $\xrightarrow[(-KX)]{R—X}$ ***N*-Alkylphthalimide**

Step 3: $\xrightarrow[\text{reflux (several steps)}]{\ddot{N}H_2\ddot{N}H_2,\ \text{ethanol}}$

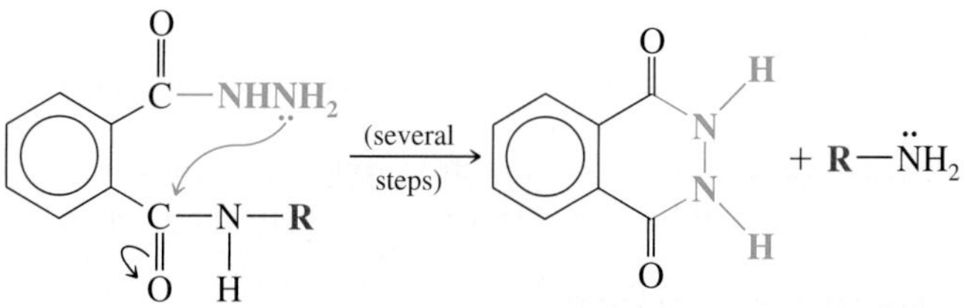

C(=O)—NHNH₂ / C(=O)—N(H)—R intermediate $\xrightarrow{\text{(several steps)}}$ **Phthalazine-1,4-dione** + $R—\ddot{N}H_2$ **Primary amine**

Phthalimide is quite acidic ($pK_a = 9$); it can be converted to potassium phthalimide by potassium hydroxide (step 1). The phthalimide anion is a strong nucleophile and (in step 2) it reacts with an alkyl halide by an S_N2 mechanism to give an *N*-alkylphthalimide. At this point, the *N*-alkylphthalimide can be hydrolyzed with aqueous acid or base, but the hydrolysis is often difficult. It is often more convenient to treat the *N*-alkylphthalimide with hydrazine (NH_2NH_2) in refluxing ethanol (step 3) to give a primary amine and phthalazine-1,4-dione.

Syntheses of amines using the Gabriel synthesis are, as we might expect, restricted to the use of methyl, primary, and secondary alkyl halides. The use of tertiary halides leads almost exclusively to eliminations.

Problem 20.3 (a) Write resonance structures for the phthalimide anion that account for the acidity of phthalimide. (b) Would you expect phthalimide to be more or less acidic than benzamide? Why? (c) After step 2 of our reaction, several steps have been omitted. Propose reasonable mechanisms for these steps.

Problem 20.4 Outline a preparation of benzylamine using the Gabriel synthesis.

Alkylation of Tertiary Amines Multiple alkylations are not a problem when tertiary amines are alkylated with methyl or primary halides. Reactions such as the following take place in good yield:

$$R_3N\!: + RCH_2{-}Br \xrightarrow{S_N2} R_3\overset{+}{N}{-}CH_2R + Br^-$$

20.5B Preparation of Aromatic Amines through Reduction of Nitro Compounds

The most widely used method for preparing aromatic amines involves nitration of the ring and subsequent reduction of the nitro group to an amino group:

$$Ar{-}H \xrightarrow[H_2SO_4]{HNO_3} Ar{-}NO_2 \xrightarrow{[H]} Ar{-}NH_2$$

We studied ring nitration in Chapter 15 and saw there that it is applicable to a wide variety of aromatic compounds. Reduction of the nitro group can also be carried out in a number of ways. The most frequently used methods employ catalytic hydrogenation, or treatment of the nitro compound with acid and iron. Zinc, tin, or a metal salt such as $SnCl_2$ can also be used. Overall, this is a $6e^-$ reduction.

General Reaction

$$Ar{-}NO_2 \xrightarrow[\text{or (1) Fe, HCl (2) } OH^-]{H_2,\ \text{catalyst}} Ar{-}NH_2$$

Specific Example

$$C_6H_5{-}NO_2 \xrightarrow[\text{(2) } OH^-]{\text{(1) Fe, HCl}} C_6H_5{-}NH_2$$

(97%)

Selective reduction of one nitro group of a dinitro compound can often be achieved through the use of hydrogen sulfide in aqueous (or alcoholic) ammonia:

$$\text{m-}C_6H_4(NO_2)_2 \xrightarrow[NH_3,\ C_2H_5OH]{H_2S} \text{m-}O_2N{-}C_6H_4{-}NH_2$$

m-Dinitrobenzene → *m*-Nitroaniline (70–80%)

When this method is used, the amount of the hydrogen sulfide must be carefully measured because the use of an excess may result in the reduction of more than one nitro group.

It is not always possible to predict just which nitro group will be reduced, however. Treating 2,4-dinitrotoluene with hydrogen sulfide and ammonia results in reduction of the 4-nitro group:

$$\text{2,4-dinitrotoluene} \xrightarrow{H_2S,\ NH_3} \text{4-methyl-3-nitroaniline}$$

On the other hand, monoreduction of 2,4-dinitroaniline causes reduction of the 2-nitro group:

$$\text{2,4-dinitroaniline} \xrightarrow[50°C]{H_2S,\ NH_3} \text{4-nitro-1,2-benzenediamine (52–58\%)}$$

20.5C Preparation of Primary, Secondary, and Tertiary Amines through Reductive Amination

Aldehydes and ketones can be converted to amines through catalytic or chemical reduction in the presence of ammonia or an amine. Primary, secondary, and tertiary amines can be prepared this way:

$$RR'C{=}O \xrightarrow[\text{[H]}]{NH_3} R{-}CHR'{-}NH_2 \quad \textbf{1° Amine}$$

$$RR'C{=}O \xrightarrow[\text{[H]}]{R''NH_2} R{-}CHR'{-}NHR'' \quad \textbf{2° Amine}$$

$$RR'C{=}O \xrightarrow[\text{[H]}]{R''R'''NH} R{-}CHR'{-}NR''R''' \quad \textbf{3° Amine}$$

Aldehyde or ketone

This process, called **reductive amination** of the aldehyde or ketone (or *reductive alkylation* of the amine), appears to proceed through the following general mechanism (illustrated with a 1° amine):

A Mechanism for the Reaction

Reductive Amination

$$RR'C{=}\ddot{O}: + H_2\ddot{N}{-}R'' \underset{\text{steps}}{\overset{\text{two}}{\rightleftharpoons}} \left[R{-}CR'(OH){-}NHR''\right] \underset{}{\overset{(-H_2O)}{\rightleftharpoons}} \left[RR'C{=}NR''\right]$$

Aldehyde or ketone **1° Amine** **Hemiaminal** **Imine**

$$\xrightarrow{\text{[H]}} R{-}CHR'{-}NHR''$$

2° Amine

When ammonia or a primary amine is used, there are two possible pathways to the product—via an amino alcohol that is similar to a hemiacetal and is called a *hemiaminal* or via an imine. When secondary amines are used, an imine cannot form, and, therefore, the pathway is through the hemiaminal or through an iminium ion:

$$\underset{R}{\overset{R'}{}}\!\!>C=\overset{+}{N}\!<\!\underset{R'''}{\overset{R''}{}}$$

Iminium ion

We saw the importance of imines in "The Chemistry of... Pyridoxal Phosphate" (vitamin B_6) in Section 16.8.

The reducing agents employed include hydrogen and a catalyst (such as nickel) or $NaBH_3CN$ or $LiBH_3CN$ (sodium or lithium cyanoborohydride). The latter two reducing agents are similar to $NaBH_4$ and are especially effective in reductive aminations. Three specific examples of reductive amination follow:

$$C_6H_5CHO \xrightarrow[\substack{90\text{ atm}\\40-70°C}]{NH_3,\ H_2,\ Ni} C_6H_5CH_2NH_2$$

Benzaldehyde → **Benzylamine (89%)**

$$C_6H_5CHO \xrightarrow[(2)\ LiBH_3CN]{(1)\ CH_3CH_2NH_2} C_6H_5CH_2NHCH_2CH_3$$

Benzaldehyde → ***N*-Benzylethanamine (89%)**

$$C_6H_{10}{=}O \xrightarrow[(2)\ NaBH_3CN]{(1)\ (CH_3)_2NH} C_6H_{11}N(CH_3)_2$$

Cyclohexanone → ***N,N*-Dimethylcyclohexanamine (52–54%)**

PROBLEM 20.5 Show how you might prepare each of the following amines through reductive amination:

(a) $CH_3(CH_2)_3CH_2NH_2$

(b) $C_6H_5CH_2CH(NH_2)CH_3$ **(Amphetamine)**

(c) $CH_3(CH_2)_4CH_2NHC_6H_5$

(d) $C_6H_5CH_2N(CH_3)_2$

PROBLEM 20.6 Reductive amination of a ketone is almost always a better method for the preparation of amines of the type $RCH(R')NH_2$ than treatment of an alkyl halide with ammonia. Why would this be true?

20.5D Preparation of Primary, Secondary, or Tertiary Amines through Reduction of Nitriles, Oximes, and Amides

Nitriles, oximes, and amides can be reduced to amines. Reduction of a nitrile or an oxime yields a primary amine; reduction of an amide can yield a primary, secondary, or tertiary amine:

$$R{-}C{\equiv}N \xrightarrow{[H]} RCH_2NH_2$$
Nitrile → 1° Amine

Nitriles can be prepared from alkyl halides and CN^- (Section 18.3) or from aldehydes and ketones as cyanohydrins (Section 16.9).

$$RCH{=}NOH \xrightarrow{[H]} RCH_2NH_2$$
Oxime → 1° Amine

Oximes can be prepared from aldehydes and ketones (Section 16.8A).

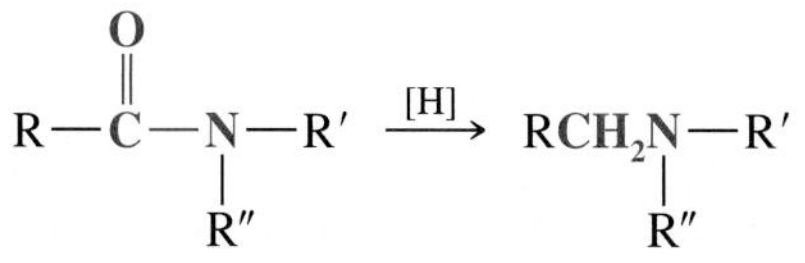

$$R{-}\overset{O}{\overset{\|}{C}}{-}\underset{R''}{\underset{|}{N}}{-}R' \xrightarrow{[H]} RCH_2\underset{R''}{\underset{|}{N}}{-}R'$$
Amide → 3° Amine

Amides can be prepared from acid chlorides, acid anhydrides, and esters (Section 18.8).

(In the last example, if R′ = H and R″ = H, the product is a 1° amine; if only R′ = H, the product is a 2° amine.)

All of these reductions can be carried out with hydrogen and a catalyst or with $LiAlH_4$. Oximes are also conveniently reduced with sodium in ethanol.

Specific examples follow:

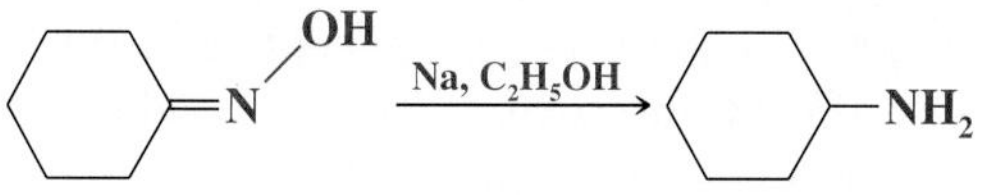

$$C_6H_{10}{=}N{-}OH \xrightarrow{Na,\ C_2H_5OH} C_6H_{11}{-}NH_2$$
(50–60%)

$$C_6H_5{-}CH_2C{\equiv}N + 2\ H_2 \xrightarrow[140°C]{Raney\ Ni} C_6H_5{-}CH_2CH_2NH_2$$

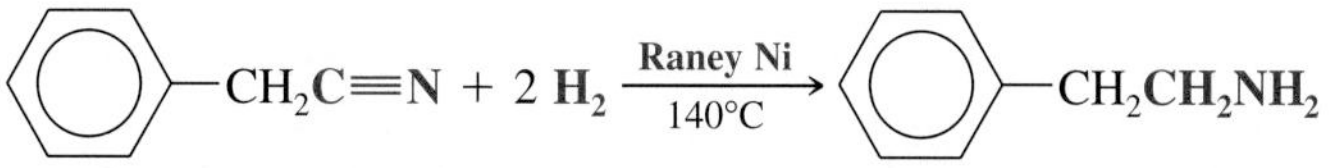

2-Phenylethanenitrile (phenylacetonitrile) → **2-Phenylethanamine (71%)**

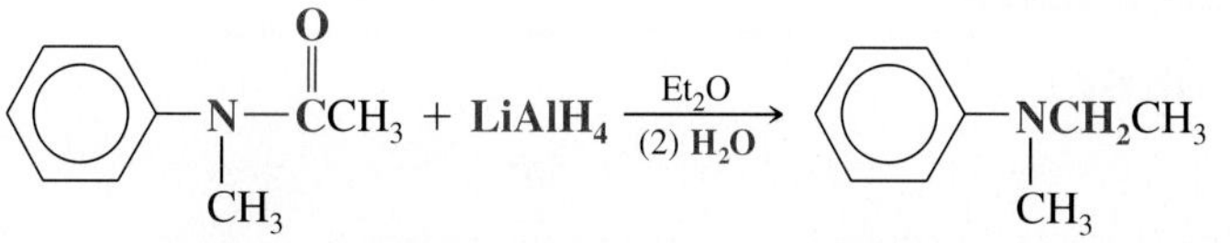

$$C_6H_5{-}\underset{CH_3}{\underset{|}{N}}{-}\overset{O}{\overset{\|}{C}}CH_3 + LiAlH_4 \xrightarrow[(2)\ H_2O]{Et_2O} C_6H_5{-}\underset{CH_3}{\underset{|}{N}}CH_2CH_3$$

***N*-Methylacetanilide** → ***N*-Ethyl-*N*-methylaniline**

Reduction of an amide is the last step in a useful procedure for **monoalkylation of an amine.** The process begins with *acylation* of the amine using an acyl chloride or acid anhydride; then the amide is reduced with lithium aluminum hydride. For example,

$$C_6H_5CH_2NH_2 \xrightarrow[base]{CH_3COCl} C_6H_5CH_2NH\overset{O}{\overset{\|}{C}}CH_3 \xrightarrow[(2)\ H_2O]{(1)\ LiAlH_4,\ Et_2O} C_6H_5CH_2NHCH_2CH_3$$

Benzylamine → **Benzylethylamine**

Problem 20.7

Show how you might utilize the reduction of an amide, oxime, or nitrile to carry out each of the following transformations:

(a) Benzoic acid ⟶ benzylethylamine
(b) 1-Bromopentane ⟶ hexylamine
(c) Propanoic acid ⟶ tripropylamine
(d) Butanone ⟶ *sec*-butylamine

20.5E Preparation of Primary Amines through the Hofmann and Curtius Rearrangements

Amides with no substituent on the nitrogen react with solutions of bromine or chlorine in sodium hydroxide to yield amines through a reaction known as the *Hofmann rearrangement* or *Hofmann degradation:*

$$R{-}\overset{\overset{\large O}{\|}}{C}{-}NH_2 + \mathbf{Br_2} + 4\ NaOH \xrightarrow{H_2O} RNH_2 + 2\ Na\mathbf{Br} + Na_2CO_3 + 2\ H_2O$$

From this equation we can see that the carbonyl carbon atom of the amide is lost (as CO_3^{2-}) and that the R group of the amide becomes attached to the nitrogen of the amine. Primary amines made this way are not contaminated by 2° or 3° amines.

The mechanism for this interesting reaction is shown in the following scheme. In the first two steps the amide undergoes a base-promoted bromination, in a manner analogous to the base-promoted halogenation of a ketone that we studied in Section 17.3B. (The electron-withdrawing acyl group of the amide makes the amido hydrogens much more acidic than those of an amine.) The *N*-bromo amide then reacts with hydroxide ion to produce an anion, which spontaneously rearranges with the loss of a bromide ion to produce an isocyanate (Section 18.9A). In the rearrangement the R— group migrates with its electrons from the acyl carbon to the nitrogen atom at the same time the bromide ion departs. The isocyanate that forms in the mixture is quickly hydrolyzed by the aqueous base to a carbamate ion, which undergoes spontaneous decarboxylation resulting in the formation of the amine.

A Mechanism for the Reaction

The Hofmann Rearrangement

$$R{-}\overset{\overset{\large O}{\|}}{C}{-}\underset{\underset{\large H}{|}}{N}{-}H + {}^{-}{:}\ddot{O}H \rightleftharpoons R{-}\overset{\overset{\large O}{\|}}{C}{-}\underset{\underset{\large H}{|}}{\ddot{N}}{:}^{-} + H_2O \xrightarrow{Br-Br} R{-}\overset{\overset{\large O}{\|}}{C}{-}\underset{\underset{\large H}{|}}{N}{-}Br + Br^{-}$$

Amide **N-Bromo amide**

Base-promoted *N*-bromination of the amide occurs.

$$R{-}\overset{\overset{\large O}{\|}}{C}{-}\underset{\underset{\large H}{|}}{N}{-}Br + {}^{-}{:}\ddot{O}H \rightleftharpoons R{-}\overset{\overset{\large O}{\|}}{C}{-}\ddot{N}^{-}{-}Br + H_2O \xrightarrow[(-Br^{-})]{} R{-}\ddot{N}{=}C{=}\ddot{O}{:}$$

***N*-Bromo amide** **Isocyanate**

Base removes a proton from the nitrogen to give a bromo amide anion.

The R— group migrates to the nitrogen as a bromide ion departs. This produces an isocyanate.

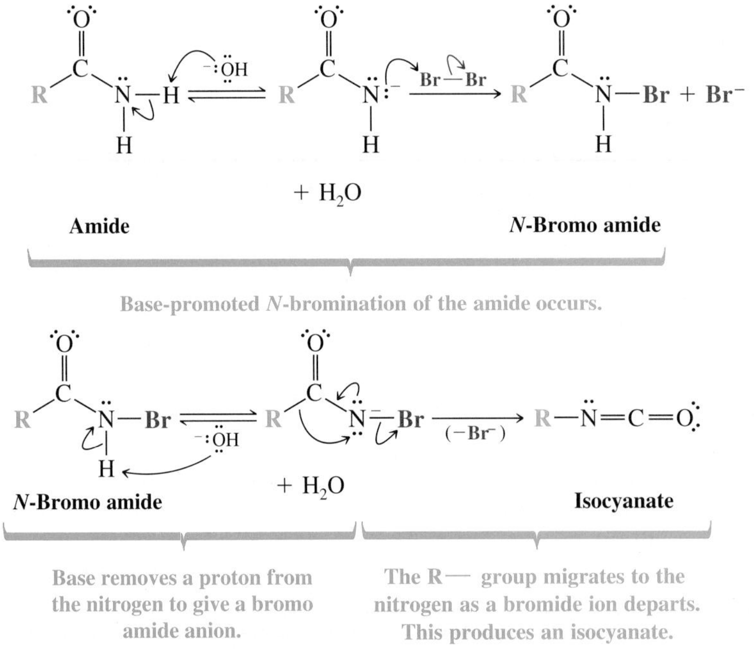

$$R-\ddot{N}=C=\ddot{O} \xrightarrow{^-:\ddot{O}H} R-\ddot{N}^{-}-C(=O)-OH \rightleftharpoons R-N(H)-C(=O)-O^- \xrightarrow{H-\ddot{O}H} R-\ddot{N}H_2 + CO_2 + OH^-$$

Isocyanate — Carbamate ion — Amine

$$\updownarrow \qquad R-\ddot{N}=C(OH)-\ddot{O}:^- \qquad\qquad OH^- + CO_2 \rightarrow HCO_3^-$$

The isocyanate undergoes hydrolysis and decarboxylation to produce the amine.

An examination of the first two steps of this mechanism shows that, initially, two hydrogen atoms must be present on the nitrogen of the amide for the reaction to occur. Consequently, the Hofmann rearrangement is limited to amides of the type $RCONH_2$.

Studies of the Hofmann rearrangement of optically active amides in which the stereogenic center is directly attached to the carbonyl group have shown that these reactions occur with *retention of configuration.* Thus, the R group migrates to nitrogen with its electrons, *but without inversion.*

The *Curtius rearrangement* is a rearrangement that occurs with acyl azides. It resembles the Hofmann rearrangement in that an R—group migrates from the acyl carbon to the nitrogen atom as the leaving group departs. In this instance the leaving group is N_2 (the best of all possible leaving groups since it is highly stable, is virtually nonbasic, and being a gas, removes itself from the medium). Acyl azides are easily prepared by allowing acyl chlorides to react with sodium azide. Heating the acyl azide brings about the rearrangement; afterward, adding water causes hydrolysis and decarboxylation of the isocyanate:

$$RC(=O)Cl \xrightarrow[(-NaCl)]{NaN_3} RC(=O)-\ddot{N}=\overset{+}{N}=N: \xrightarrow[-N_2]{heat} R-\ddot{N}=C=\ddot{O} \xrightarrow{H_2O} R-\ddot{N}H_2 + CO_2$$

Acyl chloride — Acyl azide — Isocyanate — Amine

PROBLEM 20.8

Using a different method for each part, but taking care in each case to select a *good* method, show how each of the following transformations might be accomplished:

(a) $CH_3O-C_6H_5 \longrightarrow CH_3O-C_6H_4-NH_2$

(b) $CH_3O-C_6H_5 \longrightarrow CH_3O-C_6H_4-CH(NH_2)CH_3$

(c) $C_6H_5-CH_3 \longrightarrow C_6H_5-CH_2\overset{+}{N}(CH_3)_3\ Cl^-$

(d) $O_2N-C_6H_4-CH_3 \longrightarrow O_2N-C_6H_4-NH_2$

(e) $C_6H_5-CH_3 \longrightarrow C_6H_5-CH_2CH_2NH_2$

20.6 Reactions of Amines

We have encountered a number of important reactions of amines in earlier sections. In Section 20.3 we saw reactions in which primary, secondary, and tertiary amines act *as bases.* In Section 20.5 we saw their reactions as *nucleophiles* in *alkylation reactions,* and in Chapter 18 as *nucleophiles* in *acylation reactions.* In Chapter 15 we saw that an amino group on an aromatic ring acts as a powerful *activating group* and as an *ortho–para director.*

The feature of amines that underlies all of these reactions and that forms a basis for our understanding of most of the chemistry of amines is the ability of nitrogen to share an electron pair:

Acid–Base Reactions

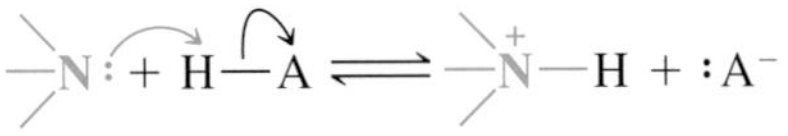

$$\text{—N: + H—A} \rightleftharpoons \text{—}\overset{+}{\text{N}}\text{—H + :A}^-$$

An amine acting as a base

Alkylation

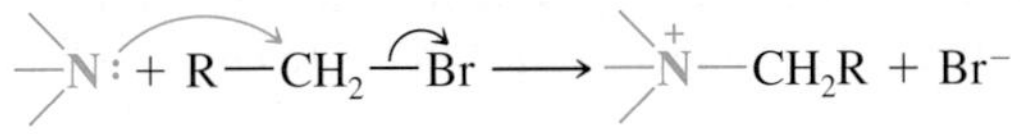

$$\text{—N: + R—CH}_2\text{—Br} \longrightarrow \text{—}\overset{+}{\text{N}}\text{—CH}_2\text{R + Br}^-$$

An amine acting as a nucleophile in an alkylation reaction

Acylation

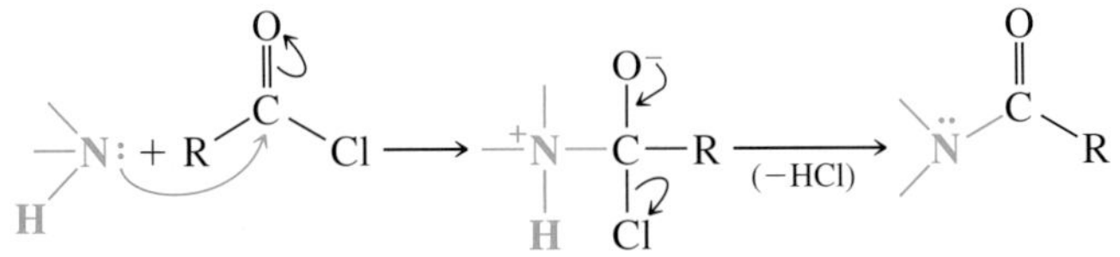

$$\text{—N(H): + R—C(=O)—Cl} \longrightarrow \text{—}\overset{+}{\text{N}}\text{(H)—C(O}^-\text{)(Cl)—R} \xrightarrow{(-\text{HCl})} \text{—}\ddot{\text{N}}\text{—C(=O)—R}$$

An amine acting as a nucleophile in an acylation reaction

In the preceding examples the amine acts as a nucleophile by donating its electron pair to an electrophilic reagent. In the following example, resonance contributions involving the nitrogen electron pair make *carbon* atoms nucleophilic:

Electrophilic Aromatic Substitution

H–N̈–H (aniline) + E—A → [ortho arenium ion, $\overset{+}{N}H_2$, H, E; A:⁻] $\xrightarrow{-\text{HA}}$ ortho-E-aniline

H–N̈–H (aniline) + A—E → [para arenium ion, $\overset{+}{N}H_2$; A:⁻, H, E] $\xrightarrow{-\text{HA}}$ para-E-aniline

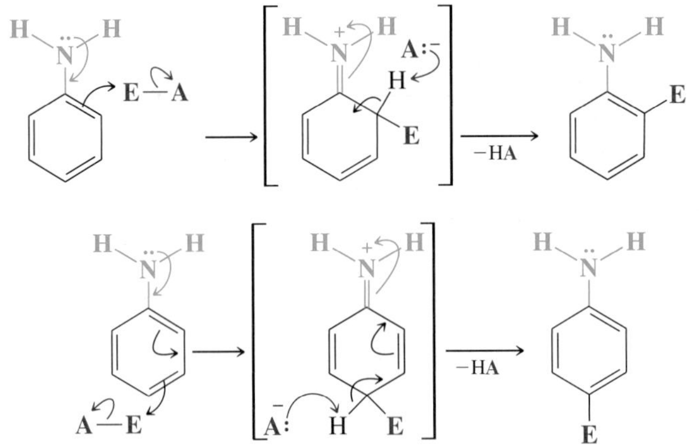

The amino group acting as an activating group and as an ortho–para director in electrophilic aromatic substitution

---

Problem 20.9 Review the chemistry of amines given in earlier sections and provide a specific example for each of the previously illustrated reactions.

20.6A Oxidation of Amines

Primary and secondary aliphatic amines are subject to oxidation, although in most instances useful products are not obtained. Complicated side reactions often occur, causing the formation of complex mixtures.

Tertiary amines can be oxidized cleanly to tertiary amine oxides. This transformation can be brought about by using hydrogen peroxide or a peroxy acid:

$$R_3N: \xrightarrow{H_2O_2 \text{ or } RCO(=O)OH} R_3\overset{+}{N}-\ddot{\underset{..}{O}}:^-$$

A tertiary amine oxide

Tertiary amine oxides undergo a useful elimination reaction to be discussed in Section 20.13B.

Arylamines are very easily oxidized by a variety of reagents, including the oxygen in air. Oxidation is not confined to the amino group but also occurs in the ring. (The amino group through its electron-donating ability makes the ring electron rich and hence especially susceptible to oxidation.) The oxidation of other functional groups on an aromatic ring cannot usually be accomplished when an amino group is present on the ring, because oxidation of the ring takes place first.

20.7 Reactions of Amines with Nitrous Acid

Nitrous acid (HONO) is a weak, unstable acid. It is always prepared *in situ,* usually by treating sodium nitrite ($NaNO_2$) with an aqueous solution of a strong acid:

$$HCl_{(aq)} + NaNO_{2(aq)} \longrightarrow HONO_{(aq)} + NaCl_{(aq)}$$
$$H_2SO_4 + 2\,NaNO_{2(aq)} \longrightarrow 2\,HONO_{(aq)} + Na_2SO_{4(aq)}$$

Nitrous acid reacts with all classes of amines. The products that we obtain from these reactions depend on whether the amine is primary, secondary, or tertiary and whether the amine is aliphatic or aromatic.

20.7A Reactions of Primary Aliphatic Amines with Nitrous Acid

Primary aliphatic amines react with nitrous acid through a reaction called *diazotization* to yield highly unstable aliphatic **diazonium salts.** Even at low temperatures, *aliphatic* diazonium salts decompose spontaneously by losing nitrogen to form carbocations. The carbocations go on to produce mixtures of alkenes, alcohols, and alkyl halides by removal of a proton, reaction with H_2O, and reaction with X^-:

General Reaction

$$\underset{\textbf{1° Aliphatic amine}}{R-NH_2} + NaNO_2 + 2\,HX \xrightarrow[H_2O]{(HONO)} \underset{\substack{\textbf{Aliphatic diazonium salt}\\\textbf{(highly unstable)}}}{[R-\overset{+}{N}\equiv N:\ X^-]} + NaX + 2\,H_2O$$

$$\big\downarrow -N_2 \text{ (i.e., } :N\equiv N:)$$

$$R^+ + X^-$$

$$\big\downarrow$$

Alkenes, alcohols, alkyl halides

Diazotizations of primary aliphatic amines are of little synthetic importance because they yield such a complex mixture of products. Diazotizations of primary aliphatic amines are used in some analytical procedures, however, because the evolution of nitrogen is quantitative. They can also be used to generate and thus study the behavior of carbocations in water, acetic acid, and other solvents.

20.7B Reactions of Primary Arylamines with Nitrous Acid

The most important reaction of amines with nitrous acid, by far, is the reaction of primary arylamines. We shall see why in Section 20.8. Primary arylamines react with nitrous acid to give arenediazonium salts. Even though arenediazonium salts are unstable, they are still far more stable than aliphatic diazonium salts; they do not decompose at an appreciable rate in solution when the temperature of the reaction mixture is kept below 5°C:

$$\underset{\textbf{Primary arylamine}}{Ar{-}NH_2} + NaNO_2 + 2\,HX \longrightarrow \underset{\substack{\textbf{Arenediazonium salt}\\ \textbf{(stable if kept below 5°C)}}}{Ar{-}\overset{+}{N}{\equiv}N{:}\ \overset{-}{X}} + NaX + 2\,H_2O$$

Primary arylamines can be converted to aryl halides, nitriles, and phenols via aryl diazonium ions (Section 20.8).

Diazotization of a primary amine takes place through a series of steps. In the presence of strong acid, nitrous acid dissociates to produce ^{+}NO ions. These ions then react with the nitrogen of the amine to form an unstable *N*-nitrosoammonium ion as an intermediate. This intermediate then loses a proton to form an *N*-nitrosoamine, which, in turn, tautomerizes to a diazohydroxide in a reaction that is similar to keto–enol tautomerization. Then, in the presence of acid, the diazohydroxide loses water to form the diazonium ion.

A Mechanism for the Reaction

Diazotization

$$HONO + H_3O^+ + A{:}^- \rightleftharpoons H_2\overset{+}{O}{-}NO + H_2O \rightleftharpoons 2\,H_2O + \overset{+}{N}{=}O$$

$$\underset{\substack{\textbf{1° Arylamine}\\ \textbf{(or alkylamine)}}}{Ar{-}NH_2{:}} + {}^{+}N{=}O \longrightarrow \underset{\substack{\textbf{\textit{N}-Nitroso-}\\ \textbf{ammonium ion}}}{Ar{-}\overset{+}{N}H_2{-}N{=}O} \xrightarrow{-H_3O^+} \underset{\textbf{\textit{N}-Nitrosoamine}}{Ar{-}NH{-}N{=}O} \quad (+\ H{-}A)$$

$$Ar{-}NH{-}N{=}\overset{+}{O}{-}H\ (A{:}^-) \underset{+HA}{\overset{-HA}{\rightleftharpoons}} \underset{\textbf{Diazohydroxide}}{Ar{-}N{=}N{-}OH} \underset{-HA}{\overset{+HA}{\rightleftharpoons}} Ar{-}N{=}N{-}OH_2^+ \rightleftharpoons$$

$$\underbrace{Ar{-}\overset{+}{N}{\equiv}N{:} \longleftrightarrow Ar{-}N{=}\overset{+}{N}}_{\textbf{Diazonium ion}} + H_2O$$

Diazotization reactions of primary arylamines are of considerable synthetic importance because the diazonium group, $-\overset{+}{N}{\equiv}N{:}$, can be replaced by a variety of other functional groups. We shall examine these reactions in Section 20.8.

The Chemistry of...

N-Nitrosoamines

N-Nitrosoamines are very powerful carcinogens which scientists fear may be present in many foods, especially in cooked meats that have been cured with sodium nitrite.

Sodium nitrite is added to many meats (e.g., bacon, ham, frankfurters, sausages, and corned beef) to inhibit the growth of *Clostridium botulinum* (the bacterium that produces botulinus toxin) and to keep red meats from turning brown. (Food poisoning by botulinus toxin is often fatal.) In the presence of acid or under the influence of heat, sodium nitrite reacts with amines always present in the meat to produce *N*-nitrosoamines. Cooked bacon, for example, has been shown to contain *N*-nitrosodimethylamine and *N*-nitrosopyrrolidine.

There is also concern that nitrites from food may produce nitrosoamines when they react with amines in the presence of the acid found in the stomach. In 1976, the FDA reduced the permissible amount of nitrite allowed in cured meats from 200 parts per million (ppm) to 50–125 ppm. Nitrites (and nitrates that can be converted to nitrites by bacteria) also occur naturally in many foods.

Cigarette smoke is known to contain *N*-nitrosodimethylamine. Someone smoking a pack of cigarettes a day inhales about 0.8 μg of *N*-nitrosodimethylamine, and even more has been shown to be present in the sidestream smoke.

A processed food preserved with sodium nitrite.

20.7C Reactions of Secondary Amines with Nitrous Acid

Secondary amines—both aryl and alkyl—react with nitrous acid to yield *N*-nitrosoamines. ***N*-Nitrosoamines** usually separate from the reaction mixture as oily yellow liquids:

Specific Examples

$$(CH_3)_2\ddot{N}H + HCl + NaNO_2 \xrightarrow[H_2O]{(HONO)} (CH_3)_2\ddot{N}{-}\ddot{N}{=}O$$

Dimethylamine → ***N*-Nitrosodimethylamine (a yellow oil)**

$$C_6H_5{-}\ddot{N}(H)CH_3 + HCl + NaNO_2 \xrightarrow[H_2O]{(HONO)} C_6H_5{-}\ddot{N}(\ddot{N}{=}O)CH_3$$

***N*-Methylaniline** → ***N*-Nitroso-*N*-methylaniline (87–93%) (a yellow oil)**

20.7D Reactions of Tertiary Amines with Nitrous Acid

When a tertiary aliphatic amine is mixed with nitrous acid, an equilibrium is established among the tertiary amine, its salt, and an *N*-nitrosoammonium compound:

$$2\ R_3N\text{:} + HX + NaNO_2 \rightleftharpoons R_3\overset{+}{N}H\ \overset{-}{X} + R_3\overset{+}{N}{-}\ddot{N}{=}O\ X^-$$

Tertiary aliphatic amine — **Amine salt** — ***N*-Nitrosoammonium compound**

While *N*-nitrosoammonium compounds are stable at low temperatures, at higher temperatures and in aqueous acid they decompose to produce aldehydes or ketones. These reactions are of little synthetic importance, however.

Tertiary arylamines react with nitrous acid to form *C*-nitroso aromatic compounds. Nitrosation takes place almost exclusively at the para position if it is open and, if not, at the ortho position. The reaction (see Problem 20.10) is another example of electrophilic aromatic substitution.

Specific Example

$$(H_3C)_2\ddot{N}{-}C_6H_5 + HCl + NaNO_2 \xrightarrow[H_2O]{8^\circ C} (H_3C)_2\ddot{N}{-}C_6H_4{-}\ddot{N}{=}O$$

***p*-Nitroso-*N*,*N*-dimethylaniline (80–90%)**

PROBLEM 20.10

Para-nitrosation of *N*,*N*-dimethylaniline (*C*-nitrosation) is believed to take place through an electrophilic attack by $\overset{+}{N}O$ ions. **(a)** Show how $\overset{+}{N}O$ ions might be formed in an aqueous solution of $NaNO_2$ and HCl. **(b)** Write a mechanism for *p*-nitrosation of *N*,*N*-dimethylaniline. **(c)** Tertiary aromatic amines and phenols undergo *C*-nitrosation reactions, whereas most other benzene derivatives do not. How can you account for this difference?

20.8 Replacement Reactions of Arenediazonium Salts

Diazonium salts are highly useful intermediates in the synthesis of aromatic compounds, because the diazonium group can be replaced by any one of a number of other atoms or groups, including —F, —Cl, —Br, —I, —CN, —OH, and —H.

Diazonium salts are almost always prepared by diazotizing primary aromatic amines. Primary arylamines can be synthesized through reduction of nitro compounds that are readily available through direct nitration reactions.

20.8A Syntheses Using Diazonium Salts

Most arenediazonium salts are unstable at temperatures above 5–10°C, and many explode when dry. Fortunately, however, most of the replacement reactions of diazonium salts do not require their isolation. We simply add another reagent (CuCl, CuBr, KI, etc.) to the mixture, gently warm the solution, and the replacement (accompanied by the evolution of nitrogen) takes place:

$$Ar{-}NH_2 \xrightarrow[0-5^\circ C]{HONO} Ar{-}\overset{+}{N}_2 \text{ (Arenediazonium salt)}$$

$$\xrightarrow{Cu_2O,\ Cu^{2+},\ H_2O} Ar{-}OH$$
$$\xrightarrow{CuCl} Ar{-}Cl$$
$$\xrightarrow{CuBr} Ar{-}Br$$
$$\xrightarrow{CuCN} Ar{-}CN$$
$$\xrightarrow{KI} Ar{-}I$$
$$\xrightarrow[(2)\ heat]{(1)\ HBF_4} Ar{-}F$$
$$\xrightarrow{H_3PO_2,\ H_2O} Ar{-}H$$

Only in the replacement of the diazonium group by —F need we isolate a diazonium salt. We do this by adding HBF_4 to the mixture, causing the sparingly soluble and reasonably stable arenediazonium fluoroborate, $ArN_2^+\ BF_4^-$, to precipitate.

20.8B The Sandmeyer Reaction: Replacement of the Diazonium Group by —Cl, —Br, or —CN

Arenediazonium salts react with cuprous chloride, cuprous bromide, and cuprous cyanide to give products in which the diazonium group has been replaced by —Cl, —Br, and —CN, respectively. These reactions are known generally as *Sandmeyer reactions.* Several specific examples follow. The mechanisms of these replacement reactions are not fully understood; the reactions appear to be radical in nature, not ionic.

$$\text{o-Toluidine} \xrightarrow[H_2O\ (0-5^\circ C)]{HCl,\ NaNO_2} \text{o-}CH_3C_6H_4\overset{+}{N}_2\ Cl^- \xrightarrow[15-60^\circ C]{CuCl} \text{o-Chlorotoluene} + N_2$$

o-Toluidine → **o-Chlorotoluene (74–79% overall)**

$$\text{m-Chloroaniline} \xrightarrow[H_2O\ (0-10^\circ C)]{HBr,\ NaNO_2} \text{m-}ClC_6H_4\overset{+}{N}_2\ Br^- \xrightarrow[100^\circ C]{CuBr} \text{m-Bromochlorobenzene} + N_2$$

m-Chloroaniline → **m-Bromochlorobenzene (70% overall)**

$$\text{o-Nitroaniline} \xrightarrow[H_2O\ (room\ temp.)]{HCl,\ NaNO_2} \text{o-}NO_2C_6H_4\overset{+}{N}_2\ Cl^- \xrightarrow[90-100^\circ C]{CuCN} \text{o-Nitrobenzonitrile} + N_2$$

o-Nitroaniline → **o-Nitrobenzonitrile (65% overall)**

20.8C Replacement by —I

Arenediazonium salts react with potassium iodide to give products in which the diazonium group has been replaced by —I. An example is the synthesis of *p*-iodonitrobenzene:

$$p\text{-}O_2N\text{-}C_6H_4\text{-}NH_2 \xrightarrow[\substack{H_2O \\ 0\text{-}5^\circ C}]{H_2SO_4,\ NaNO_2} p\text{-}O_2N\text{-}C_6H_4\text{-}N_2^+\ HSO_4^- \xrightarrow{KI} p\text{-}O_2N\text{-}C_6H_4\text{-}I + N_2$$

***p*-Nitroaniline** → ***p*-Iodonitrobenzene (81% overall)**

20.8D Replacement by —F

The diazonium group can be replaced by fluorine by treating the diazonium salt with fluoroboric acid (HBF_4). The diazonium fluoroborate that precipitates is isolated, dried, and heated until decomposition occurs. An aryl fluoride is produced:

$$m\text{-}CH_3\text{-}C_6H_4\text{-}NH_2 \xrightarrow[(2)\ HBF_4]{(1)\ HONO,\ H^+} m\text{-}CH_3\text{-}C_6H_4\text{-}N_2^+\ BF_4^- \xrightarrow{heat} m\text{-}CH_3\text{-}C_6H_4\text{-}F + N_2 + BF_3$$

***m*-Toluidine** → ***m*-Toluenediazonium fluoroborate (79%)** → ***m*-Fluorotoluene (69%)**

20.8E Replacement by —OH

The diazonium group can be replaced by a hydroxyl group by adding cuprous oxide to a dilute solution of the diazonium salt containing a large excess of cupric nitrate:

$$H_3C\text{-}C_6H_4\text{-}N_2^+\ HSO_4^- \xrightarrow[Cu^{2+},\ H_2O]{Cu_2O} H_3C\text{-}C_6H_4\text{-}OH$$

***p*-Toluenediazonium hydrogen sulfate** → ***p*-Cresol (93%)**

This variation of the Sandmeyer reaction (developed by T. Cohen of the University of Pittsburgh) is a much simpler and safer procedure than an older method for phenol preparation, which required heating the diazonium salt with concentrated aqueous acid.

PROBLEM 20.11 In the preceding examples of diazonium reactions, we have illustrated syntheses beginning with the compounds (a)–(e) here. Show how you might prepare each of the following compounds from benzene:

(a) *m*-Nitroaniline **(c)** *m*-Bromoaniline **(e)** *p*-Nitroaniline
(b) *m*-Chloroaniline **(d)** *o*-Nitroaniline

20.8F Replacement by Hydrogen: Deamination by Diazotization

Arenediazonium salts react with hypophosphorous acid (H_3PO_2) to yield products in which the diazonium group has been replaced by —H.

Since we usually begin a synthesis using diazonium salts by nitrating an aromatic compound, that is, replacing —H by —NO_2 and then by —NH_2, it may seem strange that we would ever want to replace a diazonium group by —H. However, replacement of the diazonium group by —H can be a useful reaction. We can introduce an amino group into an aromatic ring to influence the orientation of a subsequent reaction. Later we can remove the amino group (i.e., carry out a *deamination*) by diazotizing it and treating the diazonium salt with H_3PO_2.

We can see an example of the usefulness of a deamination reaction in the following synthesis of *m*-bromotoluene.

$$\text{p-Toluidine} \xrightarrow{(CH_3CO)_2O} \text{p-}CH_3C_6H_4NHCOCH_3 \xrightarrow[\text{(2) } OH^-,\ H_2O,\ \text{heat}]{\text{(1) } Br_2} \text{2-bromo-4-methylaniline (65\% from p-toluidine)} \xrightarrow[H_2O,\ 0-5^\circ C]{H_2SO_4,\ NaNO_2}$$

$$\text{2-bromo-4-methylbenzenediazonium } (N_2^+) \xrightarrow[H_2O,\ 25^\circ C]{H_3PO_2} \text{m-Bromotoluene (85\% from 2-bromo-4-methylaniline)} + N_2$$

p-Toluidine

(65% from p-toluidine)

***m*-Bromotoluene (85% from 2-bromo-4-methylaniline)**

We cannot prepare *m*-bromotoluene by direct bromination of toluene or by a Friedel–Crafts alkylation of bromobenzene because both reactions give *o*- and *p*-bromotoluene. (Both CH_3— and Br— are ortho–para directors.) However, if we begin with *p*-toluidine (prepared by nitrating toluene, separating the para isomer, and reducing the nitro group), we can carry out the sequence of reactions shown and obtain *m*-bromotoluene in good yield. The first step, synthesis of the *N*-acetyl derivative of *p*-toluidine, is done to reduce the activating effect of the amino group. (Otherwise both ortho positions would be brominated.) Later, the acetyl group is removed by hydrolysis.

PROBLEM 20.12

Suggest how you might modify the preceding synthesis in order to prepare 3,5-dibromotoluene.

PROBLEM 20.13

(a) In Section 20.8D we showed a synthesis of *m*-fluorotoluene starting with *m*-toluidine. How would you prepare *m*-toluidine from toluene? **(b)** How would you prepare *m*-chlorotoluene? **(c)** *m*-Bromotoluene? **(d)** *m*-Iodotoluene? **(e)** *m*-Tolunitrile (m-$CH_3C_6H_4CN$)? **(f)** *m*-Toluic acid?

PROBLEM 20.14

Starting with *p*-nitroaniline [Problem 20.11(e)], show how you might synthesize 1,2,3-tribromobenzene.

20.9 Coupling Reactions of Arenediazonium Salts

Arenediazonium ions are weak electrophiles; they react with highly reactive aromatic compounds—with phenols and tertiary arylamines—to yield *azo* compounds. This electrophilic aromatic substitution is often called a *diazo coupling reaction.*

General Reaction

$$\text{C}_6\text{H}_5\text{—}\overset{+}{\text{N}}\equiv\text{N}: \ \text{X}^- + \text{C}_6\text{H}_5\text{—G} \longrightarrow \left[\text{C}_6\text{H}_5\text{—N=N—}\overset{\text{H}}{\text{C}_6\text{H}_5}^{+}\text{—G}\right] \text{X}^- \xrightarrow{-\text{HX}} \text{C}_6\text{H}_5\text{—N=N—C}_6\text{H}_4\text{—G}$$

G = —NR$_2$ or —OH

An azo compound

Specific Examples

$$\text{C}_6\text{H}_5\text{—N}_2^+\ \text{Cl}^- + \text{C}_6\text{H}_5\text{—OH} \xrightarrow[\text{NaOH, H}_2\text{O}]{0°\text{C}} \text{C}_6\text{H}_5\text{—N=N—C}_6\text{H}_4\text{—OH}$$

Benzenediazonium chloride **Phenol** ***p*-(Phenylazo)phenol (orange solid)**

$$\text{C}_6\text{H}_5\text{—N}_2^+\ \text{Cl}^- + \text{C}_6\text{H}_5\text{—N(CH}_3)_2 \xrightarrow[0°\text{C, H}_2\text{O}]{\text{CH}_3\text{CO}^-\ \text{Na}^+ \ (\text{C=O})} \text{C}_6\text{H}_5\text{—N=N—C}_6\text{H}_4\text{—N(CH}_3)_2$$

Benzenediazonium chloride ***N,N*-Dimethylaniline** ***N,N*-Dimethyl-*p*-(phenylazo)aniline (yellow solid)**

Couplings between arenediazonium cations and phenols take place most rapidly in *slightly* alkaline solution. Under these conditions an appreciable amount of the phenol is present as a phenoxide ion, ArO$^-$, and phenoxide ions are even more reactive toward electrophilic substitution than are phenols themselves. (Why?) If the solution is too alkaline (pH > 10), however, the arenediazonium salt itself reacts with hydroxide ion to form a relatively unreactive diazohydroxide or diazotate ion:

$$\text{C}_6\text{H}_5\text{—}\ddot{\text{O}}\text{H} \underset{\text{HA}}{\overset{\text{OH}^-}{\rightleftharpoons}} \text{C}_6\text{H}_5\text{—}\ddot{\text{O}}:^-$$

Phenol (couples slowly) **Phenoxide ion (couples rapidly)**

$$\text{Ar—}\overset{+}{\text{N}}\equiv\text{N}: \underset{\text{HA}}{\overset{\text{OH}^-}{\rightleftharpoons}} \text{Ar—}\ddot{\text{N}}\text{=}\ddot{\text{N}}\text{—OH} \underset{\text{HA}}{\overset{\text{OH}^-}{\rightleftharpoons}} \text{Ar—}\ddot{\text{N}}\text{=}\ddot{\text{N}}\text{—}\ddot{\text{O}}:^-$$

Arenediazonium ion (couples) **Diazohydroxide (does not couple)** **Diazotate ion (does not couple)**

Couplings between arenediazonium cations and amines take place most rapidly in slightly acidic solutions (pH 5–7). Under these conditions the concentration of the arenediazonium cation is at a maximum; at the same time an excessive amount of the amine has not been converted to an unreactive aminium salt:

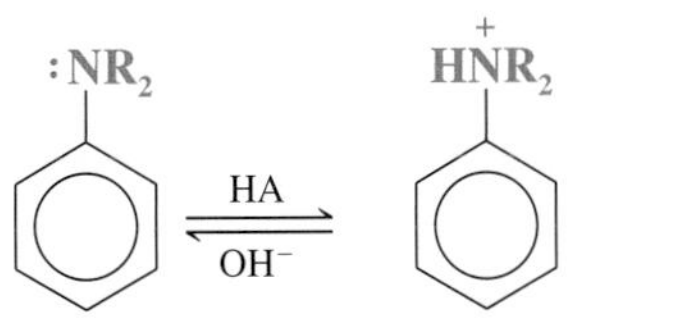

$$\text{Amine (couples)} \; \underset{OH^-}{\overset{HA}{\rightleftharpoons}} \; \text{Aminium salt (does not couple)}$$

:NR₂ → $\overset{+}{H}NR_2$

Amine
(couples)

Aminium salt
(does not couple)

If the pH of the solution is lower than 5, the rate of amine coupling is low.

With phenols and aniline derivatives, coupling takes place almost exclusively at the para position if it is open. If it is not, coupling takes place at the ortho position.

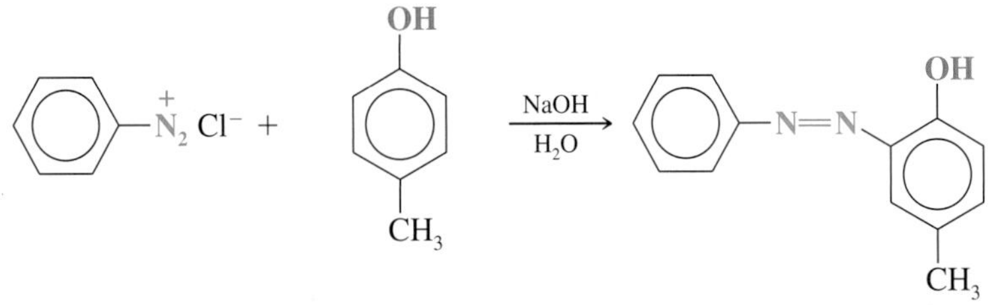

$$C_6H_5\overset{+}{N}_2\,Cl^- + \text{4-methylphenol} \xrightarrow[H_2O]{NaOH} \text{4-methyl-2-(phenylazo)phenol}$$

4-Methylphenol
(*p*-cresol)

4-Methyl-2-(phenylazo)phenol

Azo compounds are usually intensely colored because the azo (diazenediyl) linkage, —N=N—, brings the two aromatic rings into conjugation. This gives an extended system of delocalized π electrons and allows absorption of light in the visible region. Azo compounds, because of their intense colors and because they can be synthesized from relatively inexpensive compounds, are used extensively as *dyes.*

Azo dyes almost always contain one or more $—SO_3^-\ Na^+$ groups to confer water solubility on the dye and assist in binding the dye to the surfaces of polar fibers (wool, cotton, or nylon). Many dyes are made by coupling reactions of naphthylamines and naphthols.

Orange II, a dye introduced in 1876, is made from 2-naphthol:

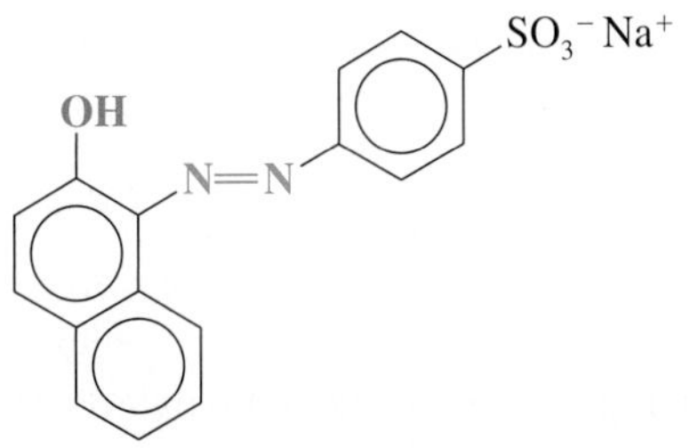

OH, N=N, $SO_3^-\ Na^+$

Orange II

PROBLEM 20.15

Outline a synthesis of orange II from 2-naphthol and *p*-aminobenzenesulfonic acid.

PROBLEM 20.16 Butter yellow is a dye once used to color margarine. It has since been shown to be carcinogenic, and its use in food is no longer permitted. Outline a synthesis of butter yellow from benzene and *N*,*N*-dimethylaniline.

$N(CH_3)_2$

N=N

Butter yellow

PROBLEM 20.17 Azo compounds can be reduced to amines by a variety of reagents including stannous chloride ($SnCl_2$):

$$\text{Ar—N=N—Ar}' \xrightarrow{SnCl_2} \text{ArNH}_2 + \text{Ar}'\text{NH}_2$$

This reduction can be useful in synthesis as the following example shows:

$$\text{4-Ethoxyaniline} \xrightarrow[\text{(2) phenol, OH}^-]{\text{(1) HONO, H}_3\text{O}^+} \textbf{A}\ (C_{14}H_{14}N_2O_2) \xrightarrow{\text{NaOH, CH}_3\text{CH}_2\text{Br}} \textbf{B}\ (C_{16}H_{18}N_2O_2) \xrightarrow{SnCl_2}$$

$$\text{two molar equivalents of } \textbf{C}\ (C_8H_{11}NO) \xrightarrow{\text{acetic anhydride}} \text{phenacetin } (C_{10}H_{13}NO_2)$$

Give a structure for phenacetin and for the intermediates **A, B,** and **C.** (Phenacetin, formerly used as an analgesic, is also the subject of Problem 18.35.)

20.10 Reactions of Amines with Sulfonyl Chlorides

Primary and secondary amines react with sulfonyl chlorides to form **sulfonamides:**

$$\text{R—NH—H} + \text{Cl—SO}_2\text{—Ar} \xrightarrow[-\text{HCl}]{} \text{R—NH—SO}_2\text{—Ar}$$

1° Amine **Sulfonyl chloride** ***N*-Substituted sulfonamide**

$$\text{R}_2\text{N—H} + \text{Cl—SO}_2\text{—Ar} \xrightarrow[-\text{HCl}]{} \text{R}_2\text{N—SO}_2\text{—Ar}$$

2° Amine ***N,N*-Disubstituted sulfonamide**

When heated with aqueous acid, sulfonamides are hydrolyzed to amines:

$$\text{R}_2\text{N—SO}_2\text{—Ar} \xrightarrow[\text{(2) OH}^-]{\text{(1) H}_3\text{O}^+\text{, heat}} \text{R}_2\text{N—H} + {}^-\text{O—SO}_2\text{—Ar}$$

This hydrolysis is much slower, however, than hydrolysis of carboxamides.

20.10A The Hinsberg Test

Sulfonamide formation is the basis for a chemical test, called the Hinsberg test, that can be used to demonstrate whether an amine is primary, secondary, or tertiary. A Hinsberg test involves two steps. First, a mixture containing a small amount of the amine and benzenesulfonyl chloride is shaken with *excess* potassium hydroxide. Next, after allowing time for a reaction to take place, the mixture is acidified. Each type of amine—primary, secondary, or tertiary—gives a different set of *visible* results after each of these two stages of the test.

Primary amines react with benzenesulfonyl chloride to form *N*-substituted benzenesulfonamides. These, in turn, undergo acid–base reactions with the excess potassium hydroxide to form water-soluble potassium salts. (These reactions take place because the hydrogen attached to nitrogen is made acidic by the strongly electron-withdrawing —SO_2— group.) At this stage our test tube contains a clear solution. Acidification of this solution will, in the next stage, cause the water-insoluble *N*-substituted sulfonamide to precipitate:

$$R{-}\ddot{N}H{-}H + Cl{-}SO_2{-}C_6H_5 \xrightarrow[(-HCl)]{OH^-} \left[R{-}\ddot{N}H{-}SO_2{-}C_6H_5\right]$$

1° Amine

Acidic hydrogen (the H on N)

$$\left[R{-}\ddot{N}H{-}SO_2{-}C_6H_5\right] \xrightarrow{KOH} R{-}\ddot{\underset{..}{N}}^{-}K^{+}{-}SO_2{-}C_6H_5$$

$$R{-}\ddot{N}H{-}SO_2{-}C_6H_5 \xleftarrow{HCl} R{-}\ddot{N}^{-}K^{+}{-}SO_2{-}C_6H_5$$

Water insoluble (precipitate) — **Water-soluble salt (clear solution)**

Secondary amines react with benzenesulfonyl chloride in aqueous potassium hydroxide to form insoluble *N*,*N*-disubstituted sulfonamides that precipitate after the first stage. *N*,*N*-Disubstituted sulfonamides do not dissolve in aqueous potassium hydroxide because they do not have an acidic hydrogen. Acidification of the mixture obtained from a secondary amine produces no visible result—the nonbasic *N*,*N*-disubstituted sulfonamide remains as a precipitate and no new precipitate forms:

$$R{-}\ddot{N}(R'){-}H + Cl{-}SO_2{-}C_6H_5 \xrightarrow[(-HCl)]{OH^-} R{-}\ddot{N}(R'){-}SO_2{-}C_6H_5$$

Water insoluble (precipitate)

If the amine is a tertiary amine and if it is water insoluble, no apparent change will take place in the mixture as we shake it with benzenesulfonyl chloride and aqueous KOH. When we acidify the mixture, the tertiary amine dissolves because it forms a water-soluble salt.

PROBLEM 20.18

An amine **A** has the molecular formula C_7H_9N. Compound **A** reacts with benzenesulfonyl chloride in aqueous potassium hydroxide to give a clear solution; acidification of the solution gives a precipitate. When **A** is treated with $NaNO_2$ and HCl at 0–5°C, and then with 2-naphthol, an intensely colored compound is formed. Compound **A** gives a single strong IR absorption peak at 815 cm^{-1}. What is the structure of **A**?

PROBLEM 20.19 Sulfonamides of primary amines are often used to synthesize *pure* secondary amines. Suggest how this synthesis is carried out.

20.11 The Sulfa Drugs: Sulfanilamide

20.11A Chemotherapy

Chemotherapy is defined as the use of chemical agents selectively to destroy infectious cells without simultaneously destroying the host. Although it may be difficult to believe (in this age of "wonder drugs"), chemotherapy is a relatively modern phenomenon. Before 1900 only three specific chemical remedies were known: mercury (for syphilis—but often with disastrous results), cinchona bark (for malaria), and ipecacuanha (for dysentery).

Paul Ehrlich won the 1908 Nobel Prize in Physiology or Medicine for his work in chemotherapy.

Modern chemotherapy began with the work of Paul Ehrlich early in the twentieth century—particularly with his discovery in 1907 of the curative properties of a dye called trypan red I when used against experimental trypanosomiasis and with his discovery in 1909 of salvarsan as a remedy for syphilis (Special Topic H). Ehrlich was awarded the Nobel Prize for Physiology or Medicine in 1908. He invented the term "chemotherapy," and in his research sought what he called "magic bullets," that is, chemicals that would be toxic to infectious microorganisms but harmless to humans.

As a medical student, Ehrlich had been impressed with the ability of certain dyes to stain tissues selectively. Working on the idea that "staining" was a result of a chemical reaction between the tissue and the dye, Ehrlich sought dyes with selective affinities for microorganisms. He hoped that in this way he might find a dye that could be modified so as to render it specifically lethal to microorganisms.

20.11B Sulfa Drugs

Gerhard Domagk won the 1939 Nobel Prize in Physiology or Medicine for discovering the antibacterial effects of prontosil.

Between 1909 and 1935, tens of thousands of chemicals, including many dyes, were tested by Ehrlich and others in a search for such "magic bullets." Very few compounds, however, were found to have any promising effect. Then, in 1935, an amazing event happened. The daughter of Gerhard Domagk, a doctor employed by a German dye manufacturer, contracted a streptococcal infection from a pin prick. As his daughter neared death, Domagk decided to give her an oral dose of a dye called prontosil. Prontosil had been developed at Domagk's firm (I. G. Farbenindustrie), and tests with mice had shown that prontosil inhibited the growth of streptococci. Within a short time the little girl recovered. Domagk's gamble not only saved his daughter's life, but it also initiated a new and spectacularly productive phase in modern chemotherapy. G. Domagk was awarded the Nobel Prize for Physiology or Medicine in 1939 but was unable to accept it until 1947.

A year later, in 1936, Ernest Fourneau of the Pasteur Institute in Paris demonstrated that prontosil breaks down in the human body to produce sulfanilamide, and that sulfanilamide is the actual active agent against streptococci. Prontosil, therefore, is a prodrug because it is converted to the active compound *in vivo*.

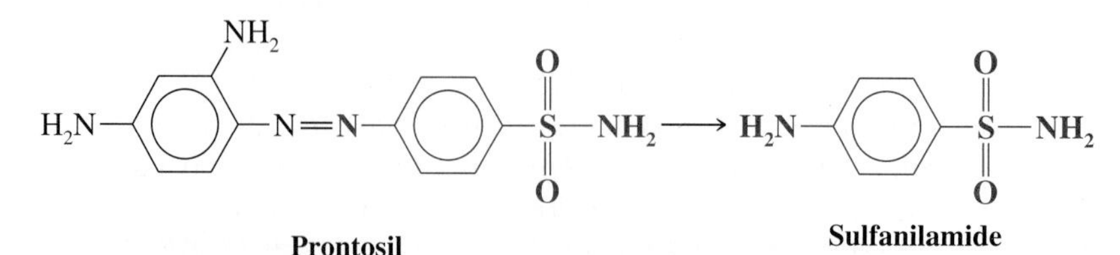

Fourneau's announcement of this result set in motion a search for other chemicals (related to sulfanilamide) that might have even better chemotherapeutic effects. Literally thousands of chemical variations were played on the sulfanilamide theme; the structure of sulfanilamide was varied in almost every imaginable way. The best therapeutic results were obtained from compounds in which one hydrogen of the $—SO_2NH_2$ group was replaced by some other group, usually a heterocyclic ring (shown in blue in the following structures). Among the most successful variations were the following compounds. Sulfanilamide itself is too toxic for general use.

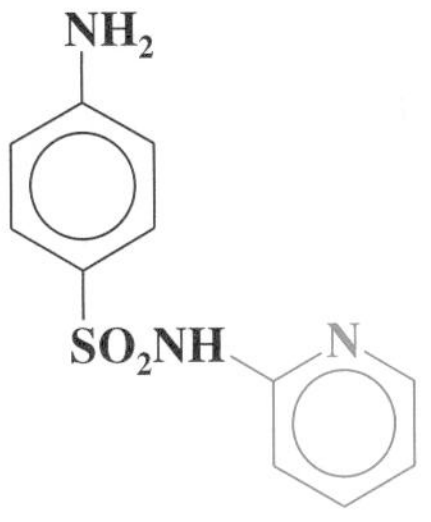

Sulfapyridine 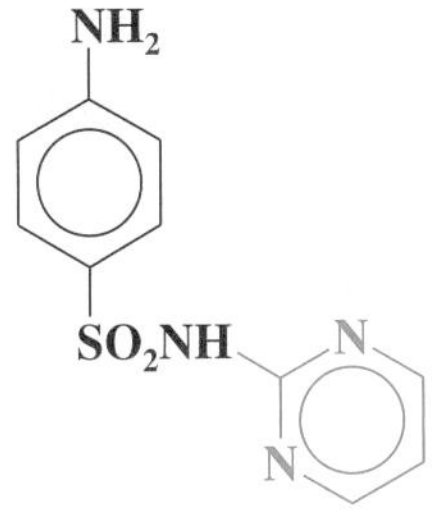 **Sulfadiazine** 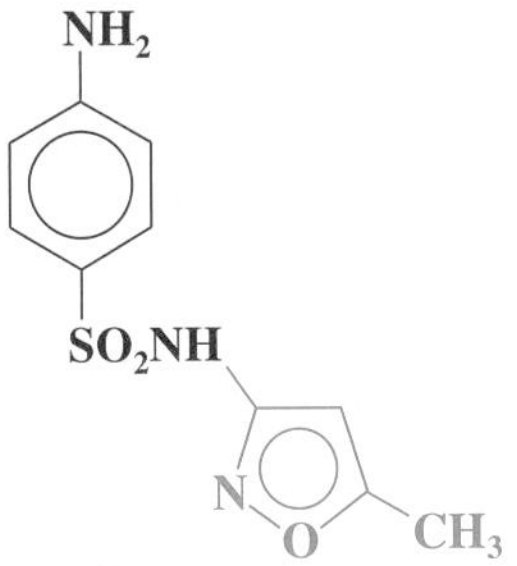 **Sulfamethoxazole**

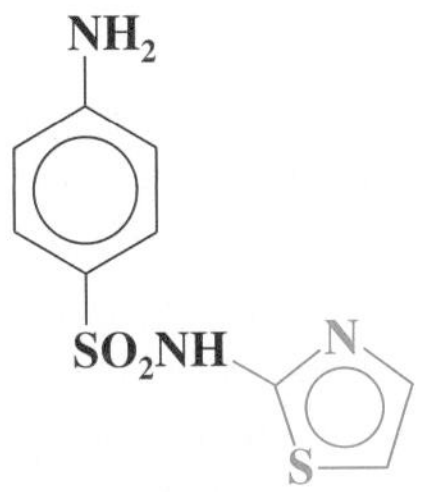

Sulfathiazole 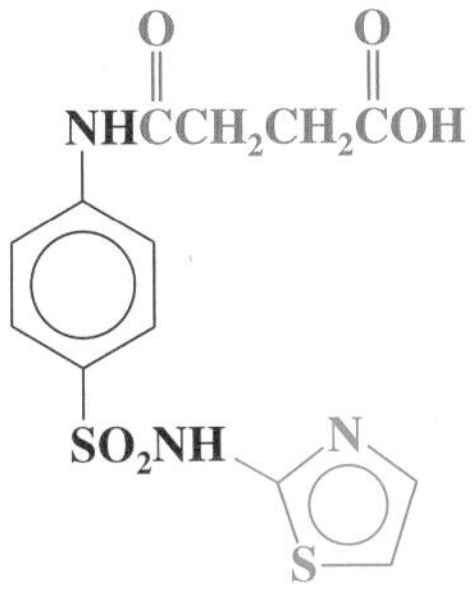 **Succinylsulfathiazole** 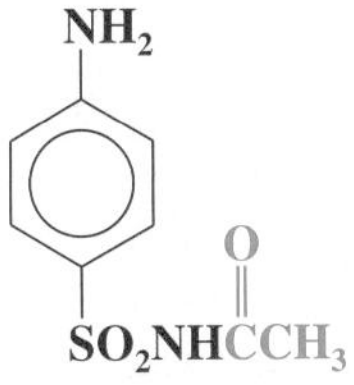 **Sulfacetamide**

Sulfapyridine was shown to be effective against pneumonia in 1938. (Before that time pneumonia epidemics had brought death to tens of thousands.) Sulfacetamide was first used successfully in treating urinary tract infections in 1941. Succinoylsulfathiazole and the related compound phthalylsulfathiazole were used as chemotherapeutic agents against infections of the gastrointestinal tract beginning in 1942. (Both compounds are slowly hydrolyzed internally to sulfathiazole.) Sulfathiazole saved the lives of countless wounded soldiers during World War II.

In 1940 a discovery by D. D. Woods laid the groundwork for our understanding of how the **sulfa drugs** work. Woods observed that the inhibition of growth of certain microorganisms by sulfanilamide is competitively overcome by *p*-aminobenzoic acid. Woods noticed the structural similarity between the two compounds (Fig. 20.4) and reasoned that the two compounds compete with each other in some essential metabolic process.

20.11C Essential Nutrients and Antimetabolites

All higher animals and many microorganisms lack the biochemical ability to synthesize certain essential organic compounds. These essential nutrients include vitamins, certain amino acids, unsaturated carboxylic acids, purines, and pyrimidines. The aromatic amine *p*-aminobenzoic acid is an essential nutrient for those bacteria that are sensitive to sulfanilamide therapy. Enzymes within these bacteria use *p*-aminobenzoic acid to synthesize another essential compound called *folic acid:*

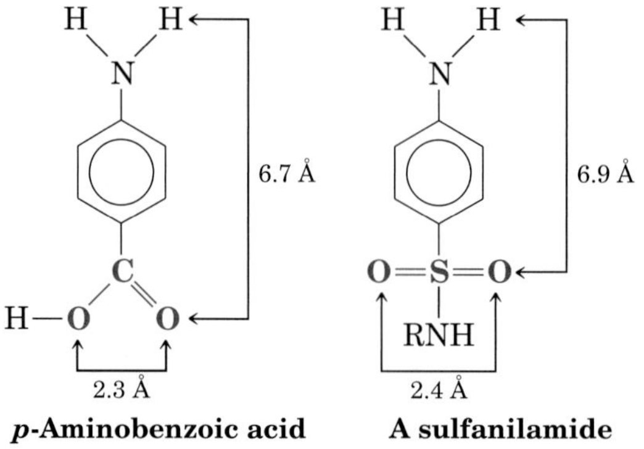

FIGURE 20.4 The structural similarity of *p*-aminobenzoic acid and a sulfanilamide. (From Korolkovas, A. *Essentials of Molecular Pharmacology;* Wiley: New York, 1970; p 105. Used with permission.)

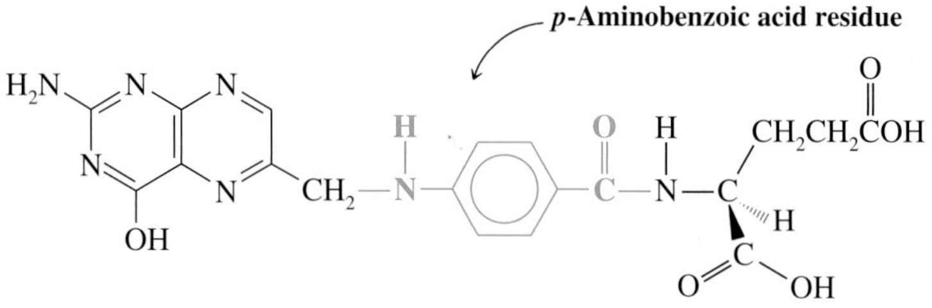

Chemicals that inhibit the growth of microbes are called *antimetabolites.* The sulfanilamides are antimetabolites for those bacteria that require *p*-aminobenzoic acid. The sulfanilamides apparently inhibit those enzymatic steps of the bacteria that are involved in the synthesis of folic acid. The bacterial enzymes are apparently unable to distinguish between a molecule of a sulfanilamide and a molecule of *p*-aminobenzoic acid; thus, sulfanilamide inhibits the bacterial enzyme. Because the microorganism is unable to synthesize enough folic acid when sulfanilamide is present, it dies. Humans are unaffected by sulfanilamide therapy because we derive our folic acid from dietary sources (folic acid is a vitamin) and do not synthesize it from *p*-aminobenzoic acid.

The discovery of the mode of action of the sulfanilamides has led to the development of many new and effective antimetabolites. One example is *methotrexate,* a derivative of folic acid that has been used successfully in treating certain carcinomas as well as rheumatoid arthritis:

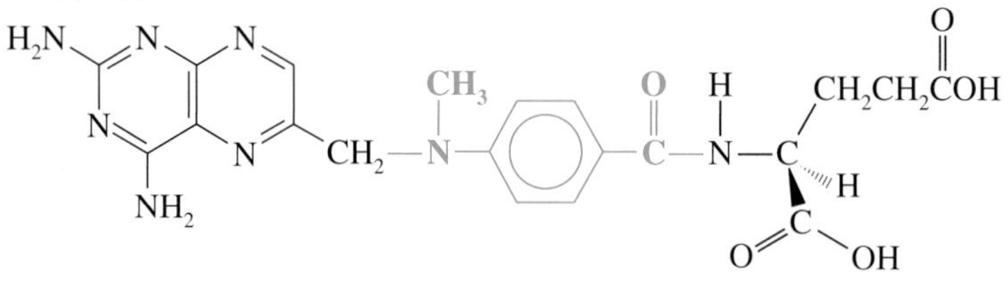

Methotrexate, by virtue of its resemblance to folic acid, can enter into some of the same reactions as folic acid, but it cannot serve the same function, particularly in important reactions involved in cell division. Although methotrexate is toxic to all dividing cells, those cells that divide most rapidly—*cancer cells*—are most vulnerable to its effect.

20.11D Synthesis of Sulfa Drugs

Sulfanilamides can be synthesized from aniline through the following sequence of reactions:

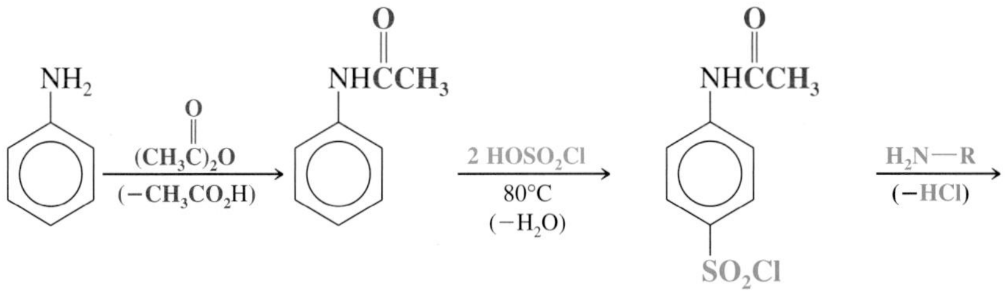

Aniline (1) $\xrightarrow[(-CH_3CO_2H)]{(CH_3C(=O))_2O}$ Acetanilide (2) $\xrightarrow[80°C\ (-H_2O)]{2\ HOSO_2Cl}$ *p*-Acetamidobenzenesulfonyl chloride (3) $\xrightarrow[(-HCl)]{H_2N—R}$

(4) $\xrightarrow[heat\ (2)\ HCO_3^-]{(1)\ dilute\ HCl}$ A sulfanilamide (5)

Aniline **(1)**

Acetanilide **(2)**

***p*-Acetamidobenzenesulfonyl chloride** **(3)**

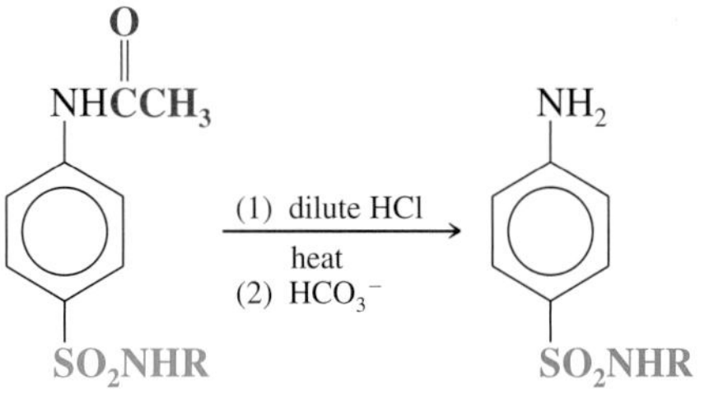

(4)

A sulfanilamide **(5)**

Acetylation of aniline produces acetanilide **(2)** and protects the amino group from the reagent to be used next. Treatment of **2** with chlorosulfonic acid brings about an electrophilic aromatic substitution reaction and yields *p*-acetamidobenzenesulfonyl chloride **(3).** Addition of ammonia or a primary amine gives the diamide, **4** (an amide of both a carboxylic acid and a sulfonic acid). Finally, refluxing **4** with dilute hydrochloric acid selectively hydrolyzes the carboxamide linkage and produces a sulfanilamide. (Hydrolysis of carboxamides is much more rapid than that of sulfonamides.)

PROBLEM 20.20

(a) Starting with aniline and assuming that you have 2-aminothiazole available, show how you would synthesize sulfathiazole. **(b)** How would you convert sulfathiazole to succinylsulfathiazole?

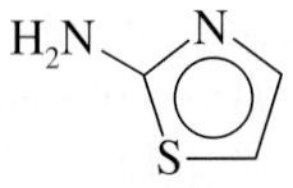

2-Aminothiazole

20.12 ANALYSIS OF AMINES

20.12A Chemical Analysis

Amines are characterized by their basicity and, thus, by their ability to dissolve in dilute aqueous acid (Section 20.3A). Moist pH paper can be used to test for the presence of an amine functional group in an unknown compound. If the compound is an amine, the pH paper shows the presence of a base. The unknown amine can then readily be classified as 1°, 2°, or 3° by IR spectroscopy (see below). Primary, secondary, and tertiary amines can also be distinguished from each other on the basis of the Hinsberg test (Section 20.10A). Primary aromatic amines are often detected through diazonium salt formation and subsequent coupling with 2-naphthol to form a brightly colored azo dye (Section 20.9).

Note these tools for characterizing amines.

Use IR Tutor to help you review infrared spectroscopy.

20.12B Spectroscopic Analysis

Infrared Spectra Primary and secondary amines are characterized by IR absorption bands in the 3300–3555-cm^{-1} region that arise from N—H stretching vibrations. Primary amines give two bands in this region (see Fig. 20.5); secondary amines generally give only one. Tertiary amines, because they have no N—H group, do not absorb in this region. Absorption bands arising from C—N stretching vibrations of aliphatic amines occur in the 1020–1220-cm^{-1} region but are usually weak and difficult to identify. Aromatic amines generally give a strong C—N stretching band in the 1250–1360-cm^{-1} region. Figure 20.5 shows an annotated IR spectrum of 4-methylaniline.

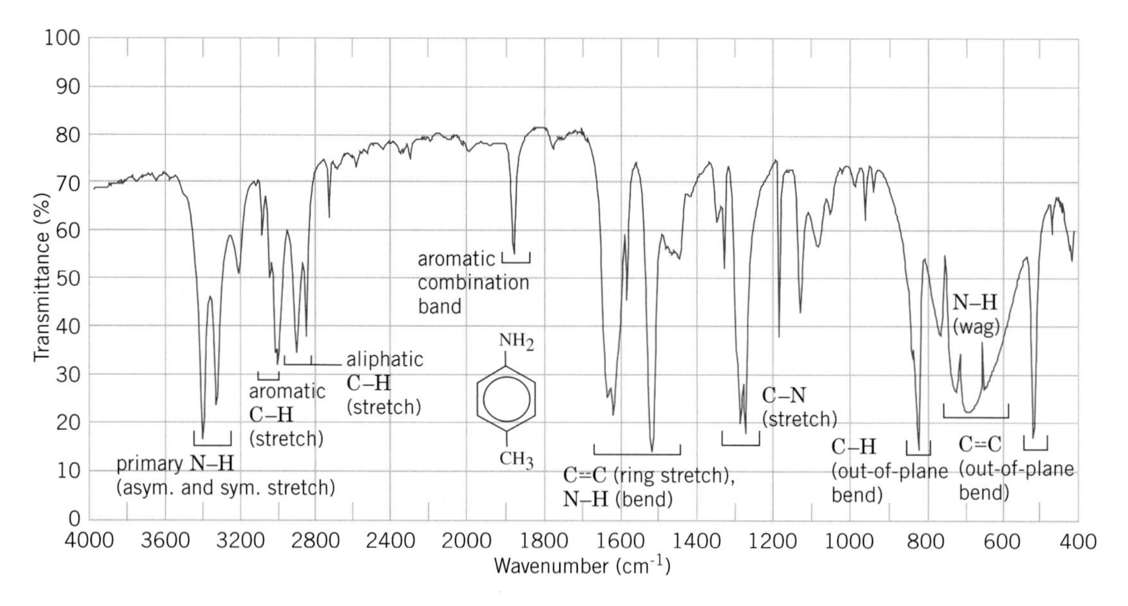

FIGURE 20.5 Annotated IR spectrum of 4-methylaniline.

^{1}H NMR Spectra Primary and secondary amines show N—H proton signals in the region δ 0.5–5. These signals are usually broad, and their exact position depends on the nature of the solvent, the purity of the sample, the concentration, and the temperature. Because of proton exchange, N—H protons are not usually coupled to protons on adjacent carbons. As such, they are difficult to identify and are best detected by proton counting or by adding a small amount of D_2O to the sample. Exchange of N—D deuterons for the N—H protons takes place, and the N—H signal disappears from the spectrum.

Protons on the α carbon of an aliphatic amine are deshielded by the electron-withdrawing effect of the nitrogen and absorb typically in the δ 2.2–2.9 region; protons on the β carbon are not deshielded as much and absorb in the range δ 1.0–1.7.

Figure 20.6 shows an annotated ^{1}H NMR spectrum of diisopropylamine.

^{13}C NMR Spectra The α carbon of an aliphatic amine experiences deshielding by the electronegative nitrogen, and its absorption is shifted downfield, typically appearing at δ 30–60. The shift is not as great as for the α carbon of an alcohol (typically δ 50–75), however, because nitrogen is less electronegative than oxygen. The downfield shift is even less for the β carbon, and so on down the chain, as the chemical shifts of the carbons of pentylamine show:

$$\begin{array}{cccccc} & H_3C— & CH_2— & CH_2— & CH_2— & CH_2—NH_2 \\ \delta & \mathbf{14.3} & \mathbf{23.0} & \mathbf{29.7} & \mathbf{34.0} & \mathbf{42.5} \end{array}$$

^{13}C NMR chemical shifts

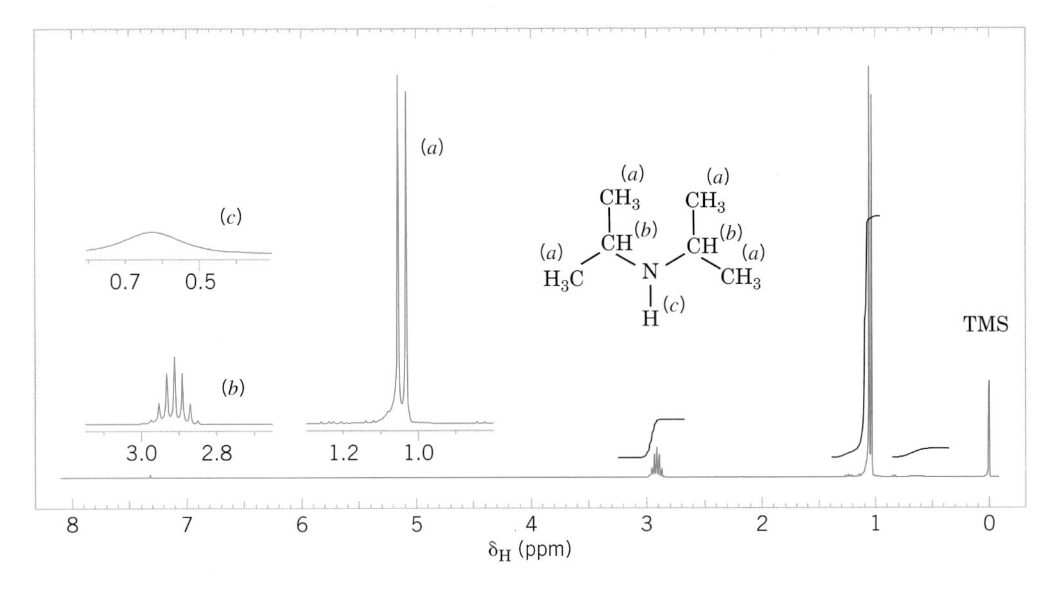

FIGURE 20.6 The 300-MHz ^{1}H NMR spectrum of diisopropylamine. Note the integral for the broad NH peak at approximately δ 0.7. Vertical expansions are not to scale.

Mass Spectra of Amines The molecular ion in the mass spectrum of an amine has an odd number mass (unless there is an even number of nitrogen atoms in the molecule). The peak for the molecular ion is usually strong for aromatic and cyclic aliphatic amines but weak for acyclic aliphatic amines. Cleavage between the α and β carbons of aliphatic amines is a common mode of fragmentation.

20.13 Eliminations Involving Ammonium Compounds

20.13A The Hofmann Elimination

All of the eliminations that we have described so far have involved electrically neutral substrates. However, eliminations are known in which the substrate bears a positive charge. One of the most important of these is the E2-type elimination that takes place when a quaternary ammonium hydroxide is heated. The products are an alkene, water, and a tertiary amine:

$$HO^{-}\ H{-}\overset{|}{\underset{|}{C}}{-}\overset{|}{\underset{|}{C}}{-}\overset{+}{N}R_3 \xrightarrow{\text{heat}} \gt C{=}C\lt + \mathbf{HOH} + :NR_3$$

A quaternary ammonium hydroxide — an alkene — a tertiary amine

This reaction was discovered in 1851 by August W. von Hofmann and has since come to bear his name.

Quaternary ammonium hydroxides can be prepared from quaternary ammonium halides in aqueous solution through the use of silver oxide or an ion exchange resin:

$$2\ RCH_2CH_2\overset{+}{N}(CH_3)_3\ X^- + Ag_2O + H_2O \longrightarrow 2\ RCH_2CH_2\overset{+}{N}(CH_3)_3\ OH^- + 2\ AgX \downarrow$$

A quaternary ammonium halide — A quaternary ammonium hydroxide

Silver halide precipitates from the solution and can be removed by filtration. The quaternary ammonium hydroxide can then be obtained by evaporation of the water.

Although most eliminations involving neutral substrates tend to follow the *Zaitsev rule* (Section 7.6A), eliminations with charged substrates tend to follow what is called the *Hofmann rule* and *yield mainly the least substituted alkene.* We can see an example of this behavior if we compare the following reactions:

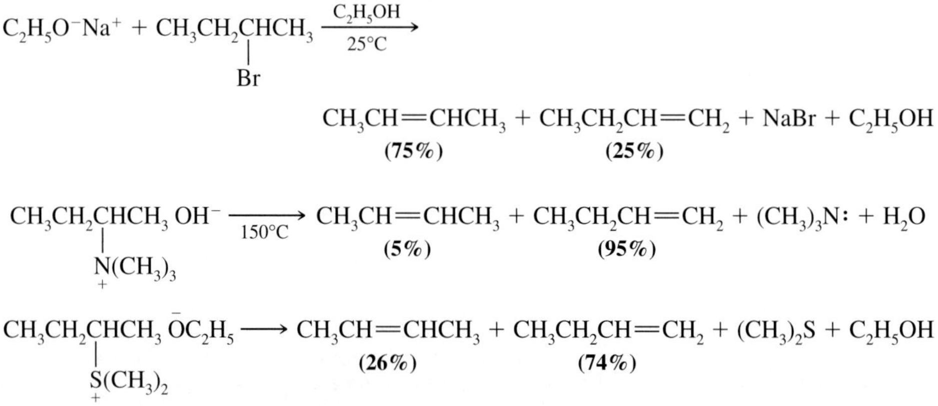

$$C_2H_5O^-Na^+ + CH_3CH_2\underset{\displaystyle Br}{CH}CH_3 \xrightarrow[25°C]{C_2H_5OH} \underset{(75\%)}{CH_3CH{=}CHCH_3} + \underset{(25\%)}{CH_3CH_2CH{=}CH_2} + NaBr + C_2H_5OH$$

$$CH_3CH_2\underset{\displaystyle \overset{+}{N}(CH_3)_3}{CH}CH_3\ OH^- \xrightarrow[150°C]{} \underset{(5\%)}{CH_3CH{=}CHCH_3} + \underset{(95\%)}{CH_3CH_2CH{=}CH_2} + (CH_3)_3N\!: + H_2O$$

$$CH_3CH_2\underset{\displaystyle \overset{+}{S}(CH_3)_2}{CH}CH_3\ \bar{O}C_2H_5 \longrightarrow \underset{(26\%)}{CH_3CH{=}CHCH_3} + \underset{(74\%)}{CH_3CH_2CH{=}CH_2} + (CH_3)_2S + C_2H_5OH$$

The precise mechanistic reasons for these differences are complex and are not yet fully understood. One possible explanation is that the transition states of elimination reactions with charged substrates have considerable carbanionic character. Therefore, these transition states show little resemblance to the final alkene product and are not stabilized appreciably by a developing double bond:

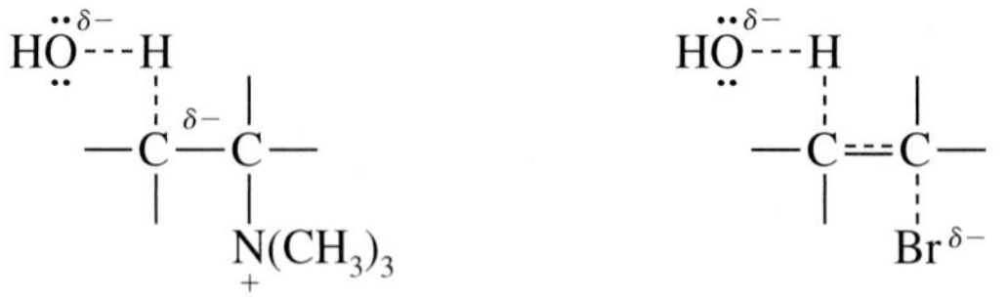

$$\begin{array}{c} \overset{\delta-}{HO}\cdots H \\ \vdots \\ -\overset{\delta-}{C}-C- \\ \quad\ \ | \\ \quad\ \ \overset{+}{N}(CH_3)_3 \end{array} \qquad\qquad \begin{array}{c} \overset{\delta-}{HO}\cdots H \\ \vdots \\ -C\overset{\cdots}{=}C- \\ \quad\ \ \vdots \\ \quad\ \ Br^{\delta-} \end{array}$$

Carbanion-like transition state (gives Hofmann orientation) — **Alkene-like transition state (gives Zaitsev orientation)**

With a charged substrate, the base attacks the most acidic hydrogen instead. A primary hydrogen atom is more acidic because its carbon atom bears only one electron-releasing group.

20.13B The Cope Elimination

Tertiary amine oxides undergo the elimination of a dialkylhydroxylamine when they are heated. This reaction is called the Cope elimination:

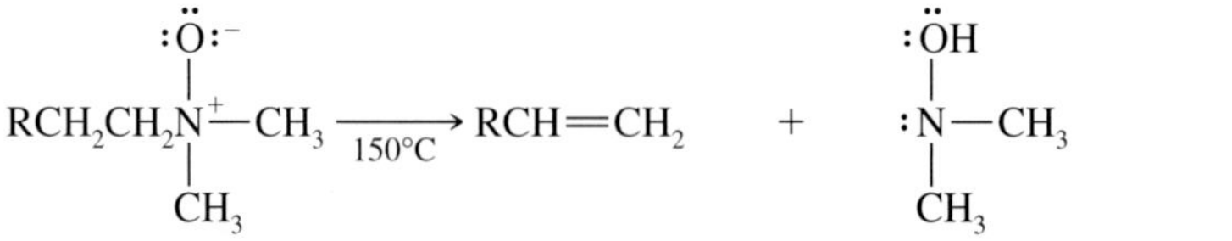

$$\underset{\text{A tertiary amine oxide}}{RCH_2CH_2\overset{\displaystyle :\ddot{O}:^-}{\underset{\displaystyle CH_3}{\overset{+}{N}}}-CH_3} \xrightarrow[150°C]{} \underset{\text{An alkene}}{RCH{=}CH_2} + \underset{N,N\text{-Dimethylhydroxylamine}}{:\overset{\displaystyle :\ddot{O}H}{\underset{\displaystyle CH_3}{N}}-CH_3}$$

The Cope elimination is a syn elimination and proceeds through a cyclic transition state:

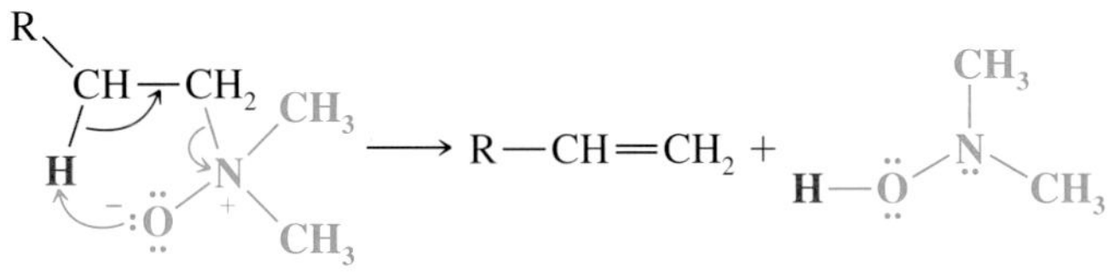

$$R-CH_2-CH_2-\overset{+}{N}(CH_3)_2-O^- \longrightarrow R-CH=CH_2 + H-\ddot{O}-\ddot{N}(CH_3)_2$$

Tertiary amine oxides are easily prepared by treating tertiary amines with hydrogen peroxide (Section 20.6A).

The Cope elimination is useful synthetically. Consider the following synthesis of methylenecyclohexane:

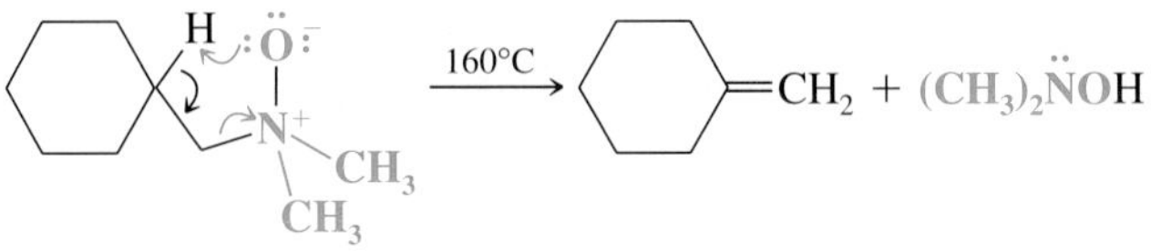

$$C_6H_{11}-CH_2-\overset{+}{N}(CH_3)_2-O^- \xrightarrow{160°C} C_6H_{10}=CH_2 + (CH_3)_2\ddot{N}OH$$

(98%)

Summary of Preparations and Reactions of Amines

Preparation of Amines

1. Gabriel synthesis (discussed in Section 20.5A):

$$\text{Phthalimide N-H} \xrightarrow[(2)\ R-X]{(1)\ KOH} \text{Phthalimide N-R} \xrightarrow[\text{ethanol reflux}]{NH_2NH_2} R-NH_2 + \text{phthalhydrazide}$$

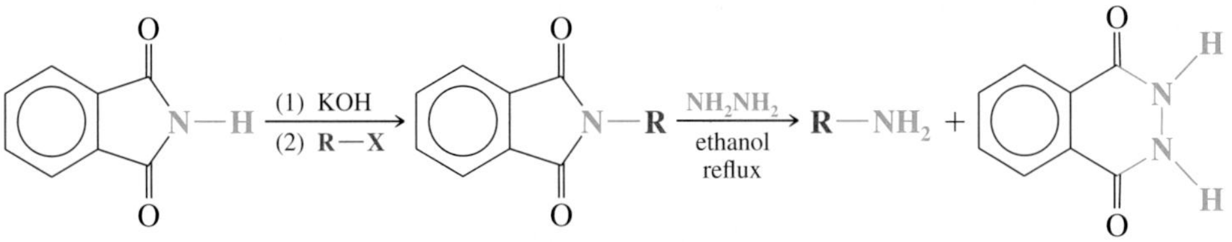

2. By reduction of alkyl azides (discussed in Section 20.5A):

$$R-Br \xrightarrow[\text{ethanol}]{NaN_3} R-N=\overset{+}{N}=\overset{-}{N} \xrightarrow[\text{or } LiAlH_4]{\text{Na/alcohol}} R-NH_2$$

3. By amination of alkyl halides (discussed in Section 20.5A):

$$R-Br + NH_3 \longrightarrow RNH_3^+\,Br^- + R_2NH_2^+\,Br^- + R_3N^+\,Br^- + R_4N^+\,Br^-$$

$$\downarrow OH^-$$

$$RNH_2 + R_2NH + R_3N + R_4N^+\,OH^-$$

(A mixture of products results.)

(**R** = a 1° alkyl group)

4. By reduction of nitroarenes (discussed in Section 20.5B):

$$Ar-NO_2 \xrightarrow[\text{or (1) Fe/HCl (2) NaOH}]{H_2,\ \text{catalyst}} Ar-NH_2$$

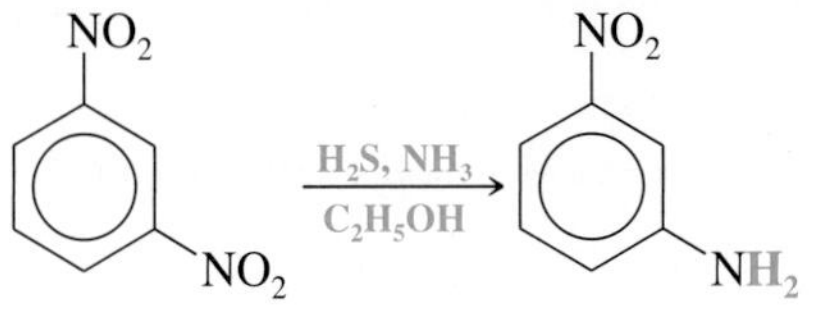

$$\text{m-}C_6H_4(NO_2)_2 \xrightarrow[C_2H_5OH]{H_2S,\ NH_3} \text{m-}O_2N-C_6H_4-NH_2$$

5. By reductive amination (discussed in Section 20.5C):

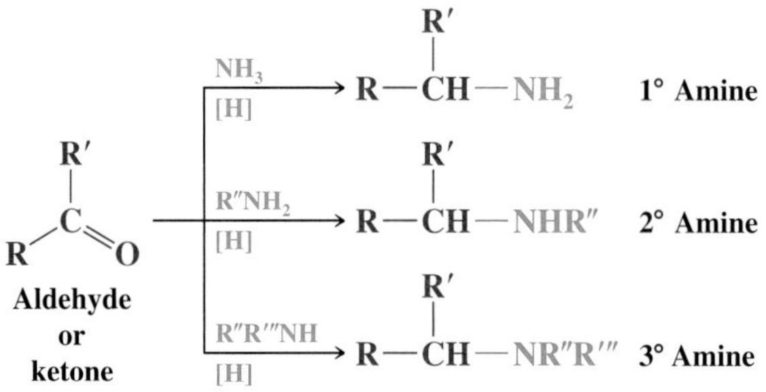

$$\underset{\text{Aldehyde or ketone}}{RC(=O)R'} \xrightarrow[\text{[H]}]{NH_3} R{-}CH(R'){-}NH_2 \quad \textbf{1° Amine}$$

$$\underset{\text{Aldehyde or ketone}}{RC(=O)R'} \xrightarrow[\text{[H]}]{R''NH_2} R{-}CH(R'){-}NHR'' \quad \textbf{2° Amine}$$

$$\underset{\text{Aldehyde or ketone}}{RC(=O)R'} \xrightarrow[\text{[H]}]{R''R'''NH} R{-}CH(R'){-}NR''R''' \quad \textbf{3° Amine}$$

6. By reduction of nitriles, oximes, and amides (discussed in Section 20.5D):

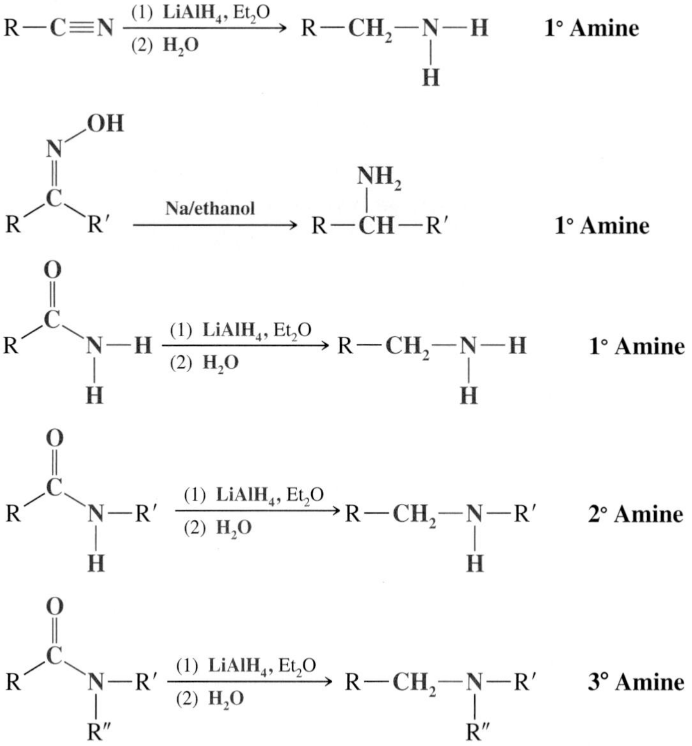

$$R{-}C{\equiv}N \xrightarrow[\text{(2) } H_2O]{\text{(1) } LiAlH_4,\ Et_2O} R{-}CH_2{-}NH{-}H \quad \textbf{1° Amine}$$

$$RC(=N{-}OH)R' \xrightarrow{\text{Na/ethanol}} R{-}CH(NH_2){-}R' \quad \textbf{1° Amine}$$

$$RC(=O){-}NH{-}H \xrightarrow[\text{(2) } H_2O]{\text{(1) } LiAlH_4,\ Et_2O} R{-}CH_2{-}NH{-}H \quad \textbf{1° Amine}$$

$$RC(=O){-}NH{-}R' \xrightarrow[\text{(2) } H_2O]{\text{(1) } LiAlH_4,\ Et_2O} R{-}CH_2{-}NH{-}R' \quad \textbf{2° Amine}$$

$$RC(=O){-}N(R''){-}R' \xrightarrow[\text{(2) } H_2O]{\text{(1) } LiAlH_4,\ Et_2O} R{-}CH_2{-}N(R''){-}R' \quad \textbf{3° Amine}$$

7. Through the Hofmann and Curtius rearrangements (discussed in Section 20.5E):

Hofmann Rearrangement

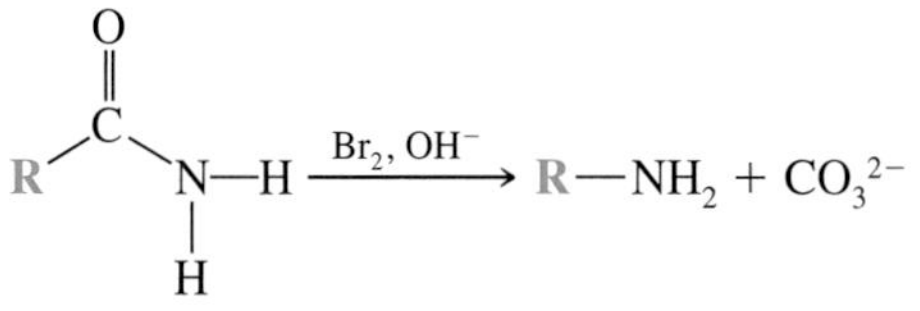

$$RC(=O){-}NH{-}H \xrightarrow{Br_2,\ OH^-} R{-}NH_2 + CO_3^{2-}$$

Curtius Rearrangement

$$RC(=O){-}Cl \xrightarrow[(-NaCl)]{NaN_3} RC(=O){-}N_3 \xrightarrow[-N_2]{\text{heat}} R{-}N{=}C{=}O \xrightarrow{H_2O} R{-}NH_2 + CO_2$$

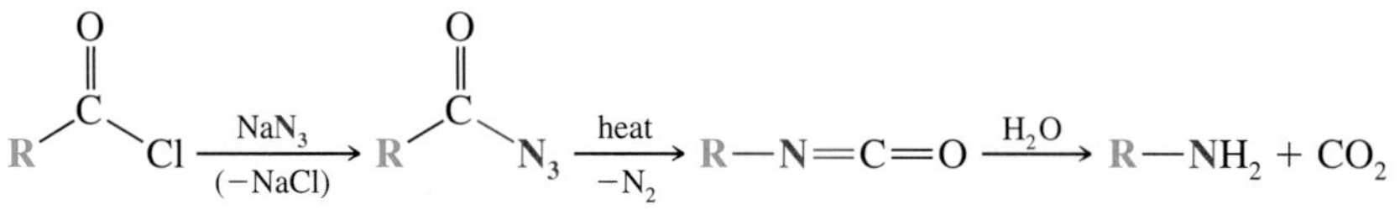

Reactions of Amines

1. As bases (discussed in Section 20.3):

$$R-\ddot{N}(R'')-R' + H-A \longrightarrow R-N^{+}H(R'')-R' \; A^-$$

(R, R′, and/or R″ may be alkyl, H, or Ar)

2. Diazotization of 1° arylamines and replacement of, or coupling with, the diazonium group (discussed in Sections 20.8 and 20.9):

$$Ar-NH_2 \xrightarrow[0\text{–}5^\circ C]{HONO} Ar-\overset{+}{N}_2$$

$$Ar-\overset{+}{N}_2 \xrightarrow{Cu_2O,\ Cu^{2+},\ H_2O} Ar-OH$$

$$Ar-\overset{+}{N}_2 \xrightarrow{CuCl} Ar-Cl$$

$$Ar-\overset{+}{N}_2 \xrightarrow{CuBr} Ar-Br$$

$$Ar-\overset{+}{N}_2 \xrightarrow{CuCN} Ar-CN$$

$$Ar-\overset{+}{N}_2 \xrightarrow{KI} Ar-I$$

$$Ar-\overset{+}{N}_2 \xrightarrow[(2)\ heat]{(1)\ HBF_4} Ar-F$$

$$Ar-\overset{+}{N}_2 \xrightarrow{H_3PO_2,\ H_2O} Ar-H$$

$$C_6H_5-\overset{+}{N}_2 \xrightarrow[G\,=\,NR_2\text{ or }OH]{C_6H_5-G} C_6H_5-N{=}N-C_6H_4-G$$

3. Conversion to sulfonamides (discussed in Section 20.10):

$$R-NH-H \xrightarrow[(2)\ HCl]{(1)\ ArSO_2Cl,\ OH^-} R-NH-SO_2-Ar$$

$$R-N(R')-H \xrightarrow{ArSO_2Cl,\ OH^-} R-N(R')-SO_2-Ar$$

4. Conversion to amides (discussed in Section 18.8):

$$R-NH-H \xrightarrow[base]{R''COCl} R-NH-\overset{O}{\overset{\|}{C}}-R'' + Cl^-$$

$$R-NH-H \xrightarrow{(R''CO)_2O} R-NH-\overset{O}{\overset{\|}{C}}-R'' + R''-\overset{O}{\overset{\|}{C}}-OH$$

$$R-N(R')-H \xrightarrow[base]{R''COCl} R-N(R')-\overset{O}{\overset{\|}{C}}-R'' + Cl^-$$

5. Hofmann and Cope eliminations (discussed in Section 20.13):

Hofmann Elimination

$$-\overset{\displaystyle H}{\underset{|}{\overset{|}{C}}}-\underset{|}{\overset{|}{C}}-\overset{+}{N}R_3\ OH^- \xrightarrow{\text{heat}} \;\rangle C{=}C\langle \; + H_2O + NR_3$$

Cope Elimination

$$-\overset{\displaystyle H}{\underset{|}{\overset{|}{C}}}-\underset{|}{\overset{{}^{+}N(CH_3)_2\,(O^-)}{C}}- \xrightarrow[\text{(syn elimination)}]{\text{heat}} \;\rangle C{=}C\langle \; + (CH_3)_2NOH$$

KEY TERMS AND CONCEPTS

Amines as resolving agents	Section 20.3F
Aminium salts	Section 20.3D
Arylamines	Section 20.1A
Basicity of amines	Section 20.3
Diazonium salts	Sections 20.7A, 20.7B, 20.8, 20.9
Heterocyclic amines	Section 20.1B
N-Nitrosoamines	Section 20.7C
Primary (1°) amines	Section 20.1
Secondary (2°) amines	Section 20.1
Tertiary (3°) amines	Section 20.1
Quaternary ammonium salts	Sections 20.2B, 20.3D
Reductive amination	Section 20.5C
Sulfa drugs	Section 20.11B
Sulfonamides	Section 20.10

Additional Problems

20.21 Write structural formulas for each of the following compounds:

(a) Benzylmethylamine
(b) Triisopropylamine
(c) *N*-Ethyl-*N*-methylaniline
(d) *m*-Toluidine
(e) 2-Methylpyrrole
(f) *N*-Ethylpiperidine
(g) *N*-Ethylpyridinium bromide
(h) 3-Pyridinecarboxylic acid
(i) Indole
(j) Acetanilide
(k) Dimethylaminium chloride
(l) 2-Methylimidazole
(m) 3-Aminopropan-1-ol
(n) Tetrapropylammonium chloride
(o) Pyrrolidine
(p) *N,N*-Dimethyl-*p*-toluidine
(q) 4-Methoxyaniline
(r) Tetramethylammonium hydroxide
(s) *p*-Aminobenzoic acid
(t) *N*-Methylaniline

20.22 Give common or systematic names for each of the following compounds:

(a) $CH_3CH_2CH_2NH_2$
(b) $C_6H_5NHCH_3$
(c) $(CH_3)_2CH\overset{+}{N}(CH_3)_3\ I^-$
(d) o-$CH_3C_6H_4NH_2$
(e) o-$CH_3OC_6H_4NH_2$
(f)  [structure: imidazole ring, N–H]
(g) 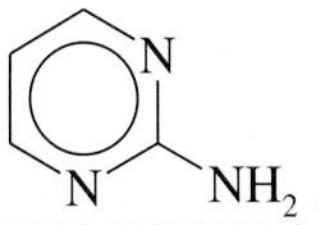 [structure: 2-aminopyrimidine ring with N, N, NH_2]
(h) $C_6H_5CH_2NH_3^+\ Cl^-$
(i) $C_6H_5N(CH_2CH_2CH_3)_2$
(j) $C_6H_5SO_2NH_2$
(k) $CH_3NH_3^+CH_3CO_2^-$
(l) $HOCH_2CH_2CH_2NH_2$
(m) 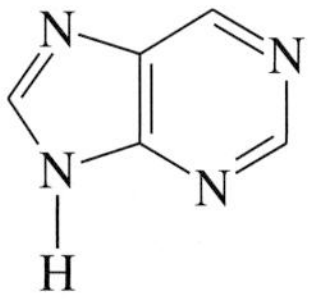 [structure: purine ring system, N–H]
(n) [structure: pyrrole ring, N–CH_3]

20.23 Show how you might prepare benzylamine from each of the following compounds:

(a) Benzonitrile
(b) Benzamide
(c) Benzyl bromide (two ways)
(d) Benzyl tosylate
(e) Benzaldehyde
(f) Phenylnitromethane
(g) Phenylacetamide

20.24 Show how you might prepare aniline from each of the following compounds:

(a) Benzene **(b)** Bromobenzene **(c)** Benzamide

20.25 Show how you might synthesize each of the following compounds from butyl alcohol:

(a) Butylamine (free of 2° and 3° amines)
(b) Pentylamine
(c) Propylamine
(d) Butylmethylamine

20.26 Show how you might convert aniline into each of the following compounds. (You need not repeat steps carried out in earlier parts of this problem.)

(a) Acetanilide
(b) *N*-Phenylphthalimide
(c) *p*-Nitroaniline
(d) Sulfanilamide
(e) *N,N*-Dimethylaniline
(f) Fluorobenzene
(g) Chlorobenzene
(h) Bromobenzene
(i) Iodobenzene
(j) Benzonitrile
(k) Benzoic acid
(l) Phenol
(m) Benzene
(n) *p*-(Phenylazo)phenol
(o) *N,N*-Dimethyl-*p*-(phenylazo)aniline

20.27 What products would you expect to be formed when each of the following amines reacts with aqueous sodium nitrite and hydrochloric acid?

(a) Propylamine
(b) Dipropylamine
(c) *N*-Propylaniline
(d) *N,N*-Dipropylaniline
(e) *p*-Propylaniline

20.28 **(a)** What products would you expect to be formed when each of the amines in the preceding problem reacts with benzenesulfonyl chloride and excess aqueous potassium hydroxide? **(b)** What would you observe in each reaction? **(c)** What would you observe when the resulting solution or mixture is acidified?

Note: Problems marked with an asterisk are "challenge problems."

20.29 **(a)** What product would you expect to obtain from the reaction of piperidine with aqueous sodium nitrite and hydrochloric acid? **(b)** From the reaction of piperidine and benzenesulfonyl chloride in excess aqueous potassium hydroxide?

20.30 Give structures for the products of each of the following reactions:

(a) Ethylamine + benzoyl chloride $\longrightarrow$
(b) Methylamine + acetic anhydride $\longrightarrow$
(c) Methylamine + succinic anhydride $\longrightarrow$
(d) Product of (c) $\xrightarrow{\text{heat}}$
(e) Pyrrolidine + phthalic anhydride $\longrightarrow$
(f) Pyrrole + acetic anhydride $\longrightarrow$
(g) Aniline + propanoyl chloride $\longrightarrow$
(h) Tetraethylammonium hydroxide $\xrightarrow{\text{heat}}$
(i) *m*-Dinitrobenzene + H_2S $\xrightarrow[C_2H_5OH]{NH_3}$
(j) *p*-Toluidine + Br_2 (excess) $\xrightarrow[H_2O]{}$

20.31 Starting with benzene or toluene, outline a synthesis of each of the following compounds using diazonium salts as intermediates. (You need not repeat syntheses carried out in earlier parts of this problem.)

(a) *p*-Fluorotoluene
(b) *o*-Iodotoluene
(c) *p*-Cresol
(d) *m*-Dichlorobenzene
(e) *m*-$C_6H_4(CN)_2$
(f) *m*-Iodophenol
(g) *m*-Bromobenzonitrile
(h) 1,3-Dibromo-5-nitrobenzene
(i) 3,5-Dibromoaniline
(j) 3,4,5-Tribromophenol
(k) 3,4,5-Tribromobenzonitrile
(l) 2,6-Dibromobenzoic acid
(m) 1,3-Dibromo-2-iodobenzene
(n) 4-Bromo-2-nitrotoluene
(o) 4-Methyl-3-nitrophenol
(p) CH_3 — (benzene ring with CN) — Br
(q) CH_3 — (benzene ring) — N=N — (benzene ring) — OH
(r) CH_3 — (benzene ring) — N=N — (benzene ring with OH and CH_3)

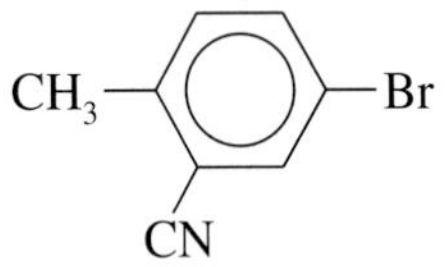

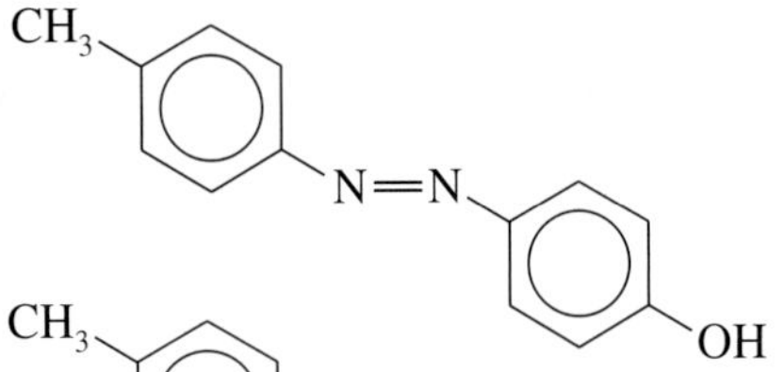

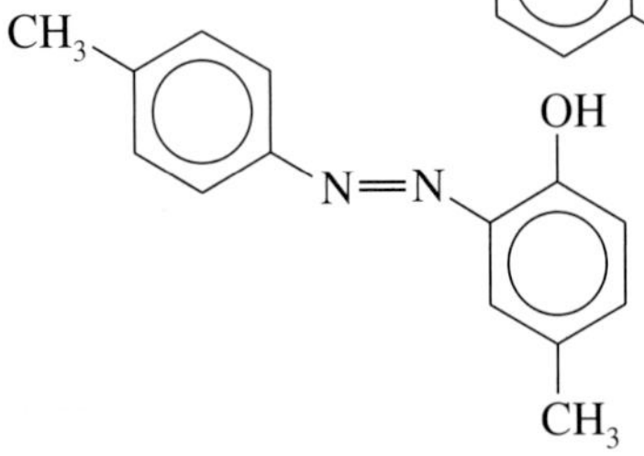

20.32 Write equations for simple chemical tests that would distinguish between

(a) Benzylamine and benzamide
(b) Allylamine and propylamine
(c) *p*-Toluidine and *N*-methylaniline
(d) Cyclohexylamine and piperidine
(e) Pyridine and benzene
(f) Cyclohexylamine and aniline
(g) Triethylamine and diethylamine
(h) Tripropylaminium chloride and tetrapropylammonium chloride
(i) Tetrapropylammonium chloride and tetrapropylammonium hydroxide

20.33 Describe with equations how you might separate a mixture of aniline, *p*-cresol, benzoic acid, and toluene using ordinary laboratory reagents.

20.34 Show how you might synthesize β-aminopropionic acid ($H_3\overset{+}{N}CH_2CH_2CO_2^-$) from succinic anhydride. (β-Aminopropionic acid is used in the synthesis of pantothenic acid; see Problem 18.34.)

20.35 Show how you might synthesize each of the following from the compounds indicated and any other needed reagents:

(a) $(CH_3)_3\overset{+}{N}(CH_2)_{10}\overset{+}{N}(CH_3)_3\ 2Br^-$ from 1,10-decanediol

(b) Succinylcholine bromide (see the chapter opening vignette) from succinic acid, 2-bromoethanol, and trimethylamine

20.36 A commercial synthesis of folic acid consists of heating the following three compounds with aqueous sodium bicarbonate. Propose reasonable mechanisms for the reactions that lead to folic acid.

(2,4,5-triaminopyrimidine with 6-OH: H_2N, N, NH_2, N, NH_2, OH) + $CHBr_2\overset{O}{\overset{\|}{C}}CH_2Br$ + $H_2N-C_6H_4-CONHCH(CO_2H)CH_2CH_2CO_2H$

$\xrightarrow{HCO_3^-,\ H_2O}$ **Folic acid** **(~10%)**

20.37 When compound **W** ($C_{15}H_{17}N$) is treated with benzenesulfonyl chloride and aqueous potassium hydroxide, no apparent change occurs. Acidification of this mixture gives a clear solution. The ^1H NMR spectrum of **W** is shown in Fig. 20.7. Propose a structure for **W**.

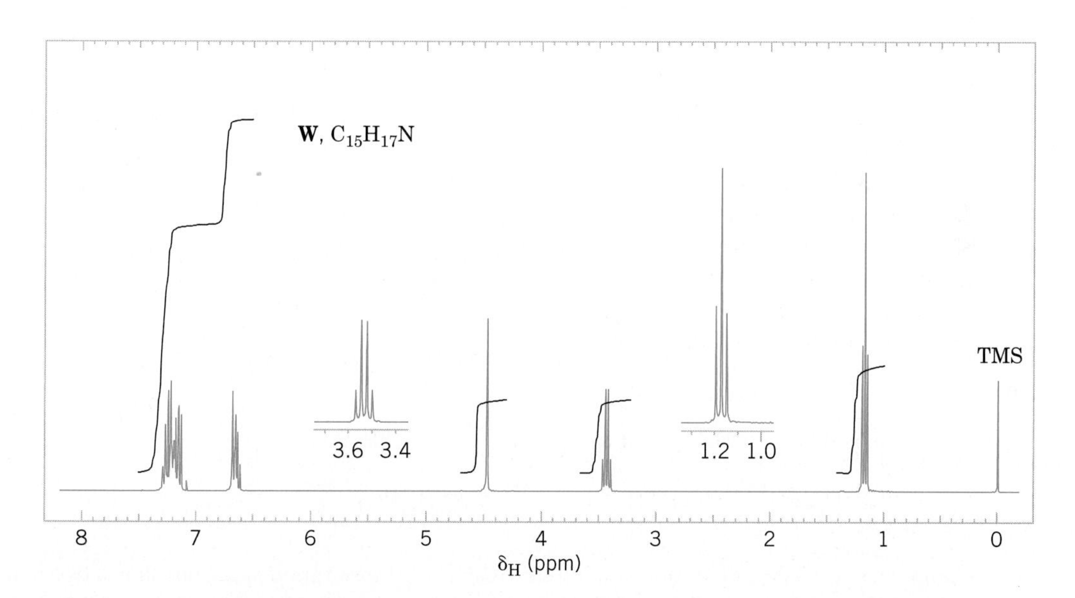

FIGURE 20.7 The 300-MHz ^1H NMR spectrum of compound **W,** Problem 20.37. Expansions of the signals are shown in the offset plots.

20.38 Propose structures for compounds **X, Y,** and **Z:**

$$\mathbf{X}\ (C_7H_7Br) \xrightarrow{NaCN} \mathbf{Y}\ (C_8H_7N) \xrightarrow{LiAlH_4} \mathbf{Z}\ (C_8H_{11}N)$$

The ^1H NMR spectrum of **X** gives two signals, a multiplet at δ 7.3 (5H) and a singlet at δ 4.25 (2H); the 680–840-cm^{-1} region of the IR spectrum of **X** shows peaks at 690 and 770 cm^{-1}. The ^1H NMR spectrum of **Y** is similar to that of **X**: multiplet at δ 7.3 (5H), singlet at δ 3.7 (2H). The ^{1}H NMR spectrum of **Z** is shown in Fig. 20.8.

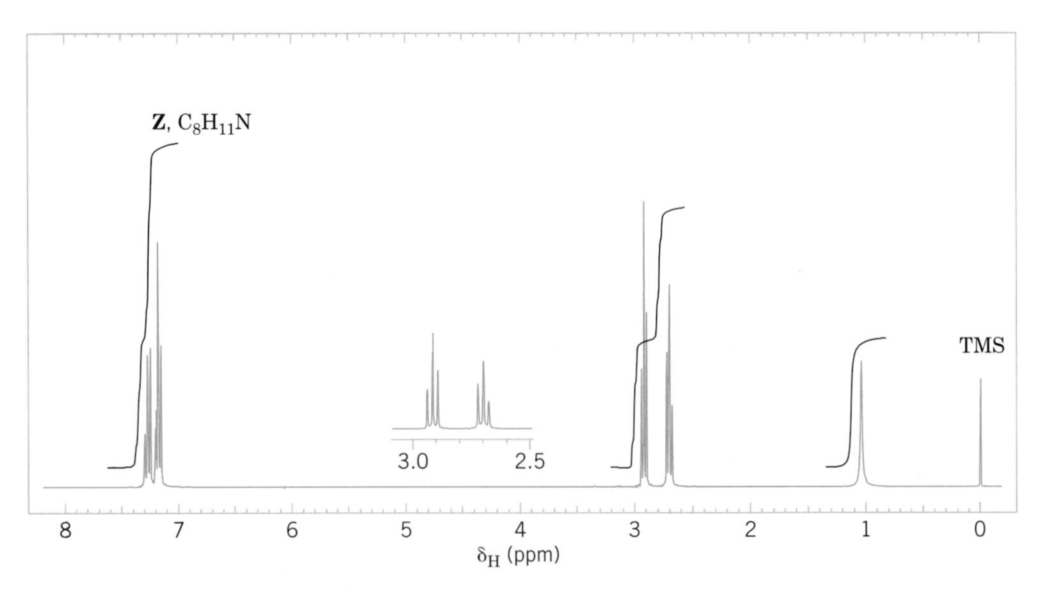

FIGURE 20.8 The 300-MHz ^{1}H NMR spectrum of compound **Z**, Problem 20.38. Expansion of the signals is shown in the offset plot.

20.39 Using reactions that we have studied in this chapter, propose a mechanism that accounts for the following reaction:

O
CH_2CH_2CN
$\xrightarrow{H_2,\ Pd}$
H
N

20.40 Give structures for compounds **R–W**:

$$N\text{-Methylpiperidine} + CH_3I \longrightarrow \mathbf{R}\ (C_7H_{16}NI) \xrightarrow[H_2O]{Ag_2O} \mathbf{S}\ (C_7H_{17}NO) \xrightarrow[(-H_2O)]{\text{heat}}$$

$$\mathbf{T}\ (C_7H_{15}N) \xrightarrow{CH_3I} \mathbf{U}\ (C_5H_{18}NI) \xrightarrow[H_2O]{Ag_2O} \mathbf{V}\ (C_8H_{19}NO) \xrightarrow{\text{heat}} \mathbf{W}\ (C_5H_8) + H_2O + (CH_3)_3N$$

20.41 Compound **A** ($C_{10}H_{15}N$) is soluble in dilute HCl. The IR absorption spectrum shows two bands in the 3300–3500-cm^{-1} region. The broadband proton-decoupled ^{13}C spectrum of **A** is given in Fig. 20.9. Propose a structure for **A.**

20.42 Compound **B,** an isomer of **A** (Problem 20.41), is also soluble in dilute HCl. The IR spectrum of **B** shows no bands in the 3300–3500-cm^{-1} region. The broadband proton-decoupled ^{13}C spectrum of **B** is given in Fig. 20.9. Propose a structure for **B.**

20.43 Compound **C** ($C_9H_{11}NO$) gives a positive Tollens test and is soluble in dilute HCl. The IR spectrum of **C** shows a strong band near 1695 cm^{-1} but shows no bands in the 3300–3500-cm^{-1} region. The broadband proton-decoupled ^{13}C NMR spectrum of **C** is shown in Fig. 20.9. Propose a structure for **C.**

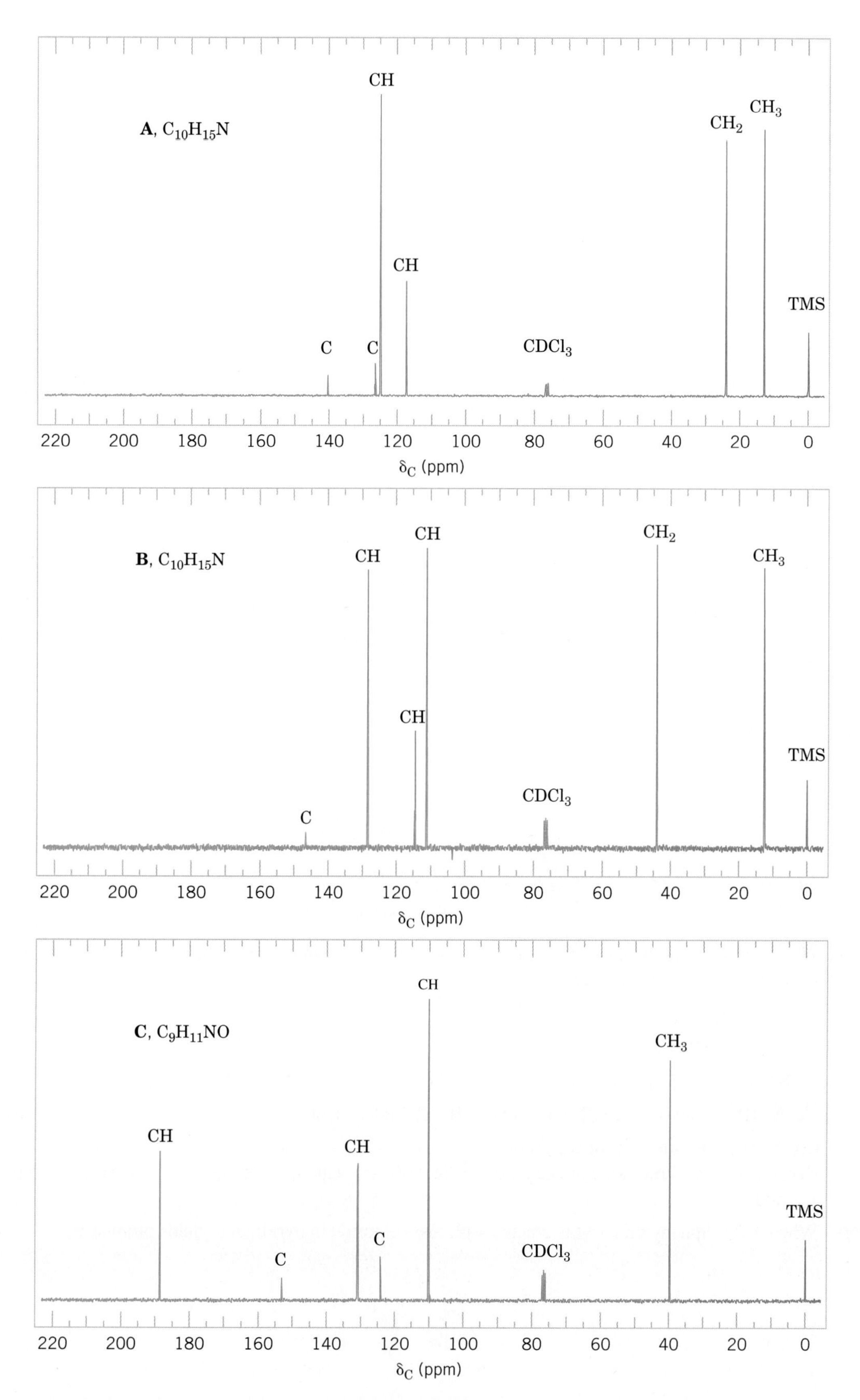

FIGURE 20.9 The broadband proton-decoupled ^{13}C NMR spectra of compounds A, B, and C, Problems 20.41–20.43. Information from the DEPT ^{13}C NMR spectra is given above each peak.

20.44 Outline a synthesis of acetylcholine iodide using as organic starting materials: dimethylamine, oxirane, iodomethane, and acetyl chloride.

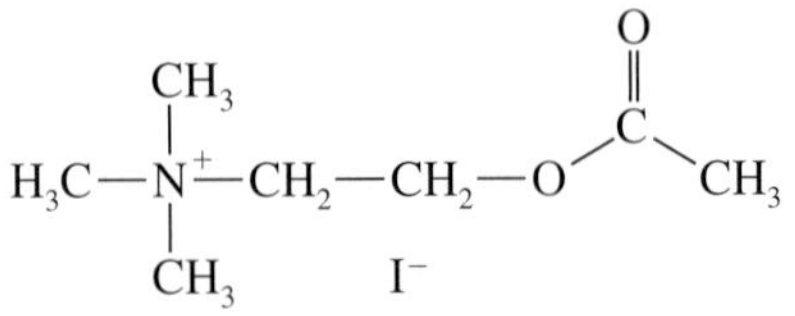

$$H_3C-\overset{CH_3}{\underset{CH_3}{\overset{|}{\underset{|}{N^+}}}}-CH_2-CH_2-O-\overset{O}{\overset{\|}{C}}-CH_3 \quad I^-$$

Acetylcholine iodide

20.45 Ethanolamine, $HOCH_2CH_2NH_2$, and diethanolamine, $(HOCH_2CH_2)_2NH$, are used commercially to form emulsifying agents and to absorb acidic gases. Propose syntheses of these two compounds.

20.46 Diethylpropion (see the following structure) is a compound used in the treatment of anorexia. Propose a synthesis of diethylpropion starting with benzene and using any other needed reagents.

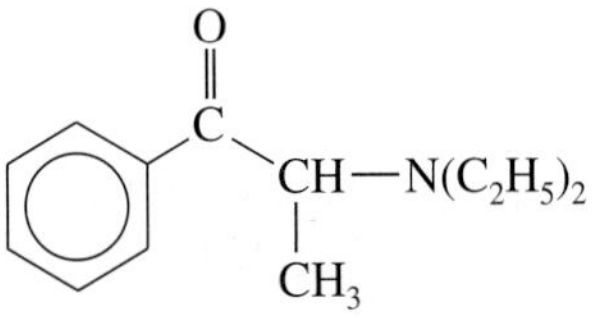

$$C_6H_5-\overset{O}{\overset{\|}{C}}-\underset{CH_3}{\underset{|}{CH}}-N(C_2H_5)_2$$

Diethylpropion

20.47 Suggest an experiment to test the proposition that the Hofmann reaction is an intramolecular rearrangement, that is, one in which the migrating R group never fully separates from the amide molecule.

20.48 Using as starting materials 2-chloropropanoic acid, aniline, and 2-naphthol, propose a synthesis of naproanilide, a herbicide used in rice paddies in Asia:

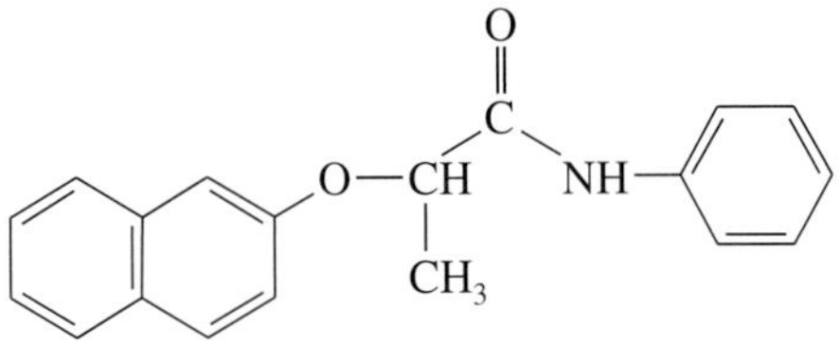

$$C_{10}H_7-O-\underset{CH_3}{\underset{|}{CH}}-\overset{O}{\overset{\|}{C}}-NH-C_6H_5$$

Naproanilide

***20.49** When phenyl isothiocyanate, $C_6H_5N{=}C{=}S$, is reduced with lithium aluminum hydride, the product formed has these spectral data:

MS (*m/z*): 107, 106

IR (cm^{-1}): 3330 (sharp), 3050, 2815, 760, 700

^{1}H NMR (δ): 2.7 (s), 3.5 (broad), 6.6 (m), 7.2 (t)

^{13}C NMR (δ): 30 (CH_3), 112 (CH), 117 (CH), 129 (CH), 150 (C)

(a) What is the structure of the product?

(b) What is the structure that accounts for the 106 *m/z* peak and how is it formed? (It is an iminium ion.)

***20.50** When *N,N′*-diphenylurea **(A)** is reacted with tosyl chloride in pyridine, it yields product **B.**

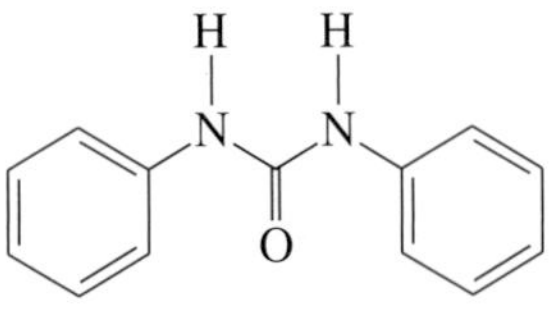

$$C_6H_5-\overset{H}{\overset{|}{N}}-\overset{}{\underset{O}{\underset{\|}{C}}}-\overset{H}{\overset{|}{N}}-C_6H_5$$

A

The spectral data for **B** include:

MS (*m/z*): 194 ($M^{\ddagger}$)

IR (cm^{-1}): 3060, 2130, 1590, 1490, 760, 700

^{1}H NMR (δ): only 6.9–7.4 (m)

^{13}C NMR (δ): 122 (CH), 127 (CH), 130 (CH), 149 (C), and 163 (C)

(a) What is the structure of **B?**

(b) Write a mechanism for the formation of **B.**

***20.51** Propose a mechanism that can explain the occurrence of this reaction:

***20.52** When acetone is treated with anhydrous ammonia in the presence of anhydrous calcium chloride (a common drying agent), crystalline product **C** is obtained on concentration of the organic liquid phase of the reaction mixture.

These are spectral data for product **C:**

MS (*m/z*): 155 ($M^{\ddagger}$), 140

IR (cm^{-1}): 3350 (sharp), 2850–2960, 1705

^{1}H NMR (δ): 2.3 (s, 4H), 1.7 (1H; disappears in D_2O), and 1.2 (s, 12H)

(a) What is the structure of **C?**

(b) Propose a mechanism for the formation of **C.**

20.53 The difference in positive-charge distribution in an amide that accepts a proton on its oxygen or its nitrogen atom can be visualized with electrostatic potential maps. Consider the electrostatic potential maps for acetamide in its O—H and N—H protonated forms shown below. On the basis of the electrostatic potential maps, which protonated form appears to delocalize, and hence stabilize, the formal positive charge more effectively? Discuss your conclusion in terms of resonance contributors for the two possible protonated forms of acetamide.

Chem3D Model

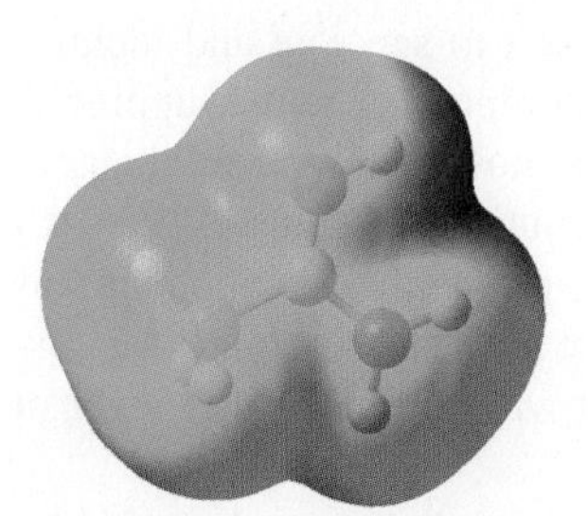

Acetamide protonated on oxygen

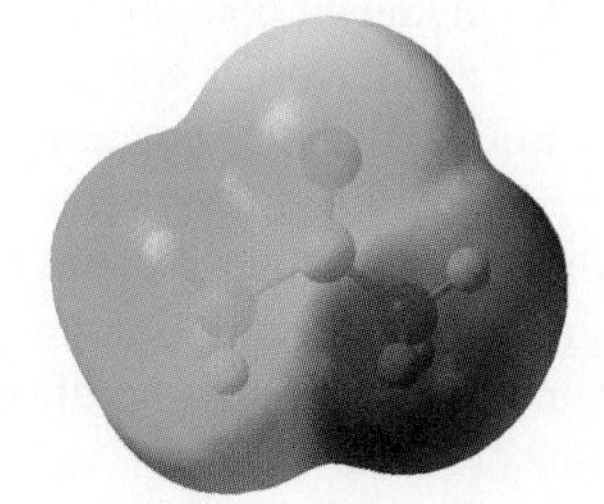

Acetamide protonated on nitrogen

LEARNING GROUP PROBLEMS

1. Reserpine is a natural product belonging to the family of alkaloids (see Special Topic E). Reserpine was isolated from the Indian snakeroot *Rauwolfia serpentina.* Clinical applications of reserpine include treatment of hypertension and nervous and mental disorders. The synthesis of reserpine, which contains six stereogenic centers, was a landmark accomplishment reported by R. B. Woodward in 1955. Incorporated in the synthesis are several reactions involving amines and related nitrogen-containing functional groups.

Reserpine

(a) The goal of the first two steps shown in the scheme that follows, prior to formation of the amide, is preparation of a secondary amine. Draw the structure of the products labeled **A** and **B** from the first and second reactions, respectively. Write a mechanism for formation of **A.**

(b) The next sequence of reactions involves formation of a tertiary amine together with closure of a new ring. Write curved arrows to show how the amide functional group reacts with phosphorus oxychloride ($POCl_3$) to place the leaving group on the bracketed intermediate.

(c) The ring closure from the bracketed intermediate involves a type of electrophilic aromatic substitution reaction characteristic of indole rings. Identify the part of the structure that contains the indole ring. Write mechanism arrows to show how the nitrogen in the indole ring, via conjugation, can cause electrons from the adjacent carbon to attack an electrophile. In this case, the attack by the indole ring in the bracketed intermediate is an addition–elimination reaction, somewhat like reactions that occur at carbonyls bearing leaving groups.

2. **(a)** A student was given a mixture of two unknown compounds and asked to separate and identify them. One of the compounds was an amine and the other was a neutral compound (neither appreciably acidic nor basic). Describe how you would go about separating the unknown amine from the neutral compound using extraction techniques involving diethyl ether and solutions of aqueous 5% HCl and 5% $NaHCO_3$. The mixture as a whole was soluble in diethyl ether, but neither component was soluble in water at pH 7. Using R groups on a generic amine, write out the reactions for any acid–base steps you propose and explain why the compound of interest will be in the ether layer or aqueous layer at any given time during the process.

(b) Once the amine was successfully isolated and purified, it was reacted with benzenesulfonyl chloride in the presence of aqueous sodium hydroxide. The reaction led to a solution that on acidification produced a precipitate. The results just described constitute a test (Hinsberg's) for the class of an amine. What class of amine was the unknown compound: primary, secondary, or tertiary? Write the reactions involved for a generic amine of the class you believe this one to be.

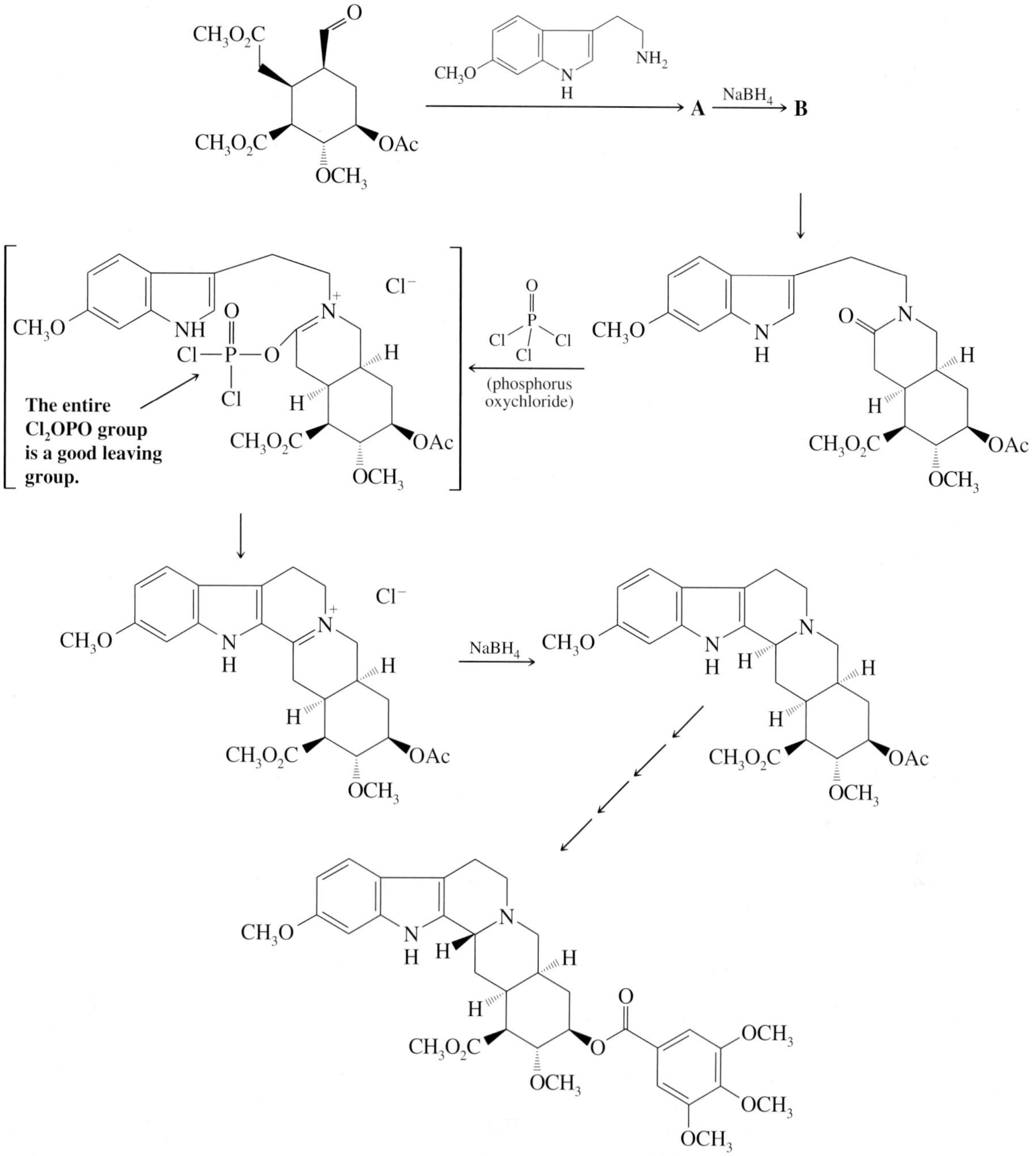

$\xrightarrow{\text{NaBH}_4}$ A $\xrightarrow{\text{NaBH}_4}$ B

(phosphorus oxychloride)

The entire Cl_2OPO group is a good leaving group.

Reserpine

(c) The unknown amine was then analyzed by IR, NMR, and MS. The following data were obtained. On the basis of this information, deduce the structure of the unknown amine. Assign the spectral data to specific aspects of the structure you propose for the amine.

IR (cm^{-1}): 3360, 3280, 3020, 2962, 1604, 1450, 1368, 1021, 855, 763, 700, 538

^{1}H NMR (δ): 1.35 (d, 3H), 1.8 (bs, 2H), 4.1 (q, 1H), 7.3 (m, 5H)

MS (*m/z*): 121, 120, 118, 106 (base peak), 79, 77, 51, 44, 42, 28, 18, 15

Alkaloids

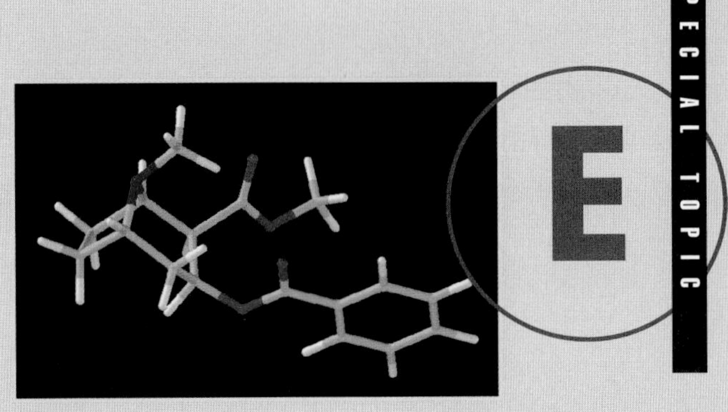

Cocaine

Erythroxylum coca, a shrub whose leaves contain about 1% cocaine.

Extracting the bark, roots, leaves, berries, and fruits of plants often yields nitrogen-containing bases called *alkaloids*. The name alkaloid comes from the fact that these substances are "alkali-like"; that is, since alkaloids are amines, they often react with acids to yield soluble salts. The nitrogen atoms of most alkaloids are present in heterocyclic rings. In a few instances, however, nitrogen may be present as a primary amine or as a quaternary ammonium group.

When administered to animals, most alkaloids produce striking physiological effects, and the effects *vary greatly* from alkaloid to alkaloid. Some alkaloids stimulate the central nervous system, others cause paralysis; some alkaloids elevate blood pressure, others lower it. Certain alkaloids act as pain relievers; others act as tranquilizers; still others act against infectious microorganisms. Most alkaloids are toxic when their dosage is large enough, and with some this dosage is very small. In spite of this, many alkaloids find use in medicine.

Systematic names are seldom used for alkaloids, and their common names have a variety of origins. In many instances the common name reflects the botanical source of the compound. The alkaloid strychnine, for example, comes from the seeds of the *Strychnos* plant. In other instances the names are more whimsical: The name of the opium alkaloid morphine comes from Morpheus, the ancient Greek god of dreams; the name of the tobacco alkaloid nicotine comes from Nicot, an early French ambassador who sent tobacco seeds to France. The one characteristic that alkaloid names have in common is the ending *-ine,* reflecting the fact that they are all amines.

Alkaloids have been of interest to chemists for centuries, and in that time thousands of alkaloids have been isolated. Most of these have had their structures determined through the application of chemical and physical methods, and in many instances these structures have been confirmed by independent synthesis. A complete account of the chemistry of the alkaloids would (and does) occupy volumes; here we have space to consider only a few representative examples.

E.1 Alkaloids Containing a Pyridine or Reduced Pyridine Ring

The predominant alkaloid of the tobacco plant is nicotine:

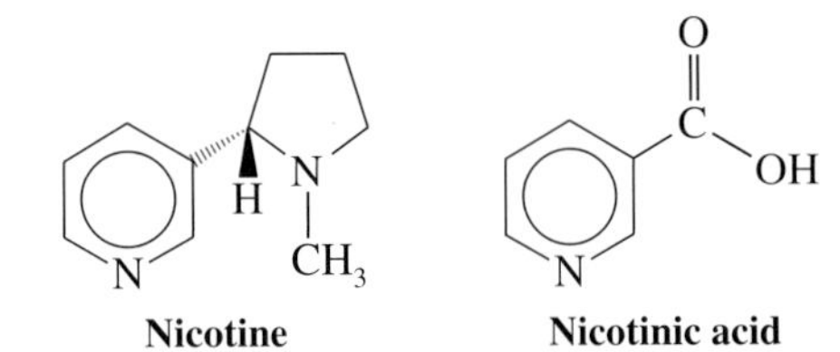

Nicotine **Nicotinic acid**

Nicotine is addictive. In very small doses nicotine acts as a stimulant, but in larger doses it causes depression, nausea, and vomiting. In still larger doses it is a violent poison. Nicotine salts are used as insecticides.

Oxidation of nicotine by concentrated nitric acid produces pyridine-3-carboxylic acid—a compound that is called *nicotinic acid.* Whereas the consumption of nicotine is of no benefit to humans, nicotinic acid is a vitamin; it is incorporated into an important coenzyme, nicotinamide adenine dinucleotide, commonly referred to as NAD^+ (oxidized form).

PROBLEM E.1

Nicotine has been synthesized by the following route. All of the steps involve reactions that we have seen before. Suggest reagents that could be used for each.

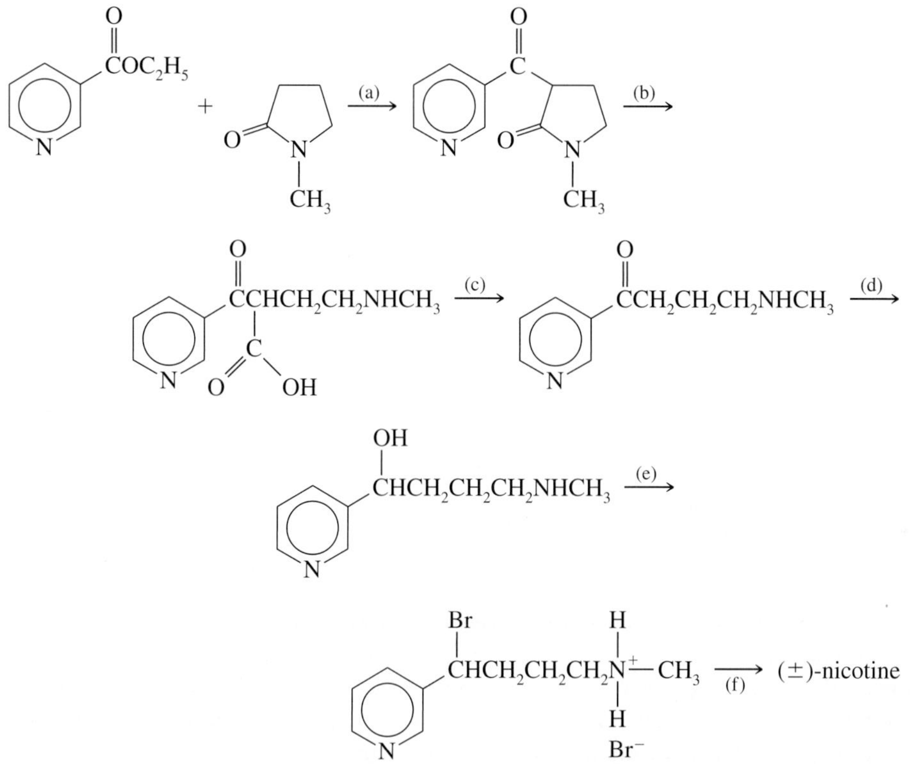

(a) (b) (c) (d) (e) (f) (±)-nicotine

A number of alkaloids contain a piperidine ring. These include coniine (from the poison hemlock, *Conium maculatum,* a member of the carrot family, Umbelliferae), atropine (from *Atropa belladonna* and other genera of the plant family, Solanaceae), and cocaine (from *Erythroxylum coca*):

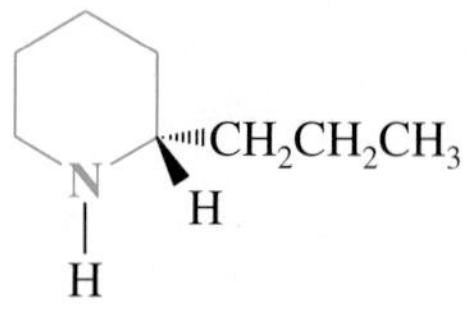

Coniine
[(+)-2-propylpiperidine]

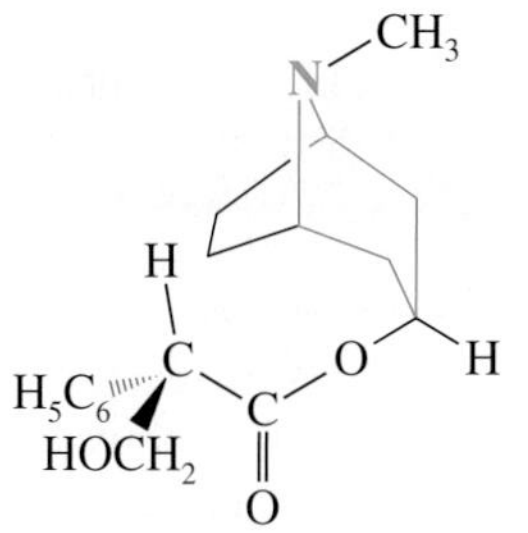

Atropine

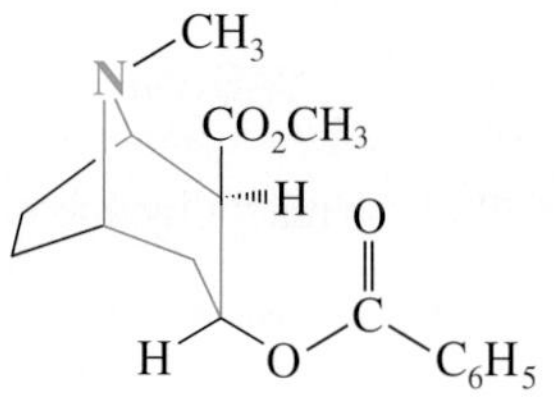

Cocaine

Coniine is toxic; its ingestion may cause weakness, drowsiness, nausea, labored respiration, paralysis, and death. Coniine was one toxic substance of the "hemlock" used in the execution of Socrates (other poisons may have been included as well).

In small doses cocaine decreases fatigue, increases mental activity, and gives a general feeling of well-being. Prolonged use of cocaine, however, leads to physical addiction and to periods of deep depression. Cocaine is also a local anesthetic, and, for a time, it was used medically in that capacity. When its tendency to cause addiction was recognized, efforts were made to develop other local anesthetics. This led, in 1905, to the synthesis of Novocain, a compound also called procaine, that has some of the same structural features as cocaine (e.g., its benzoic ester and tertiary amine groups):

$$(CH_3CH_2)_2N{-}CH_2CH_2{-}O{-}\overset{O}{\overset{\|}{C}}{-}C_6H_4{-}NH_2$$

Novocain (procaine)

Atropine is an intense poison. In dilute solutions (0.5–1.0%) it is used to dilate the pupil of the eye in ophthalmic examinations. Compounds related to atropine are contained in the 12-h continuous-release capsules used to relieve symptoms of the common cold.

PROBLEM E.2

The principal alkaloid of *Atropa belladonna* is the optically active alkaloid *hyoscyamine.* During its isolation hyoscyamine is often racemized by bases to optically inactive atropine. **(a)** What stereogenic center is likely to be involved in the racemization? **(b)** In hyoscyamine this stereogenic center has the (*S*) configuration. Write a three-dimensional structure for hyoscyamine.

PROBLEM E.3

Hydrolysis of atropine gives tropine and (±)-tropic acid. **(a)** What are their structures? **(b)** Even though tropine has a stereogenic center, it is optically inactive. Explain. **(c)** An isomeric form of tropine called ψ-tropine has also been prepared by heating tropine with base. ψ-Tropine is also optically inactive. What is its structure?

PROBLEM E.4

In 1891 G. Merling transformed tropine (see Problem E.3) into 1,3,5-cycloheptatriene (tropylidene) through the following sequence of reactions:

$$\text{Tropine } (C_8H_{15}NO) \xrightarrow{-H_2O} C_8H_{13}N \xrightarrow{CH_3I} C_9H_{16}NI \xrightarrow[\text{(2) heat}]{\text{(1) Ag}_2\text{O/H}_2\text{O}}$$

$$C_9H_{15}N \xrightarrow{CH_3I} C_{10}H_{18}NI \xrightarrow[\text{(2) heat}]{\text{(1) Ag}_2\text{O/H}_2\text{O}} \text{1,3,5-cycloheptatriene} + (CH_3)_3N + H_2O$$

Write out all of the reactions that take place.

PROBLEM E.5

Many alkaloids appear to be synthesized in plants by reactions that resemble the Mannich reaction (Section 19.10). Recognition of this (by R. Robinson in 1917) led to a synthesis of tropinone that takes place under "physiological conditions," that is, at room temperature and at pH values near neutrality. This synthesis is shown here. Propose reasonable mechanisms that account for the overall course of the reaction.

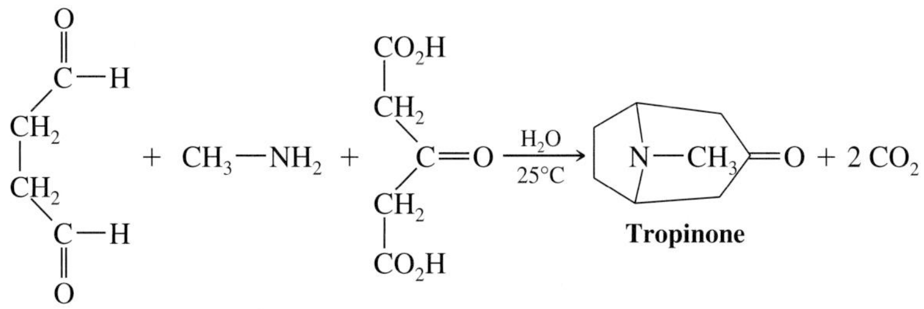

Tropinone

E.2 Alkaloids Containing an Isoquinoline or Reduced Isoquinoline Ring

Papaverine, morphine, and codeine are all alkaloids obtained from the opium poppy, *Papaver somniferum:*

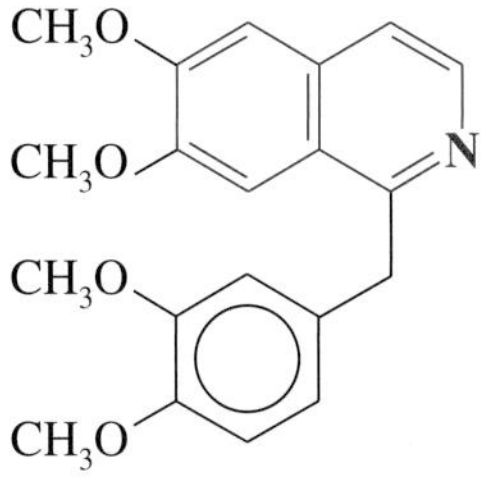

Papaverine

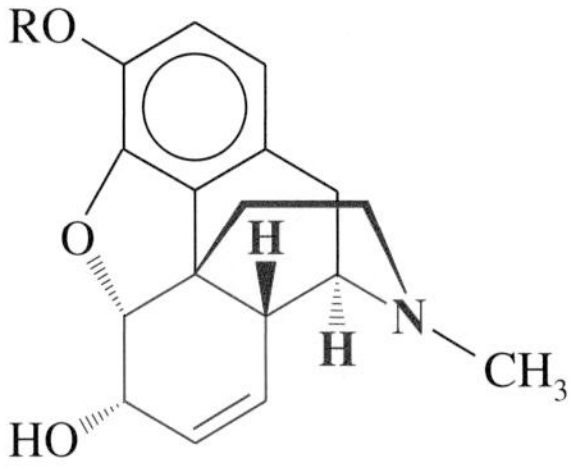

Morphine (R = H)
Codeine (R = CH_3)

Papaverine has an isoquinoline ring; in morphine and codeine the isoquinoline ring is partially hydrogenated (reduced).

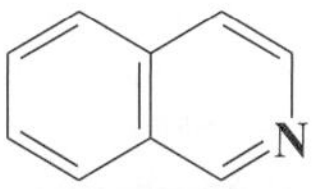

Isoquinoline

Opium has been used since earliest recorded history. Morphine was first isolated from opium in 1803, and its isolation represented one of the first instances of the purification of the active principle of a drug. One hundred twenty years were to pass, however, before the complicated structure of morphine was deduced, and its final confirmation through independent synthesis (by Marshall Gates of the University of Rochester) did not take place until 1952.

Morphine is one of the most potent analgesics known, and it is still used extensively in medicine to relieve pain, especially "deep" pain. Its greatest drawbacks, however, are its tendencies to lead to addiction and to depress respiration. These disadvantages have brought about a search for morphinelike compounds that do not have these disadvantages. One of the newest candidates is the compound pentazocine. Pentazocine is a highly effective analgesic and is nonaddictive; unfortunately, however, like morphine, it depresses respiration.

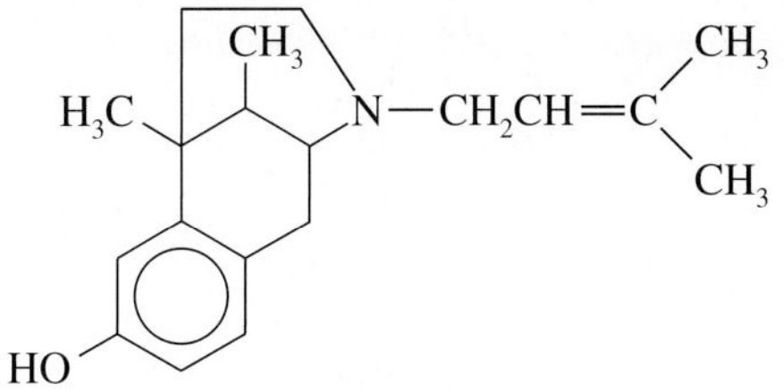

Pentazocine

PROBLEM E.6

Papaverine has been synthesized by the following route:

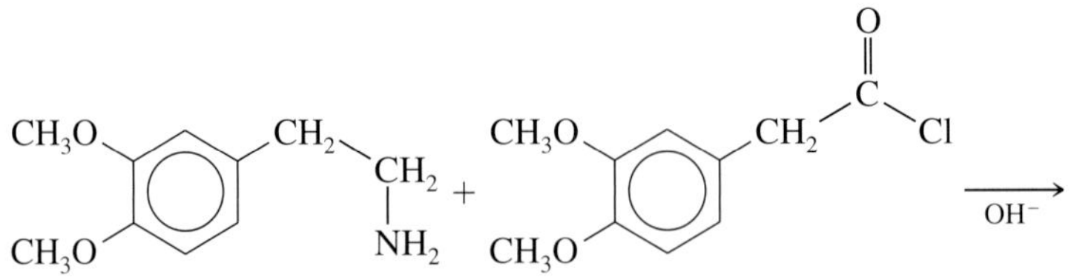

CH_3O, CH_3O — $CH_2CH_2NH_2$ + CH_3O, CH_3O — $CH_2C(=O)Cl$ $\xrightarrow{OH^-}$

$$C_{20}H_{25}NO_5 \xrightarrow[\text{heat} \; (-H_2O)]{P_4O_{10}} \text{dihydropapaverine} \xrightarrow[\text{heat} \; (-H_2)]{Pd} \text{papaverine}$$

Outline the reactions involved.

PROBLEM E.7

One of the important steps in Gates' synthesis of morphine involved the following transformation:

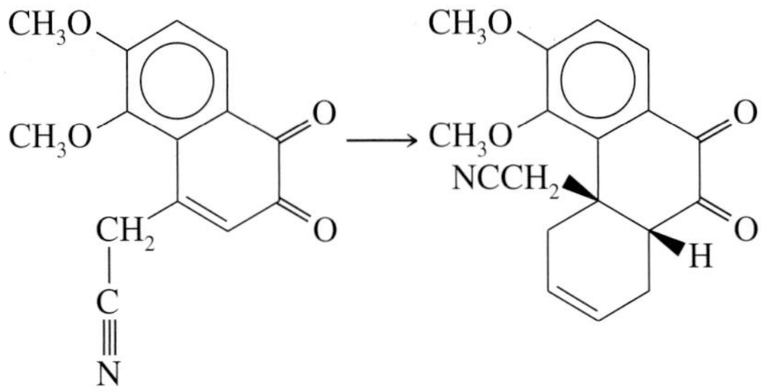

CH_3O, CH_3O, CH_2, C, N, O, O $\longrightarrow$ CH_3O, CH_3O, $NCCH_2$, O, O, H

Suggest how this step was accomplished.

PROBLEM E.8

When morphine reacts with 2 mol of acetic anhydride, it is transformed into the highly addictive narcotic, heroin. What is the structure of heroin?

E.3 Alkaloids Containing Indole or Reduced Indole Rings

A large number of alkaloids are derivatives of an indole ring system. These range from the relatively simple *gramine* to the highly complicated structures of *strychnine* and *reserpine:*

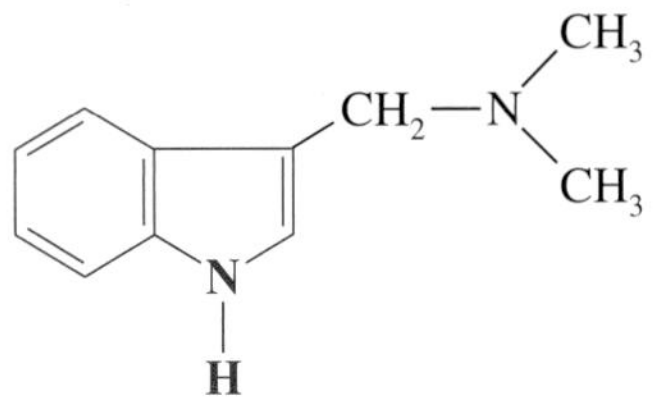

$CH_2—N(CH_3)_2$, N, H

Gramine

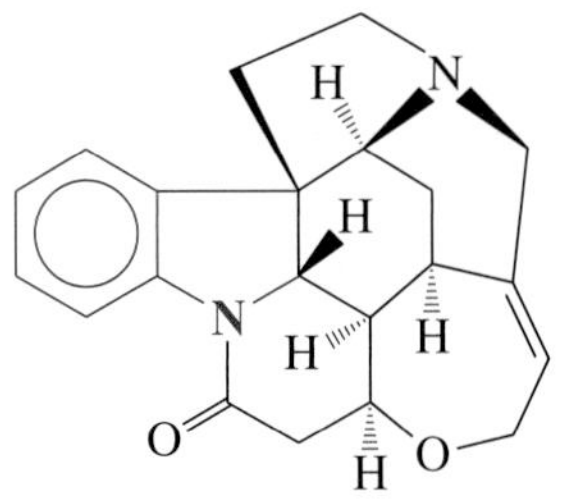

H, N, H, N, H, H, O, H, O

Strychnine

CH_3O, N, H, H, N, H, H, $CH_3OC(=O)$, OCH_3, O—C(=O)—, OCH_3, OCH_3, OCH_3

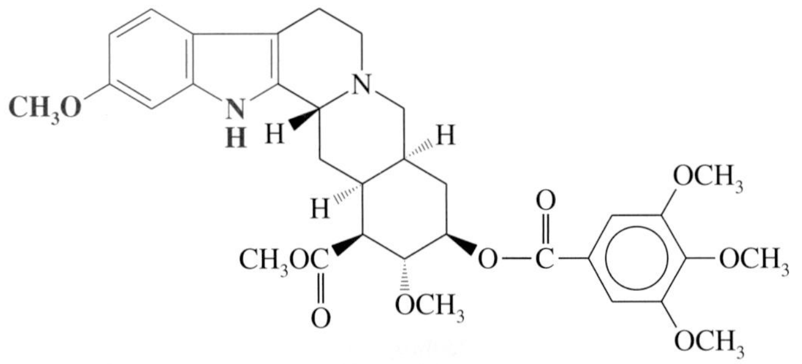

Reserpine

Gramine can be obtained from chlorophyll-deficient mutants of barley. Strychnine, a very bitter and highly poisonous compound, comes from the seeds of *Strychnos nuxvomica.* Strychnine is a central nervous system stimulant and has been used medically (in low dosage) to counteract poisoning by central nervous system depressants. Reserpine can be obtained from the Indian snakeroot *Rauwolfia serpentina,* a plant that has been used in native medicine for centuries. Reserpine is used in modern medicine as a tranquilizer and as an agent to lower blood pressure. See the Chapter 20 Learning Group Problem for an exercise related to the synthesis of reserpine.

PROBLEM E.9

Gramine has been synthesized by heating a mixture of indole, formaldehyde, and dimethylamine. **(a)** What general reaction is involved here? **(b)** Outline a reasonable mechanism for the gramine synthesis.

CHAPTER 21

Phenols and Aryl Halides: Nucleophilic Aromatic Substitution

A Silver Chalice

Calixarenes are an intriguing family of bowl-shaped molecules whose potential applications include medicinal, industrial, and analytical uses. As we shall see, the name calixarene aptly describes their structure and properties because the Greek word *calix* means chalice. An example is 4-*tert*-butylcalix[4]arene, shown below and in the molecular graphic above. 4-*tert*-Butylcalix[4]arene is formed by condensation of four 4-*tert*-butylphenol molecules with four formaldehyde molecules.

4-*tert*-Butylcalix[4]arene

Calixarenes are excellent host molecules, especially for Lewis acids. The phenolic hydroxyl groups that crown calixarenes are perfectly situated to coordinate with metal ions. Because calixarenes so readily participate in host–guest interactions, their potential as enzyme mimics, selective ionophores, and even biocidal agents has been widely investigated. Calixarenes complexed with silver ions, for example, are effective against *Escherichia coli,* herpes simplex virus, and HIV-1. Calixarene silver chalices have been tested in oral and topical antimicrobial applications, as well as in coatings for utensils, appliances, and hospital instruments. They have even shown antimicrobial activity when mixed with paint.

The structure and synthesis of calixarenes embody many aspects of the chemistry of phenols. Phenols readily undergo electrophilic aromatic substitution reactions, which is key to the synthesis of calixarenes. The hydroxyl group of phenols and calixarenes is also a locus for intermolecular interactions—the most central property of calixarenes. Furthermore, phenolic hydroxyl groups can be used to form other functional groups such as ethers, esters, and acetals. Transformations of this sort have been used to optimize calixarenes for the binding of specific metal ions and other guests. Depending on the specific ring size and substituents, calixarenes are able to form complexes with mercury, cesium, potassium, calcium, sodium, lithium, and, of course, silver.

21.1 Structure and Nomenclature of Phenols

Compounds that have a hydroxyl group directly attached to a benzene ring are called **phenols.** Thus, **phenol** is the specific name for hydroxybenzene, and it is the general name for the family of compounds derived from hydroxybenzene:

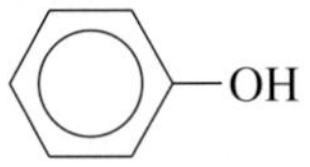

OH H_3C OH

Phenol  **4-Methylphenol (a phenol)**

Compounds that have a hydroxyl group attached to a polycyclic benzenoid ring are chemically similar to phenols, but they are called **naphthols** and **phenanthrols,** for example:

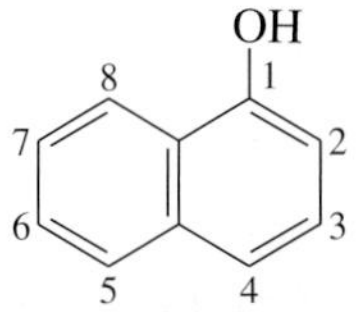

1-Naphthol (α-naphthol) 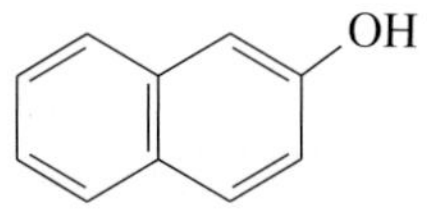 **2-Naphthol (β-naphthol)** 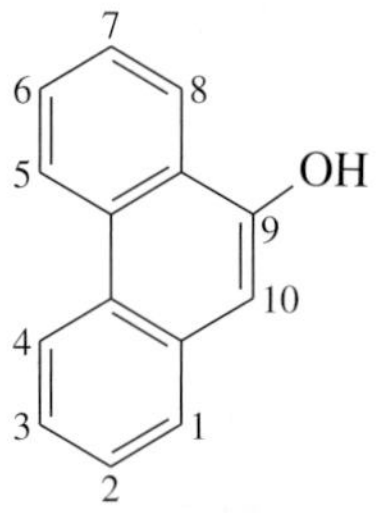 **9-Phenanthrol**

21.1A Nomenclature of Phenols

We studied the nomenclature of some of the phenols in Chapter 14. In many compounds *phenol* is the base name:

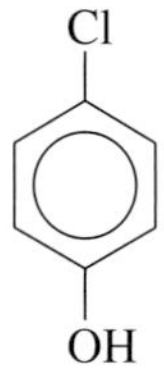

Cl, OH

4-Chlorophenol
(*p*-chlorophenol)

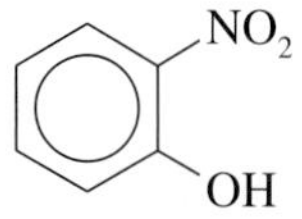

NO_2, OH

2-Nitrophenol
(*o*-nitrophenol)

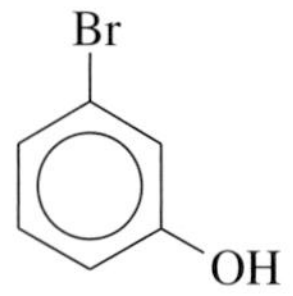

Br, OH

3-Bromophenol
(*m*-bromophenol)

The methylphenols are commonly called *cresols:*

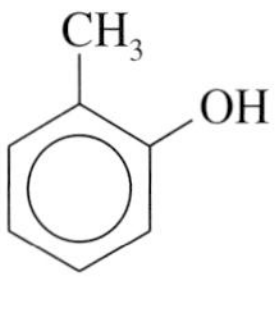

CH_3, OH

2-Methylphenol
(*o*-cresol)

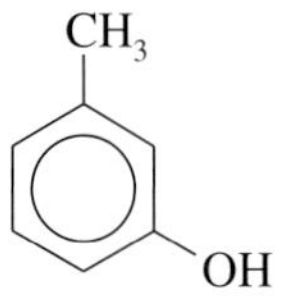

CH_3, OH

3-Methylphenol
(*m*-cresol)

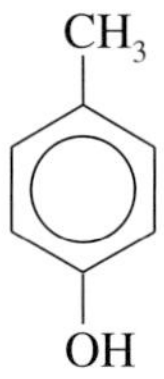

CH_3, OH

4-Methylphenol
(*p*-cresol)

The benzenediols also have common names:

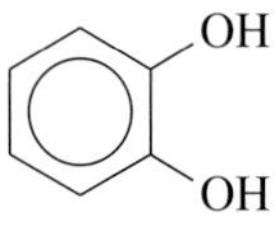

OH, OH

1,2-Benzenediol
(catechol)

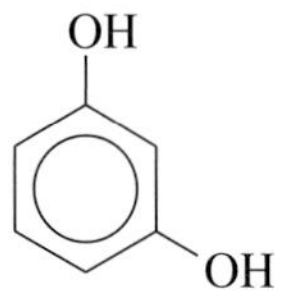

OH, OH

1,3-Benzenediol
(resorcinol)

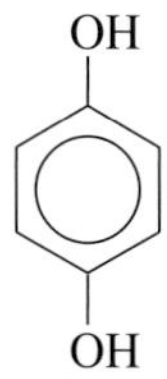

OH, OH

1,4-Benzenediol
(hydroquinone)

21.2 Naturally Occurring Phenols

Phenols and related compounds occur widely in nature. Tyrosine is an amino acid that occurs in proteins. (See the "Biosynthesis of Thyroxine" and "The Chemistry of . . . Iodine Incorporation in Thyroxine Biosynthesis" in Chapter 15.) Methyl salicylate is found in oil of wintergreen, eugenol is found in oil of cloves, and thymol is found in thyme.

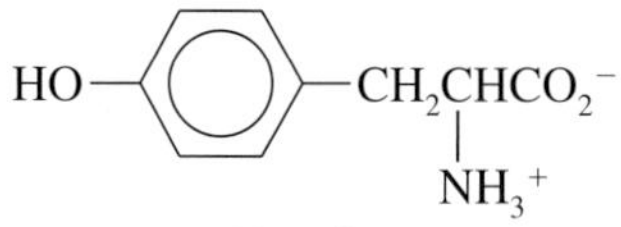

HO—C_6H_4—$CH_2CHCO_2^-$ (NH_3^+)

Tyrosine

CO_2CH_3, OH

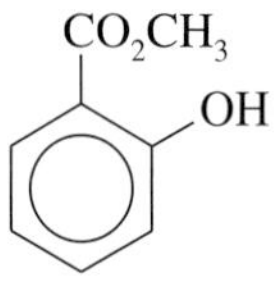

Methyl salicylate
(oil of wintergreen)

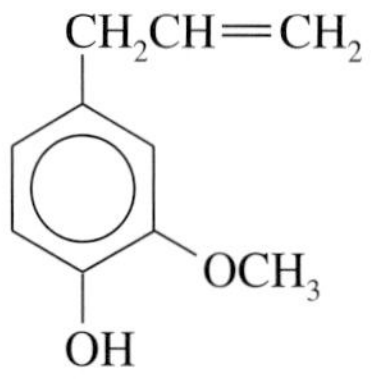

$CH_2CH{=}CH_2$, OCH_3, OH

Eugenol
(oil of cloves)

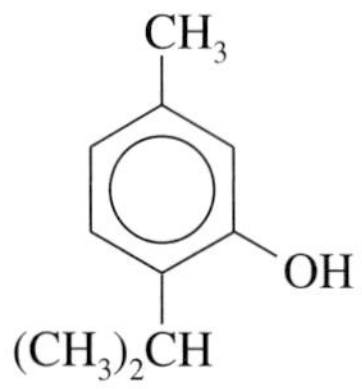

CH_3, OH, $(CH_3)_2CH$

Thymol
(thyme)

The urushiols are blistering agents (vesicants) found in poison ivy.

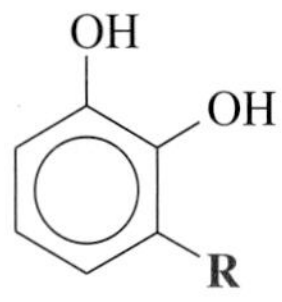

R = $-(CH_2)_{14}CH_3$,
$-(CH_2)_7CH{=}CH(CH_2)_5CH_3$, or
$-(CH_2)_7CH{=}CHCH_2CH{=}CH(CH_2)_2CH_3$, or
$-(CH_2)_7CH{=}CHCH_2CH{=}CHCH{=}CHCH_3$ or
$-(CH_2)_7CH{=}CHCH_2CH{=}CHCH_2CH{=}CH_2$

Urushiols

Estradiol is a female sex hormone, and the tetracyclines are important antibiotics.

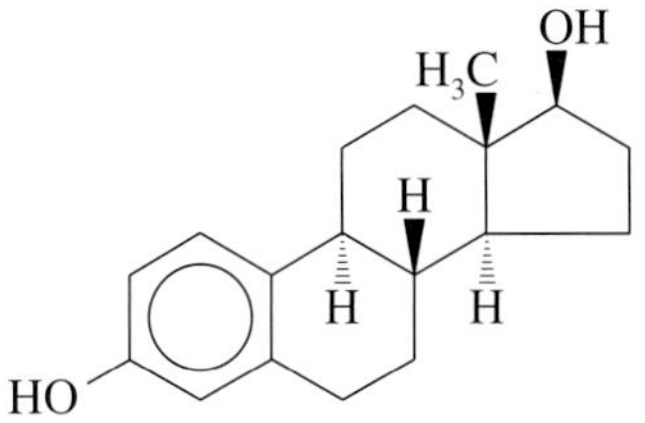

Estradiol

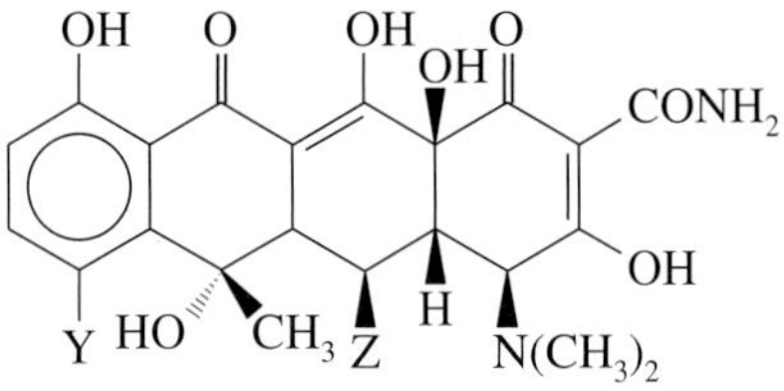

Tetracyclines
(Y = Cl, Z = H; Aureomycin)
(Y = H, Z = OH; Terramycin)

21.3 Physical Properties of Phenols

The presence of hydroxyl groups in the molecules of phenols means that phenols are like alcohols (Section 11.2) in being able to form strong intermolecular hydrogen bonds. This hydrogen bonding causes phenols to be associated and, therefore, to have higher boiling points than hydrocarbons of the same molecular weight. For example, phenol (bp 182°C) has a boiling point more than 70°C higher than toluene (bp 110.6°C), even though the two compounds have almost the same molecular weight.

The ability to form strong hydrogen bonds to molecules of water confers on phenols a modest solubility in water. Table 21.1 lists the physical properties of a number of common phenols.

21.4 Synthesis of Phenols

21.4A Laboratory Synthesis

The most important laboratory synthesis of phenols is by hydrolysis of arenediazonium salts (Section 20.8E). This method is highly versatile, and the conditions required for the diazotization step and the hydrolysis step are mild. This means that other groups present on the ring are unlikely to be affected.

General Reaction

$$Ar{-}NH_2 \xrightarrow{HONO} Ar{-}\overset{+}{N}_2 \xrightarrow[Cu^{2+},\ H_2O]{Cu_2O} Ar{-}OH$$

Specific Example

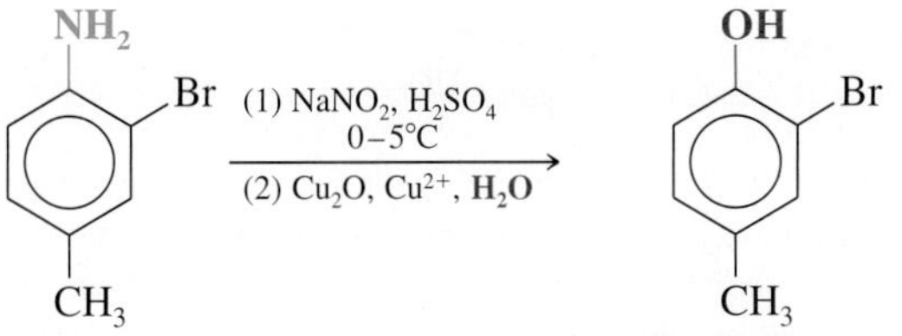

$$\text{2-Bromo-4-methylaniline} \xrightarrow[(2)\ Cu_2O,\ Cu^{2+},\ H_2O]{(1)\ NaNO_2,\ H_2SO_4,\ 0\text{–}5°C} \text{2-Bromo-4-methylphenol}$$

2-Bromo-4-methylphenol
(80–92%)

TABLE 21.1 Physical Properties of Phenols

Name	Formula	mp (°C)	bp (°C)	Water Solubility (g 100 mL^{-1} of H_2O)
Phenol	C_6H_5OH	43	182	9.3
2-Methylphenol	*o*-$CH_3C_6H_4OH$	30	191	2.5
3-Methylphenol	*m*-$CH_3C_6H_4OH$	11	201	2.6
4-Methylphenol	*p*-$CH_3C_6H_4OH$	35.5	201	2.3
2-Chlorophenol	*o*-ClC_6H_4OH	8	176	2.8
3-Chlorophenol	*m*-ClC_6H_4OH	33	214	2.6
4-Chlorophenol	*p*-ClC_6H_4OH	43	220	2.7
2-Nitrophenol	*o*-$O_2NC_6H_4OH$	45	217	0.2
3-Nitrophenol	*m*-$O_2NC_6H_4OH$	96		1.4
4-Nitrophenol	*p*-$O_2NC_6H_4OH$	114		1.7
2,4-Dinitrophenol	O_2N–C_6H_3(–NO_2)–OH	113		0.6
2,4,6-Trinitrophenol (picric acid)	O_2N–C_6H_2(NO_2)$_2$–OH	122		1.4

The Chemistry of...

Polyketide Anticancer Antibiotic Biosynthesis

Doxorubicin (also known as adriamycin) is a highly potent anticancer drug that contains phenol functional groups. It is effective against many forms of cancer, including tumors of the ovaries, breast, bladder, and lung, as well as against Hodgkin's disease and other acute leukemias. Doxorubicin is a member of the anthracycline family of antibiotics. Another member of the family is daunomycin. Both of these antibiotics are produced in strains of *Streptomyces* bacteria by a pathway called polyketide biosynthesis.

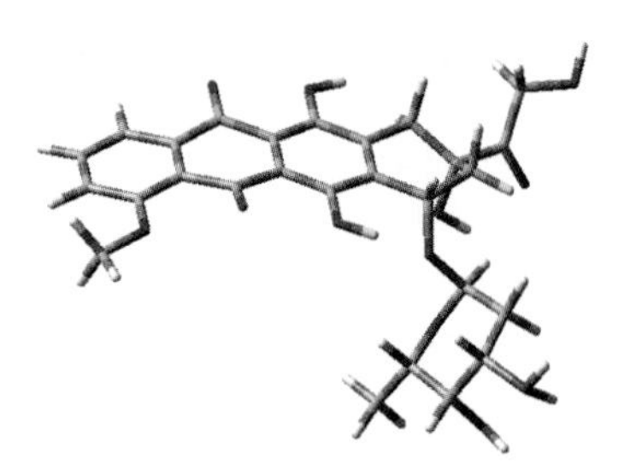

A molecular model of doxorubicin.

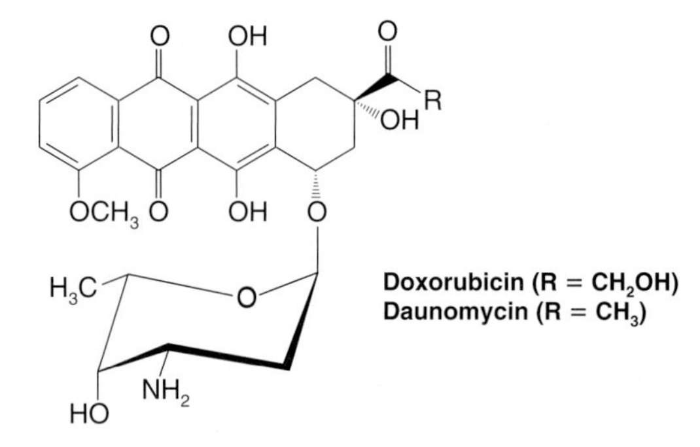

Doxorubicin (R = CH_2OH)
Daunomycin (R = CH_3)

Isotopic labeling experiments have shown that daunomycin is synthesized in *Streptomyces galilaeus* from a tetracyclic precursor called aklavinone. Aklavinone, in turn, is synthesized from acetate. When *S. galilaeus* is grown in a medium containing acetate labeled with carbon-13 and oxygen-18, the aklavinone produced has isotopic labels in the positions indicated below. Notice that oxygen atoms occur at alternate carbons in several places around the structure, consistent with the linking of acetate units in head-to-tail fashion. This is typical of aromatic polyketide biosynthesis.

Isotopically labeled acetate
■, ● = ^{13}C labels
▼ = ^{18}O labels

S. galilaeus

Aklavinone

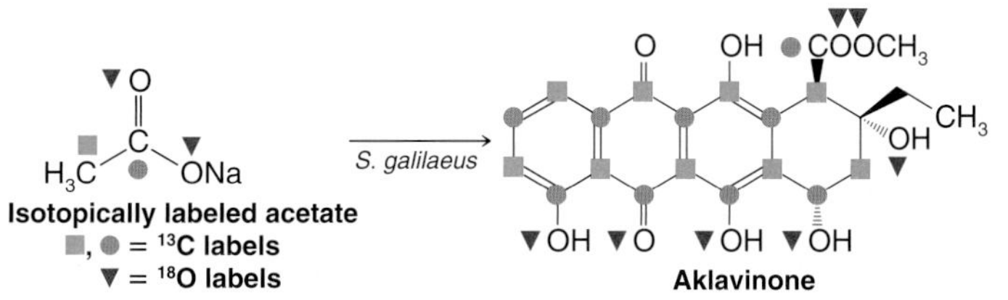

This and other information show that nine C_2 units from malonyl-coenzyme A and one C_3 unit from propionyl-coenzyme A condense to form the linear polyketide intermediate shown below. These units are joined by acylation reactions that are the biosynthetic equivalent of the *malonic ester synthesis* we studied in Section 19.4. These reactions are also similar to the acylation steps we saw in fatty acid biosynthesis (Special Topic D). Once formed, the linear polyketide cyclizes by enzymatic reactions akin to intramolecular *aldol additions and dehydrations* (Section 17.6). These steps form the tetracyclic core of aklavinone. Phenolic hydroxyl groups in aklavinone arise by enolization of ketone carbonyl groups present after the aldol condensation steps. Several other transformations ultimately lead to daunomycin:

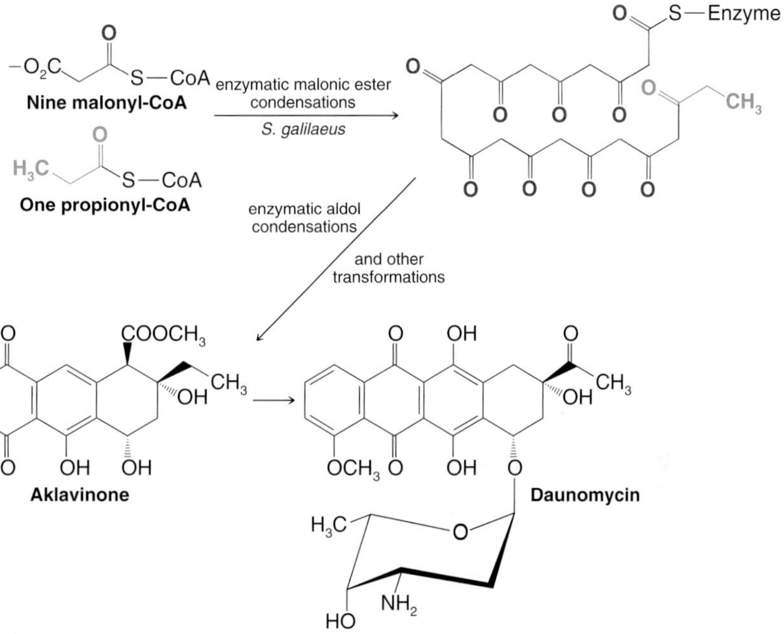

Nine malonyl-CoA

One propionyl-CoA

enzymatic malonic ester condensations
S. galilaeus

enzymatic aldol condensations
and other transformations

Aklavinone

Daunomycin

There are many examples of important biologically active molecules formed by polyketide biosynthesis. Aureomycin and Terramycin (Section 21.2) are examples of other aromatic polyketide antibiotics. Erythromycin (Section 18.7C) and aflatoxin, a carcinogen (Section 11.14), are polyketides from other pathways.

21.4B Industrial Syntheses

Phenol is a highly important industrial chemical; it serves as the raw material for a large number of commercial products ranging from aspirin to a variety of plastics. Worldwide production of phenol is more than 3 million tons per year. Several methods have been used to synthesize phenol commercially.

1. **Hydrolysis of Chlorobenzene (Dow Process).** In this process chlorobenzene is heated at 350°C (under high pressure) with aqueous sodium hydroxide. The reaction pro-

duces sodium phenoxide, which, on acidification, yields phenol. The mechanism for the reaction probably involves the formation of benzyne (Section 21.11B):

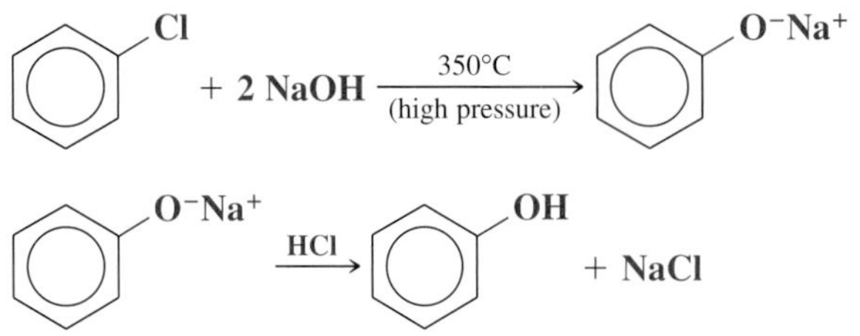

$$C_6H_5Cl + 2\,NaOH \xrightarrow[\text{(high pressure)}]{350°C} C_6H_5O^-Na^+ + NaCl + H_2O$$

$$C_6H_5O^-Na^+ \xrightarrow{HCl} C_6H_5OH + NaCl$$

2. **Alkali Fusion of Sodium Benzenesulfonate.** The first commercial process for synthesizing phenol was developed in Germany in 1890. Sodium benzenesulfonate is melted (fused) with sodium hydroxide (at 350°C) to produce sodium phenoxide. Acidification then yields phenol:

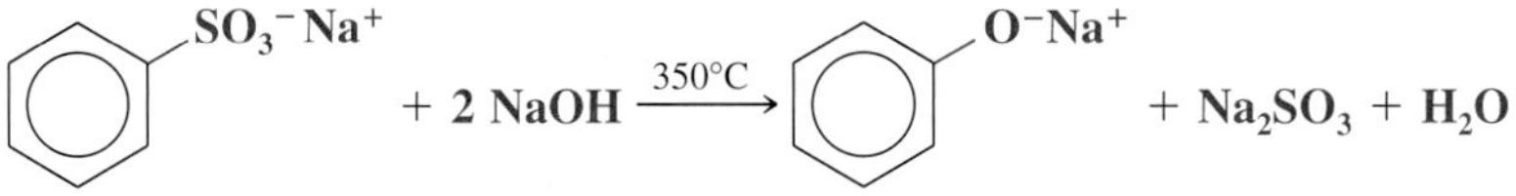

$$C_6H_5SO_3^-Na^+ + 2\,NaOH \xrightarrow{350°C} C_6H_5O^-Na^+ + Na_2SO_3 + H_2O$$

Sodium benzenesulfonate

This procedure can also be used in the laboratory and works quite well for the preparation of 4-methylphenol, as the following example shows. However, the conditions required to bring about the reaction are so vigorous that this method cannot be used for the preparation of many phenols.

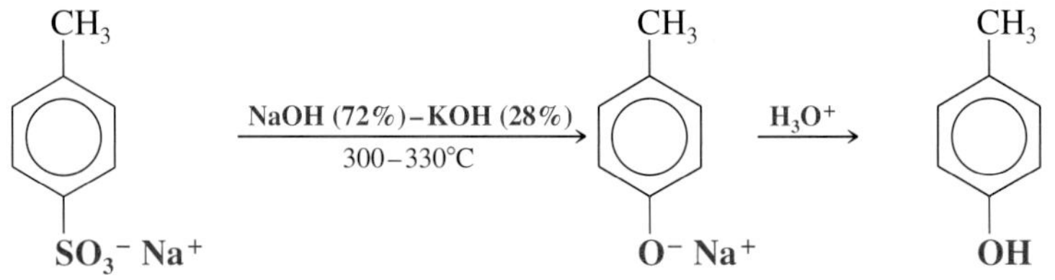

$$p\text{-}CH_3C_6H_4SO_3^-Na^+ \xrightarrow[300-330°C]{NaOH\,(72\%)-KOH\,(28\%)} p\text{-}CH_3C_6H_4O^-Na^+ \xrightarrow{H_3O^+} p\text{-}CH_3C_6H_4OH$$

Sodium *p*-toluenesulfonate → **4-Methylphenol (63–70% overall)**

3. **From Cumene Hydroperoxide.** This process illustrates industrial chemistry at its best. Overall, it is a method for converting two relatively inexpensive organic compounds—benzene and propene—into two more valuable ones—phenol and acetone. The only other substance consumed in the process is oxygen from air. Most of the worldwide production of phenol is now based on this method. The synthesis begins with the Friedel–Crafts alkylation of benzene with propene to produce cumene (isopropylbenzene):

Reaction 1

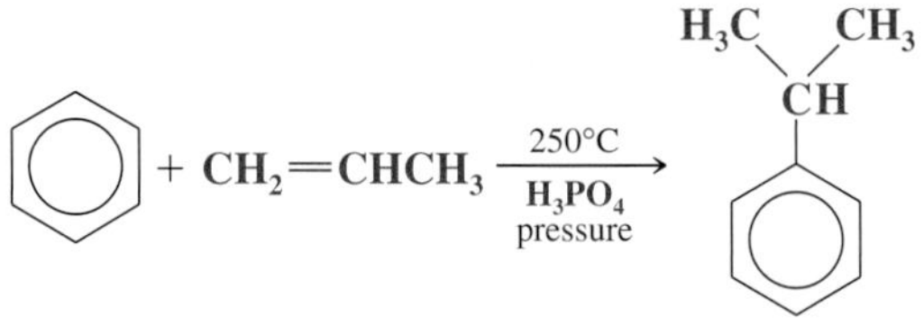

$$C_6H_6 + CH_2{=}CHCH_3 \xrightarrow[\text{H}_3\text{PO}_4,\ \text{pressure}]{250°C} C_6H_5CH(CH_3)_2$$

Cumene

Then cumene is oxidized to cumene hydroperoxide:

Reaction 2

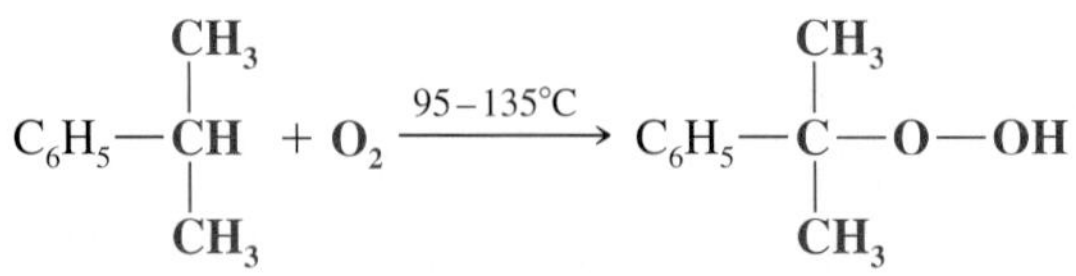

$$C_6H_5{-}CH(CH_3)_2 + O_2 \xrightarrow{95-135°C} C_6H_5{-}C(CH_3)_2{-}O{-}OH$$

Cumene hydroperoxide

Finally, when treated with 10% sulfuric acid, cumene hydroperoxide undergoes a hydrolytic rearrangement that yields phenol and acetone:

Reaction 3

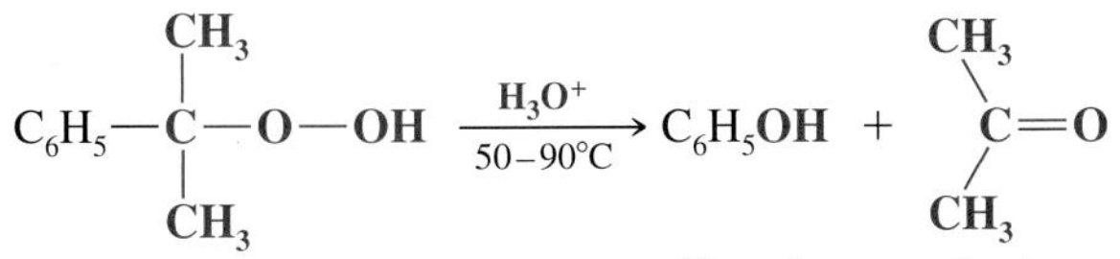

$$C_6H_5-\underset{CH_3}{\overset{CH_3}{C}}-O-OH \xrightarrow[50-90°C]{H_3O^+} C_6H_5OH + (CH_3)_2C{=}O$$

Phenol **Acetone**

The mechanism of each of the reactions in the synthesis of phenol from benzene and propene via cumene hydroperoxide requires some comment. The first reaction is a familiar one. The isopropyl cation generated by the reaction of propene with the acid (H_3PO_4) alkylates benzene in a typical electrophilic aromatic substitution:

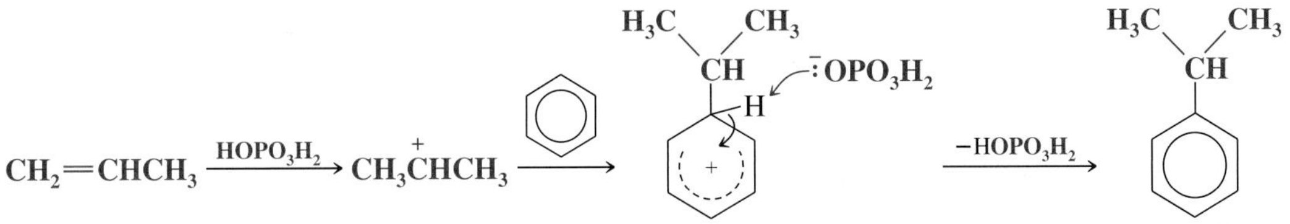

$$CH_2{=}CHCH_3 \xrightarrow{HOPO_3H_2} CH_3\overset{+}{C}HCH_3 \xrightarrow{C_6H_6} [(CH_3)_2CH-C_6H_6^+]\ \ :\overset{-}{O}PO_3H_2 \xrightarrow{-HOPO_3H_2} (CH_3)_2CH-C_6H_5$$

The second reaction is a radical chain reaction. A radical initiator abstracts the benzylic hydrogen atom of cumene, producing a 3° benzylic radical. Then a chain reaction with oxygen produces cumene hydroperoxide:

Chain Initiation

Step 1 $$C_6H_5-\underset{CH_3}{\overset{CH_3}{C}}-H + R\cdot \longrightarrow C_6H_5-\underset{CH_3}{\overset{CH_3}{C}}\cdot + R-H$$

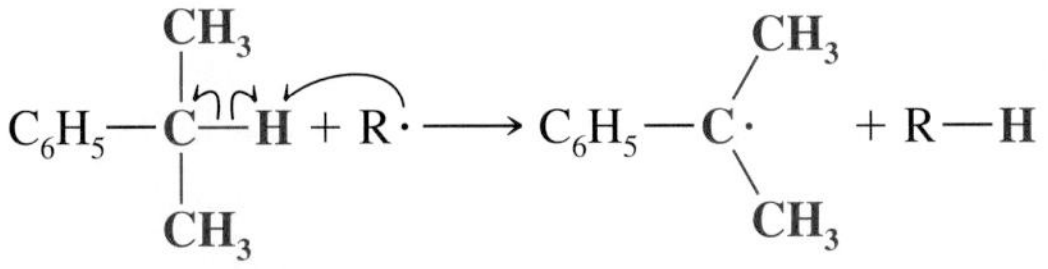

Chain Propagation

Step 2 $$C_6H_5-\underset{CH_3}{\overset{CH_3}{C}}\cdot + \cdot O-O\cdot \longrightarrow C_6H_5-\underset{CH_3}{\overset{CH_3}{C}}-O-O\cdot$$

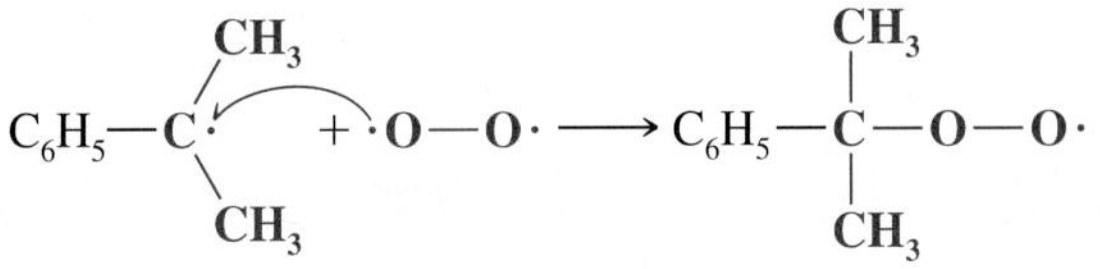

Step 3 $$C_6H_5-\underset{CH_3}{\overset{CH_3}{C}}-O-O\cdot + H-\underset{CH_3}{\overset{CH_3}{C}}-C_6H_5 \longrightarrow C_6H_5-\underset{CH_3}{\overset{CH_3}{C}}-O-O-H + C_6H_5-\underset{CH_3}{\overset{CH_3}{C}}\cdot$$

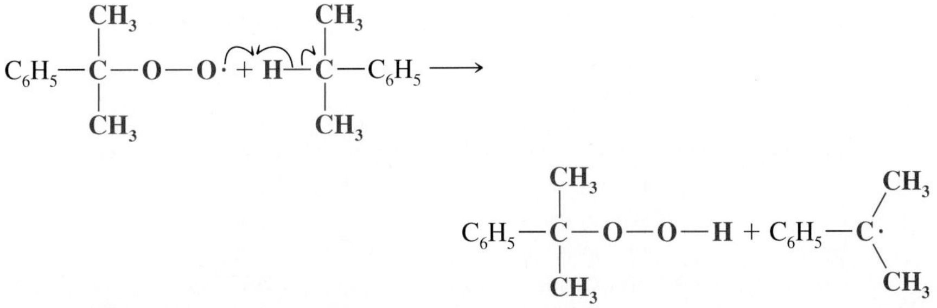

The reaction continues with steps 2, 3, 2, 3, etc.

The third reaction—the hydrolytic rearrangment—resembles the carbocation rearrangements that we have studied before. In this instance, however, the rearrangement involves the migration of a phenyl group to *a cationic oxygen atom.* Phenyl groups have a much greater tendency to migrate to a cationic center than do methyl groups (see Section 16.12A). The following equations show all the steps of the mechanism.

$$C_6H_5-C(CH_3)_2-\ddot{O}-\ddot{O}H + H-\overset{+}{O}H_2 \longrightarrow C_6H_5-C(CH_3)_2-\ddot{O}-\overset{+}{O}H_2 \xrightarrow[\text{loss of water and concurrent phenyl migration with an electron pair to oxygen}]{-H_2O} (CH_3)_2\overset{+}{C}-\ddot{O}-C_6H_5 \xrightarrow{H_2\ddot{O}:}$$

$$H-\overset{+}{\underset{H}{O}}-C(CH_3)_2-\ddot{O}-C_6H_5 \rightleftharpoons H-\ddot{O}-C(CH_3)_2-\overset{+}{\underset{H}{O}}-C_6H_5 \xrightarrow{-H_3O^+} \ddot{O}=C(CH_3)_2 + H\ddot{O}C_6H_5$$

Acetone **Phenol**

The second and third steps of the mechanism may actually take place at the same time; that is, the loss of H_2O and the migration of C_6H_5- may be concerted.

21.5 Reactions of Phenols as Acids

21.5A Strength of Phenols as Acids

Although phenols are structurally similar to alcohols, they are much stronger acids. The pK_a values of most alcohols are of the order of 18. However, as we see in Table 21.2, the pK_a values of phenols are smaller than 11.

TABLE 21.2 Acidity Constants of Phenols

Name	pK_a (in H_2O at 25°C)
Phenol	9.89
2-Methylphenol	10.20
3-Methylphenol	10.01
4-Methylphenol	10.17
2-Chlorophenol	8.11
3-Chlorophenol	8.80
4-Chlorophenol	9.20
2-Nitrophenol	7.17
3-Nitrophenol	8.28
4-Nitrophenol	7.15
2,4-Dinitrophenol	3.96
2,4,6-Trinitrophenol (picric acid)	0.38
1-Naphthol	9.31
2-Naphthol	9.55

Let us compare two *superficially* similar compounds, cyclohexanol and phenol:

Cyclohexanol
pK_a = 18

Phenol
pK_a = 9.89

Although phenol is a weak acid when compared with a carboxylic acid such as acetic acid (pK_a = 4.75), phenol is a much stronger acid than cyclohexanol (by a factor of eight pK_a units).

Experimental and theoretical results have shown that the greater acidity of phenol owes itself primarily to an electrical charge distribution in phenol that causes the —OH oxygen to be more positive; therefore, the proton is held less strongly. In effect, the benzene ring of phenol acts as if it were an electron-withdrawing group when compared with the cyclohexane ring of cyclohexanol.

We can understand this effect by noting that the carbon atom that bears the hydroxyl group in phenol is sp^2 hybridized, whereas in cyclohexane it is sp^3 hybridized. Because of their greater s character, sp^2-hybridized carbon atoms are more electronegative than sp^3-hybridized carbon atoms (Section 3.7A).

Another factor influencing the electron distribution may be the contributions to the overall resonance hybrid of phenol made by structures **2–4.** Notice that the effect of these structures is to withdraw electrons from the hydroxyl group and to make the oxygen positive:

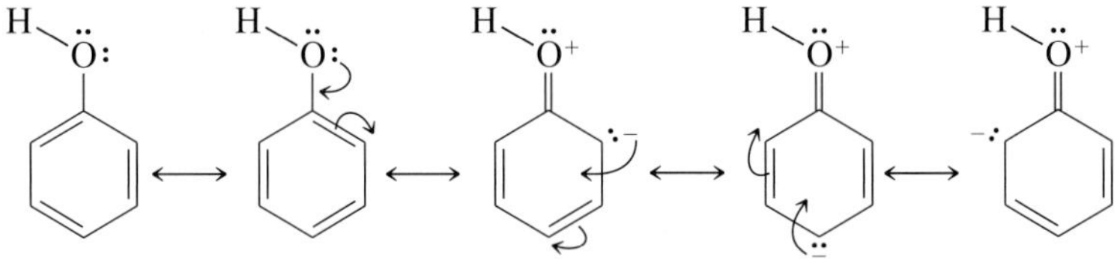

1a 1b 2 3 4

Resonance structures for phenol

An alternative explanation for the greater acidity of phenol relative to cyclohexanol can be based on similar resonance structures for the phenoxide ion. Unlike the structures for phenol, **2–4,** resonance structures for the phenoxide ion do not involve charge separation. According to resonance theory, such structures should stabilize the phenoxide ion more than structures **2–4** stabilize phenol. (No resonance structures can be written for cyclohexanol or its anion, of course.) Greater stabilization of the phenoxide ion (the conjugate base) than of phenol (the acid) has an acid-strengthening effect.*

PROBLEM 21.1

If we examine Table 21.2, we find that the methylphenols (cresols) are less acidic than phenol itself. For example,

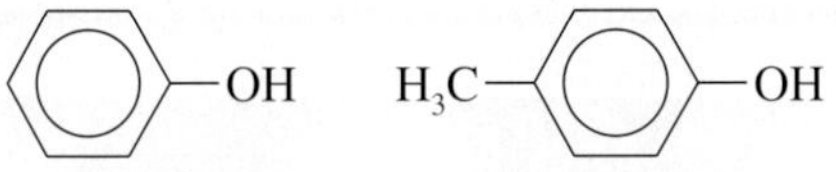

Phenol
pK_a = 9.89

4-Methylphenol
pK_a = 10.17

This behavior is characteristic of phenols bearing electron-releasing groups. Provide an explanation.

*Those who may be interested in pursuing this subject further should see the following articles: Siggel, M. R. F.; Thomas, T. D. *J. Am. Chem. Soc.* **1986,** *108,* 4360–4362; and Siggel, M. R. F.; Streitwieser, A. R.; Thomas, T. D. *J. Am. Chem. Soc.* **1988,** *110,* 8022–8028.

PROBLEM 21.2 If we examine Table 21.2, we see that phenols having electron-withdrawing groups (Cl— or O_2N—) attached to the benzene ring are more acidic than phenol itself. Account for this trend on the basis of resonance and inductive effects. Your answer should also explain the large acid-strengthening effect of nitro groups, an effect that makes 2,4,6-trinitrophenol (also called *picric acid*) so exceptionally acidic ($pK_a = 0.38$) that it is more acidic than acetic acid ($pK_a = 4.75$).

21.5B Distinguishing and Separating Phenols from Alcohols and Carboxylic Acids

Because phenols are more acidic than water, the following reaction goes essentially to completion and produces water-soluble sodium phenoxide:

$$C_6H_5\text{—}OH + NaOH \xrightleftharpoons{H_2O} C_6H_5\text{—}O^- Na^+ + H_2O$$

Stronger acid $pK_a \cong 10$ (slightly soluble)	Stronger base	Weaker base (soluble)	Weaker acid $pK_a \cong 16$

The corresponding reaction of 1-hexanol with aqueous sodium hydroxide does not occur to a significant extent because 1-hexanol is a weaker acid than water:

$$CH_3(CH_2)_4CH_2OH + NaOH \xrightleftharpoons[H_2O]{} CH_3(CH_2)_4CH_2O^- Na^+ + H_2O$$

Weaker acid $pK_a \cong 18$ (very slightly soluble)	Weaker base	Stronger base	Stronger acid $pK_a \cong 16$

The fact that phenols dissolve in aqueous sodium hydroxide, whereas most alcohols with six carbon atoms or more do not, gives us a convenient means for distinguishing and separating phenols from most alcohols. (Alcohols with five carbon atoms or fewer are quite soluble in water—some are infinitely so—and so they dissolve in aqueous sodium hydroxide even though they are not converted to sodium alkoxides in appreciable amounts.)

Most phenols, however, are not soluble in aqueous sodium bicarbonate ($NaHCO_3$), but carboxylic acids are soluble. Thus, aqueous $NaHCO_3$ provides a method for distinguishing and separating most phenols from carboxylic acids.

You will likely employ the moderate acidity of phenols in your organic chemistry laboratory work for the separation or characterization of compounds.

PROBLEM 21.3 Your laboratory instructor gives you a mixture of 4-methylphenol, benzoic acid, and toluene. Assume that you have available common laboratory acids, bases, and solvents and explain how you would proceed to separate this mixture by making use of the solubility differences of its components.

21.6 Other Reactions of the O—H Group of Phenols

Phenols react with carboxylic acid anhydrides and acid chlorides to form esters. These reactions are quite similar to those of alcohols (Section 18.7).

$$C_6H_5\text{—}OH \xrightarrow[\text{base}]{(RCO)_2O} C_6H_5\text{—}O\text{—}\overset{O}{\overset{\|}{C}}R + R\overset{O}{\overset{\|}{C}}O^-$$

$$C_6H_5\text{—}OH \xrightarrow[\text{base}]{RCOCl} C_6H_5\text{—}O\text{—}\overset{O}{\overset{\|}{C}}R + Cl^-$$

21.6A Phenols in the Williamson Synthesis

Phenols can be converted to ethers through the Williamson synthesis (Section 11.11B). Because phenols are more acidic than alcohols, they can be converted to sodium phenoxides through the use of sodium hydroxide (rather than metallic sodium, the reagent used to convert alcohols to alkoxide ions).

General Reaction

$$\text{ArOH} \xrightarrow{\text{NaOH}} \text{ArO}^-\text{Na}^+ \xrightarrow[\substack{(\text{X} = \text{Cl, Br, I,} \\ \text{OSO}_2\text{OR}' \text{ or} \\ \text{OSO}_2\text{R}')}]{\text{R}-\text{X}} \text{ArOR} + \text{NaX}$$

Specific Examples

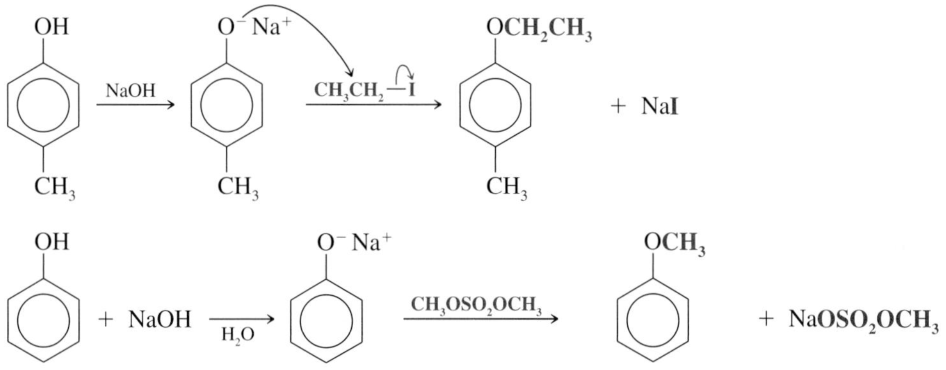

$$p\text{-}CH_3C_6H_4OH \xrightarrow{\text{NaOH}} p\text{-}CH_3C_6H_4O^-Na^+ \xrightarrow{CH_3CH_2-I} p\text{-}CH_3C_6H_4OCH_2CH_3 + \text{NaI}$$

$$C_6H_5OH + \text{NaOH} \xrightarrow[H_2O]{} C_6H_5O^-Na^+ \xrightarrow{CH_3OSO_2OCH_3} C_6H_5OCH_3 + NaOSO_2OCH_3$$

Anisole (methoxybenzene)

21.7 Cleavage of Alkyl Aryl Ethers

We learned in Section 11.12 that when dialkyl ethers are heated with excess concentrated HBr or HI, the ethers are cleaved and alkyl halides are produced from both alkyl groups:

$$\text{R}-\text{O}-\text{R}' \xrightarrow[\text{heat}]{\text{concd HX}} \text{R}-\text{X} + \text{R}'-\text{X} + H_2O$$

When alkyl aryl ethers react with strong acids such as HI and HBr, the reaction produces an alkyl halide and a phenol. The phenol does not react further to produce an aryl halide because the phenol carbon–oxygen bond is very strong (see Problem 21.1) and because phenyl cations do not form readily.

General Reaction

$$\text{Ar}-\text{O}-\text{R} \xrightarrow[\text{heat}]{\text{concd HX}} \text{Ar}-\text{OH} + \text{R}-\text{X}$$

Specific Example

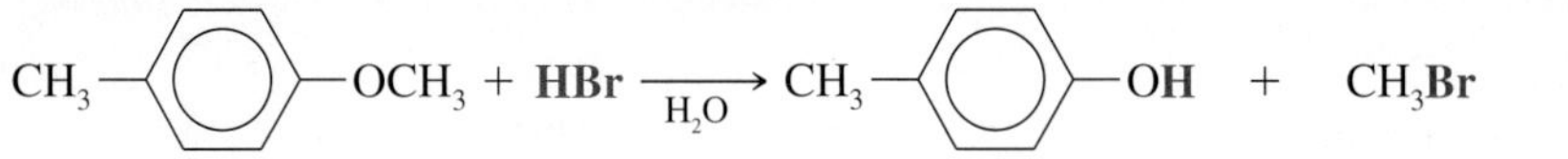

$$CH_3-C_6H_4-OCH_3 + \text{HBr} \xrightarrow[H_2O]{} CH_3-C_6H_4-OH + CH_3Br$$

p-Methylanisole → **4-Methylphenol** + **Methyl bromide**

4-Methylphenol $\xrightarrow{\text{HBr}}$ no reaction

21.8 Reactions of the Benzene Ring of Phenols

Bromination The hydroxyl group is a powerful activating group—and an ortho–para director—in **electrophilic substitutions.** Phenol itself reacts with bromine in aqueous solution to yield 2,4,6-tribromophenol in nearly quantitative yield. Note that a Lewis acid is not required for the bromination of this highly activated ring:

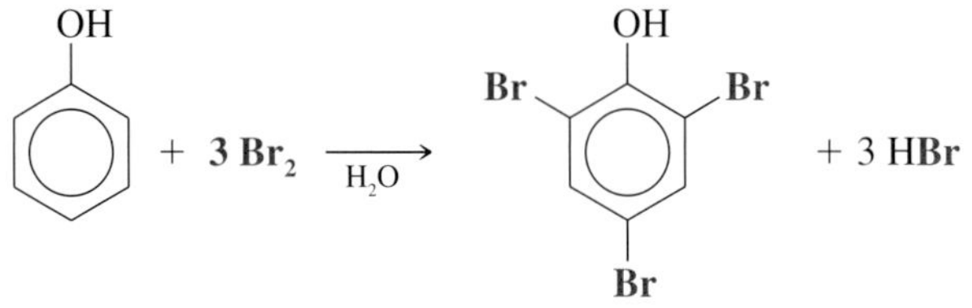

Phenol + 3 Br_2 $\xrightarrow{H_2O}$ 2,4,6-Tribromophenol (~100%) + 3 HBr

2,4,6-Tribromophenol
(~100%)

Monobromination of phenol can be achieved by carrying out the reaction in carbon disulfide at a low temperature, conditions that reduce the electrophilic reactivity of bromine. The major product is the para isomer:

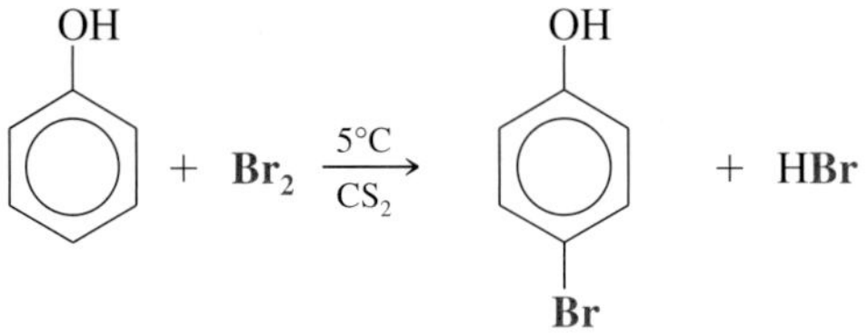

Phenol + Br_2 $\xrightarrow[CS_2]{5°C}$ *p*-Bromophenol + HBr

***p*-Bromophenol**
(80–84%)

Nitration Phenol reacts with dilute nitric acid to yield a mixture of *o*- and *p*-nitrophenol:

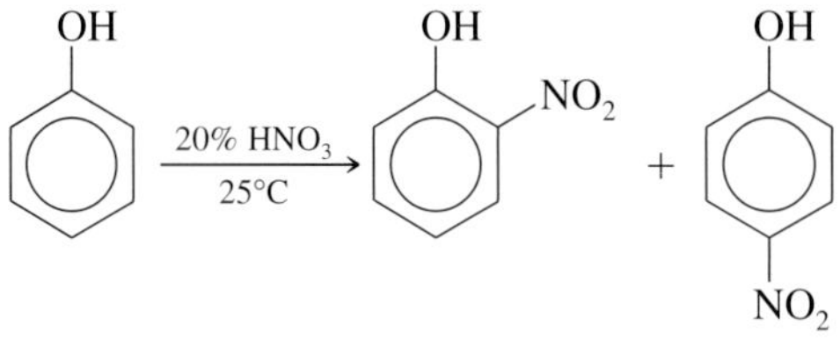

Phenol $\xrightarrow[25°C]{20\%\ HNO_3}$ *o*-nitrophenol **(30–40%)** + *p*-nitrophenol **(15%)**

Although the yield is relatively low (because of oxidation of the ring), the ortho and para isomers can be separated by steam distillation. *o*-Nitrophenol is the more volatile isomer because its hydrogen bonding (see the following structures) is *intramolecular. p*-Nitrophenol is less volatile because *intermolecular* hydrogen bonding causes association among its molecules. Thus, *o*-nitrophenol passes over with the steam, and *p*-nitrophenol remains in the distillation flask.

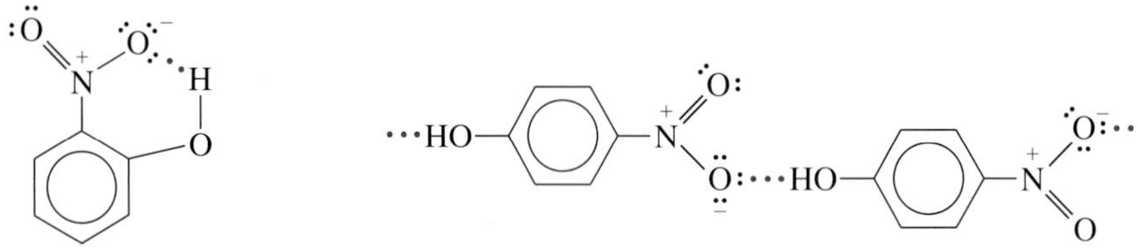

***o*-Nitrophenol**
(more volatile because of intramolecular hydrogen bonding)

***p*-Nitrophenol**
(less volatile because of intermolecular hydrogen bonding)

Sulfonation Phenol reacts with concentrated sulfuric acid to yield mainly the ortho-sulfonated product if the reaction is carried out at 25°C and mainly the para-sulfonated product at 100°C. This is another example of thermodynamic versus kinetic control of a reaction (Section 13.10A):

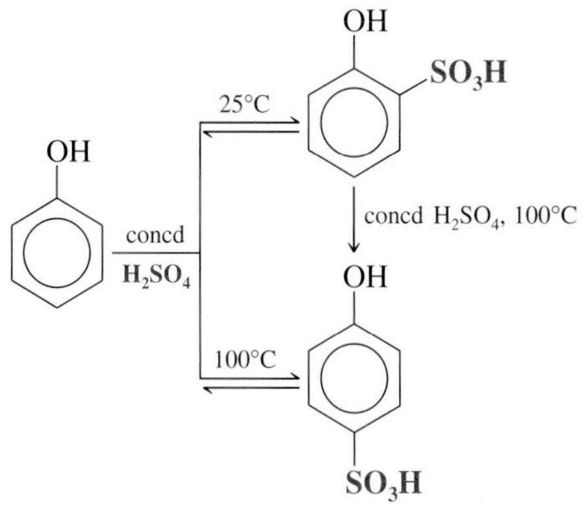

Major product under kinetic control

Major product under thermodynamic control

Problem 21.4 (a) Which sulfonic acid (see previous reactions) is more stable? (b) For which sulfonation (ortho or para) is the free energy of activation lower?

Kolbe Reaction The phenoxide ion is even more susceptible to electrophilic aromatic substitution than phenol itself. (Why?) Use is made of the high reactivity of the phenoxide ring in a reaction called the *Kolbe reaction.* In the Kolbe reaction carbon dioxide acts as the electrophile.

A Mechanism for the Reaction

The Kolbe Reaction

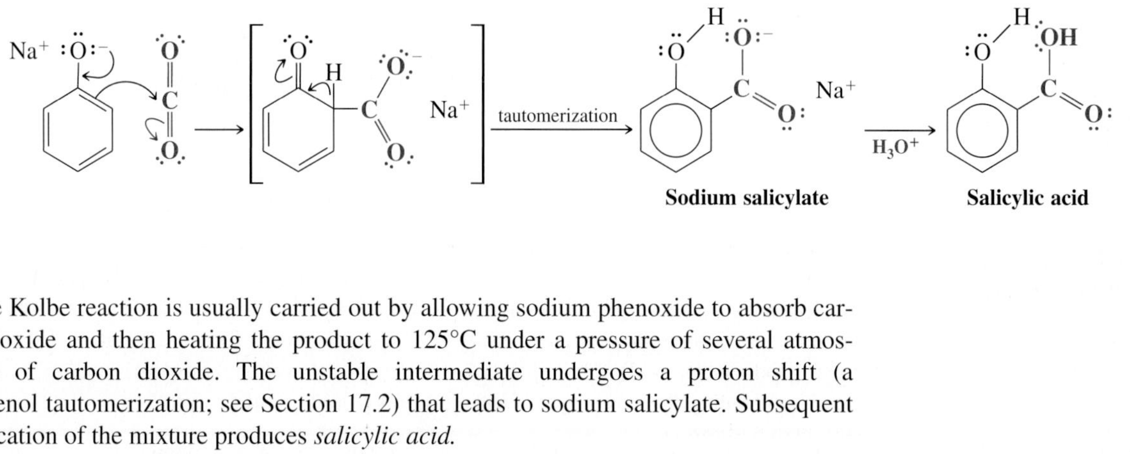

Sodium salicylate

Salicylic acid

The Kolbe reaction is usually carried out by allowing sodium phenoxide to absorb carbon dioxide and then heating the product to 125°C under a pressure of several atmospheres of carbon dioxide. The unstable intermediate undergoes a proton shift (a keto–enol tautomerization; see Section 17.2) that leads to sodium salicylate. Subsequent acidification of the mixture produces *salicylic acid.*

Reaction of salicylic acid with acetic anhydride yields the widely used pain reliever—*aspirin:*

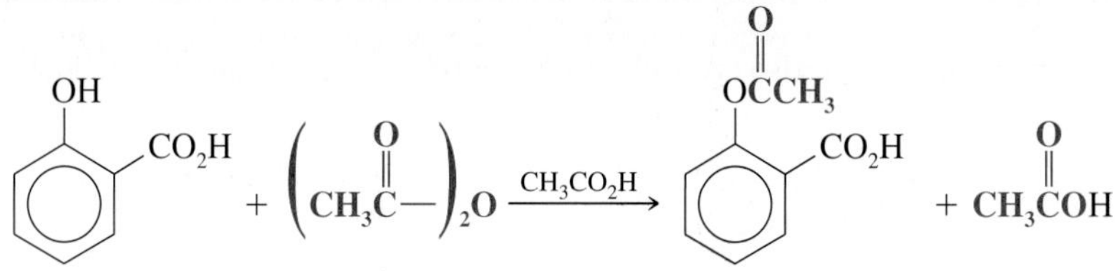

Salicylic acid + Acetic anhydride $\xrightarrow{CH_3CO_2H}$ Acetylsalicylic acid (aspirin) + CH_3COOH

21.9 The Claisen Rearrangement

Heating allyl phenyl ether to 200°C effects an intramolecular reaction called a **Claisen rearrangement.** The product of the rearrangement is *o*-allylphenol:

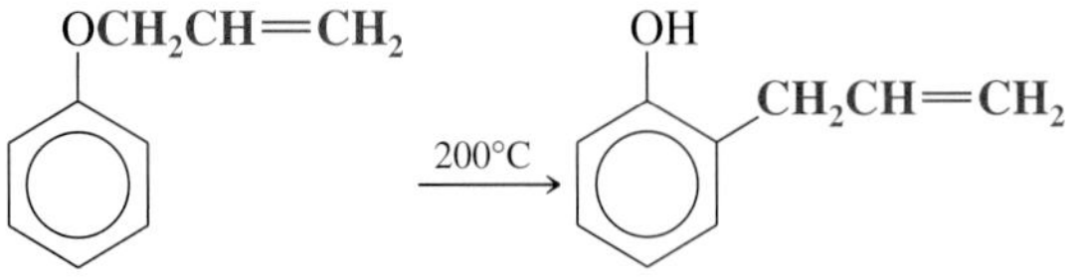

$OCH_2CH{=}CH_2$ (Allyl phenyl ether) $\xrightarrow{200°C}$ *o*-Allylphenol (OH, $CH_2CH{=}CH_2$)

Allyl phenyl ether **o-Allylphenol**

The reaction takes place through a **concerted rearrangement** in which the bond between C3 of the allyl group and the ortho position of the benzene ring forms at the same time that the carbon–oxygen bond of the allyl phenyl ether breaks. The product of this rearrangement is an unstable intermediate that, like the unstable intermediate in the Kolbe reaction (Section 21.8), undergoes a proton shift (a keto–enol tautomerization, see Section 17.2) that leads to the *o*-allylphenol:

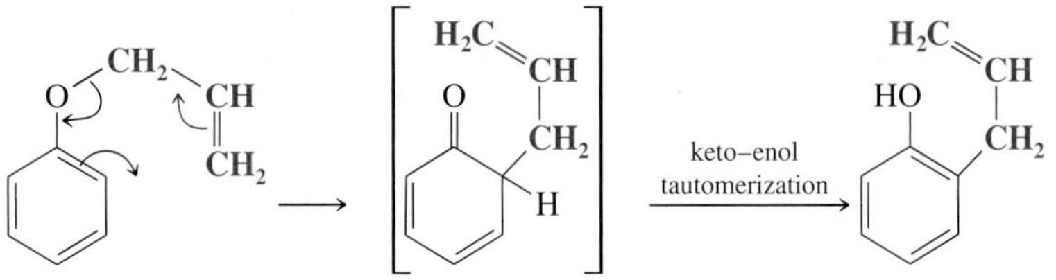

Unstable intermediate

(keto–enol tautomerization)

That only C3 of the allyl group becomes bonded to the benzene ring was demonstrated by carrying out the rearrangement with allyl phenyl ether containing ^{14}C at C3. All of the product of this reaction had the labeled carbon atom bonded to the ring:

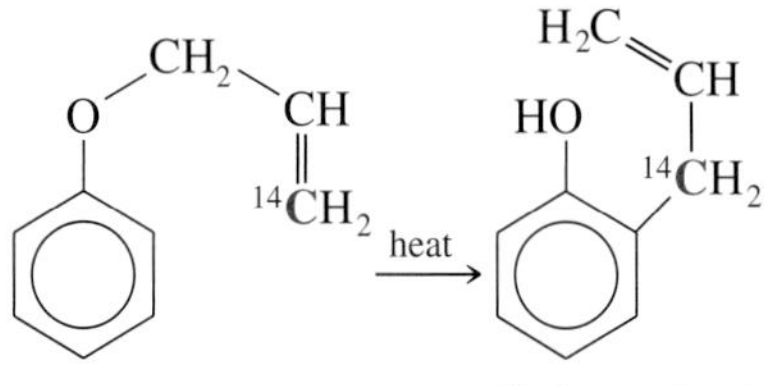

$O{-}CH_2{-}CH{=}{}^{14}CH_2$ $\xrightarrow{heat}$ HO, $H_2C{=}CH{-}{}^{14}CH_2$

Only product

Problem 21.5 The labeling experiment just described eliminates from consideration a mechanism in which the allyl phenyl ether dissociates to produce an allyl cation (Section 13.4) and a phenoxide ion, which then subsequently undergo a Friedel–Crafts alkylation (Section 15.6) to produce the *o*-allylphenol. Explain how this alternative mechanism can be discounted by showing the product (or products) that would result from it.

Problem 21.6 Show how you would synthesize allyl phenyl ether through a Williamson synthesis (Section 21.6A) starting with phenol and allyl bromide.

A Claisen rearrangement also takes place when allyl vinyl ethers are heated. For example,

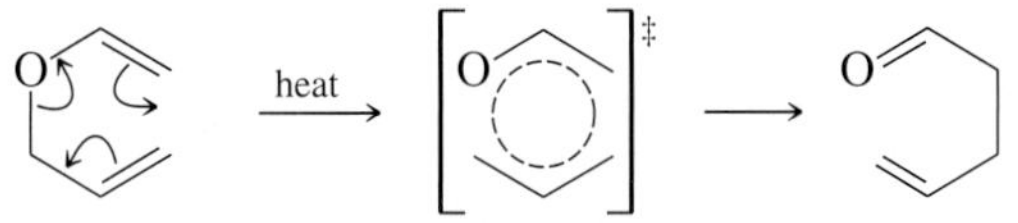

$\xrightarrow{heat}$ [ ]‡ $\longrightarrow$

Allyl vinyl ether **Aromatic transition state** **4-Pentenal**

The transition state for the Claisen rearrangement involves a cycle of six orbitals and six electrons. Having six electrons suggests that the transition state has aromatic character (Section 14.7). Other reactions of this general type are known, and they are called **pericyclic reactions.** Another similar pericyclic reaction is the **Cope rearrangement** shown here:

3,3-Dimethyl-1,5-hexadiene ⇌ [**Aromatic transition state**]‡ ⇌ **2-Methyl-2,6-heptadiene**

The Diels–Alder reaction (Section 13.11) is also a pericyclic reaction. The transition state for the Diels–Alder reaction also involves six orbitals and six electrons:

→ [**Aromatic transition state**]‡ →

The mechanism of the Diels–Alder reaction is discussed further in Special Topic F.

21.10 Quinones

Oxidation of hydroquinone (1,4-benzenediol) produces a compound known as *p*-benzoquinone. The oxidation can be brought about by mild oxidizing agents, and, overall, the oxidation amounts to the removal of a pair of electrons ($2e^-$) and two protons from hydroquinone. (Another way of visualizing the oxidation is as the loss of a hydrogen molecule, H:H, making it a dehydrogenation.)

Hydroquinone $\underset{+2e^-}{\overset{-2e^-}{\rightleftharpoons}}$ ***p*-Benzoquinone** $+ \ 2\ H^+$

This reaction is reversible; *p*-benzoquinone is easily reduced by mild reducing agents to hydroquinone.

Nature makes much use of this type of reversible oxidation–reduction to transport a pair of electrons from one substance to another in enzyme-catalyzed reactions. Important compounds in this respect are the compounds called **ubiquinones** (from *ubiquitous* + quinone—these quinones are found within the inner mitochondrial membrane of every living cell). Ubiquinones are also called coenzymes Q (CoQ).

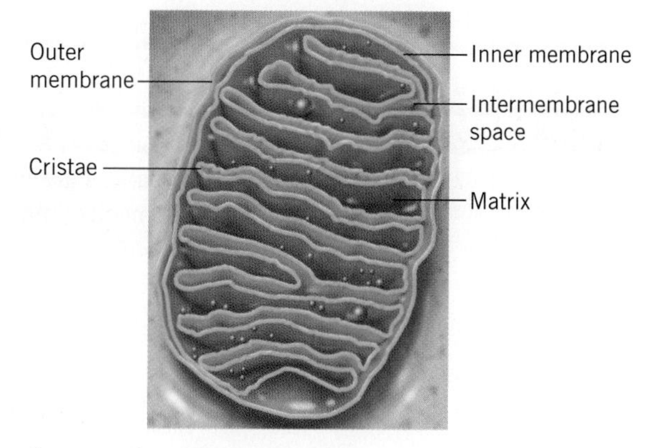

Cross-section of a mitochondrion.

Ubiquinones have a long, isoprene-derived side chain (see Special Topic D and Section 23.4). Ten isoprene units are present in the side chain of human ubiquinones. This part of their structure is highly nonpolar, and it serves to solubilize the ubiquinones within the hydrophobic bilayer of the mitochondrial inner membrane. Solubility in the membrane environment facilitates their lateral diffusion from one component of the electron transport chain to another. In the electron transport chain, ubiquinones function by accepting two electrons and two hydrogen atoms to become a hydroquinone. The hydroquinone form carries the two electrons to the next acceptor in the chain:

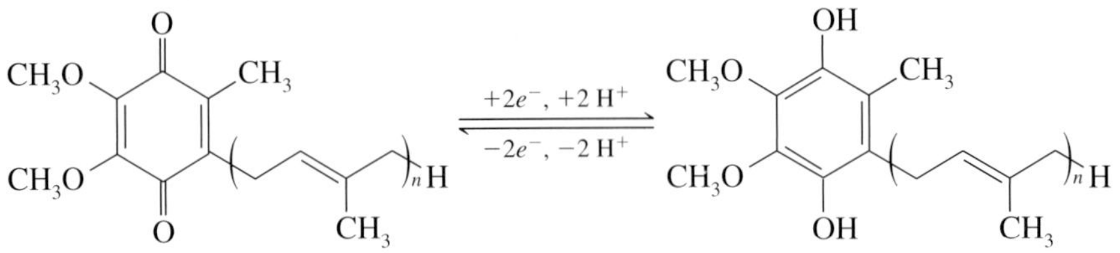

O
CH₃O
CH₃
CH₃O
O
CH₃
nH
+2e⁻, +2 H⁺
−2e⁻, −2 H⁺
OH
CH₃O
CH₃
CH₃O
OH
CH₃
nH

Ubiquinones ($n = 6-10$) (coenzymes Q) **Ubiquinol (hydroquinone form)**

Vitamin K_1, the important dietary factor that is instrumental in maintaining the coagulant properties of blood, contains a 1,4-naphthoquinone structure:

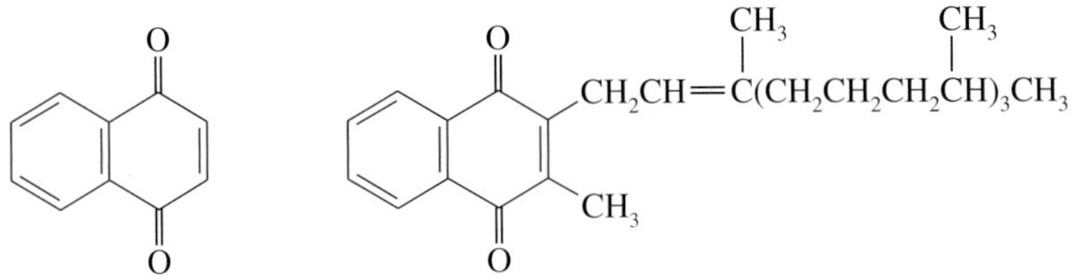

O
O
O
O
CH₃
CH₃
CH₂CH=C(CH₂CH₂CH₂CH)₃CH₃
CH₃

1,4-Naphthoquinone **Vitamin K_1**

PROBLEM 21.7

p-Benzoquinone and 1,4-naphthoquinone act as dienophiles in Diels–Alder reactions. Give the structures of the products of the following reactions:

(a) *p*-Benzoquinone + butadiene;

(b) 1,4-Naphthoquinone + butadiene;

(c) *p*-Benzoquinone + cyclopentadiene

PROBLEM 21.8

Outline a possible synthesis of the following compound.

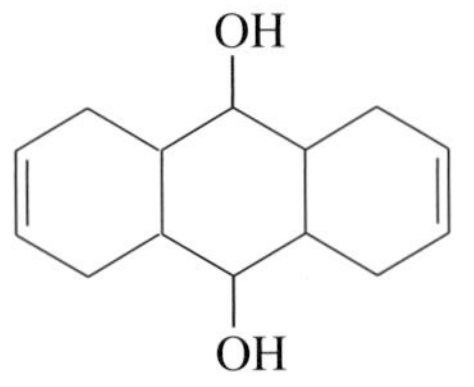

OH
OH

21.11 Aryl Halides and Nucleophilic Aromatic Substitution

Simple aryl halides are like vinylic halides (Section 6.14A) in that they are relatively unreactive toward nucleophilic substitution under conditions that give facile nucleophilic substitution with alkyl halides. Chlorobenzene, for example, can be boiled with sodium hydroxide for days without producing a detectable amount of phenol (or sodium phenoxide).* Similarly, when vinyl chloride is heated with sodium hydroxide, no substitution occurs:

*The Dow process for making phenol by substitution (Section 21.4B) requires extremely high temperature and pressure to effect the reaction. These conditions are not practical in the laboratory.

The Chemistry of...

The Bombardier Beetle's Noxious Spray

The bombardier beetle defends itself by spraying a jet stream of hot (100°C), noxious *p*-benzoquinones at an attacker. The beetle mixes *p*-hydroquinones and hydrogen peroxide from one abdominal reservoir with enzymes from another reservoir. The enzymes convert hydrogen peroxide to oxygen, which in turn oxidizes the *p*-hydroquinones to *p*-benzoquinones and explosively propels the irritating spray at the attacker. Photos by T. Eisner and D. Aneshansley (Cornell University) have shown that the amazing bombardier beetle can direct its spray in virtually any direction, even parallel over its back, to ward off a predator.

Bombardier beetle in the process of spraying. From Eisner, T.; Aneshansley, D. J. *Proc. Natl. Acad. Sci. USA* **1999,** *96,* 9705–9709.

$$C_6H_5Cl + \textbf{NaOH} \xrightarrow[\text{reflux}]{H_2O} \text{no substitution}$$

$$CH_2{=}CHCl + \textbf{NaOH} \xrightarrow[\text{reflux}]{H_2O} \text{no substitution}$$

We can understand this lack of reactivity on the basis of several factors. The benzene ring of an aryl halide prevents back-side attack in an S_N2 reaction:

$$\textbf{Nu:}^- \not\rightarrow C_6H_5{-}X \longrightarrow \text{no reaction}$$

Phenyl cations are very unstable; thus S_N1 reactions do not occur. The carbon–halogen bonds of aryl (and vinylic) halides are shorter and stronger than those of alkyl, allylic, and benzylic halides. Stronger carbon–halogen bonds mean that bond breaking by either an S_N1 or S_N2 mechanism will require more energy.

Two effects make the carbon–halogen bonds of aryl and vinylic halides shorter and stronger: (1) The carbon of either type of halide is sp^2 hybridized, and therefore the electrons of the carbon orbital are closer to the nucleus than those of an sp^3-hybridized carbon. (2) Resonance of the type shown here strengthens the carbon–halogen bond by giving it *double-bond character:*

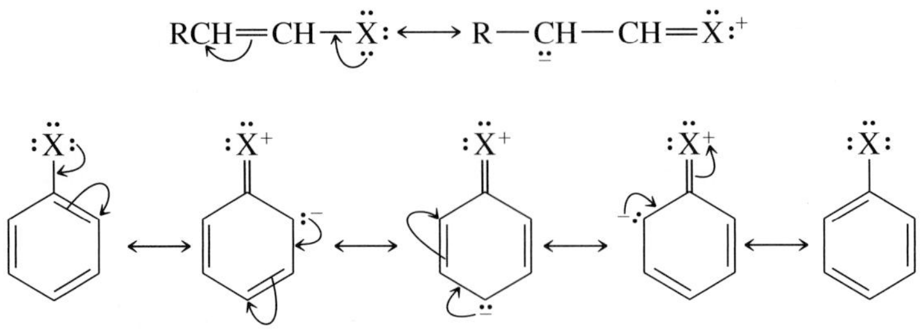

$$RCH{=}CH{-}\ddot{X}{:} \longleftrightarrow R{-}\ddot{C}H{-}CH{=}\ddot{X}{:}^{+}$$

Having said all this, we shall find in the next two subsections that *aryl halides can be remarkably reactive toward nucleophiles* if they bear certain substituents or when we allow them to react under the proper conditions.

21.11A Nucleophilic Aromatic Substitution by Addition–Elimination: The S_NAr Mechanism

Nucleophilic substitution reactions of aryl halides *do* occur readily when an electronic factor makes the aryl carbon bonded to the halogen susceptible to nucleophilic attack. ***Nucleophilic substitution can occur when strong electron-withdrawing groups are ortho or para to the halogen atom:***

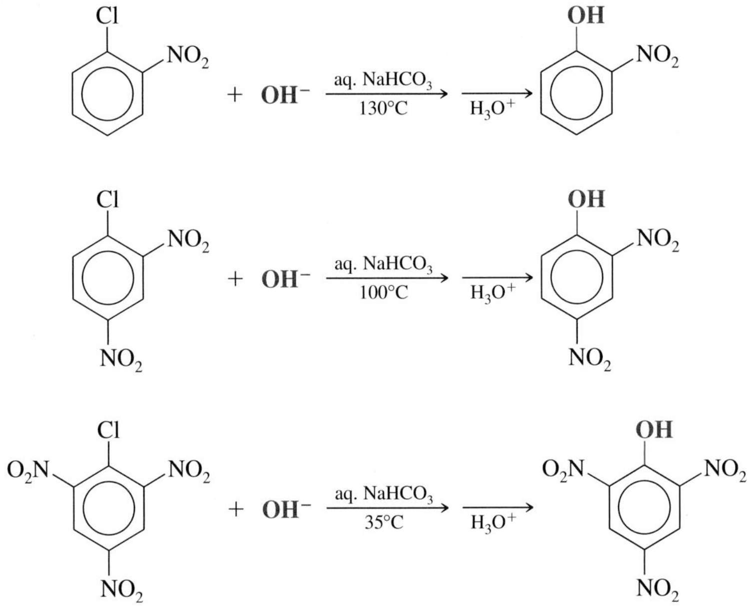

$$o\text{-}ClC_6H_4NO_2 + OH^- \xrightarrow[130°C]{aq.\ NaHCO_3} \xrightarrow{H_3O^+} o\text{-}HOC_6H_4NO_2$$

$$2,4\text{-}(NO_2)_2C_6H_3Cl + OH^- \xrightarrow[100°C]{aq.\ NaHCO_3} \xrightarrow{H_3O^+} 2,4\text{-}(NO_2)_2C_6H_3OH$$

$$2,4,6\text{-}(NO_2)_3C_6H_2Cl + OH^- \xrightarrow[35°C]{aq.\ NaHCO_3} \xrightarrow{H_3O^+} 2,4,6\text{-}(NO_2)_3C_6H_2OH$$

We also see in these examples that the temperature required to bring about the reaction is related to the number of ortho or para nitro groups. Of the three compounds, *o*-nitrochlorobenzene requires the highest temperature (*p*-nitrochlorobenzene reacts at 130°C as well) and 2,4,6-trinitrochlorobenzene requires the lowest temperature.

A meta-nitro group does not produce a similar activating effect. For example, *m*-nitrochlorobenzene gives no corresponding reaction.

The mechanism that operates in these reactions is an *addition–elimination* mechanism involving the formation of a *carbanion* with delocalized electrons called a **Meisenheimer complex** after the German chemist Jacob Meisenheimer, who proposed its correct structure. In the following first step, addition of a hydroxide ion to *p*-nitrochlorobenzene, for example, produces the carbanion; then elimination of a chloride ion yields the substitution product as the aromaticity of the ring is recovered. This mechanism is called the **nucleophilic aromatic substitution (S_NAr).**

A Mechanism for the Reaction

The S_NAr Mechanism

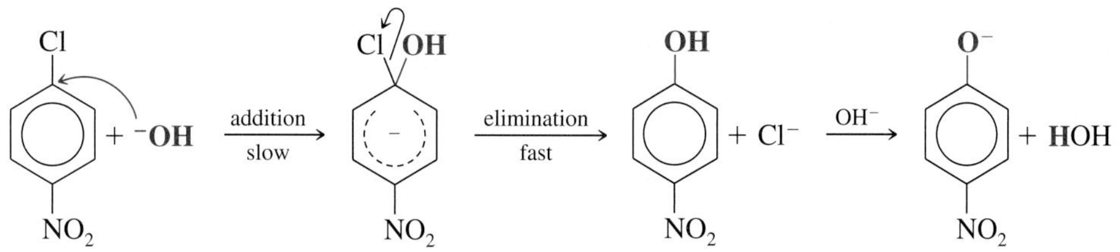

Cl, NO_2 + ^-OH —addition, slow→ Cl/OH, NO_2 —elimination, fast→ OH, NO_2 + Cl^- —OH^-→ O^-, NO_2 + HOH

Carbanion (Meisenheimer complex) Structures of the contributing resonance forms are shown below.

The carbanion is stabilized by *electron-withdrawing groups* in the positions ortho and para to the halogen atom. If we examine the following resonance structures for a Meisenheimer complex, we can see how:

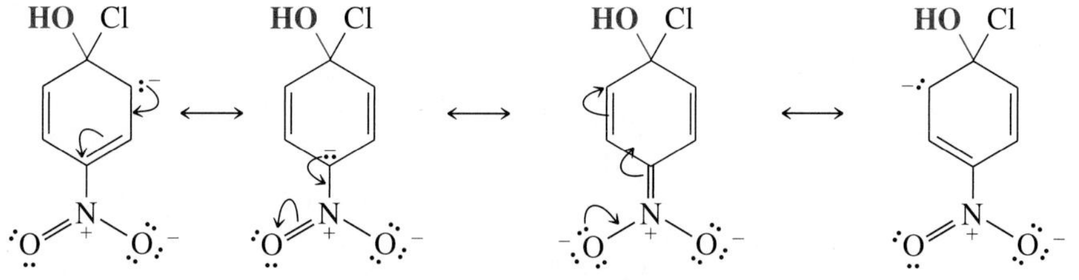

HO Cl ⟷ HO Cl ⟷ HO Cl ⟷ HO Cl

Especially stable (Negative charges are both on oxygen atoms.)

PROBLEM 21.9

1-Fluoro-2,4-dinitrobenzene is highly reactive toward nucleophilic substitution through an S_NAr mechanism. (In Section 24.5B we shall see how this reagent is used in the Sanger method for determining the structures of proteins.) What product would be formed when 1-fluoro-2,4-dinitrobenzene reacts with each of the following reagents?
(a) CH_3CH_2ONa **(b)** NH_3 **(c)** $C_6H_5NH_2$ **(d)** CH_3CH_2SNa

21.11B Nucleophilic Aromatic Substitution through an Elimination–Addition Mechanism: Benzyne

Although aryl halides such as chlorobenzene and bromobenzene do not react with most nucleophiles under ordinary circumstances, they do react under highly forcing conditions.

Chlorobenzene can be converted to phenol by heating it with aqueous sodium hydroxide in a pressurized reactor at 350°C (Section 21.4B):

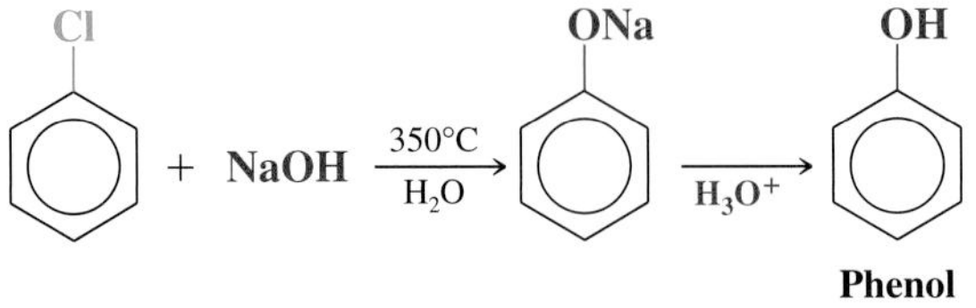

The Chemistry of...

Bacterial Dehalogenation of a PCB Derivative

Polychlorinated biphenyls (PCBs) are compounds that were once used in a variety of electrical devices, industrial applications, and polymers. Their use and production were banned in 1979, however, owing to the toxicity of PCBs and their tendency to accumulate in the food chain.

4-Chlorobenzoic acid is a degradation product of some PCBs. It is now known that certain bacteria are able to dehalogenate 4-chlorobenzoic acid by an enzymatic nucleophilic aromatic substitution reaction. The product is 4-hydroxybenzoic acid, and a mechanism for this enzyme-catalyzed process is shown here. The sequence begins with the thioester of 4-chlorobenzoic acid derived from coenzyme A (CoA):

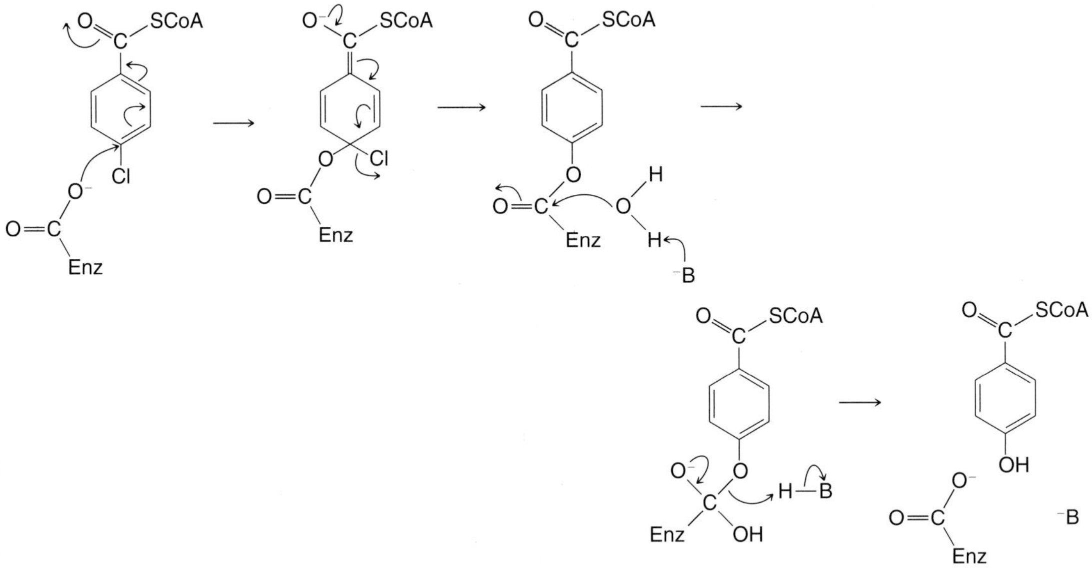

Some key features of this enzymatic S_NAr mechanism are the following. The nucleophile that attacks the chlorinated benzene ring is a carboxylate anion of the enzyme. When the carboxylate attacks, positively charged groups within the enzyme stabilize the additional electron density that develops in the thioester carbonyl group of the Meisenheimer complex. Collapse of the Meisenheimer complex, with rearomatization of the ring and loss of the chloride anion, results in an intermediate where the substrate is covalently bonded to the enzyme as an ester. Hydrolysis of this ester linkage involves a water molecule whose nucleophilicity has been enhanced by a basic site within the enzyme. Hydrolysis of the ester releases 4-hydroxybenzoic acid and leaves the enzyme ready to catalyze another reaction cycle.

Bromobenzene reacts with the very powerful base, $\bar{N}H_2$, in liquid ammonia:

$$C_6H_5Br + \overset{+}{K}\ :\ddot{\bar{N}}H_2 \xrightarrow[NH_3]{-33^{\circ}C} C_6H_5\ddot{N}H_2 + KBr$$

Aniline

These reactions take place through an **elimination–addition mechanism** that involves the formation of an interesting intermediate called *benzyne* (or *dehydrobenzene*). We can illustrate this mechanism with the reaction of bromobenzene and amide ion.

In the first step (see the following mechanism), the amide ion initiates an elimination by abstracting one of the ortho protons because they are the most acidic. The negative charge that develops on the ortho carbon is stabilized by the inductive effect of the bromine. The anion then loses a bromide ion. This elimination produces the highly unstable, and thus highly reactive, **benzyne.** Benzyne then reacts with any available nucleophile (in this case, an amide ion) by a two-step addition reaction to produce aniline.

A Mechanism for the Reaction

The Benzyne Elimination–Addition Mechanism

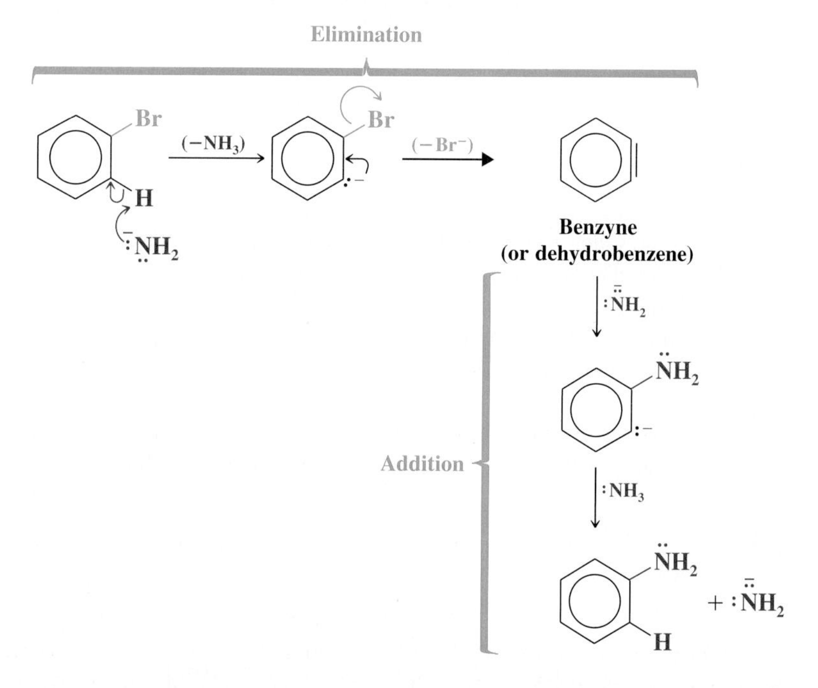

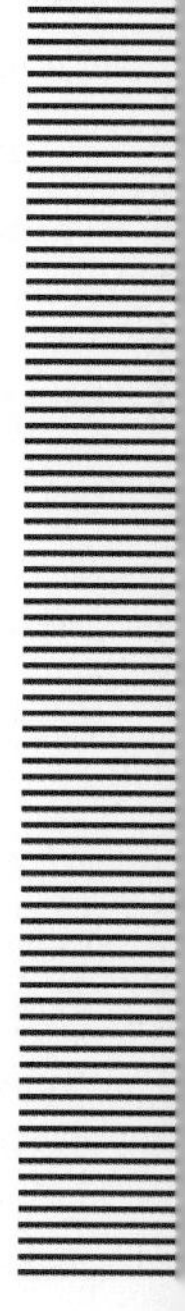

We can better understand the reactive and unstable nature of benzyne if we consider aspects of its electronic structure.

The calculated electrostatic potential map for benzyne, shown in Fig. 21.1*a*, shows the relatively greater negative charge at the edge of the ring, corresponding to the electron density from the additional π bond in benzyne. Figure 21.1*b* shows a schematic representation of the orbital associated with the additional π bond. We can see from these models that the orbitals of the additional π bond in benzyne lie in the same plane as the ring, perpendicular to the axis of the aromatic π system. We can also see in Fig. 21.1 that, because the carbon ring is not a perfect hexagon as in benzene, there is angle strain in the structure of benzyne. The distance between the carbons of the additional π bond in benzyne is shorter than between the other carbons, and the bond angles of the ring are therefore distorted from their ideal values. The result is that benzyne is highly unstable and highly reactive. Consequently, benzyne has never been isolated as a pure substance, but it has been detected and trapped in various ways (see below).

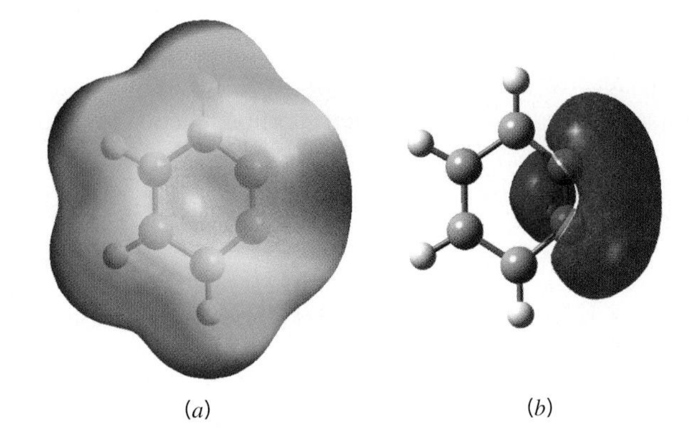

FIGURE 21.1 *(a)* A calculated electrostatic potential map for benzyne shows the relatively greater negative charge (in red) at the edge of the ring, corresponding to electron density from the additional π bond in benzyne. *(b)* A schematic representation of the molecular orbital associated with the additional π bond in benzyne. (Red and blue indicate orbital phase, not charge distribution.) Note that the orientation of this orbital is in the same plane as the ring and perpendicular to the axis of the aromatic π system.

Chem3D Model
Benzyne

What, then, is some of the evidence for an elimination–addition mechanism involving benzyne in some nucleophilic aromatic substitutions?

The first piece of clear-cut evidence was an experiment done by J. D. Roberts (Section 9.9) in 1953—one that marked the beginning of benzyne chemistry. Roberts showed that when ^{14}C-labeled (C*) chlorobenzene is treated with amide ion in liquid ammonia, the aniline that is produced has the label equally divided between the 1 and 2 positions. This result is consistent with the following elimination–addition mechanism but is, of course, not at all consistent with a direct displacement or with an addition–elimination mechanism. (Why?)

Cl
$K^+NH_2^-$
NH_2^-
NH_3
NH_2
(50%)
NH_2
Elimination **Addition** **(50%)**

An even more striking illustration can be seen in the following reaction. When the ortho derivative **1** is treated with sodium amide, the only organic product obtained is *m*-(trifluoromethyl)aniline:

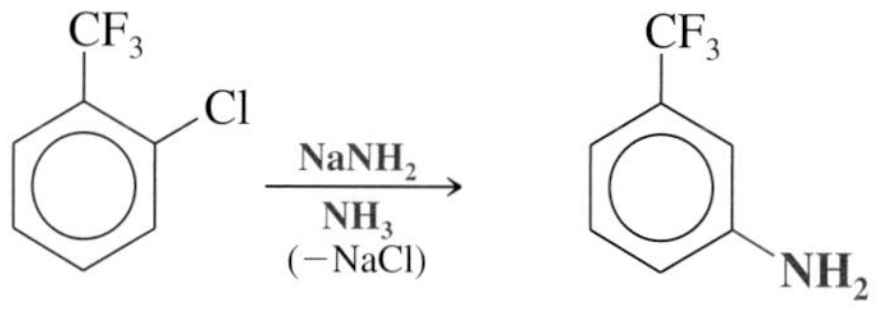

CF_3 Cl NaNH$_2$ / NH$_3$ (−NaCl) → CF_3 NH$_2$

1 ***m*-(Trifluoromethyl)aniline**

This result can also be explained by an elimination–addition mechanism. The first step produces the benzyne **2:**

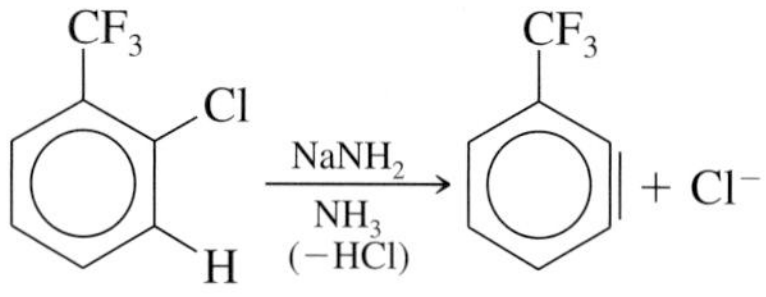

CF_3 Cl H NaNH$_2$ / NH$_3$ (−HCl) → CF_3 + Cl^-

1 **2**

This benzyne then adds an amide ion in the way that produces the more stable carbanion **3** rather than the less stable carbanion **4:**

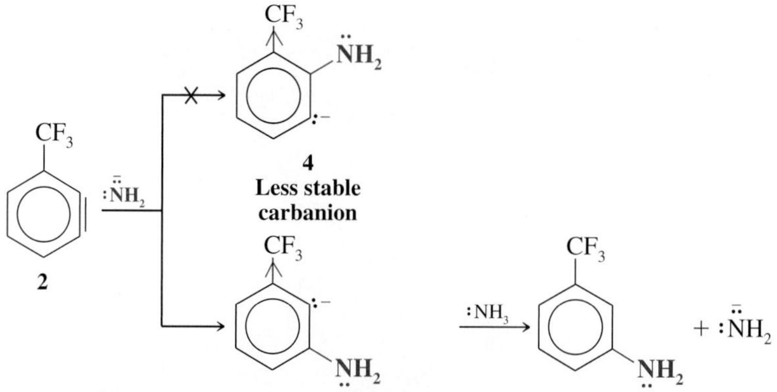

CF_3 :$\ddot{N}H_2^-$ → CF_3 $\ddot{N}H_2$:−

2

4
Less stable carbanion

CF_3 :− $\ddot{N}H_2$:NH_3 → CF_3 $\ddot{N}H_2$ + :$\ddot{N}H_2^-$

3
More stable carbanion (The negative charge is closer to the electronegative trifluoromethyl group.)

Carbanion **3** then accepts a proton from ammonia to form *m*-(trifluoromethyl)aniline.

Carbanion **3** is more stable than **4** because the carbon atom bearing the negative charge is closer to the highly electronegative trifluoromethyl group. The trifluoromethyl group stabilizes the negative charge through its inductive effect. (Resonance effects are not important here because the sp^2 orbital that contains the electron pair does not overlap with the π orbitals of the aromatic system.)

Benzyne intermediates have been "trapped" through the use of Diels–Alder reactions. One convenient method for generating benzyne is the diazotization of anthranilic acid (2-aminobenzoic acid) followed by elimination of CO_2 and N_2:

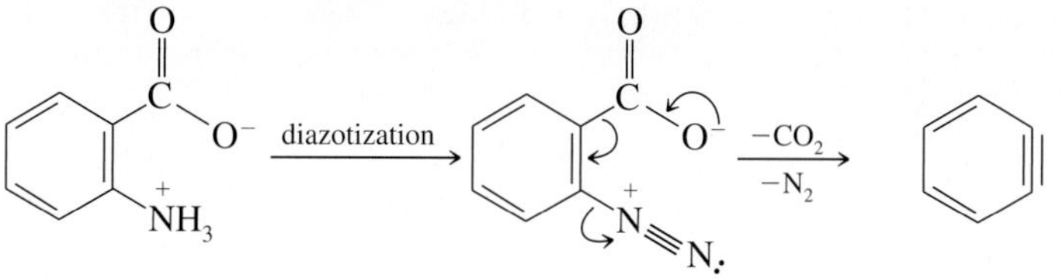

O C O⁻ $\overset{+}{N}H_3$ diazotization → O C O⁻ $\overset{+}{N}\equiv N$: −CO_2 / −N_2 →

Anthranilic acid **Benzene (trapped *in situ*)**

When benzyne is generated in the presence of the diene *furan,* the product is a Diels–Alder adduct:

Benzyne (generated by an elimination reaction) + **Furan** ⟶ **Diels–Alder adduct**

In a fascinating application of host–guest chemistry (an area founded by the late D. Cram, and for which he shared the Nobel Prize in Chemistry in 1987), benzyne itself has been trapped at very low temperature inside a molecular container called a hemicarcerand. Under these conditions, R. Warmuth and Cram found that the incarcerated benzyne was sufficiently stabilized for its ^{1}H and ^{13}C NMR spectra to be recorded (see Fig. 21.2), before it ultimately underwent a Diels–Alder reaction with the container molecule.

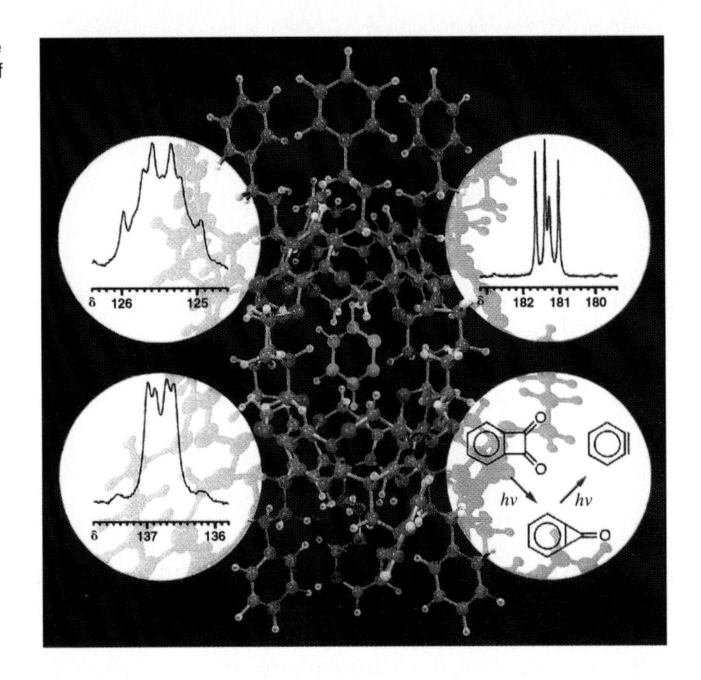

FIGURE 21.2 A molecular graphic of benzyne (green) trapped in a hemicarcerand. Images of ^{13}C NMR data from benzyne and a reaction used to synthesize it are shown in the white circles. (Image courtesy of Jan Haller; reprinted with permission of Ralf Warmuth.)

21.11C Phenylation

Reactions involving benzyne can be useful for formation of a carbon–carbon bond to a phenyl group (a process called phenylation). For example, if acetoacetic ester is treated with bromobenzene and two molar equivalents of sodium amide, phenylation of ethyl acetoacetate occurs. The overall reaction is as follows:

$$CH_3\overset{O}{\overset{\|}{C}}CH_2\overset{O}{\overset{\|}{C}}OC_2H_5 + C_6H_5Br + 2\,NaNH_2 \xrightarrow{\text{liq. } NH_3} CH_3\overset{O}{\overset{\|}{C}}\underset{C_6H_5}{\underset{|}{C}H}\overset{O}{\overset{\|}{C}}OC_2H_5$$

Malonic esters can be phenylated in an analogous way. This process is a useful complement to the alkylation reactions of acetoacetic and malonic esters that we studied in

Chapter 19 because, as you may recall, substrates like bromobenzene are not susceptible to S_N2 reactions [see Section 6.14A and Problem 19.9(c)].

PROBLEM 21.10

When *o*-chlorotoluene is subjected to the conditions used in the Dow process (i.e., aqueous NaOH at 350°C at high pressure), the products of the reaction are *o*-cresol and *m*-cresol. What does this result suggest about the mechanism of the Dow process?

PROBLEM 21.11

When 2-bromo-1,3-dimethylbenzene is treated with sodium amide in liquid ammonia, no substitution takes place. This result can be interpreted as providing evidence for the elimination–addition mechanism. Explain how this interpretation can be given.

PROBLEM 21.12

(a) Outline a step-by-step mechanism for the phenylation of acetoacetic ester by bromobenzene and two molar equivalents of sodium amide. (Why are two molar equivalents of $NaNH_2$ necessary? **(b)** What product would be obtained by hydrolysis and decarboxylation of the phenylated acetoacetic ester? **(c)** How would you prepare phenylacetic acid from malonic ester?

21.12 Spectroscopic Analysis of Phenols and Aryl Halides

Infrared Spectra Phenols show a characteristic absorption band (usually broad) arising from O—H stretching in the 3400–3600-cm^{-1} region. Phenols and aryl halides also show the characteristic absorptions that arise from their benzene rings (see Section 14.11C).

^{1}H NMR Spectra The hydroxylic proton of a phenol is more deshielded than that of an alcohol due to proximity of the benzene pi electron ring current. The exact position of the O—H signal depends on the extent of hydrogen bonding and on whether the hydrogen bonding is *intermolecular* or *intramolecular.* The extent of intermolecular hydrogen bonding depends on the concentration of the phenol, and this strongly affects the position of the O—H signal. In phenol, itself, for example, the position of the O—H signal varies from δ 2.55 for pure phenol to δ 5.63 at 1% concentration in CCl_4. Phenols with strong intramolecular hydrogen bonding, such as salicylaldehyde, show O—H signals between δ 0.5 and δ 1.0, and the position of the signal varies only slightly with concentration. As with other protons that undergo exchange (Section 9.9), the identity of the O—H proton of a phenol can be determined by adding D_2O to the sample. The O—H proton undergoes rapid exchange with deuterium and the proton signal disappears. The aromatic protons of phenols and aryl halides give signals in the δ 7–9 region.

^{13}C NMR Spectra The carbon atoms of the aromatic ring of phenols and aryl halides appear in the region δ 135–170.

Mass Spectra Mass spectra of phenols often display a prominent molecular ion peak, $M^{+\cdot}$. Phenols that have a benzylic hydrogen produce an $M^{+\cdot} - 1$ peak that can be larger than the $M^{+\cdot}$ peak.

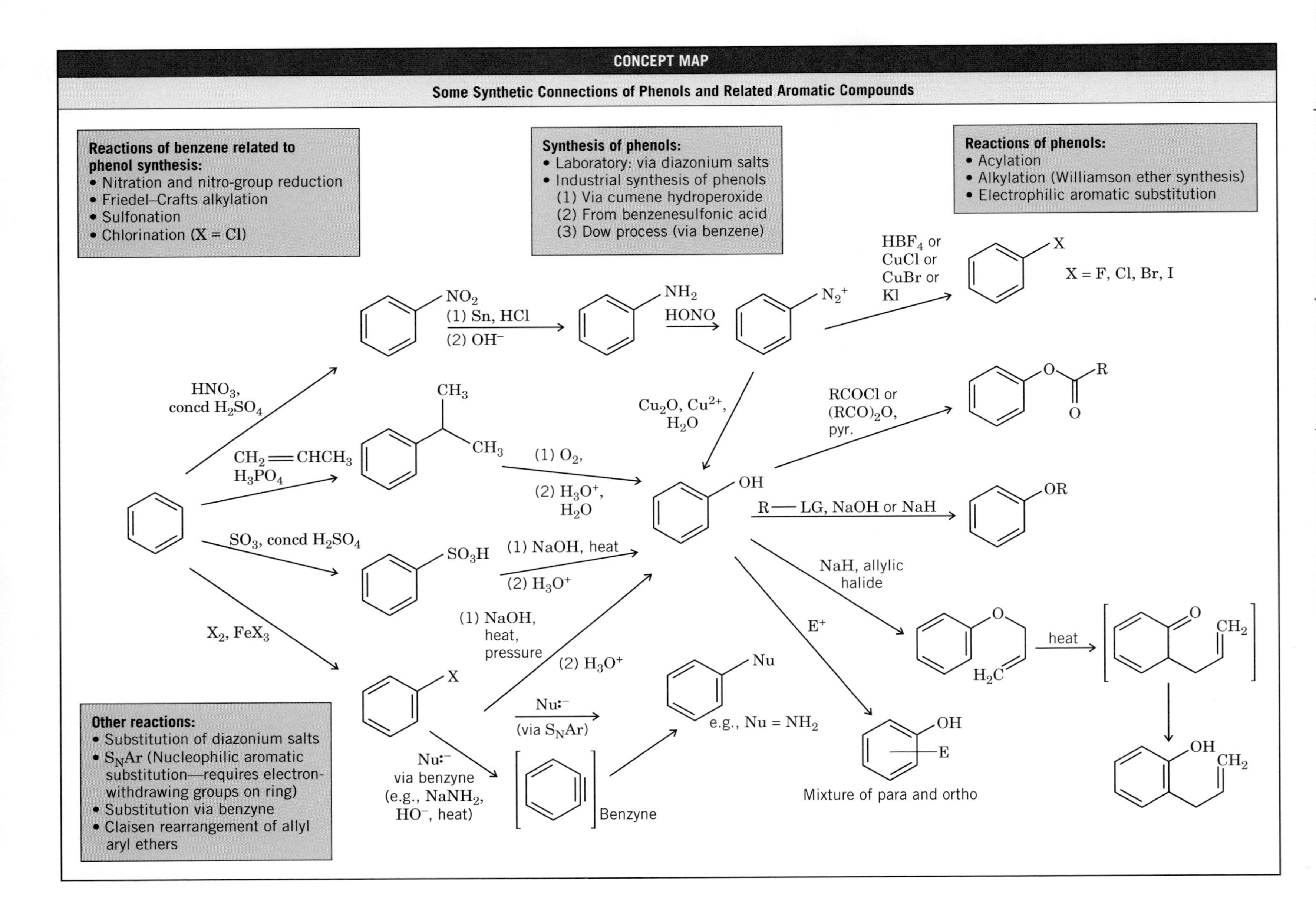
CONCEPT MAP
Some Synthetic Connections of Phenols and Related Aromatic Compounds
Reactions of benzene related to phenol synthesis:
• Nitration and nitro-group reduction
• Friedel–Crafts alkylation
• Sulfonation
• Chlorination (X = Cl)
Synthesis of phenols:
• Laboratory: via diazonium salts
• Industrial synthesis of phenols
(1) Via cumene hydroperoxide
(2) From benzenesulfonic acid
(3) Dow process (via benzene)
Reactions of phenols:
• Acylation
• Alkylation (Williamson ether synthesis)
• Electrophilic aromatic substitution
NO_2
(1) Sn, HCl
(2) OH^-
NH_2
HONO
N_2^+
HBF_4 or CuCl or CuBr or KI
X
X = F, Cl, Br, I
HNO_3, concd H_2SO_4
CH_3
CH_3
Cu_2O, Cu^{2+}, H_2O
RCOCl or $(RCO)_2O$, pyr.
O
R
O
$CH_2{=}CHCH_3$ H_3PO_4
(1) O_2,
(2) H_3O^+, H_2O
OH
R—LG, NaOH or NaH
OR
SO_3, concd H_2SO_4
SO_3H
(1) NaOH, heat
(2) H_3O^+
NaH, allylic halide
X_2, FeX_3
(1) NaOH, heat, pressure
(2) H_3O^+
E^+
O
H_2C
heat
O
CH_2
X
Nu
$Nu{:}^-$ (via S_NAr)
e.g., Nu = NH_2
OH
E
OH
CH_2
Other reactions:
• Substitution of diazonium salts
• S_NAr (Nucleophilic aromatic substitution—requires electron-withdrawing groups on ring)
• Substitution via benzyne
• Claisen rearrangement of allyl aryl ethers
$Nu{:}^-$ via benzyne (e.g., $NaNH_2$, HO^-, heat)
Benzyne
Mixture of para and ortho

Key Terms and Concepts

Claisen rearrangement	Section 21.9
Electrophilic substitution	Section 21.8
Elimination–addition (via benzyne)	Section 21.11B
Nucleophilic aromatic substitution (S_NAr)	Section 21.11A
Phenols as weak acids	Section 21.5
Synthesis and cleavage of aryl esters and ethers	Sections 21.6, 21.7

Additional Problems

21.13 What products would be obtained from each of the following acid–base reactions?

(a) Sodium ethoxide in ethanol + phenol $\longrightarrow$

(b) Phenol + aqueous sodium hydroxide $\longrightarrow$

(c) Sodium phenoxide + aqueous hydrochloric acid $\longrightarrow$

(d) Sodium phenoxide + H_2O + CO_2 $\longrightarrow$

21.14 Complete the following equations:

(a) Phenol + Br_2 $\xrightarrow{5°C,\ CS_2}$

(b) Phenol + concd H_2SO_4 $\xrightarrow{25°C}$

(c) Phenol + concd H_2SO_4 $\xrightarrow{100°C}$

(d) CH_3–C_6H_4–OH (para) + *p*-toluenesulfonyl chloride $\xrightarrow[OH^-]{}$

(e) Phenol + Br_2 $\xrightarrow{H_2O}$

(f) Phenol + phthalic anhydride (benzene ring fused to C(=O)–O–C(=O)) $\longrightarrow$

(g) *p*-Cresol + Br_2 $\xrightarrow{H_2O}$

(h) Phenol + $C_6H_5\overset{O}{\overset{\|}{C}}Cl$ $\xrightarrow[base]{}$

(i) Phenol + $\left(C_6H_5\overset{O}{\overset{\|}{C}}-\right)_2O$ $\xrightarrow[base]{}$

(j) Phenol + NaOH $\longrightarrow$

(k) Product of (j) + $CH_3OSO_2OCH_3$ $\longrightarrow$

(l) Product of (j) + CH_3I $\longrightarrow$

(m) Product of (j) + $C_6H_5CH_2Cl$ $\longrightarrow$

Note: Problems marked with an asterisk are "challenge problems."

21.15 Describe a simple chemical test that could be used to distinguish between members of each of the following pairs of compounds:

(a) 4-Chlorophenol and 4-chloro-1-methylbenzene
(b) 4-Methylphenol and 4-methylbenzoic acid
(c) Phenyl vinyl ether and ethyl phenyl ether
(d) 4-Methylphenol and 2,4,6-trinitrophenol
(e) Ethyl phenyl ether and 4-ethylphenol

21.16 When *m*-chlorotoluene is treated with sodium amide in liquid ammonia, the products of the reaction are *o*-, *m*-, and *p*-toluidine (i.e., *o*-$CH_3C_6H_4NH_2$, *m*-$CH_3C_6H_4NH_2$, and *p*-$CH_3C_6H_4NH_2$). Propose plausible mechanisms that account for the formation of each product.

21.17 Without consulting tables, select the stronger acid from each of the following pairs:

(a) 4-Methylphenol and 4-fluorophenol
(b) 4-Methylphenol and 4-nitrophenol
(c) 4-Nitrophenol and 3-nitrophenol
(d) 4-Methylphenol and benzyl alcohol
(e) 4-Fluorophenol and 4-bromophenol

21.18 Phenols are often effective antioxidants (see Problem 21.21 and Section 10.11) because they are said to "trap" radicals. The trapping occurs when phenols react with highly reactive radicals to produce less reactive (more stable) phenolic radicals. **(a)** Show how phenol itself might react with an alkoxyl radical (RO·) in a hydrogen abstraction reaction involving the phenolic —OH. **(b)** Write resonance structures for the resulting radical that account for its being relatively unreactive.

21.19 The first synthesis of a crown ether (Section 11.17A) by C. J. Pedersen (of the DuPont Company) involved treating 1,2-benzenediol with di(2-chloroethyl) ether, $(ClCH_2CH_2)_2O$, in the presence of NaOH. The product was a compound called dibenzo-18-crown-6. Give the structure of dibenzo-18-crown-6 and provide a plausible mechanism for its formation.

21.20 A compound **X** ($C_{10}H_{14}O$) dissolves in aqueous sodium hydroxide but is insoluble in aqueous sodium bicarbonate. Compound **X** reacts with bromine in water to yield a dibromo derivative, $C_{10}H_{12}Br_2O$. The 3000–4000-cm^{-1} region of the IR spectrum of **X** shows a broad peak centered at 3250 cm^{-1}; the 680–840-cm^{-1} region shows a strong peak at 830 cm^{-1}. The ^1H NMR spectrum of **X** gives the following:

Singlet	δ 1.3 (9H)
Singlet	δ 4.9 (1H)
Multiplet	δ 7.0 (4H)

What is the structure of **X**?

21.21 The widely used antioxidant and food preservative called **BHA** (**b**utylated **h**ydroxy**a**nisole) is actually a mixture of 2-*tert*-butyl-4-methoxyphenol and 3-*tert*-butyl-4-methoxyphenol. **BHA** is synthesized from *p*-methoxyphenol and 2-methylpropene. **(a)** Suggest how this is done. **(b)** Another widely used antioxidant is **BHT** (**b**utylated **h**ydroxy**t**oluene). **BHT** is actually 2,6-di-*tert*-butyl-4-methylphenol, and the raw materials used in its production are *p*-cresol and 2-methylpropene. What reaction is used here?

21.22 The herbicide **2,4-D** can be synthesized from phenol and chloroacetic acid. Outline the steps involved.

Cl—C_6H_3(Cl)—OCH_2CO_2H $ClCH_2CO_2H$

2,4-D
(2,4-dichlorophenoxyacetic acid)

Chloroacetic acid

21.23 Compound **Z** ($C_5H_{10}O$) decolorizes bromine in carbon tetrachloride. The IR spectrum of **Z** shows a broad peak in the 3200–3600-cm^{-1} region. The 300 MHz ^{1}H NMR spectrum of **Z** is given in Fig. 21.3. Propose a structure for **Z.**

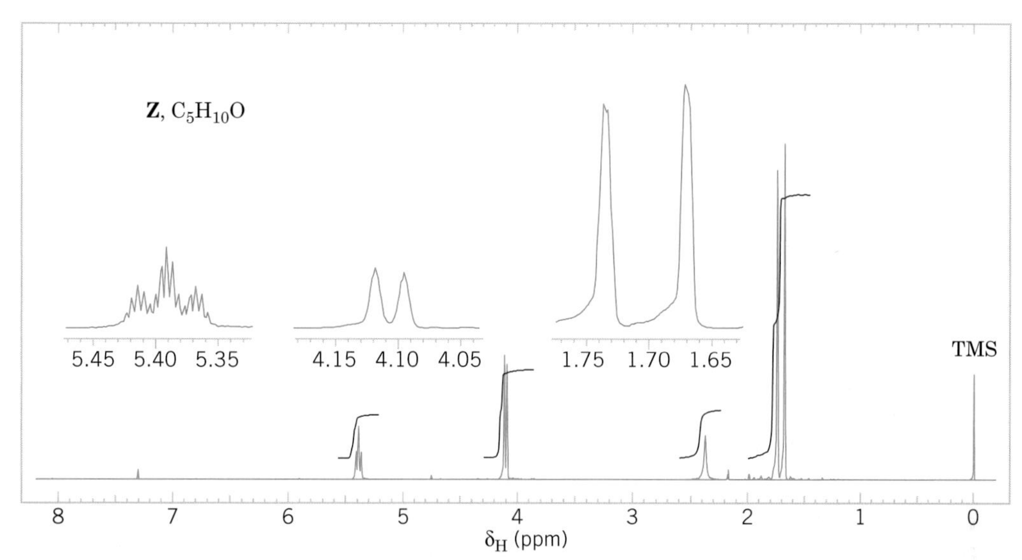

FIGURE 21.3 The 300-MHz ^{1}H NMR spectrum of compound Z (Problem 21.23). Expansions of the signals are shown in the offset plots.

21.24 A synthesis of the β-receptor blocker called toliprolol begins with a reaction between 3-methylphenol and epichlorohydrin. The synthesis is outlined below. Give the structures of the intermediates and toliprolol.

$$\text{3-Methylphenol} + \underset{\text{Epichlorohydrin}}{H_2C\overset{\text{O}}{—}CH—CH_2Cl} \longrightarrow C_{10}H_{13}O_2Cl \xrightarrow{OH^-}$$

$$C_{10}H_{12}O_2 \xrightarrow{(CH_3)_2CHNH_2} \textit{Toliprolol, } C_{13}H_{21}NO_2$$

21.25 Explain how it is possible for 2,2′-dihydroxy-1,1′-binaphthyl (below) to exist in enantiomeric forms.

OH
OH

21.26 *p*-Chloronitrobenzene was allowed to react with sodium 2,6-di-*tert*-butylphenoxide with the intention of preparing the diphenyl ether **1.** The product was not **1,** but rather was an isomer of **1** that still possessed a phenolic hydroxyl group:

$C(CH_3)_3$
O — NO_2
$C(CH_3)_3$
1

What was this product, and how can one account for its formation?

21.27 Account for the fact that the Dow process for the production of phenol produces both diphenyl ether (**1**) and 4-hydroxybiphenyl (**2**) as byproducts:

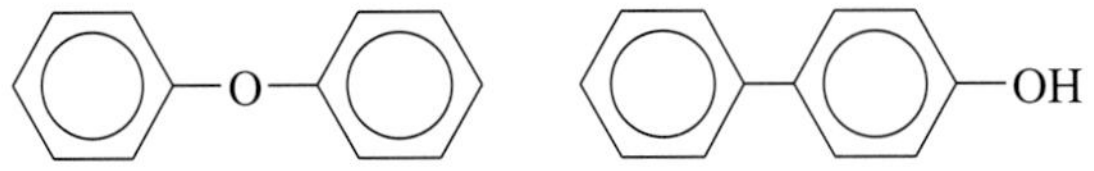

1 **2**

21.28 Predict the outcome of the following reactions:

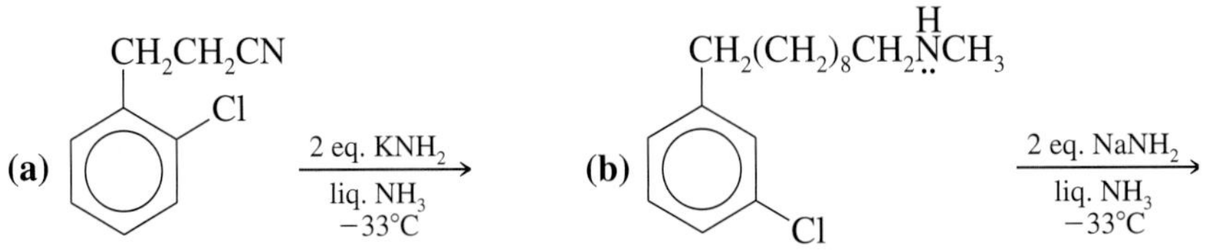

(a) CH_2CH_2CN, Cl; $\xrightarrow[\text{liq. } NH_3,\ -33°C]{\text{2 eq. } KNH_2}$

(b) $CH_2(CH_2)_8CH_2\ddot{N}HCH_3$, Cl; $\xrightarrow[\text{liq. } NH_3,\ -33°C]{\text{2 eq. } NaNH_2}$

21.29 In the case of halogen-substituted azulenes, a halogen atom on C6 can be displaced by nucleophiles while one on C1 is unreactive toward nucleophiles. Rationalize this difference in behavior.

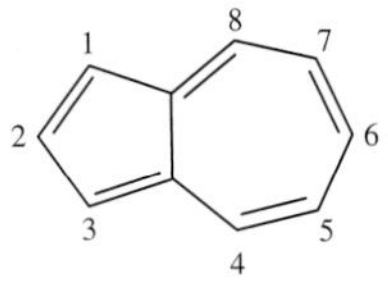

21.30 Explain why, in the case shown, the allyl group has migrated with no change having occurred in the position of the labeled carbon atom within the allyl group:

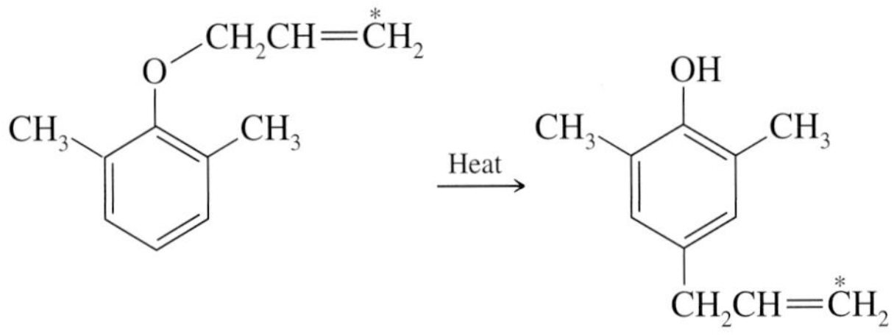

$O-CH_2CH{=}\overset{*}{C}H_2$, CH_3, CH_3 $\xrightarrow{\text{Heat}}$ OH, CH_3, CH_3, $CH_2CH{=}\overset{*}{C}H_2$

21.31 In the Sommelet–Hauser rearrangement, a benzyl quaternary ammonium salt reacts with a strong base to give a benzyl tertiary amine, as exemplified below:

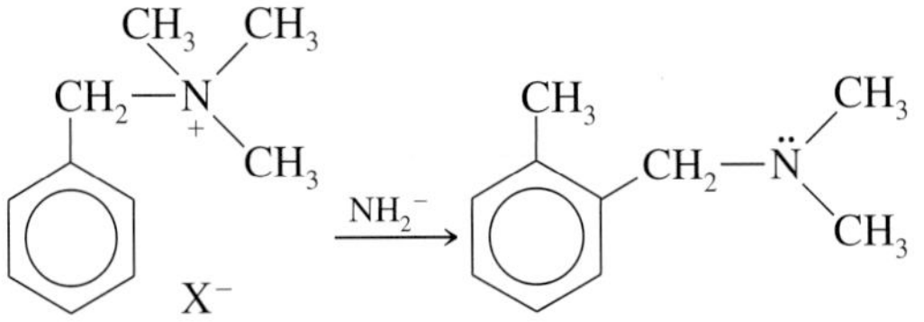

$CH_2-\overset{+}{N}(CH_3)_3$ X^- $\xrightarrow{NH_2^-}$ CH_3, $CH_2-\ddot{N}(CH_3)_2$

Suggest a mechanism for this rearrangement.

21.32 In comparing nucleophilic aromatic substitution reactions that differ only in the identity of the halogen that is the leaving group in the substrate, it is found that the fluorinated substrate reacts faster than either of the cases where bromine or chlorine is the leaving group. Explain this behavior, which is contrary to the trend among the halogens as leaving groups in S_N1 and S_N2 reactions (in protic solvents).

21.33 Hexachlorophene, [structure: OH, Cl, CH_2, OH, Cl, Cl, Cl, Cl, Cl], was, at one time, a widely used germicide (see Section G.3). Suggest how this compound might be synthesized, starting with benzene.

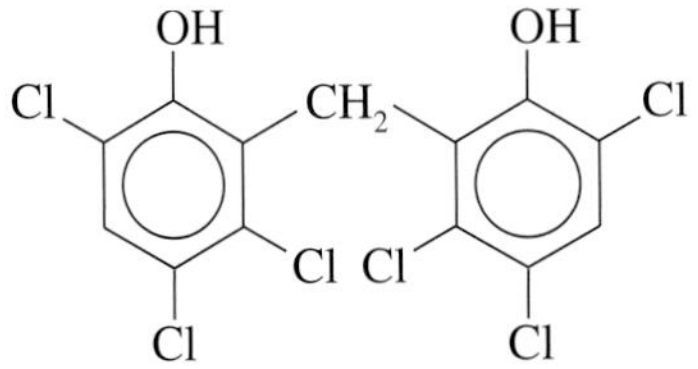

21.34 The Fries rearrangement occurs when a phenolic ester is heated with a Friedel–Crafts catalyst such as $AlCl_3$:

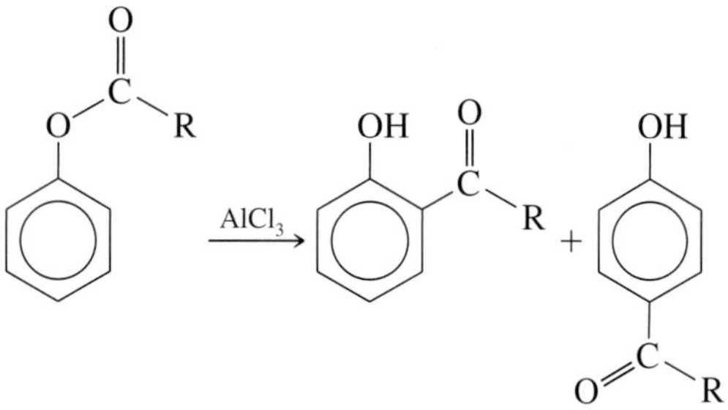

The reaction may produce both ortho and para acylated phenols, the former generally favored by high temperatures and the latter by low temperatures. **(a)** Suggest an experiment that might indicate whether the reaction is inter- or intramolecular. **(b)** Explain the temperature effect on product formation.

21.35 In protic solvents the naphthoxide ion (**I**) is alkylated primarily at position 1 (*C* alkylation) whereas in polar aprotic solvents such as DMF, the product is almost exclusively the result of a conventional Williamson ether synthesis (*O* alkylation):

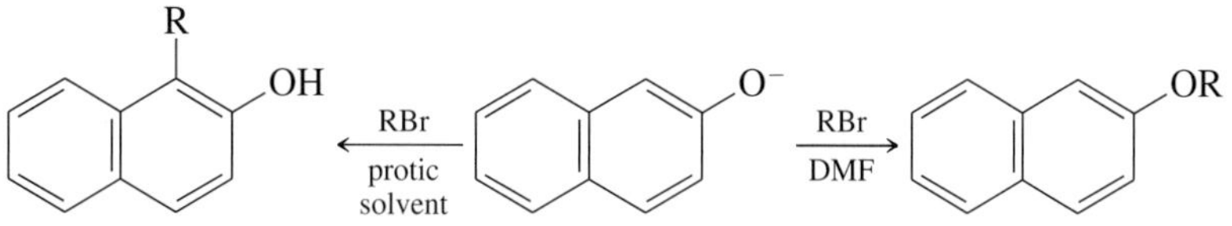

Why does the change in solvent make a difference?

***21.36** Compound **W** was isolated from a marine annelid commonly used in Japan as a fish bait, and it was shown to be the substance that gives this organism its observed toxicity to some insects that contact it.

MS (*m*/*z*): 151 (relative abundance 1.09), 149 ($\mathbf{M}^{+\cdot}$, rel. abund. 1.00), 148

IR (cm^{-1}): 2960, 2850, 2775

^{1}**H NMR** (δ): 2.3 (s, 6H), 2.6 (d, 4H), and 3.2 (m, 1H)

^{13}**C NMR** (δ): 38 (CH_3), 43 (CH_2), and 75 (CH)

These reactions were used to obtain further information about the structure of **W**:

$$\mathbf{W} \xrightarrow{NaBH_4} \mathbf{X} \xrightarrow{C_6H_5COCl} \mathbf{Y} \xrightarrow{\text{Raney Ni}} \mathbf{Z}$$

Compound **X** had a new infrared band at 2570 cm^{-1} and

^{1}**H NMR** (δ): 1.6 (t, 2H), 2.3 (s, 6H), 2.6 (m, 4H), and 3.2 (m, 1H)

^{13}**C NMR** (δ): 28 (CH_2), 38 (CH_3), and 70 (CH)

Compound **Y** had these data:

IR (cm^{-1}): 3050, 2960, 2850, 1700, 1610, 1500, 760, 690

^{1}**H NMR** (δ): 2.3 (s, 6H), 2.9 (d, 4H), 3.0 (m, 1H), 7.4 (m, 4H), 7.6 (m, 2H), 8.0 (m, 4H)

^{13}**C NMR** (δ): 34 (CH_2), 39 (CH_3), 61 (CH), 128 (CH), 129 (CH), 134 (CH), 135 (C), and 187 (C)

Compound **Z** had

MS (*m*/*z*): 87 ($M^{\ddagger}$), 86, 72
IR (cm^{-1}): 2960, 2850, 1385, 1370, 1170
^{1}H NMR (δ): 1.0 (d, 6H), 2.3 (s, 6H), and 3.0 (heptet, 1H)
^{13}C NMR (δ): 21 (CH_3), 39 (CH_3), and 55 (CH)

What are the structures of **W** and of each of its reaction products **X, Y,** and **Z**?

***21.37** Phenols generally are not changed on treatment with sodium borohydride followed by acidification to destroy the excess, unreacted hydride. For example, the 1,2-, 1,3-, and 1,4-benzenediols and 1,2,3-benzenetriol are unchanged under these conditions. However, 1,3,5-benzenetriol (phloroglucinol) gives a high yield of a product **A** that has these properties:

MS (*m*/*z*): 110
IR (cm^{-1}): 3250 (broad), 1613, 1485
^{1}H NMR (δ in DMSO): 6.15 (m, 3H), 6.89 (t, 1H), and 9.12 (s, 2H)

(a) What is the structure of **A**?
(b) Suggest a mechanism by which the above reaction occurred. (1,3,5-Benzenetriol is known to have more tendency to exist in a keto tautomeric form than do simpler phenols.)

21.38 Open the molecular model file for benzyne and examine the following molecular orbitals: the LUMO (lowest unoccupied molecular orbital), the HOMO (highest occupied molecular orbital), the HOMO-1 (next lower energy orbital), the HOMO-2 (next lower in energy), and the HOMO-3 (next lower in energy). **(a)** Which orbital best represents the region where electrons of the additional π bond in benzyne would be found? **(b)** Which orbital would accept electrons from a Lewis base upon nucleophilic addition to benzyne? **(c)** Which orbitals are associated with the six π electrons of the aromatic system. Recall that each molecular orbital can hold a maximum of two electrons.

Chem3D Model

LEARNING GROUP PROBLEMS

1. Thyroxine is a hormone produced by the thyroid gland that is involved in regulating metabolic activity. In a previous Learning Group Problem (Chapter 15) we considered reactions involved in a chemical synthesis of thyroxine. The following is a synthesis of optically pure thyroxine from the amino acid tyrosine (also see Problem 2, below). This synthesis proved to be useful on an industrial scale. (Scheme adapted from Fleming, I. *Selected Organic Syntheses,* Wiley: New York; 1973, pp 31–33.)
   (a) 1 to **2** What type of reaction is involved in the conversion of **1** to **2**? Write a detailed mechanism for this transformation. Explain why the nitro groups appear where they do in **2.**
   (b) 2 to **3** (i) Write a detailed mechanism for step (1) in the conversion of **2** to **3.**
   (ii) Write a detailed mechanism for step (2) in the conversion of **2** to **3.**
   (iii) Write a detailed mechanism for step (3) in the conversion of **2** to **3.**
   (c) 3 to **4** (i) What type of reaction mechanism is involved in the conversion of **3** to **4**?
   (ii) Write a detailed mechanism for the reaction from **3** to **4.** What key intermediate is involved?
   (d) 5 to **6** Write a detailed mechanism for conversion of the methoxyl group of **5** to the phenolic hydroxyl of **6.**

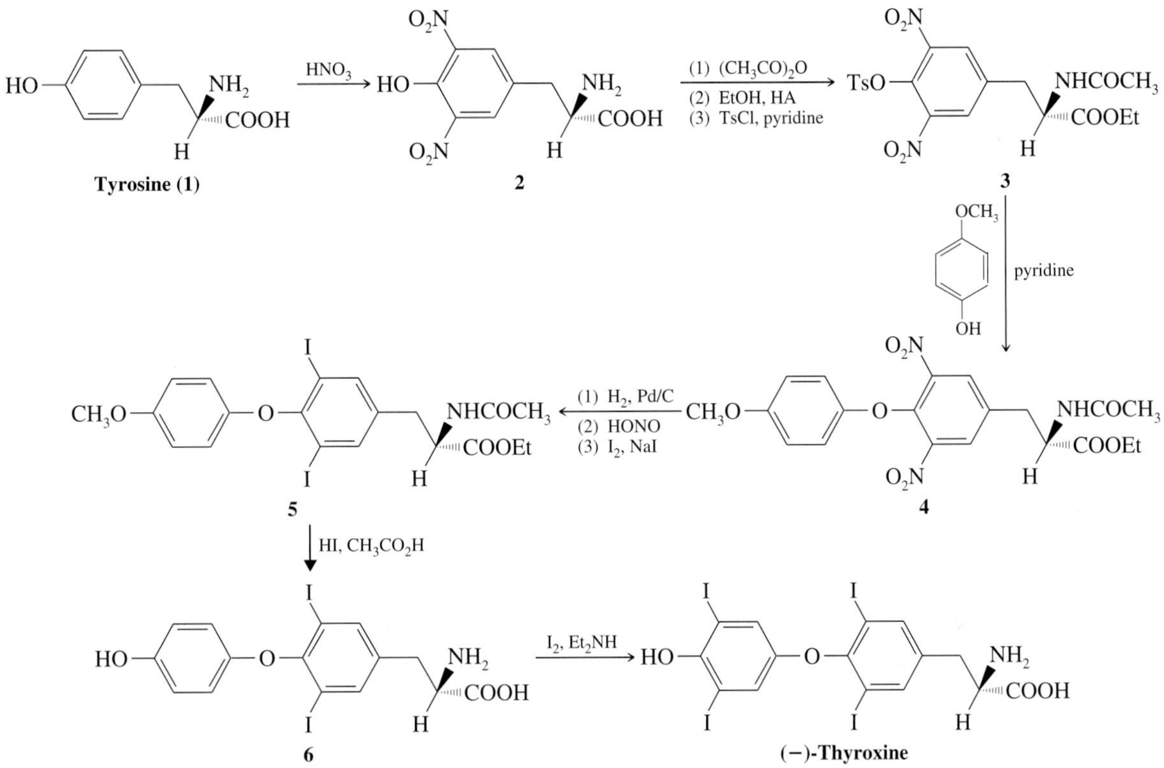

Tyrosine (1) $\xrightarrow{HNO_3}$ 2 $\xrightarrow{(1)\ (CH_3CO)_2O;\ (2)\ EtOH,\ HA;\ (3)\ TsCl,\ pyridine}$ 3

3 $\xrightarrow{p\text{-}CH_3OC_6H_4OH,\ pyridine}$ 4 $\xrightarrow{(1)\ H_2,\ Pd/C;\ (2)\ HONO;\ (3)\ I_2,\ NaI}$ 5

5 $\xrightarrow{HI,\ CH_3CO_2H}$ 6 $\xrightarrow{I_2,\ Et_2NH}$ (−)-Thyroxine

2. Tyrosine is an amino acid with a phenolic side chain. Biosynthesis in plants and microbes of tyrosine involves enzymatic conversion of chorismate to prephenate, below. Prephenate is then processed further to form tyrosine. These steps are shown here:

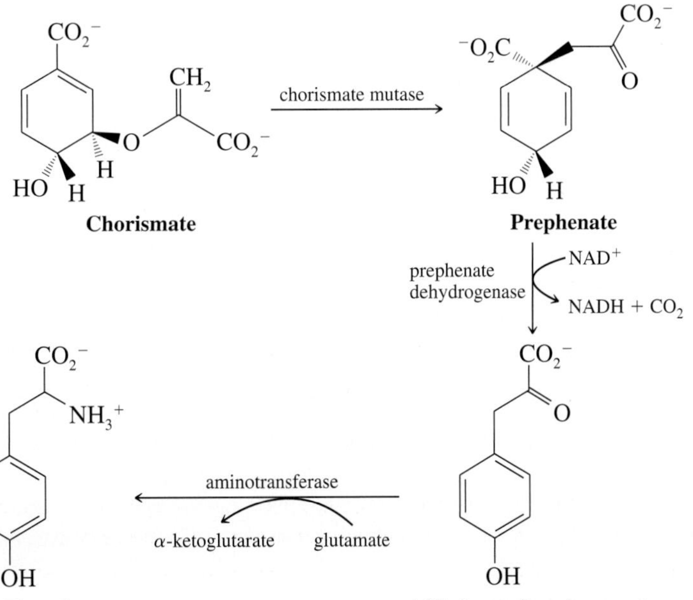

Chorismate $\xrightarrow{\text{chorismate mutase}}$ Prephenate

Prephenate $\xrightarrow{\text{prephenate dehydrogenase};\ NAD^+ \rightarrow NADH + CO_2}$ 4-Hydroxyphenylpyruvate

4-Hydroxyphenylpyruvate $\xrightarrow{\text{aminotransferase};\ \text{glutamate} \rightarrow \alpha\text{-ketoglutarate}}$ Tyrosine

(a) There has been substantial research and debate about the enzymatic conversion of chorismate to prephenate by chorismate mutase. Although the enzymatic mechanism may not be precisely analogous, what laboratory reaction have we studied in this chapter that resembles the biochemical conversion of chorismate to prephenate? Draw arrows to show the movement of electrons involved in such a reaction from chorismate to prephenate.

(b) When the type of reaction you proposed above is applied in laboratory syntheses, it is generally the case that the reaction proceeds by a concerted chair conformation transition state. Five of the atoms of the chair are carbon and one is oxygen. In both the reactant and product, the chair has one bond missing, but at the point of the bond reorganization there is roughly concerted flow of electron density throughout the atoms involved in the chair. For the reactant shown below, draw the structure of the product and the associated chair conformation transition state for this type of reaction:

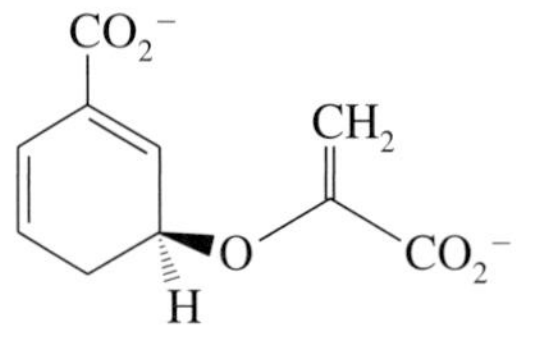

(c) Draw the structure of the nicotinamide ring of NAD^+ and draw mechanism arrows to show the decarboxylation of prephenate to 4-hydroxyphenylpyruvate with transfer of the hydride to NAD^+ (this is the type of process involved in the mechanism of prephenate dehydrogenase).

(d) Look up the structures of glutamate (glutamic acid) and α-ketoglutarate and consider the process of transamination involved in conversion of 4-hydroxyphenylpyruvate to tyrosine. Identify the source of the amino group in this transamination (i.e., what is the amino group "donor"?). What functional group is left after the amino group has been transferred from its donor? Propose a mechanism for this transamination. Note that the mechanism you propose will likely involve formation and hydrolysis of several imine intermediates—reactions similar to others we studied in Section 16.8.

Second Review Problem Set

1. Arrange the compounds of each of the following series in order of increasing acidity:

(a) CH_3CH_2OH $\quad CH_3\overset{O}{\overset{\|}{C}}OH$ $\quad CH_3O\overset{O}{\overset{\|}{C}}CH_2\overset{O}{\overset{\|}{C}}OCH_3$ $\quad CH_3\overset{O}{\overset{\|}{C}}CH_3$

(b) 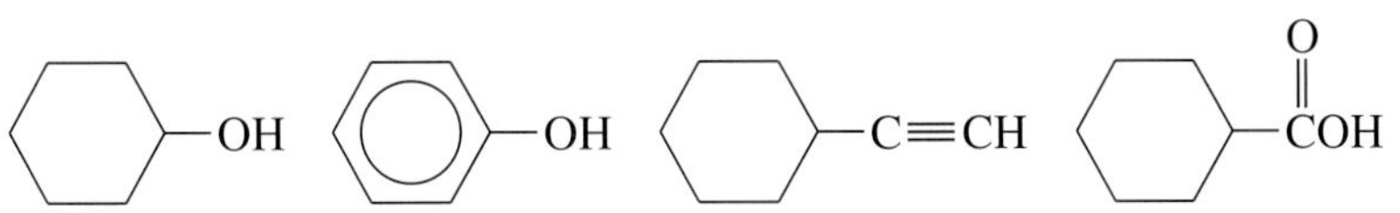 $C_6H_{11}{-}OH$ (cyclohexanol) $\quad C_6H_5{-}OH$ $\quad C_6H_{11}{-}C{\equiv}CH$ $\quad C_6H_{11}{-}\overset{O}{\overset{\|}{C}}OH$

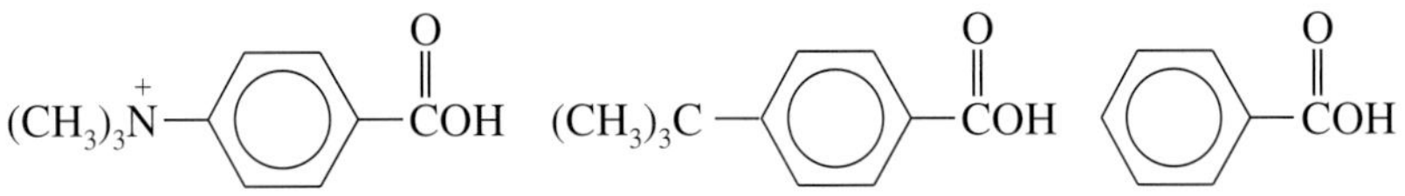

(c) $(CH_3)_3\overset{+}{N}{-}C_6H_4{-}\overset{O}{\overset{\|}{C}}OH$ $\quad (CH_3)_3C{-}C_6H_4{-}\overset{O}{\overset{\|}{C}}OH$ $\quad C_6H_5{-}\overset{O}{\overset{\|}{C}}OH$

(d) $CH_3CCl_2\overset{O}{\overset{\|}{C}}OH$ $\quad CH_3CH_2\overset{O}{\overset{\|}{C}}OH$ $\quad CH_3CHCl\overset{O}{\overset{\|}{C}}OH$

(e) 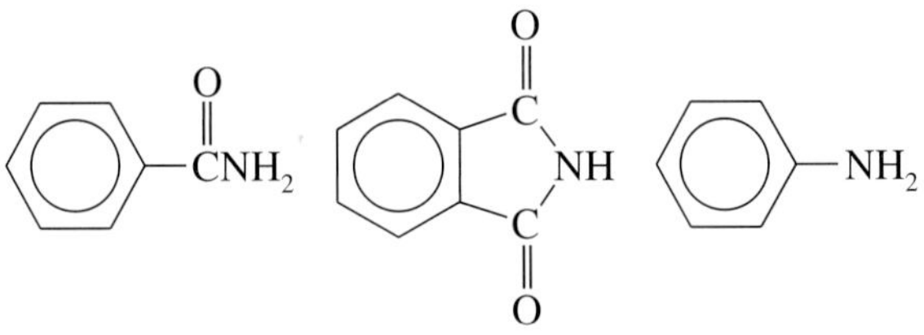 $C_6H_5{-}\overset{O}{\overset{\|}{C}}NH_2$ $\quad$ $C_6H_4(C{=}O)_2NH$ (phthalimide) $\quad C_6H_5{-}NH_2$

2. Arrange the compounds of each of the following series in order of increasing basicity:

(a) $CH_3\overset{O}{\overset{\|}{C}}NH_2$ $\quad CH_3CH_2NH_2$ $\quad NH_3$

(b) 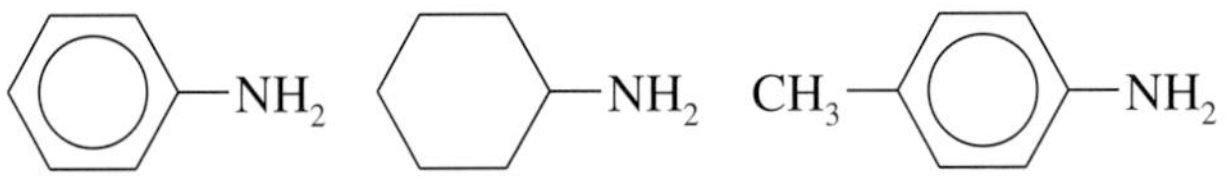 $C_6H_5{-}NH_2$ $\quad C_6H_{11}{-}NH_2$ $\quad CH_3{-}C_6H_4{-}NH_2$

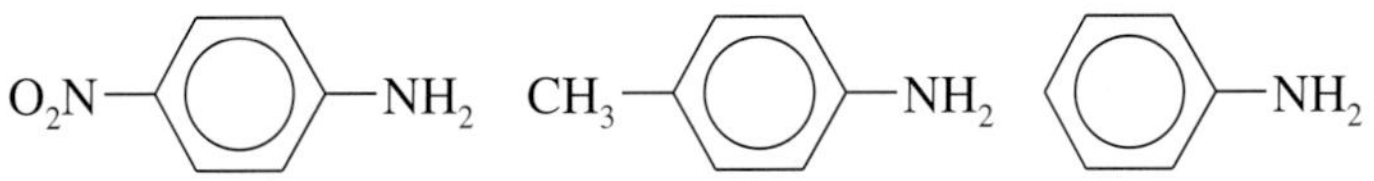

(c) $O_2N{-}C_6H_4{-}NH_2$ $\quad CH_3{-}C_6H_4{-}NH_2$ $\quad C_6H_5{-}NH_2$

(d) $CH_3CH_2CH_3$ $\quad CH_3NHCH_3$ $\quad CH_3OCH_3$

3. Starting with 1-butanol and using any other required reagents, outline a synthesis of each of the following compounds. You need not repeat steps carried out in earlier parts of this problem.

(a) Butyl bromide
(b) Butylamine
(c) Pentylamine
(d) Butanoic acid
(e) Pentanoic acid
(f) Butanoyl chloride
(g) Butanamide
(h) Butyl butanoate
(i) Propylamine
(j) Butylbenzene
(k) Butanoic anhydride
(l) Hexanoic acid

4. Starting with benzene, toluene, or aniline and any other required reagents, outline a synthesis of each of the following:

(a) 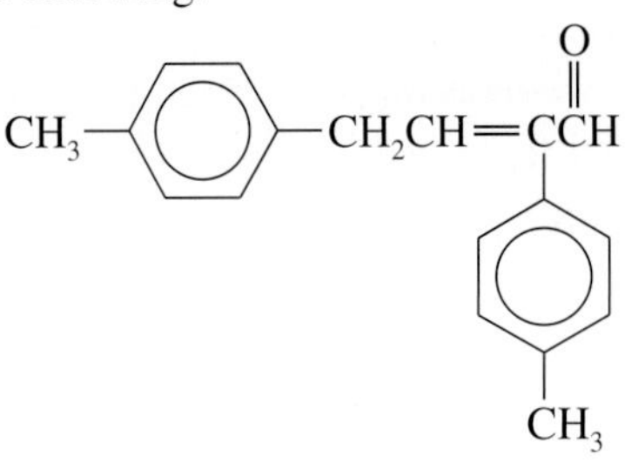 $CH_3{-}C_6H_4{-}CH_2CH{=}\overset{O}{\overset{\|}{C}}CH$, with $C_6H_4{-}CH_3$ on the carbonyl-bearing carbon

(b) 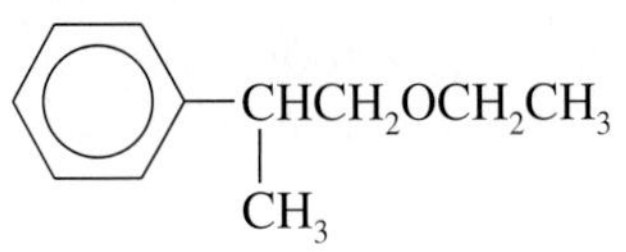 $C_6H_5{-}\underset{CH_3}{\underset{|}{C}}HCH_2OCH_2CH_3$

(c) 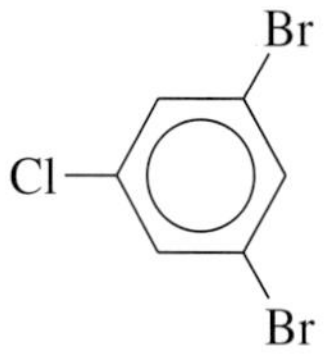 1,3-dibromo-5-chlorobenzene (Cl, Br, Br on benzene ring)

(e) $C_6H_5CH{=}C(CO_2H)CH_3$

(d) 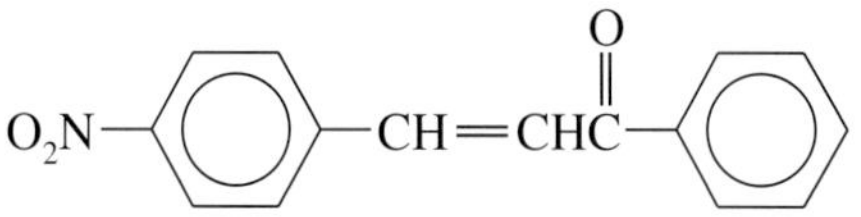 $O_2N-C_6H_4-CH{=}CHC(=O)-C_6H_5$

5. Give stereochemical structures for compounds **A–D**:

$$\text{2-Methyl-1,3-butadiene} + \text{diethyl fumarate} \longrightarrow \mathbf{A}\ (C_{13}H_{20}O_4) \xrightarrow{(1)\ LiAlH_4,\ (2)\ H_2O} \mathbf{B}\ (C_9H_{16}O_2) \xrightarrow{PBr_3} \mathbf{C}\ (C_9H_{14}Br_2) \xrightarrow{Zn,\ H_3O^+} \mathbf{D}\ (C_9H_{16})$$

6. A Grignard reagent that is a key intermediate in an industrial synthesis of vitamin A (Section 17.5B) can be prepared in the following way:

$$HC{\equiv}CLi + CH_2{=}CHCCH_3\ (\text{C}{=}\text{O}) \xrightarrow[(2)\ NH_4^+]{(1)\ \text{liq. } NH_3} \mathbf{A}\ (C_6H_8O) \xrightarrow{H_3O^+} \mathbf{B}\ HOCH_2CH{=}C(CH_3){-}C{\equiv}CH \xrightarrow{2\ C_2H_5MgBr} \mathbf{C}\ (C_6H_6Mg_2Br_2O)$$

(a) What are the structures of compounds **A** and **C**?

(b) The acid-catalyzed rearrangement of **A** to **B** takes place very readily. What two factors account for this?

7. The remaining steps in the industrial synthesis of vitamin A (as an acetate) are as follows: The Grignard reagent **C** from Problem 6 is allowed to react with the aldehyde shown here:

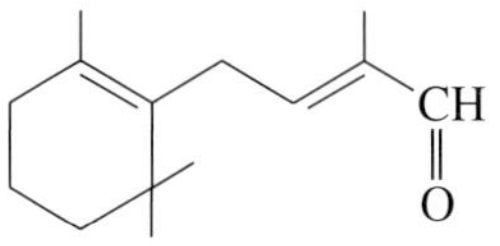

[structure: trimethylcyclohexenyl–CH₂CH=C(CH₃)–CH=O aldehyde]

After acidification, the product obtained from this step is a diol **D**. Selective hydrogenation of the triple bond of **D** using Ni_2B (P-2) catalyst yields **E** ($C_{20}H_{32}O_2$). Treating **E** with one molar equivalent of acetic anhydride yields a monoacetate (**F**), and dehydration of **F** yields vitamin A acetate. What are the structures of **D–F**?

8. Heating acetone with an excess of phenol in the presence of hydrogen chloride is the basis for an industrial process used in the manufacture of a compound called "bisphenol A." (Bisphenol A is used in the manufacture of epoxy resins and a polymer called Lexan.) Bisphenol A has the molecular formula $C_{15}H_{16}O_2$, and the reactions involved in its formation are similar to those involved in the synthesis of DDT (see Special Topic H). Write out these reactions and give the structure of bisphenol A.

9. Outlined here is a synthesis of the local anesthetic *procaine*. Provide structures for procaine and the intermediates **A–C**:

$$p\text{-Nitrotoluene} \xrightarrow[(2)\ H_3O^+]{(1)\ KMnO_4,\ OH^-,\ \text{heat}} \mathbf{A}\ (C_7H_5NO_4) \xrightarrow{SOCl_2} \mathbf{B}\ (C_7H_4ClNO_3) \xrightarrow{HOCH_2CH_2N(C_2H_5)_2} \mathbf{C}\ (C_{13}H_{18}N_2O_4) \xrightarrow{H_2,\ \text{cat.}} \text{procaine}\ (C_{13}H_{20}N_2O_2)$$

10. The sedative–hypnotic *ethinamate* can be synthesized by the following route. Provide structures for ethinamate and the intermediates **A** and **B**:

$$\text{Cyclohexanone} \xrightarrow{(1)\ HC{\equiv}CNa,\ (2)\ H_3O^+} \mathbf{A}\ (C_8H_{12}O) \xrightarrow{ClCOCl} \mathbf{B}\ (C_9H_{11}ClO_2) \xrightarrow{NH_3} \text{ethinamate}\ (C_9H_{13}NO_2)$$

11. The prototype of the antihistamines, *diphenhydramine* (also called Benadryl), can be synthesized by the following sequence of reations:

(a) Give structures for diphenhydramine and for the intermediates **A** and **B**.

(b) Comment on a possible mechanism for the last step of the synthesis.

$$\text{Benzaldehyde} \xrightarrow{(1)\ C_6H_5MgBr,\ (2)\ H_3O^+} \mathbf{A}\ (C_{13}H_{12}O) \xrightarrow{PBr_3}$$

$$\mathbf{B}\ (C_{13}H_{11}Br) \xrightarrow{(CH_3)_2NCH_2CH_2OH} \text{diphenhydramine}\ (C_{17}H_{21}NO)$$

12. Show how you would modify the synthesis given in the previous problem to synthesize the following drugs:

(a) Br–C_6H_4–$CH(C_6H_5)OCH_2CH_2N(CH_3)_2$ **Bromodiphenhydramine (an antihistamine)**

(b) (o-CH_3)C_6H_4–$CH(C_6H_5)OCH_2CH_2N(CH_3)_2$ **Orphenadrine (an antispasmodic, used in controlling Parkinson's disease)**

13. Outlined here is a synthesis of 2-methyl-3-oxocyclopentanecarboxylic acid. Give the structure of each intermediate:

$$CH_3CH(Br)CO_2C_2H_5 \xrightarrow{CH_2(CO_2C_2H_5)_2,\ EtO^-} \mathbf{A}\ (C_{12}H_{20}O_6) \xrightarrow{CH_2{=}CHCN,\ EtO^-}$$

$$\mathbf{B}\ (C_{15}H_{23}NO_6) \xrightarrow{EtOH,\ HA} \mathbf{C}\ (C_{17}H_{28}O_8) \xrightarrow{EtO^-} \mathbf{D}\ (C_{15}H_{22}O_7) \xrightarrow[(2)\ H_3O^+,\ (3)\ heat]{(1)\ OH^-,\ H_2O,\ heat} \text{2-methyl-3-oxocyclopentanecarboxylic acid (cyclopentanone ring bearing } CH_3 \text{ and } CO_2H)$$

14. Give structures for compounds **A–D.** Compound **D** gives a strong IR absorption band near 1720 cm^{-1}, and it reacts with bromine in carbon tetrachloride by a mechanism that does not involve radicals.

$$CH_3C(=O)CH_3 \xrightarrow{HCl} \mathbf{A}\ (C_6H_{10}O) \xrightarrow[base]{CH_3C(=O)CH_2C(=O)OC_2H_5} [\mathbf{B}\ (C_{12}H_{20}O_4)] \xrightarrow{base}$$

$$\mathbf{C}\ (C_{12}H_{18}O_3) \xrightarrow{HA,\ H_2O,\ heat} \mathbf{D}\ (C_9H_{14}O)$$

15. A synthesis of the broad-spectrum antibotic *chloramphenicol* is shown here. In the last step basic hydrolysis selectively hydrolyzes ester linkages in the presence of an amide group. What are the intermediates **A–E**?

$$\text{Benzaldehyde} + HOCH_2CH_2NO_2 \xrightarrow{EtO^-} \mathbf{A}\ (C_9H_{11}NO_4) \xrightarrow{H_2,\ cat.}$$

$$\mathbf{B}\ (C_9H_{13}NO_2) \xrightarrow{Cl_2CHCOCl} \mathbf{C}\ (C_{11}H_{13}Cl_2NO_3) \xrightarrow{excess\ (CH_3CO)_2O}$$

$$\mathbf{D}\ (C_{15}H_{17}Cl_2NO_5) \xrightarrow{HNO_3,\ H_2SO_4} \mathbf{E}\ (C_{15}H_{16}Cl_2N_2O_7) \xrightarrow{OH^-,\ H_2O}$$

O_2N–C_6H_4–$CH(OH)CH(NHCOCHCl_2)CH_2OH$

Chloramphenicol

16. The tranquilizing drug *meprobamate* (Equanil or Miltown) can be synthesized from 2-methylpentanal as follows. Give structures for meprobamate and for the intermediates **A–C**:

$$CH_3CH_2CH_2\underset{\underset{CH_3}{|}}{C}H\overset{O}{\overset{\|}{C}}H \xrightarrow{HCHO,\ OH^-} [\mathbf{A}\ (C_7H_{14}O_2)] \xrightarrow[OH^-]{HCHO} \mathbf{B}\ (C_7H_{16}O_2) \xrightarrow{ClCOCl}$$

$$\mathbf{C}\ (C_9H_{14}Cl_2O_4) \xrightarrow{NH_3} \text{meprobamate}\ (C_9H_{18}N_2O_4)$$

17. What are compounds **A–C?** Compound C is useful as an insect repellent.

$$\text{Succinic anhydride} \xrightarrow{CH_3CH_2CH_2OH} \mathbf{A}\ (C_7H_{12}O_4) \xrightarrow{SOCl} \mathbf{B}\ (C_7H_{11}ClO_3) \xrightarrow{(CH_3CH_2)_2NH} \mathbf{C}\ (C_{11}H_{21}NO_3)$$

18. Outlined here is the synthesis of a central nervous system stimulant called *fencamfamine*. Provide structural formulas for each intermediate and for fencamfamine itself:

$$\text{1,3-Cyclopentadiene} + (E)\text{-}C_6H_5CH{=}CHNO_2 \longrightarrow \mathbf{A}\ (C_{13}H_{13}NO_2) \xrightarrow{H_2,\ Pt}$$

$$\mathbf{B}\ (C_{13}H_{17}N) \xrightarrow{CH_3CHO} [\mathbf{C}\ (C_{15}H_{19}N)] \xrightarrow{H_2,\ Ni} \text{fencamfamine}\ (C_{15}H_{21}N)$$

19. What are compounds **A** and **B**? Compound **B** has a strong IR absorption band in the 1650–1730-cm^{-1} region and a broad strong band in the 3200–3550-cm^{-1} region.

$$\text{1-Methylcyclohexene} \xrightarrow[(2)\ NaHSO_3]{(1)\ OsO_4} \mathbf{A}\ (C_7H_{14}O_2) \xrightarrow[CH_3CO_2H]{CrO_3} \mathbf{B}\ (C_7H_{12}O_2)$$

20. Starting with phenol, outline a stereoselective synthesis of methyl *trans*-4-isopropylcyclohexanecarboxylate, that is

[Structure: cyclohexane chair with $(CH_3)_2CH$ and $COCH_3$ (C=O) substituents, each carbon bearing H]

21. Compound **Y** ($C_6H_{14}O$) shows prominent IR absorption bands at 3334 (broad), 2963, 1463, 1381, and 1053 cm^{-1}. The broadband proton-decoupled ^{13}C NMR spectrum of **Y** is given in Fig. 1. Propose a structure for **Y**.

22. Compound **Z** (C_8H_{16}) is the more stable of a pair of stereoisomers, and it reacts with bromine in carbon tetrachloride by an ionic mechanism. Ozonolysis of **Z** gives a single product. The broadband proton-decoupled ^{13}C NMR spectrum of **Z** is given in Fig. 1. Propose a structure for **Z**.

23. Consider this reaction involving peracetic acid:

[Structure: 2-allylphenol (OH) + peracetic acid ($CH_3C(=O)OOH$) ⟶ **B**]

These are spectral data for the product, **B:**

MS (*m/z*): 150 ($M^{+\cdot}$), 132

IR (cm^{-1}): 3400 (broad), 750 (no absorption in the range of 690–710)

^{1}H NMR (δ): 6.7–7.0 (m, 4H), 4.2 (m, 1H), 3.9 (d, 2H), 2.9 (d, 2H), 1.8 (1H; disappears after treatment with D_2O)

^{13}C NMR (δ): 159 (C), 129 (CH), 126 (CH), 124 (C), 120 (CH), 114 (CH), 78 (CH), 70 (CH_2), and 35 (CH_2)

Problem 23, continued:

(a) What is the structure of **B**?

(b) Propose a mechanism for formation of **B**.

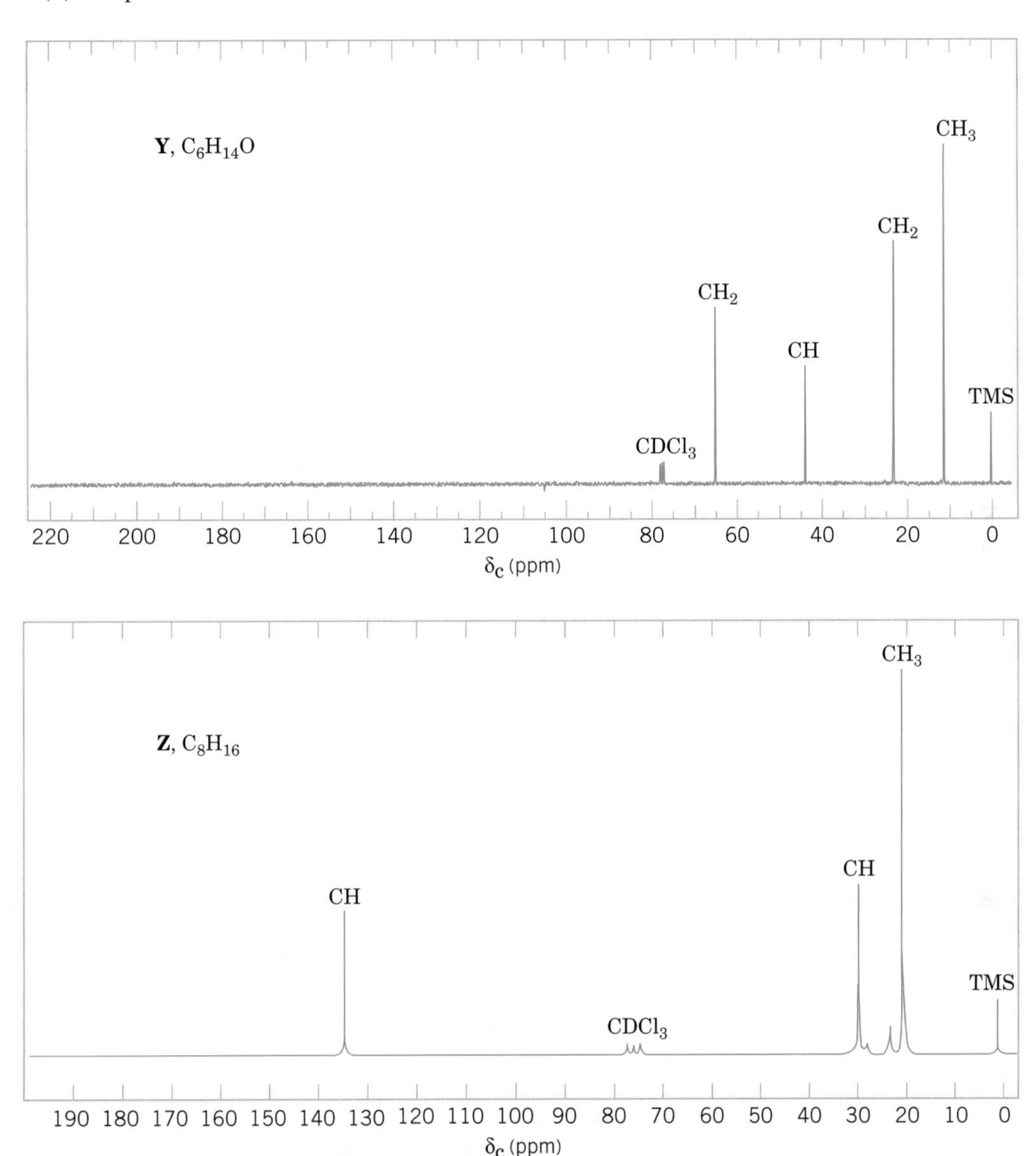

FIGURE 1 The broadband proton-decoupled ^{13}C NMR spectra of compounds Y (Problem 21) and Z (Problem 22). Information from the DEPT ^{13}C NMR spectra is given above the peaks.

24. Adult African dwarf crocodiles produce an aromatic ketone called dianeackerone, named in honor of Diane Ackerman, poet and champion of endangered species and biodiversity. Dianeackerone was isolated, characterized, and synthesized by J. Meinwald and co-workers at Cornell University, who found that dianeackerone occurs naturally in two stereoisomeric forms, 3*S*, 7*S* and 3*S*, 7*R*. An outline of the synthesis of one stereoisomer of dianeackerone is shown here. **(a)** Write structures for the intermediate compounds and necessary reagents indicated by **A–M** below. **(b)** There is an intermediate formed upon treatment of **D** with NBS and $P(C_6H_5)_3$ and before the second treatment with $P(C_6H_5)_3$. What is the structure of this intermediate? **(c)** Which stereoisomer of dianeackerone is formed by this synthesis (i.e., determine the configuration of the stereogenic centers in the formula for dianeackerone shown below).

Synthesis of 1

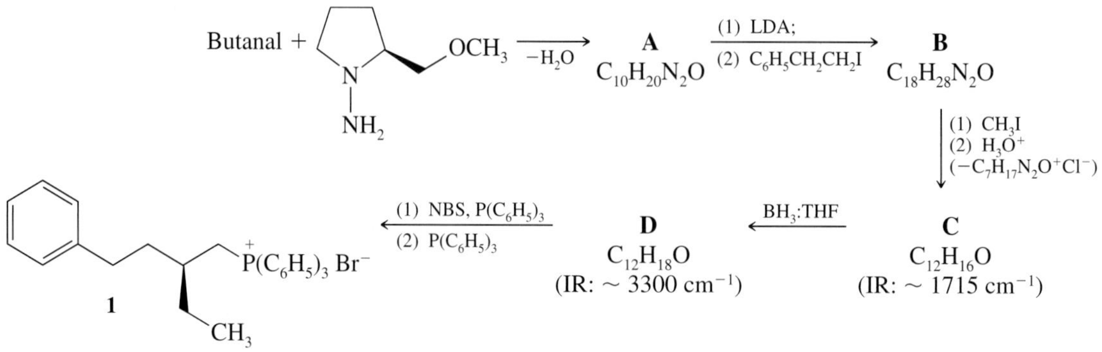

Butanal + [(S)-1-amino-2-(methoxymethyl)pyrrolidine] $\xrightarrow{-H_2O}$ **A** $C_{10}H_{20}N_2O$ $\xrightarrow{(1)\ LDA;\ (2)\ C_6H_5CH_2CH_2I}$ **B** $C_{18}H_{28}N_2O$

B $\xrightarrow{(1)\ CH_3I;\ (2)\ H_3O^+\ (-C_7H_{17}N_2O^+Cl^-)}$ **C** $C_{12}H_{16}O$ (IR: ~ 1715 cm^{-1})

C $\xrightarrow{BH_3:THF}$ **D** $C_{12}H_{18}O$ (IR: ~ 3300 cm^{-1})

D $\xrightarrow{(1)\ NBS,\ P(C_6H_5)_3;\ (2)\ P(C_6H_5)_3}$ **1** ($\overset{+}{P}(C_6H_5)_3\ Br^-$, CH_3)

Synthesis of 2

A $C_{10}H_{20}N_2O$ $\xrightarrow{(1)\ LDA;\ (2)\ \mathbf{E}}$ [hydrazone: N–N, OCH_3, *tert*-BuO_2C, CH_3] $\xrightarrow{O_3}$ **F** $C_{10}H_{18}O_3$ (IR: ~ 1715, 1740 cm^{-1})

F $\xrightarrow{(1)\ \mathbf{G};\ (2)\ \mathbf{H}}$ [*tert*-BuO_2C, O, CH_3, CH_3]

$\xrightarrow{HOCH_2CH_2OH,\ p\text{-TsOH}\ (-H_2O)}$ **I** $C_{13}H_{24}O_4$ (IR: ~ 1740 cm^{-1}) $\xrightarrow{(1)\ \mathbf{J};\ (2)\ \mathbf{K}}$ **2** [O, O, O, CH_3, CH_3]

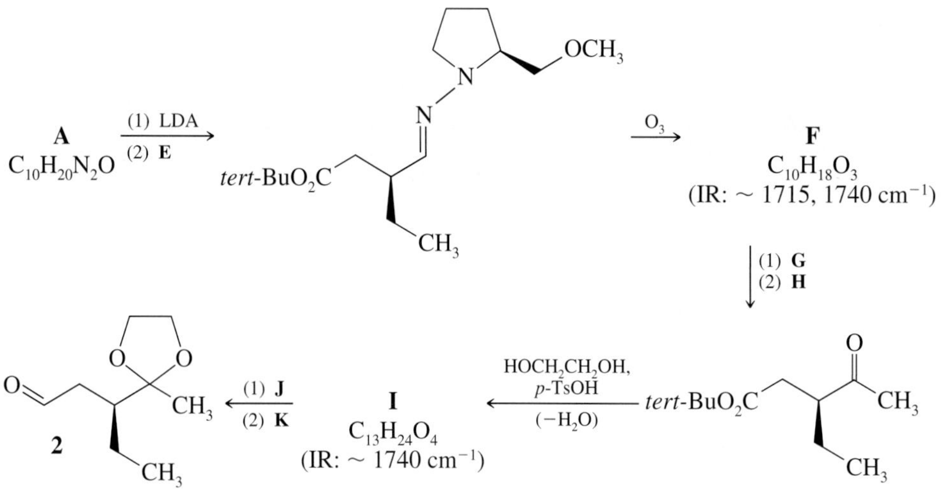

Coupling of 1 and 2 and Completion of Synthesis of Dianeackerone

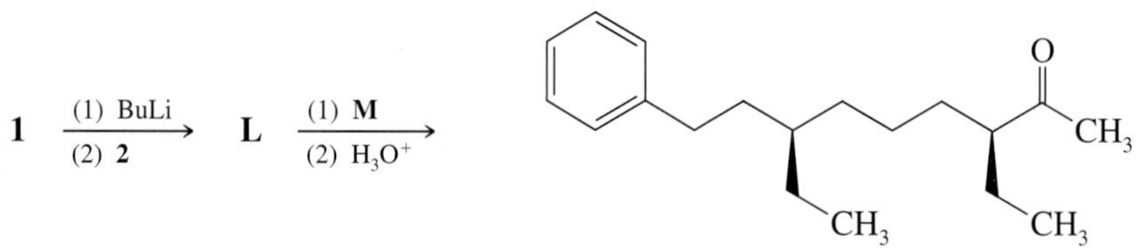

1 $\xrightarrow{(1)\ BuLi;\ (2)\ \mathbf{2}}$ **L** $\xrightarrow{(1)\ \mathbf{M};\ (2)\ H_3O^+}$ [O, CH_3, CH_3, CH_3]

Dianeackerone

SPECIAL TOPIC F

Electrocyclic and Cycloaddition Reactions

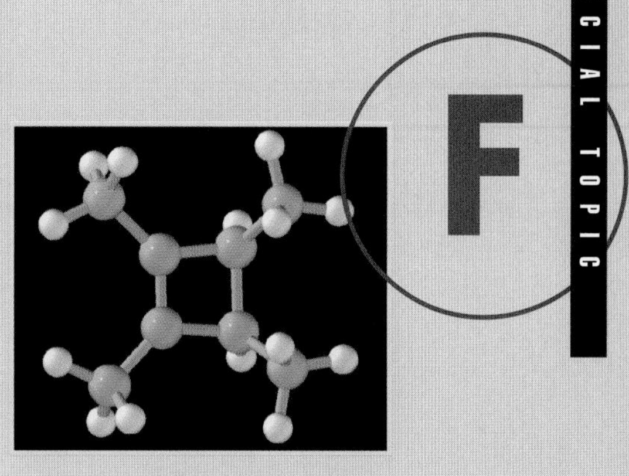

cis-Tetramethylcyclobutene

F.1 Introduction

Chemists have found that there are many reactions in which certain symmetry characteristics of molecular orbitals control the overall course of the reaction. These reactions are often called *pericyclic reactions* because they take place through cyclic transition states. Now that we have a background knowledge of molecular orbital theory—especially as it applies to conjugated polyenes (dienes, trienes, etc., see Chapter 13)—we are in a position to examine some of the intriguing aspects of these reactions. We shall look in detail at two basic types: *electrocyclic reactions* and *cycloaddition reactions.*

F.2 Electrocyclic Reactions

A number of reactions, like the one shown here, transform a conjugated polyene into a cyclic compound.

1,3-Butadiene **Cyclobutene**

In many other reactions, the ring of a cyclic compound opens and a conjugated polyene forms.

Cyclobutene **1,3-Butadiene**

Reactions of either type are called *electrocyclic reactions.*

In electrocyclic reactions, σ and π bonds are interconverted. In our first example, one π bond of 1,3-butadiene becomes a σ bond in cyclobutene. In our second example, the reverse is true: a σ bond of cyclobutene becomes a π bond in 1,3-butadiene.

Electrocyclic reactions have several characteristic features:

1. They require only heat or light for initiation.
2. Their mechanisms do not involve radical or ionic intermediates.
3. Bonds are made and broken in *a single concerted step involving a cyclic transition state.*
4. The reactions are *stereospecific.*

The examples that follow demonstrate this last characteristic of electrocyclic reactions.

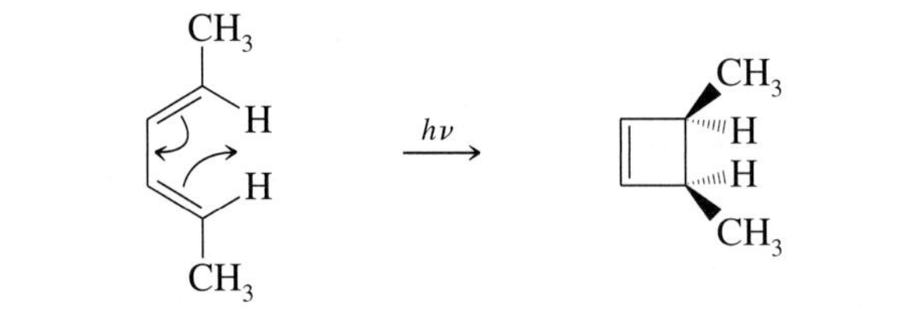

CH_3 H H CH_3 $\xrightarrow{h\nu}$ CH_3 H H CH_3

***trans,trans*-2,4-Hexadiene** **_cis_-3,4-Dimethylcyclobutene**

CH_3 H H CH_3 $\xrightarrow{\text{heat}}$ CH_3 H H CH_3

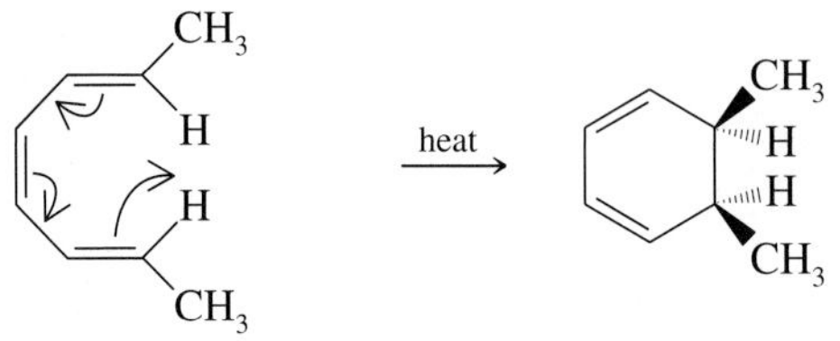

***trans,cis,trans*-2,4,6-Octatriene** **_cis_-5,6-Dimethyl-1,3-cyclohexadiene**

In each of these examples, a single stereoisomeric form of the reactant yields a single stereoisomeric form of the product. The concerted photochemical cyclization of *trans,trans*-2,4-hexadiene, for example, yields only *cis*-3,4-dimethylcyclobutene; it does not yield *trans*-3,4-dimethylcyclobutene.

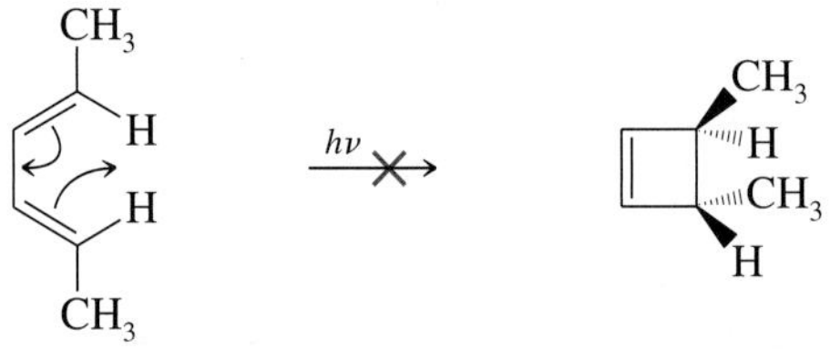

CH_3 H H CH_3 $\xrightarrow{h\nu}$ (crossed) CH_3 H CH_3 H (not formed)

***trans,trans*-2,4-Hexadiene** **_trans_-3,4-Dimethylcyclobutene**

The other two concerted reactions are characterized by the same stereospecificity.

Hoffmann and Fukui were awarded the Nobel Prize in 1981 for this work.

The electrocyclic reactions that we shall study here and the concerted cycloaddition reactions that we shall study in the next section were poorly understood by chemists before 1960. In the years that followed, several scientists, most notably K. Fukui in Japan, H. C. Longuet-Higgins in England, and R. B. Woodward and R. Hoffmann in the United States provided us with a basis for understanding how these reactions occur and why they take place with such remarkable stereospecificity.

All of these scientists worked from molecular orbital theory. In 1965, Woodward and Hoffmann formulated their theoretical insights into a set of rules that not only enabled chemists to understand reactions that were already known but that correctly predicted the outcome of many reactions that had not been attempted.

The Woodward–Hoffmann rules are formulated for concerted reactions only. Concerted reactions are reactions in which bonds are broken and formed simultaneously and, thus, no intermediates occur. The Woodward–Hoffmann rules are based on this hypothesis: *In concerted reactions molecular orbitals of the reactant are continuously converted into molecular orbitals of the product.* This conversion of molecular orbitals is not a random one, however. Molecular orbitals have symmetry characteristics. Because they do, restrictions exist on which molecular orbitals of the reactant may be transformed into particular molecular orbitals of the product.

According to Woodward and Hoffmann, certain reaction paths are said to be *symmetry allowed,* whereas others are said to be *symmetry forbidden.* To say that a particular path is symmetry forbidden does not necessarily mean, however, that the reaction will not occur. It simply means that if the reaction were to occur through a symmetry-forbidden path, the concerted reaction would have a much higher free energy of activation. The reaction may occur, but it will probably do so in a different way: through another path that is symmetry allowed or through a nonconcerted path.

A complete analysis of electrocyclic reactions using the Woodward–Hoffmann rules requires a correlation of symmetry characteristics of *all* of the molecular orbitals of the reactants and product. Such analyses are beyond the scope of our discussion here. We shall find, however, that a simplified approach can be undertaken, one that is easy to visualize and, at the same time, is accurate in most instances. In this simplified approach to electrocyclic reactions we focus our attention only on the *highest occupied molecular orbital (HOMO) of the conjugated polyene.* This approach is based on a method developed by Fukui called the *frontier orbital method.*

F.2A Electrocyclic Reactions of 4n π-Electron Systems

Let us begin with an analysis of the thermal interconversion of *cis*-3,4-dimethylcyclobutene and *cis,trans*-2,4-hexadiene shown here.

CH_3 H H CH_3 heat CH_3 H CH_3 H

***cis*-3,4-Dimethylcyclobutene** ***cis,trans*-2,4-Hexadiene**

Electrocyclic reactions are reversible, and so the path for the forward reaction is the same as that for the reverse reaction. In this example it is easier to see what happens to the orbitals if we follow the *cyclization* reaction, *cis,trans*-2,4-hexadiene $\longrightarrow$ *cis*-3,4-dimethylcyclobutene.

In this cyclization one π bond of the hexadiene is transformed into a σ bond of the cyclobutene. But which π bond? And how does the conversion occur?

Let us begin by examining the π molecular orbitals of 2,4-hexadiene, and, in particular, let us look at *the HOMO of the ground state* [Fig. F.1(a)].

The cyclization that we are concerned with now, *cis,trans*-2,4-hexadiene $\rightleftharpoons$ *cis*-3,4-dimethylcyclobutene, requires heat alone. We conclude, therefore, that excited states of the hexadiene are not involved, for these would require the absorption of light. If we focus our attention on ψ_2—the HOMO of the ground state—we can see how the *p* orbitals at C2 and C5 can be transformed into a σ bond in the cyclobutene.

A bonding σ molecular orbital between C2 and C5 is formed when the *p* orbitals *rotate in the same direction* (both clockwise, as shown, or both counterclockwise, which leads to an equivalent result). The term *conrotatory* is used to describe this type of motion of the two *p* orbitals relative to each other.

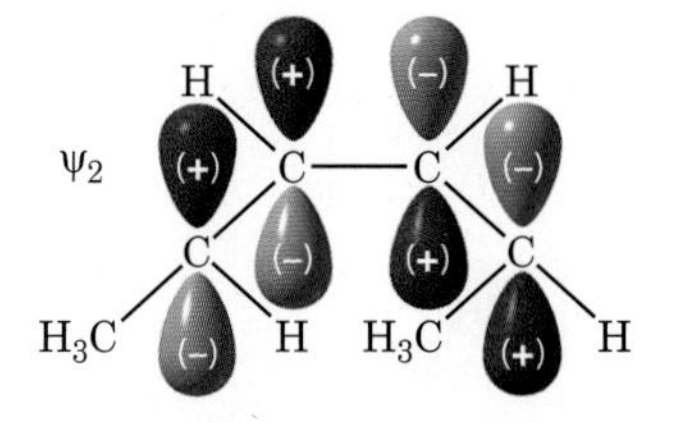

Highest occupied molecular orbital (HOMO) of the ground state

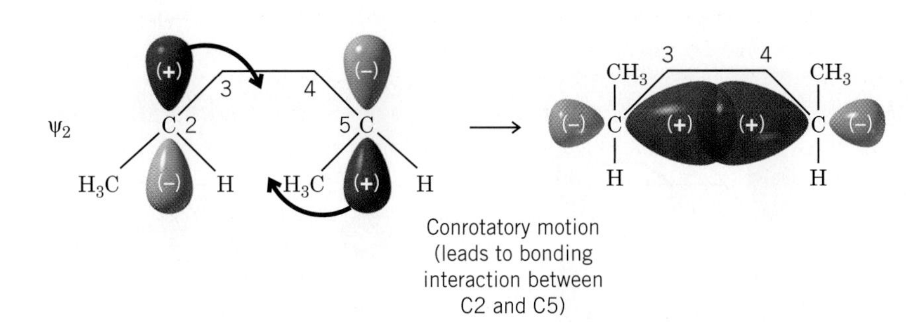

Conrotatory motion (leads to bonding interaction between C2 and C5)

FIGURE F.1 The π molecular orbitals of a 2,4-hexadiene. *(a)* The electron distribution of the ground state. *(b)* The electron distribution of the first excited state. (The first excited state is formed when the molecule absorbs a photon of light of the proper wavelength.) Notice that the orbitals of a 2,4-hexadiene are like those of 1,3-butadiene shown in Fig. 13.5.

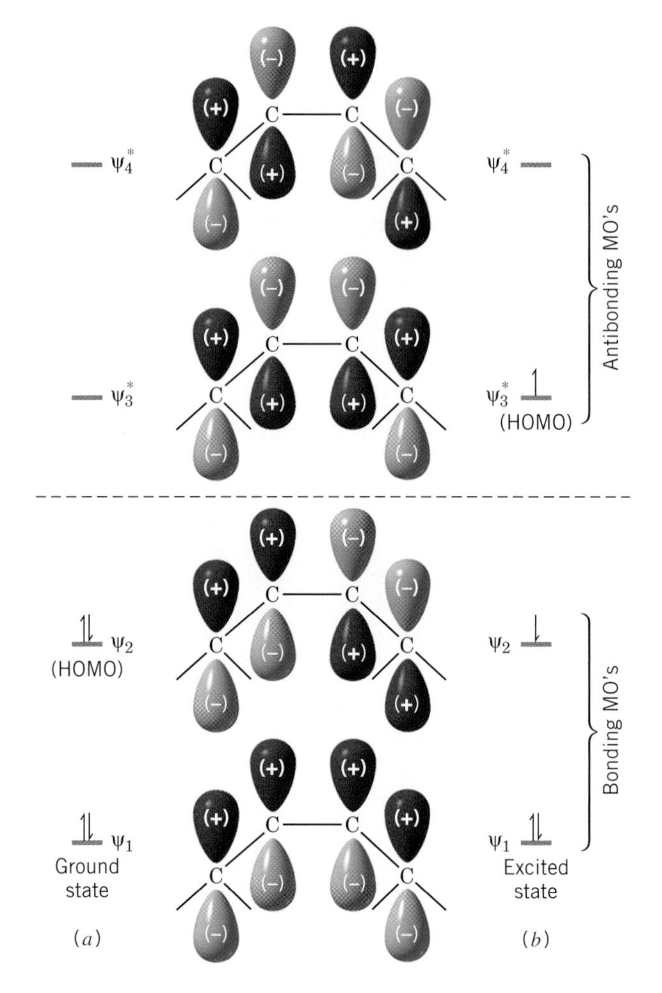

Conrotatory motion allows *p*-orbital lobes of the *same phase sign* to overlap. It also places the two methyl groups on the same side of the molecule in the product, that is, in the cis configuration.*

The pathway with conrotatory motion of the methyl groups is consistent with what we know from experiments to be true: The *thermal reaction* results in the interconversion of *cis*-3,4-dimethylcyclobutene and *cis,trans*-2,4-hexadiene.

Use hand-held molecular models to explore the stereochemistry that results from conrotatory or disrotatory motion in these and other examples.

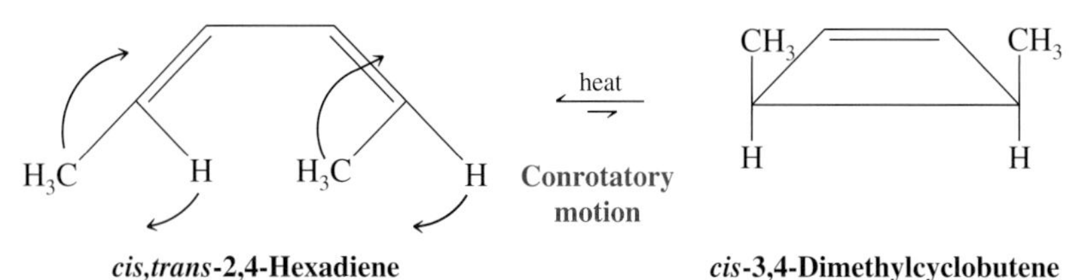

*Notice that if conrotatory motion occurs in the opposite (counterclockwise) direction, lobes of the same phase sign still overlap, and the methyl groups are still cis.

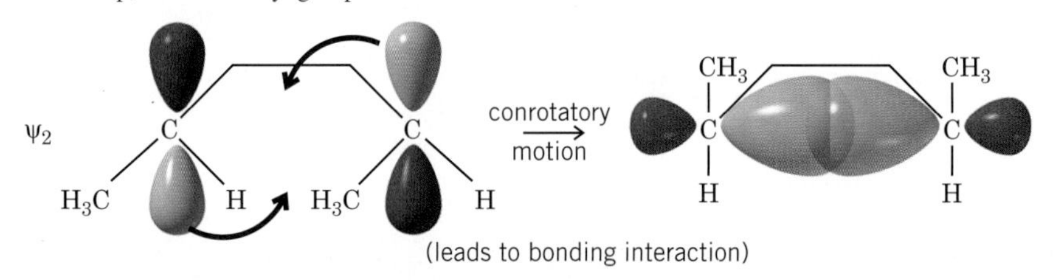

We can now examine another 2,4-hexadiene ⇌ 3,4-dimethylcyclobutene interconversion: one that takes place under the influence of light. This reaction is shown here.

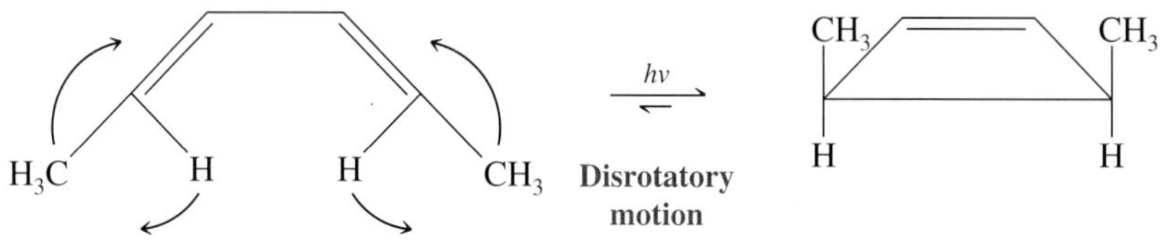

In the photochemical reaction, *cis*-3,4-dimethylcyclobutene and *trans,trans*-2,4-hexadiene are interconverted. The photochemical interconversion occurs with the methyl groups rotating in *opposite directions,* that is, with the methyl groups undergoing *disrotatory motion.*

The photochemical reaction can also be understood by considering orbitals of the 2,4-hexadiene. In this reaction, however—since the absorption of light is involved—we want to look at the first *excited state* of the hexadiene [Figure F.1(b)]. We want to examine ψ_3^*, because in the first excited state ψ_3^* *is the highest occupied molecular orbital.*

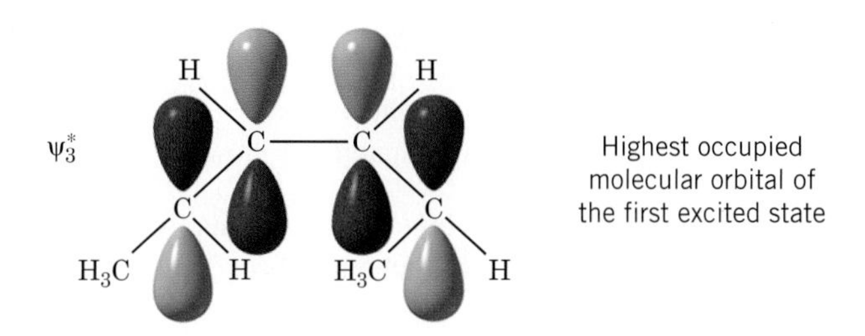

We find that disrotatory motion of the orbitals at C2 and C5 of ψ_3^* allows lobes of the same sign to overlap and form a bonding sigma molecular orbital between them. Disrotatory motion of the orbitals, of course, also requires disrotatory motion of the methyl groups, and, once again, this is consistent with what we find experimentally. The *photochemical reaction* results in the interconversion of *cis*-3,4-dimethylcyclobutene and *trans,trans*-2,4-hexadiene.

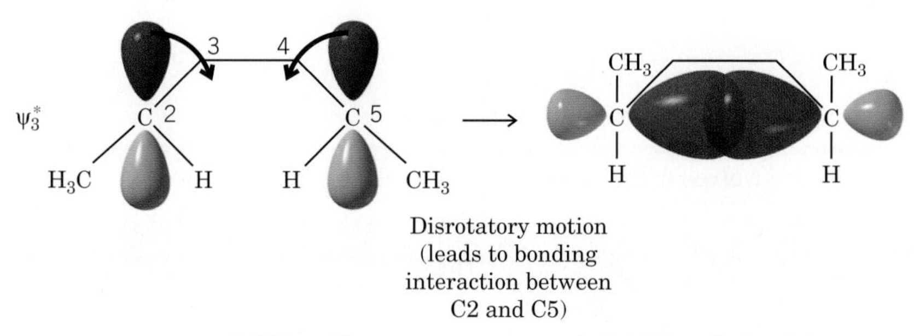

Since both of the interconversions that we have presented so far involve *cis*-3,4-dimethylcyclobutene, we can summarize them in the following way:

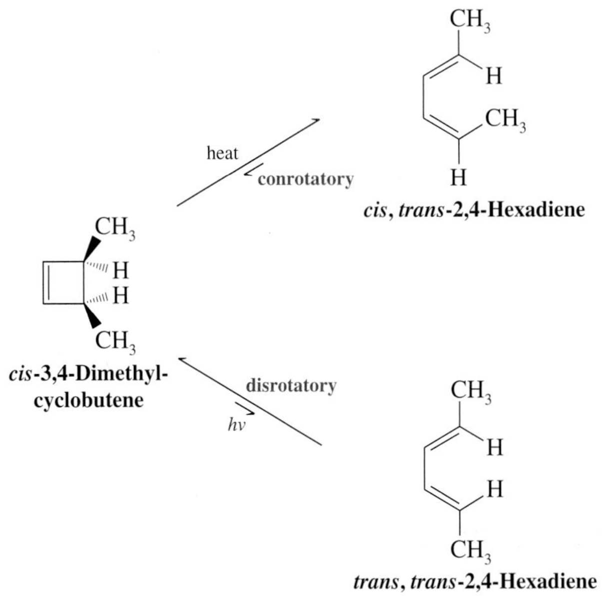

heat

conrotatory

CH_3

H

CH_3

H

cis, trans-2,4-Hexadiene

CH_3

H

H

CH_3

cis-3,4-Dimethyl-
cyclobutene

disrotatory

$h\nu$

CH_3

H

H

CH_3

trans, trans-2,4-Hexadiene

We see that these two interconversions occur with precisely opposite stereochemistry. We also see that the stereochemistry of the interconversions depends on whether the reaction is brought about by the application of heat or light.

The first Woodward–Hoffmann rule can be stated as follows:

1. **A thermal electrocyclic reaction involving 4n π electrons (where n = 1, 2, 3, . . .) proceeds with conrotatory motion; the photochemical reaction proceeds with disrotatory motion.**

Both of the interconversions that we have studied involve systems of 4 π electrons and both follow this rule. Many other 4n π-electron systems have been studied since Woodward and Hoffmann stated their rule. Virtually all have been found to follow it.

PROBLEM F.1

What product would you expect from a concerted photochemical cyclization of *cis,trans*-2,4-hexadiene?

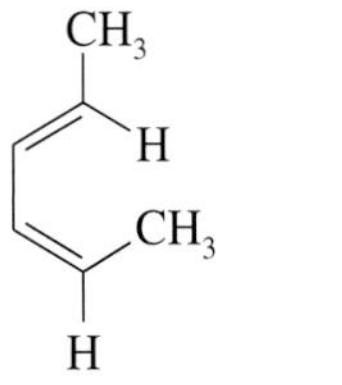

CH_3

H

CH_3

H

cis, trans-2,4-Hexadiene

PROBLEM F.2

(a) Show the orbitals involved in the following thermal electrocyclic reaction.

H_3C CH_3 H H H_3C CH_3 $\xrightarrow{200°C}$ H H_3C CH_3 H H_3C CH_3

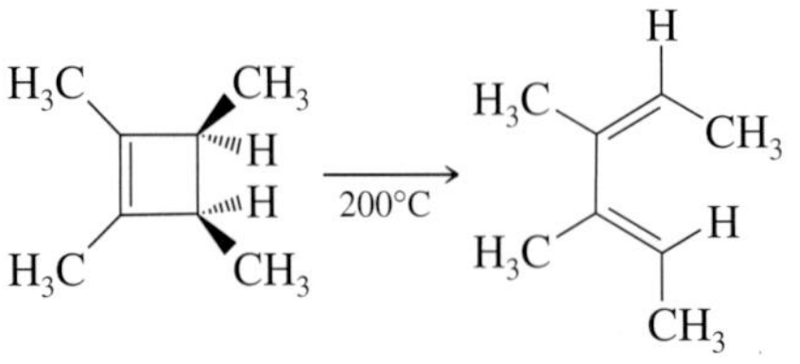

(b) Do the groups rotate in a conrotatory or disrotatory manner?

PROBLEM F.3

Can you suggest a method for carrying out a stereospecific conversion of *trans,trans*-2,4-hexadiene into *cis,trans*-2,4-hexadiene?

PROBLEM F.4

The following 2,4,6,8-decatetraenes undergo ring closure to dimethylcyclooctatrienes when heated or irradiated. What product would you expect from each reaction?

(a) 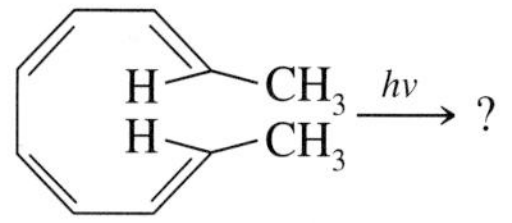 H, CH_3, H, CH_3 $\xrightarrow{h\nu}$? (b) 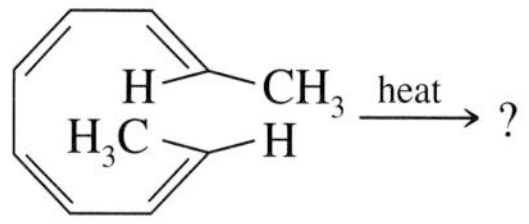 H, CH_3, H_3C, H $\xrightarrow{\text{heat}}$?

PROBLEM F.5

For each of the following reactions, (1) state whether conrotatory or disrotatory motion of the groups is involved and (2) state whether you would expect the reaction to occur under the influence of heat or of light.

(a) 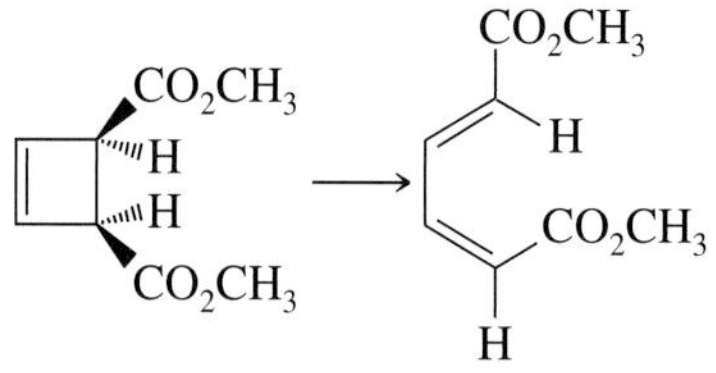 CO_2CH_3, H, H, CO_2CH_3 ⟶ CO_2CH_3, H, CO_2CH_3, H

(b) 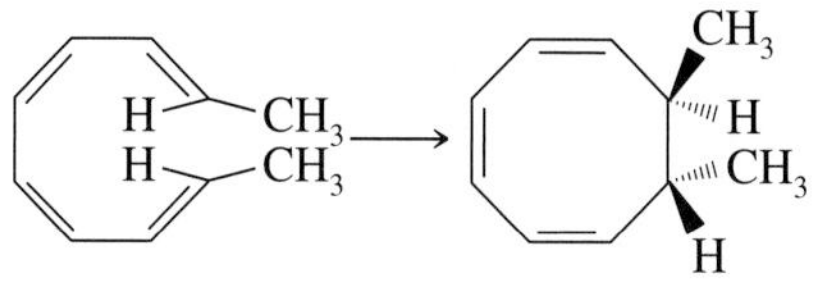 H, CH_3, H, CH_3 ⟶ CH_3, H, CH_3, H

(c) 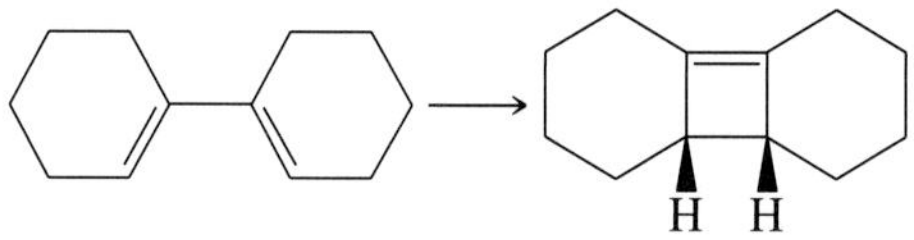 ⟶ H H

F.2B Electrocyclic Reactions of $(4n + 2)$ π-Electron Systems

The second Woodward–Hoffmann rule for electrocyclic reactions is stated as follows:

2. **A thermal electrocyclic reaction involving $(4n + 2)$ π electrons (where $n = 0, 1, 2, \ldots$) proceeds with disrotatory motion; the photochemical reaction proceeds with conrotatory motion.**

According to this rule, the direction of rotation of the thermal and photochemical reactions of $(4n + 2)$ π-electron systems is the opposite of that for corresponding $4n$ systems. Thus, we can summarize both systems in the way shown in Table F.1.

TABLE F.1 Woodward–Hoffman Rules for Electrocyclic Reactions

Number of Electrons	Motion	Rule
$4n$	Conrotatory	Thermally allowed, photochemically forbidden
$4n$	Disrotatory	Photochemically allowed, thermally forbidden
$4n + 2$	Disrotatory	Thermally allowed, photochemically forbidden
$4n + 2$	Conrotatory	Photochemically allowed, thermally forbidden

The interconversions of *trans*-5,6-dimethyl-1,3-cyclohexadiene and the two different 2,4,6-octatrienes that follow illustrate thermal and photochemical interconversions of 6 π-electron systems $(4n + 2$, where $n = 1)$.

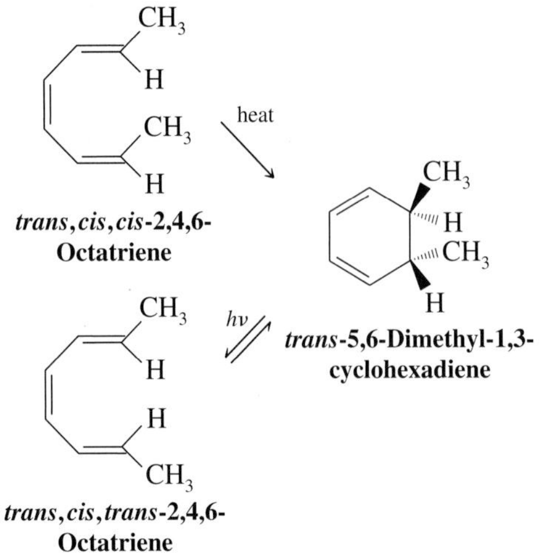

CH_3 H CH_3 H heat CH_3 H CH_3 H

***trans,cis,cis*-2,4,6-Octatriene**

CH_3 H H CH_3 $h\nu$

***trans,cis,trans*-2,4,6-Octatriene**

***trans*-5,6-Dimethyl-1,3-cyclohexadiene**

In the following thermal reaction, the methyl groups rotate in a disrotatory fashion.

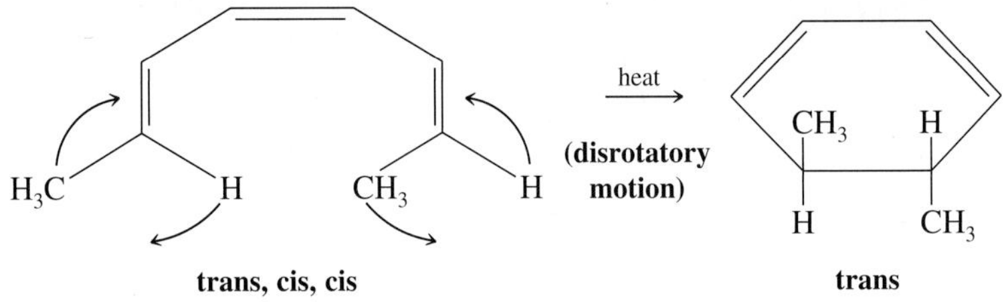

H_3C H CH_3 H heat **(disrotatory motion)** CH_3 H H CH_3

trans, cis, cis **trans**

In the photochemical reaction, the groups rotate in a conrotatory way.

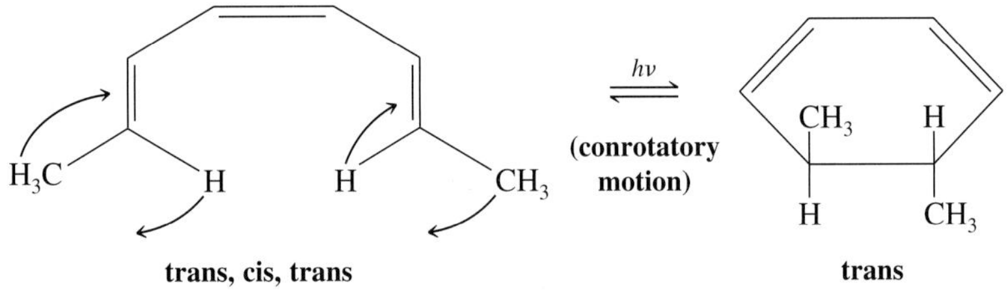

H_3C H H CH_3 $h\nu$ **(conrotatory motion)** CH_3 H H CH_3

trans, cis, trans **trans**

We can understand how these reactions occur if we examine the π molecular orbitals shown in Fig. F.2. Once again, we want to pay attention to the highest occupied molecular orbital. For the thermal reaction of a 2,4,6-octatriene, the highest occupied orbital is ψ_3 because the molecule reacts in its ground state.

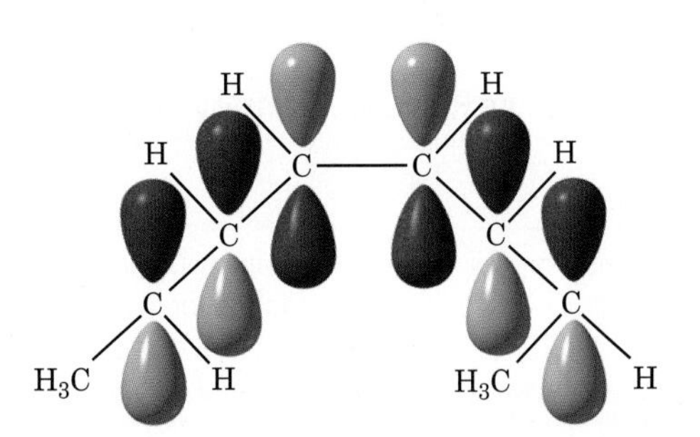

We see in the following figure that disrotatory motion of orbitals at C2 and C7 of ψ_3 allows the formation of a bonding sigma molecular orbital between them. Disrotatory

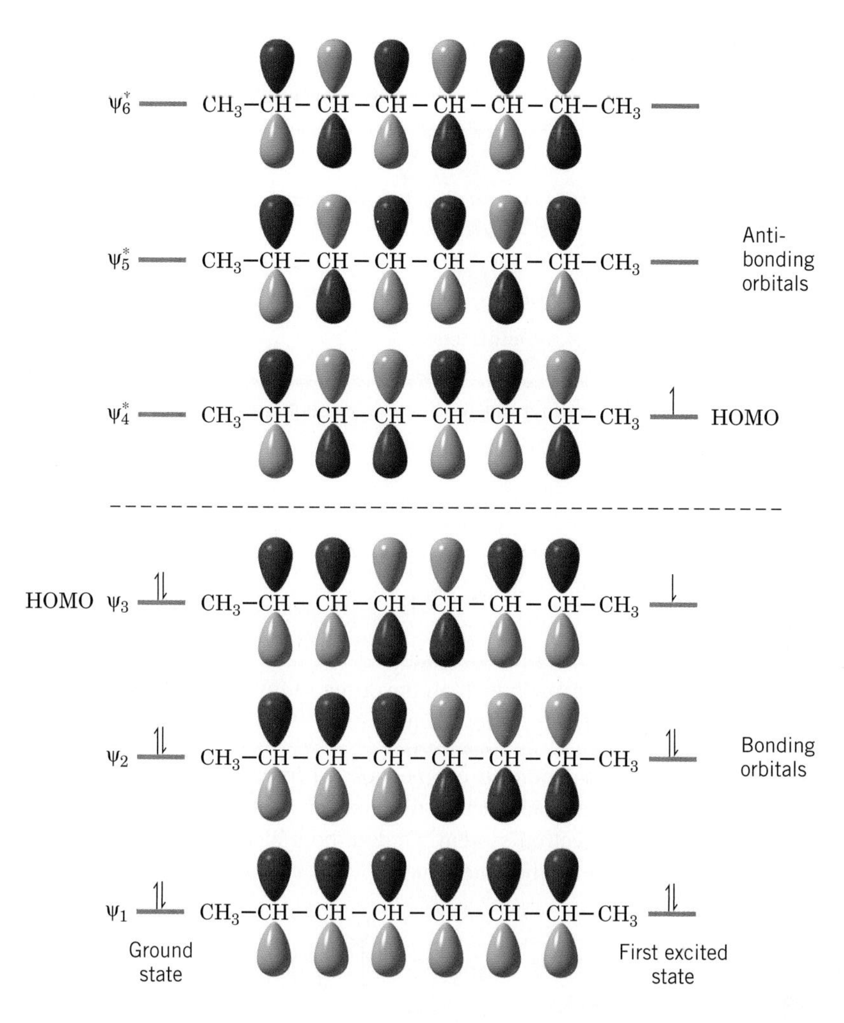

FIGURE F.2 The π molecular orbitals of a 2,4,6-octatriene. The first excited state is formed when the molecule absorbs light of the proper wavelength. (These molecular orbitals are obtained using procedures that are beyond the scope of our discussions.)

motion of the orbitals, of course, also requires disrotatory motion of the groups attached to C2 and C7. And disrotatory motion of the groups is what we observe in the thermal reaction: *trans,cis,cis*-2,4,6-octatriene $\longrightarrow$ *trans*-5,6-dimethyl-1,3-cyclohexadiene.

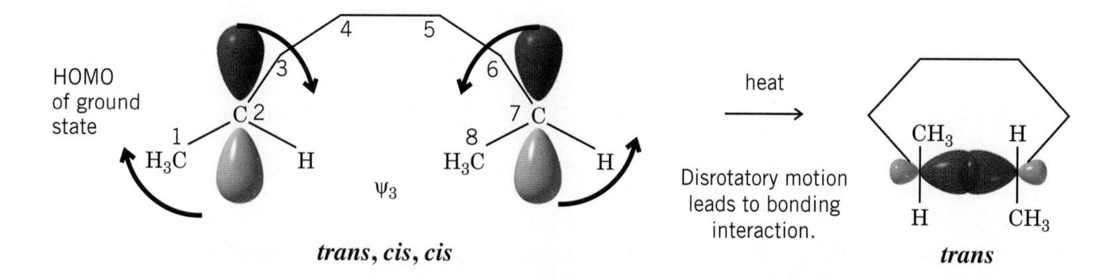

When we consider the photochemical reaction, *trans,cis,trans*-2,4,6-octatriene $\rightleftharpoons$ *trans*-5,6-dimethyl-1,3-cyclohexadiene, we want to focus our attention on ψ_4^*. In the photochemical reaction, light causes the promotion of an electron from ψ_3 to ψ_4^*, and thus ψ_4^* becomes the HOMO. We also want to look at the symmetry of the orbitals at C2 and C7 of ψ_4^*, for these are the orbitals that form a σ bond. In the interconversion shown here, conrotatory motion of the orbitals allows lobes of the same sign to overlap. Thus, we can understand why conrotatory motion of the groups is what we observe in the photochemical reaction.

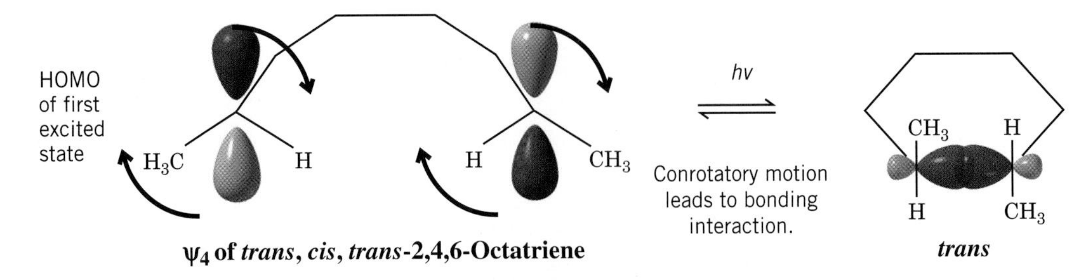

ψ_4 **of *trans, cis, trans*-2,4,6-Octatriene**

PROBLEM F.6 Give the stereochemistry of the product that you would expect from each of the following electrocyclic reactions.

(a) CH_3, H, H, CH_3 — heat ⇌

(b) CH_3, H, H, CH_3 — hv ⇌

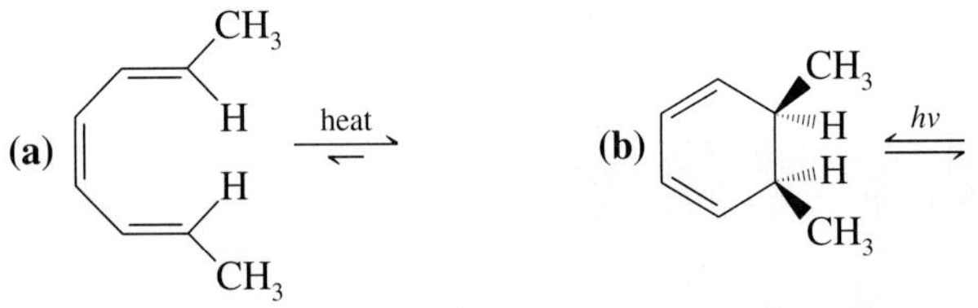

PROBLEM F.7 Can you suggest a stereospecific method for converting *trans*-5,6-dimethyl-1,3-cyclohexadiene into *cis*-5,6-dimethyl-1,3-cyclohexadiene?

PROBLEM F.8 When compound **A** is heated, compound **B** can be isolated from the reaction mixture. A sequence of two electrocyclic reactions occurs; the first involves a 4 π-electron system, and the second involves a 6 π-electron system. Outline both electrocyclic reactions and give the structure of the intermediate that intervenes.

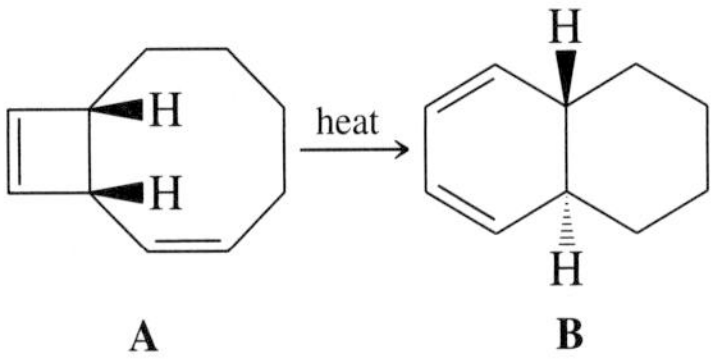

A (H, H) —heat→ **B** (H, H)

F.3 CYCLOADDITION REACTIONS

There are a number of reactions of alkenes and polyenes in which two molecules react to form a cyclic product. These reactions, called *cycloaddition* reactions, are shown next.

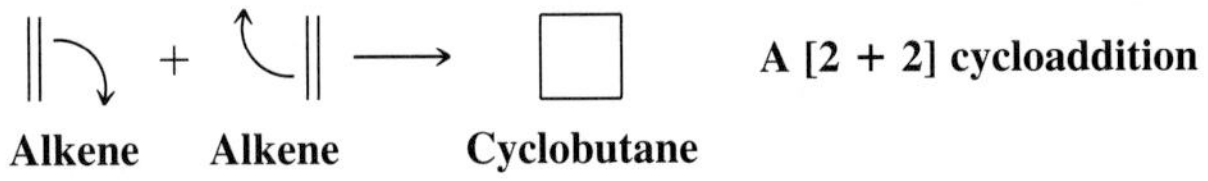

Alkene + **Alkene** → **Cyclobutane** — **A [2 + 2] cycloaddition**

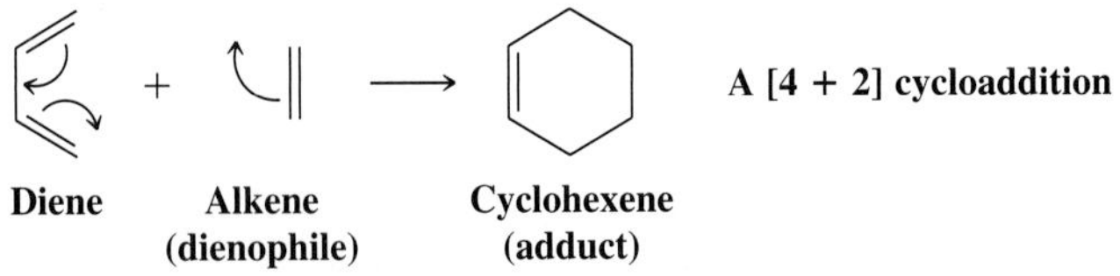

Diene + **Alkene (dienophile)** → **Cyclohexene (adduct)** — **A [4 + 2] cycloaddition**

Chemists classify cycloaddition reactions on the basis of the number of π electrons involved in each component. The reaction of two alkenes to form a cyclobutane is a [2 + 2] cycloaddition; the reaction of a diene and an alkene to form a cyclohexene is called a [4 + 2] cycloaddition. We are already familiar with the [4 + 2] cycloaddition, because it is the Diels–Alder reaction that we studied in Section 13.11.

Cycloaddition reactions resemble electrocyclic reactions in the following important ways:

1. Sigma and pi bonds are interconverted.
2. Cycloaddition reactions require only heat or light for initiation.
3. Radicals and ionic intermediates are not involved in the mechanisms for concerted cycloadditions.
4. Bonds are made and broken in a single concerted step involving a cyclic transition state.
5. Cycloaddition reactions are highly stereospecific.

As we might expect, concerted cycloaddition reactions resemble electrocyclic reactions in still another important way: The symmetry elements of the interacting molecular orbitals allow us to account for their stereochemistry. The symmetry elements of the interacting molecular orbitals also allow us to account for two other observations that have been made about cycloaddition reactions:

1. **Photochemical [2 + 2] cycloaddition reactions occur readily, whereas thermal [2 + 2] cycloadditions take place only under extreme conditions.** When thermal [2 + 2] cycloadditions do take place, they occur through radical (or ionic) mechanisms, not through a concerted process.
2. **Thermal [4 + 2] cycloaddition reactions occur readily and photochemical [4 + 2] cycloadditions are difficult.**

F.3A [2 + 2] Cycloadditions

Let us begin with an analysis of the [2 + 2] cycloaddition of two ethene molecules to form a molecule of cyclobutane.

$$2\ \begin{array}{c}CH_2\\ \|\\ CH_2\end{array} \longrightarrow \begin{array}{c}H_2C-CH_2\\ |\quad\ \ |\\ H_2C-CH_2\end{array}$$

In this reaction we see that two π bonds are converted into two σ bonds. But how does this conversion take place? One way of answering this question is by examining the frontier orbitals of the reactants. The frontier orbitals are the HOMO of one reactant and the LUMO of the other.

We can see how frontier orbital interactions come into play if we examine the possibility of a *concerted thermal* conversion of two ethene molecules into cyclobutane.

Thermal reactions involve molecules reacting in their ground states. The following is the orbital diagram for ethene in its ground state.

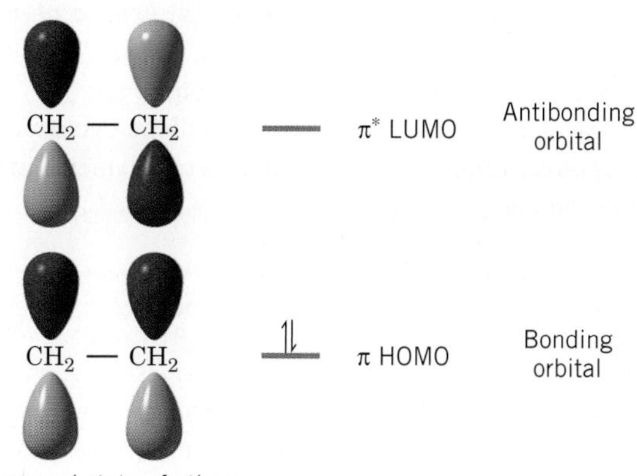

The ground state of ethene

The HOMO of ethene in its ground state is the π orbital. Since this orbital contains two electrons, it interacts with an *unoccupied* molecular orbital of another ethene molecule. The LUMO of the ground state of ethene is, of course, π^*.

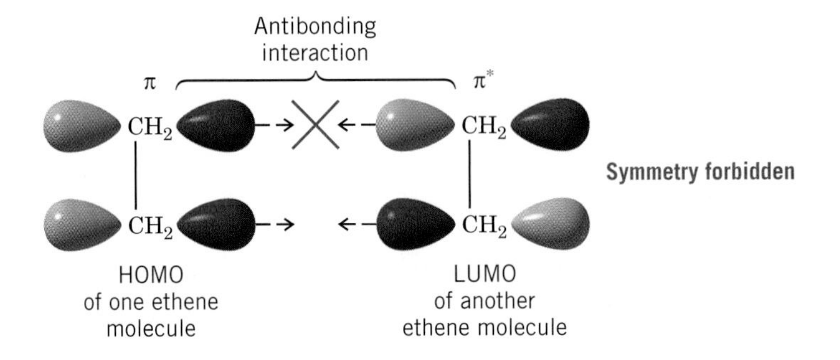

We see from the previous diagram, however, that overlapping the π orbital of one ethene molecule with the π^* orbital of another does not lead to bonding between both sets of carbon atoms because orbitals of opposite signs overlap between the top pair of carbon atoms. This reaction is said to be *symmetry forbidden.* What does this mean? It means that a thermal (or ground state) cycloaddition of ethene would be unlikely to occur in a concerted process. This is exactly what we find experimentally; thermal cycloadditions of ethene, when they occur, take place through nonconcerted, radical mechanisms.

What, then, can we decide about the other possibility—a photochemical [2 + 2] cycloaddition? If an ethene molecule absorbs a photon of light of the proper wavelength, an electron is promoted from π to π^*. In this excited state the HOMO of an ethene molecule is π^*. The following diagram shows how the HOMO of an excited state ethene molecule interacts with the LUMO of a ground state ethene molecule.

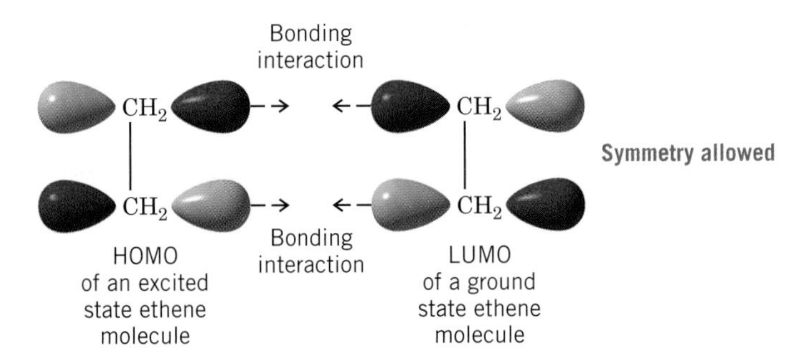

Here we find that bonding interactions occur between both CH_2 groups, that is, lobes of the same sign overlap between both sets of carbon atoms. Complete correlation diagrams also show that the photochemical reaction is *symmetry allowed* and should occur readily through a concerted process. This, moreover, is what we observe experimentally: Ethene reacts readily in a *photochemical* cycloaddition.

The analysis that we have given for the [2 + 2] ethene cycloaddition can be made for any alkene [2 + 2] cycloaddition because the symmetry elements of the π and π^* orbitals of all alkenes are the same.

PROBLEM F.9

What products would you expect from the following concerted cycloaddition reactions? (Give stereochemical formulas.)

(a) 2 *cis*-2-Butene $\xrightarrow{h\nu}$

(b) 2 *trans*-2-Butene $\xrightarrow{h\nu}$

PROBLEM F.10

Show what happens in the following reaction:

$\xrightarrow{h\nu}$

F.3B [4 + 2] Cycloadditions

Concerted [4 + 2] cycloadditions—Diels–Alder reactions—are *thermal reactions.* Considerations of orbital interactions allow us to account for this fact as well. To see how, let us consider the diagram shown in Fig. F.3.

Both modes of orbital overlap shown in Fig. F.3 lead to bonding interactions and both involve *ground states* of the reactants. The ground state of a diene has two electrons in ψ_2 (its HOMO). The overlap shown in Fig. F.3*a* allows these two electrons to flow into the LUMO, π^*, of the dienophile. The overlap shown in Fig. F.3*b* allows two electrons to flow from the HOMO of the dienophile, π, into the LUMO of the diene, ψ_3^*. This thermal reaction is said to be symmetry allowed.

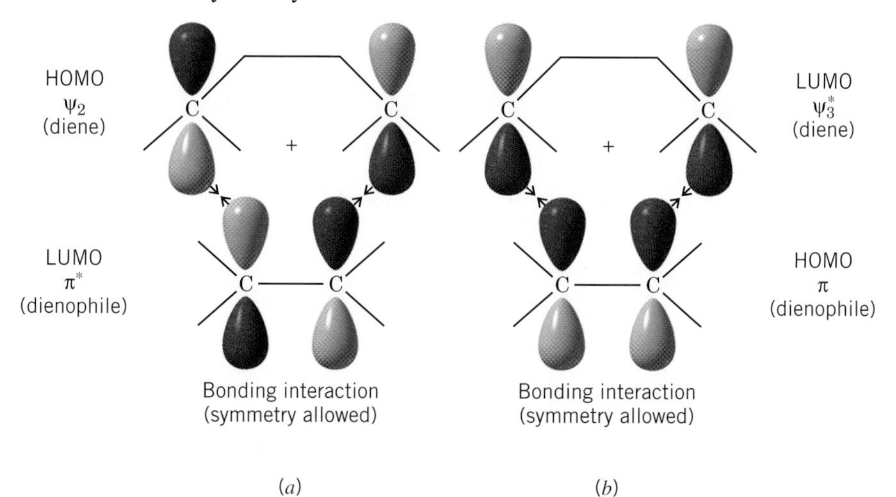

FIGURE F.3 Two symmetry-allowed interactions for a thermal [4 + 2] cycloaddition. *(a)* Bonding interaction between the HOMO of a diene and the LUMO of a dienophile. *(b)* Bonding interaction between the LUMO of the diene and the HOMO of the dienophile.

In Section 13.11 we saw that the Diels–Alder reaction proceeds with retention of configuration of the dienophile. Because the Diels–Alder reaction is usually concerted, it also proceeds with retention of configuration of the diene.

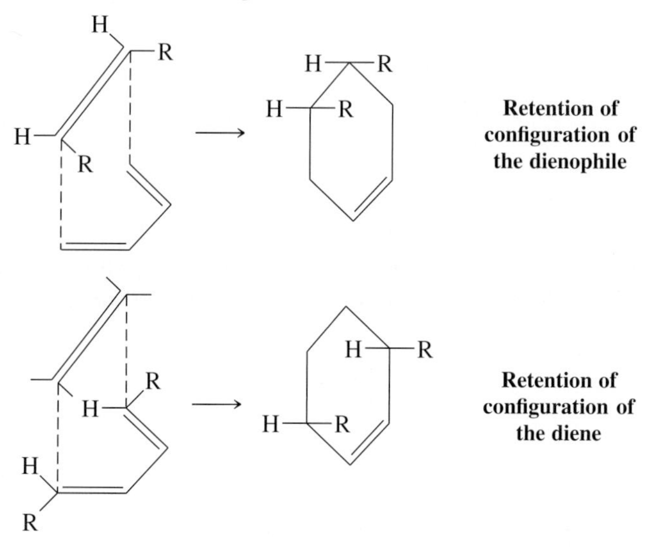

PROBLEM F.11 What products would you expect from the following reactions?

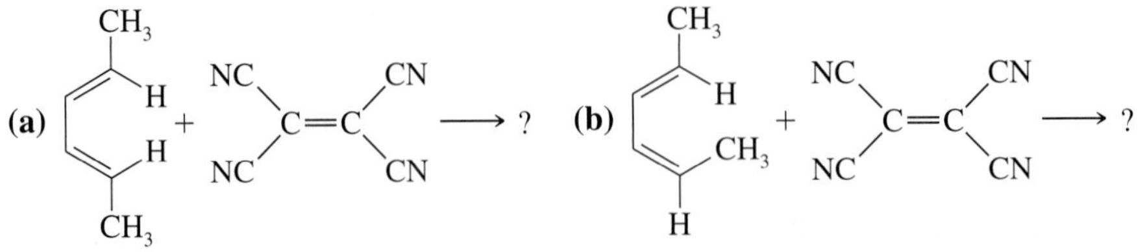

(a) CH_3, H, H, CH_3 (diene) + NC, CN, NC, CN $C{=}C$ ⟶ ?

(b) CH_3, H, CH_3, H (diene) + NC, CN, NC, CN $C{=}C$ ⟶ ?

SPECIAL TOPIC G

Transition Metal Organometallic Compounds

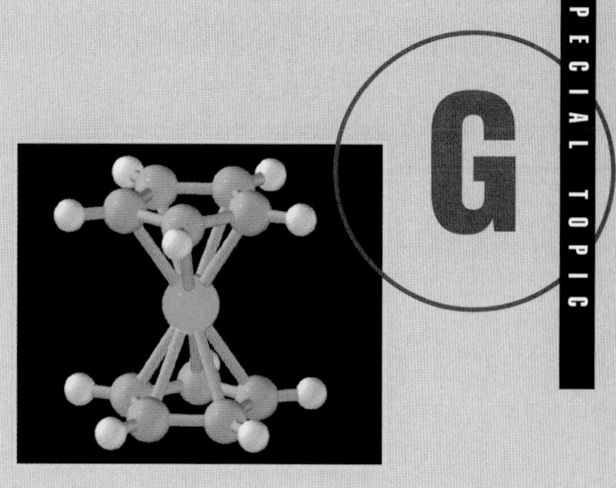

Ferrocene

G.1 Introduction

One of the most active areas of chemical research in recent years has involved studying compounds in which a bond exists between the carbon atom of an organic group and a transition metal. This field, which combines aspects of organic chemistry and inorganic chemistry and is called *organometallic chemistry*, has led to many important applications in organic synthesis. Many of these transition metal organic compounds act as catalysts of extraordinary selectivity.

The transition metals are defined as those elements that have partly filled *d* (or *f*) shells, either in the elemental state or in their important compounds. The transition metals that are of most concern to organic chemists are those shown in the green and yellow portion of the periodic table given in Fig. G.1. Transition metals react with a variety of mole-

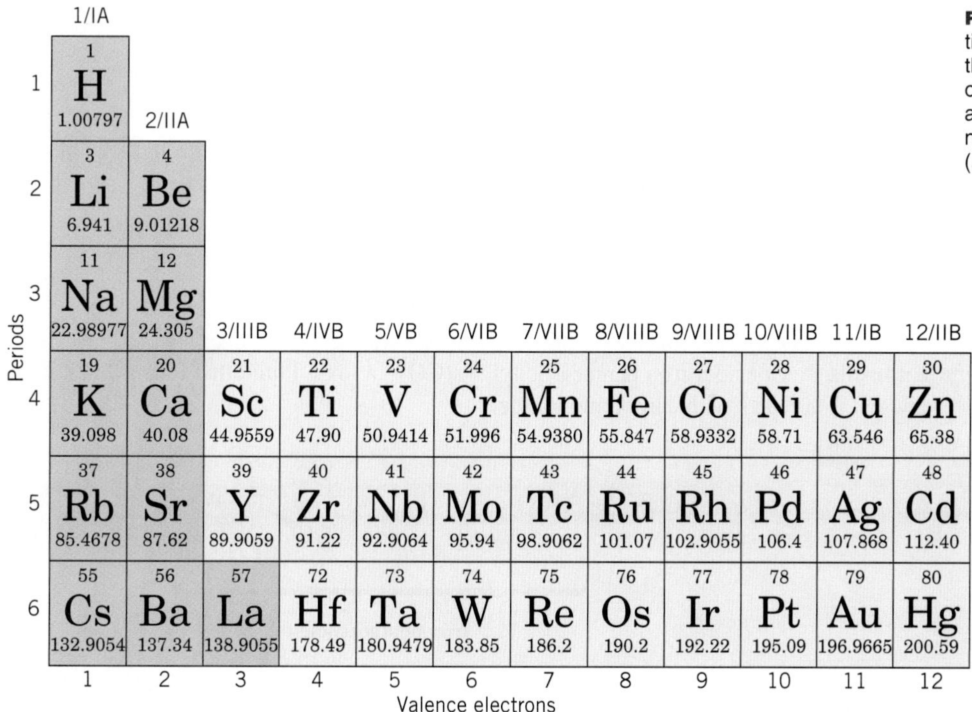

FIGURE G.1 Important transition elements are shown in the green and yellow portion of the periodic table. Given across the bottom is the total number of valence electrons (*s* and *d*) of each element.

cules or groups, called *ligands,* to form *transition metal complexes.* In forming a complex, the ligands donate electrons to vacant orbitals of the metal. The bonds between the ligand and the metal range from very weak to very strong. The bonds are covalent but often have considerable polar character.

Transition metal complexes can assume a variety of geometries depending on the metal and on the number of ligands around it. Rhodium can form complexes with four ligands, for example, that are *square planar.* On the other hand, rhodium can form complexes with five or six ligands that are trigonal bipyramidal or octahedral. These typical shapes are shown below with the letter L used to indicate a ligand.

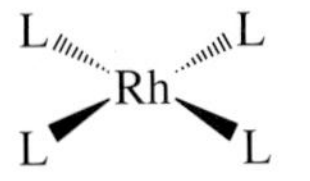

Square planar rhodium complex 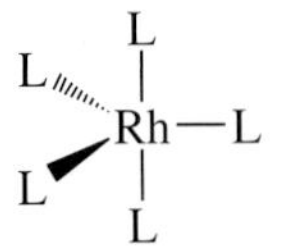 **Trigonal bipyramidal rhodium complex** 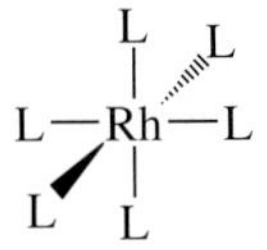 **Octahedral rhodium complex**

G.2 Electron Counting: The 18-Electron Rule

Transition metals are like the elements that we have studied earlier in that they are most stable when they have the electronic configuration of a noble gas. In addition to *s* and *p* orbitals, transition metals have five *d* orbitals (which can hold a total of 10 electrons). Therefore, the noble gas configuration for a transition metal is *18 electrons,* not 8 as with carbon, nitrogen, oxygen, and so on. When the metal of a transition metal complex has 18 valence electrons, it is said to be *coordinatively saturated.**

To determine the valence electron count of a transition metal in a complex, we take the total number of valence electrons of the metal in the elemental state (see Fig. G.1) and subtract from this number the oxidation state of the metal in the complex. This gives us what is called the *d* electron count, d^n. The oxidation state of the metal is the charge that would be left on the metal if all the ligands (Table G.1) were removed.

$$d^n = \begin{array}{c}\text{total number of valence electrons}\\ \text{of the elemental metal}\end{array} - \begin{array}{c}\text{oxidation state of}\\ \text{the metal in the complex}\end{array}$$

Then to get the total valence electron count of the metal *in the complex,* we add to d^n the number of electrons donated by all of the ligands. Table G.1 gives the number of electrons donated by several of the most common ligands.

$$\begin{array}{c}\text{total number of valence electrons}\\ \text{of the metal in the complex}\end{array} = d^n + \begin{array}{c}\text{electrons donated}\\ \text{by ligands}\end{array}$$

Let us now work out the valence electron count of two examples.

Example A Consider iron pentacarbonyl, $Fe(CO)_5$, a toxic liquid that forms when finely divided iron reacts with carbon monoxide.

$$Fe + 5\,CO \longrightarrow Fe(CO)_5 \quad \text{or}$$

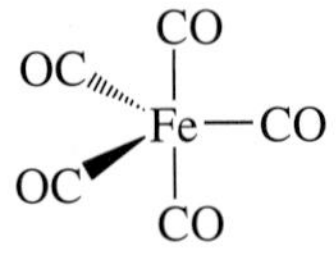

Iron pentacarbonyl

*We do not usually show the unshared electron pairs of a metal complex in our structures, because to do so would make the structure unnecessarily complicated.

TABLE G.1 Common Ligands in Transition Metal Complexes[a]

Ligand	Count as	Number of Electrons Donated
Negatively charged ligands		
H	$H:^-$	2
R	$R:^-$	2
X	$X:^-$	2
Allyl	(allyl anion)	4
Cyclopentadienyl, Cp	(cyclopentadienyl anion)	6
Electrically neutral ligands		
Carbonyl (carbon monoxide)	:C≡O:	2
Phosphine	R_3P or Ph_3P	2
Alkene	>C=C<	2
Diene	(diene)	4
Benzene	(benzene)	6

[a]Adapted from Schwartz, J.; Labinger, J. A. *J. Chem. Educ.* **1980,** *57,* 170–175.

From Fig. G.1 we find that an iron atom in the elemental state has 8 valence electrons. We arrive at the oxidation state of iron in iron pentacarbonyl by noting that the charge on the complex as a whole is zero (it is not an ion), and that the charge on each CO ligand is also zero. Therefore, the iron is in the zero oxidation state.

Using these numbers, we can now calculate d^n and, from it, the total number of valence electrons of the iron in the complex.

$$d^n = 8 - 0 = 8$$

$$\text{total number of valence electrons} = d^n + 5(\text{CO}) = 8 + 5(2) = 18$$

We find that the iron of $Fe(CO)_5$ has 18 valence electrons and is, therefore, coordinatively saturated.

Example B Consider the rhodium complex $Rh[(C_6H_5)_3P]_3H_2Cl$, a complex that, as we shall see later, is an intermediate in certain alkene hydrogenations.

(structure: Rh bearing Cl, H, H and three L ligands) **L = Ph_3P [i.e., $(C_6H_5)_3P$]**

The oxidation state of rhodium in the complex is +3. (The two hydrogen atoms and the chlorine are each counted as −1, and the charge on each of the triphenylphosphine ligands is zero. Removing all the ligands would leave a Rh^{3+} ion.) From Fig. G.1 we find that in the elemental state, rhodium has 9 valence electrons. We can now calculate d^n for the rhodium of the complex.

$$d^n = 9 - 3 = 6$$

Each of the six ligands of the complex donates two electrons to the rhodium in the complex, and, therefore, the total number of valence electrons of the rhodium is 18. The rhodium of $Rh[(C_6H_5)_3P]_3H_2Cl$ is coordinatively saturated.

$$\text{total number of valence electrons of rhodium} = d^n + 6(2) = 6 + 12 = 18$$

G.3 Metallocenes: Organometallic Sandwich Compounds

Cyclopentadiene reacts with phenylmagnesium bromide to give the Grignard reagent of cyclopentadiene. This reaction is not unusual for it is simply another acid–base reaction like those we saw earlier. The methylene hydrogen atoms of cyclopentadiene are much more acidic than the hydrogen atoms of benzene, and, therefore, the reaction goes to completion. (The methylene hydrogen atoms of cyclopentadiene are acidic relative to ordinary methylene hydrogen atoms because the cyclopentadienyl anion is aromatic; see Section 14.7C.)

$$\text{Cyclopentadiene} + C_6H_5MgBr \xrightarrow[Et_2O]{} C_5H_5^{-}\,\overset{2+}{Mg}\overset{-}{Br} + C_6H_6$$

Cyclopentadiene **Phenylmagnesium bromide** **Cyclopentadienylmagnesium bromide** **Benzene**

When the Grignard reagent of cyclopentadiene is treated with ferrous chloride, a reaction takes place that produces a product called *ferrocene.*

$$2\, C_5H_5^{-}\,\overset{2+}{Mg}\overset{-}{Br} + FeCl_2 \longrightarrow (C_5H_5)_2Fe + 2\, MgBrCl$$

Ferrocene (71% overall yield from cyclopentadiene)

Ferrocene is an orange solid with a melting point of 174°C. It is a highly stable compound; ferrocene can be sublimed at 100°C and is not damaged when heated to 400°C.

Many studies, including X-ray analysis, show that ferrocene is a compound in which the iron(II) ion is located between two cyclopentadienyl rings.

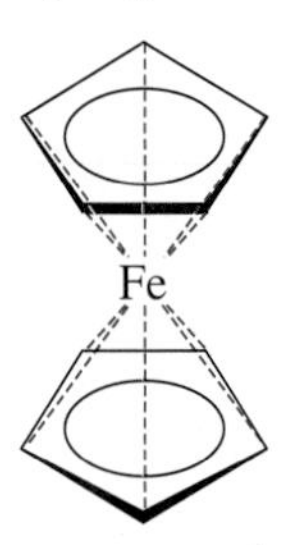

The carbon–carbon bond distances are all 1.40 Å, and the carbon–iron bond distances are all 2.04 Å. Because of their structures, molecules such as ferrocene have been called "sandwich" compounds.

The carbon–iron bonding in ferrocene results from overlap between the inner lobes of the *p* orbitals of the cyclopentadienyl anions and 3*d* orbitals of the iron atom. Studies have shown, moreover, that this bonding is such that the rings of ferrocene are capable of essentially free rotation about an axis that passes through the iron atom and that is perpendicular to the rings.

The iron of ferrocene has 18 valence electrons and is, therefore, coordinatively saturated. We calculate this number as follows:

Iron has 8 valence electrons in the elemental state and its oxidation state in ferrocene is +2. Therefore, $d^n = 6$.

$$d^n = 8 - 2 = 6$$

Each cyclopentadienyl (Cp) ligand of ferrocene donates 6 electrons to the iron. Therefore, for the iron, the valence electron count is 18.

$$\begin{array}{c}\text{total number of}\\ \text{valence electrons}\end{array} = d^n + 2(\text{Cp}) = 6 + 2(6) = 18$$

Ferrocene is an *aromatic compound.* It undergoes a number of electrophilic aromatic substitutions, including sulfonation and Friedel–Crafts acylation.

The discovery of ferrocene (in 1951) was followed by the preparation of a number of similar aromatic compounds. These compounds, as a class, are called *metallocenes.* Metallocenes with five-, six-, seven-, and even eight-membered rings have been synthesized from metals as diverse as zirconium, manganese, cobalt, nickel, chromium, and uranium.

"Half-sandwich" compounds have been prepared through the use of metal carbonyls. Several are shown here.

Ernst O. Fischer (of the Technical University, Munich) and Geoffrey Wilkinson (of Imperial College, London) received the Nobel Prize in 1973 for their pioneering work (performed independently) on the chemistry of organometallic sandwich compounds—or metallocenes.

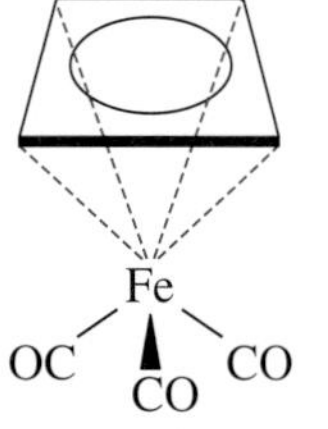

Cyclobutadiene iron tricarbonyl

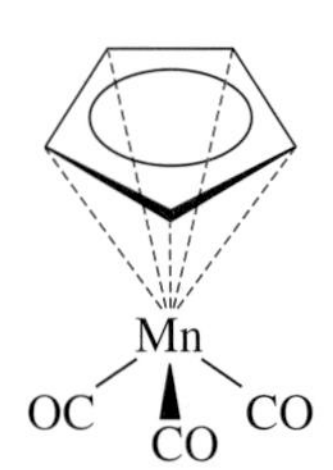

Cyclopentadienylmanganese tricarbonyl

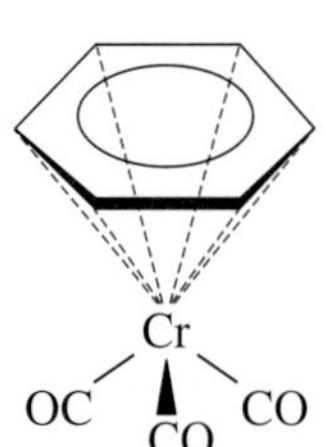

Benzene chromium tricarbonyl

Although cyclobutadiene itself is *not* stable, the cyclobutadiene iron tricarbonyl is.

PROBLEM G.1

The metal of each of the previously given half-sandwich compounds is coordinatively saturated. Show that this is true by working out the valence electron count for the metal in each complex.

G.4 REACTIONS OF TRANSITION METAL COMPLEXES

Much of the chemistry of organic transition metal compounds becomes more understandable if we are able to follow the mechanisms of the reactions that occur. These mechanisms, in most cases, amount to nothing more than a sequence of reactions, each of which represents *a fundamental reaction type that is characteristic of a transition metal complex.* Let us examine three of the fundamental reaction types now. In each instance we shall use steps that occur when an alkene is hydrogenated using a catalyst called Wilkinson's catalyst. Later (in Section G.5) we shall examine the entire hydrogenation mechanism.

1. **Ligand Dissociation–Association (Ligand Exchange).** A transition metal complex can lose a ligand (by dissociation) and combine with another ligand (by association). In the process it undergoes *ligand exchange.* For example, the rhodium complex that we encountered in Example B can react with an alkene (in this example, with ethene) as follows:

$$\mathrm{H_2RhL_3Cl} + \mathrm{CH_2{=}CH_2} \rightleftharpoons \mathrm{H_2RhL_2Cl}(\leftarrow \mathrm{CH_2{=}CH_2}) + \mathrm{L}$$

L = Ph_3P [i.e., $(C_6H_5)_3P$]

Two steps are actually involved. In the first step, one of the triphenylphosphine ligands dissociates. This leads to a complex in which the rhodium has only 16 electrons and is, therefore, coordinatively *unsaturated.*

$$\mathrm{H_2RhL_3Cl} \rightleftharpoons \mathrm{H_2RhL_2Cl} + \mathrm{L}$$

(18 electrons) **(16 electrons)**

L = Ph_3P

In the second step, the rhodium associates with the alkene to become coordinatively saturated again.

$$\mathrm{H_2RhL_2Cl} + \mathrm{CH_2{=}CH_2} \rightleftharpoons \mathrm{H_2RhL_2Cl}(\leftarrow \mathrm{CH_2{=}CH_2})$$

(16 electrons) **(18 electrons)**

The complex between the rhodium and the alkene is called a π *complex.* In it, two electrons are donated by the alkene to the rhodium. Alkenes are often called π donors to distinguish them from σ donors such as Ph_3P:, Cl^-, and so on.

In a π complex such as the one just given, there is also a donation of electrons from a populated *d* orbital of the metal back to the vacant π^* orbital of the alkene. This kind of donation is called "back-bonding."

2. Insertion–Deinsertion. An unsaturated ligand such as an alkene can undergo *insertion* into a bond between the metal of a complex and a hydrogen or a carbon. These reactions are reversible, and the reverse reaction is called *deinsertion.*

The following is an example of insertion–deinsertion.

$$\mathrm{H_2RhL_2Cl}(\leftarrow \mathrm{CH_2{=}CH_2}) \underset{\text{deinsertion}}{\overset{\text{insertion}}{\rightleftharpoons}} \mathrm{H(CH_3CH_2)RhL_2Cl}$$

(18 electrons) **(16 electrons)**

In this process, a π bond (between the rhodium and the alkene) and a σ bond (between the rhodium and the hydrogen) are exchanged for two new σ bonds (between rhodium and carbon, and between carbon and hydrogen). The valence electron count of the rhodium decreases from 18 to 16.

This insertion–deinsertion occurs in a stereospecific way, as a *syn addition* of the M—H unit to the alkene.

$$\mathrm{C{=}C} \;(\leftarrow \mathrm{M{-}H}) \rightleftharpoons \mathrm{M{-}C{-}C{-}H}$$

3. Oxidative Addition–Reductive Elimination. Coordinatively unsaturated metal complexes can undergo oxidative addition of a variety of substrates in the following way.*

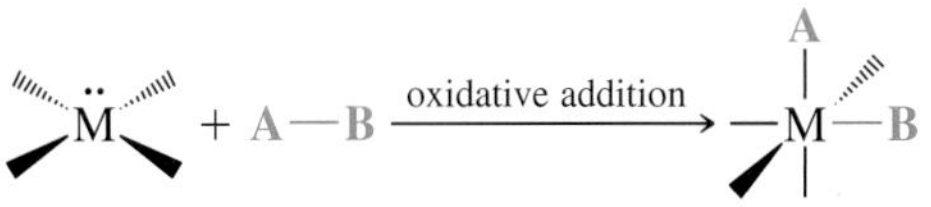

$$\ddot{M}L_n + A{-}B \xrightarrow{\text{oxidative addition}} A{-}M{-}B$$

The substrate, A—B, can be H—H, H—X, R—X, RCO—H, RCO—X, and a number of other compounds.

In this type of oxidative addition, the metal of the complex undergoes an increase in the number of its valence electrons *and in its oxidation state.* Consider, as an example, the oxidative addition of hydrogen to the rhodium complex that follows (L = Ph_3P).

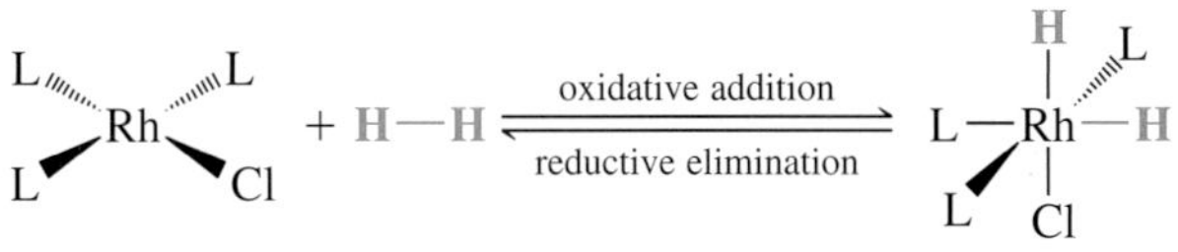

$$L_3RhCl + H{-}H \underset{\text{reductive elimination}}{\overset{\text{oxidative addition}}{\rightleftharpoons}} L_3Rh(H)_2Cl$$

(16 electrons) Rh is in +1 oxidation state. **(18 electrons) Rh is in +3 oxidation state.**

Reductive elimination is the reverse of oxidative addition. With this background, we are now in a position to examine a few interesting applications of transition metal complexes in organic synthesis.

G.5 Homogeneous Hydrogenation

Until now, all of the hydrogenations that we have examined have been heterogeneous processes. Two phases have been involved in the reaction: the solid phase of the catalyst (Pt, Pd, Ni, etc.), containing the adsorbed hydrogen, and the liquid phase of the solution, containing the unsaturated compound. In homogeneous hydrogenation using a transition metal complex such as $Rh[(C_6H_5)_3P]_3Cl$ (called Wilkinson's catalyst), hydrogenation takes place *in a single phase,* i.e., in solution.

When Wilkinson's catalyst is used to carry out the hydrogenation of an alkene, the following steps take place (L = Ph_3P).

Step 1 $L_3RhCl + H{-}H \longrightarrow L_3Rh(H)_2Cl$ **Oxidative addition**

(16 electrons, Rh^{I}) **(18 electrons, Rh^{III})**

Step 2 $L_3Rh(H)_2Cl \longrightarrow L_2Rh(H)_2Cl + L$ **Ligand dissociation**

(18 electrons) **(16 electrons)**

Step 3 $L_2Rh(H)_2Cl + CH_2{=}CH_2 \longrightarrow L_2Rh(H)_2Cl(CH_2{=}CH_2)$ **Ligand association**

(16 electrons) **(18 electrons)**

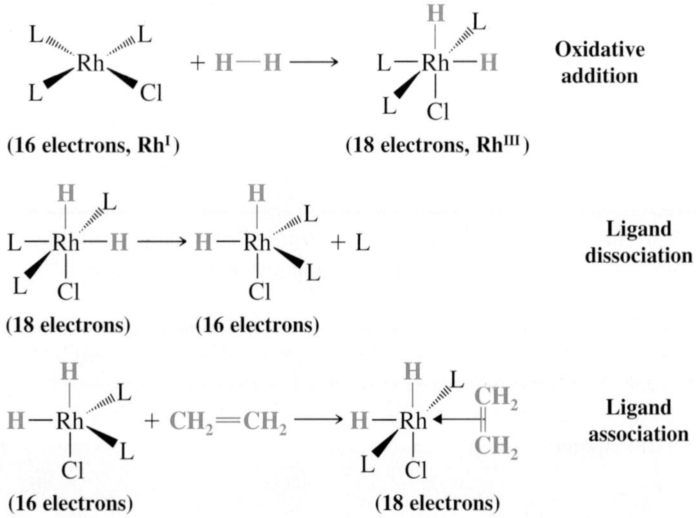

*Coordinatively saturated complexes also undergo oxidative addition.

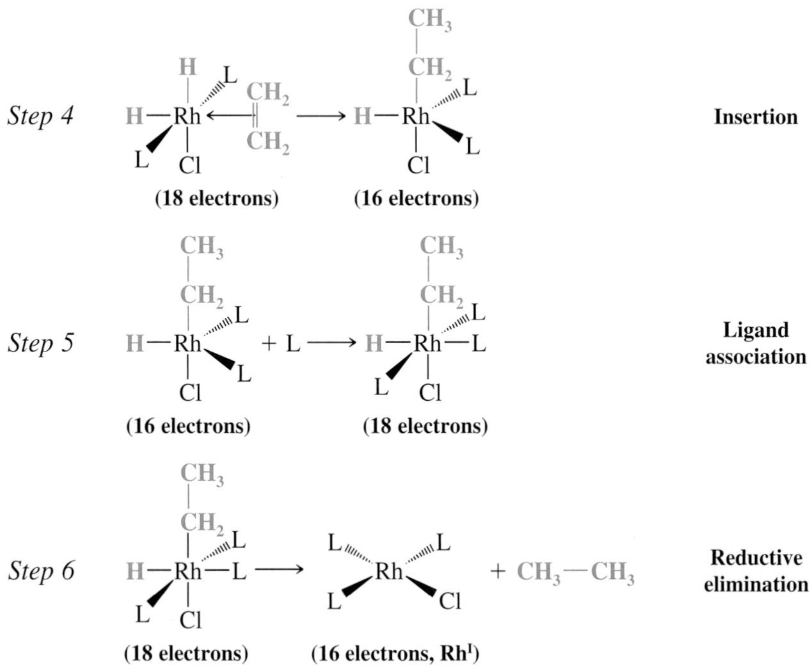

Step 4 — Insertion

$H{-}Rh(H)(L)(L)(Cl) \leftarrow CH_2{=}CH_2 \longrightarrow H{-}Rh(CH_2CH_3)(L)(L)(Cl)$

(18 electrons) (16 electrons)

Step 5 — Ligand association

$H{-}Rh(CH_2CH_3)(L)(L)(Cl) + L \longrightarrow H{-}Rh(CH_2CH_3)(L)(L)(L)(Cl)$

(16 electrons) (18 electrons)

Step 6 — Reductive elimination

$H{-}Rh(CH_2CH_3)(L)(L)(L)(Cl) \longrightarrow L_3RhCl + CH_3{-}CH_3$

(18 electrons) (16 electrons, Rh^{I})

Then steps 1, 2, 3, 4, 5, 6, and so on.

Step 6 regenerates the catalyst, which can then cause hydrogenation of another molecule of the alkene.

Because the insertion step 4 and the reductive elimination step 6 are stereospecific, the net result of the hydrogenation using Wilkinson's catalyst is a *syn addition* of hydrogen to the alkene. The following example (with D_2 in place of H_2) illustrates this aspect.

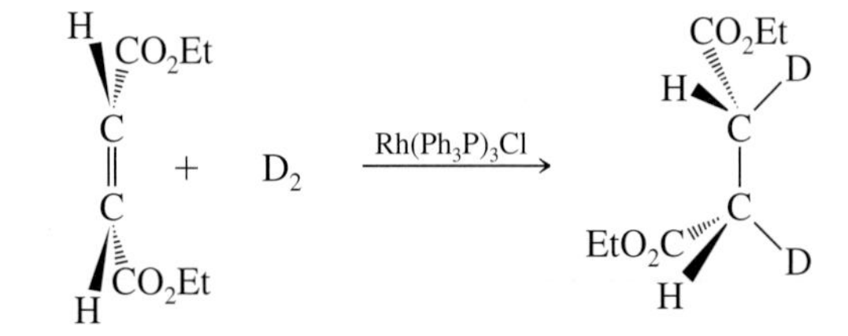

$H(CO_2Et)C{=}C(CO_2Et)H + D_2 \xrightarrow{Rh(Ph_3P)_3Cl} H(D)(CO_2Et)C{-}C(CO_2Et)(D)H$

A *cis*-alkene (diethyl maleate) **A meso compound**

PROBLEM G.2 What product (or products) would be formed if the *trans*-alkene corresponding to the *cis*-alkene (see the previous reaction) had been hydrogenated with D_2 and Wilkinson's catalyst?

G.6 Carbon–Carbon Bond-Forming Reactions Using Rhodium Complexes

Rhodium complexes have also been used to synthesize compounds in which the formation of a carbon–carbon bond is required. An example is the synthesis that follows:

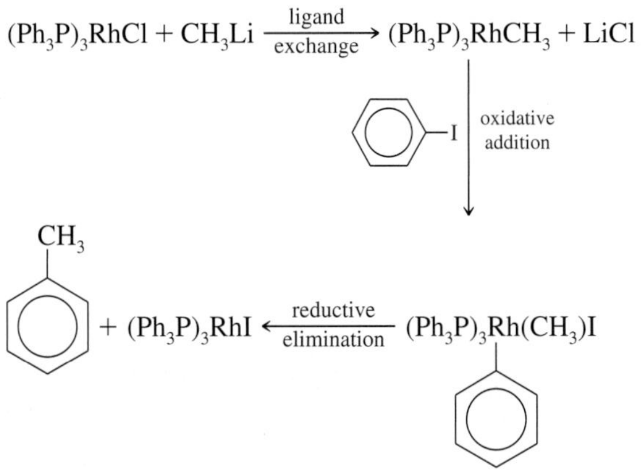

$$(Ph_3P)_3RhCl + CH_3Li \xrightarrow[\text{exchange}]{\text{ligand}} (Ph_3P)_3RhCH_3 + LiCl$$

$$\downarrow \text{oxidative addition } (C_6H_5\text{–}I)$$

$$C_6H_5CH_3 + (Ph_3P)_3RhI \xleftarrow[\text{elimination}]{\text{reductive}} (Ph_3P)_3Rh(CH_3)I(C_6H_5)$$

The first step, *a ligand exchange*, occurs by a combination of ligand association–dissociation steps and incorporates the methyl group into the coordination sphere of the rhodium. The next step, *an oxidative addition,* incorporates the phenyl group into the rhodium coordination sphere. Then, in the last step, *a reductive elimination* joins the methyl group and the benzene ring to form toluene.

PROBLEM G.3

Give the total valence electron count for rhodium in each complex in the synthesis outlined previously.

Another example is the following ketone synthesis.

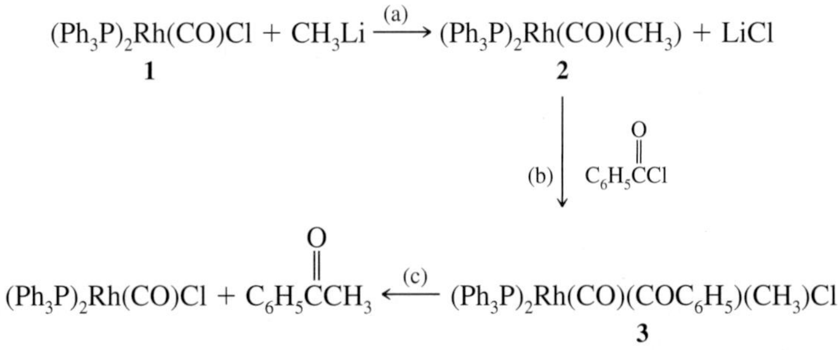

$$\underset{\mathbf{1}}{(Ph_3P)_2Rh(CO)Cl} + CH_3Li \xrightarrow{(a)} \underset{\mathbf{2}}{(Ph_3P)_2Rh(CO)(CH_3)} + LiCl$$

$$\downarrow (b)\ C_6H_5\overset{O}{\overset{\|}{C}}Cl$$

$$(Ph_3P)_2Rh(CO)Cl + C_6H_5\overset{O}{\overset{\|}{C}}CH_3 \xleftarrow{(c)} \underset{\mathbf{3}}{(Ph_3P)_2Rh(CO)(COC_6H_5)(CH_3)Cl}$$

PROBLEM G.4

Give the valence electron count and the oxidation state of rhodium in the complexes labeled **1, 2,** and **3**; then describe each step (a), (b), and (c) as to its fundamental type (oxidative addition, ligand exchange, etc.).

Still another carbon–carbon bond-forming reaction (below) illustrates the stereospecificity of these reactions.

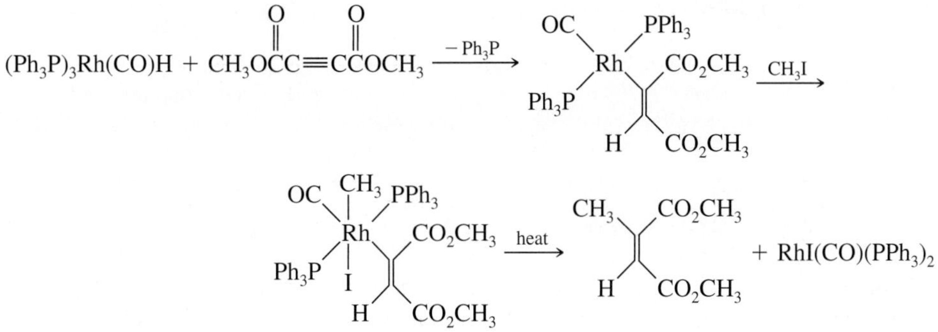

$$(Ph_3P)_3Rh(CO)H + CH_3O\overset{O}{\overset{\|}{C}}C{\equiv}C\overset{O}{\overset{\|}{C}}OCH_3 \xrightarrow{-Ph_3P} (OC)(Ph_3P)_2Rh\text{–}C(CO_2CH_3){=}C(H)(CO_2CH_3) \xrightarrow{CH_3I}$$

$$(OC)(CH_3)(I)(Ph_3P)_2Rh\text{–}C(CO_2CH_3){=}C(H)(CO_2CH_3) \xrightarrow{\text{heat}} (CH_3)(CO_2CH_3)C{=}C(H)(CO_2CH_3) + RhI(CO)(PPh_3)_2$$

PROBLEM G.5 Give, in detail, a possible mechanism for the synthesis just outlined, describing each step according to its fundamental type.

PROBLEM G.6 The actual mechanism of the Corey–Posner, Whitesides–House synthesis (Section 12.9) is not known with certainty. One possible mechanism involves the oxidative addition of R′—X or Ar—X to R_2CuLi followed by a reductive elimination to generate R—R′ or R—Ar. Outline the steps in this mechanism using $(CH_3)_2CuLi$ and C_6H_5I.

G.7 Vitamin B_{12}: A Transition Metal Biomolecule

The discovery (in 1926) that pernicious anemia can be overcome by the ingestion of large amounts of liver led ultimately to the isolation (in 1948) of the curative factor, called vitamin B_{12}. The complete three-dimensional structure of vitamin B_{12} [Fig. G.2(*a*)] was elucidated in 1956 through the X-ray studies of Dorothy Hodgkin (Nobel Prize, 1964), and in 1972 the synthesis of this complicated molecule was announced by R. B. Woodward

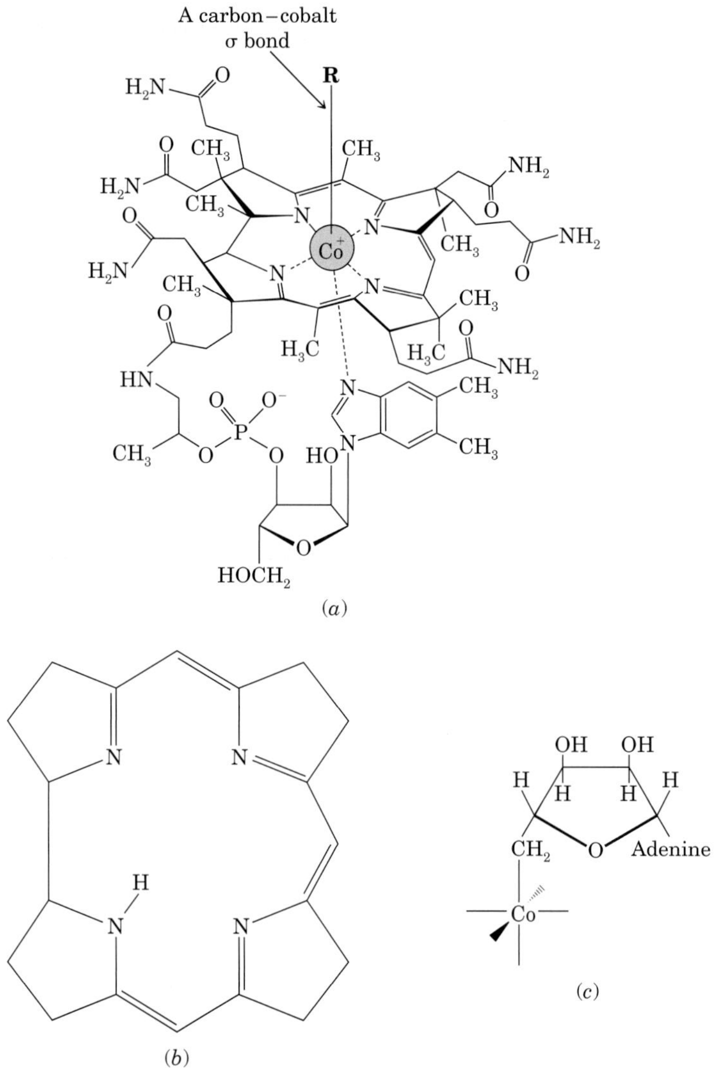

FIGURE G.2 *(a)* The structure of vitamin B_{12}. In the commercial form of the vitamin (cyanocobalamin), R = CN. *(b)* The corrin ring system. *(c)* In the biologically active form of the vitamin (5′-deoxyadenosylcobalamin), the 5′ carbon atom of 5′-deoxyadenosine is coordinated to the cobalt atom. For the structure of adenine, see Section 25.2.

A carbon–cobalt σ bond

(*a*)

(*b*)

(*c*)

(Harvard University) and A. Eschenmoser (Swiss Federal Institute of Technology). The synthesis took 11 years and involved more than 90 separate reactions. One hundred co-workers took part in the project.

Vitamin B_{12} is the only known biomolecule that possesses a carbon–metal bond. In the stable commercial form of the vitamin, a cyano group is bonded to the cobalt, and the cobalt is in the +3 oxidation state. The core of the vitamin B_{12} molecule is a *corrin ring* with various attached side groups. The corrin ring consists of four pyrrole subunits, the nitrogen of each of which is coordinated to the central cobalt. The sixth ligand (below the corrin ring in Fig. G.2(*a*) is a nitrogen of a heterocyclic molecule called 5,6-dimethylbenzimidazole.

The cobalt of vitamin B_{12} can be reduced to a +2 or a +1 oxidation state. When the cobalt is in the +1 oxidation state, vitamin B_{12} (called B_{12s}) becomes one of the most powerful nucleophiles known, being more nucleophilic than methanol by a factor of 10^{14}.

Acting as a nucleophile, vitamin B_{12s} reacts with adenosine triphosphate (Fig. 22.2) to yield the biologically active form of the vitamin [Fig. G.2(*c*)].

Organic Halides and Organometallic Compounds in the Environment

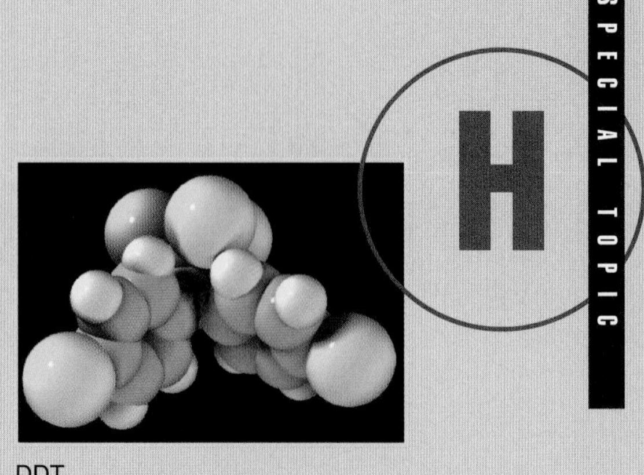

DDT

H.1 Organic Halides as Insecticides

Since the discovery of the insecticidal properties of DDT in 1942, vast quantities of chlorinated hydrocarbons have been sprayed over the surface of the earth in an effort to destroy insects. These efforts initially met with incredible success in ridding large areas of the earth of disease-carrying insects, particularly those of typhus and malaria. As time has passed, however, we have begun to understand that this prodigious use of chlorinated hydrocarbons has not been without harmful—indeed tragic—side effects. Chlorinated hydrocarbons are usually highly stable compounds and are only slowly destroyed by natural processes in the environment. As a result, many chloroorganic insecticides remain in the environment for years. These persistent pesticides are called "hard" pesticides.

Chlorohydrocarbons are also fat soluble and tend to accumulate in the fatty tissues of most animals. The food chain that runs from plankton to small fish to larger fish to birds and to larger animals, including humans, tends to magnify the concentrations of chloroorganic compounds at each step.

The chlorohydrocarbon DDT is prepared from inexpensive starting materials, chlorobenzene and trichloroacetaldehyde. The reaction is catalyzed by acid.

$$2\ \text{Cl}-\text{C}_6\text{H}_5 + \text{H}\overset{\text{O}}{\overset{\|}{\text{C}}}\text{CCl}_3 \xrightarrow{\text{H}_2\text{SO}_4} \text{Cl}-\text{C}_6\text{H}_4-\underset{\text{CCl}_3}{\underset{|}{\text{CH}}}-\text{C}_6\text{H}_4-\text{Cl}$$

DDT
[1,1,1-trichloro-2,2-bis(*p*-chlorophenyl)ethane]

In nature the principal decomposition product of DDT is DDE.

$$(p\text{-Cl}-\text{C}_6\text{H}_4)_2\text{C}=\text{CCl}_2$$

DDE
[1,1-dichloro-2,2-bis(*p*-chlorophenyl)ethene]

Estimates indicate that nearly 1 billion pounds of DDT were spread throughout the world ecosystem. One pronounced environmental effect of DDE, after conversion from DDT, has been in its action on eggshell formation in many birds. DDE inhibits the enzyme *carbonic anhydrase* that controls the calcium supply for shell formation. (See the Chapter 3 opening vignette for another important role of carbonic anhydrase.) As a consequence, the shells are often very fragile and do not survive to the time of hatching. During the late 1940s the populations of eagles, falcons, and hawks dropped dramatically. There can be little doubt that DDT was primarily responsible.

DDE also accumulates in the fatty tissues of humans. Although humans appear to have a short-range tolerance to moderate DDE levels, the long-range effects are far from certain.

Other hard insecticides are aldrin, dieldrin, and chlordan. Aldrin can be manufactured through the Diels–Alder reaction of hexachlorocyclopentadiene and norbornadiene. Dieldrin can be made by converting an aldrin double bond to an epoxide. (This reaction also takes place in nature.)

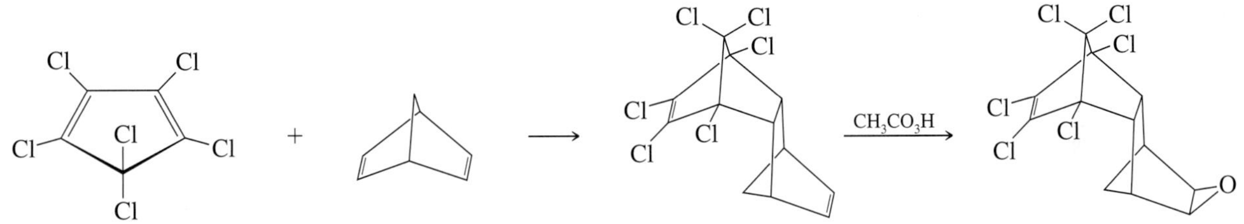

Hexachloro-cyclopentadiene + Norbornadiene ⟶ Aldrin $\xrightarrow{CH_3CO_3H}$ Dieldrin

Chlordan can be made by adding chlorine to the unsubstituted double bond of the Diels–Alder adduct obtained from hexachlorocyclopentadiene and cyclopentadiene.

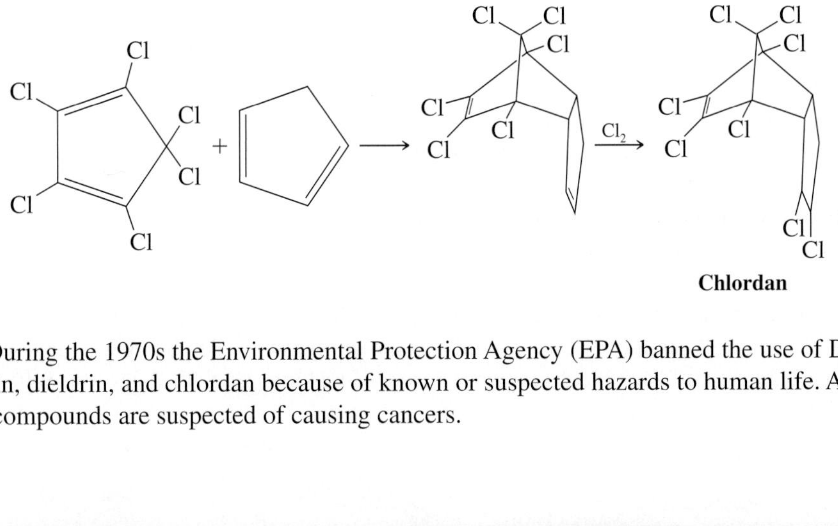

$\xrightarrow{Cl_2}$ Chlordan

During the 1970s the Environmental Protection Agency (EPA) banned the use of DDT, aldrin, dieldrin, and chlordan because of known or suspected hazards to human life. All of the compounds are suspected of causing cancers.

PROBLEM H.1

The mechanism for the formation of DDT from chlorobenzene and trichloroacetaldehyde in sulfuric acid involves two electrophilic aromatic substitution reactions. In the first electrophilic substitution reaction, the electrophile is protonated trichloroacetaldehyde. In the second, the electrophile is a carbocation. Propose a mechanism for the formation of DDT.

PROBLEM H.2

What kind of reaction is involved in the conversion of DDT to DDE?

Mirex, kepone, and lindane are also hard insecticides whose use has been banned.

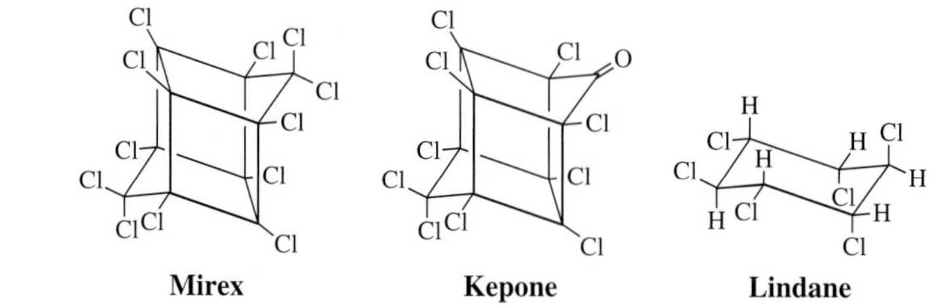

Mirex **Kepone** **Lindane**

H.2 Organic Halides as Herbicides

Other chlorinated organic compounds have been used extensively as herbicides. The following two examples are 2,4-D and 2,4,5-T.

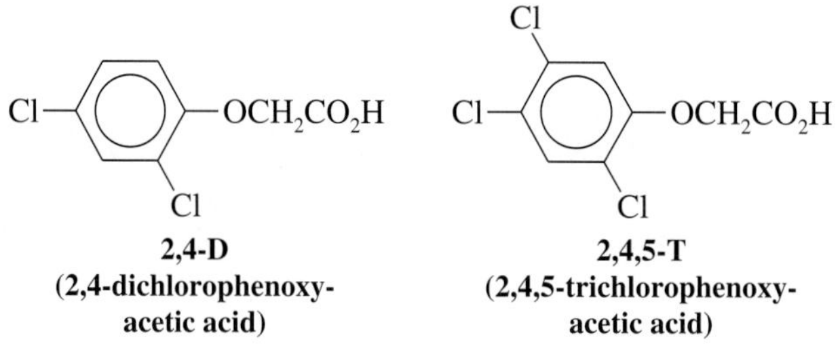

2,4-D (2,4-dichlorophenoxyacetic acid)

2,4,5-T (2,4,5-trichlorophenoxyacetic acid)

Enormous quantities of these two compounds were used as defoliants in the jungles of Indochina during the Vietnam War. Some samples of 2,4,5-T were shown to be teratogenic (a fetus-deforming agent). This teratogenic effect was the result of an impurity present in commercial 2,4,5-T, the compound 2,3,7,8-tetrachlorodibenzodioxin. 2,3,7,8-Tetrachlorodibenzodioxin is also highly toxic; it is more toxic, for example, than cyanide ion, strychnine, and the nerve gases.

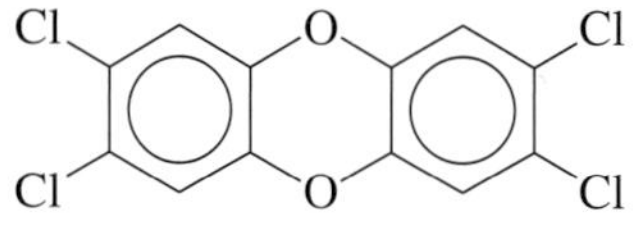

2,3,7,8-Tetrachlorodibenzodioxin (also called TCDD)

This dioxin is also highly stable; it persists in the environment and because of its fat solubility can be passed up the food chain. In sublethal amounts it can cause a disfiguring skin disease called chloracne.

In July 1976 an explosion at a chemical plant in Seveso, Italy, caused the release of between 22 and 132 pounds of this dioxin into the atmosphere. The plant was engaged in the manufacture of 2,4,5-trichlorophenol (used in making 2,4,5-T) using the following method:

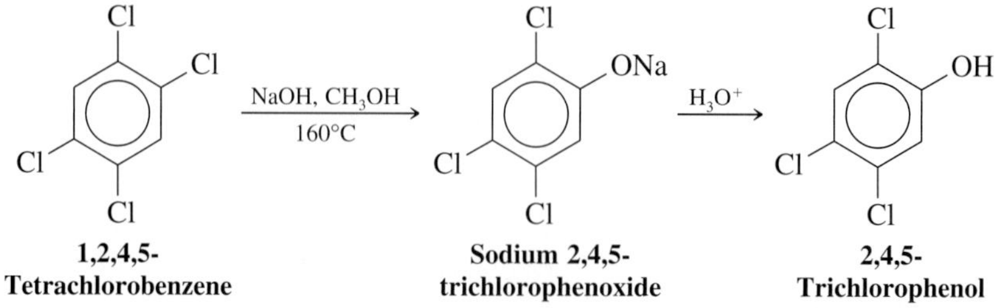

$$\text{1,2,4,5-Tetrachlorobenzene} \xrightarrow[160°C]{NaOH,\ CH_3OH} \text{Sodium 2,4,5-trichlorophenoxide} \xrightarrow{H_3O^+} \text{2,4,5-Trichlorophenol}$$

The temperature of the first reaction must be very carefully controlled; if it is not, this dioxin forms in the reaction mixture:

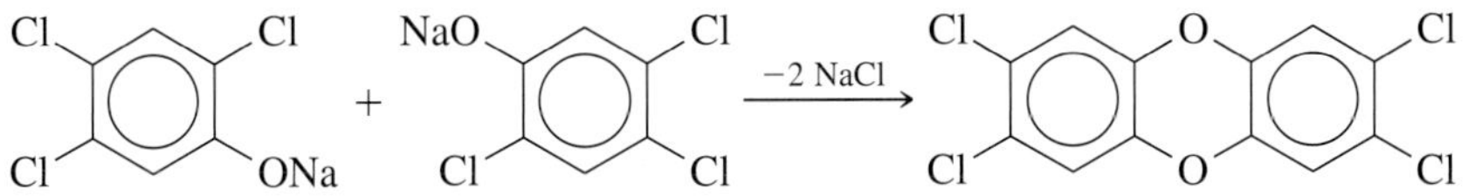

Apparently at the Italian factory, the temperature went out of control, causing the pressure to build up. Eventually, a valve opened and released a cloud of trichlorophenol and the dioxin into the atmosphere. Many wild and domestic animals were killed and many people, especially children, were afflicted with severe skin rashes.

PROBLEM H.3

(a) Assume that the ortho and para chlorine atoms provide enough activation by electron withdrawal for nucleophilic substitution to occur by an addition–elimination pathway and outline a possible mechanism for the conversion of 1,2,4,5-tetrachlorobenzene to sodium 2,4,5-trichlorophenoxide. **(b)** Do the same for the conversion of 2,4,5-trichlorophenoxide to the dioxin of Section H.2.

PROBLEM H.4

2,4,5-T is made by allowing sodium 2,4,5-trichlorophenoxide to react with sodium chloroacetate ($ClCH_2COONa$). (This produces the sodium salt of 2,4,5-T that, on acidification, gives 2,4,5-T itself.) What kind of mechanism accounts for the reaction of sodium 2,4,5-trichlorophenoxide with $ClCH_2COONa$? Write the equation.

H.3 GERMICIDES

2,4,5-Trichlorophenol is also used in the manufacture of hexachlorophene, a germicide once widely used in soaps, shampoos, deodorants, mouthwashes, aftershave lotions, and other over-the-counter products.

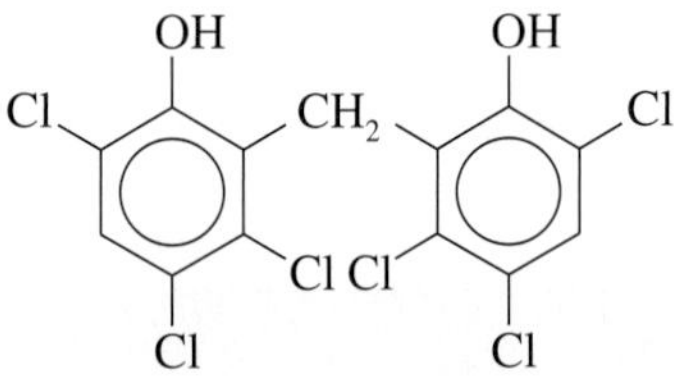

Hexachlorophene

Hexachlorophene is absorbed intact through the skin, and tests with experimental animals have shown that it causes brain damage. Since 1972, the use of hexachlorophene in cleansers and cosmetics sold over the counter has been banned by the Food and Drug Administration.

H.4 POLYCHLORINATED BIPHENYLS (PCBs)

Mixtures of polychlorinated biphenyls have been produced and used commercially since 1929. In these mixtures, biphenyls with chlorine atoms at any of the numbered positions (see the following structure) may be present. In all, there are 210 possible compounds. A typical commercial mixture may contain as many as 50 different PCBs. Mixtures are usually classified on the basis of their chlorine content, and most industrial mixtures contain from 40 to 60% chlorine.

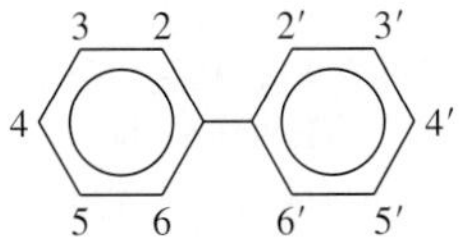

Biphenyl

Polychlorinated biphenyls have had a multitude of uses: as heat-exchange agents in transformers; in capacitors, thermostats, and hydraulic systems; as plasticizers in polystyrene coffee cups, frozen food bags, bread wrappers, and plastic liners for baby bottles. They have been used in printing inks, in carbonless carbon paper, and as waxes for making molds for metal castings. Between 1929 and 1972, about 500,000 metric tons of PCBs were manufactured.

Although they were never intended for release into the environment, PCBs have become, perhaps more than any other chemical, the most widespread pollutant. They have been found in rainwater, in many species of fish, birds, and other animals (including polar bears) all over the globe, and in human tissue.

Polychlorinated biphenyls are highly persistent and, being fat soluble, tend to accumulate in the food chain. Fish that feed in PCB-contaminated waters, for example, have PCB levels 1000–100,000 times the level of the surrounding water, and this amount is further magnified in birds that feed on the fish. The toxicity of PCBs depends on the composition of the individual mixture. The largest incident of human poisoning by PCBs occurred in Japan in 1968 when about 1000 people ingested a cooking oil accidentally contaminated with PCBs. See "The Chemistry of . . . Bacterial Dehalogenation of a PCB" (Section 21.11) for a potential method of PCB remediation.

As late as 1975, industrial concerns were legally discharging PCBs into the Hudson River. In 1977, the EPA banned the direct discharge into waterways, and since 1979 their manufacture, processing, and distribution have been prohibited. In 2000 the EPA specified certain sections of the Hudson River for cleanup of PCBs.

H.5 Polybromobiphenyls (PBBs)

Polybromobiphenyls are bromine analogs of PCBs that have been used as flame retardants. In 1973, in Michigan, a mistake at a chemical company led to PBBs being mixed into animal feeds that were sold to farmers. Before the mistake was recognized, PBBs had affected thousands of dairy cattle, hogs, chickens, and sheep, necessitating their destruction.

H.6 Organometallic Compounds

With few exceptions, organometallic compounds (see Special Topic G) are toxic. This toxicity varies greatly depending on the nature of the organometallic compound and the identity of the metal. Organic compounds of arsenic, antimony, lead, thallium, and mercury are toxic because the metal ions, themselves, are toxic. Certain organic derivatives of silicon are toxic even though silicon and most of its inorganic compounds are nontoxic.

Early in the twentieth century the recognition of the biocidal effects of organoarsenic compounds led Paul Ehrlich to his pioneering work in chemotherapy. Ehrlich sought compounds (which he called "magic bullets") that would show greater toxicity toward disease-causing microorganisms than they would toward their hosts. Ehrlich's research led to the development of Salvarsan and Neosalvarsan, two organoarsenic compounds that were used successfully in the treatment of diseases caused by spirochetes (e.g., syphilis) and trypanosomes (e.g., sleeping sickness). Salvarsan and Neosalvarsan are no longer used in the treatment of these diseases; they have been displaced by safer and more effective antibiotics. Ehrlich's research, however, initiated the field of chemotherapy (see Section 20.11).

Many microorganisms actually synthesize organometallic compounds, and this discovery has an alarming ecological aspect. Mercury metal is toxic, but mercury metal is also unreactive. In the past, untold tons of mercury metal present in industrial wastes have been disposed of by simply dumping such wastes into lakes and streams. Since mercury is toxic, many bacteria protect themselves from its effect by converting mercury metal to methylmercury ions (CH_3Hg^+) and to gaseous dimethylmercury $(CH_3)_2Hg$. These organic mercury compounds are passed up the food chain (with modification) through fish to humans, where methylmercury ions act as a deadly nerve poison. Between 1953 and 1964, 116 people in Minamata, Japan, were poisoned by eating fish containing methylmercury compounds. Arsenic is also methylated by organisms to the poisonous dimethylarsine, $(CH_3)_2AsH$.

Ironically, chlorinated hydrocarbons appear to inhibit the biological reactions that bring about mercury methylation. Lakes polluted with organochlorine pesticides show significantly lower mercury methylation. Whereas this particular interaction of two pollutants may, in a certain sense, be beneficial, it is also instructive of the complexity of the environmental problems that we face.

Tetraethyllead and other alkyllead compounds have been used as antiknock agents in gasoline since 1923. Although this use has now been phased out in the United States, more than 1 trillion pounds of lead have been introduced into the atmosphere. In the northern hemisphere, gasoline burning alone has spread about 10 mg of lead on each square meter of the earth's surface. In highly industrialized areas the amount of lead per square meter is probably several hundred times higher. Because of the well-known toxicity of lead, these facts are of great concern.

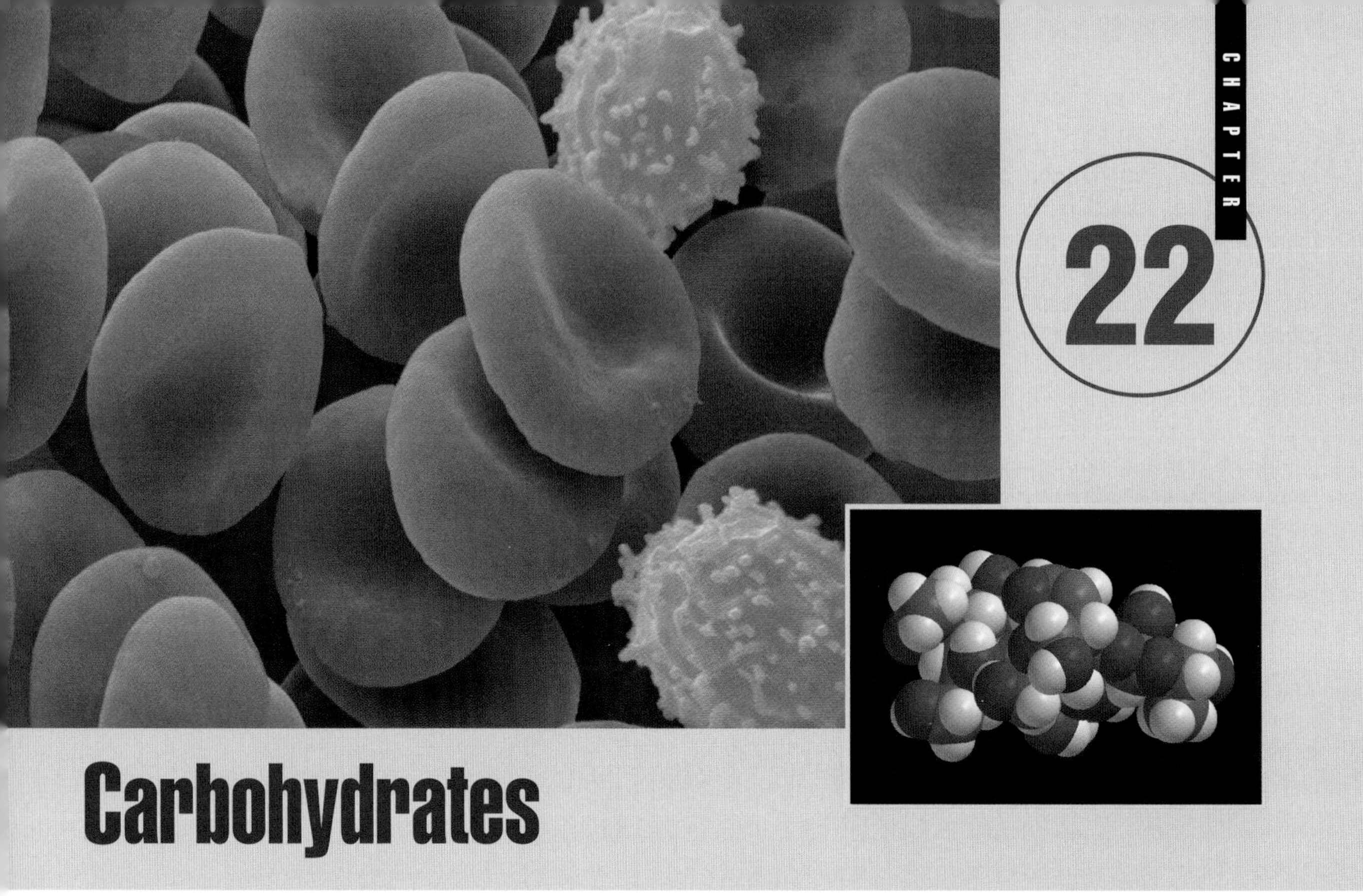

CHAPTER 22

Carbohydrates

Carbohydrate Recognition in Healing and Disease

White blood cells continually patrol the circulatory system and interstitial spaces, ready for mobilization at a site of trauma. The frontline scouts for leukocytes are carbohydrate groups on their surface called sialyl Lewis^x acids. When injury occurs, cells at the site of trauma display proteins, called selectins, that signal the site of injury and bind sialyl Lewis^x acids. Binding between selectins and the sialyl Lewis^x acids on the leukocytes causes adhesion of leukocytes at the affected area. Recruitment of leukocytes in this way is an important step in the inflammatory cascade. It is a necessary part of the healing process as well as part of our natural defense against infection. A molecular model of sialyl Lewis^x is shown above, and its structural formula is given in Section 22.16.

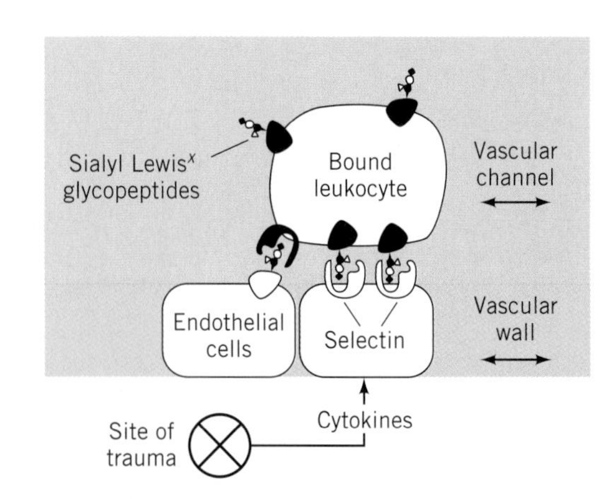

Patrolling leukocytes bind at the site of trauma by interactions between sialyl Lewis^x glycoproteins on their surface and selectin proteins on the injured cell. From Simanek, E. E.; McGarvey, G. J.; Jablonowski, J. A.; Wong, C. A. *Chem. Rev.* **1998,** *98,* 833–862 (Figure 1, p 835).

There are some maladies, however, that result from the overenthusiastic recruitment of leukocytes. Rheumatoid arthritis, strokes, and injuries related to perfusion during surgery and organ transplantation are a few examples. In these conditions, the body perceives that certain cells are under duress, and it reacts accordingly to initiate the inflammatory cascade. Unfortunately, under these circumstances the inflammatory cascade actually causes greater harm than good.

A strategy for combating undesirable initiation of the inflammatory cascade is to disrupt the adhesion of leukocytes. This can be done by blocking the selectin binding sites for sialyl Lewis^x acids. Chemists have advanced this approach by synthesizing both natural and mimetic sialyl Lewis^x acids for studies on the binding process. These compounds have helped identify key functional groups in sialyl Lewis^x acids that are required for recognition and binding. Chemists have even designed and synthesized novel compounds that have tighter binding affinities than the natural sialyl Lewis^x acids. Among them are polymers with repeating occurrences of the structural motifs essential for binding. These polymeric species presumably occupy multiple sialyl Lewis^x acid binding sites at once, thereby binding more tightly than monomeric sialyl Lewis^x acid analogs.

Efforts like these to prepare finely tuned molecular agents are typical of research in drug discovery and design. In the case of sialyl Lewis^x acid analogs, chemists hope to create new therapies for chronic inflammatory diseases by making ever-improved agents for blocking undesired leukocyte adhesion.

22.1 INTRODUCTION

22.1A Classification of Carbohydrates

The group of compounds known as carbohydrates received their general name because of early observations that they often have the formula $C_x(H_2O)_y$—that is, they appear to be "hydrates of carbon." Simple carbohydrates are also known as sugars or saccharides (Latin *saccharum,* Greek *sakcharon,* sugar) and the ending of the names of most sugars is *-ose.* Thus, we have such names as *sucrose* for ordinary table sugar, *glucose* for the principal sugar in blood, *fructose* for a sugar in fruits and honey, and *maltose* for malt sugar.

STUDY TIP

You may find it helpful now to review the chemistry of hemiacetals and acetals (Section 16.7).

Carbohydrates are usually defined as *polyhydroxy aldehydes and ketones or substances that hydrolyze to yield polyhydroxy aldehydes and ketones.* Although this definition draws attention to the important functional groups of carbohydrates, it is not entirely satisfactory. We shall later find that because carbohydrates contain $\backslash C{=}O$ groups and —OH groups, they exist, primarily, as *hemiacetals* or *acetals* (Section 16.7).

The simplest carbohydrates, those that cannot be hydrolyzed into simpler carbohydrates, are called **monosaccharides.** On a molecular basis, carbohydrates that undergo hydrolysis to produce only 2 molecules of monosaccharide are called **disaccharides;** those that yield 3 molecules of monosaccharide are called **trisaccharides;** and so on. (Carbohydrates that hydrolyze to yield 2–10 molecules of monosaccharide are sometimes called **oligosaccharides.**) Carbohydrates that yield a large number of molecules of monosaccharides (>10) are known as **polysaccharides.**

Maltose and sucrose are examples of disaccharides. On hydrolysis, 1 mol of maltose yields 2 mol of the monosaccharide glucose; sucrose undergoes hydrolysis to yield 1 mol of glucose and 1 mol of the monosaccharide fructose. Starch and cellulose are examples of polysaccharides; both are glucose polymers. Hydrolysis of either yields a large number of glucose units. The following shows these hydrolyses in a schematic way:

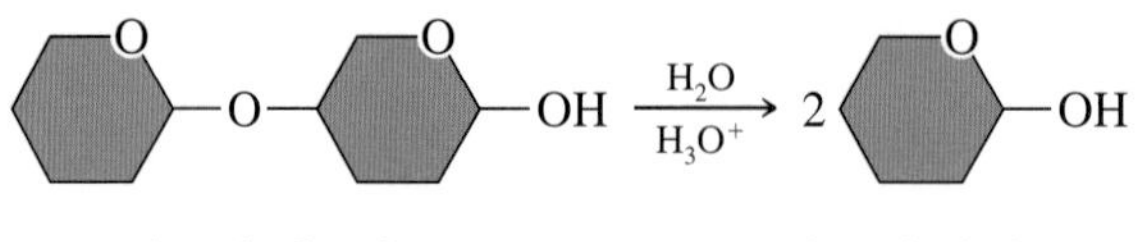

$\xrightarrow[H_3O^+]{H_2O}$ 2

1 mol of maltose
A disaccharide

2 mol of glucose
A monosaccharide

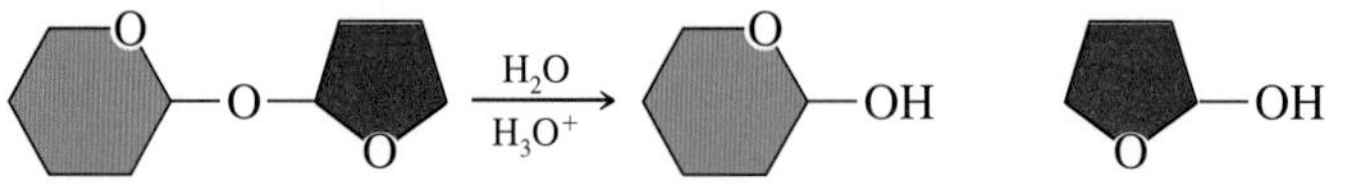

$\xrightarrow[H_3O^+]{H_2O}$

1 mol of sucrose
A disaccharide

1 mol of glucose + 1 mol of fructose
Monosaccharides

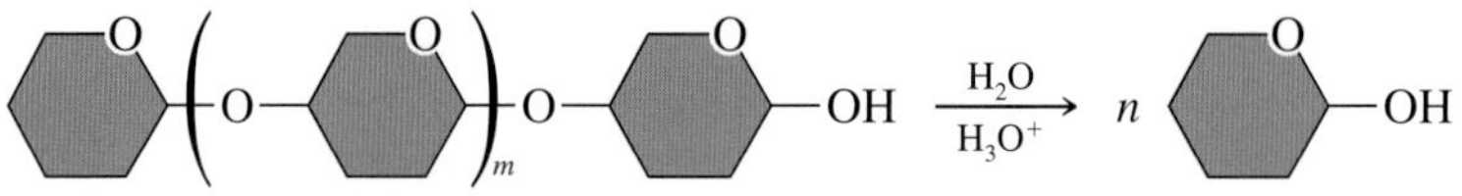

$\xrightarrow[H_3O^+]{H_2O}$ n

1 mol of starch or
1 mol of cellulose
Polysaccharides

many moles of glucose
Monosaccharides

Carbohydrates are the most abundant organic constituents of plants. They not only serve as an important source of chemical energy for living organisms (sugars and starches are important in this respect), but also in plants and in some animals they serve as important constituents of supporting tissues (this is the primary function of the cellulose found in wood, cotton, and flax, for example).

We encounter carbohydrates at almost every turn of our daily lives. The paper on which this book is printed is largely cellulose; so, too, is the cotton of our clothes and the wood of our houses. The flour from which we make bread is mainly starch, and starch is also a major constituent of many other foodstuffs, such as potatoes, rice, beans, corn, and peas. Carbohydrates are central to metabolism, and they are important for cell recognition (see the chapter opening vignette and Section 22.16).

22.1B Photosynthesis and Carbohydrate Metabolism

Carbohydrates are synthesized in green plants by *photosynthesis*—a process that uses solar energy to reduce, or "fix," carbon dioxide. Photosynthesis in algae and higher plants occurs in cell organelles called chloroplasts. The overall equation for photosynthesis can be written as follows:

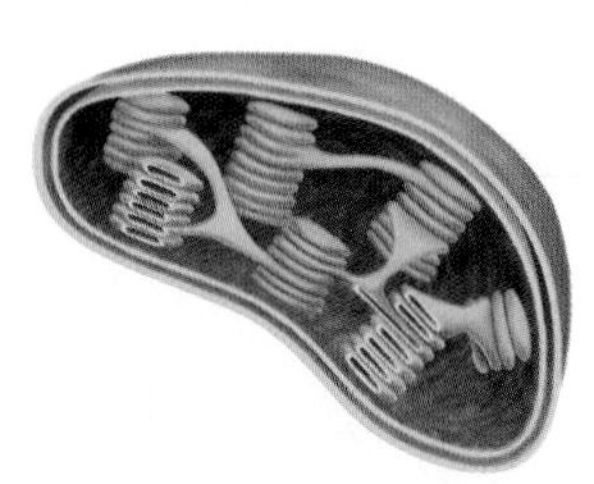

Schematic diagram of a chloroplast from corn. (From Voet, D.; Voet, J. G. *Biochemistry,* 2nd ed; Wiley: New York, 1995; p 627.)

$$x\,CO_2 + y\,H_2O + \text{solar energy} \longrightarrow \underset{\textbf{Carbohydrate}}{C_x(H_2O)_y} + x\,O_2$$

Many individual enzyme-catalyzed reactions take place in the general photosynthetic process and not all are fully understood. We know, however, that photosynthesis begins with the absorption of light by the important green pigment of plants, chlorophyll (Fig. 22.1). The green color of chlorophyll and, therefore, its ability to absorb sunlight in the visible region are due primarily to its extended conjugated system. As photons of sunlight are trapped by chlorophyll, energy becomes available to the plant in a chemical form that can be used to carry out the reactions that reduce carbon dioxide to carbohydrates and oxidize water to oxygen.

Carbohydrates act as a major chemical repository for solar energy. Their energy is released when animals or plants metabolize carbohydrates to carbon dioxide and water:

$$C_x(H_2O)_y + x\,O_2 \longrightarrow x\,CO_2 + y\,H_2O + \text{energy}$$

The metabolism of carbohydrates also takes place through a series of enzyme-catalyzed

FIGURE 22.1 Chlorophyll *a*. [The structure of chlorophyll *a* was established largely through the work of H. Fischer (Munich), R. Willstätter (Munich), and J. B. Conant (Harvard). A synthesis of chlorophyll *a* from simple organic compounds was achieved by R. B. Woodward (Harvard) in 1960, who won the Nobel Prize in 1965 for his outstanding contributions to synthetic organic chemistry.]

reactions in which each energy-yielding step is an oxidation (or the consequence of an oxidation).

Although some of the energy released in the oxidation of carbohydrates is inevitably converted to heat, much of it is conserved in a new chemical form through reactions that are coupled to the synthesis of adenosine triphosphate (ATP) from adenosine diphosphate (ADP) and inorganic phosphate (P_i) (Fig. 22.2). The phosphoric anhydride bond that

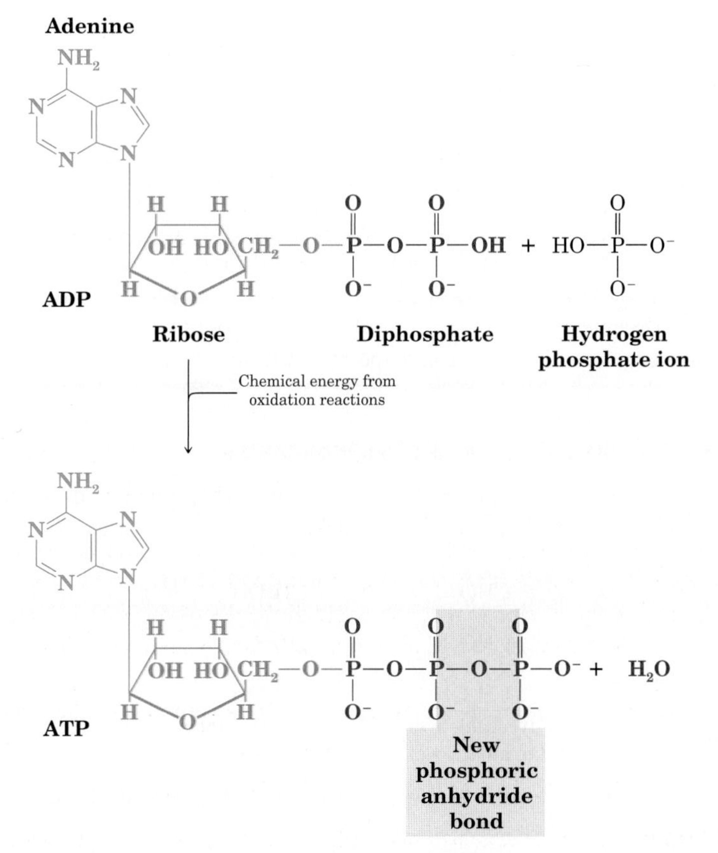

FIGURE 22.2 The synthesis of adenosine triphosphate (ATP) from adenosine diphosphate (ADP) and hydrogen phosphate ion. This reaction takes place in all living organisms, and adenosine triphosphate is the major compound into which the chemical energy released by biological oxidations is transformed.

forms between the terminal phosphate group of ADP and the phosphate ion becomes another repository of chemical energy. Plants and animals can use the conserved energy of ATP (or very similar substances) to carry out all of their energy-requiring processes: the contraction of a muscle, the synthesis of a macromolecule, and so on. When the energy in ATP is used, a coupled reaction takes place in which ATP is hydrolyzed,

$$\text{ATP} + \text{H}_2\text{O} \longrightarrow \text{ADP} + \text{P}_i + \text{energy}$$

or a new anhydride linkage is created,

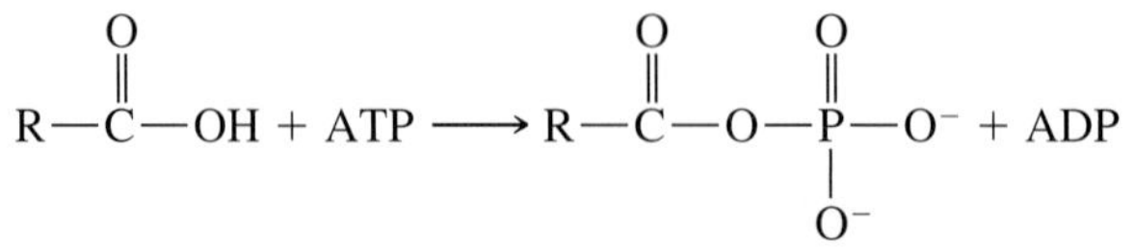

$$\text{R}-\overset{\overset{\displaystyle\text{O}}{\|}}{\text{C}}-\text{OH} + \text{ATP} \longrightarrow \text{R}-\overset{\overset{\displaystyle\text{O}}{\|}}{\text{C}}-\text{O}-\underset{\underset{\displaystyle\text{O}^-}{|}}{\overset{\overset{\displaystyle\text{O}}{\|}}{\text{P}}}-\text{O}^- + \text{ADP}$$

An acyl phosphate

22.2 Monosaccharides

22.2A Classification of Monosaccharides

Monosaccharides are classified according to (1) the number of carbon atoms present in the molecule and (2) whether they contain an aldehyde or keto group. Thus, a monosaccharide containing three carbon atoms is called a *triose;* one containing four carbon atoms is called a *tetrose;* one containing five carbon atoms is a *pentose;* and one containing six carbon atoms is a *hexose.* A monosaccharide containing an aldehyde group is called an *aldose;* one containing a keto group is called a *ketose.* These two classifications are frequently combined. A C_4 aldose, for example, is called an *aldotetrose;* a C_5 ketose is called a *ketopentose.*

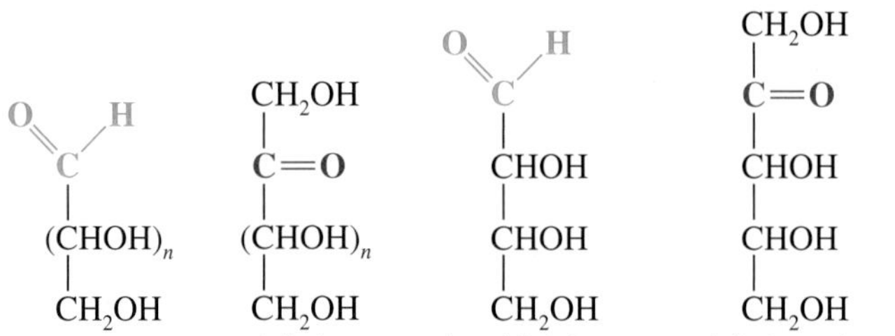

An aldose	A ketose	An aldotetrose C_4	A ketopentose C_5
O=C–H	CH_2OH	O=C–H	CH_2OH
$(CHOH)_n$	C=O	CHOH	C=O
CH_2OH	$(CHOH)_n$	CHOH	CHOH
	CH_2OH	CH_2OH	CHOH
			CH_2OH

PROBLEM 22.1

How many stereogenic centers are contained in **(a)** the aldotetrose and **(b)** the ketopentose just given? **(c)** How many stereoisomers would you expect from each general structure?

22.2B D and L Designations of Monosaccharides

The simplest monosaccharides are the compounds glyceraldehyde and dihydroxyacetone (see the following structures). Of these two compounds, only glyceraldehyde contains a stereogenic center.

Glyceraldehyde (an aldotriose)	Dihydroxyacetone (a ketotriose)
CHO	CH_2OH
*CHOH	C=O
CH_2OH	CH_2OH

Glyceraldehyde exists, therefore, in two enantiomeric forms that are known to have the absolute configurations shown here:

(+)-Glyceraldehyde: $O{=}C(H)$ over $H{-}C{-}OH$ over CH_2OH and (−)-Glyceraldehyde: $O{=}C(H)$ over $HO{-}C{-}H$ over CH_2OH

(+)-Glyceraldehyde **(−)-Glyceraldehyde**

We saw in Section 5.7 that, according to the Cahn–Ingold–Prelog convention, (+)-glyceraldehyde should be designated (*R*)-(+)-glyceraldehyde and (−)-glyceraldehyde should be designated (*S*)-(−)-glyceraldehyde.

Early in the twentieth century, before the absolute configurations of any organic compounds were known, another system of stereochemical designations was introduced. According to this system (first suggested by M. A. Rosanoff of New York University in 1906), (+)-glyceraldehyde is designated D-(+)-glyceraldehyde and (−)-glyceraldehyde is designated L-(−)-glyceraldehyde. These two compounds, moreover, serve as configurational standards for all monosaccharides. A monosaccharide *whose highest numbered stereogenic center* (the penultimate carbon) has the same configuration as D-(+)-glyceraldehyde is designated as a D sugar; one whose highest numbered stereogenic center has the same configuration as L-glyceraldehyde is designated as an L sugar. By convention, acyclic forms of monosaccharides are drawn vertically with the aldehyde or keto group at or nearest the top. When drawn in this way, D sugars have the —OH on their penultimate carbon on the right:

A D-aldopentose: 1 CHO; 2 *CHOH; 3 *CHOH; 4 H–*C–OH; 5 CH_2OH

An L-ketohexose: 1 CH_2OH; 2 C=O; 3 *CHOH; 4 *CHOH; 5 HO–*C–H; 6 CH_2OH

Highest numbered stereogenic center

A D-aldopentose **An L-ketohexose**

The D and L designations are like (*R*) and (*S*) designations in that they are not necessarily related to the optical rotations of the sugars to which they are applied. Thus, one may encounter other sugars that are D-(+)- or D-(−)- and ones that are L-(+)- or L-(−)-.

The D–L system of stereochemical designations is thoroughly entrenched in the literature of carbohydrate chemistry, and even though it has the disadvantage of specifying the configuration of only one stereogenic center—that of the highest numbered stereogenic center—we shall employ the D–L system in our designations of carbohydrates.

PROBLEM 22.2

Write three-dimensional formulas for each aldotetrose and ketopentose isomer in Problem 22.1 and designate each as a D or L sugar.

22.2C Structural Formulas for Monosaccharides

Later in this chapter we shall see how the great carbohydrate chemist Emil Fischer* was able to establish the stereochemical configuration of the aldohexose D-(+)-glucose, the

*Emil Fischer (1852–1919) was professor of organic chemistry at the University of Berlin. In addition to monumental work in the field of carbohydrate chemistry, where Fischer and his co-workers established the configuration of most of the monosaccharides, Fischer also made important contributions to studies of amino acids, proteins, purines, indoles, and stereochemistry generally. As a graduate student, Fischer discovered phenylhydrazine, a reagent that was highly important in his later work with carbohydrates. Fischer was the second recipient (in 1902) of the Nobel Prize in Chemistry.

Emil Fischer

most abundant monosaccharide. In the meantime, however, we can use D-(+)-glucose as an example illustrating the various ways of representing the structures of monosaccharides.

Fischer represented the structure of D-(+)-glucose with the cross formulation **(1)** in Fig. 22.3. This type of formulation is now called a **Fischer projection** (Section 5.13) and is still useful for carbohydrates. In Fischer projections, by convention, *horizontal lines project out toward the reader and vertical lines project behind the plane of the page. When we use Fischer projections, however, we must not* (in our mind's eye) *remove them from the plane of the page in order to test their superposability and we must not rotate them by 90°*. In terms of more familiar formulations, the Fischer projection translates into formulas **2** and **3**. In IUPAC nomenclature and with the Cahn–Ingold–Prelog system of stereochemical designations, the open-chain form of D-(+)-glucose is (2*R*,3*S*,4*R*,5*R*)-2,3,4,5,6-pentahydroxyhexanal.

The meaning of formulas **1, 2,** and **3** can be seen best through the use of molecular models: We first construct a chain of six carbon atoms with the —CHO group at the top and a —CH_2OH group at the bottom. We then bring the CH_2OH group up

FIGURE 22.3 Formulas 1–3 are used for the open-chain structure of D-(+)-glucose. Formulas 4–7 are used for the two cyclic hemiacetal forms of D-(+)-glucose.

CD Tutorial

Haworth formulas

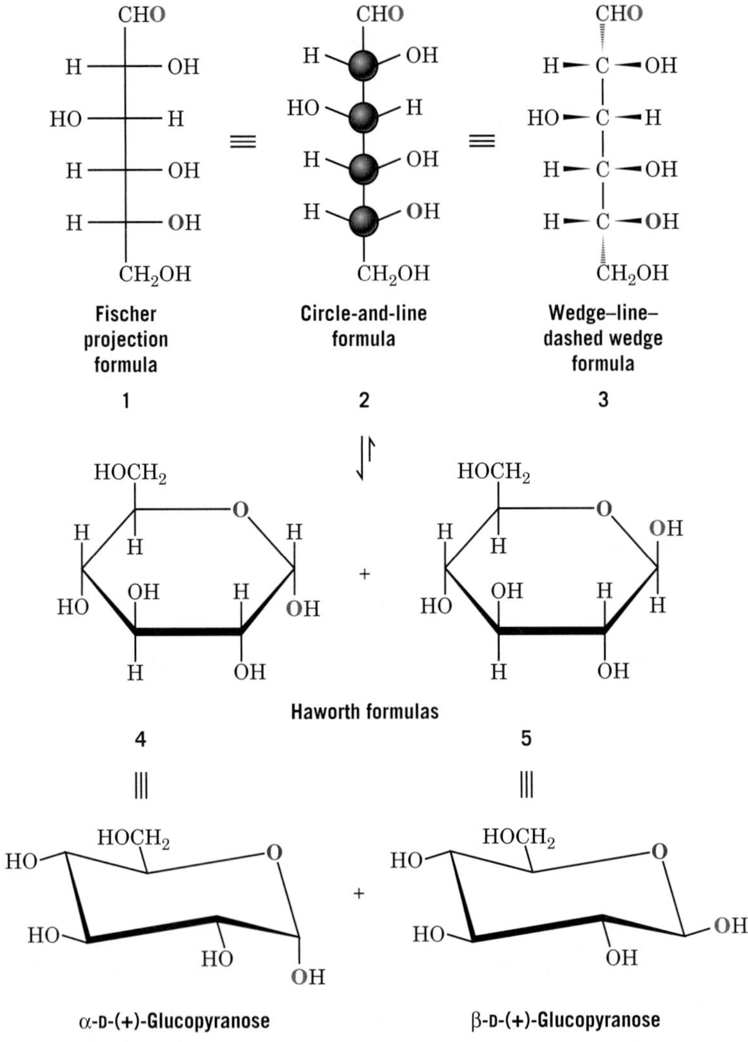

behind the chain until it almost touches the —CHO group. Holding this model so that the —CHO and —CH_2OH groups are directed generally away from us, we then begin placing —H and —OH groups on each of the four remaining carbon atoms. The —OH group of C2 is placed on the right; that of C3 on the left; and those of C4 and C5 on the right.

Use molecular models to help you learn to interpret Fischer projection formulas.

Although many of the properties of D-(+)-glucose can be explained in terms of an open-chain structure (**1, 2,** or **3**), a considerable body of evidence indicates that the open-chain structure exists, primarily, in equilibrium with two cyclic forms. These can be represented by structures **4** and **5** or **6** and **7.** The cyclic forms of D-(+)-glucose are **hemiacetals** formed by an intramolecular reaction of the —OH group at C5 with the aldehyde group (Fig. 22.4). Cyclization creates a new stereogenic center at C1, and this stereogenic center explains how two cyclic forms are possible. These two cyclic forms are *diastereomers* that differ only in the configuration of C1. In carbohydrate chemistry diastereomers of this type are called **anomers,** and the hemiacetal carbon atom is called the **anomeric carbon atom.**

Structures **4** and **5** for the glucose anomers are called **Haworth formulas*** and, although they do not give an accurate picture of the shape of the six-membered ring, they have many practical uses. Figure 22.4 demonstrates how the representation of each stereogenic center of the open-chain form can be correlated with its representation in the Haworth formula.

Each glucose anomer is designated as an **α anomer** or a **β anomer** depending on the location of the —OH group of C1. When we draw the cyclic forms of a D sugar in the orientation shown in Figs. 22.3 or 22.4, the α anomer has the —OH trans to the —CH_2OH group and the β anomer has the —OH cis to the —CH_2OH group.

α and β also find common use in steroid nomenclature (Section 23.4A).

Studies of the structures of the cyclic hemiacetal forms of D-(+)-glucose using X-ray analysis have demonstrated that the actual conformations of the rings are the chair forms represented by conformational formulas **6** and **7** in Fig. 22.3. This shape is exactly what we would expect from our studies of the conformations of cyclohexane (Chapter 4), and it is especially interesting to notice that in the β anomer of D-glucose all of the large substituents, —OH and —CH_2OH, are equatorial. In the α anomer, the only bulky axial substituent is the —OH at C1.

It is convenient at times to represent the cyclic structures of a monosaccharide without specifying whether the configuration of the anomeric carbon atom is α or β. When we do this, we shall use formulas such as the following:

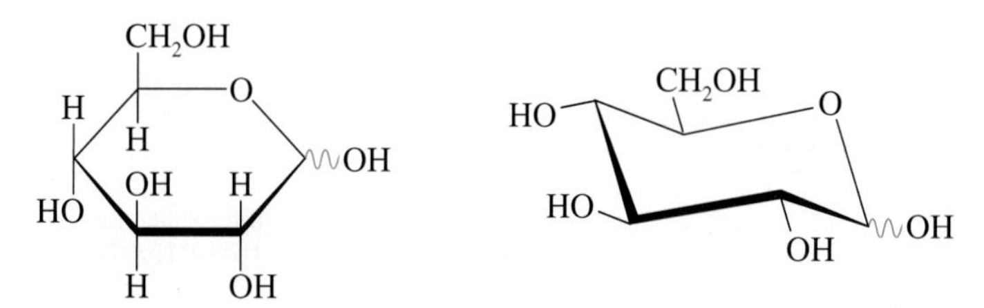

∿ **indicates α or β (three-dimensional view not specified)**

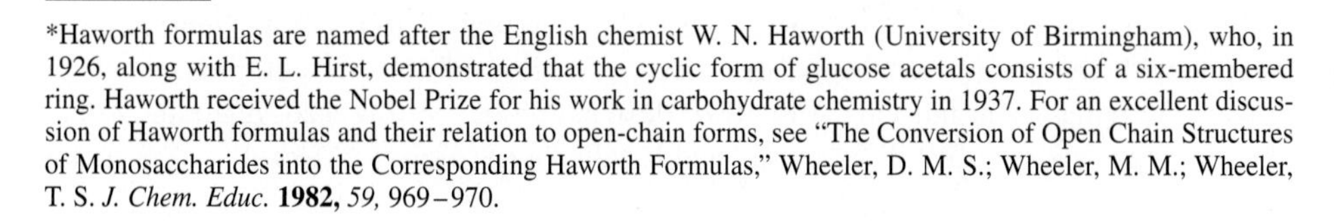

*Haworth formulas are named after the English chemist W. N. Haworth (University of Birmingham), who, in 1926, along with E. L. Hirst, demonstrated that the cyclic form of glucose acetals consists of a six-membered ring. Haworth received the Nobel Prize for his work in carbohydrate chemistry in 1937. For an excellent discussion of Haworth formulas and their relation to open-chain forms, see "The Conversion of Open Chain Structures of Monosaccharides into the Corresponding Haworth Formulas," Wheeler, D. M. S.; Wheeler, M. M.; Wheeler, T. S. *J. Chem. Educ.* **1982,** *59,* 969–970.

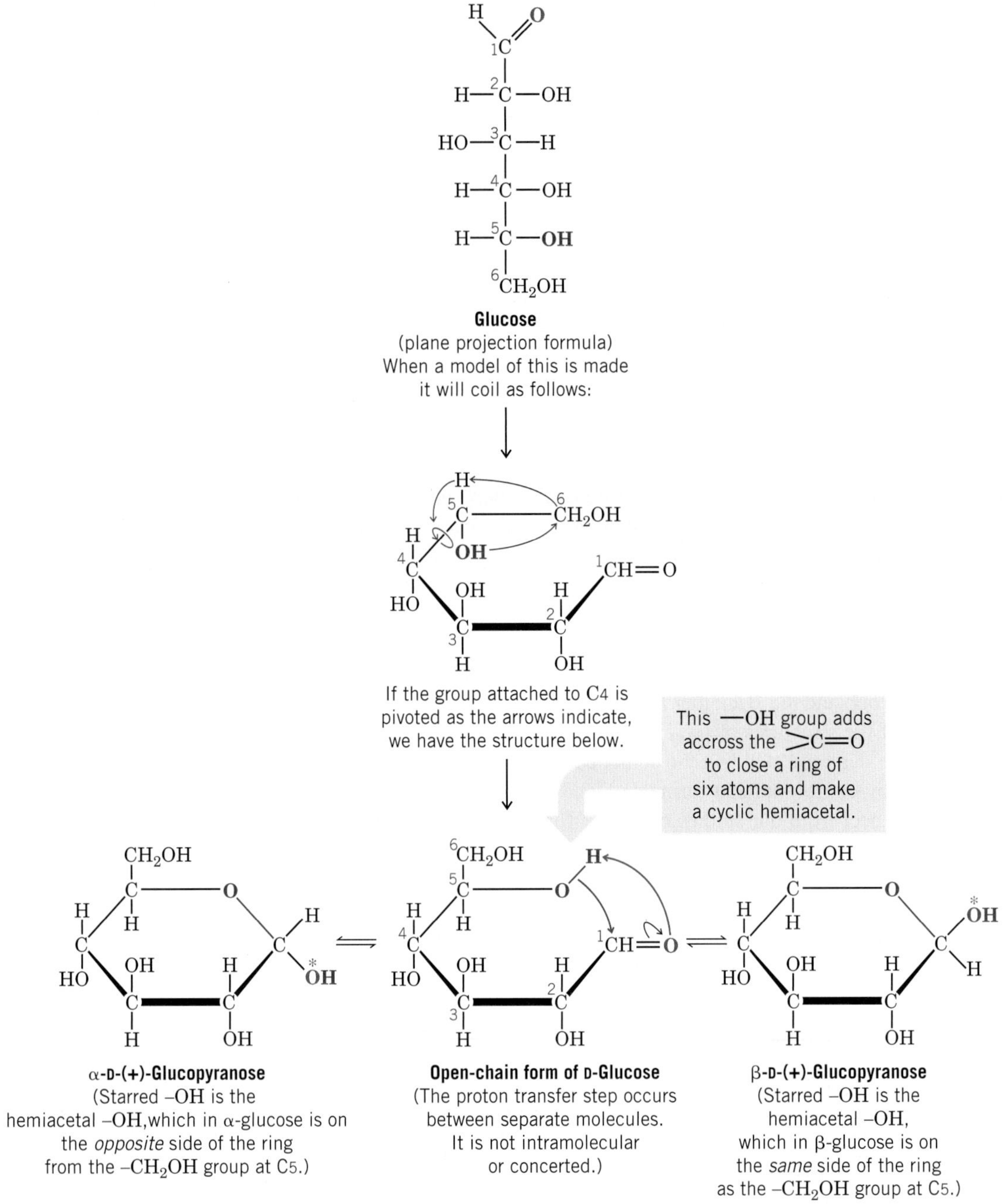

Glucose
(plane projection formula)
When a model of this is made
it will coil as follows:

If the group attached to C4 is
pivoted as the arrows indicate,
we have the structure below.

This —OH group adds
accross the >C=O
to close a ring of
six atoms and make
a cyclic hemiacetal.

α-D-(+)-Glucopyranose
(Starred –OH is the
hemiacetal –OH,which in α-glucose is on
the *opposite* side of the ring
from the $-CH_2OH$ group at C5.)

Open-chain form of D-Glucose
(The proton transfer step occurs
between separate molecules.
It is not intramolecular
or concerted.)

β-D-(+)-Glucopyranose
(Starred –OH is the
hemiacetal –OH,
which in β-glucose is on
the *same* side of the ring
as the $-CH_2OH$ group at C5.)

FIGURE 22.4 The Haworth formulas for the cyclic hemiacetal forms of D-(+)-glucose and their relation to the open-chain polyhydroxy aldehyde structure. Adapted from Holum, J. R. *Organic Chemistry: A Brief Course;* Wiley: New York, 1975; p 332. Used by permission.

Not all carbohydrates exist in equilibrium with six-membered hemiacetal rings; in several instances the ring is five membered. (Even glucose exists, to a small extent, in equilibrium with five-membered hemiacetal rings.) Because of this variation, a system of nomenclature has been introduced to allow designation of the ring size. If the monosaccharide ring is six membered, the compound is called a **pyranose;** if the ring is five membered, the compound is designated as a **furanose.*** Thus, the full name of compound **4** (or **6**) is α-D-(+)-glucopyranose, while that of **5** (or **7**) is β-D-(+)-glucopyranose.

22.3 Mutarotation

Part of the evidence for the cyclic hemiacetal structure for D-(+)-glucose comes from experiments in which both α and β forms have been isolated. Ordinary D-(+)-glucose has a melting point of 146°C. However, when D-(+)-glucose is crystallized by evaporating an aqueous solution kept above 98°C, a second form of D-(+)-glucose with a melting point of 150°C can be obtained. When the optical rotations of these two forms are measured, they are found to be significantly different, but when an aqueous solution of either form is allowed to stand, its rotation changes. The specific rotation of one form decreases and the rotation of the other increases, *until both solutions show the same value.* A solution of ordinary D-(+)-glucose (mp 146°C) has an initial specific rotation of +112°, but, ultimately, the specific rotation of this solution falls to +52.7°. A solution of the second form of D-(+)-glucose (mp 150°C) has an initial specific rotation of + 18.7°, but, slowly, the specific rotation of this solution rises to +52.7°. This change in rotation toward an equilibrium value is called **mutarotation.**

The explanation for this mutarotation lies in the existence of an equilibrium between the open-chain form of D-(+)-glucose and the α and β forms of the cyclic hemiacetals:

α-D-(+)-Glucopyranose (mp 146°C; $[\alpha]_D^{25}$ = +112°) ⇌ **Open-chain form of D-(+)-glucose** ⇌ **β-D-(+)-Glucopyranose (mp 150°C; $[\alpha]_D^{25}$ = +18.7°)**

X-ray analysis has confirmed that ordinary D-(+)-glucose has the α configuration at the anomeric carbon atom and that the higher melting form has the β configuration.

The concentration of open-chain D-(+)-glucose in solution at equilibrium is very small. Solutions of D-(+)-glucose give no observable UV or IR absorption band for a carbonyl group, and solutions of D-(+)-glucose give a negative test with Schiff's reagent—a special reagent that requires a relatively high concentration of a free aldehyde group (rather than a hemiacetal) in order to give a positive test.

Assuming that the concentration of the open-chain form is negligible, one can, by use of the specific rotations in the preceding figures, calculate the percentages of the α and β anomers present at equilibrium. These percentages, 36% α anomer and 64% β anomer, are in accord with a greater stability for β-D-(+)-glucopyranose. This preference is what we might expect on the basis of its having only equatorial groups:

---

*These names come from the names of the oxygen heterocycles *pyran* and *furan* + *ose:*

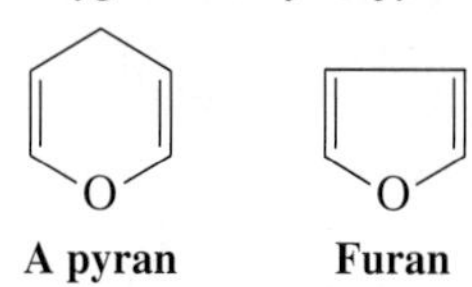

A pyran **Furan**

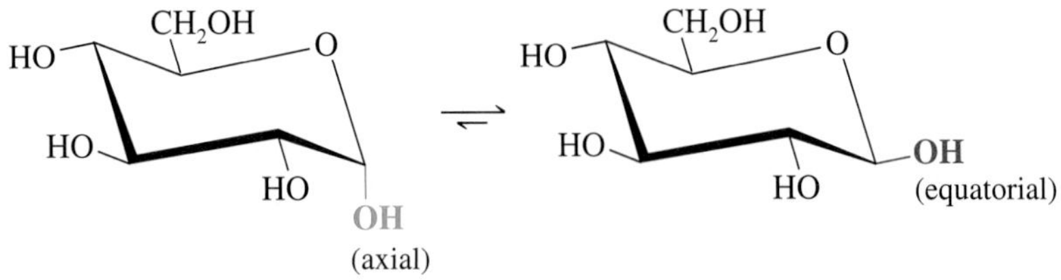

α-D-(+)-Glucopyranose
(36% at equilibrium)

β-D-(+)-Glucopyranose
(64% at equilibrium)

The β anomer of a pyranose is not always the more stable, however. With D-mannose, the equilibrium favors the α anomer, and this result is called an *anomeric effect:*

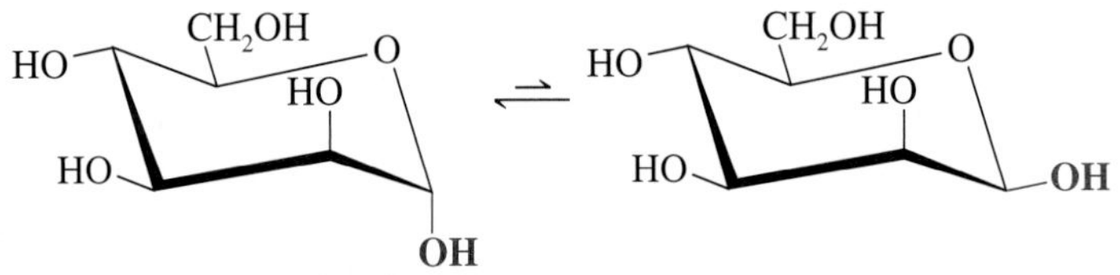

α-D-Mannopyranose
(69% at equilibrium)

β-D-Mannopyranose
(31% at equilibrium)

The anomeric effect is widely believed to be caused by hyperconjugation. An axially oriented orbital associated with nonbonding electrons of the ring oxygen can overlap with a σ* orbital of the axial exocyclic C—O hemiacetal bond. This effect is similar to that which causes the lowest energy conformation of ethane to be the anti conformation (Section 4.8). An anomeric effect will frequently cause an electronegative substituent, such as a hydroxyl or alkoxyl group, to prefer the axial orientation.

22.4 Glycoside Formation

When a small amount of gaseous hydrogen chloride is passed into a solution of D-(+)-glucose in methanol, a reaction takes place that results in the formation of anomeric methyl *acetals:*

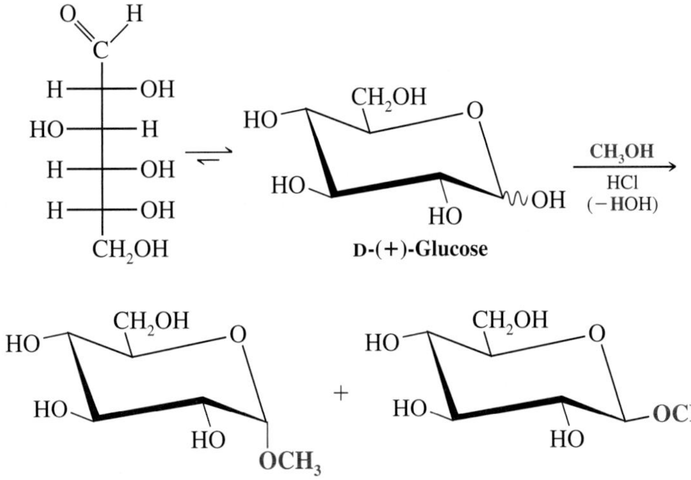

D-(+)-Glucose

CH_3OH, HCl, (−HOH)

Methyl α-D-glucopyranoside
(mp 165°C; $[\alpha]_D^{25}$ = +158°)

Methyl β-D-glucopyranoside
(mp 107°C; $[\alpha]_D^{25}$ = −33°)

Carbohydrate acetals, generally, are called **glycosides** (see the following mechanism), and an acetal of glucose is called a *glucoside.* (Acetals of mannose are *mannosides,* acetals of fructose are *fructosides,* and so on.) The methyl D-glucosides have been shown to have six-membered rings (Section 22.2C) so they are properly named methyl α-D-glucopyranoside and methyl β-D-glucopyranoside.

The mechanism for the formation of the methyl glucosides (starting arbitrarily with β-D-glucopyranose) is as follows:

A Mechanism for the Reaction

Formation of a Glycoside

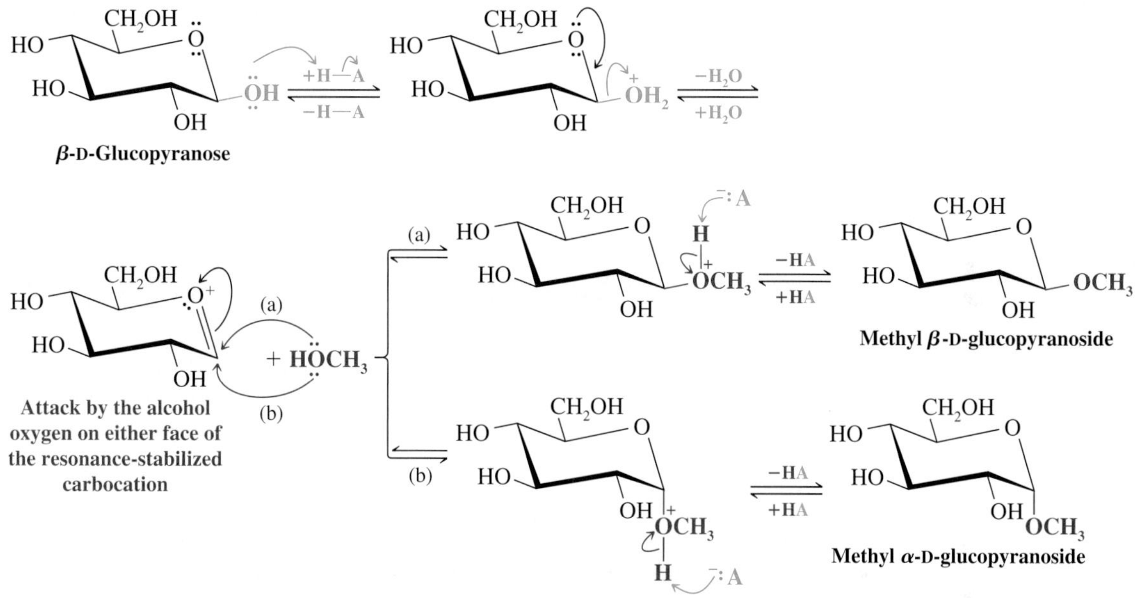

You should review the mechanism for acetal formation given in Section 16.7C and compare it with the steps given here. Notice, again, the important role played by the electron pair of the adjacent oxygen atom in stabilizing the carbocation that forms in the second step.

Glycosides are stable in basic solutions because they are acetals. In acidic solutions, however, glycosides undergo hydrolysis to produce a sugar and an alcohol (again, because they are acetals, Section 16.7). The alcohol obtained by hydrolysis of a glycoside is known as an **aglycone:**

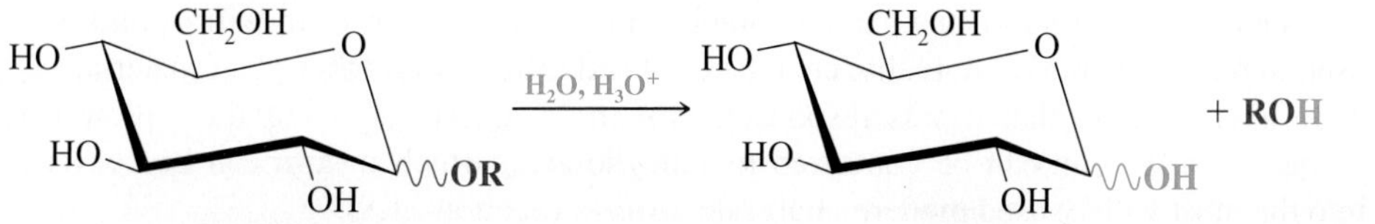

Glycoside (stable in basic solutions) **A sugar** **Aglycone**

For example, when an aqueous solution of methyl β-D-glucopyranoside is made acidic, the glycoside undergoes hydrolysis to produce D-glucose as a mixture of the two pyranose forms (in equilibrium with a small amount of the open-chain form).

A Mechanism for the Reaction

Hydrolysis of a Glycoside

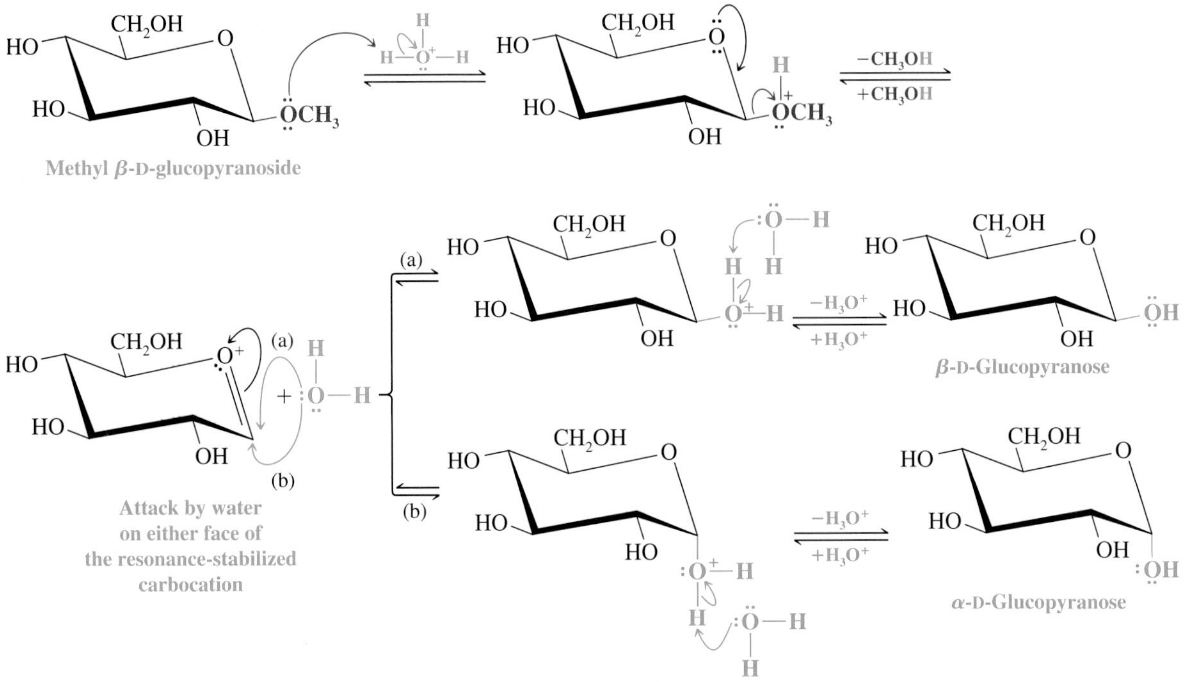

Methyl β-D-glucopyranoside

$-CH_3OH$ / $+CH_3OH$

(a) (b)

Attack by water on either face of the resonance-stabilized carbocation

$-H_3O^+$ / $+H_3O^+$

β-D-Glucopyranose

α-D-Glucopyranose

Glycosides may be as simple as the methyl glucosides that we have just studied or they may be considerably more complex. Many naturally occurring compounds are glycosides. An example is *salicin,* a compound found in the bark of willow trees:

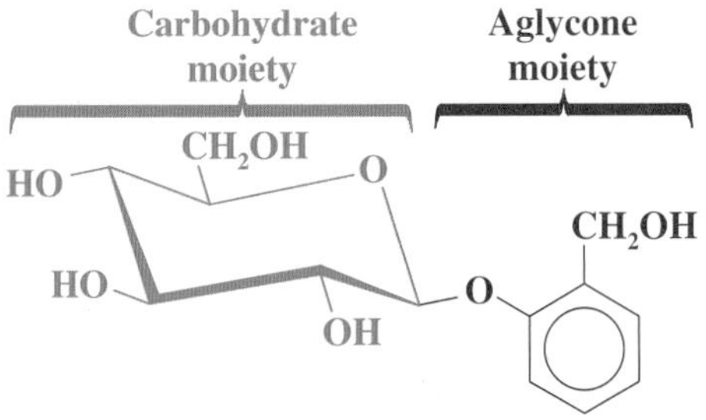

Carbohydrate moiety

Aglycone moiety

Salicin

As early as the time of the ancient Greeks, preparations made from willow bark were used in relieving pain. Eventually, chemists isolated salicin from other plant materials and were able to show that it was responsible for the analgesic effect of the willow bark preparations. Salicin can be converted to salicylic acid, which in turn can be converted into the most widely used modern analgesic, *aspirin* (Section 21.8).

Problem 22.3

(a) What products would be formed if salicin were treated with dilute aqueous HC1?
(b) Outline a mechanism for the reactions involved in their formation.

Problem 22.4

How would you convert D-glucose to a mixture of ethyl α-D-glucopyranoside and ethyl β-D-glucopyranoside? Show all steps in the mechanism for their formation.

PROBLEM 22.5 In neutral or basic aqueous solutions, glycosides do not show mutarotation. However, if the solutions are made acidic, glycosides do show mutarotation. Explain why this occurs.

22.5 Other Reactions of Monosaccharides

22.5A Enolization, Tautomerization, and Isomerization

Dissolving monosaccharides in aqueous base causes them to undergo enolizations and a series of keto–enol tautomerizations that lead to isomerizations. For example, if a solution of D-glucose containing calcium hydroxide is allowed to stand for several days, several products can be isolated, including D-fructose and D-mannose (Fig. 22.5). This type of reaction is called the **Lobry de Bruyn–Alberda van Ekenstein transformation** after the two Dutch chemists who discovered it in 1895.

When carrying out reactions with monosaccharides, it is usually important to prevent these isomerizations and thereby to preserve the stereochemistry at all of the stereogenic carbons. One way to do this is to convert the monosaccharide to the methyl glycoside first. We can then safely carry out reactions in basic media because the aldehyde group has been converted to an acetal and acetals are stable in aqueous base. Preparation of the methyl glycoside serves to "protect" the monosaccharide from undesired reactions that could occur with the anomeric carbon in its hemiacetal form.

22.5B Use of Protecting Groups in Carbohydrate Synthesis

Protecting groups are functional groups introduced selectively to block the reactivity of certain sites in a molecule while desired transformations are carried on elsewhere. After the desired transformations are accomplished, the protecting groups are removed. Laboratory reactions involving carbohydrates often require the use of protecting groups

D-Glucose (open-chain form) $\underset{H_2O}{\overset{OH^-}{\rightleftharpoons}}$ Enolate ion $\underset{OH^-}{\overset{H_2O}{\rightleftharpoons}}$ D-Mannose

Enolate ion $\underset{}{\overset{H_2O \mid OH^-}{\rightleftharpoons}}$ Enediol

D-Fructose $\underset{}{\overset{\text{tautomerization}}{\rightleftharpoons}}$ Enediol

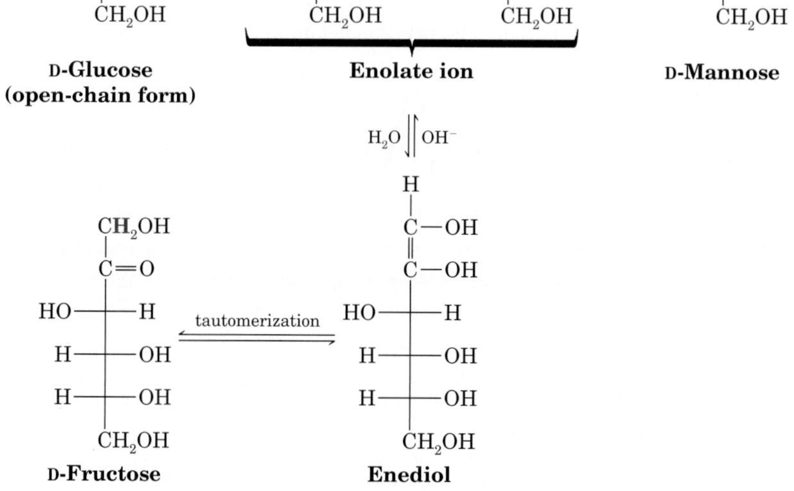

FIGURE 22.5 Monosaccharides undergo isomerizations via enolate ions and enediols when placed in aqueous base. Here we show how D-glucose isomerizes to D-mannose and to D-fructose.

due to the multiple sites of reactivity present in carbohydrates. As we have just seen, formation of a glycoside (an acetal) can be used to prevent undesired reactions that would involve the anomeric carbon in its hemiacetal form. Common protecting groups for the alcohol functional groups in carbohydrates include ethers, esters, and acetals.

22.5C Formation of Ethers

Hydroxyl groups of sugars can be converted to ethers using a base and an alkyl halide. The reaction is nothing more than an example of a multiple Williamson ether synthesis (Section 11.11B). Benzyl ethers are commonly used to protect hydroxyl groups in sugars. Benzyl halides are easily introduced because they are highly reactive in S_N2 reactions. Sodium or potassium hydride is typically used as the base in an aprotic solvent such as DMF or DMSO. The benzyl groups can later be easily removed by hydrogenolysis using a palladium catalyst.

Benzyl Ether Formation

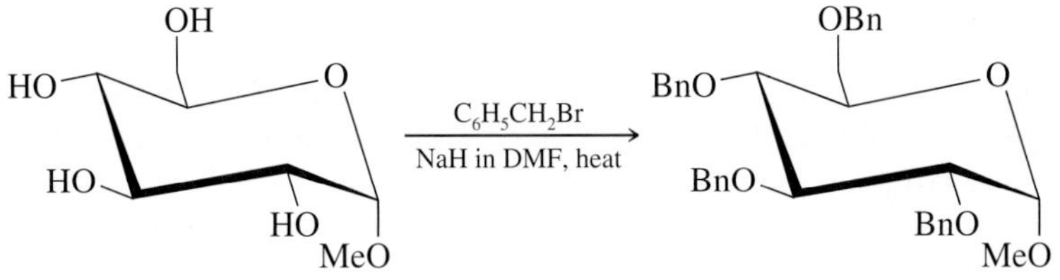

$C_6H_5CH_2Br$, NaH in DMF, heat

Bn = $C_6H_5CH_2$

Benzyl Ether Cleavage

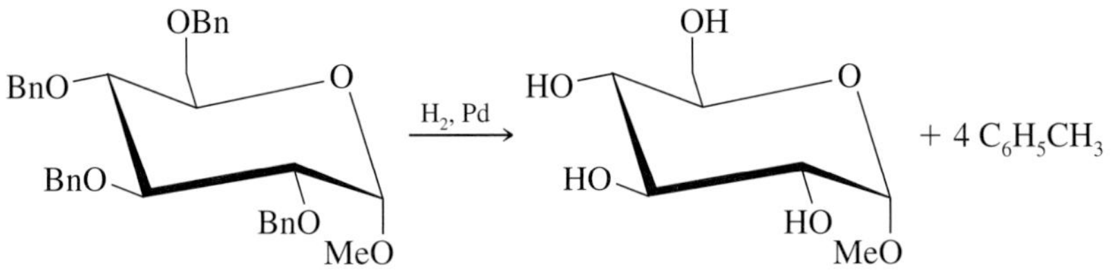

H_2, Pd

+ 4 $C_6H_5CH_3$

Methyl ethers can also be prepared. The pentamethyl derivative of methyl glucoside, for example, can be synthesized by treating it with excess dimethyl sulfate in aqueous sodium hydroxide. Sodium hydroxide is a competent base in this case because the hydroxyl groups of monosaccharides are more acidic than those of ordinary alcohols due to the many electronegative atoms in the sugar, all of which exert electron-withdrawing inductive effects on nearby hydroxyl groups. In aqueous NaOH the hydroxyl groups are all converted to alkoxide ions, and each of these, in turn, reacts with dimethyl sulfate in an S_N2 reaction to yield a methyl ether. The process is called *exhaustive methylation:*

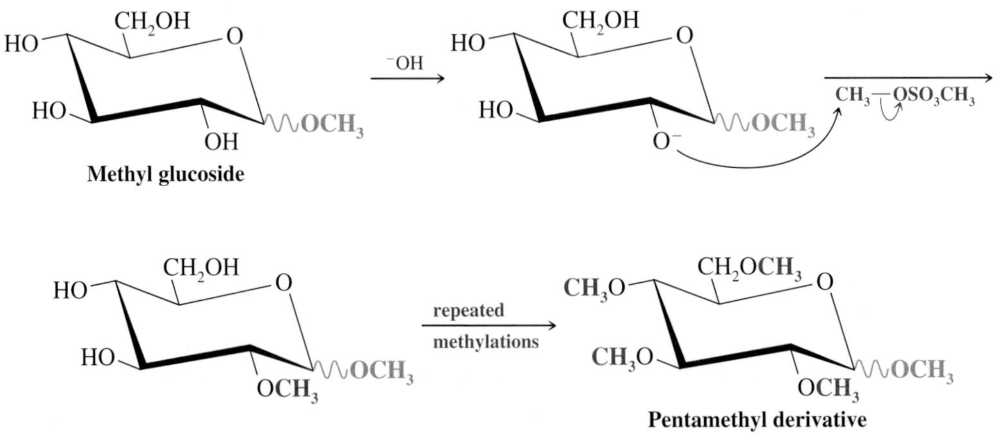

Methyl glucoside $\xrightarrow{^-OH}$ $\xrightarrow{CH_3-OSO_3CH_3}$

$\xrightarrow[\text{methylations}]{\text{repeated}}$ **Pentamethyl derivative**

Although not often used as protecting groups for alcohols in carbohydrates, methyl ethers have been useful in the structure elucidation of sugars. For example, evidence for the pyranose form of glucose can be obtained by exhaustive methylation followed by aqueous hydrolysis of the acetal linkage. Because the C2, C3, C4, and C6 methoxy groups of the pentamethyl derivative are ethers, they are not affected by aqueous hydrolysis. (To cleave them requires heating with concentrated HBr or HI, Section 11.12.) The methoxyl group at C1, however, is part of an acetal linkage, and so it is labile under the conditions of aqueous hydrolysis. Hydrolysis of the pentamethyl derivative of glucose gives evidence that the C5 oxygen was the one involved in the cyclic hemiacetal form because in the open-chain form of the product (which is in equilibrium with the cyclic hemiacetal) it is the C5 oxygen that is not methylated:

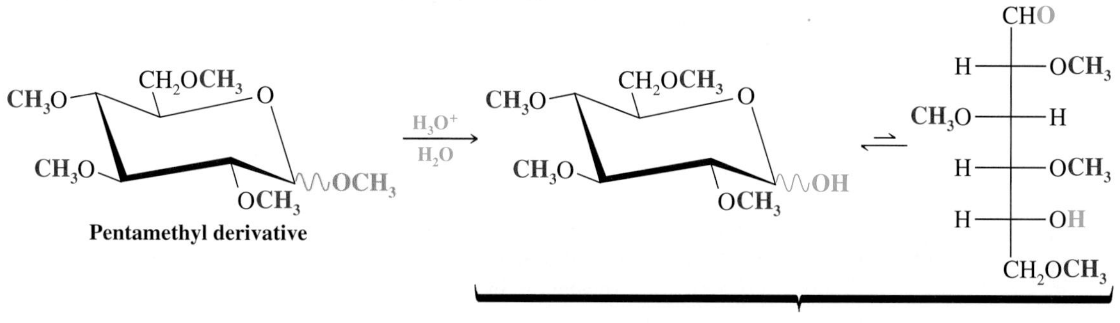

CH_2OCH_3 CH_3O CH_3O O OCH_3 OCH_3 $\xrightarrow[H_2O]{H_3O^+}$ CH_2OCH_3 CH_3O CH_3O O OCH_3 OH $\rightleftharpoons$ CHO, H—OCH_3, CH_3O—H, H—OCH_3, H—OH, CH_2OCH_3

Pentamethyl derivative

2,3,4,6-Tetra-*O*-methyl-D-glucose

Sily ethers, including *tert*-butyldimethylsilyl (TBDMS) ethers (Section 11.11D) and phenyl-substituted ethers, are also used as protecting groups in carbohydrate synthesis. *tert*-Butyldiphenylsilyl (TBDPS) ethers show excellent regioselectivity for primary hydroxyl groups in sugars, such as at C6 in a hexopyranose. (We shall see the use of some related silyl ether groups in Section 22.14.)

Regioselective TBDPS Ether Formation

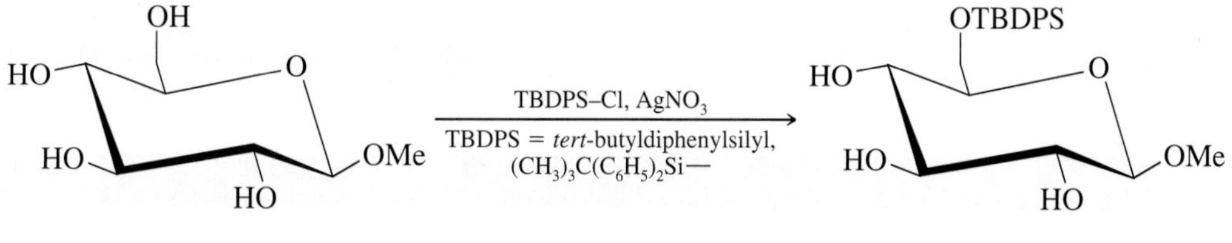

OH HO HO O OMe HO $\xrightarrow{\text{TBDPS–Cl, } AgNO_3}$ TBDPS = *tert*-butyldiphenylsilyl, $(CH_3)_3C(C_6H_5)_2Si$— OTBDPS HO HO O OMe HO

TBDPS Ether Cleavage

OTBDPS HO HO O OMe HO $\xrightarrow[\text{THF}]{Bu_4N^+F^-}$ OH HO HO O OMe HO

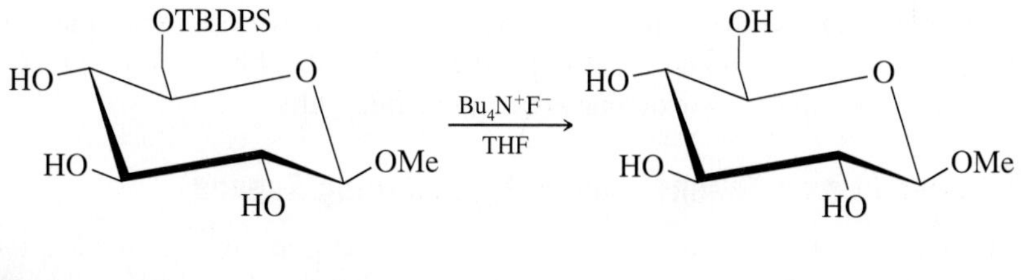

22.5D Conversion to Esters

Treating a monosaccharide with excess acetic anhydride and a weak base (such as pyridine or sodium acetate) converts all of the hydroxyl groups, including the anomeric hydroxyl, to ester groups. If the reaction is carried out at a low temperature (e.g., 0°C), the reaction occurs stereospecifically; the α anomer gives the α-acetate and the β anomer gives the β-acetate. Acetate esters are common protecting groups for carbohydrate hydroxyls.

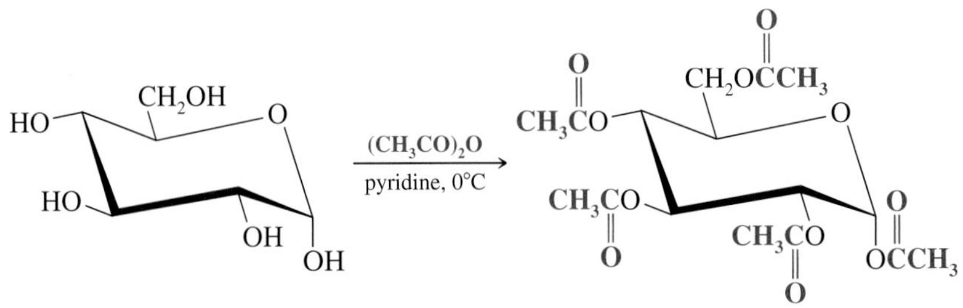

22.5E Conversion to Cyclic Acetals

In Section 16.7C we learned that aldehydes and ketones react with open-chain 1,2-diols to produce **cyclic acetals:**

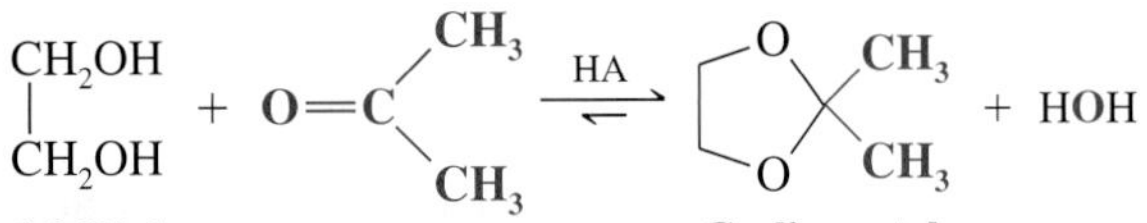

1,2-Diol

Cyclic acetal

If the 1,2-diol is attached to a ring, as in a monosaccharide, **formation of the cyclic acetals occurs only when the vicinal hydroxyl groups are cis to each other.** For example, α-D-galactopyranose reacts with acetone in the following way:

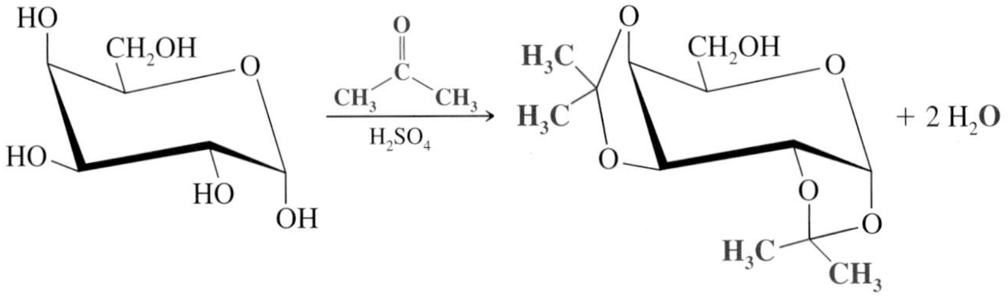

Cyclic acetals are commonly used to protect vicinal cis hydroxyl groups of a sugar while reactions are carried out on other parts of the molecule. When acetals such as these are formed from acetone, they are called **acetonides.**

22.6 Oxidation Reactions of Monosaccharides

A number of oxidizing agents are used to identify functional groups of carbohydrates, in elucidating their structures, and for syntheses. The most important are (1) Benedict's or Tollens' reagents, (2) bromine water, (3) nitric acid, and (4) periodic acid. Each of these reagents produces a different and usually specific effect when it is allowed to react with a monosaccharide. We shall now examine what these effects are.

22.6A Benedict's or Tollens' Reagents: Reducing Sugars

Benedict's reagent (an alkaline solution containing a cupric citrate complex ion) and Tollens' solution [$\overset{+}{Ag}(NH_3)_2\overset{-}{O}H$] oxidize and thus give positive tests with *aldoses and ketoses.* The tests are positive even though aldoses and ketoses exist primarily as cyclic hemiacetals.

We studied the use of Tollens' silver mirror test in Section 16.13. Benedict's solution and the related Fehling's solution (which contains a cupric tartrate complex ion) give brick-red precipitates of Cu_2O when they oxidize an aldose. [In alkaline solution ketoses are converted to aldoses (Section 22.5A), which are then oxidized by the cupric complexes.] Since the solutions of cupric tartrates and citrates are blue, the appearance of a brick-red precipitate is a vivid and unmistakable indication of a positive test.

$$\text{Cu}^{2+}\text{(complex)} + \underset{\text{Aldose}}{\text{HC(=O)}-(\text{CHOH})_n-\text{CH}_2\text{OH}} \quad \text{or} \quad \underset{\text{Ketose}}{\text{CH}_2\text{OH}-\text{C=O}-(\text{CHOH})_n-\text{CH}_2\text{OH}} \longrightarrow \text{Cu}_2\text{O}\downarrow + \text{oxidation products}$$

Benedict's solution (blue) — **Aldose** — **Ketose** — **(brick-red reduction product)**

Sugars that give positive tests with Tollens' or Benedict's solutions are known as **reducing sugars,** and all carbohydrates that contain a *hemiacetal group* give positive tests. In aqueous solution these hemiacetals exist in equilibrium with relatively small, but not insignificant, concentrations of noncyclic aldehydes or α-hydroxy ketones. It is the latter two that undergo the oxidation, perturbing the equilibrium to produce more aldehyde or α-hydroxy ketone, which then undergoes oxidation until one reactant is exhausted.

Carbohydrates that contain only acetal groups do not give positive tests with Benedict's or Tollens' solutions, and they are called *nonreducing sugars.* Acetals do not exist in equilibrium with aldehydes or α-hydroxy ketones in the basic aqueous media of the test reagents.

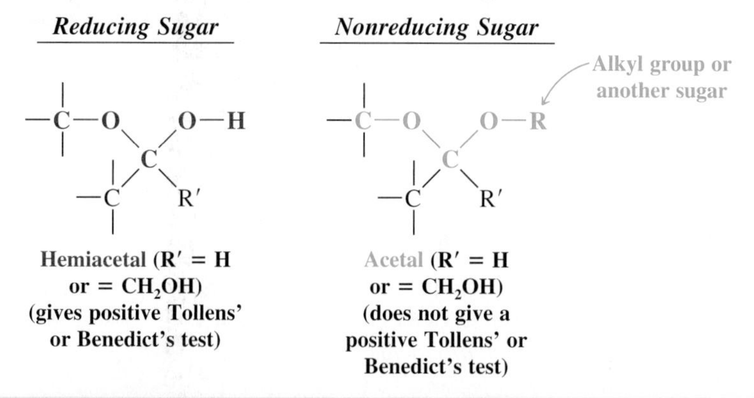

PROBLEM 22.6

How might you distinguish between α-D-glucopyranose (i.e., D-glucose) and methyl α-D-glucopyranoside?

Although Benedict's and Tollens' reagents have some use as diagnostic tools [Benedict's solution can be used in quantitative determinations of reducing sugars (reported as glucose) in blood or urine], neither of these reagents is useful as a preparative reagent in carbohydrate oxidations. Oxidations with both reagents take place in alkaline solution, *and in alkaline solutions sugars undergo a complex series of reactions that lead to isomerizations* (Section 22.5A).

22.6B Bromine Water: The Synthesis of Aldonic Acids

Monosaccharides do not undergo isomerization and fragmentation reactions in mildly acidic solution. Thus, a useful oxidizing reagent for preparative purposes is bromine in water (pH 6.0). Bromine water is a general reagent that selectively oxidizes the —CHO group to a —CO_2H group. It converts an aldose to an *aldonic acid:*

$$\underset{\textbf{Aldose}}{\text{CHO}-(\text{CHOH})_n-\text{CH}_2\text{OH}} \xrightarrow[\text{H}_2\text{O}]{\text{Br}_2} \underset{\textbf{Aldonic acid}}{\text{CO}_2\text{H}-(\text{CHOH})_n-\text{CH}_2\text{OH}}$$

Experiments with aldopyranoses have shown that the actual course of the reaction is somewhat more complex than we have indicated above. Bromine water specifically oxidizes the β anomer, and the initial product that forms is a *δ-aldonolactone.* This compound may then hydrolyze to an aldonic acid, and the aldonic acid may undergo a subsequent ring closure to form a *γ-aldonolactone:*

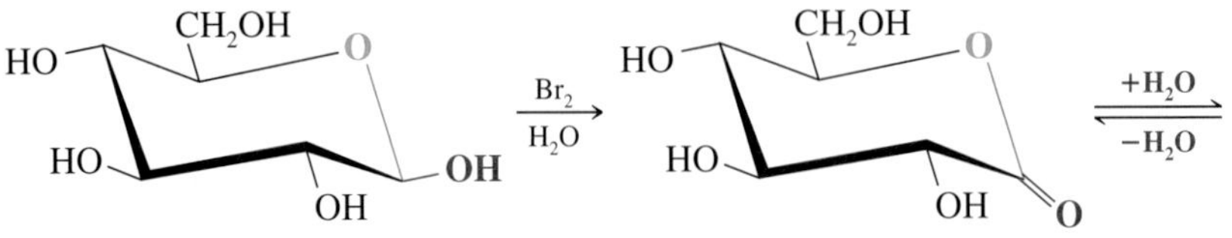

β-D-Glucopyranose $\xrightarrow[H_2O]{Br_2}$ **D-Glucono-δ-lactone** $\underset{-H_2O}{\overset{+H_2O}{\rightleftharpoons}}$ **D-Gluconic acid** $\underset{+H_2O}{\overset{-H_2O}{\rightleftharpoons}}$ **D-Gluconic-γ-lactone**

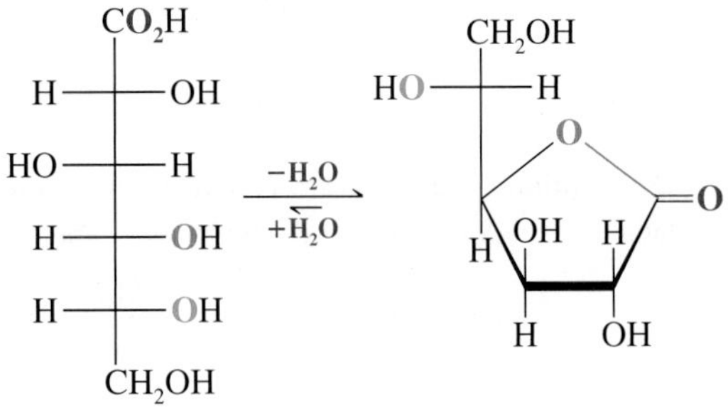

22.6C Nitric Acid Oxidation: Aldaric Acids

Dilute nitric acid—a stronger oxidizing agent than bromine water—oxidizes both the —CHO group and the terminal —CH_2OH group of an aldose to —CO_2H groups. These dicarboxylic acids are known as *aldaric acids:*

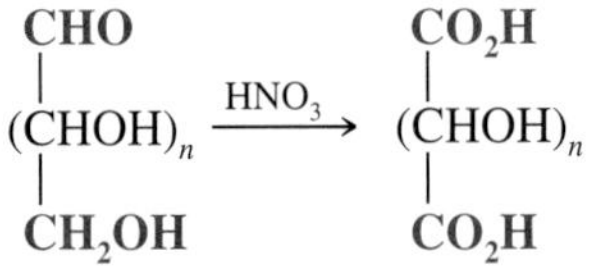

$$\underset{\textbf{Aldose}}{CHO-(CHOH)_n-CH_2OH} \xrightarrow{HNO_3} \underset{\textbf{Aldaric acid}}{CO_2H-(CHOH)_n-CO_2H}$$

It is not known whether a lactone is an intermediate in the oxidation of an aldose to an aldaric acid; however, aldaric acids form γ- and δ-lactones readily:

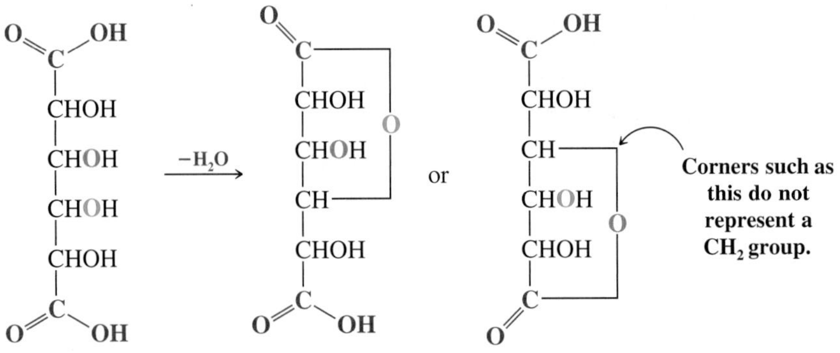

Aldaric acid (from an aldohexose) $\xrightarrow{-H_2O}$ **γ-Lactones of an aldaric acid**

Corners such as this do not represent a CH_2 group.

The aldaric acid obtained from D-glucose is called D-glucaric acid*:

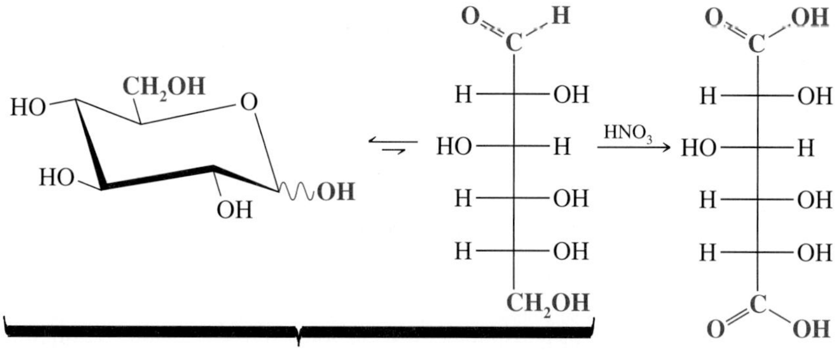

$$\text{D-Glucose} \rightleftharpoons \text{CHO–(H–C–OH)–(HO–C–H)–(H–C–OH)–(H–C–OH)–CH}_2\text{OH} \xrightarrow{HNO_3} \text{COOH–(H–C–OH)–(HO–C–H)–(H–C–OH)–(H–C–OH)–COOH}$$

D-Glucose **D-Glucaric acid**

PROBLEM 22.7

(a) Would you expect D-glucaric acid to be optically active?
(b) Write the open-chain structure for the aldaric acid (mannaric acid) that would be obtained by nitric acid oxidation of D-mannose.
(c) Would you expect mannaric acid to be optically active?
(d) What aldaric acid would you expect to obtain from D-erythrose?

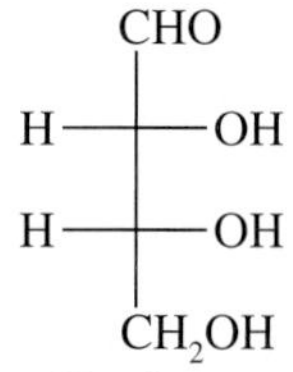

CHO
H—C—OH
H—C—OH
CH_2OH
D-Erythrose

(e) Would the aldaric acid in (d) show optical activity?
(f) D-Threose, a diastereomer of D-erythrose, yields an optically active aldaric acid when it is subjected to nitric acid oxidation. Write Fischer projection formulas for D-threose and its nitric acid oxidation product.
(g) What are the names of the aldaric acids obtained from D-erythrose and D-threose? (See Section 5.13A.)

PROBLEM 22.8

D-Glucaric acid undergoes lactonization to yield two different γ-lactones. What are their structures?

22.6D Periodate Oxidations: Oxidative Cleavage of Polyhydroxy Compounds

Compounds that have hydroxyl groups on adjacent atoms undergo oxidative cleavage when they are treated with aqueous periodic acid (HIO_4). The reaction breaks carbon–carbon bonds and produces carbonyl compounds (aldehydes, ketones, or acids). The stoichiometry of the reaction is

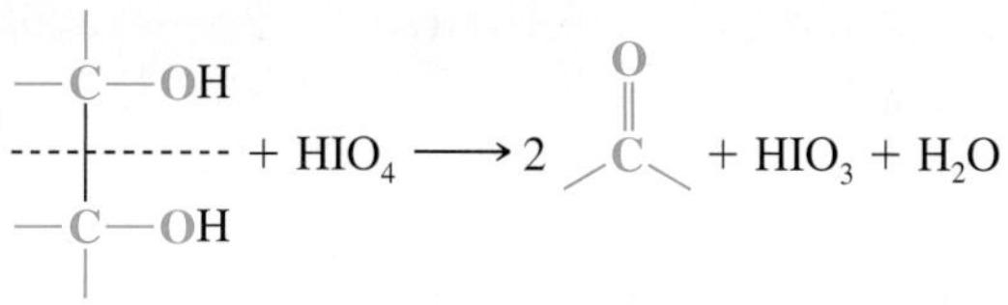

$$-\overset{|}{C}-OH \;/\; -\overset{|}{C}-OH + HIO_4 \longrightarrow 2\ \ \overset{O}{\overset{\|}{C}} + HIO_3 + H_2O$$

*Older terms for an aldaric acid are a *glycaric* acid or a *saccharic* acid.

Since the reaction usually takes place in quantitative yield, valuable information can often be gained by measuring the number of molar equivalents of periodic acid that is consumed in the reaction as well as by identifying the carbonyl products.*

Periodate oxidations are thought to take place through a cyclic intermediate:

$$\begin{array}{c}-\overset{|}{C}-OH \\ | \\ -\underset{|}{C}-OH\end{array} + IO_4^- \xrightarrow{(-H_2O)} \text{[cyclic periodate ester]} \longrightarrow \begin{array}{c}\backslash \\ C{=}O \\ / \\ \\ \backslash \\ C{=}O \\ /\end{array} + IO_3^-$$

Before we discuss the use of periodic acid in carbohydrate chemistry, we should illustrate the course of the reaction with several simple examples. Notice in these periodate oxidations that *for every C—C bond broken, a C—O bond is formed at each carbon.*

1. When three or more —CHOH groups are contiguous, the internal ones are obtained as *formic acid.* Periodate oxidation of glycerol, for example, gives two molar equivalents of formaldehyde and one molar equivalent of formic acid:

$$\text{Glycerol } (HOCH_2-CHOH-CH_2OH) + 2\,IO_4^- \longrightarrow HCHO \text{ (formaldehyde)} + HCOOH \text{ (formic acid)} + HCHO \text{ (formaldehyde)}$$

2. Oxidative cleavage also takes place when an —OH group is adjacent to the carbonyl group of an aldehyde or ketone (but not that of an acid or an ester). Glyceraldehyde yields two molar equivalents of formic acid and one molar equivalent of formaldehyde, while dihydroxyacetone gives two molar equivalents of formaldehyde and one molar equivalent of carbon dioxide:

$$\text{Glyceraldehyde } (OHC-CHOH-CH_2OH) + 2\,IO_4^- \longrightarrow HCOOH \text{ (formic acid)} + HCOOH \text{ (formic acid)} + HCHO \text{ (formaldehyde)}$$

*The reagent lead tetraacetate, $Pb(O_2CCH_3)_4$, brings about cleavage reactions similar to those of periodic acid. The two reagents are complementary; periodic acid works well in aqueous solutions and lead tetraacetate gives good results in organic solvents.

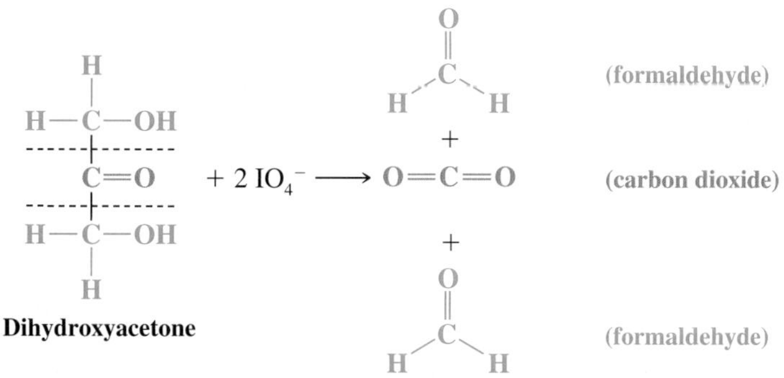

$$\begin{array}{c} H \\ | \\ H-C-OH \\ \hline C=O \\ \hline H-C-OH \\ | \\ H \end{array} + 2\,IO_4^- \longrightarrow \begin{array}{c} HCHO \text{ (formaldehyde)} \\ + \\ O=C=O \text{ (carbon dioxide)} \\ + \\ HCHO \text{ (formaldehyde)} \end{array}$$

Dihydroxyacetone

3. Periodic acid does not cleave compounds in which the hydroxyl groups are separated by an intervening $—CH_2—$ group, nor those in which a hydroxyl group is adjacent to an ether or acetal function:

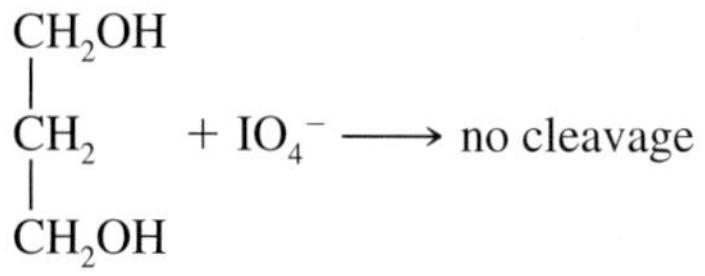

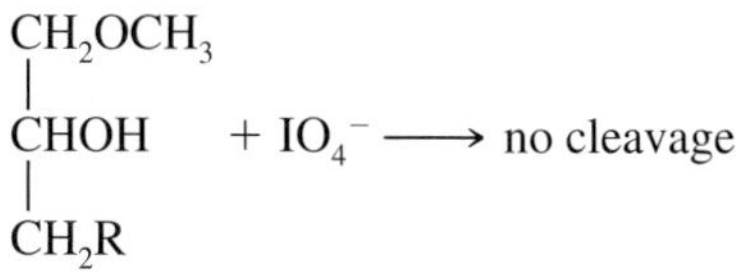

$$\begin{array}{c} CH_2OH \\ | \\ CH_2 \\ | \\ CH_2OH \end{array} + IO_4^- \longrightarrow \text{no cleavage} \qquad \begin{array}{c} CH_2OCH_3 \\ | \\ CHOH \\ | \\ CH_2R \end{array} + IO_4^- \longrightarrow \text{no cleavage}$$

PROBLEM 22.9

What products would you expect to be formed when each of the following compounds is treated with an appropriate amount of periodic acid? How many molar equivalents of HIO_4 would be consumed in each case?

(a) 2,3-Butanediol
(b) 1,2,3-Butanetriol
(c) $CH_2OHCHOHCH(OCH_3)_2$
(d) $CH_2OHCHOHCOCH_3$
(e) $CH_3COCHOHCOCH_3$
(f) *cis*-1,2-Cyclopentanediol
(g) 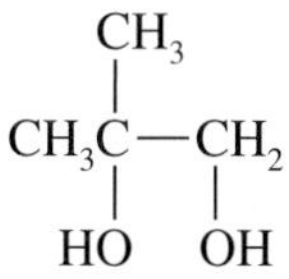 $CH_3C(CH_3)(OH)—CH_2OH$

$$\begin{array}{c} \quad CH_3 \\ \quad | \\ CH_3C—CH_2 \\ \;\;|\quad\;\; | \\ HO \quad OH \end{array}$$

(h) D-Erythrose

PROBLEM 22.10

Show how periodic acid could be used to distinguish between an aldohexose and a ketohexose. What products would you obtain from each, and how many molar equivalents of HIO_4 would be consumed?

22.7 Reduction of Monosaccharides: Alditols

Aldoses (and ketoses) can be reduced with sodium borohydride to compounds called *alditols:*

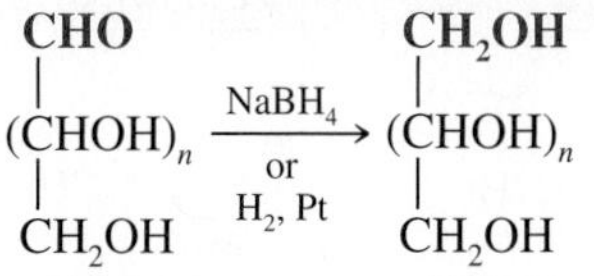

$$\begin{array}{c} CHO \\ | \\ (CHOH)_n \\ | \\ CH_2OH \\ \textbf{Aldose} \end{array} \xrightarrow[\text{or } H_2,\, Pt]{NaBH_4} \begin{array}{c} CH_2OH \\ | \\ (CHOH)_n \\ | \\ CH_2OH \\ \textbf{Alditol} \end{array}$$

Reduction of D-glucose, for example, yields D-glucitol:

$$\text{D-Glucose (pyranose)} \rightleftharpoons \begin{array}{c} \text{CHO} \\ \text{H—C—OH} \\ \text{HO—C—H} \\ \text{H—C—OH} \\ \text{H—C—OH} \\ \text{CH}_2\text{OH} \end{array} \xrightarrow{\text{NaBH}_4} \begin{array}{c} \text{CH}_2\text{OH} \\ \text{H—C—OH} \\ \text{HO—C—H} \\ \text{H—C—OH} \\ \text{H—C—OH} \\ \text{CH}_2\text{OH} \end{array}$$

D-Glucitol
(or D-sorbitol)

PROBLEM 22.11 **(a)** Would you expect D-glucitol to be optically active? **(b)** Write Fischer projection formulas for all of the D-aldohexoses that would yield *optically inactive alditols.*

22.8 Reactions of Monosaccharides with Phenylhydrazine: Osazones

The aldehyde group of an aldose reacts with such carbonyl reagents as hydroxylamine and phenylhydrazine (Section 16.8). With hydroxylamine, the product is the expected oxime. With enough phenylhydrazine, however, three molar equivalents of phenylhydrazine are consumed and a second phenylhydrazone group is introduced at C2. The product is called a *phenylosazone.* Phenylosazones crystallize readily (unlike sugars) and are useful derivatives for identifying sugars.

$$\underset{\textbf{Aldose}}{\begin{array}{c} \text{HC=O} \\ \text{CHOH} \\ (\text{CHOH})_n \\ \text{CH}_2\text{OH} \end{array}} + 3\ \text{C}_6\text{H}_5\text{NHNH}_2 \longrightarrow \underset{\textbf{Phenylosazone}}{\begin{array}{c} \text{H} \\ \text{C=NNHC}_6\text{H}_5 \\ \text{C=NNHC}_6\text{H}_5 \\ (\text{CHOH})_n \\ \text{CH}_2\text{OH} \end{array}} + \text{C}_6\text{H}_5\text{NH}_2 + \text{NH}_3 + \text{H}_2\text{O}$$

The mechanism for osazone formation probably depends on a series of reactions in which $\rangle$C=N— behaves very much like $\rangle$C=O in giving a nitrogen version of an enol.

A Mechanism for the Reaction

Phenylosazone Formation

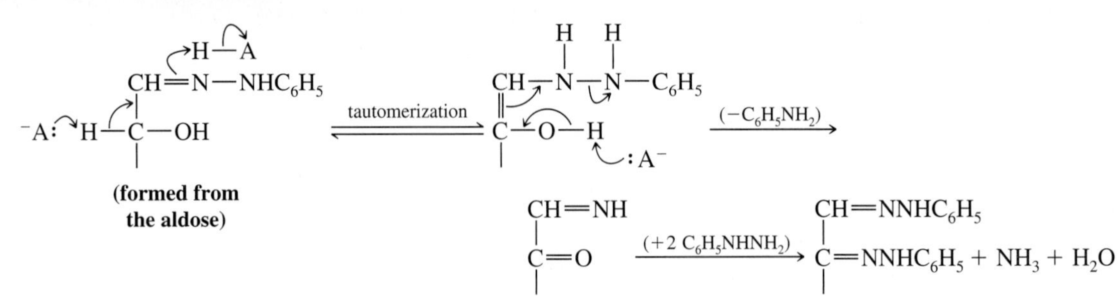

Osazone formation results in a loss of the stereogenic center at C2 but does not affect other stereogenic carbons; D-glucose and D-mannose, for example, yield the same phenylosazone:

D-Glucose (CHO; H—OH; HO—H; H—OH; H—OH; CH_2OH) $\xrightarrow{C_6H_5NHNH_2}$ Same phenylosazone ($CH{=}NNHC_6H_5$; $C{=}NNHC_6H_5$; HO—H; H—OH; H—OH; CH_2OH) $\xleftarrow{C_6H_5NHNH_2}$ D-Mannose (CHO; HO—H; HO—H; H—OH; H—OH; CH_2OH)

D-Glucose **Same phenylosazone** **D-Mannose**

This experiment, first done by Emil Fischer, established that D-glucose and D-mannose have the same configurations about C3, C4, and C5. Diastereomeric aldoses that differ in configuration at only one carbon (such as D-glucose and D-mannose) are called epimers. In general, any pair of diastereomers that differ in configuration at only a single tetrahedral stereogenic carbon can be called **epimers.**

PROBLEM 22.12

Although D-fructose is not an epimer of D-glucose or D-mannose (D-fructose is a ketohexose), all three yield the same phenylosazone. **(a)** Using Fischer projection formulas, write an equation for the reaction of fructose with phenylhydrazine. **(b)** What information about the stereochemistry of D-fructose does this experiment yield?

22.9 Synthesis and Degradation of Monosaccharides

22.9A Kiliani–Fischer Synthesis

In 1885, Heinrich Kiliani (Freiburg, Germany) discovered that an aldose can be converted to the epimeric aldonic acids having one additional carbon through the addition of hydrogen cyanide and subsequent hydrolysis of the epimeric cyanohydrins. Fischer later extended this method by showing that aldonolactones obtained from the aldonic acids can be reduced to aldoses. Today, this method for lengthening the carbon chain of an aldose is called the Kiliani–Fischer synthesis.

We can illustrate the Kiliani–Fischer synthesis with the synthesis of D-threose and D-erythrose (aldotetroses) from D-glyceraldehyde (an aldotriose) in Fig. 22.6.

Addition of hydrogen cyanide to glyceraldehyde produces two epimeric cyanohydrins because the reaction creates a new stereogenic center. The cyanohydrins can be separated easily (since they are diastereomers), and each can be converted to an aldose through hydrolysis, acidification, lactonization, and reduction with Na–Hg at pH 3–5. One cyanohydrin ultimately yields D-(−)-erythrose and the other yields D-(−)-threose.

We can be sure that the aldotetroses that we obtain from this Kiliani–Fischer synthesis are both D sugars because the starting compound is D-glyceraldehyde and its stereogenic carbon is unaffected by the synthesis. On the basis of the Kiliani–Fischer synthesis, we cannot know just which aldotetrose has both —OH groups on the right and which has the top —OH on the left in the Fischer projection. However, if we oxidize both aldotetroses to aldaric acids, one [D-(−)-erythrose] will yield an *optically inactive* (meso) product while the other [D-(−)-threose] will yield a product that is *optically active* (see Problem 22.7).

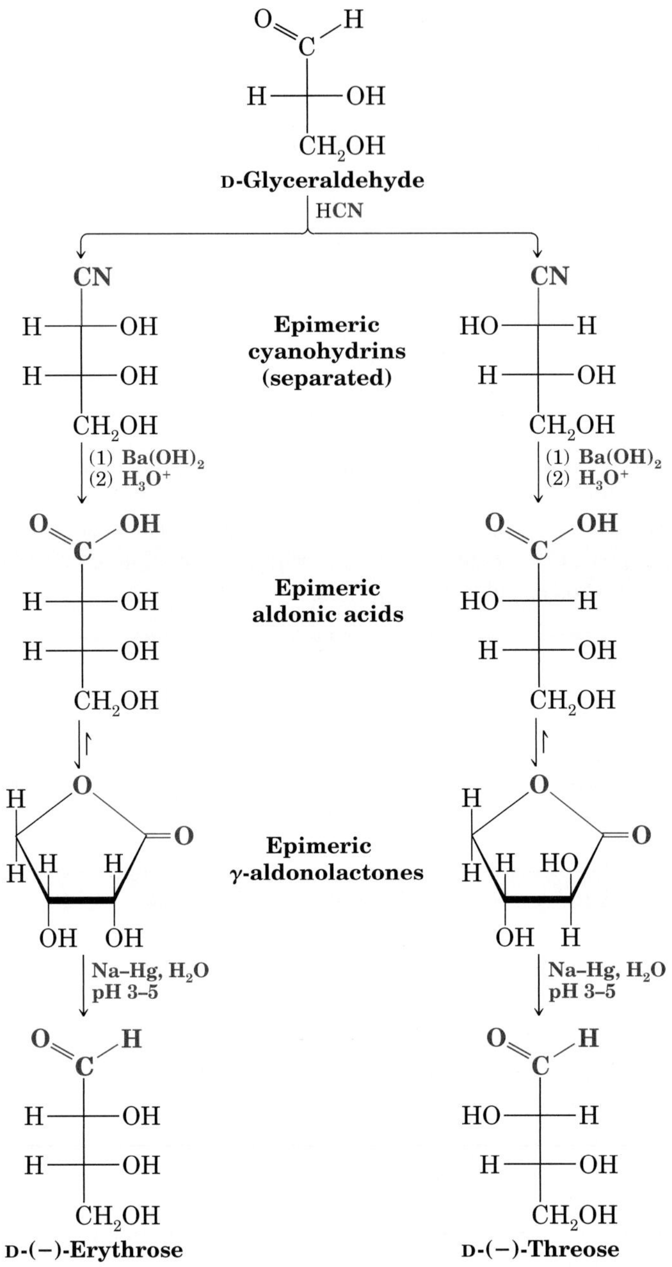

FIGURE 22.6 A Kiliani–Fischer synthesis of D-(−)-erythrose and D-(−)-threose from D-glyceraldehyde.

D-Glyceraldehyde

HCN

Epimeric cyanohydrins (separated)

(1) $Ba(OH)_2$
(2) H_3O^+

Epimeric aldonic acids

Epimeric γ-aldonolactones

Na–Hg, H_2O
pH 3–5

D-(−)-Erythrose

D-(−)-Threose

PROBLEM 22.13 (a) What are the structures of L-(+)-threose and L-(+)-erythrose? (b) What aldotriose would you use to prepare them in a Kiliani–Fischer synthesis?

PROBLEM 22.14 (a) Outline a Kiliani–Fischer synthesis of epimeric aldopentoses starting with D-(−)-erythrose (use Fischer projections). (b) The two epimeric aldopentoses that one obtains are D-(−)-arabinose and D-(−)-ribose. Nitric acid oxidation of D-(−)-ribose yields an optically inactive aldaric acid, whereas similar oxidation of D-(−)-arabinose yields an optically active product. On the basis of this information alone, which Fischer projection represents D-(−)-arabinose and which represents D-(−)-ribose?

PROBLEM 22.15

Subjecting D-(−)-threose to a Kiliani–Fischer synthesis yields two other epimeric aldopentoses, D-(+)-xylose and D-(−)-lyxose. D-(+)-Xylose can be oxidized (with nitric acid) to an optically inactive aldaric acid, while similar oxidation of D-(−)-lyxose gives an optically active product. What are the structures of D-(+)-xylose and D-(−)-lyxose?

PROBLEM 22.16

There are eight aldopentoses. In Problems 22.14 and 22.15 you have arrived at the structures of four. What are the names and structures of the four that remain?

22.9B The Ruff Degradation

Just as the Kiliani–Fischer synthesis can be used to lengthen the chain of an aldose by one carbon atom, the Ruff degradation* can be used to shorten the chain by a similar unit. The Ruff degradation involves (1) oxidation of the aldose to an aldonic acid using bromine water and (2) oxidative decarboxylation of the aldonic acid to the next lower aldose using hydrogen peroxide and ferric sulfate. D-(−)-Ribose, for example, can be degraded to D-(−)-erythrose:

O=C–H; H–C–OH; H–C–OH; H–C–OH; CH_2OH → (Br_2, H_2O) → O=C–OH; H–C–OH; H–C–OH; H–C–OH; CH_2OH → (H_2O_2, $Fe_2(SO_4)_3$) → O=C–H; H–C–OH; H–C–OH; CH_2OH + CO_2

D-(−)-Ribose **D-Ribonic acid** **D-(−)-Erythrose**

PROBLEM 22.17

The aldohexose D-(+)-galactose can be obtained by hydrolysis of *lactose,* a disaccharide found in milk. When D-(+)-galactose is treated with nitric acid, it yields an optically inactive aldaric acid. When D-(+)-galactose is subjected to Ruff degradation, it yields D-(−)-lyxose (see Problem 22.15). Using only these data, write the Fischer projection formula for D-(+)-galactose.

22.10 The D Family of Aldoses

The Ruff degradation and the Kiliani–Fischer synthesis allow us to place all of the aldoses into families or "family trees" based on their relation to D- or L-glyceraldehyde. Such a tree is constructed in Fig. 22.7 and includes the structures of the D-aldohexoses, **1–8.**

Most, but not all, of the naturally occurring aldoses belong to the D family, with D-(+)-glucose being by far the most common. D-(+)-Galactose can be obtained from milk sugar (lactose), but L-(−)-galactose occurs in a polysaccharide obtained from the vineyard snail, *Helix pomatia.* L-(+)-Arabinose is found widely, but D-(−)-arabinose is scarce, being found only in certain bacteria and sponges. Threose, lyxose, gulose, and allose do not occur naturally, but one or both forms (D or L) of each have been synthesized.

*Developed by Otto Ruff, 1871–1939, a German chemist.

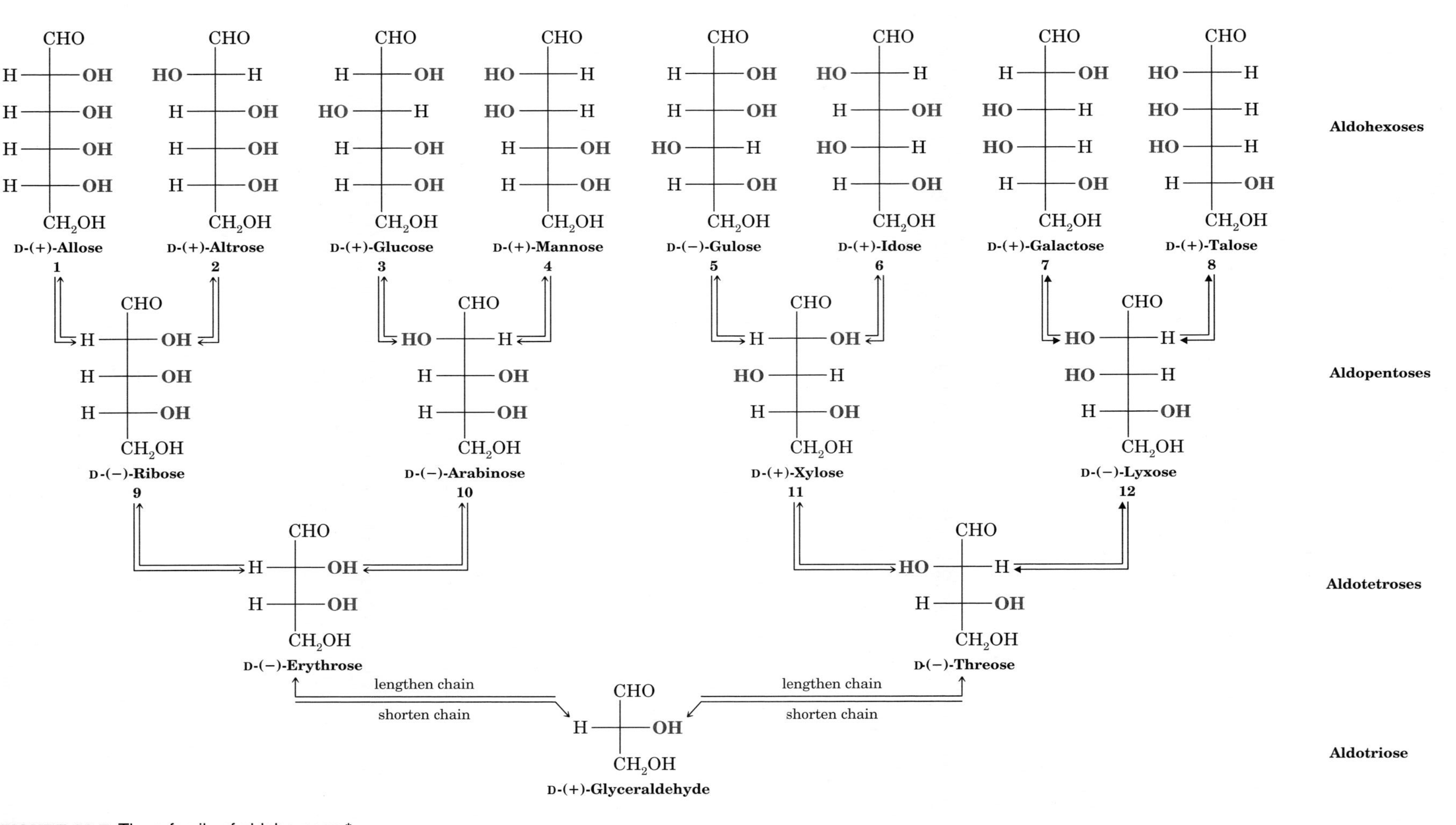

FIGURE 22.7 The D family of aldohexoses.*

*A useful mnemonic for the D-aldohexoses: All altruists gladly make gum in gallon tanks. Write the names in a line and above each write CH_2OH. Then, for C5 write OH to the right all the way across. For C4 write OH to the right four times, then four to the left; for C3, write OH twice to the right, twice to the left, and repeat; for C2, alternate OH and H to the right. (From Fieser, L. F.; Fieser, M. *Organic Chemistry;* Reinhold: New York, 1956; p 359.)

22.11 Fischer's Proof of the Configuration of D-(+)-Glucose

Emil Fischer began his work on the stereochemistry of (+)-glucose in 1888, only 12 years after van't Hoff and Le Bel had made their proposal concerning the tetrahedral structure of carbon. Only a small body of data was available to Fischer at the beginning: Only a few monosaccharides were known, including (+)-glucose, (+)-arabinose, and (+)-mannose. [(+)-Mannose had just been synthesized by Fischer.] The sugars (+)-glucose and (+)-mannose were known to be aldohexoses; (+)-arabinose was known to be an aldopentose.

Since an aldohexose has four stereogenic carbons, 2^4 (or 16) stereoisomers are possible—*one of which is (+)-glucose.* Fischer arbitrarily decided to limit his attention to the eight structures with the D configuration given in Fig. 22.7 (structures **1–8**). Fischer realized that he would be unable to differentiate between enantiomeric configurations because methods for determining the absolute configuration of organic compounds had not been developed. It was not until 1951, when Bijvoet (Section 5.15A) determined the absolute configuration of L-(+)-tartaric acid [and, hence, D-(+)-glyceraldehyde] that Fischer's arbitrary assignment of (+)-glucose to the family we call the D family was known to be correct.

Fischer's assignment of structure **3** to (+)-glucose was based on the following reasoning:

1. Nitric acid oxidation of (+)-glucose gives an optically active aldaric acid. This eliminates structures **1** and **7** from consideration because both compounds would yield *meso*-aldaric acids.

2. *Degradation of (+)-glucose gives (−)-arabinose, and nitric acid oxidation of (−)-arabinose gives an optically active aldaric acid.* This means that (−)-arabinose cannot have configuration **9** or **11** and must have either structure **10** or **12.** It also establishes that (+)-glucose cannot have configuration **2, 5,** or **6.** This leaves structures **3, 4,** and **8** as possibilities for (+)-glucose.

3. Kiliani–Fischer synthesis beginning with (−)-arabinose gives (+)-glucose and (+)-mannose; nitric acid oxidation of (+)-mannose gives an optically active aldaric acid. This, together with the fact that (+)-glucose yields a different but also optically active aldaric acid, establishes **10** as the structure of (−)-arabinose and eliminates **8** as a possible structure for (+)-glucose. Had (−)-arabinose been represented by structure **12,** a Kiliani–Fischer synthesis would have given the two aldohexoses, **7** and **8,** one of which (**7**) would yield an optically inactive aldaric acid on nitric acid oxidation.

4. Two structures now remain, **3** and **4;** one structure represents (+)-glucose and one represents (+)-mannose. Fischer realized that (+)-glucose and (+)-mannose were epimeric (at C2), but a decision as to which compound was represented by which structure was most difficult.

5. Fischer had already developed a method for effectively *interchanging the two end groups* (—CHO and —CH_2OH) *of an aldose chain.* And, with brilliant logic, Fischer realized that if (+)-glucose had structure **4,** an interchange of end groups *would yield the same aldohexose:*

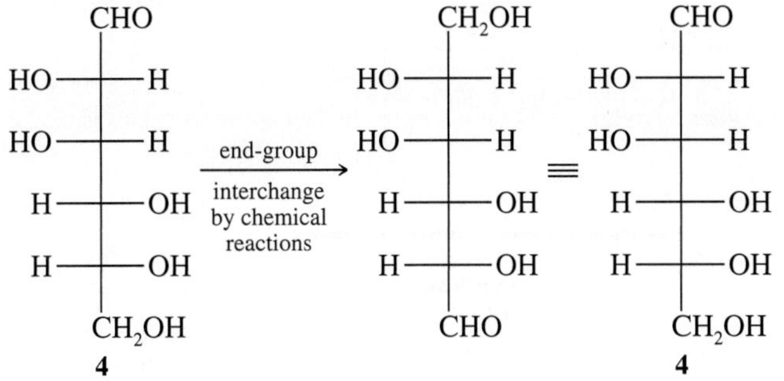

CHO; HO—H; HO—H; H—OH; H—OH; CH_2OH (**4**) —end-group interchange by chemical reactions→ CH_2OH; HO—H; HO—H; H—OH; H—OH; CHO ≡ CHO; HO—H; HO—H; H—OH; H—OH; CH_2OH (**4**)

(Recall that it is permissible to turn a Fischer projection 180° in the plane of the page.)

On the other hand, if (+)-glucose has structure **3,** *an end-group interchange will yield a different aldohexose,* **13:**

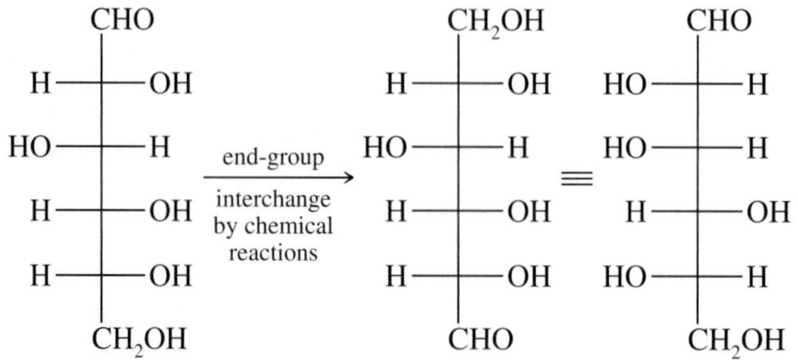

end-group interchange by chemical reactions

3

13
L-Gulose

This new aldohexose, if it were formed, would be an L sugar and it would be the mirror reflection of D-gulose. Thus its name would be L-gulose.

Fischer carried out the end-group interchange starting with (+)-glucose and *the product was the new aldohexose* **13.** This outcome proved that (+)-glucose has structure **3.** It also established **4** as the structure for (+)-mannose, and it proved the structure of L-(+)-gulose as **13.**

The procedure Fischer used for interchanging the ends of the (+)-glucose chain began with one of the γ-lactones of D-glucaric acid (see Problem 22.8) and was carried out as follows:

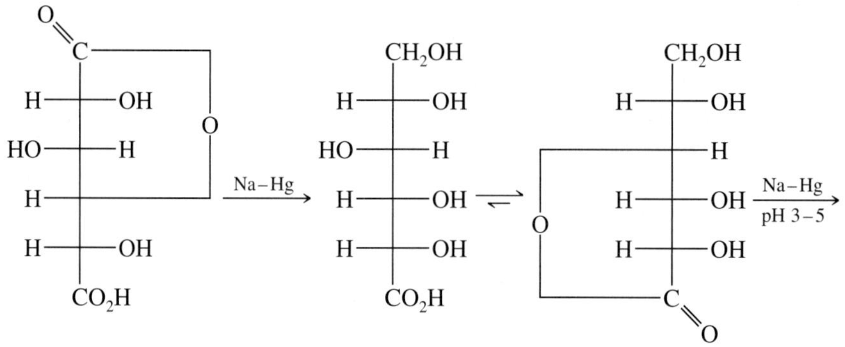

Na–Hg

Na–Hg, pH 3–5

A γ-lactone of D-glucaric acid

L-Gulonic acid

A γ-aldonolactone

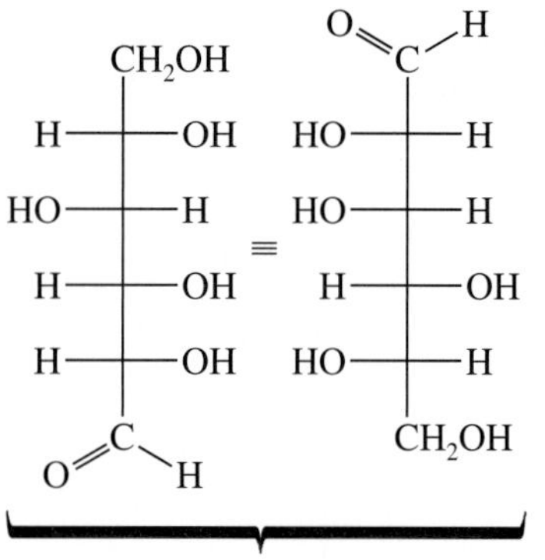

L-(+)-Gulose
13

The Chemistry of...

Stereoselective Synthesis of All the L-Aldohexoses

S. Masamune and K. B. Sharpless masterfully demonstrated the power of stereoselective reactions by designing a highly efficient synthesis of all eight stereoisomers in the family of L-aldohexoses. Using the Sharpless asymmetric epoxidation (SAE) at two points in the synthesis of each stereoisomer and including several other reactions that were themselves either stereoselective or stereospecific, they created a convergent route whose branches provided for selective assembly of all eight stereoisomers. One could say that they achieved a formal synthesis of the D-aldohexose family as well, because simply choosing the opposite stereochemical version of the Sharpless asymmetric epoxidation in the first step leads to the D stereoisomers. A summary of the Masamune–Sharpless synthesis of the L-aldohexoses is given below.

Masamune–Sharpless L-aldohexose synthesis scheme. Adapted from Nicolaou, E. J.; Sorensen, E. J. *Classics in Total Synthesis: Targets, Strategies, Methods;* VCH Publishers: New York, 1996; pp 310 and 312.

A:	B:
1. m-CPBA	1. m-CPBA
2. Ac_2O, NaOAc	2. Ac_2O, NaOAc
3. DIBAL-H	3. NaOMe, MeOH
4. TFA-H_2O	4. TFA-H_2O
5. H_2, Pd-C	5. H_2, Pd-C

The synthesis begins with application of the SAE protocol to the archiral protected *E*-2-buten-1-ol (**1**), forming chiral epoxide **2**. A stereospecific epoxide rearrangement (a Payne rearrangement) and epoxide opening leads from **2** to the chiral diol **3**, which is then protected as an acetonide. Oxidation of the thioether to a sulfoxide and acetylation sets the stage for another rearrangement (called a Pummerer rearrangement) that leads to the α-acetoxy sulfide (**4**). Basic hydrolysis of **4** with epimerization leads to **6**, while reducing conditions applied to **4** avoids epimerization and leads to **5**. Diastereomeric acetonide aldehydes **5** and **6** thus result from the first stereochemical branch point in the synthesis. Now **5** and **6** allow introduction of three carbon atoms to each by a Wittig reaction, stereoselectively producing the *E* alkene diastereomers **7** and **8** (and without epimerization of **5** and **6** as a complication under the basic conditions of the Wittig reaction). The next branch point in the synthesis occurs by application of the SAE in each of its complementary forms to both diastereomeric allylic alcohols **7** and **8** resulting in **9, 10, 11,** and **12.** Payne rearrangement with epoxide opening and protection leads to the set of four bis-acetonides **13, 14, 15,** and **16**. Oxidation and acetylation leads to a repeat of the Pummerer rearrangement with each one. Stereoselective conversion of the four intermediates at the last branch point leads to the final eight stereoisomers, L-aldohexoses **17–24.**

Contributions by Sharpless to the invention of asymmetric synthesis methods earned him half of the 2001 Nobel Prize in Chemistry. (The other half was awarded to W. Knowles and S. Noyori for their involvement in other asymmetric synthesis methods. See Section 7.12.)

Notice in this synthesis (page 1100) that the second reduction with Na–Hg is carried out at pH 3–5. Under these conditions, reduction of the lactone yields an aldehyde and not a primary alcohol.

PROBLEM 22.18 Fischer actually had to subject both γ-lactones of D-glucaric acid (Problem 22.8) to the procedure just outlined. What product does the other γ-lactone yield?

22.12 DISACCHARIDES

22.12A Sucrose

Ordinary table sugar is a disaccharide called *sucrose.* Sucrose, the most widely occurring disaccharide, is found in all photosynthetic plants and is obtained commercially from sugarcane or sugar beets. Sucrose has the structure shown in Fig. 22.8.

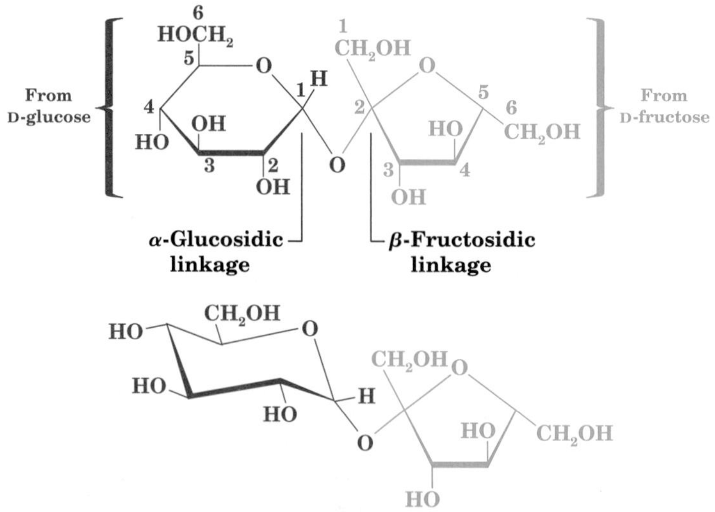

FIGURE 22.8 Two representations of the formula for (+)-sucrose (α-D-glucopyranosyl β-D-fructofuranoside).

The structure of sucrose is based on the following evidence:

1. Sucrosc has thc molccular formula $C_{12}II_{22}O_{11}$.

2. Acid-catalyzed hydrolysis of 1 mol of sucrose yields 1 mol of D-glucose and 1 mol of D-fructose.

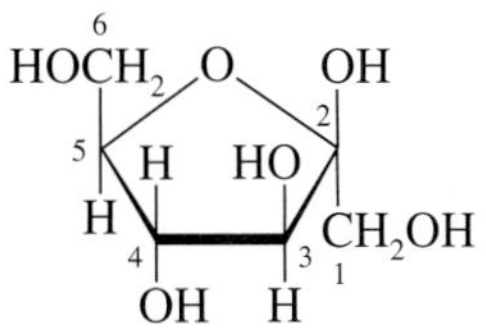

HOCH$_2$ O OH
H HO
H CH$_2$OH
OH H

Fructose
(as a β-furanose)

3. Sucrose is a nonreducing sugar; it gives negative tests with Benedict's and Tollens' solutions. Sucrose does not form an osazone and does not undergo mutarotation. These facts mean that neither the glucose nor the fructose portion of sucrose has a hemiacetal group. Thus, the two hexoses must have a glycosidic linkage that involves C1 of glucose and C2 of fructose, for only in this way will both carbonyl groups be present as full acetals (i.e., as glycosides).

4. The stereochemistry of the glycosidic linkages can be inferred from experiments done with enzymes. Sucrose is hydrolyzed by an *α-glucosidase* obtained from yeast but not by β-glucosidase enzymes. This hydrolysis indicates *an α configuration at the glucoside portion.* Sucrose is also hydrolyzed by *sucrase,* an enzyme known to hydrolyze β-fructofuranosides but not α-fructofuranosides. This hydrolysis indicates a *β configuration at the fructoside portion.*

5. Methylation of sucrose gives an octamethyl derivative that, on hydrolysis, gives 2,3,4,6-tetra-*O*-methyl-D-glucose and 1,3,4,6-tetra-*O*-methyl-D-fructose. The identities of these two products demonstrate that the glucose portion is a *pyranoside* and that the fructose portion is a *furanoside.*

The structure of sucrose has been confirmed by X-ray analysis and by an unambiguous synthesis.

22.12B Maltose

When starch (Section 22.13A) is hydrolyzed by the enzyme *diastase,* one product is a disaccharide known as *maltose* (Fig. 22.9). The structure of maltose was deduced based on the following evidence:

1. When 1 mol of maltose is subjected to acid-catalyzed hydrolysis, it yields 2 mol of D-(+)-glucose.

2. Unlike sucrose, *maltose is a reducing sugar;* it gives positive tests with Fehling's, Benedict's, and Tollens' solutions. Maltose also reacts with phenylhydrazine to form a monophenylosazone (i.e., it incorporates two molecules of phenylhydrazine).

3. Maltose exists in two anomeric forms: α-(+)-maltose, $[\alpha]_D^{25} = +168°$, and β-(+)-maltose, $[\alpha]_D^{25} = +112°$. The maltose anomers undergo mutarotation to yield an equilibrium mixture, $[\alpha]_D^{25} = +136°$.

Facts 2 and 3 demonstrate that one of the glucose residues of maltose is present in a hemiacetal form; the other, therefore, must be present as a glucoside. The configuration of this glucosidic linkage can be inferred as α, because maltose is hydrolyzed by α-glucosidase enzymes and not by β-glucosidase enzymes.

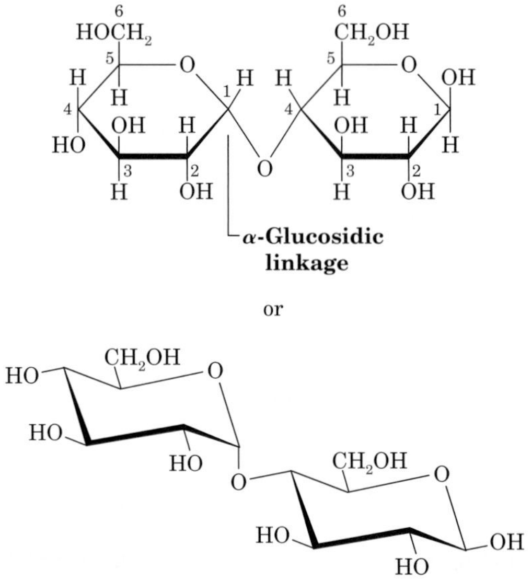

FIGURE 22.9 Two representations of the structure of the β anomer of (+)-maltose, 4-*O*-(α-D-glucopyranosyl)-β-D-glucopyranose.

4. Maltose reacts with bromine water to form a monocarboxylic acid, maltonic acid (Fig. 22.10*a*). This fact, too, is consistent with the presence of only one hemiacetal group.

5. Methylation of maltonic acid followed by hydrolysis gives 2,3,4,6-tetra-*O*-methyl-D-glucose and 2,3,5,6-tetra-*O*-methyl-D-gluconic acid. That the first product has a free —OH at C5 indicates that the nonreducing glucose portion is present as a pyranoside; that the second product, 2,3,5,6-tetra-*O*-methyl-D-gluconic acid, has a free —OH at C4 indicates that this position was involved in a glycosidic linkage with the nonreducing glucose.

Only the size of the reducing glucose ring needs to be determined.

6. Methylation of maltose itself, followed by hydrolysis (Fig. 22.10*b*), gives 2,3,4,6-tetra-*O*-methyl-D-glucose and 2,3,6-tri-*O*-methyl-D-glucose. The free —OH at C5 in the latter product indicates that it must have been involved in the oxide ring and that the reducing glucose is present as a *pyranose*.

22.12C Cellobiose

Partial hydrolysis of cellulose (Section 22.13C) gives the disaccharide cellobiose ($C_{12}H_{22}O_{11}$) (Fig. 22.11). Cellobiose resembles maltose in every respect except one: the configuration of its glycosidic linkage.

Cellobiose, like maltose, is a reducing sugar that, on acid-catalyzed hydrolysis, yields two molar equivalents of D-glucose. Cellobiose also undergoes mutarotation and forms a monophenylosazone. Methylation studies show that C1 of one glucose unit is connected in glycosidic linkage with C4 of the other and that both rings are six membered. Unlike maltose, however, cellobiose is hydrolyzed by *β-glucosidase* enzymes and not by α-glucosidase enzymes: This indicates that the glycosidic linkage in cellobiose is β (Fig. 22.11).

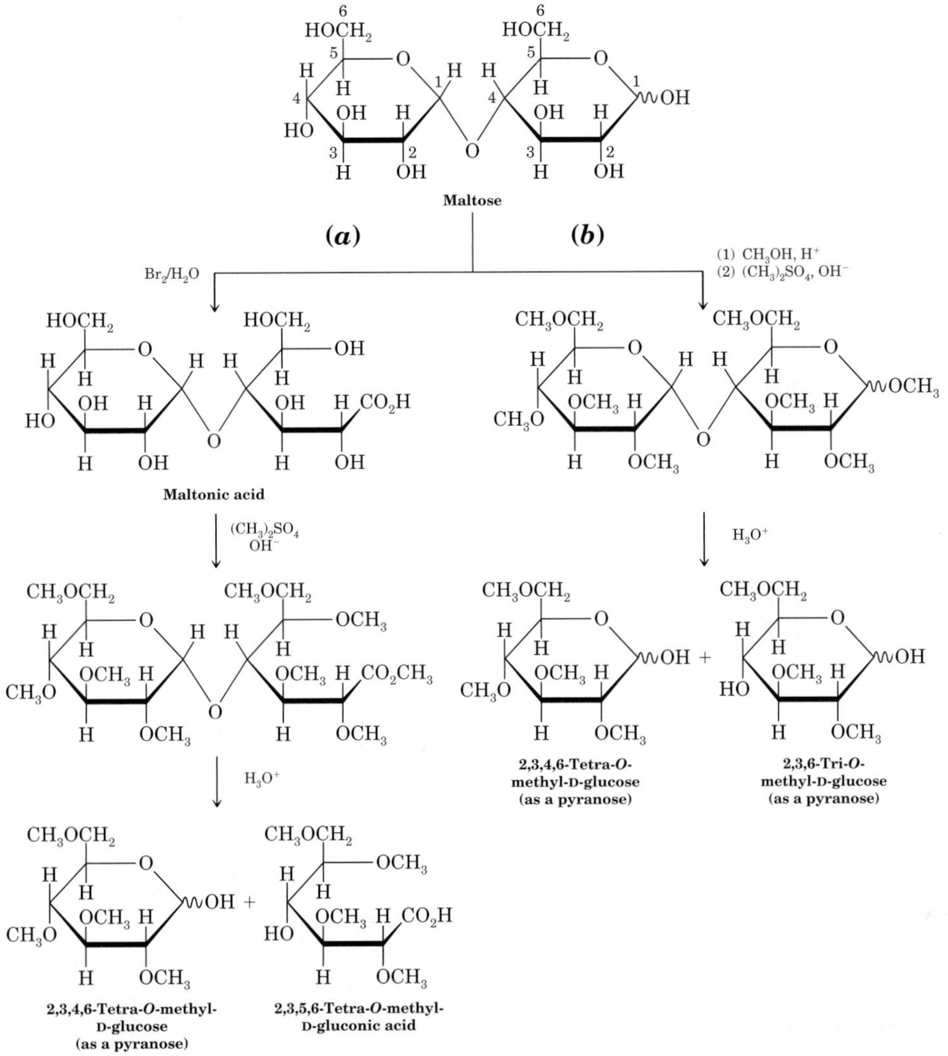

Maltose

(a) Br_2/H_2O

Maltonic acid

$(CH_3)_2SO_4$, OH^-

H_3O^+

2,3,4,6-Tetra-*O*-methyl-D-glucose (as a pyranose)

2,3,5,6-Tetra-*O*-methyl-D-gluconic acid

(b) (1) CH_3OH, H^+ (2) $(CH_3)_2SO_4$, OH^-

H_3O^+

2,3,4,6-Tetra-*O*-methyl-D-glucose (as a pyranose)

2,3,6-Tri-*O*-methyl-D-glucose (as a pyranose)

FIGURE 22.10 *(a)* Oxidation of maltose to maltonic acid followed by methylation and hydrolysis. *(b)* Methylation and subsequent hydrolysis of maltose itself.

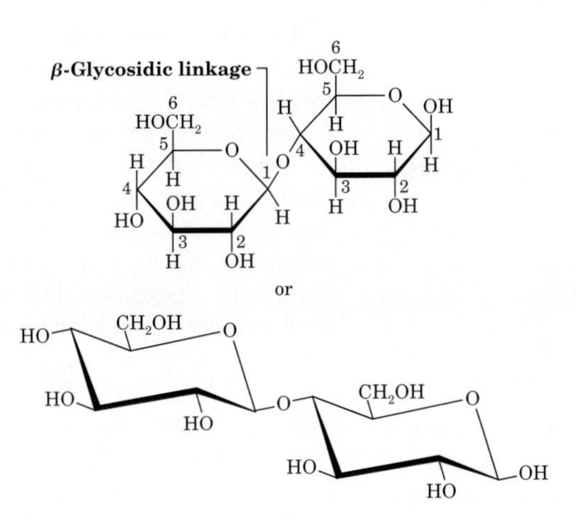

FIGURE 22.11 Two representations of the β anomer of cellobiose, 4-*O*-(β-D-glucopyranosyl)-β-D-glucopyranose.

The Chemistry of...

Artificial Sweeteners (How Sweet It Is)

Sucrose (table sugar) and fructose are the most common natural sweeteners. We all know, however, that they add to our calorie intake and promote tooth decay. For these reasons, many people find artificial sweeteners to be an attractive alternative to the natural and calorie-contributing counterparts.

Some products that contain the artificial sweetener aspartame.

Perhaps the most successful and widely used artificial sweetener is aspartame, the methyl ester of a dipeptide formed from phenylalanine and aspartic acid (Section 24.4). Aspartame is roughly 100 times as sweet as sucrose. It undergoes slow hydrolysis in solution, however, which limits its shelf life in products such as soft drinks. It also cannot be used for baking because it decomposes with heat. Furthermore, people with a genetic condition known as phenylketonuria cannot use aspartame because their metabolism causes a buildup of phenylpyruvic acid derived from aspartame. Accumulation of phenylpyruvic acid is harmful, especially to infants. Alitame, on the other hand, is a compound related to aspartame, but with improved properties. It is more stable than aspartame and roughly 2000 times as sweet as sucrose.

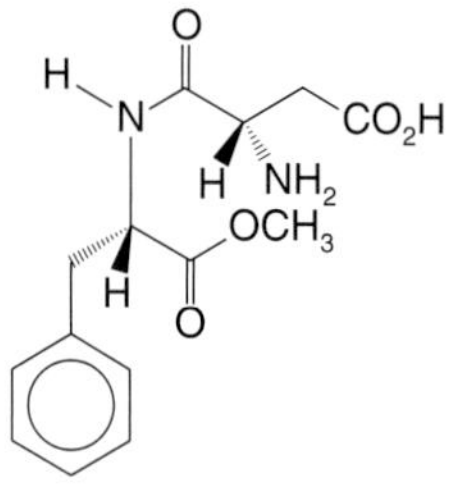

Aspartame 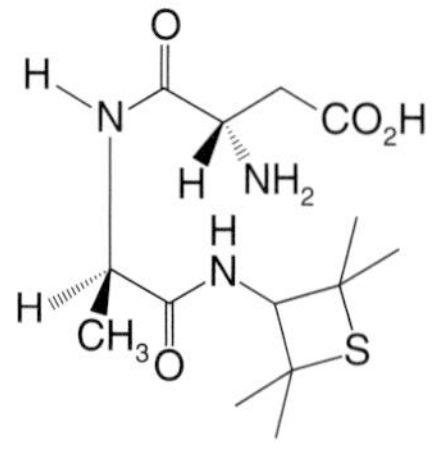 **Alitame**

Sucralose is a trichloro derivative of sucrose that is an artificial sweetener. Like aspartame, it is also approved for use by the U.S. Food and Drug Administration (FDA). Sucralose is 600 times sweeter than sucrose and has many properties desirable in an artificial sweetener. Sucralose looks and tastes like sugar, is stable at the temperatures used for cooking and baking, and it does not cause tooth decay or provide calories.

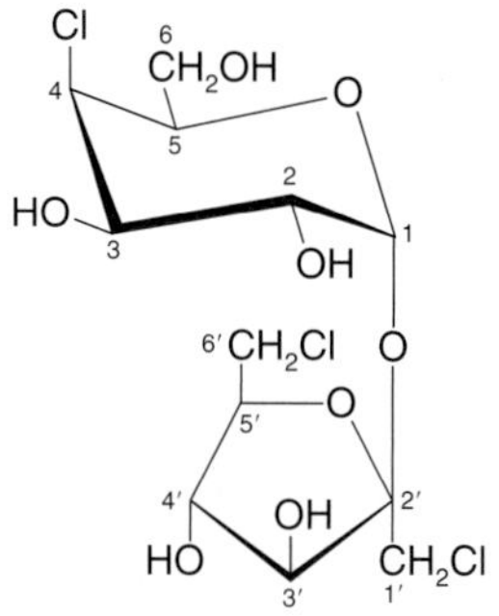

Sucralose

Cyclamate and saccharin, used as their sodium or calcium salts, were popular sweeteners at one time. A common formulation involved a 10 : 1 mixture of cyclamate and saccharin that proved sweeter than either compound individually. Tests showed, however, that this mixture produced tumors in animals, and the FDA subsequently banned it. Certain exclusions to the regulations nevertheless allow continued use of saccharin in some products.

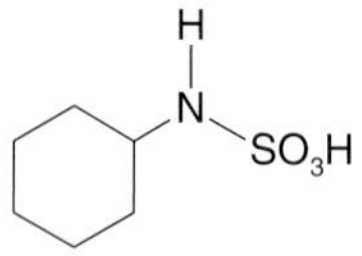

Cyclamate  **Saccharin**

Many other compounds have potential as artificial sweeteners. For example, L sugars are also sweet, and they presumably would provide either zero or very few calories because our enzymes have evolved to selectively metabolize their enantiomers instead, the D sugars. Although sources of L sugars are rare in nature, all eight L-hexoses have been synthesized by S. Masamune and K. B. Sharpless using the Sharpless asymmetric epoxidation (Section 11.13) and other enantioselective synthetic methods.

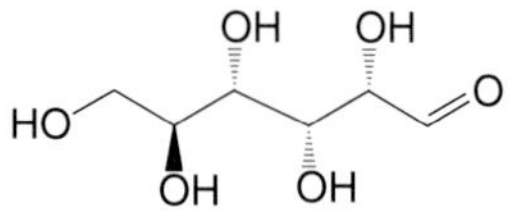

L-Glucose

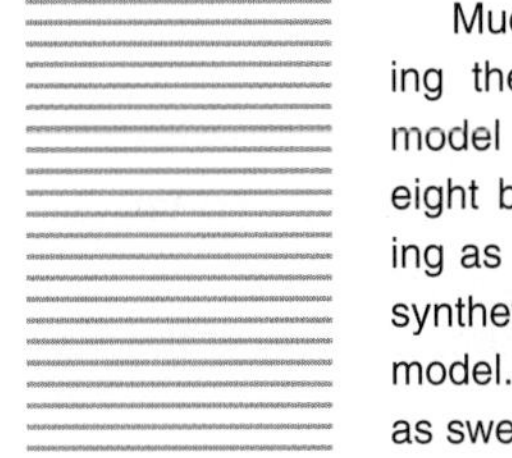

Much of the research on sweeteners involves probing the structure of sweetness receptor sites. One model proposed for a sweetness receptor incorporates eight binding interactions that involve hydrogen bonding as well as van der Waals forces. Sucronic acid is a synthetic compound designed on the basis of this model. Sucronic acid is reported to be 200,000 times as sweet as sucrose.

Sucronic acid

22.12D Lactose

Lactose (Fig. 22.12) is a disaccharide present in the milk of humans, cows, and almost all other mammals. Lactose is a reducing sugar that hydrolyzes to yield D-glucose and D-galactose; the glycosidic linkage is β.

FIGURE 22.12 Two representations of the β anomer of lactose, 4-*O*-(β-D-galactopyranosyl)-β-D-glucopyranose.

22.13 Polysaccharides

Polysaccharides, also known as **glycans,** consist of monosaccharides joined together by glycosidic linkages. Polysaccharides that are polymers of a single monosaccharide are called **homopolysaccharides;** those made up of more than one type of monosaccharide are called **heteropolysaccharides.** Homopolysaccharides are also classified on the basis of their monosaccharide units. A homopolysaccharide consisting of glucose monomeric units is called a **glucan;** one consisting of galactose units is a **galactan,** and so on.

Three important polysaccharides, all of which are glucans, are starch, glycogen, and cellulose. Starch is the principal food reserve of plants; glycogen functions as a carbohydrate reserve for animals; and cellulose serves as structural material in plants. As we examine the structures of these three polysaccharides, we shall be able to see how each is especially suited for its function.

22.13A Starch

Starch occurs as microscopic granules in the roots, tubers, and seeds of plants. Corn, potatoes, wheat, and rice are important commercial sources of starch. Heating starch with water causes the granules to swell and produce a colloidal suspension from which two major components can be isolated. One fraction is called *amylose* and the other *amylopectin.* Most starches yield 10–20% amylose and 80–90% amylopectin.

Physical measurements show that amylose typically consists of more than 1000 D-glucopyranoside units *connected in α linkages* between C1 of one unit and C4 of the next (Fig. 22.13). Thus, in the ring size of its glucose units and in the configuration of the glycosidic linkages between them, amylose resembles maltose.

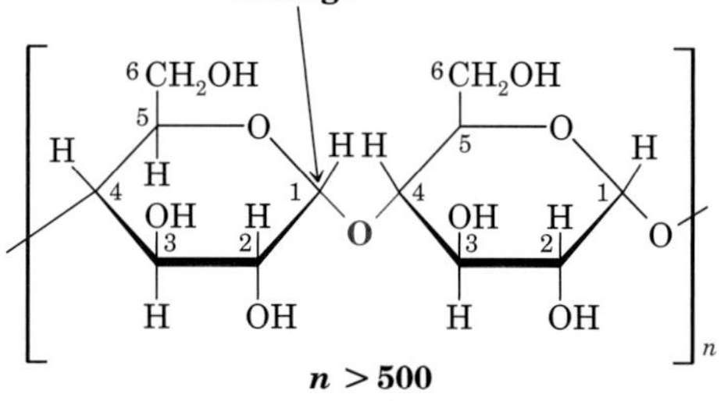

FIGURE 22.13 Partial structure of amylose, an unbranched polymer of D-glucose connected in $\alpha(1 \rightarrow 4)$ glycosidic linkages.

$\alpha(1 \rightarrow 4)$ Glucosidic linkage

$n > 500$

Chains of D-glucose units with α-glycosidic linkages such as those of amylose tend to assume a helical arrangement (Fig. 22.14). This arrangement results in a compact shape for the amylose molecule even though its molecular weight is quite large (150,000–600,000).

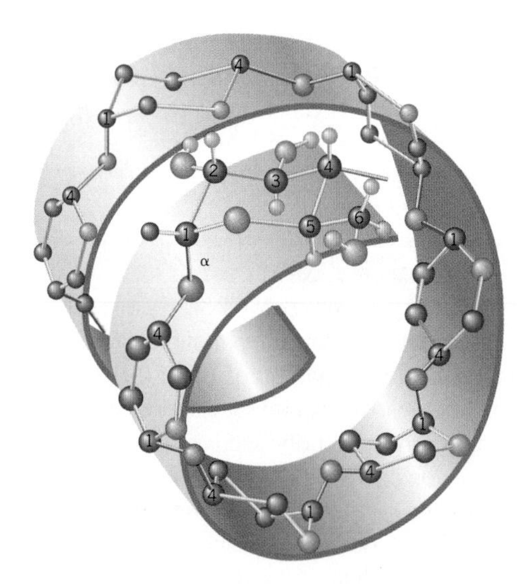

FIGURE 22.14 Amylose. The $\alpha(1 \rightarrow 4)$ linkages cause it to assume the shape of a left-handed helix. (Figure copyrighted © by Irving Geis. From Voet D.; Voet, J. G. *Biochemistry,* 2nd ed.; Wiley: New York, 1995; p 262. Used with permission.)

Amylopectin has a structure similar to that of amylose [i.e., $\alpha(1 \rightarrow 4)$ links], with the exception that in amylopectin the chains are branched. Branching takes place between C6 of one glucose unit and C1 of another and occurs at intervals of 20–25 glucose units (Fig. 22.15). Physical measurements indicate that amylopectin has a molecular weight of 1–6 million; thus amylopectin consists of hundreds of interconnecting chains of 20–25 glucose units each.

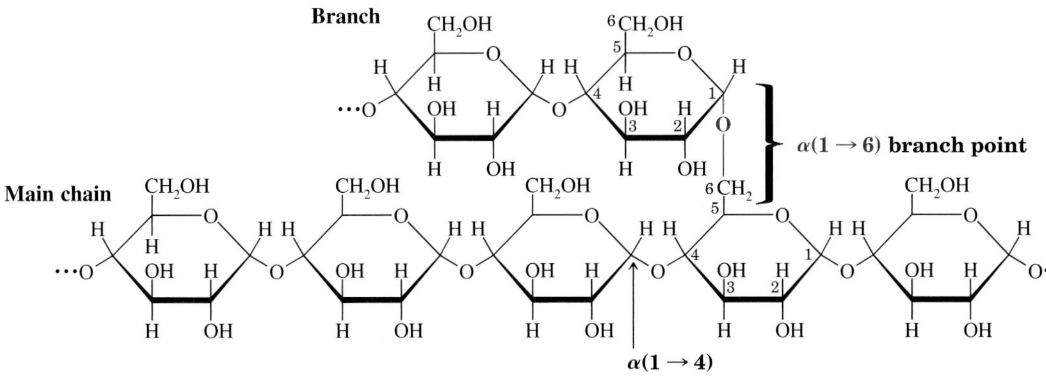

FIGURE 22.15 Partial structure of amylopectin.

22.13B Glycogen

Glycogen has a structure very much like that of amylopectin; however, in glycogen the chains are much more highly branched. Methylation and hydrolysis of glycogen indicates that there is one end group for every 10–12 glucose units; branches may occur as often as every 6 units. Glycogen has a very high molecular weight. Studies of glycogens isolated under conditions that minimize the likelihood of hydrolysis indicate molecular weights as high as 100 million.

The size and structure of glycogen beautifully suit its function as a reserve carbohydrate for animals. First, its size makes it too large to diffuse across cell membranes; thus, glycogen remains inside the cell, where it is needed as an energy source. Second, because glycogen incorporates tens of thousands of glucose units in a single molecule, it solves an important osmotic problem for the cell. Were so many glucose units present in the cell as individual molecules, the osmotic pressure within the cell would be enormous—so large that the cell membrane would almost certainly break.* Finally, the localization of glucose units within a large, highly branched structure simplifies one of the cell's logistical problems: that of having a ready source of glucose when cellular glucose concentrations are low and of being able to store glucose rapidly when cellular glucose concentrations are high. There are enzymes within the cell that catalyze the reactions by which glucose units are detached from (or attached to) glycogen. These enzymes operate at end groups by hydrolyzing (or forming) $\alpha(1 \rightarrow 4)$ glycosidic linkages. Because glycogen is so highly branched, a very large number of end groups is available at which these enzymes can operate. At the same time the overall concentration of glycogen (in moles per liter) is quite low because of its enormous molecular weight.

Amylopectin presumably serves a similar function in plants. The fact that amylopectin is less highly branched than glycogen is, however, not a serious disadvantage. Plants have a much lower metabolic rate than animals—and plants, of course, do not require sudden bursts of energy.

Animals store energy as fats (triacylglycerols) as well as glycogen. Fats, because they are more highly reduced, are capable of furnishing much more energy. The metabolism of a typical fatty acid, for example, liberates more than twice as much energy per carbon as glucose or glycogen. Why, then, we might ask, have two different energy repositories

---

*The phenomenon of osmotic pressure occurs whenever two solutions of different concentrations are separated by a membrane that allows penetration (by osmosis) of the solvent but not of the solute. The osmotic pressure (π) on one side of the membrane is related to the number of moles of solute particles (n), the volume of the solution (V), and the gas constant times the absolute temperature (RT): $\pi V = nRT$.

evolved? Glucose (from glycogen) is readily available and is highly water soluble.* Glucose, as a result, diffuses rapidly through the aqueous medium of the cell and serves as an ideal source of "ready energy." Long-chain fatty acids, by contrast, are almost insoluble in water, and their concentration inside the cell could never be very high. They would be a poor source of energy if the cell were in an energy pinch. On the other hand, fatty acids (as triacylglycerols), because of their caloric richness, are an excellent energy repository for long-term energy storage.

22.13C Cellulose

When we examine the structure of cellulose, we find another example of a polysaccharide in which nature has arranged monomeric glucose units in a manner that suits its function. Cellulose contains D-glucopyranoside units linked in $(1 \rightarrow 4)$ fashion in very long unbranched chains. Unlike starch and glycogen, however, the linkages in cellulose are *β-glycosidic linkages* (Fig. 22.16). This configuration of the anomeric carbon atoms of cellulose makes cellulose chains essentially linear; they do not tend to coil into helical structures as do glucose polymers when linked in an $\alpha(1 \rightarrow 4)$ manner.

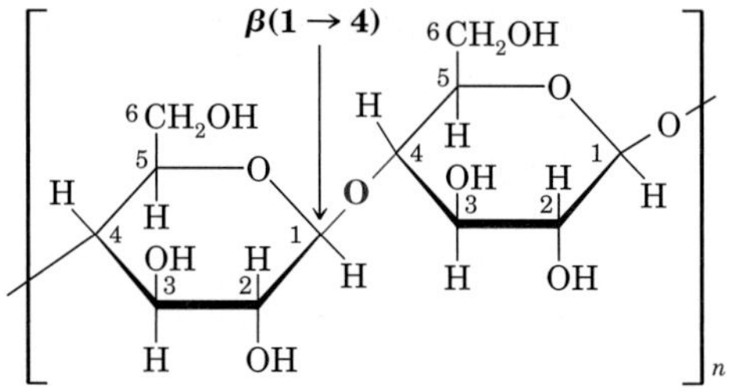

FIGURE 22.16 A portion of a cellulose chain. The glycosidic linkages are $\beta(1 \rightarrow 4)$.

The linear arrangement of β-linked glucose units in cellulose presents a uniform distribution of —OH groups on the outside of each chain. When two or more cellulose chains make contact, the hydroxyl groups are ideally situated to "zip" the chains together by forming hydrogen bonds (Fig. 22.17). Zipping many cellulose chains together in this way gives a highly insoluble, rigid, and fibrous polymer that is ideal as cell-wall material for plants.

This special property of cellulose chains, we should emphasize, is not just a result of $\beta(1 \rightarrow 4)$ glycosidic linkages; it is also a consequence of the precise stereochemistry of D-glucose at each stereogenic carbon. Were D-galactose or D-allose units linked in a similar fashion, they almost certainly would not give rise to a polymer with properties like cellulose. Thus, we get another glimpse of why D-glucose occupies such a special position in the chemistry of plants and animals. Not only is it the most stable aldohexose (because it can exist in a chair conformation that allows all of its bulky groups to occupy equatorial positions), but its special stereochemistry also allows it to form helical structures when α linked as in starches, and rigid linear structures when β linked as in cellulose.

There is another interesting and important fact about cellulose: The digestive enzymes of humans cannot attack its $\beta(1 \rightarrow 4)$ linkages. Hence, cellulose cannot serve as a food source for humans, as can starch. Cows and termites, however, can use cellulose (of grass and wood) as a food source because symbiotic bacteria in their digestive systems furnish β-glucosidase enzymes.

Perhaps we should ask ourselves one other question: Why has D-(+)-glucose been selected for its special role rather than L-(−)-glucose, its mirror image? Here an answer cannot be given with any certainty. The selection of D-(+)-glucose may simply have been a random event early in the course of the evolution of enzyme catalysts. Once this selection

*Glucose is actually liberated as glucose-6-phosphate (G6P), which is also water soluble.

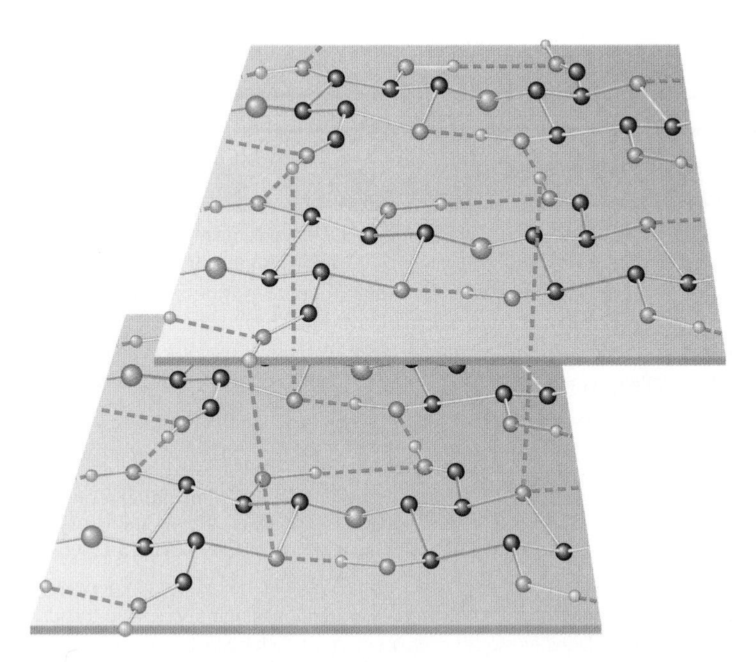

FIGURE 22.17 A proposed structure for cellulose. A fiber of cellulose may consist of about 40 parallel strands of glucose molecules linked in a $\beta(1 \rightarrow 4)$ fashion. Each glucose unit in a chain is turned over with respect to the preceding glucose unit and is held in this position by hydrogen bonds (dashed lines) between the chains. The glucan chains line up laterally to form sheets, and these sheets stack vertically so that they are staggered by one-half of a glucose unit. (Hydrogen atoms that do not participate in hydrogen bonding have been omitted for clarity.) (From Voet, D.; Voet, J. G. *Biochemistry,* 2nd ed; Wiley: New York, 1995; p 261. Used with permission.)

was made, however, the stereogenicity of the active sites of the enzymes involved would retain a bias toward D-(+)-glucose and against L-(−)-glucose (because of the improper fit of the latter). Once introduced, this bias would be perpetuated and extended to other catalysts.

Finally, when we speak about evolutionary selection of a particular molecule for a given function, we do not mean to imply that evolution operates on a molecular level. Evolution, of course, takes place at the level of organism populations, and molecules are selected only in the sense that their use gives the organism an increased likelihood of surviving and procreating.

22.13D Cellulose Derivatives

A number of derivatives of cellulose are used commercially. Most of these are compounds in which two or three of the free hydroxyl groups of each glucose unit have been converted to an ester or an ether. This conversion substantially alters the physical properties of the material, making it more soluble in organic solvents and allowing it to be made into fibers and films. Treating cellulose with acetic anhydride produces the triacetate known as "Arnel" or "acetate," used widely in the textile industry. Cellulose trinitrate, also called "gun cotton" or nitrocellulose, is used in explosives.

Rayon is made by treating cellulose (from cotton or wood pulp) with carbon disulfide in a basic solution. This reaction converts cellulose to a soluble xanthate:

$$\text{Cellulose—OH} + CS_2 \xrightarrow{\text{NaOH}} \text{cellulose—O—}\overset{\overset{\displaystyle S}{\|}}{C}\text{—}S^- \, Na^+$$

Cellulose xanthate

The solution of cellulose xanthate is then passed through a small orifice or slit into an acidic solution. This operation regenerates the —OH groups of cellulose, causing it to precipitate as a fiber or a sheet:

$$\text{Cellulose—O—}\overset{\overset{\displaystyle S}{\|}}{C}\text{—}S^- \, Na^+ \xrightarrow{H_3O^+} \text{cellulose—OH}$$

Rayon or cellophane

Cellophane on rollers at a manufacturing plant.

The fibers are *rayon;* the sheets, after softening with glycerol, are *cellophane.*

The Chemistry of...

Oligosaccharide Synthesis on a Solid Support—The Glycal Assembly Approach

Imagine being able to carry out a reaction where isolating your product involved simply filtering some plastic beads to which your product was attached and washing them with solvent. Imagine that if you needed to carry on with the synthesis, you would simply expose the beads to the next set of reagents, or if the synthesis were complete, you would retrieve your final product from the beads by a simple cleavage step. Imagine that to maximize your yields and drive reactions to completion you could use as large an excess of reagents as you wanted and that there would be no intermediate steps of recrystallization or chromatography because leftover reagents, solvents, and soluble byproducts could simply be washed away. All of this is possible with syntheses that are carried out on solid supports.

In 1984 the Nobel Prize in Chemistry was awarded to R. B. Merrifield (Rockefeller University) for his invention of a method to synthesize amino acid polymers (peptides) by "growing" the peptide from starter molecules attached to plastic beads. The method is called solid-phase peptide synthesis (see Section 24.7D), and it allows precisely the simple form of purification between each synthetic step that we described above. Merrifield's invention of solid support synthesis sparked adaptations in nucleic acid and carbohydrate syntheses as well as in general methods of organic synthesis. An elegant method for the solid-phase synthesis of oligosaccharides, called the *glycal assembly approach,* has been developed by S. J. Danishefsky and co-workers at Columbia University and Sloan Kettering Institute for Cancer Research. They have demonstrated the power of this approach by synthesizing molecules that show promise as vaccines against cancer (see "The Chemistry of . . . Vaccines Against Cancer", Section 22.16).

The glycal assembly method for solid-phase oligosaccharide synthesis begins by preparing the polystyrene solid support to bond the first carbohydrate moiety (a glycal) via a silyl ether linkage (see Section 11.11D). This is accomplished by lithiation of the polystyrene at some of its phenyl groups using butyllithium followed by reaction of those groups with diphenyldichlorosilane. The result is polystyrene with some phenyl ($—C_6H_5$) groups converted to $—C_6H_4Si(Ph)_2Cl$ groups. Reaction of a free (unprotected) hydroxyl group of the glycal moiety with a silyl chloride group of the polymer joins the initial carbohydrate group to the solid support.

Glycal Attachment to a Polystyrene Resin

(1) BuLi, TMEDA, cyclohexane; (2) R_2SiCl_2, benzene
SiR_2Cl
8
$(i\text{-Pr})_2NEt$, CH_2Cl_2
$OSiR_2$—Ⓢ
5
6: R = Ph
7: R = *i*-Pr
9: R = Ph
10: R = *i*-Pr
Ⓢ = solid support

Conversion of the double bond in the polymer-bound glycal to an epoxide (using an epoxidizing reagent called dimethyldioxirane, DMDO) prepares the polymer-bound substrate to act as an electrophile. Attack on the epoxide by a hydroxyl group of the next glycal reactant extends the oligosaccharide chain on the polymer by one group. Epoxidation of the newly bonded glycal unit prepares it for attachment of the next glycal, and so on, until the desired termination point is reached.

Epoxidation and Glycal Assembly

DMDO
3
1
2
4

Solid-Phase Synthesis of a Tetrasaccharide Using the Glycal Assembly Method

The glycal assembly method eliminates the need for tedious purifications after each step, requires no specific enzymes, and does not require complex starting materials or reagents. By varying the choice of glycal reactants and selectively protecting hydroxyl groups of these precursors, Danishefsky and co-workers have shown that linkages can be made at any position of a carbohydrate moiety desired and that complex natural and unnatural oligosaccharides can be prepared. They have already used this approach to synthesize molecules that may become vaccines against cancer (see "The Chemistry of . . . Vaccines Against Cancer"). Efforts to automate the glycal assembly method so that machines can perform the syntheses are also underway. In Section 24.7D we shall see that automated peptide synthesis is already a reality.

Figures from Seeberger, P. H.; Danishefsky, S. J. *Acc. Chem. Res.*, **1998,** *31,* p. 687

22.14 Other Biologically Important Sugars

Monosaccharide derivatives in which the —CH_2OH group at C6 has been specifically oxidized to a carboxyl group are called **uronic acids.** Their names are based on the monosaccharide from which they are derived. For example, specific oxidation of C6 of glucose to a carboxyl group converts *glucose* to **glucuronic acid.** In the same way, specific oxidation of C6 of *galactose* would yield **galacturonic acid:**

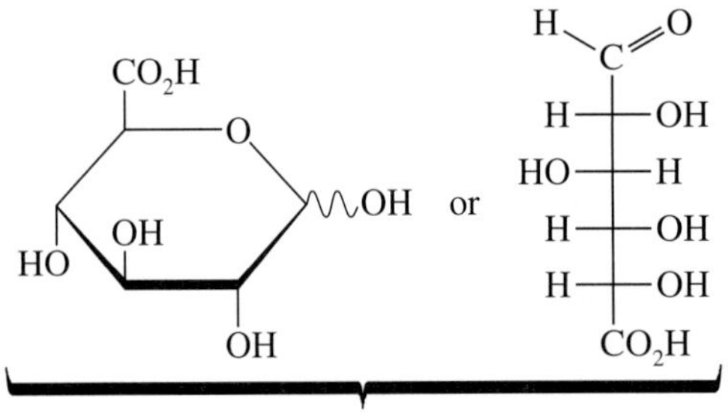

D-Glucuronic acid 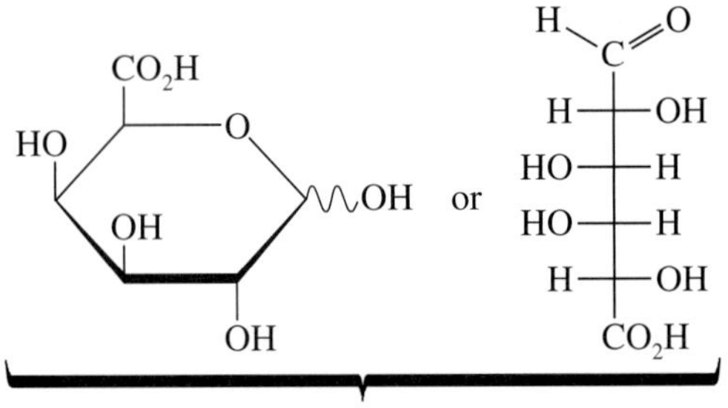 **D-Galacturonic acid**

PROBLEM 22.19 Direct oxidation of an aldose affects the aldehyde group first, converting it to a carboxylic acid (Section 22.6B), and most oxidizing agents that will attack 1° alcohol groups will also attack 2° alcohol groups. Clearly, then, a laboratory synthesis of a uronic acid from an aldose requires protecting these groups from oxidation. Keeping this in mind, suggest a method for carrying out a specific oxidation that would convert D-galactose to D-galacturonic acid. (*Hint:* See Section 22.5E.)

Monosaccharides in which an —OH group has been replaced by —H are known as **deoxy sugars.** The most important deoxy sugar, because it occurs in DNA, is **deoxyribose.** Other deoxy sugars that occur widely in polysaccharides are L-rhamnose and L-fucose:

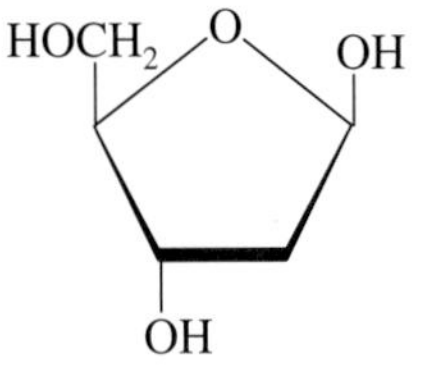

β-2-Deoxy-D-ribose 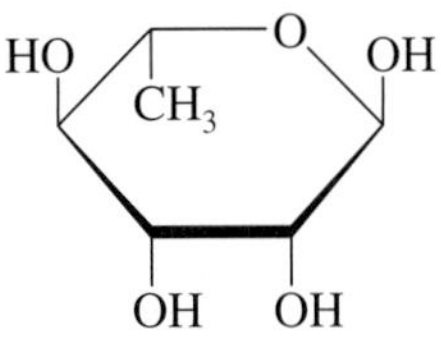 **α-L-Rhamnose (6-deoxy-L-mannose)** 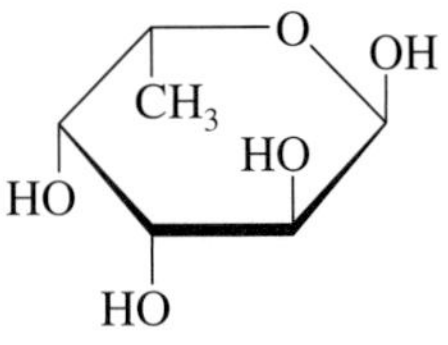 **α-L-Fucose (6-deoxy-L-galactose)**

22.15 Sugars That Contain Nitrogen

22.15A Glycosylamines

A sugar in which an amino group replaces the anomeric —OH is called a glycosylamine. Examples are β-D-glucopyranosylamine and adenosine:

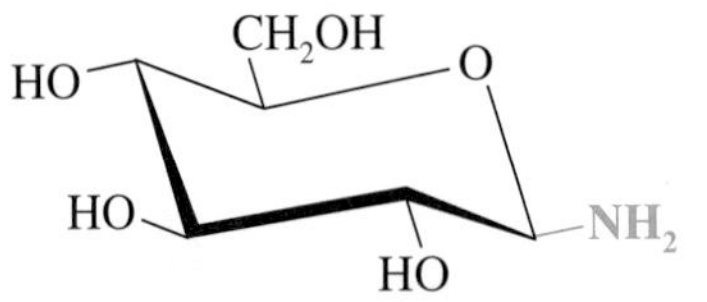

β-D-Glucopyranosylamine 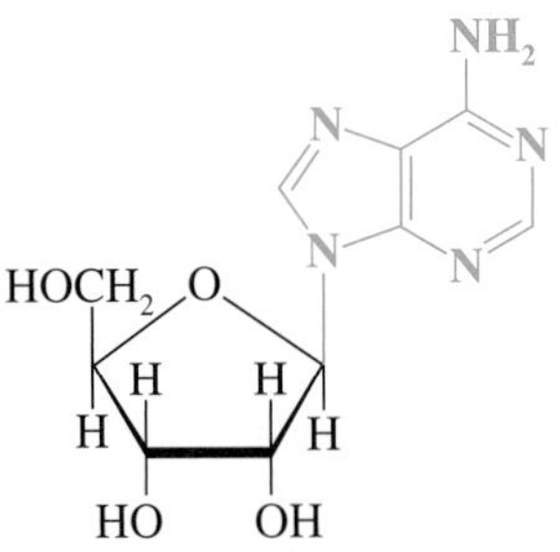 **Adenosine**

Adenosine is an example of a glycosylamine that is also called a **nucleoside.** Nucleosides are glycosylamines in which the amino component is a pyrimidine or a purine (Section 20.1B) and in which the sugar component is either D-ribose or 2-deoxy-D-ribose (i.e., D-ribose minus the oxygen at the 2 position). Nucleosides are the important components of RNA (ribonucleic acid) and DNA (deoxyribonucleic acid). We shall describe their properties in detail in Section 25.2.

22.15B Amino Sugars

A sugar in which an amino group replaces a nonanomeric —OH group is called an **amino sugar.** An example is **D-glucosamine.** In many instances the amino group is acetylated as in ***N*-acetyl-D-glucosamine.** ***N*-Acetylmuramic acid** is an important component of bacterial cell walls (Section 24.10).

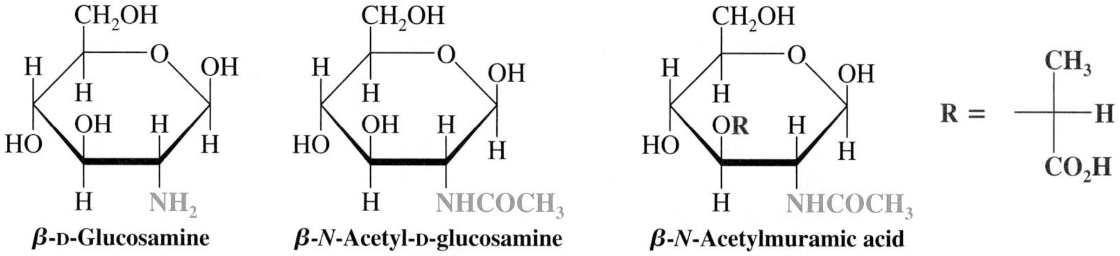

β-D-Glucosamine **β-*N*-Acetyl-D-glucosamine (NAG)** **β-*N*-Acetylmuramic acid (NAM)**

R = $CH(CH_3)CO_2H$

D-Glucosamine can be obtained by hydrolysis of **chitin,** a polysaccharide found in the shells of lobsters and crabs and in the external skeletons of insects and spiders. The amino group of D-glucosamine as it occurs in chitin, however, is acetylated; thus, the repeating unit is actually *N*-acetylglucosamine (Fig. 22.18). The glycosidic linkages in chitin are $\beta(1 \rightarrow 4)$. X-ray analysis indicates that the structure of chitin is similar to that of cellulose.

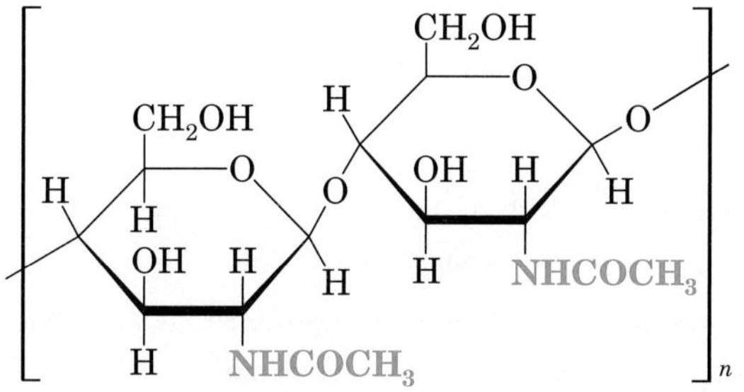

FIGURE 22.18 A partial structure of chitin. The repeating units are *N*-acetylglucosamines linked $\beta(1 \rightarrow 4)$.

D-Glucosamine can also be isolated from **heparin,** a sulfated polysaccharide that consists predominately of alternating units of D-glucuronate-2-sulfate and *N*-sulfo-D-glucosamine-6-sulfate (Fig. 22.19). Heparin occurs in intracellular granules of mast cells that line arterial walls, where, when released through injury, it inhibits the clotting of blood. Its purpose seems to be to prevent runaway clot formation. Heparin is widely used in medicine to prevent blood clotting in postsurgical patients.

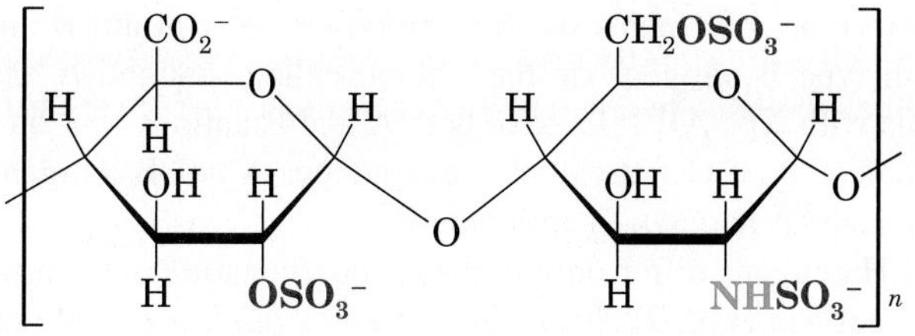

D-Glucuronate-2-sulfate **N-Sulfo-D-glucosamine-6-sulfate**

FIGURE 22.19 A partial structure of heparin, a polysaccharide that prevents blood clotting.

22.16 Glycolipids and Glycoproteins of the Cell Surface: Cell Recognition and the Immune System

Before 1960, it was thought that the biology of carbohydrates was rather uninteresting, that, in addition to being a kind of inert filler in cells, carbohydrates served only as an energy source and, in plants, as structural materials. Research has shown, however, that carbohydrates joined through glycosidic linkages to lipids (Chapter 23) and to proteins (Chapter 24), called **glycolipids** and **glycoproteins,** respectively, have functions that span the entire spectrum of activities in the cell. Indeed, most proteins are glycoproteins, of which the carbohydrate content can vary from less than 1% to greater than 90%.

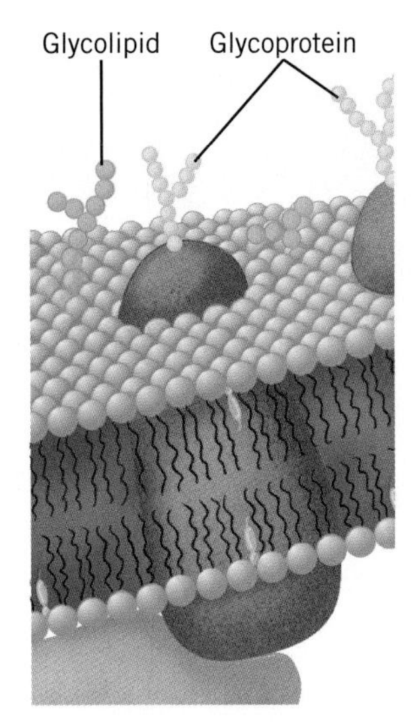

Glycolipids and glycoproteins on the cell surface (Section 23.6A) are now known to be the agents by which cells interact with other cells and with invading bacteria and viruses. The immune system's role in healing and autoimmune diseases such as rheumatoid arthritis involves cell recognition through cell surface carbohydrates. Important carbohydrates in this role are sialyl Lewis^x acids (see the chapter opening vignette). Tumor cells also have specific carbohydrate markers on their surface as well, a fact that may make it possible to develop vaccines against cancer. (See "The Chemistry of . . . Vaccines Against Cancer.")

A Sialyl Lewis^x Acid

The human blood groups offer another example of how carbohydrates, in the form of glycolipids and glycoproteins, act as biochemical markers. The A, B, and O blood types are determined, respectively, by the A, B, and H determinants on the blood cell surface. (The odd naming of the type O determinant came about for complicated historical reasons.) Type AB blood cells have both A and B determinants. These determinants are the carbohydrate portions of the A, B, and H **antigens.**

Antigens are characteristic chemical substances that cause the production of **antibodies** when injected into an animal. Each antibody can bind at least two of its corresponding antigen molecules, causing them to become linked. Linking of red blood cells causes them to agglutinate (clump together). In a transfusion this agglutination can lead to a fatal blockage of the blood vessels.

Individuals with type A antigens on their blood cells carry anti-B antibodies in their serum; those with type B antigens on their blood cells carry anti-A antibodies in their serum. Individuals with type AB cells have both A and B antigens but have neither anti-A nor anti-B antibodies. Type O individuals have neither A nor B antigens on their blood cells but have both anti-A and anti-B antibodies.

The A, B, and H antigens differ only in the monosaccharide units at their nonreducing ends. The type H antigen (Fig. 22.20) is the precursor oligosaccharide of the type A and B antigens. Individuals with blood type A have an enzyme that specifically adds an *N*-

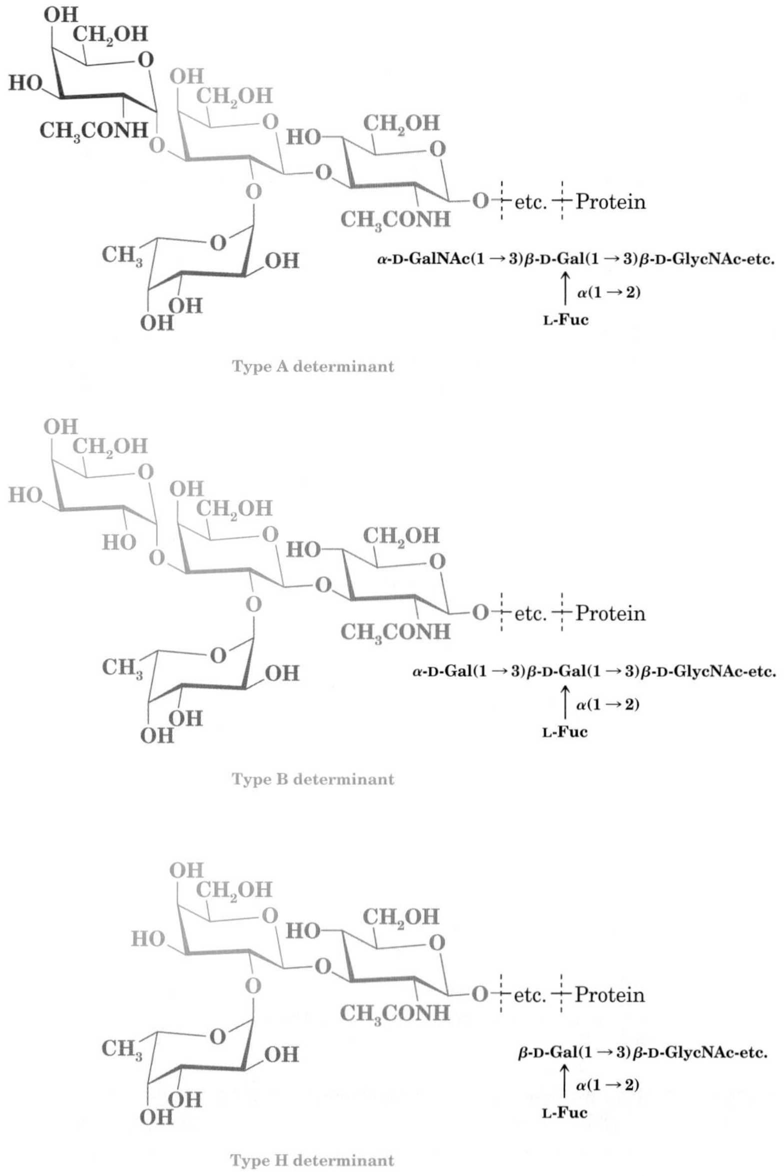

α-D-GalNAc(1 → 3)β-D-Gal(1 → 3)β-D-GlycNAc-etc.
↑ α(1 → 2)
L-Fuc

Type A determinant

α-D-Gal(1 → 3)β-D-Gal(1 → 3)β-D-GlycNAc-etc.
↑ α(1 → 2)
L-Fuc

Type B determinant

β-D-Gal(1 → 3)β-D-GlycNAc-etc.
↑ α(1 → 2)
L-Fuc

Type H determinant

FIGURE 22.20 The terminal monosaccharides of the antigenic determinants for types A, B, and O blood. The type H determinant is present in individuals with blood type O and is the precursor of the type A and B determinants. These oligosaccharide antigens are attached to carrier lipid or protein molecules that are anchored in the red blood cell membrane (see Fig. 23.8 for a depiction of a cell membrane). Ac = acetyl, Gal = D-galactose, GalNAc = *N*-acetylgalactosamine, GlycNAc = *N*-acetylglucosamine, Fuc = Fucose.

acetylgalactosamine unit to the 3-OH group of the terminal galactose unit of the H antigen. Individuals with blood type B have an enzyme that specifically adds galactose instead. In individuals with type O blood, the enzyme is inactive.

Antigen–antibody interactions like those that determine blood types are the basis of the immune system. These interactions often involve the chemical recognition of a glycolipid or glycoprotein in the antigen by a glycolipid or glycoprotein of the antibody. In "The Chemistry of . . . Antibody-Catalyzed Aldol Condensations" (Chapter 19), however, we saw a different and emerging dimension of chemistry involving antibodies. We shall explore this topic further in the Chapter 24 opening vignette, "Designer Catalysts," and "The Chemistry of . . . Some Catalytic Antibodies."

The Chemistry of...

Vaccines Against Cancer

With hopes of developing a vaccine against cancer, chemists have synthesized carbohydrate moieties corresponding to antigenic carbohydrate groups on breast and other tumor cells. They have linked these tumor antigens to peptides that promote the immune system, and they are testing them as new therapies against breast and other cancers. The hope is that injecting these antigens will stimulate a patient's immune system to produce antibodies against cancerous cells, thereby inducing an immune response that is a "natural" mode of fighting cancer. A cancer vaccine could avoid or reduce the need to use chemotherapy drugs that have toxic side effects. The structure of one cancer antigen that has been synthesized is shown here. To synthesize these complex carbohydrate antigens, chemists led by Samuel J. Danishefsky (Columbia University and the Sloan Kettering Institute for Cancer Research) used their pioneering glycal assembly approach to synthesize the oligosaccharide moiety (Section 22.13), incorporating *N*-acetylaminosugars where found in the antigen structures and adding an immunostimulant protein moiety to boost the immune response.

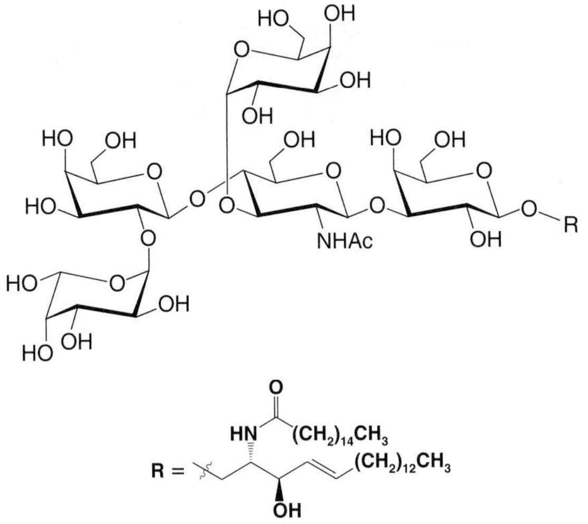

A fully synthetic glycoconjugate antigen

In something of a molecular "multiple warhead approach," Danishefsky and co-workers went on to synthesize compounds that carry *three* tumor cell antigens simultaneously in the same molecule. This approach is an ingenious effort not only to elicit a broader and even stronger immune response than might be achieved against the cancer cells by a single tumor antigen but also to avoid the need to make multiple injections to deliver each antigen as a separate vaccine.

These and other efforts of modern medicine would not be possible without the power of synthesis and new methods such as the glycal assembly approach.

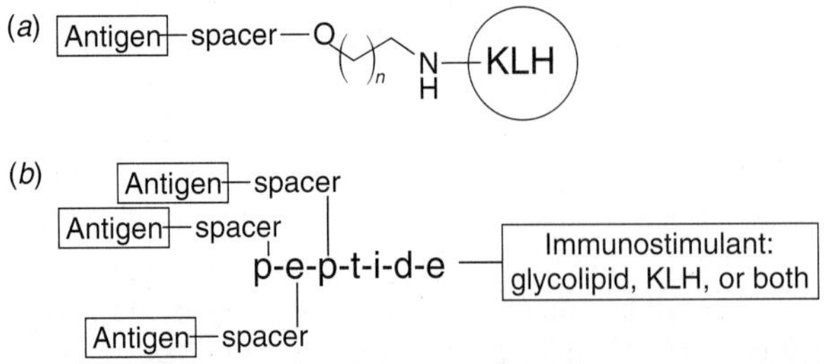

Glycoconjugate components of anticancer vaccines: (a) single tumor antigen conjugated to carrier protein KLH; (b) glycopeptide presented triple-clustered antigens.

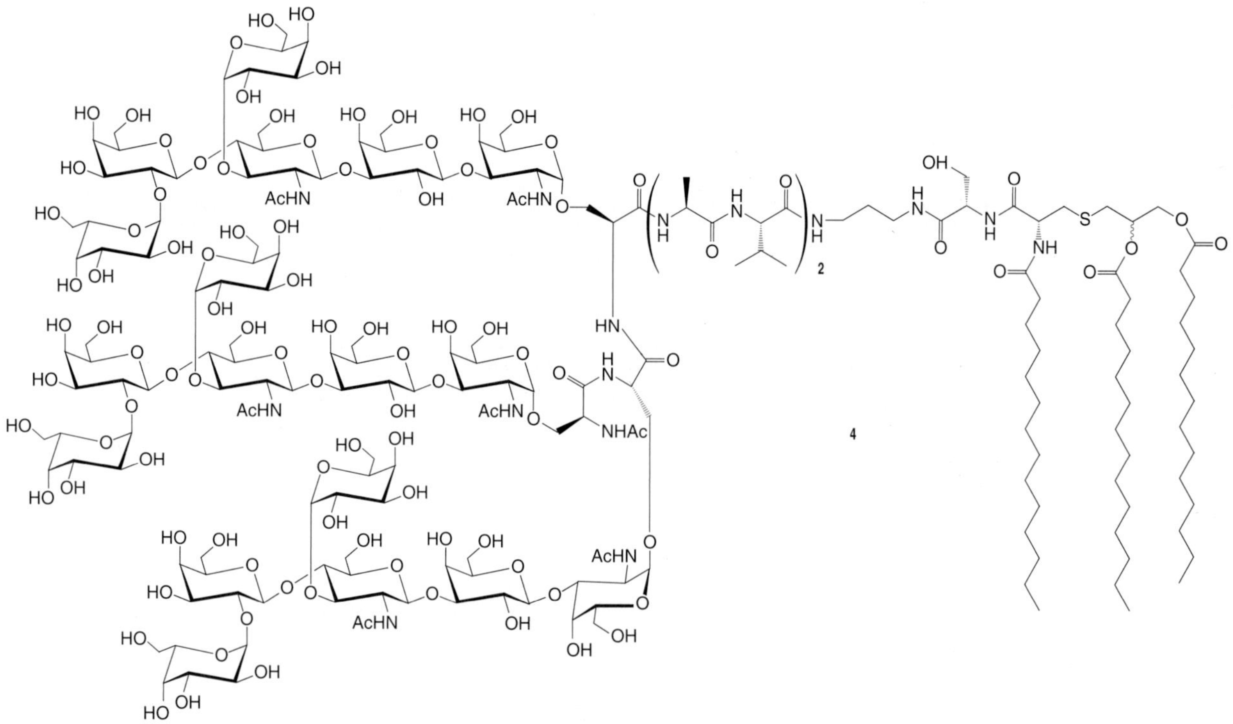

4 [$(Le^y)_3$-peptide-Pam_3Cys]
5 [$(Le^y)_3$-peptide-MBS-KLH]

Full structure of a synthetic triple-clustered anticancer vaccine.

22.17 Carbohydrate Antibiotics

One of the important discoveries in carbohydrate chemistry was the isolation (in 1944) of the carbohydrate antibiotic called *streptomycin.* Streptomycin disrupts bacterial protein synthesis. Its structure is made up of the following three subunits:

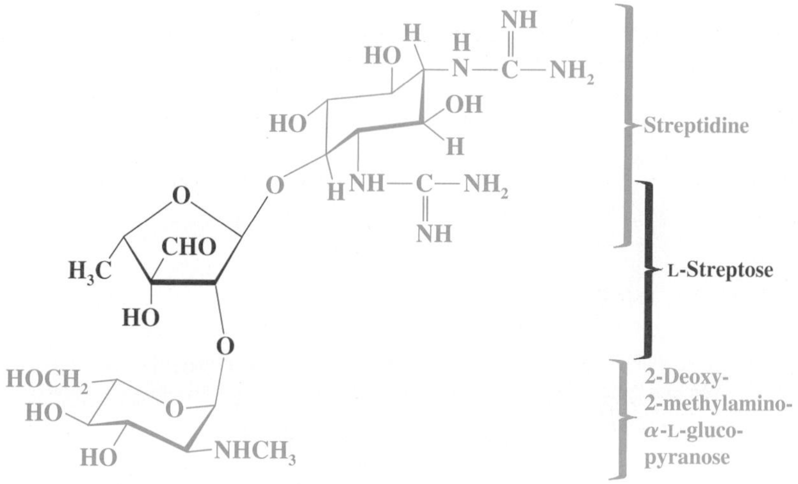

Streptidine
L-Streptose
2-Deoxy-2-methylamino-α-L-glucopyranose

All three components are unusual: The amino sugar is based on L-glucose; streptose is a branched-chain monosaccharide; and streptidine is not a sugar at all, but a cyclohexane derivative called an amino cyclitol.

Other members of this family are antibiotics called kanamycins, neomycins, and gentamicins (not shown). All are based on an amino cyclitol linked to one or more amino sugars. The glycosidic linkage is nearly always α. These antibiotics are especially useful against bacteria that are resistant to penicillins.

SUMMARY OF REACTIONS OF CARBOHYDRATES

The reactions of carbohydrates, with few exceptions, are the reactions of functional groups that we have studied in earlier chapters, especially those of aldehydes, ketones, and alcohols. The most central reactions of carbohydrates are those of hemiacetal and acetal formation and hydrolysis. Hemiacetal groups form the pyranose and furanose rings in carbohydrates, and acetal groups form glycoside derivatives and join monosaccharides together to form di-, tri-, oligo-, and polysaccharides.

Other reactions of carbohydrates include those of alcohols, carboxylic acids, and their derivatives. Alkylation of carbohydrate hydroxyl groups leads to ethers. Acylation of their hydroxyl groups produces esters. Alkylation and acylation reactions are sometimes used to protect carbohydrate hydroxyl groups from reaction while a transformation occurs elsewhere. Hydrolysis reactions are involved in converting ester and lactone derivatives of carbohydrates back to their polyhydroxyl form. Enolization of aldehydes and ketones leads to epimerization and interconversion of aldoses and ketoses. Addition reactions of aldehydes and ketones are useful, too, such as the addition of ammonia derivatives in osazone formation, and of cyanide in the Kiliani–Fischer synthesis. Hydrolysis of nitriles from the Kiliani–Fischer synthesis leads to carboxylic acids.

Oxidation and reduction reactions have their place in carbohydrate chemistry as well. Reduction reactions of aldehydes and ketones, such as borohydride reduction and catalytic hydrogenation, are used to convert aldoses and ketoses to alditols. Oxidation by Tollens' and Benedict's reagents is a test for the hemiacetal linkage in a sugar. Bromine water oxidizes the aldehyde group of an aldose to an aldonic acid. Nitric acid oxidizes both the aldehyde group and terminal hydroxymethyl group of an aldose to an aldaric acid (a dicarboxylic acid). Lastly, periodate cleavage of carbohydrates yields oxidized fragments that can be useful for structure elucidation.

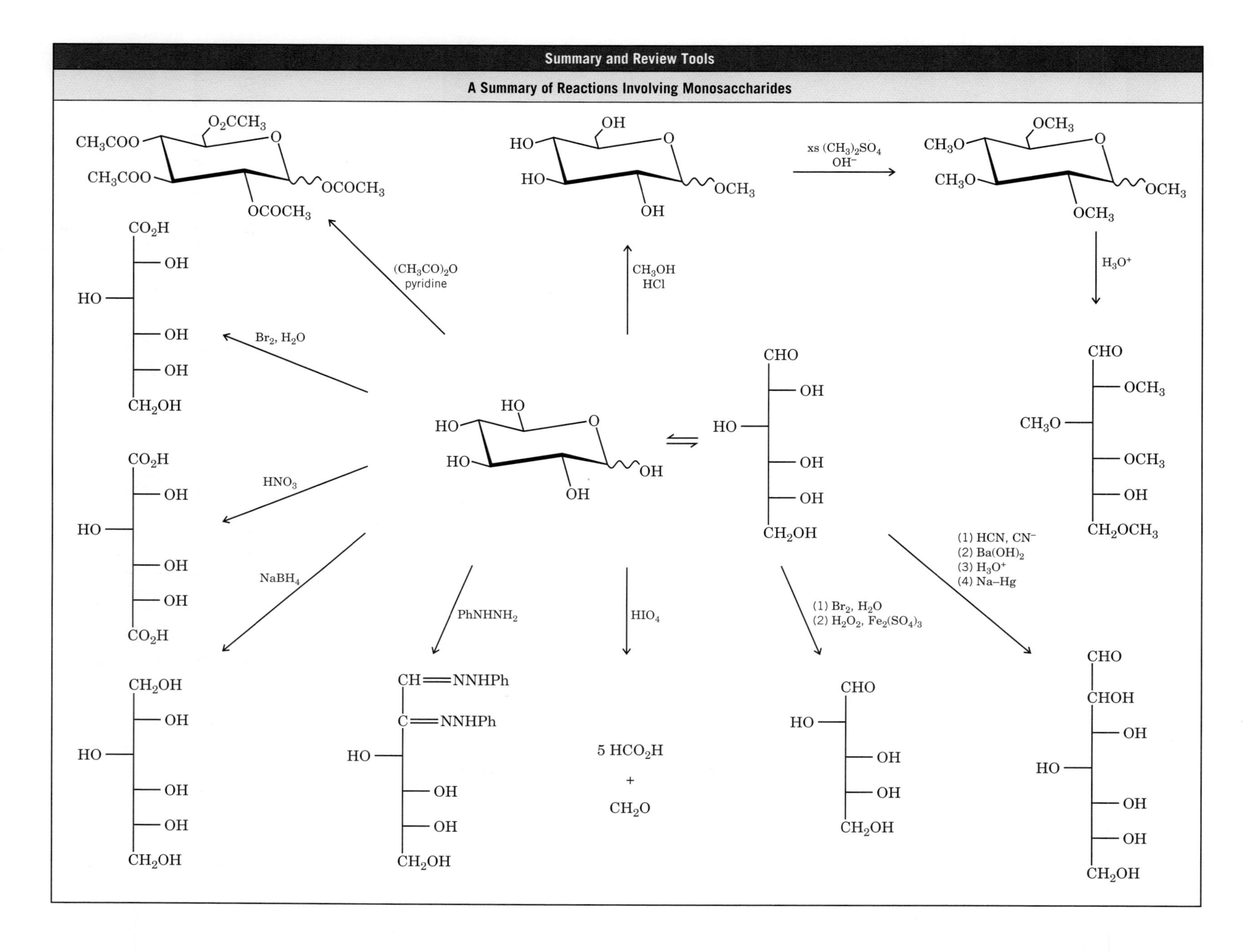
Summary and Review Tools
A Summary of Reactions Involving Monosaccharides
$(CH_3CO)_2O$
pyridine
CH_3OH
HCl
xs $(CH_3)_2SO_4$
OH^-
H_3O^+
Br_2, H_2O
HNO_3
$NaBH_4$
$PhNHNH_2$
HIO_4
(1) Br_2, H_2O
(2) H_2O_2, $Fe_2(SO_4)_3$
(1) HCN, CN^-
(2) $Ba(OH)_2$
(3) H_3O^+
(4) Na–Hg
5 HCO_2H
+
CH_2O

Key Terms and Concepts

α anomer, β anomer	Section 22.2C
Aglycone	Section 22.4
Anomer	Section 22.2C
Anomeric carbon	Section 22.2C
Cyclic acetal	Section 22.5E
D–L nomenclature	Section 22.2B
Disaccharides	Sections 22.1A, 22.12
Epimers	Sections 17.3A, 22.8
Fischer projections	Section 22.2C
Furanose	Section 22.2C
Glycolipids	Section 22.16
Glycoproteins	Section 22.16
Glycoside	Section 22.4
Haworth formula	Section 22.2C
Hemiacetal	Section 22.2C
Monosaccharides	Sections 22.1A, 22.2
Mutarotation	Section 22.3
Oligosaccharides	Section 22.1A
Osazones	Section 22.8
Polysaccharides (glycans)	Sections 22.1A, 22.13
Pyranose	Section 22.2C
Reducing sugar	Section 22.6A
Trisaccharides	Section 22.1A

Additional Problems

22.20 Give appropriate structural formulas to illustrate each of the following:

(a) An aldopentose
(b) A ketohexose
(c) An L-monosaccharide
(d) A glycoside
(e) An aldonic acid
(f) An aldaric acid
(g) An aldonolactone
(h) A pyranose
(i) A furanose
(j) A reducing sugar
(k) A pyranoside
(l) A furanoside
(m) Epimers
(n) Anomers
(o) A phenylosazone
(p) A disaccharide
(q) A polysaccharide
(r) A nonreducing sugar

22.21 Draw conformational formulas for each of the following: **(a)** α-D-Allopyranose, **(b)** Methyl β-D-allopyranoside and **(c)** Methyl 2,3,4,6-tetra-*O*-methyl-β-D-allopyranoside.

22.22 Draw structures for furanose and pyranose forms of D-ribose. Show how you could use periodate oxidation to distinguish between a methyl ribofuranoside and a methyl ribopyranoside.

22.23 One reference book lists D-mannose as being dextrorotatory; another lists it as being levorotatory. Both references are correct. Explain.

22.24 The starting material for a commercial synthesis of vitamin C is L-sorbose (see the following reaction); it can be synthesized from D-glucose through the following reaction sequence:

Note: Problems marked with an asterisk are "challenge problems."

$$\text{D-Glucose} \xrightarrow[\text{Ni}]{H_2} \text{D-Glucitol} \xrightarrow[\textit{Acetobacter suboxydans}]{O_2}$$

CH_2OH | C=O | HO—C—H | H—C—OH | HO—C—H | CH_2OH

L-Sorbose

The second step of this sequence illustrates the use of a bacterial oxidation; the microorganism *A. suboxydans* accomplishes this step in 90% yield. The overall result of the synthesis is the transformation of a D-aldohexose (D-glucose) into an L-ketohexose (L-sorbose). What does this mean about the specificity of the bacterial oxidation?

22.25 What two aldoses would yield the same phenylosazone as L-sorbose (Problem 22.24)?

22.26 In addition to fructose (Problem 22.12) and sorbose (Problem 22.24), there are two other 2-ketohexoses, *psicose* and *tagatose.* D-Psicose yields the same phenylosazone as D-allose (or D-altrose); D-tagatose yields the same osazone as D-galactose (or D-talose). What are the structures of D-psicose and D-tagatose?

22.27 **A, B,** and **C** are three aldohexoses. Compounds **A** and **B** yield the same optically active alditol when they are reduced with hydrogen and a catalyst; **A** and **B** yield different phenylosazones when treated with phenylhydrazine; **B** and **C** give the same phenylosazone but different alditols. Assuming that all are D sugars, give names and structures for **A, B,** and **C.**

22.28 Xylitol is a sweetener that is used in sugarless chewing gum. Starting with an appropriate monosaccharide, outline a possible synthesis of xylitol.

CH_2OH | H—OH | HO—H | H—OH | CH_2OH

Xylitol

22.29 Although monosaccharides undergo complex isomerizations in base (see Section 22.5), aldonic acids are epimerized specifically at C2 when they are heated with pyridine. Show how you could make use of this reaction in a synthesis of D-mannose from D-glucose.

22.30 The most stable conformation of most aldopyranoses is one in which the largest group, the $—CH_2OH$ group, is equatorial. However, D-idopyranose exists primarily in a conformation with an axial $—CH_2OH$ group. Write formulas for the two chair conformations of α-D-idopyranose (one with the $—CH_2OH$ group axial and one with the $—CH_2OH$ group equatorial) and provide an explanation.

22.31 **(a)** Heating D-altrose with dilute acid produces a nonreducing *anhydro sugar* ($C_6H_{10}O_5$). Methylation of the anhydro sugar followed by acid hydrolysis yields 2,3,4-tri-*O*-methyl-D-altrose. The formation of the anhydro sugar takes place through a chair conformation of β-D-altropyranose in which the $—CH_2OH$ group is axial. What is the structure of the anhydro sugar, and how is it formed? **(b)** D-Glucose also forms an anhydro sugar but the conditions required are much more drastic than for the corresponding reaction of D-altrose. Explain.

22.32 Show how the following experimental evidence can be used to deduce the structure of lactose (Section 22.12D):

1. Acid hydrolysis of lactose ($C_{12}H_{22}O_{11}$) gives equimolar quantities of D-glucose and D-galactose. Lactose undergoes a similar hydrolysis in the presence of a *β-galactosidase.*
2. Lactose is a reducing sugar and forms a phenylosazone; it also undergoes mutarotation.
3. Oxidation of lactose with bromine water followed by hydrolysis with dilute acid gives D-galactose and D-gluconic acid.
4. Bromine water oxidation of lactose followed by methylation and hydrolysis gives 2,3,6-tri-*O*-methylgluconolactone and 2,3,4,6-tetra-*O*-methyl-D-galactose.
5. Methylation and hydrolysis of lactose gives 2,3,6-tri-*O*-methyl-D-glucose and 2,3,4,6-tetra-*O*-methyl-D-galactose.

22.33 Deduce the structure of the disaccharide *melibiose* from the following data:

1. Melibiose is a reducing sugar that undergoes mutarotation and forms a phenylosazone.
2. Hydrolysis of melibiose with acid or with an *α-galactosidase* gives D-galactose and D-glucose.
3. Bromine water oxidation of melibiose gives *melibionic acid.* Hydrolysis of melibionic acid gives D-galactose and D-gluconic acid. Methylation of melibionic acid followed by hydrolysis gives 2,3,4,6-tetra-*O*-methyl-D-galactose and 2,3,4,5-tetra-*O*-methyl-D-gluconic acid.
4. Methylation and hydrolysis of melibiose gives 2,3,4,6-tetra-*O*-methyl-D-galactose and 2,3,4-tri-*O*-methyl-D-glucose.

22.34 Trehalose is a disaccharide that can be obtained from yeasts, fungi, sea urchins, algae, and insects. Deduce the structure of trehalose from the following information:

1. Acid hydrolysis of trehalose yields only D-glucose.
2. Trehalose is hydrolyzed by α-glucosidase but not by β-glucosidase enzymes.
3. Trehalose is a nonreducing sugar; it does not mutarotate, form a phenylosazone, or react with bromine water.
4. Methylation of trehalose followed by hydrolysis yields two molar equivalents of 2,3,4,6-tetra-*O*-methyl-D-glucose.

22.35 Outline chemical tests that will distinguish between members of each of the following pairs:

(a) D-Glucose and D-glucitol
(b) D-Glucitol and D-glucaric acid
(c) D-Glucose and D-fructose
(d) D-Glucose and D-galactose
(e) Sucrose and maltose
(f) Maltose and maltonic acid
(g) Methyl β-D-glucopyranoside and 2,3,4,6-tetra-*O*-methyl-β-D-glucopyranose
(h) Methyl α-D-ribofuranoside (**I**) and methyl 2-deoxy-α-D-ribofuranoside (**II**):

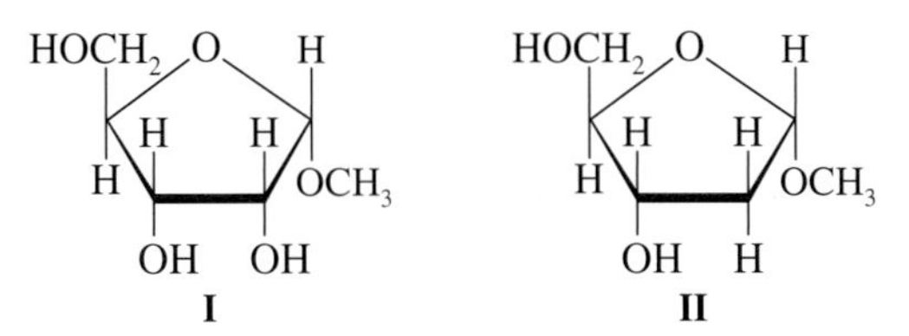

22.36 A group of oligosaccharides called *Schardinger dextrins* can be isolated from *Bacillus macerans* when the bacillus is grown on a medium rich in amylose. These oligosaccharides are all *nonreducing.* A typical Schardinger dextrin undergoes hydrolysis when treated with an acid or an α-glucosidase to yield six, seven, or eight molecules of D-glucose. Complete methylation of a Schardinger dextrin followed by acid hydrolysis yields only 2,3,6-tri-*O*-methyl-D-glucose. Propose a general structure for a Schardinger dextrin.

22.37 *Isomaltose* is a disaccharide that can be obtained by enzymatic hydrolysis of amylopectin. Deduce the structure of isomaltose from the following data:

1. Hydrolysis of 1 mol of isomaltose by acid or by an α-glucosidase gives 2 mol of D-glucose.
2. Isomaltose is a reducing sugar.
3. Isomaltose is oxidized by bromine water to isomaltonic acid. Methylation of isomaltonic acid and subsequent hydrolysis yields 2,3,4,6-tetra-*O*-methyl-D-glucose and 2,3,4,5-tetra-*O*-methyl-D-gluconic acid.

4. Methylation of isomaltose itself followed by hydrolysis gives 2,3,4,6-tetra-*O*-methyl-D-glucose and 2,3,4-tri-*O*-methyl-D-glucose.

22.38 *Stachyose* occurs in the roots of several species of plants. Deduce the structure of stachyose from the following data:

1. Acidic hydrolysis of 1 mol of stachyose yields 2 mol of D-galactose, 1 mol of D-glucose, and 1 mol of D-fructose.
2. Stachyose is a nonreducing sugar.
3. Treating stachyose with an α-galactosidase produces a mixture containing D-galactose, sucrose, and a nonreducing trisaccharide called *raffinose.*
4. Acidic hydrolysis of raffinose gives D-glucose, D-fructose, and D-galactose. Treating raffinose with an α-galactosidase yields D-galactose and sucrose. Treating raffinose with invertase (an enzyme that hydrolyzes sucrose) yields fructose and *melibiose* (see Problem 22.33).
5. Methylation of stachyose followed by hydrolysis yields 2,3,4,6-tetra-*O*-methyl-D-galactose, 2,3,4-tri-*O*-methyl-D-galactose, 2,3,4-tri-*O*-methyl-D-glucose, and 1,3,4,6-tetra-*O*-methyl-D-fructose.

22.39 *Arbutin,* a compound that can be isolated from the leaves of barberry, cranberry, and pear trees, has the molecular formula $C_{12}H_{16}O_7$. When arbutin is treated with aqueous acid or with a β-glucosidase, the reaction produces D-glucose and a compound **X** with the molecular formula $C_6H_6O_2$. The ^{1}H NMR spectrum of compound **X** consists of two singlets, one at δ 6.8 (4H) and one at δ 7.9 (2H). Methylation of arbutin followed by acidic hydrolysis yields 2,3,4,6-tetra-*O*-methyl-D-glucose and a compound **Y** ($C_7H_8O_2$). Compound **Y** is soluble in dilute aqueous NaOH but is insoluble in aqueous $NaHCO_3$. The ^{1}H NMR spectrum of **Y** shows a singlet at δ 3.9 (3H), a singlet at δ 4.8 (1H), and a multiplet (that resembles a singlet) at δ 6.8 (4H). Treating compound **Y** with aqueous NaOH and $(CH_3)_2SO_4$ produces compound **Z** ($C_8H_{10}O_2$). The ^{1}H NMR spectrum of **Z** consists of two singlets, one at δ 3.75 (6H) and one at δ 6.8 (4H). Propose structures for arbutin and for compounds **X, Y,** and **Z.**

22.40 When subjected to a Ruff degradation, a D-aldopentose, **A,** is converted to an aldotetrose, **B.** When reduced with sodium borohydride, the aldotetrose **B** forms an optically active alditol. The ^{13}C NMR spectrum of this alditol displays only two signals. The alditol obtained by direct reduction of **A** with sodium borohydride is not optically active. When **A** is used as the starting material for a Kiliani–Fischer synthesis, two diastereomeric aldohexoses, **C** and **D,** are produced. On treatment with sodium borohydride, **C** leads to an alditol **E,** and **D** leads to **F.** The ^{13}C NMR spectrum of **E** consists of three signals; that of **F** consists of six. Propose structures for **A–F.**

22.41 Figure 22.21 shows the ^{13}C NMR spectrum for the product of the reaction of D-(+)-mannose with acetone containing a trace of acid. This compound is a mannofuranose with some hydroxyl groups protected as acetone acetals (as acetonides). Use the ^{13}C NMR spectrum to determine how many acetonide groups are present in the compound.

22.42 D-(+)-Mannose can be reduced with sodium borohydride to form D-mannitol. When D-mannitol is dissolved in acetone containing a trace amount of acid and the product of this reaction subsequently oxidized with $NaIO_4$, a compound whose ^{13}C NMR spectrum consists of six signals is produced. One of these signals is near δ 200. What is the structure of this compound?

***22.43** Of the two anomers of methyl 2,3-anhydro-D-ribofuranoside, **I**, the β form has a strikingly lower boiling point. Suggest an explanation using their structural formulas.

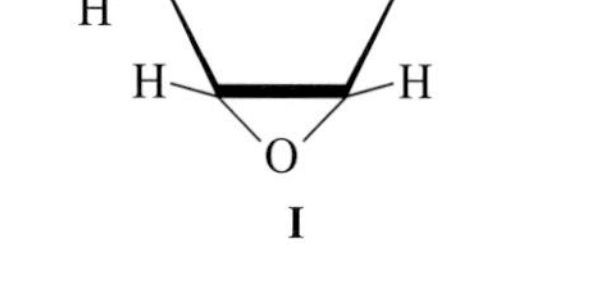

$HOCH_2$
O
OCH_3
H
H
H
O

I

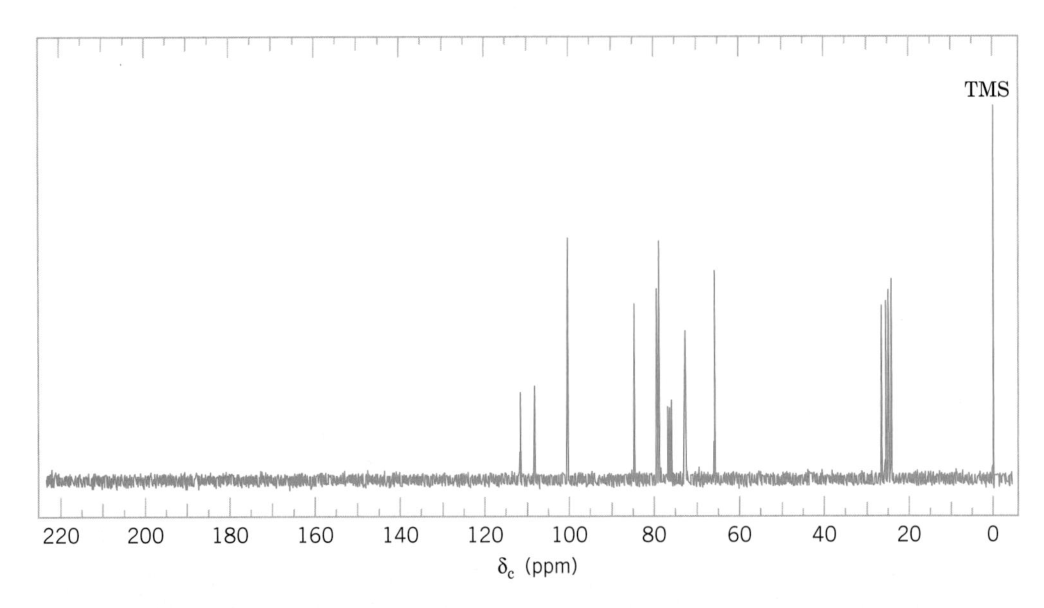

FIGURE 22.21 The broadband proton-decoupled ^{13}C NMR spectrum for the reaction product in Problem 22.41.

***22.44** The following reaction sequence represents an elegant method of synthesis of 2-deoxy-D-ribose, **IV**, published by D. C. C. Smith in 1955:

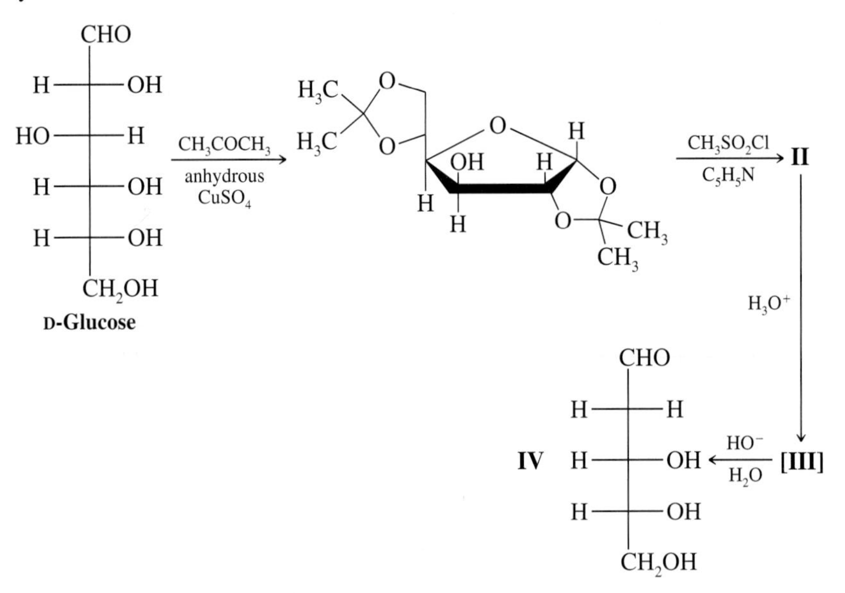

(a) What are the structures of **II** and **III**?
(b) Propose a mechanism for the conversion of **III** to **IV**.

***22.45** D-Glucose $\xrightarrow[\text{anhydrous sodium acetate}]{\text{acetic anhydride}}$ D-Glucopyranose pentaacetate, anomer **V**

D-Glucose $\xrightarrow[\text{pyridine}]{\text{acetic anhydride}}$ D-Glucopyranose pentaacetate, anomer **VI**

The ^1H NMR data for the two anomers included very comparable peaks in the δ 2.0–5.6 region but differed in that, as their highest δ peaks, anomer **V** had a doublet at δ 5.8 (1H, J = 12 Hz) while anomer **VI** had a doublet at δ 6.3 (1H, J = 4 Hz).

(a) Which proton in these anomers would be expected to have these highest δ values?

(b) Why do the signals for these protons appear as doublets?

(c) The relationship between the magnitude of the observed coupling constant and the dihedral angle (when measured using a Newman projection) between C—H bonds on the adjacent carbons of a C—C bond is given by the Karplus equation. It indicates that an axial–axial relationship results in a coupling constant of about 9 Hz (observed range is 8–14 Hz) and an equatorial–axial relationship results in a coupling constant of about 2 Hz (observed range is 1–7 Hz). Which of **V** and **VI** is the α-anomer and which is the β-anomer?

(d) Draw the most stable conformer for each of **V** and **VI**.

Learning Group Problems

1. **(a)** The members of one class of low-calorie sweeteners are called polyols. The chemical synthesis of one such polyol sweetener involves reduction of a certain disaccharide to a mixture of diastereomeric glycosides. The alcohol (actually polyol) portion of the diastereomeric glycosides derives from one of the sugar moieties in the original disaccharide. Exhaustive methylation of the sweetener (e.g., with dimethyl sulfate in the presence of hydroxide) followed by hydrolysis would be expected to produce 2,3,4,6-tetra-*O*-methyl-α-D-glucopyranose, 1,2,3,4,5-penta-*O*-methyl-D-sorbitol, and 1,2,3,4,5-penta-*O*-methyl-D-mannitol, in the ratio of 2 : 1 : 1. On the basis of this information, deduce the structure of the two disaccharide glycosides that make up the diastereomeric mixture in this polyol sweetener.

   (b) Knowing that the mixture of two disaccharide glycosides in this sweetener results from reduction of a single disaccharide starting material (e.g., reduction by sodium borohydride), what would the structure be of the disaccharide *reactant* for the reduction step? Explain how reduction of this compound would produce the two glycosides.

   (c) Write the lowest energy chair conformational structure for 2,3,4,6-tetra-*O*-methyl-α-D-glucopyranose.

2. Shikimic acid is a key biosynthetic intermediate in plants and microorganisms. In the Chapter 21 Learning Group Problem we saw that, in nature, shikimic acid is converted to chorismate, which is then converted to prephenate, ultimately leading to aromatic amino acids and other essential plant and microbial metabolites. In the course of research on biosynthetic pathways involving shikimic acid, H. Floss (University of Washington) required shikimic acid labeled with ^{13}C to trace the destiny of the labeled carbon atoms in later biochemical transformations. To synthesize the labeled shikimic acid, Floss adapted a synthesis of optically active shikimic acid from D-mannose reported earlier by G. W. J. Fleet (Oxford University). This synthesis is a prime example of how natural sugars can be excellent chiral starting materials for the chemical synthesis of optically active target molecules. It is also an excellent example of classic reactions in carbohydrate chemistry. The Fleet–Floss synthesis of D-(−)-[1,7-^{13}C]-shikimic acid **(1)** from D-mannose is shown in Scheme 1.

   (a) Comment on the several transformations that occur between D-mannose and **2.** What new functional groups are formed?

   (b) What is accomplished in the steps from **2** to **3, 3** to **4,** and **4** to **5?**

   (c) Deduce the structure of compound **9** (a reagent used to convert **5** to **6**), knowing that it was a carbanion that displaced the trifluoromethanesulfonate (triflate) group of **5.** Note that it was compound **9** that brought with it the required ^{13}C atoms for the final product.

   (d) Explain the transformation from **7** to **8.** Write out the structure of the compound in equilibrium with **7** that would be required for the process from **7** to **8** to occur. What is the name given to the reaction from this intermediate to **8?**

(e) Label the carbon atoms of D-mannose and **1** by number or letter so as to show which atoms in **1** came from which atoms of D-mannose.

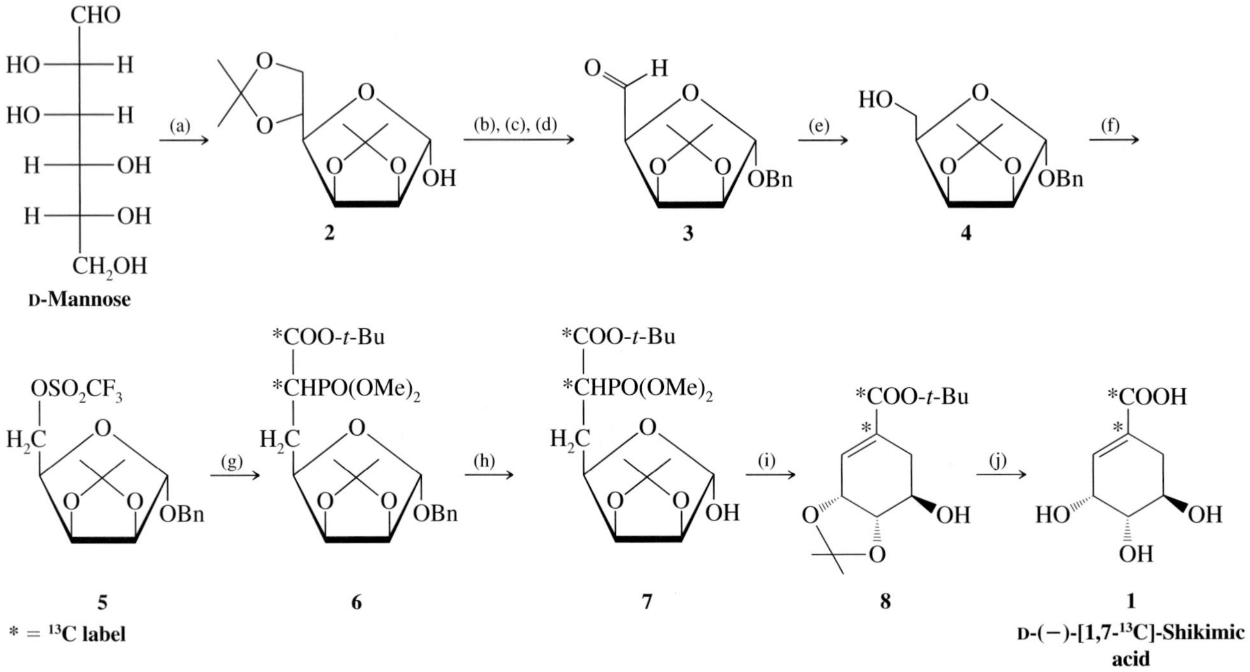

D-Mannose

2 **3** **4**

5 **6** **7** **8** **1**

* = ^{13}C label

D-(−)-[1,7-^{13}C]-Shikimic acid

SCHEME 1 The synthesis of D-(−)-[1,7-^{13}C]-shikimic acid (1) by H. G. Floss, based on the route of Fleet, et al. Conditions: (a) acetone, H^+; (b) BnCl, NaH; (c) HCl, aq. MeOH; (d) $NaIO_4$; (e) $NaBH_4$; (f) $(CF_3SO_2)_2O$, pyridine; (g) 9, NaH; (h) $HCOO^-$ NH_4^+, Pd/C; (i) NaH; (j) 60% aq. CF_3COOH.

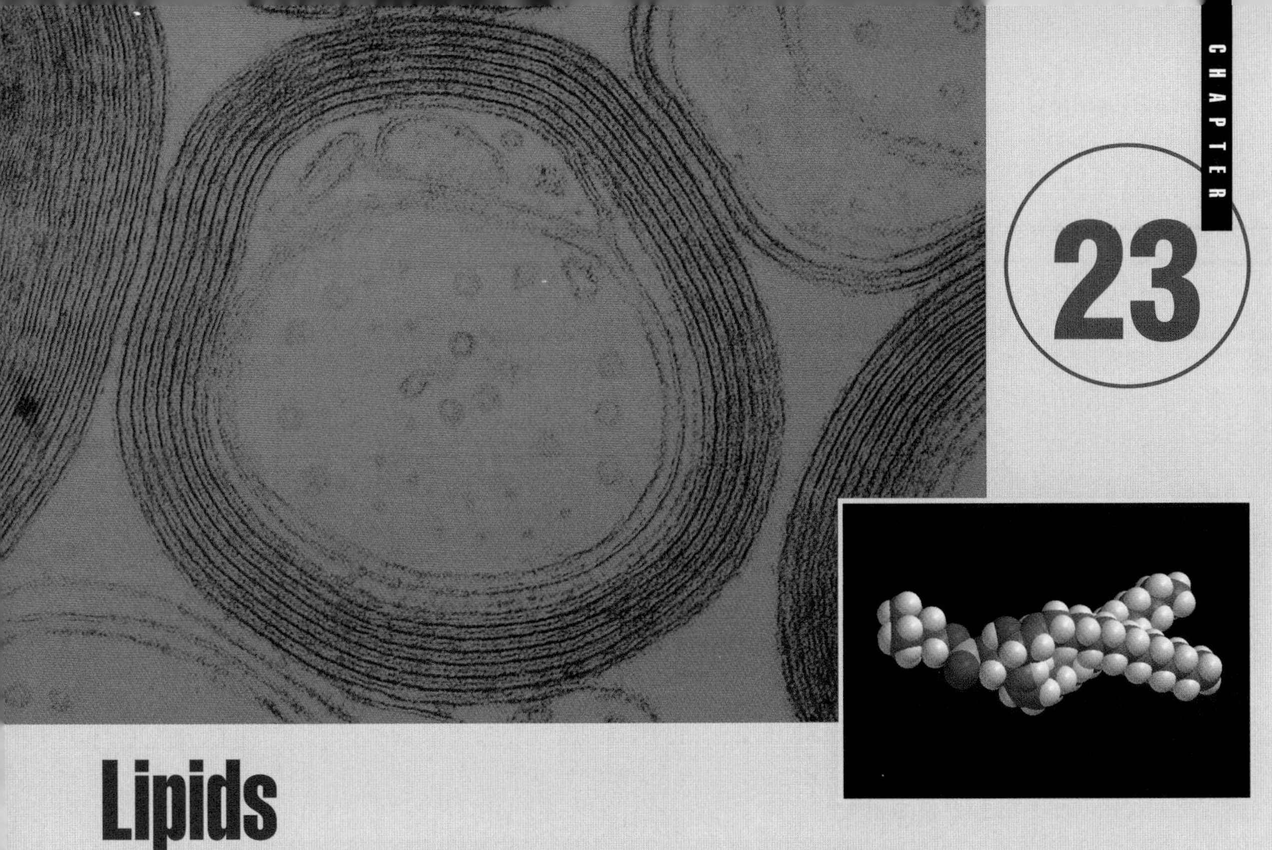

Lipids

Insulation for Nerves

A bare wire conducting electricity will form a short circuit if it touches another conductor. This, of course, is why electrical wires are insulated. The axons of large neurons, the electrical conduits of the nervous system, are also insulated. Just as in electrical wires whose covering is an insulating sheath of plastic, a feature called the myelin sheath insulates the axons of many nerve cells from their extracellular environment. The myelin sheath is formed by the membrane of specialized cells, called Schwann cells, which grow around the axon and encircle it many times. In the structure of this membrane are molecules called lipids, a major component of which in myelin is sphingomyelin. A molecular model of sphingomyelin is shown above, and its structure is given in Section 23.6B. Wrapping of the axon by the Schwann cell membrane provides layer on layer of insulation by sphingomyelin and related lipid molecules. This is the key to the insulating property of the myelin sheath.

Unlike electrical wires that require insulation from end to end, the lipid layers of the myelin sheath are not a continuous insulator for the axon. Periodic gaps in the myelin sheath create nodes (called nodes of Ranvier) between which electrical signals of the nerve impulses hop along the axon. Propagation of nerve impulses in this way occurs at velocities up to 100 m s^{-1}, much faster than propagation in unmyelinated nerve fibers where this hopping effect is not possible. Impulse propagation in unmyelinated nerves is roughly 10 times slower than in myelinated nerves. The hopping of a nerve impulse between nodes is shown schematically in the diagram on the following page.

As you might expect, myelination of nerve fibers is crucial for proper neurological function. Multiple sclerosis, for example, is an autoimmune disease that causes demyelination of nerve cells, usually with very serious neurological consequences. Other conditions called sphingolipid storage diseases cause a buildup of various sphingolipids, which has various consequences. Examples of sphingolipid storage diseases are Tay-Sachs disease and Krabbe's disease. Both of these are fatal to children under the age of 3.

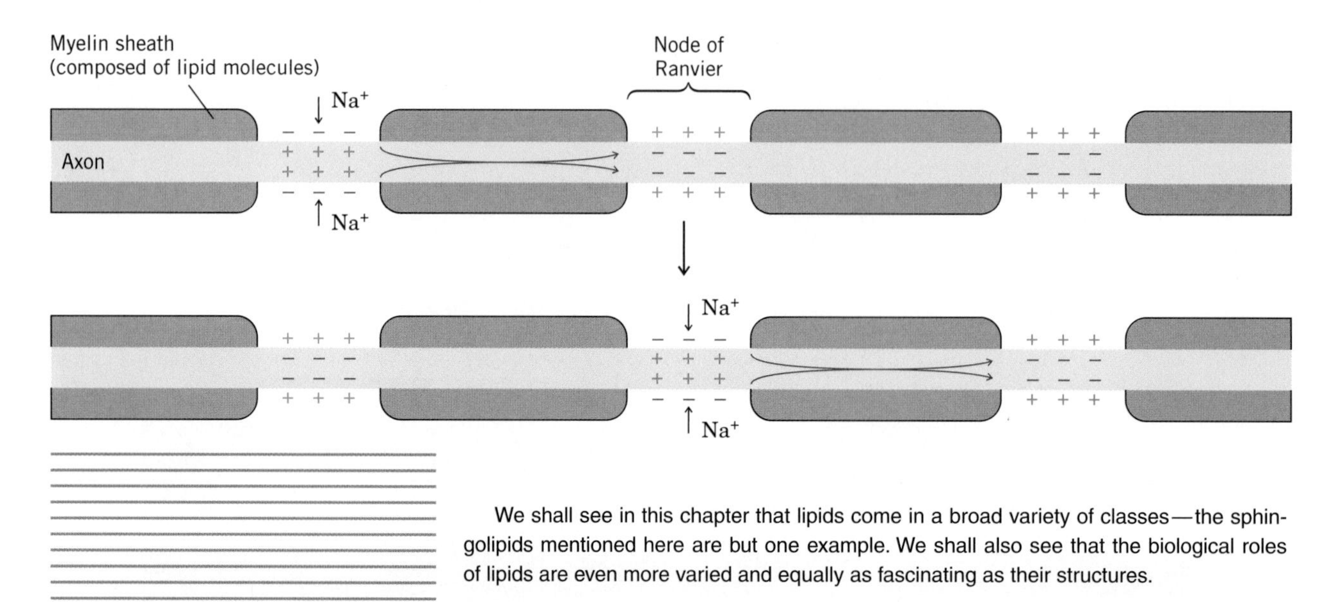

We shall see in this chapter that lipids come in a broad variety of classes—the sphingolipids mentioned here are but one example. We shall also see that the biological roles of lipids are even more varied and equally as fascinating as their structures.

23.1 Introduction

Lipids are compounds of biological origin that dissolve in nonpolar solvents, such as chloroform and diethyl ether. The name lipid comes from the Greek word *lipos,* for fat. Unlike carbohydrates and proteins, which are defined in terms of their structures, lipids are defined by the physical operation that we use to isolate them. Not surprisingly, then, lipids include a variety of structural types. Examples are the following:

$$\begin{array}{l} CH_2{-}O{-}\overset{O}{\overset{\|}{C}}{-}R \\ | \\ CH{-}O{-}\overset{O}{\overset{\|}{C}}{-}R' \\ | \\ CH_2{-}O{-}\overset{O}{\overset{\|}{C}}{-}R'' \end{array}$$

A fat or oil
(a triacylglycerol)

Menthol
(a terpenoid)

Vitamin A
(a terpenoid)

$$\begin{array}{l} CH_2{-}O{-}\overset{O}{\overset{\|}{C}}{-}R \\ | \\ CH{-}O{-}\overset{O}{\overset{\|}{C}}{-}R' \\ | \\ CH_2{-}O{-}\underset{O^-}{\underset{|}{\overset{O}{\overset{\|}{P}}}}{-}OCH_2CH_2\overset{+}{N}(CH_3)_3 \end{array}$$

A lecithin
(a phosphatide)

Cholesterol
(a steroid)

23.2 Fatty Acids and Triacylglycerols

Only a small portion of the total lipid fraction obtained by extraction with a nonpolar solvent consists of long-chain carboxylic acids. Most of the carboxylic acids of biological origin are found as *esters of glycerol,* that is, as **triacylglycerols** (Fig. 23.1).*

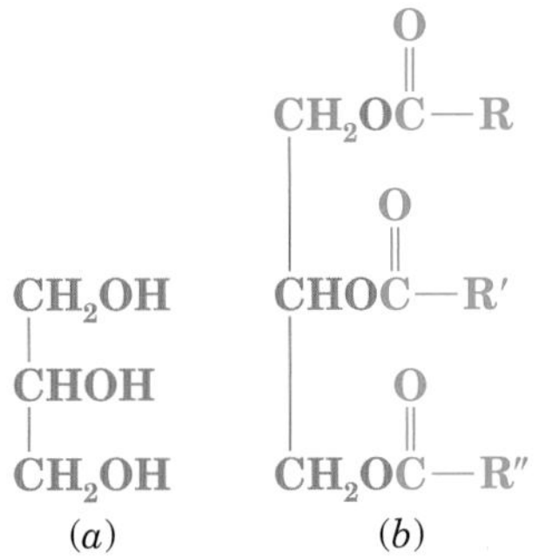

$$\begin{array}{l} CH_2OH \\ | \\ CHOH \\ | \\ CH_2OH \end{array} \qquad \begin{array}{l} CH_2OC(=O)-R \\ | \\ CHOC(=O)-R' \\ | \\ CH_2OC(=O)-R'' \end{array}$$

(*a*) (*b*)

FIGURE 23.1 *(a)* Glycerol. *(b)* A triacylglycerol. The groups R, R′, and R″ are usually long-chain alkyl groups. R, R′, and R″ may also contain one or more carbon–carbon double bonds. In a triacylglycerol R, R′, and R″ may all be different.

Triacylglycerols are the oils of plants and the fats of animal origin. They include such common substances as peanut oil, soybean oil, corn oil, sunflower oil, butter, lard, and tallow. Triacylglycerols that are liquids at room temperature are generally called **oils;** those that are solids are called **fats.** Triacylglycerols can be **simple triacylglycerols** in which all three acyl groups are the same. More commonly, however, the triacylglycerol is a **mixed triacylglycerol** in which the acyl groups are different.

Hydrolysis of a fat or oil produces a mixture of fatty acids:

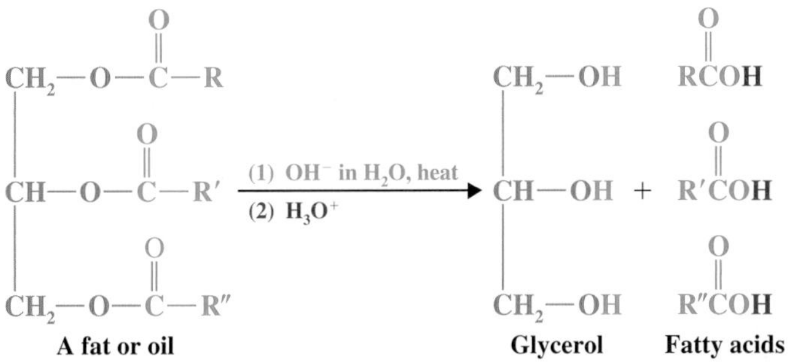

$$\begin{array}{l} CH_2-O-C(=O)-R \\ | \\ CH-O-C(=O)-R' \\ | \\ CH_2-O-C(=O)-R'' \end{array} \xrightarrow[(2)\ H_3O^+]{(1)\ OH^-\ \text{in}\ H_2O,\ \text{heat}} \begin{array}{l} CH_2-OH \\ | \\ CH-OH \\ | \\ CH_2-OH \end{array} + \begin{array}{l} RC(=O)OH \\ R'C(=O)OH \\ R''C(=O)OH \end{array}$$

A fat or oil **Glycerol** **Fatty acids**

Most natural fatty acids have **unbranched chains** and, because they are synthesized from two-carbon units, **they have an even number of carbon atoms.** Table 23.1 lists some of the most common fatty acids, and Table 23.2 gives the fatty acid composition of a number of common fats and oils. Notice that in the **unsaturated fatty acids** in Table 23.1 **the double bonds are all cis.** Many naturally occurring fatty acids contain two or three double bonds. The fats or oils that these come from are called **polyunsaturated fats or oils.** The first double bond of an unsaturated fatty acid commonly occurs between C9 and C10; the remaining double bonds tend to begin with C12 and C15 (as in linoleic acid and linolenic acid). The double bonds, therefore, *are not conjugated.* Triple bonds rarely occur in fatty acids.

We saw how fatty acids are biosynthesized in two-carbon units in Special Topic D.

The carbon chains of **saturated fatty acids** can adopt many conformations but tend to be fully extended because this minimizes steric repulsions between neighboring methyl-

*In the older literature triacylglycerols were referred to as triglycerides, or simply as glycerides. In IUPAC nomenclature, because they are esters of glycerol, they should be named as glyceryl trialkanoates, glyceryl trialkenoates, and so on.

TABLE 23.1 Common Fatty Acids

	mp (°C)
Saturated Carboxylic Acids	
$CH_3(CH_2)_{12}CO_2H$ **Myristic acid (tetradecanoic acid)**	54
$CH_3(CH_2)_{14}CO_2H$ **Palmitic acid (hexadecanoic acid)**	63
$CH_3(CH_2)_{16}CO_2H$ **Stearic acid (octadecanoic acid)**	70
Unsaturated Carboxylic Acids	
$CH_3(CH_2)_5(H)C{=}C(H)(CH_2)_7CO_2H$ **Palmitoleic acid (*cis*-9-hexadecenoic acid)**	32
$CH_3(CH_2)_7(H)C{=}C(H)(CH_2)_7CO_2H$ **Oleic acid (*cis*-9-octadecenoic acid)**	4
$CH_3(CH_2)_4(H)C{=}C(H)CH_2(H)C{=}C(H)(CH_2)_7CO_2H$ **Linoleic acid (*cis, cis*-9,12-octadecadienoic acid)**	−5
$CH_3CH_2(H)C{=}C(H)CH_2(H)C{=}C(H)CH_2(H)C{=}C(H)(CH_2)_7CO_2H$ **Linolenic acid (*cis, cis, cis*-9,12,15-octadecatrienoic acid)**	−11
(structure with repeating unit, subscript 6) COOH **DHA, an omega-3 fatty acid [(4*Z*,7*Z*,10*Z*,13*Z*,16*Z*,19*Z*)-4,7,10,13,16,19-docosahexaenoic acid]**	−44
H_3C (structure with repeating unit, subscript 4) COOH **Arachidonic acid, an omega-6 fatty acid [(5*Z*,8*Z*,11*Z*,14*Z*)-5,8,11,14-eicosatetraenoic acid]**	−49

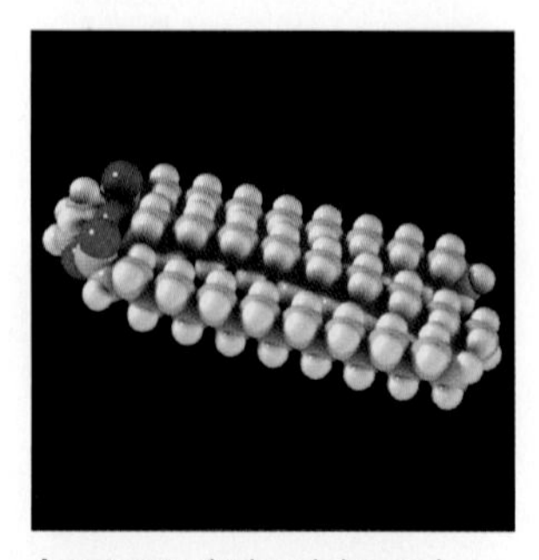
A saturated triacylglycerol.

ene groups. Saturated fatty acids pack efficiently into crystals, and because van der Waals attractions are large, they have relatively high melting points. The melting points increase with increasing molecular weight. The cis configuration of the double bond of an unsaturated fatty acid puts a rigid bend in the carbon chain that interferes with crystal packing, causing reduced van der Waals attractions between molecules. Unsaturated fatty acids, consequently, have lower melting points.

TABLE 23.2 Fatty Acid Composition Obtained by Hydrolysis of Common Fats and Oils[a]

	Average Composition of Fatty Acids (mol %)											
	Saturated								Unsaturated			
Fat or Oil	C_4 Butyric Acid	C_6 Caproic Acid	C_8 Caprylic Acid	C_{10} Capric Acid	C_{12} Lauric Acid	C_{14} Myristic Acid	C_{16} Palmitic Acid	C_{18} Stearic Acid	C_{16} Palmitoleic Acid	C_{18} Oleic Acid	C_{18} Linoleic Acid	C_{18} Linolenic Acid
Animal Fats												
Butter	3–4	1–2	0–1	2–3	2–5	8–15	25–29	9–12	4–6	18–33	2–4	
Lard						1–2	25–30	12–18	4–6	48–60	6–12	0–1
Beef tallow						2–5	24–34	15–30		35–45	1–3	0–1
Vegetable Oils												
Olive						0–1	5–15	1–4		67–84	8–12	
Peanut							7–12	2–6		30–60	20–38	
Corn						1–2	7–11	3–4	1–2	25–35	50–60	
Cottonseed						1–2	18–25	1–2	1–3	17–38	45–55	
Soybean						1–2	6–10	2–4		20–30	50–58	5–10
Linseed							4–7	2–4		14–30	14–25	45–60
Coconut		0–1	5–7	7–9	40–50	15–20	9–12	2–4	0–1	6–9	0–1	
Marine Oils												
Cod liver						5–7	8–10	0–1	18–22	27–33	27–32	

[a]Data adapted from Holum, J. R. *Organic and Biological Chemistry;* Wiley: New York, 1978; p 220 and from Altman, P. L.; Ditmer, D. S., Eds. *Biology Data Book;* Federation of American Societies for Experimental Biology: Washington, DC; 1964.

Fatty acids known as omega-3 fatty acids are those where the third to last carbon in the chain is part of a carbon–carbon double bond. Long-chain omega-3 fatty acids incorporated in the diet are believed to have beneficial effects in terms of reducing the risk of fatal heart attack and easing certain autoimmune diseases, including rheumatoid arthritis and psoriasis. Oil from fish such as tuna and salmon are good sources of omega-3 fatty acids, including the C22 omega-3 fatty acid docosahexaenoic acid [DHA, whose full IUPAC name is (4Z, 7Z, 10Z, 13Z, 16Z, 19Z)-4, 7, 10, 13, 16, 19-docosahexaenoic acid]. DHA is also found in breast milk, gray matter of the brain, and retinal tissue.

COOH

(4Z,7Z,10Z,13Z,16Z,19Z)-4,7,10,13,16,19-Docosahexaenoic acid
(DHA, an omega-3 fatty acid)

What we have just said about the fatty acids applies to the triacylglycerols as well. Triacylglycerols made up of largely saturated fatty acids have high melting points and are solids at room temperature. They are what we call *fats.* Triacylglycerols with a high proportion of unsaturated and polyunsaturated fatty acids have lower melting points. They are *oils.* Figure 23.2 shows how the introduction of a single cis double bond affects the shape of a triacylglycerol and how catalytic hydrogenation can be used to convert an unsaturated triacylglycerol into a saturated one.

23.2A Hydrogenation of Triacylglycerols

Solid commercial cooking fats are manufactured by partial hydrogenation of vegetable oils. The result is the familiar "partially hydrogenated fat" present in so many prepared foods. Complete hydrogenation of the oil is avoided because a completely saturated triacylglycerol is very hard and brittle. Typically, the vegetable oil is hydrogenated until a semisolid of appealing consistency is obtained. One commercial advantage of partial hydrogenation is to give the fat a longer shelf life. Polyunsaturated oils tend to react by au-

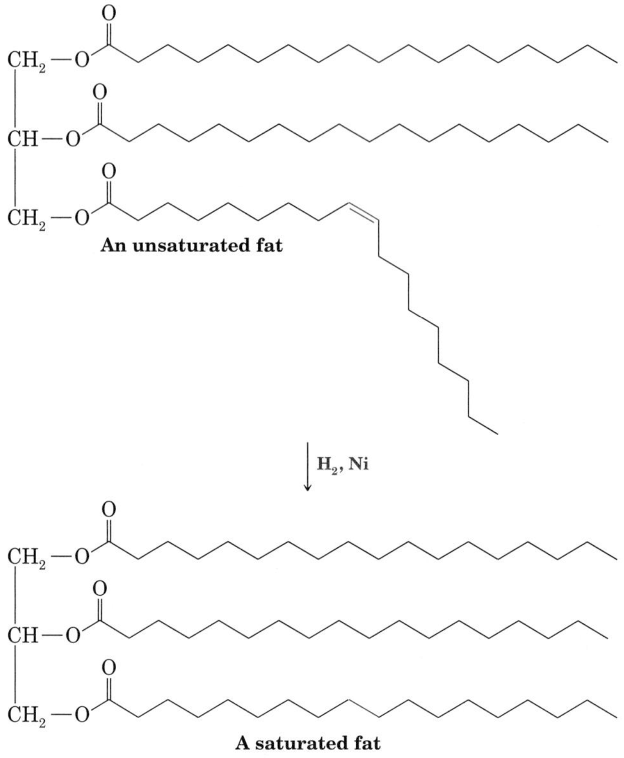

FIGURE 23.2 Two typical triacylglycerols, one unsaturated and one saturated. The cis double bond of the unsaturated triacylglycerol interferes with efficient crystal packing and causes an unsaturated fat to have a lower melting point. Hydrogenation of the double bond causes an unsaturated triacylglycerol to become saturated.

toxidation (Section 10.11D), causing them to become rancid. One problem with partial hydrogenation, however, is that the catalyst isomerizes some of the unreacted double bonds from the natural cis arrangement to the unnatural trans arrangement, and there is accumulating evidence that "trans" fats are associated with an increased risk of cardiovascular disease.

23.2B Biological Functions of Triacylglycerols

The primary function of triacylglycerols in animals is as an energy reserve. When triacylglycerols are converted to carbon dioxide and water by biochemical reactions (i.e., when triacylglycerols are *metabolized*), they yield more than twice as many kilocalories per gram as do carbohydrates or proteins. This is largely because of the high proportion of carbon–hydrogen bonds per molecule.

In animals, specialized cells called **adipocytes** (fat cells) synthesize and store triacylglycerols. The tissue containing these cells, adipose tissue, is most abundant in the abdominal cavity and in the subcutaneous layer. Men have a fat content of about 21%, women about 26%. This fat content is sufficient to enable us to survive starvation for 2–3 months. By contrast, glycogen, our carbohydrate reserve, can provide only one day's energy need.

All of the saturated triacylglycerols of the body, and some of the unsaturated ones, can be synthesized from carbohydrates and proteins. Certain polyunsaturated fatty acids, however, are essential in the diets of higher animals.

The amount of fat in the diet, especially the proportion of saturated fat, has been a health concern for many years. There is compelling evidence that too much saturated fat in the diet is a factor in the development of heart disease and cancer.

The Chemistry of...

Olestra and Other Fat Substitutes

Olestra is a zero-calorie commercial fat substitute with the look and feel of natural fats. It is a synthetic compound whose structure involves a novel combination of natural components. The core of olestra is derived from sucrose, ordinary table sugar. Six to eight of the hydroxyl groups on the sucrose framework have long-chain carboxylic acids (fatty acids) appended to them by ester linkages. These fatty acids are from C_8 to C_{22} in length. In the industrial synthesis of olestra, these fatty acids derive from cottonseed or soybean oil.

A food product made with olestra.

Olestra.

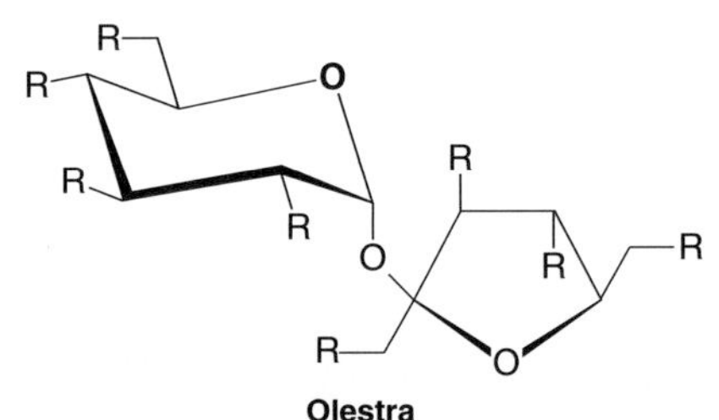

Olestra
Six to eight of the R groups are fatty acid esters, the remainder being hydroxyl groups.

The presence of fatty acid esters in olestra bestows on it the taste and culinary properties of an ordinary fat. Yet, olestra is not digestible like a typical fat. This is because the steric bulk of olestra renders it unacceptable to the enzymes that catalyze hydrolysis of ordinary fats. Olestra passes through the digestive tract unchanged and thereby adds no calories to the diet. As it does so, however, olestra associates with and carries away some of the lipid-soluble vitamins, namely, vitamins A, D, E, and K. Foods prepared with olestra are supplemented with these vitamins to compensate for any loss that may result from their extraction by olestra. Studies conducted since olestra's approval have demonstrated that people report no more bothersome digestive effects when eating Olean (the trademark name for olestra) snacks than they do when eating full-fat chips.

Many other fat substitutes have received consideration. Among these are polyglycerol esters, which presumably by their steric bulk would also be undigestible, like the polyester olestra. Another approach to low-calorie fats, already in commercial use, involves replacement of some long-chain carboxylic acids on the glycerol backbone with medium- or short-chain carboxylic acids (C_2 to C_4). These compounds provide fewer calories because each CH_2 group that is absent from the glycerol ester (as compared to long-chain fatty acids) reduces the amount of energy (calories) liberated when that compound is metabolized. The calorie content of a given glycerol ester can essentially be tailored to provide a desired calorie output, simply by adjusting the ratio of long-chain to medium- and short-chain carboxylic acids. Still other low-calorie fat substitutes are carbohydrate- and protein-based compounds. These materials act by generating a similar gustatory response to that of fat, but for various reasons produce fewer calories.

23.2C Saponification of Triacylglycerols

Alkaline hydrolysis (i.e., **saponification**) of triacylglycerols produces glycerol and a mixture of salts of long-chain carboxylic acids:

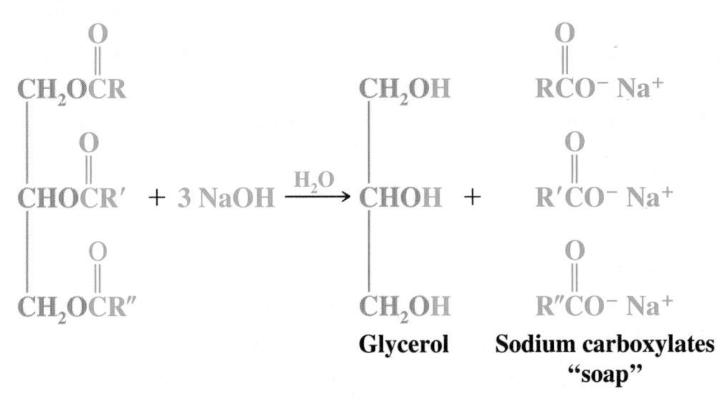

These salts of long-chain carboxylic acids are **soaps,** and this saponification reaction is the way most soaps are manufactured. Fats and oils are boiled in aqueous sodium hydroxide until hydrolysis is complete. Adding sodium chloride to the mixture then causes the soap to precipitate. (After the soap has been separated, glycerol can be isolated from the aqueous phase by distillation.) Crude soaps are usually purified by several reprecipitations. Perfumes can be added if a toilet soap is the desired product. Sand, sodium carbonate, and other fillers can be added to make a scouring soap, and air can be blown into the molten soap if the manufacturer wants to market a soap that floats.

CD TUTORIAL
Micelle formation.

The sodium salts of long-chain carboxylic acids (soaps) are almost completely miscible with water. However, they do not dissolve as we might expect, that is, as individual ions. Except in very dilute solutions, soaps exists as **micelles** (Fig. 23.3). Soap micelles are usually spherical clusters of carboxylate anions that are dispersed throughout the aqueous phase. The carboxylate anions are packed together with their negatively charged (and thus, *polar*) carboxylate groups at the surface and with their nonpolar hydrocarbon chains on the interior. The sodium ions are scattered throughout the aqueous phase as individual solvated ions.

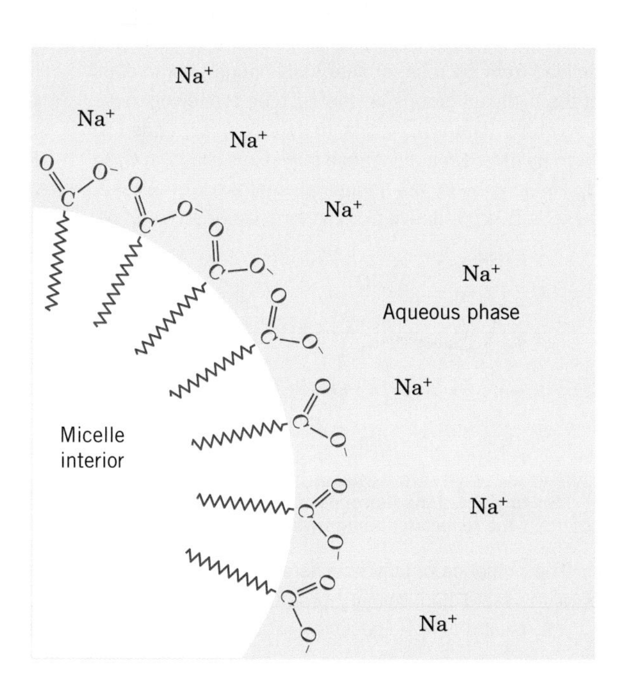

FIGURE 23.3 A portion of a soap micelle showing its interface with the polar dispersing medium.

Micelle formation accounts for the fact that soaps dissolve in water. The nonpolar (and thus **hydrophobic**) alkyl chains of the soap remain in a nonpolar environment—in the interior of the micelle. The polar (and therefore **hydrophilic**) carboxylate groups are exposed to a polar environment—that of the aqueous phase. Because the surfaces of the micelles are negatively charged, individual micelles repel each other and remain dispersed throughout the aqueous phase.

Soaps serve their function as "dirt removers" in a similar way. Most dirt particles (e.g., on the skin) become surrounded by a layer of an oil or fat. Water molecules alone are unable to disperse these greasy globules because they are unable to penetrate the oily layer and separate the individual particles from each other or from the surface to which they are stuck. Soap solutions, however, *are* able to separate the individual particles because their hydrocarbon chains can "dissolve" in the oily layer (Fig. 23.4). As this happens, each individual particle develops an outer layer of carboxylate anions and presents the aqueous

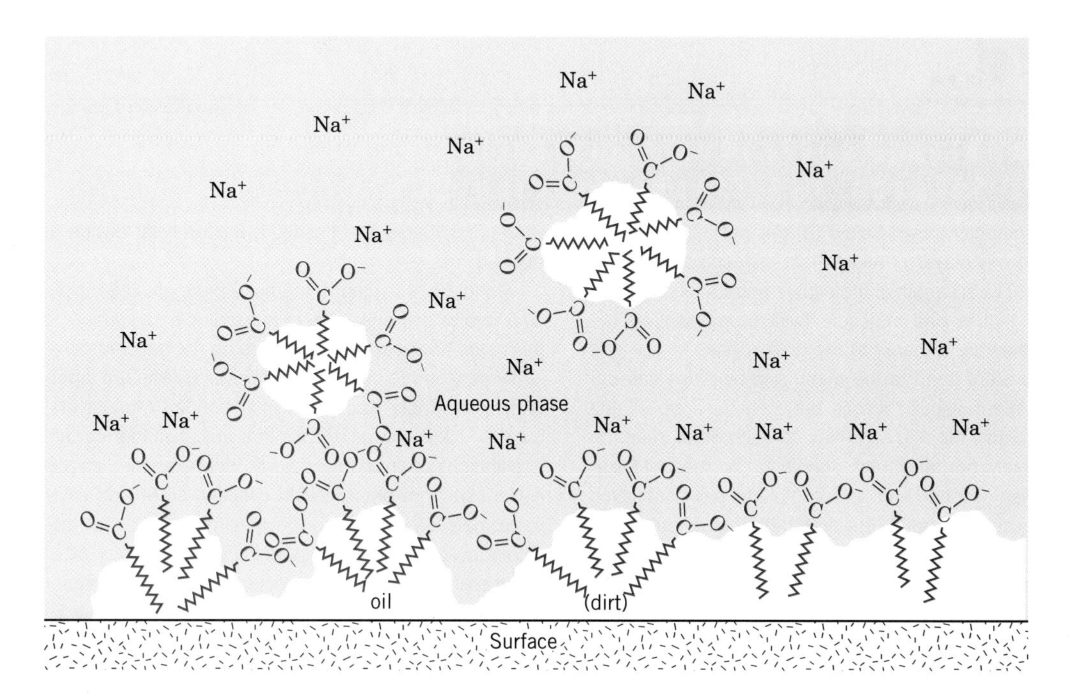

FIGURE 23.4 Dispersal of oil-coated dirt particles by a soap.

phase with a much more compatible exterior—a polar surface. The individual globules now repel each other and thus become dispersed throughout the aqueous phase. Shortly thereafter, they make their way down the drain.

Synthetic detergents (Fig. 23.5) function in the same way as soaps; they have long nonpolar alkane chains with polar groups at the end. The polar groups of most synthetic detergents are sodium sulfonates or sodium sulfates. (At one time, extensive use was made of synthetic detergents with highly branched alkyl groups. These detergents proved to be nonbiodegradable, and their use was discontinued.)

$CH_3(CH_2)_{10}CH_2SO_2O^- \, Na^+$

Sodium alkanesulfonates

$CH_3(CH_2)_{10}CH_2OSO_2O^- \, Na^+$

Sodium alkyl sulfates

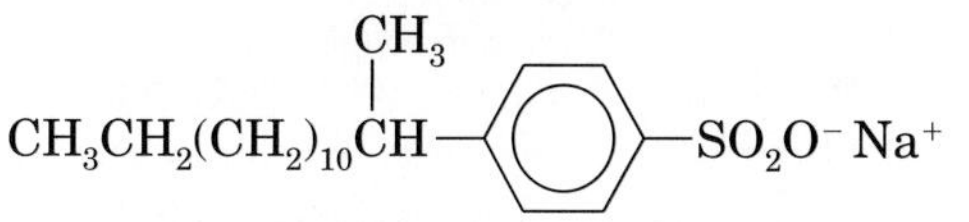

$CH_3CH_2(CH_2)_{10}CH(CH_3)-C_6H_4-SO_2O^- \, Na^+$

Sodium alkylbenzenesulfonates

FIGURE 23.5 Typical synthetic detergents.

Synthetic detergents offer an advantage over soaps; they function well in "hard" water, that is, water containing Ca^{2+}, Fe^{2+}, Fe^{3+}, and Mg^{2+} ions. Calcium, iron, and magnesium salts of alkanesulfonates and alkyl hydrogen sulfates are largely water soluble, and thus synthetic detergents remain in solution. Soaps, by contrast, form precipitates—the ring around the bathtub—when they are used in hard water.

The Chemistry of...

Self-Assembled Monolayers—Lipids in Materials Science and Bioengineering

The graphic shown below (a) depicts a self-assembled monolayer of alkanethiol molecules on a gold surface. The alkanethiol molecules spontaneously form a layer that is one molecule thick (a monolayer) because they are tethered to the gold surface at one end by a covalent bond to the metal and because van der Waals intermolecular forces between the long alkane chains cause them to align next to each other in an approximately perpendicular orientation to the gold surface. Many researchers are exploiting self-assembled monolayers (SAMs) for the preparation of surfaces that have specific uses in medicine, computing, and telecommunications. One example in biomedical engineering that may lead to advances in surgery involves testing cells for their response to SAMs with varying head groups. By varying the structure of the exposed head group of the monolayer, it may be possible to create materials that have either affinity for or resistance against cell binding (b). Such properties could be useful in organ transplants for inhibiting rejection by cells of the immune system or in prosthesis surgeries where the binding of tissue to the artificial device is desired.

Monolayers called Langmuir–Blodgett (LB) films also involve self-assembly of molecules on a surface. In this case, however, the molecules do not become covalently attached to the surface. These LB films are inherently less stable than covalently bonded monolayers, but they have characteristics that are useful for certain applications in nanotechnology. For example, an LB film made from phospholipid (Section 23.6) and catenane molecules was used in making the array of molecular switches we discussed in "The Chemistry of. . . Nanoscale Motors and Molecular Switches" (Chapter 4). This LB monolayer (c) was formed at a water–air interface where the polar phosphate head groups of the phospholipid buried themselves in water and the hydrophobic carbon tails projected out into the air. Interspersed among them were the catenane molecules. In later steps, this monolayer was lifted from the water–air surface and transferred onto a solid gold surface.

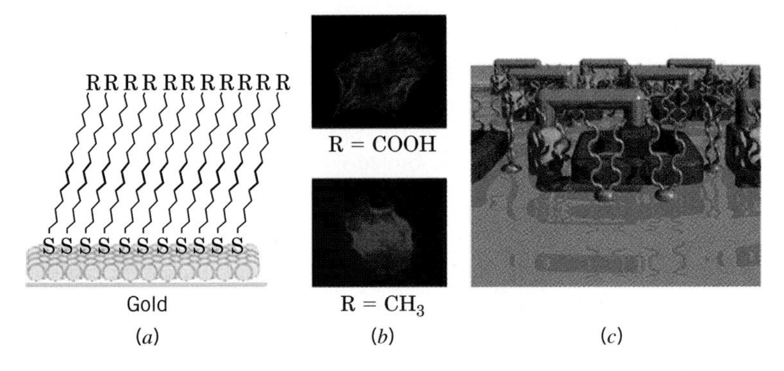

(a) A self-assembled monolayer of alkanethiol molecules on a gold surface ($R{=}CH_3$ or COOH).
(b) Spreading of a Swiss 3T3 fibroblast cell plated on a COOH-terminated self-assembled monolayer (top) indicates effective signaling on the surface. The fibroblast cell on a CH_3-terminated monolayer curls away from surface. The cells were stained with a rhodamine-tagged toxin that binds to filamentous actin and then were imaged under fluorescent light. (c) A Langmuir-Blodgett (LB) film formed from phospholipid molecules (golden color) and catenane molecules (purple and gray with green and red groups) at an air-water interface.

23.2D Reactions of the Carboxyl Group of Fatty Acids

Fatty acids, as we might expect, undergo reactions typical of carboxylic acids (see Chapter 18). They react with $LiAlH_4$ to form alcohols, with alcohols and mineral acid to form esters, and with thionyl chloride to form acyl chlorides:

$$\text{RCH}_2\text{C(=O)OH (Fatty acid)} \xrightarrow[(2)\ H_2O]{(1)\ LiAlH_4,\ Et_2O} \text{RCH}_2\text{CH}_2\text{OH (Long-chain alcohol)}$$

$$\text{RCH}_2\text{C(=O)OH} \xrightarrow{CH_3OH,\ HA} \text{RCH}_2\text{C(=O)OCH}_3 \text{ (Methyl ester)}$$

$$\text{RCH}_2\text{C(=O)OH} \xrightarrow[\text{pyridine}]{SOCl_2} \text{RCH}_2\text{C(=O)Cl (Long-chain acyl chloride)}$$

The reactions presented in Sections 23.2D and 23.2E in the context of fatty acids are the same as those we studied in earlier chapters regarding carboxylic acids and alkenes.

23.2E Reactions of the Alkenyl Chain of Unsaturated Fatty Acids

The double bonds of the carbon chains of fatty acids undergo characteristic alkene addition reactions (see Chapters 7 and 8):

$$CH_3(CH_2)_nCH{=}CH(CH_2)_mCO_2H \xrightarrow[Ni]{H_2} CH_3(CH_2)_nCH(H){-}CH(H)(CH_2)_mCO_2H$$

$$CH_3(CH_2)_nCH{=}CH(CH_2)_mCO_2H \xrightarrow[CCl_4]{Br_2} CH_3(CH_2)_nCHBrCHBr(CH_2)_mCO_2H$$

$$CH_3(CH_2)_nCH{=}CH(CH_2)_mCO_2H \xrightarrow[(2)\ NaHSO_3]{(1)\ OsO_4} CH_3(CH_2)_nCH(OH){-}CH(OH)(CH_2)_mCO_2H$$

$$CH_3(CH_2)_nCH{=}CH(CH_2)_mCO_2H \xrightarrow{HBr} CH_3(CH_2)_nCH(H)CHBr(CH_2)_mCO_2H + CH_3(CH_2)_nCHBrCH(H)(CH_2)_mCO_2H$$

PROBLEM 23.1

(a) How many stereoisomers are possible for 9,10-dibromohexadecanoic acid?
(b) The addition of bromine to palmitoleic acid yields primarily one set of enantiomers, (±)-*threo*-9,10-dibromohexadecanoic acid. The addition of bromine is an anti addition to the double bond (i.e., it apparently takes place through a bromonium ion intermediate). Taking into account the cis stereochemistry of the double bond of palmitoleic acid and the stereochemistry of the bromine addition, write three-dimensional structures for the (±)-*threo*-9,10-dibromohexadecanoic acids.

23.3 Terpenes and Terpenoids

People have isolated organic compounds from plants since antiquity. By gently heating or by steam distilling certain plant materials, one can obtain mixtures of odoriferous compounds known as *essential oils.* These compounds have had a variety of uses, particularly in early medicine and in the making of perfumes.

As the science of organic chemistry developed, chemists separated the various components of these mixtures and determined their molecular formulas and, later, their structural formulas. Even today these natural products offer challenging problems for chemists interested in structure determination and synthesis. Research in this area has also given us important information about the ways the plants themselves synthesize these compounds.

Terpene biosynthesis was described in Special Topic D.

Hydrocarbons known generally as **terpenes** and oxygen-containing compounds called **terpenoids** are the most important constituents of essential oils. Most terpenes have skeletons of 10, 15, 20, or 30 carbon atoms and are classified in the following way:

Number of Carbon Atoms	Class
10	Monoterpenes
15	Sesquiterpenes
20	Diterpenes
30	Triterpenes

One can view terpenes as being built up from two or more C_5 units known as *isoprene units.* Isoprene is 2-methyl-1,3-butadiene. Isoprene and the isoprene unit can be represented in various ways:

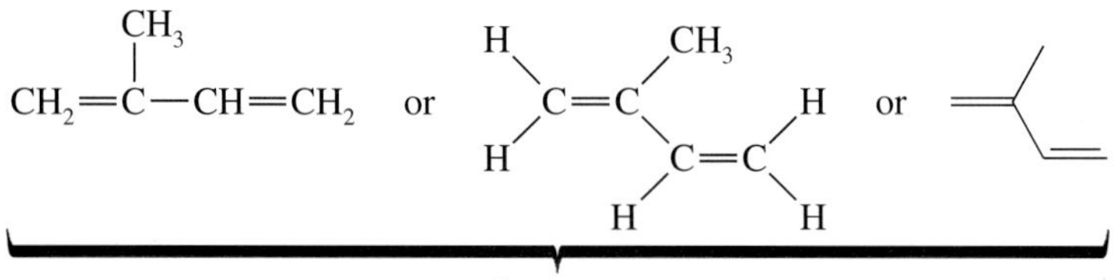

$$CH_2{=}\underset{}{\overset{CH_3}{\overset{|}{C}}}{-}CH{=}CH_2$$ or

Isoprene

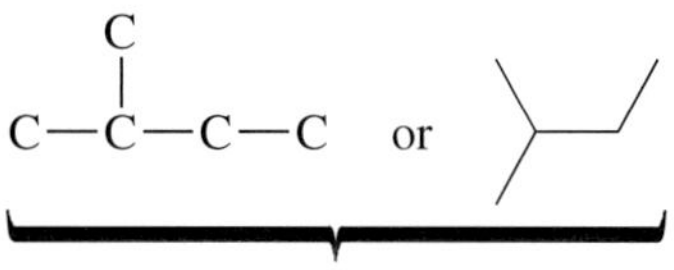

An isoprene unit

We now know that plants do not synthesize terpenes from isoprene (see Special Topic D). However, recognition of the isoprene unit as a component of the structure of terpenes has been a great aid in elucidating their structures. We can see how if we examine the following structures:

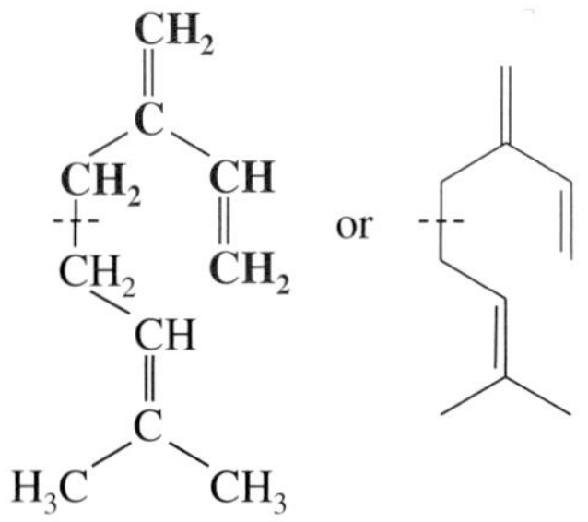

Myrcene
(isolated from bay oil)

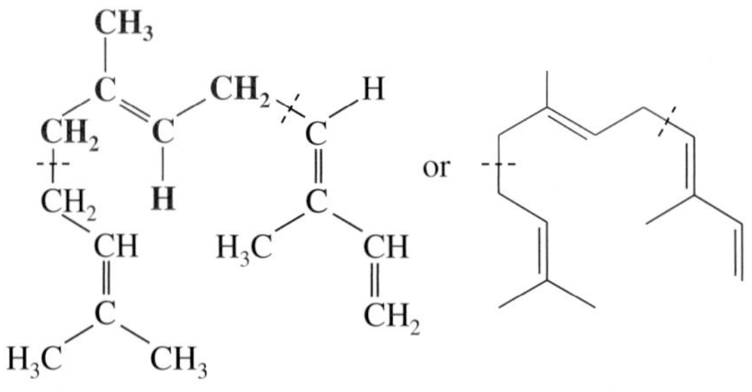

α-Farnesene
(from natural coating of apples)

Using dashed lines to separate isoprene units, we can see that the monoterpene (myrcene) has two isoprene units; the sesquiterpene (α-farnesene) has three. In both compounds the isoprene units are linked head to tail:

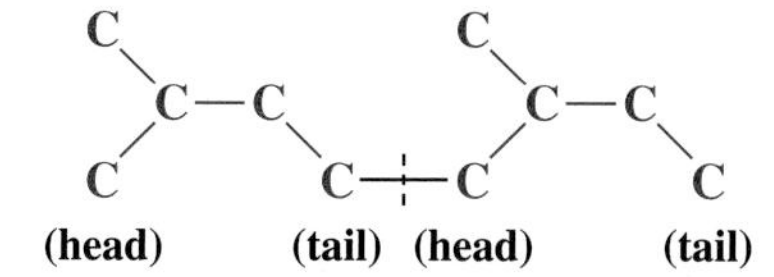

C C
C—C C—C
C C—C C
(head) (tail) (head) (tail)

Many terpenes also have isoprene units linked in rings, and others (terpenoids) contain oxygen:

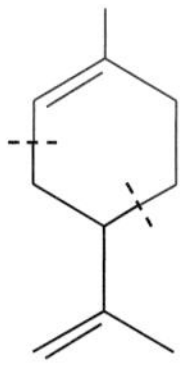

Limonene
(from oil of lemon or orange)

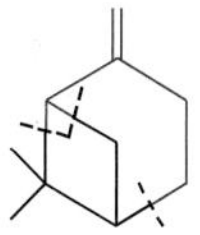

β-Pinene
(from oil of turpentine)

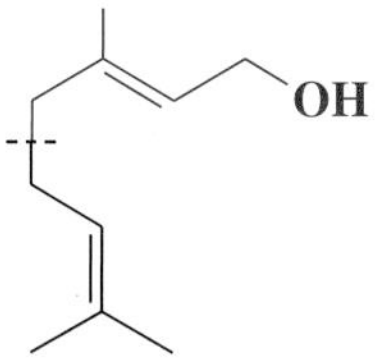

OH

Geraniol
(from roses and other flowers)

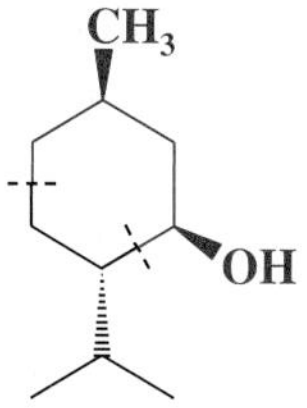

CH_3
OH

Menthol
(from peppermint)

PROBLEM 23.2

(a) Show the isoprene units in each of the following terpenes. **(b)** Classify each as a monoterpene, sesquiterpene, diterpene, and so on.

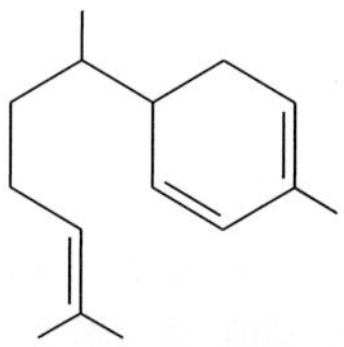

Zingiberene
(from oil of ginger)

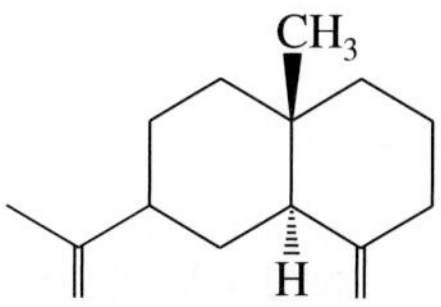

CH_3
H

β-Selinene
(from oil of celery)

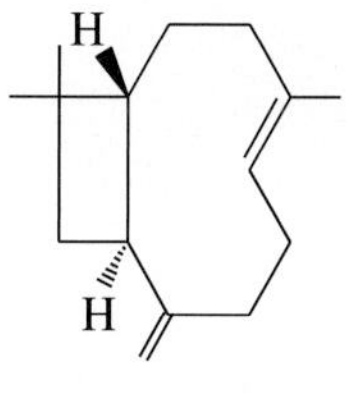

H
H

Caryophyllene
(from oil of cloves)

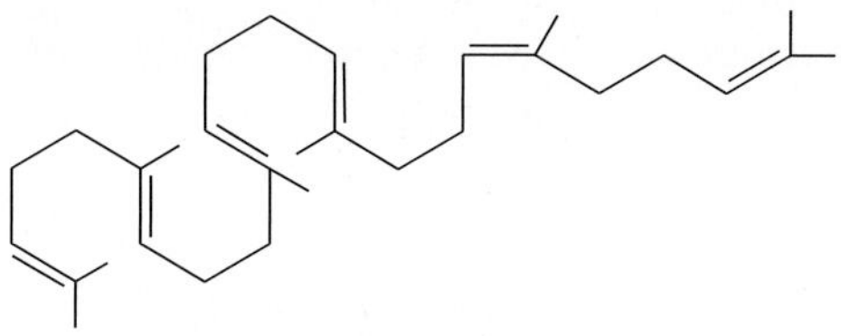

Squalene
(from shark liver oil)

PROBLEM 23.3 What products would you expect to obtain if each of the following terpenes were subjected to ozonolysis and subsequent treatment with zinc and acetic acid?

(a) Myrcene **(c)** α-Farnesene **(e)** Squalene
(b) Limonene **(d)** Geraniol

PROBLEM 23.4 Give structural formulas for the products that you would expect from the following reactions:

(a) β-Pinene + hot $KMnO_4 \longrightarrow$ **(c)** Caryophyllene + HCl $\longrightarrow$

(b) Zingiberene + $H_2 \xrightarrow{Pt}$ **(d)** β-Selinene + 2 BH_3:THF $\xrightarrow[(2)\ H_2O_2,\ OH^-]{}$

PROBLEM 23.5 What simple chemical test could you use to distinguish between geraniol and menthol?

The carotenes are tetraterpenes. They can be thought of as two diterpenes linked in tail-to-tail fashion:

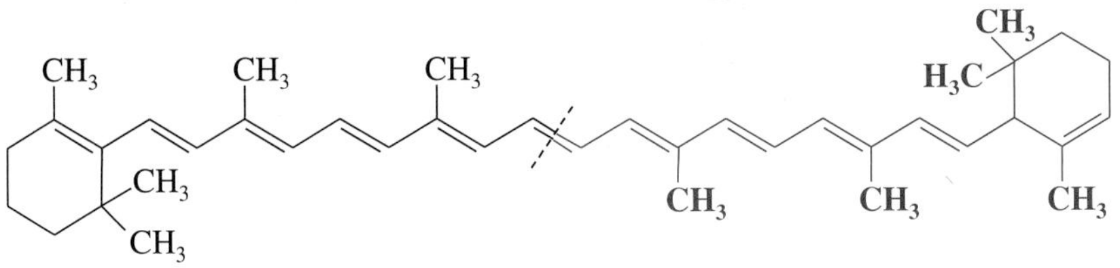

α-Carotene

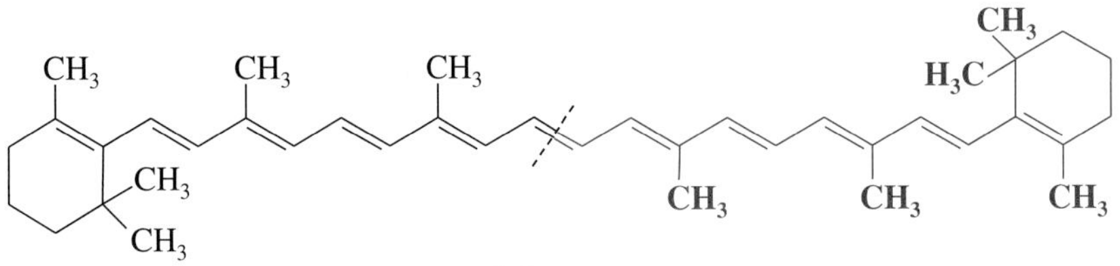

β-Carotene

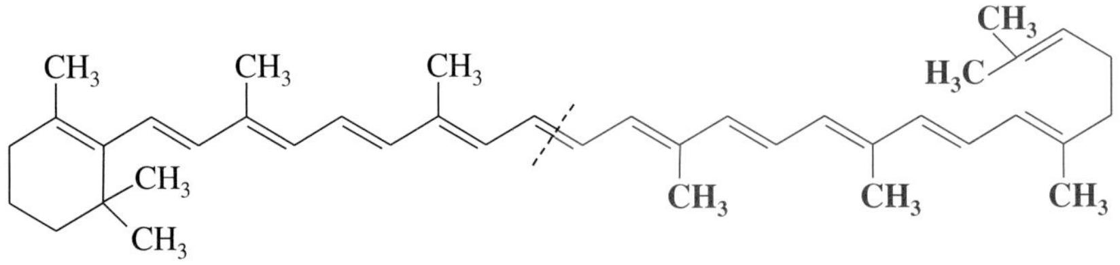

γ-Carotene

The carotenes are present in almost all green plants. In animals, all three carotenes serve as precursors for vitamin A, for they all can be converted to vitamin A by enzymes in the liver.

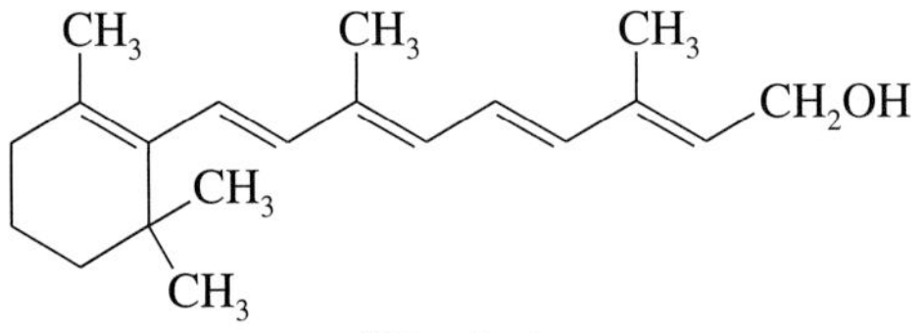

Vitamin A

In this conversion, one molecule of β-carotene yields two of vitamin A: α- and γ-carotene give only one. Vitamin A is important not only in vision but in many other ways as well. For example, young animals whose diets are deficient in vitamin A fail to grow.

Vitamin A, β-carotene, and vitamin E (Section 10.11E) are important lipid-soluble antioxidants, as well.

23.3A Natural Rubber

Natural rubber can be viewed as a 1,4-addition polymer of isoprene. In fact, pyrolysis degrades natural rubber to isoprene. Pyrolysis (Greek: *pyros,* a fire, + *lysis*) is the heating of something in the absence of air until it decomposes. The isoprene units of natural rubber are all linked in a head-to-tail fashion, and all of the double bonds are cis:

Natural rubber
(*cis*-1,4-polyisoprene)

Ziegler–Natta catalysts (see Special Topic A) make it possible to polymerize isoprene and obtain a synthetic product that is identical with the rubber obtained from natural sources.

Pure natural rubber is soft and tacky. To be useful, natural rubber has to be *vulcanized.* In vulcanization, natural rubber is heated with sulfur. A reaction takes place that produces cross-links between the *cis*-polyisoprene chains and makes the rubber much harder. Sulfur reacts both at the double bonds and at allylic hydrogen atoms:

Vulcanized rubber

23.4 Steroids

The lipid fractions obtained from plants and animals contain another important group of compounds known as **steroids.** Steroids are important "biological regulators" that nearly always show dramatic physiological effects when they are administered to living organisms. Among these important compounds are male and female sex hormones, adrenocortical hormones, D vitamins, the bile acids, and certain cardiac poisons.

23.4A Structure and Systematic Nomenclature of Steroids

Steroids are derivatives of the following perhydrocyclopentanophenanthrene ring system:

FIGURE 23.6 The basic ring systems of the 5α and 5β series of steroids.

Build hand-held molecular models of the 5α and 5β series of steroids and use them to explore the structures of steroids discussed in this chapter.

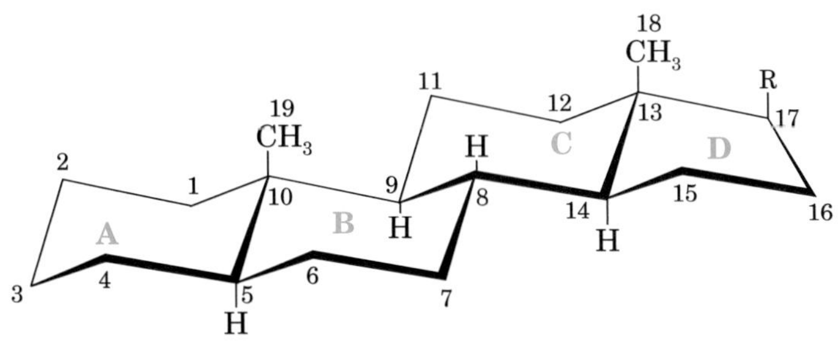

5α Series of steroids
(all ring junctions are trans.)

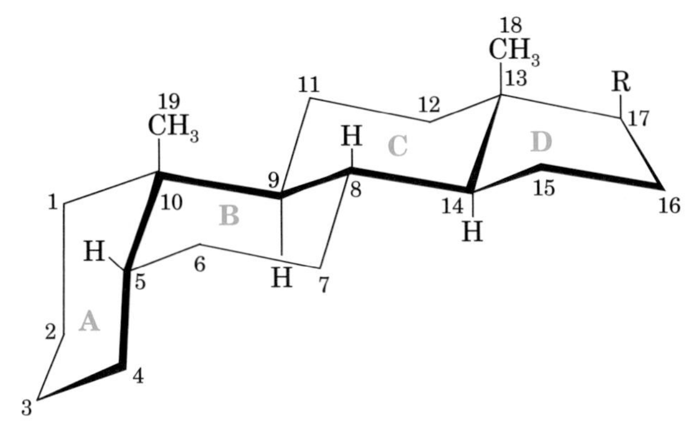

5β Series of steroids
(A,B ring junction is cis.)

The carbon atoms of this ring system are numbered as shown. The four rings are designated with letters.

In most steroids the **B,C** and **C,D** ring junctions are trans. The **A,B** ring junction, however, may be either cis or trans, and this possibility gives rise to two general groups of steroids having the three-dimensional structures shown in Fig. 23.6.

The methyl groups that are attached at points of ring junction (i.e., those numbered 18 and 19) are called **angular methyl groups,** and they serve as important reference points for stereochemical designations. The angular methyl groups protrude above the general plane of the ring system when it is written in the manner shown in Fig. 23.6. By convention, other groups that lie on the same general side of the molecule as the angular methyl groups (i.e., on the top side) are designated ***β* substituents** (these are written with a solid wedge). Groups that lie generally on the bottom (i.e., are trans to the angular methyl groups) are designated ***α* substituents** (these are written with a dashed wedge). When α and β designations are applied to the hydrogen atom at position 5, the ring system in which the **A,B** ring junction is trans becomes the 5α series; the ring system in which the **A,B** ring junction is cis becomes the 5β series.

PROBLEM 23.6

Draw the two basic ring systems given in Fig. 23.6 for the 5α and 5β series showing all hydrogen atoms of the cyclohexane rings. Label each hydrogen atom as to whether it is axial or equatorial.

In systematic nomenclature the nature of the R group at position 17 determines (primarily) the base name of an individual steroid. These names are derived from the steroid hydrocarbon names given in Table 23.3.

TABLE 23.3 Names of Steroid Hydrocarbons

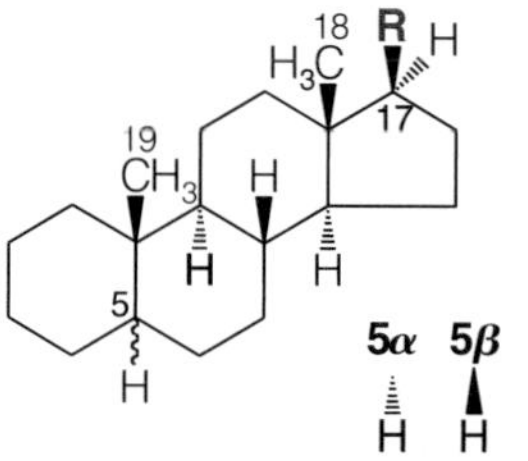

R	Name
—H	Androstane
—H (with —H also replacing —$\overset{19}{CH_3}$)	Estrane
—$\overset{20}{CH_2}\overset{21}{CH_3}$	Pregnane
—$CH(CH_3)CH_2CH_2CH_3$ (C-20, 21, 22, 23, 24)	Cholane
—$CH(CH_3)CH_2CH_2CH_2CH(CH_3)CH_3$ (C-20 to 27)	Cholestane

The following two examples illustrate the way these base names are used:

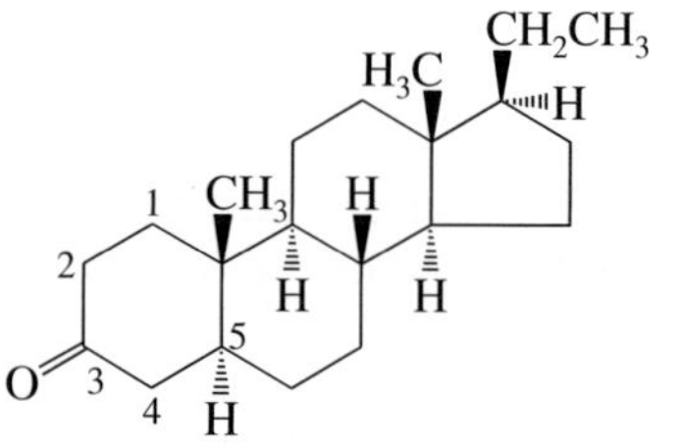

5α-Pregnan-3-one

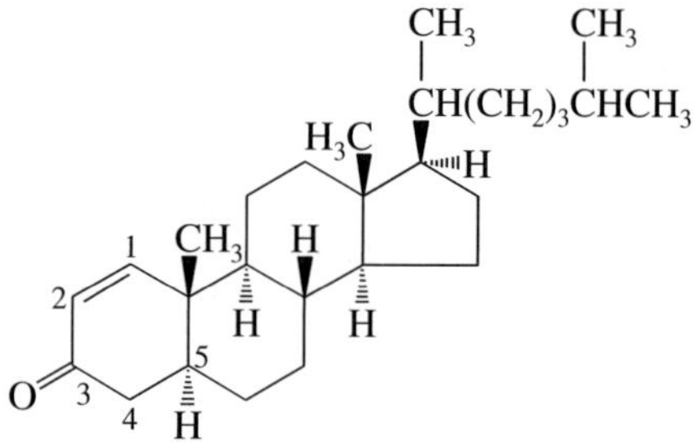

5α-Cholest-1-en-3-one

We shall see that many steroids also have common names and that the names of the steroid hydrocarbons given in Table 23.3 are derived from these common names.

PROBLEM 23.7

(a) Androsterone, a secondary male sex hormone, has the systematic name 3α-hydroxy-5α-androstan-17-one. Give a three-dimensional formula for androsterone. **(b)** Norethynodrel, a synthetic steroid that has been widely used in oral contraceptives, has the systematic name 17α-ethynyl-17β-hydroxy-5(10)-estren-3-one. Give a three-dimensional formula for norethynodrel.

23.4B Cholesterol

Cholesterol, one of the most widely occurring steroids, can be isolated by extraction of nearly all animal tissues. Human gallstones are a particularly rich source.

Cholesterol was first isolated in 1770. In the 1920s, two German chemists, Adolf Windaus (University of Göttingen) and Heinrich Wieland (University of Munich), were

We saw how cholesterol is biosynthesized in "The Chemistry of. . . Cholesterol Biosynthesis" in Chapter 8.

responsible for outlining a structure for cholesterol; they received Nobel Prizes for their work in 1927 and 1928.*

Part of the difficulty in assigning an absolute structure to cholesterol is that cholesterol contains *eight* tetrahedral stereogenic centers. This feature means that 2^8, or 256, possible stereoisomeric forms of the basic structure are possible, *only one of which is cholesterol:*

5-Cholesten-3β-ol
(absolute configuration of cholesterol)

PROBLEM 23.8 Designate with asterisks the eight stereogenic centers of cholesterol.

Cholesterol occurs widely in the human body, but not all of the biological functions of cholesterol are yet known. Cholesterol is known to serve as an intermediate in the biosynthesis of all of the steroids of the body. Cholesterol, therefore, is essential to life. We do not need to have cholesterol in our diet, however, because our body can synthesize all we need. When we ingest cholesterol, our body synthesizes less than if we ate none at all, but the total cholesterol is more than if we ate none at all. Far more cholesterol is present in the body than is necessary for steroid biosynthesis. High levels of blood cholesterol have been implicated in the development of arteriosclerosis (hardening of the arteries) and in heart attacks that occur when cholesterol-containing plaques block arteries of the heart. Considerable research is being carried out in the area of cholesterol metabolism with the hope of finding ways of minimizing cholesterol levels through the use of dietary adjustments or drugs.

It is important to note that, in common language, "cholesterol" does not necessarily refer only to the pure compound that chemists call cholesterol, but often refers instead to mixtures that contain cholesterol, other lipids, and proteins. These aggregates are called chylomicrons, high-density lipoproteins (HDLs), and low-density lipoproteins (LDLs). They have structures generally resembling globular micelles, and they are the vehicles by which cholesterol is transported through the aqueous environment of the body. Hydrophilic groups of their constituent proteins, phospholipids, and cholesterol hydroxyl groups are oriented outward toward the water medium so as to facilitate transport of the lipids through the circulatory system. HDLs (the so-called "good cholesterol") carry lipids from the tissues to the liver for degradation and excretion. LDL ("bad cholesterol") carries biosynthesized lipids from the liver to the tissues. Chylomicrons transport dietary lipids from the intestines to the tissues. (See Fig. 23.7.)

Certain compounds related to steroids and derived from plants are now known to lower total blood cholesterol when used in dietary forms approved by the FDA. Called phy-

*The original cholesterol structure proposed by Windaus and Wieland was incorrect. This became evident in 1932 as a result of X-ray diffraction studies done by the British physicist J. D. Bernal. By the end of 1932, however, English scientists, and Wieland himself, using Bernal's results, were able to outline the correct structure of cholesterol.

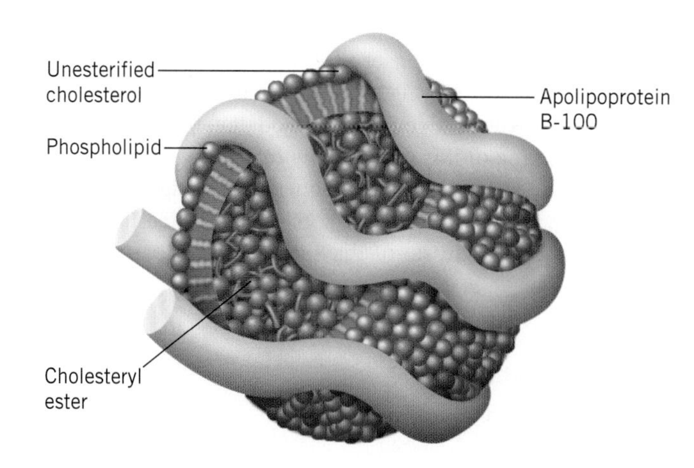

FIGURE 23.7 An LDL showing a core of cholesterol esters and a shell of phospholipids and unesterified cholesterol (hydroxyl groups exposed), wrapped in an apolipoprotein. The phospholipid head groups and hydrophilic residues of the protein support the water compatibility of the LDL particle.

tostanols and phytosterols, these patented compounds act by inhibiting intestinal absorption of dietary cholesterol. They are marketed as food in the form of edible spreads. An example of a phytostanol is shown here.

A phytostanol ester (β-sitostanol, R = fatty acid)

23.4C Sex Hormones

The sex hormones can be classified into three major groups: (1) the female sex hormones, or **estrogens;** (2) the male sex hormones, or **androgens;** and (3) the pregnancy hormones, or **progestins.**

The first sex hormone to be isolated was an estrogen, *estrone.* Working independently, Adolf Butenandt (in Germany at the University of Göttingen) and Edward Doisy (in the United States at St. Louis University) isolated estrone from the urine of pregnant women. They published their discoveries in 1929. Later, Doisy was able to isolate the much more potent estrogen, *estradiol.* In this research Doisy had to extract *4 tons* of sow ovaries in order to obtain just 12 mg of estradiol. Estradiol, it turns out, is the true female sex hormone, and estrone is a metabolized form of estradiol that is excreted.

Estrone [3-hydroxy-1,3,5(10)-estratrien-17-one]

Estradiol [1,3,5(10)-estratriene-3,17β-diol]

Estradiol is secreted by the ovaries and promotes the development of the secondary female characteristics that appear at the onset of puberty. Estrogens also stimulate the development of the mammary glands during pregnancy and induce estrus (heat) in animals.

In 1931, Butenandt and Kurt Tscherning isolated the first androgen, *androsterone.* They were able to obtain 15 mg of this hormone by extracting approximately 15,000 L of male urine. Soon afterward (in 1935), Ernest Laqueur (in Holland) isolated another male sex hormone, *testosterone,* from bull testes. It soon became clear that testosterone is the true male sex hormone and that androsterone is a metabolized form of testosterone that is excreted in the urine.

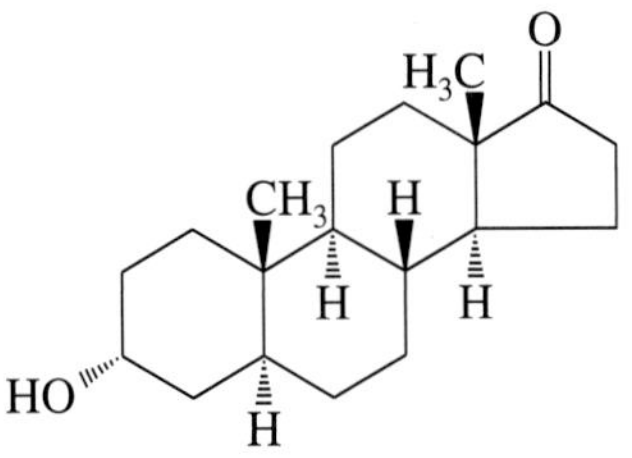

Androsterone
(3α-hydroxy-5α-androstan-17-one)

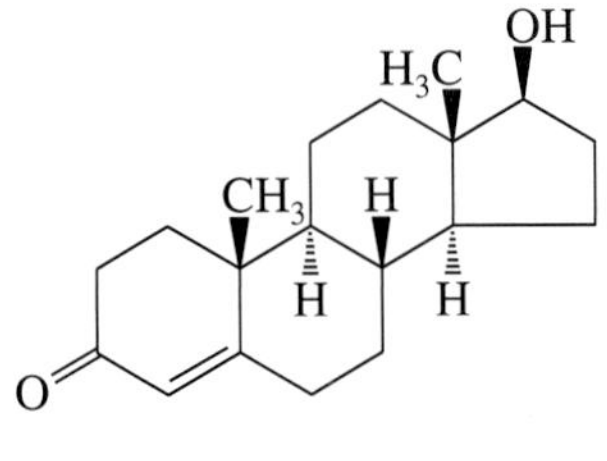

Testosterone
(17β-hydroxy-4-androsten-3-one)

Testosterone, secreted by the testes, is the hormone that promotes the development of secondary male characteristics: the growth of facial and body hair, the deepening of the voice, muscular development, and the maturation of the male sex organs.

Testosterone and estradiol, then, are the chemical compounds from which "maleness" and "femaleness" are derived. It is especially interesting to examine their structural formulas and see how very slightly these two compounds differ. Testosterone has an angular methyl group at the **A,B** ring junction that is missing in estradiol. Ring **A** of estradiol is a benzene ring and, as a result, estradiol is a phenol. Ring **A** of testosterone contains an α,β-unsaturated keto group.

PROBLEM 23.9 The estrogens (estrone and estradiol) are easily separated from the androgens (androsterone and testosterone) on the basis of one of their chemical properties. What is the property, and how could such a separation be accomplished?

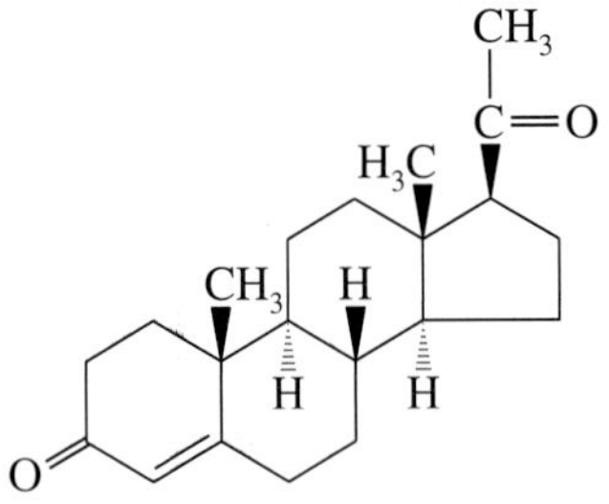

Progesterone
(4-pregnene-3,20-dione)

Progesterone is the most important *progestin* (pregnancy hormone). After ovulation occurs, the remnant of the ruptured ovarian follicle (called the *corpus luteum*) begins to secrete progesterone. This hormone prepares the lining of the uterus for implantation of the fertilized ovum, and continued progesterone secretion is necessary for the completion of pregnancy. (Progesterone is secreted by the placenta after secretion by the corpus luteum declines.)

Progesterone *also suppresses ovulation,* and it is the chemical agent that apparently accounts for the fact that pregnant women do not conceive again while pregnant. It was this observation that led to the search for synthetic progestins that could be used as oral contraceptives. (Progesterone itself requires very large doses to be effective in suppressing ovulation when taken orally because it is degraded in the intestinal tract.) A number of such compounds have been developed and are now widely used. In addition to norethynodrel (see Problem 23.7), another widely used synthetic progestin is its double-bond isomer, *norethindrone:*

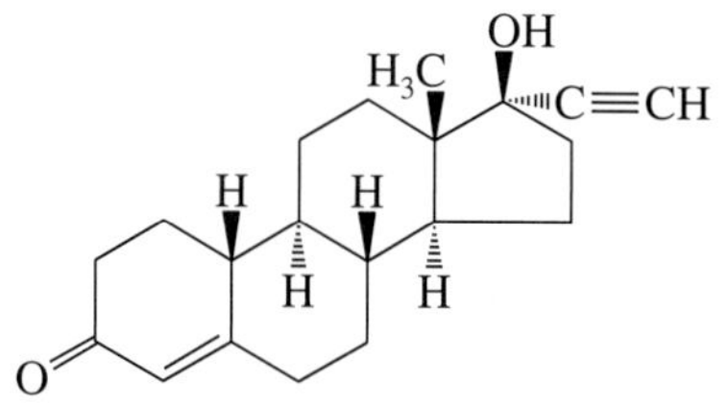

Norethindrone
(17α-ethynyl-17β-hydroxy-4-estren-3-one)

Synthetic estrogens have also been developed, and these are often used in oral contraceptives in combination with synthetic progestins. A very potent synthetic estrogen is the compound called *ethynylestradiol* or *novestrol:*

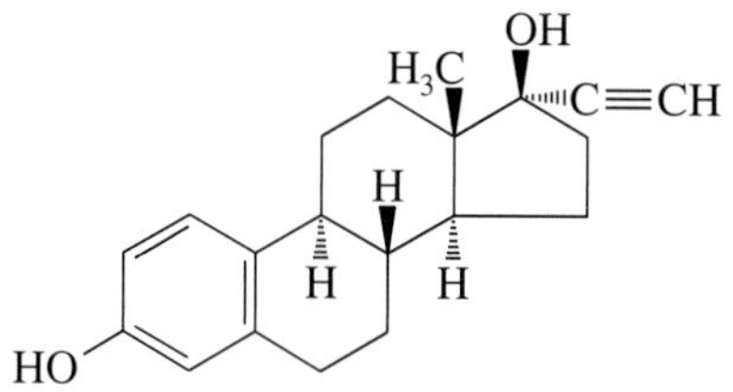

Ethynylestradiol
[17α-ethynyl-1,3,5(10)-estratriene-3,17β-diol]

23.4D Adrenocortical Hormones

At least 28 different hormones have been isolated from the adrenal cortex, part of the adrenal glands that sit on top of the kidneys. Included in this group are the following two steroids:

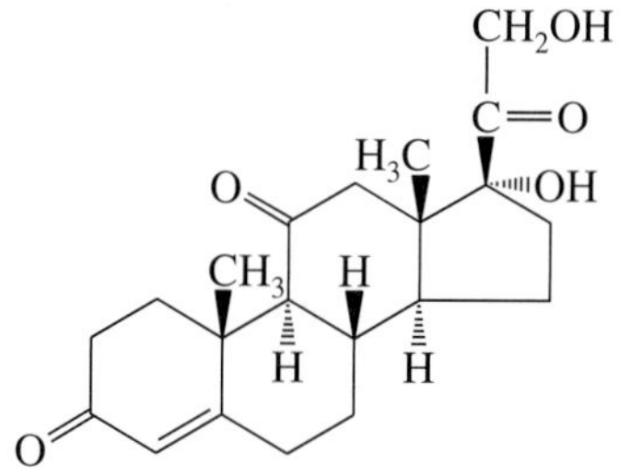

Cortisone
(17α,21-dihydroxy-4-pregnene-3,11,20-trione)

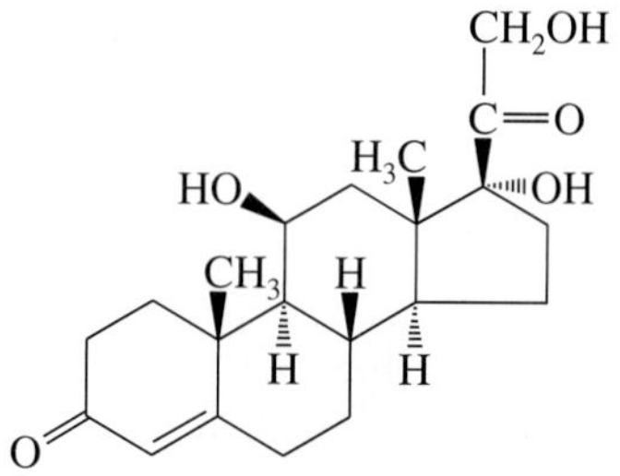

Cortisol
(11β,17α,21-trihydroxy-4-pregnene-3,20-dione)

Most of the adrenocortical steroids have an oxygen function at position 11 (a keto group in cortisone, for example, and a β-hydroxyl in cortisol). Cortisol is the major hormone synthesized by the human adrenal cortex.

The adrenocortical steroids are apparently involved in the regulation of a large number of biological activities, including carbohydrate, protein, and lipid metabolism; water and electrolyte balance; and reactions to allergic and inflammatory phenomena. Recognition, in 1949, of the anti-inflammatory effect of cortisone and its usefulness in the treatment of rheumatoid arthritis led to extensive research in this area. Many 11-oxygenated steroids are now used in the treatment of a variety of disorders ranging from Addison's disease to asthma and skin inflammations.

23.4E D Vitamins

The demonstration, in 1919, that sunlight helped cure rickets—a childhood disease characterized by poor bone growth—began a long search for a chemical explanation. Soon it was discovered that irradiation of certain foodstuffs increased their antirachitic properties, and, in 1930, the search led to a steroid that can be isolated from yeast, called *ergosterol*. Irradiation of ergosterol was found to produce a highly active material. In 1932, Windaus (Section 23.4B) and co-workers in Germany demonstrated that this highly active substance was vitamin D_2. The photochemical reaction that takes place is one in which the dienoid ring **B** of ergosterol opens to produce a conjugated triene:

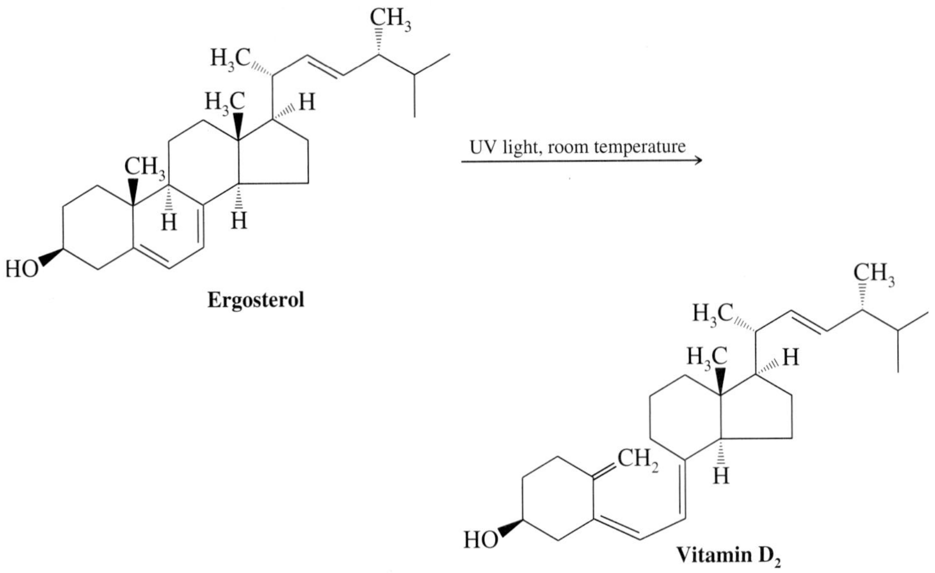

CH_3 H_3C H_3C H CH_3 H H HO

Ergosterol

UV light, room temperature →

CH_3 H_3C H_3C H CH_2 H HO

Vitamin D_2

23.4F Other Steroids

The structures, sources, and physiological properties of a number of other important steroids are given in Table 23.4.

23.4G Reactions of Steroids

Steroids undergo all of the reactions that we might expect of molecules containing double bonds, hydroxyl groups, keto groups, and so on. While the stereochemistry of steroid reactions is often quite complex, it is often strongly influenced by the steric hindrance presented at the β face of the molecule by the angular methyl groups. Many reagents react preferentially at the relatively unhindered α face, especially when the reaction takes place at a functional group very near an angular methyl group and when the attacking reagent is bulky. Examples that illustrate this tendency are shown in the reactions on page 1152:

TABLE 23.4 Other Important Steroids

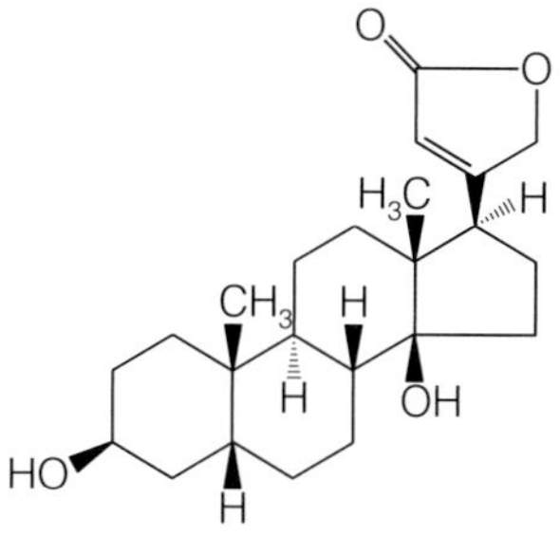 **Digitoxigenin**	Digitoxigenin is a cardiac aglycone that can be isolated by hydrolysis of digitalis, a pharmaceutical that has been used in treating heart disease since 1785. In digitalis, sugar molecules are joined in acetal linkages to the 3-OH group of the steroid. In small doses digitalis strengthens the heart muscle; in larger doses it is a powerful heart poison. The aglycone has only about one-fortieth the activity of digitalis.
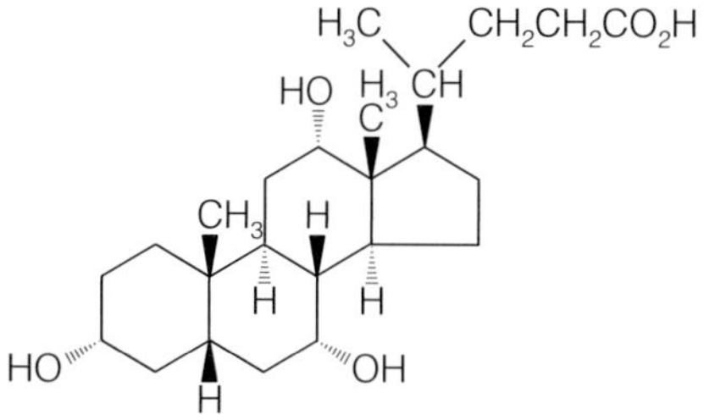 **Cholic acid**	Cholic acid is the most abundant acid obtained from the hydrolysis of human or ox bile. Bile is produced by the liver and stored in the gallbladder. When secreted into the small intestine, bile emulsifies lipids by acting as a soap. This action aids in the digestive process.
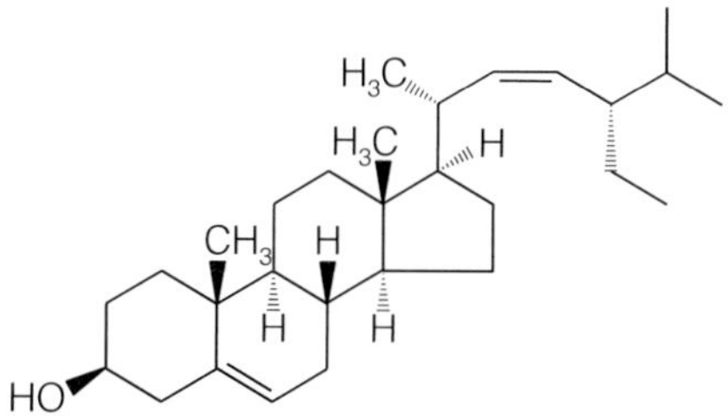 **Stigmasterol**	Stigmasterol is a widely occurring plant steroid that is obtained commercially from soybean oil. β-Sitostanol (a phytostanol, esters of which inhibit dietary cholesterol absorption) has the same formula except that it is saturated (C5 hydrogen is α).
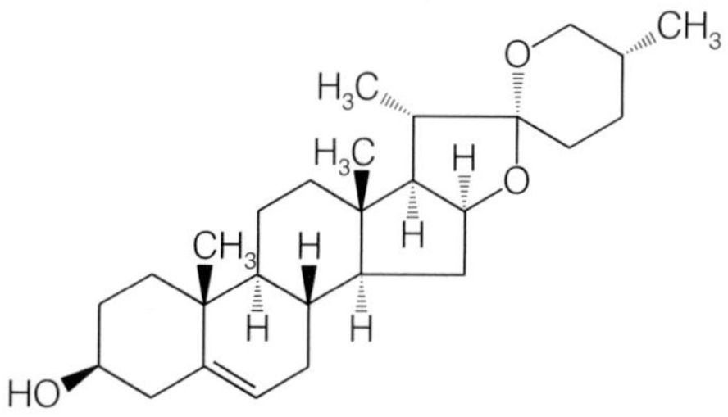 **Diosgenin**	Diosgenin is obtained from a Mexican vine, *cabeza de negro,* genus *Dioscorea.* It is used as the starting material for a commercial synthesis of cortisone and sex hormones.

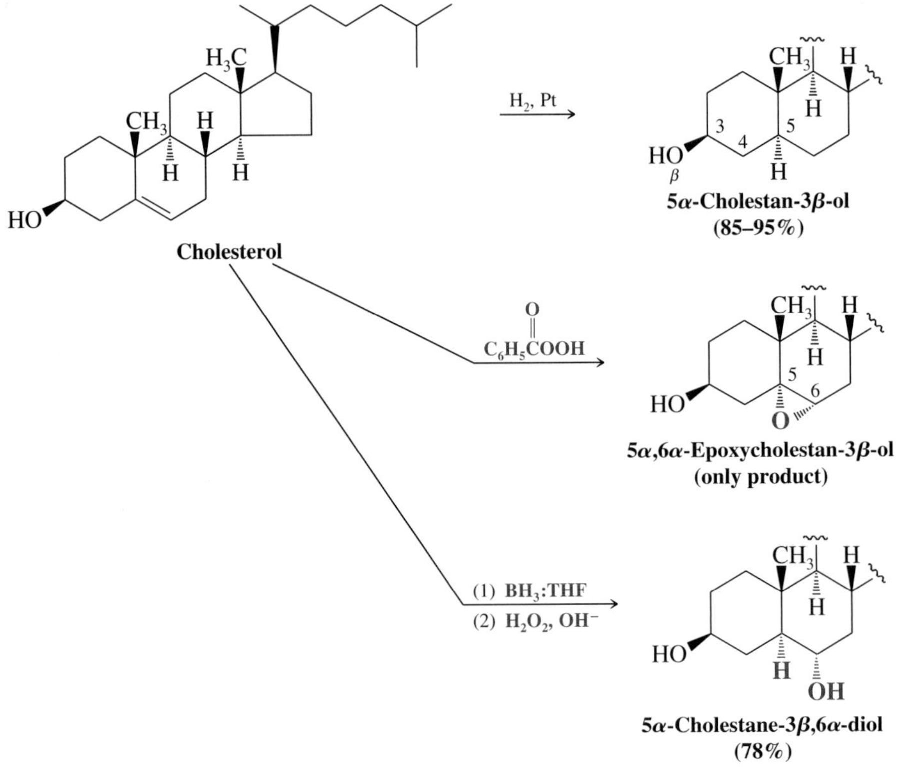

Cholesterol

$\xrightarrow{H_2, Pt}$ **5α-Cholestan-3β-ol (85–95%)**

$\xrightarrow{C_6H_5COOH \text{ (peroxy acid)}}$ **5α,6α-Epoxycholestan-3β-ol (only product)**

$\xrightarrow{(1)\ BH_3:THF \quad (2)\ H_2O_2,\ OH^-}$ **5α-Cholestane-3β,6α-diol (78%)**

When the epoxide ring of 5α,6α-epoxycholestan-3β-ol (see the following reaction) is opened, attack by chloride ion must occur from the β face, but it takes place at the more open 6 position. Notice that the 5 and 6 substituents in the product are *diaxial* (Section 8.13):

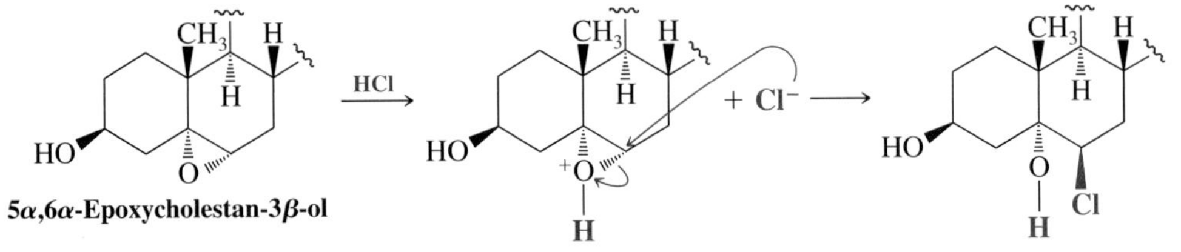

5α,6α-Epoxycholestan-3β-ol $\xrightarrow{HCl}$ + Cl^- ⟶

PROBLEM 23.10 Show how you might convert cholesterol into each of the following compounds:

(a) 5α,6β-Dibromocholestan-3β-ol
(b) Cholestane-3β,5α,6β-triol
(c) 5α-Cholestan-3-one
(d) 6α-Deuterio-5α-cholestan-3β-ol
(e) 6β-Bromocholestane-3β,5α-diol

The relative openness of equatorial groups (when compared to axial groups) also influences the stereochemical course of steroid reactions. When 5α-cholestane-3β,7α-diol (see the following reaction) is treated with excess ethyl chloroformate (C_2H_5OCOCl), only the equatorial 3β-hydroxyl becomes esterified. The axial 7α-hydroxyl is unaffected by the reaction:

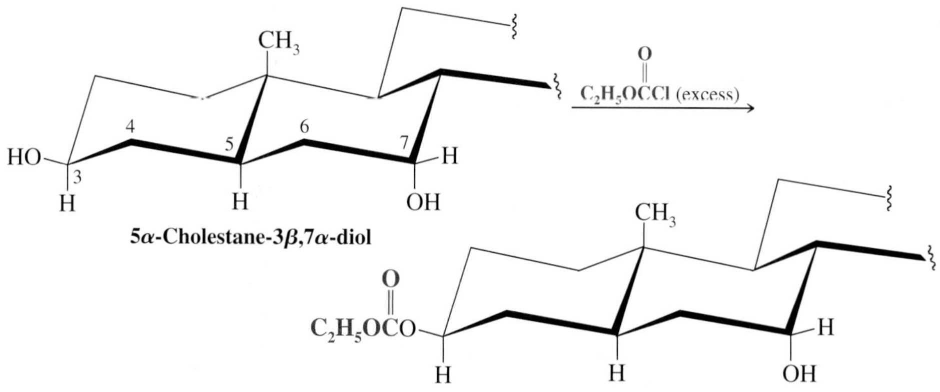

C_2H_5OCCl (excess)

5α-Cholestane-3β,7α-diol

(only product)

By contrast, treating 5α-cholestane-3β,7β-diol with excess ethyl chloroformate esterifies both hydroxyl groups. In this instance both groups are equatorial:

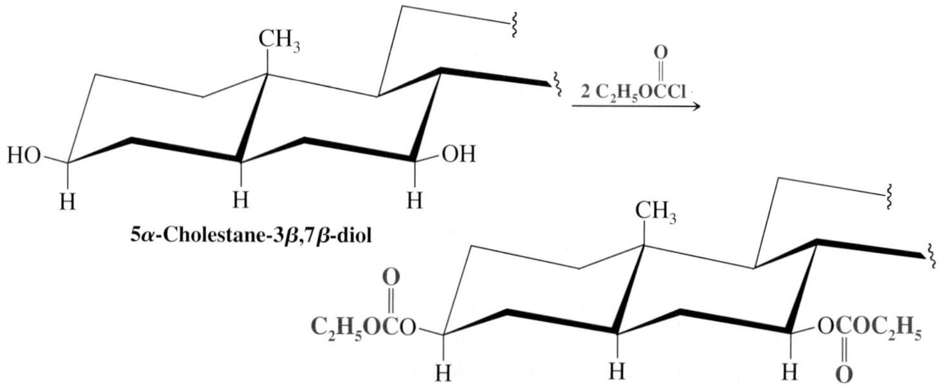

2 C_2H_5OCCl

5α-Cholestane-3β,7β-diol

23.5 PROSTAGLANDINS

One very active area of current research is concerned with a group of lipids called **prostaglandins.** Prostaglandins are C_{20} carboxylic acids that contain a five-membered ring, at least one double bond, and several oxygen-containing functional groups. Two of the most biologically active prostaglandins are prostaglandin E_2 and prostaglandin $F_{1\alpha}$:

These names for the prostaglandins are abbreviated designations used by workers in the field; systematic names are seldom used for prostaglandins.

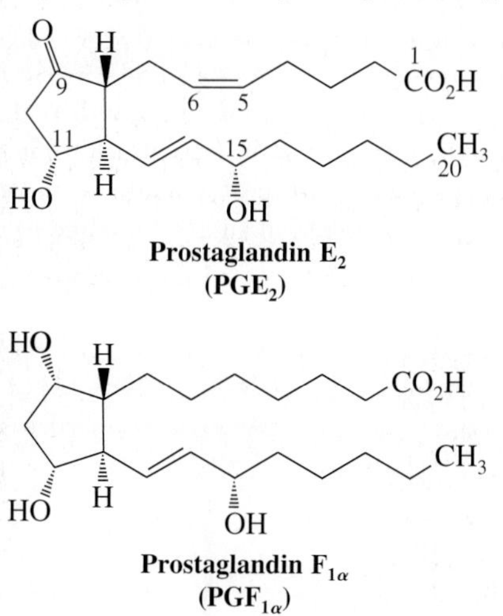

Prostaglandin E_2 (PGE_2)

Prostaglandin $F_{1\alpha}$ ($PGF_{1\alpha}$)

Prostaglandins of the E type have a carbonyl group at C9 and a hydroxyl group at C11; those of the F type have hydroxyl groups at both positions. Prostaglandins of the 2 series have a double bond between C5 and C6; in the 1 series this bond is a single bond.

The 1982 Nobel Prize in Physiology or Medicine was awarded to S. K. Bergström and B. I. Samuelsson (of the Karolinska Institute, Stockholm, Sweden) and to J. R. Vane (of the Wellcome Foundation, Beckenham, England) for their work on prostaglandins.

First isolated from seminal fluid, prostaglandins have since been found in almost all animal tissues. The amounts vary from tissue to tissue but are almost always very small. Most prostaglandins have powerful physiological activity, however, and this activity covers a broad spectrum of effects. Prostaglandins are known to affect heart rate, blood pressure, blood clotting, conception, fertility, and allergic responses.

The finding that prostaglandins can prevent formation of blood clots has great clinical significance, because heart attacks and strokes often result from the formation of abnormal clots in blood vessels. An understanding of how prostaglandins affect the formation of clots may lead to the development of drugs to prevent heart attacks and strokes.

The biosynthesis of prostaglandins of the 2 series begins with a C_{20} polyenoic acid, arachidonic acid, an omega-6 fatty acid. (Synthesis of prostaglandins of the 1 series begins with a fatty acid with one fewer double bond.) The first step requires two molecules of oxygen and is catalyzed by an enzyme called *cyclooxygenase:*

CO_2H / CH_3 —— $2\ O_2$, cyclooxygenase (inhibited by aspirin) ——→

Arachidonic acid

H, O, O, H, CO_2H, CH_3, O_2H —— several steps ——→ PGE_2 and other prostaglandins

PGG_2
(a cyclic endoperoxide)

The involvement of prostaglandins in allergic and inflammatory responses has also been of special interest. Some prostaglandins induce inflammation; others relieve it. The most widely used anti-inflammatory drug is ordinary aspirin (see Section 21.8). Aspirin blocks the synthesis of prostaglandins from arachidonic acid, apparently by acetylating the enzyme cyclooxygenase, thus rendering it inactive (see the previous reaction). This reaction may represent the origin of aspirin's anti-inflammatory properties. Another prostaglandin (PGE_1) is a potent fever-inducing agent (pyrogen), and aspirin's ability to reduce fever may also arise from its inhibition of prostaglandin synthesis.

23.6 Phospholipids and Cell Membranes

Another large class of lipids are those called **phospholipids.** Most phospholipids are structurally derived from a glycerol derivative known as a *phosphatidic acid.* In a phosphatidic acid, two hydroxyl groups of glycerol are joined in ester linkages to fatty acids and one terminal hydroxyl group is joined in an ester linkage to *phosphoric acid:*

CH_2OCR (O) / $CHOCR'$ (O) — From fatty acids

$CH_2—O—P(=O)(OH)—OH$ — From phosphoric acid

A phosphatidic acid
(a diacylglyceryl phosphate)

23.6A Phosphatides

In *phosphatides,* the phosphate group of a phosphatidic acid is bound through another phosphate ester linkage to one of the following nitrogen-containing compounds:

$HOCH_2CH_2\overset{+}{N}(CH_3)_3\ HO^-$ — **Choline**

$HOCH_2CH_2NH_2$ — **2-Aminoethanol (ethanolamine)**

$HOCH_2{-}C(\overset{+}{N}H_3)(H){-}CO_2^-$ — **L-Serine**

The most important phosphatides are the **lecithins, cephalins, phosphatidylserines,** and **plasmalogens** (a phosphatidyl derivative). Their general structures are shown in Table 23.5.

Phosphatides resemble soaps and detergents in that they are molecules having both polar and nonpolar groups (Fig. 23.8*a*). Like soaps and detergents, too, phosphatides "dissolve" in aqueous media by forming micelles. There is evidence that in biological systems the preferred micelles consist of three-dimensional arrays of "stacked" bimolecular micelles (Fig. 23.8b) that are better described as **lipid bilayers.**

The hydrophilic and hydrophobic portions of phosphatides make them perfectly suited for one of their most important biological functions: They form a portion of a structural unit that creates an interface between an organic and an aqueous environment. This struc-

TABLE 23.5 Phosphatides

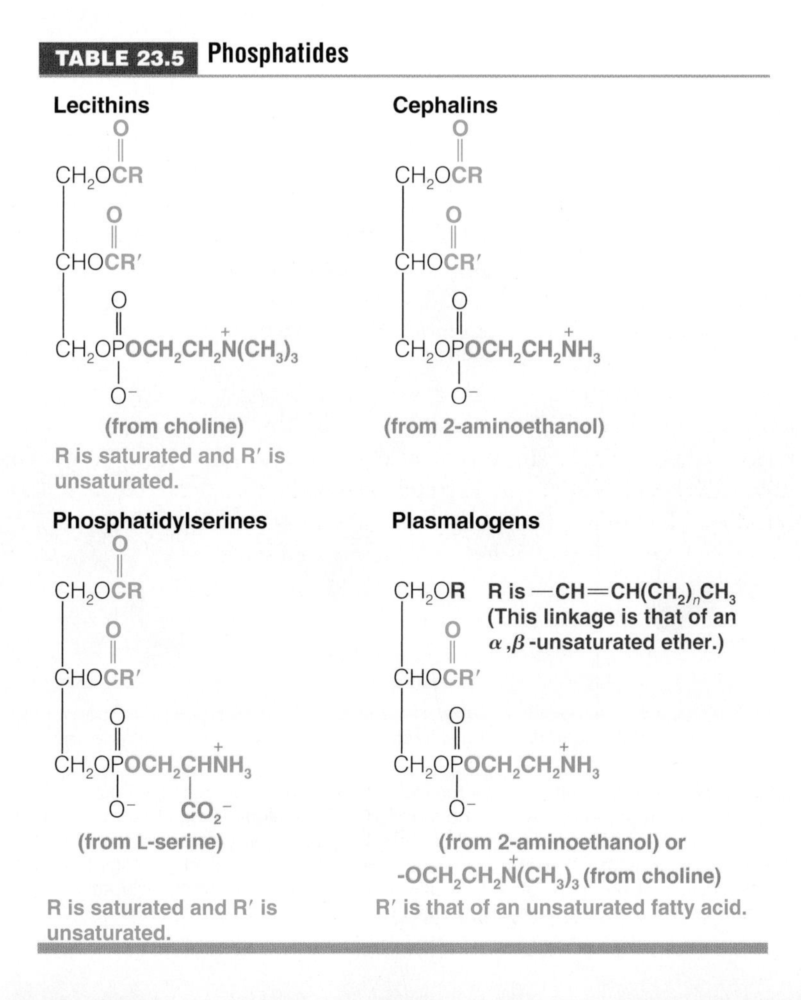

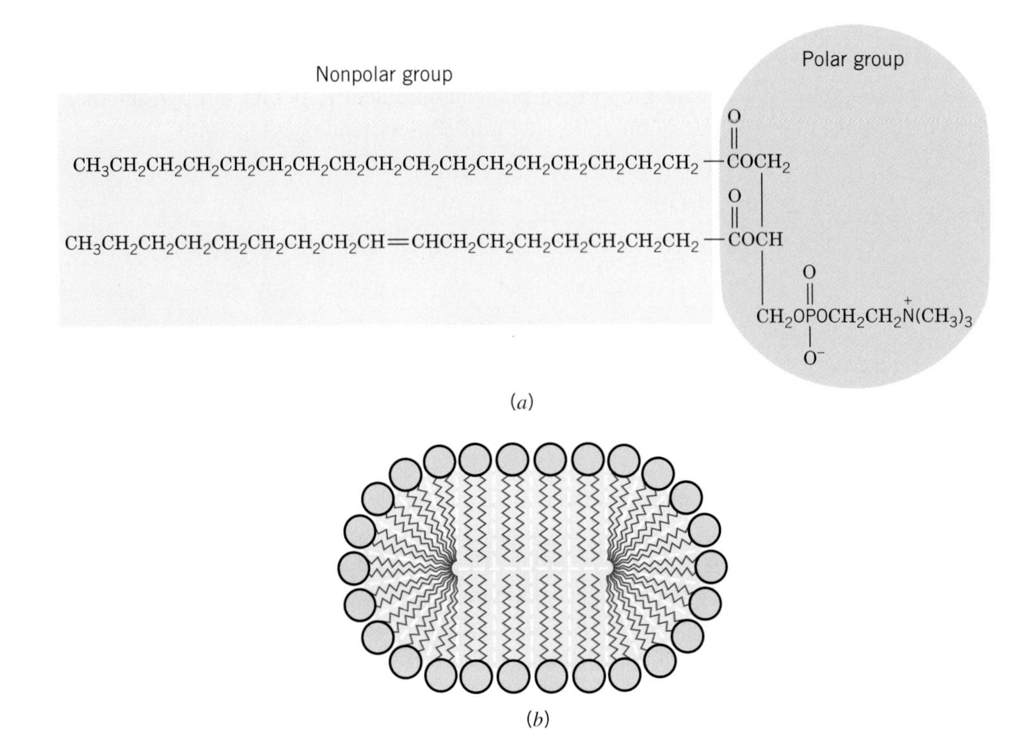

FIGURE 23.8 *(a)* Polar and nonpolar sections of a phosphatide. *(b)* A phosphatide micelle or lipid bilayer.

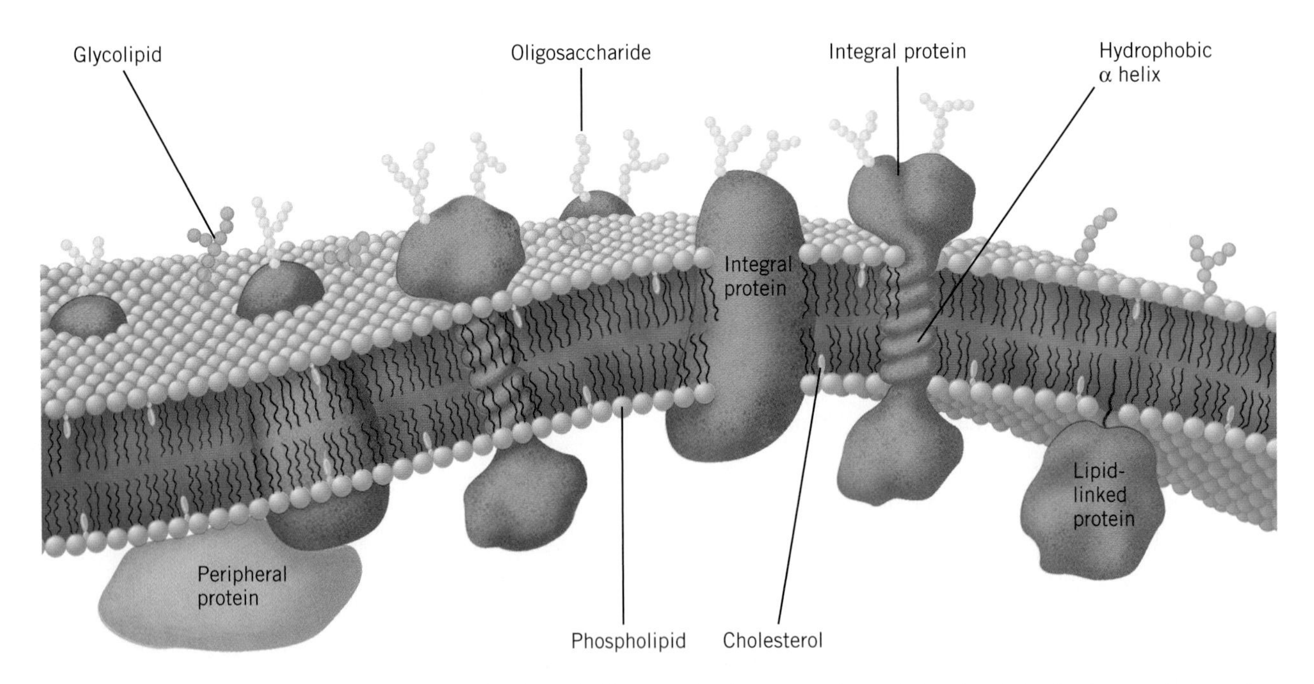

FIGURE 23.9 A schematic diagram of a plasma membrane. Integral proteins (*red-orange*), shown for clarity in much greater proportion than they are found in actual biological membranes, and cholesterol (*yellow*) are embedded in a bilayer composed of phospholipids (*blue spheres with two wiggly tails*). The carbohydrate components of glycoproteins (*yellow beaded chains*) and glycolipids (*green beaded chains*) occur only on the external face of the membrane. (From Voet, D.; Voet, J. G.; Pratt, C. W. *Fundamentals of Biochemistry;* Wiley: New York, 1999; p 248.)

ture (Fig. 23.9) is found in cell walls and membranes where phosphatides are often found associated with proteins and glycolipids (Section 23.6B).

PROBLEM 23.11

Under suitable conditions all of the ester (and ether) linkages of a phosphatide can be hydrolyzed. What organic compounds would you expect to obtain from the complete hydrolysis of **(a)** a lecithin, **(b)** a cephalin, and **(c)** a choline-based plasmalogen? [*Note:* Pay particular attention to the fate of the α,β-unsaturated ether in part (c).]

The Chemistry of...

STEALTH® Liposomes for Drug Delivery

The anticancer drug Doxil (doxorubicin) has been packaged in STEALTH® liposomes that give each dose of the drug extended action in the body. During manufacture of the drug it is ensconced in microscopic bubbles (vesicles) formed by a phospholipid bilayer and then given a special coating that masks it from the immune system. Ordinarily, a foreign particle such as this would be attacked by cells of the immune system and degraded, but a veil of polyethylene glycol oligomers on the liposome surface masks it from detection. Because of this coating, the STEALTH® liposome circulates through the body and releases its therapeutic contents over a period of time significantly greater than the lifetime for circulation of the undisguised drug. Coatings like those used for STEALTH® liposomes may also be able to reduce the toxic side effects of some drugs. Furthermore, by attaching specific cell recognition "marker" molecules to the polymer, it may be possible to focus binding of the liposomes specifically to cells of a targeted tissue. One might be tempted to call a targeted liposome a "smart stealth liposome."

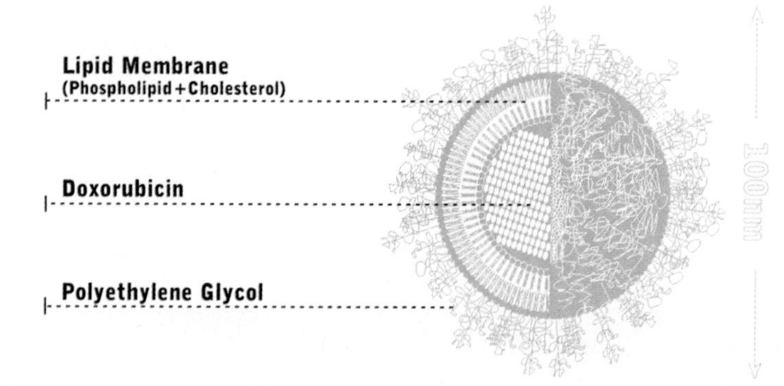

A STEALTH® liposome carrying the drug Doxorubicin. (Courtesy of Alza Corporation.)

23.6B Derivatives of Sphingosine

Another important group of lipids is derived from **sphingosine;** the derivatives are called **sphingolipids.** Two sphingolipids, a typical *sphingomyelin* and a typical *cerebroside,* are shown in Fig. 23.10.

On hydrolysis, sphingomyelins yield sphingosine, choline, phosphoric acid, and a C_{24} fatty acid called lignoceric acid. In a sphingomyelin this last component is bound to the —NH_2 group of sphingosine. The sphingolipids do not yield glycerol when they are hydrolyzed.

FIGURE 23.10 A sphingosine and two sphingolipids.

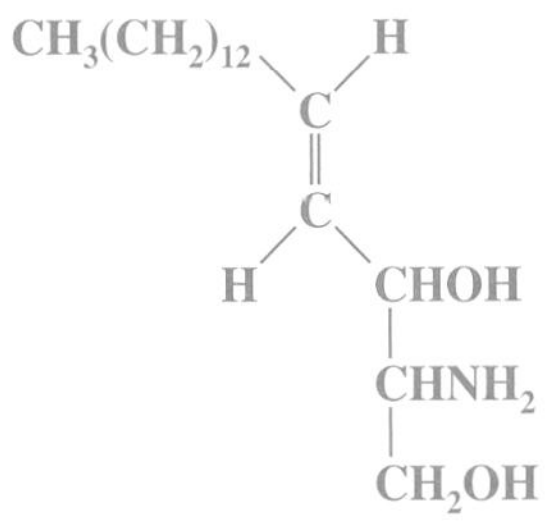

Sphingosine

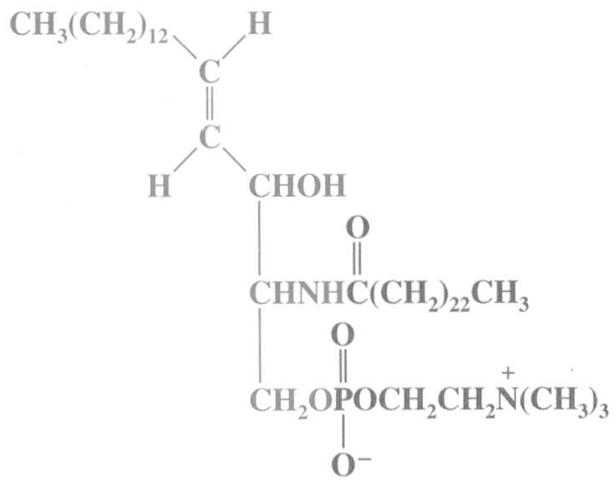

Sphingomyelin (a sphingolipid)

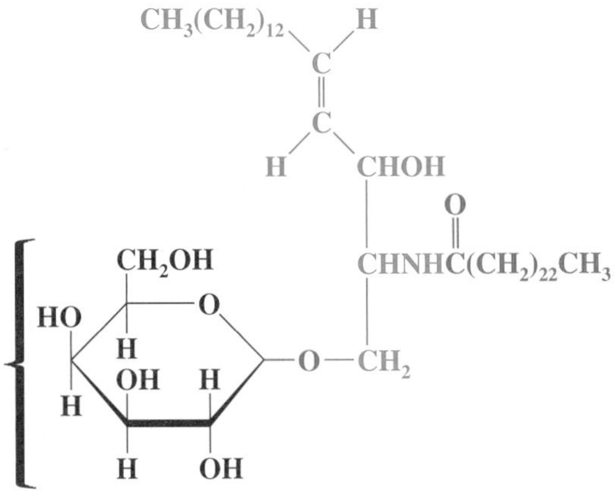

From a carbohydrate, D-galactose

Cerebroside

The cerebroside shown in Fig. 23.10 is an example of a **glycolipid.** Glycolipids have a polar group that is contributed by a *carbohydrate.* They do not yield phosphoric acid or choline when they are hydrolyzed.

The sphingolipids, together with proteins and polysaccharides, make up **myelin,** the protective coating that encloses nerve fibers or **axons.** The axons of nerve cells carry electrical nerve impulses. Myelin has a function relative to the axon similar to that of the insulation on an ordinary electric wire (see the chapter opening vignette).

23.7 WAXES

Most waxes are esters of long-chain fatty acids and long-chain alcohols. Waxes are found as protective coatings on the skin, fur, and feathers of animals and on the leaves and fruits of plants. Several esters isolated from waxes are the following:

$$CH_3(CH_2)_{14}\overset{O}{\overset{\|}{C}}OCH_2(CH_2)_{14}CH_3$$

Cetyl palmitate (from spermaceti)

$$CH_3(CH_2)_n\overset{O}{\overset{\|}{C}}OCH_2(CH_2)_mCH_3$$

n = 24 or 26; m = 28 or 30 (from beeswax)

$$HOCH_2(CH_2)_n\overset{O}{\overset{\|}{C}}{-}OCH_2(CH_2)_mCH_3$$

n = 16–28; m = 30 or 32 (from carnauba wax)

Summary of Reactions of Lipids

The reactions of lipids represent many reactions that we have studied in previous chapters, especially reactions of carboxylic acids, alkenes, and alcohols. Ester hydrolysis (e.g., saponification) liberates fatty acids and glycerol from triacylglycerols. The carboxylic acid group of a fatty acid can be reduced, converted to an activated acyl derivative such as an acyl chloride, or converted to an ester or amide. Alkene functional groups in unsaturated fatty acids can be hydrogenated, hydrated, halogenated, hydrohalogenated, converted to a vicinal diol or epoxide, or cleaved by oxidation reactions. Alcohol functional groups in lipids such as terpenes, steroids, and prostaglandins can be alkylated, acylated, oxidized, or used in elimination reactions. All of these are reactions we have studied previously in the context of smaller molecules.

Key Terms and Concepts

Fats	**Section 23.2**
Hydrophilic molecule/group	**Section 23.2C**
Hydrophobic molecule/group	**Section 23.2C**
Lipid bilayer	**Section 23.6A**
Micelles	**Section 23.2**
Oils	**Section 23.2**
Phospholipid bilayer	**Section 23.6**
Phospholipids	**Section 23.6**
Polyunsaturated fatty acid/ester	**Section 23.2**
Prostaglandins	**Section 23.5**
Saponification	**Section 23.2C**
Saturated fatty acid/ester	**Section 23.2**
Simple and mixed triacylglycerols	**Section 23.2**
Steroids	**Section 23.4**
Terpenes, Terpenoids	**Section 23.3**
Triacylglycerols	**Section 23.2**
Unsaturated fatty acid/ester	**Section 23.2**

Additional Problems

23.12 How would you convert stearic acid, $CH_3(CH_2)_{16}CO_2H$, into each of the following?

(a) Ethyl stearate, $CH_3(CH_2)_{16}CO_2C_2H_5$ (two ways)
(b) *tert*-Butyl stearate, $CH_3(CH_2)_{16}CO_2C(CH_3)_3$
(c) Stearamide, $CH_3(CH_2)_{16}CONH_2$
(d) *N,N*-Dimethylstearamide, $CH_3(CH_2)_{16}CON(CH_3)_2$
(e) Octadecylamine, $CH_3(CH_2)_{16}CH_2NH_2$
(f) Heptadecylamine, $CH_3(CH_2)_{15}CH_2NH_2$
(g) Octadecanal, $CH_3(CH_2)_{16}CHO$
(h) Octadecyl stearate, $CH_3(CH_2)_{16}\overset{O}{\overset{\|}{C}}OCH_2(CH_2)_{16}CH_3$
(i) 1-Octadecanol, $CH_3(CH_2)_{16}CH_2OH$ (two ways)
(j) 2-Nonadecanone, $CH_3(CH_2)_{16}\overset{O}{\overset{\|}{C}}CH_3$
(k) 1-Bromooctadecane, $CH_3(CH_2)_{16}CH_2Br$
(l) Nonadecanoic acid, $CH_3(CH_2)_{16}CH_2CO_2H$

23.13 How would you transform myristic acid into each of the following?

(a) $CH_3(CH_2)_{11}\underset{Br}{\underset{|}{C}}HCO_2H$
(b) $CH_3(CH_2)_{11}\underset{OH}{\underset{|}{C}}HCO_2H$
(c) $CH_3(CH_2)_{11}\underset{CN}{\underset{|}{C}}HCO_2H$
(d) $CH_3(CH_2)_{11}\underset{NH_3^+}{\underset{|}{C}}HCO_2^-$

23.14 Using palmitoleic acid as an example and neglecting stereochemistry, illustrate each of the following reactions of the double bond:

(a) Addition of bromine
(b) Addition of hydrogen
(c) Hydroxylation
(d) Addition of HCl

23.15 When oleic acid is heated to 180–200°C (in the presence of a small amount of selenium), an equilibrium is established between oleic acid (33%) and an isomeric compound called elaidic acid (67%). Suggest a possible structure for elaidic acid.

23.16 Gadoleic acid ($C_{20}H_{38}O_2$), a fatty acid that can be isolated from cod-liver oil, can be cleaved by hydroxylation and subsequent treatment with periodic acid to $CH_3(CH_2)_9CHO$ and $OHC(CH_2)_7CO_2H$. **(a)** What two stereoisomeric structures are possible for gadoleic acid? **(b)** What spectroscopic technique would make possible a decision as to the actual structure of gadoleic acid? **(c)** What peaks would you look for?

23.17 When limonene (Section 23.3) is heated strongly, it yields 2 mol of isoprene. What kind of reaction is involved here?

23.18 α-Phellandrene and β-phellandrene are isomeric compounds that are minor constituents of spearmint oil; they have the molecular formula $C_{10}H_{16}$. Each compound has a UV absorption maximum in the 230–270-nm range. On catalytic hydrogenation, each compound yields 1-isopropyl-4-methylcyclohexane. On vigorous oxidation with potassium permanganate, α-phellandrene yields $CH_3\overset{O}{\overset{\|}{C}}CO_2H$ and $CH_3\underset{CH_3}{\underset{|}{C}}HCH(CO_2H)CH_2CO_2H$. A similar oxidation of β-phellandrene yields

Note: Problems marked with an asterisk are "challenge problems."

$CH_3CH(CH_3)CH(CO_2H)CH_2CH_2C(=O)CO_2H$ as the only isolable product. Propose structures for α- and β-phellandrene.

23.19 Vaccenic acid, a constitutional isomer of oleic acid, has been synthesized through the following reaction sequence:

$$\text{1-Octyne} + NaNH_2 \xrightarrow[NH_3]{\text{liq.}} \mathbf{A}\ (C_8H_{13}Na) \xrightarrow{ICH_2(CH_2)_7CH_2Cl} \mathbf{B}\ (C_{17}H_{31}Cl) \xrightarrow{NaCN} \mathbf{C}\ (C_{18}H_{31}N) \xrightarrow{KOH,\ H_2O} \mathbf{D}\ (C_{18}H_{31}O_2K) \xrightarrow{H_3O^+} \mathbf{E}\ (C_{18}H_{32}O_2) \xrightarrow[BaSO_4]{H_2,\ Pd} \text{vaccenic acid}\ (C_{18}H_{34}O_2)$$

Propose a structure for vaccenic acid and for the intermediates **A–E.**

23.20 ω-Fluorooleic acid can be isolated from a shrub, *Dechapetalum toxicarium,* that grows in Africa. The compound is highly toxic to warm-blooded animals; it has found use as an arrow poison in tribal warfare, in poisoning enemy water supplies, and by witch doctors "for terrorizing the native population." Powdered fruit of the plant has been used as a rat poison; hence ω-fluorooleic acid has the common name "ratsbane." A synthesis of ω-fluorooleic acid is outlined here. Give structures for compounds **F–I:**

$$\text{1-Bromo-8-fluorooctane} + \text{sodium acetylide} \longrightarrow \mathbf{F}\ (C_{10}H_{17}F) \xrightarrow[(2)\ I(CH_2)_7Cl]{(1)\ NaNH_2} \mathbf{G}\ (C_{17}H_{30}FCl) \xrightarrow{NaCN} \mathbf{H}\ (C_{18}H_{30}NF) \xrightarrow[(2)\ H_3O^+]{(1)\ KOH} \mathbf{I}\ (C_{18}H_{31}O_2F) \xrightarrow[Ni_2B\ (P\text{-}2)]{H_2}$$

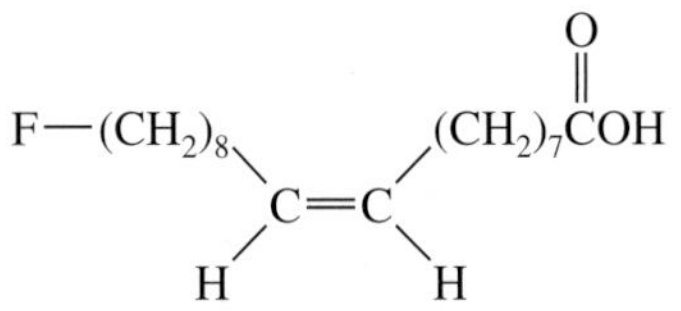

$F{-}(CH_2)_8CH{=}CH(CH_2)_7COOH$ (cis; both H on same side)

ω-Fluorooleic acid
(46% yield, overall)

23.21 Give formulas and names for compounds **A** and **B**:

$$5\alpha\text{-Cholest-2-ene} \xrightarrow{C_6H_5CO_3H} \mathbf{A}\ \text{(an epoxide)} \xrightarrow{HBr} \mathbf{B}$$

(*Hint:* **B** is not the most stable stereoisomer.)

23.22 One of the first laboratory syntheses of cholesterol was achieved by R. B. Woodward and his students at Harvard University in 1951. Many of the steps of this synthesis are outlined in the reactions that follow. Supply the missing reagents (a)–(w):

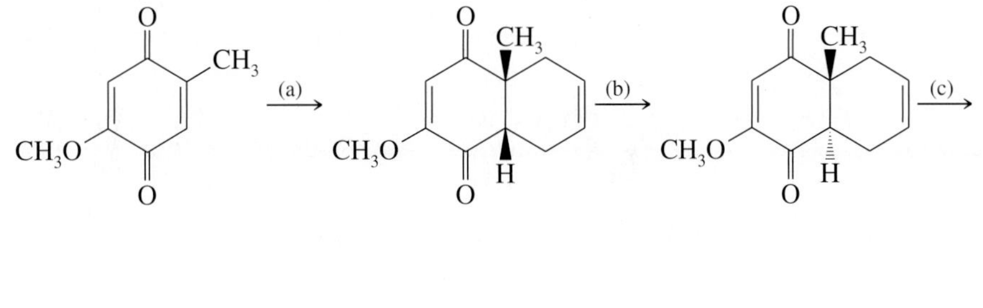

[Scheme: 2-methoxy-5-methyl-1,4-benzoquinone (CH_3O, CH_3) $\xrightarrow{(a)}$ cis-fused bicyclic dione (CH_3, CH_3O, H) $\xrightarrow{(b)}$ trans-fused bicyclic dione (CH_3, CH_3O, H) $\xrightarrow{(c)}$]

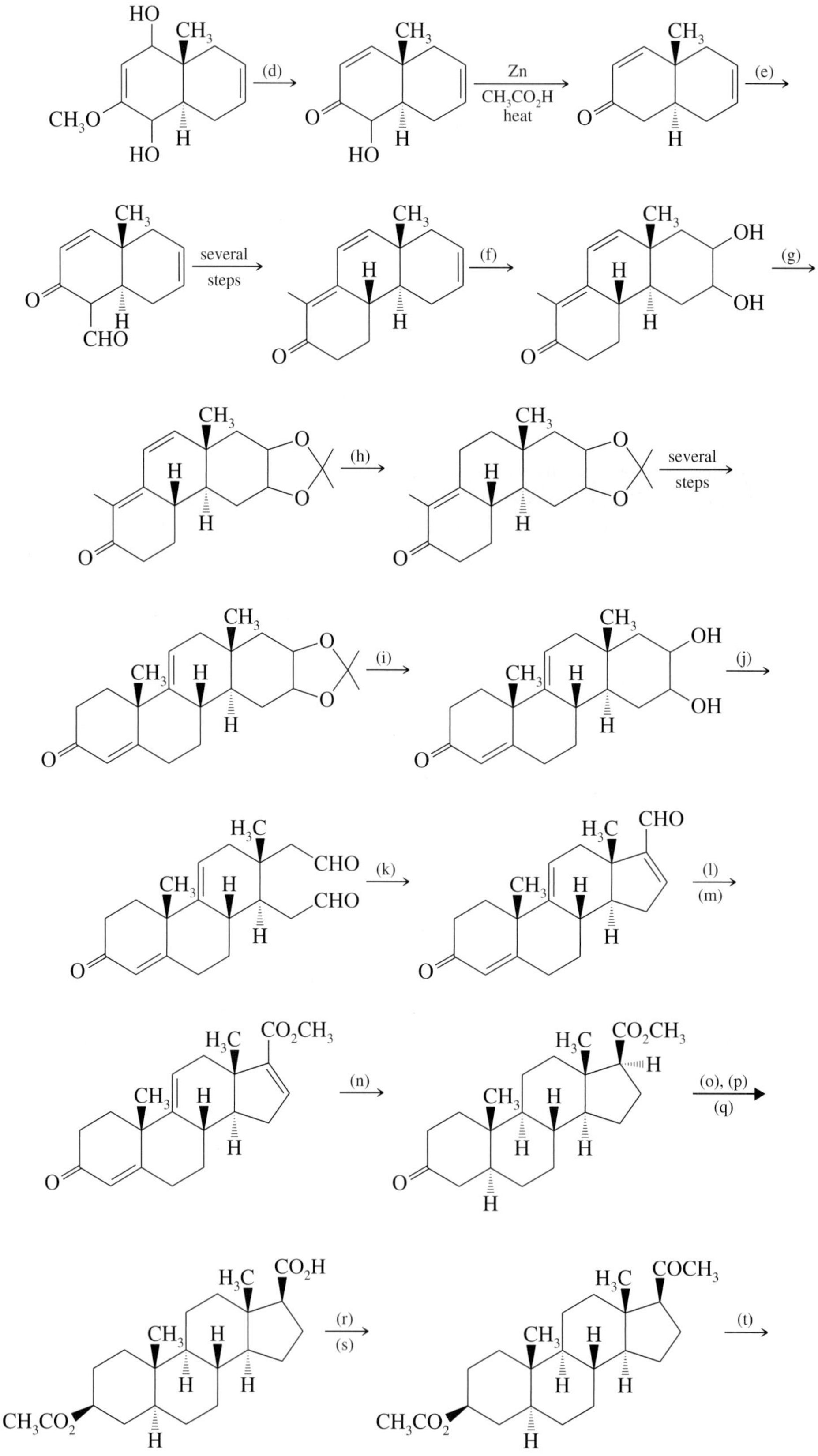

(d) → ; Zn, CH_3CO_2H, heat → ; (e) →

several steps → ; (f) → ; (g) →

(h) → ; several steps →

(i) → ; (j) →

(k) → ; (l), (m) →

(n) → ; (o), (p), (q) →

(r), (s) → ; (t) →

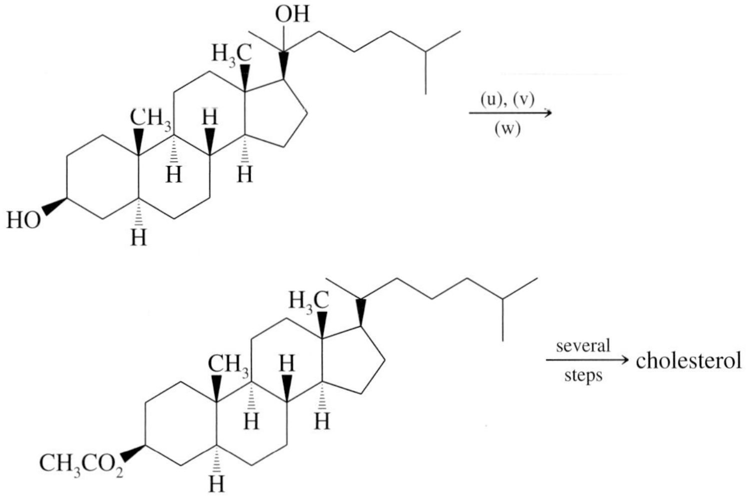

OH, H_3C, CH_3, H, H, H, H, HO, H $\xrightarrow[\text{(w)}]{\text{(u), (v)}}$

H_3C, CH_3, H, H, H, CH_3CO_2, H $\xrightarrow[\text{steps}]{\text{several}}$ cholesterol

23.23 The initial steps of a laboratory synthesis of several prostaglandins reported by E. J. Corey (Section 4.18C) and co-workers in 1968 are outlined here. Supply each of the missing reagents:

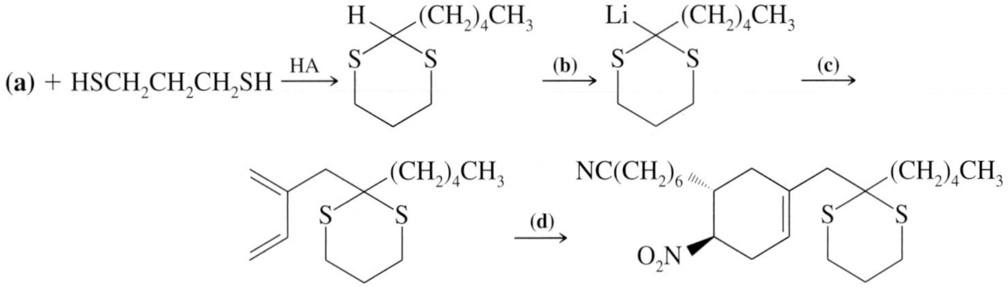

(a) + $HSCH_2CH_2CH_2SH$ $\xrightarrow{HA}$ (H, $(CH_2)_4CH_3$, S, S) $\xrightarrow{\textbf{(b)}}$ (Li, $(CH_2)_4CH_3$, S, S) $\xrightarrow{\textbf{(c)}}$

($(CH_2)_4CH_3$, S, S) $\xrightarrow{\textbf{(d)}}$ ($NC(CH_2)_6$, O_2N, $(CH_2)_4CH_3$, S, S)

(e) The initial step in another prostaglandin synthesis is shown in the following reaction. What kind of reaction—and catalyst—is needed here?

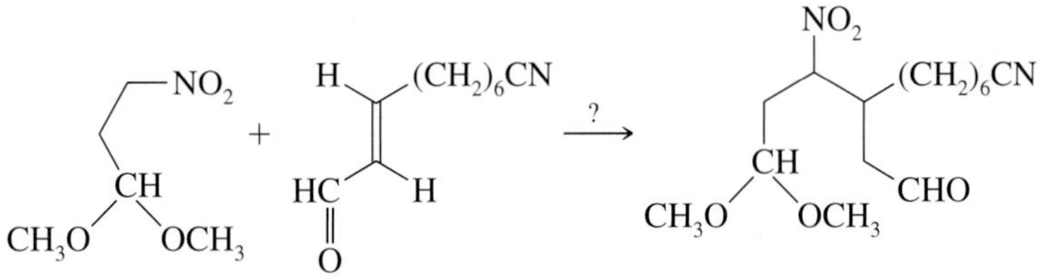

(NO_2, CH, CH_3O, OCH_3) + (H, $(CH_2)_6CN$, HC, O, H) $\xrightarrow{?}$ (NO_2, $(CH_2)_6CN$, CH, CH_3O, OCH_3, CHO)

23.24 A useful synthesis of sesquiterpene ketones, called *cyperones,* was accomplished through a modification of the following Robinson annulation procedure (Section 17.9B):

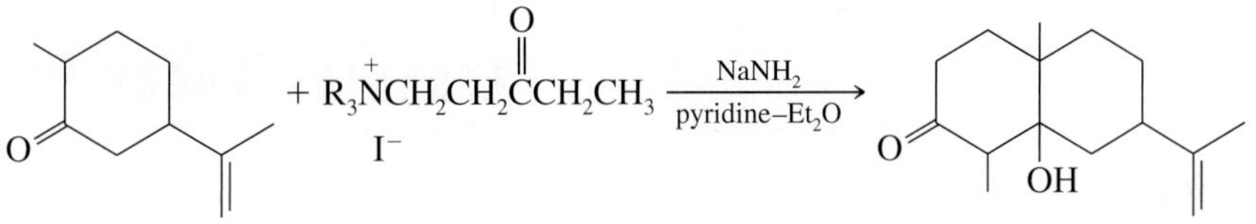

Dihydrocarvone (O) + $R_3\overset{+}{N}CH_2CH_2\overset{O}{\overset{\|}{C}}CH_2CH_3$ I^- $\xrightarrow[\text{pyridine–Et}_2\text{O}]{\text{NaNH}_2}$ (O, OH)

↓ HA, heat

(O)

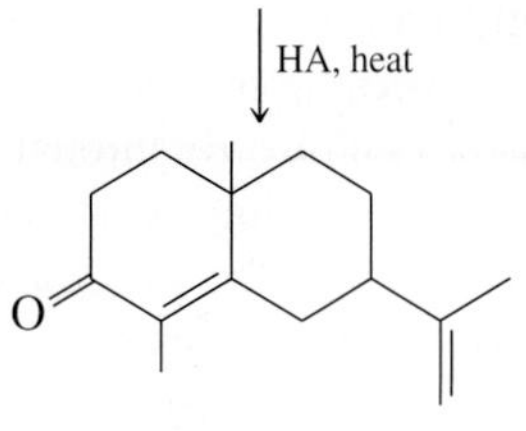

A cyperone

Write a mechanism that accounts for each step of this synthesis.

***23.25** A Hawaiian fish called the pahu or boxfish (*Ostracian lentiginosus*) secretes a toxin that kills other fish in its vicinity. The active agent in the secretion was named pahutoxin by P. J. Scheuer, and it was found by D. B. Boylan and Scheuer to contain an unusual combination of lipid moieties. To prove its structure, they synthesized it by this route:

$$CH_3(CH_2)_{12}CH_2OH \xrightarrow[\text{chlorochromate}]{\text{pyridinium}} \mathbf{A} \xrightarrow{BrCH_2CO_2Et,\ Zn} \mathbf{B} \xrightarrow[(2)\ H_3O^+]{(1)\ HO^-}$$

$$\mathbf{C} \xrightarrow[\text{pyridine}]{Ac_2O} \mathbf{D} \xrightarrow{SOCl_2} \mathbf{E} \xrightarrow[\text{pyridine}]{\text{choline chloride}} \text{Pahutoxin}$$

Compound	Selected Infrared Absorption Bands (cm^{-1})
A	1725
B	3300 (broad), 1735
C	3300–2500 (broad), 1710
D	3000–2500 (broad), 1735, 1710
E	1800, 1735
Pahutoxin	1735

What are the structures of **A**–**E** and of pahutoxin?

***23.26** The reaction illustrated by the equation below is a very general one that can be catalyzed by acid, base, and some enzymes. It therefore needs to be taken into consideration when planning syntheses that involve esters of polyhydroxy substances like glycerol and sugars:

$$HOCH_2CH(OCO(CH_2)_{14}CH_3)CH_2OH \xrightarrow[\substack{\text{10 min., room temp.} \\ \text{90\% yield}}]{\text{trace } HClO_4 \text{ in } CHCl_3} \mathbf{F}$$

Spectral data for F:

MS (*m/z*): (after trimethylsilylation): 546, 531

IR (cm^{-1}, in CCl_4 solution): 3200 (broad), 1710

^{1}H NMR (δ) (after exchange with D_2O): 4.2 (d), 3.9 (m), 3.7 (d), 2.2 (t), and others in the range 1.7 to 1

^{13}C NMR (δ): 172 (C), 74 (CH), 70 (CH_2), 67 (CH_2), 39 (CH_2), and others in the range 32 to 14

(a) What is the structure of product **F**?

(b) The reaction is intramolecular. Write a mechanism by which it probably occurs.

LEARNING GROUP PROBLEMS

1. Olestra is a fat substitute patented by Proctor and Gamble that mimics the taste and texture of triacylglycerols (see "The Chemistry of. . . Olestra and Other Fat Substitutes" in Section 23.2). It is calorie-free because it is neither hydrolyzed by digestive enzymes nor absorbed by the intestines but instead is passed directly through the body unchanged. The FDA has approved olestra for use in a variety of foods, including potato chips and other snack foods that typically have a high fat content. It can be used in both the dough and the frying process.

   (a) Olestra consists of a mixture of sucrose fatty acid esters (unlike triacylglycerols, which are glycerol esters of fatty acids). Each sucrose molecule in olestra is esterified with six to eight fatty acids. (One undesirable aspect of olestra is that it sequesters fat-soluble vitamins needed by the body, due to its high lipophilic character.) Draw the structure of a specific olestra molecule comprising six different naturally occurring fatty acids esterified to any of the available positions on sucrose. Use three saturated fatty acids and three unsaturated fatty acids.

(b) Write reaction conditions that could be used to saponify the esters of the olestra molecule you drew and give IUPAC and common names for each of the fatty acids that would be liberated on saponification.

(c) Olestra is made by sequential transesterification processes. The first transesterification involves reaction of methanol under basic conditions with natural triacylglycerols from cottonseed or soybean oil (chain lengths of C_8–C_{22}). The second transesterification involves reaction of these fatty acid methyl esters with sucrose to form olestra. Write one example reaction, including its mechanism, for each of these transesterification processes used in the synthesis of olestra. Start with any triacylglycerol having fatty acids like those incorporated into olestra.

2. The biosynthesis of fatty acids is accomplished two carbons at a time by an enzyme complex called fatty acid synthetase. The biochemical reactions involved in fatty acid synthesis are described in Special Topic D. Each of these biochemical reactions has a counterpart in synthetic reactions you have studied. Consider the biochemical reactions involved in adding each $—CH_2CH_2—$ segment during fatty acid biosynthesis (those in Special Topic D that begin with acetyl-S-ACP and malonyl-S-ACP, and end with butyryl-S-ACP). Write laboratory synthetic reactions using reagents and conditions you have studied (not biosynthetic reactions) that would accomplish the same sequence of transformations (i.e., the condensation–decarboxylation, ketone reduction, dehydration, and alkene reduction steps).

3. A certain natural terpene produced peaks in its mass spectrum at m/z 204, 111, and 93 (among others). On the basis of this and the following information, elucidate the structure of this terpene. Justify each of your conclusions.

(a) Reaction of the unknown terpene with hydrogen in the presence of platinum under pressure results in a compound with molecular formula $C_{15}H_{30}$.

(b) Reaction of the terpene with ozone followed by zinc and acetic acid produces the following mixture of compounds (1 mol of each for each mole of the unknown terpene):

O O O
O O O

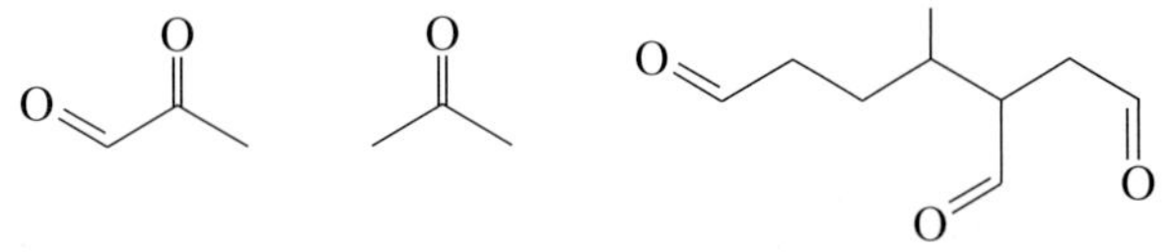

(c) After writing the structure of the unknown terpene, circle each of the isoprene units in this compound. To what class of terpenes does this compound belong (based on the number of carbons it contains)?

4. Draw the structure of a phospholipid (from any of the subclasses of phospholipids) that contains one saturated and one unsaturated fatty acid.

(a) Draw the structure of all of the products that would be formed from your phospholipid if it were subjected to complete hydrolysis (choose either acidic or basic conditions).

(b) Draw the structure of the product(s) that would be formed from reaction of the unsaturated fatty acid moiety of your phospholipid (assuming it had been released by hydrolysis from the phospholipid first) under each of the following conditions:

(i) Br_2 in CCl_4
(ii) OsO_4, followed by $NaHSO_3$
(iii) HBr
(iv) Hot alkaline $KMnO_4$, followed by H_3O^+
(v) $SOCl_2$, followed by excess CH_3NH_2

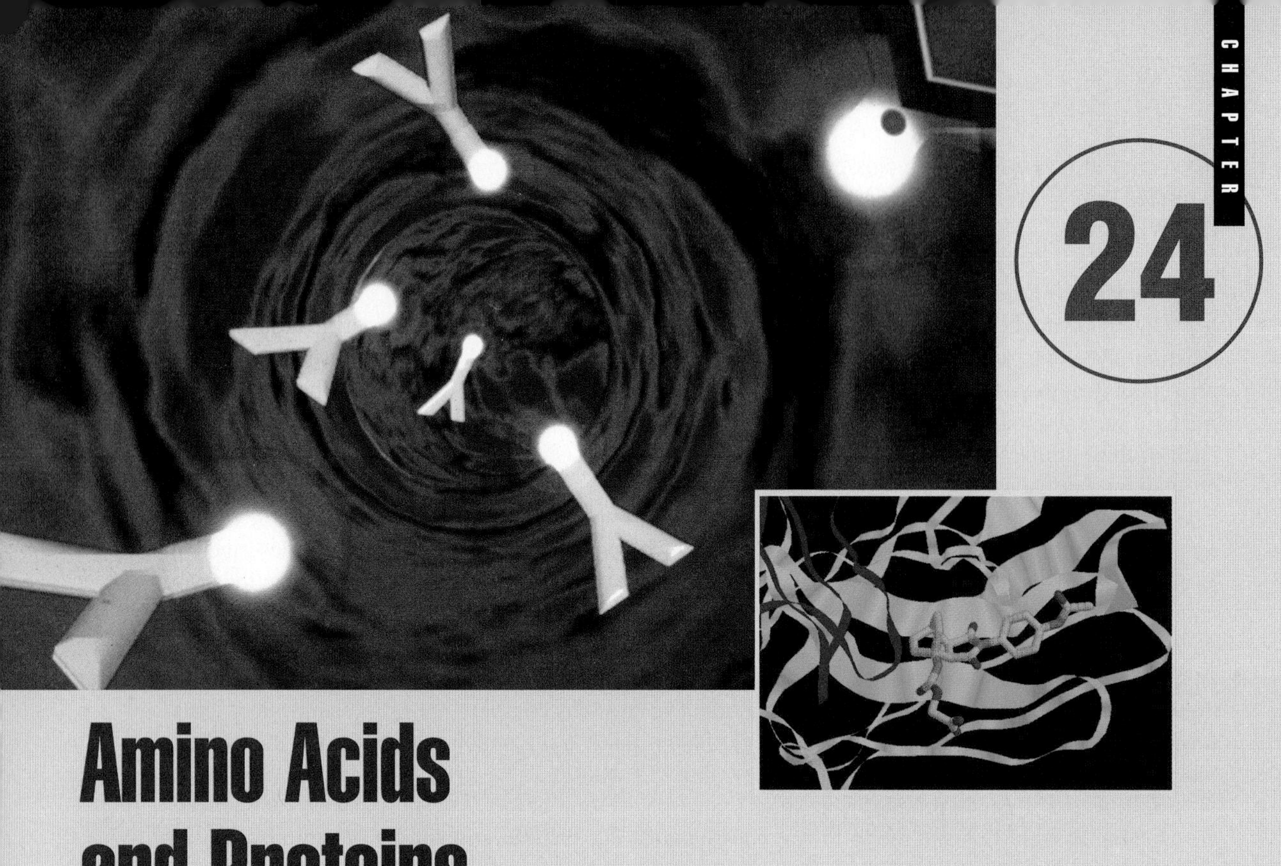

CHAPTER 24

Amino Acids and Proteins

Catalytic Antibodies: Designer Catalysts

Chemists are capitalizing on the natural adaptability of the immune system to create what we can fittingly call *designer catalysts.* These catalysts are *antibodies*—protein species usually produced by the immune system to capture and remove foreign agents but which, in this case, are elicited in a way that makes them able to catalyze chemical reactions. [The above graphic is a stylized representation of antibodies (in yellow) flowing through a blood vessel.]

The creation of the first catalytic antibodies by Richard A. Lerner and Peter G. Schultz (both of Scripps Research Institute) represented an ingenious union of principles relating to enzyme chemistry and the innate capabilities of the immune system. In some respects catalytic antibodies are like enzymes, the protein catalysts we have mentioned many times already and shall study further in this chapter. Unlike enzymes, however, catalytic antibodies can virtually be "made to order" for specific reactions by a marriage of chemistry and immunology. Examples include catalytic antibodies for Claisen rearrangements, Diels–Alder reactions (such as that shown in the molecular graphic inset above), ester hydrolyses, and aldol reactions. We shall consider how catalytic antibodies are produced in "The Chemistry of . . . Some Catalytic Antibodies" later in this chapter. Designer catalysts are indeed at hand.

24.1 Introduction

The three groups of biological polymers are polysaccharides, proteins, and nucleic acids. We studied polysaccharides in Chapter 22 and saw that they function primarily as energy reserves, as biochemical labels on cell surfaces, and, in plants, as structural materials. When we study nucleic acids in Chapter 25, we shall find that they serve two major purposes: storage and transmission of information. Of the three groups of biopolymers, proteins have the most diverse functions. As enzymes and hormones, proteins catalyze and regulate the reactions that occur in the body; as muscles and tendons they provide the body with the means for movement; as skin and hair they give it an outer covering; as hemoglobin molecules they transfer all-important oxygen to its most remote corners; as antibodies they provide it with a means of protection against disease; and in combination with other substances in bone they provide it with structural support.

Given such diversity of functions, we should not be surprised to find that proteins come in all sizes and shapes. By the standard of most of the molecules we have studied, even small proteins have very high molecular weights. Lysozyme, an enzyme, is a relatively small protein and yet its molecular weight is 14,600. The molecular weights of most proteins are much larger. Their shapes cover a range from the globular proteins such as lysozyme and hemoglobin to the helical coils of α-keratin (hair, nails, and wool) and the pleated sheets of silk fibroin.

And yet, in spite of such diversity of size, shape, and function, all proteins have common features that allow us to deduce their structures and understand their properties. Later in this chapter we shall see how this is done.

Proteins are **polyamides,** and their monomeric units are comprised of about 20 different α-amino acids:

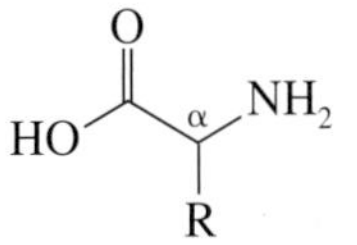

O
HO α NH$_2$
R

An α-amino acid
R is a side chain at the α carbon that determines the identity of the amino acid (Table 24.1).

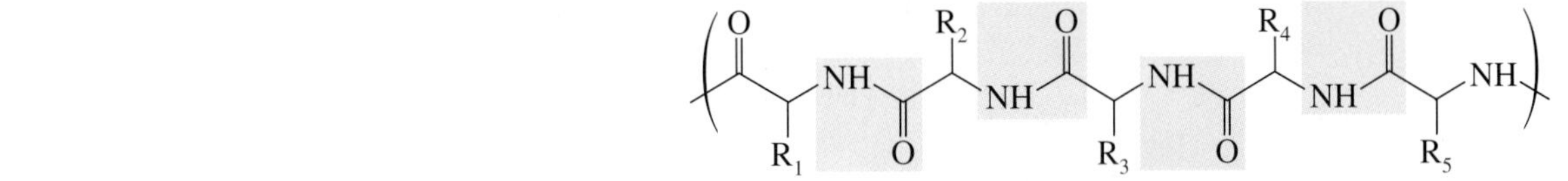

O R$_2$ O R$_4$ O
NH NH NH NH NH
R$_1$ O R$_3$ O R$_5$

A portion of a protein molecule
Amide (peptide) linkages are shaded.
R_1–R_5 may be any of the possible side chains.

Cells use α-amino acids to synthesize proteins. The exact sequence of the different α-amino acids along the protein chain is called the **primary structure** of the protein. This primary structure, as its name suggests, is of fundamental importance. For the protein to carry out its particular function, the primary structure must be correct. We shall see later that when the primary structure is correct, the polyamide chain folds in particular ways to give it the shape it needs for its particular task. This folding of the polyamide chain gives rise to higher levels of complexity called the **secondary** and **tertiary structures** of the protein. **Quaternary structure** results when a protein contains an aggregate of more than one polyamide chain.

Hydrolysis of proteins with acid or base yields a mixture of amino acids. Although hydrolysis of naturally occurring proteins may yield as many as 22 different amino acids, the amino acids have an important structural feature in common: With the exception of glycine (whose molecules are achiral), almost all naturally occurring amino acids have the L configuration at the α carbon.* That is, they have the same relative configuration as L-glyceraldehyde:

CO_2H / H_2N—C—H / R

An L-α-amino acid [usually an (S)-α-amino acid]

CHO / HO—C—H / CH_2OH

L-Glyceraldehyde [(S)-glyceraldehyde]

CO_2H / H_2N—+—H / R

CHO / HO—+—H / CH_2OH

Fischer projections for an L-α-amino acid and L-glyceraldehyde

24.2 Amino Acids

24.2A Structures and Names

The 22 α-amino acids that can be obtained from proteins can be subdivided into three different groups on the basis of the structures of their side chains, R. These are given in Table 24.1.

Only 20 of the 22 α-amino acids in Table 24.1 are actually used by cells when they synthesize proteins. Two amino acids are synthesized after the polyamide chain is intact. Hydroxyproline (present mainly in collagen) is synthesized by oxidation of proline, and cystine (present in most proteins) is synthesized from cysteine.

This conversion of cysteine to cystine requires additional comment. The —SH group of cysteine makes cysteine a *thiol.* One property of thiols is that they can be converted to disulfides by mild oxidizing agents. This conversion, moreover, can be reversed by mild reducing agents:

$$2\,\text{R—S—H} \underset{[\text{H}]}{\overset{[\text{O}]}{\rightleftharpoons}} \text{R—S—S—R}$$

Thiol **Disulfide**

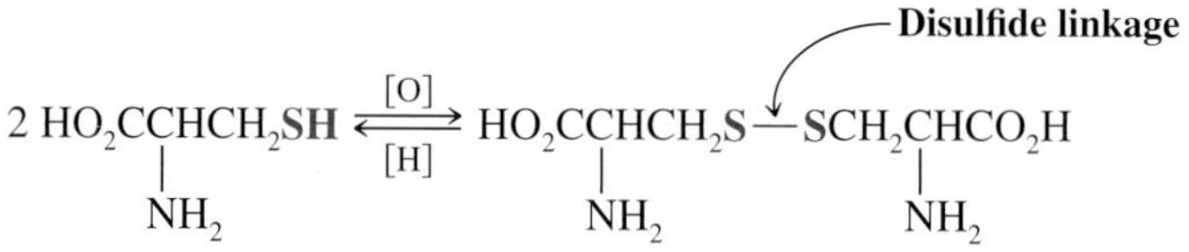

$$2\,\text{HO}_2\text{CCH(NH}_2\text{)CH}_2\text{SH} \underset{[\text{H}]}{\overset{[\text{O}]}{\rightleftharpoons}} \text{HO}_2\text{CCH(NH}_2\text{)CH}_2\text{S—SCH}_2\text{CH(NH}_2\text{)CO}_2\text{H}$$

Cysteine **Cystine** (S—S: **Disulfide linkage**)

We shall see later how the **disulfide linkage** between cysteine units in a protein chain contributes to the overall structure and shape of the protein.

24.2B Essential Amino Acids

Amino acids can be synthesized by all living organisms, plants and animals. Many higher animals, however, are deficient in their ability to synthesize all of the amino acids they need for their proteins. Thus, these higher animals require certain amino acids as a part of

*Some D-amino acids have been obtained from the material comprising the cell walls of bacteria and by hydrolysis of certain antibiotics.

TABLE 24.1 L-Amino Acids Found in Proteins

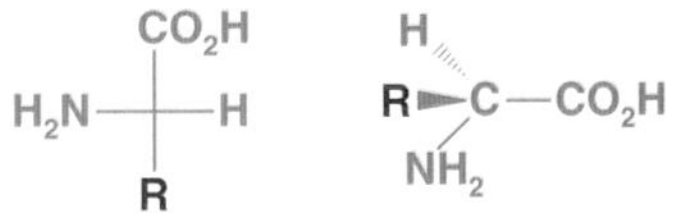

$$\begin{array}{c} CO_2H \\ H_2N-\!\!\!-\!\!\!-H \\ R \end{array} \qquad \begin{array}{c} H \\ R-C-CO_2H \\ NH_2 \end{array}$$

Structure of R	Name	Abbreviations[a]	pK_{a_1} α-CO_2H	pK_{a_2} α-NH_3^+	pK_{a_3} R group	p*I*
Neutral Amino Acids						
$—H$	Glycine	G or Gly	2.3	9.6		6.0
$—CH_3$	Alanine	A or Ala	2.3	9.7		6.0
$—CH(CH_3)_2$	Valine[b]	V or Val	2.3	9.6		6.0
$—CH_2CH(CH_3)_2$	Leucine[b]	L or Leu	2.4	9.6		6.0
$—CH(CH_3)CH_2CH_3$	Isoleucine[b]	I or Ile	2.4	9.7		6.1
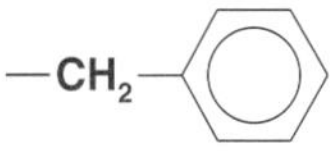 $—CH_2—C_6H_5$	Phenylalanine[b]	F or Phe	1.8	9.1		5.5
$—CH_2CONH_2$	Asparagine	N or Asn	2.0	8.8		5.4
$—CH_2CH_2CONH_2$	Glutamine	Q or Gln	2.2	9.1		5.7
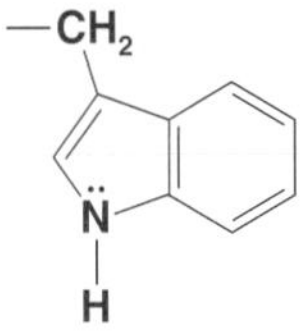 $—CH_2$–(indol-3-yl, N–H)	Tryptophan[b]	W or Trp	2.4	9.4		5.9
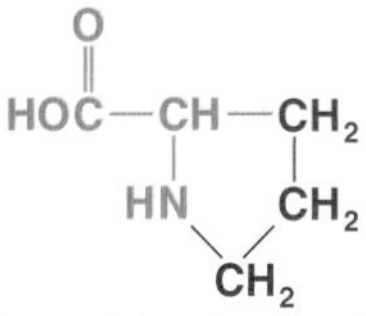 $HOC(=O)—CH—CH_2$, HN, CH_2, CH_2 (ring) (complete structure)	Proline	P or Pro	2.0	10.6		6.3
$—CH_2OH$	Serine	S or Ser	2.2	9.2		5.7
$—CH(OH)CH_3$	Threonine[b]	T or Thr	2.6	10.4		6.5
 $—CH_2—C_6H_4—OH$	Tyrosine	Y or Tyr	2.2	9.1	10.1	5.7
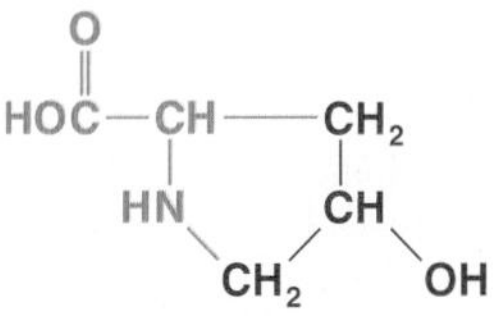 $HOC(=O)—CH—CH_2$, HN, $CH(OH)$, CH_2 (ring) (complete structure)	Hydroxyproline	Hyp	1.9	9.7		6.3
$—CH_2SH$	Cysteine	C or Cys	1.7	10.8	8.3	5.0
$—CH_2—S—S—CH_2—$	Cystine	Cys-Cys	1.6 2.3	7.9 9.9		5.1
$—CH_2CH_2SCH_3$	Methionine[b]	M or Met	2.3	9.2		5.8

TABLE 24.1 CONTINUED

Structure of R	Name	Abbreviations[a]	pK_{a_1} α-CO_2H	pK_{a_2} α-NH_3^+	pK_{a_3} R group	p*I*
R Contains an Acidic (Carboxyl) Group						
$—CH_2CO_2H$	Aspartic acid	D or Asp	2.1	9.8	3.9	3.0
$—CH_2CH_2CO_2H$	Glutamic acid	E or Glu	2.2	9.7	4.3	3.2
R Contains a Basic Group						
$—CH_2CH_2CH_2CH_2NH_2$	Lysine[b]	K or Lys	2.2	9.0	10.5[c]	9.8
$—CH_2CH_2CH_2NH—C(=NH)—NH_2$	Arginine	R or Arg	2.2	9.0	12.5[c]	10.8
$—CH_2$–(imidazolyl, N–H)	Histidine	H or His	1.8	9.2	6.0[c]	7.6

[a]Single-letter abbreviations are now the most commonly used form in current biochemical literature.
[b]An essential amino acid.
[c]pK_a is of protonated amine of R group.

their diet. For adult humans there are eight essential amino acids; these are identified in Table 24.1 by a footnote.

24.2C Amino Acids as Dipolar Ions

Amino acids contain both a basic group ($—NH_2$) and an acidic group ($—CO_2H$). In the dry solid state, amino acids exist as **dipolar ions,** a form in which the carboxyl group is present as a carboxylate ion, $—CO_2^-$, and the amino group is present as an aminium ion, $—NH_3^+$ (Dipolar ions are also called **zwitterions.**) In aqueous solution, an equilibrium exists between the dipolar ion and the anionic and cationic forms of an amino acid.

$$H_3\overset{+}{N}CH(R)CO_2H \underset{H_3O^+}{\overset{OH^-}{\rightleftharpoons}} H_3\overset{+}{N}CH(R)CO_2^- \underset{H_3O^+}{\overset{OH^-}{\rightleftharpoons}} H_2NCH(R)CO_2^-$$

Cationic form (predominant in strongly acidic solutions, e.g., at pH 0) — **Dipolar ion** — **Anionic form (predominant in strongly basic solutions, e.g., at pH 14)**

The predominant form of the amino acid present in a solution depends on the pH of the solution and on the nature of the amino acid. In strongly acidic solutions all amino acids are present primarily as cations; in strongly basic solutions they are present as anions. At some intermediate pH, called the **isoelectric point (p*I*),** the concentration of the dipolar ion is at its maximum and the concentrations of the anions and cations are equal. Each amino acid has a particular isoelectric point. These are given in Table 24.1. Proteins have isolectric points as well. As we shall see later (Section 24.13), this property of proteins is important for their separation and identification.

Let us consider first an amino acid with a side chain that contains neither acidic nor basic groups—an amino acid, for example, such as alanine.

If alanine is dissolved in a strongly acidic solution (e.g., pH 0), it is present in mainly a net cationic form. In this state the amine group is protonated (bears a formal + 1 charge) and the carboxylic acid group is neutral (has no formal charge). As is typical of α-amino acids, the pK_a for the carboxylic acid hydrogen of alanine is considerably lower (2.3) than the pK_a of an ordinary carboxylic acid (e.g., propanoic acid, pK_a 4.89):

$$\underset{\overset{+}{N}H_3}{CH_3CHCO_2H} \qquad CH_3CH_2CO_2H$$

Cationic form of alanine **Propanoic acid**
pK_{a_1} = 2.3 **pK_a = 4.89**

The reason for this enhanced acidity of the carboxyl group in an α-amino acid is the inductive effect of the neighboring aminium cation, which helps to stabilize the carboxylate anion formed when it loses a proton. Loss of a proton from the carboxyl group in a cationic α-amino acid leaves the molecule electrically neutral (in the form of a dipolar ion). This equilibrium is shown in the red-shaded portion of the equation below.

The protonated amine group of an α-amino acid is also acidic, but less so than the carboxylic acid group. The pK_a of the aminium group in alanine is 9.7. The equilibrium for loss of an aminium proton is shown in the blue-shaded portion of the equation below. The carboxylic acid proton is always lost before a proton from the aminium group in an α-amino acid.

$$\underset{\overset{+}{N}H_3}{CH_3CHCO_2H} \underset{H_3O^+}{\overset{OH^-}{\rightleftharpoons}} \underset{\overset{+}{N}H_3}{CH_3CHCO_2^-} \underset{H_3O^+}{\overset{OH^-}{\rightleftharpoons}} \underset{NH_2}{CH_3CHCO_2^-}$$

Cationic form (pK_{a_1} = 2.3) **Dipolar ion (pK_{a_2} = 9.7)** **Anionic form**

The state of an α-amino acid at any given pH is governed by a combination of two equilibria, as shown in the above equation for alanine. The isoelectric point (pI) of an amino acid such as alanine is the average of pK_{a_1} and pK_{a_2}:

$$pI = \tfrac{1}{2}(2.3 + 9.7) = 6.0 \quad \text{(isolelectric point of alanine)}$$

When a base is added to a solution of the net cationic form of alanine (initially at pH 0, for example), the first proton removed is the carboxylic acid proton, as we have said. In the case of alanine, when a pH of 2.3 is reached, the carboxylic acid proton will have been removed from half of the molecules. This pH represents the pK_a of the alanine carboxylic acid proton, as can be demonstrated using the **Henderson–Hasselbalch equation.** The Henderson–Hasselbalch equation shows that for an acid (HA) and its conjugate base (A^-),

$$pK_a = pH + \log\frac{[HA]}{[A^-]}$$

When the acid is half neutralized,

$$[HA] = [A^-] \quad \text{and} \quad \log\frac{[HA]}{[A^-]} = 0$$

thus pH = pK_a.

As more base is added to this solution, alanine reaches its isoelectric point (pI), the pH at which all of alanine's carboxylic acid protons have been removed but not its aminium protons. The molecules are therefore electrically neutral (in their dipolar ion or zwitterionic form) because the carboxylate group carries a −1 charge and the aminium group a +1 charge. The pI for alanine is 6.0.

Now, as we continue to add the base, protons from the aminium ions will begin to be removed, until at pH 9.7 half of the aminium groups will have lost a proton. This pH represents the pK_a of the aminium group. Finally, as more base is added, the remaining aminium protons will be lost until all of the alanine molecules have lost their aminium protons. At this point (e.g., pH 14) the molecules carry a net anionic charge from their carboxylate group. The amine groups are now electrically neutral.

Figure 24.1 shows a titration curve for these equilibria. The graph represents the change in pH as a function of the number of molar equivalents of base. Because alanine has two protons to lose in its net cationic form, when one molar equivalent of base has been added, the molecules will have each lost one proton and they will be electrically neutral (the dipolar ion or zwitterionic form).

FIGURE 24.1 A titration curve for $CH_3CH(\overset{+}{N}H_3)CO_2H$.

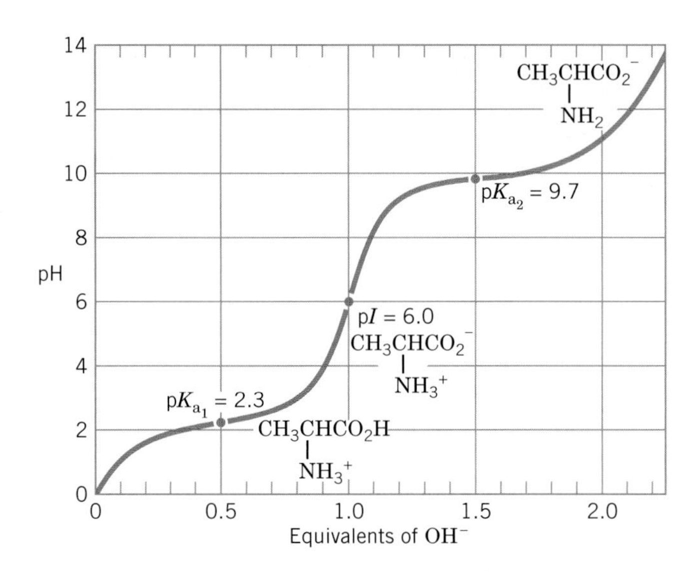

If an amino acid contains a side chain that has an acidic or basic group, the equilibria become more complex. Consider lysine, for example, an amino acid that has an additional —NH_2 group on its ε carbon. In strongly acidic solution, lysine is present as a dication because both amino groups are protonated. The first proton to be lost as the pH is raised is a proton of the carboxyl group (pK_a = 2.2), the next is from the α-aminium group (pK_a = 9.0), and the last is from the ε-aminium group:

$$\underset{\substack{\textbf{Dicationic form of}\\ \textbf{lysine}\\ (\mathbf{p}K_{a_1} = \mathbf{2.2})}}{\overset{+}{H_3N}(CH_2)_4CH(\overset{+}{N}H_3)CO_2H} \underset{H_3O^+}{\overset{OH^-}{\rightleftharpoons}} \underset{\substack{\textbf{Monocationic}\\ \textbf{form}\\ (\mathbf{p}K_{a_2} = \mathbf{9.0})}}{\overset{+}{H_3N}(CH_2)_4CH(\overset{+}{N}H_3)CO_2^-} \underset{H_3O^+}{\overset{OH^-}{\rightleftharpoons}} \underset{\substack{\textbf{Dipolar ion}\\ (\mathbf{p}K_{a_3} = \mathbf{10.5})}}{\overset{+}{H_3N}(CH_2)_4CH(NH_2)CO_2^-} \underset{H_3O^+}{\overset{OH^-}{\rightleftharpoons}} \underset{\textbf{Anionic form}}{H_2N(CH_2)_4CH(NH_2)CO_2^-}$$

The isoelectric point of lysine is the average of pK_{a_2} (the monocation) and pK_{a_3} (the dipolar ion).

$$pI = \tfrac{1}{2}(9.0 + 10.5) = 9.8 \quad \text{(isoelectric point of lysine)}$$

PROBLEM 24.1

What form of glutamic acid would you expect to predominate in **(a)** strongly acidic solution, **(b)** strongly basic solution, and **(c)** at its isoelectric point (p*I* 3.2)? **(d)** The isoelectric point of glutamine (p*I* 5.7) is considerably higher than that of glutamic acid. Explain.

PROBLEM 24.2

The guanidino group $-NH-\overset{\overset{NH}{\|}}{C}-NH_2$ of arginine is one of the most strongly basic of all organic groups. Explain.

24.3 Synthesis of α-Amino Acids

A variety of methods have been developed for the synthesis of α-amino acids. We shall begin by describing three methods that are based on reactions we have seen before. Then, in Sections 24.3D and 24.3E, we shall study methods to prepare α-amino acids in optically active form. Asymmetric synthesis is an important goal in α-amino acid synthesis due to the biological activity of the natural enantiomeric forms of α-amino acids. Several of the methods we shall study are very important in industry due to the commercial relevance of products made by these routes.

24.3A Direct Ammonolysis of an α-Halo Acid

$$\underset{\underset{X}{|}}{RCHCO_2H} \xrightarrow{NH_3\ (excess)} R-\underset{\underset{\overset{+}{N}H_3}{|}}{CHCO_2^-}$$

This method is probably used least often because yields tend to be poor (see Section 20.5A).

24.3B From Potassium Phthalimide

This method is a modification of the Gabriel synthesis of amines (Section 20.5A). The yields are usually high and the products are easily purified:

$$\text{Potassium phthalimide } (N^-K^+) + \underset{\text{Ethyl chloroacetate}}{ClCH_2CO_2C_2H_5} \longrightarrow$$

$$\underset{(97\%)}{\text{Phthalimido-}N-CH_2CO_2C_2H_5} \xrightarrow[(2)\ HCl]{(1)\ KOH/H_2O} \underset{\substack{\text{Glycine} \\ (85\%)}}{\underset{\underset{\overset{+}{N}H_3}{|}}{CH_2CO_2^-}} + \underset{\text{Phthalic acid}}{C_6H_4(CO_2H)_2} + C_2H_5OH$$

A variation of this procedure uses potassium phthalimide and diethyl α-bromomalonate to prepare an *imido* malonic ester. This method is illustrated with a synthesis of methionine:

$$\text{Potassium phthalimide } (\text{N}^- \text{K}^+) + \text{BrCH(CO}_2\text{C}_2\text{H}_5)_2 \xrightarrow{(82\text{–}85\%)} \text{Phthalimido-NCH(CO}_2\text{C}_2\text{H}_5)_2 \xrightarrow[\text{ClCH}_2\text{CH}_2\text{SCH}_3\ (96\text{–}98\%)]{\text{NaOCH}_2\text{CH}_3}$$

Diethyl α-bromomalonate **Phthalimidomalonic ester**

$$\text{Phthalimido-N—C(CO}_2\text{C}_2\text{H}_5)_2\text{CH}_2\text{CH}_2\text{SCH}_3 \xrightarrow{\text{NaOH}} o\text{-}(\text{CO}_2^-)\text{C}_6\text{H}_4\text{C(=O)—NHC(CO}_2^-)_2\text{CH}_2\text{CH}_2\text{SCH}_3 \xrightarrow[(84\text{–}85\%)]{\text{HCl}}$$

$$\text{CH}_3\text{SCH}_2\text{CH}_2\text{CH(}\overset{+}{\text{N}}\text{H}_3)\text{CO}_2^- + \text{CO}_2 + \text{phthalic acid (}o\text{-C}_6\text{H}_4(\text{COOH})_2)$$

DL-Methionine

PROBLEM 24.3 Starting with diethyl α-bromomalonate and potassium phthalimide and using any other necessary reagents, show how you might synthesize: **(a)** DL-leucine, **(b)** DL-alanine, and **(c)** DL-phenylalanine.

24.3C The Strecker Synthesis

Treating an aldehyde with ammonia and hydrogen cyanide produces an α-aminonitrile. Hydrolysis of the nitrile group (Section 18.3) of the α-aminonitrile converts the latter to an α-amino acid. This synthesis is called the Strecker synthesis:

$$\text{RCH(=O)} + \text{NH}_3 + \text{HCN} \longrightarrow \underset{\alpha\text{-Amino nitrile}}{\text{RCH(NH}_2\text{)CN}} \xrightarrow[\text{H}_2\text{O}]{\text{H}_3\text{O}^+,\ \text{heat}} \underset{\alpha\text{-Amino acid}}{\text{RCH(}\overset{+}{\text{N}}\text{H}_3\text{)CO}_2^-}$$

The first step of this synthesis probably involves the initial formation of an imine from the aldehyde and ammonia followed by the addition of hydrogen cyanide.

A Mechanism for the Reaction

Formation of an α-Aminonitrile During the Strecker Synthesis

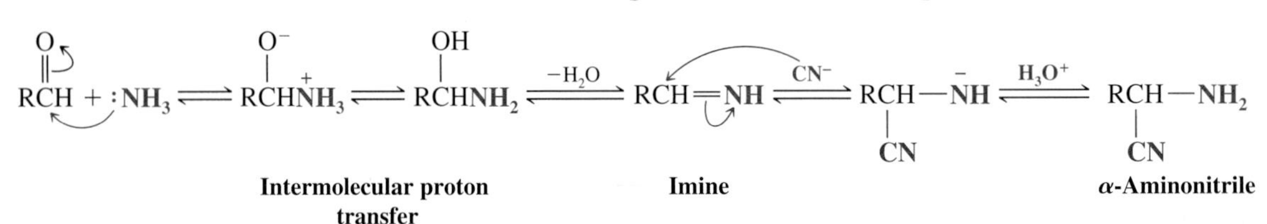

SOLVED PROBLEM

Outline a Strecker synthesis of DL-tyrosine.

ANSWER

$$HO-C_6H_4-CH_2CHO \xrightarrow{NH_3,\ HCN} HO-C_6H_4-CH_2CH(NH_2)CN \xrightarrow[\text{heat}]{H_3O^+} HO-C_6H_4-CH_2CH(\overset{+}{N}H_3)CO_2^-$$

DL-Tyrosine

PROBLEM 24.4

(a) Outline a Strecker synthesis of DL-phenylalanine. **(b)** DL-Methionine can also be synthesized by a Strecker synthesis. The required starting aldehyde can be prepared from acrolein ($CH_2{=}CHCHO$) and methanethiol (CH_3SH). Outline all steps in this synthesis of DL-methionine.

24.3D Resolution of DL-Amino Acids

With the exception of glycine, which has no stereogenic center, the amino acids that are produced by the methods we have outlined are all produced as racemic forms. To obtain the naturally occurring L-amino acid, we must, of course, resolve the racemic form. This can be done in a variety of ways, including the methods outlined in Section 20.3.

One especially interesting method for resolving amino acids is based on the use of enzymes called *deacylases.* These enzymes catalyze the hydrolysis of *N-acylamino acids* in living organisms. Since the active site of the enzyme is chiral, it hydrolyzes only *N*-acylamino acids of the L configuration. When it is exposed to a racemic mixture of *N*-acylamino acids, only the derivative of the L-amino acid is affected and the products, as a result, are separated easily:

$$\text{DL-RCH}(\overset{+}{N}H_3)CO_2^- \xrightarrow{(CH_3CO)_2O} \text{DL-RCH}(NHCOCH_3)CO_2H \xrightarrow{\text{deacylase enzyme}} CH_3CO_2H$$

(racemic form)

$$H_3\overset{+}{N}-C(CO_2^-)(R)-H \quad + \quad H-C(CO_2^-)(R)-NHCOCH_3$$

L-Amino acid **D-*N*-Acylamino acid**

Easily separated

24.3E Asymmetric Syntheses of Amino Acids

The ideal synthesis of an amino acid, of course, would be an **asymmetric synthesis** (or **enantioselective synthesis**) that produces only or predominantly the active or naturally occurring enantiomer. A variety of asymmetric synthesis methods for amino acids have been developed. One of the most important methods involves hydrogenation of an enamide using chiral transition metal catalysts.

A prime example is the enantioselective synthesis of L-DOPA developed by W. Knowles and co-workers at Monsanto Corporation. L-DOPA [(*S*)-3,4-dihydroxyphenylalanine] is a drug used against Parkinson's disease. A portion of the 2001 Nobel Prize in Chemistry was awarded to Knowles for his achievements in developing stereoselective synthesis methods, among them being the method used to synthesize L-DOPA.

The Monsanto L-DOPA synthesis involves a rhodium hydrogenation catalyst containing a chiral phosphorus ligand called (*R,R*)-DiPAMP, (*R,R*)-1,2-bis[(2-methoxyphenyl)phenylphosphino]ethane, shown below, reacting with an enamide. The hydrogenation product is obtained in 95% enantiomeric excess. Simple removal of the protecting groups leads to L-DOPA.

Asymmetric Synthesis of L-DOPA

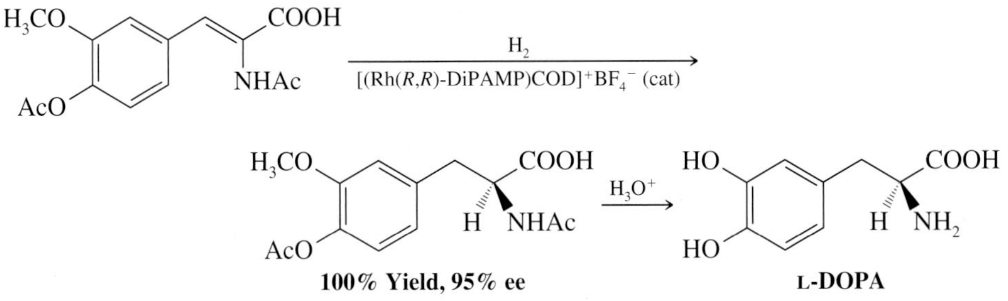

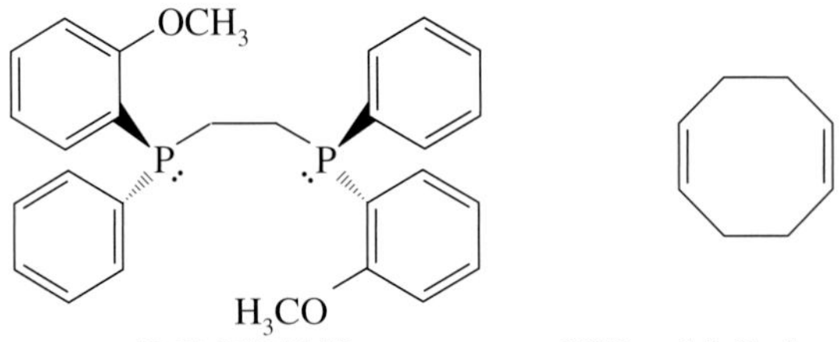

H_3CO, AcO, COOH, NHAc —[H_2; $[(Rh(R,R\text{-DiPAMP})COD]^+BF_4^-$ (cat)]→ H_3CO, AcO, COOH, H NHAc —[H_3O^+]→ HO, HO, COOH, H NH_2

100% Yield, 95% ee

L-DOPA

OCH_3, P, P, H_3CO

(*R,R*)-DiPAMP
(chiral ligand for rhodium)

COD = 1,5-Cyclooctadiene

Another important industrial synthesis of a chiral amino acid is the preparation of L-phenylalanine methyl ester, used in the synthesis of the artificial sweetener Aspartame. The chiral ligand for the hydrogenation catalyst in this case is (*R,R*)-PNNP, [*N,N′*-bis(diphenylphosphino)bis[(*R*)-1-phenylethyl]ethylenediamine], shown below.

Asymmetric Synthesis of Aspartame

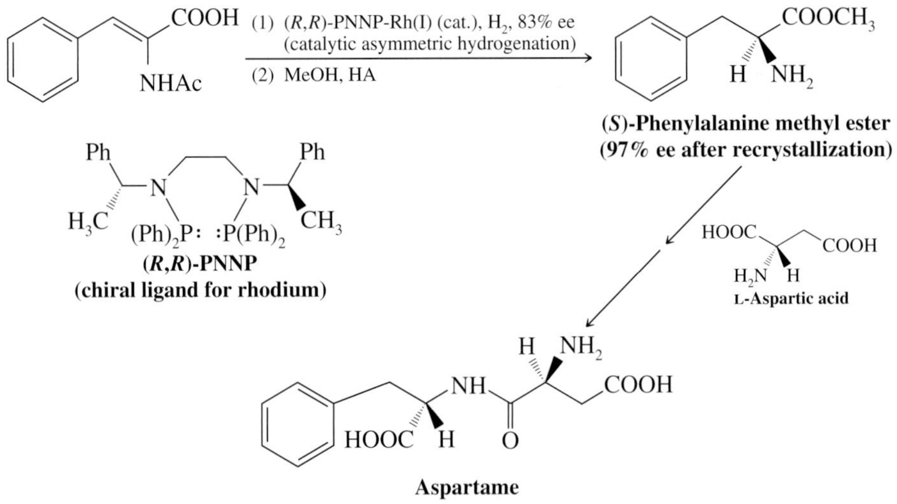

COOH, NHAc —[(1) (*R,R*)-PNNP-Rh(I) (cat.), H_2, 83% ee (catalytic asymmetric hydrogenation); (2) MeOH, HA]→ $COOCH_3$, H NH_2

(*S*)-Phenylalanine methyl ester
(97% ee after recrystallization)

Ph, Ph, N, N, H_3C, CH_3, $(Ph)_2P$: :$P(Ph)_2$

(*R,R*)-PNNP
(chiral ligand for rhodium)

HOOC, COOH, H_2N H

L-Aspartic acid

H NH_2, NH, COOH, HOOC H, O

Aspartame

An enantioselective synthesis of phenylalanine has also been accomplished using BINAP as the chiral ligand (see Section 5.16). Part of the 2001 Nobel Prize in Chemistry was awarded to R. Noyori (Nagoya University) for his development of this and other asymmetric synthesis methods.

As a final example, L-alanine and other amino acids have been synthesized using a catalyst involving (*R*)-prophos, [(*R*)-1,2-bis(diphenylphosphino)propane], developed by B. Bosnich (University of Chicago). When a rhodium complex of norbornadiene (NBD) is treated with (*R*)-prophos, the (*R*)-prophos replaces one of the norbornadiene molecules surrounding the rhodium atom to produce a chiral rhodium complex:

H_3C, H, C—CH_2, $(C_6H_5)_2P$, $P(C_6H_5)_2$

(*R*)-Prophos

$$[Rh(NBD)_2]ClO_4 + (R)\text{-prophos} \longrightarrow [Rh((R)\text{-prophos})(NBD)]ClO_4 + NBD$$

Chiral rhodium complex

Treating this rhodium complex with hydrogen in a solvent such as ethanol yields a solution containing the active chiral hydrogenation catalyst, which probably has the composition $Rh[(R)\text{-prophos}](H_2)(EtOH)_2^+$. Hydrogenation of 2-acetylaminopropenoic acid with this catalyst leads to the *N*-acetyl derivative of L-alanine in 90% enantiomeric excess. Hydrolysis of the *N*-acetyl group yields L-alanine:

$$CH_2{=}C(CO_2H)(NHCOCH_3) \xrightarrow[H_2]{[Rh((R)\text{-prophos})(H)_2(\text{solvent})_2]^+} H_3C\text{—}CH(CO_2H)(NHCOCH_3)$$

2-Acetylaminopropenoic acid → ***N*-Acetyl-L-alanine (90% ee)**

$$\xrightarrow{(1)\ OH^-,\ H_2O,\ \text{heat}\ (2)\ H_3O^+} H_3C\text{—}CH(CO_2^-)(\overset{+}{N}H_3)$$

L-Alanine

The same procedure has been used to synthesize several other L-amino acids from 2-acetylaminopropenoic acids having substituents at the C3 position. Use of the (*R*)-prophos catalyst in hydrogenation of substituted (*Z*) isomers yields the corresponding L-amino acids in 87–93% enantiomeric excess:

$$(H)(R)C{=}C(CO_2H)(NHCOCH_3) \xrightarrow[(2)\ OH^-,\ H_2O,\ \text{heat; then}\ H_3O^+]{(1)\ [Rh((R)\text{-prophos})(H)_2(\text{solvent})_2]^+,\ H_2} RCH_2\text{—}CH(CO_2^-)(\overset{+}{N}H_3)$$

(*Z*)-3-Substituted 2-acetylaminopropenoic acid → **L-Amino acid (87–93% ee)**

24.4 Polypeptides and Proteins

Amino acids are polymerized in living systems by enzymes that form amide linkages from the amino group of one amino acid to the carboxyl group of another. A molecule formed by joining amino acids together is called a **peptide,** and the amide linkages in them are called **peptide bonds** or **peptide linkages.** Each amino acid in the peptide is called an **amino acid residue.** Peptides that contain 2, 3, a few (3–10), or many amino acids are called **dipeptides, tripeptides, oligopeptides,** and **polypeptides,** respectively. **Proteins** are polypeptides consisting of one or more polypeptide chains.

$$H_3\overset{+}{N}-\underset{R}{CH}-\overset{O}{\overset{\|}{C}}-O^- + H_3\overset{+}{N}-\underset{R'}{CH}-\overset{O}{\overset{\|}{C}}-O^- \xrightarrow{[-H_2O]} H_3\overset{+}{N}-\underset{R}{CH}-\overset{O}{\overset{\|}{C}}-NH-\underset{R'}{CH}-\overset{O}{\overset{\|}{C}}-O^-$$

A dipeptide

Polypeptides are **linear polymers.** One end of a polypeptide chain terminates in an amino acid residue that has a free $-NH_3^+$ group; the other terminates in an **amino acid residue** with a free $-CO_2^-$ group. These two groups are called the **N-terminal** and the **C-terminal residues,** respectively:

$$H_3\overset{+}{N}-\underset{R}{CH}-\overset{O}{\overset{\|}{C}}\left(NH-\underset{R'}{CH}-\overset{O}{\overset{\|}{C}}\right)_n NH-\underset{R''}{CH}-\overset{O}{\overset{\|}{C}}-O^-$$

N-Terminal residue **C-Terminal residue**

By convention, we write peptide and protein structures with the N-terminal amino acid residue on the left and the C-terminal residue on the right:

$$H_3\overset{+}{N}CH_2\overset{O}{\overset{\|}{C}}-NH\underset{CH(CH_3)_2}{CH}\overset{O}{\overset{\|}{C}}O^- \qquad H_3\overset{+}{N}\underset{CH(CH_3)_2}{CH}\overset{O}{\overset{\|}{C}}-NHCH_2\overset{O}{\overset{\|}{C}}O^-$$

Glycylvaline **(Gly · Val)** **Valylglycine** **(Val · Gly)**

The tripeptide glycylvalylphenylalanine has the following structural formula:

$$H_3\overset{+}{N}CH_2\overset{O}{\overset{\|}{C}}-NH\underset{CH(CH_3)_2}{CH}\overset{O}{\overset{\|}{C}}-NH\underset{CH_2C_6H_5}{CH}\overset{O}{\overset{\|}{C}}O^-$$

Glycylvalylphenylalanine
(Gly · Val · Phe)

It becomes a significant task to write a full structural formula for a polypeptide chain that contains any more than a few amino acid residues. In this situation, use of the one-letter abbreviations is the norm for showing the sequence of amino acids. Very short peptide sequences are sometimes still represented with the three-letter abbreviations.

24.4A Hydrolysis

When a protein or polypeptide is refluxed with 6*M* hydrochloric acid for 24 h, hydrolysis of all the amide linkages usually takes place, liberating its constitutent amino acids as a mixture. Chromatographic separation and quantitative analysis of the resulting mixture can then be used to determine which amino acids comprised the intact polypeptide and their relative amounts.

One chromatographic method for separation of a mixture of amino acids is based on the use of *cation-exchange resins* (Fig. 24.2), which are insoluble polymers containing sulfonate groups.

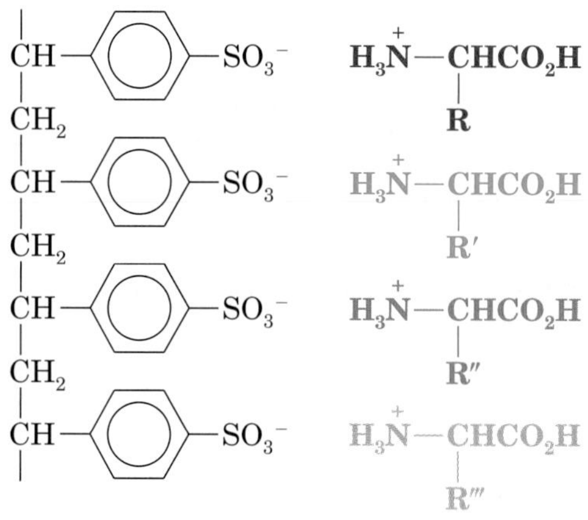

$CH-C_6H_4-SO_3^-$ $H_3\overset{+}{N}-CHCO_2H$ (R)
CH_2
$CH-C_6H_4-SO_3^-$ $H_3\overset{+}{N}-CHCO_2H$ (R′)
CH_2
$CH-C_6H_4-SO_3^-$ $H_3\overset{+}{N}-CHCO_2H$ (R″)
CH_2
$CH-C_6H_4-SO_3^-$ $H_3\overset{+}{N}-CHCO_2H$ (R‴)

FIGURE 24.2 A section of a cation-exchange resin with adsorbed amino acids.

If an acidic solution containing a mixture of amino acids is passed through a column packed with a cation-exchange resin, the amino acids will be adsorbed by the resin because of attractive forces between the negatively charged sulfonate groups and the positively charged amino acids. The strength of the adsorption varies with the basicity of the individual amino acids; those that are most basic are held most strongly. If the column is then washed with a buffered solution at a given pH, the individual amino acids move down the column at different rates and ultimately become separated. In an automated version of this analysis developed at Rockefeller University in 1950, the eluate is allowed to mix with **ninhydrin,** a reagent that reacts with most amino acids to give a derivative with an intense purple color (λ_{max} 570 nm). The amino acid analyzer is designed so that it can measure the absorbance of the eluate (at 570 nm) continuously and record this absorbance as a function of the volume of the effluent.

A typical graph obtained from an automatic amino acid analyzer is shown in Fig. 24.3. When the procedure is standardized, the positions of the peaks are characteristic of the individual amino acids, and the areas under the peaks correspond to their relative amounts.

Ninhydrin is the hydrate of indane-1,2,3-trione. With the exception of proline and hydroxyproline, all of the α-amino acids found in proteins react with ninhydrin to give the same intensely colored purple anion (λ_{max} 570 nm). We shall not go into the mechanism here, but notice that the only portion of the anion that is derived from the α-amino acid is the nitrogen:

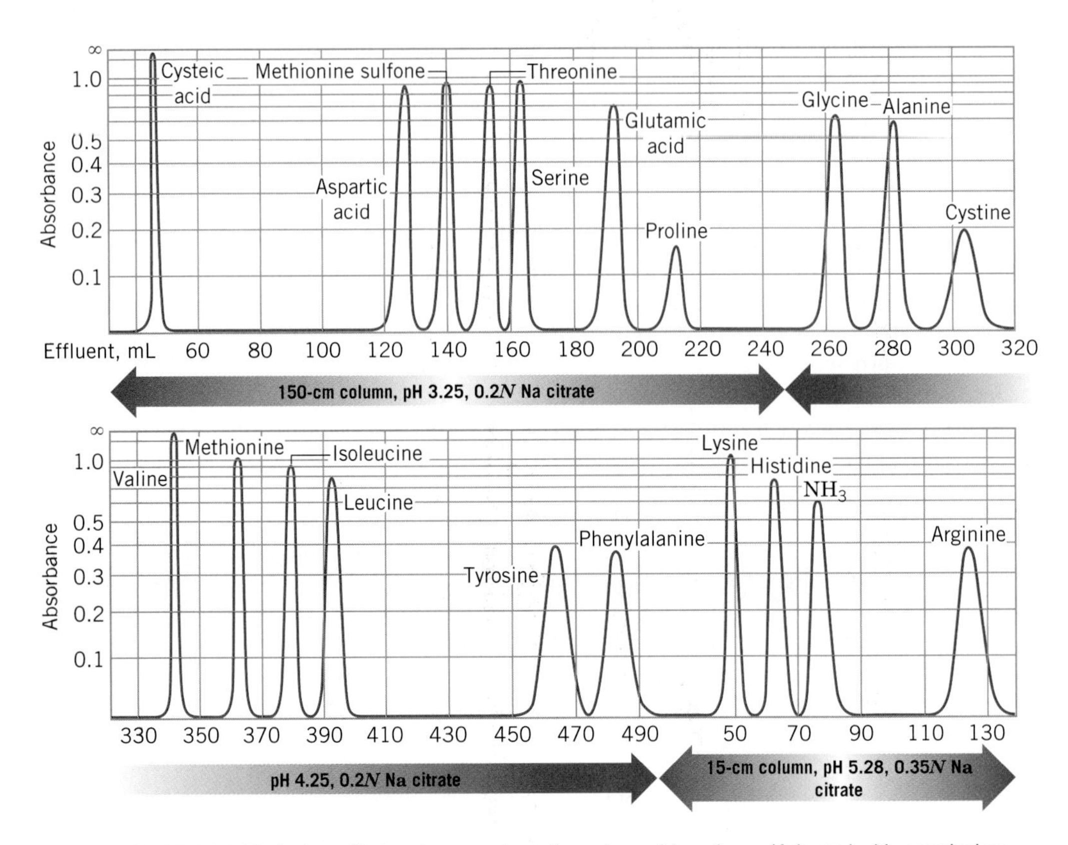

FIGURE 24.3 Typical result given by an automatic amino acid analyzer. (Adapted with permission from Spackman, D. H.; Stein, W. H.; Moore, S. *Anal. Chem.* **1958,** *30,* 1190–1206. Copyright © by the American Chemical Society.)

$$\text{Indane-1,2,3-trione} \underset{-H_2O}{\overset{+H_2O}{\rightleftharpoons}} \text{Ninhydrin}$$

Indane-1,2,3-trione **Ninhydrin**

$$2\,(\text{Indane-1,2,3-trione}) + RCH(\overset{+}{N}H_3)CO_2^- \xrightarrow[(-H_3O^+)]{} \text{Purple anion} + RCH{=}O + CO_2$$

Purple anion

Proline and hydroxyproline do not react with ninhydrin in the same way because their α-amino groups are secondary amines and part of a five-membered ring.

Analysis of amino acid mixtures can also be done very easily using high-performance liquid chromatography (HPLC), and this is now the most common method. A cation-exchange resin is used for the column packing in some HPLC analyses (See Section 24.14), while other analyses require hydrophobic (reversed-phase) column materials. Identification of amino acids separated by HPLC can be done by comparison with retention times of standard samples. Instruments that combine HPLC with mass spectrometry make direct identification possible. See Section 24.5E.

24.5 Primary Structure of Polypeptides and Proteins

The sequence of amino acid residues in a polypeptide or protein is called its **primary structure.** A simple peptide composed of three amino acids (a tripeptide) can have 6 different amino acid sequences; a tetrapeptide can have as many as 24 different sequences. For a protein composed of 20 different amino acids in a single chain of 100 residues, there are $2^{100} = 1.27 \times 10^{130}$ possible peptide sequences, a number much greater than the number of atoms estimated to be in the universe (9×10^{78})! Clearly, one of the most important things to determine about a protein is the sequence of its amino acids. Fortunately, there are a variety of methods available to determine the sequence of amino acids in a polypeptide. We shall begin with techniques used to identify the N- and C-terminal amino acids.

24.5A Edman Degradation

The most widely used procedure for identifying the N-terminal amino acid in a peptide is the **Edman degradation** method (developed by Pehr Edman of the University of Lund, Sweden). Used repetitively, the Edman degradation method can be used to sequence peptides up to about 60 residues in length. The process works so well that machines called amino acid sequencers have been developed to carry out the Edman degradation process in automated cycles.

The chemistry of the Edman degradation is based on a labeling reaction between the N-terminal amino group and phenyl isothiocyanate, $C_6H_5—N=C=S$. Phenyl isothiocyanate reacts with the N-terminal amino group to form a phenylthiocarbamyl derivative, which is then cleaved from the peptide chain by acid. The result is an unstable anilinothioazolinone (ATZ), which rearranges to a stable phenylthiohydantoin (PTH) derivative of the amino acid. In the automated process, the PTH derivative is introduced directly to a high-performance liquid chromatograph and identified by comparison of its retention time with known amino acid PTH derivatives (Fig. 24.4). The cycle is then repeated for the next N-terminal amino acid. Automated peptide sequence analyzers can perform a single iteration of the Edman degradation in approximately 30 min using only picomole amounts of the polypeptide sample.

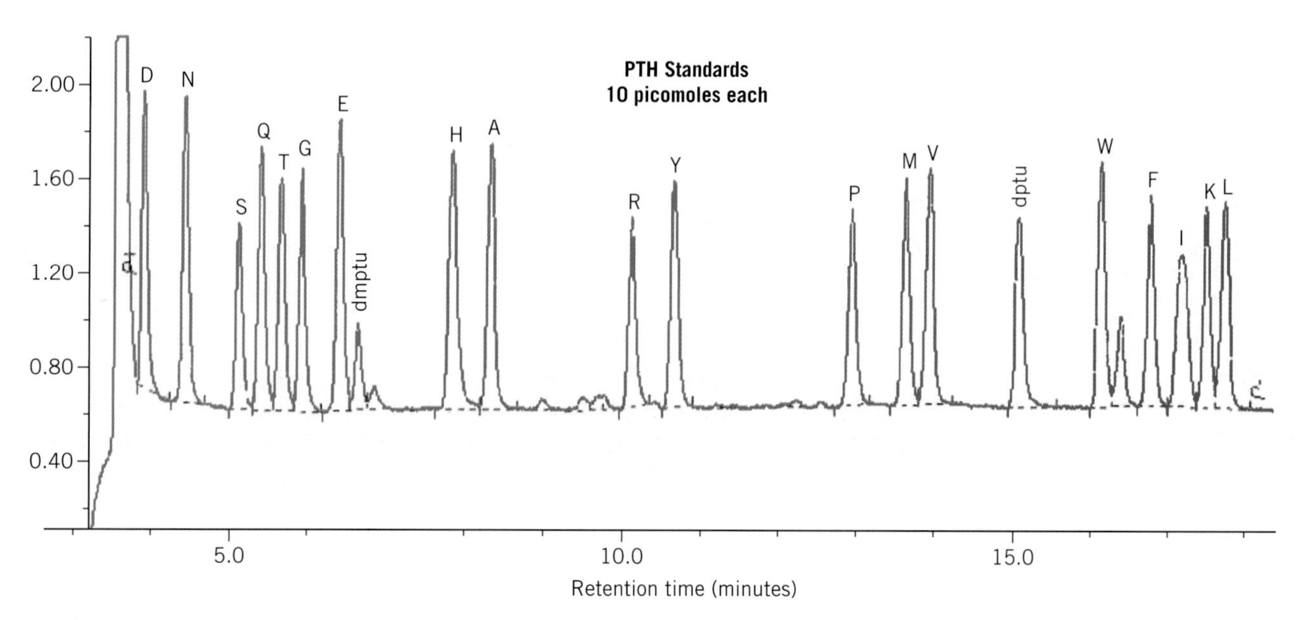

FIGURE 24.4 PTH standards run on a Procise™ instrument..

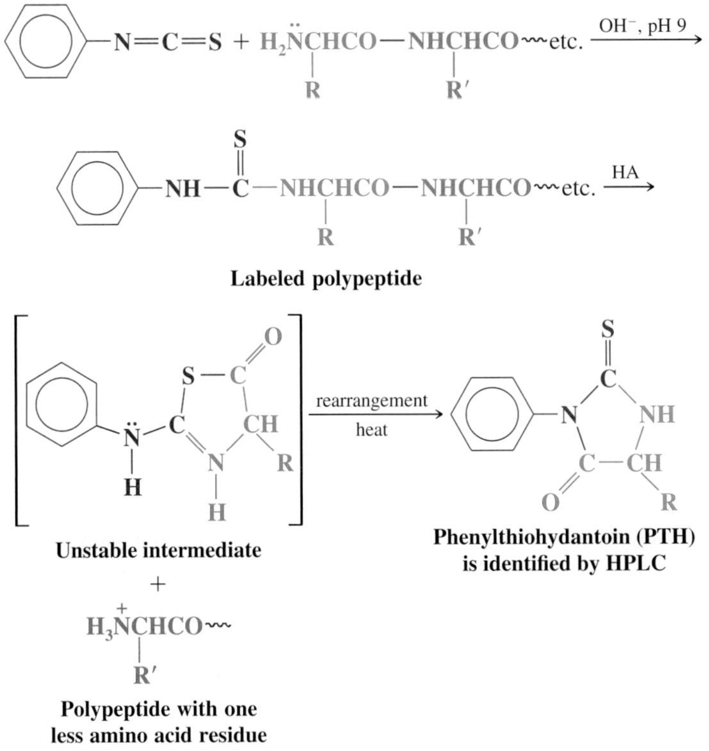

$C_6H_5-N=C=S + H_2\ddot{N}CHCO-NHCHCO\sim etc. \xrightarrow{OH^-,\ pH\ 9}$ (R, R′)

$C_6H_5-NH-C(=S)-NHCHCO-NHCHCO\sim etc. \xrightarrow{HA}$ (R, R′)

Labeled polypeptide

$\xrightarrow[\text{heat}]{\text{rearrangement}}$

Unstable intermediate

+

$H_3\overset{+}{N}CHCO\sim$ (R′)

Polypeptide with one less amino acid residue

Phenylthiohydantoin (PTH) is identified by HPLC

24.5B Sanger N-Terminal Analysis

Another method for N-terminal sequence analysis is the Sanger method, based on the use of 2,4-dinitrofluorobenzene (DNFB). When a polypeptide is treated with DNFB in mildly basic solution, a nucleophilic aromatic substitution reaction (S_NAr, Section 21.11A) takes place involving the free amino group of the N-terminal residue. Subsequent hydrolysis of the polypeptide gives a mixture of amino acids in which the N-terminal amino acid is labeled with a 2,4-dinitrophenyl group. After separating this amino acid from the mixture, it can be identified by comparison with known standards.

This method was introduced by Frederick Sanger of Cambridge University in 1945. Sanger made extensive use of this procedure in his determination of the amino acid sequence of insulin and won the Nobel Prize in Chemistry for the work in 1958.

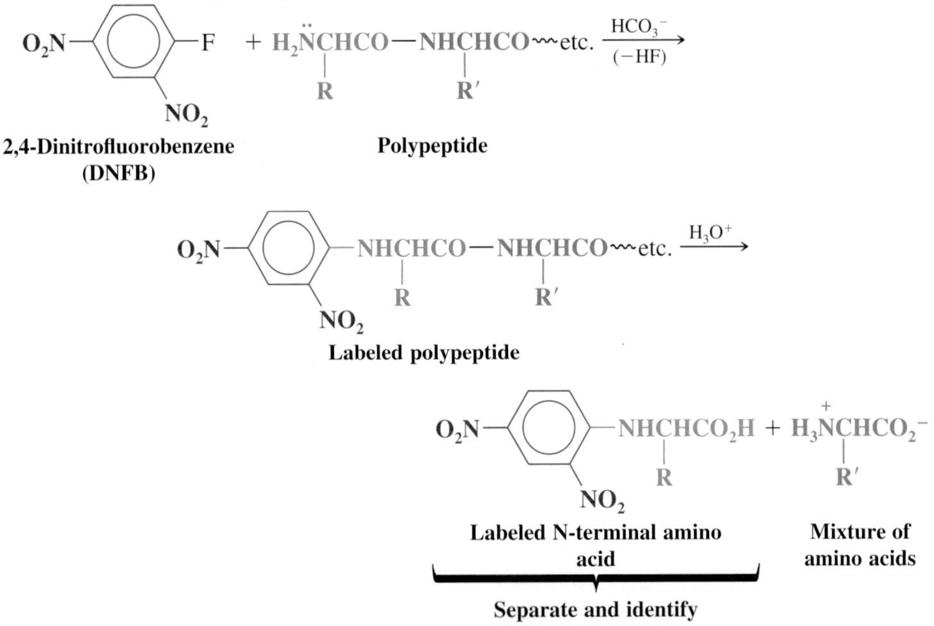

$O_2N-C_6H_3(NO_2)-F + H_2\ddot{N}CHCO-NHCHCO\sim etc. \xrightarrow[(-HF)]{HCO_3^-}$ (R, R′)

2,4-Dinitrofluorobenzene (DNFB) **Polypeptide**

$O_2N-C_6H_3(NO_2)-NHCHCO-NHCHCO\sim etc. \xrightarrow{H_3O^+}$ (R, R′)

Labeled polypeptide

$O_2N-C_6H_3(NO_2)-NHCHCO_2H + H_3\overset{+}{N}CHCO_2^-$ (R, R′)

Labeled N-terminal amino acid **Mixture of amino acids**

Separate and identify

2,4-Dinitrofluorobenzene will react with any free amino group in a polypeptide, including the ε-amino group of lysine, and this fact complicates Sanger analyses. Only the N-terminal amino acid residue of a peptide will bear the 2,4-dinitrophenyl group at its α-amino group, however. Nevertheless, the Edman method of N-terminal analysis is much more widely used.

PROBLEM 24.5

The electron-withdrawing property of the 2,4-dinitrophenyl group makes separation of the labeled amino acid very easy. Suggest how this is done.

24.5C C-Terminal Analysis

C-Terminal residues can be identified through the use of digestive enzymes called *carboxypeptidases*. These enzymes specifically catalyze the hydrolysis of the amide bond of the amino acid residue containing a free $—CO_2H$ group, liberating it as a free amino acid. A carboxypeptidase, however, will continue to attack the polypeptide chain that remains, successively lopping off C-terminal residues. As a consequence, it is necessary to follow the amino acids released as a function of time. The procedure can be applied to only a limited amino acid sequence for, at best, after a time the situation becomes too confused to sort out.

PROBLEM 24.6

(a) Write a reaction showing how 2,4-dinitrofluorobenzene could be used to identify the N-terminal amino acid of Val · Ala · Gly. **(b)** What products would you expect (after hydrolysis) when Val · Lys · Gly is treated with 2,4-dinitrofluorobenzene?

PROBLEM 24.7

Write the reactions involved in a sequential Edman degradation of Met · Ile · Arg.

24.5D Complete Sequence Analysis

Sequential analysis using the Edman degradation or other methods becomes impractical with large proteins and polypeptides. Fortunately, there are techniques to cleave peptides into fragments that are of manageable size. **Partial hydrolysis** with dilute acid, for example, generates a family of peptides cleaved in random locations and with varying lengths. Sequencing these cleavage peptides and looking for points of overlap allows the sequence of the entire peptide to be pieced together.

Consider a simple example: We are given a pentapeptide known to contain valine (two residues), leucine (one residue), histidine (one residue), and phenylalanine (one residue), as determined by hydrolysis and automatic amino acid analysis. With this information we can write the "molecular formula" of the protein in the following way, using commas to indicate that the sequence is unknown:

$$\text{Val}_2\text{, Leu, His, Phe}$$

Then, let us assume that by using DNFB and carboxypeptidase we discover that valine and leucine are the N- and C-terminal residues, respectively. So far we know the following:

Val (Val, His, Phe) Leu

But the sequence of the three nonterminal amino acids is still unknown.

We then subject the pentapeptide to partial acid hydrolysis and obtain the following dipeptides. (We also get individual amino acids and larger pieces, i.e., tripeptides and tetrapeptides.)

Val · His + His · Val + Val · Phe + Phe · Leu

The points of overlap of the dipeptides (i.e., His, Val, and Phe) tell us that the original pentapeptide must have been the following:

Val · His · Val · Phe · Leu

Site-specific cleavage of peptide bonds is possible with enzymes and specialized reagents as well, and these methods are now more widely used than partial hydrolysis. For example, the enzyme trypsin preferentially catalyzes hydrolysis of peptide bonds on the C-terminal side of arginine and lysine. Chemical cleavage at specific sites can be done with cyanogen bromide (CNBr), which cleaves peptide bonds on the C-terminal side of methionine residues. Using these site-selective cleavage methods on separate samples of a given polypeptide results in fragments that have overlapping sequences. After sequencing the individual fragments, aligning them with each other on the basis of their overlapping sections results in a sequence for the intact protein.

24.5E Peptide Sequencing Using Mass Spectrometry and Sequence Databases

Other methods for determining the sequence of a polypeptide include mass spectrometry and comparison of partial peptide sequences with databases of known complete sequences.

Mass spectrometry is especially powerful because sophisticated techniques allow mass analysis of proteins with very high precision. One mass spectrometric method is called "ladder sequencing." In this technique an enzymatic digest is prepared that yields a mixture of peptide fragments that each differ in length by one amino acid residue (e.g., by use of carboxypeptidase). The digest is a family of peptides where each one is the result of cleavage of one successive residue from the chain. Mass spectrometric analysis of this mixture yields a family of peaks corresponding to the molecular weight of each peptide. Each peak in the spectrum differs from the next by the molecular weight of the amino acid that is the difference in their structures. With these data, one can ascend the ladder of peaks from the lowest weight fragment to the highest (or vice versa), "reading" the sequence of the peptide from the difference in mass between each peak. The difference in mass between each peptide fragment and the next represents the amino acid in that spot along the sequence, and hence an entire sequence can be read from the ladder of fragment masses. This technique has also been applied to the sequencing of oligonucleotides.

Random cleavage of a peptide, similar to that from partial hydrolysis with acid, can also be accomplished with mass spectrometry. An intact protein introduced into a mass spectrometer can be cleaved into smaller fragments by collision with gas molecules deliberately leaked into the mass spectrometer vacuum chamber (a technique called collision-induced dissociation, CID). These peptide fragments can be individually selected for analysis using a technique called tandem mass spectrometry (MS/MS). The mass spectra of these random fragments can be compared with mass spectra databases to determine the protein sequence.

In some cases it is also possible to determine the sequence of an unknown polypeptide by sequencing just a few of its amino acids and comparing this partial sequence with the database of known sequences for complete polypeptides or proteins. This procedure works if the unknown peptide turns out to be one that has been studied previously. (Studies of the expression of known proteins is one dimension of the field of proteomics, Section 24.14.) Due to the many sequence permutations that are theoretically possible and the uniqueness of a given protein's structure, a sequence of just 10–25 peptide residues is usually sufficient to generate data that match only one or a small number of known polypeptides. The partial sequence can be determined by the Edman method or mass spectrometry. For example, the enzyme lysozyme with 129 amino acid residues (see Section 24.10) can be identified based on the sequence of just its first 15 amino acid residues. Structure determination based on comparison of sequences with computerized databases is part of the burgeoning field of bioinformatics.

An analogous approach using databases is to infer the *DNA sequence* that codes for a partial peptide sequence and compare this DNA sequence with the database of known DNA sequences. If a satisfactory match is found, the remaining sequence of the polypeptide can be read from the DNA sequence using the genetic code (see Section 25.5). In addition, the inferred oligonucleotide sequence for the partial peptide can be synthesized chemically (see Section 25.7) and used as a probe to find the gene that codes for the protein. This technique is part of molecular biological methods used to clone and express large quantities of a protein of interest.

PROBLEM 24.8

Glutathione is a tripeptide found in most living cells. Partial acid-catalyzed hydrolysis of glutathione yields two dipeptides, Cys · Gly and one composed of Glu and Cys. When this second dipeptide was treated with DNFB, acid hydrolysis gave *N*-labeled Glu. **(a)** On the basis of this information alone, what structures are possible for glutathione? **(b)** Synthetic experiments have shown that the second dipeptide has the following structure:

$$H_3\overset{+}{N}\underset{\substack{|\\CO_2^-}}{CH}CH_2CH_2CONH\underset{\substack{|\\CH_2SH}}{CH}CO_2^-$$

What is the structure of glutathione?

PROBLEM 24.9

Give the amino acid sequence of the following polypeptides using only the data given by partial acidic hydrolysis:

(a) Ser, Hyp, Pro, Thr $\xrightarrow[H_2O]{H_3O^+}$ Ser · Thr + Thr · Hyp + Pro · Ser

(b) Ala, Arg, Cys, Val, Leu $\xrightarrow[H_2O]{H_3O^+}$ Ala · Cys + Cys · Arg + Arg · Val + Leu · Ala

24.6 Examples of Polypeptide and Protein Primary Structure

As we discussed in the previous section, the covalent structure of a protein or polypeptide is called its **primary structure** (Fig. 24.5). Using the techniques we described, chemists have had remarkable success in determining the primary structures of polypeptides and proteins. The compounds described in the following pages are important examples.

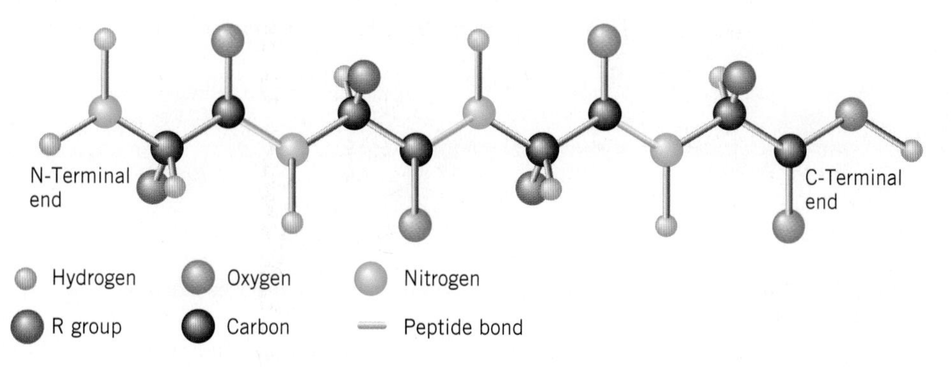

FIGURE 24.5 A representation of the primary structure of a tetrapeptide.

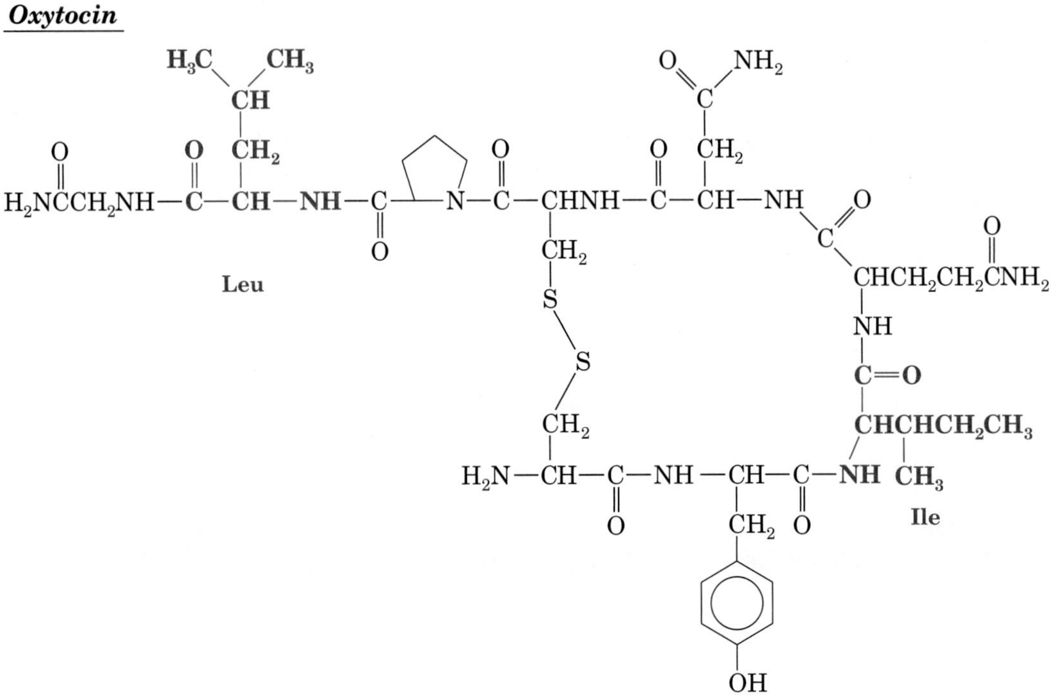

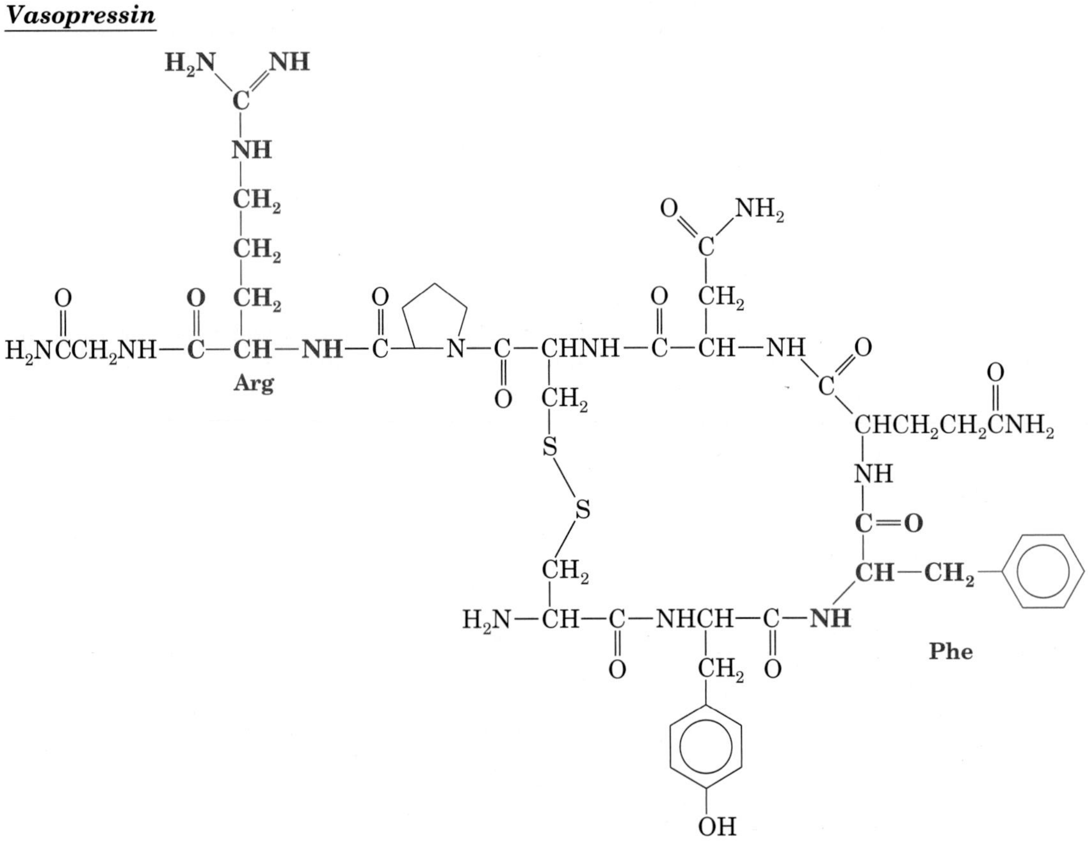

Oxytocin

Vasopressin

FIGURE 24.6 The structures of oxytocin and vasopressin. Amino acid residues that differ between them are shown in red.

24.6A Oxytocin and Vasopressin

Vincent du Vigneaud of Cornell Medical College synthesized oxytocin and vasopressin in 1953; he received the Nobel Prize in Chemistry in 1955.

Oxytocin and vasopressin (Fig. 24.6) are two rather small polypeptides with strikingly similar structures (where oxytocin has leucine, vasopressin has arginine, and where oxytocin has isoleucine, vasopressin has phenylalanine). In spite of the similarity of their amino acid sequences, these two polypeptides have quite different physiological effects. Oxytocin occurs only in the female of a species and stimulates uterine contractions during childbirth. Vasopressin occurs in males and females; it causes contraction of peripheral blood vessels and an increase in blood pressure. Its major function, however, is as an *antidiuretic;* physiologists often refer to vasopressin as an *antidiuretic hormone.*

The structures of oxytocin and vasopressin also illustrate the importance of the disulfide linkage between cysteine residues (Section 24.2A) in the overall primary structure of a polypeptide. In these two molecules this disulfide linkage leads to a cyclic structure.

PROBLEM 24.10

Treating oxytocin with certain reducing agents (e.g., sodium in liquid ammonia) brings about a single chemical change that can be reversed by air oxidation. What chemical changes are involved?

24.6B Insulin

Insulin, a hormone secreted by the pancreas, regulates glucose metabolism. Insulin deficiency in humans is the major problem in diabetes mellitus.

The amino acid sequence of bovine insulin (Fig. 24.7) was determined by Sanger in 1953 after 10 years of work. Bovine insulin has a total of 51 amino acid residues in two polypeptide chains, called the A and B chains. These chains are joined by two disulfide linkages. The A chain contains an additional disulfide linkage between cysteine residues at positions 6 and 11.

Human insulin differs from bovine insulin at only three amino acid residues: Threonine replaces alanine once in the A chain (residue 8) and once in the B chain (residue 30), and isoleucine replaces valine once in the A chain (residue 10). Insulins from most mammals have similar structures.

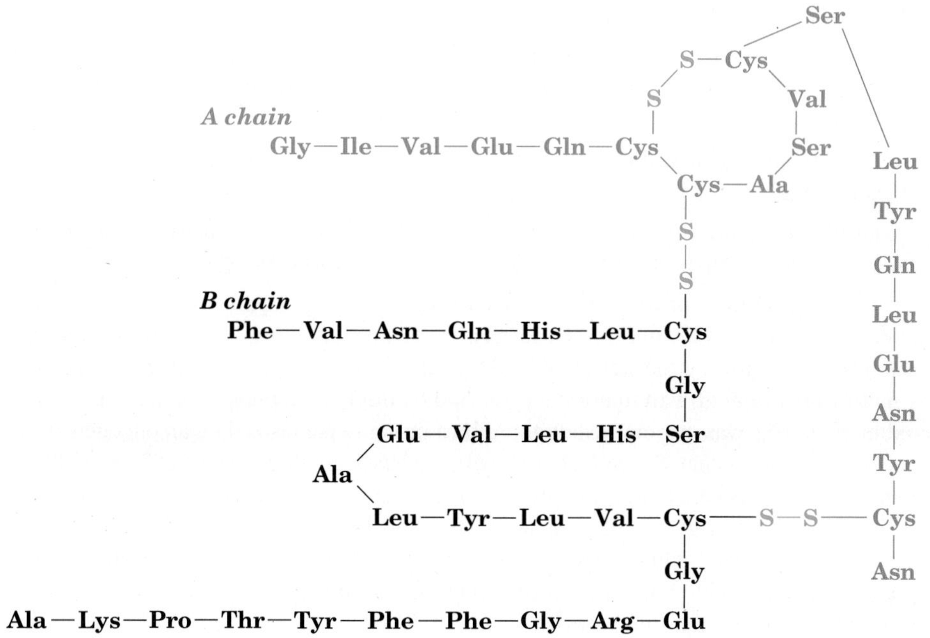

A chain
Gly—Ile—Val—Glu—Gln—Cys—(S—S)—Cys—Ser—Val—Ser—Cys—Ala—Ser—Leu—Tyr—Gln—Leu—Glu—Asn—Tyr—Cys—Asn

B chain
Phe—Val—Asn—Gln—His—Leu—Cys—Gly—Ser—His—Leu—Val—Glu—Ala—Leu—Tyr—Leu—Val—Cys—Gly—Glu—Arg—Gly—Phe—Phe—Tyr—Thr—Pro—Lys—Ala

(A chain Cys 7 —S—S— B chain Cys 7; A chain Cys 20 —S—S— B chain Cys 19)

FIGURE 24.7 The amino acid sequence of bovine insulin.

The Chemistry of...

Sickle-Cell Anemia

The genetically based disease sickle-cell anemia results from a single amino acid error in the β chain of hemoglobin. In normal hemoglobin, position 6 has a glutamic acid residue, whereas in sickle-cell hemoglobin position 6 is occupied by valine.

Red blood cells (erythrocytes) containing hemoglobin with this amino acid residue error tend to become crescent shaped ("sickle") when the partial pressure of oxygen is low, as it is in venous blood. These distorted cells are more difficult for the heart to pump through small capillaries. They may even block capillaries by clumping together; at other times the red cells may even split open. Children who inherit this genetic trait from both parents suffer from a severe form of the disease and usually do not live past the age of two. Children who inherit the disease from only one parent generally have a much milder form. Sickle-cell anemia arose among the populations of central and western Africa where, ironically, it may have had a beneficial effect. People with a mild form of the disease are far less susceptible to malaria than those with normal hemoglobin. Malaria, a disease caused by an infectious microorganism, is especially prevalent in central and western Africa. Mutational changes such as those that give rise to sickle-cell anemia are very common. Approximately 150 different types of mutant hemoglobin have been detected in humans; fortunately, most are harmless.

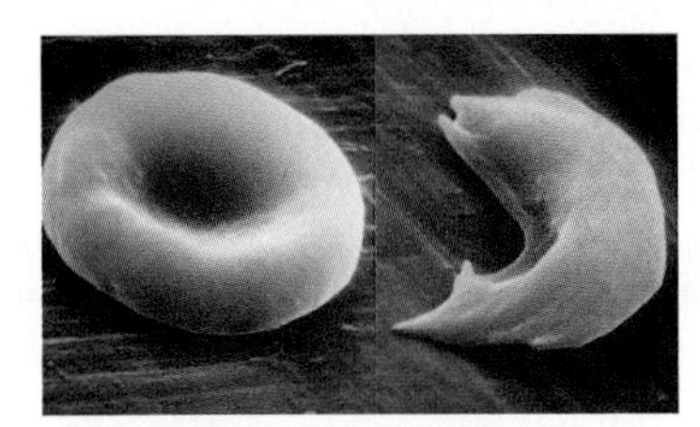

Normal (left) and sickled (right) red blood cells viewed with a scanning electron microscope at 18,000× magnification.

24.6C Other Polypeptides and Proteins

Successful sequential analyses have now been achieved with hundreds of other polypeptides and proteins, including the following:

1. **Bovine ribonuclease.** This enzyme, which catalyzes the hydrolysis of ribonucleic acid (Chapter 25), has a single chain of 124 amino acid residues and four intrachain disulfide linkages.

2. **Human hemoglobin.** There are four peptide chains in this important oxygen-carrying protein. Two identical α chains have 141 residues each, and two identical β chains have 146 residues each.

3. **Bovine trypsinogen and chymotrypsinogen.** These two enzyme precursors have single chains of 229 and 245 residues, respectively.

4. **Gamma globulin.** This immunoprotein has a total of 1320 amino acid residues in four chains. Two chains have 214 residues each; the other two have 446 each.

5. **p53, an anticancer protein.** The protein called p53 (the p stands for protein), consisting of 393 amino acid residues, has a variety of cellular functions, but the most important ones involve controlling the steps that lead to cell growth. It acts as a **tumor suppressor** by halting abnormal growth in normal cells, and by doing so it prevents cancer. Discovered in 1979, p53 was originally thought to be a protein synthesized by an oncogene (a gene that causes cancer). Research has shown, however, that the form of p53 originally thought to have this cancer-causing property was a mutant form of the normal protein. The unmutated (or *wild type*) p53 apparently coordinates a complex set of responses to changes in DNA that could otherwise lead to cancer. When p53 becomes mutated, it no longer provides the cell with its cancer-preventing role; it apparently does the opposite, by acting to increase abnormal growth.

More than half of the people diagnosed with cancer each year have a mutant form of p53 in their cancers. Different forms of cancer have been shown to result from different mutations in the protein, and the list of cancer types associated with mutant p53 includes cancers of most of the body parts: brain, breast, bladder, cervix, colon, liver, lung, ovary, pancreas, prostate, skin, stomach, and so on.

6. *Ras* proteins. *Ras* proteins are modified proteins associated with cell growth and the cell's response to insulin. They belong to a class of proteins called prenylated proteins, in which lipid groups derived from isoprenoid biosynthesis (Special Topic D) are appended as thioethers to C-terminal cysteine residues. Certain mutated forms of *ras* proteins cause oncogenic changes in various eukaryotic cell types. One effect of prenylation and other lipid modifications of proteins is to anchor these proteins to cellular membranes. Prenylation may also assist with molecular recognition of prenylated proteins by other proteins.*

24.7 Polypeptide and Protein Synthesis

We saw in Chapter 18 that the synthesis of an amide linkage is a relatively simple one. We must first "activate" the carboxyl group of an acid by converting it to an anhydride or acid chloride and then allow it to react with an amine:

$$\underset{\textbf{Anhydride}}{R{-}\overset{O}{\overset{\|}{C}}{-}O{-}\overset{O}{\overset{\|}{C}}{-}R} + \underset{\textbf{Amine}}{R'{-}NH_2} \longrightarrow \underset{\textbf{Amide}}{R{-}\overset{O}{\overset{\|}{C}}{-}NHR'} + R{-}CO_2H$$

The problem becomes somewhat more complicated, however, when both the acid group and the amino group are present in the same molecule, as they are in an amino acid, and especially when our goal is the synthesis of a naturally occurring polyamide where the sequence of different amino acids is all important. Let us consider, as an example, the synthesis of the simple dipeptide alanylglycine, Ala·Gly. We might first activate the carboxyl group of alanine by converting it to an acid chloride, and then we might allow it to react with glycine. Unfortunately, however, we cannot prevent alanyl chloride from reacting with itself. So our reaction would yield not only Ala·Gly but also Ala·Ala. It could also lead to Ala·Ala·Ala and Ala·Ala·Gly, and so on. The yield of our desired product would be low, and we would also have a difficult problem separating the dipeptides, tripeptides, and higher peptides.

$$\underset{\textbf{Ala}}{CH_3CH(\overset{+}{N}H_3)CO_2^-} \xrightarrow[\text{(2) } H_3\overset{+}{N}CH_2CO_2^- \ (\textbf{Gly})]{\text{(1) } SOCl_2} \underset{\textbf{Ala·Gly}}{CH_3CH(\overset{+}{N}H_3)CONHCH_2CO_2^-} + \underset{\textbf{Ala·Ala}}{CH_3CH(\overset{+}{N}H_3)CONHCH(CH_3)CO_2^-}$$

$$+ \underset{\textbf{Ala·Ala·Ala}}{CH_3CH(\overset{+}{N}H_3)CONHCH(CH_3)CONHCH(CH_3)CO_2^-} + \underset{\textbf{Ala·Ala·Gly}}{CH_3CH(\overset{+}{N}H_3)CONHCH(CH_3)CONHCH_2CO_2^-} + \text{others}$$

*See Gelb, M. H. Modification of Proteins by Prenyl Groups. In *Principles of Medical Biology;* Vol. 4, Bittar, E. E., Bittar, N., Eds.; JAI Press: Greenwich, CT, 1995; Chapter 14, pp 323–333.

24.7A Protecting Groups

The solution to this problem is to "protect" the amino group of the first amino acid before we activate it and allow it to react with the second. By protecting the amino group, we mean that we must convert it to some other group of low nucleophilicity—*one that will not react with a reactive acyl derivative.* The **protecting group** must be carefully chosen because after we have synthesized the amide linkage between the first amino acid and the second, we will want to be able to remove the protecting group without disturbing the new amide bond.

A number of reagents have been developed to meet these requirements. Three that are often used are *benzyl chloroformate,* di-*tert*-butyl carbonate, and 9-fluorenylmethyl chloroformate:

$C_6H_5CH_2O{-}C(=O){-}Cl$
Benzyl chloroformate

$(CH_3)_3CO{-}C(=O){-}OC(CH_3)_3$
Di-*tert*-butyl carbonate

9-Fluorenylmethyl chloroformate

All three reagents react with the amine to block it from futher acylation. These derivations, however, are types that allow removal of the protecting group under conditions that do not affect peptide bonds. The benzyloxycarbonyl group (abbreviated Z) can be removed with catalytic hydrogenation or cold HBr in acetic acid. The *tert*-butyloxycarbonyl group (Boc) can be removed with trifluoroactic acid (CF_3CO_2H) in acetic acid. The 9-fluorenylmethoxycarbonyl (Fmoc) group is stable under acid conditions but can be removed under mild basic conditions using piperidine (a secondary amine).

Benzyloxycarbonyl Group

Protection

$$C_6H_5CH_2OC(=O){-}Cl + H_2N{-}R \xrightarrow[25°C]{OH^-} C_6H_5CH_2OC(=O){-}NH{-}R + Cl^-$$

Benzyl chloroformate → **Benzyloxycarbonyl or Z group**

Deprotection

$$\xrightarrow{HBr,\ acetic\ acid\ (cold)} C_6H_5CH_2Br + CO_2 + H_2N{-}R$$

$$\xrightarrow{H_2/Pd} C_6H_5CH_3 + CO_2 + H_2N{-}R$$

tert-Butyloxycarbonyl Group

Protection

$$(CH_3)_3COC(=O)OC(CH_3)_3 + H_2N{-}R \xrightarrow[25°C]{base} (CH_3)_3COC(=O){-}NHR + (CH_3)_3COH$$

Di-*tert*-butyl carbonate → ***tert*-Butyloxycarbonyl or Boc group**

Deprotection

$$\xrightarrow{HCl\ or\ CF_3CO_2H\ in\ acetic\ acid,\ 25°C} (CH_3)_2C{=}CH_2 + CO_2 + H_2N{-}R$$

9-Fluorenylmethoxycarbonyl Group

Protection (introduction of Fmoc group)

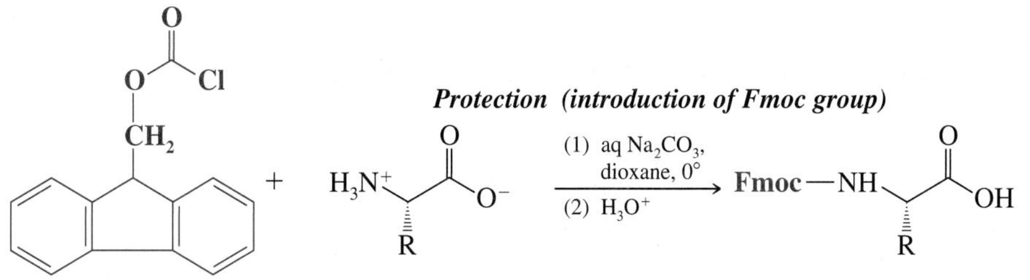

9-Fluorenylmethyl chloroformate + Amino acid (side chain protected in advance if necessary) $\xrightarrow[(2)\ H_3O^+]{(1)\ aq\ Na_2CO_3,\ dioxane,\ 0°}$ Fmoc-protected amino acid (stable in acid)

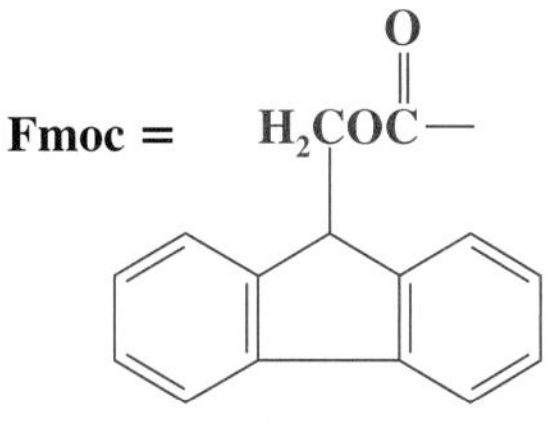

9-Fluorenylmethyl chloroformate

Amino acid (side chain protected in advance if necessary)

Fmoc-protected amino acid (stable in acid)

Fmoc = H_2COC—

(9-Fluorenylmethoxycarbonyl group)

Deprotection (removal of Fmoc group)

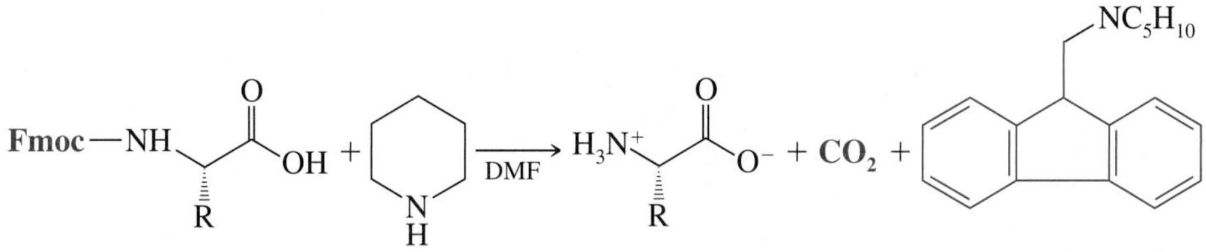

Fmoc—NH... OH + Piperidine $\xrightarrow{DMF}$ H_3N^+ ... O^- + CO_2 + Byproduct (NC_5H_{10})

Piperidine ($C_5H_{11}N$)

Unprotected amino acid

(Byproduct)

The easy removal of the Z and Boc groups in acidic media results from the exceptional stability of the carbocations that are formed initially. The benzyloxycarbonyl group gives a benzyl carbocation; the *tert*-butyloxycarbonyl group yields, initially, a *tert*-butyl cation. Removal of the benzyloxycarbonyl group with hydrogen and a catalyst depends on the fact that benzyl–oxygen bonds are weak and subject to hydrogenolysis at low temperatures, resulting in methylbenzene (toluene) as one product:

$$C_6H_5CH_2\text{—}OCR \xrightarrow[25°C]{H_2,\ Pd} C_6H_5CH_3 + HOCR$$

A benzyl ester

PROBLEM 24.11

What classes of reactions are involved in the cleavage of the Fmoc group with piperidine, leading to the unprotected amino acid and the fluorene byproduct? Write mechanisms for these reactions.

24.7B Activation of the Carboxyl Group

Perhaps the most obvious way to activate a carboxyl group is to convert it to an acyl chloride. This method was used in early peptide syntheses, but acyl chlorides are actually more reactive than necessary. As a result, their use leads to complicating side reactions. A much better method is to convert the carboxyl group of the "protected" amino acid to a mixed anhydride using ethyl chloroformate, $Cl{-}\overset{O}{\overset{\|}{C}}{-}OC_2H_5$:

$$Z{-}NHCH(R)\overset{O}{\overset{\|}{C}}{-}OH \xrightarrow[(2)\ ClCO_2C_2H_5]{(1)\ (C_2H_5)_3N} Z{-}NHCH(R){-}\overset{O}{\overset{\|}{C}}{-}O{-}\overset{O}{\overset{\|}{C}}{-}OC_2H_5$$

"Mixed anhydride"

The mixed anhydride can then be used to acylate another amino acid and form a peptide linkage:

$$Z{-}NHCH(R)\overset{O}{\overset{\|}{C}}{-}O{-}\overset{O}{\overset{\|}{C}}OC_2H_5 \xrightarrow{H_3\overset{+}{N}{-}CH(R')CO_2^-} Z{-}NHCH(R)\overset{O}{\overset{\|}{C}}{-}NHCH(R')CO_2H + CO_2 + C_2H_5OH$$

Diisopropylcarbodiimide and dicyclohexylcarbodiimide (Section 18.8E) can also be used to activate the carboxyl group of an amino acid. In Section 24.7D we shall see how it is used in an automated peptide synthesis.

24.7C Peptide Synthesis

Let us examine now how we might use these reagents in the preparation of the simple dipeptide Ala · Leu. The principles involved here can, of course, be extended to the synthesis of much longer polypeptide chains.

$$\underset{\textbf{Ala}}{CH_3CH(\overset{+}{N}H_3)CO_2^-} + \underset{\textbf{Benzyl chloroformate}}{C_6H_5CH_2O\overset{O}{\overset{\|}{C}}{-}Cl} \xrightarrow[25°C]{OH^-} \underset{\textbf{Z-Ala}}{CH_3CH(NH{-}C(=O){-}OCH_2C_6H_5){-}CO_2H} \xrightarrow[(2)\ ClCO_2C_2H_5]{(1)\ (C_2H_5)_3N}$$

$$\underset{\textbf{Mixed anhydride of Z-Ala}}{CH_3CH(NH{-}C(=O){-}OCH_2C_6H_5){-}\overset{O}{\overset{\|}{C}}{-}O\overset{O}{\overset{\|}{C}}OC_2H_5} \xrightarrow[-\ CO_2\ +\ C_2H_5OH]{\underset{\text{Leu}}{(CH_3)_2CHCH_2CH(\overset{+}{N}H_3)CO_2^-}}$$

$$\underset{\underset{\underset{C_6H_5CH_2O}{|}}{\underset{C=O}{|}}}{\underset{NH}{|}}{CH_3CH}-\overset{O}{\overset{\|}{C}}-NH\underset{\underset{\underset{H_3C\ \ CH_3}{CH}}{|}}{\underset{CH_2}{|}}{CH}CO_2H \xrightarrow{H_2/Pd}$$

Z-Ala • Leu

$$CH_3\underset{\underset{+}{NH_3}}{CH}\overset{O}{\overset{\|}{C}}NH\underset{\underset{\underset{H_3C\ \ CH_3}{CH}}{|}}{\underset{CH_2}{|}}{CH}CO_2^- + C_6H_5-CH_3 + CO_2$$

Ala • Leu

PROBLEM 24.12

Show all steps in the synthesis of Gly · Val · Ala using the *tert*-butyloxycarbonyl (Boc) group as a protecting group.

PROBLEM 24.13

The synthesis of a polypeptide containing lysine requires the protection of both amino groups. **(a)** Show how you might do this in a synthesis of Lys · Ile using the benzyloxycarbonyl group as a protecting group. **(b)** The benzyloxycarbonyl group can also be used to protect the guanidino group, $-NH\overset{NH}{\overset{\|}{C}}-NH_2$, of arginine. Show a synthesis of Arg · Ala.

PROBLEM 24.14

The terminal carboxyl groups of glutamic acid and aspartic acid are often protected through their conversion to benzyl esters. What mild method could be used for removal of this protecting group?

24.7D Automated Peptide Synthesis

The methods that we have described thus far have been used to synthesize a number of polypeptides, including ones as large as insulin. They are extremely time consuming and tedious, however. One must isolate the peptide and purify it by lengthy means at almost every stage. Furthermore, significant loss of the peptide can occur with each isolation and purification stage. The development of a procedure by R. B. Merrifield (Rockefeller University) for automating this process was therefore a breakthrough in peptide synthesis. Merrifield's method, for which he received the 1984 Nobel Prize in Chemistry, is called **solid-phase peptide synthesis (SPPS),** and it hinges upon synthesis of the peptide residue by residue while one end of the peptide remains attached to an insoluble plastic bead. Protecting groups and other reagents are still necessary, but because the peptide being synthesized is anchored to a solid support, byproducts, excess reagents, and solvents can simply be rinsed away between each synthetic step without need for intermediate purification. After the very last step the polypeptide is cleaved from the polymer support and subjected to a final purification by HPLC. The method works so well that it has been developed into an automated process.

R. B. Merrifield, 1984

Solid-phase peptide synthesis (Fig. 24.8) begins with attachment of the first amino acid by its carboxyl group to the polymer bead, usually with a linker or spacer molecule in between. Each new amino acid is then added by formation of an amide bond between

Polymer—O—C$_6$H$_4$—CH$_2$OH + HOC(=O)CH(R)NH—Fmoc

↓ Diisopropylcarbodiimide

Polymer—O—C$_6$H$_4$—CH$_2$OC(=O)CH(R)NH—Fmoc

↓ Piperidine in DMF

Polymer—O—C$_6$H$_4$—CH$_2$OC(=O)CH(R)NH$_2$

↓ HOC(=O)CH(R′)NH—Fmoc and Diisopropylcarbodiimide

Polymer—O—C$_6$H$_4$—CH$_2$OC(=O)CH(R)NHC(=O)CH(R′)NH—Fmoc

↓ Piperidine in DMF

Repetitions of steps 4–7

↓ CF_3CO_2H

Polymer—O—C$_6$H$_4$—CH$_2$OH + HOC(=O)CH(R)NHC(=O)CH(R′)NHC(=O)CH(R″)NH— etc.

Step 1	Attaches C-terminal (Fmoc protected) amino acid residue to resin
Step 2	Purifies resin with attached residue by washing
Step 3	Removes Fmoc protecting group
Step 4	Purifies by washing
Step 5	Adds next (protected) amino acid residue
Step 6	Purifies by washing
Step 7	Removes protecting group
Final Step	Detaches completed polypeptide

FIGURE 24.8 A method for automated solid-phase peptide synthesis.

the N-terminal amino group of the peptide growing on the solid support and the new amino acid's carboxyl group. Diisopropylcarbodiimide (similar in reactivity to DCC, Section 18.8E) is used as the amide bond-forming reagent. To prevent undesired reactions as each new residue is coupled, a protecting group is used to block the amino group of the residue being added. Once the new amino acid has been coupled to the growing peptide and before the next residue is added, the protecting group on the new N-terminus is removed, making the peptide ready to begin the next cycle of amide bond formation.

Although Merrifield's initial method for solid-phase peptide synthesis used the Boc group to protect the α-amino group of residues being coupled to the growing peptide, several advantages of the Fmoc group have since made it the group of choice. The reasons have mainly to do with excellent selectivity for removing the Fmoc group in the presence of other protecting groups used to block reactive side chains along the growing peptide and the ability to monitor the progress of the solid-phase synthesis by spectrophotometry as the Fmoc group is released in each cycle.

Let us discuss the choice of protecting groups further. As noted (Section 24.7A), *basic conditions* (piperidine in DMF) are used to remove the Fmoc group. On the other hand, protecting groups for the side chains of the peptide residues are generally blocked with *acid-labile* moieties. The base-labile Fmoc groups and acid-labile side-chain protecting groups are said to be **orthogonal protecting groups** because one set of protecting groups is stable under conditions for removal of the other, and vice versa. Another advantage of Fmoc as compared to Boc groups for protecting the α-amino group of each new residue is that repetitive application of the acidic conditions to remove Boc groups from each new residue slowly sabotages the synthesis by prematurely cleaving some peptide molecules from the solid support and deprotecting some of the side chains. The basic conditions for Fmoc removal avoid these problematic side reactions.

As we have said, the great advantage of solid-phase peptide synthesis is that purification of the peptide at each stage involves simply rinsing the beads of the solid support to wash away excess reagent, byproducts, and solvents. Furthermore, having the peptide attached to a tangible solid during the synthesis allows all of the steps in the synthesis to be carried out by a machine in repeated cycles. Automated peptide synthesizers are available that can complete one cycle in 40 min and carry out 45 cycles of unattended operation. Though not as efficient as protein synthesis in the body, where enzymes directed by DNA can catalyze assembly of a protein with 150 amino acids in about 1 min, automated peptide synthesis is a far cry from the tedious process of manually synthesizing a peptide step after step. A hallmark example of automated peptide synthesis was the synthesis of ribonuclease, a protein with 124 amino acid residues. The synthesis involved 369 chemical reactions and 11,930 automated steps—all carried out without isolating an intermediate. The synthetic ribonuclease not only had the same physical characteristics as the natural enzyme, it possessed the identical biological activity as well. The overall yield was 17%, which means that the average yield of each individual step was greater than 99%.

PROBLEM 24.15

One type of insoluble support used for SPPS is polymer-bound 4-benzyloxybenzyl alcohol, also known as "Wang resin," shown in Fig. 24.8. The 4-benzyloxybenzyl alcohol moiety serves as a linker between the resin backbone and the peptide. After purification, the completed polypeptide can be detached from the resin using trifluoroacetic acid under conditions that are mild enough not to affect the amide linkages. What structural features of the linker make this possible?

PROBLEM 24.16

Outline the steps in the synthesis of Lys · Phe · Ala using the SPPS procedure.

24.8 Secondary, Tertiary, and Quaternary Structures of Proteins

We have seen how amide and disulfide linkages constitute the covalent or *primary structure* of proteins. Of equal importance in understanding how proteins function is knowledge of the way in which the peptide chains are arranged in three dimensions. The secondary and tertiary structures of proteins are involved here.

24.8A Secondary Structure

The **secondary structure** of a protein is defined by the local conformation of its polypeptide backbone. These local conformations have come to be specified in terms of regular folding patterns called *helices, pleated sheets,* and *turns.* The major experimental techniques that have been used in elucidating the secondary structures of proteins are X-ray analysis and NMR (including two-dimensional NMR).

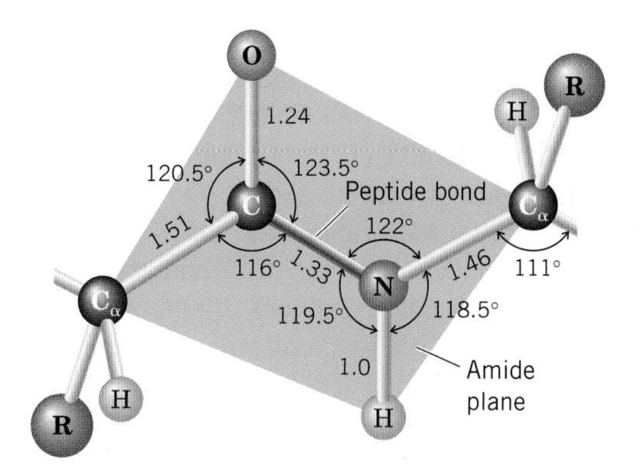

FIGURE 24.9 The geometry and bond lengths (in angstroms, Å) of the peptide linkage. The six enclosed atoms tend to be coplanar and assume a "transoid" arrangement. (From Voet, D.; Voet, J. G. *Biochemistry,* 2nd ed.; Wiley: New York, 1995; p 142. Used with permission.)

Linus Pauling
Two American scientists, Linus Pauling and Robert B. Corey, were pioneers in the X-ray analysis of proteins. Beginning in 1939, Pauling and Corey initiated a long series of studies of the conformations of peptide chains. At first, they used crystals of single amino acids, then dipeptides and tripeptides, and so on. Moving on to larger and larger molecules and using the precisely constructed molecular models, they were able to understand the secondary structures of proteins for the first time. Pauling won the 1954 Nobel Prize in Chemistry and the 1962 Nobel Peace Prize.

When X-rays pass through a crystalline substance, they produce diffraction patterns. Analysis of these patterns indicates a regular repetition of particular structural units with certain specific distances between them, called **repeat distances.** X-ray analyses have revealed that the polypeptide chain of a natural protein can interact with itself in two major ways: through formation of a **β-pleated** sheet and an **α helix.**

To understand how these interactions occur, let us look first at what X-ray analysis has revealed about the geometry at the peptide bond itself. Peptide bonds tend to assume a geometry such that six atoms of the amide linkage are coplanar (Fig. 24.9). The carbon–nitrogen bond of the amide linkage is unusually short, indicating that resonance contributions of the type shown here are important:

$$\ddot{N}{-}C(=O) \longleftrightarrow \overset{+}{N}{=}C(-O^-)$$

The carbon–nitrogen bond, consequently, has considerable double-bond character ($\sim$40%), and rotations of groups about this bond are severely hindered.

Rotations of groups attached to the amide nitrogen and the carbonyl carbon are relatively free, however, and these rotations allow peptide chains to form different conformations.

The transoid arrangement of groups around the relatively rigid amide bond would cause the R groups to alternate from side to side of a single fully extended peptide chain:

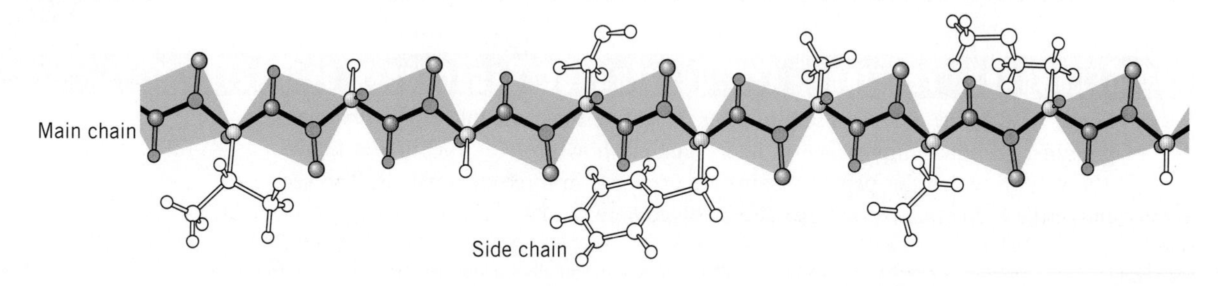

Calculations show that such a polypeptide chain would have a repeat distance (i.e., distance between alternating units) of 7.2 Å.

Fully extended polypeptide chains could conceivably form a flat-sheet structure, with each alternating amino acid in each chain forming two hydrogen bonds with an amino acid in the adjacent chain:

Hypothetical flat-sheet structure
(not formed because of steric hindrance)

This structure does not exist in naturally occurring proteins because of the crowding that would exist between R groups. If such a structure did exist, it would have the same repeat distance as the fully extended peptide chain, that is, 7.2 Å.

Slight rotations of bonds, however, can transform a flat-sheet structure into what is called the ***β*-pleated sheet** or ***β* configuration** (Fig. 24.10). The pleated-sheet structure gives small- and medium-sized R groups room enough to avoid van der Waals repulsions and is the predominant structure of silk fibroin (48% glycine and 38% serine and alanine residues). The pleated-sheet structure has a slightly shorter repeat distance, 7.0 Å, than the flat sheet.

CD Tutorial
β-pleated sheet.

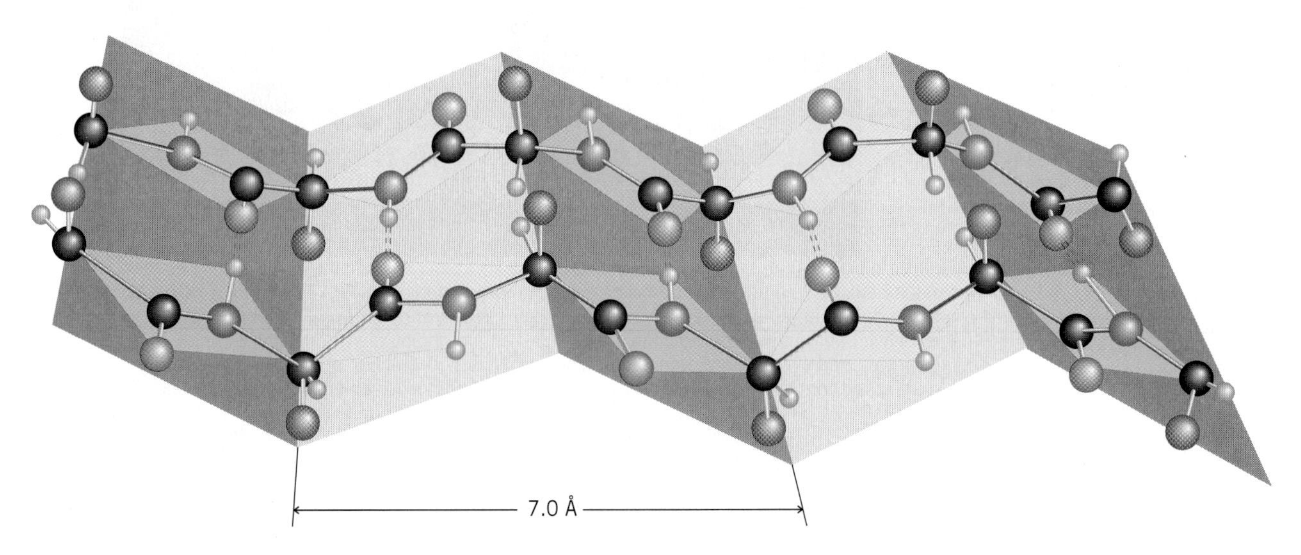

FIGURE 24.10 The *β*-pleated sheet or *β* configuration of a protein. (From Voet, D.; Voet, J. G. *Biochemistry,* 2nd ed.; Wiley: New York, 1995; p 150. Copyright © by Irving Geis. Used with permission.)

Of far more importance in naturally occurring proteins is the secondary structure called the *α* helix (Fig. 24.11). This structure is a right-handed helix with 3.6 amino acid residues per turn. Each amide group in the chain has a hydrogen bond to an amide group at a distance of three amino acid residues in either direction, and the R groups all extend away from the axis of the helix. The repeat distance of the *α* helix is 5.4 Å.

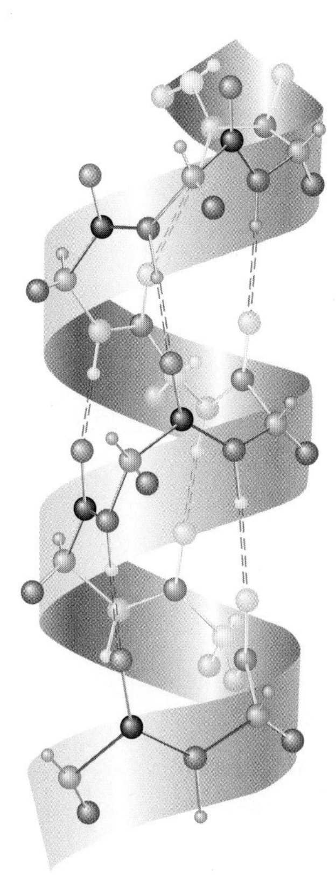

FIGURE 24.11 A representation of the α-helical structure of a polypeptide. Hydrogen bonds are denoted by dashed lines. (From Voet, D.; Voet, J. G. *Biochemistry,* 2nd ed.; Wiley: New York, 1995; p 146. Copyright © by Irving Geis. Used with permission.)

The α-helical structure is found in many proteins; it is the predominant structure of the polypeptide chains of fibrous proteins such as *myosin,* the protein of muscle, and of α-*keratin,* the protein of hair, unstretched wool, and nails.

Helices and pleated sheets account for only about one-half of the structure of the average globular protein. The remaining polypeptide segments have what is called a **coil** or **loop conformation.** These nonrepetitive structures are not random; they are just more difficult to describe. Globular proteins also have stretches, called **reverse turns** or $\boldsymbol{\beta}$ **bends,** where the polypeptide chain abruptly changes direction. These often connect successive strands of β sheets and almost always occur at the surface of proteins.

Figure 24.12 shows the structure of the enzyme human carbonic anhydrase, based on X-ray crystallographic data. Segments of α helix (magenta) and β sheets (yellow) intervene between reverse turns and nonrepetitive structures (blue and white, respectively).

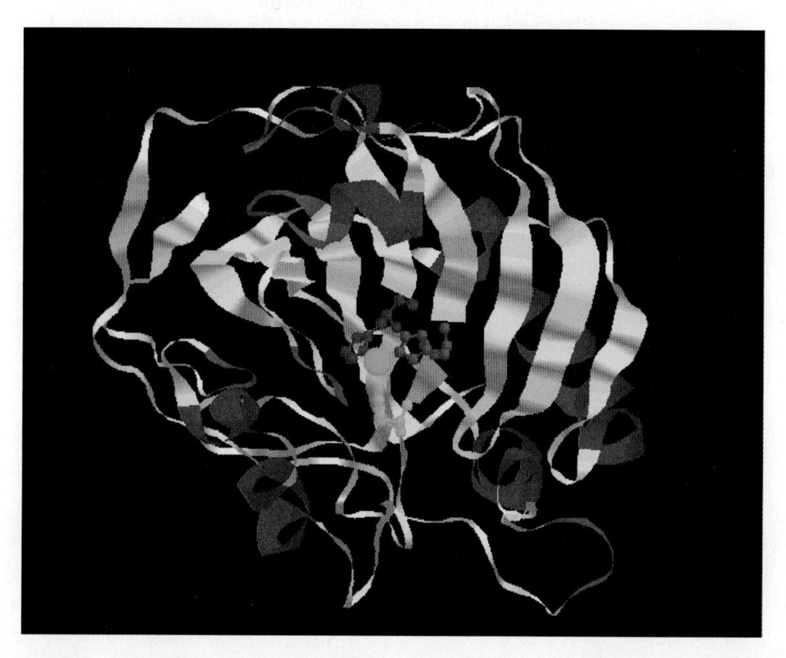

FIGURE 24.12 The structure of the enzyme human carbonic anhydrase, based on X-ray crystallographic data. Alpha helices are shown in magenta and strands of β-pleated sheets are yellow. Turns are shown in blue and random coils are white. The side chains of three histidine residues (shown in red, green, and cyan) coordinate with a zinc atom (light green). Not obvious from this image is the interesting fact that the C-terminus is tucked through a loop of the polypeptide chain, making carbonic anhydrase a rare example of a native protein in which the polypeptide chain forms a knot. (Image prepared from an X-ray crystal structure by Eriksson, A. E.; Jones, T. A.; Liljas, A. Protein Data Bank file 1CA2.pdb.)

The locations of the side chains of amino acids of globular proteins are usually those that we would expect from their polarities:

1. Residues with **nonpolar, hydrophobic side chains,** such as *valine, leucine, isoleucine, methionine, and phenylalanine,* are almost always found in the interior of the protein, out of contact with the aqueous solvent. (These hydrophobic interactions are largely responsible for the tertiary structure of proteins that we discuss in Section 24.8B.)

2. Side chains of **polar residues with positive or negative charges,** such as *arginine, lysine, aspartic acid, and glutamic acid,* are usually on the surface of the protein in contact with the aqueous solvent.

3. **Uncharged polar side chains,** such as those of *serine, threonine, asparagine, glutamine, tyrosine, and tryptophan,* are most often found on the surface, but some of these are found in the interior as well. When they are found in the interior, they are virtually all hydrogen bonded to other similar residues. Hydrogen bonding apparently helps neutralize the polarity of these groups.

Certain peptide chains assume what is called **random coil arrangement,** a structure that is flexible, changing, and statistically random. Synthetic polylysine, for example, exists as a random coil and does not normally form an α helix. At pH 7, the ε-amino groups of the lysine residues are positively charged, and, as a result, repulsive forces between them are so large that they overcome any stabilization that would be gained through hydrogen bond formation of an α helix. At pH 12, however, the ε-amino groups are uncharged and polylysine spontaneously forms an α helix.

The presence of proline or hydroxyproline residues in polypeptide chains produces another striking effect: Because the nitrogen atoms of these amino acids are part of five-membered rings, the groups attached by the nitrogen–α carbon bond cannot rotate enough to allow an α-helical structure. Wherever proline or hydroxyproline occur in a peptide chain, their presence causes a kink or bend and interrupts the α helix.

24.8B Tertiary Structure

The **tertiary structure** of a protein is its three-dimensional shape that arises from further foldings of its polypeptide chains, foldings superimposed on the coils of the α helices. These foldings do not occur randomly: Under the proper environmental conditions they occur in one particular way—a way that is characteristic of a particular protein and one that is often highly important to its function.

Various forces are involved in stabilizing tertiary structures, including the disulfide bonds of the primary structure. One characteristic of most proteins is that the folding takes place in such a way as to expose the maximum number of polar (hydrophilic) groups to the aqueous environment and enclose a maximum number of nonpolar (hydrophobic) groups within its interior.

The soluble globular proteins tend to be much more highly folded than fibrous proteins. However, fibrous proteins also have a tertiary structure; the α-helical strands of α-keratin, for example, are wound together into a "superhelix." The superhelix makes one complete turn for each 35 turns of the α helix. The tertiary structure does not end here, however. Even the superhelices can be wound together to give a ropelike structure of seven strands.

Myoglobin (Fig. 24.13) and hemoglobin (Section 24.12) were the first proteins (in 1957 and 1959, respectively) to be subjected to a completely successful X-ray analysis. This work was accomplished by J. C. Kendrew and Max Perutz at Cambridge University in England. (They received the Nobel Prize in Chemistry in 1962.) Since then many other proteins including lysozyme, ribonuclease, and α-chymotrypsin have yielded to complete structural analysis. In fact, it is now possible to retrieve data for thousands of protein X-ray crystal structures deposited by researchers in public computerized data banks.

CD Tutorial
Myoglobin

24.8C Quaternary Structure

Many proteins exist as stable and ordered noncovalent aggregates of more than one polypeptide chain. The overall structure of a protein having multiple subunits is called its **quaternary structure.** The quaternary structure of hemoglobin, for example, involves four subunits (see Section 24.12).

FIGURE 24.13 The three-dimensional structure of myoglobin. The heme ring is shown in gray. The iron atom is shown as a red sphere, and the histidine side chains that coordinate with the iron are shown in cyan. (Image prepared from an X-ray crystal structure by Phillips, S. E. V. Protein Data Bank file 1MBD.pdb.)

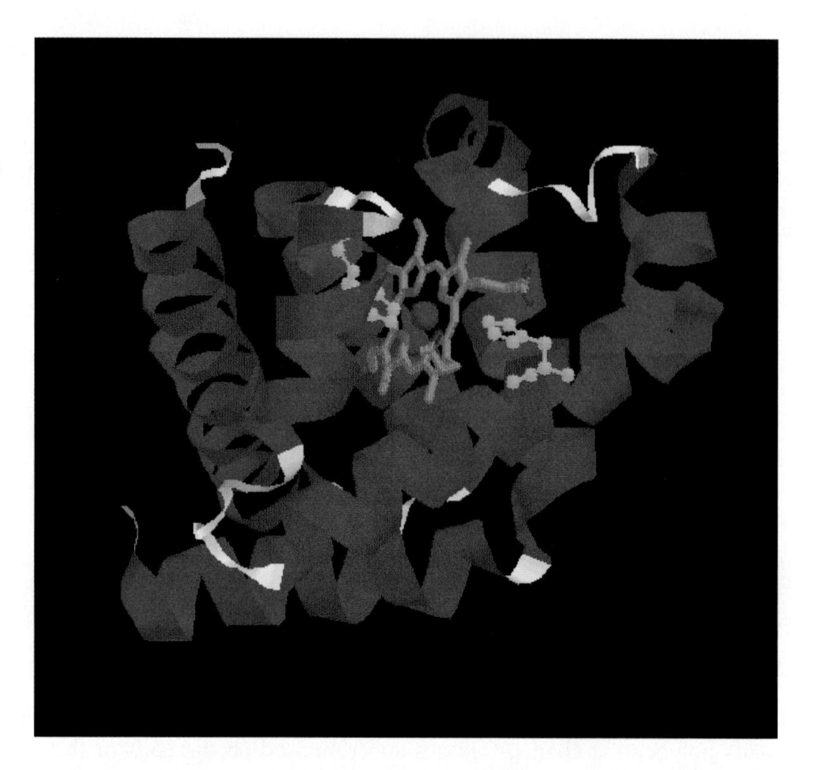

24.9 Introduction to Enzymes

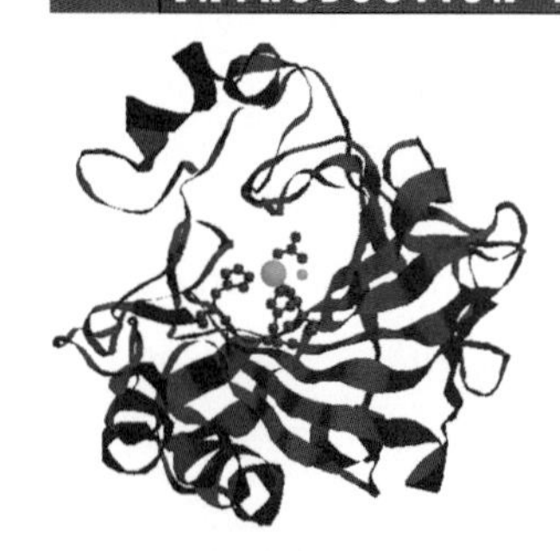

Carbonic anhydrase

Carbonic anhydrase is an enzyme that catalyzes the following reaction: $H_2O + CO_2 \rightleftharpoons H_2CO_3$. We discussed its physiological role of regulating blood pH in the Chapter 3 opening vignette.

All of the reactions that occur in living cells are mediated by remarkable biological catalysts called **enzymes.** Enzymes have the ability to bring about vast increases in the rates of reactions; in most instances, the rates of enzyme-catalyzed reactions are faster than those of uncatalyzed reactions by factors of 10^6–10^{12}. For living organisms, rate enhancements of this magnitude are important because they permit reactions to take place at reasonable rates, even under the mild conditions that exist in living cells (i.e., approximately neutral pH and a temperature of about 35°C.)

Enzymes also show remarkable **specificity** for their reactants (called **substrates**) and for their products. This specificity is far greater than that shown by most chemical catalysts. In the enzymatic synthesis of proteins, for example (through reactions that take place on ribosomes, Section 25.5D), polypeptides consisting of well over 1000 amino acid residues are synthesized virtually without error. It was Emil Fischer's discovery, in 1894, of the ability of enzymes to distinguish between α- and β-glycosidic linkages (Section 22.12) that led him to formulate his **lock-and-key hypothesis** for enzyme specificity. According to this hypothesis, the specificity of an enzyme (the lock) and its substrate (the key) comes from their geometrically complementary shapes.

The enzyme and the substrate combine to form an **enzyme–substrate complex.** Formation of the complex often induces a conformational change in the enzyme that allows it to bind the substrate more effectively. This is called an **induced fit.** Binding the substrate also often causes certain of its bonds to become strained, and therefore more easily broken. The product of the reaction usually has a different shape from the substrate, and this altered shape, or in some instances the intervention of another molecule, causes the complex to dissociate. The enzyme can then accept another molecule of the substrate, and the whole process is repeated:

$$\text{Enzyme} + \text{substrate} \rightleftharpoons \begin{matrix}\text{enzyme–substrate}\\ \text{complex}\end{matrix} \rightleftharpoons \text{enzyme} + \text{product}$$

Almost all enzymes are proteins. The substrate is bound to the protein, and the reaction takes place, at what is called the **active site.** The noncovalent forces that bind the substrate to the active site are the same forces that account for the conformations of the proteins themselves: van der Waals forces, electrostatic forces, hydrogen bonding, and hydrophobic interactions. The amino acids located in the active site are arranged so that they can interact specifically with the substrate.

Certain RNA molecules, called ribozymes, can also act as enzymes. The 1989 Nobel Prize in Chemistry went to Sidney Altman (Yale University) and to Thomas R. Cech (University of Colorado, Boulder) for the discovery of ribozymes.

Reactions catalyzed by enzymes are completely **stereospecific,** and this specificity comes from the way enzymes bind their substrates. An α-glycosidase will only bind the α form of a glycoside, not the β form. Enzymes that metabolize sugars bind only D sugars; enzymes that synthesize most proteins bind only L amino acids; and so on.

Although enzymes are absolutely stereospecific, they often vary considerably in what is called their **geometric specificity.** By geometric specificity, we mean a specificity that is related to the identities of the chemical groups of the substrates. Some enzymes will accept only one compound as their substrate. Others, however, will accept a range of compounds with similar groups. Carboxypeptidase A, for example, will hydrolyze the C-terminal peptide from all polypeptides as long as the penultimate residue is not arginine, lysine, or proline and as long as the next preceding residue is not proline. Chymotrypsin, a digestive enzyme that catalyzes the hydrolysis of peptide bonds, will also catalyze the hydrolysis of esters. We shall consider its mechanism of hydrolysis in Section 24.11.

$$\underset{\textbf{Peptide}}{R{-}C({=}O){-}NH{-}R'} + H_2O \xrightarrow{\text{chymotrypsin}} R{-}C({=}O){-}O^- + H_3\overset{+}{N}{-}R'$$

$$\underset{\textbf{Ester}}{R{-}C({=}O){-}O{-}R'} + H_2O \xrightarrow{\text{chymotrypsin}} R{-}C({=}O){-}OH + HO{-}R'$$

A compound that can negatively alter the activity of an enzyme is called an **inhibitor.** A compound that competes directly with the substrate for the active site is known as a **competitive inhibitor.** We learned in Section 20.11, for example, that sulfanilamide is a competitive inhibitor for a bacterial enzyme that incorporates *p*-aminobenzoic acid into folic acid.

We have become acquainted with several coenzymes in earlier chapters because they are the "organic chemistry machinery" for some enzymes. For example, see "Two Aspects of the Coenzyme NADH" (Chapter 12 opening vignette), "The Chemistry of . . . Pyridoxal Phosphate" (Section 16.8), and "The Chemistry of . . . Thiamine" (Section 18.10).

Some enzymes require the presence of a **cofactor.** The cofactor may be a metal ion as, for example, the zinc atom of human carbonic anhydrase (see the Chapter 3 opening vignette and Fig. 24.12). Others may require the presence of an organic molecule, such as NAD^+ (Section 14.10), called a **coenzyme.** Coenzymes become chemically changed in the course of the enzymatic reaction. NAD^+ becomes converted to **NADH.** In some enzymes the cofactor is permanently bound to the enzyme, in which case it is called a **prosthetic group.**

Many of the water-soluble vitamins are the precursors of coenzymes. Niacin (nicotinic acid) is a precursor of NAD^+, for example. Pantothenic acid is a precursor of coenzyme A.

Niacin (pyridine-3-C(=O)OH)

Pantothenic acid ($HO{-}C({=}O){-}CH_2CH_2{-}NH{-}C({=}O){-}CH(OH){-}C(CH_3)_2{-}CH_2OH$)

24.10 Lysozyme: Mode of Action of an Enzyme

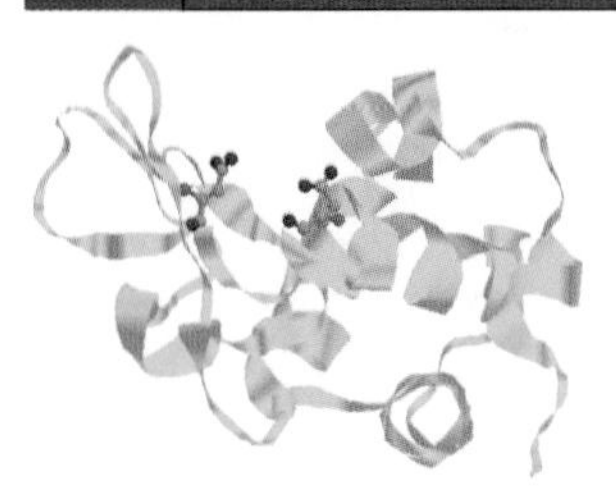

A ribbon diagram of lysozyme.

Lysozyme is made up of 129 amino acid residues (Fig. 24.14). Three short segments of the chain between residues 5 and 15, 24 and 34, and 88 and 96 have the structure of an α helix; the residues between 41 and 45 and 50 and 54 form pleated sheets, and a hairpin turn occurs at residues 46–49. The remaining polypeptide segments of lysozyme have a coil or loop conformation.

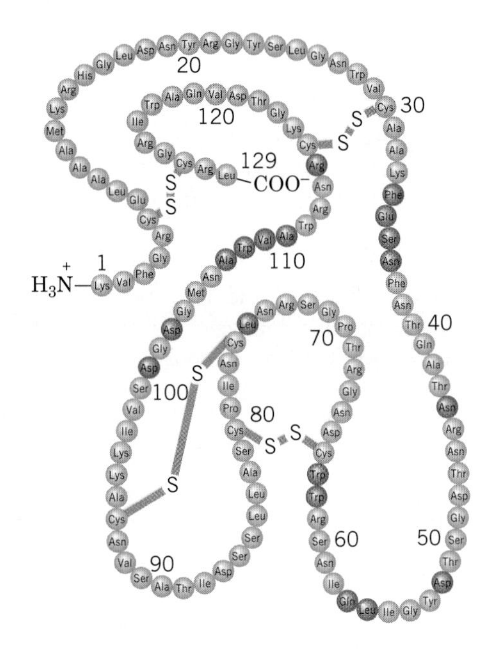

FIGURE 24.14 The primary structure of hen egg white lysozyme. The amino acids that line the substrate-binding pocket are shown in blue. (From Voet, D.; Voet, J. G. *Biochemistry,* 2nd ed.; Wiley: New York, 1995; p 382.)

The discovery of lysozyme is an interesting story in itself:

> One day in 1922 Alexander Fleming was suffering from a cold. This is not unusual in London, but Fleming was a most unusual man and he took advantage of the cold in a characteristic way. He allowed a few drops of his nasal mucus to fall on a culture of bacteria he was working with and then put the plate to one side to see what would happen. Imagine his excitement when he discovered some time later that the bacteria near the mucus had dissolved away. For a while he thought his ambition of finding a universal antibiotic had been realized. In a burst of activity he quickly established that the antibacterial action of the mucus was due to the presence of an enzyme; he called this substance lysozyme because of its capacity to lyse, or dissolve, the bacterial cells. Lysozyme was soon discovered in many tissues and secretions of the human body, in plants, and most plentifully of all in the white of an egg. Unfortunately Fleming found that it is not effective against the most harmful bacteria. He had to wait 7 years before a strangely similar experiment revealed the existence of a genuinely effective antibiotic: penicillin.

This story was related by Professor David C. Phillips of Oxford University who many years later used X-ray analysis to discover the three-dimensional structure of lysozyme.*

*Quotation from David C. Phillips, *The Three-Dimensional Structure of an Enzyme Molecule.*

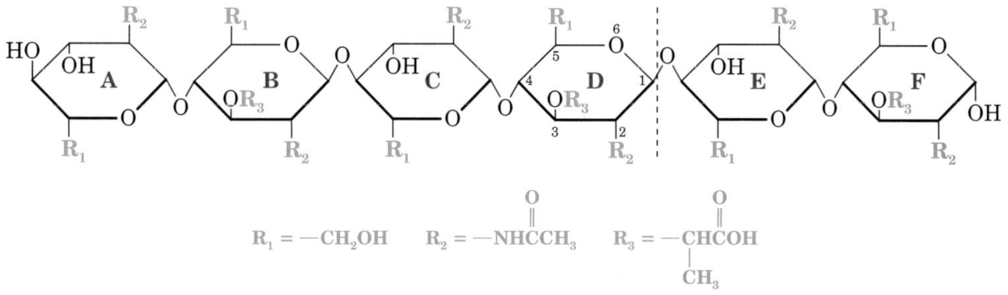

FIGURE 24.15 A hexasaccharide that has the same general structure as the cell wall polysaccharide on which lysozyme acts. Two different amino sugars are present: rings A, C, and E are derived from a monosaccharide called *N*-acetylglucosamine; rings B, D, and F are derived from a monosaccharide called *N*-acetylmuramic acid. When lysozyme acts on this oligosaccharide, hydrolysis takes place and results in cleavage at the glycosidic linkage between rings D and E.

Phillips' X-ray diffraction studies of lysozyme are especially interesting because they have also revealed important information about how this enzyme acts on its substrate. Lysozyme's substrate is a polysaccharide of amino sugars that makes up part of the bacterial cell wall. An oligosaccharide that has the same general structure as the cell wall polysaccharide is shown in Fig. 24.15.

By using oligosaccharides (made up of *N*-acetylglucosamine units only) on which lysozyme acts very slowly, Phillips and co-workers were able to discover how the substrate fits into the enzyme's active site. This site is a deep cleft in the lysozyme structure (Fig. 24.16*a*). The oligosaccharide is held in this cleft by hydrogen bonds, and, as the enzyme binds the substrate, two important changes take place: The cleft in the enzyme closes slightly and ring **D** of the oligosaccharide is "flattened" out of its stable chair conformation. This flattening causes atoms 1, 2, 5, and 6 of ring **D** to become coplanar; it also distorts ring **D** in such a way as to make the glycosidic linkage between it and ring **E** more susceptible to hydrolysis.*

Hydrolysis of the glycosidic linkage probably takes the course illustrated in Fig. 24.16*b*. The carboxyl group of glutamic acid (residue number 35) donates a proton to the oxygen between rings **D** and **E**. Protonation leads to cleavage at the glycosidic link and to the formation of a carbocation at C1 of ring **D**. This carbocation is stabilized by the negatively charged carboxylate group of aspartic acid (residue number 52), which lies in close proximity. A water molecule diffuses in and supplies an OH^- ion to the carbocation and a proton to replace that lost by glutamic acid. An X-ray crystal structure of lysozyme is shown in Fig. 24.16*c*. Glutamic acid 35 and aspartic acid 52 are highlighted in ball-and-stick format.

When the polysaccharide is a part of a bacterial cell wall, lysozyme probably first attaches itself to the cell wall by hydrogen bonds. After hydrolysis has taken place, lysozyme falls away, leaving behind a bacterium with a punctured cell wall.

24.11 Serine Proteases

Chymotrypsin, trypsin, and elastin are digestive enzymes secreted by the pancreas into the small intestine to catalyze the hydrolysis of peptide bonds. These enzymes are all called **serine proteases** because the mechanism for their proteolytic activity (one that they have in common) involves a particular serine residue that is essential for their enzymatic activity. As another example of how enzymes work, we shall examine the mechanism of action of chymotrypsin.

---

*R. H. Lemieux and G. Huber, while with the National Research Council of Canada, showed that when an aldohexose is converted to a carbocation the ring of the carbocation assumes just this flattened conformation.

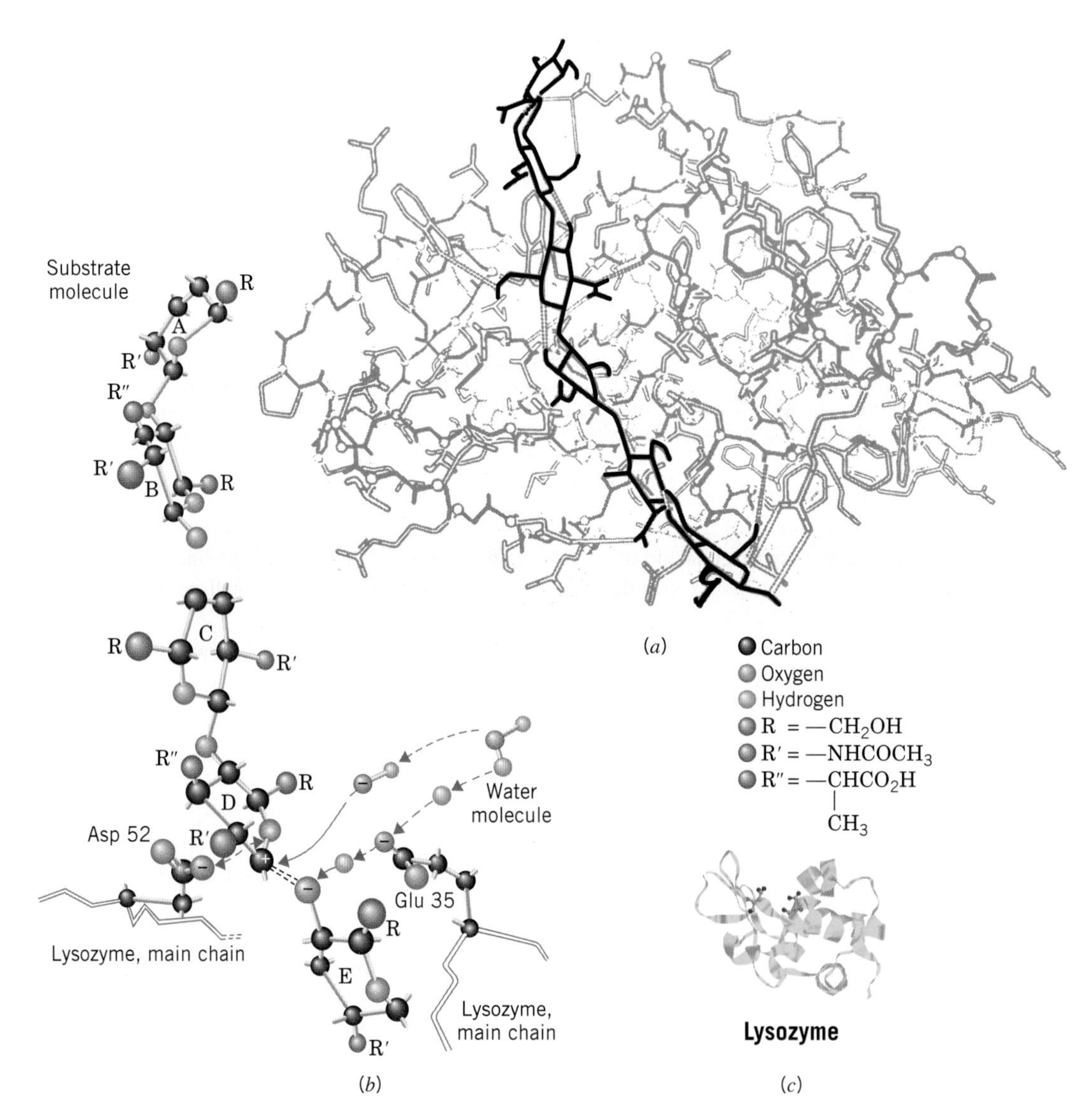

FIGURE 24.16 *(a)* The framework of the lysozyme–substrate complex. The substrate (in this drawing a hexasaccharide) fits into a cleft in the lysozyme structure and is held in place by hydrogen bonds. As lysozyme binds the oligosaccharide, the cleft in its structure closes slightly. (Adapted with permission from *Atlas of Protein Sequence and Structure, 1969;* Dayhoff, M. O., Ed.; National Biomedical Research Foundation: Washington, DC, 1969. The drawing was made by Irving Geis, based on his perspective painting of the molecule, which appeared in *Scientific American,* November 1966. The painting was made of an actual model assembled at the Royal Institution, London, by D. C. Phillips and colleagues, based on their X-ray crystallography results.) *(b)* A possible mechanism for lysozyme action. This drawing shows an expanded portion of part *(a)* and illustrates how hydrolysis of the acetal linkage between rings D and E of the substrate may occur. Glutamic acid (residue 35) donates a proton to the intervening oxygen atom. This causes the formation of a carbocation that is stabilized by the carboxylate ion of aspartic acid (residue 52). A water molecule supplies an OH^- to the carbocation and H^+ to glutamic acid. (Adapted with permission from *The Three-Dimensional Structures of an Enzyme Molecule,* by David C. Phillips. Copyright © 1966 by Scientific American, Inc. All rights reserved.) *(c)* A ribbon diagram of lysozyme highlighting aspartic acid 52 (left) and glutamic acid 35 (right) in ball-and-stick format. (This image and that in the margin by the Section 24.10 heading were created from an X-ray crystal structure by Lim, K.; Nadarajah, A.; Forsythe, E. L.; Pusey, M. L. Protein Data Bank file 1AZF.pdb.)

Chymotrypsin is formed from a precursor molecule called chymotrypsinogen, which has 245 amino acid residues. Cleavage of two dipeptide units of chymotrypsinogen produces chymotrypsin. Chymotrypsin folds in a way that brings together histidine at position 57, aspartic acid at position 102, and serine at position 195. Together, these residues constitute what is called the **catalytic triad** of the active site (Fig. 24.17). Near the active site is a hydrophobic binding site, a slitlike pocket that preferentially accommodates the nonpolar side chains of Phe, Tyr, and Trp.

A serine protease

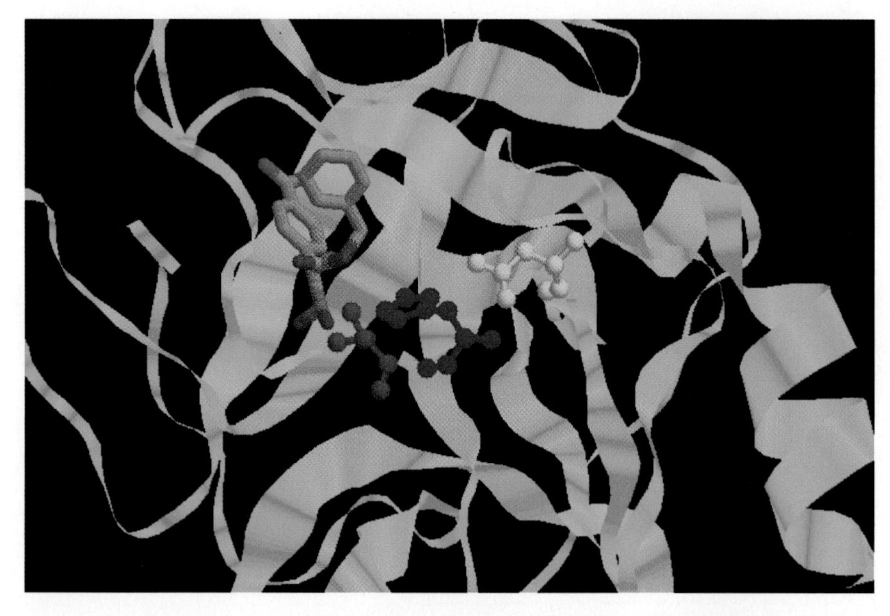

FIGURE 24.17 The catalytic triad in this serine protease (trypsin) is highlighted using ball-and-stick model format for aspartic acid 52 (yellow-green), histidine 102 (purple), and serine 195 (red). A phosphonate inhibitor bound at the active site is shown in tube format. (This image and that by the Section 24.11 heading were created from an X-ray crystal structure by Bertrand, J. A.; Oleksyszn, J.; Kam, C.-M.; Boduszek, B. Protein Data Bank file 1MAX.pdb.)

After chymotrypsin has bound its protein substrate, the serine residue at position 195 is ideally situated to attack the acyl carbon of the peptide bond (Fig. 24.18). This serine residue is made more nucleophilic by transferring its proton to the imidazole nitrogen of the histidine residue at position 57. The imidazolium ion that is formed is stabilized by the polarizing effect of the carboxylate ion of the aspartic acid residue at position 102. (Neutron diffraction studies, which show the positions of hydrogen atoms, confirm that the carboxylate ion remains as a carboxylate ion throughout and does not actually accept a proton from the imidazole.) Nucleophilic attack by the serine leads to an acylated serine through a tetrahedral intermediate. The new N-terminal end of the cleaved polypeptide chain diffuses away and is replaced by a water molecule.

Regeneration of the active site of chymotrypsin is shown in Fig. 24.19. In this process water acts as the nucleophile and, in a series of steps analogous to those in Fig. 24.18, hydrolyzes the acyl–serine bond. The enzyme is now ready to repeat the whole process.

There is much evidence for this mechanism that, for reasons of space, we shall have to ignore. One bit of evidence deserves mention, however. There are compounds such as **diisopropylphosphofluoridate (DIPF)** that irreversibly inhibit serine proteases. It has been shown that they do this by reacting only with Ser 195 (shown on page 1208):

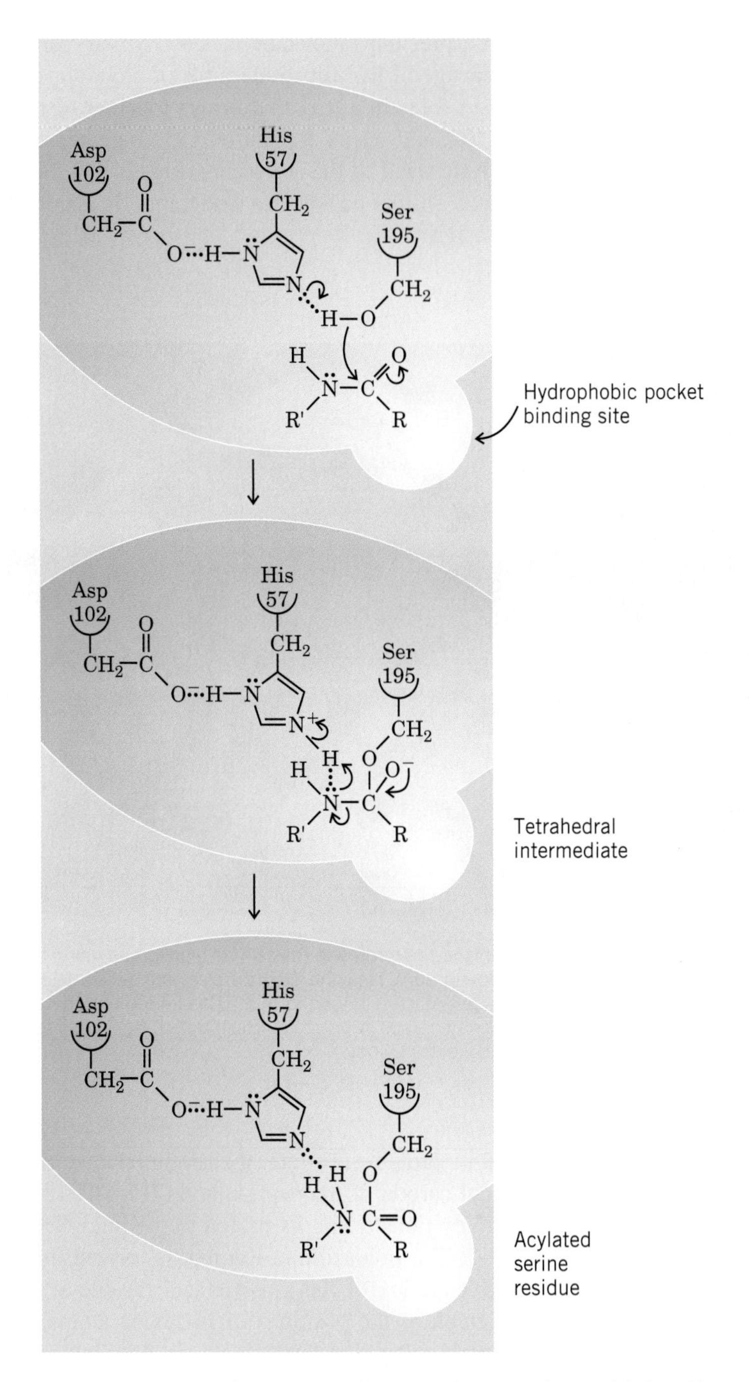

FIGURE 24.18 The catalytic triad of chymotrypsin causes cleavage of a peptide bond by acylation of the serine residue 195 of chymotrypsin. Near the active site is a hydrophobic binding site that accommodates nonpolar side chains of the protein.

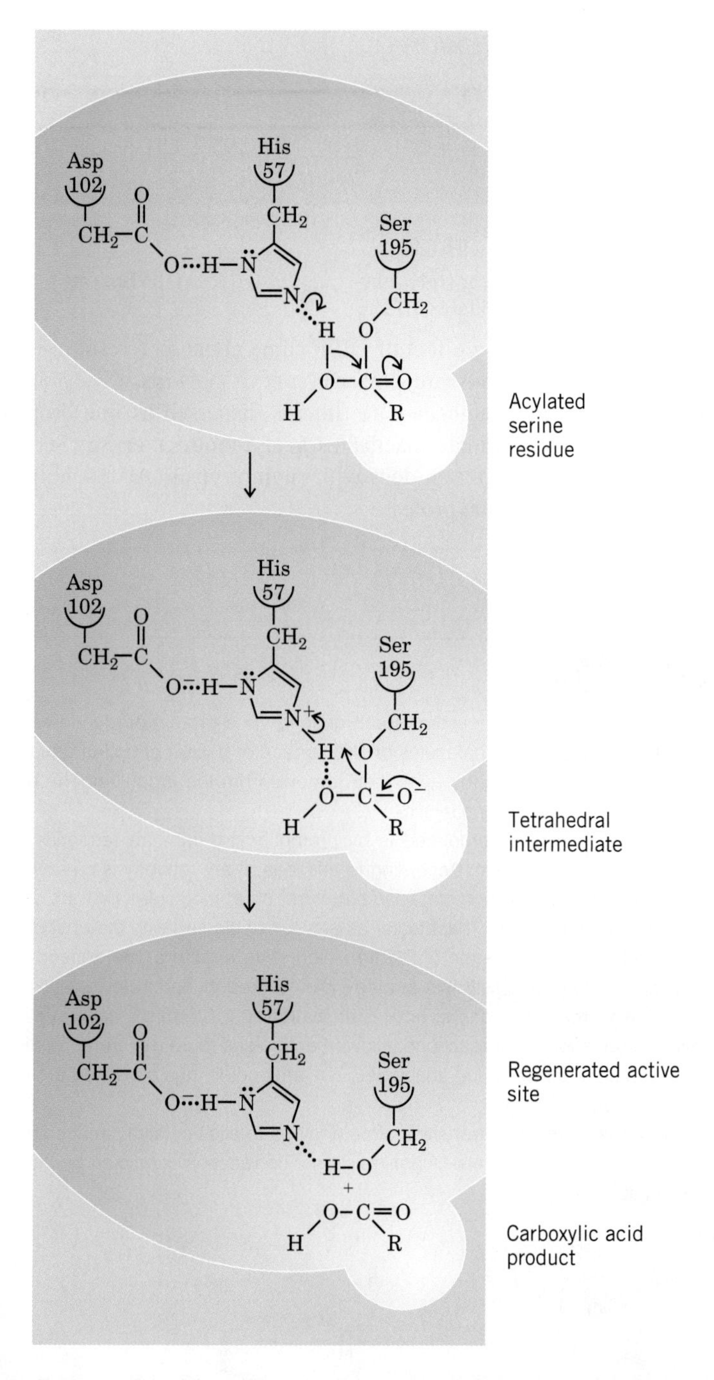

FIGURE 24.19 Regeneration of the active site of chymotrypsin. Water causes hydrolysis of the acyl–serine bond.

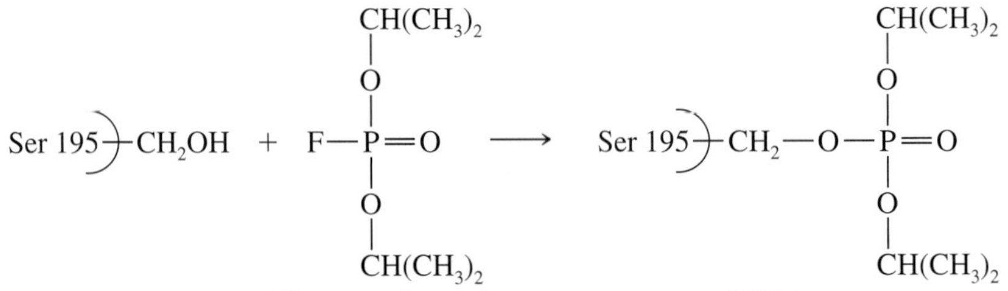

$$\text{Ser 195}\!-\!CH_2OH + F\!-\!\overset{O\!-\!CH(CH_3)_2}{\underset{O\!-\!CH(CH_3)_2}{P}}\!=\!O \longrightarrow \text{Ser 195}\!-\!CH_2\!-\!O\!-\!\overset{O\!-\!CH(CH_3)_2}{\underset{O\!-\!CH(CH_3)_2}{P}}\!=\!O$$

Diisopropylphosphofluoridate (DIPF) **DIP-Enzyme**

Recognition of the inactivating effect of DIPF came about as a result of the discovery that DIPF and related compounds are powerful **nerve poisons.** (They are the "nerve gases" of military use, even though they are liquids dispersed as fine droplets, and not gases.) Diisopropylphosphofluoridate inactivates **acetylcholinesterase** (Section 20.4) by reacting with it in the same way that it does with chymotrypsin. Acetylcholinesterase is a **serine esterase** rather than a serine protease.

The Chemistry of...

Some Catalytic Antibodies

Antibodies are chemical warriors of the immune system. Each antibody is a protein produced specifically in response to an invading chemical species (e.g., molecules on the surface of a virus or pollen grain). The purpose of antibodies is to bind with these foreign agents and cause their removal from the organism. The binding of each antibody with its target (the antigen) is usually highly specific.

One way that *catalytic* antibodies have been produced is by prompting an immune response to a chemical species resembling the transition state for a reaction. According to this idea, if an antibody is created that preferentially binds with a stable molecule that has a transition state-like structure, other molecules that are capable of reaction *through* this transition state should, in principle, react faster as a result of binding with the antibody. (By facilitating association of the reactants and favoring formation of the transition state structure, the antibody acts in a way similar to an enzyme.) In stunning fashion, precisely this strategy has worked to generate catalytic antibodies for certain Diels–Alder reactions, Claisen rearrangements, and ester hydrolyses. Chemists have synthesized stable molecules that resemble transition states for these reactions, allowed antibodies to be generated against these molecules (called haptens), and then isolated the resulting antibodies. The antibodies thus produced are catalysts when actual substrate molecules are provided.

The following are examples of haptens used as transition state analogs to elicit catalytic antibodies for a Claisen rearrangement, hydrolysis of a carbonate, and a Diels–Alder reaction. The reaction catalyzed by the antibody generated from each hapten is shown as well.

Claisen Rearrangement

Hapten: ^-O_2C, O, CO_2^-, OR; R = (amide N–H chain to aryl N_2^+)

CO_2^-, OH, O, CO_2^- → [^-O_2C, O, CO_2^-, OH]‡ → ^-O_2C, CO_2^-, O, OH

Transition state

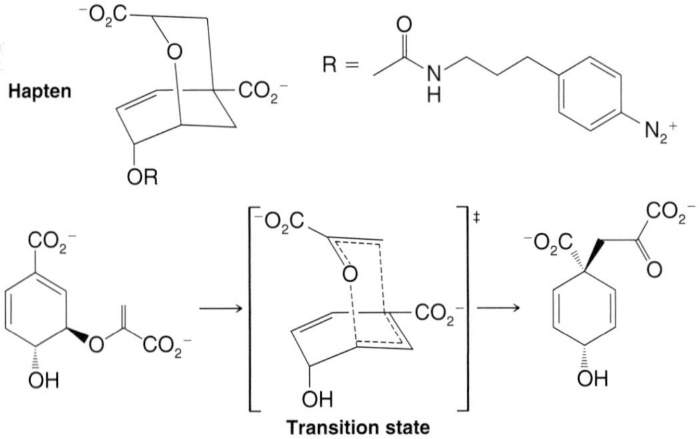

Carbonate Hydrolysis

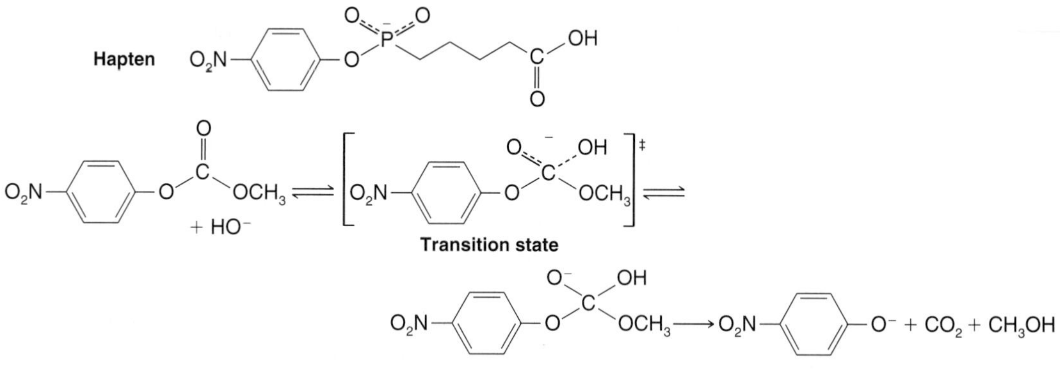

Hapten $O_2N-C_6H_4-O-P(=O)(O^-)-CH_2CH_2CH_2CH_2-C(=O)OH$

$O_2N-C_6H_4-O-C(=O)-OCH_3 + HO^- \rightleftharpoons [O_2N-C_6H_4-O-C(O^-)(OH)-OCH_3]^\ddagger \rightleftharpoons$

Transition state

$O_2N-C_6H_4-O-C(O^-)(OH)-OCH_3 \longrightarrow O_2N-C_6H_4-O^- + CO_2 + CH_3OH$

Diels–Alder Reaction

Hapten: ferrocene bearing $-C(=O)-N(CH_3)_2$ and $-NH-C(=O)-(CH_2)_3COOH$ (Fe)

Diene: $CH_2=CH-CH=CH-NH-C(=O)-O-CH_2-C_6H_4-COOH$

Dienophile: $CH_2=CH-C(=O)-N(CH_3)_2$

Diene + Dienophile $\longrightarrow$ [Transition state: RO_2CNH, H, H, $CON(CH_3)_2$]‡ $\longrightarrow$

Exo product: cyclohexene bearing $-C(=O)-N(CH_3)_2$ and $-NH-C(=O)-O-CH_2-C_6H_4-COOH$

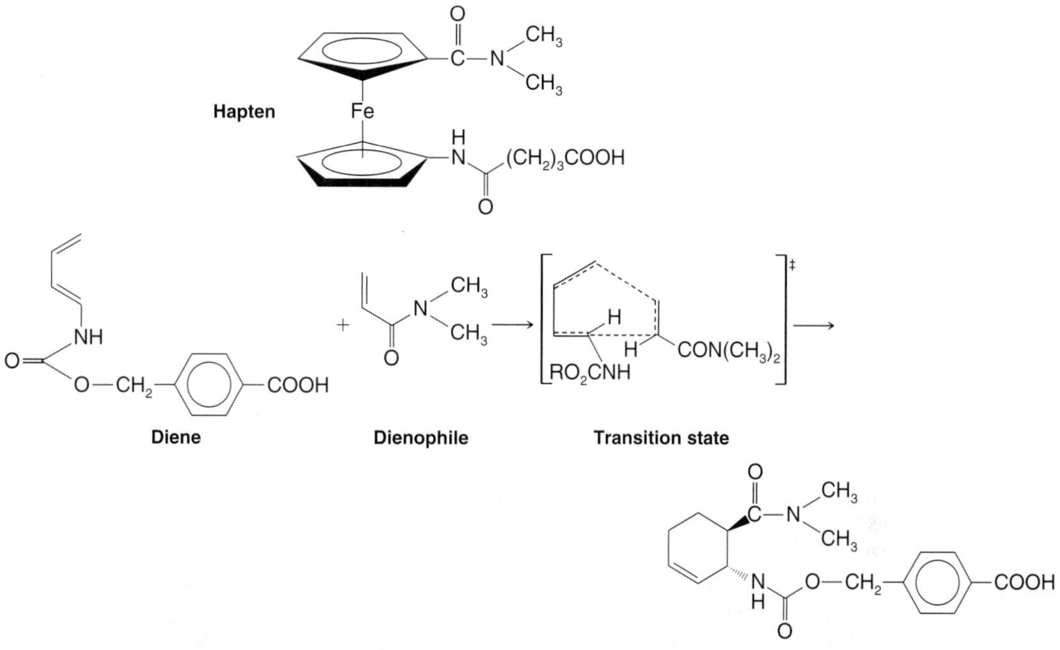

This marriage of enzymology and immunology, resulting in chemical offspring, is just one area of exciting research at the interface of chemistry and biology.

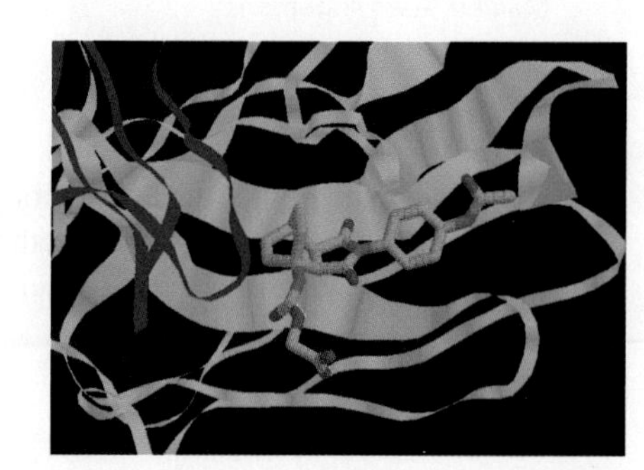

A hapten related to the Diels–Alder adduct from cyclohexadiene and maleimide, bound within a Diels–Alderase catalytic antibody. (This image and that by the opening photo for Chapter 24 were created from an X-ray crystal structure by Romesburg, F. E.; Spiller, B.; Schultz, P. G.; Stevens, R. C. Protein Data Bank file 1A4K.pdb.)

24.12 Hemoglobin: A Conjugated Protein

Some proteins, called **conjugated proteins,** contain as a part of their structure a nonprotein group called a **prosthetic group.** An example is the oxygen-carrying protein hemoglobin. Each of the four polypeptide chains of hemoglobin is bound to a prosthetic group called *heme* (Fig. 24.20). The four polypeptide chains of hemoglobin are wound in such a way as to give hemoglobin a roughly spherical shape (Fig. 24.21). Moreover, each heme group lies in a crevice with the hydrophobic vinyl groups of its porphyrin structure surrounded by hydrophobic side chains of amino acid residues. The two propanoate side chains of heme lie near positively charged amino groups of lysine and arginine residues.

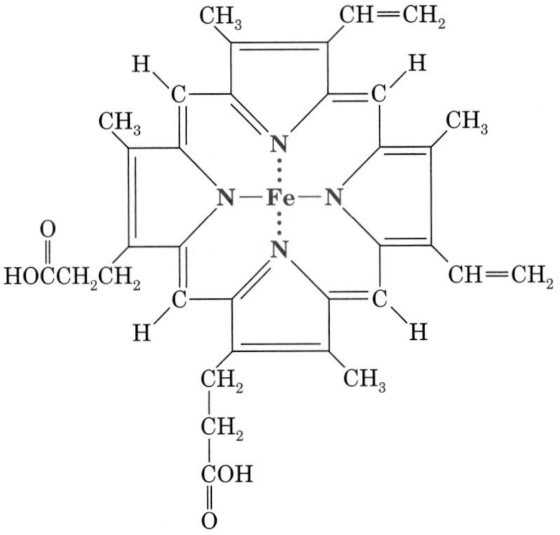

FIGURE 24.20 The structure of heme, the prosthetic group of hemoglobin. Heme has a structure similar to that of chlorophyll (Fig. 22.1) in that each is derived from the heterocyclic ring, porphyrin. The iron of heme is in the ferrous (2+) oxidation state.

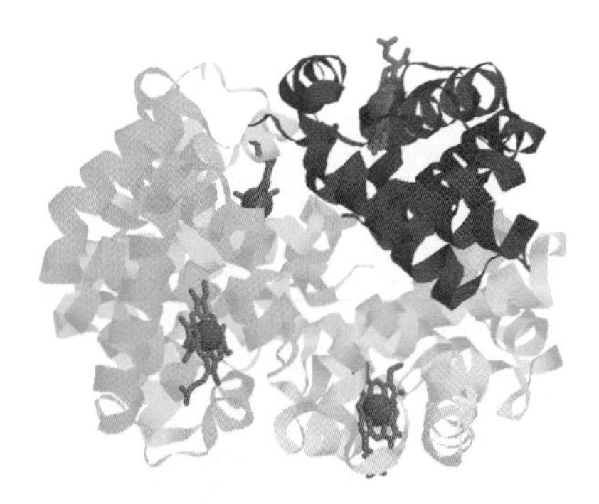

FIGURE 24.21 Hemoglobin. The two α subunits of hemoglobin are shown in blue and green. The two β subunits are shown in yellow and cyan. The four heme groups are shown in purple, and their iron atoms are in red. (Image created from an X-ray crystal structure by Tame, J. R.; Wilson, J. C.; Weber, R. E. Protein Data Bank file 1OUU.pdb.)

The iron of the heme group is in the 2+ (ferrous) oxidation state and it forms a coordinate bond to a nitrogen of the imidazole group of histidine of the polypeptide chain. This leaves one valence of the ferrous ion free to combine with oxygen as follows:

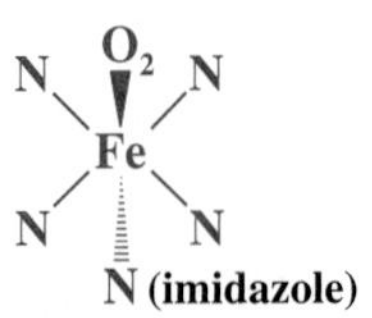

A portion of oxygenated hemoglobin

The fact that the ferrous ion of the heme group combines with oxygen is not particularly remarkable; many similar compounds do the same thing. What is remarkable about hemoglobin is that when the heme combines with oxygen the ferrous ion does not become readily oxidized to the ferric state. Studies with model heme compounds in water, for example, show that they undergo a rapid combination with oxygen but they also undergo a rapid oxidation of the iron from Fe^{2+} to Fe^{3+}. When these same compounds are embedded in the hydrophobic environment of a polystyrene resin, however, the iron is easily oxygenated and deoxygenated, and this occurs *with no change in oxidation state of iron.* In this respect, it is especially interesting to note that X-ray studies of hemoglobin have revealed that the polypeptide chains provide each heme group with a similar hydrophobic environment.

24.13 Purification and Analysis of Polypeptides and Proteins

24.13A Purification

There are many methods used to purify polypeptides and proteins. The specific methods one chooses depend on the source of the protein (isolation from a natural source or chemical synthesis), its physical properties, and the quantity of the protein on hand. Initial purification methods may involve precipitation, various forms of column chromatography, and electrophoresis. Perhaps the most important final method for peptide purification, HPLC, is used to purify peptides generated both by automated synthesis and peptides and proteins isolated from nature.

24.13B Analysis

A variety of parameters are used to characterize polypeptides and proteins. One of the most fundamental is molecular weight. Gel electrophoresis can be used to measure the approximate molecular weight of a protein. Gel electrophoresis involves migration of a peptide or protein dissolved in a buffer through a porous polymer gel under the influence of a high-voltage electric field. The buffer used (typically about pH 9) imparts an overall negative charge to the protein such that the protein migrates toward the positively charged terminal. Migration rate depends on the overall charge and size of the protein as well as the average pore size of the gel. The molecular weight of the protein is inferred by comparing the distance traveled through the gel by the protein of interest with the migration distance of proteins with known molecular weights used as internal standards. The version of this technique called SDS–PAGE (sodium dodecyl sulfate–polyacrylamide gel electrophoresis) allows protein molecular weight determinations with an accuracy of about 5–10%.

Mass spectrometry can be used to determine a peptide's molecular weight with very high accuracy and precision. Earlier we discussed mass spectrometry in the context of protein sequencing. Now we shall consider the practical aspects of how molecules with very high molecular weight, such as proteins, can be transferred to the gas phase for mass spectrometric analysis. This is necessary, of course, whether the analysis regards peptide sequencing or full molecular analysis. Small organic molecules, as we discussed in Chapter 9, can be vaporized simply with high vacuum and heat. High-molecular-weight species cannot be transferred to the gas phase solely with heat and vacuum. Fortunately, very effective techniques have been developed for generating gas-phase ions of large molecules without destruction of the sample.

One ionization method is electrospray ionization (ESI, Figure 24.22), whereby a solution of a peptide (or other analyte) in a volatile solvent containing a trace of acid is sprayed through a high-voltage nozzle into the vacuum chamber of a mass spectrometer.

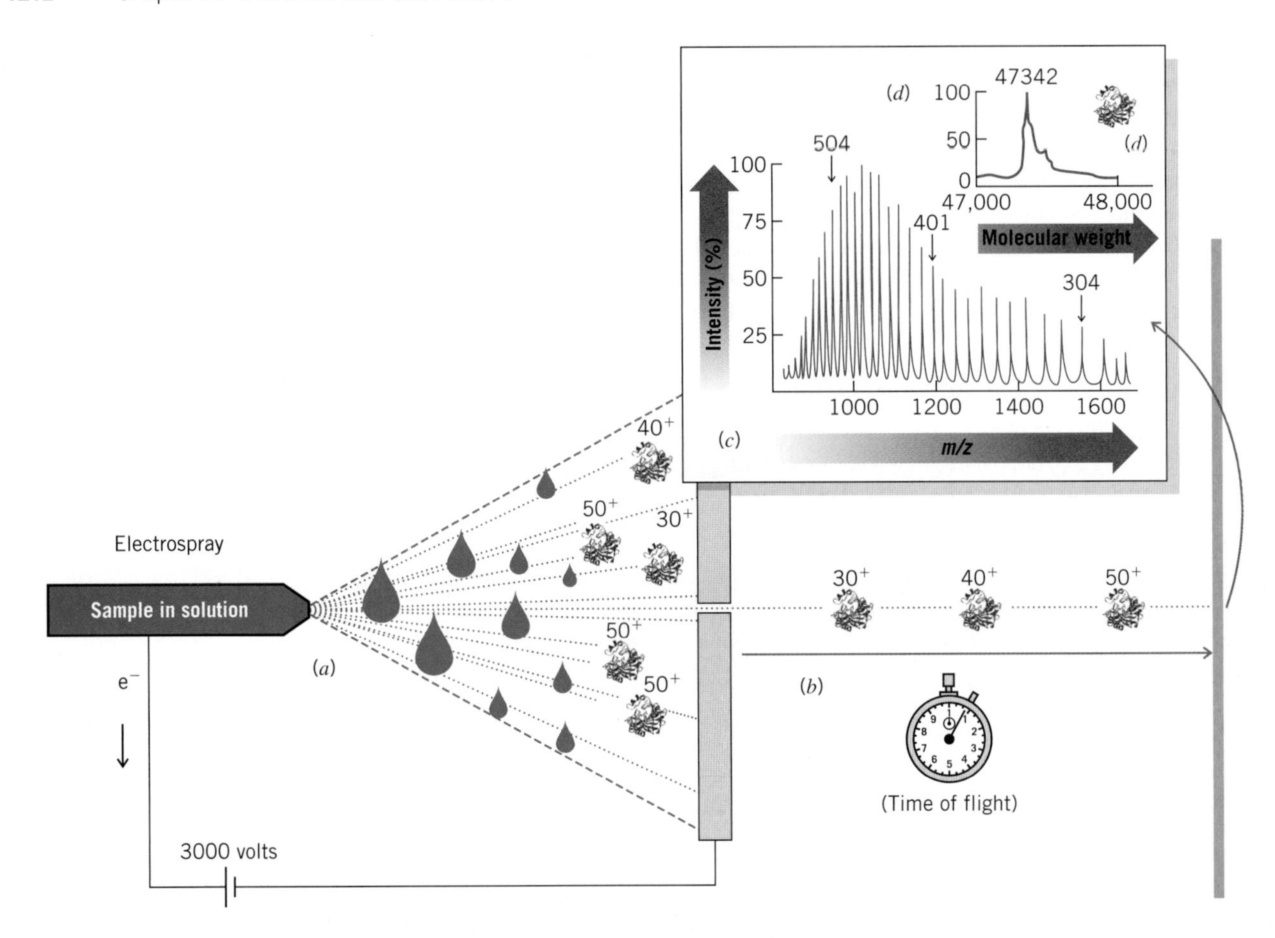

FIGURE 24.22 Electrospray ionization (ESI) mass spectrometry. *(a)* Analyte ions, protonated multiple times by an acidic solvent system, are sprayed through a high-voltage nozzle into a vacuum chamber. Molecules of the solvent evaporate. The multiply-charged analyte ions are drawn into the mass analyzer. *(b)* The analyte ions drift through the analyzer tube and are detected according to their time of flight. *(c)* The family of detected ions is displayed in a spectrum according to *m/z* ratio. *(d)* Computerized deconvolution of the *m/z* peak series leads the molecular weight of the analyte.

One-quarter of the 2002 Nobel Prize in Chemistry was awarded to John B. Fenn for his development of ESI mass spectrometry. Another quarter of the prize was awarded to Koichi Tanaku for discoveries that led to matrix-assisted laser desorption ionization (MALDI, see below).

The acid in the solvent generates ions by protonating Lewis basic sites within the analyte. Peptides are typically protonated multiple times. Once injected through the high-voltage nozzle into the vacuum chamber, solvent molecules evaporate from the analyte ions [Figure 24.22, part (a)], and the ions are drawn into the mass analyzer [Figure 24.22, part (b)]. The mass analyzer detects the analyte ions according to their time of flight, and registers their mass-to-charge (*m/z*). [Fig. 24.22, part (c)]. Each peak displayed in the mass spectrum represents the molecular weight of an ion divided by the number positive charges it carries. From this series of *m/z* peaks, the molecular weight of the analyte is calculated by a computerized process called deconvolution. An example of a deconvoluted spectrum, indicating a molecular weight of 47,342 atomic mass units (Daltons), is shown in Figure 24.22, part (d).

If fragmentation of the analyte molecules is desired, it can be caused by collision-induced dissociation (CID, Section 24.5F). In this case, tandem mass spectrometry is necessary because the first mass analyzer in the system is used to select fragments of the peptide from CID based on their overall mass, while the second mass analyzer in the system records the spectrum of the selected peptide fragment. Multiple fragments from the CID procedure can be analyzed this way. The final spectrum for each peptide fragment selected has the typical appearance of a family of ions, as shown above.

Mass spectrometry with electrospray ionization (ESI-MS) is especially powerful when combined with HPLC because the two techniques can be used in tandem. With such an instrument the effluent from the HPLC is introduced directly into an ESI mass spectrometer. Thus, chromatographic separation of peptides in a mixture and direct structural information about each of them are possible using this technique.

Another method for ionization of nonvolatile molecules is MALDI. Energy from laser bombardment of a sample adsorbed in a solid chemical matrix leads to generation of gas-phase ions that are detected by the mass spectrometer. Both MALDI and ESI are common ionization techniques for the analysis of biopolymers.

24.14 Proteomics

Proteomics and genomics are two fields that have blossomed in recent years. **Proteomics** has to do with the study of all proteins that are expressed in a cell at a given time. **Genomics** (Sections 25.1 and 25.9) focuses on the study of the complete set of genetic instructions in an organism. While the genome holds the instructions for making proteins, it is proteins that carry out the vast majority of functions in living systems. Yet, compared to the tens of thousands of proteins encoded by the genome, we know the structure and function of only a relatively small percentage of proteins in the proteome. For this reason, the field of proteomics has moved to a new level of importance since completion of sequencing the human genome. Many potential developments in health care and medicine now depend on identifying the myriad of proteins that are expressed at any given time in a cell, along with elucidation of their structures and biochemical function. New tools for medical diagnosis and targets for drug design will undoubtedly emerge at an increasing rate as the field of proteomics advances.

One of the basic challenges in proteomics is simply separation of all the proteins present in a cell extract. The next challenge is identification of those proteins that have been separated. Separation of proteins in cell extracts has classically been carried out using two-dimensional polyacrylamide gel electrophoresis (2D PAGE). In 2D PAGE the mixture of proteins extracted from an organism is separated in one dimension of the gel by the isoelectric point (a technique called isoelectric focusing) and in the second dimension by molecular weight. The result is a set of spots in the two-dimensional gel field that represents the location of separated proteins. The protein spots on the gel may then be extracted and analyzed by mass spectrometry or other methods, either as intact proteins or as enzymatic digests. Comparison of the results from mass spectrometry with protein mass spectrometry databases allows identification of many of the proteins separated by the gel.

There are limitations to protein separation by 2D PAGE, however. Not all proteins are amenable to 2D PAGE due to their size, charge, or specific properties. Furthermore, more than one protein may migrate to the same location if their isoelectric points and molecular weights are similar. Finally, 2D PAGE has inherent limits of detection that can leave some proteins of low concentration undetected.

A recent improvement over 2D PAGE involves two-dimensional microcapillary HPLC coupled with mass spectrometry. See Figure 24.23. In this technique called MudPIT (multidimensional protein identification technology, developed by John Yates and coworkers at Scripps Research Institute) a microcapillary HPLC column is used that has been packed first with a strong cation-exchange resin and then a reversed-phase (hydrophobic) material. The two packing materials used in sequence and with different resolving properties represent the two-dimensional aspect of this technique. A protein extract is introduced to the microcapillary column and eluted with pH and solvent gradients over a sequence of automated steps. As the separated proteins are eluted from the column they pass directly into a mass spectrometer. Mass spectrometric data obtained for each

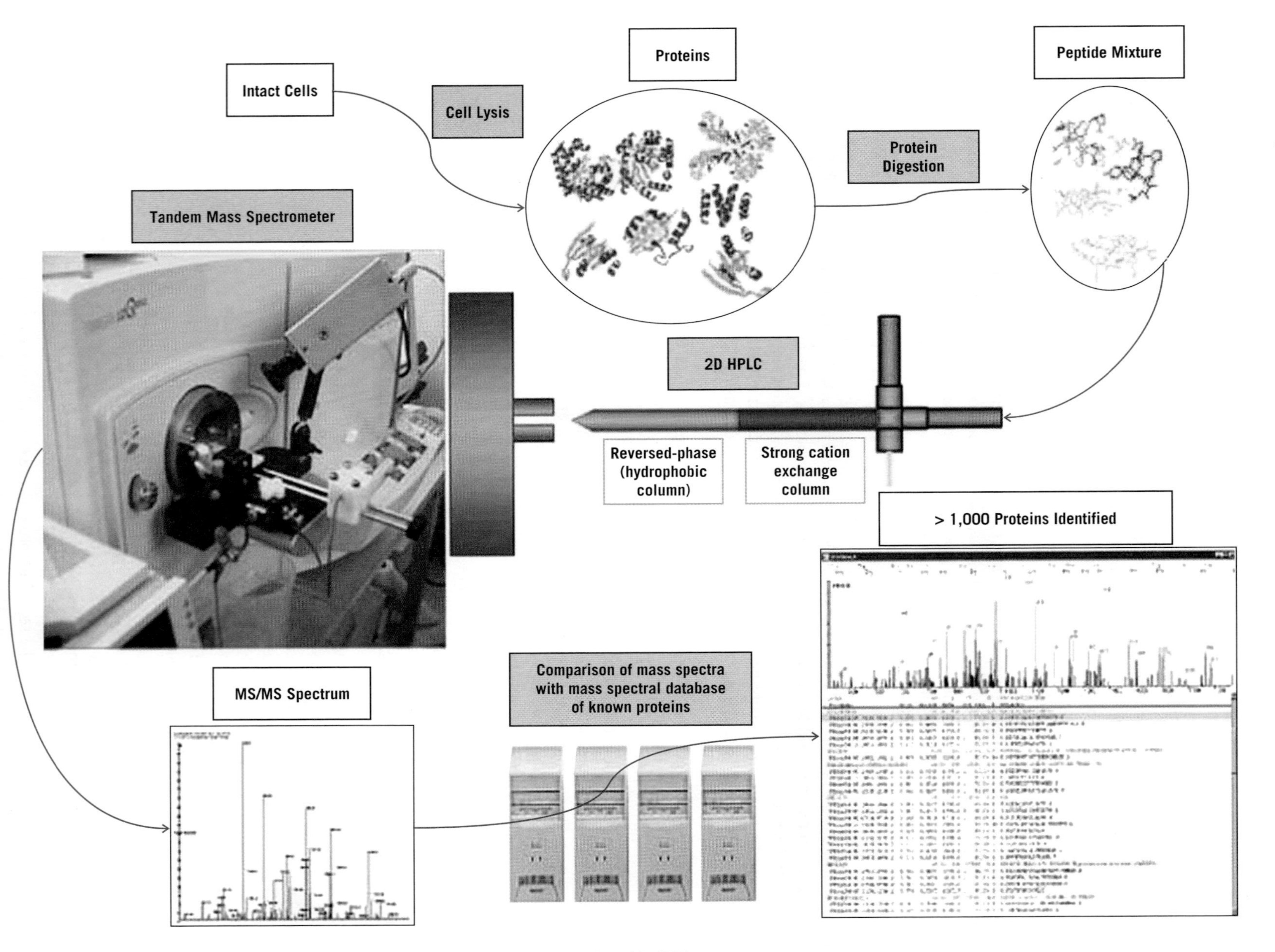

FIGURE 24.23 The high-throughput multidimensional protein identification technology MudPIT) process. (Courtesy of John Yates and Laurence Florens, Scripps Research Institute.)

protein represent a signature that allows identification of the protein by comparison with a protein mass spectrometry database. This technique of 2D HPLC coupled with mass spectrometry is inherently more sensitive and general than 2D PAGE. One powerful example of its use is the identification by Yates and co-workers of nearly 1500 proteins from the *Saccharomyces cerivisiae* (Baker's yeast) proteome in one integrated analysis.

Beyond the identification of proteins, quantitative measurement of the amounts of various proteins that are expressed is also important in proteomics. Various disease states or environmental conditions experienced by a cell may influence the amount of some proteins that are expressed. Quantitative tracking of these changes as a function of cell state could be relevant to studies of disease and the development of therapies. A technique using reagents called isotope-coded affinity tags (ICAT, developed at the University of Washington) allows quantitative analysis and identification of components in complex protein mixtures. The ICAT analysis involves mass spectrometric comparison of isotopically labeled and unlabeled protein segments that have been isolated by affinity chromatography and purified by microcapillary HPLC.

Hand in hand with identification and quantification of proteins remains the need to determine full three-dimensional protein structures. Even though thousands of proteins are encoded in the genome, only a relative handful of them have been studied in depth in terms of detailed structure and function. Full structure determination will therefore continue to be central to the field of proteomics. X-ray crystallography, NMR, and mass spectrometry are key tools that will be applied ever more fervently as the quest intensifies to elucidate as many structures in the proteome as possible.

Key Terms and Concepts

Active site	Section 24.9
α Helix, β-pleated sheet, random coil arrangement	Section 24.8A
Amino acid residue	Section 24.4
Asymmetric (enantioselective) synthesis	Sections 5.9B, 24.3*E*
Coenzyme	Section 24.9
Cofactor	Section 24.9
Conjugated proteins	Section 24.12
Dipeptide, tripeptide, oligopeptide, and polypeptide	Section 24.4
Dipolar ions, zwitterions	Section 24.2C
Disulfide linkage	Section 24.2A
Edman degradation	Section 24.5A
Enzyme	Section 24.9
Genomics	Section 24.14
Henderson–Hasselbalch equation	Section 24.2C
Inhibitor	Section 24.9
Isoelectric point (p*I*)	Section 24.2C
Lock-and-key or induced fit hypothesis	Section 24.9
Orthogonal protecting groups	Section 24.7D
Partial hydrolysis	Section 24.5D
Peptide	Section 24.4
Peptide bond, peptide linkage	Section 24.4
Primary structure	Sections 24.1, 24.5, 24.6
Prosthetic group	Section 24.9
Protecting groups	Section 24.7A
Protein	Section 24.4
Proteomics	Section 24.14
Quaternary structure	Sections 24.1, 24.8C
Secondary structure	Sections 24.1, 24.8A
Site-specific cleavage	Section 24.5D
Solid-phase peptide synthesis (SPPS)	Section 24.7D
Substrate	Section 24.9
Terminal residue analysis	Section 24.5A
Tertiary structure	Sections 24.1, 24.8B

Additional Problems

24.17 **(a)** Which amino acids in Table 24.1 have more than one stereogenic center? **(b)** Write Fischer projections for the isomers of each of these amino acids that would have the L configuration at the α carbon. **(c)** What kind of isomers have you drawn in each case?

24.18 **(a)** Which amino acid in Table 24.1 could react with nitrous acid (i.e., a solution of $NaNO_2$ and HCl) to yield lactic acid? **(b)** All of the amino acids in Table 24.1 liberate nitrogen when they are treated with nitrous acid except two; which are these? **(c)** What product would you expect to obtain from treating tyrosine with excess bromine water? **(d)** What product would you expect to be formed in the reaction of phenylalanine with ethanol in the presence of hydrogen chloride? **(e)** What product would you expect from the reaction of alanine and benzoyl chloride in aqueous base?

Note: Problems marked with an asterisk are "challenge problems."

24.19 **(a)** On the basis of the following sequence of reactions, Emil Fischer was able to show that (−)-serine and L-(+)-alanine have the same configuration. Write Fischer projections for the intermediates **A–C**:

$$(-)\text{-Serine} \xrightarrow[\text{CH}_3\text{OH}]{\text{HCl}} \mathbf{A}\ (C_4H_{10}ClNO_3) \xrightarrow{\text{PCl}_5} \mathbf{B}\ (C_4H_9Cl_2NO_2) \xrightarrow[\text{(2) OH}^-]{\text{(1) H}_3\text{O}^+,\ \text{H}_2\text{O, heat}}$$

$$\mathbf{C}\ (C_3H_6ClNO_2) \xrightarrow[\text{dilute H}_3\text{O}^+]{\text{Na–Hg}} \text{L-(+)-alanine}$$

(b) The configuration of L-(−)-cysteine can be related to that of L-(−)-serine through the following reactions. Write Fischer projections for **D** and **E**:

$$\mathbf{B}\ [\text{from part (a)}] \xrightarrow{\text{OH}^-} \mathbf{D}\ (C_4H_8ClNO_2) \xrightarrow{\text{NaSH}} \mathbf{E}\ (C_4H_9NO_2S) \xrightarrow[\text{(2) OH}^-]{\text{(1) H}_3\text{O}^+,\ \text{H}_2\text{O, heat}} \text{L-(+)-cysteine}$$

(c) The configuration of L-(−)-asparagine can be related to that of L(−)-serine in the following way. What is the structure of **F**?

$$\text{L-(−)-Asparagine} \xrightarrow[\text{Hofmann rearrangement}]{\text{NaOBr/OH}^-} \mathbf{F}\ (C_3H_7N_2O_2)$$

$$\mathbf{C}\ [\text{from part (a)}] \xrightarrow{\text{NH}_3} \mathbf{F}$$

24.20 **(a)** DL-Glutamic acid has been synthesized from diethyl acetamidomalonate in the following way. Outline the reactions involved.

$$\underset{\textbf{Diethyl acetamidomalonate}}{CH_3\overset{O}{\overset{\|}{C}}NHCH(CO_2C_2H_5)_2} + CH_2{=}CH{-}C{\equiv}N \xrightarrow[\substack{C_2H_5OH \\ (95\%\ \text{yield})}]{NaOC_2H_5}$$

$$\mathbf{G}\ (C_{12}H_{18}N_2O_5) \xrightarrow[\substack{\text{reflux 6 h} \\ (66\%\ \text{yield})}]{\text{concd HCl}} \text{DL-glutamic acid}$$

(b) Compound **G** has also been used to prepare the amino acid DL-ornithine through the following route. Outline the reactions involved here.

$$\mathbf{G}\ (C_{12}H_{18}N_2O_5) \xrightarrow[\substack{68°\text{C, 1000 psi} \\ (90\%\ \text{yield})}]{H_2,\ \text{Ni}} \mathbf{H}\ (C_{10}H_{16}N_2O_4,\ \text{a } \delta\text{-lactam}) \xrightarrow[\substack{\text{reflux 4 h} \\ (97\%\ \text{yield})}]{\text{concd HCl}}$$

$$\text{DL-ornithine hydrochloride } (C_5H_{13}ClN_2O_2)$$

(L-Ornithine is a naturally occurring amino acid but does not occur in proteins. In one metabolic pathway L-ornithine serves as a precursor for L-arginine.)

24.21 Bradykinin is a nonapeptide released by blood plasma globulins in response to a wasp sting. It is a very potent pain-causing agent. Its molecular formula is Arg_2, Gly, Phe_2, Pro_3, Ser. The use of 2,4-dinitrofluorobenzene and carboxypeptidase shows that both terminal residues are arginine. Partial acid hydrolysis of bradykinin gives the following di- and tripeptides:

Phe · Ser + Pro · Gly · Phe + Pro · Pro + Ser · Pro · Phe + Phe · Arg + Arg · Pro

What is the amino acid sequence of bradykinin?

24.22 Complete hydrolysis of a heptapeptide showed that it had the following molecular formula:

Ala_2, Glu, Leu, Lys, Phe, Val

Deduce the amino acid sequence of this heptapeptide from the following data.

1. Treatment of the heptapeptide with 2,4-dinitrofluorobenzene followed by incomplete hydrolysis gave, among other products: valine labeled at the α-amino group, lysine labeled at the ε-amino group, and a dipeptide, DNP—Val · Leu (DNP = 2,4-dinitrophenyl-).
2. Hydrolysis of the heptapeptide with carboxypeptidase gave an initial high concentration of alanine, followed by a rising concentration of glutamic acid.

3. Partial enzymatic hydrolysis of the heptapeptide gave a dipeptide (**A**) and a tripeptide (**B**).
   a. Treatment of **A** with 2,4-dinitrofluorobenzene followed by hydrolysis gave DNP-labeled leucine and lysine labeled only at the ε-amino group.
   b. Complete hydrolysis of **B** gave phenylalanine, glutamic acid, and alanine. When **B** was allowed to react with carboxypeptidase, the solution showed an initial high concentration of glutamic acid. Treatment of **B** with 2,4-dinitrofluorobenzene followed by hydrolysis gave labeled phenylalanine.

24.23 Synthetic polyglutamic acid exists as an α helix in solution at pH 2–3. When the pH of such a solution is gradually raised through the addition of a base, a dramatic change in optical rotation takes place at pH 5. This change has been associated with the unfolding of the α helix and the formation of a random coil. What structural feature of polyglutamic acid and what chemical change can you suggest as an explanation of this transformation?

24.24 Part of the evidence for restricted rotation about the carbon–nitrogen bond in a peptide linkage (see Section 24.8A) comes from ^{1}H NMR studies done with simple amides. For example, at room temperature and with the instrument operating at 60 MHz, the ^{1}H NMR spectrum of N,N-dimethylformamide, $(CH_3)_2NCHO$, shows a doublet at δ 2.80 (3H), a doublet at δ 2.95 (3H), and a multiplet at δ 8.05 (1H). When the spectrum is determined at lower magnetic field strength (i.e., with the instrument operating at 30 MHz), the doublets are found to have shifted so that the distance (in hertz) that separates one doublet from the other is smaller. When the temperature at which the spectrum is determined is raised, the doublets persist until a temperature of 111°C is reached; then the doublets coalesce to become a single signal. Explain in detail how these observations are consistent with the existence of a relatively large barrier to rotation about the carbon–nitrogen bond of DMF.

***24.25** Given this sequence of reactions, studied at SRI International during the synthesis of compounds having possible anticancer activity, consider the questions below.

$HOOCCH(NH_2)(CH_2)_nS{-}S(CH_2)_nCH(NH_2)COOH$ (**A**) $\xrightarrow{(1)\ KNCO,\ (2)\ HCl}$ bis-hydantoin $(CH_2)_n{-}S{-}S{-}(H_2C)_n$ (**B**)

B $\xrightarrow{Cl_2\ \text{in aq. HOAc}}$

$n = 2$: hydantoin–$CH_2{-}CH_2{-}SO_2Cl$ (**C**) $\xrightarrow{HNR_2}$ hydantoin–$H_2C{-}CH_2{-}SO_2NR_2$ (**E**)

$n = 1$: hydantoin–$CH_2{-}SO_2Cl$ (**D**) $\xrightarrow{HNR_2}$ [5-methylene hydantoin, $=CH_2$] ?

(a) Write a mechanism for the conversion of one of the disulfides **A** (homocystine, if $n = 2$; cystine if $n = 1$) to its bis hydantoin derivative **B**.
(b) Write a mechanism by which **B,** $n = 2$, is converted to sulfonyl chloride **C.**
(c) Since sulfonyl chloride **C** could routinely be converted to sulfonamides of type **E,** suggest a mechanism that would explain why the result was very different when sulfonyl chloride **D** was used. (Assume that the product was the one shown, even though that was not proven.)

***24.26** When reacted with ethanol in pyridine, sulfonyl chloride **C** from above gave a product that had molecular formula $C_{12}H_{17}N_3O_5S$. What is its structure?

***24.27** The Japanese company Kaneka reported this dynamic resolution method for obtaining a pure enantiomer of an amino acid:

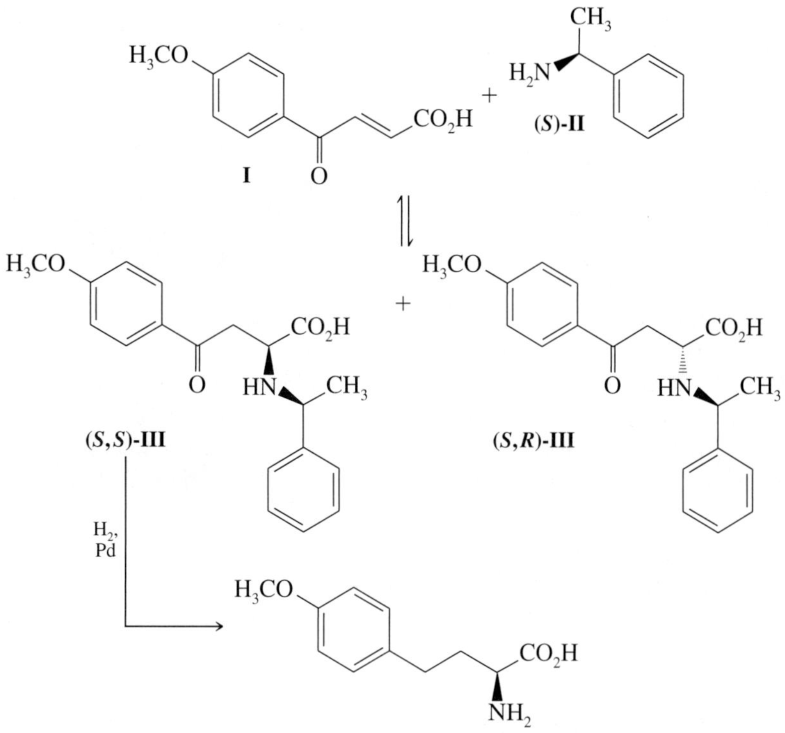

I + **(*S*)-II** ⇌ **(*S*,*S*)-III** + **(*S*,*R*)-III**

(*S*,*S*)-III $\xrightarrow{H_2, Pd}$ **(*S*)-*p*-Methoxyhomophenylalanine, IV**

In this example, the (*S,S*) form of the adduct **III** crystallizes from the above reaction in 90% yield and 97% enantiomeric excess.

(a) What explains the production of material in such high enantiomeric excess?

(b) What principle learned in one's first chemistry course explains getting such a high yield of one of the two possible products?

(c) What type of reaction is doubly illustrated in the conversion of (*S,S*)-**III** to (*S*)-**IV**?

LEARNING GROUP PROBLEMS

1. The enzyme lysozyme and its mechanism are described in Section 24.10. Using the information presented there (and perhaps with additional information from a biochemistry textbook), prepare notes for a class presentation on the mechanism of lysozyme. Consider especially the role of the enzyme in favoring formation of the carbocation intermediate, stabilizing the carbocation, and in providing or removing protons where necessary.
2. Chymotrypsin is a member of the serine protease class of enzymes. Its mechanism of action is described in Section 24.11. Using the information presented there (and perhaps supplemented by information from a biochemistry textbook), prepare notes for a class presentation on the mechanism of chymotrypsin. Consider especially the role of the "catalytic triad" with regard to acid–base catalysis and the relative propensity of various groups to act as nucleophiles or leaving groups.

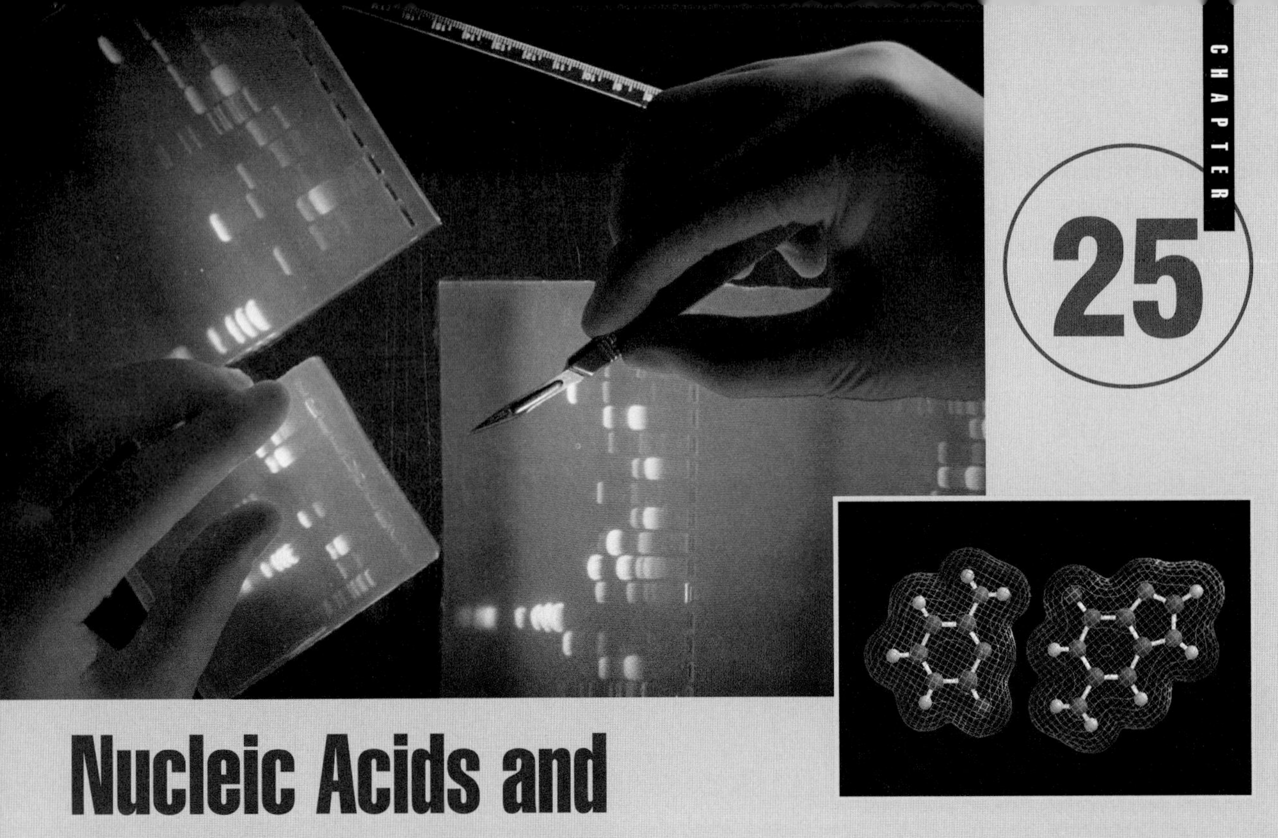

CHAPTER 25

Nucleic Acids and Protein Synthesis

Tools for Finding Families

Chemistry has long been called the central science—it is involved in every aspect of life. Much of what we have learned about chemistry is related to how things work, how diseases can be treated at the molecular level, and how materials we need in our daily lives can be improved or new ones created. Certainly not the least of chemistry's many applications, however, is an important dimension regarding work for global human rights and justice. As we are all too well aware, in many parts of the world there are situations where people have been separated from relatives because of the atrocious acts of war or terrorism. Some scientists are tracing the family connections left after these grievous events using modern tools of chemistry. Laboratories such as those of M.-C. King (University of Washington) are attempting to help families bring closure when only remains of suspected relatives have been found and to reunite people in cases where victims have survived and they or their families are searching for familial ties.

The key to this work is DNA—the chemical fingerprint present in every tissue of every individual. Although the general structure of DNA is the same from one person to another (see the molecular graphic above), evidence for familial ties is present in the detailed sequence of each person's DNA. With the use of relatively simple chemistry—involving fluorescent dyes or radioactive isotopes, enzymes from thermophilic bacteria and other sources, gel electrophoresis (see the photo on this page), and a process called the polymerase chain reaction (PCR) that earned its inventor the 1993 Chemistry Nobel Prize (Section 25.8)—it is now easy to synthesize millions of copies from a sample of DNA and to sequence it rapidly and conveniently. Application of these tools to comparison of DNA samples from victims and relatives provides hope that, at least in a few cases, the gap between family members will be closed.

25.1 INTRODUCTION

Deoxyribonucleic acid (DNA) and **ribonucleic acid (RNA)** are molecules that carry genetic information in cells. DNA is the molecular archive of instructions for protein synthesis. RNA molecules transcribe and translate the information from DNA for the mechanics of protein synthesis. The storage of genetic information, its passage from generation to generation, and the use of genetic information to create the working parts of the cell all depend on the molecular structures of DNA and RNA. For these reasons, we shall focus our attention on the structures and properties of these **nucleic acids** and of their components, nucleotides and nucleosides.

DNA is a biological polymer comprised of two molecular strands held together by hydrogen bonds. Its overall structure is that of a twisted ladder with a backbone of alternating sugar and phosphate units and rungs made of hydrogen-bonded pairs of heterocyclic amine bases (Fig. 25.1). DNA molecules are very long polymers. If the DNA from a single human cell were extracted and laid straight end-to-end, it would be roughly a meter long. To package DNA into the microscopic container of a cell's nucleus, however, it is supercoiled and bundled into the 23 chromosomes with which we are familiar from electron micrographs.

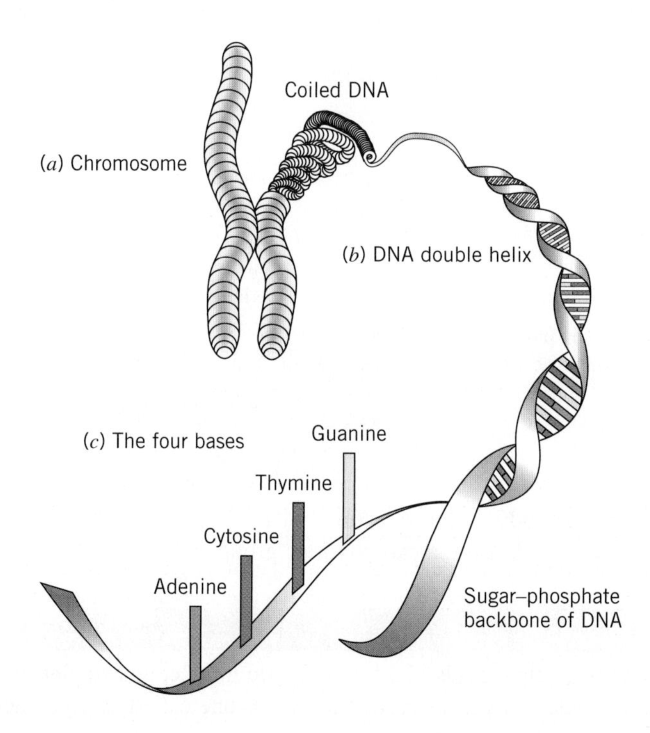

FIGURE 25.1 The basics of genetics. Each cell in the human body (except red blood cells) contains 23 pairs of chromosomes. Chromosomes are inherited: Each parent contributes one chromosome per pair to their children. *(a)* Each chromosome is made up of a tightly coiled strand of DNA. The structure of DNA in its uncoiled state reveals *(b)* the familiar double-helix shape. If we picture DNA as a twisted ladder, the sides, made of sugar and phosphate molecules, are connected by *(c)* rungs made of heterocyclic amine bases. DNA has four, and only four, bases—adenine (A), thymine (T), guanine (G), and cytosine (C)—that form interlocking pairs. The order of the bases along the length of the ladder is called the DNA sequence. Within the overall sequence are genes, which encode the structure of proteins.

Four types of heterocyclic bases are involved in the rungs of the DNA ladder, and it is the sequence of these bases that carries the information for protein synthesis. Human DNA consists of approximately 3 billion base pairs. In an effort that marks a milestone in the history of science, a working draft of the sequence of the 3 billion base pairs in the human genome was announced in 2000. A final version was announced in 2003, the 50th anniversary of the structure determination of DNA by Watson and Crick.

Each section of DNA that codes for a given protein is called a **gene.** The set of all genetic information coded by DNA in an organism is its **genome.** There are approximately 30,000–35,000 genes in the human genome. The set of all proteins encoded within the genome of an organism and expressed at any given time is called its **proteome** (Section 24.14). Some scientists estimate there could be up to one million different proteins in the cells of our various tissues—a number much greater than the number of genes in the genome due to gene splicing during protein expression and post-translational protein modification.

Hopes are very high that, having sequenced the human genome, knowledge of it will bring increased identification of genes related to disease states (Fig. 25.2) and study of these genes and the proteins encoded by them will yield a myriad of benefits for human health and longevity. Determining the structure of all of the proteins encoded in the genome, learning their functions, and creating molecular therapeutics based on this rapidly expanding store of knowledge are some of the key research challenges that lie ahead.

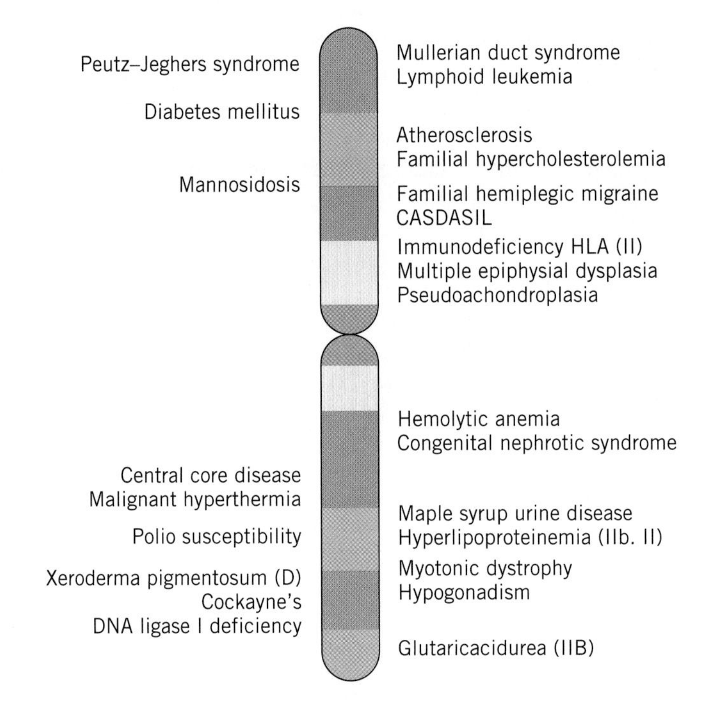

FIGURE 25.2 A schematic map of the location of genes for diseases on chromosome 19.

Let us begin with a study of the structures of nucleic acids. Each of their monomer units contains a cyclic amine base, a carbohydrate group, and a phosphate ester.

25.2 Nucleotides and Nucleosides

Mild degradations of nucleic acids yield monomeric units called **nucleotides.** A general formula for a nucleotide and the specific structure of one called adenylic acid are shown in Fig. 25.3.

Complete hydrolysis of a nucleotide furnishes:

1. A heterocyclic base from either the purine or pyrimidine family.
2. A five-carbon monosaccharide that is either D-ribose or 2-deoxy-D-ribose.
3. A phosphate ion.

The central portion of the nucleotide is the monosaccharide, and it is always present as a five-membered ring, that is, as a furanoside. The heterocyclic base of a nucleotide is attached through an *N*-glycosidic linkage to C1′ of the ribose or deoxyribose unit, and this linkage is always β. The phosphate group of a nucleotide is present as a phosphate ester

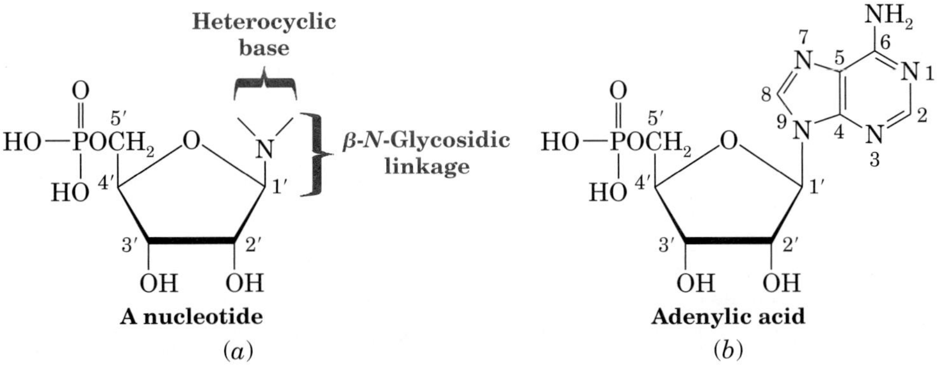

FIGURE 25.3 *(a)* General structure of a nucleotide obtained from RNA. The heterocyclic base is a purine or pyrimidine. *In nucleotides obtained from DNA, the sugar component is 2′-deoxy-D-ribose; that is, the —OH at position 2′ is replaced by —H.* The phosphate group of the nucleotide is shown attached at C5′; it may instead be attached at C3′. In DNA and RNA a phosphodiester linkage joins C5′ of one nucleotide to C3′ of another. The heterocyclic base is always attached through a *β*-*N*-glycosidic linkage at C1′. *(b)* Adenylic acid, a typical nucleotide.

and may be attached at C5′ or C3′. (In nucleotides, the carbon atoms of the monosaccharide portion are designated with primed numbers, i.e., 1′, 2′, 3′, etc.)

Removal of the phosphate group of a nucleotide converts it to a compound known as a **nucleoside** (Section 22.15A). The nucleosides that can be obtained from DNA all contain 2-deoxy-D-ribose as their sugar component and one of four heterocyclic bases, adenine, guanine, cytosine, or thymine:

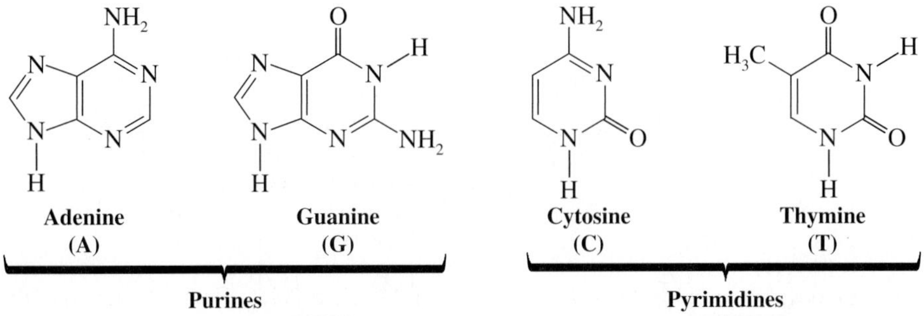

The nucleosides obtained from RNA contain D-ribose as their sugar component and adenine, guanine, cytosine, or uracil as their heterocyclic base.

Uracil replaces thymine in an RNA nucleoside (or nucleotide). (Some nucleosides obtained from specialized forms of RNA may also contain other, but similar, purines and pyrimidines.)

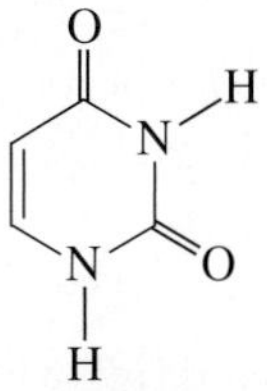

The heterocyclic bases obtained from nucleosides are capable of existing in more than one tautomeric form. The forms that we have shown are the predominant forms that the bases assume when they are present in nucleic acids.

The names and structures of the nucleosides found in DNA are shown in Fig. 25.4; those found in RNA are given in Fig. 25.5.

FIGURE 25.4 Nucleosides that can be obtained from DNA. DNA is 2′-deoxy at the position where the blue hydrogen atoms are shown. RNA has hydroxyl groups at that location. RNA has a hydrogen where there is a methyl group in thymine, which makes the base uracil (and the nucleoside uridine).

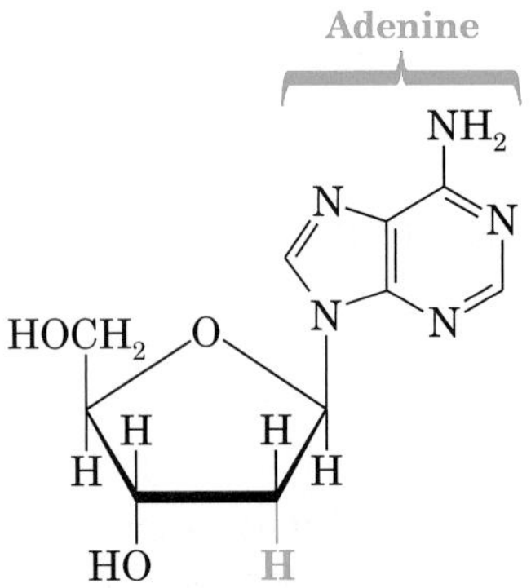

2′-Deoxyadenosine

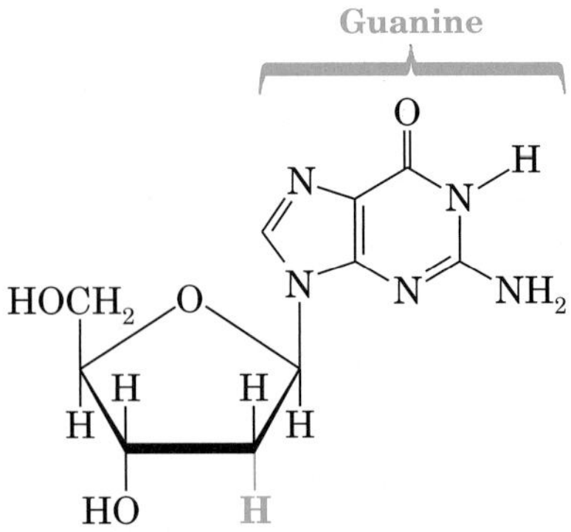

2′-Deoxyguanosine

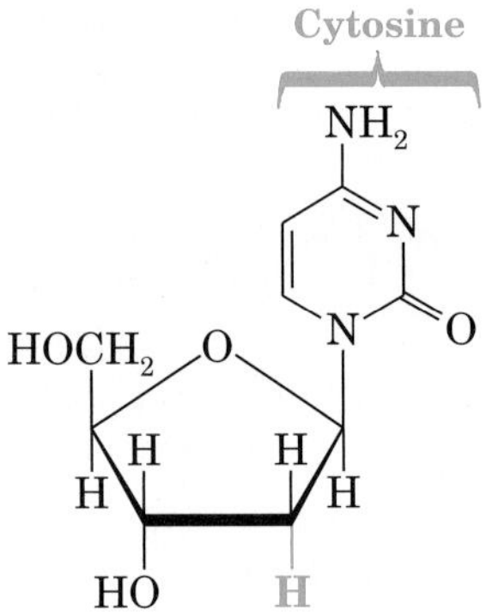

2′-Deoxycytidine

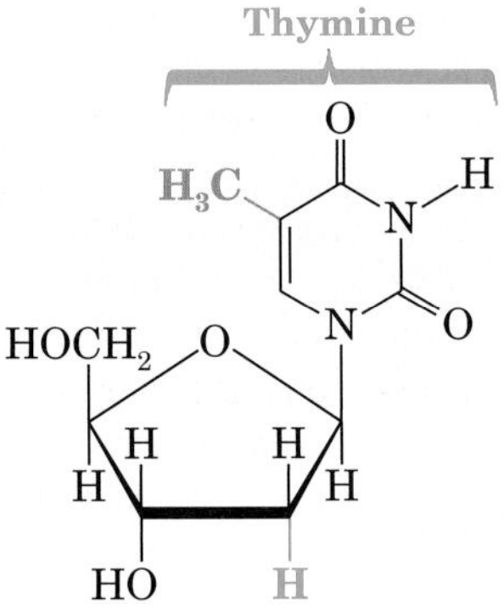

2′-Deoxythymidine

FIGURE 25.5 Nucleosides that can be obtained from RNA. DNA has hydrogen atoms where the red hydroxyl groups of ribose are shown (DNA is 2′-deoxy with respect to its ribose moiety).

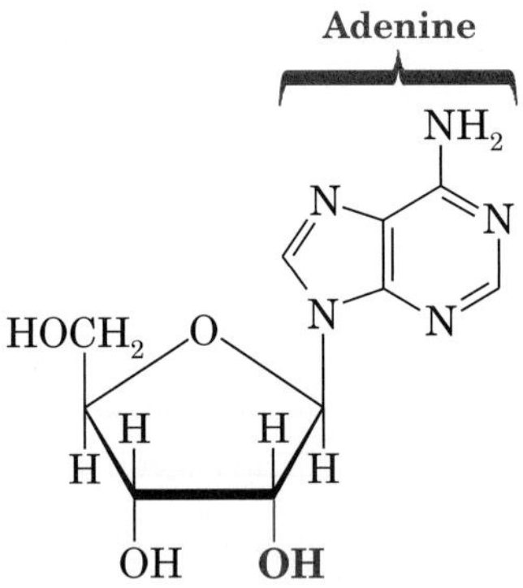

Adenosine

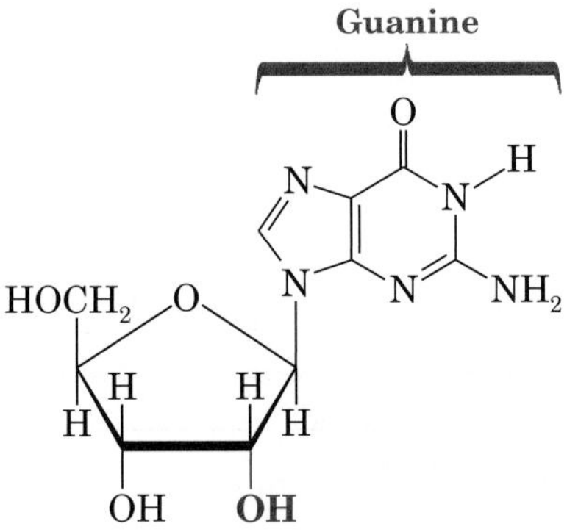

Guanosine

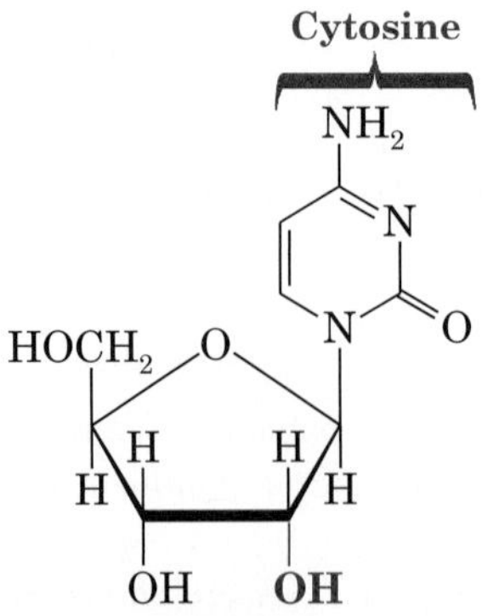

Cytidine

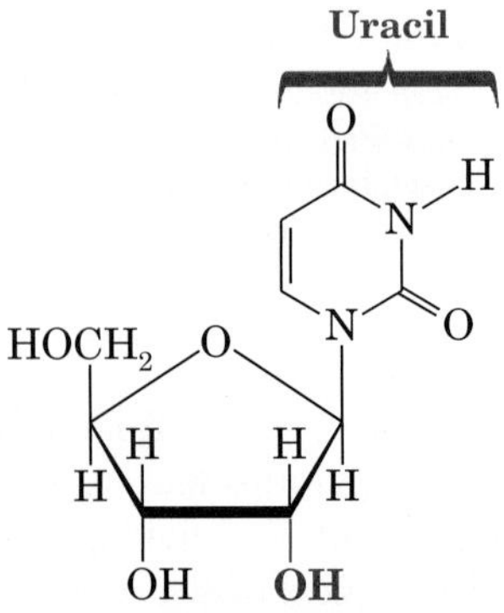

Uridine

PROBLEM 25.1

Write the structures of other tautomeric forms of adenine, guanine, cytosine, thymine, and uracil.

PROBLEM 25.2

The nucleosides shown in Figs. 25.4 and 25.5 are stable in dilute base. In dilute acid, however, they undergo rapid hydrolysis yielding a sugar (deoxyribose or ribose) and a heterocyclic base.
(a) What structural feature of the nucleoside accounts for this behavior?
(b) Propose a reasonable mechanism for the hydrolysis.

Nucleotides are named in several ways. Adenylic acid (Fig. 25.3), for example, is usually called AMP, for adenosine monophosphate. The position of the phosphate group is sometimes explicitly noted by use of the names adenosine 5′-monophosphate or 5′-adenylic acid. Uridylic acid is usually called UMP, for uridine monophosphate, although it can also be called uridine 5′-monophosphate or 5′-uridylic acid. If a nucleotide is present as a diphosphate or triphosphate, the names are adjusted accordingly, such as ADP for adenosine diphosphate or GTP for guanosine triphosphate

Nucleosides and nucleotides are found in places other than as part of the structure of DNA and RNA. We have seen, for example, that adenosine units are part of the structures of two important coenzymes, NADH and coenzyme A. The 5′-triphosphate of adenosine is, of course, the important energy source, ATP (Section 22.1B). The compound called 3′, 5′-cyclic adenylic acid (or cyclic AMP) (Fig. 25.6) is an important regulator of hormone activity. Cells synthesize this compound from ATP through the action of an enzyme, *adenylate cyclase.* In the laboratory, 3′,5′-cyclic adenylic acid can be prepared through dehydration of 5′-adenylic acid with dicyclohexylcarbodiimide.

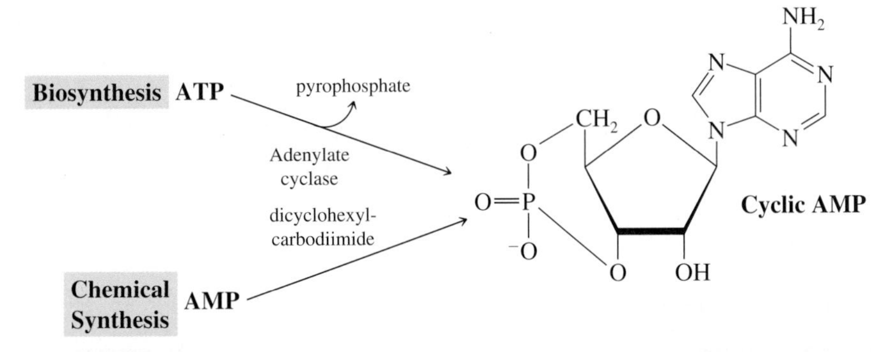

FIGURE 25.6 3′,5′-Cyclic adenylic acid (cyclic AMP) and its biosynthesis and laboratory synthesis.

PROBLEM 25.3

When 3′,5′-cyclic adenylic acid is treated with aqueous sodium hydroxide, the major product that is obtained is 3′-adenylic acid (adenosine 3′-phosphate) rather than 5′-adenylic acid. Suggest an explanation that accounts for the course of this reaction.

25.3 LABORATORY SYNTHESIS OF NUCLEOSIDES AND NUCLEOTIDES

A variety of methods have been developed for the chemical synthesis of nucleosides from the constituent sugars and bases or their precursors. The following is an example of a *Silyl–Hilbert–Johnson nucleosidation,* where a benzoyl protected sugar (D-ribose) reacts in the presence of tin chloride with an *N*-benzoyl protected base (cytidine) that is protected further by *in situ* silylation.* The trimethylsilyl protecting groups for the base are introduced using *N,O*-bis(trimethylsilyl)acetamide (BSA) and they are removed with aqueous acid in the second step. The result is a protected form of the nucleoside cytosine, from which the benzoyl groups can be removed with ease using a base:

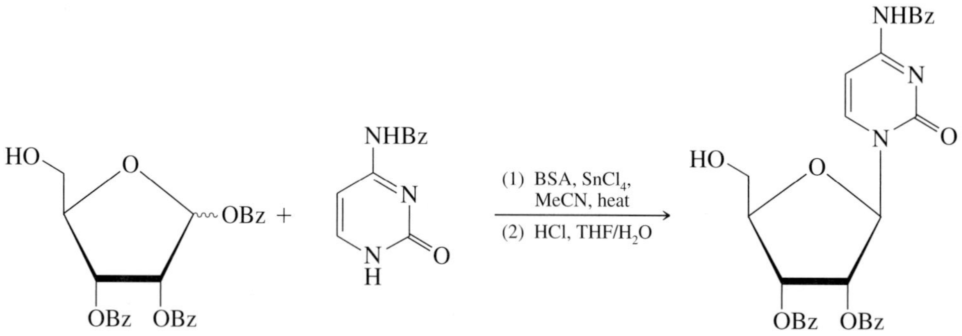

D-Ribose (Bz protected) + **N^4-Benzoylcytosine** —(1) BSA, $SnCl_4$, MeCN, heat; (2) HCl, THF/H_2O→ **Cytidine (Bz protected)**

Bz = C_6H_5CO (benzoyl)

BSA = $CH_3C[{=}NSi(CH_3)_3]OSi(CH_3)_3$,
[*N,O*-bis(trimethylsilyl)acetamide, a reagent for trimethylsilylation of nucleophilic oxygen and nitrogen atoms]

Another technique involves formation of the heterocyclic base on a protected ribosylamine derivative:

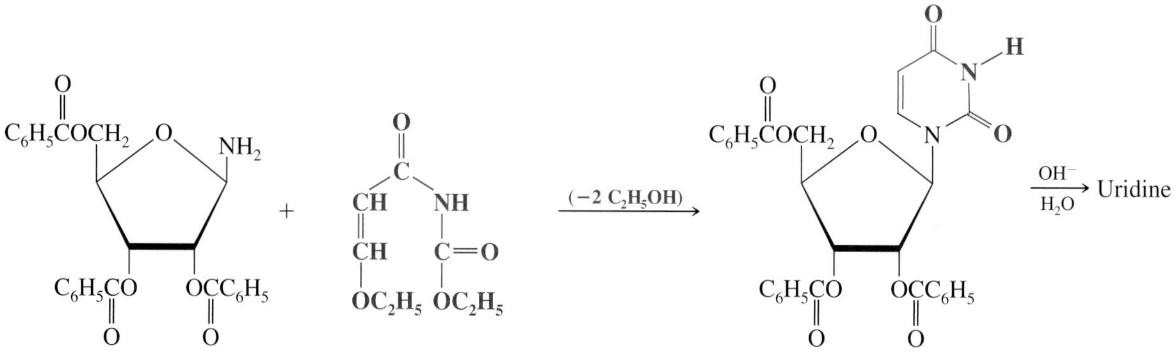

2,3,5-Tri-*O*-benzoyl-β-D-ribofuranosylamine + **β-Ethoxy-*N*-ethoxycarbonylacrylamide** —($-2\ C_2H_5OH$)→ (tri-O-benzoyl uridine) —OH^-, H_2O→ Uridine

PROBLEM 25.4 Basing your answer on reactions that you have seen before, propose a likely mechanism for the condensation reaction in the first step of the preceding uridine synthesis.

Still a third technique involves the synthesis of a nucleoside with a substituent in the heterocyclic ring that can be replaced with other groups. This method has been used extensively to synthesize unusual nucleosides that do not necessarily occur naturally. The following example makes use of a 6-chloropurine derivative obtained from the appropriate ribofuranosyl chloride and chloromercuripurine:

*These conditions were applied using L-ribose in a synthesis of the unnatural enantiomer of RNA (Pitsch, S. An Efficient Synthesis of Enantiomeric Ribonucleic Acids from D-Glucose. *Helv. Chim. Acta* **1997,** *80,* 2286–2314). The protected enantiomeric cytidine was produced in 94% yield by the above reaction. After adjusting protecting groups, solid-phase oligonucleotide synthesis methods (Section 25.7) were used with this compound and the other three nucleotide monomers (also derived from L-ribose) for preparation of the unnatural RNA enantiomer. See also Vorbrüggen, H.; Ruh-Pohlenz, C. *Handbook of Nucleoside Synthesis;* Wiley: Hoboken, NJ, 2001.

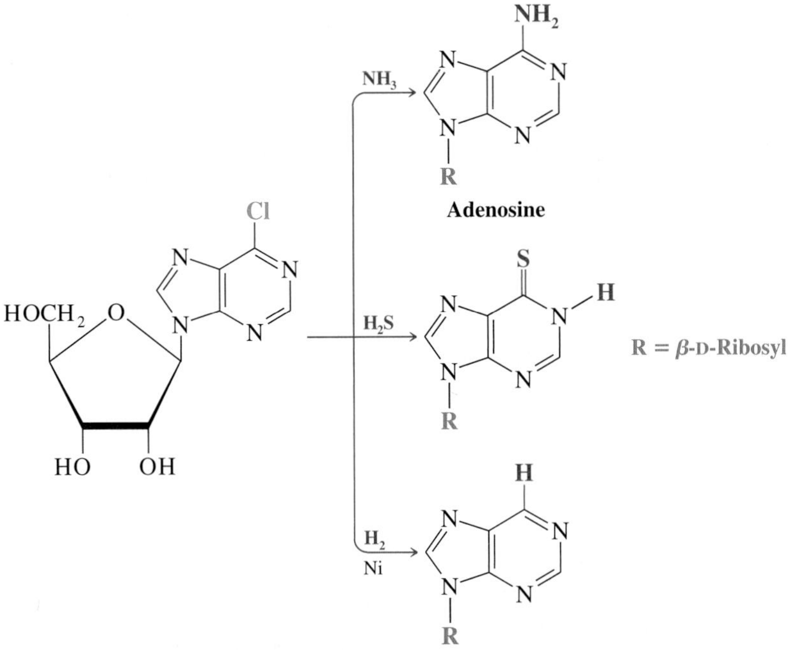

Adenosine

R = β-D-Ribosyl

Numerous phosphorylating agents have been used to convert nucleosides to nucleotides. One of the most useful is dibenzyl phosphochloridate:

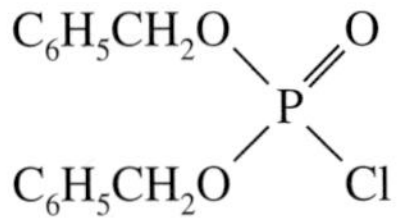

Dibenzyl phosphochloridate

Specific phosphorylation of the 5′-OH can be achieved if the 2′- and 3′-OH groups of the nucleoside are protected by an isopropylidene (acetonide) group (see the following):

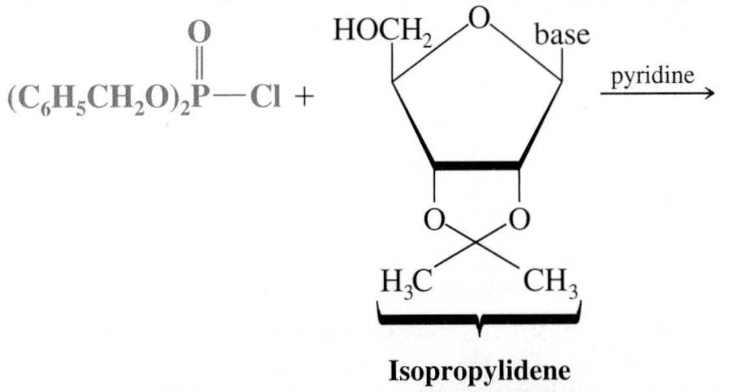

Isopropylidene protecting group

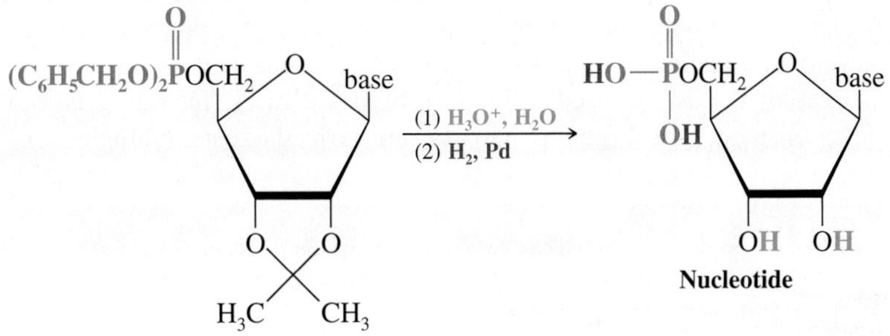

Nucleotide

Mild acid-catalyzed hydrolysis removes the isopropylidene group, and hydrogenolysis cleaves the benzyl phosphate bonds.

PROBLEM 25.5

(a) What kind of linkage is involved in the isopropylidene group of the protected nucleoside, and why is it susceptible to mild acid-catalyzed hydrolysis? **(b)** How might such a protecting group be installed?

PROBLEM 25.6

The following reaction scheme is from a synthesis of cordycepin (a nucleoside antibiotic) and the first synthesis of 2′-deoxyadenosine (reported in 1958 by C. D. Anderson, L. Goodman, and B. R. Baker, Stanford Research Institute):

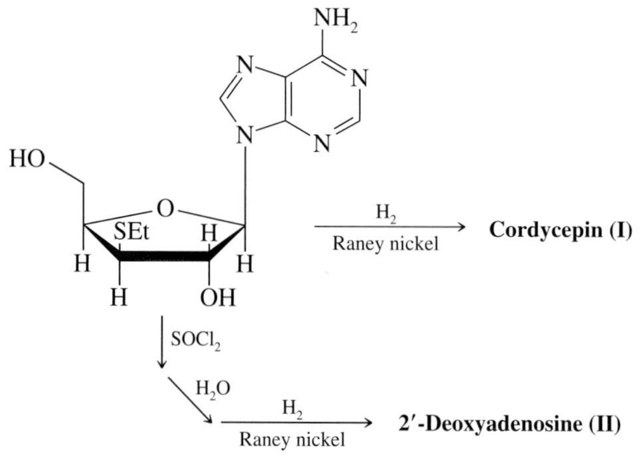

$\xrightarrow[\text{Raney nickel}]{H_2}$ **Cordycepin (I)**

$\downarrow SOCl_2$; H_2O $\xrightarrow[\text{Raney nickel}]{H_2}$ **2′-Deoxyadenosine (II)**

(a) What is the structure of cordycepin? (**I** and **II** are isomers.)
(b) Propose a mechanism that explains the formation of **II**.

25.3A Medical Applications

In the early 1950s, Gertrude Elion and George Hitchings (of the Wellcome Research Laboratories) discovered that 6-mercaptopurine had antitumor and antileukemic properties. This discovery led to the development of other purine derivatives and related compounds, including nucleosides, of considerable medical importance. Three examples are the following:

Elion and Hitchings shared the 1988 Nobel Prize in Physiology or Medicine for their work in the development of chemotherapeutic agents derived from purines.

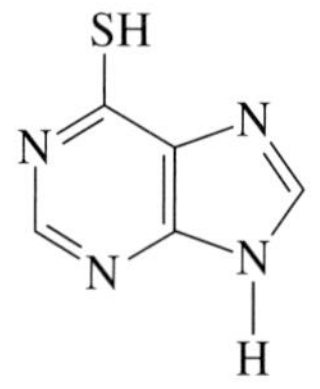

6-Mercaptopurine 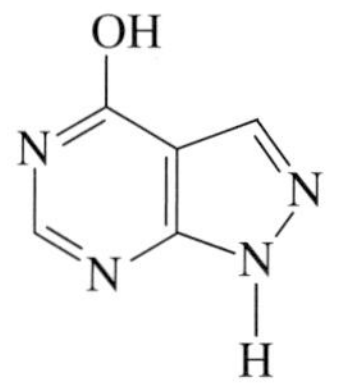 **Allopurinol** 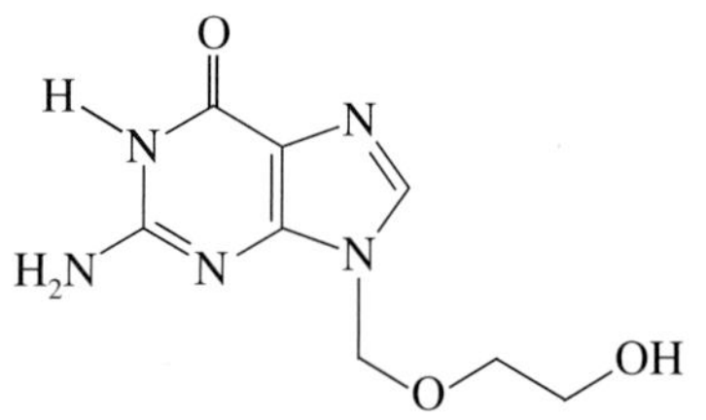 **Acyclovir**

6-Mercaptopurine is used in combination with other chemotherapeutic agents to treat acute leukemia in children, and almost 80% of the children treated are now cured. Allopurinol, another purine derivative, is a standard therapy for the treatment of gout. Acyclovir, a nucleoside that lacks two carbon atoms of its ribose ring, is highly effective in treating diseases caused by certain herpes viruses, including *herpes simplex* type 1 (fever blisters), type 2 (genital herpes), and varicella-zoster (shingles).

25.4 Deoxyribonucleic Acid: DNA

25.4A Primary Structure

Nucleotides bear the same relation to a nucleic acid that amino acids do to a protein; they are its monomeric units. The connecting links in proteins are amide groups; in nucleic acids they are phosphate ester linkages. Phosphate esters link the 3′-OH of one ribose (or deoxyribose) with the 5′-OH of another. This makes the nucleic acid a long unbranched chain with a "backbone" of sugar and phosphate units with heterocyclic bases protruding from the chain at regular intervals (Fig. 25.7). We would indicate the direction of the bases in Fig. 25.7 in the following way:

$$5' \longleftarrow \text{A—T—G—C} \longrightarrow 3'$$

It is, as we shall see, the **base sequence** along the chain of DNA that contains the encoded genetic information. The sequence of bases can be determined using enzymatic methods and chromatography (Section 25.6).

FIGURE 25.7 A segment of one DNA chain showing how phosphate ester groups link the 3′- and 5′-OH groups of deoxyribose units. RNA has a similar structure with two exceptions: A hydroxyl replaces a hydrogen atom at the 2′ position of each ribose unit and uracil replaces thymine.

25.4B Secondary Structure

It was the now-classic proposal of Watson and Crick (made in 1953 and verified shortly thereafter through the X-ray analysis by Wilkins) that gave a model for the secondary structure of DNA. This work earned Crick, Watson, and Wilkins the 1962 Nobel Prize in

I cannot help wondering whether some day an enthusiastic scientist will christen his newborn twins Adenine and Thymine.
F. H. C. Crick*

* Taken from Crick, F. H. C. The Structure of the Hereditary Material. *Sci. Am.* **1954,** *191*(10), 20, 54–61.

Physiology or Medicine. The secondary structure of DNA is especially important because it enables us to understand how the genetic information is preserved, how it can be passed on during the process of cell division, and how it can be transcribed to provide a template for protein synthesis.

Of prime importance to Watson and Crick's proposal was an earlier observation (made in the late 1940s) by E. Chargaff that certain regularities can be seen in the percentages of heterocyclic bases obtained from the DNA of a variety of species. Table 25.1 gives results that are typical of those that can be obtained.

TABLE 25.1 DNA Composition of Various Species

Species	Base Proportions (mol %)							
	G	A	C	T	$\frac{G+A}{C+T}$	$\frac{A+T}{G+C}$	$\frac{A}{T}$	$\frac{G}{C}$
Sarcina lutea	37.1	13.4	37.1	12.4	1.02	0.35	1.08	1.00
Escherichia coli K12	24.9	26.0	25.2	23.9	1.08	1.00	1.09	0.99
Wheat germ	22.7	27.3	22.8[a]	27.1	1.00	1.19	1.01	1.00
Bovine thymus	21.5	28.2	22.5[a]	27.8	0.96	1.27	1.01	0.96
Staphylococcus aureus	21.0	30.8	19.0	29.2	1.11	1.50	1.05	1.11
Human thymus	19.9	30.9	19.8	29.4	1.01	1.52	1.05	1.01
Human liver	19.5	30.3	19.9	30.3	0.98	1.54	1.00	0.98

[a]Cytosine + methylcytosine.
Source: Smith, E. L.; Hill, R. L.; Lehman, I. R.; Lefkowitz, R. J.; Handler, P.; White, A. *Principles of Biochemistry: General Aspects,* 7th ed.; McGraw-Hill: New York, 1983; p 132.

Chargaff pointed out that for all species examined:

1. The total mole percentage of purines is approximately equal to that of the pyrimidines, that is, $(\%G + \%A)/(\%C + \%T) \cong 1$.

2. The mole percentage of adenine is nearly equal to that of thymine (i.e., $\%A/\%T \cong 1$), and the mole percentage of guanine is nearly equal to that of cytosine (i.e., $\%G/\%C \cong 1$).

Chargaff also noted that the ratio which varies from species to species is the ratio $(\%A + \%T)/(\%G + \%C)$. He noted, moreover, that whereas this ratio is characteristic of the DNA of a given species, it is the same for DNA obtained from different tissues of the same animal and does not vary appreciably with the age or conditions of growth of individual organisms within the same species.

Watson and Crick also had X-ray data that gave them the bond lengths and angles of the purine and pyrimidine rings of model compounds. In addition, they had data from Wilkins that indicated an unusually long repeat distance (34 Å) in natural DNA.

Reasoning from these data, Watson and Crick proposed a double helix as a model for the secondary structure of DNA. According to this model, two nucleic acid chains are held together by hydrogen bonds between base pairs on opposite strands. This double chain is wound into a helix with both chains sharing the same axis. The base pairs are on the inside of the helix, and the sugar–phosphate backbone is on the outside (Fig. 25.8). The pitch of the helix is such that 10 successive nucleotide pairs give rise to one complete turn in 34 Å (the repeat distance). The exterior width of the spiral is about 20 Å, and the internal distance between 1′ positions of ribose units on opposite chains is about 11 Å.

Models were critical to Watson and Crick in their Nobel Prize–winning work on the three-dimensional structure of DNA.

Using molecular-scale models, Watson and Crick observed that the internal distance of the double helix is such that it allows only a purine–pyrimidine type of hydrogen bonding between base pairs. Purine–purine base pairs do not occur because they would be too

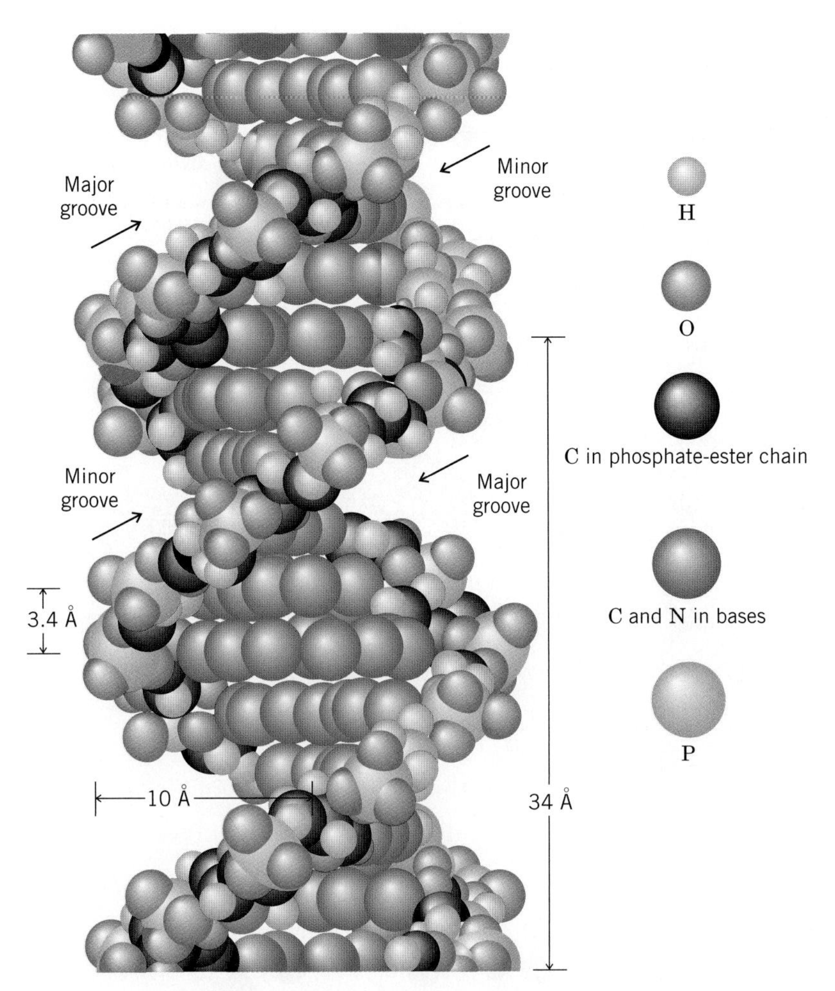

FIGURE 25.8 A molecular model of a portion of the DNA double helix. (Adapted from Neal, A. L. *Chemistry and Biochemistry: A Comprehensive Introduction;* McGraw-Hill: New York, 1971. Used with permission of McGraw-Hill Book Company, New York.)

large to fit, and pyrimidine–pyrimidine base pairs do not occur because they would be too far apart to form effective hydrogen bonds.

Watson and Crick went one crucial step further in their proposal. Assuming that the oxygen-containing heterocyclic bases existed in keto forms, they argued that base pairing through hydrogen bonds can occur in only a specific way: adenine (A) with thymine (T) and cytosine (C) with guanine (G). Dimensions of the pairs and electrostatic potential maps for the individual bases are shown in Figure 25.9.

Adenine pairs with thymine and **Guanine pairs with cytosine**

Thymine **Adenine** **Cytosine** **Guanine**

Specific base pairing of this kind is consistent with Chargaff's finding that %A/%T ≅ 1 and %G/%C ≅ 1.

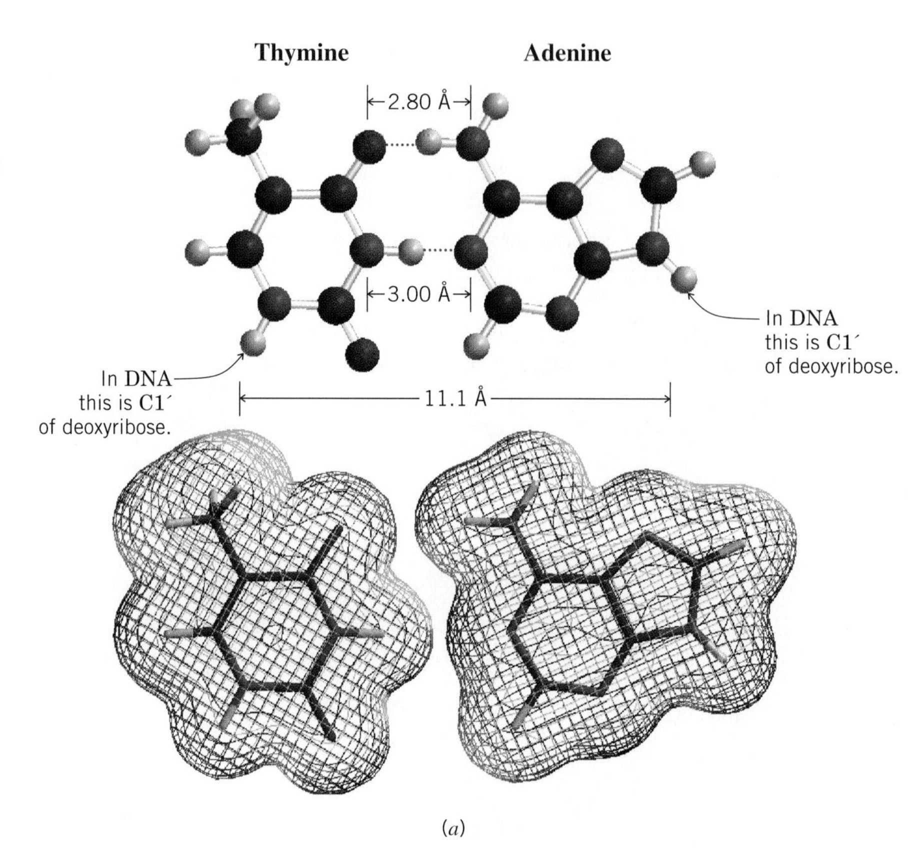

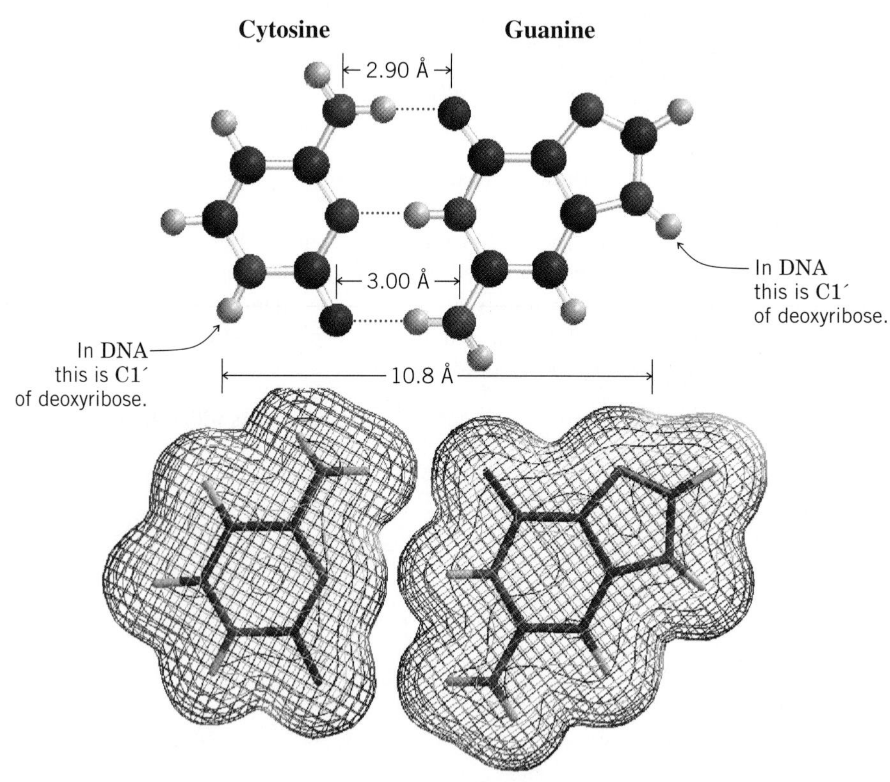

FIGURE 25.9 Base pairing of adenine with thymine *(a)* and cytosine with guanine *(b)*. The dimensions of the thymine–adenine and cytosine–guanine hydrogen-bonded pairs are such that they allow the formation of strong hydrogen bonds and also allow the base pairs to fit inside the two phosphate–ribose chains of the double helix. (Based on Pauling, L.; Corey, R. B. *Arch. Biochem. Biophys.* **1956,** *65,* 164–181.) Electrostatic potential maps calculated for the individual bases show the complementary distribution of charges that leads to hydrogen bonding.

CD Animated Graphics

Specific base pairing also means that the two chains of DNA are complementary. Wherever adenine appears in one chain, thymine must appear opposite it in the other; wherever cytosine appears in one chain, guanine must appear in the other (Fig. 25.10).

Notice that while the sugar–phosphate backbone of DNA is completely regular, the sequence of heterocyclic base pairs along the backbone can assume many different permutations. This is important because it is the precise sequence of base pairs that carries the genetic information. Notice, too, that one chain of the double strand is the complement of the other. If one knows the sequence of bases along one chain, one can write down the sequence along the other, because A always pairs with T and G always pairs with C. It is this complementarity of the two strands that explains how a DNA molecule replicates itself at the time of cell division and thereby passes on the genetic information to each of the two daughter cells.

CD Tutorial
DNA structure

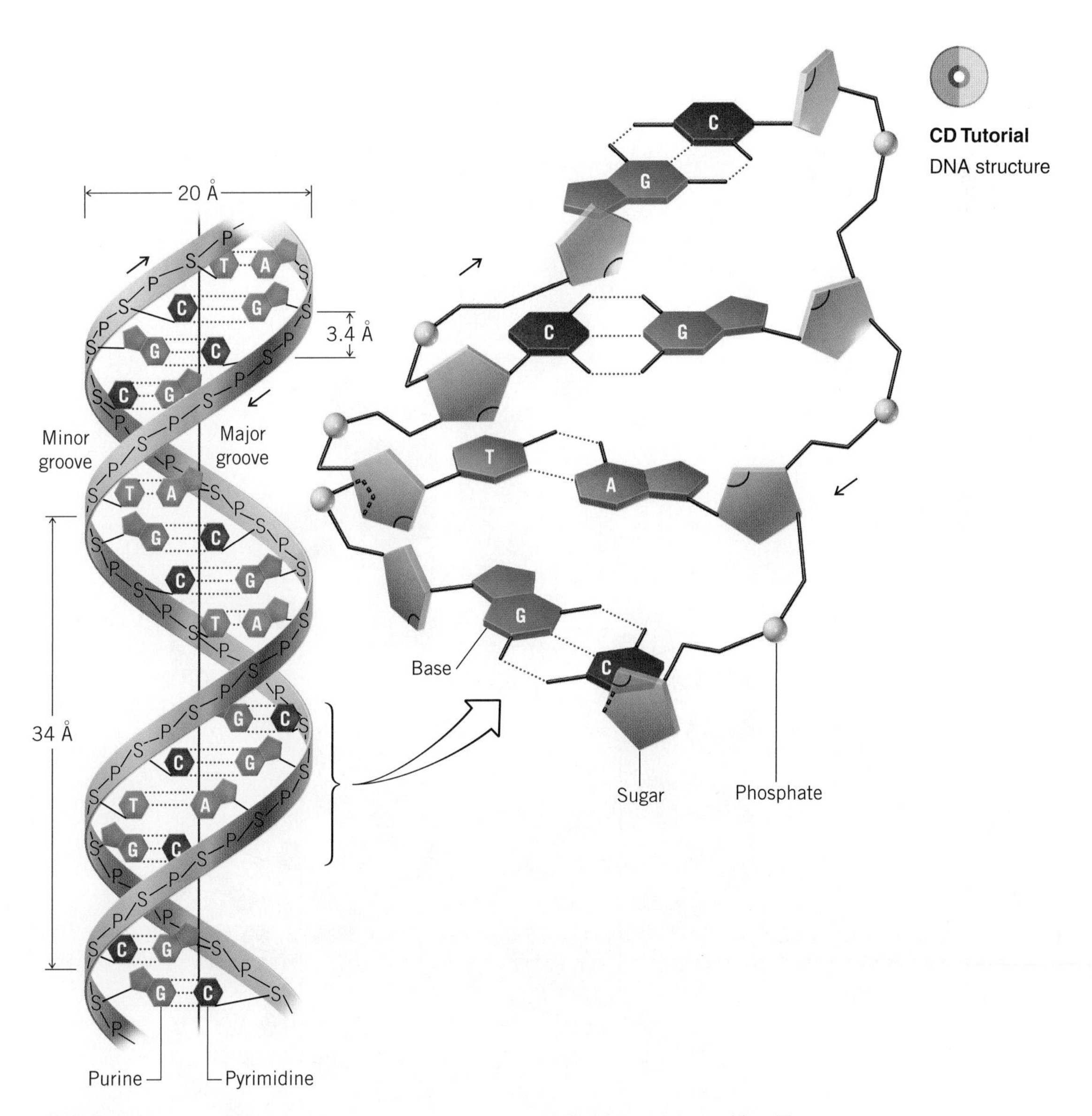

FIGURE 25.10 Diagram of the DNA double helix showing complementary base pairing. The arrows indicate the 3′ → 5′ direction.

25.4C Replication of DNA

Just prior to cell division the double strand of DNA begins to unwind. Complementary strands are formed along each chain (Fig. 25.11). Each chain acts, in effect, as a template for the formation of its complement. When unwinding and duplication are complete, there are two identical DNA molecules where only one had existed before. These two molecules can then be passed on, one to each daughter cell.

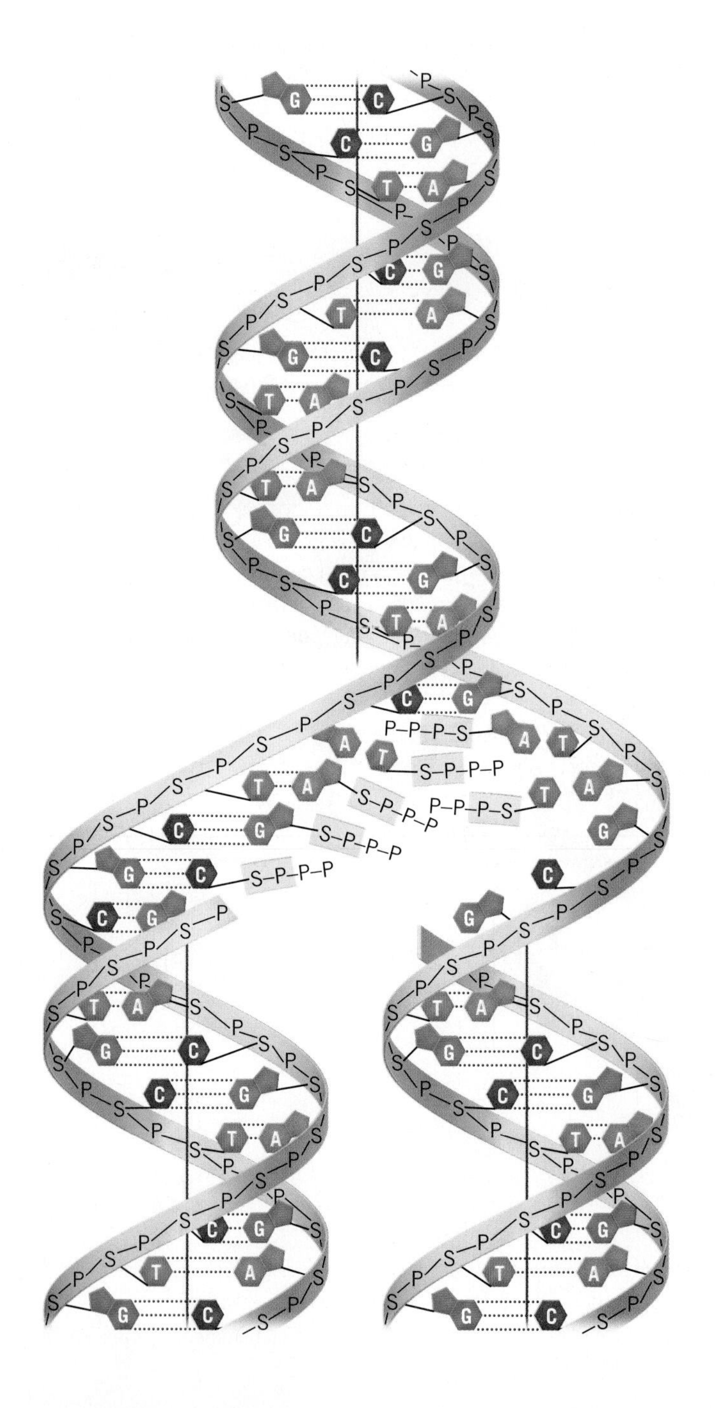

FIGURE 25.11 Replication of DNA. The double strand unwinds from one end and complementary strands are formed along each chain.

PROBLEM 25.7

(a) There are approximately 3 billion base pairs in the DNA of a single human cell. Assuming that this DNA exists as a double helix, calculate the length of all the DNA contained in a human cell. **(b)** The weight of DNA in a single human cell is 6×10^{-12} g. Assuming that Earth's population is about 5 billion, we can conclude that all of the genetic information that gave rise to all human beings now alive was once contained in the DNA of a corresponding number of fertilized ova. What is the total weight of this DNA? (The volume that this DNA would occupy is approximately that of a raindrop, yet if the individual molecules were laid end-to-end, they would stretch to the moon and back almost eight times.)

PROBLEM 25.8

(a) The most stable tautomeric form of guanine is the lactam form (or cyclic amide, see Section 18.8I). This is the form normally present in DNA, and, as we have seen, it pairs specifically with cytosine. If guanine tautomerizes (see Section 17.2) to the abnormal lactim form, it pairs with thymine instead. Write structural formulas showing the hydrogen bonds in this abnormal base pair.

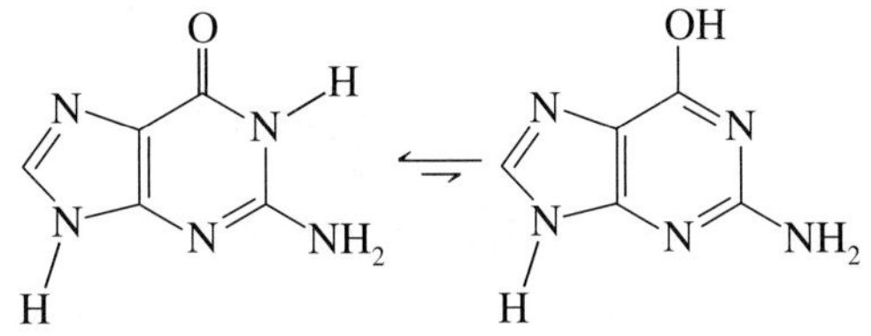

O H N N N N NH2 H ⇌ OH N N N N NH2 H

Lactam form of guanine **Lactim form of guanine**

(b) Improper base pairings that result from tautomerizations occurring during the process of DNA replication have been suggested as a source of spontaneous mutations. We saw in part (a) that if a tautomerization of guanine occurred at the proper moment, it could lead to the introduction of thymine (instead of cytosine) into its complementary DNA chain. What error would this new DNA chain introduce into *its* complementary strand during the next replication even if no further tautomerizations take place?

PROBLEM 25.9

Mutations can also be caused chemically, and nitrous acid is one of the most potent chemical **mutagens.** One explanation that has been suggested for the mutagenic effect of nitrous acid is the deamination reactions that it causes with purines and pyrimidines bearing amino groups. When, for example, an adenine-containing nucleotide is treated with nitrous acid, it is converted to a hypoxanthine derivative:

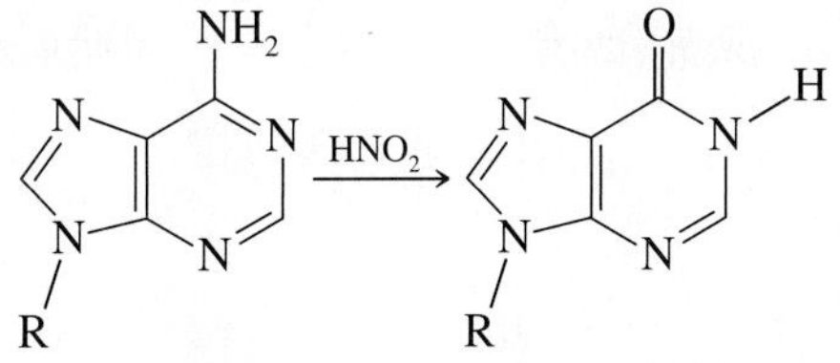

NH2 N N N N R $\xrightarrow{HNO_2}$ O H N N N N R

Adenine nucleotide **Hypoxanthine nucleotide**

(a) Basing your answer on reactions you have seen before, what are likely intermediates in the adenine $\rightarrow$ hypoxanthine interconversion? **(b)** Adenine normally pairs with thymine in DNA, but hypoxanthine pairs with cytosine. Show the hydrogen bonds of a hypoxanthine–cytosine base pair. **(c)** Show what errors an adenine $\rightarrow$ hypoxanthine interconversion would generate in DNA through two replications.

25.5 RNA AND PROTEIN SYNTHESIS

Soon after the Watson–Crick hypothesis was published, scientists began to extend it to yield what Crick called "the central dogma of molecular genetics." This dogma stated that genetic information flows as follows:

$$\text{DNA} \longrightarrow \text{RNA} \longrightarrow \text{protein}$$

The synthesis of protein is, of course, all important to a cell's function because proteins (as enzymes) catalyze its reactions. Even the very primitive cells of bacteria require as many as 3000 different enzymes. This means that the DNA molecules of these cells must contain a corresponding number of genes to direct the synthesis of these proteins. A **gene** is that segment of the DNA molecule that contains the information necessary to direct the synthesis of one protein (or one polypeptide).

DNA is found primarily in the nucleus of eukaryotic cells. Protein synthesis takes place primarily in that part of the cell called the *cytoplasm.* Protein synthesis requires that two major processes take place; the first occurs in the cell nucleus, the second in the cytoplasm. The first is **transcription,** a process in which the genetic message is transcribed onto a form of RNA called messenger RNA (mRNA). The second process involves two other forms of RNA, called ribosomal RNA (rRNA) and transfer RNA (tRNA).

There are viruses, called retroviruses, in which information flows from RNA to DNA. The virus that causes AIDS is a retrovirus.

25.5A Messenger RNA Synthesis—Transcription

The events leading to protein synthesis begin in the cell nucleus with the synthesis of mRNA. Part of the DNA double helix unwinds sufficiently to expose on a single chain a portion corresponding to at least one gene. Ribonucleotides, present in the cell nucleus, assemble along the exposed DNA chain by pairing with the bases of DNA. The pairing patterns are the same as those in DNA with the exception that in RNA uracil replaces thymine. The ribonucleotide units of mRNA are joined into a chain by an enzyme called *RNA polymerase.* This process is illustrated in Fig. 25.12.

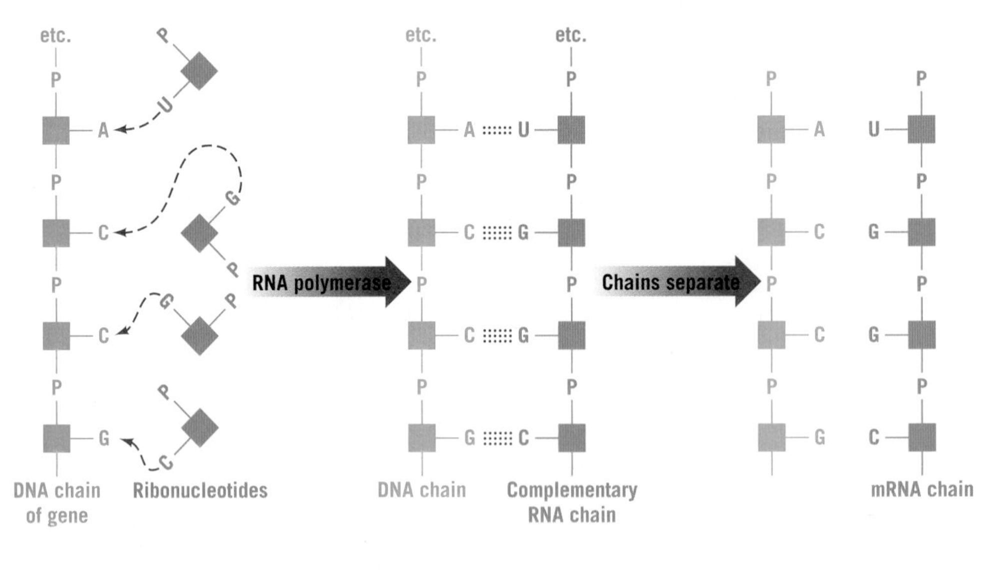

FIGURE 25.12 Transcription of the genetic code from DNA to mRNA.

Write structural formulas showing how the keto form of uracil (Section 25.2) in mRNA can pair with adenine in DNA through hydrogen bond formation.

PROBLEM 25.10

Most eukaryotic genes contain segments of DNA that are not actually used when a protein is expressed, even though they are transcribed into the initial mRNA. These segments are called **introns,** or intervening sequences. The segments of DNA within a gene that are expressed are called **exons,** or expressed sequences. Each gene usually contains a number of introns and exons. After the mRNA is transcribed from DNA, the introns in the mRNA are removed and the exons are spliced together.

After mRNA has been synthesized and processed in the cell nucleus to remove the introns, it migrates into the cytoplasm where, as we shall see, it acts as a template for protein synthesis.

25.5B Ribosomes—rRNA

Protein synthesis is catalyzed by ribosomes in the cytoplasm. Ribosomes (Fig. 25.13) are ribonucleoproteins, comprised of approximately two-thirds RNA and one-third protein. They have a very high molecular weight (about 2.6×10^6). The RNA component is present in two subunits, called the 50S and 30S subunits (classified according to their sedimentation behavior during ultracentrifugation*). The 50S subunit is roughly twice the molecular weight of the 30S subunit. Binding of RNA with mRNA is mediated by the 30S subunit. The 50S subunit carries the catalytic activity for translation that joins one amino acid by an amide bond to the next. In addition to the rRNA subunits there are approximately 30–35 proteins tightly bound to the ribosome, the entire structure resem-

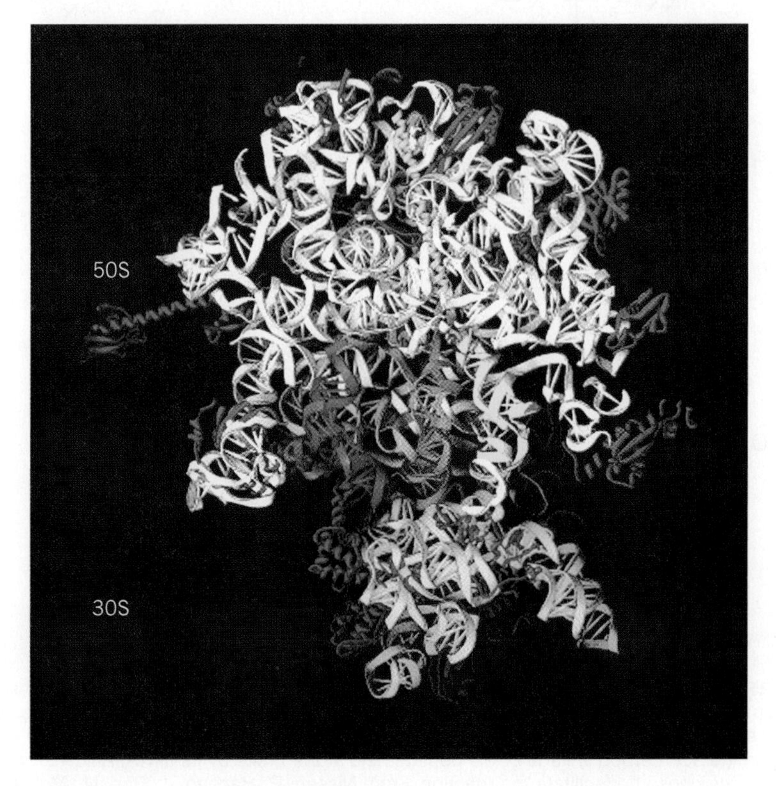

FIGURE 25.13 Structure of the *Thermus thermophilus* ribosome showing the 50S and 30S subunits and three bound transfer RNAs. The yellow tRNA is at the A-site, which would bear the new amino acid to be added to the peptide. The light orange tRNA is at the P-site, which would be the tRNA that bears the growing peptide. The red tRNA is at the E-site, which is the 'empty' tRNA after it has transferred the peptide chain to the new amino acid. Graphic courtesy of Harry F. Noller, University of California, Santa Cruz.

*S stands for svedberg unit; it is used in describing the behavior of proteins in an ultracentrifuge.

bling an exquisite three-dimensional jigsaw puzzle of RNA and protein. The mechanism for ribosome-catalyzed amide bond formation is discussed below.

Ribosomes, as reaction catalysts, are most appropriately classified as ribozymes rather than enzymes, because it is RNA that catalyzes the peptide bond formation during protein synthesis and not the protein subunits of the ribosome. The mechanism for peptide bond formation catalyzed by the 50S ribosome subunit (Fig. 25.14), proposed by Moore and co-workers based on X-ray crystal structures, suggests that attack by the α-amino group is facilitated by acid–base catalysis involving nucleotide residues along the 50S ribosome subunit chain, specifically a nearby adenine group. Full or partial removal of a proton from the α-amino group of the amino acid by N3 of the adenine group imparts greater nucleophilicity to the amino nitrogen, facilitating its attack on the acyl carbon of the adjacent peptide–tRNA moiety. A tetrahedral intermediate is formed, which collapses to form the new amide bond with release of the tRNA that had been joined to the peptide. Other moieties in the 50S ribosome subunit are believed to help stabilize the transfer of charge that occurs as N3 of the adenyl group accepts the proton from the α-amino group of the new amino acid (see Problem 25.16).

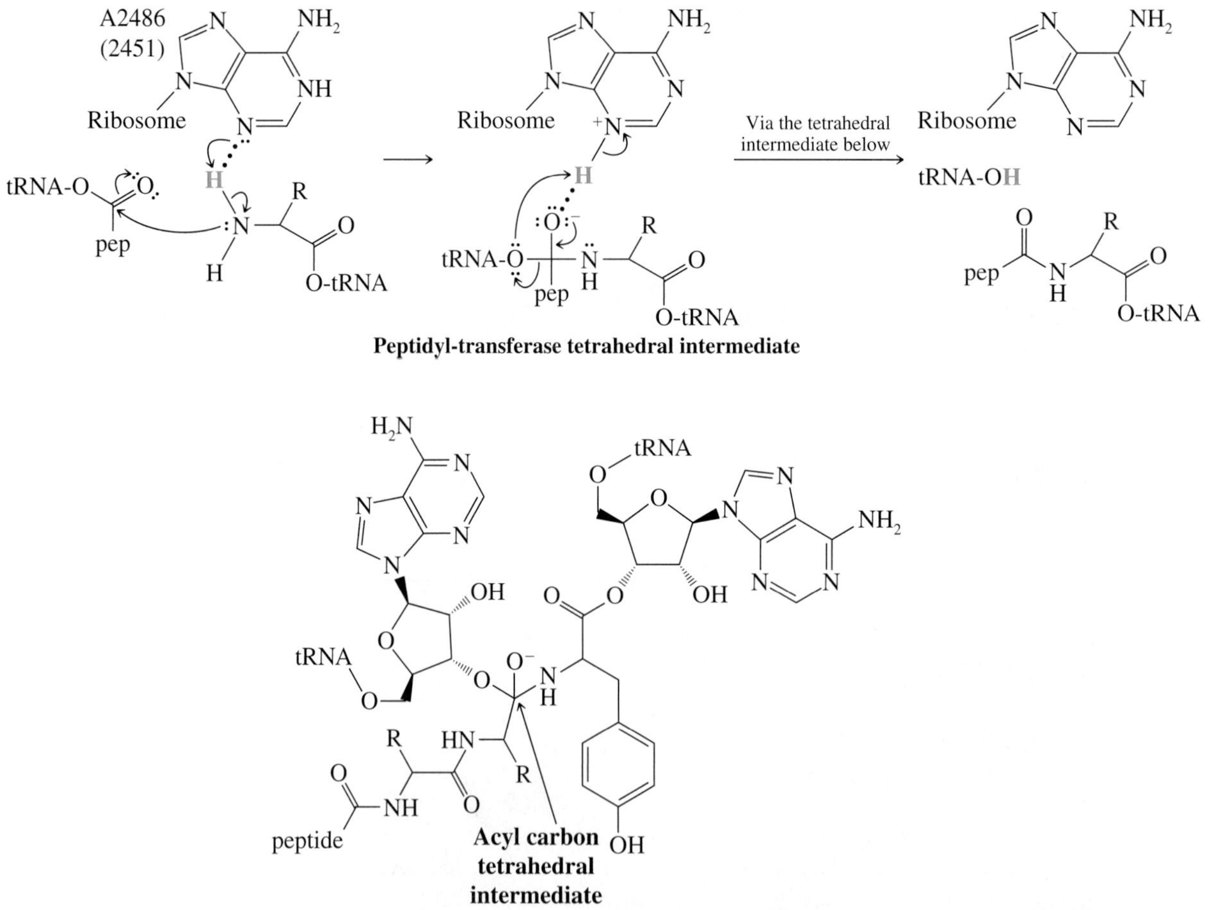

A2486
(2451)
Ribosome
tRNA-O
pep
R
O
O-tRNA
Ribosome
tRNA-O
pep
R
O
O-tRNA
Peptidyl-transferase tetrahedral intermediate
Via the tetrahedral intermediate below
Ribosome
tRNA-OH
pep
R
O-tRNA

tRNA
tRNA
peptide
Acyl carbon tetrahedral intermediate

FIGURE 25.14 A mechanism for peptide bond formation catalyzed by the 50S subunit of the ribosome (as proposed by Moore and co-workers). The new amide bond in the growing peptide chain is formed by attack of the α-amino group in the new amino acid, brought to the A site of the ribosome by its tRNA, on the acyl carbon linkage of the peptide held at the P site by its tRNA. Acid–base catalysis by groups in the ribosome facilitate the reaction. From Nissen, P.; Hansen, J.; Ban, N.; Moore, P. B.; Steitz, T. A. The Structural Basis of Ribosome Activity in Peptide Bond Synthesis, *Science* **2000,** *289,* 920–930. Also Moore, P. *Biochemistry* **2001,** *40*(11), 3243–3250.

25.5C Transfer RNA

Transfer RNA has a very low molecular weight when compared to those of mRNA and rRNA. Transfer RNA, consequently, is much more soluble than mRNA or rRNA and is sometimes referred to as soluble RNA. The function of tRNA is to transport amino acids to specific areas on the mRNA bound to the ribosome. There are, therefore, many forms of tRNA, more than one for each of the 20 amino acids that are incorporated into proteins, including the redundancies in the **genetic code** (see Table 25.2).*

TABLE 25.2 The Messenger RNA Genetic Code

Amino Acid	Base Sequence 5′ → 3′	Amino Acid	Base Sequence 5′ → 3′	Amino Acid	Base Sequence 5′ → 3′
Ala	GCA	His	CAC	Ser	AGC
	GCC		CAU		AGU
	GCG	Ile	AUA		UCA
	GCU		AUC		UCG
Arg	AGA		AUU		UCC
	AGG	Leu	CUA		UCU
	CGA		CUC	Thr	ACA
	CGC		CUG		ACC
	CGG		CUU		ACG
	CGU		UUA		ACU
Asn	AAC		UUG	Trp	UGG
	AAU	Lys	AAA	Tyr	UGG
Asp	GAC		AAG		UAC
	GAU	Met	AUG		UAU
Cys	UGC	Phe	UUU	Val	GUA
	UGU		UUC		GUG
Gln	CAA	Pro	CCA		GUC
	CAG		CCC		GUU
Glu	GAA		CCG	Chain initiation	
	GAG		CCU	fMet (*N*-formyl-methionine)	AUG
Gly	GGA			Chain termination	UAA
	GGC				UAG
	GGG				UGA
	GGU				

The structures of most tRNAs have been determined. They are composed of a relatively small number of nucleotide units (70–90 units) folded into several loops or arms through base pairing along the chain (Fig. 25.15). One arm always terminates in the sequence cytosine–cytosine–adenine (CCA). It is to this arm that a specific amino acid becomes attached *through an ester* linkage to the 3′-OH of the terminal adenosine. This attachment reaction is catalyzed by an enzyme that is specific for the tRNA and for the amino acid. The specificity may grow out of the enzyme's ability to recognize base sequences along other arms of the tRNA.

*Although proteins are composed of 22 different amino acids, protein synthesis requires only 20. Proline is converted to hydroxyproline and cysteine is converted to cystine after synthesis of the polypeptide chain has taken place.

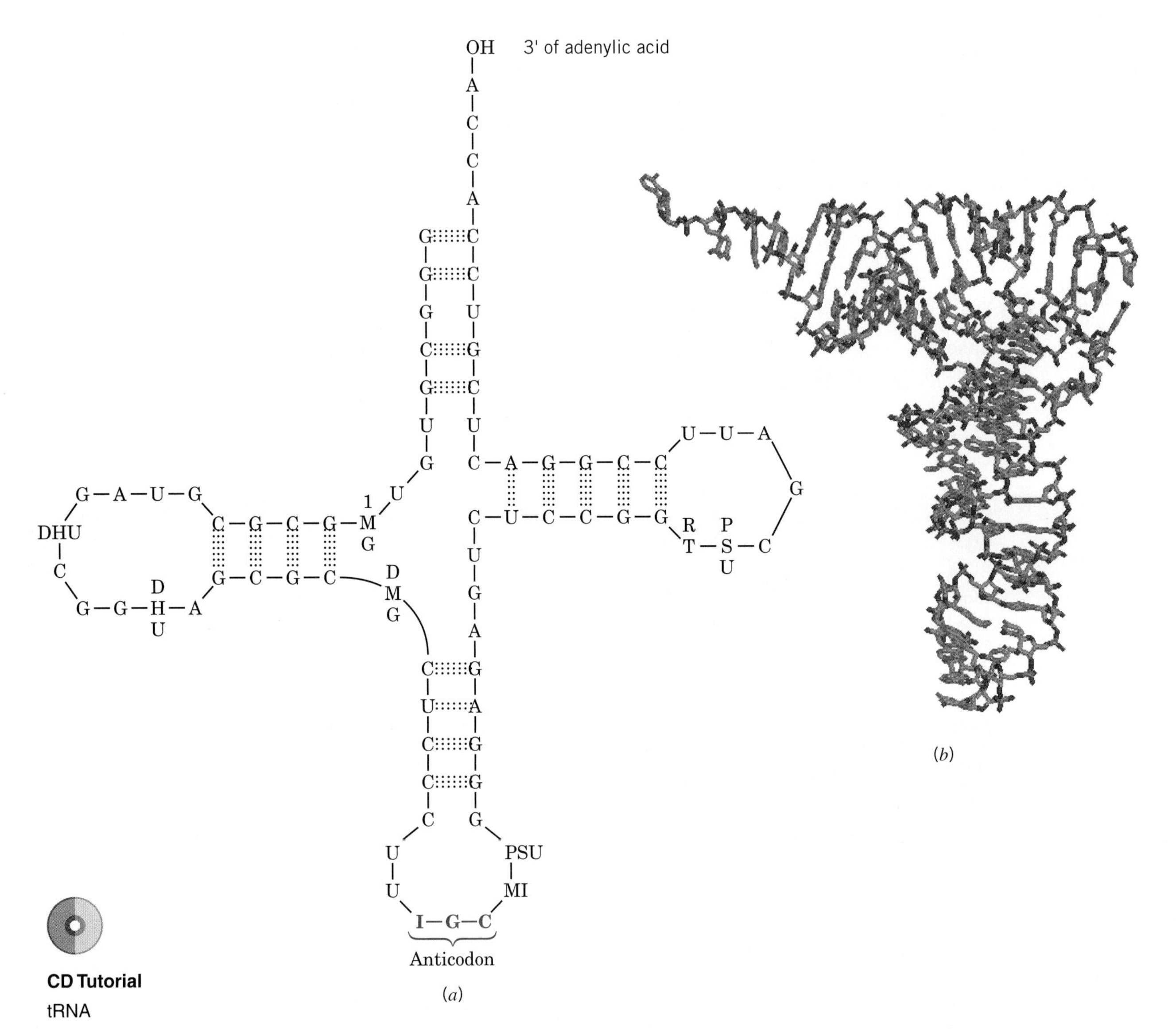

CD Tutorial
tRNA

FIGURE 25.15 *(a)* Structure of a tRNA isolated from yeast that has the specific function of transferring alanine residues. Transfer RNAs often contain unusual nucleosides. PSU = pseudouridine, RT = ribothymidine, MI = 1-methylinosine, I = inosine, DMG = N^2-methylguanosine, DHU = 4,5-dihydrouridine, 1MG = 1-methylguanosine. *(b)* The X-ray crystal structure of a phenylalanine tRNA from yeast (Hingerty, B. E.; Brown, R. S.; Jack, A. *J. Mol. Biol.* **1978,** *124,* 523; Protein Data Bank file name 4TNA.pdb).

At the loop of still another arm is a specific sequence of bases, called the **anticodon.** The anticodon is highly important because it allows the tRNA to bind with a specific site—called the **codon**—of mRNA. The order in which amino acids are brought by their tRNA units to the mRNA strand is determined by the sequence of codons. This sequence, therefore, constitutes a genetic message. Individual units of that message (the individual words, each corresponding to an amino acid) are triplets of nucleotides.

25.5D The Genetic Code

Which triplet on mRNA corresponds to which amino acid is called the genetic code (see Table 25.2). The code must be in the form of three bases, not one or two, because there are 20 different amino acids used in protein synthesis but there are only four different bases in mRNA. If only two bases were used, there would be only 4^2, or 16, possible combinations, a number too small to accommodate all of the possible amino acids. However, with a three-base code, 4^3, or 64, different sequences are possible. This is far more than are needed, and it allows for multiple ways of specifying an amino acid. It also allows for sequences that punctuate protein synthesis, sequences that say, in effect, "start here" and "end here."

Both methionine (Met) and *N*-formylmethionine (fMet) have the same mRNA code (AUG); however, *N*-formylmethionine is carried by a different tRNA from that which carries methionine. *N*-Formylmethionine appears to be the first amino acid incorporated into the chain of proteins in bacteria, and the tRNA that carries fMet appears to be the punctuation mark that says "start here." Before the polypeptide synthesis is complete, *N*-formylmethionine is removed from the protein chain by an enzymatic hydrolysis.

$$CH_3SCH_2CH_2CH(NH{-}CHO)CO_2H$$

N-Formylmethionine (fMet)

25.5E Translation

We are now in a position to see how the synthesis of a hypothetical polypeptide might take place. This process is called **translation.** Let us imagine that a long strand of mRNA is in the cytoplasm of a cell and that it is in contact with ribosomes. Also in the cytoplasm are the 20 different amino acids, each acylated to its own specific tRNA.

As shown in Fig. 25.16, a tRNA bearing fMet uses its anticodon to associate with the proper codon (AUG) on that portion of mRNA that is in contact with a ribosome. The next triplet of bases on the mRNA chain in this figure is AAA; this is the codon that specifies lysine. A lysyl-tRNA with the matching anticodon UUU attaches itself to this site. The two amino acids, fMet and Lys, are now in the proper position for the 50S ribosome subunit to catalyze the formation of an amide bond between them, as shown in Fig. 25.13 (by the mechanism in Fig. 25.14). After this happens, the ribosome moves down the chain so that it is in contact with the next codon. This one, GUA, specifies valine. A tRNA bearing valine (and with the proper anticodon) binds itself to this site. Another peptide bond-forming reaction takes place attaching valine to the polypeptide chain. Then the whole process repeats itself again and again. The ribosome moves along the mRNA chain, other tRNAs move up with their amino acids, new peptide bonds are formed, and the polypeptide chain grows. At some point an enzymatic reaction removes fMet from the beginning of the chain. Finally, when the chain is the proper length, the ribosome reaches a punctuation mark, UAA, saying "stop here." The ribosome separates from the mRNA chain and so, too, does the protein.

Even before the polypeptide chain is fully grown, it begins to form its own specific secondary and tertiary structure. This happens because its primary structure is correct—its amino acids are ordered in just the right way. Hydrogen bonds form, giving rise to specific segments of α helix, pleated sheet, and coil or loop. Then the whole chain folds and bends; enzymes install disulfide linkages, so that when the chain is fully grown, the whole protein has just the shape it needs to do its job. (Predicting 2° and 3° protein structure from amino acid sequence, however, remains a critical problem in structural biochemistry.)

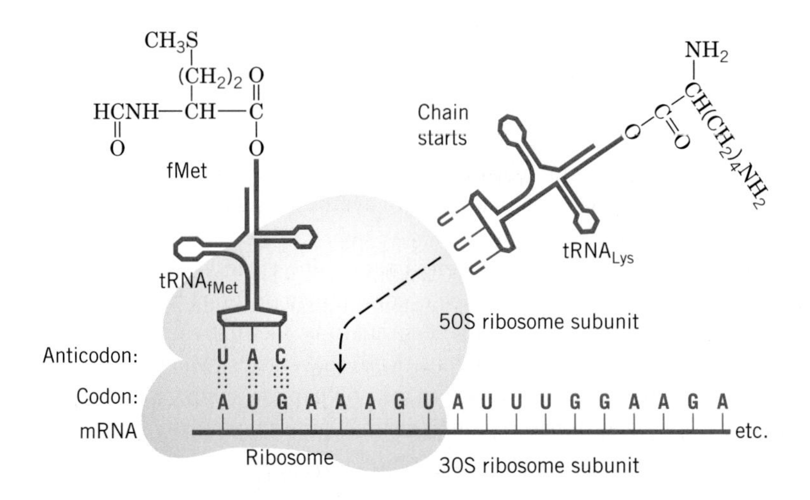

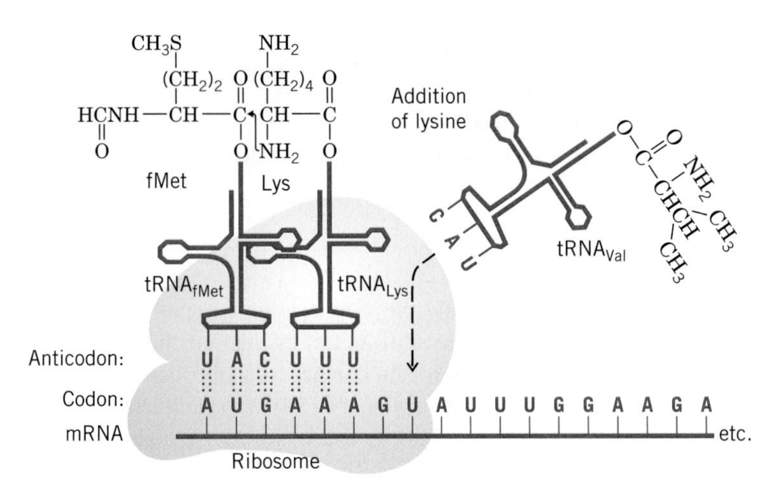

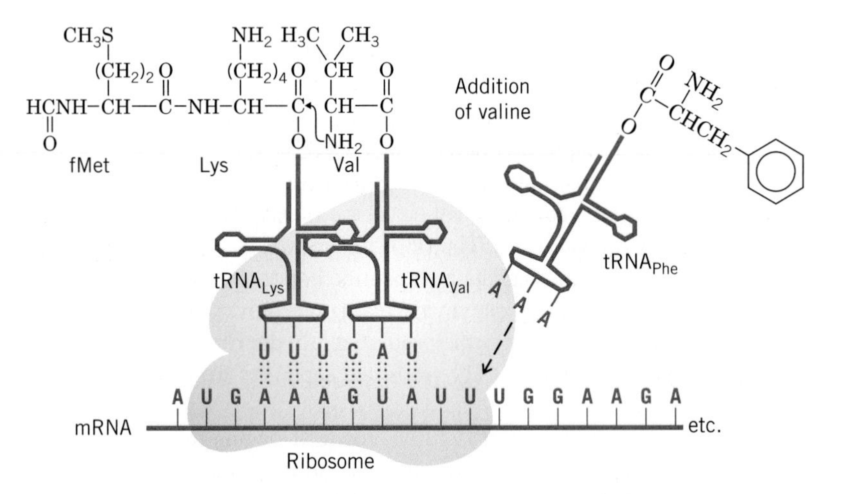

FIGURE 25.16 Step-by-step growth of a polypeptide chain with mRNA acting as a template. Transfer RNAs carry amino acid residues to the site of mRNA that is in contact with a ribosome. Codon–anticodon pairing occurs between mRNA and RNA at the ribosomal surface. An enzymatic reaction joins the amino acid residues through an amide linkage. After the first amide bond is formed, the ribosome moves to the next codon on mRNA. A new tRNA arrives, pairs, and transfers its amino acid residue to the growing peptide chain, and so on.

In the meantime, other ribosomes nearer the beginning of the mRNA chain are already moving along, each one synthesizing another molecule of the polypeptide. The time required to synthesize a protein depends, of course, on the number of amino acid residues it contains, but indications are that each ribosome can cause 150 peptide bonds to be formed each minute. Thus, a protein, such as lysozyme, with 129 amino acid residues requires less than a minute for its synthesis. However, if four ribosomes are working their way along a single mRNA chain, a protein molecule can be produced every 13 s.

But why, we might ask, is all this protein synthesis necessary—particularly in a fully grown organism? The answer is that proteins are not permanent; they are not synthesized once and then left intact in the cell for the lifetime of the organism. They are synthesized when and where they are needed. Then they are taken apart, back to amino acids; enzymes disassemble enzymes. Some amino acids are metabolized for energy; others—new ones—come in from the food that is eaten, and the whole process begins again.

PROBLEM 25.11

A segment of DNA has the following sequence of bases:

. . . A C C C C C A A A A T G T C G . . .

(a) What sequence of bases would appear in mRNA transcribed from this segment? **(b)** Assume that the first base in this mRNA is the beginning of a codon. What order of amino acids would be translated into a polypeptide synthesized along this segment? **(c)** Give anticodons for each tRNA associated with the translation in part (b).

PROBLEM 25.12

(a) Using the first codon given for each amino acid in Table 25.2, write the base sequence of mRNA that would translate the synthesis of the following pentapeptide:

Arg · Ile · Cys · Tyr · Val

(b) What base sequence in DNA would transcribe a synthesis of the mRNA? **(c)** What anticodons would appear in the tRNAs involved in the pentapeptide synthesis?

PROBLEM 25.13

Explain how an error of a single base in each strand of DNA could bring about the amino acid residue error that causes sickle-cell anemia (Section 24.6C).

25.6 Determining the Base Sequence of DNA: The Chain-Terminating (Dideoxynucleotide) Method

Certain aspects of the strategy used to sequence DNA resemble the methods used to sequence proteins. Both types of molecules require methods amenable to lengthy polymers, but in the case of DNA, a single DNA molecule is so long that it is absolutely necessary to cleave it into smaller, manageable fragments. Another similarity between DNA and proteins is that small sets of molecular building blocks comprise the structures of each, but in the case of DNA, only four nucleotide monomer units are involved instead of the 20 amino acid building blocks used to synthesize proteins. Finally, both proteins and nucleic acids are charged molecules that can be separated on the basis of size and charge using chromatography.

The first part of the process is accomplished by using enzymes called **restriction endonucleases.** These enzymes cleave double-stranded DNA at specific base sequences. Several hundred restriction endonucleases are now known. One, for example, called *Alu*I, cleaves the sequence AGCT between G and C. Another, called *Eco*R1, cleaves GAATTC between G and A. Most of the sites recognized by restriction enzymes have sequences of base pairs with the same order in both strands when read from the 5′ direction to the 3′ direction. For example:

Gilbert and Sanger shared the Nobel Prize in Chemistry in 1980 with Paul Berg for their work on nucleic acids. Sanger (Section 24.5), who pioneered the sequencing of proteins, had won an earlier Nobel Prize in 1958 for the determination of the structure of insulin.

$$5' \longleftarrow \text{G—A—A—T—T—C} \longrightarrow 3'$$
$$3' \longleftarrow \text{C—T—T—A—A—G} \longrightarrow 5'$$

Such sequences are known as **palindromes.** (Palindromes are words or sentences that read the same forward or backward. Examples are "radar" and "Madam, I'm Adam.")

Sequencing of the fragments (often called restriction fragments) can be done chemically or with the aid of enzymes. The first chemical method was introduced by A. Maxam and W. Gilbert (both of Harvard University); the **chain-terminating (dideoxynucleotide) method** was introduced in the same year by F. Sanger (Cambridge University). Essentially all DNA sequencing is currently done using an automated version of the chain-terminating method, which involves enzymatic reactions and 2′,3′-dideoxynucleotides.

25.6A DNA Sequencing by the Chain-Terminating (Dideoxynucleotide) Method

The chain-terminating method for sequencing DNA involves replicating DNA in a way that generates a family of partial copies that differ in length by one base pair. These partial copies of the parent DNA are separated according to length, and the terminal base on each strand is detected by a covalently attached fluorescent marker.

The mixture of partial copies of the target DNA is made by "poisoning" a replication reaction with a low concentration of unnatural nucleotides. The unnatural nucleotides terminate a growing DNA chain whenever an unnatural nucleotide is incorporated by chance instead of a natural nucleotide. Because a low concentration of the dideoxynucleotides is used, only occasionally is a dideoxynucleotide incorporated at random into the growing chains, and thus DNA molecules of essentially all different lengths are synthesized from the parent DNA.

The terminating nucleotides are 2′,3′-dideoxy analogues of the four natural nucleotides. Lacking the 3′-hydroxyl, the 2′,3′-dideoxynucleotide is incapable of forming a phosphodiester bond between its 3′ carbon and the next nucleotide that would be needed to continue the polymerization. Each terminating dideoxynucleotide is labeled with a fluorescent dye that gives a specific color depending on the base carried by that terminating nucleotide. (An alternate method is to label the *primers* with specific fluorescent dyes, instead of the dideoxynucleotide terminators, but the general method is the same.) One of the dye systems in use (patented by ABI) consists of a donor chromophore that is initially excited by the laser and which then transfers its energy to an acceptor moiety which produces the observed fluorescence. The donor is tethered to the dideoxynucleotide by a short linker.

A 2′3′-dideoxynucleotide, linker, and fluorescent dye moiety like those used in fluorescence-tagged dideoxynucleotide DNA sequencing reactions

2′,3′-dideoxycytosine triphosphate

Lack of 3′-hydroxyl causes chain termination.

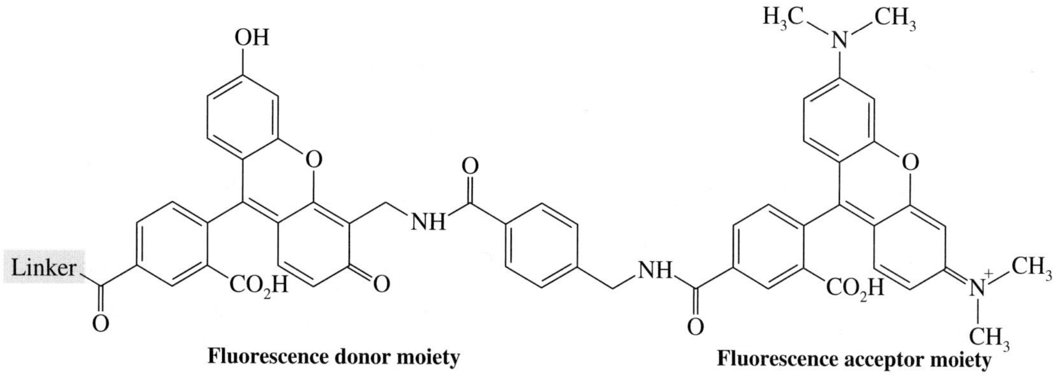

Fluorescence donor moiety **Fluorescence acceptor moiety**

Donor–acceptor fluorescent dye

The replication reaction used to generate the partial DNA copies is similar but not identical to the polymerase chain reaction (PCR) method (Section 25.8). In the dideoxy sequencing method only one primer sequence of DNA is used, and hence only one strand of the DNA is copied, whereas in the PCR, two primers are used and both strands are copied simultaneously. Furthermore, in sequencing reactions the chains are deliberately terminated by addition of the dideoxy nucleotides.

Capillary electrophoresis is the method most commonly used to separate the mixture of partial DNAs that results from a sequencing reaction. Capillary electrophoresis separates the DNAs on the basis of size and charge, allowing nucleotides that differ by only one base length to be resolved. Computerized acquisition of fluorescence data as the differently terminated DNAs pass the detector generates a four-color chromatogram, wherein each consecutive peak represents a DNA molecule one nucleotide longer than the previous one. The color of each peak represents the terminating nucleotide in that molecule. Since each of the four types of dideoxy terminating bases fluoresces a different color, the sequence of nucleotides in the DNA can be read directly. An example of sequence data from this kind of system is shown in Fig. 25.17.

Use of automated methods for DNA sequencing represents an exponential increase in speed over manual methods employing vertical slab polyacrylamide gel electrophoresis (Fig. 25.18). Only a few thousand bases per day (at most) could be sequenced by a person using the manual method. Now it is possible for a single machine running parallel and continuous analyses to sequence almost 3 million bases per day using automated capillary electrophoresis and laser fluorescence detection. As an added benefit, the ease of DNA sequencing often makes it easier to determine the sequence of a protein by the sequence of all or part of its corresponding gene, rather than by sequencing the protein itself (see Section 24.5).

The development of high-throughput methods for sequencing DNA is largely responsible for the remarkable success achieved in the Human Genome Project. Sequencing the 3 billion base pairs in the human genome could never have been completed before the 2003 fiftieth anniversary of Watson and Crick's elucidation of the structure of DNA had high-throughput sequencing methods not come into existence.*

---

*The Human Genome Project web page of the U.S. Department of Energy provides a wealth of resources for further information: www.ornl.gov/hgmis/.

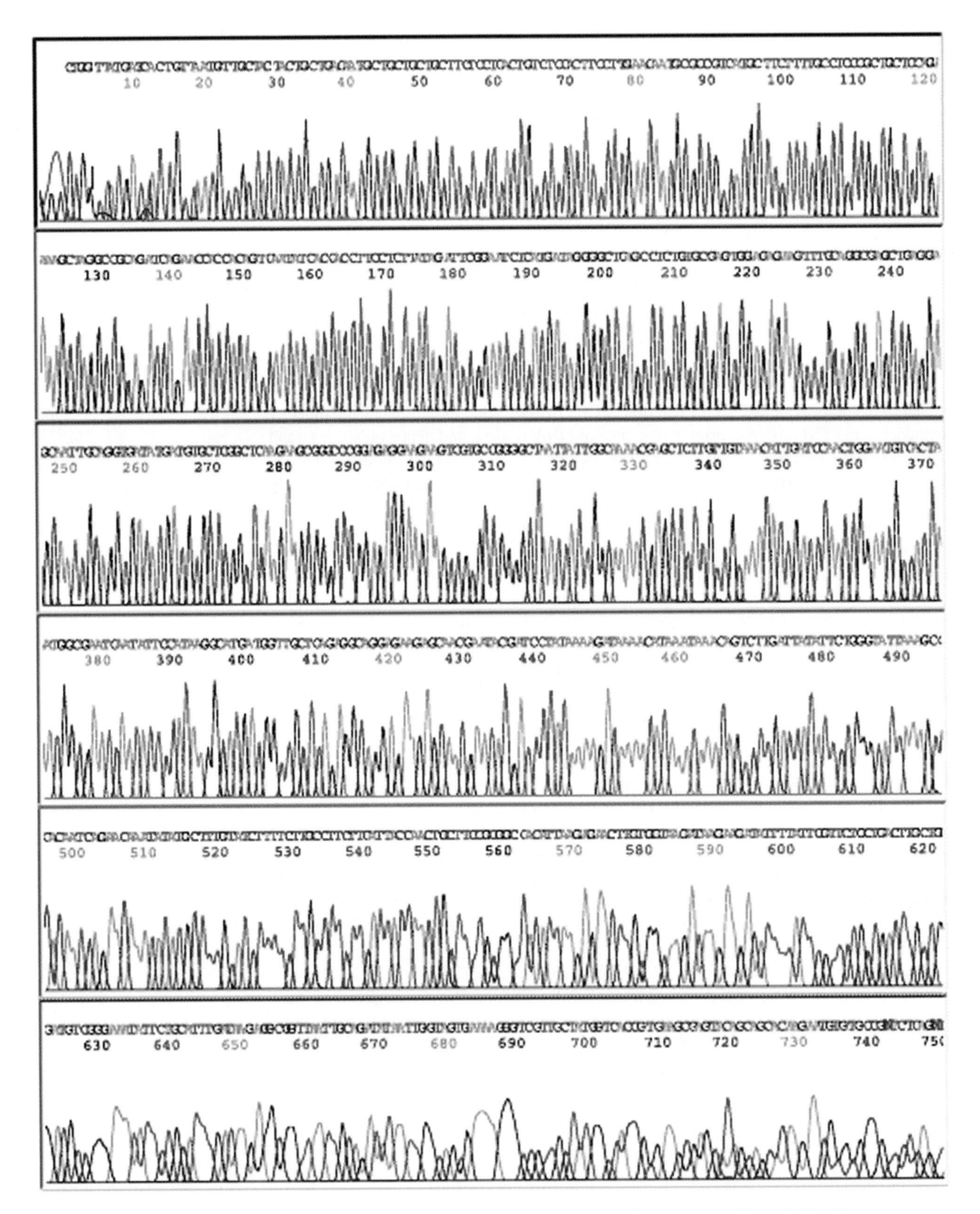

FIGURE 25.17 Example of data from an automated DNA sequencer. From Applied Biosystems, Inc.

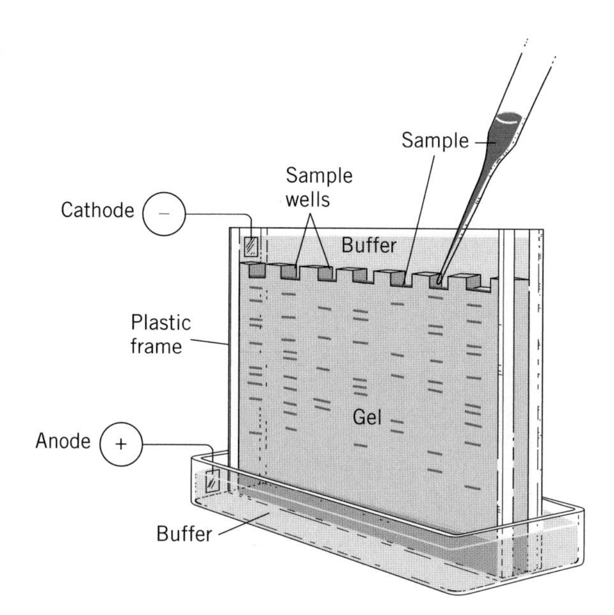

FIGURE 25.18 An apparatus for gel electrophoresis. Samples are applied in the slots at the top of the gel. Application of a voltage difference causes the samples to move. The samples move in parallel lanes. (From Voet, D.; Voet, J. G. *Biochemistry,* 2nd ed.; Wiley: New York, 1995; p 92. Used with permission.)

25.7 Laboratory Synthesis of Oligonucleotides

Synthetic oligonucleotides are needed for a variety of purposes. One of the most important and common uses of synthetic oligonucleotides is as primers for nucleic acid sequencing and the PCR (Section 25.8). Another important application is in the research and development of **antisense oligonucleotides,** which hold potential as therapies for a variety of diseases. An antisense oligonucleotide is one that has a sequence complementary to the coding sequence in a DNA or RNA molecule. Synthetic oligonucleotides that bind tightly to DNA or mRNA sequences from a virus, bacterium, or other disease condition may be able to shut down expression of the target protein associated with those conditions. For example, if the sense portion of DNA in a gene reads

A—G—A—C—C—G—T—G—G

the antisense oligonucleotide would read

T—C—T—G—G—C—A—C—C

The ability to deactivate specific genes in this way holds great medical promise. Many viruses and bacteria, during their life cycles, use a method like this to regulate some of their own genes. The hope, therefore, is to synthesize antisense oligonucleotides that will seek out and destroy viruses in a person's cells by binding with crucial sequences of the viral DNA or RNA. Synthesis of such oligonucleotides is an active area of research today and is directed at many viral diseases, including AIDS, as well as lung and other forms of cancer.

Current methods for oligonucleotide synthesis are similar to those used to synthesize proteins, including the use of automated solid-phase techniques (Section 24.7D). A suitably protected nucleotide is attached to a solid phase called a "controlled pore glass," or CPG (Fig. 25.19), through a linkage that can ultimately be cleaved. The next protected nucleotide in the form of a **phosphoramidite** is added, and coupling is brought about by a coupling agent, usually 1,2,3,4-tetrazole. The phosphite triester that results from the coupling is oxidized to phosphate triester with iodine, producing a chain that has been lengthened by one nucleotide. The **dimethoxytrityl (DMTr)** group used to protect the 5′ end of the added nucleotide is removed by treatment with acid, and the steps **coupling,**

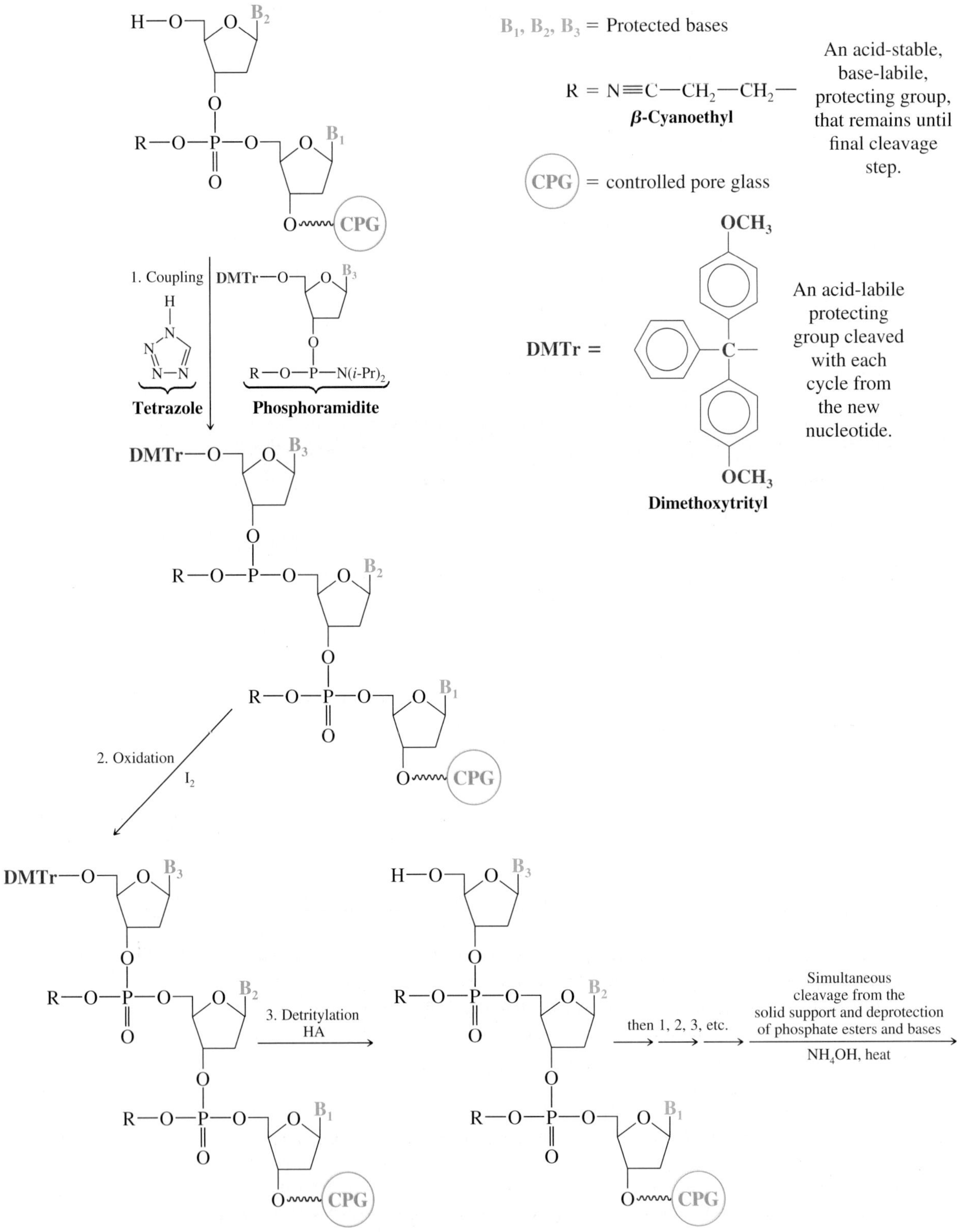

B_1, B_2, B_3 = Protected bases

R = $N{\equiv}C{-}CH_2{-}CH_2{-}$
β-Cyanoethyl

An acid-stable, base-labile, protecting group, that remains until final cleavage step.

CPG = controlled pore glass

DMTr =
Dimethoxytrityl

An acid-labile protecting group cleaved with each cycle from the new nucleotide.

1. Coupling

Tetrazole

Phosphoramidite

R—O—P—N(i-Pr)$_2$

2. Oxidation
I_2

3. Detritylation
HA

then 1, 2, 3, etc.

Simultaneous cleavage from the solid support and deprotection of phosphate esters and bases

NH_4OH, heat

FIGURE 25.19 The steps involved in automated synthesis of oligonucleotides using the phosphoramidite coupling method.

oxidation, detritylation are repeated. (All the steps are carried out in nonaqueous solvents.) With automatic synthesizers the process can be repeated at least 50 times and the time for a complete cycle is 40 min or less. The synthesis is monitored by spectrophotometric detection of the dimethoxytrityl cation as it is released in each cycle (much like the monitoring of Fmoc release in solid-phase peptide synthesis). After the desired oligonucleotide has been synthesized, it is released from the solid support and the various protecting groups, including those on the bases, are removed.

25.8 The Polymerase Chain Reaction

The **polymerase chain reaction (PCR)** is an extraordinarily simple and effective method for exponentially multiplying (amplifying) the number of copies of a DNA molecule. Beginning with even just a single molecule of DNA, the PCR can generate 100 billion copies in a single afternoon. The reaction is easy to carry out: It requires only a miniscule sample of the target DNA (picogram quantities are sufficient), a supply of nucleotide triphosphate reagents to build the new DNA, DNA polymerase to catalyze the reaction, and a device called a thermal cycler to control the reaction temperature and automatically repeat the reaction. The PCR has had a major effect on molecular biology. Perhaps its most important role has been in the sequencing of the human genome (Sections 25.6 and 25.9), but now virtually every aspect of research involving DNA involves the PCR at some point.

One of the original aims in developing the PCR was to use it in increasing the speed and effectiveness of prenatal diagnosis of sickle-cell anemia (Section 24.6C). It is now being applied to the prenatal diagnosis of a number of other genetic diseases, including muscular dystrophy and cystic fibrosis. Among infectious diseases, the PCR has been used to detect cytomegalovirus and the viruses that cause AIDS, certain cervical carcinomas, hepatitis, measles, and Epstein–Barr disease.

The PCR is a mainstay in forensic sciences as well, where it may be used to copy DNA from a trace sample of blood or semen or a hair left at the scene of a crime. It is also used in evolutionary biology and anthropology, where the DNA of interest may come from a 40,000-year-old wooly mammoth or the tissue of a mummy. It is also used to match families with lost relatives (see the chapter opening vignette). There is almost no area with biological significance that does not in some way have application for use of the PCR reaction.

The PCR was invented by Kary B. Mullis and developed by him and his co-workers at Cetus Corporation. It makes use of the enzyme DNA polymerase, discovered in 1955 by Arthur Kornberg and associates at Stanford University. In living cells, DNA polymerases help repair and replicate DNA. The PCR makes use of a particular property of DNA polymerases: their ability to attach additional nucleotides to a short oligonucleotide "primer" when the primer is bound to a complementary strand of DNA called a template. The nucleotides are attached at the 3′ end of the primer, and the nucleotide that the polymerase attaches will be the one that is complementary to the base in the adjacent position on the template strand. If the adjacent template nucleotide is G, the polymerase adds C to the primer; if the adjacent template nucleotide is A, then the polymerase adds T, and so on. Polymerase repeats this process again and again as long as the requisite nucleotides (as triphosphates) are present in the solution, until it reaches the 5′ end of the template.

Mullis was awarded the Nobel Prize in Chemistry for this work in 1993.

Figure 25.20 shows one cycle of the PCR. The target DNA, a supply of nucleotide triphosphate monomers, DNA polymerase, and the appropriate oligonucleotide primers (one primer sequence for each 5′ to 3′ direction of the target double-stranded DNA) are added to a small reaction vessel. The mixture is briefly heated to approximately 90°C to separate the DNA strands (denaturation); it is cooled to 50–60°C to allow the primer sequences and DNA polymerase to bind to each of the separated strands (annealing); and it

30–40 cycles of 3 steps:

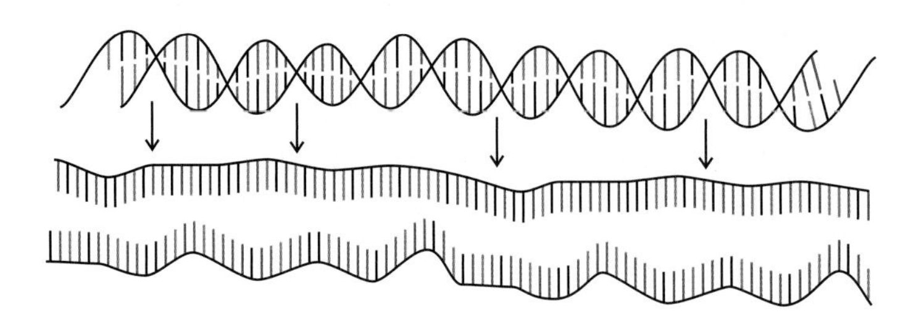

Step 1: Denaturation of double-stranded DNA to single strands.
1 minute at approximately 90°C

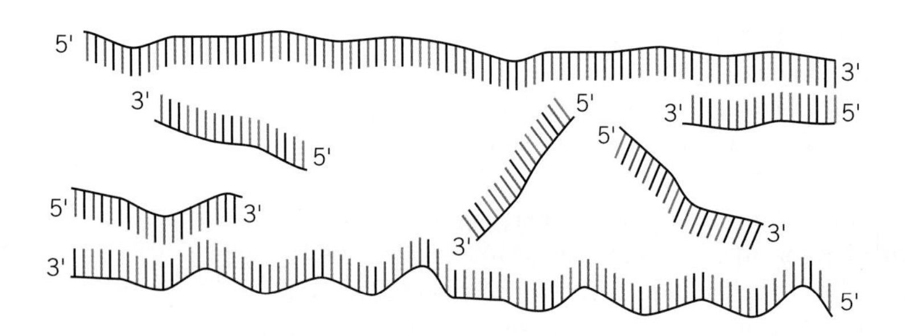

Step 2: Annealing of primers to each single-stranded DNA. Primers are needed with sequences complementary to both single strands.
45 seconds at 50–60°C

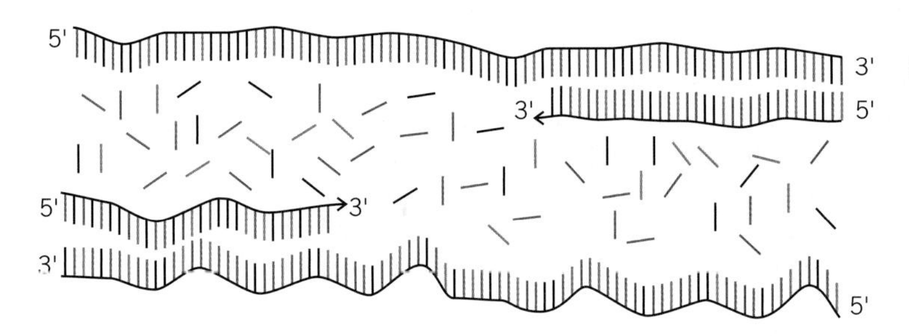

Step 3: Extension of the parent DNA strands with nucleotide triphosphate monomers from the reaction mixture.
2 minutes at approximately 70°C

FIGURE 25.20 One cycle of the PCR. Heating separates the strands of DNA of the target to give two single-stranded templates. Primers, designated to complement the nucleotide sequences flanking the targets, anneal to each strand. DNA polymerase, in the presence of nucleotide triphosphates, catalyzes the synthesis of two pieces of DNA, each identical to the original target DNA.

is warmed to about 70°C to extend each strand by polymerase-catalyzed condensation of nucleotide triphosphate monomers complementary to the parent DNA strand. Another cycle of the PCR begins by heating to separate the new collection of DNA molecules into single strands, cooling for the annealing step, and so on.

Each cycle, taking only a few minutes, doubles the amount of target DNA that existed prior to that step (Fig. 25.21). The result is an exponential increase in the amount of DNA over time. After n cycles, the DNA will have been replicated 2^n times—after 10 cycles there is roughly 1000 times as much DNA; after 20 cycles roughly 1 million times as much; and so on. Thermal cycling machines can carry out approximately 20 PCR cycles per hour, resulting in billions of DNA copies over a single afternoon.

Each application of PCR requires primers that are 10–20 nucleotides in length and whose sequences are complementary to short, conveniently located sequences flanking the target DNA sequence. The primer sequence is also chosen so that it is near sites that are cleavable with restriction enzymes. Once a researcher determines what primer sequence is needed, the primers are usually purchased from commercial suppliers who synthesize them on request using solid-phase oligonucleotide synthesis methods like that described in Section 25.7.

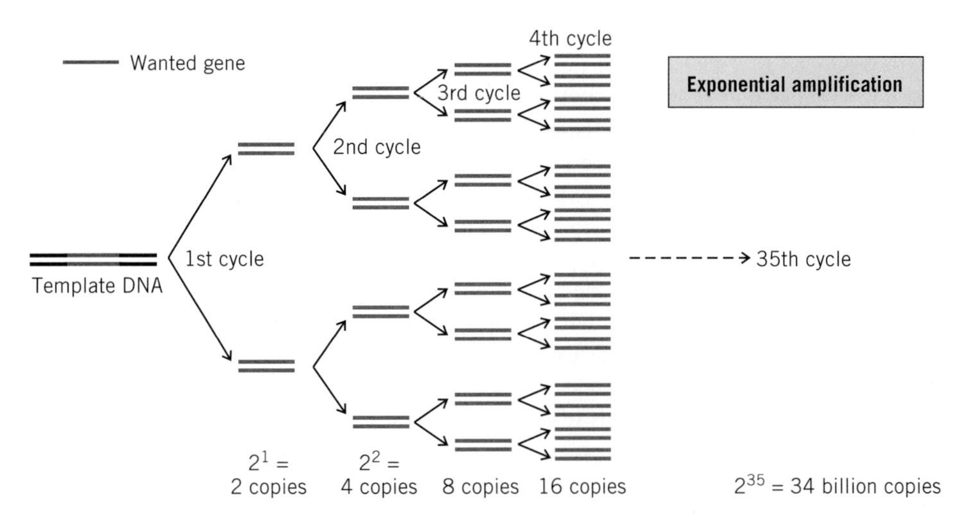

FIGURE 25.21 Each cycle of the PCR doubles the number of copies of the target area.

As an intriguing adjunct to the PCR story, it turns out that cross-fertilization between disparate research fields greatly assisted development of current PCR methods. In particular, the discovery of extremozymes, which are enzymes from organisms that live in high-temperature environments, has been of great use. DNA polymerases now typically used in PCR are heat-stable forms derived from thermophilic bacteria. Polymerases such as Taq polymerase, from the bacterium *Thermus aquaticus,* found in places such as geyser hot springs, and $Vent_R^{TM}$, from bacteria living near deep-sea thermal vents, are used. Use of extremozyme polymerases facilitates PCR by allowing elevated temperatures to be used for the DNA melting step without having to worry about denaturing the polymerase enzyme at the same time. All materials can therefore be present in the reaction mixture throughout the entire process. Furthermore, use of a higher temperature during the chain extension also leads to faster reaction rates. (See "The Chemistry . . . of Stereoselective Reductions of Carbonyl Groups", Section 12.3, for another example of the use of high-temperature enzymes.)

Thermophilic bacteria, growing in hot springs like these at Yellowstone National Park, produce heat-stable enzymes called extremozymes that have proven useful for a variety of chemical processes.

25.9 Sequencing of the Human Genome: An Instruction Book for the Molecules of Life

The announcement by scientists from the public Human Genome Project and Celera Genomics Company in June 2000 that sequencing of the approximately 3 billion base pairs in the human genome was complete marked achievement of one of the most important and ambitious scientific endeavors ever undertaken. To accomplish this feat, data were pooled from thousands of scientists working around the world using tools including the PCR reaction (Section 25.8), dideoxynucleotide sequencing reactions (Section 25.6), capillary electrophoresis, laser-induced fluorescence, and supercomputers. What was ultimately produced is a transcript of our chromosomes that could be called an instruction book for the molecules of life.

"This structure has novel features, which are of considerable biological importance." James Watson, one of the scientists who determined the structure of DNA.

But what do the instructions in the genome say? How can we best make use of the molecular instructions for life? Of the roughly 35,000 genes in our DNA, the function of only a small percentage of genes is understood. Discovering genes that can be used to benefit our human condition and the chemical means to turn them on or off presents some of the greatest opportunities and challenges for scientists of today and the future. Sequencing the genome was only the beginning of the story.

As the story unfolds, chemists will continue to add to the molecular archive of compounds used to probe our DNA. DNA microchips, with 10,000 or more short diagnostic sequences of DNA chemically bonded to their surface in predefined arrays, will be used to test DNA samples for thousands of possible genetic conditions in a single assay. With the map of our genome in hand, great libraries of potential drugs will be tested against genetic targets to discover more molecules that either promote or inhibit expression of key gene products. Sequencing of the genome will also accelerate development of molecules that interact with proteins, the products of gene expression. Knowledge of the genome sequence will expedite identification of the genes coding for interesting proteins, thus allowing these proteins to be expressed in virtually limitless quantities. With an ample supply of target proteins available, the challenges of solving three-dimensional protein structures and understanding their functions will also be overcome more easily. Optimization of the structures of small organic molecules that interact with proteins will also occur more rapidly because the protein targets for these molecules will be available faster and in greater quantity. There is no doubt that the pace of research to develop new and useful organic molecules for interaction with gene and protein targets will increase dramatically now that the genome has been sequenced. The potential to use our chemical creativity in the fields of genomics and proteomics is immense.

Key Terms and Concepts

Chain-terminating (dideoxynucleotide) method	**Section 25.6**
Codon, anticodon	**Section 25.5C**
Deoxyribonucleic acid (DNA)	**Section 25.1**
Gene	**Section 25.1**
Genetic code	**Sections 25.5C, 25.5D**
Genome	**Sections 25.1, 25.9**
Exons	**Section 25.5A**
Introns	**Section 25.5A**
Nucleic acids	**Sections 25.1, 25.4, 25.5**
Nucleosides	**Sections 25.2, 25.3**
Nucleotides	**Sections 25.2, 25.3**
Polymerase chain reaction (PCR)	**Section 25.8**
Proteome	**Sections 25.1, 25.9**
Replication	**Section 25.4C**
Restriction endonucleases	**Section 25.6**
Ribonucleic acid (RNA)	**Section 25.1**
Ribozymes	**Section 25.5B**
Transcription	**Section 25.5**
Translation	**Section 25.5E**

ADDITIONAL PROBLEMS

25.14 The example of a Silyl Hilbert-Johnson nucleosidation reaction in Section 25.3 is presumed to involve an intermediate ribosyl cation that is stabilized by intramolecular interactions involving the C2 benzoyl group. This intermediate blocks attack by the heterocyclic base from the α face of the ribose ring but allows attack on the β face, as required for formation of the desired product. Propose a structure for the ribosyl cation intermediate that explains the stereoselective bonding of the base.

25.15 **(a)** Mitomycin is a clinically used antitumor antibiotic that acts by disrupting DNA synthesis through covalent bond-forming reactions with deoxyguanosine in DNA. Maria Tomasz (Hunter College) and others have shown that alkylation of DNA by mitomycin occurs by a complex series of mechanistic steps. The process begins with reduction of the quinone ring in mitomycin to its hydroquinone form, followed by elimination of methanol from the adjacent ring to form an intermediate called leuco-aziridinomitosene, shown below. One of the paths by which leuco-aziridinomitosene alkylates DNA involves protonation and opening of the three-membered aziridine ring, resulting in an intermediate

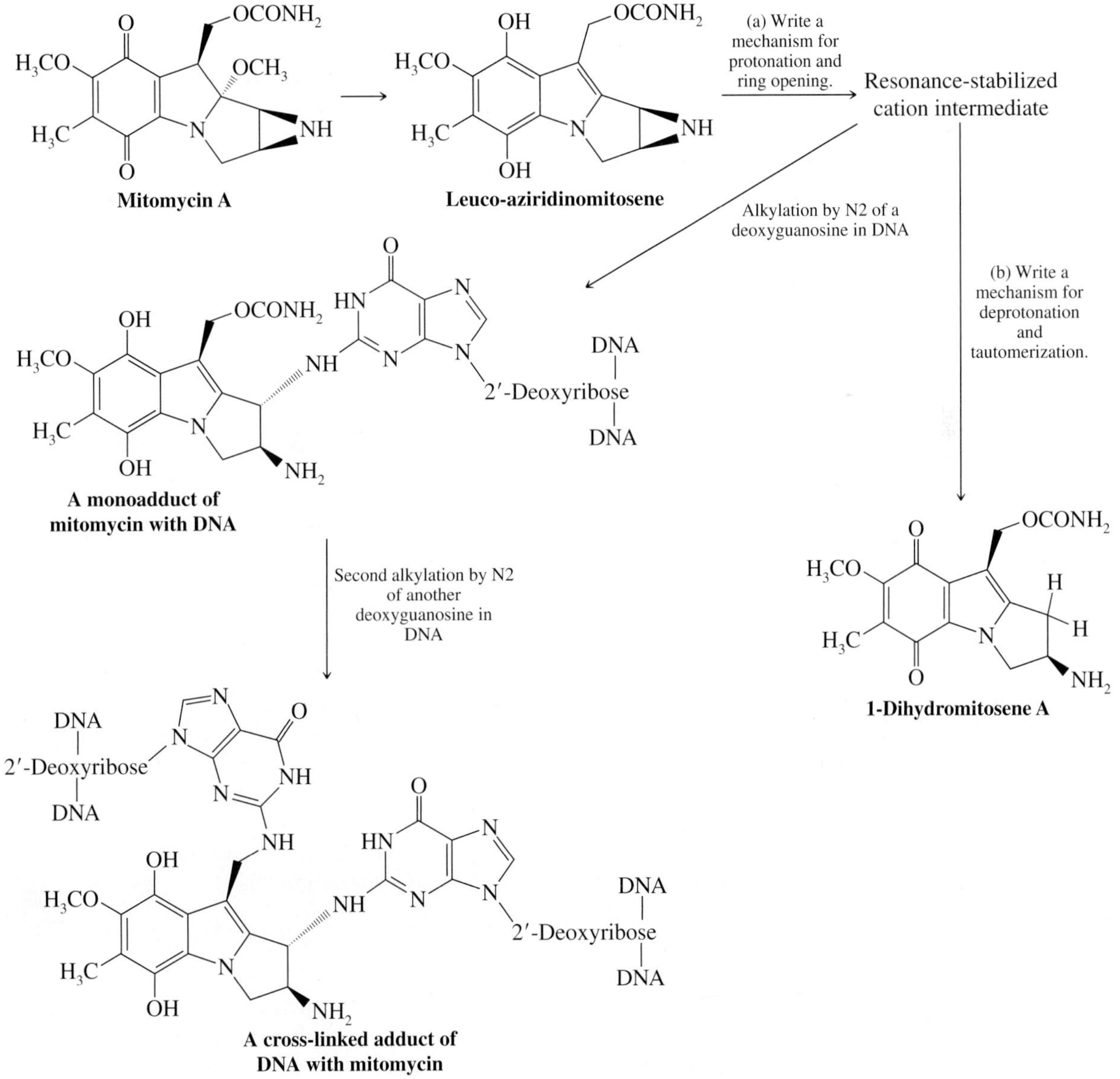

Mitomycin A → **Leuco-aziridinomitosene**

(a) Write a mechanism for protonation and ring opening. → Resonance-stabilized cation intermediate

Alkylation by N2 of a deoxyguanosine in DNA → **A monoadduct of mitomycin with DNA**

Second alkylation by N2 of another deoxyguanosine in DNA → **A cross-linked adduct of DNA with mitomycin**

(b) Write a mechanism for deprotonation and tautomerization. → **1-Dihydromitosene A** cation that is resonance stabilized by the hydroquinone group. Attack on the cation by N2 of a deoxyguanosine residue leads to a monalkylated DNA product, shown below. Write a detailed mechanism to show how the ring opening might occur, including resonance forms for the cation intermediate, followed by nucleophilic attack by DNA. (Intra- or interstrand cross-linking of DNA can further occur by reaction of another deoxyguanosine residue to displace the carbamoyl group of the initial mitosene–DNA monoadduct. A cross-linked adduct is also shown below.) **(b)** 1-Dihydromitosene A (shown below) is sometimes formed from the cation intermediate in part (a) by loss of a proton and tautomerization. Propose a detailed mechanism for the formation of 1-dihydromitosene A from the resonance-stabilized cation of part (a).

25.16 As described in Section 25.5B, acid–base catalysis is believed to be the mechanism by which ribosomes catalyze the formation of peptide bonds in the process of protein translation. Key to this proposal is assistance by the N3 nitrogen (highlighted below) of a nearby adenine in the ribosome for the removal of a proton from the α-amino group of the amino acid adding to the growing peptide chain (Fig. 25.14). The ability of this adenine group to remove the proton is, in turn, apparently facilitated by relay of charge made possible by other nearby groups in the ribosome. The constellation of these groups is shown here. Draw mechanism arrows to show formation of a resonance contributor wherein the adenine group could carry a formal negative charge, thereby facilitating its removal of the α-amino proton of the amino acid. (The true electronic structure of these groups is not accurately represented by any single resonance contributor, of course. A hybrid of the contributing resonance structures weighted according to stability would best reflect the true structure.)

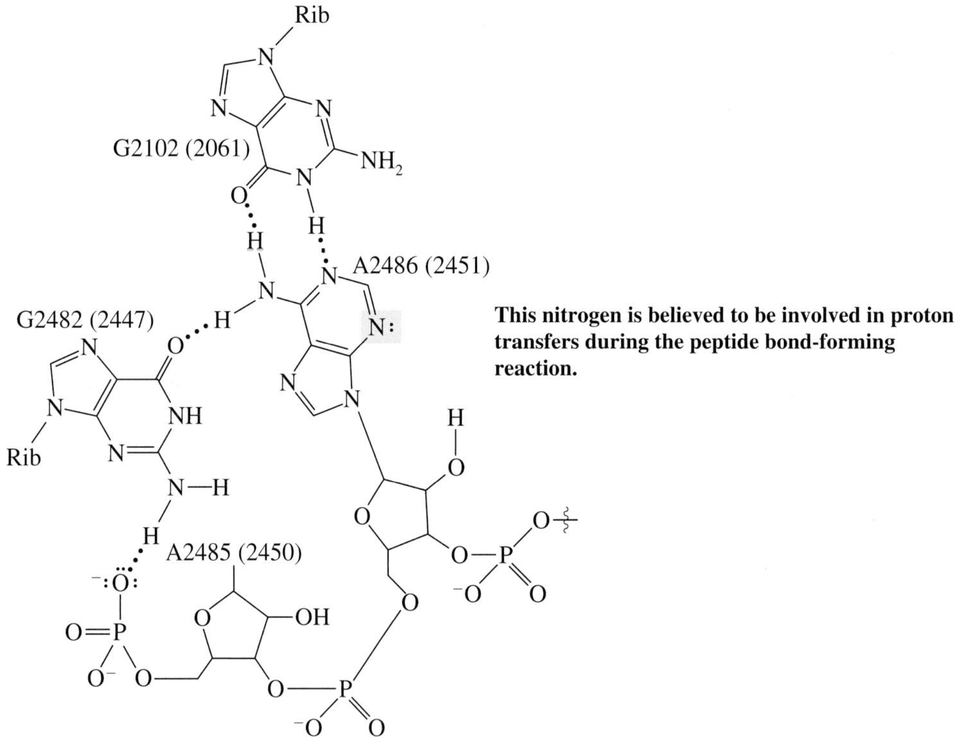

G2102 (2061)

A2486 (2451)

G2482 (2447)

A2485 (2450)

This nitrogen is believed to be involved in proton transfers during the peptide bond-forming reaction.

LEARNING GROUP PROBLEM

Research suggests that expression of certain genes is controlled by conversion of some cytosine bases in the genome to 5-methylcytosine by an enzyme called DNA methyltransferase. Cytosine methylation may be a means by which some genes are turned off as cells differentiate during growth and development. It may also play a role in some cancer processes and in defending the genome from foreign DNA such as viral genes. Measuring the level of methylation in DNA is an important analytical process. One method for measuring cytosine methylation is known as methylation-specific PCR. This technique requires that all unmethylated cytosines in a sample of DNA be converted to uracil by deamination of the C4 amino group in the unmethylated cytosines. This accomplished by treating the DNA with sodium bisulfite ($NaHSO_3$) to form a bisulfite addition product with its unmethylated cytosine residues. The cytosine sulfonates that result are then subjected to hydrolysis conditions that convert the C4 amino group to a carbonyl group, resulting in uracil sulfonate. Finally, treatment with base causes elimination of the sulfonate group to produce uracil. The modified DNA is then amplified by PCR using primers designed to distinguish DNA with methylated cytosine versus cytosine-to-uracil converted bases.

Write detailed mechanisms for the reactions used to convert cytosine to uracil by the above sequence of steps.

Answers to Selected Problems

CHAPTER 1

1.10 (a), (c), (f), (g) are tetrahedral; (e) is trigonal planar; (b) is linear; (d) is angular; (h) is trigonal pyramidal.

1.14 (a) and (d); (b) and (e); and (c) and (f).

1.22 (a), (g), (i), (l), represent different compounds that are not isomeric; (c–e), (h), (j), (m), (n), (o) represent the same compound; (b), (f), (k), (p) represent constitutional isomers.

1.27 (a) The structures differ in the positions of the nuclei.

1.29 (a) A negative charge; (b) a negative charge; (c) trigonal pyramidal.

CHAPTER 2

2.10 (c) Propyl bromide; (d) isopropyl fluoride; (e) phenyl iodide.

2.13 (a) $CH_3CH_2OCH_2CH_3$; (b) $CH_3CH_2OCH_2CH_2CH_3$; (e) diisopropyl ether.

2.17 (a) $CH_3CH_2CH_2CH_2OH$ would boil higher because its molecules can form hydrogen bonds to each other; (c) $HOCH_2CH_2CH_2OH$ would boil higher because it can form more hydrogen bonds.

2.20 (a) Ketone; (c) alcohol; (e) alcohol.

2.21 (a) 3 Alkene, and a 2° alcohol; (c) phenyl and 1° amine; (e) phenyl, ester and 3° amine; (g) alkene and 2 ester groups.

2.27 (f) $CH_3CH_2CH_2CH_2CH_2Br$;$(CH_3)_2CHCH_2CH_2Br$; $CH_3CH_2CH(CH_3)CH_2Br$; $(CH_3)_3CCH_2Br$.

2.32 Ester

CHAPTER 3

3.2 (a), (c), (d), and (f) are Lewis bases; (b) and (e) are Lewis acids.

3.4 (a) $[H_3O^+] = [HCO_2^-] = .0042\ M$; (b) Ionization $= 4.2\%$.

3.5 (a) $pK_a = 7$; (b) $pK_a = -0.7$; (c) Because the acid with a $pK_a = 5$ has a larger K_a, it is the stronger acid.

3.7 The pK_a of the methylaminium ion is equal to 10.6 (Section 3.5C). Because the pK_a of the anilinium ion is equal to 4.6, the anilinium ion is a stronger acid than the methylaminium ion, and aniline ($C_6H_5NH_2$) is a weaker base than methylamine (CH_3NH_2).

3.11 (a) $CHCl_2CO_2H$ would be the stronger acid because the electron-withdrawing inductive effect of two chlorine atoms would make its hydroxyl proton more positive. (c) CH_2FCO_2H would be the stronger acid because a fluorine atom is more electronegative than a bromine atom and would be more electron withdrawing.

3.23 (a) $pK_a = 3.752$; (b) $K_a = 10^{-13}$.

CHAPTER 4

4.5 (a) 1-(1-Methylethyl)-2-(1,1-dimethylethyl)cyclopentane or 1-*tert*-butyl-2-isopropylcyclopentane; (c) butylcyclohexane; (e) 2-chlorocyclopentanol.

4.6 (a) 2-Chlorobicyclo[1.1.0]butane; (c) bicyclo[2.1.1]hexane; (e) 2-methylbicyclo[2.2.2]octane.

4.7 (a) *trans*-3-Heptene; (c) 4-ethyl-2-methyl-1-hexene

4.8 (a) 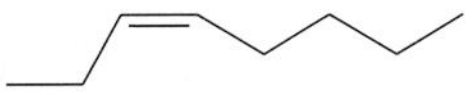 (c) 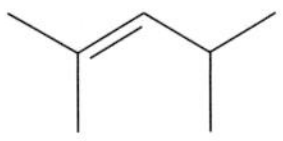 (e) 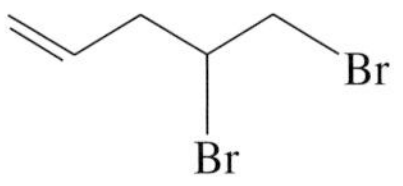 Br Br (g) 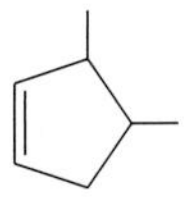 (i) 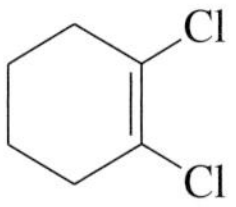 Cl Cl

4.9 1-Hexyne, 2-Hexyne, 3-Hexyne, 4-Methyl-1-pentyne, 4-Methyl-2-pentyne 3,3-Dimethyl-1-butyne

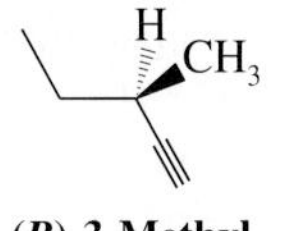

H CH_3 H_3C H

(*R*)-3-Methyl-1-pentyne 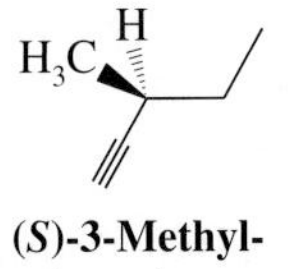 **(*S*)-3-Methyl-1-pentyne**

4.20 (a) 3,3,4-Trimethylhexane; (c) 3,5,7-Trimethylnonane; (e) 2-Bromobicyclo[3.3.1]nonane; (g) Cyclobutylcyclopentane.

4.25 The alkane is 2-methylpentane, $(CH_3)_2CHCH_2CH_2CH_3$.

4.26 The alkane is 2,3-dimethylbutane.

4.29 $(CH_3)_3CCH_3$ is the most stable isomer.

4.37 (a) Pentane would boil higher because its chain is unbranched. (c) 2-Chloropropane because it is more polar and has a higher molecular weight. (e) CH_3COCH_3 because its molecules are more polar.

4.41 (a) The trans isomer is more stable because both methyl groups can be equatorial. (c) The trans isomer is more stable because both methyl groups can be equatorial.

4.48 (a) From Table 4.8 we find that this is *trans*-1,2-dichlorocyclohexane. (b) The chlorines must have added from opposite sides of the double bond.

CHAPTER 5

5.1 (a) Achiral; (c) chiral; (e) chiral.

5.2 (a) Yes; (c) no.

5.3 (a) They are the same. (b) They are enantiomers.

5.7 The following possess a plane of symmetry and are, therefore, achiral: screwdriver, baseball bat, hammer.

5.11 (a) —Cl > —SH > —OH > —H;

(c) —OH > —CHO > —CH_3 > —H;

(e) —OCH_3 > —$N(CH_3)_2$ > —CH_3 > —H

5.13 (a) Enantiomers; (c) enantiomers.

5.17 (a) Diastereomers; (c) no; (e) no.

5.19 (a) Represents **A**; (b) represents **C**; (c) represents **B**.

5.21 **B** (2*S*,3*S*)-2,3-Dibromobutane; **C** (2*R*,3*S*)-2,3-Dibromobutane.

5.35 (a) Same; (c) diastereomers; (e) same; (g) diastereomers; (i) same; (k) diastereomers; (m) diastereomers; (o) diastereomers; (q) same.

CHAPTER 6

6.3 (a) The reaction is S_N2 and, therefore, occurs with inversion of configuration. Consequently, the configuration of (+)-2-chlorobutane is opposite [i.e., (*S*)] to that of (−)-2-butanol [i.e., (*R*)]. (b) The configuration of (−)-2-iodobutane is (*R*).

6.7 Protic solvents are formic acid, formamide, ammonia, and ethylene glycol. The others are aprotic.

6.9 (a) CH_3O^-; (c) $(CH_3)_3P$.

6.14 (a) 1-Bromopropane would react more rapidly, because, being a primary halide, it is less hindered. (c) 1-Chlorobutane, because the carbon bearing the leaving group is less hindered than in 1-chloro-2-methylpropane. (e) 1-Chlorohexane because it is a primary halide. Phenyl halides are unreactive in S_N2 reactions.

6.15 (a) Reaction (1) because ethoxide ion is a stronger nucleophile than ethanol; (c) reaction (2) because triphenylphosphine, $(C_6H_5)_3P$, is a stronger nucleophile than triphenylamine. (Phosphorus atoms are larger than nitrogen atoms.)

6.16 (a) Reaction (2) because bromide ion is a better leaving group than chloride ion; (c) reaction (2) because the concentration of the substrate is twice that of reaction (1).

6.19 The better yield is obtained by using the secondary halide, 1-bromo-1-phenylethane, because the desired reaction is E2. Using the primary halide will result in substantial S_N2 reaction as well, producing the alcohol instead of the desired alkene.

6.29 (a) You should use a strong base, such as RO^-, at a higher temperature to bring about an E2 reaction. (b) Here we want an S_N1 reaction. We use ethanol as the solvent *and as the nucleophile*, and we carry out the reaction at a low temperature so that elimination is minimized.

CHAPTER 7

7.3 (a) 2,3-Dimethyl-2-butene would be the more stable because the double bond is tetrasubstituted. (c) *cis*-3-Hexene would be more stable because its double bond is disubstituted.

7.5 (a)

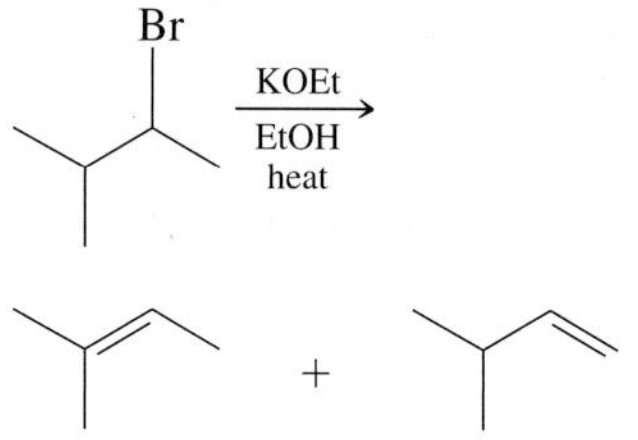

Br

KOEt / EtOH / heat

+

(trisubstituted, more stable) **(monosubstituted, less stable)**

Major product **Minor product**

7.18 (a) We designate the position of the double bond by using the *lower* of the two numbers of the doubly bonded carbon atoms, and the chain is numbered from the end nearer the double bond. The correct name is *trans*-2-pentene. (c) We use the lower number of the two doubly bonded carbon atoms to designate the position of the double bond. The correct name is 1-methylcyclohexene.

7.19 (a) 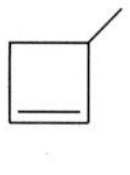 (c)

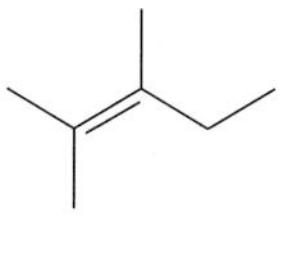

(e) 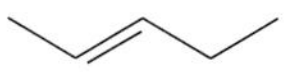 (g)

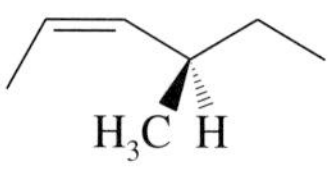

H_3C H

7.21 **(a)** (*E*)-3,5-Dimethyl-2-hexene; (c) 6-methyl-3-heptyne; (e) (3*Z*,5*R*)-5-chloro-3-hepten-6-yne.

7.35 Only the deuterium atom can assume the anti coplanar orientation necessary for an E2 reaction to occur.

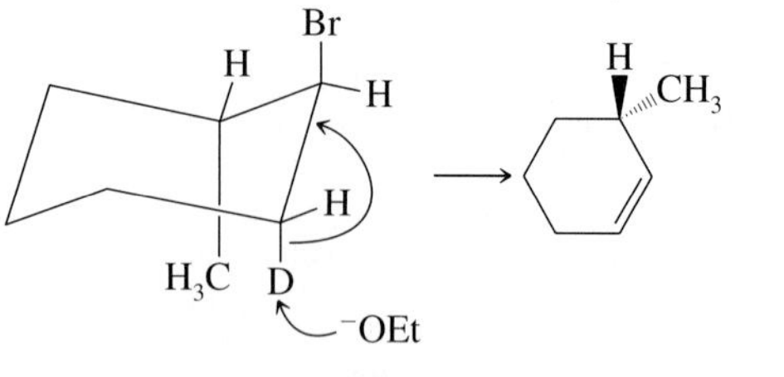

Br H H H H_3C D $^-$OEt → H CH_3

CHAPTER 8

8.1 2-Bromo-1-iodopropane.

8.7 The order reflects the relative ease with which these alkenes accept a proton and form a carbocation. 2-Methylpropene reacts fastest because it leads to a 3° cation; ethene reacts slowest because it leads to a 1° cation.

8.22 (a) syn-Hydroxylation at either face of the (*Z*)-isomer leads to the meso compound (2*R*,3*S*)-2,3-butanediol. (b) syn-Hydroxylation at one face of the (*E*)-isomer leads to (2*R*,3*R*)-2,3-butanediol; at the other face, which is equally likely, it leads to the (2*S*,3*S*)-enantiomer.

8.27 (a) $CH_3CH_2CHICH_3$; (b) $CH_3CH_2CH_2CH_3$; (e) $CH_3CH_2CH(OH)CH_3$; (h) $CH_3CH_2CH{=}CH_2$; (j) $CH_3CH_2CHClCH_3$; (l) $CH_3CH_2CHO + HCHO$.

8.29 (a) $CH_3CH_2CBr{=}CHBr$; (c) $CH_3CH_2CBr_2CH_3$; (e) $CH_3CH_2CH{=}CH_2$.

8.32

(a) 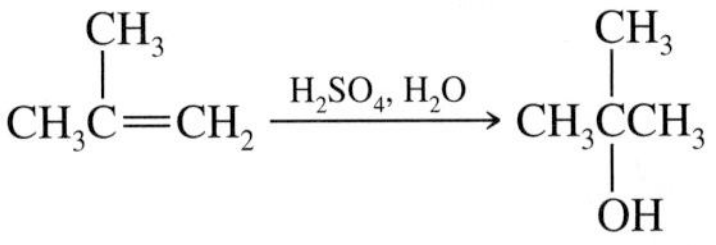 $CH_3C(CH_3){=}CH_2 \xrightarrow{H_2SO_4,\ H_2O} CH_3C(CH_3)(OH)CH_3$

(c) 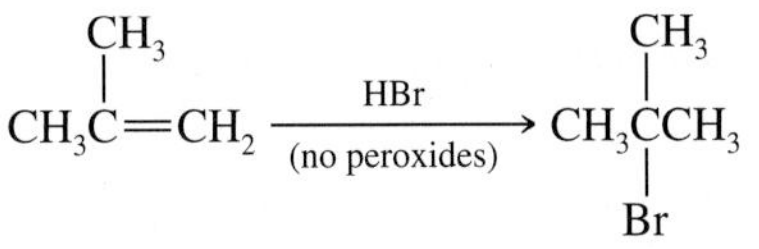 $CH_3C(CH_3){=}CH_2 \xrightarrow[\text{(no peroxides)}]{HBr} CH_3C(CH_3)(Br)CH_3$

(d) 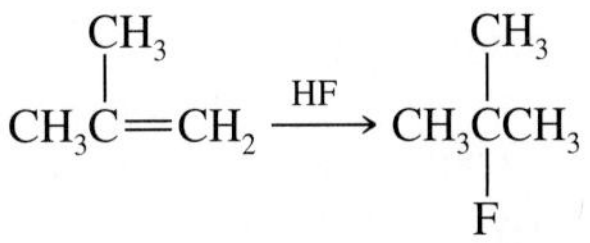 $CH_3C(CH_3){=}CH_2 \xrightarrow{HF} CH_3C(CH_3)(F)CH_3$

8.33 (a)

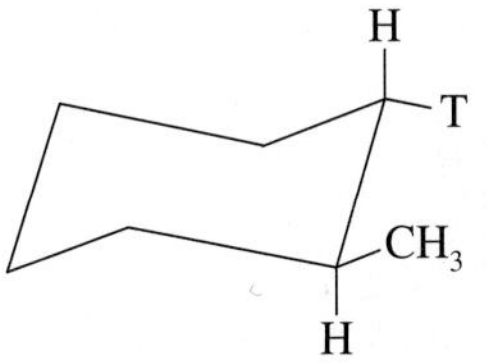

H, T, CH_3, H

(c)

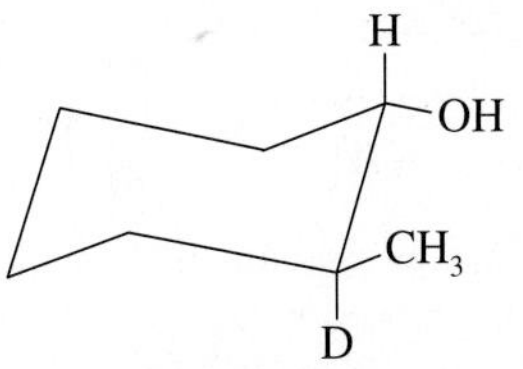

H, OH, CH_3, D

8.63

$HC{\equiv}C$, CH_3, H, C, CH_2CH_3 $\xrightarrow[Pt]{H_2}$ $CH_3CH_2CH(CH_3)CH_2CH_3$

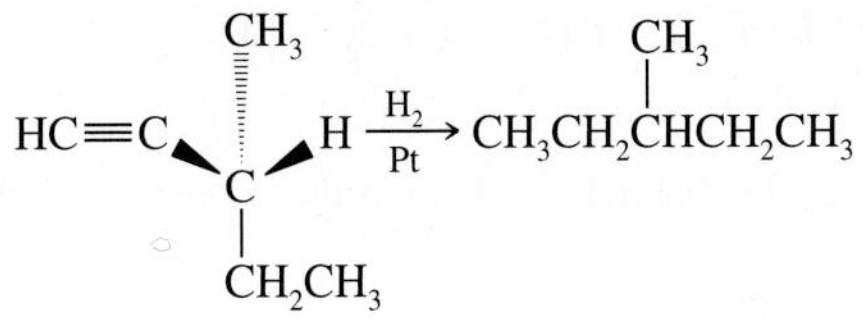

D **E**

CHAPTER 9

9.2 (a) One; (b) two; (c) two; (d) one; (e) two; (f) two.

9.7 A doublet (3H) downfield; a quartet (1H) upfield.

9.8 **A**, CH_3CHICH_3; **B**, CH_3CHCl_2; **C** $CH_2ClCH_2CH_2Cl$

9.30 Phenylacetylene.

9.31 **G**, $CH_3CH_2CHBrCH_3$
H, $CH_2{=}CBrCH_2Br$

9.36 **Q** is bicyclo[2.2.1}hepta-2,5-diene.
R is bicyclo[2.2.1]heptane.

CHAPTER 10

10.1 (a) $\Delta H° = -545$ kJ mol^{-1}; (c) $\Delta H° = -101$ kJ mol^{-1}; (e) $\Delta H° = +53$ kJ mol^{-1}; (g) $\Delta H° = -132$ kJ mol^{-1}.

10.4 A small amount of ethane is formed by the combination of two methyl radicals; the ethane then reacts with chlorine to form chloroethane.

10.14 (a) Cyclopentane; (c) 2,2-dimethylpropane.

10.16 (a)

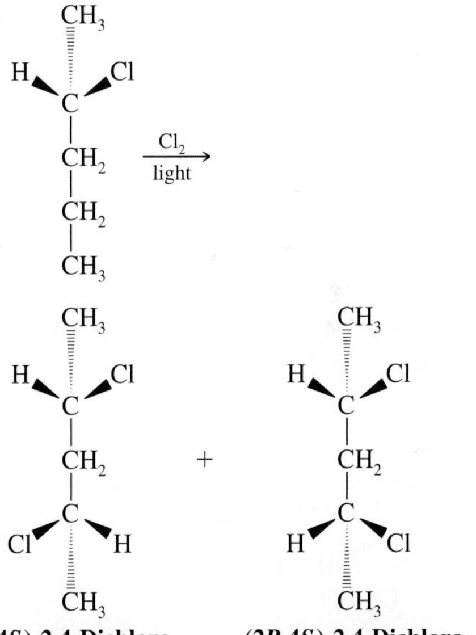

CH_3, H, Cl, C, CH_2, CH_2, CH_3 $\xrightarrow[\text{light}]{Cl_2}$

CH_3, H, Cl, C, CH_2, C, Cl, H, CH_3 + CH_3, H, Cl, C, CH_2, C, H, Cl, CH_3

(2*S*,4*S*)-2,4-Dichloropentane **(2*R*,4*S*)-2,4-Dichloropentane**

(c) No, (2*R*,4*S*)-2,4-dichloropentane is achiral because it is a meso compound. (It has a plane of symmetry passing through C3.) (e) Yes, by fractional distillation or by gas–liquid chromatography. (Diastereomers have different physical properties. Therefore, the two isomers would have different vapor pressures.)

10.17 (a) Seven fractions; (c) none of the fractions would show optical activity.

10.18 (a) No, the only fractions that would contain chiral molecules (as enantiomers) would be that containing 1-chloro-2-

methylbutane and the one containing 2-chloro-3-methylbutane. These fractions would not show optical activity, however, because they would contain racemic forms of the enantiomers. (b) Yes, the fraction containing 1-chloro-2-methylbutane and the one containing 2-chloro-3-methylbutane.

10.23 $CH_3\overset{\bullet}{C}(CH_3)CH_2CH_3$ (3°) > $CH_3\overset{\bullet}{C}HCH(CH_3)CH_3$ (2°) > $CH_3CH(CH_3)CH_2CH_2\cdot$ (1°) ~ $CH_3CH_2CH(CH_2\cdot)CH_3$ (1°)

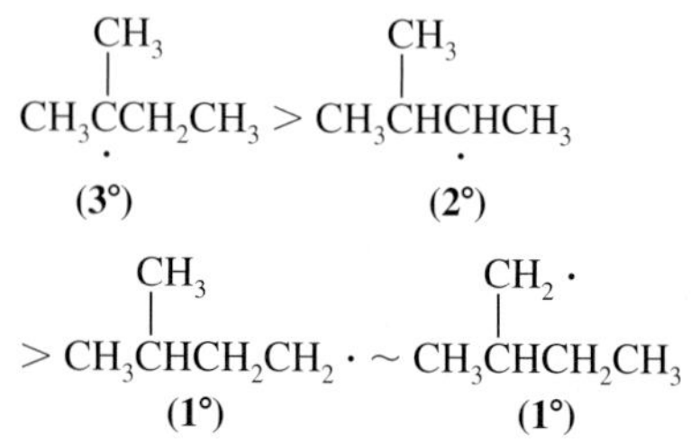

CHAPTER 11

11.3 The presence of two —OH groups in the glycols allows their molecules to form more hydrogen bonds.

11.4 (a) CH_3CH_2OH; (c) $(CH_3)_3COH$.

11.11 Use an alcohol containing labeled oxygen. If all of the labeled oxygen appears in the sulfonate ester, then it can be concluded that the alcohol C—O bond does not break during the reaction.

11.25 (a) 3,3-Dimethyl-1-butanol; (c) 2-methyl-1,4-butanediol; (e) 1-methyl-2-cyclopenten-1-ol.

11.26 (a) 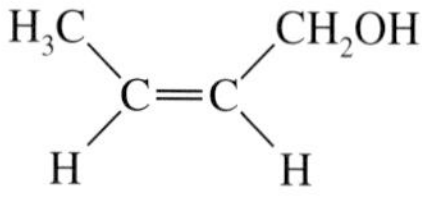 $H_3C(H)C{=}C(H)CH_2OH$ (trans)

(c) cyclopentane-1,2-diol with H, HO on one carbon and OH, H on the adjacent carbon

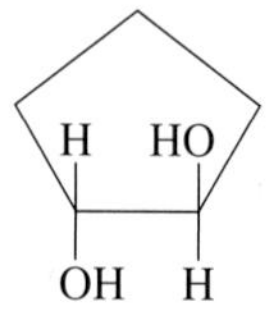

(e) 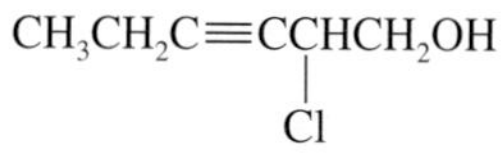 $CH_3CH_2C{\equiv}CCH(Cl)CH_2OH$

(g) 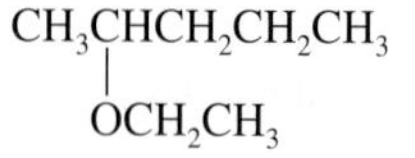 $CH_3CH(OCH_2CH_3)CH_2CH_2CH_3$

(i) 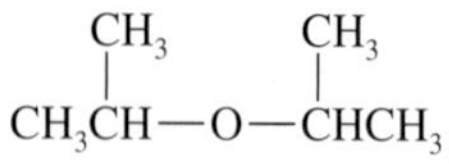 $CH_3CH(CH_3){-}O{-}CH(CH_3)CH_3$

11.32 (a) $CH_3Br + CH_3CH_2Br$; (c) $Br{-}CH_2CH_2CH_2CH_2{-}Br$

CHAPTER 12

12.4 (a) $LiAlH_4$; (c) $NaBH_4$

12.5 (a) 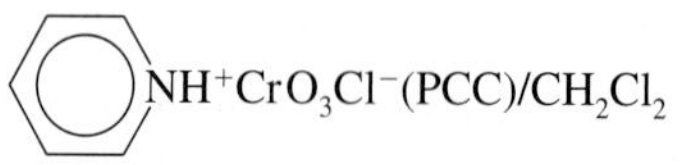 $C_5H_5NH^+CrO_3Cl^-$(PCC)/CH_2Cl_2

(c) H_2CrO_4/acetone

12.10 (a) 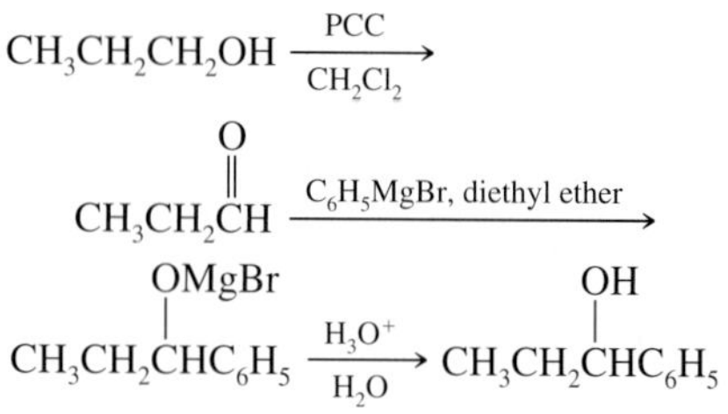 $CH_3CH_2CH_2OH \xrightarrow[CH_2Cl_2]{PCC} CH_3CH_2CH{=}O \xrightarrow{C_6H_5MgBr,\ diethyl\ ether} CH_3CH_2CH(OMgBr)C_6H_5 \xrightarrow[H_2O]{H_3O^+} CH_3CH_2CH(OH)C_6H_5$

(c) $CH_3CH_2C(=O)OCH_3 \xrightarrow[\text{[from part (a)]}]{2\ C_6H_5MgBr,\ diethyl\ ether} C_6H_5C(OMgBr)(C_6H_5)CH_2CH_3 \xrightarrow[H_2O]{NH_4^+} C_6H_5C(OH)(C_6H_5)CH_2CH_3$

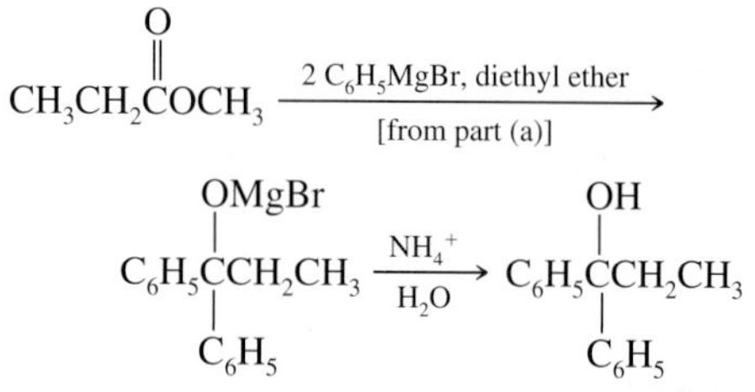

12.12 (a) CH_3CH_3; (b) CH_3CH_2D;

(c) 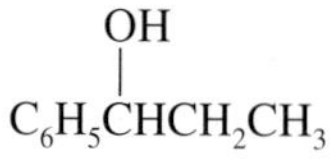 $C_6H_5CH(OH)CH_2CH_3$

(g) 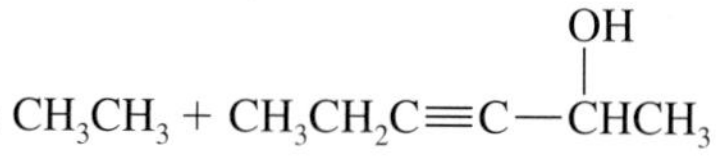 $CH_3CH_3 + CH_3CH_2C{\equiv}C{-}CH(OH)CH_3$

(h) 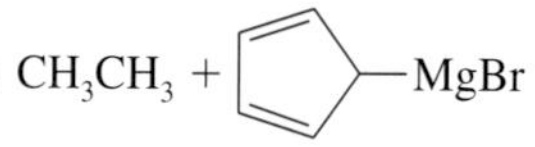 CH_3CH_3 + cyclopentadienyl—MgBr

12.13 (a) 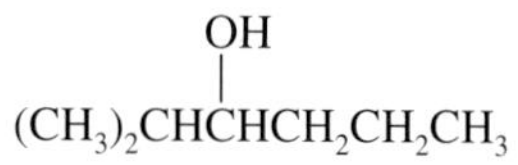 $(CH_3)_2CHCH(OH)CH_2CH_2CH_3$

(b) 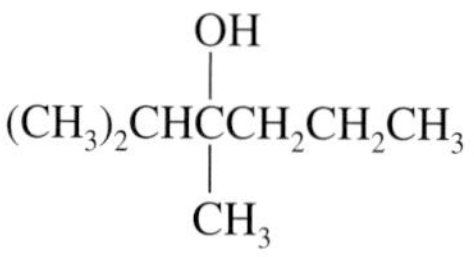 $(CH_3)_2CHC(OH)(CH_3)CH_2CH_2CH_3$

(e) $CH_3CH_2CH_2CH_2CH{=}CH_2$

(f) 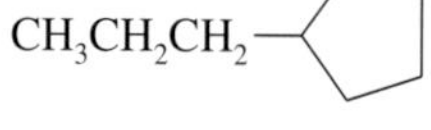 $CH_3CH_2CH_2$—cyclopentyl

12.21 γ-butyrolactone $\xrightarrow[(2)\ NH_4^+]{(1)\ 2\ CH_3MgI}$ $(H_3C)_2C(OH)CH_2CH_2CH_2OH$

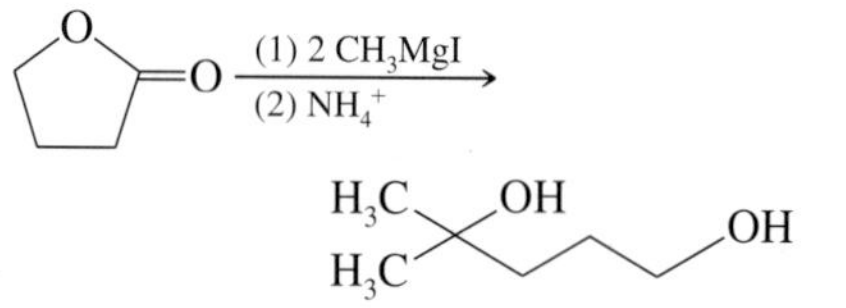

CHAPTER 13

13.1 (a) ${}^{14}CH_2{=}CH{-}CH_2{-}X + X{-}{}^{14}CH_2{-}CH{=}CH_2$; (c) in equal amounts.

13.6 (b) 1,4-Cyclohexadiene and 1,4-pentadiene are isolated dienes.

13.16 (a) 1,4-Dibromobutane + $(CH_3)_3COK$, and heat; (g) $HC{\equiv}CCH{=}CH_2 + H_2$, Ni_2B (P-2).

13.19 (a) 1-Butene + *N*-bromosuccinimide, then $(CH_3)_3COK$ and heat; (e) cyclopentane + Br_2, *hv*, then $(CH_3)_3COK$ and heat, then *N*-bromosuccinimide.

13.28 This is another example of rate versus equilibrium control of a reaction. The endo adduct, **G**, is formed faster, and at the lower temperature it is the major product. The exo adduct, **H**, is more stable, and at the higher temperature it is the major product.

CHAPTER 14

14.1 Compounds (a) and (b).

14.8 It suggests that the cyclopropenyl cation should be aromatic.

14.10 The cyclopropenyl cation.

14.15 **A**, *o*-Bromotoluene; **B**, *p*-bromotoluene; **C**, *m*-bromotoluene; **D**, benzyl bromide.

14.18 Hückel's rule should apply to both pentalene and heptalene. Pentalene's antiaromaticity can be attributed to its having 8 π electrons. Heptalene's lack of aromaticity can be attributed to its having 12 π electrons. Neither 8 nor 12 is a Hückel number.

14.20 The bridging $—CH_2—$ group causes the 10 π electron ring system (below) to become planar. This allows the ring to become aromatic.

14.23 (a) The cycloheptatrienyl anion has 8 π electrons, and does not obey Hückel's rule; the cyclononatetraenyl anion with 10 π electrons obeys Hückel's rule.

14.25 **A**, $C_6H_5CH(CH_3)_2$; **B**, $C_6H_5CH(NH_2)CH_3$;
C, (benzene ring fused with CH_2–CH_2–CH_2)

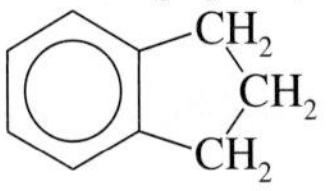

14.27 $CH(CH_3)_2$ (para-substituted benzene) CH_3
F

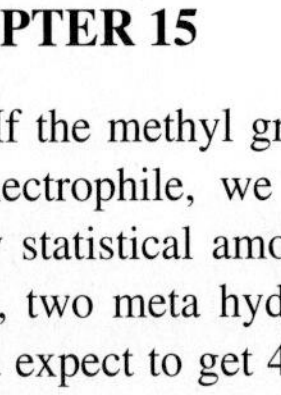

CHAPTER 15

15.8 If the methyl group had no directive effect on the incoming electrophile, we would expect to obtain the products in purely statistical amounts. Since there are two ortho hydrogen atoms, two meta hydrogen atoms, and one para hydrogen, we would expect to get 40% ortho (2/5), 40% meta (2/5), and 20% para (1/5). Thus, we would expect that only 60% of the mixture of mononitrotoluenes would have the nitro group in the ortho or para position. And, we would expect to obtain 40% of *m*-nitrotoluene. In actuality, we get 96% of combined *o*- and *p*-nitrotoluene and only 4% *m*-nitrotoluene. This result shows the ortho–para directive effect of the methyl group.

15.11 (b) Structures such as the following compete with the benzene ring for the oxygen electrons, making them less available to the benzene ring.

$C_6H_5–\ddot{O}–C(=\ddot{O}:)–CH_3 \longleftrightarrow C_6H_5–\overset{+}{O}=C(–\ddot{O}:^-)–CH_3$

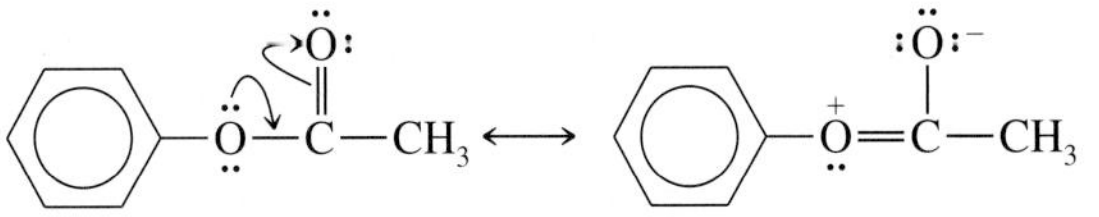

(d) Structures such as the following compete with the benzene ring for the nitrogen electrons, making them less available to the benzene ring.

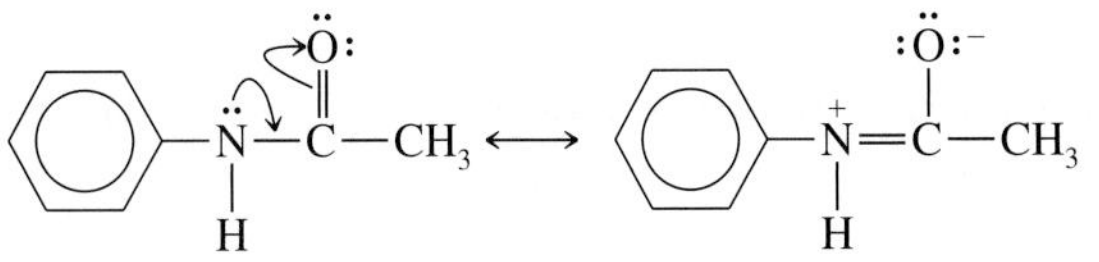

$C_6H_5–\ddot{N}(H)–C(=\ddot{O}:)–CH_3 \longleftrightarrow C_6H_5–\overset{+}{N}(H)=C(–\ddot{O}:^-)–CH_3$

15.28 (a) 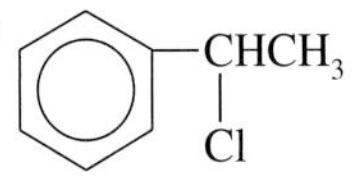 $C_6H_5–CHCH_3$ with Cl on CH

(c) $C_6H_5–CH=CHCH_2CH_3$

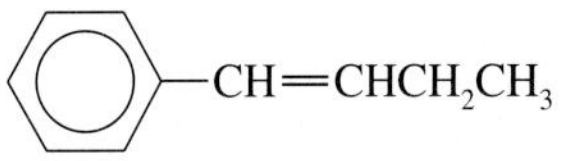

(e) $C_6H_5–CHCH_2CH_2CH_3$ with OH on CH

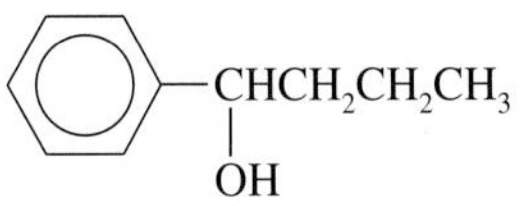

(g) $C_6H_5–C(=O)–OH$

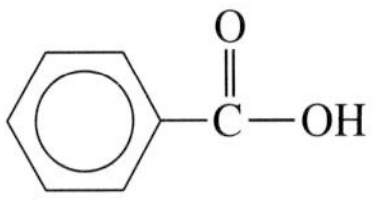

15.35

(a) $C_6H_4(o\text{-}Br)–CH_2–C(=O)–C_6H_5$ +

$Br–C_6H_4–CH_2–C(=O)–C_6H_5$ (para)

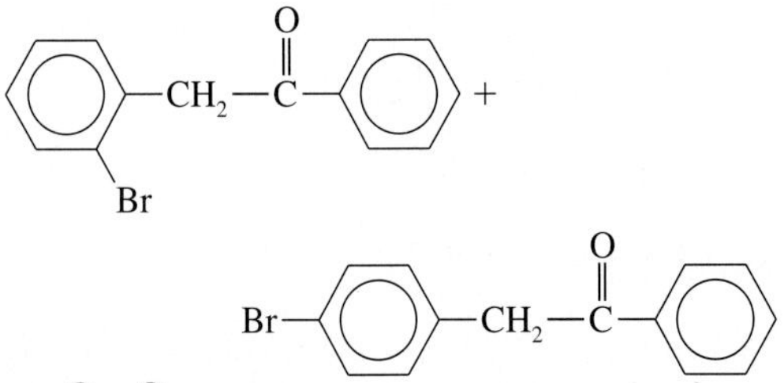

(c) $C_6H_4(m\text{-}Br)–C(=O)–C(=O)–C_6H_5$

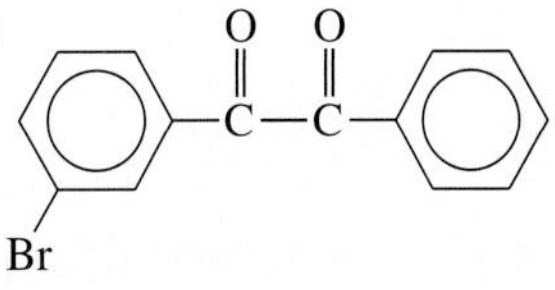

CHAPTER 16

16.2 (a) 1-Pentanol; (c) pentanal; (e) benzyl alcohol.

16.6 A hydride ion.

16.17 (b) $CH_3CH_2Br + (C_6H_5)_3P$, then strong base, then $C_6H_5COCH_3$; (d) $CH_3I + (C_6H_5)_3P$, then strong base, then cyclopentanone; (f) $CH_2=CHCH_2Br + (C_6H_5)_3P$, then strong base, then C_6H_5CHO.

16.24 (a) $CH_3CH_2CH_2OH$; (c) $CH_3CH_2CH_2OH$
(h) $CH_3CH_2CH{=}CHCH_3$; (j) $CH_3CH_2CO_2^-NH_4^+ + Ag\downarrow$
(l) $CH_3CH_2CH{=}NNHCONH_2$; (n) $CH_3CH_2CO_2^-$

16.42 (a) Tollens' reagent; (e) Br_2 in CCl_4; (f) Tollens' reagent; (h) Tollens' reagent.

16.43

X is (benzene ring fused to C(=O)–CH_2–CH_2)

16.44 **Y** is 1-phenyl-2-butanone; **Z** is 4-phenyl-2-butanone.

CHAPTER 17

17.1 The enol form is phenol. It is especially stable because it is aromatic.

17.5 Base is consumed as the reaction takes place. A catalyst, by definition, is not consumed.

17.13 $C_6H_5CHO + OH^- \xrightarrow[\text{heat}]{CH_3CHO} C_6H_5CH{=}CHCHO$

17.16 (b) $CH_3NO_2 + HCH(=O) \xrightarrow{OH^-} HOCH_2CH_2NO_2$

17.28 (a) $CH_3CH_2CH(OH)CH(CH_3)CHO$

(b) $C_6H_5CH{=}C(CH_3)CHO$

(k) $CH_3CH_2CH(OH)C_6H_5$; (l) $CH_3CH_2CH(OH)C{\equiv}CH$

17.33 **B** is $CH_3C(=O){-}C(CH_3)(OH){-}CH_3$

CHAPTER 18

18.3 (a) CH_2FCO_2H; (c) CH_2ClCO_2H; (e) $CH_3CH_2CHFCO_2H$;

(g) $CF_3{-}C_6H_4{-}CO_2H$ (para)

18.7 (a) $C_6H_5CH_2Br$ + Mg + diethyl ether, then CO_2, then H_3O^+; (c) $CH_2{=}CHCH_2Br$ + Mg + diethyl ether, then CO_2, then H_3O^+.

18.8 (a), (c), and (e).

18.10 In the carboxyl of benzoic acid.

18.15 (a) $(CH_3)_3CCO_2H + SOCl_2$, then NH_3, then P_4O_{10}, heat;

(b) $CH_2{=}C(CH_3){-}CH_3$

18.24 (a) CH_3CO_2H; (c) $CH_3CO_2CH_2(CH_2)_2CH_3$;
(e) p-$CH_3COC_6H_4CH_3$ + o-$CH_3COC_6H_4CH_3$; (g) CH_3COCH_3;
(i) $CH_3CONHCH_3$; (k) $CH_3CON(CH_3)_2$; (m) $(CH_3CO)_2O$;
(o) $CH_3CO_2C_6H_5$

18.36 (a) Diethyl succinate; (c) ethyl phenylacetate; (e) ethyl chloroacetate.

18.40 **X** is diethyl malonate.

CHAPTER 19

19.4 (a) $CH_3CH(CO_2C_2H_5)COCO_2C_2H_5$

(b) $HC(=O)CH_2CO_2C_2H_5$

19.7 *O*-alkylation that results from the oxygen of the enolate ion acting as a nucleophile.

19.9 (a) Reactivity is the same as with any S_N2 reaction. With primary halides substitution is highly favored, with secondary halides elimination competes with substitution, and with tertiary halides elimination is the exclusive course of the reaction. (b) Acetoacetic ester and 2-methylpropene. (c) Bromobenzene is unreactive toward nucleophilic substitution.

19.29 (b) **D** is racemic *trans*-1,2-cyclopentanedicarboxylic acid, **E** is *cis*-1,2-cyclopentanedicarboxylic acid, a meso compound.

19.38 (a) $CH_2{=}C(CH_3)CO_2CH_3$; (b) $KMnO_4$, OH^-; H_3O^+; (c) CH_3OH, HA; (d) CH_3ONa, then H_3O^+

(e) and (f) (cyclopentanone ring with CO_2CH_3 and CH_3 on one carbon and CO_2CH_3 on the adjacent carbon α to C=O)

and

(cyclopentanone ring with CO_2CH_3 and CH_3 on one carbon and CH_3O_2C on the carbon α to C=O)

(g) OH^-, H_2O, then H_3O^+; (h) heat ($-CO_2$); (i) CH_3OH, HA; (j) Zn, $BrCH_2CO_2CH_3$, diethyl ether, then H_3O^+

(k) (cyclopentane ring with CO_2CH_3 and CH_3 on one carbon and $=CHCO_2CH_3$ on an adjacent carbon)

(l) H_2, Pt; (m) CH_3ONa, then H_3O^+; (n) 2 $NaNH_2$ + 2 CH_3I

CHAPTER 20

20.5 (a) $CH_3(CH_2)_3CHO + NH_3 \xrightarrow{H_2, Ni} CH_3(CH_2)_3CH_2NH_2$

(c) $CH_3(CH_2)_4CHO + C_6H_5NH_2 \xrightarrow{LiBH_3CN}$

$CH_3(CH_2)_4CH_2NHC_6H_5$

20.6 The reaction of a secondary halide with ammonia is almost always accompanied by some elimination.

20.8 (a) Methoxybenzene + HNO_3 + H_2SO_4, then Fe + HCl; (b) Methoxybenzene + CH_3COCl + $AlCl_3$, then NH_3 + H_2 + Ni; (c) toluene + Cl_2 and light, then $(CH_3)_3N$; (d) *p*-nitrotoluene + $KMnO_4$ + OH^-, then H_3O^+, then $SOCl_2$ followed by NH_3, then NaOBr (Br_2 in NaOH); (e) toluene + *N*-bromosuccinimide in CCl_4, then KCN, then $LiAlH_4$.

20.14 *p*-Nitroaniline + Br_2 + Fe, followed by $H_2SO_4/NaNO_2$ followed by CuBr, then H_2/Pt, then $H_2SO_4/NaNO_2$ followed by H_3PO_2.

20.37 **W** is *N*-benzyl-*N*-ethylaniline.

CHAPTER 21

21.4 (a) The para-sulfonated phenol. (b) For ortho sulfonation.

21.9 (a) OCH_2CH_3 (benzene ring with NO_2 at positions 2 and 4) (b) NH_2 (benzene ring with NO_2 at positions 2 and 4)

21.10 That *o*-chlorotoluene leads to the formation of two products (*o*-cresol and *m*-cresol), when submitted to the conditions used in the Dow process, suggests that an elimination-addition mechanism takes place.

21.11 2-Bromo-1,3-dimethylbenzene, because it has no *o*-hydrogen atom, cannot undergo an elimination. Its lack of reactivity toward sodium amide in liquid ammonia suggests that those compounds (e.g., bromobenzene) that do react, react by a mechanism that begins with an elimination.

21.15 (a) 4-Chlorophenol will dissolve in aqueous NaOH; 4-chloro-1-methylbenzene will not. (c) Phenyl vinyl ether will react with bromine in carbon tetrachloride by addition (thus decolorizing the solution); ethyl phenyl ether will not. (e) 4-Ethylphenol will dissolve in aqueous NaOH; ethyl phenyl ether will not.

21.17 (a) 4-Fluorophenol because a fluorine substituent is more electron withdrawing than a methyl group. (e) 4-Fluorophenol because fluorine is more electronegative than bromine.

CHAPTER 22

22.1 (a) Two; (b) two; (c) four.

22.5 Acid catalyzes hydrolysis of the glycosidic (acetal) group.

22.9 (a) 2 CH_3CHO, one molar equivalent HIO_4; (b) HCHO + HCO_2H + CH_3CHO, two molar equivalents HIO_4; (c) HCHO + $OHCCH(OCH_3)_2$, one molar equivalent HIO_4; (d) HCHO + HCO_2H + CH_3CO_2H, two molar equivalents HIO_4; (e) 2 CH_3CO_2H + HCO_2H, two molar equivalents HIO_4

22.18 D-(+)-Glucose.

22.23 One anomeric form of D-mannose is dextrorotatory ($[\alpha]_D = +29.3°$), the other is levorotatory ($[\alpha]_D = -17.0°$).

22.24 The microorganism selectively oxidizes the —CHOH group of D-glucitol that corresponds to C5 of D-glucose.

22.27 **A** is D-altrose; **B** is D-talose, **C** is D-galactose

CHAPTER 23

23.5 Br_2 in CCl_4 would react with geraniol (discharging the bromine color) but would not react with menthol.

23.12 (a) C_2H_5OH, HA, heat; or $SOCl_2$, then C_2H_5OH; (d) $SOCl_2$, then $(CH_3)_2NH$; (g) $SOCl_2$, then $LiAlH[OC(CH_3)_3]_3$; (j) $SOCl_2$, then $(CH_3)_2CuLi$

23.15 Elaidic acid is *trans*-9-octadecenoic acid.

23.19 **A** is $CH_3(CH_2)_5C{\equiv}CNa$
B is $CH_3(CH_2)_5C{\equiv}CCH_2(CH_2)_7CH_2Cl$
C is $CH_3(CH_2)_5C{\equiv}CCH_2(CH_2)_7CH_2CN$
E is $CH_3(CH_2)_5C{\equiv}CCH_2(CH_2)_7CH_2CO_2H$
Vaccenic acid is

$CH_3(CH_2)_5$(H)C=C(H)$(CH_2)_9CO_2H$ (cis)

23.20 **F** is $FCH_2(CH_2)_6CH_2C{\equiv}CH$
G is $FCH_2(CH_2)_6CH_2C{\equiv}C(CH_2)_7Cl$
H is $FCH_2(CH_2)_6CH_2C{\equiv}C(CH_2)_7CN$
I is $FCH_2(CH_2)_7C{\equiv}C(CH_2)_7CO_2H$

CHAPTER 24

24.5 The labeled amino acid no longer has a basic $—NH_2$ group; it is, therefore, insoluble in aqueous acid.

24.8 Glutathione is

$H_3\overset{+}{N}CH(CO_2^-)CH_2CH_2CONHCH(CH_2SH)CONHCH_2CO_2^-$

24.21 Arg·Pro·Pro·Gly·Phe·Ser·Pro·Phe·Arg

24.22 Val·Leu·Lys·Phe·Ala·Glu·Ala

CHAPTER 25

25.2 (a) The nucleosides have an *N*-glycosidic linkage that (like an *O*-glycosidic linkage) is rapidly hydrolyzed by aqueous acid, but one that is stable in aqueous base.

25.3 The reaction appears to take place through an S_N2 mechanism. Attack occurs preferentially at the primary 5′ carbon atom rather than at the secondary 3′ carbon atom.

25.5 (a) The isopropylidene group is part of a cyclic acetal. (b) By treating the nuceloside with acetone and a trace of acid.

25.8 (b) Thymine would pair with adenine, and, therefore, adenine would be introduced into the complementary strand where guanine should occur.

25.13 A change from C-T-T to C-A-T, or a change from C-T-C to C-A-C.

Glossary

A

Absolute configuration (Section 5.15A): The actual arrangement of groups in a molecule. The absolute configuration of a molecule can be determined by X-ray analysis or by relating the configuration of a molecule, using reactions of known stereochemistry, to another molecule whose absolute configuration is known.

Absorption spectrum (Section 13.9A): A plot of the wavelength (λ) of a region of the spectrum versus the absorbance (A) at each wavelength. The absorbance at a particular wavelength (A_λ) is defined by the equation $A_\lambda = \log (I_R/I_S)$, where I_R is the intensity of the reference beam and I_S is the intensity of the sample beam.

Acetal (Section 16.7B): A functional group, consisting of a carbon bonded to alkoxy groups [i.e., $RCH(OR')_2$ or $R_2C(OR')_2$], derived by adding 2 molar equivalents of an alcohol to an aldehyde or ketone.

Acetylene (Sections 1.14, 7.11 and 7.12): A common name for ethyne. Also used as a general name for alkynes.

Acetylenic hydrogen atom (Sections 3.14, 4.18C and 7.12): A hydrogen atom attached to a carbon atom that is bonded to another carbon atom by a triple bond.

Achiral molecule (Section 5.3): A molecule that is superposable on its mirror image. Achiral molecules lack handedness and are incapable of existing as a pair of enantiomers.

Acid strength (Section 3.5): The strength of an acid is related to its acidity constant, K_a or to its pK_a. The larger the value of its K_a or the smaller the value of its pK_a, the stronger is the acid.

Acidity constant (Section 3.5A): An equilibrium constant related to the strength of an acid. For the reaction,

$$HA + H_2O \rightleftharpoons H_3O^+ + A^-$$

$$K_a = \frac{[H_3O^+][A^-]}{[HA]}$$

Activating group (Section 15.10): A group that when present on a benzene ring causes the ring to be more reactive in electrophilic substitution than benzene itself.

Activation Energy E_{act}; see **Energy of activation.**

Acyl group (Section 15.7): The general name for groups with the structure RCO— or ArCO—.

Acyl halide (Section 15.7): Also called an *acid halide*. A general name for compounds with the structure RCOX or ArCOX.

Acylation (Section 15.7): The introduction of an acyl group into a molecule.

Acylium ion (Sections 9.16 and 15.7): The resonance-stabilized cation:

$$R{-}\overset{+}{C}{=}\ddot{O}{:} \longleftrightarrow R{-}C{\equiv}\overset{+}{O}$$

Addition polymer (Section 10.10): A polymer that results from a stepwise addition of monomers to a chain (usually through a chain reaction) with no loss of other atoms or molecules in the process.

Addition reaction (Section 8.1): A reaction that results in an increase in the number of groups attached to a pair of atoms joined by a double or triple bond. An addition reaction is the opposite of an elimination reaction.

Aglycone (Section 22.4): The alcohol obtained by hydrolysis of a glycoside.

Aldaric acid (Section 22.6C): An α,ω-dicarboxylic acid that results from oxidation of the aldehyde group and the terminal 1° alcohol group of an aldose.

Alditol (Section 22.7): The alcohol that results from the reduction of the aldehyde or keto group of an aldose or ketose.

Aldonic acid (Section 22.6B): A monocarboxylic acid that results from oxidation of the aldehyde group of an aldose.

Aliphatic compound (Section 14.1): A nonaromatic compound such as an alkane, cycloalkane, alkene, or alkyne.

Alkaloid (Special Topic E): A naturally occurring basic compound that contains an amino group. Most alkaloids have profound physiological effects.

Alkylation (Sections 4.17C and 15.7): The introduction of an alkyl group into a molecule.

Allyl group (Section 4.5): The $CH_2{=}CHCH_2{-}$ group.

Allylic substituent (Section 13.2): Refers to a substituent on a carbon atom adjacent to a carbon–carbon double bond.

Ambident nucleophile (Section 17.7C): Nucleophiles that are capable of reacting at two different nucleophilic sites.

Angle strain (Section 4.11): The increased potential energy of a molecule (usually a cyclic one) caused by deformation of a bond angle away from its lowest energy value.

Annulene (Section 14.7A): Monocyclic hydrocarbons that can be represented by structures having alternating single and double bonds. The ring size of an annulene is represented by a number in brackets, e.g., benzene is [6]annulene and cyclooctatetraene is [8]annulene.

Anomers (Section 22.2C): A term used in carbohydrate chemistry. Anomers are diastereomers that differ only in configuration at the acetal or hemiacetal carbon of a sugar in its cyclic form.

Anti addition (Section 8.13): An addition that places the parts of the adding reagent on opposite faces of the reactant.

Anti conformation (Section 4.9A): An anti conformation of butane, for example, has the methyl groups at an angle of 180° to each other:

CH_3

CH_3

Anti conformation of butane

Antiaromatic compound (Section 14.7D): A cyclic conjugated system whose π electron energy is greater than that of the corresponding open-chain compound.

Antibonding molecular orbital (antibonding MO) (Sections 1.11 and 1.13): A molecular orbital whose energy is higher than that of the isolated atomic orbitals from which it is constructed. Electrons in an antibonding molecular orbital destabilize the bond between the atoms that the orbital encompasses.

Anticodon (Section 25.5C): A sequence of three bases on transfer RNA (tRNA) that associates with a codon of messenger RNA (mRNA).

Aprotic solvent (Section 6.14C): A solvent whose molecules do not have a hydrogen atom attached to a strongly electronegative element (such as oxygen). For most purposes, this means that an aprotic solvent is one whose molecules lack an —OH group.

Arene (Section 15.1): A general name for an aromatic hydrocarbon.

Aromatic compound (Sections 14.1–14.8 and 14.11): A cyclic conjugated unsaturated molecule or ion that is stabilized by π electron delocalization. Aromatic compounds are characterized by having large resonance energies, by reacting by substitution rather than addition, and by deshielding of protons exterior to the ring in their ^{1}H NMR spectra caused by the presence of an induced ring current.

Aryl group (Section 15.1): The general name for a group obtained (on paper) by the removal of a hydrogen from a ring position of an aromatic hydrocarbon. Abbreviated Ar—.

Aryl halide (Section 6.1): An organic halide in which the halogen atom is attached to an aromatic ring, such as a benzene ring.

Atactic polymer (Special Topic A): A polymer in which the configuration at the stereogenic centers along the chain is random.

Atomic orbital (AO) (Section 1.10): A volume of space about the nucleus of an atom where there is a high probability of finding an electron. An atomic orbital can be described mathematically by its **wave function.** Atomic orbitals have characteristic quantum numbers; the *principal quantum number, n,* is related to the energy of the electron in an atomic orbital and can have the values 1, 2, 3, ... The *azimuthal quantum number, l,* determines the angular momentum of the electron that results from its motion around the nucleus, and can have the values, 0, 1, 2, . . . , $(n - 1)$. The *magnetic quantum number, m,* determines the orientation in space of the angular momentum and can have values from $+l$ to $-l$. The *spin quantum number, s,* measures the intrinsic angular momentum of an electron and can have the values of $+\frac{1}{2}$ and $-\frac{1}{2}$ only.

Aufbau principle (Section 1.10): A principle that guides us in assigning electrons to orbitals of an atom or molecule in its lowest energy state or **ground state**. The aufbau principle states that electrons are added so that orbitals of lowest energy are filled first.

Autoxidation (Section 10.11C): The reaction of an organic compound with oxygen to form a hydroperoxide.

Axial bond (Section 4.13): The six bonds of a cyclohexane ring (below) that are perpendicular to the general plane of the ring, and that alternate up

and down around the ring.

B

Base peak (Section 9.14): The most intense peak in a mass spectrum.

Base strength (Section 3.5): The strength of a base is inversely related to the strength of its conjugate acid; the weaker the conjugate acid, the stronger is the base. In other words, if the conjugate acid has a large pK_a, the base will be strong.

Benzenoid aromatic compound (Section 14.8A): An aromatic compound whose molecules have one or more benzene rings.

Benzyl group (Sections 2.5B and 15.15): The $C_6H_5CH_2$— group.

Benzylic substituent (Section 15.15): Refers to a substituent on a carbon atom adjacent to a benzene ring.

Benzyne (Section 21.11B): An unstable, highly reactive intermediate consisting of a benzene ring with an additional bond resulting from sideways overlap of sp^2 orbitals on adjacent atoms of the ring.

Betaine (Section 16.10): An electrically neutral molecule that has nonadjacent cationic and anionic sites and that does not possess a hydrogen atom bonded to the cationic site.

Bimolecular reaction (Section 6.6): A reaction whose rate-determining step involves two initially separate species.

Boat conformation (Section 4.12): A conformation of cyclohexane that resembles a boat and that has eclipsed bonds along its two sides:

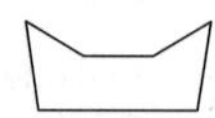

It is of higher energy than the chair conformation.

Bond angle (Sections 1.12 and 1.16): The angle between two bonds originating at the same atom.

Bond dissociation energy, see **Homolytic bond dissociation energy.**

Bond length (Sections 1.11 and 1.14A): The equilibrium distance between two bonded atoms or groups.

Bond-line formula (Section 1.17D): A formula that shows the carbon skeleton of a molecule with lines. The number of hydrogen atoms necessary to fulfill each carbon's valence is assumed to be present but not written in. Other atoms (e.g., O, Cl, N) are written in.

Bonding molecular orbital (bonding MO) (Section 1.11): The energy of a bonding molecular orbital is lower than the energy of the isolated atomic orbitals from which it arises. When electrons occupy a bonding molecular orbital they help hold together the atoms that the molecular orbital encompasses.

Bromination (Sections 8.12, 10.5C, and 10.6A): A reaction in which one or more bromine atoms are introduced into a molecule.

Bromohydrin (Section 8.14): A bromo alcohol.

Bromonium ion (Section 8.12A): An ion containing a positive bromine atom bonded to two carbon atoms.

Brønsted–Lowry theory of acids and bases (Section 3.2A): An acid is a substance that can donate (or lose) a proton; a base is a substance that can accept (or remove) a proton. The **conjugate acid** of a base is the molecule or ion that forms when a base accepts a proton. The **conjugate base** of an acid is the molecule or ion that forms when an acid loses its proton.

C

Carbanion (Section 3.3): A chemical species in which a carbon atom bears a formal negative charge.

Carbene (Section 8.15): An uncharged species in which a carbon atom is divalent. The species, $:CH_2$, called methylene, is a carbene.

Carbenoid (Section 8.15C): A carbene-like species. A species such as the reagent formed when diiodomethane reacts with a zinc–copper couple. This reagent, called the Simmons–Smith reagent, reacts with alkenes to add methylene to the double bond in a stereospecific way.

Carbocation (Section 3.3): A chemical species in which a trivalent carbon atom bears a formal positive charge.

Carbohydrate (Section 22.1A): A group of naturally occurring compounds that are usually defined as polyhydroxyaldehydes or polyhydroxyketones, or as substances that undergo hydrolysis to yield such compounds. In actuality, the aldehyde and ketone groups of carbohydrates are often present as hemiacetals and acetals. The name comes from the fact that many carbohydrates possess the empirical formula $C_x(H_2O)_y$.

Carbonyl group (Section 16.1): A functional group consisting of a carbon atom doubly bonded to an oxygen atom. The carbonyl group is found in aldehydes, ketones, esters, anhydrides, amides, acyl halides, and so on. Collectively these compounds are referred to as carbonyl compounds.

Cascade polymer (Special Topic B.5): A polymer produced from a multifunctional central core by progressively adding layers of repeating units.

CFC (see **Freon**): A chlorofluorocarbon.

Chain reaction (Section 10.4): A reaction that proceeds by a sequential, stepwise mechanism, in which each step generates the reactive intermediate that causes the next step to occur. Chain reactions have *chain initiating steps, chain propagating steps,* and *chain terminating steps.*

Chair conformation (Section 4.12): The all-staggered conformation of cyclohexane that has no angle strain or torsional strain and is, therefore, the lowest energy conformation:

Chemical shift, δ (Sections 9.3C and 9.6): The position in an NMR spectrum, relative to a reference compound, at which a nucleus absorbs. The reference compound most often used is tetramethylsilane (TMS), and its absorption point is arbitrarily designated zero. The chemical shift of a given nucleus is proportional to the strength of the magnetic field of the spectrometer. The chemical shift in delta units, δ, is determined by dividing the observed shift from TMS in hertz multiplied by 10^6 by the operating frequency of the spectrometer in hertz.

Chiral molecule (Section 5.3): A molecule that is not superposable on its mirror image. Chiral molecules have handedness and are capable of existing as a pair of enantiomers.

Chirality (Section 5.3): The property of having handedness.

Chlorination (Sections 8.12 and 10.5): A reaction in which one or more chlorine atoms are introduced into a molecule.

Chlorohydrin (Section 8.14): A chloro alcohol.

Cis–trans isomers (Sections 4.5 and 7.2): Diastereomers that differ in their stereochemistry at adjacent atoms of a double bond or on different atoms of a ring.

Codon (Section 25.5C): A sequence of three bases on messenger RNA (mRNA) that contains the genetic information for one amino acid. The codon associates, by hydrogen bonding, with an anticodon of a transfer RNA (tRNA) that carries the particular amino acid for protein synthesis on the ribosome.

Condensation polymer (Special Topic B): A polymer produced when bifunctional monomers (or potentially bifunctional monomers) react with each other through the intermolecular elimination of water or an alcohol. Polyesters, polyamides, and polyurethanes are all condensation polymers.

Condensation reaction (Section 17.4): A reaction in which molecules become joined through the intermolecular elimination of water or an alcohol.

Configuration (Section 5.7): The particular arrangement of atoms (or groups) in space that is characteristic of a given stereoisomer.

Conformation (Section 4.8): A particular temporary orientation of a molecule that results from rotations about its single bonds.

Conformational analysis (Section 4.9): An analysis of the energy changes that a molecule undergoes as its groups undergo rotation (sometimes only partial) about the single bonds that join them.

Conformer (Section 4.8): A particular staggered conformation of a molecule.

Conjugate acid (Section 3.2A): The molecule or ion that forms when a base accepts a proton.

Conjugate addition (Section 17.9): A form of nucleophilic addition to an α,β-unsaturated carbonyl compound in which the nucleophile adds to the β-carbon.

Conjugate base (Section 3.2A): The molecule or ion that forms when an acid loses its proton.

Conjugated system (Section 13.1): Molecules or ions that have an extended π system. A conjugated system has a *p* orbital on an atom adjacent to a multiple bond; the *p* orbital may be that of another multiple bond or that of a radical, carbocation, or carbanion.

Connectivity (Section 1.3): The sequence, or order, in which the atoms of a molecule are attached to each other.

Constitutional isomers (Section 1.3A): Compounds that have the same molecular formula but that differ in their connectivity (i.e., molecules that have the same molecular formula but have their atoms connected in different ways).

Coplanar (Section 7.6C): A conformation in which vicinal groups lie in the same plane.

Copolymer (Special Topic A): A polymer synthesized by polymerizing two monomers.

Coupling constant, J_{ab} (Section 9.8): The separation in frequency units (hertz) of the peaks of a multiplet caused by spin–spin coupling between atoms *a* and *b*.

Covalent bond (Section 1.4B): The type of bond that results when atoms share electrons.

Cracking (Section 4.1C): A process used in the petroleum industry for breaking down the molecules of larger alkanes into smaller ones. Cracking may be accomplished with heat (thermal cracking), or with a catalyst (catalytic cracking).

Crown ether (Section 11.16A): Cyclic polyethers that have the ability to form complexes with metal ions. Crown ethers are named as *x*-crown-*y* where *x* is the total number of atoms in the ring and *y* is the number of oxygen atoms in the ring.

Cyanohydrin (Sections 16.9 and 18.3): A functional group consisting of a carbon atom bonded to a cyano group and to a hydroxyl group, i.e., RHC(OH)(CN) or $R_2C(OH)(CN)$, derived by adding HCN to an aldehyde or ketone.

Cycloaddition (Section 13.11): A reaction, like the Diels–Alder reaction, in which two connected groups add to the end of a π system to generate a new ring.

D

D and L designations (Section 22.2B): A method for designating the configuration of monosaccarides and other similar compounds in which the reference compound is (+)- or (−)-glyceraldehyde. According to this system, (+)-glyceraldehyde is designated D-(+)-glyceraldehyde and (−)-glyceraldehyde is designated L-(−)-glyceraldehyde. Therefore, a monosaccharide whose highest numbered stereogenic center has the same general configuration as D-(+)-glyceraldehyde is designated a D sugar; one whose highest numbered stereogenic center has the same general configuration as L-(+)-glyceraldehyde is designated an L-sugar.

Deactivating group (Section 15.10): A group that when present on a benzene ring causes the ring to be less reactive in electrophilic substitution than benzene itself.

Debromination (Section 7.9): The elimination of two atoms of bromine from a *vic*-dibromide, or, more generally, the loss of bromine from a molecule.

Debye unit (Section 2.3): The unit in which dipole moments are stated. One debye, D, equals 1×10^{-18} esu cm.

Decarboxylation (Section 18.11): A reaction whereby a carboxylic acid loses CO_2.

Degenerate orbitals (Section 1.10): Orbitals of equal energy. For example, the three 2*p* orbitals are degenerate.

Dehydration reaction (Section 7.7): An elimination that involves the loss of a molecule of water from the substrate.

Dehydrohalogenation (Section 6.16): An elimination reaction that results in the loss of HX from adjacent carbons of the substrate and the formation of a π bond.

Delocalization (Section 6.12B): The dispersal of electrons (or of electrical charge). Delocalization of charge always stabilizes a system.

Dextrorotatory (Section 5.8B): A compound that rotates plane-polarized light clockwise.

Diastereomers (Section 5.2): Stereoisomers that are not mirror images of each other.

Diastereoselective reaction (See **Stereoselective reaction** and Sections 5.10B and 12.3)

Diastereotopic hydrogens (or ligands) (Section 9.7B): If replacement of each of two hydrogens (or ligands) by the same groups yields compounds that are diastereomers, the two hydrogen atoms (or ligands) are said to be diastereotopic.

Dielectric constant (Section 6.14D): A measure of a solvent's ability to insulate opposite charges from each other. The dielectric constant of a solvent roughly measures its polarity. Solvents with high dielectric constants are better solvents for ions than are solvents with low dielectric constants.

Dienophile (Section 13.11): The diene-seeking component of a Diels–Alder reaction.

Dipolar ion (Section 24.2C): The charge-separated form of an amino acid that results from the transfer of a proton from a carboxyl group to a basic group.

Dipole moment, μ (Section 2.3): A physical property associated with a polar molecule that can be measured experimentally. It is defined as the product of the charge in electrostatic units (esu) and the distance that separates them in centimeters: $\mu = e \times d$.

Dipole-dipole interaction (Section 2.14B): An interaction between molecules having permanent dipole moments.

Disaccharide (Section 22.1A): A carbohydrate that, on a molecular basis, undergoes hydrolytic cleavage to yield two molecules of a monosaccharide.

Distortionless enhanced polarization transfer (DEPT) spectra (Section 9.10E): A series of ^{13}C NMR spectra in which the signal for each type of carbon, C, CH, CH_2, and CH_3, is printed out separately. Data from DEPT spectra help identify the different types of carbon atoms in a ^{13}C NMR spectrum.

E

(E–Z) system (Section 7.2): A system for designating the stereochemistry of alkene diastereomers based on the priorities of groups in the Cahn–Ingold–Prelog convention.

E1 reaction (Section 6.18): A unimolecular elimination in which, in a slow, rate-determining step, a leaving group departs from the substrate to form a carbocation. The carbocation then in a fast step loses a proton with the resulting formation of a π bond.

E2 reaction (Section 6.17): A bimolecular 1,2 elimination in which, in a single step, a base removes a proton and a leaving group departs from the substrate, resulting in the formation of a π bond.

Eclipsed conformation (Section 4.8): A temporary orientation of groups around two atoms joined by a single bond such that the groups directly oppose each other.

An eclipsed conformation

Electromagnetic spectrum (Section 9.2): The full range of energies propagated by wave fluctuations in an electromagnetic field.

Electron isodensity surface (Section 1.12B): An electron isodensity surface shows points in space that happen to have the same electron density. An electron isodensity surface can be calculated for any chosen value of electron density. A "high" electron isodensity surface (also called a "bond" electron density surface) shows the *core* of electron density around each atomic nucleus and regions where neighboring atoms share electrons (bonding regions). A "low" electron isodensity surface roughly shows the *outline* of a molecule's electron cloud. This surface gives information about molecular shape and volume, and usually looks the same as a van der Waals or space-filling model of the molecule. (Contributed by Alan Shusterman, Reed College, and Warren Hehre, Wavefunction, Inc.)

Electronegativity (Section 1.4A): A measure of the ability of an atom to attract electrons it is sharing with another and thereby polarize the bond.

Electrophile (Sections 3.3 and 8.1): A Lewis acid, an electron-pair acceptor, an electron seeking reagent.

Electrophoresis (Section 25.6A): A technique for separating charged molecules based on their different mobilities in an electric field.

Electrostatic potential maps (Section 2.3A): Electrostatic potential maps are structures calculated by a computer that show the relative distribution of electron density at some surface of a molecule or ion. They are very useful for understanding interactions between molecules that are based on attraction of opposite charges. Usually we choose the van der Waals surface (approximately the outermost region of electron density) of a molecule to depict the electrostatic potential map because this is where the electron density of one molecule would first interact with another. In an electrostatic potential map, color trending toward red indicates a region with more negative charge, and color trending toward blue indicates a region with less negative charge (or more positive charge). An electrostatic potential map is generated by calculating the extent of charge interaction (electrostatic potential) between an imaginary positive charge and the electron density at a particular point or surface in a molecule. (Contributed by Alan Shusterman, Reed College, and Warren Hehre, Wavefunction, Inc.)

Elimination (Section 6.16): A reaction that results in the loss of two groups from the substrate and the formation of a π bond. The most common elimination is a 1,2 elimination or β elimination, in which the two groups are lost from adjacent atoms.

Empirical formula (Section 1.2B): A formula that expresses the relative proportions of atoms in a molecule as smallest whole numbers.

Enantiomeric excess or enantiomeric purity (Section 5.9B): A percentage calculated for a mixture of enantiomers by dividing the moles of one enantiomer minus the moles of the other enantiomer by the moles of both enantiomers and multiplying by 100. The enantiomeric excess equals the percentage optical purity.

Enantiomers (Section 5.2): Stereisomers that are mirror images of each other.

Enantioselective reaction (See **Stereoselective reaction** and Section 5.10B)

Enantiotopic hydrogens (or **ligands**) (Section 9.7B): If replacement of each of two hydrogens (or ligands) by the same group yields compounds that are enantiomers, the two hydrogen atoms (or ligands) are said to be enantiotopic.

Endergonic reaction (Section 6.8): A reaction that proceeds with a positive free energy change.

Endothermic reaction (Section 3.8A): A reaction that absorbs heat. For an endothermic reaction $\Delta H°$ is positive.

Energy (Section 3.8): Energy is the capacity to do work.

Energy of activation, E_{act} (Section 6.8): A measure of the difference in potential energy between the reactants and the transition state of a reaction. It is related to, but not the same as, the free energy of activation, $\Delta G^{\ddagger}$.

Enolate anion (Section 17.1): The delocalized anion formed when an enol loses its hydroxylic proton or when the carbonyl tautomer that is in equilibrium with the enol loses an α proton.

Enthalpy change (Sections 3.8A and 3.9): Also called the heat of reaction. The *standard enthalpy change,* ΔH°, is the change in enthalpy after a system in its standard state has undergone a transformation to another system, also in its standard state. For a reaction, ΔH° is a measure of the difference in the total bond energy of the reactants and products. It is one way of expressing the change in potential energy of molecules as they undergo reaction. The enthalpy change is related to the **free-energy change,** ΔG°, and to the **entropy change,** ΔS°, through the expression:

$$\Delta H^\circ = \Delta G^\circ + T\Delta S^\circ$$

Entropy change (Section 3.9): The standard entropy change, ΔS°, is the change in entropy between two systems in their standard states. Entropy changes have to do with changes in the relative order of a system. The more random a system is, the greater is its entropy. When a system becomes more disorderly its entropy change is positive.

Epoxide (Section 11.13). An oxirane. A three-membered ring containing one oxygen and two carbon atoms.

Equatorial bond (Section 4.13): The six bonds of a cyclohexane ring that lie generally around the "equator" of the molecule:

Equilibrium constant (Section 3.5A): A constant that expresses the position of an equilibrium. The equilibrium constant is calculated by multiplying the molar concentrations of the products together and then dividing this number by the number obtained by multiplying together the molar concentrations of the reactants.

Equilibrium control, see **Thermodynamic control.**

Essential oil (Section 23.3): A volatile odoriferous compound obtained by steam distillation of plant material.

Exergonic reaction (Section 6.8): A reaction that proceeds with a negative free-energy change.

Exothermic reaction (Section 3.8A): A reaction that evolves heat. For an exothermic reaction, ΔH° is negative.

F

Fat (Section 23.2): A triacylglycerol. The triester of glycerol with carboxylic acids.

Fatty acid (Section 23.2): A long-chained carboxylic acid (usually with an even number of carbon atoms) that is isolated by the hydrolysis of a fat.

Fischer projection formula (Sections 5.13 and 22.2C): A two-dimensional formula for representing the configuration of a chiral molecule. By convention, Fischer projection formulas are written with the main carbon chain extending from top to bottom with all groups eclipsed. Vertical lines represent bonds that project behind the plane of the page (or that lie in it). Horizontal lines represent bonds that project out of the plane of the page.

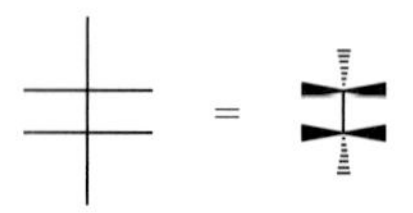

Fluorination (Section 10.5C): A reaction in which fluorine atoms are introduced into a molecule.

Formal charge (Section 1.7): The difference between the number of electrons assigned to an atom in a molecule and the number of electrons it has in its outer shell in its elemental state. Formal charge can be calculated using the formula: $F = Z - S/2 - U$, where F is the formal charge, Z is the group number of the atom (i.e., the number of electrons the atom has in its outer shell in its elemental state), S equals the number of electrons the atom is sharing with other atoms, and U is the number of unshared electrons the atom possesses.

Free energy of activation, $\Delta G^\ddagger$, (Section 6.8): The difference in free energy between the transition state and the reactants.

Free-energy change (Section 3.9): The *standard free-energy change,* ΔG°, is the change in free energy between two systems in their standard states. At constant temperature, $\Delta G^\circ = \Delta H^\circ - T\Delta S^\circ = -RT \ln K_{eq}$, where ΔH° is the standard enthalpy change, ΔS° is the standard entropy change, and K_{eq} is the equilibrium constant. A negative value of ΔG° for a reaction means that the formation of products is favored when the reaction reaches equilibrium.

Free energy diagram (Section 6.8): A plot of free energy changes that take place during a reaction versus the reaction coordinate. It displays free energy changes as a function of changes in bond orders and distances as reactants proceed through the transition state to become products.

Freon (Section 10.11E): A chlorofluorocarbon or CFC.

Frequency (abbreviated ν) (Sections 2.16 and 9.2): The number of full cycles of a wave that pass a given point in each second.

Functional class nomenclature (Section 4.3E): A system for naming compounds that uses two or more words to describe the compound. The final word corresponds to the functional group present; the preceding words, usually listed in alphabetical order, describe the remainder of the molecule. Examples are: methyl alcohol, ethyl methyl ether, and ethyl bromide.

Functional group (Section 2.5): The particular group of atoms in a molecule that primarily determines how the molecule reacts.

Functional group interconversion (Section 6.15): A process that converts one functional group into another.

Furanose (Section 22.2C): A sugar in which the cyclic acetal or hemiacetal ring is five membered.

G

Gauche conformation (Section 4.9A): A gauche conformation of butane, for example, has the methyl groups at an angle of 60° to each other:

CH_3 CH_3

A gauche conformation of butane

Geminal (gem) substituents (Section 7.9): Substituents that are on the same atom.

Glycol (Sections 4.3F and 8.16): A diol.

Glycoside (Section 22.40): A cyclic mixed acetal of a sugar with an alcohol.

Grignard reagent (Section 12.6B): An organomagnesium halide, usually written RMgX.

Ground state (Section 1.12): The lowest electronic energy state of an atom or molecule.

H

Halogenation (Section 10.3): A reaction in which one or more halogen atoms are introduced into a molecule.

Halohydrin (Section 8.14): A halo alcohol.

Halonium ion (Section 8.12A): An ion containing a positive halogen atom bonded to two carbon atoms.

Hammond–Leffler postulate (Section 6.14A): A postulate stating that the structure and geometry of the transition state of a given step will show a greater resemblance to the reactants or products of that step depending on which is closer to the transition state in energy. This means that the transition state of an endothermic step will resemble the products of that step more than the reactants, whereas the transition state of an exothermic step will resemble the reactants of that step more than the products.

Heat of combustion (Section 4.10A): The standard enthalpy change for the complete combustion of 1 mol of a compound.

Heat of hydrogenation (Section 7.3A): The standard enthalpy change that accompanies the hydrogenation of 1 mol of a compound to form a particular product.

Heisenberg uncertainty principle (Section 1.11): A fundamental principle that states that both the position and momentum of an electron (or of any object) cannot be exactly measured simultaneously.

Hemiacetal (Section 16.7A): A functional group, consisting of a carbon atom bonded to an alkoxy group and to a hydroxyl group, [i.e., RCH(OH)(OR′) or $R_2C(OH)(OR')$]. Hemiacetals are synthesized by adding one molar equivalent of an alcohol to an aldehyde or a ketone.

Hemiketal (Section 16.7A): Also called a hemiacetal. A functional group, consisting of a carbon atom bonded to an alkoxy group and to a hydroxyl group, [i.e., $R_2C(OH)(OR')$]. Hemiketals are synthesized by adding one molar equivalent of an alcohol to a ketone.

Hertz (abbreviated Hz) (Section 9.2): Now used instead of the equivalent cycles per second as a measure of the frequency of a wave.

Heterocyclic compound (Section 14.9): A compound whose molecules have a ring containing an element other than carbon.

Heterolysis (Section 3.1A): The cleavage of a covalent bond so that one fragment departs with both of the electrons of the covalent bond that joined them. Heterolysis of a bond normally produces positive and negative ions.

Hofmann rule (Sections 7.6B and 20.13A). When an elimination yields the alkene with the less substituted double bond, it is said to follow the Hofmann rule.

HOMO (Sections 3.2C and 13.9B): The highest occupied molecular orbital.

Homologous series (Section 4.7): A series of compounds in which each member differs from the next member by a constant unit.

Homolysis (Section 3.1A): The cleavage of a covalent bond so that each fragment departs with one of the electrons of the covalent bond that joined them.

Homolytic bond dissociation energy, $DH°$, (Section 10.2): The enthalpy change that accompanies the homolytic cleavage of a covalent bond.

Hückel's rule (Section 14.7): A rule stating that planar monocyclic rings with $(4n + 2)$ delocalized π electrons (i.e., with 2, 6, 10, 14, . . . , delocalized π electrons) will be aromatic.

Hund's rule (Section 1.10): A rule used in applying the **aufbau principle.** When orbitals are of equal energy (i.e., when they are **degenerate**), electrons are added to each orbital with their spins unpaired, until each degenerate orbital contains one electron. Then electrons are added to the orbitals so that the spins are paired.

Hybridization of atomic orbitals (Section 1.12): A mathematical (and theoretical) mixing of two or more atomic orbitals to give the same number of new orbitals, called *hybrid orbitals,* each of which has some of the character of the original atomic orbitals.

Hydration (Sections 8.5–8.10 and 11.4): The addition of water to a molecule, such as the addition of water to an alkene to form an alcohol.

Hydroboration (Sections 8.8 and 11.4): The addition of a boron hydride (either BH_3 or an alkylborane) to a multiple bond.

Hydrogen bond (Section 2.14C): A strong dipole–dipole interaction (4–38 kJ mol^{-1}) that occurs between hydrogen atoms bonded to small strongly electronegative atoms (O, N, or F) and the nonbonding electron pairs on other such electronegative atoms.

Hydrogenation (Sections 4.18A and 7.13–7.15): A reaction in which hydrogen adds to a double or triple bond.

Hydrogenation is often accomplished through the use of a metal catalyst such as platinum, palladium, rhodium, or ruthenium.

Hydrophilic group (Section 2.14E): A polar group that seeks an aqueous environment.

Hydrophobic group (also called a **lipophilic group**) (Sections 2.14E and 11.20): A nonpolar group that avoids an aqueous surrounding and seeks a nonpolar environment.

Hydroxylation (Sections 8.16 and 11.15): The addition of hydroxyl groups to each carbon or atom of a double bond.

I

Index of hydrogen deficiency (Section 7.16): The index of hydrogen deficiency (or IHD) equals the number of pairs of hydrogen atoms that must be subtracted from the molecular formula of the corresponding alkane to give the molecular formula of the compound under consideration.

Inductive effect (Sections 3.7B and 15.11B): An intrinsic electron-attracting or -releasing effect that results from a nearby dipole in the molecule and that is transmitted through space and through the bonds of a molecule.

Infrared (IR) spectroscopy (Section 2.16): A type of optical spectroscopy that measures the absorption of infrared radiation. Infrared spectroscopy provides structural information about functional groups present in the compound being analyzed.

Intermediate (Sections 3.1, 6.10, and 6.11): A transient species that exists between reactants and products in a state corresponding to a local energy minimum on a potential energy diagram.

Iodination (Section 10.5C): A reaction in which one or more iodine atoms are introduced into a molecule.

Ion (Sections 1.4A and 6.3): A chemical species that bears an electrical charge.

Ion–dipole interaction (Section 2.14E): The interaction of an ion with a permanent dipole. Such interactions (resulting in solvation) occur between ions and the molecules of polar solvents.

Ionic bond (Section 1.4A): A bond formed by the transfer of electrons from one atom to another resulting in the creation of oppositely charged ions.

Ionic reaction (Sections 3.1A and 6.3): A reaction involving ions as reactants, intermediates, or products. Ionic reactions occur through the heterolysis of covalent bonds.

Isoelectric point (Section 24.2C): The pH at which the number of positive and negative charges on an amino acid or protein are equal.

Isomers (Sections 1.3A and 5.1): Different molecules that have the same molecular formula.

Isoprene unit (Section 23.3): A name for the structural unit:

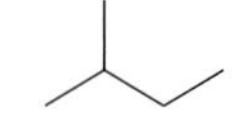

found in all terpenes.

Isotactic polymer (Special Topic A): A polymer in which the configuration at each stereogenic center along the chain is the same.

K

Kekulé structure (Sections 2.2D and 14.4): A structure in which lines are used to represent bonds. The Kekulé structure for benzene is a hexagon of carbon atoms with alternating single and double bonds around the ring, and with one hydrogen atom attached to each carbon.

Ketal (Section 16.7B): Properly called an acetal. A functional group, consisting of a carbon bonded to alkoxyl groups, [i.e., $R_2C(OR')_2$], derived by adding two molar equivalents of an alcohol to a ketone.

Kinetic control (Section 13.10A): A principle stating that when the ratio of products of a reaction is determined by relative rates of reaction, the most abundant product will be the one that is formed fastest.

Kinetic energy (Section 3.8): Energy that results from the motion of an object. Kinetic energy $(KE) = \frac{1}{2} mv^2$, where m is the mass of the object, and v is its velocity.

Kinetics (Section 6.6): A term that refers to rates of reactions.

L

Lactam (Section 18.8I): A cyclic amide.

Lactone (Section 18.7C): A cyclic ester.

Leaving group (Section 6.5): The substituent with an unshared electron pair that departs from the substrate in a nucleophilic substitution reaction.

Leveling effect of a solvent (Section 3.14): An effect that restricts the use of certain solvents with strong acids and bases. In principle, no acid stronger than the conjugate acid of a particular solvent can exist to an appreciable extent in that solvent, and no base stronger than the conjugate base of the solvent can exist to an appreciable extent in that solvent.

Levorotatory (Section 5.8B): A compound that rotates plane-polarized light in a counterclockwise direction.

Lewis structure (or electron-dot structure) (Section 1.4B): A representation of a molecule showing electron pairs as a pair of dots or as a dash.

Lewis theory of acids and bases (Section 3.2B): An acid is an electron pair acceptor, and a base is an electron pair donor.

Lipid (Section 23.1): A substance of biological origin that is soluble in nonpolar solvents. Lipids include fatty acids, triacylglycerols (fats and oils), steroids, prostaglandins, terpenes and terpenoids, and waxes.

Lipophilic group (or **hydrophopic group**) (Sections 2.14E and 11.20): A nonpolar group that avoids an aqueous surrounding and seeks a nonpolar environment.

LUMO (Sections 3.2C and 13.9B): The lowest unoccupied molecular orbital.

M

Macromolecule (Section 10.10): A very large molecule.

Magnetic resonance imaging (Section 9.11B): A technique based on NMR spectroscopy that is used in medicine.

Markovnikov's rule (Section 8.2): A rule for predicting the regiochemistry of electrophilic additions to alkenes and alkynes that can be stated in various ways. As originally stated (in 1870) by Vladimir Markovnikov, the rule provides that "if an unsymmetrical alkene combines with a hydrogen halide, the halide ion adds to the carbon with the fewer hydrogen atoms." More commonly the rule has been stated in reverse: that in the addition of HX to an alkene or alkyne the hydrogen atom adds to the carbon atom that already has the greater number of hydrogen atoms. A modern expression of Markovnikov's rule is: *In the ionic addition of an unsymmetrical reagent to a multiple bond, the positive portion of the reagent (the electrophile) attaches itself to a carbon atom of the reagent in the way that leads to the formation of the more stable intermediate carbocation.*

Mass spectrometry (Section 9.12): A technique, useful in structure elucidation, that is based on generating ions of a molecule in a magnetic field, then instrumentally determining the mass/charge ratio and relative amounts of the ions that result.

Mechanism, see **Reaction mechanism**.

Meso compound (Section 5.12A): An optically inactive compound whose molecules are achiral even though they contain tetrahedral atoms with four different attached groups.

Mesylate (Section 11.10): A methanesulfonate ester.

Methylene (Section 8.15A): The carbene with the formula CH_2.

Methylene group (Section 6.2): The $—CH_2—$ group.

Micelle (Section 23.2C): A spherical cluster of ions in aqueous solution (such as those from a soap) in which the nonpolar groups are in the interior and the ionic (or polar) groups are at the surface.

Molar absorptivity (abbreviated ε) (Section 13.9A): A proportionality constant that relates the observed absorbance (A) at a particular wavelength (λ) to the molar concentration of the sample (C) and the length (l) (in centimeters) of the path of the light beam through the sample cell:

$$\varepsilon = A/C \times l$$

Molecular formula (Section 1.2B): A formula that gives the total number of each kind of atom in a molecule. The molecular formula is a whole number multiple of the empirical formula. For example the molecular formula for benzene is C_6H_6; the empirical formula is CH.

Molecular ion (Section 9.13): The cation produced in a mass spectrometer when one electron is dislodged from the parent molecule.

Molecular orbital (MO) (Section 1.11): Orbitals that encompass more than one atom of a molecule. When atomic orbitals combine to form molecular orbitals, the number of molecular orbitals that results always equals the number of atomic orbitals that combine.

Molecularity (Section 6.6): The number of species involved in a single step of a reaction (usually the rate determining step).

Monomer (Section 10.10): The simple starting compound from which a polymer is made. For example, the polymer polyethylene is made from the monomer, ethylene.

Monosaccharide (Section 22.1A): The simplest type of carbohydrate, one that does not undergo hydrolytic cleavage to a simpler carbohydrate.

Mutarotation (Section 22.3): The spontaneous change that takes place in the optical rotation of α and β anomers of a sugar when they are dissolved in water. The optical rotations of the sugars change until they reach the same value.

N

Neighboring group participation (Problem 6.42): The effect on the course or rate of a reaction brought about by another group near the functional group undergoing reaction.

Newman projection formula (Section 4.8): A means of representing the spatial relationships of groups attached to two atoms of a molecule. In writing a Newman projection formula we imagine ourselves viewing the molecule from one end directly along the bond axis joining the two atoms. Bonds that are attached to the front atom are shown as radiating from the center of a circle; those attached to the rear atom are shown as radiating from the edge of the circle:

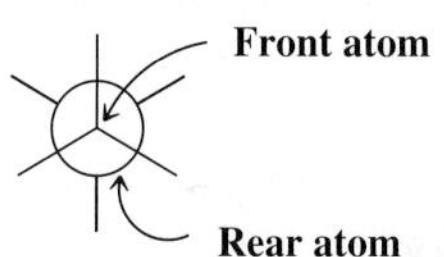

Nitrogen rule (Section 9.15A): A rule that states that if the mass of the molecular ion in a mass spectrum is an even number, the parent compound contains an even number of nitrogen atoms, and conversely.

Node (Section 1.9): A place where a wave function (Ψ) is equal to zero. The greater the number of nodes in an orbital, the greater is the energy of the orbital.

Nonbenzenoid aromatic compound (Section 14.8B): An aromatic compound, such as azulene, that does not contain benzene rings.

Nuclear magnetic resonance (NMR) spectroscopy (Section 9.3): A spectroscopic method for measuring the absorption of radiofrequency radiation by certain nuclei when the nuclei are in a strong magnetic field. The most important NMR spectra for organic chemists are ^1H NMR spectra and ^{13}C NMR spectra. These two types of spectra provide structural information about the carbon framework of the molecule, and about the number and environment of hydrogen atoms attached to each carbon atom.

Nucleic acids (Sections 25.1 and 25.2): Biological polymers of nucleotides. DNA and RNA are, respectively, nucleic acids that preserve and transcribe hereditary information within cells.

Nucleophile (Section 3.3): A Lewis base, an electron pair donor that seeks a positive center in a molecule.

Nucleophilic substitution reaction (Section 6.3): A reaction initiated by a nucleophile (a species with an unshared electron pair) in which the nucleophile reacts with a substrate to replace a substituent (called the leaving group), that departs with an unshared electron pair.

Nucleophilicity (Section 6.14B): The relative reactivity of a nucleophile in an S_N2 reaction as measured by relative rates of reaction.

Nucleoside (Section 25.2): A five-carbon monosaccharide bonded at the 1′ position to a purine or pyrimidine.

Nucleotide (Section 25.2): A five-carbon monosaccharide bonded at the 1′ position to a purine or pyrimidine and at the 3′ or 5′ position to a phosphate group.

O

Olefin (Section 7.1): An old name for an alkene.

Optical purity (Section 5.9B): A percentage calculated for a mixture of enantiomers by dividing the observed specific rotation for the mixture by the specific rotation of the pure enantiomer and multiplying by 100. The optical purity equals the enantiomeric purity or enantiomeric excess.

Optically active compound (Section 5.8): A compound that rotates the plane of polarization of plane polarized light.

Orbital (Section 1.10): A volume of space in which there is a high probability of finding an electron. Orbitals are described mathematically by the squaring of wave functions, and each orbital has a characteristic energy. An orbital can hold two electrons when their spins are paired.

Organometallic compound (Section 12.5): A compound that contains a carbon–metal bond.

Oxidation (Section 12.2): A reaction that increases the oxidation state of atoms in a molecule or ion. For an organic substrate, oxidation usually involves increasing its oxygen content or decreasing its hydrogen content. Oxidation also accompanies any reaction in which a less electronegative substituent is replaced by a more electronegative one.

Oxonium ion (Section 3.12): A chemical species with an oxygen atom that bears a formal positive charge.

Oxymercuration (Sections 8.6 and 11.4): The addition of —OH and —HgO_2CR to a multiple bond.

Ozonolysis (Section 8.17A): The cleavage of a multiple bond that makes use of the reagent O_3, called ozone. The reaction leads to the formation of a cyclic compound called an *ozonide*, which is then reduced to carbonyl compounds by treatment with zinc and acetic acid.

P

***p* orbitals** (Section 1.10): A set of three degenerate (equal energy) atomic orbitals shaped like two tangent spheres with a nodal plane at the nucleus. For *p* orbitals, the principal quantum number, n (see **atomic orbital**), is 2; the azimuthal quantum number, l, = 1; and the magnetic quantum numbers, m are +1, 0, or −1.

Paraffin (Section 4.17): An old name for an alkane.

Pauli exclusion principle (Section 1.10): A principle that states that no two electrons of an atom or molecule may have the same set of four quantum numbers. It means that only two electrons can occupy the same orbital, and then only when their spin quantum numbers are opposite. When this is true, we say that the spins of the electrons are paired.

Periplanar (See **coplanar**, Section 7.6C).

Peroxide (Section 10.1A): A compound with an oxygen–oxygen single bond.

Peroxy acid (Section 11.13): An acid with the general formula RCO_3H, containing an oxygen–oxygen single bond.

Phase sign (Section 1.9): Signs, either + or − , that are characteristic of all equations that describe the amplitudes of waves.

Phase-transfer catalyst (Section 11.20): A reagent that transports an ion from an aqueous phase into a nonpolar phase where reaction takes place more rapidly. Tetraalkylammonium ions and crown ethers are phase-transfer catalysts.

Phospholipid (Section 23.6): Compounds that are structurally derived from *phosphatidic acids*. Phosphatidic acids are derivatives of glycerol in which two hydroxyl groups are joined to fatty acids, and one terminal hydroxyl group is joined in an ester linkage to phosphoric acid. In a phospholipid the phosphate group of the phosphatidic acid is joined in ester linkage to a nitrogen-containing compound such as choline, 2-aminoethanol, or L-serine.

Pi (π) bond (Section 1.13): A bond formed when electrons occupy a bonding π molecular orbital (i.e., the lower energy molecular orbital that results from overlap of parallel *p* orbitals on adjacent atoms).

Pi (π) molecular orbital (Section 1.13): A molecular orbital formed when parallel *p* orbitals on adjacent atoms overlap. Pi molecular orbitals may be *bonding* (*p* lobes of the same phase sign overlap) or *antibonding* (*p* orbitals of opposite phase sign overlap).

pK_a (Section 3.5): The pK_a is the negative logarithm of the acidity constant, K_a. p$K_a = -\log K_a$.

Plane of symmetry (Section 5.6): An imaginary plane that bisects a molecule in a way such that the two halves of the molecule are mirror images of each other. Any molecule with a plane of symmetry will be achiral.

Plane-polarized light (Section 5.8A): Ordinary light in which the oscillations of the electrical field occur only in one plane.

Polar covalent bond (Section 2.3): A covalent bond in which the electrons are not equally shared because of differing electronegativities of the bonded atoms.

Polar molecule (Section 2.4): A molecule with a dipole moment.

Polarimeter (Section 5.8B): A device used for measuring optical activity.

Polarizability (Section 6.14C): The susceptibility of the electron cloud of an uncharged molecule to distortion by the influence of an electric charge.

Polymer (Section 10.10): A large molecule made up of many repeating subunits. For example, the polymer polyethylene is made up of the repeating subunit $—(CH_2CH_2)_n—$.

Polysaccharide (Section 22.1A): A carbohydrate that, on a molecular basis, undergoes hydrolytic cleavage to yield many molecules of a monosaccharide.

Potential energy (Section 3.8): Potential energy is stored energy; it exists when attractive or repulsive forces exist between objects.

Primary carbon (Section 2.6): A carbon atom that has only one other carbon atom attached to it.

Primary structure (Section 24.5): The covalent structure of a polypeptide or protein. This structure is determined, in large part, by determining the sequence of amino acids in the protein.

Prochiral (Section 12.3): A group is prochiral if replacement of one of two identical groups at a tetrahedral atom, or if addition of a group to a trigonal planar atom, leads to a new stereogenic center. At a tetrahedral atom where there are two identical groups, the identical groups can be designated pro-R and pro-S depending on what configuration would result when it is imagined that each is replaced by a group of next higher priority (but not higher than another existing group).

Protecting group (Sections 11.11D, 11.11E, 12.10, 16.7C, and 24.7A): A group that is introduced into a molecule to protect a sensitive group from reaction while a reaction is carried out at some other location in the molecule. Later, the protecting group is removed.

Protein (Section 24.1): A large biological polymer of α-amino acids joined by amide linkages.

Protic solvent (Sections 3.11 and 6.14C): A solvent whose molecules have a hydrogen atom attached to a strongly electronegative element such as oxygen or nitrogen. Molecules of a protic solvent can therefore form hydrogen bonds to unshared electron pairs of oxygen or nitrogen atoms of solute molecules or ions, thereby stabilizing them. Water, methanol, ethanol, formic acid, and acetic acid are typical protic solvents.

Proton decoupling (Section 9.10B): An electronic technique used in ^{13}C NMR spectroscopy that allows decoupling of spin–spin interactions between ^{13}C nuclei and ^{1}H nuclei. In spectra obtained in this mode of operation all carbon resonances appear as singlets.

Proton off-resonance decoupling (Section 9.10D): An electronic technique used in ^{13}C NMR spectroscopy that allows one-bond couplings between ^{13}C nuclei and ^{1}H nuclei. In spectra obtained in this mode of operation, CH_3 groups appear as quartets, CH_2 groups appear as triplets, CH groups appear as doublets, and carbon atoms with no attached hydrogen atoms appear as singlets.

Psi function (Ψ function or wave function) (Section 1.9): A mathematical expression derived from *quantum mechanics* corresponding to an energy state for an electron. The square of the Ψ function, Ψ^2, gives the probability of finding the electron in a particular location in space.

Pyranose (Section 22.2C): A sugar in which the cyclic acetal or hemiacetal ring is six membered.

R

R (Section 2.5A): A symbol used to designate an alkyl group. Oftentimes it is taken to symbolize any organic group.

(*R*–*S*) System (Section 5.7): A method for designating the configuration of tetrahedral stereogenic centers.

Racemic form (racemate or racemic mixture) (Section 5.9A): An equimolar mixture of enantiomers. A racemic form is optically inactive.

Racemization (Section 6.13A): A reaction that transforms an optically active compound into a racemic form is said to proceed with racemization. Racemization takes place whenever a reaction causes chiral molecules to be converted to an achiral intermediate.

Radical (or **free radical**) (Section 3.1A): An uncharged chemical species that contains an unpaired electron.

Radical reaction (Section 10.1): A reaction involving radicals. Homolysis of covalent bonds occurs in radical reactions.

Rate control, see **Kinetic control.**

Rate-determining step (Section 6.10A): If a reaction takes place in a series of steps, and if the first step is intrinsically slower than all of the others, then the rate of the overall reaction will be the same as (will be determined by) the rate of this slow step.

Reaction coordinate (Section 6.8): The abscissa in a potential energy diagram that represents the progress of the reaction. It represents the changes in bond orders and bond distances that must take place as reactants are converted to products.

Reaction mechanism (Section 3.1): A step-by-step description of the events that are postulated to take place at the molecular level as reactants are converted to products. A mechanism will include a description of all intermediates and transition states. Any mechanism proposed for a reaction must be consistent with all experimental data obtained for the reaction.

Rearrangement (Sections 3.1 and 7.8A): A reaction that results in a product with a different carbon skeleton from the reactant. The type of rearrangement called a 1,2 shift involves the migration of an organic group (with its electrons) from one atom to the atom next to it.

Reducing sugar (Section 22.6A): Sugars that reduce Tollens' or Benedict's reagents. All sugars that contain hemiacetal or hemiketal groups (and therefore are in equilibrium with aldehydes or α-hydroxyketones) are reducing sugars. Sugars in which only acetal or ketal groups are present are nonreducing sugars.

Reduction (Section 12.2): A reaction that lowers the oxidation state of atoms in a molecule or ion. Reduction of an organic compound usually involves increasing its hydrogen content or decreasing its oxygen content. Reduction also accompanies any reaction that results in replacement of a more electronegative substituent by a less electronegative one.

Regioselective reaction (Section 8.2C): A reaction that yields only one (or a predominance of one) constitutional isomer as the product when two or more constitutional isomers are possible products.

Relative configuration (Section 5.15A): The relationship between the configurations of two chiral molecules. Molecules are said to have the same relative configuration when similar or identical groups in each occupy the same position in space. The configurations of molecules can be related to each other through reactions of known stereochemistry, for example, through reactions that cause no bonds to a stereogenic center to be broken.

Resolution (Sections 5.16 and 20.3E): The process by which the enantiomers of a racemic form are separated.

Resonance effect (Sections 3.10A, 13.5 and 15.11B): An effect by which a substituent exerts either an electron-releasing or -withdrawing effect through the pi system of the molecule.

Resonance energy (Section 14.5): An energy of stabilization that represents the difference in energy between the actual compound and that calculated for a single resonance structure. The resonance energy arises from delocalization of electrons in a conjugated system.

Resonance structures (or **resonance contributors**) (Sections 1.8 and 13.5): Lewis structures that differ from one another only in the position of their electrons. A single resonance structure will not adequately represent a molecule. The molecule is better represented as a *hybrid* of all of the resonance structures.

Retrosynthetic analysis (Section 4.20A): A method for planning syntheses that involves reasoning backward from the target molecule through various levels of precursors and thus finally to the starting materials.

Ring flip (Sections 4.12 and 4.13): The change in a cyclohexane ring (resulting from partial bond rotations) that converts one ring conformation to another. A chair–chair ring flip converts any equatorial substitutent to an axial substituent and vice versa.

Ring strain (Section 4.11): The increased potential energy of the cyclic form of a molecule (usually measured by heats of combustion) when compared to its acyclic form.

S

***s* orbital** (Section 1.10): A spherical atomic orbital. For *s* orbitals the azimuthal quantum number $l = 0$ (see atomic orbital).

Saponification (Section 18.7B): Base-promoted hydrolysis of an ester.

Saturated compound (Section 2.2): A compound that does not contain any multiple bonds.

Secondary carbon (Section 2.6): A carbon atom that has two other carbon atoms attached to it.

Secondary structure (Section 24.8): The local conformation of a polypeptide backbone. These local conformations are specified in terms of regular folding patterns such as pleated sheets, α-helixes, and turns.

Shielding and deshielding (Section 9.5): Effects observed in NMR spectra caused by the circulation of sigma and pi electrons within the molecule. Shielding causes signals to appear at higher magnetic fields (upfield), deshielding causes signals to appear at lower magnetic fields (downfield).

Sigma (σ) bond (Section 1.12): A single bond. A bond formed when electrons occupy the bonding σ orbital formed by the end-on overlap of atomic orbitals (or hybrid orbitals) on adjacent atoms. In a sigma bond the electron density has circular symmetry when viewed along the bond axis.

Sigma (σ) orbital (Section 1.12): A molecular orbital formed by end-on overlap of orbitals (or lobes of orbitals) on adjacent atoms. Sigma orbitals may be *bonding* (orbitals or lobes of the same phase sign overlap) or *antibonding* (orbitals or lobes of opposite phase sign overlap).

S_N1 reaction (Sections 6.10 and 6.14): Literally, substitution nucleophilic unimolecular. A multistep nucleophilic substitution in which the leaving group departs in a unimolecular step before the attack of the nucleophile. The rate equation is first order in substrate but zero order in the attacking nucleophile.

S_N2 reaction (Sections 6.6, 6.7 and 6.14): Literally, substitution nucleophilic bimolecular. A bimolecular nucleophilic substitution reaction that takes place in a single step in which a nucleophile attacks a carbon bearing a leaving group from the backside, causing an inversion of configuration at this carbon and displacement of the leaving group.

Solvent effect (Section 6.14D): An effect on relative rates of reaction caused by the solvent. For example, the use of a polar solvent will increase the rate of reaction of an alkyl halide in an S_N1 reaction.

Solvolysis (Section 6.13B): Literally, cleavage by the solvent. A nucleophilic substitution reaction in which the nucleophile is a molecule of the solvent.

***sp* orbital** (Section 1.14): A hybrid orbital that is derived by mathematically combining one *s* atomic orbital and one *p* atomic orbital. Two *sp* hybrid orbitals are obtained by this process, and they are oriented in opposite directions with an angle of 180° between them.

sp^2 orbital (Section 1.13): A hybrid orbital that is derived by mathematically combining one *s* atomic orbital and two *p* atomic orbitals. Three sp^2 hybrid orbitals are obtained by this process, and they are directed toward the corners of an equilateral triangle with angles of 120° between them.

sp^3 orbital (Section 1.12): A hybrid orbital that is derived by mathematically combining one *s* atomic orbital and three *p* atomic orbitals. Four sp^3 hybrid orbitals are obtained by this process, and they are directed toward the corners of a regular tetrahedron with angles of 109.5° between them.

Specific rotation (Section 5.8C): A physical constant calculated from the observed rotation of a compound using the following equation:

$$[\alpha]_D = \frac{\alpha}{c \times l}$$

where α is the observed rotation using the D line of a sodium lamp, c is the concentration of the solution in grams per milliliter or the density of a neat liquid in g mL^{-1}, and l is the length of the tube in decimeters.

Spin decoupling (Section 9.9): An effect that causes spin–spin splitting not to be observed in NMR spectra.

Spin–spin splitting (Section 9.8): An effect observed in NMR spectra. Spin–spin splittings result in a signal appearing as a multiplet (i.e., doublet, triplet, quartet, etc.) and are caused by magnetic couplings of the nucleus being observed with nuclei of nearby atoms.

Staggered conformation (Section 4.8): A temporary orientation of groups around two atoms joined by a single bond such that the bonds of the back atom exactly bisect the angles formed by the bonds of the front atom in a Newman projection formula:

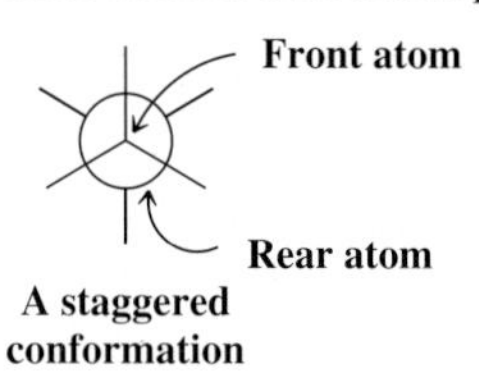

Stereogenic center (Section 5.3): An atom bearing groups of such nature that an interchange of any two groups will produce a stereoisomer.

Stereochemistry (Section 5.5): Chemical studies that take into account the spatial aspects of molecules.

Stereoisomers (Sections 1.13B, 5.2, and 5.3): Compounds with the same molecular formula that differ *only* in the arrangement of their atoms in space. Stereoisomers have the same connectivity and, therefore, are not constitutional isomers. Stereoisomers are classified further as being either enantiomers or diastereomers.

Stereoselective reaction (Sections 5.10B, 8.21, and 12.3): In reactions where stereogenic centers are altered or created, a stereoselective reaction produces a preponderance of one stereoisomer. Furthermore, a stereoselective reaction can be either enantioselective, in which case the reaction produces a preponderance of one enantiomer, or diastereoselective, in which case the reaction produces a preponderance of one diastereomer.

Stereospecific reaction (Section 8.13A): A reaction in which a particular stereoisomeric form of the reactant reacts in such a way that it leads to a specific stereoisomeric form of the product.

Steric effect (Section 6.14A): An effect on relative reaction rates caused by the space-filling properties of those parts of a molecule attached at or near the reacting site.

Steric hindrance (Section 6.14A): An effect on relative reaction rates caused when the spatial arrangement of atoms or groups at or near the reacting site hinders or retards a reaction.

Steroid (Section 23.4): Steroids are lipids that are derived from the following perhydrocyclopentanophenanthrene ring system:

Structural formula (Sections 1.2B and 1.17): A formula that shows how the atoms of a molecule are attached to each other.

Substituent effect (Sections 3.7B and 15.10): An effect on the rate of reaction (or on the equilibrium constant) caused by the replacement of a hydrogen atom by another atom or group. Substituent effects include those effects caused by the size of the atom or group, called steric effects, and those effects caused by the ability of the group to release or withdraw electrons, called electronic effects. Electronic effects are further classified as being inductive effects or resonance effects.

Substitution reaction (Sections 6.3 and 10.3): A reaction in which one group replaces another in a molecule.

Substitutive nomenclature (Section 4.3F): A system for naming compounds in which each atom or group, called a substituent, is cited as a prefix or suffix to a parent compound. In the IUPAC system only one group may be cited as a suffix. Locants (usually numbers) are used to tell where the group occurs.

Substrate (Section 6.3): The molecule or ion that undergoes reaction.

Sugar (Section 21.1A): A carbohydrate.

Superposable (Section 5.1): Two objects are superposable if, when one object is placed on top of the other, all parts of each coincide. To be superposable is different than to be superimposable. Any two objects can be superimposed simply by putting one object on top of the other, whether or not all parts coincide. The condition of superposability must be met for two things to be identical.

Syn addition (Section 7.14A): An addition that places both parts of the adding reagent on the same face of the reactant.

Syndiotactic polymer (Special Topic A): A polymer in which the configuration at the stereogenic centers along the chain alternate regularly: (*R*), (*S*), (*R*), (*S*), etc.

Synthon (Section 8.21): The fragments that result (on paper) from the disconnection of a bond. The actual reagent that will, in a synthetic step, provide the synthon is called the *synthetic equivalent*.

T

Tautomers (Section 17.2): Constitutional isomers that are easily interconverted. Keto and enol tautomers, for example, are rapidly interconverted in the presence of acids and bases.

Terpene (Section 23.3): Terpenes are lipids that have a structure that can be derived on paper by linking isoprene units.

Tertiary carbon (Section 2.6): A carbon atom that has three other carbon atoms attached to it.

Tertiary structure (Section 24.8B): The three-dimensional shape of a protein that arises from folding of its polypeptide chains superimposed on its α helixes and pleated sheets.

Thermodynamic control (Section 13.10A): A principle stating that the ratio of products of a reaction that reaches equilibrium is determined by the relative stabilities of the products (as measured by their standard free energies, ΔG°). The most abundant product will be the one that is the most stable.

Torsional barrier (Section 4.8): The barrier to rotation of groups joined by a single bond caused by repulsions between the aligned electron pairs in the eclipsed form.

Torsional strain (Section 4.11): The strain associated with an eclipsed conformation of a molecule; it is caused by repulsions between the aligned electron pairs of the eclipsed bonds.

Tosylate (Section 11.10): A *p*-toluenesulfonate ester.

Transition state (Sections 6.8 and 6.9): A state on a potential energy diagram corresponding to an energy maximum (i.e., characterized by having higher potential energy than immediately adjacent states). The term transition state is also used to refer to the species that occurs at this state of maximum potential energy; another term used for this species is *the activated complex.*

U

Ultraviolet–visible (UV–vis) spectroscopy (Section 13.9): A type of optical spectroscopy that measures the absorption of light in the visible and ultraviolet regions of the spectrum. Visible–UV spectra primarily provide structural information about the kind and extent of conjugation of multiple bonds in the compound being analyzed.

Unimolecular reaction (Section 6.10): A reaction whose rate-determining step involves only one species.

Unsaturated compound (Section 2.2): A compound that contains multiple bonds.

V

van der Waals force (or **London force**) (Sections 2.14D and 4.7): Weak forces that act between nonpolar molecules or between parts of the same molecule. Bringing two groups (or molecules) together first results in an attractive force between them because a temporary unsymmetrical distribution of electrons in one group induces an opposite polarity in the other. When groups are brought closer than their *van der Waals radii*, the force between them becomes repulsive because their electron clouds begin to interpenetrate each other.

Vicinal (*vic*) substituents (Section 7.9): Substituents that are on adjacent atoms.

Vinyl group (Section 4.5): The $CH_2{=}CH{-}$ group.

Vinylic halide (Section 6.1): An organic halide in which the halogen atom is attached to a carbon atom of a double bond.

Vinylic substituent (Section 6.1): Refers to a substituent on a carbon atom that participates in a carbon–carbon double bond.

W

Wave function (or **Ψ function**) (Section 1.9): A mathematical expression derived from *quantum mechanics* corresponding to an energy state for an electron, i.e., for an orbital. The square of the Ψ function, Ψ^2, gives the probability of finding the electron in a particular place in space.

Wavelength (abbreviated λ) (Sections 2.16 and 9.2): The distance between consecutive crests (or troughs) of a wave.

Wavenumber (Section 2.16): A way to express the frequency of a wave. The wavenumber is the number of waves per centimeter, expressed as cm^{-1}.

Y

Ylide (Section 16.10): An electrically neutral molecule that has a negative carbon with an unshared electron pair adjacent to a positive heteroatom.

Z

Zaitsev's rule (Section 7.6A): A rule stating that an elimination will give as the major product the most stable alkene (i.e., the alkene with the most highly substituted double bond).

Zwitterion (see **Dipolar ion**): Another name for a dipolar ion.

Photo and Illustration Credits

Protein Data Bank (http://www.rcsb.org/pdb/): A number of images depicting molecular structure in this book were generated using data from the Protein Data Bank (PDB). Citations to specific data files from the PDB used to depict molecular structures in this book are listed below by page number under their respective chapter. The following is a general citation for the Protein Data Bank.

H.M. Berman; J. Westbrook, Z. Feng; G. Gilliland, T.N. Bhat; H. Weissig; I.N. Shindyalov; P.E. Bourne. The Protein Data Bank. *Nucleic Acids Research*, **2000**, *28*, pp. 235–242.

CHAPTER 1

Chapter opener, p. 1: Pascale Ehrenfreund & Steven B. Charnley, *Annual Review of Astronomy & Astrophysics,* 38: 4427–83, 2000. Image courtesy of Pascale Ehrenfreund. **Chapter opener, p. 1:** From "Organic Molecules in the Interstellar Medium, Comets and Meteorites: A Voyage from Dark Clouds to the Earth," by Pascale Ehrenfreund and Steven B. Charnley, p. 453. *Annual Review of Astronomy Astrophysics,* 38: 427–83, Volume 38 © 2000 by Annual Reviews. www.annualreviews.org **P. 4**: Photo from Corbis Images. **P. 6**: Photo of G.N. Lewis. John Hagemeyer/Library of Congress.

CHAPTER 2

Chapter opener, p. 52: Reprinted with permission from *Accounts of Chemical Research,* June 2001, 34 (6), Copyright 2001 American Chemical Society. Image courtesy of Anthony Pease. **Pp. 52 & 79:** Courtesy Samuel I. Stupp and Jeffrey Hartgerink, Northwestern University. **P. 64**: Photo courtesy Alan & Linda Detrick/ Photo Researchers. **P. 72**: Melting point apparatus, and **P.75** microscale distillation apparatus from *Introduction to Organic Laboratory Techniques: A Microscale Approach, 3rd* edition by Pavia / Lampan / Kriz / Engel. © 1999. Reprinted with permission of Brooks/Cole, a division of Thomson Learning: www.thomsonrights.com **P. 77:** A representation of the α-helical structure of a polypeptide. From *Biochemistry,* Second Edition, by Voet, D., & Voet, J.G. © 1995 by John Wiley & Sons, Inc. Illustration by Irving Geis. **P. 79, The Chemistry of box:** Molecular graphic & photos from *Chemical. & Engineering News,* July 9, 2001, pp. 58–62. **P. 84, Figs. 2.13 & 2.14:** Courtesy Sadtler Research Laboratories. **P. 85, Fig. 2.15**: From *Introduction to Organic Laboratory Techniques, A Microscale Approach, 3rd* edition by Pavia / Lampan / Kriz / Engel. © 1999. Reprinted with permission of Brooks/Cole, a division of Thomson Learning: www.thomsonrights.com **P. 97, Figs. 2.16 & 2.17:** Adapted from Silverstein, R. & Webster, F.X. *Spectrometric Identification of Organic Compounds,* Sixth Edition, Copyright 1999 by John Wiley & Sons, Inc.

CHAPTER 3

Chapter opener, p. 94: Photo from Corbis Stock Market. **P. 103**: Photo of Sir Robert Robinson. © The Nobel Foundation. **P. 121, Table 3.1:** Adapted from *Advanced Organic Chemistry,* Third Edition, by March, J. Copyright ©1985 by John Wiley & Sons, Inc.

CHAPTER 4

Chapter opener, p. 134: Color enhanced transmission electron micrograph of healthy heart muscle. Dr. Gopal Murti/Photo Researchers. **P. 135**: Photo courtesy Natural History Museum of Los Angeles County. **P. 136**: Photo courtesy Richard During/Tony Stone Images/New York, Inc. **P. 136, Table 4.1**: Adapted with permission from *Elements of General, Organic, and Biological Chemistry,* Ninth Edition, by Holum, J.R. © 1995 by John Wiley & Sons, Inc. **P. 140**: © Lisa Gee. **P. 155:** Photo of Melvin S. Newman. Courtesy Ohio State University Archives. **P. 157:** Photo of J.H. van't Hoff. © The Nobel Foundation. **P. 160**: Photo of laboratory 'bomb' calorimeter. Courtesy Parr Instrument Company. **P.166**: Photos of Derek Barton and Odd Hassel. © The Nobel Foundation. **P. 167, The Chemistry of box**: Reprinted with permission from Stoddart et al., *Accounts of Chemical Research,* Vol. 34, no.6, pp. 433–444 ©2001 American Chemical Society, and from Balzani et al. *Accounts of Chemical Research,* Vol. 34, no.6, pp. 445–455 © 2001 American Chemical Society. **P. 183:** Photos of E. J. Corey © The Nobel Foundation.

CHAPTER 5

Chapter opener, p. 193: A spiral galaxy. David Malin/Anglo Australian Telescope Board. **P. 195**: Photo of bindweed. Courtesy Photo Researchers. **P. 204**: Photo of Cahn, Ingold, Prelog. Courtesy Dr. Keith Ingold. **P. 210, Fig. 5.12:** From *Organic Chemistry: A Brief Course,* by Holum, J.R., © 1975 by John Wiley & Sons. **P. 213, Fig. 5.13 a & b**: from *Physical Chemistry: Principles and Applications in Biological Sciences,* 2nd ed. By Tinco/ Wang /Sauer, © Reprinted by permission of Pearson Education, Inc., Upper Saddle River, NJ. **P. 213: Fig. 5.13 c & d**: from Adamson, A.W., *A Textbook of Physical Chemistry,* 2nd ed. Academic Press, 1973, p. 836, Fig. 19–27. **P. 218, The Chemistry of box:** Graphic courtesy John O. Trent, Ph.D., Brown Cancer Center, Dept. of Medicine, University of Louisville, KY. Based on work from Qu X, Trent JO, Fokt I, Priebe W, and Chaires JB. *Allosteric, Chiral-Selective Drug Building to DNA.* Pro. Natl. Acad. Sci. USA. 2000, Oct. 24: 97 (22), 12032–7.

CHAPTER 6

Chapter opener, p. 238: A wrecking ball demolishing a wall. Andy Whale/Tony Stone Images/New York, Inc. **P. 254**: Photo of

George Olah, © The Nobel Prize Foundation. **P. 261, Fig. 6.11**: Adapted from *Introduction to Free Radical Chemistry,* by Pyror, W.A.© 1966 by Prentice Hall.

CHAPTER 7

Chapter Opener, p. 287: Caribou in Alaska. Daniel J. Cox/ Tony Stone Images/New York, Inc. **P. 311**: Photo of margarine. © Lisa Gee. **P. 313**: Diagram of general Wilkinson hydrogenation catalyst method. Adapted from Noyoi, *Asymmetric Catalysis in Organic Synthesis,* Wiley Interscience, 1994, p.17.

CHAPTER 8

Chapter opener, p. 328: Giant barrel sponge, Cayman Islands. Andrew J. Martinex/ Photo Researchers. **P. 343**: Courtesy Herbert C. Brown. **P. 362:** Adaptations of Sharpless taxol example and catalytic asymmetric dihydroxylaton from *Journal of the American Chemical Society,* Vol. 119, 1840–1858 With permission from K Barry Sharpless (Scripps Research Institute). Sharpless taxol example in Nicoloau, p. 688, scheme 13. Used with permission of K.B. Sharpless.

CHAPTER 9

Chapter opener, p. 383: An NMR spectrometer with a superconducting magnet. Courtesy Varian Associates, Inc. **P. 417**: Magnetic resonance image. Harry Sieplinga/The Image Bank/Getty Images. **P. 418, Fig. 933 and p. 422, Table 9.1**: Adapted from Silverstein, R.M.; Webster, F.X. *Spectrometric Identification of Organic Compounds, 6th* ed. Wiley, 1998. Used with permission. **P. 419, Fig. 9.34:** From Holum, J.R. *Organic Chemistry: A Brief Course.* Wiley, 1975. Used with permission. **P. 442, Fig. 9.48:** The IR Spectrum of compound E © BioRad Laboratories, Inc, Information Division, Sadtler Software & Databases. All rights reserved. Permission for the publication herein of Sadtler Spectra has been granted by BioRAd Laboratories, Inc., Informatics Division.

CHAPTER 10

Chapter opener, p. 447: Propelled by the radical reactions of combustion, a projectile cleaves twisted strands of string resembling the helix of DNA. C.E. Miller, The Massachusetts Institute of Technology. **P. 448**: Calicheamicin $\gamma_1{}^1$ bound to DNA. PDB ID: 2PIK. Kumar, R.A.; Ikemoto, N.; Patel, D.J. Solution Structure of the Calicheamicin $\gamma_1{}^1$-DNA Complex, *J. Mol. Biol.* 1997, *265, 187.* **P. 448:** Calicheamicin $\gamma_1{}^1$ structure. From *Chemistry and Biology,* 1994, I(1), 26. Reprinted by permission of Current Biology, Ltd., London. **P. 477, Fig. 10.7:** Graphic of relaxation of smooth muscle in vascular tissues. With permission from The Nobel Committee for Physiology or Medicine at Karolinska Institutet, SE-171 77 Stockholm, Sweden.

CHAPTER 11

Chapter opener, p. 492: An industrial robot passing boxes from one side to another. Gerard Fritz/Taxi/Getty Images. **P. 498**: Grapes fermenting in stone vats from Rod Westwood/The Image Bank/Getty Images.

CHAPTER 12

Chapter opener, p. 537: Photo courtesy Jim Foster/Corbis Images. **P. 542:** Photo courtesy Aldrich Chemical Co. **P. 545**: Photo courtesy Harvey Lloyd / Corbis Stock Market. **P. 552**: Photo of Victor Grignard © The Nobel Foundation.

CHAPTER 13

Chapter opener, p. 577: The Nobel Ceremony, Stockholm, Sweden. AP/Wide World Photos. **P. 598, Fig. 13.8**: UV absorption spectrum of 2,5–dimethyl-2,4–hexadiene © Bio-Rad Laboratories, Inc., Informatics Division. Sadtler Software * Databases. All Rights Reserved. Permission for the publication herein of Sadtler Sepctra has been granted by Bio-Rad Laboratories, Inc., Informatics Division. **P. 600, Fig. 13.A:** Reprinted from Teller, D. et al, *Biochemistry, 2001*, Vol. 40, No. 26, p. 776, Figure 1. Copyright (2001) American Chemical Society. Courtesy David Teller, Biochem, Washington University. **P. 601, Fig. 13.C**: Comparison of visible absorption spectrum of rhodopsin. Reproduced from Hecht, S., Shaler, S., Pirenne, M.H., *Journal of General Physiology, 1942,* 25, 819–840 by copyright permission of The Rockefeller University Press.

CHAPTER 14

Chapter opener, p. 622: Plastic for recycling (left) and reclaimed hydrocarbon feedstock (right). James King-Holmes/Photo Researchers. **P. 643, Fig. 14.19:** Structures of C_{60} and C_{70} adapted from Diederich, F., Whetten, R.L., *Accounts of Chemical Research,* 1992, 25, 119–126. Copyright (1992) American Chemical Society. **P. 643**: Photo of nanotube. Courtesy Professor Charles M. Lieber, Harvard University. **P. 644**: Photos of Robert Curl, Harold Kroto, Richard Smalley. AP/Wide World Photos. **P. 652**: Photo of sun screen. © Lisa Gee. **P. 659, Figs. 14.30 and 14.32**: Spectrum courtesy of Aldrich Chemical Co., Milwaukee, WI. **P. 662**: Diagram of synthesis of callistephin chloride. Reprinted with permission from *Journal of Chemical Society,* 1455–1472. Copyright 1928 American Chemical Society.

CHAPTER 15

Chapter opener, p. 664: Giant kelp, a natural source of iodine. Darryl Torckler/Tony Stone Images/New York, Inc. **P. 672**: Courtesy Edgar Fahs Smith Collection, Van Pelt Library, University of Pennsylvania. **P. 691**: The biosynthesis of thyroxine. Adapted from *Biochemistry.* Second Edition, by Voet, D. & Voet, J.G. © 1995 by John Wiley & Sons, Inc.

CHAPTER 16

Chapter opener, p. 716: Wheat, a source of vitamin B_6, growing in the Palouse region of Washington state. Grant Heilman/Grant Heilman Photography. **P. 768:** Diagram of ascorbic acid. Reprinted with permission from *Journal of Chemical Society,* 1419. Copyright 1933 American Chenical Society.

CHAPTER 17

Chapter opener, P. 769: A crew team expending great amounts of energy. David Madison/David Madison Photography. **P. 801**: Diagram of calicheamicin $\gamma_1{}^1$ structure. Reprinted from *Chemistry and Biology, 1 (1),* 26, by permission of Current Biology Ltd., London. **P. 809**: Diagram of cholesterol. Reprinted from R.B. Woodward, F. Sondheimer, et al., *Journal of American Chemical Society,* 1952, 74, 4223. Copyright (1952) American Chemical Society.

CHAPTER 18

Chapter opener, p. 813: The manufacturing of nylon carpet fibers. Ted Horowitz/Corbis Stock Market. **P. 815**: Valerian plant in bloom. Jeanne White/Photo Researchers. **P. 822, Figure 18.2**: Frequency ranges based on Silverstein, R.M., Webster, F.X. *Spectrometric Identification of Organic Compounds, 6th ed.;* John Wiley & Sons, 1998. **P. 845:** Industrial-scale reactor. Michael Rosenfeld/Tony Stone Images/ New York, Inc. **P. 849:** Whole-grain breads. David Ball / Corbis Stock Market. **P. 861, Fig. 18.7:** © Bio-Rad Laboratories, Inc., Informatics Division. Sadtler Software * Databases. All Rights Reserved. Permission for the publication herein of Sadtler Sepctra has been granted by Bio-Rad Laboratories, Inc., Informatics Division. **P. 872:** Plastic for recycling (left) and reclaimed hydrocarbon feedstock (right). James King-Holmes/Photo Researchers. **P. 875 & 876, Figs. B1 & B2:** Reprinted with permission from Newkome, G.R.; Lin, X., *Macromolecules, 24,* 1443–1444. Copyright 1991 American Chemical Society.

CHAPTER 19

Chapter opener, p. 878: A Venetian carnival masquerade, Italy. Grant V. Faint/ The Image Bank/Getty Images. **P. 906:** Lys H93 image. Reprinted from Barbas III, et al., *Science,* (December 1997) Vol. 278, No. 5346, fig. 6, p. 2085. Copyright © 2000 American Association for the Advancement of Science. **Pp. 920–921**: Dehyroabietic acid structure. From *Selected Organic Synthesis* by Fleming, I. © 1973 by John Wiley & Sons, Ltd. **P. 932, Fig. D.1:** Adapted from *Biochemistry* by Albert A. Lehninger © 1970, 1975 by Worth Publishers: Used with permission of W.H. Freeman and Company.

CHAPTER 20

Chapter opener, p. 940: A poison arrow frog. Stephen J. Kraseman/ Photo Researchers. **P. 965**: Meat processed with nitrites. Charles D. Winters/Photo Reserachers. **P. 976, Fig. 20.4:** From *Essentials of Molecular Pharmacology* by Korolkovas, A. © 1970 by John Wiley & Sons, Inc. **P. 994**: Erythroxylum coca. Dr. Morley Read/ Photo Researchers.

CHAPTER 21

Chapter opener, p. 1000: A silver chalice from the 1st century. Art Resource. **P. 1017**: Bombardier beetle from Eisner & Aneshansley, *Proc. Natl. Acad. Sci. USA.,* 1999, 96, 9705–9709. Courtesy Thomas Eisner and Daniel Aneshansley, Cornell University. **P. 1024**: Molecular graphic of benzyne. Image courtesy of Jan Haller; reprinted with permission of Ralf Warmuth. **P. 1033**: Scheme of synthesis of optically pure thyroxine. Adapted from *Selected Organic Synthesis,* by Fleiming, I. ©1973 by John Wiley & Sons, Ltd. Reprinted by permission of John Wiley & Sons, Ltd.

CHAPTER 22

Chapter opener, p. 1072: A color-enhanced scanning electron micrograph of red and white blood cells. Andrew Syred/Tony Stone Images/ New York, Inc. **P. 1072:** Graphic of inflammatory cascade. Reprinted with permission from *Chemical Reviews, 98,* 833–862. Copyright 1998 American Chemical Society. **P. 1074** Schematic diagram of plasma membrane, and **P. 1111, Fig. 22.17**, proposed structure for cellulose**.** From *Biochemistry,* Second Edition, by Voet, D., Voet, J.G. ©1995 by John Wiley & Sons, Inc. **P. 1078**: Photo of Emil Fischer. Courtesy Corbis Images. **P. 1080, Fig. 22.4**: The Haworth formulas. From *Organic Chemistry: A Brief course* by Holum, J.R. © 1975 by John Wiley & Sons, Inc. **P. 1098, Fig. 22.7:** The D family of Aldohexoses. From *Organic Chemistry* by Fieser, L.F. , & Fieser, M. Copyright © 1956 by International Thompson. **P. 1101:** Masamune-Sharpless *L*-aldohexose synthesis scheme, adapted from Nicolaou, *Classics in Total Synthesis,* Jan. 1996, pp. 310 & 312, Wiley-VCH. **P. 1106**: Photo of products with artificial sweeteners. The Photo Works/Photo Researchers. **P. 1108, Fig. 22.14:** Amylose. Figure copyrighted © by Irving Geis. From Voet, D., & Voet, J.G. *Biochemistry*, Second Edition, Wiley, 1995, p. 262. **P. 1111:** Cellophane on rollers. Vince Streano /Corbis Images. **P. 1112:** Structure of glycal assembly method from Danishefsky et al., *Accounts of Chemical Research,* Vol. 31, p. 687. Copyright 1998 American Chemical Society. **P. 1116:** Structure of fully synthetic glycoconjugate antigen from Glunz, Hintermann, et al., *Journal of the American Chemical Society,* 2000, Vol. 122, pp. 7273–7279. Copyright 2000 American Chemical Society. **P. 1118:** Structure of glycoconjugate components of anticancer vaccines from *Journal of American Chemical Society* (2001) Vol. 123, p. 1890. Copyright 2001 American Chemical Society. **P. 1119:** Full structure of synthetic triple-clustered anticancer vaccine from Kudryashov, Glunz, et al., *PNAS,* March 13, 2001, Vol. 98, no. 6, pp. 3264–3269. Copyright 2001 National Academy of Sciences, U.S.A.

CHAPTER 23

Chapter opener, p. 1129: A nerve axon with a myelin sheath. C. Raines/Visuals Unlimited. **P. 1130:** Schematic diagram of myelinated axon. Adapted from *Biochemistry,* Second Edition, by Voet, D. & Voet, J.G. © 1995 by John Wiley & Sons, Inc. **P. 1133, Table 23.2:** Table of fatty acid composition obtained by hydrolysis. Reprinted from *Biology Data Book* by permission of the Federation of American Societies for Experimental Biology, Bethesda, MD. In *Organic and Biological Chemistry* by Holum, J.R. © 1978 by John Wiley & Sons, Inc. **P. 1135**: Photo (left) Courtesy of Procter & Gamble. Structure of Olestra; adapted by permission from the *Journal of Chemical Education, Vol. 74,* No. 4, 1997, pp. 370–372; copyright 1997, Division of

Chemical Education, Inc. **P. 1138:** Image of switching devices based on interlocked molecules from Pease, A.R., Jeppensen, J.D. et al., *Accounts of Chemical Research*, June 2001, Vol. 34, no. 6, p. 439. Copyright 2001 American Chemical Society. **P. 1147, Fig. 23.7:** Image of cholesterol from Voet, D., Voet, J.G. *Biochemistry,* p. 318, Fig. 11–50. © Wiley 1995. This material is used by permission of John Wiley & Sons, Inc. **P. 1156, Fig. 23.9:** Diagram of plasma membrane from Voet, Voet, Pratt, *Fundamentals of Biochemistry,* p. 248. © Wiley 1999. This material is used by permission of John Wiley & Sons, Inc. **P. 1157**: Stealth Liposome. Courtesy of ALZA Corporation.

CHAPTER 24

Chapter opener, p. 1166: Computer graphic of antibodies moving along an artery. Alfred Pasieka/Photo Researchers. **P. 1180, Fig. 24.3**: Graphic of typical result of automatic amino acid analyzer. Reprinted with permission from Spackman, D.H., Stein, W.H., and Moore, S. *Analytical Chemistry, 30* , 1190. Copyright 1958 by the American Chemical Society. **P. 1181, Fig. 24.4:** Overlay of PTH standards from Perkin Elmer Biosystems brochure for Precise Protein Sequencing System. **P. 1188:** Normal and sickle-cell anemia cells. Stan Flegler/ Visuals Unlimited. **P. 1196**: Photo of Linus Pauling. Courtesy Kenneth Dunmire, Pacific Lutheran University. **P. 1196, Fig. 24.9** and **P. 1202, Fig. 24.14**: From *Biochemistry, 2nd* ed., by Voet, D. Voet, J.G. © 1995. Reprinted by permission of John Wiley & Sons, Inc. **P. 1197, Fig. 24.10 & p. 1198, Fig. 24.11:** From *Biochemistry, 2nd* ed., by Voet, D. Voet, J.G. Copyight © 1995 by John Wiley & Sons. Illustration © Irving Geis. **P. 1200, Fig. 24.13:** Three-dimensional structure of myoglobin. Image prepared from x-ray crystal structure by Phillips, S.E.V. Structure and Refinement of Oxymyoglobin at 1.6 Angstroms Resolution. *J. Mol. Biol.*1980, *142,* 531. **P. 1200:** Carbonic anhydrase. PDB ID: 1CA2. Eriksson A.E., Jones T.A., Liljas, A. Refined Structure of Human Carbonic Anhydrase at 2.0 Angstroms Resolution. *Proteins Struct., Funct.* 1988, *4,* 274. **P. 1202 (margin)** and **P. 1204, Fig. 24.16 (c):** Lysozyme, PFB ID: 1AZF. Lim, K., Nadarahah, A., Forsythe, E.L, Pusey, M.L. Location of Halide Ions in Tetragonal Lysozyme Crystals. **P. 1202:** Quotation from Phillips, David C., *The Three-Dimensional Structure of an Enzyme Molecule.* Copyright © 1966 by Scientific American, Inc. All rights reserved. **P. 1204, Fig. 24.16 (a):** Adapted with permission from *Atlas of Protein in Sequence and Structure,* Margaret O. Dayhoff, Editor (1969). Figure reproduced with permission from the National Biomedical Research Foundation. **Fig. 24.17 (b):** From David C. Phillips, "The Three-Dimensional Structure of an Enzyme Molecule," Copyright © 1966 by Scientific American, Inc. All rights reserved. **P. 1205, Fig. 24.17:** Trypsin. PDB ID: 1 MAX. Bertrand, J.A., Oleksyszyn, J., Kam, C.-M., Boduszek, B., Presnell, S., Plaskon, R.R., Suddath, F.L., Powers, J.C., Williams, L.D. Inhibition of Trypsin and Thrombin by Amino(4–amidinopheny)-methanephophonate Diphenyl Ester Derivatives: X-Ray Structures and Molecular Models. **P. 1209:** Diels-Alderase catalytic antibody from Romesburg, F.E., Spiller, B., Schultz, P.G., Stevens, R.C., Protein Data Bank file 1A4K. **P. 1210, Fig. 24.21:** Hemoglobin from Tame, J.R., Wilson, J.C., Weber, R.E., The Crystal Structures of Trout HB 1 in the Deoxy and Carbonmonoxy Forms, *J. Mol. Biol.* 1996, *259,* 749. **P. 1214, Fig. 24.23:** Cutaway view of the T. thermophilus 70S ribosome containing three bound tRNAs. Image courtesy of Professor Harry Noller, University of California, Santa Cruz, and John Yates, The Scripps Research Institute.

CHAPTER 25

Chapter opener, p. 1220: An electrophoresis gel used for sequencing DNA. Ted Horowitz/ Corbis Stock Market. **P. 1221, Fig. 25.1:** The basics of genetics. DNA Sequencing: The next step in the search for genes, *Science and Technology Review,* November 1996. Credit to the University of California, Lawrence Livermore National Laboratory and the Department of Energy under whose auspices the work was performed. **P. 1222, Fig. 25.2**: Schematic of chromosome 19. From Dept. of Energy Joint Genome Institute web page (http://www.jgi.doe.gov/whoweare/llnl_jgi_decoding.htm#top). Credit to the University of California, Lawrence Livermore National Laboratory and the Department of Energy under whose auspices the work was performed. **P. 1230**, **Table 25.1**: From E.L. Smith, R.L. Hill, et al., *Principles of Biochemistry: General Aspects,* p. 132. Copyright (c) 1983 by McGraw-Hill Inc. Reproduced with permission of McGraw-Hill Companies. **P. 1231, Fig. 25.8**: From A.L. Neal, *Chemistry and Biochemistry: A Comprehensive Introduction.* Copyright © 1971 by McGraw-Hill Inc. Reproduced with permission of the McGraw-Hill Companies. **P. 1232, Fig. 25.9**: Reprinted from L. Pauling and R. Corey, *Archives of Biochemistry and Biophysics, 65,* 164, by permission of Academic Press, Orlando, FL. **P. 1237, Fig. 25.13:** Reprinted from Cate, J.H., Yusupov, M.M., Yusupov, G.Z., Earnest, T.N., Noller, H.F., "X-ray crystal structure of a ribosome with bound t-RNAs," *Science,* 1999, 285, 2095–2104. Copyright (1999) American Association for the Advancement of Science. **P. 1238, Fig. 25.14:** From Nissen, P., Hansen, J., Ban, N., Moore, P.B., Steitz, T.A., "The Structural Basis of Ribosome Activity in Peptide Bond Synthesis," *Science, 2000,* 289, 920–930. Copyright (2000) American Association for the Advancement of Science. **P. 1240, Fig. 25.15:** A transfer RNA, PDB ID: 4TNA. Hingerty, E. , Brown, R.S., Jack, A. Further Refinement of the Structure of Yeast $tRNA_{Phe}$, *J. Mol. Biol.* 1978, *124,* 523. **P. 1246, Fig. 25.17:** DNA automated sequencer from Applied Biosystems, ABI Prism 3100 Genetic Analyzer. **P. 1247, Fig. 25.18**: From *Biochemistry, 2nd ed.,* by Voet, D. Voet, J.G. © 1995. Reprinted by permission of John Wiley & Sons, Inc. **P. 1250, Fig. 25.20:** From one cycle of PCR flow diagram, and Fig. 25.33, cycles of polymerase chain reaction. Used with permission from Andy Vierstraete, University of Ghent. **P. 1251:** Photo of thermophilic bacteria. Harvey Lloyd / Corbis Stock Market.

Index

Page numbers followed by t indicate tables, n indicate footnotes, f indicate figures

B

D

H

I

N

O

P

Q

R

T

U

V

Periodic Table of the Elements

Atomic number → 6; Symbol → C; Name (IUPAC) → Carbon; Atomic mass → 12.011

IUPAC recommendations → (top line of group headings)
Chemical Abstracts Service group notation → (bottom line of group headings)

1 IA	2 IIA	3 IIIB	4 IVB	5 VB	6 VIB	7 VIIB	8 VIIIB	9 VIIIB	10 VIIIB	11 IB	12 IIB	13 IIIA	14 IVA	15 VA	16 VIA	17 VIIA	18 VIIIA
1 H Hydrogen 1.0079																	2 He Helium 4.0026
3 LI Lithium 6.941	4 Be Berylium 9.0122											5 B Boron 10.811	6 C Carbon 12.011	7 N Nitrogen 14.007	8 O Oxygen 15.999	9 F Fluorine 18.998	10 Ne Neon 20.180
11 Na Sodium 22,990	12 Mg Magnesium 24.305											13 Al Aluminum 26.982	14 Si Silicon 28.086	15 P Phosphorus 30.974	16 S Sulfur 32.065	17 Cl Chlorine 35.453	18 Ar Argon 39.948
19 K Potassium 39.098	20 Ca Calcium 40.078	21 Sc Scandium 44.956	22 Ti Titanium 47.867	23 V Vanadium 50.942	24 Cr Chromium 51.996	25 Mn Manganese 54.938	26 Fe Iron 55.845	27 Co Cobalt 58.933	28 Ni Nickel 58.693	29 Cu Copper 63.546	30 Zn Zinc 65.409	31 Ga Gallium 69.723	32 Ge Germanium 72.64	33 As Arsenic 74.922	34 Se Selenium 78.96	35 Br Bromine 79.904	36 Kr Krypton 83.798
37 Rb Rubidium 85.468	38 Sr Strontium 87.62	39 Y Yttrium 88.906	40 Zr Zirconium 91.224	41 Nb Niobium 92.906	42 Mo Molybdenum 95.94	43 Tc Technetium (98)	44 Ru Ruthenium 101.07	45 Rh Rhodium 102.91	46 Pd Palladium 106.42	47 Ag Silver 107.87	48 Cd Cadmium 112.41	49 In Indium 114.82	50 Sn Tin 118.71	51 Sb Antimony 121.76	52 Te Tellurium 127.60	53 I Iodine 126.90	54 Xe Xeno 131.29
55 Cs Caesium 132.91	56 Ba Barium 137.33	57 *La Lanthanum 138.91	72 Hf Hafnium 178.49	73 Ta Tantalum 180.95	74 W Tungsten 183.84	75 Re Rhenium 186.21	76 Os Osmium 190.23	77 Ir Iridium 192.22	78 Pt Platinum 195.08	79 Au Gold 196.97	80 Hg Mercury 200.59	81 Tl Thallium 204.38	82 Pb Lead 207.2	83 Bi Bismuth 208.98	84 Po Polonium (209)	85 At Astatine (210)	86 Rn Radon (222)
87 Fr Francium (223)	88 Ra Radium (226)	89 #Ac Actinium (227)	104 Rf Rutherfordium (261)	105 Db Dubnium (262)	106 Sg Seaborgium (266)	107 Bh Bohrium (264)	108 Hs Hassium (277)	109 Mt Meitnerium (268)	110 Uun (281)	111 Uuu (272)	112 Uub (285)		114 Uuq (289)				

*Lanthanide Series	58 Ce Cerium 140.12	59 Pr Praseodymium 140.91	60 Nd Neodymium 144.24	61 Pm Promethium (145)	62 Sm Samarium 150.36	63 Eu Europium 151.96	64 Gd Gadolinium 157.25	65 Tb Terbium 158.93	66 Dy Dysprosium 162.50	67 Ho Holmium 164.93	68 Er Erbium 167.26	69 Tm Thulium 168.93	70 Yb Ytterbium 173.04	71 Lu Lutetium 174.97
# Actinide Series	90 Th Thorium 232.04	91 Pa Protactinium 231.04	92 U Uranium 238.03	93 Np Neptunium (237)	94 Pu Plutonium (244)	95 Am Americium (243)	96 Cm Curium (247)	97 Bk Berkelium (247)	98 Cf Californium (251)	99 Es Einsteinium (252)	100 Fm Fermium (257)	101 Md Mendelevium (258)	102 No Nobelium (259)	103 Lr Lawrencium (262)

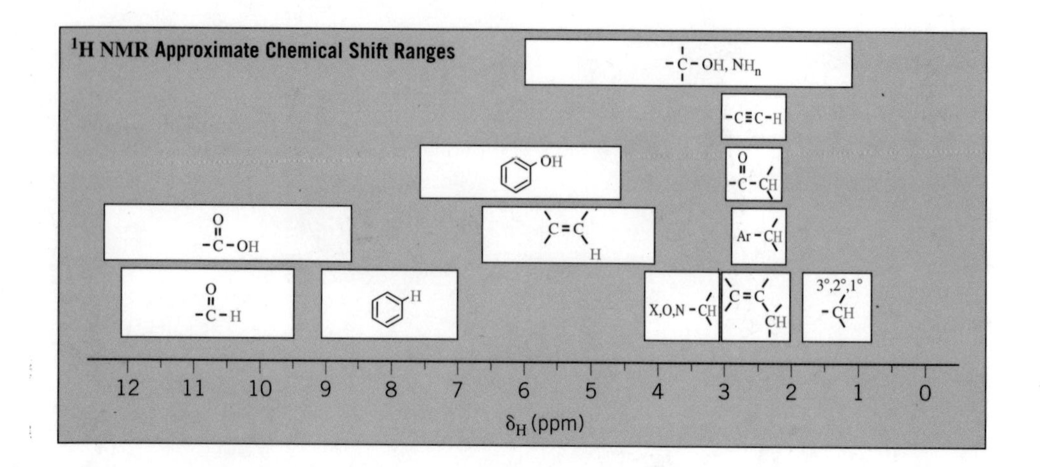

¹H NMR Approximate Chemical Shift Ranges
-C-OH, NHn
-C≡C-H
OH
O
-C-CH
C=C
H
Ar-CH
O
-C-OH
O
-C-H
H
X,O,N-CH
C=C
CH
3°,2°,1°
-CH
12
11
10
9
8
7
6
5
4
3
2
1
0
δH (ppm)

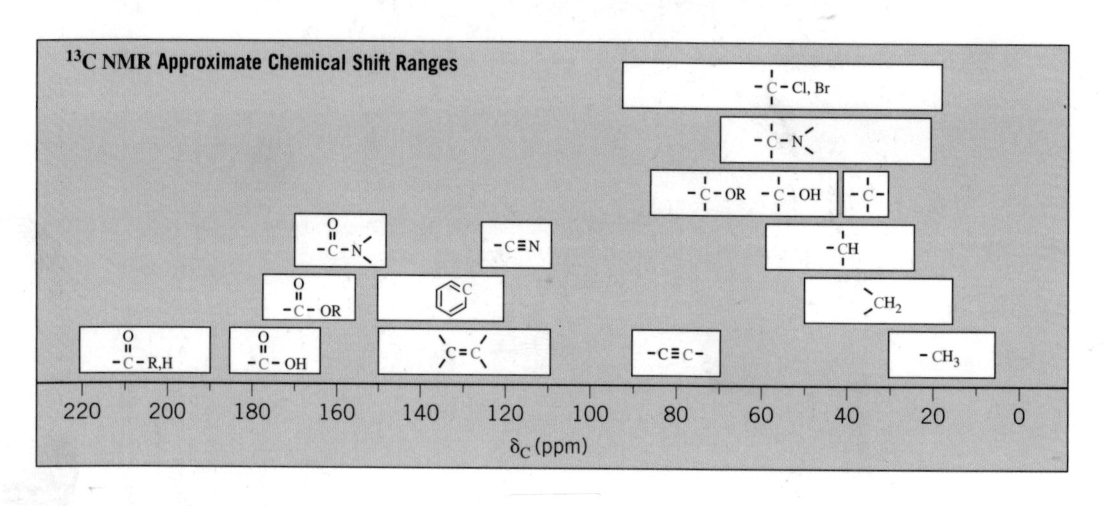

¹³C NMR Approximate Chemical Shift Ranges
-C-Cl, Br
-C-N
-C-OR
-C-OH
-C-
O
-C-N
-C≡N
-CH
O
-C-OR
C
CH2
O
-C-R,H
O
-C-OH
C=C
-C≡C-
-CH3
220
200
180
160
140
120
100
80
60
40
20
0
δC (ppm)